CONTENTS—1991 HVAC APPLICATIONS

CONTENTS—1990 REFRIGERATION

1993 ASHRAE HANDBOOK

FUNDAMENTALS

I-P Edition

American Society of Heating, Refrigerating and Air-Conditioning Engineers, Inc.
1791 Tullie Circle, N.E., Atlanta, GA 30329
404-636-8400

ISBN 0-910110-96-4
ISSN 1041-2344

CONTENTS

LOAD AND ENERGY CALCULATIONS

DUCT AND PIPE SIZING

GENERAL

ERRATA

INDEX

Composite index to the 1990 Refrigeration Systems and Applications, 1991 HVAC Applications, 1992 HVAC Systems and Equipment, 1993 Fundamentals, and the HVAC chapters in the 1988 Equipment volumes.

CONTRIBUTORS

In addition to the Technical Committees, the following individuals contributed significantly to this volume. The appropriate chapter numbers follow each contributor's name.

S.A. Sherif (2)
University of Florida

A.E. Bergles (3)
Rensselaer Polytechnic Institute

Michael M. Ohadi (3,5)
University of Maryland, College Park

Naim Z. Azer (4)
Kansas State University

Creighton A. Depew (6)
University of Washington

Bruce Herbert (6)

Anthony Jacobi (6)
University of Illinois, Urbana

Allen Walton, Jr. (6)

Howard F. Kingsbury (7)

Mark E. Schaffer (7)
McKay Conant Brook Inc.

Neil T. Shade (7)
Acoustical Design Collaborative Ltd.

John R. Breckenridge (8)

Byron W. Jones (8)
Kansas State University

Bjarne W. Olesen (8, 13)
Virginia Polytechnic Institute and State University

Louis D. Albright (9)
Cornell University

Kifle G. Gebremedhin (9)
Cornell University

Kevin A. Janni (9)
University of Minnesota

Jack A. Nienaber (9)
USDA/ARS

Larry W. Turner (9)
University of Kentucky

Garrett L. van Wicklen (9)
University of Georgia

Gerald H. Brusewitz (10)
Oklahoma State University

Terry J. Siebenmorgen (10)
University of Arkansas

Shahab Sokhansanj (10)
University of Saskatchewan

Richard L. Stroshine (10)
Purdue University

Demetrios Moschandreas (12)
Illinois Institute of Technology

Richard D. Rivers (12, 13)
Environmental Quality Sciences

Douglas M. Burch (13)
NIST

Douglas W. Dewerth (13)
American Gas Association Labs

David J. Evans (13)
NIST

James E. Hill (13)
NIST

Peter H. Huang (13)
NIST

Robert M. Judish (13)
NIST

Owen B. Laug (13)
NIST

Billy W. Magnum (13)
NIST

George E. Mattingly (13)
NIST

Norman E. Mease (13)
NIST

Phil Naughton (13)
Motorola, Inc.

Nile M. Oldham (13)
NIST

Andrew K. Persily (13)
NIST

Robert F. Richards (13)
NIST

Luis M. Solarte (13)
ALNOR Instrument Company

Charles R. Tilford (13)
NIST

Stephen J. Treado (13, 27)
NIST

Eugene L. Valerio (13)
Technical Marketing Services

Robert R. Zarr (13)
NIST

Mark P. Modera (14)
Lawrence Berkeley Laboratory

David J. Wilson (14)
University of Alberta

Hall Virgil (15)
Carrier Corporation

EdWard B. Wilson (15)
Kilborn Inc.

Earl M. Clark (16)
E.I. du Pont de Nemours and Company

Mark O. McLinden (16, 17)
NIST

Steven G. Penoncello (17)
University of Idaho

John B. Cuthbert (18)
The Dow Chemical Company

Tushar K. Ghosh (19)
University of Missouri, Columbia

Lew Harriman (19)
Mason-Grant Company

Charles Gilbo (20)
Journal of Thermal Insulation

Charles P. Hedlin (20)
Hedlin Consulting Inc.

Ev Shuman (20)

Jack D. Verschoor (20)
Verschoor Associates

Erv L. Bales (21)
New Jersey Institute of Technology

Jeffrey E. Christian (22)
Oak Ridge National Laboratory

William R. Strzepek (22)
Dow U.S.A.

Harold A. Trethowen (22)
Building Research Association of New Zealand

Martha G. van Geem (22)
Construction Technology Laboratories, Inc.

Donald G. Colliver (23)
University of Kentucky

William Fisk (23)
Lawrence Berkeley Laboratory

William Jones (23)
Ontario Hydro

Drury B. Crawley (24)
National Research Center, Canada

Robert J. Morris (24)
Environment Canada

Marc S. Plantico (24)
National Climatic Data Center

Oscar Richard (24)

Jack F. Roberts (24)
Fanning, Fanning and Associates

Hilda Snelling (24)
USAF Environmental Technical Applications Center

Raymond G. Alvine (25)
Alvine Associates, Inc.

Charles F. Tvrdik (25)
Leo A. Daly Co.

Thomas B. Romine, Jr. (26)
Romine, Romine & Burgess

Jan F. Kreider (28)
University of Colorado

Ari Rabl (28)
Ecole des Mines, France

George N. Walton (28)
NIST

Grenville K. Yuill (28)
Pennsylvania State University

Kan-Ichi Hayakawa (29)

CONTRIBUTORS *(Concluded)*

Roger W. Dickerson, Jr. (30)

Chris Abbey (31)
Anemostat Products

Leslie L. Christianson (31)
University of Illinois

Daniel Int-Hout III (31)
Carrier Corporation

Paul Miller (31)
Kansas State University

Harold Straub (31)
Titus

William J. Waeldner (31)
Anemostat Products

Jianshun Zhang (31)
University of Illinois

Alexander M. Zhivov (31)
University of Illinois

Herman F. Behls (32)
Sargent & Lundy

John D. Sofra (32)
United McGill Corporation

Robert J. Tsal (32)

Albert W. Black (33)
McClure Engineering Associates

Edwin I. Griggs (36)
Tennessee Technical University

Eleanor R. Adair (37)
The John B. Pierce Laboratory, Inc.

Brigitta Berglund (37)
University of Stockholm

Larry G. Berglund (37)
The John B. Pierce Laboratory, Inc.

Michael J. Hodgson (37)
University of Connecticut

R.J.M. Horton (37)

Preston E. McNall, Jr. (37)
Phoenix Engineers

Lars Molhave (37)
Aarhus University, Denmark

Philip R. Morey (37)
Clayton Environmental Consultants

Francis J. (Bud) Offerman (37)
Indoor Environmental Engineering

Kenneth Parsons (37)
Loughborough University of Technology,
United Kingdom

Susan L. Rose (37)
U.S. Department of Energy

ASHRAE TECHNICAL COMMITTEES AND TASK GROUPS

SECTION 1.0—FUNDAMENTALS AND GENERAL
1.1   Thermodynamics and Psychrometrics
1.2   Instruments and Measurements
1.3   Heat Transfer and Fluid Flow
1.4   Control Theory and Application
1.5   Computer Applications
1.6   Terminology
1.7   Operation and Maintenance Management
1.8   Owning and Operating Costs
1.9   Electrical Systems
1.10 Energy Resources

SECTION 2.0—ENVIRONMENTAL QUALITY
2.1   Physiology and Human Environment
2.2   Plant and Animal Environment
2.3   Gaseous Air Contaminants and Gas Contaminant Removal Equipment
2.4   Particulate Air Contaminants and Particulate Contaminant Removal Equipment
2.5   Air Flow Around Buildings
2.6   Sound and Vibration Control
TG   Global Climate Change
TG   Halocarbon Emissions
TG   Safety
TG   Seismic Restraint Design

SECTION 3.0—MATERIALS AND PROCESSES
3.1   Refrigerants and Brines
3.2   Refrigerant System Chemistry
3.3   Contaminant Control in Refrigerating Systems
3.4   Lubrication
3.5   Desiccant and Sorption Technology
3.6   Corrosion and Water Treatment
3.7   Fuels and Combustion

SECTION 4.0—LOAD CALCULATIONS AND ENERGY REQUIREMENTS
4.1   Load Calculation Data and Procedures
4.2   Weather Information
4.3   Ventilation Requirements and Infiltration
4.4   Thermal Insulation and Moisture Retarders
4.5   Fenestration
4.6   Building Operation Dynamics
4.7   Energy Calculations
4.9   Building Envelope Systems
4.10 Indoor Environmental Modeling
TG   Cold Climate Design

SECTION 5.0—VENTILATION AND AIR DISTRIBUTION
5.1   Fans
5.2   Duct Design
5.3   Room Air Distribution
5.4   Industrial Process Air Cleaning (Air Pollution Control)
5.5   Air-to-Air Energy Recovery
5.6   Control of Fire and Smoke
5.7   Evaporative Cooling
5.8   Industrial Ventilation
5.9   Enclosed Vehicular Facilities
TG   Kitchen Ventilation

SECTION 6.0—HEATING EQUIPMENT, HEATING AND COOLING SYSTEMS AND APPLICATIONS
6.1   Hydronic and Steam Equipment and Systems
6.2   District Heating and Cooling
6.3   Central Forced Air Heating and Cooling Systems
6.4   In Space Convection Heating
6.5   Radiant Space Heating and Cooling
6.6   Service Water Heating
6.7   Solar Energy Utilization
6.8   Geothermal Energy Utilization
6.9   Thermal Storage

SECTION 7.0—PACKAGED AIR-CONDITIONING AND REFRIGERATION EQUIPMENT
7.1   Residential Refrigerators and Food Freezers
7.4   Unitary Combustion-Engine-Driven Heat Pumps
7.5   Room Air Conditioners and Dehumidifiers
7.6   Unitary Air Conditioners and Heat Pumps

SECTION 8.0—AIR-CONDITIONING AND REFRIGERATION SYSTEM COMPONENTS
8.1   Positive Displacement Compressors
8.2   Centrifugal Machines
8.3   Absorption and Heat Operated Machines
8.4   Air-to-Refrigerant Heat Transfer Equipment
8.5   Liquid-to-Refrigerant Heat Exchangers
8.6   Cooling Towers and Evaporative Condensers
8.7   Humidifying Equipment
8.8   Refrigerant System Controls and Accessories
8.10 Pumps and Hydronic Piping
8.11 Electric Motors—Open and Hermetic

SECTION 9.0—AIR-CONDITIONING SYSTEMS AND APPLICATIONS
9.1   Large Building Air-Conditioning Systems
9.2   Industrial Air Conditioning
9.3   Transportation Air Conditioning
9.4   Applied Heat Pump/Heat Recovery Systems
9.5   Cogeneration Systems
9.6   Systems Energy Utilization
9.7   Testing and Balancing
9.8   Large Building Air-Conditioning Applications
9.9   Building Commissioning
9.10 Laboratory Systems
TG   Clean Spaces
TG   Tall Buildings

SECTION 10.0—REFRIGERATION SYSTEMS
10.1 Custom Engineered Refrigeration Systems
10.2 Automatic Icemaking Plants and Skating Rinks
10.3 Refrigerant Piping, Controls, and Accessories
10.4 Ultra-Low Temperature Systems and Cryogenics
10.5 Refrigerated Distribution and Storage Facilities
10.6 Transport Refrigeration
10.7 Commercial Food and Beverage Cooling, Display and Storage
10.8 Refrigeration Load Calculations

SECTION 11.0—REFRIGERATED FOOD TECHNOLOGY AND PROCESSING
11.2 Foods and Beverages
11.5 Fruits, Vegetables and Other Products

PREFACE

While much of the information in the Fundamentals Handbook does not change, research supported by ASHRAE and other organizations continues to generate new information for many chapters. In this 1993 volume, several changes and additions are worth noting. One of the important additions is the new Chapter 37 on enviromental health, which introduces the field of environmental health as it pertains to buildings. Chapter 8 on physiological principles and thermal comfort has been revised to complement this new chapter. For example, Chapter 8 includes a revised comfort chart that accounts for the effect of high humidity on comfort. Chapter 27 on fenestration contains new data for a variety of fenestration products, including quadruple glazing, operable and fixed units, double doors, and sloped skylights.

Chapters 16 and 17 now have design data and pressure-enthalpy charts for six new refrigerants—R-32, R-123, R-124, R-125, R-134a, and R-141b. Also, Chapter 18 includes data on the physical and thermal properties of several new secondary coolants.

Chapter 10 on physiological factors in drying and storing farm crops and Chapter 19 on sorbents and desiccants were substantially rewritten for the 1989 volume, but minor changes and corrections are included in this volume. Chapter 26 on nonresidential air-conditioning cooling and heating loads was rewritten for 1989 and revised again in 1993. Residential cooling load information is incorporated into Chapter 25 on residential cooling and heating load calculations. Compared to its 1989 counterpart, Chapter 26 now more clearly differentiates and compares the transfer function (TFM), cooling load temperature difference/cooling load factor (CLTD/CLF), and the total equivalent temperature difference/time averaging (TETD/TA) methods for calculating air-conditioning loads.

Chapter 31 on space air diffusion now includes information on displacement ventilation, localized ventilation, and standards for satisfactory conditions of comfort. Chapter 32 on duct design reflects recent research that developed a database for duct fitting loss coefficients. This database, which is available from ASHRAE in an electronic format for personal computers, includes coefficients for over 200 fittings in a wide range of configurations. Finally, Chapter 38, which is updated every year, lists most of the current United States and Canadian standards that relate to the industry.

Additions and corrections for the 1991 and 1992 Handbooks precede the index. Errors found in this volume will be reported in the 1994 ASHRAE *Handbook—Refrigeration*. The Handbook Committee welcomes reader input. If you have suggestions or comments on improving a chapter, write to: Handbook Editor, ASHRAE, 1791 Tullie Circle, Atlanta, GA 30329.

Robert A. Parsons
Handbook Editor

ASHRAE HANDBOOK COMMITTEE

Charles J. Procell, Chairman

1993 Fundamentals Volume Subcommittee: **Thomas B. Romine, Jr.**, Chairman

Leslie L. Christianson   **Matt R. Hargan**   **Byron W. Jones**   **John F. Klouda**   **Harold G. Lorsch**

ASHRAE HANDBOOK STAFF

Robert A. Parsons, Editor   **Claudia Forman**, Associate Editor

Laura M. Reisinger, Editorial Assistant

Ron Baker, Production Manager

Nancy F. Thysell, **Christopher R. Hart**, Typography

Lawrence H. Darrow, **Susan M. Boughadou**, Graphics

Frank M. Coda, Publisher
W. Stephen Comstock, Communications and Publications Director

CHAPTER 1

THERMODYNAMICS AND REFRIGERATION CYCLES

THERMODYNAMICS is the study of energy, its transformations, and its relation to states of matter. This chapter covers the application of thermodynamics to refrigeration cycles. The first part reviews the first and second laws of thermodynamics and presents methods for calculating thermodynamic properties. The second and third parts address compression and absorption refrigeration cycles, the two most common methods of thermal energy transfer.

THERMODYNAMICS

A *thermodynamic system* is a region in space or a quantity of matter bounded by a closed surface. The surroundings include everything external to the system, and the system is separated from the surroundings by the system boundaries. These boundaries can be movable or fixed, real or imaginary.

The concepts that operate in any thermodynamic system are *entropy* and *energy*. Entropy(s) measures the molecular disorder of a given system. The more shuffled a system, the greater its entropy; conversely, an orderly or unmixed configuration is one of low entropy.

Energy has the capacity for producing an effect, and can be categorized into either stored or transient forms. Stored forms of energy include the following:

Thermal (internal) energy u is the energy possessed by a system caused by the motion of the molecules and/or intermolecular forces.

Potential energy P.E. is the energy possessed by a system caused by the attractive forces existing between molecules, or the elevation of the system.

$$P.E. = mgz \tag{1}$$

where

 m = mass
 g = local acceleration of gravity
 z = elevation above horizontal reference plane

The preparation of the first and second parts of this chapter is assigned to TC 1.1, Thermodynamics and Psychrometrics. The third part is assigned to TC 8.3, Absorption and Heat Operated Machines.

Kinetic energy K.E. is the energy possessed by a system caused by the velocity of the molecules.

$$K.E. = mV^2/2 \tag{2}$$

where

 m = mass
 V = velocity of fluid streams crossing system boundaries

Chemical energy E_c is energy possessed by the system caused by the arrangement of atoms composing the molecules.

Nuclear (atomic) energy E_a is energy possessed by the system from the cohesive forces holding protons and neutrons together as the atom's nucleus.

Transient energy forms include the following:

Heat Q is the mechanism that transfers energy across the boundary of systems with differing temperatures, always in the direction of the lower temperature.

Work is the mechanism that transfers energy across the boundary of systems with differing pressures (or force of any kind), always in the direction of the lower pressure. If the total effect produced in the system can be reduced to the raising of a weight, then nothing but work has crossed the boundary.

Mechanical or *shaft work W* is the energy delivered or absorbed by a mechanism, such as a turbine, air compressor, or internal combustion engine.

Flow work is energy carried into or transmitted across the system boundary because a pumping process occurs somewhere outside the system, causing fluid to enter the system. It can be more easily understood as the work done by the fluid just outside the system on the adjacent fluid entering the system to force or push it into the system. Flow work also occurs as fluid leaves the system.

$$\text{Flow Work (per unit mass)} = pv \tag{3}$$

where p is the pressure and v is the specific volume, or the volume displaced per unit mass.

A *property* of a system is any observable characteristic of the system. The *state* of a system is defined by listing its properties. The most common thermodynamic properties are temperature T, pressure p, and specific volume v or density ρ. Additional thermodynamic properties include entropy, stored forms of energy, and enthalpy.

Frequently, thermodynamic properties combine to form new properties. *Enthalpy h*, a result of combining properties, is defined as:

$$h = u + pv \qquad (4)$$

where

u = internal energy
p = pressure
v = specific volume

Each property in a given state has only one definite value, and any property always has the same value for a given state, regardless of how the substance arrived at that state.

A *process* is a change in state that can be defined as any change in the properties of a system. A process is described by specifying the initial and final equilibrium states, the path (if identifiable), and the interactions that take place across system boundaries during the process.

A *cycle* is a process or a series of processes wherein the initial and final states of the system are identical. Therefore, at the conclusion of a cycle, all the properties have the same value they had at the beginning.

A *pure substance* has a homogeneous and invariable chemical composition. It can exist in more than one phase, but the chemical composition is the same in all phases.

If a substance exists as liquid at the saturation temperature and pressure, it is called *saturated liquid*. If the temperature of the liquid is lower than the saturation temperature for the existing pressure, it is called either a *subcooled liquid* (the temperature is lower than the saturation temperature for the given pressure) or a *compressed liquid* (the pressure is greater than the saturation pressure for the given temperature).

When a substance exists as part liquid and part vapor at the saturation temperature, its quality is defined as the ratio of the mass of vapor to the total mass. Quality has meaning only when the substance is in a saturated state, *i.e.*, at saturation pressure and temperature.

If a substance exists as vapor at the saturation temperature, it is called *saturated vapor*. (Sometimes the term *dry saturated vapor* is used to emphasize that the quality is 100%.) When the vapor is at a temperature greater than the saturation temperature, it is *superheated vapor*. The pressure and temperature of superheated vapor are independent properties, since the temperature can increase while the pressure remains constant. Gases are highly superheated vapors. Figure 1 depicts these thermodynamic fluid states.

FIRST LAW OF THERMODYNAMICS

The first law of thermodynamics is often called the *law of the conservation of energy*. The following form of the first law equation is valid only in the absence of a nuclear or chemical reaction.

Based on the first law or the law of conservation of energy for any system, open or closed, there is an energy balance as:

$$\begin{bmatrix} \text{Net Amount of Energy} \\ \text{Added to System} \end{bmatrix} = \begin{bmatrix} \text{Net Increase in Stored} \\ \text{Energy of System} \end{bmatrix}$$

or

Energy In − Energy Out = Increase in Energy in System

Referring to Figure 2, Δm_1 is the mass entering the system, and Δm_2 is the mass leaving. The first law in differential or incremental form becomes:

$$[\Delta m(e + pv)]_{in} - [\Delta m(e + pv)]_{out} + \Delta Q - \Delta W = dE \quad (5)$$

$$\Delta m_1 \left(u_1 + p_1 v_1 + \frac{V_1^2}{2} + z_1 g\right) - \Delta m_2$$
$$\times \left(u_2 + p_2 v_2 + \frac{V_2^2}{2} + z_2 g\right) + \Delta Q - \Delta W = dE \quad (6)$$

where ΔQ and ΔW are the increments of work and heat, and dE is the differential change in the energy of the system; if expressed as a rate equation:

$$\left(\frac{\Delta m_1}{\Delta t}\right) \left(h_1 + \frac{V_1^2}{2} + z_1 g\right) - \left(\frac{\Delta m_2}{\Delta t}\right)$$
$$\times \left(h_2 + \frac{V_2^2}{2} + z_2 g\right) + \frac{\Delta Q}{\Delta t} - \frac{\Delta W}{\Delta t} = \frac{dE}{\Delta t} \quad (7)$$

as

$$\Delta t \to 0, \qquad \frac{\Delta Q}{\Delta t} \to \dot{Q}, \qquad \frac{\Delta W}{\Delta t} \to \dot{W},$$

$$\frac{\Delta m_1}{\Delta t} \to \dot{m}_1, \qquad \frac{\Delta m_2}{\Delta t} \to \dot{m}_2, \qquad \frac{dE}{\Delta t} \to \frac{dE}{dt}$$

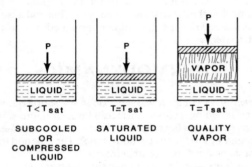

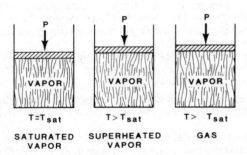

Fig. 1   Thermodynamic Fluid States

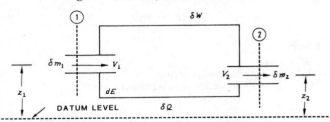

Fig. 2   Energy Flows in General Thermodynamic System

where

$\Delta Q/\Delta t = \dot{Q}$ = rate of heat transfer to system
$\Delta W/\Delta t = \dot{W}$ = rate at which work is performed by system, other than flow work pv
$\Delta m/\Delta t = \dot{m}$ = rate of mass flow crossing system boundary
$dE/\Delta t$ = rate of change of system energy
V = velocity of fluid streams crossing system boundary
g = local acceleration of gravity
z = elevation above horizontal reference plane
u = internal energy of fluid per unit mass
v = volume of fluid per unit mass

In the most general case that allows integration of the first law equation:

1. The properties of the fluids crossing the boundary remain constant at each point on the boundary.
2. The flow rate is constant at each section where mass crosses the boundary. (The flow rate cannot change as long as all properties, including velocity, remain constant at each point.)
3. All interactions with the surroundings occur at a steady rate.

This integration yields:

$$\sum m_{in} \left(u + pv + \frac{V^2}{2} + gz\right)_{in}$$
$$- \sum m_{out} \left(u + pv + \frac{V^2}{2} + gz\right)_{out} + Q - W$$
$$= \left[m_f \left(u + \frac{V^2}{2} + gz\right)_f - m_i \left(u + \frac{V^2}{2} + gz\right)_i\right]_{system} \quad (8)$$

The steady-flow process is important in engineering applications. Steady flow signifies that all quantities associated with the system do not vary with time. Consequently,

$$\sum_{\substack{all\ streams \\ leaving}} \dot{m}\left(h + \frac{V^2}{2} + gz\right)$$
$$- \sum_{\substack{all\ streams \\ entering}} \dot{m}\left(h + \frac{V^2}{2} + gz\right) + \dot{Q} - \dot{W} = 0 \quad (9)$$

where $h = u + pv$.

A second common application is the closed stationary system for which the first law equation reduces to:

$$Q - W = [m(u_f - u_i)]_{system} \quad (10)$$

SECOND LAW OF THERMODYNAMICS

In an open system, the second law of thermodynamics can be written in terms of entropy as:

$$dS_{system} = \left(\frac{\Delta Q}{T}\right) + \Delta m_i s_i - \Delta m_e s_e + dS_{irr} \quad (11)$$

where

dS_{system} = total change within system in time dt during process
$\Delta m_i s_i$ = entropy increase caused by mass entering
$\Delta m_e s_e$ = entropy decrease caused by mass leaving
$(\Delta Q/T)$ = entropy change caused by reversible heat transfer between system and surroundings
dS_{irr} = entropy created caused by irreversibilities

Equation (11) accounts for all entropy changes in the system. Rearranging, this equation becomes:

$$\Delta Q = T\left[(\Delta m_e s_e - \Delta m_i s_i) + dS_{sys} - dS_{irr}\right] \quad (12)$$

In integrated form, if inlet and outlet properties, mass flow rates, and interactions with the surroundings do not vary with time, the general equation for the second law is:

$$(S_f - S_i)_{system} = \int_{rev} \frac{\Delta Q}{T} + \sum (ms)_{in}$$
$$- \sum (ms)_{out} + \Delta S_{produced} \quad (13)$$

Available energy is energy in the form of shaft work or in a form completely convertible to shaft work by ideal processes. Energy that is in part convertible and in part nonconvertible is made up of an available part (the availability of energy) and an unavailable part (the unavailability of energy).

In thermodynamics, availability of energy is restricted to availability of energy for performing work (mechanical or electrical). The second law of thermodynamics limits transformation of heat into work, because only a portion of the heat available for heating can perform work. Therefore, the availability of energy is less than the quantity of heat under consideration.

Thermodynamic availability is useful for evaluating system performance and showing direction for improvement. The availability depends not only on system properties and interactions, but on conditions of the appropriate, naturally occurring thermal sink or source, denoted by the subscript o.

To determine availability, a reference condition that has no available energy must be specified. This reference condition is selected when the system is (1) at rest (zero velocity); (2) at the reference elevation (zero potential energy); (3) at the same pressure p_o and temperature T_o as the atmosphere; and (4) at chemical equilibrium (for reactive systems).

Irreversibility or *available energy degraded* is the decrease in the available energy due to irreversibilities and is equal to the reversible work minus the actual work for a process.

Reversible work refers to the maximum useful work obtained for a given change of state. It includes heat supplied from other systems, but excludes work done on the surroundings.

For the steady-flow system considered above, the availability formulation of the second law of thermodynamics is:

$$\dot{A}_Q + \dot{m}(a_f)_1 = \dot{m}(a_f)_2 + \dot{A}_W + \dot{I} \quad (14)$$

where

$\dot{A}_Q$ = rate of availability of heat transfer to system
$\dot{A}_W$ = rate of availability of work transfer from system
a_f = availability per unit flowing mass (flow availability)
$\dot{I}$ = irreversibility rate (always greater than zero)

Each of the terms $\dot{A}_Q$, $\dot{A}_W$, and a_f is evaluated by considering the maximum work achievable from the associated energy form, while $\dot{I}$ must be evaluated from Equation (14).

The availability of a heat transfer rate $\dot{Q}$ that occurs at temperature T is:

$$\dot{A}_Q = (1 - T_o/T)\,\dot{Q} \quad (15)$$

Often, $\dot{A}_Q$ is denoted by E_A and its converse, the unavailable energy, by E_U:

$$E_A = Q(1 - T_o/T) \quad \text{and} \quad E_U = Q - E_A = Q(T_o/T)$$

The availability of a steady work transfer rate $\dot{W}$ equals the work rate.

Flow availability a_f is the maximum work that can be obtained by allowing a unit mass of the flowing fluid to come into equilibrium with surroundings:

$$a_f = h + gZ + \frac{V^2}{2} - T_o s - \left(h_o + gZ_o + \frac{V_o{}^2}{2} - T_o s \right) \quad (16)$$

FIRST AND SECOND LAW ANALYSES OF REFRIGERATION CYCLES

Refrigeration cycles continually transfer thermal energy from a region of low temperature to one of higher temperature. The higher temperature heat sink is usually provided by ambient air or cooling water. This temperature, designated T_o, is the temperature of the surroundings.

The energy balance of the first law allows analysis and design of individual system components. The first law performance parameter for refrigeration systems is the coefficient of performance (COP), defined as:

$$\text{COP} = \frac{\text{Refrigerant Effect}}{\text{Net Work Input}} = \frac{|Q_2|}{|Q_1| - |Q_2|} \quad (17)$$

where $|Q_1|$ is the absolute value of heat transferred to the higher temperature sink, and $|Q_2|$ is the absolute value of heat transferred from the lower temperature source.

A second law analysis can demonstrate that (1) the reversed Carnot cycle is an ideal standard and the criterion for minimum requirement of work input to a compression refrigeration cycle; (2) departures from this ideal standard, such as fluid friction or heat transfer through a finite temperature difference, degrade available energy; and (3) such departures require that work input to the refrigeration cycle be greater than the ideal minimum. The increased work input equals the irreversibility evaluated from Equation (14). This represents total available energy, degraded by cyclic operation of the refrigerant and by heat transfer through finite temperature differences. An isolated system that includes the source, sink, and refrigeration unit performs as follows:

$$I = T_o(\Delta S_{sink} + \Delta S_{source}) = T_o(\Delta S_{total}) \quad (18)$$

EQUATIONS OF STATE

The equation of state of a pure substance is a mathematical relation between pressure, specific volume, and temperature, when the system is in thermodynamic equilibrium:

$$f(p,v,T) = 0 \quad (19)$$

The principles of statistical mechanics are used to (1) explore the fundamental properties of matter; (2) predict an equation of state based on the statistical nature of a particulate system; or (3) propose a functional form for an equation of state with unknown parameters that are determined by measuring thermodynamic properties of a substance. A fundamental equation with this basis is the *virial equation*.

The virial equation is expressed as an expansion in pressure p or in reciprocal values of volume per unit mass v:

$$\frac{pv}{RT} = 1 + B'p + C'p^2 + D'p^3 + \ldots \quad (20)$$

$$\frac{pv}{RT} = 1 + (B/v) + (C/v^2) + (D/v^3) + \ldots \quad (21)$$

where coefficients B', C', D', etc., and B, C, D, etc., are the virial coefficients. B' and B are second virial coefficients; C' and C are third virial coefficients, etc. The virial coefficients are functions of temperature only, and values of the respective coefficients in Equations (20) and (21) are related. For example, $B' = B/RT$ and $C' = (C - B^2)/(RT)^2$.

The quantity R is the ideal gas constant defined as:

$$R = \lim_{P \to 0} (pv)_T / T_{tp} \quad (22)$$

where $(pv)_T$ is the product of pressure and volume along an isotherm, and T_{tp} is the defined temperature of the triple point of water $T_{tp} = 491.69\,°\text{R}$. The current best value of R is 1545.32 ft·lb$_f$/lb mole·°R.

The quantity pv/RT is also referred to as the compressibility factor $Z = pv/RT$ or:

$$Z = 1 + (B/v) + (C/v^2) + (D/v^3) + \ldots \quad (23)$$

An advantage of the virial form is that statistical mechanics can be used to predict the lower order coefficients and provide physical significance to the virial coefficients. For example, in Equation (23), the term B/v is a function of interactions between two molecules, C/v^2 between three molecules, etc. Since the lower order interactions are common, the contributions of the higher order terms are successively less. Thermodynamicists use the partition or distribution function to determine virial coefficients; however, experimental values of the second and third coefficients are preferred. For dense fluid, many higher order terms are necessary that can neither be satisfactorily predicted from theory nor determined from experimental measurements. In general, a truncated virial expansion of four terms is valid for densities of less than one-half the value at the critical point. For higher densities, additional terms can be used and determined empirically.

Digital computers allow use of far more complex equations of state in calculating p-v-T values, even to high densities. The Benedict-Webb-Rubin (B-W-R) equation of state (Benedict *et al.* 1940) and the Martin-Hou equation (1955) have had considerable use, but should generally be limited to densities less than the critical value. Strobridge (1962) suggested a modified Benedict-Webb-Rubin relation that gives excellent results at higher densities and can be used for a p-v-T surface that extends into the liquid phase.

The B-W-R equation has been used extensively for hydrocarbons (Cooper and Goldfrank 1967). It is:

$$P = (RT/v) + (B_o RT - A_o - C_o/T^2)/v^2 + (bRT - a)/v^3$$
$$+ a\alpha/v^6 + [c(1 + \gamma/v^2)e^{(-\gamma/v^2)}]/v^3 T^2 \quad (24)$$

where the constant coefficients are A_o, B_o, C_o, a, b, c, α, γ.

The Martin-Hou equation, developed for fluorinated hydrocarbon properties, has been used to calculate the thermodynamic property tables in Chapter 17 and in ASHRAE *Thermodynamic Properties of Refrigerants* (Stewart *et al.* 1986). The Martin-Hou equation is as follows:

$$p = \frac{RT}{v - b} + \frac{A_2 + B_2 T + C_2 e^{(-kT/T_c)}}{(v - b)^2}$$
$$+ \frac{A_3 + B_3 T + C_3 e^{(-kT/T_c)}}{(v - b)^3} + \frac{A_4 + B_4 T}{(v - b)^4}$$
$$+ \frac{A_5 + B_5 T + C_5 e^{(-kT/T_c)}}{(v - b)^5} + (A_6 + B_6 T)e^{av} \quad (25)$$

where the constant coefficients are A_i, B_i, C_i, k, b, and α.

Strobridge (1962) suggested an equation of state that was developed for nitrogen properties and used for most cryogenic fluids. This equation combines the B-W-R equation of state with an equation for high density nitrogen suggested by Benedict (1937). These equations have been used successfully for liquid and vapor phases, extending in the liquid phase to the triple-point temperature and the freezing line, and in the vapor phase from 18 to 1800°R, with pressures to 150,000 psi. The equation suggested by Strobridge is accurate within the uncertainty of the measured p-v-T data. This equation, as originally reported by Strobridge, is:

$$p = RT\rho + \left[Rn_1 T + n_2 + \frac{n_3}{T} + \frac{n_4}{T^2} + \frac{n_5}{T^4} \right] \rho^2$$
$$+ (Rn_6 T + n_7)\rho^3 + n_8 T\rho^4$$
$$+ \rho^3 \left[\frac{n_9}{T^2} + \frac{n_{10}}{T^3} + \frac{n_{11}}{T^4} \right] \exp(-n_{16}\rho^2)$$
$$+ \rho^5 \left[\frac{n_{12}}{T^2} + \frac{n_{13}}{T^3} + \frac{n_{14}}{T^4} \right] \exp(-n_{16}\rho^2) + n_{15}\rho^6 \quad (26)$$

The 15 coefficients of this equation's linear terms are determined by a least-square fit to experimental data. Hust and Stewart (1966) and Hust and McCarty (1967) give further information on methods and techniques for determining equations of state.

In the absence of experimental data, Van der Waals' principle of corresponding states can predict fluid properties. This principle relates properties of similar substances by suitable reducing factors, *i.e.*, the p-v-T surfaces of similar fluids in a given region are assumed to be of similar shape. The critical point can be used to define reducing parameters to scale the surface of one fluid to the dimensions of another. Modifications of this principle, as suggested by Kamerlingh Onnes, a Dutch cryogenic researcher, have been used to improve correspondence at low pressures. The principle of corresponding states provides useful approximations, and numerous modifications have been reported. More complex treatments for predicting property values, which recognize similarity of fluid properties, are by generalized equations of state. These equations ordinarily allow for adjustment of the p-v-T surface by introduction of parameters. One example (Hirschfelder *et al.* 1958) allows for departures from the principle of corresponding states by adding two correlating parameters.

CALCULATING THERMODYNAMIC PROPERTY TABLES

The primary data used for these calculations are (1) p-v-T values over the entire range of application; (2) heat capacity values (c_p or c_v) along one isobar or isometric for the required temperature range; (3) vapor pressure values, where the application range includes phase changes; and (4) critical point and triple point data, where appropriate. Other measured values of thermodynamic properties can supplement these primary data or confirm the accuracy of the resulting property tables. These supplementary data include (1) heat capacities for additional isobars or isometrics, and of the saturated liquid; (2) Joule-Thomson data; and (3)

velocity of sound measurements. This discussion is limited to calculating property tables from primary data.

Following are methods for calculation of thermodynamic property tables for the validity range of the equation of state. Many tables for the fluorinated hydrocarbons use Equation (25), which is not valid for the liquid phase. The following calculations limit application of Equation (25) to the gaseous phase. However, Equation (31) can be used for calculating saturated liquid properties, provided supplementary information is available for values of saturated volume v_f. The equations of state used for most cryogenic fluids are valid for gaseous and liquid phases; therefore, the following equations can be applied for both.

Calculating Entropy

Differences in entropy can be calculated along isotherms using p-v-T data and the Maxwell relations:

$$(\partial s/\partial p)_T = -(\partial v/\partial T)_p$$

or

$$_T\!\!\int_{s_1}^{s_2} ds = -_T\!\!\int_{p_1}^{p_2} (\partial v/\partial T)_p \, dp \quad (27)$$

and

$$(\partial s/\partial v)_T = (\partial p/\partial T)_v$$

or

$$_T\!\!\int_{s_1}^{s_2} ds = _T\!\!\int_{v_1}^{v_2} (\partial p/\partial T)_v \, dv \quad (28)$$

Differences in entropy can be calculated from heat capacity data. By definition,

$$c_p \equiv (\partial h/\partial T)_p \quad \text{and} \quad c_v \equiv (\partial u/\partial T)_v$$

However, $dh = Tds + vdp$, and at constant pressure $dp = 0$. Therefore:

$$c_p = T(\partial s/\partial T)_p \quad \text{and} \quad _p\!\!\int_{s_1}^{s_2} ds = _p\!\!\int_{T_1}^{T_2} (c_p/T) dT \quad (29)$$

Similarly, with:

$$c_v \equiv (\partial u/\partial T) \text{ and } du = Tds - pdv, \, c_v = T(\partial s/\partial T)_v \text{ and}$$
$$_v\!\!\int_{s_1}^{s_2} ds = _v\!\!\int_{T_1}^{T_2} (c_v/T) dT \quad (30)$$

Therefore, values of entropy along isobars or isometrics can be obtained from c_p or c_v data.

Changes in entropy during a phase change can be calculated from the vapor pressure data, and the p-v-T data at the phase boundaries using the Clapeyron equation:

$$(dp/dT) = (s_g - s_f)/(v_g - v_f) \quad (31)$$

where subscript g refers to the saturated vapor phase, subscript f refers to the saturated liquid phase, and dp/dT is the slope of the vapor pressure curve. This relation is valid for all first order changes.

The five relations given by Equations (27), (28), (29), (30), and (31) are functions for calculating an entropy table. Choice of function depends on the nature of the tabulation, p-v-T representation (equation of state), and availability of c_p or c_v data. The choice between $ds_{T=C} = -(\partial v/\partial T)_p dp$ or $ds_{T=C} = (\partial p/\partial T)_v dv$ is determined by the relative simplicity of the derivatives $(\partial v/\partial T)_p$ or $(\partial p/\partial T)_v$; the desire to calculate entropy values along isobars or isometrics; and the choice of either c_p or c_v data.

For example, if the heat capacity data to be used are c_p values along an isobar, then entropy differences are usually calculated along isotherms using the isobar for which c_p is known and the relation $ds_{T=C} = -(\partial v/\partial T)_p dp$ (see Figure 3).

Consider calculation of $S_2 - S_1$, where available information includes (1) an equation of state for the entire region of vapor and liquid of the form $p = p(T, v)$; (2) a vapor pressure function for values from the triple point to the critical point; and (3) c_p data along an isobar of standard pressure for this temperature range, represented by a function $c_p = c_p(T)$. The path for integration between 1 and 2 in Figure 3 is as noted. The entropy difference can be calculated as:

$$s_2 - s_1 = (s_a - s_1) + (s_b - s_a)$$
$$+ (s_c - s_b) + (s_d - s_c) + (s_2 - s_d) \quad (32)$$

$$s_2 - s_1 = {}_{T_1}\!\!\int_{p_1}^{p_a} - (\partial v/T)_p \, dp$$

$$+ {}_{p_a}\!\!\int_{T_a}^{T_b} (c_p/T) dT$$

$$+ {}_{T_2}\!\!\int_{p_b}^{p_c} - (\partial v/\partial T)_p \, dp$$

$$+ (dp/dT)(v_d - v_c)$$

$$+ {}_{T_2 p_d}\!\!\int^{p_2} - (\partial v/\partial T)_p (dp) \quad (33)$$

where the relation $(\partial v/\partial T)_p$ as a function of p along an isotherm is derived from the equation of state, and dp/dT is the slope of the vapor pressure curve at T_2. Specified states are identified by p and T, and heat capacity is known along an isobar. Although the equation of state is of form $p = p(T, v)$ and the derivative $(\partial p/\partial T)_v$ is simpler to determine, the relation is more simply expressed in terms of $(\partial v/\partial T)_p$.

Calculating Internal Energy and Enthalpy

To calculate internal energy and enthalpy, the techniques for calculating entropy are combined with the relations:

$$ds = (du/T) + (p/T)dv \quad (34)$$

$$ds = (dh/T) - (v/T)dp \quad (35)$$

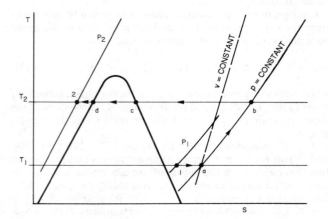

Fig. 3   Path of Integration for Entropy Calculation

These equations must be used with functions defining heat capacities $c_p \equiv (\partial h/\partial T)_p$ and $c_v \equiv (\partial u/\partial T)_v$, and the Clapeyron equation:

$$(dp/dT) = \frac{(h_g - h_f)}{T(v_g - v_f)}$$

For example, from Equation (34), $du = Tds - pdv$. Along an isotherm,

$${}_T\!\!\int_{s_1}^{s_2} ds$$

can be evaluated from Equation (27) or (28), with $T =$ constant. The remaining problem for the isothermal variation of u is calculating

$${}_T\!\!\int_{v_1}^{v_2} pdv$$

along the isotherm. If the equation of state is of form $p = p(T,v)$, this calculation is usually simple. This suggests that the equation for internal energy changes along an isotherm that has limits of integration most consistent with the paths of the parameters of the problem is:

$$(u_2 - u_1)_{T=C} = {}_T\!\!\int_{v_1}^{v_2} [T(\partial p/\partial T)_v - p]dv \quad (36)$$

Enthalpy can then be calculated from $h = u + pv$, i.e:

$$(h_2 - h_1)_{T=C} = {}_T\!\!\int_{v_1}^{v_2} [T(\partial p/\partial T)_v - p]dv + p_2 v_2 - p_1 v_1 \quad (37)$$

An alternate method for calculating internal energy and enthalpy differences along an isotherm is from Equation (35). Therefore, $dh = Tds + vdp$, and:

$$(h_2 - h_1)_{T=C} = {}_T\!\!\int_{p_1}^{p_2} [-T(\partial v/\partial T)_p + v]dp \quad (38)$$

and

$$(u_2 - u_1)_{T=C} = {}_T\!\!\int_{p_1}^{p_2} [-T(\partial v/\partial T)_P + v]dp - p_2 v_2 + p_1 v_1 \quad (39)$$

Equations (36) and (39) can be extended for internal energy and enthalpy changes along any combination of isothermal, isobaric, and isometric paths by combining with relations for the heat capacities. For example, from Equation (36), combining with the change in internal energy along a constant volume path:

$$du = c_v dT + [T(\partial p/\partial T)_v - p]dv \quad (40)$$

and from Equation (38), combining with the change of enthalpy along a constant pressure path:

$$dh = c_p dT + [v - T(\partial v/\partial T)_P]dp \quad (41)$$

Thermodynamic Property Calculations Based on p-v-T Relations and Zero-Pressure Specific Heats

Inaccurate heat data can cause significant errors in thermodynamic properties calculated as described previously. These errors are reduced by using zero-pressure specific heats calculated by methods of statistical mechanics from spectroscopic data, since ideal gas properties are more accurately known than real gas properties determined by calorimetric methods.

To use the zero-pressure specific heats, extend the isothermal path of integration to $p = 0$, as shown in Figure 3, *i.e.*, so that $a - b$ lies on the zero-pressure isobar. (The value of pv from the real gas equation of state must approach RT as p approaches zero.) Consequently, a reference state on the ideal gas surface is the most convenient choice. Since a reference point at zero pressure would result in infinite entropies at any finite pressure on the real or ideal gas surface, the standard reference state is usually chosen at 14.696 psi and T_o on the ideal gas surface. This is equivalent to choosing the standard reference values of enthalpy and internal energy at zero pressure and T_o, since h_o and u_o are functions of temperature alone for the ideal gas.

In the following discussion, superscript o denotes an ideal gas property, superscript * denotes a property (real or ideal) at zero pressure, and subscript o denotes a reference state property.

Figure 4 illustrates the path of integration for the following derivation. Paths $o - a$ and $b - c$ are defined by Equations (28) and (37). Path $a - b$ is defined by Equation (29) and the following relation:

$$\int_p\int_{h_1}^{h_2} dh = \int_p\int_{T_1}^{T_2} c_p dT$$

Equations (42) and (43) are the result of these integration paths.

$$s_c = s_{T_o^o} + \int_{T_o}\int_{v_o}^{v_a^*} (\partial p/\partial T)_v dv$$
$$+ \int_T\int_{v_b^*}^{v_c} (\partial p/\partial T)_v dv + \int_{p^*}\int_{T_o}^{T_c} (c_p^*/T) dT \qquad (42)$$

$$h_c = h_{T_o^o} + \int_{T_o}\int_{v_o^*}^{v_a} [T(\partial P/\partial T)_v - p] dv$$
$$+ \int_{T_c}\int_{v_b^*}^{v_c} [T(\partial p/\partial T)_v - p] dv$$
$$+ \int_{p^*}\int_{T_o}^{T_c} c_p^o dT + p^* v_a^* - p_o v_o + p_c v_c - p^* v_b^* \qquad (43)$$

The limit must be introduced as p^* approaches zero, since only then do the real and ideal gas surfaces coincide. For the ideal gas, $(\partial p/\partial T)_v$ is independent of temperature [$(\partial p/\partial T)_v = R/v$], and the first integral in Equation (42) can be evaluated at any temperature T:

$$\int_{T_o}\int_{v_o}^{v_a^*} (\partial p/\partial T)_v dv = \int_T\int_{v_o}^{v_a^*} (\partial p/\partial T)_v dv$$

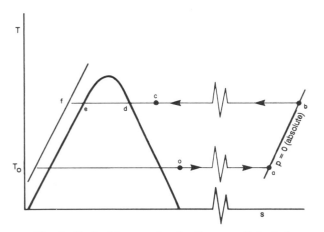

This integral is then replaced by three integrals:

$$\int_{T_o}\int_{v_o}^{v_a^*} (\partial p/\partial T)_v dv = \int_T\int_{v_o}^{v_c^*} (\partial p/\partial T)_v dv$$
$$+ \int_T\int_{v_c}^{v_b^*} (\partial p/\partial T)_v dv + \int_T\int_{v_b^*}^{v_a^*} (\partial p/\partial T)_v dv$$

and is finally evaluated, using $v_b^* = RT_c/p^*$ and $v_a^* = RT_o/p^*$, to obtain:

$$\int_{T_o}\int_{v_o}^{v_a^*} (\partial p/\partial T)_v dv = R \left(\ln \frac{v_c T_o}{v_o T_c} \right) - \int_{T_c}\int_{v_b^*}^{v_c} (R/v) dv$$

Substitution into Equation (42) yields in the limit:

$$s = s_{T_o^o} + R \left(\ln \frac{p_o v_c}{RT_c} \right) + \int_{T_c}\int_{v_c}^{\infty} [(R/v) - (\partial p/\partial T)_v] dv$$
$$+ \int_{p^*}\int_{T_o}^{T_c} (c_p^o/T) dT \qquad (44)$$

A similar simplification of Equation (43) yields:

$$h_c = (h_{T_o})^0 + \int_{T_c}\int_{v_c}^{\infty} [p - T(\partial p/\partial T)_v] dv$$
$$+ (p_c v_c - RT_c) + \int_{p^*}\int_{T_o}^{T_c} c_p^o dT \qquad (45)$$

Equations (46) and (47) are obtained by substituting $v = 1/\rho$ and $dv = - (d\rho/\rho^2)$ into Equations (44) and (45), respectively:

$$s = (s_{T_o})^0 - R \left(\ln \frac{\rho_c RT_c}{p_o} \right) + \int_{T_c}\int_{\rho=0}^{\rho_c} [(R/\rho)$$
$$- (1/\rho^2)(\partial p/\partial T)_\rho] d\rho + \int_{p^*}\int_{T_o}^{T_c} (c_p^o/T) dT \qquad (46)$$

$$h = (h_{T_o})^0 + \int_{T_c}\int_{\rho=0}^{\rho_c} [(p/\rho^2) - (T/\rho^2)(\partial p/\partial T)\rho] d\rho$$
$$+ (p_c - \rho_c RT)/\rho_c + \int_{p^*}\int_{T_o}^{T_c} c_p^o dT \qquad (47)$$

The low-pressure (large volume) behavior of the real gas equation of state must be such that $pv \rightarrow RT$, *i.e.*, $(\partial P/\partial T)_v$ for the real gas must approach R/v to cancel the existing R/v term in Equation (44). If this is not the case, the remaining R/v term will integrate to $\ln(v)$, which is infinite at the upper limit.

Derivations similar to those above can be obtained from equations in which p and T are independent variables:

$$s = (s_{T_o})^0 + R \left(\ln \frac{p_o}{p} \right) + \int_T\int_{p=0}^{p} [(R/p) - (\partial v/\partial T)_p] dp$$
$$+ \int_{T_o}^{T} (c_p^o/T) dT \qquad (48)$$

$$h = (h_{T_o})^0 + \int_T\int_{p=0}^{p} [v - T(\partial v/\partial T)_p] dp + \int_{T_o}^{T} c_p^o dT \qquad (49)$$

COMPRESSION REFRIGERATION CYCLES

Thermodynamic analysis can generally be applied to any vapor compression system. This section gives examples of thermo-

Fig. 4   Path of Integration for Property Calculation

dynamic analyses for the basic vapor compression system and for a two-stage vapor compression system with subcooling and intercooling. Thermodynamic considerations for a two-stage vapor compressor are introduced.

The discussion emphasizes second law analysis of the cycles, using the concepts introduced in the preceding section. Although pressure-enthalpy (*p-h*) coordinates are used most often to illustrate refrigeration cycles, temperature-entropy (*T-S*) coordinates better express second law concepts and are used in the following examples. Tripp (1966) and Tripp and Sun (1965) give additional examples of *T-S* coordinates for examining refrigeration cycles and compressors. Tripp and Sun (1965) introduce pressure-volume coordinates for thermodynamic analysis.

IDEAL BASIC VAPOR COMPRESSION REFRIGERATION CYCLE

Figure 5 illustrates the equipment diagram for the basic vapor compression cycle. Figures 6 and 7 show the *p-h* and *T-S* diagrams. Minimum components of this cycle include compressor, condenser, expansion valve, and evaporator. The ideal cycle considers heat transfer in the condenser and evaporator without pressure

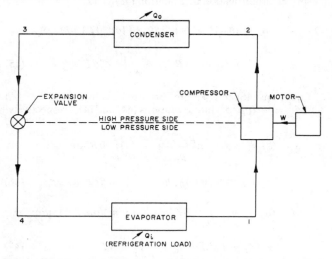

Fig. 5   Equipment Diagram for Basic Vapor Compression Cycle

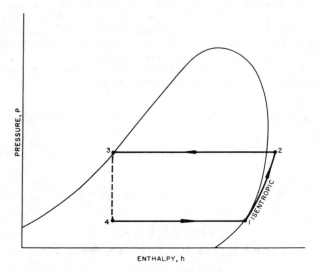

**Fig. 6   Pressure-Enthalpy Diagram for Basic
Vapor Compression Cycle**

losses, a reversible adiabatic (isentropic) compressor, and an adiabatic expansion valve, connected by piping that has neither pressure loss nor heat transfer with the surroundings. The refrigerant leaves the evaporator at point 1 as a low-pressure, low-temperature, saturated vapor and enters the compressor, where it is compressed reversibly and adiabatically (isentropic). At point 2, it leaves the compressor as a high-temperature, high-pressure, superheated vapor and enters the condenser, where it is first desuperheated and then condensed at constant pressure. At point 3, the refrigerant leaves the condenser as a high-pressure, medium-temperature, saturated liquid and enters the expansion valve, where it expands irreversibly and adiabatically (constant enthalpy). At point 4, it leaves the expansion valve as a low-pressure, low-temperature, low quality vapor and enters the evaporator, where it is evaporated reversibly at constant pressure to the saturated state at point 1. Heat transfer to the evaporator and from the condenser occurs without a finite temperature difference between the fluid emitting the heat and the fluid that absorbs the heat, except during the desuperheating process in the condenser.

Energy Balance and First Law Analysis

An energy balance and certain performance parameters can be derived from the first law of thermodynamics. Applying the steady-flow equation for the first law [Equation (9)] to each of the components of the basic vapor compression cycle, the following relationships are derived:

1-2 Compression $\quad _1\dot{W}_2 = -(h_2 - h_1)\dot{m}$ (50)

(This quantity is negative, indicating work is done by the compressor on the refrigerant.)

2-3 Condensing $\quad _2\dot{Q}_3 = -(h_2 - h_3)\dot{m}$ (51)

3-4 Expansion valve $\quad h_3 = h_4$

4-1 Evaporator $\quad _4\dot{Q}_1 = (h_1 - h_4)\dot{m}$ (52)

In applying the steady-flow equation, kinetic energy and potential energy terms were omitted; because flow velocities are low to avoid fluid friction and undesirable pressure losses, and height variation within a given refrigeration system is usually small, these terms are numerically insignificant. Since the system is cyclic, the

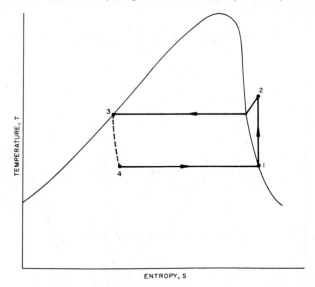

**Fig. 7   Temperature-Entropy Diagram for Basic
Vapor Compression Cycle**

heat rejected in the condenser must equal the sum of the heat absorbed in the evaporator and the work of compression.

Coefficient of performance (COP) indicates the performance of a refrigeration system. The COP was introduced as Equation (17), *i.e.*, COP = refrigeration effect/net work input.

For the basic vapor compression cycle, from Equations (50) and (52), the COP is:

$$COP = (h_1 - h_4)/(h_2 - h_1)$$

Thermodynamic Analysis of the Compressor

The change in state between the inlet and outlet is (1) reversible and adiabatic (isentropic) for the ideal compressor, or (2) adiabatic and irreversible (with an increase in entropy in the fluid passing through the compressor). The variation from the ideal compressor is then described by the adiabatic compressor efficiency.

The effect of the clearance volume, *i.e.*, the volume the refrigerant occupies within the compressor that is not displaced by the moving member, is an important thermodynamic consideration for the positive displacement compressor. For the piston compressor, the gas in clearance volume between piston and cylinder head when the piston is in a top, center position reexpands to a larger volume as the cylinder discharges and the pressure falls to the inlet pressure (see Figure 8). Consequently, the compressor discharges a refrigerant mass less than the mass that would occupy the volume swept by the piston, measured at the inlet pressure and temperature. This effect is quantitatively expressed by the volumetric efficiency η_v:

$$\eta_v = m_a/m_t \qquad (53)$$

where

m_a = actual mass of new gas entering compressor per stroke
m_t = theoretical mass of gas represented by displacement volume and determined at pressure and temperature at compressor inlet

For the basic vapor compression cycle, volumetric efficiency becomes:

$$\eta_v = \frac{V_a - V_d}{V_a - V_c} = 1 + C - C\frac{v_1}{v_2} \qquad (54)$$

where

$C = V_c/(V_a - V_c)$ = clearance ratio
v_1 = specific volume of refrigerant at beginning of compression
v_2 = specific volume at end of compression

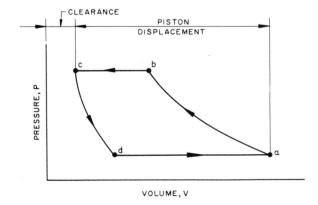

VOLUME, V

Fig. 8   Machine Cycle for Idealized Piston Compressor

Table 1   Effect of Interstage Pressure

Interstage Pressure, psia	Work Done per Unit of Cooling, Btu/min·ton		
	L.P. Compressor	H.P. Compressor	Total
18	0.0	79.0	79.0
28	15.0	61.6	76.6
38	25.7	49.3	75.0
48	34.1	39.7	73.8
58	41.0	31.8	72.8
68	46.9	25.1	72.0
78	52.1	19.2	71.3
88	56.6	14.0	70.6
98	60.7	9.4	70.1
115	66.8	2.4	69.2

Volumetric efficiency measures the effectiveness of the compressor's piston displacement (size) in moving the refrigerant vapor through the cycle. Since refrigerants differ in their specific volumes v_1, choice of refrigerant can affect the mass flow delivered by compressor displacement.

One of the design parameters of a multistage compressor is selection of the interstage pressure at which the refrigerant temperature is reduced by an intercooler. At optimum interstage pressure, total work is minimum. For two-stage compression of an ideal gas ($pv = RT$), this occurs at the geometric mean of the suction and discharge pressures and results in equal work for the stages. The application of multistage compressors to refrigeration systems, however, differs from gas compressors, since the interstage pressure is usually cooled by refrigerant diverted from some other part of the cycle. Table 1 lists the results of an analysis for a two-stage R-12 cycle (Qvale 1971a). The refrigerant at the low-pressure compressor inlet is at $-3\,°F$ and 18 psia, and the high-pressure discharge is at $117.90\,°F$ and 120 psia. Adiabatic efficiency of the compressor is assumed to be 80%, pressure loss in the desuperheater is 1.5 psia, and the refrigerant leaves the desuperheater at $5\,°F$ of superheat. (Also see Figure 19 and Tables 2 and 4.)

In most refrigeration multistage cycles, minimum work is obtained at an interstage pressure other than the geometric mean pressure. In the case considered in Table 1, however, the condition of minimum work is not a valid criterion for selecting the interstage pressure. In such cases, the geometric mean pressure is suggested for the interstage pressure as it yields approximately equal volumetric efficiencies for the two stages. Therefore, it is important to evaluate each specific cycle.

Other thermodynamic studies of compressor performance are detailed by Qvale (1971b) and Haseltine and Qvale (1971).

SECOND LAW THERMODYNAMIC ANALYSIS

In analyzing compression refrigeration cycles, the concepts of the reversible process and available energy are basic to the application of the second law of thermodynamics. From these basic concepts, the performance of a process or system can be expressed in terms of degradation (or loss) of available energy (irreversibility). The following examples (Tripp 1966) apply these concepts to compression refrigeration systems, relate the principles of irreversibility $I = T_o \Delta S_{total}$ and the degradation of available energy $(\Sigma E_A)_d$, and emphasize use of the *T-S* coordinates to illustrate the effects of losses in the refrigeration system.

REVERSED CARNOT CYCLE

For given temperatures of the refrigerated space and the atmosphere (usually ambient air or economically available water), a thermodynamically reversible cycle, such as the reversed Carnot cycle, has the highest possible coefficient of performance.

As shown on the *T-S* diagram of Figure 9, this cycle has four reversible processes: (1) isothermal expansion, from state 4 to state 1, during which heat (the refrigeration load) flows from refrigerated space to refrigerant; (2) adiabatic compression, from 1-2; (3) isothermal compression, from 2-3, in which heat flows from refrigerant to atmosphere; and (4) adiabatic expansion, from 3-4. Area (b) represents refrigeration load; area (a), net work input to the cycle; and area (a+b), heat rejection to the atmosphere. The ratio of areas (b/a) represents COP. Area (a+b) also represents unavailable heat flow both to and from the cycle, while the negative of area (a) represents the available portion of refrigeration load.

In Figure 9, the COP of the reversed Carnot cycle is a function of the absolute temperatures of refrigerated space T_R and atmosphere T_o.

$$
\begin{aligned}
\text{COP} &= \text{Refrigeration Load/Net Work Input} \\
&= Q_i/(Q_o - Q_i) = Q_i/W = T_R\,\Delta S/(T_o\Delta S - T_R\Delta S) \\
&= 1/[(T_o/T_R) - 1] \tag{55}
\end{aligned}
$$

In this cycle, departures from ideal negligible temperature differences during heat transfer and reversible processes require a greater work input to the refrigeration cycle than the ideal minimum. Such departures degrade available energy and increase entropy of the system and its surroundings ΔS_{total}. Increased net work input equals the sum of all available energies degraded and also equals the product of absolute temperature of the atmosphere and increase in entropy $T_o\Delta S_{total}$. Ideal conditions are illustrated in Example 1, and departures from this ideal are shown in Examples 2 and 3.

Example 1. Determine entropy change, work, and coefficient of performance for the cycle shown in Figure 9. Temperature of the refrigerated space T_R is 400°R and that of the atmosphere T_o is 500°R. Refrigeration load is 200 Btu.

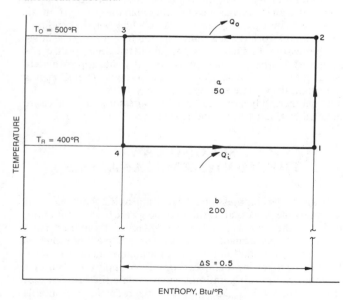

Fig. 9 Temperature-Entropy Diagram for Ideal Reversed Carnot Cycle of Example 1

Solution:

$$
\begin{aligned}
\Delta S &= S_1 - S_4 = Q_i/T_R = 200/400 = 0.500 \text{ Btu/°R} \\
W &= \Delta S(T_o - T_R) = 0.5\,(500 - 400) = 50 \text{ Btu} \\
\text{COP} &= Q_i/(Q_o - Q_i) = Q_i/W = 200/50 = 4
\end{aligned}
$$

Flow of energy, available energy E_A, unavailable energy E_U, and their area representations in Figure 9 are:

Energy	Btu	Area
Q_i	200	b
Q_o	250	a + b
W	50	a
$(E_A)_i$	−50	−a
$(E_U)_i$	250	a + b
$(E_A)_o$	0	—
$(E_U)_o$	250	a + b

In Example 1, the entropy change of the isolated system as a result of the refrigeration cycle is zero. The net change of entropy of any refrigerant in any cycle is always zero. The change in entropy of the refrigerated space in Example 1 is $\Delta S_R = 200/400 = 0.5$ Btu/°R, and that of the atmosphere is $\Delta S_o = -250/500 = -0.5$ Btu/°R, giving a change in entropy of the isolated system of $\Delta S_{total} = \Delta S_R + \Delta S_o = 0$.

In Example 1, there is no degradation of available energy, as all processes in the cycle are assumed reversible and all heat flows occur with negligible temperature difference.

In actual cycles, heat transfer occurs with finite temperature differences, and all processes of the refrigerant involve friction. These processes are irreversible and cause degradation of available energy, an increase in entropy of the isolated system and an augmented work input to the cycle.

These concepts are expressed as:

$$ \Delta W = W - W_{id} \tag{56} $$

$$ W = Q_o - Q_i \tag{57} $$

$$ W_{id} = (T_o - T_R)(-\Delta S_R) \tag{58} $$

$$ Q_o = T_o\Delta S_o \tag{59} $$

$$ Q_i = T_R(-\Delta S_R) \tag{60} $$

Therefore:

$$
\begin{aligned}
\Delta W &= T_o\Delta S_o - T_R(-\Delta S_R) - (T_o - T_R)(-\Delta S_R) \\
&= T_o(\Delta S_o + \Delta S_R) = T_o\Delta S_{total} \tag{61}
\end{aligned}
$$

Also:

$$ Q_o = (E_A)_o + (E_U)_o \tag{62} $$

$$ (E_U)_o = (E_U)_R + (\Sigma E_{Ad})_r + (E_{Ad})_i \tag{63} $$

$$ (E_U)_R = T_o(-\Delta S_R) \tag{64} $$

Therefore:

$$
\begin{aligned}
W &= (E_A)_o + T_o(-\Delta S_R) + (\Sigma E_{Ad})_r + (E_{Ad})_i \\
&\quad - T_R\,(-\Delta S_R) - (T_o - T_r)(-\Delta S_R) \\
&= (E_A)_o + (\Sigma E_{Ad})_r + (E_{Ad})_i = \Sigma E_{Ad} \tag{65}
\end{aligned}
$$

where

$(E_U)_R$ = unavailable portion of energy removed from refrigerated space at temperature T_R

$(\Sigma E_{Ad})_r$ = sum of all available energy degraded by refrigerant in cyclical operation (see Examples 2 and 3)

$(E_A)_o$ = available energy outflow from refrigerant during process of heat rejection; $(E_A)_o$ is rendered unavailable when absorbed by atmosphere (see Example 3)

$(E_{Ad})_i$ = available energy degraded by heat transfer from refrigerated space to refrigerant when process is accomplished through finite temperature difference (see Example 3)

ACTUAL BASIC VAPOR COMPRESSION REFRIGERATION CYCLE

Although the basic vapor compression refrigeration cycle shown in Figures 6 and 7 has been called ideal, the cycle, in interaction with the atmosphere, entails two irreversibilities: (1) the inherent irreversibility in the expansion valve process and (2) the process of heat transfer through a finite temperature difference during the desuperheating of the vapor in the condenser. To remove the first irreversibility, a reversible adiabatic expansion engine should be substituted for the expansion valve, and work output from this engine would supply part of the work input to the cycle. The process for this engine is shown as process 3 to 4′ in Figure 10. To remove the second irreversibility, the compression process 1 to 2 of Figure 10 should be replaced by a reversible two-part compression, the isentropic process 1 to 2′ and the isothermal process 2′ to 2″ of Figure 10. These improvements produce a reversed Carnot cycle equal to the one in Figure 9, except that, due to increased entropy change during heat influx to the refrigerant, the refrigeration load, net work input, and heat rejection would all be increased by 4.52%.

The complexity of the equipment needed for these improvements precludes its application, and a feasible arrangement is shown in Figure 5.

In the following demonstrations, Example 2 presents a second law analysis of an ideal vapor cycle, as depicted in Figure 7, and Example 3 analyzes a vapor cycle that represents actual irreversible conditions of friction and heat transfer found in practice.

Example 2. The data are those of Example 1. The refrigerant is R-12. All processes are reversible, except those through the expansion valve. Heat transfer is accomplished with negligible temperature differences, except for the desuperheating process in the condenser. Figure 10 shows the cycle.

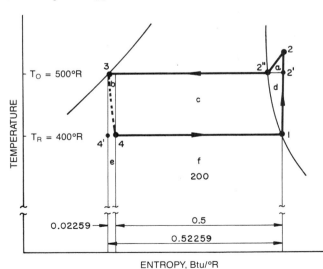

Fig. 10 Temperature-Entropy Diagram for Ideal Basic Vapor Cycle of Example 2

State-Point Properties

State	p, psia	t, °F	h, Btu/lb	s, Btu/lb·°R
1	5.409	−59.69	70.727	0.17708
2	51.94	77.31	87.295	0.17708
3	51.94	40.31	17.342	0.03759
4	5.409	−59.69	17.342	0.04362

$$M = 200/(70.727 - 17.342) = 3.7465 \text{ lb}$$
$$Q_o = 3.7465 (87.295 - 17.342) = 262.07 \text{ Btu}$$
$$W = 262.07 - 200 = 62.07 \text{ Btu}$$
$$\text{COP} = 200/62.07 = 3.222$$
$$\Delta W = 62.07 - 50 = 12.07 \text{ Btu}$$
$$\Delta S_R = -200/400 = -0.5 \text{ Btu/°R}$$
$$\Delta S_o = 262.07/500 = 0.5241 \text{ Btu/°R}$$
$$\Delta S_{total} = 0.5241 - 0.5 = 0.0241 \text{ Btu/°R}$$
$$T_o \Delta S_{total} = 500 (0.0241) = 12.07 \text{ Btu}$$
$$(E_U)_o = 500 (0.52259) = 261.29 \text{ Btu}$$
$$(E_A)_o = 262.07 - 261.29 = 0.78 \text{ Btu}$$
$$(E_{Ad})_r = 500 (0.02259) = 11.29 \text{ Btu}$$
$$\Sigma E_{Ad} = 11.29 + 0.78 = 12.07 \text{ Btu}$$

Energy	Btu	Area
Q_i	200	f
Q_o	262.07	(a→f)
W	62.07	(a→e)
$(E_A)_i$	−50	−(c + d)
$(E_U)_i$	250	c + d + f
$(E_A)_o$	0.78	a
$(E_U)_o$	261.29	(b→f)
$(E_{Ad})_r$	11.29	b + e

Example 3. In this case, deviations from the ideal vapor cycle of Example 2 are a combination of heat transfer through finite temperature differences, irreversible adiabatic compression, and pressure losses in the evaporator and condenser. The cooling is 200 Btu, and temperatures of the refrigerated space and the atmosphere are 400 and 500 °R, respectively. The cycle is shown in Figure 11, in which the constant pressure process from state 2 to 3, is the equivalent reversible process of heat emission from the vapor as it is desuperheated and condensed. Figure 12 shows an enlargement of part of Figure 11.

State-Point Properties

State	p, psia	t, °F	h, Btu/lb	s, Btu/lb·°R
1	4.0053	−69.59	69.625	0.17907
1′	3.0	−69.59	69.615	0.18373
2	64.0	150	98.427	0.19328
3	64.0	50.31	19.581	0.04196
3′	61.717	—	—	—
4	4.0053	−69.59	19.577	0.05080

$$\Delta S_{Qi} = 200/390 = 0.51282 \text{ Btu/°R}$$
$$M = 200/(69.615 - 19.577) = 3.997 \text{ lb}$$
$$Q_o = 3.997 (98.427 - 19.577) = 315.14 \text{ Btu}$$
$$W = 315.14 - 200 = 115.14 \text{ Btu}$$
$$\text{COP} = 200/115.14 = 1.737$$
$$\Delta W = 115.14 - 50 = 65.14 \text{ Btu}$$
$$\Delta S_R = -200/400 = -0.5 \text{ Btu/°R}$$
$$\Delta S_o = 315.14/500 = 0.6303 \text{ Btu/°R}$$
$$\Delta S_{total} = 0.6303 - 0.5 = 0.1303 \text{ Btu/°R}$$
$$T_o \Delta S_{total} = 500 \times 0.1303 = 65.14 \text{ Btu}$$
$$(E_U)_o = 500 (0.60499) = 302.49 \text{ Btu}$$
$$(E_A)_o = 315.14 - 302.50 = 12.64 \text{ Btu}$$
$$(E_U)_i = 500 \times 0.51282 = 256.41 \text{ Btu}$$
$$(E_A)_i = 200 - 256.41 = 56.41 \text{ Btu}$$
$$(E_{Ad})_i = -[-56.41 - (-50)] = 6.41 \text{ Btu}$$
$$(E_{Ad})_o = (E_A)_o = 12.64 \text{ Btu}$$
$$(E_{Ad})_r = 500 (0.03527 + 0.01864 + 0.03820 + 0.00005)$$
$$= 46.09 \text{ Btu}$$
$$\Sigma E_{Ad} = 6.41 + 12.64 + 46.09 = 65.14 \text{ Btu}$$

Energy	Btu	Area
Q_i	200	$q + t$
Q_o	315.14	$(a \rightarrow v)$
W	115.14	$(a \rightarrow v) - (q + t)$
$(E_A)_i$	−56.41	$-(1 + m)$
$(E_U)_i$	256.41	$1 + m + q + t$
$(E_A)_o$	12.64	$(a \rightarrow h)$
$(E_U)_o$	302.50	$(i \rightarrow v)$
$(E_{Ad})_r$	46.09	$i + (j + k) + (n + r + s + u) + (o + p + v)$

$$
\begin{aligned}
Q_i + W &= Q_o \\
200 + 115.14 &= 315.14 \\
Q_i &= (E_A)_i + (E_U)_i \\
200 &= -56.41 + 256.41 \\
Q_o &= (E_A)_o + (E_U)_o \\
315.14 &= 12.64 + 302.50 \\
W + (E_A)_i &= (E_A)_o + (E_{Ad})_r \\
115.14 - 56.41 &= 12.64 + 46.09 \\
58.73 &= 58.73 \\
(E_U)_i + (E_{Ad})r &= (E_U)_o \\
256.41 + 46.09 &= 302.50
\end{aligned}
$$

COMPLEX VAPOR COMPRESSION REFRIGERATION CYCLES

In large refrigeration systems, more equipment than needed for the basic vapor cycle shown in Figure 5 is sometimes advantageous. The design for a system can include many compressors, evaporators, expansion valves, and flash chambers. However, an optimum operational and economical system is the objective of any design.

A more complex design shown in Figure 13 includes two evaporators, two compressors, two expansion valves, and a single condenser. Evaporator II could have a higher vapor temperature (35 °F for the storage of fruits and vegetables) than Evaporator I (5 °F for frozen fish or poultry). Each evaporator could also carry a different load.

In the next example, second law analysis shows how introducing a flash chamber and doubling other equipment improves thermodynamic performance.

Example 4. In this cycle (see Figure 14), the data are those of Example 2. The equipment provides compound compression in a low-stage compressor (I) and a high-stage compressor (II), a flash intercooler, and two expansion valves. All processes of the refrigerant are reversible, except those in the flash intercooler and expansion valves.

When compared with the ideal basic vapor cycle of Example 2, the cycle in this example reduces (1) the irreversibilities caused by the refrigerant, (2) the available portion of heat rejected to the atmosphere, and (3) net work input to the cycle.

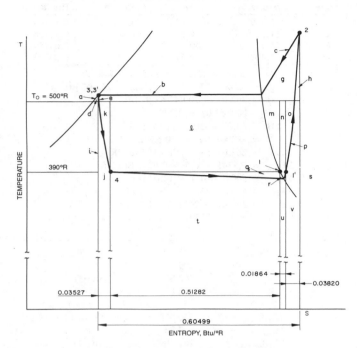

Fig. 11 Temperature-Entropy Diagram for Basic Vapor Cycle of Example 3

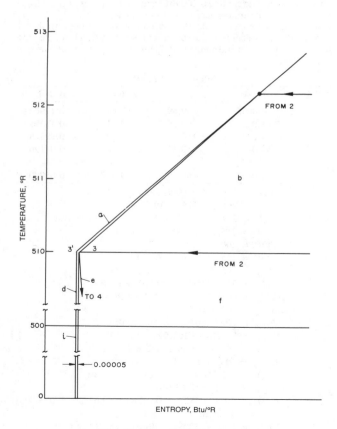

Fig. 12 Detail of Figure 11

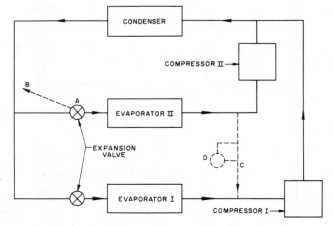

Fig. 13 Equipment Diagram for Two-Evaporator, Two-Compressor System

The *T-S* diagram is given in Figure 15. As the flow of refrigerant through each compressor is different, the liquid and vapor saturation lines are omitted.

State-Point Properties

State	p, psia	t, °F	h, Btu/lb	s, Btu/lb·°R
1	5.409	−59.69	70.727	0.17708
2	16.0	2.46	78.178	0.17708
3	16.0	−17.99	75.329	0.17078
4	51.94	56.23	83.977	0.17078
5	51.94	40.31	17.342	0.03759
6	16.0	−17.99	17.342	0.03950
7	16.0	−17.99	4.664	0.01080
8	5.409	−59.69	4.664	0.01192

$$M_I = 200/(70.727 - 4.664) = 3.0276 \text{ lb}$$
$$M_{II} = M_I(h_2 - h_7)/(h_3 - h_6)$$
$$= 3.0276 (78.178 - 4.664)/(75.329 - 17.342)$$
$$= 3.8384 \text{ lb}$$
$$Q_o = 3.8384 (83.977 - 17.342) = 255.75 \text{ Btu}$$
$$W = (255.75 - 200) = 55.75 \text{ Btu}$$
$$COP = 200/55.75 = 3.587$$

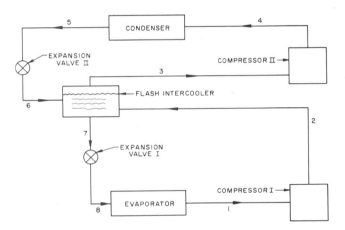

Fig. 14   Equipment Diagram for Dual-Compression, Dual-Expansion Cycle of Example 4

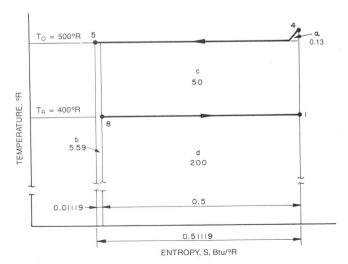

Fig. 15   Temperature-Entropy Diagram for Dual-Compression, Dual-Expansion Cycle of Example 4

$$\Delta W = 55.75 - 50 = 5.75 \text{ Btu}$$
$$\Delta S_R = -200/400 = -0.50 \text{ Btu/°R}$$
$$\Delta S_o = 255.75/500 = 0.5115 \text{ Btu/°R}$$
$$\Delta S_{total} = 0.5115 - 0.50 = 0.0115 \text{ Btu/°R}$$
$$T_o \Delta S_{total} = 500 (0.0115) = 5.75 \text{ Btu}$$
$$(E_U)_o = 500 \times 3.8384 (0.17078 - 0.03759) = 255.62 \text{ Btu}$$
$$(E_A)_o = 255.75 - 255.62 = 0.13 \text{ Btu}$$
$$(E_{Ad})_r = 500 (0.51119 - 0.50) = 5.60 \text{ Btu}$$
$$\Sigma E_{Ad} = 5.60 + 0.13 = 5.73 \text{ Btu}$$

Energy	Btu	Area
Q_i	200	d
Q_o	255.75	(a→d)
W	55.75	a+b+c
$(E_A)_i$	−50	−c
$(E_U)_i$	250	c+d
$(E_A)_o$	0.13	a
$(E_U)_o$	255.62	b+c+d
$(E_{Ad})_r$	5.60	b

SECOND LAW ANALYSIS FOR OPTIMUM CYCLE DESIGN

The concepts of available and unavailable energy, degradation of available energy, and irreversibility are used by Patel (1969), Swers (1968), and Swers *et al.* (1972) to demonstrate a logical and systematic method for selecting a refrigeration cycle's optimum parameters. This method is described for both the basic vapor compression system and a two-stage compression system with desuperheating between compressor stages (intercooling) and subcooling of the liquid before the main expansion valve.

These examples describe the effects of irreversibilities in the refrigeration systems for a range of operating conditions, list the irreversibility of each cycle component, and compare irreversibilities and parameters of the two systems.

Figures 16 and 17 show the equipment designs of the cycles; *T-S* and *P-h* diagrams appear in Figures 18 and 19. To compare operations of the two systems, various points in the operations are identified by numbers consistent for all four figures. These numbers are used as subscripts in identifying pressures and temperatures throughout the system. M_1, M_2, M_3, and M_4 identify rates of refrigerant flow through the various circuits of the two-stage cycle.

In preparing these comparisons, the thermodynamic properties of R-12 were programmed for the computer from equations using the thermodynamic property relations given in the Thermodynamics section of this chapter. A computer program for the cycle analysis was used to calculate the results. The number of significant figures in the tabulated results is not justified on the basis of the data's uncertainty, but is presented to maintain consistency.

Refrigeration Cycle Specifications

The cycles analyzed here are defined in terms of refrigerant temperatures and pressures at various points in the cycles, and temperatures of refrigeration load and of the environment. These temperatures and pressures are specified to be independent of each other, *i.e.*, changes in any one specification are independent of other input parameters. Standard state operating conditions for these cycles are specified in Table 2. Effects of each of these parameters are identified by analyzing the cycle over a range of values for each specification. The determination of heat transfer and fluid flow characteristics in the system components are not considered in this analysis, but temperatures and pressures are given as input parameters that reflect the results of heat transfer and fluid flow characteristics.

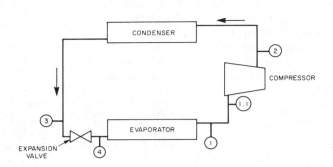

Fig. 16   Basic Vapor Compression System

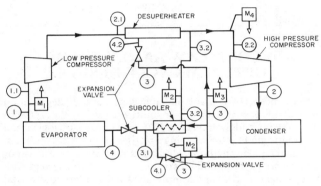

Fig. 17   Two-Stage Compression System

Table 2   Standard Operating Conditions

Item	Basic Vapor Compression System	Two-Stage Compression System
Refrigeration load temp. T_R, °F	0	0
Environment temp. T_o, °F	70	80
Efficiency of compressors, %	70	80
Temp. difference between refrigerant at condenser outlet and environment temp. $T_3 - T_o$, °F	5	5
Temp. difference between refrigeration load and refrigerant at evaporator outlet $T_R - T_1$, °F	5	5
Pressure drop between condenser inlet and outlet $p_2 - p_3$, psi	5	5
Pressure drop between evaporator inlet and outlet $p_4 - p_1$, psi	2	5
Pressure at condenser outlet p_3, psia	105	115
Pressure at evaporator outlet p_1, psia	17	19
Temp. increase of refrigerant between evaporator outlet and (low-pressure) compressor inlet $T_{1.1} - T_1$, °F	3	2
Pressure drop between evaporator outlet and (low-pressure) compressor inlet $p_1 - p_{1.1}$, psi	3	1
Pressure drop between inlet and outlet of desuperheater $p_{2.1} - p_{2.2}$, psi		1.5
Degrees of superheat at outlet of desuperheater $T_{sat} - T_{2.2}$, °F		5
Degrees of subcooling at subcooler outlet $T_{sat} - T_{3.1}$, °F		20
Temp. difference between subcooler vapor outlet and inlet $T_{3.2} - T_{3.1}$, °F		10

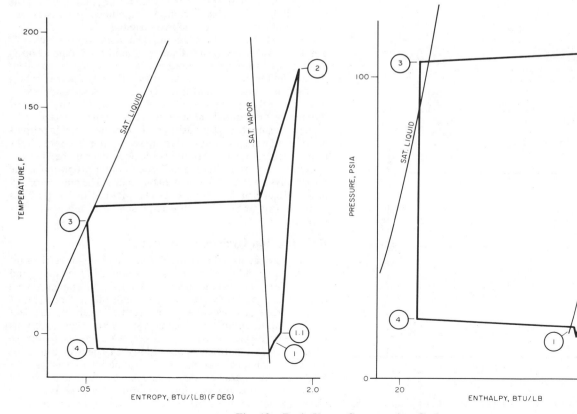

Fig. 18   Basic Vapor Compression Cycle

Selecting Intercooler Pressure for Two-Stage Compression Systems

As previously discussed, one design parameter of the two-stage compression system is selecting the interstage pressure. The total compressor work, given in Table 1, illustrates the effect of interstage pressure. However, the irreversibilities of the entire system may be a more valid basis for selecting this pressure, since a change in intercooler pressure affects conditions throughout the cycle. Table 3 presents values of irreversibilities for the system as a function of interstage pressure. In addition, it gives irreversibility of low- and high-pressure compressors and the work of compression (from Table 1) for comparison. In this example, the inlet pressure to the first stage compressor is 18 psia and the discharge pressure from the second stage is 120 psia, with the remaining cycle conditions as shown in Table 2. The values in Table 3 show that for this specific example, the minimum total work and the minimum total system irreversibility occur at a maximum intercooler pressure (*i.e.*, with single-stage compression).

This case differs from the case of two compressors in series where intercooling is provided by a separate coolant, since the refrigerant used for intercooling in this system causes irreversibilities. In addition, mass flow in the high-pressure compressor increases with decreasing intercooler pressure, since the refrigerant used for desuperheating is recycled to the second-stage compressor. These results also indicate that the low-pressure and high-pressure compressor work are equal at the interstage pressure of approximately 50 psia, whereas the geometric mean of the system's suction and discharge pressures is 46.47 psia. Under conditions in this example, the criterion of minimum work cannot be used to select the interstage pressure. Therefore, the following equations use geometric mean pressure, 46.47 psia, because it provides nearly equal volumetric efficiencies of the two stages.

Comparing Irreversibilities of Refrigeration Systems

The effects of changes in operating conditions of the basic compression refrigeration system and of the two-stage compression

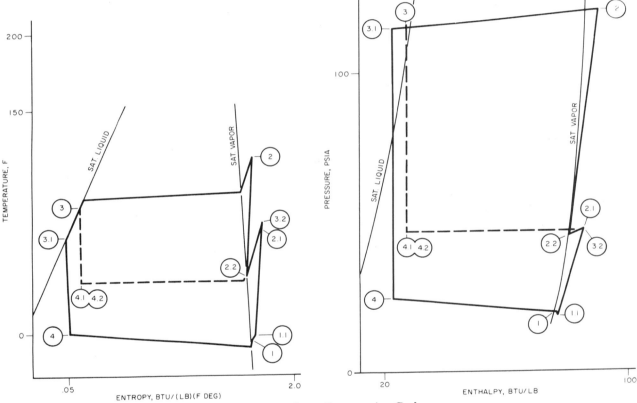

Fig. 19   Two-Stage Compression Cycle

Table 3   Effect of Interstage Pressure

Interstage Pressure, psia	Work Done per Unit of Cooling, Btu/min·ton			Irreversibility per Unit of Cooling, Btu/min·ton			
	Total L.P. Comp.	Total H.P. Comp.	Total for Compressor	Total L.P. Comp.	Total H.P. Comp.	Total for Compressor	Total for System
18	0.0	79.0	79.0	0.00	14.41	14.41	44.19
28	15.0	61.6	76.6	3.32	11.43	14.75	41.79
38	25.7	49.3	75.0	5.46	9.24	14.70	40.22
48	34.1	39.7	73.8	7.50	7.06	14.56	39.02
58	41.0	31.8	72.8	8.23	6.04	14.27	38.04
68	46.9	25.1	72.0	9.21	4.78	13.99	37.21
78	52.1	19.2	71.3	10.02	3.68	13.70	36.49
88	56.6	14.0	70.6	10.72	2.69	13.41	35.85
98	60.7	9.4	70.1	11.33	1.92	13.25	35.28
115	66.8	2.4	69.2	12.20	0.46	12.66	34.42

refrigeration system were calculated for standard state conditions in Table 2 and then for operating conditions in which each specification of Table 2 was varied, one at a time, over a range of values.

These analyses made the following assumptions:
1. The refrigeration load temperature is constant.
2. The heat transferred from the condenser is discharged to the environment at temperature T_o.
3. In the two-stage cycle, the compressor efficiencies of the two stages are equal.

4. In both systems, the piping connecting the cycle components is adiabatic with negligible fluid friction, except for the pipe connecting the evaporator outlet to the low-pressure compressor inlet. (In many cases, these piping losses are considered to be combined with losses in adjacent components.)

Table 4 lists refrigerant properties at various points of the cycle for standard state operating conditions. Table 5 gives a sample calculation of the irreversibilities at standard state conditions.

Table 4   Refrigerant Properties

Basic Vapor Compression System Refrigeration Load of 0°F, Surroundings at 70°F							Two-Stage Compression System Refrigeration Load at 0°F, Surroundings at 80°F						
State[a] Point	Temp., °F	Pressure, psia	Density, lb/ft³	Enthalpy, Btu/lb	Entropy, Btu/lb·°R	Quality, %	State[a] Point	Temp., °F	Pressure, psia	Density, lb/ft[a]	Enthalpy, Btu/lb	Entropy, Btu/lb·°R	Quality, %
1	−5	17	0.44044	77.061	0.17369	—	1	−5	19	0.49502	76.915	0.17163	—
1.1	−2	14	0.35706	77.694	0.17815	—	1.1	−3	18	0.46523	77.269	0.17325	—
							2.1	68.379	46.476	1.0675	86.168	0.17666	—
							2.2	37.264	44.976	1.1165	81.425	0.16790	—
2	176.06	110	2.1411	101.28	0.18966	—	2	117.90	120	2.7309	90.866	0.17120	—
3	75	105	82.089	25.204	0.05262	—	3	85	115	80.802	27.535	0.05688	—
							3.1	65	115	83.335	22.905	0.04834	—
							3.2	75	46.476	1.0505	87.198	0.17860	—
4	−10.444	19	1.8271	25.204	0.05654	27.09	4	0.297	24	2.9238	22.905	0.05059	20.84
							4.1	34.070	46.476	6.1653	27.535	0.05827	17.85
							4.2	34.070	46.476	6.1653	27.535	0.05827	17.85

[a]State points are identified in Figures 16 through 19.

Table 5   Sample Calculation of Second Law Analysis for Standard State Conditions*

Basic Vapor Compression System (T_o = 529.7°R)
Mass flow per unit of refrigeration is $1.0/(h_1 - h_4) = [1.0/(77.061 - 25.204)](12,000/60) = 3.857$ lb/min·ton

PROCESS

Pipe (1-1.1)
(heat transfer to atmosphere)

$$I = T_o \Delta s_{total} = T_o \left[(s_{1.1} - s_1) + \frac{-_1 Q_{1.1}}{T_o} \right] = T_o \left[(s_{1.1} - s_1) + \frac{h_1 - h_{1.1}}{T_o} \right]$$

$$I = 529.7 \left[(0.17815 - 0.17369) + \frac{77.061 - 77.694}{529.7} \right] (3.857) = 6.67 \text{ Btu/min·ton}$$

Compressor (1.1-2)
(adiabatic)

$$I = T_o(s_2 - s_{1.1})$$

$$I = 529.7(0.18966 - 0.17815)(3.857) = 23.52 \text{ Btu/min·ton}$$

Condenser (2-3)
(heat transfer to atmosphere)

$$I = T_o \left[(s_3 - s_2) + \frac{-_2 Q_3}{T_o} \right] = T_o \left[(s_3 - s_2) + \frac{h_2 - h_3}{T_o} \right]$$

$$I = 529.7 \left[(0.05262 - 0.18966) + \frac{101.28 - 25.204}{459.72} \right] (3.857) = 13.45 \text{ Btu/min·ton}$$

Expansion valve (3-4)
(adiabatic)

$$I = T_o(s_4 - s_3)$$

$$I = 529.7 (0.05654 - 0.05262)(3.857) = 8.01 \text{ Btu/min·ton}$$

Evaporator (4-1)
(heat transfer from refrigeration load at 459.7°R)

$$I = T_o \left[(s_1 - s_4) + \frac{-_4 Q_1}{T_R} \right] = T_o \left[(s_1 - s_4) + \frac{h_4 - h_1}{T_R} \right]$$

$$I = 529.7 \left[(0.17369 - 0.05654) + \frac{25.204 - 77.061}{459.72} \right] (3.857) = 8.88 \text{ Btu/min·ton}$$

Mass flow per unit of refrigeration
a. through the evaporator

$$M_1 = 1.0/(h_1 - h_4) = [1.0/(76.915 - 22.905)](12,000/60) = 3.703 \text{ lb/min·ton}$$

b. for the subcooler

$$M_2 = \frac{M_1(h_3 - h_{3.1})}{(h_{3.2} - h_{4.1})} = \frac{3.703(27.535 - 22.905)}{(87.198 - 27.535)} = 0.287 \text{ lb/min·ton}$$

c. for the desuperheater

$$M_3 = \frac{M_1(h_{2.1} - h_{2.2}) + M_2(h_{3.2} - h_{2.2})}{(h_{2.2} - h_{4.2})}$$

$$M_3 = \frac{3.703 (86.168 - 81.425) + 0.287 (87.198 - 81.425)}{(81.425 - 27.535)} = 0.357 \text{ lb/min·ton}$$

d. through the condenser

$$M_4 = M_1 + M_2 + M_3 = 4.347 \text{ lb/min·ton}$$

*See Table 2. Properties are from Table 4.

Table 5 Sample Calculation of Second Law Analysis for Standard State Conditions* (*Concluded*)

Two-Stage Compression System (T_o = 539.7°R)

PROCESS	
Pipe (1-1.1) (heat transfer to atmosphere)	$I = T_o \left[(s_{1.1} - s_1) + \dfrac{-_1Q_{1.1}}{T_o} \right] M_1 = T_o \left[(s_{1.1} - s_1) + \dfrac{h_1 - h_{1.1}}{T_o} \right] M_1$ $I = 539.7 \left[(0.17325 - 0.17163) + \dfrac{(76.915 - 77.269)}{539.7} \right] = (3.703) = 1.93 \text{ Btu/min·ton}$
Evaporator (4-1) (heat transfer from refrigeration load at 459.7°R)	$I = T_o \left[(s_1 - s_4) + \dfrac{-_4Q_1}{T_R} \right] M_1 = T_o \left[(s_1 - s_4) + \dfrac{h_4 - h_1}{T_R} \right]$ $I = 539.7 \left[(0.17163 - 0.05059) + \dfrac{(22.905 - 76.915)}{459.72} \right] (3.703) = 7.09 \text{ Btu/min·ton}$
L.P. Compressor (1.1-2.1) (adiabatic)	$I = T_o(s_{2.1} - s_{1.1})M_1$ $I = 539.7 (0.17666 - 0.17325) (3.703) = 6.81 \text{ Btu/min·ton}$
Desuperheater (adiabatic)	$I = T_o(s_{2.2} - s_{2.1}) M_1 + T_o(s_{2.2} - s_{4.2}) M_3 + T_o(s_{2.2} - s_{3.2})M_2$ $I = 539.7 [(0.16790 - 0.17666) (3.703) + (0.16790 - 0.05827) (0.357)$ $+ (0.16790 - 0.17860) (0.287)] = 1.96 \text{ Btu/min·ton}$
H.P. Compressor (2.2-2) (adiabatic)	$I = T_o(s_2 - s_{2.2})M_4$ $I = 539.7 (0.17120 - 0.16790) (4.347) = 7.74 \text{ Btu/min·ton}$
Condenser (2-3) (heat transfer to atmosphere)	$I = T_o \left[(s_3 - s_2) + \dfrac{-_2Q_3}{T_o} \right] M_4 = T_o \left[(s_3 - s_2) + \dfrac{h_2 - h_3}{T_o} \right] M_4$ $I = 539.7 \left[(0.05688 - 0.17120) + \dfrac{90.866 - 27.535}{539.7} \right] (4.347) = 7.10 \text{ Btu/min·ton}$
Subcooler (3-3.1) (adiabatic)	$I = T_o [(s_{3.1} - s_3) M_1 + (s_{3.2} - s_{4.1}) M_2]$ $I = 539.7 [(0.04834 - 0.05688) (3.703) + (0.17860 - 0.05827) (0.287)] = 1.57 \text{ Btu/min·ton}$
Expansion valve A (3.1-4) (adiabatic)	$I = T_o(s_4 - s_{3.1})M_1$ $I = 539.7 (0.05059 - 0.04834) (3.703) = 4.50 \text{ Btu/min·ton}$
Expansion valve B (3-4.2) (adiabatic)	$I = T_o(s_{4.2} - s_3)M_3$ $I = 539.7 (0.05827 - 0.05688) (0.357) = 0.27 \text{ Btu/min·ton}$
Expansion valve C (3-4.1) (adiabatic)	$I = T_o(s_{4.1} - s_3)M_2$ $I = 539.7 (0.05827 - 0.05688) (0.287) = 0.22 \text{ Btu/min·ton}$

Table 6   System Irreversibilities

Component	Basic System* (T_o = 529.7°R)	Two-Stage System* (T_o = 529.7°R)	Two-Stage System* (T_o = 539.7°R)
Evaporator	8.89	6.96	7.09
Pipe (1 to 1.1)	6.65	1.87	1.93
L.P. Compressor	—	6.68	6.81
H.P. Compressor	—	7.28	7.74
Total compressor	23.51	13.96	14.55
Desuperheater	—	1.80	1.94
Condenser	13.46	11.87	7.09
Expansion valve A	—	4.43	4.51
Expansion valve B	—	0.16	0.27
Expansion valve C	—	0.07	0.22
Total expansion valves	8.00	4.66	5.00
Subcooler	—	0.71	1.59
Total irreversibilities	60.51	41.83	39.19
Carnot work	30.46	30.46	34.81

*All units are Btu/min per ton of cooling.

The results of these second law analyses are expressed in terms of the irreversibilities various system components. Irreversibilities for both basic and two-stage systems are tabulated in Table 6 for standard state conditions. However, since operating conditions of the two systems are not identical, results for the two-stage compression system are given at (1) standard operating condition, where the environment temperature is 80°F and (2) at 70°F, which is the environment temperature for the standard operating condition of the basic vapor compression cycle.

Results given in Table 6 indicate that irreversibilities for each component of the two-stage compression system are smaller than for the basic cycle, and that total cycle irreversibility is about one-third less. In both systems, the largest irreversibilities occur in the compressor, and the next largest occur in the condenser. Losses in the expansion valve are significant but are the smallest of the major components. Piping loss for the basic system is large for the operating conditions chosen, whereas the pressure loss and temperature rise specified for the two-stage system result in smaller piping losses. Results such as those given in Table 6 should be weighed with the cost of larger heat exchangers, or improved compressors, to obtain an optimum economic design. This type of analysis for several different cycle configurations is useful in selecting an optimum configuration.

Table 7 lists the energy balance for these systems and includes mass flows for the two-stage system. (Note that total work listed in Table 7 is equal to the sum of the total irreversibility and the Carnot work given in Table 6.)

Tables 8 and 9 illustrate relative effects of the design parameters for a range of operating conditions in the basic compression cycle and the two-stage cycle, respectively. These tables show irreversibilities of cycle components for Table 1 parameter values that differ from standard operating conditions. The following paragraphs address effects of some of the operating parameters in Tables 8 and 9. These paragraphs are numbered to correspond to the columns in Tables 8 and 9, which also correspond to the items given in Table 2.

Column 1. In both systems, changes in refrigeration load temperature affect the irreversibility of the evaporator. This change in irreversibility results from a change in temperature gradient across this heat exchanger (*i.e.*, the difference between temperatures of the refrigerant and of the refrigeration load). In these calculations, the refrigerant at the evaporator outlet is at a constant pressure, with the temperature 5 °F below load temperature.

Table 7 Energy Balance

	Basic System* $(T_o = 529.7 °R)$	Two-Stage System* $(T_o = 529.7 °R)$	$(T_o = 539.7 °R)$
Energy			
Heat from condenser	293.4	273.6	275.3
Work from L.P. compressor	—	33.0	33.0
Work from H.P. compressor	—	39.3	41.0
Total work	91.0	72.3	74.0
Desuperheater heat transfer by			
M_1	—	17.6	17.6
M_2	—	0.6	1.7
M_3	—	18.2	19.2
Subcooler heat transfer	—	8.5	17.1
Heat into pipe (1 to 1.1)	2.4	1.3	1.3
Mass Flow			
Flow through evaporator, M_1	—	3.703	3.703
Flow through subcooler, M_2	—	0.140	0.287
Flow for desuperheating, M_3	—	0.322	0.357
Flow through condenser, M_4	3.857	4.165	4.347

*All units are Btu/min per ton of cooling.

(The evaporator outlet pressure for the basic cycle is 17 psia; for the two-stage cycle, 19 psia.) Refrigerant temperature at the evaporator outlet is above the saturation temperature, and the degree of superheat depends on load temperature. Therefore, changes in load temperature affect the degree of superheat at the evaporator outlet. The inlet state to the evaporator is not affected by this change.

Column 1A. The change in load temperature with the refrigerant at the outlet state of the evaporator at a constant temperature of −5 °F affects only the irreversibility of the evaporator. With refrigerant properties throughout the system held constant, a change in load temperature does not affect irreversibilities of the other components.

Column 2. Changes in environment temperature affect the irreversibility of the condenser. In the basic cycle, the irreversibility of the expansion valve also changes with this variable. (This is not the case for the two-stage cycle.) The irreversibility of the condenser shows large variations with changes in environment temperature, because the inlet state of the condenser remains constant: therefore, the average temperature gradient across the heat exchanger varies with changes in environment temperature. In this calculation, the refrigerant at the condenser outlet is at a constant pressure and a temperature 5 °F above environment temperature. Outlet pressure for the basic cycle was 105 psia; for the two-stage cycle, 115 psia. In both cases, refrigerant temperatures at the condenser outlet are below their saturation temperatures. The change in irreversibility of the expansion valve for the basic cycle results from this temperature change at the inlet to the valve (which varies with environment temperature). This difference, due to liquid subcooling in the condenser, is insignificant for the two-stage cycle, since the separate subcooler eliminates much of this irreversibility. Note that at temperatures below 65 °F, mass flow from the subcooler outlet is zero, and the irreversibility of the subcooler is zero.

Column 2A. The effect of change in environment temperature is also given in Column 2A, but the condenser outlet temperature is held constant (at 75 °F for the basic cycle, and 85 °F for the two-stage cycle). The major difference in performance of both systems is the variation in irreversibility of the condenser. With no changes

Table 8 System Irreversibilities for Range of Design Parameters in Basic Compression Refrigeration System[a,b]

Irreversibilities are in Btu/min for a refrigeration rate of 200 Btu/min; temperatures are in °F and pressures in psia.

Variable Parameter	Standard Condition	Load Temperature (1)		Load Temperature (1A)		Environment Temperature (2)		Environment Temperature (2A)		Compressor Efficiency (3)		$T_3 − T_o$ (4)	
Variable value		−10	10	−5	5	50	79	50	75	55%	85%	0	10
Irreversibilities													
Evaporator	8.89	3.87	13.56	6.35	11.37	8.28	9.19	8.55	8.97	a	a	8.81	8.97
Piping	6.65	6.89	6.43	a	a	5.80	7.10	6.31	6.74	a	a	6.51	6.81
Compressor	23.51	23.99	23.05	a	a	20.79	24.92	22.63	23.74	43.59	9.88	23.00	24.05
Condenser	13.46	12.58	14.40	a	a	22.74	8.94	24.03	10.82	18.19	11.04	13.24	13.64
Expansion valve	8.00	8.22	7.79	a	a	4.24	10.29	7.70	8.08	a	a	6.97	9.13
TOTAL	60.51	55.55	65.23	57.97	62.99	61.85	60.44	69.22	58.35	85.32	44.46	58.53	62.60

Variable Parameter	$T_R − T_1$ (5)		$p_2 − p_3$ (6)		$p_4 − p_1$ (7)		p_3 (8)		p_1 (9)		$T_{1.1} − T_1$ (10)		$p_1 − p_{1.1}$ (11)	
Variable value	0	10	2	10	0	4	98	112	15	19	0	5	0	5
Irreversibilities														
Evaporator	8.79	8.96	a	a	7.89	9.74	a	a	11.89	6.20	a	a	a	a
Piping	6.54	6.77	a	a	a	a	a	a	7.60	5.92	6.40	6.82	0.26	11.74
Compressor	23.28	23.75	23.26	23.92	a	a	22.91	24.07	25.12	22.10	23.46	23.55	21.36	25.19
Condenser	13.92	13.01	12.40	15.17	a	a	10.94	15.84	14.97	12.28	13.08	13.72	11.73	15.01
Expansion valve	7.89	8.11	a	a	9.00	7.15	a	a	8.97	7.17	a	a	a	a
TOTAL	60.42	60.60	59.20	62.63	a	a	57.39	63.45	68.55	53.67	59.83	60.98	50.24	68.83

[a]Irreversibilities are in Btu/min; for a refrigeration rate of 200 Btu/min; temperatures are in °F; and pressures are in psia.
[b]"a" indicates that the irreversibility is the same as the standard condition.

in liquid subcooling in the condenser, irreversibility in the expansion valve for the basic system does not undergo large variation (see Column 2).

Column 3. Compressor irreversibility is directly proportional to compressor efficiency, and changes in efficiency affect irreversibility. An increase in compressor efficiency also causes a decrease in condenser irreversibility, since the temperature of the compressor outlet decreases with increases in efficiency, which changes the temperature gradient between the superheated vapor in the condenser and the surroundings. A similar effect is noted in performance of the desuperheater for the two-stage cycle. In this calculation, compressors are assumed to operate adiabatically, and compressor loss is identified by assuming an adiabatic efficiency.

Columns 4, 5, 6, 10, and 12. The effect on system performance from variation of the steady-state parameters numbered 4, 5, 6, 10, and 12 is small for the range of conditions shown in Tables 8 and 9. In some cases, the difference in the irreversibility of one component is offset by a similar difference in the opposite direction in another component.

Column 7. An increase in the evaporator pressure loss increases irreversibility of the evaporator but decreases irreversibility of the expansion valve ahead of the evaporator by the same amount; therefore, the total cycle irreversibility remains the same. Because the inlet state to the expansion valve ahead of the evaporator and the outlet state to the evaporator remain fixed, the same heat is removed from the evaporator, and there is no overall change in system performance. However, evaporator heat exchanger surface designs are affected, since the average temperature gradient between refrigerant and refrigeration load is changing.

Column 8. Varying condenser pressure primarily affects irreversibility of the condenser. An increase in compressor discharge pressure also increases irreversibility of the compressor, since work of compression is increased. This change in irreversibility results from change in condensation temperature, a function of condenser pressure.

Column 9. An increase in evaporator pressure decreases the irreversibility of every component in both of these systems. Evaporator pressure affects pressure difference across the expansion valve, compression required in the compressor, and discharge temperature from the compressor (the inlet temperature to the condenser). All changes result in differences in the irreversibilities of these units. There are similar variations in the additional components of the two-stage system. In the basic system, the irreversibility change in the evaporator is larger than for other components. In the two-stage system, the evaporator irreversibility is the largest of the components, but variations in other units are relatively small by comparison. The change in irreversibility of the evaporator is attributed to change in the temperature gradient between refrigerant and refrigeration load. In these calculations, evaporator outlet temperature is constant at $-5\,°F$.

Column 11. The effect of pressure drop in the compressor inlet piping is large compared to the effect of temperature rise (Column 10). Increased pressure loss increases piping irreversibility; large changes in compressor irreversibility are also noted.

Table 9 System Irreversibilities for Range of Design Parameters in Two-Stage Compression Refrigeration System[a,b]

Variable Parameter	Standard Condition	Load Temperature (1)		Environment Temperature (1A)		Load Temperature (2)		Environment Temperature (2A)		Compressor Efficiency (3)		T_3-T_0 (4)		T_R-T_1 (5)	
Variable value	—	−5.	10.	−10	5	60.	84.	60	84.5	70%	90%	0	10	0	10
Irreversibilities															
Evaporator	7.09	4.55	11.91	1.87	9.62	6.83	7.14	6.83	7.14	a	a	a	a	7.02	7.13
Piping (1-1.1)	1.93	1.97	1.86	a	a	1.81	1.96	1.81	1.96	a	a	a	a	1.89	1.97
L. P. Compressor	6.81	6.88	6.67	a	a	6.56	6.86	6.56	6.86	11.58	3.04	a	a	6.74	6.88
Desuperheater	1.94	1.75	2.40	a	a	1.70	2.01	1.87	1.95	2.33	1.68	1.88	2.01	2.15	1.75
H. P. Compressor	7.74	7.74	7.74	a	a	6.86	7.94	7.45	7.80	13.45	3.40	7.58	7.92	7.74	7.74
Condenser	7.09	7.09	7.09	a	a	16.44	5.10	17.03	5.10	7.66	6.70	7.03	7.09	7.09	7.09
Expansion valve (C)	0.22	0.22	0.21	a	a	0.00	0.30	0.21	0.22	a	a	0.13	0.33	0.21	0.22
Expansion valve (B)	0.27	0.23	0.34	a	a	0.09	0.33	0.26	0.27	0.33	0.22	0.21	0.34	0.31	0.23
Expansion valve (A)	4.51	4.57	4.39	a	a	4.34	4.45	4.34	4.54	a	a	a	a	4.45	4.57
Subcooler	1.59	1.61	1.55	a	a	0.00	1.99	1.53	1.60	a	a	1.14	2.08	1.57	1.61
TOTAL	39.19	36.61	44.16	33.97	41.72	44.63	38.08	47.89	37.44	50.69	30.38	38.31	39.28	39.17	39.19

Variable Parameter	p_2-p_3 (6)		p_4-p_1 (7)		p_3 (8)		p_1 (9)		$T_{1.1}-T_1$ (10)		$p_1-p_{1.1}$ (11)		$p_{2.1}-p_{2.2}$ (12)		$T_3-T_{3.1}$ (13)	
Variable value	2	7	0	7.5	107	120	17	21	0	2.5	0	3	0	2	0	30
Irreversibilities																
Evaporator	a	a	5.44	7.72	a	a	9.96	4.50	a	a	a	a	a	a	7.77	6.80
Piping (1-1.1)	a	a	a	a	a	a	2.14	1.77	1.72	1.94	0.19	5.73	a	a	2.11	1.86
L.P. Compressor	6.73	6.86	a	a	6.58	6.94	7.23	6.43	6.73	6.74	6.61	7.25	a	a	7.45	6.53
Desuperheater	1.97	1.93	a	a	1.96	1.93	2.39	1.60	2.05	2.18	1.75	2.39	0.78	2.33	1.94	1.93
H.P. Compressor	7.66	7.79	a	a	7.52	7.87	8.34	7.22	7.70	7.75	7.46	8.36	7.46	7.84	7.85	7.70
Condenser	6.19	7.67	a	a	4.65	8.95	7.27	6.93	7.05	7.10	6.99	7.29	7.03	7.11	7.18	7.05
Expansion valve (C)	0.22	0.21	a	a	0.23	0.21	0.24	0.20	a	a	0.21	0.24	a	a	0.00	0.31
Expansion valve (B)	0.27	0.26	a	a	0.28	0.26	0.37	0.19	0.29	0.31	0.23	0.37	0.26	0.27	0.27	0.27
Expansion valve (A)	a	a	6.16	3.88	a	a	5.09	4.01	a	a	a	a	a	a	8.26	3.14
Subcooler	1.62	1.58	a	a	1.66	1.55	1.70	1.50	a	a	1.54	1.70	a	a	0.00	2.03
TOTAL	38.19	39.83	a	a	36.41	41.24	44.73	34.35	38.95	39.43	36.58	44.93	37.68	39.70	42.83	37.62

[a]Irreversibilities are in Btu/min for a refrigeration rate of 200 Btu/min; temperatures are in °F; and pressures are in psia.

[b]"a" indicates that the irreversibility is the same as the standard condition.

Column 13. The effect of subcooling in the subcooler on system performance is primarily on the irreversibility of the subcooler and on the main expansion valve. Increased heat transfer in the subcooler results in a corresponding increase in its irreversibility. This is offset by the decrease in irreversibility of the expansion valve. Increasing the amount of subcooling decreases the total irreversibility of the system and changes irreversibility in other components.

Recommendations

These analyses illustrate a method of systems analysis; identifying sources of irreversibilities in a system provides a basis for selecting design parameters. However, optimum system design considers the cost of system components and then compares the cost of improving the performance of a given component with the cost of improving all other components in the system.

An extension of this work, the mathematical modeling of the system components, allows study of a given system for a range of steady-state operating conditions. An additional extension, the consideration of thermal properties of the system components, allows calculation of transient operating conditions.

ABSORPTION REFRIGERATION CYCLES

Absorption refrigeration cycles are heat-operated cycles in which a secondary fluid (the absorbent) absorbs the primary fluid (gaseous refrigerant) that has been vaporized in the evaporator.

In the basic absorption cycle, low-pressure refrigerant vapor is converted to a liquid phase (solution) while still at low pressure. Conversion is made possible by the vapor being absorbed by a secondary fluid—the absorbent. Absorption proceeds because of the mixing tendency of miscible substances, and because of an affinity between absorbent and refrigerant molecules. Thermal energy released during the absorption process must be released to a sink. This energy arises from the heat of condensation, sensible heats, and heat of dilution.

The refrigerant-absorbent solution is pressurized in the solution pump and conveyed via a heat exchanger to the generator where refrigerant and absorbent are separated, *i.e.*, regenerated, by a distillation process. A simple still is adequate for the separation when the pure absorbent material is nonvolatile, as in the water-lithium bromide system. However, fractional distillation equipment is required when the pure absorbent material is volatile, as in the ammonia-water system. Refrigerant that is not essentially free of absorbent hampers vaporization in the evaporator (Threlkeld 1970). The regenerated absorbent normally contains a substantial amount of refrigerant. If the absorbent material tends to become solid, as in the water-lithium bromide system, enough refrigerant must be present to keep the pure absorbent material in a dissolved state. Certain considerations, particularly avoiding excessively high temperatures in the generator, make it desirable to leave a moderate amount of refrigerant in the regenerated absorbent.

Figure 20a diagrams a basic absorption cycle that uses a lithium bromide and water solution as an absorbent and water as a refrigerant. Figure 20b shows state points of the pressure-temperature diagram for this pair in the basic absorption cycle. The flow path for these schematics is as follows:

Path 1–2. Hot, concentrated solution (1), in equilibrium with the condenser pressure, leaves the generator; this solution is cooled in the heat exchanger by the incoming solution and throttled to the absorber (2).

Path 2–4. Cold, concentrated solution (2) absorbs low-pressure refrigerant (8 and 9) in the absorber in equilibrium with the evaporator pressure.

Path 4–5. Solution (4) is pumped to the generator via the heat exchanger, where it is heated (5) by the solution leaving the generator.

Path 5–1. Hot, dilute solution (5) enters the generator, where heat is added to distill refrigerant (6). Hot, concentrated solution leaves the generator (1).

Path 6–7. Hot, high-pressure refrigerant vapor (6) condenses (7).

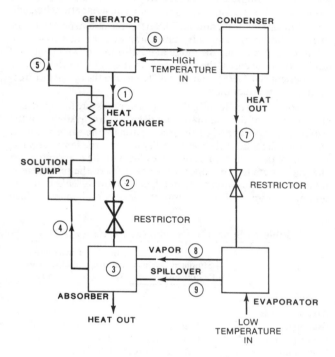

Fig. 20a   Lithium Bromide-Water Single-Stage Absorption Refrigeration Cycle

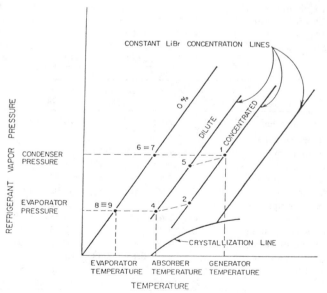

Fig. 20b   Pressure-Temperature State Points for Lithium Bromide-Water Single-Stage Absorption Refrigeration Cycle

Path 7–8. Hot, liquid refrigerant is expanded into the evaporator, where it is evaporated at low pressure and temperature with heat from the cooled space. Cold, low-pressure refrigerant vapor (8) is absorbed by the solution in the absorber (3).

ABSORPTION CYCLES IN PRACTICE

Inefficiencies in the basic absorption cycle are caused by sensible heat lossees, heats of solution, and vaporization characteristics of the absorbing fluid. Conveying hot absorbent from generator into absorber wastes considerable thermal energy. A liquid-to-liquid heat exchanger transfers energy from this stream to the refrigerant-absorbent solution being pumped back to the generator, saving a major portion of the energy. Use of this liquid heat exchanger is shown in the flow diagram for a water-lithium bromide cycle (Figure 20a) and for an ammonia-water cycle (Figure 21).

Modifications for the basic cycle do not bring the coefficient of performance over a threshold of unity, *e.g.*, heat required to generate one pound of refrigerant is not less than the heat taken up when this pound evaporates in the evaporator. Performance can be improved by using the double-effect evaporation principle and a double-effect generator (Whitlow and Swearingen 1958). With the water-lithium bromide pair, two generators can be used: one, at high temperature and pressure, heated by an external source of thermal energy; a second, at lower pressure and temperature, heated by condensation of the vapor from the first generator. Condensate from both generators moves to the evaporator.

CHARACTERISTICS OF REFRIGERANT-ABSORBENT PAIR

The materials that make up the refrigerant-absorbent pair should meet the following requirements to be suitable for absorption refrigeration:

Absence of solid phase. The refrigerant-absorbent pair should not form a solid phase over the range of composition and temper-

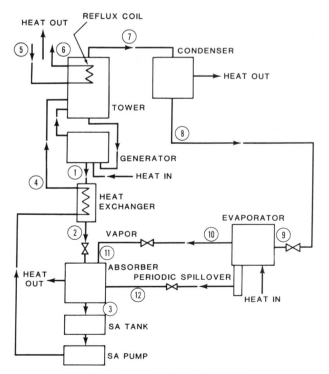

Fig. 21 Ammonia-Water Single-Stage Absorption Refrigeration Cycle

ature to which it might be subjected. If a solid forms, it presumably would stop flow and cause equipment shutdown.

Volatility ratio. The refrigerant should be much more volatile than the absorbent so the two can be easily separated. Otherwise, cost and heat requirements can prohibit separation.

Affinity. The absorbent should have a strong affinity for the refrigerant under conditions in which absorption takes place (Buffington 1949). This affinity (1) causes a negative deviation from Raoult's law and results in an activity coefficient of less than unity for the refrigerant; (2) reduces the amount of absorbent to be circulated and, consequently, the waste of thermal energy from sensible heat effects; and (3) reduces the size of the liquid heat exchanger that transfers heat from absorbent to pressurized refrigerant-absorbent solution in practical cycles. Calculations by Jacob *et al.* (1969) indicate that strong affinity has disadvantages. This affinity is associated with a high heat of dilution; consequently, extra heat is required in the generator to separate refrigerant from absorbent.

Pressure. Operating pressures, largely established by physical properties of the refrigerant, should be moderate. High pressures necessitate use of heavy-walled equipment, and significant electrical power may be required to pump the fluids from low side to high side. Low pressures (vacuum) necessitate use of large volume equipment and special means of reducing pressure drop in refrigerant vapor flow.

Stability. Almost absolute chemical stability is required, because fluids are subjected to severe conditions over many years of service. Instability could cause the undesirable formation of gases, solids, or corrosive substances.

Corrosion. Since the fluids or substances created by instability can corrode materials used in constructing equipment, corrosion inhibitors should be used.

Safety. Fluids must be nontoxic and nonflammable if they are in an occupied dwelling. Industrial process refrigeration is less critical in this respect.

Transport properties. Viscosity, surface tension, thermal diffusivity, and mass diffusivity are important characteristics of the refrigerant and absorbent pair. For example, low viscosity of the fluid promotes heat and mass transfer and reduces pumping problems.

Latent heat. The refrigerant's latent heat should be high so the circulation rate of the refrigerant and absorbent can be kept at a minimum.

No known refrigerant-absorbent pair meets all requirements listed. Ammonia-water and water-lithium bromide are two pairs in extensive commercial use. The ammonia-water pair meets most requirements, but its volatility ratio is too low, and it requires high operating pressures. Furthermore, ammonia is a Safety Code Group 2 fluid (ASHRAE *Standard* 15-1989), restricting its indoor use.

Advantages of the water-lithium bromide pair include high safety, high volatility ratio, high affinity, high stability, and high latent heat. However, this pair tends to form solids. Since the refrigerant turns to ice at 32 °F, the pair cannot be used for low-temperature refrigeration. Lithium bromide crystallizes at moderate concentrations, especially when it is air cooled, limiting the pair to applications where the absorber is water cooled. However, using a combination of salts as the absorbent can reduce this crystallizing tendency enough to permit air cooling (Macriss 1968, Weil 1968, and Rush 1968). Other disadvantages of the water-lithium bromide pair include the low operating pressures it requires and the lithium bromide solution's high viscosity. Proper equipment design can overcome these disadvantages.

Other important refrigerant-absorbent pairs include the following:

- Ammonia-salts (Blytas and Daniels 1962, Robertson *et al.* 1966)
- Methylamine-salts (Robertson *et al.* 1966, Macriss *et al.* 1969)

- Alcohols-salts (Aker *et al.* 1965)
- Ammonia-organic solvents (Robertson *et al.* 1966)
- Sulfur dioxide-organic solvents (Albright *et al.* 1963)
- Halogenated hydrocarbons-organic solvents (Albright *et al.* 1962, Albright *et al.* 1960, Albright and Thieme 1961, Hesselberth and Albright 1966)

Several pairs of refrigerant-absorbents appear suitable for a relatively simple cycle and may not create as much of a crystallization problem as the water-lithium bromide pair. However, stability and corrosion information on most of them is limited. Also, the refrigerants, except for fluororefrigerants in the last type, are somewhat hazardous. Adding corrosion inhibitors, crystallization retardants, or heat transfer enhancers may correct these problems.

THERMODYNAMIC ANALYSIS

First law analysis is useful for investigating new refrigerant-solvent pairs, improving cycles, finding effects of operating conditions, troubleshooting test equipment, and related studies. The method is illustrated in the following data on the water-lithium bromide and the ammonia-water machines. For convenience in making first law analyses and other calculations, see appropriate charts in Chapter 17.

Second law analysis is useful for quantifying the inefficiencies, or thermodynamic irreversibilities, of each process in a cycle. In such an analysis, the concept of availability—a measure of the work that could be produced from a flow stream if it were carried in an ideal reversible manner to an equilibrium condition with the environment—is used. For instance, the availability of a hot fluid stream is work that could be produced by cooling and expanding that fluid to atmospheric temperature and pressure while converting the resulting heat flow to work by an ideal Carnot cycle and the fluid expansion to work by an ideal expansion turbine. The loss of availability by irreversibilities in a process can never be recovered. The principal irreversibilities leading to this loss are heat transfer through a large temperature difference, friction, and unrestrained expansion. Availability is a state function or property. If kinetic and potential energy are negligible, availability is calculated from:

$$B = h - h_o - T_o (S - S_o) \qquad (66)$$

where

B = availability per unit mass
h = enthalpy per unit mass
h_o = enthalpy at ambient conditions
T_o = ambient absolute temperature
S = entropy per unit mass
S_o = entropy at ambient conditions

Second law analysis of absorption cycle machines generally shows that the generator produces the largest loss of availability.

WATER-LITHIUM BROMIDE MACHINE

The determination of essential fluid flow rates is based on the following relationships (Figure 20):

$$R_E = Q_E/(h_v - h_l) \qquad (67)$$

where (in consistent units)

R_E = mass flow of refrigerant from evaporator
Q_E = heat load at evaporator
h_v = enthalpy of refrigerant vapor from evaporator
h_l = enthalpy of refrigerant liquid from condenser

$$R_A X - R_G (X - 1) = 1 \qquad (68)$$

where

R_A = refrigerant mass fraction in solution leaving absorber
R_G = refrigerant mass fraction in solution leaving generator
X = mass of solution flow from absorber/unit mass flow rate of refrigerant
$X - 1$ = mass of solution flow from generator/unit mass flow rate of refrigerant
l = unit mass flow of refrigerant

In this equation, the refrigerant in the dilute solution (from the absorber) minus the refrigerant in the concentrated solution from the generator equals the unit mass flow of refrigerant being considered.

This equation can be rewritten:

$$(1 - WFS_A) X - (1 - WFS_G) (X - 1) = 1 \qquad (69)$$

where

WFS_A = mass fraction of lithium bromide in solution from absorber
WFS_G = mass fraction of lithium bromide in solution from generator

Therefore, if the concentrations of lithium bromide are $WFS_A = 0.595$ and $WFS_G = 0.646$, Equation (69) becomes:

$$(1 - 0.595) X - (1 - 0.646) (X - 1) = 1$$
$$0.405 X - 0.354 (X - 1) = 1$$
$$\text{and } X = 12.67$$

Flow rates can be developed for a hypothetical system with certain assumptions or prior knowledge of typical system conditions as in the following example:

Example 5. A large lithium bromide machine operating according to the flow diagrams of Figures 20a and 20b (including identifying numbers) has the following conditions:

1. Refrigeration load, 500 tons
2. Evaporator temperature (point 8), 41.1°F
3. Absorber equilibrium temperature (point 3), 107.2°F
4. Actual solution temperature (point 4), 100.9°F
5. Solution temperature (point 5), 170.3°F
6. Solution temperature (point 1), 209.6°F
7. Solution temperature (point 2), 128.1°F
8. Refrigerant vapor temperature (point 6), 200°F
9. Refrigerant temperature (point 7), 110°F
10. Refrigerant spillover rate (point 9), 2.5% of (point 8)
11. Concentrations of solution per above
12. Chilled water temperatures, 54 − 44°F
13. Cooling water temperature entering, 85°F
14. Assume no inerts present
15. Cooling tower water flow rate, 1800 gpm

The lithium bromide charts in Chapter 17 and the steam tables in Keenan *et al.* (1969) may be used to find fluid flow rates, heat loads for the various components, and the COP for the system.

Enthalpy	Btu/lb
Of (8), h_v =	1079.8
Of (7), h_l =	77.94
Difference, ΔH =	1001.86

$$1.025 \frac{500 \text{ tons} \times 200}{1001.86} = 102.3 \text{ lb/min refrigerant}$$

$$
\begin{array}{rll}
12.67 \times 102.3 = & 1296.14 & \text{lb/min dilute solution} \\
& - \quad 102.3 & \text{lb/min refrigerant} \\
\hline
& 1193.84 & \text{lb/min concentrated solution}
\end{array}
$$

Approximate absolute pressure at generator = 65.9 mm Hg

Enthalpy	Btu/lb
Of (9) h_l	9.15
Of (2) h_l	71.7
Of (4) h_l	47.2
Of (5) h_l	79
Of (1) h_l	107
Of (6) h_v	1150.3

Material and heat balances at each component follow:

Absorber

Heat in		Btu/min
(2) 1193.84 × 71.7	=	85,598.3
(8) (102.3/1.025) (1079.8)	=	107,769.3
(9) (102.3/1.025) (0.025) (9.15)	=	22.8
	Subtotal =	193,390.4

Heat out		
(4) 1296.14 × 47.2	=	61,177.8
Absorber load (difference)	=	132,212.6

Generator

Heat out		Btu/min
(1) 1193.84 × 107	=	127,741
(6) 102.3 × 1150.3	=	117,676
	Subtotal =	245,417

Heat in		
(5) 1296.14 × 79	=	102,395
Generator load (difference)	=	143,022

Condenser

Heat in		Btu/min
(6) 102.3 × 1150.3	=	117,676

Heat out		
(7) 102.3 × 77.94	=	7,973
Condenser load (difference)	=	109,703

Evaporator

500 tons × 200 = 100,000 Btu/min

Heat Balance

Heat in		Btu/min
Evaporator load	=	100,000
Generator load	=	143,022
	Subtotal =	243,022

Heat out		
Absorber load	=	132,213
Condenser load	=	109,703
		241,916
Balance within	=	0.46%

$$\text{COP} = \frac{\text{Evaporator Load}}{\text{Generator Load}} = \frac{100,000}{143,022} = 0.699$$

With 2% heat loss to ambient, COP = 0.685

The rate of steam flow or hot water flow to the generator can be determined by the enthalpies of steam and hot water available with assumed condensate and hot water temperatures off the generator. For saturated steam at 24.7 psia and 4°F, subcooling of steam condensate in the generator, the steam rate is:

$$\frac{143,022 \times 1.02 \times 60}{(1160.3 - 203.7) \times 500} = 18.3 \text{ lb/h per ton}$$

and for hot water at 240°F with a 10°F water range, the flow rate of hot water is:

$$\frac{143,022 \times 1.02 \times 0.01692 \text{ ft}^3/\text{lb} \times 7.48 \text{ gal/ft}^3}{(208.34 - 198.23) \times 500 \text{ tons}} = \frac{3.66 \text{ gal/min per ton}}{\text{measured at } 240°F}$$

The actual temperatures of the generator's heat source vary with the design heat transfer surface available and the application. For lower temperatures of steam or hot water, the concentrations of lithium bromide-water solutions and the COP are reduced for the same delivered chilled water temperature and available cooling tower water temperature. At lower cooling tower water temperatures, the concentrations and COP can be maintained, even though the available heat source temperature for the generator is reduced, within the practical limits for a given design.

AMMONIA-WATER CYCLE

Figure 21 diagrams an ammonia-water single-stage refrigeration cycle. This resembles a lithium bromide-water refrigeration cycle except, in large systems, the relative volatilities of ammonia and water require a fractional distillation accomplished with a packed tower, a bubble tower, or a tower with sieve trays. Reflux purifies the ammonia vapors coming off the top of the tower and assures the least possible water content of the refrigerant. The concentrated solution of ammonia-in-water (aqua), after being heated by the heat exchanger, is fed at some intermediate point generally near the lower portion of the tower (or rectifier/analyzer). The same required evaporator duty should be assumed when analyzing the system and comparing it with a lithium bromide-water cycle.

In smaller systems, the degree of ammonia vapor purity off the top of the tower (or rectifier/analyzer) is generally less than that in the following example. As a result, tower and condenser pressure is less, but water contamination of the refrigerant must be constantly bled by liquid spillover from the evaporator to the absorber. (In larger systems, a high purity of ammonia vapor off the tower is maintained, and spillover from the absorber can be a periodic occurrence.) To bleed refrigerant, the valve in the vapor line off the evaporator is throttled to establish a slight pressure differential between the evaporator and the absorber. Then, the valve in the liquid spillover line is cracked open until the ammonia in the evaporator is purified sufficiently (detected by checking the pressure in the evaporator versus the temperature of the liquid in the evaporator).

In large systems, a vertical liquid leg under the evaporator provides a relatively inactive area and accumulates ammonia rich in water. The spillover line is tapped into this liquid leg. At an evaporator pressure of 75 psia, a 10% by mass water content increases the refrigerant temperature from 41.1 to 45.8°F for a 4.7°F penalty (Jennings and Shannon 1938). For pure ammonia off the top of the tower (attained by keeping the superheat near 7°F), the operating pressure and temperatures for the tower, generator, and condenser are relatively high, established by the coolant temperature available at the condenser.

In lithium bromide-water systems, the cooling tower water is fed to the absorber and then to the condenser. In ammonia-water systems, the cooling tower water is fed first to the condenser and then to the absorber. In both cases, coolant can be in parallel to improve efficiencies; however, this requires high coolant flow rates and excessively large cooling towers.

In the ammonia-water cycle, the reflux can be created by a separate condenser or by the main condenser. Reflux can flow by gravity or pump to the tower. Ammonia-water systems do not have the potential crystallization of solution problem of lithium bromide-water systems, and the controls can be simpler. Also, the corrosion characteristics of ammonia-water solutions are less severe, although inhibitors are generally used for both types of systems. Whereas lithium bromide-water systems use combinations of steel, copper, and copper-nickel materials for shells and heat transfer surfaces, no copper-bearing materials can be used in ammonia-water systems.

For the ammonia-water cycle, Equation (67) determines the refrigerant flow rate, and Equation (70) develops the solution flow rate per unit refrigerant rate:

$$WFS_A (X) - WFS_G (X - 1) = 1 \qquad (70)$$

where

WFS_A = mass fraction of ammonia in solution from absorber
WFS_G = mass fraction of ammonia in solution from generator
X = mass of solution from absorber per unit mass of refrigerant flow
$X - 1$ = mass of solution from generator per unit mass of refrigerant flow

For large systems, a reasonable pressure drop between the evaporator and absorber is 1.5 psi.

Example 6. A large ammonia-water absorption plant operating according to the flow diagram of Figure 21 has the following conditions:

1. Refrigeration load, 500 tons
2. Evaporator temperature (point 10), 41.1°F
3. Evaporator pressure (point 10), 75 psia
4. Absorber pressure (point 11), 73.5 psia
5. Strong aqua solution temperature (point 3), 105°F
6. Condenser temperature (point 8), 100°F
7. Condenser and tower pressure (point 7), 211.9 psia
8. Concentration split ($WFS_G - WFS_A$), 6% by mass
9. Cooling tower water temperature, 85°F

Chapter 17 shows the enthalpy-concentration diagram for ammonia-water mixture and the ammonia (R-717) properties table. Assume a 3% increase in the theoretical refrigerant flow rate to accommodate heat losses from the high-temperature shells and heat gain through insulation for the evaporator.

From the ammonia-water diagram at 73.5 psia and 105°F from the absorber, assuming no subcooling of solution, the strong aqua (SA) has a concentration of 49% by mass of ammonia. A 6 to 8% concentration split allows sufficient flow and adequate wetting of plain horizontal tubes up to 1 in. in diameter in optimum arranged absorbers when using gravity feed for large systems. It also ensures reasonable maximum liquid flows for cost-effective exchangers and towers and a practical minimum temperature of heat source for the generator. Large splits reduce the flow rate, efficiency, and cost-effectiveness of absorbers and exchangers and raise the required temperature of the heat source, unless a device minimizing these effects (a solution-cooled absorber) is used. With a 6% split, the concentration of ammonia in the weak aqua (WA) from the generator or tower will be 43% by mass. At a 30% split, the concentration of WA will be 19% by mass ammonia. Use of Equation (70) develops the solution flow rates:

For 6% split:

$$0.49\,X - 0.43\,(X - 1) = 1$$
$$0.49\,X - 0.43\,(X - 1) = 1$$
$$0.06\,X + 0.43 = 1$$
$$\text{or } X = 0.57/0.06 = 9.5$$
$$X - 1 = 8.5$$

For 30% split:

$$0.49\,X - 0.19\,(X - 1) = 1$$
$$0.30\,X + 0.19 = 1$$
$$\text{or } X = 0.81/0.30 = 2.7$$
$$X - 1 = 1.7$$

The SA is heated to within 3°F of equilibrium in the tower to avoid flashing in the feed valve to the generator. Consequently, overpressurizing the SA is avoided, reducing the energy and head requirements for the SA pump. The equilibrium temperature of 49% aqua at 211.9 psia is 176.9°F. Therefore, the SA temperature to the tower is 173.9°F. For the 6% split, the 43% WA leaves the tower or generator at an equilibium temperature of 194.9°F. For the 30% split, the 19% WA leaves the tower or generator at an equilibrium temperature of 292°F. Enthalpy values to the aqua are as follows:

Enthalpy	6% Split, Btu/lb	30% Split, Btu/lb
of (1) h_l	71.4	218.7
of (4) h_l	48.3	48.3
of (3) h_l	−28.3	−28.3

The enthalpy of (2) h_l is determined by a mass-enthalpy flow rate balance, as follows:

For 6% split (43% WA):

$$71.4 - 9.5\,[48.3 - (-28.3)]/8.5 = -14.21 \text{ Btu/lb}$$

For 30% split (19% WA):

$$218.7 - 2.7\,[48.3 - (-28.3)]/1.7 = 97.04 \text{ Btu/lb}$$

These enthalpy values correspond to the following temperatures for solution entering the absorbers: for 6% split—118°F; for 30% split—181.5°F.

For a given strong ammonia (SA) concentration, tower pressure and feed temperature, and purity of ammonia produced in the main condenser at a given tower pressure, there is a minimum reflux rate to the top of the tower that requires an infinite number of trays or infinite tower height. For ammonia water mixtures using the enthalpy-concentration diagram of IGT (1964) and the procedure of Brown and Associates (1956), this minimum reflux ratio for 49% SA at the feed condition of 173.9°F for the tower pressure of 211.9 psia is 0.145 lb/lb of 99.95% ammonia refrigerant feed to the evaporator. The practical reflux rate for a reasonable tower height is at least 1.15 times this ratio or 0.167 lb/lb refrigerant.

By constructing flow lines for fluid streams entering and leaving the tower, generator, and condenser, the combined condenser and reflux coil loads, as well as the generator load, can be determined with a mass enthalpy balance calculation (Figure 22).

Enthalpy	Btu/lb
of (point 7) h_v	638.7
of (point 10) h_v	623.02
of (point 8) h_l	155.20
Difference (10 − 8) Δh	467.82

Refrigerant flow rate:

$$1.03 \times 500 \text{ tons} \times 12,000/467.82 = 13210.2 \text{ lb/h}$$

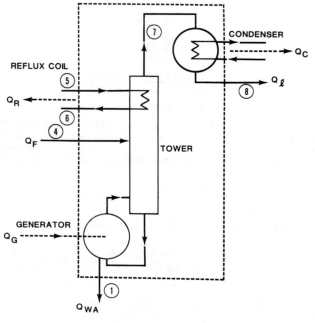

Fig. 22   Enthalpy Balance Method

Generator Heat

For 6% split:

Heat Out		Btu/h
Q_l	$13,210.2 \times 155.2$	2,050,223
Q_c	$13,210.2 \times (638.7 - 155.2)$	6,387,132
Q_r	$0.167 \times 13,210.2 \times (638.7 - 155.2)$	1,066,651
	Subtotal of $Q_l + Q_c + Q_r$	9,504,006
Q_{wa}	$8.5 \times 13\,210.2 \times 71.4$	8,017,270
	Total heat out	17,521,276
Heat In		
Q_f	$9.5 \times 13,210.2 \times 48.3$	6,061,500
Q_g	Difference, generator heat	11,459,776

For 30% split:

Heat Out		Btu/h
	Subtotal of $Q_l + Q_c + Q_r$	9,504,006
Q_{wa}	$1.7 \times 13,210.2 \times 218.7$	4,911,420
	Total heat out	14,415,426
Heat In		
Q_f	$2.7 \times 13,210.2 \times 48.3$	1,722,742
Q_g	Difference, generator heat	12,692,684

Absorber Load (see Figure 21)

For 6% split:

Heat In		Btu/h
(point 11)	$13,210.2 \times 623.02$	8,230,219
(point 2)	$8.5 \times 13,210.2 \times (-14.21)$	−1,595,594
	Subtotal heat in	6,634,625
Heat Out		
(point 3)	$9.5 \times 13,210.2 \times (-28.3)$	−3,551,562
Q_a	Difference, absorber load	10,186,187

For 30% split:

Heat In		Btu/h
(point 11)	$13,210.2 \times 623.02$	8,230,219
(point 2)	$1.7 \times 13,210.2 \times 97.04$	2,179,260
	Subtotal heat in	10,409,479
Heat Out		
(point 3)	$2.7 \times 13,210.2 \times (-28.3)$	−1,009,391
Q_a	Difference, absorber load	11,418,870

Heat Balance

	6% Split	30% Split
Heat In	Btu/h	Btu/h
Evaporator	6,180,000	6,180,000
Generator	11,459,776	12,692,684
	17,639,776	18,872,684
Heat Out		
Condenser	6,387,132	6,387,132
Reflux coil	1,066,651	1,066,651
Absorber	10,186,167	11,418,870
	17,639,950	18,872,653
COP =	$\dfrac{6,180,000}{1.03 \times 11,459,776}$	$\dfrac{6,180,000}{1.03 \times 12,692,684}$
or	0.524	0.473

The most cost-effective cooling water system, condensers, and absorbers determine the cooling tower water flow rate. In the ammonia-water cycle, an 1800 gpm cooling water rate raises the temperature 19.6°F for a 6% split and 21.0°F for a 30% split in concentration. This is about a 25% greater range than for the lithium bromide-water cycle operating at the same evaporator temperature.

A large conventional flooded water cooler in the ammonia system of the example using plain steel tubes produces chilled water with a temperature about 5°F above that for the lithium bromide-water cycle described earlier. For 44°F chilled water and a lower ammonia evaporator temperature, the procedure in the example should be used to adjust the cycle's operating temperatures and COP.

REFERENCES

Aker, J.E., R.G. Squires, and L.F. Albright. 1965. An evaluation of alcohol-salt mixtures as absorption refrigeration solutions. ASHRAE *Transactions* 71(2):14.

Albright, L.F. *et al.* 1960. Solubility of R-11, R-21, and R-22 in organic solvents containing an oxygen atom. ASHRAE *Transactions* 66:423.

Albright, L.F. *et al.* 1962. Solubility of chlorfluormethanes in nonvolatile polar organic solvents. AICHE *Journal* 8:668.

Albright, L.F. and A.A. Thieme. 1961. Solubility of refrigerants 11, 21, and 22 in organic solvents containing a nitrogen atom and in mixtures of liquids. ASHRAE *Journal* 3:71.

Benedict, M., G.B. Webb, and L.C. Rubin. 1940. An empirical equation for thermodynamic properties of light hydrocarbons and their mixtures. *Journal of Chemistry and Physics* 4:334.

Benedict, M. 1937. Pressure, volume, temperature properties of nitrogen at high density, I and II. *Journal of American Chemists Society* 59(11):2224.

Blytas, G.C. and F. Daniels. 1962. Concentrated solutions of NaSCN in liquid ammonia. *Journal of American Chemical Society* 84:1075.

Briggs, S.W. 1971. Second law analysis of absorption refrigeration. AGA and IGT Conference on Natural Gas Research and Technology, Chicago, IL.

Brown and Associates. 1956. *Unit operations*, 6th ed. John Wiley and Sons, Inc., New York, 325.

Buffington, R.M. 1949. Qualitative requirements for absorbent-refrigerant combinations. *Refrigerating Engineering* 4:343.

Cooper, H.W. and J.C. Goldfrank. 1967. B-W-R Constants and new correlations. *Hydrocarbon Processing* 46(12):141.

Haseltine, J.D. and E.B. Qvale. 1971. Comparison of power and efficiency of constant-speed compressors using three different capacity reduction methods. ASHRAE *Transactions* 77(1):158.

Hesselberth, J.F. and L.F. Albright. 1966. Solubility of mixtures of R-12 and R-22 in organic solvents of low volatility. ASHRAE *Transactions* 72(1):198.

Hirschfelder, J.O. *et al.* 1958. Generalized equation of state for gases and liquids. *Industrial and Engineering Chemistry* 50:375.

Hust, J.G. and R.D. McCarty. 1967. Curve-fitting techniques and applications to thermodynamics. *Cryogenics* 8:200.

Hust, J.G. and R.B. Stewart. 1966. Thermodynamic property computations for system analysis. ASHRAE *Journal* 2:64.

IGT. 1964. Physical and thermodynamic properties of ammonia-water. Institute of Gas Technology, Research Bulletin No. 34, Chicago, IL.

Jacob, X., L.F. Albright, and W.H. Tucker. 1969. Factors affecting the coefficient of performance for absorption air-conditioning systems. ASHRAE *Transactions* 75(1):103.

Jennings, B.H. and F.P. Shannon. 1938. The thermodynamics of absorption refrigeration. *Refrigerating Engineering* 35(5):338,

Keenan, J.H., F.G. Keyes, P.G. Hill, and J.G. Moore. 1969. *Steam tables: Thermodynamic properties of water including vapor, liquid, and solid phases.* John Wiley and Sons, Inc., New York.

Macriss, R.A. 1968. Physical properties of modified LiBr solutions. AGA Symposium on Absorption Air-Conditioning Systems, February.

Macriss, R.A., S.A. Weil, and W.F. Rush. 1969. Absorption refrigeration system containing solutions of monomethylamine with thiocyanate. U.S. Patent 3,458,445, July 29.

Martin, J.J. and Y. Hou. 1955. Development of an equation of state for gases. AICHE *Journal* 1:142.

Patel, Y.P. 1969. A thermodynamic analysis of a two-stage compression refrigeration cycle. Worcester Polytechnic Institute, Worcester, MA.

Qvale, E.B. 1971a. The development of a mathematical model for the study of rotary-vane compressors. ASHRAE *Transactions* 77(1):225.

Qvale, E.B. 1971b. Parametric study of the effect of heat exchange between the discharge and suction sides of refrigerating compressors. ASHRAE *Transactions* 77(1):152-57.

Roberson, J.P. *et al.* 1966. Vapor pressure of ammonia and methylamine in solution for absorption refrigeration system. ASHRAE *Transactions* 72(1):198.

Rush, W.F. 1968. The stability of LiBr-LiSCN solutions in water. AGA Symposium on Absorption Air-Conditioning Systems, Chicago, February.

Stewart, R.B., R.T. Jacobsen, and S.G. Penoncello. 1986. ASHRAE Thermodynamic properties of refrigerants. ASHRAE, Atlanta, GA.

Strobridge, T.R. 1962. The thermodynamic properties of nitrogen from 64 to 300 K, between 0.1 and 200 atmospheres. National Bureau of Standards Technical Note 129.

Swers, R. 1968. A thermodynamic analysis of a basic compression refrigeration cycle. Worcester Polytechnic Institute, Worcester, MA.

Swers, R., Y.P. Patel, and R.B. Stewart. 1972. Thermodynamic analysis of compression refrigeration systems. ASHRAE, New Orleans, LA, January.

Threlkeld, J.L. 1970. *Thermal environmental engineering*, 2nd ed. Prentice Hall, Inc., Englewood, CA.

Tripp, W. 1966. Second law analysis of compression refrigeration systems. ASHRAE *Journal* 1:49.

Tripp, W. and H.T. Sun. 1965. T-S and P-V Representation of refrigeration cycle rotary shaft work. ASHRAE *Journal* 7(11):70.

Weil, S.A. 1968. Correlation of the LiSCN-LiBr-H-O Thermodynamic properties. AGA Symposium on Absorption Air-Conditioning Systems, Chicago, February.

Whitlow, E.P. and J.S. Swearingen. 1958. An improved absorption refrigeration cycle. *Gas Age* 122(9):19.

BIBLIOGRAPHY

Bogart, M. 1981. *Ammonia absorption refrigeration in industrial processes*. Gulf Publishing Co., Houston, TX.

Briggs, S.W. 1971. Concurrent, crosscurrent, and countercurrent absorption in ammonia-water absorption refrigeration. ASHRAE *Transactions* 77(1):171.

Doolittle, J.S. and F.J. Hale. 1983. *Thermodynamics for engineers*. John Wiley and Sons, Inc., New York.

Jain, P.C. and G.K. Gable. 1971. Equilibrium property data for aqua-ammonia mixture. ASHRAE *Transactions* 77(1):149.

Stoecker, W.F. and L.D. Reed. 1971. Effect of operating temperatures on the coefficient of performance of aqua-ammonia refrigerating systems. ASHRAE *Transactions* 77(1):163.

Van Wylen, C.J. and R.E. Sonntag. 1985. *Fundamentals of classical thermodynamics*, 3rd ed. John Wiley and Sons, Inc., New York.

Wark, K. 1983. *Thermodynamics*. McGraw-Hill Book Co., New York.

CHAPTER 2

FLUID FLOW

FLOWING fluids in heating, ventilating, air-conditioning, and refrigeration systems transfer heat or mass (as particles). This chapter introduces those basic fluid mechanics that are related to HVAC processes, reviews pertinent flow processes, and presents a general discussion of single-phase fluid flow analysis.

FLUID PROPERTIES

Fluids differ from solids in their reaction to shearing. When placed in shear stress, a solid deforms only a finite amount, whereas a fluid deforms continuously for as long as the shear is applied. A fluid is either a liquid or a gas. Although liquids and gases differ strongly in the nature of molecular actions, their primary mechanical differences are in the degree of compressibility and liquid formation of a free surface (interface).

Fluid motion can usually be described by one of several simplified modes of action or models. The simplest is the ideal-fluid model that assumes no resistance to shearing; flow analysis is well developed (Baker 1983, Schlichting 1979, Streeter and Wylie 1979) and when properly interpreted, is valid for a wide range of applications. Nevertheless, the effects of viscous action may need to be considered. Most fluids in HVAC applications can be treated as Newtonian, where the deformation is directly proportional to the shearing stress. Turbulence, which complicates fluid behavior, does not depend on the viscous nature of a fluid; although viscosity does tend to influence turbulence.

Density

Density ρ, or the mass per unit volume, is involved in most fluid models. The densities of air and water at standard conditions of 68 °F and 14.696 psi (sea level atmospheric pressure) are:

$$\rho_{water} = 62.3 \ \text{lb}_m/\text{ft}^3$$

$$\rho_{air} = 0.075 \ \text{lb}_m/\text{ft}^3$$

Viscosity

Viscosity is the resistance of adjacent fluid layers to shear. For shearing between two parallel planes, each of area A and separated by distance Y, the tangential force F per unit area required to slide one plate with velocity V parallel to the other is proportional to V/Y.

$$F/A = \mu \ (V/Y)$$

where the proportionality factor μ is the viscosity of the fluid. The ratio of the tangential force F to area A is the shearing stress τ;

<hr>

The preparation of this chapter is assigned to TC 1.3, Heat Transfer and Fluid Flow.

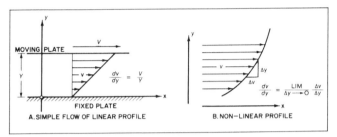

Fig. 1 Velocity Profiles and Gradients in Shear Flows

while V/Y is the lateral velocity gradient (Figure 1A). In complex flows, velocity and shear stress may vary across the flow field; this concept is expressed by the following differential equation:

$$\tau = \mu(dv/dy) \tag{1}$$

The velocity gradient associated with viscous shear for a simple case involving flow velocity in the x direction but of varying magnitude in the y direction is illustrated in Figure 1B.

Viscosity μ depends primarily on temperature. For gases (except near the critical point), viscosity increases with the square root of the absolute temperature, as predicted by the kinetic theory. Liquid viscosity decreases with increasing temperature. Viscosities of various fluids are given in Chapter 37.

At standard conditions, the viscosities of water and air are:

$$\mu_{water} = 6.7 \times 10^{-4} \ \text{lb}_m/\text{ft} \cdot \text{s or } 2.1 \times 10^{-5} \ \text{lb}_f \cdot \text{s/ft}^2$$

$$\mu_{air} = 1.2 \times 10^{-5} \ \text{lb}_m/\text{ft} \cdot \text{s or } 3.7 \times 10^{-7} \ \text{lb}_f \cdot \text{s/ft}^2$$

The centipoise is another common unit of viscosity (1 centipoise = 1 g/(s·m) = 1 mPa/s). Water at standard conditions has a viscosity close to 1.0 centipoise.

In fluid dynamics, *kinematic viscosity* ν is the ratio of viscosity to density, or $\nu = \mu/\rho$. The stoke (1 cm²/s) or centistoke (1 mm²/s) are common units for kinematic viscosity. At standard conditions:

$$\nu_{water} = 1.08 \times 10^{-5} \ \text{ft}^2/\text{s}$$

$$\nu_{air} = 1.7 \times 10^{-4} \ \text{ft}^2/\text{s}$$

BASIC RELATIONS OF FLUID DYNAMICS

This section considers homogeneous, constant-property, incompressible fluids and introduces fluid dynamic considerations of most analyses.

Continuity

Conservation of matter applied to fluid flow in a conduit requires that:

$$\int \rho v \, dA = \text{constant}$$

where v, the velocity normal to the differential area dA and ρ, the density, may vary over the cross section A of a conduit. If both ρ and v are constant over the cross-sectional area normal to the flow, then:

$$\dot{m} = \rho V A = \text{constant} \qquad (2a)$$

where $\dot{m}$ = mass flow rate across the area normal to the flow. When flow is effectively incompressible, ρ = constant; in pipeline and duct flow analyses, the average velocity is then $V = (1/A)\int v \, dA$. The continuity relation is:

$$Q = AV = \text{constant} \qquad (2b)$$

Except when branches occur, the flow rate Q is the same at all sections along the conduit.

For the ideal-fluid model, flow patterns around bodies (or in conduit section changes) result from displacement effects. An obstruction in a fluid stream, such as a strut in a flow or bump on the conduit wall, pushes the flow smoothly out of the way, so that behind the obstruction, the flow becomes uniform again. The effect of fluid inertia (density) appears only in pressure changes.

Pressure Variation across Flow

Pressure variation in fluid flow is important and can be easily measured. Variation across streamlines involves fluid rotation (vorticity). Lateral pressure variation across streamlines is given by the relation (Bober and Kenyon 1980, Olson 1980, Robertson 1965):

$$\frac{\delta}{\delta r}\left(\frac{p}{\rho} + gz\right) = \frac{v^2}{r} \qquad (3)$$

where r is the radius of curvature of the streamline. This relation explains the pressure difference found between the inside and outside walls of a bend and near other regions of conduit section change. It also states that pressure variation is hydrostatic ($p + \rho gz$ = constant) across any conduit where streamlines are parallel.

Bernoulli Equation and Pressure Variation along Flow

A basic tool of fluid flow analysis is the Bernoulli relation involving the principle of energy conservation along a streamline. Generally, the Bernoulli equation is not applicable across streamlines. The first law of thermodynamics is applied to mechanical flow energies (kinetic and potential) and thermal energies: heat is a form of energy and energy is conserved.

The change in energy content ΔE per unit mass of flowing material results from work W done on the system plus absorbed heat:

$$\Delta E = W + Q$$

Fluid energy is composed of kinetic, potential (as due to elevation z), and internal (u) energies. Per unit mass of fluid, the above energy change relation between two sections of the system is then:

$$\Delta\left(\frac{v^2}{2} + gz + u\right) = E_M - \Delta\left(\frac{p}{\rho}\right) + Q$$

The work terms are: (1) external work from a fluid machine E_M (positive for a pump or blower) and (2) pressure or flow work p/ρ. Rearranged, the energy equation can be written as the *generalized Bernoulli equation*:

$$\Delta\left(\frac{v^2}{2} + gz + \frac{p}{\rho}\right) + \Delta u = E_M + Q \qquad (4)$$

The bracketed factor in Equation (4) is the Bernoulli *constant*:

$$\frac{p}{\rho} + \frac{v^2}{2} + gz = \pi \qquad (5a)$$

In cases with no viscous action and no work interaction, π is constant; more generally its change (or lack thereof) is considered in applying the Bernoulli equation. The terms making up π are fluid energies (pressure, kinetic, and potential) per mass rate of fluid flow. Alternative forms of this relation are obtained through multiplication by ρ or division by g, thus:

$$p + \frac{\rho v^2}{2} + \rho gz = p + \frac{\rho v^2}{2} + \gamma z = \rho\pi \qquad (5b)$$

$$\frac{p}{\rho g} + \frac{v^2}{2g} + z = \frac{p}{\gamma} + \frac{v^2}{2g} + z = \frac{\pi}{g} \qquad (5c)$$

The first form involves energies per volume flow rate or pressures; the second involves energies per mass flow rate or heads. In gas flow analysis, Equation (5b) is often used with the γz term dropped as negligible. Equation (5a) should be used when density variations occur. For liquid flows, Equation (5c) is commonly used. Identical results are obtained with the three forms if units and fluids are homogeneous.

Many systems of pipes or ducts and pumps or blowers can be considered as one-dimensional flow. The Bernoulli equation is considered as velocity and pressure vary along the conduit. Analysis is adequate in terms of the section-average velocity V of Equation (2a) or (2b). The Bernoulli relation [Equations (4) and (5)] has v replaced by V, and variation across streamlines can be ignored; the whole conduit is now taken as one streamline. Two- and three-dimensional details of local flow occurrences are still significant, but their effect is accounted for in lumped factors.

The kinetic energy term of the Bernoulli constant is expressed as $\alpha V^2/2$ where the kinetic energy factor ($\alpha > 1$) expresses the true kinetic energy of the velocity profile in ratio to that of the mean flow velocity.

Heat transfer may often be ignored. The change of mechanical energy into internal energy Δu may be expressed as E_L. Flow analysis involves the change in the Bernoulli constant ($\Delta\pi = \pi_2 - \pi_1$) between stations 1 and 2 along the conduit, and the Bernoulli equation is expressed as:

$$\left(\frac{p}{\rho} + \alpha\frac{V^2}{2} + gz\right)_1 + E_M = \left(\frac{p}{\rho} + \alpha\frac{V^2}{2} + gz\right)_2 + E_L \qquad (6a)$$

or in the head form

$$\left(\frac{p}{\gamma} + \alpha\frac{V^2}{2g} + z\right)_1 + H_M = \left(\frac{p}{\gamma} + \alpha\frac{V^2}{2g} + z\right)_2 + H_L \qquad (6b)$$

The factors $gH_M = E_M$, representing energy added to the conduit flow by pumps or blowers; and $gH_L = E_L$, energy dissipated (converted into heat, as mechanically nonrecoverable energy), are defined as positive. A turbine or fluid motor thus has a negative

Fluid Flow

H_M or E_M. For conduit systems with branches involving inflow or outflow, the total energies must be treated; analysis is in terms of $\dot{m} \pi$ and not π.

When real-fluid effects of viscosity or turbulence are included, the continuity relation [Equation (2)] is not changed, but V must be evaluated from the integral of the velocity profile, using time-averaged local velocities.

In fluid flow past fixed boundaries, the velocity at the boundary is zero and shear stresses are produced. The equations of motion then become complex and exact solutions are difficult to find, except in simple cases.

Laminar Flow

For steady, fully developed laminar flow in a parallel-walled conduit, the shear stress varies linearly with distance y from the centerline. For a wide rectangular channel:

$$\tau = (y/b)\tau_w = m(dv/dy)$$

where τ_w = wall shear stress = bdp/ds.

Here the wall spacing is $2b$ and s is the flow direction. Since the velocity is zero at the wall ($y = b$), the integrated result is:

$$v = [(b_2 - y_2)/2\mu]dp/ds$$

This is the *Poiseuille-flow* parabolic velocity profile for a wide rectangular channel. The average velocity V is two-thirds the maximum velocity (at $y = 0$) and the longitudinal pressure drop in terms of conduit flow velocity is:

$$dp/ds = -(3\mu V/b^2) \tag{7}$$

The parabolic velocity profile can also be derived for the axisymmetric conduit (pipe) of radius R but with a different constant. The average velocity is then half the maximum, and the pressure drop relation is:

$$dp/ds = -(8\mu V/R^2) \tag{8}$$

In Equation (6), for laminar flow in a wide rectangular channel, $\alpha = 1.54$, while for a pipe, $\alpha = 2.0$.

Turbulence

Fluid flows are generally turbulent, which involves random perturbations or fluctuations of the flow (velocity and pressure) characterized by an extensive hierarchy of scales or frequencies (Robertson 1963). Flow disturbances that are not random, but have some degree of periodicity, such as the oscillating vortex trail behind bodies, have been erroneously identified as turbulence. Only flows involving random perturbations without any order or periodicity are turbulent; the velocity in such a flow varies with time or locale of measurement (Figure 2).

Turbulence can be quantified by statistical factors. Thus, the velocity most often used in velocity profiles is the temporal average velocity $\bar{v}$, and the strength of the turbulence is characterized by the root-mean-square of the instantaneous variation in velocity about this mean. The effects of turbulence cause the fluid to diffuse momentum, heat, and mass very rapidly across the flow.

The Reynolds number Re, a dimensionless quantity, gives the relative ratio of inertia to viscous forces.

$$\text{Re} = VL/\nu$$

where L is a characteristic length and ν is the kinematic viscosity. In flow through round pipes and tubes, the characteristic length L is the diameter D. Generally, laminar flow in pipes can be expected if the Reynolds number, based on the pipe diameter, is less than 2000. Fully turbulent flow exists when $\text{Re}_D > 10,000$. Between 2000 and 10,000, the flow is in a transition state and predictions are unreliable. In other geometries, different Re criteria exist.

BASIC FLOW PROCESSES

Wall Friction

At the boundary of real fluid flow, the relative tangential velocity at the fluid surface is zero. Sometimes in turbulent flow studies, velocity at the wall may appear finite, implying a fluid slip at the wall. However, this is not the case; the difficulty is in velocity measurement (Goldstein 1938). Zero wall velocity leads to a high shear stress near the wall boundary and a slowing down of adjacent fluid layers. A velocity profile develops near a wall, with the velocity increasing from zero at the wall to an exterior value within a finite lateral distance.

Laminar and turbulent flow differ significantly in their velocity profiles. Turbulent flow profiles are flat compared with more pointed laminar profiles (Figure 3). Near the wall, velocities of the turbulent profile must drop to zero more rapidly than those of the laminar profile, so the shear stress and friction are much greater in the turbulent flow case. Fully developed conduit flow may be characterized by the *pipe factor*, the ratio of average to maximum (centerline) velocity. Viscous velocity profiles result in pipe factors of 0.667 and 0.50 for wide rectangular and axisymmetric conduits. Figure 4 indicates much higher values for rectangular and circular conduits for turbulent flow. Due to the flat velocity profiles, the kinetic energy factor α for Equation (6) ranges from 1.01 to 1.10 in fully developed turbulent pipe flow.

Boundary-Layer Occurrence

In most flows, the friction of a bounding wall on the fluid flow is evidenced by a boundary layer. For flow around bodies, this layer (quite thin relative to distances in the flow direction) encompasses all viscous or turbulent actions, causing the velocity in it to vary rapidly from zero at the wall to that of the outer flow at its edge. Boundary layers are generally laminar near the start of their formation but may become turbulent downstream of the transition

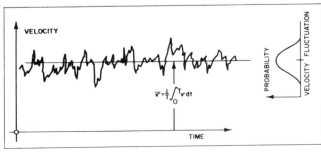

Fig. 2   Velocity Fluctuation at Point in Turbulent Flow

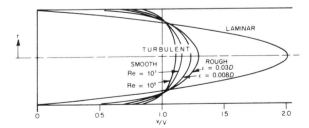

Fig. 3   Velocity Profiles of Flow in Pipes

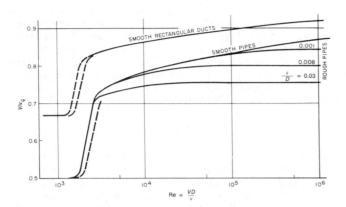

Fig. 4   Pipe Factor for Flow in Conduits

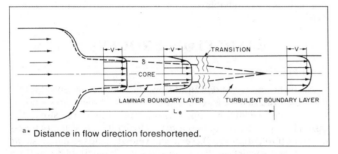

a* Distance in flow direction foreshortened.

Fig. 5   Flow in Conduit Entrance Region[a]

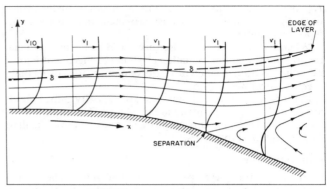

Fig. 6   Boundary Layer Flow to Separation

Flow Patterns with Separation

In technical applications, flow with separation is common and often accepted if it is too expensive to avoid. Flow separation may be geometric or dynamic. Dynamic separation is shown in Figure 6. Geometric separation (Figures 7 and 8) results when a fluid stream passes over a very sharp corner, as with an orifice; the fluid generally leaves the corner irrespective of how much its velocity has been reduced by friction.

For geometric separation in orifice flow (Figure 7), the outer streamlines separate from the sharp corners and because of fluid inertia, contract to a section smaller than the orifice opening, the *vena contracta*, with a limiting area of about six-tenths of the orifice opening. After the vena contracta, the fluid stream expands

point (Figure 5). For conduit flows, spacing between adjacent walls is generally small compared with distances in the flow direction. As a result, layers from the walls meet at the centerline to fill the conduit.

A significant boundary-layer occurrence is in a pipeline or conduit following a well-rounded entrance (Figure 5). Layers grow from the walls until they meet at the center of the pipe. Near the start of the straight conduit, the layer is very thin (and laminar in all probability), so the uniform velocity core outside has a velocity only slightly greater than the average velocity. As the layer grows in thickness, the slower velocity near the wall requires a velocity increase in the uniform core to satisfy continuity. As the flow proceeds, the wall layers grow (and the centerline velocity increases) until they join, after an entrance length L_e. Application of the Bernoulli relation of Equation (5) to the core flow indicates a decrease in pressure along the layer. Ross (1956) shows that although the entrance length L_e is many diameters, the length in which the pressure drop and head loss rates significantly exceed those for fully developed flow is on the order of 10 diameters for turbulent flow in smooth pipes.

In more general boundary-layer flows, as with wall layer development in a diffuser or for the layer developing along the surface of a strut or turning vane, pressure gradient effects can be severe and even lead to separation. The development of a layer in an adverse-pressure gradient situation (velocity v_1 at edge $y = \delta$ of layer decreasing in flow direction) with separation is shown in Figure 6. Downstream from the separation point, fluid backflows near the wall. Separation is due to frictional velocity (thus local kinetic energy) reduction near the wall. Flow near the wall no longer has energy to move into the higher pressure imposed by decrease in v_1 at the edge of the layer. The locale of this separation is difficult to predict, especially for the turbulent boundary layer. Analyses verify the experimental observation that a turbulent boundary layer is less subject to separation than a laminar one because of its greater kinetic energy.

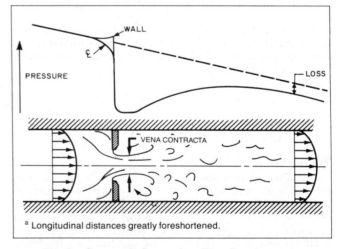

a Longitudinal distances greatly foreshortened.

Fig. 7   Geometric Separation, Flow Development, and Loss in Flow through Orifice[a]

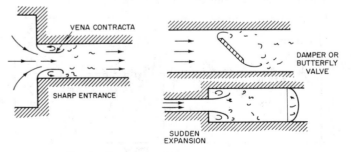

Fig. 8   Examples of Geometric Separation Encountered in Flows in Conduits

rather slowly through turbulent or laminar interaction with the fluid along its sides. Outside the jet, fluid velocity is small compared with that in the jet and is very disturbed. Strong turbulence vorticity helps spread out the jet, increases the losses, and brings velocity distribution back to a more uniform profile. Finally, at a considerable distance downstream, the velocity profile returns to the fully developed flow of Figure 2.

Other geometric separations (Figure 8) occur at a sharp entrance to a conduit, at an inclined plate or damper in a conduit, and at a sudden expansion. For these, a vena contracta can be identified; for sudden expansion, its area is that of the upstream contraction. Ideal fluid theory, using free streamlines, provides insight and predicts contraction coefficients for valves, orifices, and vanes (Robertson 1965). These geometric flow separations are large loss-producing devices. To expand a flow efficiently or to have an entrance with minimum losses, the device should be designed with gradual contours, a diffuser, or a rounded entrance.

Flow devices with gradual contours are subject to separation that is more difficult to predict, since it involves the dynamics of boundary layer growth under an adverse pressure gradient rather than flow over a sharp corner. In a diffuser, used to reduce the loss in expansion, it is possible to expand the fluid some distance at a gentle angle without difficulty (particularly if the boundary layer is turbulent). Eventually, separation may occur (Figure 9), which is frequently asymmetrical because of irregularities. Downstream flow involves flow reversal (backflow) and excess losses exist. Such separation is termed *stall* (Kline 1959). Larger area expansions may use splitters that divide the diffuser into smaller divisions less likely to have separations (Moore and Kline 1958). Another technique for controlling separation is to bleed some low velocity fluid near the wall (Furuya *et al.* 1976). Alternatively, Heskested (1965, 1970) shows that suction at the corner of a sudden expansion has a strong positive effect on geometric separation.

Drag Forces on Bodies or Struts

Bodies in moving fluid streams are subjected to appreciable fluid forces or drag. Conventionally expressed in coefficient form, drag forces on bodies can be expressed as:

$$D = C_D(\rho/2g)V^2 A \tag{9}$$

where A is the projected (normal to flow) area of the body. The drag coefficient C_D depends on the body's shape and angularity and the Reynolds number of the relative flow in terms of the body's characteristic dimension.

For Reynolds numbers of 10^3 to above 10^5, the C_D of most bodies is constant due to flow separation, but above 10^5, the C_D of rounded bodies suddenly drops as the surface boundary layer undergoes transition to turbulence. Typical C_D values are given in Table 1; Hoerner (1965) gives expanded values.

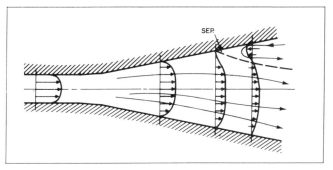

Fig. 9   Separation in Flow in Diffuser

Table 1   Drag Coefficients

Body Shape	$10^3 < Re < 2 \times 10^5$	$Re > 3 \times 10^5$
Sphere	0.36 to 0.47	~ 0.1
Disc	1.12	1.12
Streamlined strut	0.1 to 0.3	< 0.1
Circular cylinder	1.0 to 1.1	0.35
Elongated rectangular strut	1.0 to 1.2	1.0 to 1.2
Square strut	~ 2.0	~ 2.0

For a strut crossing a conduit, the contribution to the head loss of Equation (6) is:

$$H_L = C_D(A/A_c)(V^2/2g) \tag{10}$$

where A_c is the conduit cross-sectional area and A is the area of the strut facing the flow.

Cavitation

Liquid flow with gas- or vapor-filled pockets can occur if the absolute pressure is reduced to vapor pressure or less. In this case a cavity or series of cavities will form, since liquids are rarely pure enough to withstand any tensile stressing or pressures less than vapor pressure for any length of time (John and Haberman 1980, Knapp *et al.* 1970, Robertson and Wislicenus 1969). Robertson and Wislicenus (1969) indicate significant occurrences in various technical fields, chiefly in hydraulic equipment and turbomachines.

Initial evidence of cavitation is the collapse noise of many small bubbles that appear initially as they are carried by the flow into regions of higher pressure. The noise is not deleterious and serves as a warning of the occurrence. As flow velocity further increases or pressure decreases, severity of cavitation increases. More bubbles appear and may join to form large fixed cavities. The space they occupy becomes large enough to modify the flow pattern and alter performance of the flow device. Collapse of the cavities on or near solid boundaries becomes so frequent that the cumulative impact in time results in damage in the form of cavitational erosion of the surface or excessive vibration. As a result, pumps can lose efficiency or their parts may erode locally. Control valves may be noisy or seriously damaged by cavitation.

Cavitation in orifice and valve flow is indicated in Figure 10. With high upstream pressure and a low flow rate, no cavitation occurs. As pressure is reduced or flow rate increased, the minimum pressure in the flow (in the shear layer leaving the edge of the orifice) eventually approaches vapor pressure. Turbulence in this layer causes fluctuating pressures below the mean (as in vortex cores) and small bubble-like cavities. These are carried downstream into the region of pressure regain where they collapse, either in the fluid or on the wall (Figure 10A). As the pressure is reduced, more vapor- or gas-filled bubbles result and coalesce into larger ones. Eventually, a single large cavity results that collapses further downstream (Figure 10B). The region of wall damage is then as many as 20 diameters downstream from the valve or orifice plate.

Sensitivity of a device to cavitation occurrence is measured by the *cavitation index* or *number*, which is the ratio of available pressure above vapor pressure to the dynamic pressure of the reference flow, and can be expressed as:

$$\sigma = [p_o - p_v]/[(\rho/2g)V_o^2] \tag{11}$$

where the subscript o refers to appropriate reference conditions and p_v is the vapor pressure. Valve analyses use such an index in the specific form $(H_d - H_v)/(H - H_d)$ where u and d refer to upstream and downstream locales and H is the head (or pressure

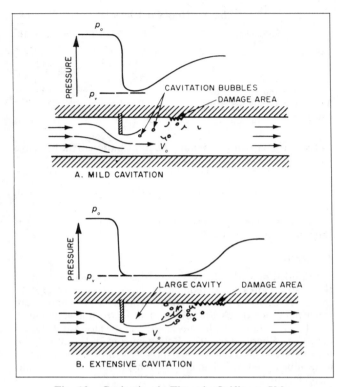

Fig. 10 Cavitation in Flows in Orifice or Valve

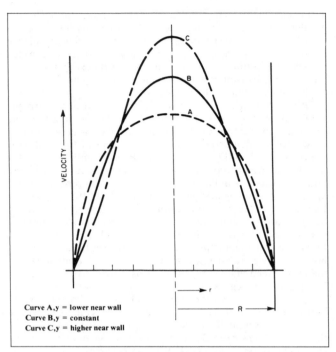

Curve A,y = lower near wall
Curve B,y = constant
Curve C,y = higher near wall

**Fig. 11 Effect of Viscosity Variation on Velocity
Profile of Laminar Flow in Pipe**

divided by liquid specific weight). For the globe valve, the discharge coefficient changes at index values below 0.8 (Ball 1957). For the gate valve, the discharge coefficient begins to be reduced for index values below 1.5 at 10% opening and below 2.0 at 40% opening. With flow-metering devices such as orifices, venturis, and flow nozzles, there is little cavitation, since it occurs mostly downstream of the flow regions involved in establishing the metering action.

The detrimental effect of cavitation can be avoided by operating the liquid-flow device at high enough pressures. When this is not possible, the flow must be changed or the device must be built to withstand cavitation effects. Some materials or surface coatings are more resistant to cavitation erosion than others, but none is immune. Surface contours can be designed to delay the advent of cavitation.

Nonisothermal Effects

When appreciable temperature variations exist, the primary fluid properties (density and viscosity) are no longer constant, as usually assumed, but vary across or along the flow. The Bernoulli equation in the form of Equations (5a) through (5c) must be used, since volumetric flow is not constant. With gas flows, the thermodynamic process involved must be considered. In general, this is assessed in applying Equation (5a), written in the form:

$$\int \frac{dp}{\rho} + \frac{V^2}{2} + gz = \pi \tag{12}$$

Effects of viscosity variations also appear. With nonisothermal laminar flow, the parabolic velocity profile (Figure 3) is no longer valid. For gases, viscosity increases as the square root of absolute temperature, while for liquids, it decreases with increasing temperature. This results in opposite effects.

For fully developed pipe flow, the linear variation in shear stress from the wall value τ_w to zero at the centerline is independent of the temperature gradient. Since $y = R - r$, where R is the pipe

radius $D/2$ and y is the distance from the wall, the relation for the change in velocity from Equation (1) becomes:

$$dv = \frac{\tau_w (R - y) dy}{R \mu} = - \frac{\tau_w}{R} \frac{1}{\mu} r dr \tag{13}$$

When the fluid has a lower viscosity near the wall than at the center (due to external heating of liquid or cooling of gas via heat transfer through the pipe wall), the velocity gradient is steeper near the wall and flatter near the center, so the profile is generally flattened. When liquid is cooled or gas is heated, the velocity profile becomes more pointed for laminar flow (Figure 11). Calculations are made for such flows of gases and liquid metals in pipes (Deissler 1951). Occurrences in turbulent flow are less apparent. If enough heating is applied to gaseous flows, the viscosity increase can cause reversion to laminar flow.

Buoyancy effects and gradual approach of the fluid temperature to equilibrium with that outside the pipe can cause considerable variation in the velocity profile along the conduit. Thus, Colborne and Drobitch (1966) found the pipe factor for upward vertical flow of hot air at a Reynolds number less than 2000 reduced to about 0.6 at 40 diameters from the entrance, then increased to about 0.8 at 210 diameters, and finally decreased to the isothermal value of 0.5 at the end of 320 diameters.

Compressibility

All fluids are compressible to some degree; their density depends on pressure. Steady liquid flow may ordinarily be treated as incompressible, and incompressible flow analysis is satisfactory for gases and vapors at velocities below about a few thousand feet per minute, except in long conduits.

For liquids in pipelines, if flow is suddenly stopped, a severe pressure surge or water hammer that travels along the pipe at the speed of sound in the liquid is produced. This pressure surge alternately compresses and decompresses the liquid. For steady gas flows in long conduits, a decrease in pressure along the conduit can reduce the specific mass (or density) of the gas significantly enough to cause the velocity to increase. If the conduit is long enough, velocities approaching the speed of sound are possible at the discharge end, and the Mach number (ratio of the flow velocity to the speed of sound) must be considered.

Some compressible flows occur without heat gain or loss (adiabatically). If there is no friction (conversion of flow mechanical energy into internal energy), the process is reversible as well. Such a reversible adiabatic process is called isentropic, and follows the relationship:

$$p/\rho^k = \text{constant}$$

where k is the ratio of specific heats at constant pressure and volume with a value of 1.4 for air and diatomic gases.

The Bernoulli equation of steady flow, Equation (12), as an integral of the ideal-fluid equation of motion along a streamline, then becomes:

$$\int \frac{dp}{\rho} + \frac{V^2}{2} = \text{constant} \qquad (14)$$

For a frictionless adiabatic process, the pressure term has the form:

$$\int_1^2 \frac{dp}{\rho} = \frac{k}{k-1}\left(\frac{p_2}{\rho_2} - \frac{p_1}{\rho_1}\right) \qquad (15)$$

As in most compressible flow analyses, the elevation terms are insignificant and are dropped. Then, between stations 1 and 2 for the isentropic process:

$$\frac{p_1}{\rho_1}\left(\frac{k}{k-1}\right)\left[\left(\frac{p_2}{p_1}\right)^{(k-1)/k} - 1\right] + 0.5(V_2^2 - V_1^2) = 0 \quad (16)$$

Equation (16) replaces the Bernoulli equation for compressible flows and may be applied to the stagnation point at the front of a body. With this point as station 2 and the upstream reference flow ahead of the influence of the body as station 1, $V_2 = 0$ and:

$$p_s = p_2 = p_1\left[1 + \frac{k-1}{2}\frac{\rho_1 V_1^2}{kp_1}\right]^{k/(k-1)} \qquad (17)$$

Since kp/ρ is the square of the acoustic velocity a and the Mach number $M = V/a$, the stagnation pressure relation becomes:

$$p_s = p_1\left[1 + \frac{k-1}{2}M_1^2\right]^{k(k-1)} \qquad (18)$$

For Mach numbers less than one:

$$p_s = p_1 + \frac{\rho_1}{2}V_1^2\left[1 + \frac{M_1}{4} + \frac{2-k}{24}M_1^4 + \dots\right] \qquad (19)$$

Equation (19) reduces to the incompressible flow result obtained from Equation (5a) when $M = 0$. Appreciable differences appear when the Mach number of the approaching flow exceeds 0.2. Thus a pitot tube in air is influenced by compressibility at velocities over 13,000 fpm.

Flows through a converging conduit, as in a flow nozzle, venturi, or orifice meter, also may be considered isentropic. Velocity at the upstream station 1 is negligible. From Equation (16), velocity at the downstream station is:

$$V_2 = \sqrt{[2k/(k-1)](p_1/\rho_1)[1 - (p_2/p_1)^{(k-1)/k}]} \qquad (20)$$

The mass flow rate is:

$$w = V_2 A_2 \rho_2 =$$
$$A_2\sqrt{[(2k/(k-1)](p_1\rho_1)[(p_2/p_1)^{2/k} - (p_2/p_1)^{(k+1)/k}]} \quad (21)$$

The corresponding incompressible flow relation is:

$$w_{in} = A_2\rho\sqrt{2\Delta p/\rho} = A_2\sqrt{2\rho(p_1 - p_2)} \qquad (22)$$

The compressibility effect is often accounted for in the expansion factor Y:

$$w = Yw_{in} = A_2 Y\sqrt{2\rho\,(p_1 - p_2)} \qquad (23)$$

Y is 1.00 for the incompressible case, while for air ($k = 1.4$), a value of 0.95 is reached with orifices at p_2/p_1 of 0.83 and with venturis at about 0.90, when these devices are of relatively small diameter ($D_2/D_1 = 0$ to 0.5). Further information on the compressibility correction factor Y is presented in the section of Chapter 13 that discusses orifice flowmeters.

As p_2/p_1 decreases, the flow rate increases, but more slowly than for the incompressible case because of the nearly linear decrease in Y. However, the downstream velocity reaches the local acoustic value and the discharge levels off at a value fixed by upstream pressure and density at the critical ratio:

$$\left.\frac{p_2}{p_1}\right|_c = \left(\frac{2}{k+1}\right)^{k/(k-1)} = 0.53 \text{ for air} \qquad (24)$$

This *choking* (no increase in flow with lowering of downstream pressure) is used in some flow control devices to avoid flow dependence on downstream conditions.

FLOW ANALYSIS

Fluid flow analysis is used to correlate pressure changes with flow rates and nature of the conduit. For a given pipeline, either the pressure drop for a certain flow rate or the flow rate for a certain pressure difference between the ends of the conduit is needed. Flow analysis ultimately involves comparing a pump or blower to a conduit piping system for evaluating the expected flow rate.

Generalized Bernoulli Equation

Internal energy differences are generally small and usually the only significant effect of heat transfer is to change ρ. For gas or vapor flows, use the generalized Bernoulli equation in the pressure-over-density form of Equation (6a), allowing for the thermodynamic process in the pressure-density relation:

$$\int_1^2 \frac{dp}{\rho} + \alpha_1\frac{V_1^2}{2} + E_M = \alpha_2\frac{V_2^2}{2} + E_L \qquad (25a)$$

where elevation changes are negligible. The pressure form of Equation (5b) is generally unacceptable when appreciable density variations occur, since the volumetric flow rate differs at the two stations. This is particularly serious in friction-loss evaluations where the density usually varies over considerable lengths of conduit (Benedict and Carlucci 1966). When the flow is essentially incompressible, Equation (25a) is satisfactory.

As an example, consider specifying the blower to produce an isothermal airflow of 400 cfm through a ducting system (Figure 12). Accounting for intake and fitting losses, the equivalent con-

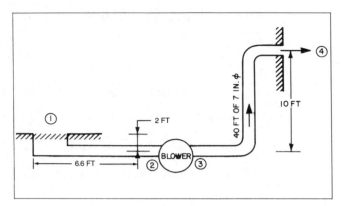

Fig. 12   Blower and Duct System

duit lengths are 60 and 165 ft and the flow is isothermal. The pressure head of the inlet (station 1) and that following the discharge (station 4) (where the velocity is zero) are the same. The frictional losses are evaluated as 24.5 ft of air between stations 1 and 2, and 237 ft between stations 3 and 4. The head form of the generalized Bernoulli relation is used in place of Equation (25a) (which could be used):

$$(p_1/\rho_1 g) + \alpha_1 (V_1^2/2g) + z_1 + H_M$$
$$= (p_2/\rho_2 g) + \alpha_2(V_2^2/2g) + z_2 + H_L \qquad (25b)$$

H_M is evaluated by applying this relation between any two points on opposite sides of the blower. Since conditions at stations 1 and 4 are fully known, they are used, and the location-specifying subscripts on the right side of Equation (25b) are changed to 4. Note that $p_1 = p_4 = p$, $\rho_1 = \rho_4 = \rho$, and $V_1 = V_4 = 0$; then:

$$(p/\rho g) + 0 + 2 + H_M = (p/\rho g) + 0 + 10 + (24.5 + 237)$$

so $H_M = 269.5$ ft of air. For standard air this corresponds to 3.88 in. of water.

The pressure (or head) difference measured across the blower (between stations 2 and 3), is often taken as the head H_M. The results are obtained by calculating the static pressure at stations 2 and 3. Applying Equation (25b) successively between stations 1 and 2 and between 3 and 4 gives:

$$(p_1/\rho g) + 0 + 2 + 0 = (p_2/\rho g) + 1.06 \times 3.54 + 0 + 24.5$$

$$(p_3/\rho g) + 1.03 \times 9.70 + 0 + 0 = (p_4/\rho g) + 0 + 10 + 237$$

Here, α just ahead of the blower has been taken as 1.06 and just after it as 1.03; the latter value is uncertain because of the possible uneven discharge from the blower. Static pressures p_1 and p_4 may be taken as zero gage; thus:

$$(p_2/\rho g) = -26.2 \text{ ft of air}$$

$$(p_3/\rho g) = 237 \text{ ft of air}$$

The static head across the blower, the difference between these two numbers, is 263.2 ft, which is *not* H_M. The apparent discrepancy results from ignoring the velocity heads at stations 2 and 3. Actually, H_M is the change in total head (pressure head plus true kinetic energy) across the machine; thus:

$$H_M = (p_3/\rho g) + \alpha_3(V_3^2/2g) - [(p_2/\rho g) + \alpha_2(V_2^2/2g)]$$
$$= 237.0 + 1.03 \times 9.7 - (-26.2 + 1.06 \times 3.54)$$
$$= 247 - (-22.5) = 269.5 \text{ ft of air}$$

The required blower head is the same, no matter how it is evaluated. It is the specific energy added to the system by the machine. Only when the conduit size and velocity profiles on both sides of the machine are the same is E_M or H_M simply found from $\Delta p = p_3 - p_2$.

Conduit Friction

The loss term E_L or H_L of Equation (6a) or (6b) accounts for friction caused by conduit-wall shearing stresses and losses from conduit-section changes. H_L is the loss of energy (ft-lb$_f$) per lb$_m$ of flowing fluid.

In real fluid flow, a frictional shear occurs at bounding walls, gradually influencing the flow further away from the boundary. A lateral velocity profile is produced and flow energy is converted into heat (fluid internal energy), generally unrecoverable (a loss). This loss in fully developed conduit flow is evaluated through the Darcy-Weisbach relation:

$$(H_L)_f = f (L/D)(V^2/2g) \qquad (26)$$

where L is the length of conduit of diameter D and f is the *friction factor*. Sometimes a numerically different relation is used with the *Fanning friction factor* (one-quarter of f). The value of f is nearly constant for turbulent flow, varying only from about 0.01 to 0.05.

For fully developed laminar-viscous flow in a pipe, the loss is evaluated from Equation (8) as follows:

$$H_L = L \frac{8\mu V/\rho}{R^2 g} = \frac{32L\nu V}{D^2 g} = \frac{64}{(VD/\nu)} \frac{L}{D} \frac{V^2}{2g}$$

so that:

$$f = 64/\text{Re} \quad \text{where} \quad \text{Re} = (VD/\nu) \qquad (27)$$

Thus with laminar flow, the friction factor varies inversely with the Reynolds number.

With turbulent flow, friction loss depends not only on flow conditions, as characterized by the Reynolds number, but also on the nature of the conduit wall surface. With smooth conduit walls, empirical correlations give:

$$f = (0.3164/\text{Re}^{0.25}) \qquad \text{for Re up to } 10^5 \qquad (28a)$$

$$= 0.0032 + \frac{0.221}{\text{Re}^{0.237}} \qquad \text{for } 10^5 < \text{Re} < 3 \times 10^6 \qquad (28b)$$

Generally, f also depends on the wall roughness ϵ. The mode of variation is complex and best expressed in chart form (Moody 1944) (Figure 13). Inspection indicates that, for high Reynolds numbers and relative roughness, the friction factor becomes independent of the Reynolds number in a *fully-rough flow regime*. Then:

$$f^{-0.5} = 1.14 + 2 \log (D/\epsilon) \qquad (29a)$$

Values of f between the values for smooth tubes and those for the fully-rough regime are represented by Colebrook's natural roughness function:

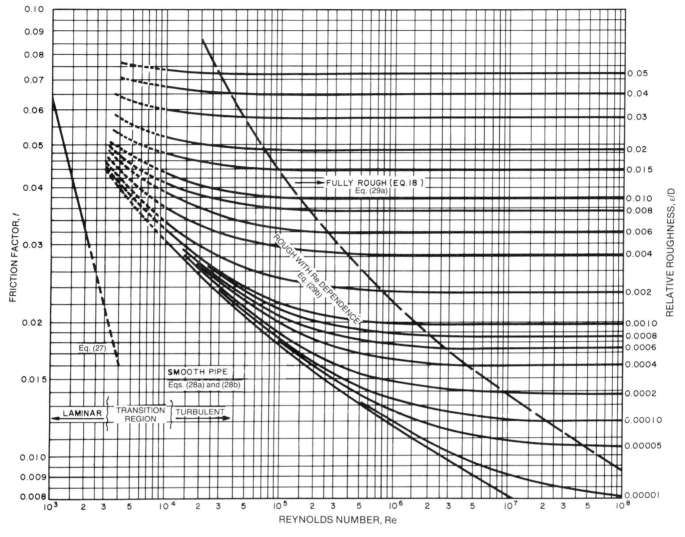

Fig. 13   Relation between Friction Factor and Reynolds Number
(Moody 1944)

$$f^{-0.5} = 1.14 + 2 \log (D/\epsilon) - 2 \log \left[1 + \frac{9.3}{\text{Re}(\epsilon/D)f^{0.5}} \right] \quad (29b)$$

A transition region appears in Figure 13 for Reynolds numbers between 2000 and 10,000. Below this critical condition, for smooth walls, Equation (27) is used to determine f; above the critical condition, Equation (28b) is used. For rough walls, Figure 13 or Equation (29b) must be used to assess the friction factor in turbulent flow. To do this, the roughness height ϵ, which may increase with conduit use or aging, must be evaluated from the conduit surface (Table 2).

While the preceding discussion has focused on circular pipes or ducts, air ducts are often rectangular in cross section. The equivalent circular conduit corresponding to the noncircular conduit

must be found before Figure 13 or Equations (28) or (29) can be entered. Based on turbulent flow concepts, the equivalent diameter is determined by:

$$D_{eq} = (4A/P_w) \quad (30)$$

where A is the flow area and P_w the wetted perimeter of the cross section. For turbulent flow, D_{eq} is substituted for D in Equation (24) and the Reynolds number definition. Noncircular duct friction can be evaluated to within 5% for all except very extreme cross sections. A more refined method for finding the equivalent circular duct diameter is given in Chapter 32. With laminar flow, the loss predictions may be off by a factor as large as two.

Section Change Effects and Losses

Valve and section changes (contractions, expansions and diffusers, elbows or bends, tees), as well as entrances, distort the fully developed velocity profiles (Figure 3) and introduce extra flow losses (dissipated as heat) into pipelines or duct systems. Valves produce such extra losses to control flow rate. In contractions and expansions, flow separation as shown in Figures 8 and 9 causes the extra loss. The loss at rounded entrances develops as

Table 2   Effective Roughness of Conduit Surfaces

Material	ϵ, ft
Commercially smooth brass, lead, copper, or plastic pipe	0.000005
Steel and wrought iron	0.00015
Galvanized iron or steel	0.0005
Cast iron	0.00085

the flow accelerates to higher velocities. The resulting higher velocity near the wall leads to wall shear stresses greater than those of fully developed flow (Figure 5). In flow around bends, the velocity increases along the inner wall near the start of the bend. This increased velocity creates a secondary motion, which is a double helical vortex pattern of flow downstream from the bend. In all these devices, the disturbance produced locally is converted into turbulence and appears as a loss in the downstream region.

The return of disturbed flow to a fully developed velocity profile is quite slow. Ito (1962) showed that the secondary motion following a bend takes up to 100 diameters of conduit to die out, but the pressure gradient settles out after 50 diameters.

With laminar flow following a rounded entrance (Figure 8), the entrance length depends on the Reynolds number:

$$(L_e/D) \cong 0.06 \text{ Re} \qquad (31)$$

At Re = 2000, a length of 120 diameters is needed to establish the parabolic profile. The pressure gradient reaches the developed value of Equation (26) much sooner. The extra head drop is 1.2 velocity heads; 1.0 velocity heads result from the change in profile from uniform to parabolic (since $\alpha = 2.0$) and the rest is due to excess friction. With turbulent flow, 80 to 100 diameters following the rounded entrance are needed for the velocity profile to become fully developed, but the friction loss per unit length reaches a value close to that of the fully developed flow value more quickly. After six diameters, the loss rate at a Reynolds number of 10^5, is only 14% above that of fully developed flow in the same length, while at 10^7 it is only 10% higher (Robertson 1963). For a sharp entrance, the flow separation (Figure 8) causes a greater disturbance, but fully developed flow is achieved in about half the length required for a rounded entrance. With sudden expansion, the pressure change settles out in about eight times the diameter change $(D_2 - D_1)$, while the velocity profile takes at least a 50% greater distance to return to fully developed pipe flow (Lipstein 1962).

These disturbance effects are assumed compressed (in the flow direction) into a point, and the losses are treated as locally occurring. Such losses are related to the velocity head by:

$$\text{Loss of section} = K(V^2/2g) \qquad (32)$$

Chapter 33 and the *Pipe Friction Manual* (HI 1961) have information for pipe applications. Chapter 32 gives information for airflow. The same type of fitting in pipes and ducts may give a different loss, since flow disturbances are controlled by the detailed geometry of the fitting. The elbow of a small pipe may be a threaded fitting that differs from a bend in a circular duct. For 90° screw-fitting elbows, K is about 0.8 (Ito 1962), whereas smooth flanged elbows have a K as low as 0.2 at the optimum curvature.

Table 3 gives a list of fitting loss coefficients. These values indicate the losses, but there is considerable variance. Not included are expansion flows, as from one conduit size to another or at exit into a room or reservoir. For such occurrences, the Borda loss prediction (from impulse-momentum considerations) is appropriate:

$$\text{Loss at expansion} = (V_1 - V_2)^2/2g = \left(1 - \frac{A_1}{A_2}\right)^2 \frac{V_1^2}{2g} \qquad (33)$$

Such expansion loss is reduced by avoiding or delaying separation using a gradual diffuser (Figure 9). For a diffuser of about 7° total angle, the loss is minimal, about one-sixth that given by Equation (21). The diffuser loss for total angles above 45 to 60° exceeds that of the sudden expansion, depending somewhat on the diameter ratio of the expansion. Optimum design of diffusers involves many factors; excellent performance can be achieved in short diffusers with splitter vanes or suction. Turning vanes in miter

Table 3 Fitting Loss Coefficients of Turbulent Flow

Fitting	Geometry	$K = \dfrac{\Delta p/\rho g}{V^2/2g}$
Entrance	Sharp	0.50
	Well-rounded	0.05
Contraction	Sharp $(D_2/D_1 = 0.5)$	0.38
90° Elbow	Miter	1.3
	Short radius	0.90
	Long radius	0.60
	Miter with turning vanes	0.2
Globe valve	Open	10
Angle valve	Open	5
Gate valve	Open	0.19 to 0.22
	75% open	1.10
	50% open	3.6
	25% open	28.8
Any valve	Closed	∞
Tee	Straight through flow	0.5
	Flow through branch	1.8

bends produce the least disturbance and loss for elbows; with careful design, the loss coefficient can be reduced to as low as 0.1.

For losses in smooth elbows, Ito (1962) found a Reynolds number effect (K slowly decreasing with Re) and the minimum loss at a bend curvature of 2.5 (bend radius to diameter ratio). At this optimum curvature, a 45° turn had 63%, and a 180° turn approximately 120%, of the loss of a 90° bend. The loss does not vary linearly with the turning angle because secondary motion occurs.

Use of coefficient K presumes its independence of the Reynolds number. Crane Co. (1976) found a variation with the Reynolds number similar to that of the friction factor; Kittridge and Rowley (1957) observed it only with laminar flow. Assuming that K varies with Re similarly to f, it is convenient to represent fitting losses as adding to the effective length of uniform conduit. The effective length of a fitting is then:

$$(L_{eff}/D) = (K/f_{ref}) \qquad (34)$$

where f_{ref} is an appropriate reference value of the friction factor. Deissler (1951) uses 0.028, while the air duct values in Chapter 32 are based on an f_{ref} of about 0.02. For rough conduits, appreciable errors can occur if the relative roughness does not correspond to that used when f_{ref} was fixed. It is unlikely that the fitting losses involving separation are affected by pipe roughness. The effective length method for fitting loss evaluation is still useful.

When a conduit contains a number of section changes or fittings, the values of K are added to the fL/D friction loss, or the L_{eff}/D of the fittings are added to the conduit length L/D for evaluating the total loss H_L. This assumes that each fitting loss is fully developed and its disturbance fully smoothed out before the next section change. Such an assumption is frequently wrong, and the total loss can be overestimated. For elbow flows, the total loss of adjacent bends may be over- or underestimated. The secondary flow pattern following a radius elbow is such that when one elbow follows another, perhaps in a different plane, the secondary flow production of the second elbow may reinforce or partially cancel that of the first. Moving the second elbow a few diameters can reduce the total loss (from more than twice the amount) to less than the loss from one elbow. Screens or perforated plates can be used for smoothing velocity profiles (Wile 1947) and flow spreading. Their effectiveness and loss coefficients depend on their amount of open area (Baines and Peterson 1951).

Compressible Conduit Flow

When friction loss is included, as it must be except for a very short conduit, the incompressible flow analysis previously considered applies until the pressure drop exceeds about 10% of the initial pressure. The possibility of sonic velocities at the end of relatively long conduits limits the amount of pressure reduction achieved. For an inlet Mach number of 0.2, the discharge pressure can be reduced to about 0.2 of the initial pressure; for an inflow at $M = 0.5$, the discharge pressure cannot be less than about $0.45p_1$ in the adiabatic case and about $0.6p_1$ in isothermal flow.

Analysis of such conduit flow must treat density change, as evaluated from the continuity relation [Equation (2)], with the frictional occurrences evaluated from wall roughness and Reynolds number correlations of incompressible flow (Binder 1944). In evaluating valve and fitting losses, consider the reduction in K caused by compressibility (Benedict and Carlucci 1966). Although the analysis differs significantly, isothermal and adiabatic flows involve essentially the same pressure variation along the conduit, up to the limiting conditions.

Control Valve Characterization

Control valves are characterized by a discharge coefficient C_d. As long as the Reynolds number is above 250, the orifice equation holds for liquids.

$$Q = C_d A_o \sqrt{2\Delta P/\rho} \qquad (35)$$

where A_o is the area of orifice opening.

The discharge coefficient is about 0.63 for sharp-edged configurations and 0.8 to 0.9 for chamfered or rounded configurations. For gas flows at pressure ratios below the choking critical [Equation (24)], the flow is:

$$w = C_d A_o C_1 \times (P_u/\sqrt{T_u})(\sqrt{P_d/P_u}) \sqrt{1 - (P_d/P_u)^{(k-1)/k}} \quad (36)$$

where

- w = mass rate of flow
- A_o = area of orifice
- $C_1 = g\sqrt{2k/R(k-1)}$
- k = ratio of specific heat at constant pressure and volume
- R = gas content
- P = absolute pressure
- T = absolute temperature

Subscripts u and d refer to upstream and downstream positions.

Incompressible Flow in Systems

Flow devices must be evaluated in terms of their interaction with other elements of the system, *e.g.*, the action of valves in modifying flow rate and in matching the flow-producing device (pump or blower) with system loss. Analysis is via the general Bernoulli equation and the loss evaluations noted previously.

A valve regulates or stops the flow of fluid by throttling. The change in flow is not proportional to the change in area of the valve opening. Figures 14 and 15 indicate the nonlinear action of valves in controlling flow. A gate valve opening controls the flow in a pipeline discharging water from a tank (Figure 14). The K values are those of Table 3; f is 0.027. The degree of control also depends on the conduit L/D. For a relatively long conduit, the valve must be nearly closed before its high K value becomes a significant portion of the loss. Figure 15 shows a control damper (essentially a butterfly valve) in a duct discharging air from a plenum held at constant pressure. With a long duct, the damper does not affect the flow rate until it is about one-quarter closed. Duct length has little effect when the damper is more than half closed. The damper closes the duct totally at the 90° position ($K = \infty$).

Flow in a system (pump or blower and conduit with fittings) involves interaction between the characteristics of the flow-producing device (pump or blower) and the loss characteristics of the pipeline or duct system. Often the devices are centrifugal,

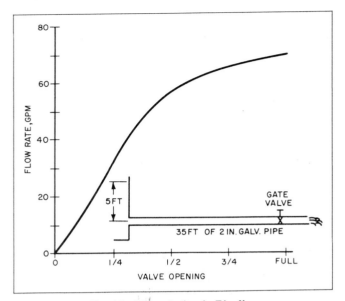

Fig. 14 Valve Action in Pipeline

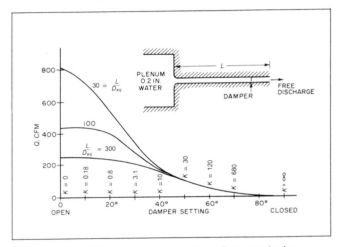

Fig. 15 Effect of Duct Length on Damper Action

in which case the head produced decreases as flow rate increases, except for the lowest flow rates. System head required to overcome losses increases roughly as the square of the flow rate. The flow rate of a given system is that for which the two heads match, where the two curves of head versus flow rate cross (point 1 in Figure 16). When a control valve (or damper) is partially closed, it increases the losses and reduces the flow (point 2 in Figure 16). For cases of constant head, the flow decrease due to valving is not as great as that indicated in Figures 14 and 15.

Flow Measurement

The general principles noted (continuity and Bernoulli equations) are basic to most fluid-metering devices. Chapter 13 has further details.

The pressure difference between the stagnation point (total pressure) and that in the ambient fluid stream (static pressure) is used to give a point velocity measurement. Flow rate in a conduit is measured with a pitot device by placing it at various locations in the cross section and spatially integrating the velocity profile found. A single point measurement may be used for approximate flow rate evaluation. When the flow is fully developed, the pipe-factor information of Figure 4 can be used to estimate the flow rate from a centerline measurement. Measurements can be made in one of two modes. With the pitot-static tube, the ambient (static) pressure is found from pressure taps along the side of the forward-facing portion of the tube. When this portion is not long and slender, static pressure indication will be low and velocity indication high; as a result, a tube coefficient less than unity must be employed. For parallel conduit flow, wall piezometers (taps) may take the ambient pressure, and the pitot tube indicates the impact (total pressure).

The venturi meter, flow nozzle, and orifice meter are flow rate metering devices based on the pressure change associated with relatively sudden changes in conduit section area (Figure 17). The elbow meter is another differential pressure flow meter (also shown in Figure 17). The flow nozzle is similar to the venturi in action, but does not have the downstream diffuser. For all these, flow rate is proportional to the square root of the pressure difference resulting from fluid flow. With the area change devices (venturi, flow nozzle, and orifice meter), a theoretical flow rate relation is found by applying the Bernoulli and continuity equations [Equations (6) and (2)] between stations 1 and 2, so that:

$$Q_{theor} = (\pi d^2/4)\sqrt{2g\Delta h/(1 - \beta^4)} \quad (37)$$

where $\Delta h = h_1 - h_2 = \Delta p/\rho g$ (h is static pressure head), and $\beta = d/D$, the ratio of throat (or orifice) diameter to conduit diameter. The actual flow rate through the device can differ because the approach flow kinetic energy factor α deviates from unity and because of small losses. More significantly, the jet contraction of orifice flow is neglected in deriving Equation (37), to the extent that it can reduce the effective flow area by a factor of 0.6. The effect of all these factors is lumped into the discharge coefficient C_d, so that:

$$Q = C_d Q_{theor} = C_d (\pi d^2/4)\sqrt{2g\Delta h/(1 - \beta^4)} \quad (38)$$

Sometimes an alternate coefficient is used in the form

$$C_d/\sqrt{1 - \beta^4}$$

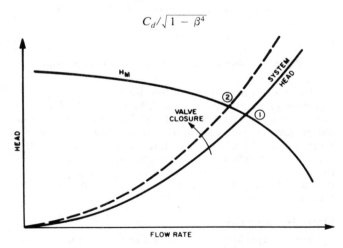

Fig. 16   Matching of Pump or Blower to System Characteristics

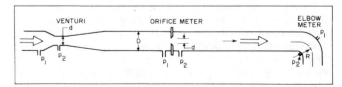

Fig. 17   Differential Head Flow Meters

For compressible fluid metering, the expansion factor Y as described by Equation (23) must be included, and the mass flow rate calculated:

$$w = C_d Y\rho Q_{theor} = C_d Y (\pi d^2/4)\sqrt{2\rho\Delta p/(1 - \beta^4)} \quad (39)$$

Values of Y depend primarily on the pressure ratio p_2/p_1, but also on the metering device and k value of the particular gas.

The general mode of variation in C_d for orifices and venturis is indicated in Figure 18 as a function of Reynolds number and, to a lesser extent, diameter ratio β. For Reynolds numbers less than 10, the coefficient varies as $\sqrt{Re}$.

The elbow meter (Murdock *et al.* 1964) employs the pressure difference between inside and outside the bend as the metering signal. A momentum analysis gives the flow rate as:

$$Q = (\pi D^2/4)\sqrt{R/2D}\sqrt{(2g\Delta h)} \quad (40)$$

where R is the radius of curvature of the bend. Again, a discharge coefficient C_d is needed; as in Figure 18, this drops off for the lower Reynolds numbers (below 10^5). These devices are calibrated in pipes with fully developed velocity profiles, so they must be located far enough downstream of sections that modify the approach velocity.

Unsteady Flow

Conduit flows are not always steady. In a compressible fluid, the acoustic velocity is usually high and conduit length is rather short, so the time of signal travel is negligibly small. Even in the incompressible approximation, system response is not instantaneous. If a pressure difference Δp is applied between the conduit ends, the fluid mass must be accelerated and wall friction overcome, so a finite time passes before the steady flow rate corresponding to the pressure drop is achieved.

The time it takes for an incompressible fluid in a horizontal constant-area conduit of length L to achieve steady flow may be estimated by using the unsteady flow equation of motion with wall friction effects included. On the quasi-steady assumption, friction is given by Equation (26); also by continuity, V is constant along the conduit. The occurrences are characterized by the relation:

$$(dV/d\theta) + (dp/\rho ds) + (fV^2/2D) = 0 \quad (41)$$

where θ is time and s is distance in the flow direction.

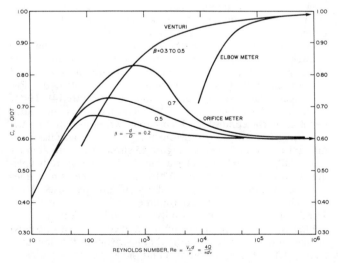

Fig. 18   Flow Meter Coefficients

Since a certain Δp is applied over the conduit length L:

$$(dV/d\theta) = (\Delta p/\rho L) - (fV^2/2D) \qquad (42)$$

For laminar flow, f is given by Equation (27), and then:

$$(dV/d\theta) = (\Delta p/\rho L) - (32\mu V/\rho D^2) = A - BV \qquad (43)$$

This is rearranged and integrated to yield the time to reach a certain velocity:

$$\theta = \int d\theta = \int \frac{dV}{(A - BV)} = -\frac{1}{B} \ln (A - BV) \qquad (44)$$

and

$$V = (\Delta p/L)(D^2/32\mu) [1 - \exp (- 32\nu\theta/D)] \qquad (45a)$$

For large times ($\theta \rightarrow \infty$), this indicates steady velocity as:

$$V_\infty = (\Delta p/L)(D^2/32\mu) = (\Delta p/L)(RZ^2/8\mu) \qquad (45b)$$

as by Equation (8). Then:

$$V = V_\infty [1 - \exp(- f_\infty V_\infty \theta/2D)] \qquad (46)$$

The general nature of velocity development for starting-up flow is derived by more complex techniques; however, the temporal variation is as given above. For shutdown flow (steady flow with $\Delta p = 0$ at $\theta > 0$), the flow decays exponentially as $e^{-\theta}$.

Turbulent flow analysis of Equation (41) also must be based on the quasi-steady approximation, with less justification. Daily *et al.* (1956) indicate frictional resistance to be slightly greater than the steady-state result for accelerating flows, but appreciably less for decelerating ones. If the friction factor is approximated as constant:

$$dV/d\theta = (\Delta p/\rho L) - f(V^2/2D) = A - BV^2$$

and, for the accelerating flow:

$$\theta = \frac{1}{\sqrt{AB}} \tanh^{-1} (V \sqrt{B/A})$$

or:

$$V = \sqrt{A/B} \tanh (\theta \sqrt{AB})$$

Since the hyperbolic tangent is zero when the independent variable is zero and unity when the variable is infinity, the initial ($V = 0$ at $\theta = 0$) and final conditions are verified. Thus, for large times ($\theta \rightarrow \infty$):

$$V_\infty = \sqrt{A/B} = \sqrt{(\Delta p/\rho L)(2D/f)}$$

which is in accord with Equation (26) when f is constant (the flow regime is the fully rough one of Figure 13). The temporal velocity variation then is:

$$V = V_\infty \tanh [(f V_\infty/2D)/\theta] \qquad (47)$$

The turbulent velocity start-up result is compared with the laminar one in Figure 19, where initially it appears somewhat faster but of the same general form, increasing rapidly at the start but reaching V_∞ asymptotically.

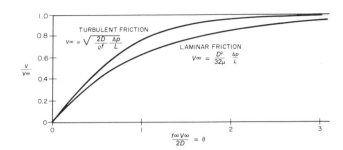

Fig. 19   Temporal Increase in Velocity Following Sudden Application of Pressure

NOISE FROM FLUID FLOW

Noise in flowing fluids results from unsteady flow fields and can be at discreet frequencies or broadly distributed over the audible range. With liquid flow, cavitation results in noise through the collapse of vapor bubbles. The noise in pumps or fittings (such as valves) can be a rattling or sharp hissing sound. It is easily eliminated by raising system pressure. With severe cavitation, the resulting unsteady flow can produce indirect noise from induced vibration of adjacent parts.

The disturbed laminar flow behind cylinders can be an oscillating motion. The shedding frequency of these vortexes is characterized by a Strouhal number St $= fd/V$ of about 0.21 for the circular cylinder of diameter d, over a considerable range in Reynolds numbers. Chapter 42 of the 1991 ASHRAE *Handbook—Applications* has further details. This oscillating flow can be a powerful noise source, particularly when f is close to the natural frequency of the cylinder or some nearby structural member, so that resonance occurs. With cylinders of another shape, as impeller blades of a pump or blower, the characterizing Strouhal number involves the trailing edge thickness of the member. The strength of the vortex wake, with its resulting vibrations and noise potential, can be reduced by breaking up the flow with downstream splitter plates or boundary-layer trip devices (wires) on the cylinder surface.

Noise produced in pipes and ducts, especially from valves and fittings, is associated with the loss through such elements. The sound pressure of noise in water pipe flow increases linearly with the head loss; the broad-band noise increases but only in the lower frequency range. Fitting-produced noise levels also increase with their loss (even without cavitation) and significantly exceed that of the pipe flow. The relation between noise and loss is not surprising since they both involve excessive flow perturbations. A valve's pressure-flow characteristics and structural elasticity may be such that for some operating point it will oscillate, perhaps in resonance with part of the piping system, to produce excessive noise. A change in the operating point conditions or details of the valve geometry can result in significant noise reduction.

Pumps and blowers are strong potential noise sources. Turbomachinery noise is associated with blade-flow occurrences. Broad-band noise appears from vortex and turbulence interaction with walls and is primarily a function of the operating point of the machine. For blowers, it has a minimum at the peak efficiency point (Groff *et al.* 1967). Narrow-band noise also appears at the blade-crossing frequency and its harmonics. Such noise can be very annoying because it stands out from the background. To reduce this noise, increase clearances between impeller and housing, and space the impeller blades unevenly around the circumference.

REFERENCES

Baines, W.D. and E.G. Peterson. 1951. An investigation of flow through screens. ASME *Transactions* 73:467.

Baker, A.J. 1983. *Finite element computational fluid mechanics*. McGraw-Hill Books Co., Inc., New York.

Ball, J.W. 1957. Cavitation characteristics of gate valves and globe values used as flow regulators under heads up to about 125 ft. ASME *Transactions* 79:1275.

Benedict, R.P. and N.A. Carlucci. 1966. *Handbook of specific losses in flow systems*. Plenum Press Data Division, New York.

Binder, R.C. 1944. Limiting isothermal flow in pipes. ASME *Transactions* 66:221.

Bober, W. and R.A. Kenyon. 1980. *Fluid mechanics*. John Wiley and Sons, Inc., New York.

Colborne, W.G. and A.J. Drobitch. 1966. An experimental study of non-isothermal flow in a vertical circular tube. ASHRAE *Transactions* 72(4):5.

Crane Co. 1976. Flow of Fluids. Technical Paper No. 410, New York.

Daily, J.W., *et al.* 1956. Resistance coefficients for accelerated and decelerated flows through smooth tubes and orifices. ASME *Transactions* 78:1071.

Deissler, R.G. 1951. Laminar flow in tubes with heat transfer. National Advisory Technical Note 2410, Committee for Aeronautics.

Furuya, Y., T. Sate, and T. Kushida. 1976. The loss of flow in the conical with suction at the entrance. *Bulletin of the Japan Society of Mechanical Engineers* 19:131.

Goldstein, S. ed. 1938. *Modern developments in fluid mechanics*. Oxford University Press, London. Reprinted by Dover Publications, New York.

Heskested, G. 1965. An edge suction effect. AIAA *Journal* 3:1958.

Heskested, G. 1970. Further experiments with suction at a sudden enlargement. *Journal of Basic Engineering*, ASME *Transactions* 92D:437.

Hoerner, S.F. 1965. *Fluid dynamic drag*, 3rd ed. Published by author, Midland Park, NJ.

Hydraulic Institute. 1961. *Pipe friction manual*. Hydraulic Institute, New York.

Ito, H. 1962. Pressure losses in smooth pipe bends. *Journal of Basic Engineering*, ASME *Transactions* 4(7):43.

John, J.E.A. and W.L. Haberman. 1980. *Introduction to fluid mechanics*, 2nd ed. Prentice Hall, Inc., Englewood Cliffs, NJ.

Kittridge, C.P. and D.S. Rowley. 1957. Resistance coefficients for laminar and turbulent flow through one-half inch valves and fittings. ASME *Transactions* 79:759.

Kline, S.J. 1959. On the nature of stall. *Journal of Basic Engineering*, ASME *Transactions* 81D:305.

Knapp, R.T., J.W. Daily, and F.G. Hammitt. 1970. *Cavitation*. McGraw-Hill Book Co., New York.

Lipstein, N.J. 1962. Low velocity sudden expansion pipe flow. ASHRAE *Journal* 4(7):43.

Moody, L.F. 1944. Friction factors for pipe flow. ASME *Transactions* 66:672.

Moore, C.A. and S.J. Kline. 1958. Some effects of vanes and turbulence in two-dimensional wide-angle subsonic diffusers. National Advisory Committee for Aeronautics, Technical Memo 4080.

Murdock, J.W., C.J. Foltz, and C. Gregory. 1964. Performance characteristics of elbow flow meters. *Journal of Basic Engineering*, ASME *Transactions* 86D:498.

Olson, R.M. 1980. *Essentials of engineering fluid mechanics*, 4th ed. Harper and Row, New York.

Robertson, J.M. 1963. *A turbulence primer*. University of Illinois, Engineering Experiment Station Circular 79, Urbana, IL.

Robertson, J.M. 1965. *Hydrodynamics in theory and application*. Prentice-Hall, Inc., Englewood Cliffs, NJ.

Robertson, J.M. and G.F. Wislicenus, ed. 1969 (discussion 1970). *Cavitation state of knowledge*. ASME, New York.

Ross, D. 1956. Turbulent flow in the entrance region of a pipe. ASME *Transactions* 78:915.

Schlichting, H. 1979. *Boundary layer theory*, 7th ed. McGraw-Hill Book Co., New York.

Streeter, V.L. and E.B. Wylie. 1979. *Fluid mechanics*, 7th ed. McGraw-Hill Book Co., New York.

Wile, D.D. 1947. Air flow measurement in the laboratory. *Refrigerating Engineering*: 515.

BIBLIOGRAPHY

ASME. 1959. *Fluid meters, their theory and application*. American Society of Mechanical Engineers, New York.

ASME. 1961. *Flow meter computation handbook*. American Society of Mechanical Engineers, New York.

Bertin, J.J. 1984. *Engineering fluid mechanics*. Prentice Hall, Inc., Englewood Cliffs, NJ.

Cebeci, T.C. and P. Bradshaw. 1977. *Momentum transfer in boundary layers*. Hemisphere Publishing Corporation, Washington, D.C.

Cusik, C.F., ed. 1961. *Flow meter engineering handbook*. Brown Instruments, Division of Honeywell, Philadelphia, PA.

Eagleson, P.S., G.K. Noutsopoulos, and J.M. Daily. 1964. The nature of self-excitation in the flow-induced vibration of flat plates. *Journal of Basic Engineering*, ASME *Transactions* 86D:599.

Ingard, U., A. Oppenheim, and M. Hirschorn. 1968. Noise generation in ducts. ASHRAE *Transactions* 74(I):V.1.1.

Kline, S.J., D.E. Abbott, and R.W. Fox. 1959. Optimum design of straight walled diffusers. *Journal of Basic Engineering*, ASME *Transactions* 81D:321.

Miller, R.W. 1982. *Flow measurement engineering handbook*. McGraw-Hill Book Co., Inc., New York.

Pigott, J.S. 1957. Losses in pipe and fittings. ASME *Transactions* 79:1767.

Protos, A., V.W. Goldschmidt, and G.H. Toebes. 1968. Hydroelastic forces on bluff cylinders. *Journal of Basic Engineering*, ASME *Transactions* 90D:378.

Rogers, W.L. 1958. Noise production and damping in water piping. *Heating, Piping and Air Conditioning* 1:181.

Shames, I.H. 1982. *Mechanics of fluids*, 2nd ed. McGraw-Hill Book Co., Inc., New York.

Sovran, G. and E.D. Klomp. 1967. Experimentally determined optimum geometries for rectilinear diffusers with rectangular, conical or annular cross-section. *Fluid mechanics of internal flow*, G. Sovran, ed., Elsevier Publishing Co.

White, F.M. 1979. *Fluid mechanics*. McGraw-Hill Book Co., Inc., New York.

CHAPTER 3

HEAT TRANSFER

HEAT is energy in transition caused by a temperature difference. The process and study thereof is called heat transfer. The thermal energy is transferred from one region to another by three modes: conduction, convection, and radiation. Heat transfer is among the transport phenomena that include mass transfer (see Chapter 5), momentum transfer or fluid friction (see Chapter 2), and electrical conduction. Transport phenomena have similar rate equations, and flux is proportional to a potential difference. In heat transfer by conduction and convection, the potential difference is the temperature difference. Heat, mass, and momentum transfer are often considered together because of their similarities and interrelationship in many common physical processes.

This chapter presents the elementary principles of single-phase heat transfer with emphasis on heating, refrigerating, and air conditioning. Boiling and condensation are discussed in Chapter 4. More specific information on heat transfer to or from buildings or refrigerated spaces can be found in Chapters 22 through 28. Physical properties of substances can be found in Chapters 16, 20, 22, 30, and 36. Heat transfer equipment, including evaporators, condensers, heating and cooling coils, furnaces, and radiators, are covered in the 1992 ASHRAE *Handbook—Systems and Equipment*. For further information on heat transfer, see the bibliography.

HEAT TRANSFER PROCESSES

Thermal conduction is the mechanism of heat transfer whereby energy is transported between parts of a continuum by the transfer of kinetic energy between particles or groups of particles at the atomic level. In gases, conduction is a result of elastic collision of molecules; in liquids and electrically nonconducting solids, it is believed to be caused by longitudinal oscillations of the lattice structure. Thermal conduction in metals occurs like electrical conduction, through motions of free electrons. Thermal energy transfer occurs in the direction of decreasing temperature, a consequence of the *second law of thermodynamics*. In solid opaque bodies, the significant heat transfer mechanism is thermal conduction, since no net material flows in the process. With flowing fluids, thermal conduction dominates in the region very close to a solid boundary where the flow is *laminar* and parallel to the surface, and where there is no eddy motion.

Thermal convection may involve energy transfer by eddy mixing and diffusion in addition to conduction (Burmeister 1983, Kays and Crawford 1980). Consider heat transfer to a fluid flowing inside a pipe. If the Reynolds number is large enough, three different flow regions exist. Immediately adjacent to the wall is a *laminar sublayer* where heat transfer occurs by thermal conduction; outside the laminar sublayer is a transition region called the *buffer layer*, where both eddy mixing and conduction effects are significant; beyond the buffer layer and extending to the center of

---

The preparation of this chapter is assigned to TC 1.3, Heat Transfer and Fluid Flow.

the pipe is the *turbulent region*, where the dominant mechanism of transfer is eddy mixing.

In most equipment, the main body of fluid is in turbulent flow, and the laminar layer exists at the solid walls only. In cases of low-velocity flow in small tubes, or with viscous liquids such as oil (*i.e.*, at low Reynolds numbers), the entire flow may be laminar with no transition or eddy region.

When fluid currents are produced by external sources (for example, a blower or pump), the solid-to-fluid heat transfer is termed *forced convection*. If the fluid flow is generated internally by nonhomogeneous densities caused by temperature variation, the heat transfer is termed *free* or *natural convection*.

In conduction and convection, heat transfer takes place through matter. For *thermal radiation*, there is a change in energy form; from internal energy at the source to electromagnetic energy for transmission, then back to internal energy at the receiver. Whereas conduction and convection are affected primarily by temperature difference and somewhat by temperature level, the heat transferred by radiation increases rapidly as the temperature increases.

Although some generalized heat transfer equations have been mathematically derived from fundamentals, they are usually obtained from correlations of experimental data. Normally, the correlations employ certain dimensionless numbers from analyses such as dimensional analysis or analogy. Table 1 lists some important dimensionless numbers.

STEADY-STATE CONDUCTION

For steady-state heat conduction in one dimension, the Fourier equation applies:

$$q = -kA(dt/dx) \qquad (1)$$

where

q = heat flow rate, Btu/h
k = thermal conductivity, Btu/h·ft·°F
A = cross-sectional area normal to flow, ft^2
dt/dx = temperature gradient, °F/ft

Equation (1) states that the heat flow rate q in the x direction is directly proportional to the temperature gradient dt/dx and the cross-sectional area A normal to the flow. The proportionality factor is the thermal conductivity k. The minus sign indicates that the heat flow is positive in the direction of decreasing temperature. Since conductivity values are sometimes given in other units, consistent units must be used.

Equation (1) may be integrated along a path of constant heat flow rate to obtain:

$$q = k(A_m/L_m)\Delta t = \Delta t/R \qquad (2)$$

Table 1   Dimensionless Numbers Commonly Used in Heat Transfer[a]

Name	Symbol	Value	Application
Nusselt number	Nu	hD/k or hL/k	Natural or forced convection, Boiling or condensing
Reynolds number	Re	GD/μ or $\rho VL/\mu$	Forced convection
Prandtl number	Pr	$\mu c_p/k$	Natural or forced convection, Boiling or condensing
Stanton number	St	h/Gc_p	Forced convection
Grashof number	Gr	$L^3\rho^2\beta g\,\Delta t/\mu^2$ or $L^3\rho^2 g\,\Delta t/T\mu^2$	Natural convection (for ideal gases)
Fourier number	Fo	$\alpha\tau/L^2$	Unsteady state conduction
Peclet number	Pe	GDc_p/k or Re Pr	Forced convection (small Pr)
Graetz number	Gz	GD^2c_p/kL or Re Pr D/L	Laminar convection

[a]A complete list of symbols appears at the end of this chapter.

where

A_m = mean cross-sectional area normal to flow, ft^2
L_m = mean length of heat flow path, ft
Δt = overall temperature difference, °F
R = thermal resistance, °F·h/Btu

The *thermal resistance R* is directly proportional to the mean length of the heat flow path L_m and inversely proportional to the conductivity k, and the mean cross-sectional area normal to the flow A_m. Equations for thermal resistances of a few common shapes are given in Table 2. Mathematical solutions to many heat conduction problems are addressed by Carslaw and Jaeger (1959). Complicated problems can be solved by graphical or numerical methods such as described by Croft and Lilley (1977), Adams and Rogers (1973), and Patankar (1980).

Equation (2) is analogous to Ohm's law for electrical circuits: thermal current (heat flow) in a thermal circuit is directly proportional to the thermal potential (temperature difference) and inversely proportional to the thermal resistance. This electrical-thermal analogy can be used for heat conduction in complex shapes that resist solution by exact analytical means. The thermal circuit concept is also useful for problems involving combined conduction, convection, and radiation.

OVERALL HEAT TRANSFER

In most steady-state heat transfer problems, more than one heat transfer mode may be involved. The various heat transfer coefficients may be combined into an overall coefficient so that the total heat transfer can be calculated from the terminal temperatures. The solution to this problem is much simpler if the concept of *thermal circuit* and *thermal resistance* is employed.

Local Overall Coefficient of Heat Transfer—Resistance Method

Consider heat transfer from one fluid to another by a three-step steady-state process: from a warmer fluid to a solid wall, through the wall, then to a colder fluid. An *overall coefficient* of heat transfer U based on the difference between the bulk temperatures of the two fluids $t_1 - t_2$ is defined:

$$q = U A(t_1 - t_2) \tag{3}$$

where A is the surface area. Since Equation (3) is a definition of U, the surface area A on which U is based is arbitrary; it should always be specified in referring to U.

The temperature drops across each part of the heat flow path are:

$$t_1 - t_{s1} = qR_1$$

$$t_{s1} - t_{s2} = qR_2$$

$$t_{s2} - t_2 = qR_3$$

where t_{s1} and t_{s2} are the warm and cold surface temperatures of the wall respectively, and R_1, R_2, and R_3 are the thermal resistances. Since the same quantity of heat flows through each thermal resistance, these equations combined give:

$$(t_1 - t_2)/q = (1/UA) = R_1 + R_2 + R_3 \tag{4}$$

Table 2   Solutions for Some Steady-State Thermal Conduction Problems

System	R in Equation $q = \Delta t/R$
Flat wall or curved wall if curvature is small (wall thickness less than 0.1 of inside diameter)	$R = \dfrac{L}{kA}$
Radial flow through a right circular cylinder	$R = \dfrac{\ln(r_o/r_i)}{2\pi kL}$
Buried cylinder	$R = \dfrac{\ln[(a + \sqrt{a^2 - r^2})/r]}{2\pi kL}$ $= \dfrac{\cosh^{-1}(a/r)}{2\pi kL} \; (L \gg 2r)$
Radial flow in a hollow sphere	$R = \dfrac{(1/r_i - 1/r_o)}{4\pi k}$

L, r, a = dimensions, ft
k = thermal conductivity (at average material temperature, Btu/h·ft^2·°F)
h = heat transfer coefficient, Btu/h·ft^2·°F
A = surface area, ft^2

As shown above, the equations are analogous to those for electrical circuits; for thermal current flowing through n resistances in *series*, the resistances are additive.

$$R_o = R_1 + R_2 + R_3 + \ldots + R_n \qquad (5)$$

Similarly, *conductance* is the reciprocal of resistance, and for heat flow through resistances in *parallel*, the conductances are additive:

$$C = (1/R_o) = (1/R_1) + (1/R_2) + (1/R_3) + \ldots + (1/R_n) \quad (6)$$

For convection, the thermal resistance is inversely proportional to the convection coefficient h_c and the applicable surface area:

$$R_c = 1/(h_c A) \qquad (7)$$

The thermal resistance for radiation is written similarly to that for convection:

$$R_r = 1/(h_r A) \qquad (8)$$

The term *radiation coefficient* h_r has no physical significance but is useful in computations. It is a function of the temperatures, radiation properties, and geometrical arrangement of the enclosure and the body in question.

Analysis by the resistance method can be illustrated by considering heat transfer from air outside to cold water inside an insulated pipe. The temperature gradients and the nature of the resistance analysis are shown in Figure 1.

Since air is sensibly transparent to radiation, some heat transfer occurs by both radiation and convection to the outer insulation surface. The mechanisms act in parallel on the air side. The total transfer then passes through the insulating layer and the pipe wall by thermal conduction, and then by convection and radiation into the cold water stream. (Radiation is not significant on the water side as liquids are sensibly opaque to radiation, although water transmits energy in the visible region.) The contact resistance between the insulation and the pipe wall is assumed negligible.

The heat transfer rate for a given length L of pipe q_{rc} may be thought of as the sum of the rates q_r and q_c flowing through the parallel resistances R_r and R_c associated with the surface radiation and convection coefficients. The total flow then proceeds through the resistance offered to thermal conduction by the insulation R_3, through the pipe wall resistance R_2, and into the water stream through the convection resistance R_1. Note the analogy to direct current electricity. A temperature (potential) drop is required to overcome resistances to the flow of thermal current. The total resistance to heat transfer R_o is the sum of the individual resistances:

$$R_o = R_1 + R_2 + R_3 + R_4 \qquad (9)$$

where the resultant parallel resistance R_4 is obtained from:

$$(1/R_4) = (1/R_r) + (1/R_c) \qquad (10)$$

If the individual resistances can be evaluated, the total resistance can be obtained from this relation. The heat transfer rate for the length of pipe L can be established by:

$$q_{rc} = (t_e - t)/R_o \qquad (11)$$

For a unit length of the pipe, the heat transfer rate is:

$$q_{rc}/L = (t_e - t)/R_o L \qquad (12)$$

The temperature drop Δt through each individual resistance may then be calculated from the relation:

$$\Delta t_n = R_n q_{rc} \qquad (13)$$

where $n = 1, 2,$ and 3.

Mean Temperature Difference

When heat is exchanged between two fluids flowing through a heat exchanger, the local temperature difference Δt varies along the flow path. Heat transfer may be calculated using:

$$q = UA\Delta t_m \qquad (14)$$

where U is the overall coefficient of heat transfer from fluid to fluid, A is an area associated with the coefficient U, and Δt_m is the appropriate mean temperature difference.

For parallel flow or counterflow exchangers and for any exchanger in which one fluid temperature is substantially constant, the mean temperature difference is:

$$\Delta t_m = \frac{\Delta t_1 - \Delta t_2}{\ln(\Delta t_1/\Delta t_2)} = \frac{\Delta t_1 - \Delta t_2}{2.3 \log (\Delta t_1/\Delta t_2)} \qquad (15)$$

where Δt_1 and Δt_2 are the temperature differences between the fluids at each end of the heat exchanger. Δt_m is called the *logarithmic mean temperature difference*. For the special case of $\Delta t_1 = \Delta t_2$, which leads to an indeterminate form of Equation (15), $\Delta t_m = \Delta t_1 = \Delta t_2$.

Equation (15) for Δt_m is true only if the overall coefficient and the specific heat of the fluids are constant through the heat exchanger, and no heat losses occur (often well-approximated in practice). Parker *et al.* (1969) give a procedure for cases with variable overall coefficient U.

Calculations using Equation (14) and Δt_m are convenient when terminal temperatures are known. In many cases, however, the temperatures of the fluids leaving the exchanger are not known. To avoid trial-and-error calculations, an alternate method involves the use of three nondimensional parameters, defined as follows:

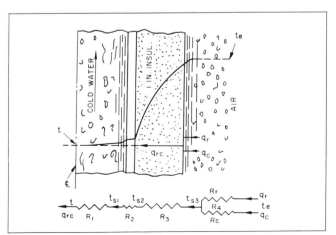

Fig. 1   Thermal Circuit Diagram for Insulated Cold Water Line

1. *Exchanger Heat Transfer Effectiveness ϵ*

$$\epsilon = \frac{(t_{hi} - t_{ho})}{(t_{hi} - t_{ci})} \quad \text{when } C_h = C_{min}$$

$$\epsilon = \frac{(t_{co} - t_{ci})}{(t_{hi} - t_{ci})} \quad \text{when } C_c = C_{min} \qquad (16)$$

where

$C_h = (\dot{m} c_p)_h$ = hot fluid capacity rate, Btu/h·°F
$C_c = (\dot{m} c_p)_c$ = cold fluid capacity rate, Btu/h·°F
C_{min} = smaller of two capacity rates
t_h = terminal temperature of hot fluid, °F. Subscript i indicates entering condition; subscript o indicates leaving condition.
t_c = terminal temperature of cold fluid, °F. Subscripts i and o are the same as for t_h.

2. *Number of Exchanger Heat Transfer Units* (NTU)

$$\text{NTU} = AU_{avg}/C_{min} = (1/C_{min}) \int_A U(dA) \qquad (17)$$

where A is the same area used to define overall coefficient U.

3. *Capacity Rate Ratio Z*

$$Z = C_{min}/C_{max} \qquad (18)$$

Generally, the heat transfer effectiveness can be expressed for a given exchanger as a function of the number of transfer units and the capacity rate ratio:

$$\epsilon = f(\text{NTU}, Z, \text{flow arrangement}) \qquad (19)$$

The effectiveness is independent of the temperatures in the exchanger. For any exchanger in which the capacity rate ratio Z is zero (where one fluid undergoes a phase change, *e.g.*, in a condenser or evaporator), the effectiveness is:

$$\epsilon = 1 - \exp(-\text{NTU}) \qquad (20)$$

Heat transferred can be determined from:

$$q = C_h(t_{hi} - t_{ho}) = C_c(t_{co} - t_{ci}) \qquad (21)$$

Combining Equations (16) and (21) produces an expression for heat transfer rate in terms of entering fluid temperatures:

$$q = \epsilon C_{min}(t_{hi} - t_{ci}) \qquad (22)$$

The proper mean temperature difference for the rate equation [Equation (14)] is then given by:

$$\Delta t_m = (t_{hi} - t_{ci}) \epsilon/\text{NTU} \qquad (23)$$

The effectiveness for parallel flow exchangers is:

$$\epsilon = \frac{1 - \exp[-\text{NTU}(1 + Z)]}{1 + Z} \qquad (24)$$

When, for a parallel flow exchanger, $Z = 1$:

$$\epsilon = \frac{1 - \exp(-2\,\text{NTU})}{2} \qquad (25)$$

The effectiveness for counterflow exchangers is:

$$\epsilon = \frac{1 - \exp[-\text{NTU}(1 - Z)]}{1 - Z \exp[-\text{NTU}(1 - Z)]} \qquad (26)$$

$$\epsilon = \frac{\text{NTU}}{1 + \text{NTU}} \quad \text{for } Z = 1 \qquad (27)$$

Incropera and DeWitt (1985) and Kays and London (1964) show the relations of ϵ, NTU, and Z for other flow arrangements. These authors and Afgan and Schlunder (1974) present graphical representations for convenience.

TRANSIENT HEAT FLOW

Often, the heat transfer and temperature distribution under unsteady state (varying with time) conditions must be known. Examples are: cold storage temperature variations on starting or stopping a refrigeration unit; variation of external air temperature, and solar irradiation affecting the heat load of a cold storage room or wall temperatures; the time required to freeze a given material under certain conditions in a storage room; quick freezing of objects by direct immersion in brines; sudden heating or cooling of fluids and solids from one temperature to a different temperature.

The fundamental equation for unsteady state conduction in solids or fluids in which there is no substantial motion is:

$$\frac{\delta t}{\delta \tau} = \alpha \left(\frac{\delta^2 t}{\delta x^2} + \frac{\delta^2 t}{\delta y^2} + \frac{\delta^2 t}{\delta z^2} \right) \qquad (28)$$

where the thermal diffusivity α is the ratio $k/\rho c_p$; k is the thermal conductivity; ρ, the density; and c_p, the specific heat. If α is large (high conductivity, low density, and specific heat or both), heat will diffuse faster.

To predict the rate of temperature change of a body or material being held at constant volume with uniform temperature, such as a well-stirred reservoir of fluid, whose temperature is changing because of a net rate of heat gain or loss, the applicable equation can be written:

$$q_{net} = Mc_v dt/d\tau \qquad (29)$$

where M is the mass of the body and c_v is its specific heat at constant volume. q_{net} is algebraic with positive being into the body and negative being out of the body. If the heating occurs at constant pressure, c_v should be replaced by c_p; however, for liquids and solids, c_v and c_p are nearly equal, and c_p can be used with negligible error. The term q_{net} may include heat transfer by conduction, convection, or radiation and is the difference between the rate of heat transfer to and away from the body.

From Equation (28), it is possible to derive expressions for temperature and heat flow variations at different instants and different locations. Most common cases have been solved and presented in graphical forms (Jakob 1957, Schneider 1964, Myers 1971). In other cases, it is simpler to use numerical methods (Croft and Lilley 1977, Patankar 1980). When convective boundary conditions are required in the solution of Equation (28), h values based on steady-state correlations are often used. However, this approach may not be valid when rapid transients are involved.

Estimating Cooling Times

Cooling times for materials can be estimated (McAdams 1954) by Gurnie-Lurie charts (Figures 2, 3, and 4), which are graphical solutions for the heating or cooling of infinite slabs, infinite cylinders, and spheres. These charts assume an initial uniform temperature distribution and no change of phase. They apply to a body exposed to a constant temperature fluid with a constant surface convection coefficient of h. The charts cannot be used to calculate freezing times because of the change of phase that occurs in freezing.

By using Figures 2, 3, and 4, it is possible to estimate the temperature at any point and the average temperature in a homogeneous mass of material as a function of time in a cooling process. In cooling, it is assumed that no freezing occurs. It is possible to estimate the cooling times for rectangular-shaped solids, cubes, cylinders, and spheres.

From a heat transfer point of view, a cylinder insulated on its ends behaves as a cylinder of infinite length, and a rectangular solid insulated so that only two parallel faces allow heat transfer behaves as an infinite slab. Also, a thin slab or a long, thin cylinder may be considered infinite objects.

Consider a slab of material with insulated edges being cooled. If the cooling time is the time required for the center of the slab to reach a temperature of t_2, the cooling time can be calculated as follows:

1. Evaluate the temperature ratio $(t_c - t_2)/(t_c - t_1)$.

where

t_c = temperature of cooling medium
t_1 = initial product temperature
t_2 = final product center temperature

Note that on Figures 2, 3, and 4, the temperature ratio $(t_c - t_2)/(t_c - t_1)$ is designated as Y to simplify the equations.

2. Determine the radius ratio r/r_m designated as n in Figures 2, 3, and 4.

where

r = distance from centerline
r_m = half thickness of slab

3. Evaluate the resistance ratio k/hr_m designated as m in Figures 2, 3, and 4.

where

k = thermal conductivity of material
h = heat transfer coefficient
r_m = half thickness of slab

4. From Figure 2 for infinite slabs, select the appropriate value of $k\tau/\rho c_p r_m^2$ designated as F_o on Figures 2, 3, and 4.

where

τ = time elapsed
c_p = specific heat
ρ = density

5. Determine τ from the value of $k\tau/\rho c_p r_m^2$.

Temperature Distribution in Finite Objects

Finite objects can be formed from the intersection of infinite objects. For example, the solid of intersection of an infinite cylinder and an infinite slab is a finite cylinder with a length equal to the thickness of the slab and a radius equal to that of the cylinder (Figure 5). Intersection of three infinite slabs with the same thickness produces a cube; intersection of three dissimilar slabs forms a finite rectangular solid.

The temperature in the finite object can be calculated from the temperature ratio Y of the infinite objects that are intersected to form the finite object. The product of temperature ratios of the infinite object is the temperature ratio of the finite object; *e.g.*, for the finite cylinder of Figure 5:

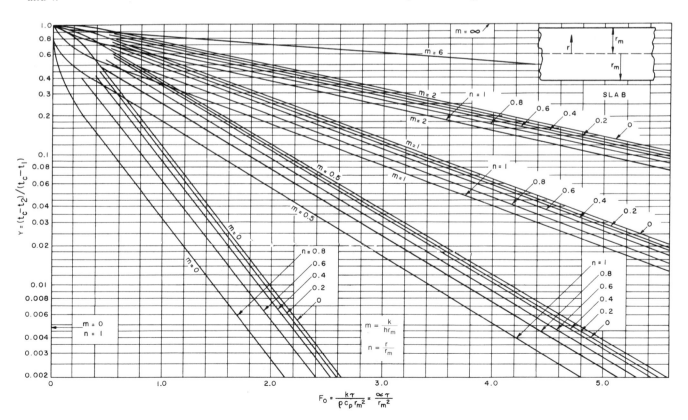

Fig. 2   Transient Temperatures for Infinite Slab

$$Y_{is} \; Y_{ic} = Y_{fc} \qquad (30)$$

where

Y_{is} = temperature ratio of infinite slab
Y_{ic} = temperature ratio of infinite cylinder
Y_{fc} = temperature ratio of finite cylinder

For a finite rectangular solid:

$$(Y_{is})_1 (Y_{is})_2 (Y_{is})_3 = Y_{frs} \qquad (31)$$

where

Y_{frs} = temperature ratio of finite rectangular solid, and subscripts 1, 2, and 3 designate three infinite slabs

Heat Exchanger Transients

Determination of the transient behavior of heat exchangers is becoming increasingly important in evaluating the dynamic behavior of heating and air-conditioning systems. Many studies have been made of the transient behavior of counterflow and parallel flow heat exchangers; some are listed in the bibliography.

THERMAL RADIATION

Radiation, one of the basic mechanisms for energy transfer between different temperature regions, is distinguished from conduction and convection in that it does not depend on an intermediate material as a carrier of energy, but is impeded by the presence of material between the regions. The radiation energy transfer process is the consequence of energy-carrying electromagnetic waves, emitted by atoms and molecules resulting from changes in their energy content. The amount and characteristics of radiant energy emitted by a quantity of material depend on the nature of the material, its microscopic arrangement, and its absolute temperature. Although rate of energy emission is independent of the surroundings, the *net* energy transfer rate depends on the temperatures and spatial relationships of the surface and its surroundings.

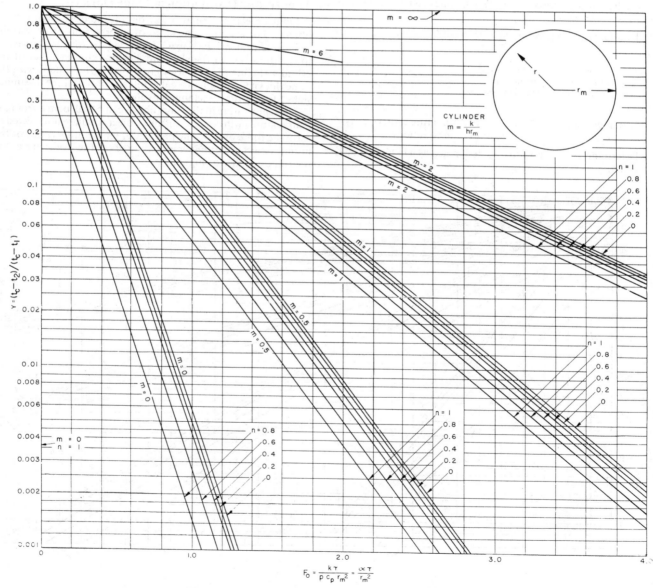

Fig. 3   Transient Temperatures for Infinite Cylinder

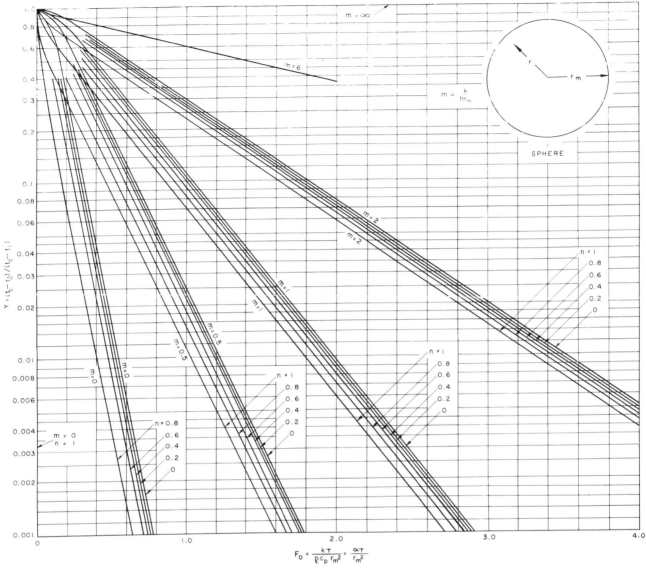

Fig. 4 Transient Temperatures for Spheres

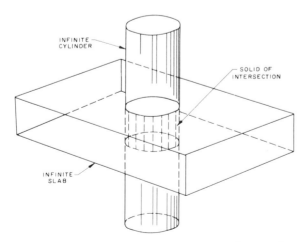

Fig. 5 Finite Cylinder of Intersection from Intersection of Infinite Cylinder and Infinite Slab

Blackbody Radiation

The rate of thermal radiant energy emitted by a surface depends on its absolute temperature. A surface is called *black* if it can absorb all incident radiation. The total energy emitted per unit time per unit area of black surface W_b to the hemispherical region above it is given by the *Stefan-Boltzmann law*.

$$W_b = \sigma T^4 \qquad (32)$$

The constant σ is 0.1714×10^{-8} Btu/h·ft²·°R⁴.

The heat radiated by a body is comprised of electromagnetic waves of many different frequencies or wavelengths. Planck showed that the spectral distribution of the energy radiated by a blackbody is given by:

$$W_{b\lambda} = \frac{C_1 \lambda^{-5}}{e^{C_2/\lambda T} - 1} \qquad (33)$$

where

$W_{b\lambda}$ = monochromatic emissive power of blackbody, Btu/(h·ft²·μm)
λ_1 = wavelength, μm
T = temperature, °R
C_1 = first Planck's law constant = 1.1870×10^8 Btu·μm⁴/h·ft²
C_2 = second Planck's law constant = 2.5896×10^4 μm·°R

$W_{b\lambda}$ is the *monochromatic emissive power*, defined as the energy emitted per unit time per unit surface area at wavelength λ per unit wavelength interval around λ; *i.e.*, the energy emitted per unit time per unit surface area in the interval $d\lambda$ is equal to $W_{b\lambda}d\lambda$.

The Stefan-Boltzmann equation can be obtained by integrating Planck's equation:

$$W_b = \sigma T^4 = \int_0^\infty W_{b\lambda}d\lambda \qquad (34)$$

Wien showed that the wavelength of maximum emissive power multiplied by the absolute temperature is a constant:

$$\lambda_{max} T = 5216 \ \mu\text{m} \cdot °\text{R} \qquad (35)$$

This useful relationship gives a rough idea of the spectral energy distribution from a black surface; 25% of the energy is radiated at wavelengths shorter than the maximum, and 75% is radiated at wavelengths longer than the maximum.

Actual Radiation

Substances and surfaces diverge variously from the Stefan-Boltzmann and Planck laws. W_b and $W_{b\lambda}$ are the maximum emissive powers at a surface temperature. Actual surfaces emit and absorb less readily and are called *nonblack*. The emissive power of a nonblack surface, at temperature T, radiating to the hemispherical region above it, is written as:

$$W = \epsilon W_b = \epsilon \sigma T^4 \qquad (36)$$

where ϵ is known as the *hemispherical emittance*. The term emittance conforms to physical and electrical terminology; the suffix "ance" denotes a property of a piece of material as it exists. The ending "ivity" denotes a property of the bulk material independent of geometry or surface condition. Thus, emittance, reflec-

tance, absorptance, and transmittance refer to actual pieces of material. Emissivity, reflectivity, absorptivity, and transmissivity refer to the properties of materials that are optically smooth and thick enough to be opaque.

The emittance is a function of the material, the condition of its surface, and the temperature of the surface. Table 3 lists selected values; Siegel and Howell (1981) have more extensive lists.

The monochromatic emissive power of a nonblack surface is similarly written as:

$$W_\lambda = \epsilon_\lambda W_{b\lambda} = \epsilon_\lambda(C_1\lambda^{-5}/e^{C_2/\lambda T} - 1) \qquad (37)$$

where ϵ_λ is the monochromatic hemispherical emittance. The relationship between ϵ and ϵ_λ is given by:

$$W = \epsilon \sigma T^4 = \int_0^\infty W_\lambda d\lambda = \int_0^\infty \epsilon_\lambda W_{b\lambda}d\lambda$$

or

$$\epsilon = (1/\sigma T^4)\int_0^\infty \epsilon_\lambda W_{b\lambda}d\lambda \qquad (38)$$

If ϵ_λ does not depend on λ, then, from Equation (38), $\epsilon = \epsilon_\lambda$. Surfaces with this characteristic are called *gray*. Gray surface characteristics are often assumed in calculations. Several classes of surfaces approximate this condition in some regions of the spectrum. The simplicity is desirable, but care must be exercised, especially if temperatures are high. Assumption of grayness is sometimes made because of the absence of information relating ϵ_λ and λ.

When radiant energy falls on a surface, it can be absorbed, reflected, or transmitted through the material. Therefore, from the first law of thermodynamics:

$$\alpha + \tau + \rho = 1 \qquad (39)$$

where

α = fraction of incident radiation absorbed or *absorptance*
τ = fraction of incident radiation transmitted or *transmittance*
ρ = fraction of incident radiation reflected or *reflectance*

If the material is opaque, as most solids are in the infrared, $\tau = 0$ and $\alpha + \rho = 1$. For a black surface, $\alpha = 1$, $\rho = 0$, and $\tau = 0$.

Table 3 Emittances and Absorptances for Some Surfaces[a]

Class	Surfaces	Total Normal Emittance[b] At 50 to 100 °F	At 1000 °F	Absorptance for Solar Radiation
1	A small hole in a large box, sphere, furnace, or enclosure	0.97 to 0.99	0.97 to 0.99	0.97 to 0.99
2	Black nonmetallic surfaces such as asphalt, carbon, slate, paint, paper	0.90 to 0.98	0.90 to 0.98	0.85 to 0.98
3	Red brick and tile, concrete and stone, rusty steel and iron, dark paints (red, brown, green, etc.)	0.85 to 0.95	0.75 to 0.90	0.65 to 0.80
4	Yellow and buff, brick and stone, firebrick, fireclay	0.85 to 0.95	0.70 to 0.85	0.50 to 0.70
5	White or light cream brick, tile, paint or paper, plaster, whitewash	0.85 to 0.95	0.60 to 0.75	0.30 to 0.50
6	Window glass	0.90	—	[c]
7	Bright aluminum paint; gilt or bronze paint	0.40 to 0.60	—	0.30 to 0.50
8	Dull brass, copper, or aluminum; galvanized steel; polished iron	0.20 to 0.30	0.30 to 0.50	0.40 to 0.65
9	Polished brass, copper, monel metal	0.02 to 0.05	0.05 to 0.15	0.30 to 0.50
10	Highly polished aluminum, tin plate, nickel, chromium	0.02 to 0.04	0.05 to 0.10	0.10 to 0.40
11	Selective surfaces			
	Stainless steel wire mesh	0.23 to 0.28	—	0.63 to 0.86
	White painted surface	0.92	—	0.23 to 0.49
	Copper treated with solution of NaClO₂ and NaOH	0.13	—	0.87
	Copper, nickel, and aluminum plate with CuO coating	0.09 to 0.21	—	0.08 to 0.93

[a] See also Chapter 37, McAdams (1954), and Siegel and Howell (1981).
[b] Hemispherical and normal emittance are not equal in many cases. The hemispherical emittance may be as much as 30% greater for polished reflectors to 7% lower for nonconductors.
[c] Absorbs 4 to 40% depending on its transmittance.

Platinum black and gold black have absorptances of about 98% in the infrared, which is as black as any actual surface is. Any desired degree of blackness can be simulated by a small hole in a large enclosure. Consider a ray of radiant energy entering the opening. It will undergo many internal reflections and be almost completely absorbed before it has a reasonable probability of passing back out of the opening.

Certain flat black paints also exhibit emittances of 98% over a wide range of conditions. They provide a much more durable surface than gold or platinum black and are frequently used on radiation instruments and as standard reference in emittance or reflectance measurements.

Kirchhoff's law, relating emittance and absorptance of any opaque surface from thermodynamic considerations, states that for any surface where the incident radiation is independent of angle or where the surface is diffuse, $\epsilon_\lambda = \alpha_\lambda$. If the surface is gray, or the incident radiation is from a black surface at the same temperature, then also $\epsilon = \alpha$, but many surfaces are not gray. For most surfaces listed in Table 3, absorptance for solar radiation is different than emittance for low-temperature radiation. This is because the wavelength distributions are different in the two cases, and ϵ_λ varies with wavelength.

The foregoing discussion relates to total hemispherical radiation from surfaces. Energy distribution over the hemispherical region above the surface also has an important effect on the rate of heat transfer in various geometric arrangements.

Lambert's law states that the emissive power of radiant energy over a hemispherical surface above the emitting surface varies as the cosine of the angle between the normal to the radiating surface and the line joining the radiating surface to the point of the hemispherical surface. This radiation is *diffuse* radiation. The Lambert emissive power variation is equivalent to assuming that radiation from a surface in a direction other than normal occurs as if it came from an equivalent area with the same emissive power (per unit area) as the original surface. The equivalent area is obtained by projecting the original area onto a plane normal to the direction of radiation. Black surfaces obey the Lambert law exactly. It is approximate for many actual radiation and reflection processes, especially those involving rough surfaces and nonmetallic materials. Most radiation analyses are based on the assumption of gray diffuse radiation and reflection.

In estimating heat transfer rates between surfaces of different geometries, radiation characteristics, and orientations, it is usually assumed that: (1) all surfaces are gray or black, (2) radiation and reflection are diffuse, (3) properties are uniform over the surfaces, (4) absorptance equals emittance and is independent of the temperature of the source of incident radiation, and (5) the material in the space between the radiating surfaces neither emits nor absorbs radiation. These assumptions simplify problems greatly, although results must be considered approximate.

Angle Factor

The distribution of radiation from a surface among the surfaces it irradiates is indicated by a quantity variously called an interception, view, configuration, or angle factor. In terms of two surfaces, i and j, the *angle factor* from surface i to surface j, F_{ij}, is defined as the fraction of diffuse radiant energy leaving surface i that falls directly on j (*i.e.*, is intercepted by j). The angle factor from j to i is similarly defined, merely by interchanging the roles of i and j. This second angle factor will not, in general, be numerically equal to the first. However, $F_{ij}A_i = F_{ji}A_j$, where A is the surface area. Note that a concave surface may *see* itself, $F_{ii} \neq 0$, and that if n surfaces form an enclosure:

$$\sum_{j=1}^{n} F_{ij} = 1 \qquad (40)$$

The angle factor F_{12} between two surfaces is:

$$F_{12} = \frac{1}{A_1} \int_{A_1} \int_{A_2} \frac{\cos \phi_1 \cos \phi_2}{\pi r^2} dA_1 dA_2 \qquad (41)$$

where dA_1 and dA_2 are elemental areas of the two surfaces, r is the distance between dA_1 and dA_2, and ϕ_1 and ϕ_2 are the angles between the respective normals to dA_1 and dA_2 and the connecting line r. Numerical, graphical, and mechanical techniques can solve this equation (Siegel and Howell 1981). Numerical values of angle factor for common geometries are given in Figure 6.

Calculation of Radiant Exchange between Surfaces Separated by Nonabsorbing Media

A surface radiates energy at a rate independent of its surroundings and absorbs and reflects incident energy at a rate dependent on its surface condition. The net energy exchange per unit area is denoted by q or q_j for unit area of A_j. It is the rate of emission of the surface minus the total rate of absorption at the surface from all radiant effects in its surroundings, possibly including the return of some of its own emission by reflection off its surroundings. The rate at which energy must be supplied to the surface by other exchange processes if its temperature is to remain constant is q; therefore, to define q, the total radiant surroundings (in effect, an enclosure) must be specified.

Several methods have been developed to solve certain problems. To calculate the radiation exchange at each surface of an n opaque surface enclosure by simple, general equations convenient for machine calculation, two terms must be defined:

G = irradiation; total radiation incident on surface per unit time and per unit area
J = radiosity; total radiation that leaves surface per unit time and per unit area

The radiosity is the sum of the energy emitted and the energy reflected:

$$J = \epsilon W_b + \rho G \qquad (42)$$

Since the transmittance is zero, the reflectance is:

$$\rho = 1 - \alpha = 1 - \epsilon$$

Thus:

$$J = \epsilon W_b + (1 - \epsilon)G \qquad (43)$$

The net energy lost by a surface is the difference between the radiosity and the irradiation:

$$q/A = J - G = \epsilon W_b + (1 - \epsilon)G - G \qquad (44)$$

Substituting for G in terms of J from Equation (43):

$$q = \frac{W_b - J}{(1 - \epsilon)/\epsilon \, A} \qquad (45)$$

Consider an enclosure of n isothermal surfaces with areas of A_1, A_2, $\cdots$, A_n, emittances of ϵ_1, ϵ_2, $\cdots$, ϵ_n, and reflectances of ρ_1, ρ_2, $\cdots$, ρ_n, respectively.

The irradiation of surface i is the sum of the radiation incident on it from all n surfaces, or

$$G_i A_i = \sum_{j=1}^{n} F_{ji} J_j A_j = \sum_{j=1}^{n} F_{ij} J_j A_i$$

or

$$G_i = \sum_{j=1}^{n} F_{ij} J_j$$

Substituting in Equation (44) yields the following simultaneous equations when each of the n surfaces is considered:

$$J_i = \epsilon_i W_{bi} + (1 - \epsilon_i) \sum_{j=1}^{n} F_{ij} J_j \qquad i = 1, 2, \cdots n \qquad (46)$$

Equation (46) can be solved manually for the unknown Js if the number of surfaces is small. More complex enclosures require a computer.

Once the radiosities (Js) are known, the net radiant energy lost by each surface is determined from Equation (45) as:

$$q_i = \frac{W_{bi} - J_i}{(1 - \epsilon_i)/\epsilon_i A_i}$$

If the surface is black, Equation (45) becomes indeterminate and an alternate expression must be used, such as:

$$q_i = \sum_{j=1}^{n} (J_i A_i F_{ij} - J_j A_j F_{ji})$$

or

$$q_i = \sum_{j=1}^{n} F_{ij} A_i (J_i - J_j) \qquad (47)$$

since

$$F_{ij} A_i = F_{ji} A_j$$

All diffuse radiation processes are included in the aforementioned enclosure method, and surfaces with special characteristics are assigned consistent properties. An opening is treated as an equivalent surface area A_e, with a reflectance of zero. If energy enters the enclosure diffusely through the opening, A_e is assigned

PERPENDICULAR RECTANGES WITH COMMON EDGE

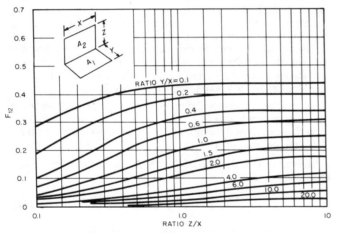

ALIGNED PARALLEL RECTANGLES

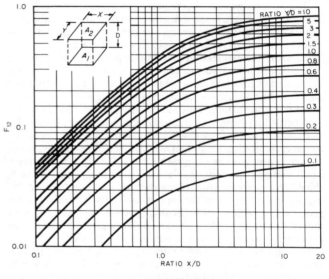

CONCENTRIC CYLINDERS OF FINITE LENGTH

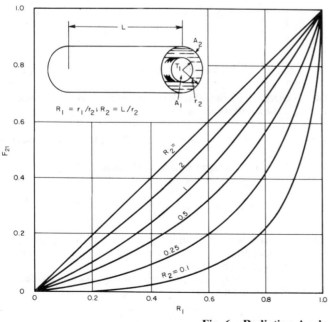

COAXIAL DISKS

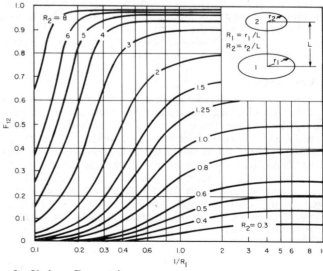

Fig. 6   Radiation Angle Factor for Various Geometries

an equivalent temperature; otherwise, its temperature is taken as zero. If the loss through the opening is desired, q_2 is found. A window in the enclosure is assigned its actual properties.

A surface in *radiant balance* is one for which radiant emission is balanced by radiant absorption; heat is neither removed from nor supplied to a surface. Reradiating surfaces, insulated surfaces with $q_{net} = 0$, can be treated in Equation (46) as being perfectly reflective, $\epsilon = 0$. The equilibrium temperature of such a surface can be found from:

$$T_k = (J_k/\sigma)^{0.25}$$

once Equation (46) has been solved for the radiosities.

Use of angle factors and radiation properties as defined assumes that the surfaces are diffuse radiators—a good assumption for most nonmetals in the infrared region, but a poor assumption for highly polished metals. Subdividing the surfaces and considering the variation of radiation properties with angle of incidence improves the approximation but increases the work required for a solution.

Radiation in Gases

Elementary gases such as oxygen, nitrogen, hydrogen, and helium are essentially transparent to thermal radiation. Their absorption and emission bands are confined mainly to the ultraviolet region of the spectrum. The gaseous vapors of most compounds, however, have absorption bands in the infrared region. Carbon monoxide, carbon dioxide, water vapor, sulfur dioxide, ammonia, acid vapors, and organic vapors absorb and emit significant amounts of energy.

Radiation exchange by opaque solids is considered a surface phenomenon. Radiant energy does, however, penetrate the surface of all materials. The absorption coefficient gives the rate of exponential attenuation of the energy. Metals have large absorption coefficients, and radiant energy penetrates only a few hundred angstroms at most. Absorption coefficients for nonmetals are lower. Radiation may be considered a surface phenomenon unless the material is transparent. Gases have small absorption coefficients, so the path length of radiation through gas becomes very significant.

Beer's law states that the attenuation of radiant energy in a gas is a function of the product of the partial pressure of the gas times the path length (p_gL). The monochromatic absorptance of a body of gas of thickness L is then given by:

$$\alpha_{\lambda L} = 1 - e^{-\alpha\lambda L} \tag{48}$$

Since absorption occurs in discrete wavelengths, the absorptances must be summed over the spectral region corresponding to the temperature of the blackbody radiation passing through the gas. The monochromatic absorption coefficient α_λ is also a function of temperature and pressure of the gas; therefore, detailed treatment of gas radiation is quite complex.

Estimated emittance for carbon dioxide and water vapor in air at 75 °F is a function of concentration and path length (Table 4). The values are for a hemispherically shaped body of gas radiating to an element of area at the center of the hemisphere. Among other, Siegel and Howell (1981) and Hottel and Sarofim (1967)

describe geometrical calculations in their texts on radiation transfer. Generally, at low values of p_gL, the mean path length L, or equivalent hemispherical radius for a gas body radiating to its surrounding surfaces, is four times the mean hydraulic radius of the enclosure. A room with a dimensional ratio of 1:1:4 has a mean path length of 0.89 times the shortest dimension when considering radiation to all walls. For a room with a dimensional ratio of 1:2:6, the mean path length for the gas radiating to all surfaces is 1.2 times the shortest dimension. The mean path length for radiation to the 2 by 6 face is 1.18 times the shortest dimension. These values are for cases where the partial pressure of the gas times the mean path length approaches zero ($p_gL \approx 0$). The factor decreases with increasing values of p_gL. For an average room with approximately 8-ft ceilings and relative humidity ranging from 10 to 75% at 75 °F, the effective path length for carbon dioxide radiation is about 85% of the ceiling height, or 6.8 ft. The effective path length for water vapor would be about 93% of the ceiling height, or 7.4 ft. The effective emittance of the water vapor and carbon dioxide radiating to the walls, ceiling, and floor of a room 16 ft by 48 ft with 8-ft walls is in the following tabulation.

Relative Humidity, %	ϵ_g
10	0.10
50	0.19
75	0.22

The radiation heat transfer from the gas to the walls is then:

$$q = \sigma A_w \epsilon_g (T_g^4 - T_w^4) \tag{49}$$

The examples in Table 4 and the preceding text indicate the importance of gas radiation in environmental heat transfer problems. Gas radiation in large furnaces is the dominant mode of heat transfer, and many additional factors must be considered. Increased pressure broadens the spectral bands, and interaction of different radiating species prohibits simple summation of the emittance factors for the individual species. Departures from blackbody conditions necessitate separate calculations of the emittance and absorptance. McAdams (1954) and Hottel and Sarofim (1967) give more complete treatments of gas radiation.

NATURAL CONVECTION

Heat transfer involving motion in a fluid caused by the difference in density and the action of gravity is called *natural* or *free convection*. Heat transfer coefficients for natural convection are generally much lower than those for forced convection, and it is therefore important not to ignore radiation in calculating the total heat loss or gain. Radiant transfer may be of the same order of magnitude as natural convection, even at room temperatures, since wall temperatures in a room can affect human comfort (Chapter 8).

Natural convection is important in a variety of heating and refrigeration equipment: (1) gravity coils used in high humidity cold storage rooms and in roof-mounted refrigerant condensers, (2) the evaporator and condenser of household refrigerators, (3) baseboard radiators and convectors for space heating, and (4) cooling panels for air conditioning. Natural convection is also involved in heat loss or gain to equipment casings and interconnecting ducts and pipes.

Consider heat transfer by natural convection between a cold fluid and a hot surface. The fluid in immediate contact with the surface is heated by conduction, becomes lighter and rises because of the difference in density of the adjacent fluid. The motion is resisted by the viscosity of the fluid. The heat transfer is influenced

Table 4 Emittance of CO₂ and Water Vapor in Air at 75 °F

Path Length, ft	CO₂, % by Volume		Relative Humidity, %			
	0.1	0.3	1.0	10	50	100
10	0.03	0.06	0.09	0.06	0.17	0.22
100	0.09	0.12	0.16	0.22	0.39	0.47
1000	0.16	0.19	0.23	0.47	0.64	0.70

by: (1) gravitational force due to thermal expansion, (2) viscous drag, and (3) thermal diffusion. Gravitational acceleration g, the coefficient of thermal expansion β, the kinematic viscosity $\nu = \mu/\sigma$, and the thermal diffusivity $\alpha = k/\rho c_p$ affect natural convection. These variables are included in the dimensionless numbers, given in Equation (1) of Table 5: the Nusselt number Nu is a function of the product of the Prandtl number Pr and Grashof number Gr. These numbers, when combined, depend on the fluid properties, the temperature difference between the surface and the fluid Δt, and the characteristic length of the surface L. The constant c and the exponent n depend on the physical configuration and the nature of flow.

Natural convection cannot be represented by a single value of exponent n, but can be divided into three regions: (1) *turbulent natural convection*, for which n equals 0.33; (2) *laminar natural convection*, for which n equals 0.25; and (3) a region that has Gr Pr less than for laminar natural convection, for which the exponent n gradually diminishes from 0.25 to lower values. Note that, for wires, the Gr Pr is likely to be very small, so that the exponent n is 0.1 [Equation (5), Table 5].

To calculate the natural convection heat transfer coefficient, determine Gr Pr to find whether the boundary layer is laminar or turbulent; then apply the appropriate equation from Table 5. The correct characteristic length indicated in the table must be used. Since the exponent n is 0.33 for a turbulent boundary layer, the characteristic length cancels out [in Equation (2), Table 5], and the heat transfer coefficient is independent of the characteristic length, as seen in Equations (7), (9), and (11) of Table 5. Turbulence occurs when length or temperature difference is large. Since the length of a pipe is generally greater than its diameter, the heat transfer coefficient for vertical pipes is larger than for horizontal pipes.

Convection from horizontal plates facing downward when heated (or upward when cooled) is a special case. Since the hot air is above the colder air, theoretically no convection should occur. Some convection is caused, however, by secondary influences such as temperature differences on the edges of the plate. As an approximation, a coefficient of somewhat less than half of the coefficient for a heated horizontal plate facing upward can be used.

Since air is often the heat transport fluid, simplified equations for air are given in Table 5. Other information on natural convection is available in the general heat transfer references.

Observed differences in the comparisons of recent experimental and numerical results with existing correlations for natural convective heat transfer coefficients indicate that caution should be taken when applying coefficients for (isolated) vertical plates to vertical surfaces in enclosed spaces (buildings). Bauman *et al.* (1983) and Altmayer *et al.* (1983) developed improved correlations for calculating natural convective heat transfer from vertical surfaces in rooms under certain temperature boundary conditions.

Natural convection can affect the heat transfer coefficient in the presence of weak forced convection. As the forced convection effect (*i.e.*, the Reynolds number) increases, the "mixed convection" (superimposed forced-on-free convection) gives way to the pure forced convection regime. In these cases, other sources describing combined free and forced convection should be consulted, since the heat transfer coefficient in the mixed convection region is often larger than that calculated based on the natural or forced convection calculation alone. Metais and Eckert (1964) summarize natural, mixed, and forced convection regimes for vertical and horizontal tubes. Figure 7 shows the approximate limits for horizontal tubes. Other studies are described by Grigull *et al.* (1982). Local conditions influence the values of the convection coefficient in a mixed convection regime, but the references permit locating the pertinent regime and approximating the convection coefficient.

Table 5   Natural Convection Heat Transfer Coefficients

I. General relationships	$Nu = c(Gr\,Pr)^n$ (1)
	$h = c\,\dfrac{k}{L}\left(\dfrac{L^3\rho^2\beta g(\Delta t)}{\mu_2}\right)^n_f \left(\dfrac{\mu c_p}{k}\right)^n_f$ (2)
Characteristic length L for vertical plates, or pipes	L = height
Horizontal plates	L = length
Horizontal pipes	L = diameter
Spheres	L = 0.5 diameter
Rectangular block, with horizontal length L_h and vertical length L_v	$1/L = (1/L_h) + (1/L_v)$

II. Planes and pipes

Horizontal or vertical planes, pipes, rectangular blocks, and spheres (excluding horizontal plates facing downward for heating and facing upward for cooling)

(a) Laminar range, when Gr Pr is between 10^4 and 10^8	$Nu = 0.56\,(Gr\,Pr)^{0.25}$	(3)
(b) Turbulent range, when Gr Pr is between 10^8 and 10^{12}	$Nu = 0.13\,(Gr\,Pr)^{0.33}$	(4)

III. Wires

For horizontal or vertical wires, use L = diameter, for Gr Pr between 10^{-7} and 1	$Nu = (Gr\,Pr)^{0.1}$	(5)

IV. With air

Gr Pr $= 1.6 \times 10^6\,L^3(\Delta t)$ (at 70 °F, L in ft, Δt in °F)

(a) Cylinders		
Small cylinder, Laminar range	$h = 0.27\,(\Delta t/L)^{0.25}$	(6)
Large cylinder, Turbulent range	$h = 0.18\,(\Delta t)^{0.33}$	(7)
(b) Vertical plates		
Small plates, Laminar range	$h = 0.29\,(\Delta t/L)^{0.25}$	(8)
Large plates, Turbulent range	$h = 0.19\,(\Delta t)^{0.33}$	(9)
(c) Horizontal plates, facing upward when heated or downward when cooled		
Small plates, Laminar range	$h = 0.27\,(\Delta t/L)^{0.25}$	(10)
Large plates, Turbulent range	$h = 0.22\,(\Delta t)^{0.33}$	(11)
(d) Horizontal plates, facing downward when heated or upward when cooled		
Small plates	$h = 0.12\,(\Delta t/L)^{0.25}$	(12)

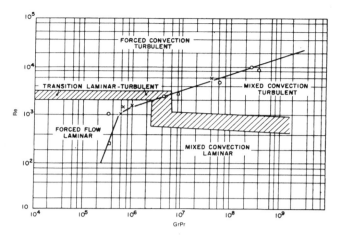

Fig. 7   Regimes of Free, Forced, and Mixed Convection for Flow-through Horizontal Tubes

FORCED CONVECTION

Forced air coolers and heaters, forced air- or water-cooled condensers and evaporators, and liquid suction heat exchangers are examples of equipment that transfer heat primarily by forced convection.

When fluid flows over a flat plate, a *boundary layer* forms adjacent to the plate. The velocity of the fluid at the plate surface is zero and increases to its maximum *free stream* value just past the edge of the boundary layer (Figure 8). Boundary layer formation is important, since the temperature change from plate to fluid (thermal resistance) is concentrated here. Where the boundary layer is thick, thermal resistance is great and the heat transfer coefficient is small. At the leading edge of the plate, boundary layer thickness is theoretically zero, and the heat transfer coefficient is infinite. Flow within the boundary layer immediately downstream from the leading edge is laminar, and is known as *laminar forced convection*. As flow proceeds along the plate, the laminar boundary layer increases in thickness to a critical value. Then, turbulent eddies develop within the boundary layer, except for a thin *laminar sublayer* adjacent to the plate. The boundary layer beyond this point is a *turbulent boundary layer*, and the flow is *turbulent forced convection*. The region between the breakdown of the laminar boundary layer and the establishment of the turbulent boundary layer is the *transition region*. Since the turbulent eddies greatly enhance heat transport into the main stream, the heat transfer coefficient begins to increase rapidly through the transition region. For a flat plate with a smooth leading edge, the turbulent boundary layer starts at Reynolds numbers, based on distance from the leading edge, of about 300,000 to 500,000. In blunt-edged plates, it can start at much smaller Reynolds numbers.

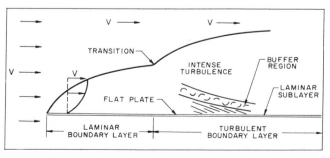

**Fig. 8   Boundary Layer Buildup on Flat Plate
(Vertical Scale Magnified)**

For flow in long tubes or channels of small hydraulic diameter, the laminar boundary layers on each wall grow until they meet, if the velocity is sufficiently low. Beyond this point, the velocity distribution does not change, and no transition to turbulent flow takes place. This is called *fully developed laminar flow.* For tubes of large diameter or at higher velocities, transition to turbulence takes place and *fully developed turbulent flow* is established (Figure 9). Therefore, the length dimension that determines the critical Reynolds number is the hydraulic diameter of the channel. For smooth circular tubes, flow is laminar for Reynolds numbers below 2100 and turbulent above 10,000.

Table 6 lists various forced convection correlations. In the generalized, dimensionless formula of Equation (1), Table 6, heat transfer is determined by flow conditions and by the fluid properties, as indicated by the Reynolds number and the Prandtl number. This equation can be modified to Equation (4), Table 6 to get the *heat transfer factor j.* The heat transfer factor is related to the *friction factor f* by the interrelationship of the transport of momentum and heat, and is approximately $f/2$ for turbulent flow in straight ducts. These factors are plotted in Figure 10.

The characteristic length D is the diameter of the tube, outside or inside, or the length of the plane plate. For other shapes, the hydraulic diameter D_h is used, where:

$$D_h = 2\,r_h = 4\,\frac{\text{Cross-sectional area for flow}}{\text{Total wetted perimeter}}$$

This reduces to twice the distance between surfaces for parallel plates or an annulus.

Simplified equations applicable to common fluids under normal operating conditions appear in Equations (8) through (25) of Table 6. Figure 11 gives graphical solutions for water.

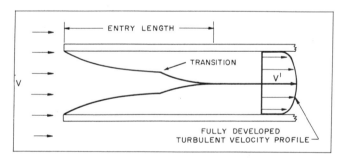

**Fig. 9   Boundary Layer Buildup in Entry Length
of Tube or Channel**

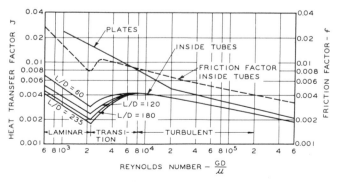

**Fig. 10   Typical Dimensionless Representation of
Forced Convection Heat Transfer**

Table 6   Equations for Forced Convection

Description	Reference Author	Reference Page	Reference Eq. No.	Equation	
I. Generalized correlations					
(a) Turbulent flow inside tubes	Jakob	491	(23–36)	$\dfrac{hD}{k} = c \left(\dfrac{GD}{\mu}\right)^{m} \left(\dfrac{\mu c_p}{k}\right)^{n}$	(1)
(1) Using fluid properties based on bulk temperature t	McAdams	219	(9–10a)	$\dfrac{hD}{k} = 0.023 \left(\dfrac{GD}{\mu}\right)^{0.8} \left(\dfrac{\mu c_p}{k}\right)^{0.4}$ (See Note a)	(2)
(2) Same as (1), except μ at surface temperature t_s	McAdams	219	(9–10c)	$\dfrac{h}{c_p G} \left(\dfrac{c_p \mu}{k}\right)^{2/3}_{f} = \dfrac{0.023}{(GD/\mu)^{0.2}}$	(3)
(3) Using fluid properties based on film temperature t_f, except c_p in Stanton modulus	McAdams	219	(9–10b)	$\dfrac{h}{c_p G} \left(\dfrac{c_p \mu}{k}\right)^{2/3}_{f} = \dfrac{0.023}{(GD/\mu_f)^{0.2}} = j$	(4)
where $t_f = 1/2(t_s + t)$					
(4) For viscous fluids (viscosities higher than twice water), using viscosity μ at bulk temperature t and μ_s at surface temperature t_s	Jakob	547	(26–12)	$\dfrac{hD}{k} = 0.027 \left(\dfrac{GD}{\mu}\right)^{0.8} \left(\dfrac{\mu c_p}{k}\right)^{1/3} \left(\dfrac{\mu}{\mu_s}\right)^{0.14}$	(5)
(b) Laminar flow inside tubes					
(1) For large D or high Δt, the effect of natural convection should be included	Jakob	544	(26–5)	$\dfrac{hD}{k} = 1.86 \left[\left(\dfrac{GD}{\mu}\right) \left(\dfrac{c_p \mu}{k}\right) \left(\dfrac{D}{L}\right) \right]^{1/3} \left(\dfrac{\mu}{\mu_s}\right)^{0.14}$	(6)
(2) For very long tubes				when $\left(\dfrac{GD}{\mu}\right) \left(\dfrac{c_p \mu}{k}\right) \left(\dfrac{D}{L}\right) < 20$, Eq. (6) should not be used	
(c) Annular spaces, Turbulent flow All fluid properties at bulk temperature except μ_s at surface temperature t_s	McAdams	242	(9–32c)	$\dfrac{h}{c_p G} \left(\dfrac{c_p \mu}{k}\right)^{2/3} \left(\dfrac{\mu_s}{\mu}\right)^{0.14} = \dfrac{0.023}{(D_e G/\mu)^{0.2}}$	(7)
II. Simplified equations for gases, Turbulent flow inside tubes (Units are in lb_m, h, ft, °F, and Btu.)					
(a) Most common gases, Turbulent flow (assuming $\mu = 0.0455\ lb_m/ft \cdot h$ and $\mu c_p/k = 0.78$)	Obtained from Eq. (2)			$h = 0.0144\ (c_p G^{0.8}/D^{0.2})$	(8)
(b) Air at ordinary temperatures	Obtained from Eq. (2)			$h = c\ (G^{0.8}/D^{0.2})$ (See Note b)	(9)
(c) Fluorinated hydrocarbon refrigerant gas at ordinary pressures	Obtained from Eq. (2)			$h = c\ (G^{0.8}/D^{0.2})$ (See Note b)	(10)
(d) Ammonia gas at approximately 150°F, 300 psi	Obtained from Eq. (2)			$h = 0.00756\ (G^{0.8}/D^{0.2})$	(11)
at 0°F, 24 psi	Obtained from Eq. (2)			$h = 0.00604\ (G^{0.8}/D^{0.2})$	(12)
III. Simplified equations for liquids, Turbulent flow inside tubes (Units are in lb_m, h, ft, °F, and Btu.)					
(a) Water at ordinary temperatures, 40 to 200°F. V is velocity in ft/s, D is tube ID in inches.	McAdams	228	(9–19)	$h = \dfrac{150\ (1 + 0.011t)\ V^{0.8}}{D^{0.2}}$	(13)
(b) Fluorinated hydrocarbon refrigerant liquid	Obtained from Eq. (2)			$h = c\ (G^{0.8}/D^{0.2})$ (See Note b)	(14)
(c) Ammonia liquid at approximately 100°F	Obtained from Eq. (2)			$h = 0.0156\ (G^{0.8}/D^{0.2})$	(15)
(d) Oil heating, Approximate equation	Brown and Marco	146	(7–15)	$h = 0.034\ V/\mu_f^{0.63}$	(16)
(e) Oil cooling, Approximate equation	Brown and Marco	146	(7–15)	$h = 0.0255\ V/\mu_f^{0.63}$	(17)
IV. Simplified equations for air					
(a) Vertical plane surfaces, V of 16 to 100 fps (room temperature)[c]	McAdams	249	(9–42)	$h' = 0.5\ (V)^{0.78}$	(18)
(b) Vertical plane surfaces, $V < 16$ fps (room temperature)[c]	McAdams	249	(9–42)	$h' = 0.99 + 0.21\ V$	(19)
(c) Single cylinder cross flow (film temperature = 200°F) $1000 < GD/\mu_f < 50,000$	McAdams	261	(10–3c)	$h = 0.026\ (G^{0.6}/D^{0.4})$	(20)
(d) Single sphere $17 < GD/\mu_f < 70,000$	McAdams	265	(10–6)	$h = 0.37\ \dfrac{k_f}{D} \left(\dfrac{GD}{\mu_f}\right)^{0.6}$	(21)
V. Gases flowing normal to pipes (dimensionless)					
(a) Single cylinder Re from 0.1 to 1000	McAdams	260	(10–3)	$\dfrac{hD}{k_f} = 0.32 + 0.43 \left(\dfrac{GD}{\mu}\right)^{0.52}$	(22)
Re from 1000 to 50,000	McAdams	260	(10–3)	$\dfrac{hD}{k_f} = 0.24 \left(\dfrac{GD}{\mu_f}\right)^{0.6}$	(23)
(b) Unbaffled staggered tubes, 10 rows, Approximate equation for turbulent flow[d]	McAdams	272	(10–11a)	$\dfrac{hD}{k_f} = 0.33 \left(\dfrac{G_{max} D}{\mu_f}\right)^{0.6} \left(\dfrac{\mu c_p}{k}\right)^{1/3}_{f}$	(24)
(c) Unbaffled in-line tubes, 10 rows, Approximate equation for turbulent flow[d] $(G_{max} D/\mu_f)$ from 2000 to 32,000	McAdams	272	(10–11a)	$\dfrac{hD}{k_f} = 0.26 \left(\dfrac{G_{max} D}{\mu_f}\right)^{0.6} \left(\dfrac{\mu c_p}{k}\right)^{1/3}_{f}$	(25)

[a]McAdams (1954) recommends this equation for heating and cooling. Others recommend an exponent of 0.4 for heating and 0.3 for cooling.
[b]Table 7, 1981 ASHRAE *Handbook*, lists values for c.

[c]h' is expressed in Btu/h·ft²·°F based on initial temperature difference.
[d]G_{max} is based on *minimum* free area. Coefficients for tube banks depend greatly on geometrical details. These values approximate only.

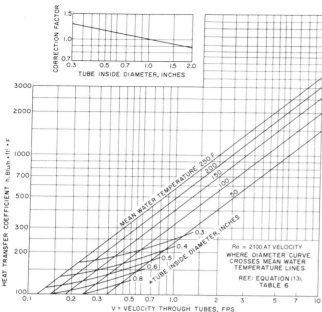

Fig. 11   Heat Transfer Coefficient for Turbulent Flow of Water Inside Tubes

Techniques to Augment Forced Convection

As discussed by Bergles (1985), techniques applied to augment heat transfer can be classified as passive methods, which require no direct application of external power, or as active schemes, which require external power. Examples of passive techniques include rough surfaces, extended surfaces, displaced promoters, and vortex flow devices. Examples of active techniques include mechanical aids, surface vibration, fluid vibration, and electrostatic fields. The effectiveness of a given augmentation technique depends largely on the mode of heat transfer or the type of heat exchanger to which it is applied.

When augmentation is used, the dominant thermal resistances in Equation (9) should be considered, *i.e.*, do not invest in a reduction of already low thermal resistance (increase of an already high heat transfer coefficient). Additionally, heat exchangers with a large number of heat transfer units (NTU) show relatively small gains in effectiveness with augmentation [see Equations (26) and (27)]. Finally, the increased friction factor that accompanies the heat transfer augmentation must be considered.

Several examples of tubes with internal roughness or fins are shown in Figure 12. Rough surfaces of the spiral repeated rib variety are widely used to improve in-tube heat transfer with water, as in flooded chillers. The roughness may be produced by spirally indenting the outer wall, forming the inner wall, or inserting coils. Internal fins in tubes, longitudinal or spiral, can be produced by extrusion or forming, with a substantial increase in the surface area. The fin efficiency, discussed *in the* next section, can usually be taken as unity. *Twisted strips* can be inserted as original equipment or as *retrofit* devices.

The increased friction factor may not require increased heat loss or pumping power if the flow rate can be adjusted or if the length of the heat exchanger can be reduced. Nelson and Bergles (1986) discuss this issue of performance evaluation criteria, especially for HVAC applications.

Of concern in chilled water systems is the fouling that in some cases may seriously reduce the overall heat transfer coefficient U. In general, fouled enhanced tubes still perform better than fouled plain tubes, as shown in recent studies of scaling of cooling tower water (Knudsen and Roy 1983) and particulate fouling (Somerscales *et al.* 1991).

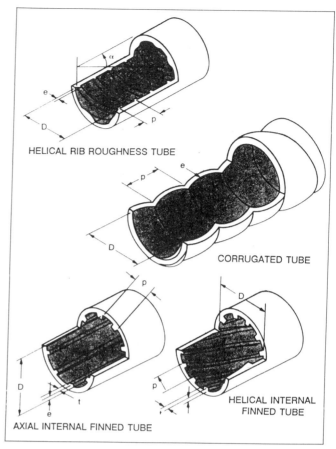

Fig. 12   Typical Tube-Side Enhancements

Fire-tube boilers are frequently fitted with turbulators to improve the turbulent convective heat transfer coefficient constituting the dominant thermal resistance. Also, due to the high gas temperatures, radiation from the connectively heated insert to the tube wall can represent as much as 50% of the total heat transfer. (Note, however, that the magnitude of the convective contribution decreases as the radiative contribution increases because of the reduced temperature difference.) Two commercial bent strip inserts, a twisted strip insert, and a simple bent tab insert are depicted in Figure 13. Design equations, for convection only, are included in Table 7. Beckermann and Goldschmidt (1986) present procedures to include radiation, and Junkhan *et al.* (1985, 1988) give friction factor data and performance evaluations.

Several enhanced surfaces for gases are depicted in Figure 14. The offset strip fin is an example of an interrupted fin that is often found in compact plate fin heat exchangers used for heat recovery from exhaust air. Design equations are included in Table 7. These equations are comprehensive in that they apply to laminar and transitional flow as well as to turbulent flow, which is a necessary feature because the small hydraulic diameter of these surfaces drives the Reynolds number down. Data for other surfaces (wavy, spine, louvered, etc.) are given in the section References.

Among the various active augmentation techniques several mechanical aids, including stirring of the fluid by mechanical means, rotating the heat transfer surface, and use of electrostatic fields have significantly increased the forced convective heat transfer. While mechanical aids are used in appropriate applications such as surface scraping, baking, and drying processes, the electrostatic technique has only been demonstrated on prototype heat exchangers. The electrostatic (also called the electrohydrodynamic

Table 7   Equations for Augmented Forced Convection

Description	Equation
I. Turbulent in-tube flow of liquids (a) Spiral repeated rib[a]	$\dfrac{h_a}{h_s} = \left[1 + \left[2.64 \left(\dfrac{GD}{\mu}\right)^{0.036} \left(\dfrac{e}{d}\right)^{0.212} \left(\dfrac{p}{d}\right)^{-0.21} \left(\dfrac{\alpha}{90}\right)^{0.29} \left(\dfrac{c_p \mu}{k}\right)^{-0.024} \right]^7 \right]^{1/7}$

$$\dfrac{f_a}{f_s} = \left\{ 1 + \left[29.1 \left(\dfrac{GD}{\mu}\right)^w \right] \right\} \left(\dfrac{e}{d}\right)^x \left(\dfrac{p}{d}\right)^y \left\{ \left[\left(\dfrac{\alpha}{90}\right)^z \left(1 + \dfrac{2.94}{n}\right) \sin\beta \right]^{15/16} \right\}^{16/15}$$

where

$w = 0.67 - 0.06\,(p/d) - 0.49\,(\alpha/90)$

$x = 0.37 - 0.157\,(p/d)$

$y = -1.66 \times 10^{-6}\,(GD/\mu) - 0.33\,(\alpha/90)$

$z = 4.59 + 4.11 \times 10^{-6}\,(GD/\mu) - 0.15\,(p/d)$

(b) Fins[b]	$\dfrac{hD_h}{k} = 0.023 \left(\dfrac{c_p \mu}{k}\right)^{0.4} \left(\dfrac{GD_h}{\mu}\right)^{0.8} \left(\dfrac{A_F}{A_{Fi}}\right)^{0.1} \left(\dfrac{A_i}{A}\right)^{0.5} (\sec \alpha)^3$
	$f_h = 0.046 \left(\dfrac{GD_h}{\mu}\right)^{-0.2} \left(\dfrac{A_F}{A_{Fi}}\right)^{0.5} (\sec \alpha)^{0.75}$
(c) Twisted strip inserts[c]	$\dfrac{hD_h}{k} = F \left[0.023 \left[1 + \left(\dfrac{\pi}{2y}\right)^2 \right]^{0.4} \left(\dfrac{GD_h}{\mu}\right)^{0.8} \left(\dfrac{c_p \mu}{k}\right)^{0.4} + 0.193 \left[\left(\dfrac{GD_h}{\mu y}\right) \dfrac{2D_h}{D_i} \dfrac{\Delta\rho}{\rho} \left(\dfrac{c_p \mu}{k}\right) \right]^{1/3} \right]$
	$f_{h,iso} = 0.127 y^{-0.406} \left(\dfrac{GD_h}{\mu}\right)^{-0.2}$
(d) Twisted strip inserts for an evaporator (cooling)[c]	$\dfrac{hD_h}{k} = 0.023\, F \left[1 + \left(\dfrac{\pi}{2y}\right)^2 \right]^{0.4} \left(\dfrac{GH_h}{\mu}\right)^{0.8} \left(\dfrac{c_p \mu}{k}\right)^{0.4}$
II. Turbulent in-tube flow of gases (a),(b) Bent strip inserts[d]	$\dfrac{hD}{k} \left(\dfrac{T_w}{T_b}\right)^{0.45} = 0.258 \left(\dfrac{GD}{\mu}\right)^{0.6}$
	$\dfrac{hD}{k} \left(\dfrac{T_w}{T_b}\right)^{0.45} = 0.208 \left(\dfrac{GD}{\mu}\right)^{0.63}$
(c) Twisted strip inserts[d]	$\dfrac{hD}{k} \left(\dfrac{T_w}{T_b}\right)^{0.45} = 0.122 \left(\dfrac{GD}{\mu}\right)^{0.65}$
(d) Bent tab inserts[d]	$\dfrac{hD}{k} \left(\dfrac{T_w}{T_b}\right)^{0.45} = 0.406 \left(\dfrac{GD}{\mu}\right)^{0.54}$
III. Offset strip fins for plate-fin heat exchangers[e]	$\dfrac{h}{c_p G} = 0.6522 \left(\dfrac{GD_h}{\mu}\right)^{-0.5403} a^{-0.1541} \delta^{0.1499} \gamma^{-0.0678} \left[1 + 5.269 \times 10^{-5} \left(\dfrac{GD_h}{\mu}\right)^{1.340} a^{0.504} \delta^{0.456} \gamma^{-1.055} \right]^{0.1}$
	$f_h = 9.6243 \left(\dfrac{GD_h}{\mu}\right)^{-0.7422} a^{-0.1856} \delta^{0.3053} \gamma^{-0.2659} \left[1 + 7.669 \times 10^{-8} \left(\dfrac{GD_h}{\mu}\right)^{4.429} a^{0.920} \delta^{3.767} \gamma^{0.236} \right]^{0.1}$

where $h/c_p G$, f_h, and GD_h/μ are based on the hydraulic diameter given by

$D_h = 4shl/[2(sl + hl + th) + ts]$

References:
[a] Ravigururajan and Bergles (1985)
[b] Carnavos (1979)
[c] Lopina and Bergles (1969)
[d] Junkhan *et al.* (1985)
[e] Manglik and Bergles (1990)

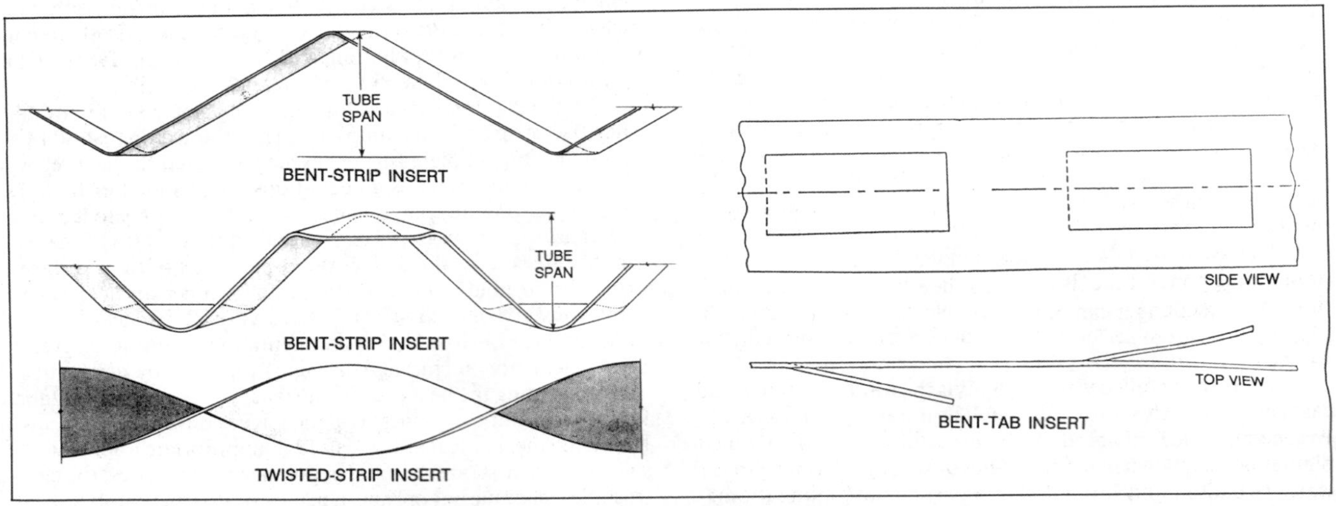

Fig. 13   Turbulators for Firetube Boilers

or EHD) augmentation technique uses electrically induced secondary motions to destabilize the thermal boundary layer near the heat transfer surface, thereby substantially increasing the heat transfer coefficients at the wall. The magnitude and nature of enhancements are a function of (1) electric field parameters (such as field potential, field polarity, pulse versus steady discharge, electrode geometry and spacing); (2) flow field parameters (such as mass flow rate, temperature, density, and electrical permittivity of the working fluid); and (3) the heat transfer surface type (such as smooth, porous, or integrally finned/grooved configurations).

The EHD effect is generally applied by placing wire or plate electrodes parallel and adjacent to the heat transfer surface. Figure 15 presents four electrode configurations for augmentation of forced convection heat transfer in tube flows. A high-voltage electric field charges the electrode and establishes the electrical body forces required to initiate and sustain augmentation.

Several important advantages have contributed to the progress of the EHD technique in recent years. The mechanically complex rotation, injection, and vibration forms of promoters are generally cumbersome to manufacture. Furthermore, these systems often require a significant fraction of the power needed to pump the fluid. In contrast, even though high voltages are employed, the amount of electrical power consumed by the EHD process is extremely small (a few watts or less) due to the very small currents (1 mA or less). The additional manufacturing costs are minimal as the EHD process requires a small transformer and simple wire or plate electrodes. As is the case with most augmentation techniques, the increase in heat transfer coefficients is associated with a corresponding increase in pressure drop coefficients. However, unlike most other techniques where the pressure drop is substantial, the rise in pressure drops in the EHD technique is usually much less than the corresponding increase in heat transfer coefficients.

The EHD technique is limited in that it is only effective for fluids with low electrical conductivity. This includes air, certain industrial fluids, and many refrigerants. Table 8 provides a summary of selected studies involving single-phase and phase-change processes, and demonstrates that the EHD technique is particularly effective for heat transfer enhancement of refrigerants, including the ozone-safe refrigerant substitutes. Ohadi (1991b) gives further details on the fundamentals, applicability, and limitations of the EHD technique.

EXTENDED SURFACE

Heat transfer from a prime surface can be increased by attaching *fins* or *extended surfaces* to increase the area available for heat transfer. Fins provide a more compact heat exchanger with lower material costs for a given performance. To achieve optimum design, fins are generally located on the side of the heat exchanger where the heat transfer coefficients are low (such as the air side of an air-to-water coil). Equipment with an extended surface includes natural and forced convection coils and shell-and-tube evaporators and condensers. Fins are also used inside tubes in condensers and dry expansion evaporators.

Fin Efficiency

As heat flows from the root of a fin to its tip, temperature drops because of the thermal resistance of the fin material. The temperature difference between the fin and the surrounding fluid is therefore greater at the root than at the tip, causing a corresponding variation in the heat flux. Therefore, increases in fin length result in proportionately less additional heat transfer. To account for this effect, *fin efficiency* is defined by Equation (50) as the ratio of the actual heat transferred from the fin to the heat that would be transferred if the entire fin were at its root or base temperature.

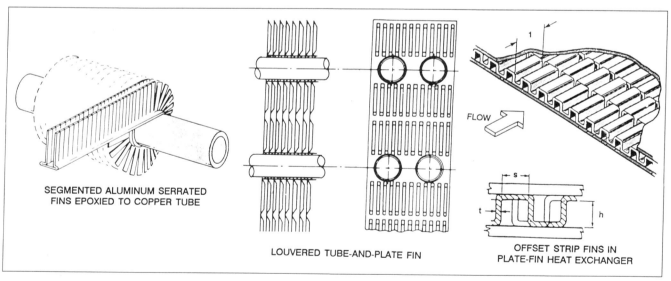

SEGMENTED ALUMINUM SERRATED FINS EPOXIED TO COPPER TUBE

LOUVERED TUBE-AND-PLATE FIN

OFFSET STRIP FINS IN PLATE-FIN HEAT EXCHANGER

FLOW

Fig. 14   Enhanced Surfaces for Gases

Table 8   EHD Heat Transfer Enhancement in Heat Exchangers

Source	Maximum Reported Enhancement, %	Test Fluid	Heat Transfer Wall/ Electrode Configuration	Process
Fernandez (1987)	2300	Transformer oil	Tube/wire	Forced convection
Ohadi *et al.* (1991)	320	Air	Tube/wire or rod	Forced convection
Ohadi *et al.* (1992)	480	R-123	Tube/wire	Boiling
Sunada *et al.* (1991)	600	R-123	Vertical wall/plate	Condensation
Uemura *et al.* (1990)	1400	R-113	Plate/wire mesh	Film boiling
Yabe and Maki (1988)	10,000	96% (by mass) R-113, 4% ethanol	Plate/ring	Natural convection

$$\phi = \frac{\int h(t - t_e)dA}{\int h(t_r - t_e)dA} \qquad (50)$$

where ϕ is the fin efficiency, t_e is the temperature of the surrounding environment, and t_r is the temperature at the fin root. Fin efficiency is low for long fins, thin fins, or fins of low thermal conductivity material. Fin efficiency decreases as the heat transfer coefficient increases because of the increased heat flow. For natural convection in air-cooled condensers and evaporators, where h for the air side is low, fins can be fairly large and of low conductivity materials, such as steel, instead of copper or aluminum. For condensing and boiling, where large heat transfer coefficients are involved, fins must be very short for optimum use of material.

The heat transfer from a finned surface, such as a tube, which includes both finned or secondary area A_s and unfinned or prime area A_p is given by Equation (51):

$$q = (h_p A_p + \phi h_s A_s)(t_r - t_e) \qquad (51)$$

Assuming the heat transfer coefficients for the finned surface and prime surface are equal, a *surface efficiency* ϕ_s can be derived for use in Equation (52).

$$\phi_s = 1 - (A_s/A)(1 - \phi) \qquad (52)$$

$$q = \phi_s h A (t_r - t_e) \qquad (53)$$

where A is the total surface area, equal to the sum of the finned and prime areas.

Temperature distribution and fin efficiencies for various fin shapes are derived in most heat transfer texts. Figures 16 through

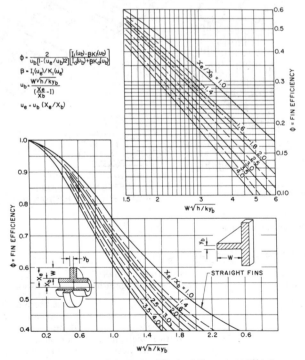

Fig. 15   Electrode Configurations for Internal Forced-Convection Flow

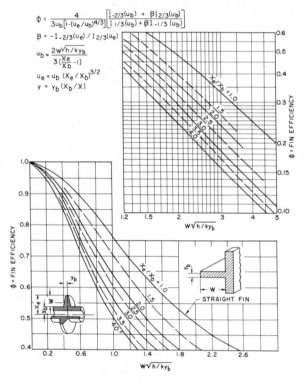

Fig. 17   Efficiency of Annular Fins with Constant Metal Area for Heat Flow

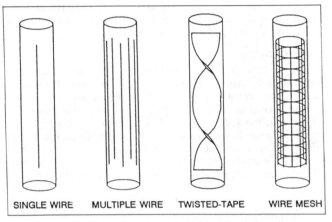

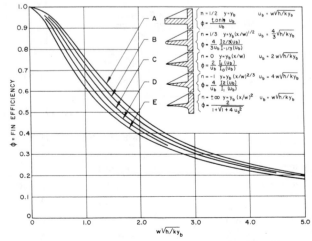

Fig. 16   Efficiency of Annular Fins of Constant Thickness

Fig. 18   Efficiency of Several Types of Straight Fin

19 show curves and equations for annular fins, straight fins, and spines. For constant thickness square fins, the efficiency of a constant thickness annular fin of the same area can be used. More accuracy, particularly with rectangular fins of large aspect ratio, can be obtained by dividing the fin into circular sectors (Rich 1966).

By defining a dimensionless thermal resistance Φ by Equations (54) and (54a) and developing expressions for its maximum limiting value Φ_{max}, Rich (1966) presents results for a wide range of geometries in a compact form for equipment designers.

$$\Phi = R_f t_o k / l^2 \tag{54}$$

$$R_f = (1/h)(1/\phi - 1) \tag{54a}$$

where

Φ = dimensionless thermal resistance
ϕ = fin efficiency

t_o = fin thickness at fin base
l = length dimension = $r_t - r_o$ for annular fins, and
= W for rectangular fins

Figure 20 gives Φ_{max} for annular fins of constant and tapered cross section as a function of $R = r_t/r_o$, i.e., the ratio of the fin tip-to-root radii. Figure 21 gives Φ_{max} for rectangular fins of a given geometry as determined by the sector method. Figure 22 gives correction factors for the determination of Φ from Φ_{max} for both annular and rectangular fins.

Example. This example illustrates the use of the fin resistance number for a rectangular fin typical of that for an air-conditioning coil.

Given: $L = 0.75$ in. $t_o = 0.006$ in.
$W = 0.50$ in. $h = 10$ Btu/h·ft²·°F
$r_o = 0.25$ in. $k = 100$ Btu/h·ft·°F

Solution: From Figure 21 at $W/r_o = 2.0$ and $L/W = 1.5$

$$\Phi_{max} = R_{f(max)} t_o k / W^2 = 1.12$$

$$R_{f(max)} = \frac{1.12 \times 0.50^2}{0.006 \times 100 \times 12} = 0.0389 \; \frac{\text{h·ft}^2\text{·°F}}{\text{Btu}}$$

The correction factor, which is multiplied by $R_{f(max)}$ to give R_f, is given in Figure 22 as a function of the fin efficiency. As a first approximation, the fin efficiency is calculated from Equation (54a) assuming $R_f = R_{f(max)}$.

$$\phi = 1/(1 + hR_f) \approx 0.72$$

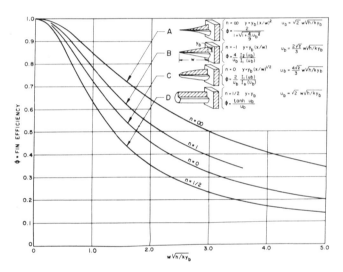

Fig. 19   Efficiency of Four Types of Spine

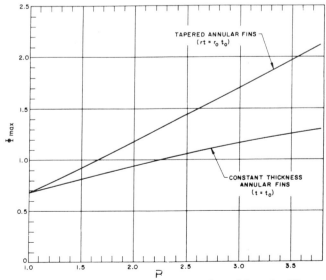

Fig. 20   Maximum Fin Resistance Number of Annular Fins
(Gardner 1945)

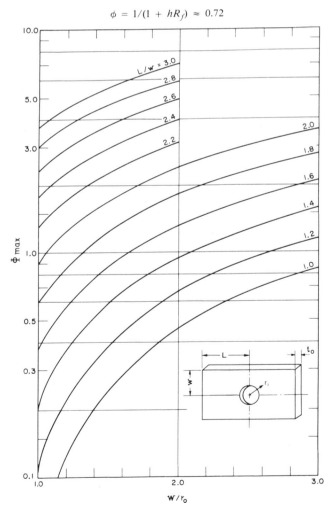

Fig. 21   Maximum Fin Resistance Number of Rectangular Fins Determined by Sector Method

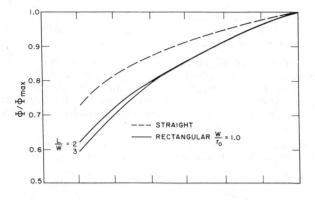

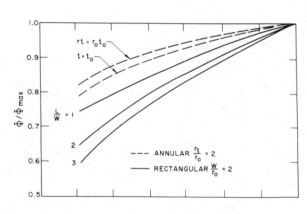

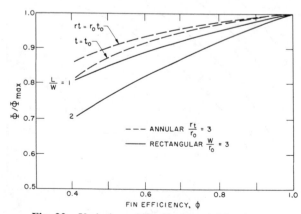

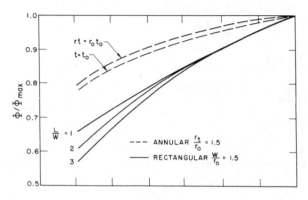

**Fig. 22   Variation of Fin Resistance Number with
Efficiency for Annular and Rectangular Fins**
(Gardner 1945)

Interpolating between $L/W = 1$ and $L/W = 2$ at $W/r_o = 2$ gives:

$$\Phi/\Phi_{max} = 0.88$$

Therefore:

$$R_f = 0.88 \times 0.0389 = 0.0342 \text{ ft}^2 \cdot {}^\circ\text{F} \cdot \text{h/Btu}$$

The above steps may now be repeated using the corrected value of fin resistance.

$$\phi = 0.745$$
$$\Phi/\Phi_{max} = 0.9$$
$$R_f = 0.035 \text{ ft}^2 \cdot {}^\circ\text{F} \cdot \text{h/Btu}$$

Note that the improvement in accuracy by reevaluating Φ/Φ_{max} is less than 1% of the overall thermal resistance (environment to fin base). The error produced by using $R_{f(max)}$ without correction is less than 3%. For many practical cases where greater accuracy is not warranted, a single value of R_f, obtained by estimating Φ/Φ_{max}, can be used over a range of heat transfer coefficients for a given fin. For approximate calculations, the fin resistance for other values of k and t_o can be obtained by simple proportion if the range covered is not excessive.

Schmidt (1949) presented approximate, but reasonably accurate, analytical expressions (for computer use) for circular, rectangular, and hexagonal fins. Hexagonal fins are the representative fin shape for the common staggered tube arrangement in finned tube heat exchangers.

Schmidt's empirical solution is given by:

$$\phi = \tanh (mr_i\phi) / mr_i$$

where $m = (2h/kt)^{0.5}$ and Φ is given by:

$$\phi = [(r_e/r_i) - 1] [1 + 0.35 \ln (r_e/r_i)]$$

For *circular fins* $r_e/r_i = r_o/r_i$

and for *rectangular fins*

$$r_e/r_i = 1.28 \, \psi \, \sqrt{(\beta - 0.2)}, \quad \psi = M/r_i, \quad \beta = L/M \geq 1,$$

where M and L are defined by Figure 23 as $a/2$ or $b/2$, depending on which is greater.

For *hexagonal fins* $r_e/r_i = 1.27 \, \psi \, \sqrt{(\beta - 0.3)}$

where ψ and β are defined as above and where M and L are defined by Figure 24 as $a/2$ or b (whichever is less) and $0.5 \sqrt{(a^2/2)^2 + b^2}$, or vice versa.

The bibliography lists other sources of information on finned surfaces.

Thermal Contact Resistance

Fins can be extruded from the prime surface—for example, the short fins on the tubes in flooded evaporators or water-cooled con-

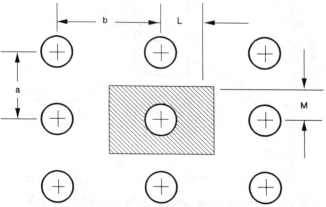

Fig. 23   Rectangular Tube Array

densers; or they can be fabricated separately, sometimes of a different material, and bonded to the prime surface. Metallurgical bonds are achieved by furnace-brazing, dip-brazing, or soldering. Nonmetallic bonding materials, such as epoxy resin, are also used. Mechanical bonds are obtained by tension winding fins around tubes (spiral fins) or expanding the tubes into the fins (plate fin). Metallurgical bonding, properly done, leaves negligible thermal resistance at the joint, but is not always economical. Thermal resistance of a mechanical bond may or may not be negligible, depending on the application, care taken in manufacture, and materials and temperatures involved. Tests of plate fin coils with expanded tubes have indicated that substantial losses in performance can occur with fins that have cracked collars; but negligible thermal resistance was found in coils with continuous collars and properly expanded tubes (Dart 1959).

Thermal resistance at an interface between two solid materials is largely a function of the surface properties and characteristics of the solids, the contact pressure, and the fluid in the interface, if any. Eckels (1977) modeled the influence of fin density, fin thickness, and tube diameter on contact pressure and compared it to data for wet and dry coils. Shlykov (1964) showed that the range of attainable contact resistances is large. Sonokama (1964) presents data on the effects of contact pressure, surface roughness, hardness, void material, and the pressure of the gas in the voids. Lewis and Sauer (1965) show the resistance of adhesive bonds, and Kaspareck (1964) and Clausing (1964) give data on the contact resistance in a vacuum environment.

Finned Tube Heat Transfer

The heat transfer coefficients for finned coils follow the basic equations of convection, condensation, and evaporation. The arrangement of the fins affects the values of constants and the exponential powers in the equations. It is generally necessary to refer to test data for the exact coefficients.

For natural convection finned coils (gravity coils), approximate coefficients can be obtained by considering the coil to be made of tubular and vertical fin surfaces at different temperatures and then applying the natural convection equations to each. This calculation is difficult because the natural convection coefficient depends on the temperature difference, which varies at different points on the fin.

Fin efficiency should be high (80 to 90%) for optimum natural convection heat transfer. A low fin efficiency reduces the temperature near the tip. This reduces Δt near the tip, and also the coefficient, h, which in natural convection depends on Δt. The

coefficient of heat transfer also decreases as the fin spacing decreases, because of interfering convection currents from adjacent fins and reduced free-flow passage; 2 to 4 in. spacing is common. Generally, high coefficients result from large temperature differences and small flow restriction.

Edwards and Chaddock (1963) give coefficients for several circular fin-on-tube arrangements, using fin spacing δ as the characteristic length and in the form $Nu = f(Gr Pr \delta / D_o)$ where D_o is the fin diameter.

Forced convection finned coils are used extensively in a wide variety of equipment. The fin efficiency for optimum performance is smaller than that for gravity coils, since the forced convection coefficient is almost independent of the temperature difference between the surface and the fluid. Very low fin efficiencies should be avoided, since inefficient surface gives a high (uneconomical) pressure drop. An efficiency of 70 to 90% is often used.

As fin spacing is decreased to obtain a large surface area for heat transfer, the coefficient generally increases because of higher air velocity between fins at the same face velocity and reduced equivalent diameter. The limit is reached when the boundary layer formed on one fin surface (Figure 8) begins to interfere with the boundary layer formed on the adjacent fin surface, resulting in a decrease of the heat transfer coefficient, which may offset the advantage of larger surface area.

Selection of the fin spacing for forced convection finned coils usually depends on economic and practical considerations, such as fouling, frost formation, condensate drainage, cost, weight, and volume. Fins for conventional coils generally are spaced 14 to 6 per inch, except where factors such as frost formation necessitate wider spacing.

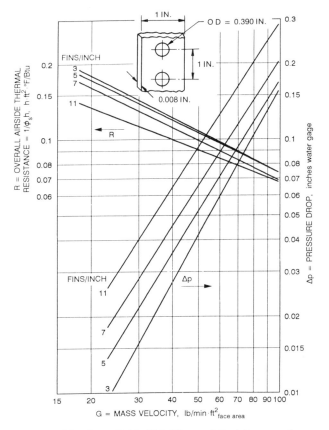

Fig. 25 Overall Air-Side Thermal Resistance and Pressure Drop for 1-Row Coils

(Shepherd 1946)

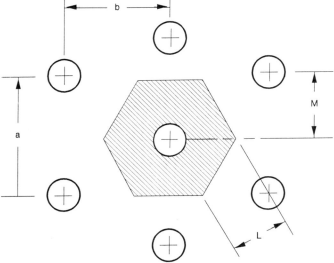

Fig. 24 Hexagonal Tube Array

Several means are used to obtain higher coefficients with a given air velocity and surface, usually by creating air turbulence, generally with a higher pressure drop: (1) staggered tubes instead of in-line tubes for multiple row coils; (2) artificial additional tubes, or collars or fingers made by suitably forming the fin materials; (3) corrugated fins instead of plane fins; and (4) louvered or interrupted fins.

Figure 25 shows data for one-row coils. The thermal resistances plotted include the temperature drop through the fins, based on one square foot of total external surface area.

The bibliography lists other sources of information on fins.

SYMBOLS

A = surface area for heat transfer
A_F = cross-sectional flow area
C = conductance; or fluid capacity rate
c = coefficient or constant
c_p = specific heat at constant pressure
c_v = specific heat at constant volume
C_1, C_2 = Planck's law constants [see Eq. (33)]
D = diameter
d = diameter; or prefix meaning differential
e = emissivity; protuberance height
F = angle factor [see Eqs. (40) and (41)]; fin factor
Fo = Fourier number (see Table 1 and Figs. 2, 3, and 4)
f = friction factor for single-phase flow
G = mass velocity; or irradiation
g = gravitational acceleration
h = heat transfer coefficient; offset strip fin height
J = mechanical equivalent of heat; or radiosity
j = heat transfer factor j [see Eq. (4), Table 6]
k = thermal conductivity
l = length; length of one module of offset strip fins
M = mass; or molecular weight
m = general exponent
$\dot{m}$ = mass rate of flow
N = number of tubes in vertical tier
n = general number [see Eq. (2), Table 5]; or ratio r/r_m (see Figs. 2, 3, and 4)
NTU = number of exchanger heat transfer units [see Eq. (17)]
p = pressure; fin pitch; repeated rib pitch
Q = total heat transfer
q = rate of heat transfer
R = thermal resistance
r = radius
s = lateral spacing of offset fin strips
T = absolute temperature
t = temperature; fin thickness at base
U = overall heat transfer coefficient
V = linear velocity
W = work; or total rate of energy emission
W_λ = monochromatic emissive power
x, y, z = lengths along principal coordinate axes
Y = temperature ratio (see Figs. 2, 3, and 4)
Y_v = mole fraction of vapor
Z = ratio of fluid capacity rates [see Eq. (18)]
α = thermal diffusivity = $k/\rho c_p$ [Eq. (28)]; absorptance; spiral angle for helical fins; aspect ratio of offset strip fins
β = coefficient of thermal expansion; contact angle of rib profile
γ = ratio, l/s
Δ = difference between values
δ = distance between fins; ratio, t/s
ϵ = hemispherical emittance; or exchanger heat transfer effectiveness [see Eq. (16)]
λ = wavelength
μ = absolute viscosity
ν = kinematic viscosity
ρ = density; or reflectance
σ = Stefan-Boltzmann constant
τ = time; or transmittance [see Eq. (39)]
Φ = fin resistance number defined by Eq. (54); Φ_{max} is maximum limiting value of Φ
ϕ = fin efficiency [see Eq. (50)]; or angle [see Eq. (41)]

Subscripts

a = augmented
b = blackbody; based on bulk fluid temperature
c = convection; or critical; or cold (fluid)
e = equivalent; or environment
f = film; or fin
fc = finite cylinder
frs = finite rectangular solid
h = horizontal; or hot (fluid); or hydraulic
i = inlet; or inside; particular surface (radiation); based on maximum inside (envelope) diameter
ic = infinite cylinder
if = interface
is = infinite slab
iso = isothermal conditions
j = particular surface (radiation)
l = liquid
m = mean
max = maximum
min = minimum
o = outside; or outlet; or overall; or fin diameter
p = prime heat transfer surface
r = radiation; root (fin); or reduced
s = surface; secondary heat transfer surface; straight or plain; accounting for flow blockage of twisted tape
st = static (pressure)
t = temperature; or terminal temperature; or tip (fin)
v = vapor; or vertical
w = wall; or wafer
λ = monochromatic
∞ = bulk

REFERENCES

Adams, J.A. and D.F. Rogers. 1973. *Computer aided heat transfer analysis.* McGraw-Hill Book Co., Inc., New York.

Afgan, N.H. and E.U. Schlunder. 1974. *Heat exchangers: Design and theory sourcebook.* McGraw-Hill Book Co., New York.

Altmayer, E.F. *et al.* 1983. Correlations for convective heat transfer from room surfaces. ASHRAE *Transactions* 89:2A.

Bauman, F. *et al.* 1983. Convective heat transfer in buildings. ASHRAE *Transactions* 89:1.

Beckerman, C. and V. Goldschmidt. 1986. Heat transfer augmentation in the flueway of a water heater. ASHRAE *Transactions* 92(2B):485-95.

Bergles, A.E. 1985. "Techniques to augment heat transfer." In *Handbook of heat transfer applications.* McGraw-Hill, New York, 3-1-3-80.

Brown, A.I. and S.M. Marco. 1958. *Introduction to heat transfer*, 3rd ed. McGraw-Hill Book Co., Inc., New York.

Burmeister, L.C. 1983. *Convective heat transfer.* John Wiley and Sons, Inc., New York.

Carnavos, T.C. 1979. Heat transfer performance of internally finned tubes in turbulent flow. Advances in enhanced heat transfer, ASME, New York, 61-67.

Carslaw, H.S. and J.C. Jaeger. 1959. *Conduction of heat in solids.* Oxford University Press, England.

Clausing, A.M. 1964. Thermal contact resistance in a vacuum environment. ASME Paper 64-HT-16, Seventh National Heat Transfer Conference.

Collicott, H.E., W.E. Fontaine, and O.W. Witzell. 1963. Radiation and free convection heat transfer from wire and tube heat exchangers. ASHRAE *Journal* 12:79.

Croft, D.R. and D.G. Lilley. 1977. *Heat transfer calculations using finite difference equations.* Applied Science Publishers, Ltd., London, England.

Cyphers, J.A., R.D. Cess, and E.V. Somers. 1959. Heat transfer characteristics of wire-and-tube heat exchangers. ASHRAE *Journal* 5:86.

Dart, D.M. 1959. Effect of fin bond on heat transfer. ASHRAE *Journal* 5:67.

Eckels, P.W. 1977. Contact conductance of mechanically expanded plate finned tube heat exchangers. AICHE-ASME Heat Transfer Conference, Salt Lake City, UT.

Edwards, J.A. and J.B. Chaddock. 1963. An experimental investigation of the radiation and free-convection heat transfer from a cylindrical disk extended surface. ASHRAE *Transactions* 69:313.

Fernandez, J. and R. Poulter. 1987. Radial mass flow in electrohydrodynamically-enhanced forced heat transfer in tubes. *International Journal of Heat and Mass Transfer* 80:2125-36.

Gardner, K.A. 1945. Efficiency of extended surface. ASME *Transactions* 67:621.

Grigull, U. *et al.* 1982. Heat transfer. Proceedings of the Seventh International Heat Transfer Conference, Munchen, Vol. 3, Hemisphere Publishing Co., New York.

Hottel, H.C. and A.F. Sarofim. 1967. *Radiation transfer.* McGraw-Hill Book Co., Inc., New York.

Incropera, F.P. and D.P. DeWitt. 1985. *Fundamentals of heat transfer.* John Wiley and Sons, Inc., New York.

Jakob, M. 1949, 1957. *Heat transfer*, Vols. I and II. John Wiley and Sons, Inc., New York.

Junkhan, G.N. *et al.* 1985. Investigation of turbulence for tube flow boilers. *Journal of Heat Transfer* 107:354-60.

Junkhan, G.N. *et al.* 1988. Performance evaluation of the effects of a group of turbulator inserts on heat transfer from gases in tubes. ASHRAE *Transactions* 94(2).

Kaspareck, W.E. 1964. Measurement of thermal contact conductance between dissimilar metals in a vacuum. ASME Paper 64-HT-38, Seventh National Heat Transfer Conference.

Kays, W.M. and A.L. London. 1964. *Compact heat exchangers*, 2nd ed. McGraw-Hill Book Co., Inc., New York.

Kays, W.M. and M. Crawford. 1980. *Convective heat and mass transfer*, 2nd ed. McGraw-Hill Book Co., Inc., New York.

Knudsen, J.G. and B.V. Roy. 1983. Studies on scaling of cooling tower water. Fouling of heat enhancement surfaces, Engineering Foundation, New York, 517-30.

Lewis, D.M. and H.J. Sauer, Jr. 1965. The thermal resistance of adhesive bonds. ASME *Journal of Heat Transfer* 5:310.

Lopina, R.F. and A.E. Bergles. 1969. Heat transfer and pressure drop in tape generated swirl flow of single-phase water. *Journal of Heat Transfer* 91:434-42.

Manglik, R.M. and A.E. Bergles. 1990. "The thermal-hydraulic design of the rectangular offset-strip-fin-compact heat exchanger." In *Compact heat exchangers*. Hemisphere, New York, 123-49.

McAdams, W.H. 1954. *Heat transmission*, 3rd ed. McGraw-Hill Book Co., Inc., New York.

Metais, B. and E.R.G. Eckert. 1964. Forced, mixed and free convection regimes. ASME *Journal of Heat Transfer* 86(C2)(5):295.

Myers, G.E. 1971. *Analytical methods in conduction heat transfer.* McGraw-Hill Book Co., Inc., New York.

Nelson, R.M. and A.E. Bergles. 1986. Performance evaluation for tubeside heat transfer enhancement of a flooded evaporative water chiller. ASHRAE *Transactions* 92(1B):739-55.

Ohadi, M.M. *et al.* 1991a. Electrohydrodynamic enhancement of heat transfer in a shell-and-tube heat exchanger. *Experimental Heat Transfer* 4(1):19-39.

Ohadi, M.M. 1991b. Heat transfer enhancement in heat exchangers. ASHRAE *Journal* (December):42-50.

Ohadi, M.M. *et al.* 1992. EHD Enhancement of shell-side boiling heat transfer coefficients of R-123/oil mixture. ASHRAE *Transactions* 98(2).

Ohadi, M.M., R. Papar, M. Fanni-Tabrizi, and R. Radermacher. 1992. Electrohydrodynamic boiling heat transfer enhancement of R-123 outside of a smooth tube. ASHRAE *Transactions* 92(2).

Parker, J.D., J.H. Boggs, and E.F. Blick. 1969. *Introduction to fluid mechanics and heat transfer.* Addison Wesley Publishing Co., Reading, MA.

Patankar, S.V. 1980. *Numerical heat transfer and fluid flow.* McGraw-Hill Book Co., Inc., New York.

Ravigururajan, T.S. and A.E. Bergles. 1985. General correlations for pressure drop and heat transfer for single-phase turbulent flow in internally ribbed tubes. *Augmentation of heat transfer in energy systems*, Vol. 52. ASME, New York, 9-20.

Rich, D.G. 1966. The efficiency and thermal resistance of annular and rectangular fins. Proceedings of the Third International Heat Transfer Conference, AICHE 111:281-89.

Schmidt, T.E. 1949. Heat transfer calculations for extended surfaces. *Refrigerating Engineering* 4:351-57.

Schneider, P.J. 1964. *Temperature response charges.* John Wiley and Sons, Inc., New York.

Shepherd, D.G. 1946. Performance of one-row tube coils with thin plate fins, low velocity forced convection. *Heating, Piping, and Air Conditioning*, April.

Shlykov, Y.P. 1964. Thermal resistance of metallic contacts. *International Journal of Heat and Mass Transfer* 7(8):921.

Siegel, R. and J.R. Howell. 1981. *Thermal radiation heat transfer.* McGraw-Hill Book Co., Inc., New York.

Somerscales, E.F.C. *et al.* 1991. Particulate fouling of heat transfer tubes enhanced on their inner surface, fouling and enhancement interactions. HTD Vol. 164, ASME, New York, 17-28.

Sonokama, K. 1964. Contact thermal resistance. *Journal of the Japan Society of Mechanical Engineers* 63(505):240. English translation in RSIC-215, AD-443429.

Sunada, K., A. Yabe, T. Taketani, and Y. Yoshizawa. 1991. Experimental study of EHD pseudo-dropwise condensation. Proceedings of the ASME-JSME Thermal Engineering Joint Conference. 3:47-53.

Uemura, M., S. Nishio, and I. Tanasawa. 1990. Enhancement of pool boiling heat transfer by static electric field. 9th International Heat Transfer Conference, 75-80.

Yabe, A. and H. Maki. 1988. Augmentation of convective and boiling heat transfer by applying an electrohydrodynamical liquid jet. International Journal of Heat Mass Transfer. 31(2):407-17.

BIBLIOGRAPHY

Fins

General

Gunter, A.Y. and A.W. Shaw. 1945. A general correlation of friction factors for various types of surfaces in cross flow. ASME *Transactions* 11:643.

Shah, R.K. and R.L. Webb. 1981. *Compact and enhanced heat exchangers, heat exchangers, theory and practice.* J. Taborek *et al.*, eds. Hemisphere, WA, 425-68.

Webb, R.L. 1980. Air-side heat transfer in finned tube heat exchangers. *Heat Transfer Engineering* 1(3):33-49.

Smooth

Clarke, L. and R.E. Winston. 195. Calculation of finside coefficients in longitudinal finned heat exchangers. *Chemical Engineering Progress* 3:147.

Elmahdy, A.H. and R.C. Biggs. 1979. Finned tube heat exchanger: Correlation of dry surface heat transfer data. ASHRAE *Transactions* 85:2.

Ghai, M.L. 1951. Heat transfer in straight fins. General discussion on heat transfer. London Conference, September.

Gray, D.L. and R.L. Webb. 1986. Heat transfer and friction correlations for plate finned-tube heat exchangers having plain fins. Proceedings of Eighth International Heat Transfer Conference, San Francisco, CA.

Wavy

Beecher, D.T. and T.J. Fagan. 1987. Fin patternization effects in plate finned tube heat exchangers. ASHRAE *Transactions* 93:2.

Yashu, T. 1972. Transient testing technique for heat exchanger fin. *Reito* 47(531):23-29.

Spine

Abbott, R.W., R.H. Norris, and W.A. Spofford. 1980. Compact heat exchangers for general electric products—Sixty years of advances in design and manufacturing technologies, compact heat exchangers—History, technological advancement and mechanical design problems. R.K. Shah, C.F. McDonald, and C.P. Howard, eds. Book No. G00183. ASME, 37-55.

Moore, F.K. 1975. Analysis of large dry cooling towers with spine-fin heat exchanger elements. ASME Paper No. 75-WA/HT-46.

Rabas, T.J. and P.W. Eckels. 1975. Heat transfer and pressure drop performance of segmented surface tube bundles. ASME Paper No. 75-HT-45.

Weierman, C. 1976. Correlations ease the selection of finned tubes. *Oil and Gas Journal* 9:94-100.

Louvered

Hosoda, T. *et al.* 1977. Louver fin type heat exchangers. *Heat Transfer Japanese Research* 6(2):69-77.

Mahaymam, W. and L.P. Xu. 1983. Enhanced fins for air-cooled heat exchangers—Heat transfer and friction factor correlations. Y. Mori and W. Yang, eds. Proceedings of the ASME-JSME Thermal Engineering Joint Conference, Hawaii.

Senshu, T. *et al.* 1979. Surface heat transfer coefficient of fins utilized in air-cooled heat exchangers. *Reito* 54(615):11-17.

Circular

Jameson, S.L. 1945. Tube spacing in finned tube banks. ASME *Transactions* 11:633.

Katz, D.L. and Associates. 1954-55. Finned tubes in heat exchangers; Cooling liquids with finned coils; Condensing vapors on finned coils; and Boiling outside finned tubes. Bulletin reprinted from Petroleum Refiner.

Heat Exchangers

Gartner, J.R. and H.L. Harrison. 1963. Frequency response transfer functions for a tube in crossflow. ASHRAE *Transactions* 69:323.

Gartner, J.R. and H.L. Harrison. 1965. Dynamic characteristics of water-to-air crossflow heat exchangers. ASHRAE *Transactions* 71:212.

McQuiston, F.C. 1981. Finned tube heat exchangers: State of the art for the air side. ASHRAE *Transactions* 87:1.

Myers, G.E., J.W. Mitchell, and R. Nagaoka. 1965. A method of estimating crossflow heat exchangers transients. ASHRAE *Transactions* 71:225.

Stermole, F.J. and M.H. Carson. 1964. Dynamics of flow forced distributed parameter heat exchangers. AICHE *Journal* 10(5):9.

Thomasson, R.K. 1964. Frequency response of linear counterflow heat exchangers. *Journal of Mechanical Engineering Science* 6(1):3.

Wyngaard, J.C. and F.W. Schmidt. Comparison of methods for determining transient response of shell and tube heat exchangers. ASME Paper 64-WA HT-20.

Yang, W.J. Frequency response of multipass shell and tube heat exchangers to timewise variant flow perturbance. ASME Paper 64HT-18.

Heat Transfer, General

Bennet, C.O. and J.E. Myers. 1984. *Momentum, heat and mass transfer*, 3rd ed. McGraw-Hill Book Co., Inc., New York.

Chapman, A.J. 1981. *Heat transfer*, 4th ed. Macmillan Publishing Co., Inc., New York.

Holman, J.D. 1981. *Heat transfer*, 5th ed. McGraw-Hill Book Co., Inc., New York.

Kern, D.Q. and A.D. Kraus. 1972. *Extended surface heat transfer*. McGraw-Hill Book Co., Inc., New York.

Kreith, F. and W.Z. Black. 1980. *Basic heat transfer*. Harper and Row, New York.

Lienhard, J.H. 1981. *A heat transfer textbook*. Prentice Hall, Englewood Cliffs, NJ.

McQuiston, F.C. and J.D. Parker. 1988. *Heating, ventilating and air-conditioning, analysis and design*, 3rd ed. John Wiley and Sons, New York.

Rohsenow, W.M. and J.P. Hartnett, eds. 1973. *Handbook of heat transfer*. McGraw-Hill Book Co., Inc., New York.

Sissom, L.E. and D.R. Pitts. 1972. *Elements of transport phenomena*. McGraw-Hill Book Co., Inc., New York.

Todd, J.P. and H.B. Ellis. 1982. *Applied heat transfer*. Harper and Row, New York.

Webb, R.L. and A.E. Bergles. 1983. Heat transfer enhancement, second generation technology. *Mechanical Engineering* 6:60-67.

Welty, J.R. 1974. *Engineering heat transfer*. John Wiley and Sons, Inc., New York.

Welty, J.R., C.E. Wicks, and R.E. Wilson. 1972. *Fundamentals of momentum, heat and mass transfer*. John Wiley and Sons, Inc., New York.

Wolf, H. 1983. *Heat transfer*. Harper and Row, New York.

CHAPTER 4

TWO-PHASE FLOW

TWO-PHASE flow is encountered extensively in the air-conditioning, heating, and refrigeration industries. Mixtures of liquid and vapor refrigerants exist in flooded coolers, direct expansion coolers, thermosiphon coolers, brazed and gasketed plate evaporators and condensers, tube-in-tube evaporators and condensers, as well as in air-cooled evaporators and condensers. A mixture of steam and water is found in the pipes of heating systems. Since the hydrodynamic and heat transfer aspects of two-phase flow are not as well understood as those of single-phase flow, no single set of correlations can be used to predict pressure drops or heat transfer rates. Instead, the correlations are for specific thermal and hydrodynamic operating conditions.

This chapter presents the basic principles of two-phase flow and provides information on the vast number of correlations that have been developed to predict heat transfer coefficients and pressure drops in these systems.

Breber *et al*. (1980) present a method of using a flow regime map to predict heat transfer coefficients for condensation of pure components in a horizontal tube.

BOILING

Commonly used refrigeration evaporators are: (1) flooded evaporators, where refrigerants boil outside or inside tubes with low fluid velocities; and (2) dry expansion shell-and-tube evaporators with boiling outside or inside tubes at substantial fluid velocities.

Flow of a two-phase mixture is characterized by various flow and thermal regimes, whether vaporization takes place under natural convection or in forced flow. As in single-phase flow systems, the heat transfer coefficient for a two-phase mixture depends on the flow regime. The thermodynamic and transport properties of the vapor and the liquid, the conditions (roughness and wettability) of the heating surface, and other parameters influence the heat transfer coefficients in different ways. Therefore, it is necessary to consider each flow and boiling regime separately to determine the coefficient.

Accurate data defining limits of regimes and determining the effects of various parameters are not available. The accuracy of previously proposed correlations to predict the heat transfer coefficient for two-phase flow is not known beyond the range of the test data.

Boiling and Pool Boiling in Natural Convection Systems

Regimes of boiling. The different regimes of pool boiling described by Farber and Scorah (1948) are illustrated in Figure 1. When the temperature of the heating surface is near the fluid saturation temperature, heat is transferred by convection currents to the free surface where evaporation occurs (Region I). Transition to nucleate boiling occurs when the surface temperature exceeds saturation by a few degrees (Region II).

In *nucleate boiling* (Region III), a thin layer of superheated liquid is formed adjacent to the heating surface. In this layer,

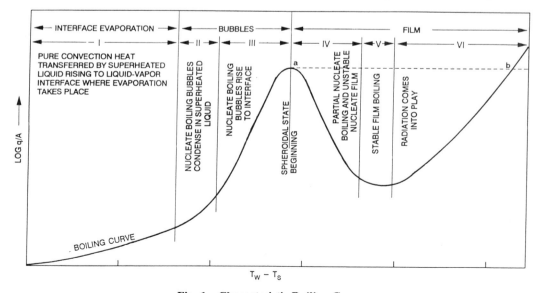

Fig. 1 Characteristic Boiling Curve

The preparation of this chapter is assigned to TC 1.3, Heat Transfer and Fluid Flow.

bubbles nucleate and grow from spots on the surface. The thermal resistance of the superheated liquid film is greatly reduced by bubble-induced agitation and vaporization. Increased wall temperature increases bubble population, causing a large increase in heat flux.

As heat flux or temperature difference increases further and as more vapor forms, the flow of the liquid toward the surface is interrupted and a vapor blanket forms. This gives the *maximum heat flux* or *peak heat flux* in nucleate boiling (point *a*, Figure 1). This flux is often termed the *burnout heat flux* because, for constant power-generating systems, an increase of heat flux beyond this point results in a jump of the heater temperature (to point *b*, Figure 1), often beyond the melting point of a metal heating surface.

In systems with controllable surface temperature, an increase beyond the temperature for peak heat flux causes a decrease of heat flux density. This is the *transitional boiling regime* (Region IV); liquid alternately falls onto the surface and is repulsed by an explosive burst of vapor.

At sufficiently high surface temperature, a stable vapor film forms at the heater surface; this is the *film boiling regime* (Regions V and VI). Since heat transfer is by conduction (and some radiation) across the vapor film, the heater temperature is much higher than for comparable heat flux densities in the nucleate boiling regime.

Free surface evaporation. In Region I, where surface temperature exceeds liquid saturation temperature by less than a few degrees, no bubbles form. Evaporation occurs at the free surface by convection of superheated liquid from the heated surface. Correlations of heat transfer coefficients for this region are similar to those for fluids under ordinary natural convection [Equations (1) through (4), Table 1].

Nucleate boiling. Much information is available on boiling heat transfer coefficients, but no universally reliable method is available for correlating this data. In the nucleate boiling regime, heat flux density is not a single, valued function of the temperature but depends also on the nucleating characteristics of the surface, as illustrated by Figure 2 (Berenson 1962).

The equations proposed for correlating nucleate boiling data can be put in a form that relates heat transfer coefficient *h* to temperature difference ($t_w - t_{sat}$):

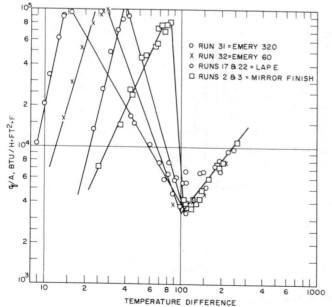

Fig. 2 Effect of Surface Roughness on Temperature in Pool Boiling of Pentane

$$h = \text{constant } (t_w - t_{sat})^a \qquad (1)$$

Exponent *a* normally is between 1 and 3; the constant depends on the thermodynamic and transport properties of the vapor and the liquid. Nucleating characteristics of the surface (size distribution of surface cavities and wettability) affect the value of the multiplying constant and the value of the exponent *a* in Equation (1). For example, variations in exponent *a* from 1 to 25 can be produced by polishing the surface with different grades of emery paper.

A generalized correlation cannot be expected without considering the nucleating characteristics of the heating surface. A statistical analysis of data for 25 liquids by Hughmark (1962) shows that in a correlation not considering surface condition, deviations of more than 100% are common.

In the absence of quantitative nucleating characteristics, Rohsenow (1951) devised a test that evaluated surface effects at atmospheric pressure with a given surface-liquid combination. The effect of pressure can be sealed by using the dimensionless groups in Equation (5), Table 1. Values of the coefficient c_{sf} for some liquid-solid combinations found by Blatt and Adt (1963) are presented in Section II, Table 1.

The nomographs of Figures 3 and 4 (Stephan 1963) can be used to estimate the heat transfer coefficient for R-11, R-12, R-13, R-21, R-22, R-113, R-114, R-142, CO_2, CH_3Cl, NH_3, and SO_2 in nucleate boiling from a horizontal plate (Figure 3) and from the outside of a horizontal cylinder OD = 1.18 in. (Figure 4). Pressures range from 1 to 3 atm.

Stephan's correlation (1963a) is subject to the aforementioned limitations, since its form is that of Equation (1), with exponent *a* equal to 4 for horizontal plates, and 2.33 for horizontal cylinders. Data show variations of *a* from 2 to 25, depending on surface conditions. The nomographs of Figures 3 and 4 are based on experimental data and can be used for estimating the heat transfer coefficient within the range tested (Stephan 1963b).

Equation (6), Table 1, presents an extensively used correlation (Kutateladze 1963). It includes the effect of the diameter of the heating surface (Gilmour 1958), the last term on the right side. This equation predicts the heat transfer coefficients in nucleate boiling from horizontal and vertical plates and cylinders.

In addition to correlations dependent on thermodynamic and transport properties of the vapor and the liquid, Borishansky *et al.* (1962) and Lienhard and Schrock (1963) documented a correlating method based on the law of corresponding states. The properties can be expressed in terms of fundamental molecular parameters, leading to scaling criteria based on the reduced pressure, $p_r = p/p_c$, where p_c is the critical thermodynamic pressure for the coolant. An example of this method of correlation is shown in Figure 5. The reference pressure p^* was chosen as $p^* = 0.029 p_c$. This correlation provides a simple method for scaling the effect of pressure if data are available for one pressure level. It also has an advantage if the thermodynamic and particularly the transport properties used in several equations in Table 1 are not accurately known. In its present form, this correlation gives a value of $a = 2.33$ for the exponent in Equation (1) and consequently should apply for typical aged metal surfaces.

There are explicit correlations for the heat transfer coefficient based on the law of corresponding states for various substances (Borishansky and Kosyrev 1966), specifically for halogenated refrigerants (Danilova 1965), including application to design of flooded evaporators (Starczewski 1965). Other investigations include the effect of oil on heat transfer from a flat plate during pool boiling (Stephan 1963); oil-R-12 mixtures boiling from a 0.55-in. OD horizontal tube (Tschernobyiski and Ratiani 1955); effect of oil on boiling R-12 from inside horizontal tubes (Breber *et al.* 1980, Worsoe-Schmidt 1959, Green and Furse 1963); R-11

Table 1   Equations for Boiling Heat Transfer

Description	Reference(s)	Equation
Free convection		
Free convection boiling, or boiling without bubbles for low Δt and Gr Pr $< 10^8$ (all properties to be based on liquid state)	Jakob (1949, 1957)	$\text{Nu} = c(\text{Gr})^m(\text{Pr})^n$ (1)
Vertical submerged surface		$\text{Nu} = 0.61\,(\text{Gr})^{0.25}\,(\text{Pr})^{0.25}$ (2)
Horizontal submerged surface		$\text{Nu} = 0.16\,(\text{Gr})^{\frac{1}{3}}\,(\text{Pr})^{\frac{1}{3}}$ (3)
Simplified equation for water		$h \approx 80\,(\Delta t)^{\frac{1}{3}}$, where h is in Btu/h·ft^2·°F, Δt in °F (4)
Nucleate boiling	Rohsenow (1951)	$(\text{Nu})_b = (\text{Re})_b{}^{\frac{2}{3}}(\text{Pr})_l{}^{-0.7}/C_{sf}$ (5)
	Kutateladze (1963) Gilmour (1958)	$(\text{Nu})_b = C_k \left[\dfrac{\rho_l}{\rho_v}\,\text{Re}\right]_b^{0.7}(\text{Pr})_l^{0.4}N_k{}^m\left[\dfrac{D_b}{D}\right]^n$ (6)

where $(\text{Nu})_b = hD_b/k_l$, $(\text{Pr})_l = [\mu c_p/k]_l$

$(\text{Re})_b = q\,D_b/Ah_{fg}\,\mu_l$

$D_b = \sqrt{\sigma_t/(\rho_l - \rho_v)g}$

$N_k = p/[\sigma_t g(\rho_l - \rho_v)]^{0.5}$

Blatt and Adt (1963) C_{sf} = constant which depends on solid-liquid combination

Refrigerant	Stainless Steel	Copper
11	0.016	0.022
113	0.09	0.013

Jakob (1949, 1957) $C_k = 7.0 \times 10^{-4}, m = 0.7, n = 0$

$C_k = 4.37 \times 10^{-3}, m = 0.95, n = 1/3$

Description	Reference(s)	Equation
Peak heat flux	Kutateladze (1951) Zuber *et al.* (1962)	$\dfrac{q/A}{\rho_v h_{fg}}\left[\dfrac{\rho^2 v}{\sigma_t g(\rho_l - \rho_v)}\right]^{0.25} = K$ (7)

For many liquids, K varies from 0.12 to 0.16. Recommended average value is 0.13.

Minimum heat flux in film boiling horizontal plates	Zuber (1959)	$\dfrac{q/A}{\rho_v h_{fg}}\left[\dfrac{(\rho_l + \rho_v)}{\sigma_t g(\rho_l - \rho_v)}\right]^{0.25} = 0.09$ (8)
Horizontal cylinders	Lienhard and Wong (1963)	$\dfrac{q/A}{\rho_v h_{fg}}\left[\dfrac{(\rho_l + \rho_v)^2}{\sigma_t g(\rho_l - \rho_v)}\right]^{0.25} = 0.114\,\dfrac{\left[\dfrac{2\sigma_t}{g(\rho_l - \rho_v)D^2}\right]^{0.5}}{\left[1 + \dfrac{2\sigma_t}{g(\rho_l - \rho_v)D^2}\right]^{0.25}}$ (9)
Minimum temperature difference for film boiling from horizontal plate	Berenson (1961)	$(t_w - t_{sat}) = 0.127\,\dfrac{\rho_v h_{fg}}{k_v}\left[\dfrac{g(\rho_l - \rho_v)}{\rho_l + \rho_v}\right]^{\frac{2}{3}}$
		$\times \left[\dfrac{\sigma_t}{g(\rho_l - \rho_v)}\right]^{0.5}\left[\dfrac{\mu_v}{\rho_l - \rho_v}\right]^{\frac{1}{3}}$ (10)
Film boiling from horizontal plate	Berenson (1961)	$h = 0.425\left[\dfrac{k_v^3 \rho_v h_{fg} g(\rho_l - \rho_v)}{\mu_v(t_w - t_{sat})\sqrt{\phi_t/g(\rho_l - \rho_v)}}\right]^{0.25}$ (11)
Film boiling from horizontal cylinders	Anderson *et al.* (1966)	$h = 0.62\left[\dfrac{k_v^3 \rho_v g(\rho_l - \rho_v)h_{fg}}{D\mu_v(t_w - t_{sat})}\right]^{0.25}$ (12)
Effect of radiation	Anderson *et al.* (1966)	Substitute $h_{fg}' = h_{fg}\left[1 + 0.4c_p\dfrac{t_w - t_b}{h_{fg}}\right]$
Effect of surface tension and of pipe diameter	Breen and Westwater (1962)	$\Lambda/D < 0.8$: $h(\Lambda)^{0.25}/F = 0.60$ (13)
		$0.8 < \Lambda/D < 8$: $hD^{0.25}/F = 0.62$ (14)
		$8 < \Lambda/D$: $h(\Lambda)^{0.25}/F = 0.016\,(\Lambda/D)^{0.83}$ (15)

where $\Lambda = 2\pi\left[\dfrac{\sigma_t}{g(\rho_l - \rho_v)}\right]^{0.25}$

$F = \left[\dfrac{\rho_v h_{fg} g(\rho_l - \rho_v)k_v^3}{\mu_v(t_w - t_{sat})}\right]^{0.25}$

Turbulent film	Frederking and Clark (1962)	$\text{Nu} = 0.15\,(\text{Ra})^{\frac{1}{3}}$ (16) for $\text{Ra} > 5 \times 10^7$

$\text{Ra} = \left[\dfrac{D^3 g(\rho_l - \rho_v)}{v_v^2 \rho_v}\left(\dfrac{c_p\mu}{k}\right)_v\left(\dfrac{h_{fg}}{c_p(t_w - t_{sat})} + 0.4\right)\dfrac{a}{g}\right]^{\frac{1}{3}}$

a = local acceleration

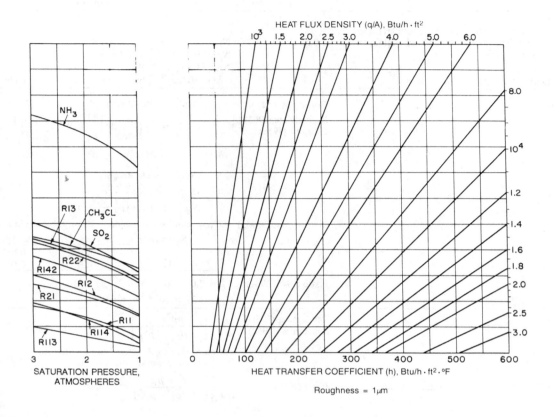

Fig. 3   Heat Transfer Coefficient for Pool Boiling from Horizontal Plate

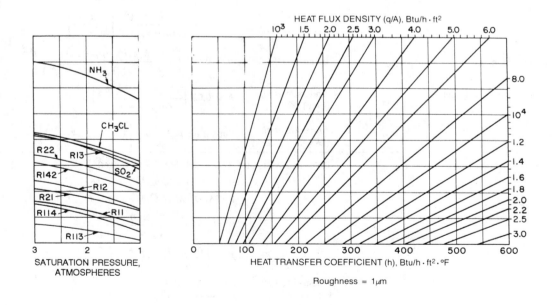

Fig. 4   Heat Transfer Coefficient for Pool Boiling from Horizontal Cylinder

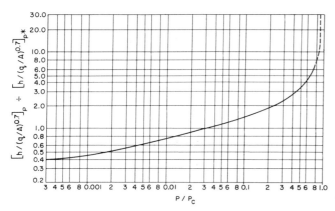

Fig. 5   Correlation of Pool Boiling Data in Terms of
Reduced Pressure

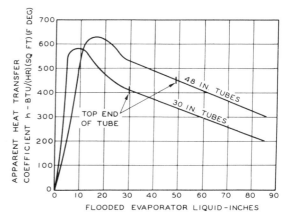

Fig. 6   Boiling Heat Transfer Coefficients for Flooded Evaporator

and R-12 boiling over a flat horizontal copper surface (Furse 1965); and R-11 and R-113 with oil content to 10% boiling from commercial copper tubing (Dougherty and Sauer 1974).

Maximum Heat Flux and Film Boiling

Maximum heat flux and the film boiling region are not as strongly affected by conditions of the heating surface as the heat flux in the nucleate boiling region, making analysis of peak heat flux and of film boiling more tractable.

When peak heat flux is assumed to be a hydrodynamic instability phenomenon, a simple relation, Equation (7), Table 1, can be derived to predict this flux for pure, wetting liquids (Kutateladze 1951, Zuber *et al.* 1962). The dimensionless constant varies from approximately 0.12 to 0.16 for a large variety of liquids. The effect of wettability is still in question. Van Stralen (1959) found that for liquid mixtures, peak heat flux is a function of the concentration.

The minimum heat flux density in film boiling from a horizontal surface and a horizontal cylinder can be predicted by Equations (8) and (9), Table 1. The numerical factors 0.09 and 0.114 were adjusted to fit experimental data; values predicted by two analyses were approximately 30% higher. Equation (10), Table 1, predicts the temperature difference at minimum heat flux of film boiling.

The heat transfer coefficient in film boiling from a horizontal surface can be predicted by Equation (11), Table 1; from a horizontal cylinder by Equation (12), Table 1 (Bromley 1950), which has been generalized to include the effect of surface tension and cylinder diameter, as shown in Equations (13), (14), and (15), Table 1 (Breen and Westwater 1962).

Frederking and Clark (1962) found that for turbulent film boiling, Equation (16), Table 1, agrees with data from experiments at reduced gravity (Rohsenow 1963, Westwater 1963, Kutateladze 1963, Jakob 1949, 1957).

Flooded Evaporators

Equations in Table 1 merely approximate heat transfer rates in flooded evaporators. One reason is that vapor entering the evaporator, combined with vapor generated within the evaporator, can produce significant forced convection effects superimposed on those caused by nucleation. Nonuniform distribution of the two-phase, vapor-liquid flow within the tube bundle of shell-and-tube evaporators or the tubes of vertical-tube flooded evaporators, is also important.

Myers and Katz (1952) investigated the effect of vapor generated by the bottom rows of a tube bundle on the heat transfer coefficient for the upper rows. Improvement in coefficients for the upper tube rows is greatest at low temperature differences where nucleation effects are less pronounced. Hofmann (1956) summarizes other data for flooded tube bundles.

Typical performance of vertical tube natural circulation evaporators, based on data for water, is shown in Figure 6 (Perry 1950). Low coefficients are at low liquid levels because insufficient liquid covers the heating surface. The lower coefficient at high levels is the result of an adverse effect of hydrostatic head on temperature difference and circulation rate. Perry (1950) noted similar effects in horizontal shell-and-tube evaporators.

Forced Convection Evaporation in Tubes

Flow mechanics. When a mixture of liquid and vapor flows inside a tube, a number of flow patterns occur, depending on the mass fraction of liquid, the fluid properties of each phase, and the flow rate. In an evaporator tube, the mass fraction of liquid decreases along the circuit length resulting in a series of changing gas-liquid flow patterns. If the fluid enters as a subcooled liquid, the first indications of vapor generation are bubbles forming at the heated tube wall (nucleation). Subsequently, bubble, plug, churn (or semiannular), annular, spray-annular, and mist flows can occur as the vapor content increases. Idealized flow patterns are illustrated in Figure 7a for a horizontal tube evaporator.

Since nucleation occurs at the heated surface in a thin sublayer of superheated liquid, boiling in forced convection may begin while the bulk of the liquid is subcooled. Depending on the nature of the fluid and the amount of subcooling, the bubbles formed can collapse, or continue to grow and coalesce (Figure 7a), as Gouse and Coumou (1965) observed for R-113. Bergles and Rohsenow (1964) developed a method to determine the point of incipient surface boiling.

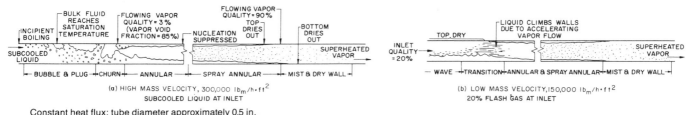

Constant heat flux; tube diameter approximately 0.5 in.

Fig. 7   Flow Regimes in Horizontal Tube Evaporator

After nucleation begins, bubbles quickly agglomerate to form vapor slugs at the center of a vertical tube, or, as shown in Figure 7a, vapor plugs form along the top surface of a horizontal tube. At the point where the bulk of the fluid reaches saturation temperature, corresponding to local static pressure, there will be up to 1% vapor quality because of the preceding surface boiling (Guerrieri and Talby 1956).

Further coalescence of vapor bubbles and plugs results in churn or semiannular flow. If the fluid velocity is high enough, a continuous vapor core surrounded by a liquid annulus at the tube wall soon forms. This annular flow occurs when the ratio of the tube cross section filled with vapor to the total cross section is approximately 85%. With common refrigerants, this equals a vapor quality of about 3 to 5%. The usual flowing vapor quality or vapor fraction is referred to throughout this discussion. Static vapor quality is smaller, since the vapor in the core flows at a higher average velocity than the liquid at the walls (see Chapter 2).

If two-phase mass velocity is high, greater than 150,000 $lb_m/h \cdot ft^2$ for a 0.5-in. tube, annular flow with small drops of entrained liquid in the vapor core (spray) can persist over a vapor quality range from a few percentage points to more than 90%. Refrigerant evaporators are fed from an expansion device at vapor qualities of approximately 20%, so that annular and spray-annular flow predominate in most tube lengths. In a vertical tube, the liquid annulus is distributed uniformly over the periphery, but it is somewhat asymmetric in a horizontal tube (Figure 7a). As vapor quality reaches about 90%, the surface dries out, although there are still entrained droplets of liquid in the vapor (mist). Chaddock and Noerager (1966) found that in a horizontal tube, dryout occurs first at the top of the tube, and later at the bottom (Figure 7a).

If two-phase mass velocity is low, less than 150,000 $lb_m/h \cdot ft^2$ for a 0.5-in. horizontal tube, liquid occupies only the lower cross section of the tube. This causes a wavy type of flow at vapor qualities above about 5%. As the vapor accelerates with increasing evaporation, the interface is disturbed sufficiently to develop annular flow (Figure 7b). Liquid slugging is superimposed on the flow configurations illustrated; the liquid forms a continuous, or nearly continuous, sheet over the tube cross section. The slugs move rapidly and at irregular intervals.

Heat transfer. It is difficult to develop a single relation to describe the heat transfer performance for evaporation in a tube over the full quality range. For refrigerant evaporators with several percentage points of flash gas at entrance, it is less difficult, since annular flow occurs in most of the tube length. The reported data are accurate only within geometry, flow, and refrigerant conditions tested; therefore, a large number of methods for calculating heat transfer coefficients for evaporation in tubes are presented in Table 2 (also see Figures 8 through 11).

Figure 8 gives data obtained for R-12 evaporating in a 0.575-in. copper tube (Ashley 1942). The curves for other tube diameters

are approximations based on an assumed dependence, as in Table 2. Heat transfer coefficient dependence can be understood better from the data in Figure 9 (Gouse and Coumou 1965). At low mass velocities, below 150,000 $lb_m/h \cdot ft^2$, the wavy flow regime of Figure 7b probably exists, and the heat transfer coefficient is nearly constant along the tube length, dropping at the tube exit as complete vaporization occurs. At higher mass velocities, the

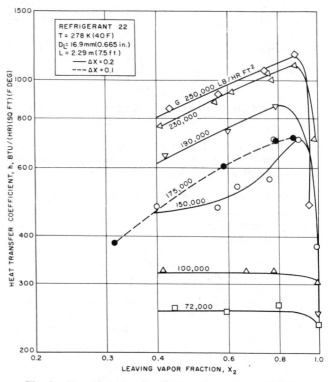

Fig. 9 Heat Transfer Coefficient versus Vapor Fraction for Partial Evaporation

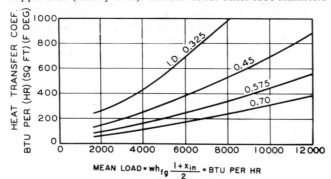

Fig. 8 Boiling Heat Transfer Coefficients for R-12 Inside Horizontal Tubes

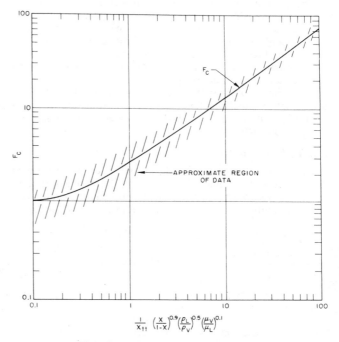

Fig. 10 Reynolds Number Factor F_c

Table 2   Equations for Forced Convection Evaporation in Tubes

Equations	Comments and References
HORIZONTAL TUBES	

HORIZONTAL TUBES

Graphical presentation in Figure 8

h versus q_m

where $q_m = wh_{fg}(1 + x)/2$

> Average coefficients for complete evaporation of R-12 at 40 °F in a 0.575-in. ID copper tube, 33 ft long; the curves for other diameters in Figure 4 are based on the assumption that h varies inversely as the square of the tube diameter (Ashley 1942).

Graphical presentation in Figure 9

h versus x_2

where $x_2 = $ leaving vapor fraction

> Average coefficients for R-22 evaporating at 40 °F in a 0.665-in. ID copper tube, 7.5 ft long. Vapor fraction varied from 20% to 100%. Average coefficients plotted are for vapor fraction changes of 0.20 (or 0.10). For average coefficients (at the same heat flux) over larger vapor fraction ranges, the curves can be integrated (Anderson *et al.* 1966).

$$h = c_1 \left(\frac{k_l}{d}\right)\left[\left(\frac{GD}{\mu_l}\right)^2\left(\frac{J\Delta xh_{fg}}{L}\right)\right]^n \quad (1)$$

where

$c_1 = 0.0009$ and $n = 0.5$

for exit qualities $\leqslant 90\%$; and

$c_1 = 0.0082$ and $n = 0.4$

for 11 °F superheat at exit

> Average coefficients for R-12 and R-22 evaporating in copper tubes of 0.472 and 0.709-in. ID, from 13.4 to 31.2 ft long, and at evaporating temperatures from −4 to 32 °F. Vapor fraction varied from 0.15 to 11 °F superheat. Note that the term $J\Delta xh_{fg}/L$ is not dimensionless, but has the units of pound force per pound mass (Pierre 1955, 1957).

Equation (1) of Pierre with

$c_1 = 0.0225$ and $n = 0.375$

> Average coefficients for R-22 evaporating at temperatures from 40 to 80 °F in a 0.343-in. ID tube, 8 ft long. Coefficients were determined for approximately 15% vapor quality changes. The range investigated was $x = 0.20$ to superheat (Altman *et al.* 1960b).

$$h = c_2 h_l \left(\frac{i+x}{1-x}\right)^{1.16}\left(\frac{q}{Gh_{fg}}\right)^{0.1} \quad (2)$$

where

$c_2 = 6.59$

$$h_l = \frac{0.023\,k_l}{d}\left[\frac{DG(1-x)}{\mu_l}\right]^{0.8}(\text{Pr})_l^{0.4} \quad (3)$$

> Local coefficients for R-12 and R-22 evaporating in a 0.732-in. ID tube 1 ft long at saturation temperatures from 75 to 90 °F. Location of transition from annular to mist flow is established, and a heat transfer equation for the mist flow regime is presented (Lavin and Young 1964).

$$h = 1.85\,h_L[B_o \times 10^4 + (1/X_{tt})^{0.67}]^{0.6} \quad (4)$$

where

$$B_o = q/Gh_{fg} \quad (5)$$

$$h_L = \frac{0.023 k_l}{d}\left(\frac{DG}{\mu_l}\right)^{0.8}(\text{Pr})_l^{0.4} \quad (6)$$

$$X_{tt} = \left(\frac{1-x}{x}\right)^{0.9}\left(\frac{\rho_v}{\rho_l}\right)^{0.5}\left(\frac{\mu_l}{\mu_v}\right)^{0.1} \quad (7)$$

> Local coefficients for R-12 evaporating in a 0.460-in. ID stainless steel tube with a uniform wall heat flux (electric heating) over a length of 6.344 ft, and an evaporating temperature of 53 °F. Vapor fraction range was 0.20 to 0.88. Equation (4) is a modified form of the Schrock and Grossman equation for vertical tube evaporation [Equation (10)] (Chaddock and Noerager 1966).

Best agreement was with Equation (11) for vertical tubes.

> Local coefficients for R-113 evaporating in a 0.430-in. ID transparent tube with a uniform wall heat flux over a length of 12.5 ft; evaporating temperature approximately 120 °F. Report includes photographs of subcooled surface boiling, bubble, plug, and annular flow evaporation regimes (Gouse and Coumou 1965).

VERTICAL TUBES

$$h = 3.4\,h_l(1/X_{tt})^{0.45} \quad (8)$$

$$h = 3.5\,h_L(1/X_{tt})^{0.5} \quad (9)$$

where

h_l is from (3), X_{tt} from (7), h_L from (6)

> Equations (8) and (9) were fitted to experimental data for vertical upflow in tubes. Both relate to forced convection evaporation regions where nucleate boiling is suppressed (Guerrieri and Talty 1956, Dengler and Addoms 1956). (A multiplying factor is recommended when nucleation is present.)

$$h = 0.74\,h_L[B_o \times 10^4 + (1/X_{tt})^{0.67}] \quad (10)$$

where

B_o is from (5), h_L from (6), X_{tt} from (7)

> Local coefficients for water in vertical upflow in tubes with diameters from 0.1162 to 0.4317 in. and lengths of 15 to 40 in. The boiling number B_o accounts for nucleation effects, and the Martinelli parameter x_{tt} for forced convection effects (Schrock and Grossman 1962).

$$h = h_{mic} + h_{mac} \quad (11)$$

where

$$h_{mac} = h_l F_c \quad (12)$$

$$h_{mic} = 0.00122\,(S_c)(E)(\Delta t)^{0.24}(\Delta P)^{0.75} \quad (13)$$

F_c and S_c from Figures 10 and 11

> Chen developed this correlation reasoning that the nucleation transfer mechanism (represented by h_{mic}) and the convective transfer mechanism (represented by h_{mac}) are additive. h_{mac} is expressed as a function of the two-phase Reynolds number after Martinelli, and h_{mic} is obtained from the nucleate boiling correlation of Forster and Zuber (1955). S_c is a suppression factor for nucleate boiling (Chen 1963).

$$E = \frac{k_l^{0.79}(c_p)_l^{0.45}\rho_l^{0.49}g_c^{0.25}}{\sigma_t^{0.50}\mu_l^{0.29}h_{fg}^{0.24}\rho_v^{0.24}} \quad (14)$$

Equation (2) with $c_2 = 3.79$

> See comments for Equation (2). Note the superior performance of the horizontal versus vertical configuration ($c_2 = 6.59$ versus 3.79) from this investigation which used the same apparatus and test techniques for both orientations (Lavin and Young 1964).

Note: Except for dimensionless equations, units are lb_m, h, ft, °F, and Btu.

Table 3 Heat Transfer Coefficients for Film-Type Condensation

$$t_f = \text{liquid film temperature} = t_{sat} - 0.75\,\Delta t$$

Description	Reference(s)	Equation	
1. Vertical surfaces, height L			
Laminar condensate flow, Re $= 4\Gamma/\mu_f < 1800$	McAdams (1954)	$h = 1.13\,F_1(h_{fg}/L\Delta t)^{0.25}$	(1)
	McAdams (1954)	$h = 1.11\,F_2(b/w_l)^{1/3}$	(2)
	Grigull (1952)	$h = 0.003\,(F_1)^2(\Delta tL/\mu_f^2 h_{fg})^{0.5}$	(3)
Turbulent flow, Re $= 4\Gamma/\mu_f > 1800$	McAdams (1954)	$h = 0.0077\,F_2(\text{Re})^{0.4}(1/\mu_f)^{1/3}$	(4)
2. Outside horizontal tubes, N rows in a vertical plane, length L, laminar flow	McAdams (1954)	$h = 0.79\,F_1\,(h_{fg}/Nd\Delta t)^{0.25}$	(5)
	McAdams (1954)	$h = 1.05\,F_2(L/w_l)^{1/3}$	(6)
Finned tubes	Beatty and Katz (1948)	$h = 0.689\,F_1(h_{fg}/\Delta tD_e)^{0.25}$ where D_e is determined from $\dfrac{1}{(D_e)^{0.25}} = 1.30\,\dfrac{A_s\phi}{A_{ef}(L_{mf})^{0.25}} + \dfrac{A_p}{A_{ef}(D)^{0.25}}$ with $A_{ef} = A_{s\phi} + A_p$ and $L_{mf} = a_f/D_o$	(7)
3. Simplified equations for steam			
Outside vertical tubes, Re $= 4\Gamma/\mu_f < 2100$	McAdams (1954)	$h = 4000/(L)^{0.25}(\Delta t)^{1/3}$	(8)
Outside horizontal tubes, Re $= 4\Gamma/\mu_f < 1800$			
Single tube	McAdams (1954)	$h = 3100/(d')^{0.25}(\Delta t)^{1/3}$	(9a)
Multiple tubes		$h = 3100/(Nd')^{0.25}(\Delta t)^{1/3}$	(9b)
4. Inside vertical tubes	Carpenter and Colburn (1949)	$h = 0.065\left(\dfrac{c_{pf}k_f\rho_f f'}{2\mu_f\rho_v}\right)$ where $G_m = \left(\dfrac{G_i^2 + G_iG_o + G_o^2}{3}\right)^{0.5}$	(10)
5. Inside horizontal tubes, $\dfrac{DG_l}{\mu_l} < 5000$			
$1000 < \dfrac{DG_v}{\mu_l}\left(\dfrac{\rho_l}{\rho_v}\right)^{0.5} < 20{,}000$	Ackers and Rosson (1960)	$\dfrac{hD}{k_l} = 13.8\left(\dfrac{c_p\mu_l}{k_l}\right)^{1/3}\left(\dfrac{h_{fg}}{c_p\Delta t}\right)^{1/6}\left[\dfrac{DG_v}{\mu_l}\left(\dfrac{\rho_l}{\rho_v}\right)^{0.5}\right]^{0.2}$	(11)
$20{,}000 < \dfrac{DG_v}{\mu_l}\left(\dfrac{\rho_l}{\rho_v}\right)^{0.5} < 100{,}000$	Ackers and Rosson (1960)	$\dfrac{hD}{k_l} = 0.1\left(\dfrac{c_p\mu_l}{k_l}\right)^{1/3}\left(\dfrac{h_{fg}}{c_p\Delta t}\right)^{1/6}\left[\dfrac{DG_v}{\mu_l}\left(\dfrac{\rho_l}{\rho_v}\right)^{0.5}\right]^{2/3}$	(12)
For $\dfrac{DG_l}{\mu_l} > 5000\ \dfrac{DG_v}{\mu_l}\left(\dfrac{\rho_l}{\rho_v}\right)^{0.5} > 20{,}000$	Ackers et al. (1959)	$\dfrac{hD}{k_l} = 0.026\left(\dfrac{c_p\mu_l}{k_l}\right)^{1/3}\left(\dfrac{DG_E}{\mu_l}\right)^{0.8}$ where $G_E = G_v(\rho_l/\rho_v)^{0.5} + G_l$	(13)
	Altman et al. (1960a)	$h = 0.057\left(\dfrac{c_pk_f\rho_f}{\mu_f}\right)^{0.5}F^{0.5}$ where $F = \Delta p_{TPF}\left(\dfrac{g_oD}{4L}\right)$	(14)
	Forster and Zuber (1955)	$\Delta p_{TPF} = $ frictional two-phase pressure drop	

Note: Equations (1) through (10) and Equation (14) are dimensional with units of Btu, h, ft, °F, and lb_m.

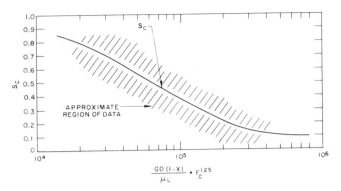

Fig. 11 Suppression Factor S_c

flow pattern is usually annular, and the coefficient increases as vapor accelerates. As the surface dries at a 90% vapor quality, the coefficient drops sharply.

Equation (1), Table 2, is recommended for broadest application to refrigerant evaporation in tubes (Pierre 1955, 1957). It fits a wide range of R-12 and R-22 data.

Equations (2), (10), and (11) in Table 2 include terms for velocity effect (convection) and heat flux (nucleation), and produce local heat transfer coefficients as a function of the local vapor quality x and heat transfer rate q. Local rather than average coefficients are used for accurate design.

The effect of oil on forced convection evaporation has not been clearly determined. Increases occur in the *average* heat transfer coefficient for R-12 up to 10% oil by weight, with a maximum of about 4% (Green and Furse 1963, Worsoe-Schmidt 1960). Oil quantities greater than 10% cause reduction in heat transfer. Oil can increase the pressure drop, offsetting possible gains in the heat transfer coefficient.

CONDENSING

Condensation occurs when a saturated vapor contacts a surface at a lower temperature than the saturation temperature of the vapor. The vapor loses its latent heat of vaporization and condenses on the surface.

Film condensation occurs in most applications. The liquid condensate covers the condensing surface with a continuous film and flows off the surface by gravity. *Dropwise condensation* is seen as steam on highly polished surfaces or on surfaces contaminated with fatty acids.

The rate of heat flow depends on the condensate film thickness, which depends on the rate of vapor condensation and the rate of condensate removal. At high reduced pressures, the heat transfer coefficients for dropwise condensation, at the same surface loading, are higher than those available in the presence of film condensation. At low reduced pressures, the reverse is true. For example, there is a reduction of 6 to 1 in the dropwise condensation coefficient of steam when saturation pressure is decreased from 0.9 to 0.16 atm. One method for correlating the dropwise condensation heat transfer coefficient employs nondimensional parameters, including the effect of surface tension gradient, temperature difference, and fluid properties.

When condensation occurs on horizontal tubes and short vertical plates, the condensate film motion is laminar. On vertical tubes and long vertical plates, the film can become turbulent. Grober *et al.* (1961) suggest using a Reynolds number of 1600 as the critical point where the flow pattern changes from laminar to turbulent. This Reynolds number is based on condensate flow rate divided by the breadth of the condensing surface. For a vertical tube, the breadth is the circumference of the tube; for a horizontal tube,

the breadth is twice the length of the tube. The Reynolds number = $4w/\mu_f$, where w is the mass flow of condensate per unit of breadth, and μ_f is the viscosity of the condensate at the film temperature t_f. In practice, condensation is usually laminar in shell-and-tube condensers with the vapor outside horizontal tubes.

Vapor velocity also affects the condensing coefficient. When this is small, condensate flows primarily by gravity and is resisted by the viscosity of the liquid. When vapor velocity is high relative to the condensate film, there is appreciable drag at the vapor-liquid interface. The thickness of the condensate film, and hence the heat transfer coefficient, is affected. When vapor flow is upward, a retarding force is added to the viscous shear, increasing the film thickness. When vapor flow is downward, the film thickness decreases and the heat transfer coefficient increases. For condensation inside horizontal tubes, the force of the vapor velocity causes the condensate to flow. When the vapor velocity is high, the transition from laminar to turbulent flow occurs at lower Reynolds numbers.

When *superheated* vapor is condensed, the heat transfer coefficient depends on the surface temperature. When the surface temperature is *below* saturation temperature, there is little error if the value of h for condensation of saturated vapor is used with the difference between the *saturation* temperature and the surface temperature (McAdams 1954). If the surface temperature is *above* the saturation temperature, there is no condensation and the equations for gas convection apply.

Correlation equations for condensing heat transfer are given in Table 3. Factors F_1 and F_2, which depend only on the physical properties of the refrigerant and which occur often in these equations, have been computed for some commonly used refrigerants in Table 4. Refrigerant properties used in the calculations may be found in Chapter 16.

Table 4 Values of Condensing Coefficient Factors for Different Refrigerants (from Chapter 16)

Refrigerant	Film Temperature, °F $t_f = t_{sat} - 0.75\,(\Delta t)$	F_1	F_2
Refrigerant 11	75	154	822
	100	153	815
	125	151	803
Refrigerant 12	75	133	672
	100	122	608
	125	112	538
Refrigerant 22	75	153	822
	100	144	755
	125	132	675
Sulfur Dioxide	75	290	1920
	100	299	2000
	125	318	2170
Ammonia	75	409	3040
	100	408	3035
	125	408	3030
Propane	75	159	850
	100	157	845
	125	154	836
Butane	75	156	840
	100	156	843
	125	157	845

$$F_1 = \left(\frac{k_f^3 \rho_f^2 g}{\mu f}\right)^{0.25} \qquad \text{Units} = \left(\frac{(\text{Btu})^3 (\text{lb}_m)}{(\text{h})^4 (\text{ft})^7 (°\text{F})^3}\right)^{0.25}$$

$$F_2 = \left(\frac{k_f^3 \rho_f^2 g}{\mu f}\right)^{1/3} \qquad \text{Units} = \left(\frac{(\text{Btu})^3 (\text{lb}_m)}{(\text{h})^4 (\text{ft})^7 (°\text{F})^3}\right)^{1/3}$$

In some cases, the equations are given in two forms: one is convenient when the amount of refrigerant to be condensed or the condensing load is known; the second is useful when the difference between the vapor temperature and the condensing surface temperature is known.

Condensation Outside Vertical Tubes

For film-type condensation outside vertical tubes and on vertical surfaces, Equations (1) and (2), Table 3, are recommended when $4w/\mu_f$ is less than 1800 (McAdams 1954). Fluid properties are evaluated at the mean film temperature. When $4w/\mu_f$ is greater than 1800 (tall vertical plates or tubes), use Equations (3) or (4), Table 3. Equations (2) and (4), Table 3, are plotted in Figure 12. The theoretical curve for laminar film-type condensation is shown for comparison. A semitheoretical relationship for turbulent film-type condensation is also shown for Pr values of 1.0 and 5.0 (Colburn 1933–34).

Condensation Outside Horizontal Tubes

For a bank of N tubes, Nusselt's equations, increased by 10% (Jakob 1949, 1957), are given in Equations (5) and (6), Table 3. Experiments by Short and Brown (1951) with R-11 suggest that drops of condensation falling from row to row cause local turbulence and increase heat transfer.

For condensation outside horizontal finned tubes, Equation (7), Table 3, is used for liquids that drain readily from the surface (Beatty and Katz 1948). For condensing steam outside finned tubes, where liquid is retained in the spaces between the tubes, coefficients substantially lower than those given by Equation (7), Table 3, were reported. For additional data on condensation outside finned tubes, see Katz *et al.* (1947).

Simplified Equations for Steam

For film-type steam condensation at atmospheric pressure and film temperature drops of 10 to 150 °F, McAdams (1954) recommends Equations (8) and (9), Table 3.

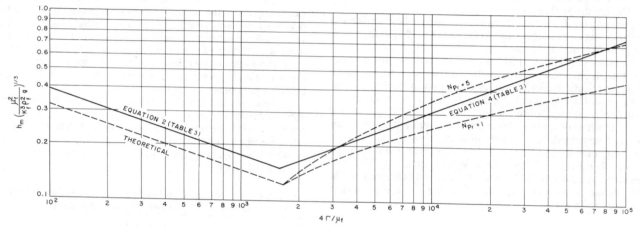

Fig. 12   Film-Type Condensation

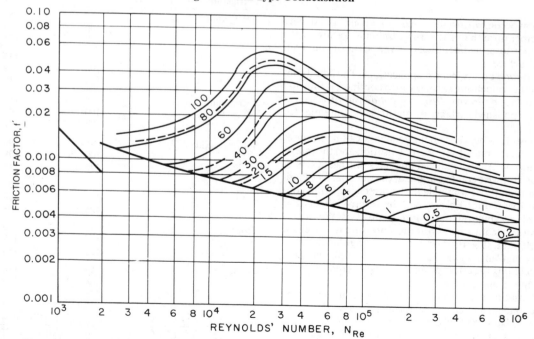

Curve parameter = $\Gamma/\rho s$, where Γ = liquid flow rate, ρ = liquid density, and s = surface tension of liquid relative to water; values of gas velocity used in calculating f and Re are calculated as though no liquid was present.

Fig. 13   Friction Factors for Gas Flow Inside Pipes with Wetted Walls

Condensation Inside Vertical Tubes

Condensation inside tubes is generally affected by appreciable vapor velocity. The measured heat transfer coefficients are as much as 10 times those predicted by Equation (4), Table 3. For vertical tubes, Jakob (1949, 1957) gives theoretical derivations for upward and downward vapor flow. For downward vapor flow, Carpenter and Colburn (1949) suggest Equation (10), Table 3. The friction factor f' for vapor in a pipe containing condensate should be taken from Figure 13.

Condensation Inside Horizontal Tubes

For condensation inside horizontal tubes (as in air-cooled condensers, evaporative condensers, and some shell-and-tube condensers), the vapor velocity and resulting shear at the vapor-liquid interface are major factors in analyzing heat transfer. Hoogendoorn (1959) identified seven types of two-phase flow patterns. For semistratified and laminar annular flow, use Equations (11) and (12), Table 3 (Ackers and Rosson 1960). Ackers *et al.* (1959) recommend Equation (13), Table 3, for turbulent annular flow (vapor Reynolds number greater than 20,000, and liquid Reynolds number greater than 5000). Equation (14), Table 3, correlates the local heat transfer coefficients for R-22 condensing inside pipes (Altman *et al.* 1960b); R-22 and several other fluids take the same form. The two-phase pressure drop in Equation (14) is determined by the method proposed by Martinelli and Nelson (1948); see also Altman *et al.* (1960a). A method for using a flow regime map to predict the heat transfer coefficient for condensation of pure components in a horizontal tube is presented in Breber *et al.* (1980).

Noncondensable Gases

Condensation heat transfer rates reduce drastically if one or more noncondensable gases are present in the condensing vapor/gas mixture. The decrease in the heat transfer coefficient is approximately linear with the weight fraction of the noncondensable gas present. The condensable component is termed *vapor*, and the noncondensable component, *gas*. In a steam chest with 2.89% air by volume, Othmer (1929) found that the heat transfer coefficient dropped from about 2000 to about 600 Btu/h·ft²·°F. Consider a surface cooled to some temperature t_s below the saturation temperature of the vapor (Figure 14). Accumulated condensate falls or is driven across the condenser surface. At a finite heat transfer rate, a temperature profile develops across the condensate that can be estimated from Table 3. Therefore, the interface of the condensate is at a temperature $t_{if} > t_s$. In the absence of gas, the interface temperature is the vapor saturation temperature at the pressure of the condenser.

The presence of noncondensable gas lowers the vapor partial pressure and, hence, the saturation temperature of the vapor in equilibrium with the condensate. Further, the movement of the vapor toward the cooled surface implies similar bulk motion of the gas. At the condensing interface, the vapor is condensed at temperature t_{if} and is then swept out of the system as a liquid. The gas concentration rises to ultimately diffuse away from the cooled surface at the same rate as it is convected toward the surface (Figure 14). If gas (mole fraction) concentration is Y_g and total pressure of the system is p, the partial pressure of the bulk gas is:

$$p_{g\infty} = Y_{g\infty}p \tag{2}$$

The partial pressure of the bulk vapor is:

$$p_{v\infty} = (1 - Y_{g\infty})p = Y_{v\infty}p \tag{3}$$

As opposing fluxes of convection and diffusion of the gas increase, the partial pressure of gas at the condensing interface is $p_{gif} > p_{g\infty}$. By Dalton's law, assuming isobaric condition:

$$p_{gif} + p_{vif} = p \tag{4}$$

Hence, $p_{vif} < p_{v\infty}$.

Sparrow *et al.* (1967) noted that thermodynamic equilibrium exists at the interface, except in the case of very low pressures or liquid metal condensation, so that:

$$p_{vif} = p_{sat}(t_{if}) \tag{5}$$

where $p_{sat}(t)$ is the saturation pressure of the vapor at temperature t. The available Δt for condensation across the condensate film is reduced from $(t_\infty - t_s)$ to $(t_{if} - t_s)$, where t_∞ is the bulk temperature of the condensing vapor-gas mixture, caused by the additional noncondensable resistance.

The equations in Table 3 are still valid for the condensate resistance, but the interface temperature t_{if} must be found. The noncondensable resistance, which accounts for the temperature difference $(t_\infty - t_{if})$, depends on the heat flux (through the convecting flow to the interface) and the diffusion of gas away from the interface.

In simple cases, Sparrow *et al.* (1967), Rose (1969), and Sparrow and Lin (1964) found solutions to the combined energy, diffusion, and momentum problem of noncondensables, but they are cumbersome.

A general method given by Colburn and Hougen (1934) can be used over a wide range if correct expressions are provided for the rate equations—add the contributions of the sensible heat transport through the noncondensable gas film and the latent heat transport via condensation:

$$h_g(t_\infty - t_{if}) + KM_v h_{lv}(p_{v\infty} - p_{vif}) = h(t_{if} - t_s) = U(t_{if} - t_c) \tag{6}$$

where h is from the appropriate equation in Table 3.

The value of the heat transfer coefficient for the stagnant gas depends on the geometry and flow conditions. For flow parallel to a condenser tube, for example:

$$j = \frac{h_g}{(c_p)_g G}\left(\frac{(c_p)_g \mu_{gv}}{K_g}\right)^{2/3} \tag{7}$$

where j is a known function of $Re = GD/\mu_{gv}$.

The mass transfer coefficient K is:

$$\frac{K}{M_m}\left[\frac{p_{g\infty} - p_{gif}}{\ln(p_{g\infty}/p_{gif})}\right]\left(\frac{\mu_{gv}}{\rho_g D}\right)^{2/3} = j \tag{8}$$

The calculation method requires substitution of Equation (8) into Equation (6). For a given flow condition, G, Re, j, M_m, $p_{g\infty}$, h_g, and h (or U) are known. Assume values of t_{if}; calculate $p_{sat}(t_{if}) = p_{vif}$ and hence p_{gif}. If t_s is not known, use the overall coef-

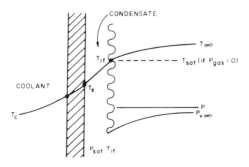

Fig. 14   Origin of Noncondensable Resistance

ficient U to the coolant at t_c in place of h and t_s in Equation (6). For either case, at each location in the condenser, iterate Equation (6) until it balances, giving the condensing interface temperature and, hence, the thermal load to that point (Colburn and Hougen 1934, Colburn 1951).

Other Impurities

Vapor entering the condenser often contains a small percentage of impurities such as oil. Oil forms a film on the condensing surfaces, creating additional resistance to heat transfer. Some allowance should be made for this, especially in the absence of an oil separator or when the discharge line from the compressor to the condenser is short.

Pressure Drop

Total pressure drop for two-phase flow in tubes consists of friction, acceleration, and gravitational components. It is necessary to know the *void fraction* (the ratio of gas flow area to total flow area) to compute the acceleration and gravitational components. To compute the frictional component or pressure drop, either the *two-phase friction factor* or the *two-phase frictional multiplier* must be known.

The homogeneous model provides a simple method for computing the acceleration and gravitational components of pressure drop. The homogeneous model assumes that the flow can be characterized by average fluid properties and that the velocities of the liquid and vapor phases are equal (Collier 1972, Wallis 1969).

Martinelli and Nelson (1948) developed a method for predicting the void fraction and two-phase frictional multiplier to use with a separated flow model. This method predicts the pressure drops of boiling refrigerants reasonably well. Other methods of computing the void fraction and two-phase frictional multiplier used in a separated flow model are given in Collier (1972) and Wallis (1969).

The general nature of annular gas-liquid flow in vertical, and to some extent horizontal, pipe is indicated (Wallis 1970) in Figure 15, which plots effective gas friction factor versus the liquid fraction $(1 - a)$. Here a is the void fraction, or fraction of the pipe cross section taken up by the gas or vapor.

The effective gas friction factor is defined as:

$$f_{eff} = \left[\frac{a^{5/2}D}{2\rho_g(4Q_g/\pi D^2)^2} \right] \left(-\frac{dp}{ds} \right) \quad (9)$$

where D is the pipe diameter, ρ_g the gas density, and Q_g the gas volume flow rate. The friction factor of gas flowing by itself in the pipe (presumed smooth) is denoted by f_g. Wallis' analysis of the flow occurrences is based on interfacial friction between the gas and liquid. The wavy film corresponds to a conduit of relative roughness (ϵ/D), about four times the liquid film thickness. Thus, the pressure drop relation of vertical flow is:

$$\left(-\frac{dp}{ds} + \rho_g g \right) = 0.01 \, (\rho_g/D^5)(4Q_g/\pi)^2 \frac{1 + 75(1 - a)}{a^{5/2}} \quad (10)$$

This corresponds to the Martinelli-type analysis with:

$$f_{two\text{-}phase} = \phi_g^2 f_g$$

when

$$\phi_g^2 = \frac{1 + 75(1 - a)}{a^{5/2}} \quad (11)$$

where the friction factor f_g (of the gas alone) is taken as 0.02, an appropriate turbulent flow value. This calculation can be modified for more detailed consideration of factors, such as Reynolds number variation in friction, gas compressibility, and entrainment (Wallis 1970).

In two-phase flow inside horizontal tubes, the pressure gradient is written as the sum of frictional and momentum terms. Thus:

$$\left(\frac{dp}{dz} \right) = \left(\frac{dp}{dz} \right)_f + \left(\frac{dp}{dz} \right)_m \quad (12)$$

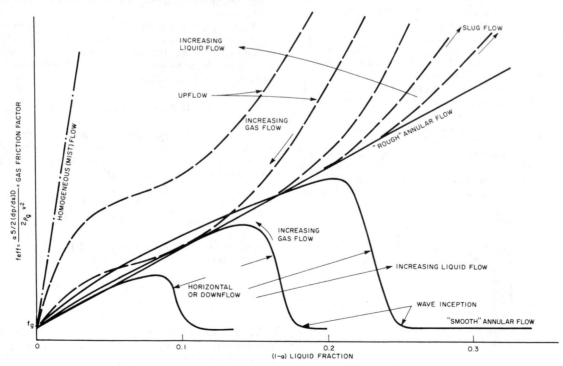

Fig. 15   Qualitative Pressure Drop Characteristics of Two-Phase Flow Regime

In adiabatic two-phase flow, the contribution of the momentum transfer to the overall pressure drop is negligibly small; theoretically, it is nonexistent if the flow is fully developed. In condensation heat transfer, the momentum transfer term contributes to the overall pressure drop due to the mass transfer that occurs at the liquid-vapor interface.

Two basic models were used in developing frictional pressure drop correlations for two-phase adiabatic flow. In the first, the flow of both phases is assumed to be homogeneous; the gas and liquid velocities are assumed equal. The frictional pressure drop is computed as if the flow were single phase, except for introducing modifiers to the single-phase friction coefficient. In the second model, the two phases are considered separate and the velocities may differ. Two correlations used to predict the frictional pressure drop are those of Lockhart-Martinelli (1949) and Dukler *et al.* (1964).

In the Lockhart-Martinelli correlation, a parameter X was defined as:

$$X = \left[\left(\frac{dp}{dz} \right)_l \div \left(\frac{dp}{dz} \right)_v \right]^{0.05} \tag{13}$$

where

$\left(\frac{dp}{dz} \right)_l$ = frictional pressure gradient, assuming that liquid alone flows in pipe

$\left(\frac{dp}{dz} \right)_v$ = frictional pressure gradient, assuming that gas (or vapor in case of condensation) alone flows in pipe

The frictional pressure gradient due to the single-phase flow of the liquid or vapor depends on the type of flow of each phase, laminar or turbulent. For turbulent flow during condensation, replace X by X_{tt}. Thus:

$$X_{tt} = \left(\frac{1-x}{x} \right)^{0.9} \left(\frac{\mu_l}{\mu_v} \right)^{0.1} \left(\frac{\rho_v}{\rho_l} \right)^{0.5} \tag{14}$$

Lockhart and Martinelli (1949) also defined ϕ_v as

$$\phi_v = \left[\left(\frac{dp}{dz} \right)_f \div \left(\frac{dp}{dz} \right)_v \right]^{0.5} \tag{15}$$

For condensation

$$\left(\frac{dp}{dz} \right)_v = -\frac{2 f_o (x\, G)^2}{\rho_v D_i} \tag{16}$$

where

$$f_o = 0.045 \Big/ \left(\frac{GxD_i}{\mu_v} \right)^{0.2} \tag{17}$$

Here f_o is the friction factor for adiabatic two-phase flow.

By analyzing the pressure drop data of simultaneous adiabatic flow of air and various liquids, Lockhart and Martinelli (1949) correlated the parameters ϕ_v and X and reported the results graphically. Soliman *et al.* (1968) approximated the graphical results of ϕ_v versus X_{tt} by:

$$\phi_v = 1 + 2.85\, X_{tt}^{0.523} \tag{18}$$

In the correlation of Dukler (1964), the frictional pressure gradient is given by:

$$\left(\frac{dp}{dz} \right)_f = -\frac{2\, G^2 f_o \alpha(\lambda) \beta}{D_i \rho_{NS}} \tag{19}$$

where

f_o = single-phase friction coefficient evaluated at two-phase Reynolds number

$$= 0.0014 + 0.125 \left(\frac{4\, \dot{M}_T \beta}{\pi D_i \mu_{NS}} \right)^{-0.32} \tag{20}$$

$$\alpha(\lambda) = 1 - (\ln\lambda)/[1.281 + 0.478 \ln\lambda + 0.444\, (\ln\lambda)^2 + 0.094\, (\ln\lambda)^3 + 0.00843\, (\ln\lambda)^4] \tag{21}$$

$$\beta = \left[\left(\frac{\rho_l}{\rho_{NS}} \right) \frac{\lambda^2}{(1-\psi)} + \left(\frac{\rho_v}{\rho_{NS}} \right) \frac{(1-\lambda)^2}{\psi} \right] \tag{22}$$

$$\rho_{NS} = [\rho_l \lambda + \rho_v (1-\lambda)] \tag{23}$$

$$\mu_{NS} = [\mu_l \lambda + \mu_v (1-\lambda)] \tag{24}$$

$$\lambda = 1 \Big/ \left(1 + \frac{x}{(1-x)} \frac{\rho_v}{\rho_e} \right) \tag{25}$$

Since the correlations mentioned here were originally developed for adiabatic two-phase flow, Luu and Bergles (1980) modified the friction coefficients in Equations (16) and (19), using the modifier suggested by Silver and Wallis (1965). The modification replaced the friction coefficient f_o with the friction coefficient f_{co}; f_o and f_{co} are related by

$$(f_{co}/f_o) = \exp(\epsilon/2 f_o) - (\xi/f_o) \tag{26}$$

where $\qquad \xi = (D_i \psi/2x) \left(\dfrac{dx}{dz} \right) \tag{27}$

Since the Lockhart-Martinelli and Dukler correlations for the frictional pressure gradient were based on the separated flow model, so should the momentum pressure gradient. Thus:

$$\left(\frac{dp}{dz} \right)_m = -G^2 \left(\frac{dx}{dz} \right) \left\{ \frac{2x}{\rho_v \psi} - \frac{2(1-x)}{\rho_l(1-\psi)} \right.$$
$$\left. + q_l \left[\frac{\psi(1-x)}{x(1-\psi)\rho_l} - \frac{x(1-\psi)}{\psi(1-x)\rho_v} \right] \right\} \tag{28}$$

To determine $(dp/dz)_m$, the void fraction ψ and the quality gradient must be known. A generalized expression for ψ was suggested by Butterworth (1975):

$$\psi = \frac{1}{1 + A_l\, [(1-x)/x]^{q_l}\, (\rho_v/\rho_l)^{r_l}\, (\mu_l/\mu_v)^{S_l}} \tag{29}$$

where A_l, q_l, r_l, and S_l are constants and are listed for the various correlations in Table 5.

Table 5   Constants in Equation (29) for Different Void Fraction Correlations

Model	A_l	q_l	r_l	S_l
Homogeneous (Collier 1972)	1.0	1.0	1.0	0
Lockhart-Martinelli (1949)	0.28	0.64	0.36	0.07
Baroczy (1963)	1.0	0.74	0.65	0.13
Thom (1964)	1.0	1.0	0.89	0.18
Zivi (1964)	1.0	1.0	0.67	0
Turner-Wallis (1965)	1.0	0.72	0.40	0.08

The quality gradient dx/dz in Equation (28) can be estimated by assuming a constant rate of cooling. In the case of complete condensation, its value is $-1/L$, where L is the length of the condenser tube.

Evaporators and condensers often have valves, tees, bends, and other fittings that contribute to the overall pressure drop of the heat exchanger. Collier (1972) summarizes methods predicting the two-phase pressure drop in these fittings.

ENHANCED SURFACES

Enhanced heat transfer surfaces are used in heat exchangers to improve performance and decrease cost. Condensing heat transfer is often enhanced with circular fins attached to the external surfaces of tubes to increase the area of heat transfer. Other enhancement methods, such as porous coatings, integral fins, and reentrant cavities, are used to augment boiling heat transfer on the external surfaces of evaporator tubes. Webb (1981) surveys external boiling surfaces and compares the performances of several enhanced surfaces with the performances of smooth tubes. Since the heat transfer coefficient for the refrigerant side is often smaller than the coefficient for the water side, enhancing the refrigerant-side surface can reduce the size of the heat exchanger and improve performance.

Internal fins increase the heat transfer coefficients during evaporation or condensation in tubes. However, internal fins increase the refrigerant pressure drop and reduce the heat transfer rate by decreasing the available temperature difference between hot and cold fluids. Designers should carefully determine the number of parallel refrigerant passes that give optimum loading for best overall heat transfer.

For additional information on enhancement methods in two-phase flow, consult Bergles' comprehensive surveys (1976, 1985).

SYMBOLS

A = area
A_{ef} = total effective area [Eq. (7a), Table 3]
a = local acceleration [Eq. (16), Table 1], void fraction [Eqs. (9) and (10)]
a_f = area of one side of one film
B = boiling number [Eq. (10), Table 2]
b = breadth of a condensing surface. For vertical tube, pd; for horizontal tube, $b = 2L$
c = a coefficient or constant
c_p = specific heat at constant pressure
c_v = specific heat at constant volume
c_1, c_2 = special constants (see Table 2)
c_{sf}, c_r = special constants (see Table 1)
D = diameter
D_i = inside tube diameter
D_o = outside tube diameter
d = diameter; or prefix meaning differential
(dp/dz) = pressure gradient
$(dp/dz)_f$ = frictional pressure gradient
$(dp/dz)_l$ = frictional pressure gradient, assuming that liquid alone is flowing in pipe
$(dp/dz)_m$ = momentum pressure gradient
$(dp/dz)_v$ = frictional pressure gradient, assuming that gas (or vapor) alone is flowing in pipe
F = special coefficient [Eq. (14), Table 3]
F_c = Reynolds number factor (Table 2 and Figure 10)
F_1, F_2 = condensing coefficient factors (Table 4)
f = friction factor for single-phase flow
f' = friction factor for gas flow inside pipes with wetted walls (Figure 13)
f_{co} = friction factor in presence of condensation [Eq. (26)]
f_o = friction factor [Eqs. (17) and (19)]
G = mass velocity
Gr = Grashof number
g = gravitational acceleration
g_o = gravitational constant

h = heat transfer coefficient
h_{fg} = latent heat of vaporization or of condensation
j = Colburn j factor
K = mass transfer coefficient
k = thermal conductivity
L = length
L_{mf} = mean length of fin [Eq. (7a), Table 3]
ln = natural logarithm
M = mass; or molecular weight
M_m = mean molecular weight of vapor-gas mixture
M_v = molecular weight of condensing vapor
m = general exponent [Eq. (1) or (6), Table 1]
$\dot{m}$ = mass rate of flow
N = number of tubes in vertical tier
Nu = Nusselt number
n = general exponent [Eq. (1) or (6), Table 1 and Eq. (1), Table 2]
p = pressure
p_c = critical thermodynamic pressure for coolant
Pr = Prandtl number
Q = total heat transfer
q = rate of heat transfer
r = radius
Ra = Rayleigh number
Re = Reynolds number
S = distance along flow direction
S_c = suppression factor (Table 2 and Figure 11)
T = absolute temperature
t = temperature
U = overall heat transfer coefficient
V = linear velocity
w = mass rate of flow of condensate, per unit of breadth (see Condensing)
x = quality, i.e., vapor fraction = w_v/w; or distance in dt/dx
X_{tt} = parameter [Figure 10, Table 2, and Eq. (14)]
x,y,z = lengths along principal coordinate axes
Y_g = mole fraction of gas [Eqs. (2) and (3)]
Y_v = mole fraction of vapor [Eq. (3)]
α = thermal diffusivity = $k/\rho c_p$
$\alpha(\lambda)$ = ratio of two-phase friction factor to single-phase friction factor at two-phase Reynolds number [Eq. (21)]
β = ratio of two-phase density to no-slip density [Eq. (22)]
Δ = difference between values
ϵ = roughness of interface
Λ = special coefficient [Eq. (13) through (15), Table 1]
λ = ratio of liquid volumetric flow rate to total volumetric flow rate [Eq. (25)]
μ = absolute viscosity
μ_l = dynamic viscosity of saturated liquid
μ_{NS} = dynamic viscosity of two-phase homogeneous mixture [Eq. (24)]
μ_v = dynamic viscosity of saturated vapor
ν = kinematic viscosity
ρ = density
ρ_l = density of saturated liquid
ρ_{NS} = density of two-phase homogeneous mixture [Eq. (23)]
ρ_v = density of saturated vapor phase
σ = surface tension
ϕ = fin efficiency, Martinelli factor [Eq. (11)]
ϕ_v = Lockhart-Martinelli parameter [Eq. (15)]
ψ = void fraction

Subscripts and Superscripts

a = exponent in Eq. (1)
b = bubble
c = critical or cold (fluid)
cg = condensing
e = equivalent
eff = effective
f = film or fin
g = gas
h = horizontal or hot (fluid) or hydraulic
i = inlet or inside
if = interface
L = liquid
l = liquid
m = mean
mac = convective mechanism [Eq. (2), Table 2]

$$
\begin{aligned}
max &= \text{maximum} \\
mic &= \text{nucleation mechanism [Eq. (2), Table 2]} \\
min &= \text{minimum} \\
o &= \text{outside or outlet or overall} \\
r &= \text{root (fin) or reduced pressure} \\
s &= \text{surface or secondary heat transfer surface} \\
sat &= \text{saturation (pressure)} \\
t &= \text{temperature or terminal temperature of tip (fin)} \\
v &= \text{vapor or vertical} \\
w &= \text{wall} \\
\infty &= \text{bulk} \\
* &= \text{reference}
\end{aligned}
$$

REFERENCES

Ackers, W.W., H.A. Deans, and O.K. Crosser. 1959. Condensing heat transfer within horizontal tubes. *Chemical Engineering Progress Symposium Series* 55(29).

Ackers, W.W. and H.F. Rosson. 1960. Condensation inside a horizontal tube. *Chemical Engineering Progress Symposium Series* 56(30).

Altman, M., R.H. Norris, and F.W. Staub. 1960a. Local and average heat transfer and pressure drop for refrigerants. ASHRAE *Transactions* (August):189.

Altman, M., F.W. Staub, and R.H. Norris. 1960b. Local heat transfer and pressure drop for Refrigerant-22 condensing to horizontal tubes. *Chemical Engineering Progress Symposium* 56(30).

Anderson, W., D.G. Rich, and D.F. Geary. 1966. Evaporation of Refrigerant 22 in a horizontal 3/4-in. OD tube. ASHRAE *Transactions* 72(1):28.

Ashley, C.M. 1942. The heat transfer of evaporating Freon. *Refrigerating Engineering* (February):89.

Baroczy, C.J. 1963. Correlation of liquid fraction in two-phase flow with application to liquid metals. *North American Aviation Report* SR-8171, El-Segundo, CA.

Beatty, K.O. and D.L. Katz. 1948. Condensation of vapors on outside of finned tubes. *Chemical Engineering Progress* 44(1):55.

Berenson, P.J. 1961. Film boiling heat transfer from a horizontal surface. ASME *Journal of Heat Transfer* 85:351.

Berenson, P.J. 1962. Experiments on pool boiling heat transfer. *International Journal of Heat and Mass Transfer* 5:985.

Bergles, A.E. 1976. Survey and augmentation of two-phase heat transfer. ASHRAE *Transactions* 82(1):891-905.

Bergles, A.E. 1985. "Techniques to augment heat transfer." In *Handbook of heat transfer application*, 2nd ed. McGraw-Hill, New York.

Bergles, A.E. and W.M. Rohsenow. 1964. The determination of forced convection surface-boiling heat transfer. ASME *Journal of Heat Transfer*, Series C, 86(August):365.

Blatt, T.A. and R.R. Adt. 1963. Boiling heat transfer and pressure drop characteristics of Freon 11 and Freon 113 refrigerants. AIChE Buffalo Meeting, Paper No. 132, Chemical Engineering Progress Monograph and Symposium Series.

Borishansky, V.M., I.I. Novikov, and S.S. Kutateladze. 1962. Use of thermodynamic similarity in generalizing experimental data on heat transfer. Proceedings of the International Heat Transfer Conference.

Borishansky, W., and A. Kosyrev. 1966. Generalization of experimental data for the heat transfer coefficient in nucleate boiling. ASHRAE *Journal* (May):74.

Breber, G., J.W. Palen, and J. Taborek. 1980. Prediction of the horizontal tubeside condensation of pure components using flow regime criteria. ASME *Journal of Heat Transfer* 102(3):471-76.

Breen, B.P. and J.W. Westwater. 1962. Effects of diameter of horizontal tubes on film boiling heat transfer. AIChE Preprint No. 19, Fifth National Heat Transfer Conference, Houston, TX. Chemical Engineering Progress, Monograph and Symposium Series, Vol. 59.

Bromley, L.A. 1950. Heat transfer in stable film boiling. *Chemical Engineering Progress* (46):221.

Butterworth, D. 1975. A comparison of some void-fraction relationships for co-current gas-liquid flow. *International Journal of Multiphase Flow* 1:845-50.

Carpenter, E.F. and A.P. Colburn. 1949. The effect of vapor velocity on condensation inside tubes. General discussion on Heat Transfer and Fluid Mechanics Institute, ASME, New York.

Chaddock, J.B. and J.A. Noerager. 1966. Evaporation of Refrigerant 12 in a horizontal tube with constant wall heat flux. ASHRAE *Transactions* 72(1):90.

Chen, J.C. 1963. A correlation for boiling heat transfer to saturated fluids on convective flow. ASME Paper 63-HT-34.

Colburn, A.P. 1933-34. Note on the calculation of condensation when a portion of the condensate layer is in turbulent motion. AIChE *Transactions* No. 30, London.

Colburn, A.P. 1951. Problems in design and research on condensers of vapours and vapour mixtures. Proceedings of the Institute of Mechanical Engineers, 164:448, London.

Colburn, A.P. and O.A. Hougen. 1934. Design of cooler condensers for mixtures of vapors with noncondensing gases. *Industrial and Engineering Chemistry* 26(November):1178.

Collier, J.G. 1972. *Convective boiling and condensation.* McGraw-Hill Book Co. (UK), Ltd., London.

Danilova, G. 1965. Influence of pressure and temperature on heat exchange in the boiling of halogenated hydrocarbons. Kholodilnaya Teknika, No. 2. English abstract, *Modern refrigeration* (December).

Dengler, C.E. and J.N. Addoms. 1956. Heat transfer mechanism for vaporization of water in a vertical tube. *Chemical Engineering Progress Symposium Series* 52(18):95.

Dougherty, R.L. and H.J. Sauer, Jr. 1974. Nucleate pool boiling of refrigerant-oil mixtures from tubes. ASHRAE *Transactions* 80(2):175.

Dukler, A.E., M. Wicks, III, and R.G. Cleveland. 1964. Frictional pressure drop in two-phase flow: An approach through similarity analysis. AIChE *Journal* 10(January):44-51.

Farber, E.A. and R.L. Scorah. 1948. Heat transfer to water boiling under pressure. ASME *Transactions* (May):373.

Forster, H.K. and N. Zuber. 1955. Dynamics of vapor bubbles and boiling heat transfer. AIChE *Journal* 1(4):531-35.

Frederking, T.H.K. and J.A. Clark. 1962. "Natural convection film boiling on a sphere." In *Advances in cryogenic engineering*, ed. K.D. Timmerhouse, Plenum Press, New York.

Furse, F.G. 1965. Heat transfer to refrigerants 11 and 12 boiling over a horizontal copper surface. ASHRAE *Transactions* 71(1):231.

Gilmour, C.H. 1958. Nucleate boiling—A correlation. *Chemical Progress* 54 (October):77.

Gouse, S.W., Jr. and K.G. Coumou. 1965. Heat transfer and fluid flow inside a horizontal tube evaporator, Phase I. ASHRAE *Transactions* 71(2):152.

Green, G.H. and F.G. Furse. 1963. Effect of oil on heat transfer from a horizontal tube to boiling refrigerant 12-oil mixtures. ASHRAE *Journal* (October):63.

Grigull, U. 1952. Warmeubergang bei filmkondensation. *Forsch, Gebiete Ingenieurw*, 18.

Grober, H., S. Erk, and U. Grigull. 1961. *Fundamentals of heat transfer.* McGraw-Hill Book Co., Inc., New York.

Guerrieri, S.A. and R.D. Talty. 1956. A study of heat transfer to organic liquids in single tube boilers. *Chemical Engineering Progress Symposium Series* 52(18):69.

Hofmann, E. 1957. Heat transfer coefficients for evaporating refrigerants. Kaltetechnik 9 Jahrgang, Heft 1/1957. Argonne National Laboratory Translation, Lemont, IL (September) 1958.

Hoogendoorn, C.J. 1959. Gas-liquid flow in horizontal pipes. *Chemical Engineering Sciences* IX(1).

Hughmark, G.A. 1962. A statistical analysis of nucleate pool boiling data. *International Journal of Heat and Mass Transfer* 5:667.

Jakob, M. 1949 and 1957. *Heat transfer*, Vols. I and II. John Wiley and Sons, Inc., New York.

Katz, D.L., P.E. Hope, S.C. Datsko, and D.B. Robinson. 1947. Condensation of Freon-12 with finned tubes. Part I, Single horizontal tubes; Part II, Multitube condensers. *Refrigerating Engineering* (March):211, (April):315.

Kutateladze, S.S. 1951. A hydrodynamic theory of changes in the boiling process under free convection. Izvestia Akademii Nauk, USSR, Otdelenie Tekhnicheski Nauk 4:529.

Kutateladze, S.S. 1963. *Fundamentals of heat transfer.* E. Arnold Press Ltd., London, England.

Lavin, J.G. and E.H. Young. 1964. Heat transfer to evaporating refrigerants in two-phase flow. AIChE Preprint 21e (February), Symposium on Two-Phase Flow and Heat Transfer.

Lienhard, J.H. and V.E. Schrock. 1963. The effect of pressure, geometry and the equation of state upon peak and minimum boiling heat flux. ASME *Journal of Heat Transfer* 85:261.

Lienhard, J.H. and P.T.Y. Wong. 1963. The dominant unstable wave length and minimum heat flux during film boiling on a horizontal cylinder. ASME Paper No. 63-HT-3, ASME-AIChE Heat Transfer Conference, Boston, MA, August.

Lockhart, R.W. and R.C. Martinelli. 1949. Proposed correlation of data for isothermal two-phase, two-component flow in pipes. *Chemical Engineering Progress* 45(1):39-48.

Luu, M. and A.E. Bergles. 1980. Augmentation of in-tube condensation of R-113. ASHRAE Project RP-219.

Martinelli, R.C. and D.B. Nelson. 1948. Prediction of pressure drops during forced circulation boiling of water. ASME *Transactions* 70:695.

McAdams, W.H. 1954. *Heat transmission*, 3rd ed. McGraw-Hill Book Co., Inc., New York.

Myers, J.E. and D.L. Katz. 1952. Boiling coefficients outside horizontal plain, and finned tubes. *Refrigerating Engineering* (January):56.

Pierre, B. 1955. S.F. Review. *A.B. Svenska Flaktafabriken*, Stockholm, Sweden 2(1):55.

Pierre, B. 1957. *Kylteknisk Tidskrift* 3 (May):129.

Pierre, B. 1964. Flow resistance with boiling refrigerant. ASHRAE *Journal* (September through October).

Rohsenow, W.M. 1951. A method of correlating heat transfer for surface boiling of liquids. ASME *Transactions* 73:609.

Rohsenow, W.M. 1963. "Boiling heat transfer." In *Modern developments in heat transfer*, ed. W. Ibele, Academic Press, New York.

Rose, J.W. 1969. Condensation of a vapour in the presence of a non-condensable gas. *International Journal of Heat and Mass Transfer* 12:233.

Schrock, V.E. and L.M. Grossman. 1962. Forced convection boiling in tubes. *Nuclear Science and Engineering* 12:474.

Short, B.E. and H.E. Brown. 1951. Condensation of vapors on vertical banks of horizontal tubes. ASME, New York.

Silver, R.S. and G.B. Wallis. 1980. A simple theory for longitudinal pressure drop in the presence of lateral condensation. Proceedings of Institute of Mechanical Engineering, 36-42.

Soliman, M., J.R. Schuster, and P.J. Berenson. 1968. A general heat transfer correlation for annular flow condensation. *Journal of Heat Transfer* 90:267-76.

Sparrow, E.M. and S.H. Lin. 1964. Condensation in the presence of a non-condensable gas. ASME *Transactions, Journal of Heat Transfer* 86C:430.

Sparrow, E.M., W.J. Minkowycz, and M. Saddy. 1967. Forced convection condensation in the presence of noncondensables and interfacial resistance. *International Journal of Heat and Mass Transfer* 10:1829.

Starczewski, J. 1965. Generalized design of evaporation heat transfer to nucleate boiling liquids. *British Chemical Engineering* (August).

Stephan, K. 1963. Influence of oil on heat transfer of boiling Freon-12 and Freon-22. Eleventh International Congress of Refrigeration, I.I.R. Bulletin No. 3.

Stephan, K. 1963a. A mechanism and picture of the processes involved in heat transfer during bubble evaporation. *Chemic Ingenieur Technik* 35:775.

Stephan, K. 1963b. The computation of heat transfer to boiling refrigerants. *Kaltetechnik* 15:231.

Thom, J.R.S. 1964. Prediction of pressure drop during forced circulation boiling water. *International Journal of Heat and Mass Transfer* 7:709-24.

Tschernobyiski, I. and G. Ratiani. 1955. *Kholodilnaya Teknika* 32.

Turner, J.M. and G.B. Wallis. 1965. The separate-cylinders model of two-phase flow. Report No. NYO-3114-6. Thayer's School of Engineering, Dartmouth College, Hanover, NH.

Van Stralen, S.J. 1959. Heat transfer to boiling binary liquid mixtures. *Chemical Engineering* (British) 4(January):78.

Wallis, G.B. 1969. *One-dimensional two-phase flow*. McGraw-Hill Book Co., New York.

Wallis, G.C. 1970. Annular two-phase flow; Part I: A simple theory, Part II: Additional effect. ASME *Transactions, Journal of Basic Engineering* 92D:59 and 73.

Webb, R.L. 1981. The evolution of enhanced surface geometries for nucleate boiling. *Heat Transfer Engineering* 2(3-4):46-69.

Westwater, J.W. 1963. "Things we don't know about boiling". In *Research in Heat Transfer*, ed. J. Clark. Pergamon Press, New York.

Worsoe-Schmidt, P. 1959. Some characteristics of flow-pattern and heat transfer of Freon-12 evaporating in horizontal tubes. *Ingenieren*, International edition 3(3).

Worsoe-Schmidt, P. 1960. ASME *Transactions* (August):197.

Zivi, S.M. 1964. Estimation of steady-state steam void-fraction by means of the principle of minimum entropy production. *Journal of Heat Transfer* 86:247-52.

Zuber, N. 1959. Hydrodynamic aspects of boiling heat transfer. U.S. Atomic Energy Commission, Technical Information Service, Report AECU 4439, Oak Ridge, TN.

Zuber, N., M. Tribus, and J.W. Westwater. 1962. The hydrodynamic crisis in pool boiling of saturated and subcooled liquids. Proceedings of the International Heat Transfer Conference 2:230, and discussion of the papers, Vol. 6.

CHAPTER 5

MASS TRANSFER

MASS transfer by either molecular diffusion or convection is the transport of one component of a mixture, relative to the motion of the mixture, and is the result of a *concentration gradient*. In an air-conditioning process, water vapor is added or removed from the air, with a simultaneous transfer of heat and mass (water vapor) between the airstream and a wetted surface. The wetted surface can be water droplets in an air washer, wetted slats of a cooling tower, condensate on the surface of a dehumidifying coil, surface presented by a spray of liquid absorbent, or wetted surfaces of an evaporative condenser. The performance of equipment with these phenomena must be calculated carefully because of the simultaneous heat and mass transfer processes.

This chapter addresses the principles of mass transfer and provides methods of solving a simultaneous heat and mass transfer problem involving air and water vapor. Emphasis is on the air-conditioning processes involving mass transfer. The formulations presented can help in analyzing the performance of specific equipment. For a discussion on the performance of air washers, cooling coils, evaporative condensers, and cooling towers, see the 1992 ASHRAE *Handbook—Systems and Equipment*.

MOLECULAR DIFFUSION

Most mass transfer problems can be analyzed by considering the diffusion of a gas into a second gas, a liquid, or a solid. In this chapter, the diffusing or dilute component is designated as component B and the other component, component A. For example, when water vapor diffuses into air, the water vapor is component B and dry air is component A. Properties with subscripts A or B are local properties of that component. Properties without subscripts are local properties of the mixture.

This chapter is divided into (1) the principles of molecular diffusion, (2) a discussion on the convection of mass, and (3) simultaneous heat and mass transfer and its application to specific equipment.

The primary mechanism of mass diffusion at ordinary temperature and pressure conditions is *molecular diffusion*, a result of density gradient. In a binary gas mixture, the presence of a concentration gradient causes transport of matter by molecular diffusion, *i.e.*, because of random molecular motion, gas B diffuses through the mixture of gases A and B in a direction that reduces the concentration gradient.

Fick's Law

The basic equation for molecular diffusion is Fick's law. Expressing the concentration of component B of a binary mixture in terms of the mass fraction (ρ_B/ρ) or mole fraction (C_B/C), Fick's law is:

$$J_B = -\rho D_v \frac{d(\rho_B/\rho)}{dy} \tag{1a}$$

The preparation of this chapter is assigned to TC 1.3, Heat Transfer and Fluid Flow.

$$J_B^* = -CD_v \frac{d(C_B/C)}{dy} \tag{1b}$$

The minus sign indicates that the concentration gradient is negative in the direction of diffusion. The proportionality factor D_v is the *mass diffusivity* or the *diffusion coefficient*. The diffusive mass flux J_B and the diffusive molar flux J_B^* are:

$$J_B \equiv \rho_B(v_B - v) \tag{2a}$$

$$J_B^* \equiv C_B(v_B - v) \tag{2b}$$

where $(v_B - v)$ is the velocity of component B relative to the velocity of the mixture. Bird *et al.* (1960) presented an analysis of Equations (1a) and (1b).

Equations (1a) and (1b) are equivalent forms of Fick's law. The equation used depends on the problem and individual preference. This chapter emphasizes mass analysis rather than molar analysis. However, all results can be converted to the molar form using the relation $C_B \equiv \rho_B/M_B$.

Frick's Law for Dilute Mixtures

In many mass diffusion problems, component B is dilute. The density of component B is small compared with the density of the mixture, and the variation in the density of the mixture throughout the problem is about ρ_B or less. In this case, Equation (1a) can be written as

$$J_B = -D_v(d\rho_B/dy) \tag{3}$$

when $\rho_B << \rho$, $\Delta\rho < \Delta\rho_B$.

Equation (3) can be used without significant error for water vapor diffusing through air at atmospheric pressure and a temperature less than 27 °C. In this case, $\rho_B < 0.02\rho$, where ρ_B is the density of water vapor and ρ is the density of moist air (air and water vapor mixture). The error in J_B caused by replacing ρ $[d(\rho_B/\rho)/dy]$ with $d\rho_B/dy$ is less than 2%. At temperatures below 60 °C where $\rho_B < 0.10\rho$, Equation (3) can still be used if errors in J_B as great as 10% are tolerable.

Fick's Law for Mass Diffusion through Solids or Stagnant Fluids

Fick's law can be simplified for cases of dilute mass diffusion in solids, stagnant liquids, or stagnant gases. In these cases, $\rho_B << \rho$ and $v = 0$, which yields the approximate result, $J_B = \dot{m}_B''$. That is,

$$J_B = \rho_B(v_B - v) = \rho_B(v_B - \rho_B v_B/\rho) \cong \rho_B v_B = \dot{m}_B'' \tag{4}$$

Therefore, Fick's law reduces to:

$$\dot{m}_B'' = -D_v(\rho_B/dy) \tag{5}$$

when $\rho_B << \rho$, $\Delta\rho < \Delta\rho_B$, $v_A = 0$.

Fick's Law for Ideal Gases with Negligible Temperature Gradient

For cases of dilute mass diffusion, Fick's law can be written in terms of pressure gradient instead of concentration gradient when gas B can be approximated as ideal

$$p_B = \rho_B R_U T / M_B \tag{6}$$

and when the gradient in T is small. Under these conditions, Equation (3) can be written as

$$J_B = - M_B D_v / R_U T \, (dp_B/dy) \tag{7a}$$

or

$$J_B^* = - D_v / R_U T \, (dp_B/dy) \tag{7b}$$

when B is dilute and ideal and $T = $ is constant. Also, under the above conditions, Equation (5) may be written as

or

$$\dot{m}_B'' = - M_B D_v / R_U T \, (d\rho_B/dy) \tag{8a}$$

$$\dot{m}_B''^* = - D_v / R_U T \, (d\rho_B/dy) \tag{8b}$$

when B is dilute and ideal, $v_A = 0$, and $T = $ constant.

The pressure gradient formulation for mass transfer analysis has been used extensively; this is unfortunate, since the pressure formulation, Equations (7) and (8), applies only to cases where one component is dilute, the fluid closely approximates an ideal gas, and the temperature gradient has a negligible effect. The density (or concentration) gradient formulation expressed in Equations (2) through (5) is the more general formulation and can be applied to a wider range of mass transfer problems, including cases where neither component is dilute [Equation (1)]. The gases need not be ideal, nor the temperature gradient negligible. Consequently, this chapter emphasizes the density formulation.

Diffusion Coefficient

For a binary mixture, the diffusion coefficient D_v is a function of temperature, pressure, and composition. Experimental measurements of D_v for most binary mixtures are limited in range and accuracy. Table 1 gives a few experimental values for diffusion of some gases in air. For more detailed tables, see the references at the end of this chapter.

Without data, use equations developed from (1) theory or (2) theory with constants adjusted from limited experimental data. For binary gas mixtures at low pressure, D_v is inversely proportional to pressure, increases with increasing temperature, and is almost independent of composition for a given gas pair. The following equation for estimating D_v at pressures less than $0.1 \, p_{c\,min}$ was developed by Bird et al. (1960) from kinetic theory and corresponding-states arguments.

Table 1 Mass Diffusivities for Gases in Air*

Gas	D, ft²/h
Ammonia	1.08
Benzene	0.34
Carbon dioxide	0.64
Ethanol	0.46
Hydrogen	1.60
Oxygen	0.80
Water vapor	0.99

*Gases at 77°F and 14.696 psi.

$$D_v = a \left(\frac{T}{\sqrt{T_{cA} + T_{cB}}} \right)^b \left(\frac{1}{M_A} + \frac{1}{M_B} \right)^{0.5}$$
$$\times \left(\frac{(p_{cA} p_{cB})^{0.33} (T_{cA} T_{cB})^{0.42}}{p} \right) \tag{9}$$

where

D_v = diffusion coefficient, ft²/h
a = constant
b = constant, dimensionless
T = absolute temperature, K
p = pressure, psi
M = molecular weight, lb_m/lb mol

The subscripts cA and cB refer to the critical state of the two gases. Analysis of experimental data gave the following values of the constants a and b:

For nonpolar gas pairs:

$$a = 7.266 \times 10^{-3}$$
$$b = 1.823$$

For water vapor with a nonpolar gas:

$$a = 9.635 \times 10^{-3}$$
$$b = 2.334$$

A *nonpolar gas* is one for which the intermolecular forces are independent of the relative orientation of molecules, depending only on the separation distance from each other. Air, composed of nonpolar gases O_2 and N_2, is nonpolar.

Equation (9) is stated to agree with experimental data at atmospheric pressure to within about 8% (Bird et al. 1960).

The mass diffusivity D_v for binary mixtures at low pressure is predictable within about 10% by kinetic theory (Reid et al. 1987).

$$D_v = 0.722 \frac{T^{1.5}}{(\sigma_{AB})^2 \Omega_{D,AB} p} \left(\frac{1}{M_A} + \frac{1}{M_B} \right)^{0.5} \tag{10}$$

where

σ_{AB} = characteristic molecular diameter, nm
$\Omega_{D, AB}$ = temperature function, dimensionless

D_v is in ft²/h, p in atmospheres, and T in kelvins. If the gas molecules of A and B are considered rigid spheres having diameters σ_A and σ_B [and $\sigma_{AB} = (\sigma_A/2) + (\sigma_B/2)$, all expressed in nanometers], the dimensionless function $\Omega_{D,AB}$ equals unity. More realistic models for the molecules having intermolecular forces of attraction and repulsion lead to values of $\Omega_{D,AB}$ that are functions of temperature. Reid et al. (1987) present tabulations of this quantity. These results show that D_v increases as the 2.0 power of T at low temperatures and as the 1.65 power of T at very high temperatures.

The diffusion coefficient of moist air has been calculated for Equation (9), using a simplified intermolecular potential field function for water vapor and air (Mason and Monchick 1965).

An empirical equation for mass diffusivity of water vapor in air up to 2000°F is (Sherwood and Pigford 1952):

$$D_v = (0.00215/p) [T^{2.5} / (T + 441)] \tag{11}$$

where D_v is in ft²/h, p in psi, and T in °R.

Analogy between Heat and Mass Diffusion

Molecular diffusion is, in some cases, directly analogous to conduction heat transfer. Both result from random molecular mixing in a stagnant fluid, or in a fluid in laminar (streamline) flow. The equation governing each process can be expressed in the form:

$$\text{Flux} = \text{Diffusivity} \times \text{Concentration gradient}$$

Examination of Equation (5) reveals that Fick's law applied to mass diffusion of a dilute gas through solids and stagnant fluids is analogous to Fourier's law:

$$q'' = -k\,dt/dx \tag{12}$$

The usefulness of the analogy can be proven as follows. Transient mass diffusion in a dilute mixture region where D_v is constant and component A is stagnant is governed by the following system of equations:

$$\frac{\partial \rho_B}{\partial \tau} = D_v\left(\frac{\partial^2 \rho_B}{\partial x^2} + \frac{\partial^2 \rho_B}{\partial y^2} + \frac{\partial^2 \rho_B}{\partial z^2}\right) \tag{13a}$$

IC: $\rho_B = \rho_{Bi}(x, y, z)$ at $\tau = 0$ \hfill (13b)

BC: $\rho_B = f(x, y, z, \tau)$; on boundary \hfill (13c)

Transient heat diffusion in a constant k and constant ρc_p region is governed by:

$$\frac{\partial t}{\partial \bar{\tau}} = \alpha\left(\frac{\partial^2 t}{\partial \bar{x}^2} + \frac{\partial^2 t}{\partial \bar{y}^2} + \frac{\partial^2 t}{\partial \bar{z}^2}\right) \tag{14a}$$

IC: $t = t_i(\bar{x}, \bar{y}, \bar{z})$ at $\tau = 0$ \hfill (14b)

BC: $t = f(\bar{x}, \bar{y}, \bar{z}, \bar{\tau})$; on boundary \hfill (14c)

These two systems of equations can be put in identical form by selecting the following dimensionless parameters:

$$\theta \equiv \frac{t - t_l}{t_h - t_l}, \qquad \bar{\theta} \equiv \frac{\rho_B - \rho_{Bl}}{\rho_{Bh} - \rho_{Bl}} \tag{15a}\dots$$

$$X \equiv x/L,\ Y \equiv y/L,\ Z \equiv z/L,\ \bar{X} \equiv \bar{x}/L,\ \bar{Y} \equiv \bar{y}/\bar{L},\ \bar{Z} \equiv \bar{z}/\bar{L}$$

$$\text{Fo} \equiv \frac{\alpha\bar{\tau}}{L^2}, \qquad \bar{\tau} \equiv \frac{D_v\tau}{L^2} \qquad \dots (15j)$$

Using these parameters, the heat and mass diffusion equations become identical.

Heat diffusion:

$$\frac{\partial \theta}{\partial \text{Fo}} = \frac{\partial^2 \theta}{\partial \bar{x}^2} + \frac{\partial^2 \theta}{\partial \bar{y}^2} + \frac{\partial^2 \theta}{\partial \bar{z}^2} \tag{16a}$$

IC: $\theta = \theta_i(X, Y, Z)$; Fo $= 0$ \hfill (16b)

BC: $\theta = F(X, Y, Z, \text{Fo})$; on boundary \hfill (16c)

Mass diffusion:

$$\frac{\partial \bar{\theta}}{\partial \tau} = \frac{\partial^2 \bar{\theta}}{\partial x^2} + \frac{\partial^2 \bar{\theta}}{\partial y^2} + \frac{\partial^2 \bar{\theta}}{\partial z^2} \tag{17a}$$

IB: $\bar{\theta} = \bar{\theta}_i(\bar{X}, \bar{Y}, \bar{Z})$; $\bar{\tau} = 0$ \hfill (17b)

BC: $\bar{\theta} = \bar{F}(\bar{X}, \bar{Y}, \bar{Z}, \text{Fo}_m)$; on boundary \hfill (17c)

Therefore, if $\theta_i = \bar{\theta}_i$ and $F = \bar{F}$, then $\theta = \bar{\theta}$. That is, if the two geometries are of the same shape and if the initial potential distributions θ_i and $\bar{\theta}_i$ are the same and the boundary conditions F and $\bar{F}$ are the same, then the solutions θ and $\bar{\theta}$ are the same. Therefore, all of the heat transfer solutions for steady-state and transient conduction are available for solving analogous steady-state and transient mass transfer problems when $\rho_B \ll \rho$, $\Delta\rho < \rho_B$, and $v_B = 0$. For further details on the analogy between heat and mass transfer, refer to Eckert and Drake (1972).

Diffusion of One Gas through Second Stagnant Gas

Figure 1 shows diffusion of one gas through a second stagnant gas. Water vapor diffuses from the liquid surface into surrounding stationary air. It is assumed that local equilibrium exists through the gas mixture, that the gases are ideal, and that the Gibbs-Dalton law is valid, which implies that the temperature gradient has a negligible effect. Water vapor diffuses because of concentration gradient and is given by Equation (7a). There is a continuous gas phase, so the mixture pressure p is constant, and the Gibbs-Dalton law yields:

$$p_A + p_B = p = \text{constant} \tag{18a}$$

or

$$\rho_A/M_A + \rho_B/M_B = p/R_U T = \text{constant} \tag{18b}$$

The partial pressure gradient of the water vapor causes a partial pressure gradient of the air such that:

$$dp_A/dy = -dp_B/dy$$

or

$$\frac{1}{M_A}\left(\frac{d\rho_A}{dy}\right) = -\frac{1}{M_B}\left(\frac{d\rho_B}{dy}\right) \tag{19}$$

Air, then, diffuses toward the liquid water interface. Since it cannot be absorbed there, a bulk velocity v of the gas mixture is established in a direction away from the liquid surface, so that the net transport of air is zero, *i.e.*, the air is stagnant.

$$\dot{m}_A'' = -D_v\left(\frac{d\rho_A}{dy}\right) + \rho_A v = 0 \tag{20}$$

The bulk velocity v not only transports air but also water vapor away from the interface. Therefore, total rate of water vapor diffusion is:

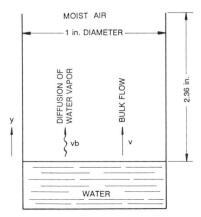

Fig. 1   Diffusion of Water Vapor through Stagnant Air

$$\dot{m}_B'' = -D_v \left(\frac{d\rho_B}{dy} \right) + \rho_B v \qquad (21)$$

Substituting for the velocity from Equation (20) and using Equations (18b) and (19) gives:

$$\dot{m}_B'' = \frac{D_v M_B p}{\rho_A R_U T} \left(\frac{d\rho_A}{dy} \right) \qquad (22)$$

Integration yields:

$$\dot{m}_B'' = \frac{D_v M_B p}{R_U T} \left(\frac{\ln (\rho_{AL}/\rho_{AO})}{y_L - y_O} \right) \qquad (23a)$$

or

$$\dot{m}_B'' = -D_v P_{Am} \left(\frac{\rho_{BL} - \rho_{BO}}{y_L - y_O} \right) \qquad (23b)$$

where

$$P_{AM} \equiv \frac{p}{p_{AL}} \rho_{AL} \left(\frac{\ln (\rho_{AL}/\rho_{AO})}{\rho_{AL} - \rho_{AO}} \right) \qquad (24)$$

P_{AM} is the logarithmic mean density factor of the stagnant air. The pressure distribution for this type of diffusion is illustrated in Figure 2. *Stagnant* refers to the net behavior of the air; it does not move because the bulk flow exactly offsets diffusion. The term P_{AM} in Equation (23b) approximately equals unity for dilute mixtures, such as water vapor in air at near atmospheric conditions. This condition makes it possible to simplify Equation (23) and implies that in the case of dilute mixtures, the partial pressure distribution curves in Figure 2 are straight lines.

Example 1. A vertical tube of 1-in. diameter is partially filled with water so that the distance from the water surface to the open end of the tube is 2.362 in., as shown in Figure 1. Perfectly dried air is blown over the open tube end, and the complete system is at a constant temperature of 59 °F. In 200 h of steady operation, 0.00474 lb of water evaporates from the tube. The total pressure of the system is 14.696 psia. Using these data, (a) cal-

culate the mass diffusivity of water vapor in air and (b) compare this experimental result with that from Equation (11).
   Solution:
(a) The mass diffusion flux of water vapor from the water surface is:

$$\dot{m}_B = 0.00474/200 = 0.0000237$$

The cross-sectional area of a 1-in. diameter tube is $\pi (1)^2/(4 \times 144) = 0.005454$ ft^2. Therefore, $\dot{m}_B'' = 1.207 \times 10^6$ lb/ft$^2 \cdot$ s. The partial densities are determined with the aid of the psychrometric tables.

$$\rho_{BL} = 0; \; \rho_{BO} = 0.000801 \text{ lb/ft}^3$$
$$\rho_{AL} = 0.0765 \text{ lb/ft}^3; \; \rho_{AO} = 0.0752 \text{ lb/ft}^3$$

Since $p = p_{AL} = 1$ atm, the logarithmic mean density factor [Equation (24)] is:

$$P_{AM} = 0.0765 \left(\frac{\ln (0.0765/0.0752)}{(0.0765 - 0.0752)} \right) = 1.009$$

The mass diffusivity is now computed from Equation (23b) as:

$$D_v = \frac{-\dot{m}_B'' (y_L - y_O)}{P_{AM} (\rho_{BL} - \rho_{BO})} = \frac{-(0.000001207)(2.362)(3600)}{(1.009)(0 - 0.000801)}$$
$$= 1.058 \text{ ft}^2/\text{h}$$

(b) By Equation (11) with $p = 14.696$ psi; $T = 59 + 460 = 519$°R

$$D_v = \frac{0.00215}{14.696} \left(\frac{519^{2.5}}{519 + 441} \right) = 0.935 \text{ ft}^2/\text{h}$$

Neglecting the correction factor P_{AM} for this example results in only a 1% change in the calculated experimental value of D_v.

Molecular Diffusion in Liquids and Solids

Because of the greater density, diffusion is slower in liquids than in gases. No satisfactory molecular theories have been developed for calculating diffusion coefficients. The limited measured values of D_v show that, unlike gas mixtures at low pressures, the diffusion coefficient for liquids varies appreciably with concentration.

Reasoning largely from analogy to the case of one-dimensional diffusion in gases and employing Fick's law as expressed by Equation (5):

$$\dot{m}_B''^* = D_v (C_{B1} - C_{B2})/(y_1 - y_2) \qquad (25)$$

where C_B = molal concentration of solute in solvent, lb·mol/ft^3.

Equation (25) expresses the steady-state diffusion of the solute B through the solvent A, in terms of the molal concentration difference of the solute at two locations separated by the distance $\Delta y = y_1 - y_2$. Bird *et al.* (1960), Hirschfelder *et al.* (1954), Sherwood and Pigford (1952), Reid and Sherwood (1966), Treybal (1952), and Eckert and Drake (1972) provide equations and tables for evaluating D_v. Hirschfelder *et al.* (1954) provide comprehensive treatment of the molecular developments.

Diffusion through a solid when the solute is dissolved to form a homogeneous solid solution is known as *structure-insensitive* diffusion (Treybal 1952). This solid diffusion closely equates diffusion through fluids, and Equation (25) can be applied to one-dimensional steady-state problems. Values of mass diffusivity are generally lower than they are for liquids and vary with temperature.

The flow of a liquid or gas through the interstices and capillaries of a porous or granular solid is a concern. The fundamental mechanism of transport differs for gaseous diffusion. It is

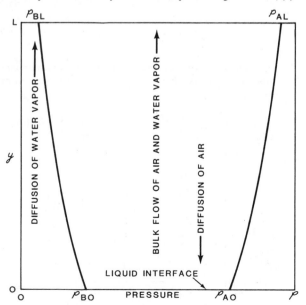

Fig. 2   Pressure Profiles for Diffusion of Water Vapor through Stagnant Air

considered a form of molecular diffusion called *structure sensitive*. Experimental measurements are important because of the complex geometry of the flow passages. Generally, a factor $\bar{\mu}$ called permeability, is defined by the following equation:

$$\dot{m}_B'' = -D_v d\rho_B/(dRT) = \bar{\mu}(\Delta p_B/\Delta y) \qquad (26)$$

For moisture transfer through a porous building material of thickness Δy, the water vapor pressure gradient $\Delta p_B/\Delta y$ is expressed as in. Hg per inch of thickness, and the mass flux in grains/h·ft², so that the permeability $\bar{\mu}$ has the units of grains in/h·ft²·in. Hg. Chapter 21 has further information.

CONVECTION OF MASS

Convection of mass involves the mass transfer mechanisms of molecular diffusion and bulk fluid motion. Fluid motion in the region adjacent to a mass transfer surface is laminar or turbulent, depending on geometry and flow conditions.

Mass Transfer Coefficient

Convective mass transfer is analogous to convective heat transfer where geometry and boundary conditions are similar. The analogy holds for both laminar and turbulent flows and applies to both external and internal flow problems.

Mass Transfer Coefficients for External Flows

Most convective mass transfer problems can be solved with an appropriate relationship that relates the mass transfer flux (to or from an interfacial surface) to the concentration difference across the boundary layer illustrated in Figure 3. This formulation gives rise to the convective mass transfer coefficient defined by:

$$h_m \equiv \dot{m}_B''/(\rho_{Bi} - \rho_{B\infty}) \qquad (27)$$

where

h_m = local external mass transfer coefficient, ft/h
$\dot{m}_B''$ = mass flux of gas B from surface, $lb_m/ft^2 \cdot h$
ρ_{Bi} = concentration or density of gas B at interface, lb_m/ft^3
$\rho_{B\infty}$ = density of component B outside boundary layer

If ρ_{Bi} and $\rho_{B\infty}$ are constant over the entire interfacial surface, the mass transfer rate from the surface can be expressed as:

$$\dot{m}_B'' = \bar{h}_m(\rho_{Bi} - \rho_{B\infty}) \qquad (28)$$

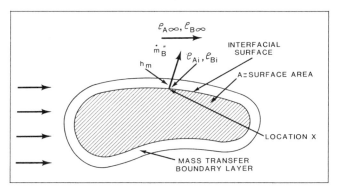

Fig. 3 Nomenclature for Convective Mass Transfer from External Surface at Location X where Surface is Impermeable to Gas A

where $\bar{h}_m$ is the average mass transfer coefficient defined as:

$$\bar{h}_m \equiv \frac{1}{A}\int_A h_m \, dA \qquad (29a)$$

Mass Transfer Coefficients for Internal Flows

Most internal convective mass transfer problems, such as those that occur in channels or in the cores of dehumidification coils, can be solved if an appropriate relationship is available to relate the mass transfer flux (to or from the interfacial surface) to the difference between the concentration at the surface and the bulk concentration in the channel, as illustrated in Figure 4. This formulation leads to the definition of the mass transfer coefficient for internal flows:

$$h_m \equiv \dot{m}_B''/(\rho_{Bi} - \rho_{Bb}) \qquad (29b)$$

where

h_m = internal mass transfer coefficient, ft/h
$\dot{m}_B''$ = mass flux of gas B at interfacial surface, $lb_m/ft^2 \cdot h$
ρ_{Bi} = density of gas B at interfacial surface, lb_m/ft^3
$\rho_{Bb} \equiv 1/(\bar{u}_B A_{cs}\int_{A_{cs}} u_B\rho_B dA_{cs})$ = bulk density of gas B at location x
$\bar{u}_B \equiv 1/(A_{cs}\int_A u_B \, dA_{cs})$ = average velocity of gas B at location x, ft/min
A_{cs} = cross-sectional area of channel at station x, ft²
u_B = velocity of component B in x direction, ft/min
ρ_B = density distribution of component B at station x, lb_m/ft^3

Often, it is easier to obtain the bulk density of gas B from:

$$\rho_{Bb} = (\dot{m}_{BO} + \int_A \dot{m}_B'' \, dA)/(\bar{u}_B A_{cs}) \qquad (30)$$

where

$\dot{m}_{BO}$ = mass flow rate of component B at station x = 0, lb_m/h
A = interfacial area of channel between station x = 0 and station x = x, ft²

Equation (30) can be derived from the preceding definitions. The major problem is the determination of $\bar{u}_B$. If, however, the analysis is restricted to cases where B is dilute and concentration gradients of B in the x direction are negligibly small, $\bar{u}_B \cong \bar{u}$. Component B is swept along in the x direction with an average velocity equal to the average velocity of the dilute mixture.

Analogy between Convective Heat and Mass Transfer

Most expressions for the convective mass transfer coefficient h_m are determined from expressions for the convective heat transfer coefficient h.

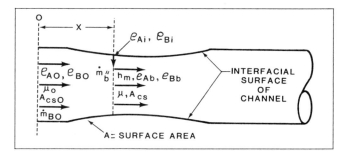

Fig. 4 Nomenclature for Convective Mass Transfer from Internal Surface Impermeable to Gas A

For problems in internal and external flow where mass transfer occurs at the convective surface and where component B is dilute, it was shown by Bird *et al.* (1960) and Incropera and DeWitt (1990) that:

$$\text{Nu} = f(X, Y, Z, \text{Pr}, \text{Re}) \text{ and } \overline{\text{Nu}} = g(\text{Pr, Re}) \qquad (31), (32)$$

and

$$\text{Sh} = f(X, Y, Z, \text{Sc}, \text{Re}) \text{ and } \overline{\text{Sh}} = g(\text{Sc, Re}) \qquad (33), (34)$$

where the function f is the same in Equations (31) and (33), and the function g is the same in Equations (32) and (34). The quantities Pr and Sc are dimensionless Prandlt and Schmidt numbers, respectively, as defined in the Symbols section. The primary restrictions on the analogy are that the surface shapes are the same and that the dimensionless temperature boundary conditions are analogous to the dimensionless density distribution boundary conditions for component B, as indicated in Equations (16) and (17). Several primary factors prevent the analogy from being perfect. In some cases, the Nusselt number was derived for smooth surfaces. Many mass transfer problems involve wavy, droplet-like, or roughened surfaces. Many Nusselt number relations are obtained for constant temperature surfaces. Sometimes ρ_{Bi} is not constant over the entire surface because of varying saturation conditions and the possibility of surface dryout.

In all mass transfer problems, there is some blowing or suction at the surface because of the condensation, evaporation, or transpiration of component B. In most cases, this blowing/suction phenomenon has little effect on the Sherwood number, but the analogy should be examined closely if $v_i/u_\infty > 0.01$ or $v_i/\bar{u} > 0.01$, especially if the Reynolds number is large.

Example 2. Use the analogy expressed in Equations (32) and (34) to solve the following problem. An expression for heat transfer from a constant-temperature flat plate in laminar flow is:

$$\overline{\text{Nu}}_L = 0.644 \, \text{Pr}^{1/3} \, \text{Re}_L^{1/2} \qquad (35)$$

Using the heat/mass transfer analogy, determine the mass transfer rate and temperature of the water-wetted flat plate in Figure 5.

Solution: To solve the problem, properties should be evaluated at film conditions. However, since the plate temperature and the interfacial water vapor density are not known, a first estimate will be obtained assuming the plate T_{i1} to be at 77 °F. The plate Reynolds number is:

$$\text{Re}_{L1} = \frac{\rho u_\infty L}{\mu} = \frac{(0.0728 \text{ lb/ft}^3) \, (1970 \text{ ft/min}) \, (0.328 \text{ ft})}{(1.32 \times 10^{-5} \text{ lb} \cdot \text{ft/s}) \, (60 \text{ s/min})} = 59{,}340$$

The plate is entirely in laminar flow, since the transitional Reynolds number is about 5×10^5. Using the mass transfer analogy, Equation (35) yields:

$$\begin{aligned} \overline{\text{Sh}}_{L1} &= 0.664 \, \text{Sc}^{1/3} \, \text{Re}_L^{1/2} \\ &= 0.664 \, (0.35)^{1/3} \, (59{,}340)^{1/2} = 114 \end{aligned}$$

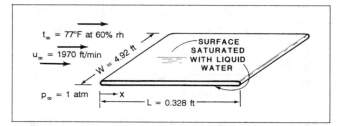

Fig. 5   Water Saturated Flat Plate in Flowing Airstream

From the definition of the Sherwood number:

$$\bar{h}_{m1} = \overline{\text{SL}}_L \, D/L = (114) \, (3.87 \times 10^{-4} \text{ ft}^2/\text{s})/0.328 \text{ ft} = 0.135 \text{ ft/s}$$

The psychrometric tables give a humidity ratio of 0.0121 at 77 °F and 60% rh. Therefore,

$$\rho_{B\infty} = (0.0121) \, \rho_{A\infty} = (0.0121)(0.0728 \text{ lb/ft}^3) = 0.000881 \text{ lb/ft}^3$$

Psychrometric tables give the saturation density for water at 77 °F as

$$\rho_{Bil} = (0.02016) \, \rho_{A\infty} = (0.02016) \, (0.0728 \text{ lb/ft}^3) = 0.001468 \text{ lb/ft}^3$$

Therefore, the mass transfer rate from the double-sided plate is

$$\begin{aligned} \dot{m}_{B1} &= \bar{h}_m A (\rho_{Bi} - \rho_{B\infty}) \\ &= (0.135 \text{ ft/s}) \, (0.328 \text{ ft} \times 4.92 \text{ ft}) \, (2) \, (0.001468 - 0.000881) \text{ lb/ft}^3 \\ &= 0.000255 \text{ lb/s} \end{aligned}$$

This mass rate, transformed from the liquid state to the vapor state, requires the following heat rate to the plate to maintain the evaporation:

$$q_{il} = \dot{m}_B'' \, h_{fg} = (0.000255 \text{ lb/s}) \, (1050 \text{ Btu/lb}) = 0.268 \text{ Btu/s}$$

To obtain a second estimate of the wetted plate temperature in this type of problem, the following criteria are used. Calculate the T_i necessary to provide a heat rate of q_{i1}. If this temperature T_{iq1} is above the dew-point temperature T_{id}, set the second estimate at $T_{i2} = (T_{iq1} + T_{i1})/2$. If T_{iq1} is below the dew-point temperature, set $T_{i2} = (T_{id} + T_{i1})/2$. For this problem, the dew point is $T_{id} = 57$ °F.

Obtaining the second estimate of the plate temperature requires an approximate value of the heat transfer coefficient.

$$\begin{aligned} \overline{\text{Nu}}_{L1} &= 0.664 \, \text{Pr}^{1/3} \, \text{Re}_L^{1/2} = 0.664 \, (0.708)^{1/3} \, (59{,}340)^{1/2} \\ &= 144.2 \end{aligned}$$

From the definition of the Nusselt number:

$$\begin{aligned} \bar{h}_1 &= \overline{\text{Nu}}_L \, k/L = (144.2) \, (0.0151 \text{ Btu/h} \cdot \text{ft} \cdot {}^\circ\text{F}) \, (0.328 \text{ ft}) \\ &= 6.63 \text{ Btu/h} \cdot \text{ft}^2 \cdot {}^\circ\text{F} \end{aligned}$$

Therefore, the second estimate for the plate temperature is

$$\begin{aligned} T_{iq1} &= T_\infty - q_i/(\bar{h}A) \\ &= 77 \, ^\circ\text{F} - (0.268 \text{ Btu/s}) \, (3600 \text{ s/h}) \, (663 \text{ Btu/h} \cdot \text{ft}^2 \cdot {}^\circ\text{F}) \, (3.228 \text{ ft}^2) \\ &= 77 \, ^\circ\text{F} - 45 \, ^\circ\text{F} = 32 \, ^\circ\text{F} \end{aligned}$$

This temperature is below the dew-point temperature; therefore,

$$T_{i2} = (57 \, ^\circ\text{F} + 77 \, ^\circ\text{F})/2 = 67 \, ^\circ\text{F}$$

The second estimate of the film temperature is

$$T_{f2} = (T_{i2} + T_\infty)/2 = (67 \, ^\circ\text{F} + 77 \, ^\circ\text{F})/2 = 72 \, ^\circ\text{F}$$

The next iteration on the solution is as follows:

$$\begin{aligned} \text{Re}_{L2} &= 61{,}010 \\ \overline{\text{Sh}}_{L2} &= 0.664 \, (0.393)^{1/3} \, (61{,}010)^{1/2} = 120 \\ \bar{h}_{m2} &= (120) \, (3.63 \times 10^{-4})/(0.328) = 0.133 \text{ ft/s} \end{aligned}$$

The free stream density of the water vapor has been evaluated. The density of the water vapor at the plate surface is the saturation density at 67 °F.

$$\rho_{Bi2} = (0.01374)(0.0739 \text{ lb/ft}^3) = 0.00101 \text{ lb/ft}^3$$
$$A = 2 \times 4.92 \times 0.328 = 3.228 \text{ ft}^2$$
$$\dot{m}_{B2} = (0.133 \text{ ft/s})(3.228 \text{ ft}^2)(0.00101 \text{ lb/ft}^3 - 0.000881 \text{ lb/ft}^3)$$
$$= 0.0000554 \text{ lb/s}$$
$$\dot{q}_{i2} = (0.0000554)(1056) = 0.0585 \text{ Btu/s}$$
$$\overline{Nu}_{L2} = 0.644(0.709)^{1/3}(61,000)^{1/2} = 146$$
$$\bar{h}_2 = (146)(0.01493)/(0.328) = 6.65 \text{ Btu/h} \cdot \text{ft}^2 \cdot {}^\circ\text{F}$$
$$T_{iq2} = 77{}^\circ\text{F} - (0.0585)(3600)/(6.65)(3.228) = 67.2{}^\circ\text{F}$$

This temperature is above the dew-point temperature; therefore

$$T_{i3} = (T_{iq2} + T_{i2})/2 = (67 + 67.2)/2 = 67.1{}^\circ\text{F}$$

This is approximately the same result as that obtained in the previous iteration. Therefore, the problem solution is:

$$T_i = 67{}^\circ\text{F}$$
$$\dot{m}_B = 0.199 \text{ lb/h}$$

Eddy Diffusivity

Turbulent flow is characterized by random velocity fluctuations superimposed on the time-average velocity. The fluctuations occur in the direction of flow and normal to it. Small mixing actions or *eddy currents* are established within the turbulent flow field. The eddies cause an exchange of momentum between different layers of the moving fluid. If the fluid has a temperature gradient, energy is exchanged by this mixing action in much the same way. If a mass concentration gradient exists, a similar mass exchange known as *eddy diffusion* occurs, which is relatively faster than molecular diffusion and depends on the *intensity* of the velocity fluctuations, or turbulence. Since intensity of turbulence is determined by the Reynolds number of the flow, the rate of eddy diffusion depends on the Reynolds number (Sherwood and Pigford 1952). The *eddy diffusivity* ϵ_D is defined by the same form of equation as that in molecular diffusion [Equation (5)]:

$$\dot{m}_B'' = \epsilon_D (d\rho_B/dy)$$

where ϵ_D = eddy diffusivity, ft²/h.

Because data on eddy diffusivities are difficult to obtain, the mass transfer coefficient, analogous to the heat transfer coefficient in convective heat transfer, is usually defined and determined experimentally.

Application to Turbulent Flow

Consider an airstream in steady turbulent flow over a wetted surface (Figure 6). The liquid-vapor interface is assumed to be at zero velocity, which results in a slow-moving layer of fluid in laminar flow next to the surface. Between this *laminar sublayer* and the main body of the turbulent stream, a transition region or buffer layer exists, in which the fluid may be alternately in

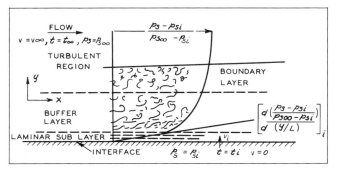

Fig. 6 Turbulent Diffusion Boundary Layer on Flat Surface

laminar flow and in turbulent flow. Within the laminar sublayer, only molecular diffusion can occur; in the buffer layer, both molecular and eddy diffusion contribute to mass transfer. In the turbulent region, eddy diffusion predominates and is so rapid that it almost equalizes the concentration gradient.

Because of the presence of the laminar sublayer, the rate of molecular mass diffusion from the wetted surface to the airstream, from Equation (6), is:

$$J_B = -D_v (d\rho_B/dy)_i$$

where $(d\rho_B/dy)_i$ is the partial pressure gradient at the interface. Assuming that the gases are ideal and obey the Gibbs-Dalton law, and that the total pressure is constant, then a partial pressure gradient must also exist in the air. The wetted surface is impermeable to air, so a convective or bulk velocity must be established to counter the air diffusion rate. The total mass transfer from the wetted surface to the airstream must then be given by:

$$\dot{m}_B'' = -D_v (d\rho_B/dy)_i + \rho_{Bi} v_i \qquad (37)$$

where v_i is the convective velocity of the component B vapor at the interface (Figure 6), and the partial mass density of the water vapor at the interface is:

$$\rho_{Bi} = M_B p_{Bi}/R_U T_i$$

For the diffusion of one gas through a second stagnant gas, the simple molecular diffusion equation corrected by the factor P_{AM} [Equation (24)] accounts for the mass transfer contribution of the convective velocity. For dilute mixtures, as in Example 1, this contribution is small. The same correction factor can be used for this forced convection mass transfer process—at least at low mass transfer rates where v_i is small. The total mass transfer rate from the interface is:

$$\dot{m}_B'' = -D_v P_{AM} (d\rho_B/dy)_i \qquad (38)$$

The concentration gradient $(d\rho_B/dy)_i$ must be evaluated experimentally. Rather than work with this gradient, a mass transfer coefficient h_m may be defined as in Equation (27). It follows that:

$$\dot{m}_B'' = (h_m M_B/R_U)(p_{Bi} - p_{B\infty})$$
$$= -(M_B D_v P_{AM}/R_U T_i)(dp_B/dy)_i$$

By mathematical rearrangement, this becomes:

$$\text{Sh}_L = h_m L/D$$
$$= P_{AM}\left[d\left(\frac{p_B - p_{Bi}}{p_{B\infty} - p_{Bi}}\right) / d(y/L)\right]_i \qquad (39)$$

where Sh_L is the dimensionless Sherwood number and L is some characteristic dimension of the mass transfer surface, such as the length of a plate or diameter of a cylinder. Equation (39) gives a simple, physical interpretation of the dimensionless mass transfer coefficient, i.e., the Sherwood number. It is a measure of the dimensionless concentration gradient at the interface or mass transfer boundary (Figure 6).

Bird *et al.* (1960), Hirschfelder *et al.* (1954), Sherwood and Pigford (1952), Incropera and DeWitt (1990), and Reid *et al.* (1987) experimentally established mass transfer coefficients for a number of flow geometries. Because of the analogy between these transfer processes, heat transfer data have been used to predict mass transfer coefficients. The reliability of such similar relations has been well-established at low mass transfer rates for some flow geometries.

Analogy Relations for Convective Mass Transfer

Heat transfer from the wetted surface of Figure 6 to the air-stream can be expressed in an analogous form to that of mass transfer. Therefore, the heat flux per unit area is:

$$\dot{q}'' = -k\,(dt/dy)_i = h\,(t_i - t_\infty)$$

or, by mathematical rearrangement:

$$\mathrm{Nu} = hL/k = d\left(\frac{t - t_i}{t_\infty - t_i}\right) / d(y/L) \qquad (40)$$

The dimensionless Nusselt number Nu of heat transfer is the dimensionless temperature gradient at the heat transfer boundary (interface), and the Sherwood and Nusselt numbers are analogous expressions for dimensionless mass transfer and heat transfer coefficients, respectively.

For momentum transfer, a friction factor f is defined so that the momentum flux at the interface τ_i per unit area is:

$$\tau_i = \mu(du/dy)_i = f\rho\bar{u}^2/2$$

or

$$f/2(\rho\bar{u}L/\mu) = \left[\frac{d(u/\bar{u})}{d(y/L)}\right]_i \qquad (41)$$

where $\bar{u}$ = mean free stream velocity of gas, ft/h.

Equation (41) states that one-half the friction factor times the Reynolds number (Re = $\rho\bar{u}L/\mu$) equals the dimensionless velocity gradient at the stationary surface or interface.

Equations (39), (40), and (41) show the similarity of mass, heat, and momentum transport in a turbulent boundary layer (with both eddy and molecular diffusion), and Equations (11), (6), and (7) demonstrate the similarity for a laminar boundary layer (with only molecular diffusion). Again, assume a gaseous system in which, by chance, $D_v = \alpha = \nu$ (i.e., the molecular diffusion constants for mass, energy, and momentum are identical). Theoretically, the eddy diffusivities for these three rate processes are identical, since experiments verify they rely on the same particle mixing action for the transfer operation. Therefore, the dimensionless partial pressure, temperature, and velocity profiles must be identical. This permits equating the left side of Equations (39), (40), and (41) to obtain

$$\mathrm{Sh}_L = \mathrm{Nu}_L = (f/2)(\mathrm{Re}_L)$$

or

$$h_mL/DP_{AM} = hL/k = (f/2)(\bar{u}L/\nu) \qquad (42)$$

This is equivalent to the statement that the Prandtl number (Pr = $c_p\mu/k$) and Schmidt number (Sc = $\mu/\rho D_v$) are unity. Using this statement, Equation (42) can be rearranged to give:

$$\frac{\mathrm{Sh}}{\mathrm{Re}\,\mathrm{Sc}} = \frac{\mathrm{Nu}}{\mathrm{Re}\,\mathrm{Pr}} = \frac{f}{2} \qquad (43a)$$

or

$$h_m/\bar{u}P_{AM} = h/\rho c_p\bar{u} = f/2 \qquad (43b)$$

The two right-hand terms of Equation (43) are known as the *Reynolds analogy*. The left-hand term is an extension of the Reynolds analogy to include mass transfer. The Reynolds analogy gives a plausible correlation between friction factors and heat transfer coefficients for common gases (where Pr $\cong$ 1) with moderate temperature potential. Since Reynolds developed this analogy, at least five modified analogies have been suggested to account for the effect of the Prandtl number. One of the simplest and most often used is the one suggested by Chilton and Colburn (1934), who showed that the correlation of heat transfer data with friction data could be improved by substituting $(\mathrm{Pr})^{1/3}$ for Pr in Equation (43a), which leads to:

$$j_H = \left(\frac{h}{\rho c_p\bar{u}}\right)\left(\frac{c_p\mu}{K}\right)^{2/3}$$
$$= \mathrm{St}\,\mathrm{Pr}^{2/3} = f/2 \qquad (44)$$

By analogy, Chilton and Colburn suggested mass transfer data could be represented with good accuracy by a similar change in the Schmidt number parameter:

$$j_D = \left(\frac{h_m P_{AM}}{\bar{u}}\right)\left(\frac{\mu}{\rho D_v}\right)^{2/3}$$
$$= \mathrm{St}_m\mathrm{Sc}^{2/3} = f/2 \qquad (45)$$

where St and St_m refer to the dimensionless Stanton numbers for heat and mass transfer, respectively (see Symbols section for definition of St).

Equations (44) and (45) are known as the *j-factor analogy* and are widely used for plotting and predicting heat and mass transfer data.

The power of the Chilton-Colburn j-factor analogy is represented in Figures 7 through 10. Figure 7 plots various experimental values of j_D from a flat plate with flow parallel to the plate surface. The solid line, which represents the data to near perfection, is actually $f/2$ from Blasius' solution of laminar flow on a flat plate (left-hand portion of the solid line) and Goldstein's solution for a turbulent boundary layer (right-hand portion). The right-hand portion of the solid line also represents McAdams' (1954) correlation of turbulent flow heat transfer coefficient for a flat plate.

A *wetted-wall column* is a vertical tube in which a thin liquid film adheres to the tube surface and exchanges mass by evaporation or absorption with a gas flowing through the tube. Figure 8

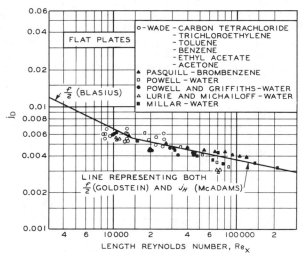

Fig. 7   Mass Transfer from Flat Plate

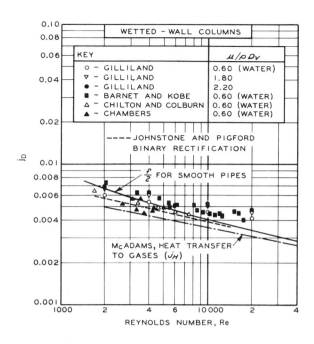

Fig. 8　Vaporization and Absorption in Wetted Wall Column

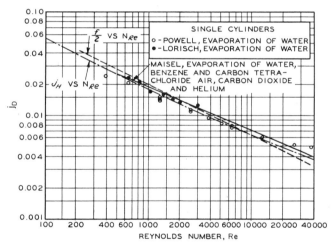

Fig. 9　Mass Transfer from Single Cylinders in Crossflow

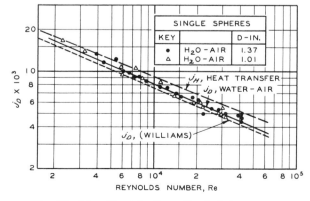

Fig. 10　Mass Transfer from Single Spheres

illustrates typical data on vaporization in wetted-wall columns, plotted as j_D versus Re. The spread of the points with variation in $\mu/\rho D_v$ results from Gilliland's finding of an exponent of 0.56, not 2/3, representing the effect of the Schmidt number. Gilliland's equation can be written as follows:

$$j_D = 0.023 \, \mathrm{Re}^{-0.17} \, (\mu/\rho D_v)^{0.11} \qquad (46)$$

Similarly, McAdams' (1954) equation for heat transfer in pipes can be expressed as:

$$j_H = 0.023 \, \mathrm{Re}^{-0.20} \, (c_p \mu/k)^{0.07} \qquad (47)$$

This is represented by the dashed curve in Figure 8, which falls below the mass transfer data. The curve $f/2$ representing friction in smooth tubes is the upper solid curve.

Data for the evaporation of liquids from single cylinders into gas streams flowing transversely to the cylinders' axes are shown in Figure 9. Although the dash-dot line on Figure 9 represents the data, it is actually taken from McAdams (1954) as representative of a large collection of data on heat transfer to single cylinders placed transverse to airstreams. To compare these data with friction, it is necessary to distinguish between total drag and skin friction. Since the analogies are based on skin friction, the normal pressure drag must be subtracted from the measured total drag. At Re = 1000, the skin friction is 12.6% of the total drag; at Re = 31,600, it is only 1.9%. Consequently, the values of $f/2$ at a high Reynolds number, obtained by the difference, are subject to considerable error.

In Figure 10, data on the evaporation of water into air for single spheres are presented. The solid line, which best represents these data, agrees with the dashed line representing McAdams' correlation for heat transfer to spheres. These results cannot be compared with friction or momentum transfer, since total drag has not been allocated to skin friction and normal pressure drag. Application of these data to air-water contacting devices such as air washers and spray cooling towers is well substantiated.

When the temperature of the heat exchanger surface in contact with moist air is below the dew-point temperature of the air, vapor condensation occurs. Typically, the air dry-bulb temperature and humidity ratio both decrease as the air flows through the exchanger. Therefore, sensible and latent heat transfer occur simultaneously. This process is similar to one that occurs in a spray dehumidifier and can be analyzed using the same procedure; however, this is not generally done.

Cooling coil analysis and design is complicated by the problem of determining transport coefficients h, h_m, and f. It would be convenient if heat transfer and friction data for dry heating coils could be used with the Colburn analogy to obtain the mass transfer coefficients. However, this approach is not always reliable, and work by Guillory and McQuiston (1973) and Helmer (1974) shows that the analogy is not consistently true. Figure 11 shows j factors for a simple parallel plate exchanger for different surface conditions with sensible heat transfer. Mass transfer j factors and the friction factors exhibit the same behavior. Dry surface j factors fall below those obtained under dehumidifying conditions with the surface wet. At low Reynolds numbers, the boundary layer grows quickly; the droplets are soon covered and have little effect on the flow field. As the Reynolds number is increased, the boundary layer becomes thin and more of the total flow field is exposed to the droplets. The roughness caused by the droplets induces mixing and larger j factors. The data in Figure 11 cannot be applied to all surfaces, because the length of the flow channel is also an important variable. However, the water collecting on the surface is mainly responsible for breakdown of the j factor analogy. The j factor analogy is approximately true when the surface

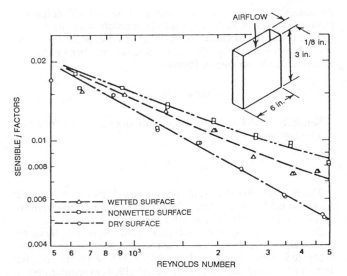

Fig. 11 Sensible Heat Transfer j-Factors for Parallel Plate Exchanger

conditions are identical. Under some conditions, it is possible to obtain a film of condensate on the surface instead of droplets. Guillory and McQuiston (1973) and Helmer (1974) related dry sensible j and f factors to those for wetted dehumidifying surfaces.

The equality of j_H, j_D, and $f/2$ for certain streamline shapes at low mass transfer rates has experimental verification. For flow past bluff objects, j_H and j_D are much smaller than $f/2$, based on total pressure drag. The heat and mass transfer, however, still relate in a useful way by equating j_H and j_D.

Example 3. Using solid cylinders of volatile solids (*e.g.*, naphthalene, camphor, dichlorobenzene) with airflow normal to these cylinders, Bedingfield and Drew (1950) found that the ratio between the heat and mass transfer coefficients could be closely correlated by the relation:

$$h/\rho h_m = [(0.294 \text{ Btu/lb}_m \cdot °F)(\mu/\rho D_v)]^{0.56}$$

For completely dry air at 70°F flowing at a velocity of 31 fps over a wet-bulb thermometer of diameter $d = 0.300$ in., determine the heat and mass transfer coefficients from Figure 9 and compare their ratio with the Bedingfield-Drew relation.

Solution: For dry air at 70°F, $\rho = 0.075$ lb/ft³, $\mu = 0.0441$ lb/ft · h, $k = 0.0150$ Btu/h · ft · °F, and $c_p = 0.2398$ Btu/lb$_m$ · °F. From Equation (11), $D_v = 0.971$ ft²/h. Therefore:

$\text{Re}_{da} = \rho u_\infty d/\mu = 0.075 \times 31 \times 3600 \times 0.300/(12 \times 0.0441) = 4740$

$\text{Pr} = c_p\mu/k = 0.2398 \times 0.0441/0.0150 = 0.706$

$\text{Sc} = \mu/\rho D_v = 0.0441/(0.075 \times 0.971) = 0.605$

From Figure 9 at $\text{Re}_{da} = 4740$, $j_H = 0.0088$, $j_D = 0.0099$, and

$h = j_H\rho c_p U_\infty/(\text{Pr})^{2/3}$
$h = 0.0088 \times 0.075 \times 0.2398 \times 31 \times 3600/(0.706)^{2/3}$
$\quad = 22.3 \text{ Btu/h} \cdot \text{ft}^2 \cdot °F$
$h_m = j_D U_\infty (\text{Sc})^{2/3} = 0.0099 \times 31/(0.605)^{2/3}$
$\quad = 0.429 \text{ ft/s}$
$h/\rho h_m = 22.3/(0.075 \times 0.429 \times 3600) = 0.193 \text{ Btu/lb}_m \cdot °F$

From the Bedingfield and Drew relation:

$h/\rho h_m = 0.294 (0.605)^{0.56} = 0.222 \text{ Btu/lb}_m \cdot °F$

The Reynolds analogy, Equation (43b), suggests that $h/\rho h_m = c_p = 0.2398$ Btu/lb$_m$ · °F. This close agreement is because the ratio Sc/Pr is 0.605/0.706 or 0.86, so that the exponent of these numbers has little effect on the ratio of the transfer coefficients.

The extensive developments for calculating heat transfer coefficients can be applied to calculate mass transfer coefficients under similar geometrical and flow conditions using the j-factor analogy. For example, Chapter 3 lists equations for calculating heat transfer coefficients for flow inside and normal to pipes. Each equation can be used for mass transfer coefficient calculations by equating j_H and j_D, and imposing the same restriction to each stated in Chapter 3. Similarly, mass transfer experiments often replace corresponding heat transfer experiments with complex geometries where exact boundary conditions are difficult to model (Sparrow and Ohadi 1987a, 1987b).

The j-factor analogy is useful only at low mass transfer rates. As the rate of mass transfer increases, the movement of matter normal to the transfer surface increases the convective velocity [v_i in Equation (37) and Figure 6]. For example, if a gas is blown from many small holes in a flat plate placed parallel to an airstream, the boundary layer thickens, and resistance to both mass and heat transfer increases with the increasing blowing rate. Heat transfer data are usually collected at zero or, at least, insignificant mass transfer rates. Therefore, if such data are to be valid for a mass transfer process, the mass transfer rate, *i.e.*, the blowing, must be low.

The j-factor relationship, $j_H = j_D$, can still be valid at high mass transfer rates, but neither j_H nor j_D can be represented by data at zero mass transfer conditions. Chapter 18 of Eckert and Drake (1972) and Chapter 24 of Bird *et al.* (1960) have detailed information on high mass transfer rates.

Lewis Relation

Heat and mass transfer coefficients are satisfactorily related, at the same Reynolds number, by equating the Chilton-Colburn j-factors. This leads to:

$$h_m/\bar{u} P_{AM} (\mu/\rho D_v)^{2/3} = (h/\rho c_p \bar{u})(c_p\mu/k)^{2/3}$$

or

$$h/h_m\rho c_p = (P_{AM})[\mu/\rho D_v/(c_p\mu/k)]^{2/3}$$
$$= (P_{AM})(\alpha/D_v) \tag{48}$$

The quantity α/D_v is the Lewis number Le. Its magnitude expresses relative rates of propagation of energy and mass within a system. It is fairly insensitive to temperature variation. For air and water vapor mixtures, the ratio is (0.60/0.71) or 0.845, and $(0.845)^{2/3}$ is 0.894. At low diffusion rates, where the heat-mass transfer analogy is valid, P_{AM} is essentially unity. Therefore, for air and water vapor mixtures:

$$h/h_m\rho c_p = 1 \tag{49}$$

The ratio of the heat transfer coefficient to the mass transfer coefficient is equal to the specific heat per unit volume at constant pressure of the mixture. This relation is usually called the Lewis relation and is nearly true for air and water vapor at low mass transfer rates. It is generally not true for other gas mixtures because the ratio of thermal to vapor diffusivity Le can differ from unity. The agreement between wet-bulb temperature and adiabatic saturation temperature is a direct result of the nearness of Lewis number to unity for air and water vapor.

The Lewis relation is valid in turbulent flow whether or not the ratio of α/D_v equals 1, because eddy diffusion in turbulent flow involves the same mixing action for heat exchange as for mass exchange, and this action overwhelms any molecular diffusion. Deviations from the Lewis relation are, therefore, due to a laminar boundary layer or a laminar sublayer and buffer zone, as in

Figure 6, where molecular transport phenomena are the controlling factors.

SIMULTANEOUS HEAT AND MASS TRANSFER BETWEEN WATER-WETTED SURFACES AND AIR

A simplified method used to solve simultaneous heat and mass transfer problems was developed using the Lewis relation and gives satisfactory results for most air-conditioning processes. Extrapolation to very high mass transfer rates, where the simple heat-mass transfer analogy is not valid, will lead to erroneous results.

Enthalpy Potential

The water vapor concentration in the air is the humidity ratio W defined as:

$$W \equiv \rho_B / \rho_A \tag{50}$$

A mass transfer coefficient is defined using W as the driving potential:

$$\dot{m}_B'' = K_m(W_i - W_\infty) \tag{51}$$

where the coefficient K_m has the units $lb_m/h \cdot ft^2$. For dilute mixtures $\rho_{Ai} \cong \rho_{A\infty}$ (i.e., the partial mass density of dry air changes by only a small percentage between interface and free stream conditions). Therefore:

$$\dot{m}_B'' = (K_m/\rho_{Am})(\rho_{Bi} - \rho_\infty) \tag{52}$$

where ρ_{Am} is the *mean* density of dry air, lb_m/ft^3. Comparing this equation with Equation (27) shows that:

$$h_m = K_m/\rho_{Am} \tag{53}$$

The *humid* specific heat c_{pm} of the airstream is, by definition (Mason and Monchick 1965),

$$c_{pm} = (1 + W_\infty)C_p \quad Btu \cdot lb(dry \ air) \cdot °F$$

or

$$c_{pm} = (\rho/\rho_{A\infty})c_p \tag{54}$$

Substituting from these expressions into the Lewis relation [Equation (49)] gives:

$$(h\rho_{Am}/K_m\rho_{A\infty}c_{pm}) = 1 \cong (h/K_mc_{pm}) \tag{55}$$

since $\rho_{Am} \cong \rho_{A\infty}$, because of the small change in dry-air density. Using a mass transfer coefficient with humidity ratio as the driving force, the Lewis relation becomes: ratio of heat to mass transfer coefficient equals humid specific heat.

For the plate humidifier illustrated in Figure 5, the total heat transfer from liquid to interface is:

$$\dot{q}'' = \dot{q}_A'' + \dot{m}_B'' h_{fg}$$

Using the definitions of the transfer coefficients:

$$\dot{q}'' = h(t_i - t_\infty) + K_m(W_i - W_\infty)h_{fg}$$

Assuming Equation (55) is valid:

$$\dot{q}'' = K_m[c_{pm}(t_i - t_\infty) + (W_i - W_\infty)h_{fg}] \tag{56}$$

The enthalpy of the air is approximately:

$$h = c_{pa}t + Wh_s \tag{57}$$

The enthalpy of the water vapor h_s by the ideal gas law can be expressed as:

$$h_s = c_{ps}(t - t_o) + h_{fgo}$$

where the base of enthalpy is taken as saturated water at temperature t_o. Choosing $t_o = 0°F$ to correspond with the base of the dry-air enthalpy gives:

$$h = (c_{pa} + Wc_{ps})t + Wh_{fgo} = c_{pm}t + Wh_{fgo} \tag{58}$$

If small changes in the latent heat of vaporization of water with temperature are neglected when comparing Equations (57) and (58), the total heat transfer can be written as:

$$\dot{q}'' = K_m(h_i - h_\infty) \tag{59}$$

Where the driving potential for heat transfer is temperature difference and the driving potential for mass transfer is mass concentration or partial pressure, the driving potential for simultaneous transfer of heat and mass in an air water-vapor mixture is, to a close approximation, enthalpy.

Basic Equations for Direct Contact Equipment

Air-conditioning equipment can be classified by (1) whether there is direct contact between air and water used as a cooling or heating fluid, or (2) whether the heating or cooling fluid is separated from the airstream by a solid wall. Examples of the former are air washers and cooling towers; an example of the latter is a direct expansion refrigerant (or water) cooling and dehumidifying coil. In both cases, the airstream is in contact with a water surface. *Direct contact* implies a direct contact with the cooling (or heating) fluid. In the dehumidifying coil, the contact is directly with the condensate removed from the airstream, but indirectly with the refrigerant flowing inside the tubes of the coil. These two cases are treated separately because the surface areas of direct contact equipment cannot be evaluated.

For the direct contact spray chamber air washer of cross-sectional area A_{cs} and length l (Figure 12), the steady mass flow rate of dry air per unit of cross-sectional areas is:

$$\dot{m}_a/A_{cs} = G_a \tag{60}$$

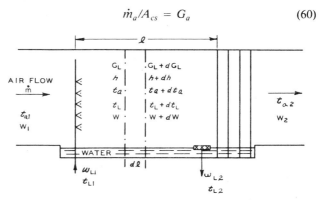

Fig. 12   Air Washer Spray Chamber

and the corresponding mass velocity of water flowing parallel with the air is:

$$\dot{m}_L/A_{cs} = G_L \qquad (61)$$

where

$\dot{m}_a$ = mass flux, mass flow rate of air, lb/h
G_a = mass flux, mass velocity, or flow rate per unit cross-sectional area for air, lb/h·ft²
$\dot{m}_L$ = mass flow rate of liquid, lb/h
G_L = mass velocity, or flow rate per unit cross-sectional area, for liquid, lb/h·ft²

Since water is evaporating or condensing, G_L changes by an amount dG_L in a differential length dl of the chamber. Similar changes occur in temperature, humidity ratio, enthalpy, and other properties.

Since evaluating the true surface area in direct contact equipment is difficult, it is common to work on a unit volume basis. If a_H and a_M are the area of heat transfer and mass transfer surface per unit of chamber volume, respectively, the total surface areas for heat and mass transfer are:

$$A_H = a_H A_{cs} l \quad \text{and} \quad A_M = a_M A_{cs} l \qquad (62)$$

The basic equations for the process occurring in the differential length dl can be written for:
1. *Mass transfer*

$$- dG_L = G_a dW = K_m a_M (W_i - W) dl \qquad (63)$$

That is, the water evaporated, the moisture increase of the air, and the mass transfer rate are all equal.
2. *Heat transfer to air*

$$G_a c_{pm} dt_a = h_a a_H (t_i - t_a) dl \qquad (64)$$

3. *Total energy transfer to air*

$$G_a(c_{pm} dt_a + h_{fgo} dW) = [K_m a_M (W_i - W) h_{fg} + h_a a_H (t_i - t_a)] dl \quad (65)$$

As shown in the preceding section, assuming $a_H = a_M$, Le = 1, and neglecting small variations in h_{fg}, Equation (65) reduces to:

$$G_a dh = K_m a_M (h_i - h) dl \qquad (66)$$

The heat and mass transfer areas of spray chambers are assumed to be identical ($a_H = a_M$). Where packing materials, such as wood slats or Raschig rings, are used, the two areas may be considerably different because the packing may not be wet uniformly. The validity of the Lewis relation was discussed previously. It is not necessary to account for the small changes in latent heat h_{fg} after making the two previous assumptions.
4. *Energy balance*

$$G_a dh = \pm G_L c_L dt_L \qquad (67)$$

A minus sign refers to parallel flow of air and water; a plus sign refers to counterflow (water flow in the opposite direction from airflow).

The water flow rate changes between inlet and outlet as a result of the mass transfer. For exact energy balance, the term ($c_L t_L dG_L$) should be added to the right side of Equation (67). The percentage change in G_L is quite small in usual applications of air-conditioning equipment and, therefore, can be ignored.

5. *Heat transfer to water*

$$G_L c_L dt_L = h_L a_H (t_L - t_i) dl \qquad (68)$$

Equations (63) to (68) are the basic relations for solution of simultaneous heat and mass transfer processes in direct contact air-conditioning equipment.

To facilitate the use of these relations in equipment design or performance, three other equations can be extracted from the above set. Combining Equations (66), (67), and (68) gives:

$$(h - h_i)/(t_L - t_i) = -(h_L a_H/K_m a_M) = -(h_L/K_m) \quad (69)$$

Equation (69) relates the enthalpy potential for the total heat transfer through the gas film to the temperature potential for this same transfer through the liquid film. Physical reasoning leads to the conclusion that this ratio is proportional to the ratio of gas film resistance ($1/K_m$) to the liquid film resistance ($1/h_L$). Combining Equations (64), (66), and the Lewis relation [Equation (55)] gives:

$$dh/dt_a = (h - h_i)/(t_a - t_i) \qquad (70)$$

Similarly, combining Equations (63), (64), and the Lewis relation gives:

$$dW/dt_a = (W - W_i)/(t_a - t_i) \qquad (71)$$

Equation (71) indicates that at any cross section in the spray chamber, the instantaneous slope of the air path dW/dt_a on a psychrometric chart is determined by a straight line connecting the air state with the interface saturation state at that cross section. In Figure 13, state 1 represents the state of the air entering the parallel flow air washer chamber of Figure 12. The washer is operating as a heating and humidifying apparatus so that the interface saturation state of the water at air inlet is the state designated 1_i. Therefore, the initial slope of the air path is along a line directed from state 1 to state 1_i. As the air is heated, the water cools and the interface temperature drops. Corresponding air states and interface saturation states are indicated by the letters a, b, c, and d in

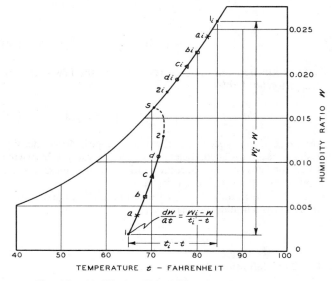

Fig. 13   Air Washer Humidification Process on Psychrometric Chart

Figure 13. In each instance, the air path is directed toward the associated interface state. The interface states are derived from Equations (67) and (69). Equation (67) describes how the air enthalpy changes with water temperature; Equation (69) describes how the interface saturation state changes to accommodate this change in air and water conditions. The solution for the interface state on the normal psychrometric chart of Figure 13 can be determined either by trial and error from Equations (67) and (69) or by a complex graphical procedure (Kusuda 1957).

Air Washers

Air washers are direct contact apparatus used to simultaneously change the temperature and humidity content of air passing through the chamber and to remove air contaminants such as dust and odors. Adiabatic spray washers, which have no external heating or chilling source, are used to cool and humidify air. Chilled spray air washers have an external chiller to cool and dehumidify air. Heated spray air washers, whose external heating source provides additional energy for evaporation of water, are used to humidify and possibly heat air.

Example 4. A parallel flow air washer with the following design conditions is to be designed (see Figure 12).

Water temperature at inlet $t_{L1} = 95\,°F$
Water temperature at outlet $t_{L2} = 75\,°F$
Air temperature at inlet $t_{a1} = 65\,°F$
Air wet-bulb at inlet $t_{a1} = 45\,°F$
Air mass flow rate per unit area $G_a = 1200\ lb/h \cdot ft^2$
Spray ratio $G_L/G_a = 0.70$
Air heat transfer coefficient per cubic foot of chamber volume
$\quad h_a a_H = 72\ Btu/h \cdot °F \cdot ft^3$
Liquid heat transfer coefficient per cubic foot of chamber volume
$\quad h_L a_H = 900\ Btu/h \cdot °F \cdot ft^3$
Air volume flow rate $Q = 6500\ ft^3/min$

Solution: The air mass flow rate $w_a = (6500/13.25) = 490\ lb/min$, and the required spray chamber cross-sectional area is, then, $A_{cs} = w_a/G_a = 490\,(60)/1200 = 24.5\ ft^2$. The mass transfer coefficient is given by the Lewis relation as:

$$K_m a_M = (h_a a_H)/c_{pm} = 72/0.24 = 300\ lb/h \cdot ft^3$$

Figure 14 shows the enthalpy-temperature psychrometric chart with the graphical solution for the interface states and the air path through the washer spray chamber. The solution proceeds as follows:

1. Enter bottom of chart with t_{ai}' of 45 °F, follow up to saturation curve to establish air enthalpy h_1 of 17.65 Btu/lb. Extend this enthalpy line to intersect initial air temperature t_{ai}' of 65 °F (state 1 of air), and initial water temperature t_{L1} of 95 °F at point A. (Note that the temperature scale is used for both air and water temperatures.)
2. Through point A, construct the *energy balance* AB with a slope of:

$$(dh/dt_L) = -G_L/G_a = -0.7$$

Point B is determined by intersection with the leaving water temperature $t_{L2} = 75\,°F$. The negative slope here is a consequence of the parallel flow, which results in the air-water mixture approaching, but not reaching, the common saturation state *s*. (The line AB has no physical significance in representing any *air state* on the psychrometric chart. It is merely a construction line in the graphical solution.)
3. Through point A, construct the *tie-line* A1$_i$ having a slope of:

$$(h - h_i)(t_L - t_i) = -(h_L a_H/K_m a_M) = -(900/300) = -3$$

The intersection of this line with the saturation curve gives the initial interface state 1$_i$ at the chamber inlet. (Note how the energy balance line and tie-line, representing Equations (67) and (69), combine for a simple graphical solution on the $h - t$ chart for the interface state.)
4. The initial slope of the air path can now be constructed, according to Equation (70), drawing line 1a toward the initial interface state 1$_i$. (The

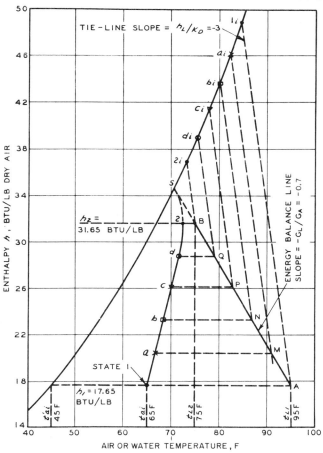

Fig. 14 Graphical Solution for Air-State Path in Parallel Flow Air Washer

length of the line 1a will depend on the degree of accuracy required in the solution and the rate at which the slope of the air path is changing.)
5. Construct the horizontal line a-M locating the point M on the energy-balance line. Draw a new tie-line (slope of -3 as before) from M to a_i locating interface state a_i. Continue the air path from a to b by directing it toward the new interface state a_i. (Note that the change in slope of the air path from 1a to ab is quite small, justifying the path incremental lengths used.)
6. Continue in the manner of step 5 until point 2, when the final state of the air leaving the chamber is reached. In this example, six steps are used in the graphical construction with the following results:

State	1	a	b	c	d	2
t_L	95.0	91.0	87.0	83.0	79.0	75.0
h	17.65	20.45	23.25	26.05	28.85	31.65
t_i	84.5	82.3	80.1	77.8	75.6	73.2
h_i	49.00	46.25	43.80	41.50	39.10	37.00
t_a	65.0	66.8	68.5	70.0	71.4	72.4

The final state of the air leaving the washer is $t_{a2} = 72.4\,°F$ and $h_2 = 31.65\ Btu/lb$ (wet-bulb temperature $t_{a2}' = 67\,°F$).
7. The final step involves calculating the required length of the spray chamber. From Equation (66):

$$l = \frac{G_a}{K_m a_M} \int_1^2 \frac{dh}{(h_i - h)}$$

The integral is evaluated graphically by plotting $1/(h_i - h)$ versus h as shown in Figure 15. Any satisfactory graphical method can be used to evaluate the area under the curve. Simpson's rule with four equal increments of Δh equal to 3.5 gives:

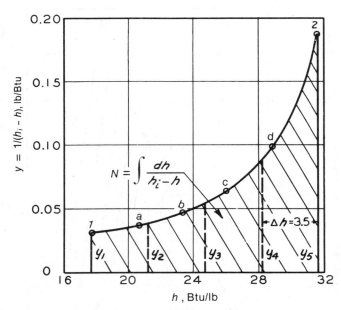

Fig. 15   Graphical Solution of $\int dh/(h_i - h)$

$$N = \int_1^2 \frac{dh}{(h_i - h)} \approx (\Delta h/3)(y_1 + 4y_2 + 2y_3 + 4y_4 + y_5)$$

$$N = (3.5/3)[(1 \times 0.0319) + (4 \times 0.0400) + (2 \times 0.0553) +$$
$$(4 \times 0.0865) + (1 \times 0.1870)] = 0.975$$

The design length is, therefore, $l = (1200/300)(0.975) = 3.9$ ft.

The method used in Example 4 can also be used to predict the performance of existing direct contact equipment and can determine the transfer coefficients when performance data from test runs are available. By knowing the water and air temperatures entering and leaving the chamber and the spray ratio, it is possible, by trial and error, to determine the proper slope of the tie-line necessary to achieve the measured final air state. The tie-line slope gives the ratio $h_L a_H/K_m a_M$; $K_m a_M$ is found from the integral relationship in Example 4 from the known chamber length l.

Additional descriptions of air spray washers and general performance criteria are given in Chapter 19 of the 1992 ASHRAE *Handbook—Systems and Equipment.*

Cooling Towers

A cooling tower is a direct contact heat exchanger in which waste heat picked up by the cooling water from a refrigerator, air conditioner, or industrial process is transferred to atmospheric air by cooling the water. Cooling is achieved by breaking up the water flow to provide a large water surface for air, moving by natural or forced convection through the tower, to contact the water. Cooling towers may be counterflow, crossflow, or a combination of both.

The temperature of the water leaving the tower and the packing depth needed to achieve the desired leaving water temperature are of primary interest for design. Therefore, the mass and energy balance equations are based on an overall coefficient K, which is based on (1) the enthalpy driving force due to h at the bulk water temperature and (2) the neglecting of the film resistance. Combining Equations (66) and (67) and using the parameters described above, yields

$$G_L C_L dt = K_M a_M (h_i - h)dl = G_a dh$$
$$= K_a dV(h' - h_a)/A_{cs} \tag{72}$$

or

$$K_a V/\dot{m}_L = \int_{t_1}^{t_2} C_L dt/(h' - h_a) \tag{73}$$

Chapter 37 of the 1992 ASHRAE *Handbook—Systems and Equipment* covers cooling tower design in detail.

Cooling and Dehumidifying Coils

When water vapor is condensed out of an airstream onto an extended surface (finned) cooling coil, the simultaneous heat and mass transfer problem can be solved by the same procedure set forth for direct contact equipment. The basic equations are the same, except that the true surface area of the coil A is known and the problem does not have to be solved on a unit volume basis. Therefore, if in Equations (63), (64), and (66) $a_M dl$ or $a_H dl$ is replaced by A_{cs}/dA, these equations become the basic heat, mass, and total energy transfer equations for indirect contact equipment such as dehumidifying coils. The energy balance shown by Equation (67) remains unchanged. The heat transfer from the interface to the refrigerant now encounters the combined resistances of the condensate film ($R_L = 1/h_L$), the metal wall and fins (if any) R_m, and the refrigerant film ($R_r = A/h_r A_r$). If this combined resistance is designated as $R_i = R_L + R_m + R_r = 1/U_i$, Equation (68) becomes, for a coil dehumidifier:

$$\dot{m}_L c_L dt_L = U_i (t_L - t_i) dA \tag{74}$$

(plus sign for counterflows, minus sign for parallel flow).

The tie-line slope is then:

$$(h - h_i)/(t_L - t_i) = -(U_i/K_m) \tag{75}$$

Figure 16 illustrates the graphical solution on a temperature-enthalpy (psychrometric) chart for the air path through a dehumidifying coil with a constant refrigerant temperature. Since the tie-line slope is infinite in this case, the energy balance line is

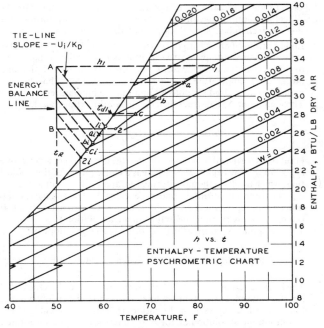

Fig. 16   Graphical Solution for Air-State Path in Dehumidifying Coil with Constant Refrigerant Temperature

vertical. The corresponding interface states and air states are denoted by the same letter symbol, and the solution follows the same procedure as in Example 4.

If the problem is to determine the required coil surface area for a given performance, the area is computed by the relation:

$$A = (\dot{m}_a/K_m)\int_1^2 dh/(h_i - h) \qquad (76)$$

This graphical solution on the psychrometric chart automatically determines whether any part of the coil is dry. Thus, in the example illustrated in Figure 16, the entering air at state 1 initially encounters an interface saturation state 1_i, clearly below its dew-point temperature t_{d1}, so the coil immediately becomes wet. Had the graphical technique resulted in an initial interface state above the dew-point temperature of the entering air, the coil would then be initially dry. The air would then follow a constant humidity ratio line (the sloping W = constant lines on the chart) until the interface state reached the air dew-point temperature.

Mizushina *et al.* (1959) developed this method not only for water vapor and air, but also for other vapor-gas mixtures. Chapter 21 of the 1992 ASHRAE *Handbook—Systems and Equipment*, based on ARI *Standard* 410, shows another related method of determining air cooling and dehumidifying coil performance.

SYMBOLS

a = constant, dimensionless; or surface area per unit volume, ft^2/ft^3

A = surface area, ft^2

A_{cs} = cross-sectional area, ft^2

b = exponent, dimensionless

C = molal concentration of solute in solvent, lb mol/ft^2

c_L = specific heat of liquid, Btu/lb·°F

c_p = specific heat at constant pressure, Btu/lb·°F

c_{pm} = specific heat of moist air at constant pressure, Btu/lb (da)·°F (Note: da = dry air)

d = diameter, ft

D_v = diffusion coefficient (mass diffusivity), ft^2/h

f = Fanning friction factor, dimensionless

Fo = Fourier number = $\alpha\tau/L^2$, dimensionless

Fo_m = mass transfer Fourier number $D_v\tau/L^2$

G = mass flux, flow rate per unit of cross-sectional area, $lb_m/h·ft^2$

h = enthalpy, Btu/lb; or heat transfer coefficient, Btu/h·ft^2·°F

h_{fg} = enthalpy of vaporization, Btu/lb_m

h_m = mass transfer coefficient, ft/h

J = diffusive mass flux, $lb_m/h·ft^2$

J^* = diffusive molar flux, lb mol/h·ft^2

j_D = Colburn mass transfer group = $Sh/(Re·Pr^{1/3})$, dimensionless

j_H = Colburn heat transfer group = $Nu/(Re·Pr^{1/3})$, dimensionless

k = thermal conductivity, Btu/h·ft^2·°F

K_m = mass transfer coefficient, lb/h·ft^2

K_{mo} = overall transfer coefficient, lb/h·ft^2

l = length, ft

L = characteristic length, ft

L/G = liquid-to-air mass flow ratio

Le = Lewis number = a/D_v, dimensionless

$\dot{m}$ = rate of mass transfer, lb/h

$\dot{m}''$ = mass flux, lb/h·ft^2

$\dot{m}''^*$ = molar flux, lb mol/h·ft^2

M = molecular weight, lb/lb mol

N = number of transfer units, dimensionless

N_o = number of overall transfer units, dimensionless

Nu = Nusselt number = hD/k, dimensionless

p = pressure, atmospheres or lb_f/ft^2

P_{AM} = logarithmic mean density factor, Eq. (24)

Pr = Prandtl number = $c_p\mu/k$, dimensionless

$\dot{q}$ = rate of heat transfer, Btu/h

$\dot{q}''$ = heat flux per unit area, Btu/h·ft^2

Re = Reynolds number = $\rho VD/\mu$, dimensionless

R_U = universal gas constant = 1545 lb_f·ft/lb mol·R

R_i = combined thermal resistance, ft^2·°F·h/Btu

R_L = thermal resistance of condensate film, ft^2·°F·h/Btu

R_m = thermal resistance across metal wall and fins, ft^2·°F·h/Btu

R_r = thermal resistance of refrigerant film, ft^2·°F·h/Btu

Sc = Schmidt number = $\mu/\rho D_v$, dimensionless

Sh = Sherwood number = $h_D L/D_v$, dimensionless

t = temperature, °F

T = absolute temperature, °R

u = velocity in x direction, ft/min

U_i = overall conductance from refrigerant to air-water interface for dehumidifying coil, Btu/h·ft^2·°F

v = velocity in y direction, ft/min

v_i = velocity normal to mass transfer surface for component i, ft/h

V = fluid stream velocity, ft/min

W = humidity ratio, lb of water vapor per lb of dry air, lb(w)/lb(da)

x,y,z = coordinate direction, ft

X,Y,Z = coordinate direction, dimensionless

α = thermal diffusivity = $k/\rho c_p$, ft^2/h

ϵ_D = eddy mass diffusivity, ft^2/h

μ = absolute viscosity, $lb_m/ft·h$

$\bar{\mu}$ = permeability, grains/h·ft^2·in. Hg/in.

ν = kinematic viscosity, ft^2/h

σ = characteristic molecular diameter, angstroms

ρ = mass density or concentration, lb_m/ft^3

$\bar{\rho}$ = molal density or concentration, lb mol/ft^3

τ = time

ω = mass fraction, lb/lb

$\Omega_{D,AB}$ = temperature function in Eq. (10)

Subscripts

a = air property

da = dry air property or air-side transfer quantity

Am = logarithmic mean

A = gas component of binary mixture

B = the more dilute gas component of binary mixture

c = critical state

H = heat transfer quantity

i = air-water interface value

L = liquid

m = mean value

M = mass transfer quantity

o = property evaluated at 0°F

s = water vapor property or transport quantity

t = total

w = water vapor

∞ = property of main fluid stream

Superscripts

j^* = on molar basis

$\bar{u}$ = average value

REFERENCES

Bedingfield, G.H., Jr. and T.B. Drew. 1950. Analogy between heat transfer and mass transfer—A psychrometric study. *Industrial and Engineering Chemistry* 42:1164.

Bird, R.B., W.E. Stewart, and E.N. Lightfoot. 1960. *Transport phenomena*. John Wiley & Sons, New York.

Chilton, T.H. and A.P. Colburn. 1934. Mass transfer (absorption) coefficients—Prediction from data on heat transfer and fluid friction. *Industrial and Engineering Chemistry* 26 November:1183.

Eckert, E.R.G. and R.M. Drake, Jr. 1972. *Analysis of heat and mass transfer.* McGraw-Hill Book Company, New York.

Guillory, J.L. and F.C. McQuiston. 1973. An experimental investigation of air dehumidification in a parallel plate heat exchanger. ASHRAE *Transactions* 79(2):146.

Helmer, W.A. 1974. *Condensing water vapor—Airflow in a parallel plate heat exchanger.* Ph.D. Thesis, Purdue University, West Lafayette, IN.

Hirschfelder, J.O., C.F. Curtiss, and R.B. Bird. 1954. *Molecular theory of gases and liquids.* John Wiley & Sons, New York.

Incropera, F.P. and D.P. DeWitt. 1990. *Fundamentals of heat and mass transfer.* John Wiley & Sons, New York, 715-32.

Kusuda, T. 1957. Graphical method simplifies determination of aircoil, wet-heat-transfer surface temperature. *Refrigerating Engineering* 65:41.

Mason, E.A. and L. Monchick. 1965. Survey of the equation of state and transport properties of moist gases. *Humidity and Moisture* 3. Reinhold Publishing Corp., New York.

McAdams, W.H. 1954. *Heat transmission,* 3rd ed. McGraw-Hill Book Company, New York.

Mizushina, T., N. Hashimoto, and M. Nakajima. 1959. Design of cooler condensers for gas-vapour mixtures. *Chemical Engineering Science* 9:195.

Reid, R.C. and T.K. Sherwood. 1966. *The properties of gases and liquids: Their estimation and correlation,* 2nd ed. McGraw-Hill Book Company, New York, 520-43.

Reid, R.C., J.M. Prausnitz, and B.E. Poling. 1987. *The properties of gases and liquids,* 4th ed. McGraw-Hill Book Company, New York. 21-78.

Sherwood, T.K. and R.L. Pigford. 1952. *Absorption and extraction.* McGraw-Hill Book Company, New York, 1-28.

Sparrow, E.M. and M.M. Ohadi. 1987a. Comparison of turbulent thermal entrance regions for pipe flows with developed velocity and velocity developing from a sharp-edged inlet. ASME *Transactions, Journal of Heat Transfer* 109:1028-30.

Sparrow, E.M. and M.M. Ohadi. 1987b. Numerical and experimental studies of turbulent flow in a tube. *Numerical Heat Transfer* 11:461-76.

BIBLIOGRAPHY

Bennett, C.O. and J.E. Myers. 1982. *Momentum, heat and mass transfer,* 3rd ed. McGraw-Hill Book Company, New York.

Slattery, J.C. 1972. *Momentum, energy and mass transfer in continua.* McGraw-Hill Book Company, New York.

PSYCHROMETRICS

PSYCHROMETRICS deals with thermodynamic properties of moist air and uses these properties to analyze conditions and processes involving moist air. Hyland and Wexler (1983a, 1983b) developed formulas for thermodynamic properties of moist air and water. Perfect gas relations can be used in most air-conditioning problems instead of these formulas. Threlkeld (1970) showed that errors are less than 0.7% in calculating humidity ratio, enthalpy, and specific volume of saturated air at standard atmospheric pressure for a temperature range of -60 to $120\,°F$. Furthermore, these errors decrease with decreasing pressure.

This chapter discusses perfect gas relations and describes their use in common air-conditioning problems. The formulas developed by Hyland and Wexler (1983a) may be used where greater precision is required.

COMPOSITION OF DRY AND MOIST AIR

Atmospheric air contains many gaseous components as well as water vapor and miscellaneous contaminants (*e.g.*, smoke, pollen, and gaseous pollutants not normally present in free air far from pollution sources).

Dry air exists when all water vapor and contaminants have been removed from atmospheric air. The composition of dry air is relatively constant, but small variations in the amounts of individual components occur with time, geographic location, and altitude. Harrison (1965) lists the approximate percentage composition of dry air by volume as: nitrogen, 78.084; oxygen, 20.9476; argon, 0.934; carbon dioxide, 0.0314; neon, 0.001818; helium, 0.000524; methane, 0.00015; sulfur dioxide, 0 to 0.0001; hydrogen, 0.00005; and minor components such as krypton, xenon, and ozone, 0.0002. The apparent molecular mass or weighted average molecular weight of all components, for *dry air* is 28.9645, based on the carbon-12 scale (Harrison 1965). The gas constant for dry air, based on the carbon-12 scale, is:

$$R_a = 1545.32/28.9645 = 53.352 \text{ ft} \cdot \text{lb}_f/\text{lb}_m \cdot °R \qquad (1)$$

Moist air is a binary (or two-component) mixture of dry air and water vapor. The amount of water vapor in moist air varies from zero (dry air) to a maximum that depends on temperature and pressure. The latter condition refers to *saturation*, a state of neutral equilibrium between moist air and the condensed water phase (liquid or solid). Unless otherwise stated, saturation refers to a flat interface surface between the moist air and the condensed phase. The molecular weight of water is 18.01528 on the carbon-12 scale. The gas constant for water vapor is:

$$R_w = 1545.32/18.01528 = 85.778 \text{ ft} \cdot \text{lb}_f/\text{lb}_m \cdot °R \qquad (2)$$

UNITED STATES STANDARD ATMOSPHERE

The temperature and barometric pressure of atmospheric air vary considerably with altitude as well as with local geographic and weather conditions. The standard atmosphere gives a standard of reference for estimating properties at various altitudes. At sea level, standard temperature is $59\,°F$; standard barometric pressure is 29.921 in. Hg. The temperature is assumed to decrease linearly with increasing altitude throughout the troposphere (lower atmosphere), and to be constant in the lower reaches of the stratosphere. The lower atmosphere is assumed to consist of dry air that behaves as a perfect gas. Gravity is also assumed constant at the standard value, 32.1740 ft/s^2. Table 1 summarizes property data for altitudes to 60,000 ft.

Table 1 Standard Atmospheric Data for Altitudes to 60,000 ft

Altitude, ft	Temperature, °F	Pressure in. Hg	Pressure psia
-1000	62.6	31.02	15.236
-500	60.8	30.47	14.966
0	59.0	29.921	14.696
500	57.2	29.38	14.430
1000	55.4	28.86	14.175
2000	51.9	27.82	13.664
3000	48.3	26.82	13.173
4000	44.7	25.82	12.682
5000	41.2	24.90	12.230
6000	37.6	23.98	11.778
7000	34.0	23.09	11.341
8000	30.5	22.22	10.914
9000	26.9	21.39	10.506
10,000	23.4	20.58	10.108
15,000	5.5	16.89	8.296
20,000	-12.3	13.76	6.758
30,000	-47.8	8.90	4.371
40,000	-69.7	5.56	2.731
50,000	-69.7	3.44	1.690
60,000	-69.7	2.14	1.051

[a] Data adapted from NASA (1976).

The preparation of this chapter is assigned to TC 1.1, Thermodynamics and Psychrometrics.

Table 2   Thermodynamic Properties of Moist Air, Standard Atmospheric Pressure, 14.696 psi (29.921 in. Hg)

Temp. t, °F	Humidity Ratio, lb_w/lb_{da} W_s	Volume, ft³/lb dry air v_a	v_{as}	v_s	Enthalpy, Btu/lb dry air h_a	h_{as}	h_s	Entropy, Btu/(lb dry air)·°F s_a	s_{as}	s_s	Condensed Water Enthalpy, Btu/lb h_w	Entropy, Btu/lb·°F s_w	Vapor Press., in. Hg p_s	Temp., °F
−80	0.0000049	9.553	0.000	9.553	−19.221	0.005	−19.215	−0.04594	0.00001	−0.04592	−193.45	−0.4067	0.000236	−80
−79	0.0000053	9.579	0.000	9.579	−18.980	0.005	−18.975	−0.04531	0.00002	−0.04529	−193.06	−0.4056	0.000255	−79
−78	0.0000057	9.604	0.000	9.604	−18.740	0.006	−18.734	−0.04468	0.00002	−0.04466	−192.66	−0.4046	0.000275	−78
−77	0.0000062	9.629	0.000	9.629	−18.500	0.007	−18.493	−0.04405	0.00002	−0.04403	−192.27	−0.4036	0.000296	−77
−76	0.0000067	9.655	0.000	9.655	−18.259	0.007	−18.252	−0.04342	0.00002	−0.04340	−191.87	−0.4025	0.000319	−76
−75	0.0000072	9.680	0.000	9.680	−18.019	0.007	−18.011	−0.04279	0.00002	−0.04277	−191.47	−0.4015	0.000344	−75
−74	0.0000078	9.705	0.000	9.705	−17.778	0.008	−17.770	−0.04217	0.00002	−0.04215	−191.07	−0.4005	0.000371	−74
−73	0.0000084	9.731	0.000	9.731	−17.538	0.009	−17.529	−0.04155	0.00002	−0.04152	−190.68	−0.3994	0.000400	−73
−72	0.0000090	9.756	0.000	9.756	−17.298	0.010	−17.288	−0.04093	0.00003	−0.04090	−190.27	−0.3984	0.000430	−72
−71	0.0000097	9.781	0.000	9.782	−17.057	0.010	−17.047	−0.04031	0.00003	−0.04028	−189.87	−0.3974	0.000463	−71
−70	0.0000104	9.807	0.000	9.807	−16.806	0.011	−16.817	−0.03969	0.00003	0.03966	−189.47	−0.3963	0.000498	−70
−69	0.0000112	9.832	0.000	9.832	−16.577	0.012	−16.565	−0.03907	0.00003	−0.03904	−189.07	−0.3953	0.000536	−69
−68	0.0000120	9.857	0.000	9.858	−16.336	0.013	−16.324	−0.03846	0.00003	−0.03843	−188.66	−0.3943	0.000576	−68
−67	0.0000129	9.883	0.000	9.883	−16.096	0.013	−16.083	−0.03785	0.00004	−0.03781	−188.26	−0.3932	0.000619	−67
−66	0.0000139	9.908	0.000	9.908	−15.856	0.015	−15.841	−0.03724	0.00004	−0.03720	−187.85	−0.3922	0.000665	−66
−65	0.0000149	9.933	0.000	9.934	−15.616	0.015	−15.600	−0.03663	0.00004	−0.03659	−187.44	−0.3912	0.000714	−65
−64	0.0000160	9.959	0.000	9.959	−15.375	0.017	−15.359	−0.03602	0.00005	−0.03597	−187.04	−0.3901	0.000766	−64
−63	0.0000172	9.984	0.000	9.984	−15.117	0.018	−15.135	−0.03541	0.00005	−0.03536	−186.63	−0.3891	0.000822	−63
−62	0.0000184	10.009	0.000	10.010	−14.895	0.019	−14.876	−0.03481	0.00005	−0.03476	−186.22	−0.3881	0.000882	−62
−61	0.0000198	10.035	0.000	10.035	−14.654	0.021	−14.634	−0.03420	0.00006	−0.03415	−185.81	−0.3870	0.000945	−61
−60	0.0000212	10.060	0.000	10.060	−14.414	0.022	−14.392	−0.03360	0.00006	−0.03354	−185.39	−0.3860	0.001013	−60
−59	0.0000227	10.085	0.000	10.086	−14.174	0.024	−14.150	−0.03300	0.00006	−0.03294	−184.98	−0.3850	0.001086	−59
−58	0.0000243	10.111	0.000	10.111	−13.933	0.025	−13.908	−0.03240	0.00007	−0.03233	−184.57	−0.3839	0.001163	−58
−57	0.0000260	10.136	0.000	10.137	−13.693	0.027	−13.666	−0.03180	0.00007	−0.03173	−184.15	−0.3829	0.001246	−57
−56	0.0000279	10.161	0.000	10.162	−13.453	0.029	−13.424	−0.03121	0.00008	−0.03113	−183.74	−0.3819	0.001333	−56
−55	0.0000298	10.187	0.000	10.187	−13.213	0.031	−13.182	−0.03061	0.00008	−0.03053	−183.32	−0.3808	0.001427	−55
−54	0.0000319	10.212	0.001	10.213	−12.972	0.033	−12.939	−0.03002	0.00009	−0.02993	−182.90	−0.3798	0.001526	−54
−53	0.0000341	10.237	0.001	10.238	−12.732	0.035	−12.697	−0.02943	0.00009	−0.02934	−182.48	−0.3788	0.001632	−53
−52	0.0000365	10.263	0.001	10.263	−12.492	0.038	−12.454	−0.02884	0.00010	−0.02874	−182.06	−0.3778	0.001745	−52
−51	0.0000390	10.288	0.001	10.289	−12.251	0.041	−12.211	−0.02825	0.00011	−0.02814	−181.64	−0.3767	0.001865	−51
−50	0.0000416	10.313	0.001	10.314	−12.011	0.043	−11.968	−0.02766	0.00011	−0.02755	−181.22	−0.3757	0.001992	−50
−49	0.0000445	10.339	0.001	10.340	−11.771	0.046	−11.725	−0.02708	0.00012	−0.02696	−180.80	−0.3747	0.002128	−49
−48	0.0000475	10.364	0.001	10.365	−11.531	0.050	−11.481	−0.02649	0.00013	−0.02636	−180.37	−0.3736	0.002272	−48
−47	0.0000507	10.389	0.001	10.390	−11.290	0.053	−11.237	−0.02591	0.00014	−0.02577	−179.95	−0.3726	0.002425	−47
−46	0.0000541	10.415	0.001	10.416	−11.050	0.056	−10.994	−0.02533	0.00015	−0.02518	−179.52	−0.3716	0.002587	−46
−45	0.0000577	10.440	0.001	10.441	−10.810	0.060	−10.750	−0.02475	0.00016	−0.02459	−179.10	−0.3705	0.002760	−45
−44	0.0000615	10.465	0.001	10.466	−10.570	0.064	−10.505	−0.02417	0.00017	−0.02400	−178.67	−0.3695	0.002943	−44
−43	0.0000656	10.491	0.001	10.492	−10.329	0.068	−10.261	−0.02359	0.00018	−0.02342	−178.24	−0.3685	0.003137	−43
−42	0.0000699	10.516	0.001	10.517	−10.089	0.073	−10.016	−0.02302	0.00019	−0.02283	−177.81	−0.3675	0.003343	−42
−41	0.0000744	10.541	0.001	10.543	−9.849	0.078	−9.771	−0.02244	0.00020	−0.02224	−177.38	−0.3664	0.003562	−41
−40	0.0000793	10.567	0.001	10.568	−9.609	0.083	−9.526	−0.02187	0.00021	−0.02166	−176.95	−0.3654	0.003793	−40
−39	0.0000844	10.592	0.001	10.593	−9.368	0.088	−9.280	−0.02130	0.00022	−0.02107	−176.52	−0.3644	0.004039	−39
−38	0.0000898	10.617	0.002	10.619	−9.128	0.094	−9.034	−0.02073	0.00024	−0.02049	−176.08	−0.3633	0.004299	−38
−37	0.0000956	10.643	0.002	10.644	−8.888	0.100	−8.788	−0.02016	0.00025	−0.01991	−175.65	−0.3623	0.004575	−37
−36	0.0001017	10.668	0.002	10.670	−8.648	0.106	−8.541	−0.01959	0.00027	−0.01932	−175.21	−0.3613	0.004866	−36
−35	0.0001081	10.693	0.002	10.695	−8.407	0.113	−8.294	−0.01902	0.00028	−0.01874	−174.78	−0.3603	0.005175	−35
−34	0.0001150	10.719	0.002	10.721	−8.167	0.120	−8.047	−0.01846	0.00030	−0.01816	−174.34	−0.3529	0.005502	−34
−33	0.0001222	10.744	0.002	10.746	−7.927	0.128	−7.799	−0.01790	0.00032	−0.01758	−173.90	−0.3582	0.005848	−33
−32	0.0001298	10.769	0.002	10.772	−7.687	0.136	−7.551	−0.01733	0.00034	−0.01699	−173.46	−0.3572	0.006214	−32
−31	0.0001379	10.795	0.002	10.797	−7.447	0.145	−7.302	−0.01677	0.00036	−0.01641	−173.02	−0.3561	0.006601	−31
−30	0.0001465	10.820	0.003	10.822	−7.206	0.154	−7.053	−0.01621	0.00038	−0.01583	−172.58	−0.3551	0.007009	−30
−29	0.0001555	10.845	0.003	10.848	−6.966	0.163	−6.803	−0.01565	0.00040	−0.01525	−172.14	−0.3541	0.007442	−29
−28	0.0001650	10.871	0.003	10.873	−6.726	0.173	−6.553	−0.01510	0.00043	−0.01467	−171.70	−0.3531	0.007898	−28
−27	0.0001751	10.896	0.003	10.899	−6.486	0.184	−6.302	−0.01454	0.00045	−0.01409	−171.25	−0.3520	0.008381	−27
−26	0.0001858	10.921	0.003	10.924	−6.245	0.195	−6.051	−0.01399	0.00048	−0.01351	−170.81	−0.3510	0.008890	−26
−25	0.0001970	10.947	0.003	10.950	−6.005	0.207	−5.798	−0.01343	0.00051	−0.01293	−170.36	−0.3500	0.009428	−25
−24	0.0002088	10.972	0.004	10.976	−5.765	0.220	−5.545	−0.01288	0.00054	−0.01235	−169.92	−0.3489	0.009995	−24
−23	0.0002214	10.997	0.004	11.001	−5.525	0.233	−5.292	−0.01233	0.00057	−0.01176	−169.47	−0.3479	0.010594	−23
−22	0.0002346	11.022	0.004	11.027	−5.284	0.247	−5.038	−0.01178	0.00060	−0.01118	−169.02	−0.3469	0.011226	−22
−21	0.0002485	11.048	0.004	11.052	−5.044	0.261	−4.783	−0.01123	0.00063	−0.01060	−168.57	−0.3459	0.011893	−21
−20	0.0002632	11.073	0.005	11.078	−4.804	0.277	−4.527	−0.01069	0.00067	−0.01002	−168.12	−0.3448	0.012595	−20
−19	0.0002786	11.098	0.005	11.103	−4.564	0.293	−4.271	−0.01014	0.00071	−0.00943	−167.67	−0.3438	0.013336	−19
−18	0.0002950	11.124	0.005	11.129	−4.324	0.311	−4.013	−0.00960	0.00075	−0.00885	−167.21	−0.3428	0.014117	−18
−17	0.0003121	11.149	0.006	11.155	−4.084	0.329	−3.754	−0.00905	0.00079	−0.00826	−166.76	−0.3418	0.014939	−17
−16	0.0003303	11.174	0.006	11.180	−3.843	0.348	−3.495	−0.00851	0.00083	−0.00768	−166.30	−0.3407	0.015806	−16
−15	0.0003493	11.200	0.006	11.206	−3.603	0.368	−3.235	−0.00797	0.00088	−0.00709	−165.85	−0.3397	0.016718	−15
−14	0.0003694	11.225	0.007	11.232	−3.363	0.390	−2.973	−0.00743	0.00093	−0.00650	−165.39	−0.3387	0.017679	−14
−13	0.0003905	11.250	0.007	11.257	−3.123	0.412	−2.710	−0.00689	0.00098	−0.00591	−164.93	−0.3377	0.018690	−13
−12	0.0004128	11.276	0.007	11.283	−2.882	0.436	−2.447	−0.00635	0.00103	−0.00532	−164.47	−0.3366	0.019754	−12
−11	0.0004362	11.301	0.008	11.309	−2.642	0.460	−2.182	−0.00582	0.00109	−0.00473	−164.01	−0.3356	0.020873	−11

Table 2 Thermodynamic Properties of Moist Air, Standard Atmospheric Pressure, 14.696 psi (29.921 in. Hg) (*Continued*)

Temp. t, °F	Humidity Ratio, lb_w/lb_da W_s	Volume, ft³/lb dry air			Enthalpy, Btu/lb dry air			Entropy, Btu/(lb dry air)·°F			Condensed Water			Temp., °F
											Enthalpy, Btu/lb h_w	Entropy, Btu/lb·°F s_w	Vapor Press., in. Hg p_s	
		v_a	v_as	v_s	h_a	h_as	h_s	s_a	s_as	s_s				
-10	0.0004608	11.326	0.008	11.335	-2.402	0.487	-1.915	-0.00528	0.00115	-0.00414	-163.55	-0.3346	0.022050	-10
-9	0.0004867	11.351	0.009	11.360	-2.162	0.514	-1.647	-0.00475	0.00121	-0.00354	-163.09	-0.3335	0.023289	-9
-8	0.0005139	11.377	0.009	11.386	-1.922	0.543	-1.378	-0.00422	0.00127	-0.00294	-162.63	-0.3325	0.024591	-8
-7	0.0005425	11.402	0.010	11.412	-1.681	0.574	-1.108	-0.00369	0.00134	-0.00234	-162.17	-0.3315	0.025959	-7
-6	0.0005726	11.427	0.010	11.438	-1.441	0.606	-0.835	-0.00316	0.00141	-0.00174	-161.70	-0.3305	0.027397	-6
-5	0.0006041	11.453	0.011	11.464	-1.201	0.640	-0.561	-0.00263	0.00149	-0.00114	-161.23	-0.3294	0.028907	-5
-4	0.0006373	11.478	0.012	11.490	-0.961	0.675	-0.286	-0.00210	0.00157	-0.00053	-160.77	-0.3284	0.030494	-4
-3	0.0006722	11.503	0.012	11.516	-0.721	0.712	-0.008	-0.00157	0.00165	0.00008	-160.30	-0.3274	0.032160	-3
-2	0.0007088	11.529	0.013	11.542	-0.480	0.751	0.271	-0.00105	0.00174	0.00069	-159.83	-0.3264	0.033909	-2
-1	0.0007472	11.554	0.014	11.568	-0.240	0.792	0.552	-0.00052	0.00183	0.00130	-159.36	-0.3253	0.035744	-1
0	0.0007875	11.579	0.015	11.594	0.0	0.835	0.835	0.00000	0.00192	0.00192	-158.89	-0.3243	0.037671	0
1	0.0008298	11.604	0.015	11.620	0.240	0.880	1.121	0.00052	0.00202	0.00254	-158.42	-0.3233	0.039694	1
2	0.0008742	11.630	0.016	11.646	0.480	0.928	1.408	0.00104	0.00212	0.00317	-157.95	-0.3223	0.041814	2
3	0.0009207	11.655	0.017	11.672	0.721	0.978	1.699	0.00156	0.00223	0.00380	-157.47	-0.3212	0.044037	3
4	0.0009695	11.680	0.018	11.699	0.961	1.030	1.991	0.00208	0.00235	0.00443	-157.00	-0.3202	0.046370	4
5	0.0010207	11.706	0.019	11.725	1.201	1.085	2.286	0.00260	0.00247	0.00506	-156.52	-0.3192	0.048814	5
6	0.0010743	11.731	0.020	11.751	1.441	1.143	2.584	0.00311	0.00259	0.00570	-156.05	-0.3182	0.051375	6
7	0.0011306	11.756	0.021	11.778	1.681	1.203	2.884	0.00363	0.00635	0.00272	-155.57	-0.3171	0.054060	7
8	0.0011895	11.782	0.022	11.804	1.922	1.266	3.188	0.00414	0.00286	0.00700	-155.09	-0.3161	0.056872	8
9	0.0012512	11.807	0.024	11.831	2.162	1.332	3.494	0.00466	0.00300	0.00766	-154.61	-0.3151	0.059819	9
10	0.0013158	11.832	0.025	11.857	2.402	1.402	3.804	0.00517	0.00315	0.00832	-154.13	-0.3141	0.062901	10
11	0.0013835	11.857	0.026	11.884	2.642	1.474	4.117	0.00568	0.00330	0.00898	-153.65	-0.3130	0.066131	11
12	0.0014544	11.883	0.028	11.910	2.882	1.550	4.433	0.00619	0.00347	0.00966	-153.17	-0.3120	0.069511	12
13	0.0015286	11.908	0.029	11.937	3.123	1.630	4.753	0.00670	0.00364	0.01033	-152.68	-0.3110	0.073049	13
14	0.0016062	11.933	0.031	11.964	3.363	1.714	5.077	0.00721	0.00381	0.01102	-152.20	-0.3100	0.076751	14
15	0.0016874	11.959	0.032	11.991	3.603	1.801	5.404	0.00771	0.00400	0.01171	-151.71	-0.3089	0.080623	15
16	0.0017724	11.984	0.034	12.018	3.843	1.892	5.736	0.00822	0.00419	0.01241	-151.22	-0.3079	0.084673	16
17	0.0018613	12.009	0.036	12.045	4.084	1.988	6.072	0.00872	0.00439	0.01312	-150.74	-0.3069	0.088907	17
18	0.0019543	12.035	0.038	12.072	4.324	2.088	6.412	0.00923	0.00460	0.01383	-150.25	-0.3059	0.093334	18
19	0.0020515	12.060	0.040	12.099	4.564	2.193	6.757	0.00973	0.00482	0.01455	-149.76	-0.3049	0.097962	19
20	0.0021531	12.085	0.042	12.127	4.804	2.303	7.107	0.01023	0.00505	0.01528	-149.27	-0.3038	0.102798	20
21	0.0022592	12.110	0.044	12.154	5.044	2.417	7.462	0.01073	0.00529	0.01602	-148.78	-0.3028	0.107849	21
22	0.0023703	12.136	0.046	12.182	5.285	2.537	7.822	0.01123	0.00554	0.01677	-148.28	-0.3018	0.113130	22
23	0.0024863	12.161	0.048	12.209	5.525	2.662	8.187	0.01173	0.00580	0.01753	-147.79	-0.3008	0.118645	23
24	0.0026073	12.186	0.051	12.237	5.765	2.793	8.558	0.01223	0.00607	0.01830	-147.30	-0.2997	0.124396	24
25	0.0027339	12.212	0.054	12.265	6.005	2.930	8.935	0.01272	0.00636	0.01908	-146.80	-0.2987	0.130413	25
26	0.0028660	12.237	0.056	12.293	6.246	3.073	9.318	0.01322	0.00665	0.01987	-146.30	-0.2977	0.136684	26
27	0.0030039	12.262	0.059	12.321	6.486	3.222	9.708	0.01371	0.00696	0.02067	-145.81	-0.2967	0.143233	27
28	0.0031480	12.287	0.062	12.349	6.726	3.378	10.104	0.01420	0.00728	0.02148	-145.31	-0.2956	0.150066	28
29	0.0032984	12.313	0.065	12.378	6.966	3.541	10.507	0.01470	0.00761	0.02231	-144.81	-0.2946	0.157198	29
30	0.0034552	12.338	0.068	12.406	7.206	3.711	10.917	0.01519	0.00796	0.02315	-144.31	-0.2936	0.164631	30
31	0.0036190	12.363	0.072	12.435	7.447	3.888	11.335	0.01568	0.00832	0.02400	-143.80	-0.2926	0.172390	31
32	0.0037895	12.389	0.075	12.464	7.687	4.073	11.760	0.01617	0.00870	0.02487	-143.30	-0.2915	0.180479	32
32*	0.003790	12.389	0.075	12.464	7.687	4.073	11.760	0.01617	0.00870	0.02487	0.02	0.0000	0.18050	32
33	0.003947	12.414	0.079	12.492	7.927	4.243	12.170	0.01665	0.00905	0.02570	1.03	0.0020	0.18791	33
34	0.004109	12.439	0.082	12.521	8.167	4.420	12.587	0.01714	0.00940	0.02655	2.04	0.0041	0.19559	34
35	0.004277	12.464	0.085	12.550	8.408	4.603	13.010	0.01763	0.00977	0.02740	3.05	0.0061	0.20356	35
36	0.004452	12.490	0.089	12.579	8.648	4.793	13.441	0.01811	0.01016	0.02827	4.05	0.0081	0.21181	36
37	0.004633	12.515	0.093	12.608	8.888	4.990	13.878	0.01860	0.01055	0.02915	5.06	0.0102	0.22035	37
38	0.004820	12.540	0.097	12.637	9.128	5.194	14.322	0.01908	0.01096	0.03004	6.06	0.0122	0.22920	38
39	0.005014	12.566	0.101	12.667	9.369	5.405	14.773	0.01956	0.01139	0.03095	7.07	0.0142	0.23835	39
40	0.005216	12.591	0.105	12.696	9.609	5.624	15.233	0.02004	0.01183	0.03187	8.07	0.0162	0.24784	40
41	0.005424	12.616	0.110	12.726	9.849	5.851	15.700	0.02052	0.01228	0.03281	9.08	0.0182	0.25765	41
42	0.005640	12.641	0.114	12.756	10.089	6.086	16.175	0.02100	0.01275	0.03375	10.08	0.0202	0.26781	42
43	0.005863	12.667	0.119	12.786	10.330	6.330	16.660	0.02148	0.01324	0.03472	11.09	0.0222	0.27831	43
44	0.006094	12.692	0.124	12.816	10.570	6.582	17.152	0.02196	0.01374	0.03570	12.09	0.0242	0.28918	44
45	0.006334	12.717	0.129	12.846	10.810	6.843	17.653	0.02244	0.01426	0.03669	13.09	0.0262	0.30042	45
46	0.006581	12.743	0.134	12.877	11.050	7.114	18.164	0.02291	0.01479	0.03770	14.10	0.0282	0.31206	46
47	0.006838	12.768	0.140	12.908	11.291	7.394	18.685	0.02339	0.01534	0.03873	15.10	0.0302	0.32408	47
48	0.007103	12.793	0.146	12.939	11.531	7.684	19.215	0.02386	0.01592	0.03978	16.10	0.0321	0.33651	48
49	0.007378	12.818	0.152	12.970	11.771	7.984	19.756	0.02433	0.01651	0.04084	17.10	0.0341	0.34937	49
50	0.007661	12.844	0.158	13.001	12.012	8.295	20.306	0.02480	0.01712	0.04192	18.11	0.0361	0.36264	50
51	0.007955	12.869	0.164	13.033	12.252	8.616	20.868	0.02528	0.01775	0.04302	19.11	0.0381	0.37636	51
52	0.008259	12.894	0.171	13.065	12.492	8.949	21.441	0.02575	0.01840	0.04415	20.11	0.0400	0.39054	52
53	0.008573	12.920	0.178	13.097	12.732	9.293	22.025	0.02622	0.01907	0.04529	21.11	0.0420	0.40518	53
54	0.008897	12.945	0.185	13.129	12.973	9.648	22.621	0.02668	0.01976	0.04645	22.11	0.0439	0.42030	54
55	0.009233	12.970	0.192	13.162	13.213	10.016	23.229	0.02715	0.02048	0.04763	23.11	0.0459	0.43592	55
56	0.009580	12.995	0.200	13.195	13.453	10.397	23.850	0.02762	0.02122	0.04884	24.11	0.0478	0.45205	56
57	0.009938	13.021	0.207	13.228	13.694	10.790	24.484	0.02808	0.02198	0.05006	25.11	0.0497	0.46870	57
58	0.010309	13.046	0.216	13.262	13.934	11.197	25.131	0.02855	0.02277	0.05132	26.11	0.0517	0.48589	58

*Extrapolated to represent metastable equilibrium with undercooled liquid.

Table 2   Thermodynamic Properties of Moist Air, Standard Atmospheric Pressure, 14.696 psi (29.921 in. Hg) (Continued)

Temp. t, °F	Humidity Ratio, lb_w/lb_{da} W_s	Volume, ft³/lb dry air v_a	v_{as}	v_s	Enthalpy, Btu/lb dry air h_a	h_{as}	h_s	Entropy, Btu/(lb dry air)·°F s_a	s_{as}	s_s	Condensed Water Enthalpy, Btu/lb h_w	Entropy, Btu/lb·°F s_w	Vapor Press., in. Hg p_s	Temp., °F
59	0.010692	13.071	0.224	13.295	14.174	11.618	25.792	0.02901	0.02358	0.05259	27.11	0.0536	0.50363	59
60	0.011087	13.096	0.233	13.329	14.415	12.052	26.467	0.02947	0.02442	0.05389	28.11	0.0555	0.52193	60
61	0.011496	13.122	0.242	13.364	14.655	12.502	27.157	0.02994	0.02528	0.05522	29.12	0.0575	0.54082	61
62	0.011919	13.147	0.251	13.398	14.895	12.966	27.862	0.03040	0.02617	0.05657	30.11	0.0594	0.56032	62
63	0.012355	13.172	0.261	13.433	15.135	13.446	28.582	0.03086	0.02709	0.05795	31.11	0.0613	0.58041	63
64	0.012805	13.198	0.271	13.468	15.376	13.942	29.318	0.03132	0.02804	0.05936	32.11	0.0632	0.60113	64
65	0.013270	13.223	0.281	13.504	15.616	14.454	30.071	0.03178	0.02902	0.06080	33.11	0.0651	0.62252	65
66	0.013750	13.248	0.292	13.540	15.856	14.983	30.840	0.03223	0.03003	0.06226	34.11	0.0670	0.64454	66
67	0.014246	13.273	0.303	13.577	16.097	15.530	31.626	0.03269	0.03107	0.06376	35.11	0.0689	0.66725	67
68	0.014758	13.299	0.315	13.613	16.337	16.094	32.431	0.03315	0.03214	0.06529	36.11	0.0708	0.69065	68
69	0.015286	13.324	0.326	13.650	16.577	16.677	33.254	0.03360	0.03325	0.06685	37.11	0.0727	0.71479	69
70	0.015832	13.349	0.339	13.688	16.818	17.279	34.097	0.03406	0.03438	0.06844	38.11	0.0746	0.73966	70
71	0.016395	13.375	0.351	13.726	17.058	17.901	34.959	0.03451	0.03556	0.07007	39.11	0.0765	0.76567	71
72	0.016976	13.400	0.365	13.764	17.299	18.543	35.841	0.03496	0.03677	0.07173	40.11	0.0783	0.79167	72
73	0.017575	13.425	0.378	13.803	17.539	19.204	36.743	0.03541	0.03801	0.07343	41.11	0.0802	0.81882	73
74	0.018194	13.450	0.392	13.843	17.779	19.889	37.668	0.03586	0.03930	0.07516	42.11	0.0821	0.84684	74
75	0.018833	13.476	0.407	13.882	18.020	20.595	38.615	0.03631	0.04062	0.07694	43.11	0.0840	0.87567	75
76	0.019491	13.501	0.422	13.923	18.260	21.323	39.583	0.03676	0.04199	0.07875	44.10	0.0858	0.90533	76
77	0.020170	13.526	0.437	13.963	18.500	22.075	40.576	0.03721	0.04339	0.08060	45.10	0.0877	0.93589	77
78	0.020871	13.551	0.453	14.005	18.741	22.851	41.592	0.03766	0.04484	0.08250	46.10	0.0896	0.96733	78
79	0.021594	13.577	0.470	14.046	18.981	23.652	42.633	0.03811	0.04633	0.08444	47.10	0.0914	0.99970	79
80	0.022340	13.602	0.487	14.089	19.222	24.479	43.701	0.03855	0.04787	0.08642	48.10	0.0933	1.03302	80
81	0.023109	13.627	0.505	14.132	19.462	25.332	44.794	0.03900	0.04945	0.08844	49.10	0.0951	1.06728	81
82	0.023902	13.653	0.523	14.175	19.702	26.211	45.913	0.03944	0.05108	0.09052	50.10	0.0970	1.10252	82
83	0.024720	13.678	0.542	14.220	19.943	27.120	47.062	0.03988	0.05276	0.09264	51.09	0.0988	1.13882	83
84	0.025563	13.703	0.561	14.264	20.183	28.055	48.238	0.04033	0.05448	0.09481	52.09	0.1006	1.17608	84
85	0.026433	13.728	0.581	14.310	20.424	29.021	49.445	0.04077	0.05626	0.09703	53.09	0.1025	1.21445	85
86	0.027329	13.754	0.602	14.356	20.664	30.017	50.681	0.04121	0.05809	0.09930	54.09	0.1043	1.25388	86
87	0.028254	13.779	0.624	14.403	20.905	31.045	51.949	0.04165	0.05998	0.10163	55.09	0.1061	1.29443	87
88	0.029208	13.804	0.646	14.450	21.145	32.105	53.250	0.04209	0.06192	0.10401	56.09	0.1080	1.33613	88
89	0.030189	13.829	0.669	14.498	21.385	33.197	54.582	0.04253	0.06392	0.10645	57.09	0.1098	1.37893	89
90	0.031203	13.855	0.692	14.547	21.626	34.325	55.951	0.04297	0.06598	0.10895	58.08	0.1116	1.42298	90
91	0.032247	13.880	0.717	14.597	21.866	35.489	57.355	0.06810	0.04340	0.11150	59.08	0.1134	1.46824	91
92	0.033323	13.905	0.742	14.647	22.107	36.687	58.794	0.04384	0.07028	0.11412	60.08	0.1152	1.51471	92
93	0.034433	13.930	0.768	14.699	22.347	37.924	60.271	0.04427	0.07253	0.11680	61.08	0.1170	1.56248	93
94	0.035577	13.956	0.795	14.751	22.588	39.199	61.787	0.04471	0.07484	0.11955	62.08	0.1188	1.61154	94
95	0.036757	13.981	0.823	14.804	22.828	40.515	63.343	0.04514	0.07722	0.12237	63.08	0.1206	1.66196	95
96	0.037972	14.006	0.852	14.858	23.069	41.871	64.940	0.04558	0.07968	0.12525	64.07	0.1224	1.71372	96
97	0.039225	14.032	0.881	14.913	23.309	43.269	66.578	0.04601	0.08220	0.12821	65.07	0.1242	1.76685	97
98	0.040516	14.057	0.912	14.969	23.550	44.711	68.260	0.04644	0.08480	0.13124	66.07	0.1260	1.82141	98
99	0.041848	14.082	0.944	15.026	23.790	46.198	69.988	0.04687	0.08747	0.13434	67.07	0.1278	1.87745	99
100	0.043219	14.107	0.976	15.084	24.031	47.730	71.761	0.04730	0.09022	0.13752	68.07	0.1296	1.93492	100
101	0.044634	14.133	1.010	15.143	24.271	49.312	73.583	0.04773	0.09306	0.14079	69.07	0.1314	1.99396	101
102	0.046090	14.158	1.045	15.203	24.512	50.940	75.452	0.04816	0.09597	0.14413	70.06	0.1332	2.05447	102
103	0.047592	14.183	1.081	15.264	24.752	52.621	77.373	0.04859	0.09897	0.14756	71.06	0.1349	2.11661	103
104	0.049140	14.208	1.118	15.326	24.993	54.354	79.346	0.04901	0.10206	0.15108	72.06	0.1367	2.18037	104
105	0.050737	14.234	1.156	15.390	25.233	56.142	81.375	0.04944	0.10525	0.15469	73.06	0.1385	2.24581	105
106	0.052383	14.259	1.196	15.455	25.474	57.986	83.460	0.04987	0.10852	0.15839	74.06	0.1402	2.31297	106
107	0.054077	14.284	1.236	15.521	25.714	59.884	85.599	0.05029	0.11189	0.16218	75.06	0.1420	2.38173	107
108	0.055826	14.309	1.279	15.588	25.955	61.844	87.799	0.05071	0.11537	0.16608	76.05	0.1438	2.45232	108
109	0.057628	14.335	1.322	15.657	26.195	63.866	90.061	0.05114	0.11894	0.17008	77.05	0.1455	2.52473	109
110	0.059486	14.360	1.367	15.727	26.436	65.950	92.386	0.05156	0.12262	0.17418	78.05	0.1473	2.59891	110
111	0.061401	14.385	1.414	15.799	26.677	68.099	94.776	0.05198	0.12641	0.17839	79.05	0.1490	2.67500	111
112	0.063378	14.411	1.462	15.872	26.917	70.319	97.237	0.05240	0.13032	0.18272	80.05	0.1508	2.75310	112
113	0.065411	14.436	1.511	15.947	27.158	72.603	99.760	0.05282	0.13434	0.18716	81.05	0.1525	2.83291	113
114	0.067512	14.461	1.562	16.023	27.398	74.964	102.362	0.05324	0.13847	0.19172	82.04	0.1543	2.91491	114
115	0.069676	14.486	1.615	16.101	27.639	77.396	105.035	0.05366	0.14274	0.19640	83.04	0.1560	2.99883	115
116	0.071908	14.512	1.670	16.181	27.879	79.906	107.786	0.05408	0.14713	0.20121	84.04	0.1577	3.08488	116
117	0.074211	14.537	1.726	16.263	28.120	82.497	110.617	0.05450	0.15165	0.20615	85.04	0.1595	3.17305	117
118	0.076586	14.562	1.784	16.346	28.361	85.169	113.530	0.05492	0.15631	0.21122	86.04	0.1612	3.26335	118
119	0.079036	14.587	1.844	16.432	28.601	87.927	116.528	0.05533	0.16111	0.21644	87.04	0.1629	3.35586	119
120	0.081560	14.613	1.906	16.519	28.842	90.770	119.612	0.05575	0.16605	0.22180	88.04	0.1647	3.45052	120
121	0.084169	14.638	1.971	16.609	29.083	93.709	122.792	0.05616	0.17115	0.22731	89.04	0.1664	3.54764	121
122	0.086860	14.663	2.037	16.700	29.323	96.742	126.065	0.05658	0.17640	0.23298	90.03	0.1681	3.64704	122
123	0.089633	14.688	2.106	16.794	29.564	99.868	129.432	0.05699	0.18181	0.23880	91.03	0.1698	3.74871	123
124	0.092500	14.714	2.176	16.890	29.805	103.102	132.907	0.05740	0.18739	0.24480	92.03	0.1715	3.85298	124
125	0.095456	14.739	2.250	16.989	30.045	106.437	136.482	0.05781	0.19314	0.25096	93.03	0.1732	3.95961	125
126	0.098504	14.764	2.325	17.090	30.286	109.877	140.163	0.05823	0.19907	0.25729	94.03	0.1749	4.06863	126
127	0.101657	14.789	2.404	17.193	30.527	113.438	143.965	0.05864	0.20519	0.26382	95.03	0.1766	4.18046	127
128	0.104910	14.815	2.485	17.299	30.767	117.111	147.878	0.05905	0.21149	0.27054	96.03	0.1783	4.29477	128
129	0.108270	14.840	2.569	17.409	31.008	120.908	151.916	0.21800	0.21810	0.27745	97.03	0.1800	4.41181	129

Table 2   Thermodynamic Properties of Moist Air, Standard Atmospheric Pressure, 14.696 psi (29.921 in. Hg) (Concluded)

Temp., t, °F	Humidity Ratio, lb_w/lb_da, W_s	Volume, ft³/lb dry air v_a	v_{as}	v_s	Enthalpy, Btu/lb dry air h_a	h_{as}	h_s	Entropy, Btu/(lb dry air)·°F s_a	s_{as}	s_s	Condensed Water Enthalpy, Btu/lb h_w	Entropy, Btu/lb·°F s_w	Vapor Press., in. Hg p_s	Temp., °F
130	0.111738	14.865	2.655	17.520	31.249	124.828	156.076	0.05986	0.22470	0.28457	98.03	0.1817	4.53148	130
131	0.115322	14.891	2.745	17.635	31.489	128.880	160.370	0.06027	0.23162	0.29190	99.02	0.1834	4.65397	131
132	0.119023	14.916	2.837	17.753	31.730	133.066	164.796	0.06068	0.23876	0.29944	100.02	0.1851	4.77919	132
133	0.122855	14.941	2.934	17.875	31.971	137.403	169.374	0.06109	0.24615	0.30723	101.02	0.1868	4.90755	133
134	0.126804	14.966	3.033	17.999	32.212	141.873	174.084	0.06149	0.25375	0.31524	102.02	0.1885	5.03844	134
135	0.130895	14.992	3.136	18.127	32.452	146.504	178.957	0.06190	0.26161	0.32351	103.02	0.1902	5.17258	135
136	0.135124	15.017	3.242	18.259	32.693	151.294	183.987	0.06230	0.26973	0.33203	104.02	0.1919	5.30973	136
137	0.139494	15.042	3.352	18.394	32.934	156.245	189.179	0.06271	0.27811	0.34082	105.02	0.1935	5.44985	137
138	0.144019	15.067	3.467	18.534	33.175	161.374	194.548	0.06311	0.28707	0.35018	106.02	0.1952	5.59324	138
139	0.148696	15.093	3.585	18.678	33.415	166.677	200.092	0.06351	0.29602	0.35954	107.02	0.1969	5.73970	139
140	0.153538	15.118	3.708	18.825	33.656	172.168	205.824	0.06391	0.30498	0.36890	108.02	0.1985	5.88945	140
141	0.158643	15.143	3.835	18.978	33.897	177.857	211.754	0.06431	0.31456	0.37887	109.02	0.2002	6.04256	141
142	0.163748	15.168	3.967	19.135	34.138	183.754	217.892	0.06471	0.32446	0.38918	110.02	0.2019	6.19918	142
143	0.169122	15.194	4.103	19.297	34.379	189.855	244.233	0.06511	0.33470	0.39981	111.02	0.2035	6.35898	143
144	0.174694	15.219	4.245	19.464	34.620	196.183	230.802	0.06551	0.34530	0.41081	112.02	0.2052	6.52241	144
145	0.180467	15.244	4.392	19.637	34.860	202.740	237.600	0.06591	0.35626	0.42218	113.02	0.2068	6.68932	145
146	0.186460	15.269	4.545	19.815	35.101	209.550	244.651	0.06631	0.36764	0.43395	114.02	0.2085	6.86009	146
147	0.192668	15.295	4.704	19.999	35.342	216.607	251.949	0.06671	0.37941	0.44611	115.02	0.2101	7.03435	147
148	0.199110	15.320	4.869	20.189	35.583	223.932	259.514	0.06710	0.39160	0.45871	116.02	0.2118	7.21239	148
149	0.205792	15.345	5.040	20.385	35.824	231.533	267.356	0.06750	0.40424	0.47174	117.02	0.2134	7.39413	149
150	0.212730	15.370	5.218	20.589	36.064	239.426	275.490	0.06790	0.41735	0.48524	118.02	0.2151	7.57977	150
151	0.219945	15.396	5.404	20.799	36.305	247.638	283.943	0.06829	0.43096	0.49925	119.02	0.2167	7.76958	151
152	0.227429	15.421	5.596	21.017	36.546	256.158	292.705	0.06868	0.44507	0.51375	120.02	0.2184	7.96306	152
153	0.235218	15.446	5.797	21.243	36.787	265.028	301.816	0.06908	0.45973	0.52881	121.02	0.2200	8.16087	153
154	0.243309	15.471	6.005	21.477	37.028	274.245	311.273	0.06947	0.47494	0.54441	122.02	0.2216	8.36256	154
155	0.251738	15.497	6.223	21.720	37.269	283.849	321.118	0.06986	0.49077	0.56064	123.02	0.2233	8.56871	155
156	0.260512	15.522	6.450	21.972	37.510	293.849	331.359	0.07025	0.50723	0.57749	124.02	0.2249	8.77915	156
157	0.269644	15.547	6.686	22.233	37.751	304.261	342.012	0.07065	0.52434	0.59499	125.02	0.2265	8.99378	157
158	0.279166	15.572	6.933	22.505	37.992	315.120	353.112	0.07104	0.54217	0.61320	126.02	0.2281	9.21297	158
159	0.289101	15.598	7.190	22.788	38.233	326.452	364.685	0.07143	0.56074	0.63216	127.02	0.2297	9.43677	159
160	0.29945	15.623	7.459	23.082	38.474	338.263	376.737	0.07181	0.58007	0.65188	128.02	0.2314	9.6648	160
161	0.31027	15.648	7.740	23.388	38.715	350.610	389.325	0.07220	0.60025	0.67245	129.02	0.2330	9.8978	161
162	0.32156	15.673	8.034	23.707	38.956	363.501	402.457	0.07259	0.62128	0.69388	130.03	0.2346	10.1353	162
163	0.33336	15.699	8.341	24.040	39.197	376.979	416.175	0.07298	0.64325	0.71623	131.03	0.2362	10.3776	163
164	0.34572	15.724	8.664	24.388	39.438	391.095	430.533	0.07337	0.66622	0.73959	132.03	0.2378	10.6250	164
165	0.35865	15.749	9.001	24.750	39.679	405.865	445.544	0.07375	0.69022	0.76397	133.03	0.2394	10.8771	165
166	0.37220	15.774	9.355	25.129	39.920	421.352	461.271	0.07414	0.71535	0.78949	134.03	0.2410	11.1343	166
167	0.38639	15.800	9.726	25.526	40.161	437.578	477.739	0.07452	0.74165	0.81617	135.03	0.2426	11.3965	167
168	0.40131	15.825	10.117	25.942	40.402	454.630	495.032	0.07491	0.76925	0.84415	136.03	0.2442	11.6641	168
169	0.41698	15.850	10.527	26.377	40.643	472.554	513.197	0.07529	0.79821	0.87350	137.04	0.2458	11.9370	169
170	0.43343	15.875	10.959	26.834	40.884	491.372	532.256	0.07567	0.82858	0.90425	138.04	0.2474	12.2149	170
171	0.45079	15.901	11.414	27.315	41.125	511.231	552.356	0.07606	0.86058	0.93664	139.04	0.2490	12.4988	171
172	0.46905	15.926	11.894	27.820	41.366	532.138	573.504	0.07644	0.89423	0.97067	140.04	0.2506	12.7880	172
173	0.48829	15.951	12.400	28.352	41.607	554.160	595.767	0.07682	0.92962	1.00644	141.04	0.2521	13.0823	173
174	0.50867	15.976	12.937	28.913	41.848	577.489	619.337	0.07720	0.96707	1.04427	142.04	0.2537	13.3831	174
175	0.53019	16.002	13.504	29.505	42.089	602.139	644.229	0.07758	1.00657	1.08416	143.05	0.2553	13.6894	175
176	0.55294	16.027	14.103	30.130	42.331	628.197	670.528	0.07796	1.04828	1.12624	144.05	0.2569	14.0010	176
177	0.57710	16.052	14.741	30.793	42.572	655.876	698.448	0.07834	1.09253	1.17087	145.05	0.2585	14.3191	177
178	0.60274	16.078	15.418	31.496	42.813	685.260	728.073	0.07872	1.13943	1.21815	146.05	0.2600	14.6430	178
179	0.63002	16.103	16.139	32.242	43.054	716.524	759.579	0.07910	1.18927	1.26837	147.06	0.2616	14.9731	179
180	0.65911	16.128	16.909	33.037	43.295	749.871	793.166	0.07947	1.24236	1.32183	148.06	0.2632	15.3097	180
181	0.69012	16.153	17.730	33.883	43.536	785.426	828.962	0.07985	1.29888	1.37873	149.06	0.2647	15.6522	181
182	0.72331	16.178	18.609	34.787	43.778	823.487	867.265	0.08023	1.35932	1.43954	150.06	0.2663	16.0014	182
183	0.75885	16.204	19.551	35.755	44.019	864.259	908.278	0.08060	1.42396	1.50457	151.07	0.2679	16.3569	183
184	0.79703	16.229	20.564	36.793	44.260	908.061	952.321	0.08098	1.49332	1.57430	152.07	0.2694	16.7190	184
185	0.83817	16.254	21.656	37.910	44.501	955.261	999.763	0.08135	1.56797	1.64932	153.07	0.2710	17.0880	185
186	0.88251	16.280	22.834	39.113	44.742	1006.149	1050.892	0.08172	1.64834	1.73006	154.08	0.2725	17.4634	186
187	0.93057	16.305	24.111	40.416	44.984	1061.314	1106.298	0.08210	1.73534	1.81744	155.08	0.2741	17.8462	187
188	0.98272	16.330	25.498	41.828	45.225	1121.174	1166.399	0.08247	1.82963	1.91210	156.08	0.2756	18.2357	188
189	1.03951	16.355	27.010	43.365	45.466	1186.382	1231.848	0.08284	1.93221	2.01505	157.09	0.2772	18.6323	189
190	1.10154	16.381	28.661	45.042	45.707	1257.614	1303.321	0.08321	2.04412	2.12733	158.09	0.2787	19.0358	190
191	1.16965	16.406	30.476	46.882	45.949	1335.834	1381.783	0.08359	2.16684	2.25043	159.09	0.2803	19.4465	191
192	1.24471	16.431	32.477	48.908	46.190	1422.047	1468.238	0.08396	2.30193	2.38589	160.10	0.2818	19.8652	192
193	1.32788	16.456	34.695	51.151	46.431	1517.581	1564.013	0.08433	2.45144	2.53576	161.10	0.2834	20.2913	193
194	1.42029	16.481	37.161	53.642	46.673	1623.758	1670.430	0.08470	2.61738	2.70208	162.11	0.2849	20.7244	194
195	1.52396	16.507	39.928	56.435	46.914	1742.879	1789.793	0.08506	2.80332	2.88838	163.11	0.2864	21.1661	195
196	1.64070	16.532	43.046	59.578	47.155	1877.032	1924.188	0.08543	3.01244	3.09787	164.12	0.2880	21.6152	196
197	1.77299	16.557	46.580	63.137	47.397	2029.069	2076.466	0.08580	3.24914	3.33494	165.12	0.2895	22.0714	197
198	1.92472	16.583	50.636	67.218	47.638	2203.464	2251.102	0.08617	3.52030	3.60647	166.13	0.2910	22.5367	198
199	2.09975	16.608	55.316	71.923	47.879	2404.668	2452.547	0.08653	3.83275	3.91929	167.13	0.2926	23.0092	199
200	2.30454	16.633	60.793	77.426	48.121	2640.084	2688.205	0.08690	4.19787	4.28477	168.13	0.2941	23.4906	200

Table 3 Thermodynamic Properties of Water at Saturation

Temp. t, °F	Absolute Pressure p psi	Absolute Pressure p in. Hg	Sat. Solid v_i	Evap. v_{ig}	Sat. Vapor v_g	Sat. Solid h_i	Evap. h_{ig}	Sat. Vapor h_g	Sat. Solid s_i	Evap. s_{ig}	Sat. Vapor s_g	Temp., °F
−80	0.000116	0.000236	0.01732	1953234	1953234	−193.50	1219.19	1025.69	−0.4067	3.2112	2.8045	−80
−79	0.000125	0.000254	0.01732	1814052	1814052	−193.11	1219.24	1026.13	−0.4056	3.2029	2.7972	−79
−78	0.000135	0.000275	0.01732	1685445	1685445	−192.71	1219.28	1026.57	−0.4046	3.1946	2.7900	−78
−77	0.000145	0.000296	0.01732	1566663	1566663	−192.31	1219.33	1027.02	−0.4036	3.1964	2.7828	−77
−76	0.000157	0.000319	0.01732	1456752	1456752	−191.92	1219.38	1027.46	−0.4025	3.1782	2.7757	−76
−75	0.000169	0.000344	0.01733	1355059	1355059	−191.52	1219.42	1027.90	−0.4015	3.1701	2.7685	−75
−74	0.000182	0.000371	0.01733	1260977	1260977	−191.12	1219.47	1028.34	−0.4005	3.1619	2.7615	−74
−73	0.000196	0.000399	0.01733	1173848	1173848	−190.72	1219.51	1028.79	−0.3994	3.1539	2.7544	−73
−72	0.000211	0.000430	0.01733	1093149	1093149	−190.32	1219.55	1029.23	−0.3984	3.1459	2.7475	−72
−71	0.000227	0.000463	0.01733	1018381	1018381	−189.92	1219.59	1029.67	−0.3974	3.1379	2.7405	−71
−70	0.000245	0.000498	0.01733	949067	949067	−189.52	1219.63	1030.11	−0.3963	3.1299	2.7336	−70
−69	0.000263	0.000536	0.01733	884803	884803	−189.11	1219.67	1030.55	−0.3953	3.1220	2.7267	−69
−68	0.000283	0.000576	0.01733	825187	825187	−188.71	1219.71	1031.00	−0.3943	3.1141	2.7199	−68
−67	0.000304	0.000619	0.01734	769864	769864	−188.30	1219.74	1031.44	−0.3932	3.1063	2.7131	−67
−66	0.000326	0.000664	0.01734	718508	718508	−187.90	1219.78	1031.88	−0.3922	3.0985	2.7063	−66
−65	0.000350	0.000714	0.01734	670800	670800	−187.49	1219.82	1032.32	−0.3912	3.0907	2.6996	−65
−64	0.000376	0.000766	0.01734	626503	626503	−187.08	1219.85	1032.77	−0.3901	3.0830	2.6929	−64
−63	0.000404	0.000822	0.01734	585316	585316	−186.67	1219.88	1033.21	−0.3891	3.0753	2.6862	−63
−62	0.000433	0.000882	0.01734	548041	547041	−186.26	1219.91	1033.65	−0.3881	3.0677	2.6730	−62
−61	0.000464	0.000945	0.01734	511446	511446	−185.85	1219.95	1034.09	−0.3870	3.0601	2.6730	−61
−60	0.000498	0.001013	0.01734	478317	478317	−185.44	1219.98	1034.54	−0.3860	3.0525	2.6665	−60
−59	0.000533	0.001086	0.01735	447495	447495	−185.03	1220.01	1034.98	−0.3850	3.0449	2.6600	−59
−58	0.000571	0.001163	0.01735	418803	418803	−184.61	1220.03	1035.42	−0.3839	3.0374	2.6535	−58
−57	0.000612	0.001246	0.01735	392068	392068	−184.20	1220.06	1035.86	−0.3829	3.0299	2.6470	−57
−56	0.000655	0.001333	0.01735	367172	367172	−183.78	1220.09	1036.30	−0.3819	3.0225	2.6406	−56
−55	0.000701	0.001427	0.01735	343970	343970	−183.37	1220.11	1036.75	−0.3808	3.0151	2.6342	−55
−54	0.000750	0.001526	0.01735	322336	322336	−182.95	1220.14	1037.19	−0.3798	3.0077	2.6279	−54
−53	0.000802	0.001632	0.01735	302157	302157	−182.53	1220.16	1037.63	−0.3788	3.0004	2.6216	−53
−52	0.000857	0.001745	0.01735	283335	283335	−182.11	1220.18	1038.07	−0.3778	2.9931	2.6153	−52
−51	0.000916	0.001865	0.01736	265773	265773	−181.69	1220.21	1038.52	−0.3767	2.9858	2.6091	−51
−50	0.000979	0.001992	0.01736	249381	249381	−181.27	1220.23	1038.96	−0.3757	2.9786	2.6029	−50
−49	0.001045	0.002128	0.01736	234067	234067	−180.85	1220.25	1039.40	−0.3747	2.9714	2.5967	−49
−48	0.001116	0.002272	0.01736	219766	219766	−180.42	1220.26	1039.84	−0.3736	2.9642	2.5906	−48
−47	0.001191	0.002425	0.01736	206398	206398	−180.00	1220.28	1040.28	−0.3726	2.9570	2.5844	−47
−46	0.001271	0.002587	0.01736	193909	193909	−179.57	1220.30	1040.73	−0.3716	2.9499	2.5784	−46
−45	0.001355	0.002760	0.01736	182231	182231	−179.14	1220.31	1041.17	−0.3705	2.9429	2.5723	−45
−44	0.001445	0.002943	0.01736	171304	171304	−178.72	1220.33	1041.61	−0.3695	2.9358	2.5663	−44
−43	0.001541	0.003137	0.01737	161084	161084	−178.79	1220.34	1042.05	−0.3685	2.9288	2.5603	−43
−42	0.001642	0.003343	0.01737	151518	151518	−177.86	1220.36	1042.50	−0.3675	2.9218	2.5544	−42
−41	0.001749	0.003562	0.01737	142566	142566	−177.43	1220.37	1042.94	−0.3664	2.9149	2.5485	−41
−40	0.001863	0.003793	0.01737	134176	134176	−177.00	1220.38	1043.38	−0.3654	2.9080	2.5426	−40
−39	0.001984	0.004039	0.01737	126322	126322	−176.57	1220.39	1043.82	−0.3644	2.9011	2.5367	−39
−38	0.002111	0.004299	0.01737	118959	118959	−176.13	1220.40	1044.27	−0.3633	2.8942	2.5309	−38
−37	0.002247	0.004574	0.01737	112058	112058	−175.70	1220.40	1044.71	−0.3623	2.8874	2.5251	−37
−36	0.002390	0.004866	0.01738	105592	105592	−175.26	1220.41	1045.15	−0.3613	2.8806	2.5193	−36
−35	0.002542	0.005175	0.01738	99522	99522	−174.83	1220.42	1045.59	−0.3603	2.8738	2.5136	−35
−34	0.002702	0.005502	0.01738	93828	93828	−174.39	1220.42	1046.03	−0.3592	2.8671	2.5078	−34
−33	0.002872	0.005848	0.01738	88489	88489	−173.95	1220.43	1046.48	−0.3582	2.8604	2.5022	−33
−32	0.003052	0.006213	0.01738	83474	83474	−173.51	1220.43	1046.92	−0.3572	2.8537	2.4965	−32
−31	0.003242	0.006600	0.01738	78763	78763	−173.07	1220.43	1047.36	−0.3561	2.8470	2.4909	−31
−30	0.003443	0.007009	0.01738	74341	74341	−172.63	1220.43	1047.80	−0.3551	2.8404	2.4853	−30
−29	0.003655	0.007441	0.01738	70187	70187	−172.19	1220.43	1048.25	−0.3541	2.8338	2.4797	−29
−28	0.003879	0.007898	0.01739	66282	66282	−171.74	1220.43	1048.69	−0.3531	2.8272	2.4742	−28
−27	0.004116	0.008380	0.01739	62613	62613	−171.30	1220.43	1049.13	−0.3520	2.8207	2.4687	−27
−26	0.004366	0.008890	0.01739	59161	59161	−170.86	1220.43	1049.57	−0.3510	2.8142	2.4632	−26
−25	0.004630	0.009428	0.01739	55915	55915	−170.41	1220.42	1050.01	−0.3500	2.8077	2.4577	−25
−24	0.004909	0.009995	0.01739	52861	52861	−169.96	1220.42	1050.46	−0.3489	2.8013	2.4523	−24
−23	0.005203	0.010594	0.01739	49986	49986	−169.51	1220.41	1050.90	−0.3479	2.7948	2.4469	−23
−22	0.005514	0.011226	0.01739	47281	47281	−169.07	1220.41	1051.34	−0.3469	2.7884	2.4415	−22
−21	0.005841	0.011892	0.01740	44733	44733	−168.62	1220.40	1051.78	−0.3459	2.7820	2.4362	−21
−20	0.006186	0.012595	0.01740	42333	42333	−168.16	1220.39	1052.22	−0.3448	2.7757	2.4309	−20
−19	0.006550	0.013336	0.01740	40073	40073	−167.71	1220.38	1052.67	−0.3438	2.7694	2.4256	−19
−18	0.006933	0.014117	0.01740	37943	37943	−167.26	1220.37	1053.11	−0.3428	2.7631	2.4203	−18
−17	0.007337	0.014939	0.01740	35934	35934	−166.81	1220.36	1053.55	−0.3418	2.7568	2.4151	−17
−16	0.007763	0.015806	0.01740	34041	34041	−166.35	1220.34	1053.99	−0.3407	2.7506	2.4098	−16
−15	0.008211	0.016718	0.01740	32256	32256	−165.90	1220.33	1054.43	−0.3397	2.7444	2.4046	−15
−14	0.008683	0.017678	0.01741	30572	30572	−165.44	1220.31	1054.87	−0.3387	2.7382	2.3995	−14

Table 3   Thermodynamic Properties of Water at Saturation (*Continued*)

Temp. t, °F	Absolute Pressure p		Specific Volume, ft³/lb			Enthalpy, Btu/lb			Entropy, Btu/lb·°F			Temp., °F
	psi	in. Hg	Sat. Solid/Liq. v_i	Evap. v_{ig}	Sat. Vapor v_g	Sat. Solid/Liq. h_i	Evap. h_{ig}	Sat. Vapor h_g	Sat. Solid/Liq. s_i	Evap. s_{ig}	Sat. Vapor s_g	
−13	0.009179	0.018689	0.01741	28983	28983	−164.98	1220.30	1055.32	−0.3377	2.7320	2.3943	−13
−12	0.009702	0.019753	0.01741	27483	27483	−164.52	1220.28	1055.76	−0.3366	2.7259	2.3892	−12
−11	0.010252	0.020873	0.01741	26067	26067	−164.06	1220.26	1056.20	−0.3356	2.7197	2.3841	−11
−10	0.010830	0.022050	0.01741	24730	24730	−163.60	1220.24	1056.64	−0.3346	2.7136	2.3791	−10
−9	0.011438	0.023288	0.01741	23467	23467	−163.14	1220.22	1057.08	−0.3335	2.7076	2.3740	−9
−8	0.012077	0.024590	0.01741	22274	22274	−162.68	1220.20	1057.53	−0.3325	2.7015	2.3690	−8
−7	0.012749	0.025958	0.01742	21147	21147	−162.21	1220.18	1057.97	−0.3315	2.6955	2.3640	−7
−6	0.013456	0.027396	0.01742	20081	20081	−162.75	1220.16	1058.41	−0.3305	2.6895	2.3591	−6
−5	0.014197	0.028906	0.01742	19074	19074	−161.28	1220.13	1058.85	−0.3294	2.6836	2.3541	−5
−4	0.014977	0.030493	0.01742	18121	18121	−160.82	1220.11	1059.29	−0.3284	2.6776	2.3492	−4
−3	0.015795	0.032159	0.01742	17220	17220	−160.35	1220.08	1059.73	−0.3274	2.6717	2.3443	−3
−2	0.016654	0.033908	0.01742	16367	16367	−159.88	1220.05	1060.17	−0.3264	2.6658	2.3394	−2
−1	0.017556	0.035744	0.01742	15561	15561	−159.41	1220.02	1060.62	−0.3253	2.6599	2.3346	−1
0	0.018502	0.037671	0.01743	14797	14797	−158.94	1220.00	1061.06	−0.3243	2.6541	2.3298	0
1	0.019495	0.039693	0.01743	14073	14073	−158.47	1219.96	1061.50	−0.3233	2.6482	2.3249	1
2	0.020537	0.041813	0.01743	13388	13388	−157.99	1219.93	1061.94	−0.3223	2.6424	2.3202	2
3	0.021629	0.044037	0.01743	12740	12740	−157.52	1219.90	1062.38	−0.3212	2.6367	2.3154	3
4	0.022774	0.046369	0.01743	12125	12125	−157.05	1219.87	1062.82	−0.3202	2.6309	2.3107	4
5	0.023975	0.048813	0.01743	11543	11543	−156.57	1219.83	1063.26	−0.3192	2.6252	2.3060	5
6	0.025233	0.051375	0.01743	10991	10991	−156.09	1219.80	1063.70	−0.3182	2.6194	2.3013	6
7	0.026552	0.054059	0.01744	10468	10468	−155.62	1219.76	1064.14	−0.3171	2.6138	2.2966	7
8	0.027933	0.056872	0.01744	9971	9971	−155.14	1219.72	1064.58	−0.3161	2.6081	2.2920	8
9	0.029379	0.059817	0.01744	9500	9500	−154.66	1219.68	1065.03	−0.3151	2.6024	2.2873	9
10	0.030894	0.062901	0.01744	9054	9054	−154.18	1219.64	1065.47	−0.3141	2.5968	2.2827	10
11	0.032480	0.066131	0.01744	8630	8630	−153.70	1219.60	1065.91	−0.3130	2.5912	2.2782	11
12	0.034140	0.069511	0.01744	8228	8228	−153.21	1219.56	1066.35	−0.3120	2.5856	2.2736	12
13	0.035878	0.073047	0.01745	7846	7846	−152.73	1219.52	1066.79	−0.3110	2.5801	2.2691	13
14	0.037696	0.076748	0.01745	7483	7483	−152.24	1219.47	1067.23	−0.3100	2.5745	2.2645	14
15	0.039597	0.080621	0.01745	7139	7139	−151.76	1219.43	1067.67	−0.3089	2.5690	2.2600	15
16	0.041586	0.084671	0.01745	6811	6811	−151.27	1219.38	1068.11	−0.3079	2.5635	2.2556	16
17	0.043666	0.088905	0.01745	6501	6501	−150.78	1219.33	1068.55	−0.3069	2.5580	2.2511	17
18	0.045841	0.093332	0.01745	6205	6205	−150.30	1219.28	1068.99	−0.3059	2.5526	2.2467	18
19	0.048113	0.097960	0.01745	5924	5924	−149.81	1219.23	1069.43	−0.3049	2.5471	2.2423	19
20	0.050489	0.102796	0.01746	5657	5657	−149.32	1219.18	1069.87	−0.3038	2.5417	2.2379	20
21	0.052970	0.107849	0.01746	5404	5404	−148.82	1219.13	1070.31	−0.3028	2.5363	2.2335	21
22	0.055563	0.113128	0.01746	5162	5162	−148.33	1219.08	1070.75	−0.3018	2.5309	2.2292	22
23	0.058271	0.118641	0.01746	4932	4932	−147.84	1219.02	1071.19	−0.3008	2.5256	2.2248	23
24	0.061099	0.124398	0.01746	4714	4714	−147.34	1218.97	1071.63	−0.2997	2.5203	2.2205	24
25	0.064051	0.130408	0.01746	4506	4506	−146.85	1218.91	1072.07	−0.2987	2.5149	2.2162	25
26	0.067133	0.136684	0.01747	4308	4308	−146.35	1218.85	1072.50	−0.2977	2.5096	2.2119	26
27	0.070349	0.143233	0.01747	4119	4119	−145.85	1218.80	1072.94	−0.2967	2.5044	2.2077	27
28	0.073706	0.150066	0.01747	3940	3940	−145.35	1218.74	1073.38	−0.2956	2.4991	2.2035	28
29	0.077207	0.157195	0.01747	3769	3769	−144.85	1218.68	1073.82	−0.2946	2.4939	2.1992	29
30	0.080860	0.164632	0.01747	3606	3606	−144.35	1218.61	1074.26	−0.2936	2.4886	2.1951	30
31	0.084669	0.172387	0.01747	3450	3450	−143.85	1218.55	1074.70	−0.2926	2.4834	2.1909	31
32	0.088640	0.180474	0.01747	3302	3302	−143.35	1218.49	1075.14	−0.2915	2.4783	2.1867	32
32*	0.08865	0.18049	0.01602	3302.07	3302.09	−0.02	1075.14	1075.14	0.0000	2.1867	2.1867	32
33	0.09229	0.18791	0.01602	3178.15	3178.16	0.99	1074.59	1075.58	0.0020	2.1811	2.1832	33
34	0.09607	0.19559	0.01602	3059.47	3059.49	2.00	1074.02	1076.01	0.0041	2.1756	2.1796	34
35	0.09998	0.20355	0.01602	2945.66	2945.68	3.00	1073.45	1076.45	0.0061	2.1700	2.1761	35
36	0.10403	0.21180	0.01602	2836.60	2836.61	4.01	1072.88	1076.89	0.0081	2.1645	2.1726	36
37	0.10822	0.22035	0.01602	2732.12	2732.15	5.02	1072.32	1077.33	0.0102	2.1590	2.1692	37
38	0.11257	0.22919	0.01602	2631.88	2631.89	6.02	1071.75	1077.77	0.0122	2.1535	2.1657	38
39	0.11707	0.23835	0.01602	2535.86	2535.88	7.03	1071.18	1078.21	0.0142	2.1481	2.1623	39
40	0.12172	0.24783	0.01602	2443.67	2443.69	8.03	1070.62	1078.65	0.0162	2.1426	2.1589	40
41	0.12654	0.25765	0.01602	2355.22	2355.24	9.04	1070.05	1079.09	0.0182	2.1372	2.1554	41
42	0.13153	0.26780	0.01602	2270.42	2270.43	10.04	1069.48	1079.52	0.0202	2.1318	2.1521	42
43	0.13669	0.27831	0.01602	2189.02	2189.04	11.04	1068.92	1079.96	0.0222	2.1265	2.1487	43
44	0.14203	0.28918	0.01602	2110.92	2110.94	12.05	1068.35	1080.40	0.0242	2.1211	2.1454	44
45	0.14755	0.30042	0.01602	2035.91	2035.92	13.05	1067.79	1080.84	0.0262	2.1158	2.1420	45
46	0.15326	0.31205	0.01602	1963.85	1963.87	14.05	1067.22	1081.28	0.0282	2.1105	2.1387	46
47	0.15917	0.32407	0.01602	1894.71	1894.73	15.06	1066.66	1081.71	0.0302	2.1052	2.1354	47
48	0.16527	0.33650	0.01602	1828.28	1828.30	16.06	1066.09	1082.15	0.0321	2.1000	2.1321	48
49	0.17158	0.34935	0.01602	1764.44	1764.46	17.06	1065.53	1082.59	0.0341	2.0947	2.1288	49
50	0.17811	0.36263	0.01602	1703.18	1703.20	18.06	1064.96	1083.03	0.0361	2.0895	2.1256	50
51	0.18484	0.37635	0.01602	1644.25	1644.26	19.06	1064.40	1083.46	0.0381	2.0843	2.1224	51
52	0.19181	0.39053	0.01603	1587.64	1587.65	20.07	1063.83	1083.90	0.0400	2.0791	2.1191	52

*Extrapolated to represent metastable equilibrium with undercooled liquid.

Table 3   Thermodynamic Properties of Water at Saturation (Continued)

Temp. t, °F	Absolute Pressure p		Specific Volume, ft³/lb			Enthalpy, Btu/lb			Entropy, Btu/lb·°F			Temp., °F
	psi	in. Hg	Sat. Liquid v_f	Evap. v_{fg}	Sat. Vapor v_g	Sat. Liquid h_f	Evap. h_{fg}	Sat. Vapor h_g	Sat. Liquid s_f	Evap. s_{fg}	Sat. Vapor s_g	
53	0.19900	0.40516	0.01603	1533.22	1533.24	21.07	1063.27	1084.34	0.0420	2.0740	2.1159	53
54	0.20643	0.42029	0.01603	1480.89	1480.91	22.07	1062.71	1084.77	0.0439	2.0689	2.1128	54
55	0.21410	0.43591	0.01603	1430.61	1430.62	23.07	1062.14	1085.21	0.0459	2.0637	2.1096	55
56	0.22202	0.45204	0.01603	1382.19	1382.21	24.07	1061.58	1085.65	0.0478	2.0586	2.1064	56
57	0.23020	0.46869	0.01603	1335.65	1335.67	25.07	1061.01	1086.08	0.0497	2.0536	2.1033	57
58	0.23864	0.48588	0.01603	1290.85	1290.87	26.07	1060.45	1086.52	0.0517	2.0485	2.0002	58
59	0.24735	0.50362	0.01603	1247.76	1247.78	27.07	1059.89	1086.96	0.0536	2.0435	2.0971	59
60	0.25635	0.52192	0.01604	1206.30	1206.32	28.07	1059.32	1087.39	0.0555	2.0385	2.0940	60
61	0.26562	0.54081	0.01604	1166.38	1166.40	29.07	1058.76	1087.83	0.0575	2.0334	2.0909	61
62	0.27519	0.56029	0.01604	1127.93	1127.95	30.07	1058.19	1088.27	0.0594	2.0285	2.0878	62
63	0.28506	0.58039	0.01604	1090.94	1090.96	31.07	1057.63	1088.70	0.0613	2.0235	2.0848	63
64	0.29524	0.60112	0.01604	1055.32	1055.33	32.07	1057.07	1089.14	0.0632	2.0186	2.0818	64
65	0.30574	0.62249	0.01604	1020.98	1021.00	33.07	1056.50	1089.57	0.0651	2.0136	2.0787	65
66	0.31656	0.64452	0.01604	987.95	987.97	34.07	1055.94	1090.01	0.0670	2.0087	2.0758	66
67	0.32772	0.66724	0.01605	956.11	956.12	35.07	1055.37	1090.44	0.0689	2.0039	2.0728	67
68	0.33921	0.69065	0.01605	925.44	925.45	36.07	1054.81	1090.88	0.0708	1.9990	2.0698	68
69	0.35107	0.71478	0.01605	895.86	895.87	37.07	1054.24	1091.31	0.0727	1.9941	2.0668	69
70	0.36328	0.73964	0.01605	867.34	867.36	38.07	1053.68	1091.75	0.0746	1.9893	2.0639	70
71	0.37586	0.76526	0.01605	839.87	839.88	39.07	1053.11	1092.18	0.0765	1.9845	2.0610	71
72	0.38882	0.79164	0.01606	813.37	813.39	40.07	1052.55	1092.61	0.0783	1.9797	2.0580	72
73	0.40217	0.81883	0.01606	787.85	787.87	41.07	1051.98	1093.05	0.0802	1.9749	2.0552	73
74	0.41592	0.84682	0.01606	763.19	763.21	42.06	1051.42	1093.48	0.0821	1.9702	2.0523	74
75	0.43008	0.87564	0.01606	739.42	739.44	43.06	1050.85	1093.92	0.0840	1.9654	2.0494	75
76	0.44465	0.90532	0.01606	716.51	726.53	44.06	1050.29	1094.35	0.0858	1.9607	2.0465	76
77	0.45966	0.93587	0.01607	694.38	794.40	45.06	1049.72	1094.78	0.0877	1.9560	2.0437	77
78	0.47510	0.96732	0.01607	673.05	673.06	46.06	1049.16	1095.22	0.0896	1.9513	2.0409	78
79	0.49100	0.99968	0.01607	652.44	652.46	47.06	1048.59	1095.65	0.0914	1.9466	2.0380	79
80	0.50736	1.03298	0.01607	632.54	632.56	48.06	1048.03	1096.08	0.0933	1.9420	2.0352	80
81	0.52419	1.06725	0.01608	613.35	613.37	49.06	1047.46	1096.51	0.0951	1.9373	2.0324	81
82	0.54150	1.10250	0.01608	594.82	594.84	50.05	1046.89	1096.95	0.0970	1.9327	2.0297	82
83	0.55931	1.13877	0.01608	576.90	576.92	51.05	1046.33	1097.38	0.0988	1.9281	2.0269	83
84	0.57763	1.17606	0.01608	559.63	559.65	52.05	1045.76	1097.81	0.1006	1.9235	2.0242	84
85	0.59647	1.21442	0.01609	542.93	542.94	53.05	1045.19	1098.24	0.1025	1.9189	2.0214	85
86	0.61584	1.25385	0.01609	526.80	526.81	54.05	1044.63	1098.67	0.1043	1.9144	2.0187	86
87	0.63575	1.29440	0.01609	511.21	511.22	55.05	1044.06	1099.11	0.1061	1.9098	2.0160	87
88	0.65622	1.33608	0.01609	496.14	496.15	56.05	1043.49	1099.54	0.1080	1.9053	2.0133	88
89	0.67726	1.37892	0.01610	481.60	481.61	57.04	1042.92	1099.97	0.1098	1.9008	2.0106	89
90	0.69889	1.42295	0.01610	467.52	467.53	58.04	1042.36	1100.40	0.1116	1.8963	2.0079	90
91	0.72111	1.46820	0.01610	453.91	453.93	59.04	1041.79	1100.83	0.1134	1.8918	2.0053	91
92	0.74394	1.51468	0.01611	440.76	440.78	60.04	1041.22	1101.26	0.1152	1.8874	2.0026	92
93	0.76740	1.56244	0.01611	428.04	428.06	61.04	1040.65	1101.69	0.1170	1.8829	2.0000	93
94	0.79150	1.61151	0.01611	415.74	415.76	62.04	1040.08	1102.12	0.1188	1.8785	1.9973	94
95	0.81625	1.66189	0.01612	403.84	403.86	63.03	1039.51	1102.55	0.1206	1.8741	1.9947	95
96	0.84166	1.71364	0.01612	392.33	392.34	64.03	1038.95	1102.98	0.1224	1.8697	1.9921	96
97	0.86776	1.76678	0.01612	381.20	381.21	65.03	1038.38	1103.41	0.1242	1.8653	1.9895	97
98	0.89456	1.82134	0.01612	370.42	370.44	66.03	1037.81	1103.84	0.1260	1.8610	1.9870	98
99	0.92207	1.87736	0.01613	359.99	360.01	67.03	1037.24	1104.26	0.1278	1.8566	1.9844	99
100	0.95031	1.93485	0.01613	349.91	349.92	68.03	1036.67	1104.69	0.1296	1.8523	1.9819	100
101	0.97930	1.99387	0.01613	340.14	340.15	69.03	1036.10	1105.12	0.1314	1.8479	1.9793	101
102	1.00904	2.05443	0.01614	330.69	330.71	70.02	1035.53	1105.55	0.1332	1.8436	1.9768	102
103	1.03956	2.11667	0.01614	321.53	321.55	71.02	1034.95	1105.98	0.1349	1.8393	1.9743	103
104	1.07088	2.18034	0.01614	312.67	312.69	72.02	1034.38	1106.40	0.1367	1.8351	1.9718	104
105	1.10301	2.24575	0.01615	304.08	304.10	73.02	1033.81	1106.83	0.1385	1.8308	1.9693	105
106	1.13597	2.31285	0.01615	295.76	295.77	74.02	1033.24	1107.26	0.1402	1.8266	1.9668	106
107	1.16977	2.38168	0.01616	287.71	287.73	75.01	1032.67	1107.68	0.1420	1.8223	1.9643	107
108	1.20444	2.45226	0.01616	279.91	279.92	76.01	1032.10	1108.11	0.1438	1.8181	1.9619	108
109	1.23999	2.52464	0.01616	272.34	272.36	77.01	1031.52	1108.54	0.1455	1.8139	1.9594	109
110	1.27644	2.59885	0.01617	265.02	265.03	78.01	1030.95	1108.96	0.1473	1.8097	1.9570	110
111	1.31381	2.67494	0.01617	257.91	257.93	79.01	1030.38	1109.39	0.1490	1.8055	1.9546	111
112	1.35212	2.75293	0.01617	251.02	251.04	80.01	1029.80	1109.81	0.1508	1.8014	1.9521	112
113	1.39138	2.83288	0.01618	244.36	244.38	81.01	1029.23	1110.24	0.1525	1.7972	1.9497	113
114	1.43162	2.91481	0.01618	237.89	237.90	82.00	1028.66	1110.66	0.1543	1.7931	1.9474	114
115	1.47286	2.99878	0.01619	231.62	231.63	83.00	1028.08	1111.09	0.1560	1.7890	1.9450	115
116	1.51512	3.08481	0.01619	225.53	225.55	84.00	1027.51	1111.51	0.1577	1.7849	1.9426	116
117	1.55842	3.17296	0.01619	219.63	219.65	85.00	1026.93	1111.93	0.1595	1.7808	1.9402	117
118	1.60277	3.26327	0.01620	213.91	213.93	86.00	1026.36	1112.36	0.1612	1.7767	1.9379	118
119	1.64820	3.35577	0.01620	208.36	208.37	87.00	1025.78	1112.78	0.1629	1.7726	1.9356	119
120	1.69474	3.45052	0.01620	202.98	202.99	88.00	1025.20	1113.20	0.1647	1.7686	1.9332	120

Table 3 Thermodynamic Properties of Water at Saturation (*Continued*)

Temp. t, °F	Absolute Pressure p		Specific Volume, ft³/lb			Enthalpy, Btu/lb			Entropy, Btu/lb·°F			Temp., °F
	psi	in. Hg	Sat. Liquid v_f	Evap. v_{fg}	Sat. Vapor v_g	Sat. Liquid h_f	Evap. h_{fg}	Sat. Vapor h_g	Sat. Liquid s_f	Evap. s_{fg}	Sat. Vapor s_g	
121	1.74240	3.54755	0.01621	197.76	197.76	89.00	1023.62	1113.62	0.1664	1.7645	1.9309	121
122	1.79117	3.64691	0.01621	192.69	192.69	90.00	1024.05	1114.05	0.1681	1.7605	1.9286	122
123	1.84117	3.74863	0.01622	187.78	187.78	90.99	1024.47	1114.47	0.1698	1.7565	1.9263	123
124	1.89233	3.85282	0.01622	182.98	182.99	91.99	1022.90	1114.89	0.1715	1.7525	1.9240	124
125	1.94470	3.95945	0.01623	178.34	178.36	92.99	1022.32	1115.31	0.1732	1.7485	1.9217	125
126	1.99831	4.06860	0.01623	173.85	173.86	93.99	1021.74	1115.73	0.1749	1.7445	1.9195	126
127	2.05318	4.18032	0.01623	169.47	169.49	94.99	1021.16	1116.15	0.1766	1.7406	1.9172	127
128	2.10934	4.29465	0.01624	165.23	165.25	95.99	1020.58	1116.57	0.1783	1.7366	1.9150	128
129	2.16680	4.41165	0.01624	161.11	161.12	96.99	1020.00	1116.99	0.1800	1.7327	1.9127	129
130	2.22560	4.53136	0.01625	157.11	157.12	97.99	1019.42	1117.41	0.1817	1.7288	1.9105	130
131	2.28576	4.65384	0.01625	153.22	153.23	98.99	1018.84	1117.83	0.1834	1.7249	1.9083	131
132	2.34730	4.77914	0.01626	149.44	149.46	99.99	1018.26	1118.25	0.1851	1.7210	1.9061	132
133	2.41025	4.90730	0.01626	145.77	145.78	100.99	1017.68	1118.67	0.1868	1.7171	1.9039	133
134	2.47463	5.03839	0.01627	142.21	142.23	101.99	1017.10	1119.08	0.1885	1.7132	1.9017	134
135	2.54048	5.17246	0.01627	138.74	138.76	102.99	1016.52	1119.50	0.1902	1.7093	1.8995	135
136	2.60782	5.30956	0.01627	135.37	135.39	103.98	1015.93	1119.92	0.1919	1.7055	1.8974	136
137	2.67667	5.44975	0.01628	132.10	132.12	104.98	1015.35	1120.34	0.1935	1.7017	1.8952	137
138	2.74707	5.59308	0.01628	128.92	128.94	105.98	1014.77	1120.75	0.1952	1.6978	1.8930	138
139	2.81903	5.73961	0.01629	125.83	125.85	106.98	1014.18	1121.17	0.1969	1.6940	1.8909	139
140	2.89260	5.88939	0.01629	122.82	122.84	107.98	1013.60	1121.58	0.1985	1.6902	1.8888	140
141	2.96780	6.04250	0.01630	119.90	119.92	108.98	1013.01	1122.00	0.2002	1.6864	1.8867	141
142	3.04465	6.19897	0.01630	117.05	117.07	109.98	1012.43	1122.41	0.2019	1.6827	1.8845	142
143	3.12320	6.35888	0.01631	114.29	114.31	110.98	1011.84	1122.83	0.2035	1.6789	1.8824	143
144	3.20345	6.52229	0.01631	111.60	111.62	111.98	1011.26	1123.24	0.2052	1.6752	1.8803	144
145	3.28546	6.68926	0.01632	108.99	109.00	112.98	1010.67	1123.66	0.2068	1.6714	1.8783	145
146	3.36924	6.85984	0.01632	106.44	106.45	113.98	1010.09	1124.07	0.2085	1.6677	1.8762	146
147	3.45483	7.03410	0.01633	103.96	103.98	114.98	1009.50	1124.48	0.2101	1.6640	1.8741	147
148	3.54226	7.21211	0.01633	101.55	101.57	115.98	1008.91	1124.89	0.2118	1.6603	1.8721	148
149	3.63156	7.39393	0.01634	99.21	99.22	116.98	1008.32	1125.31	0.2134	1.6566	1.8700	149
150	3.72277	7.57962	0.01634	96.93	96.94	117.98	1007.73	1125.72	0.2151	1.6529	1.8680	150
151	3.81591	7.76925	0.01635	94.70	94.72	118.99	1007.14	1126.13	0.2167	1.6492	1.8659	151
152	3.91101	7.96289	0.01635	92.54	92.56	119.99	1006.55	1126.54	0.2184	1.6455	1.8639	152
153	4.00812	8.16061	0.01636	90.44	90.46	120.99	1005.96	1126.95	0.2200	1.6419	1.8619	153
154	4.10727	8.36247	0.01636	88.39	88.41	121.99	1005.37	1127.36	0.2216	1.6383	1.8599	154
155	4.20848	8.56854	0.01637	86.40	86.41	122.99	1004.78	1127.77	0.2233	1.6346	1.8579	155
156	4.31180	8.77890	0.01637	84.45	84.47	123.99	1004.19	1128.18	0.2249	1.6310	1.8559	156
157	4.41725	8.99360	0.01638	82.56	82.58	124.99	1003.60	1128.59	0.2265	1.6274	1.8539	157
158	4.52488	9.21274	0.01638	80.72	80.73	125.99	1003.00	1128.99	0.2281	1.6238	1.8519	158
159	4.63472	9.43637	0.01639	78.92	78.94	126.99	1002.41	1129.40	0.2297	1.6202	1.8500	159
160	4.7468	9.6646	0.01639	77.175	77.192	127.99	1001.82	1129.81	0.2314	1.6167	1.8480	160
161	4.8612	9.8974	0.01640	75.471	75.488	128.99	1001.22	1130.22	0.2330	1.6131	1.8461	161
162	4.9778	10.1350	0.01640	73.812	73.829	130.00	1000.63	1130.62	0.2346	1.6095	1.8441	162
163	5.0969	10.3774	0.01641	72.196	72.213	131.00	1000.03	1131.03	0.2362	1.6060	1.8422	163
164	5.2183	10.6246	0.01642	70.619	70.636	132.00	999.43	1131.43	0.2378	1.6025	1.8403	164
165	5.3422	10.8768	0.01642	69.084	69.101	133.00	998.84	1131.84	0.2394	1.5989	1.8383	165
166	5.4685	11.1340	0.01643	67.587	67.604	134.00	998.24	1132.24	0.2410	1.5954	1.8364	166
167	5.5974	11.3963	0.01643	66.130	66.146	135.00	997.64	1132.64	0.2426	1.5919	1.8345	167
168	5.7287	11.6638	0.01644	64.707	64.723	136.01	997.04	1133.05	0.2442	1.5884	1.8326	168
169	5.8627	11.9366	0.01644	63.320	63.336	137.01	996.44	1133.45	0.2458	1.5850	1.8308	169
170	5.9993	12.2148	0.01645	61.969	61.986	138.01	995.84	1133.85	0.2474	1.5815	1.8289	170
171	6.1386	12.4983	0.01646	60.649	60.666	139.01	995.24	1134.25	0.2490	1.5780	1.8270	171
172	6.2806	12.7874	0.01646	59.363	59.380	140.01	994.64	1134.66	0.2506	1.5746	1.8251	172
173	6.4253	13.0821	0.01647	58.112	58.128	141.02	994.04	1135.06	0.2521	1.5711	1.8233	173
174	6.5729	13.3825	0.01647	56.887	56.904	142.02	993.44	1135.46	0.2537	1.5677	1.8214	174
175	6.7232	13.6886	0.01648	55.694	55.711	143.02	992.83	1135.86	0.2553	1.5643	1.8196	175
176	6.8765	14.0006	0.01648	54.532	54.549	144.02	992.23	1136.26	0.2569	1.5609	1.8178	176
177	7.0327	14.3186	0.01649	53.397	53.414	145.03	991.63	1136.65	0.2585	1.5575	1.8159	177
178	7.1918	14.6426	0.01650	52.290	52.307	146.03	991.02	1137.05	0.2600	1.5541	1.8141	178
179	7.3539	14.9727	0.01650	51.210	51.226	147.03	990.42	1137.45	0.2616	1.5507	1.8123	179
180	7.5191	15.3091	0.01651	50.155	50.171	148.04	989.81	1137.85	0.2632	1.5473	1.8105	180
181	7.6874	15.6518	0.01651	49.126	49.143	149.04	989.20	1138.24	0.2647	1.5440	1.8087	181
182	7.8589	16.0008	0.01652	48.122	48.138	150.04	988.60	1138.64	0.2663	1.5406	1.8069	182
183	8.0335	16.3564	0.01653	47.142	47.158	151.05	987.99	1139.03	0.2679	1.5373	1.8051	183
184	8.2114	16.7185	0.01653	46.185	46.202	152.05	987.38	1139.43	0.2694	1.5339	1.8034	184
185	8.3926	17.0874	0.01654	45.251	45.267	153.05	986.77	1139.82	0.2710	1.5306	1.8016	185
186	8.5770	17.4630	0.01654	44.339	44.356	154.06	986.16	1140.22	0.2725	1.5273	1.7998	186
187	8.7649	17.8455	0.01655	43.448	43.465	155.06	985.55	1140.61	0.2741	1.5240	1.7981	187
188	8.9562	18.2350	0.01656	42.579	42.595	156.07	984.94	1141.00	0.2756	1.5207	1.7963	188

Table 3 Thermodynamic Properties of Water at Saturation (*Concluded*)

Temp. t, °F	Absolute Pressure p psi	in. Hg	Specific Volume, ft³/lb Sat. Liquid v_f	Evap. v_{fg}	Sat. Vapor v_g	Enthalpy, Btu/lb Sat. Liquid h_f	Evap. h_{fg}	Sat. Vapor h_g	Entropy, Btu/lb·°F Sat. Liquid s_f	Evap. s_{fg}	Sat. Vapor s_g	Temp., °F
189	9.1510	18.6316	0.01656	41.730	41.746	157.07	984.32	1141.39	0.2772	1.5174	1.7946	189
190	9.3493	19.0353	0.01657	40.901	40.918	158.07	983.71	1141.78	0.2787	1.5141	1.7929	190
191	9.5512	19.4464	0.01658	40.092	40.108	159.08	983.10	1142.18	0.2803	1.5109	1.7911	191
192	9.7567	19.8648	0.01658	39.301	39.317	160.08	982.48	1142.57	0.2818	1.5076	1.7894	192
193	9.9659	20.2907	0.01659	38.528	38.544	161.09	981.87	1142.95	0.2834	1.5043	1.7877	193
194	10.1788	20.7242	0.01659	37.774	37.790	162.09	981.25	1143.34	0.2849	1.5011	1.7860	194
195	10.3955	21.1653	0.01660	37.035	37.052	163.10	980.63	1143.73	0.2864	1.4979	1.7843	195
196	10.6160	21.6143	0.01661	36.314	36.331	164.10	980.02	1144.12	0.2880	1.4946	1.7826	196
197	10.8404	22.0712	0.01661	35.611	35.628	165.11	979.40	1144.51	0.2895	1.4914	1.7809	197
198	11.0687	22.5361	0.01662	34.923	34.940	166.11	978.78	1144.89	0.2910	1.4882	1.7792	198
199	11.3010	23.0091	0.01663	34.251	34.268	167.12	978.16	1145.28	0.2926	1.4850	1.7776	199
200	11.5374	23.4904	0.01663	33.594	33.610	168.13	977.54	1145.66	0.2941	1.4818	1.7759	200
201	11.7779	23.9800	0.01664	32.951	32.968	169.13	976.92	1146.05	0.2956	1.4786	1.7742	201
202	12.0225	24.4780	0.01665	32.324	32.340	170.14	976.29	1146.43	0.2971	1.4755	1.7726	202
203	12.2713	24.9847	0.01665	31.710	31.726	171.14	975.67	1146.81	0.2986	1.4723	1.7709	203
204	12.5244	25.5000	0.01666	31.110	31.127	172.15	975.05	1147.20	0.3002	1.4691	1.7693	204
205	12.7819	26.0241	0.01667	30.523	30.540	173.16	974.42	1147.58	0.3017	1.4660	1.7677	205
206	13.0436	26.5571	0.01667	29.949	29.965	174.16	973.80	1147.96	0.3032	1.4628	1.7660	206
207	13.3099	27.0991	0.01668	29.388	29.404	175.17	973.17	1148.34	0.3047	1.4597	1.7644	207
208	13.5806	27.6503	0.01669	28.839	28.856	176.18	972.54	1148.72	0.3062	1.4566	1.7628	208
209	13.8558	28.2108	0.01669	28.303	28.319	177.18	971.92	1149.10	0.3077	1.4535	1.7612	209
210	14.1357	28.7806	0.01670	27.778	27.795	178.19	971.29	1149.48	0.3092	1.4503	1.7596	210
212	14.7096	29.9489	0.01671	26.763	26.780	180.20	970.03	1150.23	0.3122	1.4442	1.7564	212
214	15.3025	31.1563	0.01673	25.790	25.807	182.22	968.76	1150.98	0.3152	1.4380	1.7532	214
216	15.9152	32.4036	0.01674	24.861	24.878	184.24	967.50	1151.73	0.3182	1.4319	1.7501	216
218	16.5479	33.6919	0.01676	23.970	23.987	186.25	966.23	1152.48	0.3212	1.4258	1.7469	218
220	17.2013	35.0218	0.01677	23.118	23.134	188.27	964.95	1153.22	0.3241	1.4197	1.7438	220
222	17.8759	36.3956	0.01679	22.299	22.316	190.29	963.67	1153.96	0.3271	1.4136	1.7407	222
224	18.5721	37.8131	0.01680	21.516	21.533	192.31	962.39	1154.70	0.3301	1.4076	1.7377	224
226	19.2905	39.2758	0.01682	20.765	20.782	194.33	961.11	1155.43	0.3330	1.4016	1.7347	226
228	20.0316	40.7848	0.01683	20.045	20.062	196.35	959.82	1156.16	0.3359	1.3957	1.7316	228
230	20.7961	42.3412	0.01684	19.355	19.372	198.37	958.52	1156.89	0.3389	1.3898	1.7287	230
232	21.5843	43.9461	0.01686	18.692	18.709	200.39	957.22	1157.62	0.3418	1.3839	1.7257	232
234	22.3970	45.6006	0.01688	18.056	18.073	202.41	955.92	1158.34	0.3447	1.3780	1.7227	234
236	23.2345	47.3060	0.01689	17.446	17.463	204.44	954.62	1159.06	0.3476	1.3722	1.7198	236
238	24.0977	49.0633	0.01691	16.860	16.877	206.46	953.31	1159.77	0.3505	1.3664	1.7169	238
240	24.9869	50.8738	0.01692	16.298	16.314	208.49	952.00	1160.48	0.3534	1.3606	1.7140	240
242	25.9028	52.7386	0.01694	15.757	15.774	210.51	950.68	1161.19	0.3563	1.3548	1.7111	242
244	26.8461	54.6591	0.01695	15.238	15.255	212.54	949.35	1161.90	0.3592	1.3491	1.7083	244
246	27.8172	56.6364	0.01697	14.739	14.756	214.57	948.03	1162.60	0.3621	1.3434	1.7055	246
248	28.8169	58.6717	0.01698	14.259	14.276	216.60	946.70	1163.29	0.3649	1.3377	1.7026	248
250	29.8457	60.7664	0.01700	13.798	13.815	218.63	945.36	1163.99	0.3678	1.3321	1.6998	250
252	30.9043	62.9218	0.01702	13.355	13.372	220.66	944.02	1164.68	0.3706	1.3264	1.6971	252
254	31.9934	65.1391	0.01703	12.928	12.945	222.69	942.68	1165.37	0.3735	1.3208	1.6943	254
256	33.1135	67.4197	0.01705	12.526	12.147	226.73	939.99	1166.72	0.3764	1.3153	1.6691	256
258	34.2653	69.7649	0.01707	12.123	12.140	226.76	939.97	1166.73	0.3792	1.3097	1.6889	258
260	35.4496	72.1760	0.01708	11.742	11.759	228.79	938.61	1167.40	0.3820	1.3042	1.6862	260
262	36.6669	74.6545	0.01710	11.376	11.393	230.83	937.25	1168.08	0.3848	1.2987	1.6835	262
264	37.9180	77.2017	0.01712	11.024	11.041	232.87	935.88	1168.74	0.3876	1.2932	1.6808	264
266	39.2035	79.8190	0.01714	10.684	10.701	234.90	934.50	1169.41	0.3904	1.2877	1.6781	266
268	40.5241	82.5078	0.01715	10.357	10.374	236.94	933.12	1170.07	0.3932	1.2823	1.6755	268
270	41.8806	85.2697	0.01717	10.042	10.059	238.98	931.74	1170.72	0.3960	1.2769	1.6729	270
272	43.2736	88.1059	0.01719	9.737	9.755	241.03	930.35	1171.38	0.3988	1.2715	1.6703	272
274	44.7040	91.0181	0.01721	9.445	9.462	243.07	928.95	1172.02	0.4016	1.2661	1.6677	274
276	46.1723	94.0076	0.01722	9.162	9.179	245.11	927.55	1172.67	0.4044	1.2608	1.6651	276
278	47.6794	97.0761	0.01724	8.890	8.907	247.16	926.15	1173.31	0.4071	1.2554	1.6626	278
280	49.2260	100.2250	0.01726	8.627	8.644	249.20	924.74	1173.94	0.4099	1.2501	1.6600	280
282	50.8128	103.4558	0.01728	8.373	8.390	251.25	923.32	1174.57	0.4127	1.2448	1.6575	282
284	52.4406	106.7701	0.01730	8.128	8.146	253.30	921.90	1175.20	0.4154	1.2396	1.6550	284
286	54.1103	110.1695	0.01731	7.892	7.910	255.35	920.47	1175.82	0.4182	1.2343	1.6525	286
288	55.8225	113.6556	0.01733	7.664	7.681	257.40	919.03	1176.44	0.4209	1.2291	1.6500	288
290	57.5780	117.2299	0.01735	7.444	7.461	259.45	917.59	1177.05	0.4236	1.2239	1.6476	290
292	59.3777	120.8941	0.01737	7.231	7.248	261.51	916.15	1177.66	0.4264	1.2187	1.6451	292
294	61.2224	124.6498	0.01739	7.026	7.043	263.56	914.69	1178.26	0.4291	1.2136	1.6427	294
296	63.1128	128.4987	0.01741	6.827	6.844	265.62	913.24	1178.86	0.4318	1.2084	1.6402	296
298	65.0498	132.4425	0.01743	6.635	6.652	267.68	911.77	1179.45	0.4345	1.2033	1.6378	298
300	67.0341	136.4827	0.01745	6.450	6.467	269.74	910.30	1180.04	0.4372	1.1982	1.6354	300

THERMODYNAMIC PROPERTIES OF MOIST AIR

Table 2, developed from formulas by Hyland and Wexler (1983a, 1983b), shows values of thermodynamic properties based on the *thermodynamic temperature scale*. This ideal scale differs slightly from practical temperature scales used for physical measurements. For example, the standard boiling point for water (at 14.696 psia or 29.921 in. Hg) occurs at 211.95 °F on this scale rather than at the traditional value of 212 °F. Most measurements are currently based on the International Practical Temperature Scale of 1968 (IPTS-68) (Preston-Thomas 1976). The following paragraphs briefly describe each column of Table 2.

t = Fahrenheit temperature, based on thermodynamic temperature scale and expressed relative to absolute temperature T in degrees Rankine (°R) by the relation:

$$T = t + 459.67$$

W_s = humidity ratio *at saturation*, condition at which gaseous phase (moist air) exists in equilibrium with condensed phase (liquid or solid) at given temperature and pressure (standard atmospheric pressure). At given values of temperature and pressure, humidity ratio W can have any value from zero to W_s.

v_a = specific volume of dry air, ft³/lb.

v_{as} = $v_s - v_a$, difference between volume of moist air *at saturation*, per pound of dry air, and specific volume of dry air itself, ft³/lb of dry air, at same pressure and temperature.

v_s = volume of moist air *at saturation* per pound of dry air, ft³/lb of dry air.

h_a = specific enthalpy of dry air, Btu/lb of dry air. Specific enthalpy of dry air has been assigned a value of zero at 0 °F and standard atmospheric pressure in Table 2.

h_{as} = $h_s - h_a$, difference between enthalpy of moist air *at saturation*, per pound of dry air, and specific enthalpy of dry air itself, Btu/lb of dry air, at same pressure and temperature.

h_s = enthalpy of moist air *at saturation* per pound of dry air, Btu/lb of dry air.

h_w = specific enthalpy of condensed water (liquid or solid) in equilibrium with saturated air at specified temperature and pressure, Btu per pound of water. Specific enthalpy of liquid water is assigned a value of zero at its triple point (32.018 °F) and saturation pressure.

Note that h_w is greater than the steam-table enthalpy of saturated pure condensed phase by the amount of enthalpy increase governed by the pressure increase from saturation pressure to 1 atmosphere plus influences from presence of air.

s_a = specific entropy of dry air, Btu/lb·°R. In Table 2, specific entropy of dry air has been assigned a value of zero at 0 °F and standard atmospheric pressure.

s_{as} = $s_s - s_a$, difference between entropy of moist air *at saturation*, per pound of dry air, and specific entropy of dry air itself, Btu/lb of dry air·°R, at same pressure and temperature.

s_s = entropy of moist air *at saturation* per pound of dry air, Btu/lb of dry air·°R.

s_w = specific entropy of condensed water (liquid or solid) in equilibrium with saturated air, Btu/lb of water·°R; s_w differs from entropy of pure water at saturation pressure, similar to h_w.

p_s = vapor pressure of water in saturated moist air, in. Hg. Pressure p_s differs negligibly from saturation vapor pressure of pure water p_{ws} at least for conditions shown. Consequently, values of p_s can be used at same pressure and temperature in equations where p_{ws} appears. Pressure p_s is defined as $p_s = x_{ws}p$, where x_{ws} is mole fraction of water vapor in moist air saturated with water at temperature t and pressure p, and where p is total barometric pressure of moist air.

THERMODYNAMIC PROPERTIES OF WATER AT SATURATION

Table 3 shows thermodynamic properties of water at saturation for temperatures from −80 to 300 °F, calculated by the formulations described by Hyland and Wexler (1983b). Symbols in the table follow standard steam table nomenclature. These properties are based on the thermodynamic temperature scale. The enthalpy and entropy of saturated liquid water are both assigned the value zero at the triple point, 32.018 °F. Between the triple-point and critical-point temperatures of water, two states—liquid and vapor—may coexist in equilibrium. These states are called saturated liquid and saturated vapor.

In determining a number of moist air properties, principally the saturation humidity ratio, the *water vapor saturation pressure* is required. Values may be obtained from Table 3 or calculated from the following formulas (Hyland and Wexler 1983b).

The saturation pressure over *ice* for the temperature range of −148 to 32 °F is given by:

$$\ln(p_{ws}) = C_1/T + C_2 + C_3T + C_4T^2 + C_5T^3 \\ + C_6T^4 + C_7\ln(T) \tag{3}$$

where

$C_1 = -1.021\ 416\ 5\ E+04$
$C_2 = -4.893\ 242\ 8\ E+00$
$C_3 = -5.376\ 579\ 4\ E-03$
$C_4 = 1.920\ 237\ 7\ E-07$
$C_5 = 3.557\ 583\ 2\ E-10$
$C_6 = -9.034\ 468\ 8\ E-14$
$C_7 = 4.163\ 501\ 9\ E+00$

The saturation pressure over *liquid water* for the temperature range of 32 to 392 °F is given by:

$$\ln(p_{ws}) = C_8/T + C_9 + C_{10}T + C_{11}T^2 \\ + C_{12}T^3 + C_{13}\ln(T) \tag{4}$$

where

$C_8 = -1.044\ 039\ 7\ E+04$
$C_9 = -1.129\ 465\ 0\ E+01$
$C_{10} = -2.702\ 235\ 5\ E-02$
$C_{11} = 1.289\ 036\ 0\ E-05$
$C_{12} = -2.478\ 068\ 1\ E-09$
$C_{13} = 6.545\ 967\ 3$

In both Equations (3) and (4),

$\ln$ = natural logarithm
p_{ws} = saturation pressure, psia
T = absolute temperature, °R = °F + 459.67

The coefficients of Equations (3) and (4) have been derived from the Hyland-Wexler equations, which are given in SI units. Due to rounding errors in the derivations and in some computers' calculating precision, the results obtained from Equations (3) and (4) may not agree precisely with Table 3 values.

HUMIDITY PARAMETERS

Humidity ratio (alternatively, the moisture content or mixing ratio) W of a given moist air sample is defined as the ratio of the mass of water vapor to the mass of dry air contained in the sample:

$$W = M_w/M_a \tag{5}$$

The humidity ratio W is equal to the mole fraction ratio x_w/x_a multiplied by the ratio of molecular masses; namely, 18.01528/ 28.9645 = 0.62198, *i.e.*:

$$W = 0.62198 x_w/x_a \tag{6}$$

Specific humidity q is the ratio of the mass of water vapor to the total mass of the moist air sample:

$$q = M_w/(M_w + M_a) \qquad (7a)$$

In terms of the humidity ratio:

$$q = W/(1 + W) \qquad (7b)$$

Absolute humidity (alternatively, *water vapor density*) d_v is the ratio of the mass of water vapor to the total volume of the sample:

$$d_v = M_w/V \qquad (8)$$

The *density* ρ of a moist air mixture is the ratio of the total mass to the total volume:

$$\rho = (M_a + M_w)/V = (1/v)(1 + W) \qquad (9)$$

where v is the moist air specific volume, ft^3/lb dry air, as defined by Equation (25).

HUMIDITY PARAMETERS INVOLVING SATURATION

The following definitions of humidity parameters involve the concept of moist air saturation:

Saturation humidity ratio $W_s(t, p)$ is the humidity ratio of moist air saturated with respect to water (or ice) at the same temperature t and pressure p.

Degree of saturation μ is the ratio of the air humidity ratio W to the humidity ratio W_s of saturated air at the same temperature and pressure:

$$\mu = \left.\frac{W}{W_s}\right|_{t,p} \qquad (10)$$

Relative humidity ϕ is the ratio of the mole fraction of water vapor x_w in a given moist air sample to the mole fraction x_{ws} in an air sample, saturated at the same temperature and pressure:

$$\phi = \left.\frac{x_w}{x_{ws}}\right|_{t,p} \qquad (11)$$

Combining Equations (6), (10), and (11):

$$\mu = \frac{\phi}{1 + (1 - \phi)W_s/0.62198} \qquad (12)$$

Dew-point temperature t_d is the temperature of moist air saturated at the same pressure p, with the same humidity ratio W as that of the given sample of moist air. It is defined as the solution $t_d(p, W)$ of the equation:

$$W_s(p, t_d) = W \qquad (13)$$

Thermodynamic wet-bulb temperature t^* is the temperature at which water (liquid or solid), by evaporating into moist air at a given dry-bulb temperature t and humidity ratio W, can bring air to saturation adiabatically at the same temperature t^* while the pressure p is maintained constant. This parameter is considered separately in a later section.

PERFECT GAS RELATIONSHIPS FOR DRY AND MOIST AIR

When moist air is considered a mixture of independent perfect gases, dry air, and water vapor, each is assumed to obey the perfect gas equation of state as follows:

$$\text{Dry air } p_a V = n_a RT \qquad (14)$$
$$\text{Water vapor } p_w V = n_w RT \qquad (15)$$

where

p_a = partial pressure of dry air
p_w = partial pressure of water vapor
V = total mixture volume
n_a = number of moles of dry air
n_w = number of moles of water vapor
R = universal gas constant 1545.32 ft·lb_f/lb mol·°R
T = absolute temperature, °R

The mixture also obeys the perfect gas equation:

$$pV = nRT \qquad (16)$$

or

$$(p_a + p_w)V = (n_a + n_w)RT \qquad (17)$$

where $p = p_a + p_w$ is the total mixture pressure and $n = n_a + n_w$ is the total number of moles in the mixture. From Equations (14) through (17), the mole fractions of dry air and water vapor are, respectively:

$$x_a = p_a/(p_a + p_w) = p_a/p \qquad (18)$$

and

$$x_w = p_w/(p_a + p_w) = p_w/p \qquad (19)$$

From Equations (6), (18), and (19), the *humidity ratio W* is given by:

$$W = 0.62198\frac{p_w}{p - p_w} \qquad (20)$$

The degree of saturation μ is, by definition, Equation (10):

$$\mu = \left.\frac{W}{W_s}\right|_{t,p}$$

where

$$W_s = 0.62198\frac{p_{ws}}{p - p_{ws}} \qquad (21)$$

The term p_{ws} represents the saturation pressure of water vapor in the absence of air at the given temperature t. This pressure p_{ws} is a function only of temperature and differs slightly from the vapor pressure of water in saturated moist air.

The *relative humidity* ϕ is, by definition, Equation (11):

$$\phi = \left.\frac{x_w}{x_{ws}}\right|_{t,p}$$

Substituting Equation (19) for x_w and x_{ws}:

$$\phi = \left.\frac{p_w}{p_{ws}}\right|_{t,p} \qquad (22)$$

Substituting Equation (19) for x_{ws} into Equation (12):

$$\phi = \frac{\mu}{1 - (1 - \mu)(p_{ws}/p)} \qquad (23)$$

Both ϕ and μ are zero for dry air and unity for saturated moist air. At intermediate states their values differ, substantially so at higher temperatures.

The *specific volume v* of a moist air mixture is expressed in terms of a unit mass of dry air, *i.e.*:

$$v = V/M_a = V/(28.9645n_a) \qquad (24)$$

where V is the total volume of the mixture, M_a is the total mass of dry air, and n_a is the number of moles of dry air. By Equations (14) and (24), with the relation $p = p_a + p_w$:

$$v = \frac{RT}{28.9645(p - p_w)} = \frac{R_a T}{(p - p_w)} \qquad (25)$$

Using Equation (20):

$$v = \frac{RT(1 + 1.6078W)}{28.9645p} = \frac{R_a T(1 + 1.6078W)}{p} \qquad (26)$$

In Equations (25) and (26), v is specific volume, T is absolute temperature, p is total pressure, p_w is the partial pressure of water vapor, and W is the humidity ratio.

The *enthalpy* of a mixture of perfect gases equals the sum of the individual partial enthalpies of the components. Therefore, the enthalpy of moist air can be written:

$$h = h_a + Wh_g \qquad (27)$$

where h_a is the specific enthalpy for dry air and h_g is the specific enthalpy for saturated water vapor at the temperature of the mixture. Approximately:

$$h_a = 0.240t \qquad \text{(Btu/lb)} \qquad (28)$$
$$h_g = 1061 + 0.444t \qquad \text{(Btu/lb)} \qquad (29)$$

where t is the dry-bulb temperature, °F. The moist air enthalpy then becomes:

$$h = 0.240t + W(1061 + 0.444t) \qquad \text{(Btu/lb)} \qquad (30)$$

THERMODYNAMIC WET-BULB TEMPERATURE AND DEW-POINT TEMPERATURE

For any state of moist air, a temperature t^* exists at which liquid (or solid) water evaporates into the air to bring it to saturation at exactly this same temperature and pressure (Harrison 1965). During the adiabatic saturation process, the saturated air is expelled at a temperature equal to that of the injected water (Figures 8 and 9). In the constant pressure process, the humidity ratio is increased from a given initial value W to the value W_s^*, corresponding to saturation at the temperature t^*; the enthalpy is increased from a given initial value h to the value h_s^*, corresponding to satu-

ration at the temperature t^*; the mass of water added per unit mass of dry air is $(W_s^* - W)$, which adds energy to the moist air of amount $(W_s^* - W)h_w^*$, where h_w^* denotes the specific enthalpy of the water added at the temperature t^*. Therefore, if the process is strictly adiabatic, conservation of enthalpy at constant pressure requires that:

$$h + (W_s^* - W) h_w^* = h_s^* \qquad (31)$$

The properties W_s^*, h_w^*, and h_s^* are functions only of the temperature t^* for a fixed value of pressure. The value of t^*, which satisfies Equation (31) for given values of h, W, and p, is the *thermodynamic wet-bulb temperature*.

The *psychrometer* consists of two thermometers; one thermometer's bulb is covered by a wick that has been thoroughly wetted with water. When the wet bulb is placed in an airstream, water evaporates from the wick, eventually reaching an equilibrium temperature called the *wet-bulb temperature*. This process is not one of adiabatic saturation, which defines the thermodynamic wet-bulb temperature, but is one of simultaneous heat and mass transfer from the wet bulb. The fundamental mechanism of this process is described by the Lewis relation (Chapter 5). Fortunately, only small corrections must be applied to wet-bulb thermometer readings to obtain the thermodynamic wet-bulb temperature.

As defined, thermodynamic wet-bulb temperature is a unique property of a given moist air sample independent of measurement techniques. Chapter 5 of the ASHRAE *Brochure on Psychrometry* (1977) discusses the concept in more detail.

Equation (31) is exact since it defines the thermodynamic wet-bulb temperature t^*. Substituting the approximate perfect gas relation [Equation (30)] for h, the corresponding expression for h_s^*, and the approximate relation

$$h_w^* = t^* - 32 \qquad \text{(Btu/lb)} \qquad (32)$$

into Equation (31), and solving for the humidity ratio:

$$W = \frac{(1093 - 0.556t^*)W_s^* - 0.240(t - t^*)}{1093 + 0.444t - t^*} \qquad (33)$$

where t and t^* are in °F.

The *dew-point temperature* t_d of moist air with humidity ratio W and pressure p was defined earlier as the solution $t_d(p, w)$ of $W_s(p, t_d) = W$. For perfect gases, this reduces to:

$$p_{ws}(t_d) = p_w = (pW)/(0.62198 + W) \qquad (34)$$

where p_w is the water vapor partial pressure for the moist air sample and $p_{ws}(t_d)$ is the saturation vapor pressure at temperature t_d. The saturation vapor pressure is derived from Table 3 or from Equations (3) or (4). Alternatively, the dew-point temperature can be calculated directly by one of the following equations (Peppers 1988):

For the dew-point temperature range of 32 to 200°F:

$$t_d = C_{14} + C_{15}\alpha + C_{16}\alpha^2 + C_{17}\alpha^3 + C_{18}(p_w)^{0.1984} \qquad (35)$$

and for temperatures below 32°F:

$$t_d = 90.12 + 26.412\alpha + 0.8927\alpha^2 \qquad (36)$$

where

t_d = dew-point temperature, °F
α = ln (p_w)
p_w = water vapor partial pressure, psia
C_{14} = 100.45
C_{15} = 33.193
C_{16} = 2.319
C_{17} = 0.17074
C_{18} = 1.2063

NUMERICAL CALCULATION OF MOIST AIR PROPERTIES

The following are outlines, citing equations and tables already presented, for calculating moist air properties using perfect gas relations. These relations are sufficiently accurate for most engineering calculations in air-conditioning practice, and are readily adapted to either hand or computer calculating methods. Graphical procedures are discussed in the section on psychrometric charts.

Situation 1.   Given: Dry-bulb temperature t
Wet-bulb temperature t^*
Pressure p

To Obtain	Use	Comments
$p_{ws}(t^*)$	Table 3 or Eq. (3) or (4)	Sat. press. for temp. t^*
$W_s{}^*$	Eq. (21)	Using $p_{ws}(t^*)$
W	Eq. (33)	
$p_{ws}(t)$	Table 3 or Eq. (3) or (4)	Sat. press. for temp. t
W_s	Eq. (21)	Using $p_{ws}(t)$
μ	Eq. (10)	Using W_s
ϕ	Eq. (23)	Using $p_{ws}(t)$
v	Eq. (26)	
h	Eq. (30)	
p_w	Eq. (34)	
t_d	Table 3 with Eq. (34), (35), or (36)	

Situation 2.   Given: Dry-bulb temperature t
Dew-point temperature t_d
Pressure p

To Obtain	Use	Comments
$p_w = p_{ws}(t_d)$	Table 3 or Eq. (3) or (4)	Sat. press. for temp. t_d
W	Eq. (20)	
$p_{ws}(t)$	Table 3 or Eq. (3) or (4)	Sat. press. for temp. t_d
W_s	Eq. (21)	Using $p_{ws}(t)$
μ	Eq. (10)	Using W_s
ϕ	Eq. (23)	Using $p_{ws}(t)$
v	Eq. (26)	
h	Eq. (30)	
t^*	Eq. (21) and (33) with Table 3 or with Eq. (3) or (4)	Requires trial-and-error or numerical solution method

Situation 3.   Given: Dry-bulb temperature t
Relative humidity ϕ
Pressure p

To Obtain	Use	Comments
$p_{ws}(t)$	Table 3 or Eq. (3) or (4)	Sat. press. for temp. t
p_w	Eq. (22)	
W	Eq. (20)	
W_s	Eq. (21)	Using $p_{ws}(t)$
μ	Eq. (10)	Using W_s
v	Eq. (26)	
h	Eq. (30)	
t_d	Table 3 with Eq. (34), (35), or (36)	
t^*	Eq. (21) and (33) with Table 3 or with Eq. (3) or (4)	Requires trial-and-error or numerical solution method

EXACT RELATIONS FOR COMPUTING W_s AND ϕ

Corrections that account for (1) the effect of dissolved gases on properties of condensed phase; (2) the effect of pressure on properties of condensed phase; and (3) the effect of intermolecular force on properties of moisture itself, can be applied to Equations (21) or (23):

$$W_s = 0.62198[f_s p_{ws}/(p - f_s p_{ws})] \qquad (21a)$$

$$\phi = \frac{\mu}{1 - (1 - \mu)(f_s p_{ws}/p)} \qquad (23a)$$

Table 4 lists f_s values for a number of pressure and temperature combinations. Hyland and Wexler (1983a) give additional values.

Table 4   Values of f_s and Estimated Maximum Uncertainties (EMU)

	14.50 psia		72.52 psia		145.04 psia	
T, °R	f	EMU E+04	f	EMU E+04	f	EMU E+04
311.67	1.0105	134	1.0540	66	1.1130	136
491.67	1.0039	2	1.0177	10	1.0353	19
671.67	1.0039	0.1	1.0180	4	1.0284	11

MOIST AIR PROPERTY TABLES FOR STANDARD PRESSURE

Table 2 shows values of thermodynamic properties for standard atmospheric pressure at temperatures from −80 to 200°F. The properties of intermediate moist air states can be calculated using the degree of saturation μ:

Volume	$v = v_a + \mu v_{as}$	(37)
Enthalpy	$h = h_a + \mu h_{as}$	(38)
Entropy	$s = s_a + \mu s_{as}$	(39)

These equations are accurate to about 160°F. At higher temperatures, the errors can be significant. Hyland and Wexler (1983a) include charts that can be used to estimate errors for v, h, and s for standard barometric pressure.

PSYCHROMETRIC CHARTS

A psychrometric chart graphically represents the thermodynamic properties of moist air.

The choice of coordinates for a psychrometric chart is arbitrary. A chart with coordinates of enthalpy and humidity ratio provides convenient graphical solutions of many moist air problems with a minimum of thermodynamic approximations. ASHRAE developed seven such psychrometric charts.

Charts 1, 2, and 3 are for sea level pressure. Chart 4 is for 5000-ft altitude (24.89 in. Hg). Chart 5 is for 7500-ft altitude (22.65 in. Hg). All charts use oblique-angle coordinates of enthalpy and humidity ratio, and are consistent with the data of Table 2 and the properties computation methods of Goff and Gratch (1945, 1949) as well as Hyland and Wexler (1983a). Palmatier (1963) describes the geometry of chart construction applying specifically to Charts 1 and 4.

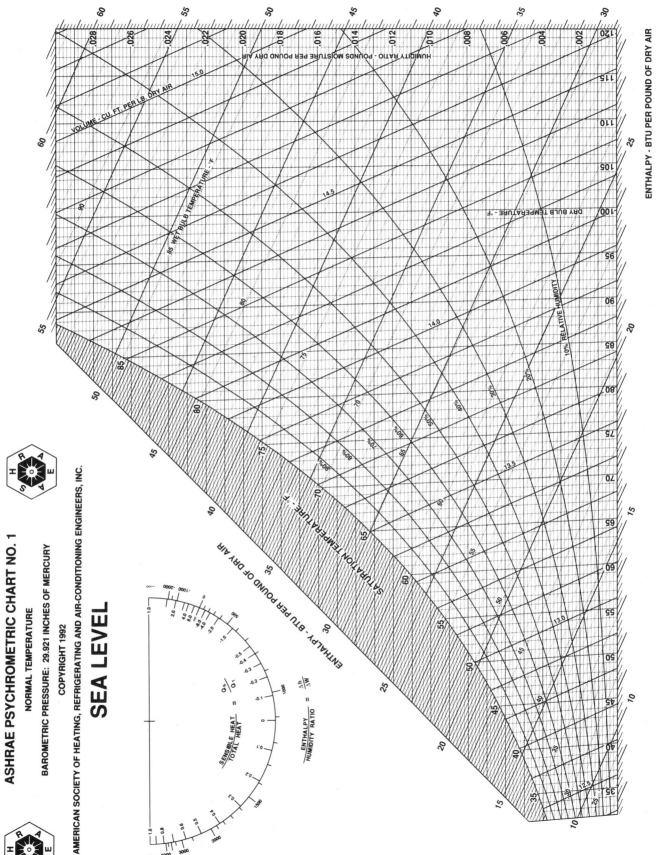

Fig. 1   ASHRAE Psychrometric Chart No. 1

The dry-bulb temperature ranges covered by the charts are:

Charts 1, 4, 5	Normal temperature	32 to 120 °F
Chart 2	Low temperature	−40 to 50 °F
Chart 3	High temperature	60 to 250 °F

Charts 1 and 4 are constructed on the identical system of coordinates. Psychrometric properties or charts for other barometric pressures can be derived by interpolation. Sufficiently exact values for most purposes can be derived by methods described in the section on perfect gas relations. The construction of charts for altitude conditions has been treated by Haines (1961), Rohsenow (1946), and Karig (1946).

Comparison of Charts 1 and 4 by overlay reveals:

1. The dry-bulb lines coincide.
2. Wet-bulb lines for a given temperature originate at the intersections of the corresponding dry-bulb line and the two saturation curves, and they have the same slope.
3. Humidity ratio and enthalpy for a given dry- and wet-bulb increase with altitude, but there is little change in relative humidity.
4. Volume changes rapidly; for a given dry-bulb and humidity ratio, it is practically inversely proportional to barometric pressure.

The following table compares properties at sea level (Chart 1) and 5000-ft (Chart 4):

Chart No.	db	wb	h	W	rh	v
1	100	81	44.6	0.0186	45	14.5
4	100	81	49.8	0.0234	46	17.6

Figure 1, which is Chart 1 of the ASHRAE psychrometric charts, shows humidity ratio lines (horizontal) for the range from 0 (dry air) to 0.03 lb water/lb dry air. Enthalpy lines are oblique lines drawn across the chart precisely parallel to each other.

Dry-bulb temperature lines are drawn straight, not precisely parallel to each other, and inclined slightly from the vertical position. Thermodynamic wet-bulb temperature lines are oblique lines that differ slightly in direction from that of enthalpy lines. They are identically straight but are not precisely parallel to each other.

Relative humidity (rh) lines are shown in intervals of 10%. The saturation curve is the line of 100% rh, while the horizontal line for $W = 0$ (dry air) is the line for 0% rh.

Specific volume lines are straight but are not precisely parallel to each other.

A narrow region above the saturation curve has been developed for fog conditions of moist air. This two-phase region represents a mechanical mixture of saturated moist air and liquid water, with the two components in thermal equilibrium. Isothermal lines in the fog region coincide with extensions of thermodynamic wet-bulb temperature lines. If required, the fog region can be further expanded by extension of humidity ratio, enthalpy, and thermodynamic wet-bulb temperature lines.

The protractor to the left of the chart shows two scales—one for sensible-total heat ratio, and one for the ratio of enthalpy difference to humidity ratio difference. The protractor is used to establish the direction of a condition line on the psychrometric chart. The full-size charts available from ASHRAE include a nomograph that provides an alternate method for determining enthalpy from the psychrometric chart, and allows a direct reading of the enthalpy $\Delta W h_w$ of liquid water added or rejected in a process.

Example 1 illustrates use of the ASHRAE psychrometric chart to determine moist air properties.

Example 1. Moist air exists at 100 °F dry-bulb temperature, 65 °F thermodynamic wet-bulb temperature, and 29.921 in. Hg pressure. Determine the humidity ratio, enthalpy, dew-point temperature, relative humidity, and volume.

Solution: Locate state point on Chart 1 (Figure 1) at the intersection of 100 °F dry-bulb temperature and 65 °F thermodynamic wet-bulb temperature lines. Read $W = 0.00523$ lb water/lb dry air.

The enthalpy can be found by two methods. Using two triangles, draw a line parallel to the nearest enthalpy line (30 Btu/lb dry air) through the state point to the nearest edge scale. Read $h = 29.80$ Btu/lb dry air.

Enthalpy can also be derived from Equation (31):

$$h = h_s^* - (W_s^* - W)h_w^*$$

where $(W_s^* - W)h_w^* = D$, the enthalpy deviation given by the nomograph, and h_s^* is the enthalpy of saturated moist air at the thermodynamic wet-bulb temperature at 65 °F, read $h_s^* = 30.06$ Btu/lb of dry air. By nomograph, at $W = 0.00523$ lb water/dry air and $t^* = 65$ °F, read $D = -0.26$ Btu/lb dry air. Thus, $h = 30.06 - 0.26 = 29.80$ Btu/lb dry air.

Dew-point temperature can be read at the intersection of $W = 0.00523$ lb water/lb dry air with the saturation curve. Thus, $t_d = 40$ °F.

Relative humidity ϕ can be estimated directly. Thus, $\phi = 13\%$.

Volume can be found by linear interpolation between the volume lines for 14.0 and 14.5 ft^3/lb dry air. Thus, $v = 14.22$ ft^3/lb dry air.

TYPICAL AIR-CONDITIONING PROCESSES

The ASHRAE psychrometric chart can be used to solve numerous process problems with moist air. Its use is best explained through illustrative examples. In each of the following examples, the process takes place at a constant pressure of 29.921 in. Hg.

Moist Air Heating

The process of adding heat alone to moist air is represented by a horizontal line on the ASHRAE chart, since the humidity ratio remains unchanged.

Figure 2 shows a device that adds heat to a stream of moist air. For steady flow conditions, the required rate of heat addition is:

$$_1q_2 = m_a(h_2 - h_1) \qquad (40)$$

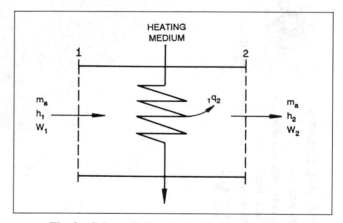

Fig. 2   Schematic Device for Heating Moist Air

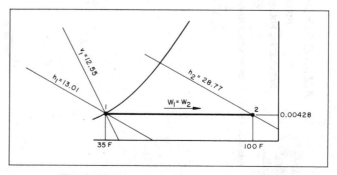

Fig. 3   Schematic Solution for *Example 2*

Example 2. Moist air, saturated at 35 °F, enters a heating coil at a rate of 20,000 cfm. Air leaves the coil at 100 °F. Find the required rate of heat addition in Btu/h.

Solution: Figure 3 schematically shows the solution. State 1 is located on the saturation curve at 35 °F. Thus, h_1 = 13.01 Btu/lb dry air, W_1 = 0.00428 lb water/lb dry air, and v_1 = 12.55 ft^3/lb dry air. State 2 is located at the intersection of t = 100 °F and W_2 = W_1 = 0.00428 lb water/lb dry air. Thus, h_2 = 28.77 Btu/lb dry air. The mass flow of dry air is:

$$m_a = [(20,000)(60)/12.55] = 95,620 \text{ lb dry air/h}$$

From Equation (40):

$$_1q_2 = (95,620)(28.77 - 13.01) = 1,507,000 \text{ Btu/h}$$

Moist Air Cooling

Moisture separation occurs when moist air is cooled to a temperature below its initial dew point. Figure 4 shows a schematic cooling coil where moist air is assumed to be uniformly processed. Although water can be separated at various temperatures ranging from the initial dew point to the final saturation temperature, it is assumed that condensed water is cooled to the final air temperature t_2 before it drains from the system.

For the system of Figure 4, the steady flow energy and material balance equations are:

$$m_a h_1 = m_a h_2 + {}_1q_2 + m_w h_{w2}$$
$$m_a W_1 = m_a W_2 + m_w$$

Thus:

$$m_w = m_a(W_1 - W_2) \qquad (41)$$
$$_1q_2 = m_a[(h_1 - h_2) - (W_1 - W_2)h_{w2}] \quad (42)$$

Example 3. Moist air at 85 °F dry-bulb temperature and 50% rh enters a cooling coil at 10,000 cfm and is processed to a final saturation condition at 50 °F. Find the tons of refrigeration required.

Solution: Figure 5 shows the schematic solution. State 1 is located at the intersection of t = 85 °F and ϕ = 50%. Thus, h_1 = 34.62 Btu/lb dry air, W_1 = 0.01292 lb water/lb dry air, and v_1 = 14.01 ft^3/lb dry air. State 2 is located on the saturation curve at 50 °F. Thus, h_2 = 20.30 Btu/lb dry air and W_2 = 0.00766 lb water/lb dry air. From Table 2, h_{w2} = 18.11 Btu/lb water. The mass flow of dry air is:

$$m_a = 10,000/14.01 = 713.8 \text{ lb dry air/min}$$

From Equation (42):

$$_1q_2 = 713.8[(34.62 - 20.30) - (0.01292 - 0.00766)(18.11)]$$
$$= 10,150 \text{ Btu/min}$$

Since one ton of refrigeration equals a heat withdrawal of 200 Btu/min, the required refrigerating capacity is 50.75 tons.

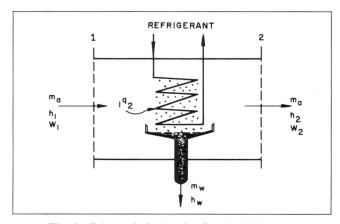

Fig. 4   Schematic Device for Cooling Moist Air

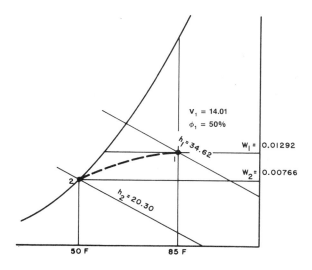

Fig. 5   Schematic Solution for *Example 3*

Adiabatic Mixing of Two Moist Airstreams

A common process in air-conditioning systems is the adiabatic mixing of two moist airstreams. Figure 6 schematically shows the problem. Adiabatic mixing is governed by three equations:

$$m_{a1} h_1 + m_{a2} h_2 = m_{a3} h_3$$
$$m_{a1} + m_{a2} = m_{a3}$$
$$m_{a1} W_1 + m_{a2} W_2 = m_{a3} W_3$$

Eliminating m_{a3} gives:

$$\frac{h_2 - h_3}{h_3 - h_1} = \frac{W_2 - W_3}{W_3 - W_1} = \frac{m_{a1}}{m_{a2}} \qquad (43)$$

according to which, on the ASHRAE chart, the state point of the resulting mixture lies on the straight line connecting the state points of the two streams being mixed, and divides the line into two segments, in the same ratio as the masses of dry air in the two streams.

Example 4. A stream of 5000 cfm of outdoor air at 40 °F dry-bulb temperature and 35 °F thermodynamic wet-bulb temperature is adiabatically mixed with 15,000 cfm of recirculated air at 75 °F dry-bulb temperature and 50% rh. Find the dry-bulb temperature and thermodynamic wet-bulb temperature of the resulting mixture.

Solution: Figure 7 shows the schematic solution. States 1 and 2 are located on the ASHRAE chart, revealing that v_1 = 12.65 ft^3/lb dry air, and v_2 = 13.68 ft^3/lb dry air. Therefore:

$$m_{a1} = 5000/12.65 = 395 \text{ lb dry air/min}$$
$$m_{a2} = 15,000/13.68 = 1096 \text{ lb dry air/min}$$

According to Equation (43):

$$\frac{\text{Line } 3—2}{\text{Line } 1—3} = \frac{m_{a1}}{m_{a2}} \quad \text{or} \quad \frac{\text{Line } 1—3}{\text{Line } 1—2} = \frac{m_{a2}}{m_{a3}} = \frac{1096}{1491} = 0.735$$

Consequently, the length of line segment 1—3 is 0.735 times the length of entire line 1—2. Using a ruler, State 3 is located, and the values t_3 = 65.9 °F and t_3^* = 56.6 °F found.

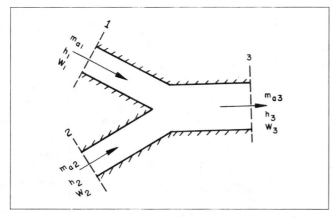

Fig. 6   Adiabatic Mixing of Two Moist Airstreams

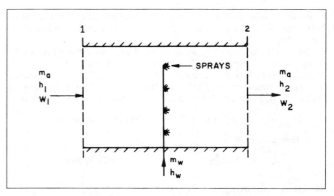

Fig. 8   Schematic Injection of Water into Moist Air

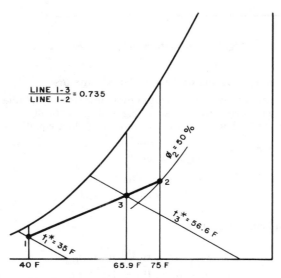

Fig. 7   Schematic Solution for *Example 4*

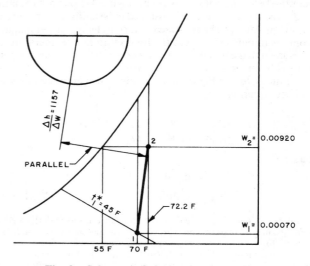

Fig. 9   Schematic Solution for *Example 5*

Adiabatic Mixing of Water Injected into Moist Air

Steam or liquid water can be injected into a moist airstream to raise its humidity. Figure 8 represents a diagram of this common air-conditioning process. If the mixing is adiabatic, the following equations apply:

$$m_a h_1 + m_w h_w = m_a h_2$$
$$m_a W_1 + m_w = m_a W_2$$

Therefore,

$$\frac{h_2 - h_1}{W_2 - W_1} = h_w \tag{44}$$

according to which, on the ASHRAE chart, the final state point of the moist air lies on a straight line whose direction is fixed by the specific enthalpy of the injected water, drawn through the initial state point of the moist air.

Example 5. Moist air at 70°F dry-bulb and 45°F thermodynamic wetbulb temperature is to be processed to a final dew-point temperature of 55°F by adiabatic injection of saturated steam at 230°F. The rate of dry airflow is 200 lb/min. Find the final dry-bulb temperature of the moist air and the rate of steam flow required in lb/h.

Solution: Figure 9 shows the schematic solution. By Table 3, the enthalpy of the steam h_g = 1157 Btu/lb water. Therefore, according to Equation (44), the condition line on the ASHRAE chart connecting States 1 and 2 must have a direction:

$$\Delta h / \Delta W = 1157 \text{ Btu/lb water}$$

The condition line can be drawn with the $\Delta h / \Delta W$ protractor. First, establish the reference line on the protractor by connecting the origin with the value $\Delta h / \Delta W$ = 1157. Draw a second line parallel to the reference line and through the initial state point of the moist air. This second line is the condition line. State 2 is established at the intersection of the condition line with the horizontal line extended from the saturation curve at 55°F (t_{d2} = 55°F). Thus, t_2 = 72.2°F.

Values of W_2 and W_1 can be read from the chart. The required steam flow is:

$$m_w = m_a(W_2 - W_1) = (200)(60)(0.00920 - 0.00070)$$
$$= 102 \text{ lb steam/h}$$

Space Heat Absorption and Moist Air Moisture Gains

The problem of air conditioning a space is usually determined by (1) the quantity of moist air to be supplied, and (2) the air condition necessary to remove given amounts of energy and water from the space and be withdrawn at a specified condition.

Figure 10 schematically shows a space with incident rates of energy and moisture gains. The quantity q_s denotes the net sum of all rates of heat gain in the space, arising from transfers through

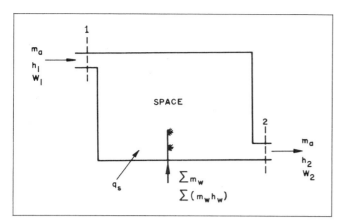

Fig. 10   Schematic Problem of Air Conditioning a Space

boundaries and from sources within the space. This heat gain involves addition of energy alone and does not include energy contributions due to addition of water (or water vapor). It is usually called the *sensible heat gain*. The quantity Σm_w denotes the net sum of all rates of moisture gain on the space arising from transfers through boundaries and from sources within the space. Each pound of moisture injected into the space adds an amount of energy equal to its specific enthalpy.

Assuming steady-state conditions, governing equations are:

$$m_a h_1 + q_s + \Sigma(m_w h_w) = m_a h_2$$
$$m_a W_1 + \Sigma m_w = m_a W_2$$

or

$$q_s + \Sigma(m_w h_w) = m_a(h_2 - h_1) \qquad (45)$$
$$\Sigma m_w = m_a(W_2 - W_1) \qquad (46)$$

The left side of Equation (45) represents the total rate of energy addition to the space from all sources. By Equations (45) and (46):

$$\frac{h_2 - h_1}{W_2 - W_1} = \frac{q_s + \Sigma(m_w h_w)}{\Sigma m_w} \qquad (47)$$

according to which, on the ASHRAE chart and for a given state of the withdrawn air, all possible states (conditions) for the supply air must lie on a straight line drawn through the state point of the withdrawn air, that has a direction specified by the numerical value of $[q_s + \Sigma(m_w h_w)]/\Sigma m_w$. This line is the condition line for the given problem.

Example 6. Moist air is withdrawn from a room at 80 °F dry-bulb temperature and 66 °F thermodynamic wet-bulb temperature. The sensible rate of heat gain for the space is 30,000 Btu/h. A rate of moisture gain of 10 lb/h occurs from the space occupants. This moisture is assumed as saturated water vapor at 90 °F. Moist air is introduced into the room at a dry-bulb temperature of 60 °F. Find the required thermodynamic wet-bulb temperature and volume flow rate of the supply air.

Solution: Figure 11 shows the schematic solution. State 2 is located on the ASHRAE chart. From Table 3, specific enthalpy of added water vapor is $h_g = 1100.40$ Btu/lb. From Equation (47):

$$\frac{\Delta h}{\Delta W} = \frac{30,000 + (10)(1100.40)}{10} = 4100 \text{ Btu/lb water}$$

With the $\Delta h/\Delta W$ protractor, establish a reference line of direction $\Delta h/\Delta W = 4100$ Btu/lb water. Parallel to this reference line, draw a straight line on the chart through State 2. The intersection of this line with the 60 °F dry-bulb temperature line is State 1. Thus, $t_1^* = 56.4$ °F.

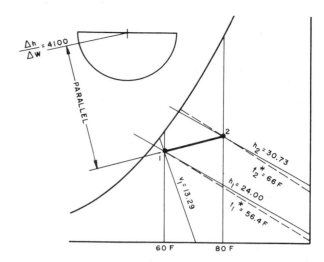

Fig. 11   Schematic Solution for *Example 6*

An alternate (and approximately correct) procedure in establishing the condition line is to use the protractor's sensible-total heat ratio scale instead of the $\Delta h/\Delta W$ scale. The quantity $\Delta H_s/\Delta H_T$ is the ratio of the rate of sensible heat gain for the space to the rate of total energy gain for the space. Therefore:

$$\frac{\Delta H_s}{\Delta H_T} = \frac{q_s}{q_s + \Sigma(m_w h_w)} = \frac{30,000}{30,000 + (10 \times 1100.44)} = 0.732$$

Note that $\Delta H_s/\Delta H_T = 0.732$ on the protractor coincides closely with $\Delta h/\Delta W = 4100$ Btu/lb water.

The flow rate of dry air can be calculated from either Equation (45) or (46). From Equation (45):

$$m_a = \frac{q_s + \Sigma(m_w h_w)}{h_2 - h_1} = \frac{30,000 + (10 \times 1100.44)}{(60)(30.73 - 24.00)}$$

$$= 101.5 \text{ lb dry air/min}$$

At State 1, $v_1 = 13.29$ ft^3/lb dry air.

Therefore, supply volume $= m_a v_1 = 101.5 \times 13.29 = 1349$ cfm

TRANSPORT PROPERTIES OF MOIST AIR

For certain scientific and experimental work, particularly in the heat transfer field, many other moist air properties are important. Generally classified as transport properties, these include diffusion coefficient, viscosity, thermal conductivity, and thermal diffusion factor. Mason and Monchick (1965) derive these properties by calculation. Table 5 and Figures 12 and 13 summarize the

Table 5   Calculated Diffusion Coefficients for Water-Air at 29.921 in. Hg Barometric Pressure

Temp., °F	ft²/h	Temp., °F	ft²/h	Temp., °F	ft²/h
−100	0.504	40	0.884	140	1.205
−50	0.6	50	0.915	150	1.240
−40	0.655	60	0.942	200	1.414
−30	0.682	70	0.973	250	1.600
−20	0.709	80	1.008	300	1.794
−10	0.736	90	1.042	350	1.996
0	0.767	100	1.073	400	2.205
10	0.794	110	1.104	450	2.422
20	0.825	120	1.139	500	2.647
30	0.853	130	1.170		

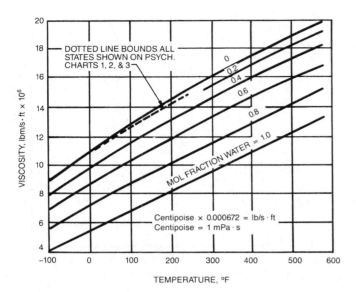

Fig. 12   Viscosity of Moist Air

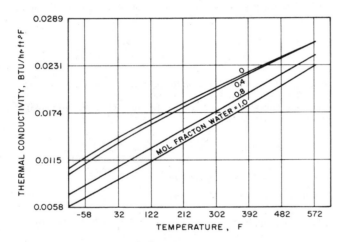

Fig. 13   Thermal Conductivity of Moist Air

author's results on the first three properties listed. Note that, within the boundaries of Charts 1, 2, and 3, the viscosity varies little from that of dry air at normal atmospheric pressure, and the thermal conductivity is essentially identical.

AIR, WATER, AND STEAM PROPERTIES

Coefficient f_w (or f_s) (over water) at pressures from 0.145 to 31.799 in. Hg for temperatures from −58 to 140°F (Smithsonian Institution).

Coefficient f_i (over ice) at pressures from 0.145 to 31.799 in. Hg for temperatures from 32 to 212°F (Smithsonian Institution).

Compressibility factor of dry air at pressures from 0.147 to 1470 psi and at temperatures from 90 to 5400°R (Hilsenrath *et al.* 1960).

Compressibility factor of moist air at pressures from 25 to 31 in. Hg at humidity ratios from 0.006 to 0.30 lb of water per lb of dry air, and for temperatures from 40 to 100°F. At pressures from 0 to 31.799 in. Hg, at values of degree of saturation from 0 to 100, and for temperatures from 32 to 140 °F (Smithsonian

Institution). [Note: At the time the Smithsonian Meteorological Tables were published, the value $\mu = W/W_s$ was known as relative humidity, in terms of a percentage. Since that time, there has been general agreement to designate the value μ as degree of saturation, usually expressed as a decimal and sometimes as a percentage. See Goff (1949) for more recent data and formulations.]

Compressibility factor for steam at pressures from 14.696 to 440 psi and at temperatures from 684 to 1530°R (Hilsenrath *et al.* 1960).

Density, enthalpy, entropy, Prandtl number, specific heat, specific heat ratio, and viscosity of dry air (Hilsenrath *et al.* 1960).

Density, enthalpy, entropy, specific heat, viscosity, thermal conductivity, and free energy of steam (Hilsenrath *et al.* 1960).

Dry air. Thermodynamic properties over a wide range of temperature (Keenan and Kaye 1945).

Enthalpy of saturated steam (Osborne *et al.* 1939).

Ideal-gas thermodynamic functions of dry air at temperatures from 18 to 5400°R (Hilsenrath *et al.* 1960).

Ideal-gas thermodynamic functions of steam at temperatures from 90 to 9000°R. Functions included are specific heat, enthalpy, free energy, and entropy (Hilsenrath *et al.* 1960).

Moist air properties from tabulated virial coefficients (Chaddock 1965).

Saturation humidity ratio over ice at pressures from 8.672 to 28.908 in. Hg and for temperatures from −128 to 32°F (Smithsonian Institution).

Saturation humidity ratio over water at pressures from 1.73 to 30.35 and for temperatures from −58 to 138.2°F (Smithsonian Institution).

Saturation vapor pressure over water in in. Hg and for temperatures from −60 to 212°F (Smithsonian Institution).

Speed of sound in dry air at pressures from 0.147 to 1470 psi for temperatures from 90 to 5400°R (Hilsenrath *et al.* 1960). At atmospheric pressure for temperatures from −130 to 140°F (Smithsonian Institution).

Speed of sound in moist air. Relations using the formulation of Goff and Gratch and studies by Hardy *et al.* (1942) give methods for calculating this speed (Smithsonian Institution).

Steam tables covering the range from 32 to 1472°F.

Transport properties of moist air. Diffusion coefficient, viscosity, thermal conductivity, and thermal diffusion factor of moist air are listed (Mason and Monchick 1965). The author's results are summarized in Table 5 and Figures 11 and 12.

Virial coefficients and other information for use with Goff and Gratch formulation (Goff 1949).

Volume of water in cubic feet for temperatures from −14 to 482°F (Smithsonian Institution 1954).

Water properties. Includes properties of ordinary water substance for the gaseous, liquid, and solid phases (Dorsey 1940).

SYMBOLS

α = $\ln(p_w)$, parameter used in Equations (35) and (36)
μ = degree of saturation W/W_s, dimensionless
ρ = moist air density, lb/ft^3
ϕ = relative humidity, dimensionless
C_1 to C_{18} = constants in Equations (3), (4), and (35)
d_v = absolute humidity of moist air, mass of water per unit volume of mixture
D = enthalpy deviation, Btu/lb dry air
f_s = enhancement factor, used in Equations (21a) and (23a)
h = enthalpy of moist air, Btu/lb dry air
h_a = specific enthalpy of dry air, Btu/lb
h_{as} = $h_s - h_a$
h_f = specific enthalpy of saturated liquid water
h_{fg} = $h_g - h_f$
h_g = specific enthalpy of saturated water vapor

h_s = enthalpy of moist air at saturation per unit mass of dry air

h_s^* = enthalpy of moist air at saturation at thermodynamic wet-bulb temperature per unit mass of dry air

h_w = specific enthalpy of water (any phase) added to or removed from moist air in a process

h_w^* = specific enthalpy of condensed water (liquid or solid) at thermodynamic wet-bulb temperature and pressure of 29.921 in. Hg

H_s = rate of sensible heat gain for space

H_t = rate of total energy gain for space

m_a = mass flow of dry air, per unit time

m_w = mass flow of water (any phase), per unit time

M_a = mass of dry air in moist air sample

M_w = mass of water vapor in moist air sample

$n = n_a + n_w$, total number of moles in moist air sample

n_a = moles of dry air

n_w = moles of water vapor

p = total pressure of moist air

p_a = partial pressure of dry air

p_s = vapor pressure of water in moist air at saturation. Differs from saturation pressure of pure water because of presence of air.

p_w = partial pressure of water vapor in moist air

p_{ws} = pressure of saturated pure water

q = specific humidity of moist air, mass of water per unit mass of mixture

q_s = rate of addition (or withdrawal) of sensible heat

R = universal gas constant, 1545.32 $(lb_f/ft^2)ft^3/(lb\ mole \cdot °R)$

R_a = gas constant for dry air

R_w = gas constant for water vapor

s = entropy of moist air per unit mass of dry air

s_a = specific entropy of dry air

$s_{as} = s_s - s_a$

s_f = specific entropy of saturated liquid water

$s_{fg} = s_g - s_f$

s_g = specific entropy of saturated water vapor

s_s = specific entropy of moist air at saturation per unit mass of dry air

s_w = specific entropy of condensed water (liquid or solid) at pressure of 29.921 in. Hg

t = dry-bulb temperature of moist air, °F

t_d = dew-point temperature of moist air, °F

t^* = thermodynamic wet-bulb temperature of moist air, °F

T = absolute temperature, °R

v = volume of moist air, per unit mass of dry air

v_a = specific volume of dry air

$v_{as} = v_s - v_a$

v_f = specific volume of saturated liquid water

$v_{fg} = v_g - v_f$

v_g = specific volume of saturated water vapor

v_s = volume of moist air at saturation, per unit mass of dry air

v_T = total gas volume

V = total volume of moist air sample

W = humidity ratio of moist air, mass of water per unit mass of dry air

W_s = humidity ratio of moist air at saturation

W_s^* = humidity ratio of moist air at saturation at thermodynamic wet-bulb temperature

x_a = mole-fraction of dry air, moles of dry air per mole of mixture

x_w = mole-fraction of water, moles of water per mole of mixture

x_{ws} = mole-fraction of water vapor under saturated conditions, moles of vapor per mole of saturated mixture

REFERENCES

ASHRAE. 1977. Brochure on psychrometry.

Chaddock, J.B. 1965. Moist air properties from tabulated virial coefficients. *Humidity and moisture measurement and control in science and industry* 3:273. A. Wexler and W.A. Wildhack, eds. Reinhold Publishing Corp., New York.

Dorsey, N.E. 1940. *Properties of ordinary water substance*. Reinhold Publishing Corp., New York.

Goff, J.A. 1949. Standardization of thermodynamic properties of moist air. *Heating, Piping, and Air Conditioning* 21(11):118.

Goff, J.A. and S. Gratch. 1945. Thermodynamic properties of moist air. ASHVE *Transactions* 51:125.

Goff, J.A., J.R. Anderson, and S. Gratch. 1943. Final values of the interaction constant for moist air. ASHVE *Transactions* 49:269.

Haines, R.W. 1961. How to construct high altitude psychrometric charts. *Heating, Piping, and Air Conditioning* 33(10):144.

Hardy, H.C., D. Telfair, and W.H. Pielemeier. 1942. The velocity of sound in air. *Journal of the Acoustical Society of America* 13:226.

Harrison, L.P. 1965. Fundamental concepts and definitions relating to humidity. In *Humidity and moisture measurement and control in science and industry* 3:289. A. Wexler and W.H. Wildhack, eds. Reinhold Publishing Corp., New York.

Hilsenrath, J. *et al.* 1960. Tables of thermodynamic and transport properties of air, argon, carbon dioxide, carbon monoxide, hydrogen, nitrogen, oxygen, and steam. National Bureau of Standards. *Circular* 564, Pergamon Press, New York.

Hyland, R.W. and A. Wexler. 1983a. Formulations for the thermodynamic properties of dry air from 173.15 K to 473.15 K, and of saturated moist air from 173.15 K to 372.15 K, at pressures to 5 MPa. ASHRAE *Transactions* 89(2A):520-35.

Hyland, R.W. and A. Wexler. 1983b. Formulations for the thermodynamic properties of the saturated phases of H_2O from 173.15 K to 473.15 K. ASHRAE *Transactions* 89(2A):500-519.

Karig, H.E. 1946. Psychrometric charts for high altitude calculations. *Refrigerating Engineering* 52(11):433.

Keenan, J.H. and J. Kaye. 1945. *Gas tables*. John Wiley and Sons, New York.

Kusuda, T. 1970. Algorithms for psychrometric calculations. NBS Publication BSS21 (January) for sale by Superintendent of Documents, U.S. Government Printing Office, Washington, D.C.

Mason, E.A. and L. Monchick. 1965. Humidity and moisture measurement and control in science and industry. *Survey of the Equation of State and Transport Properties of Moist Gases* 3:257. Reinhold Publishing Corp., New York.

NASA. 1976. U.S. Standard atmosphere, 1976. National Oceanic and Atmospheric Administration, National Aeronautics and Space Administration, and the United States Air Force, Superintendent of Documents. U.S. Government Printing Office, Washington, D.C.

NIST. 1990. Guidelines for realizing the international temperature scale of 1990 (ITS-90). NIST Technical Note 1265. National Institute of Technology and Standards, Gaithersburg, MD.

Osborne, N.S. 1939. Stimson and Ginnings. Thermal properties of saturated steam. *Journal of Research*, National Bureau of Standards, 23(8):261.

Palmatier, E.P. 1963. Construction of the normal temperature. ASHRAE psychrometric chart. ASHRAE *Journal* 5:55.

Peppers, V.W. 1988. Unpublished paper. Available from ASHRAE.

Preston-Thomas, H. 1976. The international practical temperature scale of 1968, amended edition of 1975. *Metrologia* 12:7-17.

Rohsenow, W.M. 1946. Psychrometric determination of absolute humidity at elevated pressures. *Refrigerating Engineering* 51(5):423.

Smithsonian Institution. 1954. *Smithsonian physical tables*, 9th rev. ed. Available from the Smithsonian Institution, Washington, D.C.

Smithsonian Institution. *Smithsonian meteorological tables*, 6th rev. ed. Out of print, but available in many libraries. Washington, D.C.

The international temperature scale of 1990 (ITS-90). Metrologia 27:3-10.

Threlkeld, J.L. 1970. *Thermal environmental engineering*, 2nd ed. Prentice-Hall, New York, 175.

SOUND AND VIBRATION

THE design, installation, and use of HVAC and refrigeration systems, without attention to sound and vibration control, can result in complaints due to an unacceptable acoustical environment. These problems can be avoided if fundamental principles of sound and vibration control are applied. This chapter introduces these principles, including characteristics of sound; basic definitions and terminology; human response to sound; acoustical design goals; the source/path/receiver concept in sound and vibration control; and vibration isolation fundamentals. Chapter 42 of the 1991 ASHRAE *Handbook—Applications* and the references listed at the end of this chapter provide further information on the subject.

ACOUSTICAL DESIGN

The primary objective of HVAC system and equipment acoustical design is to create an appropriate acoustical environment for a given space. To achieve appropriate sound levels, the source/path/receiver concept must be applied; *i.e.*, sound and vibration are created by a *source*, transmitted along one or more *paths*, and reach a *receiver*. Treatments and modifications can be applied to any or all of these elements to achieve a proper acoustical or vibration environment.

CHARACTERISTICS OF SOUND

Sound is a travelling oscillation in a medium exhibiting the properties of both elasticity and inertia. In fluid media (air or water), the disturbance travels as a longitudinal wave. Sound is generated by a vibrating surface or a turbulent fluid stream. In HVAC system design, both air- and structureborne sound propagation are of concern.

Speed

The speed of a longitudinal wave in a fluid medium is a function of the medium's density and modulus of elasticity. In air, at room temperature, the speed of sound is about 1100 ft/s; in water, about 5000 ft/s.

Frequency

Frequency is the number of oscillations (or cycles) per second completed by a vibrating object. The international unit for frequency is hertz (Hz).

Wavelength

Wavelength is the distance between successive rarefactions or compressions of the propagation medium. Wavelength, speed, and frequency are interrelated by the following equation:

$$\lambda = c/f \tag{1}$$

where

λ = wavelength, ft
c = speed of sound, ft/s
f = frequency, Hz

Frequency Spectrum and Bandwidths

The audible frequency range extends from about 20 Hz to 20 kHz. In some cases, infrasound (< 20 Hz) or ultrasound (> 20 kHz) are important, but methods and instrumentation for these frequency regions are specialized and are not considered here. Within the frequency range of interest, a sound source is characterized by its sound power output in octave or 1/3 octave bands, although narrower bandwidths may be appropriate for certain analyses. An octave is a frequency band with its upper band limit twice the frequency of its lower band limit. Table 1 lists the preferred series of octave bands and the upper and lower band limit

Table 1   Center Approximate Cutoff Frequencies for Octave and 1/3 Octave Band Series (ANSI *Standard* S1.6)

Octave Bands, Hz			1/3 Octave Bands, Hz		
Lower	Center	Upper	Lower	Center	Upper
			22.4	25	28
22.4	31.5	45	28	31.5	35.5
			35.5	40	45
			45	50	56
45	63	90	56	63	71
			71	80	90
			90	100	112
90	125	180	112	125	140
			140	160	180
			180	200	224
180	250	355	224	250	280
			280	315	355
			355	400	450
355	500	710	450	500	560
			560	630	710
			710	800	900
710	1000	1400	900	1000	1120
			1120	1250	1400
			1400	1600	1800
1400	2000	2800	1800	2000	2240
			2240	2500	2800
			2800	3150	3550
2800	4000	5600	3550	4000	4500
			4500	5000	5600
			5600	6300	7100
5600	8000	11 200	7100	8000	9000
			9000	10,000	11,200
			11,200	12,500	14,000
11 200	16 000	22 400	14,000	16,000	18,000
			18,000	20,000	22,400

The preparation of this chapter is assigned to TC 2.6, Sound and Vibration Control.

frequencies. An octave band can, on a logarithmic frequency scale, be divided into three equally wide 1/3 octave bands with upper and lower frequency limits in the ratio of the cube root of two to one. One-third octave band center frequencies and upper and lower band limits are also listed in Table 1. The center frequency of an octave or 1/3 octave band is the geometric mean of its upper and lower band limits. Octave and 1/3 octave bands are identified by their center frequencies, not by their upper and lower band limit frequencies (ANSI *Standard* S1.11). While analysis in octave bands is usually acceptable for rating acoustical environments in rooms, 1/3 octave band analysis is often useful in product development and troubleshooting investigations.

Noise

The first and simplest definition of noise is any unwanted sound. The second definition is that noise is broadband sound with no distinguishable frequency characteristics, such as the sound of a waterfall. This definition is appropriate when one sound is used to mask another, as when controlled sound radiated into a room from a well-designed air-conditioning system is used to mask or hide low-level intrusive sounds from adjacent spaces to increase privacy. This controlled sound is called noise, but not in the context of unwanted sound; rather, it is a broadband, bland sound that is frequently unobtrusive. Three types of noise in the second context are frequently encountered in acoustics:

1. Random noise is an oscillation, the instantaneous magnitude of which is not specified for any given instant. The instantaneous magnitudes of a random noise are specified only by probability distributions, giving the fraction of the total time that the magnitude, or some sequence of magnitudes, lies within a specified range.
2. White noise has a continuous frequency spectrum with equal energy/Hz over a specified frequency range. In this sense, it is like white light. White noise is not necessarily random.
3. Pink noise also has a continuous frequency spectrum, but has equal energy per constant-percentage bandwidth, such as per octave or 1/3 octave band. The need for pink noise arose primarily in acoustical laboratory testing. Since the octave bands in white noise double in width for each successive band, the energy in the band also doubles in each successive band. Pink noise, a constant energy per bandwidth noise, is obtained by sloping the energy profile by 3 dB per octave.

BASIC DEFINITIONS

Decibel

The decibel (dB) is a basic unit of measurement in acoustics. Its use is often confusing because it is used to quantify many different descriptors relating to sound source strength, sound level, and sound attenuation. For this reason, it is important to be aware of the context in which the term is used.

Numerically, the decibel is ten times the base 10 logarithm of the ratio of two like quantities proportional to acoustical power or energy. The term level, when used in relation to sound power, sound intensity, or sound pressure, connotes that dB notation is being used. A reference quantity is always implied if it does not appear.

Sound Intensity and Sound Intensity Level

If a sphere is circumscribed around an arbitrary distance from a source, all the energy radiated by the source must pass through the sphere. Power flow through a unit area of the sphere is intensity, expressed in watts per square meter. (The SI system units are used here rather than I-P system units because of international agreement on the definition.) Sound intensity follows the inverse

square law; *i.e.*, sound intensity varies inversely as the square of distance from the source. This is true for acoustical sources outdoors and, to a limited extent, indoors. Sound intensity level is expressed in dB with a reference quantity of 10^{-12} W/m^2.

Sound Pressure and Sound Pressure Level

Sound intensity is difficult to measure directly. On the other hand, sound pressure is relatively easy to measure; the human ear and microphones are pressure-sensitive devices. It can be shown that pressure squared is proportional to intensity. Using the same techniques, a decibel scale for sound pressure can be created in a manner analogous to the decibel scale for sound intensity. In this case, the reference pressure is 20 μPa, which corresponds to the approximate threshold of hearing. Since pressure squared is proportional to intensity, sound pressure level is:

$$L_p = 10 \log (p/p_{ref})^2 \text{ re } p_{ref} \tag{2}$$

Since p_{ref} is 20 μPa, which is 2×10^{-5} Pa, and since $10 \log p^2 = 20 \log p$:

$$L_p = 20 \log (p/2 \times 10^{-5}) \text{ re } 20 \mu\text{Pa} \tag{3}$$

where p is the root mean square (rms) value of pressure in micropascals. Or

$$L_p = 20 \log p + 94 \text{ dB re } 20 \mu\text{Pa} \tag{4}$$

The human ear responds across a broad range of sound pressures; threshold of hearing to threshold of pain covers a range of approximately 10^{14}:1. The linear range scale for sound pressure in Table 2 is awkward in this form; therefore, the equivalent logarithmic notations in the third column of Table 2 should be used. Table 2 also gives the sound pressures and sound pressure levels of various typical sources.

Since the decibel is a logarithmic unit, two sound levels cannot be added arithmetically. Two common methods for adding decibels are used. For the more accurate method of adding sound pressure levels, divide each by 10 and take the antilogarithm. Then

Table 2 Typical Sound Pressures and Sound Pressure Levels

Source	Sound Pressure, Pa	Sound Pressure Level, dB re 20μPa	Subjective Reaction
Military jet takeoff at 100 ft	200	140	Extreme
Artillery fire at 10 ft	63.2	130	danger
Passenger's ramp at jet airliner (peak)	20	120	Threshold of pain
Loud rock band[a]	6.3	110	Threshold of discomfort
Platform of subway station (steel wheels)	2	100	
Unmuffled large diesel engine at 130 ft	0.6	90	Very loud
Computer printout room[a]	0.2	80	
Freight train at 100 ft	0.06	70	
Conversational speech at 3 ft	0.02	60	
Window air conditioner[a]	0.006	50	Moderate
Quiet residential area[a]	0.002	40	
Whispered conversation at 6 ft	0.0006	30	
Buzzing insect at 3 ft	0.0002	20	
Threshold of good hearing	0.00006	10	Faint
Threshold of excellent youthful hearing	0.00002	0	Threshold of hearing

[a]Ambient.

Table 3 Combining Decibels to Determine Overall Sound Pressure Level

Octave Band Frequency, Hz	Octave Band Level, L_p	Antilog		
63	85	3.2×10^8	$= 0.32$	$\times 10^9$
125	90	1.0×10^9	$= 1.0$	$\times 10^9$
250	92	1.6×10^9	$= 1.6$	$\times 10^9$
500	87	5.0×10^8	$= 0.5$	$\times 10^9$
1000	82	1.6×10^8	$= 0.16$	$\times 10^9$
2000	78	6.3×10^7	$= 0.06$	$\times 10^9$
4000	65	3.2×10^6	$= 0.003$	$\times 10^9$
8000	54	2.5×10^5	$= 0.0002$	$\times 10^9$
			3.6432×10^9	

$$10 \log (3.6 \times 10^9) = 96 \text{ dB}$$

add the antilogarithms—take the logarithm, and multiply by 10 to obtain the combined level. The process can be used to add a series of decibel levels, such as the octave band analysis of a noise (Table 3), to reach the overall level. As previously noted, this can be done by converting all the levels to mixed notations, adding them, and converting back to logarithmic notations. Alternatively, add the largest and next largest, and repeat this process until the next addition has little or no influence.

The second method is simpler and slightly less accurate. In this method, simply refer to Table 4 to perform the desired addition. This method, although not exact, results in errors of 1 dB or less.

Table 4 Combining Two Sound Levels

Difference between Two Levels to be Combined, dB	0 to 1	2 to 4	5 to 9	10 and More
Number of decibels to be added to higher level to obtain combined level	3	2	1	0

Sound Power and Sound Power Level

Referring back to the definition for sound intensity, note the provision for a sound source. A fundamental characteristic of an acoustic source is its ability to radiate energy, whether weak and small in size (a cricket) or strong and large (a compressor). Some energy input excites the source, which radiates some fraction of this energy in the form of sound. Since unit power radiated through a unit sphere yields unit intensity, the power reference base, established by international agreement, is 1 picowatt (pW) (10^{-12} W). The reference quantity used should be stated explicitly. A definition of sound power level is, therefore:

$$L_w = \log (w/10^{-12}\text{W}) \text{ dB re 1 pW}$$

or

$$L_w = 10 \log w + 120 \text{ dB re 1 pW} \qquad (5)$$

Sound power outputs for common sources are shown in Table 5.

HUMAN RESPONSE TO SOUND

Objective Determination

Objectively determining the subjective reaction to sound in terms of loudness is complex; the primary method for determining this reaction is a statistically representative sample of human observers. To determine the loudness of a sound, a standard sound is chosen; a sampling of people compares an unknown sound with the standard sound. The accepted standard sound is a pure tone of 1000 Hz or a narrow band of random noise centered on 1000 Hz. The loudness level of sound is defined as the sound pressure level of a standard sound that seems to be as loud as the unknown to the sample observers. Loudness level is expressed in phons, and the loudness level of any sound in phons is equal to the sound

Table 5 Typical Sound Power Outputs and Sound Power Levels

Source	Approximate Power Output	
	Watts	Decibel re: 10^{-12} W
Saturn rocket	10^8	200
Turbojet engine[a]	10^5	170
Jet aircraft at takeoff[b]	10^4	160
Turboprop at takeoff	1000	150
Prop aircraft at takeoff[c]	100	140
Large pipe organ	10	130
Small aircraft engine	1	120
Blaring radio	0.1	110
Automobile at highway speed	0.01	100
Voice, shouting	0.001	90
Garbage disposal unit	10^{-4}	80
Voice, conversation level	10^{-5}	70
Electronic equipment ventilation fan	10^{-6}	60
Office air diffuser	10^{-7}	50
Small electric clock	10^{-8}	40
Voice, soft whisper	10^{-9}	30
Rustling leaves	10^{-10}	20
Human breath	10^{-11}	10
Threshold of hearing	10^{-12}	0

[a]With afterburner.
[b]Four jet engines.
[c]Four propeller engines.

pressure level in decibels of an equally loud standard sound. Therefore, a sound that is judged to be as loud as a 40-dB, 1000-Hz tone has a loudness level of 40 phons. Statistical reactions of humans to pure tones are shown in Figure 1 (Robinson and Dadson 1956). The reaction changes when the sound is a band of random noise (Pollack 1952), rather than a pure tone (Figure 2). These curves are equal loudness contours and were developed by asking the subjects to match the loudness of tones or bands of random noise presented at other frequencies and levels to the loudness of a tone or band centered on 1000 Hz. The figures indicate that the human perception of sound is most sensitive in the mid-frequency range.

The designer's fundamental concern is how humans respond to sound; frequently this is only generally related to the variation of sound energy. Humans subjectively say that one sound is louder than another, higher or lower pitched, etc. Under carefully controlled experimental conditions, humans can detect small changes in sound level. However, the human reaction describing the halving or doubling of loudness requires changes in a sound pressure level of about 10 dB. For broadband sounds, 3 dB is the minimum perceptible change. This means that halving the power output of

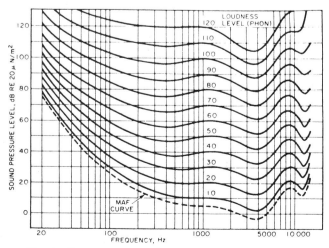

Fig. 1 Free-Field Loudness Contours for Pure Tones

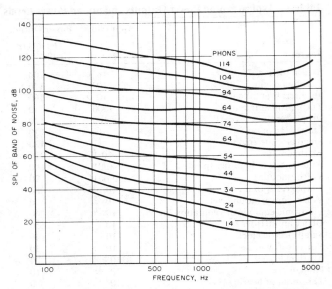

Fig. 2   Equal Loudness Contours for Relatively Narrow Bands of Random Noise

Table 6   Subjective Effect of Changes in Sound Pressure Level, Broadband Sounds

Change in Sound Pressure Level	Apparent Change in Loudness
3 dB	Just noticeable
5 dB	Clearly noticeable
10 dB	Twice (or half) as loud

the source causes a barely noticeable change in sound pressure level, and the sound power output must be reduced by a factor of 10 before humans determine that loudness has been halved. Subjective changes are shown in Table 6.

Quality of Sound

The way in which humans react to the quality of sound is as important to the designer as their reaction to the loudness. Sound quality is a function of the relative intensities of sound levels in each region of the audible frequency spectrum and the presence of tones or level fluctuations. Figure 3 shows an indoor sound spectrum and the manner in which fan and diffuser noise contributes to that spectrum at various frequencies. The fan noise has been attenuated to a degree that it approaches the sound criterion (in this case an RC-35 contour) only in the lower octave bands. If this were the only noise present in the space, it would be considered rumbly by most listeners. However, the diffusers have been selected to balance the spectrum by filling in the higher frequencies so that the quality of the sound is more pleasant. Unfortunately, achieving a balanced sound spectrum is not usually this easy—there may be a multiplicity of sound sources to consider. As a guide to the designer, Figure 4 shows the more common mechanical and electrical noise sources and frequency regions that control the indoor noise spectrum. Chapter 42 of the 1991 ASHRAE *Handbook—Applications* provides more detailed information on treating some of these noise sources.

MEASURING SOUND AND VIBRATION

Instrumentation

The great majority of HVAC sound level measurements can be made using a battery-operated, handheld sound level meter with an attached filter set. The meter uses a microphone, internal

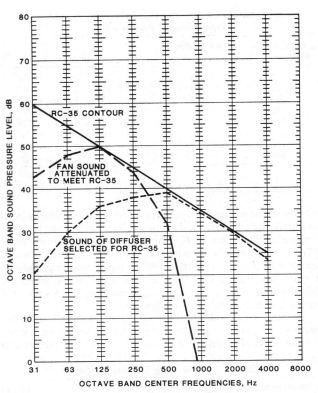

Fig. 3   Well-Balanced Sound Spectrum Resulting from Proper Selection of Air Outlets and Adequate Fan Noise Attenuation

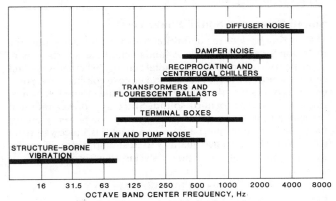

Fig. 4   Frequencies at which Various Types of Mechanical and Electrical Equipment Generally Control Sound Spectra

electronic circuits, and a readout display to measure and display the L_p at the desired location. The most useful meters have high (Type 1) or moderate (Type 2) precision; Linear, A, and C weighting networks; octave band filters with center frequencies from 31 to 8000 Hz; and a moving needle or LCD thermometer-style display to show sound level fluctuations. Some manufacturers also provide 1/3 octave band filters for their handheld meters. Sound level meters without octave or 1/3 octave band filters are much less valuable because, in a troubleshooting situation, they provide only limited help in determining the frequency content of the offending noise, a necessary step in determining noise reduction strategies.

Measurement Procedures

Determining the sound spectrum in a room or troubleshooting a noise complaint usually requires measuring the octave band

Table 7 Guidelines for Determining Equipment Sound Levels in Presence of Contaminating Background Noise

Measurement A minus Measurement B	Correction to Measurement A to Obtain Equipment Sound Level
10 dB or more	0 dB
6 to 9 dB	−1 dB
4 to 5 dB	−2 dB
3 dB	−3 dB
2 dB	−4 dB
1 dB	−7 dB
0 dB	Equipment sound level is at least 9 dB below Measurement A

Measurement A = Tested equipment plus background noise
Measurement B = Background noise alone

L_p values in the octave bands from 31 to 4000 Hz. In cases where tonal noise or rumble is the complaint, 1/3 octave band measurements are recommended because of their ability to discriminate narrower frequency bands of noise.

Measuring noise emissions from a particular piece of equipment or group of equipment requires a measurement plan specific to the situation. The Air-Conditioning and Refrigeration Institute (ARI), the Air Movement and Control Association (AMCA), the American Society of Testing and Materials (ASTM), the American National Standards Institute (ANSI), the Acoustical Society of America (ASA), and ASHRAE all publish sound level measurement procedures for various laboratory and field noise measurement situations.

Sometimes the noise due to a particular piece of HVAC equipment must be measured in the presence of background noise from other sources that cannot be turned off, such as automobile traffic or certain business machines. Determining the sound level due to the selected equipment alone requires making two sets of measurements, one set with the HVAC equipment operating with the background noise and another set with only the background noise (with the equipment turned off). This situation might occur, for example, when determining whether or not the property line noise exposure from a cooling tower meets a local noise ordinance. The guidelines given in Table 7 will help to determine the sound level of a particular machine in the presence of background noise.

Vibration Measurements

While the control of HVAC system noise and vibration are of equal importance, the measurement of vibration is not usually necessary for determining the sources or transmission paths of disturbing noise. In addition, the techniques and instrumentation used for vibration measurement and analysis are specialized and beyond the scope of this handbook. Therefore, designers should consult other sources (*e.g.*, Harris 1991) for descriptions of vibration measurement and analysis methods.

DETERMINING SOUND POWER

Sound power cannot be determined directly. Rather, it is determined by measuring the sound pressure or sound intensity created by a source under one of several circumstances and then calculating the sound power. Sound power is determined by measuring one of four different quantities—sound pressure in a free field, sound pressure in a reverberant field, sound pressure in a progressive wave field, or sound intensity in a nonreactive field.

Free-Field Method

A sound field in which the effects of the boundaries are negligible over the frequency range of interest is a free field. In a free field or in spaces where free-field conditions can be approximated (*e.g.*, by sound-absorbing walls, floor, and ceiling), the sound power of a sound source can be determined from a number of sound pressure level measurements around the source and equally spaced from it. This method is based on the fact that, since absorption of sound in air can be practically neglected at small distances from the sound source, the sound power generated by a source must flow through an imagined sphere with the source at its center. The intensity of the sound is determined at each of the measuring points around the source and multiplied by the area of the imagined sphere associated with the measuring points. Total sound power is the sum of these products for each point.

Using numerical values of acoustic velocity c for air at 68 °F and 14.7 psi (standard atmosphere) and using sound power level L_w (dB re 10^{-12} W) and sound pressure level L_p (dB re 20 μPa), yields the following relationships between sound power level and sound pressure level for a nondirectional sound source, measured in a free field at a distance r (in feet):

$$L_w = L_p + 20 \log r + 0.5 \text{ dB} \qquad (6)$$

Actual sound sources radiate different amounts of sound power in different directions, because the source is a certain size and the various areas of its surface do not necessarily vibrate at the same level or in phase. A directivity pattern can be established by measuring sound pressure under free-field conditions, either in an anechoic room or over a reflecting plane. Directivity is described by the directivity factor Q, which is a function of frequency and direction. Directivity is defined as the ratio of sound pressure at a given angle from the sound source to the sound pressure that would be produced by the same source radiating uniformly in all directions. Chapter 42 of the 1991 ASHRAE *Handbook—Applications* provides more detailed information on sound source directivity.

Since most sound sources have more or less pronounced directional characteristics, measurements have to be made at several points. ANSI *Standard* S12.35 describes various methods used to calculate the sound power level under free-field conditions.

In many cases, a fully free field is not available, and measurements have to be made in a free field over a reflecting plane. This means that the sound source is placed on a hard floor (in an otherwise sound-absorbing room) or on pavement outdoors. Since the sound is then radiated into a hemisphere rather than a full sphere, the relationship for L_w and L_p for a nondirectional sound source is:

$$L_w = L_p + 20 \log r - 2.5 \text{ dB} \qquad (7)$$

The free-field method, with measurements made in a room with sound-absorbing walls, floor, and ceiling, is limited in the lower frequencies by the difficulty of obtaining room surface treatments that have high sound absorption coefficients at low frequencies. For example, a glass fiber wedge structure useful at 70 Hz must be at least 4 ft long.

Reverberant Field Method

A sound field in which the sound intensity is the same in all directions and at every point due to reflections from the room surfaces is called a reverberant field. Another widely used method to determine sound power places the sound source in a reverberation room that has sound-reflecting walls, floor, and ceiling. The sound pressure level is then measured at some distance from the source and the surfaces of the room. The sound power level is calculated from the sound pressure level, if the rate of decay of sound (in dB/s) and the volume of the reverberation room are known (direct method). Alternately, the sound pressure level of a reference sound source (RSS) with known sound power output in the reverberation room may be measured, and then the unknown sound source may be measured to obtain its sound power output through comparison (substitution method).

Standardized methods for determining the sound power of HVAC equipment in reverberation rooms are given in ANSI *Standard* 12.31-90, when the sound source contains mostly broadband noise; ANSI *Standard* 12.32-90, when tonal noise is prominent; and AMCA *Standard* 300-85 for testing fans.

The relationship between sound power level and sound pressure level in a reverberation room is given by:

$$L_w = L_p + 10 \log V + 10 \log D - 47.3 \text{ dB} \qquad \text{(direct method)} \quad (8)$$

and

$$L_w = L_p + (L_w - L_p)_{ref} \qquad \text{(substitution method)} \quad (9)$$

where

$$
\begin{aligned}
L_p &= \text{sound pressure level averaged over room} \\
V &= \text{volume of room, ft}^3 \\
D &= \text{decay rate, dB/s} \\
(L_w - L_p)_{ref} &= \text{difference between sound power level and sound} \\
&\quad\ \text{pressure level of reference sound source}
\end{aligned}
$$

Some sound sources that can be measured by these methods are room air conditioners, refrigeration compressors, components of central HVAC systems, and air terminal devices. AMCA *Standard* 300-85 and ARI *Standard* 880-89 establish special measuring procedures for some of these units. Large equipment that can operate on a large paved area, such as a parking lot, can also be measured under free-field conditions on a reflecting plane. Determining the sound power of large equipment is difficult; however, data may be available from some manufacturers.

Progressive Wave Method

One form of sound propagation is the progressive movement of a sound wave along a duct. Fan sound power can be determined in a progressive wave field by measuring the sound pressure level inside a duct. The method is described in detail in ASHRAE *Standard* 68-1986 (AMCA *Standard* 330-86) for in-duct testing of fans.

Sound Intensity Method

Recent advances in acoustical instrumentation now permit the direct determination of sound intensity, previously defined as the sound power per unit area flowing through a control surface. By measuring the sound intensity over the sphere or hemisphere surrounding a sound source, the spatially averaged sound power radiated by the source can be determined. One of the advantages of this method is that, with a few limitations, sound intensity (and, therefore, sound power) measurements can be made in the presence of background noise in ordinary rooms, thereby eliminating the need for a special testing environment. Another advantage is that by measuring sound intensity over restricted areas around a sound source, sound directivity can be determined. This procedure can be particularly useful in reducing noise of products during their development.

Sound intensity measurement procedures are still experimental and as yet no national or international standards exist. One particular concern is that small test rooms or those having somewhat flexible boundaries (sheet metal or thin drywall) can permit a reactive sound field to exist, one in which the room's acoustical characteristics cause it to affect the sound power output of the source.

Measurement Bandwidths

Sound power is normally determined in octave or 1/3 octave bands. Occasionally, a more exact determination of the sound source spectrum is required. In these cases, narrowband analysis, using either constant relative bandwidth (*e.g.*, 1/12 octave) or constant absolute bandwidth (*e.g.*, 5 Hz) can be applied. The most frequently used analyzer types are digital filter analyzers for constant relative bandwidth measurements and fast Fourier transform (FFT) analyzers for constant bandwidth measurements. Narrowband analysis results are used to determine the exact frequencies of pure tones and their harmonics in a sound spectrum.

CONVERTING FROM SOUND POWER TO SOUND PRESSURE

Provided with sound power level information on a source, the designer may be required to convert the sound power level to predict the sound pressure level at a given location.

The sound pressure level at a given location in a room relative to a source of known sound power level depends on: (1) room volume, (2) room furnishings and surface treatments, (3) magnitude of the sound source(s), and (4) distance from the sound source(s) to the point of observation.

In most typical rooms, the presence of acoustically absorbent surfaces and sound-scattering elements, such as furniture, creates a relationship between sound power and sound pressure level that is independent of the absorptive properties of the space. For example, hospital rooms, which have only a small amount of absorption, and executive offices, which have substantial absorption, are similar when the comparison is based on the same room volume and distance between the source and point of observation.

Equation (10) can be used to estimate the sound pressure level at a chosen observation point in a normally furnished room. The estimate is accurate to ±2 dB (Schultz 1985).

$$L_p = L_w - 5 \log V - 3 \log f - 10 \log r + 25 \text{ dB} \quad (10)$$

where

$$
\begin{aligned}
L_p &= \text{room sound pressure level at chosen reference point,} \\
&\quad\ \text{dB re 20 } \mu\text{Pa} \\
L_w &= \text{source sound power level, dB re } 10^{-12} \text{ W} \\
V &= \text{room volume, ft}^3 \\
f &= \text{octave band center frequency, Hz} \\
r &= \text{distance from source to observation point, ft}
\end{aligned}
$$

Equation (10) applies to a single sound source in the room. With more than one source, total sound pressure level at the observation point is obtained by adding (on an energy basis) the individual contribution of each source, using the corresponding L_w and r for each source.

TYPICAL SOURCES OF SOUND

Whenever mechanical power is generated or transmitted, a fraction of this power is converted into sound power and radiated into the air. Sound source characteristics depend on the form of mechanical power and the type of medium or structure of the mechanical power generation or transmission system. The most important sound source characteristics include total sound power output L_w, frequency spectrum, and radiation directivity Q.

Sound sources are so numerous that it is impossible to provide a complete listing. All mechanical and most electrical equipment is capable of generating noise and vibration. Typical sources of noise and vibration in building mechanical systems are:

- Rotating and reciprocating equipment such as fans, motors, pumps, and chillers. A minor unbalance in this equipment can vibrate machine surfaces to produce airborne noise.
- Air and fluid noises such as those associated with ductwork, piping systems, grilles, diffusers, terminal boxes, manifolds, and pressure-reducing stations.

- Excitation of surfaces, for example, friction; movement of mechanical linkages, ducts, and pipes; and impact within mechanisms, such as cams and valve slap.
- Magnetostriction (transformer hum), which becomes significant in motor laminations, transformers, switchgear, lighting ballasts, and dimmers. A characteristic of magnetostrictive oscillations is that their fundamental frequency is twice the line frequency (120 Hz in a 60 Hz system).

SOUND TRANSMISSION PATHS

Sound from a source is transmitted via a path to a receiver. Airborne and structureborne transmission paths are of principal concern for the HVAC system designer. Airborne paths can be via the atmosphere or through ductwork; structureborne paths are via solid materials. Room-to-room paths comprise both airborne and structureborne transmission paths. Chapter 42 of the 1991 ASHRAE *Handbook—Applications* has additional information on transmission paths.

Airborne Transmission

Atmospheric transmission. Sound transmits readily in the atmosphere, both indoors and outdoors. Indoor sound transmission differs from outdoor transmission due to the presence of room boundaries, which allow sound reflections. Thus, in most rooms, sound can be transmitted via a straight-line direct path from the source to the receiver, by one or more reflected paths (from the source to the receiver via boundary reflections), or by a combination of both. Sound propagation indoors does not follow the inverse square law. Outdoors, the effects of the reflections are small, provided that the source is not located near large reflecting surfaces. However, sound outdoors can refract and change propagation direction due to the presence of wind and temperature gradient effects. Sound propagation outdoors follows the inverse square law. Therefore, Equations (6) and (7) can be used to calculate the relationship between sound power level and sound pressure level for fully free-field and hemispherical free-field conditions, respectively.

Ductborne transmission. Ductwork can provide an effective sound transmission path since the sound is readily contained within the boundaries of the ductwork. Sound can transmit both upstream and downstream from the source. A special case of ductborne transmission is crosstalk, where sound is transmitted from one room to another via the duct path.

Room-to-room transmission. Room-to-room sound transmission generally involves both airborne and structureborne sound paths. The sound power incident on a room surface element undergoes three transmission phenomena: (1) some of the sound energy is reflected from the surface element back into the room, (2) a portion of the sound energy is lost as energy transfer to the material comprising the element, and (3) the remainder of the sound energy is transmitted through the element to the other room. Airborne sound is radiated as the element vibrates, and structureborne sound can be transmitted via interconnecting studs or the floor. Some of the sound from the source room can be radiated around the primary separating element and into the receiving room via sound flanking. A common sound flanking path is the return air plenum.

Structureborne Transmission

Solid structures are efficient transmission paths for sound, which frequently originates as a vibration imposed on the transmitting structure. The vibration can be either a single impulse or a steady-state energy input. Typically, only a small amount of the input energy is radiated by the structure as airborne sound. A lightweight structure with little inherent damping radiates more sound than a massive structure with greater damping.

NOISE CONTROL

Terminology

The following noninterchangeable terms are used to describe the acoustical performance of many system components. ASTM *Standard* C634 defines additional terms used to describe acoustical performance parameters.

Sound attenuation is a general term describing the reduction of the intensity of sound as it travels from a source to a receiver.

Insertion loss (IL) of a silencer or other sound-reducing element is expressed in dB and is defined as the decrease in sound pressure level or sound intensity level, measured at the location of the receiver, when the silencer or a sound-reducing element is inserted into the transmission path between the source and the receiver. For example, a duct silencer would be replaced by a section of straight unlined duct to determine the silencer's insertion loss. Measurements are typically made in octave bands.

Sound transmission loss (TL) of a partition or other building element is expressed in dB and is the loss equal to 10 times the common logarithm (base 10) of the ratio of the airborne sound power incident on the partition to the sound power transmitted by the partition and radiated on the other side. Measurements are typically made in octave or 1/3 octave bands. Chapter 42 of the 1991 ASHRAE *Handbook—Applications* defines the special case of breakout transmission loss through duct walls.

Noise reduction (NR) is also expressed in dB and is the difference between the space-time average sound pressure levels produced in two enclosed spaces or rooms by one or more sound sources in the rooms. An alternate, non-ASTM definition of NR is the difference in sound pressure levels measured upstream and downstream of the silencer or sound-reducing element. Measurements are typically made in octave or 1/3 octave bands.

Sound absorption coefficient (α) is the fraction of the incident sound energy that is absorbed by a construction material or assembly. It is measured using 1/3 octave bands of broadband noise and is commonly reported at the full octave band center frequencies. The α value of a material in a specific octave band depends on the material's thickness, flow resistivity, stiffness, and method of attachment to the supporting structure. Duct and plenum liners are usually fibrous materials with high sound absorption values at mid to high frequencies.

Spherical spreading is the decrease in sound intensity due to the effects of inverse square law propagation. It occurs when the sound source is located either in free space or positioned on a reflecting plane.

Scattering is the change in direction of sound propagation due to an obstacle or inhomogeneity in the transmission medium. It results in the incident sound energy being dispersed in many directions.

Enclosures and Barriers

Enclosing the noise source is a common means of controlling airborne sound transmission. A homogeneous, single panel has a certain weight or mass per unit area and a certain stiffness and internal resistance; the sound transmission loss of such a panel has a frequency characteristic like the one shown in Figure 5. For most common materials, audible frequencies are in the mass-controlled part of the curve. The transmission loss (TL) of a limp panel without stiffness, with sound impinging perpendicular to the panel, is modeled by the mass law equation for sound at normal incidence, giving:

$$TL_o = 20 \log w_s f - 28 \text{ dB} \qquad (11)$$

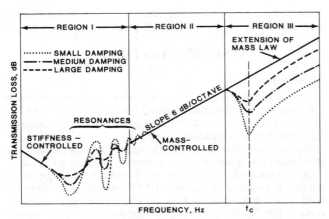

Fig. 5 Typical Transmission Loss of Panel

where w_s is the surface mass of the panel in lb/ft² and f is the frequency in Hz. For homogeneous panels, heavy materials attenuate sound better than light materials. Transmission loss in a diffuse sound field is averaged over all angles of incidence to yield:

$$TL_{random} = TL_o - 10 \log (0.23\ TL_o)\ \text{dB} \qquad (12)$$

Thus, the random incidence transmission loss of common building materials is often 5 to 10 dB less than the normal incidence transmission loss.

If the sound fields on both sides of a panel are diffuse, the panel's noise reduction is a function of its area S_p and the total sound absorption A_r in the receiving space, according to the following equation:

$$NR = TL + 10 \log A_r - 10 \log S_p\ \text{dB} \qquad (13)$$

Since the total sound absorption in a room is expressed as the equivalent area of perfect sound absorption, both S_p and A_r are expressed in consistent units, usually square feet.

The sound transmission class (STC) rating of a partition or assembly is often used in architecture to classify sound isolation performance. However, the STC rating should not be used as an indicator of an assembly's ability to contain equipment noise or any source rich in low frequencies. The reason is that the STC rating system was developed to deal effectively with speech frequency sound sources (250 to 4000 Hz). It is possible for a drywall partition to have a higher STC rating than a masonry wall, yet the masonry wall is likely to isolate most HVAC noise sources more effectively. Because of the limited frequency range and selective frequency discrimination of most single number rating systems, the designer should work with octave band sound transmission loss values rather than single number ratings.

In many applications, noise reduction of an enclosure is severely compromised by openings or leaks in the enclosure. Ducts that lead into or through a noisy space can carry sound to many areas of a building. Designers should consider this when designing duct, piping, and electrical systems.

Attenuation of Noise in Ducts and Plenums

Most ductwork, even a sheet metal duct without acoustical lining or silencers, attenuates sound. The natural attenuation of unlined ductwork is minimal, but can, for long duct runs, significantly reduce ductborne sound. Acoustic lining of ductwork can greatly attenuate the propagation of sound through ducts, particularly at mid to high frequencies. Chapter 42 of the 1991 ASHRAE *Handbook—Applications* has a detailed discussion of lined and unlined ductwork attenuation.

When lined ductwork cannot reduce sound propagation adequately, commercially available sound attenuators (also known as sound traps or duct silencers) can be used. There are three types: dissipative, reactive, and active. The first two are commonly known as passive silencers.

Dissipative silencers contain thick, perforated sheet metal baffles that restrict the air passage width within the silencer housing. The baffles are filled with low-density mineral fiber insulation. This type of silencer is most effective in reducing mid- and high-frequency sound energy.

Reactive silencers are similar to dissipative silencers except that the baffles are not filled with fibrous materials. This silencer type is typically used in HVAC systems serving hospitals, laboratories, or other areas with strict air quality standards. The reactive silencer acts as a complex of resonators that absorb incident sound energy via the Helmhotz resonator concept. Since reactive silencers do not contain fibrous materials, they are not as effective as dissipative silencers. Therefore, for equal performance, a reactive silencer must be longer than a dissipative silencer.

The first active silencers for commercial HVAC systems are currently being installed. Controlled laboratory arrangements have shown that active silencers effectively reduce both broadband and tonal noise in the 31 through 250 Hz octave bands. Active silencer systems use microphones, loudspeakers, and appropriate electronics to reduce in-duct noise by generating inverse phase sound waves that destructively interfere with the incident sound energy. Since the system's microphones and loudspeakers are mounted flush with the duct wall, there is no obstruction to airflow and, therefore, a negligible pressure drop. However, since active silencers are not effective in the presence of excessively turbulent airflow, their use is limited to relatively long, straight duct sections with an air velocity less than about 1500 ft/min.

Silencers and duct liner materials are tested according to ASTM *Standard* E477, which defines acoustical and aerodynamic performance in terms of insertion loss, self-generated noise (or self-noise), and airflow pressure drop. Insertion loss performance is measured in the presence of both forward and reverse flow. Forward flow occurs when the air and sound move in the same direction, as in a supply air or fan discharge system; reverse flow occurs when the air and sound travel in opposite directions, as in the case of a return air or fan intake system. Self-noise can limit a silencer's effective insertion loss for air velocities in excess of about 2000 ft/min. Use extreme caution when reviewing manufacturer's performance data for silencers and duct liner materials to be sure that the test conditions are representative of the specific design conditions.

End reflection losses due to abrupt area changes in duct cross section are sometimes useful in controlling low frequencies. The end reflection effect can be maximized at the end of a duct run by designing the last few feet of duct with the characteristic dimension of less than 15 in.

Where space is available, a lined plenum can provide excellent attenuation across a broad frequency range. The combination of end reflection at the plenum's entrance and exit, a large distance between the entrance and exit, and sound-absorbing lining on the plenum walls can perform as effectively as a duct silencer, but with less pressure drop.

Chapter 42 of the 1991 ASHRAE *Handbook—Applications* has additional information on the control of noise.

ACOUSTICAL DESIGN GOALS AND RATING SYSTEMS

Establishing Design Goals

The primary acoustical design goal for air-conditioning systems is the achievement of a level of background sound that is unobtrusive in quality and low enough in level that it does not interfere with the occupancy requirements of the space being served.

For example, large conference rooms, auditoriums, and recording studios can tolerate only a low level of background sound before interference problems develop. On the other hand, higher levels of background sound are acceptable and even desirable in certain situations, such as in open-plan offices or music practice rooms, where a certain amount of speech and activity masking is essential. Therefore, the system noise control goal is a variable that depends on the required use of the space.

The degree of occupancy satisfaction achieved with a given level of background sound is multidimensional. To be unobtrusive, the background sound should be steady in level, bland in character, and free of identifiable machinery noises. Background sound should have the following properties:

- A balanced distribution of sound energy over a broad frequency range.
- No audible tonal characteristics such as a whine, whistle, hum, or rumble.
- No noticeable time-varying levels from beats or other system-induced aerodynamic instability.

Four types of acoustical design criteria are used by the air-conditioning industry: A-weighted sound level (dB), loudness (sones), noise criteria (NC) curves, and room criteria (RC) curves. A more detailed discussion of acoustical design goals, as well as a table of recommended design goals, can be found in Chapter 42 of the 1991 ASHRAE *Handbook—Applications*.

A-Weighted Sound Level

The A-weighted sound level (L_A) is widely used to state design goals as a single number, but its usefulness is limited because it gives no information on spectrum content. The measuring method is simple, since the L_A can be obtained from a single reading on a handheld instrument. The standard sound level meter includes an electronic weighting network that deemphasizes the low-frequency portions of a noise spectrum, automatically compensating for the lower sensitivity of the human ear to low-frequency sounds. Figure 6 shows the weighting characteristic of the A-weighting and C-weighting networks.

The A-weighted sound level has the advantage of identifying the desirable level as a single-valued number that correlates well with human judgment of relative loudness. However, it has the disadvantage of not correlating well with human judgment of relative noisiness or the subjective quality of a sound.

The A-weighted level comparison is best used with noises that sound alike but differ in level. It should not be used to compare sounds with distinctly different spectral characteristics; *i.e.*, two sounds at the same sound level, but with different spectral

content, may be judged differently by the listener for an acceptable background sound. One of these noises might be completely acceptable, while the other could be objectionable because its spectrum shape was rumbly, hissy, or tonal in character.

NC Curves

The noise criteria (NC) curves shown in Figure 7 (Beranek 1957) are widely used. These curves define the limits that the octave-band spectrum of a noise source must not exceed to achieve a level of occupant acceptance. For example, an NC-35 design goal is commonly used for private offices; the background noise level meets this goal, provided no portion of its spectrum lies above the designated NC-35 curve.

Two problems occur in using the NC design goal: (1) if the NC-level is determined by a singular tangent peak, the actual level of resulting background sound may be quieter than that desired for masking unwanted speech and activity noises because the spectrum on either side of the tangent peak drops off too rapidly, and (2) if the shape of the NC-curve is closely matched, the resulting sound will be either rumbly or hissy.

Because the shape of the NC-curve is not that of a well-balanced, bland-sounding noise, these curves should be used with caution in critical noise situations where the background sound of an air-conditioning system is required to mask speech and activity noise.

RC Curves

Room criteria (RC) curves (Figure 8) are designed especially for establishing HVAC system design goals. The shape of these curves differs from that of the NC curves at both low and high frequencies.

The shape of the RC curve is a close approximation to a well-balanced, bland-sounding spectrum. It provides guidance whenever the space requirements dictate that a certain level of background sound be maintained for masking or other purposes.

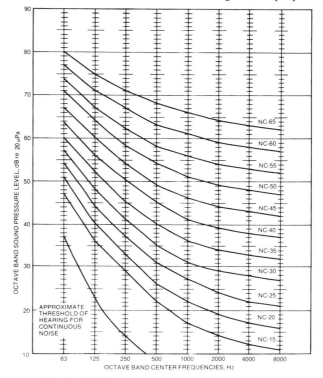

Fig. 7   NC Curves for Specifying Design Level in Terms of Maximum Permissible Sound Pressure Level for Each Frequency Band

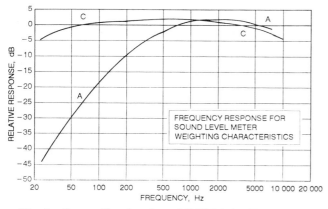

Fig. 6   Curves Showing A- and C-Weighting Responses

Sound that approximates the shape of the curve within 2 dB over the entire frequency range achieves an optimum balance in quality. If the low-frequency levels (31.5 to 500 Hz) exceed the design curve by more than 5 dB, the sound is likely to be judged rumbly; exceeding the design curve by more than 3 dB at high frequencies (1000 to 4000 Hz) causes the sound to be judged hissy.

Another feature of the RC rating system is the presence of the shaded areas at high levels of low frequency noise. These regions predict conditions of frequency and level under which noise may induce lightweight constructions (gypsum walls or ceilings) and their attached services (electrical or HVAC system components) to vibrate or rattle.

Loudness (Sones) and Loudness Level (Phons)

Although the logarithmic phon scale covers the large dynamic range of the ear, it does not fit a subjective linear loudness scale. A sound that is twice as loud as another sound does not double the number of phons; as previously noted, over most of the audible range, a doubling of loudness corresponds to a change of approximately 10 phons. To obtain a quantity proportional to the loudness sensation, a loudness scale is defined in which the unit of loudness is known as a sone. One sone is equal to a loudness level of 40 phons.

ANSI *Standard* S3.4-80 calculates loudness or loudness level by using octave-band sound pressure level data as a starting point. These data are then plotted on a graph constructed as shown in Figure 9. After plotting this graph, the number corresponding to the loudness index for each octave band is obtained. Then total loudness is calculated:

$$S = S_m + 0.3 (\Sigma S - S_m) \qquad (14)$$

where

S = total loudness in sones
S_m = greatest loudness index
ΣS = sum of all loudness indices

After determining the loudness in sones, the loudness level in phons can be found on the right side of Figure 9. In interpreting loudness data, different authors have reported values based on differing loudness contours; also, several methods are available to calculate the single-number loudness or loudness level of a complex sound. For these reasons, this tool is not widely used in engineering practice because of its complexity and uncertainty. AMCA *Standard* 301 describes how the sone method is applied to rating the relative loudness of fans and ventilators.

This calculation method is usually acceptable when there are no strong tonal components in the measured sound spectrum. A more complex calculation method using 1/3 octave band sound pressure levels by Zwicker (ISO *Standard* 532) is more accurate in predicting loudness of sound spectra with tones.

A detailed discussion on the selection of sound design goals, as well as a table of recommended design goals, can be found in Chapter 42 of the 1991 ASHRAE *Handbook—Applications*. Sound levels below NC-35 or RC-35 contribute to good speech intelligibility, and those at or above NC-35 or RC-35 interfere with or mask speech. In mechanical systems with variable air volume, it may not be possible to fill in the higher frequencies when the quantity of supply air is moderate to low. In these cases, if acoustic privacy is important, it may be necessary to provide controlled amounts of electronic masking noise or to request the building designer to improve the sound-isolating construction.

Even if the occupancy noise is significantly higher than the ambient noise level, ambient noise levels should not necessarily be raised to levels approaching occupancy noise. Where speech contributes heavily to occupancy noise, raising the ambient noise levels raises the occupancy levels to the point of physical discomfort.

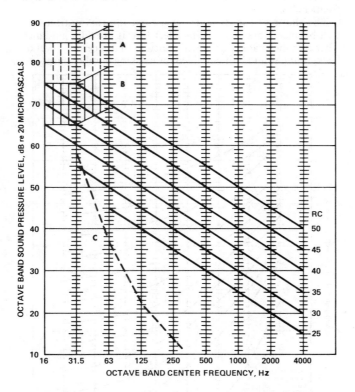

Region A: High probability that noise-induced vibration levels in lightweight wall and ceiling constructions will be felt; anticipate audible rattles in light fixtures, doors, windows, etc.
Region B: Noise-induced vibration levels in lightweight wall and ceiling constructions may be felt; slight possibility of rattles in light fixtures, doors, windows, etc.
Region C: Below threshold of hearing for continuous noise.

**Fig. 8   RC Curves for Specifying Design Level
in Terms of Balanced Spectrum Shape**

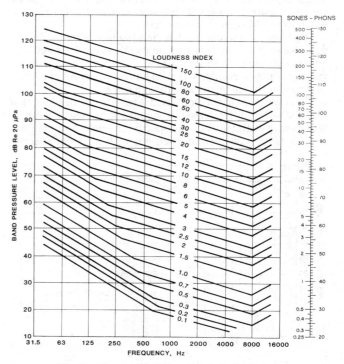

Fig. 9   Contours of Equal Loudness Index

FUNDAMENTALS OF VIBRATION

A rigidly mounted machine transmits all vibratory forces to its supporting structure. This vibration can be isolated or, more realistically, reduced to a fraction of the original force by inserting resilient mounts between the equipment and the building structure. The following sections on vibration isolation and information in Chapter 42 of the 1991 ASHRAE *Handbook—Applications* are useful in selecting vibration isolators and analyzing and correcting field vibration problems.

The following terms apply to vibration isolation:

M = mass of equipment, $\text{lb}_f \cdot \text{s}^2/\text{in}.$
M_f = mass of floor, lb
k = stiffness of isolator, $\text{lb}_f/\text{in}.$
k_f = stiffness of floor, $\text{lb}_f/\text{in}.$
F = vibratory force, lb_f
f = frequency of vibratory force (disturbing frequency), Hz
f_n = natural frequency of vibration isolation, Hz
f_f = natural frequency of floor, Hz
∂_{st} = static deflection of vibration isolator, in.
T = transmissibility, dimensionless ratio
$X''(\theta)$ = acceleration of system at time θ, in/s^2
X = displacement of equipment at time θ, in.
X_f = displacement of floor at time θ, in.

Vibration evaluation requires the study of system motion. Most applications can be treated as single- or two-degree of freedom systems where only motion along the vertical axis is considered, and damping is disregarded. Figure 10 schematically shows a single degree of freedom system where the equation of motion is:

$$M \ddot{X} + k X = F \sin (f_d)\theta \qquad (15)$$

Displacement (absolute value) is given by:

$$X = \left| \frac{F/k}{1 - (f_d/f_n)^2} \right| \qquad (16)$$

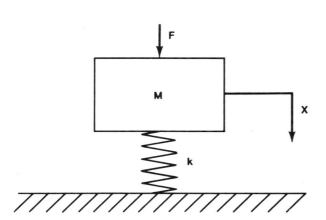

Fig. 10   Single Degree of Freedom System

where

$$f_n = \frac{1}{2\pi} \left(\frac{k}{M} \right)^{0.5}$$

and transmission (absolute value) to building structure is given by:

$$T = \left| \frac{1}{1 - (f_d/f_n)^2} \right| \qquad (17)$$

The disturbing vibration can be caused by unbalance, misalignment, eccentricities in rotating components, or defective bearings; if the frequency of the disturbing vibration is not known, it can be considered to be the lowest rotating frequency. Transmissibility T is the ratio of force transmitted by the isolator (to the floor) to force applied by the equipment. For spring isolators commonly used with HVAC equipment, the natural frequency of the isolator f_n is a function of the isolator static deflection (the distance the spring compresses under the supported equipment), as shown in Figure 11.

Equation (17) indicates that the amount of transmitted energy is a function of disturbing frequency f_d to isolator natural frequency f_n, and varies as the square of this ratio. As shown in Figure 12, when $f_d = f_n$, the denominator of Equation (17) equals zero and theoretically infinite transmission exists—a condition known as resonance. In actual resonant conditions,

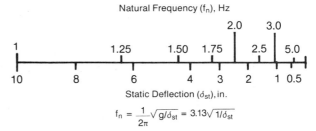

Fig. 11   Natural Frequency of Vibration Isolators

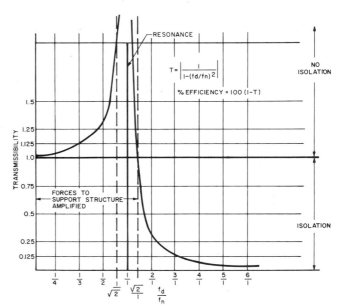

Fig. 12   Vibration Transmissibility as Function of f_d/f_n

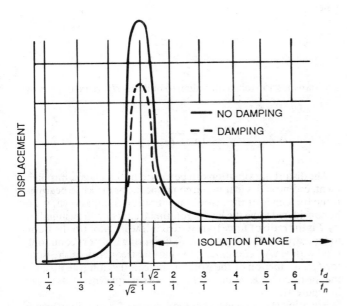

Fig. 13   Displacement of Equipment on Vibration Isolators

some limit to the transmission always occurs because of damping characteristics of isolators and structures. When the isolator natural frequency f_n is lower than the disturbing frequency f_d, isolation occurs (when $f_d/f_n \geqslant \sqrt{2}$). The transmissibility curve becomes nearly asymptotic above a f_d/f_n ratio of about 5:1, and a large increase in the f_d/f_n ratio is required to reduce the transmissibility further.

Figure 13 plots equipment displacement in accordance with Equation (16). This is similar to the transmissibility curve (Figure 12) in that equipment displacement is very great at resonance and decreases when f_d is greater than f_n and isolation occurs. Note the following:

1. Equipment displacement is greater with isolators than without isolators. Without isolators, displacement is a function of vibratory forces acting against the stiffness of the structure; with isolators, the vibratory forces act against the isolator stiffness, and the isolators are generally not as stiff as the structure.

2. Once in the effective isolation range, the curve flattens out, indicating that the stiffness of the isolator has little effect on equipment displacement. This is because as isolator stiffness decreases, the isolator deflection increases, resulting in a lower natural frequency f_n (see Figure 11). Therefore, as isolator stiffness k decreases, the denominator of Equation (16) increases faster than the numerator, and equipment displacement actually decreases.

The mass of equipment has no effect on transmission through isolators to the structure (Figure 14). As shown, a 1000-lb piece of equipment installed on isolators with stiffness k of 1000 lb_f/in. provides a 1-in. deflection for an isolator natural frequency f_n of 3.13 Hz. If equipment is operated at 564 rpm (9.4 Hz) and develops a force of 100 lb_f, 12.5 lb_f of force is transmitted to the structure. If the mass of the equipment is increased to 10,000 lb and spring stiffness k increased to 10,000 lb_f/in., isolator deflection is still 1 in., resulting in the same 12.5 lb_f force transmitted to the structure. However, the increased mass affects displacement of equipment X. Solving Equation (16) for both sets of conditions results in equipment displacement of 0.0125 in. for a 1000-lb mass, and 0.00125 in. for a 10,000-lb mass. Once in the isolation range, a mass-controlled system is created, where equipment displacement varies linearly with total system mass.

The foregoing discussion assumes an infinitely rigid supporting structure, which is not the case for most upper-floor locations. These can be represented as a two-degree of freedom system shown schematically in Figure 15, where equipment displacement is in accordance with Equation (18).

$$X = \left| \frac{(F/k)\,[1\,+\,(k/k_f)\,-\,(f_d/f_f)^2]}{[1\,-\,(f_d/f_n)^2]\,[1\,+\,(k/k_f)\,-\,(f_d/f_f)^2]\,-\,k/k_f} \right| \quad (18)$$

Floor displacement is in accordance with Equation (19).

$$X_f = \left| \frac{(F/k_f)}{[1\,-\,(f_d/f_n)^2]\,[1\,+\,(k/k_f)\,-\,(f_d/f_f)^2]\,-\,k/k_f} \right| \quad (19)$$

Transmissibility at the column supports is shown in Equation (20).

$$T_c = \left| \frac{1}{[1\,-\,(f_d/f_n)^2]\,[1\,+\,(k/k_f)\,-\,(f_d/f_f)^2]\,-\,k/k_f} \right| \quad (20)$$

The denominator of Equations (18), (19), and (20) is the same; as it approaches zero, resonance occurs, resulting in excessive transmissibility as well as excessive equipment and floor displacement.

Floor stiffness is included in the ratio of isolator stiffness to floor stiffness k/k_f. If total floor deflection and isolator deflection are the same, it does not mean that the floor and the isolators are of equal stiffness. Isolator deflection is a function of equipment mass and isolator stiffness, while floor deflection is a

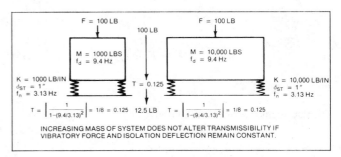

Fig. 14   Effect of Mass on Transmissibility

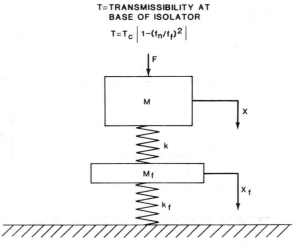

Fig. 15   Two Degrees of Freedom System

function of its own mass and floor spring rate. If the equipment room floor is ten times or more stiff than the spring isolators used to support the equipment, then the ratio of k/k_f becomes so small that it is not significant. Often, this is not the case. For large equipment, refer to Table 34 in Chapter 42 of the 1991 ASHRAE *Handbook—Applications.*

The natural frequency of the floor f_f is critical, because if it is equal or close to the disturbing frequency f_d, a resonant condition exists, resulting in excessive motion of both the equipment and the floor, as well as vibration transmission to the building structure. The natural frequency of the floor is a function of its mass and flexural stiffness, which depends on the material, type of construction, geometry, and support conditions. Floor deflection can indicate potential problems; as floor deflection increases, the fundamental frequency of floor vibration decreases.

If the natural frequency of the floor f_f is not at least three times the disturbing frequency f_d, the possibility of a resonant condition exists, and the following approaches should be considered:

1. Relocate the equipment.
2. Change the operating speed of the equipment to achieve a f_d/f_f ratio of at least 3:1.
3. Stiffen the structure to increase f_f.
4. Increase isolator deflection, thereby decreasing f_n.

Of these options, increasing the isolator deflection is usually the most economical, in which case the selected isolator should provide an f_d/f_n ratio of about 10:1. For example, if f_d is 15 Hz, an isolator with an f_n of about 1.5 Hz is desired, which can be provided by a 4-in. deflection isolation system. In certain situations, especially for large slow-speed equipment operating below 10 Hz, one of the other options might be more useful or cost-effective.

REFERENCES

AMCA. 1985. Reverberant room method for sound testing of fans. *Standard* 300-85. Air Movement and Control Association, Inc., Arlington Heights, IL.

ANSI. 1980. Procedure for computation of loudness of noise. *Standard* S3.4-80 (Revised 1986; ASA *Standard* 37-80). American National Standards Institute, New York.

ANSI. 1983. Specifications for sound level meters. *Standard* S1.4-83 (Amendment S1.4a-85; ASA *Standard* 47-83). American National Standards Institute, New York.

ANSI. 1984. Preferred frequencies, frequency levels, and band numbers for acoustical measurements. *Standard* S1.6-84 (Revised 1990; ASA *Standard* 53-84). American National Standards Institute, New York.

ANSI. 1986. Specifications for octave-band and fractional-octave-band analog and digital filters. *Standard* S1.11-86 (ASA *Standard* 65-86). American National Standards Institute, New York.

ANSI. 1990. Precision methods for the determination of sound power levels of broad-band noise sources in reverberation rooms. *Standard* S12.31-90 (Revision of ANSI S1.31-80 and ASA 11-80). American National Standards Institute, New York.

ANSI. 1990. Precision methods for the determination of sound power levels of discrete-frequency and narrow-band noise sources in reverberation rooms. *Standard* S12.32-90 (Revision of ANSI S1.32-80 and ASA 12-80). American National Standards Institute, New York.

ANSI. 1990. Precision methods for the determination of sound power levels of noise sources in anechoic and hemi-anechoic rooms. *Standard* S12.35-90 (Revision of ANSI S1.35-79 and ASA 15-79). American National Standards Institute, New York.

ASHRAE. 1986. Laboratory method of testing in-duct sound power measurement procudure for fans. *Standard* 68-86 (AMCA 330-86). American Society for Heating, Refrigerating and Air-Conditioning Engineers, Inc., Atlanta, GA.

ASTM. 1987. Classification for rating sound insulation. *Standard* E413-87. American Society for Testing and Materials, Philadelphia, PA.

ASTM. 1989. Standard terminology relating to environmental acoustics. *Standard* C634-89. American Society for Testing and Materials, Philadelphia, PA.

ASTM. 1990. Standard test method for measuring acoustical and airflow performance of duct liner materials and prefabricated silencers. *Standard* E477-90. American Society for Testing and Materials, Philadelphia, PA.

Beranek, L.L. 1957. Revised criteria for noise in buildings. *Noise Control* 1:19.

ISO. 1975. Methods for calculating loudness level. *Standard* 532-1975. American National Standards Institute, New York.

Pollack, I. 1952. The loudness of bands of noise. *Journal of the Acoustical Society of America* 24(9):533.

Robinson, D.W. and R.S. Dadson. 1956. A redetermination of the equal loudness relations for pure tones. *British Journal of Applied Physics* 7(5):166.

Schultz, T.J. 1985. Relationship between sound power level and sound pressure level in dwellings and offices. ASHRAE *Transactions* 91(1):124-53.

BIBLIOGRAPHY

Beranek, L.L. 1960. *Noise reduction.* McGraw-Hill Book Co., New York.

Beranek, L.L. 1988. *Noise and vibration control,* Revised ed. Institute of Noise Control Engineering, Washington, D.C.

Crede, C.E. 1951. *Vibration and shock isolation.* John Wiley & Sons, New York.

Harris, C.M. 1991. *Handbook of acoustical measurements and noise control.* McGraw-Hill, Inc., New York.

Harris, C.M. and C.E. Crede. 1979. *Shock and vibration handbook,* 2nd ed. McGraw-Hill Book Co., New York.

Peterson, A.P.G. and E.E. Gross, Jr. 1974. *Handbook of noise measurement.* GenRad, Inc., Concord, MA.

Schaffer, M.E. 1991. *A practical guide to noise and vibration control for HVAC systems.* American Society for Heating, Refrigerating and Air-Conditioning Engineers, Inc., Atlanta, GA.

CHAPTER 8

PHYSIOLOGICAL PRINCIPLES AND THERMAL COMFORT

THE human body continuously generates heat, with an output varying from about 340 Btu/h for a sedentary person to 3400 Btu/h for a person exercising strenuously. Body temperature must be maintained within a narrow temperature range, to avoid discomfort, and within a somewhat wider range, to avoid danger from heat or cold stress. Consequently, heat must be dissipated in a carefully controlled manner for a body to stay within this range. Heat is not generated uniformly throughout the body, nor is it dissipated uniformly. However, for most engineering applications, it is sufficient to consider the body as a uniform cylinder when describing heat dissipation to the environment (Figure 1).

The total metabolic energy M produced within the body is the metabolic energy required for the person's activity M_{act} plus the metabolic energy required for shivering M_{shiv} (should shivering occur). A portion of the body's energy production may be expended as external work done by the muscles W; the net heat production $M - W$ is either stored, causing the body's temperature to rise, or is dissipated to the environment through the skin surface and respiratory tract. Heat dissipation from the body to the immediate surroundings occurs by several modes of heat exchange: sensible heat flow from the skin $C + R$; latent heat flow from the evaporation of sweat E_{rsw} and from evaporation of moisture diffused through the skin E_{dif}; sensible heat flow during respiration C_{res}; and latent heat flow due to evaporation of moisture during respiration E_{res}. Sensible heat flow from the skin may be a complex mixture of conduction, convection, and radiation for a clothed person; however, it is equal to the sum of the convection C and radiation R heat transfer at the outer clothing surface (or exposed skin).

ENERGY BALANCE AND THERMAL COMFORT

Most models of thermal exchanges between the body and the environment and the subsequent measures of the physiological strain or thermal sensation are similar in that they use classic heat transfer theory as a rational starting basis and introduce empirical equations describing the effects of known physiological regulatory controls. Sensible and latent heat losses from the skin are expressed in terms of environmental factors, skin temperature t_{sk}, and skin wettedness w. The expressions also incorporate factors that account for the thermal insulation and moisture permeability of clothing. The independent environmental variables can be summarized as: air temperature t_a, mean radiant temperature $\bar{t}_r$, relative air velocity v, and ambient water vapor pressure p_a. The independent personal variables that influence thermal comfort are activity and clothing.

The preparation of this chapter is assigned to TC 2.1, Physiology and Human Environment.

Two commonly used models are briefly described here, and more details are given in the section Prediction of Thermal Comfort and Thermal Sensation. Both models assume that skin temperature and skin surface heat transfer are uniform across the body. All terms in Equations (1) and (2) have units of energy per unit area and are referenced to the surface area of the nude body A_D. A correction factor is introduced later to account for the increased surface area of the clothed body A_{cl}. An equation for A_D, referred to as the DuBois surface area, is presented in the Engineering Data and Measurements section.

Steady-State Energy Balance

The steady-state model developed by Fanger (1970, 1982) assumes that the body is in a state of thermal equilibrium with negligible heat storage. The body is assumed to be near thermal neutrality—there is no shivering and vasoregulation is not considered because the core and skin are modeled as one compartment. At steady state, the rate of heat generation equals the rate of heat loss, and the energy balance is:

$$M - W = Q_{sk} + Q_{res}$$
$$= (C + R + E_{sk}) + (C_{res} + E_{res}) \quad (1)$$

where

M = rate of metabolic heat production, Btu/(h·ft²)
W = rate of mechanical work accomplished, Btu/(h·ft²)
Q_{res} = total rate of heat loss through respiration, Btu/(h·ft²)
Q_{sk} = total rate of heat loss from skin, Btu/(h·ft²)

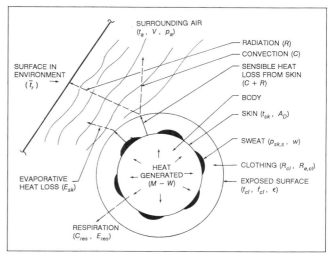

Fig. 1 Cylindrical Model of Thermal Interaction of Human Body and Environment

C_{res} = rate of convective heat loss from respiration, Btu/(h·ft^2)
E_{res} = rate of evaporative heat loss from respiration, Btu/(h·ft^2)
$C + R$ = sensible heat loss from skin, Btu/(h·ft^2)
E_{sk} = rate of total evaporative heat loss from skin [see Equation (13)]

Two-Node, Transient Energy Balance

A two-compartment (or two-node) model developed by Gagge *et al.* (1971, 1986) represents the body as two concentric cylinders—the inner cylinder represents the body core (skeleton, muscle, internal organs), and the outer cylinder represents the skin layer. The model is based on the following assumptions: conductive heat exchange from the skin is negligible; the temperature in each compartment is uniform (core = t_{cr} and skin = t_{sk}); metabolic heat production, external work, and respiratory losses are associated with the core compartment; and the core and skin compartments exchange energy passively through direct contact and through the thermoregulatory controlled peripheral blood flow. A transient energy balance states that the rate of heat storage equals the net rate of heat gain minus the heat loss. The thermal model is described by two coupled heat balance equations, one applied to each compartment:

$$S_{cr} = M - W - (C_{res} + E_{res}) - Q_{cr,sk} \qquad (2)$$

$$S_{sk} = Q_{cr,sk} - (C + R + E_{sk}) \qquad (3)$$

where

S_{cr} = rate of heat storage in core compartment, Btu/(h·ft^2)
S_{sk} = rate of heat storage in skin compartment, Btu/(h·ft^2)
$Q_{cr,sk}$ = rate of heat transport from core to skin (includes both conduction through body tissue and convection through blood flow), Btu/(h·ft^2)

The rate of heat storage in the body equals the rate of increase in internal energy. The rate of storage can be written separately for each compartment in terms of thermal capacity and time rate of change of temperature in each compartment:

$$S_{cr} = (1 - \alpha)mc_{p,b}(dt_{cr}/d\theta)/A_D \qquad (4)$$

$$S_{sk} = \alpha mc_{p,b}(dt_{sk}/d\theta)/A_D \qquad (5)$$

where

α = fraction of body mass concentrated in skin compartment, n.d.
m = body mass, lb
$c_{p,b}$ = specific heat capacity of body = 0.834 Btu/(lb·°F)
θ = time, h

The fractional skin mass α depends on the rate of blood flowing to the skin surface $\dot{m}_{bl}$. An empirical equation relating α to $\dot{m}_{bl}$ is presented in the Thermoregulatory Control Mechanisms section.

THERMAL EXCHANGES WITH THE ENVIRONMENT

Fanger (1967, 1970), Hardy (1949), Rapp and Gagge (1967), and Gagge and Hardy (1967) give quantitative information on calculating the heat exchange between people and the environment. A summary of the mathematical statements for various terms of heat exchange used in the heat balance equations (C, R, E_{sk}, C_{res}, E_{res}) follows. Terms describing the heat exchanges associated with the thermoregulatory control mechanisms ($Q_{cr,sk}$, M_{shiv}, E_{rsw}) and values for the coefficients and appropriate equations for M_{act} and A_D are presented in later sections. The mathematical description of the energy balance of the human body represents a combined rational/empirical approach to describing the thermal exchanges with the environment. Fundamental heat transfer theory is used to describe the various mechanisms of sensible and latent heat exchange, while empirical expressions are used to determine the values of the coefficients describing these rates of heat exchange. Empirical equations are also used to describe the thermophysiological control mechanisms as a function of skin and core temperatures in the body.

Sensible Heat Loss from Skin

Sensible heat exchange from the skin surface must pass through clothing to the surrounding environment. These paths are treated in series and can be described in terms of: (1) heat transfer from the skin surface, through the clothing insulation, to the outer clothing surface, and (2) from the outer clothing surface to the environment.

Both convective C and radiative R heat losses from the outer surface of a clothed body can be expressed in terms of a heat transfer coefficient, and the difference between the mean temperature of the outer surface of the clothed body t_{cl} and the appropriate environmental temperature:

$$C = f_{cl}h_c(t_{cl} - t_a) \qquad (6)$$

$$R = f_{cl}h_r(t_{cl} - \bar{t}_r) \qquad (7)$$

where

h_c = convective heat transfer coefficient, Btu/(h·ft^2·°F)
h_r = linear radiative heat transfer coefficient, Btu/(h·ft^2·°F)
f_{cl} = clothing area factor = A_{cl}/A_D

The coefficients h_c and h_r are both evaluated at the clothing surface. Equations (6) and (7) are commonly combined to describe the total sensible heat exchange by these two mechanisms in terms of an operative temperature t_o and a combined heat transfer coefficient h:

$$(C + R) = f_{cl}h(t_{cl} - t_o) \qquad (8)$$

where

$$t_o = (h_r\bar{t}_r + h_ct_a)/(h_r + h_c) \qquad (9)$$

$$h = h_r + h_c \qquad (10)$$

Based on Equation (9), operative temperature t_o can be defined as the average of the mean radiant and ambient air temperatures, weighted by their respective heat transfer coefficients.

The actual transport of sensible heat through clothing involves conduction, convection, and radiation. It is usually most convenient to combine these into a single thermal resistance value R_{cl}, defined by:

$$(C + R) = (t_{sk} - t_{cl})/R_{cl} \qquad (11)$$

where R_{cl} = thermal resistance of clothing, h·ft^2·°F/Btu.

Since it is often inconvenient to include the clothing surface temperature in calculations, Equations (8) and (11) can be combined to eliminate t_{cl}:

$$(C + R) = (t_{sk} - t_o)/[R_{cl} + 1/(f_{cl}h)] \qquad (12)$$

where t_o is defined in Equation (9).

Evaporative Heat Loss from Skin

Evaporative heat loss from skin E_{sk} depends on the difference between the water vapor pressure at the skin and in the ambient environment, and the amount of moisture on the skin:

$$E_{sk} = w(p_{sk,s} - p_a)/[R_{e,cl} + 1/(f_{cl}h_e)] \qquad (13)$$

where

p_a = water vapor pressure in ambient air, psi
$p_{sk,s}$ = water vapor pressure at skin, normally assumed to be that of saturated water vapor at t_{sk}, psi
$R_{e,cl}$ = evaporative heat transfer resistance of clothing layer (analogous to R_{cl}), ft^2·psi·h/Btu
h_e = evaporative heat transfer coefficient (analogous to h_c), Btu/(h·ft^2·psi)
w = skin wettedness, dimensionless

Procedures for calculating $R_{e,cl}$ and h_e are given in the Engineering Data and Measurements section. Skin wettedness is the ratio of the actual evaporative heat loss to the maximum possible evaporative heat loss E_{max} with the same conditions and a completely wetted skin ($w = 1$). The value of skin wettedness is important in determining evaporative heat loss. Maximum evaporative potential E_{max} occurs when the skin surface is completely wetted, or $w = 1.0$.

Evaporative heat loss from the skin is a combination of the evaporation of sweat secreted due to thermoregulatory control mechanisms E_{rsw} and the natural diffusion of water through the skin E_{dif}:

$$E_{sk} = E_{rsw} + E_{dif} \qquad (14)$$

Evaporative heat loss by regulatory sweating is directly proportional to the regulatory sweat generated:

$$E_{rsw} = \dot{m}_{rsw} h_{fg} \qquad (15)$$

where

h_{fg} = heat of vaporization of water = 1045 Btu/lb at 86°F
$\dot{m}_{rsw}$ = rate at which regulatory sweat is generated, lb/h·ft^2 [see Equation (45)]

The portion of a body that must be wetted to evaporate the regulatory sweat w_{rsw} is:

$$w_{rsw} = E_{rsw}/E_{max} \qquad (16)$$

With no regulatory sweating, skin wettedness due to diffusion is approximately 0.06 for normal conditions. For large values of E_{max} or long exposures to low humidities, the value may drop to as low as 0.02, since dehydration of the outer skin layers alters its diffusive characteristics. With regulatory sweating, the 0.06 value applies only to the portion of skin not covered with sweat $(1 - w_{rsw})$; the diffusion evaporative heat loss is:

$$E_{dif} = (1 - w_{rsw})\, 0.06\, E_{max} \qquad (17)$$

These equations can be solved to give a value of w, given the maximum evaporative potential E_{max} and the regulatory sweat generation E_{rsw}:

$$w = w_{rsw} + 0.06(1 - w_{rsw}) = 0.06 + 0.94 E_{rsw}/E_{max} \qquad (18)$$

Once skin wettedness is determined, evaporative heat loss from the skin is calculated from Equation (13), or by:

$$E_{sk} = wE_{max} \qquad (19)$$

To summarize, the following calculations determine w and E_{sk}:

E_{max} Equation (13), with $w = 1.0$
E_{rsw} Equation (15)
w Equation (18)
E_{sk} Equation (19) or (13)

Although evaporation from the skin E_{sk} as described in Equation (13) depends on w, the body does not directly regulate skin wettedness but, rather, regulates sweat rate $\dot{m}_{rsw}$ [Equation (15)]. Skin wettedness is then an indirect result of the relative activity of the sweat glands and the evaporative potential of the environment. Skin wettedness of 1.0 is the upper theoretical limit. If the aforementioned calculations yield a wettedness of more than 1.0, then Equation (15) is no longer valid since all of the sweat is not evaporated. In this case, $E_{sk} = E_{max}$.

Skin wettedness is strongly correlated with warm discomfort and is also a good measure of thermal stress. Theoretically, skin wettedness can approach 1.0 while the body still maintains thermoregulatory control. In most situations, it is difficult to exceed 0.8. Azer (1982) recommends 0.5 as a practical upper limit for sustained activity for a healthy acclimatized person.

Respiratory Losses

During respiration, the body loses both sensible and latent heat by convection and evaporation of heat and water vapor from the respiratory tract to the inhaled air. A significant amount of heat can be associated with respiration because the air is inspired at ambient conditions and expired nearly saturated at a temperature only slightly cooler than t_{cr}.

Sensible C_{res} and latent E_{res} heat losses due to respiration are:

$$C_{res} = \dot{m}_{res}c_{p,a}(t_{ex} - t_a)/A_D \qquad (20)$$

$$E_{res} = \dot{m}_{res}h_{fg}(W_{ex} - W_a)/A_D \qquad (21)$$

where

$\dot{m}_{res}$ = pulmonary ventilation rate, lb/h
t_{ex} = temperature of exhaled air, °F
W_{ex} = humidity ratio of exhaled air, lb H$_2$O/lb d.a.
W_a = humidity ratio of inhaled (ambient) air, lb H$_2$O/lb d.a.
$c_{p,a}$ = specific heat of air, Btu/(lb·°F)

These equations can be reduced by using approximations and empirical relationships to estimate the values of the parameters. Under normal circumstances, pulmonary ventilation rate is primarily a function of metabolic rate (Fanger 1970):

$$\dot{m}_{res} = K_{res}M \qquad (22)$$

where K_{res} is a proportionality constant (0.0645 lb·ft^2/Btu).

Respiratory air is nearly saturated and near the body temperature when exhaled. The following empirical equations developed by Fanger (1970) can be used to estimate these conditions for typical indoor environments:

$$t_{ex} = 88.6 + 0.066\, t_a + 57.6\, W_a \qquad (23)$$

$$W_{ex} - W_a = 0.0265 + 0.000036\, t_a - 0.80\, W_a \qquad (24)$$

where t_a and t_{ex} are in °F. For more extreme conditions, such as outdoor winter environments, different relationships may be required (see Holmer 1984).

The humidity ratio of ambient air can be expressed in terms of the total or barometric pressure p_t and ambient partial vapor pressure p_a:

$$W_a = 0.622\, p_a/(p_t - p_a) \qquad (25)$$

Two approximations are commonly used to simplify Equations (20) and (21). First, because the dry respiratory heat loss is relatively small compared to the other terms in the heat balance, an average value for t_{ex} is determined by evaluating Equation (23) at standard conditions (68 °F, 50% rh, sea level). Second, noting in Equation (24) that there is only a weak dependence on ambient air temperature, the second term in Equation (24) and the denominator in Equation (25) are evaluated at standard conditions. Using these approximations, and substituting values of h_{fg} and $c_{p,a}$ at standard conditions, Equations (20) and (21) can be combined and written as:

$$C_{res} + E_{res} = [0.0084M(93.2 - t_a) + 1.28M(0.851 - p_a)]/A_D \qquad (26)$$

where p_a is expressed in psi and t_a is in °F.

Alternative Formulations

Equations (12) and (13) describe heat loss from skin for clothed people in terms of clothing parameters R_{cl}, $R_{e,cl}$, and f_{cl}; parameters h and h_e describe outer surface resistances. Other parameters and definitions are also used. Although these alternate parameters and definitions may be confusing, note that the information presented in one form can be converted to another form. Common parameters and their qualitative descriptions are presented in Table 1. Equations showing their relation to each other are presented in Table 2. Generally, parameters related to dry or evaporative heat flows are not independent because they both rely, in part, on the same physical processes. The Lewis relation describes the relations between convective heat transfer and mass transfer coefficients for a surface (see Chapter 5). The Lewis relation can be used to relate convective and evaporative heat transfer coefficients defined in Equations (4) and (11) according to:

$$h_e/h_c = LR \qquad (27)$$

where LR is referred to as the "Lewis ratio" and, at typical indoor conditions, equals approximately 205 °F/psi. The Lewis relation applies to surface convection coefficients. Heat transfer coefficients that include the effects of insulation layers and/or radiation will still be coupled, but the relationship may deviate significantly from that for a surface. The i terms in Tables 1 and 2 describe how the actual ratios of these parameters deviate from the ideal Lewis ratio (Woodcock 1962, Oohori 1985).

Depending on the combination of parameters used, heat transfer from the skin can be calculated using several different formulations (see Table 3). If the parameters are used correctly, the end result will be the same regardless of the formulation used.

Total Skin Heat Loss

Total skin heat loss—sensible heat plus evaporative heat—can be calculated from any combination of the equations presented in Table 3. Total skin heat loss is used as a measure of the thermal environment; two combinations of parameters that yield the same total heat loss for a given set of body conditions (t_{sk} and w) are considered to be approximately equivalent. The fully expanded skin heat loss equation, showing each parameter that must be known or specified, is as follows:

$$Q_{sk} = \frac{t_{sk} - \dfrac{\bar{t}_r h_r + t_a h_c}{h_r + h_c}}{R_{cl} + \dfrac{1}{[(h_r + h_c)f_{cl}]}} + w\frac{p_{sk,s} - p_a}{R_{e,cl} + \dfrac{1}{(LRh_c f_{cl})}} \qquad (28)$$

or

$$Q_{sk} = F_{cl}f_{cl}\,h + w\text{LR}\,F_{pcl}h_c \qquad (29)$$

This equation allows the trade-off between any two, or more, parameters to be evaluated under given conditions. If the trade-off between two specific variables is to be examined, then a simplified form of the equation suffices. The trade-off between operative temperature and humidity is often of interest. Equation (28) can be written in a much simpler form for this purpose (Fobelets and Gagge 1988):

$$Q_{sk} = h'\,[(t_{sk} + wi_m\text{LR}p_{sk,s}) - (t_o + wi_m\text{LR}p_a)] \qquad (30)$$

Equation (30) can be used to define a combined temperature t_{com}, which reflects the combined effect of operative temperature and humidity for an actual environment:

Table 1   Parameters Used to Describe Clothing

Sensible Heat Flow

R_{cl} = instrinsic clothing insulation, the thermal resistance of a uniform layer of insulation covering the entire body that has the same effect on sensible heat flow as the actual clothing.

R_t = total insulation, the total equivalent uniform thermal resistance between the body and the environment: clothing and boundary resistance.

R_{cle} = effective clothing insulation, the increased body insulation due to clothing as compared to the nude state.

R_a = boundary insulation, the thermal resistance at the skin boundary for a nude body.

$R_{a,cl}$ = outer boundary insulation, the thermal resistance at the outer boundary (skin or clothing).

h' = overall sensible heat transfer coefficient, the overall equivalent uniform conductance between the body and the environment.

h'_{cl} = clothing conductance, the thermal conductance of a uniform layer of insulation covering the entire body that has the same effect on sensible heat flow as the actual clothing.

F_{cle} = effective clothing thermal efficiency, the ratio of the actual sensible heat loss to that of a nude body at the same conditions.

F_{cl} = intrinsic clothing thermal efficiency, the ratio of the actual sensible heat loss to that of a nude body at the same conditions including an adjustment for the increase in surface area due to the clothing.

Evaporative Heat Flow

$R_{e,cl}$ = evaporative heat transfer resistance of the clothing, the impedance to transport of water vapor of a uniform layer of insulation covering the entire body that has the same effect on evaporative heat flow as the actual clothing.

$R_{e,t}$ = total evaporative resistance, the total equivalent uniform impedance to the transport of water vapor from the skin to the environment.

F_{pcl} = permeation efficiency, the ratio of the actual evaporative heat loss to that of a nude body at the same conditions, including an adjustment for the increase in surface area due to the clothing.

Parameters Relating Sensible and Evaporative Heat Flow

i_{cl} = clothing vapor permeation efficiency, the ratio of the actual evaporative heat flow capability through the clothing to the sensible heat flow capability as compared to the Lewis ratio.

i_m = moisture permeability index, the ratio of the actual evaporative heat flow capability between the skin and the environment to the sensible heat flow capability as compared to the Lewis ratio.

i_a = air layer vapor permeation efficiency, the ratio of the actual evaporative heat flow capability through the outer air layer to the sensible heat flow capability as compared to the Lewis ratio.

Table 2 Relationships between Clothing Parameters

Sensible Heat Flow

$$R_t = R_{cl} + 1/(h f_{cl}) = R_{cl} + R_a/f_{cl}$$
$$R_t = R_{cle} + 1/h = R_{cle} + R_a$$
$$h'_{cl} = 1/R_{cl}$$
$$h' = 1/R_t$$
$$h = 1/R_a$$
$$R_{a,cl} = R_a/f_{cl}$$
$$F_{cl} = h'/(h f_{cl}) = 1/(1 + f_{cl} h R_{cl})$$
$$F_{cle} = h'/h = f_{cl}/(1 + f_{cl} h R_{cl}) = f_{cl} F_{cl}$$

Evaporative Heat Flow

$$R_{e,t} = R_{e,cl} + 1/(h_e f_{cl}) = R_{e,cl} + R_{e,a}/f_{cl}$$
$$h_e = 1/R_{e,a}$$
$$h'_{e,cl} = 1/R_{e,cl}$$
$$h'_e = 1/R_{e,t} = f_{cl} F_{pcl} h_e$$
$$F_{pcl} = 1/(1 + f_{cl} h_e R_{e,cl})$$

Parameters Relating Sensible and Evaporative Heat Flows

$$i_{cl}LR = h'_{e,cl}/h'_{cl} = R_{cl}/R_{e,cl}$$
$$i_m LR = h'_e/h' = R_t/R_{e,t}$$
$$i_m = (R_{cl} + R_{a,cl})/[(R_{cl}/i_{cl}) + (R_{a,cl}/i_a)]$$
$$i_a LR = h_e/h$$
$$i_a = h_c/(h_c + h_r)$$

Table 3 Skin Heat Loss Equations

Sensible Heat Loss

$$C + R = (t_{sk} - t_o)/[R_{cl} + 1/(f_{cl} h)]$$
$$C + R = (t_{sk} - t_o)/R_t$$
$$C + R = F_{cle} h (t_{sk} - t_o)$$
$$C + R = F_{cl} f_{cl} h (t_{sk} - t_o)$$
$$C + R = h' (t_{sk} - t_o)$$

Evaporative Heat Loss

$$E_{sk} = w (p_{sk,s} - p_a)/[R_{e,cl} + 1/(f_{cl} h_e)]$$
$$E_{sk} = w (p_{sk,s} - p_a)/R_{e,t}$$
$$E_{sk} = w F_{pcl} f_{cl} h_e (p_{sk,s} - p_a)$$
$$E_{sk} = h'_e w (p_{sk,s} - p_a)$$
$$E_{sk} = h' w_{im} LR (p_{sk,s} - p_a)$$

$$t_{com} + w i_m LR p (t_{com}) = t_o + w i_m LR p_a$$

or

$$t_{com} = t_o + w i_m LR p_a - w i_m LR p(t_{com}) \qquad (31)$$

where $p(t_{com})$ is a vapor pressure related in some fixed way to t_{com} and is analogous to $p_{wb,s}$ for t_{wb}. The term $w i_m LR p(t_{com})$ is constant, to the extent that i_m is constant and any combination of t_o and p_a that gives the same t_{com} will result in the same total heat loss.

Environmental indices are discussed in a later section of this chapter. Two of these, the humid operative temperature t_{oh} and the effective temperature ET*, can be represented in terms of Equation (31). The humid operative temperature is that temperature which at 100% rh yields the same total heat loss as for the actual environment:

$$t_{oh} = t_o + w i_m LR (p_a - p_{oh,s}) \qquad (32)$$

where $p_{oh,s}$ = saturated vapor pressure at t_{oh}, psi.

The effective temperature is the temperature at 50% rh that yields the same total heat loss as for the actual environment:

$$ET* = t_o + w i_m LR (p_a - 0.5 p_{ET*,s}) \qquad (33)$$

where $p_{ET*,s}$ = saturated vapor pressure at ET*, psi.

The psychrometric chart in Figure 2 shows a constant total heat loss line and the relationship between these indices. This line represents only one specific skin wettedness and moisture permeability index. The relationship between indices depends on these two parameters (see the Environmental Indices section).

THERMOREGULATORY CONTROL MECHANISMS

Zones of Physiological and Behavioral Response

Heat exchange exists in relation to zones of different physiological and behavioral response. In the operative temperature range from 84 to 88°F for unclothed, and 73 to 81°F for normally clothed, sedentary people (0.6 clo), there is no body cooling or heating and no increase in evaporative heat loss. Within this zone, each individual has a neutral temperature where the environment is neither hot nor cold, and where no action from the physiological control system is required to maintain normal body temperature. If the rate of heat loss from skin to environment increases, the body decreases blood flow to the skin. This cools the skin and subjacent tissues and maintains the temperature of deep tissues. The external conditions over which this occurs fall in a narrow range called the *zone of vasomotor regulation against cold*. Below this range, temperatures of superficial and deep tissues fall, unless another control reaction occurs. Typically, the body generates heat through muscular tension, shivering, or spontaneous activity. If the heat generated balances the increased heat loss to the environment, deep body temperature is maintained. An alternative control reaction in the *zone of behavioral regulation against cold* involves donning more clothing or increasing activity (*e.g.*, walking faster). If all control reactions prove inadequate, the body enters the *zone of body cooling*. When core temperature falls below 95°F, people suffer major losses in efficiency (*e.g.*, in manual dexterity under arctic conditions); core temperatures below 87.8°F can be lethal.

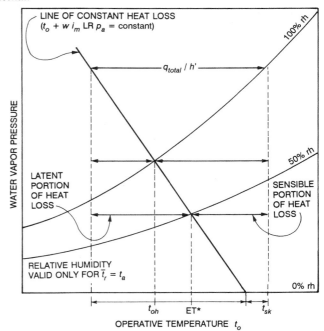

Fig. 2 Constant Heat Loss Line and Its Relationship to t_{oh} and ET*

Over a wide range of cold external conditions, the body adjusts by providing life-sustaining conditions for crucial internal regions at the expense of an energy loss from and possible damage to peripheral tissues. The farther the superficial tissues (*e.g.*, hands and feet) are from the central body mass, the more readily their temperatures fall. The larger the surface area per unit of thermal mass (*e.g.*, ears versus torso), the more rapidly its temperature falls.

On the hot side of the neutral midpoint is a narrow *zone of vasomotor regulation against heat*. Physiological control mechanisms for this zone correspond to those used against cold. Blood flow to the skin increases when desirable heat loss to the environment is restricted. This increase in blood flow can double or triple the conductance of heat to superficial tissues over that characteristic of the neutral point, causing skin surface temperature to come closer to the temperature of the deep tissues. If, in spite of this increase in blood flow, the core temperature rises above 98.6 °F, the body enters its second line of defense, releasing water from sweat glands for evaporative cooling. As long as this cooling can maintain required heat loss, the body is in the *zone of evaporative regulation*. Therefore, any environmental factor that affects evaporation of water from the skin affects temperature regulation. For example, increased atmospheric water vapor pressure, reduced air movement, and added clothing all affect the upper limit of E_{sk}. When this cooling is inadequate, the body is in the *zone of body heating*. When the core temperature rises above 102.2 °F, people suffer major losses in efficiency (become sluggish or refuse to contine their present exertion level). Deep body temperatures above 109 °F can be lethal.

Regulatory Signals and Responses

The following discussion is based on the two-node model (Gagge *et al.* 1971); other, more complex sets of control relationships have also been developed for other models (see Hardy 1961, for example). When the body is able to maintain its thermal equilibrium with the environment with minimal regulatory effort, it is in a state of physiological thermal neutrality, *i.e.*, the average core and skin temperatures are at their neutral values:

$$t_{sk,n} = 92.7\,°\text{F} \tag{34}$$

$$t_{cr,n} = 98.2\,°\text{F} \tag{35}$$

Empirical relationships derived from laboratory experiments describe how the thermoregulatory control processes (vasomotor regulation, sweating, and shivering) are probably stimulated and governed by temperature signals from the skin and core (deviations from their respective neutral set points). Five signals trigger these processes: warm signal from the core (WSIG_{cr}), cold signal from the core (CSIG_{cr}), warm signal from the skin (WSIG_{sk}), cold signal from the skin (CSIG_{sk}), and warm signal from the body (WSIG_b). The signals are written in terms of actual temperature t, and neutral temperature t_n, such that they only take on positive values:

$$\text{WSIG}_{cr} = \begin{cases} 0 & t_{cr} \leqslant t_{cr,n} \\ t_{cr} - t_{cr,n} & t_{cr} > t_{cr,n} \end{cases} \tag{36}$$

$$\text{CSIG}_{cr} = \begin{cases} t_{cr,n} - t_{cr} & t_{cr} < t_{cr,n} \\ 0 & t_{cr} \geqslant t_{cr,n} \end{cases} \tag{37}$$

$$\text{WSIG}_{sk} = \begin{cases} 0 & t_{sk} \leqslant t_{sk,n} \\ t_{sk} - t_{sk,n} & t_{sk} > t_{sk,n} \end{cases} \tag{38}$$

$$\text{CSIG}_{sk} = \begin{cases} t_{sk,n} - t_{sk} & t_{sk} < t_{sk,n} \\ 0 & t_{sk} \geqslant t_{sk,n} \end{cases} \tag{39}$$

$$\text{WSIG}_b = \begin{cases} 0 & t_b \leqslant t_{b,n} \\ t_b - t_{b,n} & t_b > t_{b,n} \end{cases} \tag{40}$$

The average temperature of the human body t_b can be predicted by the weighted average of the skin and core temperatures:

$$t_b = \alpha\, t_{sk} + (1 - \alpha)\, t_{cr} \tag{41}$$

The neutral body temperature $t_{b,n}$ is calculated from the neutral skin and core temperature in the same manner. The value of the weighting coefficient α depends on the rate of blood flow to the skin; an empirical relationship between α and $\dot{m}_{bl}$ is presented in Equation (44). In the two-compartment thermal model, α represents the fraction of the total body mass concentrated in the skin compartment.

The following empirical equations are introduced into the Pierce two-node heat balance model of the human body. They are not used in the Fanger whole-body steady-state model, because that model does not independently calculate t_{sk} and t_{cr}.

Skin blood flow and heat transport. Vasomotor activity is controlled by warm and cold temperature signals from the core and skin, respectively. Normal blood flow for sedentary activity at thermally neutral body conditions is approximately 1.29 lb/(h·ft²), expressed per unit DuBois surface area. Vasodilation is governed by warm signals from the core. For each 1 °F increase above $t_{cr,n}$, blood flow increases by 41 lb/(h·ft²). Warm signals from the skin play a more important role in body temperature regulation by governing sweating rather than vasodilation, and hence its effect on dilation is neglected. Vasoconstriction is governed by cold signals from the skin, and for each one degree decrease below $t_{sk,n}$, blood flow encounters a proportional increase in resistance. This resistance factor varies throughout the body (*i.e.*, it is high in the hands and feet, negligible in the trunk) and its effect is given an average value. A cold signal from the core causes vasoconstriction, but not as rapidly or effectively as a cold signal from the skin, and its effect is consequently neglected. The effects of core and skin temperature deviations on blood flow can be expressed mathematically as:

$$\dot{m}_{bl} = 0.205\,[(11.3 + 200\,\text{WSIG}_{cr})/(1.8 + 0.5\,\text{CSIG}_{sk})] \tag{42}$$

Blood flow is limited to the range: $0.10 < \dot{m}_{bl} < 18$ lb/(h·ft²).

In the two-compartment model, heat is transferred from the core to the skin passively (by direct contact), and through the skin blood flow, which is controllable. The combined thermal exchange between the core and skin can be written as:

$$Q_{cr,sk} = (K + c_{p,bl}\,\dot{m}_{bl})\,(t_{cr} - t_{sk}) \tag{43}$$

where

K = effective conductance between core and skin
  = 0.930 Btu/(h·ft²·°F)
$c_{p,bl}$ = specific heat capacity of blood = 1.00 Btu/(lb·°F)

Fractional skin mass. Changes in $\dot{m}_{bl}$ influence the effective masses of the skin and core compartments by effectively changing the thermal resistance of the skin layers and the depth of the temperature gradient from the skin surface to the internal core. Since the Pierce two-compartment model assumes that each compartment is of a uniform temperature, this is equivalent to changing the apparent fractional mass of the skin compartment. For example, in cold environments, blood flow to the skin lowers and the skin and fat become isolated from the core. The actual body mass approximated by t_{cr} is therefore smaller and α increases. The effect of blood flow on the relative masses of the skin and core compartments can be written as:

$$\alpha = 0.0418 + 0.745/(4.89 \dot{m}_{bl} + 0.585) \qquad (44)$$

During thermal equilibrium and while sedentary, $\alpha \cong 0.2$. With exercise or overheating, circulation to the extremities increases and the skin layer becomes more closely coupled to the core, producing a value of $\alpha \cong 0.1$. When the body is exposed to cold, circulation to the extremities decreases and the skin layer becomes less closely coupled to the core, resulting in a higher value of $\alpha \cong 0.33$. As indicated in Equation (42), $\dot{m}_{bl}$ and, consequently, α are both determined by the deviations of t_{cr} and t_{sk} from their neutral values.

Regulatory sweating. The activity of the sweat glands is activated by warm signals from both the core and the skin, and is expressed in terms of the warm signal from the weighted average body temperature and a local control factor at the skin:

$$\dot{m}_{rsw} = 0.019 \, WSIG_b \exp (WSIG_{sk}/19.3) \qquad (45)$$

The value of $\dot{m}_{rsw}$ [calculated in Equation (45)] is then used to determine E_{rsw} in Equation (15).

Shivering. Generating additional metabolic heat through shivering and muscle tension is a more effective mechanism for maintaining the body's heat balance in extremely cold weather than vasoconstriction of the blood vessels. Shivering can raise M as much as three times its normal sedentary value. Metabolic energy production due to shivering requires simultaneous cold signals from both the skin and core, and is related to the two signals by the expression:

$$M_{shiv} = 1.90 \, CSIG_{sk} \, CSIG_{cr} \qquad (46)$$

The total metabolic energy M is then the sum of the shivering energy M_{shiv} and the metabolic heat generated due to activity M_{act}:

$$M = M_{act} + M_{shiv} \qquad (47)$$

The following section describes the estimation of M_{act} in more detail.

ENGINEERING DATA AND MEASUREMENTS

Applying the preceding basic equations to practical problems of the thermal environment requires quantitative estimates of the body's surface area, metabolic requirements for a given activity and the mechanical efficiency for the work accomplished, evaluation of the heat transfer coefficient h_r and h_c, and the general nature of the clothing insulation used. This section provides the necessary data and describes methods used to measure the parameters of the heat balance equation.

Body Surface Area

The terms in the heat balance equations previously presented have units of energy per unit area and refer to the surface area of the nude body. The most useful measure of nude body surface area, originally proposed by DuBois (1916), is described by:

$$A_D = 0.108 \, m^{0.425} \, l^{0.725} \qquad (48)$$

where

A_D = DuBois surface area, ft^2
m = mass, lb
l = height, in.

A correction factor $f_{cl} = A_{cl}/A_D$ must be applied to the heat transfer terms from the skin (C, R, and E_{sk}) to account for the actual surface area of the clothed body A_{cl}. This factor can be found in Table 7 for various clothing ensembles. For an average sized man, 68 in. and 154 lb, $A_D = 19.6$ ft^2. All terms in the basic heat balance equations are expressed per unit DuBois surface area.

Metabolic Rate and Mechanical Efficiency

Maximum energy capacity. In choosing optimal conditions for comfort and health, the energy expended during routine physical activities must be known, because metabolic energy production increases in proportion to exercise intensity. Metabolic rate varies over a wide range, depending on the activity, the person, and the conditions under which the activity is performed. Table 4 lists typical metabolic rates for an average adult ($A_D = 19.6$ ft^2) for activities performed continuously. The highest energy level a

Table 4   Typical Metabolic Heat Generation for Various Activities

	Btu/(h·ft^3)	met[a]
Resting		
Sleeping	13	0.7
Reclining	15	0.8
Seated, quiet	18	1.0
Standing, relaxed	22	1.2
Walking (on level surface)		
2.9 ft/s (2 mph)	37	2.0
4.4 ft/s (3 mph)	48	2.6
5.9 ft/s (4 mph)	70	3.8
Office Activities		
Reading, seated	18	1.0
Writing	18	1.0
Typing	20	1.1
Filing, seated	22	1.2
Filing, standing	26	1.4
Walking about	31	1.7
Lifting/packing	39	2.1
Driving/Flying		
Car	18 to 37	1.0 to 2.0
Aircraft, routine	22	1.2
Aircraft, instrument landing	33	1.8
Aircraft, combat	44	2.4
Heavy vehicle	59	3.2
Miscellaneous Occupational Activities		
Cooking	29 to 37	1.6 to 2.0
Housecleaning	37 to 63	2.0 to 3.4
Seated, heavy limb movement	41	2.2
Machine work		
sawing (table saw)	33	1.8
light (electrical industry)	37 to 44	2.0 to 2.4
heavy	74	4.0
Handling 110 lb bags	74	4.0
Pick and shovel work	74 to 88	4.0 to 4.8
Miscellaneous Leisure Activities		
Dancing, social	44 to 81	2.4 to 4.4
Calisthenics/exercise	55 to 74	3.0 to 4.0
Tennis, singles	66 to 74	3.6 to 4.0
Basketball	90 to 140	5.0 to 7.6
Wrestling, competitive	130 to 160	7.0 to 8.7

Compiled from various sources. For additional information, see Buskirk (1960), Passmore and Durnin (1967), and Webb (1964).
[a] 1 met = 18.43 Btu/(h·ft^2)

person can maintain for any continuous length of time is approximately 50% of the maximal capacity to use oxygen (maximum energy capacity).

A unit used to express the metabolic rate per unit DuBois area is the met, defined as the metabolic rate of a sedentary person (seated, quiet): 1 met = 18.43 Btu/(h·ft²). A normal, healthy man has a maximum energy capacity of approximately M_{act} = 12 met at age 20, which drops to 7 met at age 70. Maximum rates for women are about 30% lower. Long-distance runners and trained athletes have maximum rates as high as 20 met. An average 35 year-old that does not exercise has a maximum rate of about 10 met and activities with $M_{act} > 5$ met are likely to prove exhausting.

Intermittent activity. The activity of many people consists of a mixture of activities or a combination of work-rest periods. A weighted average metabolic rate is generally satisfactory, provided that activities alternate frequently (several times per hour). For example, a person typing 50% of the time, filing while seated 25% of the time, and walking about 25% of the time would have an average metabolic rate of $0.50 \times 20 + 0.25 \times 22 + 0.25 \times 31 = 23$ Btu/(h·ft²) (see Table 4).

Accuracy. Estimating metabolic rates is difficult. The values given in Table 4 only indicate metabolic rates for the specific activities listed. Some entries give a range and some a single value, depending on the source of the data. The level of accuracy depends on the value of M_{act} and how well the activity can be defined. For well-defined activities with $M_{act} < 1.5$ met (e.g., reading), Table 4 is sufficiently accurate for most engineering purposes. For values of $M_{act} > 3$, where a task is poorly defined or where there are a variety of ways of performing a task (e.g., heavy machine work), the values may be in error by as much as ±50% for a given application. Engineering calculations should thus allow for potential variations.

Measurement. When metabolic rates must be determined more accurately than is possible with tabulated data, physiological measurements with human subjects may be necessary. The rate of metabolic heat produced by the body is most accurately measured by the rate of respiratory oxygen consumption and carbon dioxide production. An empirical equation for metabolic rate is given by Nishi (1981):

$$M = 567 (0.23 \, RQ + 0.77) \, V_{O_2}/A_D, \text{ in Btu/(h·ft}^2) \quad (49)$$

where

RQ = respiratory quotient; molar ratio of V_{CO_2} exhaled to V_{O_2} inhaled, dimensionless
V_{O_2} = volumetric rate of oxygen consumption at conditions (STPD) of 32°F, 14.7 psi, ft³/h

The exact value of the respiratory quotient used in Equation (49) RQ depends on a person's activity, diet, and physical condition. It can be determined by measuring both carbon dioxide and oxygen in the respiratory airflows, or it can be estimated with reasonable accuracy. A good estimate for the average adult is RQ = 0.83 for light or sedentary activities (M < 1.5 met), increasing proportionally to RQ = 1.0 for heavy exertion (M = 5.0 met). In extreme cases, the maximum range is 0.7 < RQ < 1.0. Estimation of RQ is generally sufficient for all except precision laboratory measurements since it does not strongly affect the value of the metabolic rate. A 10% error in estimating the respiratory quotient results in an error of less than 3% in the metabolic rate.

A second, much less accurate, method of estimating metabolic rate physiologically is to measure the heart rate. Table 5 shows the relationship between heart rate and oxygen consumption at different levels of physical exertion for a typical person. Once oxygen consumption is estimated from heart rate information, Equation (49) can be used to estimate the metabolic rate. A number of fac-

Table 5 Heart Rate and Oxygen Consumption at Different Activity Levels[a]

Level of Exertion	Oxygen Consumed, ft³/h	Heart Rate, beat/min
Light work	< 0.1	< 90
Moderate work	1.0 to 2.0	90 to 110
Heavy work	2.0 to 3.0	110 to 130
Very heavy work	3.0 to 4.0	130 to 150
Extremely heavy work	> 4.0	150 to 170

[a]Astrand and Rodahl (1977).

tors other than metabolic rate affect heart rate, such as physical condition, heat, emotional factors, muscles used, etc. Astrand and Rodahl (1977) show that heart rate is only a very approximate measure of metabolic rate and should not be the only source of information where accuracy is required.

Mechanical efficiency. In the heat balance equation, the rate of work accomplished W must be in the same units as metabolism M and expressed in terms of A_D in Btu/(h·ft²). The mechanical work done by the muscles for a given task is often expressed in terms of the body's mechanical efficiency $\mu = W/M$. It is unusual for μ to be more than 5 to 10%; for most activities, it is close to zero. The maximum value under optimal conditions (e.g., bicycle ergometer) is $\mu = 20$ to 24% (Nishi 1981). It is common to assume that mechanical work is zero for several reasons: (1) the mechanical work produced is small compared to metabolic rate, especially for office activities; (2) estimates for metabolic rates can often be inaccurate; and (3) this assumption results in a more conservative estimate when designing air-conditioning equipment for upper comfort and health limits. More accurate calculation of heat generation may require estimation of the mechanical work produced for activities where it is significant (walking on a grade, climbing a ladder, bicycling, lifting, etc.). In some cases, it is possible to either estimate or measure the mechanical work. For example, a 200-lb person walking up a 5% grade at 4.4 ft/s (3 mph) would lift a 200-lb weight a height of 0.22 ft every second, for a work rate of 44 ft·lb$_f$/s, or 204 Btu/h. This rate of mechanical work would then be subtracted from M to determine the net heat generated.

Heat Transfer Coefficients

Values for the linearized radiative heat transfer coefficient, convective heat transfer coefficient, and evaporative heat transfer coefficient are required to solve the equations describing heat transfer from the body.

Radiative heat transfer coefficient. The linearized radiative heat transfer coefficient can be calculated by:

$$h_r = 4 \epsilon \sigma (A_r/A_D) [459.7 + (t_{cl} + \bar{t}_r)/2]^3 \quad (50)$$

where

ϵ = average emissivity of clothing or body surface, dimensionless
σ = Stefan-Boltzmann constant, 0.1714×10^{-8} Btu/(h·ft²·°R⁴)
A_r = effective radiation area of body, ft²

The ratio A_r/A_D is 0.70 for a sitting person and 0.73 for a standing person (Fanger 1967). The emissivity is close to unity (typically 0.95), unless special reflective materials are used or high-temperature sources are involved. It is not always possible to solve Equation (50) explicitly for h_r, since t_{cl} may also be an unknown. Some form of iteration may be required if a precise solution is required. Fortunately, h_r is nearly constant for typical indoor temperatures, and a value of 0.83 Btu/(h·ft²·°F) suffices for most calculations. If the emissivity is significantly less than unity, the value should be adjusted by:

$$h_r = 0.83 \, \epsilon \; \text{Btu}/(\text{h} \cdot \text{ft}^2 \cdot {}^\circ\text{F}) \qquad (51)$$

where ϵ represents the area-weighted average emissivity for the clothing/body surface.

Convective heat transfer coefficient. Heat transfer by convection is usually caused by air movement within the living space or by body movements. Equations for estimating h_c under various conditions are presented in Table 6. An additional relationship is presented in Equation (75). Where two conditions apply (e.g., walking in moving air), a reasonable estimate can be obtained by taking the larger of the two values for h_c. Limits have been given to all equations. If no limits were given in the source, reasonable limits have been estimated. Care should be exercised in using these values for seated and reclining persons. The heat transfer coefficients may be accurate, but the effective heat transfer area may be substantially reduced due to body contact with a padded chair or bed.

Quantitative values of h_c are important, not only in estimating convection loss, but in evaluating (1) operative temperature t_o, (2) clothing parameters I_t and i_m, and (3) rational effective temperatures t_{oh} and ET*. All heat transfer coefficients in Table 6 were evaluated at or near 14.7 psia. These coefficients should be corrected as follows for atmospheric pressure:

$$h_{cc} = h_c(p_t/14.7)^{0.55} \qquad (52)$$

where

h_{cc} = corrected convective heat transfer coefficient, $\text{Btu}/(\text{h} \cdot \text{ft}^2 \cdot {}^\circ\text{F})$
p_t = local atmospheric pressure, psia

The combined coefficient h is the sum of h_r and h_c described in Equation (50) and Table 6, respectively. The coefficient h governs exchange by radiation and convection from the exposed body surface to the surrounding environment.

Evaporative heat transfer coefficient. The evaporative heat transfer coefficient h_e for the outer air layer of a nude or clothed person can be estimated from the convective heat transfer coefficient using the Lewis relationship given in Equation (27). If the atmospheric pressure is significantly different from standard (14.7 psia), the correction to the value obtained from Equation (27) is:

$$h_{ec} = h_e (14.7/p_t)^{0.45} \qquad (53)$$

Table 6   Equations for Convection Heat Transfer Coefficients

Equation	Limits	Condition	Remarks/Sources
$h_c = 0.061 \, V^{0.6}$ $h_c = 0.55$	$40 < V < 800$ $0 < V < 40$	Seated with moving air	Mitchell (1974)
$h_c = 0.475 + 0.044 \, V^{0.67}$ $h_c = 0.90$	$30 < V < 300$ $0 < V < 30$	Reclining with moving air	Colin and Houdas (1967)
$h_c = 0.092 \, V^{0.53}$	$100 < V < 400$	Walking in still air	V is walking speed (Nishi and Gagge 1970)
$h_c = (M - 0.85)^{0.39}$	$1.1 < M < 3.0$	Active in still air	Gagge et al. (1976)
$h_c = 0.146 \, V^{0.39}$	$100 < V < 400$	Walking on treadmill in still air	V is treadmill speed (Nishi and Gagge 1970)
$h_c = 0.068 \, V^{0.69}$ $h_c = 0.70$	$30 < V < 300$ $0 < V < 30$	Standing person in moving air	Developed from data presented by Seppenan et al. (1972)

Note: h_c in $\text{Btu}/(\text{h} \cdot \text{ft}^2 \cdot {}^\circ\text{F})$, V in fpm, and M in met units, where 1 met = 18.43 $\text{Btu/h} \cdot \text{ft}^2$.

where h_{ec} is the corrected evaporative heat transfer coefficient, $\text{Btu}/(\text{h} \cdot \text{ft}^2 \cdot {}^\circ\text{F})$.

Clothing Insulation and Moisture Permeability

Thermal insulation. The most accurate methods for determining clothing insulation are: (1) measurements on heated manikins (McCullough and Jones 1984, Olesen and Nielsen 1983) and (2) measurements on active subjects (Nishi et al. 1975). For most routine engineering work, estimates based on tables and equations presented in this section are sufficient. Thermal manikins can measure the sensible heat loss from the "skin" $(C + R)$ in a given environment. Equation (12) can then be used to evaluate R_{cl} if the environmental conditions are well defined and f_{cl} is measured. Evaluation of clothing insulation on subjects requires measurement of t_{sk}, t_{cl}, and t_o. The clothing thermal efficiency is calculated by:

$$F_{cl} = (t_{cl} - t_o)/(t_{sk} - t_o) \qquad (54)$$

The intrinsic clothing insulation can then be calculated from manikin measurements by the following relationship, provided f_{cl} is measured and conditions are sufficiently well-defined to make an accurate determination of h:

$$R_{cl} = (t_{sk} - t_o)/q - 1/(hf_{cl}) \qquad (55)$$

where q = heat loss from the manikin, $\text{Btu}/(\text{h} \cdot \text{ft}^2)$.

Traditionally, clothing insulation value is expressed in clo units. In order to avoid confusion, the symbol I is used with the clo unit instead of the symbol R. The relationship between the two is:

Table 7   Typical Insulation and Permeability Values for Clothing Ensembles[a]

Ensemble Description[b]	I_{cl} (clo)	I_t[c] (clo)	f_{cl}	i_{cl}	i_m[c]
Walking shorts, short-sleeve shirt	0.36	1.02	1.10	0.34	0.42
Trousers, short-sleeve shirt	0.57	1.20	1.15	0.36	0.43
Trousers, long-sleeve shirt	0.61	1.21	1.20	0.41	0.45
Same as above, plus suit jacket	0.96	1.54	1.23		
Same as above, plus vest and t-shirt	1.14	1.69	1.32	0.32	0.37
Trousers, long-sleeve shirt, long-sleeve sweater, t-shirt	1.01	1.56	1.28		
Same as above, plus suit jacket and long underwear bottoms	1.30	1.83	1.33		
Sweat pants, sweat shirt	0.74	1.35	1.19	0.41	0.45
Long-sleeve pajama top, long pajama trousers, short 3/4 sleeve robe, slippers (no socks)	0.96	1.50	1.32	0.37	0.41
Knee-length skirt, short-sleeve shirt, panty hose, sandals	0.54	1.10	1.26		
Knee-length skirt, long-sleeve shirt, full slip, panty hose	0.67	1.22	1.29		
Knee-length skirt, long-sleeve shirt, half slip, panty hose, long-sleeve sweater	1.10	1.59	1.46		
Same as above, replace sweater with suit jacket	1.04	1.60	1.30	0.35	0.40
Ankle-length skirt, long-sleeve shirt, suit jacket, panty hose	1.10	1.59	1.46		
Long-sleeve coveralls, t-shirt	0.72	1.30	1.23		
Overalls, long-sleeve shirt, t-shirt	0.89	1.46	1.27	0.35	0.40
Insulated coveralls, long-sleeve thermal underwear, long underwear bottoms	1.37	1.94	1.26	0.35	0.39

[a] From McCullough and Jones (1984) and McCullough et al. (1989)
[b] All ensembles include shoes and briefs or panties. All ensembles except those with panty hose include socks unless otherwise noted.
[c] For $t_r = t_a$ and air velocity less than 40 ft/min ($I_a = 0.72$ clo and $i_m = 0.48$ when nude)

$$R = 0.88 I \qquad (56)$$

or 1.0 clo is equivalent to 0.88 ft² · h · °F/Btu.

Since clothing insulation cannot be measured for most routine engineering applications, tables of measured values for various clothing ensembles can be used to select an ensemble comparable to the one(s) in question. Table 7 gives values for typical indoor clothing ensembles. More detailed tables are presented by McCullough and Jones (1984) and Olesen and Nielsen (1983). Accuracies for I_{cl} on the order of ±20% are typical if good matches between ensembles are found.

Often it is not possible to find an already measured clothing ensemble that matches the one in question. In this case, the ensemble insulation can be estimated from the insulation of individual garments. Table 8 gives a list of individual garments commonly worn. The insulation of an ensemble is estimated from the individual values using a summation formula (McCullough and Jones 1984):

$$I_{cl} = 0.835 \sum_i I_{clu,i} + 0.161 \qquad (57)$$

where $I_{clu,i}$ is the effective insulation of garment i, and I_{cl}, as before, is the insulation for the entire ensemble. A simpler and nearly as accurate summation formula is (Olesen 1985):

$$I_{cl} = \sum_i I_{clu,i} \qquad (58)$$

Either Equation (57) or (58) gives acceptable accuracy for typical indoor clothing. The main source of inaccuracy is in determining the appropriate values for individual garments. Overall accuracies are on the order of ±25% if the tables are used carefully. If it is important to include a specific garment that is not included in Table 8, its insulation can be estimated by (McCullough and Jones 1984):

$$I_{clu,i} = (0.534 + 3.43 x_f)(A_G/A_D) - 0.0549 \qquad (59)$$

where

x_f = fabric thickness, in.
A_G = body surface area covered by garment, ft²

Values in Table 7 may be adjusted by information in Table 8 and a summation formula. Using this method, values of $I_{clu,i}$ for the selected items in Table 8 are then added to or subtracted from the ensemble value of I_{cl} in Table 7.

Moisture permeability. Moisture permeability data for some clothing ensembles are presented in terms of i_{cl} and i_m in Table 7. The values of i_m can be used to calculate $R_{e,t}$ using the relationships in Table 2. Ensembles worn indoors generally fall in the range $0.3 < i_m < 0.5$ and assuming $i_m = 0.4$ is reasonably accurate (McCullough et al. 1989). This latter value may be used if a good match to ensembles in Table 7 cannot be made. The value of i_m or $R_{e,t}$ may be substituted directly into equations for body heat loss calculations (see Table 3). However, i_m for a given clothing ensemble is a function of the environment as well as the clothing properties. Unless i_m is evaluated at conditions very similar to the intended application, it is more rigorous to use i_{cl} to describe the moisture permeability of the clothing. The value of i_{cl} is not as sensitive to environmental conditions; thus, given data are more accurate over a wider range of air velocity and radiant and air temperature combinations for i_{cl} than for i_m. The relationships in Table 2 can be used to determine $R_{e,cl}$ from i_{cl}, and i_{cl} or $R_{e,cl}$ can then be used for body heat loss calculations (see Table 3). McCullough et al. (1989) found an average value of $i_{cl} = 0.34$ for common indoor clothing; this value can be used when other data are not available.

Measurements of i_m or i_{cl} may be necessary if unusual clothing (e.g., impermeable or metalized) and/or extreme environments (e.g., high radiant temperatures or high air velocities) are to be addressed. There are three different methods for measuring the moisture permeability of clothing: the first uses a wet manikin to measure the effect of sweat evaporation on heat loss (McCullough 1986); the second uses moisture permeability measurements on component fabrics as well as dry manikin measurements (Umbach 1980); the third uses measurements from sweating subjects (Nishi et al. 1975, Holmer 1984).

Table 8 Garment Insulation Values

Garment Description[a]	$I_{clu,i}$, clo[b]	Garment Description[a]	$I_{clu,i}$, clo[b]	Garment Description[a]	$I_{clu,i}$, clo[b]
Underwear		Short-sleeve, knit sport shirt	0.17	Sleeveless vest (thick)	0.22
Men's briefs	0.04	Long-sleeve, sweat shirt	0.34	Long-sleeve (thin)	0.25
Panties	0.03			Long-sleeve (thick)	0.36
Bra	0.01	**Trousers and Coveralls**			
T-shirt	0.08	Short shorts	0.06	**Suit jackets and vests (lined)**	
Full slip	0.16	Walking shorts	0.08	Single-breasted (thin)	0.36
Half slip	0.14	Straight trousers (thin)	0.15	Single-breasted (thick)	0.44
Long underwear top	0.20	Straight trousers (thick)	0.24	Double-breasted (thin)	0.42
Long underwear bottoms	0.15	Sweatpants	0.28	Double-breasted (thick)	0.48
		Overalls	0.30	Sleeveless vest (thin)	0.10
Footwear		Coveralls	0.49	Sleeveless vest (thick)	0.17
Ankle-length athletic socks	0.02				
Calf-length socks	0.03	**Dresses and skirts[c]**		**Sleepwear and Robes**	
Knee socks (thick)	0.06	Skirt (thin)	0.14	Sleeveless, short gown (thin)	0.18
Panty hose	0.02	Skirt (thick)	0.23	Sleeveless, long gown (thin)	0.20
Sandals/thongs	0.02	Long-sleeve shirtdress (thin)	0.33	Short-sleeve hospital gown	0.31
Slippers (quilted, pile-lined)	0.03	Long-sleeve shirtdress (thick)	0.47	Long-sleeve, long gown (thick)	0.46
Boots	0.10	Short-sleeve shirtdress (thin)	0.29	Long-sleeve pajamas (thick)	0.57
		Sleeveless, scoop neck (thin)	0.23	Short-sleeve pajamas (thin)	0.42
Shirts and Blouses		Sleeveless, scoop neck (thick),		Long-sleeve, long wrap robe	
Sleeveless, scoop-neck blouse	0.12	i.e., jumper	0.27	(thick)	0.69
Short-sleeve, dress shirt	0.19			Long-sleeve, short wrap robe	
Long-sleeve, dress shirt	0.25	**Sweaters**		(thick)	0.48
Long-sleeve, flannel shirt	0.34	Sleeveless vest (thin)	0.13	Short-sleeve, short robe (thin)	0.34

[a] "Thin" garments are made of light, thin fabrics worn in the summer; "thick" garments are made of heavy, thick fabrics worn in the winter.

[b] 1 clo = 0.880°F · ft² · h/Btu
[c] Knee-length

Clothing surface area. Many clothing heat transfer calculations require the clothing area factor f_{cl} to be known. The most reliable approach is to measure it using photographic methods (Olesen *et al.* 1982). Other than actual measurements, the best method is to use previously tabulated data for similar clothing ensembles. Table 7 is adequate for most indoor clothing ensembles. No good method of estimating f_{cl} for a clothing ensemble from other information is available, although a rough estimate can be made by (McCullough and Jones 1984):

$$f_{cl} = 1.0 + 0.3\, I_{cl} \qquad (60)$$

Total Evaporative Heat Loss

The total evaporative heat loss (latent heat) from the human body due to both respirative losses and skin losses, $E_{sk} + E_{res}$, can be measured directly from the body's rate of mass loss as observed by a sensitive scale:

$$E_{sk} + E_{res} = h_{fg}\,(dm/d\theta)/A_D \qquad (61)$$

where

h_{fg} = latent heat of vaporization of water, Btu/lb
m = body mass, lb
θ = time, h

When using Equation (61), adjustments should be made for any materials consumed (*e.g.*, food and drink) and body effluents (*e.g.*, wastes). The fuel burned by the body also contributes slightly to weight loss and can be accounted for with the following relationship (Astrand and Rodahl 1977):

$$dm_{ge}/d\theta = 2.2\, V_{O_2}(0.1225\ \mathrm{RQ} - 0.0891) \qquad (62)$$

where

$dm_{ge}/d\theta$ = mass loss due to respiratory gas exchange, lb/h
V_{O_2} = O_2 uptake at STPD, ft^3/h
RQ = respiratory quotient
0.1225 = density of CO_2 at STPD, lb/ft^3
0.0891 = density of O_2 STPD, lb/ft^3
STPD = standard temperature and pressure dry, 32 °F, 14.7 psia

Environmental Parameters

The parameters describing the thermal environment that must be measured or otherwise quantified if accurate estimates of human thermal response are to be made are divided into two groups—those that can be measured directly and those that are calculated from other measurements.

Directly measured. Seven of the parameters frequently used to describe the thermal environment are psychrometric and include (1) air temperature t_a; (2) wet-bulb temperature t_{wb}; (3) dew-point temperature t_{dp}; (4) water vapor pressure p_a; (5) total atmospheric pressure p_t; (6) relative humidity (rh); and (7) humidity ratio W. These parameters are discussed in detail in Chapter 6, and methods for measuring them are discussed in Chapter 13. Two other important parameters that can be measured directly include air velocity V and mean radiant temperature $\bar{t}_r$. Air velocity measurements are also discussed in Chapter 13. The radiant temperature is the temperature of an exposed surface in the environment. The temperatures of individual surfaces are usually combined into a mean radiant temperature $\bar{t}_r$. Finally, globe temperature t_g, which can also be measured directly, is a good approximation of the operative temperature t_o and is also used with other measurements to calculate the mean radiant temperature (see Chapter 13).

Calculated parameters. The *mean radiant temperature* $\bar{t}_r$ is a key variable in making thermal calculations for the human body.

It is the uniform temperature of an imaginary enclosure in which radiant heat transfer from the human body equals the radiant heat transfer in the actual nonuniform enclosure. Measurements of the globe temperature, air temperature, and the air velocity can be combined to estimate the mean radiant temperature (see Chapter 13). The accuracy of the mean radiant temperature determined this way varies considerably depending on the type of environment and the accuracy of the individual measurements. Since the mean radiant temperature is defined with respect to the human body, the shape of the sensor is also a factor. The spherical shape of the globe thermometer gives a reasonable approximation for a seated person; an ellipsoid-shaped sensor gives a better approximation to the shape of a human, both upright and seated.

The mean radiant temperature can also be calculated from measured values of the temperature of the surrounding walls and surfaces and their positions with respect to the person. As most building materials have a high emittance ϵ, all the surfaces in the room can be assumed to be black. The following equation is then used:

$$\bar{T}_r^4 = T_1^4 F_{p-1} + T_2^4 F_{p-2} + \ldots + T_N^4 F_{p-N} \qquad (63)$$

where

$\bar{T}_r$ = mean radiant temperature, °R
T_N = surface temperature of surface N, °R
F_{p-N} = angle factor between a person and surface N

As the sum of the angle factors is unity, the fourth power of mean radiant temperature equals the mean value of the surrounding surface temperatures to the fourth power, weighted by the respective angle factors. In general, angle factors are difficult to determine, although Figures 3a and 3b may be used to estimate them for rectangular surfaces. The angle factor normally depends on the position and orientation of the person (Fanger 1982).

If relatively small temperature differences exist between the surfaces of the enclosure, Equation (63) can be simplified to a linear form:

$$\bar{t}_r = t_1 F_{p-1} + t_2 F_{p-2} + \ldots + t_N F_{p-N} \qquad (64)$$

Equation (64) always gives a slightly lower mean radiant temperature than Equation (63), but in many cases the difference is small. If, for example, half the surroundings ($F_{p-N} = 0.5$) has a temperature 10 °F higher than the other half, the difference between the calculated mean radiant temperatures—according to Equations (63) and (64)—is only 0.4 °F. If, however, this difference is 200 °F, the mean radiant temperature calculated by Equation (64) is 20 °F too low.

The mean radiant temperature may also be calculated from the plane radiant temperature t_{pr} (defined later) in six directions (up, down, left, right, front, back) and for the projected area factors of a person in the same six directions. For a standing person, the mean radiant temperature may be estimated as:

$$\begin{aligned}
\bar{t}_r = \{ & 0.08\,[t_{pr}\,(\text{up}) + t_{pr}\,(\text{down})] + 0.23\,[t_{pr}\,(\text{right}) \\
& + t_{pr}\,(\text{left})] + 0.35\,[t_{pr}\,(\text{front}) + t_{pr}\,(\text{back})]\} \\
& / [2\,(0.08 + 0.23 + 0.35)] \qquad (65)
\end{aligned}$$

For a seated person, the mean radiant temperature may be estimated as:

$$\begin{aligned}
\bar{t}_r = \{ & 0.18\,[t_{pr}\,(\text{up}) + t_{pr}(\text{down})] + 0.22\,[t_{pr}\,(\text{right}) \\
& + t_{pr}\,(\text{left})] + 0.30\,[t_{pr}\,(\text{front}) + t_{pr}\,(\text{back})]\} \\
& / [2\,(0.18 + 0.22 + 0.30)] \qquad (66)
\end{aligned}$$

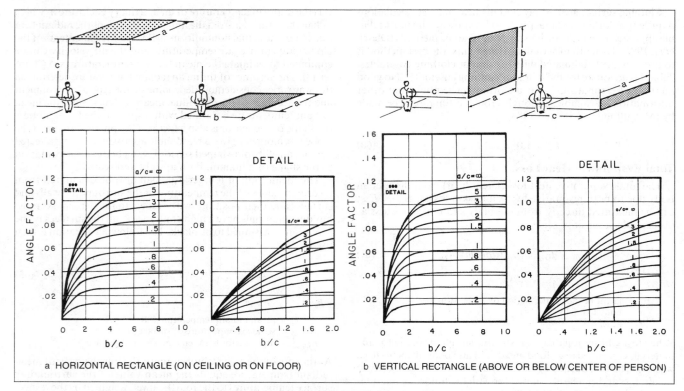

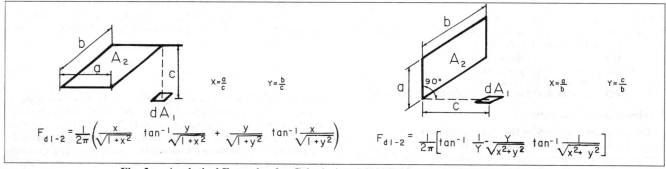

Fig. 3 **Mean Value of Angle Factor between Seated Person and Horizontal or Vertical Rectangle when Person is Rotated around Vertical Axis**
(Fanger 1982)

Fig. 3c **Analytical Formulae for Calculating Angle Factor for Small Plane Element**

The *plane radiant temperature* t_{pr}, first introduced by McIntyre (1974), is the uniform temperature of an enclosure in which the incident radiant flux on one side of a small plane element is the same as that in the actual environment. The plane radiant temperature describes the thermal radiation in one direction and its value thus depends on the direction. In comparison, the mean radiant temperature describes the thermal radiation for the human body from all directions. The plane radiant temperature can be calculated using Equations (63) and (64) with the same limitations. Area factors are determined from Figure 3c.

The *radiant temperature asymmetry* Δt_{pr} is the difference between the plane radiant temperature of the opposite sides of a small plane element. This parameter describes the asymmetry of the radiant environment, and is especially important in comfort conditions. Because it is defined with respect to a plane element, its value depends on the orientation of that plane. This orientation may be specified in some situations (*e.g.*, floor to ceiling asymmetry) and not in others. If direction is not specified, the radiant asymmetry should be for the orientation that gives the maximum value.

ENVIRONMENTAL INDICES

A number of indices simplify the description of the thermal environment and the stress imposed by an environment. An environmental index combines two or more parameters, such as air temperature, mean radiant temperature, humidity, or air velocity, into a single variable. Environmental indices may be classified according to how they are developed. Rational indices are based on the theoretical concepts presented earlier. Empirical indices are based on measurements with subjects or on simplified relationships that do not necessarily follow theory. Indices may also be classified according to their application, generally either heat stress or cold stress.

Effective Temperature

The *effective temperature* ET* is probably the most common environmental index and has the widest range of application. It combines temperature and humidity into a single index, so two environments with the same ET* should evoke the same thermal response even though they have different temperatures and humidities; but they must have the same air velocities. The original empirical effective temperature was developed by Houghton and Yaglou (1923). Gagge *et al.* (1971) defined a new effective temperature using a rational approach. Defined mathematically in Equation (33), this is the temperature of an environment at 50% rh that results in the same total heat loss from the skin E_{sk} as in the actual environment. Since the index is defined in terms of operative temperature t_o, it combines the effect of three parameters—$\bar{t}_r$, t_a, and p_a—into a single index. Skin wettedness w and the permeability index i_m must be specified and are constant for a given ET* line for a particular situation. The two-node model is used to determine skin wettedness in the zone of evaporative regulation. At the upper limit of regulation, w approaches 1.0, and at the lower limit, w approaches 0.06; skin wettedness equals one of these values when the body is outside the zone of evaporative regulation. Since the slope of a constant ET* line depends on skin wettedness and clothing moisture permeability, effective temperature for a given temperature and humidity may depend on the clothing of the person and their activity. This difference is shown in Figure 4. At low skin wettedness, the air humidity has little influence, and lines of constant ET* are nearly vertical. As skin wettedness increases due to activity and/or heat stress, the lines become more horizontal and the influence of humidity is much more pronounced. The ASHRAE comfort envelope shown in Figure 5 is described in terms of ET*.

Since ET* depends on clothing and activity, it is not possible to generate a universal ET* chart. Calculation of ET* can also be tedious, requiring the solution of multiple coupled equations to determine skin wettedness. These equations can be solved with the appropriate computer routines (Gagge *et al.* 1971), but not always with hand calculations. A standard set of conditions representative of typical indoor applications is used to define a *standard effective temperature* (SET*). The standard effective temperature is then defined as the equivalent air temperature of an isothermal environment at 50% rh in which a subject, while wearing clothing standardized for the activity concerned, has the same heat stress (skin temperature t_{sk}) and thermoregulatory strain (skin wettedness w) as in the actual environment.

Humid Operative Temperature

The *humid operative temperature* t_{oh} is the temperature of a uniform environment at 100% rh in which a person loses the same total amount of heat from the skin as in the actual environment. This index is defined mathematically in Equation (32). It is analogous to ET*, the only difference being that it is defined at 100% rh and 0% rh rather than at 50% rh. Figures 2 and 4 indicate that lines of constant ET* are also lines of constant t_{oh}. However, the values of these two indices differ for a given environment.

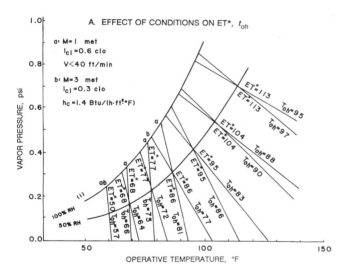

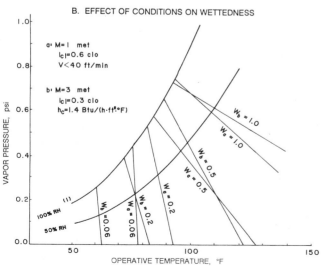

Fig. 4 Effective Temperature (ET*) and Skin Wettedness (W)
(Adapted from Nishi *et al.* 1975 and Gonzalez *et al.* 1978)

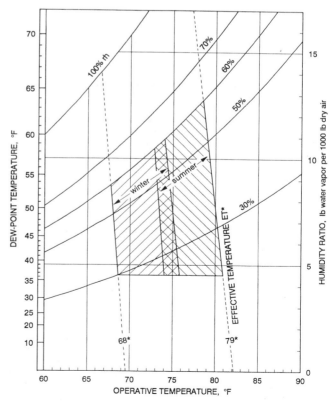

Fig. 5 Standard Effective Temperature and ASHRAE Comfort Zones

Heat Stress Index

Originally proposed by Belding and Hatch (1955), this rational index is the ratio of the total evaporative heat loss E_{sk} required for thermal equilibrium (the sum of metabolism plus dry heat load) to the maximum evaporative heat loss E_{max} possible for the environment, multiplied by 100, for steady-state conditions (S_{sk} and S_{cr} are zero), and with t_{sk} held constant at 95 °F. The ratio E_{sk}/E_{max} equals skin wettedness as is shown in Equation (19). When HSI > 100, body heating occurs; when HSI < 0, body cooling occurs. Belding and Hatch (1955) limited E_{max} to 220 Btu/(h·ft²), which corresponds to a sweat rate of approximately 0.21 lb/(h·ft²). When t_{sk} is constant, loci of constant HSI coincide with lines of constant ET* on a psychrometric chart. Other indices based on wettedness of the original HSI concept have the same practical applications (Gonzalez *et al.* 1978, Belding 1970, ISO *Standard* 7933), but differ in their treatment of E_{max} and the effect of clothing. Table 9 describes physiological factors associated with HSI values.

Index of Skin Wettedness

This index is the ratio of observed skin sweating E_{sk} to the E_{max} of the environment as defined by t_{sk}, t_a, humidity, air movement, and clothing in Equation (13). Except for the factor of 100, it is essentially the same as the heat stress index. Skin wettedness is more closely related to the sense of discomfort or unpleasantness than to temperature sensation (Gagge *et al.* 1969a, 1969b; Gonzalez *et al.* 1978).

Wet-Bulb Globe Temperature

The WBGT is an environmental heat stress index that combines dry-bulb temperature t_{db}, a *naturally ventilated* (not aspirated) wet-bulb temperature t_{nwb}, and black globe temperature t_g, according to the relation (Dukes-Dobos and Henschel 1971, 1973):

$$\text{WBGT} = 0.7\,t_{nwb} + 0.2\,t_g + 0.1\,t_a \qquad (67)$$

Table 9   Evaluation of Heat Stress Index

Heat Stress Index	Physiological and Hygienic Implications of 8-h Exposures to Various Heat Stresses
0	No thermal strain.
10 20 30	Mild to moderate heat strain. If job involves higher intellectual functions, dexterity, or alertness, subtle to substantial decrements in performance may be expected. In performing heavy physical work, little decrement is expected, unless ability of individuals to perform such work under no thermal stress is marginal.
40 50 60	Severe heat strain involving a threat to health unless men are physically fit. Break-in period required for men not previously acclimatized. Some decrement in performance of physical work is to be expected. Medical selection of personnel desirable, because these conditions are unsuitable for those with cardiovascular or respiratory impairment or with chronic dermatitis. These working conditions are also unsuitable for activities requiring sustained mental effort.
70 80 90	Very severe heat strain. Only a small percentage of the population may be expected to qualify for this work. Personnel should be selected: (a) by medical examination, and (b) by trial on the job (after acclimatization). Special measures are needed to assure adequate water and salt intake. Amelioration of working conditions by any feasible means is highly desirable, and may be expected to decrease the health hazard while increasing job efficiency. Slight "indisposition", which in most jobs would be insufficient to affect performance, may render workers unfit for this exposure.
100	The maximum strain tolerated daily by fit, acclimatized young men.

This form of the equation is usually used where solar radiation is present. The naturally ventilated wet-bulb thermometer is left exposed to the sunlight, but the air temperature t_a sensor is shaded. In enclosed environments, Equation (67) is simplified by dropping the t_a term and using a 0.3 weighting factor for t_g.

The black globe thermometer is responsive to air temperature, mean radiant temperature, and air movement, while the naturally ventilated wet-bulb thermometer responds to air humidity, air movement, radiant temperature, and air temperature. Thus, WBGT is a function of all four environmental factors affecting human environmental heat stress.

The WBGT is a better index of heat stress than the old ET; it shows almost as good a correlation with sweat rate as do the later Corrected Effective Temperature (CET) and the Effective Temperature with Radiation (ETR) indices (Minard 1961); the CET and ETR both require direct measurement of wind velocity which, for accuracy, requires special instruments and trained technicians.

The WBGT index is widely used for estimating the heat stress potential of industrial environments (Davis 1976). In the United States, the National Institute of Occupational Safety and Health (NIOSH) developed a criteria document for a heat-stress limiting standard (NIOSH 1986). ISO *Standard* 7243 (ISO 1982) also uses the WBGT. Figure 6 is a graphical summary of the permissible heat exposure limits, expressed as working time per hour, for a fit individual, as specified for various WBGT levels. Values apply for normal permeable clothing (0.6 clo) and must be adjusted for heavy or partly vapor-permeable clothing. USAF (1980) recommends adjusting the measured WBGT upwards by 10 °F for personnel wearing chemical protective clothing or body armor. This type of clothing increases the resistance to sweat evaporation about threefold (higher if it is totally impermeable), requiring an adjustment in WBGT level to compensate for reduced evaporative cooling at the skin.

Several mathematical models are available for predicting WBGT from the environmental factors: air temperature, psychrometric wet-bulb temperature, mean radiant temperature, and air motion (Azer and Hsu 1977, Sullivan and Gorton 1976). A simpler approach, involving plotting WBGT lines on a psychromet-

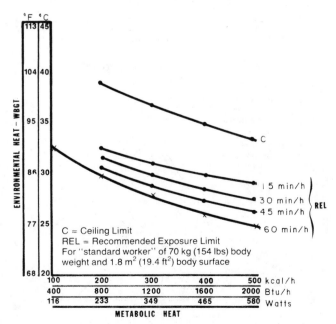

Fig. 6   Recommended Heat Stress Exposure Limits for Heat Acclimatized Workers
(From United States DHHS (NIOSH) Publication 86-113, 1986)

ric chart, is recommended. Isotherms of WBGT are parallel and have negative slopes varying from 0.014 psi/°F for still air to 0.016 psi/°F for air motion greater than 200 fpm. By comparison, psychrometric wet-bulb lines have negative slopes of about 0.0056 psi/°F, or are about 35% as steep.

Wet-Globe Temperature

The WGT, introduced by Botsford (1971), is a simpler approach to measuring environmental heat stress than the WBGT. The measurement is made with a wetted globe thermometer called a Botsball, which consists of a 2.5 in. black copper sphere covered with a fitted wet black mesh fabric, into which the sensor of a dial thermometer is inserted. A polished stem attached to the sphere supports the thermometer and contains a water reservoir for keeping the sphere covering wet. This instrument is suspended by the stem at the indoor (or outdoor) site to be measured.

Onkaram (1980) has shown that WBGT can be predicted with reasonable accuracy from WGT for temperate to warm environments with medium to high humidities. With air temperatures between 68 and 95 °F, dew points ranging from 45 to 77 °F (relative humidities above 30%), and wind speeds of 15 mph or less, the experimental regression equation ($r = 0.98$) in °F for an outdoor environment is:

$$WBGT = 1.044\ WGT - 1.745 \quad (68)$$

This equation should not be used outside the experimental range just given, since data from hot-dry desert environments show differences between WBGT and WGT that are too large (10 °F and above) to be adjusted by Equation (68) (Matthew 1986). With very low humidity combined with high wind, WGT approaches the psychrometric wet-bulb temperature, which is greatly depressed below t_a. However, in the WBGT, t_{nwb} accounts for only 70% of the index value, with the remaining 30% at or above t_a.

Ciriello and Snook (1977) handle the problem by providing a series of regression equations, the choice depending on the levels of wind speed, humidity, and radiant heat. They report an accuracy of conversion from WGT to WBGT within 0.7 °F (90% confidence level), if good estimates of wind speed, humidity, and radiation level are available.

Wind Chill Index

The wind chill index (WCI) is an empirical index developed from cooling measurements obtained in Antarctica on a cylindrical flask partly filled with water (Siple and Passel 1945). The index describes the rate of heat loss from the cylinder by radiation and convection for a surface temperature of 91.4 °F, as a function of ambient temperature and wind velocity. As originally proposed:

$$WCI = (10.45 - 0.447\ V + 6.686\ \sqrt{V}\,)\,(91.4 - t_a)/1.8 \quad (69)$$

where V and t_a are in mph and °F, respectively. The 91.4 °F surface temperature was chosen to be representative of the mean skin temperature of a resting human in comfortable surroundings.

A number of valid objections have been raised about this formulation. Cooling rate data from which it was derived were measured on a 2.24 in. diameter plastic cylinder, making it unlikely that WCI would be an accurate measure of heat loss from exposed flesh, which has different characteristics than the plastic (curvature, roughness, and radiation exchange properties) and is invariably below 91.4 °F in a cold environment. Moreover, values given by the equation peak at 56 mph, then decrease with increasing velocity.

Nevertheless, for velocities below 50 mph, this index reliably expresses combined effects of temperature and wind on subjective discomfort. For example, if the calculated WCI is less than 1400 and actual air temperature is above 14 °F, there is little risk of frostbite during brief exposures (1 h or less), even for bare skin. However, at a WCI of 2000 or more, the probability is high that exposed flesh will begin to freeze in 1 min or less unless preventive measures are taken to shield the exposed skin (such as a fur ruff to break up the wind around the face).

Rather than using the WCI to express the severity of a cold environment, meteorologists use an index derived from the WCI called the *equivalent wind chill temperature*. This is the ambient temperature that would produce, in a calm wind (defined for this application as 4 mph), the same WCI as the actual combination of air temperature and wind velocity. Equivalent wind chill temperature $t_{eq,wc}$ in °F can be calculated by:

$$t_{eq,wc} = -0.0818\ (WCI) + 91.4 \quad (70)$$

where $t_{eq,wc}$ is expressed as a temperature (and frequently referred to as a wind chill factor), thus distinguishing it from WCI, which is given either as a cooling rate or as a plain number with no units. For velocities less than 4 mph, Equation (70) does not apply, and the wind chill temperature is equal to the air temperature.

Equation (70) does not imply cooling to below ambient temperature, but recognizes that, because of wind, the cooling rate is increased as though it were occurring at the lower equivalent wind chill temperature. Wind accelerates the rate of heat loss, so that

Table 10 Equivalent Wind Chill Temperatures of Cold Environments[a]

Wind Speed, mph	Actual Thermometer Reading, °F											
	50	40	30	20	10	0	−10	−20	−30	−40	−50	−60
	Equivalent Chill Temperature, °F											
0	50	40	30	20	10	0	−10	−20	−30	−40	−50	−60
5	48	37	27	16	6	−5	−15	−26	−36	−47	−57	−68
10	40	28	16	3	−9	−21	−34	−46	−58	−71	−83	−95
15	36	22	9	−5	−18	−32	−45	−59	−72	−86	−99	−113
20	32	18	4	−11	−25	−39	−53	−68	−82	−96	−110	−125
25	30	15	0	−15	−30	−44	−59	−74	−89	−104	−119	−134
30	28	13	−3	−18	−33	−48	−64	−79	−94	−110	−125	−140
35	27	11	−4	−20	−36	−51	−67	−83	−98	−114	−129	−145
40	26	10	−6	−22	−38	−53	−69	−85	−101	−117	−133	−148

Little danger: In less than 5 h, with dry skin. Maximum danger from false sense of security. (WCI less than 1400)	**Increasing danger:** Danger of freezing exposed flesh within one minute. (WCI between 1400 and 2000)	**Great danger:** Flesh may freeze within 30 seconds. (WCI greater than 2000)

[a] Cooling power of environment expressed as an equivalent temperature under calm conditions [Equation (70)].

[b] Winds greater than 43 mph have little added chilling effect.
Source: U.S. Army Research Institute of Environmental Medicine.

the skin surface is cooling faster toward the ambient temperature. Table 10 shows a typical wind chill chart, expressed in equivalent wind chill temperature.

PREDICTION OF THERMAL COMFORT AND THERMAL SENSATION

Thermal comfort is defined as "that condition of mind in which satisfaction is expressed with the thermal environment." Because comfort is a "condition of mind," empirical equations must be used to relate comfort perceptions to specific physiological responses. In addition to the previously discussed independent environmental and personal variables influencing thermal response and comfort, other factors such as nonuniformity of the environment, visual stimuli, age, outdoor climate, etc. may also have some effect, but are generally considered to be secondary factors.

Conditions for Thermal Comfort

Studies by Rohles and Nevins (1971) and Rohles (1973) on 1600 college-age students revealed statistical correlations between comfort level, temperature, humidity, sex, and length of exposure. The equations from this study for predicting thermal sensations from air temperature and atmospheric water vapor pressure for men and women for different exposure periods are shown in Table 11. The thermal sensation scale used in these equations is referred to as the ASHRAE thermal sensation scale and is the same as the PMV scale described later. This scale is as follows:

    +3 hot
    +2 warm
    +1 slightly warm
     0 neutral
    −1 slightly cool
    −2 cool
    −3 cold

Vapor pressure rather than conventional relative humidity was used in these equations. Plotting the regression equation results in sets of observed lines of *equal thermal sensation* for lightly clothed, sedentary young adults.

Design data given in the equations (Table 11) define conditions that maximize thermal acceptability of the environment for a large group of similarly clothed, active adults and minimize the fraction of expected complaints.

Steady-State Energy Balance

Fanger (1982) related the comfort data to physiological variables. At a given level of metabolic activity M, and when the body

Table 11 Equations for Predicting Thermal Sensation (Y)[a] of Men, Women, and Men and Women Combined[b]

Exposure Period, h	Sex	Regression Equations t = dry-bulb temperature, °F p = vapor pressure, psi
1.0	Male	$Y = 0.122\,t + 1.61\,p - 9.584$
	Female	$Y = 0.151\,t + 1.71\,p - 12.080$
	Combined	$Y = 0.136\,t + 1.71\,p - 10.880$
2.0	Male	$Y = 0.123\,t + 1.86\,p - 9.953$
	Female	$Y = 0.157\,t + 1.45\,p - 12.725$
	Combined	$Y = 0.140\,t + 1.65\,p - 11.339$
3.0	Male	$Y = 0.118\,t + 2.02\,p - 9.718$
	Female	$Y = 0.153\,t + 1.76\,p - 13.511$
	Combined	$Y = 0.135\,t + 1.92\,p - 11.122$

[a] Y values refer to the ASHRAE thermal sensation scale.
[b] For young adult subjects with sedentary activity and wearing clothing with a thermal resistance of approximately 0.5 clo, $\bar{t}_r \cong \bar{t}_a$ and air velocities are < 40 fpm.

is not far from thermal neutrality, the mean skin temperature t_{sk} and sweat rate E_{rsw} are the only physiological parameters influencing the heat balance. However, heat balance alone is not sufficient to establish thermal comfort. In the wide range of environmental conditions where heat balance can be obtained, only a narrow range provides thermal comfort. The following linear regression equations based on Rohles' and Nevins' data indicate values of t_{sk} and E_{rsw} that provide thermal comfort.

$$t_{sk,req} = 96.3 - 0.156\,(M - W)\ \text{in °F} \tag{71}$$

$$E_{rsw,req} = 0.42\,(M - W - 18.43)\ \text{in Btu/(h·ft}^2) \tag{72}$$

The mean skin temperature decreases at higher activities and the sweat loss increases. Both reactions increase the heat loss from the body core to the environment. These two empirical relationships link the physiological and heat flow equations and thermal comfort perceptions. By substituting these values into Equation (12) for $(C + R)$, and into Equations (18) and (19) for E_{sk}, the energy balance Equation (1) can be used to determine combinations of the six environmental and personal parameters that optimize comfort for steady-state conditions.

Fanger (1982) reduced these relationships to a single equation, which assumed all sweat generated is evaporated, eliminating clothing moisture permeability as a factor in the equation. This assumption is valid for normal indoor clothing worn in typical indoor environments with low or moderate activity levels. At higher activity levels ($M_{act} > 3$ met), where a significant amount of sweating occurs even at optimum comfort conditions, this assumption may limit accuracy. The reduced equation is slightly different from the heat transfer equations developed here. The radiant heat exchange is expressed in terms of the Stefan-Boltzmann law (instead of using h_r), and diffusion of water vapor through the skin is expressed in terms of a diffusivity coefficient and a linear approximation for saturated vapor pressure evaluated at t_{sk}. The combination of environmental and personal variables that produce a neutral sensation may be expressed as follows:

$$
\begin{aligned}
(M - W) = {}& 1.196 \times 10^{-9} f_{cl}[(t_{cl} + 460)^4 - (\bar{t}_r + 460)^4] \\
& + f_{cl}\,h_c\,(t_{cl} - t_a) \\
& + 0.97\,[5.73 - 0.022\,(M - W) - 6.9\,p_a] \\
& + 0.42\,[(M - W) - 18.43] \\
& + 0.0173\,M\,(5.87 - 6.9\,p_a) \\
& + 0.00077\,M\,(93.2 - t_a) \tag{73}
\end{aligned}
$$

where

$$
\begin{aligned}
t_{cl} = {}& 96.3 - 0.156\,(M - W) \\
& - R_{cl}[(M - W) \\
& - 0.97\,[5.73 - 0.022\,(M - W) - 6.9\,p_a] \\
& - 0.42\,[(M - W) - 18.43] - 0.0173\,M(5.87 - 6.9\,p_a) \\
& - 0.00077\,M\,(93.2 - t_a)] \tag{74}
\end{aligned}
$$

The values of h_c and f_{cl} can be estimated from tables and equations given in the Engineering Data and Measurements section. Fanger used the following relationships:

$$
h_c = \begin{cases}
0.361\,(t_{cl} - t_a)^{0.25} & 0.361\,(t_{cl} - t_a)^{0.25} > 0.151\,\sqrt{V} \\
0.151\,\sqrt{V} & 0.361\,(t_{cl} - t_a)^{0.25} < 0.151\,\sqrt{V}
\end{cases} \tag{75}
$$

$$f_{cl} = \begin{cases} 1.0 + 0.2\,I_{cl} & I_{cl} < 0.5 \text{ clo} \\ 1.05 + 0.1\,I_{cl} & I_{cl} > 0.5 \text{ clo} \end{cases} \qquad (76)$$

Figures 7 and 8 show examples of how Equation (73) can be used.

Equation (73) is expanded to include a range of thermal sensations by using a *predicted mean vote* (PMV) index. The PMV index predicts the mean response of a large group of people according to the ASHRAE thermal sensation scale. Fanger (1970) related PMV to the imbalance between the actual heat flow from the body in a given environment and the heat flow required for optimum comfort at the specified activity by the following equation:

$$\text{PMV} = [0.303\exp(-0.036\,M) + 0.028]\,L \qquad (77)$$

where L is the thermal load on the body, defined as the difference between the internal heat production and the heat loss to the actual environment for a person hypothetically kept at comfort values of t_{sk} and E_{rsw} at the actual activity level. Thermal load L is then the difference between the left and right sides of Equation (73) when evaluated at the actual environmental conditions.

Two-Node Transient Energy Balance

After replacing the terms in Equations (2) and (3) with the expressions presented in the previous sections, the equations can be rearranged to calculate t_{sk} and t_{cr} at any time θ, using numerical integration for a specified time of exposure. At each time step, values for the individual energy flows and the thermoregulatory responses have to be recalculated since they are functions of t_{sk} and/or t_{cr}.

After calculating values of t_{sk}, t_{cr}, and w, the model uses empirical expressions to predict thermal sensation (TSENS) and thermal discomfort (DISC). These indices are based on 11-point numerical scales, where positive values represent the warm side of neutral sensation or comfort, and negative values represent the cool side. TSENS is based on the same scale as PMV, but with extra terms for ± 4 (very hot/cold) and ± 5 (intolerably hot/cold). Recognizing the same positive/negative convention for warm/cold discomfort, DISC is defined as:

5 intolerable
4 limited tolerance
3 very uncomfortable
2 uncomfortable and unpleasant
1 slightly uncomfortable but acceptable
0 comfortable

TSENS is defined in terms of deviations of mean body temperature t_b from cold and hot set points representing the lower and upper limits for the zone of evaporative regulation: $t_{b,c}$ and $t_{b,h}$, respectively. The values of these set points depend on the net rate of internal heat production and are calculated by:

$$t_{b,c} = (0.194/58.15)(M - W) + 36.301 \qquad (78)$$

$$t_{b,h} = (0.347/58.15)(M - W) + 36.669 \qquad (79)$$

TSENS is then determined by:

$$\text{TSENS} = \begin{cases} 0.4685\,(t_b - t_{b,c}) & t_b < t_{b,c} \\ 4.7\,\eta_{ev}\,(t_b - t_{b,c})/(t_{b,h} - t_{b,c}) & t_{b,c} \leqslant t_b \leqslant t_{b,h} \\ 4.7\,\eta_{ev} + 0.4685\,(t_b - t_{b,h}) & t_{b,h} < t_b \end{cases} \qquad (80)$$

where η_{ev} is the evaporative efficiency (assumed to be 0.85).

Thermal discomfort is numerically equal to TSENS when t_b is below its cold set point $t_{b,c}$ and is related to skin wettedness when body temperature is regulated by sweating:

$$\text{DISC} = \begin{cases} 0.4685\,(t_b - t_{b,c}) & t_b < t_{b,c} \\ \dfrac{4.7\,(E_{rsw} - E_{rsw,req})}{(E_{max} - E_{rsw,req} - E_{dif})} & t_{b,c} \leqslant t_b \end{cases} \qquad (81)$$

where $E_{rsw,req}$ is calculated as in Fanger's model, using Equation (72).

Zones of Comfort and Discomfort

The preceding section shows that comfort and thermal sensation are not necessarily the same variable, especially for a person in the zone of evaporative thermal regulation. Figures 9 and 10 show this difference for the standard combination of met-clo-air movement used in the standard effective temperature. Figure 9 demonstrates that practically all basic physiological variables predicted by the two-node model are functions of ambient temperature and are relatively independent of vapor pressure. All exceptions occur at relative humidities above 80% and as the isotherms reach the ET* = 41.4°C line, where regulation by

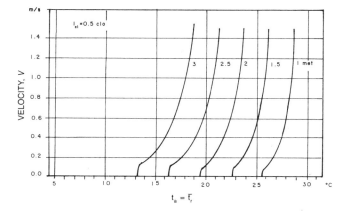

Fig. 7 Interaction between Air Velocity, Temperature, and Activity

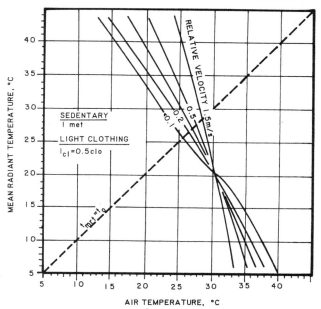

Fig. 8 Interaction between Air Temperature, Mean Radiant Temperature, and Velocity

evaporation fails. Figure 10 shows that lines of constant ET* and wettedness are functions of both ambient temperature and vapor pressure. Thus, human thermal responses are divided into two classes—those in Figure 9, which respond only to heat stress from the environment, and those in Figure 10, which respond to both the heat stress from the environment and the resultant heat strain (Stolwijk *et al.* 1968).

For warm environments, any index with isotherms parallel to skin temperature is a reliable index of thermal sensation alone, and not of discomfort caused by increased humidity. Indices with isotherms parallel to ET* are reliable indicators of discomfort or dissatisfaction with thermal environments. For a fixed exposure time to cold, lines of constant t_{sk}, ET*, and t_o are essentially identical, and cold sensation is no different from cold discomfort. For a state of comfort with sedentary or light activity, lines of constant t_{sk} and ET* coincide. Thus comfort and thermal sensations coincide in this region as well. The upper and lower temperature limits for comfort at these levels can be specified either by thermal sen-

sation (Fanger 1982) or by ET*, as is done in ASHRAE *Standard* 55-92, since lines of constant comfort and lines of constant thermal sensation should be identical. Figure 5 shows comfort zones for summer and winter; they are intended to provide acceptable conditions for occupants wearing typical indoor clothing and at or near sedentary activity.

Individual Variations

No single environment is judged satisfactory by everybody, even if they are wearing identical clothing and performing the same activity. The comfort zone specified in ASHRAE *Standard* 55-92 is based on 90% acceptance, or 10% dissatisfied. Fanger (1982) related the predicted percent dissatisfied (PPD) to the PMV as follows:

$$PPD = 100 - 95 \exp[-(0.03353\,PMV^4 + 0.2179\,PMV^2)] \quad (82)$$

where dissatisfied is defined as anybody not voting either −1, +1, or 0. This relationship is shown in Figure 11. A PPD of 10% corresponds to the PMV range of ±0.5, and even with PMV = 0, about 5% of the people are dissatisfied.

Day-to-Day Variations

Fanger (1973) conducted an experiment with a group of subjects, where the preferred ambient temperature for each subject under identical conditions was determined on four different days. Since the standard deviation was only 1.0°F, Fanger concluded that the comfort conditions for the individual can be reproduced and will vary only slightly from day to day.

Age

Because metabolism decreases slightly with age, many have stated comfort conditions based on experiments with young and healthy subjects cannot be used for other age groups. Fanger (1982), Fanger and Langkilde (1975), Langkilde (1979), Nevins *et al.* (1966), and Rohles and Johnson (1972) conducted comfort studies in Denmark and the United States on different age groups (mean age 21 to 84). The studies revealed that the thermal environments preferred by older people do not differ from those preferred by younger people. The lower metabolism in older people is compensated for by a lower evaporative loss. Collins and Hoinville (1980) confirmed these results.

The fact that young and old people prefer the same thermal environment does not necessarily mean they are equally sensitive when exposed to cold or heat. In practice, the ambient temperature level in the homes of older people is often higher than that for younger people. This may be explained by the lower activity of elderly people, who are normally sedentary for a greater part of the day.

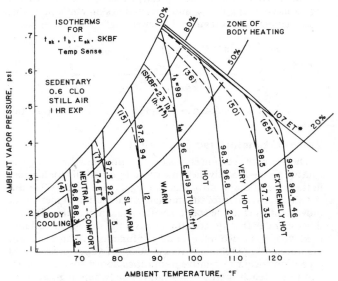

Fig. 9   Effect of Environmental Conditions on Physiological Variables

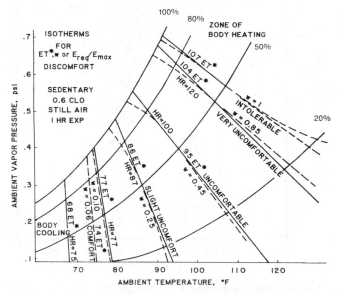

Fig. 10   Effect of Thermal Environment on Discomfort

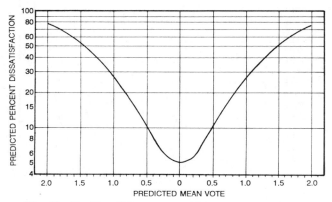

Fig. 11   Predicted Percentage of Dissatisfied (PPD) as Function of Predicted Mean Vote (PMV)

Adaptation

Many believe that people can acclimatize themselves by exposure to hot or cold surroundings, so that they prefer other thermal environments. Fanger (1982) conducted experiments involving subjects from the United States, Denmark, and tropical countries. The latter group was tested in Copenhagen immediately after their arrival by plane from the tropics where they had lived all their lives. Other experiments were conducted for two groups exposed to cold daily. One group comprised subjects who for 8 h daily for at least one year were doing sedentary work in cold surroundings (in the meat-packing industry). The other group consisted of winter swimmers who bathed in the sea daily.

Only slight differences regarding both the preferred ambient temperature and the physiological parameters in the comfort conditions were reported for the various groups. These results indicate that people cannot adapt to preferring warmer or colder environments. It is therefore likely that the same comfort conditions can be applied throughout the world. However, in determining the preferred ambient temperature from the comfort equations, a clo-value that corresponds to the local clothing habits should be used. A comparison of field comfort studies from different parts of the world shows significant differences in clothing habits depending on, among other things, the outdoor climate (Nichol and Humphreys 1972). According to these results, adaptation has little influence on the preferred ambient temperatures. In uncomfortable warm or cold environments, there will, however, often be an influence of adaptation. People used to working and living in warm climates can more easily accept and maintain a higher work performance in hot environments than people from colder climates.

Sex

Previously cited experiments by Fanger (1982), Fanger and Langkilde (1975), and Nevins et al. (1966) use equal numbers of male and female subjects, so comfort conditions for the two sexes can be compared. The experiments show that men and women prefer almost the same thermal environments. Women's skin temperature and evaporative loss are slightly lower than those for men, and this balances the somewhat lower metabolism of women. The reason that women often prefer higher ambient temperatures than men may be explained by the lighter clothing normally worn by women.

Seasonal and Circadian Rhythms

Since people cannot adapt to prefer warmer or colder environments, it follows that there is no difference between comfort conditions in winter and in summer. McNall et al. (1968) confirmed this in an investigation where results of winter and summer experiments showed no difference. On the other hand, it is reasonable to expect the comfort conditions to alter during the day as the internal body temperature has a daily rhythm—a maximum occuring late in the afternoon, and a minimum early in the morning.

In determining the preferred ambient temperature for each of 16 subjects both in the morning and in the evening, Fanger et al. (1974) and Ostberg and McNicholl (1973) observed no difference. Furthermore, Fanger et al. (1973) found only small fluctuations in the preferred ambient temperature during a simulated 8-h workday (sedentary work). There is a slight tendency to prefer somewhat warmer surroundings before lunch, but none of the fluctuations is significant.

Local Thermal Discomfort

As predicted by the comfort equation (PMV = 0), thermal neutrality is not the only condition for thermal comfort. A person may feel thermally neutral for the body as a whole, but might not be comfortable if one part of the body is warm and another cold.

Therefore, thermal comfort also requires that no local warm or cold discomfort exists at any part of the human body. Such local discomfort may be caused by an asymmetric radiant field (cold windows, warm heaters); by local convective cooling (draft); by contact with a hot or cold floor; or by a vertical air temperature difference between the feet and the head.

Asymmetric Thermal Radiation

Asymmetric or nonuniform thermal radiation in a space may be caused by cold windows, uninsulated walls, cold products, cold or warm machinery, or by improperly sized panels on the wall or ceiling. In residential buildings, offices, restaurants, etc., the most common reasons for discomfort due to asymmetric thermal radiation are large windows in the winter or improperly sized or installed ceiling heating panels. At industrial workplaces, the reasons include cold or warm products, cold or warm equipment, etc.

Among the studies conducted on the influence of asymmetric thermal radiation are those by McIntyre (1974, 1976), McIntyre and Griffiths (1975), Fanger and Langkilde (1975), McNall and Biddison (1970), and Olesen et al. (1972). These studies all used seated subjects. The recommendations in the existing comfort standards (ISO 7730, ASHRAE 55-92) are based on studies reported by Fanger et al. (1980) and Fanger and Christensen (1985), and include guidelines regarding the radiant temperature asymmetry from an overhead warm surface (heated ceiling) and a vertical surface (cold window).

To establish a relationship between the radiant temperature asymmetry and the sensation of comfort or discomfort, human subjects were seated, dressed in standard clothing (approximately 0.6 clo), and exposed to either an overhead warm surface or a vertical cold surface from one side. During the experiments, the radiant temperature asymmetry was increased by increasing the temperature of the overhead warm surface or decreasing the temperature of the vertical cold surface. The temperature of all other surfaces in the climatic chamber was kept equal to the air temperature. Changing the temperature on the warm/cold surface also influenced the mean radiant temperature and then the general thermal sensation of the subjects. This was, however, compensated for by changing the air temperature according to the subjects' wishes. In this way, they were always in thermal neutrality and exposed only to the discomfort resulting from excessive asymmetry.

The subjects gave their reactions on their comfort sensation, and a relationship between the radiant temperature asymmetry and the number of subjects feeling dissatisfied was established. Note that the percentages of dissatisfied have nothing to do with the PPD-index, which predicts the percentage of dissatisfied due to a general warm or cold sensation. In all the experiments, the subjects were in thermal neutrality so the percentage of people feeling generally uncomfortably warm or cold (PPD-value) cannot be added and assumed to be equal to the total number of people feeling dissatisfied. Figure 12 shows that people are more sensitive to the asymmetry caused by an overhead warm surface than by a vertical cold surface. The influence of an overhead cold surface and a vertical warm surface is much less. These data are particularly important when applying radiant panels to provide comfort in spaces with large cold surfaces or cold windows.

Draft

Draft is an undesired local cooling of the human body caused by air movement. This is a serious problem, not only in many ventilated buildings but also in automobiles, trains, and aircraft. Draft has been identified as one of the most annoying factors in offices. When people sense draft, it often results in a demand for higher air temperatures in the room or for stopping ventilation systems.

Rohles et al. (1974), Ostergaard et al. (1974), and Burton et al. (1975) investigated the convective heat loss for the entire body as a function of air velocity. However, this information is of limited value for predicting a local cooling felt as a draft. A person may feel thermally neutral for the body as a whole, yet may not be comfortable if air movements cause an unwanted cooling of a particular part of the body.

Houghton (1938) studied ten male subjects exposed to constant local velocities at the back of the neck and at the ankles; McIntyre (1979) exposed the head to constant local velocities; and Fanger and Pedersen (1977) exposed subjects to a periodically fluctuating airflow directed towards the back of the neck or the ankles. The latter showed that fluctuating airflow is more uncomfortable than a constant flow. In real spaces, occupants are not exposed to a well-defined, periodically fluctuating airflow. Rather, in practice, fluctuations are stochastical. In field studies by Thorshauge (1982), Hanzawa *et al.* (1987), and Melikow *et al.* (1988), fluctuations were determined in ventilated spaces.

To establish the scientific basis required to predict human response to fluctuating air velocities as they occur in practice, Fanger and Christensen (1985) aimed to establish the percentage of the population feeling draft when exposed to a given mean velocity. Figure 13 shows the percentage of subjects who felt draft

on the head region (the dissatisfied) as a function of the mean velocity at the neck. The head region comprises head, neck, shoulders, and back. The air temperature had a significant influence on the percentage of dissatisfied. There was no significant difference between responses of men and women to draft. The data in Figure 13 applies, therefore, only to persons wearing normal indoor clothing and performing light, mainly sedentary work. Persons with higher activity levels are not so sensitive to draft (Jones *et al.* 1986).

In another study, Fanger *et al.* (1987) investigated the effect of turbulence intensity on sensation of draft. The turbulence intensity had a significant effect on the occurrence of draft sensation. The following model predicts the percentage of people dissatisfied because of draft intensity. The model can be used for quantifying draft risk in spaces and for developing air distribution systems with a low draft risk.

$$PD = 0.06533 (93.2 - t_a)(V - 9.8)^{0.622}$$
$$+ 3.90 \times 10^{-5} (93.2 - t_a)(V - 9.8)^{0.622} V \, Tu \qquad (83)$$

For $V < 9.8$ fpm, insert $V = 9.8$ and for PD $> 100\%$, insert PD $= 100\%$ where Tu is the turbulence intensity defined by:

$$Tu = 100 \frac{V_{sd}}{V} \qquad (84)$$

and V_{sd} is the standard deviation of the velocity measured with an omnidirectional anemometer having a 0.2-s time constant.

The model extends the Fanger and Christensen draft chart model to include turbulence intensity. In this study, Tu decreases when V increases. This means that the effect of V for the experimental data to which the model is fitted are: $68 < t_a < 79\,°F$, $10 < V < 100$ fpm, and $0 < Tu < 70\%$. Figure 14 gives more precisely the curves that result from intersections between planes of constant Tu and the surfaces of PD $= 15\%$.

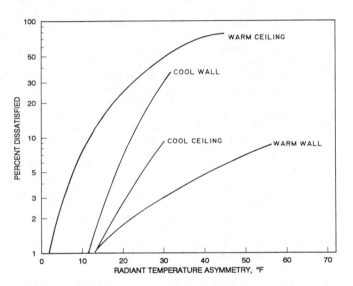

Fig. 12　Percentage of People Expressing Discomfort Due to Asymmetric Radiation

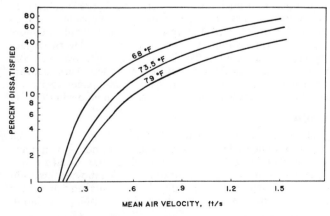

Fig. 13　Percentage of People Dissatisfied as Function of Mean Air Velocity

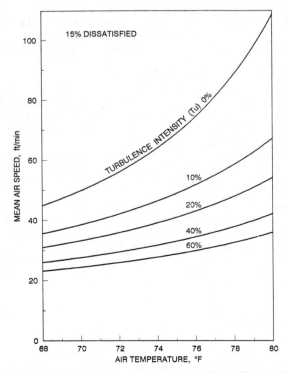

Fig. 14　Draft Conditions Dissatisfying 15% of Population

Vertical Air Temperature Difference

In most spaces in buildings, the air temperature normally increases with height above the floor. If the gradient is sufficiently large, local warm discomfort can occur at the head, and/or cold discomfort can occur at the feet, although the body as a whole is thermally neutral. Among the few studies of vertical air temperature differences and the influence of thermal comfort reported are Olesen *et al.* (1979), McNair (1973), McNair and Fishman (1974), and Eriksson *et al.* (1975). Subjects were seated in a climatic chamber so they were individually exposed to different air temperature differences between head and ankles (Olesen *et al.* 1979). During the tests, the subjects were in thermal neutrality because they were allowed to change the temperature level in the test room whenever they desired; the vertical temperature difference, however, was kept unchanged. The subjects gave subjective reactions to their thermal sensation, and Figure 15 shows the percentage of dissatisfied as a function of the vertical air temperature difference between head (43 in. above the floor) and ankles (4 in. above the floor).

The case where the air temperature at head level is lower than that at ankle level will not be as critical for the occupants. Eriksson (1975) indicated that his subjects could tolerate much greater differences if the head were cooler. This observation is verified in the experiments with asymmetric thermal radiation from a cooled ceiling (Fanger *et al.* 1985).

Warm or Cold Floors

Due to the direct contact between the feet and the floor, local discomfort of the feet can often be caused by a too-high or too-low floor temperature. Also, the floor temperature has a significant influence on the mean radiant temperature in a room. The floor temperature is greatly influenced by the way a building is constructed (*e.g.*, insulation of the floor, above a cellar, directly on the ground, above another room, use of floor heating, floors in radiant heated areas). If a floor is too cold and the occupants feel cold discomfort in their feet, a common reaction is to increase the temperature level in the room; in the heating season, this also increases energy consumption. A radiant system, which will radiantly heat the floor, is another method used to avoid discomfort from cold floors.

The influence of floor temperature on feet comfort has been reported by Nevins *et al.* (1958, 1964, 1967), Olesen (1977a, 1977b), Frank (1959), Cammerer and Schule (1960), Schule and Monroe (1971), Schule (1954), and Missenard (1955). The most extensive studies were performed by Olesen (1977a, 1977b), who, based on his own experiments and reanalysis of the data from Nevins *et al.*, recorded the following results. For floors occupied by people with

bare feet (in swimming halls, gymnasiums, dressing rooms, bathrooms, and bedrooms), flooring material is important. Ranges for some typical floor materials are as follows:

Textiles (rugs)	70 to 82 °F
Pinewood floor	72.5 to 82 °F
Oakwood floor	76 to 82 °F
Hard linoleum wood	75 to 82 °F
Concrete	79 to 83 °F

To save energy, flooring materials with a low contact coefficient (cork, wood, carpets), radiant heated floors, or floor heating systems can be used to eliminate a desire for higher ambient temperatures caused by cold feet. These recommendations should also be followed in schools, where the children often play directly on the floor.

For floors occupied by people with normal indoor footwear, flooring material is insignificant. Olesen (1977b) found an optimal temperature of 77 °F for sedentary and 73.5 °F for standing or walking persons. At the optimal temperature, 6% of the occupants felt warm or cold discomfort in the feet. Figure 16 shows the relationship between floor temperature and percentage of dissatisfied, combining data from experiments with seated and standing subjects. In all experiments, the subjects were in thermal neutrality; thus, the percentage of dissatisfied is only related to the discomfort due to cold or warm feet. Again, no significant difference in floor temperature was preferred by females and males.

SPECIAL ENVIRONMENTS

Infrared Heating

Optical and thermal properties of skin must be considered in studies concerning the effects of infrared radiation in (1) producing changes in skin temperature and skin blood flow, and (2) evoking sensations of temperature and comfort (Hardy 1961). Although the body can be considered to have the properties of water, thermal sensation and heat transfer with the environment require a study of the skin and its interaction with visible and infrared radiation.

Figure 17 shows how skin reflectance and absorptance vary for a blackbody heat source at the temperature (in °R) indicated. These curves show that darkly pigmented skin is heated more by direct radiation from a high-intensity heater at 4500 °R than is lightly pigmented skin. With low-temperature and low-intensity heating equipment used for total area heating, there is minimal, if any, difference. Also, in practice, clothing minimizes differences.

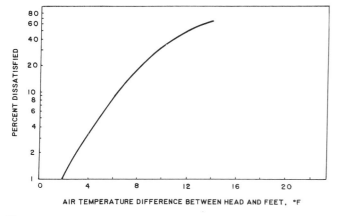

Fig. 15   Percentage of People Dissatisfied as Function of Vertical Air Temperature Difference between Head and Ankles

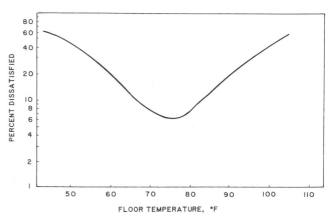

Fig. 16   Percentage of People Dissatisfied as Function of Floor Temperature

Changes in skin temperature caused by high-intensity infrared radiation depend on the thermal conductivity, density, and specific heat of the living skin (Lipkin and Hardy 1954). Modeling of skin heating with the heat transfer theory yields a parabolic relation between exposure time and skin temperature rise for nonpenetrating radiation:

$$t_{sf} - t_{si} = \Delta t = 2J\epsilon \sqrt{\theta/(\pi k\rho c_p)} \qquad (85)$$

where

t_{sf} = final skin temperature, °F
t_{si} = initial skin temperature, °F
J = irradiation intensity from source at °R, Btu/(h·ft²)
θ = time, h
k = specific thermal conductivity of tissue, Btu/(h·ft·°R)
ρ = density, lb/ft³
c_p = specific heat, Btu/(lb·°R)
ϵ = skin absorptance for radiation at °R, dimensionless

Product $k\rho c_p$ is the physiologically important quantity that determines temperature elevation of skin or other tissue on exposure to nonpenetrating radiation. Fatty tissue, because of its relatively low specific heat, is heated more rapidly than moist skin or bone. Experimentally, $k\rho c_p$ values can be determined by plotting Δt^2 against $1.13\,J^2\theta$ (Figure 18). Lines are linear and their slopes are inversely proportional to the $k\rho c_p$ of the specimen. Comparing leather and water with body tissues suggests that thermal inertia values depend largely on tissue water content.

Living tissues do not conform strictly to this simple mathematical formula. Figure 19 compares excised skin with living skin with normal blood flow, and skin with blood flow occluded. For short exposure times, the $k\rho c_p$ of normal skin is the same as that in which blood flow has been stopped; excised skin heats more rapidly due to unavoidable dehydration that occurs postmortem. However, with longer exposure to thermal radiation, vasodilatation increases blood flow, cooling the skin. For the first 20 s of irradiation, skin with normally constricted blood vessels has a $k\rho c_p$ value of one-fourth that for skin with fully dilated vessels.

Skin temperature is the best single index of thermal comfort. The most rapid changes in skin temperature occur during the first 60 s of exposure to infrared radiation. During this initial period, thermal sensation and the heating rate of the skin vary with the quality of infrared radiation (color temperature in degrees Rankine). Since radiant heat from a gas-fired heater is absorbed at the skin surface, the same unit level of absorbed radiation during the first 60 s of exposure can cause an even warmer initial sensation than penetrating solar radiation. Because skin heating curves tend to level off after a 60 s exposure (Figure 19), a relative balance is quickly created between heat absorbed, heat flow to the skin surface, and heat loss to the ambient environment. Therefore, the effects of radiant heating on thermal comfort should be examined for conditions approaching thermal equilibrium.

Stolwijk and Hardy (1966) described an unclothed subject's response for a 2-h exposure to temperatures of 41 to 95 °F. Nevins *et al.* (1966) showed a relation between ambient temperatures and thermal comfort of clothed, resting subjects. For any given uniform environmental temperature, both initial physiological response and degree of comfort can be determined for a subject at rest.

Physiological implications for radiant heating can be defined by two environmental temperatures: (1) mean radiant temperature or $\bar{t}_r$, and (2) ambient air temperature t_a. For this discussion on radiant heat, assume that (1) relative humidity is less than 50%, and (2) air movement is low and constant, with an equivalent convection coefficient of 0.51 Btu/(h·ft²·°F).

The equilibrium equation, describing heat exchange between skin surface at mean temperature t_{sk} and the radiant environment, is given in Equation (28), and can be transformed to give (see Table 2):

$$M' - E_{sk} - F_{cle}[h_r(t_{sk} - \bar{t}_r) + h_c(t_{sk} - t_a)] = 0 \qquad (86)$$

where M' is the net heat production ($M - W$) less respiratory losses.

By algebraic transformation, Equation (86) can be rewritten:

$$M' + h_r(\bar{t}_r - t_a)F_{cle} = E_{sk} + (h_r + h_c)(t_{sk} - t_a)F_{cle} \qquad (87)$$

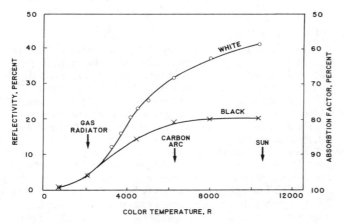

Fig. 17 Variation in Skin Reflection and Absorptivity for Blackbody Heat Sources

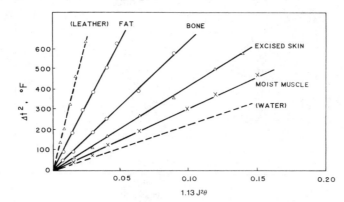

Fig. 18 Comparing Thermal Inertia of Fat, Bone, Moist Muscle, and Excised Skin to that of Leather and Water

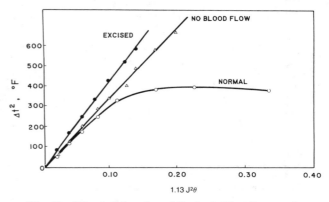

Fig. 19 Thermal Inertias of Excised, Bloodless, and Normal Living Skin

In Equation (87), the factor common to both terms describing dry heat exchange is the ambient temperature t_a. In the left-hand term, $\bar{t}_r$ is related to t_a independent of t_{sk}; in the right, t_{sk} is related to t_a independent of $\bar{t}_r$.

The last term in Equation (87) describes heat exchange with an environment uniformly heated to temperature t_a. The term h_r, evaluated in Equation (50), is also a function of posture, for which factor A_r/A_D can vary from 0.67 for crouching to 0.73 for standing. For preliminary analysis, a useful value for h_r is 0.83 Btu/ (h·ft²·°F), which corresponds to a normally clothed (at 75 °F) sedentary subject. Ambient air movement affects h_c, which appears only in the right-hand term of Equation (87).

The radiation term on the left in Equation (87) describes the net radiation exchange caused by the presence of a warm panel; an infrared electric or gas heater; a cold wall, floor, or window; or any surface with a temperature different from the air temperature. When $\bar{t}_r$ equals t_a, the result is the same as inside an enclosure at uniform temperature t_a. The radiation term on the left in Equation (87) is the basis for defining *effective radiant field* (ERF), which is the net radiant energy absorbed on the exposed body surface of a human-shaped object, with a surface temperature of uniform value t_a. This field energy term (1) is a function of body position, (2) is independent of air movement, (3) as an energy, adds to metabolism M, and (4) is modified by factor F_{cl} with clothing, and by absorptivity of the clothing surface.

Although linear radiation coefficient h_r is used in Equations (86) and (87), the same definition of ERF follows if the fourth power radiation law is used. By this law, assuming emissivity of the body surface is unity, the ERF term in Equation (87) is now:

$$ERF = \sigma(A_r/A_D)[(\bar{t}_r + 460)^4 - (t_a + 460)^4]F_{cle} \qquad (88)$$

Since $\bar{t}_r$ equals the radiation of several surfaces at different temperatures $(T_1, T_2, \cdots, T_j)$; then:

$$\begin{aligned}
ERF = \sigma(A_r/A_D)[&\epsilon_1 F_{m-1}(T_1^4 - T_a^4) \\
&+ \epsilon_2 F_{m-2}(T_2^4 - T_a^4) \\
&+ \cdots + \epsilon_j F_{m-j}(T_j^4 - T_a^4)]F_{cle} \qquad (89)
\end{aligned}$$

where

σ = Stefan-Boltzmann constant, 0.1714×10^{-8} Btu/(h·ft²·°R⁴)
ϵ_j = emissivity of skin or clothing surface for source radiating at temperature T_j
F_{m-j} = angle factor to subject m from source j
T_a = ambient air temperature, °R

$$ERF = (ERF)_1 + (ERF)_2 + \cdots + (ERF)_j \qquad (90)$$

where any $(ERF)_j$ is given by:

$$(ERF)_j = \sigma(A_r/A_D)\epsilon_j F_{m-j}(T_j^4 - T_a^4)F_{cle} \qquad$$

ERF is the sum of the fields caused by each surface T_j [e.g., T_1 may be an infrared beam heater; T_2, a heated floor; T_3, a warm ceiling; T_4, a cold plate glass window $(T_4 < T_a)$; etc.]. Only surfaces with temperature T_j differing from T_a contribute to the ERF.

Comfort Equations for Radiant Heating

The *comfort equation for radiant heat* (Gagge *et al.* 1967a, 1967b) follows from Equation (87):

$$t_o \text{ (for comfort)} = t_a + ERF \text{ (for comfort)}/h \qquad (91)$$

Thus, operative temperature for comfort is the temperature of the ambient air plus a temperature increment ERF/h, a ratio that measures the effectiveness of the incident radiant heating on occupants. Higher air movement (greater values of h or h_c) reduces the effectiveness of radiant heating systems. Clothing lowers t_o for comfort and for thermal neutrality.

Values for ERF and h must be determined to apply the comfort equation for radiant heating. Table 3 may be used to estimate h. One method of determining ERF is to calculate it directly from radiometric data that give, (1) radiation emission spectrum of the source, (2) concentration of the beam, (3) radiation from the floor, ceiling, and windows, and (4) corresponding angle factors involved. This analytical approach is described in Chapter 48 of the 1991 ASHRAE *Handbook—Applications*.

For direct measurement, a skin-colored or black globe, 6 in. in diameter, can measure the radiant field ERF for comfort, in terms of the uncorrected globe temperature t_g in °F and air movement in fpm, by the following relation:

$$ERF = (A_r/A_D)[1.07 + 0.169\sqrt{V}](t_g - t_a) \text{ in Btu/(h·ft²)} \qquad (92)$$

The average value of A_r/A_D is 0.7. For a skin-colored globe, no correction is needed for the quality of radiation. For a black globe, ERF must be multiplied by ϵ for the exposed clothing/skin surface. For a subject with 0.6 to 1.0 clo, t_o for comfort should agree numerically with t_a for comfort in Figure 5. When t_o replaces t_a in Figure 5, humidity is measured in psi vapor pressure rather than relative humidity, which refers only to air temperature.

Other methods may be used to measure ERF. The most accurate is by physiological means. In Equation (87), when M, $t_{sk} - t_a$, and the associated transfer coefficients are experimentally held constant:

$$\Delta E = \Delta ERF \qquad (93)$$

The variation in evaporative heat loss (rate of weight loss) caused by changing the wattage of two T-3 infrared lamps is a measure in absolute terms of the radiant heat received by the body.

A third method uses a directional radiometer to measure ERF directly. For example, radiation absorbed at the body surface [in Btu/(h·ft²·°F)] is:

$$ERF = \epsilon(A_i/A_D)J \qquad (94)$$

where irradiance J can be measured by a directional (Hardy-type) radiometer; ϵ is the surface absorptance effective for the source used; and A_i is the projection area of the body normal to the directional irradiance. Equation (94) can be used to calculate ERF only for the simplest geometrical arrangements. For a human subject lying supine and irradiated uniformly from above, A_i/A_D is 0.3. Figure 17 shows variance of ϵ for human skin with blackbody temperature (in °R) of the radiating source. When irradiance J is uneven and coming from many directions, as is usually the case, the previous physiological method can be used to obtain an effective A_i/A_D from the observed ΔE and $\Delta(\epsilon J)$.

Hot and Humid Environments

Tolerance limits to high temperature vary with the ability to (1) sense temperature, (2) lose heat by regulatory sweating, and (3) move heat from the body core by blood flow to the skin surface, where cooling is the most effective. Many interrelating processes are involved in heat stress (Figure 20).

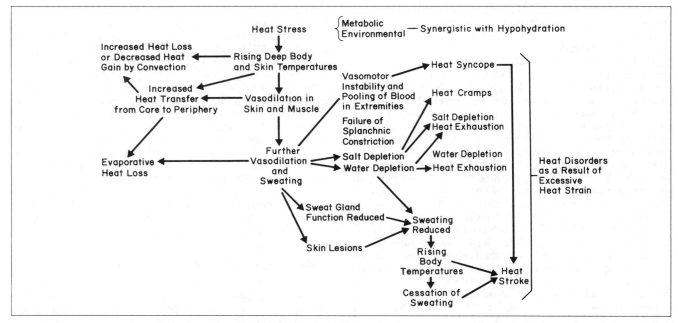

Fig. 20   Schematic Design of Heat Stress and Heat Disorders
[Modified by Buskirk from scale diagram by Belding (1967) and Leithead and Lind (1964)]

Skin surface temperatures of 115 °F trigger pain receptors in the skin; direct contact with metal at this temperature is painful. However, since thermal insulation of the air layer around the skin is high, much higher dry air temperatures can be tolerated. For lightly clothed subjects at rest, tolerance times of nearly 50 min have been reported at 180 °F dry-bulb temperature; 33 min at 200 °F; 26 min at 220 °F; and 24 min at 240 °F. In each case, dew points were lower than 86 °F. Many individuals are stimulated by brief periods of exposure to 185 °F dry air in a sauna. Short exposures to these extremely hot environments are tolerable because of cooling by sweat evaporation. However, when ambient vapor pressure approaches 0.87 psi (97 °F dew point, typically found on sweating skin), tolerance is drastically reduced. Temperatures of 122 °F can be intolerable if the dew-point temperature is greater than 77 °F and both deep body temperature and heart rate rise rapidly within minutes (Gonzalez *et al.* 1978).

The rate and length of time a body can sweat are limited. The maximum rate of sweating for an average man is about 4 lb/h. If all this sweat evaporates from the skin surface under conditions of low humidity and air movement, maximum cooling is about 214 Btu/(h·ft²·°F). However, this value does not normally occur because sweat rolls off the skin surface without evaporative cooling or is absorbed by or evaporated within clothing. A more typical cooling limit is 6 mets, 110 Btu/(h·ft²·°F), representing approximately 2.4 lb/h of sweating for the average man.

Thermal equilibrium is maintained by dissipation of resting heat production (1 met) plus any radiant and convective load. If the environment does not limit heat loss from the body during heavy activity, decreasing skin temperature compensates for the core temperature rise. Therefore, mean body temperature is maintained, although the gradient from core to skin is increased. Blood flow through the skin is reduced, but muscle blood flow necessary for exercise is preserved. The upper limit of skin blood flow ($\dot{m}_{bl}$) is about 200 lb/h (Burton and Bazett 1936).

Body heat storage of 318 Btu (or a rise in t_b of 2.5 °F) for an average-sized man represents an average voluntary tolerance limit. Continuing work beyond this limit increases the risk of heat exhaustion. Collapse can occur at about 635 Btu of storage (5 °F); few individuals can tolerate heat storage of 872 Btu (6.8 °F above normal).

The cardiovascular system affects tolerance limits. In normal, healthy subjects exposed to extreme heat, heart rate and cardiac output increase in an attempt to maintain blood pressure and supply of blood to the brain. At a heart rate of about 180 bpm, the short time between contractions prevents adequate blood supply to the heart chambers. As heart rate continues to increase, cardiac output drops, causing inadequate convective blood exchange with the skin and, perhaps more important, inadequate blood supply to the brain. Victims of this heat exhaustion faint or black out. Accelerated heart rate can also result from inadequate venous return to the heart caused by pooling of blood in the skin and lower extremities. In this case, cardiac output is limited because not enough blood is available to refill the heart between beats. This occurs most frequently when an overheated individual, having worked hard in the heat, suddenly stops working. The muscles no longer massage the blood back past the valves in the veins toward the heart. Dehydration compounds the problem, since fluid volume in the vascular system is reduced.

If core temperature increases above 106 °F, critical hypothalamic proteins can be damaged, resulting in inappropriate vasoconstriction, cessation of sweating, increased heat production by shivering, or some combination of these. Heat stroke damage is frequently irreversible and carries a high risk of fatality.

A final problem, hyperventilation, occurs predominantly in hot-wet conditions, when too much CO_2 is washed from the blood. This can lead to tingling sensations, skin numbness, and vasoconstriction in the brain with occasional loss of consciousness.

Since a rise in heart rate or rectal temperature is essentially linear with ambient vapor pressure above a dew-point temperature of 77 °F, these two changes can measure severe heat stress. Although individual heart rate and rectal temperature responses to mild heat stress vary, severe heat stress saturates physiological regulating systems, producing uniform increases in heart rate and rectal temperature. In contrast, sweat production measures stress under milder conditions but becomes less useful under more severe stress. The maximal sweat rate compatible with body cooling varies with (1) degree of heat acclimatization, (2) duration of sweating, and (3) whether the sweat evaporates or merely saturates the skin and drips off. Total sweat rates in excess of 4.4 lb/h can occur in short exposures, but about 2.2 lb/h is an average maximum level sustainable for an acclimatized man.

Figure 21 illustrates the decline in heart rate, rectal temperature, and skin temperature when exercising subjects are exposed to 104°F over a period of days. Acclimatization can be achieved by working in the heat for 100 min each day—30% improvement occurs after the first day, 50% after 3 days, and 95% after 6 or 7 days. Increased sweat secretion while working in the heat can be induced by rest. Although reducing salt intake during the first few days in the heat can conserve sodium, heat cramps may result. Working regularly in the heat improves cardiovascular efficiency, sweat secretion, and sodium conservation. Once induced, heat acclimatization can be maintained by as few as once-a-week workouts in the heat; otherwise, it diminishes slowly over a 2- to 3-week period and disappears.

Extreme Cold Environments

Human performance in extreme cold ultimately depends on maintaining thermal balance. Whether at work or at rest, a body exposed to cold can lose heat faster than it produces heat for only a limited time. Subjective discomfort is reported by an average 154-lb man with 2800 in² of body surface area, when a heat debt of about 100 Btu is incurred. A heat debt of about 600 Btu is acutely uncomfortable; this represents a drop of approximately 4.7°F (or about 7% of total heat content) in mean body temperature.

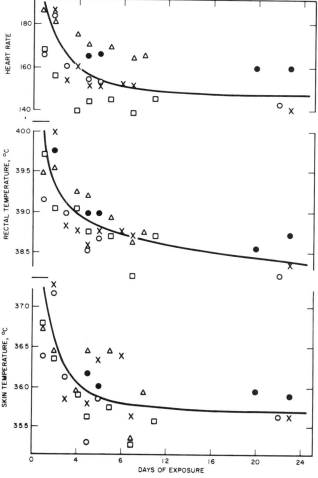

Fig. 21 Acclimatization to Heat Resulting from Daily Exposure of Five Subjects to Room Maintained at 40°C with 23% rh
(Robinson *et al.* 1943)

This loss can occur during 1 to 2 h of sedentary activity outdoors. A sleeping individual will awake after losing about 300 Btu, decreasing mean skin temperature by about 5.5°F and deep body temperature by about 1°F, using the Burton relationship [$\alpha = 0.67$, in Equation (41)]. A drop in deep body temperature (*e.g.*, rectal temperature) below 95°F threatens a loss of body temperature regulation, while 82.4°F is considered critical for survival, despite recorded survival from a deep body temperature of 64.4°F. Temperature is more crucial than rate of temperature change; Witherspoon *et al.* (1971) observed a rate of fall in the core temperature of 5.4°F per hour in subjects immersed in 50°F water, without residual effect.

Activity level also affects human performance. Subjective sensations, reported by sendentary subjects at a mean skin temperature of 92°F, are *comfortable*; at 88°F, *uncomfortably cold*; at 86°F, *shivering cold*; and at 84°F, *extremely cold*. The critical subjective tolerance limit (without numbing) for mean skin temperature appears to be about 77°F. However, during moderate to heavy activity, subjects reported the same skin temperatures as comfortable. Although mean skin temperature is significant, the temperature of the extremities is more frequently the critical factor for comfort in the cold. Consistent with this, one of the first responses to cold exposure is vasoconstriction, which reduces circulatory heat input to the hands and feet. A hand-skin temperature of 68°F causes a report of *uncomfortably cold*; 59°F, *extremely cold*; and 41°F, *painful*. Identical verbal responses for the foot surface occur at approximately 2.7 to 3.5°F warmer temperatures.

An ambient temperature of −31°F is the lower limit for useful outdoor activity, even with adequate insulative clothing. At −58°F, almost all outdoor effort becomes exceedingly difficult; even with appropriate protective equipment, only limited exposure is possible. Reported exposures of 30 min at −103°F have occurred in the Antarctic without injury.

In response to extreme heat loss, maximal heat production becomes very important. When the less efficient vasoconstriction cannot prevent body heat loss, shivering is an automatic, more efficient defense against cold. This can be triggered by low deep body temperature, low skin temperature, rapid change of skin temperature, or some combination of all three. Shivering is usually preceded by an imperceptible increase in muscle tension and by noticeable gooseflesh produced by muscle contraction in the skin. It begins slowly in small muscle groups, initially increasing total heat production by 1.5 to 2 times resting levels. As body cooling increases, the reaction spreads to additional body segments. Ultimately violent, whole body shivering causes maximum heat production of about 6 times resting levels, rendering the individual totally ineffective.

Given sufficient cold exposure, the body undergoes changes that indicate cold acclimatization. These physiological changes include, (1) endocrine changes (*e.g.*, sensitivity to norepinephrine), causing nonshivering heat production by metabolism of free fatty acids released from adipose tissue; (2) improved circulatory heat flow to skin, causing an overall sensation of greater comfort; and (3) improved circulatory heat flow to the extremities, reducing the risk of injury and permitting activities at what ordinarily would be severely uncomfortable temperatures in the extremities. Generally, these physiological changes are minor and are induced only by repeated extreme exposures. Nonphysiological factors, including training, experience, and selection of adequate protective clothing, are more useful and may be safer than dependence on physiological changes.

The caloric requirement for adequately clothed subjects in extreme cold is only slightly greater than that for subjects living and working in temperate climates. This greater requirement results from added caloric work caused by two factors: (1) carrying the weight of heavy clothing, since energy cost is a linear function of total weight (energy cost for heavy protective footwear may be six times that of an equivalent weight on the torso); and (2) the inefficiency of walking in snow, snowshoeing, or skiing, which can increase energy cost up to 300%.

To achieve proper protection in low temperatures, a person must either maintain high metabolic heat production by activity or reduce heat loss by controlling the body's microclimate with clothing. Other protective measures include spot radiant heating, showers of hot air for work at a fixed site, and warm air ventilated or electrically heated clothing. The extremities, such as fingers and toes, pose more of a problem than the torso because, as thin cylinders, they are particularly susceptible to heat loss and difficult to insulate without increasing the surface for heat loss. Vasoconstriction can reduce circulatory heat input to extremities by over 90%.

Although there is no ideal insulating material for protective clothing, radiation-reflective materials are promising. Insulation is primarily a function of clothing thickness; the thickness of trapped air, rather than fibers used, determines insulation effectiveness.

Protection for the respiratory tract seems unnecessary in healthy individuals, even at $-49\,°F$. However, asthmatics or individuals with mild cardiovascular problems may benefit from a face mask that warms inspired air. Masks are unnecessary for protecting the face since heat to facial skin is not reduced by local vasoconstriction, as it is for hands. If wind chill is great, there is always a risk of cold injury caused by freezing of exposed skin. Using properly designed torso clothing, such as a parka with fur-lined hood to minimize wind penetration to the face, and 34 Btu/h of auxiliary heat to each hand and foot, inactive people can tolerate $-67\,°F$ with a 10 mph wind for more than 6 h. As long as skin temperature of fingers remains above $60\,°F$, manual dexterity can be maintained and useful work performed without difficulty.

SYMBOLS

A = area, ft^2
    A_{cl} = surface area of clothed body
    A_D = DuBois surface area of nude body
    A_G = body surface area covered by garment
    A_r = effective radiation area of body
c_p = constant pressure specific heat, Btu/(lb·°F)
    $c_{p,a}$ = of air
    $c_{p,b}$ = of body tissue
    $c_{p,bl}$ = of blood
C_{res} = sensible heat loss due to respiration, Btu/(h·ft²)
$C+R$ = total sensible heat loss from skin, Btu/(h·ft²)
E = evaporative heat loss, Btu/(h·ft²)
    E_{dif} = due to moisture diffusion through skin
    E_{max} = maximum possible
    E_{rsw} = due to regulatory sweating
    E_{res} = due to respiration
    $E_{rsw,req}$ = required for comfort
    E_{sk} = total from skin
ERF = effective radiant field, Btu/(h·ft²)
ET* = effective temperature, °F
f_{cl} = clothing area factor, A_{cl}/A_D, dimensionless
F_{cl} = intrinsic clothing thermal efficiency, dimensionless
F_{cle} = effective clothing thermal efficiency, dimensionless
F_{pcl} = permeation efficiency, dimensionless
F_{m-j} = angle factor to person from source j, dimensionless
F_{p-N} = angle factor from person to source N, dimensionless
h = sensible heat transfer coefficient, Btu/(h·ft²·°F)
    h = total at surface
    h' = overall including clothing
    h_c = convection at surface
    h_{cc} = corrected convection at surface

h_r = radiation
h_e = evaporative heat transfer coefficient, Btu/(h·ft²·psi)
    h_e = at surface
    h_{ec} = at surface, corrected for atmospheric pressure
    $h_{e'}$ = overall including clothing
h_{fg} = heat of vaporization of water, Btu/lb
i = vapor permeation efficiency, dimensionless
    i_a = air layer
    i_{cl} = clothing
    i_m = total
I = thermal resistance in clo units, clo
    All subscripts given for symbol R apply to symbol I
J = irradiance, Btu/(h·ft²)
k = thermal conductivity of body tissue, Btu/(h·ft·°F)
K = effective conductance between core and skin, Btu/(h·ft²·°F)
l = height, ft
L = thermal load on body, Btu/(h·ft²)
LR = Lewis ratio, °F/psi
m = body mass, lb
m_{ge} = mass to gas exchange, lb
$\dot{m}$ = mass flow, lb/(h·ft²)
    $\dot{m}_{bl}$ = blood circulation between core and skin
    $\dot{m}_{rsw}$ = rate of regulatory sweat generation
$\dot{m}_{res}$ = pulmonary ventilation rate flow, lb/h
M = metabolic heat production, Btu/(h·ft²)
    M = total
    M' = net
    M_{act} = due to activity
    M_{shiv} = due to shivering
p = water vapor pressure, psi
    p_a = in ambient air
    $p_{ET*,s}$ = saturated at ET*
    $p_{oh,s}$ = saturated at t_{oh}
    $p_{sk,s}$ = saturated at t_{sk}
p_t = atmospheric pressure, psi
Q = heat flow, Btu/(h·ft²)
    Q_{crsk} = from core to skin
    Q_{res} = total due to respiration
    Q_{sk} = total from the skin
R = thermal insulation, h·ft²·°F/Btu
    R_a = air layer on nude skin
    $R_{a,cl}$ = air layer at outer surface
    R_{cl} = clothing
    R_{cle} = change due to clothing
    R_t = total
R_e = evaporative resistance, h·ft²·psi/Btu
    $R_{e,cl}$ = clothing
    $R_{e,t}$ = total
RQ = respiratory quotient, dimensionless
S = heat storage, Btu/(h·ft²)
    S_{cr} = in core compartment
    S_{sk} = in skin compartment
SET* = standard effective temperature, °F
t = temperature, °F
    t_a = ambient air
    t_b = average of body
        $t_{b,c}$ = lower limit for evaporative regulation zone
        $t_{b,h}$ = upper limit for evaporative regulation zone
        $t_{b,n}$ = at neutrality
    t_{com} = combined temperature
    t_{cl} = clothing surface
    t_{cr} = core
        $t_{cr,n}$ = at neutrality

t_{db} = dry bulb

$t_{eq,wc}$ = equivalent wind chill temperature

t_{ex} = of exhaled air

t_g = globe

t_N = of surface N

t_{nwb} = naturally ventilated wet bulb

t_o = operative

t_{pr} = plane radiant

t_s = surface

t_{sf} = final skin

t_{si} = initial skin

t_{sk} = skin

$t_{sk,n}$ = at neutrality

$t_{sk,req}$ = required for comfort

$\bar{t}_r$ = mean radiant

T = absolute temperature, °R

All subscripts that apply to symbol t may apply to symbol T.

Tu = turbulence intensity, %

V = air velocity, fpm

V_{CO_2} = volume rate of CO_2 produced, ft^3/h

V_{O_2} = volume rate of O_2 consumed, ft^3/h

w = skin wettedness, dimensionless

w_{rsw} = required to evaporate regulatory sweat

W = external work accomplished, $Btu/(h \cdot ft^2)$

W_a = humidity ratio of ambient air, lb H_2O/lb d.a.

W_{ex} = humidity ratio of exhaled air, lb H_2O/lb d.a.

α = fraction of total body mass concentrated in skin compartment, dimensionless

σ = Stefan-Boltzman = n constant, $Btu/(h \cdot ft^2 \cdot °R^4)$

ϵ = emissivity, dimensionless

η = evaporative efficiency, dimensionless

ρ = density, lb/ft^3

θ = time, h

REFERENCES

ASHRAE. 1992. Thermal environmental conditions for human occupancy. ANSI/ASHRAE *Standard* 55-1992.

Astrand, P. and K. Rodahl. 1977. *Textbook of work physiology: Physiological bases of exercise.* McGraw-Hill, New York.

Azer, N.Z. 1982. Design guidelines for spot cooling systems: Part I—Assessing the acceptability of the environment. ASHRAE *Transactions* 88:1.

Azer, N.Z. and S. Hsu. 1977. OSHA heat stress standards and the WBGT index. ASHRAE *Transactions* 83(2):30.

Belding, H.S. and T.F. Hatch. 1955. Index for evaluating heat stress in terms of resulting physiological strains. *Heating, Piping and Air Conditioning* 207:239.

Belding, H.D. 1967. "Heat stress." In *Thermobiology*, ed. A.H. Rose. Academic Press, New York.

Belding, H.S. 1970. The search for a universal heat stress index. *Physiological and Behavioral Temperature Regulation*, eds. J.D. Hardy *et al.* Springfield, IL.

Botsford, J.H. 1971. A wet globe thermometer for environmental heat measurement. *American Industrial Hygiene Association Journal* 32:1-10.

Burton, A.C. and H.C. Bazett. 1936. A study of the average temperature of the tissues, the exchange of heat and vasomotor responses in man, by a bath calorimeter. *American Journal of Physiology* 117:36.

Burton, D.R., K.A. Robeson, and R.G. Nevins. 1975. The effect of temperature on preferred air velocity for sedentary subjects dressed in shorts. ASHRAE *Transactions* 2:157-68.

Buskirk, E.R. 1960. Problems related to the caloric cost of living. Bulletin of the New York Academy of Medicine 26:365.

Cammerer, J.S. and W. Schule. 1960. Erprobung verschiedener mess methoden zur bestimmung der warmeableitung von fussboden. *Gesundheits Ingenieur* 82:1-8.

Carlton-Foss, J.A. 1983. Tight building syndrome: Diagnosis and cure. ASHRAE *Journal* 25:38.

Ciriello, V.M. and S.H. Snook. 1977. The prediction of WBGT from the Botsball. *American Industrial Hygiene Association Journal* 38:264.

Colin, J. and Y. Houdas. 1967. Experimental determination of coefficient of heat exchange by convection of the human body. *Journal of Applied Physiology* 22:31.

Collins, K.J. and E. Hoinville. 1980. Temperature requirements in old age. *Building Services Engineering Research and Technology* 1(4):165-72.

Davis, W.J. 1976. Typical WBGT indexes in various industrial environments. ASHRAE *Transactions* 82(2):303.

DuBois, D. and E.F. DuBois. 1916. A formula to estimate approximate surface area, if height and weight are known. *Archives of Internal Medicine* 17:863-71.

Dukes-Dobos, F. and A. Henschel. 1971. The modification of the WBGT index for establishing permissible heat exposure limits in occupational work. HEW, USPHE, NIOSH, TR-69.

Dukes-Dobos, F. and A. Henschel. 1973. Development of permissible heat exposure limits for occupational work. ASHRAE *Journal* 9:57.

Eriksson, H.A. 1975. Heating and ventilating of tractor cabs. Presented at the 1975 Winter Meeting, American Society of Agricultural Engineers, Chicago.

Fanger, P.O. 1967. Calculation of thermal comfort: introduction of a basic comfort equation. ASHRAE *Transactions* 73(2):III.4.1.

Fanger, P.O. 1970. *Thermal comfort analysis and applications in environmental engineering.* McGraw-Hill, New York.

Fanger, P.O. 1973. The variability of man's preferred ambient temperature from day to day. *Archives des Sciences Physiologiques* 27(4):A403.

Fanger, P.O. 1982. *Thermal comfort.* Robert E. Krieger Publishing Company, Malabar, FL.

Fanger, P.O., L. Banhidi, B.W. Olesen, and G. Langkilde. 1980. Comfort limits for heated ceilings. ASHRAE *Transactions* 86.

Fanger, P.O. and N.K. Christensen. 1985. Perception of draught in ventilated spaces. *Ergonomics.*

Fanger, P.O., J. Hojbjerre, and J.O.B. Thomsen. 1974. Thermal comfort conditions in the morning and the evening. *International Journal of Biometeorology* 18(1):16.

Fanger, P.O., J. Hojbjerre, and J.O.B. Thomsen. 1973. Man's preferred ambient temperature during the day. *Arch. Science Physiology* 27(4):A395-A402.

Fanger, P.O., B.M. Ipsen, G. Langkilde, B.W. Olesen, N.K. Christensen, and S. Tanabe. 1985. Comfort limits for asymmetric thermal radiation. *Energy and Buildings.*

Fanger, P.O. and G. Langkilde. 1975. Interindividual differences in ambient temperature preferred by seated persons. ASHRAE *Transactions* 81(2):140-47.

Fanger, P.O., A. Melikov, H. Hanzawa, and J. Ring. 1987. Air turbulence and sensation of draught. *Energy and Buildings.*

Fanger, P.O., O. Ostberg, A.G.M. Nichell, N.O. Breum, and E. Jerking. 1974. Thermal comfort conditions during day and night. *European Journal of Physiology* 33:225-63.

Fanger, P.O. and C.J.K. Pedersen. 1977. Discomfort due to air velocities in spaces. Proceedings of the meeting of Commissions B1, B2, E1 of the IIR, Belgrade, 289-96.

Fobelets, A.P.R. and A.P. Gagge. 1988. Rationalization of the ET* as a measure of the enthalpy of the human enviroment. ASHRAE *Transactions* 94:1.

Frank, W. 1959. Fusswarmeuntersuchungen am bekleideten fuss. *Gesundheits Ingenieur* 80:193-201.

Gagge, A.P., A.C. Burton, and H.D. Bazett. 1971. A practical system of units for the description of heat exchange of man with his environment. *Science* 94:428-30.

Gagge, A.P. and J.D. Hardy. 1967. Thermal radiation exchange of the human by partitional calorimetry. *Journal of Applied Physiology.* 23:248.

Gagge, A.P., A.P. Fobelets, and L.G. Berglund. 1986. A standard predictive index of human response to the thermal environment. ASHRAE *Transactions* 92(1).

Gagge, A.P., Y. Nishi, and R.G. Nevins. 1976. The role of clothing in meeting FEA energy conservation guidelines. ASHRAE *Transactions* 82(2):234.

Gagge, A.P., G.M. Rapp, and J.D. Hardy. 1967a. The effective radiant field and operative temperature necessary for comfort with radiant heating. ASHRAE *Transactions* 73(1):I.2.1.

Gagge, A.P., G.M. Rapp, and J.D. Hardy. 1967b. The effective radiant field and operative temperature necessary for comfort with radiant heating. ASHRAE *Journal* 9(5):63.

Gagge, A.P., J.A.J. Stolwijk, and Y. Nishi. 1971. An effective temperature scale based on a simple model of human physiological regulatory response. ASHRAE *Transactions* 77(1):247.

Gagge, A.P., J.A.J. Stolwijk, and B. Saltin. 1969a. Comfort and thermal sensation and associated physiological responses during exercise at various ambient temperatures *Environmental Research* 2:209.

Gagge, A.P., J.A.J. Stolwijk, and Y. Nishi. 1969b. The prediction of thermal comfort when thermal equilibrium is maintained by sweating. ASHRAE *Transactions* 75(2):108.

Gonzalez, R.R., L.G. Berglund, and A.P. Gagge. 1978. Indices of thermoregulatory strain for moderate exercise in the heat. *Journal of Applied Physiology* 44:889.

Hanzawa, H., A.K. Melikov, and P.O. Fanger. 1987. Airflow characteristics in the occupied zone of ventilated spaces. ASHRAE *Transactions* 93(1).

Hardy, J.D. 1949. Heat transfer, In *Physiology of heat regulation and science of clothing*. eds. L.H. Newburgh and W.B. Saunders Ltd., London, 78.

Hardy, J.D. 1961. Physiological effects of high intensity infrared heating. ASHRAE *Journal* 4:11.

Hardy, J.D. 1961. Physiology of temperature regulation. *Physiological Reviews* 41:521-606.

Holmer, I. 1984. Required clothing insulation (IREQ) as an analytical index of cold stress. ASHRAE *Transactions* 90(1).

Horton, R.J.M. 1976. Help wanted. ASHRAE *Journal* 18:75.

Houghten, F.C. and C.P. Yaglou. 1923. ASHVE Research Report No. 673, Determination of the Comfort Zone. ASHVE *Transactions* 29:361.

Houghten, F.C. 1938. Draft temperatures and velocities in relation to skin temperature and feeling of warmth. ASHVE *Transactions* 44:289.

ISO. 1982. Hot environments—Estimation of the heat stress on working man, based on the WBGT-index (wet-bulb globe temperature). ISO (International Organization for Standardization) *Standard* 7243, 1st ed.

ISO. 1984. Moderate thermal environments—Determination of the PMV and PPD indices and specification of the conditions for thermal comfort. ISO *Standard* 7730.

ISO. 1989. Hot environments—analytical determination and interpretation of thermal stress using calculations of required sweat rate. ISO *Standard* 7933. International Organization for Standardization, Geneva.

Jones, B.W., K. Hsieh, and M. Hashinaga. 1986. The effect of air velocity on thermal comfort at moderate activity levels. ASHRAE *Transactions* 92:2.

Langkilde, G. 1979. Thermal comfort for people of high age. In *Comfort thermique: Aspects physiologiques et psychologiques*, INSERM, Paris 75:187-93.

Leithead, C.S. and A.R. Lind. 1964. *Heat stress and heat disorders*. Cassell & Co., London, England, 108. (F.A. Davis, Philadelphia, 1964).

Lipkin, M. and J.D. Hardy 1954. Measurement of some thermal properties of human tissues. *Journal of Applied Physiology* 7:212.

Matthew, W.H., *et al.* 1986. Botsball (WGT) performance characteristics and their impact on the implementation of existing military hot weather doctrine. U.S. Army Reserves Institute of Environmental Medicine Technical Report T 9/86, April.

McCullough, E.A. 1986. An insulation data base for military clothing. Institute for Environmental Research Report 86-01, Kansas State University, Manhattan, KS.

McCullough, E.A. and B.W. Jones. 1984. A comprehensive data base for estimating clothing insulation. IER Technical Report 84-01, Institute for Environmental Research, Kansas State University, Manhattan, KS. (Final report to ASHRAE research project 411-RP).

McCullough, E.A., B.W. Jones, and T. Tamura. 1989. A data base for determining the evaporative resistance of clothing. ASHRAE *Transactions* 95(2).

McIntyre, D.A. 1974. The thermal radiation field. *Building Science* 9:247-62.

McIntyre, D.A. 1976. Overhead radiation and comfort. *The Building Services Engineer* 44:226-32.

McIntyre, D.A. 1979. The effect of air movement on thermal comfort and sensation, P.O. Fanger and O. Valbjorn eds. In *Indoor Climate*, Danish Building Research Institute, Copenhagen.

McIntyre, D.A. and I.D. Griffiths. 1975. The effect of uniform and asymmetric thermal radiation on comfort. Proceedings of the Sixth International Congress of Climatistics CLIMA 2000, Milan.

McNall, P.E., Jr. and R.E. Biddison. 1970. Thermal and comfort sensations of sedentary persons exposed to asymmetric radiant fields. ASHRAE *Transactions* 76(1):123.

McNall, P.E., P.W. Ryan, and J. Jaax. 1968. Seasonal variation in comfort conditions for college-age persons in the Middle West. ASHRAE *Transactions* 74(1):IV.2.1—IV.2.9.

McNair, H.P. 1973. A preliminary study of the subjective effects of vertical air temperature gradients. British Gas Corporation Report No. WH/T/R&D/73/94, London.

McNair, H.P. and D.S. Fishman. 1974. A further study of the subjective effects of vertical air temperature gradients. British Gas Corporation Report No. WH/T/R&D/73/94, London.

Melikov, A.K., H. Hanzawa, and P.O. Fanger. 1988. Air flow characteristics in the occupied zone of nonventilated spaces. ASHRAE *Transactions* 95:1.

Minard, D. 1961. Prevention of heat casualties in marine corps recruits. *Military Medicine* 126:261.

Missenard, A. 1955. Chauffrage par rayonnement temperature limite du sol. *Chaleur et Industrie* 155:37.

Mitchell, D. 1974. *Convective heat transfer in man and other animals, heat loss from animals and man.* Butterworth Publishing Inc., London, 59.

Nevins, R.G. and A.M. Feyerherm. 1967. Effect of floor surface temperature on comfort: Part IV, Cold floors. ASHRAE *Transactions* 73(2):III.2.1.

Nevins, R.G. and A.O. Flinner. 1958. Effect of heated-floor temperatures on comfort. ASHRAE *Transactions* 64:175.

Nevins, R.G., K.B. Michaels, and A.M. Feyerherm. 1964. The effect of floor surface temperature on comfort: Part 1, College age males; Part II, College age females. ASHRAE *Transactions* 70:29.

Nevins, R.G., F.H. Rohles, Jr., W.E. Springer, and A.M. Feyerherm. 1966. Temperature-humidity chart for thermal comfort of seated persons. ASHRAE *Transactions* 72(1):283.

Nicol, J.F. and M.A. Humphreys. 1972. Thermal comfort as part of a self-regulating system. Proceedings of CIB symposium on thermal comfort, Building Research Station, London.

NIOSH. 1986. Criteria for a recommended standard—Occupational exposure to hot environments, revised criteria. U.S. Dept. of Health and Human Services, USDHHS (NIOSH) Publication 86-113.

Nishi, Y. 1981. Measurement of thermal balance of man. *Bioengineering Thermal Physiology and Comfort*, K. Cena and J.A. Clark, eds. Elsevier, New York.

Nishi, Y. and A.P. Gagge. 1970. Direct evaluation of convective heat transfer coefficient by naphthalene sublimation. *Journal of Applied Physiology* 29:830.

Nishi, Y., R.R. Gonzalez, and A.P. Gagge. 1975. Direct measurement of clothing heat transfer properties during sensible and insensible heat exchange with thermal environment. ASHRAE *Transactions* 81(2):183.

Olesen, B.W. 1977a. Thermal comfort requirements for floors. Proceedings of Commissions B1, B2, E1 of the IIR, Belgrade, 337-43.

Olesen, B.W. 1977b. Thermal comfort requirements for floors occupied by people with bare feet. ASHRAE *Transactions* 83(2).

Olesen, B.W. 1985. A new and simpler method for estimating the thermal insulation of a clothing ensemble. ASHRAE *Transactions* 92(1).

Olesen, B.W. and R. Nielsen. 1983. Thermal insulation of clothing measured on a moveable manikin and on human subjects. Technical University of Denmark, Lyngby, Denmark.

Olesen, B.W., M. Scholer, and P.O. Fanger. 1979. Vertical air temperature differences and comfort. In *Indoor climate*, P.O. Fanger and O. Valbjorn, eds. Danish Building Research Institute, Copenhagen 561-79.

Olesen, B.W., E. Sliwinska, T.L. Madsen, and P.O. Fanger. 1982. Effect of body posture and activity on the thermal insulation of clothing, measurements by a moveable thermal manikin. ASHRAE *Transactions* 88(2):791-805.

Olesen, S., J.J. Bassing, and P.O. Fanger. 1972. Physiological comfort conditions at sixteen combinations of activity, clothing, air velocity and ambient temperature. ASHRAE *Transactions* 78(2)199.

Onkaram, B. Stroschein, and R.F. Goldman. 1980. Three instruments for assessment of WBGT and a comparison with WGT (Botsball). American Industrial Hygiene Association 41:634-41.

Oohori, T., L.G. Berglund, and A.P. Gagge. 1984. Comparison of current two-parameter indices of vapor permeation of clothing—As factors governing thermal equilibrium and human comfort. ASHRAE *Transactions* 90(2).

Ostberg, O. and A.G. McNicholl. 1973. The preferred thermal conditions for "morning" and "evening" types of subjects during day and night—Preliminary results. *Build International* 6(1):147-57.

Ostergaard, J., P.O. Fanger, S. Olesen, and T.L. Madsen. 1974. The effect of man's comfort of a uniform air flow different directions. ASHRAE *Transactions* 82(2):142-157.

Passmore, R. and J.V.G. Durnin. 1967. *Energy, work and leisure*. Heinemann Educational Books, Ltd., London.

Rapp, G. and A.P. Gagge. 1967. Configuration factors and comfort design in radiant beam heating of man by high temperature infrared sources. ASHRAE *Transactions* 73(2):III.1.1.

Robinson. *et al.* 1943. American Journal of Physiology 140:168.

Rohles, F.H., Jr. 1973. The revised modal comfort envelope. ASHRAE *Transactions* 79(2):52.

Rohles, F.H., Jr. 1980. The preferred indoor comfort temperatures. Report No. 80-02, Institute for Environmental Research, Kansas State University, Manhattan, KS.

Rohles, F.H., Jr. and M.A. Johnson. 1972. Thermal comfort in the elderly. ASHRAE *Transactions* 78(1):131.

Rohles, F.H., Jr. and R.G. Nevins. 1971. The nature of thermal comfort for sedentary man. ASHRAE *Transactions* 77(1):239.

Rohles, F.H., Jr., J.E. Woods, and R.G. Nevins. 1974. The effects of air movement and temperature on the thermal sensations of sedentary man. ASHRAE *Transactions* 81:1.

Schule, W. 1954. Untersuchungen uber die Hauttemperatur des Fusses stehen auf verschiedenartigen Fussboden. *Gesundheits Ingenieur* 75:380.

Schule, W. and J.L. Monroe. 1971. Automatische Messeinrichtungen zur Bestimmung der Warmeableitung von Fussboden. *Gesundheits Ingenieur* 92:164-66.

Seppanen, O., P.E. McNall, D.M. Munson, and C.H. Sprague. 1972. Thermal insulating values for typical indoor clothing ensembles. ASHRAE *Transactions* 78(1):120-30.

Siple, P.A. and C.F. Passel. 1945. Measurements of dry atmospheric cooling in subfreezing temperatures. Proceedings of the American Philosophical Society 89:177.

Stolwijk, J.A.J., A.P. Gagge, and B. Saltin. 1968. Physiological factors associated with sweating during exercise. *Journal of Aerospace Medicine* 39:1101.

Stolwijk, J.A.J. and J.D. Hardy. 1966. Partitional calorimetric studies of response of man to thermal transients. *Journal of Applied Physiology* 21:967.

Sullivan, C.D. and R.L. Gorton. 1976. A method of calculating WBGT from environmental factors. ASHRAE *Transactions* 82(2):279.

Thorshauge, J. 1982. Air velocity fluctuations in the occupied zone of ventilated spaces. ASHRAE *Transactions* 88:2.

Umbach, K.H. 1980. Measuring the physiological properties of textiles for clothing. *Melliand Textilberichte* (English Edition) G1:543-48.

Webb, P. 1964. *Bioastronautics Data Base*, NASA.

Witherspoon, J.M., R.F. Goldman, and J.R. Breckenridge. 1971. Heat transfer coefficients of humans in cold water. *Journal de Physiologie*, Paris, 63:459.

Woodcock, A.H. 1962. Moisture transfer in textile systems. *Textile Research Journal* 8:628-33.

CHAPTER 9

ENVIRONMENTAL CONTROL FOR ANIMALS AND PLANTS

THERMAL conditions, air quality, lighting, noise, ion concentration, and crowding are important in designing structures for animals and plants. Thermal environment influences heat dissipation by animals and chemical process rates in plants. Lighting influences photoperiodism in animals and plants, and photosynthesis and regulation in plants. Air quality, noise, ion concentrations, and crowding can affect the health and/or productivity of animals or plants. This chapter summarizes the published results from various research projects and provides a concept of the physiological factors involved in controlling the environment.

ANIMALS

Animal performance (growth, egg or milk production, wool growth, and reproduction) and their conversion of feed to useful products are closely tied to the thermal environment. For each homeothermic species, an optimum thermal environment permits necessary and desirable body functions with minimum energetic input (Figure 1A). The optimal thermal environment—in terms of an effective temperature that integrates the effects of dry-bulb temperature, humidity, air movement, and radiation—is less important to the designer than the range of conditions that provides acceptable animal performance, efficiency, well-being, and economic return for a given species. Figure 1A depicts this range as the *zone of nominal losses*, selected to limit losses in performance to a level acceptable to the livestock manager. Researchers have found that the zone of nominal losses corresponds to the welfare plateau (*i.e.*, welfare is enhanced by maintaining environmental conditions within the zone of nominal losses). Milk and egg production by mature animals also shows an optional thermal environment zone, or zone of nominal losses (Figure 2).

Developed from actual measurements of swine growth, Figure 1B shows the relationships of energy, growth, and efficiency with air temperature. In the case of growing pigs in Figure 1B, the range of temperatures from 60 to 72 °F, which includes both optimal productivity and efficiency levels, represents acceptable design conditions to achieve maximum performance and efficiency. Even beyond that temperature range, performance and efficiency do not markedly decline in the growing pig until near the lower critical temperature (LCT) or upper critical temperature (UCT), and

potential performance losses within the temperature range from 50 to 76 °F may be acceptable. Response relationships, as shown in Figure 1B, allow environmental selection and design criteria to be based on penalties to performance (*i.e.*, economic costs) and animal well-being—particularly when used with climatological information to evaluate risks for a particular situation (Hahn *et al.* 1983). Choosing housing requires caution, because research indicates that factors such as group versus individual penning, feed intake, and floor type can affect the LCT by 10 °F.

The limits of acceptable values of the LCT and UCT depend on such effects as the species, breed, genetic characteristics of an individual animal's age, weight, sex, level of feeding and type of feed, prior conditioning, parasites, disease, social factors such as space allocation, lactation or gestation, and physical features of the environment. The LCT and UCT vary among individuals; data reported are for group means. As a result, the limits become statistical values based on animal population and altered by time-dependent factors.

Acceptable conditions are most commonly established based on temperature because an animal's sensible heat dissipation is largely influenced by the temperature difference between the animal's surface and ambient air. Humidity and air movement are sometimes included as modifiers for an effective temperature. This has been a logical development. Air movement is a secondary but influential factor in sensible heat dissipation. The importance of air velocities is species and age dependent (*e.g.*, swine under 8 weeks of age experience slower gains and increased disease susceptibility when air velocity is increased from 25 to 50 ft/min).

With warm or hot ambient temperature, elevated humidity can restrict performance severely. Relative humidity has little effect on the animal's heat dissipation during cold temperatures, and it is usually only moderately important to thermal comfort during moderate temperatures. (Information is limited on such interactions, as well as on the effects of barometric pressure, air composition, and thermal radiation.)

Animals housed in a closed environment alter air composition by reducing oxygen content and increasing carbon dioxide and vapor content. Decomposing waste products add methane, hydrogen sulfide, and ammonia. Animal activities and air movement add microscopic particles of dust from feed and bedding. Generally, a ventilation rate sufficient to remove water vapor adequately controls dusts and gases that can irritate the animal's respiratory tract. However, improper air movement patterns, certain waste-handling methods, and special circumstances (*e.g.*, disease outbreak) may indicate that more ventilation is necessary.

The preparation of this chapter is assigned to TC 2.2, Plant and Animal Environment, with cooperation of Committee SE-300 of the American Society of Agricultural Engineers.

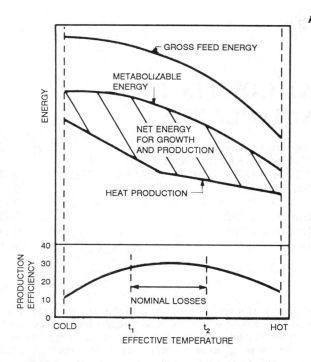

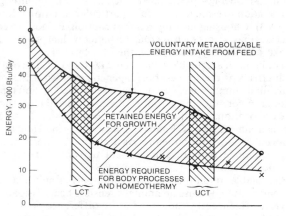

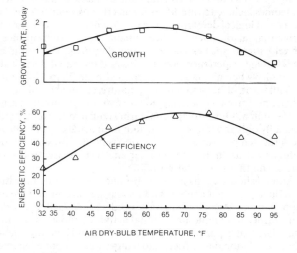

Fig. 1 Energetic and Performance Relationships Typical for Animals as Affected by Effective Environmental Temperatures
(Hahn *et al.* 1983)

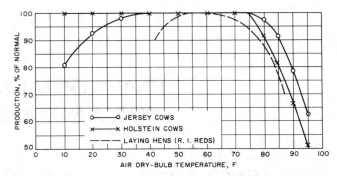

Fig. 2 Comparative Effect of Air Temperature in Mature Animal Production
(Hahn *et al.* 1983, Tienhoven *et al.* 1979)

ANIMAL CARE/WELFARE

Animal facilities that facilitate good animal care and welfare must be designed considering a wide range of environmental factors beyond thermal conditions. These include space requirements, flooring type, lighting, feed and water requirements, animal handling; and waste management. Facilities that meet animal care and welfare needs vary considerably by purpose, animal species, and geographic location. For additional information and guidelines for designing animal facilities to meet animal care and welfare needs, refer to the bibliography.

PHYSIOLOGICAL CONTROL SYSTEMS

An animal has a phenomenally stable control system (Scott *et al.* 1974, Tienhoven *et al.* 1979). Despite wide variations in environmental, nutritional, or pathological conditions, animals can control blood pressure and composition, body temperature, respiration, and cardiac output without conscious effort. Physiological control systems react to unfavorable conditions to ensure survival of the animal at the expense of production or reproduction.

One physiological control system important for air-conditioning design is the homeothermic system—the means by which an animal adapts to its thermal environment. The animal strives to control body temperature by adjusting both heat production in the tissues and transfer of heat to the environment from exposed surfaces. The homeothermic system in domestic animals is a closed-loop system and can be analyzed like any closed-loop control system.

Heat Production

Heat production data have been measured for many farm animals (Yeck and Stewart 1959, Longhouse *et al.* 1960, 1968, Bond *et al.* 1959). Much of the early data were obtained for a basal condition (*i.e.*, all life processes at a minimum level). Heat production under conditions of normal metabolic activity is more useful to the engineer. Such data are now available for most farm animals (Figures 3 and 4). Where possible, the data show total or sensible and latent heat for animals fed in a typical housing situation. Thus, the sensible heat used to evaporate urine moisture (assuming no other moisture) appears as latent heat. For design purposes, such data are better reflections of total heat partitioning than metabolic heat production obtained by calorimetry. Scott *et al.* (1983) provide additional information on specific situations.

The rate of heat production is primarily a function of temperature, animal species, and animal size (Brody *et al.* 1945). Heat production varies diurnally, depending on animal activity and eating times, and may change dramatically under special circumstances (*e.g.*, when animals are disease-stressed). Building type and waste-handling system can affect the conversion of sensible to latent heat by more than 50%. Therefore, the ranges of ventilation require careful analysis.

Heat Transfer to Environment

An animal can control, to some extent, the amount of heat transferred from its body by chemical and physical regulations. In a cold environment, the animal's metabolism increases, which increases the amount of heat production, offsetting heat transfer to the environment (Brody 1945). (The principal result of such chemical regulation is inefficient feed utilization.) Physical regulation is also used in a cold environment; blood circulation to subcutaneous capillaries decreases, hair or feathers are erected, and animals huddle together in an attempt to reduce sensible heat loss. In a hot environment, opposite physical and physiological responses generally occur to enable transfer to the environment of heat associated with necessary and productive life processes. In addition, as sensible heat dissipation becomes more difficult,

evaporative heat loss increases and air moisture content becomes a factor in heat loss. Since the production of heat is a necessary by-product of growth and useful production, environmental limits to dissipating this heat cause a decrease in feed consumption and subsequent decrease in growth and production.

Most sensible heat of domestic animals is dissipated through the skin. Birds, sheep, and swine transfer most of their latent heat through the respiratory tract; cattle and horses transfer most of their latent heat through the skin. Blood transfers the heat produced by metabolism to the skin or to the lung surfaces, where it is dissipated by evaporation from the mucous layer coating the inside of the alveoli. Inspired air reaching the alveoli is heated almost to body temperature and may become nearly saturated with moisture. Expired air may not be saturated at body temperature— especially during periods of heat stress—because it is a combination of air, some of which has not reached deeply into the lungs. Minor amounts of heat are also transferred by ingestion of feed and water and through excretion.

Design factors affecting animal heat loss are (1) air temperature; (2) air vapor pressure; (3) air movement; (4) configuration, emissivity, absorptivity, and surface temperature of the surrounding shelter; and (5) temperature and conductivity of surfaces (*e.g.*, floors with which the animal may be in contact).

Cyclic Conditions

The physiologic sensing elements respond both to environmental conditions and to changes in those conditions. Cycles of temperature, pressure, light, nutrients, parasites, magnetic fields, ionization, and other factors frequently occur with little engineering control.

Light is perhaps the earliest discovered and most important controlled environmental variable affecting reproductive processes (Farner 1961). However, the effects vary widely among animal species and age. Some animals grow and remain healthy with or without light (*e.g.*, growing swine), while for others, lighting management is important. Studies have shown that (1) short photoperiods induce or accelerate estrus development in sheep; (2) day length affects semen production in sheep and horses (Farner 1961); (3) continuous white incandescent light during incubation of White Leghorn eggs caused eggs to hatch from 16 to 24 h earlier than eggs incubated in darkness (Shutze *et al.* 1962); and (4) red, yellow, or blue lights gave comparable results. Use of lights for only one of the three weeks also reduced the time required for hatching, but differences were not as marked. Percent hatchability was not affected by lighting treatments.

Light is used to delay sexual maturity in hens, which enhances subsequent production. This is done by gradually decreasing day length from hatching to 22 weeks of age or by abruptly decreasing day length to 9 h at 14 to 16 weeks of age. If pullets have previously been exposed to an increasing day length, the change should take place at 14 weeks of age; if exposed to a constant or decreasing schedule, a change at 16 weeks is adequate. Light intensities of 1 to 2 footcandles measured at bird height were adequate in all cases. Light can then be abruptly increased at 21 weeks of age to 14 or 16 h of light. The economic value of increasing day length beyond 14 h in a windowless poultry house, and 16 h in an open poultry house, has not been proved. Recently, photoperiods such as 8 h of light (L), 10 h of darkness (D), 2 L and 4 D, and other cycles have improved feed utilization, egg production, and poultry growth (Buckland 1975).

Continuous light from hatching through 20 to 21 weeks of age markedly depressed subsequent egg production and caused a severe eye abnormality, but did not depress egg mass. (Light intensity was 1 to 3 footcandles, measured at bird height.)

Temperature cycles have been studied in cattle (Brody *et al.* 1955, Kibler and Brody 1956), swine (Bond *et al.* 1963, Nienaber *et al.* 1987), and poultry (Squibb 1959). The results differ somewhat

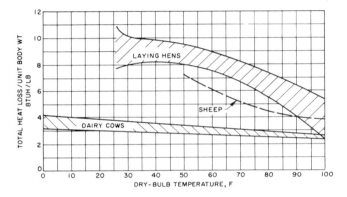

Fig. 3 Comparative Heat Loss of Mature, Producing Animals
(Tienhoven *et al.* 1979)

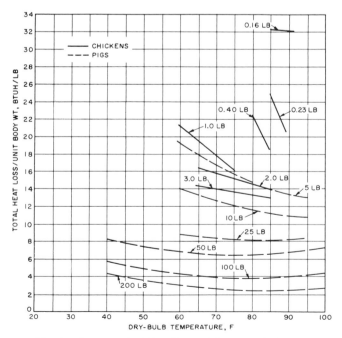

Fig. 4 Comparative Heat Loss of Growing Animals
(Bond *et al.* 1959)

among animals. Heat loss from cattle and swine can usually be calculated from average daily environmental temperatures with sufficient accuracy for design load calculation.

Productivity is only slightly different for averaged temperatures when cyclic conditions with a range less than 20°F are experienced (Squibb 1959). Above that range, productivity is depressed below that expected from the averaged temperature as determined in constant temperature tests, but under diurnally varying air conditions (68 to 77°F temperature and widely varying humidities), near-normal egg production is maintained.

Air Composition and Contaminants

The major contaminants in livestock housing are (1) respirable dusts from feed; manure; and animal skins, hair, and feathers; (2) microbes, both pathogenic and nonpathogenic, hosted in the respiratory tracts or animal wastes; and (3) gases of many types. Respirable dust particles have diameters between approximately 0.5 and 5 μm. The gases of most concern are ammonia, hydrogen sulfide, and carbon monoxide.

Dust is generated primarily by increases in animal activity and air movement causing reentrainment of settled dust. Airborne microbes in a calf nursery are generated through defecation, urination, and coughing by the calves (Van Wicklen and Albright 1987). Gases are produced from the metabolic processes of animals and from the anaerobic microbial degradation of wastes (Muehling 1970).

Contaminants are a concern because they predispose animals to disease and poorer performance and affect operator health. Common agricultural animals can tolerate higher levels of most air contaminants than most humans without adverse health effects. However, animals experiencing stress (*e.g.*, newborns, hot or cold, nutritionally limited, and sick animals) are more sensitive, and the presence of low levels of contaminants can have adverse effects on them.

The Occupational Safety and Health Administration has established minimum standards for exposure to air contaminants that are indicators of the maximum safe level for the operators (Table 1) (OSHA 1985). These are time-dependent—people experiencing contaminants for less than a 40-h week can tolerate higher levels, while those exposed for longer periods can tolerate less. They are also additive; for example, someone exposed to 75% of the maximum dust level may be exposed to no more than 25% of the maximum for any other contaminant or combination of contaminants.

The main adverse effects of dust are that it is a carrier for microbial and contaminant gases and is an irritant that increases animal susceptibility to other contaminants. High levels of dust overload the lung clearance ability of animals. Dust may have several types of microbes that can affect animal performance. Pneumonia and other respiratory tract pathogenic bacteria can be transferred between animals from the air. Nonpathogenic bacteria may cause health problems if present in sufficient quantities to overload the body's immune system. Sick animals are more susceptible to pathogenic or nonpathogenic bacteria. Normally, when ventilation is adequate for moisture removal (Figure 5), that rate is sufficient to prevent problems due to airborne microbes; however, contaminants reach undesirable levels in confinement housing that is underventilated during cold weather (Bundy 1984). Air

Table 1   OSHA Standards for 8-h Day Worker Exposure

Dust	Total:	15 mg/m³
	Respirable:	5 mg/m³ (<5μm)
Ammonia (NH)₄	25 ppm	
Hydrogen sulfide (H₂S)	10 ppm	
Carbon monoxide (CO)	50 ppm	

Source: Squibb (1959).

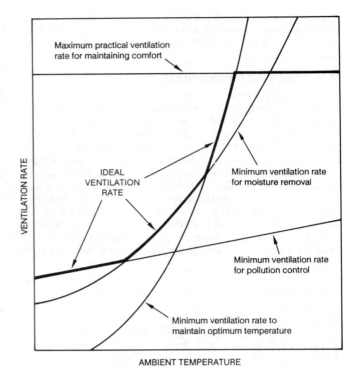

Fig. 5   Logic for Selecting the Appropriate Ventilation Rate in Livestock Buildings
(Christianson and Fehr 1983)

filtration is suggested as a means of reducing the incidence of pneumonia in an intensive calf housing (Hillman *et al.* 1992).

Airborne dust and microbes have been linked to respiratory disease in cattle, poultry, swine, horses, and laboratory animals. Infectious diseases, such as viral pneumonia and diarrhea, in calf barns have caused mortality rates between 20 and 80%, with the highest mortality occurring in enclosed barns that were underventilated during cold weather. Many respiratory diseases of poultry have been found to be transmitted via pathogenic microbes that can travel over 31 miles and remain infectious for months (Siegmund 1979). Over 1.2-million broiler chickens in Georgia were condemned postmortem at processing by inspectors during 1985 as the result of respiratory-related diseases (NASS 1986). Respiratory diseases are also costly to the swine industry—the incidence of enzootic pneumonia in pigs ranged from 30 to 75% with an estimated cost of $200-million per year (Armstrong 1982).

Gaseous ammonia is frequently a contaminant causing serious problems in swine and poultry housing. Ammonia levels of 50 ppm have been shown to reduce body mass gain rates of swine (Curtis 1983), and many researchers suspect that adverse health effects begin at levels below 25 ppm. Some physiologists and veterinary science researchers suggest that the level is between 5 and 10 ppm (Donham 1987). Ammonia at concentrations greater than 60 ppm has been implicated in reduced body weight and feed consumption, and in increased ocular and respiratory problems with broiler chickens (Carr and Nicholson 1980). Ammonia levels are influenced by the ventilation rate, sanitation practices, and the waste-handling system. Ventilating for moisture control is usually adequate to control ammonia below 10 mg in buildings sanitized monthly, where wastes are not allowed to accumulate for more than one month and where manure solids are covered by water.

Hydrogen sulfide is produced mainly from the wastes and is present at toxic levels that usually cause problems only when liquid wastes are agitated and gases from the waste handling can filter into the animal housing area. Hydrogen sulfide production can rise to lethal levels when manure in a pit below the animals is agitated, so extra precautions are necessary to ensure that ventilation is sufficient to remove the contamination.

Carbon monoxide is usually produced by malfunctioning heating or ventilation equipment. Many heaters are unventilated, which, when combined with the corrosive environment in livestock housing, makes this equipment susceptible to failure. Heaters should be vented to the outdoors or sensors should be installed to ensure complete combustion.

Control of air contaminants is achieved primarily through proper ventilation system design and management. High contaminant levels are a greater problem during cold weather when ventilation rates are lowered to conserve heat (Sutton *et al.* 1987). Dust removal systems such as settling systems, cyclones, fibrous filters, and electrostatic precipitation have been considered (Bundy 1986). Proper design and management of waste-handling facilities, adding fat to feed, and reducing animal and human activity have also been used to control contaminants. Respirators have been recommended to provide minimal protection for operators.

Air contaminants, the thermal environment, or both can reach lethal levels if the ventilation equipment is stopped by a power outage or equipment failure. Emergency power generation equipment should be available and alarm systems installed on important ventilation equipment.

Air Ionization

Charging the atmosphere with negative ions shows the beneficial results of reduced air contaminant levels and bacteria levels. Although negative ions kill bacteria, research results are inconclusive in determining whether the mode of action is air purification, a physiological effect on blood pH, ciliary stimulation, or a tranquilizing effect. Some researchers suggest that negative ionization is not beneficial because it can increase particle retention in the respiratory tract (Janni *et al.* 1984).

Of the various methods of ion generation investigated, those most used in the environmental field are thermionic, alpha particle emission, and high-voltage corona discharge. The latter operates more satisfactorily when discharge points are distributed over the ionized area, which helps overcome (1) ozone and nitrogen oxide production, which can occur when sufficient ion production for a large space is generated at one source and (2) much of the neutralizing effect caused by ions attached to dust particles, due to the short life of air ions. The entire enclosure acts as a low-level air cleaner because of electrostatic dust precipitation to grounded surfaces. Air bacteria counts drop about 50%, and even further reduction may be possible.

Interactive Stressors

Thermal (air temperature, air humidity, air velocity, and surrounding surface temperatures) and air quality (relative absence of dust, microbes, and contaminant gases) are the two stressors of most concern to ventilating a livestock structure. However, health status, animal age and stage of production, nutrition, and social conditions interact with the thermal and air quality conditions to determine animal health and performance.

Whether stressors are linearly additive or otherwise is an important research area. McFarlane (1987) found that ammonia, heat, accoustic noise, disease, and beak-trimming stressor effects on chickens were linearly additive in effect on the feed efficiencies and daily gain rates.

Comparing Individual Animal and Room Heat Production Data

Liquid wastes from urine, manure, waterer spillage, and cleaning affect the sensible and latent heat fraction in a room (ASAE 1987). More water on the floors requires more thermal energy from

the animals or the building heating system to evaporate the moisture. This evaporated moisture must then be removed by ventilation to prevent condensation on building surfaces and equipment and minimize adverse health effects, which can occur when humidities exceed 80%.

Floor flushing or excessive water spillage by animals can increase latent heat production by one-third from animal moisture production data, with a corresponding reduction in sensible heat production on a per-animal basis (ASAE 1987). Partially slatted floors with underfloor storage of wastes reduce the room moisture production by approximately 35%, compared to solid floor systems. Rooms with slatted floors throughout the building may have moisture production rates as much as 50% lower than solid floors. Therefore, slatted floors reduce the ventilation rate required for moisture removal, while excessive water spillage increases the moisture removal ventilation rate.

Influence of Genetic Change and Breed on Heat Production

Researchers have noted significant genetic correlations with performance. For example, swine breeds with high fat content characteristics outgained low fat content breeds, but the low fat content lines used feed more efficiently (Bereskin *et al.* 1975). Barrows (male hogs castrated before sexual maturity) ate 6% more than gilts (young sows that have not farrowed) to sustain a 7% faster growth rate.

Genetic improvements within breeds suggest that performance data should be updated periodically. Between 1962 and 1977, for example, the average gain rate for swine increased 0.15 lb per day, and feed efficiency improved 15% for Missouri hogs (Thomeczek *et al.* 1977).

CATTLE

Growth

Figure 6 shows the general growth rate for both beef and dairy breeds. Efficiency of beef calf growth is of economic importance; the dairy calf is developed for adult productive and reproductive capacity. Fattening calves and yearlings could be fed increasing amounts of grain and hay, with an expected gain of about 2 lb per day. Figure 7 shows the effect of temperature on the growth rate of several breeds of beef calves.

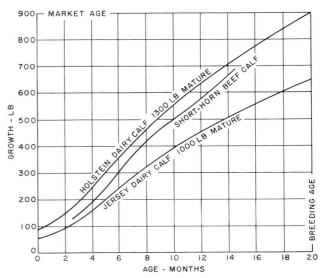

Fig. 6   General Growth Curves for Calves Fed Grain
(Johnson *et al.* 1958, Ragsdale 1960)

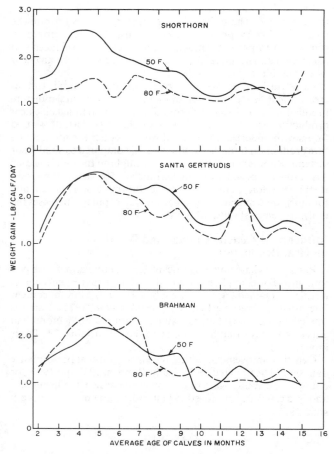

Fig. 7   Daily Weight Gain of Beef Calves
(Longhouse *et al.* 1960)

Food and energy requirements for growth can be computed by taking the difference between the energy in the available ration consumed and the energy required for maintenance and growth. Energy requirements for maintenance at an air temperature of 68°F expressed per pound body weight has been reported to be 8.65 Btu/h (Blaxter and Wood 1951), 7.98 Btu/h (Bryant *et al.* 1967), and 8.34 Btu/h (Gebremedhin *et al.* 1983). Using 370 kcal body tissue for conversion of calories to body tissue (Brody 1945), the growth rate of a calf at 70°F (5.38×10^{-4} lb gain per hour per pound of body weight) can be nearly 30 times that of a calf being kept at 37°F (0.18×10^{-4} lb/h per pound of body weight) for the same investment in food (10% of body weight). Alternatively, to get the same growth rate at 37°F as at 70°F, the same ration must be fed at the rate of 11.6% of body weight each day, a 16% increase (Gebremedhin *et al.* 1981).

Lactation

Figure 8 illustrates the effect of high temperature on the milk production of one Holstein in a test designed to study the combined effects of temperature and humidity. The dotted line represents the normal decline in milk production (persistency) from an advancing stage of lactation.

The ideal environment for Holstein cattle should not exceed 75°F. Jerseys are somewhat more heat tolerant, and their limit can be 80°F (Yeck and Stewart 1959). At the lower end of the temperature scale, production decreases may be expected at 30°F for Jerseys and below 10°F for Holsteins and Brown Swiss.

A temperature-humidity index (THI) expresses the relationship of temperature and humidity to milk production. Temperature-

humidity indexes have a greater effect on cows with a genetic potential for high milk production than for those with a lower potential (Figure 9).

Under high temperatures, a cooling system can help maintain milk production and reproductive function. In field studies conducted by Bucklin and Turner (1991) in both warm and hot, humid climates, a system providing evaporative cooling by forced air movement and direct sprinkling of lactating cows increased milk production by a range of 7 to 15% over cows not exposed to cooling.

Reproduction

Prolonged low temperatures, even those well below freezing, do not affect the reproductive performance of farm livestock, but breeding efficiency of both sexes decreases under summer conditions. Sustained temperatures above 85°F may decrease fertility, sperm production, and semen quality of males, and increase anestrus and embryonic death in females. In bulls, temperatures above 75°F decrease spermatogenesis, and long exposures at 85°F or above cause temporary sterility. The extent of the reaction depends on temperature rise and exposure duration.

In females, Guazdauskas (1985) observed that conception rates decreased from a range of 40 to 80% conception in thermoneutral environments (50 to 72°F) to a range of 10 to 51% conception in hot environments (> 81.5°F). A cooling system involving a cooled

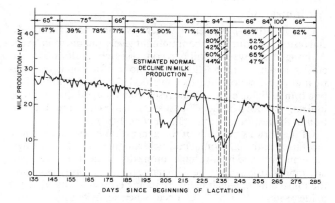

**Fig. 8   Milk Production Decline of One Holstein during
Temperature-Humidity Test**
(Yeck and Stewart 1959)

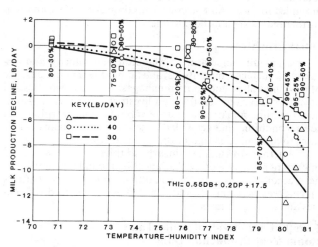

**Fig. 9   Effect of Milk Production Level on Average Decline
in Milk Production, versus Temperature-Humidity Index**
(Johnson *et al.* 1962)

shade and artificially induced air movement increased breeding efficiency by approximately 100% in a hot climate (Wiersma and Stott 1969).

Heat and Moisture Production

Sensible and latent heat production from individual animals (Figure 10) differ from the *stable heat and moisture production* of animals (Figures 11 and 12). The data in Figures 11 and 12 were obtained while ambient conditions were at constant temperatures and relative humidities were between 55 and 70%. The effects of evaporation from feces and urine are included in these data.

The rate of cutaneous water loss is very small at colder temperatures but rises sharply above 64°F. Cutaneous evaporation as a means of heat loss in calves becomes increasingly important as the air temperature rises above 75°F. As the air temperature rises, the proportion of nonevaporative cooling decreases. Above 86°F, about 80% of the heat transferred is by evaporative cooling. Gebremedhin *et al.* (1981) observed that cutaneous water loss from calves varied between 1.3×10^{-4} and 8.8×10^{-4} lb/h per pound of body weight for air temperatures between 32 and 64°F and increases steadily beyond 64°F (Figure 13). Water loss by respiration, the other avenue of loss, is shown in Figure 14.

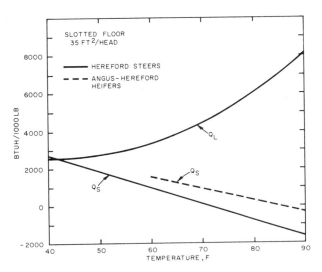

Fig. 10 Sensible and Latent Heat Production per Unit Livestock Mass (Weight)
(Hellickson *et al.* 1974)

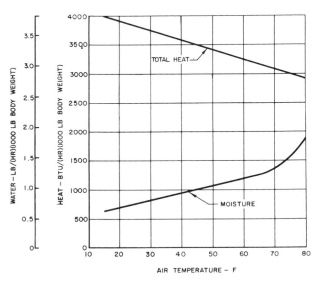

Fig. 11 Stable Heat and Moisture Dissipation Rates Dairy Cattle Stanchioned in Enclosed Stables
(Yeck and Stewart 1959)

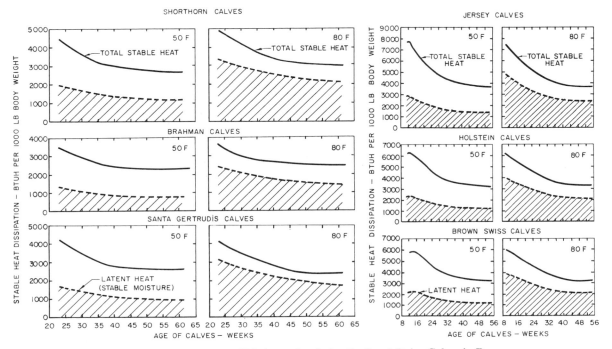

Fig. 12 Stable Heat and Moisture Loads for Beef and Dairy Calves in Pens
(Yeck 1957, Yeck and Stewart 1960)

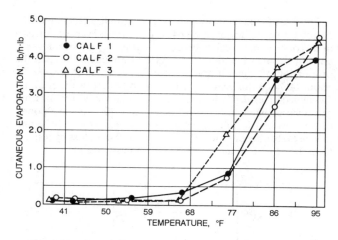

**Fig. 13   Cutaneous Water Loss per Unit Body Weight
of Holstein Calves**
(Gebremedhin *et al.* 1981)

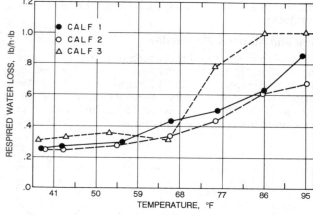

**Fig. 14   Respiratory Water Loss per Unit Body Weight
of Holstein Calves**
(Gebremedhin *et al.* 1981)

SHEEP

Growth

In normal environments, an average daily gain of 0.5 lb per day can be expected of lambs marketed at 80 to 120 lb, depending on breed. Hampshires often gain more than 1 lb per day, and lambs can be fed to weigh over 200 lb while still under a year old.

Variations exist among breeds and strains of sheep in their ability to adjust to environmental changes. Temperature effects on sheep growth suggest a lower rate at elevated temperatures. A South African study shows that lack of shade in warm climates reduces growth rate. Australian studies indicate that Merino lambs survive for only about 2 h in an air temperature of 100 °F.

Wool Production

The amount and quality of wool produced varies considerably among breeds, ranging from about 3 lb of poor quality wool from Dorset ewes (a breed developed primarily for mutton) to well over 10 lb of high quality fleece from dual-purpose breeds. Rambouillet, Merino, and Columbia-Southdale breeds supplied 7 lb per year, and about 4.5 lb was grown by a Hampshire, where monthly shearing increased wool production by about 1 lb per year.

Environmental factors such as photoperiod, nutritional level, and temperature also affect wool growth. Skin temperature is considered a dominant factor; high wool growth is associated with high skin temperatures. In a related study, low subcutaneous blood circulation limited wool growth. A thick fleece appears to limit radiant heat loads.

Reproduction

Sheep mate only during certain periods of the year, but they can sometimes be induced to mate outside the normal season if the natural environment is modified. Ewes may breed if exposed to an air temperature of 45 °F before mating is attempted.

In studies on the effects of high temperatures on sheep reproduction, air temperature was reported to affect the spermatogenesis rate. Rams kept at 90 °F environmental temperature showed a reduced rate of spermatogenesis, although a 3-week period at 70 °F effected recovery to the normal rate. Later reports suggest that high air temperatures may cause lowered fertility for

other reasons. Subjecting ewes to 100 °F just before mating causes fertilization failure because of some form of ovum structure degeneration.

Under practical conditions, early embryonic death appears to be the greatest loss in potential offspring conceived in a high-temperature environment. Degree of susceptibility to high temperature is greatest near mating time but generally decreases as the length of time after mating increases. A temperature of 90 °F and 65% rh at mating kills most embryos at an early age. The same conditions applied later in gestation do not cause death, but do cause the birth of small, weak lambs. The longer the high-temperature period, especially during the last third of gestation, the greater the number of weak lambs born. All reproductive processes appear to be more adversely affected by high air temperature when accompanied by high relative humidity.

Heat Production

Only a few heat-production tests have been conducted, mostly for correlation with other physiological data. A lamb's ability to generate heat is important because it is normally born during the most severe climatic season. One study indicated newborn lambs were limited in heat production to 30 Btu/h·lb—five times the basal level. Test conditions were from 14 to 32 °F with a 12 mph wind, but lambs are capable of withstanding temperatures as low as −40 °F. Table 2 provides calorimeter heat production data for sheep; while the total heat production data are reliable, the latent heat proportions do not reflect the portions of sensible heat used for water and urine evaporation from flooring and bedding.

Table 2   Heat Production of Sheep

Fleece Length, in.	At 46 °F		At 68 °F		At 90 °F	
	Total[a]	Latent[b]	Total[a]	Latent[b]	Total[a]	Latent[b]
Mature, maintenance-fed						
Shorn	4.0	8%	2.7	12%	2.0	38%
1.2	2.2	29%	2.0	28%	2.0	65%
2.4	2.0	23%	1.9	43%	1.9	76%
Lambs, 1 to 14 days						
Normal	10.5	—	8.0	—	—	—

Source: Scott *et al.* (1983).
[a]Btu/h·lb of body weight.
[b]Percent of total heat.

SWINE

Ambient air temperature affects the feed conversion and daily weight of growing swine. As shown in Figure 1B, a temperature range of 60 to 72 °F produces maximum rates of gain and feed use for 155- to 220-lb hogs, while a broader range of 50 to 76 °F reduces performance only slightly. For 45- to 130-lb animals, the optimal and nominal loss ranges are about 62 to 73 °F and 55 to 76 °F, respectively (Kibler and Brady 1956). Younger animals require temperatures of 73 to 82 °F for best performance, and piglets from 3 days to 2 weeks of age should have 86 to 90 °F conditions (Hahn 1983).

Daily air temperature cycles of more than 10 °F on either side of 70 °F result in reduced daily gain by pigs and an increased feed requirement per unit of gain (Nienaber et al. 1987). Reasonably constant air temperatures are desirable.

The level of air temperature in which swine grow affects deposition and retention of protein (carcass quality). Lean meat formation is reportedly highest in pigs raised between 60 and 70 °F (Mount 1963). However, the ratio of protein to fat decreases at air temperatures above 59 °F.

Swine gains and feed conversion rates are highly sensitive to air velocities and, in many cases, are affected adversely at velocities as low as 50 fpm. Swine less than 8 weeks of age should not be exposed to velocities greater than 50 fpm; a velocity lower than 25 fpm is preferred when temperatures are in the recommended range. Even in hot conditions, pigs are affected adversely by air velocities greater than 200 fpm (Bond et al. 1965, Gunmarson et al. 1967, Riskowski and Bundy 1986). Feed utilization and gain are much better at low air velocities (35 fpm) than at high (300 fpm) air velocities when temperatures are optimal (Figure 15).

Reproduction

Merkle and Hazen (1967) and Heard et al. (1986) have shown that sows benefit from some type of cooling in hot weather. In these studies, cool, dry air was directed at the sow to relieve heat stress by increasing evaporative and convective heat dissipation. Field studies of breeding problems and resultant small litters have established that both sexes suffer losses from high temperature. Sprinkling the sow during hot weather at breeding time and shortly after resulted in more live births than with unsprinkled sows. Sprinkling boars before mating also increased the number of live births per litter. Conception rate varied from about 100% of normal at 70 °F to only 70% at 90 °F. Breeding difficulties and a decrease in live embryos were observed in tests with controlled temperatures and relative humidities (Roller and Teague 1966). In some species, spontaneous abortion under severe heat stress may save the mother's life; however, sows appear to die of heat prostration, due to extra metabolic heat generation in late pregnancy, before spontaneous abortion occurs.

Little information is available on the value of temperature control for sows during cold weather. Forcing sows to produce the necessary heat to maintain body temperature may require rigid nutritional management to avoid the animals from becoming overweight—a condition resulting in poorer conception and smaller litters.

Heat and Moisture Production

Table 3 (Butchbaker and Shanklin 1964, Ota et al. 1982) shows direct calorimetry heat production for piglets. Figures 16 and 17 (Bond et al. 1959) show heat and moisture that must be accounted for in ventilating or air conditioning older swine housing for swine between 50 and 400 lb, housed at temperatures from 50 to 90 °F. These data are the sensible and latent heat levels measured in a room containing hogs. For design purposes, the room heat data reflect swine housing conditions rather than metabolic heat production from an individual animal.

The same data can be applied even where building temperatures are cycling, if the average air temperature during the cycle is used for design. The total animal heat load and the latent load that the ventilation system must remove are greater than the values in Figures 15 and 16, if the air velocity around the animals is more than about 50 fpm.

The lower critical temperature (the ambient temperature below which heat production increases) is about 77 to 86 °F for a group of newborn pigs; for a single piglet, the critical temperature is about 93 to 95 °F. As the pigs grow, the critical temperature falls—for 4.5- to 9-lb pigs, it is between 86 and 95 °F; for 9- to 18-lb pigs, between 77 and 86 °F. The floor bedding materials for growing piglets influence these critical temperatures. For a group of nine pigs averaging 88 lb, the critical temperature is 53 to 55 °F on straw, 57 to 59 °F on asphalt, and 66 to 68 °F on concrete slabs (Butchbaker and Shanklin 1964).

Table 3   Heat Production of Grouped Nursery Pigs

Weight Range, lb	Temperature, °F	Total Heat Production, Btu/h·lb	Latent Heat, % of Total
9 to 13	85	5.1	33
13 to 24	75	7.0	31
24 to 37	65	7.8	30

Source: Ota et al. (1982).

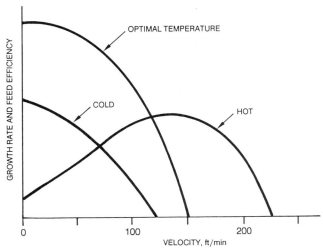

Fig. 15   Swine Response to Air Velocity
(Riskowski and Bundy 1986, Yao et al. 1986)

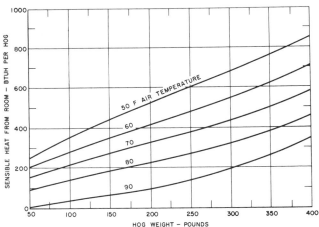

Fig. 16   Room Sensible Heat in Hog House
(Bond et al. 1959)

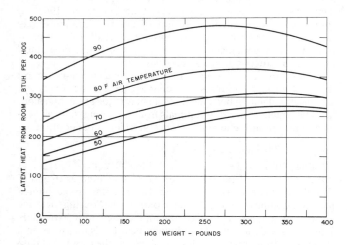

Fig. 17   Room Latent Heat in Hog House
(Bond *et al*. 1959)

CHICKENS

The tremendous worldwide production of broiler chickens is a result of improvements in genetics, nutrition, and housing. Four-pound broilers are produced in about 8 weeks. Table 4 indicates the feed consumption and growth of one strain of broilers (North 1984).

Research has shown that intermittent light and darkness (photoperiod) improve growth and conserve feed and energy (McDaniel and Brewer 1975).

Supplemental heat is generally needed for newly hatched chicks. Modern broiler strains respond well to brooder temperatures of 92 °F under the heaters. Thereafter, air temperatures may be decreased at a rate of 7 °F per week until 70 °F is reached. Relative humidities of 65 to 70% promote good feathering and market-quality broilers. Air velocity effects have received limited attention. Some benefit from increased velocity (up to 500 fpm) is obtained at air temperatures from 75 to 95 °F (Drury 1966). High-velocity air, at a temperature above the feather temperatures, causes more, not less, heat stress. Observations showed that 28-day-old broilers rested in areas with airflow temperature combinations of 55 fpm at 60 °F, 100 fpm at 70 °F, and 150 fpm at 75 °F.

Table 4   Weight Gain and Feed Conversion for Meat-Type Growing Pullets

Age, Weeks	Restricted Feeding		Full Feeding	
	Gain in Weight for Week, %	Feed Conversion for Week	Gain in Weight for Week, %	Feed Conversion for Week
4	22.2	3.32	44.4	2.21
6	15.4	3.85	25.7	2.32
8	11.8	4.24	20.0	2.71
10	9.5	4.66	13.2	3.54
12	8.0	5.15	9.5	4.88
14	6.9	5.64	7.6	6.50
16	6.1	6.13	5.9	7.56
18	5.4	6.62	4.3	9.03
20	4.9	7.11	3.3	11.00
22	4.4	6.04	2.5	14.10
24	3.9	5.39	1.8	14.44

Source: North (1984).

Reproduction

Adverse effects of high thermal environments (above 85 °F) on egg production include fewer eggs, reduced egg weight, and thinner shells. Hens over 1 year of age are more adversely affected by high thermal environments than younger hens. Larger hens, such as Rhode Island Reds, are more adversely affected than smaller hens, such as White Leghorns. Hatchability also declines as temperatures increase. Although the fertility rate is good at 70 °F, it declines at 86 °F. Feed requirements increase markedly below 45 °F; activity and productivity decline below 32 °F. The suggested ideal environment is between 55 and 75 °F.

Increasing relative humidity above 79%, at 85 °F dry-bulb temperature, produces an increased respiration rate and drooping wings. After 7 to 10 days, acclimatization may reduce the respiration rate, but production will still be less than optimal. Generally, 70 to 75% rh is recommended to maintain conditions that are neither too dusty nor too restrictive of latent heat dissipation.

Heat and Moisture Production

Figure 18 shows the total sensible and latent heat produced by laying hens during day and night at various temperatures. Figure 19 shows the same data for chicks. Figures 20 and 21 show sensible and latent heat production for broilers grown on litter at typical stocking rates at two temperatures. The effects of heat and moisture absorption or release by the litter are included in these figures.

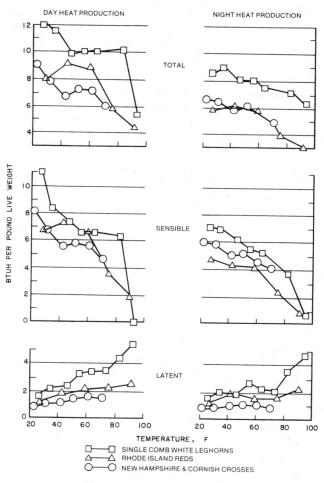

Fig. 18   Heat and Moisture Loads for Caged Laying Hens at Various Air Temperatures
(Ota and McNally 1961)

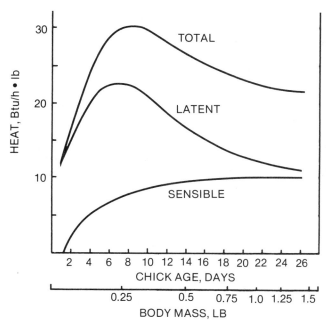

Fig. 19 Sensible, Latent, and Total Heat for Chicks Brooded on Litter
(Reece and Lott 1982)

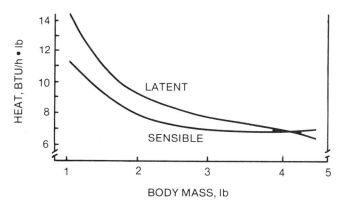

Fig. 20 Sensible and Latent Heat for Broilers Raised at 60 °F on Litter
(Reece and Lott 1982)

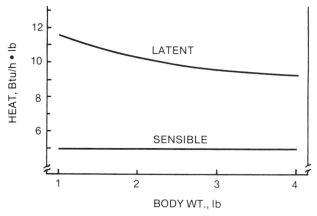

Fig. 21 Sensible and Latent Heat for Broilers Raised at 80 °F on Litter
(Reece and Lott 1982)

The sensible heat loss of a layer pullet is similar to that of a broiler at the same weight (Zulovich *et al.* 1987). However, the latent heat loss of the pullet is approximately 50% that of the broiler. Layer pullets consume 20 to 25% less feed energy than broilers. Thus, the total heat production is decreased with the reduction being in latent heat loss.

TURKEYS

Growth

The turkey poult has the best rate and efficiency of gain at a brooder room temperature of 70 to 75 °F for the first 2 weeks, and 65 °F thereafter, with about 70% rh. Initial brooder temperature should be 100 °F, reduced 5 °F per week until room temperature is reached. Proper control of photoperiod stimulates the growth rate in turkeys. Table 5 shows weekly cumulative average live weight for both sexes (Sell 1990). Figure 22 provides calorimetric heat production data for medium breed turkeys.

Reproduction

Mature turkeys tolerate temperature and humidity over a range of at least 20 to 90 °F and 35 to 85% rh. Since mature birds are kept only for fertile egg production, lighting for off-season egg production becomes very important. The young stock are raised to 20 weeks on natural light. At 20 weeks, females are placed in a totally darkened pen and given 8 h per day illumination at an intensity of 2 to 5 footcandles at bird's-eye level. At 30 weeks, the sexes are mixed, and the lighting period is increased to 13 to 15 h. The breeding season then continues for 12 to 26 weeks.

Heat and Moisture Production

Table 6 gives limited calorimetric heat loss data on large breed growing turkeys (DeShazer *et al.* 1974); however, an estimate from data on heavy chickens, applied to turkeys on a live-weight basis, should be satisfactory for design purposes. The reduction in heat loss during the dark period of the day is 25% for large breeds (Buffington *et al.* 1974) and between 5 and 40% for small breeds.

Table 5 Weekly Average Live Weights of Male and Female Turkeys

Age, Weeks	Male, lb	Female, lb
0	0.07	0.07
1	0.33	0.31
2	0.66	0.64
3	1.15	1.10
4	1.85	1.76
5	2.80	2.58
6	3.92	3.55
7	5.27	4.65
8	6.75	5.89
9	8.33	7.19
10	10.10	8.56
11	11.95	9.97
12	14.00	11.40
13	16.07	12.81
14	18.26	14.18
15	20.46	15.52
16	22.67	16.80
17	24.87	17.99
18	27.06	19.14
19	29.24	20.20
20	31.40	21.17
21	33.52	
22	35.54	
23	37.51	
24	39.42	

Source: Sell (1990).

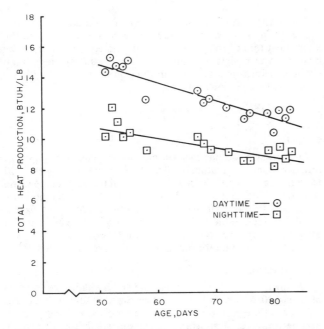

Fig. 22   Total Heat Production of Medium Breed Turkeys versus Age during Daytime and Nighttime
(Buffington *et al.* 1974)

Table 6   Heat Loss of Growing Turkeys at Various Air Temperatures and Relative Humidities

Age, Days	Weight of Poultry, lb	Dry Bulb, °F	Relative Humidity, %	Heat Loss, Btu/h·lb Sensible	Latent	Total Live Weight, lb
15	0.487	100	23	2.3	16.5	18.8
6	0.233	95	26	7.7	18.2	25.9
19	0.802	95	26	3.3	10.1	13.4
14	0.518	90	31	9.2	10.9	20.1
29	1.629	90	31	5.6	6.1	11.7
7	0.244	85	36	12.2	14.1	26.3
21	0.923	85	36	7.7	8.0	15.7
36	2.000	85	36	6.8	4.3	11.1
28	1.385	80	42	9.0	4.1	13.1
23	0.963	75	50	10.8	6.2	17.0
27	1.250	75	50	11.7	3.1	14.8
35	2.120	75	50	9.5	2.6	12.1

Source: DeShazer *et al.* (1974).

The latent heat production of turkeys may be significantly higher than the 1970s data show if the growth rate or feed consumption is significantly higher in the 1990s than it was in the 1970s (see Chickens, Heat and Moisture Production).

LABORATORY ANIMALS

Significant environmental conditions for facilities that house laboratory animals include temperature, humidity, air motion, illumination, noise, and gaseous and viable particulate contaminants (Moreland 1975). Design conditions vary widely, depending on whether the animals are experiencing disease-induced stress (which alters environmental needs), subjected to test environments, or simply housed (Besch 1975, Murakami 1971, Nienaber and Hahn 1983). This fact reflects differing housing and ventilation guidelines for animals used in research (NIH 1985), compared to recommendations by the American Society of Agricultural Engineers (ASAE 1987) and Midwest Plan Service (MWPS 1983) for production agriculture. Little is known about the influence of

disease on environmental requirements, animal performance, and well-being. Since significantly different conditions may exist between animal cage and animal room (macroenvironment), control of cage microenvironments is essential to ensure the animal's physiological well-being.

Heat and Moisture Production

Table 7 approximates heat released by laboratory animals at rest and during normal activity. Temperature and velocity gradient controls require low supply air-to-room air temperature differential, overhead high induction diffusion, uniform horizontal and vertical air distribution, and low return outlets. For load calculation purposes, heat gain from all laboratory animal species can be estimated (Wood *et al.* 1972, Gorton *et al.* 1976) with an acceptable level of error from:

$$ATHG = 2.5M$$
$$M = 6.6\,W^{0.75}$$

where

ATHG = average total heat gain, Btu/h per animal
M = metabolic rate of animal, Btu/h per animal
W = mass of animal, lb

Conditions in animal rooms must be maintained continuously. This requires year-round availability of refrigeration and, in some cases, dual air-conditioning facilities and emergency power for motor drives and control instruments. Chapter 14 of the 1991 ASHRAE *Handbook—Applications* has additional information on laboratory animal facilities.

PLANTS: GREENHOUSES, GROWTH CHAMBERS, AND OTHER FACILITIES

Most agronomically important plant crops are produced outdoors in favorable climates and seasons. Greenhouses and other indoor facilities are used for the out-of-season production of horticultural crops for both commercial sales and research purposes, and for producing food, floricultural, and other crops

Table 7   Heat Generated by Laboratory Animals

Animal	Weight, lb	Heat Generation, Btu/h per Animal Basal[a]	Normally Active[b,c] Sensible	Latent	Total
Mouse	0.046	0.66	1.11	0.54	1.65
Hamster	0.260	2.40	4.02	1.98	6.00
Rat	0.62	4.64	7.77	3.83	11.60
Guinea pig	0.90	6.10	10.22	5.03	15.25
Rabbit	5.41	23.41	39.22	19.31	58.53
Cat	6.61	27.21	45.57	22.45	68.02
Primate	12	42.55	71.27	35.10	106.38
Dog	22.7	68.64	104.8	56.4	161.2
Dog	50.0	124.10	230.7	124.2	354.9
Goat	79.3	175.39	293.78	144.70	438.48
Sheep	99	207.14	346.96	170.89	517.85
Pig	150	282.09	371	292	663
Chicken	4.00	18.67	12.9	21.9	34.7

[a] Based on standard metabolic rate $M = 6.6W^{0.75}$ watt per animal (Kleiber 1961) or appropriate reference (W = animal mass, lb).
[b] Referenced according to availability of heat generation data. Otherwise, heat generation is calculated on basis of ATHG = 2.5M (Gorton *et al.* 1976). Latent heat is assumed to be approximately 33% of total; sensible heat, 67% of total heat (Besch 1973, Woods *et al.* 1972).
[c] Data taken from Runkle (1964), Kleiber (1961), Besch (1973), Woods and Besch (1974), Woods *et al.* (1972), Bond *et al.* (1959), and Ota and McNally (1961).

in conditions that permit the highest quality by buffering the crops from the vagaries of weather. The industry that produces crops in greenhouses may be termed *controlled environment agriculture* (CEA).

Historically, many cold-climate commercial greenhouses were operated only from late winter into early summer, and during autumn. Greenhouses were too warm during midsummer; during winter in some cold-climate locations, light levels were too low and the day length inadequate for many crops. Mechanical ventilation, evaporative cooling, centralized heating systems, movable insulations, carbon dioxide enrichment, and supplemental lighting have extended the use of greenhouses to year-round cropping on a relatively large scale.

Growth chambers, growth rooms, and propagation units are environmentally controlled spaces used for either research or commercial crop production. Environmentally controlled chambers may include highly sophisticated facilities used for micropropagation (*e.g.*, tissue culture), or may be simple boxes in which air temperature and lights are controlled. Indoor facilities having controlled temperature and humidity environments may be used as warehouses to hold plants and plant products prior to commercial sale. Often these are simple refrigerated storage rooms or chambers.

Primary atmospheric requirements for plant production include: (1) favorable temperatures, (2) adequate light intensity and suitable radiation spectrum, and (3) favorable air composition and circulation. Engineering design to meet these requirements typically is based on steady-state assumptions. The thermal and ventilation time constants of most greenhouses are sufficiently short that transient conditions are seldom considered.

TEMPERATURE

Plant Requirements

Leaf and root temperatures are dominant environmental factors for plant growth and flowering. Factors in the energy balance of a plant canopy include air temperature, relative humidity, air movement, thermal radiation exchange, and convective exchange coefficients for sensible and latent heat. Therefore, leaf temperature is affected by environmental influences such as the type of heating and ventilating systems, supplemental lighting, light transmittance characteristics of the greenhouse cover, misting or evaporative cooling, location of the leaf on the plant, and the geometry of the surrounding leaf canopy.

Most information on plant responses to temperature is based on air temperature rather than plant temperature. Leaf temperature is difficult to measure, and one or several leaves represent neither the average nor the extreme temperatures of the plant. Since plants cannot actively regulate their cell and tissue temperatures in response to changing ambient conditions (passive regulation by opening and closing leaf stomata, which controls evapotranspiration, provides a small degree of control), their leaves and stems are usually within a few degrees of the surrounding air temperature (above during times of solar insolation, below at other times due to thermal reradiation and evapotranspiration).

All plants have minimum, optimum, and maximum temperatures for growth. Optimum temperature depends on the physiological process desired. Thermoperiodic species have different optimum day and night temperatures for each stage of growth, and each stage of plant growth may have its own unique optimum temperature influenced by radiant flux density, the ambient carbon dioxide level, and water and nutrient availability.

Historically, plants have been grown with night temperatures lower than day temperatures. In practice, many greenhouse crops are grown at standard (blueprint) night temperatures. Day temperatures are increased from 10 to 20°F (depending on solar

intensity) above night temperatures. Table 8 presents recommended ranges of night temperatures for a selection of greenhouse crops. New practices have been developed in which night temperature is kept higher than day temperature, providing a nonchemical means of height control for some plant species (*e.g.*, Easter lilies). However, thus far this technique has had limited application.

Heating Greenhouses

Heat loss from greenhouses is caused primarily by conduction through the structural cover and infiltration of outdoor air. The heating system is designed to meet the sum of the two. Perimeter heat loss is generally only a few percent of the total and is often neglected in design. When movable night insulation is used to conserve energy, heating systems are still designed to match conduction and infiltration losses without insulation because movable insulation may be opened early in the morning when outdoor air temperature is near its minimum, and excess heating capacity may be useful to melt snow in climatic regions where this occurs. Greenhouses are not designed to carry heavy snow loads.

Energy Balance

Radiation energy exchange. Solar gain can be estimated using procedures presented in Chapter 27. Not all insolation appears as sensible heat, however. As a general rule, from two-thirds to three-quarters of ambient insolation is available inside a typical commercial greenhouse. (Highly detailed models for calculating solar transmittance may be found in the literature.) If a greenhouse is filled with mature plants, approximately one-half of the available insolation (transmitted) may be converted to latent heat, one-quarter to one-third released as sensible heat, and the rest either reflected back outdoors or converted through photosynthesis (perhaps 3%).

Supplemental lighting can add significantly to the thermal load in a greenhouse. If movable night insulation is used, venting may be required during times of lighting, even during cold weather. The components of heat addition from supplemental lighting are divided between sensible and latent loads, with approximately one-quarter to one-third of the total heat load in latent form.

Reradiation heat loss from greenhouses comprises complex processes and may involve both reradiation from the structural cover and reradiation from inside the greenhouse if the cover is not thermally opaque. Glass is nearly thermally opaque, but many plastics are not. Newer plastic films may contain IR-inhibiting substances and can save a significant amount of heating energy, while adding only slightly to summer ventilation needs. Condensation on plastic films also reduces transmittance of thermal radiation, while diffusing but not seriously impairing transmittance of solar radiation. Generally, heat loss coefficients used in greenhouse design include the effects of thermal radiation exchange by the structural cover.

Structural heat loss. Conduction q_c plus infiltration q_i determine total heating requirements q_t. While infiltration heat loss is most accurately calculated using enthalpy differences, in practice only air temperature changes are considered, but with the apparent specific heat of air adjusted upward to account for the latent heat component.

$$q_t = q_c + q_i$$
$$q_c = \Sigma UA\Delta t$$
$$q_i = c_p VN\Delta t$$

where

U = heat loss coefficient, Btu/h·ft^2·°F (Table 9)
A = exposed surface area, ft^2

Table 8 Recommended Night Temperatures for Greenhouse Crops

Crop Species	Night Temperatures, °F	Remarks	Crop Species	Night Temperatures, °F	Remarks
Aster *Callistephus chinensis*	50–55	Long days during early stages of growth	Gloxinia *Sinningia speciosa*	64–70	Lower temperatures increase bud brittleness
Azalea *Rhododendron* spp.	61–64	Vegetative growth and forcing specific temperatures required for flower initiation and development	Hydrangea *H. macrophylla*	55–61 61–63 (forcing)	Specific temperature for flower initiation and development
Calceolaria *C. herbeohydrida*	61 50	Vegetative growth Flower initiation and development; initiation also occurs with long days and high temperatures if photon flux density is high	Iris *I. tingitana* (Wedgewood)	45–61 (forcing)	Forcing temperature 55 to 57 °F for 10/11 bulbs; 50 to 54 °F for 9/10 bulbs
Calendula *C. officinallis*	39–45		Kalanchoe *K. Blossfeldiana*	61	Temperatures influence rate of flower development and incidence of powdery mildew
Calla *Zantedeschia* spp.	55–61	Decrease to 55 °F as plants bloom	Lily *Lilium longiflorum*	61	Temperatures manipulated to alter rate of flower development; specific temperatures for flower initiation
Carnation *Dianthus caryophyllus*	50–52 winter 55 spring 55–61 summer	Night temperatures adjusted seasonally in relation to photon energy flux density	Orchida *Cattleya* spp.	61	Temperature requirement of hybrids related in parental species
Chrysanthemum *C. morifolium*	61 cut flowers 63–64 pot plants	Temperatures during flower initiation especially critical; uniform initiation very important for pot mums; cultivars classified on basis of temperature for development	Orchids *Phalaenopsis* spp. *Cymbidium* spp. *Cypripedium*	64 50 50–55	
Cineraria *Senecio cruentus*	61 48–52	Vegetative growth Flower initiation and development; plant quality best at low temperatures	Poinsettia *Euphorbia pulcherrima*	64 61–63	Vegetative growth Photoperiod requirement changes with temperature; bract development influenced by temperature
Crossandra *C. infundibuliformis*	75–81 64	Germination Growth and flowering	Roses *Rosa* spp.	61–63	
Cyclamen *C. indicum*	61–64 55 50–52	Germination Seedlings Growth and flowering	Saintpaulia *S. ionantha*	64–70	Below 61 °F, growth is slow, hard, and brittle
Foliage plants	64–70	Species differ in their temperature and radiant energy requirements	Snapdragon *Antirrhinum maias*	48–50 55–61	Winter Spring and Fall seedlings benefit from 61 to 64 °F temperatures
Fuchsia *F. hydrida*	52–61	Long days for flower initiation	Stock *Matthiola incana*	45–50	Buds fail to set if temperatures are above 64 °F for 6 h or more per day. Grown mainly as field crop in California and Arizona
Geranium *Pelargonium x hortorum*	55–61	61 to 64 °F for fast crops at high radiant energy flux	Tomato	61–66	Dry temperatures from 70 to 81 °F on sunny days
Gardenia *G. grandiflora* *G. jasminoides*	61–63 61–63	Lower temperatures result in iron chlorosis; higher temperatures increase bud abscission	Lettuce	55	63 to 64 °F on cloudy days 70 to 79 °F on sunny days
			Cucumber	64	75 °F on cloudy days 81 °F on sunny days

Δt = inside minus outside air temperature, °F
c_p = volumetric specific heat of air (adjusted upward to account for latent heat component), 0.03 Btu/ft$^3 \cdot$ °F
V = greenhouse internal volume, ft^3
N = number of air exchanges per hour (Table 10)
Σ = summation over all exposed surfaces of the greenhouse, and perimeter heat losses may be added to conduction loss equation for completeness

When design conditions are assumed for indoor and outdoor air temperatures and air exchange rate, the resulting heat loss may be assumed equal to the peak heating requirement for the greenhouse.

No universally accepted method exists to determine season-long heating needs for greenhouses. The heating degree-day method may be applied, but heating degree-day data must be adjusted to a base lower than 65 °F because of the significant passive solar heating effect in greenhouses. The proper base must be determined locally to reflect the expected solar climate of the region and the expected greenhouse operating temperature. These difficulties often lead designers to obtain season-long heating data from comparable, existing greenhouses in the region, and apply them to new designs.

LIGHT AND RADIATION

Plant Requirements

Light (400 to 700 nm) is essential for plant vegetative growth and reproduction (Figures 23 and 24). Intensity integrated over time provides the energy for growth and development, while duration (either long or short, depending on species) may be essential for certain physiological processes such as flowering. High light intensity may exceed the ability of individual leaves to photosynthesize. However, if there is a dense canopy, excess light may be beneficial to lower leaves even when upper leaves are light saturated. The intensity at which light saturates a leaf depends on various environmental factors, such as the concentration of carbon dioxide in the ambient air, as well as biological factors (Figure 25).

Table 9 Suggested Heat Transmission Coefficients

		U, Btu/h $\cdot$ ft$^2 \cdot$ °F
Glass		
	Single-glazing	1.13
	Double-glazing	0.70
	Insulating	Manufacturers Data
Plastic film		
	Single film[a]	1.20
	Double film, inflated	0.70
	Single film over glass	0.85
	Double film over glass	0.60
Corrugated glass fiber		
	Reinforced panels	1.20
Plastic structured sheet[b]		
	16 mm thick	0.58
	8 mm thick	0.65
	6 mm thick	0.72

[a] Infrared barrier polyethylene films reduce heat loss; however, use this coefficient when designing heating systems because the structure could occasionally be covered with non-IR materials.
[b] Plastic structured sheets are double-walled, rigid plastic panels.

Table 10 Construction U-Value Multipliers

Metal frame and glazing system, 16 to 24 in. spacing	1.08
Metal frame and glazing system, 48 in. spacing	1.05
Fiberglass on metal frame	1.03
Film plastic on metal frame	1.02
Film or fiberglass on wood	1.00

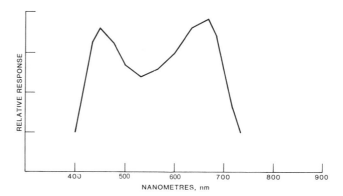

Fig. 23 Traditional Photosynthesis Action Spectra Based on Chlorophyll Absorption

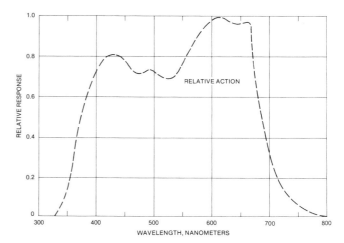

Fig. 24 Relative Photosynthetic Response

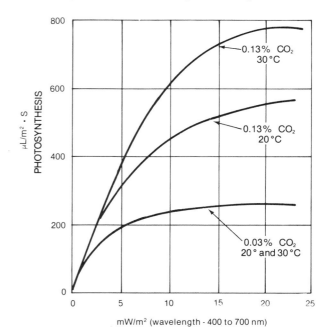

Fig. 25 Photosynthesis of Cucumber Leaf at Limiting and Saturating Carbon Dioxide Concentrations under Incandescent Light

Spectral distribution of light can affect plant development, but sunlight's spectral distribution need not be duplicated by artificial lighting to have suitable growth and development. Certain reproductive changes are initiated by red (660 nm) and far red (730 nm) light (Figure 26), and excessive ultraviolet light (290 to 390 nm) may be detrimental to growth.

Plants that respond to the durations of light and dark periods are termed photoperiodic (photoperiodic effects generally relate to flowering). Some plant species are long-day obligates, some are short-day obligates, some are day length-intermediate, and others are day-neutral. Such responses are usually (relatively) independent of light intensity. Photoperiodic effects can be initiated by very low light levels (less than 0.0929 W/ft^2), such as that provided to chrysanthemums by incandescent lights for a short period during the middle of the night to promote vegetative growth and inhibit flowering (until a suitable size has been attained) during the winter. Some plant species can tolerate continuous light, but others require some period of darkness for proper growth and development.

Sunlight is the most common source of photosynthetically active radiation (PAR, 400 to 700 nm). Although specially designed lamp sources may provide light similar to sunlight, no single source or combination of sources has spectral radiation exactly like the emission of the sun from 300 to 2700 nm. Table 11 summarizes the spectral distribution of various light sources. Three systems of measuring radiation are as follows:

1. Radiometric units (irradiance) in watts per square foot (W/ft^2), with specified wavelength intervals.
2. Quantum units as photon flux density in μmol/s·m^2 (400 to 700 nm unless otherwise specified). A mole of photons delivered in one second over a square meter may be referred to as an einstein.
3. Photometric units (illuminance) as one lumen per square meter, or lux (lx). One lumen per square foot is equivalent to one foot-candle (fc).

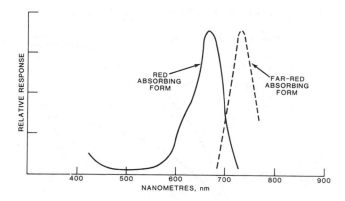

Fig. 26   Phytochrome Action Spectra

Table 11   Radiation Power Distribution of Light Sources

Light Sources	UV 300-400 nm	PAR + FR 400-850 nm	IR 850-2700 nm	Thermal 2700 + nm	Total Radiation
FCW	2	36	1	61	100
HG/DX	3	19	18	60	100
MH	4	41	8	47	100
HPS	0.4	50	12	38	100
LPS	0.1	56	3	41	100
INC	0.2	17	74	9	100
SUN	6	59	33	2	100

Note: Values are in watts per 100 W of total radiation.

Plant scientists use photosynthetic photon flux density (PPFD) in μmol/s·m^2 (400 to 700 nm). Engineering organizations and manufacturers of light sources use photometric and radiometric units. Because of the variation in spectral power distribution, conversion from one system of units to another must be made individually for each light source for the wavelength interval included (Table 12). To obtain comparable plant growth from different light sources, the same radiation levels (PAR) and red/far-red ratios must be maintained.

Radiation Levels for Plant Growth

Display 0.03 W/ft^2. For display purposes, plants can exist at an irradiance of 0.03 W/ft^2. The preferred lamp has changed with technological advances in efficiency and distribution. The emphasis, however, has always been on color rendering and the type of atmosphere created in the display space. Low-wattage incandescent and fluorescent lamps are preferred. At this irradiance, plants can be displayed (seen), but little or no significant positive effect on plants can be expected. Extended holding in such low light conditions will have a negative effect on many plant species. Timing (light-dark durations) and temperature interaction are not a concern.

Photoperiod response 0.08 W/ft^2 (4 to 6 h). For a photoperiod response, plant growth can be regulated at an irradiance of 0.08 W/ft^2 for as little time as 1 h. This irradiance is called a low light intensity system. The range of plant responses (promote or delay flowering, promote growth) that can be regulated is extensive, and this lighting is widely used by commercial growers.

Survival 0.3 W/ft^2 (8 h). Plants can survive at an irradiance of 0.3 W/ft^2 for 8 or more hours daily. This level enables many green plants to maintain their color. However, stem lengthening (etiolation) and reduction in new leaf size and thickness occur under this irradiance level. In time, the overall development of the plants falls behind that of other plants grown under higher radiation levels. Photoperiod responses do not function well at this irradiance. However, strong interactions occur between this irradiance and temperature, watering frequency, and nutrition. Cooler temperatures (less than 63 °F) help conserve previously stored material, while frequent watering and fertilization aggravate stem lengthening and senescence of older foliage.

Growth maintenance 0.8 W/ft^2 (12 h). Plants maintain growth over many months when exposed to an irradiance of 0.8 W/ft^2 of 12-h duration daily. This is the intensity at which many indoor gardeners (professional or hobbyists) grow their plants when starting them from seeds, cuttings, or meristems. Interactions with the environment (temperature, airflow, relative humidity, and pollutants) can vary among installations. Simple facilities with good air exchange and limited lamp concentration can grow a wide

Table 12   Light Conversion Factors

Light Source	Multiply W/ft^2 (400-850 nm) to Obtain: μmol/(s·ft^2) (400-700 nm)	Divide lux by Constant to Obtain: μmol/(s·ft^2) (400-700 nm)	μmol/(s·ft^2) (400-700 nm)
Sun and sky, daylight	4.57	581	388
Blue sky only	4.24	560	441
High-pressure sodium	4.98	883	581
Metal halide	4.59	764	657
Mercury deluxe	4.52	904	829
Warm white fluorescent	4.67	818	797
Cool white fluorescent	4.59	797	775
Plant growth fluorescent A	4.80	355	334
Plant growth fluorescent B	4.69	581	506
Incandescent	5.00	538	215
Low-pressure sodium	4.92	1141	958

Adapted from Thimijan and Heins (1983).

range of plant species. The rate of development, particularly as the plants grow in size, can be slow compared to plants grown at higher irradiances.

Propagation 1.7 W/ft² (6 to 8 h). Plants propagate rapidly when exposed to an irradiance of 1.7 W/ft² for a minimum of 6 to 8 h daily, but they prefer 12 h. Above this level, many propagators attempt to shade their greenhouses with one or several layers of neutral filters (painted films on glazing, or movable or semipermanent plastic or other fabric shade cloth materials) to restrict light (and heat) in the propagation area.

Cuttings rooted at this intensity maintain a growth rate much like that of similar tissue on a stock plant. Stem length, branching, and leaf color, however, can be regulated by manipulating temperature, moisture, stress, and nutrients. Most plants grown for their flowers and fruits can be brought to maturity by increasing the day length to 16 to 18 h for flower initiation (or rapid growth) and then reducing the day length to 8 to 12 h for development. The growth rate, however, is relatively slow. For quickest development (leaf number, number of branches, and early flower initiation), the plants must be transferred to a higher lighting regime—2.2 to 4.6 W/ft².

Greenhouse supplemental light 2.2 W/ft² (8 to 16 h). When natural light is inadequate, it may be supplemented up to approximately 2.2 W/ft² for 8 to 16 h daily. When coupled with the ambient sunlight (shaded by clouds, greenhouse structures, and lamp fixtures), this irradiance simulates many of the growth responses and rates associated with growth chamber studies. Plants grown in greenhouses without supplemental lighting grow slower and flower later than lighted ones in cloudy regions or in northern areas during winter. Duration (in hours) and timing (day-night) of lighting is critical.

Supplemental lighting for 8 h, particularly during the day (0800 to 1600) may not be as cost-effective as lighting at night (2000 to 0400) if off-peak electric rates are available. Neither of these lighting regimes, however, is as effective as lighting for 16 h from morning to midnight.

Lighting short-day plants, such as chrysanthemums and poinsettias, is relatively inefficient because they can be lighted only during the 8- to 12-h day, followed by an obligatory 12- to 16-h daily dark period.

Growth chambers 4.6 W/ft² (8 to 24 h). Plants grow in growth chambers or growth rooms if the light irradiance is a minimum of 4.6 W/ft² for 8 to 24 h daily. This irradiance is approximately one-fourth that of outdoor sunlight. Cool, white fluorescent lamps, combined with incandescent lamps, are widely used. More recently, HID lamps have been substituted for fluorescent lamps. For consistent results, all require a barrier of glass or other material between the lamp and the plants, and a separate ventilating system to remove the heat from such enclosed spaces.

Since filters cannot remove infrared completely, chambers are difficult to standardize. This often leads to confusing information on plant growth and flowering of plants grown in greenhouses and outdoors. When the total irradiance is 4.6 W/ft² and 10 to 20% of the total radiation is provided by incandescent lamps, most kinds of plants can be grown. In typical plant forms, flowering and fruiting responses occur when the plants are subjected to the following parameters:

- Day length, 8 to 24 h
- Temperature, 50 to 95 °F
- Carbon dioxide, 300 to 2000 ppm
- Relative humidity, 20 to 80%

Photoperiod

Day length affects the performance of some plants. There are four basic day length plant groups:

1. *Short-day plants* flower only when the length of the daily light period is less than the critical number of hours. Daily light periods longer than the critical length inhibit flowering.

2. *Long-day plants* flower only when the daily light period is longer than the critical number of hours. They become dormant or remain vegetative when the daily light period is shorter than the critical length.

3. *Day length-intermediate plants* flower only within a narrow range of day length, usually between about 10 and 14 h. If the day length is shorter than the optimum day length for flowering, the plants stop growing.

4. *Day-neutral plants* continue in vegetative growth or flower regardless of the day length.

Continuous light inhibits flowering and promotes vegetative growth of short-day plants, but encourages continued vegetative growth and early flowering of long-day plants, blocks the flowering of day length-intermediate plants, and in many instances, increases the stem length of day-neutral plants. Plants vary in their responsiveness to light source, duration, and intensity. The technology that has evolved to control the photoperiod of plants is based primarily on the incandescent-filament lamp. Of all the light sources available, this lamp creates the regulating mechanism most similar to that of sunlight. This is because the red/far-red wavelength ratio of light from an incandescent lamp is similar to the ratio of sunlight.

The effectiveness for photoperiod response in plants peaks at wavelengths of 660 nm (red) and 730 nm (far-red). The relative order of activity in regulating photoperiod responses by lamp type is as follows: Incandescent (INC) > High-pressure sodium (HPS) ≫ Metal halide (MN) = Cool white fluorescent (F) = Low-pressure sodium (LPS) ≫ Clear mercury (Hg). Photoperiod lighting is always used in combination with daylight or another main light source. Short days (less than normal day length) are created in the greenhouse with opaque materials that surround the plants.

RELATIVE HUMIDITY

Relative humidity affects the rate at which plants take water up, the rate of latent heat transfer, and certain diseases. Normal plant growth generally occurs at relative humidities between 20 and 80% if the plants have a well-developed root system, although relative humidities above 40% are preferred to avoid water stress conditions.

Transpiration, the movement of water vapor and gases from the plant to its surroundings, is controlled by the plant's stomatal openings. It is a function of air velocity and the vapor pressure difference between water at saturation at the leaf temperature and the actual water vapor partial pressure in the air. Generally, as relative humidity decreases, transpiration increases. Very low relative humidities (less than 20%) can cause wilting, since evaporation losses may be higher than the plant can replace, especially when light intensity is high.

High humidity provides a good environment for pathogenic organisms. Many pathogenic spores do not germinate unless relative humidity is 96% or more and many require a film of water on the leaves. Somewhat lower relative humidities may support other pathogen growth stages.

Still air surrounding a plant may be much wetter than the general atmosphere because evapotranspiration from the leaves raises the relative humidity in interfoliage air. The lower leaves, which stay moist longer, are more susceptible to disease. The upper leaves are dried by radiation and air currents.

AIR COMPOSITION

Carbon dioxide, which comprises about 0.035% of ambient air, is essential for plant growth. There are basically three ways to obtain carbon dioxide for greenhouse enrichment: pure in solid, liquid, or gaseous form; from burning fuels such as propane, natural gas, or kerosene; and by the aerobic breakdown of organic matter. The three ways are listed in order of purity and reliability. Carbon dioxide enters plants through stomata and is converted to carbohydrates through photosynthesis. The carbon dioxide concentration in air surrounding a plant, as well as light level, affects the rate of photosynthesis. The concentration for maximum growth depends on many factors, including the stage of growth, leaf area, light intensity, temperature, and air velocity past the stomatal openings.

An important relationship exists between light level and carbon dioxide uptake (Figure 25). As light level increases, carbon dioxide concentration must increase concurrently to take maximum advantage of the greater photosynthetic potential. In plastic greenhouses, and in glass greenhouses sealed against infiltration, the carbon dioxide level can drop below 200 ppm when the weather is cold, light levels are moderate, and the greenhouse is not ventilated. Carbon dioxide enrichment just to maintain normal levels can then be beneficial. During times of high light levels, carbon dioxide enrichment gives maximum benefit from the available light and may even be economically desirable when greenhouse ventilation is modest. However, carbon dioxide concentrations above 1500 ppm are seldom recommended; levels between 800 and 1200 ppm are typically used.

The effects of enrichment are not always positive. Without proper crop management, the yield, quality, or both may decrease, and timing of crop maturity may change.

Pollutants

Plants are sensitive to atmospheric pollutants such as ethylene, ammonia, gaseous fuels, ozone, fluorides, photochemical smog, and oxidants (from nitrogen and sulfur). Pollution damage can range from small spots on leaves, to yellowing of leaves, to severe foliage burn, and, ultimately, to plant death in extreme but not rare situations. The effect occurs both outdoors and in greenhouses; however, this is more common in greenhouses, because of their closed nature. Pollutants indoors can be removed by activated charcoal filters in the ventilation system; however, these are seldom used in commercial greenhouses. Economically, the more feasible approach is to limit pollutant production within, or introduction into, the greenhouse air space.

Ethylene is produced naturally by plants and leads to flower and whole plant senescence. It is also produced by combustion of gaseous and liquid fuels and can rapidly cause plant damage. Concentrations above 0.2 ppm can have a detrimental effect on plant growth. Unvented heaters, air currents that bring vented combustion products back into the greenhouse, and burners for carbon dioxide production are common sources of ethylene injury. Liquefied carbon dioxide may be used to supplement natural levels rather than combustion, specifically to avoid introducing ethylene into the greenhouse air, but even liquefied carbon dioxide should be carefully selected to avoid residual amounts of ethylene that may be contained within it.

Nitrogen oxides, common components of air pollution, can cause serious plant damage. Greenhouse locations near highways, nearby industrial complexes, and even a truck left running for an extended time near a greenhouse air intake vent may lead to leaf damage from NO and NO_2.

Sulfur dioxide, produced by the burning of sulfur containing fuels, causes injury to many plants within a short time. Sources of sulfur dioxide may be nearby, such as an industrial area, or may be within the greenhouse complex, such as the vented combustion products from a central heating facility, combustion products from carbon dioxide burners (using kerosene as a fuel, for example), and sulfur burned for mildew control.

Ozone is widely recognized as a serious pollutant affecting the production of many agronomic crops. Although few research results exist to quantify the effect of ozone on greenhouse crops, damage is likely to occur when greenhouses are located near ozone sources.

Phenolics and certain other organic vapors are phytotoxic. Phenolics, as volatiles from certain wood preservatives (creosote and pentachlorophenol), can cause leaf and petal damage. Vapors from some paints can also be damaging. Misuse of herbicides and pesticides can lead to plant injury, either through spray drift or volatilization.

Covering and sealing greenhouses for energy conservation can increase concentrations of ethylene and other air pollutants if their sources are within the air space, since infiltration and ventilation are decreased. Sealing to reduce infiltration can also lead to rapid carbon dioxide depletion and inhibited plant growth during cold temperatures when, even with bright light, ventilation is not required.

AIR MOVEMENT

Air movement influences transpiration, evaporation, and the availability of carbon dioxide. Air speed affects the thickness of the boundary layer at the leaf surface, which in turn influences the transport resistance between the ambient air and the leaf stomatal cavities. Air speed of 100 to 150 ft/min is commonly accepted as suitable for plant growth under CEA conditions. Air speeds across the leaf of 6 to 20 ft/min are needed to facilitate carbon dioxide uptake. Air speeds above 200 ft/min can induce excessive transpiration, cause the stomatal guard cells to close, reduce carbon dioxide uptake, and inhibit plant growth. Air speeds above 1000 ft/min may cause physical damage to plants. Generally, if plants within a greenhouse move noticeably due to ventilation, air speed is excessive.

Air circulation within greenhouses may be created to reduce thermal stratification and maintain suitable levels of carbon dioxide within the leaf canopy. Horizontal air flow, produced by small propeller fans that move air around the greenhouse in a racetrack pattern, has been found to be effective. Such fans are approximately 14 in. in diameter, with approximately one-sixth horsepower (0.2 kW) motors, spaced at approximately 50-ft intervals. Total fan capacity in cfm should equal approximately 25% of the greenhouse volume in cubic feet.

REFERENCES

Animals

Armstrong, C.H. 1982. Mycoplasmal pneumonia of swine. *International Swine Update* (Squibb) 1:1.

ASAE. 1987. Design of ventilation systems for poultry and livestock shelters. ASAE *Standard* D270.4, American Society of Agricultural Engineers, St. Joseph, MI.

Bereskin, B., R.J. Davey, W.H. Peters, and H.O. Hetzer. 1975. Genetic and environmental effects and interactions in swine growth and feed utilization. *Journal of Animal Science* 40(1):53.

Besch, E.L. 1973. Final report to Animal Resources Branch, Division of Research Resources, National Institutes of Health. NIH Contract 71-2511.

Besch, E.L. 1975. Animal cage room dry-bulb and dew point temperature differentials. ASHRAE *Transactions* 81(2):549.

Bond, T.E., H. Heitman, Jr., and C.F. Kelly. 1965. Effects of increased air velocities on heat and moisture loss and growth of swine. *Transactions of ASAE* 8:167. American Society of Agricultural Engineers, St. Joseph, MI.

Bond, T.E., C.F. Kelly, and H. Heitman, Jr. 1959. Hog house air conditioning and ventilation data. *Transactions of* ASAE 2:1.

Bond, T.E., C.F. Kelly, and H. Heitman, Jr. 1963. Effect of diurnal temperature upon swine heat loss and well-being. *Transactions of* ASAE 6:132.

Brody, S. 1945. *Bioenergetics and growth*. Reinhold Publishing Co., New York.

Brody, S., A.C. Ragsdale, R.G. Yeck, and D. Worstell. 1955. Milk production, feed and water consumption, and body weight of Jersey and Holstein cows in relation to several diurnal temperature rhythms. University of Missouri Agricultural Experiment Station Research Bulletin No. 578.

Buckland, R.B. 1975. The effect of intermittent lighting programmes on the production of market chickens and turkeys. *World Poultry Science Journal* 31(4):262.

Bucklin, R.A. and L.W. Turner. 1991. Methods to relieve heat stress in hot, humid climates. *Applied Engineering in Agriculture* 7. American Society of Agricultural Engineers, St. Joseph, MI.

Buffington, D.E., K.A. Jordan, W.A. Junnila, and L.L. Boyd. 1974. Heat production of active, growing turkeys. *Transactions of* ASAE 17:542.

Bundy, D.S. 1984. Rate of dust decay as affected by relative humidity, ionization and air movement. *Transactions of* ASAE 27(3):865-70.

Bundy, D.S. 1986. Sound preventive measures to follow when working in confinement buildings. Presented at the American Pork Congress, St. Louis, MO.

Butchbaker, A.F. and M.D. Shanklin. 1964. Partitional heat losses of newborn pigs as affected by air temperature, absolute humidity, age and body weight. *Transactions of* ASAE 7(4):380.

Carr, L.E. and J.L. Nicholson. 1980. Broiler response to three ventilation ranges. *Transactions of* ASAE 22(2):414-18.

Christianson, L.L. and R.L. Fehr. 1983. Ventilation energy and economics. In *Ventilation of agricultural structures*, M.A. Hellickson and J.N. Walker, eds. ASAE Nomograph No. 6, 336.

Consortium for Developing a Guide for the Care and Use of Agricultural Animals in Agricultural Research and Teaching. 1988. *Guide for the care and use of agricultural animals in agricultural research and teaching*. Agricultural Animal Care Guide Division of Agriculture, NASULGC, Washington, D.C. 20036-1191.

Curtis, S.E. 1983. *Environmental management in animal agriculture*. Iowa State University Press, Ames, IA.

DeShazer, J.A., L.L. Olson, and F.B. Mather. 1974. Heat losses of large white turkeys—6 to 36 days of age. *Poultry Science* 53(6):2047.

Donham, K.J. 1987. Human health and safety for workers in livestock housing. CIGR Proceedings. ASAE *Special Publication* 6-87.

Drury, L.N. 1966. The effect of air velocity of broiler growth in a diurnally cycling hot humid environment. *Transactions of* ASAE 9:329.

Farner, D.S. 1961. Comparative physiology: Photoperiodicity. *Annual Review of Physiology* 23:71.

Gebremedhin, K.G., C.O. Cramer, and W.P. Porter. 1981. Predictions and measurements of heat production and food and water requirements of Holstein calves in different environments. *Transactions of* ASAE 24(3):715-20.

Gebremedhin, K G., W.P. Porter, and C.O. Cramer. 1983. Quantitative analysis of the heat exchange through the fur layer of Holstein calves. *Transactions of* ASAE 26(1):188-93.

Gordon, R.L., J.E. Woods, and E.L. Besch. 1976. System load characteristics and estimation of animal heat loads for laboratory animal facilities. ASHRAE *Transactions* 82(1):107.

Guazdauskas, F.C. 1985. Effects of climate on reproduction in cattle. *Journal of Dairy Science* 68:1568-78.

Gunnarson, H.J. *et al.* 1967. Effect of air velocity, air temperatures and mean radiant temperature on performance of growing-finishing swine. *Transactions of* ASAE 10:715.

Hahn, G.L. 1983. Management and housing of farm animals environments. *Stress physiology in livestock*. CRC Press, Boca Raton, FL.

Hahn, G.L., A. Nygaard, and E. Simensen. 1983. Toward establishing rational criteria for selection and design of livestock environments. ASAE Paper No. 83-4517.

Heard, L., D. Froehlich, L. Christianson, R. Woerman, and W. Witmer. 1986. Snout cooling effects on sows and litters. *Transactions of* ASAE 29(4):1097.

Hellickson, M.A., H.G. Young, and W.B. Witmer. 1974. Ventilation design for closed beef buildings. Proceedings of the International Livestock Environment Symposium, SP-0174, 123.

Hillman, P.E., K.G. Gebremedhin, and R.G. Warner. 1992. Ventilation system to minimize airborne bacteria, dust, humidity, and ammonia in calf nurseries. *Journal of Dairy Science* 75:1305-12.

Hubbard Broiler. 1974. *Management guide for the Hubbard broiler*. Hubbard Farms, Walpole, NH.

Janni, K.A., P.T. Redig, J. Newmen, and J. Mulhausen. 1984. Respirable aerosol concentration in turkey grower building. ASAE Paper No. 84-4522.

Jensen, L. 1975. Growth standards feature changes, one new standard. *Turkey World* 50(1):26.

Johnson, H.D., A.C. Ragsdale, and R.G. Yeck. 1958. Effects of constant environmental temperature of 50°F and 80°F on the feed and water consumption of Brahman, Santa Gertrudis and Shorthorn calves during growth. University of Missouri Research Bulletin 683.

Kibler, H.H. and S. Brody. 1956. Influence of diurnal temperature cycles on heat production and cardiorespirativities in Holstein and Jersey cows. University of Missouri Agricultural Experiment Station Research Bulletin No. 601.

Kleiber, M. 1961. *The fire of life: An introduction to animal energetics*. John Wiley and Sons, New York.

Longhouse, A.D. *et al.* 1968. Heat and moisture design data for broiler houses. *Transactions of* ASAE 41(5):694.

Longhouse, A.D., H. Ota, and W. Ashby. 1960. Heat and moisture design data for poultry housing. *Agricultural Engineering* 41(9):567.

McDaniel, G.R. and R.N. Brewer. 1975. Intermittent light speeds broiler growth and improves efficiency. *Highlights of Agricultural Research* 4:9. Auburn University, Auburn, AL.

McFarlane, J. 1987. Linear additivity of multiple concurrent environmental stressors effects on chick performance, physiology, histopathology and behavior. Unpublished PhD thesis, Animal Sciences Department, University of Illinois, Urbana, IL.

Merkle, J.A. and T.E. Hazen. 1967. Zone cooling for lactating sows. *Transactions of* ASAE 10:444.

Moreland, A.F. 1975. Characteristics of the research animal bioenvironment. ASHRAE *Transactions* 81(2):542.

Mount, L.E. 1963. Food, meat, and heat conservation. Pig Industry Development Authority Conference Circulat. Buxton, Derbyshire, England.

Muehling, A.J. 1970. Gases and odors from stored swine wastes. *Journal of Animal Science* 30:526-31.

Murakami, H. 1971. Differences between internal and external environments of the mouse cage. *Laboratory Animal Science* 21:680.

MWPS. 1983. *Structures and environment handbook*. Midwest Plan Service, Ames, IA.

NASS. 1986. *Poultry slaughter*. Agricultural Statistics Board, National Agricultural Statistics Service, USDA, Washington, D.C. 20250.

Nienaber, J.A. and G.L. Hahn. 1983. Temperature distribution within controlled-environment animal rooms. *Transactions of* ASAE 26:895.

Nienaber, J.A., G.L. Hahn, H.G. Klencke, B.A. Becker, and F. Blecha. 1987. Cyclic temperature effects on growing-finishing swine. CIGR Proceedings. ASAE *Special Publication* 687:312.

NIH. 1985. Guide for the care and use of laboratory animals. National Institutes of Health Publication 85-23. Bethesda, MD.

North, M.O. 1984. *Commercial chicken production manual*, 3rd ed. AVI Publishing, Westport, CT.

OSHA. 1985. OSHA Safety and Health Standard. U.S. Department of Labor Code 1910.1000, 653-59.

Ota, H.J. and E.H. McNally. 1961. Poultry respiration calorimetric studies of laying hens. ARS-USDA 42-13, June.

Ota, H., J.A. Whitehead, and R.J. Davey. 1982. Heat production of male and female piglets. Proceedings of Second International Livestock Environment Symposium, SP-03-82. ASAE, St. Joseph, MI.

Reece, F.N. and B.D. Lott. 1982. Heat and moisture production of broiler chickens. Proceedings of the Second International Livestock Environment Symposium, SP-03-82. ASAE, St. Joseph, MI.

Riskowski, G.L. and D.S. Bundy. 1986. The effect of air velocity and temperature on growth performance and stress indicators of weanling pigs. ASAE Paper No. 86-4531. ASAE, St. Joseph, MI.

Roller, W.L. and H.S. Teague. 1966. Effect of controlled thermal environment on reproductive performance of swine. Proceedings of the Fourth International Biometeorological Congress. Rutgers University, New Brunswick, NJ.

Runkle, R.S. 1964. Laboratory animal housing, Part II. AIA *Journal* 4:73.

Scott, N.R., J.A. DeShazer, and W.L. Roller. 1983. Effects of the thermal and gaseous environment on livestock. ASAE Monograph No. 6. ASAE, St. Joseph, MI.

Scott, N.R., A. van Tienhoven, and C.A. Pettibone. 1974. Thermoregulation in poultry housing. Proceedings of the International Livestock Environment Symposium, SP-0174. ASAE, St. Joseph, MI, 211.

Sell, J.L. 1990. Faster growing, more efficient turkeys in 1989. *Turkey World* 66(1):12.

Shutze, J.V., J.K. Lauber, M. Kato, and W.O. Wilson. 1962. Influence of incandescent and colored light on chicken embryos during incubation. *Nature* 196(4854):594.

Siegmund, H., ed. 1979. *The Merck veterinary manual, A handbook of diagnosis and therapy for the veterinarian.* Merck & Co., Inc., Rahway, NJ.

Squibb, R.L. 1959. Relation of diurnal temperature and humidity ranges to egg production and feed efficiency of New Hampshire hens. *Journal of Agricultural Science* 52(2):217.

Sutton, A.L., S.R. Nichols, D.D. Jones, D.T. Kelley, and A.B. Scheidt. 1987. Survey of seasonal atmospheric changes in confinement farrowing houses. In Latest developments in livestock housing. International Commission of Agricultural Engineering (CIGR) Meeting, Urbana-Champaign, IL, 106-17.

Thomeczek, F.J., M.R. Ellersieck, R.K. Leavitt, and J.F. Lasley. 1977. Trends in economic traits of production tested boars in the Missouri Evaluation Station. University of Missouri Research Bulletin No. 1021.

Tienhoven, A.V., N.R. Scott, and P.E. Hillman. 1979. The hypothalamus and thermoregulation: A review. *Poultry Science* 52(6):1633.

Van Wicklen, G.L. and L.D. Albright. 1987. Removal mechanisms for calf barn aerosol particles. *Transactions of* ASAE 30(6):1758-63.

Verstegen, M.W.A. and W. van der Hel. 1974. The effects of temperature and type of floor on metabolic rate effective critical temperature in groups of growing pigs. *Animal Production* 18:1.

Wiersma, F. and G.H. Stott. 1969. New concepts in the physiology of heat stress in dairy cattle of interest of engineers. *Transactions of* ASAE 12(1):130-32.

Woods, J.E. and E.L. Besch. 1974. Influence of group size on heat dissipation from dogs in a controlled environment. *Laboratory Animal Science* 24:72.

Woods, J.E., E.L. Besch, and R.G. Nevins. 1972. A direct calorimetric analysis of heat and moisture dissipated from dogs. ASHRAE *Transactions* 78(2):170-83.

Yao, W.Z., L.L. Christianson, and A.J. Muehling. 1986. Air movement in neutral pressure swine buildings similitude theory and test results. ASAE Paper No. 86-4532.

Yeck, R.G. and R.E. Stewart. 1959. A ten-year summary of the psychro-energetic laboratory dairy cattle research at the University of Missouri. *Transactions of* ASAE 2(1):71.

Yeck, R.G. and R.E. Stewart. 1960. Stable heat and moisture dissipation with dairy calves at temperatures of 50 and 80°F. University of Missouri Research Bulletin No. 759.

Zulovich, J.M., M.B. Manbeck, and W.B. Roush. 1987. Whole-house heat and moisture production of young floor brood layer pullets. *Transactions of* ASAE 30(2)455-58.

BIBLIOGRAPHY

Animals

ASAE. 1986. Design of ventilation systems for poultry and livestock shelters. ASAE *Standard* D270.4. American Society of Agricultural Engineers, St. Joseph, MI.

CIGR. 1987. Latest development in livestock housing. Proceedings of CIGR Conference. American Society of Agricultural Engineers, St. Joseph, MI.

Curtis, S.E. 1983. *Environmental management in animal agriculture.* Iowa State University Press, Ames, IA.

Esmay, M.E. and J.E. Dixon. 1986. *Environmental control for agricultural buildings.* AVI Publications, Westport, CT.

Hahn, G.L., J.A. Nienaber, and J.A. DeShazer. 1987. Air temperature influences on swine performance and behavior. Applied Engineering in Agriculture. American Society of Agricultural Engineers, St. Joseph, MI.

Hellickson, M.A. and J.N. Walker, eds. 1983. Ventilation of agricultural structures. ASAE Monograph No. 6. American Society of Agricultural Engineers, St. Joseph, MI.

MWPS. 1987. The Midwest Plan Service handbooks and plans series. Midwest Plan Service, Ames, IA.

Yousef, M., ed. *Stress physiology in livestock* (3 volumes). CRC Press, Boca Raton, FL.

Plants

Cathey, H.M. and L.E. Campbell. 1975. Effectiveness of five vision lighting sources on photo-regulation of 22 species of ornamental plants. *Journal of American Society of Horticultural Science* 100(1):65.

Downs, R.J. 1975. *Controlled environments for plant research.* Columbia University Press, New York.

Gaastra, P. 1963. Climatic control of photosynthesis and respiration. In *Environmental control of plant growth,* L.T. Evans, ed. Academic Press, New York.

Hellickson, M.A. and J.N. Walker, eds. 1983. *Ventilation of agricultural structures.* American Society of Agricultural Engineers, St. Joseph, MI.

Hendricks, S.B. and H.A. Borthwik. 1963. Control of plant growth by light. In *Environmental control of plant growth,* L.T. Evans, ed. Academic Press, New York.

Mastalerz. J.W. 1977. *The greenhouse environment.* John Wiley & Sons, New York.

Tibbits, T.W. and T.T. Kozlowski. 1979. *Controlled environment guidelines for plant research.* Academic Press, New York.

Light and Radiation

Austin, R.B. and J.A. Edrich. 1974. A comparison of six sources of supplementary light for growing cereals in glasshouses during winter time. *Journal of Agricultural Research* 19:339.

Biran, I. and A.M. Kofranek. 1976. Evaluation of fluorescent lamps as an energy source for plant growth. *Journal of American Society of Horticultural Science* 101(6):625.

Campbell, L.E., R.W. Thimijan, and H.M. Cathey. 1975. Spectral radiant power of lamps used in horticulture. *Transactions of* ASAE 18(5):952.

Carpenter, W.J. and G.A. Anderson. 1972. High intensity supplementary lighting increases yields of greenhouse roses. *Journal of American Society of Horticultural Science* 101:331.

Cathey, H.M. and L.E. Campbell. 1977a. Lamps and lighting: A horticultural view. *Lighting Design & Application* 4(2):41.

Cathey, H.M. and L.E. Campbell. 1977b. Plant productivity: New approaches to efficient sources and environmental control. *Transactions of* ASAE 20(2):360.

Cathey, H.M. and L.E. Campbell. 1980. *Horticultural reviews,* Vol. II. AVI Publishing Co., Westport, CT, Chapter 10, 491.

Cathey, H.M., L.E. Campbell, and R.W. Thimijan. 1978. Comparative development of 11 plants grown under various fluorescent lamps and different duration of irradiation with and without additional incandescent lighting. *Journal of American Society of Horticultural Science* 103:781.

Duke, W.B. *et al.* 1975. Metal halide lamps for supplemental lighting in greenhouses, crop response and spectral distribution. *Agronomics Journal* 67:49.

Gates, D.M. 1968. Transpiration and leaf temperature. *Annual Review of Plant Physiology* 19:211.

McCree, K.J. 1972a. Significance of enhancement for calculation based on the action spectrum for photosynthesis. *Plant Physiology* 49:704.

McCree, K.J. 1972b. Test of current definitions of photosynthetically active radiation against leaf photosynthesis data. *Agricultural Metrord* 10:443.

Meijer, G. 1971. Some aspects of plant irradiation. *Acta Horticulture* 22:103.

Parker, M.W. and H.A. Borthwick. 1950. Influence of light on plant growth. *Annual Review of Plant Physiology,* 43.

Stoutmeyer, V.T. and A.W. Close. 1946. Rooting cuttings and germinating seeds under fluorescent and cold cathode lighting. Proceedings of American Society of Horticultural Science 48:309.

Thimijam, R.W. and R.D. Heins. 1983. Photometric, radiometric, and quantum light units of measure: A review of procedures for interconversion. *HortScience* 18(6):818-22.

Photoperiod

Cathey, H.M. and H.A. Borthwick. 1961. Cyclic lighting for controlling flowering of chrysanthemums. Proceedings of ASAE 78:545.

Downs, R.J. and A.A. Piringer, Jr. 1958. Effects of photoperiod and kind of supplemental light on vegetative growth of pines. *Forest Science* 4(3):185.

Downs, R.J., H.A. Borthwick, and A.A. Piringer, Jr. 1958. Comparison of incandescent and fluorescent lamps for lengthening photoperiods. Proceedings of American Society of Horticultural Science 71:568.

Hillman, W.S. 1962. *The physiology of flowering.* Holt, Rinehart & Winston, Inc., New York.

Jose, A.M. and D. Vince-Prue. 1978. Phytochrome action. A reappraisal. *Photochemistry & Photobiology* 27:209.

Lane, H.C., H.M. Cathey, and L.T. Evans. 1965. The dependence of flowering in several long-day plants on the spectral composition of light extending the photoperiod. *American Journal of Botany* 52:1006.

Whalley, D.N. and K.E. Cockshull. 1976. The photoperiodic control of rooting, growth and dormancy in Cornus Alba L. *Science Horticulture* 5:127.

Withrow, A.P. 1958. Artificial lighting for forcing greenhouse crops. Purdue University Agricultural Experiment Station Bulletin No. 533.

Temperature

Gates, D.M. 1968. Transpiration and leaf temperature. *Annual Review of Plant Physiology* 19:211.

Joffe, A. 1962. An evaluation of controlled temperature environments for plant growth investigations. *Nature* 195:1043.

Humidity, Carbon Dioxide, and Air Composition

Bailey, W.A. *et al.* 1970. CO_2 systems for growing plants. *Transactions of* ASAE 13(2):263.

Holley, W.D. 1970. CO_2 enrichment for flower production. *Transactions of* ASAE 13(3):257.

Kretchman, J. and F.S. Howlett. 1970. CO_2 enrichment for vegetable production. *Transactions of* ASAE 13(2):22.

Pettibone, C.A. *et al.* 1970. The control and effects of supplemental carbon dioxide in air-supported plastic greenhouses. *Transactions of* ASAE 13(2):259.

Tibbits, T.W., J.C. McFarlane, D.T. Krizek, W.L. Berry, P.A. Hammer, R.H. Hodgsen, and R.W. Langhans. 1977. Contaminants in plant growth chambers. *Horticulture Science* 12:310.

Wittwer, S.H. 1970. Aspects of CO_2 enrichment for crop production. *Transactions of* ASAE 13(2):249.

CHAPTER 10

PHYSIOLOGICAL FACTORS IN DRYING AND STORING FARM CROPS

THIS chapter focuses on the drying and storage of grains, oil-seeds, hay, cotton, and tobacco. However, the primary focus is on grains and oilseeds (collectively referred to as *grain*). Major causes of postharvest losses in these products are fungi, insects, and rodents. Substantial deterioration of grain can occur in storage. However, where the principles of good grain storage are applied, losses are minimal.

Preharvest invasion of grains by storage insects is usually not a problem in the midwestern United States. Field infestations can occur in grains when they are dried in the field at warm temperatures during harvest. Preharvest invasion by storage fungi is possible and does occur if certain weather conditions prevail when the grain is ripening. For example, preharvest invasion of corn by *Aspergillus flavus* occurs when hot weather is prevalent during grain ripening; it is, therefore, more common in the southeastern United States (McMillan *et al.* 1985). Invasion of wheat, soybeans, and corn by other fungi can occur when high ambient relative humidities prevail during grain ripening (Christensen and Meronuck 1986). However, the great majority of damage occurs during storage due to improper conditions that permit storage fungi to develop.

Deterioration from fungi during storage is prevented or minimized by (1) reducing grain moisture content below limits for growth of fungi, (2) maintaining low grain temperatures throughout the storage period to prevent fungal growth, (3) chemical treatment to prevent development of fungi or to reduce rate of fungal growth while the grain is being lowered to a safe moisture content, and (4) airtight storage in which initial microbial and seed respiration reduces the oxygen level so that further activity by potentially harmful aerobic fungi is prevented. Reduction of moisture by artificial drying is the most commonly used technique. The longer grain is stored, the lower its storage moisture should be. Some of the basic principles of grain drying and a summary of methods for predicting grain drying rate are included in the section Drying Theory.

Reduction of grain temperature by aeration is practical in temperate climates and for grains that are harvested during cooler seasons. Often, grain can be temporarily stored by using aeration to reduce its temperature to near freezing. Fans are operated when ambient temperatures are low. Basic information on aeration is summarized in the section Drying Theory. Use of refrigeration systems to reduce temperature is not usually cost-effective. Chemical treatment of grain is becoming more common and is briefly described in the section Prevention of Deterioration. When grain is placed in airtight silos, the oxygen level is rapidly reduced and carbon dioxide increases.

Although many fungi will not grow under ideal hermetic conditions in imperfectly sealed bins, some will grow initially, and this growth can reduce the feeding value of the grain for some animals. Partially emptied bins may support harmful mold growth. Sealed storage, even if successful, makes the grain unsuitable for human consumption because of yeast and bacterial growth. Airtight storage is briefly addressed in the section Factors Determining Safe Storage.

Deterioration from insects can also be prevented by a combination of reducing moistures and lowering temperatures. Lowering of temperatures is best achieved by aeration with cool ambient air during cool nights and periods of cool weather. Use of clean metal storage structures and segregation of new crop grain from carryover grain or grain contaminated with insects are important. In cases where insect infestation has already occurred, fumigation is often used to kill the insects. Prevention and control of insect infestations are addressed in the section Prevention of Deterioration.

Moisture content is the most important factor determining successful storage. Although some grains are harvested at safe storage moistures, other grains, notably corn, rice, and most oilseeds, must usually be artificially dried prior to storage. During some harvest seasons, wheat and soybeans are harvested at moistures above those safe for storage and, therefore, also require drying.

Christensen (1982), Brooker *et al.* (1974), Hall (1980), Christensen and Meronuck (1986), and Gunasekaran (1986) summarize the basic aspects of grain storage and grain drying. Chapter 22 of the 1991 ASHRAE *Handbook—Applications* covers crop drying equipment and aeration systems.

FACTORS DETERMINING SAFE STORAGE

Moisture Content

Grain is bought and sold on the basis of characteristics of representative samples. Probes or mechanical samplers, such as diverters, are used to obtain representative subsamples. Often a representative 2-lb sample must be taken from a large quantity (several tons) of grain. Parker *et al.* (1982) summarizes sampling procedures and equipment. For safe storage, it is necessary to know the range in moisture content within a given bulk and whether any of the grain in the bulk has a moisture content high enough to permit damaging fungal growth. This range can be determined by taking probe samples from different portions of the bulk. Commonly, in large quantities of bulk-stored grain, portions have moisture contents from 2 to 3% higher than the average (Brusewitz 1987). If the moisture content anywhere in the bulk is too high, fungi will grow, regardless of the average. Therefore, the moisture content of a single representative sample is not a reliable measure of storage risk or spoilage hazard. Measurement of moisture content and the precision of various moisture-measuring methods are covered in the section Moisture Measurement.

The preparation of this chapter is assigned to TC 2.2, Plant and Animal Environment.

Table 1 summarizes recommended safe storage moistures for several of the more common grains. Note that for long-term storage, lower moistures are recommended. Most storage fungi will not grow in environments where the relative humidity of the air between kernels is less than 70%. The relationship between grain moisture and the relative humidity of air between kernels is addressed in the section Equilibrium Moisture. Table 2 summarizes the minimum relative humidities and temperatures that permit the growth of common storage fungi.

Moisture Transfer

If temperatures vary within bulk-stored grain, moisture migrates from the warmer to the cooler portions. The rate of movement depends on the gradients in moisture content and temperature. Sellam and Christensen (1976) studied moisture transfer in a sample of one cubic foot of shelled corn initially at 15.5% moisture. They used heat lamps to produce a temperature differential of 18°F along the length of a sealed plastic container 1.22 ft long. After 2 days, this gradient (approximately 15°F/ft) caused

the moisture content at the cool end to increase by 1.2% and the moisture content at the warm end to decrease by 1.1%.

Thorpe (1982) developed an equation to describe moisture transfer caused by a temperature gradient. The equation was solved numerically, and laboratory experiments of moisture transfer in wheat were successfully modeled initially at 12% moisture content. In the experiments, an 18°F temperature gradient was developed across a column of wheat 0.66 ft thick (equivalent to a gradient of 27°F/ft). After one month, the moisture of the warmest grain dropped to 10.6%, while the moisture of the coolest grain increased to 14%.

Smith and Sokhansanj (1990a) provided a method of approximate analysis of the energy and velocity equations of the natural convection in grain bins. They showed that for small cereals, such as wheat, the influence of convection on temperature gradients may not be significant, whereas for larger cereals, such as corn, the effect of convection is more noticeable. Smith and Sokhansanj (1990b) also showed that convection flows in a grain bin are significant if the radius of the storage bin is approximately equal to the height of the bin.

Christensen and Meronuck (1986) cite an example of heating that developed in wheat initially at 13.2% in a nonaerated bin. Specially prepared samples had been placed at various positions in the bin at the time the bin was filled. After 3 months, the grain began to heat from fungal activity. Moistures in some of the samples had increased to 18%, while in others moistures had decreased to 10%.

These examples illustrate the importance of aeration in long-term storage. Aeration is generally required for storage structures with capacities exceeding 2000 bu or 50 tons. Moisture migration can initiate fungal and insect growth, and the heat of respiration generated by these organisms accelerates their growth and leads to spoilage. Studies suggest that temperature gradients could promote spoilage of grain loaded into a ship or barge—even if the grain is initially at a uniform moisture. Most shipments do not spoil because they remain in the ship or barge for a short time and because large temperature gradients do not develop. Christensen and Meronuck (1986) report studies of grain quality in barges and ships.

Table 1   Safe Storage Moisture for Aerated Good Quality Grain
(McKenzie 1980)

Grain	Maximum Safe Moisture Content (% Wet Basis), %
Shelled corn and sorghum	
To be sold as #2 grain or equivalent by spring	15.5
To be stored up to 1 year	14
To be stored more than 1 year	13
Soybeans	
To be sold by spring	14
To be stored up to 1 year	12
Wheat	13
Small grain (oats, barley, etc.)	13
Sunflower	
To be stored up to 6 months	10
To be stored up to 1 year	8

Table 2   Approximate Temperature and Relative Humidity Requirements for Spore Germination and Growth of Fungi Common on Corn Kernels
(Stroshine et al. 1984)

Fungus	Minimum Percent Relative Humidity for Spore Germination[b]	Grain Moisture[a], % w.b.	Growth Temperature, °F Min.	Growth Temperature, °F Optimum	Growth Temperature, °F Max.
Alternaria	91	19	25	68	97 to 104
Aspergillus glaucus	70 to 72	13.5 to 14	46	75	100
Aspergillus flavus	82	16 to 17	42 to 43	97 to 100	111 to 115
Aspergillus fumigatus	82	16 to 17	54	104 to 108	131
Cephalosporium acremonium[c]	97	22	46	75	104
Cladosporium	88	18	23	75 to 77	86 to 90
Epicoccum	91	19	25	75	82
Fusarium moniliforme	91	19	39	82	97
Fusarium graminearum, Fusarium roseum, (Gibberella zeae)	94	20 to 21	39	75	90
Mucor	91	19	25	82	97
Nigrospora oryzae[c]	91	19	39	82	90
Penicillium funiculosum[c] (field)	91	19	46	86	97
Penicillium oxalicum[c] (field)	86	17 to 18	46	86	97
Penicillium brevicompactum (storage)	81[b]	16	28	73	86
Penicillium cyclopium (storage)	81[b]	16	28	73	86
Penicillium viridicatum (storage)	81[b]	16	28	73	97

[a]Approximate corn moisture content at 77°F, which gives an interseed relative humidity equal to the minimum at which fungus can germinate. It is probably *below* the moisture content at which the fungus would be able to compete with other fungi on grain, except for *Aspergillus glaucus*. The latter has no real competitor at 72% rh, except occasionally *Aspergillus restrictus*.

[b]Approximately 5% or more of the population can germinate at this relative humidity.

[c]Rarely found growing in stored grain, regardless of moisture and temperature.

Temperature

Most processes that cause spoilage in stored grains are accompanied by a temperature rise. Therefore, temperatures should be monitored throughout the bulk. Temperature monitoring is commonly done by attaching thermocouples to cables that extend through the bulk from the top to the bottom, with thermocouples about 6 ft apart on each cable. The cables are 20 to 25 ft apart, with usually one thermocouple for each 2500 to 4000 bu of grain. Relatively dry grain is a good insulator, so a *hot spot* can develop without being detected immediately. However, when these thermocouple spacings are used, extensive spoilage can usually be detected by a temperature rise at a nearby thermocouple. A temperature rise of even a few degrees is evidence that grain has spoiled or is spoiling. Forced aeration maintains a uniform and preferably low temperature throughout the bulk.

Most grain-infesting insects become inactive below about 55 °F. Mites remain active but cannot develop rapidly below about 40 °F. Table 2 summarizes minimum, optimum, and maximum temperatures for the growth of some common storage fungi. Storage molds grow slowly at 32 to 40 °F. However, at higher moisture contents, some species of *Penicillium* will grow when the temperature is slightly below freezing. Grains with a moisture content high enough for invasion by *Aspergillus glaucus* will deteriorate rapidly at temperatures of 75 to 85 °F but can be kept for months without damage at 40 to 50 °F. Control of fungi and insects is described further in the sections Prevention of Deterioration and Aeration of Grain.

Oxygen and Carbon Dioxide

Only a few fungi that cause stored grain deterioration can grow in an atmosphere containing only 0.1 to 0.2% oxygen or in an atmosphere containing more than 60% carbon dioxide. Some yeasts can grow in grain stored in airtight storage at moisture contents above 18 to 19% and temperatures above 40 °F, producing flavors that make the grain unsuitable as food. However, the grain remains suitable feed for cattle and swine (Hyde and Burrell 1982), and its nutritional value may be enhanced (Beeson and Perry 1958).

Airtight storage of dry grain in underground or earth-sheltered structures is increasing in popularity in many parts of the world (Dunkel 1985, Hyde and Burrell 1982). Insects present when dry grain is put into storage usually die when oxygen has been depleted and will not usually reproduce if grain is sufficiently dry and in good condition. Hyde and Burrell (1982) and Shejbal (1980) cover controlled atmosphere storage in more detail.

Insects can also be controlled in conventional storage structures by forcing carbon dioxide or other gases such as nitrogen through the grain (Jay 1980, Ripp 1984). However, the costs of controlled atmosphere storage may be high, unless the structure can be inexpensively sealed or the gases can be easily generated or purchased at a low price.

Grain Condition

Old grain or grain already invaded by storage fungi is partly deteriorated, whether or not this is evident to the naked eye. Molding occurs more rapidly in partially deteriorated grain than in sound grain when the grain is exposed to conditions favorable to mold growth. Microscopic examination, and plating techniques can often reveal the fungal infection of grain in its early stages (Christensen and Sauer 1982, Christensen and Meronuck 1986, Stroshine *et al.* 1984). Accelerated storage tests, in which samples of grain are stored at a moisture content in equilibrium with air at 80% rh and 85 °F, and examined periodically, are useful in evaluating storability. These tests enable a manager to estimate the risk of spoilage during storage and to take appropriate action.

Equilibrium Moisture

If air remains in contact with a product for sufficient time, the partial pressure of the water vapor in the air reaches an equilibrium with the partial pressure of the water vapor in the material. The relative humidity of the air at equilibrium with the material of a given moisture is the *equilibrium relative humidity*. The moisture content of the hygroscopic material in equilibrium with air of a given relative humidity is the *equilibrium moisture content M_e*.

Several theoretical, semitheoretical, and empirical models have been proposed for calculating the M_e of grains. Morey *et al.* (1978) report that the modified Henderson equation is among the best equations available:

$$M_e = \frac{1}{100}\left[\frac{\ln(1.0 - \phi)}{-K\,(T + C)}\right]^{1/N} \tag{1}$$

where

M_e = equilibrium moisture content, decimal, dry basis
T = temperature, °C = (°F − 32)/1.8
ϕ = relative humidity, decimal equivalent
K, N, C = empirical constant

Table 3 lists values for K, N, and C for various crops. ASAE (1990) also gives the Chung equation, another equation used to predict M_e. Figure 1, based on the Chung equation, shows equilibrium moisture content curves for shelled corn, wheat, soybeans, and rice. Note that equilibrium moisture depends highly on temperature. ASAE (1990) gives additional curves drawn from the Chung equation and tabulated experimental data. Locklair *et al.* (1957) give data for tobacco.

Both the modified Henderson and the Chung equations give only approximate values of M_e and are for desorption. When grain is rewetted after it has been dried to a low moisture, the value of M_e is generally lower for a given relative humidity. Variations of as much as 0.5 to 1.0% can result from differences in variety; maturity; and relative starch, protein, and oil content. High-temperature drying can decrease the M_e of shelled corn by 0.5 to 1.0% for a given relative humidity (Tuite and Foster 1963). Therefore, to prevent mold development, shelled corn dried at high temperatures should be stored at moistures 0.5 to 1.0% below naturally dried corn. Pfost *et al.* (1976) summarize variations in reported values of M_e for several grains.

MOISTURE MEASUREMENT

Rapid and accurate determination of moisture of grains, seeds, and other farm crops determines whether they can be safely stored. Allowable upper limits for moisture are set by the marketing

Table 3   Desorption M_e Constants for Modified Henderson Equation for Various Crops
(ASAE 1990)

Product	K	N	C
Barley	0.000022919	2.0123	195.267
Beans, edible	0.000020899	1.8812	254.230
Canola (rapeseed)*	0.000505600	1.5702	40.1205
Corn, yellow dent	0.000086541	1.8634	49.810
Peanut, kernel	0.000650413	1.4984	50.561
Peanut, pod	0.000066587	2.5362	23.318
Rice, rough	0.000019187	2.4451	51.161
Sorghum	0.000085320	2.4757	113.725
Soybean	0.000305327	1.2164	134.136
Wheat, durum	0.000025738	2.2110	70.318
Wheat, hard	0.000023007	2.2857	55.815
Wheat, soft	0.000012299	2.5558	64.346

*Sokhansanj *et al.* (1986) from 41 to 86 °F.

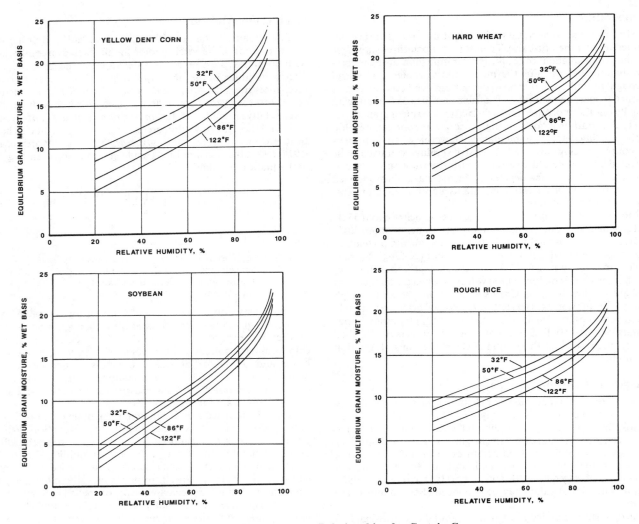

Fig. 1 Equilibrium Moisture Relationships for Certain Crops
(ASAE 1990)

system, and discounts and/or drying charges are usually imposed for higher moistures. For example, the accepted market limit on shelled corn is 15.5% moisture. If it will be stored for 4 or 5 months, it should be dried to 14.5%, and for longer storage times, 13% moisture is desirable (Table 1). Drying of the grain to moistures below the accepted market limit or the limit for safe storage moisture results in additional drying expense and may actually decrease the value of the grain.

If shelled corn is dried to 12% moisture, it becomes brittle and breaks more easily during handling. The moisture removal also reduces the total weight of grain. After drying one bushel (56.0 lbs) of shelled corn at 14% moisture to 12%, only 0.977 bushels (54.7 lbs) remain, and the market value has decreased by 2.3%. However, at high moistures, seed respiration and fungal growth can cause greater loss in dry weight.

Moisture content can be expressed on a wet or dry basis. The wet basis is used by farmers and the grain trade, while dry-basis moistures are often used by engineers and scientists to describe drying rates. Unless otherwise noted, moisture contents in this chapter are on a *wet basis* and are calculated by dividing the mass of water in the material by the total mass. The dry basis is calculated by dividing the mass of water by the mass of dry matter.

% Moisture (wet basis): $M_w = 100\, W_w/(W_w + W_d)$ (2)

% Moisture (dry basis): $M_d = 100\, W_w/W_d$ (3)

where

W_w = mass of water
W_d = mass of dry matter

Percent moisture on a wet basis M_w can be converted to percent moisture on a dry basis M_d and vice versa by the following formulas:

$$M_d = 100M_w/(100 - M_w) \qquad (4)$$

$$M_w = 100M_d/(100 + M_d) \qquad (5)$$

The weight change resulting from a change in moisture can be determined by assuming that the weight of dry matter is constant. The dry matter is calculated by multiplying the weight of grain by the quantity $(1 - M_w/100)$. For example, if 1000 lbs of wheat at 15% moisture is dried to 13%, the new weight W_n can be calculated by equating the amount of dry matter at 15% to that at 13% as follows:

$$1000\,(1 - 0.15) = W_n\,(1 - 0.13)$$

$$W_n = 1000\,(0.85/0.87) = 977 \text{ lbs}$$

If the final weight were known, the final moisture could be calculated by making it the unknown value in the above calculation. If two quantities of grain at differing moistures are mixed, the final moisture of the mixture can be determined by calculating the water in each, adding these together, and dividing by the total weight. The weight of water is the product of the decimal equivalent of wet-basis moisture and the weight. For example, if 500 lbs of shelled corn at 16% moisture is mixed with 1000 lbs of shelled corn at 14% moisture, the weight of water in each sample is:

$$(500 \text{ lbs}) (0.16) = 80 \text{ lbs}$$

$$(1000 \text{ lbs}) (0.14) = 140 \text{ lbs}$$

The moisture after mixing will be:

$$\left[\frac{80 + 140}{500 + 1000} \right] (100\%) = 14.7\%$$

Methods for determining moisture content are either *direct* or *indirect*. In direct methods involving the use of an oven, water is driven from the product, and the loss in product mass caused by evaporation of the water determines the moisture content. The Karl Fischer method, a basic reference method involving a chemical reaction of water and a reagent, is also classified as a direct method.

Indirect methods, such as electric or electronic moisture meters, measure the properties of the material that are functions of the moisture content. Moisture meters are used in commercial practice, while direct methods are used in research and for calibration of the indirect methods. Christensen *et al.* (1982) summarize approved methods used in Europe and the United States.

Direct Methods

Christensen *et al.* (1982) describe fundamental or basic methods of moisture determination as (1) drying in a vacuum with a desiccant and (2) titration with a Karl Fischer reagent. It is assumed that these methods measure the true water content and can be used to verify measurements obtained with routine reference methods, including oven drying and the Brown-Duval distillation method. The Brown-Duval method, not commonly used, involves heating the grain in a special apparatus and condensing and collecting the vaporized water.

Oven techniques use either forced convection air ovens or vacuum ovens and either ground or whole kernels. Drying times and temperatures vary considerably, and the different techniques can give significantly different results. Oven techniques are used to calibrate moisture meters (see Indirect Methods). As a result, during the export of grain, the meter moisture measurements can vary between arrival and destination if the importing country uses a different standard oven technique than the exporting country.

ASAE (1990) has developed a widely used standard which recommends temperatures and heating times of various grains. Temperatures may be either 217 °F (shelled corn, soybeans, sunflower) or 266 °F (wheat, barley, onion). Heating times vary between 50 min (onion seeds) and 72 h (soybeans, shelled corn).

Grinding of samples and use of a vacuum oven reduce heating time. When initial moistures are high, two-stage methods may be used (USDA 1971). A weighed sample of whole grain is partially dried to a moisture content of 13% or below, weighed, and then ground and completely dried, as in the one-stage method. The moisture lost in both stages is used to calculate moisture content.

Indirect Methods

Electronic moisture meters are simple to operate and give readings within minutes. Therefore, they are common in commerce and the grain processing industry. Direct methods of moisture measurement are used to calibrate the meters for each type of grain. Meters are sensitive to grain temperature, and calibration must include a correction factor. The newer automatic meters or *moisture computers* sense and correct for sample temperature and print or display the corrected moisture.

Near infrared transmittance (NIR) instruments have been developed, which measure moisture, protein, starch, and oil content of ground samples (Butler 1983, Cooper 1983, Watson 1977). However, such devices are relatively expensive and require careful calibration.

Conductance-type meters measure resistance, which varies with grain moisture. The practical range of moisture content measurable by conductive meters is approximately 7 to 23%. For up to 72 h after moisture addition or removal, the moisture at the surface of the kernels differs from the moisture in the interior. Therefore, recently dried grain reads low and recently wetted grain reads high. Mixing of wet and dry grain and mixing of good grain with partially deteriorated grain also results in erroneous readings. Martin *et al.* (1986) measured the signal from the conductance-type Tag-Heppenstall meter, analyzed the standard deviation of the AC component generated during measurement, and related this to individual kernel moisture variations in mixtures of wet and dry corn.

The dielectric properties of products depend largely on moisture content. The *capacitance meter* uses this relationship by introducing grain as the dielectric in a capacitor in a high-frequency electrical circuit. Although the capacitive reactance is the primary portion of the overall impedance measured, the resistive component is also significant in many capacitance meters. At higher frequencies and in instruments with insulated electrodes, the relative effect of the resistance is reduced, which is important in reducing errors introduced by unusual product surface conditions. Capacitance meters are affected less than conductance meters by uneven moisture distribution within kernels. Sokhansanj and Nelson (1988a) showed that the capacitance-type meters give low and high readings, respectively, on recently dried or rewetted grain. The range of measurable moisture content is slightly wider than that for conductance meters.

Moisture measurement by capacitance meters is sensitive to temperature, product mass, and product density (Sokhansanj and Nelson 1988b). To reduce these sources of error, a weighed sample is introduced into the measuring cell by reproducible mechanical means. Calibration, including temperature correction, is required. At least one commercially available unit measures bulk density and corrects for this factor as well as temperature. Tests of moisture meter accuracy have been reported by Hurburgh *et al.* (1985, 1986). Accuracy of moisture readings can be improved by taking multiple samples from a grain lot and averaging the meter measurements. Equipment for continuous measurement of moisture in flowing grain is available commercially, but this equipment is not widely used in the grain trade, and its accuracy is lowered by the difficulty in controlling the density of the grain as it flows through the detector.

Equilibrium relative humidity (described in the section Equilibrium Moisture) can be used to indicate moisture content. It also indicates storability independent of the actual moisture content because the equilibrium relative humidity of the air surrounding the grain, to a large extent, determines whether mold growth can occur (see Table 2). Measurement of equilibrium relative humidity at specific points within a grain mass requires specialized sampling equipment and has been used primarily for research.

Determination of hay moisture content does not receive the consideration devoted to grains. Oven methods (ASAE 1991) are used extensively, but several conductance-type moisture meters are available for both hay and forages. The extreme variability of the moisture and density of the material tested lead to great variability in the readings obtained. A reasonable indication of the average moisture content of a mass of hay can be obtained if many (25 or more) measurements are taken and averaged.

PREVENTION OF DETERIORATION

Fungal Growth and Mycotoxins

Fungal growth is the most important limitation on the successful storage of grain. In cases where high-temperature drying is used, it is sometimes impossible to dry grain immediately because the harvesting capacity often exceeds the capacity of the drier. Molding may occur before the grain can be dried, so the allowable storage time at the harvest moisture and temperature must be known. Low-temperature or ambient drying techniques may also be used as an alternative to high-temperature drying. In this case, wet grain is placed in the bin immediately after harvest, and the drying air is blown into a plenum in the bottom of the bin.

Drying in most farm bins begins in the lower layers and proceeds upward through the grain. The layer of grain that is drying is called the *drying zone*, and the upper boundary of the zone, where drying is just beginning, is called the *drying front*. The drying front may not reach the top of the bin for several days or weeks after drying begins. A risk associated with this type of drying is molding of the upper grain layers. Increasing the airflow increases the rate of drying but also increases costs. Therefore, the designer of low-temperature drying systems must know the maximum time that the grain at the top can be held at its harvest moisture and storage temperature before molding is significant.

In the United States, corn is one of the major crops that must be harvested above safe storage moistures. Shelled corn can be held at these higher moistures for a limited time before it must be dried. Mold growth produces carbon dioxide (CO_2). Allowable storage time at moistures above those for safe storage can be estimated by measuring CO_2 production of samples. By assuming that a simple sugar is being oxidized by microbial respiration, CO_2 production can be expressed in terms of dry matter loss in percent by mass.

Saul and Steele (1966) and Steele *et al.* (1969) studied the production of CO_2 in shelled corn, mostly on samples above 18%. Based on changes in the official grade of shelled corn, Saul and Steele (1966) established a criterion for acceptable deterioration of quality as 0.5% dry matter loss. This is equivalent to the production of 0.00735 lbs of CO_2 per pound of dry matter. Curves were developed showing grain temperature versus acceptable storage time for moistures of 18 to 30%, which are included in Chapter 22 of the 1991 ASHRAE *Handbook—Applications*. Thompson (1972) expressed Saul's data on dry matter loss per pound of dry matter as a function of moisture, time, and temperature using the following mathematical expression:

$$D_m = 1.3 \left[\exp \left(\frac{0.006t}{K_m K_t} \right) - 1.0 \right] + \frac{0.015t}{K_m K_t} \quad (6)$$

where

$$K_m = 0.103[\exp(455/M_d^{1.53}) - 0.00845 M_d + 1.558] \quad (7)$$

$$K_t = A \exp(BT/60) + C \exp[0.61(T - 60)/60] \quad (8)$$

and

- t = time in storage, h
- T = grain temperature, °F
- M_d = moisture content, % dry basis

Table 4 lists values for A, B, and C. According to Steele (1967), the damage level effect can be determined for dry matter losses

(DML) of 0.1, 0.5, and 1.0% by multiplying t from Equation (6) by K_d, where K_d is calculated as follows:

$$0.1\% \text{ DML: } K_d = 1.82 \exp(-0.0143 d) \quad (9a)$$

$$0.5\% \text{ DML: } K_d = 2.08 \exp(-0.0239 d) \quad (9b)$$

$$1.0\% \text{ DML: } K_d = 2.17 \exp(-0.0254 d) \quad (9c)$$

where d = mechanical damage, % by weight.

Based on a simulation, Thompson (1972) concluded that for airflow rates between 0.5 and 2.0 cfm/bu grain deterioration in the top layer during low drying is doubled when the airflow rate is halved. Thompson also concluded that weather variations during harvest and storage seasons can cause up to a two-fold difference in deterioration.

Seitz *et al.* (1982a, 1982b) found unacceptable levels of aflatoxin production prior to the time when 0.5% dry matter loss occurred. Nevertheless, the equations give approximate predictions of mold activity, and they have been used in several computer simulation studies (Pierce and Thompson 1979 and Brooker and Duggal 1982). Pierce and Thompson (1979) give recommended airflow rates for several common low-temperature drying systems and for various Midwest locations.

Acceptable dry-matter losses for wheat and barley are much lower than those for shelled corn, 0.085% and 0.104%, respectively (Brook 1987). Brook found reasonable agreement with published experimental data for the following equation (Frazer and Muir 1981) for days of allowable storage time D as a function of percent wet basis moisture M_w and temperature T in °F.

$$\log D = A + B M_w + C T + G \quad (10)$$

where $A = 6.234$, $B = -0.2118$, $C = -0.0293$, and $G = 0.937$ for $12.0 < M_w < 19.0$ and $A = 4.129$, $B = -0.0997$, $C = -0.03150$, and $G = 1.008$ for $19.0 < M_w < 24.0$.

Equation (10) was based on the development of visible mold. Brook also found that an adaptation of Saul's Equation (6) by Morey *et al.* (1981) gave reasonable results for storage time of wheat. Morey's method predicts dry matter loss by adjusting M_d for differences between corn and wheat equilibrium relative humidities.

Table 2 can be used to gain insight into the deterioration of stored grain. *Aspergillus* and *Penicillium* sp. are primarily responsible for deterioration because some of their species can grow at storage moistures and temperatures frequently encountered in commercial storage. In temperate climates, shelled corn is often harvested at relatively high moistures; during the harvest and storage season, ambient temperatures can be relatively low. Aeration of the grain during cold weather and cool nights can reduce the temperature of the grain to 40 to 60°F. This is below the optimum temperature for growth of *Aspergillus* sp. (Table 2). However, *Penicillium* sp. can still grow if grain moisture is above 16 to 17%; therefore, its growth is a persistent problem in such climates. If hot weather prevails prior to harvest, *Aspergillus*

Table 4 Constants for Dry Matter Loss of Shelled Corn [Equation (8)]

A	B	C	Temperature Range	Moisture Range, % w.b.
128.76	−4.68	0	T < 60	all moistures
32.3	−3.48	0	T ≥ 60	M_w ≤ 19
32.3	−3.48	$\dfrac{M_w - 19}{100}$	T ≥ 60	$19 < M_w$ ≤ 28
32.3	−3.48	0.09	T ≥ 60	M_w > 28

flavus, which competes effectively at warmer temperatures and higher moistures, can begin to grow in the field and continue to grow in stored shelled corn. In growing seasons when shelled corn must be harvested at moistures above 22%, *Fusarium, Alternaria, Epiccocum*, and *Mucor* can compete with *Penicillium* sp.

Application of chemicals slows deterioration until grain can be either dried or fed to animals. Preservatives include propionic acid, acetic acid, isobutyric acid, butyric acid (Sauer and Burroughs 1974), a combination of sorbic acid and carbon dioxide (Danziger *et al.* 1973), ammonia (Peplinski *et al.* 1978), and sulfur dioxide (Eckhoff *et al.* 1984). Propionic acid (Hall *et al.* 1974) or propionic-acetic acid mixtures, although not extensively used, are perhaps the most popular in the United States with high-moisture corn. Acetic acid and formic acid are most popular in Europe. Grain treated with propionic acid can only be used as animal feed. Inert carriers have been used to lessen its corrosiveness.

Hertung and Drury (1974) summarize fungicidal levels needed to preserve grain at various moistures. Both ammonia (Nofsinger *et al.* 1979, Nofsinger 1982) and sulfur dioxide (Eckhoff *et al.* 1984, Tuite *et al.* 1986) treatments require considerable management. Attention must be given to uniform application of the chemicals to the entire quantity of stored grain.

Insect Infestation

Insects cause major losses of stored grain. Grain containing live insects or insect fragments in sufficient numbers is unsuitable for human food. When grain is stored for long periods (a year or more), insects can infest the grain and cause significant amounts of deterioration. Traps and chemical attractants have been developed which monitor insects in storage facilities (Barak and Harein 1982, Barak and Burkholder 1985, Burkholder and Ma 1985). Detection in samples of grain taken for grading and inspection is often difficult. Many of the insects are relatively small and can only be seen easily with a magnifying lens. Many of the insect larvae develop within the kernels and cannot be detected without the use of staining techniques or grinding of the grain sample. Infested grain mixed with good grain in marketing channels compounds the infestation problem.

Sanitation is one of the most effective methods of insect control. Cleaning of bins after removal of old-crop grain and prior to filling with new-crop grain is essential. In bins containing perforated floors, fine material that collects beneath the floors can harbor insects, which infest new-crop grain when it is added. Control by aeration is feasible in temperate climates, because insect activity is reduced greatly at temperatures below 50°F. The effectiveness of temperature control has been documented by Bloome and Cuperus (1984) and Epperly *et al.* (1987). Chemicals have frequently been used to control live insects in grain, and methods are described by Harein (1982). Recently, thermal treatments have also been investigated (Lapp *et al.* 1986). Cotton and Wilbur (1982) summarize the types of grain insects, the ecology of insect growth, and the methods of detecting insects in samples of grain. Control of insects in farm-stored grain is detailed by Storey *et al.* (1979).

Rodents

The shift from ear corn harvesting and storage to field shelling and the introduction of metal bins have helped to reduce rodent problems. However, significant problems can arise when rodents consume grain and contaminate it with their hair and droppings. Storage structures should be made rodent-proof whenever possible. Rats can reach 13 in. up a wall, so storage structures should have concrete foundations and metal sides that resist gnawing.

In some countries, smaller on-farm storage structures are often elevated 18 in. to give protection from rodents. Double-wall construction and false ceilings should be avoided, and vents and holes should be covered with wire grates. Proper sanitation can help

prevent rodent problems by eliminating areas where rodents can nest and hide. Rodents need water to survive, so elimination of available water is also effective. Techniques for killing rodents include trapping, poisoning with bait, and fumigation. Harris and Bauer (1982) address rodent problems and control in more detail.

DRYING THEORY

In ordinary applications, drying is a heat and mass transfer process that vaporizes liquid water, mixes the vapor with the drying air, and removes the vapor by carrying away the mixture mechanically. In forced convection drying, sufficient heat for vaporization of product moisture (about 1100 Btu/lb of water) comes from the sensible heat in the drying air. A few types of driers, mostly experimental types, have been developed to apply heat directly to the product by conduction, radiation, or dielectric heating.

The most common mode of drying uses the sensible heat content of the air. The method can be diagrammed on the psychrometric chart by locating the state points for the air as it is first heated from ambient temperature to plenum temperature and then exhausted from the grain. The process is assumed to be adiabatic, *i.e.*, all the sensible heat lost by the air is used for moisture vaporization and converted to latent heat of the water vapor in the drying air. Therefore, the state point of the air can be considered to move along adiabatic saturation lines on the psychrometric chart. In the simplified psychrometric chart in Figure 2, the ambient air at dry-bulb temperature t_a and dew-point temperature t_{dp} is heated to drying air temperature t_d, where it has a relative humidity ϕ_1. As the air passes through the grain, its sensible heat provides the latent heat of vaporization of the water. When the air exits from the grain, its temperature has dropped to t_e, and its relative humidity has increased to ϕ_2. The moisture gained by each pound of drying air is the difference in humidity ratio $W_2 - W_1$. If the air has sufficient contact time with the grain, the value for ϕ_2 will be the equilibrium relative humidity of the grain at that moisture and temperature t_e.

The following example illustrates a grain drying process. Shelled corn at 20% moisture is to be dried with air heated to 140°F from an ambient temperature of 68°F and a dew point of 50°F. The humidity ratio of the ambient air remains constant at 0.0076 lb/lb

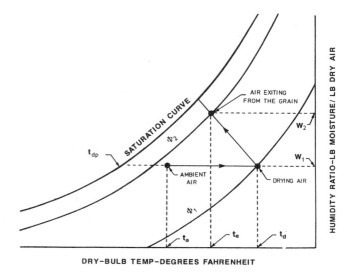

Fig. 2   Drying Process Diagrammed on Psychrometric Chart Showing Adiabatic Evaporation of Moisture from Grain

as it is heated. However, the relative humidity drops from 53% at t_a to 6% at t_d. The amount of heat added to each pound of dry air can be estimated as 18.5 Btu, the difference in enthalpy between the two state points (43 Btu/lb minus 24.5 Btu/lb). As the heated air passes through the grain and picks up moisture, it will reach 91% rh and its temperature drops to approximately 82°F.

The exhaust air relative humidity, 91%, can be estimated by reading the equilibrium relative humidity from the curve for shelled corn shown in Figure 1. Enter the curve with a grain moisture of 20% and a temperature of 82°F, which is slightly above the wet-bulb temperature of the drying air. As the air passes through the grain, the humidity ratio increases from 0.0076 lb/lb to 0.0215 lb/lb, and, therefore, each pound of air carries with it 0.0139 lbs (0.0215 lb/lb minus 0.0076 lb/lb) of water from the grain.

If the grain had been near 23% moisture, the exhaust air may have been nearly saturated, and the exhaust air temperature would be the wet-bulb temperature of the drying air, 79°F. The exhaust air would have a humidity ratio of 0.0220 lb/lb, and each pound of air would carry 0.0144 lbs of water from the grain.

For a grain moisture of 16%, the equilibrium moisture content curve from Figure 1 can be used to estimate the exhaust air relative humidity as 80%. For this grain moisture, the exhaust air would have a temperature of 86°F and a humidity ratio of 0.0210 lb/lb. Each pound of air would carry 0.0132 lbs of water from the grain.

Thin Layer Drying

A *thin layer* of grain is a layer of grain no more than several kernels deep. The ratio of grain to air is such that there is only a small change in temperature and relative humidity of the drying air when it exits the grain. The maximum rate (dM/dt) at which a thin layer of a granular hygroscopic material (such as grain) transfers moisture to or from air can be approximated by the following equation (Hukill 1947):

$$dM/dt = -C(p_g - p_a) \tag{11}$$

where

C = constant, representing vapor conductivity of kernel and surrounding air film
p_g = partial pressure of water vapor in grain
p_a = partial pressure of water vapor in drying air

If $p_g > p_a$, drying takes place. If $p_g = p_a$, moisture equilibrium exists and no moisture transfer occurs. If $p_g < p_a$, wetting occurs. The assumptions of a linear relationship between water vapor pressure and equilibrium relative humidity, and between equilibrium relative humidity and moisture content, over the range in which drying occurs, lead to the following equation:

$$dM/dt = -k(M - M_e) \tag{12}$$

where

M = moisture content (dry basis) of material at time t
M_e = equilibrium moisture content (dry basis) of material in reference to drying air
k = constant, dependent on material

The solution to this differential equation is:

$$(M - M_e)/(M_o - M_e) = \exp(-kt) \tag{13}$$

where

M_o = moisture content, dry basis, when $t = 0$

In later work (Hukill and Schmidt 1960, Troeger and Hukill 1971), Hukill recognized that Equation (13) did not describe

the drying rate of grain adequately. Misra and Brooker (1980) identified the following model as more promising for shelled corn:

$$(M - M_e)/(M_o - M_e) = \exp(-Kt^N) \tag{14}$$

They give an equation for K, which is a function of drying air temperature and velocity, and another equation for N as a function of drying air relative humidity and initial grain moisture. Their equations are valid for drying air temperatures from 36 to 160°F, drying air relative humidities of 3 to 83%, drying air velocities of 5 to 459 ft/min, and initial moistures of 18 to 60% (dry basis).

Li and Morey (1984) also fit their data to Equation (14) and found that within the limits of drying airflow rates and air relative humidities used, K and N can be expressed as functions of air temperature and initial grain moisture only. Their equations for K and N apply to air temperatures ranging from 80 to 240°F, initial grain moistures in the range of 23 to 36% dry basis, air velocities of 20 to 100 cfm/bu, and air relative humidities of 5 to 40%.

Other forms of the thin layer drying equation have also been proposed. Thompson *et al.* (1968) fit data for shelled corn to the following equation, which is applicable in the range of 140 to 300°F:

$$t = A \ln MR + B(\ln MR)^2 \tag{15}$$

where

A = $-1.86178 + 0.00488\,T$
B = $427.3740 \exp(-0.03301\,T)$
MR = $(M - M_e)/(M_o - M_e)$
t = time, h
T = temperature, °F

Martins and Stroshine (1987) describe the effects of hybrid and damage on the thin layer drying rate and give values for constants A and B in Equation (15) for several hybrids and damage levels.

Equations (13) through (15) do not describe the usual drying process where grain is in a deep bed and where drying air changes condition but does not necessarily reach moisture equilibrium with the grain. Those models, which are formulated using thin layer drying equations such as these, are summarized in the section Deep Bed Drying.

Results of thin layer drying tests for other grains have also been reported. Data are available for wheat (Watson and Bhargava 1974, Sokhansanj *et al.* 1984, Bruce and Sykes 1983), soybeans (Hukill and Schmidt 1960, Overhults *et al.* 1973, Sabbah *et al.* 1976), barley (O'Callaghan *et al.* 1971, Sokhansanj *et al.* 1984, Bruce 1985), sorghum (Hukill and Schmidt 1960, Paulsen and Thompson 1973), rice (Agrawal and Singh 1977, Noomhorm and Verma 1986, Banaszak and Siebenmorgen 1989), sunflower (Syarief *et al.* 1984, Li *et al.* 1987), canola (Sokhansanj *et al.* 1984), oats (Hukill and Schmidt 1960), and lentil seeds (Tang *et al.* 1989). Sokhansanj and Bruce (1987) developed more rigorous thin layer drying equations based on simultaneous heat and mass transfer through a single kernel and demonstrated that such a model accurately predicts the temperature and moisture content of the grain throughout the drying process.

Airflow Resistance

Data on resistance of grain to airflow are used for a variety of design calculations such as selecting fans, determining optimum depths for drying bins, predicting airflow paths in bins with aeration ducts, and determining the practical limitations on airflow caused by fan power requirements. For a given fan and drier or bin, airflow resistance can change with the type of grain being dried, the depth of grain, and the amount of fine material in the grain. In many grain-drying applications, such as when air is forced through a grain bin that has a uniform grain depth and a full perforated floor, airflow is one-dimensional and the pressure drop per unit depth of grain can be assumed to be constant. The data on pressure drop per foot versus airflow in cfm per ft^2 for a number of grains and seeds were determined by Shedd (1953) and summarized by plotting them on logarithmic axes. These curves are commonly referred to as *Shedd's curves* and are included in ASAE *Standard* D272.1 (ASAE 1990). They can also be calculated from the following equation (ASAE 1990):

$$\frac{\Delta p}{L} = \frac{aQ^2}{\ln(1 + bQ)} \qquad (16)$$

Table 5 summarizes the constants for Equation (16) for some of the more common grains. Constants for grass seeds and some vegetables are included in ASAE (1990).

Equation (16) gives the airflow resistance for clean, dry grain when the bin is loaded by allowing the grain to flow into the bin through a chute from a relatively low height. In the case of shelled corn, predictions of Equation (16) can be corrected for fine material using the following equation (ASAE 1990, Haque *et al.* 1978), which is valid for airflows at 15 to 40 cfm/ft^2 and on fine material fractions of 0.0 to 0.2.

$$(\Delta p/L)_{ctd} = (\Delta p/L)_{\text{Eq. (16)}} [1 + (14.5566 - 0.1342Q)W_{fm}] \qquad (17)$$

where

$$
\begin{aligned}
\Delta p &= \text{pressure, in. of water} \\
L &= \text{bed depth, ft} \\
Q &= \text{airflow rate, cfm/ft}^2 \\
W_{fm} &= \text{decimal fraction of fine material, by weight. Fine material} \\
&\quad \text{is defined as broken grain and other matter that will} \\
&\quad \text{pass through a 12/64-in. round hole sieve.}
\end{aligned}
$$

Grama *et al.* (1984) reported the effect of fine material particle size distribution on resistance. They also report the effect of the increased resistance from fines on fan power requirements. Kumar and Muir (1986) report the effects of fines in wheat and barley.

Bulk density can have a significant effect on airflow resistance. For moderate heights of 14 to 24 ft, drop height does not affect

bulk density in bins filled with a spout (Chang *et al.* 1986). Bern *et al.* (1982) reported that auger stirring can decrease the bulk density of bins filled with a grain spreader but has no effect or increases bulk density in bins filled by gravity. The magnitudes of the increases in bulk density caused by grain spreaders have been reported by Stephens and Foster (1976b, 1978) and Chang *et al.* (1983). In such cases, the pressure drop per foot is routinely corrected by multiplying the value from Equation (16) by a packing factor. A value of 1.5 is frequently used. If the bulk density is known or can be accurately estimated, the pressure drop for shelled corn can be estimated using the following equation (ASAE 1990):

$$\frac{\Delta p}{L} = X_1 + X_2 \frac{(\rho_b/\rho_k)^2 Q}{[1-(\rho_b/\rho_k)]^3} + X_3 \frac{(\rho_\beta/\rho_k) Q^2}{[1-(\rho_b/\rho_k)]^3} \qquad (18)$$

where

$$
\begin{aligned}
\Delta p &= \text{pressure drop, in. of water} \\
L &= \text{bed depth, ft} \\
\rho_b &= \text{corn bulk density, lb/ft}^3 \\
\rho_k &= \text{corn kernel density, lb/ft}^3 \\
Q &= \text{airflow, cfm/ft}^2 \\
X_1, X_2, X_3 &= \text{constants (Table 6)}
\end{aligned}
$$

Kumar and Muir (1986) report on the effect of filling method on the airflow resistance of wheat and barley. Jayas *et al.* (1987) showed that the resistance of canola to airflow in a horizontal direction was 0.5 to 0.7 times the resistance to airflow for the vertical direction.

Moisture content also affects airflow resistance. Its effect may, in part, be caused by its influence on bulk density. Shedd's curves include a footnote recommending that for loose fill of clean grain, airflow resistance should be multiplied by 0.80 if the grain is in equilibrium with air at relative humidities greater than 85% (ASAE 1990). At 70 °F, this corresponds to a moisture of 18% or more for shelled corn (Figure 1). Haque *et al.* (1982) give equations that correct for the effects of moisture content of shelled corn, sorghum, and wheat.

When the flow lines are parallel and airflow is linear (as is the case in a drying bin with a full perforated floor), calculation of the airflow is a straightforward application of Equation (16). For a given fan attached to a particular bin filled to a uniform depth with grain, the operating point of the fan can be determined as follows. A curve is plotted showing the total static pressure in the bin plenum versus airflow to the bin. Airflow rate is calculated by dividing the total air volume supplied to the plenum by the cross-sectional area of the bin. Using Equation (16), the pressure drop per unit depth can be calculated and multiplied by the total depth of grain in the bin to give total static pressure in the plenum. The fan curve showing air delivery volume versus static pressure can be plotted on the same axes. The intersection of the curves is the operating point for the fan. These calculations can also be done on a computer, and the point of intersection of the curves can be determined using appropriate numerical methods. McKenzie *et al.* (1980) and Hellevang (1983) summarize airflow resistances for various bin and fan combinations in tabular and graphical form.

Table 5   Constants for Airflow Resistance [Equation (16)]
(ASAE 1990)

Material	Value of a	Value of b, ft^2/cfm	Range of Q, cfm/ft^2
Barley	6.76×10^{-4}	6.71×10^{-2}	1.1 – 40
Canola (rapeseed)[1]	16.48×10^{-4}	3.69×10^{-2}	4.66– 52
Ear corn	3.29×10^{-4}	1.65	10 – 69
Lentils[2]	17.15×10^{-4}	18.7×10^{-2}	0.55–116
Oats	7.62×10^{-4}	7.06×10^{-2}	1.1 – 40
Peanuts	1.20×10^{-4}	5.64×10^{-1}	6 – 60
Popcorn, White	6.92×10^{-4}	5.99×10^{-2}	1.1 – 40
Popcorn, Yellow	5.63×10^{-4}	8.94×10^{-2}	1.1 – 40
Rice, Rough	8.12×10^{-4}	6.71×10^{-2}	1.1 – 30
Rice, Long brown	6.48×10^{-4}	3.93×10^{-2}	1.1 – 32
Rice, Long milled	6.89×10^{-4}	4.24×10^{-2}	1.1 – 32
Rice, Medium brown	1.10×10^{-3}	5.53×10^{-2}	1.1 – 32
Rice, Medium milled	9.16×10^{-4}	5.38×10^{-2}	1.1 – 32
Shelled corn	6.54×10^{-4}	1.54×10^{-1}	1.1 – 60
Shelled corn, Low airflow	3.09×10^{-4}	4.34×10^{-2}	0.05– 4
Sorghum	6.70×10^{-4}	4.09×10^{-2}	1.1 – 40
Soybeans	3.22×10^{-4}	8.13×10^{-2}	1.1 – 60
Sunflower, Confectionary	3.48×10^{-4}	9.19×10^{-2}	1.1 – 35
Sunflower, Oil	7.87×10^{-4}	1.20×10^{-1}	5 –112
Wheat	8.53×10^{-4}	4.45×10^{-2}	1.1 – 40
Wheat, Low airflow	2.66×10^{-4}	1.38×10^{-2}	0.05– 4

Notes: Units for a are in. of water·min^2/ft^2.
[1] Jayas and Sokhansanj (1989).
[2] Sokhansanj *et al.* (1990).

Table 6   Constants for Equation (18) for Airflow Resistance in Shelled Corn as Function of Bulk Density, Kernel Density, and Airflow

Airflow Range, cfm/ft^2	X_1	X_2	X_3
$5.3 < Q < 26.3$	-0.0012	5.53×10^{-4}	1.62×10^{-5}
$26.3 < Q < 52.3$	-0.013	6.94×10^{-4}	1.39×10^{-5}
$52.3 < Q \leqslant 117$	-0.094	10.2×10^{-4}	1.23×10^{-5}

Range of applicability: corn bulk density of 45.7 to 49.9 lb/ft^3 and airflow of 5.3 to 117 cfm/ft^2.
Source: ASAE (1990), Bern and Charity (1975).

Sokhansanj and Woodward (1991) developed a design procedure for use on personal computers to select fans for near-ambient drying of grain.

In cases where airflow is nonlinear, as in conical piles or systems with air ducts, computation is complex (Miketinac and Sokhansanj 1985). Numerical methods for predicting airflow patterns have been developed and applied to bins aerated with ducts (Brooker 1969, Segerlind 1982, Khompos *et al.* 1984), conical-shaped piles (Jindal and Thompson 1972), and bins in which porosity varies within the bed (Lai 1980). Lai's study applies to bins in which filling methods have created differences in bulk density within the bin or where fine material is unevenly distributed.

Analysis of Deep Bed Drying

Ability to predict the rate at which grain dries in a given type of drier operating in specific weather conditions with a specified airflow and air temperature can assist designers in developing driers for maximum efficiency. It can also guide operators in finding the optimum way to operate their particular driers for given weather conditions. Computer simulations have helped researchers understand the mechanisms and processes involved in drying. This section references two methods for making hand calculations of drying time and then summarizes computer simulations that have been developed.

Two relatively simple prediction equations can be solved on a hand calculator. Hukill (1947, 1972) developed a widely known and used method which predicts the moisture distribution in a bed of grain during drying. A graphical presentation of one of the equations, which further simplifies calculations, is available. Hukill's method is summarized by Brooker *et al.* (1974), who give an example calculation for shelled corn drying. Barre *et al.* (1971) made further adaptations of Hukill's method, and Foster (1986) gives a historical perspective on its development and utility. Brooker *et al.* also present a technique called the *heat balance equation*, which equates the heat available in the air for drying with the amount of heat needed to evaporate the desired amount of water from the grain. Both of the above methods take into account airflow, drying air temperature and relative humidity, exit air conditions, grain moisture, and the amount of grain to be dried.

Thompson *et al.* (1968) considered a deep bed of grain to be a series of thin layers of grain stacked one on top of another. Algebraic heat and mass balances were applied to each layer, with the resulting exit air conditions of one layer becoming the input conditions of the next layer. Thompson *et al.* (1969) used the model to predict concurrent flow, crossflow, and counterflow drying of shelled corn. Paulsen and Thompson (1973) used it to evaluate crossflow drying of sorghum. Stevens and Thompson (1976) and Pierce and Thompson (1981) used the model to make recommendations about optimum design of high-temperature grain driers.

Bakker-Arkema *et al.* (1978) used simultaneous heat and mass transfer equations in a series of coupled partial differential equations to describe deep bed drying. The equations, solved using a finite difference technique, predict grain temperature, grain moisture content, and air temperature and humidity ratio. Bakker-Arkema *et al.* (1979, 1984) give solutions for in-bin, batch, continuous crossflow, and continuous concurrent flow driers. Morey *et al.* (1976) used the model to evaluate energy requirements for drying. Morey and Li (1984) and Bakker-Arkema *et al.* (1983) demonstrated the effect of thin layer drying rate on the model predictions.

Other researchers have also developed simulation models. Hamdy and Barre (1970) developed a hybrid computer simulation of high-temperature drying and compared predictions to data on moisture and temperature measurements in a laboratory cross-flow drier. Bridges *et al.* (1980) used the Thompson model for simulation of batch-in-bin drying. Morey *et al.* (1978) and Parry

(1985) review many of the mathematical models used for high-temperature grain drying.

Computer simulations have also been developed for low-temperature and solar drying. Some of these models have been referenced in the section Fungal Growth and Mycotoxins under Prevention of Deterioration. Thompson (1972) developed a model that was later used by Pierce and Thompson (1979) to make recommendations on airflow in solar grain drying and by Pierce (1986) to evaluate natural air drying. Sabbah *et al.* (1979) used the logarithmic model of Barre *et al.* (1971) for simulation of solar grain drying. Bridges *et al.* (1984) used a model to evaluate the economics of stirring devices in in-bin drying systems. Morey *et al.* (1979), Fraser and Muir (1981), Bowden *et al.* (1983), and Smith and Bailey (1983) have also modeled low-temperature drying. Sharp (1982) reviewed low-temperature drying simulation models.

Aeration of Grain

Aeration involves forcing small amounts of air through the stored grain to maintain a uniform temperature. Prior to the development of this concept, grain was turned by moving it from one storage bin to another. Foster (1986) credits Hukill *et al.* (1953) with developing the concept. As mentioned in the previous sections Fungal Growth and Mycotoxins and Insect Infestation, lowering of the grain temperature during winter in temperate climates can reduce the rate of deterioration from molds and insects. Aeration can also prevent temperature gradients from developing within the grain mass. Such gradients can cause moisture migration, which results in unacceptably high moistures in certain portions of the bin.

Aeration is used to cool stored grain in the fall. A typical practice is to aerate the grain when the difference between grain temperature and average daily outside temperature exceeds 10 °F. In the United States, grain is usually not warmed in the spring unless it is to be stored past July. Foster and McKenzie (1979) and McKenzie (1980) give practical recommendations for aeration of grain. Airflow rates of 0.025 to 0.5 cfm per bushel are normally used. Air is usually distributed through the bottom of the bin using ducts. Duct spacing and fan selection are related to bin size and shape and the airflow rate. Foster and Tuite (1982) give an overview of the topic and include information and charts used for design of such systems. Peterson (1982) gives recommendations for duct spacing in flat storages.

Several computer simulations have been developed to study the effects of heat buildup from microbial activity with and without aeration (Thompson 1972, Brooker and Duggal 1982, Metzger and Muir 1983, Lissik 1986). Aldis and Foster (1977) and Schultz *et al.* (1984) studied the effect of aeration on grain moisture changes.

DRYING SPECIFIC CROPS

Hay

Forage crops can be harvested, dried, and stored as hay, or harvested and stored under anaerobic conditions as grass silage. Hay quality can be judged by its color, leafiness, and appearance. Laboratory tests and feeding trials give a more detailed picture of hay quality. The traditional method of making hay is to mow the forage and allow it to field cure or dry in the swath and windrow. More recently, harvesting at higher moistures with subsequent artificial drying has become feasible.

Basic principles of hay drying and storage are covered by Hall (1980) and Schuler *et al.* (1986). Forage must be harvested in the proper stage of maturity to attain maximum feeding value. Leaf loss from alfalfa is high when it is handled at moistures below 39%. Therefore, if it is baled at 40% moisture and dried artificially to the recommended storage moisture of 20% (Schuler *et al.*

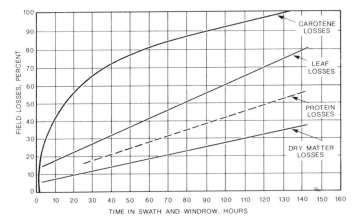

Fig. 3   Relation between Time in Swath and Windrow and Field Losses of Leaves, Dry Matter, Protein, and Carotene

(1986), a significantly higher feeding value can be achieved. Both Schuler *et al.* (1986) and Hall (1980) give sketches for batch and in-storage hay driers. They recommend airflows of 15 to 20 cfm per square foot of mow floor area or 200 to 500 cfm per ton.

Dehydrated alfalfa meal supplies provitamin A (carotene), vitamin E, xanthophylls (poultry pigmenting factors), vitamin K, vitamin C, and B vitamins. Figure 3 shows losses from field drying of hay found in tests conducted by Shepherd (1954). The rapid loss of carotene immediately after the forage is cut indicates the need for rapid transport to the dehydrator when alfalfa meal with high vitamin content is desired.

Several factors influence retention of vitamins during storage, such as the starting plant material, dehydration conditions, addition of stabilizers, and storage conditions. Lowering the temperature reduces the loss rate. Inert gas atmosphere in storage also reduces losses (Hoffman *et al.* 1945). According to Shepherd (1954), blanching of fresh alfalfa before drying does not alter the storage stability of carotene. Table 7 shows Shepherd's results on the effect of prolonged heating at 212°F. The alfalfa was dried after 45 min; heating beyond this time represented excessive exposure to this temperature. Carotene retention in the intact meal at 150°F storage temperature was considered a measure of storage stability. Normal storage moisture is 8 to 9%. Thompson *et al.* (1960) summarize the effects of over- and underdrying on carotene stability.

Drying and handling of large round bales has been researched. These bales may weigh from 850 to 1500 lbs and are handled individually with forklifts. Verma and Nelson (1983) studied storage of large round bales and found that dry matter loss was the primary component of the total storage losses. Bales stored so that they were protected from the weather had lower losses of dry matter than bales exposed to the weather. They were also higher in total protein.

Table 7   Effect of Heating Chopped Alfalfa on Carotene Loss during Subsequent Storage of Meal
(Thompson *et al.* 1960)

Hours in Oven at 212°F	Initial Carotene, ppm	Carotene Retained 7 Days at 149°F, %
0.75	229	37
1	228	37
2	197	37
3	176	28
4	149	21
5.5	112	18
7.5	86	15

Note: Alfalfa is fresh frozen from Ryer Island, CA.

Jones *et al.* (1985) found significant dry matter losses in large round bales of mature fescue hay. Harrison (1985) found that addition of sulfur dioxide at the rate of 1% of dry matter had little effect on dry matter loss and nutrient contents for a mixture of alfalfa and brome grass. However, bales protected with plastic bags did have significantly lower dry matter losses. Jones *et al.* (1985) found that bales of mature fescue hay stored inside and bales treated with ammonia had less dry matter loss and higher in-vitro dry matter digestibility. Henry *et al.* (1977) and Frisby *et al.* (1985) developed and tested solar driers for large round bales.

Grain

The physiological factors involved in drying and storing grain are different from those of forages. Grain is the end product of plant growth, and most physiological activity within the grain or seed is approaching a low level when harvested. With forage, the biological activity within the plant is at or near its peak at the time of harvest.

Deterioration of grain harvested at moistures above those safe for storage and the chemical preservation of grain is addressed in the section Fungal Growth and Mycotoxins. Preservation by ensiling or airtight storage is addressed in the section Oxygen and Carbon Dioxide.

Corn

Shelled field corn is used primarily as livestock feed, but some is used by milling or processing industries for manufacturing starch, corn oil, and other products. Little information is available on the relationship between the drying method and the feed value of corn. Market grade, as established by the USDA Agricultural Marketing Service, is the primary criterion for determining corn value. Tests by Cabell *et al.* (1958) indicated that shelled corn with a moisture content of 29 to 32% can be dried without loss of protein nutritive value by air with temperatures as high as 240°F, provided the airflow rate is approximately 110 cfm/bu.

The market grade of dry corn is affected more by the amount of fine material than by other grading factors. Fine material is defined as the broken grain and other material that passes through a 12/64-in. round-holed sieve. The brittleness imparted to the corn during drying or the physical damage done to wet corn causes it to break each time it is handled. The propensity of corn to break during subsequent handling, called *breakage susceptibility*, can be measured with a multiple-impact device called the *Stein breakage tester*. Stephens and Foster (1976a) demonstrated that corn breakage in the tester was correlated with damage during handling. Watson *et al.* (1986) give a standardized procedure for using the Stein breakage tester, and Watson and Herum (1986) describe and compare other devices developed for measurement of breakage susceptibility. They concluded that a device developed by Singh and Finner (1983) offers great potential for testing of grain for breakage susceptibility in commercial situations.

Paulsen *et al.* (1983) found significant variations in breakage susceptibility among hybrids. Corn dried with air at high temperatures (140°F) was two to six times more susceptible to breakage than corn dried at near-ambient temperatures. Gustafson and Morey (1979) found that delayed cooling (maintaining the corn at or near its temperature at the end of drying for 6 to 12 h) reduced breakage susceptibility and improved the test weight.

In a study of combination drying, Gustafson *et al.* (1978) found that combination drying (high-temperature drying to 18% followed by low-temperature drying to 16.6% moisture or below) significantly decreased the increase in breakage susceptibility normally caused by high-temperature drying. Morey and Cloud (1980) summarize principles, energy savings, and advantages of combination drying.

Both drying temperature and corn hybrid can affect the quality of shelled corn for specific end uses. Brekke *et al.* (1973) found that drying at temperatures above 140 °F decreased the quality of the corn for dry milling. Peplinski *et al.* (1982) found that optimum dry milling quality could be achieved by harvesting corn at moistures below 25%, minimizing machinery-induced damage to the kernels, and drying at air temperatures below 180 °F. Paulsen and Hill (1985) found that the yield of flaking grits from dry milling of corn was significantly greater for corn that had a high test weight and relatively low breakage susceptibility. Weller *et al.* (1987) found that variety affected wet milling quality. At drying temperatures between 120 and 160 °F, protein conformational changes occurred and decreased the ethanol soluble protein. Hybrids differ in resistance to storage mold (Tuite and Foster 1979), thin layer drying rate, and dry milling quality (Stroshine *et al.* 1986). Watson (1987) gives an extensive summary of measurement and maintenance of quality of corn, and Foster (1975) summarizes approaches to reducing damage during harvesting, handling, and drying.

Cotton

The lint moisture content for best results in ginning cotton appears to be 5 to 7%, with an optimum moisture content of 6% (Franks and Shaw 1962). Cotton, like grain, is hygroscopic and should be dried just prior to ginning. The wide variation in incoming moisture content usually requires different amounts of drying for each load. Rapid changes in the amount of drying required can best be handled by using a multipath drying tower in which the cotton is exposed for various lengths of time (2 to 10 s) at temperatures not exceeding 350 °F. The air-to-cotton ratio can range from 40 to 100 cfm/lb of cotton (Franks and Shaw 1962). Laird and Baker (1983) found that substantial amounts of heat could be reclaimed and used for drying in commercial cotton gin plants. Equilibrium moisture content data for newly harvested cotton fibers is given by Griffin (1974). Anthony (1982) studied moisture gain of cotton bales during storage.

Cottonseed removed from the fibers is also dried. The germination of cottonseed is unimpaired by drying if the internal cottonseed temperatures do not exceed 140 °F (Franks and Shaw 1962). This temperature is not exceeded in the tower drier described previously. However, the moisture content of the seed can be above the recommended level of 12% following the multipath tower drying. Drying seed in a triple-pass drum at 250 to 300 °F with an exposure time of 4 min, followed by cooling, reduces moisture content, inhibiting formation of free fatty acids, and improving germination over undried seed. Anthony (1983) dried cottonseed in a microwave drier. The cottonseed would not germinate, but its oil properties were not harmed as long as lower temperatures were used. The drying rate was increased by reducing pressure below atmospheric. Rayburn *et al.* (1978) studied preservation of high-moisture cottonseed with propionic acid.

Peanuts

Peanuts in the shell normally have a moisture content of about 50% at the time of digging. Allowing peanuts to dry on the vines in the windrows for a few days removes much of this water. However, peanuts normally contain 20 to 30% moisture when removed from the vines, and some artificial drying of peanuts in the shell is necessary. Drying should begin within 6 h after harvesting to keep peanuts from self-heating. The maximum temperature and rate of drying must be controlled to maintain quality. High temperatures result in off-flavor or bitterness. Overly rapid drying without high temperatures results in blandness or inability to develop flavor when roasted (Bailey *et al.* 1954). High temperatures and rapid or excessive drying also cause the skin to slip easily and the kernels to become brittle. These conditions result in damage in the shelling operation and can be avoided if the moisture removal rate does not exceed 0.5% per hour. Because of these limitations, continuous-flow drying is not usually recommended.

Young (1984) found energy savings up to 26% when comparing recirculating driers with conventional peanut driers. Smith and Davidson (1982) and Smith *et al.* (1985) address the aeration of peanuts during warehouse storage.

Rice

Of all grains, rice is possibly the most difficult to process without quality loss. Rice containing more than 12.5% moisture cannot be stored safely for long periods, yet the recommended harvest moisture content for best milling and germination ranges from 20 to 26% (Kramer 1951). If harvested at this moisture content, drying must begin promptly to prevent heat-related damage which can result in "stack-burn," a yellowing of the kernel. To prevent excessive internal fissuring, which results in broken kernels during milling, multiple-pass drying is usually necessary (Calderwood and Webb 1971). Kunze and Calderwood (1980) summarize rice drying techniques.

Since the market demands polished whole kernels of rice, it is necessary to prevent damage in the form of fissures. Rapid moisture removal or addition can create moisture gradients within kernels. According to Kunze (1984), gradients can develop in the field on a humid night before harvest, in a hopper containing a mixture of rice kernels at varying moistures, and in certain types of driers. Banaszek and Siebenmorgen (1990) quantified the rate at which moisture adsorption reduces head rice yields. Velupillai and Verma (1986) report that drying at 200 °F followed by tempering in a sealed container for 24 h gave good kernel strength and head rice yields. They also found that storing the rice after drying for 3 weeks gave optimum grain quality. Bakker-Arkema *et al.* (1984) achieved good rice quality with concurrent flow drying of rice.

Soybean, Sunflower, and Edible Beans

Prolonged periods of extremely wet weather during the harvest season can make artificial drying of soybeans necessary. Like peanuts and other oilseeds, soybeans cannot be dried satisfactorily with the high-temperature, high-speed methods used for cereal grains. Because of the different seed structure, rapid drying splits the seed coat and reduces quality and storage life. Overhults *et al.* (1975) report a significant decrease in the quality of oil extracted from soybeans dried at temperatures above 160 °F. Soybeans have one of the slowest thin layer drying rates of commonly grown cereals and oilseeds (Bakker-Arkema *et al.* 1983). Therefore, they dry slower and require more energy when dried in continuous-flow driers.

Sunflower is a major crop in some areas of the United States. Hellevang (1987) recommends maximum drying temperatures of up to 200 °F for continuous-flow drying of oil sunflower and maximum drying temperatures of 180 °F for non-oil sunflowers, where scorching of the seed meat should be prevented. Schuler (1974) gives data on equilibrium moisture, airflow resistance, and specific heat of sunflower seeds. Because sunflower is about half the density of shelled corn, moisture can be removed more rapidly, and there is a tendency to overdry. This factor, along with accumulation of foreign material when drying, causes an increased fire hazard (Hellevang 1982). Schmidt and Backer (1980) attribute most of the problems encountered with storage of sunflower seed to improper drying or aeration.

Edible beans, a major crop in several states, should be dried with air at relative humidities above 40% to prevent stress cracking. Natural air or low-temperature drying is best (Hellevang 1987). If dried at high rates, seed coats may crack, and beans may split during subsequent handling. Broken beans can develop a bitter or undesirable flavor and spoil more easily during storage.

Wheat and Barley

In northern regions of the United States, wheat and barley may be harvested above safe storage moistures to prevent excessive field losses. Moilanen *et al.* (1973) recommended that hard red spring wheat be dried at temperatures below 160, 140, and 120 °F, respectively, for harvest moistures of 16, 20, and 24% wet basis. These data assumed airflows of 100 to 150 cfm/ft². For airflows of 50 cfm/ft², they recommended that the drying air temperature be reduced by 10 to 15 °F. In the case of barley used for malting, the seed must be able to germinate. Therefore, the maximum recommended drying air temperature is 110 to 120 °F (Hellevang 1987). Watson *et al.* (1962) studied the effects of harvest moisture and drying temperature on barley malting quality and recommended harvesting below 20% moisture. If wheat or barley is used for seed, the maximum recommended drying air temperature is 110 °F.

In regions where soft wheat is grown, it may be economical to harvest at 20 to 24% moisture to allow double cropping with soybeans; this allows wheat harvest to begin 5 to 7 days earlier than normal and increases the yield of the soybeans (Swearingen 1979). In areas where double cropping is feasible, soft wheat can be dried using low-temperature solar drying or ambient drying with intermittent fan operation (Barrett *et al.* 1981). High-speed and continuous-flow systems with reduced drying air temperatures can also be used (Parsons *et al.* 1979). Kirleis *et al.* (1982) harvested soft red winter wheat at moistures of 25% or below and dried with air temperatures of 150 °F or below without adverse effects on milling or cookie baking quality.

In high-temperature continuous-flow driers, wheat or barley reduce airflow because they have a high airflow resistance. Bakker-Arkema *et al.* (1983) report that thin layer drying rates of barley and wheat are much faster than for corn. Barley dries more slowly than wheat, presumably because the kernels are larger. In their computer simulations of a concurrent-flow drier, wet bushel capacity for wheat was about 80% of the capacity for shelled corn when 4.7 percentage points of moisture were removed. The drying capacity difference was probably caused by a decrease in airflow.

Tobacco (Curing)

Tobacco leaves normally have a moisture content of about 85% at harvest. The major methods of tobacco drying are *air curing* and *flue curing* (Johnson *et al.* 1960). For air curing, whole plants are cut and allowed to wilt in the field until the leaves reach about 70% moisture. Plants are then hung in open barns, where temperatures range from 60 to 90 °F and humidities from 65 to 70%. The curing period is 28 to 56 days (Jefries 1940). The desired end product is a tan leaf. Overdrying at low temperatures results in green color and low sugar content; overdrying at high temperatures results in yellow color (Walton and Henson 1971). Both conditions are undesirable, since the normal chemical changes are arrested prematurely. Subsequent drying at optimum rates can reverse some damage. Underdrying at all temperatures results in undesirable dark color and damage from mold and bacterial growth (Walton *et al.* 1973).

Flue curing uses artificial heat. The leaves are harvested and hung in closed barns where temperatures are increased gradually during the curing period. Normally, 3 days of drying at temperatures of 90 to 120 °F bring about yellowing. For the next 2 days, temperatures of 120 to 140 °F are used for leaf drying; then, stems are dried using temperatures of 170 °F for 1 to 2 days. A bright yellow to orange color is desirable in flue-cured or bright-leaf tobacco.

REFERENCES

Agrawal, Y.C. and R.P. Singh. 1977. Thin-layer drying studies on short grain rice. Paper 77-3531. American Society of Agricultural Engineers, St. Joseph, MI.

Aldis, D.F. and G.H. Foster. 1977. Moisture changes in grain from exposure to ambient air. Paper 77-3524. American Society of Agricultural Engineers, St. Joseph, MI.

Anthony, W.S. 1982. Moisture gain and resilient forces of cotton bales during equilibration. *Transactions of* ASAE 25(4):1066-70.

Anthony, W.S. 1983. Vacuum microwave drying of cotton: Effect on cottonseed. *Transactions of* ASAE 26(1):275-78.

ASAE. 1990. *Standard* 245.4, Moisture relationships of grains. In *Standards Engineering Practice and Data* adopted by the American Society of Agricultural Engineers. American Society of Agricultural Engineers, St. Joseph, MI.

Bailey, W.K., T.A. Pickett, and J.G. Futral. 1954. Rapid curing adversely affects quality of peanuts. *Peanut Journal and Nut World* 33(8):37-39.

Bakker-Arkema, F.W., R.C. Brook, and L.E. Lerew. 1978. Cereal grain drying. In *Advances in cereal science and technology*, Y. Pomeranz, ed. American Association of Cereal Chemists, St. Paul, MN, 1-90.

Bakker-Arkema, F.W., S. Fosdick, and J. Naylor. 1979. Testing of commercial crossflow dryers. Paper 79-3521. American Society of Agricultural Engineers, St. Joseph, MI.

Bakker-Arkema, F.W., C. Fontana, R.C. Brook, and C.W. Westlake. 1984. Concurrent flow rice drying. *Drying Technology* 1(2):171-91.

Bakker-Arkema, F.W., C. Fontana, G.L. Fedewa, and I.P. Schisler. 1983. A comparison of drying rates of different grains. Paper 83-3009. American Society of Agricultural Engineers, St. Joseph, MI.

Barak, A.V. and W.E. Burkholder. 1985. A versatile and effective trap for detecting and monitoring stored-product coleoptera. *Agricultural Ecosystems and Environment* 12:207-18.

Barak, A.V. and P.K. Harein. 1982. Trap detection of stored-grain insects in farm-stored shelled corn. *Journal of Economic Entomology* 75(1):108-11.

Barre, H.J., G.R. Baughman, and M.Y. Hamdy. 1971. Application of the logarithmic model to cross-flow deep-bed grain drying. *Transactions of* ASAE 14(6):1061-64.

Barrett, Jr., J.R., M.R. Okos, and J.B. Stevens. 1981. Simulation of low temperature wheat drying. *Transactions of* ASAE 24(4):1042-46.

Beeson, W.M. and T.W. Perry. 1958. The comparative feeding value of high moisture corn and low moisture corn with different feed additives for fattening beef cattle. *Journal of Animal Science* 17(2):368-73.

Bern, C.J. and L.F. Charity. 1975. Airflow resistance characteristics of corn as influenced by bulk density. Paper 75-3510. American Society of Agricultural Engineers, St. Joseph, MI.

Bern, C.J., M.E. Anderson, W.F. Wilcke, and C.R. Hurburgh. 1982. Auger-stirring wet and dry corn—Airflow resistance and bulk density effects. *Transactions of* ASAE 25(1):217-20.

Bloome, P.D. and G.W. Cuperus. 1984. Aeration for management of stored grain insects in wheat. Paper 84-3517. American Society of Agricultural Engineers, St. Joseph, MI.

Bowden, P.J., W.J. Lamond, and E.A. Smith. 1983. Simulation of near-ambient grain drying. I, Comparison of simulations with experimental results. *Journal of Agricultural Engineering Research* 28:279-300.

Brekke, O.L., E.L. Griffin, Jr., and G.C. Shove. 1973. Dry milling of corn artificially dried at various temperatures. *Transactions of* ASAE 16(4):761-65.

Bridges, T.C., D.G. Colliver, G.M. White, and O.J. Loewer. 1984. A computer aid for evaluation of on-farm stir drying systems. *Transactions of* ASAE 27(5):1549-55.

Bridges, T.C., I.J. Ross, G.M. White, and O.J. Loewer. 1980. Determination of optimum drying depth for batch-in bin corn drying systems. *Transactions of* ASAE 23(1):228-33.

Brook, R.C. 1987. Modelling grain spoilage during near-ambient grain drying. *Divisional Note* DN. 1388, AFRC Institute of Engineering Research, Wrest Park, Silsoe, Bedford, MK45 4HS, England, 20 p.

Brooker, D.B. 1969. Computing air pressure and velocity distribution when air flows through a porous medium and nonlinear velocity-pressure relationships exist. *Transactions of* ASAE 12(1):118-20.

Brooker, D.B. and A.K. Duggal. 1982. Allowable storage time of corn as affected by heat buildup, natural convection and aeration. *Transactions of* ASAE 25(3):806-10.

Brooker, D.B., F.W. Bakker-Arkema, and C.W. Hall. 1974. *Drying cereal grains.* AVI Publishing Co., Westport, CT.

Bruce, D.M. 1985. Exposed-layer barley drying: Three models fitted to new data up to 150 °C. *Journal of Agricultural Engineering Research* 32:337-47.

Bruce, D.M. and R.A. Sykes. 1983. Apparatus for determining mass transfer coefficients at high temperatures for exposed particulate crops, with initial results for wheat and hops. *Journal of Agricultural Engineering Research* 28:385-400.

Brusewitz, G.H. 1987. Corn moisture variability during drying, mixing and storage. *Journal of Agricultural Engineering Research* 38:281-88.

Burkholder, W.E. and M. Ma. 1985. Pheromones for monitoring and control of stored-product insects. *Annual Review of Entomology* 30:257-72.

Butler, L.A. 1983. The history and background of NIR. *Cereal Foods World* 28(4):238-40.

Cabell, C.A., R.E. Davis, and R.A. Saul. 1958. Relation of drying air temperature, time and air flow rate to the nutritive value of field-shelled corn. *Technical Progress Report* 1957-58 ARS 44-41, USDA, Washington, D.C.

Calderwood, D.L. and B.D. Webb. 1971. Effect of the method of dryer operation on performance and on the milling and cooking characteristics of rice. *Transactions of* ASAE 14(1):142-46.

Chang, C.S., H.H. Converse, and F.S. Lai. 1986. Technical Notes: Distribution of fines and bulk density of corn as affected by choke-flow, spout-flow, and drop-height. *Transactions of* ASAE 29(2):618-20.

Chang, C.S., H.H. Converse, and C.R. Martin. 1983. Bulk properties of grain as affected by self-propelled rotational type grain spreaders. *Transactions of* ASAE 26(5):1543-50.

Christensen, C.M., ed. 1982. *Storage of cereal grains and their products.* American Association of Cereal Chemists, St. Paul, MN.

Christensen, C.M. and R.A. Meronuck. 1986. *Quality maintenance in stored grains and seeds.* University of Minnesota Press, Minneapolis, MN.

Christensen, C.M. and D.R. Sauer. 1982. Microflora. In *Storage of cereal grains and their products,* C.M. Christenson, ed.

Christensen, C.M., B.S. Miller, and J.A. Johnston. 1982. Moisture and its measurement. In *Storage of cereal grains and their products,* C.M. Christenson, ed.

Colliver, D.G., R.M. Peart, R.C. Brook, and J.R. Barrett, Jr. 1983. Energy usage for low temperature grain drying with optimized management. *Transactions of* ASAE 26(2):594-600.

Cooper, P.J. 1983. NIR analysis for process control. *Cereal Foods World* 28(4):241-45.

Cotton, R.T. and D.A. Wilbur. 1982. Insects. In *Storage of cereal grains and their products,* C.M. Christenson, ed. 281-318.

Danziger, M.T., M.P. Steinberg, and A.I. Nelson. 1973. Effect of CO_2, moisture content, and sorbate on safe storage of wet corn. *Transactions of* ASAE 16(4):679-82.

Dunkel, F.V. 1985. Underground and earth sheltered food storage: Historical, geographic, and economic considerations. *Underground Space* 9:310-15.

Eckhoff, S.R., J. Tuite, G.H. Foster, R.A. Anderson, and M.R. Okos. 1984. Inhibition of microbial growth during ambient air corn drying using sulfur dioxide. *Transactions of* ASAE 27(3):907-14.

Epperly, D.R., R.T. Noyes, G.W. Cuperus, and B.L. Clary. 1987. Control stored grain insects by grain temperature management. Paper 87-6035. American Society of Agricultural Engineers, St. Joseph, MI.

Foster, G.H. 1975. Causes and cures of physical damage to corn. In *Corn quality in world markets,* L.D. Hill, ed. Interstate Printers and Publishers, Inc., Danville, IL.

Foster, G.H. 1986. William V. Hukill, a pioneer in crop drying and storage. *Drying Technology* 4(3):461-71.

Foster, G.H. and B.A. McKenzie. 1979. Managing grain for year-round storage. AE-90. Cooperative Extension Service, Purdue University, West Lafayette, IN.

Foster, G.H. and J.F. Tuite. 1982. Aeration and stored grain management. In *Storage of cereal grains and their products,* C.M. Christenson, ed., 117-43.

Franks, G.N. and C.S. Shaw. 1962. Multipath drying for controlling moisture in cotton. ARS 42-69, USDA, Washington, D.C.

Frazer, B.M. and W.E. Muir. 1981. Airflow requirements for drying grain with ambient and solar-heated air in Canada. *Transactions of* ASAE 24(1):208-10.

Frisby, J.C., J.T. Everett, and R.M. George. 1985. A solar dryer for large, round alfalfa bales. *Applied Engineering in Agriculture* 1(2):50-52.

Grama, S.N., C.J. Bern, and C.R. Hurburgh, Jr. 1984. Airflow resistance of moistures of shelled corn and fines. *Transactions of* ASAE 27(1):268-72.

Griffin, A.C., Jr. 1974. The equilibrium moisture content of newly harvested cotton fibers. *Transactions of* ASAE 17(2):327-28.

Gunasekaran, S. 1986. Optimal energy management in grain drying. *Critical Reviews in Food Science and Nutrition* 25(1):1-48.

Gustafson, R.J. and R.V. Morey. 1979. Study of factors affecting quality changes during high-temperature drying. *Transactions of* ASAE 22(4):926-32.

Gustafson, R.J., R.V. Morey, C.M. Christensen, and R.A. Meronuck. 1978. Quality changes during high-low temperature drying. *Transactions of* ASAE 21(1):162-69.

Hall, C.W. 1980. *Drying and storage of agricultural crops.* AVI Publishing Company, Westport, CT.

Hall, G.E., L.D. Hill, E.E. Hatfield, and A.H. Jenson. 1974. Propionic-acetic acid for high-moisture preservation. *Transactions of* ASAE 17(2):379-82, 387.

Hamdy, M.Y. and H.J. Barre. 1970. Analysis and hybrid simulation of deep-bed drying of grain. *Transactions of* ASAE 13(6):752.

Haque, E., Y.N. Ahmed, and C.W. Deyoe. 1982. Static pressure drop in a fixed bed of grain as affected by grain moisture content. *Transactions of* ASAE 25(4):1095-98.

Haque, E., G.H. Foster, D.S. Chung, and F.S. Lai. 1978. Static pressure drop across a bed of corn mixed with fines. *Transactions of* ASAE 21(5):997-1000.

Harein, P.K. 1982. Chemical control alternatives for stored-grain insects. In *Storage of cereal grains and their products,* C.M. Christensen, ed. 319-62.

Harris, K.L. and F.J. Bauer. 1982. Rodents. In *Storage of cereal grains and their products,* C.M. Christensen, ed. 363-405.

Harrison, H.P. 1985. Preservation of large round bales at high moisture. *Transactions of* ASAE 28(3):675-79, 686.

Hellevang, K.J. 1982. Crop dryer fires while drying sunflower. Paper 82-3563. American Society of Agricultural Engineers, St. Joseph, MI.

Hellevang, K.J. 1983. Natural air/low temperature crop drying. Bulletin 35. Cooperative Extension Service, North Dakota State University, Fargo, ND.

Hellevang, K.J. 1987. Grain drying. Publication AE-701. Cooperative Extension Service, North Dakota State University, Fargo, ND.

Henry, Z.A., B.L. Bledsoe, and D.D. Eller. 1977. Drying of large hay packages with solar heated air. Paper 77-3001. American Society of Agricultural Engineers, St. Joseph, MI.

Hertung, D.C. and E.E. Drury. 1974. Antifungal activity of volatile fatty acids on grains. *Cereal Chemistry* 51(1):74-83.

Hoffman, E.J., G.F. Lum, and A.L. Pitman. 1945. Retention of carotene in alfalfa stored in atmospheres of low oxygen content. *Journal of Agricultural Research* 71:361-73.

Hukill, W.V. 1947. Basic principles in drying corn and grain sorghum. *Agricultural Engineering* 28(8):335-38, 340.

Hukill, W.V. 1953. Grain cooling by air. *Agricultural Engineering* 34(7):456-58.

Hukill, W.V. 1972. Grain drying. In *Storage of cereal grains and their products,* C.M. Christensen, ed. 481-508.

Hukill, W.V. and J.L. Schmidt. 1960. Drying rate of fully exposed grain kernels. *Transactions of* ASAE 3(2): 71-77, 80.

Hurburgh, C.R., T.E. Hazen, and C.J. Bern. 1985. Corn moisture measurement accuracy. *Transactions of* ASAE 28(2):634-40.

Hurburgh, C.R., L.N. Paynter, S.G. Schmitt, and C.J. Bern. 1986. Performance of farm-type moisture meters. *Transactions of* ASAE 29(4):1118-23.

Hyde, M.B. and N.J. Burrell. 1982. Controlled atmosphere storage. In *Storage of cereal grains and their products,* C.M. Christensen, ed. 443-78.

Jay, E. 1980. Methods of applying carbon dioxide for insect control in stored grain. Science and Education Administation, Advances in Agricultural Technology, Southern Series, AAT-S-13, Agricultural Research (Southern Region), SEA, USDA, P.O. Box 53326, New Orleans, LA 70153.

Jayas, D.S. and S. Sokhansanj. 1989. Design data on the airflow resistance to canola (rapeseed). *Transactions of* ASAE 32(1):295-96.

Jayas, D.S., S. Sokhansanj, E.B. Moysey, and E.M. Barber. 1987. The effect of airflow direction on the resistance of canola (rapeseed) to airflow. *Canadian Agricultural Engineering* 29(2):189-92.

Jefries, R.N. 1940. The effect of temperature and relative humidity during curing upon the quality of white burley tobacco. Bulletin No. 407, Kentucky Agricultural Experiment Station, Lexington, KY.

Jindal, V.K. and T.L. Thompson. 1972. Air pressure patterns and flow paths in two-dimensional triangular-shaped piles of sorghum using forced convection. *Transactions of* ASAE 15(4):737-44.

Johnson, W.H., W.H. Henson, Jr., F.J. Hassler, and R.W. Watkins. 1960. Bulk curing of bright-leaf tobacco. *Agricultural Engineering* 41(8):511-15, 517.

Jones, A.L., R.E. Morrow, W.G. Hires, G.B. Garner, and J.E. Williams. 1985. Quality evaluation of large round bales treated with sodium diacetate or anhyrdous ammonia. *Transactions of* ASAE 28(4):1043-45.

Khompos, V., L.J. Segerlind, and R.C. Brook. 1984. Pressure patterns in cylindrical grain storages. Paper 84-3011. American Society of Agricultural Engineers, St. Joseph, MI.

Kirleis, A.W., T.L. Housley, A.M. Emam, F.L. Patterson, and M.R. Okos. 1982. Effect of preripe harvest and artificial drying on the milling and baking quality of soft red winter wheat. *Crop Science* 22:871-76.

Kramer, H.A. 1951. Engineering aspects of rice drying. *Agricultural Engineering* 32(1):44-45, 50.

Kumar, A. and W.E. Muir. 1986. Airflow resistance of wheat and barley affected by airflow direction, filling method and dockage. *Transactions of* ASAE 29(5):1423-26.

Kunze, O.R. 1984. Physical properties of rice related to drying the grain. *Drying Technology* 2(3):369-87.

Kunze, O.R. and D.L. Calderwood. 1980. Systems for drying of rice. In *Drying and storage of agriculture crops*, C.W. Hall. AVI Publishing Company, Inc., Westport, CT.

Lai, F.S. 1980. Three dimensional flow of air through nonuniform grain beds. *Transactions of* ASAE 23(3):729-34.

Laird, W. and R.V. Baker. 1983. Heat recapture for cotton gin drying systems. *Transactions of* ASAE 26(3):912-17.

Lapp, H.M., F.J. Madrid, and L.B. Smith. 1986. A continuous thermal treatment to eradicate insects from stored wheat. Paper 86-3008. American Society of Agricultural Engineers, St. Joseph, MI.

Li, H. and R.V. Morey. 1984. Thin-layer drying of yellow dent corn. *Transactions of* ASAE 27(2):581-85.

Li, Y., R.V. Morey, and M. Afinrud. 1987. Thin-layer drying rates of oilseed sunflower. *Transactions of* ASAE 30(4):1172-75, 1180.

Lissik, E.A. 1986. A model for the removal of heat in respiring grains. Paper 86-6509. American Society of Agricultural Engineers, St. Joseph, MI.

Locklair, E.E., L.G. Veasey, and M. Samfield. 1957. Equilibrium desorption of water vapor on tobacco. *Journal of Agricultural and Food Chemistry* 5:294-98.

Martin, C.R., Z. Czuchajowska, and Y. Pomeranz. 1986. Aquagram standard deviations of moisture in mixtures of wet and dry corn. *Cereal Chemistry* 63(5):442-45.

Martins, J. and R.L. Stroshine. 1987. Difference in drying efficiencies among corn hybrids dried in a high-temperature column-batch dryer. Paper 87-6559. American Society of Agricultural Engineers, St. Joseph, MI.

McKenzie, B.A. 1980. Managing dry grain in storage. AED-20, Midwest Plan Service, Iowa State University, Ames, IA.

McKenzie, B.A., G.H. Foster, and S.S. DeForest. 1980. Fan sizing and application for bin drying/cooling of grain. AE-106, Cooperative Extension Service, Purdue University, West Lafayette, IN.

McMillan, W.W., D.M. Wilson, and N.W. Widstrom. 1985. Aflatoxin contamination of preharvest corn in Georgia—A six-year study of insect damage and visible *Aspergillus flavus. Journal of Environmental Quality* 14:200-202.

Metzger, J.F. and W.E. Muir. 1983. Computer model of two-dimensional conduction and forced convection in stored grain. *Canadian Agricultural Engineering* 25:119-25.

Miketinac, M.J. and S. Sokhansanj. 1985. Velocity-pressure distribution in grain bins—Brooker model. *International Journal of Applied Numerical Analysis in Engineering* 21:1067-75.

Misra, M.K. and D.B. Brooker. 1980. Thin-layer drying and rewetting equations for shelled yellow corn. *Transactions of* ASAE 23(5):1254-60.

Moilanen, C.W., R.T. Schuler, and E.R. Miller. 1973. Effect on wheat quality of air flow and temperatures in mechanical dryers. *North Dakota Farm Research* 30(6):15-19.

Morey, R.V. and H.A. Cloud. 1980. Combination high-speed, natural-air corn drying. M-163, Agricultural Extension Service, University of Minnesota, St. Paul, MN.

Morey, R.V. and H. Li. 1984. Thin-layer equation effects on deep-bed drying prediction. *Transactions of* ASAE 27(6):1924-28.

Morey, R.V., H.A. Cloud, and D.J. Hansen. 1981. Ambient air wheat drying. *Transactions of* ASAE 24(5):1312-16.

Morey, R.V., H.A. Cloud, and W.E. Lueschen. 1976. Practices for the efficient utilization of energy from drying corn. *Transactions of* ASAE 19(1):151-55.

Morey, R.V., H.A. Cloud, R.J. Gustafson, and D.W. Peterson. 1979. Management of ambient air drying systems. *Transactions of* ASAE 22(6):1418-25.

Nofsinger, G.W. 1982. The trickle ammonia process—An update. Grain Conditioning Conference Proceedings, Agricultural Engineering Department, University of Illinois, Champaign-Urbana, IL.

Nofsinger, G.W., R.J. Bothast, and R.A. Anderson. 1979. Field trials using extenders for ambient-conditioning high-moisture corn. *Transactions of* ASAE 22(5):1208-13.

Noomhorm, A. and L.R. Verma. 1986. Generalized single-layer rice drying models. *Transactions of* ASAE 29(2):587-91.

O'Callaghan, J.R., D.J. Menzies, and P.H. Bailey. 1971. Digital simulation of agricultural dryer performance. *Journal of Agricultural Engineering Research* 16:223-44.

Overhults, D.D., G.M. White, M.E. Hamilton, and I.J. Ross. 1973. Drying soybeans with heated air. *Transactions of* ASAE 16(1):112-13.

Overhults, D.G., G.M. White, M.E. Hamilton, I.J. Ross, and J.D. Fox. 1975. Effect of heated air drying on soybean oil quality. *Transactions of* ASAE 16(1):112-13.

Parker, P.E., G.R. Bauwin, and H.L. Ryan. 1982. Sampling, inspection, and grading of grain. In *Storage of cereal grains and their products*, C.M. Christensen, ed. 1-35.

Parry, J.L. 1985. Mathematical modelling and computer simulation of heat and mass transfer in agricultural grain drying: A review. *Journal of Agricultural Engineering Research* 32:1-29.

Parsons, S.D., B.A. McKenzie, and J.R. Barrett, Jr. 1979. Harvesting and drying high-moisture wheat. In *Double cropping winter wheat and soybeans in Indiana*. ID 96, Cooperative Extension Service, Purdue University, West Lafayette, IN.

Paulsen, M.R. and L.D. Hill. 1985. Corn quality factors affecting dry milling performance. *Journal of Agricultural Engineering Research* 31:255-63.

Paulsen, M.R. and T.L. Thompson. 1973. Drying analysis of grain sorghum. *Transactions of* ASAE 16(3):537-40.

Paulsen, M.R., L.D. Hill, D.G. White, and G.F. Sprague. 1983. Breakage susceptibility of corn-belt genotypes. *Transactions of* ASAE 26(6):1830-36.

Peplinski, A.J., R.A. Anderson, and O.L. Brekke. 1982. Corn dry milling as influenced by harvest and drying conditions. *Transactions of* ASAE 25(4):1114-17.

Peplinski, A.J., O.L. Brekke, R.J. Bothast, and L.T. Black. 1978. High moisture corn—An extended preservation trial with ammonia. *Transactions of* ASAE 21(4): 773-76, 781.

Peterson, W.H. 1982. Design principles for grain aeration in flat storages. Illinois Farm Electrification Council Fact Sheet No. 9. University of Illinois, Agricultural Engineering Department, Urbana, IL.

Pfost, H.B., S.G. Mauer, D.S. Chung, and G.A. Milliken. 1976. Summarizing and reporting equilibrium moisture data for grains. Paper 76-3520. American Society of Agricultural Engineers, St. Joseph, MI.

Pierce, R.O. 1986. Economic consideration for natural air corn drying in Nebraska. *Transactions of* ASAE 29(4):1131-35.

Pierce, R.O. and T.L. Thompson. 1979. Solar grain drying in the North Central Region—Simulation results. *Transactions of* ASAE 15(1):178-87.

Pierce, R.O. and T.L. Thompson. 1981. Energy use and performance related to crossflow dryer design. *Transactions of* ASAE 24 (1):216-20.

Rayburn, S.T., A.C. Griffin, Jr., and M.E. Whitten. 1978. Storing cottonseed with propionic acid. *Transactions of* ASAE 21(5):990-92.

Ripp, B.E., ed. 1984. Controlled atmosphere and fumigation in grain storages. Proceedings of an International Symposium, Practical Aspects of Controlled Atmosphere and Fumigation in Grain Storages, in Perth, Western Australia. Elsevier Science Publishing Company, New York.

Sabbah, M.A., H.M. Keener, and G.E. Meyer. 1979. Simulation of solar drying of shelled corn using the logarithmic model. *Transactions of* ASAE 22(3):637-43.

Sabbah, M.A., G.E. Meyer, H.M. Keener, and W.L. Roller. 1976. Reversed-air drying for fixed bed of soybean seed. Paper 76-3023. American Society of Agricultural Engineers, St. Joseph, MI.

Sauer, D.B., and R. Burroughs. 1974. Efficacy of various chemicals as grain mold inhibitors. *Transactions of ASAE* 17(3):557-59.

Saul, R.A. and J.L. Steele. 1966. Why damaged shelled corn costs more to dry. *Agricultural Engineering* 47:326-29, 337.

Schmidt, B.J. and L.F. Backer. 1980. Results of a sunflower storage monitoring program in North Dakota. Paper 80-6033, American Society of Agricultural Engineers, St. Joseph, MI.

Schuler, R.T. 1974. Drying related properties of sunflower seeds. Paper 74-3534. American Society of Agricultural Engineers, St. Joseph, MI.

Schuler, R.T., B.J. Holmes, R.J. Straub, and D.A. Rohweder. 1986. Hay Drying. Publication A3380. Cooperative Extension Service, University of Wisconsin, Madison, WI.

Schultz, L.J., M.L. Stone, and P.D. Bloome. 1984. A comparison of simulation techniques for wheat aeration. Paper 84-3012. American Society of Agricultural Engineers, St. Joseph, MI.

Segerlind, L.J. 1982. Solving the nonlinear airflow equation. Paper 82-3017. American Society of Agricultural Engineers, St. Joseph, MI.

Seitz, L.M., D.B. Sauer, and H.E. Mohr. 1982a. Storage of high-moisture corn: Fungal growth and dry matter loss. *Cereal Chemistry* 59(2):100-105.

Seitz, L.M., D.B. Sauer, H.E. Mohr, and D.F. Aldis. 1982b. Fungal growth and dry matter loss during bin storage of high-moisture corn. *Cereal Chemistry* 59(1):9-14.

Sellam, M.A. and C.M. Christensen. 1976. Temperature differences, moisture transfer and spoilage in stored corn. *Feedstuffs* 48(36):28, 33.

Singh, S.S. and M.F. Finner. 1983. A centrifugal impacter for damage susceptibility evaluation of shelled corn. *Transactions of ASAE* 26(6):1858-63.

Sharp, J.R. 1982. A review of low-temperature drying simulation models. *Journal of Agricultural Engineering Research* 27(3):169-90.

Shaw, C.S. and G.N. Franks. 1962. Cottonseed drying and storage at cotton gins. *Technical Bulletin* 1262, USDA, ARS, Washington, D.C.

Shedd, C.K. 1953. Resistance of grains and seeds to air flow. *Agricultural Engineering* 34(9):616-18.

Shejbal, J., ed. 1980. *Controlled atmosphere storage of grains.* Elsevier Science Publishing Company, New York.

Shepherd, J.B. 1954. Experiments in harvesting and preserving alfalfa for dairy cattle feed. *Technical Bulletin* 1079, USDA, Washington, D.C.

Smith, E.A. and P.H. Bailey. 1983. Simulation of near-ambient grain drying. II, Control strategies for drying barley in Northern Britain. *Journal of Agricultural Engineering Research* 28:301-17.

Smith, E.A. and S. Sokhansanj. 1990a. Moisture transport due to natural convection in grain stores. *Journal of Agricultural Engineering Research* 47:23-34.

Smith, E.A. and S. Sokhansanj. 1990b. Natural convection and temperature of stored products—A theoretical analysis. *Canadian Agricultural Engineering* 32(1):91-97.

Smith, J.S., Jr. and J.I. Davidson, Jr. 1982. Psychrometrics and kernel moisture content as related to peanut storage. *Transactions of ASAE* 25(1):231-36.

Smith, J.S., Jr., J.I. Davidson, Jr., T.H. Sanders, and R.J. Cole. 1985. Storage environment in a mechanically ventilated peanut warehouse. *Transactions of ASAE* 28(4):1248-52.

Sokhansanj, S. and D.M. Bruce. 1987. A conduction model to predict grain drying simulation. *Transactions of ASAE* 30(4):1181-84.

Sokhansanj, S. and S.O. Nelson. 1988a. Dependence of dielectric properties of whole-grain wheat on bulk density. *Journal Agricultural Engineering Research* 39:173-79.

Sokhansanj, S. and S.O. Nelson. 1988b. Transient dielectric properties of wheat associated with non-equilibrium kernel moisture conditions. *Transactions of ASAE* 31(4):1251-54.

Sokhansanj, S. and G.E. Woodward. 1990. Computer assisted fan selection for natural grain drying—A teaching and extension tool. *Applied Engineering in Agriculture* 6(6):782-84.

Sokhansanj, S., D. Singh, and J.D. Wasserman. 1984. Drying characteristics of wheat, barley and canola subjected to repetitive wetting and drying cycles. *Transactions of ASAE* 27(3):903-906, 914.

Sokhansanj, S., W. Zhijie, D.S. Jayas, and T. Kameoka. 1986. Equilibrium relative humidity moisture content of rapeseed from 5 to 25 °C. *Transactions of ASAE* 29(3):837-39.

Sokhansanj, S., A.A. Falacinski, F.W. Sosulski, D.S. Jayas, and J. Tang. 1990. Resistance of bulk lentils to airflow. *Transactions of ASAE* 33(4):1281-85.

Steele, J.L. 1967. Deterioration of damaged shelled corn as measured by carbon dioxide production. Unpublished PhD Thesis. Department of Agricultural Engineering, Iowa State University, Ames, IA.

Steele, J.L., R.A. Saul, and W.V. Hukill. 1969. Deterioration of shelled corn as measured by carbon dioxide production. *Transactions of ASAE* 12(5):685-89.

Stephens, G.R. and T.L. Thompson. 1976. Improving crossflow grain dryer design using simulation. *Transactions of ASAE* 19(4):778-81.

Stephens, L.B. and G.H. Foster. 1976a. Breakage tester predicts handling damage in corn. ARS-NC-49. ARS, USDA, Washington, D.C.

Stephens, L.E. and G.H. Foster. 1976b. Grain bulk properties as affected by mechanical grain spreaders. *Transactions of ASAE* 19(2):354-58.

Stephens, L.E. and G.H. Foster. 1978. Bulk properties of wheat and grain sorghum as affected by a mechanical grain spreader. *Transactions of ASAE* 21(6):1217-18.

Storey, C.L., R.D. Speirs, and L.S. Henderson. 1979. Insect control in farm-stored grain. *Farmers Bulletin* 2269. USDA-SEA (Available for sale from Superintendent of Documents, U.S. Government Printing Office, Washington, D.C. 20402).

Stroshine, R.L., A.W. Kirleis, J.F. Tuite, L.F. Bauman, and A. Emam. 1986. Differences in grain quality among selected corn hybrids. *Cereal Foods World* 31(4):311-16.

Stroshine, R.L., J. Tuite, G.H. Foster, and K. Baker. 1984. Self-study guide for grain drying and storage. Department of Agricultural Engineering, Purdue University, West Lafayette, IN.

Swearingen, M.L. 1979. A practical guide to no-till double cropping. In *Double cropping winter wheat and soybeans in Indiana.* ID 96. Cooperative Extension Service, Purdue University, West Lafayette, IN.

Syarief, A.M., R.V. Morey, and R.J. Gustafson. 1984. Thin-layer drying rates of sunflower seed. *Transactions of ASAE* 27(1):195-200.

Tang, J., S. Sokhansanj, and F.W. Sosulski. 1989. Thin-layer drying of lentil, Paper 89-6607. American Society of Agricultural Engineers, St. Joseph, MI.

Thompson, C.R., E.M. Bickoff, G.R. VanAtta, G.O. Kohler, J. Guggolz, and A.L. Livingston. 1960. Carotene stability in alfalfa as affected by laboratory- and industrial-scale processing. *Technical Bulletin* 1232, ARS, USDA, Washington, D.C.

Thompson, T.L. 1972. Temporary storage of high-moisture shelled corn using continuous aeration. *Transactions of ASAE* 15(2):333-37.

Thompson, T.L., G.H. Foster, and R.M. Peart. 1969. Comparison of concurrent-flow, crossflow and counterflow grain drying methods. *Marketing Research Report* 841. USDA-ARS, Washington, D.C.

Thompson, T.L., R.M. Peart, and G.H. Foster. 1968. Mathematical simulation of corn drying—A new model. *Transactions of ASAE* 11(4):582-86.

Thorpe, G.R. 1982. Moisture diffusion through bulk grain subjected to a temperature gradient. *Journal of Stored Products Research* 18:9-12.

Troeger, J.M. and W.V. Hukill. 1971. Mathematical description of the drying rate of fully exposed corn. *Transactions of ASAE* 14(6):1153-56, 1162.

Tuite, J. and G.H. Foster. 1963. Effect of artificial drying on the hygroscopic properties of corn. *Cereal Chemistry* 40:630-37.

Tuite, J. and G.H. Foster. 1979. Control of storage diseases of grain. *Annual Review of Phytopathology* 17:343.

Tuite, J., G.H. Foster, S.R. Eckhoff, and O.L. Shotwell. 1986. Sulfur dioxide treatment to extend corn drying time (note). *Cereal Chemistry* 63(5):462-64.

U.S. Department of Agriculture. 1971. Oven methods for determining moisture content of grain and related agricultural commodities, Chapter 12. *Equipment manual*, GR Instruction 916-6, Consumer and Marketing Service, Grain Division, Hyattville, MD.

Velupillai, L. and L.R. Verma. 1986. Drying and tempering effects on par-boiled rice quality. *Transactions of ASAE* 29(1):312-19.

Verma, L.R. and B.D. Nelson. 1983. Changes in round bales during storage. *Transactions of ASAE* 26(2):328-32.

Walton, L.R. and W.H. Henson, Jr. 1971. Effect of environment during curing on the quality of burley tobacco: Effect of low humidity curing on support price. *Tobacco Science* 15:54-57.

Walton, L.R., W.H. Henson, Jr., and J.M. Bunn. 1973. Effect of environment during curing on the quality of burley tobacco: effect of high humidity curing on support price. *Tobacco Science* 17:25-27.

Watson, C.A. 1977. Near infrared reflectance spectro-photometric analysis of agricultural products. *Analytical Chemistry* 49(9):835A-40A.

Watson, C.A., O.J. Banasick, and G.L. Pratt. 1962. Effect of drying temperature on barley malting quality. *Brewers Digest* 37(7):44-48.

Watson, E.L. and V.K. Bhargava. 1974. Thin-layer drying studies on wheat. *Canadian Agricultural Engineering* 16(1):18-22.

Watson, S.A. 1987. Measurement and maintenance of quality. In *Corn: Chemistry and technology*, S.A. Watson and P.E. Ramstad, eds. American Association of Cereal Chemists, St. Paul, MN, 125-83.

Watson, S.A. and F.L. Herum. 1986. Comparison of eight devices for measuring breakage susceptibility of shelled corn. *Cereal Chemistry* 63(2):139-42.

Watson, S.A. and P.E. Ramstad. 1987. *Corn: Chemistry and technology.* American Association of Cereal Chemists, Inc., St. Paul, MN.

Watson, S.A., L.L. Darrah, and F.L. Herum. 1986. Measurement of corn breakage susceptibility with the Stein breakage tester: A collaborative study. *Cereal Foods World* 31(5):366-72.

Weller, C.L., M.R. Paulsen, and M.P. Steinberg. 1987. Varietal, harvest moisture and drying air temperature effects on quality factors affecting corn wet milling. Paper 87-6046. American Society of Agricultural Engineers, St. Joseph, MI.

Young, J.H. 1984. Energy conservation by partial recirculation of peanut drying air. *Transactions of* ASAE 27(3):928-34.

AIR CONTAMINANTS

AIR is composed of many gases. The gaseous components of clean, dry air near sea level are approximately 21% oxygen, 78% nitrogen, 1% argon, and 0.03% carbon dioxide. Also included are trace amounts of hydrogen, neon, krypton, helium, ozone, and xenon, in addition to varying amounts of water vapor and small quantities of microscopic and submicroscopic solid matter called *permanent atmospheric impurities.*

Air composition may be changed accidentally or deliberately. In sewers, sewage treatment plants, tunnels, and mines, the oxygen content of air can become so low that people cannot remain conscious or survive. Concentrations of people in confined spaces (theaters, survival shelters, submarines) require that carbon dioxide given off by normal respiratory functions be removed and replaced with oxygen. Pilots of high-altitude aircraft, breathing at greatly reduced pressure, require systems that increase oxygen concentration. Conversely, for divers working at extreme depths, it is common to increase the percentage of helium in the atmosphere and reduce nitrogen and sometimes oxygen concentrations.

At atmospheric pressure, oxygen concentrations of less than 12% or carbon dioxide concentrations greater than 5% are dangerous, even for short periods.

Lesser deviations from normal composition can be hazardous under prolonged exposures. Chapter 11 of the 1991 ASHRAE *Handbook—Applications* details acceptable variations in composition. Chapter 37 further details environmental health issues.

AIR CONTAMINANTS

Normal air contains varying amounts of foreign materials (*permanent atmospheric impurities*). These materials can arise from natural processes such as wind erosion, sea spray evaporation, and volcanic eruption. The concentrations in the air vary but are usually less than the concentrations caused by human activity.

Man-made contaminants are many and varied, originating from numerous areas of human activity. Electric power-generating plants, various modes of transportation, industrial processes, mining and smelting, construction, and agriculture generate large amounts of contaminants. Contaminants that present particular problems in the indoor environment include, among others, tobacco smoke, radon, and formaldehyde.

Air contaminants can be classified as follows:

- Particulate or gaseous
- Organic or inorganic
- Visible or invisible
- Submicroscopic, microscopic, or macroscopic
- Toxic or harmless
- Stable or unstable

The preparation of this chapter is assigned to TC 2.3, Gaseous Air Contaminants and Gas Contaminant Removal Equipment, in conjunction with TC 2.4, Particulate Air Contaminants and Particulate Contaminant Removal Equipment.

Loose classifications based on the phase of the suspended contaminant (solid, liquid, or gas) and method of formation of the contaminant are as follows:

- Dusts, fumes, and smokes that are solid particulate matter, although smoke often contains liquid particulates.
- Mists, fogs, and smokes that are suspended liquid particulates.
- Vapors and gases.

Dusts, Fumes, and Smokes

Dusts are solid particles projected into the air by natural forces such as wind, volcanic eruption, or earthquakes, or by mechanical processes including crushing, grinding, demolition, blasting, drilling, shoveling, screening, and sweeping. Some of these forces produce dusts by reducing larger masses, while others disperse materials that have already been reduced.

Particles are not dust unless they are smaller than about 100 μm. Dusts can be mineral, such as rock, metal, or clay; vegetable, such as grain, flour, wood, cotton, or pollen; or animal, including wool, hair, silk, feathers, and leather.

Fumes are solid particles formed by condensation of vapors of solid materials. Metallic fumes are generated from molten metals and usually occur as oxides because of the highly reactive nature of finely divided matter. Fumes can also be formed by sublimation, distillation, or chemical reaction. Such processes create airborne particles smaller than 1 μm. Fumes permitted to age may agglomerate into larger clusters.

Smokes are small solid and/or liquid particles produced by incomplete combustion of organic substances such as tobacco, wood, coal, oil, and other carbonaceous materials. The term smoke is applied to a mixture of solid, liquid, and gaseous products, although technical literature distinguishes between such components as soot or carbon particles, flyash, cinders, tarry matter, unburned gases, and gaseous combustion products. Smoke particles vary in size—the smallest being much less than 1 μm. The average is often in the range of 0.1 to 0.3 μm.

Environmental tobacco smoke consists of a suspension of small 0.01 to 1.0 μm (mass median diameter of 0.5 μm) liquid particles that form as the superheated vapors leaving the burning tobacco condense. Also produced are numerous gaseous contaminants including carbon monoxide.

Viruses range in size from 0.003 to 0.06 μm, although they usually occur in colonies or attached to other particles. Most *bacteria* range between 0.4 and 5 μm and are usually associated with large particles. *Fungus* spores are usually from 10 to 30 μm, while *pollen* grains are from 10 to 100 μm, with many common varieties in the 20 to 40 μm range. Airborne viruses, bacteria, pollen, and fungus spores are sometimes referred to as *bioaerosols.*

Mists and Fogs

Mists are small airborne droplets of materials that are ordinarily liquid at normal temperatures and pressure. They can be formed by atomizing, spraying, mixing, violent chemical reactions, evolution of gas from liquid, or escape as a dissolved gas when pressure is released. Small droplets expelled or atomized by sneezing constitute mists containing microorganisms that become air contaminants.

Fogs are fine airborne droplets usually formed by condensation of vapor. Fog nozzles are named for their ability to produce extra fine droplets, as compared with mists from ordinary spray devices. Many droplets in fogs or clouds are microscopic and submicroscopic and serve as a transition stage between larger mists and vapors.

The volatile nature of most liquids reduces the size of their airborne droplets from the mist to the fog range and eventually to the vapor phase, until the air becomes saturated with that liquid. If solid material is suspended or dissolved in the liquid droplet, it remains in the air as particulate contamination. For example, sea spray evaporates fairly rapidly, generating a large number of fine salt particles that remain suspended in the atmosphere.

Smog commonly refers to air pollution; it implies an air mixture of smoke particles, mists, and fog droplets of such concentration and composition as to impair visibility, in addition to being irritating or harmful. The composition varies among different locations and at different times. The term is often applied to haze caused by a sunlight-induced photochemical reaction involving the materials in automobile exhausts. Smog is often associated with temperature inversions in the atmosphere that prevent normal dispersion of contaminants.

Vapors and Gases

The terms gas and vapor are often used to describe a common state of a substance. Gas is normally used to describe any mixture, except atmospheric air, that exists in the gaseous state at normal atmospheric conditions. Examples are oxygen, helium, and nitrogen. Vapor is used to describe a substance in the gaseous state that can also exist as a liquid or solid at normal atmospheric conditions. Examples include gasoline, benzine, carbon tetrachloride, and water.

NATURE OF AIRBORNE CONTAMINANTS

Sizes of Airborne Particles

Figure 1 shows the sizes and characteristics of airborne solids and liquids. Particles less than 0.1 μm begin to behave similarly to gas molecules, traveling with Brownian movement and with no predictable or measurable settling velocity. Particles in the range from 0.1 to 1 μm have settling velocities that can be calculated but are so low that settling is usually negligible, since normal air currents counteract any settling. On a particle count basis, over 99% of the particles in a typical atmosphere are below 1 μm.

Particles in the 1 to 10 μm range settle in still air at constant and appreciable velocity. However, normal air currents keep them in suspension for appreciable periods.

Industrial hygienists are primarily concerned with particles less than 2 μm, since this is the range of sizes most likely to be retained in the lungs (Morrow 1964). Particles larger than 8 to 10 μm in diameter are separated and retained by the upper respiratory tract. Intermediate sizes are deposited mainly in the conducting airways of the lungs, from which they are rapidly cleared and swallowed or coughed out. About 50% or less of the particles in inhaled air settle in the respiratory tract.

Particles larger than 10 μm settle fairly rapidly and can be found suspended in air only near their source or under strong wind conditions. Exceptions are lint and other light fibrous materials, such as portions of certain weed seeds, which remain suspended longer. Most individual particles 10 μm or larger are visible to the naked

Table 1   Relation of Screen Mesh to Particle Size

U.S. Standard sieve mesh	400	325	200	140	100	60	35	18
Nominal sieve opening in μm	37	44	74	105	149	250	500	1000

eye under favorable conditions of lighting and contrast. Smaller particles are visible in high concentrations. Cigarette smoke (with an average particle size less than 0.5 μm) and clouds are common examples.

Direct fallout in the vicinity of the dispersing stack or flue and other nuisance problems of air pollution involve larger particles. Smaller particles, as well as mists, fogs, and fumes generated, remain in suspension longer. In this size range, meteorology and topography are more important than physical characteristics of the particles. Since settling velocities are small, the ability of the atmosphere to disperse these small particles depends largely on local weather conditions.

Comparison is often made to screen sizes used for grading useful industrial dusts and granular materials. Table 1 illustrates the relation of U.S. Standard sieve mesh to particle size in micrometres. Particles above 40 μm are the screen sizes, and those below are the subscreen or microscopic sizes.

Particle Size Distribution

The particle size distribution in any sample can be expressed as the percentage of the number of particles smaller than a specified size. The upper curve of Figure 2 shows these data plotted for typical atmospheric contamination (Whitby *et al.* 1955, 1957). The middle curve shows the percentage of the total projected area of the particles contributed by particles less than a specified size. The lower curve shows the percentage of the total particle weight contributed by those particles less than a given size.

The differences among values presented by the three curves should be noted. For example, particles 0.1 μm or less in diameter (but still above electron microscope minimum detection size of about 0.005 μm) make up 80% of the number of particles in the atmosphere but contribute only 1% of the mass. Also, the 0.1% of particles by number larger than 1 μm carry 70% of the total mass, which is the direct result of the mass of a spherical particle increasing as the cube of its diameter. Although most of the mass is contributed by intermediate and larger particles, about 80% of the contamination is supplied by particles less than 5 μm in diameter. Most of the staining effect results from particles less than 1 μm in diameter. Suspended particles in urban air are predominantly smaller than 1 μm (expressed as equivalent spheres of unit density) and have a distribution that is approximately lognormal.

SUSPENDED PARTICULATES

The total amount of suspended particulate matter in the atmosphere can influence the loading rate of air filters and their selection. The amount of soot that falls in U.S. cities ranges from 20 to 200 ton/mile^2 per month. Soot fall data indicate effectiveness of smoke abatement and proper combustion methods and serve as comparative indices of such control programs. These values have little significance to the ventilating and air-conditioning engineer, since they do not express any measure of the suspended material that must be cleaned from the ventilation air before it can be used.

In particle count, the number of suspended particulates is enormous. A room with heavy cigarette smoke has a total particle concentration of 30 × 10^6/ft^3; even clean air contains over 10^6 particles/ft^3. If smaller particles detectable by other means, such an electron microscope or condensation nuclei counter, are also included, the total particle count would be greater than the above concentrations by a factor of 10 to 100.

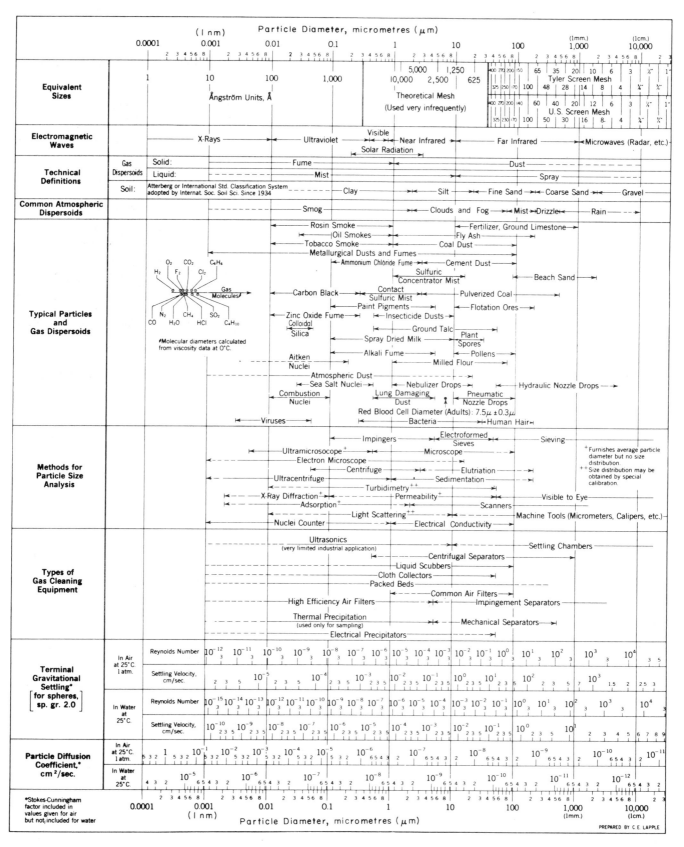

Fig. 1   Characteristics of Particles and Particle Dispersoids
(Courtesy of Stanford Research Institute)

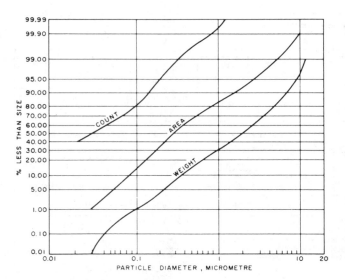

Count curve: Based on measurements by electron microscope.
Area curve: Calculated.
Weight curve: Solid section based on measurements by sedimentation.

Fig. 2   Particle Size Distribution of Atmospheric Dust
[Reproduced from Whitby *et al.* (1955) by permission.]

Extensive measurements have been made of outdoor pollution, but limited data have been gathered on indoor pollution not associated with specific industrial processes. Indoor levels are influenced by the number of people and their activities, building materials and construction, outside conditions, ventilation rate, and the air-conditioning and filtration system. For further information, see the section Indoor Air Quality, Spengler *et al.* (1982), and NRC (1981).

Measurement of Airborne Particles

The quantity of dust can be determined by particle count or total mass. Another indirect method measures optical density of the collected dust based on the projected area of the particles. Dust particles can be sized with graduated scales or optical comparisons, using a standard microscope. The lower limit for sizing with the light field microscope is approximately 0.9 μm, depending on the vision of the observer, the dust color, and the contrast available. This size can be reduced to about 0.4 μm by using oil immersion objective techniques. Dark field microscopic techniques reveal particles smaller than these, to a limit of approximately 0.1 μm. Smaller submicroscopic dusts can be sized and compared with the aid of an electron microscope. Other sizing techniques may take into account velocity of samplings in calibrated devices and actual settlement measurements in laboratory equipment. The electron microscope and various sampling instruments, such as the cascade impacter, have been successful in sizing particulates, including fogs and mists.

Instruments have been developed and test methods devised for continuous counting and sizing of airborne contaminants. These instruments use the light-scattering principle and, depending on the model, can detect particles down to 0.3 μm. Instruments using laser light sources can detect particles down to 0.1 μm (ASTM 1983). A condensation nucleus counter can count submicrometric particles to below 0.01 μm. These particles, present in great numbers in the atmosphere, serve as nuclei for condensation of water vapor (Scala 1963).

Each of the various methods of measuring particle size distribution gives a different value for the same size particle, since different properties are actually measured. For example, a microscopic technique may measure longest dimension, while impacter results are based on aerodynamic behavior (ACGIH 1983).

INDUSTRIAL AIR CONTAMINANTS

Many industrial processes produce air contaminants in the form of dusts, fumes, smokes, mists, vapors, and gases. Contaminants should be controlled at the source so that they are not dispersed through the factory nor allowed to increase to toxic concentration levels. Control methods are covered in Chapter 10 of the 1992 ASHRAE *Handbook—Systems and Equipment* and Chapters 25 and 27 of the 1991 *Applications* volume.

Zero concentration of all contaminants is not economically feasible. Absolute control of all contaminants cannot be maintained, and workers can assimilate small quantities of various toxic materials without injury. Industrial hygiene science is based on the fact that most air contaminants become toxic only if their concentration exceeds a maximum allowable limit for a specified period.

Mandatory U.S. standards promulgated by the Occupational Safety and Health Administration (OSHA) are published initially in the Federal Register and then in the Code of Federal Regulations (29 CFR 1910.1000), which is revised annually. Most of the standards promulgated initially by OSHA were those previously recommended by the American Conference of Governmental Industrial Hygienists (ACGIH) and the American Industrial Hygiene Association. The health effects on which these standards were based can be found in their publications (ACGIH 1987). Revisions of the initial OSHA standards are now recommended by the National Institute of Occupational Safety and Health (NIOSH) and are published in a series of individual criteria documents. For protection of nonindustrial occupants such as office workers, which may include sensitive persons, indoor concentrations should be kept below the ACGIH threshold limit values (see ASHRAE *Standard* 62-1989).

FLAMMABLE GASES AND VAPORS

The use of flammable materials is widespread. Flammable gases and vapors (NFPA 1984) can be found in sewage treatment plants, sewage and utility tunnels, dry-cleaning plants, automobile garages, and industrial finishing process plants. Adequate ventilation minimizes or prevents fires and explosions and is necessary, regardless of other precautions, such as elimination of the ignition sources, safe building construction, and the use of automatic alarm and extinguisher systems.

A flammable liquid's vapor pressure and volatility or rate of evaporation determines its ability to form an explosive mixture. These properties can be expressed by the flash point, which is the temperature to which a combustible liquid must be heated to produce a flash when a small flame is passed across the surface of the liquid. Depending on the test methods, either the open cup or closed cup flash point may be listed. The higher the flash point, the more safely the liquid can be handled. Liquids with flash points under 70°F should be regarded as highly flammable.

In practice, the air-vapor or air-gas mixture must be in the *explosive range* before it can be ignited. The explosive range is the range between the upper and lower explosive limits, expressed as percent by volume in air. Concentrations of material above the higher range or below the lower range will not explode. The range for many chemicals is found in National Fire Code bulletins published by the National Fire Protection Association (NFPA). Some representative limits of flammability are listed in Table 2.

Chapter 27 of the 1991 ASHRAE *Handbook—Applications* covers equipment for control of combustible materials. In designing ventilation systems to control flammable gases and vapors, the engineer must consider the following:

Table 2 Flammable Limits of Some Gases and Vapors

Gas or Vapor	Flash Point, °F	Flammable Limits, % by Volume	
		Lower	Upper
Acetone	0	2.6	12.8
Ammonia	Gas	16	25
Benzene (benzol)	12	1.3	7.1
n-Butane	Gas	1.9	8.5
Carbon disulfide	−22	1.3	44
Carbon monoxide	Gas	12.5	74
1,2-Dichloroethylene	43	9.7	12
Diethyl ether	−49	1.9	48
Ethyl alcohol	55	4.3	19
Ethylene	Gas	3.1	32
Gasoline	−45	1.4	7.6
Hydrogen	Gas	4.0	75
Hydrogen sulfide	Gas	4.3	45
Isopropyl alcohol	53	2.0	12
Methyl alcohol	52	7.3	36
Methyl ethyl ketone	21	1.8	10
Natural gas (variable)	Gas	3.8	17
Naphtha (benzine)	50	0.9	6.7
Propane	Gas	2.2	9.5
Toluene (toluol)	40	1.2	7.1
o-xylene	90	1.0	6.0

1. Most safety authorities and fire underwriters prefer to limit concentrations to 20 to 25% of the lower explosive limit of a material. The resulting safety factor of 4 or 5 allows latitude for imperfections in air distribution and variations of temperature or mixture, and guards against unpredictable or unrecognized sources of ignition. Operation at concentrations above the upper explosive limit should be resorted to only in rare instances. To reach the upper explosive limit, the flammable gas or vapor must pass through the active explosive range, in which any source of ignition causes an explosion. In addition, a drop in gas concentration due to unforeseen dilution or reduced evaporation rate results in operation in the dangerous explosive range.

2. In occupied places where ventilation is applied for proper health control, the danger of an explosion is minimized. In most instances, flammable gases and vapors are also toxic, and the maximum allowable concentrations are far below the lower explosive limit of the material. For example, proper ventilation for acetone vapors keeps the concentration below 1000 ppm. This is equivalent to 0.1% by mass. The lower explosive limit for acetone is 2.5% by volume.

3. Proper location of exhaust and supply ventilation equipment depends primarily on how the contaminant is given off and on other problems of the process, and secondarily on the relative density of flammable vapor.

The specific density of the explosive mixture is the same as that of air. Cross drafts, moving equipment, and temperature differentials may cause sufficient mixing to produce explosive concentrations and disperse these throughout the atmosphere, without regard to heavier-than-air vapors. Therefore, the engineer must either provide proper exhaust and supply air patterns to control the hazardous material, preferably at its source, or offset the effects of drafts, equipment movement, and convective forces by providing good distribution of exhaust and supply air for general dilution and exhaust. The intake duct should be positioned so that it does not bring in exhaust gases or emissions from ambient sources.

COMBUSTIBLE DUSTS

Many organic and some mineral dusts can produce dust explosions (Hartmann 1958). Often, a primary explosion results from a small amount of dust in suspension that has been exposed to a source of ignition; the pressure and vibration created can dislodge large accumulations of dust on horizontal surfaces, creating a larger secondary explosion.

For ignition, dust clouds require high temperatures and sufficient dust concentration. These temperatures and concentrations and the minimum spark energy can be found in Avallone and Baumeister (1987).

Explosive dusts are potential hazards whenever uncontrolled dust escapes, dispersing in the atmosphere or settling on horizontal surfaces such as beams and ledges. Proper exhaust ventilation design involves the principles covered in Chapter 27 of the 1991 ASHRAE *Handbook—Applications*. The ventilation systems and equipment chosen must prevent the pocketing of dust inside the equipment. When local exhaust ventilation is used, separation equipment should be installed as close to the dust source as possible to prevent transport of dust in the exhaust system.

AIR POLLUTION

Concentration levels of outdoor pollutants vary. Table 3 presents a sample of the annual mean of several pollutants in selected urban centers. The U.S. Environmental Protection Agency has set primary and secondary standards for several pollutants known as *criteria pollutants* (Table 4).

Table 3 Annual Median Concentrations for TSP, NO_2, O_3, and CO for 1979[a]

Location	Concentration			
	μg/m³			mg/m³
	TSP (Annual Average)[b]	NO_2 (1-h Average)	O_3 (1-h Average)	CO (1-h Average)
Baltimore	43–102	45	20	1.5
Boston	67	75	—	3.5
Burbank, CA	—	124	39	3.5
Charleston, WV	43–70	37	14	1.2
Chicago	56–125	63	29	2.9
Cincinnati	47–87	60	24	1.0
Cleveland	58–155	89[c]	26	2.0
Dallas	43–73	59[c]	39	1.4
Denver	80–194	89	37	4.6
Detroit	52–135	68	14	1.8
Houston	51–147	90[c]	39[d]	1.0
Indianapolis	48–81	91[c]	33	2.7
Los Angeles	90	85	117	2.6
Louisville	60–102	70[c]	31	1.5
Milwaukee	47–105	86[c]	41	1.4
Minneapolis	45–87	65[c]	—	1.8
Nashville	41–82	62[c]	49[d]	2.6
New York	40–77	57	35	5.5
Philadelphia	51–109	85	39	3.2
Pittsburgh	88–162	—	29[d]	3.9
St. Louis	63–107	90[d]	22[d]	2.3[d]
San Diego	57–75	69	39	1.1
San Francisco	51	46	20[e]	2.1
Washington, D.C.	47–70	52	29	1.6

[a]EPA (1980)
[b]Annual geometric mean of 24-h averages
[c]24-h averages
[d]Not a full year
[e]Total oxidants
Source: Wadden and Schiff (1983).

Table 4   U.S. Ambient Air Quality Standards

Pollutant	Averaging Time	Primary Standard Levels	Secondary Standard Levels
Particulate matter	Annual (geometric mean)	75 μg/m^3	60 μg/m^3
	24 h[b]	260 μg/m^3	150 μg/m^3
Sulfur oxides	Annual (arithmetic mean)	80 μg/m^3 (0.03 ppm)	—
	24 h[b]	365 μg/m^3 (0.14 ppm)	—
	3 h[b]	—	1300 μg/m^3 (0.5 ppm)
Carbon monoxide	8 h[b]	10 mg/m^3 (9 ppm)	10 mg/m^3 (9 ppm)
	1 h[b]	40 mg/m^{3c} (35 ppm)	40 mg/m^3 (35 ppm)
Nitrogen dioxide	Annual (arithmetic mean)	100 μg/m^3 (0.05 ppm)	100 μg/m^3 (0.05 ppm)
Ozone	1 h[b]	240 μg/m^3 (0.12 ppm)	240 μg/m^3 (0.12 ppm)
Hydrocarbons (nonmethane)[a]	3 h (6 to 9 A.M.)	160 μg/m^3 (0.24 ppm)	160 μg/m^3 (0.24 ppm)
Lead	3 mos.	1.5 μg/m^3	1.5 μg/m^3

[a] A nonhealth-related standard used as a guide for ozone control
[b] Not to be exceeded more than once a year
[c] EPA has proposed a reduction of the standard to 25 ppm (29 mg/m^3)
Source: U.S. Environmental Protection Agency

Odorants are also objectionable in occupied spaces. In many cases (see Chapter 40 of the 1991 ASHRAE *Handbook—Applications*), equipment that removes odorants also removes some nonodorous gases. The engineer must cope with existing pollution by selecting equipment that provides clean air, and reduce pollution by providing adequate controls to accompany the design. This includes selecting the appropriate equipment for satisfactory combustion, as well as controls that enable the operator to run the equipment according to good practice and applicable air pollution ordinances. When dealing with industrial contaminants, the engineer must design local exhaust systems that minimize the removal of useful materials and provide collection equipment that reduces the amount of material discharged to the atmosphere as required by good practice and existing codes and standards.

RADIOACTIVE AIR CONTAMINANTS

Radioactive contaminants (Jacobson and Morris 1976-77) can be particulate or gaseous and are similar to ordinary industrial contaminants. A major source of airborne radioactive exposure to the population is radon from soil gas. Many radioactive materials would be chemically toxic if present in high concentrations; however, in most cases, the radioactivity necessitates limiting their concentration in air.

Most radioactive air contaminants affect the body when they are absorbed and retained. This is known as the *internal radiation hazard*. Inert gases may be hazardous because the whole body comes in contact with radiation from the gas in the air that surrounds it. These gases are, therefore, *external radiation hazards*. Particulate contaminants may settle to the ground, where they contaminate plants and eventually enter the food chain and the human body. Deposited material on the ground increases external radiation exposure. However, except for fallout from nuclear weapons or a serious reactor accident, such exposure is insignificant.

Radioactive air contaminants can emit alpha, beta, or gamma rays. The alpha rays have very low penetrating power and present

no hazard, except when the material is deposited inside or on the body. Beta rays are somewhat more penetrating and can be both an internal and an external hazard. The penetrating ability of gamma rays depends on their energy, which varies from one type of radioactive element or isotope to another. A distinction should be made between the properties of the radioactive material and the radiation given off by this material. Radioactive particles and gases can be removed from air by devices such as filters and absorption traps, but the gamma radiation from such material is capable of penetrating solid walls. This distinction is frequently overlooked. The amount of radioactive material in air is measured in becquerels per cubic metre (1 becquerel equals 2.702702×10^{-11} curies), while the dose of radiation from deposited material is measured in rads.

Radioactive materials present problems that make them distinctive. The contaminants can generate enough heat to damage filtration equipment or ignite spontaneously. The concentrations at which most radioactive materials are hazardous are much lower than those of ordinary materials; as a result, special electronic instruments that respond to radioactivity must be used to detect these hazardous levels.

Radon is a colorless, odorless, radioactive gas found in all soils in various concentrations. Radon gas primarily enters a house or building through leakage paths in the foundation. Radon can then decay through a succession of decay products, producing metallic ions. These decay products then attach to particles suspended in the air or deposit onto surfaces within an occupied space; they can then be inhaled and deposited in the respiratory tract. The subsequent decay produces alpha particles, which may increase the health risk.

The ventilation engineer faces difficulty in dealing with radioactive air contamination because of the extremely low permissible concentrations for radioactive materials. For certain sensitive industrial plants, such as those in the photographic industry, contaminants must be kept from entering the plant. If radioactive materials are handled inside the plant, the problem is to collect the contaminated air as close to the source as possible, and then remove the contaminant from the air with a high degree of efficiency, before releasing it to the outdoors. Filters are generally used for particulate materials, but venturi scrubbers, wet washers, and other devices can be used as prefilters to meet special needs.

The design of equipment and systems for control of radioactive particulates and gases in nuclear laboratories, power plants, and fuel-processing facilities is a highly specialized technology. Careful attention must be given to the reliability, as well as the contaminant-removal ability of the equipment under the special environmental stresses involved. Various publications of the U.S. Department of Energy can provide guidance in this field.

The basic standards for permissible air concentrations are those of the National Committee on Radiation Protection, published by the National Bureau of Standards as Handbook No. 69. Industries operating under licenses from the U.S. Nuclear Regulatory Commission or state licensing agencies must meet requirements of the Code of Federal Regulations, Title 10, Part 20. Some states have additional requirements.

ATMOSPHERIC POLLEN

Pollen grains discharged by weeds, grasses, and trees (Hewson *et al.* 1967, Jacobson and Morris 1976-77, Solomon and Mathews 1978) (and cause hay fever) have properties of special interest to air-cleaning equipment designers (see Chapter 25 of the 1991 ASHRAE *Handbook—Applications*). Whole grains and fragments transported by air range between 10 and 50 μm; however, some measure as small as 5 μm, and others measure over 100 μm in diameter. Ragweed pollen grains are fairly uniform in size, ranging from 15 to 25 μm. Pollen grains can be removed from the air more readily than the dust particles prevalent in outdoor air or those produced by dusty processes, since the latter predominate in the size range from 0.1 to 10 μm.

Most grains are hygroscopic and, therefore, vary in weight with the humidity. Illustrations and data on pollen grains are available in botanical literature. Geographical distribution of plants that produce hay fever is also recorded.

The quantity of pollen grains in the air is generally estimated by exposing an adhesive-coated glass plate outdoors for 24 h, then counting calibrated areas under the microscope. Methods are available for determining the number of grains in a measured volume of air, but, despite their greater accuracy, they have not replaced the simpler gravity slide method used for most pollen counts. Counting techniques vary, but daily pollen counts reported in local newspapers during hay fever season usually represent the number of grains found on 0.28 in^2 of a 24-h gravity slide.

Hay fever sufferers may experience the first symptoms when the pollen count is 10 to 25; in some localities, maximum figures for the seasonal peak may approach 1000 for a 24-h period, depending on the sampling and reporting methods used by the laboratory. Translation of gravity counts by special formulas to a volumetric basis (*i.e.*, number of grains per unit volume of air) is unreliable, due to the complexity of the modifying factors. When such information is important, it should be obtained directly by a volumetric instrument. The number of pollen grains per cubic yard of air varies from 2 to 20 times the number found on 1 cm^2 of a 24-h gravity slide, depending on grain diameter, shape, specific gravity, wind velocity, humidity, and physical placement of the collecting plate.

Whole grain pollens are easily removed from the outside air entering a ventilation system with medium efficiency filters selected to remove 99% of particles 10μm and greater. Once they have entered a building, the rapid settling rate of whole grain pollens makes it difficult for further air cleaning to reduce the concentrations of these particles. On resuspension from occupant activities, the whole grain pollens may be broken into particles, which may then be controlled effectively with a high-efficiency filter capable of removing a high percentage of particles a few microns in diameter.

AIRBORNE MICROORGANISMS

Interest in airborne microorganisms and their ability to subsist in different environments increased with the need to assemble and launch space probes under sterile conditions, as well as to prevent cross-infection in hospitals. Methods for contamination control are similar to those used in clean rooms to protect critical manufacturing areas from dust contamination.

Public interest focuses on airborne microorganisms, primarily bacteria, responsible for diseases and infections. However, a variety of airborne microorganisms besides bacteria are of economic significance and can cause product contamination or loss. In the food processing industry, yeast and mold can reduce the shelf life of some products. Refined syrups can be damaged by mold scums. Wild yeast can destroy a batch of beer. Antibiotic yields can be reduced by foreign organisms in the culture mix.

Most microorganisms become airborne by attachment to dust particles. Bacteria from the soil are likely to be spore-formers and are capable of surviving in hostile environments. Other airborne bacteria, especially within closed occupied spaces, often originate from droplet nuclei caused by actions such as sneezing. Concentration of microorganisms in the atmosphere vary from a few to several hundred per cubic foot, depending on many factors. The

Table 5 Sources, Possible Concentrations, and Indoor-to-Outdoor Concentration Ratios of Some Indoor Pollutants

Pollutant	Sources of Indoor Pollution	Possible Indoor Concentration[a]	I/O Concentration Ratio	Location
Carbon monoxide	Combustion equipment, engines, faulty heating systems	100 ppm	>>1	Skating rinks, offices, homes, cars, shops
Respirable particles	Stoves, fireplaces, cigarettes, condensation of volatiles, aerosol sprays, resuspension, cooking	100 to 500 μg/m^3	>>1	Homes, offices, cars, public facilities, bars, restaurants
Organic vapors	Combustion, solvents, resin products, pesticides, aerosol sprays	NA	>1	Homes, restaurants, public facilities, offices, hospitals
Nitrogen dioxide	Combustion, gas stoves, water heaters, driers, cigarettes, engines	200 to 1000 μg/m^3	>>1	Homes, skating rinks
Sulfur dioxide	Heating system	20 μg/m^3	<1	Removal inside
Total suspended particles without smoking	Combustion, resuspension, heating system	100 μg/m^3	1	Homes, offices, transportation, restaurants
Sulfate	Matches, gas stoves	5 μg/m^3	<1	Removal inside
Formaldehyde	Insulation, product binders, particleboard	0.05 to 1.0 ppm	>1	Homes, offices
Radon and progeny	Building materials, groundwater, soil	0.1 to 200 nCi/m^3	>>1	Homes, buildings
Asbestos	Fireproofing	< 10^6 fiber/m^3	1	Homes, schools, offices
Mineral and synthetic fibers	Products, cloth, rugs, wallboard	NA	—	Homes, schools, offices
Carbon dioxide	Combustion, humans, pets	3000 ppm	>>1	Homes, schools, offices
Viable organisms	Humans, pets, rodents, insects, plants, fungi, humidifiers, air conditioners	NA	>1	Homes, hospitals, schools, offices, public
Ozone	Electric arcing UV light sources	20 ppb 200 ppb	<1 >1	Airplanes Offices

[a]Concentrations listed are only those reported indoors. Both higher and lower concentrations have been measured. No averaging times are given. NA indicates it is not appropriate to list a concentration.

Source: NRC (1981).

sampling method for microorganisms has an effect on the measured count. Collection on dry filter paper can cause count degradation because of the dehydration loss of some organisms. Glass impingers may give high counts because agitation can cause clusters to break up into smaller individual organisms. Slit samples may give a more accurate colony count.

When maximum removal of airborne microorganisms is either necessary or desirable, super-interception, high-efficiency particulate air (HEPA) filters are used. These filters create essentially sterile atmospheres and are preferred over chemical scrubbers and ultraviolet radiation that are also used to control airborne microorganisms.

In many situations, total control of airborne microorganisms is not required. For these situations, there are different types of high-efficiency dry media extended surface filters that provide the desired efficiency necessary for certain applications. These filters have lower pressure drops than HEPA filters and can manage or filter out the concerned contaminant.

INDOOR AIR QUALITY

Indoor air quality in residences, offices, and other indoor, nonindustrial environments is a concern (Spengler *et al.* 1982, NRC 1981). Exposure to indoor pollutants can be as important as exposure to outdoor pollutants because a large portion of the population spends up to 90% of their time indoors and because indoor pollutant concentrations are frequently higher than corresponding outdoor pollutant levels.

Characterization of the indoor air quality has been the subject of numerous recent studies. ASHRAE Indoor Air Quality (IAQ) Conference proceedings discuss indoor air quality problems and some practical controls. ASHRAE *Standard* 62-1989 addresses many indoor air quality concerns. Table 5 illustrates the sources, levels, and the indoor-to-outdoor concentration ratio of several pollutants found in indoor environments. Chapter 37 has further information.

REFERENCES

ACGIH. 1983. *Air sampling instruments*, 6th ed. American Conference of Governmental Industrial Hygienists, Cincinnati, OH.

ASHRAE. 1989. Ventilation for acceptable indoor air quality. ASHRAE *Standard* 62-1989.

ASTM. 1983. Practice for continuous sizing and counting of airborne particles in dust-controlled areas using instruments based upon light-scattering principles. ASTM *Standard* F50-83.

Avallone, E.A. and T. Baumeister. 1987. *Marks' standard handbook for mechanical engineers*. McGraw-Hill Book Company, New York.

Hartmann, I. 1958. Explosion and fire hazards of combustible dusts. *Industrial Hygiene and Toxicology* 1:2. Interscience Publishers, Inc., New York.

Hewson, E.W., *et al.* 1967. Air pollution by ragweed pollen. *Journal of the Air Pollution Control Association* 17(10):651.

Jacobson, A.R. and S.C. Morris. 1977. "The primary pollutants, viable particulates. Their occurrence, sources and effects." In *Air pollution*, 3rd ed., Academic Press, New York, 169.

Morrow, P.E. 1964. Evaluation of inhalation hazards based upon the respirable dust concept and the philosophy and application of selective sampling. AIHA *Journal* 25:213.

National Primary and Secondary Ambient Air Quality Standards. Code of Federal Regulations, Title 40, Part 50, 40CFR50.

NFPA. 1984. Fire hazard properties of flammable liquids, gases and volatile solids. NFPA *Standard* 325M, National Fire Protection Association, Quincy, MA.

NRC. 1981. *Indoor pollutants*. 1981. National Research Council, National Academy Press, Washington, D.C.

Scala, G.F. 1963. A new instrument for the continuous measurement of condensation nuclei. *Analytical Chemistry* 35(5):702.

Solomon, W.R. and K.P. Mathews. 1978. Aerobiology and inhalant allergens. *Allergy, principles and practices*, St. Louis, MO.

Spengler, J.C., Hallowell, D. Moschandreas, and O. Fanger. 1982. Environmental international. *Indoor air pollution*. Pergamon Press, Ltd, Oxford, England.

Wadden, A. and P.A. Schiff. 1983. *Indoor air pollution: Characterization, prediction and control*. John Wiley and Sons, New York.

Whitby, K.T., A.B. Algren, and R.C. Jordan. 1955. Size distribution and concentration of airborne dust. *Heating, Piping and Air Conditioning* 27:121.

Whitby, K.T., A.B. Algren, and R.C. Jordan. 1957. The ASHRAE airborne dust survey. *Heating, Piping and Air Conditioning* 29(11):185.

CHAPTER 12

ODORS

VARIOUS factors make odor control a primary consideration in ventilation engineering: (1) modern buildings permit less air infiltration through walls and have more indoor sources of odors associated with building materials, furnishings, and office equipment; (2) outdoor air, previously clean and odor-free, is often polluted; and (3) high energy costs have caused a reduction in ventilating rates at a time when requirements for a relatively odor-free environment are greater than ever.

This chapter reviews how odoriferous substances are perceived and controlled. Chapter 40 of the 1991 ASHRAE *Handbook—Applications* has more information on control methods.

SENSE OF SMELL

Olfactory Stimuli

Among organic substances, those with molecular weights greater than 300 are generally odorless. However, some substances with molecular weights less than 300 are such potent olfactory stimuli that they can be perceived at concentrations too low to be detected directly by existing instruments. Trimethylamine, for example, can be recognized by a human observer at a concentration of about 10^{-4} ppm. Table 1 shows threshold concentrations for selected compounds. These threshold values are not precise numbers and may vary by more than one order of magnitude. The American Industrial Hygiene Association (1989) provides a large, critically evaluated list of threshold values, which documents this point.

Olfactory sensitivity often makes it possible to detect and eliminate potentially harmful substances at concentrations below dangerous levels. Often foul-smelling air is also assumed to be unhealthy. When symptoms such as nausea, headache, and loss of appetite are caused by an unpleasant odor, it may not matter whether the air is toxic. The magnitude of the symptoms is related to the magnitude of the odor. Even a room with a low but recognizable odor can arouse uneasiness among occupants. Cometto *et al.* (1987) review sensory irritation and its relation to indoor air pollution.

Anatomy and Physiology

The olfactory receptors lie in the *olfactory cleft*, which is high in the nasal passages. The surrounding tissue contains other diffusely distributed receptors—free nerve endings that also respond to airborne vapors. These receptors of common chemical sense mediate the tickling, burning, cooling, and occasionally, painful sensations that accompany olfactory sensations. Ammonia stimulates these free nerve endings, as do other substances that have any degree of pungency. Thus, olfaction and the common chemical sense operate as a single perceptual system (Cain 1976).

About 5-million olfactory neurons, a small cluster of nerve cells inside the nasal cavity above the bridge of the nose, determine whether the air is odoriferous. These olfactory receptors are connected by a nerve fiber into the olfactory bulb of the brain. The bulb passes information it receives from the receptors to various central structures of the brain, including an area called the *seat of emotion*. One sniff of an odorant can often evoke a complex, emotion-laden memory, such as a scene from childhood.

Hormonal factors, which often mediate emotional states, can modulate olfactory sensitivity. Although the evidence is not uniformly compelling, research has found that (1) the sensitivity of women varies during the menstrual cycle, reaching a peak just before and during ovulation (Schneider 1974); (2) women are generally more sensitive than men, but this difference only emerges around the time of sexual maturity (Koelega and Koster 1974); (3) sensitivity is altered by certain diseases (Schneider 1974); and (4) various hormones and drugs (*e.g.*, estrogen and alcohol) alter sensitivity (Schneider 1974, Engen *et al.* 1975). Other factors that may affect olfactory perception include the olfactory acuity of an individual, the magnitude of the flow rate toward the olfactory receptors, the temperature, and the relative humidity. Humans are able to perceive over 4000 different odors, yet individuals are able to name only a small number of odors (Ruth 1986).

Table 1 Odor Threshold of Selected Gaseous Air Pollutants

Pollutant	Odor Threshold, mg/m^3	Pollutant	Odor Threshold, mg/m^3
Acetaldehyde	1.2	Hydrogen cyanide	1
Acetone	47	Hydrogen fluoride	2.7
Acetonitrile	>0	Hydrogen sulfide	0.007
Acrolein	0.35	Mercury	∞
Acrylonitrile	50	Methanol	130
Allyl chloride	1.4	Methyl chloride	595
Ammonia	33	Methylene chloride	750
Benzene	15	Nitric oxide	>0
Benzyl chloride	0.2	Nitrogen dioxide	51
2-Butanone (MEK)	30	Ozone	0.2
Carbon dioxide	∞	Phenol	0.18
Carbon monoxide	∞	Phosgene	4
Carbon disulfide	0.6	Propane	1800
Carbon tetrachloride	130	Sulfur dioxide	1.2
Chlorine	0.007	Sulfuric acid	1
Chloroform	1.5	Tetrachloroethane	24
p-Cresol	0.056	Tetrachloroethylene	140
Dichlorodifluromethane	5400	o-Toluidene	24
Dioxane	304	Toluene	8
Ethylene dichloride	25	Toluene diisocyanate	15
Ethylene oxide	196	1,1,1 Trichloroethane	1.1
Formaldehyde	1.2	Trichloroethylene	120
n-Heptane	2.4	Vinyl chloride monomer	1400
Hydrogen chloride	12	Xylene	2

The preparation of this chapter is assigned to TC 2.3, Gaseous Air Contaminants and Gas Contaminant Removal Equipment.

SENSORY MEASUREMENT

Sensory measurements yield data under one of the following categories:

- Nominal data that indicate different categories, but not in any order of magnitude; for example, the numbers used in a team or panel of odor sniffers.
- Ordinal data that fall into two or more classes belonging to an ordered series, *i.e.*, the odor is either slight, moderate, or strong.
- Interval data that are placed in ordered classes separated by a meaningful measure or interval between them, for example, first, second, third.
- Ratio data that indicate how many times a quantity is larger than a reference quantity, for example, temperature scales (Meilgaard *et al.* 1987).

The odor sensation has four odor components or attributes: detectability, intensity, character, and hedonic note. Detectability or threshold is the minimum concentration of an odorant that provoked detection of the odorant to some predetermined segment of the population. Two types of thresholds exist: the *detection* threshold and the *recognition* threshold. The detection threshold is the lowest level that elicits response by a segment of the population. If that segment is 50%, the detection threshold is denoted by ED_{50}. Such thresholds can, however, be attributed to 100%, which includes all olfactory sensitivities, or to 10%, which refers only to the most sensitive segment of the population. Threshold values are not physical constants; rather they are statistical measurements of best estimates.

Intensity is a quantitative aspect of a descriptive analysis, stating the degree to which a characteristic odor is present. Intensity of the perceived odor is, therefore, the strength of the odoriferous sensation. Detection threshold values and, most often, odor intensity determine the need for indoor odor controls.

Odor character defines the odor as similar to some familiar smell, *e.g.*, fishy, sewer, flowery, and the like. Hedonics, or the hedonic tone of an odor, is the degree to which an odor is perceived as pleasant or unpleasant. Hedonic judgments include both *category* judgments (pleasant, neutral, unpleasant) and a *magnitude* judgment (very unpleasant, slightly pleasant) and so on.

Important questions are (1) What is the minimum concentration of odorant necessary to produce a just-detectable odor? and (2) How does perceived odor magnitude grow with concentration above the threshold? No universal way has been accepted to measure either the threshold or suprathreshold magnitude. However, certain guidelines and conventions simplify the choice of methods.

Olfactory acuity. The olfactory acuity of the population is normally distributed—that is, most people have an average ability to smell substances or to respond to odoriferous stimuli, a few people are very sensitive or hypersensitive, and a few others are insensitive, including some who are totally unable to smell, *i.e.*, they are anosmic. The segment of the population with average ability to smell varies between 90% and 96%. In addition, the olfactory acuity of an individual varies as a function of the odorant. For instance, repeated exposure to an odorant may lead to anosmia (absence of the sense of smell) only to that odorant.

Threshold. The perception of weak odoriferous signals is probabilistic; at one moment the signal may make a discernible impact, at the next moment it may not. One reason for this phenomenon is moment-to-moment variability in the number of molecules striking the receptors; another is the variability in which receptors are contacted. Other important sources are a person's state of readiness or fatigue. The combined effect of these various sources prevents an individual from perceiving an odor during the entire time of the stimulus. The objective of dilution to detection (recognition) threshold values is to find the largest number of dilutions that would cause half of the panelists to detect (recognize) the odor.

Measurement of odor attributes is complex. Several methods, such as those described by Dravnieks (1979) and NAS (1979), are used to measure odor threshold numbers. ASTM STP 434 (1982) describes the forced choice dynamic olfactometer method. The most frequently used dynamic olfactometer has six ports or sets. Within each port, three glass sniffers are arranged in a circular pattern to achieve blind presentation to the subject. Two of the sniffers supply nonodoriferous air, and the third sniffer provides the diluted odorous air. The subject is asked to identify or guess the sniffer with the odoriferous air.

The odoriferous air is supplied to each port at various levels of dilution. Samples are diluted either by compressed air or by room air using a peristaltic pump and are presented in decreasing levels of dilution. The dilution factors from port one to port six are 3600x, 1200x, 400x, 45x, and 15x (IITRI 1980). The odorant is presented in increasing rather than random order because exposure to one of the higher concentrations can temporarily impair the ability to detect weaker concentrations.

The subject can be tested with additional ascending series. The concentration of interest in each series is the lowest in the string of three correct detections. The geometric mean taken across the various series provides a good index of the subject's threshold. In a similar manner, the geometric mean taken across several subjects provides a good index of the threshold for a group.

Three issues of technical and practical importance are: (1) rate of airflow to the sniffers, (2) size of panel, and (3) design of the sniffers. Presently, the flow rate varies from 8 to 50 mL/s, and, in some cases, much greater. Present test practices use as few as five panelists or as many as ten. While a small panel is representative only of itself, too large a panel is difficult to manage. A panel of ten provides statistically sound data.

Finally, the sniffers used with current dynamic olfactometers are small glass ports that allow the panelist to sniff as close or as far as desired. An alternative design is a cup arrangement that requires the panelist to insert the nose into the cup and eliminates variation in the dilution between the delivered air and the air actually sniffed by the panelist.

In all sensory testing, it is necessary to guard against subtle biases and clues that may influence results. Also, adequate statistical procedures should be used to detect biases such as whether there is a reliable psychophysical difference between odorants, between means of air sampling, etc.

For more details regarding psychophysical procedures, ways to sample odoriferous air, handling samples, means of stimulus presentation, and statistical procedures, consult ASTM STP 434.

Suprathreshold intensity. The relation between *perceived odor magnitude S* and *concentration C* conforms to a power function

$$S = kC^n \tag{1}$$

This exemplifies the psychophysical power law, also called Stevens' law (1957). The law applies throughout the sensory realm and has almost completely replaced the Weber-Fechner logarithmic law in modern psychophysics. In the olfactory realm, the exponent n of the power function is less than 1.0. Accordingly, a given percentage change in concentration causes a smaller percentage change in perceived magnitude. The size of the exponent varies from one odorant to another, ranging from less than 0.2 to about 0.7 (Cain and Moskowitz 1974), which has important consequences for malodor control. An exponent of 0.7 implies that in order to reduce perceived intensity by a factor of 5, for instance, concentration must be reduced by a factor of 10. An exponent of 0.2 implies that a fivefold reduction in perceived magnitude can be achieved by slightly more than a three-thousandfold reduction in concentration.

Of the various ways to scale perceived magnitude, a *category scale*, which can be either number or word categorized, is commonly used. Numerical values on this scale do not reflect ratio relations among odor magnitudes; *i.e.*, a value of 2 does not represent a perceived magnitude twice as great as a value of 1. Table 2 gives four examples of category scales (Meilgaard *et al.* 1988).

Although category scaling procedures can be advantageous in field situations, *ratio scaling* techniques are used frequently in the laboratory (Cain and Moskowitz 1974). Ratio scaling procedures require observers to assign numbers proportional to perceived magnitude. If the observer is instructed to assign the number 10 to one concentration and a subsequently presented concentration seems three times as strong, the observer calls it 30; if it seems one-half as strong, the observer assigns it 5, etc. This ratio scaling procedure, called *magnitude estimation*, was used to derive the power function for butanol (Figure 1). Equation (1) in this case becomes $\log S = n \log C + \log k$, which has the form $y = ax + b$. By means of this transformation, the exponent becomes the slope of the line. The function for butanol conforms to the equation $S = 0.26 \, C^{0.66}$, which represents a consensus of results obtained in various laboratories (Moskowitz *et al.* 1974). According to this equation, a concentration of 250 ppm has a perceived magnitude of 10. This corresponds to *moderate intensity*. A concentration near threshold (2 to 5 ppm) has a perceived magnitude of about 0.5.

A third way to measure suprathreshold odor intensity is to match the intensity of odorants. An observer can be given a concentration series of a matching odorant (*e.g.*, 1-butanol) to choose the member that matches most closely the intensity of an *unknown* odorant. The matching odorant can be generated by a relatively inexpensive olfactometer such as that shown in Figure 2 (Dravnieks 1975). Figure 3 shows, in logarithmic coordinates, functions for various odorants obtained by the matching method (Dravnieks and Laffort 1972). The left-hand ordinate expresses intensity in terms of concentration of butanol, and the right-hand ordinate expresses intensity in terms of perceived magnitude. The two ordinates are related by the function in Figure 1, the standardized function for butanol. The matching method illustrated here has been incorporated into ASTM E544-75, *Practice for Referencing Suprathreshold Odor Intensity.*

A National Academy of Sciences document (NAS 1979) elaborates on attitudes regarding odor units. The objection to the odor unit is associated with the fact that odor is a sensation not a substance. Here is an example of the concept of odor concentration: if 1 ft³ of odorous air were diluted with N ft³ of fresh air to reach the threshold value, then the odoriferous sample has an odor concentration of N units/ft³. Confusion occurs when other dimensions for volume are used. In response to this confusion, ASTM defined the dimensionless factor Z for the dilution to detection threshold: Z equals the ratio of the odorant concentration C divided by its concentration at threshold C_{thr} (ASTM 1978). Note: Z numerically equals the odor units per cubic foot.

The following example from the NAS document illustrates the concepts: $Z = 1000$ when one volume of odorous sample requires 1000 volumes to reach the dilution threshold. Furthermore, if the

Table 2 Examples of Category Scales

| Number Category | | Word Category | |
Scale I	Scale II	Scale I	Scale II
0	0	None	None at all
1	1	Threshold	Just detectable
2	2.5	Very slight	Very mild
3	5	Slight	Mild
4	7.5	Slight-moderate	Mild-distinct
5	10	Moderate	Distinct
6	12.5	Moderate-strong	Distinct-strong
7	15	Strong	Strong

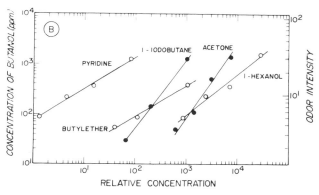

Fig. 2   Panelist Using Dravnieks Binary Dilution Olfactometer

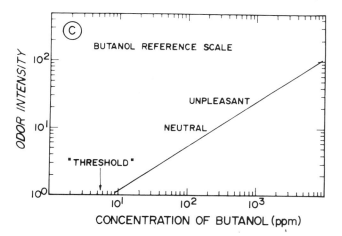

Fig. 1   Standardized Function Relating Perceived Magnitude to Concentration of 1-Butanol
(Moskowitz *et al.* 1974)

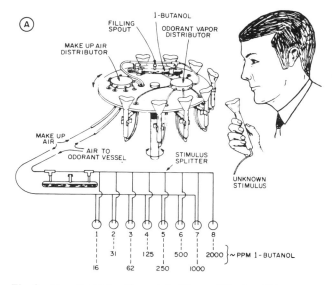

Fig. 3   Matching Functions Obtained with Dravnieks Olfactometer
(Dravnieks and Laffort 1972, Cain 1978)

sampled air is emitted at 50 m³/min, the odorous emission rate is 1000 × 50 = 50,000 m³/min, which is the clean air demand of the odorous emission. This value is equivalent to odor units per minute after conversion from cubic meters to cubic feet. The engineer should be aware of the present understanding on this unit and deal accordingly with existing and proposed regulations on odor measurement and removal.

Odor units. According to ASTM D135, Standard Definitions of Terms Relating to Atmospheric Sampling and Analysis, odor concentration can be expressed as the number of unit volumes that a unit volume of odorous sample occupies when diluted to the odor threshold with nonodorous air. If a sample of odorous air can be reduced to threshold by a tenfold dilution with pure air, the concentration of the original sample is said to be 10 odor units. Hence, odor units are equivalent to multiples of threshold concentrations.

Odor units are used widely to express legal limits for emission of odoriferous materials. For example, the law may state that a factory operation may not cause the ambient odor level to exceed 15 odor units. It should be recognized that odor units are *not* units of perceived magnitude. For every odorant, odor units and parts-per-million are proportional. The proportionality constant varies from one odorant to another, depending on the number of ppm needed to evoke a threshold response. Since perceived odor magnitude does not grow proportionally with concentration expressed in ppm, it cannot grow proportionally with concentration expressed in odor units. According to the results of psychophysical scaling, a sample of 20 odor units is always perceived as less than twice as strong as a sample of 10 odor units. Moreover, since the rate of growth of the psychophysical function varies from one odorant to another, samples of two odorants, each at 20 odor units, may have unequal perceived intensities.

Although odor units are not equivalent to units of perceived magnitude, they can be useful. Most indoor and outdoor contaminants are complex mixtures; the actual concentration of the odoriferous portion of a sample cannot be expressed with certainty. Thus, the odor unit is a useful measure of *concentration* when evaluating, for example, the efficiency of a filter or ventilation system.

Odor acceptability. The acceptability or pleasantness of an odor can be measured psychophysically in the same way as odor intensity. Either ratio or category scaling procedures can be adapted to this task.

Odors do not always cause adverse reactions. Products are manufactured to elicit favorable responses. Acceptance tests may involve product comparison (frequently used in the perfume industry) or a hedonic scale (Figure 4). The premise of acceptance tests is that the larger the segment of subjects accepting the odor, the better the odor. A hedonic scale that allows for negative as well as positive responses is likely to be more responsive to the question of how acceptable the odor is.

A given odor is not likely to be perceived as acceptable or unacceptable by all those potentially exposed. Acceptability of a given odor by a given person is due to a complex combination of associations and is not simply a characteristic of the odor itself (Beck and Day 1991).

Odor quality. The quality or character of an odor is difficult to assess quantitatively. A primary difficulty is that odors can vary along many dimensions. One way to assess quality is to ask panelists to judge the similarity between a test sample and various reference samples, using a 5-point category scale. For certain applications, reference odorants can be chosen to represent only that portion of the qualitative range relevant to the odor problem under investigation (*e.g.*, animal odors). Another procedure is to ask panelists to assess the degree of association between a test sample's quality and certain verbal descriptors (*e.g.*, sweaty, woody, chalky, sour).

The number of odorant descriptors and the descriptors to be used have been subjects of disagreement (Harper *et al.* 1968). The number of descriptors varies from a minimum of seven (Amoore 1962) to as many as 830 used by an ASTM subcommittee. An atlas of odor characters, containing 146 descriptors, was compiled for 180 chemicals by ASTM (1985).

An odor can be characterized either by an open-ended word description or by multidimensional scaling. Multidimensional scaling is based on similarity/dissimilarity judgments to a set of standard odors or to various descriptors.

In some instances, the interest merely may be whether an odor's quality has changed as a result of some treatment (*e.g.*, use of a bacteriostat). Under these circumstances, samples of air taken before and after treatment can be compared directly (using a simple scale of similarity) or indirectly (with appropriate verbal descriptors).

Olf unit. Indoor air quality scientists have not successfully resolved complaints associated with pollution in offices, schools, and other nonindustrial environments. Customarily, complaints are attributed to elevated pollutant concentrations; frequently, however, such high concentrations are not found indoors, yet complaints persist. The lack of an apparent cause and effect relationship between complaint and elevated indoor pollutant concentrations is sometimes known as the *sick building syndrome*.

Assuming that the inability to find a difference between air pollutant levels in buildings with registered complaints and in buildings without any complaints is due to inadequacies of prevailing

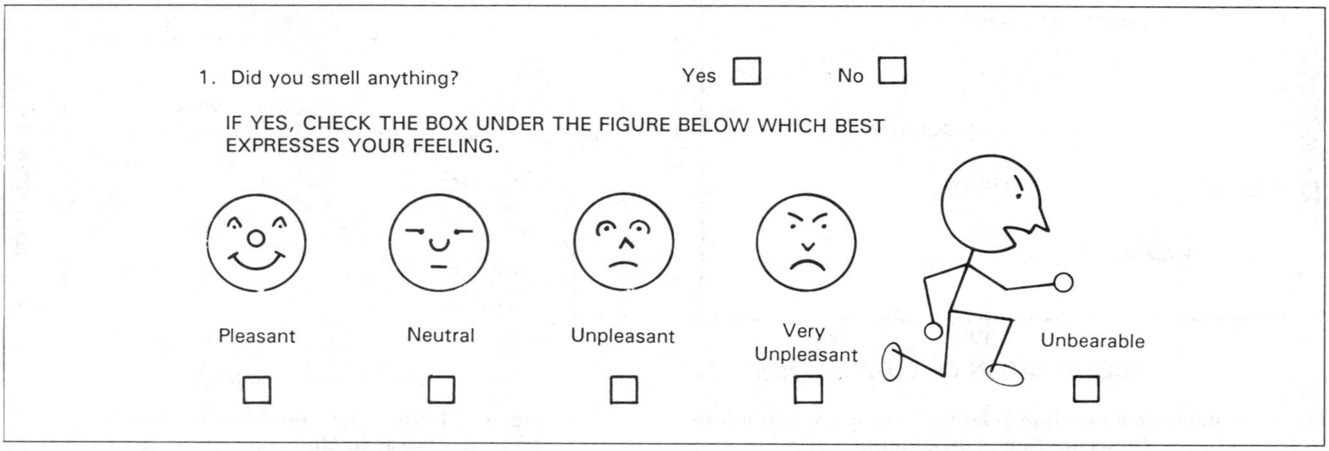

1. Did you smell anything?   Yes ☐   No ☐

IF YES, CHECK THE BOX UNDER THE FIGURE BELOW WHICH BEST EXPRESSES YOUR FEELING.

Pleasant   Neutral   Unpleasant   Very Unpleasant   Unbearable

☐   ☐   ☐   ☐   ☐

Fig. 4 Hedonic Scale Used for Evaluation of Odors in Population Surveys

measurement techniques, Fanger and others changed the focus from chemical analysis to sensory analysis (Fanger 1987, 1988). Fanger quantified air pollution sources by comparing them with a well-known source—a sedentary person in thermal comfort. A new unit, the olf, was introduced. An olf is defined as the emission rate of air pollutants (bioeffluents) from a standard person.

To use this unit, Fanger generated curves that relate the percentage of persons dissatisfied with the emissions of one person in a chamber as a function of fresh air ventilation rate, obtaining the following expression:

$$D = 395 \exp(-1.83\, q^{0.26}) \text{ for } q \geqslant 0.332$$

$$D = 100 \text{ for } q < 0.322$$

where

D = percentage dissatisfied persons
q = ventilation-emission ratio, L/s × olf

Such curves can be used in the study of indoor nonindustrial environments. A large number of judges (over 50) is used in a study. They visit the test environment under three different conditions: when the test indoor environment is (1) unoccupied and unventilated; (2) unoccupied but ventilated; and (3) occupied and ventilated. These measurements help in quantifying the sensory impact from such sources as building materials, the ventilation system, and occupant activity. Results are obtained by analyzing the responses to a questionnaire (Fanger *et al.* 1988).

ANALYTICAL MEASUREMENTS

More exact performance data on the control of specific odorants can be obtained using suitable analytical methods (Kaiser 1963). Detectors exist that permit detection of substances in amounts of about one nanogram. Since air usually contains many minor components, gas-chromatographic separation of the components must precede detection. Since odor thresholds for some compounds are low, a preconcentration of the minor components is necessary. Preconcentration is usually achieved by adsorption or absorption by a stable, sufficiently nonvolatile material, followed by thermal desorption or extraction.

Mass spectrometry can be used with gas chromatography to identify constituents of complex mixtures. The chromatograph resolves a mixture into its constituents, and the spectrometer yields data (mass spectra) that permit specification of chemical structure.

Several detectors are sufficiently sensitive and specific to detect resolved components. Hydrogen flame ionization detectors respond adequately and almost mass-proportionally to almost all organic compounds. Flame-photometric detectors can pinpoint, with equal sensitivity, compounds that contain sulfur; many sulfur compounds are strongly odorous and are of interest in odor work. A Coulson conductometric detector is specifically and adequately sensitive to ammonia and organic nitrogen compounds. Thermal conductivity detectors are generally not sensitive enough for analytical work on odors.

Airborne volatile organic compounds (VOCs) cause odors, but the correlation between indoor VOC concentrations and odor complaints in indoor environments is poor. An ACGIH document (1991), and others review the state of the art for sampling and analysis of VOCs in indoor air. The nose is the most sensitive instrument, however, and it can perceive odors from complex mixtures of VOCs at low indoor concentrations.

Frequently, a sniffing port (Dravnieks and O'Donnell 1971, Dravnieks and Krotoszynski 1969) is installed in parallel with the detector(s). Part of the resolved effluent exhausts through the port and allows the components that are particularly odorous or carry some relevant odor quality to be annotated. Usually, only a fraction of all observed components exhibit odors—particularly those of interest in odor evaluation. A change in odorant concentration indicated by detector responses may result from odorant removal treatment by ventilation or another process (Dravnieks *et al.* 1971). Odor control may not influence all odorant concentrations to the same extent, and new odorants may have appeared.

Since the physicochemical correlates of olfaction are poorly understood, no simple analytical means to predict the perceived quality and intensity of an odorant exists. Moreover, since acceptability of an odorant depends strongly on context, it is unlikely that analytical instruments will supplant human evaluation.

ODOR SOURCES

Outdoor sources of odors include automotive exhaust (emitting, among others, organic compounds); hazardous waste sites (organic compounds); wastewater treatment, composting, and refuse facilities (hydrogen sulfide, ammonia, and organic compounds); chemicals emitted from printing, paint, synthetics (organic compounds); and other stationary and mobile sites. Odors emitted by sources in the ambient environment eventually enter the indoor environment.

Indoor sources also emit odors. Such sources include cigarette and other tobacco product smoke, which includes particulate matter and volatile organic compounds; bathrooms (ammonia and other odors); building materials, such as adhesives, paints, processed wood, carpets, caulks, and other volatile organic compounds; consumer products, such as food, bathroom products, hobbies materials, paint sprays, personal care products, drapes, upholstery, and other furnishings (VOCs); reproduction and print shops in office buildings (ozone and organics).

Odors may also be emitted from wetted air-conditioning coils. Putrefaction of animal and vegetable matter also contribute to indoor odors. Several studies have been performed to measure the emission rates of organic compounds from indoor sources (Moschandreas and Gordon 1991), but no such efforts have been made to measure indoor odor emissions. Chapter 40 in the 1991 ASHRAE *Handbook—Applications* gives further information on contaminant sources and generation rates.

FACTORS AFFECTING ODORS

Humidity and Temperature

Odor perception of cigarette smoke (suspension of tobacco tar droplets plus vapor) and pure vapors is affected by temperature and humidity. An increase in humidity, at constant dry-bulb temperature, lowers the intensity level of cigarette smoke odor, as well as that of pure vapors (Kerka and Humphreys 1956, Kuehner 1956). This effect is more pronounced for some odorants than for others. An increase in temperature at constant specific humidity lowers the odor level of cigarette smoke slightly. Adaptation to odors takes place more rapidly during the initial stages of exposure. While the perceptible odor level of cigarette smoke decreases with exposure time, irritation to the eyes and nose generally increases. The irritation is greatest at low relative humidities.

To keep odor perception and irritation at a minimum, the air-conditioned space should be operated at about 45 to 60% rh. Since temperature has only a slight effect on odor level at constant specific humidity, it generally can be ignored; temperature should be maintained at conditions desired for comfort or economy.

Sorption and Release of Odors

Frequently, spaces develop normal occupancy odor levels long after occupancy has ceased. This results from sorption of odors on the furnishings during occupancy, with later desorption. This is observed when furnaces or radiators, after a long shutdown, are heated at winter startup. It also can be observed when evaporator coils warm up. Hubbard *et al.* (1955) tested the odorant sorption-release of cotton, wool, nylon, rayon, and fir wood. As a guide to correcting the sorbed odor situation, the rate of desorption can be decreased by decreasing temperature and relative humidity, and increased (as for cleaning) by the reverse.

Smoke, cooking, and body odor perception decreases as humidity increases. Where the odor source is intrinsic with the materials, as in linoleum, paint, rubber, and upholstery, reducing relative humidity is beneficial (Kuehner 1956) and decreases the rate of odor release. Quantitative values should not be used without considering the sorption-desorption phenomenon.

Dilution of Odors by Ventilation

Cain *et al.* (1983, 1987) compared occupancy odor with smoke odor by relating perceived odor intensity and odor acceptability during smoking and nonsmoking under controlled chamber conditions. When smoking took place, they found that the odor intensity was nearly five times as high as the odor intensity perceived under severe occupancy conditions (hot, humid, and low ventilation) but with no smoking. This study also determined that the ventilation rate required to control occupancy odor to a criterion of 80% acceptance is 17 cfm; for tobacco smoke odor, it is 53 cfm.

REFERENCES

ACGIH. 1988. *Advances in air sampling.* American Conference of Government Industrial Hygienists, Cincinnati, OH.

Amoore, J.E. 1962. The stereochemical theory of olfaction 1, Identification of seven primary odors. *Proc. Sci. Sect. Toilet Goods Assoc.* 37:1-12.

ASTM. 1975. Standard recommended practices for referencing suprathreshold odor intensity. ASTM E 544-75. ASTM, Philadelphia.

ASTM. 1976. Standard method for measurement of odor in atmosphere (dilution method).

ASTM. 1982. Manual on sensory testing methods. STP 434.

ASTM. 1985. Atlas of odor character profiles. DS 61.

Beck, L. and V. Day. 1991. New Jersey's approach to odor problems. From *Transactions: Recent developments and current practices and odor regulations control and technology,* D.R. Derenzo and A. Gnyp, eds. Air and Waste Management Association, Pittsburgh, PA.

Cain, W.S. 1978. The odoriferous environment and the application of olfactory research. Carterette and Friedman, eds. *Handbook of perception,* Vol. 6, Tasting, smelling, feeling, and hurting. Academic Press, Inc., New York.

Cain, W.S. 1976. Olfaction and the common chemical sense; Some psychophysical contrasts. *Sensory Processes* 1:57.

Cain, W.S. and H.R. Moskowitz. 1974. Psychophysical scaling of odor. Turk, Johnson, and Moulton, eds. *Human responses to environmental odors.* Academic Press, Inc, New York, 1.

Cain, W.S., T. Tosun, L.C. See, and B.P. Leaderer. 1987. Environmental tobacco smoke: Sensory reactions of occupants. *Atmospheric Environment* 21:347-53.

Cain, W.S., B.P. Leaderer, R. Isseroff, L.G. Berglund, R.I. Huey, E.D. Lipsitt, and D. Perlman. 1983. Ventilation requirements in buildings I. Control of occupancy odor and tobacco smoke odor. *Atmospheric Environment* 17:1183-97.

Clemens, J.B. and Lewis, R.G. 1988. Sampling for organic compounds. In *Principles of environmental sampling* 20:147-57. American Chemical Society, Washington, D.C.

Cometto-Muniz, J.E. and W.S. Cain. 1992. Sensory irritation, relation to indoor air pollution in sources of indoor air contaminants—characterizing emissions and health effects. *Annals of the New York Academy of Sciences,* Vol. 641.

Dravnieks, A. 1975. Evaluation of human body odors, methods and interpretations. *Journal of the Society of Cosmetic Chemists* 26:551.

Dravnieks, A. and B. Krotoszynski. 1969. Analysis and systematization of data for odorous compounds in air. ASHRAE Bulletin, Odor and odorants. The engineering view. New York.

Dravnieks, A. and P. Laffort. 1972. Physicochemical basis of quantitative and qualitative odor discrimination in humans. Olfaction and Taste IV, MBH, 142.

Dravnieks, A. and A. O'Donnell. 1971. Principles and some techniques of high resolution headspace analysis. *Journal of Agricultural and Food Chemistry* 19:1049.

Engen, T., R.A. Kilduff, and N.J. Rummo. 1975. The influence of alcohol on odor detection. *Chemical Senses and Flavor* 1:323.

Fanger, P.O. 1987. A solution to the sick building mystery. Indoor Air '87, Proceedings of the International Conference on Indoor Air and Climate. Institute of Water, Soil and Air Hygiene, Berlin.

Fanger, P.O., J. Lautidsen, P. Bluyssen, and G. Clausen. 1988. Air pollution sources in offices and assembly halls quantified by the olf unit. *Energy and Buildings* 12:7-19.

Hanna, G.F. *et al.* 1963. A chemical method for odor control. Conference on Recent Advances in Odor. Theory, Measurement, and Control. New York Academy of Sciences, November.

Harper, R., E.C. Bate Smith, and D.G. Land. 1968. *Odour description and odour classification.* American Elsevier Publishing Co, New York, distributors for Churchill Livingston Publishing Co, Edinburgh, Scotland.

Hubbard, A.B., N. Deininger, and F. Sullivan. 1955. Air conditioning coil odors. ASHAE Report No. 1547, ASHAE *Transactions* 61:431.

IITRI. 1980. *Instruction manual: Dynamic dilution forced-choice triangle olfactometer, model 101 for emission odors.* IITRI, Chicago.

Johnson, C.A. 1974. Electrolytic odor abatement process. State of the art of odor control technology. Publication SP-5, Air Pollution Control Association, Pittsburgh, PA.

Kerka, W.F. and C.M. Humphreys. 1956. Temperature and humidity effect on odor perception. ASHAE Report No. 1587, ASHAE *Transactions* 62:531.

Koelega, H.S. and E.P. Koster. 1974. Some experiments on sex differences in odor perception. *Annals of the New York Academy of Sciences* 237:234.

Koster, E.P. 1969. Intensity in mixtures of odorous substances. Pfaffmann, ed. Olfaction and taste III. Rockefeller University, New York, 142.

Kuehner, R.L. 1956. Humidity effects on the odor problem. ASHAE Report No. 1569, ASHAE *Transactions* 62:249.

Leonardos, G., D. Kendall, and N. Bernard. 1969. Odor threshold determinations of 53 odorant chemicals. *Journal of the Air Pollution Control Association* 19.91.

Leopold, C.S. 1945. Tobacco smoke control—A preliminary study. ASHVE Report No. 1277, ASHVE *Transactions* 51:255.

Meilgaard, M., G.V. Civille, and B.T. Carr. 1987. *Sensory evaluation techniques.* CRC Press, Boca Raton, FL.

Moschandreas, D.J. and Gordon S.M. 1991. Volatile organic compounds in the indoor environment: review of characterization methods and indoor air quality studies. In *Organic chemistry of the atmosphere,* CRC Press, Boca Raton, FL.

Moskowitz, H.R., A. Dravnieks, W.S. Cain, and A. Turk. 1974. Standardized procedure for expressing odor intensity. *Chemical Senses and Flavor* 1:235.

NAS. 1979. *Odors from stationary and mobile sources.* National Academy of Sciences, Washington, D.C.

Ruth, J.H. 1986. Odor thresholds and irritation levels of several chemical substances: A review. *American Industrial Hygiene Association Journal* 47:142-51.

Schneider, R.A. 1974. Newer insights into the role and modifications of olfaction in man through clinical studies. *Annals of the New York Academy of Sciences* 237:217.

Stevens, S.S. 1957. On the psychophysical law. *Psychological Review* 64:153.

Von Bergen, J. 1957. Industrial odor control. *Chemical Engineering* 64(8):239.

CHAPTER 13

MEASUREMENT AND INSTRUMENTS

HEATING, refrigerating, and air-conditioning engineers and technicians require instruments for laboratory and field work. Precision is more essential in the laboratory (where research and development are undertaken) than in the field (where acceptance and adjustment tests are conducted). This chapter describes some of these instruments.

TERMINOLOGY

The following definitions are generally accepted.

Accuracy. The capability of an instrument to indicate the true value of measured quantity. This is often confused with inaccuracy, which is the departure from the true value into which all causes of error are lumped, *e.g.*, hysteresis, nonlinearity, drift, temperature effect, and other sources of error.

Amplitude. The magnitude of variation in an alternating quantity from its zero value.

Average. The sum of a number of values divided by the number of values.

Band width. The range of frequencies over which a given device is designed to operate within specified limits.

Bias. The tendency of an estimate to deviate in one direction from a true value (a systematic error).

Calibration. (1) The process of comparing a set of discreet magnitudes or the characteristic curve of a continuously varying magnitude with another set or curve previously established as a standard. Deviation between indicated values and their corresponding standard values constitutes the correction (or calibration curve) for inferring true magnitude from indicated magnitude thereafter; (2) the process of adjusting an instrument to fix, reduce, or eliminate the deviation defined in (1).

Calibration curve. (1) The path or locus of a point that moves so that its coordinates on a graph correspond to values of input signals and output deflections; (2) the plot of error versus input (or output).

Confidence. The degree to which a statement (measurement) is believed to be true.

Deviate. Any item of a statistical distribution that differs from the selected measure of control tendency (average, median, mode).

Deviation, standard. The square root of the average of the squares of the deviations from the mean (root-mean-square) deviation. A measure of dispersion of a population.

Distortion. An unwanted change in wave form. Principal forms of distortion are inherent nonlinearity of the device, nonuniform response at different frequencies, and lack of constant proportionality between phase-shift and frequency. (A wanted or intentional change might be identical, but is called *modulation*.)

Drift. A gradual, undesired change in output over a period of time, unrelated to input, environment, or load. Drift is gradual. If variation is rapid, the fluctuation is referred to as cycling.

Dynamic error band. Band refers to the spread or band of output-amplitude deviation incurred by a constant amplitude sine wave as its frequency is varied over a specified portion of the frequency spectrum (see Static Error Band).

Error. The difference between the true or actual value to be measured (input signal) and the indicated value (output) from the measuring system. Errors can be systematic or random.

Error, accuracy. See Error, Systematic.

Error, fixed. See Error, Systematic.

Error, instrument. An instrument reading that includes random or systematic errors.

Error, mean square, or **RMS.** An accuracy statement of a system comprised of several items. For example, a laboratory potentiometer, volt box, null detector, and reference voltage source have individual accuracy statements assigned to them. These errors are generally independent of one another, so a system comprised of these units displays an accuracy given by the square root of the sum of the squares of the individual limits of error. For example, four individual errors of 0.1% yield a calibrated accuracy of 0.4% but an RMS error of only 0.2%.

Error, measurement. The result of a correction that is applied for any system error.

Error, precision. See Error, Random.

Error, probable. An error with a 50% chance of occurrence. A statement of probable error is of little value.

Error, random. A statistical error caused by chance and not recurring. This term is a general category for errors that can take values on either side of an average value. To describe a random error, its distribution must be known.

Error, systematic. A persistent error not caused by chance. Systematic errors are causal. A systematic error is likely to have the same magnitude and sign for every instrument constructed with the same components and procedures. Errors in calibrating equipment cause systematic errors, because all instruments calibrated are biased in the direction of the calibrating equipment error. Voltage and resistance drifts with time are generally in one direction and are classed as systematic errors.

Frequency response (flat). The portion of the frequency spectrum over which the measuring system has a constant value of amplitude response and a constant value of time lag. Input signals that have frequency components within this range are indicated by the measuring system (without distortion).

The preparation of this chapter is assigned to TC 1.2, Instruments and Measurements.

Hysteresis. The summation of all effects, under constant environmental conditions, which causes output of an instrument to assume different values at a given stimulus point when that point is approached first with increasing stimulus and then with decreasing stimulus. Hysteresis includes backlash. It is usually measured as a percent of full scale when input varies over the full increasing and decreasing range. In instrumentation, hysteresis and dead band are similar quantities.

Linearity. The straight-lineness of the transfer curve between an input and an output—that condition prevailing when output is directly proportional to input (see Nonlinearity).

Loading error. A loss of output signal from a device caused by a current drawn from its output. It increases the voltage drop across the internal impedance where no voltage drop is desired.

Mean. See Average.

Median. The middle value in a distribution, above and below which lie an equal number of values.

Noise. Any unwanted disturbance or spurious signal that modifies the transmission, measurement, or recording of desired data.

Nonlinearity. The prevailing condition (and the extent of its measurement) under which the input-output relationship fails to be a straight line, which is known as the input-output curve, transfer characteristic, calibration curve, or response curve. Nonlinearity is measured and reported in several ways, and the way, along with the magnitude, must be stated in any specification.

Minimum-deviation-based nonlinearity. The maximum departure between the calibration curve and a straight line drawn to give the greatest accuracy; expressed as a percent of full-scale deflection.

Slope-based nonlinearity. The ratio of maximum slope error anywhere on the calibration curve to the slope of the nominal sensitivity line; usually expressed as a percent of nominal slope.

Most variations beyond these two definitions result from the many ways in which the straight line can be arbitrarily drawn. All are valid as long as construction of the straight line is explicit.

Population. A group of individual persons, objects, or items from which samples may be taken for statistical measurement.

Precision. The repeatability of measurements of the same quantity under the same conditions, not a measure of absolute accuracy. The precision of a measurement is used here to describe the relative tightness of the distribution of measurements of a quantity about their mean value. Therefore, precision of a measurement is associated more with its repeatability than its accuracy. It combines uncertainty caused by random differences in a number of identical measurements and the smallest readable increment of the scale or chart. Precision is given in terms of deviation from a mean value.

Primary calibration. A calibration procedure in which the instrument output is observed and recorded, while the input stimulus is applied under precise conditions—usually from a primary external standard traceable directly to the National Institute of Standards and Technology.

Range. A statement of the upper and lower limits between which an instrument's input can be received and for which the instrument is calibrated.

Reliability. The probability that an instrument's repeatability and accuracy will continue to fall within specified limits.

Repeatability. See Precision.

Reproducibility. In instrumentation, the closeness of agreement among repeated measurements of the output for the same value of input made under the same operating conditions over a period of time, approaching from both directions; it is usually measured as a nonreproducibility and expressed as reproducibility in percent of span for a specified time period. Normally, this implies a long period of time, but under certain conditions, the period may be a short time during which drift may not be included. Reproducibility includes hysteresis, dead bank, drift,

and repeatability. Between repeated measurements, the input may vary over the range, and operating conditions may vary within normal operating conditions.

Resolution. The smallest change in input that produces a detectable change in instrument output. Resolution differs from precision in that it is a psychophysical term referring to the smallest increment of humanly perceptible output (rated in terms of the corresponding increment of input). Either or both the precision and resolution may be better than the accuracy. An ordinary six-digit (or dial) instrument has a resolution of one part per million (ppm) of full scale; however, it is possible that the accuracy is no better than 25 ppm (0.0025%). Note that the practical resolution of an instrument cannot be any better than the resolution of the indicator or detector—whether it is an internal or external device.

Scale factor. (1) The amount by which a measured quantity must change to produce unity output; (2) the ratio of real to analog values.

Sensitivity. The property of an instrument that determines scale factor. The word is often short for maximum sensitivity or the minimum scale factor with which an instrument can respond. The minimum input signal strength required to produce a desired value of output signal (*e.g.*, full scale or unit output or the ratio of output to input values).

Sensitivity inaccuracy. The maximum error in sensitivity displayed as a result of the summation of the following: frequency response; attenuator inaccuracy; hysteresis or dead-band; amplitude distortion (sensitivity nonlinearity); phase distortion (change in phase relationship between input signal and output deflection); and gain instability. Only by taking into account all these factors can nominal sensitivity, as indicated by the numeral on the attenuator readout, be discounted for accurate interpretation.

Stability. (1) Independence or freedom from changes in one quantity as the result of a change in another; (2) the absence of drift.

Static error band. (1) The spread of error present if the indicator (pen, needle) stopped at some value, *e.g.*, at one-half of full scale. It is normally reported as a percent of full scale; (2) a specification or rating of maximum departure from the point where the indicator must be when an on-scale signal is stopped and held at a given signal level. This definition stipulates that the stopped position can be approached from either direction in following any random waveform. Therefore, it is a quantity that includes hysteresis and nonlinearity but excludes items such as chart paper accuracy or electrical drift (see Dynamic Error Band).

Step-function response. The characteristic curve or output plotted against time resulting from the input application of a step function (a function that is zero for all values of time before a certain instant, and a constant for all values of time thereafter).

Threshold. The smallest stimulus or signal that results in a detectable output.

Time constant. The time required for an exponential quantity to change by an amount equal to 0.632 times the total change required to reach steady state for first-order systems.

Transducer. A device for translating the changing magnitude of one kind of quantity into corresponding changes of another kind of quantity. The second quantity often has dimensions different from the first and serves as the source of a useful signal. The first quantity may be considered an input and the second an output. Significant energy may or may not transfer from the transducer's input to output.

Uncertainty. An estimated value for the error, *i.e.*, what an error might be if it were measured by calibration. Although uncertainty may be the result of both accuracy and precision errors, only precision error can be treated by statistical methods.

Zero shift. Drift in zero indication of an instrument without any change in measured variable.

ERROR ANALYSIS

The measurement of any physical quantity such as length, mass, time, etc. is rarely made without error. The only measurement that has any possibility of being without error is one that involves the counting of discrete items. The objective of a measurement is to determine the value of the specific quantity to be measured (measurand). In general, the result of a measurement is only an estimate of the value of the measurand; thus it is complete only when an uncertainty for that estimate is also given.

The concepts of *error* and *uncertainty* are often confusing, and, in many cases, the two are not clearly distinguished. The error of a measurement is the difference between the measured value of a quantity and its "true value." The uncertainty expresses incomplete knowledge about the effects of errors in measurements. (*Note*: The word uncertainty is used in two senses. It refers to both the general concept and to specific quantitative measures, such as the standard deviation.) The terms error and true value are ideal concepts based on unknowable quantities.

The error in a measurement result is often composed of a number of individual errors from many sources. Error analysis is the study and evaluation of the effects these individual errors have on the total error of a measurement result. The error of a measurement can never be completely known; at best, a range of values within which the true value of a measurand is believed to lie can be determined. This range of values is the uncertainty, an expression of the possible magnitude of the error. In arriving at a realistic measure of uncertainty, the effects of all potential sources of error should be evaluated.

Error Sources

Measurement generally consists of a sequence of operations or steps. Virtually every step introduces a conceivable source of error the effect of which must be assessed. The following list is representative of the most common, but not all, sources of error.

- Inaccuracy in the mathematical model that describes the physical quantity
- Inherent stochastic variability of the measurement process
- Uncertainties in measurement standards and calibrated instrumentation
- Time-dependent instabilities due to gradual changes in standards and instrumentation
- Effects of environmental factors such as temperature, humidity, and pressure
- Values of constants and other parameters obtained from outside sources
- Uncertainties arising from interferences, impurities, inhomogeneity, inadequate resolution, and incomplete discrimination
- Computational uncertainties and data analysis
- Incorrect specifications and procedural errors
- Laboratory practice including handling techniques, cleanliness, and operator techniques, etc.

Error Classifications

Measurement includes two properties: (1) repeated measurements of the same quantity will disagree, and (2) the limiting mean might differ from the *true value* of the quantity of interest. (The limiting mean μ is the value the sample average converges to as the number of measurements are increased indefinitely. If $\bar{x}_n$ is the sample average based on n measurements, as $n \rightarrow \infty$, $\bar{x}_n \rightarrow \mu$.) These properties are the result of two kinds of errors—*random* and *systematic*. This classification of error is based on the way it behaves under repeated measurements of the same quantity under nearly identical measurement conditions.

Random errors are associated with effects that produce variations in repeated observations of the measurand. The variations are assumed to arise from the inability to hold completely constant all the variables and factors that can affect the measurement. Each measurement is a single realization of a hypothetical infinite population of similar measurements, and for that measurement, the random error is not known in sign or magnitude. The random error cannot be corrected for. Its effect can be reduced by increasing the number of measurements. What is known about the error is expressed as a limit based on statistical confidence intervals.

An error that remains unchanged when the measurement is repeated under substantially the same conditions is termed a *systematic error*. They cause all measurements to be in error by the same amount and, as such, cannot be detected by replicate measurements. For example, any error in a calibrated instrument will affect all measurements by the same amount. Some systematic errors may be estimated and compensated for by applying appropriate corrections. However, since systematic error cannot be known completely, some uncertainty is associated with the correction. Knowledge concerning systematic error is expressed by bounds that cover the error.

A graphical depiction of random and systematic errors is given in Figure 1. The (total) error in a measurement is simply the difference between the measured value x_i and the true value τ. This error is given by

$$\epsilon_i = x_i - \tau \tag{1}$$

The curve is the *normal* or *Guassian* distribution and is the theoretical distribution function for the conceptual infinite population of measurements that generated x_i. This distribution is characterized by a limiting mean μ and standard deviation σ. The random error associated with taking x_i as a measure of μ is given by

$$\epsilon_i = x_i - \mu \tag{2}$$

while the systematic error is given by

$$\delta = \mu - \tau \tag{3}$$

The total error ϵ_i is composed of two components, or

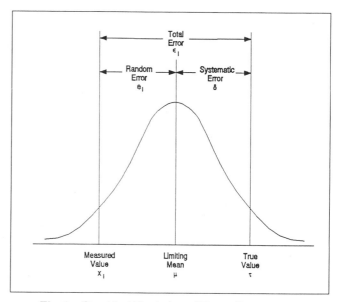

Fig. 1 Graphical Depiction of Error Components

$$\begin{aligned}
\epsilon_i &= x_i - \tau \\
&= (x_i - \mu) + (\mu - \tau) \\
&= \text{Random} + \text{Systematic}
\end{aligned} \tag{4}$$

Another measurement, *e.g.*, x_j, will give a different value as an estimate for τ. The systematic error will remain the same, but the random error $(x_j - \mu)$ will, in general, be different.

Assessment of Random Errors

The magnitude of a specific random error cannot be established because it is a chance event from a distribution of possible outcomes. The distribution is characterized by a standard deviation σ, which describes the spread of possible values. This standard deviation is estimated from repeated measurements, $x_1, x_2, \cdots$, x_n and is given by s, where

$$s = \sqrt{\frac{\sum\limits_{i=1}^{n}(x_i - \bar{x})^2}{(n-1)}} \text{ and } \bar{x} = \sum\limits_{i=1}^{n}\frac{x_i}{n} \tag{5}$$

Usually, the mean of a number of measurements $\bar{x}$ is taken as the measurement result. In this case, the effects of random error are reduced, and a measure of the uncertainty in $\bar{x}$ is given by the standard deviation of the mean (or the standard error) and computed by

$$s_{\bar{x}} = s/\sqrt{n} \tag{6}$$

The limit to random error for $\bar{x}$ at the $(1 - \alpha)$ confidence level is

$$t_{\alpha/2}(\nu)\,s_{\bar{x}} \tag{7}$$

where $t_{\alpha/2}(\nu)$ is the upper $\alpha/2$ percentage point of a t-distribution with ν degrees of freedom. Common values for α are 0.05 and 0.01, corresponding to 95 and 99% confidence intervals for μ, respectively. Often, instead of using the t-statistic, the random uncertainty is given by two or three times the standard deviation.

If there are several sources of random error affecting the measured quantity, a total standard deviation is computed; this accounts for the combined effect of the individual sources of error. A discussion of the statistical methods for more complex error models can be found in Croakin (1984), Mandel (1986), Ku (1966), and Eisenhard (1963).

Assessment of Systematic Errors

The evaluation of systematic errors is more troublesome and problematic and generally is not amenable to the use of statistical methods. Contributions to systematic error are from auxiliary quantities, constants, corrections, and so forth. The effects of these errors are often determined by subjective methods and are largely based on a subject matter expertise. Their magnitudes may be based on experimental verification, but judgment and experience on the part of the experimentalist may have to be used.

Since the systematic errors are not reliably known, credible bounds must be estimated. Credible bounds are analogous to limits to random error, in that they should be large enough to guarantee coverage for the systematic error with high reliability. Each systematic uncertainty component is considered a quasi-absolute upper bound for the effect of the error. Its magnitude is typically estimated in terms of an interval from $-\delta$ to $+\delta$. Technical literature should be consulted for specific applications of credible bounds.

Where several sources of systematic error exist, estimated bounds $\pm\delta_i$ are obtained for each component. The question of how to combine these into an overall systematic uncertainty is much debated, and there is generally no agreement on how these various bounds should be combined. The two most common methods of combining the components are by linear addition and quadrature addition (root-sum-square).

Overall Uncertainty

The result of a measurement is only an estimate of the value of the measurand, because of uncertainty due to random error and with corrections for both the known and unknown systematic errors. The uncertainty of a measurement is, therefore, a measure of the extent to which it *may* be in error, *i.e.*, how well it estimates the value of the measurand.

While it is important to identify and place limits on individual error sources, an overall uncertainty that accounts for the combined effect of all the sources of error is often needed. Several procedures combine the individual uncertainty components into an overall uncertainty.

One method is to linearly add all the systematic uncertainty components to the random uncertainty component. Thus, if there are k sources of systematic error, each having bounds given by δ_i, and s is the total standard deviation from all sources of random error, then the overall uncertainty U is given by

$$U = ts + \sum_{i=1}^{k}\delta_i \tag{8}$$

where t is the experimentalist's t-value.

Two other methods have also been used. These add in quadrature the systematic uncertainty components, and either add the resulting quantity linearly to the random uncertainty

$$U = ts + \sqrt{\sum_{i=1}^{k}\delta_i^2} \tag{9}$$

or add in quadrature to the random uncertainty

$$U = \sqrt{(ts)^2 + \sum_{i=1}^{k}\delta_i^2} \tag{10}$$

The procedures in Equations (9) and (10) are recommended in the ANSI/ASME standard on measurement uncertainty (ANSI/ASME 1985). Another guideline for the classical treatment of uncertainties can be found in Eisenhart *et al.* (1983).

TEMPERATURE MEASUREMENT

Instruments for measuring temperatures are listed in Table 1. Temperature sensor output must be related to an accepted temperature scale. This is achieved by manufacturing the instrument according to certain specifications or by calibrating it against a temperature standard. To help users conform to standard temperatures and temperature measurements, the International Committee of Weights and Measures (CIPM) has adopted the International Temperature Scale of 1990 (ITS-90).

The unit of temperature of the ITS-90 is the kelvin (K) and has a size equal to the fraction 1/273.16 of the thermodynamic temperature of the triple point of water.

The ITS-90 is maintained in the United States by the National Institute of Standards and Technology (NIST), and any laboratory may obtain calibrations from NIST based on this scale.

Benedict (1984), Considine (1985), Quinn (1990), Schooley (1986), DeWitt and Nutter (1988), and the American Institute of Physics' publication *Temperature: Its Measurement and Control in Science and Industry* (1992) cover temperature measurement in more detail.

Table 1 Temperature Measurement

Measurement Means	Application	Approximate Range, °F	Uncertainty, °F	Limitations
Glass-stem thermometers				
Mercury-glass thermometer	Temperature of gases and liquids by contact	−36/1000	0.05 to 3.6	In gases, accuracy affected by radiation
Organic thermometer	Temperature of gases and liquids by contact	−330/400	0.05 to 3.6	In gases, accuracy affected by radiation
Gas thermometer	Primary standard	−456/1200	Less than 0.02	Requires considerable skill to use
Resistance thermometers				
Platinum-resistance thermometer	Precision; remote readings; temperature of fluids or solids by contact	−430/1800	Less than 0.0002 to 0.2	High cost; accuracy affected by radiation in gases
Rhodium-iron resistance thermometer		−460/−400	0.0002 to 0.2	High cost
Nickel-resistance thermometer	Remote readings; temperature by contact	−420/400	0.02 to 2	Accuracy affected by radiation in gases
Germanium-resistance thermometer	Remote readings; temperature by contact	−460/−400	0.0002 to 0.2	
Thermistors	Remote readings; temperature by contact	Up to 400	0.0002 to 0.2	
Thermocouples				
Pt-Rh/Pt (type S) thermocouple	Standard for thermocouples on IPTS-68, not on ITS-90	32/2650	0.2 to 5	High cost
Au/Pt thermocouple		−60/1800	0.1 to 2	High cost
Types K and N thermocouples	General testing of high temperature; remote rapid readings by direct contact	Up to 2300	0.2 to 18	Less accurate than listed above thermocouples
Iron/constantan (type J) thermocouple		Up to 1400	0.2 to 10	Subject to oxidation
Copper/constantan (type T) thermocouple	Same as above, especially suited for low temperature	Up to 660	0.2 to 5	
Ni-Cr/constantan (type E) thermocouple	Same as above, especially suited for low temperature	Up to 1650	0.2 to 13	
Beckman thermometers (metastatic)	For differential temperature in same applications as in glass-stem thermometer	10°F scale, used 32–212°F	0.01	Must be set for temperature to be measured
Bimetallic thermometers	For approximate temperature	−4/1200	2, usually much more	Time lag; unsuitable for remote use
Pressure-bulb thermometers				
Gas-filled bulb	Remote-testing	−100/1200	4	Caution must be exercised so that installation is correct
Vapor-filled bulb	Remote-testing	−25/500	4	Caution must be exercised so that installation is correct
Liquid-filled bulb	Remote-testing	−60/2100	4	Caution must be exercised so that installation is correct
Optical pyrometers	For intensity of narrow spectral band of high-temperature radiation (remote)	1500 upward	30	
Radiation pyrometers	For intensity of total high-temperature radiation (remote)	Any range		
Seger cones (fusion pyrometers)	Approximate temperature (within temperature source)	1200/3600	90	
Triple points, freezing/melting and boiling points of materials	Standards	All except extremely high temperatures	Extremely precise	For laboratory use only

Thermometers

Any device that changes monotonically with temperature is a thermometer; however, the term usually signifies an ordinary liquid-in-glass temperature-indicating device. Mercury-filled thermometers have a useful range from $-37.8\,°F$, the freezing point of mercury, to about $1000\,°F$, near which the glass usually softens. Lower temperatures can be measured with organic-liquid-filled thermometers, e.g., alcohol, with ranges of -330 to $400\,°F$. During manufacture, thermometers are roughly calibrated for at least two temperatures, often the freezing and boiling points of water; space between the calibration points is divided into desired scale divisions. Thermometers that are intended for precise measurement applications have scales etched into the glass that forms their stems. The probable error for as-manufactured, etched-stem thermometers is ± 1 scale division. The highest quality mercury thermometers may have uncertainties of ± 0.06 to $\pm 4\,°F$, if they have been calibrated by comparison against primary reference standards.

Liquid-in-glass thermometers are used for many applications within the HVAC industry. Some of these uses include local indication of process fluids related to HVAC systems, such as cooling and heating fluid temperatures, and air temperatures.

The use of mercury-in-glass thermometers as temperature measurement standards is fairly common because of their relatively high accuracy and low cost. Such thermometers used as references must be calibrated on the ITS-90 by comparison in a uniform bath with a standard platinum resistance thermometer that has been calibrated either by the appropriate standards agency or a laboratory that has direct traceability to the standards agency and the ITS-90. Such a calibration is necessary in order to determine the proper corrections to be applied to the scale readings. For application and calibration of liquid-in-glass thermometers, refer to NIST (1976, 1986).

Liquid-in-glass thermometers are calibrated by the manufacturer for total or partial stem immersion. If a thermometer calibrated for total immersion is used at partial immersion, i.e., with a portion of the liquid column at a temperature different from that of the bath, an emergent stem correction must be made. This correction can be calculated as follows:

$$\text{Stem correction} = K_n (t_b - t_s) \tag{11}$$

where

K = differential expansion coefficient of mercury or other liquid in glass. K is 0.00016 for Celsius mercurial thermometers. For K values for other liquids and specific glasses, refer to AIP (1962).
n = number of degrees that liquid column emerges from bath
t_b = temperature of bath
t_s = average temperature of emergent liquid column of n degrees

Sources of Thermometer Errors

A thermometer measuring gas temperatures can be affected by radiation from surrounding surfaces. If the temperature is approximately the same as that of the surrounding surfaces, radiation effects can be ignored. If the temperature differs considerably from that of the surroundings, radiation effects should be minimized by *shielding* or *aspiration* (ASME 1985a). Shielding may be provided by highly reflective surfaces placed between the thermometer bulb and the surrounding surfaces, so that air movement around the bulb is not appreciably restricted (Parmelee and Huebscher 1946). Improper shielding can increase errors. Aspiration results from passing a high-velocity stream of air or gas over the thermometer bulb.

When a *thermometer well* within a container or pipe under pressure is required, the thermometer should fit snugly and be surrounded with a high thermal conductivity material (oil, water,

or mercury, if suitable). Liquid in a long, thin-walled well is advantageous for rapid response to temperature changes. The surface of the pipe or container around the well should be insulated to eliminate heat transfer to or from the well.

Industrial thermometers are available for permanent installation in pipes or ducts. These instruments are fitted with metal guards to prevent breakage and are useful for many other purposes. The considerable heat capacity and conductance of the guards or shields can cause errors, however.

Ample time for the thermometer to attain temperature equilibrium with the surrounding fluid prevents excessive errors in temperature measurements. When reading a liquid-in-glass thermometer, the eye should be kept at the same level as the top of the liquid column to avoid parallax.

Resistance Thermometers

Resistance thermometers depend on a change of the electrical resistance of a sensing element (usually metal) with a change in temperature; resistance increases with increasing temperature. The use of resistance thermometers largely parallels that of thermocouples, although readings are usually unstable above about $1000\,°F$. Two-lead temperature elements are not recommended, since they do not permit correction for lead resistance. Three leads to each resistor are necessary to obtain consistent readings, and four leads are preferred. Wheatstone bridge circuits or 6-1/2-digit multimeters can be used for measurements.

A typical circuit used by several manufacturers is shown in Figure 2. In this design, a differential galvanometer is used, in which coils L and H exert opposing forces on the indicating needle. Coil L is in series with the thermometer resistance AB, and coil H is in series with the constant resistance R. As the temperature falls, the resistance of AB decreases, allowing more current to flow through coil L than coil H. This causes an increase in the force exerted by coil L, pulling the needle down to a lower reading. Likewise, as the temperature rises, the resistance of AB increases, causing less current to flow through coil L than coil H. This forces the indicating needle to a higher reading. Rheostat S must be adjusted occasionally to maintain a constant flow of current.

The resistance thermometer is more costly to make and more likely to have considerably longer response times than thermocouples. A resistance thermometer gives best results when used to measure steady or slowly changing temperature.

Resistance Temperature Devices

Resistance temperature devices or RTDs are typically constructed from platinum, rhodium-iron, nickel, nickel-iron, tungsten, or copper. These devices are further characterized by their simple circuit designs, a high degree of linearity, good sensitivity,

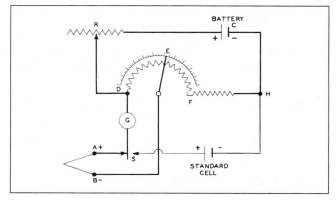

Fig. 2   Basic Thermocouple and Potentiometer Circuit

and excellent stability. The choice of materials for an RTD usually depends on the intended application, where temperature range, corrosion protection, mechanical stability, and cost are some of the selection criteria.

Platinum RTDs. Presently, for HVAC applications, RTDs constructed of platinum are the most widely used. Platinum is extremely stable and corrosion resistant. Platinum RTDs possess high malleability, and can thus be drawn into fine wires. They have a high melting point and can be refined to a high degree of purity, thus attaining highly reproducible properties. Due to these properties, platinum RTDs are used to define the ITS-90 for the range of 13.8033 K (triple point of equilibrium hydrogen) to 1234.93 K (freezing-point temperature of silver).

Platinum resistance temperature devices (PRTs) can measure the widest range of temperatures and are the most accurate and stable temperature sensors. Their resistance-temperature relationship is one of the most linear. The higher the purity of platinum, the more stable and accurate the sensor. With high-purity platinum, primary grade PRTs are capable of achieving reproducibilities of ±0.00002°F, whereas the minimum uncertainty of a recently calibrated thermocouple is only ±0.4°F.

Platinum RTD design. The most widely used RTD is designed with a resistance of 100 Ω at 32°F ($R_0 = 100 \, \Omega$). Other resistance values are available where lower resistances are used at temperatures above 1100°F. The lower the resistance value, the faster the response time for sensors of the same size.

Thin-film 1000 Ω and higher platinum RTDs are readily available. They have the excellent linear properties of lower resistance platinum RTDs and are more cost-effective since they are mass produced and have lower platinum purity. However, the problem with many platinum RTDs with R_0 values of greater than 100 Ω is the difficulty in obtaining transmitters or electronic interface boards from sources other than the platinum RTD manufacturer. In addition to a *nonstandard* interface, higher R_0 value platinum RTDs may have higher self-heating losses if the excitation current is not controlled properly.

Thin-film RTDs have the advantages of lower cost and smaller sensor size. They are specifically adapted to surface mounting. Thin-film sensors tend to have an accuracy limitation of ±0.1% or 0.2°F. This may prove to be adequate for most HVAC applications; only in tightly controlled facilities may users wish to install the standard wire-wound platinum RTDs with accuracies of ±0.01% or 0.02°F (these are available upon special request for certain temperature ranges).

Regardless of the R_0 resistance value of RTDs, their assembly and construction are relatively simple. The electrical connections come in three basic types, depending on the number of wires to be connected to the resistance measurement circuitry. Either two, three, or four wires are used for electrical connection using a Wheatstone bridge or a variation of it (Figure 3). In the basic two-wire configuration, the resistance of the RTD is measured through the two connecting wires. Since the connecting wires extend from the site of the temperature measurement, any additional changes in resistivity due to a different temperature may change the measured resistance.

Three- and four-wire assemblies are built to compensate for the connecting lead resistance values. The original three-wire circuit improved the resistance measurement by adding a compensating wire to the voltage side of the circuit. This helps reduce self-heating and part of the connecting wire resistance. When more accurate measurements are required, better than ±0.2°F, the four-wire bridge is recommended. The four-wire bridge eliminates all connecting wire resistance errors.

All the bridges discussed here are dc circuits and were used extensively until the advent of precision ac circuits using microprocessor-controlled ratio transformers, dedicated analog-to-digital convertors, and other solid-state devices that measure resistance with uncertainties of less than 1 ppm. The newer resistance measurement devices allow for more portable thermometers, lower cost, ease of use, and the application of high-precision temperature measurement in industrial uses.

Thermistors

Certain semiconductor compounds (usually sintered metallic oxides) exhibit large changes in resistance with temperature, usually decreasing as the temperature increases. For use, the thermistor element may be connected by lead wires into a galvanometer bridge circuit and calibrated. A 6-1/2-digit multimeter and a constant-current source, with a means for reversing the current to eliminate thermal electromotive force (emf) effects, may also be used. This method of measurement is easier, faster, and may be more precise and accurate. Thermistors are usually applied to electronic temperature compensation circuits, such as thermocouple reference junction compensation, or to other applications where high resolution and limited operating temperature ranges exist. Figure 4 illustrates a typical thermistor circuit.

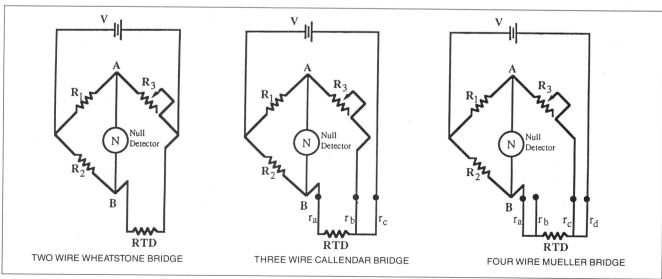

Fig. 3 Typical Resistance Temperature Bridge Circuits

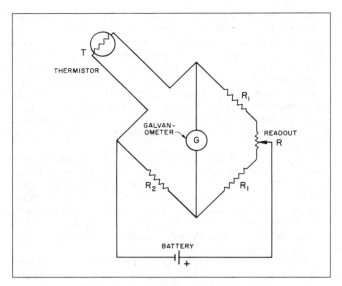

Fig. 4   Basic Thermistor Circuit

Thermocouples

When two wires of dissimilar metals are joined by soldering, welding, or twisting, they form a thermocouple junction or thermojunction. An emf that depends on the wire materials and the junction temperature exists between the wires.

Thermocouples for temperature measurement yield less precise results than those for platinum resistance thermometers; but, except for glass thermometers, thermocouples are the most common instruments of temperature measurement for the range of 32 to 1800 °F. Due to their low cost, moderate reliability, and ease of use, thermocouples continue to maintain widespread acceptance.

The most commonly used thermocouples in industrial applications are assigned letter designations. The tolerances of such commercially available thermocouples are given in Table 2.

Because the measured emf is a function of the *difference* in temperature and the type of dissimilar metals used, a known temperature at one junction is required, whereas the remaining junction temperature may be calculated. Common practice refers to the *reference* junction as the one with a known temperature and the *measured* junction as the unknown temperature junction. The reference junction is typically kept at a reproducible temperature, such as the ice point of water.

Various systems are used to maintain the reference junction temperature—a mixture of ice and water contained in an insulated flask, or commercially available thermoelectric coolers to maintain the ice point temperature automatically within a reference chamber. When these systems cannot be used in an application, measuring instruments with automatic reference junction temperature compensation may be used.

As previously described, the principle for measuring temperature with a thermocouple is based on the accurate measurement of the Seebeck voltage. The acceptable dc voltage measurements methods are (1) millivoltmeter, (2) millivolt potentiometer, and (3) a high-input impedance digital voltmeter. Many digital voltmeters include built-in software routines for the direct calculation and display of temperature. Regardless of the method selected, many options to simplify the measurement process are available.

Solid-state digital readout devices in combination with a millivolt or microvoltmeter, as well as packaged thermocouple readouts with built-in cold junction and linearization circuits, are available. The latter requires a proper thermocouple to provide direct meter reading of temperature. Accuracies approaching or surpassing that of potentiometers can be attained, depending on the instrument quality. This method is popular because it eliminates the null balancing requirement and reads temperature directly in a digital readout.

Thermocouple wire is selected by considering the temperature to be measured, the corrosion protection afforded the thermocouple, and the precision and service life required. Type T thermocouples are suitable for temperatures up to 220 °F; type J, up to 1400 °F; and types K and N, up to 2300 °F. Higher temperatures require noble metal thermocouples (types S, R, or B), which have

Table 2   Tolerances on Initial Values of Electromotive Force versus Temperature

Thermocouple Type	Material Identification	Temperature Range, °F	Reference Junction Tolerance, 32 °F	
			Standard Tolerance (whichever is greater)	Special Tolerance (whichever is greater)
T	Copper versus Constantan	32 to 700	±1.8° or ±0.75%	±0.9°F or ±0.4%
J	Iron versus Constantan	32 to 1400	±4°F or ±0.75%	±2°F to ±0.4%
E	Nickel-10% Chromium versus Constantan	32 to 1600	±3.1°F or ±0.5%	±1.8°F or ±0.4%
K	Nickel-10% Chromium versus 5% Aluminum, Silicon	32 to 2300	±4°F or ±0.75%	±2°F or ±0.4%
N	Nickel-14% Chromium, 1.5% Silicon versus Nickel-4.5% Silicon, 0.1% Magnesium	32 to 2300	±4°F or ±0.75%	±2°F or ±0.4%
R	Platinum-13% Rhodium versus Platinum	32 to 2700	±2.7°F or ±0.25%	±1.1°F or ±0.1%
S	Platinum-10% Rhodium versus Platinum	32 to 2700	±2.7°F or ±0.25%	±1.1°F or ±0.1%
B	Platinum-30% Rhodium versus Platinum-6% Rhodium	1600 to 3100	±0.5%	±0.25%
T[a]	Copper versus Constantan	−328 to 32	±1.8°F or ±1.5%	b
E[a]	Nickel-10% Chromium versus Constantan	−328 to 32	±3.1°F or ±1%	b
K[a]	Nickel-10% Chromium versus 5% Aluminum, Silicon	−328 to 32	±4°F or ±2%	b

Source: ASTM E230-87, Temperature-Electromotive Force (EMF) Tables For Standardized Thermocouples.

Notes:

Tolerances in this table apply to new thermocouple wire, normally in the size range of 0.01 to 0.1 in. diameter and used at temperatures not exceeding the recommended limits.

[a]Thermocouples and thermocouple materials are normally supplied to meet the tolerance specified in the table for temperatures above 32 °F. The same materials, however, may not fall within the tolerances given in the second section of the table when operated below freezing (32 °F). If materials are required to meet tolerances

at low temperatures, the purchase order must state so. Selection of material usually will be required.

[b]Little information is available to justify establishing special tolerances for below-freezing temperatures. Limitied experience suggests the following special tolerances for types E and T thermocouples:

Type E   −330 to 32 °F; ±2 °F or ±0.5% (whichever is greater)
Type T   −330 to 32 °F; ±1 °F or ±0.8% (whichever is greater)

These tolerances are given only as a guide for discussion between purchaser and supplier.

a higher initial cost and do not develop as high an emf as the base metal thermocouples. Thermocouple wires of the same type have small compositional variation from lot to lot from the same manufacturer and especially from different manufacturers. Consequently, calibrating samples from each wire spool is essential for precision. Calibration data on wire may be obtained from the manufacturer.

Reference functions are available for relating temperature and emf of letter-designated thermocouple types. Such functions are easy to use with computers. The functions depend on thermocouple type and on the range of temperature; they are used to generate reference tables of emf as a function of temperature but are not well suited for calculating temperatures directly from values of emf. Approximate inverse functions are available, however, for calculating temperature and are of the form

$$t_{90} = \sum_{i=0}^{n} a_i E^i \tag{12}$$

where t_{90} = temperature, and E = voltage. Burns et al. (1992) give the reference functions and approximate inverses for all letter-designated thermocouples.

The emf of a thermocouple, as measured with a high-input impedance device, is independent of the diameters of its constituent wires. Thermocouples with small-diameter wires respond faster to temperature changes and are less affected by radiation than larger ones. Large-diameter wire thermocouples, however, are necessary for high-temperature work when wire corrosion is a problem. For use in heated air or gases, thermocouples are often shielded and sometimes aspirated. An arrangement for avoiding error because of radiation involves using several thermocouples of different wire sizes and estimating the true temperature by extrapolating readings to zero diameter.

By using thermocouples, temperatures can be indicated or recorded remotely on conveniently located instruments. Since thermocouples can be made of small-diameter wire, they can be used to measure temperatures within thin materials, within narrow spaces, or in otherwise inaccessible locations.

Thermocouples in series, with alternate junctions maintained at a common temperature, produce an emf that, when divided by the number of thermocouples, gives the average emf corresponding to the temperature difference between two sets of junctions. This series arrangement of thermocouples, often called a thermopile, is used to increase the sensitivity and is often used for measuring small temperature changes and differences.

A number of thermocouples of the same type, connected in parallel and having a common reference junction, is useful to obtain an average temperature of an object or volume. In such measurements, however, it is important that the electrical resistances of the individual thermocouples be the same. The use of thermocouples in series and parallel arrangements is discussed in ASTM STP 470B (1981).

The thermocouple is useful in determining surface temperature. It can be attached to a metal surface in several ways. For permanent installations, soldering, brazing, or peening is suggested. For the latter, a small hole is drilled and the thermocouple measuring junction is driven into it. For temporary arrangements, thermocouples can be attached by tape, adhesive, or putty-like material. For boiler or furnace surfaces, furnace cement should be used. To minimize the possibility of error caused by heat conduction along wires, a surface thermocouple should be made of fine wires placed in close contact with the surface being measured for about an inch from the junction to ensure good thermal contact. The wires must be insulated electrically from each other and from the metal surface (except at the junction).

Thermocouple construction. The thermocouple wires are typically insulated with fibrous glass, fluorocarbon resin, or ceramic insulators. Another form of thermocouple consists of the thermocouple wires being insulated with compacted ceramic insulation inside a metal sheath. This form of thermocouple provides both mechanical protection and protection from stray electromagnetic fields. The measuring junction may be exposed or enclosed within the metal sheath. The enclosed junction may be either grounded or ungrounded to the metal sheath.

For the exposed junction type, the measuring junction is in direct contact with the process stream; thus, it is subject to corrosion or contamination but provides a fast temperature response. The enclosed junction type, in which the thermocouple wires are welded to the metal sheath, provides for electrical grounding, as well as for mechanical and corrosion protection. This type, however, has a slower response time than the exposed junction type. With the ungrounded enclosed junction construction, the response time is even slower, but the thermocouple wires are isolated electrically and are less susceptible to some forms of mechanical strain than those of the grounded construction.

Semiconductor devices. In addition to the positive resistance coefficient RTDs and the negative resistance coefficient thermistor, there are two other types of devices that vary resistance or impedance with temperature. While the principle of their operation has long been known, their reliability was questioned because of imprecise manufacturing techniques. Improvements in silicon microelectronics manufacturing techniques during the 1980s have improved to the point where low-cost, precise temperature sensors are now commercially available.

Elemental semiconductors. Through the use of controlled doping of impurities into elemental germanium, a germanium semiconductor is a reliable temperature sensor for cryogenic temperature measurement in the range from 1.8 to 150 R.

Junction semiconductors. The first simple junction semiconductor device consisted of a single diode or transistor, in which the forward-connected base emitter voltage is very sensitive to temperature. Today the more common form is a pair of diode-connected transistors, which make the device suitable for ambient temperature measurement. Applications include thermocouple reference junction compensation.

The primary advantage of silicon transistor temperature sensors are their extreme linearity and exact R_0 value. Another advantage is the ability to incorporate signal conditioning circuitry into the same device as the sensor element. As with thermocouples, these semiconductors require highly precise manufacturing techniques, extremely precise voltage measurements, multiple point calibration, and temperature compensation to achieve an accuracy as high as $\pm 0.02\,°F$, but with a much higher cost. Lower cost devices will achieve accuracies of $\pm 0.2\,°F$, using mass manufacturing techniques and single point calibration. A mass-produced silicon temperature sensor can be easily interchanged. If one device fails, only the sensor element need be changed. Electronic circuitry can be used to recalibrate the new device.

Winding temperature. The winding temperature of electrical operating equipment is usually determined from the resistance change of these windings in operation. With copper windings, the relation between these parameters is

$$(R_1/R_2) = (148.1 + t_1)/(148.1 + t_2) \tag{13}$$

where

R_1 = winding resistance at temperature t_1, °F
R_2 = winding resistance at temperature t_2, °F

The classical method of determining winding temperature is to measure the equipment inoperative and temperature-stabilized at room temperature. After the equipment has operated sufficiently to cause temperature stabilization under load conditions, the winding resistance should be measured again. The latter value is obtained by taking resistance measurements at known short time intervals after shutdown. These values may be extrapolated to zero time to indicate the winding resistance at the time of shutdown. The obvious disadvantage of this method is that the device must be shut down to determine winding temperature. A circuit described by Seely (1955), however, makes it possible to measure resistances while the device is operating.

Sampling and Averaging

Although temperature is usually measured within, and is associated with, a relatively small volume (depending on the size of the thermometer), it can also be associated with an area such as on a surface or in a flowing stream. To determine average stream temperature, the cross section must be divided into smaller areas and the temperature of each area measured. The temperatures measured are then combined into a weighted mass flow average either by (1) using equal areas and multiplying each temperature by the fraction of total mass flow in its area, or (2) using areas of size inversely proportional to mass flow and taking a simple arithmetic average of the temperatures in each. A means of mixing or selective sampling may be preferable to these cumbersome procedures. While mixing can occur from turbulence alone, transposition is much more effective. In transposition, the stream is divided into parts determined by the type of stratification, and alternate parts pass through one another.

Static Temperature versus Total Temperature

When a fluid stream impinges on a temperature-sensing element such as a thermometer or thermocouple, the element is at a temperature greater than the true stream temperature. The difference is a fraction of the temperature equivalent of the stream velocity t_e.

$$t_e = V^2/(2c_p) \tag{14}$$

where

t_e = temperature equivalent of stream velocity, °F
V = velocity of stream, fpm
c_p = specific heat at constant pressure, Btu/lb · °F

This fraction of the temperature equivalent of the velocity is the *recovery factor*, and it varies from 0.6 to 0.8 °F (for bare thermometers) to 1.0 °F (for aerodynamically shielded thermocouples). For precise temperature measurement, each temperature sensor must be calibrated to determine its recovery factor. However, for most applications where air velocities are below 2000 fpm, this factor can be omitted.

Various temperature sensors are available for temperature measurement in fluid streams. The principal sensors are the *static temperature thermometer*, which indicates true stream temperature but is cumbersome, and the *thermistor*, used for accurate temperature measurement within a limited range.

Infrared Radiometers

Infrared radiation thermometers, also known as remote temperature sensors (Hudson 1969), permit noncontact measurement of surface temperature over a wide range. In these instruments, radiant flux from the observed object is focused by an optical system onto an infrared detector that generates an output signal, proportional to the incident radiation, which can be read from a meter or display unit. Point and scanning radiometers are available; the latter are able to display the temperature variation existing within the field of view.

Radiometers are usually classified according to the detector used—either thermal or photon. In thermal detectors, a change in electrical property is caused by the heating effect of the incident radiation. Examples of thermal detectors are the thermocouple, thermopile, and metallic and semiconductor bolometers. In photon detectors, a change in electrical property is caused by the surface absorption of incident photons. Because these detectors do not require an increase in temperature for activation, their response time is much shorter than that of thermal detectors. Scanning radiometers usually use photon detectors.

A radiometer only measures the power level of the radiation incident on the detector; this incident radiation is a combination of the thermal radiation emitted by the object and the surrounding background radiation reflected from the surface of the object. An accurate measurement of the temperature, therefore, requires knowledge of the long-wavelength emissivity of the object as well as the effective temperature of the thermal radiation field surrounding the object. Calibration against an internal or external source of known temperature and emissivity is required in order to obtain true surface temperature from the radiation measurements.

The temperature resolution of a radiometer decreases as the object temperature decreases. For example, a radiometer that can resolve a temperature difference of 0.5 °F on an object near 70 °F may only resolve a difference of 2 °F on an object at 32 °F.

Infrared Thermography

Infrared thermography is the discipline concerned with the acquisition and analysis of thermal information in the form of images from an infrared imaging system. An infrared imaging system consists of (1) an infrared television camera and (2) a display unit. The infrared television camera scans a surface and senses the self-emitted and reflected radiation viewed from the surface. The display unit contains either a cathode-ray tube (CRT) that displays a gray-tone or color-coded thermal image of the surface or a color LCD screen. A photograph of the image on the CRT is called a thermogram. An introductory treatise on infrared thermography is given by Paljak and Pettersson (1972).

Thermography has been used successfully to detect missing insulation and air infiltration paths in building envelopes (Burch and Hunt 1978). Standard practices for conducting thermographic inspections of buildings are given in ASTM (1989a). A technique to quantitatively map the heat loss in building envelopes is given by Mack (1986).

Aerial infrared thermography of buildings is effective in identifying regions of an individual built-up roof that have wet insulation (Tobiasson and Korhonen 1985), but it is ineffective in ranking a group of roofs according to their thermal resistance (Goldstein 1978, Burch 1980). In this latter application, the emittances of the separate roofs and outdoor climate (*i.e.*, temperature and wind speed) throughout the microclimate often produce changes in the thermal image that may be incorrectly attributed to differences in thermal resistance.

Industrial applications include locating defective or missing pipe insulation in buried heat distribution systems, surveys of manufacturing plants to quantify energy loss from equipment, and locating defects in coatings (Bentz and Martin 1987).

PRESSURE MEASUREMENTS

Definitions

Pressure is the force exerted per unit area by a pressure media, generally a liquid or gas. Pressure so defined is also sometimes referred to as **absolute pressure**. Thermodynamic and material properties are expressed in terms of absolute pressures; thus, the properties of a refrigerant will be given in terms of absolute pressures. **Vacuum** refers to low pressures, less than atmospheric.

Differential pressure is the difference between two absolute pressures. In many cases, the differential pressure can be very small compared to either of the absolute pressures (these are often referred to as low-range, high-line differential pressures). A common example of differential pressures is the pressure drop, or difference between inlet and outlet pressures, across a filter or flow element.

Gage pressure is a special case of differential pressure where one of the pressures (reference pressures) is atmospheric pressure. Many pressure gages, including most refrigeration test sets, are designed to make gage pressure measurements, and there are probably more gage pressure measurements made than any other. Gage pressure measurements are often used as surrogates for absolute pressures. However, because of variations in atmospheric pressure, due to elevation (atmospheric pressure in Denver, CO is about 81% of sea-level pressure) and weather changes, the measurement of gage pressures to determine absolute pressures can significantly restrict the accuracy of the measured pressure, unless corrections are made for the local atmospheric pressure at the time of the measurement.

Pressures can be further classified as **static** or **dynamic**. Static pressures have a small or undetectable change with time; dynamic pressures include a significant pulsed, oscillatory, or other time-dependent component. Static pressure measurements are the most common, but equipment such as blowers and compressors can generate significant oscillatory pressures at discrete frequencies. Flow in pipes and ducts can generate resonant pressure changes, as well as turbulent "noise" that can span a wide range of frequencies.

Units

A plethora of pressure units, many of them poorly defined, are in common use. The internationally accepted pressure unit (SI unit) is the newton/square meter, called the pascal (Pa). The bar is an acceptable alternate unit, but it should not be introduced where it is not used at present. Although not internationally recognized, the pound/square inch (psi) is a widely used and properly defined unit. Also widely used, but not as rigorously defined and, therefore, a potential source of error, are units based on the length of liquid columns, including inches of mercury (in. Hg), mm of mercury (mm Hg), and inches of water (in. H_2O). The latter is often used for low-range differential pressure measurements. In the case of pounds per square inch, the type of pressure measurement is often indicated by a modification of the unit, *i.e.*, both psi and psia are used to indicate absolute pressure measurements, psid indicates a differential measurement, and psig indicates a gage measurement. No such standard convention exists for other units, and unless explicitly stated, reported values are assumed to be absolute pressures. Conversion units for different pressure units can be found in Chapter 35.

The differences between the conversion factors for inches of mercury and inches of water at the different temperatures is indicative of the errors that can arise from uncertainties as to how these units are defined.

Types of Pressure-Measuring Instruments

Broadly speaking, pressure instruments can be divided into three different categories—standards, mechanical gages, and electromechanical transducers. Standards instruments are used for most-accurate calibrations. The most common, and potentially the most accurate standard, the liquid-column manometer, is used for a variety of applications, including field applications. Mechanical pressure gages are generally the least expensive and the most common pressure instruments. However, electromechanical transducers have become much less expensive and are easier to use, so that they are being used more often.

Pressure Standards

Liquid-column manometers measure pressure by determining the vertical displacement of a liquid of known density in a known gravitational field. Typically they are constructed as a "U" tube of transparent material (glass or plastic). The pressure to be measured is applied to one side of the U tube. If the other (reference) side is evacuated (zero pressure), the manometer measures absolute pressure, and if the reference side is open to the atmosphere, it measures gage pressures; if the reference side is connected to some other pressure, the manometer measures the differential between the two pressures. Manometers filled with water and different oils are often used to measure low-range differential pressures. In some low-range instruments, one tube of the manometer is inclined to enhance the readability. Mercury-filled manometers are used for higher range differential and absolute pressure measurements. In the latter case, the reference side is evacuated, generally with a mechanical vacuum pump. Typical full-scale ranges for manometers vary from 10 in. of water to 3 atm.

For pressures above the range of manometers, standards are generally of the piston-gage, pressure-balance, or deadweight-tester type. These instruments apply pressure to the bottom of a vertical piston, which is surrounded by a close-fitting cylinder (typical clearances are millionths of an inch). The pressure generates a force approximately equal to the pressure times the area of the piston. This force is balanced by weights stacked on the top of the piston. If the mass of the weights, the local acceleration of gravity, and the area of the piston (or more properly, the "effective area" of the piston and cylinder assembly) are known, the applied pressure can be calculated. Piston gages generally generate gage pressures with respect to the atmospheric pressure above the piston. They can be used to measure absolute pressures by separately measuring the atmospheric pressure and adding it to the gage pressure determined by the piston gage, or directly by surrounding the top of the piston and weights with an evacuated bell jar. Piston gage full-scale ranges vary from 5 psi to 200,000 psi.

At the other extreme, very low absolute pressures (below about 0.4 in. of water), a number of different types of standards are used. These tend to be specialized and expensive instruments found only in major standards laboratories. However, one low-pressure standard, the McLeod gage, has been used for field applications. Unfortunately, although the theory of the McLeod gage is simple and straightforward, it is difficult to make accurate measurements with this instrument, and major errors can occur when it is used to measure gases that condense or are adsorbed (*e.g.*, water). In general, gages other than the McLeod gage should be used for most low-pressure or vacuum applications.

Mechanical Pressure Gages

Mechanical pressure gages couple a pressure sensor to a mechanical readout, typically a pointer and dial. The most common type employs a Bourdon tube sensor, which is essentially a coiled metal tube of circular or elliptical cross section. Pressure applied to the inside of the tube causes it to uncoil with increasing pressure. A mechanical linkage translates the motion of the end of the tube to the rotation of a pointer. In most cases, the Bourdon tube is surrounded by atmospheric pressure so that the gages measure gage pressure. A few instruments surround the Bourdon tube with a sealed enclosure that can be evacuated for absolute measurements or connected to another pressure for differential measurements. Available instruments vary widely in cost, size, pressure range, and accuracy. Full-scale ranges can vary from 5 to 100,000 psi. Accuracy of properly calibrated and used instruments can vary from 0.1 to 10% of full scale. Generally there is a strong correlation between size, price, and accuracy; larger instruments are more accurate and expensive.

To achieve better sensitivity, some low-range mechanical gages, sometimes called aneroid gages, employ corrugated diaphragms or capsules as sensors. The capsule is basically a short bellows sealed with end caps. These sensors are more compliant than a Bourdon tube and a given applied pressure will cause a larger deflection of the sensor. The inside of a capsule can be evacuated and sealed in order to measure absolute pressures, or connected to an external fitting to allow differential pressures to be measured. Typically, these gages are used for low-range measurements of 1 atm or less. In instruments of better quality, accuracies of 0.1% of reading or better can be achieved.

Electromechanical Transducers

Mechanical pressure gages are generally limited by inelastic behavior of the sensing element, friction in the readout mechanism, and limited resolution of the pointer and dial. These effects can be eliminated or reduced by using electronic techniques to sense the distortion or stress of a mechanical sensing element and electronically convert that stress or distortion to a pressure reading. A wide variety of sensors is used, including Bourdon tubes, capsules, diaphragms, and different resonant structures whose vibration frequency varies with the applied pressure. Capacitive, inductive, and optical lever sensors are used to measure the displacement of the sensor element. In some cases, feedback techniques may be used to constrain the sensor in a null position, minimizing distortion and hysteresis of the sensing element. Temperature control or compensation is often included. Readout may be in the form of a digital display, analog voltage or current, or a digital code. Size varies, but in the case of transducers employing a diaphragm fabricated as part of a silicon chip, the sensor and signal conditioning electronics can be contained in a small transistor package, and the largest part of the device is the pressure fitting. The best of these instruments achieve long-term instabilities of 0.01% of full scale or better, and corresponding accuracies when properly calibrated. Performance of the less expensive instruments can be more on the order of several percent.

While the dynamic response of most mechanical gages is limited by the sensor and readout, the response of some electromechanical transducers can be much faster, allowing measurements of dynamic pressures at frequencies up to 1 kHz and beyond in the case of transducers specifically designed for dynamic measurements. Manufacturers' literature should be consulted as a guide to the dynamic response of specific instruments.

As the measured pressure is reduced below about 1.5 psi, it becomes increasingly difficult to mechanically sense the lower pressures. A variety of gages have been developed that measure some other property of the gas that is related to the pressure. In particular, thermal conductivity gages, known as thermocouple, thermistor, pirani, and convection gages, are used for pressures down to about 0.0004 in. of water. These gages have a sensor tube with a small heated element and a temperature sensor; the temperature of the heated element is determined by the thermal conductivity of the gas, and the output of the temperature sensor is displayed on an analog or digital electrical meter contained in an attached electronics unit. The accuracy of these gages is limited by their nonlinearity, dependence on gas species, and tendency to read high when contaminated. Oil contamination is a particular problem. However, these gages are small, reasonably rugged, and relatively inexpensive; in the hands of a typical user, they will give far more reliable results than a McLeod gage. They can be used to check the base pressure in a system that is being evacuated prior to being filled with refrigerant. They should be periodically checked for contamination by comparing the reading with that from a new, clean sensor tube.

General Considerations

Accurate values of atmospheric or barometric pressure are required for weather prediction and aircraft altimetry. In the United States, a network of calibrated instruments, generally accurate to within 0.1% of reading and located at airports, are maintained by the National Weather Service, the Federal Aviation Administration, and local airport operating authorities. These agencies are generally cooperative in providing current values of atmospheric pressure that can be used to check the calibration of absolute pressure gages or to correct gage pressure readings to absolute pressures. However, the pressure readings generally reported for weather and altimetry purposes are not the true atmospheric pressure, but rather a value adjusted to an equivalent sea level pressure. Therefore, unless located near sea level, it is important to ask for the station or true atmospheric pressure and not to use the values broadcast by radio stations as they are always adjusted values. Further, the atmospheric pressure decreases with increasing elevation at a rate (near sea level) of about 0.001 in. Hg/ft, and corresponding corrections should be made to account for the difference in elevation between the instruments being compared.

As noted before, gage-pressure instruments are sometimes used to measure absolute pressures, and the accuracy of these measurements can be compromised by uncertainties in the atmospheric pressure. This error can be particularly serious when gage-pressure instruments are used to measure a vacuum (negative gage pressures). For all but the most crude measurements, absolute-pressure gages should be used for vacuum measurements; for pressures below about 0.4 in. of water, a thermal conductivity gage should be used.

All pressure gages are susceptible to temperature errors. Several techniques are used to minimize these errors—sensor materials are generally chosen to minimize temperature effects, mechanical readouts can include temperature compensation elements, electromechanical transducers may include a temperature sensor and compensation circuit, and some transducers are operated at a controlled temperature. Clearly, temperature effects are of greater concern for field applications, and it is prudent to check the manufacturers' literature for the temperature range over which the specified accuracy can be maintained. Abrupt temperature changes can also cause large transient errors that may take some time to decay.

The readings of some electromechanical transducers with a resonant or vibrating sensor can depend on the gas species. Although some of these units can achieve calibrated accuracies of the order of 0.01% of reading, they are typically calibrated with dry air or nitrogen, and the readings for other gases can be in error by several percent, quite possibly much more for refrigerants and other high-density gases. High accuracy readings can be maintained by calibrating these devices with the gas to be measured. Manufacturer's literature should be consulted.

The measurement of dynamic pressures is limited not just by the frequency response of the pressure gage, but also by the hydraulic or pneumatic time constant of the connection between the gage and the system to be monitored. As a general rule, the longer the connecting lines and the smaller their diameter, the lower the frequency response of the system. Further, even if only the static component of the pressure is of interest and a gage with a low frequency response is used, a significant pulsating or oscillating pressure component can cause significant errors in pressure gage readings and, in some cases, can damage the gage, particularly gages with a mechanical readout mechanism. In these cases, a filter or snubber should be used to reduce the higher frequency components.

VELOCITY MEASUREMENT

Heating and air-conditioning engineers measure the flow of air more often than any other gases, and the air is usually measured at or near atmospheric pressure. Under this condition, the air can be treated as an incompressible fluid, and simple formulas give

Table 3 Velocity Measurement

Measurement Means	Application	Range, fpm	Precision	Limitations
Smoke puff or airborne solid tracer	Low air velocities in rooms; highly directional	5 to 50	10 to 20%	Awkward to use but valuable in tracing air movement
Deflecting-vane anemometer	Air velocities in rooms, at outlets, etc., directional	30 to 24,000	5%	Needs periodic check calibration
Revolving-vane anemometer	Moderate air velocities in ducts and rooms; somewhat directional	100 to 3000	2 to 5%	Extremely subject to error with variations in velocities with space or time; easily damaged; needs periodic calibration
Pitot tube	Standard instrument for measuring duct velocities	180 to 10,000 with micromanometer; 600 to 10,000 with draft gages; 10,000 up with manometer	1 to 5%	Accuracy falls off at low end of range
Impact tube and sidewall or other static tap	High velocities, small tubes and where air direction may be variable	120 to 10,000 with micromanometer; 600 to 10,000 with draft gages; 10,000 up with manometer	1 to 5%	Accuracy depends on constancy of static pressure across stream section
Heated thermocouple anemometer	Air velocities in ducts, velocity distributions	10 to 2000	3 to 20%	Accuracy of some types not good at lower end of range; steady-state measurements only
Hot-wire anemometer	a. Low air velocities; directional and nondirectional available	1 to 1000	2 to 5%	Requires accurate calibration at frequent intervals. Some are relatively costly.
	b. High air velocities	Up to 60,000	0.2 to 5%	
	c. Transient velocity and turbulence			

sufficient precision to solve many problems. Instruments that measure fluid velocity and their application range and precision are listed in Table 3.

Airborne Tracer Techniques

Tracer techniques are suitable for measuring velocity in an open space. Typical tracers include smoke, feathers, pieces of lint, and radioactive or nonradioactive gases. Measurements are made by timing the rate of movement of solid tracers or by monitoring the change in concentration level of gas tracers.

Smoke is a useful qualitative tool in studying air movements. Smoke can be obtained from titanium tetrachloride (irritating to nasal membranes) or by mixing potassium chlorate and powdered sugar (a nonirritating smoke) and firing the mixture with a match. The latter process produces considerable heat and should be confined to a pan away from flammable materials. The titanium tetrachloride smoke works well for spot tests, particularly for leakage through casings and ducts, because it can be handled easily in a pistol-like ejector.

The fumes of aqua ammonia and sulfuric acid, if permitted to mix, form a white precipitate. Two bottles, one containing ammonia water and the other containing acid, are connected to a common nozzle by rubber tubing. A syringe forces air over the liquid surfaces in the bottles; the two streams mix at the nozzle and form a white cloud.

A satisfactory test smoke also can be made by bubbling an airstream through ammonium hydroxide and then muriatic acid (Nottage *et al.* 1952). Smoke tubes, smoke candles, and smoke bombs are available for studying airflow patterns.

Deflecting Vane Anemometers

The deflecting vane anemometer consists of a pivoted vane enclosed in a case. Air exerts pressure on the vane as it passes through the instrument from an upstream to a downstream opening. A hair spring and a damping magnet resist vane movement. The instrument gives instantaneous readings of directional velocities on an indicating scale. With fluctuating velocities, it is necessary to average the needle swings visually to obtain average velocities. This instrument is useful for studying air motion in a room and in locating objectionable drafts; for measuring air velocities at supply and return diffusers and grilles; and for measuring laboratory hood face velocities.

Propeller or Revolving Vane Anemometers

The propeller anemometer consists of a light, revolving wind-driven wheel connected through a gear train to a set of recording dials that read linear feet of air passing in a measured length of time. It is made in various sizes—3, 4, and 6 in. are the most common. Each instrument requires individual calibration. At low velocities, the friction drag of the mechanism is considerable. To compensate for this, a gear train that overspeeds is commonly used. For this reason, the correction is often additive at the lower range and subtractive at the upper range, with the least correction in the middle range of velocities. The best of these instruments have starting speeds of 50 fpm or higher; therefore, they cannot be used below that air speed. Electronic revolving vane anemometers, with optical or magnetic pick-ups to sense the rotation of the vane, are available. Sizes for the vanes range as small as 1 in. in diameter for the electronic versions.

Cup Anemometers

The cup-type anemometer is primarily used to measure outdoor, meteorological wind speeds. It consists of three or four hemispherical cups mounted radially from a vertical shaft. Wind from any direction with a vector component in the plane of cup rotation causes the cups and shaft to rotate. Since the primary use of this type of instrument is to make meteorological wind speed measurements, the instrument is usually constructed so that wind speeds can be recorded or indicated electrically at a remote point.

Pitot-Static Tubes

The pitot-static tube, in conjunction with a suitable manometer or differential pressure transducer, provides a simple method of determining air velocity at a point in a flow field. Figure 5 shows the construction of a standard pitot tube (ASHRAE 1985) and the method of connecting it to a draft gage. The equation for determining air velocity from measured velocity pressure is

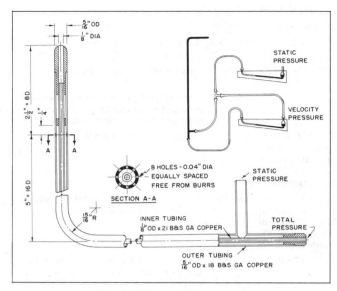

Fig. 5   Standard Pitot Tube

$$V = C\sqrt{p_w/\rho} \qquad (15)$$

where

V = velocity, fpm
p_w = velocity pressure (pitot-tube manometer reading), in. of water
ρ = density of air, lb/ft^3
C = 1096.5

The type of manometer or differential pressure transducer used with a pitot-static tube depends on the magnitude of velocity pressure being measured and the desired accuracy. At velocities greater than 1500 fpm, a draft gage of appropriate range is usually satisfactory. If the pitot-static tube is used to measure air velocities lower than 1500 fpm, a precision manometer or comparable pressure differential transducer is essential.

Other pitot-static tubes have been used and calibrated. To meet special conditions, various sizes of pitot-static tubes geometrically similar to the standard tube can be used. For relatively high velocities in ducts of small cross-sectional area, total pressure readings can be obtained with an impact (pitot) tube. Where static pressure across the stream is relatively constant, as in turbulent flow in a straight duct, a sidewall tap to obtain static pressure can be used with the impact tube to obtain the velocity pressure head. One form of impact tube is a small streamlined tube with a fine hole in its upstream end and with its axis placed parallel to the stream.

If the Mach number of the flow is greater than about 0.3, the effects of compressibility should be included in computing the air speed from pitot-static and impact (stagnation or pitot) tube measurements (Mease *et al.* 1992).

Thermal Anemometers

The thermal or hot-wire anemometer consists of a heated RTD, thermocouple junction, or thermistor sensor constructed at the end of a probe; it is designed to provide a direct, simple method of determining air velocity at a point in the flow field. The probe is placed into an airstream, and the movement of air past the electrically heated velocity sensor tends to cool the sensor in proportion to the speed of the airflow. The electronics and sensor are commonly combined into a portable, hand-held device that interprets the sensor signal and provides a direct reading of air velocity in either an analog or digital display format. Often the sensor probe also incorporates an ambient temperature-sensing RTD or

thermistor, in which case the indicated air velocity is "temperature compensated" to "standard" air density conditions (typically 0.0748 lbs/ft^3).

Hot-wire anemometers have long been used in the fluid flow research field. Research anemometer sensors have been constructed using very fine wires in configurations that allow the researcher to characterize fluid flows in one, two, and three dimensions with sensor/electronics response rates up to several hundred kilohertz. This technology has been incorporated into more ruggedized sensors suitable for measurements in the HVAC field, primarily for unidirectional airflow measurement. Omnidirectional sensing instruments suitable for thermal comfort studies are also available.

The principal advantages of thermal anemometers are their wide dynamic range and their ability to sense extremely low velocities. The same instruments can measure air velocities up to 10,000 fpm and as low as 20 fpm. Typical accuracy (including repeatability) of 2 to 5% of reading over the entire velocity range is often achieved in commercially available portable instruments.

Among the limitations of thermal anemometers are the following: (1) the unidirectional sensor must be carefully aligned in the airstream (typically to within ±20° rotation) to achieve accurate results; (2) the velocity sensor must be kept clean because contaminant buildup will cause the calibration to change; and (3) due to the inherent high speed of response of thermal anemometers, measurements in turbulent flows can yield fluctuating velocity measurements. Electronically controlled time-integrated functions are now available in many digital air velocity meters to help smooth these turbulent flow measurements.

In the HVAC field, thermal anemometers are suitable for use in a variety of applications. They are particularly well-suited to the low velocities associated with laboratory fume hood face velocity measurements (typically in the 60 to 200 fpm range). Thermal anemometers can also be used for taking multipoint traverse measurements in ventilation duct work.

Measuring Flow in Ducts

Since velocity in a duct is seldom uniform across any section, and since a pitot tube reading or thermal anemometer indicates velocity at only one location, a traverse is usually made to determine average velocity. Generally, velocity is lowest near the edges or corners and greatest at or near the center.

To determine the velocity in the traverse plane, a straight average of individual point velocities will give satisfactory results when point velocities are determined by the log-Tchebycheff rule (ISO 3966). Figure 6 shows suggested sensor locations for traversing round and rectangular ducts. The log-Tchebycheff rule provides the greatest accuracy because its location of traverse points accounts for the effect of wall friction and the fall-off of velocity near the duct walls. For circular ducts, the log-Tchebycheff and log-linear traverse methods are similar. Log-Tchebycheff is now recommended for rectangular ducts as well. It minimizes the positive error (measured greater than actual) caused by the failure to account for losses at the duct wall. This error can occur when using the older method of equal subareas to traverse rectangular ducts.

For a rectangular duct traverse, a minimum of 25 points should be measured. For a duct side less than 18 in., locate the points at the center of equal areas not more than 6 in. apart, and use a minimum of 2 points per side. For a duct side greater than 56 in., the maximum distance between points is 8 in. For a circular duct traverse, the log-linear rule and three symmetrically disposed diameters may be used (Figure 6). Points on two perpendicular diameters may be used where access is limited.

If possible, measuring points should be located at least 7.5 diameters downstream and 3 diameters upstream from a disturbance (*i.e.*, caused by a turn). Compromised traverses as close as 2 diameters downstream and 1 diameter upstream can be per-

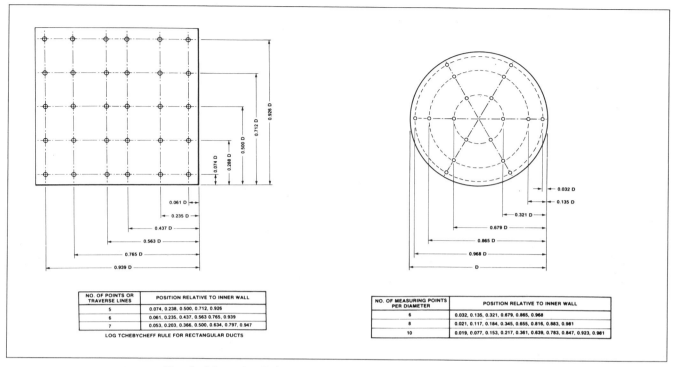

NO. OF POINTS OR TRAVERSE LINES	POSITION RELATIVE TO INNER WALL
5	0.074, 0.238, 0.500, 0.712, 0.926
6	0.061, 0.235, 0.437, 0.563 0.765, 0.939
7	0.053, 0.203, 0.366, 0.500, 0.634, 0.797, 0.947

LOG TCHEBYCHEFF RULE FOR RECTANGULAR DUCTS

NO. OF MEASURING POINTS PER DIAMETER	POSITION RELATIVE TO INNER WALL
6	0.032, 0.135, 0.321, 0.679, 0.865, 0.968
8	0.021, 0.117, 0.184, 0.345, 0.655, 0.816, 0.883, 0.981
10	0.019, 0.077, 0.153, 0.217, 0.361, 0.639, 0.783, 0.847, 0.923, 0.981

Fig. 6   Measuring Points for Round and Rectangular Duct Traverse

formed with an increase in measurement error. Since field-measured airflows are rarely steady and uniform, particularly near disturbances, accuracy can be improved by increasing the number of measuring points. Straightening vanes (ASHRAE 1985) located 1.5 duct diameters ahead of the traverse plane improve measurement precision.

When velocities at a traverse plane fluctuate, the readings should be averaged on a time-weighted basis. Two traverse readings in short succession also help to average out velocity variations that occur with time. If negative velocity pressure readings are encountered, they are considered a measurement value of zero and calculated in the average velocity pressure. ASHRAE *Standard* 111 (1988) has further information on measuring flow in ducts.

Airflow-Measuring Hoods

Flow-measuring hoods are portable instruments designed to measure supply or exhaust airflow through diffusers and grilles in HVAC systems. A flow-measuring hood assembly typically consists of a fabric hood section, a plastic or metal base, an airflow-measuring manifold, a meter, and handles for carrying and holding the hood in place.

For volumetric airflow measurements, the flow-measuring hood is placed over a diffuser or grille. The fabric hood captures and directs airflow from the outlet or inlet across the flow-sensing manifold in the base of the instrument. The manifold consists of a number of tubes containing upstream and downstream holes in a grid pattern designed to simultaneously sense and average multiple velocity points across the base of the flow-measuring hood. Air from the upstream holes flows through the tubes, past a sensor, and then exits through the downstream holes. Sensors employed by different manufacturers include swinging vane anemometers, electronic micromanometers, and thermal anemometers. In the case of the electronic micromanometer sensor, air does not actually flow through the manifold, but the airtight sensor senses the pressure differential from the upstream to downstream series of holes. The meter on the base of the flow-measuring hood

interprets the signal from the sensor and provides a direct reading of volumetric flow in either an analog or digital display format.

As a performance check in the field, the indicated flow of a measuring hood can be compared to a duct traverse flow measurement (using a pitot-tube or a thermal anemometer). All flow-measuring hoods induce some back pressure on the air-handling system, because the hood restricts the flow coming out of the diffuser. This added resistance alters the true amount of air coming out of the diffuser. In most cases, this error is negligible and is less than the accuracy of the instrument. When doing proportional balancing, this error need not be taken into account since all similar diffusers will have about the same amount of back pressure. To determine if back pressure is significant, a velocity traverse can be made in the duct ahead of the diffuser with and without the flow-measuring hood in place. The difference in the average velocity of the traverse indicates the degree of back-pressure compensation required on similar diffusers in the system. For example, if the average velocity is 800 fpm with the hood in place, and 820 fpm without the hood, the indicated flow reading can be multiplied by 1.025 on similar diffusers in the system (820/800 = 1.025). As an alternative, the designer of the air-handling system can predict the amount of airflow that is reduced because of the additional pressure of the hood by using a curve supplied by the flow-measuring hood manufacturer. This curve indicates the pressure drop through the hood for different flow rates.

Laser Doppler Velocimeter (or Anemometer)

The laser Doppler velocimeter (LDV) or laser Doppler anemometer (LDA) is an extremely complex system that collects scattered light produced by a particle passing through the intersection volume of two intersecting laser beams of the same light frequency (Mease *et al.* 1992). The scattered light consists of bursts containing a regularly spaced fringe pattern whose frequency is linearly proportional to the speed of the particle. Due to the cost and complexity of these systems, they are usually not suitable for in situ field measurements. Rather, the primary application of LDV systems in the HVAC industry is to calibrate the calibration systems used to calibrate other air velocity instruments.

The greatest advantage of an LDV is its low-speed performance. It is capable of reading air speeds as low as 15 fpm with uncertainty levels of 1% or less (Mease *et al.* 1992). In addition, it is nonintrusive in the flow—only optical access is required. It can be used to measure fluctuating components as well as mean speeds and is available in one, two, and even three-dimensional configurations. Its biggest disadvantage is its high cost and extreme technological complexity, requiring highly skilled operators. Modern fiber optic systems require less skilled operators but at a considerable increase in cost.

FLOW RATE MEASUREMENT

Various means of measuring fluid flow rate are listed in Table 4. The values for volume or mass flow rate measurement (ASME 1971, Benedict 1984) are often determined by measuring pressure difference across an orifice, nozzle, or venturi tube. These types of meter have different advantages and disadvantages. For exam-

ple, the orifice plate is more easily changed than the complete nozzle or venturi tube assembly. However, the nozzle is often preferred to the orifice because its discharge coefficient is more precise. The venturi tube is a nozzle followed by an expanding recovery section to reduce net pressure loss. Differential pressure-type flow measurement has benefited recently through workshops addressing fundamental issues, new textbooks, research, and improved standards (Miller 1983, DeCarlo 1984, Mattingly 1984, ASME 1985b, ASME 1988a, ASME 1988b, ASME 1991).

Fluid meters use a wide variety of physical techniques to make flow measurements (ASME 1971, Miller 1983, DeCarlo 1984); those more prevalently used are described in the following section. The search for high-accuracy flow measurement includes the arrangement of appropriate calibration procedures. While these used to be available only in calibration laboratories, they are now frequently purchased along with flowmeters so that flow measurements can be efficiently and effectively assured and validated at high levels of performance. To assure and validate calibration facilities and procedures, realistic traceability should be established and maintained for the calibration facilities and procedures. This is best done using round-robin testing programs (Mattingly 1985, Mattingly 1988, Benson *et al.* 1986, Olsen 1974).

Table 4   Volume or Mass Flow Rate Measurement

Measurement Means	Application	Range	Precision	Limitations
Orifice and differential pressure measurement system	Flow through pipes, ducts, and plenums for all fluids	Above Reynolds number of 5000	1 to 5%	Discharge coefficient and accuracy influenced by installation conditions
Nozzle and differential pressure measurement system	Flow through pipes, ducts, and plenums for all fluids	Above Reynolds number of 5000	0.5 to 2.0%	Discharge coefficient and accuracy influenced by installation conditions
Venturi tube and differential pressure measurement system	Flow through pipes, ducts, and plenums for all fluids	Above Reynolds number of 5000	0.5 to 2.0%	Discharge coefficient and accuracy influenced by installation conditions
Rotameters	Used for liquids or gases	Any	0.5 to 5.0%	Should be calibrated for fluid with which used
Turbine flowmeters	Used for liquids or gases	Any	0.25 to 2.0%	Uses electronic readout
Timing given mass flow	Liquids or gases; used to calibrate other flowmeters	Any	0.1 to 0.5%	System is bulky and slow
Displacement meter	Relatively small volume flow with high pressure loss	As high as 1000 cfm depending on type	0.1 to 2.0% depending on type	Most types require calibration with fluid being metered
Gasometer or volume displacement	Short duration tests; used for calibrating other flowmeters	Total flow limited by available volume of containers	0.5 to 1.0%	—
Element of resistance to flow and differential pressure measurement system	Used for check where system has calibrated resistance element	Lower limit set by readable pressure drop	1 to 5%	Secondary reading depends on accuracy of calibration
Thomas meter (temperature rise of stream due to electrical heating)	Elaborate setup justified by need for good accuracy	Any	1%	Uniform velocity; usually used with gases
Heat input and temperature changes with steam and water coil	Check value in heater or cooler tests	Any	1 to 3%	—
Instrument for measuring velocity at point in flow	Primarily for installed systems where no special provision for low measurement has been made	Lower limit set by accuracy of velocity measurement	2 to 4%	Accuracy depends on uniformity of flow and completeness of traverse
Laminar flow element and differential pressure measurement system	Measure liquid or gas volume flow rate; nearly linear relationship with pressure drop; simple and easy to use	0.0001 to 2000 cfm	1%	Fluid must be free of dirt, oil, or other impurities, which could plug meter or affect its calibration
Magnetohydrodynamic flowmeter (electromagnetic)	Measures electrically conductive fluids, slurries; meter does not obstruct flow; no moving parts	0.1 to 10,0000 gpm	1%	At present state-of-the-art, conductivity of fluid must be greater than 5 μmho/cm
Swirl flowmeter and vortex shedding meter	Measure liquid or gas flow in pipe; no moving parts	Above Reynolds number of 10^4	1%	—

Direct and Indirect Flow Measurement Methods

Both gas and liquid flow can be measured quite accurately by timing a collected amount of fluid that is determined gravimetrically or volumetrically. While this method is commonly used for calibrating other metering devices, it is particularly useful where the flow rate is low or intermittent and where a high degree of accuracy is required. These systems are generally large and slow, but in their simplicity, they can be considered primary devices.

The variable area meter or rotameter is a convenient direct reading flowmeter for liquids and gases. This is a vertical, tapered tube in which the flow rate is indicated by the position of a float suspended in the upward flow. The position of the float is determined by its buoyancy and the upwardly directed fluid drag.

Displacement meters measure total liquid or gas flow over time. The two major types of displacement meters used for gases are the conventional gas meter, which uses a set of bellows, and the wet test meter, which uses a water displacement principle.

The Thomas meter is used in laboratories to measure high gas flow rates with low pressure losses. The gas is heated by electric heaters, and the temperature rise is measured by two resistance thermometer grids. When the heat input and temperature rise are known, the mass flow of gas is calculated as the quantity of gas that will remove the equivalent heat at the same temperature rise.

A velocity traverse (made using a pitot tube or other velocity-measuring instrument) measures airflow rates in the field or calibrates large nozzles. This method can be imprecise at low velocities and impracticable where many test runs are in progress.

Another field-estimating method measures the pressure drop across elements with known pressure drop characteristics, such as heating and cooling coils or fans. If the pressure drop/flow rate relationship has been calibrated, the results can be precise. If the method depends on rating data, it should be used for check purposes only.

Venturi, Nozzle, and Orifice Flowmeters

Flow in a pipeline can be measured by a venturi meter (Figure 7), flow nozzle (Figure 8), or orifice plate (Figure 9). The American Society of Mechanical Engineers publication MFC-3M (ASME 1989) describes measurement of fluid flow in pipes using the orifice, nozzle, and venturi; the ASME Performance Test Code PTC 19.5-72 (ASME 1972) specifies their construction.

Assuming an incompressible fluid (liquid or slow-moving gas), uniform velocity profile, frictionless flow, and no gravitational effects, the principle of conservation of mass and energy can be applied to the venturi and nozzle geometries to give

$$w = \rho V_1 A_1 = \rho V_2 A_2 = A_2 \sqrt{\frac{2g\rho (p_1 - p_2)}{1 - \beta^4}} \qquad (16)$$

where

w = flow rate, lb/s
V = velocity of stream, ft/s
A = flow area, ft^2
g = gravitational factor = 32.174 ft/s^2
ρ = density of fluid, lb/ft^3
p = absolute pressure, lb/ft^2
β = (D_2/D_1) for venturi and sharp edge orifice and d/D for flow nozzle

Note: Subscript 1 refers to the entering conditions; subscript 2 refers to the throat conditions.

Since the flow through the meter is not frictionless, a correction factor C is defined to account for friction losses. If the fluid is at a high temperature, an additional correction factor F_a should be included to account for thermal expansion of the primary element. Since this amounts to less than 1% at 500 °F, it can usually be omitted. Equation (16) then becomes

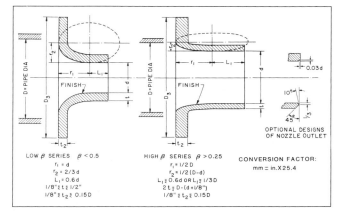

Fig. 8   Dimensions of ASME Long Radius Flow Nozzles
(Use of either recess or bevel at nozzle outlet is optional)
From ASME PTC 19.5. Reprinted with permission of ASME.

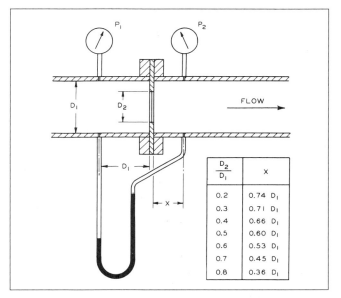

Fig. 9   Sharp Edge Orifice with Pressure Tap Locations
From ASME PTC 19.5. Reprinted with permission of ASME.

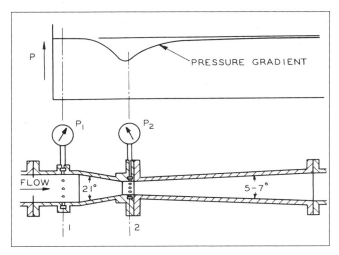

Fig. 7   Typical Herschel Type Venturi Meter

$$w = CA_2\sqrt{2g\rho\,(p_1 - p_2)/(1 - \beta^4)} \qquad (17)$$

where Q is the discharge rate in ft^3/s.

The factor C is a function of geometry and Reynolds number. Values of C are given in ASME PTC 19.5-72. The approach factor, $1/\sqrt{1 - \beta^4}$, can be combined with the discharge coefficient, as described later. The jet passing through an orifice plate contracts to a minimum area at the vena contracta located a short distance downstream from the orifice plate. The contraction coefficient, energy loss coefficient, and approach factor can be combined into a single constant K, which is a function of geometry and Reynolds number. The orifice flow rate equations then become

$$Q = KA_2\sqrt{2g\rho\,(p_1 - p_2)} \qquad (18)$$

where

$$
\begin{aligned}
Q &= \text{discharge, ft}^3/\text{s} \\
A_2 &= \text{orifice area, ft}^2 \\
p_1 - p_2 &= \text{pressure drop in lb/ft}^2 \text{ as obtained by pressure taps}
\end{aligned}
$$

Values of K are shown in ASME PTC 19.5-72.

Valves, bends, and fittings upstream from the flowmeter can cause errors. Long, straight pipes should be installed upstream and downstream from the flow devices to assure fully developed flow for proper measurement. ASHRAE *Standard* 41.8 specifies upstream and downstream pipe lengths for measuring flow of liquids with an orifice plate. ASME PTC 19.5-72 gives the piping requirements between various fittings and valves and the venturi, nozzle, and orifice. If these conditions cannot be met, flow conditioners or straightening banes can be used (ASME 1972, ASME 1988b, Mattingly 1984, Miller 1983).

Compressibility effects must be considered for gas flow if the pressure drop across the measuring device is more than a few percent of the initial pressure.

Nozzles are sometimes arranged in parallel pipes from a common manifold; thus, the capacity of the testing equipment can be changed by shutting off the flow through one or more nozzles. An apparatus designed for testing airflow and capacity of air-conditioning equipment is described by Wile (1947), who also presents pertinent information on nozzle discharge coefficients, Reynolds numbers, and resistance of perforated plates. Some laboratories refer to this apparatus as a code tester.

Variable Area Flowmeters (Rotameters)

In permanent installations where high precision, ruggedness, and operational ease are important, the variable area flowmeter is satisfactory. It is frequently used to measure liquids or gases in small-diameter pipes. For ducts or pipes over 6 in. in diameter, the expense of this meter may not be warranted. In larger systems, however, the meter can be placed in a bypass line and used with an orifice.

The variable area meter (Figure 10) commonly consists of a float that is free to move vertically in a transparent tapered tube. The fluid to be metered enters at the narrow bottom end of the tube and moves upward, passing at some point through the annulus formed between the float and the inside wall of the tube. At any particular flow rate, the float assumes a definite position in the tube; a calibrated scale on the tube shows the float's location and the fluid flow rate.

The position of the float is established by a balance between the fluid pressure forces across the annulus and gravity on the float. The buoyant force supporting the float, $\nu_f(\rho_f - \rho)$, is balanced by the pressure difference acting on the cross-sectional area of the float $A_f\Delta p$, where ρ_f, A_f, ν_f, are, respectively, the float density,

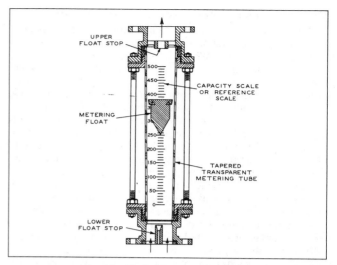

Fig. 10   Variable Area Flowmeter

float cross-sectional area, and float volume. The pressure difference across the annulus is

$$\Delta p = \nu_f(\rho_f - \rho)/A_f \qquad (19)$$

The mass flow follows from Equation (14) as

$$w = KA_2\sqrt{2g\nu_f(\rho_f - \rho)\,\rho/A_f} \qquad (20)$$

The flow for any selected fluid is nearly proportional to the area, so that calibration of the tube is convenient. To use the meter for different fluids, the flow coefficient variation for any float must be known. Float design can reduce variation of the flow coefficient with Reynolds number; float materials can reduce the dependence of mass flow calibration on fluid density.

Turbine Flowmeters

Turbine flowmeters are volumetric flow rate sensing meters with a magnetic stainless steel turbine rotor suspended in the flow stream of a nonmagnetic meter body. The fluid stream exerts a force on the blades of the turbine rotor, setting it in motion and converting the fluid's linear velocity to an angular velocity. Design motivation for turbine meters is to have the rotational speed of the turbine proportional to the average fluid velocity and thus to the volume rate of fluid flow (Miller 1983, DeCarlo 1984, Mattingly 1992).

The rotational speed of the rotor is monitored by an externally mounted pickoff assembly. Magnetic and radio frequency are the most commonly used pickoffs. The magnetic pickoff contains a permanent magnet and coil. As the turbine rotor blades pass through the field produced by the permanent magnet, a shunting action induces ac voltage in the winding of the coil wrapped around the magnet. A sine wave with a frequency proportional to the flow rate develops. With the radio frequency pickoff, an oscillator applies a high-frequency carrier signal to a coil in the pickoff assembly. The rotor blades pass through the field generated by the coil and modulate the carrier signal by shunting action on the field shape. The carrier signal is modulated at a rate corresponding to the rotor speed, which is proportional to the flow rate. With both pickoffs, frequency of the pulses generated becomes a measure of flow rate, and the total number of pulses measures total volume (Woodring 1969, Shafer 1961, Mattingly 1992).

Since output frequency of the turbine flowmeter is proportional to flow rate, every pulse from the turbine meter is equivalent to a known volume of fluid that has passed through the meter; the sum of these pulses yields total volumetric flow. Totalization is accomplished by electronic counters designed for use with turbine flowmeters; they combine a mechanical or electronic register with the basic electronic counter.

Turbine flowmeters should be installed with straight lengths of pipe upstream and downstream from the meter. The length of the inlet and outlet pipes should be according to manufacturers' recommendations or to pertinent standards. Where recommendations on standards cannot be accommodated, the meter installation should be calibrated. Some turbine flowmeters can be used in bidirectional flow applications. A fluid strainer, used with liquids of poor or marginal lubricity, minimizes bearing wear.

The lubricity of the process fluid and the type and quality of rotor bearings determine whether the meter is satisfactory for the particular application. When choosing turbine flowmeters for use with fluorocarbon refrigerants, attention must be paid to the type of bearings used in the meter and to the oil content of the refrigerant. For these applicants, sleeve-type rather than standard ball bearings are recommended. The amount of oil in the refrigerant can severely affect calibration and bearing life.

In metering liquid fluorocarbon refrigerants, the liquid must not flash to a vapor (cavitate). This would cause a tremendous increase in flow volume. Flashing results in erroneous measurements and rotor speeds that can damage the bearings or cause a failure. Flashing can be avoided by maintaining an adequate back pressure on the downstream side of the meter (Liptak 1972).

Positive Displacement Meters

Many positive displacement meters are available for measuring total liquid or gas volume flow rates. The fluid measured in these meters flows progressively into compartments of definite size. As the compartments are filled, they are rotated so that the fluid discharges from the meter. The flow rate through the meter is equal to the product of the compartment size, the number of compartments, and the rotation rate of the rotor. Most of these meters have a mechanical register, calibrated to show total flow.

ELECTRIC MEASUREMENT

Ammeters

Ammeters are low-resistance instruments for measuring current. They should be connected in series with the circuit being measured (Figure 11). Ideally, they have the appearance of a short circuit, but in practice, all ammeters have a nonzero input impedance that influences the measurement to some extent.

Ammeters often have several ranges, and it is good practice to start with the highest range when measuring unknown currents and then reduce the range to the appropriate value to obtain the most sensitive reading. Ammeters with range switches maintain circuit continuity during switching. On some older instruments, it may be necessary to short-circuit the ammeter terminals when changing the range.

Current transformers are often used to increase the operating range of ammeters. They may also provide isolation and thus protection from a high-voltage line. Current transformers have at least two separate windings on a magnetic core (Figure 12). The primary winding is connected in series with the circuit in which the current is measured. In the case of a "clamp-on" probe, the transformer core is actually opened and then connected around a single conductor carrying the current to be measured. That conductor serves as the primary winding. The secondary winding carries a scaled down version of the primary current, which is connected to an ammeter. Depending on the type of instrument, the ammeter reading may have to be multiplied by the ratio of the transformer.

When using an auxiliary current transformer, the secondary circuit must not be open when current is flowing in the primary winding; dangerous high voltage may exist across the secondary terminals. A short-circuiting blade between the secondary terminals should be closed before the secondary circuit is opened at any point.

Transformer accuracy can be impaired by the residual magnetism in the core when the primary circuit is opened at an instant when the flux is large. The transformer core may be left magnetized, resulting in ratio and phase angle errors. The primary and secondary windings should be short-circuited before making changes.

Voltmeters

Voltmeters are high-resistance instruments that should be connected across the load (in parallel), as shown in Figure 13. Ideally they have the appearance of an open circuit, but in practice, all voltmeters have some finite impedance that influences the measurement to some extent.

Voltage transformers are often used to increase the operating range of a voltmeter (Figure 14). They also provide isolation from high voltages and prevent injury to the operator. Like current transformers, voltage transformers consist of two or more windings on a magnetic core. The primary winding is generally connected across the high voltage to be measured, and the secondary winding is connected to the voltmeter. Unlike current transformers, it is important not to short-circuit the secondary winding of a voltage transformer.

Wattmeters

Wattmeters are instruments that measure the active power of an ac circuit, which equals the voltage multiplied by that part of the current in phase with the voltage. There are generally two sets of terminals—one to connect the load voltage and the other to connect in series with the load current. Current and voltage transformers can be used to extend the range of a wattmeter or to isolate it from high voltage. Figures 15 and 16 show connections for single-phase wattmeters, and Figure 17 shows use of current and voltage transformers with a single-phase wattmeter.

Wattmeters with multiple current and voltage elements are available to measure polyphase power. Polyphase wattmeter connections are shown in Figures 18 and 19.

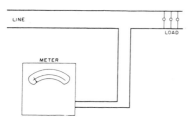

Fig. 11   Ammeter Connected in Power Circuit

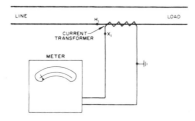

Fig. 12   Ammeter with Current Transformer

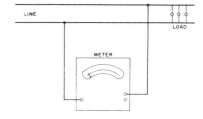

Fig. 13   Voltmeter Connected Across Load

Power-Factor Meters

Power-factor meters measure the ratio of the active power to the apparent power (product of the voltage and current). The connections for power-factor meters and wattmeters are similar, and current and voltage transformers can be used to extend their range. Connections for single-phase and polyphase power-factor meters are shown in Figures 20 and 21, respectively.

HUMIDITY MEASUREMENT

Many instruments are available for measuring the moisture content of air. The indication sensors used on the instruments respond to different moisture property contents. These responses are related to factors such as wet-bulb temperature, relative humidity, humidity (mixing) ratio, dew point, and frost point.

Table 5 lists instruments for measuring humidity. Each is capable of accurate measurement under certain conditions and within specific limitations. The following sections describe various instruments used to measure humidity.

Psychrometers

Any instrument capable of measuring the humidity or psychrometric state of air is a hygrometer. A typical industrial psy-

chrometer consists of a pair of matched electrical or mechanical temperature sensors, one of which is kept wet with a moistened wick. A blower aspirates the sensor, which lowers the temperature at the moistened temperature sensor. The lowest temperature depression occurs when the evaporation rate required to saturate the moist air adjacent to the wick is constant. This is a steady-state, open-loop, nonequilibrium process, which depends on the purity of the water, cleanliness of the wick, ventilation rate, radiation effects, size and accuracy of the temperature sensors, and the transport properties of the gas.

ASHRAE *Standard* 41.6-1992R recommends an airflow over both the wet and dry bulbs of 600 to 1000 fpm for transverse ventilation and 300 to 500 fpm for axial ventilation.

The sling psychrometer consists of two thermometers mounted side by side in a frame fitted with a handle for whirling the device through the air. The thermometers are spun until their readings become steady. In the ventilated or aspirated psychrometer, the thermometers remain stationary, and a small fan, blower, or syringe moves the air across the thermometer bulbs. Various designs are used in the laboratory, and commercial models are available.

Other temperature sensors, such as thermocouples and thermistors, are also used and can be adapted for recording temperatures

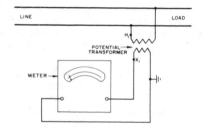

Fig. 14   Voltmeter with Potential Transformer

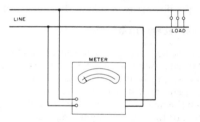

Fig. 15   Wattmeter in Single-Phase Circuit (Measuring Power Load Plus Loss in Its Current-Coil Circuit)

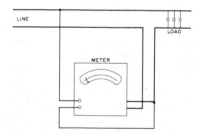

Fig. 16   Wattmeter in Single-Phase Circuit (Measuring Power Load Plus Loss in Its Potential-Coil Circuit)

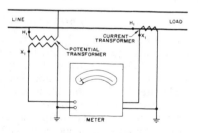

Fig. 17   Wattmeter with Current and Potential Transformer

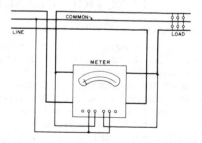

Fig. 18   Polyphase Wattmeter in Two-Phase, Three-Wire Circuit with Balanced or Unbalanced Voltage or Load

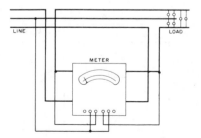

Fig. 19   Polyphase Wattmeter in Three-Phase, Three-Wire Circuit

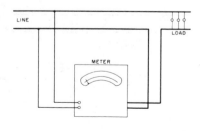

Fig. 20   Single-Phase Power-Factor Meter

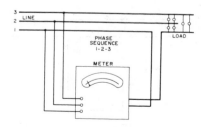

Fig. 21   Three-Wire, Three-Phase Power-Factor Meter

Table 5 Humidity Sensor Properties

Type of Sensor	Sensor Category	Method of Operation	Approximate Range	Some Uses	Approximate Accuracy
Dunmore type	Electrical	Impedance	7 to 98% rh at 40 to 140°F	Measurement, control	±1.5% rh
Surface acoustic wave	Electrical	SAW Attenuation	85 to 98% rh	Measurement, control	±1% rh
Ion exchange resin	Electrical	Impedance or capacitance	10 to 100% rh at −40 to 190°F	Measurement, control	±5% rh
Aluminum oxide	Electrical	Capacitance	5 to 100% rh	Measurement, control	±3% rh
Aluminum oxide	Electrical	Capacitance	−110 to 140°F dp	Trace moisture measurement, control	±2°F dp
Porous ceramic	Electrical	Impedance or capacitance			
Electrolytic hygrometer	Electrical	Capacitance			
Infrared laser diode	Electrical	Optical diodes	0.1 to 100 ppm	Trace moisture measurement	±0.1 ppm
Hair	Mechanical	Dimensional change	5 to 100% rh	Measurement, control	±5% rh
Cellulosic materials	Mechanical	Dimensional change	5 to 100% rh	Measurement, control	±5% rh
Nylon	Mechanical	Dimensional change	5 to 100% rh	Measurement, control	±5% rh
Dacron thread	Mechanical	Dimensional change	5 to 100% rh	Measurement	±7% rh
Goldbeaters skin	Mechanical	Dimensional change	5 to 100% rh	Measurement	±7% rh
Carbon sensor	Mechanical	Dimensional change	5 to 100% rh	Measurement	±5% rh
Piezoelectric	Mass sensitive	Mass changes due to adsorbed moisture	−100 to 0°F	Trace moisture measurement, control	±2 to ±10°F dp
Color change	Physical	Color changes	10 to 80% rh	Warning device	±10% rh
Chilled mirror	Dew point	Optical determination of moisture formation	−110 to 200°F dp	Measurement, control, meteorology	±0.4 to ±4°F
Heated saturated salt solution	Water vapor pressure	Vapor pressure depression in salt solution	−20 to 160°F dp	Measurement, control, meteorology	±3°F
Gravimetric	Direct measurement of mixing ratio	Comparison of sample gas with dry airstream	0.12 g/kg to 20g/kg mixing ratio	Primary standard, research and laboratory	±0.13% of reading
Coulometric	Electrolytic cell	Electrolyses due to adsorbed moisture	1 to 1000 ppm	Measurement	
Psychrometer	Evaporative cooling	Temperature measurement of wet bulb	32 to 180°F	Measurement, standard	±3% to ±7% rh
Adiabatic saturation psychrometer	Evaporative cooling	Temperature measurement of thermodynamic wet bulb	40 to 85°F	Measurement, standard	±0.2% to ±2% rh
Radiation absoption	Moisture absorption	Moisture adsorption by UV or IR radiation	0 to 180°F dp	Measurement, control, meteorology	±4°F dp, ±5%

Notes:
1. This table does not encompass all of the available technology for the measurement of humidity.
2. The approximate range for the device types listed is based on surveys of device manufacturers.

3. The approximate accuracy is based on manufacturers' data.
4. Presently the National Institute of Standards and Technology (NIST) will only certify instruments whose operating range is within −75 to 100°C dew point.

or for use where a small instrument is required. Small-diameter wet-bulb sensors operate with low ventilation rates.

Charts and tables showing the relationship between the temperatures and humidity are available. Data are usually based on a barometric pressure equal to one standard atmosphere. To meet special needs, charts can be produced that apply to nonstandard pressure, *e.g.*, the ASHRAE 7500-ft chart. Alternatively, mathematical calculations can be made (Kusuda 1965). Uncertainties of ±3 to ±7% rh are typical for psychrometer-based derivation. The degree of uncertainty is a function of the accuracy of the temperature measurements, wet and dry bulb, knowledge of the barometric pressure, and conformance to accepted operational procedures such as those outlined in ASHRAE *Standard* 41.6.

In air temperatures below 32°F, the water on the wick may either freeze or supercool. Since the wet-bulb temperature is different for ice and water, its state must be known and a proper chart or table used. Some operators remove the wick from the wet-bulb for freezing conditions and dip the bulb in water a few times; this allows water to freeze on the bulb between dips, forming a film of ice.

Since the wet-bulb depression is slight at low temperatures, precise temperature readings are essential. A psychrometer can be used at high temperatures, but, if the wet-bulb depression is large, the wick must remain wet and water supplied to the wick must be cooled so as not to influence the wet-bulb temperature by carrying sensible heat to it (Richardson 1965, Worrall 1965).

Greenspan and Wexler (1968) and Wentzel (1961) have developed devices to measure adiabatic saturation temperature.

Dew-Point Hygrometers

Chilled mirror dew-point hygrometer. The condensation-type (chilled mirror) dew-point hygrometer is an accurate and reliable instrument, with a wide humidity range. However, these features are obtained through an increase in complexity and cost when compared to the psychrometer. In the condensation-type hygrometer, a surface is cooled (thermoelectrically, mechanically, or chemically) until dew or frost begins to condense out. The condensate surface is maintained electronically in vapor pressure equilibrium with the surrounding gas, while surface condensation is detected by optical, electrical, or nuclear techniques. The measured surface temperature is then the dew-point temperature.

The largest source of error in a condensation hygrometer stems from the difficulty in measuring condensate surface temperature accurately. Typical industrial versions of the instrument are accurate to $\pm 1.0\,°F$ over wide temperature spans. With proper attention to the condensate surface temperature measuring system, errors can be reduced to about $\pm 0.4\,°F$. Condensation-type hygrometers can be made surprisingly compact using solid-state optics and thermoelectric cooling.

Wide-span and minimal errors are two of the main features of this instrument. A properly designed condensation hygrometer can measure dew points from $200\,°F$ down to frost points of $-100\,°F$. Typical condensation hygrometers can cool to $150\,°F$ below the ambient temperature, establishing lower limits of the instrument to dew points corresponding to approximately 0.5% rh. Accuracies for measurements above $-40\,°F$ can be $\pm 2\,°F$ or better, deteriorating to $\pm 4\,°F$ at lower temperatures.

The response time of a condensation dew-point hygrometer is usually specified in terms of its cooling/heating rate, typically $4\,°F/s$ for thermoelectric cooled mirrors. This makes it somewhat faster than a heated salt hygrometer. Perhaps the most significant feature of the condensation hygrometer is its fundamental measuring technique, which essentially renders the instrument self-calibrating. For calibration, it is only necessary to manually override the surface cooling control loop, causing the surface to heat, and witness that the instrument recools to the same dew point when the loop is closed. Assuming that the surface temperature measuring system is correct, this is a reasonable check on the instrument's performance.

Although condensation hygrometers can become contaminated, they can easily be cleaned and returned to service with no impairment in performance.

Salt-phase heated hygrometer. Another instrument in which the temperature varies with ambient dew-point temperature is variously designated as a self-heating, salt-phase transition or heated electrical hygrometer. This device usually consists of a tubular substrate covered by glass fiber fabric, with a spiral bifilar winding for electrodes. The surface is covered with a salt solution—usually lithium chloride. The sensor is connected in series with a ballast and a 24-V (ac) supply. When in operation, electrical current flowing through the salt film heats the sensor. The electrical resistance characteristics of the salt are such that a balance is reached with the salt and its critical moisture content, corresponding to a saturated solution. The sensor temperature adjusts automatically so that the water vapor pressures of the salt film and ambient atmosphere are equal. With lithium chloride, this sensor cannot be used to measure relative humidity below approximately 12% (the equilibrium relative humidity of this salt), and it has an upper dew-point limit of about $160\,°F$. The regions of highest precision are between -10 and $93\,°F$, and above $105\,°F$ dew point. Another problem is that of the lithium chloride solution being washed off when exposed to water. In addition, this type of sensor is subject to contamination problems, which limits its accuracy. Its response time is also very slow; it takes approximately 2 min for a 67% step change.

Mechanical Hygrometers

Many organic materials change in dimension with changes in humidity; this action is used in a number of simple and effective humidity indicators, recorders, and controllers (see Chapter 41 of the 1992 ASHRAE *Handbook—Systems and Equipment*). They are coupled to pneumatic leakports, mechanical linkages, or electrical transduction elements to form hygrometers.

Commonly used organic materials are human hair, nylon, dacron, animal membrane, animal horn, wood, and paper. Their inherent nonlinearity and hysteresis must be compensated for

within the hygrometer. These devices are generally unreliable below $32\,°F$. The response is generally inadequate for monitoring a changing process. Responses can be affected significantly by exposure to extremes of humidity. Such devices require initial calibration and frequent recalibration; however, they are useful, because they can be arranged to read directly in terms of relative humidity, and they are simpler and less expensive than most other types.

Electrical Impedance Hygrometers

Many substances adsorb or lose moisture with changing relative humidity and exhibit corresponding changes in electrical impedance.

Dunmore hygrometers. This sensor consists of dual electrodes on a tubular or flat substrate; it is coated with a film containing salt, such as lithium chloride, in a binder to form an electrical connection between windings. The relation of sensor resistance to humidity is usually represented by graphs. Since the sensor is highly sensitive, the graphs are a series of curves, each for a given temperature, with intermediate values found by interpolation. Several resistance elements, called Dunmore elements, cover a standard range. Systematic calibration is essential, since the resistance grid varies with time and contamination as well as with exposure to temperature and humidity extremes.

Polymer film electronic hygrometers. These devices have a hygroscopic organic polymer, which is deposited with thin or thick film-processing technology on a water permeable substrate. Both capacitance and impedance sensors are available. The impedance devices are composed of either ionic or electronic conduction types. These hygrometers typically have integrated circuits that provide temperature correction and signal conditioning. The primary advantages of this sensor technology are small size, low cost, and fast response times in the order of 1 to 120 s for 64% change in relative humidity.

Ion exchange resin electric hygrometers. A conventional ion exchange resin consists of a polymer with a high relative molecular mass and having polar groups of positive or negative charge in cross-link structure. Associated with these polar groups are ions of opposite charge that are held by electrostatic forces to the fixed polar groups. In the presence of water or water vapor, the electrostatically held ions become mobile; thus, when a voltage is impressed across the resin, the ions are capable of electrolytic conduction. The Pope cell is one example of an ion exchange element. It is a wide-range sensor, typically covering 15 to 95% rh. Therefore, one sensor can be used where several Dunmore elements would be required. The Pope cell, however, has a nonlinear characteristic from approximately 1000 ohms at 100% rh to several megohms at 10% rh.

Impedance-based porous ceramic electronic hygrometers. Utilizing the adsorption characteristics of oxides, humidity-sensitive ceramic oxide devices use either ionic or electronic measurement techniques to relate adsorbed water to relative humidity. Ionic conduction is produced by dissociation of water molecules forming surface hydroxyls. The dissociation causes migration of protons such that device impedance decreases with increasing water content. The ceramic oxide is sandwiched between porous metal electrodes that connect the device to an impedance-measuring circuit for linearizing and signal conditioning. These sensors have excellent sensitivity, resistance to contamination, high temperature (up to $400\,°F$), and may get fully wet without sensor degradation. These sensors are accurate to about $\pm 1.5\%$ and $\pm 1\%$ rh when temperature compensated. These sensors have a moderate cost.

Aluminum oxide capacitive sensor. This sensor consists of an aluminum strip that is anodized by a process that forms a porous oxide layer. A very thin coating of gold is then evaporated over this structure. The aluminum base and the gold layer form the two electrodes of what is essentially an aluminum oxide capacitor.

Water vapor is rapidly transported through the gold layer and equilibrates on the pore walls in a manner functionally related to the vapor pressure of water in the atmosphere surrounding the sensor. The number of water molecules adsorbed on the oxide structure determines the capacitance between the two electrodes.

Electrolytic Hygrometers

In electrolytic hygrometers, air is passed through a tube where moisture is adsorbed by a highly effective desiccant, usually phosphorous pentoxide, and electrolyzed. The airflow is regulated to 0.0035 cfm at a standard temperature and pressure. As the incoming water vapor is absorbed by the desiccant and electrolyzed into hydrogen and oxygen, the current of electrolysis determines the mass of water vapor entering the sensor. The flow rate of the entering gas is controlled precisely to maintain a standard sample mass flow rate into the sensor. The instrument is usually designed for use with moisture-air ratios in the range of less than 1 to 1000 ppm, but can be used with higher humidities.

Piezoelectric sorption. This hygrometer compares the changes in frequency of two hygroscopically coated quartz crystal oscillators. As the mass of the crystal changes due to the absorption of water vapor, the frequency changes. The amount of water sorbed on the sensor is a function of relative humidity, *i.e.*, partial pressure of water as well as the ambient temperature.

A commercial instrument uses a hygroscopic polymer coating on the crystal. The humidity is measured by monitoring the vibration frequency change of the quartz crystal when the crystal is alternately exposed to wet and dry gas.

Radiation Absorption (Infrared/Ultraviolet)

Radiation absorption devices operate on the principle that selective absorption of radiation is a function of frequency for different mediums. Water vapor absorbs infrared radiation at 2 to 3 μm wavelengths and ultraviolet radiation centered about the Lyman-alpha line at 0.122 μm. The amount of absorbed radiation is directly related to the absolute humidity or water vapor content in the gas mixture according to Beer's law. The basic unit consists of an energy source and optical system for isolating wavelengths in the spectral region of interest, and a measurement system for determining the attenuation of radiant energy caused by the water vapor in the optical path. The adsorbed radiation is measured extremely fast and independent of the degree of saturation of the gas mixture. Response times of 0.1 to 1 s for 90% change in moisture content is common. Spectroscopic hygrometers are primarily used where a noncontact application is required; this may include atmospheric studies, industrial drying ovens, and harsh environments. The primary disadvantages of this device are its high cost and relatively large size.

Gravimetric Hygrometers

Humidity levels can be measured by extracting and weighing water vapor in a known quantity or atmosphere. For precise laboratory work, powerful desiccants, such as phosphorous pentoxide and magnesium perchlorate, are used for the extraction process; for other purposes, calcium chloride or silica gel are satisfactory.

When the highest level of accuracy is required, the gravimetric hygrometer, developed and maintained by NIST, is the ultimate in the measurement hierarchy. The gravimetric hygrometer yields and determines the absolute water vapor content, where the mass of the absorbed water and the precise measurement of the gas volume associated with the water vapor determine the mixing ratio or absolute humidity of the sample. This system has been chosen as the primary standard, because the required measurement of mass, temperature, pressure, and volume can be made with extreme precision. However, its complexity and required attention to detail limit its usefulness.

Calibration

The need for recalibration for many hygrometers depends on the accuracy required, on the stability of the sensor, and on the conditions to which the sensor is being subjected. Many hygrometers should be calibrated regularly by exposure to an atmosphere maintained at a known humidity and temperature, or by comparison with a transfer standard hygrometer. Complete calibration usually requires observation of a series of temperatures and humidities. Methods for producing known humidities include saturated salt solutions (Greenspan 1976, Huang and Whetstone 1985), sulfuric acid solutions, and mechanical systems, such as the divided flow, two-pressure (Amdur 1965), two-temperature (Till and Handegord 1960), and the NIST two-pressure humidity generator (Hasegawa 1976). All these systems rely on precise methods of temperature and pressure control within a controlled environment to produce a known humidity, usually with accuracies of ±0.5 to ±1.0%. The operating range for the precision generator is typically 5 to 95% rh.

SOUND AND VIBRATION MEASUREMENT

Measurement systems for determining sound pressure, sound intensity, and mechanical vibration generally involve the use of transducers to convert mechanical signals into electrical signals, which are then processed electronically in order to characterize the measured mechanical signals. These measurement systems contain one or more of the following elements which may, or may not, be contained in a single instrument:

1. A transducer, or an assembly of transducers, to convert sound pressure, sound intensity, or mechanical vibration (either time-varying strain, displacement, velocity, acceleration, or force) into an electrical signal that is quantitatively related to the mechanical quantity being measured.
2. Amplifiers and networks to provide such functions as electrical impedance matching, signal conditioning, integration, differentiation, frequency weighting, and gain.
3. Signal-processing equipment to detect and quantify those aspects of the signal that are being measured (peak value, rms value, time-weighted average level, power spectral density, or magnitude or phase of a complex linear spectrum or transfer function).
4. A device, such as a meter, oscilloscope, digital display, or level recorder, to display the signal or the aspects of it that are being quantified.

The relevant range of sound and vibration signals can vary over more than 12 orders of magnitude in amplitude and more than 8 orders of magnitude in frequency, depending on the application. References on instrumentation, measurement procedures, and signal analysis are given in the bibliography. Product and application notes, technical reviews, and books published by instrumentation manufacturers are an excellent source of additional reference material. See Chapter 7 for further information on sound and vibration.

Sound Measurement

A microphone is an electroacoustical transducer that transforms an acoustical signal into an electrical signal. The two predominant transduction principles used to measure sound (as opposed to broadcast or record) are the electrostatic and piezoelectric. Electrostatic (capacitor) microphones are available either as electric microphones, which do not require an external polarizing voltage, or as condenser microphones, which do require an external polarizing voltage, typically in the range of 28 to 200 V (dc). Piezoelectric microphones may be manufactured using either natural piezoelectric crystals or poled ferroelectric crystals. The types of response characteristics of measuring microphones are pressure, free field, and random incidence (diffuse field). A microphone with a uniform pressure response characteristic maintains uniform sensitivity over its operating frequency range when exposed to a sound pressure that is uniform over the surface of the sensing element. A microphone with a uniform free-field response characteristic maintains uniform sensitivity over its operating frequency range when exposed to a plane progressive sound wave at a specified angle of incidence to the surface of the sensing element. A microphone with a uniform random-incidence response characteristic maintains uniform sensitivity over its operating frequency range when exposed to a diffuse sound field. The sensitivity and the frequency range over which the microphone has uniform sensitivity (flat frequency response) vary with the diameter (surface area) of the sensing element and the microphone type. Other factors that may critically affect the performance or response of a measuring microphone and preamplifier in a given measurement application are atmospheric pressure, temperature, relative humidity, external magnetic and electrostatic fields, mechanical vibration, and radiation. A microphone should be selected based on its long- and short-term stability; its performance characteristics (sensitivity, frequency response, amplitude linearity, self-noise, etc.) versus the expected amplitude of sound pressure, frequency, range of analysis, and expected environmental conditions of measurement; and any other pertinent considerations, such as size and directional characteristics.

Microphone preamplifiers, amplifiers, weighting networks (see Chapter 7), filters, and displays are available as separate components, or as an integral part of a measuring instrument, such as a sound level meter, personal noise exposure meter, measuring amplifier, or real-time fractional octave or Fourier (*e.g.*, FFT) signal analyzer. Which instrument(s) to include in a sound measurement system depends on the purpose of the measurement and the frequency range and resolution of signal analysis. In the case of community and industrial noise measurements for regulatory purposes, the instrument, signal processing, and quantity to be measured are usually dictated by the pertinent regulation. Acoustical measurements made to quantify other parameters, characteristics, or criteria [*e.g.*, sound power in HVAC ducts, sound power emitted by machinery, noise criteria (NC) numbers, sound absorption coefficients, sound transmission loss of building partitions, reverberation times (T_{60}), etc.] will generally each dictate a different set of optimal instrumentation.

Measurement criteria often dictate the use of filters to perform signal analysis in the frequency domain in order to indicate the spectrum of the sound being measured. Filters of different bandwidths for different purposes include fractional octave band (one, one-third, one-twelfth, etc.), constant percentage bandwidth, and constant (typically narrow) bandwidth. The filters may be analog or digital and, if digital, may, or may not, be capable of real-time data acquisition during the measurement period, depending on the bandwidth of frequency analysis. FFT signal analyzers are generally used in situations that require very narrow-band signal analysis when the amplitudes of the sound spectra vary significantly with respect to frequency. This may occur in regions of resonance or when it is necessary to identify narrow-band or discrete sine-wave signal components of a spectrum in the presence of other such components or of broad-band noise.

Special rooms and procedures are required to characterize and calibrate sound sources and receivers. These rooms are generally classified into three types—anechoic, semianechoic, and reverberant. The ideal anechoic room or chamber would have boundary surfaces that completely absorb sound energy at all frequencies. The ideal semianechoic room or chamber would be identical to the ideal anechoic room, except that one surface would totally reflect sound energy at all frequencies. The ideal reverberant room

or chamber would have boundary surfaces that totally reflect sound energy at all frequencies.

Anechoic chambers are used to perform measurements under conditions approximating those of a free sound field. They can be used in calibrating and characterizing individual microphones, microphone arrays, acoustic intensity probes, reference sound power sources, loudspeakers, sirens, and other individual or complex sources of sound.

Semianechoic chambers are built with a hard reflecting floor in order to accommodate heavy machinery or to simulate large factory-floor or outdoor conditions. They can be used in calibrating and characterizing reference sound power sources, obtaining sound power levels of noise sources, and characterizing the sound output of emergency vehicle sirens when mounted on an emergency motor vehicle.

Reverberation chambers are used to perform measurements under conditions approximating those of a diffuse sound field. They can be used in calibrating and characterizing random incidence microphones and reference sound power sources, obtaining sound power ratings of equipment and sound power levels of noise sources, measuring sound absorption coefficients of building materials and panels, and measuring the transmission loss of building partitions and components such as doors and windows.

Prior to using a measurement system to perform absolute measurements of sound, it should be calibrated as a system from microphone, or probe, to indicating device. Acoustic calibrators and piston phones of fixed or variable frequency and amplitude are available for this purpose. These calibrators should be used at a frequency low enough that the pressure, free-field, and random-incidence response characteristics of the measuring microphone(s) are, for practical purposes, equivalent, or at least related in a known quantitative manner for that specific measurement system. In general, the sound pressure produced by these calibrators may vary, depending on the microphone type, whether or not the microphone has a protective grid, atmospheric pressure, temperature, and relative humidity. Correction factors and coefficients are required for conditions of use that differ from those existing during the calibration of the acoustic calibrator or piston phone. For demanding applications, precision sound sources and measuring microphones should periodically be sent to the manufacturer, a private testing laboratory, or a national standards laboratory for calibration.

Vibration Measurement

With the exception of seismic instruments that record or indicate mechanical motion directly via a mechanical or optomechanical mechanism connected to the test surface, vibration measurements involve the use of an electromechanical or interferometric vibration transducer. Here, the term vibration transducer refers to a generic mechanical vibration transducer. Electromechanical and interferometric vibration transducers belong to a large and varied group of transducers that detect mechanical motion and furnish an electrical signal that is quantitatively related to a particular physical characteristic of the motion. The electrical signal may be related to mechanical strain, displacement, velocity, acceleration, or force depending on the design of the transducer. The principles of operation of vibration transducers may involve optical interference; electrodynamic coupling; piezoelectric (including poled ferroelectric) or piezoresistive crystals; or variable capacitance, inductance, reluctance, or resistance. A considerable variety of vibration transducers with a wide range of sensitivities and bandwidths is commercially available. Vibration transducers may be contacting (*e.g.*, seismic transducers) or noncontacting (*e.g.*, interferometric or capacitive). Seismic transducers use a spring mass resonator within the transducer. At frequencies much greater than the fundamental natural frequency of the mechanical resonator, the relative displacement between the base and the seismic mass of the transducer is nearly proportional to the displacement of the transducer base. At frequencies much lower than the fundamental resonant frequency, the relative displacement between the base and seismic mass of the transducer is nearly proportional to the acceleration of the transducer base. Therefore, seismic displacement and seismic electrodynamic velocity transducers tend to have a relatively compliant suspension with a low resonant frequency; piezoelectric accelerometers and force transducers have a relatively stiff suspension with a high resonant frequency.

Strain transducers include the metallic resistance and the piezoresistive strain gage. For dynamic strain measurements, these are usually of the bonded type, where the gages are bonded directly to the test surface. How accurately a bonded strain gage replicates strain occurring in the test structure is largely a result of how well the strain gage was oriented and bonded to the test surface.

Displacement transducers include the capacitance gage, fringe-counting interferometer, seismic displacement, and linear variable differential transformer (LVDT). Velocity transducers include the reluctance (magnetic) gage, laser Doppler interferometer, and seismic electrodynamic velocity transducer. Accelerometers and force transducers include the piezoelectric, piezoresistive, and force-balance servo.

The sensitivity, frequency limitations, bandwidth, and amplitude linearity of vibration transducers vary greatly with the transduction mechanism and the manner in which the transducer is applied in a given measurement apparatus. The performance of contacting-type transducers can be significantly affected by the mechanical mounting methods and points of attachment of the transducer and connecting cable, and by the mechanical impedance of the structure loading the transducer. Amplitude linearity varies significantly over the operating range of the transducer, with some transducer types or configurations being inherently more linear than others. Other factors that may critically affect the performance or response of a vibration transducer in a given measurement application are temperature; relative humidity; external acoustic, magnetic, and electrostatic fields; transverse vibration; base strain; chemicals; and radiation. A vibration transducer should be selected based on its long- and short-term stability; its performance characteristics (*e.g.*, sensitivity, frequency response, amplitude linearity, self-noise) versus the expected amplitude of vibration, frequency range of analysis, and expected environmental conditions of measurement; and any other pertinent considerations (*e.g.*, size, mass, and resonant frequency).

Vibration exciters, or shakers, are used in structural analysis, vibration analysis of machinery, fatigue testing, mechanical impedance measurements, and vibration calibration systems. Vibration exciters have a table or moving element with a drive mechanism that may be mechanical, electrodynamic, piezoelectric, or hydraulic. They range from relatively small low-power units for calibrating transducers, such as accelerometers, to relatively large high-power units for structural and fatigue testing.

Conditioning amplifiers, power supplies, preamplifiers, charge amplifiers, voltage amplifiers, power amplifiers, filters, controllers, and displays are available as separate components, or as an integral part of a measuring instrument or system, such as a structural analysis system, vibration analyzer, vibration monitoring system, vibration meter, measuring amplifier, multichannel data-acquisition and modal analysis system, or real-time fractional-octave or FFT signal analyzer. Which instrument(s) to include in a vibration measurement system depends on the mechanical quantity to be determined, the purpose of the measurement, and the frequency range and resolution of signal analysis. In the case of vibration measurements, the signal analysis is relatively narrow in bandwidth and may be relatively low in frequency in order to accurately characterize structural resonances. Accelerometers with internal integrated circuitry are available to provide impedance matching or servo control for measuring very low-frequency acceleration (servo accelerometers). Analog and digital integration and differentiation of vibration signals are available through integrating and differentiating networks and amplifiers, and through FFT analyzers, respectively. Vibration measurements made for different purposes (*e.g.*, machinery diagnostics and health monitoring, balancing rotating machinery, analysis of torsional vibration, analysis of machine-tool vibration, modal analysis, analysis of vibration isolation, stress monitoring, industrial control) will generally each dictate different mechanical measurement requirements and a different set of optimal intrumentation.

Because of their inherent long- and short-term stability, amplitude linearity, wide bandwidth, wide dynamic range, low noise, and wide range of sensitivities, seismic accelerometers have traditionally been used as a reference standard for dynamic mechanical measurements. Prior to using a measurement system to perform absolute dynamic measurements of mechanical quantities, it should be calibrated as a system from transducer to indicating device. Calibrated reference vibration exciters, standard reference accelerometers, precision conditioning amplifiers, and precision calibration exciters are available for this purpose. These exciters and standard reference accelerometers can be used to transfer a calibration to another transducer. For demanding applications, a calibrated exciter, or a standard reference accelerometer with connecting cable and conditioning amplifier should periodically be sent to the manufacturer, a private testing laboratory, or a national standards laboratory for calibration.

COMBUSTION ANALYSIS

Two approaches are used to measure the thermal output or capacity of a boiler, furnace, or other fuel-burning device. The direct or calorimetric test measures change in enthalpy or heat content of the fluid, air, or water heated by the device, and multiplies this by the flow rate to arrive at the unit's capacity. The indirect test or flue gas analysis method determines the heat losses in the flue gases and the jacket and deducts them from the heat content (higher heating value) of the measured fuel input to the appliance. A heat balance simultaneously applies both tests to the same device. The indirect test usually indicates the greater capacity, and the difference is credited to radiation from the casing or jacket and unaccounted-for losses.

With small equipment, the expense of the direct test is usually not justified, and the indirect test is used with an arbitrary radiation and unaccounted-for loss factor.

Flue Gas Analysis

The flue gases from burning fossil fuels generally contain CO_2, H_2O, and H_2 with some small amounts of CO, NO_X, SO_X, and unburned hydrocarbons. However, to determine completeness of combustion and efficiency, generally only CO_2 (or O_2) and CO are measured.

In the laboratory, the instruments most commonly used to measure CO and CO_2 are nondispersive infrared (NDIR) analyzers. The NDIR instruments have several advantages: (1) they are not very sensitive to flow rate, (2) no wet chemicals are required, (3) they have a relatively fast response, (4) measurements can be made over a wide range of concentrations, and (5) they are not sensitive to the presence of contaminants in the ambient air.

In the laboratory, oxygen is generally measured with an instrument based on its paramagnetic properties. The paramagnetic instruments are generally used because of their excellent accuracy and because they can be made specific to the measurement of oxygen.

For field testing and burner adjustment, protable combustion testing equipment is available. These instruments generally measure O_2 and CO with electrochemical cells. The CO_2 is then calculated by an on-board microprocessor and, together with temperature, is used to calculate thermal efficiency. If a less expensive approach is required, a portable orsat system can be used to measure CO_2, and a length-of-stain tube to measure CO.

ROTATIVE SPEED MEASUREMENT

Tachometers

Tachometers, or direct-measuring rpm counters, vary from hand-held mechanical or electric meters to shaft-driven and electronic pulse counters. They are used in general laboratory and shop work to check the rotative speeds of motors, engines, and turbines.

Stroboscopes

Optical rpm counters work by producing a controlled high-speed electronic flashing light. The operator directs the light on a rotating member and increases the rate of flashes-per-minute until the optical effect of stopping rotation of the member is achieved. At this point, the rpm measured is equal to the flashes per minute emitted by the strobe unit. Care must be taken to start at the bottom of the instrument scale and work up, since multiples of the rpm produce almost the same optical effect as true synchronism. Multiples can be indicated by suitable marks positioned on the shaft, such as a bar on one side and a circle on the opposite side. If, for example, the two are seen superimposed, then the strobe light is flashing at an even multiple of the true rpm.

AC Tachometer-Generators

A tachometer-generator consists of a rotor and a stator. The rotor is a permanent magnet driven by the equipment. The stator is a winding with a hole through the center for the rotor. Concentricity is not critical; bearings are not required between rotor and stator. The output can be a single cycle-per-revolution signal whose voltage is a linear function of rotor speed. The polypole configuration that generates 10 cycles per revolution permits measurement of speeds as low as 20 rpm without causing the indicating needle to flutter. The output of the ac tachometer-generator is rectified and connected to a dc voltmeter.

THERMAL COMFORT MEASUREMENT

Thermal comfort depends on the combined influence of clothing, activity, air temperature, air velocity, mean radiant temperature, and air humidity. Thermal comfort is influenced by heating or cooling of particular body parts. This is due to radiant temperature asymmetry (plane radiant temperature), draft (air temperature, air velocity, turbulence), vertical air temperature differences, and floor temperature (surface temperature).

A general description of thermal comfort is given in Chapter 8, and guidelines for an acceptable thermal environment are given in ASHRAE *Standard* 55 (1992b) and ISO 7730 (1984). ASHRAE *Standard* 55 also includes required measuring accuracy. In addition to specified accuracy, ISO 7726 (1985) includes recommended measuring locations and a detailed description of instruments and methods.

Clothing and activity. These are estimated from tables (Chapter 8, ISO/DIS 9920 1990, ISO 8996 1990). The thermal insulation of clothing (clo-value) can be measured on a thermal mannequin (McCullough *et al.* 1985, Olesen 1985). The activity (met-value) can be estimated from measuring CO_2 and O_2 in a person's expired air.

Air temperature. Various types of thermometers may be used to measure air temperature. Placed in a room, the sensor registers a temperature between air temperature and mean radiant temper-

ature. One way of reducing the radiant error is to make the sensor as small as possible, because the convective heat transfer coefficient increases as the size decreases while the radiant heat transfer coefficient is constant. A sensor with a smaller dimension also provides a favorably low time constant. The radiant error can also be reduced by using a shield (an open, polished aluminum cylinder) around the sensor, by using a sensor with a low-emittance surface, or by increasing the air velocity around the sensor artificially (aspirating air through a tube in which the sensor is placed).

Air velocities. In occupied zones, air velocities are usually small (0 to 100 fpm) but have an effect on human thermal sensation. Since the velocity fluctuates, the mean value should be measured over a suitable period, typically 3 min. Velocity fluctuations with frequencies up to 1 Hz significantly increase human discomfort due to draft, which is a function of air temperature, mean air velocity, and turbulence (Chapter 8). The fluctuations can be given as the standard deviation of the air velocity over the measuring period (3 min) or as the turbulence intensity (standard deviation/mean air velocity). Velocity directions may change and are difficult to identify at low air velocities. An omnidirectional sensor with a short response time should be used. A thermal anemometer is suitable. If a hot-wire anemometer is used, the direction of the flow being measured must be perpendicular to the hot wire. Smoke puffs can be used to identify the direction.

Plane radiant temperature. This refers to the uniform temperature of an enclosure in which the radiant flux on one side of a small plane element is the same as in the actual nonuniform environment. It describes the radiation in one direction. The plane radiant temperature can be calculated from the surface temperatures of the environment (half-room) and the angle factors between the surfaces and a plane element (Chapter 8, ASHRAE *Standard* 55-92). The plane radiant temperature may also be measured by a net-radiometer (Chapter 8) or a radiometer with a sensor consisting of a reflective disc (polished) and an absorbent disc (painted black) (Olesen *et al.* 1989).

Mean radiant temperature. This is the uniform temperature of an imaginary black enclosure in which an occupant would exchange the same amount of radiant heat as in the actual nonuniform enclosure. The mean radiant temperature can be calculated from measured surface temperatures and the corresponding angle factors between the person and the surfaces (Chapter 8).

The mean radiant temperature can also be determined from the plane radiant temperature in six opposite directions weighted according to the projected area factors for a person (Chapter 8).

Because of its simplicity, the most commonly used instrument to determine the mean radiant temperature is a black globe thermometer (Vernon 1932, Bedford and Warmer 1935). This thermometer consists of a hollow sphere usually 6 in. in diameter, a flat black paint coating, and a thermocouple or thermometer bulb at its center. The temperature assumed by the globe at equilibrium results from a balance between heat gained or lost by radiation and by convection.

The mean radiant temperatures are calculated from

$$\bar{t}_r = \left[(t_g + 460)^4 + \frac{4.74 \times 10^7 \, V_a^{0.6}}{\epsilon D^{0.4}} (t_g - t_a) \right]^{1/4} - 460 \quad (21)$$

where

$\bar{t}_r$ = mean radiant temperature, °F
t_g = globe temperature, °F
V_a = air velocity, fpm
t_a = air temperature, °F
D = globe diameter, ft
ϵ = emissivity (0.95 for black globe)

According to Equation (21), air temperature, and air velocity around the globe must also be determined. The globe thermometer is spherical, while mean radiant temperature is defined in relation to the human body. For sedentary people, the globe represents a good approximation. For people who are standing, the globe, in a radiant nonuniform environment, overestimates the radiation from floor or ceiling. An ellipsoid-shaped sensor gives a closer approximation to the human shape. A black globe will also overestimate the influence of short-wave radiation, *e.g.*, sunshine. A flat grey color better represents the radiant characteristic of normal clothing (Olesen 1989). The hollow sphere is usually made of copper, which results in an undesirable high time constant. This can be overcome by using lighter materials, *e.g.*, a thin plastic bubble.

Air Humidity

The water vapor pressure (absolute humidity) is usually uniform in the occupied zone of a space; therefore, it is sufficient to measure absolute humidity at one location. Many of the instruments listed in Table 5 are applicable. At ambient temperatures that provide comfort or slight discomfort, the thermal effect of humidity is only moderate, and highly accurate humidity measurements are unnecessary.

Calculating Thermal Comfort

When the thermal parameters have been measured, their combined effect can be calculated by the thermal indices in Chapter 8. For example, the effective temperature (Gagge *et al.* 1971) can be determined from air temperature and humidity. Based on the four environmental parameters and an estimation of clothing and activity, the predicted mean vote (PMV) can be determined with the aid of tables (Fanger 1982, ISO 1984, Chapter 8). The PMV is an index predicting the average thermal sensation that a group of occupants may experience in a given space.

For certain types of normal activity and clothing, the environmental parameters measured can be compared directly with those described in ASHRAE *Standard* 55 or ISO *Standard* 7730.

Integrating Instruments

Several instruments have been developed to evaluate the combined effect of two or more thermal parameters on human comfort. Madsen (1976) developed an instrument that gives information on the occupants' expected thermal sensation by direct measurement of the PMV value. The comfort meter has a heated ellipsoid-shaped sensor that simulates the body (Figure 22). The estimated clothing (clo), activity in the actual space, and humidity are set on the instrument. The sensor then integrates the thermal effect of the air temperature, mean radiant temperature, and air velocity in approximately the same way the body does. The electronic instrument gives the measured operative and equivalent temperature and calculated PMV and predicted percentage of dissatisfied (PPD).

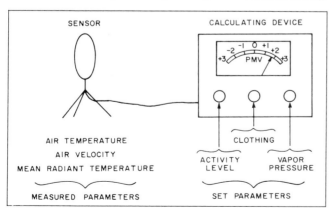

Fig. 22   Madsen's Comfort Meter
(Madsen 1976)

MOISTURE CONTENT AND TRANSFER MEASUREMENT

Little off-the-shelf instrumentation exists to measure the moisture content of porous materials or the moisture transfer through those materials. However, many measurements can be set up with a small investment of time and money. Three moisture properties are most commonly desired—(1) the sorption isotherm, a measure of the amount of water vapor a hygroscopic material will adsorb from humid air; (2) vapor permeability, a measure of the rate at which water vapor will pass through a given material; and (3) liquid diffusivity, a measure of the rate at which liquid water will pass through a porous material.

Sorption Isotherm

A sorption isotherm relates the equilibrium moisture content (EMC) of a hygroscopic material to the ambient relative humidity under conditions of constant temperature. Moisture content is the ratio of the total mass of water in a sample to the dry mass of the sample. Determining a sorption isotherm involves exposing a sample of material to a known relative humidity at a known temperature and then measuring the sample's moisture content after a sufficient period of time for the sample to reach equilibrium with its surroundings. Hysteresis in the sorption behavior of most hygroscopic materials requires that measurements be made both for increasing relative humidity (the adsorption isotherm) and for decreasing relative humidity (the desorption isotherm).

Controlling the ambient relative humidity can be accomplished using saturated salt solutions or mechanical refrigeration equipment (Tveit 1966, Cunningham and Spratt 1984, Carotenulo *et al.* 1991). Precise measurements of the relative humidity produced by various salt solutions have been reported by Greenspan (1977). ASTM *Standard* E104 describes the use of saturated salt solutions (ASTM 1991a). The equilibrium moisture content of a sample is usually determined gravimetrically using a percision balance. The sample dry mass, necessary to calculate moisture content, can be found by oven drying or desiccant drying. Oven dry mass may be less than desiccant dry mass because of the loss of volatiles other than water in the oven (Richards *et al.* 1992).

A major difficulty in the measurement of the sorption isotherms of engineering materials is the long time required for many materials to reach equilibrium—often as long as weeks or months. The rate-limiting mechanism for these measurements is usually the slow process of vapor diffusion into the pores of the material. The use of smaller samples can help reduce the diffusion time.

Vapor Permeability

The diffusive transfer of water vapor through porous materials is often described by a modified form of Fick's law

$$w_v = -\mu \, (dp/dx) \qquad (22)$$

where

w_v = mass of vapor diffusing through unit area in unit time
dp/dx = vapor pressure gradient
μ = vapor permeability

In engineering practice, permeance may be used instead of permeability. Permeance is simply permeability divided by the material thickness in the direction of the flow of vapor; thus, while permeability is a material property, permeance depends on thickness.

Measurement of permeability is made with wet cup, dry cup, or modified cup tests. Wet and dry cup tests are described in Chapter 20.

For many engineering materials, vapor permeability is a strong function of mean relative humidity. Wet and dry cups cannot adequately characterize this dependence on relative humidity. Instead, a modified cup method can be used (McLean *et al.* 1990, Burch *et al.* 1992). The modified cup method replaces the pure water or desiccant within a cup with a saturated salt solution. A second saturated salt solution is used to condition the environment external to the cup. With such an arrangement, the relative humidities on both sides of the sample material can be varied from 0 to 100%. Several cups with a range of mean relative humidities are used to map out the dependence of vapor permeability on relative humidity.

In measuring materials of high permeability, the finite rate of vapor diffusion through air in the cup may become a factor. The air-film resistance could then be a significant fraction of the resistance to vapor flow presented by the sample material. An accurate measurement of high-permeability materials may require an accounting of diffusive rates across all air gaps (Fanney *et al.* 1991).

Liquid Diffusivity

The transfer of liquid water through porous materials may be characterized as a diffusion-like process

$$w_l = - \rho D_1 \, d\gamma/dx \qquad (23)$$

where

w_l = mass of liquid transferred through unit area per unit time
ρ = liquid density
$d\gamma/dx$ = moisture content gradient

and where the liquid diffusivity D_1 typically shows a strong dependence on moisture content.

Transient measurement methods deduce the functional form of $D_1\gamma$ by observing the evolution of a one-dimensional moisture content profile over time. An initially dry specimen is brought into contact with liquid water. The free water will migrate into the specimen, drawn in by surface tension. The resulting moisture content profile, which changes with time, must be differentiated to find the liquid diffusivity of the material (Bruce and Klute 1956).

Determining the transient moisture content profile typically involves the use of a noninvasive and nondestructive method of measuring local moisture content. Gamma ray absorption (Freitas *et al.* 1991, Kumaran and Bomberg 1985, Quenard and Sallee 1989), X-ray radiography (Ambrose *et al.* 1990), neutron radiography (Prazak *et al.* 1990), and nuclear magnetic resonance (NMR) (Gummerson *et al.* 1979) have all been employed.

Uncertainty in the resulting measurement of the liquid diffusivity is often large because of the necessity to differentiate noisy experimental data.

HEAT TRANSFER THROUGH BUILDING MATERIALS

Thermal Conductivity

The thermal conductivity of a heat insulator, as defined in Chapter 20, is a unit heat transfer factor. Two methods of determining the thermal conductivity of flat insulation are the guarded hot plate and the heat-flowmeter apparatus, according to ASTM *Standards* C 177 and C 518, respectively (ASTM 1985, ASTM 1991b). Both methods use parallel, isothermal plates to induce a steady temperature gradient across the thickness of the specimen(s). The guarded hot plate is considered an absolute method for determining thermal conductivity. The heat-flowmeter apparatus requires calibration, with a specimen having a known thermal conductivity usually determined in the guarded hot plate. The heat-flowmeter apparatus is calibrated by determining the voltage output of its heat flux transducer(s) as a function of the heat flux through the transducer(s).

The basic design of the guarded hot plate consists of an electrically heated plate and two liquid-cooled plates. Two similar specimens of a material are required for a test; one is mounted on each side of the hot plate. A cold plate is then pressed against the outside of each specimen by a clamp screw. The heated plate consists of two sections separated by a small gap. During tests, the central (metering) section and the outer (guard) section are maintained at the same temperature to minimize errors caused by edge effects. The electric energy required to heat the metering section is measured carefully and converted to heat flow. The thermal conductivity of the material can be calculated under steady-state conditions using this heat flow quantity, the area of the metering section, the temperature gradient, and the specimen thickness. The thermal conductivity of cylindrical or pipe insulation (Chapter 20) is determined in a similar manner, but an equivalent thickness must be calculated to account for the cylindrical shape (ASTM 1989b). Transient methods have been developed by D'Eustachio and Schreiner (1952), Hooper and Lepper (1950), and Hooper and Chang (1953) using a line heat source within a slender probe. These instruments are available commercially and have the advantages of rapidity and a small test specimen requirement. The probe is a useful research and development tool, but it has not been as accepted as the guarded hot plate, heat-flowmeter apparatus, or pipe insulation apparatus.

Thermal Conductances and Resistances

Thermal conductances (C-values) and resistances (R-values) of many building assemblies can be calculated from the conductivities and dimensions of their components, as described in Chapter 22. Test values can also be determined experimentally using large specimens representative of the building assemblies tested in the hot box apparatus described in ASTM *Standards* C 236 and C 976 (ASTM 1989c, ASTM 1990a). This laboratory apparatus allows measurement of heat transfer through a specimen under controlled air temperature, air velocity, and radiation conditions. It is specially suited for large nonhomogeneous specimens.

For in-situ measurements, heat flux and temperature transducers are useful in measuring the dynamic or steady-state behavior of opaque building components (ASTM 1991c). A heat flux transducer is simply a differential thermopile within a core or substrate material. There are two types of construction—(1) multiple thermocouple junctions wrapped around a core material, or (2) printed circuits with a uniform array of thermocouple junctions. The transducer is calibrated by determining its voltage output as a function of the heat flux through the transducer. For in-situ measurements, the transducer is installed in either the wall or roof construction, or mounted on an exterior surface with tape or glue. The data obtained from the heat flux and temperature transducers can be used to compute the thermal conductance or resistance of the building component (ASTM 1990b).

AIR INFILTRATION AND AIRTIGHTNESS MEASUREMENT

Two major characteristics describe air infiltration in buildings—air change rates and envelope air leakage. The measurement approaches used to determine these factors are described in Chapter 23. The air change rate of a building refers to the rate at which outdoor air enters the building under normal conditions of weather and ventilation system operation. In general, the air change rate includes both outdoor air taken in through the air-handling system and air leakage through the building envelope (infiltration). The outdoor air intake rate is determined by the design, installation, and operation of the mechanical ventilation system. Infiltration is determined by the extent and distribution of leaks over the building envelope and the pressure differences across these leaks. These pressure differences are induced by wind, inside-outside temperature differences, and the operation of building mechanical equipment. Building air change rates can only be measured by injecting a tracer gas into a building and monitoring and analyzing the tracer gas concentration response. The instrumentation required for tracer gas testing includes a means of injecting the tracer gas and sampling the air in the building and a tracer gas monitor. There is a variety of tracer gas techniques, distinguished by the injection strategy and analysis approach employed. To fully characterize the air change performance of a building, the air change rate must be measured over a range of weather and equipment operation.

The airtightness of a building envelope can be measured relatively quickly using building pressurization techniques, which are described in Chapter 23. In the pressurization technique, a large fan or blower mounted in a door or window induces a large and roughly uniform pressure difference across the building shell. The airflow required to maintain this pressure difference is then measured. The more leakage in the building, the more airflow is required to induce a specific indoor-outdoor pressure difference. The building airtightness is characterized by the airflow rate at a reference pressure, normalized by the building volume or surface area. Under proper test conditions, the results of a pressurization test are independent of weather conditions. The instrumentation requirements for pressurization testing include air-moving equipment, a device to measure airflow, and a differential pressure gage.

AIR CONTAMINANT MEASUREMENT

Three measures of particulate air contamination include the number, projected area, and mass of particles per unit volume of air (ASTM 1991d, ASTM 1987b, ASHRAE 1971). Each requires an appropriate sampling technique.

Particles are counted by capturing them in impinges, impactors, membrane filters, and thermal or electrostatic precipitators. Counting is done by microscope, using stagecounts if the sample covers a broad range of sizes.

Electronic particle counters give rapid data of particle size distribution and concentration depending on the design of the instrument. However, their accuracy depends on careful calibration, appropriate maintenance, and proper application. Particle counters have been used in indoor office environments as well as clean rooms.

Projected area determinations are usually made by sampling onto a filter paper and comparing the light transmitted or scattered by this filter to a standard filter. The staining ability of dusts depends on the projected area and refractive index per unit volume. For sampling, filters must collect the minimum size particle of interest. In this respect, membrane or glass fiber filters are recommended.

To determine particle mass, a measured quantity of air is drawn through filters, preferably of membrane or glass fiber, and the filter mass is compared to the mass before sampling. Electrostatic or thermal precipitators and various impactors have also been used. For further information, see ACGIH (1983), Lundgren *et al.* (1979), and Lodge (1989).

Chapter 40 of the 1991 ASHRAE *Handbook—Applications* presents information on measuring and monitoring gaseous contaminants. Relatively costly analytical equipment, which must be calibrated and operated carefully by experienced personnel, is needed. Numerous methods of sampling the contaminants as well as the laboratory analysis techniques used after sampling are specified. Some of the analytical methods are specific to a single pollutant; others are capable of presenting a concentration spectrum for many compounds simultaneously.

LIGHTING MEASUREMENT

Light level, or illuminance, is usually measured with a photocell made from a semiconductor such as silicon or selenium. Such photocells produce an output current proportional to incident luminous flux, and when linked with a microammeter, color- and cosine-corrected filters and multirange switches are used in inexpensive hand-held light meters and more precise instruments. Different cell heads allow multirange use in precision meters.

Cadmium sulfide photocells, in which the resistance varies with illumination, are also used in light meters. Both gas-filled and vacuum photoelectric cells are in use.

Small survey-type meters are not as accurate as laboratory meters; their readings should be considered approximate, although consistent, for a given condition. Their range is usually from 5 to 5000 footcandles. Precision low-level meters have cell heads with ranges down to 0 to 2 footcandles.

A photometer installed in a revolving head is called a goniophotometer and is used to measure the distribution of light sources or luminaires. To measure total luminous flux, the luminaire is placed in the center of a sphere painted inside with a high-reflectance white with a near perfect diffusing matte surface. Total light output is measured through a small baffled window in the sphere wall.

To measure irradiation from germicidal lamps, a filter of fused quartz with fluorescent phosphor is placed over the light meter cell.

If meters are used to measure the number of lumens per unit area diffusely leaving a surface, luminance (cd/in^2) instead of illumination (footcandles) is read. Light meters can be used to measure luminance; or electronic lux meters containing a photo tube, amplifier, and microammeter can read luminance directly. In the case of a perfectly diffuse reflecting surface, which has a constant luminance regardless of viewing angle, the unit of foot-lamberts in $lumens/ft^2$ is sometimes applied.

DATA RECORDING

The trend toward computer entry and processing of data, greater automation, and improved data accuracy have led to the pervasive use of digital data acquisition systems for dynamic as well as static data. Almost every type of transducer and sensor is available with the necessary interface system to make it computer compatible. The transducer itself begins to lose its identity when integrated into a system that incorporates such features as linearization, offset correction, self-calibration, and so forth. This trend has eliminated the concern regarding the details of signal conditioning and amplification of basic transducer outputs. The personal computer is integrated into every aspect of data recording, including sophisticated graphics, acquisition and control, and analysis.

The event duration for a specific test may last from microseconds to years. For short-term transient testing, data are often recorded on analog tape or stored in a digital memory for subsequent playback, processing, and display. Long-term data are usually recorded on digital cassette tapes at timed intervals and collected on a periodic (*e.g.*, quarterly) basis for processing. Between these two extremes, the more typical application involves real-time monitoring and quick-look display of events lasting up to an hour, with off-line processing and analysis performed as needed.

Other means of recording, such as chart recorders, which can either be multipurpose or specifically designed for a given sensor, are available. Chart recorders provide a visual indication and a hard copy record of the data. Rarely is the output of a chart recorder used to process data. Simple indicators and readouts are used mostly to monitor the output of a sensor visually. In most situations, analog indicators such as d'Arsonval movement meters have been rendered obsolete by modern digital indicators. Industrial environments commonly employ signal transmitters for control or computer data-handling systems to convert the signal output of the primary sensor into a compatible common signal span, for example, the standard 4-20 mA current loop. All signal conditioning (ranging, zero suppression, reference-junction compensation) is provided at the transmitter. Thus, all recorders and controllers in the system can have an identical electrical span, with variations only in charts and scales offering the advantages of interchangeability and economy in equipment cost. Long signal transmission lines can be used, and receiving devices can be added to the loop without degrading performance.

The vast selection of available hardware, an often confusing set of terminology, and the challenge of optimizing the performance/cost ratio for a specific application makes the task of configuring a data acquisition system difficult. Two major factors influence the selection of recording systems—(1) the need for a number of measurements of the desired parameters, and (2) the need for acquired data to be processed. A system specifically configured to meet a particular measurement need can quickly become obsolete if it has inadequate flexibility. Other considerations, such as thermal, mechanical, electromagnetic interference, portability, and meteorological factors, also influence the selection of the recording system.

Digital Recording

A digital data acquisition system must contain an interface, *i.e.*, a system involving one or several analog-to-digital converters, and in the case of multichannel inputs, circuitry for multiplexing. The interface may also provide excitation for transducers, calibration, and conversion of units. The digital data are arranged into one or several standard digital bus formats. Many data acquisition systems are designed to acquire data rapidly and store large records of data for later recording and analysis. Once the input signals have been digitized, the digital data is essentially immune from noise and can be transmitted over great distances.

Information is transferred to a computer/recorder from the interface as a pulse train, which can be transmitted as 4-, 8-, 12-, 16-, or 32-bit words. An 8-bit word or byte and many communications methods are rated according to their bytes per second transfer rate. Digital data is transferred in either a serial or parallel mode. Serial transmission means that the data is sent as a series of pulses, one bit at a time. Although slower than a parallel system, serial interfaces require only two wires, which lowers their cabling cost. The speed of serial transmissions are rated according to their symbols/second rate, or baud rate. In parallel transmission, the entire data word is transmitted at one time. To do this, each bit of a data word has to have its own transmission line; other lines are needed for clocking and control. Parallel transmission is used for short distances or when high data rates are required. Serial transmission must be used for long-distance communications where wiring costs are prohibitive.

The two most popular interface bus standards currently used for data transmission are the IEEE 488 or the GPIB (general purpose interface bus) and the RS232 serial interface. The IEEE 488 bus system feeds data down 8 parallel wires, one 8-bit data byte at a time. This parallel operation allows it to transfer data rapidly with up to 1-million characters per second. However, the IEEE 488 bus is limited to a cable length of 65 ft and requires an interface connection on every meter for proper termination. The RS232 system feeds data serially down two wires, one bit at a time. The RS232 system can operate up to 9600 baud, which amounts to only 1200 8-bit bytes per second. An RS232 line may be over 1000 ft long. For longer distances, it may feed a modem to send data over standard telephone lines. A local area network (LAN) may be available in a facility for transmitting information. With appropriate interfacing, transducer data is available to any computer connected to the network.

Bus measurements can greatly simplify three basic applications—data gathering, automated limit testing, and computer-controlled processes. In data gathering applications, readings are collected over time. The most common applications include aging tests in quality control, temperature tests in quality assurance, and testing for intermittents in service. A controller can monitor any output indefinitely and then display the data directly on its screen or record it on magnetic tape or disks for future use.

In automated limit testing, the computer simply compares each measurement with limits programmed ahead of time. The controller converts the readings to a good/bad readout. Automatic limit testing becomes highly cost-effective when working with large number of parameters of a particular unit under test.

In computer-controlled processes, the IEEE 488 bus system becomes a permanent part of a larger, completely automated system. For example, a large industrial process may require many electrical sensors that feed a central computer controlling many parts of the manufacturing process. For example, an IEEE 488 bus controller collects readings from several sensors and saves the data until asked to dump an entire batch of readings to a larger central computer at one time. Used in this manner, the IEEE 488 bus controller serves as a slave of the central processor unit.

Dynamic range and accuracy are two important parameters that must be considered in a digital recording system. *Dynamic range* refers to the ratio of the maximum input signal for which the system is useful to the noise floor of the system. The *accuracy* figure for a system is impacted by the noise floor, nonlinearity, temperature, time, crosstalk, and so forth. In selecting an 8-, 10-, or 12-bit analog-to-digital converter, the designer cannot assume that the system accuracy will necessarily be determined by the resolution of the encoders, *i.e.*, 0.4%, 0.1%, and 0.25%, respectively. If the sensor preceding the converter is limited to 1% full-scale accuracy, for example, no significant benefits are gained by using a 10- or 12-bit system over an 8-bit system and suppressing the least significant bit. However, a greater number of bits may be required to cover a larger dynamic range.

Chart Recorders

Chart recorders convert electrical signals (analog or digital) to records of the data against time on a hard copy, usually paper. Mechanical styluses use ink, hot wire, pressure, or electrically sensitive paper to provide a continuous trace. They are useful up to a few hundred hertz. Thermal and ink recorders are confined to chart speeds of a few inches per second for recording relatively slow processes. Newer advances in portable recorders provide multichannel inputs and up to 25 kHz real time frequency response without using a pen or pen motor. XY recorders and plotters allow two variables to be recorded with respect to one another. Their response times are generally limited to that of thermal and ink recorders. Oscillographic recorders have largely been made obsolete by digital oscilloscopes.

Tape Recorders

A tape recorder records and reproduces information tape coated with a magnetizable material. Tapes can be of the open-reel type, which permit long-time recordings and multichannel data recording; or tapes can be housed in compact cassettes, which enhances portability. A number of information channels are usually recorded side by side on tape, *e.g.*, 7 on 0.5-in. tape or 14 on 1-in. tape. Instrumentation tape recorders are classified into three general types that refer to the coding method of the signal—direct or analog recording, FM recording, and digital or pulse code modulation.

Direct recording tape recorders, similar to general-purpose audio tape recorders, generally have a frequency response from 4 Hz at a tape speed of 1.5 in/s and up to about 60 kHz at 15 in/s. Their reproduction accuracy is about ±5%, and their dynamic range is limited, unless a compressor/expander (compander) unit is employed which can increase the dynamic range to about 70 dB. The main virtue of direct recorders is their low cost.

Frequency modulation (FM) instrumentation recorders convert the input signal proportional to frequency. Their main advantage is that signal frequencies down to dc can be recorded. Their upper frequency response is about 20 kHz at 15 in/s. Both the FM and direct-type recorders have a signal to noise ratio of about 40 dB; FM recorders are accurate to about ±1%.

Digital tape recorders, sometimes referred to as rotary storage recorders (RSR) or rotary digital audio tape (R-DAT) recorders, are the newest recording devices in the field of magnetic tape recording. The term rotary refers to the rotating recording head, which increases the head-to-tape speed. (Greater bandwidth is achieved in any recorder by increasing head-to-tape speed.) Usually, the rotating head is arranged to create a helical scan. Thus, while a longitudinal recording has continuously recorded signals on tracks running the length of the tape, each swipe on a helical scan recording provides discrete blocks of recorded signal.

A digital tape recorder first requires an analog-to-digital conversion of the signal. The resultant bit stream of binary numbers, pulse code modulation (PCM), are, in turn, recorded on the tape in terms of high and low values (saturated positive and negative magnetizations). During playback, the pulse train that is first read and then reconstructed permits the signal's sequence of binary numbers to be reconstituted. This, in turn, leads to the reconstruction of an analog signal after an appropriate digital-to-analog conversion. Most digital tape recorders have standard interfaces allowing data transfer directly to other equipment, such as computers.

An R-DAT recorder has a bandwidth comparable to direct or FM recording, but the most common recorders have a sampling frequency of 48 kHz giving a frequency response of 20 kHz. Multiple recording channels are usually available, but the frequency response drops proportionally to the number of channels. However, their main feature is superior signal-to-noise ratio and dynamic range, which is at least 70 dB. Typical data quantization levels are 14 to 16 bits with error correction codes.

REFERENCES

ACGIH. 1983. *Air sampling instruments for evaluation of atmospheric contaminants*, 6th ed. American Conference of Government Industrial Hygienists, Cincinnati, OH.

AIP. 1992. *Temperature, Its measurement and control in science and in industry*, J.F. Schooley, ed. Vol. 6. American Institute of Physics, New York.

Ambrose, J.H., L.C. Chow, and J.E. Beam. 1990. Capillary flow properties of mesh wicks. AIAA *Journal of Thermophysics* 4:318-24.

Amdur, E.J. 1965. Two-pressure relative humidity standards. In *Humidity and Moisture* 3:445. Reinhold Publishing Corporation, New York.

ANSI/ASME. 1985a. Performance test codes. *Standard* 19.1-1985, Part 1, Measurement uncertainty, February.

ANSI/ASME. 1985b. Gauges-pressure and vacuum, indicating dial type-elastic elements. *Standard* B40.1-1985. American Society of Mechanical Engineers, New York.

ASHRAE. 1971. Symposium on field testing for air pollution emissions, Philadelphia, PA.

ASHRAE. 1985. Laboratory methods of testing fans for rating. ASHRAE *Standard* 51-85, also AMCA *Standard* 210-85.

ANSI/ASHRAE. 1988. Practices for measurement, testing, adjusting, and balancing of building heating, ventilation, air-conditioning and refrigeration systems. ANSI/ASHRAE *Standard* 111-1988.

ASHRAE. 1992a. Standard method for measurement of moist air properties. ASHRAE *Standard* 41.6-1992R.

ASHRAE. 1992b. Thermal environmental conditions for human occupancy. ASHRAE *Standard* 55-1992.

ASME. 1972. Application of fluid meters, Part II, 6th ed. (Interim Supplement 19.5 on Instruments and Apparatus.) ASME Performance Test Code PTC 19.5-72. American Society of Mechanical Engineers, New York.

ASME. 1986. Temperature measurement instruments and apparatus, ASME Performance Test Code PTC 19.3-74 (revised 1986).

ASME. 1989. Measurement of fluid flow in pipes using orifice, nozzle, and venturi. MFC-3M-1985, ASME, New York.

ASME. 1988a. Measurement of liquid flow in closed conduits by weighing methods. MFC-9M-1988, ASME, New York.

ASME. 1988b. Method for establishing installation effects on flowmeters. MFC-10M 1988, ASME, New York.

ASME. 1991. Glossary of terms used in the measurement of fluid flow in pipes. MFC-1M-1991, ASME, New York.

ASTM. 1980. Standard test methods for water vapor transmission of materials. ASTM *Designation* E96-80.

ASTM. 1981. *Manual on the use of thermocouples in temperature measurement*. American Society for Testing and Materials, Philadelphia, PA.

ASTM. 1985. Standard test method for steady-state heat flux measurements and thermal transmission properties by means of the guarded-hot-plate apparatus. ASTM *Designation* C 177-85.

ASTM. 1987a. Standard temperature-electromotive force (EMF) tables for standardized thermocouples. ASTM *Designation* E 23-087.

ASTM. 1987b. Sampling and calibration for atmospheric measurements. ASTM STP 957.

ASTM. 1989. Standard metric practice guide, Publication E380-89a. American Society for Testing and Materials, Philadelphia, PA.

ASTM. 1989a. Standard practice for thermographic inspection of insulation installations in envelope cavities of wood frame buildings. ASTM *Designation* C 1060-86.

ASTM. 1989b. Standard test method for steady-state heat transfer properties of horizontal pipe insulation. ASTM *Designation* C 355-89.

ASTM. 1989c. Standard test method of steady-state thermal performance of building assemblies by means of a guarded hot box. ASTM *Designation* C 236-87.

ASTM. 1990a. Standard practice for determining thermal resistance of building envelope components from in-situ data. ASTM *Designation* C 1155-90.

ASTM. 1990b. Standard test method for thermal performance of building assemblies by means of a calibrated hot box. ASTM *Designation* C 976-90.

ASTM. 1991a. Standard recommended practice for maintaining constant relative humidity by means of aqueous solutions. ASTM *Standard* E 104 1-85.

ASTM. 1991b. Standard test method for steady state heat flux measurements and thermal transmission properties by means of the heat flow meter apparatus. ASTM *Standard* C 518-91.

ASTM. 1991c. Standard practice for in-situ measurement of heat flux and temperature on building envelope components. ASTM *Standard* C 1046-91.

ASTM. 1991d. *Atmospheric analysis: Occupational health and safety* 11.03.

ASTM. 1991e. Standard practice for the use of the international system of units (SI)(the modernized metric system). *Standard* E380-91a. Philadelphia, PA.

Bedford, T. and C.G. Warmer. 1935. The globe thermometer in studies of heating and ventilating. *Journal of the Institution of Heating and Ventilating Engineers* 2:544.

Benedict, R.P. 1984. *Fundamentals of temperature, pressure and flow measurements*, 3rd ed. John Wiley and Sons, Inc., New York.

Benson, K.R., *et al.* 1986. NBS Primary calibration facilities for air flow rate, air speed, and slurry flow. Proceedings of the National Flow Measurement Symposium, Washington D.C., published by the American Gas Association, Arlington, VA.

Bentz, D.P. and J.W. Martin. 1987. Using the computer to analyze coating defects. *Journal of Protective Coatings & Linings* 4(5).

BIPM. 1981. Report of the BIPM Working Group on the Statement of Uncertainties to the Comité International des Poids et Mesures, Kaarls R. Rapporteur.

Birge, R.T. 1939. The propagation of errors. In *The American physics teacher* 7(6).

Bruce, R.R. and A. Klute. 1956. The measurement of soil moisture diffusivity. Proceedings of the Soil Science Society of America 20:458-62.

Burch, D.M., W.C. Thomas, and A.H. Fanney. 1992. Water vapor permeability measurements of common building materials. ASHRAE *Transactions* 98(1).

Burch, D.M. 1980. Infrared audits of roof heat loss. ASHRAE *Transactions* 86(2).

Burch, D.M. and C.M. Hunt. 1978. Retrofitting an existing residence for energy conservation—An experimental study. Building Science Series 105, National Institute of Standards and Technology (formerly National Bureau of Standards).

Burns, G.W., M.G. Scroger, G.F. Strouse, M.C. Croarkin, and W.F. Guthrie. 1992. Temperature-electromotive force reference functions and tables for the letter-designated thermocouple types based on the ITS-90, NIST Monograph 175, U.S. Government Printing Office, Washington, D.C.

Carotenuto, A., F. Fucci, and G. LaFianzi. 1991. Adsorption phenomena in porous media in the presence of moist air. *International Journal of Heat and Mass Transfer* 18:71-81.

Cohen, E.R. 1990. The expression of uncertainty in physical measurements. 1990 Measurement Science Conference Proceedings, Anaheim, CA.

Colclough, A.R. 1987. Two theories of experimental error. *Journal of Research of the National Bureau of Standards* 92(3).

Considine, D.M. 1985. *Process instruments and controls handbook*, 3rd ed. McGraw-Hill Book Company, New York.

Croarkin, C. 1984. Measurement assurance programs, Part II: Development and implementation. NBS Special Publication 676-II.

Cunningham, M.J. and T.J. Sprott. 1984. Sorption properties of New Zealand building materials. Building Research Association of New Zealand Research Report No. 45, Judgeford.

R.S. Dadson, S.L. Lewis, and G.N. Peggs. 1982. *The pressure balance: Theory and practice*. Her Majesty's Stationary Office, London.

D'Eustachio, D. and R.E. Schreiner. 1952. A study of transient heat method for measuring thermal conductivity. ASHVE *Transactions* 58:331.

DeCarlo, J.P. 1984. *Fundamentals of flow measurement*. Instrumentation Society of America, Research Triangle Park, NC.

DeWitt, D.P. and G.D. Nutter. 1988. *Theory and practice of radiation thermometry*. John Wiley and Sons, Inc., New York.

Dring, R.P. and B. Gebhart. 1969. Hot-wire anemometer calibration for measurements at very low velocity. *Journal of Heat Transfer*, ASME *Transactions* 5:241.

Eisenhart, C. 1963. Realistic evaluation of the precision and accuracy of instrument calibration systems. *Journal of Research of the National Bureau of Standards* 67C(2). Reprinted as paper 1.2 in NBS Special Publication 300-1.

Eisenhart, C., H.H. Ku, R. Colle. 1983. Expression of the uncertainties of final measurement results. NBS Special Publication 644.

Fanger, P.O. 1982. *Thermal comfort*. Robert E. Krieger, Malabar, FL.

Fanney, A.H., W.C. Thomas, D.M. Burch, and L.R. Mathena. 1991. Measurements of moisture diffusion in building materials. ASHRAE *Transactions* 97:99-113.

Freitas, V., P. Crausse, and V. Abrantes. 1991. Moisture diffusion in thermal insulating materials. In *Insulation materials: Testing and applications*, Vol. 2, ASTM STP 1116.

Gagge, A.P., J.A.J. Stolwijk, and Y. Nishi. 1971. An effective temperature scale based on a simple model of human physiological regulatory response. ASHRAE *Transactions* 77(1).

Goldstein, R.J. 1978. Application of aerial infrared thermography. ASHRAE *Transactions* 84(1).

Grant, H.P. and R.E. Kronauer. 1962. Fundamentals of hot wire anemometry. ASME Symposium on Measurement in Unsteady Flow, May.

Greenspan, L. 1976. Humidity fixed points of binary saturated aqueous solutions. *Journal of Research of the National Bureau of Standards* 81A:89.

Greenspan, L. and A. Wexler. 1968. An adiabatic saturation psychrometer. *Journal of Research of the National Bureau of Standards* 72C(1):33.

Greenspan, L. 1977. Humidity fixed points of binary saturated aqueous solutions. *Journal of Research of the National Bureau of Standards* 81A:89-95.

Gummerson, R.J., C. Hall, W.D. Hoff, R. Hawkes, G.N. Holland, and W.S. Moore. 1979. Unsaturated water flow within porous materials observed by NMR imaging. *Nature* 281:56-57.

Hasegawa, S. 1976. The NBS two-pressure humidity generator, Mark 2, *Journal of Research of the National Bureau of Standards* 81A:81.

Hooper, F.C. and F.C. Lepper. 1950. Transient heat flow apparatus for the determination of thermal conductivity. ASHVE *Transactions* 56:309.

Hooper, F.C. and S.C. Chang. 1953. Development of thermal conductivity probe. ASHVE *Transactions* 59:463.

Huang, P. and J.R. Whetstone. 1985. Evaluation of relative humidity values for saturated aqueous salt solutions using osmotic coefficients between 50 and 100°C. *Moisture and Humidity* I:577. Instrumentation Society of America, Research Triangle Park, NC.

Hudson, R.D., Jr. 1969. *Infrared system engineering*. John Wiley and Sons, New York.

ISO. 1985. Thermal environments—Instruments and methods for measuring physical quantities, *Standard* 8990. International Standards Organization, Geneva.

ISO. 1990. Ergonomics—Determination of metabolic heat production, *Standard* 8996. International Standards Organization, Geneva.

ISO/DIS. 1990. Ergonomics of the thermal environment—Estimation of the thermal insulation and evaporative resistance of a clothing ensemble, *Standard* 9920. International Standards Organization, Geneva.

Madsen, T.L. 1976. Thermal comfort measurements. ASHRAE *Transactions* 82(1).

ISO. 1984. Moderate thermal environments—Determination of the PMV and PPD indices and specification of the conditions for thermal comfort, *Standard* 7730. International Standards Organization, Geneva.

Kovasznay, L.S.G. 1959. Turbulence measurements. *Applied Mechanics Reviews* 12(6):375.

Ku, H.H. 1965. Notes on the use of propagation of error formulas. *Journal of Research of the National Bureau of Standards* 70C(4).

Kumaran, M.K. and M. Bomberg. 1985. A gamma-spectrometer for determination of density distribution and moisture distribution in building materials. Proceedings of the International Symposium on Moisture and Humidity, Washington, D.C., 485-90.

Kusuda, T. 1965. Calculation of the temperature of a flat-plate wet surface under adiabatic conditions with respect to the Lewis relation. *Humidity and Moisture* I:16. Reinhold Publishing Corporation, New York.

Liptak, B.G., ed. 1972. *Instrument engineers handbook*, Vol. 1. Chilton Book Company, Philadelphia, PA.

Lodge, J.P., ed. 1989. *Methods of air sampling and analysis*, 3rd ed. Lewis Publishers, MI.

Lundgren, D.A., et al., eds. 1979. *Aerosol measurement*. University Press of Florida.

Mack, R.T. 1986. Energy loss profiles: Foundation for future profit in thermal imager, sales, and service. Proceedings of the 5th Infrared Information Exchange, Book 1, AGEMA Infrared Systems, Secaucus, NJ.

Mandel, J. and L.F. Nanni. 1986. Measurement evaluation. NBS Special Publication 700-2.

Mattingly, G.E. 1985. Gas flow measurement calibration facilities and fluid metering traceability at the National Bureau of Standards. Proceedings of the Institute for Gas Technology Conference on Energy Measurement, Chicago, IL.

Mattingly, G.E. 1988. A round robin flow measurement testing program using hydrocarbon liquids: Results for first phase testing. NISTIR 88-4013, NIST, Gaithersburg, MD.

Mattingly, G.E. 1992. The characterization of a piston displacement-type flowmeter calibration facility and the calibration and use of pulsed output type flowmeters. *Journal of Research of the National Institute of Standards and Technology*. 97(5):509.

Mattingly, G.E. 1984. Workshop on fundamental research issues in orifice metering. Gas Research Institute Report 84/0190, Gas Research Institute, Chicago, IL.

McCullough, E.A., B.W. Jones, and J. Huck. 1985. A comprehensive data base for estimating clothing insulation. ASHRAE *Transactions* 92:29-47.

McLean, R.C., G.H. Galbraith, and C.H. Sanders. 1990. Moisture transmission testing of building materials and the presentation of vapour permeability values. *Building Research and Practice*, 82-103.

Mease, N.E., W.G. Cleveland, Jr., G.E. Mattingly, and J.M. Hall. 1992. Air speed calibrations at the National Institute of Standards and Technology. Proceedings of the 1992 Measurement Science Conference, Anaheim, CA.

Miller, R.W. 1983. *Measurement engineering handbook*. McGraw-Hill Book Company, New York.

NIST. 1976. Liquid-in-glass thermometry. NIST *Monograph* 150.

NIST. 1986. Thermometer calibrations. NIST *Monograph* 174.

Nottage, H.B., J.G. Slaby, and W.P. Gojsza. 1952. A smoke-filament technique for experimental research in room air distribution. ASHVE *Transactions* 58:399.

Olesen, B.W. 1985. A new and simpler method for estimating the thermal insulation of a clothing ensemble. ASHRAE *Transactions* 92:478-92.

Olesen, B.W., J. Rosendahl, L.N. Kalisperis, L.H. Summers, and M. Steinman. 1989. Methods for measuring and evaluating the thermal radiation in a room. ASHRAE *Transactions* 95(1).

Olsen, L.O. 1974. Introduction to liquid flow metering and calibration of liquid flowmeters. NBS Technical Note 831.

Paljak, I. and B. Pettersson. 1972. *Thermography of buildings*. National Swedish Institute for Materials Testing, Stockholm.

Parmelee, G.V. and R.G. Huebscher. 1946. The shielding of thermocouples from the effects of radiation. ASHVE Research Report No. 1292, ASHVE *Transactions* 52:183.

Prazak, J., J. Tywoniak, F. Peterka, and T. Slonc. 1990. Description of transport of liquid in porous media—A study based on neutron radiography data. *International Journal of Heat and Mass Transfer* 33:1105-20.

Quenard, D. and H. Sallee. 1989. A gamma-ray spectrometer for measurement of the water diffusivity of cementitious materials. Proceedings of the Materials Research Society Symposium, Vol. 137.

Quinn, T.J. 1990. *Temperature*, 2nd ed. Academic Press, Inc., New York.

Richards, R.F., D.M. Burch, and W.C. Thomas. 1992. Water vapor sorption measurements of common building materials. ASHRAE *Transactions* 98(1).

Richardson, L. 1965. A thermocouple recording psychrometer for measurement of relative humidity in hot, arid atmosphere. *Humidity and Moisture* I:101. Reinhold Publishing Corporation, New York.

Schooley, J.F. 1986. *Thermometry*. CRC Press, Inc. Boca Raton, FL.

Seely, R.E. 1955. A circuit for measuring the resistance of energized A-C windings. AIEE *Transactions*, 214.

Shafer, M.R. 1961. Performance characteristics of turbine flowmeters. Proceedings of the Winter Annual Meeting, Paper No. 61-WA-25, ASME, New York.

Tilford, C.R. 1992. Pressure and vacuum measurements. In *Physical methods of chemistry*, 2nd ed., Vol. 6. B.W. Rossiter and J.F. Hamilton, eds. John Wiley and Sons, New York, 101-73.

Till, C.E. and G.E. Handegord. 1960. Proposed humidity standard. ASHRAE *Transactions* 66:288.

Tobiasson, W. and C. Korhonen. 1985. Roofing moisture surveys: Yesterday, today, and tomorrow. Proceedings of the Second International Symposium on Roofing Technology, Gaithersburg, MD.

Tveit, A. 1966. Measurement of moisture sorption and moisture permeability of porous materials. Norwegian Building Research Institute Report, No. 45, Oslo.

Tuve, G.L., G.B. Priester, and D.K. Wright, Jr. 1942. Entrainment and jet-pump action of air streams. ASHVE Research Report No. 1204, ASHVE *Transactions* 48:241.

Vernon, H.M. 1932. The globe thermometer. Proceedings of the Institution of Heating and Ventilating Engineers 39:100.

Wentzel, J.D. 1961. An instrument for measurement of the humidity of air. ASHRAE *Journal* 11:67.

Wile, D.D. 1947. Air flow measurement in the laboratory. *Refrigerating Engineering* 6:515.

Woodring, E.D. 1969. Magnetic turbine flowmeters. *Instruments and Control Systems* 6:133.

Worrall, R.W. 1965. Psychrometric determination of relative humidities in air with dry-bulb temperatures exceeding 212°F. *Humidity and Moisture* I:105. Reinhold Publishing Corporation, New York.

BIBLIOGRAPHY

ANSI. 1983. Specification for sound level meters. ANSI *Standard* S1.4-1983 (ASA 47) including ANSI *Standard* S1.4A-1985 amendment to ANSI S1.4-1983. American National Standards Institute, New York.

ANSI. 1986. Design response of weighting networks for acoustical measurements. ANSI *Standard* S1.42-1986 (ASA 64).

ANSI. 1986. Method for the calibration of microphones. ANSI *Standard* S1.10-1966 (R 1986).

ANSI. 1986. Methods for the measurement of sound pressure levels. ANSI *Standard* S1.13-1971 (R 1986).

ANSI. 1986. Selection of calibrations and tests for electrical transducers used for measuring shock and vibration. ANSI *Standard* S2.11-1969 (R 1986).

ANSI. 1986. Specifications for laboratory standard microphones. ANSI *Standard* S1.12-1967 (R 1986).

ANSI. 1986. Specification for octave-band and fractional-octave-band analog and digital filters. ANSI *Standard* S1.11-1986 (ASA 65).

ANSI. 1986. Techniques of machinery vibration measurement. ANSI *Standard* S2.17-1980 (R 1986) (ASA 24).

ANSI. 1987. Methods for determination of insertion loss of outdoor noise barriers. ANSI *Standard* S12.8-1987 (ASA 73).

ANSI. 1988. Engineering methods for the determination of sound power levels of noise sources for essentially free-field conditions over a reflecting plane. ANSI *Standard* S12.34-1988 (ASA 77).

ANSI. 1989. Guide to the mechanical mounting of accelerometers. ANSI *Standard* S2.61-1989 (ASA 78).

ANSI. 1989. Mechanical vibration-balance quality requirements of rigid rotors, Part 1, Determination of permissible residual unbalance. ANSI *Standard* S2.19-1989 (ASA 86).

ANSI. 1989. Method for the designation of sound power emitted by machinery and equipment. ANSI *Standard* S12.23-1989 (ASA 83).

ANSI. 1989. Reference quantities for acoustical levels. ANSI *Standard* S1.8-1989 (ASA 84).

ANSI. 1990. Engineering methods for the determination of sound power levels of noise sources in a special reverberation test room. ANSI *Standard* S12.33-1990 (ASA 91).

ANSI. 1990. Mechanical vibration of rotating and reciprocating machinery—Requirements for instruments for measuring vibration severity. ANSI *Standard* S2.40-1984 (R 1990) (ASA 50).

ANSI. 1990. Methods for analysis and presentation of shock and vibration data. ANSI *Standard* S2.10-1971 (R 1990).

ANSI. 1990. Statistical methods for determining and verifying stated noise emission values of machinery and equipment. ANSI *Standard* S12.3-1985 (R 1990) (ASA 57).

ANSI. 1990. Preferred frequencies, frequency levels, and band numbers for acoustical measurements. ANSI *Standard* S1.6-1984 (R 1990) (ASA 53).

ANSI. 1990. Requirements for the performance and calibration of reference sound sources. ANSI *Standard* S12.5-1990 (ASA 87).

ANSI. 1990. Specification for acoustical calibrators. ANSI *Standard* S1.40-1984 (R 1990) (ASA 40).

ANSI. 1990. Survey methods for the determination of sound power levels of noise sources. ANSI *Standard* S12.36-1990 (ASA 89).

ANSI. 1990. Vibrations of buildings—Guidelines for the measurements of vibrations and evaluation of their effects on buildings. ANSI *Standard* S2.47-1990 (ASA 95).

ANSI. 1991. Specification for personal noise dosimeters. ANSI *Standard* S1.25-1991 (ASA 98).

ANSI/EIA. 1986. Interface between data terminal equipment and data circuit-terminating equipment employing serial binary data interchange. *Standard* 232-D-1986.

ANSI/IEEE. 1987. Codes, formats, protocols, and common commands for use with IEEE 488.1-1987. *Standard* 488.2-1987.

ANSI/IEEE. 1987. IEEE Standard digital interface for programmable instrumentation. *Standard* 488.1-1987.

ASHRAE. 1977. ASHRAE Brochure on psychrometrics.

ASHRAE. 1984. Standard method for measurement of flow of gas. *Standard* 41.7-1984.

ASHRAE. 1984. Standard method for measurement of proportion of oil in liquid refrigerant. *Standard* 41.4-1984.

ASHRAE. 1986. Engineering analysis of experimental data. *Guideline* 2-1986.

ASHRAE. 1986. Laboratory method of testing in-duct sound power measurement procedure for fans. ASHRAE *Standard* 68-86, AMCA *Standard* 330-86.

ASHRAE. 1987. Standard methods for laboratory air-flow measurement. *Standard* 41.2-1987.

ASHRAE. 1988. A standard calorimeter test method for flow measurement of a volatile refrigerant. *Standard* 41.9-1988.

ASHRAE. 1989. Standard method for pressure measurement. *Standard* 41.3-1989.

ASHRAE. 1989. Standard methods of measurement of flow of liquids in pipes using orifice flowmeters. *Standard* 41.8-1989.

ASHRAE. 1991. Standard method for temperature measurement. *Standard* 41.1-1986.

ASTM. 1986. Standard specifications for ASTM thermometers. ASTM Designation E 1-86. American Society for Testing and Materials, Philadelphia, PA.

ASTM. 1992. *American Society for Testing and Materials Annual Book of ASTM Standards*, Section 4, Volume 04.06.

Beckwith, T.G., N.L. Buck, and R.D. Marangoni. 1982. *Mechanical measurements*, 3rd ed. Addison-Wesley Publishing Company, Reading, MA.

Bendat, J.S. and A.G. Piersol. 1980. *Engineering applications of correlation and spectral analysis*. John Wiley & Sons, New York.

Beranek, L.L. 1988. *Acoustical measurements*. Published for the Acoustical Society of America by the American Institute of Physics, New York.

Beranek, L.L. 1989. *Noise and vibration control*. Institute of Noise Control Engineering, Poughkeepsie, NY.

Blackman, R.B. and J.W. Tukey. 1959. *The measurement of power spectra from the point of view of communications engineering*. Dover Publications, Inc., New York.

Dally, J.W., W.F. Riley, and K.G. McConnell. 1984. *Instrumentation for engineering measurements*. John Wiley & Sons, Inc., New York.

Den Hartog, J.P. 1985. *Mechanical vibrations*. Dover Publications, Inc., New York.

DeWerth, D.W. 1975. Flue gas analysis of gas-fired appliances. *Appliance Engineer*, February.

EPA. 1991. Introduction to indoor air quality: A self-paced learning module, EPA/400/3-91/002, United States Environmental Protection Agency, Washington, D.C.

Fingerson, L.M. 1967. Practical extensions of anemometer techniques. Proceedings of the International Symposium on Hot-Wire Anemometry, University of Maryland, College Park, MD.

Gray, M.H. 1971. Modern anemometers vs. traditional flow meters. ASHRAE Symposium Bulletin, *Flow Measurement*.

Harris, C.M. 1979. *Handbook of noise control*, 2nd ed. McGraw-Hill Book Company, New York.

Harris, C.M. 1987. *Shock and vibration handbook*, 3rd ed. McGraw-Hill Book Company, New York.

Humphrey, R.L. 1967. Anemometry in research and industry. Proceedings of the International Symposium on Hot-Wire Anemometry, University of Maryland, College Park, MD.

Hunt, F.V. 1982. *Electroacoustics*. Published for the Acoustical Society of America by the American Institute of Physics, New York.

IEC. 1979. Values for the difference between free-field and pressure sensitivity levels for one-inch standard condenser microphones. IEC *Standard* 655-1979. International Electrotechnical Commission, Geneva, Switzerland.

IEC. 1984. Advance edition of the International electrotechnical vocabulary, Chapter 801: Acoustics and electro-acoustics, IEC *Standard* 50(801)-1984. International Electrotechnical Commission, Geneva, Switzerland.

IEC. 1990. Rotating electrical machines, Part 9: Noise limits. IEC *Standard* 34/9-1990 including Corrigendum 1991, 2nd ed. International Electrotechnical Commission, Geneva, Switzerland.

IES. 1984. *Lighting handbook*, Reference volume. Illuminating Engineering Society of North America, New York.

International Symposium on Humidity and Moisture. 1963. Humidity and moisture: Measurement and control in science and industry. International Symposium on Humidity and Moisture, Washington, D.C. Reinhold Publishing Corporation, New York.

Kemp, J.F. 1957. Stable hot-wire anemometer for low speeds. *Heating, Piping, & Air Conditioning* 10:137.

Knight, W.H. 1965. The piezoelectric sorption hygrometer. *Humidity and Moisture* I:578. Reinhold Publishing Corp., New York.

Lederer, P.S. 1981. Sensor handbook for automatic test, monitoring, diagnostic, and control systems applications to military vehicles and machinery. NBS Special Publication 615.

Lewis, R.G. and L. Wallace. 1989. Workshop: Instrumentation and methods for measurement of indoor air quality and related factors. Design and protocol for monitoring indoor air quality. ASTM STP 1002, American Society for Testing and Materials, Philadelphia, PA, 219-33.

Lord, H.W. *et al.* 1987. *Noise control for engineers*. Krieger Publishing Co. Melbourne, FL.

Morrison, R. 1986. *Grounding and shielding techniques in instrumentation*, 3rd ed. John Wiley and Sons, New York.

Nottage, H.B. 1950. A simple heated-thermocouple anemometer. ASHVE *Transactions* 56:431.

Papoulis, A. 1977. *Signal analysis*. McGraw-Hill Book Company, New York.

Papoulis, A. 1987. *The Fourier integral and its applications*. McGraw-Hill Book Company, Inc., New York.

Scott, R.W.W. 1978. *Developments in flow measurement*. Alyn-Bacon Co., New York.

Spitzer, D.W. 1983. *Industrial flow measurement*. Instrumentation Society of America, Research Triangle Park, NC.

Spitzer, D.W., ed. 1991. *Flow measurement*. Instrumentation Society of America, Research Triangle Park, NC.

Tsoufanidis, N. 1983. *Measurement and detection of radiation*. McGraw-Hill Book Company, New York.

Tuve, G.L., D.K. Wright, Jr., and L.J. Siegel. 1939. The use of air velocity meters. ASHVE Research Report No. 1140, ASHVE *Transactions* 45:645.

Wexler, A. and R.W. Hyland. 1965. The NBS standard hygrometer. *Humidity and Moisture* IV:389. Reinhold Publishing Corporation, New York.

Wylie, R.G., D.K. Davies, and W.A. Caw. 1965. The basic process of the dewpoint hygrometer. *Humidity and Moisture* I:125. Reinhold Publishing Corporation, New York.

CHAPTER 14

AIRFLOW AROUND BUILDINGS

AIRFLOW around buildings affects worker safety, process and building equipment operation, weather and pollution protection at inlets, and the ability to control environmental factors of temperature, humidity, air motion, and contaminants. Wind causes surface pressures that vary around buildings, changing intake and exhaust system flow rates, natural ventilation, infiltration and exfiltration, and interior pressure. The mean flow patterns and turbulence of wind passing over a building can cause a recirculation of exhaust gases to air intakes. This chapter contains information for evaluating flow patterns, estimating wind pressures and air intake contamination, and solving problems caused by the effects of wind on intakes, exhausts, and equipment. Related information can be found in Chapters 11, 13, 23, and 24 of this volume; in Chapters 25, 27, and 47 of the 1991 Applications volume; and in Chapters 26, 31, 36, and 37 of the 1992 Systems and Equipment volume.

FLOW PATTERNS

Buildings of an even moderately complex shape, such as L- or U-shaped structures formed by two or three rectangular blocks, can generate flow patterns too complex to generalize for design. To determine flow conditions influenced by surrounding buildings or topography, a wind tunnel or water channel test of scale models or tests of existing buildings are required. However, if a building is oriented perpendicular to the wind, it can be considered as consisting of several independent rectangular blocks. Only isolated rectangular block buildings will be discussed here. Hosker (1984, 1985) reviews the effects of nearby buildings.

The mean speed of wind approaching a building increases with height above the ground (Figure 1). Both the upwind velocity profile shape and its turbulence level strongly influence flow patterns and surface pressures. A stagnation zone exists on the upwind wall. The flow separates at the sharp edges to generate recirculating flow zones that cover the downwind surfaces of the building (roof, sides, and leeward walls) and extend for some distance into the wake. If the building has sufficient length L in the windward direction, the flow will reattach to the building (Figure 2) and may generate two distinct regions of separated recirculating flow—on the building and in its wake.

Surface flow patterns on the upwind wall are largely influenced by approach wind characteristics. Higher wind speed at roof level causes a larger stagnation pressure on the upper part of the wall than near the ground, which leads to downwash on the lower one-half to two-thirds of the building (Figure 1). On the upper one-quarter to one-third of the building, the surface flow is directed upward over the roof. For a building whose height H is three or more times the width W of the upwind face, an intermediate zone can exist between the upwash and downwash regions, where the surface streamlines pass horizontally around the building. The

downwash on the lower surface of the upwind face separates from the building before it reaches ground level and moves upwind to form a vortex that can generate high velocities close to the ground. This ground level upwind vortex is carried around the sides of the building in a U shape (Figure 1b) and is responsible for the suspension of dust and debris that can contaminate air intakes close to ground level.

Recirculation and High Turbulence Regions

For wind perpendicular to a building wall, the height H and width W of the upwind building face determine the flow patterns shown in Figure 3. According to Wilson (1979), the scaling length R which combines these dimensions is:

$$R = B_s^{0.67} B_L^{0.33} \tag{1}$$

where B_s is the smaller and B_L the larger of the dimensions H and W. When B_L is larger than $8B_s$, use $B_L = 8B_s$ in Equation (1). For buildings with varying roof levels or with wings separated by at least a distance B_s, only the height and width of the building face below the portion of the roof in question should be used to calculate R. Wilson (1976) indicates that for a flat-roofed building, the recirculation region maximum height H_c, at location X_c, and reattachment lengths L_c and L_r shown in Figures 3 and 17 are:

$$H_c = 0.22R \tag{2}$$
$$X_c = 0.5R \tag{3}$$
$$L_c = 0.9R \tag{4}$$
$$L_r = 1.0R \tag{5}$$

The downwind boundary of the rooftop recirculation region may be approximated by a straight line sloping downward from H_c to the roof at L_c. The dimensions of the recirculation zones are somewhat sensitive to the intensity and scale of turbulence in the approaching wind. High levels of turbulence from upwind obstacles may decrease the coefficients in Equations (2) through (5) by up to a factor of 2. Turbulence in the recirculation region and in the approaching wind also causes the reattachment locations on Figure 2 to fluctuate.

To account for changes in roof level, penthouses, and equipment housings and enclosures, the scaling length R of each of these obstacles should be calculated from Equation (1) using the dimensions of the upwind face of the obstacle. The recirculation region for each obstacle may be calculated from Equations (2), (3), and (4). The length L_r of the recirculation region downwind from the obstacle, or from the entire building, is given by Equation (5), with R based on the dimensions of the downwind face of the obstacle. The high turbulence region boundary Z_2 in Figure 17 follows a 1:10 (5.7°) downward slope from the top of the recirculation regions at X_c or L_r. When an obstacle is close to the

---

The preparation of this chapter is assigned to TC 2.5, Air Flow Around Buildings

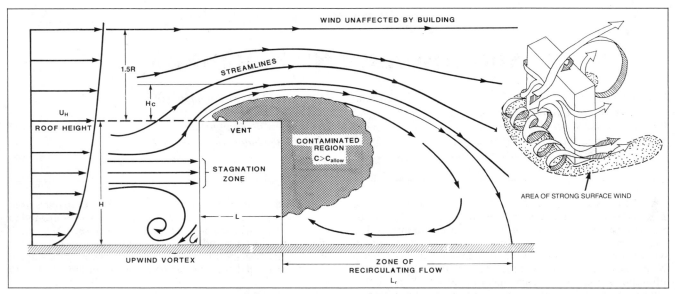

Fig. 1   Flow Patterns around Rectangular Building

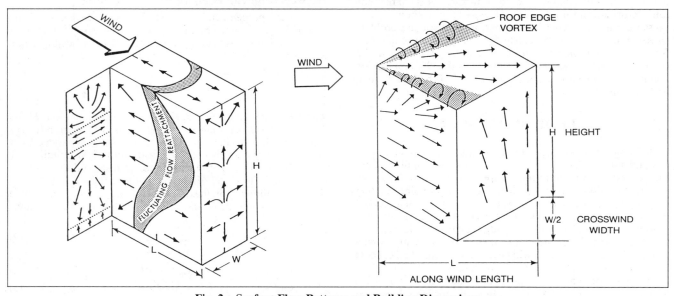

Fig. 2   Surface Flow Patterns and Building Dimensions

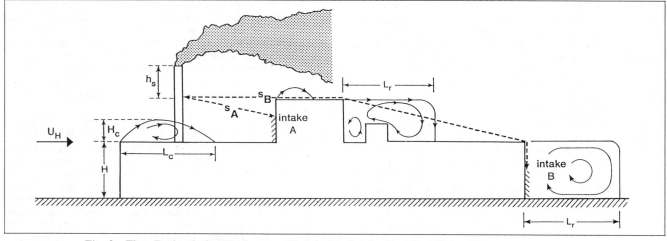

Fig. 3   Flow Recirculation Regions and Exhaust to Intake Stretched String Distances (Wilson 1982)

upwind edge of a roof or near another obstacle, the flow recirculation zones interact. Wilson (1976) gives methods of dealing with these situations.

Building-generated turbulence is confined to the roof wake region, whose upper boundary Z_3 in Figure 17, is given by

$$Z_3/R = 0.28(X/R)^{0.33} \qquad (6)$$

where X is the distance from the upwind roof edge where the recirculation region forms. Building-generated turbulence decreases with height above roof level. At the edge of the rooftop wake boundary Z_3, the turbulence intensity is close to the background level in the approach wind. High levels of turbulence exist below the boundary Z_2 (Figure 17).

Streamline patterns are independent of wind speed and depend mainly on building shape and upwind conditions. Because of the three-dimensional flow around a building, the shape and size of the recirculation airflow is not constant over the surface; instead, it reattaches closer to the upwind face of the building along the edges at the roof and at ground level (Figure 2). The height of the recirculation region also decreases near roof edges.

The wind above the roof recirculation region is affected by the presence of the building; the flow accelerates as the streamlines curve upward over the roof and decelerates as they curve downward into the wake on the lee side of the building. The distance above roof level where a building influences the flow is approximately $1.5R$ (Figure 1). The roof pitch begins to affect flow when it exceeds about $15°$ (1:4). When roof pitch reaches $20°$ (1:3), the flow remains attached to the upwind pitched roof and produces a recirculation region downwind from the roof ridge that is larger than that for a flat roof. The extent of these flow recirculation zones can be estimated by using the roof ridge height for the dimension H in Equations (1) to (5).

The downwind wall of a building faces a region of low average velocity and high turbulence. Velocities near this wall are typically one-quarter of those at the corresponding upwind wall location. Figures 1 and 2 show that an upward flow exists over most of the downwind walls. A flow recirculation region extends for an approximate distance $L_r = 1.0R$ downwind.

If the angle of the approach wind is not perpendicular to the upwind face, complex flow patterns result. Strong vortices develop from the upwind edges of a roof, causing a strong downwash into the building wake above the roof. High speeds in these vortices cause large negative pressures near roof corners that can be a hazard to roof-mounted equipment during high winds. When the angle between the wind direction and the upwind face of the building is less than about $70°$, the downwash-upwash patterns on the upwind face of the building are less pronounced, as is the ground level vortex. For an approach flow angle of $45°$, streamlines remain close to the horizontal in their passage around the sides of the building (Figures 1 and 2), except near roof level where the flow is sucked upward into the roof edge vortices.

WIND PRESSURES ON BUILDINGS

In addition to the flow patterns described previously, the turbulence or gustiness of the approaching wind and the unsteady character of the separated flows cause surface pressures to fluctuate. The pressures discussed here are time-averaged values, with an averaging period of about 600 s. Instantaneous pressures may vary significantly above and below these averages, and peak pressures two or three times the mean values are possible. While peak pressures are important with regard to structural loads, mean values are more appropriate for computing infiltration and ventilation rates. The time-averaged surface pressures are proportional to the wind velocity pressure p_v in lb/ft^2 given by Bernoulli's equation

$$p_v = \rho U_H^2/2_g \qquad (7)$$

where U_H is the approach wind speed (fps) at upwind wall height (Figures 1 and 17), and ρ is the outdoor air density (lb/ft^3). The difference p_s between the pressure on the building surface and the local outdoor atmospheric pressure at the same level in an undisturbed wind approaching the building is given by

$$p_s = C_p p_v \qquad (8)$$

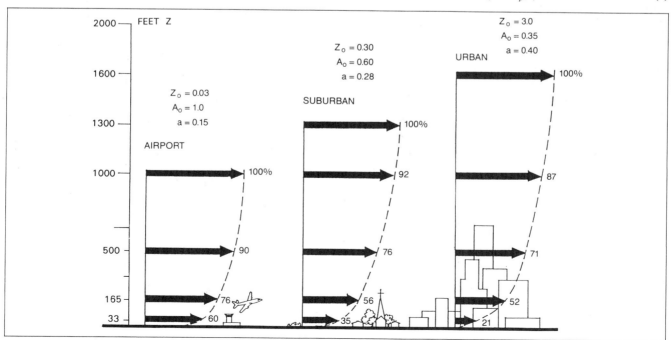

Fig. 4   Typical Mean Wind Speed Profiles over Different Terrain Roughness

The local wind speed U_H at the top of the wall required for Equation (7) is estimated by applying terrain and height corrections to the speed U_{met} from a nearby meteorological station. The wind speed U_{met} from airport meteorological stations is measured at a height H_{ref}, usually about 33 ft, in open flat terrain. U_{met} is related to the reference wind speed U_{ref} at the same height in the local building terrain by

$$U_{ref} = A_o U_{met} \qquad (9)$$

where the constant A_o in Figure 4 depends on the local terrain roughness. Figure 1 shows that the wind speed U_H at wall height depends on the velocity profile in the approach wind. Thus:

$$U_H = U_{ref}(H/H_{ref})^a \qquad (10)$$

where a is the velocity profile exponent given in Figure 4. The scaling length Z_o for terrain roughness in Figure 4 is based on an alternate logarithmic variation of wind speed with height. Values of Z_o are typically 5 to 10% of the average height of the small-scale terrain roughness features, such as buildings, obstacles, trees, and other vegetation. This average roughness height includes roughness features for a distance of about 3 mi upwind of the site. Equation (10) gives the wind speed at height H above the plan area-weighted average height of local obstacles, such as buildings and vegetation. At heights at or below this average obstacle height, *e.g.*, at roof height in densely built-up suburbs, the speed depends on the geometrical arrangement of the buildings, and Equations (9) and (10) are less reliable.

Local Wind Pressure Coefficients

Values of the pressure coefficient C_p depend on building shape, wind direction, and the influence of nearby buildings, vegetation, and terrain features. Accurate determination of C_p can be obtained only from wind tunnel model tests of the specific site and building. Ventilation rate calculations for single, unshielded rectangular buildings can be reasonably estimated using existing wind tunnel data.

Figure 5 shows wall pressure coefficients on a tall building with a square cross section for simulated urban terrain (Davenport and Hui 1982), and Figure 6 for a low-rise building (Holmes 1986). Generally, high-rise buildings are those where the height H is more than three times the crosswind width W. For $H > 3W$, use Figure 5, and for $H < W$, use Figure 6. At a wind angle $\theta = 0°$, with the wind perpendicular to the face in question, the pressure coefficients are positive, and the magnitude reduces near the sides and the top as the flow velocities increase near the edges. There is a general increase of C_p with height, which reflects the increasing velocity pressure in the approach flow. As the wind direction moves off normal, the region of maximum pressure occurs closer to the upwind edge of the building. At a wind angle of 45°, the pressures become negative at the downwind edge of the front face; at some angle between $\theta = 60°$ and $\theta = 75°$, they become negative over the whole face. Maximum suction (negative) pressure occurs near the upwind edge of the side and for $\theta = 90°$, the pressures then recover towards zero away from the leading edge. The degree of this recovery depends on the length of the side in relation to the width of the structure. For wind angles larger than 100°, the side is completely within the separated flow of the wake, and the spatial variations in pressure over the face are not as great. The average pressure on a face is above zero for wind angles from 0 to

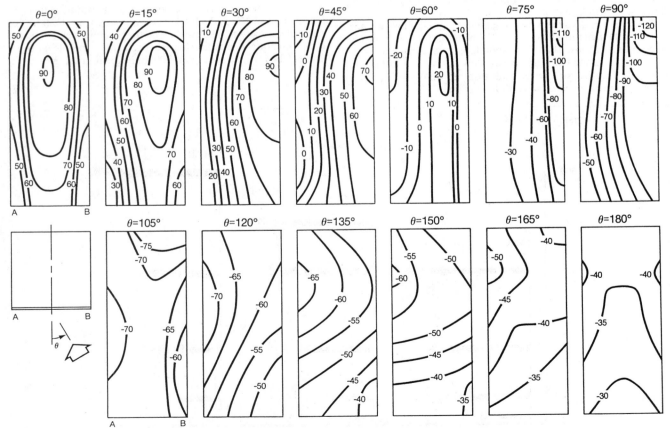

Fig. 5 Mean Pressure Coefficients ($C_p \times 100$) on Tall Building for Varying Wind Direction
(Davenport and Hui 1982)

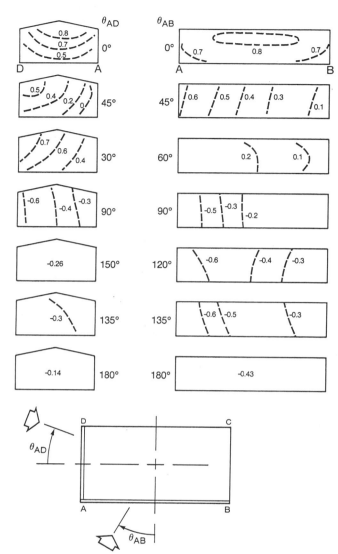

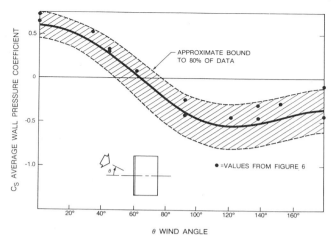

**Fig. 7 Variation of Wall Averaged Pressure Coefficients
for Low-Rise Building**
(Swami and Chandra 1987)

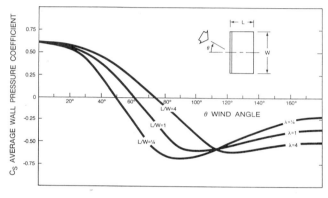

**Fig. 8 Wall Averaged Pressure Coefficients
for Tall Building, $\lambda = L/W$**

**Fig. 6 Mean Pressure Coefficients on Walls
of Low-Rise Building**
(Holmes 1986)

about 60° and below zero for 60 to 180°. A similar pattern of behavior in the wall pressure coefficients for the low-rise building is shown in Figure 6. Here the recovery from the strong suction with distance from the upwind edge is more rapid.

Surface Averaged Wall Pressures

Figure 7 shows the surface pressure coefficient averaged over a complete wall C_s of a low-rise building (Swami and Chandra 1987). The figure also includes the values calculated from the pressure distributions shown in Figure 6. Similar results for a tall building are shown in Figure 8 (Akins *et al.* 1979).

The wind-induced indoor-outdoor pressure difference is found using the coefficient $C_{p(in-out)}$ which is

$$C_{p(in-out)} = C_p - C_{in} \qquad (11)$$

The mean of the spatially averaged wall pressure coefficients C_s in Figures 7 and 8 for the four building walls is about -0.2 for most wind directions. This implies that for uniformly distributed air leakage sites in all the walls, the internal wind-induced pres-

sure coefficient will be about $C_{in} = -0.2$. Chapter 22 explains how stack effect pressures induced by the indoor-outdoor temperature difference combine with wind-induced pressures.

Roof Pressures

Surface pressures on the roof of a low-rise building depend strongly on roof slope. Figure 9 shows typical distributions for a wind direction normal to a side of the building. For very low slopes, the pressures are negative over the whole roof surface. The magnitude is greatest within the separated flow zone near the leading edge and recovers toward the free stream pressure away from the edge. For steeper slopes, the pressures are weakly positive on the windward slope and negative within the separated flow over the leeward slope. With a wind angle of about 45°, the vortices originating at the leading corner of a roof with a low slope can induce very large localized negative pressures. Figure 10 shows the average pressure coefficient over the roof of a tall building (Akins *et al.* 1979).

Interference and Shielding Effects on Pressures

Nearby structures strongly influence surface pressures on both high- and low-rise buildings. These effects are very strong for spacing to height ratios less than five, where the distributions of pressure shown in Figures 6 through 9 do not apply. While the effect of shielding is still significant at larger spacings, for low-rise buildings it is largely accounted for by the reduction in P_v with the increased terrain roughness shown in Figure 4 and defined in Equation (7). Chapter 23 gives shielding coefficients for air infiltration and ventilation applications.

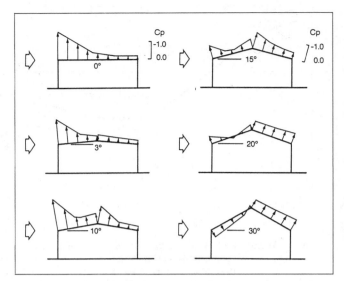

Fig. 9 Pressure Coefficients on Roof of Low-Rise Building
(Holmes 1986)

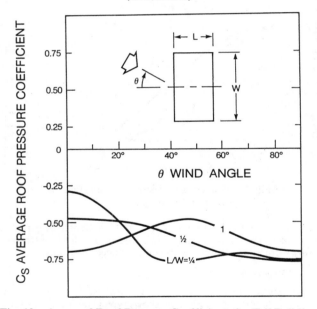

Fig. 10 Averaged Roof Pressure Coefficients for Tall Building

Sources of Wind Data

To design for the effects of airflow around buildings, wind speed and direction frequency data must be obtained. These data provide the number of calms and occurrences of wind speeds within several wind speed intervals, for each of 16 wind directions. Frequency data for mean hourly wind speeds nominally at 30 ft above ground level can be found in a summarized printed pamphlet form or in full digital records on magnetic tape for 1945 through 1985 from the National Climatic Center in Asheville, North Carolina, for numerous sites within the United States and worldwide, and from the Atmospheric Environment Service in Downsview, Ontario, for Canadian sites. ASHRAE (1986) also has limited wind frequency data available for microcomputers.

Using a single prevailing wind direction for design may cause serious errors. For any set of wind direction frequencies, one direction will always have a somewhat higher frequency of occurrence. Thus, it is often called the prevailing wind, even when winds from different directions may be almost as frequent.

Winds aloft data measured at heights of 1300 ft and higher above ground level can also be obtained from the National Climatic Center for fewer sites. These data are often used in boundary layer wind tunnel studies to relate actual wind probabilities to the velocities measured around model buildings. The mean wind speeds aloft are not influenced by individual obstructions at ground level and are more representative of a larger region around a meteorological station. The variation of wind speed with height causes wind changes from those associated with winds aloft to those at ground level. These anticlockwise (*e.g.*, north toward northwest) directional changes are up to 30° in the roughest urban terrain and 15° in smooth terrain.

When using long-term meteorological records, check the history of mounting conditions for the anemometer, because the instrument may have been relocated and its height varied. This can affect its directional exposure and recorded wind speed. Equation (10) can be used to correct wind data collected at different mounting heights. Poor anemometer exposure due to obstructions or mounting on top of a building cannot be easily corrected, and the records for that period should be deleted.

If an estimate of the probability of an extreme wind speed outside the range of the recorded values at a site is required, the observations may be fitted to an appropriate probability distribution and particular probabilities calculated from the fitted curve function, as in Figure 11. This process is usually repeated for each of the 16 wind directions. Where estimates of extreme probabilities are required, curve fitting at the tail of the distribution is extremely important.

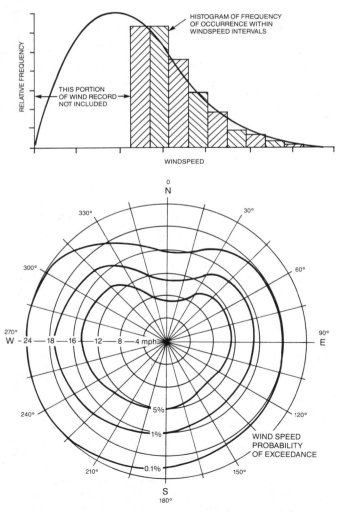

Fig. 11 Probability Distributions of Wind Speed and Direction

Accurate extrapolation of frequency distributions to rarely occurring high wind speeds requires special statistical techniques. Building codes for wind loading on structures contain information on estimating extreme wind conditions. For ventilation applications, extreme winds are usually not required, and the 99 percentile limit can be accurately estimated from airport data averaged over less than 10 years.

Estimating Wind at Sites Remote from Recording Stations

Many building sites are located far from the nearest long-term wind recording site, usually an airport. To estimate wind conditions at such sites, the terrain surrounding both the anemometer site and the building site should be checked. In the simplest case of flat or slightly undulating terrain with few obstructions extending for large distances around and between the anemometer site and building site, recorded wind data can be assumed to be representative of that at the building site.

In cases where the only significant difference between the airport recording site terrain and the terrain at the building site is surface roughness, the mean wind speed profile can be adjusted to yield approximate wind velocities at the building site (Figure 12). The constant A_o in Equation (9) corrects for the changes in a fully developed wind profile from an airport site in open terrain to the roughness of the building site. Swami and Chandra (1987) give a general procedure that can be applied to other values of the wind speed power law exponent a and boundary layer thickness δ.

If the building is near the upwind edge of the change in roughness, it may not be completely covered by the new internal boundary layer. The approximate height δ_2 of this internal boundary layer is given in Table 1 (Elliot 1958).

In the process of development, the mean hourly wind speed profile over the new terrain roughness Z_{o2} has approximately the same characteristics as the fully developed profile from Figure 4 below δ_2 and the same characteristics as the fully developed profile over Z_{o1} for heights above δ_2. The height δ_2 of the new internal boundary layer will become constant when it equals the fully developed heights shown in Figure 4.

In addition to changes in surface roughness, several other factors are important in causing the wind speed and direction at a building site to differ from values recorded at a nearby airport. Wind speeds can be 50% higher than airport data for buildings on hill crests or in valleys where the wind is channeled along a valley. Shelter in the lee of hills and escarpments can reduce speeds by a factor of 2 below values at nearby flat airport terrain.

Table 1   Depth of Internal Boundary Layer δ_2 at Transition in Roughness (see Figure 12)

	Z_{o1}, in.	Z_{o2}, in.	Travel Distance over Z_{o2}		
			160 ft	1600 ft	16000 ft
Open country to suburb	0.8	8	36	230	140
Suburb to open country	8	0.8	30	180	150
Suburb to city center	8	79	62	360	2300
City center to suburb	79	8	43	295	1640

Solar heating of valley slopes can cause light winds of 200 to 800 fpm to occur as warm air flows upslope. At night, cooling of the ground by heat radiation can produce these speeds as cold air drains downslope. In general, rolling terrain will experience a smaller fraction of low speeds than nearly flat terrain.

When the wind is calm or light in the rural area surrounding a city, urban air tends to rise in a buoyant plume over the city center. This rising air, heated by man-made sources and higher solar absorption in the city, is replaced by air sucked toward the city center from the edges. In this way, the urban heat island can produce light wind speeds and direction frequencies significantly different than measurements at an airport.

A rough guideline is that only wind speeds of 9 mph or greater at the building site can be transferred to a nearby airport using Equations (9), (10), and Figure 3 to determine their frequency of occurrence. Wind direction occurrence frequency at a building site should be inferred from airport data only if the two locations are on the same terrain, with no terrain features that could alter wind direction between them.

Where the only change in terrain between the building site and the airport anemometer is a topographic feature, such as a hill, ridge, or escarpment, adjustments can be made to the mean hourly wind speed profile using amplification factors from data collected by Simiu and Scanlan (1986) and Cook (1986). In more complex terrain with more than one hill, both wind speed and direction may be significantly different from those at the distant recording site. Here, building site wind conditions should not be estimated from airport data. Options in these cases are either to establish an on-site wind recording station or to commission a detailed wind tunnel correlation study between the building site and long-term airport wind observations.

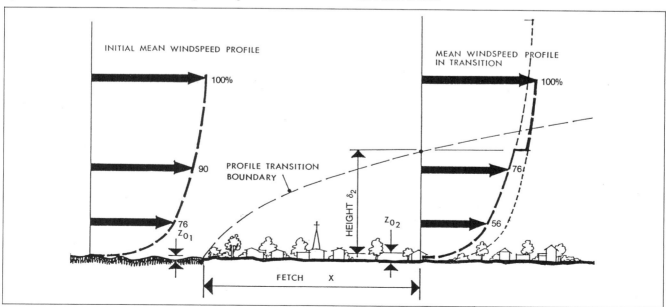

Fig. 12   Development of Internal Boundary Layer Following Change in Terrain Roughness

WIND EFFECTS ON SYSTEM OPERATION

With few exceptions, building intakes and exhausts cannot be located or oriented for a prevailing wind to assure ventilation and air-conditioning system operation. Wind can assist or hinder inlet and exhaust fans depending on their position on the building, but even in locations with a predominant wind direction, the ventilating system must perform adequately for all other directions. To avoid variability in system flow rates, use Figures 5, 6, 9, and 10 as a guide to placing inlets and exhausts in locations where the surface pressure coefficients do not vary greatly with the wind direction.

Cooling towers and similar equipment should be oriented to take advantage of prevailing wind directions, based on careful study of the meteorological data and flow patterns on the building for the area and time of year involved.

A building with only upwind openings is under a positive pressure (Figure 13). Building pressures are negative when there are only downwind openings. A building with internal partitions and openings is under various pressures depending on the relative sizes of the openings and the wind direction. With larger openings on the windward face, the building interior tends to remain under positive pressure; the reverse is also true (see Figures 5 through 8, and Chapter 23).

Airflow through a wall opening results from positive or negative external and internal pressures. Such differential pressures may exceed 0.5 in. of water during high winds. Supply and exhaust systems, and openings, dampers, louvers, doors, and windows make the building flow conditions too complex for most calculation. The opening and closing of doors and windows by building occupants add further complications. In determining $C_{p(in-out)}$ from Equation (11), the wind direction is more important than the position of an opening on a wall, as shown in Figures 5 and 6.

Natural and Mechanical Ventilation

With natural ventilation, wind may augment, impede, or sometimes reverse the airflow through a building. For large roof areas (Figure 2), the wind can reattach to the roof downstream from the cavity and thus reverse natural ventilation discharging out of monitor or similar windows. These reversals can be avoided by using stacks, continuous roof ventilators, or other exhaust devices in which the flow is augmented by the wind.

Mechanical ventilation is also affected by wind conditions. A low-pressure wall exhaust fan, 0.05 to 0.1 in. of water, can suffer drastic reduction in capacity. Flow can be reversed by wind pressures on windward walls, or its rate can be increased substantially when subjected to negative pressures on the lee and other sides. Clarke (1967), when measuring medium pressure air-conditioning systems, 1 to 1.5 in. of water, found flow rate changes of 25% for wind blowing into intakes on an L-shaped building compared to the reverse condition. Such changes in flow rate can cause noise at the supply outlets and drafts in the space served.

For mechanical systems, the wind can be thought of as producing a pressure in series with a system fan, either assisting or opposing it (Houlihan 1965). Where system stability is essential, the supply and exhaust systems must be designed for high pressures (about 3 to 4 in. of water) to minimize unacceptable variations in flow rate. To conserve energy, the system pressure selected should be consistent with system needs.

Building Pressure Balance

Proper building pressure balance avoids flow conditions that make doors hard to open, cause drafts, and prevent confining of contaminants to specific areas. Although the supply and exhaust systems in an area may be in nominal balance, wind can upset this balance, not only because of the changes in fan capacity but also by superimposing infiltrated or exfiltrated air or both on the area. The effects can make it impossible to control environmental conditions. Where building balance and minimum infiltration are important, consider the following:

- Fan system design with pressure adequate to minimize wind effects.
- Controls to regulate flow rate or pressure or both.
- Separate supply and exhaust systems to serve each building area requiring control or balance.
- Doors (possibly self-closing) or double-door air locks to non-controlled adjacent areas, particularly outside doors.
- Sealing windows and other leakage sources and closing natural vent openings.

System volume and pressure control is described in Chapter 41 of the 1991 ASHRAE *Handbook—Applications*. Such control is not possible without adequate system pressure for both the supply and exhaust systems to overcome wind effects. Such a control system may require fan inlet or discharge dampers, fan speed or pitch control, or both.

Fume Hood Operation

Wind effects can interfere with safe hood operation. Supply (makeup) volume variations can cause disturbances at hood faces or a lack of adequate makeup air. Volume surges, caused by fluctuating wind pressures acting on the exhaust system, can cause momentary inadequate hood exhaust. If highly toxic contaminants are involved, such surging is unacceptable. The system should be designed to eliminate this condition. On low-pressure exhaust systems, it is impossible to test the hoods under wind-induced, surging conditions. These systems should be tested during calm conditions for safe flow into the hood faces; they should be rechecked by smoke tests during high wind conditions.

Minimizing Wind Effect on System Volume

Wind effect can be reduced by careful selection of inlet and outlet locations. Because wall surfaces are subject to a wide variety of positive and negative pressures, wall openings should be avoided when possible. When they are required, wall openings should be away from corners formed by building wings (Figure 13). Mechanical ventilation systems should operate at a pressure

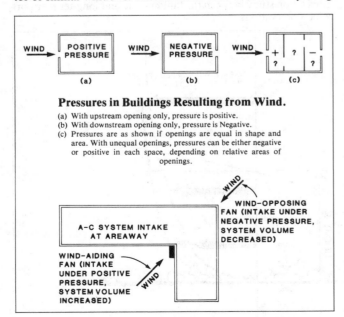

Pressures in Buildings Resulting from Wind.

(a) With upstream opening only, pressure is positive.
(b) With downstream opening only, pressure is Negative.
(c) Pressures are as shown if openings are equal in shape and area. With unequal openings, pressures can be either negative or positive in each space, depending on relative areas of openings.

Fig. 13   Sensitivity of System Volume to Locations of Building Openings, Intakes, and Exhausts

high enough to minimize wind effect. Low-pressure systems and propeller exhaust fans should not be used with wall openings unless their ventilation rates are small or used in noncritical services such as storage areas.

Although roof air intakes within flow recirculation zones minimize wind effect on system flow rates best, current and future air quality in these zones must be considered. These locations should be avoided if a source of contamination exists or may appear in the future. The best area is near the middle of the roof because the negative pressure there is small and least affected by changes in wind direction (Figure 9). Avoid the edges of the roof and walls, where large pressure fluctuations occur. Either vertical or horizontal (mushroom) openings can be used. On large roof areas, where the intake may be outside the roof recirculation zone, mushroom or 180° gooseneck designs minimize impact pressure from wind flow. The frequently used 135° gooseneck or vertical louvered openings are undesirable for this purpose or for rain protection.

Exhausts discharging heated air or industrial contaminants should exhaust vertically through stacks, above the roof recirculation zone. Horizontal, louvered (45° down), and 135° gooseneck discharges are undesirable, even for heat removal systems because of their sensitivity to wind effect. A 180° gooseneck for systems handling hot air may be undesirable because of air impingement on tar and felt roofs. Vertically discharging stacks located in a recirculation region (except near a wall) have the advantage of being subjected only to negative pressure created by wind flow over the tip of the stack.

BUILDING INTERNAL PRESSURE AND FLOW CONTROL

In air-conditioning and ventilation systems for a building containing airborne contaminants, the correct airflow is toward the contaminated areas. Airflow direction is maintained by controlling pressure differentials between spaces. In a laboratory building, for example, peripheral rooms such as offices and conference rooms are maintained at a positive pressure, and laboratories at a negative pressure, both with reference to corridor pressure. Pressure differentials between spaces are normally obtained by balancing the air-conditioning and ventilation supply system airflows in the spaces in conjunction with the exhaust systems in the laboratories, with differential pressure instrumentation to control the airflow. Chapter 41 of the 1991 ASHRAE *Handbook—Applications* has further information on controls.

Airflow in corridors is sometimes controlled by an outdoor reference probe that senses static pressure at doorways and air intakes. The differential pressure measured between the corridor and the outside may then signal a controller to increase or decrease airflow to the corridor. Unfortunately, it is difficult to locate an external probe where it will sense the proper external static pressure. High wind velocity and resulting pressure changes around entrances can cause great variations in pressure. Care must be taken to ensure that the probe is unaffected by wind pressure.

The pressure differential for a room adjacent to a corridor can be controlled by using the corridor pressure as the reference. Outdoor pressure cannot control pressure differentials within rooms, even during periods of relatively constant wind velocity and pressure. A single pressure sensor can measure the outside pressure at one point only and may not be representative of pressures elsewhere.

ATMOSPHERIC DISPERSION OF BUILDING EXHAUST

A building exhaust system is frequently used to release a mixture of building air and pollutant gas at concentration C_e (mass of pollutant per volume of air) into the atmosphere through a

stack or vent on the building. The exhaust mixes with atmospheric air to produce a pollutant concentration C, which may contaminate an air intake or receptor if the concentration is larger than some specified allowable value C_{allow} (Figure 1). The dilution D between source and receptor is defined as

$$D = C_e/C \qquad (12)$$

The dilution increases with distance from the source, starting from its initial value of unity. If C is replaced by C_{allow} in Equation (12), the required atmospheric dilution D_{req} to meet the allowable concentration at the intake (receptor) is C_e/C_{allow}. The exhaust (source) concentration is given by

$$C_e = m/Q_e = m/(A_e V_e) \qquad (13)$$

where

m = pollutant release rate, lb/s
$Q_e = A_e V_e$ = total exhaust volume flow rate, ft³/s
A_e = exhaust face area, ft²
V_e = exhaust face velocity, ft/s

The concentration units of mass/mixture volume are appropriate for gaseous pollutants, aerosols, dusts, and vapors. The concentration of gaseous pollutants is usually stated as a volume fraction f (contaminant volume/mixture volume), or ppm (parts per million) if the volume fraction is multiplied by 10^6. The pollutant volume fraction f_e in the exhaust is

$$f_e = Q/Q_e \qquad (14)$$

where Q is the volume release rate of the contaminant gas, and Q_e is the total exhaust mixture volume flow rate, both calculated at the exhaust temperature T_e.

If the exhaust gas mixture has a relative molecular mass close to that of air, the volume concentration dilution factor $D_v = f_e/f$ may be calculated from the mass concentration dilution by

$$D_v = (T_e/T_a)D \qquad (15)$$

where T_e is the exhaust air absolute temperature, T_a is the outdoor ambient air absolute temperature, and f_e is the pollutant volume fraction in the exhaust air. Many building exhausts are close enough to ambient temperature to assume volume fraction and mass concentration dilutions D_v and D are equal.

Exhaust Gas Concentrations at Air Intakes

The dispersion of pollutants from a building exhaust depends on the combined effect of atmospheric turbulence in the wind approaching the building and turbulence generated by the building itself. This building-generated turbulence is most intense in and near the flow recirculation zones that occur on the upwind edges of the building (Figures 2 and 3). Because of turbulence and distortion of the wind streamlines by the building, the concentration caused by a source near a building cannot be estimated accurately by the design procedures developed for tall isolated stacks. Meroney (1982), Wilson and Britter (1982), Hosker (1984), and Halitsky (1982) review gas diffusion near buildings.

When the exhaust source is in or near a flow recirculation zone, contours of the normalized concentration coefficient $K_c = CU_HHW/Q$, developed from wind tunnel model studies on buildings of similar shape, as shown in Figure 14, may be used. Halitsky (1985) gives details of this technique. Surface concentrations for block buildings with uncapped exhaust vents and short stacks in a uniform nonturbulent airstream are reported by Halitsky (1963).

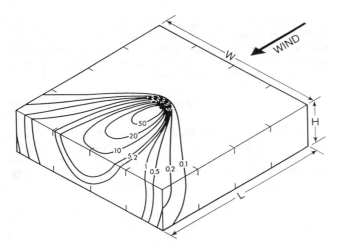

Fig. 14   Normalized Concentration Contour for Central Vent on Roof of Low-Rise Building
(Wilson 1976)

Jet diameters and emission velocities were large enough to project the exhaust plume above the roof and sometimes through the recirculation cavity.

ESTIMATING INTAKE CONTAMINATION

If a representative configuration with published surface concentration coefficients K_c cannot be found to match the existing building, available data may be used to estimate the minimum dilution observed at the same receptor distance from the source.

The most important variables determining minimum dilution are the exhaust to intake stretched-string distance S and the effective stack height h_s, both of which are defined in Figure 3. Only the stack height extending above large rooftop obstacles such as penthouses and architectural barriers should be used to define the effective stack height.

The stretched-string distance S is defined as the shortest length of string connecting the point on a stack where $h_s = 0$ to the nearest point on an air intake. The point where $h_s = 0$ is defined as the height of obstacles close to the stack, as shown in Figure 3, or the boundary of any roof recirculation region through which the stack passes.

All the following equations for D_{min} were developed from wind tunnel simulations on low-rise buildings for exposures equivalent to 10-min averaging times in the atmosphere, and with nonbuoyant exhaust jets from roof vents. If the exhaust gases are hot, buoyancy increases the rise of the exhaust gas mixture and produces lower concentrations at roof level. By neglecting buoyant plume rise, the D_{min} equations have an inherent factor of safety, particularly at low wind speed where buoyancy is most important.

The averaging time t_a over which exhaust gas concentration exposures are measured is also important in determining minimum dilution. As the averaging time increases, the exhaust gas plume meanders more from side to side, thus reducing the time-averaged concentration at an intake location. Wilson (1985) describes the effect of changing the averaging time over a range of about 3 min to 3 h, estimated by adjusting the 10-min values given in Equations (17) to (25) using

$$D_{min,1}/D_{min,2} = (t_{a1}/t_{a2})^{0.33} \qquad (16)$$

The ratio of minimum dilution at any two averaging times t_{a1} and t_{a2} applies for $S/A^{0.5} > 10$ to allow the exhaust jet to begin meandering. Where the exhaust and intake are both located in the same flow recirculation region, dilution will be less sensitive to averaging

time than predicted by Equation (16). In this case, assume the D_{min} values for 3-min averages also apply for averaging times from 3 to 60 min. Use Equation (16) to adjust the 10-min values in Equations (17) to (25) to 3-min values for this flow recirculation situation.

Strong Jets in Flow Recirculation Cavity

For block buildings with surface vents or short stacks that produce significant jet rise because of their large diameter, high emission velocity, and free discharge, the equation for minimum dilution at a surface intake on the same building or on the jet centerline above the building roof is given by Halitsky (1963)

$$D_{min} = [\alpha + 0.11(1 + 0.2\alpha)S/A_e^{0.5}]^2 \qquad (17)$$

where α is a numerical constant related to building shape, emission velocity ratio V_e/U_H, building orientation to the wind, and stack height. Values of α can be found in Halitsky (1963). The smallest D_{min} occurs along the elevated centerline of the jet plume, where $\alpha = 1.0$ is the appropriate value. Larger values of D_{min} occur on the building surfaces, where the range of α is 2.0 to 20. The larger values are associated with greater stack height and buoyant exhaust. Halitsky (1963) gives the tested configuration that most closely resembles the one under consideration.

Strong Jets on Multiwinged Buildings

For a strong vertical exhaust with $V_e/U_H \cong 2$ from surface vents or short stacks on the roof of a multiwinged building with different wing roof heights, Halitsky (1962, 1982) recommends

$$D_{min} = M(3.16 + 0.1S/A_e^{0.5})^2 \qquad (18)$$

where M is an intake location factor with values of $M = 1.5$ when the intake is on the same roof as the source, $M = 2.0$ when the source and intake are on different wings separated by an air space, and $M = 4.0$ when the intake is substantially lower than the source.

Surface Vents on Flat-Roofed Buildings

For buildings with exhausts of zero stack height on a flat roof, the minimum dilution at a roof or wall intake is given by Wilson and Chui (1985, 1987) and Chui and Wilson (1988) as

$$D_{min,o} = [D_o^{0.5} + D_s^{0.5}]^2 \qquad (19)$$

where

$$D_o = 1 + 7.0\,\beta(V_e/U_H)^2 \qquad (20)$$

and

$$D_s = B_1(U_H/V_e)(S^2/A_e) \qquad (21)$$

$D_{min,o}$ is the minimum dilution for a roof level exhaust, D_o is the apparent initial dilution at roof level, and D_s is the distance dilution caused by the combined action of building and atmospheric turbulence. The initial dilution D_o is caused by internal turbulence in the exhaust jet (Halitsky 1962, 1966). The wind speed U_H at wall height is calculated from local airport weather data using Equations (9) and (10).

Equation (20) was developed for exhaust velocity to wind speed ratios from $V_e/U_H = 0$ to 2, and should be used cautiously above this value. For vertically directed uncapped exhausts, the capping factor $\beta = 1.0$; for capped, louvered, or downward-facing exhausts, it is $\beta = 0$. The distance dilution parameter B_1 depends on the exhaust jet trajectory, and the intensity of turbulence in the approach wind and generated by the building. For flat-roofed buildings in low-rise urban surroundings, $B_1 = 0.0625$ for all wind directions, when the exhaust is uncapped ($\beta = 1$) and $V_e/U_H > 0.5$. A value of $B_1 = 0.0204$ should be used if $V_e/U_H < 0.5$ or the stack is capped ($\beta = 1$) and the wind is at 45° to the upwind wall, or if there is no significant atmospheric turbulence, for example, at roof level of high-rise buildings, or in flat rural surroundings [Wilson and Chui (1987), Chui and Wilson (1988)].

Equations (19), (20), and (21) imply that minimum dilution does not depend on the location of either the exhaust or intake, only on the distance S between them. This is true when exhaust and intake locations are on the same building wall or on the roof. The dilution may increase if the intake and exhaust are located on different faces, as indicated by the M factor in Equation (18). For roof exhausts with wall intakes, the results of Li and Meroney (1983) suggest that $B_1 \cong 0.20$ in Equation (21).

For buildings less than about 330 ft high and also less than twice as high as the surrounding buildings, atmospheric turbulence makes a significant contribution to exhaust gas dilution. Wilson (1976, 1977) gives surface concentration contours for flat-roofed buildings in a simulated approach wind typical of an urban area. Flush vents with small exhaust velocity make these results suitable for estimates for capped exhaust stacks or louvered exhaust vents.

The effect of atmospheric turbulence is relatively insignificant for high-rise buildings taller than 330 ft and also twice the average building height for 3000 ft upwind. On these high-rise buildings, where the effects of atmospheric turbulence are small, Wilson and Chui (1987) found that maximum surface concentrations for 10-min exposures were two to ten times higher than on an equivalent low-rise building. A dilution coefficient of $B_1 = 0.02$ should be used for high-rise buildings.

When exhaust from several collecting stations is combined in a single vent or in a tight cluster of stacks, the effective exhaust flow area A_e will increase, causing the minimum dilution in Equation (19) to decrease. To qualify as a cluster, the stacks must all lie within a two-stack diameter radius of the middle of the group. Stacks lined up in a row do not act as a single stack, as shown by Gregoric *et al.* (1982). However, the exhaust concentration C_e of each contaminant will decrease by mixing with other exhaust streams, and the plume rise will increase due to the higher momentum in the combined jets. For combined vertical exhaust jets, the roof level intake concentration C in Equation (12) will almost always be lower than the intake concentration caused by separate exhausts. Where possible, exhausts should be combined before release to take advantage of this increase in overall dilution.

Critical Wind Speed and Dilution

At very low wind speed, the exhaust jet from an uncapped stack will rise high above roof level, producing a large exhaust dilution D_{min} at a given intake location. Likewise, at high wind speed, the dilution will also be large because of the longitudinal stretching of the plume by the wind. Between these extremes, a critical wind speed exists at which the least dilution will occur for a given exhaust and intake location. This critical, absolute minimum dilution D_{crit} may be used to determine if an exhaust vent will be safe under all wind conditions. The critical wind speed for an uncapped vertical exhaust ($\beta = 1.0$) can be evaluated by finding the absolute minimum in Equations (19), (20), and (21). It is closely approximated by

$$U_{crit,o}/V_e = 2.9B_1{}^{-0.33}(S/A_e{}^{0.5})^{-0.67} \qquad (22)$$

where $U_{crit,o}$ is the critical wind speed producing the smallest minimum dilution for an uncapped vertical exhaust with negligible stack height. This critical dilution $D_{crit,o}$ may be found by using Equation (22) in Equation (19). For $S/A_e{}^{0.5} > 5$, this minimum is closely approximated by

$$D_{crit,o} = 1 + 7.0B_1{}^{0.67}(S/A_e{}^{0.5})^{1.33} \qquad (23)$$

The critical dilution in Equation (23) depends only on distance from the exhaust and not on the exhaust velocity V_e. However, increasing the exhaust velocity increases the critical wind speed in Equation (22), usually causing this worst-case critical dilution to occur less frequently.

To assess the severity of the hazard caused by intake contamination, it is useful to know how often the worst case D_{crit} is likely to occur. The number of hours per year during which the dilution is no more than a factor of 2 higher than the critical minimum value may be estimated from weather records by finding the fraction of time that the wind speed lies in the range from $0.5\,U_{crit,o}$ to $3.0\,U_{crit,o}$ (Wilson 1982, 1983). This fraction is then multiplied by the fraction of time the local wind direction lies in a sector ± 22.5° on each side of the line joining the exhaust and intake location.

EXHAUST STACK DESIGN

Before discharge, exhaust contamination should be reduced by filters, collectors, and scrubbers. Central exhaust systems that combine flows from many collecting stations should always be used where safe and practical. By combining several exhaust streams, central systems dilute intermittent bursts of contamination from a single station. However, in some cases, separate exhaust systems are mandatory. The nature of the contaminants to be combined, the recommended industrial hygiene practice, and the applicable safety codes need to be considered. Halitsky (1966) and Briggs (1984) present methods for estimating the trajectory of jets and the subsequent dispersion of jet plumes.

Separate exhaust stacks should be grouped in a tight cluster to take advantage of the larger plume rise of the resulting combined jet. In addition, a single stack from a central exhaust system or a tight cluster of stacks allows building air intakes to be placed as far as possible from the exhaust location. As shown in Figure 3, the effective stack height h_s is the portion of the exhaust stack that extends above local recirculation zones and upwind and downwind obstacles. Wilson and Winkel (1982) demonstrated that stacks terminating below the level of adjacent walls and architectural enclosures do not effectively reduce roof-level exhaust contamination. To take full advantage of their height, stacks should be located on the highest roof of a building. Where architectural enclosures are used to mask rooftop equipment, stacks must extend above the height H_c of the flow recirculation zone over the enclosure to prevent exhaust contamination of equipment within the enclosure.

Required Stack Exhaust Velocity

High stack discharge velocity and temperature increase plume rise and reduce intake contamination by increased jet dilution and by the elevated plume trajectory. However, high discharge velocity is a poor substitute for increased stack height.

As shown in Figure 15, stacks should have vertically directed uncapped exhaust jets. Stack caps which deflect the exhaust jet have a detrimental effect on both minimum dilution and critical wind speed. In any case, conical stack caps often do not eliminate rain, because rain does not usually fall straight down. Changnon (1966) shows that periods of heavy rainfall are often accompanied by high winds that deflect the raindrops under the cap and into the stack. A stack velocity of about 2500 fpm prevents condensed moisture from draining down the stack and keeps rain from entering the stack. Even when there are drains in the stack, the exhaust velocity should be maintained above 2000 fpm to provide adequate plume rise and jet dilution. Where stack condensate is corrosive, the body of the stack should be sized for a velocity of 1000 fpm or less, and a drain provided for the condensate (Anonymous 1964). The stack tip should have a converging cone (Figure 15B) to provide the required high-velocity discharge of 2000 to 3000 fpm. For intermittently operated systems, protection from rain and snow should be provided by stack drains as shown in Figures 15F through 15J.

Stack Height to Avoid Exhaust Entrainment

To avoid entrainment of exhaust gases into the wake, stacks must terminate above the flow recirculation height H_c. Where stacks or exhaust vents discharge within this recirculation region,

gases rapidly diffuse to the roof and may enter ventilation intakes or other openings. Figure 1 shows that this effluent will flow into the zone of recirculating flow behind the downwind face and will, in some cases, be brought back up onto the roof.

A high velocity exhaust with V_e at least 1.5 times as large as the wind speed U_H at roof height is essential not only to provide good initial dilution near the stack, but also to avoid stack wake downwash, which can reduce or eliminate plume rise. Downwash of the exhaust into the stack wake, shown in Figure 16, is caused by the low-pressure region which develops in the wake on the lee side of the stack. In situations where exhaust velocity cannot be maintained at a value larger than 1.5 times the wind speed, an additional downwash height h_d (see Figure 16) should be added to the stack height h_s. For a vertically directed jet from an uncapped stack ($\beta = 1.0$), Briggs (1973) recommends

$$h_d = 2.0d(1.5\beta V_e / U_H) \qquad (24)$$

for $V_e / U < 1.5$, where $d = (4A_e / \pi)^{0.5}$ is the effective stack diameter. Rain caps are frequently used on stacks of gas-and oil-fired furnaces and package ventilation units. These units will have $\beta = 0$, so $h_d = 3.0d$, and this should be added to the nominal height h_s to avoid flue gas contamination of roof-mounted equipment and air intakes.

The design procedure for selecting an appropriate stack height starts by calculating the height h_{sc} of a stack with a rain cap and, therefore, no plume rise. For an uncapped vertical exhaust, the minimum rise h_r of the bent-over exhaust jet is estimated, and the

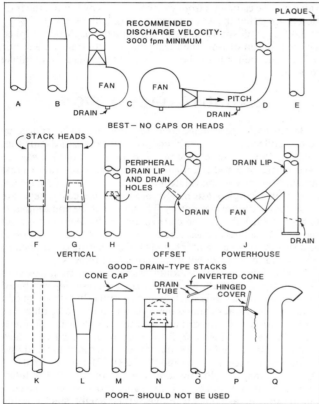

Fig. 15   Stack Designs Providing Vertical Discharge and Rain Protection

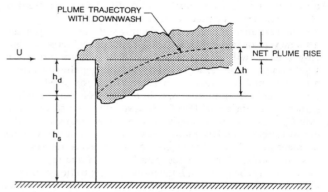

Fig. 16   Reduction of Effective Stack Height by Stack Wake Downwash

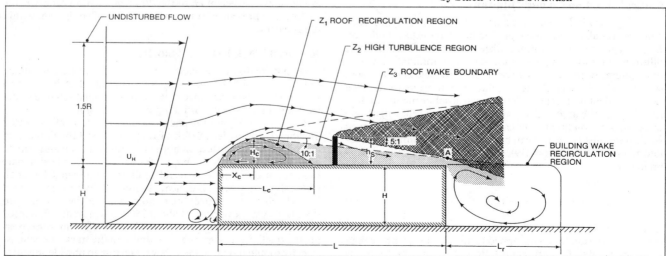

Fig. 17   Design Procedure for Required Stack Height to Avoid Contamination
(Wilson 1979)

capped height h_{sc} is lowered by an amount h_r to give credit for plume rise (see Figure 16).

The capped stack height h_{sc} required to avoid excessive exhaust gas reentry is estimated by assuming that the plume spreads upward and downward from h_{sc} with a 1:5 slope (11.3°), as shown in Figure 17. The first step is to raise the capped height h_{sc} until the lower edge of the 1:5 sloping plume avoids contact with all recirculation (zone 1) boundaries on rooftop obstacles such as air intake housings, architectural screens, or penthouses. The size of these recirculation zones, shown in Figures 3 and 17, are calculated using Equations (2), (3), and (4).

If air intakes are located on the downwind wall, the lower edge of the plume, sloping down at 1:5, must lie above the downwind edge of the roof when a nontoxic exhaust contaminant, such as an odor or water vapor, is being dealt with. For a toxic contaminant that requires a large dilution factor at the wall intake, the lower edge of the plume should lie above the flow recirculation zone in the wake downwind of the building. The boundary of the building wake recirculation, shown in Figures 1, 3, and 17, is defined by a horizontal line extending a distance L_r from the downwind edge of the roof. The recirculation length L_r is calculated from Equation (5).

For an uncapped stack, the plume rise h_r due to the vertical momentum of the exhaust is estimated from Briggs (1984) as

$$h_r = 3.0(V_e/U_H)d \tag{25}$$

where the wind speed U_H is the maximum design wind speed for which air intake contamination must be avoided. The required height h_s of the uncapped stack extending above local recirculation zones and obstacles is

$$H_S = h_{SC} - h_r + h_d \tag{26}$$

If the minimum recommended exhaust velocity of $V_e = 1.5\ U_H$ is maintained, plume downwash $h_d = 0$, and $h_r = 4.5d$; thus, an uncapped stack can be made $4.5d$ shorter than a capped one.

The largest flow recirculation, high turbulence, and wake regions occur when wind is normal to the upwind wall of the building. Required stack heights should be the largest of the heights determined for all four directions for which the wind is normal to a building wall.

Estimating Critical Dilution for Exhaust Stacks

The geometric design for avoiding excessive contamination does not give any estimate of the worst case critical dilution factor D_{crit} between the stack and an air intake. In this section, D_{crit} will be estimated for a predetermined stack height.

An increase in stack height or in exhaust velocity ratio V_e/U_H reduces roof-level contamination by keeping the high concentrations on the plume centerline far enough above the roof so that the intakes see only intermittent concentrations in the fringes of the plume. In addition, stack height or high exhaust velocity increases the critical wind speed at which the absolute minimum dilution occurs. This higher critical wind speed often reduces significantly the number of hours per year that high intake contamination (i.e., low dilution) will be observed.

Using a Gaussian plume dispersion equation, with a plume spread standard deviation of $0.14S$, and an uncapped vertical exhaust jet with no buoyancy and with plume rise inversely proportional to wind speed, the critical wind speed U_{crit} at which the smallest minimum dilution D_{crit} observed is

$$\frac{U_{crit,o}}{U_{crit}} = (Y + 1)^{0.5} - Y^{0.5} \tag{27}$$

where $U_{crit,o}$ is the critical wind speed for a flush (zero stack height) vertical exhaust, computed from Equation (22). The influence of stack height on the worst case critical dilution for the standard 10-min exposure time may be calculated from

$$\frac{D_{crit}}{D_{crit,o}} = \frac{U_{crit}}{U_{crit,o}} \exp[\,Y + Y^{0.5}(Y +1)^{0.5}\,] \tag{28}$$

where $Y = 12.6(h_s/S)^2$, and $D_{crit,o}$ is the dilution at critical wind speed for a flush vertical roof exhaust with no stack height, from Equation (23). Equations (27) and (28) are reliable only for $Y < 2.0$. Close to the stack, where $Y > 2.0$, use $Y = 2.0$ in Equations (27) and (28). Because both wind speed and turbulence intensity vary strongly with height above the building roof, the plume rise of the exhaust jet may not be inversely proportional to wind speed; normally its behavior is between $\Delta h \propto U^{-0.4}$ and $U^{-1.0}$. Thus, Equations (27) and (28) are only approximations. Because buoyancy is not included, the added rise due to buoyancy provides a factor of safety, particularly at low wind speed.

Because Equations (27) and (28) give the effect of a stack relative to a flush exhaust with $h_s = 0$, they are useful for assessing the advantages of increasing stack height as a remedial measure. By comparing two different heights, this calculation allows the relative benefits of a stack to be estimated without knowing any details of the contaminant concentrations or exhaust velocity in the existing stack. For example, the stack height required using the simple geometrical design procedure in the following section will have h_s/S of at least 0.2. Equations (27) and (28) show that the critical wind speed U_{crit} for this stack height will be about a factor of 2 larger, and the critical dilution D_{crit}, about eight times more than for the vertical jet from an uncapped exhaust with zero effective stack height.

Example 1. The stack height h_s of the uncapped vertical exhaust on the building shown in Figure 3 must be specified to avoid excessive contamination of air intakes A and B by stack gases. The stack has a diameter d of 1.64 ft and an exhaust velocity V_e of 1770 fpm. It is located 52.5 ft from the upwind edge of the roof. The penthouse has its upwind wall (with intake A) located 98.4 ft from the upwind edge of the roof, a height of 13.1 ft, and a length of 23.0 ft in the wind direction. The top of intake A is 6.56 ft below the penthouse roof. The building has a height H of 49.2 ft and a length of 203 ft. The top of intake B is 19.7 ft below roof level. The width (measured into the page) of the building is 164 ft, and the penthouse is 29.5 ft wide. What are the required stack heights h_s for both nontoxic and highly toxic exhaust contaminants for a design wind speed specified at a factor of 2 higher than the annual average hourly wind speed of 9.32 mph at a nearby airport with anemometer height H_{ref} of 32.8 ft? The building is located in unshielded suburban terrain.

Solution: The first step is to set the height h_{sc} of a capped stack by projecting lines with 5:1 slopes upwind from points of potential plume impact. For intake A, the highest point of impact is the top of the recirculation zone on the roof of the penthouse. To find the height of this recirculation zone, start with Equation (1).

$$R = (13.1)^{0.67}(29.5)^{0.33} = 17.2\text{ ft}$$

Then use Equations (2) and (3):

$$H_c = 0.22(17.2) = 3.77\text{ ft}$$
$$X_c = 0.5(17.2) = 8.60\text{ ft}$$

With the 5:1 slope of the lower plume boundary shown in Figure 17, the capped stack in Figure 3 must be

$$h_{sc} = 0.2(98.4 - 52.5 + 8.60) + 3.77 = 14.7\text{ ft}$$

above the penthouse roof to avoid intake A. For intake B on the downwind wall, the plume boundary from the stack in Figure 3 must lie above the end of the roof for nontoxic exhaust gas or the end of the building flow recirculation zone for highly toxic exhaust gas. For this recirculation zone, from Equation (1):

$$R = (49.2)^{0.67}(164)^{0.33} = 73.2 \text{ ft}$$

Then, from Equation (5)

$$L_r = 1.0(73.2) = 73.2 \text{ ft}$$

Thus, for a highly toxic exhaust, the capped stack height to avoid contamination at intake B is

$$h_{sc} = 0.2(203 - 52.5 + 73.2) - 13.1 = 31.7 \text{ ft}$$

above the penthouse roof. For a nontoxic exhaust

$$h_{sc} = 0.2(203 - 52.5) - 13.1 = 17.1 \text{ ft}$$

The required stack height is set by the condition of avoiding contamination of intake B, because intake A requires only a 14.7 ft capped stack. Credit for plume rise h_r from the uncapped stack requires the building wind speed U_H at $H = 49.2$ ft to be calculated. The design airport wind speed $U_{met} = 18.6$ mph $= 1640$ fpm at $H_{ref} = 33$ ft has been specified. From Figure 4, $A_o = 0.60$ and $a = 0.28$ at the suburban building site, so from Equations (9) and (10)

$$U_{ref} = 0.60(1640) = 984 \text{ fpm}$$

and

$$U_H = 984(49.2/32.8)^{0.28} = 1102 \text{ fpm}$$

Because $V_e/U_H = 1770/1102 = 1.61$ is greater than 1.5, there is no plume downwash, and $h_d = 0$ from Equation (24). Using Equation (25), the plume rise at the design wind speed is

$$h_r = 3.0(1.61)(1.64) = 7.92 \text{ ft}$$

Deducting this rise from the uncapped height h_{sc} using Equation (25) for intake B

$$h_s = 17.1 - 7.92 = 9.18 \text{ ft}$$

for nontoxic exhaust gas, and

$$h_s = 31.7 - 7.92 = 23.8 \text{ ft}$$

for highly toxic exhaust gas. As shown in Figure 3, both these stack heights are measured above the roof of the nearby penthouse. Adding the 13.1 ft penthouse height sets the required stack heights at 22.2 ft and 37.1 ft above roof level.

Example 2. The uncapped stack from Example 1, with $h_s = 9.18$ ft above the penthouse roof is used to exhaust a toxic contaminant, the stack gas concentration of which must be diluted by 1000:1 to be considered safe for 60-min exposures. Calculate the critical minimum dilution factor D_{crit} and the critical wind speed U_{crit} for contamination for intakes A and B. Is the stack height of 9.18 ft sized in Example 1 for nontoxic exhaust sufficient to handle this toxic substance?

Solution: At intake A, the stretched-string distance is:

$$S_A = [(98.4 - 52.5)^2 + (6.56)^2]^{0.5} = 46.3 \text{ ft}$$

and the exhaust area of the 1.64 ft diameter stack is $A_e = 2.11$ ft^2. The first step is to determine the critical wind speed and dilution for an uncapped stack with $h_s = 0$. Assuming that V_e/U_H will be greater than 0.5, a dilution constant $B_1 = 0.0625$ is chosen from the recommended values in the text following Equation (21). The normalized exhaust to intake distance is

$$S_A/A_e^{0.5} = 46.3/(2.11)^{0.5} = 31.8$$

and $V_e = 1770$ fpm is specified, so Equation (22) yields

$$U_{crit,o} = 1770(2.9)(0.0625)^{-0.33}(31.8)^{-0.67} = 1263 \text{ fpm}$$

so that the ratio $V_e/U_H = 1770/1263 = 1.40$ meets the requirement of being greater than 0.5, used to set $B_1 = 0.0625$. From Equation (23), the critical dilution for zero stack height is

$$D_{crit,o} = 1 + 7.0(0.0625)^{0.67}(31.8)^{1.33} = 110$$

at intake A. The effect of stack height is calculated from Equations (27) and (28) for the 9.18 ft height, at intake A

$$Y = 12.6(9.18/46.3)^2 = 0.49$$
$$U_{crit} = 1260/[(0.49 + 1)^{0.5} - (0.49)^{0.5}] = 2420 \text{ fpm}$$
$$D_{crit} = 110(2420/1260) \exp[0.49 + (0.49)^{0.5}(0.49 + 1)^{0.5}] = 808$$

The stack has increased the critical wind speed to a very high, infrequently occurring value of 12.3 m/s or 44.3 km/h, and increased the critical worst case dilution by more than a factor of 7. To convert the dilution from a 10-min exposure time (which applies to all the equations in this chapter), to a 60-min exposure, use Equation (16)

$$D_{crit} = 808(60/10)^{0.33} = 1460$$

showing that the stack produces adequate dilution at intake A.

Repeating the calculations for intake B, the stretched-string distance S_B in Figure 3 is

$$S_B = 45.9 + 23.0 + [(82.0)^2 + (13.1)^2]^{0.5} + 19.7 = 172 \text{ ft}$$

and the normalized distance is

$$S_B/A_e^{0.5} = 172/(2.11)^{0.5} = 118$$

Using Equations (22) and (23) for intake B

$$U_{crit,o} = 1770(2.9)(0.0625)^{-0.33}(118)^{-0.67} = 525 \text{ fpm}$$
$$D_{crit,o} = 1 + 7.0(0.0625)^{0.67}(118)^{1.33} = 623$$

For the stack with $h_s = 9.18$ ft, at intake B,

$$Y = 12.6(9.18/172)^2 = 0.0356$$

and Equations (27) and (28) yield

$$U_{crit} = 525/[(0.0356 + 1)^{0.5} - (0.0356)^{0.5}] = 633 \text{ fpm}$$
$$D_{crit} = 808(633/525) \exp[0.0356 + (0.0356)^{0.5}(0.0356 + 1)^{0.5}] = 1220$$

Intake B, further from the stack than intake A, has its critical wind speed increased by a factor of 1.2 and its critical dilution increased by a factor of 1.5 due to stack height. For a 60-min exposure, Equation (16) predicts that for the 9.1 ft stack height

$$D_{crit} = 1220(60/10)^{0.33} = 2200$$

which also meets the required 1000:1 dilution for the exhaust contaminant. This example shows that detailed dilution calculations may allow shorter stacks than the simpler but more conservative estimates using the 5:1 slope method of Example 1.

Intake and Exhaust Locations to Minimize Contamination

Inlets should be located to minimize reentry from contaminated sources, including heat, humidity, and dust that may affect system performance. In warm weather, the entry of water vapor from a cooling tower can significantly increase air-conditioning load and reduce the effectiveness of heat relief ventilation systems. In cold weather, freezing water vapor can damage equipment, and ice can block intake grilles and filters, substantially reducing performance. Inlets should be located away from hot exhausts that discharge horizontally or are deflected downward. Do not locate inlets near plumbing vents. Clarke (1965) notes that adequate exhaust dilution often requires larger separation distances between intake and exhaust than are specified in local building codes.

Changes and additions are often made to industrial and laboratory buildings. Initially satisfactory inlet and exhaust conditions can be the source of future unacceptable contaminant reentry. Exhaust outlets should not be located within enclosures or architectural screens because they hold building exhaust in their recirculation zones, increasing reentry of contaminated air and causing plugging and corrosion of heat exchangers. Restaurant kitchen hood exhausts are particularly destructive in this respect. Intakes and contaminated exhaust outlets should never be located in the same enclosure.

When the wind is normal to the upwind wall, the streamline patterns in Figures 1 and 2 show that air flows up and down the wall, dividing at about half the wall height. The downward flow creates turbulence that stirs up dust and debris. Changnon (1966), Clarke (1967), and Houlihan (1965) have shown that the upward flow on the top half of the wall can carry raindrops up the wall during high winds. To take advantage of the natural separation of wind flow on the upper and lower half of a building, toxic or nuisance exhaust should be located on the upper one-third, and intakes on the lower one-third of the building, high enough above ground to avoid wind-blown dust, vegetation, and vehicle exhaust. In particular, intakes near vehicle loading zones should be avoided. Overhanging canopies on vehicle docks do not effectively prevent hot vehicle exhaust from rising to intakes above the canopy.

To protect intakes from contamination by wind-blown dust and debris, they are best located on the roof. If contaminated exhaust is discharged from several locations on a roof, the intakes should be sited to minimize contamination by exhaust gases. In this case, intakes should be placed as far away from exhausts as possible, except when most of the exhaust is discharged from a single location. In this case, the most effective location for intakes on the roof is at the base of a single high exhaust stack that discharges the contaminated exhaust high above the intake location. If possible, air intakes and exhausts should not be located near the edges of a wall or roof because of the pressure fluctuations occurring there.

HEAT REJECTION EQUIPMENT

Cooling towers and similar heat rejection devices are very sensitive to airflow around buildings. This equipment is frequently roof-mounted, with intakes close to the roof where air is considerably hotter and at a higher wet-bulb temperature than air that is not influenced by the roof. This can reduce the capacity of cooling towers and air-cooled condensers.

Heat exchangers often take in air on one side and discharge heated, moist air horizontally from the other side. For these horizontal flow cooling towers, changes in wind direction and velocity caused by immediately adjacent obstructions can drastically reduce equipment performance by reducing the airflow rate. Recirculation is even more serious than reduction in airflow rate to such devices. Recirculation of warm moist exhaust raises the inlet wet-bulb temperature, which reduces performance. Recirculation can be caused by local disturbance of the airflow by upwind obstructions or by a close downwind obstruction. The vertical discharge ducts may need to be extended to improve the effectiveness of this equipment.

SCALE MODEL SIMULATION AND TESTING

For many routine design applications, flow patterns, wind pressures, and exhaust dilution can be estimated using the data and equations set out in the previous sections. However, in critical applications, such as when health and safety are of concern, physical modeling or full-scale field evaluations may be required to obtain more accurate estimates. Measurements on small-scale models in wind tunnels or water channels can provide information for design prior to construction and an economical method of performance evaluation for existing facilities. Full-scale testing is not generally useful in the initial design process because of the time and expense required to obtain meaningful information. On the other hand, full-scale testing is useful for verifying data

derived from physical modeling and for planning remedial changes to improve existing facilities.

Detailed accounts of physical modeling, field measurements and applications, and engineering problems resulting from atmospheric flow around buildings are available in the proceedings of conferences on wind engineering (see Bibliography).

The wind tunnel is the main tool used to assess and understand the airflow around buildings. A water channel can also be used. Models of buildings, complexes, and the local surrounding topography are constructed and tested in a simulated turbulent atmospheric boundary layer. The airflow, wind pressures, snow loads, structural response, or pollutant concentrations can then be measured directly by properly scaling the wind and exhaust flow characteristics. Weil *et al.* (1981), Petersen (1987a), and Dagliesh (1975) found generally good agreement between the results of wind tunnel simulations and corresponding full-scale data.

Similarity Requirements

Physical modeling is most appropriate for applications involving small-scale atmospheric motions, such as recirculation of exhaust downwind of a laboratory, wind loads on structures, wind speeds around building clusters, snow loads on roofs, and airflow over hills. At present, winds associated with tornadoes, thunderstorms, and large-scale atmospheric motion cannot be simulated.

Snyder (1981) gives guidelines for fluid modeling of atmospheric diffusion. This report contains explicit directions and should be used whenever designing wind tunnel studies to assess concentration levels due to air pollutants. ANSI *Standard* A58.1 (1982) also provides guidance when wind tunnels are used for evaluating wind effects on structures.

A complete and exact simulation of the airflow over buildings and the resulting concentration or pressure distributions cannot be achieved in a physical model. This is not a serious limitation, however. Cermak (1971, 1976a,b), Snyder (1981), and Petersen (1987a,b) found that an accurate simulation of the transport and dispersion of laboratory exhaust can be achieved if the following criteria are used:

1. Match (equal in model and full scale) the exhaust velocity to wind speed ratio, V_e/U_H.
2. Match exhaust to ambient air density ratio, ρ_e/ρ_a.
3. Match exhaust Froude number, $\text{Fr}^2 = \rho_r V_e{}^2/[(\rho_e - \rho_a)gd]$.
4. Ensure a fully turbulent stack gas flow; stack Reynolds number $Re_s = V_e d/\nu$ greater than 2000, or place an obstruction inside stack to enhance turbulence.
5. Ensure a fully turbulent wind flow with roughness Reynolds number $Re_z = U*Z_o/\nu$ greater than 2.5.
6. Scale all dimensions and roughness by a common factor.
7. Match atmospheric stability by the bulk Richardson number (see Cermak 1975). For most applications related to airflow around buildings, neutral stratification is assumed, and no Richardson number matching is required.
8. Match mean velocity and turbulence distributions in the wind approaching the model.
9. Building wind Reynolds number $U_H R/\mu$ must be greater than 11,000 for sharp-edged structures, or greater than 90,000 for round-edged structures.

For wind speeds, flow patterns, or pressure distributions around buildings, only conditions 5 through 9 are necessary. Usually, each wind tunnel study requires a detailed assessment to determine which appropriate parameters to match in model and full scale.

In wind tunnel simulation of exhaust gas recirculation, the buoyancy of the exhaust gas (condition 3) is often not modeled. This allows using a high wind tunnel speed or a smaller model to achieve a high enough Reynolds number (conditions 4, 5, and 9). Neglecting buoyancy is justified if the density of building exhaust air is within 10% of the outdoor ambient air. Also, critical minimum dilution D_{crit} occurs at wind speeds high enough to produce a well-mixed, neutrally stable atmosphere, allowing stability matching (condition 7) to be neglected. Omission of conditions 3 and 7 simplifies the test procedure considerably, reducing both testing time and cost.

Wind Simulation Facilities

Boundary layer wind tunnels are required for conducting most wind studies. The wind tunnel test section should be long enough so that a deep boundary layer that slowly changes with downwind distance can be established upwind of the model building. Boundary layer wind tunnels can be either open- or closed-circuit; an open-circuit tunnel is open to the ambient air at both ends, and a closed-circuit tunnel is attached at both ends so that the air continually circulates within the tunnel.

Other important wind tunnel characteristics include the width and height of the test section, range of wind speeds, roof adjustability, and temperature control. Larger models can be used in tunnels that are wider and taller which, in turn, gives better measurement resolution. Model blockage effects can be minimized by an adjustable roof height. Temperature control is required when atmospheric conditions other than neutral stability are to be simulated. Stable atmospheric conditions are simulated by heating the air and cooling the ground surface; unstable atmospheric conditions are simulated by heating the surface and cooling the air. Boundary layer characteristics appropriate for the site are established by using roughness elements on the tunnel floor that produce mean velocity and turbulence intensity profiles characteristic of the full scale.

Water as well as air can be used for the modeling fluid if an appropriate flow facility is available. Flow facilities may be in the form of a tunnel, tank, or open channel. Water tanks with a free surface ranging in size up to that of a wind tunnel test section have been used by towing a model (upside down) through the nonflowing fluid. Stable stratification can be obtained by adding a salt solution at the tank bottom. This technique does not permit development of a boundary layer and yields only approximate qualitative information on flow around buildings. Water channels can be designed to develop thick turbulent boundary layers similar to those developed in the wind tunnel. One advantage of such a flow system is ease of flow visualization, but this is offset by a greater difficulty in measurement of flow variables and concentrations.

Designing Model Test Programs

The first step in planning a test program is the selection of the model length scale. Choice of this scale depends on cross-sectional dimensions of the test section, dimensions of the building to be studied together with nearby neighboring buildings, and/or topographic features and thickness of the simulated atmospheric boundary layer. For tall buildings, the scale is often determined by the requirement that the ratio of boundary layer thickness to building height (δ/H) should be approximately equal for model and prototype. Strong wind boundary layers range in depth from 800 to 2000 ft for surroundings ranging from flat open terrain to suburbs of urban centers (see Figure 4). Boundary layer thicknesses in wind tunnel and water channel flow facilities have nominal depths ranging from 20 to 60 in; therefore, typical geometric scales range from about 150:1 to 1000:1.

For complexes of low to moderately tall buildings extending over a wide area, such as an industrial plant or an urban center, the scale is usually controlled by test section width. Most flow facilities available for testing purposes range in width from 40 to 160 in. If interest were focused on an area of 1600 ft radius, the scale could range from 1:1000 to 1:250 depending on the test section width. A large model size is desirable to meet minimum Reynolds and Froude number requirements; therefore, a wide test section provides advantages for testing purposes. In general, the model

at any section should be small compared to the test section area so that blockage is less than 5%.

The test program must include specifications of the meteorological variables to be considered. These include wind direction, wind speed, and thermal stability. Data taken at the nearest meteorological station should be reviewed to obtain a realistic assessment of wind climate for a particular site. Ordinarily, local winds around a building, pressures, and/or concentrations are measured for 16 wind directions in 22.5° intervals. This is easily accomplished by mounting a building and its near surroundings on a turntable. If only local wind information and pressures are of interest, testing at one wind speed with neutral stability is sufficient.

In addition to source geometry, specifications of source characteristics must include the exhaust discharge face velocity V_e, the exhaust temperature, and the composition components of the effluent to determine exhaust density ρ_e. This information and meteorological data for the site determine a realistic range of values for the velocity ratio and the densimetric Froude number. Equality of model and prototype Froude numbers requires wind speeds of less than 100 fpm for testing. However, greater wind speeds may be needed to meet the minimum building Reynolds number requirement. Larger wind speeds are possible by using a density difference for the model effluent larger than the prototype difference. For effluent with positive buoyancy, it is convenient to use a mixture of air and helium plus a tracer. A mixture of air and carbon dioxide plus a tracer may be used to simulate a negatively buoyant effluent.

The following elements should be considered as part of a quality assurance plan for testing:

- Accuracy of scale models?
- Boundary layer similarity compared to full-scale wind and turbulence profiles?
- Wind tunnel blockage less than 5%?
- Tunnel speed monitored during testing?
- Sensors calibrated periodically against standard?
- Model scale adequate to give good measurement resolution?
- Model scale adequate to reproduce sufficient depth of approach wind boundary layer?
- Testing conducted at sufficient speeds to achieve Reynolds number independence?
- Wind tunnel flow characteristics uniform in crosswind direction?
- Regular checks of data collection procedures and repeatability?

SYMBOLS

a = exponent in power law wind speed profile, Equation (10)

A_e = stack or exhaust exit face area, ft^2

A_o = terrain roughness correction factor for wind speeds measured at standard 33-ft height, Equation (9)

B_L = larger of two upwind building face dimensions, H and W, ft

B_s = smaller of two upwind building face dimensions, H and W, ft

B_1 = air entrainment parameter in distance dilution D_s, Equation (21)

C_e = contaminant mass concentration in exhaust at exhaust temperature T_e, Equation (13), lb/ft^3

C = contaminant mass concentration at receptor at ambient air temperature T_a, lb/ft^3

C_p = local wind pressure coefficient on building surface, Equation (8)

d = $(4A_e/\pi)^{0.5}$, effective exhaust stack diameter, ft

D = C_e/C dilution factor between source and receptor mass concentrations, Equation (12)

D_{min} = minimum dilution factor D at given wind speed for all exhaust locations at same fixed distance S from intake

$D_{min,0}$ = minimum dilution factor D_{min} at roof level for flush vent with zero stack height, $h_s = 0$, Equation (19)

D_{crit} = critical dilution factor at roof level for uncapped vertical exhaust at critical wind speed U_{crit} that produces smallest value of D_{min} for given exhaust to intake distance S and stack height h_s, Equations (23), (27), and (28)

$D_{crit,0}$ = critical dilution factor D_{crit}, at roof level for uncapped vertical exhaust with stack height $h_s = 0$, Equation (23)

D_0 = apparent initial dilution factor for exhaust jet, Equation (20)

D_s = distance dilution factor at fixed wind speed, Equation (21)

D_v = dilution factor for volume fraction concentration, Equation (15)

f = contaminant volume concentration fraction at receptor; ratio of contaminant gas volume to total mixture volume, ppm $\times 10^6$

f_e = contaminant volume concentration fraction in exhaust gas; ratio of contaminant gas volume to total mixture volume, Equation (14), ppm $\times 10^{-6}$

g = acceleration of gravity, ft/s^2

H = wall height aboveground on upwind building face, Figures 2 and 14, ft

H_c = maximum height above roof level of upwind roof edge flow recirculation zone, Figure 3, ft

H_{ref} = height of wind anemometer at meteorological station, ft

h_d = downwash correction to be subtracted from stack height, Equation (24), ft

h_s = effective exhaust stack height above rooftop obstacles and enclosures, Figure 3, ft

h_{sc} = required height of capped exhaust stack to avoid excessive intake contamination, ft

h_r = plume rise of uncapped vertical exhaust jet, ft

K_c = normalized concentration coefficient $CU_H H W/Q$

L = length of building in wind direction, Figures 2 and 14, ft

L_c = length of upwind roof edge recirculation zone, Figure 3, ft

L_r = length of flow recirculation zone behind rooftop obstacle or building, Figure 3, ft

m = contaminant mass release rate, Equation (13), lb/s

M = configuration factor, Equation (18)

p_s = wind pressure difference between exterior building surface location and local outdoor atmospheric pressure at same level in undisturbed approach wind, Equation (8), lb/ft^2

p_v = velocity pressure of wind at roof level, Equation (7), lb/ft^2

Q_e = $A_e V_e$, total exhaust gas mixture flow rate, ft^3/s

R = scaling length for roof flow patterns, Equation (1), ft

S = stretched string distance; the shortest distance from exhaust to intake over and along building surface, Figure 3, ft

T_a = outdoor ambient air absolute temperature, °R

T_e = exhaust air mixture absolute temperature, °R

t_a = time interval over which receptor (intake) concentrations are averaged in computing dilution, Equation (16), s

U^* = ground surface friction velocity, ft/s

U_{ref} = wind speed at height of meteorological station anemometer in undisturbed wind in same terrain roughness in which building is located, Equation (9), ft/s

U_{crit} = critical wind speed that produces smallest minimum dilution factor D_{crit} for uncapped vertical exhaust at given S and h_s, Equation (27), ft/s

$U_{crit,o}$ = critical wind speed for smallest minimum dilution factor $D_{crit,o}$ for *flush uncapped exhaust with zero stack height $h_s = 0$*, Equation (22), ft/s

U_H = mean wind speed for 10-min averaging at height of upwind wall in undisturbed flow approaching building, Figure 1, ft/s

U_{met} = meteorological station wind speed, measured (or corrected to) height of $Z = 33$ ft aboveground in smooth terrain, ft/s

V_e = exhaust face velocity, Q_e/A_e, ft/s

W = width of upwind building face, Figures 2 and 14, ft

X = distance from upwind roof edge, ft

X_c = distance from upwind roof edge to H_c, Figure 17, ft

Y = $12.6(h_s/s)^2$ in Equations (25, 26)

Z = height above local ground level, ft

Z_1 = height of flow recirculation zone boundary above roof, Figure 17, ft

Z_2 = height of high turbulence zone boundary above roof, Figure 17, ft

Z_3 = height of roof edge wake boundary above the roof, Figure 17, ft

Z_o = log-law terrain roughness scaling length, Figure 4, ft

α = configuration parameter, Equation (17)

β = capping factor; $\beta = 1.0$ for vertical uncapped roof exhaust; $\beta = 0$ for capped, louvered, or downward-facing exhaust, Equation (20)

δ = fully developed strong wind atmospheric boundary layer thickness, Figure 4, ft

δ_2 = developing internal boundary layer thickness from change in terrain roughness, Figure 12, Table 1, ft

θ = wind angle between perpendicular line from the upwind building face to wind direction, Figure 6, degrees

ρ_a = density of outdoor air, lb/ft^3

ρ_e = density of exhaust gas mixture, lb/ft^3

ν = kinematic viscosity of outdoor air, ft^2/s

REFERENCES

Akins, R.E., J.A. Peterka, and J.E. Cermak. 1979. "Averaged pressure coefficients for rectangular buildings." In *Wind engineering*. Proceedings of the Fifth International Conference, Fort Collins, CO 7:369-80.

Anonymous. 1964. How to design drain-type stacks. *Heating Piping and Air Conditioning* 36(6):143.

ANSI. 1982. Minimum design loads for buildings and other structures. ANSI *Standard* A58.1-82. American National Standards Institute, NY.

ASHRAE. 1986. Bin and degree hour weather data for simplified energy calculations.

Briggs. 1973. Diffusion estimates for small emissions. Oak Ridge Atmospheric Turbulence and Diffusion Laboratory, Draft Report No. 79.

Briggs. 1984. "Plume rise and buoyancy effects." In *Atmospheric science and power production*, D. Randerson, ed. U.S. Department of Energy DOE/TIC-27601 (DE 84005177).

Cermak, J.E. 1971. Laboratory simulation of the atmospheric boundary layer. AIAA *Journal* 9(9):1746.

Cermak, J.E. 1975. Applications of fluid mechanics to wind engineering. *Journal of Fluid Engineering*. Transactions of ASME 97:9.

Cermak, J.E. 1976a. Nature of airflow around buildings. ASHRAE *Transactions* 82(2):1044.

Cermak, J.E. 1976b. Aerodynamics of buildings. *Annual review of fluid mechanics* 8:75.

Cermak, J.E. 1977. Wind-tunnel testing of structures. *Journal of the Engineering Mechanics Division*, ASCE 103, EM6:1125.

Changnon, S.A. 1966. Selected rain-wind relations applicable to stack design. *Heating Piping and Air Conditioning* 38(3):93.

Chui, E.H. and D.J. Wilson. 1988. Effects of varying wind direction on exhaust gas dilution. *Journal of Wind Engineering and Industrial Aerodynamics* 31:87-104.

Clarke, J.H. 1965. The design and location of building inlets and outlets to minimize wind effect and building reentry of exhaust fumes. *Journal of American Industrial Hygiene Association* 26:242.

Clarke, J.H. 1967. Airflow around buildings. *Heating Piping and Air Conditioning* 39(5):145.

Cook, N.J. 1986. *The designer's guide to wind loading of building structures*, Chapter 9, Butterworths, London.

Dagliesh, W.A. 1975. Comparison of model/full-scale wind pressures on a high-rise building. *Journal of Industrial Aerodynamics* 1:55-66.

Davenport, A.G. and H.Y.L. Hui. 1982. External and internal wind pressures on cladding of buildings. Boundary Layer Wind Tunnel Laboratory, The University of Western Ontario, London, Ontario. BLWT-820133.

Elliot, W.P. 1958. The growth of the atmospheric internal boundary layer. Transactions of the American Geophysical Union 39:1048-54.

Gregoric, M., L.R. Davis, and D.J. Bushnell. 1982. An experimental investigation of merging buoyant jets in a crossflow. *Journal of Heat Transfer*, Transactions of ASME 104:236-40.

Halitsky, J. 1962. Diffusion of vented gas around buildings. *Journal of the Air Pollution Control Association* 12:74-80.

Halitsky, J. 1963. Gas diffusion near buildings. ASHRAE *Transactions* 69:464-84.

Halitsky, J. 1966. A method of estimating concentrations in transverse jet plumes. *International Journal of Air and Water Pollution* 10:821-43.

Halitsky, J. 1982. Atmospheric dilution of fume hood exhaust gases. *American Industrial Hygiene Association Journal* 43(3):185-89.

Halitsky, J. 1985. Concentration coefficients in atmospheric dispersion calculations. ASHRAE *Transactions* 91(2B):1722-36.

Holmes, J.D. 1986. *Wind loads on low-rise buildings: The structural and environmental effects of wind on buildings and structures*, Chapter 12. Faculty of Engineering, Monash University, Melbourne, Australia.

Hosker, R.P. 1984. "Flow and diffusion near obstacles." In *Atmospheric science and power production*, D. Randerson, ed. U.S. Department of Energy DOE/TIC-27601 (DE 84005177).

Hosker, R.P. 1985. Flow around isolated structures and building clusters: A review. ASHRAE *Transactions* 91(2b):1671-92.

Houlihan, T.F. 1965. Effects of relative wind on supply air systems. ASHRAE *Journal* 7(7):28.

Li, W.W. and R.N. Meroney. 1983. Gas dispersion near a cubical building. *Journal of Wind Engineering and Industrial Aerodynamics* 12:15-33.

Meroney, R.N. 1982. "Turbulent diffusion near buildings." In *Engineering meteorology*, E.J. Plate, ed. Elsevier, Amsterdam 48:525.

Petersen, R.L. 1987a. Wind tunnel investigation of the effect of platform-type structures on dispersion of effluents from short stacks. *Journal of Air Pollution Control Association* 36:1347-52.

Petersen, R.L. 1987b. Designing building exhausts to achieve acceptable concentrations of toxic effluent. ASHRAE *Transactions* 93:2.

Simiu, V. and R. Scanlan. 1986. *Wind effects on structures: An introduction to wind engineering*, 2nd ed. Wiley Interscience.

Snyder, W.H. 1981. Guideline for fluid modelling of atmospheric diffusion. Environmental Protection Agency Report, EPA-600/881-009.

Swami, H.V. and S. Chandra. 1987. Procedures for calculating natural ventilation airflow rates in buildings. Final Report FSEC-CR-163-86. Florida Solar Energy Center, Cape Canaveral, Florida.

Weil, J.C., J.E. Cermak, and R.L. Petersen. 1981. Plume dispersion about the windward side of a hill at short range: Wind tunnel versus field measurements. Paper presented at Fifth American Meteorological Society Symposium, Atlanta, GA.

Wilson, D.J. 1976. Contamination of air intakes from roof exhaust vents. ASHRAE *Transactions* 82:1024-38.

Wilson, D.J. 1977. Dilution of exhaust gases from building surface vents. ASHRAE *Transactions* 83(1):168-76.

Wilson, D.J. 1979. Flow patterns over flat roofed buildings and application to exhaust stack design. ASHRAE *Transactions* 85:284-95.

Wilson, D.J. 1982. Critical wind speeds for maximum exhaust gas reentry from flush vents at roof level intakes. ASHRAE *Transactions* 88(1):503-13.

Wilson, D.J. 1983. A design procedure for estimating air intake contamination from nearby exhaust vents. ASHRAE *Transactions* 89(2):136-52.

Wilson, D.J. 1985. Ventilation intake air contamination by nearby exhausts. Proceedings of the Air Pollution Control Association Conference, Ottowa, Canada.

Wilson, D.J. and R.E. Britter. 1982. Estimates of building surface concentrations from nearby point sources. *Atmospheric Environment* 16:2631-46.

Wilson, D.J. and E.H. Chui. 1985. Influence of exhaust velocity and wind incidence angle on dilution from roof vents. ASHRAE *Transactions* 91(2b):1693-1706.

Wilson, D.J. and E.H. Chui. 1987. Effect of turbulence from upwind buildings on exhaust gas dilution. ASHRAE *Transactions* 93(2).

Wilson, D.J. and G. Winkel. 1982. The effect of varying exhaust stack height on contaminant concentration at roof level. ASHRAE *Transactions* 88(1):513-33.

BIBLIOGRAPHY

Cermak, J.E., ed. 1979. *Wind engineering*. Proceedings of Fifth International Conference, Fort Collins, CO. Pergamon Press, New York.

Defant, F. 1951. "Local winds." In *Compendium of meteorology*. American Meteorology Society, Boston, 655-72.

Geiger, R. 1966. *The climate near the ground*. Harvard University Press, Cambridge, MA.

Houghton, E.L. and N.B. Carruthers. 1976. *Wind forces on buildings and structures: An introduction*. Edward Arnold, London.

Landsberg, H. 1981. *The urban climate*. Academic Press, New York.

Panofsky, H.A. and J.A. Dutton. 1984. *Atmospheric turbulence: Models and methods for engineering applications*. John Wiley and Sons, New York.

Pasquill, F. and F. Smith. 1983. *Atmospheric diffusion*, 3rd ed. Halstead Press, New York.

Proceedings of the Fifth National Conference on Wind Engineering. 1985. Texas Tech University, Lubbock, TX.

CHAPTER 15

COMBUSTION AND FUELS

PRINCIPLES OF COMBUSTION

COMBUSTION is the chemical process in which an oxidant is reacted rapidly with a fuel to liberate stored energy as thermal energy, generally in the form of high-temperature gases. Small amounts of electromagnetic energy (light), electric energy (free ions and electrons), and mechanical energy (noise) are also released during the combustion process. Except in special applications, the oxidant for combustion is oxygen in the air.

Conventional hydrocarbon fuels contain primarily hydrogen and carbon, in elemental form or in various compounds. Their complete combustion produces mainly carbon dioxide and water; however, small quantities of carbon monoxide and partially reacted flue gas constituents (gases and liquid or solid aerosols) may form. Most conventional fuels also contain small amounts of sulfur, oxidized to SO_2 or SO_3 during combustion, and noncombustible substances, i.e., mineral matter (ash), water, and inert gases. Flue gas is the product of complete or incomplete combustion including excess air (if present), but not dilution air.

Fuel combustion rate depends on: (1) the chemical reaction rate of combustible fuel constituents with oxygen, (2) the rate at which oxygen is supplied to fuel (mixing of air and fuel), and (3) the temperature in the combustion region. The reaction rate is fixed by fuel selection. Increasing the mixing rate or temperature increases the combustion rate.

With *complete combustion* of hydrocarbon fuels, all hydrogen and carbon in the fuel are oxidized to H_2O and CO_2. Generally, for complete combustion, excess oxygen or excess air must be supplied beyond the amount theoretically required to oxidize the fuel; this is usually expressed as a percentage of the air required to completely oxidize the fuel.

In *stoichiometric combustion* of a hydrocarbon fuel, fuel is reacted with the exact amount of oxygen required to oxidize all carbon, hydrogen, and sulfur in the fuel to CO_2, H_2O, and SO_2. Therefore, exhaust gas from stoichiometric combustion theoretically contains no incompletely oxidized fuel constituents and no unreacted oxygen; e.g., no excess air or oxygen, and no carbon monoxide. The percentage of CO_2 contained in products of stoichiometric combustion is the maximum attainable, referred to as *stoichiometric CO_2*, *ultimate CO_2*, or *maximum theoretical percentage of CO_2*.

Stoichiometric combustion is seldom realized in practice because of imperfect mixing and finite reaction rates. Consequently, for economy and safety, most combustion equipment should operate with some excess air. This ensures that fuel is not wasted and that combustion equipment is sufficiently flexible to provide complete combustion despite variations in fuel properties and in the supply rates of fuel and air. Combustion equipment is designed and operated to ensure complete, not stoichiometric, combustion. The exact amount of excess air supplied to any particular combustion equipment depends on such factors as (1) expected variations in fuel properties and in fuel and air supply rates, (2) equipment

application, (3) degree of operator supervision required or available, and (4) control requirements. For maximum efficiency, combustion at low excess air is desirable.

Incomplete combustion occurs when a fuel element is not completely oxidized in the combustion process. For example, a hydrocarbon may not completely oxidize to carbon dioxide and water but may form partially oxidized compounds, such as carbon monoxide, aldehydes, and ketones. Conditions that promote incomplete combustion include (1) insufficient air and fuel mixing (causing local fuel-rich and fuel-lean zones), (2) insufficient air supply to the flame (providing less than the required quantity of oxygen), (3) insufficient reactant residence time in the flame (preventing completion of combustion reactions), (4) flame impingement on a cold surface (quenching combustion reactions), or (5) a too-low flame temperature (slowing combustion reactions).

Incomplete combustion uses fuel inefficiently, can be hazardous because of carbon monoxide production, and contributes to air pollution.

Combustion Reactions

Combustion of oxygen with the combustible elements and compounds in fuels occurs according to fixed chemical principles, including:

- Chemical reaction equations.
- Law of matter conservation. The mass of each element in the reaction products must equal the mass of that element in the reactants.
- Law of combining weights. Chemical compounds are formed by elements combining in fixed weight relationships.
- Chemical reaction rates.

Oxygen for combustion is normally obtained from air, a physical mixture of nitrogen, oxygen, small amounts of water vapor, carbon dioxide, and inert gases. For practical combustion calculations, dry air consists of 20.95% oxygen and 79.05% inert gases (nitrogen, argon, and so forth) by volume, or 23.15% oxygen and 76.85% inert gases by weight. For calculation purposes, nitrogen is assumed to pass through the combustion process unchanged (although small quantities of nitrogen oxides are known to form). Table 1 lists oxygen and air requirements for stoichiometric combustion of some pure combustible materials (or constituents) found in common fuels. Table 2 lists the products of combustion for stoichiometric combustion of the same pure combustible materials.

Flammability Limits

Fuel will burn in a self-sustained reaction only when the volume percentages of fuel and air in a mixture at standard temperature and pressure are within specific limits: the upper and lower flammability limits or explosive limits (UEL and LEL). Both temperature and pressure affect these limits. As the temperature of the mixture increases, the upper limit increases and the lower limit decreases. As the pressure of the mixture decreases below atmospheric pressure, the upper limit decreases and the lower limit

---

The preparation of this chapter is assigned to TC 3.7, Fuels and Combustion.

Table 1   Combustion Reactions of Common Fuel Constituents

| Constituent | Molecular Symbol | Combustion Reactions | Stoichiometric Oxygen and Air Requirements | | | |
| | | | lb/lb Fuel[a] | | ft^3/ft^3 Fuel | |
			O_2	Air	O_2	Air
Carbon (to CO)	C	$C + 0.5\alpha_2 \rightarrow CO$	1.33	5.75	—	—
Carbon (to CO_2)	C	$C + O_2 \rightarrow CO_2$	2.66	11.51	—	—
Carbon monoxide	CO	$CO + 0.5\alpha_2 \rightarrow CO_2$	0.57	2.47	0.50	2.39
Hydrogen	H_2	$H_2 + 0.5\alpha_2 \rightarrow H_2O$	7.94	34.28	0.50	2.39
Methane	CH_4	$CH_4 + 2O_2 \rightarrow CO_2 + 2H_2O$	3.99	17.24	2.00	9.57
Ethane	C_2H_6	$C_2H_6 + 3.5O_2 \rightarrow 2CO_2 + 3H_2O$	3.72	16.09	3.50	16.75
Propane	C_3H_8	$C_3H_8 + 5O_2 \rightarrow 3CO_2 + 4H_2O$	3.63	15.68	5.00	23.95
Butane	C_4H_{10}	$C_4H_{10} + 6.5O_2 \rightarrow 4CO_2 + 5H_2O$	3.58	15.47	6.50	31.14
—	C_nH_{2n+2}	$C_nH_{2n+2} + (1.5n+0.5)O_2 \rightarrow nCO_2 + (n+1)H_2O$	—	—	$1.5n + 0.5$	$7.18n + 2.39$
Ethylene	C_2H_4	$C_2H_4 + 3O_2 \rightarrow 2CO_2 + 2H_2O$	3.42	14.78	3.00	14.38
Propylene	C_3H_6	$C_3H_6 + 4.5O_2 \rightarrow 3CO_2 + 3H_2O$	3.42	14.78	4.50	21.53
—	C_nH_{2n}	$C_nH_{2n} + 1.5nO_2 \rightarrow nCO_2 + nH_2O$	3.42	14.78	$1.50n$	$7.18n$
Acetylene	C_2H_2	$C_2H_2 + 2.5\alpha_2 \rightarrow 2CO_2 + H_2O$	3.07	13.27	2.50	11.96
—	C_nH_{2m}	$C_nH_{2m} + (n+0.5m)O_2 \rightarrow nCO_2 + mH_2O$	—	—	$n + 0.5m$	$4.78n + 2.39m$
Sulfur (to SO_2)	S	$S + O_2 \rightarrow SO_2$	1.00	4.31	—	—
Sulfur (to SO_3)	S	$S + 1.5O_2 \rightarrow SO_3$	1.50	6.47	—	—
Hydrogen sulfide	H_2S	$H_2S + 1.5O_2 \rightarrow SO_2 + H_2O$	1.41	6.08	1.50	7.18

[a] Atomic masses: H = 1.008; C = 12.01; O = 16.00; S = 32.06.

increases. However, as pressure increases above atmospheric pressure, the upper limit increases and the lower limit is relatively constant (see Table 3).

Ignition Temperature

Ignition temperature is the lowest temperature at which heat is generated by combustion faster than heat is lost to the surroundings and combustion becomes self-propagating. The fuel-air mixture will not burn freely and continuously below the ignition temperature unless heat is supplied, but chemical reaction between the fuel and air may occur. Ignition temperature is affected by a large number of factors in varying degrees.

The ignition temperature and flammability limits of a fuel-air mixture, together, are a measure of the potential for ignition (*Gas Engineers Handbook* 1965).

Combustion Modes

Combustion reactions occur in either continuous or pulse flame modes. *Continuous combustion* burns fuel in a sustained manner as long as fuel and air are continuously fed to the combustion zone and the fuel-air mixture is within the flammability limits. Continuous combustion is more common than pulse combustion and is used in most fuel-burning equipment.

Pulse combustion is an acoustically resonant process that burns various fuels in small, discrete fuel-air mixture volumes in a very rapid series of combustions.

The introduction of fuel and air into the pulse combustor is controlled by mechanical or aerodynamic valves. Typical combustors consist of one or more valves, a combustion chamber, an exit pipe, and a control system (ignition means, fuel metering devices, etc.). Typically, warm air furnace, hot water boiler, and commercial cooking equipment combustors use mechanical valves. Aerodynamic valves are usually used in higher pressure applications, such as thrust engines. Separate valves for air and fuel, a single valve for premixed air and fuel, or multiple valves of either type can be used. Premix valve systems may require a flame trap at the combustion chamber entrance to prevent flashback.

In a mechanically valved pulse combustor, air and fuel are forced into the combustion chamber through the valves under less than 0.5 psi pressure. An ignition source, such as a spark, ignites the fuel-air mixture, causing a positive pressure build-up in the combustion chamber. The positive pressure causes the valves to close, leaving only the exit pipe of the combustion chamber as a pressure relief opening. The combustion chamber and exit pipe geometry determine the resonant frequency of the combustor. The pressure wave from the initial combustion travels down the exit pipe at sonic velocity. As this wave exits the combustion chamber, most of the flue gases present in the chamber will be carried with it into the exit pipe. Flue gases remaining in the combustion chamber immediately begin to cool. The contraction of the cooling gases and the momentum of gases in the exit pipe create a vacuum inside the chamber that opens the valves and allows more fuel and air into the chamber. While the fresh charge of fuel-air enters the chamber, the pressure wave reaches the end of the exit pipe and is partially reflected from the open end of the pipe. The fresh fuel-air charge is ignited by residual combustion and/or heat. The resulting combustion starts another cycle.

Typical pulse combustors operate at 30 to 100 cycles per second and emit resonant sound, which must be considered in their application. Pulse combustion produces heat, as does any other combustion process. The pulses produce high convective heat transfer rates.

Heating Values

Combustion releases thermal energy or heat. The quantity of heat generated by complete combustion of a unit of specific fuel is constant and is termed the *heating value, heat of combustion*, or *caloric value* of that fuel. The heating value of a fuel can be determined by measuring the heat evolved during combustion of a known quantity of the fuel in a calorimeter, or it can be estimated from chemical analysis of the fuel and the heating value of the various chemical elements in the fuel.

Higher heating value, gross heating value, or *total heating value* is determined when water vapor in fuel combustion products is condensed and the latent heat of vaporization is included in the fuel's heating value. Conversely, *lower heating value* or *net heating value* is obtained when latent heat of vaporization is not included. When the heating value of a fuel is specified without designating higher or lower, it generally means the higher heating value in the United States. (Lower heating value is mainly used for internal combustion engine fuels.)

Heating values are usually expressed in Btu/ft³ for gaseous fuels, Btu/gal for liquid fuels, and Btu/lb for solid fuels. Heating values are always given in relation to a certain reference temperature and pressure, usually 60, 68, or 77 °F and 14.696 psia, depending on the particular industry practice. Heating values of several substances in common fuels are listed in Table 4.

Table 2 Combustion Reactions of Common Fuel Constituents

Constituent	Molecular Symbol	Ultimate CO_2, %	Dew Point, °F	Flue Gas from Stoichiometric Combustion			
				Unit Volume		Unit Mass	
				Unit Volume of Fuel		Unit Mass of Fuel	
				CO_2	H_2O	CO_2	H_2O
Carbon (to CO)	C	—	—	—	—	—	—
Carbon (to CO_2)	C	29.30	—	—	—	3.664	—
Carbon monoxide	CO	34.70	—	1.0	—	1.571	—
Hydrogen	H_2	—	162	—	1.0	—	8.937
Methane	CH_4	11.73	139	1.0	2.0	2.744	2.246
Ethane	C_2H_6	13.18	134	2.0	3.0	2.927	1.798
Propane	C_3H_8	13.75	131	3.0	4.0	2.994	1.634
Butane	C_4H_{10}	14.05	129	4.0	5.0	3.029	1.550
—	C_nH_{2n+2}	—	128 to 127	n	$n+1$	$\dfrac{44.01n}{14.026n + 2.016}$	$\dfrac{18.016(n+1)}{14.026n + 2.016}$
Ethylene	C_2H_4	15.05	125	2.0	2.0	3.138	1.285
Propylene	C_3H_6	15.05	125	3.0	3.0	3.138	1.285
—	C_nH_{2n}	15.05	125	n	n	3.138	1.285
Acetylene	C_2H_2	17.53	103	2.0	1.0	3.384	0.692
—	C_nH_{2m}	—	—	n	$m/2$	$\dfrac{6.005n}{6.005n + 1.008m}$	$\dfrac{9.008m}{6.005n + 1.008m}$
				SO_x	H_2O	SO_2	H_2O
Sulfur (to SO_2)	S	—	—	1.0 SO_2	—	1.998	—
Sulfur (to SO_3)	S	—	—	1.0 SO_3	—	2.497	—
Hydrogen sulfide	H_2S	—	125	1.0 SO_2	1.0	1.880	0.528

Adapted, in part, from *Gas Engineers Handbook* (1965).
Note: Dew point is obtained from Figure 2.

Table 3 Flammability Limits and Ignition Temperatures of Common Fuels in Fuel-Air Mixtures

Substance	Molecular Symbol	Lower Flammability Limit, %	Upper Flammability Limit, %	Ignition Temperature, °F	References
Carbon (activated coke)	C			1220	Hartman (1958)
Carbon monoxide	CO	12.5	74	1128	Scott *et al.* (1948)
Hydrogen	H_2	4.0	75.0	968	Zabetakis (1956)
Methane	CH_4	5.0	15.0	1301	*Gas Engineers Handbook* (1965)
Ethane	C_2H_6	3.0	12.5	968 to 1166	Trinks (1947)
Propane	C_3H_8	2.1	10.1	871	NFPA (1962)
Butane, n	C_4H_{10}	1.86	8.41	761	NFPA (1962)
Ethylene	C_2H_4	2.75	28.6	914	Scott *et al.* (1948)
Propylene	C_3H_6	2.00	11.1	856	Scott *et al.* (1948)
Acetylene	C_2H_2	2.50	81	763 to 824	Trinks (1947)
Sulfur	S	—	—	374	Hartman (1958)
Hydrogen sulfide	H_2S	4.3	45.50	558	Scott *et al.* (1948)

Flammability limits adapted from Coward and Jones (1952). All values corrected to 60°F, 30 in. Hg, dry.

With incomplete combustion, not all fuel is completely oxidized, and the heat released is less than the heating value of the fuel. Therefore, the quantity of heat produced per unit of fuel consumed decreases, implying lower combustion efficiency.

Not all heat released during combustion can be used effectively. The greatest heat loss is in the form of increased temperature (thermal energy) of hot exhaust gases above the temperature of incoming air and fuel. Other heat losses include radiation and convection heat transfer from the outer walls of combustion equipment to the environment.

Altitude Compensation

Air at altitudes above sea level has less density and, therefore, less oxygen per unit volume. Combustion at altitudes above sea level will have less available oxygen to burn with fuel unless compensation is made for the altitude. Combustion will occur, but the excess air will be reduced. If excess air is reduced excessively by an increase in altitude, combustion will be incomplete or cease.

Altitude compensation is achieved by matching the fuel and air supply rates to attain complete combustion without too much excess air or too much fuel. Fuel and air supply rates can be matched by increasing the air supply rate to the combustion zone or by decreasing the fuel supply rate to the combustion zone. The air supply rate can be increased with a combustion air blower, and the fuel supply rate can be reduced by decreasing the fuel input (derating).

Power burners use combustion air blowers and can increase the air supply rate to compensate for altitude. The combustion zone can be pressurized to attain the same air density in the combustion chamber as that attained at sea level.

Derating can be used as an alternative to power combustion. In the United States, the fuel gas codes generally do not require derating of nonpower burners at altitudes up to 2000 ft. At altitudes above 2000 ft, burners should be derated 4% for each 1000 ft above sea level (NFPA and AGA 1992). Chimney or vent operation also must be considered at high altitudes (see Chapter 31 of the 1992 ASHRAE *Handbook—Systems and Equipment*).

FUEL CLASSIFICATION

Generally, hydrocarbon fuels are classified according to physical state (gaseous, liquid, or solid). Different types of combustion

Table 4 Heating Values of Substances Occurring in Common Fuels

Substance	Molecular Symbol	Higher Heating Values[a], Btu/lb	Lower Heating Values[a], Btu/lb	Specific Volume[b], ft³/lb
Carbon (to CO)	C	3950	3950	—
Carbon (to CO_2)	C	14,093	14,093	—
Carbon monoxide	CO	4347	4347	13.5
Hydrogen	H_2	61,095	51,623	188.0
Methane	CH_4	23,875	21,495	23.6
Ethane	C_2H_6	22,323	20,418	12.5
Propane	C_3H_8	21,669	19,937	8.36
Butane	C_4H_{10}	21,321	19,678	6.32
Ethylene	C_2H_4	21,636	20,275	—
Propylene	C_3H_6	21,048	19,687	9.01
Acetylene	C_2H_2	21,502	20,769	14.3
Sulfur (to SO_2)	S	3980	3980	—
Sulfur (to SO_3)	S	5940	5940	—
Hydrogen sulfide	H_2S	7097	6537	11.0

Adapted from *Gas Engineers Handbook* (1965).
[a] All values corrected to 60°F, 30 in. Hg, dry. For gases saturated with water vapor at 60°F, deduct 1.74% of the value to adjust for gas volume displaced by water vapor.
[b] At 32°F and 29.92 in. Hg.

Table 5 Propane-Air and Butane-Air Gas Mixtures

Btu/ft³	Propane-Air[a]			Butane-Air[b]		
	%Gas	%Air	Sp Gr	%Gas	%Air	Sp Gr
500	19.8	80.2	1.103	15.3	84.7	1.155
600	23.8	76.2	1.124	18.4	81.6	1.186
700	27.8	72.2	1.144	21.5	78.5	1.216
800	31.7	68.3	1.165	24.5	75.5	1.248
900	35.7	64.3	1.185	27.6	72.4	1.278
1000	39.7	60.3	1.206	30.7	69.3	1.310
1100	43.6	56.4	1.227	33.7	66.3	1.341
1200	47.5	52.5	1.248	36.8	63.2	1.372
1300	51.5	48.5	1.268	39.8	60.2	1.402
1400	55.5	44.5	1.288	42.9	57.1	1.433
1500	59.4	40.6	1.309	46.0	54.0	1.464
1600	63.4	36.6	1.330	49.0	51.0	1.495
1700	67.4	32.6	1.350	52.1	47.9	1.526
1800	71.3	28.7	1.371	55.2	44.8	1.557

Adapted from *Gas Engineers Handbook* (1965).
[a] Values used for calculation: 2522 Btu/ft³; 1.52 specific gravity.
[b] Values used for calculation: 3261 Btu/ft³; 2.01 specific gravity.

equipment are usually needed to burn fuels in different physical states. Gaseous fuels can be burned in premix or diffusion burners that take advantage of the gaseous state. Liquid fuel burners must include a means for atomizing or vaporizing fuel into small droplets or a vapor for burning and must provide adequate mixing of fuel and air. Finally, solid fuel combustion equipment must (1) heat fuel to vaporize sufficient volatiles to initiate and sustain combustion, (2) provide residence time to complete combustion, and (3) provide space for ash containment.

Principal fuel applications include space heating and cooling of residential, commercial, industrial, and institutional buildings; service water heating; steam generation; and refrigeration. Major fuels for these applications are natural and liquefied petroleum gases, fuel oils, diesel and gas turbine fuels (for total energy applications), and coal.

Fuels of limited use, such as manufactured gases, kerosene, briquettes, wood, and coke, are not discussed here. Fuel choice is based on one or more of the following:

1. Fuel factors
   - Availability, including dependability of supply
   - Convenience of use and storage
   - Economy
   - Cleanliness
2. Combustion equipment factors
   - Operating requirements
   - Cost
   - Service requirements
   - Ease of control

GASEOUS FUELS

Although various gaseous fuels have been used as energy sources in the past, heating and cooling applications are presently limited to natural and liquefied petroleum gases.

Types and Properties

Natural gas is a nearly odorless and colorless gas that accumulates in the upper parts of oil and gas wells. Raw natural gas is a mixture of methane (55 to 98%), higher hydrocarbons (primarily ethane), and noncombustible gases. Some constituents, principally water vapor, hydrogen sulfide, helium, liquefied petroleum gases, and gasoline are removed prior to distribution.

Typical compositions of natural gas used as fuel include: methane, CH_4 (70 to 96%); ethane, C_2H_6 (1 to 14%); propane, C_3H_8 (0 to 4%); butane, C_4H_{10} (0 to 2%); pentane, C_5H_{12} (0 to 0.5%); hexane, C_6H_{14} (0 to 2%); carbon dioxide, CO_2 (0 to 2%); oxygen, O_2 (0 to 1.2%); and nitrogen, N_2 (0.4 to 17%).

The composition of a natural gas depends on its geographical source. The composition of gas distributed in a given location can vary slightly since the gas is drawn from various sources, but a fairly constant heating value is usually maintained for control and safety. Local gas utilities are the best sources of current gas composition data for a particular area.

Heating values of natural gases vary from 900 to 1200 Btu/ft³; the usual range is 1000 to 1050 Btu/ft³ at sea level. Unknown heating values for particular gases can be calculated from composition data and Table 4.

For safety purposes, odorants (such as mercaptans) are added to natural gas and LPG to give them noticeable odors.

Liquefied petroleum gases (LPG) consist primarily of propane and butane, usually obtained as a byproduct of oil refinery operations or by stripping natural gas. Propane and butane are gaseous under usual atmospheric conditions, but can be liquefied by moderate pressures at normal temperatures.

Three liquefied petroleum gases—butane, propane, and a mixture of the two—are commercially available as fuels.

Commercial propane consists primarily of propane but generally contains about 5 to 10% propylene. It has a heating value of about 21,560 Btu/lb or about 2500 Btu/ft³ of gas. At atmospheric pressure, commercial propane has a boiling point of about −40°F. The low boiling point of propane makes it usable during winter in the northern United States and in Canada. Tank heaters and vaporizers permit its use in colder climates and where high fuel flow rates are required. It is available in cylinders, bottles, tank trucks, or tank cars.

Commercial butane consists primarily of butane but may contain up to 5% butylene. It has a heating value of about 21,180 Btu/lb or about 3200 Btu/ft³. At atmospheric pressure, commercial butane has a relatively high boiling point of about 32°F. Therefore, butane cannot be used in cold weather unless the gas temperature is maintained above 32°F or butane partial pressure is decreased by dilution with lower boiling point gases. Butane is usually available in bottles, tank trucks, or tank cars, but not in cylinders.

Commercial propane-butane mixtures with varying ratios of propane and butane are available. Their properties generally fall between those of the unmixed fuels. Propane-air and butane-air mixtures are used in place of natural gas in small communities and by natural gas companies at peak loads. Table 5 lists heating values and specific gravities for various fuel-air ratios.

Manufactured gases are combustible gases produced from coal, coke, oil, liquefied petroleum gases, or natural gas. For more detailed information, see *Gas Engineers Handbook* (1965). These fuels are used primarily for industrial in-plant operations or as specialty fuels (*e.g.*, acetylene for welding).

LIQUID FUELS

Significant liquid fuels include various fuel oils for firing combustion equipment and engine fuels for total energy systems. Liquid fuels, with few exceptions, are mixtures of hydrocarbons derived by refining crude petroleum. In addition to hydrocarbons, crude petroleum usually contains small quantities of sulfur, oxygen, nitrogen, vanadium, other trace metals, and impurities such as water and sediment. Refining produces a variety of fuels and other products. Nearly all lighter hydrocarbons are refined into fuels (*e.g.*, liquefied petroleum gases, gasoline, kerosene, jet fuels, diesel fuels, and light heating oils). Heavy hydrocarbons are refined into residual fuel oils and other products (*e.g.*, lubricating oils, waxes, petroleum coke, and asphalt).

Crude petroleums from different oil fields vary in hydrocarbon molecular structure. Crude is paraffin-base (principally chain-structured paraffin hydrocarbons), naphthene- or asphaltic-base (containing relatively large quantities of saturated ring-structural naphthenes), aromatic-base (containing relatively large quantities of unsaturated, ring-structural aromatics), or mixed- or intermediate-base (between paraffin- and naphthene-base crudes). Except for heavy fuel oils, crude type has little significant effect on resultant products and combustion applications.

Types of Fuel Oils

Fuel oils for heating are broadly classified as *distillate* fuel oils (lighter oils) or *residual* fuel oils (heavier oils). ASTM has established specifications for fuel oil properties which subdivide the oils into various grades. Grades No. 1 and 2 are distillate fuel oils. Grades 4, 5 (Light), 5 (Heavy), and 6 are residual fuel oils. Specifications for the grades are based on required characteristics of fuel oils for use in different types of burners. The ANSI standard specification for fuel oils is ASTM *Standard* D396-86.

Grade No. 1 is a light distillate intended for vaporizing-type burners. High volatility is essential to continued evaporation of the fuel oil with minimum residue.

Grade No. 2 is a heavier (API Gravity) distillate than No. 1. It is used primarily with pressure-atomizing (gun) burners that spray the oil into a combustion chamber. The atomized oil vapor mixes with air and burns. This grade is used in most domestic burners and many medium capacity commercial-industrial burners.

Grade No. 4 is an intermediate fuel that is considered either a light residual or a heavy distillate. Intended for burners that atomize oils of higher viscosity than domestic burners can handle, its permissible viscosity range allows it to be pumped and atomized at relatively low storage temperatures. A dewaxed No. 2 oil with a pour point of $-58\,°F$ is supplied only to areas where regular No. 2 oil would jell.

Grade No. 5 (Light) is a residual fuel of intermediate viscosity for burners that handle fuel more viscous than No. 4 without preheating. Preheating may be necessary in some equipment for burning and, in colder climates, for handling.

Grade No. 5 (Heavy) is a residual fuel more viscous than No. 5 (Light), but intended for similar purposes. Preheating is usually necessary for burning and, in colder climates, for handling.

Grade No. 6, sometimes referred to as Bunker C, is a high viscosity oil used mostly in commercial and industrial heating. It requires preheating in the storage tank to permit pumping, and additional preheating at the burner to permit atomizing.

Low sulfur residual oils are marketed in many areas to permit users to meet sulfur dioxide emission regulations. These fuel oils are produced (1) by refinery processes that remove sulfur from the oil (hydrodesulfurization), (2) by blending high sulfur residual oils with low sulfur distillate oils, or (3) by a combination of these methods. These oils have significantly different characteristics than regular residual oils. For example, the viscosity-temperature relationship can be such that low sulfur fuel oils have viscosities of No. 6 fuel oils when cold, and of No. 4 when heated. Therefore, normal guidelines for fuel handling and burning can be altered when using these fuels.

Fuel oil grade selection for a particular application is usually based on availability and economic factors, including fuel cost, clean air requirements, preheating and handling costs, and equipment cost. Installations with low firing rates and low annual fuel consumption cannot justify the cost of preheating and other methods that use residual fuel oils. Large installations with high annual fuel consumption cannot justify the premium cost of distillate fuel oils. Disagreements on economy occur somewhere in between.

Characteristics of Fuel Oils

Characteristics that determine grade classification and suitability for given applications are: (1) flash point, (2) viscosity, (3) pour point, (4) water and sediment content, (5) carbon residue, (6) ash, (7) distillation qualities, (8) specific gravity, (9) sulfur, (10) carbon-hydrogen content, and (11) heating value. Not all of these are included in ASTM *Standard* D396-86.

Viscosity is an oil's resistance to flow. It is significant because it indicates the ease at which oil will flow or be pumped and the ease of atomization. Approximate viscosities of fuel oils are shown in Figure 1.

Flash point is the maximum temperature at which an oil can be stored and handled. Minimum permissible flash point is usually prescribed by state and municipal laws.

Pour point is the lowest temperature at which a fuel can be stored and handled. Higher pour point fuels can be used when heated storage and piping facilities are provided.

Water and *sediment* should be low to prevent fouling the facilities. Sediment accumulates on filter screens and burner parts. Water in distillate fuels can cause tanks to corrode and emulsions to form in residual oil.

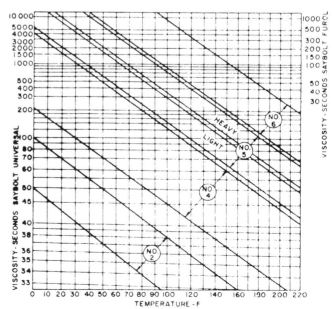

Fig. 1   Approximate Viscosity of Fuel Oils

Carbon residue is obtained by a test in which the oil sample is destructively distilled in the absence of air. When commercial fuels are used in proper burners, this residue has almost no relationship to soot deposits, except indirectly when deposits are formed by vaporizing burners.

Ash is the noncombustible material in an oil. An excessive amount indicates the presence of materials that cause high wear on burner pumps.

The *distillation* test shows the volatility and ease of vaporization of a fuel.

Specific gravity is the ratio of a fuel oil density to the density of water at a specific temperature. Differences in fuel oil viscosities are caused by variations in the concentrations of fuel oil constituents and different refining methods. Specific gravities cover a range in each grade, with some overlap between distillate and residual grades. API gravity, a parameter widely used in place of specific gravity, is obtained by the following formula, in which specific gravity at 60/60°F indicates the ratio of the mass of a given volume of oil at 60°F to the same volume of water at 60°F:

$$\text{deg API} = \frac{141.5}{\text{Sp Gr at } 60/60°F} - 131.5 \qquad (1)$$

The API gravity of water at 60°F is 10.0.

Air pollution considerations are important in determining the allowable sulfur content of fuel oils. Sulfur content is frequently limited by legislation aimed at reducing sulfur oxide emissions from combustion equipment. These laws require sulfur content to be below a certain level, usually 1.0, 0.5, or 0.3%.

Sulfur in fuel oils is also undesirable because of the corrosiveness of sulfur compounds in the flue gas. While low-temperature corrosion can be minimized by maintaining the stack at temperatures above the dew point of the flue gas, this limits the overall thermal efficiency of combustion equipment.

For certain industrial applications, the sulfur content of a fuel must be limited because of adverse effects on product quality. These include direct-fired metallurgical applications where work is in the combustion zone.

Table 6 lists the sulfur levels of some marketed fuel oils.

Heating value is an important property, although ASTM *Standard* D396-86 does not list it as one of the criteria for fuel oil classification. Heating value can generally be correlated with the API gravity. Table 7 shows the relationship between heating value, API gravity, and density for several oil grades. In the absence of more specific data, heating values can be calculated as shown in the *North American Combustion Handbook* (1965):

Higher Heating Value, Btu/lb
$$= 22{,}320 - 3780 \text{ (Specific Gravity)}^2 \qquad (2)$$

Distillate fuel oils (Grades 1 and 2) have a carbon-hydrogen content of 84 to 86% carbon, with the remainder predominantly hydrogen. The heavier grades (4, 5, and 6) may contain up to 88% carbon and as little as 11% hydrogen. An approximate relationship for determining the hydrogen content of fuel oils is:

$$\text{Hydrogen, \%} = 26 - (15 \times \text{Specific Gravity}) \qquad (3)$$

ASTM *Standard* D396-86 is more a classification than a specification, distinguishing between six generally nonoverlapping grades, one of which characterizes any commercial fuel oil. Quality is not defined, as a refiner might control it; for example, the standard lists the distillation temperature 90% point for Grade No. 2 as having a maximum of 640°F, whereas commercial practice rarely exceeds 600°F.

Types and Properties of Liquid Fuels for Engines

The primary stationary engine fuels are diesel and gas turbine oils, natural gases, and liquefied petroleum gases. Other fuels include sewage gas, manufactured gas, and gas mixtures. Gasoline and the JP series of gas turbine fuels are rarely used for stationary engines.

Only properties of diesel and gas turbine fuel oils are covered here; properties of natural and liquefied petroleum gases are found in the section Gaseous Fuels. For properties of gasolines and JP turbine fuel, consult texts on internal combustion engines and gas turbines (Bibliography). Properties of currently marketed gasolines can be found in the latest volumes of *Mineral Industry Surveys, Motor Gasolines*, issued semiannually by the Bureau of Mines.

Properties of the three grades of diesel fuel oils are listed in ASTM *Specification* D975.

Grade No. 1-D includes the class of volatile fuel oils from kerosene to intermediate distillates. These fuels are used in high-speed engines with frequent and relatively wide variations in loads and speeds and where abnormally low fuel temperatures are encountered.

Grade No. 2-D includes the class of lower volatility distillate gas oils. These fuels are used in high-speed engines with relatively high loads and uniform speeds, or in engines not requiring fuels with the higher volatility or other properties specified for Grade No. 1-D.

Grade No. 4-D covers the class of more viscous distillates and blends of these distillates with residual fuel oils. These fuels are used in low- and medium-speed engines involving sustained loads at essentially constant speed.

Property specifications and test methods for No. 1-D, 2-D, and 4-D diesel fuel oils are essentially identical to specifications of No. 1, 2, and 4 fuel oils, respectively. However, diesel fuel oils have an additional specification for cetane number, which measures ignition quality and influences combustion roughness. Cetane number requirements depend on engine design, size, speed and load variations, and starting and atmospheric conditions. The increase in cetane number over values actually required does not materially improve engine performance. Thus, the cetane number should be as low as possible to assure maximum fuel availability. ASTM *Standard* D975 states several methods for estimating cetane number from other fuel oil properties.

Table 6 Sulfur Content of Marketed Fuel Oils

Grade of Oil	No. 1	No. 2	No. 4	No. 5 (Light)	No. 5 (Heavy)	No. 6
Total fuel samples	123	158	13	15	16	96
Sulfur content,						
% Wt. Min.	0.002	0.03	0.46	0.90	0.57	0.32
Max.	0.380	0.64	1.44	3.50	2.92	4.00
No. samples with S						
over 0.3%	1	32	13	15	16	96
over 0.5%	0	1	11	15	16	93
over 1.0%	0	0	3	9	11	60
over 3.0%	0	0	0	2	0	8

Derived from *Burner Oil Fuels* (1974).

Table 7 Typical Gravity and Heating Value of Standard Grades of Fuel Oil

Grade No.	Gravity API	Density, lb/gal	Heating Value, Btu/gal
1	38 to 45	6.950 to 6.675	137,000 to 132,900
2	30 to 38	7.296 to 6.960	141,800 to 137,000
4	20 to 28	7.787 to 7.396	148,100 to 143,100
5L	17 to 22	7.940 to 7.686	150,000 to 146,800
5H	14 to 18	8.080 to 7.890	152,000 to 149,400
6	8 to 15	8.448 to 8.053	155,900 to 151,300

ASTM *Specification* D2880 for gas turbine fuel oils relates gas turbine fuel oil grades to fuel and diesel fuel oil grades. Test methods for determining properties of gas turbine fuel oils are essentially identical to those for fuel oils. However, gas turbine specifications contain quantity limits on some trace elements that may be present. These limits are intended to prevent excessive corrosion in gas turbine engines. For a detailed discussion of fuels for gas turbines and combustion in gas turbines, see Chapters 5 and 9, respectively (Hazard 1971).

SOLID FUELS

Solid fuels include coal, coke, wood, and waste products of industrial and agricultural operations. Of these, only coal is widely used for heating and cooling applications.

The complex composition of coal makes classification difficult. Chemically, coal consists of carbon, hydrogen, oxygen, nitrogen, sulfur, and a mineral residue, ash. Chemical analysis provides some indication of coal quality, but does not define its burning characteristics sufficiently. The coal user is principally interested in the available heat per unit mass of coal and the amount of ash and dust produced, but is also interested in handling and storing properties and burning characteristics. A description of coal qualities and their characteristics can be obtained from the U.S. Bureau of Mines.

Types of Coals

Commonly accepted definitions for classifying coals are listed in Table 8. However, this classification is arbitrary, since there are no distinct demarcation lines between coal types.

Anthracite is a clean, dense, hard coal that creates little dust in handling. It is comparatively hard to ignite, burning freely once started. It is noncaking and burns uniformly and smokelessly with a short flame.

Semianthracite has a higher volatile content than anthracite. It is not as hard and ignites more easily. Otherwise, its properties are similar to those of anthracite.

Bituminous coal includes many types of coal with distinctly different compositions, properties, and burning characteristics. Coals range from high-grade bituminous of the East to low-rank coals of the West. Caking properties range from coals that melt or become fully plastic, to those from which volatiles and tars are distilled without changing form (classed as noncaking or free-burning). Most bituminous coals are strong and nonfriable enough to permit screened sizes to be delivered free of fines. Generally, they ignite easily and burn freely. Flame length is long and varies with different coals. If improperly fired, much smoke and soot are possible, especially at low burning rates.

Semibituminous coal is soft and friable, and handling creates fines and dust. It ignites slowly and burns with a medium flame length. Its caking properties increase as volatile matter increases, but the coke formed is weak. With only half the volatile matter content of bituminous coals, burning produces less smoke; hence, it is sometimes called smokeless coal.

Subbituminous coals, found in the West, are high in moisture when mined and tend to break up as they dry or are exposed to the weather; they are likely to ignite spontaneously when piled or stored. They ignite easily and quickly, have a medium-length flame, and are noncaking and free-burning. The lumps tend to break into small pieces if poked. Very little smoke and soot are formed.

Lignite is woody in structure, very high in moisture when mined, of low heating value, and clean to handle. It has a greater tendency than subbituminous coals to disintegrate as it dries and is also more liable to ignite spontaneously. Because of its high moisture, freshly mined lignite ignites slowly and is noncaking. The char left after moisture and volatile matter are driven off burns very easily, like charcoal. The lumps tend to break up in the fuel bed and pieces of char that fall into the ashpit continue to burn. Very little smoke or soot forms.

Table 8 Classification of Coals by Rank[a]

	Class	Group	Limits of Fixed Carbon or Energy Content, Mineral-Matter-Free Basis	Requisite Physical Properties
I	Anthracite	1. Metaanthracite	Dry FC, 98% or more (Dry VM, 2% or less)	Nonagglomerating
		2. Anthracite	Dry FC, 92% or more, and less than 98% (Dry VM, 8% or less, and more than 2%)	
		3. Semianthracite	Dry FC, 86% or more, and less than 92% (Dry VM, 14% or less, and more than 8%)	
II	Bituminous[d]	1. Low-volatile bituminous coal	Dry FC, 78% or more, and less than 86% (Dry VM, 22% or less, and more than 14%)	Either agglomerating[b] or nonweathering[f]
		2. Medium-volatile bituminous coal	Dry FC, 69% or more, and less than 78% (Dry VM, 31% or less, and more than 22%)	
		3. High-volatile A bituminous coal	Dry FC, less than 69% (Dry VM, more than 31%), and moist[c], about 14,000 Btu[e] or more	
		4. High-volatile B bituminous coal	Moist[c], about 13,000 Btu or more, and less than 14,000[e]	
		5. High-volatile C bituminous coal	Moist, about 11,000 Btu or more, and less than 13,000[e]	
III	Subbituminous	1. Subbituminous A coal	Moist, about 11,000 Btu or more, and less than 13,000[e]	Both weathering and nonagglomerating[b]
		2. Subbituminous B coal	Moist, about 9500 Btu or more, and less than 11,000[e]	
		3. Subbituminous C coal	Moist, about 8300 Btu or more, and less than 95000[e]	
IV	Lignitic	1. Lignite	Moist, less than 8300 Btu	Consolidated
		2. Brown coal	Moist, less than 8300 Btu	Unconsolidated

Adapted from ASTM *Standard* D388 Rev A-91, Standard Classification of Coals by Rank.

Legend: FC = Fixed Carbon; VM = Volatile Matter.

[a] This classification does not include a few coals of unusual physical and chemical properties which come within the limits of fixed carbon or Btu of high-volatile bituminous and subbituminous ranks. All these coals either contain less than 48% dry, mineral-matter-free fixed carbon, or have more than about 15,500 moist, mineral-matter-free Btu.

[b] If agglomerating, classify in low-volatile group of the bituminous class.

[c] Moist (Btu) refers to coal containing its natural bed moisture but not including visible water on the coal surface.

[d] There may be noncaking varieties in each group of the bituminous class.

[e] Coals having 69% or more fixed carbon on the dry, mineral-matter-free basis shall be classified according to fixed carbon, regardless of Btu.

[f] There are three varieties of coal in the high-volatile C bituminous coal group: Variety 1, agglomerating and nonweathering; Variety 2, agglomerating and weathering; Variety 3, nonagglomerating and nonweathering.

Characteristics of Coals

The characteristics of coals that determine classification and suitability for given applications are the proportions of (1) volatile matter, (2) fixed carbon, (3) moisture, (4) sulfur, and (5) ash. Each of these is reported in the proximate analysis. Coal analyses can be reported on several bases: as-received, moisture-free (dry) or dry, and mineral-matter-free (or ash-free). As-received is applicable for combustion calculations; dry and mineral-matter-free, for classification purposes.

Volatile matter is driven off as gas or vapor when the coal is heated according to a standardized temperature test. It consists of a variety of organic gases, generally resulting from distillation and decomposition. Volatile products given off by coals when heated differ materially in the ratios by weight of the gases to oils and tars. No heavy oils or tars are given off by anthracite, and very small quantities are given off by semianthracite. As volatile matter in the coal increases to as much as 40% of the coal (dry and ash-free basis), increasing amounts of oils and tars are released. However, for coals of higher volatile content, the quantity of oils and tars decreases and is relatively low in subbituminous coals and in lignite.

Fixed carbon is the combustible residue left after the volatile matter is driven off. It is not all carbon. Its form and hardness are an indication of fuel coking properties and, therefore, guide the choice of combustion equipment. Generally, fixed carbon represents that portion of fuel that must be burned in the solid state.

Moisture is difficult to determine accurately because a sample can lose moisture on exposure to the atmosphere, particularly when reducing the sample size for analysis. To correct for this loss, total moisture content of a sample is customarily determined by adding the moisture loss obtained when air-drying the sample and the measured moisture content of the dried sample. Moisture does not represent all of the water present in coal; water of decomposition (combined water) and of hydration are not given off under standardized test conditions.

Ash is the noncombustible residue remaining after complete coal combustion. Generally, the weight of ash is slightly less than that of mineral matter before burning.

Sulfur is an undesirable constituent in coal, since the sulfur oxides formed when it burns contribute to air pollution and cause combustion system corrosion. Table 9 lists the sulfur content of typical coals. Legislation has limited the sulfur content of coals burned in certain locations.

Heating values may be reported on an as-received, dry, dry and mineral-matter-free, or moist and mineral-matter-free basis. Higher heating values of coals are frequently reported with their proximate analysis. When more specific data are lacking, the higher heating value of higher quality coals can be calculated by the Dulong formula:

Higher Heating Value, Btu/lb
$$= 14{,}544C + 62{,}028[H - (O/8)] + 4050S \qquad (4)$$

where C, H, O, and S are the mass fractions of carbon, hydrogen, oxygen, and sulfur in the coal.

Other important parameters in judging coal suitability include: (1) ultimate analysis, (2) ash-fusion temperature, (3) grindability, and (4) free-swelling index.

Ultimate analysis is another method of reporting coal composition.

Percentages of C, H, O, N, S, and ash in the coal sample are reported. Ultimate analysis is used for detailed fuel studies and for computing a heat balance when required in heating device testing. Typical ultimate analyses of various coals are shown in Table 9.

Ash-fusion temperature indicates the fluidity of the ash at elevated temperatures. It is helpful in selecting coal to be burned in a particular furnace and in estimating the possibility of ash handling and slagging problems.

The *grindability index* indicates the ease with which a coal can be pulverized and is helpful in estimating ball mill capacity with various coals. These are two common methods for determining the index—hardgrove and ball mill.

The *free swelling index* denotes the extent of coal swelling on combustion on a fuel bed and indicates the coking characteristics of coal.

COMBUSTION CALCULATIONS

Calculations of the quantity of air required for combustion and the quantity of flue gas products generated during combustion are frequently needed for sizing system components and as input to efficiency calculations. Other calculations, such as values for excess air and theoretical CO_2, are useful in estimating combustion system performance.

Frequently, combustion calculations can be simplified by using molecular mass as the basis for calculations. Molecular mass can be expressed in any mass units. The pound molecular weight or pound mole is frequently used where the pound molecular weight of a compound equals the molecular weight of the compound expressed in pounds. (The molecular mass of a compound equals the sum of the atomic mass of each element in the compound.) A molecular mass of any substance contains the same number of molecules as a molecular mass of any other substance.

Corresponding to measurement standards common to the industries, calculations involving gaseous fuels are generally based on volume, while calculations involving liquid and solid fuels are generally based on mass.

Some calculations described here require data on concentrations of carbon dioxide, carbon monoxide, and oxygen in the flue gas. Gas analyses for CO_2, CO, and O_2 can be obtained by volumetric chemical analysis and other analytical techniques including electromechanical cells used in portable electronic flue gas analyzers.

Table 9   Typical Ultimate Analyses for Coals

Rank	As Received, Btu/lb	Constituents, Percent by Weight					
		Oxygen	Hydrogen	Carbon	Nitrogen	Sulfur	Ash
Anthracite	12,700	5.0	2.9	80.0	0.9	0.7	10.5
Semianthracite	13,600	5.0	3.9	80.4	1.1	1.1	8.5
Low-volatile bituminous	14,350	5.0	4.7	81.7	1.4	1.2	6.0
Medium-volatile bituminous	14,000	5.0	5.0	81.4	1.4	1.5	6.0
High-volatile bituminous *A*	13,800	9.3	5.3	75.9	1.5	1.5	6.5
High-volatile bituminous *B*	12,500	13.8	5.5	67.8	1.4	3.0	8.5
High-volatile bituminous *C*	11,000	20.6	5.8	59.6	1.1	3.5	9.4
Subbituminous *B*	9000	29.5	6.2	52.5	1.0	1.0	9.8
Subbituminous *C*	8500	35.7	6.5	46.4	0.8	1.0	9.6
Lignite	6900	44.0	6.9	40.1	0.7	1.0	7.3

Air Required for Combustion

Stoichiometric or theoretical air is the exact quantity of air required to provide oxygen for complete combustion.

The three most prevalent components in hydrocarbon fuels are completely combusted by the following reactions:

$$C + O_2 \rightarrow CO_2$$
$$H_2 + 0.5 O_2 \rightarrow H_2O$$
$$S + O_2 \rightarrow SO_2$$

In the above reactions, C, H_2, S, and O_2 can be taken to represent 1 lb mole of carbon, hydrogen, sulfur, and oxygen, respectively. Using approximate atomic mass (C = 12, H = 1, S = 32, and O = 16), 12 lb of C are oxidized by 32 lb of O to form 44 lb of CO_2, 2 lb of H are oxidized by 16 lb of O to form 18 lb of H_2O, and 32 lb of S are oxidized by 32 lb of O to form 64 lb of SO_2. These relationships can be extended to include other hydrocarbon compounds.

The mass of dry air required to supply a given quantity of the oxygen is 4.32 times the mass of the oxygen. The mass of oxygen or air required to oxidize fuel constituents listed in Table 1 were calculated on this basis. Oxygen contained in the fuel, except that contained in ash, should be deducted from the amount of oxygen required, since this oxygen is already combined with fuel components. Also, water vapor is always present in atmospheric air, and when the mass of air to be supplied for combustion is calculated, allowance should be made for the water vapor.

As stated previously, combustion calculations are sometimes based on volume. Avogadro's law shows that, for any gas, one mole occupies the same volume at a given temperature and pressure. Therefore, in reactions involving gaseous compounds, the gases react in volume ratios identical to the pound mole ratios. That is, for the oxidation of hydrogen in the above reaction, one volume (or one lb mole) of hydrogen reacts with one-half volume (or one-half lb mole) of oxygen to form one volume (or one lb mole) of water vapor.

The volume of air required to supply a given volume of oxygen is 4.79 times the volume of oxygen. The volumes of oxygen or of dry air required to oxidize the fuel constituents listed in Table 1 were calculated on this basis. Volume ratios are not given for fuels that do not exist in vapor form at reasonable temperatures or pressures. Again, oxygen contained in the fuel should be deducted from the quantity of oxygen required, since this oxygen is already combined with fuel components. Allowance should be made for the water vapor that increases the volume of dry air by 1 to 3%.

From the relationships just described, the theoretical mass of dry air required for stoichiometric combustion of a unit mass of any hydrocarbon fuel is:

$$W_a = 0.0144 (8 C + 24 H + 3 S - 3 O) \qquad (5)$$

where C, H, S, and O are the mass percentages of carbon, hydrogen, sulfur, and oxygen in the fuel, respectively.

Analyses of gaseous fuels are generally based on hydrocarbon components rather than elemental content.

If the fuel analysis is based on mass, the theoretical mass of dry air required for stoichiometric combustion of a unit mass of gaseous fuels is:

$$W_a = 2.47 \, CO + 34.28 \, H_2 + 17.24 \, CH_4 + 16.09 \, C_2H_6$$
$$+ 15.68 \, C_3H_8 + 15.47 \, C_4H_{10} + 13.27 \, C_2H_2$$
$$+ 14.78 \, C_2H_4 + 6.08 \, H_2S - 4.32 \, O_2 \qquad (6)$$

If the fuel analysis is reported on a volumetric or molecular basis, it is simplest to calculate air requirements based on volume and, if necessary, convert to mass. The theoretical volume of air required for stoichiometric combustion of a unit volume of gaseous fuels is:

$$V_a = 2.39 \, CO + 2.39 \, H_2 + 9.57 \, CH_4 + 16.75 \, C_2H_6$$
$$+ 23.95 \, C_3H_8 + 31.14 \, C_4H_{10} + 11.96 \, C_2H_2$$
$$+ 14.38 \, C_2H_4 + 7.18 \, H_2S - 4.78 \, O_2$$
$$+ 30.47 \, illuminants \qquad (7)$$

where CO, H_2, and so forth are the volumetric fractions of each constituent in the fuel gas.

Illuminants include a variety of compounds not separated by usual gas analysis. The principal illuminants, in addition to ethylene (C_2H_4) and acetylene (C_2H_2), included in Equation (7), and the dry air required for combustion, per unit volume of each gas, are: propylene (C_3H_6), 21.44; butylene (C_4H_8), 28.58; pentene (C_5H_{10}), 35.73; benzene (C_6H_6), 35.73; toluene (C_7H_8), 42.88; and xylene (C_8H_{10}), 50.02. Since toluene and xylene are normally scrubbed from the gas before distribution, they can be disregarded in computing air required for combustion of gaseous fuels. Since the percentage of illuminants present in gaseous fuels is small, the values can be lumped together, and an approximate value of 30 unit volume of dry air per unit volume of gas can be used. If ethylene and acetylene are included as illuminants, use the suggested value of 20 unit volume of dry air/unit volume of gaseous illuminants.

For many combustion calculations, only approximate values of air requirements are necessary. If approximate values for theoretical air are sufficient, or if complete information on the fuel is not available, the values in Tables 10 and 11 can be used. Another frequently used value for estimating air requirements is that 0.9 ft^3 of air is required for 100 Btu of fuel.

Excess air, in addition to the amount theoretically required for combustion, must be supplied to most practical combustion systems to ensure complete combustion.

$$\text{Excess air, } \% = 100 \left(\frac{\text{Air supplied} - \text{Theoretical air}}{\text{Theoretical air}} \right) \qquad (8)$$

The excess air level at which a combustion process operates significantly affects its overall efficiency. Too much excess air dilutes flue gas excessively, lowering its heat transfer temperature and increasing sensible flue gas loss. Conversely, if the level of excess air is too low, incomplete combustion and loss of unburned combustible gases from the equipment can result. Highest combustion efficiency is usually obtained when just enough excess air is supplied and properly mixed with combustible gases to ensure complete combustion. The general practice is to supply from 5 to

Table 10   Approximate Air Requirements for Stoichiometric Combustion of Fuels

Type of Fuel	Air Required for Stoichiometric Combustion		Approx. Precision, %	Exceptions
	lb/lb Fuel	ft^3/Unit Fuel[a]		
Solid	Btu/lb × 0.00073	Btu/lb × 0.0097	3	Fuels containing more than 30% water
Liquid	Btu/lb × 0.00071	Btu/lb × 0.0094	3	Results low for gasoline and kerosene
Gas	Btu/lb × 0.00067	Btu/ft^3 × 0.0089	5	300 Btu/ft^3 or less

Data based on Shnidman (1954).
[a] Units for solid and liquid fuels in Btu/lb; for gas in Btu/ft^3.

**Table 11 Approximate Air Requirements
for Stoichiometric Combustion of Various Fuels**

Type of Fuel	Theoretical Air Required for Combustion
Solid fuels	lb/lb fuel
Anthracite	9.6
Semibituminous	11.2
Bituminous	10.3
Lignite	6.2
Coke	11.2
Liquid fuels	lb/gal fuel
No. 1 fuel oil	103
No. 2 fuel oil	106
No. 5 fuel oil	112
No. 6 fuel oil	114
Gaseous fuels	ft^3/ft^3 fuel
Natural gas	9.6
Butane	31.1
Propane	24.0

50% excess air, the exact amount depending on the type of fuel burned, combustion equipment, and other factors.

The amount of dry air supplied per unit mass of fuel burned can be obtained from Equation (9), which is reasonably precise for most solid and liquid fuels. Values for CO_2, CO, and N_2 are percentages by volume from the flue gas analysis, and C is the mass of carbon burned per unit mass of fuel, corrected for carbon in the ash.

Dry air supplied, unit mass per unit mass of fuel

$$= \frac{C\,(3.04\,N_2)}{(CO_2 + CO)} \qquad (9)$$

This value, together with the value for theoretical air, can be used in Equation (8) to determine excess air.

Equation (10) can be used to calculate excess air on a percentage basis for fuel gas from unit volumes of stoichiometric combustion products and air, and volumetric analysis of the flue gas.

$$\text{Excess air, \%} = [(U - CO_2)/(CO_2)] \times 100\,(P/A) \qquad (10)$$

where

U = ultimate carbon dioxide of flue gases resulting from stoichiometric combustion, %
CO_2 = carbon dioxide content of flue gases, %
P = dry products from stoichiometric combustion, unit volume per unit volume of gas burned
A = air required for stoichiometric combustion, unit volume per unit volume of gas burned

As the ratio P/A is approximately 0.9 for most natural gases, a value of 90 can be substituted for 100 (P/A) in Equation (10) for rough calculation.

Because excess air calculations are almost invariably made from flue gas analysis results and theoretical air requirements are not always known, another convenient method of expressing the relation of Equation (8) is:

$$\text{Excess air, \%} = \frac{100[O_2 - (CO/2)]}{0.264\,N_2 - [O_2 - (CO/2)]} \qquad (11)$$

where O_2, CO, and N_2 are percentages (by volume) from the flue gas analysis.

Theoretical CO_2

The theoretical CO_2, ultimate CO_2, or maximum CO_2 concentration attainable in combustion products of a hydrocarbon fuel with air is the CO_2 concentration obtained when the fuel was completely burned with the theoretical quantity of air, or zero excess air. Theoretical CO_2 varies with the carbon-hydrogen ratio of the fuel. For combustion with excess air present, theoretical CO_2 values can be calculated from a flue-gas analysis:

Theoretical CO_2, % =

$$U = \frac{\%\ CO_2 \text{ in flue gas sample}}{1 - (\%\ O_2 \text{ in flue gas sample}/20.95)} \qquad (12)$$

Table 12 gives approximate theoretical CO_2 values for stoichiometric combustion of several common types of fuel, as well as CO_2 values attained with different amounts of excess air. In practice, desirable CO_2 values depend on the excess air, fuel, firing method, and other considerations.

Quantity of Flue Gas Produced

The mass of dry flue gas produced per pound of fuel burned is required in heat loss and efficiency calculations. This mass is equal to the sum of the mass of: (1) fuel (minus ash retained in the furnace), (2) air theoretically required for combustion, and (3) excess air. For solid fuels, this mass, determined from the flue gas analysis, is:

$$\frac{11\,CO_2 + 8\,O_2 + 7(CO + N_2)}{3(CO_2 + CO)}\ C \text{ lb/lb of fuel} \qquad (13)$$

where CO_2, O_2, CO, and N_2 are percentages by volume from the flue gas analysis, and C is the mass of carbon burned per pound of fuel, corrected for carbon in the ash.

Total dry gas volume of flue gases resulting from combustion of one unit volume of gaseous fuels for various percentages of CO_2 is:

Dry flue gas, unit volume per unit volume of fuel gas =

Unit volume of CO_2 produced per unit volume of gas
burned × 100/(Percent CO_2 by analysis) (14)

**Table 12 Approximate Maximum Theoretical
(Stoichiometric) CO_2 Values, and CO_2 Values
of Various Fuels with Different Percentages of Excess Air**

Type of Fuel	Theoretical or Maximum CO_2, %	Percent CO_2 at Given Excess Air Values		
		20	40	60
Gaseous fuels				
Natural gas	12.1	9.9	8.4	7.3
Propane gas (Commercial)	13.9	11.4	9.6	8.4
Butane gas (Commercial)	14.1	11.6	9.8	8.5
Mixed gas (Natural and carbureted water gas)	11.2	12.5	10.5	9.1
Carbureted water gas	17.2	14.2	12.1	10.6
Coke oven gas	11.2	9.2	7.8	6.8
Liquid fuels				
No. 1 and 2 fuel oil	15.0	12.3	10.5	9.1
No. 6 fuel oil	16.5	13.6	11.6	10.1
Solid fuels				
Bituminous coal	18.2	15.1	12.9	11.3
Anthracite	20.2	16.8	14.4	12.6
Coke	21.0	17.5	15.0	13.0

Excess air quantity can be estimated by subtracting the quantity of dry flue gases resulting from stoichiometric combustion from the total volume of flue gas.

Water Vapor and Dew Point of Flue Gas

Water vapor in flue gas is the total of the water contained in the fuel; water contained in the stoichiometric, excess, and dilution air admitted; and water produced from the combustion of hydrogen or hydrocarbons in the fuel. The amount of water vapor in the stoichiometric combustion products may be calculated from the fuel burned by using the water data in Table 2.

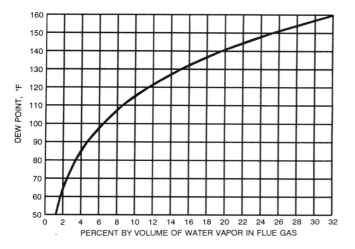

Fig. 2   Water Vapor and Dew Point of Flue Gas
Adapted from *Gas Engineers Handbook* (1965). Printed with permission of Industrial Press and American Gas Association.

The dew point is the temperature at which condensation begins and can be determined from Figure 2. The volume percentage of water vapor P_{wv} in the flue gas can be determined as follows:

$$P_{wv} = \frac{V_w}{\dfrac{100\,V_c}{P_c} + V_w} \qquad (15)$$

where

V_w = total water vapor volume (from fuel; stoichiometric, excess, and dilution air; and combustion)
V_c = unit volume of CO_2 produced per volume of fuel gas
P_c = percent CO_2 in flue gas

Using Figure 3, the dew points of solid, liquid, or gaseous fuels may be estimated. For example, to find the dew point of flue gas resulting from the combustion of a solid fuel with a weight ratio (hydrogen to carbon-plus-sulfur) of 0.088 and sufficient excess air to produce 11.4% oxygen in the flue gas, start with the weight ratio of 0.088. Proceed vertically to the intersection of the solid fuels curve and then to the theoretical dew point of 115 °F on the dew-point scale. See the dotted lines. Follow the curve fixed by this point (down and to the right) to 11.4% oxygen in the flue gas (on the abscissa). The actual dew point is 93 °F and is found on the dew-point scale.

The dew point of the flue gas from natural gas of 1020 Btu/ft³ higher heating value (HHV) with 6.3% oxygen or 31.5% air can be estimated. Proceed vertically to the intersection of the gaseous fuels curve and then to the theoretical dew point of 139 °F on the dew-point scale. Follow the curve fixed by this point to 6.3% oxygen or 31.5% air in the flue gas. The actual dew point is 127 °F.

The presence of sulfur dioxide, and particularly sulfur trioxide, influences the vapor pressure of condensate in flue gas, and the dew point can be raised by as much as 25 to 75 °F, as shown in

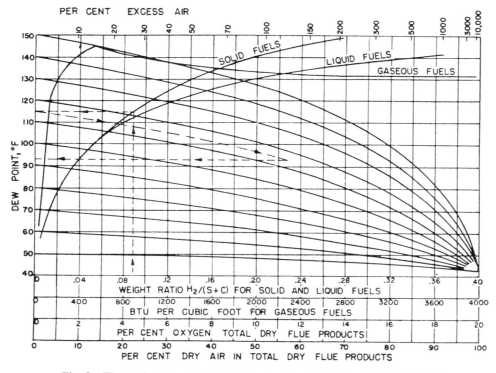

Fig. 3   Theoretical Dew Points of Combustion Products of Industrial Fuels
Adapted from *Gas Engineers Handbook* (1965). Printed with permission of Industrial Press and American Gas Association.

Figure 4. To illustrate the use of Figure 4, for a manufactured gas with a HHV of 550 Btu/ft³ containing 15 grains of sulfur per 100 ft³ being burned with 40% excess air, the proper curve in Figure 4 is determined as follows:

$$\frac{\text{Grains S per 100 ft}^3}{\text{Btu per cubic foot}} \times 100 \quad \text{or} \quad \frac{15}{500} \times 100 = 2.73 \quad (16)$$

This curve lies between the 0- and 3-curves and is close to the 3-curve. The dew point for any percentage of excess air from zero to 100% can be determined on this curve. For this flue gas with 40% excess air, the dew point is about 160°F, instead of 127°F for zero sulfur at 40% excess air.

Sample Combustion Calculations

Applications of the preceding equations and tables are illustrated by Examples 1 and 2.

Example 1. Analysis of flue gases from the burning of a natural gas is 10.0% CO_2, 3.1% O_2, and 86.9% N_2 by volume. Analysis of the fuel is 90% CH_4, 5% N_2, and 5% C_2H_6 by volume. Find U (maximum theoretical percent CO_2) and the percent of excess air.
Solution: From Equation (12)

$$U = \frac{10.0}{1 - (3.1/20.95)} = 11.74\% \ CO_2$$

From Equation (10), Excess air $= \dfrac{(11.74 - 10.0)\ 90}{10} = 15.66\%$

Example 2. For the same analysis as in Example 1, find, per ft³ of fuel gas, the ft³ of dry air required for combustion, ft³ of each constituent in the flue gases, and total volume of dry and wet flue gases.
Solution: From Equation (7) or Table 10, the volume of dry air required for combustion is

$$9.57\ CH_4 + 16.75\ C_2H_6 = (9.57 \times 0.90) + (16.75 \times 0.05)$$
$$= 9.45\ \text{ft}^3/\text{ft}^3 \text{ gas}$$

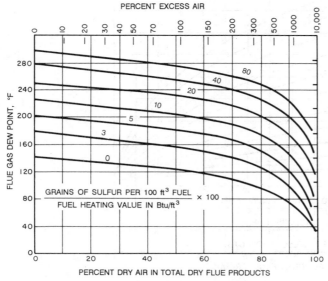

PERCENT EXCESS AIR

FLUE GAS DEW POINT, °F

PERCENT DRY AIR IN TOTAL DRY FLUE PRODUCTS

$$\frac{\text{GRAINS OF SULFUR PER 100 ft}^3 \text{ FUEL}}{\text{FUEL HEATING VALUE IN Btu/ft}^3} \times 100$$

Fig. 4   Influence of Sulfur Oxides on Flue Gas Dew Point
(Stone 1969)

From Table 1, constituents per ft³ of flue gas are:

Nitrogen, N_2

From methane	$(0.9\ CH_4)(9.57 - 2.0)$	= 6.81
From ethane	$(0.05\ C_2H_6)(16.75 - 3.5)$	= 0.61
Nitrogen in fuel		= 0.05
Nitrogen in excess air	$0.791 \times 0.157 \times 9.45$	= 1.17
Total nitrogen		8.64 ft³

Oxygen, O_2

Oxygen in excess air	$0.209 \times 0.157 \times 9.45$	= 0.31 ft³

Carbon dioxide, CO_2

From methane	$(0.9\ CH_4)(1.0)$	= 0.90
From ethane	$(0.05\ C_2H_6)(2.0)$	= 0.10
Total carbon dioxide		1.00 ft³

Water vapor, H_2O (does not appear in some flue gas analyses)

From methane	$(0.9\ CH_4)(2.0)$	= 1.8
From ethane	$(0.05\ C_2H_6)(3.0)$	= 0.15
Total water vapor		1.95 ft³

Total volume of dry gas per ft³ of gas

$$8.64 + 0.31 + 1.00 = 9.95\ \text{ft}^3$$

Total volume of wet gases per ft³ of gas (neglecting water vapor in combustion air)

$$9.95 + 1.95 = 11.90\ \text{ft}^3$$

The ft³ of dry flue gas per ft³ of fuel gas can also be computed from Equation (14)

$$(1.00)(100)/10.0 = 10.0\ \text{ft}^3$$

EFFICIENCY CALCULATIONS

Usually, in analyzing heating appliance efficiency, an energy balance is made that accounts (as much as possible) for disposition of all thermal energy released by combustion of the fuel quantity consumed. The various components of this balance are generally expressed in terms of Btu/lb of fuel burned or as a percentage of its higher heating value. Major components and their calculation methods are:

1. Useful heat, or heat transferred to the heated medium; for convection heating equipment, this value q_1 is computed as the product of the mass rate of flow and enthalpy change.

2. Heat loss as sensible heat in the dry flue gases

$$q_2 = w_g c_{pg}(t_g - t_a) \quad (17)$$

3. Heat loss in water vapor in products formed by combustion

$$q_3 = (9H_2/100)[(h)_{tg} - (h_f)_{ta}] \quad (18)$$

4. Heat loss in water vapor in the combustion air

$$q_4 = Mw_a[(h)_{tg} - (h_g)_{ta}] \quad (19)$$

5. Heat loss from incomplete combustion

$$q_5 = 10,143\ C[CO/(CO_2 + CO)] \quad (20)$$

6. Heat loss from unburned carbon in the ash or refuse

$$q_6 = 14,600[(C_u/100) - C] \quad (21)$$

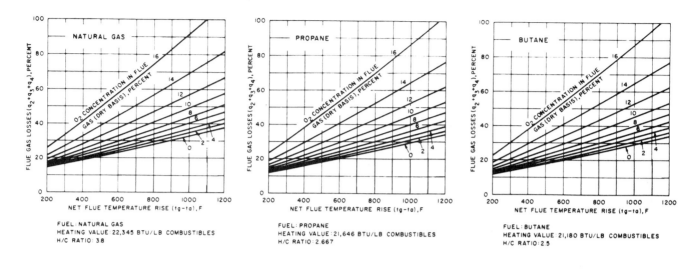

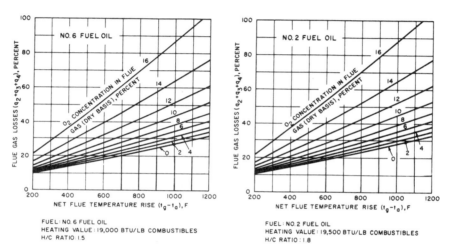

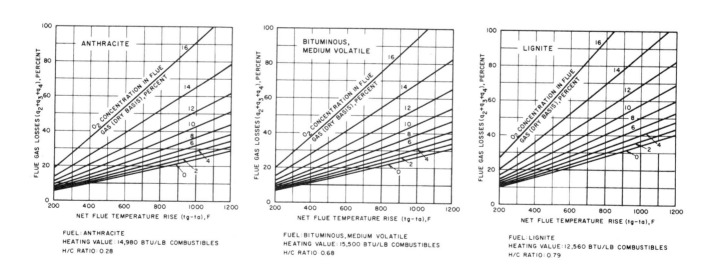

Fig. 5   Flue Gas Losses with Various Fuels
(Flue gas temperature shown. Loss based on 65°F room temperature.)

7. Unaccounted-for losses q_7.

Symbols used in Equations (17) through (21) are:

q_1 = useful heat, Btu/lb of fuel
q_2 = heat loss in dry flue gases, Btu/lb of fuel
q_3 = heat loss in water vapor from combustion of hydrogen, Btu/lb of fuel
q_4 = heat loss in water vapor in combustion air, Btu/lb of fuel
q_5 = heat loss from incomplete combustion of carbon, Btu/lb of fuel
q_6 = heat loss from unburned carbon in ash, Btu/lb of fuel
q_7 = unaccounted-for heat losses, Btu/lb of fuel
$(h_g)_{ta}$ = enthalpy of saturated steam at combustion air temperature, Btu/lb
$(h)_{tg}$ = enthalpy of superheated steam at flue gas temperature and 1 psia, Btu/lb
$(h_f)_{ta}$ = enthalpy of saturated water at air temperature, Btu/lb
t_g = temperature of flue gases at exit of heating device, °F
t_a = temperature of combustion air, °F
w_a = mass of combustion air per lb of fuel used [from Equations (5), (6), (8), (9), (10), and (11)], lb
w_g = mass of dry flue gas per lb of fuel [from Equation (13)], lb
c_{pg} = mean specific heat of flue gases at constant pressure [c_{pg} ranges from 0.242 to 0.254 Btu/lb·°F for flue gas temperatures from 300 to 1000°F], Btu/lb·°F
H_2 = hydrogen in fuel, percent by weight (from ultimate analysis of fuel)
M = specific humidity of combustion air, unit mass of water vapor per unit mass of dry air
CO, CO_2 = carbon monoxide and carbon dioxide in flue gases, percent by volume
C = mass of carbon burned, corrected for carbon in ash, lb/lb of fuel

$$C = (WC_u - W_aC_a)/(100W) \qquad (22)$$

where

C_u = percentage of carbon in fuel by mass from ultimate analysis
W_a = mass of ash and refuse
C_a = percent of combustible in ash by mass (combustible in ash is usually considered to be carbon)
W = mass of fuel used

Item 1, useful heat, is generally measured for a particular piece of combustion equipment.

Flue gas loss is the sum of Items 2 through 6. However, for clean burning gas- and oil-fired equipment, Items 5 and 6 are usually not determined, and flue gas loss is taken to be the sum of Items 2, 3, and 4.

Flue gas losses (the sum of items 2, 3, and 4 of the heat balance) can be determined with sufficient precision for most purposes from curves in Figure 5, if O_2 content and flue gas temperature are known. Values of the losses were computed from typical ultimate analyses, and assuming 1% water vapor (by mass) in the combustion air. Curves for medium-volatile bituminous coal can be used for high-volatile bituminous coal with no appreciable error.

Generally, Item 5 is negligible for modern combustion equipment in good operating condition.

Item 6 is generally negligible for gas and oil firing, but should be determined for coal-firing applications.

Item 7 consists primarily of radiation and convection losses from combustion equipment surfaces and losses caused by incomplete combustion, not included in Items 5 and 6. Incomplete combustion is determined by subtracting the sum of Items 1 through 6 from the fuel heating value.

Radiation and convection losses are not usually determined by direct measurement. But if the heating appliance is located within the heated space, radiation and convection losses can be consid-

ered useful rather than lost heat and can be omitted from heat loss calculations or added to Item 1.

If CO is present in flue gases, small amounts of unburned hydrogen and hydrocarbons may also be present. The small losses caused by incomplete combustion of these gases would be included in Item 7, if the item was determined by subtracting Items 1 through 6 from the fuel heating value.

The overall thermal efficiency of combustion equipment is defined as:

$$\text{Thermal efficiency, \% } = 100\,\frac{\text{Useful heat}}{\text{Heating value of fuel}} \qquad (23)$$

Efficiency can be estimated for equipment, where Item 7 is small or radiation and convection are useful heat, by:

Thermal efficiency, % =

$$100\,\frac{\text{Heating value of fuel} - (q_2 + q_3 + q_4 + q_5 + q_6)}{\text{Heating value of fuel}} \qquad (24)$$

The thermal efficiency of a gas appliance can be computed with sufficient precision, using heating values based on gas volume, by:

$$\eta = 100[(Q_g - Q_{fl})]/Q_g \qquad (25)$$

where

η = thermal efficiency, %
Q_g = gross heating value of fuel gas per unit volume
Q_{fl} = flue gas losses per unit volume of fuel gas

To produce heat efficiently by burning any common fuel, flue gas losses must be minimized by: (1) providing adequate heat-absorbing surface in the appliance, (2) maintaining heat transfer surfaces clean on both fire and water or air sides, and (3) reducing excess air to the minimum level consistent with complete combustion and discharge of combustion products.

Seasonal Efficiency

The method just presented is useful for calculating the steady-state efficiency of a heating system or device. Unfortunately, the seasonal efficiency of a combustion heating system can be significantly different from the steady-state efficiency. The primary factor affecting the seasonal efficiency is flue loss during the burner-off period. The warm stack that exists at the end of the firing period can cause an airflow to be maintained in the stack while the burner is off. This airflow can remove heat from furnace and heat exchanger components, from the structure itself, and from pilot flames. Also, if combustion air is drawn from the heated space within the structure, there is a loss of heated air that must be replaced with cold infiltrated air. For discussion of seasonal efficiency, see Bibliography and Chapter 29 of the 1992 ASHRAE *Handbook—Systems and Equipment*.

PRACTICAL COMBUSTION CONSIDERATIONS

Air Pollution

Combustion processes constitute the largest single source of air pollution. Pollutants can be grouped in four categories:

1. Products of incomplete fuel combustion
   - Combustible aerosols (solid and liquid), including smoke, soot, and organics, and excluding ash
   - Carbon monoxide, CO
   - Gaseous hydrocarbons, HC

2. Oxides of nitrogen (generally grouped and referred to as NO_x)
   - Nitric oxide, NO
   - Nitrogen dioxide, NO_2
3. Emissions resulting from fuel contaminants
   - Sulfur oxides, primarily sulfur dioxide, SO_2, and small quantities of sulfur trioxide, SO_3
   - Ash
   - Trace metals
4. Emissions resulting from additives
   - Combustion-controlling additives
   - Other additives

Emissions of products of incomplete combustion and nitrogen oxides are directly related to the combustion process and can be controlled to some extent by process modification. Emissions of fuel contaminants are related to fuel selection and are slightly affected by the combustion process. Additives are used to achieve specific objectives; however, possible emission problems must be considered in an overall evaluation of the merits of using additives.

Emission levels of products of incomplete fuel combustion can be reduced by ensuring adequate excess air, improving mixing of air and fuel (increasing turbulence, improving distribution, and improving liquid fuel atomization), increasing residence time in the hot combustion zone (possibly by decreasing firing rate), increasing combustion zone temperatures (to speed reactions), and avoiding quenching the flame before reactions are completed.

Nitrogen oxides form during the combustion process, either (1) by thermal fixation (reaction of nitrogen and oxygen at high combustion temperatures), or (2) from fuel nitrogen (oxidation of organic nitrogen in fuel molecules). Unfortunately, high excess air and high flame temperature techniques for ensuring complete fuel combustion and, therefore, low emissions of incomplete combustion products, tend to promote increased NO_x formation.

Some methods of reducing NO_x emissions include (Murphy and Putnam 1985):

- Burner adjustment
- Flame inserts
- Staged combustion and delayed mixing
- Secondary air baffling
- Catalytic and radiant burners

Radiation screens (flame inserts) surrounding the flame will absorb radiation to reduce flame temperature and retard NO_x formation.

Two-stage or biased firing is the only technique that reduces NO_x produced both by thermal fixation and fuel nitrogen in industrial and utility applications. The lean or air-deficient primary combustion zone retards NO_x formation early in the combustion process (when NO_x forms most readily from fuel nitrogen), and peak temperatures are avoided, reducing thermal fixation. Two-stage firing is not practical in gas appliances because of increased complexity and dangerous interstage CO levels.

Proprietary appliance burners with no flame inserts have been developed and produced to comply with California's Air Quality Management Districts' NO_x emission restrictions. NO_x control techniques have also been widely applied to large boilers.

The practical minimums for NO_x emissions differ for different types of equipment as reflected in the regulatory maximum NO_x requirements. Table 13 lists the U.S. Environmental Protection Agency's maximums for large stationary sources (Engdahl and Barrett 1977, Barrett *et al.* 1977). The EPA's automobile emission standard is 1.0 g/mile of NO_2, equivalent to 750 ng/J of NO_x emission. California's maximum is 0.4 g/mile, equivalent to 300 ng/J. California's Air Quality Management Districts for the South Coast (Los Angeles) and the San Francisco Bay Area limit NO_x emission to 40 ng/J of useful heat for some natural gas-fired residential heating appliances.

Table 13 Maximum Emission Levels for Large Fossil Fuel-Fired Steam Generators

Species	Emission Level, lb/million Btu (ppm) Type of Fuel		
	Solid	Liquid	Gaseous
Particulates	0.21 (260)	0.21 (260)	0.21 (260)
SO_2	1.23 (1550)	0.79 (1000)	0.79 (1000)
NO_x as NO_2	0.72 (600)	0.30 (250)	0.21 (170)

Group 1 (larger than 2.56×10^8 Btu/h of fuel input, which may be equivalent to an electrical output of about 30 MW). Pollutants in lb/million Btu of fuel consumption. Numbers in parentheses refer to approximate stack conditions in parts per million, assuming stoichiometric combustion, mass/mass on a mass basis, volume/volume on a volume basis (EPA 1971a, 1971b).

For further discussion of air pollution aspects of fuel combustion, see EPA (1971a, 1971b).

Condensation and Corrosion

Fuel-burning systems that cycle on and off to meet demand cool down during the off-cycle. When the appliance starts again, condensate forms briefly on surfaces until they are heated above the dew-point temperature. Low-temperature corrosion occurs in system components (heat exchangers, flues, vents, chimneys) when their surfaces remain below the dew-point temperature of flue gas constituents (water vapor, sulfides, chlorides, fluorides, etc.) long enough to cause condensation. Corrosion increases as condensate dwell time increases.

Acids in the flue gas condensate are the principle substances responsible for low-temperature corrosion in fuel-fired systems. Sulfuric, hydrochloric, and other acids are formed when acidic compounds in fuel and air combustion products combine with condensed moisture in appliance heat exchangers, flues, or vents. Corrosion can be avoided by maintaining these surfaces above the flue gas dew point.

In high-efficiency, condensing-type appliances and economizers, the flue gas temperatures are intentionally reduced below the flue gas dew-point temperatures to achieve efficiencies approaching 100%. In these systems, the surfaces subjected to condensate must be made of corrosion-resistant materials. The most corrosive conditions exist at the leading edge of the condensing region, especially those areas that experience evaporation during each cycle (Strickland *et al.* 1987). Regions in which condensate partially or completely drains away before evaporation are less severely attacked than regions from which condensate does not drain before evaporation. Drainage of condensate retards the concentration of acids on system surfaces.

The metals most resistant to condensate corrosion are stainless-steel alloys, with high chrome and molybdenum content, and nickel-chrome alloys, with high molybdenum content (Stickford *et al.* 1988). Aluminum experiences general corrosion rather than pitting when exposed to flue gas condensate. If applied in sufficiently thick cross section to allow for metal loss, aluminum can be used in condensing regions. Most ceramic and high-temperature polymer materials resist the corrosive effects of flue gas condensate. These materials may have application in the condensing regions, if they can meet the structural and temperature requirements of a particular application.

In coal-fired power plants, the rate of corrosion for carbon steel condensing surfaces by the mixed acids (primarily sulfuric and hydrochloric) is reported to be maximum at about $122 \pm 18°F$ (Davis 1987). Mitigation techniques have included (1) acid neutralization with a base, for example, NH_3 or $Ca(OH)_2$; (2) use of protective linings of glass-filled polyester or coal-tar epoxy; and (3) replacement of steel with molybdenum bearing stainless steels, nickel alloys, polymers, or other corrosion-resistant materials.

Other elements in residual fuel oils and coals that contribute to high-temperature corrosion include sodium, potassium, and vanadium. Each fuel-burning system component should be evaluated during installation, or when modified, to determine the potential for corrosion and the means to retard corrosion (Paul *et al.* 1988).

Soot

Soot deposits on flue surfaces of a boiler or heater act as an insulating layer over the surface, reducing heat transfer to the water or air. Soot can also clog flues, reduce draft and available air, and prevent proper combustion. Proper burner adjustment can minimize soot accumulation.

REFERENCES

ASTM 1986. Standard specification for fuel oils. ASTM *Standard* D396-86.

Barrett, R.E., R.B. Engdahl, and D.W. Locklin. 1977. "Space heating and steam generation." In *Air pollution*, Academic Press, New York.

Bonne, U. and A. Patani. 1982. Combustion system performance analysis and simulation study. Report to GRI-81/0093 (PB 83-161 406). Honeywell SSPL, Bloomington, MN.

Bonne, U., J.E. Janssen, and R.H. Torborg. 1977. "Efficiency and relative operating cost of central combustion heating system, IV, Oil fired residential systems." Symposium on Seasonal Operating Performance of Central Forced Air Heating and Cooling Systems. ASHRAE Semi-annual Meeting, Chicago, IL.

Bonne, U., L. Katzman, and G.E. Kelly. 1976. Effect of reducing excess firing rate on seasonal efficiency of 26 Boston oil-fired heating systems. Conference on Efficiency in HVAC Equipment and Components II, Purdue University, IN, 81, 376.

Bonne, U., J.E. Janssen, L.W. Nelson, and R.H. Torborg. 1976. "Control of overall thermal efficiency of combustion heating systems." 16th International Symposium on Combustion, MIT, Cambridge, MA, 37.

Coward, H.F. and G.W. Jones. 1952. Limits of flammability of gases and vapors. U.S. Bureau of Mines Bulletin 503, Washington, D.C.

Davis, J.R., ed. 1987. *Metals handbook*, 9th ed., Vol. 13. ASM International, Metals Park, OH.

Engdahl, R.B. and R.E. Barrett. 1977. "Fuels and their utilization." In *Air pollution*, Academic Press, New York.

EPA. 1971a. Standards of performance for new stationary sources, Group I, Federal Register 36, August 17.

EPA. 1971b. Standards of performance for new stationary sources, Group I, Part II, Federal Register 36, December 23.

Gas engineer's handbook. 1965. The Industrial Press, New York.

Hartman, I. 1958. "Dust explosions." In *Mechanical engineers' handbook*, 6th ed., Section 7, 41-48. McGraw-Hill, New York.

Hazard, H.R. 1971. "Gas turbine fuels." In *Gas turbine handbook*, Gas Turbine Publications, Inc., Stamford, CT.

Janssen, J.E. and U. Bonne. 1976. Improvement of seasonal efficiency of residential heating systems. ASME Winter Annual Meeting, Paper 76-WA/Fu-7, New York.

Murphy, M.J. and A.A. Putnam. 1985. Burner technology bulletin: Control of NO_x emissions from residential gas appliances. Report GRI-85/0132. Battelle Columbus Division for Gas Research Institute.

NFPA. 1962. Fire-hazard properties of flammable liquids, gases and volatile solids, Tables 6-126. In *Fire protection handbook*, 12th ed., Boston, 6-131 ff.

NFPA and AGA. 1992. NFPA *Standard* 54-1992/ANSI Z223.1-1992, National Fuel Gas Code 8.1.2. Quincy, MA and Arlington, VA.

North American combustion handbook. 1965. The North American Manufacturing Co., Cleveland, OH.

Paul, D.D., A.L. Rutz, S.G. Talbert, J.J. Crisafolli, G.R. Whitacre, and R.D. Fischer. 1988. User's manual for Vent-II Ver. 3.0—A dynamic microcomputer program for analyzing gas venting systems, Report GRI-88/0304. Battelle Columbus Division for Gas Research Institute.

Scott, G.S. *et al.* 1948. Determination of ignition temperatures of combustible liquids and gases. *Analytical Chemistry* 20:238-41.

Shelton, E.M. 1974. *Burner oil fuels*. Petroleum Products Survey 86, U.S. Department of the Interior, Bureau of Mines.

Shnidman, L. 1954. *Gaseous fuels*. American Gas Association, New York.

Stickford, G.H., S.G. Talbert, and D.W. Locklin. 1987. Condensate corrosivity in residential condensing appliances. Proceedings of the International Symposium on Condensing Heat Exchangers, Paper 3, Brookhaven National Laboratory Report No. BNL 52068, 1 and 2.

Stickford, G.H., S.G. Talbert, B. Hindin, and D.W. Locklin. 1988. Research on corrosion-resistant materials for condensing heat exchangers. Proceedings of the 39th Annual International Appliance Technical Conference.

Trinks, W. 1947. Simplified calculation of radiation from non-luminous furnace gases. *Industrial Heating* 14:40-46.

Zabetakis, M.G. 1956. Research on the combustion and explosion hazards of hydrogen-water vapor-air mixtures. U.S. Bureau of Mines, Division of Explosives Technology, Progress Report 1, Washington, D.C.

BIBLIOGRAPHY

Barnett, H.C. and R.R. Hibbard. 1957. Basic considerations in the combustion of hydrocarbon fuels with air. Report No. 1300, National Advisory Committee for Aeronautics, predecessor of NASA.

Combustion engineering. 1966. Combustion Engineering, Inc., New York.

Cornelius, W. and W.G. Agnew, eds. 1972. *Emissions from continuous combustion systems*. Plenum Press, New York.

Faust, F.H. and G.T. Kaufman. 1951. *Handbook of oil burning*. Oil-Heat Institute of America, now National Oil Fuel Institute, Inc., New York.

Gas Appliance Technology Center, Gas Research Institute, Manufacturer update on status of GATC research on heat-exchanger corrosion, May 1984. Battelle-Columbus Laboratories and American Gas Association Laboratories.

Gruse, W.A. 1967. *Motor fuels, Performance and testing*. Reinhold Publishing Corp., New York.

Guthrie, V.B. 1960. *Petroleum products handbook*. McGraw-Hill Book Co., Inc., New York.

Johnson, A.J. and G.H. Auth. 1951. *Fuels and combustion handbook*. McGraw-Hill Book Co., Inc., New York.

Lewis, B. and G. von Elbe. 1987. *Combustion, flames, and explosion of gases*, 3rd ed. Academic Press, Inc., New York.

McCracken, D.J. 1970. Hydrocarbon combustion and physical properties. Report No. 1496. Ballistic Research Laboratory, Aberdeen Proving Ground, MD.

Nelson, W.L. 1949. *Petroleum refinery engineering*. McGraw-Hill Book Co., Inc., New York.

Schmidt, P.F. 1958. *Fuel oil manual*. The Industrial Press, New York.

Smith, M.L. and K.W. Stinson. 1952. *Fuels and combustion*. McGraw-Hill Book Co., Inc., New York.

Solberg, H.L., O.C. Cromer, and A.R. Spalding. *Elementary heat power*, Chapter 2. John Wiley & Sons, Inc., New York.

Spiers, H.M. 1962. *Technical data on fuel*. The British National Committee, Work Power Conference.

Steam, Its generation and use. 1963. Babcock & Wilcox Co., New York.

CHAPTER 16

REFRIGERANTS

REFRIGERANTS are the vital working fluids in refrigeration, air-conditioning, and heat pumping systems. They absorb heat from one area, such as an air-conditioned space, and reject it into another, such as outdoors, usually through evaporation and condensation processes, respectively. These phase changes occur both in absorption and mechanical vapor compression systems, but they do not occur in systems operating on a gas cycle using a fluid such as air. The design of the refrigeration equipment depends strongly on the properties of the selected refrigerant. Table 1 lists ASHRAE standard refrigerant designations.

A refrigerant must satisfy many requirements, some of which do not directly relate to its ability to transfer heat. Chemical stability under conditions of use is the most important characteristic. Safety codes may require a nonflammable refrigerant of low toxicity for some applications. Cost, availability, and compatibility with compressor lubricants and materials with which the equipment is constructed are other concerns.

The environmental consequences of a refrigerant that leaks from a system must also be considered. Because of their great stability, fully halogenated compounds, called CFCs, persist in the atmosphere for many years and eventually diffuse into the stratosphere. The molecules of CFCs, such as R11 and R12, contain only carbon and the halogens chlorine, fluorine, and/or bromine. Once in the upper atmosphere, the refrigerant molecule breaks down and releases chlorine, which destroys ozone. In the lower atmosphere, these molecules absorb infrared radiation, which may contribute to the warming of the earth. Substitution of a hydrogen atom for one or more of the halogens in a CFC molecule greatly reduces its atmospheric lifetime and lessens its environmental impact.

Refrigerant selection also involves compromises between conflicting desirable thermodynamic properties. A desirable refrigeration system has an evaporator pressure as high as possible and, simultaneously, a condenser pressure as low as possible. High evaporator pressures imply high vapor densities, and thus a greater system capacity for a given compressor. However, the efficiency at the higher pressures is lower, especially as the condenser pressure approaches the critical pressure.

Latent heat of vaporization is another important property. On a molar basis, fluids with similar boiling points have almost the same latent heat. Since the compressor operates on volumes of gas, refrigerants with similar boiling points produce similar capacities in a given compressor. On a mass basis, latent heat varies widely among fluids. The maximum efficiency of a theoretical vapor compression cycle is achieved by fluids with low vapor heat capacity. This property is associated with fluids having a simple molecular structure and low molecular weight.

Transport properties of thermal conductivity and viscosity affect the performance of heat exchangers and piping. High thermal conductivity and viscosity affect the performance of heat exchangers and piping. High thermal conductivity and low viscosity are desirable.

No single fluid satisfies all the attributes desired of a refrigerant; as a result, a variety of refrigerants is used. This chapter describes the basic characteristics of various refrigerants, and Chapter 17 lists thermophysical properties.

REFRIGERANT PROPERTIES

Physical Properties

Table 2 lists some physical properties of commonly used refrigerants, a few very low-boiling cryogenic fluids, some newer refrigerants, and some older refrigerants of historical interest. These refrigerants are arranged in increasing order of atmospheric boiling point, from helium at $-452\,°F$ to water at $212\,°F$.

Table 2 also includes the boiling point, freezing point, critical properties, surface tension, and refractive index. Of these properties, the boiling point is most important because it is a direct indicator of the temperature level at which a refrigerant can be used. The freezing point must be lower than any contemplated usage. The critical properties describe a material at the point where the distinction between liquid and gas is lost. At higher temperatures, no separate liquid phase is possible. In refrigeration cycles involving condensation, a refrigerant must be chosen that allows this change of state to occur at a temperature somewhat below the critical.

Lithium bromide-water and ammonia-water solutions. Figure 1 shows specific gravity, Figure 2 specific heat, and Figure 3 viscosity of lithium bromide-water solutions. Chapter 17 has an enthalpy concentration chart and a vapor pressure diagram for lithium bromide-water solutions; it also lists equilibrium properties of water-ammonia solutions.

Electrical Properties

Tables 3 and 4 list the electrical characteristics of refrigerants that are especially important in hermetic systems.

Sound Velocity

Table 5 gives examples of the velocity of sound in the vapor phase of several fluorinated refrigerants. The velocity increases when the temperature is increased and decreases when the pressure is increased. The velocity of sound can be calculated from the equation

$$V_a = \sqrt{g(dp/d\rho)_s} = \sqrt{g\gamma(dp/d\rho)_T} \qquad (1)$$

where

V_a = sound velocity, ft/s
g = 32.1740 ft/s^2
p = pressure, lb/ft^2, absolute
ρ = density, lb/ft^2
γ = c_p/c_v = ratio of specific heats
S = entropy, Btu/lb · °R
T = temperature, °R

The preparation of this chapter is assigned to TC 3.1, Refrigerants and Brines.

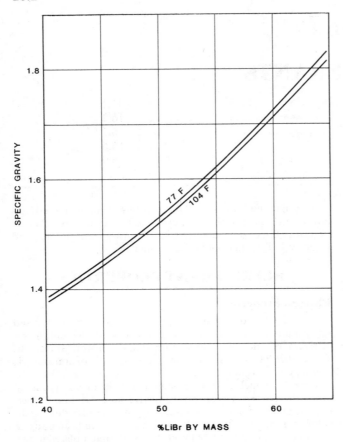

Fig. 1 Specific Gravity of Aqueous Solutions of Lithium Bromide

The sound velocity can be estimated from the tables of thermodynamic properties. The change in pressure with a change in density ($dp/d\rho$) can be estimated either at constant entropy or at constant temperature. It is simpler to estimate at constant temperature but, in this case, the heat capacity ratio must also be known as previously shown. The practical velocity of a gas in piping or through openings is limited by the sound velocity in the gas.

Latent Heat

An empirical rule of chemistry (Trouton's rule) states that the latent heat of vaporization at the boiling point on a molar basis, divided by the temperature in absolute units, is a constant for most materials. This rule is applied to refrigerants in Table 6. It applies fairly well to these refrigerants, although the result is not entirely constant. The rule helps in comparing different refrigerants and in understanding the operation of refrigeration systems.

REFRIGERANT PERFORMANCE

Chapter 1 describes several methods of calculating refrigerant performance, and Chapter 17 includes tables of thermodynamic properties of the various refrigerants.

Table 7 shows the theoretical calculated performance of a number of refrigerants for the U.S. standard cycle of 5 °F evaporation

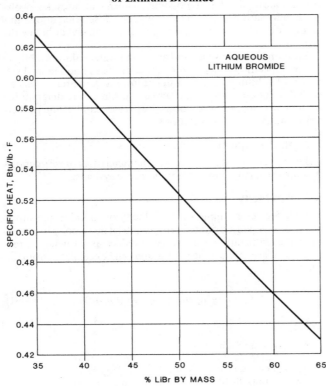

Fig. 2 Specific Heat of Aqueous Solutions of Lithium Bromide

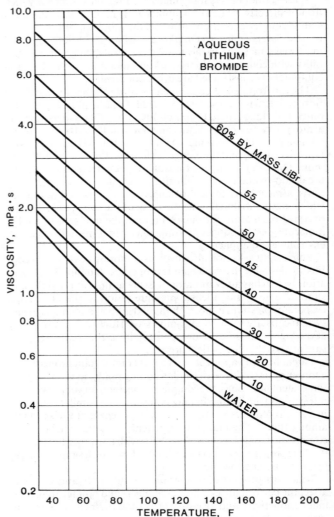

Fig. 3 Viscosities of Aqueous Solutions of Lithium Bromide

Table 1 ASHRAE Standard Designation of Refrigerants (ANSI/ASHRAE *Standard* 34-1992)

Refrigerant Number	Chemical Name or Composition	Chemical Formula
Methane Series		
10	tetrachloramethane (carbon tetrachloride)	CCl_4
11	trichlorofluoromethane	CCl_3F
12	dichlorodifluoromethane	CCl_2F_2
12B1	bromochlorodifluoromethane	$CBrClF_2$
12B2	dibromodifluoromethane	CBr_2F_2
13	chlorotrifluoromethane	$CClF_3$
13B1	bromotrifluoromethane	$CBrF_3$
14	carbon tetrafluoride	CF_4
20	trichloromethane (chloroform)	$CHCl_3$
21	dichlorofluoromethane	$CHCl_2F$
22	chlorodifluoromethane	$CHClF_2$
22B1	bromodifluoromethane	$CHBrF_2$
23	trifluoromethane	CHF_3
30	dichloromethane (methylene chloride)	CH_2Cl_2
31	chlorofluoromethane	CH_2ClF
32	difluoromethane (methylene fluoride)	CH_2F_2
40	chloromethane (methyl chloride)	CH_3Cl
41	fluoromethane (methyl fluoride)	CH_3F
50	methane	CH_4
Ethane Series		
110	hexachloroethane	CCl_3CCl_3
111	pentachlorofluoroethane	CCl_3CCl_2F
112	1,1,2,2-tetrachlorodifluoroethane	CCl_2FCCl_2F
112a	1,1,1,2-tetrachlorodifluoroethane	CCl_3CClF_2
113	1,1,2-trichlorotrifluoroethane	CCl_2FCClF_2
113a	1,1,1-trichlorotrifluoroethane	CCl_3CF_3
114	1,2-dichlorotetrafluoroethane	$CClF_2CClF_2$
114a	1,1-dichlorotetrafluoroethane	CCl_2FCF_3
114B2	1,2-dibromotetrafluoroethane	$CBrF_2CBrF_2$
115	chloropentafluoroethane	$CClF_2CF_3$
116	hexafluoroethane	CF_3CF_3
120	pentachloroethane	$CHCl_2CCl_3$
123	2,2-dichloro-1,1,1-trifluoroethane	$CHCl_2CF_3$
123a	1,2-dichloro-1,1,2-trifluoroethane	$CHClFCClF_2$
124	2-chloro-1,1,1,2-tetrafluoroethane	$CHClFCF_3$
124a	1-chloro-1,1,2,2-tetrafluoroethane	CHF_2CClF_2
125	pentafluoroethane	CHF_2CF_3
133a	2-chloro-1,1,1-trifluoroethane	CH_2ClCF_3
134a	1,1,1,2-tetrafluoroethane	CH_2FCF_3
140a	1,1,1-trichloroethane (methylchloroform)	CH_3CCl_3
141b	1,1-dichloro-1-fluoroethane	CH_3CCl_2F
142b	1-chloro-1,1-difluoroethane	CH_3CClF_2
143a	1,1,1-trifluoroethane	CH_3CF_3
T152a	1,1-difluoroethane	CH_3CHF_2
160	chloroethane (ethyl chloride)	CH_3CH_2Cl
170	ethane	CH_3CH_3
Propane Series		
216ca	1,3-dichloro-1,1,2,2,3,3-hexafluoropropane	$CClF_2CF_2CClF_2$
218	octafluoropropane	$CF_3CF_2CF_3$
245cb	1,1,1,2,2-pentafluoropropane	$CF_3CF_2CH_3$
290	propane	$CH_3CH_2CH_3$

Refrigerant Number	Chemical Name or Composition	Chemical Formula
Cyclic Organic Compounds		
C316	1,2-dichlorohexafluorocyclobutane	$C_4Cl_2F_6$
C317	chloroheptafluorocyclobutane	C_4ClF_7
C318	octafluorocyclobutane	C_4F_8
Zeotropes		
400	R-12/114 (must be specified)	
Azeotropes		
500	R-12/152a (73.8/26.2% by mass)	
501	R-22/12 (75.0/25.0)	
502	R-22/115 (48.8/51.2)	
503	R-23/13 (40.1/59.9)	
504	R-32/115 (48.2/51.8)	
505	R-12/31 (78.0/22.0)	
506	R-31/114 (55.1/44.9)	
Miscellaneous Organic Compounds		
Hydrocarbons		
600	butane	$CH_3CH_2CH_2CH_3$
600a	2-methyl propane (isobutane)	$CH(CH_3)_3$
Oxygen Compounds		
610	ethyl ether	$C_2H_5OC_2H_5$
611	methyl formate	$HCOOCH_3$
Sulfur Compounds		
620	(Reserved for future assignment)	
Nitrogen Compounds		
630	methyl amine	CH_3NH_2
631	ethyl amine	$C_2H_5NH_2$
Inorganic Compounds		
702	hydrogen	H_2
704	helium	He
717	ammonia	NH_3
718	water	H_2O
720	neon	Ne
728	nitrogen	N_2
732	oxygen	O_2
740	argon	Ar
744	carbon dioxide	CO_2
744A	nitrous oxide	N_2O
764	sulfur dioxide	SO_2
Unsaturated Organic Compounds		
1112a	1,1-dichlorodifluoroethene	$CCl_2 = CF_2$
1113	chlorotrifluoroethene	$CClF = CF_2$
1114	tetrafluoroethene	$CF_2 = CF_2$
1120	trichloroethene	$CHCl = CCl_2$
1130	1,2-dichloroethene (trans)	$CHCl = CHCl$
1132a	vinylidene fluoride	$CH_2 = CF_2$
1140	vinyl chloride	$CH_2 = CHCl$
1141	vinyl fluoride	$CH_2 = CHF$
1150	ethene (ethylene)	$CH_2 = CH_2$
1270	propene (propylene)	$CH_3CH = CH_2$

Table 2 Physical Properties of Refrigerants[a]

No.	Name	Chemical Formula	Molecular Mass	Boiling Pt. (NBP) at 14.696 psia, °F	Freezing Point, °F	Critical Temperature, °F	Critical Pressure, psia	Critical Volume, ft³/lb	Refractive Index of Liquid[b, c]
704	Helium	He	4.0026	−452.1	None	−450.3	33.21	0.2311	1.021 (NBP) 5461 Å
702n	Hydrogen (normal)	H_2	2.0159	−423.0	−434.5	−399.9	190.8	0.5320	1.097(NBP)5791 Å
702p	Hydrogen (para)	H_2	2.0159	−423.2	−434.8	−400.3	187.5	0.5097	1.09(NBP)[j]
720	Neon	Ne	20.183	−410.9	−415.5	−379.7	493.1	0.03316	—
728	Nitrogen	N_2	28.013	−320.4	−346.0	−232.4	492.9	0.05092	1.205(83K)5893 Å
729	Air	—	28.97	−317.8	—	−221.3	547.4	0.04883	
						−221.1	546.3	0.05007	
740	Argon	A	39.948	−302.6	−308.7	−188.1	710.4	0.02990	1.233(TP)5893 Å
732	Oxygen	O_2	31.9988	−297.3	−361.8	−181.1	736.9	0.0375	1.221(92K)5893 Å
50	Methane	CH_4	16.04	−258.7	−296	−116.5	673.1	0.099	
14	Tetrafluoromethane	CF_4	88.01	−198.3	−299	−50.2	543	0.0256	
1150	Ethylene	C_2H_4	28.05	−154.7	−272	48.8	742.2	0.070	1.363(−148)[1]
503	f	—	87.5	−127.6	—	67.1	607	0.0326	
170	Ethane	C_2H_6	30.07	−127.85	−297	90.0	709.8	0.0830	
744A[2]	Nitrous Oxide	N_2O	44.02	−129.1	−152	97.7	1048	0.0355	
23	Trifluoromethane	CHF_3	70.02	−115.7	−247	78.1	701.4	0.0311	
13	Chlorotrifluoromethane	$CClF_3$	104.47	−114.6	−294	83.9	561	0.0277	1.146(77)[4]
744	Carbon Dioxide	CO_2	44.01	−109.2[d]	−69.9[e]	87.9	1070.0	0.0342	1.195(59)
13B1	Bromotrifluoromethane	$CBrF_3$	148.93	−71.95	−270	152.6	575	0.0215	1.239(77)[4]
504	g	—	79.2	−71.0	—	151.5	690.5	0.0324	
32	Difluoromethane	CH_2F_2	52.02	−61.1	−213	173.14	845.6	.03726	
125	Pentafluoroethane	C_2HF_5	120.03	−55.43	−153.67	151.34	526.57	—	—
1270	Propylene	C_3H_6	42.09	−53.86	−301	197.2	670.3	0.0720	1.3640(−58)[1]
502[5]	h	—	111.63	−49.8	—	179.9	591.0	0.0286	
290	Propane	C_3H_8	44.10	−43.73	−305.8	206.3	617.4	0.0728	1.3397(−43)
22	Chlorodifluoromethane	$CHClF_2$	86.48	−41.36	−256	204.8	721.9	0.0305	1.234(77)[4]
115	Chloropentafluoroethane	$CClF_2CF_3$	154.48	−38.4	−159	175.9	457.6	0.0261	1.221(77)[4]
717	Ammonia	NH_3	17.03	−28.0	−107.9	271.4	1657	0.068[d]	1.325(61.7)
500	j	—	99.31	−28.3	−254	221.9	641.9	0.0323	
12	Dichlorodifluoromethane	CCl_2F_2	120.93	−21.62	−252	233.6	596.9	0.0287	1.288(77)[4]
134a	Tetrafluoroethane	CF_3CH_2F	102.03	−15.08	−141.9	214.0	589.8	0.029	—
152a	Difluoroethane	CH_3CHF_2	66.05	−13.0	−178.6	236.3	652	0.0439	
40[2]	Methyl Chloride	CH_3Cl	50.49	−11.6	−144	289.6	968.7	0.0454	16.2(68)
124	Chlorotetrafluoroethane	$CHFClCF_3$	136.47	8.26	−326.47	252.5	530.84	—	
600a	Isobutane	C_4H_{10}	58.13	10.89	−255.5	275.0	529.1	0.0725	1.3514(−13)[1]
764[6]	Sulfur Dioxide	SO_2	64.07	14.0	−103.9	315.5	1143	0.0306	
142b	Chlorodifluoroethane	CH_3CClF_2	100.5	14.4	−204	278.8	598	0.0368	
630[6]	Methyl Amine	CH_3NH_2	31.06	19.9	−134.5	314.4	1082		1.432(63.5)
C318	Octafluorocyclobutane	C_4F_8	200.04	21.5	−42.5	239.6	403.6	0.0258	
600	Butane	C_4H_{10}	58.13	31.1	−217.3	305.6	550.7	0.0702	1.3562(5)[1]
114	Dichlorotetrafluoroethane	$CClF_2CClF_2$	170.94	38.8	−137	294.3	473	0.0275	1.294(77)
21[7]	Dichlorofluoromethane	$CHCl_2F$	102.92	47.8	−211	353.3	750	0.0307	1.332(77)[4]
160[2]	Ethyl Chloride	C_2H_5Cl	64.52	54.32	−216.9	369.0	764.4	0.0485	
631[6]	Ethyl Amine	$C_2H_5NH_2$	45.08	61.88	−113	361.4	815.6		
11	Trichlorofluoromethane	CCl_3F	137.38	74.87	−168	388.4	639.5	0.0289	1.362(77)[4]
123	Dichlorotrifluoroethane	$CHCl_2CF_3$	152.93	82.17	−160.87	362.82	532.87	—	—
611[6]	Methyl Formate	$C_2H_4O_2$	60.05	89.2	−146	417.2	870	0.0459	
141b	Dichlorofluoroethane	CCl_2FCH_3	116.95	89.6	—	399.6	616.4	—	
610[6]	Ethyl Ether	$C_4H_{10}O$	74.12	94.3	−177.3	381.2	523	0.0607	1.3526(68)
216	Dichlorohexafluoropropane	$C_3Cl_2F_6$	220.93	96.24	−193.7	356.0	399.5	0.0279	
30[6]	Methylene Chloride	CH_2Cl_2	84.93	104.4	−142	458.6	882		1.4244(68)[i]
113	Trichlorotrifluoroethane	CCl_2FCClF_2	187.39	117.63	−31	417.4	498.9	0.0278	1.357(77)[4]
1130[8]	Dichloroethylene	$CHCl = CHCl$	96.95	118	−58	470	795		
1120[6]	Trichloroethylene	$CHCl = CCl_2$	131.39	189.0	−99	520	728		1.4782(68)[j]
718[6]	Water	H_2O	18.02	212	32	705.6	3208	0.0501	

Notes

[a] Data are from *Thermodynamic Properties of Refrigerants* (Stewart *et al.* 1987) or from McLinden (1990).
[b] The temperature of measurement (Fahrenheit) is shown in parentheses. The data are from *Handbook of Chemistry and Physics* (CRC 1987) unless otherwise noted.
[c] For the sodium D line.
[d] Sublimes.
[e] At 76.4 psia.
[f] Refrigerants 23 and 13 (40.1/59.9% by mass).
[g] Refrigerants 32 and 115 (48.2/51.82% by mass).
[h] Refrigerants 22 and 115 (48.8/51.2% by mass).
[i] Data from Electrochemicals Department, E.I. duPont de Nemours & Co.

[j] Refrigerants 12 and 152a (73.8/26.2% by weight).
[k] Dielectric constant data.

References:
[1] Kirk and Othmer (1956)
[2] *Matheson Gas Data Book* (1966)
[3] *Bulletin* D-27 (duPont 1967)
[4] *Bulletin* B-32A (duPont)
[5] *Bulletin* T-502 (duPont 1980)
[6] *Handbook of Chemistry* (1967)
[7] *Bulletin* G-1 (duPont)
[8] CRC *Handbook of Chemistry and Physics* (CRC 1987)

Table 3 Electrical Properties of Liquid Refrigerants

No.	Name	Temp., °F	Dielectric Constant	Volume Resistivity, MΩ·m	Ref.
11	Trichlorofluoromethane	84	2.28		1
		b	1.92	63680	2
		77	2.5	90	3
12	Dichlorodifluoromethane	84	2.13		1
		b	1.74	53900	2
		77	2.1	> 120	3
		77	2.100		4
13	Chlorotrifluoromethane	−22	2.3	120	4
		68	1.64		
22	Chlorodifluoromethane	75	6.11		1
		b	6.12	0.83	2
		77	6.6	75	3
22	Trifluoromethane	−22	6.3		3
		68	5.51		4
113	Trichlorotrifluoromethane	86	2.44		1
		b	1.68	45490	2
		77	2.6	> 120	3
114	Dichlorotetrafluoroethane	88	2.17		1
		b	1.83	66470	2
		77	2.2	> 70	3
124a	Chlorotetrafluoroethane	77	4.0	50	3
290	Propane	b	1.27	73840	2
500	...[a]	b	1.80	55750	2
717	Ammonia	69	15.5		5
744	Carbon Dioxide	32	1.59		5

[a]Azeotrope of Refrigerants 12 and 152a (73.8/26.2% by mass)
[b]Ambient temperature

References
1 Data from E. I. duPont de Nemours & Co., Inc. Used by permission.
2 Beacham and Divers (1955)
3 Eiseman (1955)
4 Makita *et al.* (1976)
5 CRC *Handbook of Chemistry and Physics* (CRC 1987)

Table 4 Electrical Properties of Refrigerant Vapors

No.	Name	Pressure, atm.	Temperature, °F	Dielectric Constant	Relative Dielectric Strength (Nitrogen=1)	Volume Resistivity, GΩ·m	Ref.
11	Trichlorofluoromethane	0.5	79	1.0019			3
		3	b	1.009		74.35	2
		1.0	73		3.1		4
12	Dichlorodifluoromethane	0.5	84	1.0016			3
		a	b	1.012	452[c]	72.77	2
		1.0	73		2.4		4
		4.9	68	1.019			6
13	Chlorotrifluoromethane	0.5	84	1.0013			3
		1.0	73		1.4		4
		4.9	68	1.013			6
		19.5	90	1.055			7
14	Tetrafluoromethane	0.5	76	1.0006			3
		1.0	73		1.0		4
22	Chlorodifluoromethane	0.5	78	1.0035			3
		a	b	1.004	460[c]	2113	2
		1.0	73		1.3		4
		4.9	68	1.033			6
23	Trifluoromethane	4.9	68	1.042			6
113	Trichlorotrifluoroethane	a	b	1.010	440[c]	94.18	2
		0.4	73		2.6		4
114	Dichlorotetrafluoroethane	0.5	80	1.0021			3
		a	b	1.002	295[c]	148.3	2
		1.0	73		2.8		4
116	Hexafluoroethane	0.94	73	1.002			3
133b	Chlorotrifluoroethane	0.94	80	1.010			3
142b	Chlorodifluoroethane	0.93	81	1.013			3
143a	Trifluoroethane	0.85	77	1.013			3
170	Ethane	1.0	32	1.0015			1
290	Propane	a	b	1.009	440[c]	105.3	2
500	...[d]	a	b	1.024	470[c]	76.45	2
717	Ammonia	1.0	32	1.0072			1
		a	b		0.82		4
729	Air	1.0	32	1.00059			1
744	Carbon Dioxide	1.0	32	1.00099			1
		1.0	b		0.88		4
1150	Ethylene	1.0	32	1.00144			1
		1.0	73		1.21		4

Notes
[a]Saturation vapor pressure
[b]Ambient temperature
[c]Measured breakdown voltage, volts/mil
[d]Azeotrope of Refrigerants 12 and 152a (73.8/26.2% by mass)

References
1 CRC *Handbook of Chemistry and Physics* (CRC 1987)
2 Beacham and Divers (1955)
3 Fuoss (1938)
4 Charlton and Cooper (1937)
5 Eiseman (1955)
6 Makita *et al.* (1976)
7 Hess *et al.* (1962)

and 86°F condensation. Calculated data for other conditions are given in Table 8. The tables can be used to compare the properties of different refrigerants, but actual operating conditions are somewhat different from the calculated data. In most cases, the suction vapor is assumed to be saturated, and the compression is assumed adiabatic or at constant entropy. For R-113 and R-114, these assumptions would cause some liquid in the discharge vapor. In these cases, it is assumed that the discharge vapor is saturated and that the suction vapor is slightly superheated. In Section F of Table 8, the temperature of the suction gas is assumed to be 65°F. Comparison with Section E illustrates the effect of suction gas superheating on the refrigerant properties.

The comparisons shown in Tables 9 and 10 are based on constant compressor displacement. They illustrate the effect of different evaporating and condensing temperatures on theoretical capacity and horsepower of compression.

SAFETY

Table 11 summarizes the toxicity and flammability characteristics of many refrigerants. In ANSI/ASHRAE *Standard* 15-1992, refrigerants are classified according to the hazard involved in their use. Group A1 refrigerants are the least hazardous, Group B3 the most hazardous.

Table 5 Velocity of Sound in Refrigerant Vapors[a]

Refrigerant	Pressure, psia	Temperature, °F		
		50	100	150
		Velocity of Sound, ft/s		
11	10	b	469	490
12	10	480	503	525
22	10	583	610	635
113	10	b	435	456
114	10	391	411	430
502	10	501	525	547
123	10	b	435	456
124	10	443	465	486
125	10	477	500	521
134a	10	517	543	566
12	100	b	457	490
22	100	b	574	607
502	100	450	488	519
124	100	b	b	440
125	100	b	466	497
134a	100	b	490	528
12	200	b	b	442
22	200	b	523	572
502	200	b	435	483
125	200	b	420	467
134a	200	b	b	476

[a]Data in I-P units from E. I. duPont de Nemours & Co., Inc. Used by permission, or NIST Standard Reference Database 23.
[b]Below saturation temperature.

LEAK DETECTION

Leak detection in refrigeration equipment is a major problem for manufacturers and service engineers. The following sections describe several leak detection methods.

Electronic Detector

The electronic detector is widely used in the manufacture and assembly of refrigeration equipment. Instrument operation depends on the variation in current flow caused by ionization of decomposed refrigerant between two oppositely charged platinum electrodes. This instrument can detect any of the halogenated refrigerants except R-14. *It is not recommended for use in atmospheres containing explosive or flammable vapors.* Other vapors, such as alcohol and carbon monoxide, may interfere with the test.

The electronic detector is the most sensitive of the various leak detection methods, reportedly capable of sensing a leak of 1/100 oz per year of R-12.

A portable model is available for field testing. Other models are available with automatic balancing systems that correct for refrigerant vapors that might be present in the atmosphere around the test area.

Halide Torch

The halide torch is a fast and reliable method of detecting leaks of halogenated refrigerants. Air is drawn over a copper element heated by a methyl alcohol or hydrocarbon flame. If halogenated vapors are present, they decompose, and the color of the flame changes to bluish-green. Although not as sensitive as the electronic detector, this method is suitable for most purposes.

Bubble Method

The object to be tested is pressurized with air or nitrogen. A pressure corresponding to operating conditions is generally used. The object is immersed in water, and any leaks are detected by

Table 6 Latent Heat of Vaporization versus Boiling Point

Refrigerant		Normal Boiling Pt., °F	Latent Heat at Boiling Pt., λ Btu/lb · mol	Trouton Constant, λ/°R[b]	Ref.
No.	Name				
630	Methyl amine[a]	23.0	11141	23.086	4
717	Ammonia	−28.0	10036	23.256	1
764	Sulfur dioxide	13.6	10705	22.626	2
631	Ethyl amine	68.0	11645	22.076	4
611	Methyl formate[a]	100.0	12094	21.616	4
504	—	−71.0	8282	21.316	1
23	Trifluoromethane	−115.7	7325	21.29	1
21	Dichlorofluoromethane	47.8	10557	20.80	3
30	Methylene chloride[a]	120.0	11398	19.66	4
C318	Octafluorocyclobutane	21.5	10017	20.81	1
22	Chlorodifluoromethane	−41.4	8687	20.76	1
40	Methyl chloride	−10.8	9305	20.73	3
506	—	9.9	9644	21.44	3
113	Trichlorotrifluoroethane	117.6	11828	20.49	1
152a	Difluoroethane	−13.0	9045	20.25	1
502	—	−49.9	8280	20.21	3
114	Dichlorotetrafluoroethane	38.8	10005	20.07	1
216	Dichlorohexafluoropropane	96.2	11154	20.07	1
11	Trichlorofluoromethane	74.9	10648	19.92	1
505	—	−21.8	8735	19.95	3
500	—	−28.3	8588	19.91	1
290	Propane	−43.7	8026	19.29	1
14	Tetrafluoromethane	−198.3	5146	19.69	1
600	Butane	31.1	9641	19.64	1
13B1	Bromotrifluoromethane	−72.0	7607	19.62	1
12	Dichlorodifluoromethane	−21.6	8591	19.61	1
142b	Chlorodifluoroethane	14.4	9297	19.61	1
115	Chloropentafluoroethane	−38.4	8245	19.57	1
503	—	−126.1	6483	19.43	1
1270	Propylene	−53.9	7931	19.55	1
600a	Isobutane	10.9	9103	19.34	1
13	Chlorotrifluoromethane	−114.6	6670	19.33	1
1150	Ethylene	−154.7	5793	19.00	1
170	Ethane	−127.9	6296	18.98	1
50	Methane	−258.7	3521	17.52	1
123	Dichlorotrifluoroethane	82.17	11215	20.70	5
124	Chlorotetrafluoroethane	8.26	9742	20.82	5
125	Pentafluoroethane	−55.43	8295	20.52	5
134a	Tetrafluoroethane	−15.07	9531	21.44	5

Notes

[a]Note at normal atmospheric pressure
[b]Normal boiling temperatures

References

1 Thermodynamic Properties of Refrigerants (Stewart *et al.* 1986)
2 *Handbook of Chemistry and Physics* (CRC 1987)
3 ASHRAE (1977)
4 *Chemical Engineer's Handbook* (1973)
5 NIST Standard Reference Database 23

observing the formation of bubbles in the liquid. Adding a detergent to the water decreases the surface tension, prevents escaping gas from clinging to the side of the object, and promotes the formation of a regular stream of small bubbles. Kerosene or other organic liquids are sometimes used for the same reason. A solution of soap or detergent can be brushed or poured onto joints or other spots where leakage is suspected. Leaking gas forms soap bubbles that can be readily detected.

Detection of Ammonia and Sulfur Dioxide Leaks

Ammonia can be detected by burning a sulfur candle in the vicinity of the suspected leak or by bringing a solution of hydrochloric acid near the object. If ammonia vapor is present, a white cloud or smoke of ammonium sulfite or ammonium chloride forms. Ammonia can also be detected with indicating paper that changes color in the presence of a base.

Sulfur dioxide can be detected by the appearance of white smoke when aqueous ammonia is brought near the leak.

Table 7 Comparative Refrigerant Performance per Ton of Refrigeration

No.	Name	Evaporator Pressure, psia	Condensing Pressure, psia	Compression Ratio	Net Refrigerating Effect, Btu/lb	Refrigerant Circulated, lb/min	Liquid Circulated, in³/min	Specific Volume of Suction Gas, ft³/lb	Compressor Displacement, cfm	Horsepower hp	Coefficient of Performance	Comp. Discharge Temp., °F
170	Ethane	236.410	674.710	2.85	69.27	2.88704	289.1266	0.5344	1.543	1.733	2.72	123
744	Carbon Dioxide	332.375	1045.360	3.15	57.75	3.46320	158.5272	0.2639	0.914	1.678	2.81	156
13B1	Bromotrifluoromethane	77.820	264.128	3.39	28.45	7.02901	129.7814	0.3798	2.669	1.134	4.16	104
125	Pentafluoroethane	58.870	228.110	3.87	37.69	5.30645	126.8148	0.6281	3.333	1.283	3.67	108
1270	Propylene	52.704	189.440	3.59	123.15	1.62401	90.7048	2.0487	3.327	1.035	4.56	108
290	Propane	42.37	156.820	3.70	120.30	1.66251	95.0386	2.4589	4.088	1.031	4.57	98
502	22/115 Azeotrope[a]	50.561	191.290	3.78	44.91	4.45305	103.3499	0.8015	3.569	1.067	4.42	98
22	Chlorodifluoromethane	42.963	172.899	4.02	69.90	2.86144	67.6465	1.2394	3.546	1.011	4.67	128
717	Ammonia	34.170	168.795	4.94	474.20	0.42177	19.6087	8.1790	3.450	0.989	4.77	210
500	12/152a Azeotrope[a]	31.064	127.504	4.10	60.64	3.29834	80.1925	1.5022	4.955	1.005	4.69	105
12	Dichlorodifluoromethane	26.505	107.991	4.07	50.25	3.97981	85.2280	1.4649	5.830	0.992	4.75	100
134a	Tetrafluoroethane	23.790	111.630	4.69	64.77	3.08785	71.8199	1.9500	6.021	1.070	4.41	108
124	Chlorotetrafluoroethane	12.960	64.590	4.98	50.93	3.92696	81.1580	2.7140	10.658	1.054	4.47	90
600a	Isobutane	12.924	59.286	4.59	113.00	1.76991	90.0059	6.4189	11.361	1.070	4.41	80
600	Butane	8.176	41.191	5.04	125.55	1.59299	77.7772	10.2058	16.258	0.952	4.95	88
114	Dichlorotetrafluoroethane[b]	6.747	36.493	5.41	43.02	4.64889	89.5631	4.3400	20.176	1.015	4.65	86
11	Trichlorofluoromethane	2.937	18.318	6.24	67.21	2.97592	56.2578	12.2400	36.425	0.939	5.02	110
123	Dichlorotrifluoroethane	2.290	15.900	6.94	61.19	3.26829	62.3495	14.0800	46.018	0.974	4.84	94
113	Trichlorotrifluoroethane[b]	1.006	7.884	7.83	52.08	3.84047	68.5997	26.2845	100.945	1.105	4.27	86

Note: Based on 5°F Evaporation and 86°F Condensation.
[a]See Table 1 for composition.
[b]Saturated suction except R-113, 114. Enough superheat was added to give saturated discharge.

Table 8 Comparative Refrigerant Performance per Ton at Various Evaporating and Condensing Temperatures

No.	Name	Suction Temp., °F	Evaporator Pressure, psia	Condensing Pressure, psia	Compression Ratio	Net Refrigerating Effect, Btu/lb	Refrigerant Circulated, lb/min	Volume of Suction Gas, ft³/lb	Compressor Displacement, cfm	Horsepower hp
A. −130°F Evaporating, −40°F Condensing										
1150	Ethylene	−130	30.887	210.670	6.82	142.01	1.40835	3.8529	5.426	1.756
170	Ethane	−130	13.620	112.790	8.28	156.58	1.27730	8.3575	10.675	1.633
13	Monochlorotrifluoromethane	−130	9.059	88.037	9.72	45.82	4.36529	3.6245	15.822	1.685
23	Trifluoromethane	−130	9.06	103.03	11.37	79.38	2.51953	5.4580	13.752	1.753
B. −100°F Evaporating, −30°F Condensing										
170	Ethane	−100	31.267	134.730	4.31	157.76	1.26775	3.8671	4.903	1.118
13	Monochlorotrifluoromethane	−100	22.276	106.290	4.77	46.23	4.32581	1.5631	6.762	1.153
125	Pentafluoroethane	−100	3.780	27.760	7.34	56.43	3.54403	8.3900	29.734	1.101
22	Chlorodifluoromethane	−100	2.380	19.629	8.25	90.75	2.20397	18.5580	40.901	1.074
23	Trifluoromethane	−100	23.74	125.99	5.31	79.37	2.51984	2.219	5.592	1.178
C. −76°F Evaporating, 5°F Condensing										
1150	Ethylene	−76	109.370	416.235	3.81	116.95	1.71021	1.1617	1.987	1.478
170	Ethane	−76	54.634	235.440	4.31	322.65	0.61987	2.2906	1.420	0.566
23	Trifluoromethane	−76	45.410	237.180	5.22	69.60	2.87356	1.2030	3.457	1.394
13	Monochlorotrifluoromethane	−76	40.872	192.135	4.70	39.42	5.07389	0.8801	4.465	1.382
13B1	Bromotrifluoromethane	−76	13.173	77.820	5.91	37.80	5.29128	2.0339	10.757	1.253
125	Pentafluoroethane	−76	8.210	58.870	7.17	50.62	3.95101	4.0720	16.089	1.277
290	Propane	−76	6.150	42.367	6.89	147.39	1.35699	14.8560	20.159	1.196
22	Chlorodifluoromethane	−76	5.438	42.963	7.90	84.24	2.37425	8.5925	20.401	1.195
717	Ammonia	−76	3.169	34.162	10.78	535.08	0.37378	75.5710	28.247	1.237
12	Dichlorodifluoromethane	−76	3.277	26.501	8.09	58.61	3.41219	10.2448	34.957	1.191
134a	Tetrafluoroethane	−76	2.360	23.790	10.08	77.94	2.56621	16.9300	43.446	1.198
D. −40°F Evaporating, 68°F Condensing										
744	Carbon Dioxide	−40	145.770	830.530	5.70	77.22	2.59000	0.6128	1.587	2.208
23	Trifluoromethane	−40	103.030	597.900	5.80	45.67	4.37924	0.5448	2.386	2.442
13B1	Bromotrifluoromethane	−40	31.855	207.854	6.53	28.81	6.94155	0.8915	6.189	1.855
125	Pentafluoroethane	−40	21.840	175.100	8.02	37.44	5.34188	1.6250	8.681	1.962
290	Propane	−40	16.099	121.560	7.55	119.33	1.67602	6.0829	10.195	1.670
22	Chlorodifluoromethane	−40	15.268	131.997	8.65	70.65	2.83106	3.2805	9.287	1.606
717	Ammonia	−40	10.376	124.009	11.95	479.49	0.41711	24.9350	10.401	1.577
500	12/152a Azeotrope	−40	10.959	96.948	8.85	60.24	3.31989	3.9895	13.245	1.583
12	Dichlorodifluoromethane	−40	9.304	82.295	8.84	49.44	4.04572	3.8868	15.725	1.596
134a	Tetrafluoroethane	−40	7.500	82.820	11.04	63.77	3.13627	5.7630	18.074	1.652

Table 8 Comparative Refrigerant Performance per Ton at Various Evaporating and Condensing Temperatures (*Concluded*)

No.	Refrigerant Name	Suction Temp., °F	Evaporator Pressure, psia	Condensing Pressure, psia	Compression Ratio	Net Refrigerating Effect, Btu/lb	Refrigerant Circulated, lb/min	Volume of Suction Gas, ft³/lb	Compressor Displacement, cfm	Horsepower, hp
				E. –10°F Evaporating, 100°F Condensing						
123	Dichlorotrifluoroethane	–10	1.460	20.810	14.25	55.64	3.59473	21.4300	77.035	1.477
11	Trichlorofluoromethane	–10	1.921	23.637	12.31	62.34	3.20837	18.1590	58.261	1.436
114	Dichlorotetrafluroethane	5	4.560	46.148	10.12	47.22	4.23550	6.2274	26.376	1.339
124	Chlorotetrafluoroethane	–10	8.950	80.920	9.04	44.99	4.44543	3.8410	17.075	1.649
134a	Tetrafluoroethane	–10	16.680	138.840	8.32	57.66	3.46861	2.7250	9.452	1.671
12	Dichlorodifluoromethane	–10	19.197	131.720	6.86	44.89	4.45563	1.9803	8.823	1.606
717	Ammonia	–10	23.664	211.400	8.93	453.05	0.44145	11.5260	5.088	1.497
22	Chlorodifluoromethane	–10	31.231	210.670	6.75	64.07	3.12173	1.6757	5.231	1.602
502	22/115 Azeotrope	–10	37.256	230.890	6.20	39.05	5.12177	1.0727	5.494	1.904
125	Pentafluoroethane	–10	43.320	276.950	6.39	31.09	6.43294	0.8459	5.442	2.172
				F. –10°F Evaporating, 100°F Condensing						
123	Dichlorotrifluoroethane	65	1.460	20.810	14.25	55.64	3.59473	25.1100	90.264	1.710
11	Trichlorofluoromethane	65	1.921	23.637	12.31	62.34	3.20837	21.2388	68.142	1.679
114	Dichlorotetrafluroethane	65	4.560	46.148	10.12	47.22	4.23550	7.1379	30.233	1.534
124	Chlorotetrafluoroethane	65	8.950	80.920	9.04	44.99	4.44543	4.5310	20.142	1.919
134a	Tetrafluoroethane	65	16.680	138.840	8.32	57.66	3.46861	3.2310	11.207	1.945
12	Dichlorodifluoromethane	65	19.197	131.720	6.86	44.89	4.45563	2.3597	10.514	1.914
717	Ammonia	65	23.664	211.400	8.93	453.05	0.44145	13.7053	6.050	1.780
22	Chlorodifluoromethane	65	31.231	210.670	6.75	64.07	3.12173	2.0121	6.281	1.924
502	22/115 Azeotrope	65	37.256	230.890	6.20	39.05	5.12177	1.3015	6.666	2.310
125	Pentafluoroethane	65	43.320	276.950	6.39	31.09	6.43294	1.0280	6.613	2.573
				G. –10°F Evaporating plus Refrigeration Effect to 65°F, 100°F Condensing						
123	Dichlorotrifluoroethane	65	1.460	20.810	14.25	67.05	2.98298	25.1100	74.903	1.419
11	Trichlorofluoromethane	65	1.921	23.637	12.31	72.15	2.77204	21.2388	58.875	1.451
114	Dichlorotetrafluroethane	65	4.560	46.148	10.12	47.22	4.23550	7.1379	30.233	1.534
124	Chlorotetrafluoroethane	65	8.950	80.920	9.04	57.33	3.48857	4.5310	15.807	1.506
134a	Tetrafluoroethane	65	16.680	138.840	8.32	71.95	2.77971	3.2310	8.981	1.558
12	Dichlorodifluoromethane	65	19.197	131.720	6.86	55.83	3.58251	2.3597	8.454	1.539
717	Ammonia	65	23.664	211.400	8.93	493.76	0.40506	13.7053	5.551	1.633
22	Chlorodifluoromethane	65	31.231	210.670	6.75	75.95	2.63326	2.0121	5.298	1.623
502	22/115 Azeotrope	65	37.256	230.890	6.20	51.23	3.90362	1.3015	5.081	1.761
125	Pentafluoroethane	65	43.320	276.950	6.39	45.13	4.43164	1.0280	4.556	1.773
				H. 20°F Evaporating, 80°F Condensing						
125	Pentafluoroethane	20	78.400	209.270	2.67	41.47	4.82276	0.4735	2.284	0.831
290	Propane	20	55.931	144.330	2.58	128.39	1.55775	1.8873	2.940	0.721
22	Chlorodifluoromethane	20	57.786	158.360	2.74	73.12	2.73512	0.9334	2.553	0.707
717	Ammonia	20	48.062	152.690	3.18	485.52	0.41193	5.9202	2.439	0.678
500	12/152a Azeotrope	20	41.936	116.620	2.78	64.15	3.11784	1.1294	3.521	0.702
12	Dichlorodifluoromethane	20	35.765	98.850	2.76	53.22	3.75827	1.1045	4.151	0.701
134a	Tetrafluoroethane	20	33.110	101.300	3.06	69.06	2.89603	1.4250	4.127	0.719
124	Chlorotetrafluoroethane	20	18.290	58.410	3.19	54.67	3.65831	1.9640	7.185	0.710
600a	Isobutane	20	17.916	53.907	3.01	121.45	1.64677	4.7361	7.799	0.706
600	Butane	20	11.557	37.225	3.22	134.18	1.49054	7.3947	11.022	0.686
114	Dichlorotetrafluroethane	30	9.699	32.859	3.39	44.42	4.50207	3.0678	13.812	0.769
123	Dichlorotrifluoroethane	20	3.480	14.100	4.05	64.80	3.08634	9.5420	29.450	0.630
				I. 40°F Evaporating, 100°F Condensing						
125	Pentafluoroethane	40	111.710	276.950	2.48	37.10	5.39084	0.3312	1.785	0.860
290	Propane	40	78.782	189.040	2.40	114.96	1.73974	1.3563	2.360	0.750
22	Chlorodifluoromethane	40	83.246	210.670	2.53	68.71	2.91091	0.6557	1.909	0.696
717	Ammonia	40	73.114	211.400	2.89	467.39	0.42791	3.9783	1.702	0.653
500	12/152a Azeotrope	40	60.722	155.790	2.57	60.54	3.30344	0.7920	2.616	0.692
12	Dichlorodifluoromethane	40	51.705	131.720	2.55	50.50	3.96024	0.7784	3.083	0.689
134a	Tetrafluoroethane	40	49.670	138.840	2.80	65.15	3.06984	0.9663	2.966	0.712
124	Chlorotetrafluoroethane	40	27.890	80.920	2.90	52.06	3.84172	1.3180	5.063	0.698
600a	Isobutane	40	26.750	73.364	2.74	115.83	1.72667	3.2564	5.623	0.693
600	Butane	40	17.679	51.683	2.92	129.22	1.54775	4.9754	7.701	0.669
114	Dichlorotetrafluroethane	52	15.134	46.148	3.05	42.46	4.70998	2.0321	9.571	0.738
11	Trichlorofluoromethane	40	7.050	23.637	3.35	68.50	2.91984	5.4311	15.858	0.636
123	Dichlorotrifluoroethane	40	5.800	20.810	3.59	62.86	3.18183	5.9290	18.865	0.663
113	Trichlorotrifluoroethane	47	2.695	10.494	3.89	54.14	3.69433	10.7059	39.551	0.710

Table 9 Effect of Temperature on Capacity

Evap. Temp., °F	Condensing Temperature, °F			
	60	80	100	120
Refrigerant 12 Capacity, Btu/h				
−20	5241	4780	4307	3820
0	8280	7582	6864	6124
20	12585	11563	10513	9431
40	18511	17062	15571	14035
Refrigerant 22 Capacity, Btu/h				
−20	8694	8010	7297	6550
0	13539	12503	11422	10288
20	20320	18801	17218	15557
40	29563	27401	25148	22783

Note: Constant compressor displacement = 4 cfm

Table 10 Effect of Temperature on Theoretical Horsepower per Ton of Refrigeration

Evap. Temp., °F	Condensing Temperature, °F			
	60	80	100	120
Refrigerant 12 Theoretical Horsepower per Ton of Refrigeration				
−20	1.03	1.38	1.80	2.32
0	0.71	1.01	1.37	1.80
20	0.44	0.70	1.00	1.36
40	0.20	0.43	0.68	0.99
Refrigerant 22 Theoretical Horsepower per Ton of Refrigeration				
−20	1.02	1.35	1.75	2.22
0	0.71	1.00	1.34	1.75
20	0.44	0.69	0.99	1.34
40	0.20	0.43	0.68	0.98

Leaks can also be determined by pressurizing or evacuating and observing the change in pressure or vacuum over a period of time. This is effective in checking the tightness of the system but does not locate the point of leakage.

EFFECT ON CONSTRUCTION MATERIALS

Metals

Halogenated refrigerants can be used satisfactorily under normal conditions with most common metals, such as steel, cast iron, brass, copper, tin, lead, and aluminum. Under more severe conditions, various metals affect such properties as hydrolysis and thermal decomposition in varying degrees. The tendency of metals to promote thermal decomposition of halogenated compounds is in the following order:

(Least decomposition) Inconel < 18-8 stainless steel < nickel < copper < 1340 steel < aluminum < bronze < brass < zinc < silver (most decomposition)

This order is only approximate, and exceptions may be found for individual compounds or for special use conditions. The effect of metals on hydrolysis is probably similar.

Magnesium, zinc, and aluminum alloys containing more than 2% magnesium are not recommended for use with halogenated compounds where even trace amounts of water may be present.

Warning: Never use methyl chloride with aluminum in any form. A highly flammable gas is formed and the explosion hazard is great.

Ammonia should never be used with copper, brass, or other alloys containing copper. When water is present in sulfur dioxide systems, sulfurous acid is formed and can attack iron or steel rapidly and other metals at a slower rate.

Table 11 Comparison of Safety Group Classifications to Those under ASHRAE *Standard* 34-1989

Refrigerant Number	Chemical Formula	Safety Group	
		Old	New
10	CCl_4	2	B1
11	CCl_3F	1	A1
12	CCl_2F_2	1	A1
13	$CClF_3$	1	A1
13B1	$CBrF_3$	1	A1
14	CF_4	1	A1
21	$CHCl_2F$	2	B1
22	$CHClF_2$	1	A1
30	CH_2Cl_2	2	B2
40	CH_3Cl	2	B2
50	CH_4	3a	A3
113	CCl_2FCClF_2	1	A1
114	$CClF_2CClF_2$	1	A1
115	$CClF_2CF_3$	1	A1
123	$CHCl_2CF_3$		B1
134a	CH_2FCF_3		A1
142b	CH_3CClF_2	3b	A2
152a	CH_3CHF_2	3b	A2
170	CH_3CH_3	3a	A3
290	$CH_3CH_2CH_3$	3a	A3
C318	C_4F_8	1	A1
400	R-12/114	1	A1/A1
500	R-12/152a	1	A1
501	R-22/12	1	A1
502	R-22/115	1	A1
600	$CH_3CH_2CH_2CH_3$	3a	A3
600a	$CH(CH_3)_3$	3a	A3
611	$HCOOCH_3$	2	B2
702	H_2		A3
704	He		A1
717	NH_3	2	B2
718	H_2O		A1
720	Ne		A1
728	N_2		A1
740	Ar		A1
744	CO_2	1	A1
764	SO_2	2	B1
1140	$CH_2 = CHCl$		B3
1150	$CH_2 = CH_2$	3a	A3
1270	$CH_3CH = CH_2$	3a	A3

Elastomers

The linear swelling of some elastomers in the liquid phase of various refrigerants is shown in Table 12. Swelling data can be used to a limited extent in comparing the effect of refrigerants on elastomers. However, other factors, such as the amount of extraction, tensile strength, and degree of hardness of the exposed elastomer must be considered. When other fluids are present in addition to the refrigerant, the combined effect on elastomers should be determined. In some instances, somewhat higher swelling of elastomers is found in mixtures of R-22 and lubricating oil than in either fluid alone. Table 13 shows the diffusion of water through elastomers.

Plastics

The effect of a refrigerant on a plastic material should be thoroughly examined under the conditions of intended use. Plastics are often mixtures of two or more basic types, and it is difficult to predict the effect of the refrigerant. The linear swelling of some plastic materials in refrigerants is shown in Table 14. Swelling data can be used as a guide but, as with elastomers, the effect on the properties of the plastic should also be examined. Comparable

Table 12   Swelling of Elastomers in Liquid Refrigerants at Room Temperature

Refrigerant		Linear Swell, %							
No.	Name	Buna N	Buna S (GR-S)	Butyl (GR-1)	Natural Rubber	Neoprene GN	Thiokol FA	Viton B	Silicone
11	Trichlorofluoromethane	6	21	41	23	17	2	6	38
12	Dichlorodifluoromethane	2	3	6	6	0	1	9	—
13	Chlorotrifluoromethane	1	1	0	1	0	0	4	—
13B1	Bromotrifluoromethane	1	1	2	1	2	—	7	—
21	Dichlorofluoromethane	48	49	24	34	28	28	22	—
22	Chlorodifluoromethane	26	4	1	6	2	4	20	20
30	Methylene chloride	52	26	23	34	37	59	—	—
40	Methyl chloride	35	20	16	26	22	11	—	—
113	Trichlorotrifluoroethane	1	9	21	17	3	1	7	34
114	Dichlorotetrafluoroethane	0	2	2	2	0	0	9	—
502	R22 and 115 (48.8/51.2% by mass)	7	3	—	4	1	—	—	—
600	Butane	1	8	20	16	3	0	—	—

Adapted from Eiseman (1949)

Table 13   Diffusion of Water, R12, and R-22 through Elastomers

Elastomer	Diffusion Rate		
	Water[a]	R-12[b]	R-22[b]
Neoprene	0.717	0.21	1.31
Buna N	0.109	0.066	19.7
Hypalon 40	0.457	0.16	0.52
Butyl	0.043	1.48	0.30
Viton	—	0.98	3.61
Polyethylene	0.123	—	—
Natural	1.428	—	—

Adapted from Eiseman (1966)

[a]0.003-in. film, 100% rh at 100°F. Diffusion rate per hour in pounds of water per 1000 ft^2 of elastomer.

[b]Film thickness = 0.001 in; temperature = 77°F. Gas at 1 atm and 32°F. Diffusion rate per day in ft^3 of gas per ft^2 of elastomer.

Table 14   Swelling of Plastics in Liquid Refrigerants at Room Temperature

Plastic	Linear Swell, %						
	Refrigerant						
	11	12	21	30	113	114a	22
Phenol formaldehyde resin	0	0	0	0	−0.2	−0.2	a
Cellulose acetate	0.4	0	b	b	0	−0.1	a
Cellulose nitrate	0.6	0	b	b	0	−0.1	a
Nylon	0	0	0	0	0	−0.2	1
Methyl methacrylate resin	0	−0.1	b	b	−0.2	−0.2	b
Polyethylene	6.7	0.4	4.5	4.6	2.3	0.6	2
Polystyrene	b	−0.1	b	b	−0.2	−0.2	a
Polyvinyl alcohol	0.3	−0.7	12.9	9.1	−0.1	0.2	a
Polyvinyl chloride	0	0	15.1	b	0	0.1	a
Polyvinylidene chloride	−0.2	0	1.0	2.4	−0.1	0	4
Polytetrafluoroethylene	0	−0.7	0.1	0	0	−0.3	1

Adapted from Brown (1960)   [a]Data not available.

[b]Sample completely disintegrated.

data for R-22 is limited, but the effect on plastics is generally more severe than that of R-12, but not as severe as that of R-21. The effect of R-114 is very similar to that of R-114a.

REFERENCES

ASHRAE. 1986. *Thermodynamic properties of refrigerants.*

Beacham, E.A. and R.T. Divers. 1955. Some aspects of the dielectric properties of refrigerants. *Refrigerating Engineering* 7:33.

Brown, J.A. 1960. Effect of propellants on plastic valve components. *Soap and Chemical Specialties* 3:87.

DuPont. 1967. Bulletin D-27 Freon Products Division. E.I. duPont de Nemours & Co., Inc., Wilmington, DE.

DuPont. 1980. Bulletin T-502 Freon Products Division. E.I. duPont de Nemours & Co., Inc., Wilmington, DE.

Charlton, E.E. and F.S. Cooper. 1937. Dielectric strengths of insulating fluids. *General Electric Review* 865(9):438.

Chemical engineer's handbook, 5th ed. 1973. McGraw-Hill Book Co., Inc., New York.

Eiseman, B.J., Jr. 1949. Effect on elastomers of Freon compounds and other halohydrocarbons. *Refrigerating Engineering* 12:1171.

Eiseman, B.J., Jr. 1955. How electrical properties of Freon compounds affect hermetic system's insulation. *Refrigerating Engineering* 4:61.

Fellows, B.R., R.G. Richard, and I.R. Shankland. 1991. Electrical characterization of alternate refrigerants. Proceedings of the 18th International Congress of Refrigeration. International Institute of Refrigeration, Paris, France.

Fuoss, R.M. 1938. Dielectric constants of some fluorine compounds. *Journal of the American Chemical Society*, 1633.

Handbook of chemistry, 10th ed. 1967. McGraw-Hill Book Co., Inc., New York.

Handbook of chemistry and physics, 41st ed. 1959-60. The Chemical Rubber Publishing Co., Cleveland, OH.

Kirk and Othmer. 1956. *The encyclopedia of chemical technology.* The Interscience Encyclopedia, Inc., New York.

Matheson gas data book. 1966. The Matheson Company, Inc., East Rutherford, NJ.

McLinden, M.O. 1990. *International Journal of Refrigeration* 13:149-62.

National Institute of Standards and Technology Standard Reference Database 23.

CHAPTER 17

REFRIGERANT TABLES AND CHARTS

THIS chapter presents tabular data for the thermodynamic and transport properties of refrigerants arranged for the occasional user. Each refrigerant has a thermodynamic property chart on pressure-enthalpy coordinates with an abbreviated set of tabular data for the saturated liquid and vapor on the facing page. In addition, tabular data in the superheated vapor region are given for R-134a to assist in working on compression cycle examples. New for this Handbook are data for several "new" or "alternative" refrigerants, namely, R-32, R-123, R-124, R-125, R-134a, and R-141b. All of the CFC refrigerants have been retained to assist in making comparisons. Revised formulations, especially for the transport properties, have been used for a number of refrigerants; the formulations used are detailed in the References section.

The reference states used for most of the refrigerants correspond to the international convention of 200 kJ/kg for enthalpy and 1.00 kJ/(kg·K) for entropy, both for the saturated liquid at 0°C. For this I-P edition, the traditional American convention of 0 Btu/lb and 0 Btu/lb·°F for enthalpy and entropy, respectively, for the saturated liquid at −40°F is used. The exceptions are water and fluids which have very low critical temperatures, such as R-14 and the cryogens.

These data are intended to help engineers make preliminary comparisons among unfamiliar fluids. For greater detail and a wider range of data, consult the sources listed in the References section at the end of the chapter and/or ASHRAE *Thermodynamic Properties of Refrigerants* (1985) and ASHRAE *Thermophysical Properties of Refrigerants* (1993).

The preparation of this chapter is assigned to TC 3.1, Refrigerants and Brines.

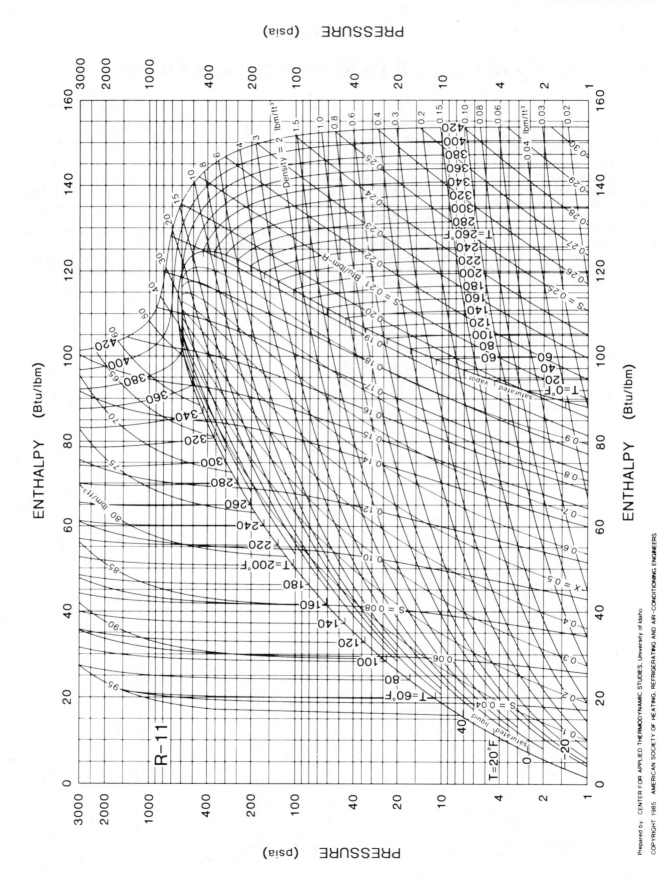

Prepared by CENTER FOR APPLIED THERMODYNAMIC STUDIES, University of Idaho
COPYRIGHT 1985   AMERICAN SOCIETY OF HEATING, REFRIGERATING AND AIR-CONDITIONING ENGINEERS

Fig. 1 Pressure-Enthalpy Diagram for Refrigerant 11

Refrigerant 11 (trichlorofluoromethane) Properties of Saturated Liquid and Saturated Vapor

Temp,* °F	Pressure, psia	Density, lb/ft³ Liquid	Volume, ft³/lb Vapor	Enthalpy, Btu/lb Liquid	Enthalpy, Btu/lb Vapor	Entropy, Btu/lb·°F Liquid	Entropy, Btu/lb·°F Vapor	Specific Heat c_p, Btu/lb·°F Liquid	Specific Heat c_p, Btu/lb·°F Vapor	c_p/c_v Vapor	Velocity of Sound, ft/s Liquid	Velocity of Sound, ft/s Vapor	Viscosity, lb_m/ft·h Liquid	Viscosity, lb_m/ft·h Vapor	Thermal Cond, Btu/h·ft·°F Liquid	Thermal Cond, Btu/h·ft·°F Vapor	Surface Tension, dyne/cm	Temp, °F
−166.85a	0.001	110.41	24230.	−24.922	73.328	−0.07085	0.26467	—	0.1005	1.1680	—	352.	—	—	—	—	—	−166.85
−160.00	0.002	109.93	14650.	−23.116	74.019	−0.06474	0.25940	0.1840	0.1019	1.1655	—	356.	—	—	—	—	—	−160.00
−150.00	0.003	109.23	7355.6	−21.267	75.045	−0.05867	0.25235	0.1860	0.1038	1.1620	3953.	361.	—	—	—	—	35.26	−150.00
−140.00	0.006	108.52	3882.0	−19.398	76.088	−0.05273	0.24597	0.1878	0.1058	1.1588	3895.	366.	—	—	—	—	34.44	−140.00
−130.00	0.012	107.81	2143.2	−17.511	77.149	−0.04692	0.24022	0.1894	0.1077	1.1558	3829.	371.	—	—	—	—	33.63	−130.00
−120.00	0.022	107.09	1232.5	−15.610	78.226	−0.04123	0.23502	0.1909	0.1095	1.1531	3758.	376.	—	—	—	—	32.82	−120.00
−110.00	0.037	106.37	735.62	−13.694	79.318	−0.03568	0.23032	0.1922	0.1114	1.1506	3684.	381.	—	—	—	—	32.01	−110.00
−100.00	0.062	105.65	454.14	−11.767	80.425	−0.03024	0.22608	0.1933	0.1133	1.1482	3609.	386.	—	—	—	—	31.21	−100.00
−90.00	0.100	104.92	289.15	−9.829	81.546	−0.02493	0.22225	0.1943	0.1151	1.1461	3533.	391.	—	—	—	—	30.41	−90.00
−80.00	0.156	104.19	189.37	−7.880	82.681	−0.01973	0.21880	0.1953	0.1170	1.1442	3457.	396.	—	—	—	—	29.62	−80.00
−70.00	0.239	103.46	127.27	−5.923	83.827	−0.01464	0.21568	0.1962	0.1188	1.1425	3382.	400.	—	—	—	—	28.83	−70.00
−60.00	0.355	102.73	87.583	−3.957	84.986	−0.00966	0.21288	0.1970	0.1206	1.1410	3308.	405.	—	—	—	—	28.04	−60.00
−50.00	0.517	101.99	61.601	−1.983	86.155	−0.00478	0.21036	0.1978	0.1225	1.1396	3235.	409.	—	—	—	—	27.26	−50.00
−40.00	0.738	101.25	44.203	0.000	87.335	0.00000	0.20810	0.1986	0.1243	1.1384	3163.	414.	—	—	—	—	26.49	−40.00
−30.00	1.032	100.50	32.310	1.991	88.524	0.00469	0.20608	0.1994	0.1262	1.1374	3093.	418.	—	—	—	—	25.72	−30.00
−20.00	1.419	99.75	24.022	3.989	89.721	0.00928	0.20427	0.2002	0.1280	1.1366	3024.	422.	1.889	0.0222	0.0595	—	24.95	−20.00
−10.00	1.918	99.00	18.143	5.997	90.927	0.01380	0.20267	0.2011	0.1298	1.1360	2955.	426.	1.782	0.0227	0.0587	—	24.19	−10.00
0.00	2.554	98.24	13.903	8.013	92.139	0.01823	0.20124	0.2020	0.1316	1.1356	2888.	430.	1.678	0.0231	0.0579	—	23.43	0.00
5.00	2.931	97.86	12.233	9.024	92.747	0.02041	0.20059	0.2024	0.1325	1.1354	2855.	432.	1.627	0.0233	0.0575	—	23.05	5.00
10.00	3.352	97.48	10.798	10.038	93.357	0.02258	0.19998	0.2029	0.1334	1.1354	2822.	433.	1.578	0.0236	0.0571	—	22.68	10.00
15.00	3.822	97.09	9.5606	11.054	93.967	0.02473	0.19941	0.2033	0.1343	1.1353	2789.	435.	1.529	0.0238	0.0567	—	22.30	15.00
20.00	4.343	96.71	8.4906	12.072	94.579	0.02686	0.19887	0.2038	0.1352	1.1354	2757.	437.	1.482	0.0240	0.0563	—	21.93	20.00
25.00	4.920	96.32	7.5621	13.093	95.192	0.02898	0.19837	0.2043	0.1361	1.1355	2725.	438.	1.436	0.0243	0.0559	—	21.56	25.00
30.00	5.557	95.93	6.7536	14.117	95.806	0.03108	0.19790	0.2048	0.1370	1.1356	2693.	440.	1.390	0.0245	0.0555	—	21.19	30.00
35.00	6.259	95.54	6.0477	15.143	96.420	0.03316	0.19747	0.2053	0.1379	1.1358	2661.	442.	1.347	0.0247	0.0551	0.00470	20.82	35.00
40.00	7.031	95.14	5.4294	16.172	97.035	0.03523	0.19706	0.2059	0.1388	1.1361	2629.	443.	1.304	0.0250	0.0548	0.00475	20.45	40.00
45.00	7.876	94.75	4.8863	17.203	97.650	0.03728	0.19668	0.2064	0.1397	1.1364	2598.	444.	1.262	0.0252	0.0544	0.00480	20.08	45.00
50.00	8.800	94.35	4.4080	18.238	98.265	0.03931	0.19633	0.2070	0.1406	1.1368	2567.	446.	1.222	0.0254	0.0540	0.00485	19.72	50.00
55.00	9.809	93.95	3.9856	19.275	98.880	0.04134	0.19601	0.2075	0.1415	1.1373	2536.	447.	1.183	0.0256	0.0536	0.00490	19.35	55.00
60.00	10.907	93.55	3.6116	20.315	99.495	0.04334	0.19571	0.2081	0.1424	1.1378	2505.	448.	1.145	0.0259	0.0532	0.00495	18.99	60.00
65.00	12.099	93.14	3.2795	21.358	100.109	0.04534	0.19543	0.2087	0.1433	1.1385	2474.	450.	1.109	0.0261	0.0528	0.00501	18.63	65.00
70.00	13.392	92.73	2.9841	22.405	100.723	0.04732	0.19518	0.2093	0.1442	1.1392	2444.	451.	1.073	0.0264	0.0524	0.00506	18.27	70.00
74.67b	14.696	92.35	2.7369	23.386	101.297	0.04916	0.19496	0.2099	0.1451	1.1399	2415.	452.	1.041	0.0266	0.0521	0.00511	17.94	74.67
75.00	14.790	92.33	2.7206	23.455	101.337	0.04928	0.19495	0.2100	0.1452	1.1400	2413.	452.	1.039	0.0266	0.0521	0.00512	17.91	75.00
80.00	16.301	91.91	2.4851	24.507	101.949	0.05124	0.19473	0.2106	0.1461	1.1409	2383.	453.	1.006	0.0268	0.0517	0.00517	17.56	80.00
85.00	17.929	91.50	2.2741	25.564	102.560	0.05318	0.19454	0.2113	0.1470	1.1419	2353.	454.	0.974	0.0271	0.0513	0.00523	17.20	85.00
90.00	19.681	91.08	2.0846	26.624	103.170	0.05511	0.19437	0.2119	0.1479	1.1429	2323.	455.	0.943	0.0273	0.0509	0.00528	16.85	90.00
95.00	21.563	90.66	1.9142	27.687	103.778	0.05703	0.19421	0.2126	0.1489	1.1441	2293.	456.	0.913	0.0276	0.0505	0.00534	16.49	95.00
100.00	23.581	90.23	1.7605	28.754	104.384	0.05894	0.19407	0.2134	0.1498	1.1454	2263.	456.	0.884	0.0278	0.0502	0.00540	16.14	100.00
105.00	25.743	89.81	1.6217	29.825	104.989	0.06083	0.19394	0.2141	0.1508	1.1468	2234.	457.	0.856	0.0281	0.0498	0.00546	15.79	105.00
110.00	28.053	89.38	1.4960	30.900	105.591	0.06272	0.19383	0.2149	0.1518	1.1483	2204.	458.	0.829	0.0283	0.0494	0.00552	15.45	110.00
115.00	30.520	88.94	1.3821	31.978	106.191	0.06459	0.19374	0.2156	0.1528	1.1500	2174.	458.	0.804	0.0286	0.0490	0.00559	—	115.00
120.00	33.150	88.51	1.2787	33.060	106.788	0.06646	0.19365	0.2164	0.1538	1.1517	2145.	459.	0.778	0.0288	0.0486	0.00565	—	120.00
125.00	35.950	88.07	1.1846	34.147	107.382	0.06832	0.19358	0.2172	0.1548	1.1536	2115.	459.	0.754	0.0291	0.0483	0.00572	—	125.00
130.00	38.926	87.62	1.0988	35.238	107.974	0.07017	0.19352	0.2181	0.1558	1.1557	2086.	459.	0.731	0.0293	0.0479	0.00578	—	130.00
135.00	42.087	87.17	1.0205	36.333	108.562	0.07200	0.19346	0.2189	0.1569	1.1579	2057.	459.	0.708	0.0296	0.0475	0.00585	—	135.00
140.00	45.439	86.72	0.9489	37.433	109.146	0.07383	0.19342	0.2198	0.1580	1.1602	2027.	459.	0.687	0.0299	0.0471	0.00592	—	140.00
145.00	48.989	86.26	0.8833	38.537	109.727	0.07565	0.19339	0.2207	0.1591	1.1628	1998.	460.	0.666	0.0302	0.0468	0.00599	—	145.00
150.00	52.745	85.80	0.8232	39.646	110.304	0.07747	0.19336	0.2217	0.1602	1.1655	1969.	459.	0.646	0.0304	0.0464	0.00606	—	150.00
160.00	60.905	84.86	0.7172	41.878	111.445	0.08107	0.19333	0.2236	0.1626	1.1715	1910.	459.	0.607	0.0310	0.0456	0.00621	—	160.00
170.00	69.979	83.91	0.6273	44.130	112.566	0.08464	0.19333	0.2257	0.1650	1.1783	1851.	458.	0.572	0.0316	0.0449	0.00637	—	170.00
180.00	80.030	82.93	0.5506	46.403	113.666	0.08819	0.19334	0.2279	0.1677	1.1861	1793.	457.	0.539	0.0322	0.0442	0.00654	—	180.00
190.00	91.120	81.93	0.4849	48.699	114.743	0.09171	0.19337	0.2303	0.1705	1.1950	1734.	456.	0.508	0.0328	0.0434	0.00671	—	190.00
200.00	103.31	80.90	0.4283	51.019	115.793	0.09521	0.19341	0.2328	0.1735	1.2052	1675.	454.	0.479	0.0334	0.0427	0.00689	—	200.00
210.00	116.68	79.85	0.3793	53.364	116.815	0.09870	0.19344	0.2356	0.1768	1.2169	1615.	451.	0.453	0.0341	0.0419	0.00707	—	210.00
220.00	131.28	78.77	0.3367	55.736	117.805	0.10216	0.19348	0.2385	0.1804	1.2302	1555.	448.	0.428	0.0347	0.0412	0.00727	—	220.00
230.00	147.18	77.65	0.2996	58.136	118.760	0.10561	0.19351	0.2418	0.1844	1.2456	1495.	445.	0.404	0.0354	0.0404	0.00748	—	230.00
240.00	164.46	76.50	0.2670	60.567	119.677	0.10905	0.19353	0.2453	0.1888	1.2634	1434.	441.	0.383	0.0362	0.0397	0.00769	—	240.00
250.00	183.19	75.30	0.2384	63.032	120.551	0.11248	0.19353	0.2493	0.1937	1.2842	1372.	437.	0.363	0.0369	0.0390	0.00792	—	250.00
260.00	203.43	74.06	0.2131	65.533	121.378	0.11591	0.19351	0.2537	0.1993	1.3086	1309.	432.	0.344	0.0377	0.0382	0.00815	—	260.00
270.00	225.26	72.78	0.1906	68.074	122.152	0.11934	0.19346	0.2587	0.2057	1.3375	1245.	426.	0.326	0.0385	0.0375	0.00840	—	270.00
280.00	248.77	71.43	0.1706	70.659	122.867	0.12278	0.19336	0.2645	0.2131	1.3721	1180.	420.	0.310	0.0393	0.0367	0.00866	—	280.00
290.00	274.03	70.01	0.1528	73.294	123.513	0.12623	0.19322	0.2713	0.2219	1.4142	1114.	413.	0.294	0.0401	0.0360	0.00893	—	290.00
300.00	301.12	68.51	0.1367	75.985	124.082	0.12970	0.19301	0.2794	0.2326	1.4663	1046.	406.	0.280	0.0410	0.0352	0.00921	—	300.00
310.00	330.14	66.92	0.1222	78.742	124.559	0.13320	0.19273	0.2894	0.2457	1.5322	976.	397.	—	—	—	—	—	310.00
320.00	361.18	65.21	0.1090	81.578	124.926	0.13675	0.19235	0.3020	0.2625	1.6179	904.	388.	—	—	—	—	—	320.00
330.00	394.36	63.35	0.0970	84.510	125.159	0.14037	0.19184	0.3188	0.2848	1.7336	830.	378.	—	—	—	—	—	330.00
340.00	429.78	61.29	0.0859	87.565	125.220	0.14408	0.19117	0.3423	0.3159	1.8976	752.	366.	—	—	—	—	—	340.00
350.00	467.60	58.97	0.0755	90.787	125.054	0.14794	0.19026	0.3778	0.3628	2.1476	671.	354.	—	—	—	—	—	350.00
360.00	507.98	56.24	0.0657	94.248	124.561	0.15203	0.18901	0.4376	0.4418	2.5730	586.	340.	—	—	—	—	—	360.00
370.00	551.15	52.86	0.0561	98.092	123.541	0.15651	0.18718	0.5570	0.6042	3.4525	499.	325.	—	—	—	—	—	370.00
380.00	597.49	48.25	0.0458	102.678	121.417	0.16180	0.18412	—	—	—	—	—	—	—	—	—	—	380.00
388.33c	639.27	34.59	0.0289	112.749	112.749	0.17350	0.17350	∞	∞	∞	0.	0.	—	—	∞	∞	0.00	388.33

*temperatures are on the ITS-90 scale a = triple point b = boiling point c = critical point

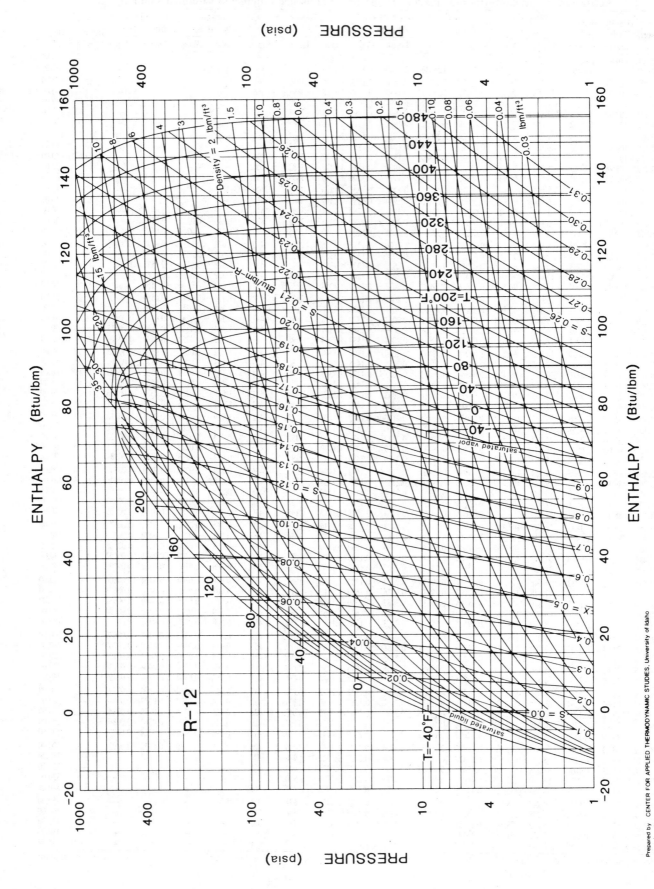

Fig. 2 Pressure-Enthalpy Diagram for Refrigerant 12

Prepared by   CENTER FOR APPLIED THERMODYNAMIC STUDIES, University of Idaho

COPYRIGHT 1985   AMERICAN SOCIETY OF HEATING, REFRIGERATING AND AIR-CONDITIONING ENGINEERS

Refrigerant 12 (dichlorodifluoromethane)   Properties of Saturated Liquid and Saturated Vapor

Temp,* °F	Pressure, psia	Density, lb/ft³ Liquid	Volume, ft³/lb Vapor	Enthalpy, Btu/lb Liquid	Vapor	Entropy, Btu/lb·°F Liquid	Vapor	Specific Heat c_p, Btu/lb·°F Liquid	Vapor	c_p/c_v Vapor	Velocity of Sound, ft/s Liquid	Vapor	Viscosity, lb$_m$/ft·h Liquid	Vapor	Thermal Cond, Btu/h·ft·°F Liquid	Vapor	Surface Tension, dyne/cm	Temp, °F
−140.00	0.260	104.06	108.98	−20.324	61.801	−0.05521	0.20169	0.1978	0.1099	1.1791	2676.	393.	—	—	—	—	25.79	−140.00
−130.00	0.417	103.16	69.954	−18.338	62.880	−0.04910	0.19727	0.1994	0.1122	1.1759	2665.	398.	—	—	—	—	24.91	−130.00
−120.00	0.648	102.24	46.326	−16.337	63.974	−0.04312	0.19332	0.2005	0.1145	1.1732	2648.	403.	—	—	—	—	24.04	−120.00
−110.00	0.979	101.32	31.556	−14.327	65.081	−0.03729	0.18981	0.2014	0.1168	1.1708	2625.	408.	—	—	—	—	23.18	−110.00
−100.00	1.438	100.40	22.049	−12.308	66.199	−0.03160	0.18668	0.2023	0.1190	1.1688	2597.	413.	—	—	—	—	22.33	−100.00
−95.00	1.727	99.93	18.594	−11.295	66.762	−0.02880	0.18525	0.2027	0.1202	1.1680	2581.	415.	—	—	—	—	21.90	−95.00
−90.00	2.063	99.46	15.766	−10.280	67.327	−0.02604	0.18390	0.2031	0.1213	1.1673	2563.	417.	1.343	0.0200	0.0594	0.00294	21.48	−90.00
−85.00	2.450	98.99	13.438	−9.263	67.894	−0.02331	0.18262	0.2035	0.1224	1.1666	2545.	420.	1.302	0.0204	0.0588	0.00303	21.06	−85.00
−80.00	2.894	98.52	11.511	−8.244	68.462	−0.02061	0.18142	0.2039	0.1235	1.1661	2526.	422.	1.262	0.0209	0.0582	0.00311	20.64	−80.00
−75.00	3.402	98.04	9.9073	−7.223	69.032	−0.01794	0.18029	0.2044	0.1245	1.1657	2505.	424.	1.223	0.0213	0.0576	0.00320	20.22	−75.00
−70.00	3.980	97.56	8.5655	−6.200	69.603	−0.01530	0.17923	0.2048	0.1256	1.1654	2484.	426.	1.185	0.0217	0.0569	0.00328	19.80	−70.00
−65.00	4.634	97.08	7.4374	−5.173	70.175	−0.01269	0.17823	0.2053	0.1267	1.1653	2462.	428.	1.148	0.0222	0.0563	0.00337	19.39	−65.00
−60.00	5.372	96.60	6.4844	−4.145	70.747	−0.01010	0.17728	0.2059	0.1278	1.1652	2440.	429.	1.112	0.0226	0.0557	0.00345	18.98	−60.00
−55.00	6.202	96.12	5.6756	−3.113	71.320	−0.00754	0.17640	0.2064	0.1289	1.1653	2416.	431.	1.077	0.0230	0.0551	0.00353	18.57	−55.00
−50.00	7.131	95.63	4.9863	−2.078	71.893	−0.00500	0.17556	0.2070	0.1300	1.1655	2392.	433.	1.043	0.0233	0.0545	0.00361	18.16	−50.00
−45.00	8.167	95.14	4.3963	−1.041	72.465	−0.00249	0.17478	0.2076	0.1311	1.1658	2367.	435.	1.010	0.0237	0.0538	0.00369	17.75	−45.00
−40.00	9.319	94.65	3.8894	0.000	73.038	0.00000	0.17404	0.2082	0.1322	1.1663	2342.	436.	0.978	0.0241	0.0532	0.00377	17.35	−40.00
−35.00	10.597	94.15	3.4520	1.044	73.609	0.00247	0.17334	0.2089	0.1334	1.1669	2316.	438.	0.947	0.0244	0.0526	0.00385	16.95	−35.00
−30.00	12.008	93.65	3.0734	2.092	74.179	0.00492	0.17269	0.2096	0.1345	1.1676	2289.	439.	0.917	0.0248	0.0520	0.00393	16.55	−30.00
−25.00	13.562	93.15	2.7443	3.144	74.749	0.00734	0.17207	0.2104	0.1357	1.1685	2262.	440.	0.888	0.0251	0.0514	0.00401	16.15	−25.00
−21.63b	14.696	92.81	2.5467	3.855	75.131	0.00897	0.17168	0.2109	0.1365	1.1692	2244.	441.	0.869	0.0254	0.0510	0.00407	15.88	−21.63
−20.00	15.270	92.64	2.4574	4.200	75.316	0.00975	0.17150	0.2112	0.1369	1.1696	2235.	442.	0.860	0.0255	0.0508	0.00409	15.75	−20.00
−15.00	17.141	92.13	2.2065	5.260	75.881	0.01214	0.17096	0.2120	0.1381	1.1708	2207.	443.	0.834	0.0258	0.0502	0.00417	15.36	−15.00
−10.00	19.186	91.62	1.9863	6.324	76.445	0.01451	0.17045	0.2129	0.1393	1.1722	2178.	444.	0.808	0.0262	0.0496	0.00426	14.97	−10.00
−5.00	21.414	91.10	1.7924	7.392	77.006	0.01686	0.16997	0.2138	0.1406	1.1737	2149.	445.	0.783	0.0265	0.0490	0.00434	14.58	−5.00
0.00	23.837	90.58	1.6213	8.466	77.564	0.01920	0.16952	0.2147	0.1419	1.1755	2120.	446.	0.758	0.0268	0.0484	0.00442	14.19	0.00
5.00	26.466	90.05	1.4698	9.544	78.119	0.02152	0.16910	0.2157	0.1432	1.1774	2091.	446.	0.735	0.0272	0.0478	0.00450	—	5.00
10.00	29.313	89.52	1.3353	10.627	78.671	0.02383	0.16870	0.2167	0.1446	1.1796	2061.	447.	0.713	0.0275	0.0472	0.00458	—	10.00
15.00	32.387	88.99	1.2156	11.716	79.219	0.02612	0.16833	0.2177	0.1460	1.1819	2030.	447.	0.691	0.0278	0.0466	0.00467	—	15.00
20.00	35.702	88.44	1.1088	12.810	79.763	0.02840	0.16798	0.2188	0.1474	1.1845	2000.	448.	0.670	0.0282	0.0460	0.00475	—	20.00
25.00	39.269	87.90	1.0133	13.910	80.303	0.03066	0.16765	0.2199	0.1489	1.1873	1969.	448.	0.650	0.0285	0.0454	0.00484	—	25.00
30.00	43.101	87.35	0.9276	15.016	80.838	0.03291	0.16734	0.2211	0.1504	1.1904	1938.	448.	0.631	0.0289	0.0448	0.00492	—	30.00
35.00	47.209	86.79	0.8506	16.127	81.369	0.03516	0.16704	0.2223	0.1520	1.1938	1906.	448.	0.612	0.0292	0.0442	0.00501	—	35.00
40.00	51.605	86.23	0.7813	17.245	81.894	0.03739	0.16677	0.2236	0.1537	1.1974	1875.	448.	0.595	0.0296	0.0436	0.00510	—	40.00
45.00	56.303	85.66	0.7187	18.370	82.413	0.03961	0.16651	0.2249	0.1554	1.2014	1843.	448.	0.577	0.0299	0.0431	0.00519	—	45.00
50.00	61.316	85.08	0.6621	19.501	82.927	0.04181	0.16626	0.2263	0.1571	1.2057	1810.	448.	0.561	0.0303	0.0425	0.00528	—	50.00
55.00	66.656	84.49	0.6109	20.640	83.434	0.04401	0.16602	0.2278	0.1589	1.2103	1778.	448.	0.545	0.0306	0.0419	0.00537	—	55.00
60.00	72.336	83.90	0.5643	21.786	83.935	0.04621	0.16580	0.2293	0.1608	1.2154	1745.	447.	0.529	0.0310	0.0413	0.00546	—	60.00
65.00	78.369	83.30	0.5219	22.939	84.428	0.04839	0.16559	0.2308	0.1628	1.2209	1712.	446.	0.514	0.0314	0.0407	0.00556	—	65.00
70.00	84.770	82.70	0.4833	24.100	84.914	0.05056	0.16538	0.2324	0.1648	1.2268	1679.	445.	0.500	0.0318	0.0401	0.00566	—	70.00
75.00	91.551	82.08	0.4481	25.269	85.392	0.05273	0.16518	0.2342	0.1670	1.2332	1645.	444.	0.486	0.0322	0.0396	0.00576	—	75.00
80.00	98.727	81.45	0.4158	26.447	85.861	0.05489	0.16499	0.2359	0.1692	1.2402	1611.	443.	0.473	0.0326	0.0390	0.00586	—	80.00
85.00	106.31	80.82	0.3863	27.633	86.321	0.05705	0.16480	0.2378	0.1716	1.2478	1577.	442.	0.460	0.0330	0.0384	0.00596	—	85.00
90.00	114.32	80.17	0.3592	28.829	86.772	0.05920	0.16462	0.2398	0.1740	1.2560	1543.	441.	0.448	0.0335	0.0378	0.00606	—	90.00
95.00	122.76	79.51	0.3342	30.034	87.212	0.06135	0.16444	0.2419	0.1766	1.2650	1508.	439.	0.436	0.0339	0.0373	0.00617	—	95.00
100.00	131.65	78.84	0.3113	31.249	87.641	0.06349	0.16425	0.2441	0.1794	1.2749	1473.	437.	0.424	0.0344	0.0367	0.00628	—	100.00
105.00	141.01	78.16	0.2901	32.474	88.059	0.06563	0.16407	0.2464	0.1823	1.2856	1438.	435.	0.413	0.0349	0.0361	0.00639	—	105.00
110.00	150.85	77.46	0.2706	33.711	88.464	0.06777	0.16389	0.2489	0.1855	1.2974	1402.	433.	0.402	0.0354	0.0356	0.00650	—	110.00
115.00	161.19	76.75	0.2525	34.959	88.856	0.06991	0.16370	0.2515	0.1888	1.3103	1367.	431.	0.392	0.0359	0.0350	0.00662	—	115.00
120.00	172.04	76.02	0.2358	36.219	89.233	0.07205	0.16351	0.2543	0.1924	1.3246	1331.	429.	0.382	0.0364	0.0345	0.00674	—	120.00
125.00	183.41	75.28	0.2203	37.492	89.596	0.07419	0.16332	0.2573	0.1963	1.3404	1294.	426.	0.372	0.0370	0.0339	0.00686	—	125.00
130.00	195.33	74.51	0.2058	38.778	89.942	0.07633	0.16309	0.2606	0.2005	1.3579	1258.	423.	0.363	0.0375	0.0333	0.00698	—	130.00
135.00	207.80	73.73	0.1924	40.079	90.270	0.07847	0.16287	0.2641	0.2050	1.3774	1220.	420.	0.354	0.0381	0.0328	0.00711	—	135.00
140.00	220.86	72.93	0.1799	41.394	90.578	0.08062	0.16264	0.2679	0.2100	1.3992	1183.	417.	0.345	0.0387	0.0322	0.00724	—	140.00
145.00	234.50	72.10	0.1682	42.727	90.866	0.08277	0.16239	0.2721	0.2155	1.4238	1145.	413.	0.337	0.0394	0.0317	0.00737	—	145.00
150.00	248.75	71.25	0.1573	44.076	91.131	0.08494	0.16212	0.2767	0.2217	1.4516	1107.	410.	0.329	0.0400	0.0311	0.00750	—	150.00
155.00	263.64	70.37	0.1471	45.444	91.371	0.08711	0.16183	0.2817	0.2285	1.4833	1069.	406.	0.321	0.0407	0.0306	0.00764	—	155.00
160.00	279.17	69.45	0.1375	46.833	91.583	0.08929	0.16151	0.2874	0.2363	1.5197	1030.	401.	—	—	—	—	—	160.00
165.00	295.37	68.51	0.1285	48.244	91.765	0.09149	0.16116	0.2937	0.2451	1.5619	990.	397.	—	—	—	—	—	165.00
170.00	312.25	67.52	0.1200	49.680	91.912	0.09371	0.16078	0.3009	0.2553	1.6113	950.	392.	—	—	—	—	—	170.00
175.00	329.85	66.50	0.1120	51.142	92.021	0.09594	0.16035	0.3092	0.2673	1.6697	910.	387.	—	—	—	—	—	175.00
180.00	348.18	65.42	0.1044	52.635	92.085	0.09820	0.15988	0.3188	0.2814	1.7397	868.	382.	—	—	—	—	—	180.00
185.00	367.27	64.28	0.0973	54.162	92.099	0.10050	0.15935	0.3302	0.2986	1.8253	826.	376.	—	—	—	—	—	185.00
190.00	387.14	63.08	0.0904	55.727	92.054	0.10283	0.15874	0.3439	0.3197	1.9317	783.	370.	—	—	—	—	—	190.00
195.00	407.83	61.80	0.0839	57.339	91.938	0.10520	0.15805	0.3607	0.3465	2.0675	739.	364.	—	—	—	—	—	195.00
200.00	429.37	60.43	0.0776	59.004	91.736	0.10764	0.15726	0.3822	0.3817	2.2464	693.	357.	—	—	—	—	—	200.00
210.00	475.17	57.29	0.0656	62.550	90.978	0.11276	0.15521	0.4508	0.4997	2.8478	595.	343.	—	—	—	—	—	210.00
220.00	524.96	53.26	0.0537	66.578	89.410	0.11848	0.15207	0.6212	0.8091	4.4129	483.	326.	—	—	—	—	—	220.00
230.00	579.55	46.30	0.0397	72.122	85.410	0.12628	0.14554	—	—	—	—	—	—	—	—	—	—	230.00
233.20c	598.27	35.25	0.0284	78.775	78.775	0.13578	0.13578	∞	∞	∞	0.	0.	—	—	∞	∞	0.00	233.20

*temperatures are on the ITS-90 scale   b = boiling point   c = critical point

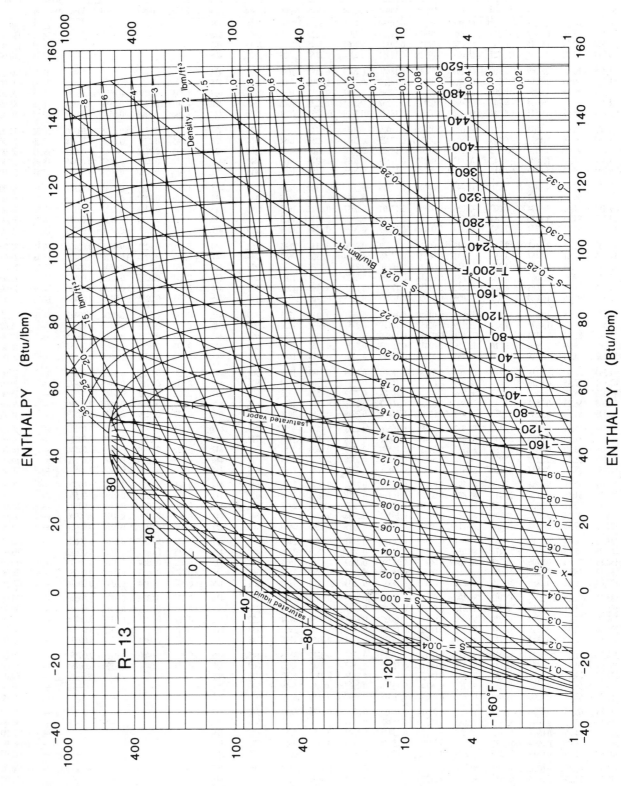

Fig. 3   Pressure-Enthalpy Diagram for Refrigerant 13

Prepared by: CENTER FOR APPLIED THERMODYNAMIC STUDIES, University of Idaho

Refrigerant 13 (chlorotrifluoromethane) Properties of Saturated Liquid and Saturated Vapor

Temp,* °F	Pressure, psia	Density, lb/ft³ Liquid	Volume, ft³/lb Vapor	Enthalpy, Btu/lb Liquid	Vapor	Entropy, Btu/lb·°F Liquid	Vapor	Specific Heat c_p, Btu/lb·°F Liquid	Vapor	c_p/c_v Vapor	Velocity of Sound, ft/s Liquid	Vapor	Viscosity, lb$_m$/ft·h Liquid	Vapor	Thermal Cond, Btu/h·ft·°F Liquid	Vapor	Surface Tension, dyne/cm	Temp, °F
−150.00	4.452	99.55	7.0160	−24.352	43.797	−0.06649	0.15358	—	0.1138	1.2216	—	417.	—	—	—	—	17.33	−150.00
−145.00	5.368	98.92	5.8959	−23.355	44.297	−0.06330	0.15169	—	0.1152	1.2216	—	419.	—	—	—	—	16.86	−145.00
−140.00	6.431	98.29	4.9849	−22.347	44.796	−0.06013	0.14991	—	0.1167	1.2218	—	421.	—	—	—	—	16.39	−140.00
−135.00	7.655	97.65	4.2390	−21.329	45.292	−0.05698	0.14822	—	0.1182	1.2223	—	423.	—	—	—	—	15.93	−135.00
−130.00	9.059	97.00	3.6245	−20.301	45.786	−0.05384	0.14662	—	0.1198	1.2232	—	425.	—	—	—	—	15.46	−130.00
−125.00	10.659	96.35	3.1150	−19.262	46.276	−0.05072	0.14511	—	0.1214	1.2244	—	426.	—	—	—	—	15.00	−125.00
−120.00	12.474	95.70	2.6902	−18.212	46.762	−0.04762	0.14367	—	0.1231	1.2259	—	428.	—	—	—	—	14.55	−120.00
−115.00	14.523	95.03	2.3340	−17.151	47.245	−0.04453	0.14230	—	0.1248	1.2277	—	429.	—	—	—	—	14.09	−115.00
−114.60b	14.696	94.98	2.3083	−17.067	47.283	−0.04429	0.14220	—	0.1250	1.2279	—	429.	—	—	—	—	14.06	−114.60
−110.00	16.826	94.36	2.0338	−16.080	47.722	−0.04146	0.14100	—	0.1266	1.2299	—	431.	—	—	—	—	13.64	−110.00
−105.00	19.403	93.69	1.7795	−14.998	48.191	−0.03840	0.13977	—	0.1285	1.2325	—	432.	—	—	—	—	13.20	−105.00
−100.00	22.276	93.01	1.5632	−13.905	48.660	−0.03536	0.13860	—	0.1304	1.2355	—	433.	—	—	—	—	12.75	−100.00
−95.00	25.466	92.32	1.3781	−12.802	49.121	−0.03233	0.13748	—	0.1324	1.2390	—	434.	—	—	—	—	12.31	−95.00
−90.00	28.995	91.62	1.2192	−11.689	49.574	−0.02932	0.13641	0.1721	0.1344	1.2428	—	434.	0.634	0.0243	0.0468	0.00383	11.87	−90.00
−85.00	32.885	90.91	1.0822	−10.565	50.021	−0.02632	0.13539	0.1761	0.1365	1.2472	—	435.	0.600	0.0247	0.0459	0.00395	11.44	−85.00
−80.00	37.160	90.20	0.9636	−9.431	50.459	−0.02333	0.13441	0.1800	0.1388	1.2520	—	435.	0.569	0.0251	0.0450	0.00407	11.01	−80.00
−75.00	41.842	89.47	0.8606	−8.287	50.890	−0.02037	0.13347	0.1839	0.1411	1.2574	—	435.	0.540	0.0255	0.0442	0.00418	10.58	−75.00
−70.00	46.955	88.74	0.7706	−7.133	51.312	−0.01741	0.13257	0.1878	0.1435	1.2634	—	436.	0.514	0.0259	0.0434	0.00430	10.16	−70.00
−68.00	49.126	88.44	0.7379	−6.668	51.478	−0.01623	0.13222	0.1893	0.1445	1.2659	—	436.	0.504	0.0261	0.0431	0.00435	9.99	−68.00
−66.00	51.372	88.15	0.7069	−6.202	51.643	−0.01506	0.13188	0.1909	0.1455	1.2686	—	435.	0.494	0.0262	0.0428	0.00440	9.82	−66.00
−64.00	53.694	87.85	0.6774	−5.734	51.806	−0.01389	0.13154	0.1924	0.1466	1.2713	—	435.	0.484	0.0264	0.0425	0.00444	9.66	−64.00
−62.00	56.092	87.55	0.6494	−5.265	51.968	−0.01272	0.13120	0.1940	0.1476	1.2742	—	435.	0.475	0.0265	0.0422	0.00449	9.49	−62.00
−60.00	58.570	87.24	0.6228	−4.794	52.128	−0.01155	0.13087	0.1956	0.1487	1.2772	—	435.	0.466	0.0267	0.0418	0.00454	9.33	−60.00
−58.00	61.129	86.94	0.5975	−4.322	52.287	−0.01038	0.13055	0.1971	0.1498	1.2803	—	435.	0.458	0.0269	0.0416	0.00459	9.16	−58.00
−56.00	63.769	86.63	0.5735	−3.848	52.443	−0.00922	0.13023	0.1987	0.1509	1.2835	—	435.	0.449	0.0270	0.0413	0.00464	9.00	−56.00
−54.00	66.493	86.32	0.5506	−3.373	52.599	−0.00806	0.12991	0.2003	0.1520	1.2869	—	435.	0.441	0.0272	0.0410	0.00468	8.83	−54.00
−52.00	69.302	86.01	0.5288	−2.895	52.752	−0.00690	0.12960	0.2019	0.1532	1.2904	—	434.	0.433	0.0273	0.0407	0.00473	8.67	−52.00
−50.00	72.199	85.70	0.5081	−2.417	52.903	−0.00575	0.12929	0.2035	0.1543	1.2940	—	434.	0.426	0.0275	0.0404	0.00478	8.51	−50.00
−48.00	75.183	85.38	0.4883	−1.937	53.053	−0.00459	0.12898	0.2051	0.1555	1.2977	—	434.	0.418	0.0276	0.0401	0.00484	8.35	−48.00
−46.00	78.258	85.06	0.4694	−1.455	53.201	−0.00344	0.12868	0.2068	0.1568	1.3016	—	433.	0.411	0.0278	0.0398	0.00489	8.18	−46.00
−44.00	81.424	84.74	0.4515	−0.972	53.347	−0.00229	0.12838	0.2085	0.1580	1.3057	—	433.	0.404	0.0279	0.0395	0.00494	8.02	−44.00
−42.00	84.683	84.42	0.4343	−0.487	53.491	−0.00114	0.12809	0.2102	0.1593	1.3099	—	433.	0.397	0.0281	0.0393	0.00499	7.86	−42.00
−40.00	88.037	84.10	0.4179	0.000	53.633	0.00000	0.12780	0.2119	0.1606	1.3143	—	432.	0.391	0.0282	0.0390	0.00505	7.70	−40.00
−38.00	91.487	83.77	0.4022	0.488	53.773	0.00114	0.12751	0.2136	0.1620	1.3188	—	432.	0.384	0.0284	0.0387	0.00510	7.55	−38.00
−36.00	95.036	83.44	0.3873	0.978	53.911	0.00228	0.12722	0.2154	0.1633	1.3236	—	431.	0.378	0.0286	0.0385	0.00516	7.39	−36.00
−34.00	98.684	83.11	0.3730	1.469	54.046	0.00342	0.12694	0.2172	0.1647	1.3285	—	431.	0.372	0.0287	0.0382	0.00521	7.23	−34.00
−32.00	102.43	82.77	0.3593	1.962	54.180	0.00456	0.12665	0.2190	0.1662	1.3336	—	430.	0.366	0.0289	0.0379	0.00527	7.08	−32.00
−30.00	106.29	82.43	0.3462	2.457	54.311	0.00569	0.12637	0.2208	0.1676	1.3389	—	429.	0.360	0.0291	0.0377	0.00533	6.92	−30.00
−28.00	110.24	82.09	0.3337	2.953	54.440	0.00682	0.12609	0.2227	0.1691	1.3445	—	429.	0.354	0.0292	0.0374	0.00539	6.77	−28.00
−26.00	114.31	81.75	0.3217	3.451	54.566	0.00795	0.12582	0.2246	0.1707	1.3502	—	428.	0.349	0.0294	0.0372	0.00545	6.61	−26.00
−24.00	118.48	81.40	0.3102	3.951	54.691	0.00908	0.12554	0.2266	0.1723	1.3562	—	428.	0.343	0.0296	0.0369	0.00551	6.46	−24.00
−22.00	122.76	81.05	0.2992	4.452	54.812	0.01020	0.12527	0.2286	0.1739	1.3625	—	427.	0.338	0.0298	0.0367	0.00558	6.31	−22.00
−20.00	127.15	80.70	0.2886	4.955	54.931	0.01133	0.12500	0.2306	0.1756	1.3690	—	426.	0.333	0.0299	0.0364	0.00564	6.16	−20.00
−18.00	131.65	80.34	0.2785	5.460	55.048	0.01245	0.12472	0.2327	0.1773	1.3758	—	425.	0.328	0.0301	0.0362	0.00571	6.00	−18.00
−16.00	136.27	79.98	0.2688	5.967	55.162	0.01357	0.12445	0.2348	0.1791	1.3829	—	424.	0.323	0.0303	0.0359	0.00578	5.85	−16.00
−14.00	141.01	79.62	0.2594	6.476	55.273	0.01469	0.12418	0.2370	0.1809	1.3903	—	423.	0.318	0.0305	0.0357	0.00585	5.71	−14.00
−12.00	145.86	79.25	0.2505	6.986	55.381	0.01581	0.12391	0.2392	0.1827	1.3980	—	423.	0.314	0.0307	0.0354	0.00592	5.56	−12.00
−10.00	150.83	78.88	0.2418	7.498	55.486	0.01692	0.12364	0.2414	0.1847	1.4061	—	422.	0.309	0.0309	0.0352	0.00599	5.41	−10.00
−8.00	155.93	78.51	0.2336	8.013	55.588	0.01804	0.12337	0.2437	0.1867	1.4145	—	421.	0.305	0.0311	0.0350	0.00606	5.26	−8.00
−6.00	161.14	78.13	0.2256	8.529	55.687	0.01915	0.12310	0.2461	0.1887	1.4234	—	420.	0.301	0.0313	0.0347	0.00614	5.12	−6.00
−4.00	166.48	77.74	0.2179	9.048	55.783	0.02027	0.12283	0.2485	0.1909	1.4326	—	419.	0.296	0.0315	0.0345	0.00622	4.97	−4.00
−2.00	171.95	77.36	0.2106	9.568	55.875	0.02138	0.12256	0.2509	0.1931	1.4423	—	417.	0.292	0.0317	0.0343	0.00629	4.83	−2.00
0.00	177.55	76.96	0.2035	10.091	55.965	0.02249	0.12228	0.2534	0.1954	1.4525	—	416.	0.288	0.0319	0.0340	0.00638	4.69	0.00
5.00	192.12	75.96	0.1869	11.408	56.172	0.02526	0.12159	0.2599	0.2015	1.4802	—	413.	0.279	0.0325	0.0335	0.00659	4.33	5.00
10.00	207.53	74.92	0.1717	12.740	56.354	0.02803	0.12089	0.2668	0.2082	1.5117	—	410.	0.270	0.0331	0.0329	0.00681	3.99	10.00
15.00	223.81	73.85	0.1579	14.089	56.510	0.03080	0.12017	0.2742	0.2157	1.5478	—	406.	0.261	0.0337	0.0324	0.00705	3.65	15.00
20.00	241.00	72.73	0.1452	15.457	56.637	0.03358	0.11943	0.2819	0.2241	1.5894	—	402.	0.253	0.0343	0.0318	0.00730	3.31	20.00
25.00	259.13	71.56	0.1335	16.846	56.730	0.03636	0.11865	0.2901	0.2337	1.6380	—	398.	0.245	0.0350	0.0313	0.00756	2.98	25.00
30.00	278.23	70.34	0.1227	18.260	56.787	0.03916	0.11784	0.2988	0.2447	1.6953	—	394.	0.238	0.0358	0.0308	0.00784	2.67	30.00
35.00	298.34	69.06	0.1127	19.702	56.800	0.04198	0.11698	—	0.2577	1.7640	—	389.	—	—	—	—	2.35	35.00
40.00	319.50	67.69	0.1035	21.177	56.765	0.04483	0.11606	—	0.2731	1.8477	—	383.	—	—	—	—	2.05	40.00
45.00	341.75	66.24	0.0948	22.691	56.671	0.04773	0.11506	—	0.2920	1.9522	—	378.	—	—	—	—	1.76	45.00
50.00	365.15	64.68	0.0867	24.256	56.506	0.05068	0.11396	—	0.3158	2.0862	—	371.	—	—	—	—	1.48	50.00
55.00	389.75	62.98	0.0791	25.883	56.254	0.05372	0.11273	—	0.3470	2.2644	—	365.	—	—	—	—	1.20	55.00
60.00	415.62	61.10	0.0718	27.593	55.889	0.05688	0.11133	—	0.3900	2.5141	—	357.	—	—	—	—	0.95	60.00
65.00	442.85	58.97	0.0648	29.418	55.370	0.06021	0.10967	—	0.4541	2.8901	—	349.	—	—	—	—	0.70	65.00
70.00	471.55	56.47	0.0579	31.418	54.624	0.06383	0.10764	—	0.5612	3.5254	—	340.	—	—	—	—	0.47	70.00
75.00	501.88	53.36	0.0507	33.723	53.494	0.06797	0.10495	—	0.7824	4.8478	—	331.	—	—	—	—	0.27	75.00
80.00	534.14	48.85	0.0426	36.745	51.495	0.07338	0.10071	—	—	—	—	—	—	—	—	—	0.09	80.00
83.93c	561.23	36.07	0.0277	44.270	44.270	0.08704	0.08704	∞	∞	∞	0.	0.	—	—	∞	∞	0.00	83.93

*temperatures are on the IPTS-68 scale   b = boiling point   c = critical point

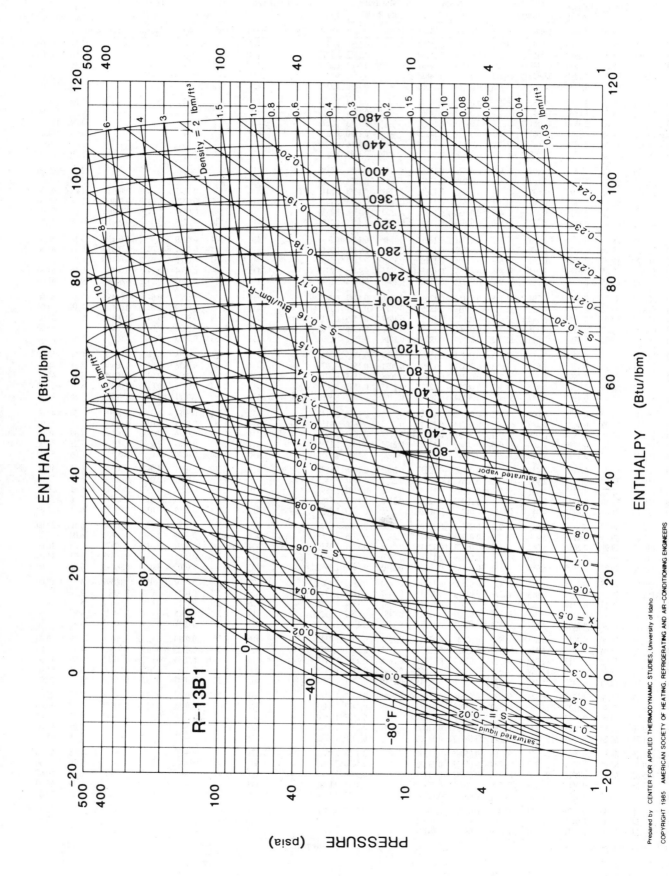

Fig. 4 Pressure-Enthalpy Diagram for Refrigerant 13B1

Refrigerant 13B1 (bromotrifluoromethane) Properties of Saturated Liquid and Saturated Vapor

Temp,* °F	Pressure, psia	Density, lb/ft³ Liquid	Volume, ft³/lb Vapor	Enthalpy, Btu/lb Liquid	Enthalpy Vapor	Entropy, Btu/lb·°F Liquid	Entropy Vapor	Specific Heat c_p, Btu/lb·°F Liquid	Specific Heat Vapor	c_p/c_v Vapor	Velocity of Sound, ft/s Liquid	Velocity of Sound Vapor	Viscosity, lb_m/ft·h Liquid	Viscosity Vapor	Thermal Cond., Btu/h·ft·°F Liquid	Thermal Cond. Vapor	Surface Tension, dyne/cm	Temp, °F
−140.00	1.549	134.07	14.7850	−15.587	40.243	−0.04219	0.13246	0.1466	—	—	—	—	1.938	0.0224	0.0517	—	20.58	−140.00
−130.00	2.305	132.72	10.2210	−14.107	41.066	−0.03763	0.12973	0.1491	—	—	—	—	1.722	0.0232	0.0505	—	19.68	−130.00
−120.00	3.340	131.34	7.2472	−12.616	41.895	−0.03318	0.12730	0.1514	—	—	—	—	1.539	0.0240	0.0493	—	18.78	−120.00
−110.00	4.724	129.94	5.2571	−11.110	42.728	−0.02882	0.12515	0.1534	0.0888	—	—	—	1.384	0.0248	0.0481	—	17.90	−110.00
−100.00	6.536	128.51	3.8924	−9.586	43.564	−0.02453	0.12325	0.1553	0.0924	—	—	—	1.252	0.0256	0.0469	—	17.03	−100.00
−90.00	8.864	127.06	2.9356	−8.044	44.399	−0.02031	0.12156	0.1570	0.0952	—	—	—	1.138	0.0265	0.0457	—	16.17	−90.00
−85.00	10.250	126.32	2.5656	−7.265	44.816	−0.01822	0.12079	0.1579	0.0964	—	—	—	1.087	0.0269	0.0451	—	15.74	−85.00
−80.00	11.803	125.58	2.2510	−6.481	45.232	−0.01614	0.12006	0.1587	0.0975	—	—	—	1.040	0.0273	0.0446	—	15.31	−80.00
−75.00	13.534	124.83	1.9825	−5.691	45.647	−0.01409	0.11937	0.1595	0.0985	—	—	—	0.995	0.0277	0.0440	—	14.89	−75.00
−71.92b	14.696	124.37	1.8365	−5.202	45.902	−0.01282	0.11897	0.1600	0.0990	—	—	—	0.969	0.0280	0.0436	—	14.63	−71.92
−70.00	15.458	124.07	1.7522	−4.896	46.061	−0.01204	0.11873	0.1603	0.0993	—	—	—	0.954	0.0281	0.0434	—	14.47	−70.00
−60.00	19.939	122.54	1.3824	−3.288	46.883	−0.00798	0.11755	0.1620	0.1008	—	—	—	0.878	0.0290	0.0423	—	13.64	−60.00
−58.00	20.944	122.23	1.3204	−2.963	47.047	−0.00717	0.11733	0.1623	0.1011	—	—	—	0.865	0.0291	0.0421	0.00329	—	−58.00
−56.00	21.989	121.91	1.2618	−2.638	47.210	−0.00637	0.11712	0.1626	0.1013	—	—	—	0.851	0.0293	0.0419	0.00332	—	−56.00
−54.00	23.073	121.60	1.2063	−2.312	47.373	−0.00557	0.11691	0.1630	0.1016	—	—	—	0.838	0.0295	0.0417	0.00336	—	−54.00
−52.00	24.197	121.28	1.1538	−1.984	47.535	−0.00477	0.11670	0.1633	0.1018	—	—	—	0.825	0.0296	0.0414	0.00339	—	−52.00
−50.00	25.364	120.96	1.1041	−1.656	47.697	−0.00397	0.11650	0.1637	0.1020	—	—	—	0.812	0.0298	0.0412	0.00342	—	−50.00
−48.00	26.572	120.65	1.0570	−1.327	47.859	−0.00317	0.11631	0.1640	0.1022	—	—	—	0.800	0.0300	0.0410	0.00345	—	−48.00
−46.00	27.825	120.33	1.0123	−0.997	48.019	−0.00238	0.11611	0.1644	0.1025	—	—	—	0.788	0.0301	0.0408	0.00349	—	−46.00
−44.00	29.122	120.00	0.9699	−0.665	48.180	−0.00158	0.11593	0.1647	0.1027	—	—	—	0.776	0.0303	0.0406	0.00352	—	−44.00
−42.00	30.465	119.68	0.9297	−0.333	48.340	−0.00079	0.11574	0.1651	0.1029	—	—	—	0.765	0.0305	0.0404	0.00355	—	−42.00
−40.00	31.855	119.36	0.8915	0.000	48.499	0.00000	0.11557	0.1655	0.1031	—	—	—	0.754	0.0306	0.0401	0.00359	—	−40.00
−38.00	33.292	119.03	0.8552	0.334	48.658	0.00079	0.11539	0.1659	0.1033	—	—	—	0.743	0.0308	0.0399	0.00362	—	−38.00
−36.00	34.778	118.70	0.8208	0.669	48.817	0.00158	0.11522	0.1662	0.1035	—	—	—	0.732	0.0310	0.0397	0.00366	—	−36.00
−34.00	36.314	118.37	0.7880	1.005	48.974	0.00236	0.11505	0.1666	0.1038	—	—	—	0.722	0.0311	0.0395	0.00369	—	−34.00
−32.00	37.901	118.04	0.7568	1.343	49.131	0.00315	0.11489	0.1670	0.1040	—	—	—	0.712	0.0313	0.0393	0.00373	—	−32.00
−30.00	39.540	117.71	0.7271	1.681	49.288	0.00393	0.11473	0.1675	0.1042	—	—	—	0.702	0.0315	0.0391	0.00376	—	−30.00
−28.00	41.233	117.37	0.6988	2.020	49.443	0.00471	0.11457	0.1679	0.1044	—	—	—	0.692	0.0316	0.0389	0.00380	—	−28.00
−26.00	42.979	117.04	0.6718	2.360	49.598	0.00549	0.11442	0.1683	0.1046	—	—	—	0.682	0.0318	0.0387	0.00383	—	−26.00
−24.00	44.780	116.70	0.6461	2.701	49.753	0.00627	0.11427	0.1687	0.1049	—	—	—	0.673	0.0320	0.0385	0.00387	—	−24.00
−22.00	46.638	116.36	0.6216	3.043	49.906	0.00705	0.11412	0.1692	0.1051	—	—	—	0.664	0.0321	0.0382	0.00391	—	−22.00
−20.00	48.554	116.02	0.5982	3.386	50.059	0.00782	0.11397	0.1696	0.1054	—	—	—	0.655	0.0323	0.0380	0.00395	—	−20.00
−18.00	50.528	115.67	0.5759	3.731	50.211	0.00860	0.11383	0.1701	0.1056	—	—	—	0.647	0.0325	0.0378	0.00399	—	−18.00
−16.00	52.562	115.33	0.5546	4.076	50.362	0.00937	0.11369	0.1706	0.1059	—	—	—	0.638	0.0326	0.0376	0.00402	—	−16.00
−14.00	54.656	114.98	0.5342	4.422	50.513	0.01014	0.11356	0.1711	0.1062	—	—	—	0.630	0.0328	0.0374	0.00406	—	−14.00
−12.00	56.813	114.63	0.5148	4.770	50.662	0.01091	0.11342	0.1716	0.1065	—	—	—	0.622	0.0329	0.0372	0.00410	—	−12.00
−10.00	59.033	114.28	0.4962	5.118	50.811	0.01168	0.11329	0.1721	0.1068	—	—	—	0.614	0.0331	0.0370	0.00415	—	−10.00
−8.00	61.317	113.92	0.4784	5.468	50.958	0.01244	0.11316	0.1726	0.1071	—	—	—	0.606	0.0333	0.0368	0.00419	—	−8.00
−6.00	63.666	113.57	0.4613	5.818	51.105	0.01321	0.11303	0.1731	0.1074	—	—	—	0.598	0.0334	0.0366	0.00423	—	−6.00
−4.00	66.082	113.21	0.4450	6.170	51.251	0.01398	0.11291	0.1737	0.1077	—	—	—	0.591	0.0336	0.0364	0.00427	—	−4.00
−2.00	68.566	112.85	0.4294	6.523	51.396	0.01474	0.11279	0.1742	0.1081	—	—	—	0.583	0.0338	0.0362	0.00431	—	−2.00
0.00	71.119	112.49	0.4144	6.876	51.539	0.01550	0.11266	0.1748	0.1085	—	—	—	0.576	0.0339	0.0360	0.00436	—	0.00
2.00	73.742	112.12	0.4001	7.231	51.682	0.01626	0.11254	0.1754	0.1088	—	—	—	0.569	0.0341	0.0358	0.00440	—	2.00
4.00	76.436	111.75	0.3864	7.587	51.823	0.01702	0.11243	0.1760	0.1092	—	—	—	0.562	0.0342	0.0356	0.00445	—	4.00
6.00	79.203	111.38	0.3732	7.944	51.964	0.01778	0.11231	0.1766	0.1097	—	—	—	0.556	0.0344	0.0354	0.00450	—	6.00
8.00	82.043	111.01	0.3605	8.303	52.103	0.01854	0.11219	0.1772	0.1101	—	—	—	0.549	0.0346	0.0352	0.00454	—	8.00
10.00	84.959	110.64	0.3484	8.662	52.241	0.01929	0.11208	0.1779	0.1106	—	—	—	0.543	0.0347	0.0350	0.00459	—	10.00
12.00	87.950	110.26	0.3367	9.023	52.378	0.02005	0.11197	0.1785	0.1111	—	—	—	0.536	0.0349	0.0348	0.00464	—	12.00
14.00	91.019	109.88	0.3255	9.385	52.513	0.02081	0.11186	0.1792	0.1116	—	—	—	0.530	0.0350	0.0346	0.00469	—	14.00
16.00	94.167	109.49	0.3147	9.748	52.647	0.02156	0.11175	0.1799	0.1121	—	—	—	0.524	0.0352	0.0344	0.00474	—	16.00
18.00	97.394	109.11	0.3044	10.112	52.780	0.02231	0.11164	0.1806	0.1127	—	—	—	0.518	0.0353	0.0342	0.00479	—	18.00
20.00	100.700	108.72	0.2945	10.477	52.911	0.02306	0.11153	0.1813	0.1133	—	—	—	0.512	0.0355	0.0340	0.00485	—	20.00
25.00	109.340	107.74	0.2712	11.396	53.234	0.02494	0.11126	0.1832	0.1148	—	—	—	0.498	0.0359	0.0336	0.00498	—	25.00
30.00	118.510	106.73	0.2501	12.323	53.547	0.02681	0.11099	0.1852	0.1166	—	—	—	0.484	0.0363	0.0331	0.00512	—	30.00
35.00	128.230	105.71	0.2308	13.259	53.849	0.02867	0.11073	0.1874	0.1185	—	—	—	0.471	0.0367	0.0326	0.00527	—	35.00
40.00	138.530	104.66	0.2133	14.204	54.141	0.03054	0.11047	0.1896	0.1207	—	—	—	0.459	0.0370	0.0321	0.00543	—	40.00
45.00	149.430	103.59	0.1972	15.157	54.420	0.03240	0.11020	0.1920	0.1230	—	—	—	0.447	0.0374	0.0317	0.00559	—	45.00
50.00	160.940	102.49	0.1825	16.122	54.687	0.03426	0.10993	0.1946	0.1256	—	—	—	0.436	0.0378	0.0312	0.00576	—	50.00
55.00	173.090	101.37	0.1689	17.096	54.941	0.03612	0.10965	0.1972	0.1285	—	—	—	0.425	0.0381	0.0308	0.00594	—	55.00
60.00	185.910	100.21	0.1565	18.083	55.179	0.03798	0.10937	0.2001	0.1316	—	—	—	0.415	0.0385	0.0303	0.00613	—	60.00
65.00	199.410	99.02	0.1451	19.081	55.401	0.03985	0.10907	0.2031	0.1350	—	—	—	0.405	0.0388	0.0299	0.00632	—	65.00
70.00	213.620	97.80	0.1345	20.094	55.606	0.04172	0.10876	0.2062	0.1387	—	—	—	0.396	0.0392	0.0294	0.00653	—	70.00
80.00	244.260	95.23	0.1156	22.164	55.956	0.04548	0.10809	0.2130	0.1470	—	—	—	0.378	0.0399	0.0285	0.00697	—	80.00
90.00	278.020	92.46	0.0994	24.306	56.214	0.04929	0.10734	—	—	—	—	—	—	—	—	—	—	90.00
100.00	315.130	89.43	0.0852	26.539	56.357	0.05318	0.10646	—	—	—	—	—	—	—	—	—	—	100.00
110.00	355.790	86.07	0.0728	28.890	56.353	0.05719	0.10540	—	—	—	—	—	—	—	—	—	—	110.00
120.00	400.260	82.23	0.0618	31.400	56.152	0.06139	0.10409	—	—	—	—	—	—	—	—	—	—	120.00
130.00	448.770	77.63	0.0518	34.151	55.664	0.06590	0.10238	—	—	—	—	—	—	—	—	—	—	130.00
140.00	501.600	71.59	0.0424	37.327	54.692	0.07102	0.09998	—	—	—	—	—	—	—	—	—	—	140.00
150.00	559.050	60.76	0.0323	41.798	52.528	0.07814	0.09574	—	—	—	—	—	—	—	—	—	—	150.00
152.60c	574.900	46.50	0.0215	46.740	46.740	0.08613	0.08613	∞	∞	∞	0	0	—	—	∞	∞	0	152.60

*temperatures are on the IPTS-68 scale   b = boiling point   c = critical point

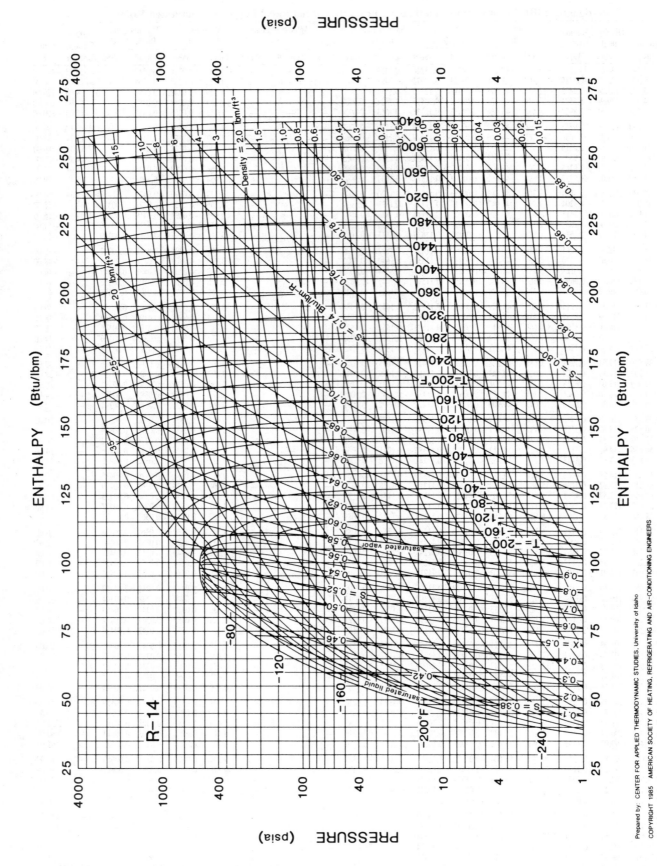

Fig. 5 Pressure-Enthalpy Diagram for Refrigerant 14

Refrigerant 14 (tetrafluoromethane) Properties of Saturated Liquid and Saturated Vapor

Temp,* °F	Pressure, psia	Density, lb/ft³ Liquid	Volume, ft³/lb Vapor	Enthalpy, Btu/lb Liquid	Enthalpy, Btu/lb Vapor	Entropy, Btu/lb·°F Liquid	Entropy, Btu/lb·°F Vapor	Specific Heat c_p, Btu/lb·°F Liquid	Specific Heat c_p, Btu/lb·°F Vapor	c_p/c_v Vapor	Velocity of Sound, ft/s Liquid	Velocity of Sound, ft/s Vapor	Viscosity, lb_m/ft·h Liquid	Viscosity, lb_m/ft·h Vapor	Thermal Cond, Btu/h·ft·°F Liquid	Thermal Cond, Btu/h·ft·°F Vapor	Surface Tension, dyne/cm	Temp, °F
−220.00	5.721	106.00	4.9883	43.772	104.335	0.36173	0.61442	—	0.1116	1.2925	—	408.	—	—	—	—	13.99	−220.00
−215.00	7.231	105.09	4.0116	44.836	104.783	0.36611	0.61113	—	0.1136	1.2939	—	411.	—	—	—	—	13.43	−215.00
−210.00	9.042	104.17	3.2572	45.919	105.224	0.37048	0.60802	—	0.1156	1.2960	—	413.	—	—	—	—	12.88	−210.00
−205.00	11.197	103.24	2.6685	47.018	105.656	0.37482	0.60508	—	0.1178	1.2988	—	416.	—	—	—	—	12.33	−205.00
−200.00	13.737	102.29	2.2045	48.134	106.079	0.37915	0.60230	—	0.1202	1.3024	—	418.	—	—	—	—	11.79	−200.00
−198.30b	14.696	101.97	2.0696	48.518	106.220	0.38061	0.60138	0.2180	0.1210	1.3037	—	418.	0.768	0.0212	0.0539	0.00265	11.61	−198.30
−198.00	14.870	101.91	2.0468	48.586	106.245	0.38087	0.60122	0.2182	0.1211	1.3040	—	418.	0.764	0.0213	0.0538	0.00265	11.58	−198.00
−196.00	16.076	101.52	1.9027	49.040	106.410	0.38259	0.60017	0.2191	0.1221	1.3058	—	419.	0.744	0.0215	0.0533	0.00271	11.37	−196.00
−194.00	17.356	101.14	1.7708	49.496	106.572	0.38431	0.59915	0.2201	0.1232	1.3077	—	420.	0.725	0.0217	0.0528	0.00277	11.15	−194.00
−192.00	18.714	100.75	1.6499	49.956	106.733	0.38602	0.59814	0.2210	0.1242	1.3097	—	420.	0.706	0.0219	0.0523	0.00283	10.94	−192.00
−190.00	20.153	100.35	1.5390	50.417	106.892	0.38773	0.59715	0.2218	0.1253	1.3119	—	421.	0.687	0.0220	0.0518	0.00288	10.73	−190.00
−188.00	21.676	99.96	1.4370	50.882	107.049	0.38943	0.59618	0.2227	0.1264	1.3142	—	422.	0.669	0.0222	0.0513	0.00294	10.52	−188.00
−186.00	23.286	99.56	1.3432	51.349	107.203	0.39114	0.59523	0.2235	0.1275	1.3167	—	422.	0.652	0.0224	0.0508	0.00300	10.32	−186.00
−184.00	24.986	99.16	1.2568	51.818	107.356	0.39283	0.59430	0.2242	0.1287	1.3193	—	423.	0.635	0.0226	0.0504	0.00305	10.11	−184.00
−182.00	26.780	98.76	1.1771	52.290	107.507	0.39453	0.59339	0.2250	0.1299	1.3221	—	423.	0.619	0.0228	0.0499	0.00311	9.91	−182.00
−180.00	28.671	98.36	1.1036	52.765	107.655	0.39622	0.59249	0.2258	0.1311	1.3250	—	423.	0.603	0.0230	0.0494	0.00316	9.70	−180.00
−178.00	30.661	97.95	1.0355	53.242	107.801	0.39790	0.59161	0.2265	0.1323	1.3282	—	424.	0.588	0.0232	0.0489	0.00322	9.50	−178.00
−176.00	32.755	97.54	0.9726	53.721	107.945	0.39959	0.59074	0.2272	0.1336	1.3314	—	424.	0.573	0.0233	0.0485	0.00328	9.30	−176.00
−174.00	34.956	97.12	0.9143	54.202	108.087	0.40126	0.58989	0.2280	0.1349	1.3349	—	424.	0.559	0.0235	0.0480	0.00334	9.10	−174.00
−172.00	37.267	96.71	0.8602	54.686	108.226	0.40293	0.58905	0.2287	0.1363	1.3385	—	425.	0.545	0.0237	0.0475	0.00339	8.90	−172.00
−170.00	39.691	96.29	0.8099	55.172	108.362	0.40460	0.58823	0.2295	0.1377	1.3424	—	425.	0.531	0.0239	0.0471	0.00345	8.70	−170.00
−168.00	42.232	95.86	0.7633	55.661	108.497	0.40626	0.58742	0.2303	0.1391	1.3464	—	425.	0.518	0.0241	0.0466	0.00351	8.50	−168.00
−166.00	44.893	95.44	0.7198	56.151	108.628	0.40792	0.58662	0.2310	0.1406	1.3506	—	425.	0.505	0.0242	0.0462	0.00357	8.30	−166.00
−164.00	47.678	95.01	0.6794	56.644	108.757	0.40958	0.58583	0.2319	0.1421	1.3551	—	425.	0.493	0.0244	0.0458	0.00363	—	−164.00
−162.00	50.591	94.58	0.6417	57.139	108.883	0.41123	0.58506	0.2327	0.1436	1.3598	—	425.	0.481	0.0246	0.0453	0.00369	—	−162.00
−160.00	53.634	94.14	0.6065	57.636	109.007	0.41287	0.58430	0.2336	0.1452	1.3647	—	425.	0.469	0.0248	0.0449	0.00375	—	−160.00
−158.00	56.811	93.70	0.5736	58.135	109.128	0.41451	0.58354	0.2345	0.1468	1.3698	—	425.	0.458	0.0250	0.0444	0.00381	—	−158.00
−156.00	60.126	93.26	0.5429	58.636	109.245	0.41614	0.58280	0.2354	0.1485	1.3752	—	425.	0.447	0.0252	0.0440	0.00387	—	−156.00
−154.00	63.581	92.81	0.5142	59.139	109.360	0.41777	0.58207	0.2365	0.1502	1.3809	—	425.	0.436	0.0254	0.0436	0.00393	—	−154.00
−152.00	67.182	92.36	0.4872	59.645	109.472	0.41940	0.58135	0.2375	0.1520	1.3868	—	425.	0.426	0.0255	0.0432	0.00400	—	−152.00
−150.00	70.931	91.90	0.4620	60.152	109.581	0.42102	0.58064	0.2386	0.1538	1.3930	—	424.	0.416	0.0257	0.0427	0.00406	—	−150.00
−148.00	74.831	91.44	0.4383	60.661	109.687	0.42263	0.57993	0.2398	0.1557	1.3996	478.	424.	0.406	0.0259	0.0423	0.00413	—	−148.00
−146.00	78.887	90.98	0.4161	61.173	109.789	0.42424	0.57923	0.2410	0.1577	1.4064	473.	424.	0.396	0.0261	0.0419	0.00419	—	−146.00
−144.00	83.102	90.51	0.3952	61.686	109.888	0.42584	0.57854	0.2423	0.1597	1.4136	467.	423.	0.387	0.0263	0.0415	0.00426	—	−144.00
−142.00	87.480	90.03	0.3756	62.201	109.984	0.42744	0.57786	0.2437	0.1617	1.4212	461.	423.	0.378	0.0265	0.0411	0.00433	—	−142.00
−140.00	92.023	89.55	0.3571	62.718	110.076	0.42904	0.57718	0.2452	0.1639	1.4291	455.	422.	0.370	0.0267	0.0407	0.00440	—	−140.00
−138.00	96.737	89.07	0.3397	63.237	110.165	0.43062	0.57651	0.2467	0.1661	1.4375	450.	422.	0.361	0.0269	0.0403	0.00447	—	−138.00
−136.00	101.63	88.58	0.3233	63.758	110.250	0.43221	0.57585	0.2483	0.1683	1.4462	444.	421.	0.353	0.0271	0.0399	0.00455	—	−136.00
−134.00	106.69	88.09	0.3078	64.282	110.331	0.43379	0.57518	0.2500	0.1707	1.4555	438.	421.	0.345	0.0273	0.0395	0.00462	—	−134.00
−132.00	111.94	87.59	0.2932	64.807	110.408	0.43536	0.57453	0.2519	0.1731	1.4652	432.	420.	0.338	0.0275	0.0391	0.00470	—	−132.00
−130.00	117.37	87.08	0.2794	65.334	110.482	0.43693	0.57388	0.2538	0.1757	1.4754	426.	419.	0.330	0.0278	0.0387	0.00477	—	−130.00
−128.00	122.98	86.57	0.2664	65.863	110.551	0.43849	0.57323	0.2558	0.1783	1.4863	420.	418.	0.323	0.0280	0.0383	0.00485	—	−128.00
−126.00	128.80	86.05	0.2540	66.395	110.616	0.44005	0.57258	0.2579	0.1810	1.4977	414.	418.	0.316	0.0282	0.0379	0.00493	—	−126.00
−124.00	134.80	85.53	0.2423	66.928	110.676	0.44161	0.57194	0.2602	0.1839	1.5097	408.	417.	0.309	0.0284	0.0375	0.00502	—	−124.00
−122.00	141.01	84.99	0.2313	67.464	110.732	0.44316	0.57130	0.2626	0.1869	1.5225	402.	416.	0.302	0.0286	0.0371	0.00510	—	−122.00
−120.00	147.42	84.46	0.2208	68.002	110.783	0.44471	0.57066	0.2650	0.1900	1.5360	396.	415.	—	0.0289	0.0367	0.00519	—	−120.00
−118.00	154.05	83.91	0.2108	68.543	110.829	0.44625	0.57002	0.2677	0.1932	1.5503	390.	413.	—	0.0291	0.0363	0.00528	—	−118.00
−116.00	160.88	83.35	0.2013	69.086	110.870	0.44779	0.56938	0.2704	0.1966	1.5656	384.	412.	—	0.0294	0.0360	0.00537	—	−116.00
−114.00	167.93	82.79	0.1923	69.631	110.906	0.44933	0.56873	0.2733	0.2001	1.5818	378.	411.	—	0.0296	0.0356	0.00546	—	−114.00
−112.00	175.20	82.22	0.1838	70.180	110.936	0.45087	0.56809	0.2763	0.2038	1.5990	371.	410.	—	0.0299	0.0352	0.00555	—	−112.00
−110.00	182.70	81.64	0.1756	70.731	110.960	0.45240	0.56745	0.2795	0.2078	1.6174	365.	408.	—	0.0301	0.0348	0.00565	—	−110.00
−108.00	190.42	81.05	0.1679	71.285	110.978	0.45393	0.56680	0.2829	0.2119	1.6372	359.	407.	—	0.0304	0.0345	0.00575	—	−108.00
−106.00	198.38	80.45	0.1605	71.842	110.989	0.45546	0.56614	0.2863	0.2163	1.6583	352.	405.	—	0.0306	0.0341	0.00585	—	−106.00
−104.00	206.58	79.84	0.1535	72.403	110.994	0.45698	0.56549	0.2900	0.2209	1.6810	346.	404.	—	0.0309	0.0337	0.00595	—	−104.00
−102.00	215.03	79.21	0.1467	72.968	110.991	0.45851	0.56482	0.2938	0.2258	1.7055	339.	402.	—	0.0312	0.0333	0.00606	—	−102.00
−100.00	223.72	78.58	0.1403	73.537	110.981	0.46004	0.56415	0.2978	0.2311	1.7319	332.	401.	—	0.0315	0.0330	0.00617	—	−100.00
−98.00	232.67	77.93	0.1342	74.109	110.962	0.46157	0.56347	0.3020	0.2367	1.7605	325.	399.	—	0.0317	0.0326	0.00628	—	−98.00
−96.00	241.87	77.26	0.1283	74.687	110.935	0.46310	0.56278	0.3063	0.2427	1.7917	319.	397.	—	0.0320	0.0322	0.00639	—	−96.00
−94.00	251.35	76.59	0.1227	75.269	110.899	0.46464	0.56207	—	0.2492	1.8256	312.	395.	—	—	—	—	—	−94.00
−92.00	261.09	75.89	0.1173	75.858	110.852	0.46618	0.56136	—	0.2562	1.8628	305.	393.	—	—	—	—	—	−92.00
−90.00	271.11	75.18	0.1121	76.452	110.795	0.46772	0.56063	—	0.2639	1.9037	297.	391.	—	—	—	—	—	−90.00
−88.00	281.42	74.45	0.1071	77.053	110.726	0.46927	0.55988	—	0.2722	1.9490	290.	389.	—	—	—	—	—	−88.00
−86.00	292.01	73.70	0.1024	77.661	110.645	0.47084	0.55911	—	0.2814	1.9992	283.	386.	—	—	—	—	—	−86.00
−84.00	302.91	72.93	0.0978	78.278	110.550	0.47241	0.55831	—	0.2916	2.0554	275.	384.	—	—	—	—	—	−84.00
−82.00	314.11	72.13	0.0933	78.905	110.440	0.47400	0.55749	—	0.3029	2.1186	268.	381.	—	—	—	—	—	−82.00
−80.00	325.63	71.30	0.0890	79.542	110.313	0.47560	0.55665	—	0.3156	2.1902	260.	379.	—	—	—	—	—	−80.00
−75.00	355.85	69.10	0.0789	81.192	109.909	0.47971	0.55436	—	0.3558	2.4199	—	372.	—	—	—	—	—	−75.00
−70.00	388.26	66.64	0.0695	82.953	109.346	0.48403	0.55176	—	0.4151	2.7661	—	364.	—	—	—	—	—	−70.00
−65.00	423.04	63.79	0.0606	84.885	108.550	0.48870	0.54867	—	0.5137	3.3495	—	356.	—	—	—	—	—	−65.00
−60.00	460.45	60.32	0.0518	87.111	107.371	0.49403	0.54472	—	0.7139	4.5478	—	347.	—	—	—	—	—	−60.00
−55.00	500.84	55.54	0.0425	89.966	105.384	0.50080	0.53890	—	—	—	—	—	—	—	—	—	—	−55.00
−50.17c	543.17	39.06	0.0256	97.671	97.671	0.51931	0.51931	∞	∞	∞	0.	0.	—	—	∞	∞	0.00	−50.17

*temperatures are on the IPTS-68 scale b = boiling point c = critical point

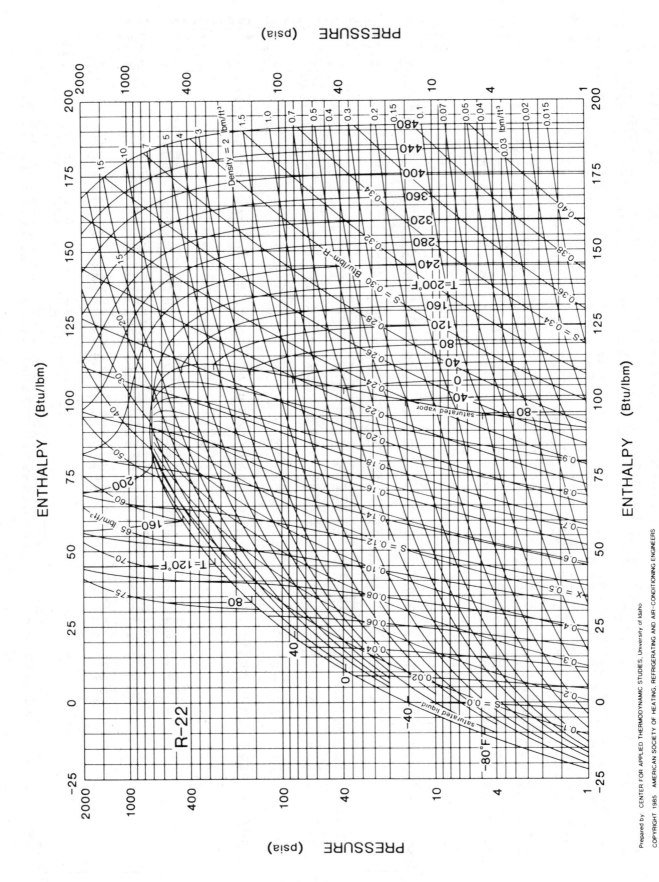

Fig. 6 Pressure-Enthalpy Diagram for Refrigerant 22

Refrigerant 22 (chlorodifluoromethane)   Properties of Saturated Liquid and Saturated Vapor

Temp,* °F	Pressure, psia	Density, lb/ft³ Liquid	Volume, ft³/lb Vapor	Enthalpy, Btu/lb Liquid	Enthalpy, Btu/lb Vapor	Entropy, Btu/lb·°F Liquid	Entropy, Btu/lb·°F Vapor	Specific Heat c_p, Btu/lb·°F Liquid	Specific Heat c_p, Btu/lb·°F Vapor	c_p/c_v Vapor	Velocity of Sound, ft/s Liquid	Velocity of Sound, ft/s Vapor	Viscosity, lb$_m$/ft·h Liquid	Viscosity, lb$_m$/ft·h Vapor	Thermal Cond, Btu/h·ft·°F Liquid	Thermal Cond, Btu/h·ft·°F Vapor	Surface Tension, dyne/cm	Temp, °F
−250.00	—	107.37	—	−63.169	76.604	−0.21914	0.44952	—	0.1018	1.2914	—	395.	—	—	—	—	—	−250.00
−240.00	—	106.41	—	−56.462	77.629	−0.18786	0.42332	—	0.1033	1.2860	—	403.	—	—	—	—	—	−240.00
−230.00	—	105.48	—	−51.569	78.669	−0.16605	0.40101	—	0.1048	1.2807	—	411.	—	—	—	—	36.75	−230.00
−220.00	0.002	104.58	16805.	−47.705	79.724	−0.14958	0.38211	—	0.1064	1.2754	—	419.	—	—	—	—	35.70	−220.00
−210.00	0.004	103.70	6982.6	−44.426	80.796	−0.13616	0.36538	—	0.1080	1.2703	—	427.	—	—	—	—	34.67	−210.00
−200.00	0.010	102.81	3151.5	−41.474	81.882	−0.12457	0.35048	—	0.1096	1.2653	—	435.	—	—	—	—	33.63	−200.00
−190.00	0.022	101.92	1527.4	−38.706	82.984	−0.11411	0.33715	—	0.1113	1.2604	—	442.	—	—	—	—	32.61	−190.00
−180.00	0.044	101.03	787.79	−36.038	84.100	−0.10439	0.32518	—	0.1130	1.2558	—	449.	—	—	—	—	31.59	−180.00
−170.00	0.084	100.12	429.22	−33.424	85.230	−0.09521	0.31441	—	0.1147	1.2515	—	456.	—	—	—	—	30.58	−170.00
−160.00	0.151	99.22	245.51	−30.839	86.373	−0.08644	0.30470	—	0.1165	1.2474	—	463.	—	—	—	—	29.57	−160.00
−150.00	0.262	98.30	146.65	−28.269	87.528	−0.07800	0.29594	—	0.1183	1.2437	—	470.	—	—	—	—	28.57	−150.00
−140.00	0.435	97.38	91.059	−25.708	88.692	−0.06986	0.28801	—	0.1201	1.2403	—	476.	—	—	—	—	27.57	−140.00
−130.00	0.696	96.46	58.544	−23.150	89.864	−0.06198	0.28082	—	0.1221	1.2374	—	482.	—	—	—	—	26.59	−130.00
−120.00	1.080	95.53	38.833	−20.594	91.040	−0.05435	0.27430	0.2555	0.1241	1.2349	3483.	488.	—	—	—	—	25.61	−120.00
−110.00	1.626	94.60	26.494	−18.038	92.218	−0.04694	0.26838	0.2555	0.1262	1.2329	3384.	494.	—	—	0.0765	—	24.64	−110.00
−100.00	2.384	93.66	18.540	−15.481	93.397	−0.03973	0.26298	0.2557	0.1285	1.2315	3290.	500.	—	—	0.0749	—	23.67	−100.00
−90.00	3.413	92.71	13.275	−12.921	94.572	−0.03271	0.25807	0.2561	0.1308	1.2307	3198.	505.	—	—	0.0734	0.00292	22.71	−90.00
−80.00	4.778	91.75	9.7044	−10.355	95.741	−0.02587	0.25357	0.2567	0.1334	1.2305	3110.	510.	—	—	0.0718	0.00315	21.76	−80.00
−70.00	6.555	90.79	7.2285	−7.783	96.901	−0.01919	0.24945	0.2574	0.1361	1.2310	3023.	514.	—	—	0.0703	0.00338	20.82	−70.00
−60.00	8.830	89.81	5.4766	−5.201	98.049	−0.01266	0.24567	0.2584	0.1389	1.2323	2937.	519.	—	—	0.0688	0.00360	19.89	−60.00
−50.00	11.696	88.83	4.2138	−2.608	99.182	−0.00627	0.24220	0.2596	0.1420	1.2344	2852.	522.	—	—	0.0673	0.00382	18.96	−50.00
−45.00	13.383	88.33	3.7160	−1.306	99.742	−0.00312	0.24056	0.2604	0.1436	1.2358	2810.	524.	—	—	0.0665	0.00393	18.50	−45.00
−41.44b	14.696	87.97	3.4048	−0.377	100.138	−0.00090	0.23944	0.2609	0.1448	1.2369	2780.	525.	—	—	0.0660	0.00401	18.18	−41.44
−40.00	15.255	87.82	3.2880	0.000	100.296	0.00000	0.23899	0.2611	0.1453	1.2374	2768.	526.	—	—	0.0658	0.00404	18.05	−40.00
−35.00	17.329	87.32	2.9185	1.310	100.847	0.00309	0.23748	0.2620	0.1471	1.2393	2725.	527.	—	—	0.0651	0.00414	17.59	−35.00
−30.00	19.617	86.81	2.5984	2.624	101.391	0.00616	0.23602	0.2629	0.1489	1.2414	2683.	529.	—	—	0.0643	0.00425	17.14	−30.00
−25.00	22.136	86.29	2.3202	3.944	101.928	0.00920	0.23462	0.2638	0.1507	1.2437	2641.	530.	—	—	0.0636	0.00435	16.69	−25.00
−20.00	24.899	85.77	2.0774	5.268	102.461	0.01222	0.23327	0.2648	0.1527	1.2463	2599.	531.	—	—	0.0629	0.00445	16.24	−20.00
−15.00	27.924	85.25	1.8650	6.598	102.986	0.01521	0.23197	0.2659	0.1547	1.2493	2557.	532.	—	—	0.0622	0.00456	15.79	−15.00
−10.00	31.226	84.72	1.6784	7.934	103.503	0.01818	0.23071	0.2671	0.1567	1.2525	2515.	533.	—	—	0.0614	0.00466	—	−10.00
−5.00	34.821	84.18	1.5142	9.276	104.013	0.02113	0.22949	0.2684	0.1589	1.2560	2473.	534.	—	—	0.0607	0.00476	—	−5.00
0.00	38.726	83.64	1.3691	10.624	104.515	0.02406	0.22832	0.2697	0.1611	1.2599	2431.	535.	0.615	0.0268	0.0600	0.00486	—	0.00
5.00	42.960	83.09	1.2406	11.979	105.009	0.02697	0.22718	0.2710	0.1634	1.2641	2389.	535.	0.597	0.0271	0.0593	0.00496	—	5.00
10.00	47.538	82.54	1.1265	13.342	105.493	0.02987	0.22607	0.2725	0.1658	1.2687	2346.	535.	0.580	0.0274	0.0586	0.00506	—	10.00
15.00	52.480	81.98	1.0250	14.712	105.968	0.03275	0.22500	0.2740	0.1683	1.2737	2304.	536.	0.563	0.0276	0.0579	0.00516	—	15.00
20.00	57.803	81.41	0.9343	16.090	106.434	0.03561	0.22395	0.2756	0.1709	1.2792	2262.	536.	0.546	0.0279	0.0572	0.00526	—	20.00
25.00	63.526	80.84	0.8532	17.476	106.891	0.03846	0.22294	0.2773	0.1737	1.2851	2219.	536.	0.530	0.0282	0.0566	0.00536	—	25.00
30.00	69.667	80.26	0.7804	18.871	107.336	0.04129	0.22195	0.2791	0.1765	1.2915	2177.	536.	0.515	0.0284	0.0559	0.00546	—	30.00
35.00	76.245	79.67	0.7150	20.275	107.769	0.04411	0.22098	0.2809	0.1794	1.2984	2134.	535.	0.499	0.0287	0.0552	0.00555	—	35.00
40.00	83.280	79.07	0.6561	21.688	108.191	0.04692	0.22004	0.2829	0.1825	1.3059	2091.	535.	0.484	0.0290	0.0545	0.00565	—	40.00
45.00	90.791	78.46	0.6029	23.111	108.600	0.04972	0.21912	0.2849	0.1857	1.3141	2048.	534.	0.470	0.0292	0.0538	0.00575	—	45.00
50.00	98.799	77.84	0.5548	24.544	108.997	0.05251	0.21821	0.2870	0.1891	1.3229	2005.	533.	0.456	0.0295	0.0532	0.00584	—	50.00
55.00	107.32	77.22	0.5111	25.988	109.379	0.05529	0.21732	0.2893	0.1927	1.3324	1962.	532.	0.442	0.0298	0.0525	0.00594	—	55.00
60.00	116.58	76.58	0.4715	27.443	109.748	0.05806	0.21644	0.2916	0.1964	1.3428	1919.	531.	0.429	0.0301	0.0518	0.00604	—	60.00
65.00	126.00	75.93	0.4355	28.909	110.103	0.06082	0.21557	0.2941	0.2003	1.3540	1876.	530.	0.416	0.0303	0.0512	0.00613	—	65.00
70.00	136.19	75.27	0.4026	30.387	110.441	0.06358	0.21472	0.2967	0.2045	1.3663	1832.	528.	0.404	—	0.0505	0.00623	—	70.00
75.00	146.98	74.60	0.3726	31.877	110.761	0.06633	0.21387	0.2994	0.2089	1.3796	1788.	527.	0.392	—	0.0499	0.00632	—	75.00
80.00	158.40	73.92	0.3451	33.381	111.066	0.06907	0.21302	0.3024	0.2135	1.3941	1744.	525.	0.380	—	0.0492	0.00642	—	80.00
85.00	170.45	73.22	0.3199	34.898	111.350	0.07182	0.21218	0.3055	0.2185	1.4100	1700.	523.	0.369	—	0.0486	0.00652	—	85.00
90.00	183.17	72.51	0.2968	36.430	111.616	0.07456	0.21134	0.3088	0.2238	1.4275	1655.	520.	0.358	—	0.0479	0.00661	—	90.00
95.00	196.57	71.79	0.2756	37.977	111.859	0.07730	0.21050	0.3123	0.2295	1.4467	1611.	518.	0.348	—	0.0473	0.00671	—	95.00
100.00	210.69	71.05	0.2560	39.538	112.081	0.08003	0.20965	0.3162	0.2356	1.4678	1566.	515.	0.338	—	0.0466	0.00680	—	100.00
105.00	225.53	70.29	0.2379	41.119	112.278	0.08277	0.20879	0.3203	0.2422	1.4912	1520.	512.	—	—	0.0460	0.00690	—	105.00
110.00	241.14	69.51	0.2212	42.717	112.448	0.08552	0.20793	0.3248	0.2495	1.5173	1474.	509.	—	—	0.0454	0.00699	—	110.00
115.00	257.52	68.71	0.2058	44.334	112.591	0.08827	0.20705	0.3298	0.2573	1.5464	1428.	506.	—	—	0.0447	0.00709	—	115.00
120.00	274.71	67.89	0.1914	45.972	112.704	0.09103	0.20615	0.3353	0.2660	1.5791	1382.	502.	—	—	0.0441	0.00719	—	120.00
125.00	292.73	67.05	0.1781	47.633	112.783	0.09379	0.20522	0.3413	0.2756	1.6160	1334.	498.	—	—	—	—	—	125.00
130.00	311.61	66.17	0.1657	49.319	112.825	0.09657	0.20427	0.3482	0.2864	1.6581	1287.	494.	—	—	—	—	—	130.00
135.00	331.38	65.27	0.1542	51.032	112.826	0.09937	0.20329	0.3559	0.2985	1.7063	1238.	489.	—	—	—	—	—	135.00
140.00	352.07	64.33	0.1434	52.775	112.784	0.10220	0.20227	0.3648	0.3123	1.7621	1189.	485.	—	—	—	—	—	140.00
145.00	373.71	63.35	0.1332	54.553	112.692	0.10504	0.20119	0.3752	0.3282	1.8275	1139.	479.	—	—	—	—	—	145.00
150.00	396.32	62.33	0.1237	56.370	112.541	0.10793	0.20006	0.3873	0.3468	1.9050	1088.	474.	—	—	—	—	—	150.00
160.00	444.65	60.12	0.1063	60.145	112.035	0.11383	0.19757	0.4198	0.3957	2.1126	983.	462.	—	—	—	—	—	160.00
170.00	497.35	57.59	0.0907	64.175	111.165	0.12001	0.19464	0.4711	0.4716	2.4409	873.	448.	—	—	—	—	—	170.00
180.00	554.82	54.57	0.0763	68.597	109.753	0.12668	0.19102	0.5657	0.6073	3.0349	752.	433.	—	—	—	—	—	180.00
190.00	617.53	50.62	0.0625	73.742	107.398	0.13432	0.18613	0.7952	0.9222	4.4150	616.	415.	—	—	—	—	—	190.00
200.00	686.11	44.44	0.0478	80.558	102.809	0.14432	0.17805	—	—	—	—	—	—	—	—	—	—	200.00
205.06c	723.74	32.70	0.0306	91.052	91.052	0.15989	0.15989	∞	∞	∞	0.	0.	—	—	∞	∞	0.00	205.06

*temperatures are on the ITS-90 scale   b = boiling point   c = critical point

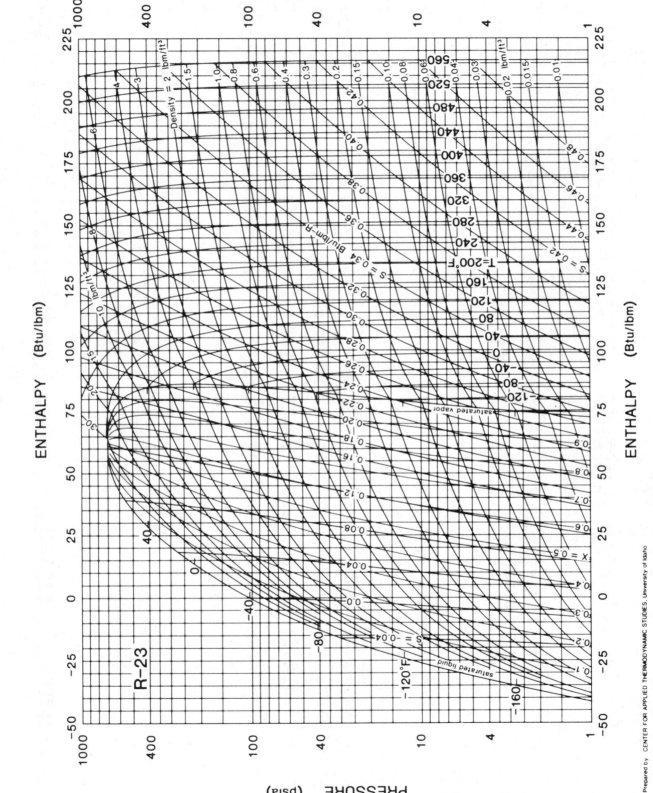

Fig. 7 Pressure-Enthalpy Diagram for Refrigerant 23

Refrigerant 23 (trifluoromethane) Properties of Saturated Liquid and Saturated Vapor

Temp,* °F	Pressure, psia	Density, lb/ft³ Liquid	Volume, ft³/lb Vapor	Enthalpy, Btu/lb Liquid	Enthalpy, Btu/lb Vapor	Entropy, Btu/lb·°F Liquid	Entropy, Btu/lb·°F Vapor	Specific Heat c_p, Btu/lb·°F Liquid	Specific Heat c_p, Btu/lb·°F Vapor	c_p/c_v Vapor	Velocity of Sound, ft/s Liquid	Velocity of Sound, ft/s Vapor	Viscosity, lb_m/ft·h Liquid	Viscosity, lb_m/ft·h Vapor	Thermal Cond, Btu/h·ft·°F Liquid	Thermal Cond, Btu/h·ft·°F Vapor	Surface Tension, dyne/cm	Temp, °F
−160.00	2.774	94.42	16.315	−35.171	75.195	−0.09766	0.27063	—	0.1363	1.2960	—	517.	—	—	—	—	—	−160.00
−155.00	3.442	93.95	13.334	−33.755	75.756	−0.09298	0.26646	—	0.1380	1.2969	—	521.	—	—	—	—	—	−155.00
−150.00	4.237	93.46	10.981	−32.354	76.309	−0.08842	0.26247	—	0.1399	1.2980	—	524.	—	—	—	—	—	−150.00
−145.00	5.177	92.97	9.1062	−30.964	76.853	−0.08398	0.25866	—	0.1419	1.2995	—	527.	—	—	—	—	—	−145.00
−140.00	6.279	92.46	7.6019	−29.581	77.388	−0.07962	0.25500	—	0.1441	1.3014	—	529.	—	—	—	—	—	−140.00
−135.00	7.566	91.95	6.3857	−28.201	77.913	−0.07535	0.25149	—	0.1464	1.3036	—	532.	—	—	—	—	—	−135.00
−130.00	9.058	91.42	5.3956	−26.822	78.427	−0.07114	0.24811	—	0.1488	1.3062	—	535.	—	—	—	—	—	−130.00
−125.00	10.778	90.88	4.5844	−25.440	78.930	−0.06699	0.24487	—	0.1514	1.3091	—	537.	—	—	—	—	—	−125.00
−120.00	12.751	90.32	3.9155	−24.052	79.421	−0.06289	0.24174	—	0.1542	1.3125	—	539.	—	—	—	—	—	−120.00
−115.64b	14.696	89.83	3.4267	−22.837	79.839	−0.05935	0.23911	—	0.1567	1.3157	—	541.	—	—	—	—	—	−115.64
−115.00	15.003	89.76	3.3607	−22.657	79.900	−0.05882	0.23873	—	0.1571	1.3162	—	541.	—	—	—	—	—	−115.00
−110.00	17.561	89.18	2.8981	−21.252	80.366	−0.05479	0.23582	—	0.1602	1.3204	—	543.	—	—	—	—	—	−110.00
−105.00	20.452	88.59	2.5102	−19.835	80.818	−0.05078	0.23301	—	0.1634	1.3251	—	545.	—	—	—	—	—	−105.00
−100.00	23.708	87.98	2.1834	−18.405	81.256	−0.04680	0.23029	—	0.1669	1.3302	—	546.	—	—	—	—	—	−100.00
−98.00	25.118	87.74	2.0672	−17.829	81.428	−0.04521	0.22923	—	0.1683	1.3324	—	547.	—	—	—	—	—	−98.00
−96.00	26.594	87.49	1.9584	−17.251	81.597	−0.04363	0.22818	—	0.1698	1.3346	—	547.	—	—	—	—	—	−96.00
−94.00	28.137	87.24	1.8565	−16.670	81.764	−0.04204	0.22714	—	0.1713	1.3370	—	548.	—	—	—	—	—	−94.00
−92.00	29.748	86.98	1.7609	−16.087	81.928	−0.04046	0.22612	—	0.1728	1.3394	—	548.	—	—	—	—	—	−92.00
−90.00	31.431	86.73	1.6712	−15.501	82.090	−0.03888	0.22511	—	0.1744	1.3419	—	549.	—	—	—	—	—	−90.00
−88.00	33.187	86.47	1.5871	−14.912	82.249	−0.03730	0.22411	—	0.1760	1.3445	—	549.	—	—	—	—	—	−88.00
−86.00	35.019	86.21	1.5079	−14.321	82.407	−0.03573	0.22313	—	0.1776	1.3472	—	550.	—	—	—	—	—	−86.00
−84.00	36.927	85.95	1.4336	−13.728	82.561	−0.03416	0.22216	—	0.1792	1.3500	—	550.	—	—	—	—	—	−84.00
−82.00	38.916	85.68	1.3636	−13.132	82.713	−0.03258	0.22119	—	0.1809	1.3529	—	550.	—	—	—	—	—	−82.00
−80.00	40.986	85.42	1.2977	−12.533	82.863	−0.03101	0.22025	—	0.1826	1.3559	—	551.	—	—	—	—	—	−80.00
−78.00	43.140	85.15	1.2356	−11.931	83.010	−0.02945	0.21930	—	0.1844	1.3589	—	551.	—	—	—	—	—	−78.00
−76.00	45.380	84.87	1.1771	−11.327	83.155	−0.02788	0.21838	—	0.1862	1.3621	—	551.	—	—	—	—	13.68	−76.00
−74.00	47.709	84.60	1.1219	−10.720	83.297	−0.02632	0.21746	—	0.1880	1.3654	—	551.	—	—	—	—	13.43	−74.00
−72.00	50.128	84.32	1.0698	−10.110	83.436	−0.02475	0.21655	—	0.1899	1.3688	—	551.	—	—	—	—	13.18	−72.00
−70.00	52.639	84.04	1.0207	−9.497	83.573	−0.02319	0.21565	—	0.1918	1.3722	—	551.	—	—	—	—	12.93	−70.00
−68.00	55.246	83.76	0.9742	−8.882	83.707	−0.02163	0.21477	—	0.1937	1.3758	—	552.	—	—	—	—	12.69	−68.00
−66.00	57.951	83.47	0.9302	−8.264	83.839	−0.02007	0.21389	—	0.1957	1.3796	—	552.	—	—	—	—	12.44	−66.00
−64.00	60.755	83.18	0.8887	−7.644	83.968	−0.01852	0.21302	—	0.1977	1.3834	—	552.	—	—	—	—	12.20	−64.00
−62.00	63.660	82.89	0.8493	−7.021	84.094	−0.01696	0.21216	—	0.1997	1.3874	—	552.	—	—	—	—	11.96	−62.00
−60.00	66.671	82.59	0.8121	−6.395	84.217	−0.01541	0.21131	—	0.2018	1.3915	—	552.	—	—	—	—	11.72	−60.00
−58.00	69.787	82.30	0.7768	−5.767	84.337	−0.01386	0.21047	—	0.2039	1.3957	—	551.	—	—	—	—	11.48	−58.00
−56.00	73.013	82.00	0.7433	−5.136	84.455	−0.01231	0.20963	—	0.2061	1.4000	—	551.	—	—	—	—	11.24	−56.00
−54.00	76.350	81.69	0.7116	−4.502	84.570	−0.01076	0.20881	—	0.2083	1.4045	—	551.	—	—	—	—	11.01	−54.00
−52.00	79.800	81.38	0.6814	−3.866	84.682	−0.00922	0.20799	—	0.2106	1.4092	—	551.	—	—	—	—	10.77	−52.00
−50.00	83.367	81.07	0.6528	−3.228	84.791	−0.00767	0.20718	—	0.2129	1.4140	—	551.	—	—	—	—	10.54	−50.00
−48.00	87.052	80.76	0.6257	−2.587	84.897	−0.00614	0.20637	—	0.2152	1.4190	—	550.	—	—	—	—	10.31	−48.00
−46.00	90.858	80.44	0.5998	−1.944	85.000	−0.00460	0.20558	—	0.2176	1.4241	—	550.	—	—	—	—	10.08	−46.00
−44.00	94.787	80.12	0.5753	−1.298	85.100	−0.00306	0.20479	—	0.2200	1.4294	—	550.	—	—	—	—	9.85	−44.00
−42.00	98.842	79.80	0.5519	−0.650	85.197	−0.00153	0.20401	—	0.2225	1.4349	—	549.	—	—	—	—	9.62	−42.00
−40.00	103.03	79.47	0.5296	0.000	85.291	0.00000	0.20323	0.1979	0.2251	1.4405	738.	549.	0.396	—	0.0595	0.00654	9.40	−40.00
−38.00	107.34	79.14	0.5085	0.652	85.382	0.00153	0.20246	0.2071	0.2277	1.4464	719.	548.	0.391	—	0.0590	0.00661	9.17	−38.00
−36.00	111.79	78.80	0.4883	1.307	85.469	0.00305	0.20170	0.2159	0.2303	1.4525	701.	548.	0.386	—	0.0585	0.00668	8.95	−36.00
−34.00	116.37	78.47	0.4690	1.964	85.553	0.00457	0.20094	0.2243	0.2330	1.4588	684.	547.	0.380	—	0.0581	0.00675	8.73	−34.00
−32.00	121.09	78.12	0.4507	2.623	85.634	0.00609	0.20019	0.2325	0.2358	1.4653	667.	547.	0.375	—	0.0576	0.00683	8.51	−32.00
−30.00	125.95	77.78	0.4332	3.284	85.711	0.00761	0.19944	0.2404	0.2386	1.4720	651.	546.	0.370	—	0.0572	0.00691	8.29	−30.00
−25.00	138.74	76.89	0.3928	4.947	85.888	0.01138	0.19760	0.2590	0.2460	1.4901	613.	544.	0.357	—	0.0561	0.00711	7.76	−25.00
−20.00	152.46	75.98	0.3567	6.622	86.042	0.01514	0.19577	0.2761	0.2538	1.5099	578.	542.	0.344	0.0299	0.0550	0.00732	7.23	−20.00
−15.00	167.17	75.04	0.3244	8.311	86.171	0.01888	0.19397	0.2921	0.2622	1.5318	546.	540.	0.331	0.0304	0.0540	0.00755	6.72	−15.00
−10.00	182.90	74.07	0.2954	10.013	86.274	0.02260	0.19219	0.3071	0.2711	1.5561	517.	537.	0.318	0.0309	0.0530	0.00778	6.22	−10.00
−5.00	199.70	73.07	0.2692	11.730	86.346	0.02630	0.19041	0.3214	0.2808	1.5833	491.	534.	0.306	0.0314	0.0519	0.00803	—	−5.00
0.00	217.62	72.03	0.2457	13.462	86.387	0.02999	0.18863	0.3352	0.2912	1.6138	466.	530.	0.293	0.0319	0.0508	0.00829	—	0.00
5.00	236.71	70.96	0.2243	15.213	86.392	0.03367	0.18685	0.3488	0.3027	1.6484	444.	527.	0.281	0.0324	0.0497	0.00856	—	5.00
10.00	257.02	69.84	0.2049	16.985	86.358	0.03735	0.18505	0.3624	0.3153	1.6879	423.	523.	0.269	0.0330	0.0484	0.00884	—	10.00
15.00	278.61	68.67	0.1872	18.783	86.278	0.04103	0.18323	—	0.3293	1.7334	403.	518.	—	0.0337	—	—	—	15.00
20.00	301.55	67.46	0.1710	20.611	86.148	0.04474	0.18136	—	0.3452	1.7866	385.	513.	—	0.0343	—	—	—	20.00
25.00	325.90	66.19	0.1562	22.479	85.959	0.04847	0.17944	—	0.3634	1.8495	367.	508.	—	0.0351	—	—	—	25.00
30.00	351.74	64.85	0.1426	24.396	85.701	0.05225	0.17745	—	0.3847	1.9250	349.	502.	—	0.0359	—	—	—	30.00
35.00	379.14	63.43	0.1300	26.375	85.361	0.05611	0.17536	—	0.4100	2.0176	332.	495.	—	0.0368	—	—	—	35.00
40.00	408.19	61.93	0.1183	28.434	84.923	0.06008	0.17314	—	0.4411	2.1338	314.	488.	—	0.0379	—	—	—	40.00
45.00	438.99	60.32	0.1074	30.598	84.365	0.06421	0.17074	—	0.4803	2.2840	296.	481.	—	0.0390	—	—	—	45.00
50.00	471.64	58.58	0.0971	32.901	83.654	0.06855	0.16813	—	0.5319	2.4859	277.	472.	—	0.0402	—	—	—	50.00
55.00	506.27	56.68	0.0874	35.389	82.745	0.07319	0.16520	—	0.6039	2.7721	257.	463.	—	0.0416	—	—	—	55.00
60.00	543.01	54.55	0.0780	38.138	81.561	0.07827	0.16182	—	0.7125	3.2100	236.	452.	—	—	—	—	—	60.00
65.00	581.99	52.08	0.0688	41.270	79.972	0.08400	0.15777	—	0.8981	3.9663	213.	441.	—	—	—	—	—	65.00
70.00	623.39	49.05	0.0594	45.032	77.695	0.09085	0.15252	—	1.2953	5.5959	—	—	—	—	—	—	—	70.00
75.00	667.39	44.71	0.0488	50.138	73.842	0.10012	0.14445	—	—	—	—	—	—	—	—	—	—	75.00
78.66c	701.40	32.78	0.0305	61.455	61.455	0.12090	0.12090	∞	∞	∞	0.	0.	—	—	∞	∞	0.00	78.66

*temperatures are on the IPTS-68 scale   b = boiling point   c = critical point

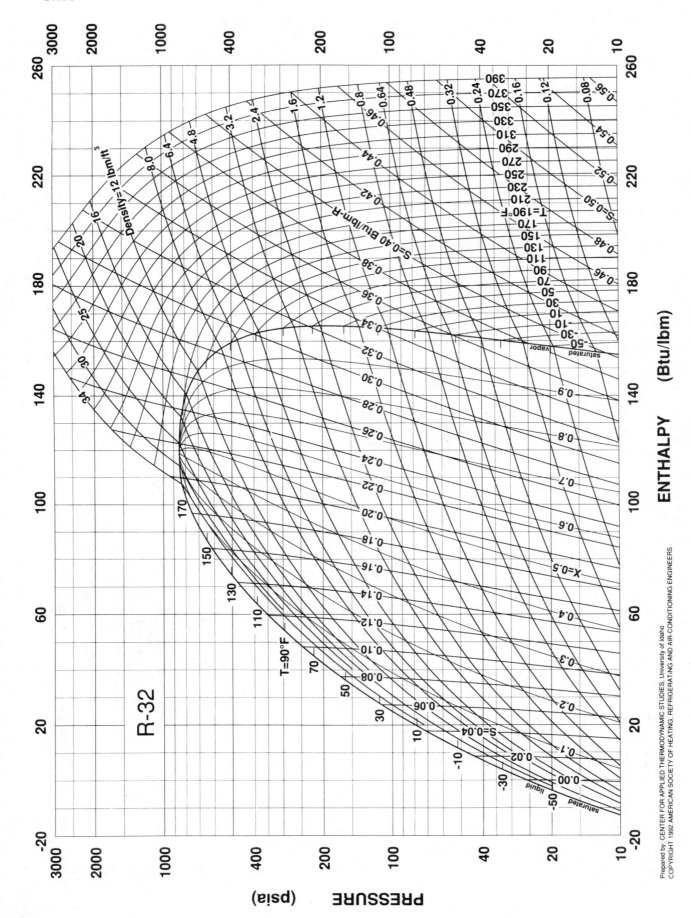

Fig. 8   Pressure-Enthalpy Diagram for Refrigerant 32

Refrigerant 32 (difluoromethane) Properties of Saturated Liquid and Saturated Vapor

Temp,* °F	Pressure, psia	Density, lb/ft³ Liquid	Volume, ft³/lb Vapor	Enthalpy, Btu/lb Liquid	Enthalpy, Btu/lb Vapor	Entropy, Btu/lb·°F Liquid	Entropy, Btu/lb·°F Vapor	Specific Heat c_p, Btu/lb·°F Liquid	Specific Heat c_p, Btu/lb·°F Vapor	c_p/c_v Vapor	Velocity of Sound, ft/s Liquid	Velocity of Sound, ft/s Vapor	Viscosity, lb$_m$/ft·h Liquid	Viscosity, lb$_m$/ft·h Vapor	Thermal Cond, Btu/h·ft·°F Liquid	Thermal Cond, Btu/h·ft·°F Vapor	Surface Tension, dyne/cm	Temp, °F
−70.00	11.318	75.87	6.8747	−11.246	155.538	−0.02771	0.40031	—	0.1862	1.3250	—	680.	0.993	—	0.1115	—	—	−70.00
−65.00	13.108	75.41	5.9913	−9.405	156.182	−0.02302	0.39653	—	0.1885	1.3278	—	682.	0.952	—	0.1102	—	—	−65.00
−61.00b	14.696	75.04	5.3818	−7.923	156.687	−0.01930	0.39360	—	0.1904	1.3302	—	684.	0.921	—	0.1092	—	—	−61.00
−60.00	15.117	74.95	5.2410	−7.551	156.812	−0.01837	0.39288	—	0.1909	1.3309	—	685.	0.914	—	0.1089	—	—	−60.00
−55.00	17.365	74.49	4.6010	−5.683	157.427	−0.01374	0.38933	—	0.1935	1.3343	—	687.	0.878	—	0.1076	—	—	−55.00
−50.00	19.872	74.02	4.0529	−3.802	158.026	−0.00913	0.38589	—	0.1962	1.3380	—	689.	0.844	—	0.1063	—	—	−50.00
−45.00	22.657	73.56	3.5817	−1.907	158.608	−0.00455	0.38254	—	0.1991	1.3421	—	691.	0.811	—	0.1051	—	—	−45.00
−40.00	25.741	73.08	3.1751	0.000	159.174	0.00000	0.37928	—	0.2021	1.3466	—	693.	0.781	—	0.1038	—	—	−40.00
−35.00	29.147	72.61	2.8229	1.920	159.721	0.00453	0.37611	—	0.2053	1.3515	—	694.	0.752	—	0.1025	—	—	−35.00
−30.00	32.897	72.12	2.5168	3.852	160.251	0.00903	0.37302	—	0.2086	1.3568	—	696.	0.725	—	0.1012	—	—	−30.00
−25.00	37.014	71.64	2.2499	5.797	160.761	0.01350	0.37001	—	0.2122	1.3626	—	697.	0.699	—	0.0999	—	—	−25.00
−20.00	41.521	71.14	2.0165	7.753	161.251	0.01795	0.36707	—	0.2159	1.3688	—	698.	0.674	—	0.0986	—	—	−20.00
−15.00	46.444	70.65	1.8116	9.721	161.721	0.02237	0.36420	—	0.2197	1.3756	—	699.	0.650	—	0.0973	—	—	−15.00
−10.00	51.807	70.14	1.6313	11.701	162.170	0.02677	0.36139	—	0.2238	1.3829	—	700.	0.627	—	0.0960	—	—	−10.00
−5.00	57.636	69.64	1.4721	13.693	162.596	0.03114	0.35864	—	0.2281	1.3908	—	700.	0.605	—	0.0948	—	—	−5.00
0.00	63.956	69.12	1.3312	15.696	163.000	0.03549	0.35594	—	0.2326	1.3993	—	700.	0.584	—	0.0935	—	—	0.00
2.00	66.629	68.91	1.2794	16.500	163.155	0.03722	0.35488	—	0.2345	1.4029	—	701.	0.576	—	0.0929	—	—	2.00
4.00	69.385	68.70	1.2300	17.306	163.306	0.03894	0.35382	—	0.2364	1.4066	—	701.	0.568	—	0.0924	—	—	4.00
6.00	72.229	68.49	1.1829	18.115	163.453	0.04066	0.35277	—	0.2384	1.4105	—	701.	0.560	—	0.0919	—	—	6.00
8.00	75.161	68.28	1.1379	18.925	163.596	0.04238	0.35173	—	0.2404	1.4145	—	700.	0.552	—	0.0914	—	—	8.00
10.00	78.183	68.07	1.0949	19.736	163.735	0.04410	0.35069	—	0.2424	1.4186	—	700.	0.545	—	0.0909	—	—	10.00
12.00	81.297	67.85	1.0538	20.550	163.870	0.04581	0.34967	—	0.2445	1.4228	—	700.	0.537	—	0.0904	—	—	12.00
14.00	84.504	67.63	1.0146	21.365	164.001	0.04752	0.34864	—	0.2466	1.4271	—	700.	0.530	—	0.0899	—	—	14.00
16.00	87.808	67.42	0.9771	22.183	164.128	0.04922	0.34763	—	0.2487	1.4316	—	700.	0.522	—	0.0893	—	—	16.00
18.00	91.208	67.20	0.9412	23.002	164.250	0.05092	0.34662	—	0.2509	1.4363	—	700.	0.515	—	0.0888	—	—	18.00
20.00	94.709	66.98	0.9069	23.823	164.368	0.05261	0.34562	—	0.2532	1.4411	—	699.	0.508	—	0.0883	—	—	20.00
22.00	98.310	66.76	0.8740	24.646	164.481	0.05430	0.34462	—	0.2555	1.4460	—	699.	0.501	—	0.0878	—	—	22.00
24.00	102.02	66.53	0.8425	25.471	164.590	0.05599	0.34362	—	0.2578	1.4512	—	699.	0.494	—	0.0873	—	—	24.00
26.00	105.82	66.31	0.8124	26.299	164.695	0.05768	0.34264	—	0.2602	1.4565	—	698.	0.487	—	0.0868	—	—	26.00
28.00	109.74	66.08	0.7835	27.128	164.794	0.05936	0.34165	—	0.2627	1.4619	—	698.	0.481	—	0.0863	—	—	28.00
30.00	113.77	65.86	0.7558	27.959	164.889	0.06104	0.34067	—	0.2652	1.4676	—	698.	0.474	—	0.0857	—	—	30.00
32.00	117.90	65.63	0.7292	28.792	164.979	0.06271	0.33970	—	0.2678	1.4734	—	697.	0.467	—	0.0852	—	—	32.00
34.00	122.15	65.40	0.7038	29.627	165.064	0.06438	0.33873	—	0.2704	1.4795	—	697.	0.461	—	0.0847	—	—	34.00
36.00	126.52	65.16	0.6793	30.465	165.144	0.06605	0.33776	—	0.2731	1.4858	—	696.	0.455	—	0.0842	—	—	36.00
38.00	131.00	64.93	0.6559	31.305	165.218	0.06771	0.33680	—	0.2758	1.4922	—	695.	0.448	—	0.0837	—	—	38.00
40.00	135.60	64.69	0.6333	32.146	165.288	0.06938	0.33583	—	0.2787	1.4989	—	695.	0.442	—	0.0832	—	—	40.00
42.00	140.32	64.45	0.6117	32.991	165.352	0.07104	0.33488	—	0.2816	1.5059	—	694.	0.436	—	0.0826	—	—	42.00
44.00	145.17	64.21	0.5909	33.837	165.410	0.07269	0.33392	—	0.2845	1.5131	—	693.	0.430	—	0.0821	—	—	44.00
46.00	150.14	63.97	0.5709	34.686	165.463	0.07435	0.33297	—	0.2876	1.5206	—	693.	0.424	—	0.0816	—	—	46.00
48.00	155.24	63.73	0.5517	35.538	165.510	0.07600	0.33201	—	0.2907	1.5283	—	692.	0.418	—	0.0811	—	—	48.00
50.00	160.46	63.48	0.5332	36.392	165.552	0.07765	0.33106	—	0.2939	1.5363	—	691.	0.412	—	0.0806	—	—	50.00
52.00	165.82	63.23	0.5154	37.248	165.587	0.07929	0.33012	—	0.2972	1.5447	—	690.	0.407	—	0.0801	—	—	52.00
54.00	171.32	62.98	0.4983	38.107	165.616	0.08094	0.32917	—	0.3006	1.5533	—	689.	0.401	—	0.0796	—	—	54.00
56.00	176.95	62.73	0.4818	38.970	165.639	0.08258	0.32822	—	0.3040	1.5623	—	688.	0.395	—	0.0790	—	—	56.00
58.00	182.72	62.48	0.4659	39.834	165.655	0.08422	0.32727	—	0.3076	1.5716	—	687.	0.390	—	0.0785	—	—	58.00
60.00	188.62	62.22	0.4506	40.702	165.665	0.08586	0.32633	—	0.3113	1.5813	—	686.	—	—	0.0780	—	—	60.00
62.00	194.68	61.96	0.4359	41.573	165.667	0.08750	0.32538	—	0.3151	1.5914	—	685.	—	—	0.0775	—	—	62.00
64.00	200.88	61.70	0.4217	42.447	165.663	0.08914	0.32443	—	0.3190	1.6020	—	684.	—	—	0.0770	—	—	64.00
66.00	207.22	61.43	0.4080	43.324	165.652	0.09077	0.32348	—	0.3231	1.6129	—	683.	—	—	0.0765	—	—	66.00
68.00	213.72	61.16	0.3948	44.205	165.633	0.09241	0.32253	—	0.3272	1.6243	—	682.	—	—	0.0760	—	—	68.00
70.00	220.37	60.89	0.3820	45.089	165.607	0.09404	0.32157	—	0.3315	1.6362	—	681.	—	—	0.0754	—	—	70.00
75.00	237.68	60.20	0.3520	47.316	165.507	0.09812	0.31918	—	0.3429	1.6683	—	677.	—	—	0.0741	—	7.11	75.00
80.00	255.99	59.49	0.3246	49.567	165.355	0.10221	0.31676	—	0.3554	1.7042	—	674.	—	—	—	—	6.68	80.00
85.00	275.36	58.76	0.2993	51.847	165.145	0.10630	0.31432	—	0.3691	1.7446	—	670.	—	—	—	—	6.26	85.00
90.00	295.83	58.01	0.2761	54.157	164.874	0.11041	0.31183	—	0.3843	1.7902	—	666.	—	—	—	—	5.84	90.00
95.00	317.43	57.23	0.2547	56.502	164.536	0.11453	0.30930	—	0.4012	1.8420	—	661.	—	—	—	—	5.42	95.00
100.00	340.22	56.42	0.2349	58.885	164.124	0.11867	0.30671	—	0.4203	1.9015	—	656.	—	—	—	—	5.00	100.00
105.00	364.24	55.59	0.2166	61.313	163.631	0.12285	0.30405	—	0.4419	1.9702	—	651.	—	—	—	—	4.59	105.00
110.00	389.54	54.71	0.1996	63.791	163.046	0.12707	0.30130	—	0.4667	2.0504	—	646.	—	—	—	—	4.19	110.00
115.00	416.19	53.80	0.1838	66.326	162.360	0.13134	0.29845	—	0.4956	2.1450	—	640.	—	—	—	—	3.79	115.00
120.00	444.23	52.84	0.1690	68.928	161.558	0.13568	0.29548	—	0.5297	2.2583	—	634.	—	—	—	—	3.39	120.00
125.00	473.74	51.83	0.1553	71.607	160.624	0.14010	0.29235	—	0.5706	2.3959	—	628.	—	—	—	—	3.01	125.00
130.00	504.76	50.75	0.1424	74.378	159.537	0.14463	0.28905	—	0.6208	2.5665	—	621.	—	—	—	—	2.63	130.00
135.00	537.39	49.61	0.1302	77.258	158.269	0.14929	0.28552	—	0.6839	2.7830	—	614.	—	—	—	—	2.26	135.00
140.00	571.69	48.37	0.1187	80.273	156.784	0.15412	0.28171	—	0.7659	3.0661	—	606.	—	—	—	—	1.90	140.00
145.00	607.76	47.02	0.1078	83.455	155.029	0.15918	0.27754	—	0.8769	3.4516	—	598.	—	—	—	—	1.55	145.00
150.00	645.68	45.52	0.0973	86.856	152.930	0.16453	0.27290	—	1.0357	4.0057	—	590.	—	—	—	—	1.21	150.00
155.00	685.57	43.82	0.0871	90.554	150.366	0.17030	0.26760	—	1.2823	4.8681	—	580.	—	—	—	—	0.89	155.00
160.00	727.57	41.81	0.0769	94.694	147.123	0.17671	0.26132	—	1.7181	6.3925	—	570.	—	—	—	—	0.59	160.00
165.00	771.85	39.28	0.0665	99.588	142.741	0.18425	0.25333	—	2.6971	9.8124	—	559.	—	—	—	—	0.32	165.00
170.00	818.67	35.45	0.0544	106.291	135.718	0.19457	0.24130	—	—	—	—	—	—	—	—	—	0.09	170.00
173.14c	849.61	26.20	0.0382	120.726	120.726	0.21713	0.21713	∞	∞	∞	0.	0.	—	—	∞	∞	0.00	173.14

*temperatures are on the IPTS-68 scale   b = boiling point   c = critical point

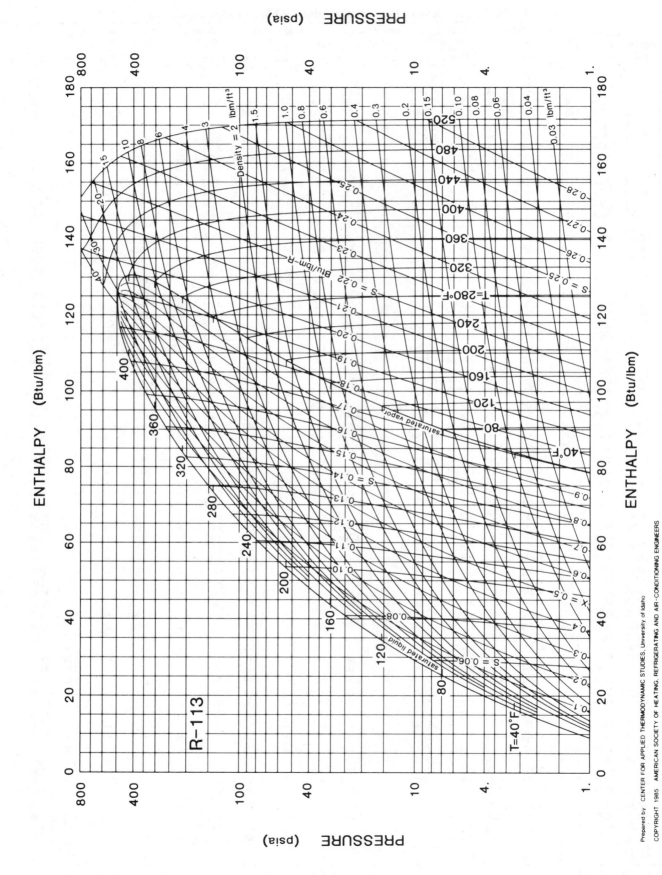

Fig. 9   Pressure-Enthalpy Diagram for Refrigerant 113

Refrigerant 113 (1,1,2-trichlorotrifluoroethane)   Properties of Saturated Liquid and Saturated Vapor

Temp,* °F	Pressure, psia	Density, lb/ft³ Liquid	Volume, ft³/lb Vapor	Enthalpy, Btu/lb Liquid	Vapor	Entropy, Btu/lb·°F Liquid	Vapor	Specific Heat c_p, Btu/lb·°F Liquid	Vapor	c_p/c_v Vapor	Velocity of Sound, ft/s Liquid	Vapor	Viscosity, lb$_m$/ft·h Liquid	Vapor	Thermal Cond, Btu/h·ft·°F Liquid	Vapor	Surface Tension, dyne/cm	Temp, °F
−20.00	0.445	105.08	56.424	4.000	75.095	0.00931	0.17101	—	—	—	—	—	—	—	—	—	22.33	−20.00
−10.00	0.624	104.32	41.061	6.035	76.554	0.01388	0.17071	—	—	—	—	—	—	—	—	—	21.69	−10.00
0.00	0.861	103.56	30.372	8.092	78.023	0.01841	0.17054	—	—	—	—	—	—	—	—	—	21.06	0.00
10.00	1.170	102.78	22.808	10.172	79.503	0.02288	0.17050	—	—	—	—	—	—	—	—	—	20.43	10.00
20.00	1.566	102.01	17.370	12.274	80.991	0.02731	0.17057	—	—	—	—	—	—	—	—	—	19.81	20.00
30.00	2.068	101.23	13.403	14.398	82.488	0.03169	0.17074	—	—	—	—	—	—	—	—	—	19.19	30.00
40.00	2.695	100.44	10.469	16.544	83.991	0.03603	0.17101	0.2220	—	—	—	—	2.173	0.0206	0.0469	0.00457	18.57	40.00
50.00	3.470	99.65	8.2704	18.712	85.500	0.04032	0.17136	0.2240	—	—	—	—	2.016	0.0215	0.0464	0.00472	17.96	50.00
60.00	4.417	98.85	6.6030	20.900	87.014	0.04457	0.17179	0.2258	—	—	—	—	1.873	0.0224	0.0458	0.00487	17.35	60.00
70.00	5.563	98.05	5.3241	23.109	88.531	0.04877	0.17229	0.2275	—	—	—	—	1.744	0.0232	0.0452	0.00503	16.74	70.00
75.00	6.220	97.64	4.7975	24.221	89.291	0.05086	0.17256	0.2283	—	—	—	—	1.683	0.0236	0.0449	0.00510	16.44	75.00
80.00	6.937	97.23	4.3326	25.338	90.051	0.05294	0.17285	0.2291	—	—	—	—	1.625	0.0239	0.0447	0.00518	16.14	80.00
85.00	7.719	96.82	3.9212	26.460	90.812	0.05500	0.17315	0.2298	—	—	—	—	1.570	0.0243	0.0444	0.00526	15.84	85.00
90.00	8.570	96.41	3.5562	27.587	91.573	0.05706	0.17347	0.2305	—	—	—	—	1.518	0.0246	0.0441	0.00534	15.55	90.00
95.00	9.494	96.00	3.2316	28.718	92.334	0.05911	0.17380	0.2312	—	—	—	—	1.467	0.0249	0.0438	0.00541	15.25	95.00
100.00	10.494	95.58	2.9424	29.854	93.095	0.06114	0.17414	0.2319	—	—	—	—	1.419	0.0252	0.0435	0.00549	14.95	100.00
102.00	10.917	95.42	2.8356	30.310	93.400	0.06195	0.17428	0.2322	—	—	—	—	1.400	0.0253	0.0434	0.00552	14.84	102.00
104.00	11.353	95.25	2.7335	30.766	93.705	0.06276	0.17442	0.2325	0.1690	—	—	—	1.382	0.0254	0.0433	0.00555	14.72	104.00
106.00	11.803	95.08	2.6358	31.223	94.009	0.06357	0.17456	0.2328	0.1697	—	—	—	1.364	0.0256	0.0432	0.00559	14.60	106.00
108.00	12.267	94.91	2.5423	31.681	94.313	0.06438	0.17471	0.2330	0.1704	—	—	—	1.346	0.0257	0.0431	0.00562	14.48	108.00
110.00	12.745	94.74	2.4529	32.140	94.618	0.06518	0.17486	0.2333	0.1710	—	—	—	1.329	0.0258	0.0430	0.00565	14.37	110.00
112.00	13.237	94.58	2.3672	32.599	94.922	0.06599	0.17500	0.2336	0.1716	—	—	—	1.312	0.0259	0.0428	0.00568	14.25	112.00
114.00	13.744	94.41	2.2852	33.059	95.227	0.06679	0.17515	0.2338	0.1722	—	—	—	1.295	0.0260	0.0427	0.00571	14.13	114.00
116.00	14.266	94.24	2.2066	33.520	95.531	0.06759	0.17531	0.2341	0.1727	—	—	—	1.279	0.0261	0.0426	0.00574	14.02	116.00
117.60b	14.696	94.10	2.1460	33.890	95.775	0.06823	0.17543	0.2343	0.1732	—	—	—	1.266	0.0262	0.0425	0.00577	13.93	117.60
118.00	14.804	94.07	2.1313	33.981	95.835	0.06838	0.17546	0.2344	0.1733	—	—	—	1.262	0.0262	0.0425	0.00577	13.90	118.00
120.00	15.357	93.90	2.0591	34.443	96.139	0.06918	0.17561	0.2346	0.1738	—	—	—	1.246	0.0263	0.0424	0.00581	13.79	120.00
122.00	15.926	93.73	1.9899	34.906	96.443	0.06998	0.17577	0.2349	0.1743	—	—	—	1.231	0.0264	0.0423	0.00584	13.67	122.00
124.00	16.512	93.56	1.9235	35.369	96.747	0.07077	0.17593	0.2352	0.1748	—	—	—	1.215	0.0265	0.0422	0.00587	13.55	124.00
126.00	17.114	93.38	1.8598	35.833	97.051	0.07156	0.17609	0.2355	0.1752	—	—	—	1.200	0.0266	0.0421	0.00590	13.44	126.00
128.00	17.733	93.21	1.7987	36.298	97.355	0.07235	0.17625	0.2357	0.1757	—	—	—	1.185	0.0266	0.0419	0.00593	13.32	128.00
130.00	18.369	93.04	1.7400	36.763	97.659	0.07314	0.17641	0.2360	0.1761	—	—	—	1.171	0.0267	0.0418	0.00596	13.21	130.00
132.00	19.023	92.87	1.6836	37.230	97.962	0.07393	0.17657	0.2363	0.1765	—	—	—	1.156	0.0268	0.0417	0.00600	13.09	132.00
134.00	19.695	92.69	1.6294	37.696	98.266	0.07471	0.17674	0.2365	0.1769	—	—	—	1.142	0.0269	0.0416	0.00603	12.98	134.00
136.00	20.385	92.52	1.5774	38.164	98.569	0.07550	0.17690	0.2368	0.1773	—	—	—	1.128	0.0270	0.0415	0.00606	12.87	136.00
138.00	21.093	92.34	1.5273	38.632	98.872	0.07628	0.17707	0.2371	0.1777	—	—	—	1.114	0.0271	0.0414	0.00609	12.75	138.00
140.00	21.820	92.17	1.4792	39.100	99.176	0.07706	0.17724	0.2374	0.1780	—	—	—	1.101	0.0272	0.0413	0.00612	12.64	140.00
142.00	22.567	91.99	1.4330	39.570	99.478	0.07784	0.17741	0.2376	0.1784	—	—	—	1.088	0.0272	0.0412	0.00616	12.52	142.00
144.00	23.333	91.82	1.3885	40.040	99.781	0.07861	0.17758	0.2379	0.1787	—	—	—	1.075	0.0273	0.0410	0.00619	12.41	144.00
146.00	24.119	91.64	1.3457	40.510	100.080	0.07939	0.17775	0.2382	0.1791	—	—	—	1.062	0.0274	0.0409	0.00622	12.30	146.00
148.00	24.925	91.47	1.3044	40.981	100.390	0.08016	0.17792	0.2385	0.1794	—	—	—	1.049	0.0275	0.0408	0.00625	12.18	148.00
150.00	25.752	91.29	1.2648	41.453	100.690	0.08094	0.17810	0.2388	0.1797	—	—	—	1.037	0.0276	0.0407	0.00629	12.07	150.00
155.00	27.910	90.84	1.1718	42.636	101.440	0.08286	0.17853	0.2395	0.1805	—	—	—	1.007	0.0278	0.0404	0.00637	11.79	155.00
160.00	30.204	90.39	1.0871	43.822	102.200	0.08477	0.17898	0.2403	0.1812	—	—	—	0.978	0.0279	0.0401	0.00645	11.51	160.00
165.00	32.640	89.94	1.0098	45.012	102.950	0.08668	0.17943	0.2411	0.1819	—	—	—	0.950	0.0281	0.0399	0.00653	11.23	165.00
170.00	35.222	89.48	0.9391	46.205	103.700	0.08857	0.17988	0.2419	0.1826	—	—	—	0.923	0.0283	0.0396	0.00661	10.95	170.00
175.00	37.956	89.03	0.8743	47.403	104.450	0.09046	0.18034	0.2427	0.1833	—	—	—	0.897	0.0285	0.0393	0.00670	10.68	175.00
180.00	40.848	88.56	0.8149	48.604	105.200	0.09233	0.18081	0.2436	0.1841	—	—	—	0.873	0.0286	0.0390	0.00678	10.40	180.00
185.00	43.903	88.09	0.7604	49.808	105.940	0.09420	0.18127	0.2445	0.1848	—	—	—	0.849	0.0288	0.0387	0.00687	—	185.00
190.00	47.127	87.62	0.7102	51.017	106.690	0.09606	0.18175	0.2455	0.1856	—	—	—	0.826	0.0290	0.0385	0.00695	—	190.00
195.00	50.527	87.14	0.6640	52.229	107.430	0.09790	0.18222	0.2465	0.1865	—	—	—	0.804	0.0291	0.0382	0.00704	—	195.00
200.00	54.107	86.66	0.6214	53.445	108.170	0.09974	0.18270	0.2475	0.1874	—	—	—	0.783	0.0293	0.0379	0.00712	—	200.00
210.00	61.835	85.68	0.5457	55.888	109.640	0.10339	0.18366	0.2498	0.1895	—	—	—	0.743	0.0297	0.0373	0.00730	—	210.00
220.00	70.358	84.68	0.4808	58.346	111.100	0.10701	0.18462	0.2522	0.1921	—	—	—	0.705	0.0300	0.0368	0.00747	—	220.00
230.00	79.726	83.66	0.4249	60.820	112.550	0.11059	0.18559	0.2550	0.1952	—	—	—	0.671	0.0304	0.0362	0.00765	—	230.00
240.00	89.990	82.61	0.3766	63.312	113.980	0.11415	0.18656	0.2579	0.1989	—	—	—	0.638	0.0308	0.0357	0.00783	—	240.00
250.00	101.20	81.53	0.3346	65.820	115.400	0.11767	0.18753	0.2612	0.2033	—	—	—	0.608	0.0312	0.0351	0.00802	—	250.00
260.00	113.42	80.42	0.2979	68.349	116.790	0.12117	0.18849	0.2648	0.2086	—	—	—	0.580	0.0317	0.0345	0.00821	—	260.00
270.00	126.69	79.27	0.2658	70.898	118.170	0.12464	0.18943	0.2687	0.2148	—	—	—	0.554	0.0322	0.0340	0.00840	—	270.00
280.00	141.08	78.08	0.2375	73.472	119.530	0.12810	0.19037	0.2730	0.2220	—	—	—	0.529	0.0327	0.0334	0.00859	—	280.00
290.00	156.64	76.84	0.2125	76.072	120.850	0.13154	0.19128	0.2777	0.2303	—	—	—	0.506	0.0333	0.0329	0.00879	—	290.00
300.00	173.45	75.55	0.1903	78.703	122.150	0.13497	0.19216	0.2828	0.2399	—	—	—	0.485	0.0340	0.0323	0.00899	—	300.00
310.00	191.55	74.20	0.1705	81.369	123.410	0.13840	0.19302	0.2883	—	—	—	—	0.465	0.0347	0.0318	0.00920	—	310.00
320.00	211.04	72.77	0.1528	84.076	124.620	0.14183	0.19383	0.2943	—	—	—	—	0.446	0.0355	0.0312	0.00941	—	320.00
330.00	231.97	71.26	0.1369	86.832	125.780	0.14527	0.19460	0.3008	—	—	—	—	0.428	0.0364	0.0307	0.00962	—	330.00
340.00	254.43	69.65	0.1225	89.647	126.880	0.14874	0.19530	0.3078	—	—	—	—	0.411	0.0373	—	—	—	340.00
350.00	278.52	67.91	0.1094	92.533	127.900	0.15224	0.19592	0.3153	—	—	—	—	0.395	0.0384	—	—	—	350.00
360.00	304.31	66.01	0.0974	95.510	128.830	0.15580	0.19645	—	—	—	—	—	—	—	—	—	—	360.00
370.00	331.93	63.90	0.0863	98.602	129.630	0.15944	0.19684	—	—	—	—	—	—	—	—	—	—	370.00
380.00	361.50	61.50	0.0760	101.850	130.260	0.16322	0.19704	—	—	—	—	—	—	—	—	—	—	380.00
390.00	393.16	58.67	0.0661	105.330	130.630	0.16720	0.19697	—	—	—	—	—	—	—	—	—	—	390.00
417.80c	494.70	35.58	0.0281	123.900	123.900	0.18710	0.18710	∞	∞	∞	0	0	—	—	∞	∞	0	417.80

*temperatures are on the IPTS-68 scale   b = boiling point   c = critical point

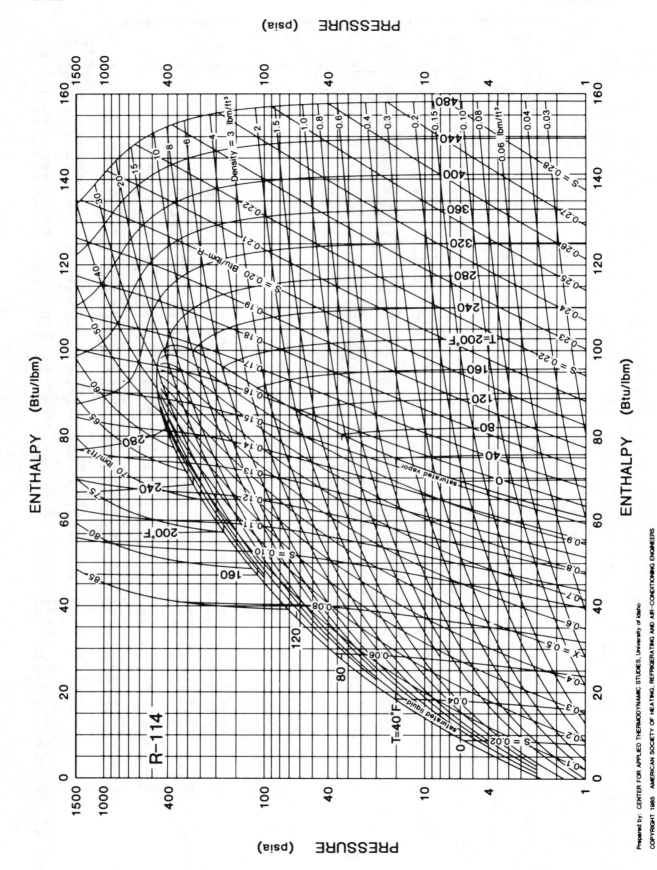

Prepared by: CENTER FOR APPLIED THERMODYNAMIC STUDIES, University of Idaho

COPYRIGHT 1985   AMERICAN SOCIETY OF HEATING, REFRIGERATING AND AIR-CONDITIONING ENGINEERS

Fig. 10   Pressure-Enthalpy Diagram for Refrigerant 114

Refrigerant 114 (1,2-dichlorotetrafluoroethane) Properties of Saturated Liquid and Saturated Vapor

Temp,* °F	Pressure, psia	Density, lb/ft³ Liquid	Volume, ft³/lb Vapor	Enthalpy, Btu/lb Liquid	Vapor	Entropy, Btu/lb·°F Liquid	Vapor	Specific Heat c_p, Btu/lb·°F Liquid	Vapor	c_p/c_v Vapor	Velocity of Sound, ft/s Liquid	Vapor	Viscosity, lb$_m$/ft·h Liquid	Vapor	Thermal Cond, Btu/h·ft·°F Liquid	Vapor	Surface Tension, dyne/cm	Temp, °F
−100.00	0.196	98.40	115.22	−14.466	55.647	−0.03723	0.15880	—	—	—	—	—	—	—	—	—	—	−100.00
−80.00	0.455	98.04	52.141	−9.506	58.230	−0.02381	0.15599	—	—	—	—	—	—	—	—	—	—	−80.00
−60.00	0.966	97.60	25.809	−4.693	60.887	−0.01145	0.15412	—	—	—	—	—	—	—	—	—	—	−60.00
−40.00	1.890	97.07	13.787	0.000	63.604	0.00000	0.15301	0.2072	0.1388	—	2617.	362.	2.012	—	0.0466	—	—	−40.00
−20.00	3.452	96.42	7.8606	4.603	66.367	0.01071	0.15253	0.2128	0.1485	—	2437.	366.	1.744	—	0.0449	—	—	−20.00
−15.00	3.975	96.24	6.8926	5.744	67.063	0.01329	0.15250	0.2140	0.1505	—	2395.	368.	1.682	—	0.0445	—	—	−15.00
−10.00	4.560	96.05	6.0640	6.881	67.761	0.01583	0.15249	0.2151	0.1523	—	2353.	369.	1.622	—	0.0441	—	—	−10.00
−5.00	5.213	95.85	5.3523	8.017	68.460	0.01834	0.15252	0.2161	0.1539	—	2313.	370.	1.563	—	0.0436	—	—	−5.00
0.00	5.939	95.64	4.7389	9.151	69.161	0.02081	0.15257	0.2171	0.1554	—	2273.	371.	1.506	—	0.0432	—	—	0.00
5.00	6.745	95.41	4.2083	10.283	69.863	0.02326	0.15265	0.2181	0.1568	—	2234.	372.	1.452	—	0.0428	—	—	5.00
10.00	7.636	95.18	3.7479	11.416	70.567	0.02568	0.15275	0.2190	0.1581	—	2196.	373.	1.399	—	0.0424	—	—	10.00
12.00	8.018	95.08	3.5810	11.869	70.848	0.02664	0.15280	0.2193	0.1585	—	2181.	374.	1.378	—	0.0423	—	—	12.00
14.00	8.415	94.98	3.4230	12.322	71.130	0.02760	0.15285	0.2196	0.1590	—	2166.	374.	1.358	—	0.0421	—	—	14.00
16.00	8.827	94.88	3.2734	12.775	71.412	0.02855	0.15291	0.2200	0.1594	—	2151.	374.	1.338	—	0.0420	—	—	16.00
18.00	9.255	94.78	3.1317	13.228	71.694	0.02950	0.15297	0.2203	0.1599	—	2136.	375.	1.318	—	0.0418	—	—	18.00
20.00	9.699	94.67	2.9973	13.682	71.976	0.03045	0.15303	0.2206	0.1603	—	2122.	375.	1.298	—	0.0417	—	—	20.00
22.00	10.160	94.56	2.8699	14.136	72.258	0.03139	0.15310	0.2209	0.1607	—	2107.	376.	1.279	—	0.0415	—	—	22.00
24.00	10.638	94.45	2.7490	14.590	72.540	0.03233	0.15317	0.2212	0.1611	—	2093.	376.	1.260	—	0.0414	—	—	24.00
26.00	11.134	94.33	2.6343	15.044	72.822	0.03326	0.15324	0.2216	0.1615	—	2078.	376.	1.242	—	0.0412	—	—	26.00
28.00	11.648	94.22	2.5253	15.499	73.104	0.03420	0.15331	0.2219	0.1618	—	2064.	377.	1.224	—	0.0410	—	—	28.00
30.00	12.180	94.10	2.4218	15.954	73.386	0.03513	0.15339	0.2222	0.1622	—	2050.	377.	1.206	—	0.0409	—	—	30.00
32.00	12.732	93.98	2.3235	16.410	73.669	0.03605	0.15347	0.2225	0.1626	—	2036.	377.	1.188	0.0254	0.0407	—	13.80	32.00
34.00	13.302	93.86	2.2299	16.866	73.951	0.03698	0.15356	0.2228	0.1629	—	2022.	378.	1.170	0.0255	0.0406	—	13.67	34.00
36.00	13.892	93.73	2.1410	17.322	74.233	0.03789	0.15365	0.2231	0.1632	—	2009.	378.	1.153	0.0256	0.0404	—	13.54	36.00
38.00	14.503	93.61	2.0563	17.780	74.516	0.03881	0.15374	0.2234	0.1636	—	1995.	378.	1.136	0.0257	0.0403	—	13.41	38.00
38.62b	14.696	93.56	2.0309	17.921	74.603	0.03910	0.15377	0.2235	0.1637	—	1991.	378.	1.131	0.0258	0.0402	—	13.37	38.62
40.00	15.134	93.47	1.9757	18.237	74.798	0.03973	0.15383	0.2237	0.1639	—	1982.	379.	1.119	0.0258	0.0401	—	13.28	40.00
42.00	15.787	93.34	1.8989	18.696	75.080	0.04064	0.15393	0.2240	0.1642	—	1968.	379.	1.103	0.0260	0.0400	—	13.15	42.00
44.00	16.461	93.20	1.8257	19.155	75.362	0.04155	0.15403	0.2242	0.1645	—	1955.	379.	1.087	0.0261	0.0398	—	13.02	44.00
46.00	17.158	93.07	1.7560	19.614	75.644	0.04246	0.15413	0.2245	0.1648	—	1941.	379.	1.071	0.0262	0.0397	—	12.89	46.00
48.00	17.877	92.92	1.6895	20.075	75.927	0.04337	0.15423	0.2248	0.1652	—	1928.	380.	1.055	0.0263	0.0395	—	12.76	48.00
50.00	18.619	92.78	1.6260	20.536	76.209	0.04427	0.15434	0.2251	0.1655	—	1915.	380.	1.040	0.0264	0.0394	—	12.63	50.00
52.00	19.384	92.63	1.5655	20.998	76.491	0.04517	0.15445	0.2254	0.1658	—	1902.	380.	1.025	0.0265	0.0392	—	12.51	52.00
54.00	20.174	92.48	1.5076	21.460	76.772	0.04607	0.15456	0.2257	0.1661	—	1889.	380.	1.010	0.0266	0.0391	—	12.38	54.00
56.00	20.988	92.33	1.4524	21.924	77.054	0.04697	0.15467	0.2260	0.1664	—	1876.	381.	0.995	0.0267	0.0389	—	12.25	56.00
58.00	21.828	92.17	1.3997	22.388	77.336	0.04786	0.15479	0.2263	0.1667	—	1863.	381.	0.981	0.0269	0.0388	—	12.12	58.00
60.00	22.693	92.02	1.3492	22.853	77.617	0.04876	0.15491	0.2266	0.1670	—	1850.	381.	0.966	0.0270	0.0386	—	12.00	60.00
62.00	23.584	91.85	1.3010	23.319	77.899	0.04965	0.15503	0.2270	0.1673	—	1838.	381.	0.952	0.0271	0.0385	—	11.87	62.00
64.00	24.501	91.69	1.2549	23.786	78.180	0.05054	0.15515	0.2273	0.1676	—	1825.	381.	0.939	0.0272	0.0383	—	11.74	64.00
66.00	25.446	91.52	1.2108	24.254	78.461	0.05143	0.15528	0.2276	0.1679	—	1812.	381.	0.925	0.0273	0.0382	—	11.62	66.00
68.00	26.418	91.35	1.1685	24.722	78.742	0.05231	0.15540	0.2279	0.1682	—	1800.	381.	0.912	0.0274	0.0380	—	11.49	68.00
70.00	27.418	91.18	1.1281	25.192	79.023	0.05320	0.15553	0.2282	0.1686	—	1787.	382.	0.899	0.0275	0.0379	—	11.36	70.00
72.00	28.447	91.00	1.0894	25.662	79.303	0.05408	0.15566	0.2286	0.1689	—	1775.	382.	0.886	0.0276	0.0377	—	11.24	72.00
74.00	29.505	90.83	1.0522	26.133	79.584	0.05496	0.15580	0.2289	0.1692	—	1763.	382.	0.873	0.0277	0.0376	—	11.11	74.00
76.00	30.593	90.64	1.0167	26.605	79.864	0.05584	0.15593	0.2292	0.1696	—	1750.	382.	0.861	0.0278	0.0374	—	10.99	76.00
78.00	31.710	90.46	0.9826	27.078	80.144	0.05672	0.15607	0.2296	0.1699	—	1738.	382.	0.848	0.0279	0.0373	—	10.86	78.00
85.00	35.868	89.79	0.8737	28.740	81.112	0.05977	0.15656	0.2309	0.1713	—	1696.	382.	0.807	0.0283	0.0367	—	10.43	85.00
90.00	39.080	89.29	0.8049	29.934	81.818	0.06194	0.15692	0.2319	0.1723	—	1666.	382.	0.779	0.0285	0.0364	—	10.12	90.00
95.00	42.504	88.78	0.7426	31.132	82.513	0.06410	0.15729	0.2329	0.1734	—	1636.	382.	0.752	0.0288	0.0360	—	9.81	95.00
100.00	46.148	88.25	0.6861	32.335	83.205	0.06624	0.15767	0.2340	0.1746	—	1606.	381.	0.726	0.0290	0.0356	—	9.51	100.00
105.00	50.021	87.70	0.6347	33.541	83.895	0.06837	0.15805	0.2352	0.1759	—	1577.	381.	0.701	0.0293	0.0353	—	9.21	105.00
110.00	54.132	87.14	0.5880	34.752	84.583	0.07049	0.15845	0.2364	0.1773	—	1547.	380.	0.677	0.0295	0.0349	—	8.90	110.00
120.00	63.103	85.98	0.5063	37.185	85.949	0.07469	0.15925	0.2391	0.1806	—	1489.	378.	0.632	0.0301	0.0341	—	8.31	120.00
130.00	73.134	84.76	0.4380	39.630	87.302	0.07884	0.16007	0.2423	0.1845	—	1431.	376.	0.591	0.0306	0.0334	—	7.72	130.00
140.00	84.299	83.50	0.3803	42.088	88.638	0.08293	0.16089	0.2458	0.1892	—	1372.	373.	0.552	0.0312	0.0326	—	7.14	140.00
150.00	96.678	82.19	0.3314	44.561	89.955	0.08697	0.16173	0.2498	0.1946	—	1313.	370.	0.517	0.0318	0.0319	—	6.57	150.00
160.00	110.35	80.84	0.2896	47.050	91.250	0.09097	0.16256	0.2543	0.2009	—	1253.	366.	0.485	0.0324	0.0311	—	6.01	160.00
170.00	125.41	79.44	0.2537	49.560	92.518	0.09494	0.16338	0.2594	0.2082	—	1192.	361.	0.455	0.0331	0.0304	—	—	170.00
180.00	141.93	77.99	0.2227	52.096	93.753	0.09887	0.16418	0.2652	0.2167	—	1130.	355.	0.427	0.0339	0.0296	—	—	180.00
190.00	160.03	76.48	0.1958	54.666	94.950	0.10279	0.16495	0.2715	0.2263	—	1067.	348.	0.401	0.0347	0.0288	—	—	190.00
200.00	179.78	74.90	0.1722	57.275	96.101	0.10670	0.16568	0.2787	0.2372	—	1001.	341.	0.378	0.0356	0.0280	—	—	200.00
210.00	201.32	73.23	0.1515	59.935	97.195	0.11062	0.16636	0.2865	0.2494	—	934.	332.	0.356	0.0366	0.0272	—	—	210.00
220.00	224.75	71.46	0.1332	62.655	98.218	0.11456	0.16696	0.2952	0.2632	—	864.	323.	0.335	—	0.0263	—	—	220.00
230.00	250.19	69.55	0.1168	65.449	99.154	0.11855	0.16747	0.3048	0.2785	—	792.	313.	0.316	—	0.0255	—	—	230.00
240.00	277.77	67.47	0.1021	68.337	99.977	0.12260	0.16785	0.3152	0.2954	—	716.	302.	0.299	—	0.0246	—	—	240.00
250.00	307.65	65.15	0.0888	71.346	100.650	0.12675	0.16806	—	—	—	638.	289.	—	—	—	—	—	250.00
260.00	339.97	62.49	0.0766	74.519	101.110	0.13106	0.16801	—	—	—	556.	276.	—	—	—	—	—	260.00
270.00	374.90	59.28	0.0650	77.941	101.240	0.13563	0.16757	—	—	—	470.	261.	—	—	—	—	—	270.00
280.00	412.62	55.02	0.0537	81.819	100.810	0.14074	0.16641	—	—	—	381.	245.	—	—	—	—	—	280.00
290.00	453.31	47.62	0.0409	87.049	98.880	0.14757	0.16335	—	—	—	—	—	—	—	—	—	—	290.00
294.60c	473.00	34.75	0.0288	93.94	93.94	0.1566	0.1566	∞	∞	∞	0	0	—	—	∞	∞	0	294.60

*temperatures are on the IPTS-68 scale   b = boiling point   c = critical point

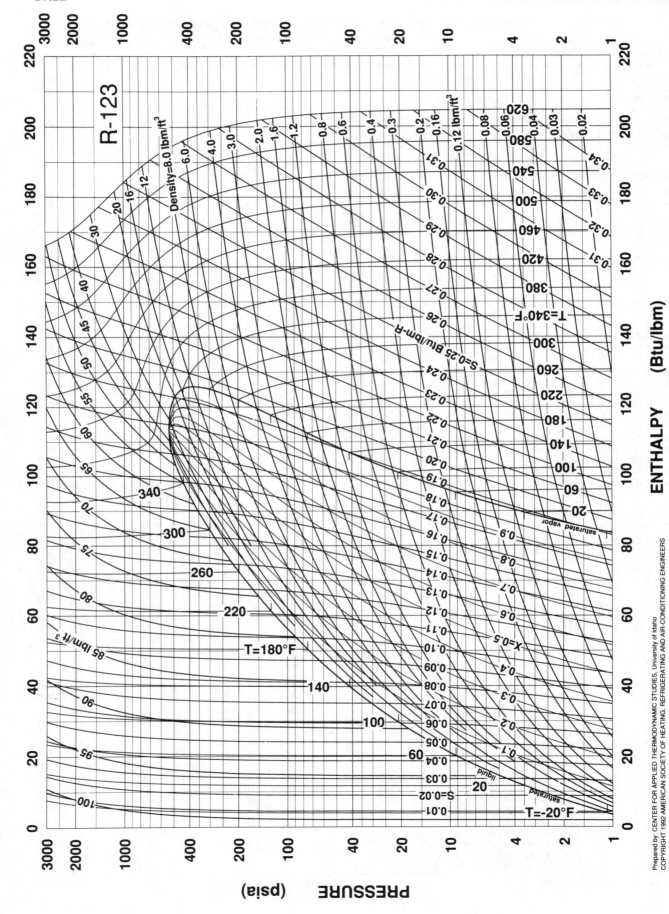

PRESSURE (psia)

ENTHALPY (Btu/lbm)

Fig. 11 Pressure-Enthalpy Diagram for Refrigerant 123

Prepared by: CENTER FOR APPLIED THERMODYNAMIC STUDIES, University of Idaho
COPYRIGHT 1992 AMERICAN SOCIETY OF HEATING, REFRIGERATING AND AIR-CONDITIONING ENGINEERS

Refrigerant 123 (1,1-dichloro-2,2,2-trifluoroethane)   Properties of Saturated Liquid and Saturated Vapor

Temp,* °F	Pressure, psia	Density, lb/ft³ Liquid	Volume, ft³/lb Vapor	Enthalpy, Btu/lb Liquid	Enthalpy, Btu/lb Vapor	Entropy, Btu/lb·°F Liquid	Entropy, Btu/lb·°F Vapor	Specific Heat c_p, Btu/lb·°F Liquid	Specific Heat c_p, Vapor	c_p/c_v Vapor	Velocity of Sound, ft/s Liquid	Velocity of Sound, ft/s Vapor	Viscosity, lb$_m$/ft·h Liquid	Viscosity, Vapor	Thermal Cond, Btu/h·ft·°F Liquid	Thermal Cond, Vapor	Surface Tension, dyne/cm	Temp, °F
−40.00	0.548	100.71	53.445	0.000	81.340	0.00000	0.19382	—	0.1419	1.1064	—	386.	2.537	—	—	—	23.49	−40.00
−30.00	0.776	99.96	38.593	2.005	82.707	0.00472	0.19254	—	0.1442	1.1057	—	390.	2.303	—	—	—	22.78	−30.00
−20.00	1.080	99.20	28.345	3.923	84.085	0.00913	0.19146	—	0.1466	1.1052	—	394.	2.100	—	—	—	22.06	−20.00
−10.00	1.478	98.44	21.141	5.812	85.476	0.01338	0.19054	—	0.1490	1.1047	—	398.	1.922	—	—	—	21.35	−10.00
0.00	1.993	97.67	15.993	7.710	86.877	0.01755	0.18978	—	0.1514	1.1045	—	401.	1.766	—	0.0515	—	20.65	0.00
5.00	2.303	97.28	13.979	8.672	87.581	0.01963	0.18945	—	0.1526	1.1044	—	403.	1.695	—	0.0510	—	20.30	5.00
10.00	2.651	96.89	12.258	9.645	88.289	0.02171	0.18916	—	0.1539	1.1043	—	405.	1.629	—	0.0506	—	19.95	10.00
15.00	3.042	96.49	10.782	10.631	88.998	0.02380	0.18890	—	0.1551	1.1043	—	406.	1.566	—	0.0501	—	19.61	15.00
20.00	3.480	96.09	9.5110	11.632	89.710	0.02590	0.18867	—	0.1563	1.1043	—	408.	1.507	—	0.0496	—	19.26	20.00
25.00	3.969	95.69	8.4141	12.647	90.424	0.02800	0.18847	0.2046	0.1576	1.1043	2636.	409.	1.451	—	0.0492	—	18.92	25.00
30.00	4.513	95.29	7.4644	13.679	91.140	0.03012	0.18831	0.2079	0.1588	1.1044	2596.	411.	1.398	—	0.0487	—	18.57	30.00
35.00	5.116	94.88	6.6397	14.727	91.859	0.03224	0.18817	0.2111	0.1601	1.1046	2558.	412.	1.348	—	0.0482	—	18.23	35.00
40.00	5.785	94.47	5.9215	15.792	92.579	0.03438	0.18806	0.2144	0.1613	1.1047	2521.	414.	1.301	—	0.0478	—	17.89	40.00
45.00	6.522	94.06	5.2943	16.873	93.301	0.03653	0.18797	0.2177	0.1626	1.1050	2486.	415.	1.257	—	0.0473	—	17.55	45.00
50.00	7.334	93.64	4.7450	17.971	94.025	0.03869	0.18791	0.2209	0.1639	1.1052	2452.	416.	1.214	—	0.0468	—	17.21	50.00
55.00	8.226	93.22	4.2628	19.084	94.750	0.04086	0.18788	0.2240	0.1651	1.1056	2419.	417.	1.174	—	0.0464	—	16.87	55.00
60.00	9.203	92.80	3.8384	20.213	95.477	0.04304	0.18787	0.2271	0.1664	1.1059	2387.	418.	1.136	—	0.0459	—	16.53	60.00
65.00	10.271	92.38	3.4639	21.356	96.205	0.04523	0.18789	0.2300	0.1676	1.1063	2355.	419.	1.099	—	0.0454	—	16.20	65.00
70.00	11.436	91.95	3.1326	22.515	96.933	0.04742	0.18792	0.2329	0.1689	1.1068	2323.	420.	1.065	—	0.0450	0.00572	15.86	70.00
75.00	12.704	91.52	2.8389	23.687	97.663	0.04962	0.18798	0.2356	0.1701	1.1074	2292.	421.	1.032	—	0.0445	0.00583	15.53	75.00
80.00	14.080	91.08	2.5779	24.873	98.394	0.05182	0.18805	0.2382	0.1714	1.1080	2261.	422.	1.000	—	0.0440	0.00594	15.20	80.00
82.11b	14.696	90.89	2.4764	25.378	98.702	0.05275	0.18809	0.2392	0.1719	1.1083	2248.	422.	0.987	—	0.0438	0.00598	15.06	82.11
85.00	15.572	90.64	2.3454	26.072	99.124	0.05403	0.18815	0.2406	0.1726	1.1087	2231.	423.	0.970	—	0.0436	0.00605	14.87	85.00
90.00	17.185	90.20	2.1379	27.282	99.855	0.05623	0.18826	0.2430	0.1738	1.1095	2200.	423.	0.941	—	0.0431	0.00616	14.54	90.00
95.00	18.927	89.75	1.9522	28.504	100.586	0.05844	0.18839	0.2452	0.1751	1.1103	2169.	424.	0.913	—	0.0426	0.00627	14.21	95.00
100.00	20.803	89.30	1.7858	29.737	101.317	0.06065	0.18854	0.2473	0.1763	1.1113	2138.	425.	0.887	—	0.0422	0.00638	13.89	100.00
105.00	22.821	88.85	1.6363	30.980	102.047	0.06285	0.18871	0.2493	0.1775	1.1124	2107.	425.	0.861	—	0.0417	0.00649	13.56	105.00
110.00	24.988	88.39	1.5018	32.233	102.777	0.06505	0.18888	0.2511	0.1788	1.1135	2075.	425.	0.837	—	0.0412	0.00659	13.24	110.00
115.00	27.310	87.92	1.3804	33.495	103.505	0.06725	0.18907	0.2529	0.1800	1.1148	2044.	425.	0.813	—	0.0408	0.00670	12.92	115.00
120.00	29.796	87.46	1.2708	34.766	104.231	0.06944	0.18928	0.2545	0.1812	1.1162	2012.	426.	0.790	—	0.0403	0.00681	12.59	120.00
125.00	32.451	86.98	1.1716	36.044	104.956	0.07163	0.18949	0.2560	0.1825	1.1177	1980.	426.	0.768	0.0285	0.0398	0.00692	12.27	125.00
130.00	35.285	86.51	1.0816	37.330	105.679	0.07381	0.18972	0.2575	0.1837	1.1194	1948.	426.	0.747	0.0287	0.0394	0.00703	11.96	130.00
135.00	38.304	86.03	0.9999	38.623	106.399	0.07598	0.18995	0.2588	0.1850	1.1212	1915.	425.	0.727	0.0289	0.0389	0.00714	11.64	135.00
140.00	41.515	85.54	0.9255	39.922	107.116	0.07814	0.19019	0.2601	0.1862	1.1232	1882.	425.	0.707	0.0292	0.0384	0.00725	11.32	140.00
145.00	44.928	85.05	0.8577	41.228	107.830	0.08030	0.19044	0.2613	0.1875	1.1254	1849.	425.	0.687	0.0294	0.0380	0.00736	11.01	145.00
150.00	48.549	84.55	0.7958	42.540	108.540	0.08245	0.19070	0.2624	0.1888	1.1277	1816.	425.	0.669	0.0296	0.0375	0.00747	10.70	150.00
155.00	52.388	84.05	0.7393	43.857	109.246	0.08458	0.19097	0.2635	0.1901	1.1303	1782.	424.	0.651	0.0298	0.0370	0.00758	10.39	155.00
160.00	56.451	83.54	0.6874	45.179	109.948	0.08671	0.19123	0.2645	0.1914	1.1331	1749.	423.	0.633	0.0301	0.0366	0.00769	10.08	160.00
165.00	60.747	83.02	0.6399	46.506	110.645	0.08883	0.19151	0.2655	0.1927	1.1361	1715.	423.	0.616	0.0303	0.0361	0.00780	9.77	165.00
170.00	65.285	82.50	0.5963	47.838	111.337	0.09094	0.19178	0.2664	0.1941	1.1393	1681.	422.	0.599	0.0305	0.0356	0.00791	9.47	170.00
175.00	70.073	81.98	0.5561	49.174	112.022	0.09303	0.19206	0.2673	0.1955	1.1429	1646.	421.	0.583	0.0307	0.0352	0.00802	9.16	175.00
180.00	75.119	81.44	0.5191	50.515	112.702	0.09512	0.19234	0.2682	0.1969	1.1467	1612.	420.	—	0.0310	0.0347	0.00813	8.86	180.00
185.00	80.434	80.90	0.4850	51.860	113.374	0.09719	0.19261	0.2691	0.1984	1.1509	1578.	419.	—	0.0312	0.0342	0.00824	8.56	185.00
190.00	86.024	80.35	0.4535	53.209	114.039	0.09926	0.19289	0.2700	0.1999	1.1554	1543.	417.	—	0.0314	0.0338	0.00835	8.26	190.00
195.00	91.900	79.79	0.4244	54.562	114.697	0.10131	0.19317	0.2709	0.2015	1.1604	1509.	416.	—	0.0316	0.0333	0.00846	7.96	195.00
200.00	98.071	79.23	0.3974	55.920	115.346	0.10336	0.19344	0.2719	0.2032	1.1657	1474.	415.	—	0.0318	0.0328	—	7.67	200.00
205.00	104.55	78.65	0.3724	57.281	115.986	0.10539	0.19371	0.2728	0.2049	1.1715	1439.	413.	—	0.0321	0.0324	—	7.38	205.00
210.00	111.33	78.07	0.3492	58.647	116.617	0.10741	0.19398	0.2738	0.2067	1.1777	1405.	411.	—	0.0323	0.0319	—	7.09	210.00
215.00	118.44	77.48	0.3277	60.018	117.237	0.10943	0.19424	0.2749	0.2086	1.1846	1370.	409.	—	0.0325	—	—	6.80	215.00
220.00	125.89	76.88	0.3076	61.393	117.847	0.11143	0.19449	0.2760	0.2106	1.1920	1336.	407.	—	0.0327	—	—	6.51	220.00
225.00	133.67	76.26	0.2889	62.774	118.445	0.11343	0.19474	0.2772	0.2127	1.2001	1301.	405.	—	0.0329	—	—	6.23	225.00
230.00	141.81	75.64	0.2715	64.159	119.032	0.11542	0.19498	0.2786	0.2150	1.2089	1267.	403.	—	0.0332	—	—	5.94	230.00
235.00	150.31	75.00	0.2552	65.551	119.605	0.11740	0.19521	0.2800	0.2174	1.2185	1233.	401.	—	0.0334	—	—	5.67	235.00
240.00	159.19	74.35	0.2400	66.948	120.165	0.11937	0.19543	0.2816	0.2200	1.2291	1198.	398.	—	0.0336	—	—	5.39	240.00
245.00	168.44	73.69	0.2258	68.353	120.711	0.12134	0.19564	0.2834	0.2228	1.2407	1164.	396.	—	0.0338	—	—	5.11	245.00
250.00	178.10	73.01	0.2125	69.764	121.241	0.12330	0.19583	0.2853	0.2258	1.2534	1130.	393.	—	0.0340	—	—	4.84	250.00
260.00	198.63	71.61	0.1882	72.613	122.250	0.12721	0.19618	0.2899	0.2327	1.2831	1061.	387.	—	—	—	—	4.31	260.00
270.00	220.87	70.13	0.1668	75.501	123.184	0.13112	0.19646	0.2957	0.2411	1.3198	992.	380.	—	—	—	—	3.79	270.00
280.00	244.92	68.56	0.1478	78.437	124.031	0.13503	0.19667	0.3031	0.2515	1.3663	922.	372.	—	—	—	—	3.28	280.00
290.00	270.89	66.88	0.1308	81.434	124.773	0.13895	0.19676	0.3127	0.2649	1.4268	851.	363.	—	—	—	—	2.79	290.00
300.00	298.88	65.06	0.1154	84.507	125.389	0.14292	0.19674	0.3254	0.2829	1.5085	778.	353.	—	—	—	—	2.31	300.00
310.00	329.01	63.06	0.1015	87.677	125.845	0.14695	0.19654	0.3431	0.3083	1.6250	701.	342.	—	—	—	—	1.85	310.00
320.00	361.45	60.82	0.0887	90.977	126.088	0.15109	0.19612	0.3691	0.3471	1.8034	621.	329.	—	—	—	—	1.42	320.00
330.00	396.37	58.22	0.0768	94.457	126.025	0.15538	0.19536	0.4109	0.4137	2.1089	534.	315.	—	—	—	—	1.01	330.00
340.00	434.02	55.08	0.0652	98.209	125.478	0.15995	0.19405	0.4911	0.5508	2.7349	441.	300.	—	—	—	—	0.63	340.00
350.00	474.80	50.88	0.0534	102.454	124.000	0.16505	0.19166	—	—	—	—	—	—	—	—	—	0.29	350.00
360.00	519.44	43.25	0.0389	108.347	119.584	0.17206	0.18577	—	—	—	—	—	—	—	—	—	0.04	360.00
362.63c	532.00	34.34	0.0291	113.806	113.806	0.17788	0.17788	∞	∞	∞	0.	0.	—	—	∞	∞	0.00	362.63

*temperatures are on the ITS-90 scale   b = boiling point   c = critical point

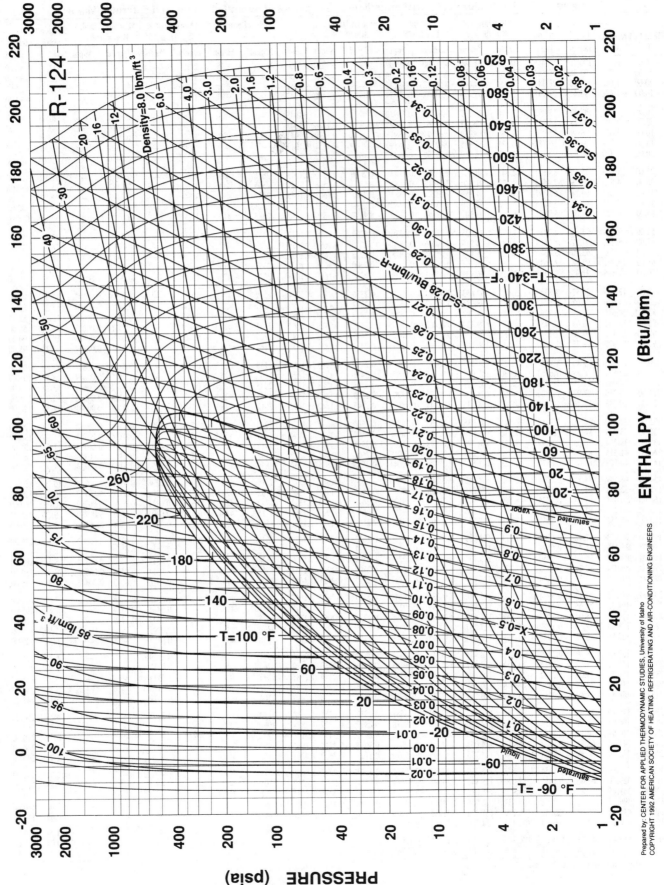

ENTHALPY (Btu/lbm)

PRESSURE (psia)

R-124

Fig. 12 Pressure-Enthalpy Diagram for Refrigerant 124

Refrigerant Properties

Refrigerant 124 (1-chloro-1,2,2,2-tetrafluoroethane)   Properties of Saturated Liquid and Saturated Vapor

Temp,* °F	Pressure, psia	Density, lb/ft³ Liquid	Volume, ft³/lb Vapor	Enthalpy, Btu/lb Liquid	Enthalpy, Btu/lb Vapor	Entropy, Btu/lb·°F Liquid	Entropy, Btu/lb·°F Vapor	Specific Heat c_p, Btu/lb·°F Liquid	Specific Heat c_p, Btu/lb·°F Vapor	c_p/c_v Vapor	Velocity of Sound, ft/s Liquid	Velocity of Sound, ft/s Vapor	Viscosity, lb$_m$/ft·h Liquid	Viscosity, lb$_m$/ft·h Vapor	Thermal Cond, Btu/h·ft·°F Liquid	Thermal Cond, Btu/h·ft·°F Vapor	Surface Tension, dyne/cm	Temp, °F
−100.00	0.438	100.79	64.170	−16.341	67.816	−0.03222	0.19140	0.2290	0.1382	1.1232	3019.	382.	2.721	—	0.0672	—	22.10	−100.00
−90.00	0.666	100.79	43.371	−12.650	69.146	−0.02797	0.18901	0.2294	0.1408	1.1219	2778.	386.	2.390	—	0.0661	—	21.53	−90.00
−80.00	0.988	100.64	29.950	−9.306	70.485	−0.02328	0.18687	0.2296	0.1436	1.1219	2654.	390.	2.118	—	0.0649	—	20.92	−80.00
−70.00	1.433	99.74	21.136	−7.002	71.832	−0.01730	0.18501	0.2311	0.1464	1.1217	2585.	394.	1.902	—	0.0638	—	20.16	−70.00
−60.00	2.034	98.83	15.228	−4.684	73.184	−0.01143	0.18340	0.2325	0.1493	1.1218	2523.	398.	1.719	—	0.0626	—	19.41	−60.00
−50.00	2.830	97.91	11.182	−2.350	74.542	−0.00566	0.18203	0.2340	0.1524	1.1223	2464.	402.	1.563	—	0.0615	—	18.66	−50.00
−45.00	3.315	97.44	9.6449	−1.177	75.223	−0.00282	0.18142	0.2349	0.1539	1.1226	2436.	404.	1.493	—	0.0609	—	18.29	−45.00
−40.00	3.865	96.97	8.3539	0.000	75.904	0.00000	0.18087	0.2357	0.1555	1.1231	2409.	405.	1.428	—	0.0603	—	17.92	−40.00
−35.00	4.487	96.50	7.2644	1.181	76.586	0.00280	0.18036	0.2366	0.1572	1.1236	2381.	407.	1.368	—	0.0598	—	17.55	−35.00
−30.00	5.189	96.02	6.3410	2.368	77.268	0.00557	0.17989	0.2375	0.1588	1.1243	2354.	408.	1.311	—	0.0592	—	17.18	−30.00
−25.00	5.977	95.54	5.5550	3.559	77.950	0.00832	0.17947	0.2385	0.1605	1.1250	2327.	410.	1.259	—	0.0586	—	16.81	−25.00
−20.00	6.858	95.06	4.8834	4.755	78.633	0.01105	0.17909	0.2395	0.1622	1.1258	2300.	411.	1.209	—	0.0580	—	16.45	−20.00
−15.00	7.842	94.57	4.3073	5.956	79.316	0.01377	0.17874	0.2406	0.1640	1.1267	2273.	412.	1.163	—	0.0575	—	16.09	−15.00
−10.00	8.935	94.08	3.8113	7.163	79.999	0.01646	0.17844	0.2417	0.1658	1.1278	2245.	414.	1.120	—	0.0569	—	15.72	−10.00
−5.00	10.146	93.58	3.3827	8.376	80.681	0.01914	0.17817	0.2429	0.1676	1.1289	2218.	415.	1.079	—	0.0563	0.00447	15.36	−5.00
0.00	11.485	93.08	3.0110	9.594	81.363	0.02180	0.17793	0.2441	0.1695	1.1301	2190.	416.	1.040	—	0.0557	0.00456	15.00	0.00
5.00	12.960	92.58	2.6877	10.819	82.045	0.02444	0.17772	0.2453	0.1713	1.1314	2162.	417.	1.004	—	0.0552	0.00466	14.65	5.00
10.00	14.580	92.07	2.4056	12.051	82.726	0.02707	0.17755	0.2466	0.1732	1.1329	2134.	418.	0.969	—	0.0546	0.00476	14.29	10.00
10.34b	14.696	92.03	2.3877	12.135	82.772	0.02725	0.17754	0.2467	0.1734	1.1330	2132.	418.	0.967	—	0.0546	0.00476	14.27	10.34
15.00	16.356	91.55	2.1586	13.289	83.406	0.02968	0.17740	0.2479	0.1752	1.1344	2105.	418.	0.937	—	0.0540	0.00486	13.94	15.00
20.00	18.298	91.03	1.9417	14.533	84.086	0.03228	0.17728	0.2493	0.1772	1.1361	2076.	419.	0.905	—	0.0535	0.00495	13.58	20.00
25.00	20.416	90.51	1.7507	15.785	84.764	0.03487	0.17719	0.2507	0.1792	1.1379	2047.	420.	0.876	—	0.0529	0.00505	13.23	25.00
30.00	22.720	89.98	1.5821	17.044	85.441	0.03745	0.17713	0.2521	0.1812	1.1398	2017.	420.	0.848	—	0.0523	0.00515	12.88	30.00
35.00	25.221	89.44	1.4328	18.310	86.116	0.04001	0.17708	0.2536	0.1833	1.1419	1987.	420.	0.821	—	0.0517	0.00524	12.53	35.00
40.00	27.930	88.90	1.3003	19.584	86.790	0.04256	0.17706	0.2551	0.1854	1.1441	1957.	421.	0.795	—	0.0512	0.00534	12.18	40.00
45.00	30.859	88.35	1.1823	20.866	87.462	0.04510	0.17706	0.2567	0.1875	1.1465	1926.	421.	0.770	—	0.0506	0.00544	11.84	45.00
50.00	34.020	87.80	1.0771	22.155	88.132	0.04763	0.17708	0.2582	0.1897	1.1491	1895.	421.	0.746	—	0.0500	0.00553	11.49	50.00
55.00	37.423	87.24	0.9831	23.453	88.800	0.05015	0.17712	0.2599	0.1919	1.1518	1863.	421.	0.723	—	0.0494	0.00563	11.15	55.00
60.00	41.081	86.67	0.8988	24.759	89.465	0.05266	0.17717	0.2616	0.1942	1.1547	1831.	420.	0.701	—	0.0489	0.00573	10.81	60.00
65.00	45.006	86.10	0.8231	26.074	90.127	0.05516	0.17724	0.2633	0.1965	1.1579	1799.	420.	0.680	—	0.0483	0.00583	10.47	65.00
70.00	49.210	85.52	0.7549	27.397	90.787	0.05765	0.17733	0.2650	0.1989	1.1612	1766.	420.	0.659	—	0.0477	0.00592	10.13	70.00
75.00	53.706	84.93	0.6935	28.729	91.443	0.06014	0.17743	0.2669	0.2013	1.1648	1733.	419.	0.639	—	0.0472	0.00602	9.79	75.00
80.00	58.506	84.33	0.6379	30.070	92.095	0.06262	0.17755	0.2687	0.2037	1.1686	1700.	418.	0.619	0.0283	0.0466	0.00611	9.46	80.00
85.00	63.624	83.72	0.5876	31.421	92.744	0.06509	0.17767	0.2706	0.2063	1.1728	1666.	417.	0.600	0.0285	0.0460	0.00621	9.13	85.00
90.00	69.073	83.11	0.5420	32.781	93.388	0.06755	0.17781	0.2726	0.2089	1.1772	1632.	416.	0.582	0.0288	0.0454	0.00631	8.80	90.00
95.00	74.865	82.49	0.5005	34.151	94.027	0.07001	0.17796	0.2746	0.2115	1.1820	1597.	415.	0.564	0.0290	0.0449	0.00640	8.47	95.00
100.00	81.015	81.85	0.4627	35.531	94.661	0.07246	0.17811	0.2768	0.2143	1.1872	1562.	414.	0.546	0.0292	0.0443	0.00649	8.14	100.00
105.00	87.535	81.21	0.4282	36.922	95.290	0.07491	0.17827	0.2789	0.2171	1.1928	1527.	412.	0.529	0.0295	0.0437	0.00659	7.82	105.00
110.00	94.441	80.55	0.3967	38.323	95.912	0.07735	0.17844	0.2812	0.2200	1.1988	1491.	411.	0.512	0.0297	0.0432	0.00668	7.49	110.00
115.00	101.75	79.88	0.3679	39.735	96.527	0.07979	0.17861	0.2836	0.2231	1.2054	1456.	409.	0.496	0.0300	0.0426	0.00677	7.17	115.00
120.00	109.46	79.20	0.3415	41.158	97.135	0.08222	0.17879	0.2860	0.2262	1.2125	1419.	407.	0.479	0.0302	0.0420	0.00687	6.85	120.00
125.00	117.61	78.51	0.3172	42.594	97.734	0.08466	0.17897	0.2886	0.2296	1.2203	1383.	405.	0.463	0.0305	0.0415	0.00696	6.54	125.00
130.00	126.20	77.80	0.2949	44.041	98.325	0.08709	0.17914	0.2913	0.2330	1.2288	1346.	403.	0.448	0.0307	0.0409	0.00705	6.22	130.00
135.00	135.24	77.08	0.2743	45.501	98.906	0.08951	0.17932	0.2942	0.2367	1.2382	1309.	400.	0.432	0.0310	0.0403	0.00713	5.91	135.00
140.00	144.76	76.34	0.2553	46.974	99.476	0.09194	0.17949	0.2972	0.2406	1.2484	1271.	397.	0.417	0.0313	0.0398	0.00722	5.61	140.00
145.00	154.77	75.59	0.2378	48.461	100.035	0.09437	0.17966	0.3004	0.2447	1.2598	1233.	394.	0.403	0.0316	0.0392	0.00731	5.30	145.00
150.00	165.28	74.81	0.2215	49.963	100.580	0.09680	0.17983	0.3038	0.2491	1.2724	1195.	391.	0.388	0.0319	0.0386	0.00739	5.00	150.00
155.00	176.31	74.01	0.2065	51.480	101.112	0.09924	0.17998	0.3075	0.2538	1.2864	1156.	388.	0.374	0.0322	0.0381	0.00747	4.70	155.00
160.00	187.88	73.20	0.1925	53.013	101.627	0.10167	0.18012	0.3114	0.2590	1.3022	1117.	384.	0.360	0.0325	0.0375	0.00755	4.40	160.00
165.00	200.00	72.35	0.1795	54.563	102.125	0.10411	0.18025	0.3157	0.2646	1.3199	1077.	381.	0.346	0.0329	0.0369	0.00763	4.11	165.00
170.00	212.70	71.48	0.1674	56.132	102.604	0.10656	0.18037	0.3204	0.2707	1.3400	1038.	376.	0.333	0.0332	0.0364	0.00770	3.82	170.00
175.00	225.99	70.59	0.1561	57.720	103.061	0.10902	0.18046	0.3256	0.2775	1.3629	997.	372.	0.319	0.0337	0.0358	0.00777	3.54	175.00
180.00	239.89	69.66	0.1455	59.329	103.494	0.11149	0.18053	0.3314	0.2852	1.3893	957.	368.	0.306	0.0341	0.0352	0.00783	3.26	180.00
185.00	254.41	68.69	0.1357	60.962	103.900	0.11397	0.18058	0.3379	0.2938	1.4200	916.	363.	0.294	0.0346	0.0347	0.00789	2.98	185.00
190.00	269.60	67.68	0.1264	62.620	104.274	0.11647	0.18059	0.3452	0.3038	1.4560	874.	357.	0.282	0.0352	0.0341	0.00795	2.71	190.00
195.00	285.45	66.63	0.1176	64.306	104.613	0.11899	0.18056	0.3537	0.3153	1.4987	832.	352.	0.270	0.0358	0.0336	0.00800	2.44	195.00
200.00	302.01	65.52	0.1094	66.024	104.910	0.12153	0.18048	0.3636	0.3289	1.5502	789.	346.	0.258	0.0364	0.0330	0.00803	2.18	200.00
205.00	319.28	64.35	0.1016	67.778	105.160	0.12410	0.18035	0.3754	0.3453	1.6133	746.	339.	0.247	0.0372	0.0325	0.00806	1.92	205.00
210.00	337.32	63.11	0.0942	69.573	105.353	0.12672	0.18015	0.3897	0.3654	1.6921	701.	333.	0.236	0.0381	0.0319	0.00808	1.67	210.00
215.00	356.13	61.78	0.0871	71.416	105.477	0.12938	0.17986	0.4076	0.3909	1.7932	656.	325.	0.226	—	0.0314	0.00808	1.43	215.00
220.00	375.76	60.34	0.0804	73.319	105.518	0.13210	0.17947	0.4307	0.4243	1.9270	610.	318.	0.216	—	0.0309	0.00806	1.19	220.00
225.00	396.25	58.77	0.0738	75.295	105.451	0.13490	0.17895	0.4619	0.4698	2.1119	563.	310.	0.207	—	0.0303	0.00801	0.97	225.00
230.00	417.64	57.01	0.0675	77.367	105.245	0.13782	0.17824	0.5069	0.5360	2.3825	514.	301.	0.199	—	0.0298	0.00793	0.75	230.00
235.00	440.00	55.01	0.0612	79.572	104.843	0.14090	0.17727	0.5779	0.6410	2.8142	463.	292.	—	—	—	—	0.55	235.00
240.00	463.38	52.61	0.0547	81.981	104.143	0.14424	0.17591	0.7078	0.8330	3.6065	410.	281.	—	—	—	—	0.36	240.00
245.00	487.91	49.49	0.0479	84.761	102.909	0.14807	0.17382	1.0242	1.2942	5.5136	353.	271.	—	—	—	—	0.19	245.00
250.00	513.78	44.38	0.0392	88.581	100.240	0.15333	0.16975	—	—	—	—	—	—	—	—	—	0.04	250.00
252.45c	527.07	34.57	0.0289	94.596	94.596	0.16087	0.16087	∞	∞	∞	0.	0.	—	—	∞	∞	0.00	252.45

*temperatures are on the ITS-90 scale   b = boiling point   c = critical point

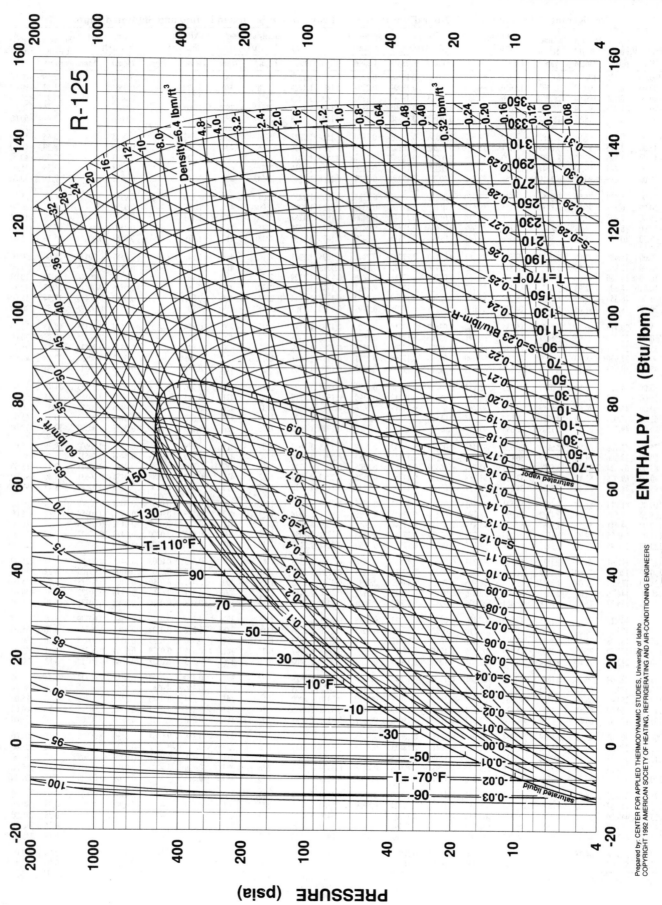

ENTHALPY (Btu/lbm)

Fig. 13 Pressure-Enthalpy Diagram for Refrigerant 125

Prepared by: CENTER FOR APPLIED THERMODYNAMIC STUDIES, University of Idaho
COPYRIGHT 1992 AMERICAN SOCIETY OF HEATING, REFRIGERATING AND AIR-CONDITIONING ENGINEERS

PRESSURE (psia)

R-125

Refrigerant 125 (pentafluoroethane)   Properties of Saturated Liquid and Saturated Vapor

Temp,* °F	Pressure, psia	Density, lb/ft³ Liquid	Volume, ft³/lb Vapor	Enthalpy, Btu/lb Liquid	Vapor	Entropy, Btu/lb·°F Liquid	Vapor	Specific Heat c_p, Btu/lb·°F Liquid	Vapor	c_p/c_v Vapor	Velocity of Sound, ft/s Liquid	Vapor	Viscosity, lb$_m$/ft·h Liquid	Vapor	Thermal Cond, Btu/h·ft·°F Liquid	Vapor	Surface Tension, dyne/cm	Temp, °F
−100.00	3.691	99.92	8.5830	−13.439	60.483	−0.03294	0.17155	—	0.1511	1.1333	—	405.	1.557	—	—	—	—	−100.00
−95.00	4.376	99.34	7.3248	−12.593	61.186	−0.03088	0.17071	—	0.1526	1.1333	—	407.	1.474	—	—	—	—	−95.00
−90.00	5.164	98.76	6.2782	−11.675	61.890	−0.02861	0.16994	—	0.1542	1.1333	—	409.	1.398	—	—	—	—	−90.00
−85.00	6.067	98.17	5.4034	−10.694	62.595	−0.02617	0.16922	—	0.1558	1.1335	—	410.	1.327	—	—	—	—	−85.00
−80.00	7.096	97.58	4.6691	−9.658	63.301	−0.02359	0.16856	—	0.1574	1.1339	—	412.	1.262	—	—	—	—	−80.00
−75.00	8.264	96.99	4.0501	−8.573	64.006	−0.02089	0.16795	—	0.1590	1.1344	—	414.	1.201	—	—	—	—	−75.00
−70.00	9.586	96.39	3.5261	−7.444	64.710	−0.01809	0.16738	—	0.1607	1.1350	—	415.	1.144	—	—	—	—	−70.00
−65.00	11.075	95.78	3.0809	−6.278	65.414	−0.01521	0.16687	—	0.1624	1.1359	—	416.	1.092	—	—	—	—	−65.00
−60.00	12.746	95.17	2.7010	−5.077	66.116	−0.01226	0.16639	—	0.1642	1.1368	—	418.	1.042	—	—	—	—	−60.00
−55.00	14.615	94.55	2.3758	−3.846	66.817	−0.00925	0.16595	0.2473	0.1660	1.1380	2413.	419.	0.997	—	—	—	—	−55.00
−54.79b	14.696	94.52	2.3634	−3.794	66.846	−0.00912	0.16594	0.2476	0.1661	1.1381	2409.	419.	0.995	—	—	—	—	−54.79
−50.00	16.698	93.92	2.0963	−2.587	67.515	−0.00620	0.16556	0.2530	0.1678	1.1394	2338.	420.	0.953	—	—	—	—	−50.00
−45.00	19.012	93.29	1.8553	−1.305	68.211	−0.00311	0.16520	0.2581	0.1697	1.1409	2269.	421.	0.913	—	—	—	—	−45.00
−40.00	21.574	92.65	1.6468	0.000	68.904	0.00000	0.16487	0.2626	0.1716	1.1427	2204.	421.	0.875	—	—	—	—	−40.00
−35.00	24.402	92.01	1.4658	1.325	69.593	0.00313	0.16458	0.2667	0.1736	1.1447	2141.	422.	0.839	—	—	—	—	−35.00
−30.00	27.514	91.35	1.3083	2.668	70.278	0.00628	0.16432	0.2705	0.1757	1.1469	2082.	423.	0.805	—	—	—	—	−30.00
−25.00	30.928	90.69	1.1706	4.027	70.959	0.00943	0.16409	0.2739	0.1778	1.1494	2024.	423.	0.773	—	—	—	—	−25.00
−20.00	34.664	90.02	1.0500	5.403	71.636	0.01259	0.16388	0.2770	0.1799	1.1522	1969.	423.	0.742	—	—	—	—	−20.00
−15.00	38.742	89.35	0.9441	6.793	72.307	0.01575	0.16370	0.2799	0.1822	1.1552	1914.	423.	0.713	—	—	—	—	−15.00
−10.00	43.179	88.66	0.8508	8.197	72.972	0.01890	0.16354	0.2826	0.1845	1.1586	1861.	423.	0.686	—	—	—	—	−10.00
−5.00	47.997	87.96	0.7684	9.614	73.632	0.02205	0.16341	0.2852	0.1868	1.1623	1810.	423.	0.659	—	—	—	—	−5.00
0.00	53.215	87.25	0.6954	11.044	74.284	0.02520	0.16329	0.2877	0.1893	1.1663	1759.	422.	0.634	—	—	—	—	0.00
2.00	55.419	86.97	0.6685	11.620	74.543	0.02645	0.16325	0.2886	0.1903	1.1680	1739.	422.	0.624	—	—	—	—	2.00
4.00	57.691	86.68	0.6429	12.197	74.801	0.02771	0.16321	0.2896	0.1913	1.1698	1719.	422.	0.615	—	—	—	—	4.00
6.00	60.034	86.39	0.6184	12.777	75.058	0.02896	0.16318	0.2906	0.1924	1.1717	1699.	421.	0.605	—	—	—	—	6.00
8.00	62.448	86.10	0.5951	13.358	75.314	0.03021	0.16315	0.2915	0.1934	1.1736	1679.	421.	0.596	—	—	—	—	8.00
10.00	64.934	85.80	0.5728	13.942	75.568	0.03147	0.16311	0.2925	0.1945	1.1756	1659.	421.	0.587	—	—	—	—	10.00
12.00	67.495	85.51	0.5515	14.527	75.821	0.03271	0.16309	0.2934	0.1956	1.1776	1640.	420.	0.578	—	—	—	—	12.00
14.00	70.131	85.21	0.5311	15.114	76.073	0.03396	0.16306	0.2944	0.1967	1.1798	1621.	420.	0.569	—	—	—	—	14.00
16.00	72.843	84.91	0.5116	15.704	76.323	0.03521	0.16304	0.2953	0.1978	1.1820	1601.	420.	0.560	—	—	—	—	16.00
18.00	75.633	84.61	0.4930	16.295	76.572	0.03645	0.16301	0.2963	0.1989	1.1843	1582.	419.	0.552	—	—	—	—	18.00
20.00	78.503	84.30	0.4752	16.888	76.819	0.03769	0.16299	0.2972	0.2001	1.1867	1563.	419.	0.543	—	—	—	—	20.00
22.00	81.454	84.00	0.4581	17.483	77.065	0.03893	0.16298	0.2982	0.2013	1.1892	1544.	418.	0.535	—	—	—	—	22.00
24.00	84.487	83.69	0.4418	18.081	77.309	0.04017	0.16296	0.2992	0.2025	1.1918	1525.	418.	0.526	—	—	—	—	24.00
26.00	87.603	83.38	0.4261	18.680	77.552	0.04141	0.16294	0.3002	0.2037	1.1944	1506.	417.	0.518	—	—	—	—	26.00
28.00	90.805	83.06	0.4111	19.281	77.793	0.04265	0.16293	0.3012	0.2049	1.1972	1487.	416.	0.510	—	—	—	—	28.00
30.00	94.092	82.75	0.3967	19.885	78.033	0.04388	0.16292	0.3023	0.2062	1.2001	1468.	416.	0.502	—	—	—	—	30.00
32.00	97.468	82.43	0.3829	20.491	78.270	0.04512	0.16291	0.3033	0.2075	1.2030	1450.	415.	0.494	—	—	—	—	32.00
34.00	100.93	82.11	0.3696	21.098	78.506	0.04635	0.16289	0.3044	0.2088	1.2061	1431.	414.	0.487	—	—	—	—	34.00
36.00	104.49	81.78	0.3569	21.708	78.741	0.04758	0.16288	0.3055	0.2102	1.2094	1412.	414.	0.479	—	—	—	—	36.00
38.00	108.14	81.45	0.3447	22.320	78.973	0.04881	0.16288	0.3066	0.2116	1.2127	1394.	413.	0.472	—	—	—	—	38.00
40.00	111.88	81.12	0.3330	22.935	79.203	0.05003	0.16287	0.3078	0.2130	1.2162	1375.	412.	0.464	—	—	—	—	40.00
42.00	115.72	80.79	0.3217	23.551	79.431	0.05126	0.16286	0.3090	0.2144	1.2198	1357.	411.	0.457	—	—	—	—	42.00
44.00	119.65	80.45	0.3109	24.170	79.658	0.05248	0.16285	0.3102	0.2159	1.2236	1338.	410.	0.449	—	—	—	—	44.00
46.00	123.68	80.11	0.3005	24.792	79.882	0.05371	0.16284	0.3115	0.2174	1.2275	1320.	409.	0.442	—	—	—	—	46.00
48.00	127.81	79.77	0.2905	25.415	80.104	0.05493	0.16284	0.3128	0.2189	1.2316	1302.	408.	0.435	—	—	—	—	48.00
50.00	132.05	79.42	0.2808	26.042	80.323	0.05615	0.16283	0.3141	0.2205	1.2358	1283.	407.	0.428	—	—	—	—	50.00
55.00	143.09	78.54	0.2583	27.618	80.862	0.05920	0.16281	0.3177	0.2247	1.2473	1238.	405.	0.411	—	—	—	—	55.00
60.00	154.80	77.63	0.2377	29.212	81.385	0.06225	0.16278	0.3217	0.2291	1.2600	1192.	402.	0.394	—	—	—	—	60.00
65.00	167.20	76.69	0.2190	30.824	81.889	0.06530	0.16274	0.3260	0.2339	1.2743	1147.	399.	0.378	—	—	—	—	65.00
70.00	180.34	75.73	0.2018	32.456	82.375	0.06835	0.16269	0.3309	0.2391	1.2904	1102.	395.	0.362	—	—	—	—	70.00
75.00	194.23	74.73	0.1861	34.108	82.839	0.07140	0.16263	0.3364	0.2447	1.3087	1057.	392.	0.346	—	—	—	3.99	75.00
80.00	208.91	73.69	0.1716	35.784	83.279	0.07445	0.16255	0.3426	0.2510	1.3295	1012.	387.	0.331	—	—	—	3.65	80.00
85.00	224.41	72.61	0.1582	37.485	83.692	0.07752	0.16244	0.3497	0.2579	1.3535	967.	383.	0.316	—	—	—	3.31	85.00
90.00	240.78	71.49	0.1459	39.215	84.075	0.08061	0.16229	0.3579	0.2657	1.3814	922.	378.	0.301	—	—	—	2.99	90.00
95.00	258.04	70.31	0.1345	40.976	84.423	0.08371	0.16211	0.3674	0.2746	1.4142	878.	373.	0.287	—	—	—	2.67	95.00
100.00	276.24	69.07	0.1239	42.772	84.730	0.08684	0.16189	0.3787	0.2850	1.4534	833.	368.	0.273	—	—	—	2.36	100.00
105.00	295.41	67.75	0.1140	44.610	84.991	0.09001	0.16160	0.3922	0.2971	1.5009	788.	362.	0.260	—	—	—	2.05	105.00
110.00	315.62	66.35	0.1048	46.495	85.198	0.09323	0.16124	0.4085	0.3118	1.5598	743.	355.	0.247	—	—	—	1.76	110.00
115.00	336.89	64.84	0.0961	48.438	85.338	0.09652	0.16080	0.4288	0.3300	1.6345	698.	348.	0.234	—	—	—	1.48	115.00
120.00	359.28	63.21	0.0879	50.450	85.397	0.09988	0.16024	0.4547	0.3534	1.7326	653.	340.	0.222	—	—	—	1.21	120.00
125.00	382.82	61.41	0.0801	52.549	85.356	0.10337	0.15954	0.4887	0.3846	1.8667	607.	332.	0.210	—	—	—	0.95	125.00
130.00	407.55	59.39	0.0726	54.765	85.184	0.10701	0.15866	0.5353	0.4292	2.0609	561.	323.	0.198	—	—	—	0.71	130.00
135.00	433.46	57.06	0.0654	57.146	84.837	0.11090	0.15751	0.6037	0.4984	2.3666	515.	312.	—	—	—	—	0.48	135.00
140.00	460.51	54.23	0.0583	59.789	84.238	0.11519	0.15599	0.7158	0.6211	2.9150	466.	302.	—	—	—	—	0.27	140.00
145.00	488.50	50.45	0.0511	62.944	83.233	0.12030	0.15385	—	—	—	—	—	—	—	—	—	0.10	145.00
150.00	516.72	43.11	0.0344	68.072	76.834	0.12864	0.14293	—	—	—	—	—	—	—	—	—	—	150.00
150.87c	521.41	35.66	0.0280	72.653	72.653	0.13611	0.13611	∞	∞	∞	0.	0.	—	—	∞	∞	0.00	150.87

*temperatures are on the IPTS-68 scale   b = boiling point   c = critical point

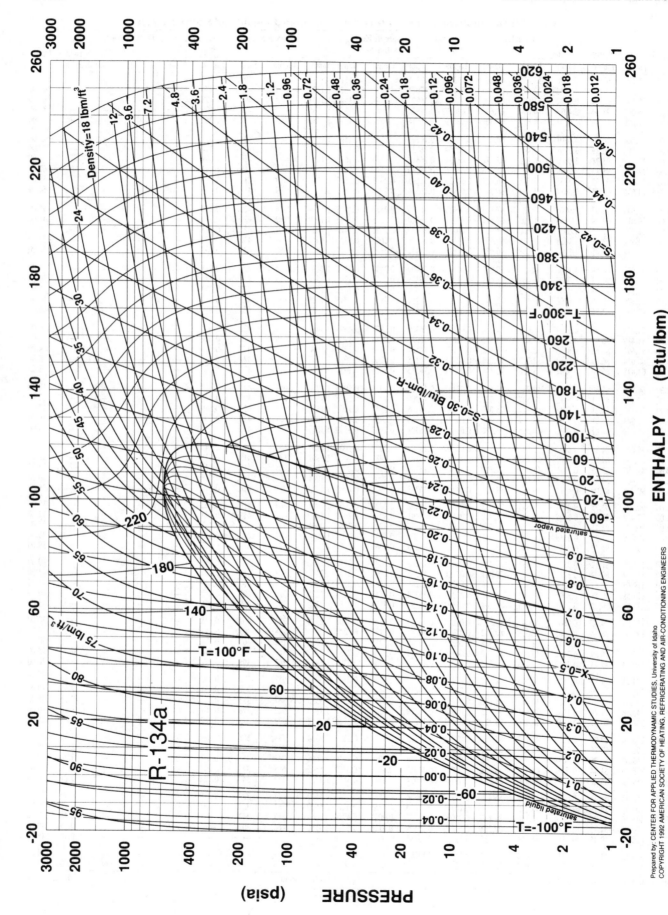

ENTHALPY (Btu/lbm)

PRESSURE (psia)

R-134a

Fig. 14 Pressure-Enthalpy Diagram for Refrigerant 134a

Prepared by: CENTER FOR APPLIED THERMODYNAMIC STUDIES, University of Idaho
COPYRIGHT 1992 AMERICAN SOCIETY OF HEATING, REFRIGERATING AND AIR-CONDITIONING ENGINEERS

Refrigerant 134a (1,1,1,2-tetrafluoroethane) Properties of Saturated Liquid and Saturated Vapor

Temp,* °F	Pressure, psia	Density, lb/ft³ Liquid	Volume, ft³/lb Vapor	Enthalpy, Btu/lb Liquid	Enthalpy, Btu/lb Vapor	Entropy, Btu/lb·°F Liquid	Entropy, Btu/lb·°F Vapor	Specific Heat c_p, Btu/lb·°F Liquid	Specific Heat c_p, Btu/lb·°F Vapor	c_p/c_v Vapor	Velocity of Sound, ft/s Liquid	Velocity of Sound, ft/s Vapor	Viscosity, lb_m/ft·h Liquid	Viscosity, lb_m/ft·h Vapor	Thermal Cond, Btu/h·ft·°F Liquid	Thermal Cond, Btu/h·ft·°F Vapor	Surface Tension, dyne/cm	Temp, °F
−153.94a	0.057	99.34	564.85	−32.989	80.235	−0.09154	0.27880	0.2740	0.1397	1.1628	3723.	416.	5.289	0.0160	—	—	28.15	−153.94
−150.00	0.072	98.95	449.29	−31.902	80.783	−0.08801	0.27588	0.2776	0.1409	1.1615	3670.	418.	4.913	0.0163	—	—	27.76	−150.00
−140.00	0.130	97.98	259.15	−29.093	82.190	−0.07908	0.26903	0.2837	0.1438	1.1583	3552.	424.	4.117	0.0168	—	—	26.78	−140.00
−130.00	0.222	97.01	155.69	−26.238	83.618	−0.07029	0.26294	0.2870	0.1467	1.1554	3448.	430.	3.497	0.0173	—	—	25.81	−130.00
−120.00	0.367	96.05	97.027	−23.359	85.066	−0.06169	0.25752	0.2886	0.1497	1.1528	3352.	436.	3.006	0.0179	—	—	24.86	−120.00
−110.00	0.586	95.09	62.509	−20.467	86.531	−0.05330	0.25270	0.2895	0.1527	1.1507	3261.	441.	2.613	0.0184	—	—	23.92	−110.00
−100.00	0.906	94.13	41.496	−17.569	88.011	−0.04513	0.24842	0.2900	0.1559	1.1489	3173.	446.	2.292	0.0190	—	—	22.99	−100.00
−90.00	1.363	93.17	28.303	−14.665	89.504	−0.03717	0.24462	0.2906	0.1591	1.1475	3087.	451.	2.028	0.0195	0.0721	—	22.08	−90.00
−80.00	1.997	92.21	19.783	−11.755	91.005	−0.02940	0.24125	0.2913	0.1624	1.1466	3001.	456.	1.808	0.0200	0.0706	—	21.17	−80.00
−75.00	2.396	91.73	16.680	−10.297	91.759	−0.02559	0.23972	0.2917	0.1641	1.1463	2959.	458.	1.712	0.0203	0.0699	—	20.73	−75.00
−70.00	2.859	91.25	14.138	−8.837	92.514	−0.02182	0.23827	0.2922	0.1658	1.1462	2916.	460.	1.623	0.0206	0.0691	—	20.28	−70.00
−65.00	3.393	90.77	12.045	−7.374	93.270	−0.01809	0.23691	0.2928	0.1676	1.1461	2874.	462.	1.542	0.0209	0.0684	—	19.84	−65.00
−60.00	4.006	90.28	10.310	−5.907	94.026	−0.01440	0.23563	0.2935	0.1694	1.1462	2832.	464.	1.466	0.0212	0.0677	0.00405	19.40	−60.00
−55.00	4.707	89.80	8.8656	−4.437	94.783	−0.01075	0.23443	0.2943	0.1712	1.1465	2790.	466.	1.396	0.0214	0.0669	0.00423	18.96	−55.00
−50.00	5.505	89.31	7.6569	−2.963	95.539	−0.00713	0.23331	0.2951	0.1731	1.1468	2747.	468.	1.331	0.0217	0.0662	0.00440	18.52	−50.00
−45.00	6.409	88.82	6.6405	−1.484	96.295	−0.00355	0.23225	0.2960	0.1750	1.1473	2705.	470.	1.271	0.0220	0.0654	0.00457	18.09	−45.00
−40.00	7.429	88.32	5.7819	0.000	97.050	0.00000	0.23125	0.2970	0.1769	1.1479	2663.	471.	1.215	0.0223	0.0647	0.00473	17.66	−40.00
−35.00	8.577	87.83	5.0533	1.489	97.804	0.00352	0.23032	0.2981	0.1789	1.1487	2621.	473.	1.163	0.0225	0.0639	0.00489	17.23	−35.00
−30.00	9.862	87.33	4.4325	2.984	98.556	0.00701	0.22945	0.2992	0.1810	1.1497	2579.	474.	1.113	0.0228	0.0632	0.00505	16.81	−30.00
−25.00	11.297	86.82	3.9014	4.484	99.306	0.01048	0.22863	0.3004	0.1831	1.1508	2536.	476.	1.068	0.0231	0.0625	0.00521	16.38	−25.00
−20.00	12.895	86.32	3.4452	5.991	100.054	0.01392	0.22786	0.3017	0.1852	1.1521	2494.	477.	1.024	0.0234	0.0617	0.00536	15.96	−20.00
−15.00	14.667	85.81	3.0519	7.505	100.799	0.01733	0.22714	0.3031	0.1874	1.1535	2452.	478.	0.984	0.0237	0.0610	0.00550	15.54	−15.00
−14.92b	14.696	85.80	3.0462	7.529	100.811	0.01739	0.22713	0.3031	0.1874	1.1535	2451.	478.	0.983	0.0237	0.0610	0.00551	15.54	−14.92
−10.00	16.626	85.29	2.7116	9.026	101.542	0.02073	0.22647	0.3045	0.1897	1.1552	2410.	479.	0.946	0.0240	0.0602	0.00565	15.13	−10.00
−5.00	18.787	84.77	2.4161	10.554	102.280	0.02409	0.22584	0.3060	0.1920	1.1570	2367.	480.	0.910	0.0243	0.0595	0.00579	14.71	−5.00
0.00	21.162	84.25	2.1587	12.090	103.015	0.02744	0.22525	0.3075	0.1943	1.1590	2325.	481.	0.876	0.0245	0.0588	0.00593	14.30	0.00
5.00	23.767	83.72	1.9337	13.634	103.745	0.03077	0.22470	0.3091	0.1968	1.1613	2283.	481.	0.843	0.0248	0.0580	0.00607	13.89	5.00
10.00	26.617	83.18	1.7365	15.187	104.471	0.03408	0.22418	0.3108	0.1993	1.1637	2240.	482.	0.813	0.0251	0.0573	0.00621	13.48	10.00
15.00	29.726	82.64	1.5630	16.748	105.192	0.03737	0.22370	0.3126	0.2018	1.1664	2198.	482.	0.784	0.0254	0.0565	0.00635	13.08	15.00
20.00	33.110	82.10	1.4101	18.318	105.907	0.04065	0.22325	0.3144	0.2045	1.1694	2155.	482.	0.756	0.0257	0.0558	0.00649	12.67	20.00
25.00	36.785	81.55	1.2749	19.897	106.617	0.04391	0.22283	0.3162	0.2072	1.1726	2113.	482.	0.730	0.0260	0.0550	0.00662	12.27	25.00
30.00	40.768	80.99	1.1550	21.486	107.320	0.04715	0.22244	0.3182	0.2100	1.1761	2070.	482.	0.705	0.0263	0.0543	0.00676	11.87	30.00
35.00	45.075	80.42	1.0484	23.085	108.016	0.05038	0.22207	0.3202	0.2129	1.1799	2027.	482.	0.681	0.0267	0.0536	0.00690	11.48	35.00
40.00	49.724	79.85	0.9534	24.694	108.705	0.05359	0.22172	0.3223	0.2159	1.1841	1985.	482.	0.658	0.0270	0.0528	0.00704	11.08	40.00
45.00	54.732	79.26	0.8685	26.314	109.386	0.05679	0.22140	0.3244	0.2190	1.1886	1942.	481.	0.636	0.0273	0.0521	0.00718	10.69	45.00
50.00	60.116	78.67	0.7925	27.944	110.058	0.05998	0.22110	0.3267	0.2222	1.1935	1899.	481.	0.615	0.0276	0.0513	0.00732	10.30	50.00
55.00	65.895	78.07	0.7243	29.586	110.722	0.06316	0.22081	0.3290	0.2255	1.1988	1856.	480.	0.595	0.0280	0.0506	0.00746	9.91	55.00
60.00	72.087	77.46	0.6630	31.239	111.376	0.06633	0.22054	0.3314	0.2289	1.2046	1813.	479.	0.576	0.0283	0.0499	0.00761	9.53	60.00
65.00	78.712	76.84	0.6077	32.905	112.019	0.06949	0.22028	0.3339	0.2325	1.2109	1770.	477.	0.557	0.0286	0.0491	0.00776	9.15	65.00
70.00	85.787	76.21	0.5577	34.583	112.652	0.07264	0.22003	0.3366	0.2363	1.2178	1726.	476.	0.539	0.0290	0.0484	0.00791	8.77	70.00
75.00	93.333	75.57	0.5125	36.274	113.272	0.07578	0.21979	0.3393	0.2402	1.2252	1683.	474.	0.522	0.0294	0.0476	0.00806	8.39	75.00
80.00	101.37	74.91	0.4715	37.978	113.880	0.07892	0.21957	0.3422	0.2444	1.2334	1640.	472.	0.505	0.0297	0.0469	0.00822	8.02	80.00
85.00	109.92	74.25	0.4343	39.697	114.475	0.08205	0.21934	0.3453	0.2487	1.2424	1596.	470.	0.489	0.0301	0.0462	0.00838	7.65	85.00
90.00	119.00	73.57	0.4004	41.430	115.055	0.08518	0.21912	0.3485	0.2533	1.2522	1552.	468.	0.473	0.0305	0.0454	0.00855	7.28	90.00
95.00	128.63	72.87	0.3694	43.179	115.619	0.08830	0.21890	0.3519	0.2582	1.2630	1509.	466.	0.458	0.0309	0.0447	0.00872	6.91	95.00
100.00	138.83	72.16	0.3411	44.943	116.166	0.09142	0.21868	0.3555	0.2633	1.2748	1465.	463.	0.443	0.0313	0.0439	0.00890	6.55	100.00
105.00	149.63	71.43	0.3153	46.725	116.694	0.09454	0.21845	0.3594	0.2689	1.2880	1421.	460.	0.428	0.0318	0.0432	0.00908	6.20	105.00
110.00	161.05	70.68	0.2915	48.524	117.203	0.09766	0.21822	0.3635	0.2748	1.3026	1376.	457.	0.414	0.0322	0.0425	0.00926	5.84	110.00
115.00	173.11	69.91	0.2697	50.343	117.690	0.10078	0.21797	0.3680	0.2812	1.3189	1332.	454.	0.400	0.0327	0.0417	0.00946	5.49	115.00
120.00	185.84	69.12	0.2497	52.181	118.153	0.10391	0.21772	0.3728	0.2881	1.3372	1288.	450.	0.387	0.0332	0.0410	0.00965	5.15	120.00
125.00	199.25	68.31	0.2312	54.040	118.591	0.10704	0.21744	0.3781	0.2957	1.3577	1243.	446.	0.374	0.0338	0.0403	0.00986	4.80	125.00
130.00	213.38	67.47	0.2141	55.923	119.000	0.11018	0.21715	0.3839	0.3040	1.3810	1198.	442.	0.361	0.0343	0.0395	0.01007	4.47	130.00
135.00	228.25	66.60	0.1983	57.830	119.377	0.11333	0.21683	0.3903	0.3133	1.4075	1153.	437.	0.348	0.0349	0.0388	0.01029	4.13	135.00
140.00	243.88	65.70	0.1836	59.764	119.720	0.11650	0.21648	0.3974	0.3236	1.4379	1108.	432.	0.335	0.0356	0.0380	0.01052	3.81	140.00
145.00	260.31	64.77	0.1700	61.727	120.024	0.11968	0.21609	0.4053	0.3353	1.4731	1062.	427.	0.323	0.0363	0.0373	0.01075	3.48	145.00
150.00	277.57	63.80	0.1574	63.722	120.284	0.12288	0.21566	0.4144	0.3486	1.5143	1017.	421.	0.311	0.0370	0.0366	0.01100	3.17	150.00
155.00	295.69	62.78	0.1455	65.752	120.495	0.12611	0.21517	0.4247	0.3640	1.5630	971.	416.	0.298	0.0378	0.0358	0.01125	2.86	155.00
160.00	314.69	61.72	0.1345	67.823	120.650	0.12938	0.21463	0.4368	0.3821	1.6213	924.	409.	0.286	0.0387	0.0351	0.01151	2.55	160.00
165.00	334.62	60.60	0.1241	69.939	120.739	0.13268	0.21400	0.4511	0.4036	1.6921	877.	403.	0.274	0.0397	0.0343	0.01178	2.26	165.00
170.00	355.51	59.42	0.1144	72.106	120.753	0.13603	0.21329	0.4683	0.4299	1.7798	829.	396.	0.262	0.0407	0.0336	0.01206	1.97	170.00
175.00	377.40	58.16	0.1052	74.335	120.677	0.13945	0.21247	0.4896	0.4627	1.8908	781.	388.	0.249	0.0420	0.0329	0.01235	1.69	175.00
180.00	400.34	56.80	0.0965	76.636	120.493	0.14295	0.21151	0.5168	0.5048	2.0354	731.	380.	0.237	0.0434	0.0321	0.01265	1.41	180.00
185.00	424.37	55.33	0.0881	79.027	120.175	0.14655	0.21037	0.5527	0.5612	2.2309	680.	372.	0.224	0.0450	0.0314	0.01296	1.15	185.00
190.00	449.55	53.70	0.0801	81.534	119.684	0.15029	0.20901	0.6031	0.6406	2.5087	627.	363.	0.211	0.0469	—	—	0.90	190.00
195.00	475.95	51.86	0.0723	84.196	118.963	0.15423	0.20733	0.6794	0.7612	2.9338	572.	353.	0.197	0.0493	—	—	0.67	195.00
200.00	503.64	49.70	0.0646	87.088	117.906	0.15847	0.20519	0.8100	0.9673	3.6640	514.	343.	0.182	0.0523	—	—	0.45	200.00
205.00	532.72	47.00	0.0566	90.368	116.289	0.16326	0.20226	1.0906	1.4042	5.2174	450.	331.	0.164	0.0566	—	—	0.25	205.00
210.00	563.34	43.03	0.0474	94.548	113.411	0.16933	0.19750	—	—	—	378.	316.	0.142	0.0639	—	—	0.09	210.00
213.85c	588.27	32.04	0.0312	103.775	103.775	0.18128	0.18128	∞	∞	∞	0.	0.	—	—	∞	∞	0.00	213.85

*temperatures are on the ITS-90 scale   a = triple point   b = boiling point   c = critical point

Refrigerant 134a   Properties of Superheated Vapor

Pressure = 14.696 psia — Saturation temperature = −14.92 °F

Temp., °F	Density, lb/ft³	Enthalpy, Btu/lb	Entropy, Btu/lb·°F	Vel. Sound, ft/s
Saturated Liquid	85.7972	7.53	0.01739	2451.2
Vapor	0.3283	100.81	0.22713	478.0
0.00	0.3158	103.62	0.23335	487.2
20.00	0.3008	107.45	0.24149	499.0
40.00	0.2874	111.34	0.24944	510.2
60.00	0.2753	115.31	0.25723	521.0
80.00	0.2642	119.35	0.26486	531.5
100.00	0.2541	123.47	0.27236	541.6
120.00	0.2448	127.68	0.27974	551.4
140.00	0.2362	131.96	0.28700	561.0
160.00	0.2282	136.32	0.29416	570.4
180.00	0.2208	140.77	0.30122	579.5
200.00	0.2139	145.30	0.30819	588.5
220.00	0.2074	149.90	0.31507	597.3
240.00	0.2013	154.59	0.32187	606.0
260.00	0.1955	159.36	0.32858	614.5
280.00	0.1901	164.20	0.33522	622.8
300.00	0.1850	169.12	0.34178	631.1

Pressure = 25.00 psia — Saturation temperature = 7.22 °F

Temp., °F	Density, lb/ft³	Enthalpy, Btu/lb	Entropy, Btu/lb·°F	Vel. Sound, ft/s
Saturated Liquid	83.4823	14.32	0.03224	2263.9
Vapor	0.5426	104.07	0.22446	481.5
20.00	0.5245	106.60	0.22982	489.9
40.00	0.4991	110.61	0.23800	502.4
60.00	0.4765	114.66	0.24596	514.1
80.00	0.4563	118.78	0.25373	525.4
100.00	0.4379	122.96	0.26135	536.2
120.00	0.4212	127.22	0.26881	546.6
140.00	0.4058	131.55	0.27615	556.7
160.00	0.3916	135.95	0.28337	566.5
180.00	0.3786	140.43	0.29048	576.0
200.00	0.3663	144.98	0.29750	585.3
220.00	0.3549	149.61	0.30441	594.4
240.00	0.3443	154.32	0.31124	603.3
260.00	0.3343	159.10	0.31798	612.0
280.00	0.3248	163.96	0.32464	620.6
300.00	0.3160	168.90	0.33122	629.0

Pressure = 50.00 psia — Saturation temperature = 40.29 °F

Temp., °F	Density, lb/ft³	Enthalpy, Btu/lb	Entropy, Btu/lb·°F	Vel. Sound, ft/s
Saturated Liquid	79.8125	24.79	0.05377	1982.3
Vapor	1.0545	108.74	0.22170	481.7
60.00	0.9982	113.00	0.23005	496.2
80.00	0.9489	117.32	0.23822	509.8
100.00	0.9055	121.68	0.24614	522.5
120.00	0.8670	126.07	0.25385	534.5
140.00	0.8322	130.51	0.26139	545.8
160.00	0.8008	135.01	0.26877	556.7
180.00	0.7718	139.57	0.27601	567.2
200.00	0.7454	144.20	0.28313	577.4
220.00	0.7208	148.89	0.29014	587.2
240.00	0.6980	153.65	0.29704	596.8
260.00	0.6768	158.48	0.30385	606.1
280.00	0.6569	163.38	0.31056	615.2
300.00	0.6383	168.35	0.31719	624.1

Pressure = 75.00 psia — Saturation temperature = 62.24 °F

Temp., °F	Density, lb/ft³	Enthalpy, Btu/lb	Entropy, Btu/lb·°F	Vel. Sound, ft/s
Saturated Liquid	77.1862	31.98	0.06775	1793.6
Vapor	1.5686	111.67	0.22042	478.1
80.00	1.4873	115.74	0.22809	492.7
100.00	1.4092	120.30	0.23639	507.7
120.00	1.3416	124.85	0.24439	521.6
140.00	1.2822	129.43	0.25215	534.5
160.00	1.2294	134.04	0.25971	546.6
180.00	1.1817	138.69	0.26710	558.2
200.00	1.1383	143.39	0.27434	569.2
220.00	1.0984	148.15	0.28145	579.8
240.00	1.0620	152.97	0.28843	590.1
260.00	1.0280	157.85	0.29531	600.0
280.00	0.9966	162.79	0.30208	609.7
300.00	0.9671	167.80	0.30876	619.0
320.00	0.9398	172.87	0.31535	628.2
340.00	0.9138	178.01	0.32186	637.1
360.00	0.8895	183.21	0.32828	645.9
380.00	0.8665	188.48	0.33463	654.5
400.00	0.8448	193.82	0.34091	662.9

Pressure = 100.00 psia — Saturation temperature = 79.17 °F

Temp., °F	Density, lb/ft³	Enthalpy, Btu/lb	Entropy, Btu/lb·°F	Vel. Sound, ft/s
Saturated Liquid	75.0245	37.69	0.07840	1646.8
Vapor	2.0917	113.78	0.21960	472.8
80.00	2.0858	113.98	0.21998	473.6
100.00	1.9576	118.80	0.22874	491.6
120.00	1.8509	123.55	0.23709	507.8
140.00	1.7597	128.29	0.24512	522.4
160.00	1.6800	133.02	0.25288	536.0
180.00	1.6094	137.78	0.26044	548.8
200.00	1.5463	142.57	0.26781	560.8
220.00	1.4891	147.40	0.27502	572.3
240.00	1.4368	152.27	0.28210	583.3
260.00	1.3886	157.21	0.28905	593.9
280.00	1.3444	162.19	0.29588	604.1
300.00	1.3031	167.24	0.30261	614.0
320.00	1.2647	172.35	0.30925	623.6
340.00	1.2287	177.52	0.31579	632.9
360.00	1.1950	182.75	0.32225	642.1
380.00	1.1633	188.04	0.32863	651.0
400.00	1.1334	193.39	0.33494	659.8

Pressure = 125.00 psia — Saturation temperature = 93.15 °F

Temp., °F	Density, lb/ft³	Enthalpy, Btu/lb	Entropy, Btu/lb·°F	Vel. Sound, ft/s
Saturated Liquid	73.1279	42.53	0.08715	1524.7
Vapor	2.6279	115.41	0.21898	466.7
100.00	2.5638	117.16	0.22212	473.9
120.00	2.4025	122.16	0.23090	492.9
140.00	2.2694	127.08	0.23924	509.7
160.00	2.1561	131.96	0.24725	525.0
180.00	2.0577	136.83	0.25498	539.0
200.00	1.9710	141.71	0.26250	552.2
220.00	1.8935	146.62	0.26983	564.6
240.00	1.8233	151.56	0.27700	576.4
260.00	1.7592	156.55	0.28402	587.6
280.00	1.7006	161.59	0.29093	598.4
300.00	1.6463	166.67	0.29771	608.8
320.00	1.5959	171.82	0.30440	618.9
340.00	1.5492	177.02	0.31098	628.7
360.00	1.5055	182.27	0.31747	638.2
380.00	1.4644	187.59	0.32388	647.5
400.00	1.4258	192.97	0.33021	656.6

Pressure = 150.00 psia — Saturation temperature = 105.17 °F

Temp., °F	Density, lb/ft³	Enthalpy, Btu/lb	Entropy, Btu/lb·°F	Vel. Sound, ft/s
Saturated Liquid	71.4013	46.78	0.09464	1419.1
Vapor	3.1801	116.71	0.21844	460.0
120.00	3.0077	120.64	0.22530	476.6
140.00	2.8181	125.78	0.23403	496.0
160.00	2.6620	130.83	0.24231	513.3
180.00	2.5295	135.83	0.25026	528.9
200.00	2.4146	140.82	0.25794	543.3
220.00	2.3132	145.82	0.26539	556.7
240.00	2.2223	150.83	0.27267	569.3
260.00	2.1401	155.88	0.27978	581.3
280.00	2.0658	160.97	0.28675	592.7
300.00	1.9971	166.10	0.29360	603.7
320.00	1.9338	171.28	0.30033	614.2
340.00	1.8751	176.51	0.30696	624.5
360.00	1.8208	181.80	0.31349	634.4
380.00	1.7695	187.14	0.31993	644.0
400.00	1.7216	192.54	0.32628	653.4
420.00	1.6766	198.00	0.33256	662.6
440.00	1.6341	203.51	0.33876	671.6
460.00	1.5940	209.08	0.34488	680.4
480.00	1.5558	214.71	0.35094	689.0
500.00	1.5197	220.40	0.35692	697.4

Pressure = 175.00 psia — Saturation temperature = 115.76 °F

Temp., °F	Density, lb/ft³	Enthalpy, Btu/lb	Entropy, Btu/lb·°F	Vel. Sound, ft/s
Saturated Liquid	69.7902	50.62	0.10126	1325.3
Vapor	3.7511	117.76	0.21794	453.0
120.00	3.6836	118.95	0.21999	458.4
140.00	3.4148	124.38	0.22921	481.3
160.00	3.2025	129.64	0.23783	500.9
180.00	3.0271	134.79	0.24602	518.3
200.00	2.8785	139.90	0.25388	534.0
220.00	2.7494	144.99	0.26148	548.5
240.00	2.6349	150.08	0.26887	562.1
260.00	2.5328	155.19	0.27607	574.8
280.00	2.4403	160.34	0.28312	586.9
300.00	2.3558	165.51	0.29003	598.5
320.00	2.2785	170.73	0.29681	609.5
340.00	2.2071	176.00	0.30348	620.2
360.00	2.1411	181.32	0.31004	630.5
380.00	2.0795	186.69	0.31651	640.5
400.00	2.0216	192.11	0.32290	650.3
420.00	1.9675	197.59	0.32920	659.7
440.00	1.9164	203.12	0.33542	669.0
460.00	1.8683	208.71	0.34156	678.0
480.00	1.8228	214.36	0.34763	686.9
500.00	1.7797	220.05	0.35363	695.6

Pressure = 200.00 psia — Saturation temperature = 125.27 °F

Temp., °F	Density, lb/ft³	Enthalpy, Btu/lb	Entropy, Btu/lb·°F	Vel. Sound, ft/s
Saturated Liquid	68.2602	54.14	0.10721	1240.5
Vapor	4.3437	118.61	0.21743	445.6
140.00	4.0726	122.86	0.22460	465.2
160.00	3.7850	128.36	0.23363	487.8
180.00	3.5561	133.70	0.24210	507.2
200.00	3.3656	138.94	0.25018	524.5
220.00	3.2036	144.14	0.25793	540.2
240.00	3.0623	149.31	0.26544	554.7
260.00	2.9371	154.50	0.27274	568.3
280.00	2.8247	159.69	0.27987	581.1
300.00	2.7234	164.92	0.28684	593.2
320.00	2.6305	170.18	0.29368	604.8
340.00	2.5455	175.49	0.30039	615.9
360.00	2.4668	180.83	0.30700	626.7
380.00	2.3934	186.23	0.31350	637.0
400.00	2.3254	191.68	0.31991	647.1
420.00	2.2614	197.18	0.32624	656.9
440.00	2.2017	202.73	0.33248	666.4
460.00	2.1453	208.34	0.33864	675.7
480.00	2.0920	214.00	0.34473	684.8
500.00	2.0417	219.71	0.35075	693.7

Refrigerant 134a   Properties of Superheated Vapor (*Concluded*)

Pressure = 225.00 psia Saturation temperature = 133.93 °F					Pressure = 250.00 psia Saturation temperature = 141.89 °F					Pressure = 275.00 psia Saturation temperature = 149.27 °F				
Temp., °F	Density, lb/ft³	Enthalpy, Btu/lb	Entropy, Btu/lb·°F	Vel. Sound, ft/s	Temp., °F	Density, lb/ft³	Enthalpy, Btu/lb	Entropy, Btu/lb·°F	Vel. Sound, ft/s	Temp., °F	Density, lb/ft³	Enthalpy, Btu/lb	Entropy, Btu/lb·°F	Vel. Sound, ft/s
Saturated					Saturated					Saturated				
Liquid	66.7870	57.42	0.11266	1162.8	Liquid	65.3526	60.50	0.11770	1090.7	Liquid	63.9423	63.43	0.12241	1023.4
Vapor	4.9609	119.30	0.21690	438.1	Vapor	5.6060	119.84	0.21634	430.3	Vapor	6.2831	120.25	0.21572	422.3
140.00	4.8123	121.16	0.22002	447.3										
160.00	4.4191	126.99	0.22959	473.6	160.00	5.1189	125.49	0.22560	458.2	160.00	5.9060	123.82	0.22155	441.2
180.00	4.1206	132.54	0.23840	495.5	180.00	4.7275	131.31	0.23484	483.1	180.00	5.3869	129.98	0.23133	469.9
200.00	3.8796	137.94	0.24671	514.6	200.00	4.4239	136.89	0.24343	504.2	200.00	5.0031	135.78	0.24026	493.4
220.00	3.6784	143.25	0.25465	531.6	220.00	4.1756	142.34	0.25156	522.8	220.00	4.6978	141.38	0.24862	513.7
240.00	3.5058	148.52	0.26229	547.2	240.00	3.9664	147.71	0.25935	539.5	240.00	4.4465	146.87	0.25658	531.7
260.00	3.3542	153.78	0.26970	561.6	260.00	3.7854	153.05	0.26688	554.9	260.00	4.2314	152.30	0.26423	548.0
280.00	3.2202	159.04	0.27691	575.1	280.00	3.6265	158.37	0.27418	569.2	280.00	4.0446	157.69	0.27162	563.1
300.00	3.0995	164.32	0.28395	587.9	300.00	3.4847	163.71	0.28129	582.6	300.00	3.8803	163.08	0.27881	577.2
320.00	2.9899	169.62	0.29084	600.0	320.00	3.3571	169.06	0.28824	595.3	320.00	3.7317	168.49	0.28583	590.5
340.00	2.8897	174.97	0.29761	611.7	340.00	3.2408	174.44	0.29506	607.4	340.00	3.5987	173.91	0.29270	603.1
360.00	2.7978	180.35	0.30425	622.8	360.00	3.1342	179.85	0.30175	618.9	360.00	3.4764	179.36	0.29943	615.1
380.00	2.7122	185.77	0.31079	633.6	380.00	3.0359	185.31	0.30832	630.1	380.00	3.3646	184.84	0.30604	626.6
400.00	2.6330	191.24	0.31723	644.0	400.00	2.9451	190.81	0.31479	640.8	400.00	3.2612	190.37	0.31254	637.7
420.00	2.5592	196.77	0.32358	654.0	420.00	2.8604	196.35	0.32117	651.2	420.00	3.1653	195.93	0.31894	648.4
440.00	2.4900	202.34	0.32984	663.9	440.00	2.7813	201.94	0.32745	661.3	440.00	3.0758	201.55	0.32525	658.8
460.00	2.4249	207.96	0.33603	673.4	460.00	2.7072	207.58	0.33365	671.1	460.00	2.9922	207.20	0.33147	668.9
480.00	2.3636	213.64	0.34213	682.8	480.00	2.6374	213.27	0.33977	680.7	480.00	2.9136	212.91	0.33761	678.7
500.00	2.3057	219.36	0.34816	691.9	500.00	2.5717	219.02	0.34582	690.1	500.00	2.8397	218.67	0.34368	688.3

Pressure = 300.00 psia Saturation temperature = 156.16 °F					Pressure = 325.00 psia Saturation temperature = 162.62 °F					Pressure = 350.00 psia Saturation temperature = 168.71 °F				
Temp., °F	Density, lb/ft³	Enthalpy, Btu/lb	Entropy, Btu/lb·°F	Vel. Sound, ft/s	Temp., °F	Density, lb/ft³	Enthalpy, Btu/lb	Entropy, Btu/lb·°F	Vel. Sound, ft/s	Temp., °F	Density, lb/ft³	Enthalpy, Btu/lb	Entropy, Btu/lb·°F	Vel. Sound, ft/s
Saturated					Saturated					Saturated				
Liquid	62.5436	66.23	0.12686	959.8	Liquid	61.1446	68.92	0.13110	899.5	Liquid	59.7334	71.54	0.13516	841.7
Vapor	6.9967	120.54	0.21505	414.2	Vapor	7.7526	120.71	0.21431	405.9	Vapor	8.5577	120.76	0.21349	397.5
160.00	6.8168	121.92	0.21730	422.0										
180.00	6.1118	128.55	0.22782	455.8	180.00	6.9220	126.96	0.22423	440.3	180.00	7.8491	125.18	0.22046	423.2
200.00	5.6239	134.61	0.23715	482.1	200.00	6.2928	133.36	0.23408	470.2	200.00	7.0242	132.01	0.23098	457.6
220.00	5.2494	140.39	0.24578	504.3	220.00	5.8341	139.34	0.24301	494.5	220.00	6.4561	138.24	0.24029	484.4
240.00	4.9472	146.00	0.25393	523.7	240.00	5.4723	145.10	0.25136	515.5	240.00	6.0219	144.17	0.24888	507.1
260.00	4.6939	151.52	0.26171	541.1	260.00	5.1741	150.73	0.25930	534.0	260.00	5.6728	149.91	0.25698	526.9
280.00	4.4758	157.00	0.26921	557.0	280.00	4.9208	156.29	0.26692	550.9	280.00	5.3805	155.56	0.26472	544.7
300.00	4.2852	162.45	0.27649	571.8	300.00	4.7017	161.81	0.27428	566.4	300.00	5.1295	161.15	0.27218	561.0
320.00	4.1160	167.90	0.28357	585.7	320.00	4.5082	167.31	0.28144	580.9	320.00	4.9098	166.72	0.27941	576.1
340.00	3.9631	173.37	0.29049	598.8	340.00	4.3352	172.82	0.28841	594.5	340.00	4.7148	172.27	0.28644	590.2
360.00	3.8247	178.85	0.29727	611.2	360.00	4.1790	178.35	0.29524	607.4	360.00	4.5396	177.84	0.29332	603.6
380.00	3.6981	184.37	0.30392	623.1	380.00	4.0368	183.90	0.30193	619.7	380.00	4.3807	183.42	0.30005	616.3
400.00	3.5816	189.92	0.31045	634.6	400.00	3.9063	189.48	0.30849	631.5	400.00	4.2355	189.03	0.30665	628.4
420.00	3.4737	195.51	0.31688	645.6	420.00	3.7859	195.09	0.31495	642.8	420.00	4.1019	194.67	0.31313	640.1
440.00	3.3735	201.15	0.32321	656.3	440.00	3.6743	200.75	0.32131	653.8	440.00	3.9784	200.35	0.31952	651.4
460.00	3.2799	206.83	0.32945	666.6	460.00	3.5703	206.44	0.32757	664.4	460.00	3.8636	206.06	0.32580	662.2
480.00	3.1922	212.55	0.33561	676.7	480.00	3.4731	212.19	0.33375	674.7	480.00	3.7564	211.82	0.33199	672.8
500.00	3.1098	218.32	0.34169	686.5	500.00	3.3819	217.97	0.33984	684.7	500.00	3.6561	217.62	0.33811	683.0

Pressure = 375.00 psia Saturation temperature = 174.46 °F					Pressure = 400.00 psia Saturation temperature = 197.93 °F					Pressure = 600.00 psia Saturation temperature = n/a (supercritical)				
Temp., °F	Density, lb/ft³	Enthalpy, Btu/lb	Entropy, Btu/lb·°F	Vel. Sound, ft/s	Temp., °F	Density, lb/ft³	Enthalpy, Btu/lb	Entropy, Btu/lb·°F	Vel. Sound, ft/s	Temp., °F	Density, lb/ft³	Enthalpy, Btu/lb	Entropy, Btu/lb·°F	Vel. Sound, ft/s
Saturated					Saturated									
Liquid	58.2974	74.09	0.13908	785.9	Liquid	56.8213	76.60	0.14289	731.8					
Vapor	9.4209	120.69	0.21256	389.0	Vapor	10.3541	120.50	0.21152	380.4					
180.00	8.9498	123.10	0.21634	403.8	180.00	10.3454	120.53	0.21158	380.6					
200.00	7.8311	130.54	0.22781	444.1	200.00	8.7370	128.93	0.22451	429.5					
220.00	7.1211	137.08	0.23758	474.0	220.00	7.8399	135.85	0.23484	463.0	220.00	19.6784	118.27	0.20421	340.3
240.00	6.6028	143.19	0.24644	498.5	240.00	7.2145	142.18	0.24403	489.7	240.00	14.2159	131.50	0.22343	409.1
260.00	6.1926	149.07	0.25473	519.6	260.00	6.7351	148.21	0.25252	512.3	260.00	12.2674	139.92	0.23530	449.7
280.00	5.8555	154.82	0.26260	538.4	280.00	6.3472	154.06	0.26055	532.1	280.00	11.0672	147.15	0.24522	480.8
300.00	5.5694	160.49	0.27016	555.5	300.00	6.0221	159.81	0.26821	550.0	300.00	10.2049	153.83	0.25413	506.7
320.00	5.3212	166.11	0.27747	571.3	320.00	5.7425	165.49	0.27560	566.5	320.00	9.5351	160.21	0.26241	529.2
340.00	5.1022	171.72	0.28457	586.0	340.00	5.4977	171.15	0.28277	581.7	340.00	8.9895	166.39	0.27024	549.2
360.00	4.9066	177.32	0.29149	599.8	360.00	5.2802	176.80	0.28975	596.0	360.00	8.5305	172.45	0.27774	567.5
380.00	4.7300	182.94	0.29826	612.9	380.00	5.0848	182.45	0.29656	609.5	380.00	8.1351	178.45	0.28496	584.4
400.00	4.5692	188.58	0.30490	625.4	400.00	4.9075	188.12	0.30323	622.4	400.00	7.7885	184.40	0.29197	600.1
420.00	4.4217	194.24	0.31141	637.4	420.00	4.7454	193.82	0.30978	634.7	420.00	7.4804	190.34	0.29879	615.0
440.00	4.2857	199.94	0.31782	648.9	440.00	4.5964	199.54	0.31621	646.5	440.00	7.2035	196.26	0.30546	629.0
460.00	4.1596	205.68	0.32413	660.1	460.00	4.4584	205.30	0.32254	657.9	460.00	6.9523	202.20	0.31198	642.4
480.00	4.0421	211.46	0.33034	670.8	480.00	4.3303	211.09	0.32878	669.0	480.00	6.7229	208.15	0.31838	655.3
500.00	3.9323	217.28	0.33647	681.3	500.00	4.2107	216.93	0.33492	679.6	500.00	6.5118	214.12	0.32467	667.6

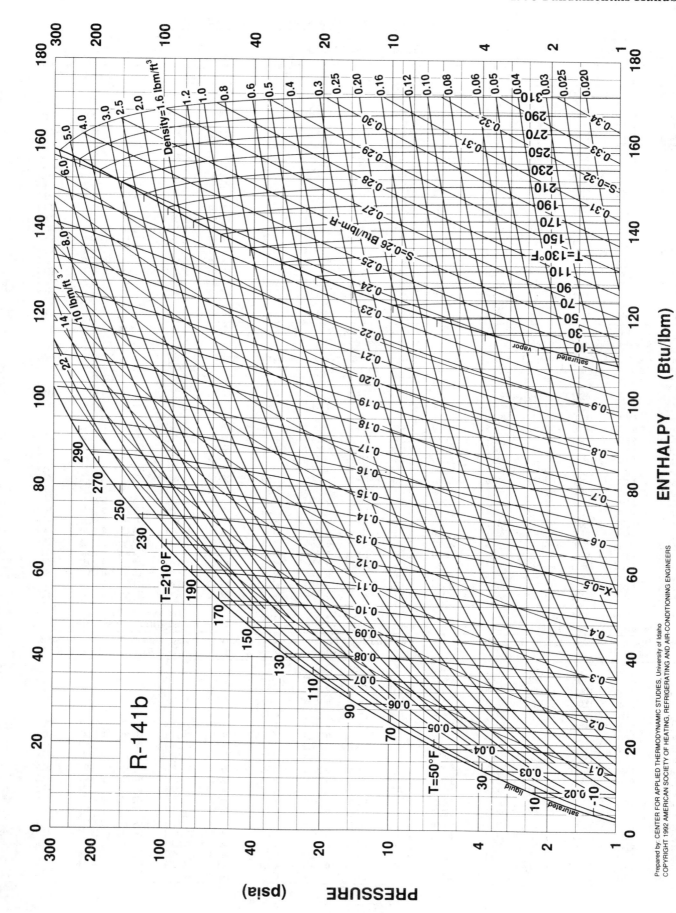

PRESSURE (psia)

ENTHALPY (Btu/lbm)

Fig. 15 Pressure-Enthalpy Diagram for Refrigerant 141b

Prepared by: CENTER FOR APPLIED THERMODYNAMIC STUDIES, University of Idaho
COPYRIGHT 1992 AMERICAN SOCIETY OF HEATING, REFRIGERATING AND AIR-CONDITIONING ENGINEERS

Refrigerant Properties

Refrigerant 141b (1,1-dichloro-1-fluoroethane)   Properties of Saturated Liquid and Saturated Vapor

Temp,* °F	Pressure, psia	Density, lb/ft³ Liquid	Volume, ft³/lb Vapor	Enthalpy, Btu/lb Liquid	Vapor	Entropy, Btu/lb·°F Liquid	Vapor	Specific Heat c_p, Btu/lb·°F Liquid	Vapor	c_p/c_v Vapor	Velocity of Sound, ft/s Liquid	Vapor	Viscosity, lb$_m$/ft·h Liquid	Vapor	Thermal Cond, Btu/h·ft·°F Liquid	Vapor	Surface Tension, dyne/cm	Temp, °F
0.00	1.695	81.68	24.691	6.101	111.614	0.01372	0.24326	—	0.1655	1.1197	—	464.	1.712	—	0.0623	—	—	0.00
5.00	1.962	81.36	21.538	7.331	112.410	0.01638	0.24252	—	0.1667	1.1193	—	466.	1.645	—	0.0617	—	—	5.00
10.00	2.263	81.05	18.852	8.606	113.208	0.01911	0.24182	—	0.1679	1.1191	—	468.	1.582	—	0.0610	—	—	10.00
15.00	2.602	80.73	16.555	9.915	114.008	0.02188	0.24118	—	0.1691	1.1188	—	470.	1.524	—	0.0604	—	—	15.00
20.00	2.981	80.41	14.586	11.248	114.811	0.02467	0.24058	—	0.1703	1.1187	—	472.	1.469	—	0.0598	—	—	20.00
25.00	3.404	80.10	12.890	12.599	115.615	0.02747	0.24002	—	0.1716	1.1186	—	474.	1.417	—	0.0591	—	—	25.00
30.00	3.874	79.78	11.427	13.963	116.422	0.03027	0.23951	—	0.1728	1.1185	—	475.	1.369	—	0.0585	—	—	30.00
35.00	4.396	79.46	10.159	15.333	117.230	0.03305	0.23904	—	0.1741	1.1186	—	477.	1.323	—	0.0579	—	21.12	35.00
40.00	4.972	79.13	9.0571	16.708	118.039	0.03581	0.23861	—	0.1753	1.1186	—	479.	1.280	—	0.0572	—	20.77	40.00
45.00	5.609	78.81	8.0968	18.084	118.850	0.03855	0.23822	—	0.1766	1.1188	—	480.	1.239	—	0.0566	—	20.41	45.00
50.00	6.309	78.49	7.2572	19.460	119.662	0.04126	0.23786	—	0.1779	1.1190	—	482.	1.200	—	0.0560	—	20.06	50.00
55.00	7.077	78.16	6.5209	20.836	120.475	0.04394	0.23754	—	0.1792	1.1193	—	483.	1.163	—	0.0554	—	19.70	55.00
60.00	7.918	77.84	5.8735	22.210	121.288	0.04660	0.23725	—	0.1805	1.1196	—	485.	1.128	—	0.0548	—	19.35	60.00
65.00	8.836	77.51	5.3026	23.583	122.102	0.04922	0.23700	—	0.1819	1.1200	—	486.	1.095	—	0.0542	—	19.00	65.00
70.00	9.838	77.18	4.7979	24.955	122.916	0.05182	0.23677	—	0.1832	1.1205	—	488.	1.063	—	0.0536	—	18.65	70.00
75.00	10.927	76.85	4.3506	26.327	123.731	0.05439	0.23657	—	0.1846	1.1210	—	489.	1.033	—	0.0530	—	18.30	75.00
80.00	12.109	76.52	3.9531	27.700	124.546	0.05694	0.23640	—	0.1860	1.1216	—	490.	1.003	—	0.0524	—	17.95	80.00
85.00	13.390	76.18	3.5990	29.073	125.360	0.05947	0.23625	—	0.1874	1.1223	—	491.	0.975	—	0.0518	—	17.61	85.00
89.72b	14.696	75.86	3.2994	30.374	126.130	0.06184	0.23614	—	0.1888	1.1230	—	492.	0.949	—	0.0512	—	17.28	89.72
90.00	14.775	75.84	3.2829	30.450	126.174	0.06198	0.23613	—	0.1888	1.1231	—	492.	0.948	—	0.0512	—	17.26	90.00
95.00	16.270	75.51	3.0001	31.830	126.988	0.06447	0.23603	—	0.1903	1.1239	—	493.	0.922	—	0.0506	—	16.92	95.00
100.00	17.881	75.17	2.7465	33.214	127.801	0.06695	0.23596	—	0.1918	1.1248	—	494.	0.896	—	0.0500	—	16.58	100.00
105.00	19.614	74.82	2.5187	34.603	128.614	0.06941	0.23590	—	0.1933	1.1258	—	495.	0.872	—	0.0495	—	16.24	105.00
110.00	21.474	74.48	2.3136	35.998	129.425	0.07187	0.23587	—	0.1948	1.1269	—	496.	0.848	—	0.0489	—	15.90	110.00
115.00	23.469	74.13	2.1285	37.400	130.236	0.07431	0.23585	—	0.1963	1.1281	—	497.	0.825	—	0.0483	—	15.56	115.00
120.00	25.604	73.78	1.9613	38.808	131.045	0.07674	0.23586	—	0.1979	1.1293	—	497.	0.802	—	0.0477	—	15.22	120.00
125.00	27.886	73.43	1.8099	40.225	131.853	0.07916	0.23588	—	0.1995	1.1306	—	498.	0.780	—	0.0472	—	14.89	125.00
130.00	30.321	73.08	1.6726	41.648	132.660	0.08158	0.23592	—	0.2011	1.1321	—	498.	0.759	—	0.0466	—	14.55	130.00
135.00	32.916	72.72	1.5479	43.080	133.464	0.08398	0.23597	—	0.2028	1.1336	—	499.	0.737	—	0.0461	—	14.22	135.00
140.00	35.677	72.36	1.4343	44.520	134.267	0.08638	0.23604	—	0.2045	1.1352	—	499.	0.717	—	0.0455	—	13.89	140.00
145.00	38.611	72.00	1.3308	45.968	135.069	0.08877	0.23613	—	0.2062	1.1369	—	499.	0.696	—	0.0450	—	13.56	145.00
150.00	41.726	71.63	1.2364	47.423	135.868	0.09116	0.23623	—	0.2079	1.1387	—	499.	0.676	—	0.0444	—	13.23	150.00
155.00	45.027	71.26	1.1500	48.886	136.665	0.09353	0.23634	—	0.2097	1.1407	—	499.	0.656	—	0.0439	—	12.90	155.00
160.00	48.522	70.89	1.0708	50.357	137.459	0.09590	0.23646	—	0.2116	1.1427	—	499.	0.637	—	0.0433	—	12.58	160.00
165.00	52.218	70.51	0.9983	51.834	138.252	0.09826	0.23660	—	0.2134	1.1449	—	499.	0.618	—	0.0428	—	12.26	165.00
170.00	56.121	70.13	0.9316	53.319	139.041	0.10061	0.23675	—	0.2153	1.1472	—	499.	0.599	—	0.0422	—	—	170.00
175.00	60.240	69.75	0.8703	54.810	139.828	0.10295	0.23691	—	0.2173	1.1496	—	498.	0.580	—	0.0417	—	—	175.00
180.00	64.581	69.36	0.8138	56.307	140.612	0.10528	0.23708	—	0.2193	1.1521	—	498.	—	—	0.0412	—	—	180.00
185.00	69.152	68.97	0.7617	57.811	141.392	0.10761	0.23726	—	0.2214	1.1548	—	497.	—	—	0.0407	—	—	185.00
190.00	73.960	68.57	0.7136	59.321	142.169	0.10992	0.23744	—	0.2235	1.1577	—	496.	—	—	0.0401	—	—	190.00
195.00	79.013	68.17	0.6691	60.837	142.943	0.11222	0.23764	—	0.2256	1.1607	—	496.	—	—	0.0396	—	—	195.00
200.00	84.320	67.76	0.6278	62.359	143.713	0.11452	0.23784	—	0.2279	1.1639	—	494.	—	—	0.0391	—	—	200.00
205.00	89.887	67.35	0.5896	63.888	144.479	0.11680	0.23805	—	0.2302	1.1673	—	493.	—	—	0.0386	—	—	205.00
210.00	95.723	66.93	0.5541	65.424	145.240	0.11908	0.23827	—	0.2325	1.1709	—	492.	—	—	0.0381	—	—	210.00
215.00	101.84	66.51	0.5211	66.968	145.997	0.12135	0.23849	—	0.2350	1.1747	—	491.	—	—	0.0376	—	—	215.00
220.00	108.24	66.08	0.4903	68.519	146.748	0.12362	0.23872	—	0.2375	1.1788	—	489.	—	—	0.0371	—	—	220.00
225.00	114.93	65.65	0.4616	70.080	147.494	0.12588	0.23895	—	0.2401	1.1832	—	487.	—	—	0.0366	—	—	225.00
230.00	121.93	65.21	0.4349	71.651	148.234	0.12813	0.23918	—	0.2428	1.1878	—	485.	—	—	0.0361	—	—	230.00
235.00	129.24	64.76	0.4099	73.233	148.968	0.13039	0.23941	—	0.2457	1.1928	—	483.	—	—	0.0356	—	—	235.00
240.00	136.87	64.31	0.3865	74.829	149.694	0.13265	0.23965	—	0.2486	1.1982	—	480.	—	—	0.0351	—	—	240.00
245.00	144.83	63.84	0.3645	76.439	150.413	0.13491	0.23989	—	0.2517	1.2040	—	478.	—	—	0.0346	—	—	245.00
250.00	153.13	63.37	0.3439	78.067	151.123	0.13718	0.24012	—	0.2550	1.2104	—	475.	—	—	0.0341	—	—	250.00
255.00	161.78	62.89	0.3246	79.713	151.823	0.13945	0.24035	—	0.2584	1.2172	—	472.	—	—	—	—	—	255.00
260.00	170.80	62.41	0.3064	81.382	152.513	0.14174	0.24058	—	0.2620	1.2248	—	468.	—	—	—	—	—	260.00
265.00	180.18	61.91	0.2892	83.075	153.191	0.14405	0.24080	—	0.2659	1.2331	—	464.	—	—	—	—	—	265.00
270.00	189.95	61.40	0.2730	84.797	153.855	0.14638	0.24102	—	0.2701	1.2424	—	460.	—	—	—	—	—	270.00
275.00	200.11	60.88	0.2577	86.551	154.505	0.14873	0.24122	—	0.2746	1.2528	—	455.	—	—	—	—	—	275.00
280.00	210.67	60.35	0.2432	88.341	155.136	0.15111	0.24142	—	0.2795	1.2645	—	450.	—	—	—	—	—	280.00
285.00	221.65	59.81	0.2294	90.174	155.748	0.15354	0.24160	—	0.2849	1.2779	—	444.	—	—	—	—	—	285.00
290.00	233.05	59.26	0.2163	92.055	156.337	0.15601	0.24175	—	0.2909	1.2935	—	438.	—	—	—	—	—	290.00
295.00	244.90	58.69	0.2037	93.993	156.897	0.15853	0.24189	—	0.2977	1.3120	—	431.	—	—	—	—	—	295.00
300.00	257.19	58.10	0.1917	95.996	157.424	0.16113	0.24199	—	0.3056	1.3342	—	423.	—	—	—	—	—	300.00

*temperatures are on the IPTS-68 scale   b = boiling point

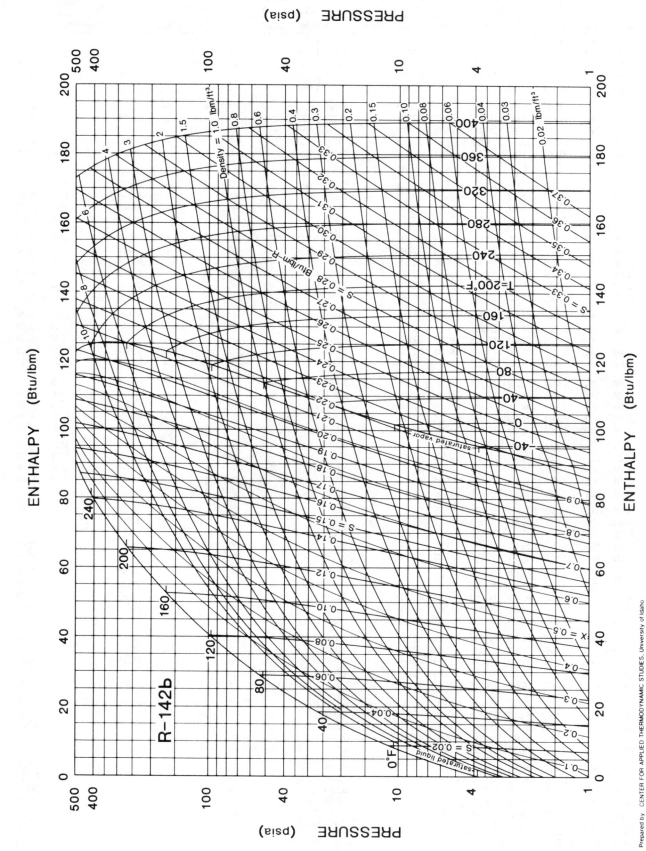

Fig. 16   Pressure-Enthalpy Diagram for Refrigerant 142b

Refrigerant 142b (1-chloro-1,1-difluoroethane)   Properties of Saturated Liquid and Saturated Vapor

Temp,* °F	Pressure, psia	Density, lb/ft³ Liquid	Volume, ft³/lb Vapor	Enthalpy, Btu/lb		Entropy, Btu/lb·°F		Specific Heat c_p, Btu/lb·°F		c_p/c_v Vapor	Velocity of Sound, ft/s		Viscosity, lb$_m$/ft·h		Thermal Cond, Btu/h·ft·°F		Surface Tension, dyne/cm	Temp, °F
				Liquid	Vapor	Liquid	Vapor	Liquid	Vapor		Liquid	Vapor	Liquid	Vapor	Liquid	Vapor		
−60.00	2.135	80.26	19.796	−3.732	92.263	−0.00910	0.23109	—	0.1623	1.1441	—	471.	—	—	—	—	—	−60.00
−55.00	2.485	79.89	17.199	−2.822	93.043	−0.00684	0.23006	—	0.1636	1.1435	—	473.	—	—	—	—	—	−55.00
−50.00	2.881	79.52	14.995	−1.897	93.826	−0.00457	0.22909	—	0.1649	1.1430	—	475.	—	—	—	—	—	−50.00
−45.00	3.329	79.15	13.116	−0.956	94.611	−0.00229	0.22818	—	0.1663	1.1425	—	477.	—	—	—	—	—	−45.00
−40.00	3.833	78.77	11.509	0.000	95.399	0.00000	0.22732	—	0.1676	1.1422	—	479.	—	—	—	—	—	−40.00
−35.00	4.399	78.40	10.130	0.972	96.187	0.00230	0.22651	—	0.1690	1.1420	—	481.	—	—	—	—	—	−35.00
−30.00	5.033	78.02	8.9418	1.961	96.977	0.00461	0.22575	—	0.1704	1.1419	—	483.	—	—	—	—	—	−30.00
−25.00	5.740	77.64	7.9154	2.967	97.768	0.00693	0.22503	—	0.1718	1.1419	—	485.	—	—	—	—	—	−25.00
−20.00	6.526	77.26	7.0257	3.990	98.560	0.00927	0.22436	—	0.1732	1.1420	—	487.	—	—	—	—	—	−20.00
−15.00	7.400	76.88	6.2523	5.030	99.351	0.01162	0.22373	—	0.1746	1.1422	—	488.	—	—	—	—	—	−15.00
−10.00	8.366	76.49	5.5778	6.088	100.143	0.01398	0.22314	—	0.1760	1.1425	—	490.	—	—	—	—	—	−10.00
−5.00	9.433	76.11	4.9880	7.163	100.933	0.01635	0.22259	—	0.1775	1.1430	—	491.	—	—	0.0561	—	—	−5.00
0.00	10.609	75.72	4.4709	8.257	101.722	0.01874	0.22207	—	0.1789	1.1436	—	493.	—	—	0.0555	—	—	0.00
5.00	11.901	75.33	4.0163	9.370	102.510	0.02114	0.22158	—	0.1804	1.1444	—	494.	—	—	0.0548	—	—	5.00
10.00	13.318	74.93	3.6156	10.501	103.295	0.02355	0.22113	—	0.1819	1.1453	—	495.	—	—	0.0542	—	16.17	10.00
14.47b	14.696	74.58	3.2973	11.527	103.995	0.02572	0.22074	—	0.1833	1.1463	—	496.	—	—	0.0537	—	15.85	14.47
15.00	14.868	74.53	3.2616	11.651	104.078	0.02598	0.22070	—	0.1834	1.1464	—	496.	—	—	0.0536	—	15.81	15.00
20.00	16.560	74.13	2.9480	12.819	104.857	0.02842	0.22030	—	0.1850	1.1477	—	497.	—	—	0.0530	—	15.44	20.00
25.00	18.404	73.73	2.6696	14.007	105.633	0.03087	0.21992	—	0.1866	1.1492	—	498.	—	—	0.0524	—	15.08	25.00
30.00	20.410	73.32	2.4219	15.213	106.404	0.03334	0.21957	—	0.1882	1.1509	—	499.	—	—	0.0519	—	14.72	30.00
35.00	22.587	72.91	2.2010	16.439	107.171	0.03582	0.21924	0.2993	0.1898	1.1528	—	499.	—	0.0236	0.0513	—	14.37	35.00
40.00	24.945	72.50	2.0037	17.684	107.932	0.03831	0.21892	0.3009	0.1915	1.1549	—	500.	—	0.0238	0.0507	—	14.01	40.00
45.00	27.495	72.08	1.8270	18.948	108.687	0.04081	0.21863	0.3026	0.1932	1.1572	—	500.	—	0.0241	0.0501	—	13.65	45.00
50.00	30.248	71.66	1.6684	20.230	109.436	0.04333	0.21835	0.3043	0.1950	1.1599	—	500.	—	0.0243	0.0495	—	13.30	50.00
55.00	33.216	71.24	1.5260	21.531	110.177	0.04585	0.21809	0.3059	0.1968	1.1628	—	500.	—	0.0245	0.0489	—	12.95	55.00
60.00	36.408	70.80	1.3976	22.852	110.910	0.04839	0.21784	0.3076	0.1986	1.1660	—	501.	—	0.0247	0.0483	—	12.60	60.00
65.00	39.838	70.37	1.2819	24.190	111.635	0.05094	0.21760	0.3093	0.2005	1.1696	—	500.	—	0.0250	0.0478	—	12.25	65.00
70.00	43.517	69.93	1.1773	25.547	112.350	0.05349	0.21737	0.3111	0.2025	1.1735	—	500.	—	0.0252	0.0472	—	11.91	70.00
75.00	47.458	69.48	1.0826	26.922	113.055	0.05606	0.21715	0.3128	0.2046	1.1778	—	500.	—	0.0254	0.0466	—	11.56	75.00
80.00	51.672	69.03	0.9967	28.314	113.750	0.05863	0.21694	0.3146	0.2067	1.1825	—	500.	—	0.0256	0.0460	—	11.22	80.00
85.00	56.172	68.57	0.9187	29.724	114.433	0.06121	0.21673	0.3164	0.2089	1.1876	—	499.	—	0.0258	0.0455	—	10.88	85.00
90.00	60.971	68.11	0.8478	31.151	115.103	0.06379	0.21652	0.3183	0.2112	1.1933	—	498.	—	0.0260	0.0449	—	10.54	90.00
95.00	66.083	67.64	0.7832	32.595	115.761	0.06638	0.21632	0.3202	0.2136	1.1994	—	498.	—	0.0262	0.0443	—	10.20	95.00
100.00	71.521	67.16	0.7242	34.054	116.405	0.06897	0.21611	0.3222	0.2161	1.2062	—	497.	—	0.0265	0.0438	—	9.86	100.00
105.00	77.297	66.67	0.6704	35.530	117.034	0.07157	0.21591	0.3242	0.2187	1.2135	—	496.	—	0.0267	0.0432	—	9.53	105.00
110.00	83.427	66.18	0.6211	37.021	117.647	0.07416	0.21570	0.3263	0.2214	1.2216	—	495.	—	0.0269	0.0427	—	9.20	110.00
115.00	89.923	65.68	0.5760	38.526	118.245	0.07676	0.21548	0.3284	0.2243	1.2304	—	493.	—	0.0271	0.0421	—	8.87	115.00
120.00	96.799	65.17	0.5346	40.046	118.824	0.07936	0.21527	0.3307	0.2274	1.2400	—	492.	—	0.0274	0.0415	—	8.54	120.00
125.00	104.07	64.64	0.4966	41.579	119.386	0.08196	0.21504	0.3330	0.2306	1.2505	—	491.	—	0.0276	0.0410	—	8.22	125.00
130.00	111.75	64.11	0.4617	43.126	119.929	0.08456	0.21481	0.3354	0.2340	1.2620	—	489.	—	0.0279	0.0404	—	7.90	130.00
135.00	119.85	63.57	0.4295	44.685	120.451	0.08715	0.21456	0.3378	0.2377	1.2746	—	487.	—	0.0281	0.0399	—	7.58	135.00
140.00	128.39	63.01	0.3999	46.256	120.952	0.08974	0.21430	0.3404	0.2415	1.2883	—	486.	—	0.0284	0.0394	—	7.26	140.00
145.00	137.39	62.44	0.3726	47.838	121.432	0.09232	0.21403	0.3431	0.2457	1.3033	—	484.	—	0.0287	0.0388	—	6.94	145.00
150.00	146.85	61.86	0.3473	49.432	121.888	0.09490	0.21375	0.3458	0.2501	1.3198	—	482.	—	0.0290	0.0383	—	6.63	150.00
155.00	156.79	61.26	0.3239	51.036	122.321	0.09747	0.21345	0.3487	0.2549	1.3379	—	480.	—	0.0293	0.0377	—	6.32	155.00
160.00	167.22	60.65	0.3023	52.650	122.728	0.10004	0.21313	0.3517	0.2600	1.3576	—	478.	—	0.0296	0.0372	—	6.01	160.00
165.00	178.17	60.02	0.2823	54.275	123.109	0.10259	0.21279	0.3548	0.2655	1.3794	—	476.	—	0.0299	0.0367	—	5.70	165.00
170.00	189.64	59.37	0.2637	55.909	123.463	0.10514	0.21243	0.3580	0.2714	1.4032	—	473.	—	0.0303	0.0361	—	5.40	170.00
175.00	201.66	58.71	0.2465	57.552	123.789	0.10768	0.21205	—	0.2779	1.4294	—	471.	—	0.0307	—	—	5.10	175.00
180.00	214.23	58.02	0.2305	59.206	124.085	0.11022	0.21164	—	0.2849	1.4582	—	469.	—	0.0311	—	—	4.81	180.00
185.00	227.37	57.30	0.2156	60.871	124.351	0.11274	0.21121	—	0.2925	1.4900	—	466.	—	0.0315	—	—	4.51	185.00
190.00	241.09	56.56	0.2017	62.547	124.585	0.11527	0.21076	—	0.3009	1.5251	—	463.	—	0.0319	—	—	4.22	190.00
195.00	255.41	55.80	0.1888	64.236	124.786	0.11778	0.21027	—	0.3100	1.5638	—	461.	—	0.0324	—	—	3.94	195.00
200.00	270.35	55.00	0.1767	65.940	124.952	0.12030	0.20976	—	0.3200	1.6067	—	458.	—	0.0329	—	—	3.65	200.00
205.00	285.92	54.16	0.1654	67.661	125.082	0.12282	0.20921	—	0.3310	1.6544	—	455.	—	0.0334	—	—	3.38	205.00
210.00	302.13	53.29	0.1548	69.404	125.174	0.12535	0.20863	—	0.3432	1.7076	—	451.	—	0.0340	—	—	3.10	210.00
215.00	319.00	52.38	0.1449	71.173	125.225	0.12789	0.20801	—	0.3568	1.7673	—	448.	—	0.0345	—	—	2.83	215.00
220.00	336.54	51.41	0.1356	72.975	125.232	0.13046	0.20735	—	0.3721	1.8347	—	444.	—	0.0352	—	—	2.56	220.00
225.00	354.77	50.40	0.1269	74.819	125.192	0.13307	0.20664	—	0.3894	1.9116	—	439.	—	0.0358	—	—	2.30	225.00
230.00	373.70	49.31	0.1186	76.717	125.099	0.13573	0.20588	—	0.4093	2.0006	—	434.	—	0.0365	—	—	2.05	230.00
235.00	393.34	48.16	0.1107	78.686	124.943	0.13846	0.20505	—	0.4327	2.1059	—	428.	—	—	—	—	1.80	235.00
240.00	413.72	46.92	0.1032	80.750	124.714	0.14131	0.20415	—	0.4611	2.2343	—	421.	—	—	—	—	1.55	240.00
245.00	434.84	45.57	0.0959	82.943	124.391	0.14431	0.20313	—	0.4970	2.3984	—	411.	—	—	—	—	1.31	245.00
250.00	456.71	44.10	0.0887	85.320	123.937	0.14755	0.20196	—	0.5463	2.6252	—	399.	—	—	—	—	1.08	250.00

*temperatures are on the IPTS-68 scale   b = boiling point

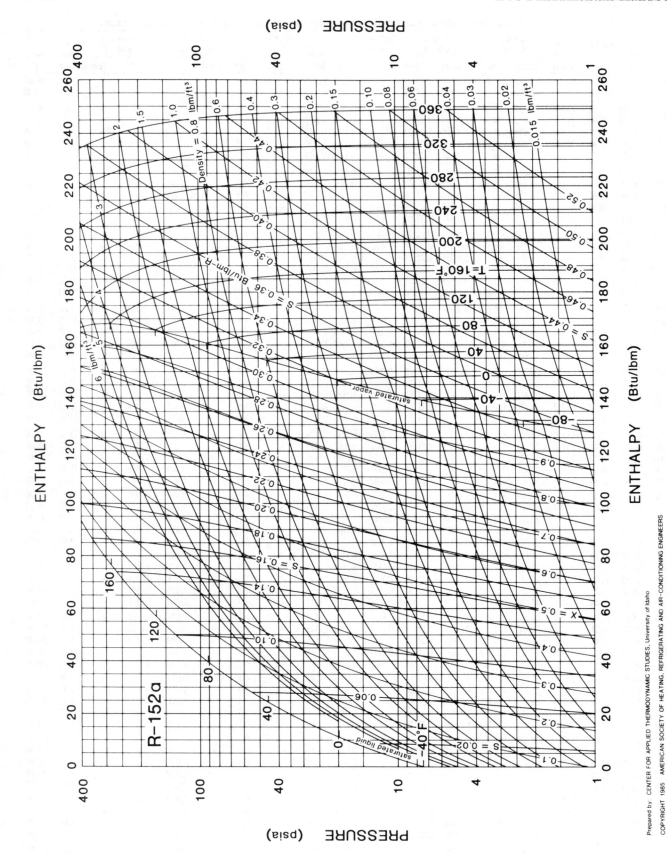

Fig. 17 Pressure-Enthalpy Diagram for Refrigerant 152a

Prepared by: CENTER FOR APPLIED THERMODYNAMIC STUDIES, University of Idaho

COPYRIGHT 1985   AMERICAN SOCIETY OF HEATING, REFRIGERATING AND AIR-CONDITIONING ENGINEERS

R152a (1,1-difluoroethane) Properties of Saturated Liquid and Saturated Vapor

Temp,* °F	Pressure, psia	Density, lb/ft³ Liquid	Volume, ft³/lb Vapor	Enthalpy, Btu/lb Liquid	Vapor	Entropy, Btu/lb·°F Liquid	Vapor	Specific Heat c_p, Btu/lb·°F Liquid	Vapor	c_p/c_v Vapor	Velocity of Sound, ft/s Liquid	Vapor	Viscosity, lb$_m$/ft·h Liquid	Vapor	Thermal Cond, Btu/h·ft·°F Liquid	Vapor	Surface Tension, dyne/cm	Temp, °F
−181.46a	0.012	74.28	3784.5	—	122.659	—	0.46931	—	0.1626	1.2274	—	507.	—	—	—	—	—	−181.46
−180.00	0.013	74.19	3483.2	—	122.896	—	0.46751	—	0.1632	1.2266	—	508.	—	—	—	—	—	−180.00
−170.00	0.024	73.58	1962.1	—	124.536	—	0.45497	—	0.1668	1.2208	—	516.	—	—	—	—	—	−170.00
−160.00	0.044	72.96	1113.9	—	126.207	—	0.44262	—	0.1705	1.2156	—	523.	—	—	—	—	—	−160.00
−150.00	0.078	72.34	646.96	—	127.907	—	0.43090	—	0.1742	1.2107	—	531.	—	—	—	—	—	−150.00
−140.00	0.134	71.72	386.82	—	129.635	—	0.42001	—	0.1780	1.2063	—	538.	—	—	—	—	—	−140.00
−130.00	0.224	71.09	238.52	—	131.387	—	0.41000	—	0.1819	1.2023	—	545.	—	—	—	—	—	−130.00
−120.00	0.363	70.45	151.60	—	133.160	—	0.40084	—	0.1858	1.1988	—	552.	—	—	—	—	—	−120.00
−110.00	0.570	69.81	99.179	—	134.952	—	0.39250	—	0.1899	1.1958	—	558.	—	—	—	—	—	−110.00
−100.00	0.871	69.16	66.660	−21.865	136.759	−0.05611	0.38492	—	0.1942	1.1932	—	565.	—	—	—	—	—	−100.00
−90.00	1.297	68.51	45.934	−18.413	138.576	−0.04664	0.37803	—	0.1987	1.1912	—	571.	1.301	—	—	—	—	−90.00
−80.00	1.884	67.85	32.386	−14.852	140.402	−0.03714	0.37178	—	0.2035	1.1898	—	577.	1.187	—	—	—	—	−80.00
−70.00	2.678	67.19	23.317	−11.216	142.230	−0.02769	0.36609	—	0.2086	1.1890	—	582.	1.088	—	—	—	—	−70.00
−60.00	3.729	66.52	17.113	−7.522	144.058	−0.01834	0.36092	—	0.2140	1.1889	—	587.	1.000	—	—	—	—	−60.00
−50.00	5.096	65.84	12.781	−3.781	145.882	−0.00911	0.35622	0.3759	0.2199	1.1894	3248.	592.	0.921	—	—	—	—	−50.00
−45.00	5.918	65.49	11.113	−1.895	146.790	−0.00454	0.35403	0.3778	0.2229	1.1900	3202.	594.	0.886	—	—	—	—	−45.00
−40.00	6.845	65.15	9.7001	0.000	147.696	0.00000	0.35193	0.3797	0.2261	1.1907	3157.	597.	0.852	—	—	—	—	−40.00
−35.00	7.885	64.80	8.4974	1.905	148.599	0.00450	0.34994	0.3815	0.2294	1.1916	3112.	599.	0.820	—	—	—	—	−35.00
−30.00	9.049	64.45	7.4699	3.819	149.499	0.00898	0.34803	0.3833	0.2328	1.1928	3068.	601.	0.790	—	—	—	—	−30.00
−25.00	10.347	64.10	6.5885	5.742	150.394	0.01342	0.34621	0.3851	0.2363	1.1941	3023.	602.	0.761	—	—	—	—	−25.00
−20.00	11.789	63.74	5.8297	7.674	151.285	0.01783	0.34446	0.3869	0.2399	1.1956	2979.	604.	0.734	—	—	—	—	−20.00
−15.00	13.387	63.38	5.1740	9.616	152.171	0.02221	0.34280	0.3887	0.2436	1.1973	2935.	606.	0.708	—	—	—	—	−15.00
−11.24b	14.696	63.11	4.7396	11.080	152.833	0.02548	0.34159	0.3901	0.2464	1.1988	2901.	607.	0.689	—	—	—	—	−11.24
−10.00	15.151	63.02	4.6055	11.566	153.052	0.02656	0.34120	0.3906	0.2474	1.1993	2890.	607.	0.683	—	—	—	—	−10.00
−5.00	17.095	62.66	4.1108	13.526	153.927	0.03088	0.33968	0.3924	0.2513	1.2015	2846.	608.	0.660	—	0.0737	—	—	−5.00
0.00	19.231	62.29	3.6791	15.496	154.796	0.03518	0.33822	0.3943	0.2553	1.2039	2802.	609.	0.638	—	0.0728	—	—	0.00
5.00	21.570	61.92	3.3010	17.476	155.658	0.03944	0.33682	0.3963	0.2594	1.2066	2758.	611.	0.617	—	0.0718	—	—	5.00
10.00	24.127	61.55	2.9691	19.465	156.513	0.04369	0.33548	0.3982	0.2636	1.2094	2714.	611.	0.597	—	0.0709	—	14.98	10.00
15.00	26.915	61.17	2.6767	21.465	157.361	0.04790	0.33420	0.4003	0.2679	1.2126	2670.	612.	0.578	—	0.0700	—	14.57	15.00
20.00	29.948	60.79	2.4184	23.475	158.201	0.05210	0.33297	0.4024	0.2723	1.2160	2626.	613.	0.559	—	0.0690	—	14.17	20.00
25.00	33.239	60.41	2.1898	25.496	159.033	0.05627	0.33179	0.4045	0.2768	1.2197	2582.	613.	0.542	—	0.0681	—	13.77	25.00
30.00	36.804	60.02	1.9867	27.529	159.856	0.06042	0.33066	0.4068	0.2814	1.2237	2537.	613.	0.525	—	0.0672	—	13.37	30.00
35.00	40.657	59.62	1.8060	29.572	160.669	0.06455	0.32957	0.4091	0.2860	1.2280	2493.	614.	0.509	—	0.0662	—	12.98	35.00
40.00	44.813	59.23	1.6447	31.628	161.473	0.06865	0.32852	0.4114	0.2908	1.2326	2448.	613.	0.494	—	0.0653	—	12.59	40.00
45.00	49.289	58.83	1.5005	33.695	162.266	0.07274	0.32751	0.4139	0.2957	1.2375	2404.	613.	0.479	—	0.0643	—	12.20	45.00
50.00	54.099	58.42	1.3712	35.775	163.048	0.07681	0.32653	0.4164	0.3007	1.2429	2359.	613.	0.464	0.0232	0.0634	—	11.81	50.00
55.00	59.260	58.01	1.2550	37.868	163.818	0.08087	0.32559	0.4190	0.3058	1.2486	2314.	612.	0.451	0.0235	0.0625	—	11.42	55.00
60.00	64.788	57.59	1.1505	39.974	164.576	0.08491	0.32468	0.4218	0.3110	1.2547	2269.	612.	0.438	0.0237	0.0615	—	11.04	60.00
65.00	70.700	57.17	1.0561	42.094	165.320	0.08893	0.32379	0.4246	0.3163	1.2613	2224.	611.	0.425	0.0239	0.0606	—	10.65	65.00
70.00	77.014	56.74	0.9708	44.228	166.051	0.09294	0.32294	0.4276	0.3218	1.2684	2179.	609.	0.413	0.0241	0.0597	—	10.28	70.00
75.00	83.746	56.31	0.8936	46.377	166.767	0.09694	0.32210	0.4306	0.3275	1.2760	2134.	608.	0.401	0.0244	0.0588	—	9.90	75.00
80.00	90.913	55.87	0.8235	48.542	167.468	0.10092	0.32129	0.4339	0.3333	1.2842	2088.	607.	0.390	0.0246	0.0579	—	9.52	80.00
85.00	98.535	55.42	0.7597	50.722	168.151	0.10490	0.32049	0.4372	0.3394	1.2931	2042.	605.	0.379	0.0248	0.0570	—	9.15	85.00
90.00	106.63	54.97	0.7016	52.919	168.818	0.10886	0.31971	0.4408	0.3456	1.3026	1996.	603.	0.368	0.0251	0.0561	—	8.78	90.00
95.00	115.21	54.51	0.6487	55.133	169.465	0.11282	0.31895	0.4445	0.3522	1.3129	1950.	601.	0.358	0.0253	0.0552	—	8.42	95.00
100.00	124.31	54.04	0.6002	57.366	170.092	0.11677	0.31819	0.4485	0.3590	1.3241	1904.	598.	0.348	0.0256	0.0543	—	8.06	100.00
105.00	133.93	53.56	0.5559	59.617	170.698	0.12072	0.31744	0.4526	0.3661	1.3363	1857.	596.	0.338	0.0259	0.0534	—	7.70	105.00
110.00	144.10	53.07	0.5153	61.887	171.281	0.12466	0.31669	0.4570	0.3736	1.3496	1811.	593.	0.328	0.0261	0.0525	—	7.34	110.00
115.00	154.84	52.57	0.4779	64.179	171.839	0.12860	0.31594	0.4617	0.3815	1.3640	1763.	590.	0.319	0.0264	0.0516	—	6.98	115.00
120.00	166.17	52.06	0.4436	66.493	172.371	0.13254	0.31519	0.4667	0.3900	1.3799	1716.	587.	0.310	0.0267	0.0508	—	6.63	120.00
125.00	178.11	51.54	0.4119	68.829	172.874	0.13648	0.31443	0.4721	0.3989	1.3973	1668.	583.	0.302	0.0270	0.0499	—	6.29	125.00
130.00	190.68	51.00	0.3827	71.190	173.347	0.14042	0.31367	0.4779	0.4085	1.4165	1620.	579.	0.293	0.0273	0.0491	—	5.94	130.00
135.00	203.90	50.46	0.3557	73.577	173.787	0.14437	0.31288	0.4841	0.4189	1.4377	1572.	575.	0.285	0.0276	0.0483	—	5.60	135.00
140.00	217.79	49.90	0.3307	75.991	174.192	0.14833	0.31208	0.4909	0.4301	1.4613	1523.	571.	0.277	0.0280	0.0474	—	5.26	140.00
145.00	232.39	49.32	0.3075	78.435	174.557	0.15230	0.31126	0.4983	0.4424	1.4877	1474.	566.	0.269	0.0283	0.0466	—	4.93	145.00
150.00	247.71	48.72	0.2860	80.911	174.880	0.15628	0.31041	0.5064	0.4559	1.5174	1425.	561.	0.261	0.0287	0.0458	—	4.60	150.00
155.00	263.77	48.11	0.2660	83.421	175.156	0.16028	0.30952	0.5154	0.4708	1.5509	1375.	555.	0.254	0.0291	0.0450	—	4.28	155.00
160.00	280.61	47.48	0.2473	85.968	175.382	0.16430	0.30859	0.5254	0.4874	1.5890	1324.	550.	—	0.0296	0.0443	—	3.96	160.00
165.00	298.24	46.82	0.2299	88.555	175.551	0.16835	0.30761	0.5366	0.5062	1.6328	1273.	544.	—	0.0300	0.0435	—	3.64	165.00
170.00	316.70	46.13	0.2136	91.187	175.657	0.17243	0.30657	0.5493	0.5276	1.6834	1221.	537.	—	0.0305	0.0428	—	3.33	170.00
175.00	336.02	45.42	0.1984	93.869	175.692	0.17654	0.30547	0.5639	0.5523	1.7427	1169.	530.	—	0.0311	—	—	3.02	175.00
180.00	356.22	44.67	0.1840	96.605	175.647	0.18071	0.30427	0.5808	0.5812	1.8128	1115.	523.	—	0.0317	—	—	2.72	180.00
185.00	377.34	43.89	0.1705	99.403	175.511	0.18493	0.30299	0.6007	0.6154	1.8970	1061.	516.	—	0.0323	—	—	2.43	185.00
190.00	399.41	43.06	0.1578	102.272	175.270	0.18921	0.30158	0.6245	0.6568	1.9999	1006.	507.	—	0.0330	—	—	2.14	190.00
195.00	422.47	42.17	0.1457	105.223	174.905	0.19359	0.30002	0.6537	0.7081	2.1282	949.	499.	—	—	—	—	1.86	195.00
200.00	446.56	41.23	0.1343	108.272	174.392	0.19806	0.29829	0.6904	0.7733	2.2925	891.	490.	—	—	—	—	1.59	200.00
210.00	498.01	39.08	0.1127	114.753	172.776	0.20746	0.29410	0.8040	0.9778	2.8110	769.	469.	—	—	—	—	1.06	210.00
220.00	554.22	36.38	0.0922	122.043	169.915	0.21785	0.28829	—	—	—	636.	446.	—	—	—	—	0.59	220.00
230.00	606.14	29.31	0.0341	135.228	135.236	0.23666	0.23667	—	—	—	—	—	—	—	—	—	0.17	230.00
235.87c	655.07	22.97	0.0435	147.302	147.302	0.25359	0.25359	∞	∞	∞	0.	0.	—	—	∞	∞	0.00	235.87

*temperatures are on the ITS-90 scale   a = triple point   b = boiling point   c = critical point

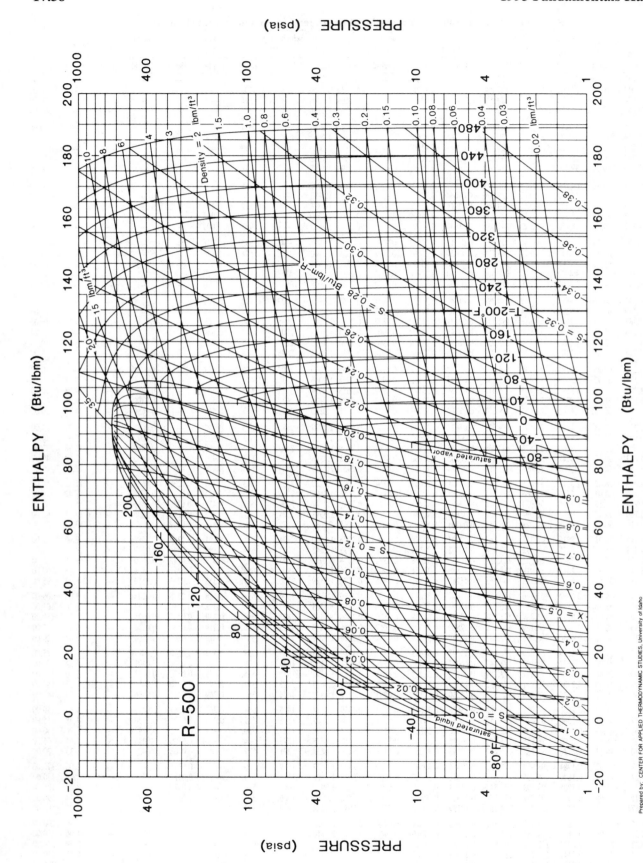

Fig. 18   Pressure-Enthalpy Diagram for Refrigerant 500

Refrigerant 500 (Azeotrope of R-12 and R-152a) Properties of Saturated Liquid and Saturated Vapor

Temp,* °F	Pressure, psia	Density, lb/ft³ Liquid	Volume, ft³/lb Vapor	Enthalpy, Btu/lb Liquid	Vapor	Entropy, Btu/lb·°F Liquid	Vapor	Specific Heat c_p, Btu/lb·°F Liquid	Vapor	c_p/c_v Vapor	Velocity of Sound, ft/s Liquid	Vapor	Viscosity, lb_m/ft·h Liquid	Vapor	Thermal Cond, Btu/h·ft·°F Liquid	Vapor	Surface Tension, dyne/cm	Temp, °F
−90.00	2.482	88.78	15.887	−10.763	81.207	−0.02723	0.22156	0.2451	0.1425	—	—	—	1.336	—	0.0643	—	—	−90.00
−80.00	3.456	87.90	11.676	−8.700	82.493	−0.02173	0.21846	0.2422	0.1456	—	—	—	1.231	—	0.0630	—	—	−80.00
−70.00	4.725	87.01	8.7298	−6.593	83.779	−0.01626	0.21566	0.2398	0.1486	—	—	—	1.137	—	0.0617	—	—	−70.00
−60.00	6.351	86.11	6.6307	−4.442	85.060	−0.01082	0.21312	0.2381	0.1517	—	—	—	1.053	—	0.0605	—	—	−60.00
−50.00	8.404	85.21	5.1094	−2.244	86.337	−0.00540	0.21083	0.2371	0.1549	—	—	—	0.978	—	0.0592	—	—	−50.00
−40.00	10.959	84.29	3.9895	0.000	87.604	0.00000	0.20875	0.2366	0.1582	—	—	—	0.911	—	0.0579	—	—	−40.00
−38.00	11.538	84.10	3.8025	0.455	87.857	0.00108	0.20835	0.2366	0.1589	—	—	—	0.898	—	0.0577	—	—	−38.00
−36.00	12.141	83.92	3.6260	0.911	88.109	0.00215	0.20797	0.2366	0.1595	—	—	—	0.886	—	0.0574	—	—	−36.00
−34.00	12.768	83.73	3.4594	1.369	88.360	0.00323	0.20759	0.2367	0.1602	—	—	—	0.874	—	0.0572	—	—	−34.00
−32.00	13.421	83.54	3.3019	1.829	88.612	0.00430	0.20722	0.2368	0.1609	—	—	—	0.862	—	0.0569	—	—	−32.00
−30.00	14.101	83.35	3.1530	2.291	88.862	0.00538	0.20686	0.2369	0.1616	—	—	—	0.850	—	0.0567	—	—	−30.00
−28.31b	14.696	83.19	3.0335	2.683	89.074	0.00629	0.20656	0.2370	0.1622	—	—	—	0.841	—	0.0565	—	—	−28.31
−28.00	14.807	83.17	3.0122	2.755	89.112	0.00645	0.20651	0.2370	0.1623	—	—	—	0.839	—	0.0564	—	—	−28.00
−26.00	15.541	82.98	2.8789	3.221	89.362	0.00753	0.20616	0.2371	0.1630	—	—	—	0.828	—	0.0562	—	—	−26.00
−24.00	16.304	82.79	2.7527	3.689	89.611	0.00860	0.20582	0.2373	0.1637	—	—	—	0.817	—	0.0559	—	—	−24.00
−22.00	17.096	82.60	2.6331	4.159	89.859	0.00967	0.20548	0.2375	0.1644	—	—	—	0.806	—	0.0557	—	—	−22.00
−20.00	17.918	82.41	2.5198	4.631	90.107	0.01074	0.20515	0.2378	0.1652	—	—	—	0.795	—	0.0554	—	—	−20.00
−18.00	18.771	82.22	2.4124	5.104	90.355	0.01181	0.20483	0.2380	0.1659	—	—	—	0.785	—	0.0552	—	—	−18.00
−16.00	19.656	82.02	2.3104	5.580	90.602	0.01288	0.20451	0.2383	0.1667	—	—	—	0.775	—	0.0550	—	—	−16.00
−14.00	20.573	81.83	2.2137	6.057	90.848	0.01395	0.20420	0.2386	0.1674	—	—	—	0.765	—	0.0547	—	—	−14.00
−12.00	21.523	81.64	2.1218	6.537	91.093	0.01502	0.20390	0.2390	0.1682	—	—	—	0.755	—	0.0545	—	—	−12.00
−10.00	22.507	81.44	2.0345	7.018	91.338	0.01609	0.20360	0.2393	0.1689	—	—	—	0.746	—	0.0542	—	—	−10.00
−8.00	23.526	81.25	1.9515	7.501	91.583	0.01716	0.20331	0.2397	0.1697	—	—	—	0.736	—	0.0540	—	—	−8.00
−6.00	24.581	81.06	1.8726	7.987	91.826	0.01822	0.20302	0.2402	0.1705	—	—	—	0.727	—	0.0537	—	—	−6.00
−4.00	25.672	80.86	1.7975	8.474	92.069	0.01929	0.20274	0.2406	0.1713	—	—	—	0.718	—	0.0535	—	—	−4.00
−2.00	26.801	80.66	1.7261	8.963	92.311	0.02035	0.20247	0.2411	0.1721	—	—	—	0.709	—	0.0532	—	—	−2.00
0.00	27.968	80.47	1.6581	9.454	92.553	0.02142	0.20220	0.2416	0.1730	—	—	—	0.701	—	0.0530	—	—	0.00
2.00	29.174	80.27	1.5933	9.947	92.794	0.02248	0.20193	0.2422	0.1738	—	—	—	0.692	—	0.0527	—	—	2.00
4.00	30.420	80.07	1.5316	10.443	93.034	0.02355	0.20167	0.2427	0.1746	—	—	—	0.684	—	0.0525	—	—	4.00
6.00	31.707	79.87	1.4727	10.940	93.273	0.02461	0.20142	0.2433	0.1755	—	—	—	0.676	—	0.0522	—	—	6.00
8.00	33.036	79.67	1.4166	11.439	93.511	0.02567	0.20116	0.2439	0.1764	—	—	—	0.667	—	0.0520	—	—	8.00
10.00	34.408	79.47	1.3631	11.940	93.749	0.02673	0.20092	0.2446	0.1773	—	—	—	0.660	—	0.0517	—	—	10.00
12.00	35.823	79.27	1.3120	12.443	93.986	0.02780	0.20068	0.2453	0.1782	—	—	—	0.652	—	0.0515	—	—	12.00
14.00	37.282	79.07	1.2632	12.948	94.221	0.02886	0.20044	0.2460	0.1791	—	—	—	0.644	—	0.0512	—	—	14.00
16.00	38.787	78.86	1.2165	13.455	94.457	0.02992	0.20021	0.2467	0.1800	—	—	—	0.637	—	0.0510	—	—	16.00
18.00	40.338	78.66	1.1720	13.964	94.691	0.03098	0.19998	0.2475	0.1809	—	—	—	0.629	—	0.0507	—	—	18.00
20.00	41.936	78.45	1.1294	14.475	94.924	0.03204	0.19976	0.2483	0.1819	—	—	—	0.622	—	0.0505	—	—	20.00
22.00	43.583	78.25	1.0887	14.988	95.157	0.03310	0.19954	0.2491	0.1828	—	—	—	0.615	—	0.0503	—	—	22.00
24.00	45.278	78.04	1.0498	15.503	95.388	0.03416	0.19932	0.2500	0.1838	—	—	—	0.608	—	0.0500	—	—	24.00
26.00	47.024	77.83	1.0125	16.020	95.619	0.03522	0.19911	0.2509	0.1848	—	—	—	0.601	—	0.0498	—	—	26.00
28.00	48.820	77.62	0.9768	16.539	95.848	0.03627	0.19890	0.2518	0.1858	—	—	—	0.595	—	0.0495	—	—	28.00
30.00	50.669	77.42	0.9426	17.060	96.077	0.03733	0.19870	0.2528	0.1868	—	—	—	0.588	—	0.0493	—	—	30.00
32.00	52.570	77.20	0.9099	17.583	96.304	0.03839	0.19850	0.2537	0.1878	—	—	—	0.581	—	0.0490	—	—	32.00
34.00	54.525	76.99	0.8785	18.109	96.531	0.03944	0.19830	0.2547	0.1889	—	—	—	0.575	—	0.0488	—	—	34.00
36.00	56.535	76.78	0.8485	18.636	96.757	0.04050	0.19811	0.2558	0.1900	—	—	—	0.569	—	0.0485	—	—	36.00
38.00	58.600	76.57	0.8196	19.165	96.981	0.04156	0.19792	0.2568	0.1910	—	—	—	0.563	—	0.0483	—	—	38.00
40.00	60.722	76.35	0.7920	19.697	97.204	0.04261	0.19773	0.2579	0.1921	—	—	—	0.557	—	0.0480	—	—	40.00
45.00	66.283	75.81	0.7276	21.034	97.758	0.04525	0.19727	0.2608	0.1950	—	—	—	0.542	—	0.0474	—	—	45.00
50.00	72.219	75.26	0.6694	22.385	98.304	0.04788	0.19684	0.2639	0.1979	—	—	—	0.528	—	0.0468	—	—	50.00
55.00	78.548	74.71	0.6168	23.749	98.843	0.05051	0.19642	0.2672	0.2009	—	—	—	0.514	—	0.0462	—	—	55.00
60.00	85.283	74.15	0.5690	25.127	99.374	0.05315	0.19602	0.2706	0.2041	—	—	—	0.501	—	0.0455	—	—	60.00
65.00	92.443	73.58	0.5255	26.519	99.895	0.05578	0.19563	0.2743	0.2074	—	—	—	0.489	—	0.0449	—	—	65.00
70.00	100.04	73.00	0.4860	27.924	100.41	0.05841	0.19526	0.2782	0.2108	—	—	—	0.477	—	0.0443	—	—	70.00
75.00	108.10	72.41	0.4499	29.344	100.91	0.06103	0.19489	0.2822	0.2143	—	—	—	0.465	—	0.0436	—	—	75.00
80.00	116.62	71.81	0.4169	30.777	101.41	0.06366	0.19454	0.2865	0.2180	—	—	—	0.454	—	0.0430	—	—	80.00
85.00	125.64	71.20	0.3867	32.225	101.89	0.06629	0.19419	0.2909	0.2218	—	—	—	0.444	—	0.0424	—	—	85.00
90.00	135.16	70.58	0.3591	33.688	102.36	0.06892	0.19385	0.2956	0.2257	—	—	—	0.433	—	0.0417	—	—	90.00
100.00	155.79	69.30	0.3103	36.661	103.27	0.07418	0.19319	0.3056	0.2341	—	—	—	0.414	—	0.0404	—	—	100.00
110.00	178.65	67.96	0.2689	39.699	104.12	0.07945	0.19253	0.3164	0.2431	—	—	—	0.396	—	0.0391	—	—	110.00
120.00	203.89	66.56	0.2335	42.807	104.90	0.08474	0.19186	0.3281	0.2527	—	—	—	0.379	—	0.0378	—	—	120.00
130.00	231.65	65.09	0.2031	45.992	105.61	0.09005	0.19116	0.3406	0.2630	—	—	—	0.364	—	0.0365	—	—	130.00
140.00	262.10	63.53	0.1768	49.263	106.23	0.09540	0.19040	0.3540	0.2741	—	—	—	—	—	0.0352	—	—	140.00
150.00	295.41	61.85	0.1539	52.633	106.75	0.10080	0.18956	—	—	—	—	—	—	—	—	—	—	150.00
160.00	331.75	60.04	0.1338	56.119	107.12	0.10629	0.18859	—	—	—	—	—	—	—	—	—	—	160.00
170.00	371.35	58.06	0.1160	59.752	107.32	0.11189	0.18744	—	—	—	—	—	—	—	—	—	—	170.00
180.00	414.45	55.84	0.1000	63.577	107.29	0.11768	0.18602	—	—	—	—	—	—	—	—	—	—	180.00
190.00	461.34	53.28	0.0855	67.676	106.94	0.12376	0.18419	—	—	—	—	—	—	—	—	—	—	190.00
200.00	512.40	50.19	0.0719	72.219	106.07	0.13036	0.18168	—	—	—	—	—	—	—	—	—	—	200.00
210.00	568.16	46.10	0.0584	77.651	104.25	0.13809	0.17781	—	—	—	—	—	—	—	—	—	—	210.00
220.00	629.50	38.52	0.0416	86.497	99.24	0.15030	0.16905	—	—	—	—	—	—	—	—	—	—	220.00
222.00c	641.9	31.10	0.0321	94.40	94.40	0.1608	0.1608	∞	∞	∞	0	0	—	—	∞	∞	0	222.00

*temperatures are on the IPTS-68 scale   b = boiling point   c = critical point

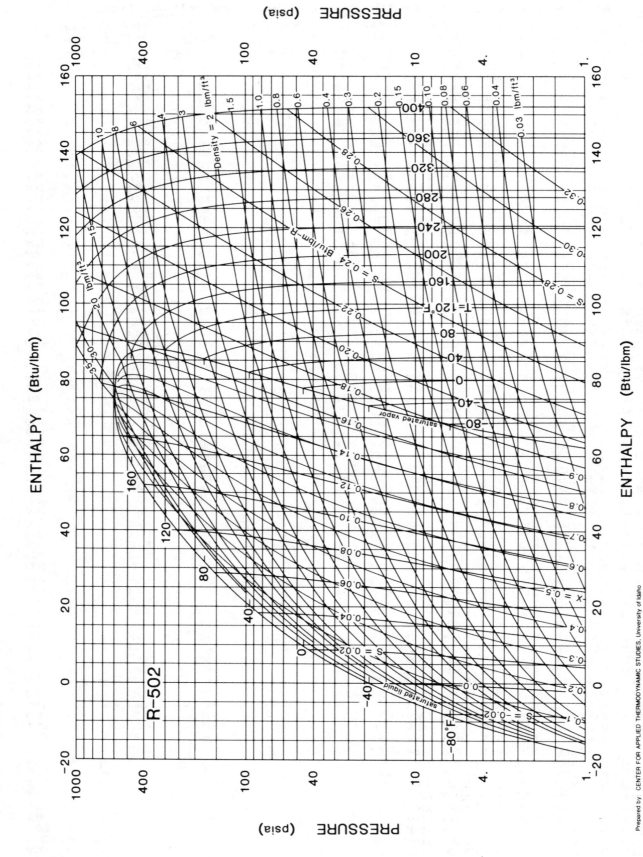

Fig. 19 Pressure-Enthalpy Diagram for Refrigerant 502

Refrigerant 502 (Azeotrope of R-22 and R-115) Properties of Saturated Liquid and Saturated Vapor

Temp,* °F	Pressure, psia	Density, lb/ft³ Liquid	Volume, ft³/lb Vapor	Enthalpy, Btu/lb Liquid	Vapor	Entropy, Btu/lb·°F Liquid	Vapor	Specific Heat c_p, Btu/lb·°F Liquid	Vapor	c_p/c_v Vapor	Velocity of Sound, ft/s Liquid	Vapor	Viscosity, lb$_m$/ft·h Liquid	Vapor	Thermal Cond, Btu/h·ft·°F Liquid	Vapor	Surface Tension, dyne/cm	Temp, °F
−70.00	8.434	94.70	4.3241	−6.564	69.570	−0.01617	0.17921	0.2506	0.1387	—	—	—	1.081	—	0.0558	—	16.89	−70.00
−65.00	9.731	94.16	3.7849	−5.511	70.174	−0.01349	0.17828	0.2518	0.1399	—	—	—	1.040	—	0.0552	—	16.46	−65.00
−60.00	11.182	93.61	3.3248	−4.441	70.777	−0.01080	0.17740	0.2532	0.1410	—	—	—	1.002	—	0.0545	—	16.03	−60.00
−55.00	12.802	93.07	2.9306	−3.355	71.377	−0.00811	0.17656	0.2545	0.1422	—	—	—	0.965	—	0.0539	—	15.61	−55.00
−50.00	14.602	92.51	2.5915	−2.252	71.975	−0.00541	0.17578	0.2559	0.1433	—	—	—	0.930	—	0.0533	—	15.19	−50.00
−49.75b	14.696	92.49	2.5761	−2.198	72.004	−0.00528	0.17574	0.2559	0.1433	—	—	—	0.929	—	0.0533	—	15.16	−49.75
−48.00	15.376	92.29	2.4693	−1.807	72.213	−0.00433	0.17547	0.2564	0.1437	—	—	—	0.917	—	0.0530	—	15.02	−48.00
−46.00	16.183	92.07	2.3541	−1.359	72.451	−0.00325	0.17518	0.2570	0.1442	—	—	—	0.904	—	0.0528	—	14.85	−46.00
−44.00	17.022	91.84	2.2452	−0.909	72.689	−0.00217	0.17489	0.2575	0.1446	—	—	—	0.891	—	0.0525	—	14.68	−44.00
−42.00	17.895	91.62	2.1425	−0.456	72.925	−0.00108	0.17461	0.2581	0.1451	—	—	—	0.879	—	0.0523	—	14.51	−42.00
−40.00	18.802	91.39	2.0453	0.000	73.162	0.00000	0.17433	0.2587	0.1456	—	—	—	0.866	—	0.0520	0.00411	14.35	−40.00
−38.00	19.746	91.17	1.9535	0.458	73.398	0.00109	0.17406	0.2592	0.1460	—	—	—	0.854	—	0.0518	0.00415	14.18	−38.00
−36.00	20.726	90.94	1.8666	0.919	73.633	0.00217	0.17380	0.2598	0.1465	—	—	—	0.843	—	0.0516	0.00420	14.01	−36.00
−34.00	21.744	90.71	1.7844	1.382	73.867	0.00326	0.17354	0.2604	0.1470	—	—	—	0.831	—	0.0513	0.00424	13.85	−34.00
−32.00	22.801	90.48	1.7066	1.848	74.101	0.00434	0.17329	0.2610	0.1474	—	—	—	0.820	—	0.0511	0.00429	13.68	−32.00
−30.00	23.897	90.25	1.6328	2.317	74.335	0.00543	0.17304	0.2616	0.1479	—	—	—	0.809	—	0.0508	0.00433	13.52	−30.00
−28.00	25.034	90.02	1.5629	2.787	74.567	0.00652	0.17280	0.2622	0.1484	—	—	—	0.798	—	0.0506	0.00438	13.35	−28.00
−26.00	26.212	89.79	1.4966	3.261	74.799	0.00761	0.17257	0.2627	0.1489	—	—	—	0.787	—	0.0503	0.00442	13.19	−26.00
−24.00	27.433	89.56	1.4336	3.737	75.031	0.00870	0.17234	0.2633	0.1494	—	—	—	0.777	—	0.0501	0.00447	13.03	−24.00
−22.00	28.697	89.32	1.3739	4.215	75.261	0.00979	0.17211	0.2639	0.1499	—	—	—	0.767	—	0.0498	0.00451	12.86	−22.00
−20.00	30.006	89.09	1.3172	4.696	75.491	0.01088	0.17189	0.2646	0.1504	—	—	—	0.757	—	0.0496	0.00455	12.70	−20.00
−18.00	31.361	88.85	1.2633	5.179	75.720	0.01197	0.17168	0.2652	0.1510	—	—	—	0.747	—	0.0494	0.00460	12.54	−18.00
−16.00	32.762	88.62	1.2120	5.665	75.948	0.01306	0.17147	0.2658	0.1515	—	—	—	0.737	—	0.0491	0.00464	12.38	−16.00
−14.00	34.211	88.38	1.1633	6.154	76.175	0.01415	0.17126	0.2664	0.1521	—	—	—	0.728	—	0.0489	0.00468	12.22	−14.00
−12.00	35.709	88.14	1.1169	6.644	76.402	0.01524	0.17106	0.2670	0.1526	—	—	—	0.719	—	0.0486	0.00473	12.05	−12.00
−10.00	37.256	87.90	1.0727	7.138	76.627	0.01633	0.17087	0.2676	0.1532	—	—	—	0.710	—	0.0484	0.00477	11.89	−10.00
−8.00	38.854	87.66	1.0307	7.633	76.852	0.01742	0.17067	0.2683	0.1538	—	—	—	0.701	—	0.0481	0.00481	11.73	−8.00
−6.00	40.504	87.42	0.9907	8.131	77.075	0.01852	0.17049	0.2689	0.1544	—	—	—	0.692	—	0.0479	0.00485	11.58	−6.00
−4.00	42.207	87.17	0.9525	8.632	77.298	0.01961	0.17030	0.2695	0.1550	—	—	—	0.684	—	0.0476	0.00490	11.42	−4.00
−2.00	43.964	86.93	0.9161	9.135	77.520	0.02070	0.17012	0.2702	0.1557	—	—	—	0.675	—	0.0474	0.00494	11.26	−2.00
0.00	45.776	86.68	0.8813	9.640	77.741	0.02180	0.16995	0.2708	0.1563	—	—	—	0.667	—	0.0472	0.00498	11.10	0.00
2.00	47.644	86.44	0.8482	10.147	77.960	0.02289	0.16978	0.2715	0.1570	—	—	—	0.659	—	0.0469	0.00502	10.94	2.00
4.00	49.569	86.19	0.8166	10.657	78.179	0.02398	0.16961	0.2721	0.1577	—	—	—	0.651	—	0.0467	0.00507	10.79	4.00
6.00	51.552	85.94	0.7864	11.169	78.397	0.02508	0.16944	0.2728	0.1584	—	—	—	0.643	—	0.0464	0.00511	10.63	6.00
8.00	53.594	85.69	0.7575	11.684	78.613	0.02617	0.16928	0.2734	0.1591	—	—	—	0.636	—	0.0462	0.00515	10.47	8.00
10.00	55.697	85.43	0.7299	12.200	78.828	0.02726	0.16912	0.2741	0.1598	—	—	—	0.628	—	0.0459	0.00519	10.32	10.00
15.00	61.226	84.80	0.6661	13.502	79.362	0.02999	0.16874	0.2758	0.1618	—	—	—	0.610	—	0.0453	0.00530	9.93	15.00
20.00	67.155	84.15	0.6088	14.818	79.887	0.03272	0.16838	0.2775	0.1638	—	—	—	0.593	—	0.0447	0.00541	9.55	20.00
25.00	73.503	83.50	0.5575	16.147	80.405	0.03545	0.16803	0.2792	0.1660	—	—	—	0.576	—	0.0441	0.00552	9.17	25.00
30.00	80.287	82.83	0.5112	17.490	80.913	0.03818	0.16770	0.2809	0.1684	—	—	—	0.561	—	0.0435	0.00562	8.79	30.00
35.00	87.523	82.16	0.4695	18.846	81.413	0.04090	0.16738	0.2827	0.1709	—	—	—	0.545	0.0285	0.0429	0.00573	8.42	35.00
40.00	95.229	81.47	0.4318	20.216	81.903	0.04362	0.16707	0.2845	0.1735	—	—	—	0.531	0.0288	0.0423	0.00584	8.05	40.00
45.00	103.42	80.77	0.3976	21.597	82.383	0.04633	0.16678	0.2863	0.1764	—	—	—	0.517	0.0292	0.0417	0.00595	7.68	45.00
50.00	112.12	80.06	0.3666	22.991	82.852	0.04904	0.16649	0.2882	0.1794	—	—	—	—	0.0296	0.0411	0.00607	—	50.00
55.00	121.34	79.33	0.3383	24.397	83.310	0.05174	0.16621	0.2900	0.1827	—	—	—	—	0.0300	0.0404	0.00618	—	55.00
60.00	131.10	78.59	0.3126	25.814	83.755	0.05444	0.16594	0.2919	0.1861	—	—	—	—	0.0304	0.0398	0.00629	—	60.00
65.00	141.42	77.83	0.2892	27.244	84.187	0.05713	0.16566	0.2938	0.1898	—	—	—	—	0.0308	0.0392	0.00641	—	65.00
70.00	152.32	77.06	0.2677	28.685	84.606	0.05982	0.16539	0.2958	0.1937	—	—	—	—	0.0312	—	0.00653	—	70.00
75.00	163.81	76.27	0.2480	30.138	85.009	0.06249	0.16512	0.2977	0.1978	—	—	—	—	0.0316	—	0.00665	—	75.00
80.00	175.92	75.46	0.2299	31.602	85.397	0.06517	0.16484	0.2997	0.2022	—	—	—	—	0.0320	—	0.00678	—	80.00
85.00	188.66	74.62	0.2133	33.078	85.767	0.06783	0.16456	0.3017	0.2068	—	—	—	—	0.0324	—	0.00690	—	85.00
90.00	202.06	73.77	0.1980	34.566	86.118	0.07049	0.16427	0.3037	0.2117	—	—	—	—	0.0328	—	0.00703	—	90.00
95.00	216.13	72.88	0.1839	36.066	86.449	0.07314	0.16397	0.3058	0.2168	—	—	—	—	0.0332	—	0.00716	—	95.00
100.00	230.89	71.97	0.1708	37.578	86.758	0.07579	0.16366	0.3079	0.2223	—	—	—	—	0.0336	—	0.00730	—	100.00
105.00	246.38	71.02	0.1587	39.104	87.042	0.07843	0.16332	—	—	—	—	—	—	—	—	—	—	105.00
110.00	262.61	70.04	0.1474	40.644	87.298	0.08107	0.16297	—	—	—	—	—	—	—	—	—	—	110.00
115.00	279.61	69.02	0.1369	42.201	87.524	0.08371	0.16258	—	—	—	—	—	—	—	—	—	—	115.00
120.00	297.41	67.96	0.1271	43.774	87.716	0.08635	0.16216	—	—	—	—	—	—	—	—	—	—	120.00
125.00	316.05	66.84	0.1179	45.369	87.869	0.08901	0.16170	—	—	—	—	—	—	—	—	—	—	125.00
130.00	335.54	65.66	0.1094	46.987	87.977	0.09167	0.16118	—	—	—	—	—	—	—	—	—	—	130.00
135.00	355.94	64.41	0.1013	48.634	88.032	0.09435	0.16060	—	—	—	—	—	—	—	—	—	—	135.00
140.00	377.30	63.08	0.0936	50.316	88.024	0.09706	0.15994	—	—	—	—	—	—	—	—	—	—	140.00
145.00	399.65	61.65	0.0863	52.045	87.941	0.09982	0.15919	—	—	—	—	—	—	—	—	—	—	145.00
150.00	423.06	60.09	0.0793	53.834	87.763	0.10265	0.15830	—	—	—	—	—	—	—	—	—	—	150.00
155.00	447.61	58.37	0.0726	55.705	87.463	0.10558	0.15724	—	—	—	—	—	—	—	—	—	—	155.00
160.00	473.29	56.43	0.0660	57.698	86.997	0.10867	0.15595	—	—	—	—	—	—	—	—	—	—	160.00
165.00	500.50	54.17	0.0595	59.878	86.293	0.11202	0.15430	—	—	—	—	—	—	—	—	—	—	165.00
170.00	529.11	51.39	0.0527	62.386	85.198	0.11584	0.15207	—	—	—	—	—	—	—	—	—	—	170.00
175.00	559.42	47.55	0.0451	65.610	83.303	0.12073	0.14861	—	—	—	—	—	—	—	—	—	—	175.00
179.90c	591.0	35.00	0.0286	74.81	74.81	0.1346	0.1346	∞	∞	∞	0	0	—	—	∞	∞	0	179.90

*temperatures are on the IPTS-68 scale   b = boiling point   c = critical point

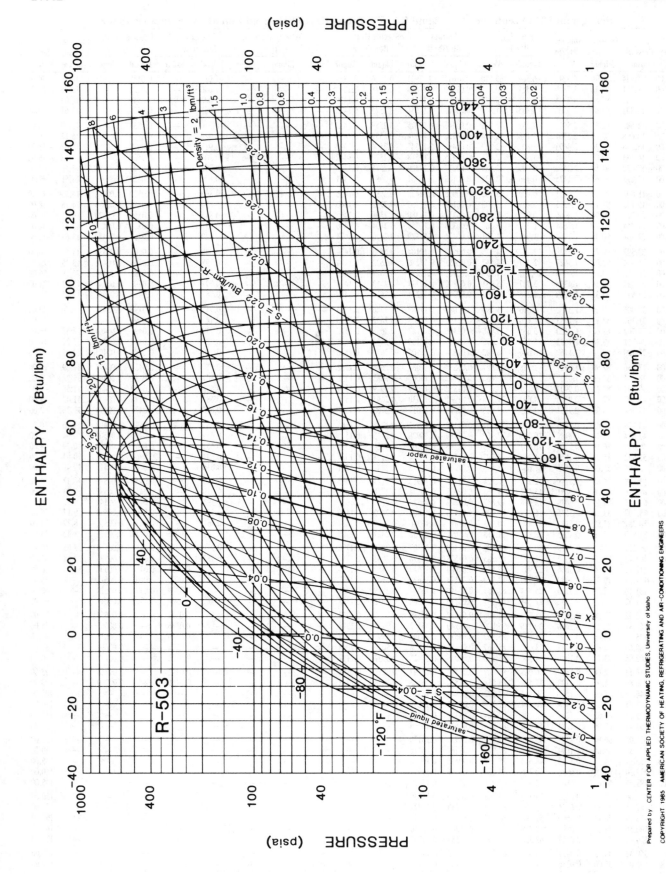

Fig. 20 Pressure-Enthalpy Diagram for Refrigerant 503

Prepared by CENTER FOR APPLIED THERMODYNAMIC STUDIES, University of Idaho

COPYRIGHT 1985 AMERICAN SOCIETY OF HEATING, REFRIGERATING AND AIR-CONDITIONING ENGINEERS

Refrigerant 503 (Azeotrope of R-23 and R-13)   Properties of Saturated Liquid and Saturated Vapor

Temp,* °F	Pressure, psia	Density, lb/ft³ Liquid	Volume, ft³/lb Vapor	Enthalpy, Btu/lb Liquid	Enthalpy, Btu/lb Vapor	Entropy, Btu/lb·°F Liquid	Entropy, Btu/lb·°F Vapor	Specific Heat c_p, Btu/lb·°F Liquid	Specific Heat c_p, Btu/lb·°F Vapor	c_p/c_v Vapor	Velocity of Sound, ft/s Liquid	Velocity of Sound, ft/s Vapor	Viscosity, lb$_m$/ft·h Liquid	Viscosity, lb$_m$/ft·h Vapor	Thermal Cond, Btu/h·ft·°F Liquid	Thermal Cond, Btu/h·ft·°F Vapor	Surface Tension, dyne/cm	Temp, °F
−190.00	1.069	97.81	30.827	−38.625	47.727	−0.11247	0.20774	—	—	—	—	—	—	—	—	—	—	−190.00
−180.00	1.768	97.04	19.265	−36.283	48.778	−0.10395	0.20020	0.1241	—	—	—	—	—	—	0.0724	—	—	−180.00
−170.00	2.809	96.23	12.506	−33.900	49.821	−0.09559	0.19344	0.1341	—	—	—	—	—	—	0.0705	—	—	−170.00
−160.00	4.307	95.37	8.3963	−31.475	50.851	−0.08737	0.18736	0.1435	0.1203	—	—	—	—	—	0.0686	—	—	−160.00
−150.00	6.396	94.46	5.8079	−29.011	51.864	−0.07929	0.18187	0.1524	0.1225	—	—	—	—	—	0.0668	—	—	−150.00
−140.00	9.227	93.49	4.1253	−26.510	52.854	−0.07136	0.17691	0.1611	0.1247	—	—	—	0.871	—	0.0649	—	—	−140.00
−138.00	9.897	93.29	3.8638	−26.006	53.050	−0.06979	0.17597	0.1628	0.1251	—	—	—	0.857	—	0.0645	—	—	−138.00
−136.00	10.605	93.09	3.6222	−25.500	53.244	−0.06823	0.17505	0.1645	0.1256	—	—	—	0.842	—	0.0642	—	—	−136.00
−134.00	11.353	92.88	3.3987	−24.993	53.437	−0.06667	0.17415	0.1662	0.1260	—	—	—	0.828	—	0.0638	—	—	−134.00
−132.00	12.141	92.68	3.1918	−24.484	53.629	−0.06512	0.17327	0.1679	0.1265	—	—	—	0.814	—	0.0634	—	—	−132.00
−130.00	12.972	92.47	3.0001	−23.975	53.819	−0.06358	0.17240	0.1697	0.1269	—	—	—	0.801	—	0.0630	—	—	−130.00
−128.00	13.847	92.26	2.8223	−23.464	54.009	−0.06204	0.17155	0.1714	0.1274	—	—	—	0.787	—	0.0627	—	—	−128.00
−126.15b	14.696	92.06	2.6693	−22.991	54.183	−0.06062	0.17077	0.1730	0.1278	—	—	—	0.775	—	0.0623	—	—	−126.15
−126.00	14.767	92.04	2.6572	−22.952	54.197	−0.06050	0.17071	0.1731	0.1279	—	—	—	0.774	—	0.0623	—	—	−126.00
−124.00	15.735	91.83	2.5037	−22.439	54.385	−0.05897	0.16989	0.1748	0.1284	—	—	—	0.761	—	0.0619	—	—	−124.00
−122.00	16.751	91.61	2.3610	−21.924	54.570	−0.05745	0.16908	0.1765	0.1289	—	—	—	0.749	—	0.0615	—	—	−122.00
−120.00	17.818	91.39	2.2281	−21.409	54.755	−0.05594	0.16829	0.1783	0.1293	—	—	—	0.737	—	0.0612	—	—	−120.00
−118.00	18.936	91.17	2.1042	−20.892	54.938	−0.05443	0.16751	0.1800	0.1298	—	—	—	0.724	—	0.0608	—	—	−118.00
−116.00	20.109	90.94	1.9887	−20.375	55.120	−0.05292	0.16675	0.1818	0.1303	—	—	—	0.713	—	0.0604	—	—	−116.00
−114.00	21.336	90.71	1.8809	−19.856	55.301	−0.05143	0.16600	0.1836	0.1309	—	—	—	0.701	—	0.0601	—	—	−114.00
−112.00	22.621	90.48	1.7801	−19.336	55.480	−0.04993	0.16526	0.1854	0.1314	—	—	—	0.689	—	0.0597	—	—	−112.00
−110.00	23.965	90.25	1.6859	−18.815	55.658	−0.04845	0.16453	0.1872	0.1319	—	—	—	0.678	—	0.0593	—	—	−110.00
−108.00	25.369	90.01	1.5977	−18.293	55.834	−0.04697	0.16382	0.1890	0.1325	—	—	—	0.667	—	0.0590	—	—	−108.00
−106.00	26.835	89.78	1.5152	−17.771	56.009	−0.04550	0.16312	0.1909	0.1330	—	—	—	0.657	—	0.0586	—	—	−106.00
−104.00	28.365	89.53	1.4378	−17.247	56.183	−0.04403	0.16243	0.1927	0.1336	—	—	—	0.646	—	0.0582	—	—	−104.00
−102.00	29.962	89.29	1.3652	−16.722	56.354	−0.04256	0.16175	0.1946	0.1341	—	—	—	0.636	—	0.0578	—	—	−102.00
−100.00	31.626	89.05	1.2971	−16.197	56.525	−0.04111	0.16108	0.1966	0.1347	—	—	—	0.626	—	0.0575	—	—	−100.00
−98.00	33.359	88.80	1.2331	−15.670	56.694	−0.03966	0.16042	0.1985	0.1353	—	—	—	0.616	—	0.0571	—	—	−98.00
−96.00	35.164	88.55	1.1729	−15.143	56.861	−0.03822	0.15978	0.2005	0.1359	—	—	—	0.606	—	0.0567	—	—	−96.00
−94.00	37.041	88.29	1.1163	−14.614	57.026	−0.03678	0.15914	0.2024	0.1365	—	—	—	0.596	—	0.0564	—	—	−94.00
−92.00	38.994	88.03	1.0630	−14.085	57.190	−0.03535	0.15851	0.2045	0.1371	—	—	—	0.587	—	0.0560	—	—	−92.00
−90.00	41.024	87.78	1.0128	−13.555	57.352	−0.03392	0.15789	0.2065	0.1378	—	—	—	0.578	—	0.0556	—	—	−90.00
−88.00	43.133	87.51	0.9654	−13.024	57.513	−0.03250	0.15728	0.2086	0.1384	—	—	—	0.569	—	0.0553	—	—	−88.00
−86.00	45.322	87.25	0.9208	−12.493	57.671	−0.03109	0.15668	0.2107	0.1391	—	—	—	0.560	—	0.0549	—	—	−86.00
−84.00	47.594	86.98	0.8786	−11.960	57.828	−0.02968	0.15609	0.2129	0.1397	—	—	—	0.551	—	0.0545	—	—	−84.00
−82.00	49.950	86.71	0.8388	−11.427	57.984	−0.02827	0.15551	0.2151	0.1404	—	—	—	0.543	—	0.0542	—	—	−82.00
−80.00	52.393	86.43	0.8012	−10.892	58.137	−0.02688	0.15494	0.2173	0.1411	—	—	—	0.535	—	0.0538	—	—	−80.00
−78.00	54.925	86.16	0.7656	−10.357	58.288	−0.02549	0.15437	0.2196	0.1419	—	—	—	0.526	—	0.0534	—	—	−78.00
−76.00	57.546	85.87	0.7320	−9.821	58.438	−0.02410	0.15381	0.2219	0.1426	—	—	—	0.518	—	0.0531	—	11.24	−76.00
−74.00	60.261	85.59	0.7001	−9.284	58.585	−0.02272	0.15326	0.2242	0.1433	—	—	—	0.511	—	0.0527	—	11.04	−74.00
−72.00	63.069	85.30	0.6699	−8.747	58.731	−0.02134	0.15271	0.2266	0.1441	—	—	—	0.503	—	0.0523	—	10.84	−72.00
−70.00	65.974	85.01	0.6412	−8.208	58.874	−0.01997	0.15218	0.2291	0.1449	—	—	—	0.495	—	0.0519	—	10.63	−70.00
−65.00	73.672	84.28	0.5759	−6.857	59.224	−0.01657	0.15086	0.2354	0.1469	—	—	—	0.477	—	0.0510	—	10.13	−65.00
−60.00	82.017	83.51	0.5184	−5.500	59.560	−0.01320	0.14958	0.2420	0.1490	—	—	—	0.460	—	0.0501	—	9.64	−60.00
−55.00	91.041	82.73	0.4677	−4.138	59.882	−0.00987	0.14834	0.2489	0.1513	—	—	—	0.443	—	0.0492	—	9.15	−55.00
−50.00	100.78	81.93	0.4228	−2.766	60.188	−0.00655	0.14712	0.2562	0.1536	—	—	—	0.428	—	0.0483	—	8.67	−50.00
−45.00	111.26	81.10	0.3830	−1.387	60.478	−0.00326	0.14593	0.2639	0.1561	—	—	—	0.413	—	0.0473	—	8.19	−45.00
−40.00	122.53	80.24	0.3475	0.000	60.751	0.00000	0.14476	0.2720	0.1588	—	—	—	0.398	—	0.0464	—	7.72	−40.00
−35.00	134.61	79.36	0.3158	1.398	61.005	0.00325	0.14361	0.2805	0.1615	—	—	—	0.385	—	0.0455	—	7.25	−35.00
−30.00	147.54	78.45	0.2873	2.808	61.238	0.00647	0.14246	0.2895	0.1644	—	—	—	0.372	—	0.0445	—	6.80	−30.00
−25.00	161.36	77.52	0.2618	4.231	61.450	0.00969	0.14133	0.2989	0.1675	—	—	—	0.360	—	0.0436	—	6.35	−25.00
−20.00	176.11	76.55	0.2388	5.670	61.638	0.01290	0.14020	0.3088	0.1707	—	—	—	0.348	—	0.0427	—	5.90	−20.00
−15.00	191.83	75.55	0.2180	7.126	61.799	0.01611	0.13906	0.3192	0.1741	—	—	—	0.336	—	0.0417	—	5.47	−15.00
−10.00	208.55	74.51	0.1992	8.601	61.932	0.01931	0.13791	0.3302	0.1776	—	—	—	0.326	—	0.0408	—	5.04	−10.00
−5.00	226.32	73.43	0.1820	10.101	62.032	0.02253	0.13674	0.3417	0.1814	—	—	—	0.315	—	0.0399	—	—	−5.00
0.00	245.19	72.31	0.1664	11.626	62.097	0.02576	0.13555	—	—	—	—	—	—	—	—	—	—	0.00
5.00	265.15	71.15	0.1522	13.183	62.121	0.02901	0.13433	—	—	—	—	—	—	—	—	—	—	5.00
10.00	286.37	69.94	0.1391	14.776	62.100	0.03230	0.13306	—	—	—	—	—	—	—	—	—	—	10.00
15.00	308.79	68.67	0.1271	16.411	62.026	0.03563	0.13173	—	—	—	—	—	—	—	—	—	—	15.00
20.00	332.49	67.33	0.1161	18.095	61.892	0.03902	0.13032	—	—	—	—	—	—	—	—	—	—	20.00
25.00	357.53	65.92	0.1058	19.838	61.687	0.04248	0.12882	—	—	—	—	—	—	—	—	—	—	25.00
30.00	383.97	64.43	0.0963	21.649	61.398	0.04603	0.12720	—	—	—	—	—	—	—	—	—	—	30.00
35.00	411.85	62.83	0.0873	23.544	61.006	0.04970	0.12543	—	—	—	—	—	—	—	—	—	—	35.00
40.00	441.25	61.10	0.0789	25.543	60.487	0.05352	0.12345	—	—	—	—	—	—	—	—	—	—	40.00
45.00	472.23	59.20	0.0710	27.673	59.804	0.05754	0.12120	—	—	—	—	—	—	—	—	—	—	45.00
50.00	504.86	57.05	0.0633	29.977	58.900	0.06182	0.11857	—	—	—	—	—	—	—	—	—	—	50.00
55.00	539.21	54.55	0.0558	32.533	57.678	0.06651	0.11537	—	—	—	—	—	—	—	—	—	—	55.00
60.00	575.35	51.35	0.0481	35.513	55.945	0.07188	0.11120	—	—	—	—	—	—	—	—	—	—	60.00
65.00	613.38	46.08	0.0397	39.596	53.255	0.07909	0.10513	—	—	—	—	—	—	—	—	—	—	65.00
67.00c	631.90	35.21	0.0284	47.41	47.41	0.0924	0.0924	∞	∞	∞	0	0	—	—	∞	∞	0	67.00

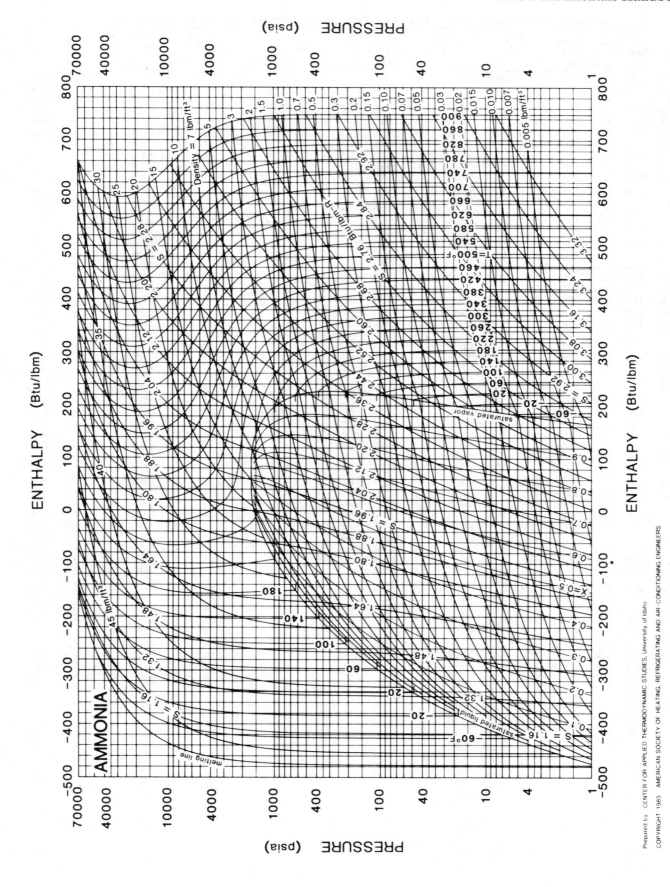

Fig. 21 Pressure-Enthalpy Diagram for Refrigerant 717 (Ammonia)
Note: the reference states for enthalpy and entropy differ from those in the table.

Prepared by CENTER FOR APPLIED THERMODYNAMIC STUDIES, University of Idaho

COPYRIGHT 1985 AMERICAN SOCIETY OF HEATING, REFRIGERATING AND AIR-CONDITIONING ENGINEERS

Refrigerant 717 (Ammonia) Properties of Saturated Liquid and Saturated Vapor

Temp,* °F	Pressure, psia	Density, lb/ft³ Liquid	Volume, ft³/lb Vapor	Enthalpy, Btu/lb Liquid	Enthalpy, Btu/lb Vapor	Entropy, Btu/lb·°F Liquid	Entropy, Btu/lb·°F Vapor	Specific Heat c_p, Btu/lb·°F Liquid	Specific Heat c_p, Btu/lb·°F Vapor	c_p/c_v Vapor	Velocity of Sound, ft/s Liquid	Velocity of Sound, ft/s Vapor	Viscosity, lb_m/ft·h Liquid	Viscosity, lb_m/ft·h Vapor	Thermal Cond, Btu/h·ft·°F Liquid	Thermal Cond, Btu/h·ft·°F Vapor	Surface Tension, dyne/cm	Temp, °F
−107.78a	0.876	45.81	252.01	−71.827	569.276	−0.18678	1.63586	—	0.4752	1.3352	—	1167.	1.224	0.0166	—	0.00742	—	−107.78
−100.00	1.230	45.51	183.31	−63.032	572.742	−0.16206	1.60621	—	0.4777	1.3362	—	1179.	1.167	0.0169	—	0.00769	—	−100.00
−90.00	1.856	45.13	124.64	−52.230	577.125	−0.13244	1.57049	—	0.4817	1.3379	—	1194.	1.076	0.0172	—	0.00802	41.94	−90.00
−80.00	2.733	44.73	86.784	−41.713	581.419	−0.10437	1.53719	—	0.4869	1.3399	—	1208.	0.983	0.0176	—	0.00836	40.67	−80.00
−70.00	3.931	44.32	61.753	−31.305	585.612	−0.07732	1.50607	—	0.4932	1.3423	—	1222.	0.897	0.0180	—	0.00869	39.41	−70.00
−60.00	5.539	43.91	44.818	−20.912	589.693	−0.05100	1.47691	—	0.5007	1.3450	—	1235.	0.820	0.0183	—	0.00902	38.16	−60.00
−50.00	7.656	43.49	33.121	−10.484	593.653	−0.02526	1.44953	—	0.5097	1.3481	—	1247.	0.753	0.0187	0.3543	0.00936	36.90	−50.00
−45.00	8.941	43.28	28.652	−5.250	595.584	−0.01257	1.43646	—	0.5146	1.3499	—	1253.	0.723	0.0189	0.3510	0.00953	36.28	−45.00
−40.00	10.396	43.07	24.885	0.000	597.482	0.00000	1.42376	1.0508	0.5200	1.3518	5046.	1259.	0.694	0.0191	0.3477	0.00971	35.65	−40.00
−35.00	12.040	42.85	21.694	5.267	599.344	0.01246	1.41144	1.0541	0.5256	1.3538	5025.	1264.	0.668	0.0193	0.3444	0.00987	35.03	−35.00
−30.00	13.890	42.64	18.981	10.551	601.171	0.02481	1.39946	1.0575	0.5317	1.3559	5006.	1269.	0.643	0.0195	0.3411	0.01004	34.41	−30.00
−27.99b	14.696	42.55	18.006	12.682	601.896	0.02975	1.39473	1.0589	0.5342	1.3568	4999.	1272.	0.634	0.0196	0.3398	0.01011	34.16	−27.99
−25.00	15.964	42.42	16.665	15.854	602.962	0.03706	1.38781	1.0609	0.5380	1.3582	4988.	1275.	0.620	0.0197	0.3379	0.01021	33.78	−25.00
−20.00	18.281	42.21	14.680	21.174	604.714	0.04921	1.37648	1.0644	0.5448	1.3607	4971.	1279.	0.598	0.0199	0.3346	0.01038	33.16	−20.00
−15.00	20.861	41.99	12.973	26.511	606.428	0.06125	1.36545	1.0679	0.5518	1.3633	4954.	1284.	0.577	0.0201	0.3313	0.01055	32.54	−15.00
−10.00	23.727	41.77	11.500	31.867	608.102	0.07320	1.35471	1.0714	0.5593	1.3661	4936.	1288.	0.558	0.0203	0.3281	0.01073	31.92	−10.00
−5.00	26.900	41.55	10.223	37.242	609.736	0.08506	1.34424	1.0750	0.5671	1.3692	4916.	1293.	0.539	0.0205	0.3248	0.01092	31.31	−5.00
0.00	30.402	41.33	9.1135	42.635	611.327	0.09682	1.33403	1.0786	0.5752	1.3724	4894.	1297.	0.521	0.0207	0.3216	0.01112	30.69	0.00
5.00	34.258	41.11	8.1464	48.046	612.876	0.10849	1.32408	1.0822	0.5837	1.3759	4870.	1300.	0.504	0.0209	0.3183	0.01133	30.07	5.00
10.00	38.492	40.88	7.3005	53.477	614.382	0.12008	1.31437	1.0860	0.5925	1.3797	4844.	1304.	0.488	0.0211	0.3151	0.01155	29.46	10.00
15.00	43.131	40.65	6.5585	58.928	615.842	0.13158	1.30488	1.0898	0.6017	1.3838	4816.	1307.	0.473	0.0213	0.3119	0.01178	28.84	15.00
20.00	48.199	40.43	5.9059	64.398	617.256	0.14299	1.29561	1.0937	0.6113	1.3881	4785.	1310.	0.458	0.0215	0.3086	0.01202	28.23	20.00
25.00	53.724	40.20	5.3303	69.888	618.623	0.15432	1.28654	1.0977	0.6212	1.3928	4751.	1313.	0.444	0.0217	0.3053	0.01227	27.62	25.00
30.00	59.734	39.96	4.8211	75.400	619.941	0.16558	1.27767	1.1018	0.6315	1.3979	4716.	1316.	0.431	0.0219	0.3021	0.01252	27.01	30.00
35.00	66.258	39.73	4.3695	80.933	621.208	0.17676	1.26898	1.1060	0.6422	1.4033	4677.	1318.	0.417	0.0221	0.2988	0.01278	26.40	35.00
40.00	73.324	39.49	3.9682	86.488	622.424	0.18787	1.26047	1.1103	0.6533	1.4092	4637.	1320.	0.405	0.0223	0.2956	0.01305	25.79	40.00
45.00	80.962	39.25	3.6105	92.066	623.585	0.19891	1.25213	1.1148	0.6648	1.4155	4594.	1322.	0.393	0.0225	0.2923	0.01333	25.18	45.00
50.00	89.204	·39.01	3.2910	97.666	624.692	0.20987	1.24394	1.1194	0.6767	1.4222	4549.	1323.	0.381	0.0227	0.2890	0.01361	24.57	50.00
55.00	98.081	38.76	3.0050	103.291	625.741	0.22077	1.23590	1.1241	0.6890	1.4295	4502.	1325.	0.370	0.0229	0.2857	0.01390	23.97	55.00
60.00	107.62	38.51	2.7484	108.941	626.731	0.23161	1.22800	1.1290	0.7018	1.4374	4452.	1326.	0.359	0.0231	0.2824	0.01419	23.36	60.00
65.00	117.87	38.26	2.5177	114.615	627.659	0.24238	1.22022	1.1341	0.7152	1.4458	4401.	1327.	0.349	0.0234	0.2791	0.01449	22.76	65.00
70.00	128.84	38.00	2.3098	120.315	628.523	0.25309	1.21257	1.1393	0.7290	1.4549	4348.	1327.	0.339	0.0236	0.2757	0.01479	22.16	70.00
75.00	140.58	37.75	2.1222	126.041	629.321	0.26374	1.20503	1.1448	0.7434	1.4647	4293.	1327.	0.329	0.0238	0.2724	0.01510	21.56	75.00
80.00	153.12	37.49	1.9525	131.795	630.051	0.27434	1.19759	1.1504	0.7584	1.4752	4236.	1327.	0.320	0.0240	0.2690	0.01541	20.96	80.00
85.00	166.50	37.22	1.7987	137.577	630.709	0.28488	1.19024	1.1563	0.7740	1.4865	4178.	1327.	0.310	0.0242	0.2657	0.01572	20.36	85.00
90.00	180.74	36.95	1.6591	143.389	631.292	0.29537	1.18298	1.1625	0.7903	1.4987	4118.	1327.	0.302	0.0244	0.2623	0.01604	19.76	90.00
95.00	195.90	36.68	1.5321	149.232	631.799	0.30581	1.17580	1.1689	0.8074	1.5117	4057.	1326.	0.293	0.0246	0.2589	0.01636	19.17	95.00
100.00	211.99	36.40	1.4165	155.106	632.224	0.31621	1.16868	1.1757	0.8253	1.5258	3994.	1325.	0.285	0.0249	0.2555	0.01669	18.57	100.00
105.00	229.07	36.12	1.3109	161.013	632.567	0.32656	1.16163	1.1829	0.8441	1.5410	3931.	1324.	0.277	0.0251	0.2520	0.01703	17.98	105.00
110.00	247.17	35.84	1.2144	166.955	632.821	0.33687	1.15462	1.1905	0.8639	1.5574	3865.	1322.	0.270	0.0253	0.2486	0.01738	17.39	110.00
115.00	266.32	35.55	1.1261	172.933	632.985	0.34715	1.14766	1.1986	0.8847	1.5750	3799.	1320.	0.262	0.0256	0.2451	0.01773	16.80	115.00
120.00	286.57	35.26	1.0451	178.950	633.053	0.35739	1.14073	1.2072	0.9067	1.5941	3731.	1318.	0.255	0.0258	0.2417	0.01809	16.22	120.00
125.00	307.96	34.96	0.9708	185.007	633.023	0.36760	1.13383	1.2165	0.9299	1.6146	3663.	1315.	0.248	0.0260	0.2382	0.01846	15.63	125.00
130.00	330.52	34.66	0.9024	191.107	632.888	0.37778	1.12694	1.2264	0.9546	1.6368	3593.	1312.	0.241	0.0263	0.2347	0.01885	15.05	130.00
135.00	354.30	34.35	0.8394	197.253	632.646	0.38795	1.12006	1.2372	0.9808	1.6609	3523.	1309.	0.235	0.0265	0.2311	0.01925	14.46	135.00
140.00	379.34	34.03	0.7813	203.447	632.289	0.39809	1.11318	1.2489	1.0088	1.6869	3451.	1306.	0.229	0.0268	0.2276	0.01966	13.88	140.00
145.00	405.67	33.71	0.7277	209.694	631.813	0.40823	1.10628	1.2616	1.0387	1.7152	3378.	1302.	0.222	0.0271	0.2240	0.02009	13.31	145.00
150.00	433.35	33.39	0.6781	215.997	631.213	0.41836	1.09937	1.2754	1.0707	1.7460	3305.	1298.	0.216	0.0274	0.2204	0.02054	12.73	150.00
155.00	462.42	33.05	0.6322	222.361	630.480	0.42849	1.09241	1.2907	1.1051	1.7796	3230.	1293.	0.211	0.0276	0.2167	0.02102	12.16	155.00
160.00	492.92	32.71	0.5896	228.792	629.609	0.43863	1.08541	1.3074	1.1423	1.8162	3154.	1288.	0.205	0.0279	0.2130	0.02151	11.58	160.00
165.00	524.91	32.36	0.5500	235.294	628.590	0.44878	1.07836	1.3259	1.1826	1.8564	3078.	1283.	0.199	0.0282	0.2093	0.02204	11.01	165.00
170.00	558.42	32.01	0.5133	241.874	627.416	0.45897	1.07123	1.3464	1.2264	1.9007	3000.	1277.	0.194	0.0286	0.2056	0.02259	10.45	170.00
175.00	593.51	31.64	0.4790	248.540	626.075	0.46919	1.06401	1.3692	1.2743	1.9495	2922.	1271.	0.189	0.0289	0.2018	0.02318	9.88	175.00
180.00	630.22	31.26	0.4471	255.298	624.558	0.47946	1.05669	1.3946	1.3268	2.0037	2842.	1264.	0.184	0.0293	0.1979	0.02380	9.32	180.00
185.00	668.61	30.87	0.4173	262.159	622.851	0.48978	1.04925	1.4232	1.3848	2.0640	2761.	1257.	0.178	0.0296	0.1940	0.02446	8.76	185.00
190.00	708.73	30.47	0.3894	269.133	620.938	0.50018	1.04167	1.4555	1.4493	2.1317	2679.	1249.	0.173	0.0300	0.1900	0.02517	8.20	190.00
195.00	750.64	30.06	0.3632	276.232	618.804	0.51067	1.03393	1.4921	1.5214	2.2081	2595.	1240.	0.168	0.0305	0.1860	0.02593	7.65	195.00
200.00	794.38	29.63	0.3387	283.469	616.428	0.52127	1.02599	1.5339	1.6028	2.2950	2511.	1231.	0.163	0.0309	0.1819	0.02674	7.10	200.00
205.00	840.03	29.18	0.3157	290.862	613.785	0.53200	1.01782	1.5821	1.6955	2.3947	2424.	1222.	0.159	0.0314	0.1777	0.02762	6.55	205.00
210.00	887.64	28.71	0.2940	298.429	610.846	0.54289	1.00939	1.6382	1.8022	2.5104	2336.	1211.	0.154	0.0320	0.1734	0.02857	6.00	210.00
215.00	937.28	28.23	0.2735	306.194	607.577	0.55396	1.00065	1.7041	1.9265	2.6462	2247.	1200.	0.149	0.0326	0.1689	0.02960	5.46	215.00
220.00	989.03	27.71	0.2541	314.185	603.934	0.56525	0.99154	1.7826	2.0737	2.8080	2155.	1188.	0.144	0.0332	0.1644	0.03073	4.93	220.00
225.00	1043.0	27.17	0.2357	322.439	599.861	0.57682	0.98199	1.8778	2.2509	3.0042	2061.	1175.	0.139	0.0339	0.1597	0.03196	4.40	225.00
230.00	1099.1	26.59	0.2182	330.998	595.287	0.58871	0.97190	1.9956	2.4691	3.2471	1966.	1162.	0.134	0.0348	—	—	3.87	230.00
235.00	1157.7	25.98	0.2015	339.924	590.119	0.60101	0.96115	2.1451	2.7450	3.5558	1867.	1147.	0.129	0.0357	—	—	3.35	235.00
240.00	1218.6	25.31	0.1854	349.296	584.228	0.61382	0.94957	2.3415	3.1056	3.9616	1766.	1130.	0.124	0.0368	—	—	2.84	240.00
245.00	1282.1	24.58	0.1698	359.229	577.434	0.62729	0.93692	2.6113	3.5985	4.5184	1662.	1113.	0.119	0.0381	—	—	2.33	245.00
250.00	1348.3	23.76	0.1545	369.899	569.465	0.64166	0.92285	3.0058	4.3134	5.3293	1555.	1094.	0.114	0.0396	—	—	1.83	250.00
255.00	1417.3	22.83	0.1393	381.593	559.887	0.65731	0.90677	3.6381	5.4427	6.6150	1443.	1073.	0.108	0.0416	—	—	1.34	255.00
260.00	1489.4	21.73	0.1239	394.854	547.928	0.67496	0.88765	4.8147	7.4812	8.9443	1327.	1049.	0.101	0.0443	—	—	0.87	260.00
269.99c	1643.7	14.67	0.0682	467.152	467.152	0.77240	0.77240	∞	∞	∞	0.	0.	—	—	∞	∞	0.00	269.99

*temperatures have been converted from the IPTS-68 scale of the original formulation to the ITS-90 scale a = triple point b = boiling point c = critical point

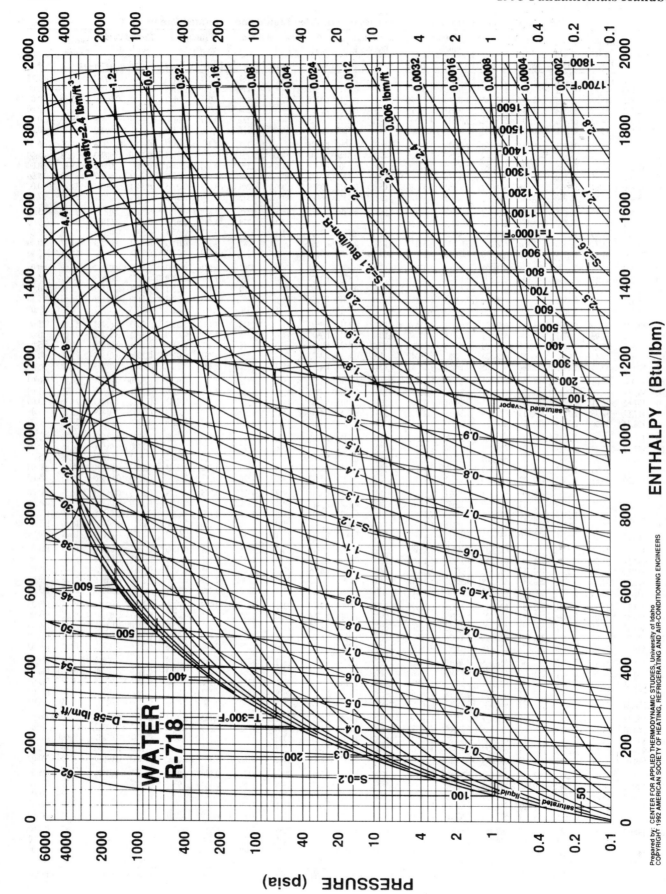

ENTHALPY (Btu/lbm)

PRESSURE (psia)

Fig. 22 Pressure-Enthalpy Diagram for Refrigerant 718 (Water/Steam)

Refrigerant 718 (Water/Steam) Properties of Saturated Liquid and Saturated Vapor

Temp,* °F	Pressure, psia	Density, lb/ft³ Liquid	Volume, ft³/lb Vapor	Enthalpy, Btu/lb Liquid	Enthalpy, Btu/lb Vapor	Entropy, Btu/lb·°F Liquid	Entropy, Btu/lb·°F Vapor	Specific Heat c_p, Btu/lb·°F Liquid	Specific Heat c_p, Btu/lb·°F Vapor	c_p/c_v Vapor	Velocity of Sound, ft/s Liquid	Velocity of Sound, ft/s Vapor	Viscosity, lb_m/ft·h Liquid	Viscosity, lb_m/ft·h Vapor	Thermal Cond, Btu/h·ft·°F Liquid	Thermal Cond, Btu/h·ft·°F Vapor	Surface Tension, dyne/cm	Temp, °F
32.02a	0.089	62.41	3299.6	0.0	1075.0	0.0000	2.1864	1.0100	0.4461	1.3305	4596.	1343.	4.336	0.0223	0.3241	0.00986	75.65	32.02
40.00	0.122	62.42	2443.5	8.0	1078.5	0.0162	2.1586	1.0037	0.4467	1.3303	4670.	1354.	3.740	0.0226	0.3290	0.01000	75.02	40.00
50.00	0.178	62.41	1703.1	18.1	1082.9	0.0361	2.1254	1.0003	0.4476	1.3302	4750.	1367.	3.161	0.0229	0.3351	0.01018	74.22	50.00
60.00	0.256	62.37	1206.3	28.0	1087.3	0.0555	2.0938	0.9993	0.4486	1.3301	4818.	1380.	2.713	0.0232	0.3411	0.01037	73.41	60.00
70.00	0.363	62.30	867.45	38.0	1091.7	0.0745	2.0637	0.9992	0.4497	1.3299	4876.	1393.	2.360	0.0236	0.3469	0.01057	72.57	70.00
80.00	0.507	62.21	632.67	48.0	1096.0	0.0932	2.0351	0.9992	0.4511	1.3299	4926.	1405.	2.075	0.0240	0.3524	0.01078	71.72	80.00
90.00	0.699	62.11	467.66	58.0	1100.3	0.1116	2.0078	0.9991	0.4525	1.3298	4967.	1418.	1.842	0.0244	0.3576	0.01100	70.85	90.00
100.00	0.950	61.99	350.05	68.0	1104.6	0.1296	1.9818	0.9990	0.4542	1.3299	5002.	1430.	1.648	0.0248	0.3625	0.01123	69.96	100.00
110.00	1.276	61.86	265.16	78.0	1108.9	0.1473	1.9569	0.9989	0.4560	1.3300	5030.	1442.	1.486	0.0252	0.3670	0.01147	69.06	110.00
120.00	1.694	61.71	203.11	88.0	1113.2	0.1646	1.9332	0.9988	0.4580	1.3303	5053.	1454.	1.348	0.0256	0.3710	0.01171	68.13	120.00
130.00	2.224	61.55	157.23	98.0	1117.4	0.1817	1.9105	0.9988	0.4602	1.3305	5070.	1465.	1.230	0.0260	0.3747	0.01197	67.20	130.00
140.00	2.891	61.38	122.93	108.0	1121.6	0.1985	1.8888	0.9990	0.4627	1.3310	5083.	1476.	1.129	0.0264	0.3781	0.01224	66.24	140.00
150.00	3.720	61.19	97.021	118.0	1125.7	0.2150	1.8680	0.9995	0.4654	1.3315	5091.	1488.	1.040	0.0269	0.3810	0.01252	65.27	150.00
160.00	4.743	61.00	77.262	128.0	1129.8	0.2313	1.8481	1.0002	0.4683	1.3322	5096.	1498.	0.963	0.0273	0.3836	0.01281	64.29	160.00
170.00	5.994	60.80	62.046	138.0	1133.9	0.2473	1.8290	1.0011	0.4715	1.3332	5096.	1509.	0.894	0.0278	0.3859	0.01311	63.29	170.00
180.00	7.513	60.58	50.224	148.0	1137.9	0.2631	1.8107	1.0023	0.4750	1.3343	5092.	1520.	0.834	0.0282	0.3879	0.01342	62.27	180.00
190.00	9.341	60.36	40.962	158.0	1141.9	0.2787	1.7930	1.0036	0.4788	1.3356	5085.	1530.	0.781	0.0287	0.3896	0.01374	61.24	190.00
200.00	11.526	60.12	33.646	168.1	1145.8	0.2940	1.7761	1.0052	0.4829	1.3372	5075.	1540.	0.733	0.0291	0.3910	0.01408	60.19	200.00
210.00	14.122	59.88	27.826	178.2	1149.6	0.3091	1.7598	1.0069	0.4873	1.3390	5061.	1549.	0.690	0.0296	0.3922	0.01443	59.13	210.00
212.00b	14.695	59.83	26.809	180.2	1150.4	0.3121	1.7566	1.0072	0.4882	1.3394	5058.	1551.	0.682	0.0297	0.3924	0.01450	58.92	212.00
220.00	17.184	59.63	23.159	188.2	1153.4	0.3241	1.7441	1.0087	0.4921	1.3411	5044.	1558.	0.651	0.0300	0.3931	0.01479	58.06	220.00
230.00	20.774	59.37	19.393	198.3	1157.0	0.3388	1.7289	1.0108	0.4973	1.3435	5025.	1568.	0.616	0.0305	0.3939	0.01516	56.97	230.00
240.00	24.960	59.10	16.332	208.5	1160.6	0.3534	1.7143	1.0129	0.5029	1.3463	5002.	1576.	0.585	0.0310	0.3944	0.01555	55.87	240.00
250.00	29.814	58.82	13.830	218.6	1164.1	0.3678	1.7001	1.0153	0.5089	1.3495	4976.	1585.	0.556	0.0314	0.3948	0.01595	54.75	250.00
260.00	35.411	58.54	11.772	228.8	1167.6	0.3820	1.6864	1.0177	0.5155	1.3530	4948.	1593.	0.530	0.0319	0.3950	0.01636	53.62	260.00
270.00	41.835	58.24	10.069	239.0	1170.9	0.3960	1.6732	1.0203	0.5225	1.3570	4918.	1600.	0.506	0.0324	0.3950	0.01679	52.48	270.00
280.00	49.173	57.94	8.6520	249.2	1174.1	0.4099	1.6603	1.0231	0.5301	1.3614	4884.	1608.	0.484	0.0328	0.3949	0.01723	51.33	280.00
290.00	57.516	57.63	7.4677	259.5	1177.2	0.4237	1.6478	1.0261	0.5383	1.3663	4849.	1615.	0.464	0.0333	0.3946	0.01769	50.16	290.00
300.00	66.963	57.31	6.4725	269.8	1180.2	0.4373	1.6356	1.0292	0.5472	1.3718	4811.	1621.	0.445	0.0338	0.3942	0.01816	48.99	300.00
310.00	77.615	56.99	5.6322	280.1	1183.0	0.4507	1.6238	1.0326	0.5567	1.3778	4771.	1628.	0.428	0.0342	0.3936	0.01864	47.80	310.00
320.00	89.580	56.65	4.9191	290.5	1185.7	0.4641	1.6123	1.0362	0.5670	1.3845	4728.	1633.	0.412	0.0347	0.3929	0.01914	46.60	320.00
330.00	102.97	56.31	4.3118	300.9	1188.3	0.4773	1.6010	1.0401	0.5780	1.3917	4684.	1639.	0.397	0.0351	0.3920	0.01965	45.39	330.00
340.00	117.90	55.96	3.7920	311.3	1190.7	0.4904	1.5900	1.0443	0.5899	1.3997	4637.	1644.	0.383	0.0356	0.3910	0.02018	44.17	340.00
350.00	134.50	55.60	3.3457	321.8	1193.0	0.5033	1.5792	1.0488	0.6028	1.4084	4588.	1648.	0.371	0.0361	0.3898	0.02072	42.94	350.00
360.00	152.89	55.23	2.9608	332.4	1195.1	0.5162	1.5687	1.0537	0.6166	1.4180	4537.	1652.	0.359	0.0365	0.3885	0.02128	41.70	360.00
370.00	173.20	54.85	2.6276	342.9	1197.0	0.5289	1.5583	1.0590	0.6314	1.4284	4484.	1656.	0.347	0.0370	0.3871	0.02185	40.46	370.00
380.00	195.57	54.46	2.3382	353.6	1198.7	0.5416	1.5481	1.0647	0.6473	1.4398	4430.	1658.	0.337	0.0375	0.3854	0.02244	39.20	380.00
390.00	220.14	54.06	2.0859	364.3	1200.3	0.5542	1.5381	1.0709	0.6645	1.4522	4373.	1661.	0.327	0.0379	0.3837	0.02305	37.94	390.00
400.00	247.05	53.66	1.8654	375.1	1201.6	0.5667	1.5282	1.0776	0.6830	1.4657	4314.	1663.	0.318	0.0384	0.3817	0.02367	36.67	400.00
410.00	276.45	53.24	1.6720	385.9	1202.8	0.5791	1.5184	1.0849	0.7028	1.4804	4253.	1664.	0.309	0.0389	0.3796	0.02431	35.39	410.00
420.00	308.50	52.81	1.5017	396.8	1203.7	0.5915	1.5087	1.0927	0.7242	1.4965	4190.	1665.	0.300	0.0393	0.3774	0.02497	34.11	420.00
430.00	343.36	52.37	1.3515	407.8	1204.5	0.6038	1.4992	1.1013	0.7473	1.5140	4126.	1665.	0.293	0.0398	0.3749	0.02565	32.82	430.00
440.00	381.18	51.92	1.2186	418.9	1204.9	0.6160	1.4897	1.1106	0.7722	1.5332	4059.	1664.	0.285	0.0403	0.3723	0.02635	31.52	440.00
450.00	422.14	51.46	1.1006	430.1	1205.2	0.6282	1.4802	1.1208	0.7991	1.5541	3990.	1663.	0.278	0.0407	0.3695	0.02708	30.23	450.00
460.00	466.41	50.98	0.9957	441.4	1205.1	0.6403	1.4708	1.1319	0.8281	1.5770	3920.	1662.	0.271	0.0412	0.3664	0.02783	28.93	460.00
470.00	514.15	50.49	0.9020	452.8	1204.8	0.6525	1.4614	1.1440	0.8596	1.6022	3847.	1659.	0.264	0.0417	0.3632	0.02861	27.62	470.00
480.00	565.56	49.99	0.8183	464.3	1204.3	0.6646	1.4521	1.1572	0.8938	1.6299	3772.	1656.	0.258	0.0422	0.3597	0.02943	26.31	480.00
490.00	620.82	49.47	0.7433	475.9	1203.4	0.6767	1.4427	1.1717	0.9311	1.6604	3695.	1652.	0.252	0.0427	0.3560	0.03029	25.01	490.00
500.00	680.11	48.93	0.6759	487.7	1202.2	0.6888	1.4333	1.1877	0.9717	1.6942	3616.	1647.	0.246	0.0432	0.3521	0.03119	23.70	500.00
510.00	743.64	48.38	0.6152	499.6	1200.6	0.7009	1.4238	1.2053	1.0163	1.7317	3534.	1641.	0.240	0.0438	0.3480	0.03215	22.39	510.00
520.00	811.60	47.81	0.5603	511.7	1198.7	0.7130	1.4143	1.2249	1.0653	1.7735	3450.	1634.	0.235	0.0443	0.3435	0.03317	21.09	520.00
530.00	884.21	47.22	0.5107	524.0	1196.4	0.7252	1.4047	1.2466	1.1196	1.8205	3364.	1627.	0.229	0.0449	0.3388	0.03427	19.78	530.00
540.00	961.69	46.60	0.4657	536.4	1193.7	0.7374	1.3949	1.2709	1.1798	1.8732	3275.	1618.	0.224	0.0455	0.3339	0.03546	18.48	540.00
550.00	1044.2	45.97	0.4248	549.1	1190.5	0.7497	1.3850	1.2982	1.2473	1.9332	3184.	1609.	0.219	0.0461	0.3287	0.03676	17.19	550.00
560.00	1132.1	45.31	0.3875	562.0	1186.9	0.7621	1.3749	1.3292	1.3233	2.0016	3089.	1598.	0.214	0.0467	0.3233	0.03820	15.90	560.00
570.00	1225.5	44.62	0.3535	575.2	1182.7	0.7745	1.3645	1.3646	1.4097	2.0803	2992.	1586.	0.209	0.0474	0.3176	0.03980	14.63	570.00
580.00	1324.8	43.89	0.3224	588.6	1177.9	0.7872	1.3539	1.4054	1.5089	2.1718	2891.	1573.	0.204	0.0481	0.3117	0.04162	13.36	580.00
590.00	1430.1	43.13	0.2938	602.4	1172.4	0.7999	1.3429	1.4531	1.6240	2.2791	2787.	1559.	0.199	0.0489	0.3056	0.04369	12.10	590.00
600.00	1541.7	42.34	0.2675	616.6	1166.2	0.8129	1.3316	1.5094	1.7594	2.4067	2679.	1543.	0.194	0.0497	0.2994	0.04609	10.86	600.00
610.00	1660.0	41.49	0.2432	631.2	1159.2	0.8261	1.3197	1.5772	1.9212	2.5606	2567.	1526.	0.189	0.0506	0.2930	0.04890	9.63	610.00
620.00	1785.3	40.59	0.2207	646.3	1151.2	0.8396	1.3073	1.6603	2.1184	2.7500	2451.	1507.	0.184	0.0516	0.2866	0.05225	8.43	620.00
630.00	1917.9	39.62	0.1998	662.0	1142.1	0.8536	1.2941	1.7650	2.3644	2.9882	2331.	1486.	0.178	0.0527	0.2800	0.05631	7.24	630.00
640.00	2058.2	38.57	0.1803	678.4	1131.6	0.8679	1.2801	1.9010	2.6805	3.2965	2204.	1462.	0.173	0.0540	0.2734	0.06133	6.09	640.00
650.00	2206.6	37.43	0.1619	695.8	1119.5	0.8830	1.2648	2.0855	3.1030	3.7114	2072.	1436.	0.167	0.0555	0.2667	0.06765	4.97	650.00
660.00	2363.6	36.15	0.1445	714.3	1105.2	0.8989	1.2481	2.3508	3.6981	4.2993	1934.	1406.	0.161	0.0572	0.2600	0.07586	3.89	660.00
670.00	2529.7	34.70	0.1279	734.5	1088.2	0.9161	1.2292	2.7658	4.6038	5.1991	1786.	1372.	0.154	0.0594	0.2532	0.08693	2.85	670.00
680.00	2705.6	32.97	0.1115	757.1	1067.1	0.9352	1.2072	3.5086	6.1610	6.7546	1629.	1332.	0.146	0.0622	0.2468	0.10285	1.89	680.00
690.00	2892.1	30.78	0.0948	784.0	1039.1	0.9577	1.1796	5.2151	9.5419	10.1841	1459.	1284.	0.137	0.0662	0.2424	0.12873	1.00	690.00
700.00	3090.8	27.42	0.0756	821.9	994.3	0.9894	1.1381	13.7960	25.9020	26.2094	1245.	1209.	0.123	0.0736	0.2543	0.19639	0.26	700.00
705.18c	3199.2	20.10	0.0497	896.8	896.8	1.0531	1.0531	∞	∞	∞	0.	0.	0.104	0.1043	∞	∞	0.00	705.02

*temperatures are on the IPTS-68 scale a = triple point b = boiling point c = critical point

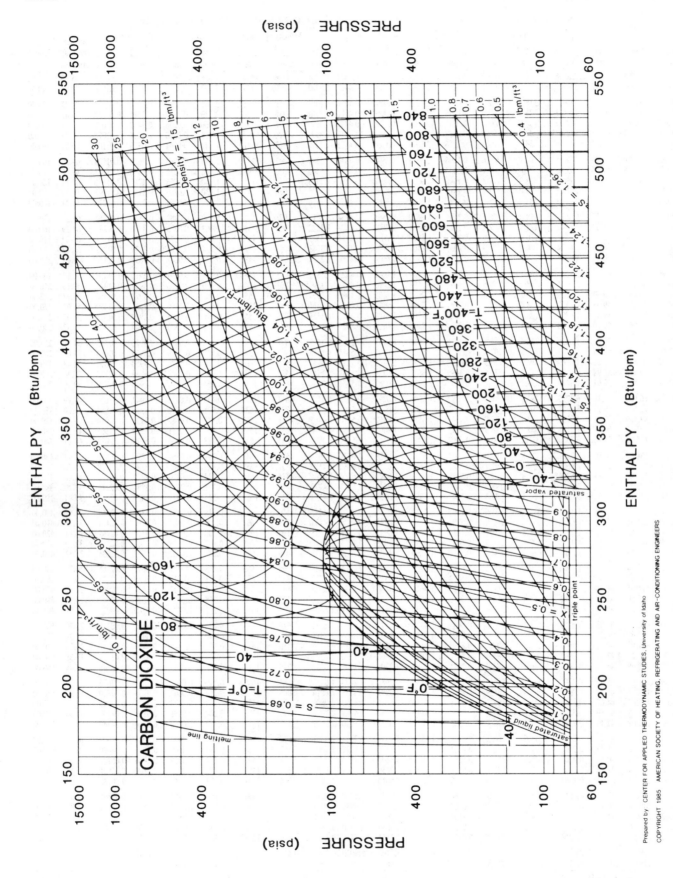

Fig. 23 Pressure-Enthalpy Diagram for Refrigerant 744 (Carbon Dioxide)
Note: the reference states for enthalpy and entropy differ from those in the table.

Prepared by CENTER FOR APPLIED THERMODYNAMIC STUDIES, University of Idaho

COPYRIGHT 1985   AMERICAN SOCIETY OF HEATING, REFRIGERATING AND AIR-CONDITIONING ENGINEERS

Refrigerant 744 (Carbon Dioxide) Properties of Saturated Liquid and Saturated Vapor

Temp,* °F	Pressure, psia	Density, lb/ft³ Liquid	Volume, ft³/lb Vapor	Enthalpy, Btu/lb Liquid	Enthalpy, Btu/lb Vapor	Entropy, Btu/lb·°F Liquid	Entropy, Btu/lb·°F Vapor	Specific Heat c_p, Btu/lb·°F Liquid	Specific Heat c_p, Btu/lb·°F Vapor	c_p/c_v Vapor	Velocity of Sound, ft/s Liquid	Velocity of Sound, ft/s Vapor	Viscosity, lb_m/ft·h Liquid	Viscosity, lb_m/ft·h Vapor	Thermal Cond, Btu/h·ft·°F Liquid	Thermal Cond, Btu/h·ft·°F Vapor	Surface Tension, dyne/cm	Temp, °F
−69.83a	75.138	73.56	1.1640	−14.225	136.515	−0.03471	0.35196	0.4764	0.2052	1.4669	3224.	737.	0.625	0.0267	0.1044	0.00654	17.09	−69.83
−65.00	84.296	72.95	1.0427	−11.922	136.902	−0.02890	0.34819	0.4751	0.2106	1.4750	3164.	737.	0.597	0.0271	0.1024	0.00667	16.41	−65.00
−60.00	94.643	72.30	0.9328	−9.541	137.274	−0.02297	0.34437	0.4745	0.2166	1.4845	3101.	738.	0.571	0.0274	0.1004	0.00681	15.72	−60.00
−55.00	105.91	71.65	0.8368	−7.161	137.616	−0.01712	0.34065	0.4745	0.2228	1.4951	3037.	738.	0.545	0.0277	0.0984	0.00694	15.03	−55.00
−50.00	118.16	70.99	0.7525	−4.780	137.927	−0.01135	0.33699	0.4752	0.2293	1.5069	2972.	738.	0.521	0.0281	0.0963	0.00708	14.35	−50.00
−48.00	123.34	70.73	0.7216	−3.826	138.041	−0.00906	0.33555	0.4757	0.2321	1.5121	2946.	737.	0.512	0.0282	0.0955	0.00714	14.08	−48.00
−46.00	128.69	70.46	0.6923	−2.871	138.151	−0.00678	0.33412	0.4762	0.2348	1.5174	2920.	737.	0.503	0.0283	0.0947	0.00719	13.81	−46.00
−44.00	134.21	70.20	0.6644	−1.915	138.255	−0.00451	0.33270	0.4769	0.2377	1.5230	2893.	737.	0.494	0.0285	0.0939	0.00725	13.54	−44.00
−42.00	139.90	69.93	0.6379	−0.958	138.353	−0.00225	0.33129	0.4776	0.2406	1.5288	2867.	737.	0.485	0.0286	0.0931	0.00731	13.28	−42.00
−40.00	145.77	69.65	0.6126	0.000	138.445	0.00000	0.32989	0.4784	0.2436	1.5349	2840.	737.	0.477	0.0287	0.0923	0.00737	13.01	−40.00
−38.00	151.82	69.38	0.5885	0.961	138.532	0.00224	0.32850	0.4794	0.2466	1.5413	2814.	736.	0.469	0.0289	0.0915	0.00743	12.75	−38.00
−36.00	158.06	69.11	0.5655	1.923	138.612	0.00448	0.32711	0.4804	0.2498	1.5479	2787.	736.	0.460	0.0290	0.0907	0.00749	12.48	−36.00
−34.00	164.48	68.83	0.5436	2.887	138.686	0.00671	0.32574	0.4816	0.2530	1.5549	2760.	735.	0.453	0.0291	0.0899	0.00756	12.22	−34.00
−32.00	171.09	68.55	0.5227	3.853	138.754	0.00893	0.32437	0.4828	0.2563	1.5621	2732.	735.	0.445	0.0293	0.0891	0.00762	11.96	−32.00
−30.00	177.90	68.27	0.5027	4.822	138.816	0.01115	0.32300	0.4841	0.2596	1.5697	2705.	734.	0.437	0.0294	0.0883	0.00769	11.70	−30.00
−28.00	184.90	67.99	0.4836	5.793	138.871	0.01336	0.32165	0.4856	0.2631	1.5776	2678.	734.	0.430	0.0295	0.0875	0.00776	11.45	−28.00
−26.00	192.11	67.71	0.4654	6.767	138.919	0.01557	0.32030	0.4871	0.2667	1.5859	2650.	733.	0.422	0.0297	0.0867	0.00783	11.19	−26.00
−24.00	199.52	67.42	0.4480	7.743	138.960	0.01777	0.31895	0.4888	0.2703	1.5945	2622.	733.	0.415	0.0298	0.0859	0.00790	10.93	−24.00
−22.00	207.14	67.13	0.4313	8.723	138.995	0.01996	0.31761	0.4905	0.2741	1.6036	2594.	732.	0.408	0.0300	0.0851	0.00797	10.68	−22.00
−20.00	214.97	66.84	0.4154	9.706	139.021	0.02215	0.31627	0.4924	0.2779	1.6130	2566.	731.	0.401	0.0301	0.0843	0.00805	10.43	−20.00
−18.00	223.02	66.54	0.4001	10.693	139.041	0.02434	0.31494	0.4944	0.2819	1.6229	2538.	730.	0.394	0.0302	0.0835	0.00813	10.18	−18.00
−16.00	231.29	66.25	0.3855	11.683	139.053	0.02653	0.31361	0.4965	0.2860	1.6333	2509.	730.	0.388	0.0304	0.0827	0.00822	9.93	−16.00
−14.00	239.78	65.95	0.3714	12.677	139.056	0.02871	0.31228	0.4987	0.2903	1.6441	2481.	729.	0.381	0.0305	0.0819	0.00830	9.68	−14.00
−12.00	248.49	65.65	0.3580	13.675	139.052	0.03089	0.31095	0.5011	0.2947	1.6554	2452.	728.	0.375	0.0307	0.0811	0.00839	9.43	−12.00
−10.00	257.44	65.34	0.3451	14.677	139.040	0.03307	0.30963	0.5036	0.2992	1.6673	2423.	727.	0.368	0.0309	0.0804	0.00849	9.19	−10.00
−8.00	266.62	65.03	0.3328	15.683	139.019	0.03524	0.30831	0.5062	0.3039	1.6798	2394.	726.	0.362	0.0310	0.0796	0.00858	8.95	−8.00
−6.00	276.04	64.72	0.3209	16.694	138.989	0.03741	0.30698	0.5090	0.3088	1.6929	2364.	725.	0.356	0.0312	0.0788	0.00869	8.70	−6.00
−4.00	285.70	64.41	0.3095	17.710	138.950	0.03959	0.30566	0.5119	0.3138	1.7067	2335.	724.	0.350	0.0314	0.0780	0.00879	8.46	−4.00
−2.00	295.60	64.09	0.2986	18.732	138.901	0.04176	0.30433	0.5150	0.3190	1.7211	2305.	722.	0.344	0.0316	0.0772	0.00890	8.22	−2.00
0.00	305.76	63.76	0.2880	19.758	138.843	0.04394	0.30300	0.5183	0.3245	1.7363	2275.	721.	0.338	0.0317	0.0764	0.00902	7.99	0.00
2.00	316.17	63.44	0.2779	20.790	138.775	0.04611	0.30167	0.5218	0.3301	1.7523	2245.	720.	0.333	0.0319	0.0756	0.00914	7.75	2.00
4.00	326.83	63.11	0.2682	21.828	138.697	0.04829	0.30034	0.5254	0.3360	1.7691	2215.	719.	0.327	0.0321	0.0749	0.00927	7.52	4.00
6.00	337.76	62.77	0.2589	22.872	138.608	0.05046	0.29900	0.5292	0.3422	1.7868	2185.	717.	0.322	0.0323	0.0741	0.00940	7.29	6.00
8.00	348.95	62.43	0.2498	23.923	138.508	0.05264	0.29766	0.5333	0.3486	1.8055	2154.	716.	0.316	0.0325	0.0733	0.00954	7.05	8.00
10.00	360.41	62.09	0.2412	24.980	138.397	0.05483	0.29631	0.5376	0.3553	1.8253	2123.	714.	0.311	0.0327	0.0725	0.00968	6.83	10.00
12.00	372.14	61.74	0.2328	26.044	138.273	0.05702	0.29495	0.5421	0.3624	1.8461	2092.	713.	0.305	0.0330	0.0718	0.00984	6.60	12.00
14.00	384.15	61.39	0.2248	27.116	138.138	0.05920	0.29359	0.5469	0.3698	1.8682	2061.	711.	0.300	0.0332	0.0710	0.01000	6.37	14.00
16.00	396.44	61.03	0.2170	28.195	137.990	0.06140	0.29222	0.5519	0.3776	1.8917	2029.	709.	0.295	0.0334	0.0702	0.01016	6.15	16.00
18.00	409.02	60.67	0.2095	29.283	137.828	0.06360	0.29084	0.5573	0.3857	1.9165	1997.	708.	0.290	0.0337	0.0694	0.01034	5.93	18.00
20.00	421.89	60.30	0.2023	30.379	137.653	0.06581	0.28945	0.5630	0.3944	1.9429	1966.	706.	0.284	0.0339	0.0687	0.01052	5.71	20.00
22.00	435.05	59.92	0.1953	31.484	137.463	0.06802	0.28805	0.5690	0.4035	1.9710	1933.	704.	0.279	0.0342	0.0679	0.01072	5.49	22.00
24.00	448.52	59.54	0.1886	32.598	137.258	0.07025	0.28663	0.5755	0.4131	2.0009	1901.	702.	0.274	0.0345	0.0671	0.01092	5.28	24.00
26.00	462.28	59.15	0.1821	33.722	137.037	0.07248	0.28520	0.5823	0.4234	2.0329	1869.	700.	0.269	0.0347	0.0663	0.01114	5.06	26.00
28.00	476.36	58.75	0.1758	34.857	136.800	0.07472	0.28376	0.5896	0.4342	2.0670	1836.	698.	0.264	0.0350	0.0656	0.01136	4.85	28.00
30.00	490.74	58.35	0.1697	36.002	136.545	0.07697	0.28230	0.5974	0.4458	2.1037	1803.	696.	0.259	0.0353	0.0648	0.01160	4.64	30.00
32.00	505.45	57.94	0.1638	37.159	136.273	0.07923	0.28082	0.6057	0.4582	2.1430	1770.	694.	0.254	0.0356	0.0640	0.01185	4.43	32.00
34.00	520.47	57.52	0.1581	38.328	135.981	0.08151	0.27932	0.6146	0.4715	2.1853	1736.	692.	0.249	0.0360	0.0633	0.01211	4.23	34.00
36.00	535.83	57.09	0.1526	39.511	135.669	0.08380	0.27779	0.6242	0.4858	2.2310	1702.	689.	0.244	0.0363	0.0625	0.01239	4.03	36.00
38.00	551.51	56.65	0.1473	40.706	135.336	0.08610	0.27625	0.6345	0.5011	2.2804	1668.	687.	0.239	0.0366	0.0617	0.01268	3.82	38.00
40.00	567.54	56.20	0.1421	41.917	134.980	0.08842	0.27467	0.6457	0.5177	2.3339	1634.	684.	0.235	0.0370	0.0610	0.01299	3.63	40.00
42.00	583.91	55.74	0.1371	43.143	134.600	0.09076	0.27307	0.6577	0.5358	2.3923	1599.	682.	0.230	0.0374	0.0602	0.01332	3.43	42.00
44.00	600.62	55.27	0.1322	44.385	134.195	0.09312	0.27144	0.6708	0.5554	2.4559	1565.	679.	0.225	0.0378	0.0595	0.01366	3.24	44.00
46.00	617.69	54.79	0.1274	45.645	133.763	0.09551	0.26977	0.6851	0.5769	2.5257	1529.	677.	0.220	0.0382	0.0587	0.01402	3.05	46.00
48.00	635.13	54.29	0.1228	46.924	133.302	0.09791	0.26806	0.7007	0.6005	2.6025	1494.	674.	0.215	0.0386	0.0579	0.01441	2.86	48.00
50.00	652.92	53.78	0.1183	48.223	132.809	0.10035	0.26631	0.7179	0.6265	2.6875	1458.	671.	0.210	0.0390	0.0572	0.01481	2.67	50.00
52.00	671.09	53.25	0.1140	49.544	132.283	0.10281	0.26451	0.7369	0.6554	2.7818	1421.	668.	0.205	0.0395	0.0564	0.01525	2.49	52.00
54.00	689.64	52.71	0.1097	50.888	131.720	0.10531	0.26267	0.7580	0.6877	2.8873	1384.	665.	0.200	0.0400	0.0557	0.01570	2.31	54.00
56.00	708.58	52.15	0.1056	52.259	131.117	0.10784	0.26077	0.7815	0.7240	3.0059	1347.	662.	0.194	0.0404	0.0549	0.01618	2.14	56.00
58.00	727.91	51.57	0.1015	53.657	130.471	0.11041	0.25880	0.8081	0.7652	3.1402	1309.	658.	0.189	0.0410	0.0542	0.01669	1.96	58.00
60.00	747.64	50.97	0.0975	55.087	129.777	0.11303	0.25676	0.8383	0.8122	3.2935	1270.	655.	0.184	0.0415	0.0534	0.01724	1.80	60.00
62.00	767.78	50.34	0.0937	56.551	129.029	0.11570	0.25464	0.8729	0.8665	3.4703	1230.	652.	0.179	0.0420	0.0526	0.01781	1.63	62.00
64.00	788.33	49.68	0.0899	58.054	128.222	0.11843	0.25243	0.9132	0.9299	3.6763	1189.	648.	0.173	0.0426	0.0519	0.01842	1.47	64.00
66.00	809.32	49.00	0.0861	59.600	127.348	0.12123	0.25011	0.9605	1.0050	3.9194	1148.	644.	0.168	0.0432	0.0511	0.01907	1.31	66.00
68.00	830.74	48.28	0.0824	61.197	126.396	0.12411	0.24767	1.0173	1.0952	4.2107	1105.	640.	0.163	0.0438	0.0504	0.01977	1.16	68.00
70.00	852.61	47.51	0.0788	62.851	125.355	0.12708	0.24508	1.0866	1.2057	4.5663	1060.	636.	0.157	0.0445	0.0496	0.02050	1.01	70.00
75.00	909.34	45.36	0.0697	67.321	122.243	0.13505	0.23777	1.3554	1.6345	5.9316	941.	625.	0.143	—	—	—	0.66	75.00
80.00	969.22	42.64	0.0604	72.565	117.967	0.14434	0.22847	1.9829	2.6237	9.0199	809.	610.	0.128	—	—	—	0.35	80.00
85.00	1032.7	38.47	0.0494	79.825	110.771	0.15720	0.21401	—	—	—	—	—	—	—	—	—	0.10	85.00
87.76c	1069.6	29.20	0.0342	94.234	94.234	0.18320	0.18320	∞	∞	∞	0.	0.	—	—	∞	∞	0.00	87.76

*temperatures are on the IPTS-68 scale   a = triple point   c = critical point

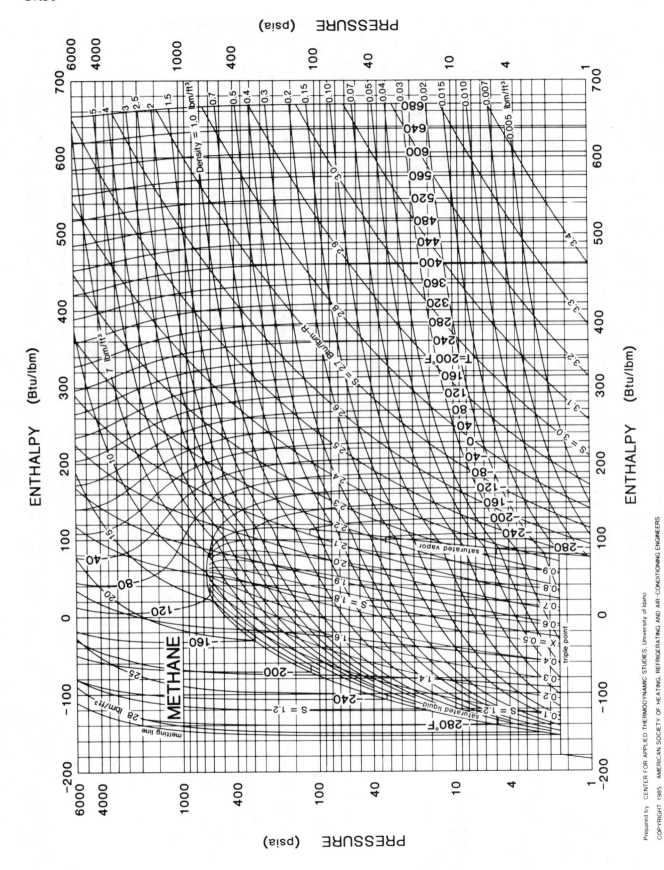

Prepared by CENTER FOR APPLIED THERMODYNAMIC STUDIES, University of Idaho
COPYRIGHT 1985 AMERICAN SOCIETY OF HEATING, REFRIGERATING AND AIR-CONDITIONING ENGINEERS

Fig. 24 Pressure-Enthalpy Diagram for Refrigerant 50 (Methane)

Refrigerant 50 (Methane)   Properties of Saturated Liquid and Saturated Vapor

Temp, °F	Pressure psia	Pressure psig	Volume, ft³/lbm Vapor	Density, lbm/ft³ Liquid	Enthalpy, Btu/lbm Liquid	Enthalpy, Btu/lbm Vapor	Entropy, Btu/lbm·°R Liquid	Entropy, Btu/lbm·°R Vapor
#−296.45	1.6997	26.461*	63.723	28.169	−153.88	79.912	1.0252	2.4575
−295	1.8831	26.087*	57.992	28.104	−152.75	80.584	1.0321	2.4490
−290	2.6448	24.536*	42.441	27.878	−148.81	82.886	1.0556	2.4211
−285	3.6375	22.515*	31.678	27.648	−144.82	85.152	1.0788	2.3953
−280	4.9082	19.928*	24.070	27.414	−140.79	87.377	1.1015	2.3713
−270	8.4934	12.629*	14.564	26.936	−132.63	91.690	1.1455	2.3282
−268	9.4086	10.765*	13.261	26.838	−130.98	92.528	1.1541	2.3202
−266	10.399	8.7491*	12.099	26.740	−129.33	93.358	1.1626	2.3125
−264	11.468	6.5716*	11.061	26.641	−127.68	94.179	1.1711	2.3049
−262	12.621	4.2244*	10.132	26.542	−126.02	94.990	1.1795	2.2976
−260	13.862	1.6988*	9.2970	26.442	−124.35	95.792	1.1878	2.2904
−258.73	14.696	0.0	8.8117	26.378	−123.30	96.296	1.1931	2.2859
−258	15.194	0.4980	8.5463	26.341	−122.68	96.584	1.1961	2.2834
−256	16.623	1.9267	7.8696	26.239	−121.01	97.366	1.2043	2.2765
−254	18.152	3.4562	7.2585	26.137	−119.33	98.138	1.2125	2.2699
−252	19.787	5.0910	6.7054	26.034	−117.65	98.898	1.2205	2.2633
−250	21.532	6.8357	6.2040	25.931	−115.97	99.648	1.2286	2.2569
−248	23.391	8.6951	5.7485	25.827	−114.27	100.39	1.2365	2.2507
−246	25.370	10.674	5.3340	25.722	−112.58	101.11	1.2444	2.2446
−244	27.472	12.776	4.9562	25.616	−110.88	101.83	1.2523	2.2386
−242	29.704	15.008	4.6113	25.510	−109.17	102.53	1.2601	2.2327
−240	32.069	17.373	4.2959	25.403	−107.46	103.22	1.2678	2.2269
−238	34.573	19.877	4.0069	25.295	−105.74	103.90	1.2755	2.2213
−236	37.221	22.525	3.7418	25.186	−104.02	104.57	1.2832	2.2158
−234	40.017	25.321	3.4983	25.076	−102.29	105.22	1.2908	2.2103
−232	42.967	28.271	3.2742	24.966	−100.56	105.85	1.2983	2.2050
−230	46.075	31.379	3.0677	24.855	−98.817	106.48	1.3059	2.1997
−228	49.348	34.652	2.8772	24.743	−97.071	107.09	1.3133	2.1946
−226	52.789	38.093	2.7012	24.630	−95.318	107.68	1.3207	2.1895
−224	56.405	41.709	2.5384	24.516	−93.558	108.26	1.3281	2.1845
−222	60.200	45.504	2.3876	24.401	−91.792	108.82	1.3355	2.1795
−220	64.179	49.483	2.2477	24.285	−90.018	109.37	1.3428	2.1747
−218	68.349	53.653	2.1178	24.168	−88.236	109.89	1.3500	2.1699
−216	72.713	58.017	1.9970	24.050	−86.447	110.41	1.3573	2.1651
−214	77.278	62.582	1.8846	23.930	−84.649	110.90	1.3645	2.1605
−212	82.049	67.353	1.7799	23.810	−82.843	111.38	1.3716	2.1558
−210	87.030	72.334	1.6822	23.688	−81.027	111.84	1.3788	2.1513
−208	92.229	77.533	1.5909	23.565	−79.203	112.28	1.3859	2.1467
−206	97.649	82.953	1.5057	23.441	−77.368	112.70	1.3930	2.1423
−204	103.30	88.601	1.4259	23.315	−75.523	113.09	1.4001	2.1378
−202	109.18	94.481	1.3511	23.188	−73.668	113.47	1.4071	2.1334
−200	115.50	100.60	1.2810	23.059	−71.801	113.83	1.4141	2.1290
−198	121.66	106.96	1.2152	22.928	−69.922	114.16	1.4212	2.1246
−196	128.27	113.58	1.1534	22.796	−68.032	114.48	1.4281	2.1203
−194	135.14	120.44	1.0953	22.662	−66.128	114.76	1.4351	2.1160
−192	142.27	127.57	1.0406	22.526	−64.211	115.03	1.4421	2.1117
−190	149.66	134.97	0.98903	22.389	−62.280	115.26	1.4491	2.1074
−188	157.33	142.63	0.94044	22.249	−60.334	115.48	1.4560	2.1031
−186	165.28	150.58	0.89458	22.107	−58.373	115.66	1.4630	2.0989
−184	173.50	158.81	0.85127	21.962	−56.395	115.82	1.4699	2.0946
−182	182.02	167.33	0.81032	21.816	−54.400	115.94	1.4769	2.0903
−180	190.84	176.14	0.77158	21.666	−52.388	116.04	1.4838	2.0860
−178	199.95	185.26	0.73489	21.514	−50.357	116.10	1.4908	2.0817
−176	209.38	194.68	0.70012	21.359	−48.305	116.13	1.4977	2.0774
−174	219.12	204.42	0.66715	21.202	−46.234	116.13	1.5047	2.0731
−172	229.18	214.48	0.63584	21.041	−44.140	116.09	1.5117	2.0687
−170	239.56	224.87	0.60610	20.877	−42.023	116.01	1.5187	2.0643
−168	250.28	235.58	0.57782	20.709	−39.882	115.89	1.5258	2.0598
−166	261.34	246.64	0.55090	20.537	−37.715	115.73	1.5328	2.0553
−164	272.75	258.05	0.52527	20.362	−35.521	115.53	1.5399	2.0508
−162	284.51	269.81	0.50083	20.182	−33.298	115.28	1.5471	2.0462
−160	296.63	281.93	0.47752	19.998	−31.045	114.98	1.5542	2.0415
−155	328.55	313.86	0.42370	19.516	−25.265	114.00	1.5724	2.0295
−150	362.90	348.21	0.37547	18.998	−19.246	112.63	1.5909	2.0168
−145	399.81	385.11	0.33196	18.437	−12.937	110.81	1.6099	2.0032
−140	439.41	424.71	0.29239	17.821	−6.2642	108.43	1.6297	1.9885
−135	481.88	467.18	0.25597	17.132	0.8920	105.30	1.6505	1.9721
−130	527.42	512.73	0.22182	16.335	8.7538	101.12	1.6730	1.9532
−125	576.50	561.60	0.18855	15.350	17.826	95.192	1.6986	1.9298
−120	628.91	614.22	0.15232	13.902	29.880	85.179	1.7323	1.8951
*−116.7	666.4	651.8	0.09877	10.12	56.91	56.91	1.810	1.810

#Triple point   *in. Hg. Vacuum   **Critical point

Temp., °F	Viscosity, lbm/ft·h Sat. Liquid	Viscosity, lbm/ft·h Sat. Vapor	Viscosity, lbm/ft·h Gas at p = 1 atm × 10⁻²‡	Thermal Conductivity, Btu/h·ft·°F Sat. Liquid	Thermal Conductivity, Btu/h·ft·°F Sat. Vapor	Thermal Conductivity, Btu/h·ft·°F Gas at p = 1 atm × 10⁻³‡	Specific Heat, c_p, Btu/lbm·°F Sat. Liquid	Specific Heat, c_p, Btu/lbm·°F Sat. Vapor	Specific Heat, c_p, Btu/lbm·°F Gas at p = 0 atm	Specific Heat, c_p, Btu/lbm·°F Gas at p = 1 atm	Temp., °F
−290	0.436	0.0094		0.127	0.0055		0.797				−290
−280	0.372	0.0099		0.122	0.0059		0.804		0.497		−280
−270	0.324	0.0103		0.117	0.0064		0.811		0.496		−270
−260	0.285	0.0108		0.112	0.0069		0.820	0.509	0.496		−260
−240	0.229	0.0117	1.18	0.102	0.0078	7.6	0.841	0.586	0.496		−240
−220	0.191	0.0127	1.27	0.093	0.0088	8.3	0.871	0.644	0.495		−220
−200	0.165	0.0139	1.37	0.083	0.0099	9.0	0.912	0.698	0.495		−200
−180	0.145	0.0156	1.47	0.074	0.0110	9.6	0.966	0.764	0.496		−180
−160	0.120	0.0177	1.57	0.064	0.0126	10.4	1.074	0.858	0.497		−160
−140	0.086	0.021	1.67	0.054	0.015	11.1	1.30	1.03	0.497	0.505	−140
−120	0.057	0.026	1.77	0.038	0.023	11.8	1.71	1.85	0.498	0.505	−120
116*	0.040	0.040	1.79	0.029	0.029	12.0			0.499	0.505	116*
−110			1.82			12.2			0.499	0.505	−110
−100			1.87			12.6			0.499	0.505	−100
−60			2.07			14.1			0.503	0.507	−60
−40			2.17			14.9			0.506	0.509	−40
−20			2.26			15.7			0.508	0.512	−20
0			2.35			16.5			0.511	0.514	0
20			2.44			17.3			0.515	0.517	20
40			2.53			18.1			0.519	0.522	40
60			2.62			19.0			0.526	0.528	60
80			2.71			19.9			0.533	0.535	80
100			2.79			20.7			0.540	0.542	100
120			2.88			21.5			0.548	0.549	120
140			2.96			22.4			0.556	0.557	140
160			3.05			23.3			0.564	0.565	160
180			3.13			24.2			0.572	0.573	180
200			3.21			25.1			0.580	0.581	200
300			3.60			30.4			0.624	0.625	300
400			3.96			36.4			0.671	0.672	400
500			4.31			42.5			0.720	0.720	500

*Critical temperature. Tabulated properties ignore critical region effects.   ‡Actual value = (Table value) × (Indicated multiplier).

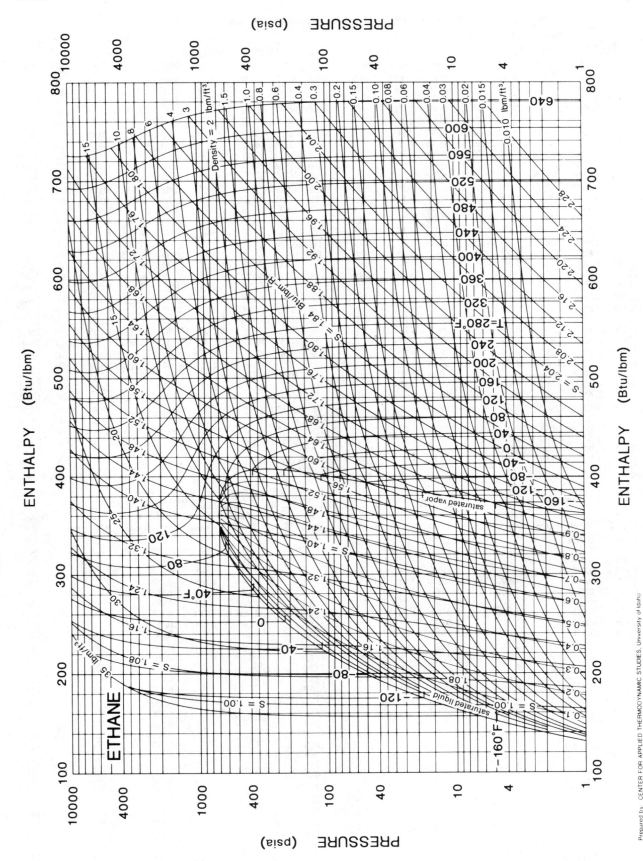

Fig. 25 Pressure-Enthalpy Diagram for Refrigerant 170 (Ethane)

Prepared by CENTER FOR APPLIED THERMODYNAMIC STUDIES, University of Idaho

Refrigerant 170 (Ethane) Properties of Saturated Liquid and Saturated Vapor

Temp, °F	Pressure psia	Pressure psig	Volume, ft³/lbm Vapor	Density, lbm/ft³ Liquid	Enthalpy, Btu/lbm Liquid	Enthalpy, Btu/lbm Vapor	Entropy, Btu/lbm·°R Liquid	Entropy, Btu/lbm·°R Vapor
#−297.04	0.00016	29.921*	351536.	40.698	76.078	332.08	0.61190	2.1861
−280	0.00155	29.918*	41296.	40.050	85.317	337.11	0.66592	2.0661
−270	0.00471	29.912*	14350.	39.670	90.758	340.06	0.69540	2.0081
−260	0.01268	29.895*	5617.1	39.288	96.212	343.01	0.72343	1.9575
−250	0.03075	29.859*	2433.0	38.904	101.68	345.97	0.75013	1.9131
−240	0.06822	29.782*	1149.3	38.520	107.15	348.93	0.77565	1.8741
−230	0.14010	29.636*	584.95	38.133	112.64	351.90	0.80008	1.8397
−220	0.26909	29.373*	317.59	37.744	118.15	354.87	0.82354	1.8091
−210	0.48753	28.929*	182.41	37.352	123.67	357.84	0.84610	1.7820
−200	0.83922	28.213*	110.04	36.958	129.21	360.80	0.86785	1.7578
−190	1.3809	27.110*	69.301	36.560	134.77	363.76	0.88887	1.7362
−180	2.1836	25.475*	45.331	36.158	140.37	366.70	0.90921	1.7169
−170	3.3327	23.136*	30.660	35.752	145.99	369.62	0.92895	1.6995
−160	4.9289	19.886*	21.360	35.341	151.65	372.51	0.94812	1.6838
−150	7.0872	15.492*	15.276	34.925	157.34	375.37	0.96679	1.6696
−140	9.9369	9.6895*	11.183	34.503	163.09	378.18	0.98499	1.6567
−130	13.620	2.1903*	8.3575	34.074	168.88	380.94	1.0028	1.6450
−128	14.470	0.4595*	7.9027	33.988	170.04	381.49	1.0063	1.6428
−127.48	14.696	0.0	7.7904	33.966	170.34	381.63	1.0072	1.6422
−126	15.361	0.6651	7.4780	33.901	171.21	382.03	1.0098	1.6406
−124	16.294	1.5980	7.0811	33.814	172.38	382.57	1.0133	1.6385
−122	17.270	2.5743	6.7099	33.726	173.55	383.11	1.0167	1.6364
−120	18.291	3.5954	6.3625	33.638	174.73	383.64	1.0202	1.6343
−118	19.358	4.6625	6.0370	33.550	175.90	384.17	1.0236	1.6323
−116	20.473	5.7771	5.7318	33.462	177.08	384.70	1.0270	1.6303
−114	21.636	6.9405	5.4455	33.373	178.27	385.23	1.0304	1.6283
−112	22.850	8.1541	5.1767	33.284	179.45	385.75	1.0338	1.6264
−110	24.115	9.4194	4.9240	33.194	180.64	386.27	1.0372	1.6245
−108	25.434	10.738	4.6865	33.104	181.83	386.79	1.0406	1.6226
−106	26.807	12.111	4.4630	33.014	183.02	387.30	1.0440	1.6208
−104	28.236	13.540	4.2525	32.923	184.22	387.81	1.0473	1.6190
−102	29.722	15.026	4.0541	32.832	185.41	388.32	1.0506	1.6172
−100	31.267	16.571	3.8671	32.741	186.61	388.82	1.0540	1.6154
−98	32.873	18.177	3.6906	32.649	187.82	389.32	1.0573	1.6137
−96	34.540	19.844	3.5240	32.556	189.03	389.82	1.0606	1.6120
−94	36.271	21.575	3.3666	32.464	190.24	390.31	1.0639	1.6103
−92	38.067	23.371	3.2178	32.370	191.45	390.80	1.0672	1.6087
−90	39.929	25.233	3.0770	32.277	192.67	391.29	1.0704	1.6071
−88	41.860	27.164	2.9437	32.183	193.89	391.77	1.0737	1.6055
−86	43.860	29.164	2.8175	32.088	195.11	392.24	1.0769	1.6039
−84	45.931	31.235	2.6980	31.993	196.34	392.72	1.0802	1.6023
−82	48.075	33.379	2.5846	31.898	197.57	393.18	1.0834	1.6008
−80	50.293	35.597	2.4770	31.802	198.80	393.65	1.0866	1.5993
−78	52.587	37.891	2.3748	31.705	200.04	394.11	1.0899	1.5978
−76	54.959	40.263	2.2778	31.608	201.28	394.56	1.0931	1.5963
−74	57.410	42.714	2.1857	31.510	202.53	395.01	1.0963	1.5948
−72	59.942	45.246	2.0980	31.412	203.78	395.46	1.0995	1.5934
−70	62.556	47.860	2.0147	31.314	205.03	395.90	1.1026	1.5919
−65	69.462	54.766	1.8233	31.064	208.18	396.97	1.1106	1.5884
−60	76.919	62.223	1.6537	30.811	211.36	398.02	1.1185	1.5850
−55	84.954	70.258	1.5029	30.554	214.56	399.03	1.1263	1.5817
−50	93.593	78.897	1.3684	30.293	217.79	400.00	1.1341	1.5785
−45	102.86	88.167	1.2482	30.027	221.06	400.93	1.1419	1.5753
−40	112.79	98.094	1.1404	29.756	224.36	401.82	1.1497	1.5722
−35	123.40	108.71	1.0435	29.479	227.69	402.67	1.1574	1.5691
−30	134.73	120.03	0.95614	29.198	231.06	403.47	1.1651	1.5661
−25	146.80	132.10	0.87726	28.910	234.47	404.22	1.1728	1.5631
−20	159.63	144.94	0.80583	28.615	237.91	404.92	1.1805	1.5601
−15	173.27	158.57	0.74102	28.314	241.41	405.56	1.1882	1.5571
−10	187.73	173.04	0.68207	28.005	246.95	406.14	1.1959	1.5542
−5	203.05	188.36	0.62833	27.688	248.54	406.65	1.2036	1.5512
0	219.26	204.57	0.57924	27.362	252.18	407.10	1.2114	1.5482
10	254.47	239.77	0.49307	26.680	259.68	407.76	1.2270	1.5422
20	293.61	278.91	0.42022	25.950	267.47	408.06	1.2428	1.5359
30	336.96	322.26	0.35806	25.161	275.58	407.93	1.2589	1.5292
40	384.81	370.11	0.30448	24.296	284.10	407.25	1.2754	1.5219
50	437.50	422.80	0.25771	23.329	293.15	405.86	1.2925	1.5137
60	495.41	480.71	0.21615	22.215	302.94	403.65	1.3106	1.5041
70	559.03	544.33	0.17813	20.862	313.88	399.45	1.3304	1.4921
80	629.07	614.37	0.14098	19.023	327.13	392.28	1.3540	1.4748
** 89.9	706.5	691.8	0.1120	8.928	384.3	384.3	1.457	1.457

#Triple point | *in. Hg. Vacuum | **Critical point

Temp., °F	Viscosity, lbm/ft·h Sat. Liquid	Viscosity Sat. Vapor	Viscosity Gas at p = 1 atm × 10⁻²‡	Thermal Conductivity, Btu/h·ft·°F Sat. Liquid	Thermal Cond. Sat. Vapor	Thermal Cond. Gas at p = 1 atm × 10⁻³‡	Specific Heat, cp, Btu/lbm·°F Sat. Liquid	Sp. Heat Sat. Vapor	Sp. Heat Gas at p = 0 atm	Sp. Heat Gas at p = 1 atm	Temp., °F
−160	0.48			0.0940	0.0039		0.5665		0.3176		−160
−140	0.41			0.0896	0.0049		0.5736		0.3245		−140
−120	0.36	0.0147	1.47	0.0852	0.0057	5.41	0.5825		0.3317	0.346	−120
−100	0.32	0.0160	1.55	0.0808	0.0064	5.99	0.5934	0.3808	0.3391	0.352	−100
−80	0.29	0.0172	1.64	0.0765	0.0070	6.61	0.6067	04147	0.3468	0.357	−80
−60	0.26	0.0180	1.72	0.0721	0.0076	7.25	0.6227	0.4340	0.3549	0.364	−60
−40	0.24	0.0187	1.80	0.0677	0.0083	7.93	0.6511	0.4508	0.3630	0.371	−40
−20	0.21	0.0196	1.88	0.0633	0.0093	8.63	0.6702	0.4771	0.3715	0.379	−20
0	0.181	0.0208	1.96	0.0589	0.0106	9.35	0.6931	0.5248	0.3803	0.387	0
20	0.158	0.023	2.05	0.0545	0.0122	10.11	0.7345	0.6060	0.3893	0.396	20
40	0.135	0.025	2.13	0.0501	0.0144	10.9	0.8091	0.7413	0.3984	0.405	40
60	0.114	0.028	2.21	0.0458	0.0169	11.7	0.9314	0.9040	0.4097	0.415	60
70	0.102	0.030	2.24	0.0428	0.0197	11.9		1.041	0.4153	0.420	70
80	0.088	0.034	2.28	0.0391	0.0228	12.6			0.4210	0.424	80
90*	0.052	0.052	2.32	0.0312	0.0312	13.0			0.4266	0.429	90*
100			2.36			13.4			0.4323	0.435	100
120			2.44			14.3			0.4436	0.445	120
140			2.52			15.2			0.4548	0.456	140
180			2.67			17.0			0.4773		180
220			2.82			18.9			0.4996		220
260			2.96			20.8			0.5218		260
300			3.11			22.7			0.5439		300
400			3.45			27.8			0.5979		400
500			3.78			33.0			0.6501		500
600			4.10			38.4			0.6998		600
700			4.40			43.9			0.7467		700
800			4.69			49.7			0.7901		800
1000			5.23			63.6			0.8662		1000

*Critical temperature. Tabulated properties ignore critical region effects.   ‡Actual value = (Table value) × (Indicated multiplier).

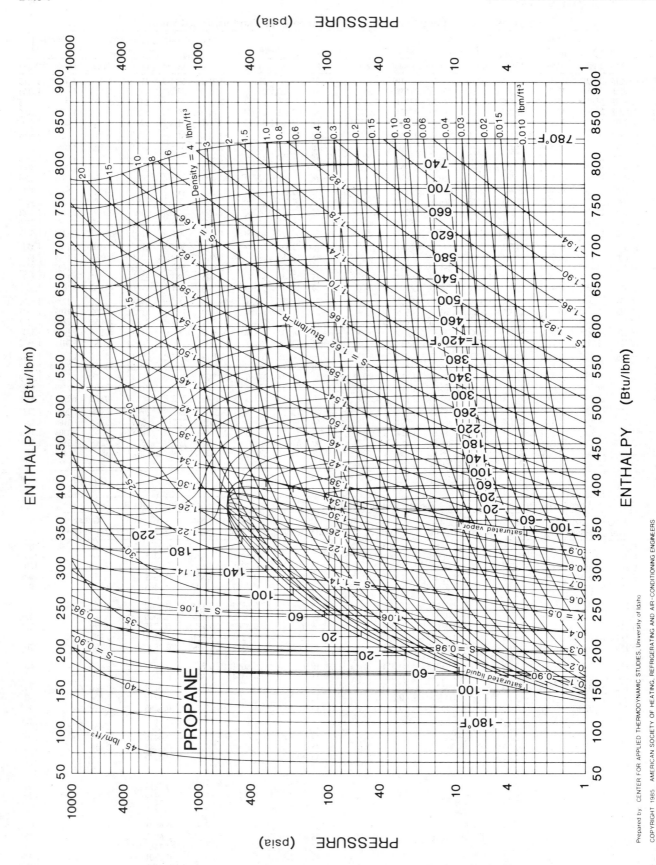

Fig. 26 Pressure-Enthalpy Diagram for Refrigerant 290 (Propane)

Prepared by   CENTER FOR APPLIED THERMODYNAMIC STUDIES, University of Idaho

COPYRIGHT 1985   AMERICAN SOCIETY OF HEATING, REFRIGERATING AND AIR-CONDITIONING ENGINEERS

Refrigerant Properties

Refrigerant 290 (Propane) Properties of Saturated Liquid and Saturated Vapor

Temp, °F	Pressure psia	Pressure psig	Volume, ft³/lb_m Vapor	Density, lb_m/ft³ Liquid	Enthalpy, Btu/lb_m Liquid	Enthalpy Vapor	Entropy, Btu/lb_m·°R Liquid	Entropy Vapor	Temp, °F	Pressure psia	Pressure psig	Volume, ft³/lb_m Vapor	Density, lb_m/ft³ Liquid	Enthalpy, Btu/lb_m Liquid	Enthalpy Vapor	Entropy, Btu/lb_m·°R Liquid	Entropy Vapor
#−305.82	0.44E-7	29.921*	860×10⁶	45.753	53.742	296.86	0.44784	1.9968	−20	25.438	10.742	3.9709	35.278	194.46	372.19	0.95639	1.3606
−300	0.14E-6	29.921*	276×10⁶	45.551	56.397	297.94	0.46478	1.9518	−18	26.560	11.864	3.8137	35.194	195.58	372.76	0.95894	1.3601
−280	0.44E-5	29.921*	985×10⁴	44.857	65.557	302.03	0.51882	1.8222	−16	27.720	13.024	3.6641	35.110	196.71	373.33	0.96148	1.3596
−260	0.69E-4	29.921*	699×10³	44.161	74.782	306.58	0.56749	1.7224	−14	28.918	14.222	3.5217	35.026	197.84	373.90	0.96401	1.3590
−240	0.00065	29.920*	82014.	43.462	84.079	311.42	0.61186	1.6444	−12	30.155	15.459	3.3861	34.941	198.98	374.46	0.96654	1.3585
−220	0.00415	29.913*	14052.	42.760	93.455	316.48	0.65270	1.5825	−10	31.432	16.736	3.2568	34.856	200.11	375.03	0.96906	1.3580
−200	0.01963	29.881*	3225.8	42.055	102.92	321.69	0.69062	1.5330	−8	32.749	18.053	3.1336	34.771	201.25	375.59	0.97158	1.3576
−180	0.07311	29.772*	932.71	41.344	112.48	327.02	0.72608	1.4933	−6	34.108	19.412	3.0161	34.686	202.39	376.15	0.97410	1.3571
−160	0.22488	29.463*	324.42	40.628	122.15	332.45	0.75947	1.4613	−4	35.509	20.813	2.9039	34.600	203.54	376.71	0.97661	1.3567
−140	0.59203	28.716*	131.07	39.905	131.94	337.97	0.79110	1.4356	−2	36.954	22.258	2.7968	34.513	204.69	377.27	0.97912	1.3562
−120	1.3716	27.129*	59.862	39.172	141.87	343.58	0.82121	1.4150	0	38.442	23.746	2.6946	34.427	205.84	377.83	0.98162	1.3558
−100	2.8586	24.101*	30.240	38.428	151.96	349.25	0.85005	1.3985	2	39.975	25.279	2.5969	34.340	206.99	378.39	0.98412	1.3554
−95	3.3856	23.028*	25.844	38.239	154.51	350.68	0.85708	1.3949	4	41.554	26.858	2.5035	34.252	208.15	378.94	0.98661	1.3550
−90	3.9888	21.800*	22.197	38.050	157.07	352.11	0.86404	1.3915	6	43.179	28.483	2.4143	34.165	209.32	379.50	0.98910	1.3546
−85	4.6757	20.401*	19.154	37.860	159.64	353.54	0.87094	1.3884	8	44.852	30.156	2.3289	34.077	210.48	380.05	0.99158	1.3542
−80	5.4548	18.815*	16.604	37.669	162.22	354.98	0.87779	1.3854	10	46.572	31.876	2.2471	33.988	211.65	380.60	0.99407	1.3538
−75	6.3343	17.025*	14.455	37.477	164.82	356.41	0.88457	1.3826	12	48.342	33.646	2.1689	33.899	212.82	381.15	0.99654	1.3534
−70	7.3233	15.011*	12.635	37.284	167.43	357.85	0.89130	1.3799	14	50.163	35.467	2.0939	33.810	214.00	381.70	0.99901	1.3531
−65	8.4310	12.756*	11.088	37.090	170.05	359.29	0.89797	1.3774	16	52.033	37.337	2.0221	33.720	215.18	382.25	1.0015	1.3527
−60	9.6670	10.239*	9.7672	36.894	172.69	360.73	0.90459	1.3750	18	53.956	39.260	1.9533	33.630	216.36	382.79	1.0039	1.3524
−55	11.041	7.4407*	8.6341	36.697	175.34	362.17	0.91117	1.3728	20	55.931	41.235	1.8873	33.539	217.54	383.34	1.0064	1.3520
−50	12.564	4.3399*	7.6583	36.499	178.00	363.61	0.91769	1.3707	25	61.106	46.410	1.7339	33.311	220.52	384.69	1.0125	1.3513
−48	13.218	3.0100*	7.3063	36.419	179.08	364.18	0.92029	1.3699	30	66.630	51.934	1.5954	33.079	223.53	386.03	1.0187	1.3505
−46	13.897	1.6269*	6.9739	36.339	180.15	364.76	0.92288	1.3692	35	72.517	57.821	1.4700	32.845	226.55	387.35	1.0248	1.3498
−44	14.603	0.1890*	6.6601	36.259	181.22	365.33	0.92547	1.3684	40	78.782	64.086	1.3563	32.607	229.60	388.66	1.0308	1.3492
−43.74	14.696	0.0	6.6210	36.249	181.36	365.41	0.92580	1.3683	50	92.506	77.810	1.1590	32.121	235.77	391.23	1.0429	1.3480
−42	15.337	0.6409	6.3634	36.179	182.31	365.91	0.92807	1.3676	60	107.92	93.226	0.99495	31.619	242.05	393.73	1.0550	1.3469
−40	16.099	1.4030	6.0829	36.098	183.40	366.48	0.93067	1.3669	70	125.16	110.46	0.85752	31.100	248.44	396.15	1.0670	1.3459
−38	16.890	2.1941	5.8174	36.017	184.49	367.06	0.93327	1.3662	80	144.33	129.64	0.74162	30.561	254.95	398.48	1.0790	1.3449
−36	17.711	3.0149	5.5661	35.936	185.59	367.63	0.93586	1.3655	90	165.58	150.89	0.64327	30.000	261.59	400.69	1.0909	1.3440
−34	18.562	3.8662	5.3281	35.855	186.69	368.20	0.93844	1.3649	100	189.04	174.34	0.55928	29.413	268.38	402.78	1.1029	1.3431
−32	19.445	4.7487	5.1025	35.773	187.79	368.77	0.94102	1.3642	120	243.13	228.43	0.42475	28.145	282.49	406.47	1.1271	1.3410
−30	20.359	5.6632	4.8886	35.691	188.89	369.35	0.94360	1.3636	140	307.74	293.04	0.32288	26.704	297.49	409.30	1.1518	1.3383
−28	21.306	6.6104	4.6857	35.609	190.00	369.92	0.94617	1.3630	160	384.12	369.42	0.24319	24.995	313.76	410.71	1.1776	1.3341
−26	22.287	7.5911	4.4930	35.527	191.11	370.49	0.94873	1.3624	180	473.82	459.12	0.17744	22.784	332.12	409.40	1.1847	1.3265
−24	23.302	8.6060	4.3101	35.444	192.22	371.06	0.95129	1.3618	**206.0	615.3	600.6	0.03947	25.34	333.4	333.4	1.274	1.274
−22	24.352	9.6560	4.1362	35.361	193.34	371.63	0.95384	1.3612									

#Triple point   *in. Hg. Vacuum   **Critical point

Temp., °F	Viscosity, lb_m/ft·h Sat. Liquid	Sat. Vapor	Gas at p = 1 atm × 10⁻²‡	Thermal Conductivity, Btu/h·ft·°F Sat. Liquid	Sat. Vapor	Gas at p = 1 atm × 10⁻³‡	Specific Heat, c_p, Btu/lb_m·°F Sat. Liquid	Sat. Vapor	Gas at p = 0 atm	Gas at p = 1 atm	Temp., °F
−80	0.63			0.0803	0.0049		0.5151	0.3246	0.3128		−80
−60	0.55			0.0772	0.0058		0.5250	0.3453	0.3230		−60
−40	0.48			0.0741	0.0067	6.50	0.5361	0.3645	0.3334	0.345	−40
−20	0.43			0.0711	0.0076	7.13	0.5484	0.3833	0.3440	0.355	−20
0	0.386			0.0680	0.0084	7.77	0.5621	0.4031	0.3548	0.365	0
20	0.348		1.78	0.0649	0.0092	8.43	0.5771	0.4251	0.3658	0.375	20
40	0.317	0.019	1.85	0.0618	0.0100	9.10	0.5932	0.4507	0.3756	0.386	40
60	0.290	0.021	1.93	0.0587	0.0108	9.79	0.6124	0.4810	0.3885	0.397	60
80	0.266	0.022	2.00	0.0557	0.0117	10.48	0.6381	0.5174	0.4013	0.408	80
100	0.222	0.022	2.07	0.0526	0.0126	11.2	0.6727	0.5602	0.4140	0.419	100
120	0.194	0.023	2.14	0.0495	0.0137	11.9	0.7187	0.6123	0.4266	0.430	120
140	0.166	0.025	2.21	0.0464	0.0148	12.6	0.7785	0.6746	0.4391	0.442	140
160	0.138	0.028	2.28	0.0434	0.0161	13.4	0.8547	0.7468	0.4515	0.454	160
180	0.111	0.032	2.35	0.0402	0.0190	14.1	0.9498		0.4637	0.465	180
200	0.076	0.041	2.42	0.035	0.025	14.8	1.07		0.4758	0.477	200
206*	0.072	0.072	2.44	0.032	0.032	15.0			0.4794	0.481	206*
220			2.49			15.6			0.4878	0.489	220
240			2.55			16.3			0.4997	0.501	240
260			2.62			17.1			0.5114	0.512	260
280			2.69			17.9			0.5231	0.524	280
300			2.76			18.6			0.5346	0.536	300
320			2.82			19.4			0.5459		320
340			2.89			20.1			0.5572		340
360			2.96			20.9			0.5683		360
380			3.02			21.7			0.5793		380
400			3.09			22.5			0.5901		400
500			3.41			26.3			0.6422		500
600									0.6908		600

*Critical temperature. Tabulated properties ignore critical region effects.   ‡Actual value = (Table value) × (Indicated multiplier).

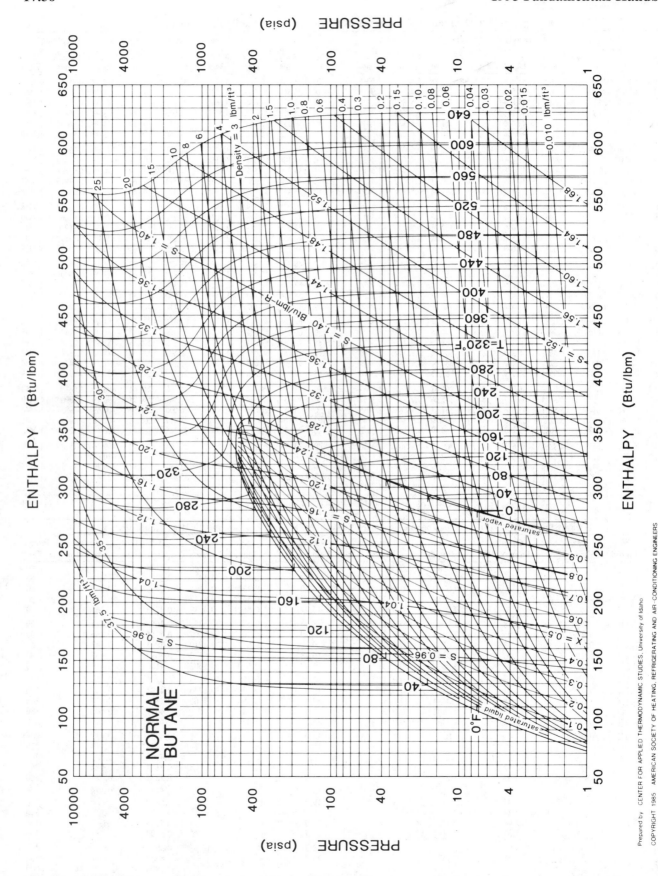

Fig. 27   Pressure-Enthalpy Diagram for Refrigerant 600 (n-Butane)

Refrigerant 600 (n-Butane) Properties of Saturated Liquid and Saturated Vapor

Temp, °F	Pressure psia	Pressure psig	Volume, ft³/lb_m Vapor	Density, lb_m/ft³ Liquid	Enthalpy, Btu/lb_m Liquid	Enthalpy, Btu/lb_m Vapor	Entropy, Btu/lb_m·°R Liquid	Entropy, Btu/lb_m·°R Vapor	Temp, °F	Pressure psia	Pressure psig	Volume, ft³/lb_m Vapor	Density, lb_m/ft³ Liquid	Enthalpy, Btu/lb_m Liquid	Enthalpy, Btu/lb_m Vapor	Entropy, Btu/lb_m·°R Liquid	Entropy, Btu/lb_m·°R Vapor
#−216.92	0.00010	29.921*	459×10⁶	45.902	0.0	212.62	0.55105	1.4269	70	31.268	16.572	2.9130	36.035	145.78	302.92	094155	1.2382
−210	0.00020	29.921*	232×10⁶	45.678	3.2017	214.46	0.56402	1.4102	72	32.397	17.701	2.8170	35.956	146.93	303.61	0.94371	1.2384
−190	0.00123	29.919*	40586.	45.031	12.513	219.92	0.59983	1.3689	74	33.556	18.860	2.7248	35.876	148.09	304.29	0.94587	1.2386
−170	0.00572	29.910*	9347.9	44.382	21.913	225.53	0.63345	1.3364	76	34.747	20.051	2.6363	35.796	149.25	304.98	0.94802	1.2387
−150	0.02132	29.878*	2680.5	43.730	31.408	231.30	0.66516	1.3107	78	35.970	21.274	2.5512	35.716	150.41	305.67	0.95017	1.2389
−130	0.06632	29.786*	916.96	43.075	41.009	237.22	0.69524	1.2904	80	37.225	22.529	2.4695	35.636	151.57	306.35	0.95232	1.2391
−110	0.17792	29.559*	362.16	42.417	50.729	243.29	0.72390	1.2746	82	38.513	23.817	2.3909	35.555	152.74	307.04	0.95447	1.2393
−90	0.42245	29.061*	160.97	41.752	60.581	249.50	0.75133	1.2624	84	39.835	25.139	2.3154	35.474	153.91	307.72	0.95661	1.2395
−70	0.90610	28.076*	78.897	41.081	70.579	255.84	0.77769	1.2531	86	41.191	26.495	2.2428	35.393	155.09	308.41	0.95875	1.2397
−50	1.7843	26.288*	41.961	40.402	80.736	262.31	0.80311	1.2463	88	42.581	27.885	2.1728	35.311	156.26	309.09	0.96088	1.2399
−30	3.2687	23.266*	23.900	39.713	91.065	268.89	0.82771	1.2416	90	44.007	29.311	2.1056	35.229	157.44	309.78	0.96302	1.2402
−20	4.3207	21.124*	18.444	39.365	96.298	272.22	0.83973	1.2399	92	45.469	30.773	2.0408	35.146	158.62	310.46	0.96515	1.2404
−10	5.6305	18.457*	14.424	39.013	101.58	275.58	0.85158	1.2385	94	46.966	32.270	1.9785	35.064	159.81	311.14	0.96728	1.2406
0	7.2414	15.178*	11.418	38.657	106.91	278.95	0.86328	1.2375	96	48.501	33.805	1.9184	34.980	161.00	311.83	0.96940	1.2408
10	9.2001	11.190*	9.1413	38.297	112.29	282.34	0.87483	1.2369	98	50.073	35.377	1.8605	34.897	162.19	312.51	0.97152	1.2411
20	11.557	6.3919*	7.3947	37.933	117.73	285.75	0.88624	1.2365	105	55.879	41.183	1.6739	34.602	166.38	314.89	0.97894	1.2419
30	14.364	0.6758*	6.0391	37.565	123.22	289.17	0.89753	1.2364	110	60.325	45.629	1.5543	34.388	169.40	316.58	0.98421	1.2426
31.08	14.696	0.0	5.9118	37.525	123.82	289.54	0.89874	1.2364	115	65.030	50.334	1.4449	34.171	172.44	318.27	0.98948	1.2432
32	14.985	0.2886	5.8056	37.491	124.33	289.86	0.89977	1.2364	120	70.004	55.308	1.3446	33.952	175.50	319.95	0.99474	1.2439
34	15.626	0.9299	5.5831	37.416	125.43	290.54	0.90201	1.2365	125	75.256	60.560	1.2525	33.730	178.58	321.62	0.99998	1.2446
36	16.288	1.5924	5.3709	37.341	126.54	291.23	0.90424	1.2365	130	80.795	66.099	1.1678	33.505	181.69	323.29	1.0052	1.2454
38	16.973	2.2765	5.1685	37.266	127.65	291.92	0.90647	1.2365	140	92.774	78.078	1.0179	33.044	187.97	326.59	1.0157	1.2468
40	17.679	2.9829	4.9754	37.191	128.77	292.60	0.90869	1.2366	150	106.02	91.320	0.89011	32.570	194.35	329.86	1.0261	1.2484
42	18.408	3.7118	4.7911	37.116	129.89	293.29	0.91091	1.2366	160	120.60	105.91	0.78063	32.080	200.85	333.07	1.0366	1.2500
44	19.160	4.4639	4.6150	37.040	131.00	293.98	0.91312	1.2367	170	136.61	121.92	0.68634	31.572	207.46	336.23	1.0470	1.2515
46	19.936	5.2396	4.4468	36.964	132.13	294.66	0.91533	1.2368	180	154.13	139.44	0.60472	31.044	214.20	339.33	1.0575	1.2531
48	20.735	6.0394	4.2861	36.888	133.25	295.35	0.91754	1.2368	190	173.24	158.55	0.53371	30.493	221.09	342.34	1.0680	1.2547
50	21.560	6.8638	4.1324	36.812	134.38	296.04	0.91974	1.2369	200	194.04	179.35	0.47164	29.916	228.14	345.26	1.0786	1.2562
52	22.409	7.7133	3.9854	36.736	135.50	296.73	0.92194	1.2370	210	216.63	201.93	0.41708	29.309	235.36	348.07	1.0893	1.2576
54	23.284	8.5884	3.8447	36.659	136.64	297.42	0.92414	1.2371	220	241.10	226.40	0.36888	28.665	242.78	350.75	1.1001	1.2589
56	24.186	9.4896	3.7101	36.582	137.77	298.10	0.92633	1.2372	230	267.57	252.87	0.32603	27.979	250.41	353.25	1.1110	1.2601
58	25.113	10.417	3.5812	36.504	138.91	298.79	0.92851	1.2374	250	327.03	312.34	0.25312	26.433	266.47	357.53	1.1334	1.2617
60	26.068	11.372	3.4577	36.427	140.04	299.48	0.93069	1.2375	270	396.28	381.59	0.19253	24.524	283.93	360.25	1.1570	1.2616
62	27.051	12.355	3.3394	36.349	141.19	300.17	0.93287	1.2376	290	477.16	462.46	0.13812	21.795	303.90	359.40	1.1831	1.2571
64	28.062	13.366	3.2260	36.271	142.33	300.86	0.93505	1.2378	** 305.6	550.6	535.9	0.07064	14.16	337.1	337.1	1.226	1.226
66	29.101	14.405	3.1173	36.193	143.48	301.54	0.93722	1.2379									
68	30.170	15.474	3.0130	36.114	144.63	302.23	0.93939	1.2381									

#Triple point

*in. Hg. Vacuum

**Critical point

Temp., °F	Viscosity, lb_m/ft·h Sat. Liquid	Viscosity Sat. Vapor	Viscosity Gas at p = 1 atm × 10⁻²‡	Thermal Conductivity, Btu/h·ft·°F Sat. Liquid	Thermal Conductivity Sat. Vapor	Thermal Conductivity Gas at p = 1 atm × 10⁻³‡	Specific Heat, c_p, Btu/lb_m·°F Sat. Liquid	Specific Heat Sat. Vapor	Specific Heat Gas at p = 0 atm	Specific Heat Gas at p = 1 atm	Temp., °F
−80	0.99			0.088			0.497				−80
−40	0.74			0.083			0.511				−40
−20	0.65			0.080			0.519				−20
0	0.583			0.077	0.0069		0.529	0.365	0.357		0
20	0.525	0.0157		0.075	0.0076		0.540	0.386	0.370		20
40	0.477	0.0172	1.70	0.073	0.0082	8.2	0.552	0.404	0.382		40
60	0.431	0.0184	1.77	0.070	0.0089	8.7	0.566	0.420	0.394		60
80	0.391	0.0192	1.83	0.067	0.0096	9.3	0.582	0.438	0.406		80
100	0.354	0.0198	1.90	0.065	0.0103	9.9	0.600	0.456	0.418		100
120	0.318	0.0203	1.96	0.062	0,0112	10.5	0.620	0.477	0.430		120
140	0.283	0.0209	2.02	0.060	0.0121	11.1	0.642	0.503	0.422		140
160	0.251	0.0217	2.09	0.057	0.0132	11.8	0.666	0.534	0.454	0.459	160
180	0.222	0.0226	2.15	0.055	0.0144	12.4	0.693	0.571	0.465	0.471	180
200	0.195	0.0240	2.22	0.052	0.0157	13.1	0.722	0.616	0.477	0.482	200
220	0.172	0.026	2.28	0.050	0.017	13.9		0.677	0.488	0.494	220
240	0.152	0.028	2.34	0.047	0.019	14.5		0.750	0.500	0.506	240
260	0.134	0.032	2.41	0.045	0.021	15.3		0.860	0.511	0.517	260
280	0.122	0.036	2.47	0.042	0.024	16.0		1.06	0.522	0.529	280
300	0.080	0.043	2.53	0.037	0.027	16.7			0.533	0.540	300
306*	0.058	0.058	2.55	0.032	0.032	16.9			0.536	0.543	306*
320			2.60			17.4			0.544	0.551	320
340			2.66			18.1			0.555	0.562	340
360			2.72			18.9			0.566	0.573	360
380			2.78			19.6			0.576	0.584	380
400			2.85			20.3			0.587	0.594	400
500			3.15			23.8			0.637	0.645	500
600									0.685	0.691	600
700									0.728	0.733	700

*Critical temperature. Tabulated properties ignore critical region effects.

‡Actual value = (Table value) × (Indicated multiplier).

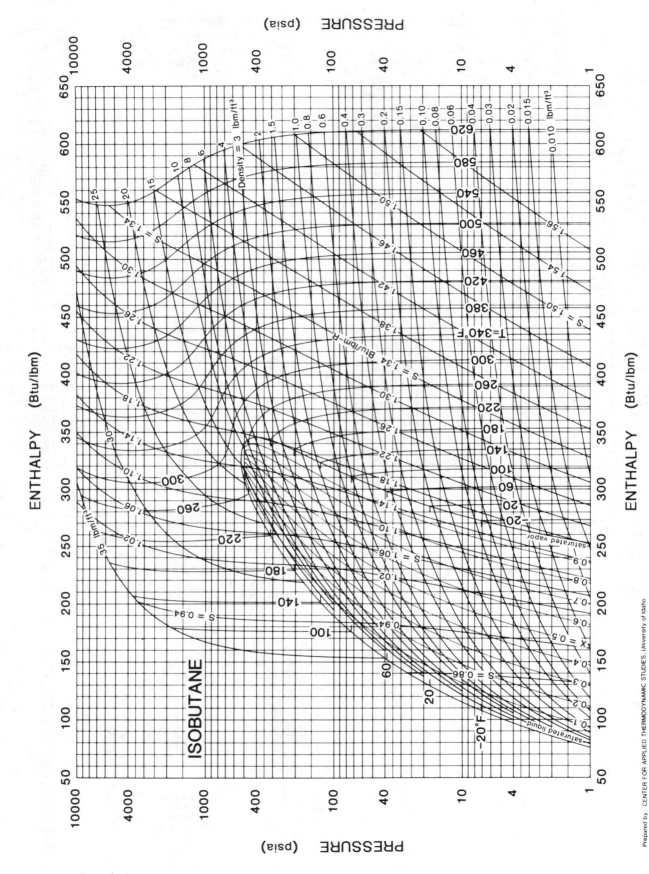

ISOBUTANE

Fig. 28   Pressure-Enthalpy Diagram for Refrigerant 600a (Isobutane)

Refrigerant 600a (Isobutane) — Properties of Saturated Liquid and Saturated Vapor

Temp, °F	Pressure psia	Pressure psig	Volume, ft³/lb$_m$ Vapor	Density, lb$_m$/ft³ Liquid	Enthalpy, Btu/lb$_m$ Liquid	Enthalpy, Btu/lb$_m$ Vapor	Entropy, Btu/lb$_m$·°R Liquid	Entropy, Btu/lb$_m$·°R Vapor
#−255.28	0.27E-5	29.921*	138E×10⁵	46.283	0.0	208.78	044514	1.4666
−250	0.58E-5	29.921*	667E×10⁴	46.108	2.1522	209.90	0.45554	1.4464
−230	0.71E-4	29.921*	599E×10³	45.445	10.392	214.27	0.49308	1.3808
−210	0.00056	29.920*	82960.	44.781	18.781	218.88	0.52810	1.3296
−190	0.00311	29.915*	16004.	44.115	27.330	223.72	0.56105	1.2893
−170	0.01335	29.894*	4003.5	43.446	36.047	228.78	0.59226	1.2576
−150	0.04638	29.827*	1231.4	42.774	44.941	234.04	0.62198	1.2326
−130	0.13579	29.645*	447.28	42.097	54.021	239.49	0.65042	1.2130
−110	0.34564	29.217*	186.05	41.415	63.298	245.12	0.67775	1.1977
−90	0.78378	28.325*	86.523	40.726	72.786	250.91	0.70414	1.1860
−70	1.6140	26.635*	44.142	40.027	82.499	256.87	0.72969	1.1772
−50	3.0650	23.681*	24.334	39.318	92.452	262.97	0.75454	1.1708
−40	4.1136	21.546*	18.525	38.958	97.524	266.07	0.76672	1.1683
−30	5.4342	18.857*	14.315	38.595	102.66	269.20	0.77876	1.1663
−20	7.0745	15.517*	11.213	38.228	107.86	272.35	0.79066	1.1648
−10	9.0862	11.422*	8.8939	37.857	113.13	275.54	0.80243	1.1636
0	11.524	6.4576*	7.1357	37.481	118.47	278.74	0.81408	1.1627
2	12.068	5.3502*	6.8370	37.406	119.55	279.39	0.81639	1.1626
4	12.632	4.2024*	6.5535	37.330	120.63	280.03	0.81870	1.1625
6	13.216	3.0132*	6.2843	37.253	121.71	280.67	0.82101	1.1624
8	13.821	1.7815*	6.0286	37.177	122.79	281.32	0.82331	1.1623
10	14.447	0.5064*	5.7855	37.100	123.88	281.96	0.82561	1.1622
10.78	14.696	0.0	5.6945	37.071	124.30	282.22	0.82650	1.1622
12	15.095	0.3994	5.5544	37.024	124.97	282.61	0.82791	1.1621
14	15.766	1.0699	5.3345	36.946	126.06	283.26	0.83020	1.1621
16	16.459	1.7632	5.1252	36.869	127.16	283.90	0.83248	1.1620
18	17.176	2.4798	4.9260	36.792	128.25	284.55	0.83476	1.1620
20	17.916	3.2204	4.7361	36.714	129.36	285.20	0.83704	1.1619
22	18.681	3.9853	4.5552	36.636	130.46	285.85	0.83932	1.1619
24	19.471	4.7751	4.3827	36.558	131.57	286.50	0.84159	1.1619
26	20.286	5.5904	4.2182	36.479	132.68	287.14	0.84385	1.1619
28	21.128	6.4317	4.0612	36.400	133.79	287.79	0.84611	1.1619
30	21.995	7.2994	3.9114	36.321	134.90	288.44	0.84837	1.1619
32	22.890	8.1942	3.7682	36.242	136.02	289.09	0.85063	1.1620
34	23.813	9.1165	3.6315	36.162	137.14	289.74	0.85288	1.1620
36	24.763	10.067	3.5008	36.082	138.27	290.39	0.85513	1.1620
38	25.742	11.046	3.3759	36.002	139.39	291.04	0.85737	1.1621

Temp, °F	Pressure psia	Pressure psig	Volume, ft³/lb$_m$ Vapor	Density, lb$_m$/ft³ Liquid	Enthalpy, Btu/lb$_m$ Liquid	Enthalpy, Btu/lb$_m$ Vapor	Entropy, Btu/lb$_m$·°R Liquid	Entropy, Btu/lb$_m$·°R Vapor
40	26.750	12.054	3.2564	35.922	140.52	291.69	0.85961	1.1621
42	27.788	13.092	3.1420	35.841	141.65	292.34	0.86185	1.1622
44	28.857	14.161	3.0326	35.760	142.79	292.99	0.86408	1.1623
46	29.956	15.260	2.9278	35.679	143.93	293.64	0.86631	1.1624
48	31.087	16.391	2.8274	35.597	145.07	294.29	0.86854	1.1625
50	32.249	17.553	2.7312	35.515	146.21	294.94	0.87076	1.1626
52	33.444	18.748	2.6390	35.433	147.36	295.59	0.87298	1.1627
54	34.672	19.976	2.5505	35.351	148.51	296.24	0.87520	1.1628
56	35.934	21.238	2.4657	35.268	149.66	296.89	0.87742	1.1629
58	37.230	22.534	2.3843	35.184	150.82	297.53	0.87963	1.1630
60	38.561	23.865	2.3062	35.101	151.98	298.18	0.88184	1.1632
62	39.927	25.231	2.2312	35.017	153.14	298.83	0.88404	1.1633
64	41.329	26.633	2.1592	34.933	154.31	299.48	0.88625	1.1635
66	42.767	28.071	2.0899	34.848	155.48	300.12	0.88845	1.1636
68	44.243	29.547	2.0234	34.763	156.65	300.77	0.89065	1.1638
70	45.756	31.060	1.9595	34.678	157.83	301.42	0.89284	1.1639
72	47.308	32.612	1.8980	34.592	159.01	302.06	0.89504	1.1641
74	48.898	34.202	1.8388	34.506	160.19	302.71	0.89723	1.1643
76	50.527	35.831	1.7819	34.419	161.37	303.35	0.89942	1.1645
78	52.197	37.501	1.7271	34.332	162.56	303.99	0.90161	1.1647
80	53.907	39.211	1.6743	34.245	163.75	304.64	0.90379	1.1648
90	63.085	48.389	1.4380	33.802	169.76	307.84	0.91469	1.1659
100	73.364	58.668	1.2409	33.347	175.86	311.02	0.92554	1.1670
110	84.818	70.122	1.0753	32.879	182.06	314.16	0.93636	1.1683
120	97.523	82.827	0.93529	32.397	188.35	317.27	0.94716	1.1696
130	111.55	96.858	0.81625	31.899	194.76	320.34	0.95795	1.1709
140	126.99	112.30	0.71439	31.383	201.27	323.35	0.96874	1.1723
150	143.92	129.22	0.62674	30.847	207.92	326.30	0.97954	1.1737
160	162.41	147.72	0.55088	30.289	214.69	329.17	0.99037	1.1751
170	182.58	167.88	0.48485	29.703	221.61	331.95	1.0012	1.1765
190	228.29	213.60	0.37612	28.435	235.96	337.13	1.0232	1.1789
210	281.99	267.30	0.29068	26.990	251.14	341.59	1.0456	1.1807
230	344.89	330.19	0.22141	25.267	267.46	344.81	1.0689	1.1811
250	418.76	404.07	0.16239	23.013	285.69	345.63	1.0941	1.1785
270	506.63	491.93	0.10380	18.937	309.91	338.88	1.1265	1.1662
** 274.7	530.1	515.4	0.07144	14.00	323.7	323.7	1.145	1.145

#Triple point   *in. Hg. Vacuum   **Critical point

Temp., °F	Viscosity, lb$_m$/ft·h Sat. Liquid	Viscosity, lb$_m$/ft·h Sat. Vapor	Viscosity Gas at p = 1 atm × 10⁻²‡	Thermal Conductivity, Btu/h·ft·°F Sat. Liquid	Thermal Conductivity Sat. Vapor	Thermal Cond. Gas at p = 1 atm × 10⁻³‡	Specific Heat, c$_p$, Btu/lb$_m$·°F Sat. Liquid	Specific Heat Sat. Vapor	Specific Heat Gas at p = 0 atm	Specific Heat Gas at p = 1 atm	Temp., °F
−80	1.19			0.080			0.499				−80
−60	0.98			0.077			0.508				−60
−40	0.83			0.075			0.518				−40
−20	0.71			0.073			0.528				−20
0	0.61			0.071			0.540				0
20	0.54	0.0164	1.64	0.068	0.0077	7.7	0.554	0.348	0.346		20
40	0.48	0.0175	1.71	0.066	0.0083	8.3	0.569	0.371	0.360	0.369	40
60	0.43	0.0183	1.77	0.064	0.0091	8.8	0.586	0.395	0.373	0.381	60
80	0.39	0.0191	1.84	0.061	0.0099	9.4	0.605	0.418	0.387	0.394	80
100	0.35	0.0198	1.90	0.059	0.0108	10.1	0.627	0.441	0.400	0.407	100
120	0.322	0.0205	1.97	0.057	0.0118	10.7	0.651	0.488	0.426	0.432	120
140	0.295	0.0214	2.03	0.054	0.0129	11.4	0.679	0.512	0.439	0.444	140
160	0.269	0.0226	2.10	0.052	0.0141	12.1	0.709	0.537	0.452	0.456	160
180	0.242	0.024	2.16	0.049	0.0154	12.8	0.743	0.567	0.464	0.468	180
200	0.215	0.026	2.23	0.047	0.0169	13.5	0.781	0.616	0.476	0.480	200
220	0.187	0.028	2.29	0.045	0.0188	14.3	0.823	0.695	0.489	0.492	220
240	0.157	0.032	2.35	0.042	0.021	15.0	0.864	0.813	0.501	0.503	240
260	0.125	0.037	2.42	0.038	0.025	15.7	0.92	1.245	0.512	0.515	260
275*	0.056	0.056	2.46	0.030	0.030	16.2			0.521	0.523	275*
280			2.48			16.4			0.524	0.526	280
300			2.54			17.1			0.536	0.537	300
320			2.60			17.8			0.547	0.549	320
340			2.67			18.5			0.558	0.560	340
360			2.73			19.3			0.569	0.570	360
380			2.79			20.0			0.580	0.581	380
400			2.86			20.8			0.591	0.592	400
500			3.16			24.4			0.642	0.643	500
600									0.689	0.690	600
700									0.733	0.734	700

*Critical temperature. Tabulated properties ignore critical region effects.   ‡Actual value = (Table value) × (Indicated multiplier).

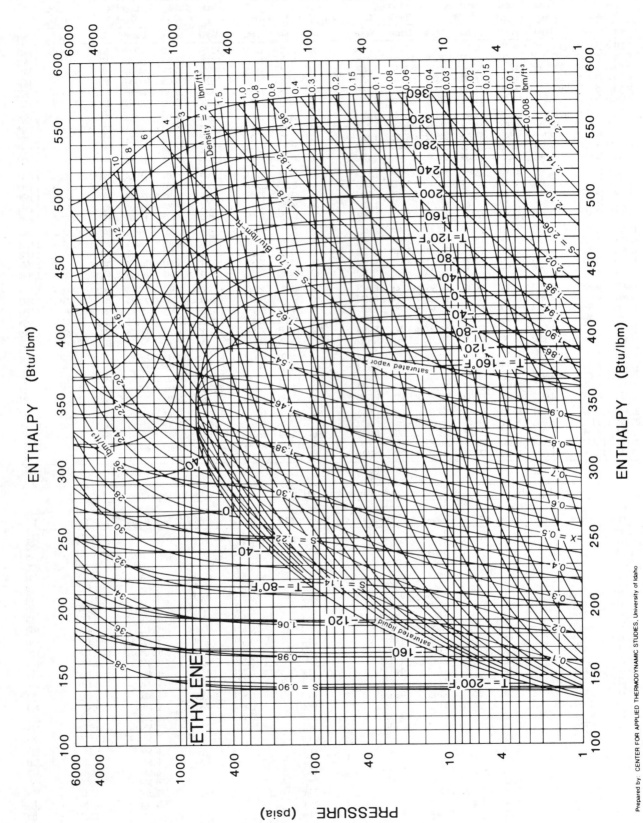

Fig. 29 Pressure-Enthalpy Diagram for Refrigerant 1150 (Ethylene)

Prepared by: CENTER FOR APPLIED THERMODYNAMIC STUDIES, University of Idaho

COPYRIGHT 1985 AMERICAN SOCIETY OF HEATING, REFRIGERATING AND AIR-CONDITIONING ENGINEERS

Refrigerant 1150 (Ethylene) Properties of Saturated Liquid and Saturated Vapor

Temp, °F	Pressure psia	Pressure psig	Volume, ft³/lbm Vapor	Density, lbm/ft³ Liquid	Enthalpy, Btu/lbm Liquid	Enthalpy, Btu/lbm Vapor	Entropy, Btu/lbm·°R Liquid	Entropy, Btu/lbm·°R Vapor	Temp, °F	Pressure psia	Pressure psig	Volume, ft³/lbm Vapor	Density, lbm/ft³ Liquid	Enthalpy, Btu/lbm Liquid	Enthalpy, Btu/lbm Vapor	Entropy, Btu/lbm·°R Liquid	Entropy, Btu/lbm·°R Vapor
−230	0.49203	28.919*	178.02	38.930	126.39	357.70	0.83928	1.8464	−110	51.797	37.101	2.3749	33.110	195.82	386.30	1.0812	1.6260
−220	0.88923	28.111*	102.63	38.493	132.09	360.44	0.86357	1.8163	−108	54.356	39.660	2.2693	32.999	197.03	386.66	1.0846	1.6239
−210	1.5232	26.820*	62.275	38.052	137.78	363.14	0.88682	1.7895	−106	57.008	42.312	2.1695	32.888	198.24	387.01	1.0879	1.6218
−200	2.4897	24.852*	39.511	37.604	143.47	365.80	0.90913	1.7654	−104	59.755	45.059	2.0751	32.776	199.45	387.36	1.0913	1.6197
−190	3.9056	21.969*	26.060	37.148	149.17	368.40	0.93063	1.7437	−102	62.599	47.903	1.9857	32.663	200.66	387.70	1.0947	1.6177
−185	4.8243	20.099*	21.442	36.916	152.02	369.68	0.94110	1.7337	−100	65.542	50.846	1.9010	32.550	201.88	388.03	1.0980	1.6157
−180	5.9086	17.891*	17.783	36.682	154.88	370.95	0.95139	1.7241	−98	68.587	53.891	1.8206	32.437	203.10	388.36	1.1014	1.6137
−175	7.1789	15.305*	14.858	36.445	157.74	372.19	0.96152	1.7150	−96	71.735	57.039	1.7445	32.323	204.32	388.68	1.1047	1.6117
−170	8.6568	12.296*	12.501	36.205	160.61	373.42	0.97149	1.7063	−94	74.988	60.292	1.6722	32.208	205.55	388.99	1.1080	1.6097
−165	10.365	8.8182*	10.587	35.962	163.49	374.63	0.98132	1.6980	−92	78.350	63.654	1.6035	32.092	206.78	389.30	1.1113	1.6078
−160	12.327	4.8232*	9.0215	35.717	166.38	375.82	0.99100	1.6900	−90	81.821	67.125	1.5382	31.976	208.01	389.60	1.1146	1.6059
−158	13.189	3.0692*	8.4759	35.618	167.54	376.28	0.99483	1.6870	−85	90.993	76.297	1.3888	31.682	211.11	390.32	1.1228	1.6011
−156	14.096	1.2210*	7.9705	35.519	168.70	376.75	0.99864	1.6839	−80	100.90	86.201	1.2566	31.384	214.24	391.00	1.1309	1.5965
−154.73	14.696	0.0	7.6695	35.456	169.43	377.04	1.0010	1.6820	−75	111.57	96.873	1.1393	31.080	217.39	391.63	1.1390	1.5920
−154	15.052	0.3559	7.5018	35.419	169.86	377.21	1.0024	1.6810	−70	123.04	108.35	1.0349	30.771	220.57	392.20	1.1470	1.5875
−152	16.057	1.3611	7.0666	35.319	171.02	377.67	1.0062	1.6780	−65	135.35	120.65	0.94175	30.456	223.78	392.73	1.1550	1.5831
−150	17.114	2.4176	6.6623	35.219	172.18	378.13	1.0100	1.6752	−60	148.53	133.83	0.85834	30.135	227.03	393.20	1.1630	1.5788
−148	18.223	3.5271	6.2862	35.118	173.34	378.58	1.0137	1.6723	−55	162.61	147.92	0.78345	29.806	230.31	393.60	1.1710	1.5745
−146	19.387	4.6913	5.9360	35.016	174.51	379.02	1.0174	1.6696	−50	177.64	162.94	0.71603	29.470	233.64	393.94	1.1789	1.5702
−144	20.608	5.9119	5.6097	34.914	175.68	379.47	1.0211	1.6668	−45	193.64	178.95	0.65516	29.124	237.02	394.21	1.1868	1.5659
−142	21.887	7.1909	5.3053	34.812	176.85	379.90	1.0248	1.6641	−40	210.67	195.97	0.60006	28.769	240.44	394.41	1.1948	1.5616
−140	23.226	8.5299	5.0212	34.709	178.02	380.34	1.0284	1.6615	−35	228.74	214.04	0.55006	28.404	243.92	394.52	1.2027	1.5574
−138	24.627	9.9308	4.7556	34.606	179.19	380.77	1.0320	1.6589	−30	247.91	233.21	0.50456	28.027	247.45	394.55	1.2107	1.5530
−136	26.091	11.395	4.5073	34.502	180.37	381.20	1.0357	1.6563	−25	268.21	253.51	0.46305	27.636	251.06	394.48	1.2187	1.5487
−134	27.622	12.926	4.2748	34.398	181.54	381.62	1.0393	1.6538	−20	289.68	274.98	0.42509	27.231	254.73	394.30	1.2268	1.5442
−132	29.219	14.523	4.0571	34.293	182.72	382.03	1.0428	1.6513	−15	312.37	297.67	0.39027	26.810	258.49	394.00	1.2350	1.5397
−130	30.887	16.191	3.8529	34.188	183.90	382.45	1.0464	1.6488	−10	336.32	321.62	0.35826	26.370	262.34	393.57	1.2432	1.5350
−128	32.625	17.929	3.6614	34.082	185.08	382.85	1.0499	1.6464	−5	361.57	346.88	0.32872	25.909	266.29	393.00	1.2515	1.5302
−126	34.436	19.740	3.4815	33.976	186.27	383.26	1.0535	1.6440	0	388.19	373.49	0.30141	25.423	270.35	392.25	1.2600	1.5252
−124	36.323	21.627	3.3124	33.870	187.45	383.66	1.0570	1.6416	5	416.20	401.51	0.27605	24.909	274.55	391.32	1.2686	1.5199
−122	38.287	23.591	3.1534	33.763	188.64	384.05	1.0605	1.6393	10	445.69	430.99	0.25242	24.362	278.90	390.16	1.2775	1.5144
−120	40.329	25.633	3.0038	33.655	189.83	384.44	1.0640	1.6370	20	509.30	494.60	0.20947	23.138	288.18	386.98	1.2960	1.5019
−118	42.453	27.757	2.8629	33.547	191.03	384.82	1.0674	1.6348	30	579.59	564.89	0.17067	21.650	298.61	382.08	1.3163	1.4868
−116	44.660	29.964	2.7300	33.438	192.22	385.20	1.0709	1.6325	40	657.38	642.68	0.13295	19.595	311.46	373.66	1.3409	1.4653
−114	46.951	32.255	2.6048	33.329	193.42	385.57	1.0743	1.6303	** 48.5	731.0	716.3	0.07480	13.37	342.2	342.2	1.400	1.400
−112	49.330	34.634	2.4865	33.220	194.62	385.94	1.0778	1.6282									

*in. Hg. Vacuum

**Critical point

Temp., °F	Viscosity, lbm/ft·h Sat. Liquid	Viscosity, lbm/ft·h Sat. Vapor	Viscosity, lbm/ft·h Gas at p = 1 atm × 10⁻²‡	Thermal Conductivity, Btu/h·ft·°F Sat. Liquid	Thermal Conductivity, Btu/h·ft·°F Sat. Vapor	Thermal Conductivity, Btu/h·ft·°F Gas at p = 1 atm × 10⁻²‡	Specific Heat, c_p, Btu/lbm·°F Sat. Liquid	Specific Heat, c_p, Btu/lbm·°F Sat. Vapor	Specific Heat, c_p, Btu/lbm·°F Gas at p = 0 atm	Specific Heat, c_p, Btu/lbm·°F Gas at p = 1 atm	Temp., °F
−180	0.487			0.119			0.572				−180
−160	0.412			0.113			0.573	0.288			−160
−140	0.356			0.107	0.0049	0.49	0.576	0.299			−140
−120	0.314	0.016	1.62	0.100	0.0055	0.55	0.582	0.313			−120
−100	0.279	0.017	1.71	0.094	0.0061	0.61	0.592	0.335			−100
−80	0.252	0.018	1.80	0.087	0.0068	0.67	0.607	0.364			−80
−60	0.230	0.019	1.89	0.081	0.0075	0.73	0.642	0.403			−60
−40	0.211	0.020	1.98	0.074	0.0085	0.79	0.705	0.454			−40
−20	0.187	0.023	2.07	0.068	0.0097	0.85	0.804	0.529			−20
0	0.157	0.026	2.16	0.061	0.0110	0.91	0.946	0.657	0.331	0.336	0
20	0.123	0.029	2.25	0.053	0.013	0.98	1.139	0.882	0.342	0.346	20
30	0.103	0.031	2.30	0.048	0.015	1.01	1.25	1.014	0.347	0.351	30
40	0.082	0.036	2.34	0.041	0.018	1.04	1.39	1.35	0.352	0.356	40
49*	0.053	0.053	2.38	0.027	0.027	1.07			0.357	0.361	49*
60			2.43			1.11			0.362	0.366	60
80			2.51			1.18			0.373	0.376	80
100			2.60			1.25			0.383	0.386	100
120			2.68			1.33			0.393	0.395	120
140			2.77			1.41			0.403	0.405	140
180			2.93			1.58			0.423	0.424	180
220			3.10			1.77			0.442	0.443	220
260			3.25			1.96			0.461	0.462	260
300			3.41			2.14			0.479	0.480	300
400			3.79			2.64			0.523	0.524	400
500			4.15			3.15			0.564	0.565	500
600			4.49			3.67			0.602	0.602	600
700			4.81			4.19			0.637		700
800			5.13			4.73			0.670		800
1000			5.72			5.87			0.726		1000

*Critical temperature. Tabulated properties ignore critical region effects.

‡Actual value = (Table value) × (Indicated multiplier).

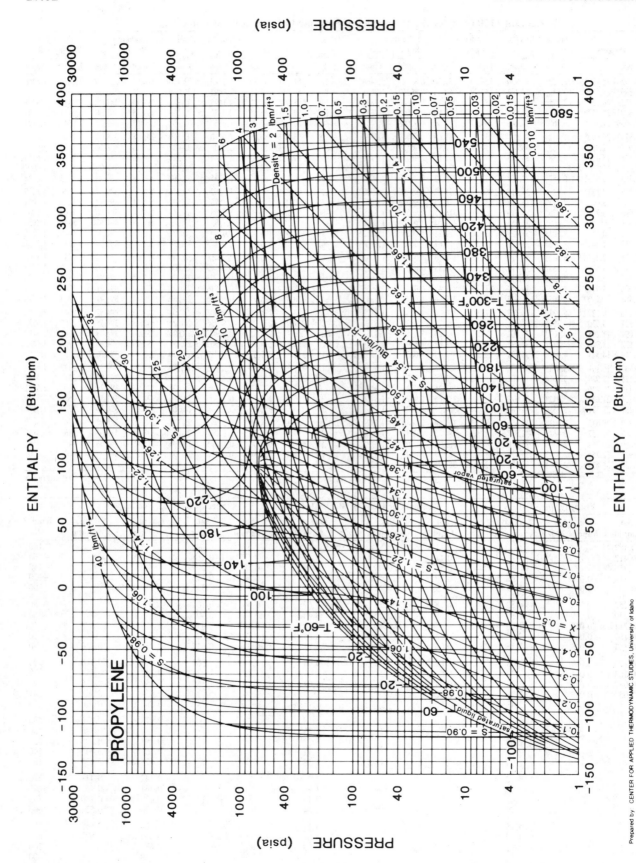

Fig. 30   Pressure-Enthalpy Diagram for Refrigerant 1270 (Propylene)

Refrigerant 1270 (Propylene) Properties of Saturated Liquid and Saturated Vapor

Temp, °F	Pressure psia	Pressure psig	Volume, ft³/lb$_m$ Vapor	Density, lb$_m$/ft³ Liquid	Enthalpy, Btu/lb$_m$ Liquid	Enthalpy, Btu/lb$_m$ Vapor	Entropy, Btu/lb$_m$·°R Liquid	Entropy, Btu/lb$_m$·°R Vapor
#−301.47	0.23E−6	29.921*	292E×10⁶	47.998	−213.80	31.094	0.51019	2.0581
−300	0.30E−6	29.921*	218E×10⁶	47.940	−212.95	31.407	0.51552	2.0459
−280	0.83E−5	29.921*	692E×10⁴	47.136	−203.33	35.763	0.57244	1.9032
−260	0.00012	29.921*	468E×10³	46.325	−194.87	40.275	0.61709	1.7948
−240	0.00108	29.919*	54132.	45.519	−186.29	44.930	0.65801	1.7106
−220	0.00663	29.908*	9340.1	44.720	−177.40	49.714	0.69676	1.6444
−200	0.03031	29.859*	2187.1	43.928	−168.22	54.619	0.73353	1.5917
−180	0.10956	29.698*	649.45	43.140	−158.83	59.637	0.76834	1.5495
−160	0.32751	29.254*	232.62	42.353	−149.29	64.755	0.80129	1.5156
−140	0.83912	28.213*	96.766	41.562	−139.62	69.956	0.83253	1.4881
−120	1.8952	26.063*	45.404	40.763	−129.81	75.215	0.86226	1.4659
−100	3.8587	22.065*	23.490	39.954	−119.87	80.504	0.89067	1.4478
−95	4.5452	20.667*	20.184	39.750	−117.36	81.827	0.89759	1.4438
−90	5.3266	19.076*	17.425	39.544	−114.85	83.149	0.90444	1.4400
−85	6.2120	17.274*	15.111	39.338	−112.32	84.470	0.91123	1.4365
−80	7.2109	15.240*	13.160	39.130	−109.78	85.789	0.91795	1.4330
−75	8.3334	12.954*	11.508	38.921	−107.22	87.105	0.92461	1.4298
−70	9.5899	10.396*	10.102	38.711	−104.66	88.418	0.93121	1.4267
−65	10.991	7.5434*	8.9010	38.499	−102.09	89.728	0.93776	1.4238
−60	12.548	4.3732*	7.8702	38.286	−99.499	91.034	0.94426	1.4210
−55	14.273	0.8620*	6.9821	38.072	−96.899	92.334	0.95070	1.4183
−53.85	14.696	0.0	6.7948	38.022	−96.297	92.634	0.95219	1.4177
−50	16.177	1.4805	6.2140	37.856	−94.285	93.630	0.95710	1.4158
−48	16.991	2.2951	5.9360	37.769	−93.236	94.146	0.95965	1.4148
−46	17.837	3.1411	5.6732	37.682	−92.184	94.662	0.96219	1.4139
−44	18.715	4.0194	5.4246	37.594	−91.130	95.176	0.96472	1.4129
−42	19.627	4.9308	5.1893	37.507	−90.074	95.689	0.96724	1.4120
−40	20.572	5.8760	4.9663	37.419	−89.015	96.201	0.96976	1.4111
−38	21.552	6.8561	4.7551	37.330	−87.954	96.712	0.97227	1.4102
−36	22.568	7.8717	4.5548	37.242	−86.891	97.222	0.97477	1.4093
−34	23.620	8.9238	4.3648	37.153	−85.825	97.731	0.97727	1.4085
−32	24.709	10.013	4.1845	37.064	−84.757	98.238	0.97976	1.4076
−30	25.837	11.141	4.0133	36.974	−83.686	98.744	0.98225	1.4068
−28	27.003	12.307	3.8506	36.884	−82.612	99.249	0.98472	1.4060
−26	28.210	13.514	3.6960	36.794	−81.536	99.752	0.98720	1.4052
−24	29.457	14.761	3.5489	36.704	−80.458	100.25	0.98967	1.4045
−22	30.746	16.050	3.4090	36.613	−79.376	100.75	0.99213	1.4037
−20	32.078	17.382	3.2759	36.521	−78.292	101.25	0.99458	1.4029
−18	33.453	18.757	3.1490	36.430	−77.206	101.75	0.99703	1.4022
−16	34.873	20.177	3.0282	36.338	−76.116	102.24	0.99948	1.4015
−14	36.338	21.642	2.9130	36.246	−75.024	102.74	1.0019	1.4008
−12	37.849	23.153	2.8032	36.153	−73.929	103.23	1.0044	1.4001
−10	39.407	24.711	2.6984	36.060	−72.831	103.72	1.0068	1.3994
−8	41.014	26.318	2.5984	35.966	−71.730	104.21	1.0092	1.3987
−6	42.670	27.974	2.5029	35.872	−70.626	104.69	1.0116	1.3981
−4	44.375	29.679	2.4117	35.778	−69.519	105.18	1.0140	1.3974
−2	46.132	31.436	2.3245	35.683	−68.409	105.66	1.0165	1.3968
0	47.941	33.245	2.2412	35.588	−67.296	106.14	1.0189	1.3962
2	49.803	35.107	2.1615	35.493	−66.180	106.62	1.0213	1.3956
4	51.718	37.022	2.0852	35.397	−65.060	107.10	1.0237	1.3949
6	53.689	38.993	2.0122	35.300	−63.938	107.57	1.0260	1.3944
8	55.716	41.020	1.9424	35.203	−62.812	108.04	1.0284	1.3938
10	57.799	43.103	1.8754	35.106	−61.683	108.51	1.0308	1.3932
12	59.940	45.244	1.8113	35.008	−60.551	108.98	1.0332	1.3926
14	62.140	47.444	1.7498	34.910	−59.415	109.44	1.0356	1.3921
16	64.401	49.705	1.6908	34.811	−58.276	109.91	1.0380	1.3915
18	66.722	52.026	1.6343	34.712	−57.133	110.37	1.0403	1.3910
25	75.340	60.644	1.4536	34.360	−53.106	111.96	1.0486	1.3892
30	81.984	67.288	1.3393	34.104	−50.202	113.07	1.0545	1.3879
35	89.055	74.359	1.2356	33.845	−47.274	114.16	1.0604	1.3867
40	96.569	81.873	1.1414	33.582	−44.321	115.24	1.0662	1.3855
45	104.54	89.845	1.0557	33.315	−41.343	116.29	1.0721	1.3844
50	112.99	98.294	0.97747	33.033	−38.338	117.33	1.0779	1.3833
60	131.38	116.69	0.84066	32.487	−32.245	119.32	1.0895	1.3812
70	151.89	137.19	0.72565	31.910	−26.033	121.20	1.1011	1.3791
80	174.64	159.95	0.62850	31.310	−19.690	122.96	1.1128	1.3771
90	199.80	185.10	0.54535	30.685	−13.206	124.57	1.1244	1.3750
100	227.50	212.80	0.47422	30.029	−6.5635	126.03	1.1361	1.3730
120	291.16	276.46	0.35950	28.606	7.2746	128.32	1.1596	1.3685
140	366.99	352.29	0.27172	26.978	22.063	129.48	1.1839	1.3630
160	456.56	441.87	0.20226	25.009	38.298	128.83	1.2094	1.3555
180	562.06	547.36	0.14370	22.290	57.444	124.53	1.2385	1.3434
**198.4	676.5	661.8	0.07171	13.95	97.42	97.42	1.298	1.298

#Triple point *in. Hg. Vacuum **Critical point

Temp., °F	Viscosity, lb$_m$/ft·h Sat. Liquid	Sat. Vapor	Gas at p = 1 atm × 10⁻²‡	Thermal Conductivity, Btu/h·ft·°F Sat. Liquid	Sat. Vapor	Gas at p = 1 atm × 10⁻³‡	Specific Heat, c_p, Btu/lb$_m$·°F Sat. Liquid	Sat. Vapor	Gas at p = 0 atm	Gas at p = 1 atm	Temp., °F
−100	0.595	0.0150		0.094	0.0035		0.485		0.287		−100
−80	0.511	0.0157		0.091	0.0044		0.483		0.295		−80
−60	0.443			0.087	0.0052		0.484		0.303		−60
−40	0.388	0.0165	1.65	0.084	0.0058	0.56	0.486	0.321	0.312		−40
−20	0.344	0.0173	1.73	0.081	0.0064	0.63	0.492	0.334	0.321		−20
0	0.307	0.0182	1.82	0.078	0.0069	0.69	0.501	0.344	0.329	0.333	0
20	0.276	0.0191	1.90	0.074	0.0076	0.76	0.513	0.359	0.338	0.343	20
40	0.250	0.0197	1.98	0.071	0.0083	0.83	0.530	0.375	0.347	0.353	40
60	0.220	0.0206	2.06	0.067	0.0091	0.90	0.551	0.408	0.356	0.363	60
80	0.191	0.0218	2.14	0.064	0.0102	0.97	0.577	0.458	0.366	0.373	80
100	0.163	0.0233	2.21	0.061	0.0116	1.05	0.609	0.540	0.375	0.383	100
120	0.135	0.026	2.29	0.057	0.0133	1.12	0.647	0.644	0.384	0.393	120
140	0.111	0.028	2.37	0.054	0.0153	1.19	0.692	0.798	0.394	0.403	140
160	0.089	0.032	2.44	0.050	0.018	1.27	0.743		0.403	0.413	160
180	0.071	0.037	2.52	0.045	0.022	1.35			0.413	0.422	180
190	0.065	0.045	2.55	0.040	0.025	1.39			0.418	0.427	190
197*	0.058	0.058	2.58	0.029	0.029	1.42			0.421	0.430	197*
200			2.59			1.42			0.423	0.432	200
220						1.50			0.432	0.441	220
240						1.58			0.442	0.450	240
260						1.67			0.452	0.459	260
280						1.75			0.461	0.468	280
300						1.84			0.471	0.477	300
320						1.92			0.480	0.486	320
340						2.01			0.490	0.495	340
400						2.29			0.518	0.521	400
500						2.75			0.563	0.563	500
600						3.25			0.605		600
700						3.77			0.642		700
800						4.27			0.674		800

*Critical temperature. Tabulated properties ignore critical region effects. ‡Actual value = (Table value) × (Indicated multiplier).

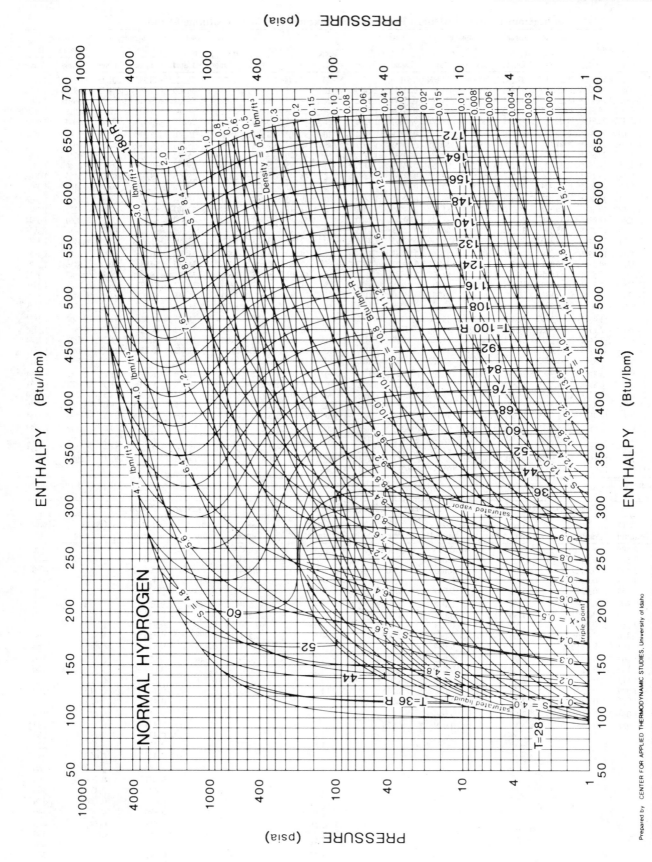

Fig. 31   Temperature-Entropy Diagram for Refrigerant 702 (Normal Hydrogen)

Prepared by   CENTER FOR APPLIED THERMODYNAMIC STUDIES, University of Idaho

COPYRIGHT 1985   AMERICAN SOCIETY OF HEATING, REFRIGERATING AND AIR-CONDITIONING ENGINEERS

Refrigerant 702 (Normal Hydrogen)   Properties of Saturated Liquid and Saturated Vapor

Temp, °R	Pressure psia	Pressure psig	Volume, ft³/lbₘ Vapor	Density, lbₘ/ft³ Liquid	Enthalpy, Btu/lbₘ Liquid	Enthalpy, Btu/lbₘ Vapor	Entropy, Btu/lbₘ·°R Liquid	Entropy, Btu/lbₘ·°R Vapor	Temp, °R	Pressure psia	Pressure psig	Volume, ft³/lbₘ Vapor	Density, lbₘ/ft³ Liquid	Enthalpy, Btu/lbₘ Liquid	Enthalpy, Btu/lbₘ Vapor	Entropy, Btu/lbₘ·°R Liquid	Entropy, Btu/lbₘ·°R Vapor
#25.11	1.0214	27.842*	128.82	4.8013	93.914	287.21	3.3650	11.1459	40.0	24.295	9.5987	7.6245	4.2721	125.07	312.18	4.2930	8.9963
26	1.4644	26.940*	92.561	4.7748	95.529	289.07	3.4275	10.8701	40.2	24.994	10.298	7.4277	4.2631	125.61	312.37	4.3059	8.9766
27	1.9346	25.982*	72.446	4.7448	97.269	291.22	3.4925	10.6828	40.4	25.707	11.011	7.2372	4.2541	126.17	312.55	4.3189	8.9571
28	2.5135	24.804*	57.548	4.7142	98.998	293.30	3.5545	10.5076	40.6	26.435	11.739	7.0526	4.2449	126.72	312.73	4.3318	8.9377
29	3.2165	23.372*	46.324	4.6830	100.75	295.33	3.6152	10.3434	40.8	27.177	12.481	6.8739	4.2357	127.28	312.90	4.3448	8.9183
30	4.0602	21.655*	37.736	4.6511	102.56	297.29	3.6752	10.1889	41.0	27.935	13.239	6.7006	4.2264	127.85	313.06	4.3578	8.8991
31	5.0618	19.615*	31.072	4.6183	104.43	299.18	3.7352	10.0431	41.2	28.707	14.011	6.5327	4.2171	128.42	313.22	4.3708	8.8799
32	6.2391	17.218*	25.834	4.5846	106.37	301.00	3.7954	9.9050	41.4	29.495	14.799	6.3699	4.2076	128.99	313.37	4.3838	8.8608
33	7.6106	14.426*	21.668	4.5499	108.39	302.74	3.8559	9.7739	41.6	30.298	15.602	6.2121	4.1981	129.57	313.52	4.3969	8.8418
34	9.1953	11.200*	18.319	4.5141	110.50	304.39	3.9168	9.6490	41.8	31.117	16.421	6.0590	4.1885	130.15	313.66	4.4100	8.8229
35.0	11.013	7.4994*	15.600	4.4772	112.69	305.96	3.9782	9.5296	42.0	31.952	17.256	5.9104	4.1789	130.74	313.80	4.4232	8.8040
35.2	11.406	6.6990*	15.119	4.4696	113.14	306.26	3.9906	9.5064	42.2	32.802	18.106	5.7663	4.1691	131.33	313.92	4.4363	8.7853
35.4	11.809	5.8777*	14.656	4.4620	113.60	306.55	4.0029	9.4833	42.4	33.669	18.973	5.6265	4.1593	131.93	314.05	4.4496	8.7666
35.6	12.223	5.0352*	14.212	4.4544	114.05	306.85	4.0153	9.4604	42.6	34.552	19.856	5.4907	4.1494	132.53	314.16	4.4628	8.7479
35.8	12.647	4.1711*	13.784	4.4467	114.51	307.14	4.0277	9.4377	42.8	35.451	33.755	5.3588	4.1394	133.14	314.27	4.4761	8.7294
36.0	13.082	3.2852*	13.372	4.4390	114.98	307.42	4.0401	9.4152	43.0	36.367	21.671	5.2308	4.1293	133.75	314.38	4.4894	8.7109
36.2	13.528	2.3770*	12.976	4.4312	115.44	307.70	4.0526	9.3928	43.2	37.300	22.604	5.1064	4.1191	134.36	314.48	4.5027	8.6924
36.4	13.986	1.4464*	12.594	4.4233	115.92	307.98	4.0650	9.3706	43.4	38.249	23.553	4.9855	4.1089	134.99	314.57	4.5161	8.6740
36.6	14.454	0.4928*	12.226	4.4154	116.39	308.25	4.0775	9.3486	43.6	39.216	24.520	4.8681	4.0985	135.61	314.65	4.5295	8.6557
36.70	14.696	0.0	12.045	4.4114	116.63	308.39	4.0839	9.3375	43.8	40.200	25.504	4.7540	4.0881	136.25	314.73	4.5419	8.6374
36.8	14.934	0.2377	11.872	4.4075	116.87	308.52	4.0900	9.3268	44	41.202	26.506	4.6430	4.0775	136.88	314.80	4.5564	8.6191
37.0	15.425	0.7290	11.531	4.3995	117.35	308.78	4.1025	9.3051	45	46.477	31.781	4.1320	4.0233	140.15	315.04	4.6244	8.5285
37.2	15.928	1.2320	11.202	4.3914	117.84	309.04	4.1151	9.2836	46	52.214	37.518	3.6857	3.9665	143.56	315.09	4.6936	8.4385
37.4	16.443	1.7469	10.884	4.3833	118.33	309.30	4.1277	9.2622	47	58.434	43.738	3.2939	3.9067	147.13	314.94	4.7640	8.3489
37.6	16.970	2.2739	10.578	4.3751	118.82	309.55	4.1402	9.2409	48	65.158	50.462	2.9484	3.8435	150.87	314.55	4.8360	8.2592
37.8	17.509	2.8131	10.283	4.3668	119.32	309.80	4.1529	9.2198	49	72.407	57.711	2.6423	3.7765	154.80	313.92	4.9098	8.1689
38.0	18.061	3.3647	9.9976	4.3585	119.82	310.04	4.1655	9.1989	50	80.202	65.506	2.3698	3.7051	158.95	313.01	4.9857	8.0775
38.2	18.625	3.9289	9.7223	4.3502	120.33	310.27	4.1781	9.1781	51	88.563	73.867	2.1260	3.6285	163.33	311.78	5.0642	7.9844
38.4	19.202	4.5057	9.4563	4.3418	120.84	310.51	4.1908	9.1574	52	97.512	82.816	1.9068	3.5456	168.00	310.19	5.1459	7.8887
38.6	19.791	5.0955	9.1994	4.3333	121.35	310.73	4.2035	9.1368	53	107.07	92.373	1.7086	3.4549	173.02	308.18	5.2318	7.7896
38.8	20.394	5.6983	8.9511	4.3248	121.87	310.96	4.2162	9.1164	54	117.25	102.56	1.5282	3.3543	178.46	305.66	5.3231	7.6856
39.0	21.010	6.3143	8.7112	4.3161	122.39	311.17	4.2290	9.0961	55	128.09	113.39	1.3626	3.2402	184.46	302.54	5.4221	7.5752
39.2	21.640	6.9437	8.4792	4.3075	122.92	311.39	4.2417	9.0759	56	139.59	124.89	1.2092	3.1060	191.27	298.64	5.5326	7.4556
39.4	22.283	7.5867	8.2548	4.2987	123.45	311.59	4.2545	9.0558	57	151.78	137.08	1.0653	2.9369	199.44	293.71	5.6640	7.3231
39.6	22.939	8.2434	8.0377	4.2899	123.99	311.79	4.2673	9.0359	58	164.68	149.98	0.92814	2.6779	211.03	287.39	5.8506	7.1718
39.8	23.610	8.9140	7.8277	4.2811	124.52	311.99	4.2802	9.0160	**59.74	190.8	176.1	0.5320	1.880	248.3	248.3	6.444	6.444

#Triple point   *in. Hg. Vacuum   **Critical point

Temp., K	Viscosity, N·s/m² Sat. Liquid × 10⁻³‡	Viscosity, N·s/m² Sat. Vapor × 10⁻³‡	Viscosity, N·s/m² Gas at p = 1 atm × 10⁻⁶‡	Thermal Conductivity, W/(m·K) Sat. Liquid	Thermal Conductivity, W/(m·K) Sat. Vapor	Thermal Conductivity, W/(m·K) Gas at p = 1 atm	Specific Heat, cₚ, kJ/(kg·K) Sat. Liquid	Specific Heat, cₚ, kJ/(kg·K) Sat. Vapor	Specific Heat, cₚ, kJ/(kg·K) Gas at p = 0 atm	Specific Heat, cₚ, kJ/(kg·K) Gas at p = 1 atm	Temp., K
15	0.0225			0.1022	0.0117		7.199	10.6			15
16	0.0200			0.1055	0.0126		7.539	10.8			16
18	0.0163			0.1121	0.0142		8.380	11.2			18
20	0.0139	0.00109	1.09	0.1184	0.0159	0.0159	9.408	11.8	10.31		20
22	0.0120	0.00121	1.20	0.1238	0.0180	0.0173	10.59	12.6	10.31		22
24	0.0103	0.00134	1.31	0.1272	0.0205	0.0186	11.91	13.8	10.31		24
26	0.00881	0.00146	1.41	0.1251	0.023	0.0200	13.33	15.7	10.31		26
28	0.00759	0.00161	1.51	0.1168	0.027	0.0213	16.01	18.8	10.31		28
30	0.00640	0.00183	1.61	0.106	0.031	0.0227	20.6	25.2	10.31	10.83	30
31	0.00585	0.00201	1.66	0.100	0.035	0.0234	25.9	31.1	10.31	10.78	31
32	0.00485	0.00227	1.71	0.091	0.040	0.0241	38.5		10.31	10.73	32
33*	0.00364	0.00364	1.76	0.060	0.060	0.0247			10.31	10.69	33*
40			2.09			0.0294			10.31	10.57	40
60			2.91			0.0426			10.39	10.49	60
80			3.60			0.0552			10.68	10.74	80
100			4.21			0.0676			11.19	11.22	100
120			4.77			0.0801			11.81	11.85	120
140			5.31			0.0926			12.36	12.39	140
160			5.82			0.1046			12.83	12.85	160
180			6.31			0.1164			13.21	13.23	180
200			6.78			0.1280			13.52	13.54	200
240			7.68			0.1506			13.95	13.99	240
280			8.53			0.1717			14.20	14.25	280
320			9.35			0.1910			14.33	14.38	320
360			10.14			0.2069			14.40	14.43	360
400			10.91			0.2212			14.48	14.49	400
450			11.84			0.2389			14.50	14.51	450
500			12.74			0.2564			14.51	14.52	500
550			13.6			0.274			14.52	14.53	550
600			14.5			0.291			14.53	14.54	600

*Critical temperature. Tabulated properties ignore critical region effects.   ‡Actual value = (Table value) × (Indicated multiplier).

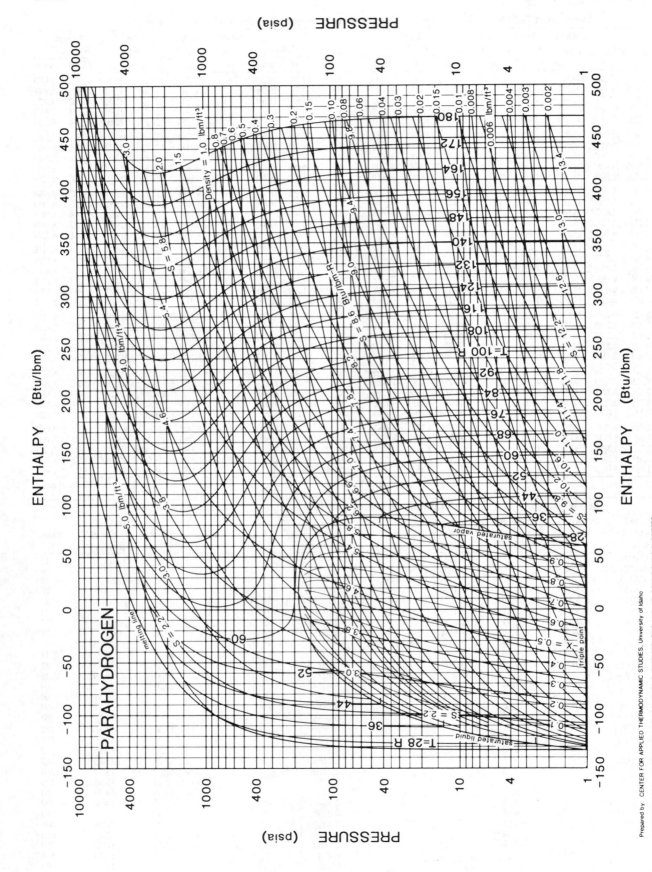

Fig. 32   Temperature-Entropy Diagram for Refrigerant 702p (Parahydrogen)

Prepared by   CENTER FOR APPLIED THERMODYNAMIC STUDIES, University of Idaho
COPYRIGHT 1985   AMERICAN SOCIETY OF HEATING, REFRIGERATING AND AIR-CONDITIONING ENGINEERS

Refrigerant 702p (Parahydrogen)   Properties of Saturated Liquid and Saturated Vapor

Temp, °R	Pressure psia	Pressure psig	Volume, ft³/lb_m Vapor	Density, lb_m/ft³ Liquid	Enthalpy, Btu/lb_m Liquid	Enthalpy, Btu/lb_m Vapor	Entropy, Btu/lb_m·°R Liquid	Entropy, Btu/lb_m·°R Vapor	Temp, °R	Pressure psia	Pressure psig	Volume, ft³/lb_m Vapor	Density, lb_m/ft³ Liquid	Enthalpy, Btu/lb_m Liquid	Enthalpy, Btu/lb_m Vapor	Entropy, Btu/lb_m·°R Liquid	Entropy, Btu/lb_m·°R Vapor
#24.84	1.0214	27.842*	127.37	4.8093	−132.98	60.124	1.1850	8.9581	40.0	25.059	10.363	7.3511	4.2727	−101.35	85.263	2.1310	6.7960
25	1.0751	27.732*	121.72	4.8046	−132.68	60.475	1.1967	8.9226	40.2	25.774	11.078	7.1622	4.2637	−100.80	85.440	2.1439	6.7764
26	1.4620	26.945*	92.720	4.7748	−130.90	62.648	1.2661	8.7104	40.4	26.504	11.808	6.9793	4.2547	−100.25	85.611	2.1568	6.7570
27	1.9480	25.955*	71.937	4.7448	−129.17	64.772	1.3308	8.5141	40.6	27.248	12.552	6.8021	4.2456	−99.693	85.777	2.1698	6.7376
28	2.5484	24.733*	56.733	4.7142	−127.44	66.840	1.3929	8.3317	40.8	28.007	13.311	6.6305	4.2364	−99.132	85.938	2.1827	6.7184
29	3.2795	23.244*	45.400	4.6831	−125.68	68.844	1.4536	8.1616	41.0	28.781	14.085	6.4643	4.2271	−98.566	86.093	2.1957	6.6992
30	4.1580	21.455*	36.809	4.6511	−123.87	70.781	1.5137	8.0023	41.2	29.571	14.875	6.3032	4.2178	−97.996	86.242	2.2088	6.6802
31	5.2012	19.331*	30.195	4.6184	−122.00	72.644	1.5737	7.8526	41.4	30.375	15.679	6.1469	4.2084	−97.422	86.385	2.2218	6.6612
32	6.4271	16.836*	25.032	4.5847	−120.05	74.428	1.6338	7.7115	41.6	31.194	16.498	5.9955	4.1989	−96.843	86.523	2.2349	6.6423
33	7.8538	13.931*	20.950	4.5500	−118.03	76.130	1.6942	7.5780	41.8	32.030	17.334	5.8486	4.1893	−96.259	86.654	2.2480	6.6235
34	9.4999	10.579*	17.684	4.5143	−115.93	77.743	1.7550	7.4512	42.0	32.880	18.184	5.7061	4.1797	−95.671	86.780	2.2612	6.6048
35.0	11.384	6.7425*	15.043	4.4774	−113.73	79.264	1.8164	7.3305	42.2	33.747	19.051	5.5678	4.1699	−95.078	86.899	2.2743	6.5861
35.2	11.792	5.9134*	14.577	4.4698	−113.28	79.556	1.8287	7.3071	42.4	34.630	19.934	5.4336	4.1601	−94.480	87.012	2.2875	6.5675
35.4	12.209	5.0631*	14.129	4.4623	−112.83	79.845	1.8410	7.2838	42.6	35.529	20.833	5.3033	4.1503	−93.877	87.119	2.3008	6.5490
35.6	12.637	4.1912*	13.698	4.4546	−112.38	80.129	1.8534	7.2607	42.8	36.444	21.748	5.1768	4.1403	−93.270	87.219	2.3140	6.5306
35.8	13.076	3.2974*	13.284	4.4470	−111.91	80.410	1.8658	7.2379	43.0	37.376	22.680	5.0539	4.1302	−92.657	87.313	2.3273	6.5122
36.0	13.526	2.3814*	12.886	4.4392	−111.45	80.686	1.8782	7.2152	43.2	38.324	23.628	4.9345	4.1201	−92.040	87.401	2.3407	6.4939
36.2	13.987	1.4428*	12.503	4.4314	−110.98	80.958	1.8906	7.1927	43.4	39.289	24.593	4.8185	4.1099	−91.418	87.481	2.3540	6.4756
36.4	14.460	0.4814*	12.135	4.4236	−110.51	81.226	1.9031	7.1704	43.6	40.272	25.576	4.7058	4.0995	−90.790	87.555	2.3674	6.4574
36.50	14.696	0.0	11.959	4.4197	−110.28	81.356	1.9092	7.1595	43.8	41.271	26.575	4.5962	4.0891	−90.158	87.622	2.3809	6.4392
36.6	14.943	0.2472	11.780	4.4157	−110.04	81.489	1.9155	7.1483	44	42.288	27.592	4.4897	4.0786	−89.520	87.682	2.3944	6.4211
36.8	15.438	0.7423	11.438	4.4078	−109.56	81.749	1.9280	7.1264	45	47.638	32.942	3.9989	4.0246	−86.250	87.872	2.4624	6.3311
37.0	15.945	1.2492	11.109	4.3998	−109.07	82.003	1.9405	7.1046	46	53.446	38.750	3.5697	3.9680	−82.837	87.869	2.5314	6.2418
37.2	16.464	1.7678	10.792	4.3917	−108.59	82.254	1.9531	7.0830	47	59.735	45.039	3.1925	3.9084	−79.270	87.653	2.6018	6.1526
37.4	16.995	2.2985	10.486	4.3836	−108.10	82.500	1.9656	7.0616	48	66.526	51.830	2.8593	3.8456	−75.533	87.203	2.6736	6.0632
37.6	17.537	2.8414	10.191	4.3755	−107.60	82.741	1.9782	7.0403	49	73.841	59.145	2.5634	3.7790	−71.610	86.492	2.7472	5.9729
37.8	18.093	3.3966	9.9066	4.3672	−107.10	82.978	1.9908	7.0192	50	81.704	67.008	2.2993	3.7080	−67.477	85.486	2.8228	5.8812
38.0	18.660	3.9643	9.6321	4.3589	−106.60	83.210	2.0035	6.9982	51	90.139	75.443	2.0623	3.6320	−63.105	84.142	2.9010	5.7872
38.2	19.241	4.5446	9.3672	4.3506	−106.09	83.438	2.0161	6.9774	52	99.172	84.476	1.8482	3.5500	−58.456	82.403	2.9822	5.6901
38.4	19.834	5.1377	9.1113	4.3422	−105.58	83.660	2.0288	6.9567	53	108.83	94.133	1.6535	3.4606	−53.477	80.189	3.0673	5.5884
38.6	20.440	5.7438	8.8643	4.3337	−105.07	83.878	2.0415	6.9362	54	119.14	104.44	1.4749	3.3619	−48.093	77.389	3.1575	5.4803
38.8	21.059	6.3630	8.6256	4.3252	−104.55	84.091	2.0542	6.9158	55	130.14	115.44	1.3091	3.2510	−42.189	73.828	3.2545	5.3630
39.0	21.691	6.9955	8.3950	4.3166	−104.03	84.299	2.0669	6.8955	56	141.86	127.16	1.1525	3.1228	−35.571	69.218	3.3614	5.2319
39.2	22.337	7.6414	8.1720	4.3080	−103.50	84.502	2.0797	6.8754	57	154.34	139.65	1.0003	2.9677	−27.863	63.002	3.4844	5.0779
39.4	22.997	8.3009	7.9564	4.2993	−102.97	84.700	2.0925	6.8554	**59.29	186.2	171.5	0.5108	1.958	16.77	16.77	4.208	4.208
39.6	23.670	8.9741	7.7479	4.2905	−102.43	84.893	2.1053	6.8355									
39.8	24.357	9.6613	7.5463	4.2816	−101.89	85.080	2.1181	6.8157									

#Triple point   *in. Hg. Vacuum   **Critical point

Temp., K	Viscosity, N·s/m² Sat. Liquid × 10⁻³‡	Viscosity, N·s/m² Sat. Vapor × 10⁻³‡	Viscosity, N·s/m² Gas at p = 1 atm × 10⁻⁶‡	Thermal Conductivity, W/(m·K) Sat. Liquid	Thermal Conductivity, W/(m·K) Sat. Vapor	Thermal Conductivity, W/(m·K) Gas at p = 1 atm	Specific Heat, c_p, kJ/(kg·K) Sat. Liquid	Specific Heat, c_p, kJ/(kg·K) Sat. Vapor	Specific Heat, c_p, kJ/(kg·K) Gas at p = 0 atm	Specific Heat, c_p, kJ/(kg·K) Gas at p = 1 atm	Temp., K
14				0.0824	0.0111		6.752	10.6			14
15				0.0855	0.0118		7.190	10.7			15
16				0.0885	0.0128		7.610	10.9			16
18				0.0933	0.0147		8.445	11.2			18
20				0.0972	0.0164	0.0159	9.358	11.7	10.31	12.15	20
22				0.0999	0.0194	0.0173	10.45	12.5	10.31	11.68	22
24				0.1006	0.0224	0.0186	11.84	13.7	10.31	11.38	24
26				0.0975	0.0260	0.0200	13.61	15.6	10.31	11.16	26
28				0.0910	0.0304	0.0213	15.88	19.4	10.31	11.00	28
30				0.0826	0.0373	0.0227	20.3	24.2	10.31	10.88	30
31				0.077	0.0428	0.0234	25.8	27.3	10.31	10.84	31
33*				0.058	0.0575	0.0247			10.31	10.76	33*
40						0.0294			10.32	10.60	40
60						0.0434			10.61	10.76	60
80						0.0601			11.72	11.77	80
100						0.0797			13.40	13.43	100
120						0.1000			14.95	15.10	120
160						0.1316			16.33	16.43	160
200						0.1512			16.07	16.09	200
250						0.1696			15.33	15.44	250
300						0.1880			14.84	15.02	300
400						0.2223			14.55	14.55	400
500						0.2565			14.52	14.54	500
600						0.291			14.54	14.55	600

*Critical temperature. Tabulated properties ignore critical region effects.   ‡Actual value = (Table value) × (Indicated multiplier).

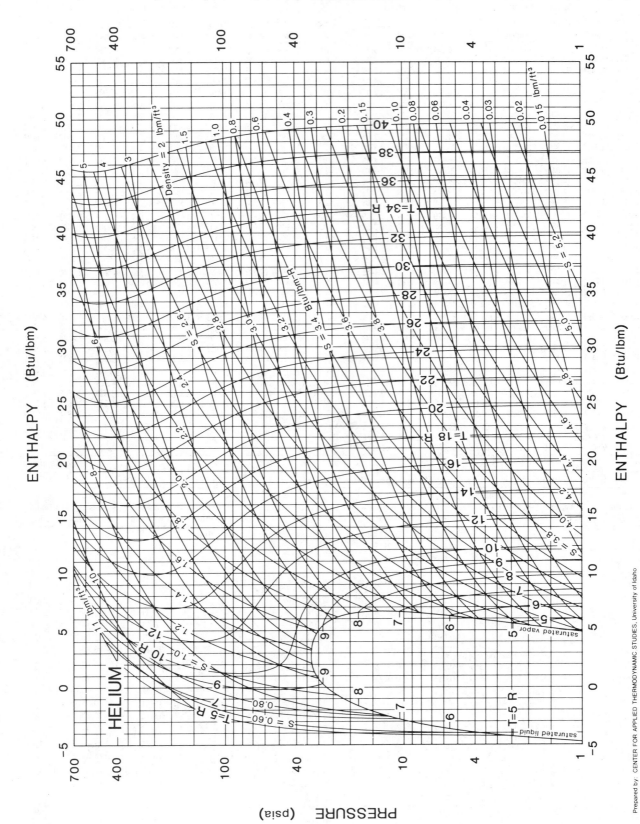

Fig. 33   Temperature-Entropy Diagram for Refrigerant 704 (Helium)

Refrigerant 704 (Helium)   Properties of Saturated Liquid and Saturated Vapor

Temp, °R	Pressure psia	Pressure psig	Volume, ft³/lb_m Vapor	Density, lb_m/ft³ Liquid	Enthalpy, Btu/lb_m Liquid	Enthalpy, Btu/lb_m Vapor	Entropy, Btu/lb_m·°R Liquid	Entropy, Btu/lb_m·°R Vapor	Temp, °R	Pressure psia	Pressure psig	Volume, ft³/lb_m Vapor	Density, lb_m/ft³ Liquid	Enthalpy, Btu/lb_m Liquid	Enthalpy, Btu/lb_m Vapor	Entropy, Btu/lb_m·°R Liquid	Entropy, Btu/lb_m·°R Vapor
#3.9191	0.73085	28.433*	13.597	9.1281	−4.9148	4.8864	0.41780	2.9187	6.0	5.5357	18.651*	2.3888	8.6095	−3.6316	6.4091	0.66001	2.3335
4.0	0.81460	28.263*	12.402	9.1191	−4.8521	4.9328	0.43322	2.8794	6.10	5.9422	17.823*	2.2420	8.5716	−3.5565	6.4585	0.67098	2.3128
4.1	0.92663	28.035*	11.120	9.1062	−4.7827	5.0121	0.44980	2.8388	6.20	6.3685	16.955*	2.1065	8.5323	−3.4792	6.5045	0.68204	2.2923
4.2	1.0487	27.786*	10.013	9.0915	−4.7198	5.1041	0.46436	2.8034	6.30	6.8149	16.046*	1.9812	8.4917	−3.3998	6.5472	0.69320	2.2721
4.3	1.1814	27.516*	9.0518	9.0751	−4.6612	5.1993	0.47751	2.7706	6.40	7.2819	15.095*	1.8651	8.4496	−3.3181	6.5863	0.70446	2.2520
4.4	1.3253	27.223*	8.2112	9.0572	−4.6054	5.2926	0.48966	2.7392	6.50	7.7700	14.101*	1.7574	8.4060	−3.2340	6.6219	0.71583	2.2321
4.5	1.4809	26.906*	7.4727	9.0379	−4.5512	5.3821	0.50113	2.7085	6.60	8.2797	13.064*	1.6552	8.3609	−3.1475	6.6539	0.72732	2.2124
4.6	1.6487	26.565*	6.8210	9.0173	−4.4976	5.4673	0.51214	2.6784	6.70	8.8113	11.981*	1.5639	8.3141	−3.0584	6.6822	0.73894	2.1928
4.7	1.8291	26.197*	6.2438	8.9955	−4.4441	5.5486	0.52286	2.6490	6.80	9.3654	10.853*	1.4769	8.2657	−2.9667	6.7068	0.75070	2.1733
4.8	2.0225	25.803*	5.7305	8.9725	−4.3901	5.6267	0.53339	2.6202	6.90	9.9423	9.6785*	1.3956	8.2154	−2.8721	6.7276	0.76261	2.1539
4.9	2.2295	25.382*	5.2725	8.9484	−4.3353	5.7022	0.34381	2.5923	7.0	10.543	8.4561*	1.3196	8.1633	−2.7746	6.7446	0.77468	2.1346
5.0	2.4505	24.932*	4.8623	8.9231	−4.2793	5.7756	0.53419	2.5652	7.10	11.167	7.1850*	1.2483	8.1092	−2.6741	6.7576	0.78693	2.1153
5.10	2.6860	24.452*	4.4938	8.8988	−4.2220	5.8473	0.56456	2.5389	7.20	11.816	5.8642*	1.1814	8.0530	−2.5703	6.7665	0.79937	2.0961
5.20	2.9364	23.943*	4.1616	8.8694	−4.1633	5.9174	0.57495	2.5136	7.30	12.489	4.4926*	1.1184	7.9947	−2.4633	6.7710	0.81201	2.0769
5.30	3.2023	23.401*	3.8613	8.8408	−4.1031	5.9861	0.58537	2.4890	7.40	13.189	3.0691*	1.0592	7.9339	−2.3522	6.7708	0.82488	2.0577
5.40	3.4840	22.828*	3.5891	8.8112	−4.0412	6.0532	0.59584	2.4652	7.50	13.914	1.5927*	1.0033	7.8707	−2.2375	6.7665	0.83800	2.0384
5.50	3.7822	22.221*	3.3416	8.7805	−3.9775	6.1184	0.60636	2.4420	7.604	14.696	0.0	0.94839	7.8021	−2.1139	6.7546	0.85192	2.0182
5.60	4.0972	21.579*	3.1160	8.7487	−3.9121	6.1817	0.61695	2.4194	8.0	17.951	3.2349	0.76482	7.5094	−1.5961	6.6463	0.90816	1.9385
5.70	4.4296	20.903*	2.9100	8.7157	−3.8449	6.2426	0.62760	2.3973	8.5	22.733	1.5927	0.57509	7.0376	−0.7977	6.2812	0.99012	1.8229
5.80	4.7798	20.189*	2.7212	8.6815	−3.7757	6.3010	0.63833	2.3757	9.0	28.348	13.652	0.40758	6.3134	0.3226	5.3685	1.1002	1.6608
5.90	5.1484	19.439*	2.5481	8.6462	−3.7047	6.3565	0.64913	2.3544	** 9.36252	32.99	18.29	0.2300	4.347	2.723	2.723	1.343	1.343

#Lower lambda point   *in. Hg. Vacuum   **Critical point

Liquid on the Melting Line					Lambda Line		
Temp., R	Pressure, psi	Density, lb_m/ft³	Enthalpy, Btu/lb_m	Entropy, Btu/lb_m·R	Temp., R	Pressure, psi	Density, lb_m/ft³
3.200	441.35	11.275	2.413	0.276	3.1820	436.9	11.264
4.0	676.61	11.850	6.308	0.317	3.2	428.8	11.238
7.0	1804.6	13.529	22.72	0.335	3.3	381.7	11.076
10.0	3289.2	14.750	42.57	0.383	3.4	331.5	10.893
13.0	5089.1	15.849	65.17	0.456	3.5	278.0	10.686
16.0	7163.1	16.863	89.43	0.512	3.6	221.0	10.445
19.0	9469.8	17.794	114.6	0.545	3.7	159.7	10.157
22.0	11968.	18.621	140.5	0.571	3.8	93.08	9.794
25.0	14616.	19.328	167.0	0.600	3.9	17.42	9.267
26.820	16278.	19.697	183.3	0.625	3.9191	0.7679	9.124

Temp., K	Viscosity, N·s/m² Sat. Liquid × 10⁻³‡	Viscosity, N·s/m² Sat. Vapor × 10⁻³‡	Viscosity, N·s/m² Gas at p = 1 atm × 10⁻⁶‡	Thermal Conductivity, W/(m·K) Sat. Liquid	Thermal Conductivity, W/(m·K) Sat. Vapor	Thermal Conductivity, W/(m·K) Gas at p = 1 atm	Specific Heat, c_p, kJ/(kg·K) Sat. Liquid	Specific Heat, c_p, kJ/(kg·K) Sat. Vapor	Specific Heat, c_p, kJ/(kg·K) Gas at p = 0 atm	Specific Heat, c_p, kJ/(kg·K) Gas at p = 1 atm	Temp., K
3.0			0.80	0.0203		0.0061	2.490				3.0
3.5			0.93	0.0232		0.0071	3.110	6.00			3.5
4.0			1.05	0.0281		0.0080	3.983	6.46			4.0
4.5			1.16	0.0348		0.0088	5.50	7.52			4.5
5.0			1.27	0.0434		0.0096	11.5	10.6	5.1931	6.629	5.0
5.4*			1.35			0.0103			5.1931		5.4*
8			1.85			0.0139			5.1931		8
10			2.18			0.0164			5.1931	5.732	10
20			3.50			0.0258			5.1931	5.407	20
40			5.48			0.0400			5.1931	5.215	40
60			7.06			0.0521			5.1931	5.200	60
80			8.41			0.0631			5.1931	5.197	80
100			9.63			0.0730			5.1931	5.195	100
120			10.75			0.0819			5.1931	5.194	120
140			11.8			0.0907			5.1931	5.194	140
160			12.9			0.0992			5.1931	5.193	160
180			13.9			0.1072			5.1931	5.1931	180
200			15.0			0.1151			5.1931	5.1931	200
220			16.0			0.123			5.1931	5.1931	220
240			17.0			0.130			5.1931	5.1931	240
260			18.0			0.137			5.1931	5.1931	260
280			19.0			0.145			5.1931	5.1931	280
300			19.9			0.152			5.1931	5.1931	300
400			24.3			0.187			5.1931	5.1931	400
500			28.3			0.220			5.1931	5.1931	500
600			32.0			0.252			5.1931	5.1931	600

*Critical temperature. Tabulated properties ignore critical region effects.   ‡Actual value = (Table value) × (Indicated multiplier).

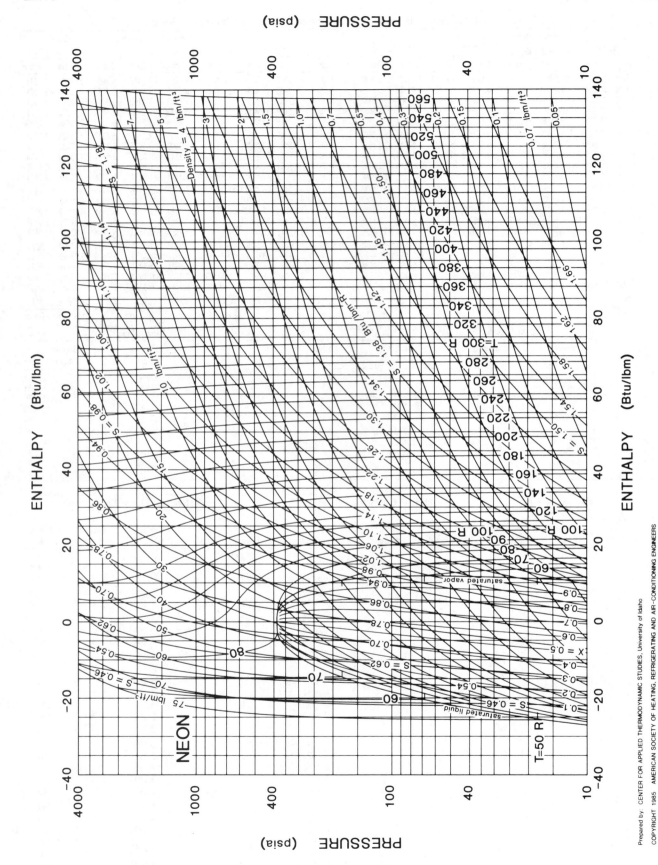

Fig. 34   Temperature-Entropy Diagram for Refrigerant 720 (Neon)

Refrigerant 720 (Neon)   Properties of Liquid and Vapor

Temp, °R	Pressure psia	Pressure psig	Volume, ft³/lb$_m$ Vapor	Density, lb$_m$/ft³ Liquid	Enthalpy, Btu/lb$_m$ Liquid	Enthalpy, Btu/lb$_m$ Vapor	Entropy, Btu/lb$_m$·°R Liquid	Entropy, Btu/lb$_m$·°R Vapor
#44.21	6.2916	17.111*	3.6302	77.994	−28.168	10.174	0.32872	1.1960
45	7.3808	14.894*	3.1400	77.552	−27.791	10.323	0.33711	1.1841
46	8.9558	11.687*	2.6316	76.987	−27.316	10.504	0.34747	1.1697
47	10.777	7.9801*	2.2224	76.414	−26.843	10.678	0.35756	1.1559
48.0	12.866	3.7257*	1.8900	75.832	−26.370	10.842	0.36741	1.1427
48.2	13.318	2.8049*	1.8312	75.715	−26.275	10.874	0.36936	1.1401
48.4	13.783	1.8599*	1.7747	75.596	−26.180	10.905	0.37129	1.1375
48.6	14.259	0.8903*	1.7203	75.478	−26.086	10.936	0.37322	1.1350
48.78	14.696	0.0	1.6734	75.371	−26.001	10.964	0.37495	1.1327
48.8	14.747	0.0512	1.6680	75.359	−25.991	10.967	0.37515	1.1325
49.0	15.248	0.5522	1.6177	75.239	−25.896	10.997	0.37706	1.1300
49.2	15.762	1.0658	1.5693	75.119	−25.801	11.027	0.37897	1.1275
49.4	16.288	1.5922	1.5227	74.998	−25.706	11.056	0.38087	1.1250
49.6	16.828	2.1317	1.4779	74.877	−25.611	11.085	0.38277	1.1226
49.8	17.381	2.6845	1.4346	74.756	−25.516	11.114	0.38465	1.1202
50.0	17.947	3.2507	1.3930	74.633	−25.420	11.142	0.38654	1.1178
50.2	18.527	3.8306	1.3528	74.511	−25.325	11.170	0.38841	1.1154
50.4	19.120	4.4242	1.3141	74.387	−25.229	11.197	0.39028	1.1130
50.6	19.728	5.0320	1.2768	74.263	−25.134	11.224	0.39215	1.1107
50.8	20.350	5.6539	1.2408	74.139	−25.038	11.250	0.39400	1.1083
51.0	20.986	6.2903	1.2061	74.014	−24.942	11.276	0.39586	1.1060
51.2	21.637	6.9413	1.1726	73.888	−24.846	11.302	0.39770	1.1037
51.4	22.303	7.6072	1.1402	73.762	−24.750	11.327	0.39955	1.1014
51.6	22.984	8.2881	1.1089	73.635	−24.653	11.351	0.40138	1.0991
51.8	23.680	8.9843	1.0788	73.508	−24.557	11.375	0.40322	1.0969
52.0	24.392	9.6958	1.0496	73.379	−24.460	11.399	0.40505	1.0946
52.2	25.119	10.423	1.0214	73.251	−24.363	11.422	0.40687	1.0924
52.4	25.862	11.166	0.99417	73.121	−24.266	11.445	0.40869	1.0902
52.6	26.621	11.925	0.96783	72.991	−24.169	11.467	0.41050	1.0880
52.8	27.397	12.701	0.94236	72.861	−24.072	11.489	0.41231	1.0858
53.0	28.189	13.493	0.91771	72.729	−23.974	11.510	0.41412	1.0836
53.2	28.997	14.301	0.89387	72.597	−23.876	11.530	0.41592	1.0815
53.4	29.823	15.127	0.87080	72.465	−23.778	11.551	0.41772	1.0793
53.6	30.665	15.969	0.84847	72.331	−23.680	11.570	0.41951	1.0772
53.8	31.525	16.829	0.82684	72.197	−23.582	11.589	0.42131	1.0750
54.0	32.402	17.706	0.80590	72.062	−23.483	11.608	0.42309	1.0729
54.2	33.298	18.602	0.78561	71.927	−23.385	11.626	0.42488	1.0708
54.4	34.211	19.515	0.76596	71.791	−23.285	11.644	0.42666	1.0687
54.6	35.142	20.446	0.74691	71.654	−23.186	11.661	0.42843	1.0667
54.8	36.091	21.395	0.72845	71.516	−23.087	11.677	0.43021	1.0646
55.0	37.059	22.363	0.71056	71.378	−22.987	11.693	0.43198	1.0625
55.2	38.046	23.350	0.69320	71.238	−22.887	11.708	0.43375	1.0605
55.4	39.052	24.356	0.67637	71.099	−22.787	11.723	0.43551	1.0584
55.6	40.077	25.381	0.66004	70.958	−22.686	11.737	0.43728	1.0564
55.8	41.122	26.426	0.64419	70.817	−22.586	11.751	0.43904	1.0544
56	42.186	27.490	0.62882	70.674	−22.485	11.764	0.44079	1.0524
57	47.809	33.113	0.55839	69.952	−21.976	11.820	0.44954	1.0424
58	53.956	39.260	0.49740	69.210	−21.459	11.861	0.45823	1.0327
59	60.655	45.959	0.44435	68.448	−20.936	11.885	0.46687	1.0232
60	67.934	53.238	0.39801	67.664	−20.404	11.893	0.47548	1.0138
61	75.822	61.126	0.35754	66.858	−19.864	11.882	0.48405	1.0045
62	84.347	69.651	0.32152	66.030	−19.314	11.851	0.49260	0.99527
63	93.541	78.845	0.28983	65.177	−18.755	11.800	0.50113	0.98613
64	103.43	88.738	0.26170	64.299	−18.185	11.727	0.50966	0.97704
65	114.06	99.360	0.23663	63.393	−17.603	11.631	0.51821	0.96795
66	125.44	110.74	0.21421	62.457	−17.008	11.508	0.52678	0.95884
67	137.62	122.92	0.19409	61.488	−16.399	11.358	0.53540	0.94967
68	150.62	135.93	0.17596	60.482	−15.772	11.177	0.54409	0.94040
69	164.49	149.80	0.15956	59.432	−15.127	10.962	0.55289	0.93098
70	179.26	164.56	0.14469	58.330	−14.459	10.709	0.56183	0.92137
71	194.96	180.27	0.13113	57.168	−13.763	10.414	0.57098	0.91150
72	211.64	196.95	0.11873	55.929	−13.035	10.070	0.58040	0.90130
73	229.34	214.64	0.10733	54.596	−12.266	9.6700	0.59019	0.89068
74	248.10	233.40	0.096813	53.139	−11.445	9.2033	0.60049	0.87951
75	267.97	253.28	0.087050	51.520	−10.555	8.6579	0.61149	0.86766
76	289.01	274.32	0.077931	49.675	−9.5711	8.0158	0.62349	0.85490
77	311.28	296.59	0.069344	47.507	−8.4539	7.2510	0.63699	0.84095
78	334.86	320.16	0.061161	44.858	−7.1363	6.3226	0.65278	0.82533
79	359.83	345.14	0.053179	41.466	−5.5031	5.1511	0.67224	0.80710
**80.006	386.4	371.7	0.03317	30.15	−0.4476	0.4476	0.7341	0.7341

#Triple point   *in. Hg. Vacuum   **Critical point

Temp., K	Viscosity, N·s/m² Sat. Liquid × 10⁻³‡	Viscosity, N·s/m² Sat. Vapor × 10⁻³‡	Viscosity, N·s/m² Gas at p = 1 atm × 10⁻⁶‡	Thermal Conductivity, W/(m·K) Sat. Liquid	Thermal Conductivity, W/(m·K) Sat. Vapor	Thermal Conductivity, W/(m·K) Gas at p = 1 atm	Specific Heat, c_p, kJ/(kg·K) Sat. Liquid	Specific Heat, c_p, kJ/(kg·K) Sat. Vapor	Specific Heat, c_p, kJ/(kg·K) Gas at p = 0 atm	Specific Heat, c_p, kJ/(kg·K) Gas at p = 1 atm	Temp., K
26	0.139			0.115			1.839	1.13	1.0301		26
28	0.116	0.00485	4.80	0.112	0.0082	0.0082	1.910	1.19	1.0301		28
30	0.098	0.00524	5.15	0.108	0.0089	0.0086	1.978	1.25	1.0301	1.157	30
32	0.084	0.00564	5.48	0.104	0.0097	0.0090	2.062	1.32	1.0301	1.136	32
34	0.072	0.00604	5.81	0.099	0.0107	0.0094	2.183	1.42	1.0301	1.119	34
36	0.0619	0.00644	6.13	0.092	0.0118	0.0098	2.359	1.59	1.0301	1.105	36
38	0.0517	0.00703	6.44	0.084	0.0131	0.0103	2.608	1.91	1.0301	1.093	38
40	0.0427	0.00781	6.75	0.073	0.0147	0.0107	3.00	2.41	1.0301	1.084	40
42	0.0343	0.00913	7.06	0.061	0.017	0.0111	3.88	3.23	1.0301	1.077	42
44*	0.0167	0.0167	7.37	0.033	0.033	0.0115			1.0301	1.071	44*
50			8.24			0.0128			1.0301	1.058	50
60			9.63			0.0148			1.0301	1.046	60
80			12.1			0.0186			1.0301	1.037	80
90			13.3			0.0204			1.0301	1.035	90
100			14.4			0.0222			1.0301	1.034	100
120			16.5			0.0256			1.0301	1.032	120
140			18.5			0.0288			1.0301	1.031	140
160			20.4			0.0318			1.0301	1.031	160
180			22.2			0.0347			1.0301	1.031	180
200			23.9			0.0375			1.0301	1.031	200
240			27.2			0.0426			1.0301	1.030	240
280			30.3			0.0472			1.0301	1.030	280
300			31.7			0.0493			1.0301	1.030	300
400			38.5			0.0590			1.0301	1.030	400
500			44.5			0.0685			1.0301		500
600			50.0			0.0771			1.0301		600

*Critical temperature. Tabulated properties ignore critical region effects.   ‡Actual value = (Table value) × (Indicated multiplier).

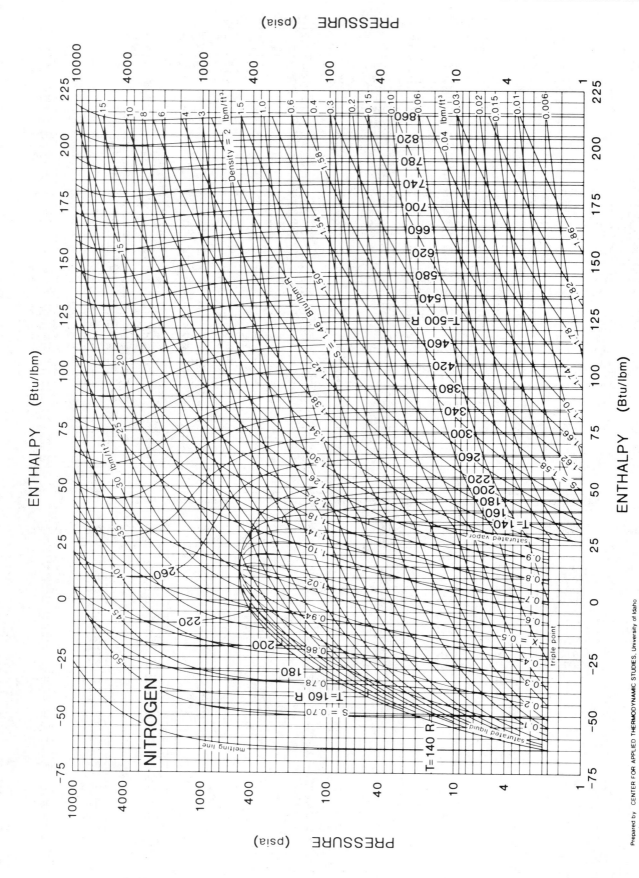

Fig. 35 Temperature-Entropy Diagram for Refrigerant 728 (Nitrogen)

Refrigerant 728 (Nitrogen)   Properties of Liquid and Vapor

Temp, °R	Pressure psia	Pressure psig	Volume, ft³/lbm Vapor	Density, lbm/ft³ Liquid	Enthalpy, Btu/lbm Liquid	Enthalpy, Btu/lbm Vapor	Entropy, Btu/lbm·°R Liquid	Entropy, Btu/lbm·°R Vapor
#113.67	1.8173	26.221*	23.735	54.174	-64.725	27.851	0.58010	1.3953
114	1.8787	26.096*	23.021	54.131	-64.569	27.928	0.58147	1.3937
116	2.2931	25.252*	19.161	53.869	-63.627	28.383	0.58965	1.3836
118	2.7781	24.265*	16.061	53.603	-62.674	28.832	0.59778	1.3740
120	3.3419	23.117*	13.552	53.331	-61.713	29.276	0.60584	1.3648
121	3.6561	22.477*	12.478	53.194	-61.230	29.496	0.60984	1.3604
122	3.9933	21.791*	11.506	53.055	-60.745	29.714	0.61382	1.3560
123	4.3547	21.055*	10.626	52.915	-60.260	29.930	0.61777	1.3518
124	4.7415	20.268*	9.8270	52.773	-59.773	30.145	0.62170	1.3476
125	5.1549	19.426*	9.1007	52.631	-59.285	30.358	0.62561	1.3435
126	5.5961	18.527*	8.4394	52.487	-58.797	30.570	0.62949	1.3394
127	6.0665	17.570*	7.8363	52.342	-58.307	30.779	0.63334	1.3355
128	6.5674	16.550*	7.2857	52.196	-57.817	30.987	0.63717	1.3316
129	7.1000	15.466*	6.7820	52.049	-57.327	31.193	0.64098	1.3278
130	7.6657	14.314*	6.3208	51.901	-56.835	31.397	0.64475	1.3241
131	8.2658	13.092*	5.8978	51.751	-56.344	31.598	0.64851	1.3205
132	8.9019	11.797*	5.5093	51.601	-55.852	31.798	0.65223	1.3169
133	9.5753	10.426*	5.1520	51.449	-55.359	31.996	0.65593	1.3134
134	10.287	8.9761*	4.8231	51.296	-54.866	32.191	0.65960	1.3099
135	11.040	7.4444*	4.5199	51.142	-54.373	32.384	0.66325	1.3065
136	11.834	5.8279*	4.2401	50.987	-53.879	32.575	0.66687	1.3032
137	12.671	4.1236*	3.9814	50.831	-53.385	32.764	0.67047	1.2999
138	13.552	2.3283*	3.7421	50.674	-52.891	32.950	0.67404	1.2967
139	14.480	0.4389*	3.5205	50.515	-52.396	33.134	0.67759	1.2935
139.23	14.696	0.0	3.4729	50.480	-52.284	33.175	0.67839	1.2928
140	15.456	0.7602	3.3150	50.356	-51.900	33.315	0.68112	1.2904
142	17.557	2.8612	2.9470	50.034	-50.909	33.670	0.68809	1.2843
144	19.568	5.1723	2.6285	49.708	-49.915	34.014	0.69498	1.2783
146	22.402	7.7063	2.3519	49.377	-48.919	34.348	0.70179	1.2726
148	25.172	10.476	2.1106	49.043	-47.921	34.670	0.70851	1.2670
150	28.192	13.496	1.8995	48.703	-46.920	34.979	0.71515	1.2616
152	31.475	16.779	1.7140	48.360	-45.916	35.277	0.72171	1.2563
154	35.034	20.338	1.5505	48.012	-44.908	35.562	0.72821	1.2511
156	38.884	24.188	1.4060	47.659	-43.896	35.833	0.73464	1.2461
158	43.038	28.342	1.2779	47.301	-42.880	36.090	0.74101	1.2412
160	47.511	32.815	1.1639	46.939	-41.860	36.333	0.74732	1.2363
162	52.317	37.621	1.0622	46.571	-40.833	36.560	0.75358	1.2316
164	57.470	42.774	0.97118	46.198	-39.801	36.772	0.75978	1.2270
166	62.985	48.289	0.88959	45.819	-38.762	36.968	0.76594	1.2224
168	68.876	54.180	0.81621	45.435	-37.717	37.147	0.77206	1.2179
170	75.159	60.463	0.75008	45.045	-36.663	37.307	0.77814	1.2135
172	81.848	67.152	0.69031	44.648	-35.602	37.450	0.78419	1.2091
174	88.957	74.261	0.63619	44.245	-34.532	37.573	079020	1.2048
176	96.503	81.807	0.58706	43.834	-33.452	37.676	0.79619	1.2005
178	104.50	89.804	0.54237	43.416	-32.362	37.757	0.80216	1.1962
180	112.96	98.268	0.50162	42.991	-31.262	37.816	0.80810	1.1920
182	121.91	107.21	0.46440	42.557	-30.149	37.851	0.81404	1.1878
184	131.35	116.66	0.43033	42.114	-29.024	37.862	0.81996	1.1835
186	141.31	126.62	0.39908	41.662	-27.886	37.846	0.82587	1.1793
188	151.80	137.11	0.37036	41.200	-26.733	37.802	0.83179	1.1751
190	162.84	148.14	0.34391	40.727	-25.564	37.728	0.83771	1.1709
192	174.44	159.74	0.31950	40.243	-24.378	37.622	0.84364	1.1666
194	186.62	171.92	0.29694	39.745	-23.173	37.481	0.84959	1.1622
196	199.40	184.70	0.27603	39.234	-21.949	37.304	0.85556	1.1579
198	212.79	198.10	0.25663	38.708	-20.701	37.086	0.56157	1.1534
200	226.82	212.13	0.23857	38.164	-19.429	36.823	0.86762	1.1489
202	241.51	226.81	0.22174	37.601	-18.130	36.512	0.87373	1.1442
204	256.87	242.18	0.20600	37.017	-16.799	36.146	0.87991	1.1394
206	272.93	258.23	0.19124	36.407	-15.432	35.719	0.88618	1.1344
208	289.71	275.01	0.17737	35.769	-14.024	35.222	0.89257	1.1293
210	307.23	292.53	0.16428	35.096	-12.567	34.646	0.89910	1.1239
212	325.52	310.82	0.15188	34.381	-11.052	33.974	0.90582	1.1181
214	344.61	329.91	0.14006	33.615	-9.4659	33.190	0.91278	1.1120
216	364.52	349.83	0.12871	32.782	-7.7891	32.264	0.92006	1.1054
218	385.30	370.61	0.11771	31.861	-5.9927	31.157	0.92779	1.0981
220	406.99	392.29	0.10689	30.814	-4.0281	29.801	0.93618	1.0899
222	429.63	414.94	0.095971	29.573	-1.8048	28.073	0.94561	1.0802
224	453.30	438.61	0.084380	27.974	0.8886	25.696	0.95700	1.0677
226	478.11	463.41	0.070205	24.820	5.6946	21.686	0.97747	1.0483
**227.2	493.1	478.4	0.05101	19.60	13.21	13.21	1.010	1.010

#Triple point   *in. Hg. Vacuum   **Critical point

Temp., K	Viscosity, N·s/m² Sat. Liquid ×10⁻³‡	Viscosity, N·s/m² Sat. Vapor ×10⁻³‡	Viscosity, N·s/m² Gas at p = 1 atm ×10⁻⁶‡	Thermal Conductivity, W/(m·K) Sat. Liquid	Thermal Conductivity, W/(m·K) Sat. Vapor	Thermal Conductivity, W/(m·K) Gas at p = 1 atm	Specific Heat, c_p, kJ/(kg·K) Sat. Liquid	Specific Heat, c_p, kJ/(kg·K) Sat. Vapor	Specific Heat, c_p, kJ/(kg·K) Gas at p = 0 atm	Specific Heat, c_p, kJ/(kg·K) Gas at p = 1 atm	Temp., K
70	0.217			0.151	0.0066		2.024	1.06			70
80	0.1480	0.00560	5.59	0.1322	0.0077	0.0076	2.059	1.10	1.039		80
90	0.1101	0.00636	6.22	0.1142	0.0091	0.0085	2.115	1.18	1.039		90
100	0.0869	0.00728	6.87	0.0966	0.0111	0.0094	2.247	1.41	1.039	1.072	100
110	0.0708	0.00842	7.52	0.0795	0.0138	0.0103	2.510	1.87	1.039	1.059	110
120	0.0484	0.01068	8.15	0.0628	0.0195	0.0112	2.96		1.039	1.054	120
122	0.0429	0.0115	8.28			0.0114			1.039	1.053	122
124	0.0361	0.0129	8.40			0.0116			1.039	1.053	124
125	0.0316	0.0144	8.53	0.0520	0.0265	0.0117			1.039	1.053	125
126*	0.0191	0.0191	8.65	0.037	0.037	0.0119			1.039	1.052	126*
130			8.78			0.0121			1.039	1.051	130
140			9.40			0.0130			1.039	1.050	140
160			10.6			0.0147			1.039	1.046	160
180			11.8			0.0165			1.039	1.044	180
200			12.9			0.0183			1.039	1.043	200
240			15.0			0.0215			1.039	1.042	240
260			16.0			0.0230			1.039	1.041	260
280			16.9			0.0245			1.039	1.041	280
300			17.9			0.0260			1.039	1.040	300
320			18.8			0.0274			1.039	1.040	320
340			19.7			0.0287			1.040	1.041	340
360			20.5			0.0300			1.041	1.042	360
380			21.3			0.0313			1.042	1.043	380
400			22.1			0.0325			1.044	1.044	400
500			25.9			0.0386			1.056	1.057	500
600			29.3			0.0441			1.075	1.075	600

*Critical temperature. Tabulated properties ignore critical region effects.   ‡Actual value = (Table value) × (Indicated multiplier).

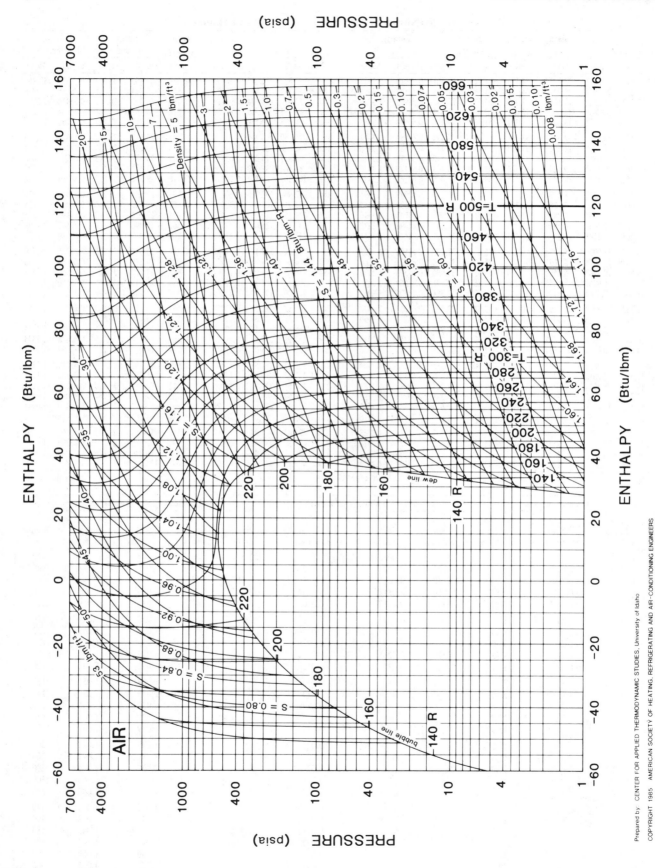

Fig. 36   Temperature-Entropy Diagram for Refrigerant 729 (Air)

Prepared by   CENTER FOR APPLIED THERMODYNAMIC STUDIES, University of Idaho

COPYRIGHT 1985   AMERICAN SOCIETY OF HEATING, REFRIGERATING AND AIR-CONDITIONING ENGINEERS

Refrigerant 729 (Air) Properties of Liquid on the Bubble Line and Vapor on the Dew Line

Temp, °R	Pressure, psia Liquid	Pressure, psia Vapor	Volume, ft³/lbm Vapor	Density, lbm/ft³ Liquid	Enthalpy, Btu/lbm Liquid	Enthalpy, Btu/lbm Vapor	Entropy, Btu/lbm·°R Liquid	Entropy, Btu/lbm·°R Vapor
110	1.1716	0.4746	85.644	58.874	-62.598	26.158	0.64918	1.4951
112	1.4338	0.6140	67.362	58.601	-62.685	26.620	0.64838	1.4817
114	1.7423	0.7854	53.555	58.323	-62.623	27.079	0.64892	1.4689
116	2.1031	0.9943	43.009	58.042	-62.436	27.534	0.65054	1.4567
118	2.5226	1.2463	34.866	57.758	-62.144	27.986	0.65302	1.4452
120	3.0075	1.5479	28.515	57.471	-61.763	28.433	0.65620	1.4342
122	3.5653	1.9058	23.515	57.180	-61.308	28.875	0.65995	1.4237
124	4.2036	2.3273	19.542	56.887	-60.788	29.313	0.66416	1.4136
126	4.9305	2.8203	16.360	56.591	-60.213	29.745	0.66874	1.4041
128	5.7546	3.3929	13.789	56.293	-59.591	30.171	0.67362	1.3949
130	6.6849	4.0540	11.698	55.992	-58.928	30.591	0.67873	1.3861
132	7.7307	4.8126	9.9836	55.689	-58.230	31.004	0.68404	1.3777
134	8.9017	5.6783	8.5692	55.383	-57.500	31.411	0.68949	1.3696
136	10.208	6.6613	7.3947	55.075	-56.744	31.810	0.69506	1.3618
138	11.660	7.7718	6.4135	54.764	-55.964	32.201	0.70072	1.3544
140	13.269	9.0206	5.5889	54.450	-55.162	32.585	0.70645	1.3472
142	15.046	10.419	4.8922	54.134	-54.342	32.959	0.71222	1.3402
144	17.002	11.978	4.3005	53.816	-53.504	33.325	0.71803	1.3336
146	19.148	13.710	3.7955	53.494	-52.651	33.682	0.72386	1.3271
148	21.497	15.627	3.3624	53.170	-51.784	34.029	0.72971	1.3208
150	24.061	17.742	2.9893	52.843	-50.904	34.367	0.73555	1.3148
152	26.853	20.066	2.6666	52.513	-50.012	34.693	0.74140	1.3089
154	29.883	22.613	2.3862	52.179	-49.109	35.010	0.74723	1.3032
156	33.167	25.397	2.1418	51.843	-48.195	35.315	0.75305	1.2976
158	36.715	28.430	1.9279	51.503	-47.271	35.608	0.75886	1.2922
160	40.542	31.726	1.7400	51.159	-46.337	35.890	0.76464	1.2870
162	44.658	35.300	1.5744	50.811	-45.394	36.160	0.77040	1.2819
164	49.079	39.165	1.4279	50.460	-44.442	36.416	0.77615	1.2768
166	53.818	43.335	1.2981	50.104	-43.481	36.660	0.78186	1.2720
168	58.888	47.825	1.1826	49.744	-42.511	36.890	0.78756	1.2672
170	64.301	52.650	1.0795	49.379	-41.532	37.106	0.79323	1.2625
172	70.071	57.825	0.98734	49.009	-40.545	37.307	0.79888	1.2579
174	76.212	63.363	0.90467	48.634	-39.548	37.494	0.80451	1.2533
176	82.735	69.281	0.83034	48.254	-38.542	37.665	0.81012	1.2489
178	89.655	75.592	0.76334	47.867	-37.526	37.819	0.81571	1.2445
180	96.985	82.313	0.70282	47.475	-36.500	37.957	0.82128	1.2402
182	104.74	89.459	0.64802	47.075	-35.465	38.077	0.82683	1.2359
184	112.93	97.045	0.59828	46.669	-34.418	38.179	0.83237	1.2317
186	121.56	105.09	0.55305	46.255	-33.360	38.263	0.83790	1.2275
188	130.66	113.60	0.51184	45.834	-32.291	38.326	0.84343	1.2234
190	140.23	122.60	0.47421	45.403	-31.209	38.369	0.84894	1.2193
192	150.28	132.10	0.43978	44.963	-30.114	38.389	0.85446	1.2152
194	160.84	142.12	0.40822	44.513	-29.005	38.387	0.85998	1.2111
196	171.90	152.68	0.37923	44.053	-27.881	38.361	0.86551	1.2070
198	183.49	163.79	0.35256	43.580	-26.740	38.310	0.87105	1.2029
200	195.61	175.46	0.32798	43.095	-25.582	38.231	0.87661	1.1988
202	208.28	187.71	0.30527	42.596	-24.405	38.124	0.88219	1.1947
204	221.51	200.57	0.28427	42.081	-23.207	37.985	0.88781	1.1906
206	235.30	214.04	0.26479	41.549	-21.987	37.814	0.89347	1.1864
208	249.68	228.14	0.24671	40.998	-20.741	37.607	0.89917	1.1822
210	264.65	242.90	0.22987	40.426	-19.467	37.361	0.90494	1.1779
212	280.22	258.32	0.21417	39.830	-18.162	37.072	0.91079	1.1735
214	296.41	274.43	0.19949	39.206	-16.821	36.736	0.91672	1.1691
216	313.22	291.25	0.18573	38.551	-15.440	36.347	0.92278	1.1645
218	330.67	308.79	0.17280	37.859	-14.012	35.899	0.92897	1.1597
220	348.77	327.09	0.16060	37.124	-12.529	35.384	0.93533	1.1548
222	367.53	346.17	0.14906	36.337	-10.979	34.789	0.94191	1.1496
224	386.96	366.06	0.13807	35.485	-9.3489	34.101	0.94878	1.1442
226	407.08	386.79	0.12756	34.551	-7.6163	33.299	0.95600	1.1383
228	427.88	408.41	0.11743	33.511	-5.7502	32.355	0.96373	1.1320
230	449.40	430.98	0.10753	32.322	-3.6998	31.222	0.97215	1.1250
232	471.64	454.59	0.097693	30.916	-1.3770	29.826	0.98164	1.1170
234	494.62	479.41	0.087610	29.152	1.3948	28.019	0.99293	1.1075
236	518.37	505.77	0.076556	26.726	5.0229	25.435	1.0077	1.0948
238	542.91	535.08	0.061042	23.292	10.080	20.202	1.0283	1.0711
**238.36	547.37	541.4	0.05568	22.72	10.96	17.79	1.032	1.061
*238.54		546.2	0.04944		14.46			1.047

**Maximum pressure   *Maximum temperature

Temp., K	Viscosity, N·s/m² Sat. Liquid × 10⁻³‡	Viscosity Sat. Vapor × 10⁻³‡	Viscosity Gas at p = 1 atm × 10⁻⁶‡	Thermal Conductivity, W/(m·K) Sat. Liquid	Thermal Cond. Sat. Vapor	Thermal Cond. Gas at p = 1 atm	Specific Heat, cp, kJ/(kg·K) Sat. Liquid	Sp. Heat Sat. Vapor	Sp. Heat Gas at p = 0 atm	Sp. Heat Gas at p = 1 atm	Temp., K
60	0.325			0.180	0.0054						60
70	0.221			0.163	0.0064						70
80	0.165	0.0055		0.145	0.0075		1.963		i.002		80
90	0.132	0.0065	6.35	0.128	0.0086	0.0083	2.077	1.12	1.002		90
100	0.1101	0.0075	7.06	0.110	0.0101	0.0092	2.205	1.21	1.002	1.028	100
110	0.0949	0.0086	7.75	0.093	0.0122	0.0102	2.421	1.44	1.002	1.022	110
120	0.0750	0.0102	8.43	0.076	0.0154	0.0111	2.80	2.00	1.002	1.017	120
130	0.0420	0.0143	9.09	0.054	0.021	0.0120			1.002	1.014	130
133*	0.0207	0.0207	9.29	0.034	0.034	0.0123			1.002	1.013	133*
140			9.74			0.0129			1.002	1.012	140
150			10.38			0.0138			1.002	1.011	150
160			11.0			0.0146			1.002	1.009	160
180			12.2			0.0164			1.002	1.007	180
200			13.4			0.0181			1.002	1.006	200
220			14.5			0.0198			1.003	1.006	220
240			15.5			0.0215			1.003	1.005	240
260			16.6			0.0231			1.003	1.005	260
280			17.6			0.0246			1.004	1.006	280
300			18.5			0.0261			1.005	1.006	300
320			19.5			0.0276			1.006	1.007	320
340			20.4			0.0290			1.007	1.008	340
360			21.3			0.0304			1.009	1.010	360
380			22.1			0.0317			1.011	1.012	380
400			22.9			0.0331			1.013	1.014	400
500			26.8			0.0395			1.029	1.030	500
600			30.3			0.0456			1.051	1.052	600

*Critical temperature. Tabulated properties ignore critical region effects.   ‡Actual value = (Table value) × (Indicated multiplier).

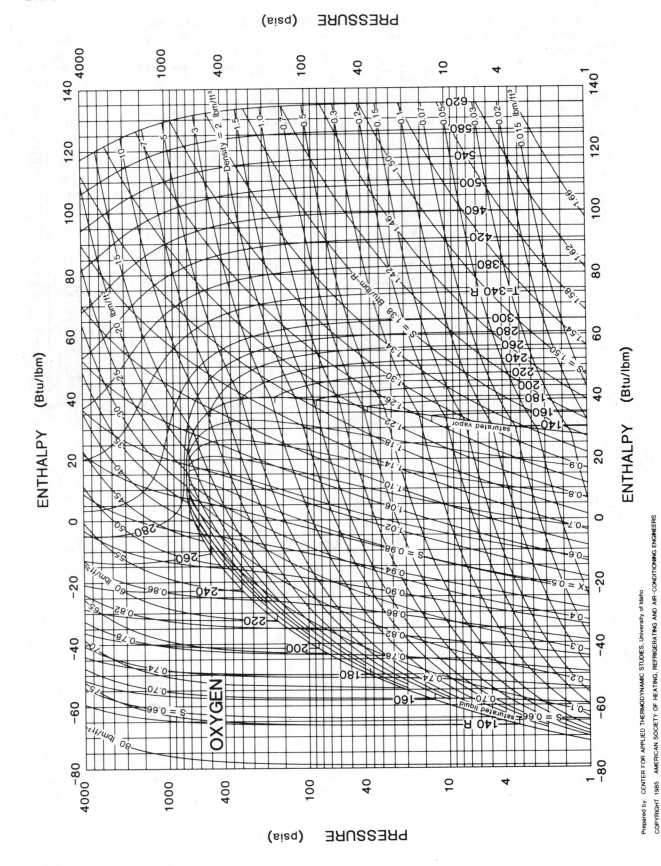

Prepared by: CENTER FOR APPLIED THERMODYNAMIC STUDIES, University of Idaho
COPYRIGHT 1985 AMERICAN SOCIETY OF HEATING, REFRIGERATING AND AIR-CONDITIONING ENGINEERS

Fig. 37   Temperature-Entropy Diagram for Refrigerant 732 (Oxygen)

Refrigerant 732 (Oxygen) Properties of Liquid and Vapor

Temp, °R	Pressure psia	Pressure psig	Volume, ft³/lbm Vapor	Density, lbm/ft³ Liquid	Enthalpy, Btu/lbm Liquid	Enthalpy, Btu/lbm Vapor	Entropy, Btu/lbm·°R Liquid	Entropy, Btu/lbm·°R Vapor
# 97.85	0.02122	29.878*	1545.8	81.581	-83.245	21.155	0.50026	1.5672
100	0.03069	29.859*	1092.5	81.243	-82.349	21.621	0.50932	1.5490
105	0.06791	29.783*	518.19	80.466	-80.321	22.700	0.52912	1.5102
110	0.13897	29.638*	265.14	79.693	-78.331	23.775	0.54762	1.4758
115	0.26576	29.380*	144.85	78.919	-76.354	24.843	0.56520	1.4451
120	0.47911	28.946*	83.756	78.141	-74.377	25.902	0.58203	1.4176
125	0.82027	28.251*	50.887	77.356	-72.394	26.950	0.59821	1.3929
130	1.3421	27.189*	32.284	76.564	-70.404	27.984	0.61381	1.3705
135	2.1100	25.625*	21.274	75.765	-68.408	29.000	0.62886	1.3503
140	3.2018	23.402*	14.495	74.957	-66.404	29.997	0.64342	1.3319
145	4.7084	20.335*	10.171	74.140	-64.394	30.970	0.65749	1.3151
150	6.7325	16.214*	7.3257	73.314	-62.378	31.917	0.67113	1.2997
152	7.7124	14.219*	6.4671	72.981	-61.570	32.287	0.67647	1.2939
154	8.8010	12.002*	5.7292	72.646	-60.760	32.653	0.68174	1.2883
156	10.007	9.5476*	5.0926	72.309	-59.950	33.013	0.68695	1.2829
158	11.337	6.8380*	4.5413	71.970	-59.138	33.368	0.69210	1.2776
160	12.802	3.8560*	4.0620	71.629	-58.324	33.717	0.69719	1.2724
162	14.409	0.5837*	3.6440	71.286	-57.509	34.060	0.70223	1.2675
162.34	14.696	0.0	3.5786	71.228	-57.371	34.111	0.70308	1.2666
164	16.168	1.4721	3.2780	70.941	-56.693	34.397	0.70721	1.2626
166	18.088	3.3918	2.9567	70.594	-55.875	34.727	0.71214	1.2579
168	20.178	5.4817	2.6736	70.244	-55.055	35.051	0.71702	1.2534
170	22.447	7.7514	2.4236	69.892	-54.233	35.367	0.72184	1.2489
172	24.907	10.211	2.2020	69.537	-53.409	35.677	0.72662	1.2446
174	27.565	12.869	2.0051	69.180	-52.583	35.979	0.73136	1.2403
176	30.434	15.738	1.8297	68.820	-51.754	36.274	0.73605	1.2362
178	33.532	18.826	1.6730	68.457	-50.923	36.560	0.74070	1.2322
180	36.840	22.144	1.5327	68.091	-50.090	36.838	0.74530	1.2282
182	40.399	25.703	1.4068	67.721	-49.253	37.108	0.74987	1.2244
184	44.209	29.513	1.2935	67.349	-48.413	37.369	0.75440	1.2206
186	48.280	33.584	1.1913	66.973	-47.571	37.621	0.75890	1.2169
188	52.624	37.928	1.0990	66.593	-46.724	37.863	0.76336	1.2133
190	57.252	42.556	1.0154	66.210	-45.874	38.096	0.76779	1.2097
192	62.174	47.478	0.93952	65.823	-45.020	38.319	0.77219	1.2062
194	67.401	52.705	0.87056	65.431	-44.162	38.532	0.77656	1.2028
196	72.945	58.249	0.80774	65.035	-43.300	38.734	0.78090	1.1994
198	78.816	64.120	0.75042	64.635	-42.433	38.925	0.78521	1.1961
200	85.026	70.330	0.69801	64.230	-41.561	39.105	0.78950	1.1928
202	91.587	76.891	0.65003	63.820	-40.685	39.273	0.79377	1.1896
204	98.508	83.812	0.60601	63.405	-39.802	39.429	0.79802	1.1864
206	105.80	91.107	0.56557	62.985	-38.914	39.573	0.80225	1.1832
208	113.48	98.786	0.52835	62.559	-38.021	39.704	0.80645	1.1801
210	121.56	106.86	0.49406	62.126	-37.120	39.821	0.81065	1.1770
212	130.04	115.34	0.46240	61.688	-36.213	39.925	0.81483	1.1740
214	138.94	124.25	0.43314	61.243	-35.299	40.014	0.81899	1.1709
216	148.27	133.58	0.40605	60.791	-34.378	40.089	0.82314	1.1679
218	158.05	143.35	0.38094	60.332	-33.449	40.148	0.82729	1.1649
220	168.28	153.59	0.35763	59.864	-32.511	40.191	0.83143	1.1619
222	178.98	164.28	0.33596	59.389	-31.564	40.217	0.83556	1.1589
224	190.16	175.46	0.31580	58.905	-30.609	40.226	0.83969	1.1559
226	201.83	187.13	0.29701	58.412	-29.643	40.217	0.84382	1.1529
228	214.00	199.30	0.27947	57.909	-28.667	40.189	0.84795	1.1499
230	226.69	212.00	0.26309	57.396	-27.679	40.141	0.85208	1.1470
232	239.91	225.22	0.24776	56.871	-26.680	40.072	0.85622	1.1439
234	253.68	238.98	0.23340	56.334	-25.668	39.980	0.86037	1.1409
236	268.00	253.30	0.21993	55.785	-24.642	39.866	0.86454	1.1379
238	282.89	268.19	0.20729	55.221	-23.601	39.726	0.86872	1.1348
240	298.36	283.67	0.19539	54.643	-22.544	39.560	0.87292	1.1317
242	314.44	299.74	0.18420	54.048	-21.471	39.366	0.87715	1.1286
244	331.12	316.43	0.17363	53.435	-20.378	39.142	0.88141	1.1254
246	348.44	333.74	0.16366	52.803	-19.265	38.885	0.88570	1.1221
248	366.40	351.70	0.15423	52.149	-18.130	38.592	0.89004	1.1188
250	385.02	370.32	0.14529	51.471	-16.970	38.261	0.89444	1.1154
252	404.32	389.62	0.13681	50.766	-15.782	37.887	0.89889	1.1119
254	424.31	409.62	0.12875	50.031	-14.563	37.466	0.90342	1.1083
256	445.02	430.33	0.12106	49.262	-13.310	36.992	0.90803	1.1045
258	466.47	451.77	0.11372	48.453	-12.017	36.459	0.91274	1.1006
260	488.67	473.97	0.10668	47.597	-10.677	35.857	0.91758	1.0966
265	547.63	532.94	0.090170	45.194	-7.0664	33.970	0.93044	1.0853
270	611.90	597.20	0.074640	42.194	-2.8705	31.255	0.94511	1.0715
275	682.09	667.39	0.058708	37.824	2.6584	26.764	0.96420	1.0519
** 278.2	731.4	716.7	0.03673	27.23	14.23	14.23	1.005	1.005

#Triple point *in. Hg. Vacuum **Critical point

Temp., K	Viscosity, N·s/m² Sat. Liquid × 10⁻³‡	Sat. Vapor × 10⁻³‡	Gas at p = 1 atm × 10⁻⁶‡	Thermal Conductivity, W/(m·K) Sat. Liquid	Sat. Vapor	Gas at p = 1 atm	Specific Heat, c_p, kJ/(kg·K) Sat. Liquid	Sat. Vapor	Gas at p = 0 atm	Gas at p = 1 atm	Temp., K
80	0.257	0.00627	6.27	0.1623			1.680	0.943	0.9098		80
100	0.1560	0.00772	7.68	0.1372	0.0093	0.0091	1.719	0.982	0.9098		100
120	0.1117	0.00949	9.12	0.1096	0.0124	0.0109	1.78	1.25	0.9097	0.9265	120
130	0.0960	0.01057	9.85	0.0949	0.015	0.0119		1.58	0.9097	0.9227	130
140	0.0780	0.01231	10.6	0.0796	0.018	0.0128		2.10	0.9097	0.9203	140
145	0.0665	0.0139	10.9	0.0712	0.021	0.0133			0.9097	0.9193	145
150	0.0510	0.0161	11.3	0.0610	0.025	0.0138			0.9097	0.9184	150
155*	0.0259	0.0259	11.6	0.041	0.041	0.0142			0.9097	0.9176	155*
160			12.0			0.0147			0.9098	0.9170	160
170			12.7			0.0156			0.9098	0.9160	170
180			13.3			0.0165			0.9099	0.9153	180
200			14.7			0.0182			0.9102	0.9143	200
220			15.9			0.0200			0.9109	0.9141	220
240			17.2			0.0217			0.9119	0.9144	240
260			18.4			0.0234			0.9128	0.9151	260
280			19.5			0.0251			0.9149	0.9168	280
300			20.7			0.0267			0.9184	0.9198	300
320			21.8			0.0283			0.9223	0.9233	320
340			22.8			0.0298			0.9266	0.9273	340
360			23.9			0.0313			0.9313	0.9317	360
380			24.9			0.0328			0.9363	0.9365	380
400			25.9			0.0342			0.9415	0.9416	400
500			30.5			0.0412			0.9709	0.9710	500
600			34.7			0.0480			1.002	1.003	600
700			38.5			0.0544			1.031	1.032	700

*Critical temperature. Tabulated properties ignore critical region effects. ‡Actual value = (Table value) × (Indicated multiplier).

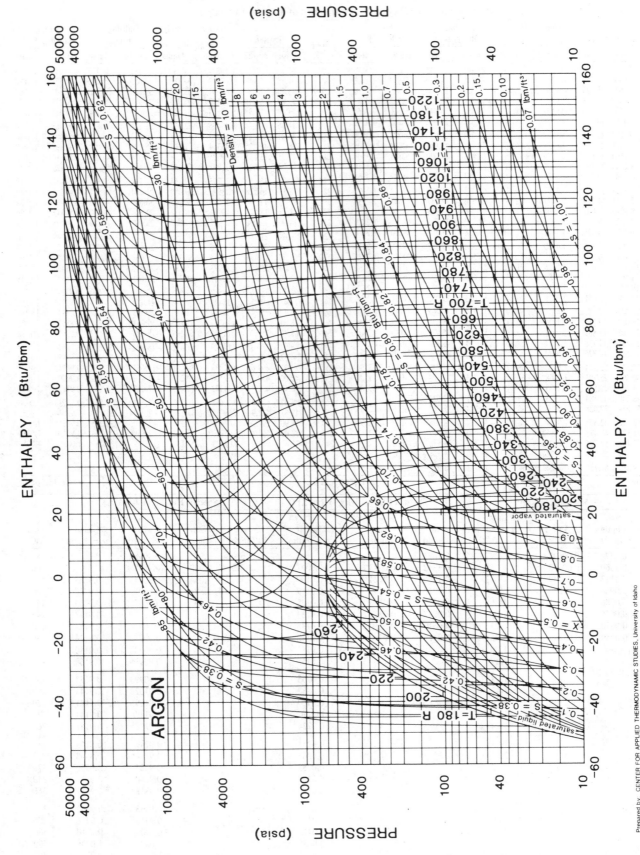

Fig. 38   Temperature-Entropy Diagram for Refrigerant 740 (Argon)

Refrigerant 740 (Argon)   Properties of Liquid and Vapor

Temp, °R	Pressure psia	Pressure psig	Volume, ft³/lbm Vapor	Density, lbm/ft³ Liquid	Enthalpy, Btu/lbm Liquid	Enthalpy, Btu/lbm Vapor	Entropy, Btu/lbm·°R Liquid	Entropy, Btu/lbm·°R Vapor
#150.85	10.000	9.5604*	3.9510	88.336	−52.410	18.246	0.31667	0.78491
151	10.098	9.3609*	3.9159	88.304	−52.368	18.261	0.31695	0.78454
152	10.760	8.0141*	3.6950	88.094	−52.095	18.357	0.31874	0.78209
153	11.454	6.5999*	3.4893	87.884	−51.822	18.451	0.32052	0.77968
154	12.183	5.1160*	3.2976	87.674	−51.550	18.544	0.32228	0.77731
155	12.947	3.5603*	3.1189	87.462	−51.277	18.636	0.32404	0.77497
156	13.748	1.9304*	2.9521	87.250	−51.004	18.727	0.32579	0.77266
157	14.586	0.2241*	2.7963	87.038	−50.730	18.816	0.32752	0.77038
157.13	14.696	0.0	2.7771	87.010	−50.695	18.827	0.32774	0.77009
158	15.463	0.7667	2.6506	86.824	−50.457	18.904	0.32924	0.76813
159	16.379	1.6834	2.5142	86.610	−50.184	18.991	0.33096	0.76592
160	17.337	2.6410	2.3866	86.395	−49.910	19.077	0.33266	0.76373
161	18.337	3.6409	2.2669	86.179	−49.637	19.161	0.33435	0.76157
162	19.380	4.6842	2.1546	85.963	−49.363	19.244	0.33603	0.75944
163	20.468	5.7721	2.0492	85.745	−49.089	19.325	0.33770	0.75734
164	21.602	6.9058	1.9501	85.527	−48.814	19.405	0.33937	0.75526
165	22.783	8.0865	1.8570	85.308	−48.540	19.484	0.34102	0.75321
166	24.012	9.3155	1.7694	85.088	−48.265	19.561	0.34266	0.75119
167	25.290	10.594	1.6868	84.868	−47.990	19.637	0.34430	0.74919
168	26.619	11.923	1.6091	84.646	−47.714	19.711	0.34593	0.74721
169	28.001	13.305	1.5358	84.423	−47.438	19.784	0.34755	0.74525
170	29.435	14.739	1.4666	84.200	−47.162	19.855	0.34916	0.74332
172	32.469	17.773	1.3395	83.750	−46.609	19.992	0.35235	0.73952
174	35.732	21.036	1.2260	83.296	−46.054	20.123	0.35552	0.73580
176	39.233	24.537	1.1242	82.838	−45.497	20.247	0.35866	0.73216
178	42.983	28.287	1.0328	82.375	−44.937	20.364	0.36177	0.72860
180	46.994	32.298	0.95049	81.907	−44.376	20.474	0.36486	0.72510
182	51.274	36.578	0.87620	81.435	−43.812	20.576	0.36792	0.72167
184	55.837	41.141	0.80902	80.958	−43.246	20.671	0.37096	0.71830
186	60.691	45.995	0.74814	80.476	−42.677	20.758	0.37397	0.71499
188	65.847	51.151	0.69285	79.988	−42.105	20.836	0.37697	0.71174
190	71.318	56.622	0.64253	79.495	−41.529	20.906	0.37995	0.70853
192	77.114	62.418	0.59667	78.996	−40.950	20.968	0.38291	0.70538
194	83.245	68.549	0.55477	78.490	−40.368	21.020	0.38585	0.70227
196	89.723	75.027	0.51644	77.979	−39.782	21.063	0.38878	0.69919
198	96.559	81.863	0.48131	77.460	−39.192	21.097	0.39169	0.69616

Temp, °R	Pressure psia	Pressure psig	Volume, ft³/lbm Vapor	Density, lbm/ft³ Liquid	Enthalpy, Btu/lbm Liquid	Enthalpy, Btu/lbm Vapor	Entropy, Btu/lbm·°R Liquid	Entropy, Btu/lbm·°R Vapor
200	103.76	89.069	0.44905	76.935	−38.597	21.121	0.39459	0.69317
202	111.35	96.655	0.41939	76.403	−37.998	21.135	0.39748	0.69020
204	119.33	104.63	0.39207	75.863	−37.395	21.138	0.40036	0.68727
206	127.71	113.01	0.36686	75.315	−36.786	21.130	0.40322	0.68436
208	136.51	121.81	0.34357	74.758	−36.172	21.110	0.40609	0.68147
210	145.73	131.03	0.32203	74.194	−35.553	21.079	0.40894	0.67860
212	155.39	140.69	0.30206	73.620	−34.927	21.036	0.41179	0.67575
214	165.50	150.80	0.28353	73.036	−34.296	20.980	0.41464	0.67292
216	176.07	161.37	0.26632	72.443	−33.657	20.910	0.41748	0.67009
218	187.11	172.41	0.25030	71.839	−33.012	20.827	0.42032	0.66728
220	198.64	183.94	0.23537	71.223	−32.359	20.729	0.42317	0.66447
222	210.67	195.97	0.22144	70.595	−31.698	20.616	0.42601	0.66165
224	223.20	208.51	0.20843	69.955	−31.029	20.487	0.42887	0.65884
226	236.26	221.57	0.19626	69.301	−30.351	20.341	0.43173	0.65601
228	249.86	235.16	0.18485	68.632	−29.663	20.177	0.43460	0.65318
230	264.00	249.30	0.17416	67.948	−28.964	19.994	0.43748	0.65033
232	278.70	264.01	0.16411	67.246	−28.255	19.791	0.44038	0.64746
234	293.98	279.29	0.15466	66.526	−27.533	19.567	0.44330	0.64456
236	309.85	295.15	0.14576	65.786	−26.797	19.319	0.44624	0.64163
238	326.32	311.62	0.13737	65.023	−26.047	19.047	0.44920	0.63866
240	343.41	328.71	0.12943	64.236	−25.382	18.748	0.45220	0.63565
242	361.13	346.44	0.12193	63.422	−24.498	18.420	0.45524	0.63258
244	379.50	364.81	0.11481	62.578	−23.695	18.060	0.45832	0.62944
246	398.54	383.84	0.10805	61.700	−22.869	17.665	0.46146	0.62622
248	418.26	403.56	0.10162	60.782	−22.019	17.230	0.46466	0.62291
250	438.68	423.99	0.095479	59.821	−21.140	16.752	0.46794	0.61950
252	459.83	445.13	0.089606	58.807	−20.228	16.224	0.47131	0.61595
254	481.71	467.01	0.083969	57.732	−19.278	15.638	0.47479	0.61224
256	504.36	489.66	0.078537	56.584	−18.281	14.984	0.47841	0.60834
258	527.80	513.10	0.073277	55.346	−17.228	14.251	0.48221	0.60421
260	552.05	537.36	0.068151	53.995	−16.104	13.418	0.48623	0.59977
262	577.15	562.46	0.063114	52.477	−14.889	12.462	0.49055	0.59494
264	603.14	588.45	0.058101	50.797	−13.548	11.341	0.49530	0.58957
266	630.06	615.37	0.053105	48.804	−12.024	9.9869	0.50067	0.58341
268	657.97	643.28	0.047686	46.333	−10.201	8.2669	0.50709	0.57600
**271.2	705.7	691.0	0.02988	33.46	−1.419	−1.419	0.5388	0.5388

#Triple point   *in. Hg. Vacuum   **Critical point

Temp., K	Viscosity, N·s/m² Sat. Liquid ×10⁻³‡	Viscosity, N·s/m² Sat. Vapor ×10⁻³‡	Viscosity, N·s/m² Gas at p = 1 atm ×10⁻⁶‡	Thermal Conductivity, W/(m·K) Sat. Liquid	Thermal Conductivity, W/(m·K) Sat. Vapor	Thermal Conductivity, W/(m·K) Gas at p = 1 atm	Specific Heat, c_p, kJ/(kg·K) Sat. Liquid	Specific Heat, c_p, kJ/(kg·K) Sat. Vapor	Specific Heat, c_p, kJ/(kg·K) Gas at p = 0 atm	Specific Heat, c_p, kJ/(kg·K) Gas at p = 1 atm	Temp., K
90	0.2396	0.00765	7.57	0.1201	0.0059	0.00587	1.163	0.555			90
100	0.1823	0.00855	8.34	0.1083	0.0068	0.00652	1.200	0.591	0.52030	0.54271	100
110	0.1458	0.00955	9.11	0.0963	0.0077	0.00716	1.247	0.627	0.52030	0.53686	110
120	0.1210	0.01070	9.91	0.0842	0.0088	0.00779	1.361	0.731	0.52030	0.53299	120
130	0.1010	0.0122	10.7	0.0718	0.0103	0.00839	1.595	0.971	0.52030	0.53039	130
140	0.0750	0.0145	11.5	0.0592	0.0120	0.00898	2.01	1.42	0.52030	0.52843	140
145	0.0603	0.0162	11.9	0.0518	0.0140	0.00928			0.52030	0.52774	145
150	0.0447	0.0226	12.3	0.0404	0.019	0.00957			0.52030	0.52704	150
151*	0.0279	0.0279	12.4	0.025	0.025	0.00963			0.52030	0.52650	151*
160			13.0			0.01016			0.52030	0.52595	160
170			13.8			0.0107			0.52030	0.52513	170
180			14.6			0.0113			0.52030	0.52444	180
200			16.0			0.0124			0.52030	0.52350	200
220			17.4			0.0136			0.52030	0.52285	220
240			18.8			0.0146			0.52030	0.52231	240
260			20.2			0.0157			0.52030	0.52196	260
280			21.5			0.0167			0.52030	0.52169	280
300			22.7			0.0177			0.52030	0.52148	300
320			23.9			0.0187			0.52030	0.52131	320
340			25.1			0.0197			0.52030	0.52119	340
360			26.3			0.0206			0.52030	0.52107	360
380			27.4			0.0215			0.52030	0.52098	380
400			28.5			0.0223			0.52030	0.52090	400
500			33.7			0.0264			0.52030	0.52067	500
600			38.3			0.0301			0.52030	0.52054	600

*Critical temperature. Tabulated properties ignore critical region effects.   ‡Actual value = (Table value) × (Indicated multiplier).

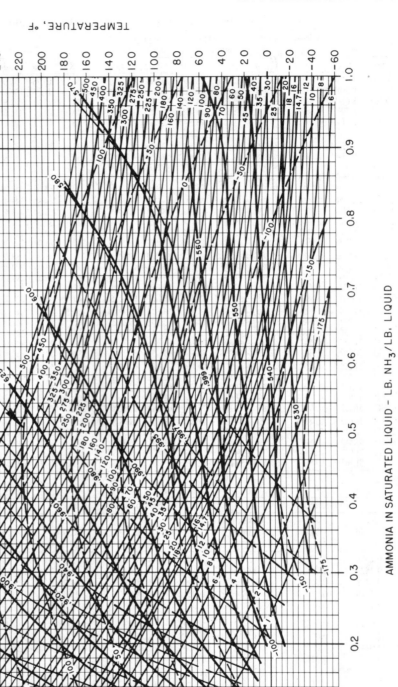

Fig. 39   Enthalpy-Concentration Diagram for Ammonia-Water Solution

Specific Volume of Saturated Ammonia Solutions, ft³/lb

Temp., °F	0	10	20	30	40	50	60	70	80	90	100	Temp., °F
20	0.0160	0.0165	0.0170	0.0176	0.0182	0.0190	0.0197	0.0207	0.0217	0.0230	0.0245	20
40	0.0160	0.0165	0.0171	0.0177	0.0184	0.0191	0.0200	0.0209	0.0221	0.0236	0.0253	40
60	0.0160	0.0166	0.0172	0.0178	0.0186	0.0193	0.0202	0.0212	0.0225	0.0241	0.0260	60
80	0.0161	0.0167	0.0173	0.0180	0.0188	0.0196	0.0205	0.0216	0.0230	0.0247	0.0267	80
100	0.0161	0.0168	0.0174	0.0182	0.0190	0.0198	0.0208	0.0220	0.0235	0.0254	0.0275	100
120	0.0162	0.0169	0.0176	0.0184	0.0192	0.0201	0.0211	0.0224	0.0241	0.0261	0.0284	120
140	0.0163	0.0170	0.0177	0.0185	0.0194	0.0203	0.0215	0.0229	0.0247	0.0268	0.0294	140
160	0.0164	0.0172	0.0179	0.0187	0.0196	0.0206	0.0219	0.0235	0.0254	0.0277	0.0306	160
180	0.0165	0.0173	0.0181	0.0190	0.0199	0.0210	0.0223	0.0241	0.0262	0.0286	0.0320	180
200	0.0166	0.0175	0.0183	0.0192	0.0202	0.0213	0.0228	0.0247	0.0270	0.0298	0.0338	200
220	0.0168	0.0176	0.0185	0.0194	0.0205	0.0217	0.0234	0.0255	0.0279	0.0312	0.0361	220

Refrigerant Temperature ($t' = °F$) and Enthalpy ($h = Btu/lb$) of Lithium Bromide Solutions

Temp. ($t = °F$)		0	10	20	30	40	45	50	55	60	65	70
80	t'	80.0	78.2	75.6	70.5	60.9	53.5	42.1	28.6	13.8	−0.2#	−11.6#
	h	48.0	39.2	31.8	25.6	21.6	21.2	23.0	28.7	38.9	52.7#	67.1#
100	t'	100.0	98.1	95.3	89.9	79.6	71.8	60.0	46.1	30.9	16.2#	3.8#
	h	68.0	56.6	47.0	38.7	33.2	32.1	33.2	38.2	47.8	61.1#	75.1#
120	t'	120.0	117.9	114.9	109.2	98.3	90.1	77.9	63.6	48.1	32.7	19.1#
	h	87.9	73.6	61.7	51.7	44.7	43.0	43.6	48.0	56.9	69.4	83.0#
140	t'	140.0	137.8	134.6	128.5	117.1	108.5	95.8	81.2	65.2	49.1	34.4#
	h	107.9	91.0	77.0	65.1	56.5	54.1	54.1	57.9	66.1	78.0	91.1#
160	t'	160.0	157.7	154.3	147.9	135.8	126.8	113.8	98.7	82.3	65.6	49.7#
	h	127.9	108.2	92.0	78.2	68.1	65.1	64.7	67.9	75.4	86.6	99.2#
180	t'	180.0	177.5	173.9	167.2	154.5	145.1	131.7	116.2	99.5	82.0	65.1#
	h	147.9	125.4	107.9	91.9	80.4	76.6	75.3	77.7	84.6	95.1	107.2#
200	t'	200.0	197.4	193.6	186.5	173.3	163.5	149.6	133.7	116.6	98.5	80.4#
	h	168.0	143.4	123.3	105.3	92.1	87.4	85.9	87.8	94.1	104.0	115.6#
220	t'	220.0	217.2	213.3	205.8	192.0	181.8	167.5	151.3	133.7	114.9	95.7
	h	188.1	160.7	138.2	119.0	104.1	99.0	96.5	97.8	103.3	112.5	123.6
240	t'	240.0*	237.1*	232.9	225.2	210.7	200.2	185.4	168.8	150.9	131.4	111.0
	h	208.3*	178.4*	154.0	132.6	116.0	110.3	107.1	107.7	112.5	121.1	131.6
260	h	260.0*	256.9*	252.6*	244.5*	229.4	218.5	203.3	186.3	168.0	147.9	126.4
	h	228.6*	195.7*	169.1*	146.2*	128.1	121.6	117.6	117.6	121.6	129.5	139.5
280	t'	280.0*	276.8*	272.3*	263.8*	248.2*	236.8*	221.2	203.9	185.1	164.3	141.7
	h	249.1*	213.8*	185.1*	159.7*	140.0*	132.8*	128.1	127.5	130.6	137.9	147.6
300	t'	300.0*	296.7*	291.9*	283.1*	266.9*	255.2*	239.2*	221.4	202.3	180.8	157.0
	h	269.6*	231.6*	200.7*	173.5*	152.1*	144.1*	138.9*	137.3	139.8	146.5	155.5
320	t'	320.0*	316.5*	311.6*	302.5*	285.6*	273.5*	257.1*	238.9*	219.4	197.2	172.4
	h	290.3*	249.7*	216.3*	187.2*	164.2*	155.3*	149.5*	147.1*	148.8	154.9	163.4
340	t'	340.0*	336.4*	331.3*	321.8*	304.4*	291.9*	275.0*	256.4*	236.5*	213.7	187.7
	h	311.1*	267.9*	232.1*	201.0*	176.1*	166.6*	160.1*	157.0*	158.0*	163.5	171.0
360	t'	360.0*	356.2*	350.9*	341.1*	323.1*	310.2*	292.9*	274.0*	253.7*	230.1	203.0
	h	332.2*	286.1*	248.0*	214.9*	188.2*	178.0*	170.6*	166.8*	167.0*	171.9	178.3

*Extensions of data above 235°F are well above the original data and should be used with care.
#Supersaturated solution.

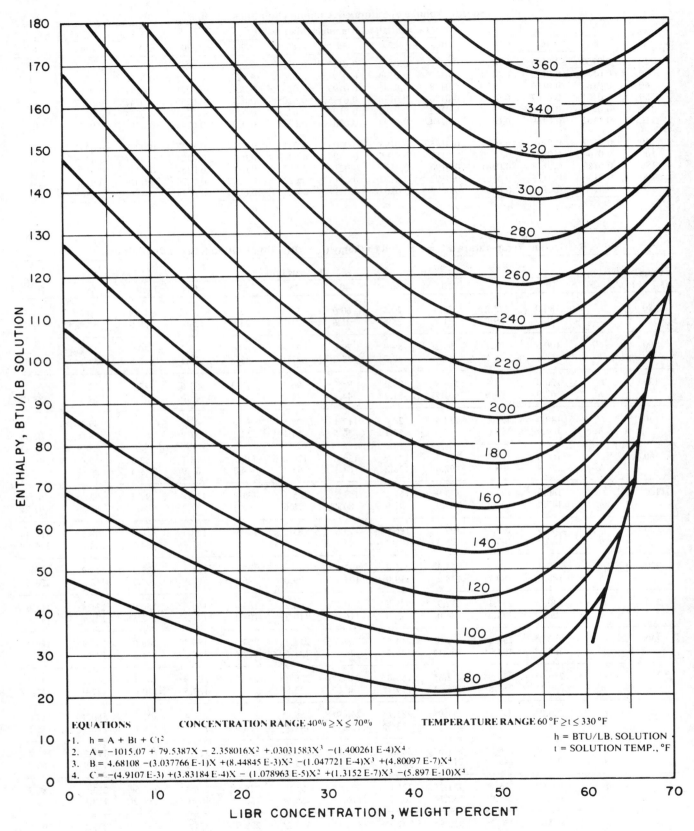

Fig. 39 Enthalpy-Concentration Diagram for Lithium Bromide-Water Solutions

EQUATIONS   CONCENTRATION RANGE 40% ≥ X ≤ 70%   TEMPERATURE RANGE 60 °F ≥ t ≤ 330 °F

1. $h = A + Bt + Ct^2$

2. $A = -1015.07 + 79.5387X - 2.358016X^2 + .03031583X^3 - (1.400261 \text{ E-4})X^4$

3. $B = 4.68108 - (3.037766 \text{ E-1})X + (8.44845 \text{ E-3})X^2 - (1.047721 \text{ E-4})X^3 + (4.80097 \text{ E-7})X^4$

4. $C = -(4.9107 \text{ E-3}) + (3.83184 \text{ E-4})X - (1.078963 \text{ E-5})X^2 + (1.3152 \text{ E-7})X^3 - (5.897 \text{ E-10})X^4$

h = BTU/LB. SOLUTION
t = SOLUTION TEMP., °F

ENTHALPY, BTU/LB SOLUTION

LIBR CONCENTRATION, WEIGHT PERCENT

EQUATIONS

1. $t = At' + B$
2. $t' = (t - B)/A$
3. $A = -2.00755 + .16976X - (3.133362 \text{ E-3})X^2 + (1.97668 \text{ E-5})X^3$
4. $B = 321.128 - 19.322X + .374382X^2 - (2.0637 \text{ E-3})X^3$
5. $\log_{10} P = C + D/(t' + 459.72) + E/(t' + 459.72)^2$
6. $t' = \dfrac{-2E}{D + [^2 - 4E(C - \log_{10} P)]^{0.5}} - 459.72$

TEMP. RANGE (REFRIGERANT)	$0 \geq t' \leq 230 \text{ F}$
TEMP. RANGE (SOLUTION)	$40 \geq t \leq 350 \text{ °F}$
CONCENTRATION RANGE	$45\% \geq X \leq 70\%$

$C = 6.21147$
$D = -2886.373$
$E = -337269.46$
$t' = \text{REFRIGERANT TEMP. °F}$
$t = \text{SOLUTION TEMP. °F}$
$X = \text{PERCENT LIBR}$
$P = \text{PSIA}$

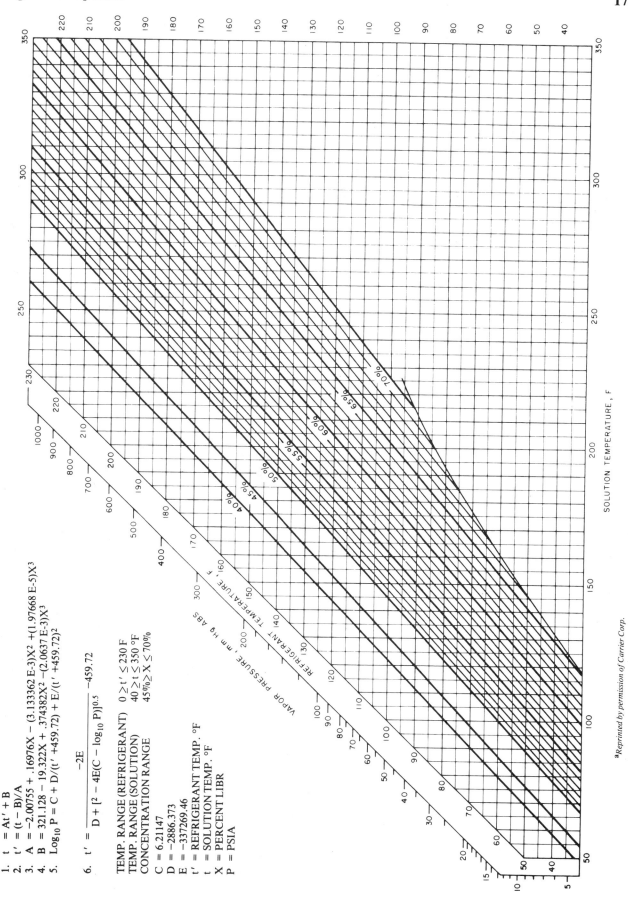

SOLUTION TEMPERATURE, F

REFRIGERANT TEMPERATURE, F

VAPOR PRESSURE, m m Hg ABS

Fig. 40 Equilibrium Chart for Aqueous Lithium Bromide Solutions

[a]*Reprinted by permission of Carrier Corp.*

REFERENCES

The sources for the tabular data in this chapter are listed by fluid and property. The reference listed under "equation of state" is the source for vapor pressure, liquid density, vapor volume, enthalpy, entropy, specific heat, and the velocity of sound, unless indicated otherwise by additional entries for one or more of these properties. Transport properties for many of the refrigerants have been updated as part of ASHRAE research project 558RP (Liley 1993); the values in the tables in this Handbook were calculated from the equations given by Liley (1993) and may differ slightly from the tabular values given by Liley. Data for the hydrocarbon refrigerants, cryogens, absorption solutions, and several of the CFC refrigerants have been copied directly from the 1989 Handbook; these are indicated by a (*) preceding the citation.

R-11

Equation of state:
Jacobsen, R.T., S.G. Penoncello, and E.W. Lemmon. 1992. A fundamental equation for trichlorofluoromethane (R-11). *Fluid Phase Equilibria* 80(11):45-56.
Viscosity, thermal conductivity, and surface tension:
Liley, P.E. 1993. ASHRAE *Thermophysical properties of refrigerants*. ASHRAE, Atlanta.

R-12

Equation of state:
Penoncello, S.G., R.T. Jacobsen, and E.W. Lemmon. 1992. A fundamental equation for dichlorodifluoromethane (R-12). *Fluid Phase Equilibria* 80(11):57-70.
Viscosity, thermal conductivity, and surface tension:
Liley, P.E. 1993. ASHRAE *Thermophysical properties of refrigerants*. ASHRAE, Atlanta.

R-13

Equation of state:
Downing, R.C. 1974. Refrigerant equations. ASHRAE *Transactions* 80(2):159.
Liquid heat capacity, viscosity, thermal conductivity, and surface tension:
Liley, P.E. 1993. ASHRAE *Thermophysical Properties of Refrigerants*. ASHRAE, Atlanta.

R-13B1

Equation of state:
*Martin, J.J., L.M. Welshans, C.H. Chou, and G.E. Gryka. 1953. Data and equations for some thermodynamic properties of Freon-13B1 [CBrF(3)]. Engineering Research Institute, Report on Project M777, University of Michigan, Ann Arbor, MI.
Heat capacity, viscosity, thermal conductivity, and surface tension:
Liley, P.E. 1993. ASHRAE *Thermophysical properties of refrigerants*. ASHRAE, Atlanta.

R-14

Equation of state:
Downing, R.C. 1974. Refrigerant equations. ASHRAE *Transactions* 80(2):159.
Liquid heat capacity, liquid sound speed, viscosity, thermal conductivity, and surface tension:
Liley, P.E. 1993. ASHRAE *Thermophysical properties of refrigerants*. ASHRAE, Atlanta.

R-22

Equation of state:
Kamei, A. and S.W. Beyerlein. 1992. A fundamental equation for chlorodifluoromethane (R-22). *Fluid Phase Equilibria* 80(11):71-86.
Viscosity, thermal conductivity, and surface tension:
Liley, P.E. 1993. ASHRAE *Thermophysical properties of refrigerants*. ASHRAE, Atlanta.

R-23

Equation of state:
Hou, Y.C. and J.J. Martin. 1959. Physical and thermodynamic properties of trifluoromethane. AIChE *Journal* 5:125-29.
Liquid heat capacity, liquid sound speed, viscosity, thermal conductivity, and surface tension:
Liley, P.E. 1993. ASHRAE *Thermophysical properties of refrigerants*. ASHRAE, Atlanta.

R-32

Equation of state:
Malbrunot, P.F., P.A. Meunier, G.M. Scatena, W.H. Mears, K.P. Murphy, and J.V. Sinka. 1968. Pressure-volume-temperature behavior of difluoromethane. *Journal of Chemical and Engineering Data* 13:16-21.
Viscosity:
Phillips, T.W. and K.P. Murphy. 1970. Liquid viscosity of halocarbons. *Journal of Chemical and Engineering Data* 15:304-307.
Thermal conductivity:
Tauscher, W. 1969. Measurement of the thermal conductivity of liquid refrigerants by an unsteady-state hot wire method. ASHRAE *Journal* 11(1):97-104.
Surface tension:
Schmidt, J.W. and M.R. Moldover. 1992. Alternative refrigerants R32 and R125: Critical temperature, refractive index, surface tension, and estimates of liquid, vapor and critical densities. National Institute of Standards and Technology, Gaithersburg, MD. Personal communication to M.O. McLinden, November 2, 1992.

R-113

Equation of state:
*Mastroianni, M.J., R.F. Stahl, and P.N. Sheldon. 1978. Physical and thermodynamic properties of 1,1,2-trifluorotrichloroethane (R113). *Journal of Chemical and Engineering Data* 23:113-18.
Heat capacity, viscosity, thermal conductivity, and surface tension:
Liley, P.E. 1993. ASHRAE *Thermophysical properties of refrigerants*. ASHRAE, Atlanta.

R-114

Equation of state:
*Hules, K.R. and D.P. Wilson. 1982. An interim engineering model of the thermodynamic properties of 1,2-dichlorotetrafluoroethane, Refrigerant 114. Proceedings of the 8th ASME Thermophysical Properties Symposium 2:370-79.
Heat capacity, velocity of sound, viscosity, thermal conductivity, and surface tension:
Liley, P.E. 1993. ASHRAE *Thermophysical properties of refrigerants*. ASHRAE, Atlanta.

R-123

Equation of state:
Younglove, B.A. 1992. National Institute of Standards and Technology, Boulder, CO. Personal communication to M.O. McLinden, January 16, 1992.
Liquid thermal conductivity:
Yata, J. 1990. Thermal conductivity of alternatives to CFCs. Proceedings of the 11th Japan Symposium on Thermophysical Properties, 143-46.
Vapor thermal conductivity:
Gross, U., Y.W. Song, J. Kallweit, and E. Hahne. 1990. Thermal conductivity of saturated R123 and R134a—Transient hot wire measurements. Proceedings of the meeting of IIR Commission B1, Tel Aviv.
Liquid viscosity:
Diller, D.E., A.S. Aragon, and A. Laesecke. 1993. Measurements of the viscosities of saturated and compressed liquid 1,1,1,2-tetrafluoroethane (R134a), 2,2-dichloro-1,1,1-trifluoroethane (R123) and 1,1-dichloro-1-fluoroethane (R141b). *Fluid Phase Equilibria*.
Vapor viscosity:
Takahashi, M., C. Yokoyama, and S. Takahashi. 1990. Gas viscosities of HCFC-123 and HCFC-123a. Proceedings of the 11th Japan Symposium on Thermophysical Properties, 115-18.
Surface tension:
Chae, H.B., J.W. Schmidt, and M.R. Moldover. 1990. Surface Tension of Refrigerants R123 and R134a. *Journal of Chemical and Engineering Data* 35:6-8.

R-124

Equation of state:
Sandarusi, J., B.A. Younglove, and M.O. McLinden. 1992. Submitted to the *International Journal of Refrigeration*.
Thermal conductivity:
Perkins, R.A., L.J. van Poolen, J.B. Howley, and M.L. Huber. 1993. Thermal conductivity of 2-chloro-1,1,1,2-tetrafluoroethane (R124). *International Journal of Thermophysics*.
Liquid viscosity:
Diller, D.E. and S.M. Peterson. 1993. Measurements of the viscosities of saturated and compressed fluid 1-chloro-1,2,2,2-tetrafluoroethane (R124) and pentafluoroethane (R125) at temperatures between 120 and 420 K. *International Journal of Thermophysics*.

Vapor viscosity:
Takahashi, M., C. Yokoyama, and S. Takahashi. 1992. Gas viscosity of HFC-124 [sic] at high pressures. Proceedings of the 13th Japan Symposium on Thermophysical Properties, 347-50.
Surface tension:
Okada, M. and Y. Higashi. 1992. Measurements of the surface tension for HCFC-124 and HCFC-141b. Proceedings of the 13th Japan Symposium on Thermophysical Properties, 73-76.

R-125

Equation of state:
Wilson, L.C., W.V. Wilding, G.M. Wilson, R.L. Rowley, V.M. Felix, and T. Chisolm-Carter. 1992. Thermophysical properties of HFC-125. *Fluid Phase Equilibria* 80(11):167-78.
Liquid viscosity:
Diller, D.E. and S.M. Peterson. 1993. Measurements of the viscosities of saturated and compressed fluid 1-chloro-1,2,2,2-tetrafluoroethane (R124) and pentafluoroethane (R125) at temperatures between 120 and 420 K. *International Journal of Thermophysics*.
Surface tension:
Schmidt, J.W. and M.R. Moldover. 1992. Alternative refrigerants R32 and R125: Critical temperature, refractive index, surface tension, and estimates of liquid, vapor and critical densities. National Institute of Standards and Technology, Gaithersburg, MD. Personal communication to M.O. McLinden, November 2, 1992.

R-134a

Equation of state:
Huber, M.L. and M.O. McLinden. 1992. Thermodynamic properties of R134a (1,1,1,2-tetrafluoroethane). Proceedings of the International Refrigeration Conference, Purdue University, West Layfayette, Indiana. July 14-17, 453-62.
Thermal conductivity:
Laesecke, A., R.A. Perkins, and C.A. Nieto de Castro. 1992. Thermal conductivity of R134a. *Fluid Phase Equilibria* 80(11):263-74.
Viscosity:
Huber, M.L. and J.F. Ely. 1992. Prediction of viscosity of refrigerants and refrigerant mixtures. *Fluid Phase Equilibria 80(11):249-62.*
Surface tension:
Chae, H.B., J.W. Schmidt, and M.R. Moldover. 1990. Surface tension of refrigerants R123 and R134a. Journal of Chemical and Engineering Data 35:6-8.

R-141b

Equation of state:
Weber, L.A. 1991. PVT and thermodynamic properties of R141b in the gas phase. Proceedings of the 18th International Congress of Refrigeration. Paper no. 69.
Liquid thermal conductivity:
Yata, J., M. Hori, T. Kurahashi, and T. Minamiyama. 1992. Thermal conductivity of alternative fluorocarbons in liquid phase. *Fluid Phase Equilibria* 80(11):287-96.
Liquid viscosity:
Diller, D.E., A.S. Aragon, and A. Laesecke. 1993. Measurements of the viscosities of saturated and compressed liquid 1,1,1,2-tetrafluoroethane (R134a), 2,2-dichloro-1,1,1-trifluoroethane (R123) and 1,1-dichloro-1-fluoroethane (R141b). *Fluid Phase Equilibria*.
Surface tension:
Okada, M. and Y. Higashi. 1992. Measurements of the surface tension for HCFC-124 and HCFC-141b. Proceedings of the 13th Japan Symposium on Thermophysical Properties, 73-76.

R-142b

Equation of state:
Mears, W.H., R.F. Stahl, S.R. Orfeo, R.C. Shair, L.F. Kells, W. Thompson, and H. McCann. 1955. Thermodynamic properties of halogenated ethanes and ethylenes. *Industrial and Engineering Chemistry* 47:1449-53.
Liquid heat capacity:
Nakagawa, S., H. Sato, and K. Watanabe. 1990. Measurements of isobaric heat capacity of liquid HCFC-142b. Proceedings of the 11th Japan Symposium on Thermophysical Properties, 103-106.
Liquid thermal conductivity:
Yata, J., M. Hori, T. Kurahashi, and T. Minamiyama. 1992. Thermal conductivity of alternative fluorocarbons in liquid phase. *Fluid Phase Equilibria* 80(11):287-96.

Vapor viscosity:
Takahashi, M., C. Yokoyama, and S. Takahashi. 1987. Viscosities of gaseous R13B1, R142b, and R152a. *Journal of Chemical and Engineering Data* 32:98-103.
Surface tension:
Okada, M., Y. Higashi, T. Ikeda, and T. Kuwana. 1991. Measurements of the surface tension for HCFC-142b and HFC-152a. Proceedings of the 12th Japan Symposium on Thermophysical Properties, 105-108.

R-152a

Equation of state:
Tillner-Roth, R. 1992. Universität Hannover, Germany, Personal communication to M.O. McLinden, June 26, 1992.
Liquid thermal conductivity:
Yata, J., M. Hori, T. Kurahashi, and T. Minamiyama. 1992. Thermal conductivity of alternative fluorocarbons in liquid phase. *Fluid Phase Equilibria* 80(11):287-96.
Liquid viscosity:
Correlation based on data of Phillips and Murphy (1970), ASHRAE *Transactions* 76(2):146-56, and Kumagai and Takahashi (1991), *International Journal of Thermophysics* 12:105-17.
Vapor viscosity:
Takahashi, M., C. Yokoyama, and S. Takahashi. 1987. Viscosities of gaseous R13B1, R142b, and R152a. *Journal of Chemical and Engineering Data* 32:98-103.
Surface tension:
Okada, M., Y. Higashi, T. Ikeda, and T. Kuwana. 1991. Measurements of the surface tension for HCFC-142b and HFC-152a. Proceedings of the 12th Japan Symposium on Thermophysical Properties, 105-108.

R-500

Equation of state:
*Sinka, J.V. and K.P. Murphy. 1967. Pressure-volume-temperature relationship for a mixture of difluorodichloromethane and 1,1-difluoroethane. *Journal of Chemical and Engineering Data* 12:315-16.
Heat capacity, viscosity, and thermal conductivity:
Liley, P.E. 1993. ASHRAE *Thermophysical properties of refrigerants.* ASHRAE, Atlanta.

R-502

Equation of state:
*Martin, J.J. and R.C. Downing. 1970. Thermodynamic properties of refrigerant 502. ASHRAE *Transactions* 76(2):129-39.
Heat capacity, viscosity, thermal conductivity, and surface tension:
Liley, P.E. 1993. ASHRAE *Thermophysical properties of refrigerants.* ASHRAE, Atlanta.

R-503

Equation of state:
*Sinka, J.V., E. Rosenthal, and R.P. Dixon. 1970. Pressure-volume-temperature relationship for a mixture of monochlorotrifluoromethane and trifluoromethane. *Journal of Chemical and Engineering Data* 15:73-74.
Heat capacity, viscosity, thermal conductivity, and surface tension:
Liley, P.E. 1993. ASHRAE *Thermophysical properties of refrigerants.* ASHRAE, Atlanta.

R-717 (ammonia)

Equation of state:
Haar, L. and J.S. Gallagher. 1978. Thermodynamic properties of ammonia. *Journal of Physical and Chemical Reference Data* 7:635-792.
Viscosity:
Watson, J.T.R. 1983. The dynamic viscosity of ammonia. Paper presented to Transport Formulations Subcommittee of International Union of Pure and Applied Chemistry.
Thermal conductivity:
Krauss, R. 1991. Section in *VDI-Warmeatlas* (Thermal Atlas of the Association of German Engineers), 6th ed., Vol. D, Section b, 59-71. Dusseldorf: VDI-Verlag.
Surface tension:
Liley, P.E. 1993. ASHRAE *Thermophysical properties of refrigerants.* ASHRAE, Atlanta.

R-718 (water/steam)

Equation of state:
Haar, L., J.S. Gallagher, and G.S. Kell. 1984. NBS/NRC Steam Tables. Hemisphere Publishing Corporation, Washington, D.C.

Viscosity and thermal conductivity:

Sengers, J.V. and J.T.R. Watson. 1986. Improved international formulations for the viscosity and thermal conductivity of water substance. *Journal of Physical and Chemical Reference Data* 15:1291-1322.

Surface tension:

International Association for the Properties of Steam. 1975. Release on surface tension of water substance.

R-744 (carbon dioxide)

Equation of state:

Ely, J.F., J.W. Magee, and W.M. Haynes. 1987. Thermophysical properties for special high CO_2 content mixtures. Research Report RR-110. Gas Processors Association, Tulsa, OK.

Viscosity and thermal conductivity:

Vesovic, V., W.A. Wakeham, G.A. Olchowy, J.V. Sengers, J.T.R. Watson, and J. Millat. 1990. The transport properties of carbon dioxide. *Journal of Physical and Chemical Reference Data* 19:763-808.

Surface tension:

Liley, P.E. 1993. ASHRAE *Thermophysical properties of refrigerants*. ASHRAE, Atlanta.

R-50 (methane)

Equation of state:

*Angus, S., B. Armstrong, and K.M. de Reuck. 1978. International Thermodynamic Tables of the Fluid State—5: Methane. International Union of Pure and Applied Chemistry, Chemical Data Series, No. 16. Pergammon Press, Oxford, England.

Heat capacity, viscosity, and thermal conductivity:

*ASHRAE *Thermophysical properties of refrigerants*. 1976. ASHRAE, Atlanta.

R-170 (ethane)

Equation of state:

*Goodwin, R.D., H.M. Roder, and G.C. Straty. 1976. Thermodynamic properties of ethane, from 90 to 600 K at pressures to 700 bar. National Bureau of Standards Technical Note 684. Government Printing Office, Washington, D.C.

Heat capacity, viscosity, and thermal conductivity:

*ASHRAE *Thermophysical properties of refrigerants*. 1976. ASHRAE, Atlanta.

R-290 (ethane)

Equation of state:

*Goodwin, R.D. 1977. Provisional thermodynamic functions of propane, from 85 to 700 K at pressures to 700 bar. National Bureau of Standards NBSIR 77-860. Government Printing Office, Washington, D.C.

Heat capacity, viscosity, and thermal conductivity:

*ASHRAE *Thermophysical properties of refrigerants*. 1976. ASHRAE, Atlanta.

R-600 (n butane)

Equation of state:

*Goodwin, R.D. 1979. Normal butane: Provisional thermodynamic functions from 135 to 700 K at pressures to 700 bar. National Bureau of Standards NBSIR 79-1621. Government Printing Office, Washington, D.C.

Heat capacity, viscosity, and thermal conductivity:

*ASHRAE *Thermophysical properties of refrigerants*. 1976. ASHRAE, Atlanta. ʼ

R-600a (isobutane)

Equation of state:

*Goodwin, R.D. 1979. Isobutane: Provisional thermodynamic functions from 114 to 700 K at pressures to 700 bar. National Bureau of Standards NBSIR 79-1612. Government Printing Office, Washington, D.C.

Heat capacity, viscosity, and thermal conductivity:

*ASHRAE *Thermophysical properties of refrigerants*. 1976. ASHRAE, Atlanta.

R-702 (hydrogen)

Equation of state:

*McCarty, R.D. (1975). Hydrogen: Technology survey—Thermophysical properties. National Aeornautics and Space Administration, NASA SP-3089. Government Printing Office, Washington, D.C.

Heat capacity, viscosity, and thermal conductivity:

*ASHRAE *Thermophysical properties of refrigerants*. 1976. ASHRAE, Atlanta.

R-702p (parahydrogen)

Equation of state:

*Roder, H.M. and R.D. McCarty. 1975. Modified Benedict-Webb-Rubin equation of state for parahydrogen-II. National Bureau of Standards NBSIR 75-814. Government Printing Office, Washington, D.C.

Heat capacity and thermal conductivity:

*ASHRAE *Thermophysical properties of refrigerants*. 1976. ASHRAE, Atlanta.

R-704 (helium)

Equation of state:

*Angus, S., K.M. de Reuck, and R.D. McCarty. 1977. International Thermodynamic Tables of the Fluid State—4: Helium. International Union of Pure and Applied Chemistry, Pergammon Press, Oxford, England.

Heat capacity, viscosity, and thermal conductivity:

*ASHRAE *Thermophysical properties of refrigerants*. 1976. ASHRAE, Atlanta.

R-720 (neon)

Equation of state:

*Teng, C.J. 1981. Thermodynamic properties of neon from the triple point to 700 K with pressures to 700 MPa, M.S. Thesis, University of Idaho, Moscow, ID.

Heat capacity, viscosity, and thermal conductivity:

*ASHRAE *Thermophysical properties of refrigerants*. 1976. ASHRAE, Atlanta.

R-728 (nitrogen)

Equation of state:

*Angus, S., K.M. de Reuck, B. Armstrong, and R.T. Jacobsen. 1979. International Thermodynamic Tables of the Fluid State—6: Nitrogen. International Union of Pure and Applied Chemistry, Chemical Data Series, No. 20. Pergammon Press, Oxford, England.

Heat capacity, viscosity, and thermal conductivity:

*ASHRAE *Thermophysical properties of refrigerants*. 1976. ASHRAE, Atlanta.

R-729 (air)

Equation of state:

*Bender, R. 1973. The calculation of phase equilibria from a thermal equation of state applied to the pure fluids argon, nitrogen, oxygen and their mixtures. C.F. Muller, Karlsruhe.

Phase equilibria:

*Blanke, W. 1973. Messung der thermischen Zustandsgrossen von Luft im Zweiphasengeheit und seiner Umgebung. Dr. Ing. Dissertation, Ruhr-Universität, Bochum, Germany.

Heat capacity, viscosity, and thermal conductivity:

*ASHRAE *Thermophysical properties of refrigerants*. 1976. ASHRAE, Atlanta.

R-732 (oxygen)

Equation of state:

*Stewart, R.B. and R.T. Jacobsen. 1977. A thermodynamic property formulation for oxygen for temperatures from 55 K to 3000 K and pressures to 34 MPa. Proceedings of the 7th Symposium on Thermophysical Properties, American Society of Mechanical Engineers, 549-63.

Heat capacity, viscosity, and thermal conductivity:

*ASHRAE *Thermophysical properties of refrigerants*. 1976. ASHRAE, Atlanta.

R-740 (argon)

Equation of state:

*Jacobsen, R.T., R.B. Stewart, and C.J. Teng. 1981. The thermodynamic properties of argon from the triple point to 1200 K with pressures to 1000 MPa. Center for Applied Thermodynamic Studies, University of Idaho, Moscow, ID.

Heat capacity, viscosity, and thermal conductivity:

*ASHRAE *Thermophysical properties of refrigerants*. 1976. ASHRAE, Atlanta.

R-1150 (ethylene)

Equation of state:

*McCarty, R.D. and R.T. Jacobsen. 1981. An equation of state for fluid ethylene. National Bureau of Standards Technical Note. Government Printing Office, Washington, D.C.

Heat capacity, viscosity, and thermal conductivity:

*ASHRAE *Thermophysical properties of refrigerants*. 1976. ASHRAE, Atlanta.

R-1270 (propylene)

Equation of state:

*Angus, S., B. Armstrong, and K.M. de Reuck. 1980. International Thermodynamic Tables of the Fluid State—7: Propylene. International Union of Pure and Applied Chemistry. Pergammon Press, Oxford, England.

Heat capacity, viscosity, and thermal conductivity:

*ASHRAE *Thermophysical properties of refrigerants*. 1976. ASHRAE, Atlanta.

---

*Indicates data reprinted from the 1989 Handbook.

CHAPTER 18

SECONDARY COOLANTS (BRINES)

IN many refrigeration applications, heat is transferred to a secondary coolant, which can be any liquid cooled by the refrigerant and used to transfer heat without changing state. These liquids are known as *heat transfer fluids*, *brines*, or *secondary refrigerants*.

Other ASHRAE Handbooks describe various applications for secondary coolants. In the 1990 ASHRAE *Handbook—Refrigeration*, refrigeration systems are discussed in Chapter 5; their uses in food processing are found in Chapters 9, 12, and 14; ice rinks are discussed in Chapter 34; and environmental test facilities are covered in Chapter 37. In the 1991 ASHRAE *Handbook—Applications*, solar energy utilization is discussed in Chapter 39, thermal storage in Chapter 39, and snow melting in Chapter 45.

This chapter describes the physical properties of several secondary coolants and provides information on their use. The chapter also includes information on corrosion protection. Additional information on corrosion inhibition can be found in Chapter 43 of the 1991 ASHRAE *Handbook—Applications* and Chapter 5 of the 1990 ASHRAE *Handbook—Refrigeration*.

BRINES

Properties

Water solutions of calcium and sodium chloride are the most common refrigeration brines. Tables 1 and 2 list the properties of pure calcium chloride brine and sodium chloride brine. For commercial grades, use the formulae indicated in the footnotes to these tables. Figures 1 and 5 give the specific heats for liquid brines and are used for computation of heat loads with ordinary brine (Carrier 1959). Figures 2 and 6 show the ratio of the mass of the solution to that of water, which is commonly used as the measure of salt concentration. Viscosities are given in Figures 3 and 7. Figures 4 and 8 show thermal conductivity of calcium and sodium brines at varying temperatures and concentrations.

Brine applications in refrigeration are mainly in the industrial machinery field and in skating rinks. Corrosion is the principal problem for calcium chloride brines, especially in ice-making tanks where galvanized iron cans are immersed.

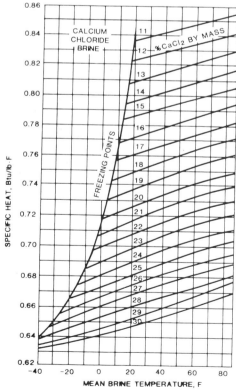

Fig. 1   Specific Heat of Calcium Chloride Brine

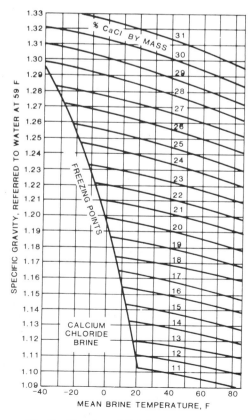

Fig. 2   Density of Calcium Chloride Brine

The preparation of this chapter is assigned to TC 3.1, Refrigerants and Brines.

Table 1 Properties of Pure Calcium Chloride[a] Brine

Pure CaCl$_2$, % by Mass	Ratio of Mass to Water at 60°F	Baume Density at 60°F	Specific Heat Btu/lb·°F	Crystallization Starts, °F	Mass per Unit Volume[b] at 60°F				Ratio of Mass at Various Temperatures to Water at 60°F			
					CaCl$_2$, lb/gal	Brine, lb/gal	CaCl$_2$, lb/ft^3	Brine, lb/ft^3	−4°F	14°F	32°F	50°F
0	1.000	0.0	1.000	32.0	0.000	8.34	0.00	62.40				
5	1.044	6.1	0.924	27.7	0.436	8.717	3.26	65.15			1.043	1.042
6	1.050	7.0	0.914	26.8	0.526	8.760	3.93	65.52			1.052	1.051
7	1.060	8.2	0.898	25.9	0.620	8.851	4.63	66.14			1.061	1.060
8	1.069	9.3	0.884	24.6	0.714	8.926	5.34	66.70			1.071	1.069
9	1.078	10.4	0.869	23.5	0.810	9.001	6.05	67.27			1.080	1.078
10	1.087	11.6	0.855	22.3	0.908	9.076	6.78	67.83			1.089	1.087
11	1.096	12.6	0.842	20.8	1.006	9.143	7.52	68.33			1.098	1.096
12	1.105	13.8	0.828	19.3	1.107	9.227	8.27	68.95			1.108	1.105
13	1.114	14.8	0.816	17.6	1.209	9.302	9.04	69.51			1.117	1.115
14	1.124	15.9	0.804	15.5	1.313	9.377	9.81	70.08			1.127	1.124
15	1.133	16.9	0.793	13.5	1.418	9.452	10.60	70.64		1.139	1.137	1.134
16	1.143	18.0	0.779	11.2	1.526	9.536	11.40	71.26		1.149	1.146	1.143
17	1.152	19.1	0.767	8.6	1.635	9.619	12.22	71.89		1.159	1.156	1.153
18	1.162	20.2	0.756	5.9	1.747	9.703	13.05	72.51		1.169	1.166	1.163
19	1.172	21.3	0.746	2.8	1.859	9.786	13.90	73.13		1.180	1.176	1.173
20	1.182	22.1	0.737	− 0.4	1.970	9.853	14.73	73.63		1.190	1.186	1.183
21	1.192	23.0	0.729	− 3.9	2.085	9.928	15.58	74.19				
22	1.202	24.4	0.716	− 7.8	2.208	10.037	16.50	75.00	1.215	1.211	1.207	1.203
23	1.212	25.5	0.707	− 11.9	2.328	10.120	17.40	75.63				
24	1.223	26.4	0.697	− 16.2	2.451	10.212	18.32	76.32	1.236	1.232	1.228	1.224
25	1.233	27.4	0.689	− 21.0	2.574	10.295	19.24	76.94				
26	1.244	28.3	0.682	− 25.8	2.699	10.379	20.17	77.56				
27	1.254	29.3	0.673	− 31.2	2.827	10.471	21.13	78.25				
28	1.265	30.4	0.665	− 37.8	2.958	10.563	22.10	78.94				
29	1.276	31.4	0.658	− 49.4	3.090	10.655	23.09	79.62				
29.87	1.290	32.6	0.655	− 67.0	3.16	10.75	23.65	80.45				
30	1.295	33.0	0.653	− 50.8	3.22	10.80	24.06	80.76				
32	1.317	34.9	0.640	− 19.5	3.49	10.98	26.10	82.14				
34	1.340	36.8	0.630	+ 4.3	3.77	11.17	28.22	83.57				

[a]Mass of Type 1 (77% min) CaCl$_2$ = (mass of pure CaCl$_2$)/(0.77). Mass of Type 2 (94% min) CaCl$_2$ = (mass of pure CaCl$_2$)/(0.94).
[b]Mass of water per unit volume = Brine mass minus CaCl$_2$ mass.

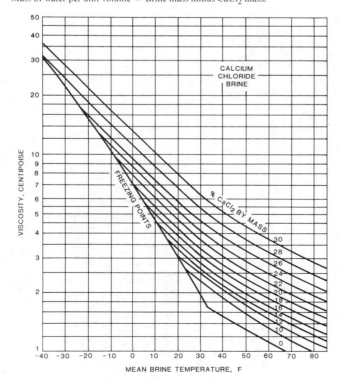

Fig. 3 Viscosity of Calcium Chloride Brine

Ordinary salt (sodium chloride) is used where contact with calcium chloride is intolerable, *e.g.*, the brine fog method of freezing fish and other foods. It is used as a spray in air cooling of unit coolers to prevent frost formation on coils. In most refrigerating work, the lower freezing point of calcium chloride solution makes it more convenient to use.

Commercial calcium chloride, available as Type 1 (77% min) and Type 2 (94% min), is marketed in flake, solid, and solution forms; flake form is used most extensively. Commercial sodium chloride is available both in crude (rock salt) and refined grades. Because magnesium salts tend to form sludge, their presence in sodium or calcium chloride is undesirable.

Corrosion

Brine systems must be treated to control corrosion and deposits. The standard chromate treatment program is the most effective. Calcium chloride brines require a minimum of 1800 ppm of sodium chromate with pH 6.5 to 8.5. Sodium chloride brines require a minimum of 3600 ppm of sodium chromate and a pH of 6.5 to 8.5. Sodium nitrite at 3000 ppm in calcium brines or 4000 ppm in sodium brines will control pH between 7.0 and 8.5, and should provide adequate protection. Organic inhibitors are available that may provide adequate protection where neither chromates nor nitrites can be used.

Before using any chromate-based inhibitor package, review federal, state, and local regulations concerning the use and disposal of chromate-containing fluids. If the regulations prove too restrictive, an alternative inhibition system should be considered.

Table 2 Properties of Pure Sodium Chloride[a] Brine

Pure NaCl, % by Mass	Ratio of Mass to Water at 59°F	Baume Density at 60°F	Specific Heat at 59°F, Btu/lb·°F	Crystalli- zation Starts, °F	Mass per Unit Volume[b] at 60°F				Ratio of Mass at Various Temperatures to Water at 60°F			
					NaCl, lb/gal	Brine, lb/gal	NaCl, lb/ft³	Brine, lb/ft³	14°F	32°F	50°F	68°F
0	1.000	0.0	1.000	32.0	0.000	8.34	0.000	62.4				
5	1.035	5.1	0.938	26.7	0.432	8.65	3.230	64.6		1.0382	1.0366	1.0341
6	1.043	6.1	0.927	25.5	0.523	8.71	3.906	65.1		1.0459	1.0440	1.0413
7	1.050	7.0	0.917	24.3	0.613	8.76	4.585	65.5		1.0536	1.0515	1.0486
8	1.057	8.0	0.907	23.0	0.706	8.82	5.280	66.0		1.0613	1.0590	1.0559
9	1.065	9.0	0.897	21.6	0.800	8.89	5.985	66.5		1.0691	1.0665	1.0633
10	1.072	10.1	0.888	20.2	0.895	8.95	6.690	66.9		1.0769	1.0741	1.0707
11	1.080	10.8	0.879	18.8	0.992	9.02	7.414	67.4		1.0849	1.0817	1.0782
12	1.087	11.8	0.870	17.3	1.090	9.08	8.136	67.8		1.0925	1.0897	1.0857
13	1.095	12.7	0.862	15.7	1.188	9.14	8.879	68.3		1.1004	1.0933	1.0971
14	1.103	13.6	0.854	14.0	1.291	9.22	9.632	68.8		1.1083	1.1048	1.1009
15	1.111	14.5	0.847	12.3	1.392	9.28	10.395	69.3	1.1195	1.1163	1.1126	1.1086
16	1.118	15.4	0.840	10.5	1.493	9.33	11.168	69.8	1.1277	1.1243	1.1205	1.1163
17	1.126	16.3	0.833	8.6	1.598	9.40	11.951	70.3	1.1359	1.1323	1.1284	1.1241
18	1.134	17.2	0.826	6.6	1.705	9.47	12.744	70.8	1.1442	1.1404	1.1363	1.1319
19	1.142	18.1	0.819	4.5	1.813	9.54	13.547	71.3	1.1535	1.1486	1.1444	1.1398
20	1.150	19.0	0.813	+ 2.3	1.920	9.60	14.360	71.8	1.1608	1.1568	1.1542	1.1478
21	1.158	19.9	0.807	− 0.0	2.031	9.67	15.183	72.3	1.1692	1.1651	1.1606	1.1559
22	1.166	20.8	0.802	− 2.3	2.143	9.74	16.016	72.8	1.1777	1.1734	1.1688	1.1640
23	1.175	21.7	0.796	− 5.1	2.256	9.81	16.854	73.3	1.1862	1.1818	1.1771	1.1721
24	1.183	22.5	0.791	+ 3.8	2.371	9.88	17.712	73.8	1.1948	1.1902	1.1854	1.1804
25	1.191	23.4	0.786	+ 16.1	2.488	9.95	18.575	74.3				
25.2	1.200			+ 32.0								

[a] Mass of commercial NaCl required = (mass of pure NaCl required)/(% purity).
[b] Mass of water per unit volume = Brine mass minus NaCl mass.

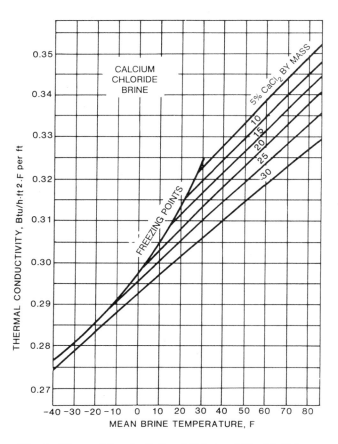

Fig. 4 Thermal Conductivity of Calcium Chloride Brine

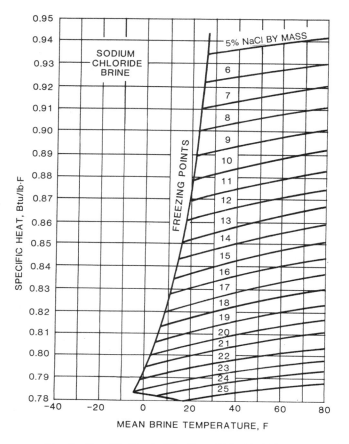

Fig. 5 Specific Heat of Sodium Chloride Brine

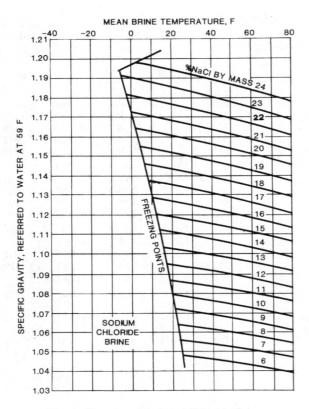

Fig. 6 Density of Sodium Chloride Brine

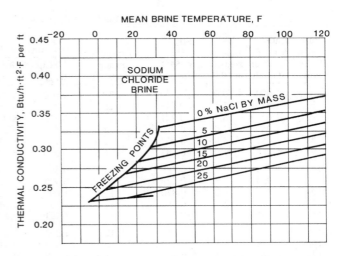

Fig. 8 Thermal Conductivity of Sodium Chloride Brine
(Carrier 1959)

INHIBITED GLYCOLS

Ethylene glycol and propylene glycol, inhibited for corrosion control, are used as aqueous freezing point depressants and heat transfer media in heating and cooling systems. Their chief attributes are their ability to lower the freezing point of water, low volatility, and relatively low corrosivity when properly inhibited.

Inhibited ethylene glycol solutions have better physical properties than propylene glycol solutions, especially at lower temperatures. However, the less toxic propylene glycol is preferred for applications involving possible human contact or where mandated by regulations.

Physical Properties

Ethylene and propylene glycol are colorless, practically odorless liquids that are miscible with water and many organic compounds. Table 3 shows properties of the pure materials.

Table 3 Summary of Physical Properties

Property	Ethylene Glycol	Propylene Glycol
Molecular weight	62.07	76.10
Ratio of mass to water at 68/68°F	1.1155	1.0381
Density at 68°F		
lb/ft^3	69.50	64.68
lb/gal	9.29	8.65
Boiling point, °F		
at 760 mm Hg	388	369
at 50 mm Hg	253	241
at 10 mm Hg	192	185
Vapor pressure, at 68°F, mm Hg	0.05	0.07
Freezing point, °F	9.1	Sets to glass below −76°F
Viscosity, centipoise		
at 32°F	57.4	243
at 68°F	20.9	60.5
at 104°F	9.5	18.0
Refractive index, ν_D, at 68°F	1.4319	1.4329
Specific heat at 68°F, Btu/lb·°F	0.561	0.593
Heat of fusion at 9.1°F, Btu/lb	80.5	—
Heat of vaporization at 1 atm, Btu/lb	364	296
Heat of combustion at 68°F, Btu/lb	8280	10,312

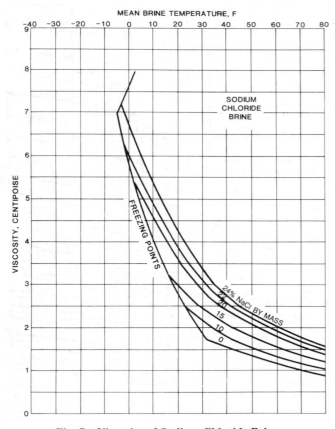

Fig. 7 Viscosity of Sodium Chloride Brine

The freezing and boiling points of aqueous solutions of ethylene glycol and propylene glycol are given in Tables 4 and 5. Note that increasing the concentration of ethylene glycol above 60% by mass causes the freezing point of the solution to increase. Propylene glycol solutions above 60% by mass do not have freezing points. Instead of freezing, propylene glycol solutions become a glass (glass being an amorphous, undercooled liquid of extremely high viscosities that has all the appearances of a solid). On the dilute side of the eutectic, ice forms on freezing; on the concentrated side, solid glycol separates from solution on freezing. Freezing velocity of such solutions is often quite slow; but, in time, they set to a hard, solid mass.

Physical properties, i.e., specific heat, specific gravity, viscosity, and thermal conductivity, for aqueous solutions of ethylene glycol can be found in Tables 6 through 9 and Figures 9 through 12; data for aqueous solutions of propylene glycol can be found in Tables 10 through 13 and Figures 13 through 16. Densities are for aqueous solutions of industrially inhibited glycols. These densities are somewhat higher than those for pure glycol and water alone. Typical corrosion inhibitor packages do not significantly affect the other physical properties. The physical properties for the two fluids are similar, with the exception of viscosity. Aqueous solutions of propylene glycol are more viscous than similar concentration solutions of ethylene glycol-water. This higher viscosity

Table 4 Freezing and Boiling Points of Aqueous Solutions of Ethylene Glycol

Percent Ethylene Glycol		Freezing Point, °F	Boiling Point, °F at 14.6 psia
By Mass	By Volume		
0.0	0.0	32.0	212
5.0	4.4	29.4	213
10.0	8.9	26.2	214
15.0	13.6	22.2	215
20.0	18.1	17.9	216
21.0	19.2	16.8	216
22.0	20.1	15.9	216
23.0	21.0	14.9	217
24.0	22.0	13.7	217
25.0	22.9	12.7	218
26.0	23.9	11.4	218
27.0	24.8	10.4	218
28.0	25.8	9.2	219
29.0	26.7	8.0	219
30.0	27.7	6.7	220
31.0	28.7	5.4	220
32.0	29.6	4.2	220
33.0	30.6	2.9	220
34.0	31.6	1.4	220
35.0	32.6	−0.2	221
36.0	33.5	−1.5	221
37.0	34.5	−3.0	221
38.0	35.5	−4.5	221
39.0	36.5	−6.4	221
40.0	37.5	−8.1	222
41.0	38.5	−9.8	222
42.0	39.5	−11.7	222
43.0	40.5	−13.5	223
44.0	41.5	−15.5	223
45.0	42.5	−17.5	224
46.0	43.5	−19.8	224
47.0	44.5	−21.6	224
48.0	45.5	−23.9	224
49.0	46.6	−26.7	224
50.0	47.6	−28.9	225
51.0	48.6	−31.2	225
52.0	49.6	−33.6	225
53.0	50.6	−36.2	226
54.0	51.6	−38.8	226
55.0	52.7	−42.0	227
56.0	53.7	−44.7	227
57.0	54.7	−47.5	228
58.0	55.7	−50.0	228
59.0	56.8	−52.7	229
60.0	57.8	−54.9	230
65.0	62.8	a	235
70.0	68.3	a	242
75.0	73.6	a	248
80.0	78.9	−52.2	255
85.0	84.3	−34.5	273
90.0	89.7	−21.6	285
95.0	95.0	−3.0	317

aFreezing points are below −60°F.

Table 5 Freezing and Boiling Points of Aqueous Solutions of Propylene Glycol

Percent Propylene Glycol		Freezing Point, °F	Boiling Point, °F at 14.6 psia
By Mass	By Volume		
0.0	0.0	32.0	212
5.0	4.8	29.1	212
10.0	9.6	26.1	212
15.0	14.5	22.9	212
20.0	19.4	19.2	213
21.0	20.4	18.3	213
22.0	21.4	17.6	213
23.0	22.4	16.6	213
24.0	23.4	15.6	213
25.0	24.4	14.7	214
26.0	25.3	13.7	214
27.0	26.4	12.6	214
28.0	27.4	11.5	215
29.0	28.4	10.4	215
30.0	29.4	9.2	216
31.0	30.4	7.9	216
32.0	31.4	6.6	216
33.0	32.4	5.3	216
34.0	33.5	3.9	216
35.0	34.4	2.4	217
36.0	35.5	0.8	217
37.0	36.5	−0.8	217
38.0	37.5	−2.4	218
39.0	38.5	−4.2	218
40.0	39.6	−6.0	219
41.0	40.6	−7.8	219
42.0	41.6	−9.8	219
43.0	42.6	−11.8	219
44.0	43.7	−13.9	219
45.0	44.7	−16.1	220
46.0	45.7	−18.3	220
47.0	46.8	−20.7	220
48.0	47.8	−23.1	221
49.0	48.9	−25.7	221
50.0	49.9	−28.3	222
51.0	50.9	−31.0	222
52.0	51.9	−33.8	222
53.0	53.0	−36.7	223
54.0	54.0	−39.7	223
55.0	55.0	−42.8	223
56.0	56.0	−46.0	223
57.0	57.0	−49.3	224
58.0	58.0	−52.7	224
59.0	59.0	−56.2	224
60.0	60.0	−59.9	225
65.0	65.0	a	227
70.0	70.0	a	230
75.0	75.0	a	237
80.0	80.0	a	245
85.0	85.0	a	257
90.0	90.0	a	270
95.0	95.0	a	310

aFreezing points are below −60°F.

Table 6   Density of Aqueous Solutions of Ethylene Glycol

Concentrations in Volume Percent Ethylene Glycol

Temperature, °F	10%	20%	30%	40%	50%	60%	70%	80%	90%
−30					68.12	69.03	69.90	70.75	
−20					68.05	68.96	69.82	70.65	71.45
−10				67.04	67.98	68.87	69.72	70.54	71.33
0				66.97	67.90	68.78	69.62	70.43	71.20
10			65.93	66.89	67.80	68.67	69.50	70.30	71.06
20		64.83	65.85	66.80	67.70	68.56	69.38	70.16	70.92
30	63.69	64.75	65.76	66.70	67.59	68.44	69.25	70.02	70.76
40	63.61	64.66	65.66	66.59	67.47	68.31	69.10	69.86	70.59
50	63.52	64.56	65.55	66.47	67.34	68.17	68.95	69.70	70.42
60	63.42	64.45	65.43	66.34	67.20	68.02	68.79	69.53	70.23
70	63.31	64.33	65.30	66.20	67.05	67.86	68.62	69.35	70.04
80	63.19	64.21	65.17	66.05	66.90	67.69	68.44	69.15	69.83
90	63.07	64.07	65.02	65.90	66.73	67.51	68.25	68.95	69.62
100	62.93	63.93	64.86	65.73	66.55	67.32	68.05	68.74	69.40
110	62.79	63.77	64.70	65.56	66.37	67.13	67.84	68.52	69.17
120	62.63	63.61	64.52	65.37	66.17	66.92	67.63	68.29	68.92
130	62.47	63.43	64.34	65.18	65.97	66.71	67.40	68.05	68.67
140	62.30	63.25	64.15	64.98	65.75	66.48	67.16	67.81	68.41
150	62.11	63.06	63.95	64.76	65.53	66.25	66.92	67.55	68.14
160	61.92	62.86	63.73	64.54	65.30	66.00	66.66	67.28	67.86
170	61.72	62.64	63.51	64.31	65.05	65.75	66.40	67.01	67.58
180	61.51	62.42	63.28	64.07	64.80	65.49	66.12	66.72	67.28
190	61.29	62.19	63.04	63.82	64.54	65.21	65.84	66.42	66.97
200	61.06	61.95	62.79	63.56	64.27	64.93	65.55	66.12	66.65
210	60.82	61.71	62.53	63.29	63.99	64.64	65.24	65.81	66.33
220	60.57	61.45	62.27	63.01	63.70	64.34	64.93	65.48	65.99
230	60.31	61.18	61.99	62.72	63.40	64.03	64.61	65.15	65.65
240	60.05	60.90	61.70	62.43	63.10	63.71	64.28	64.81	65.29
250	59.77	60.62	61.40	62.12	62.78	63.39	63.94	64.46	64.93

Note: Density in lb/ft^3.

Table 7   Specific Heat of Aqueous Solutions of Ethylene Glycol

Concentrations in Volume Percent Ethylene Glycol

Temperature, °F	10%	20%	30%	40%	50%	60%	70%	80%	90%
−30					0.734	0.680	0.625	0.567	
−20					0.739	0.686	0.631	0.574	0.515
−10				0.794	0.744	0.692	0.638	0.581	0.523
0				0.799	0.749	0.698	0.644	0.588	0.530
10			0.849	0.803	0.754	0.703	0.651	0.595	0.538
20		0.897	0.853	0.808	0.759	0.709	0.657	0.603	0.546
30	0.940	0.900	0.857	0.812	0.765	0.715	0.664	0.610	0.553
40	0.943	0.903	0.861	0.816	0.770	0.721	0.670	0.617	0.561
50	0.945	0.906	0.864	0.821	0.775	0.727	0.676	0.624	0.569
60	0.947	0.909	0.868	0.825	0.780	0.732	0.683	0.631	0.576
70	0.950	0.912	0.872	0.830	0.785	0.738	0.689	0.638	0.584
80	0.952	0.915	0.876	0.834	0.790	0.744	0.696	0.645	0.592
90	0.954	0.918	0.880	0.839	0.795	0.750	0.702	0.652	0.600
100	0.957	0.922	0.883	0.843	0.800	0.756	0.709	0.659	0.607
110	0.959	0.925	0.887	0.848	0.806	0.761	0.715	0.666	0.615
120	0.961	0.928	0.891	0.852	0.811	0.767	0.721	0.673	0.623
130	0.964	0.931	0.895	0.857	0.816	0.773	0.728	0.680	0.630
140	0.966	0.934	0.898	0.861	0.821	0.779	0.734	0.687	0.638
150	0.968	0.937	0.902	0.865	0.826	0.785	0.741	0.694	0.646
160	0.971	0.940	0.906	0.870	0.831	0.790	0.747	0.702	0.654
170	0.973	0.943	0.910	0.874	0.836	0.796	0.754	0.709	0.661
180	0.975	0.946	0.913	0.879	0.842	0.802	0.760	0.716	0.669
190	0.978	0.949	0.917	0.883	0.847	0.808	0.766	0.723	0.677
200	0.980	0.952	0.921	0.888	0.852	0.813	0.773	0.730	0.684
210	0.982	0.955	0.925	0.892	0.857	0.819	0.779	0.737	0.692
220	0.985	0.958	0.929	0.897	0.862	0.825	0.786	0.744	0.700
230	0.987	0.961	0.932	0.901	0.867	0.831	0.792	0.751	0.708
240	0.989	0.964	0.936	0.905	0.872	0.837	0.799	0.758	0.715
250	0.992	0.967	0.940	0.910	0.877	0.842	0.805	0.765	0.723

Note: Specific heat in Btu/(lb·°F).

Table 8   Thermal Conductivity of Aqueous Solutions of Ethylene Glycol

	Concentrations in Volume Percent Ethylene Glycol								
Temperature, °F	10%	20%	30%	40%	50%	60%	70%	80%	90%
−30					0.190	0.178	0.167	0.158	
−20					0.193	0.181	0.170	0.160	0.151
−10				0.212	0.197	0.184	0.172	0.161	0.152
0				0.216	0.200	0.186	0.174	0.163	0.153
10			0.238	0.220	0.204	0.189	0.176	0.164	0.154
20		0.264	0.243	0.224	0.207	0.191	0.178	0.166	0.155
30	0.294	0.269	0.247	0.227	0.210	0.194	0.180	0.167	0.156
40	0.300	0.274	0.251	0.231	0.212	0.196	0.182	0.169	0.157
50	0.305	0.279	0.255	0.234	0.215	0.198	0.183	0.170	0.158
60	0.311	0.284	0.259	0.237	0.218	0.200	0.185	0.171	0.159
70	0.316	0.288	0.263	0.240	0.220	0.202	0.186	0.172	0.160
80	0.320	0.292	0.266	0.243	0.223	0.204	0.188	0.173	0.161
90	0.325	0.296	0.269	0.246	0.225	0.206	0.189	0.174	0.161
100	0.329	0.299	0.272	0.248	0.227	0.208	0.190	0.175	0.162
110	0.333	0.302	0.275	0.251	0.229	0.209	0.192	0.176	0.163
120	0.336	0.305	0.277	0.253	0.230	0.210	0.193	0.177	0.163
130	0.339	0.308	0.280	0.255	0.232	0.212	0.194	0.178	0.164
140	0.342	0.311	0.282	0.256	0.233	0.213	0.195	0.179	0.165
150	0.345	0.313	0.284	0.258	0.235	0.214	0.196	0.180	0.165
160	0.347	0.315	0.285	0.259	0.236	0.215	0.197	0.180	0.166
170	0.349	0.316	0.287	0.261	0.237	0.216	0.197	0.181	0.166
180	0.351	0.318	0.288	0.262	0.238	0.217	0.198	0.181	0.167
190	0.352	0.319	0.289	0.263	0.239	0.218	0.199	0.182	0.167
200	0.353	0.320	0.290	0.263	0.240	0.218	0.199	0.182	0.168
210	0.354	0.321	0.291	0.264	0.240	0.219	0.200	0.183	0.168
220	0.355	0.321	0.291	0.265	0.240	0.219	0.200	0.183	0.168
230	0.355	0.322	0.291	0.265	0.241	0.219	0.200	0.183	0.169
240	0.355	0.322	0.291	0.265	0.241	0.219	0.200	0.184	0.169
250	0.354	0.321	0.291	0.265	0.241	0.220	0.201	0.184	0.169

Note: Thermal conductivity in Btu · ft/(h · ft^2 · °F).

Table 9   Viscosity of Aqueous Solutions of Ethylene Glycol

	Concentrations in Volume Percent Ethylene Glycol								
Temperature, °F	10%	20%	30%	40%	50%	60%	70%	80%	90%
−30					63.69	89.67	128.79	185.22	
−20					40.38	60.46	89.93	131.32	284.48
−10				19.58	27.27	42.05	63.50	91.88	169.83
0				13.76	19.34	30.08	45.58	65.04	107.77
10			6.83	10.13	14.26	22.06	33.31	46.89	71.87
20		3.90	5.38	7.74	10.85	16.56	24.79	34.48	49.94
30	2.16	3.14	4.33	6.09	8.48	12.68	18.77	25.84	35.91
40	1.82	2.59	3.54	4.91	6.77	9.90	14.45	19.71	26.59
50	1.56	2.18	2.95	4.04	5.50	7.85	11.31	15.29	20.18
60	1.35	1.86	2.49	3.38	4.55	6.33	8.97	12.05	15.65
70	1.18	1.61	2.13	2.87	3.81	5.17	7.22	9.62	12.37
80	1.04	1.41	1.84	2.46	3.23	4.28	5.88	7.79	9.93
90	0.93	1.24	1.60	2.13	2.76	3.58	4.85	6.38	8.10
100	0.83	1.11	1.41	1.87	2.39	3.03	4.04	5.28	6.68
110	0.75	0.99	1.25	1.64	2.08	2.58	3.40	4.41	5.58
120	0.68	0.90	1.11	1.46	1.82	2.23	2.88	3.73	4.71
130	0.62	0.81	1.00	1.30	1.61	1.93	2.47	3.17	4.01
140	0.57	0.74	0.90	1.17	1.43	1.69	2.13	2.72	3.45
150	0.53	0.68	0.82	1.05	1.28	1.49	1.86	2.35	2.98
160	0.49	0.63	0.75	0.95	1.15	1.32	1.63	2.05	2.60
170	0.46	0.58	0.68	0.87	1.04	1.18	1.43	1.80	2.28
180	0.43	0.54	0.63	0.79	0.94	1.06	1.27	1.58	2.01
190	0.40	0.50	0.58	0.73	0.85	0.95	1.14	1.40	1.79
200	0.37	0.47	0.54	0.67	0.78	0.86	1.02	1.25	1.60
210	0.35	0.43	0.50	0.61	0.71	0.78	0.92	1.12	1.43
220	0.33	0.41	0.46	0.57	0.66	0.72	0.83	1.01	1.29
230	0.32	0.38	0.43	0.53	0.60	0.66	0.76	0.91	1.16
240	0.30	0.36	0.40	0.49	0.56	0.61	0.69	0.83	1.06
250	0.29	0.34	0.38	0.45	0.52	0.56	0.63	0.75	0.96

Note: Viscosity in centipoise.

Table 10   Density of Aqueous Solutions of an Industrially Inhibited Propylene Glycol

Concentrations in Volume Percent Ethylene Glycol

Temperature, °F	10%	20%	30%	40%	50%	60%	70%	80%	90%
−30						67.05	67.47	68.38	68.25
−20					66.46	66.93	67.34	68.13	68.00
−10					66.35	66.81	67.20	67.87	67.75
0				65.71	66.23	66.68	67.05	67.62	67.49
10			65.00	65.60	66.11	66.54	66.89	67.36	67.23
20		64.23	64.90	65.48	65.97	66.38	66.72	67.10	66.97
30	63.38	64.14	64.79	65.35	65.82	66.22	66.54	66.83	66.71
40	63.30	64.03	64.67	65.21	65.67	66.05	66.35	66.57	66.44
50	63.20	63.92	64.53	65.06	65.50	65.87	66.16	66.30	66.18
60	63.10	63.79	64.39	64.90	65.33	65.68	65.95	66.04	65.91
70	62.98	63.66	64.24	64.73	65.14	65.47	65.73	65.77	65.64
80	62.86	63.52	64.08	64.55	64.95	65.26	65.51	65.49	65.37
90	62.73	63.37	63.91	64.36	64.74	65.04	65.27	65.22	65.09
100	62.59	63.20	63.73	64.16	64.53	64.81	65.03	64.95	64.82
110	62.44	63.03	63.54	63.95	64.30	64.57	64.77	64.67	64.54
120	62.28	62.85	63.33	63.74	64.06	64.32	64.51	64.39	64.26
130	62.11	62.66	63.12	63.51	63.82	64.06	64.23	64.11	63.98
140	61.93	62.46	62.90	63.27	63.57	63.79	63.95	63.83	63.70
150	61.74	62.25	62.67	63.02	63.30	63.51	63.66	63.55	63.42
160	61.54	62.03	62.43	62.76	63.03	63.22	63.35	63.26	63.13
170	61.33	61.80	62.18	62.49	62.74	62.92	63.04	62.97	62.85
180	61.11	61.56	61.92	62.22	62.45	62.61	62.72	62.68	62.56
190	60.89	61.31	61.65	61.93	62.14	62.29	62.39	62.39	62.27
200	60.65	61.05	61.37	61.63	61.83	61.97	62.05	62.10	61.97
210	60.41	60.78	61.08	61.32	61.50	61.63	61.69	61.81	61.68
220	60.15	60.50	60.78	61.00	61.17	61.28	61.33	61.51	61.38
230	59.89	60.21	60.47	60.68	60.83	60.92	60.96	61.21	61.08
240	59.61	59.91	60.15	60.34	60.47	60.55	60.58	60.91	60.78
250	59.33	59.60	59.82	59.99	60.11	60.18	60.19	60.61	60.48

Note: Density in lb/ft^3.

Table 11   Specific Heat of Aqueous Solutions of Propylene Glycol

Concentrations in Volume Percent Propylene Glycol

Temperature, °F	10%	20%	30%	40%	50%	60%	70%	80%	90%
−30						0.741	0.680	0.615	0.542
−20					0.799	0.746	0.687	0.623	0.550
−10					0.804	0.752	0.693	0.630	0.558
0				0.855	0.809	0.758	0.700	0.637	0.566
10			0.898	0.859	0.814	0.764	0.707	0.645	0.574
20		0.936	0.902	0.864	0.820	0.770	0.713	0.652	0.583
30	0.966	0.938	0.906	0.868	0.825	0.776	0.720	0.660	0.591
40	0.968	0.941	0.909	0.872	0.830	0.782	0.726	0.667	0.599
50	0.970	0.944	0.913	0.877	0.835	0.787	0.733	0.674	0.607
60	0.972	0.947	0.917	0.881	0.840	0.793	0.740	0.682	0.615
70	0.974	0.950	0.920	0.886	0.845	0.799	0.746	0.689	0.623
80	0.976	0.953	0.924	0.890	0.850	0.805	0.753	0.696	0.631
90	0.979	0.956	0.928	0.894	0.855	0.811	0.760	0.704	0.639
100	0.981	0.959	0.931	0.899	0.861	0.817	0.766	0.711	0.647
110	0.983	0.962	0.935	0.903	0.866	0.823	0.773	0.718	0.656
120	0.985	0.965	0.939	0.908	0.871	0.828	0.779	0.726	0.664
130	0.987	0.967	0.942	0.912	0.876	0.834	0.786	0.733	0.672
140	0.989	0.970	0.946	0.916	0.881	0.840	0.793	0.740	0.680
150	0.991	0.973	0.950	0.921	0.886	0.846	0.799	0.748	0.688
160	0.993	0.976	0.953	0.925	0.891	0.852	0.806	0.755	0.696
170	0.996	0.979	0.957	0.929	0.896	0.858	0.812	0.762	0.704
180	0.998	0.982	0.961	0.934	0.902	0.864	0.819	0.770	0.712
190	1.000	0.985	0.964	0.938	0.907	0.869	0.826	0.777	0.720
200	1.002	0.988	0.968	0.943	0.912	0.875	0.832	0.784	0.729
210	1.004	0.991	0.971	0.947	0.917	0.881	0.839	0.792	0.737
220	1.006	0.994	0.975	0.951	0.922	0.887	0.845	0.799	0.745
230	1.008	0.996	0.979	0.956	0.927	0.893	0.852	0.806	0.753
240	1.011	0.999	0.982	0.960	0.932	0.899	0.859	0.814	0.761
250	1.013	1.002	0.986	0.965	0.937	0.905	0.865	0.821	0.769

Note: Specified heat in Btu/(lb·°F).

Table 12 Thermal Conductivity of Aqueous Solutions of Propylene Glycol

Concentrations in Volume Percent Propylene Glycol

Temperature, °F	10%	20%	30%	40%	50%	60%	70%	80%	90%
−30						0.171	0.159	0.147	0.137
−20					0.188	0.174	0.160	0.148	0.137
−10					0.191	0.176	0.161	0.148	0.136
0				0.211	0.194	0.178	0.162	0.149	0.136
10			0.235	0.215	0.196	0.179	0.163	0.149	0.136
20		0.262	0.239	0.218	0.199	0.181	0.164	0.150	0.136
30	0.293	0.267	0.243	0.222	0.201	0.183	0.165	0.150	0.135
40	0.299	0.272	0.247	0.225	0.204	0.184	0.166	0.150	0.135
50	0.304	0.277	0.251	0.227	0.206	0.186	0.167	0.150	0.135
60	0.310	0.281	0.254	0.230	0.208	0.187	0.168	0.150	0.134
70	0.315	0.285	0.258	0.233	0.210	0.188	0.168	0.151	0.134
80	0.319	0.289	0.261	0.235	0.211	0.189	0.169	0.151	0.134
90	0.323	0.292	0.263	0.237	0.213	0.190	0.169	0.151	0.133
100	0.327	0.295	0.266	0.239	0.214	0.191	0.170	0.151	0.133
110	0.331	0.298	0.268	0.241	0.215	0.192	0.170	0.151	0.132
120	0.334	0.301	0.270	0.243	0.217	0.193	0.170	0.150	0.132
130	0.338	0.304	0.272	0.244	0.218	0.193	0.170	0.150	0.131
140	0.340	0.306	0.274	0.245	0.218	0.194	0.171	0.150	0.131
150	0.343	0.308	0.276	0.246	0.219	0.194	0.171	0.150	0.130
160	0.345	0.309	0.277	0.247	0.220	0.194	0.171	0.150	0.130
170	0.347	0.311	0.278	0.248	0.220	0.195	0.171	0.149	0.129
180	0.348	0.312	0.279	0.249	0.221	0.195	0.170	0.149	0.129
190	0.350	0.313	0.280	0.249	0.221	0.195	0.170	0.148	0.128
200	0.351	0.314	0.280	0.249	0.221	0.194	0.170	0.148	0.127
210	0.351	0.314	0.280	0.249	0.221	0.194	0.169	0.147	0.127
220	0.352	0.314	0.280	0.249	0.220	0.194	0.169	0.147	0.126
230	0.352	0.314	0.280	0.249	0.220	0.193	0.168	0.146	0.125
240	0.351	0.314	0.280	0.249	0.220	0.193	0.168	0.146	0.125
250	0.351	0.314	0.279	0.248	0.219	0.192	0.167	0.145	0.124

Note: Thermal conductivity in Btu·ft/(h·ft^2·°F).

Table 13 Viscosity of Aqueous Solutions of Propylene Glycol

Concentrations in Volume Percent Propylene Glycol

Temperature, °F	10%	20%	30%	40%	50%	60%	70%	80%	90%
−30						497.57	864.87	1363.75	3555.22
−20					156.08	298.75	493.93	820.58	1819.72
−10					95.97	182.96	291.28	495.68	983.05
0				40.99	61.32	114.90	177.73	303.94	558.32
10			13.44	27.17	40.62	74.19	112.20	190.41	332.02
20		5.36	9.91	18.64	27.83	49.29	73.22	122.30	205.91
30	2.80	4.23	7.47	13.20	19.66	33.68	49.32	80.66	132.67
40	2.28	3.41	5.75	9.63	14.28	23.65	34.22	54.64	88.51
50	1.89	2.79	4.52	7.22	10.65	17.05	24.41	37.99	60.93
60	1.60	2.32	3.61	5.55	8.13	12.59	17.86	27.10	43.16
70	1.38	1.95	2.94	4.36	6.34	9.51	13.38	19.79	31.37
80	1.20	1.66	2.43	3.50	5.04	7.34	10.25	14.79	23.35
90	1.05	1.43	2.04	2.86	4.08	5.77	8.00	11.29	17.75
100	0.93	1.25	1.73	2.37	3.35	4.62	6.37	8.79	13.76
110	0.83	1.10	1.49	2.00	2.79	3.76	5.15	6.97	10.86
120	0.75	0.97	1.30	1.71	2.36	3.11	4.23	5.62	8.71
130	0.68	0.87	1.14	1.49	2.02	2.61	3.53	4.60	7.09
140	0.62	0.78	1.01	1.30	1.75	2.22	2.98	3.82	5.85
150	0.57	0.71	0.90	1.16	1.53	1.91	2.54	3.22	4.89
160	0.52	0.64	0.82	1.03	1.35	1.66	2.19	2.75	4.13
170	0.48	0.59	0.74	0.93	1.20	1.45	1.91	2.37	3.52
180	0.44	0.54	0.68	0.85	1.08	1.29	1.69	2.07	3.04
190	0.41	0.50	0.62	0.78	0.97	1.15	1.50	1.82	2.64
200	0.38	0.46	0.58	0.72	0.88	1.04	1.34	1.61	2.31
210	0.36	0.43	0.54	0.67	0.81	0.94	1.21	1.45	2.04
220	0.34	0.40	0.50	0.62	0.74	0.86	1.10	1.31	1.82
230	0.32	0.38	0.47	0.59	0.69	0.79	1.00	1.19	1.63
240	0.30	0.36	0.45	0.55	0.64	0.73	0.92	1.09	1.47
250	0.28	0.34	0.42	0.52	0.59	0.68	0.85	1.00	1.33

Note: Viscosity in centipoises.

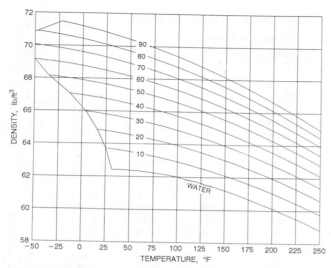

Fig. 9   Densities of Aqueous Solutions of Industrially Inhibited Ethylene Glycol

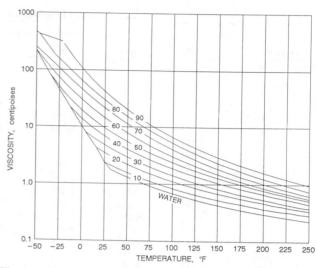

Fig. 12   Viscosities of Aqueous Solutions of Industrially Inhibited Ethylene Glycol

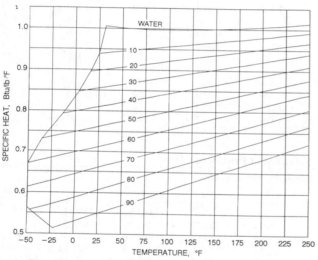

Fig. 10   Specific Heats of Aqueous Solutions of Industrially Inhibited Ethylene Glycol

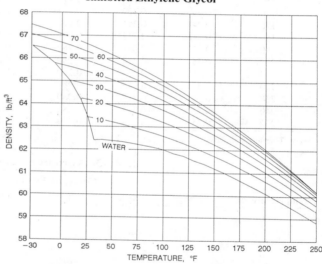

Fig. 13   Densities of Aqueous Solutions of Industrially Inhibited Propylene Glycol

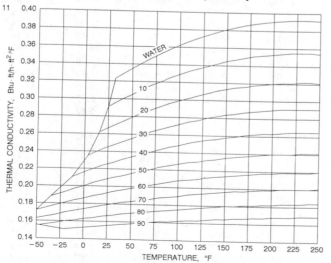

Fig. 11   Thermal Conductivities of Aqueous Solutions of Industrially Inhibited Ethylene Glycol

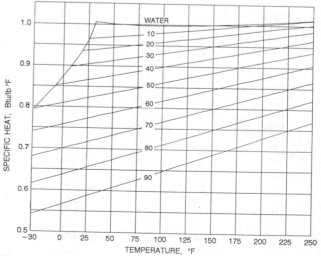

Fig. 14   Specific Heats of Aqueous Solutions of Industrially Inhibited Propylene Glycol

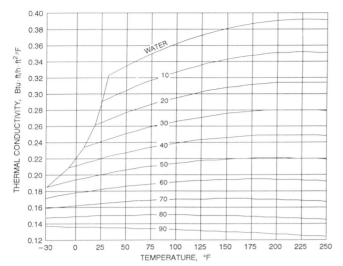

Fig. 15 Thermal Conductivities of Aqueous Solutions of Industrially Inhibited Propylene Glycol

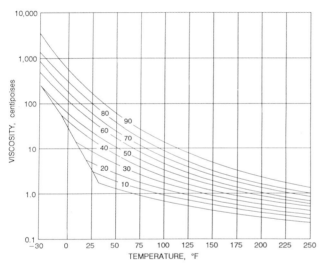

Fig. 16 Viscosities of Aqueous Solutions of Industrially Inhibited Propylene Glycol

accounts for the majority of the performance difference between the two fluids.

The choice of glycol concentration depends on the type of protection required by the application. If the fluid is being used to prevent equipment damage during idle periods in cold weather, such as winterizing coils in an HVAC system, 30% ethylene glycol or 35% propylene glycol is sufficient. These concentrations will allow the fluid to freeze. As the fluid freezes, it forms a slush that expands and flows to any available space. Therefore, expansion volume must be included with this type of protection. If the application requires that the fluid remain entirely liquid, a concentration with a freezing point 5°F below the lowest expected temperature should be chosen. Avoid excessive glycol concentration because it only increases initial cost and adversely affects the physical properties of the fluid.

Additional physical property data is available from suppliers of industrially inhibited ethylene and propylene glycol.

Corrosion Inhibition

Commercial ethylene or propylene glycol, when pure, is generally less corrosive than water to common metals used in con-

struction. However, aqueous solutions of these glycols assume the corrosivity of the water from which they are prepared and can become increasingly corrosive with use if they are not properly inhibited. Without inhibitors, glycols oxidize into acidic end products. The amount of oxidation is influenced by temperature, degree of aeration, and, to some extent, the particular combination of metal components to which the glycol solution is exposed.

Corrosion inhibition can be described by classifying additives as either (1) corrosion inhibitors, or (2) environmental stabilizers and adjusters. Corrosion inhibitors form a surface barrier that protects the metal from attack. These barriers are usually formed by adsorption of the inhibitor by the metal, by reaction of the inhibitor with the metal, or by the incipient reaction product. In most cases, metal surfaces are covered by films of their oxides that inhibitors reinforce.

Environmental stabilizers or adjusters, while not corrosion inhibitors in the strict sense, decrease corrosion by stabilizing or favorably altering the overall environment. An alkaline buffer such as borax is an example of an environmental stabilizer, since its prime purpose is to maintain an alkaline condition (pH above 7). Some chelating agents function as stabilizers by removing from the solution certain deleterious ions that accelerate the corrosion process or mechanism; however, exercise caution in their use as improper combinations of pH and concentration may lead to excessive corrosion.

Certain oxidants, such as sodium chromate, should not be used with glycol solutions, because the glycol can oxidize prematurely. Generally, combinations of the two types of additives, inhibitors, and stabilizers offer the best corrosion resistance in a given system. Commercial inhibited glycols are available from several suppliers.

Service Considerations

Design considerations. Inhibited glycols can be used at temperatures as high as 350°F. However, maximum-use temperatures vary from fluid to fluid. Therefore, the manufacturer's suggested temperature-use ranges should be followed. In systems with a high degree of aeration, the bulk fluid temperature should not exceed 180°F, but up to 350°F is permissible in a pressurized system if air intake is eliminated. Maximum film temperatures should not exceed 50°F above the bulk temperature. Nitrogen blanketing will minimize oxidation when the system operates at elevated temperatures for extended periods.

Minimum operating temperatures are typically −10°F for ethylene glycol solutions and 0°F for propylene glycol solutions. Operation below these temperatures is generally impractical, as the viscosity of the fluids builds dramatically, thus increasing pumping horsepower requirements and reducing heat transfer film coefficients.

Standard materials can be used with most inhibited glycol solutions except galvanized steel, because the galvanizing material, zinc, will react with a portion of the inhibitor package found in most formulated glycols.

Because the removal of active sludge and other contaminants is critical, install suitable filters. If inhibitors are rapidly and completely adsorbed by such contamination, the fluid is ineffective for corrosion inhibition. Consider such adsorption when selecting filters.

Storage and handling. Inhibited glycol concentrates are stable, relatively noncorrosive materials with high flash points. These fluids can be stored in mild steel, stainless steel, or aluminum vessels. However, aluminum should be used only when the fluid temperature is below 150°F. Corrosion in the vapor space of vessels may be a problem, because the fluid's inhibitor package cannot reach these surfaces to protect them. To prevent this problem, a coating may be used. Suitable coatings include novolac-based vinyl ester resins, high-bake phenolic resins, polypropylene, and polyvinylidene fluoride. To ensure the coating is suitable for a particular application and temperature, the manufacturer should be consulted. Since chemical properties of an inhibited glycol concentrate differ from those of its dilution, the effect of the concentrate on different containers should be known when selecting storage.

Choose transfer pumps only after considering temperature-viscosity data. Centrifugal pumps with electric motor drives are often used. Materials compatible with ethylene or propylene glycol should be used for pump packing material. Mechanical seals are also satisfactory. Welded mild steel transfer piping with a minimum diameter is normally used in conjunction with the piping, although flanged and gasketed joints are also satisfactory.

Preparation before application. Before an inhibited glycol is charged into a system, remove residual contaminants such as sludge, rust, brine deposits, and oil so the contained inhibitor functions properly. Avoid strong acid cleaners; if they are required, consider inhibited acids. Completely remove the cleaning agent before charging with inhibited glycol.

Use distilled, deionized, or condensate water, because water from some sources contains elements that reduce the effectiveness of the inhibited formulation. If water of this quality is unavailable, water containing less than 25 ppm chloride, less than 25 ppm sulfate, and less than 100 ppm of total hardness may be used.

Fluid maintenance. Glycol concentrations can be determined by refractive index, gas chromatography, or Karl Fischer analysis for water (assuming that the concentration of other fluid components, such as inhibitor, is known). Using specific gravity to determine glycol concentration is unsatisfactory because: (1) specific gravity measurements are temperature sensitive, (2) inhibitor concentrations can change specific gravity, (3) values for propylene glycol are close to those of water, and (4) propylene glycol values are maximum at 70 to 75% concentration.

A rigorous inhibitor monitoring and maintenance schedule is essential to maintain a glycol solution in relatively noncorrosive condition for a long period. However, a specific schedule is not always easy to establish, because inhibitor depletion rate depends on the particular conditions of use. Analysis of samples immediately after installation, after two to three months, and after six months should establish the pattern for the schedule. Visually inspecting the solution and filter residue can detect active corrosion.

Properly inhibited and maintained glycol solutions provide better corrosion protection over brine solutions in most systems. A long, though not indefinite, service life can be expected. Avoid indiscriminate mixing of inhibited formulations. Exercise caution in replacing brine systems with inhibited glycols because brine components are incompatible with glycol formulations.

HALOCARBONS

Many common refrigerants are used as secondary coolants as well as primary refrigerating media. Their favorable properties as heat transfer fluids include low freezing points, low viscosities, nonflammability, and good stability. Chapters 16 and 17 present physical and thermodynamic properties for common refrigerants. Table 14 lists four halocarbon compounds that are commonly used as secondary coolants. Table 15 gives vapor pressure, viscosity, density, specific heat, and thermal conductivity values for methylene chloride. Table 16 gives the same properties for trichlorethylene.

Volatility of R-12 and R-11 must be considered in equipment design. A pressure-tight system is essential when R-12 is used, and it is highly advisable to use R-11 in a closed system to avoid excessive material losses.

Chapter 16, Table 11, summarizes comparative safety characteristics for halocarbons. Threshold Limit Values and Biological Exposure Indices (ACGIH 1988) have more information on halocarbon toxicity.

Table 14 Freezing and Boiling Points of Halocarbon Coolants

Refrigerant	Name	Freezing Point, °F	Boiling Point, °F
12	Dichlorodifluoromethane	−252	−21.6
11	Trichloromonofluoromethane	−168	74.8
30	Methylene chloride	−142	103.6
1120	Trichloroethylene	−123	189

Table 15 Properties of Liquid Methylene Chloride

Temperature, °F	Vapor Pressure, psia	Specific Heat, Btu/lb·°F	Thermal Conductivity, Btu/h·ft·°F	Density, lb/ft³	Viscosity, Centipoise
140	25.4	0.296	0.074	78.3	0.32
122	19.9	0.293	0.076	79.4	0.34
104	14.5	0.289	0.079	80.5	0.37
86	10.2	0.286	0.081	81.6	0.40
68	6.82	0.284	0.083	82.7	0.44
50	4.39	0.282	0.085	83.8	0.48
32	2.73	0.280	0.087	84.9	0.53
14	1.64	0.278	0.089	86.0	0.59
−4	0.97	0.277	0.091	87.1	0.66
−22	0.55	0.275	0.093	88.2	0.76
−40	0.32	0.274	0.094	89.3	0.88
−58	0.18	0.273	0.096	90.4	1.05
−76	0.10	0.273	0.098	91.5	1.29
−94	0.06	0.273	0.099	92.6	1.68
−112	0.03	0.272	0.101	93.7	2.50

Table 16 Properties of Liquid Trichloroethylene

Temperature, °F	Vapor Pressure, psia	Specific Heat, Btu/lb·°F	Thermal Conductivity, Btu/h·ft·°F	Density, lb/ft³	Viscosity, Centipoise
140	5.73	0.231	0.062	86.8	0.40
122	4.21	0.228	0.063	88.0	0.44
104	2.87	0.225	0.065	89.0	0.48
86	1.86	0.223	0.066	90.1	0.52
68	1.13	0.220	0.068	91.3	0.57
50	0.667	0.218	0.069	92.4	0.63
32	0.370	0.216	0.071	93.5	0.70
14	0.199	0.213	0.073	94.6	0.78
−4	0.102	0.211	0.074	95.6	0.87
−22	0.052	0.209	0.076	96.6	0.99
−40	0.024	0.207	0.077	97.7	1.14
−58	0.011	0.206	0.079	98.7	1.33
−76	0.005	0.204	0.080	99.7	1.60
−94	0.002	0.202	0.082	100.6	1.93
−112	0.001	0.201	0.084	101.6	2.45

Table 17 Summary of Physical Properties of Polydimethylsiloxane Mixture and d-limonene

	Polydimethylsiloxane Mixture	d-limonene
Flash point, °F, closed cup	116	115
Boiling point, °F	347	310
Freeze point, °F	−168	−142
Operational temperature range, °F	−100 to 500	None published

Construction materials and stability factors in halocarbon use are discussed in Chapter 16 of this volume and Chapter 8 of the 1991 ASHRAE *Handbook—Applications*. Note particularly that methylene chloride and trichloroethylene should not be used in contact with aluminum components.

NONHALOCARBON, NONAQUEOUS FLUIDS

In addition to the aforementioned fluids, numerous other secondary refrigerants are available. These fluids have been used pri-

Table 18 Properties of a Polydimethylsiloxane Heat Transfer Fluid

Temperature, °F	Vapor Pressure, psia	Viscosity, centipoise	Density, lb/ft³	Heat Capacity, Btu/lb·°F	Thermal Conductivity, Btu/h·ft·°F
−100	.00	12.5	57.8	.337	.0651
−90	.00	10.5	57.5	.340	.0649
−80	.00	8.82	57.2	.344	.0647
−70	.00	7.50	56.9	.347	.0645
−60	.00	6.43	56.6	.350	.0643
−50	.00	5.55	56.3	.354	.0641
−40	.00	4.83	56.0	.357	.0639
−30	.00	4.22	55.7	.361	.0636
−20	.00	3.72	55.4	.364	.0634
−10	.00	3.29	55.1	.367	.0632
0	.00	2.93	54.8	.371	.0630
10	.00	2.62	54.5	.374	.0628
20	.00	2.36	54.2	.378	.0626
30	.00	2.13	53.9	.381	.0624
40	.01	1.93	53.6	.384	.0622
50	.01	1.76	53.3	.388	.0620
60	.02	1.60	53.0	.391	.0618
70	.03	1.47	52.7	.395	.0616
80	.04	1.35	52.4	.398	.0614
90	.05	1.25	52.1	.402	.0611
100	.08	1.15	51.8	.405	.0609
110	.11	1.07	51.5	.408	.0607
120	.15	.993	51.1	.412	.0605
130	.20	.926	50.8	.415	.0603
140	.27	.865	50.5	.419	.0601
150	.35	.810	50.2	.422	.0599
160	.46	.760	49.8	.425	.0597
170	.60	.715	49.5	.429	.0595
180	.76	.673	49.2	.432	.0593
190	.96	.635	48.8	.436	.0591
200	1.20	.601	48.5	.439	.0589
210	1.49	.569	48.1	.442	.0586
220	1.84	.540	47.8	.446	.0584
230	2.24	.513	47.4	.449	.0582
240	2.72	.488	47.0	.453	.0580
250	3.27	.465	46.7	.456	.0578
260	3.91	.443	46.3	.459	.0576
270	4.65	.424	45.9	.463	.0574
280	5.50	.405	45.5	.466	.0572
290	6.46	.388	45.1	.470	.0570
300	7.55	.372	44.7	.473	.0568
310	8.78	.357	44.3	.476	.0566
320	10.16	.343	43.9	.480	.0564
330	11.71	.330	43.5	.483	.0561
340	13.43	.317	43.1	.487	.0559
350	15.33	.306	42.6	.490	.0557
360	17.45	.295	42.2	.494	.0555
370	19.77	.285	41.7	.497	.0553
380	22.32	.275	41.3	.500	.0551
390	25.12	.266	40.8	.504	.0549
400	28.17	.257	40.4	.507	.0547
410	31.49	.249	39.9	.511	.0545
420	35.10	.242	39.4	.514	.0543
430	39.00	.234	38.9	.517	.0541
440	43.21	.227	38.4	.521	.0538
450	47.75	.221	37.9	.524	.0536
460	52.63	.214	37.4	.528	.0534
470	57.86	.209	36.8	.531	.0532
480	63.46	.203	36.3	.534	.0530
490	69.44	.197	35.8	.538	.0528
500	75.81	.192	35.2	.541	.0526

Table 19 Physical Properties of d-limonene

Temperature, °F	Specific Heat, Btu/lb·°F	Viscosity, Centipoise	Density, lb/ft³	Thermal Conductivity, Btu/h·ft²·°F
−100	0.3	3.8	57.1	0.0794
−50	0.34	2.8	55.8	0.0764
0	0.37	2.1	54.5	0.0734
50	0.41	1.6	53.2	0.0704
100	0.44	1.2	51.8	0.0674
150	0.48	0.9	50.4	0.0644
200	0.51	0.7	49	0.0614
250	0.54	0.6	47.6	0.0584
300	0.58	0.4	46	0.0554

Note: Properties are estimated or based on incomplete data.

Tables 17 through 19 contain physical property information on a mixture of various molecular weight dimethylsiloxane polymers (Dow Corning 1989) and d-limonene. *Note*: the information on d-limonene is limited; it is based on measurements made over small data temperature ranges or simply on standard physical property estimation techniques. The compound is an optically active terpene (molecular formula $C_{10}H_{16}$) derived as an extract from orange and lemon oils. The "d" indicates that the material is dextrorotatory, which is a physical property of the material, but does not affect the transport properties of the material significantly.

The mixture of dimethylsiloxane polymers can be used with most standard construction materials; d-limonene, however, can be quite corrosive, as it easily autooxidizes at ambient temperatures. This fact should be understood and considered before using d-limonene in a system.

REFERENCES

ACGIH. 1988. Threshold limit values and biological exposure indices for 1988-89. Published annually by the American Conference of Governmental Industrial Hygienists, Cincinnati, OH.

Carrier Air Conditioning Company. 1959. Basic data, Section 17M. Syracuse, NY.

Dow Corning USA. 1989. Syltherm® heat transfer liquids. Midland, MI: The Dow Corning Corporation.

BIBLIOGRAPHY

Born, D.W. 1989. Inhibited glycols for corrosion and freeze protection in water-based heating and cooling systems. Midland, MI.

CCI. Calcium chloride for refrigeration brine. Manual RM-1. Calcium Chloride Institute.

Dow Chemical USA. 1990. Engineering manual for DOWFROST® and DOWFROST HD heat transfer fluids. The Dow Chemical Company, Midland, MI.

Dow Chemical USA. 1990. Engineering manual for Dowtherm® SR-1 and Dowtherm 4000 heat transfer fluids. The Dow Chemical Company, Midland, MI.

Fontana, M.G. 1986. *Corrosion engineering*. McGraw-Hill Book Company, New York.

NACE. 1984. *Corrosion basics: An introduction*. National Association of Corrosion Engineers, Houston, TX.

Nathan C.C. 1973. *Corrosion inhibitors*. National Association of Corrosion Engineers, Houston, TX.

Refrigeration Engineering Application Data—Section 40. *Refrigerating Engineering* 54 (November).

Sawens, R.H. 1947. Calcium chloride and sodium chloride refrigeration brines.

Sax, N.I. and R.J. Lewis. 1987. *Hawley's condensed chemical dictionary*. Van Nostrand Reinhold Company, New York.

Union Carbide Corporation. 1971. Glycols. Tarrytown, NY.

marily by the chemical processing and pharmaceutical industries. They have been used rarely in the HVAC and allied industries due to their cost and relative novelty. Before choosing these types of fluids, consider electrical classifications, disposal, potential worker exposure, process containment, and other relevant issues.

SORBENTS AND DESICCANTS

VIRTUALLY all materials are desiccants; that is, they attract and hold water vapor. Wood, natural fibers, clays, and many synthetic materials attract and release moisture like commercial desiccants do, but they lack the holding capacity of what are known as desiccant materials. For example, woolen carpet fibers attract up to 23% of their dry weight in water vapor, and nylon can take up almost 6% of its weight in water. In contrast, a commercial desiccant takes up between 10 and 1100% of its dry weight in water vapor, depending on its type and the moisture available in the environment. Furthermore, commercial desiccants continue to attract moisture even when the surrounding air is quite dry, a characteristic that other materials do not share.

All desiccants behave in a similar way—they attract moisture until they reach equilibrium with the surrounding air. Moisture is usually removed from the desiccant by heating it to temperatures between 120 and 500 °F and exposing it to a scavenger airstream. After the desiccant dries, it must be cooled so that it can attract moisture once again. Sorption always generates sensible heat equal to the latent heat of the water vapor taken up by the desiccant, plus an additional heat of sorption that varies between 5 and 25% of the latent heat of the water vapor. This heat is transferred to the desiccant and to the surrounding air.

The process of attracting and holding moisture is described as either adsorption or absorption, depending on whether the desiccant undergoes a chemical change as it takes on moisture. *Adsorption* does not change the desiccant, except by the addition of the weight of water vapor, similar in some ways to a sponge soaking up water. *Absorption*, on the other hand, changes the desiccant. An example of this is table salt, which changes from a solid to a liquid as it absorbs moisture.

Sorption refers to the binding of one substance to another. Sorbents are materials that have an ability to attract and hold other gases or liquids. They can be used to attract gases or liquids other than water vapor, a characteristic that makes them very useful in chemical separation processes. *Desiccants* are a subset of sorbents; they have a particular affinity for water.

DESICCANT APPLICATIONS

Desiccants can dry either liquids or gases, including ambient air, and are used in many air-conditioning applications, particularly when:

- The latent load is large in comparison to the sensible load.
- The cost of energy to regenerate the desiccant is low when compared with the cost of energy to dehumidify the air by chilling it below its dew point.

The preparation of this chapter is assigned to TC 3.5, Sorption.

- The moisture control level required in the space would require chilling the air to subfreezing dew points if compression refrigeration alone were used to dehumidify the air.
- The temperature control level required by the space or process requires continuous delivery of air at subfreezing temperatures.

In any of these situations, the cost of running a vapor compression cooling system can be very high. A desiccant process may offer considerable advantages in energy, the initial cost of equipment, and maintenance.

Since desiccants are able to absorb more than simply water vapor, they can remove contaminants from airstreams to improve indoor air quality. Desiccants have been used to remove organic vapors, and in special circumstances, to control microbiological contaminants (Batelle 1971).

Desiccants are also used in drying compressed air to low dew points. In this application, moisture can be removed from the desiccant without heat. Desorption is accomplished using differences in vapor pressures compared to the total pressures of the compressed and ambient pressure airstreams.

Hines *et al.* (1991) have also confirmed the usefulness of desiccants in removing vapors that can degrade indoor air quality. Desiccant materials are capable of adsorbing hydrocarbon vapors at the same time they are collecting moisture from air. These desiccant cosorption phenomena show promise of improving indoor air quality in typical building HVAC systems.

Finally, desiccants are used to dry the refrigerant circulating in air-conditioning and refrigeration systems. This reduces corrosion in refrigerant piping and prevents valves and capillaries from becoming clogged with ice crystals. In this application, the desiccant is not regenerated; it is discarded when it has adsorbed its limit of water vapor.

This chapter discusses the water sorption characteristics of desiccant materials and explains some of the implications of those characteristics in ambient pressure air-conditioning applications. Information on other aspects of desiccants can be found in Chapters 7, 27, 32, and 34 of the 1990 ASHRAE *Handbook—Refrigeration*, and Chapters 12, 14 through 24, 28, and 40 of the 1992 ASHRAE *Handbook—Systems and Equipment*.

DESICCANT CYCLE

Practically speaking, all desiccants function by the same mechanism—transferring moisture because of a difference between the water vapor pressure at their surface and that of the surrounding air. When the vapor pressure at the desiccant surface is lower than that of the air, the desiccant attracts moisture. When the surface vapor pressure is higher than that of the surrounding air, the desiccant releases moisture.

Figure 1 shows the relationship between the moisture content of the desiccant and its surface vapor pressure. As the moisture content of the desiccant rises, so does the water vapor pressure at its surface. At some point, the vapor pressure at the desiccant is the same as that of the air—the two are in equilibrium. Then moisture cannot move in either direction until some external force changes the vapor pressure at the desiccant or in the air.

Figure 2 shows the effect of temperature on the vapor pressure at the desiccant. Both higher temperatures and increased moisture content increase the vapor pressure at the surface. When the surface vapor pressure exceeds that of the surrounding air, moisture leaves the desiccant—a process called reactivation or regeneration. After the desiccant is dried (reactivated) by the heat, its vapor pressure remains high, so that it has very little ability to absorb moisture. Cooling the desiccant reduces its surface vapor pressure so that it can absorb moisture once again. The complete cycle is illustrated in Figure 3.

The economics of desiccant operation depends on the energy cost of moving a given material through this cycle. The dehumidification of air (loading the desiccant with water vapor) generally proceeds without energy input, other than fan and pump costs.

The major portion of energy is invested in regenerating the desiccant (moving from point 2 to point 3) and cooling the desiccant (point 3 to point 1). Regeneration energy is equal to the sum of three variables:

- The heat necessary to raise the desiccant to a temperature high enough to make its surface vapor pressure higher than the surrounding air
- The heat necessary to vaporize the moisture it contains (about 1060 Btu/lb)
- The small amount of heat from desorption of the water from the desiccant

The cooling energy is proportional to the mass of the desiccant, and the difference between its temperature after regeneration and the lower temperature that allows the desiccant to remove water from the airstream once again.

The cycle is similar when desiccants are regenerated using pressure differences in a compressed air application. The desiccant is saturated in a high-pressure chamber, *i.e.*, that of the compressed air. Then valves open, isolating the compressed air from the material, and the desiccant is exposed to air at ambient pressure. The vapor pressure of the saturated desiccant is much higher than ambient air at normal pressures; thus the moisture leaves the desiccant for the surrounding air. An alternate desorption stategy uses a small portion of the dried air, returning it to the moist desiccant bed to reabsorb the moisture, then venting the air to the atmosphere at ambient pressures.

To quantify the range of vapor pressures in which the desiccant must operate in space-conditioning applications, see Table 1. It converts the relative humidity at 70 °F to dew point and, therefore, moisture content and the corresponding vapor pressure. The greater the difference between the air and desiccant surface vapor

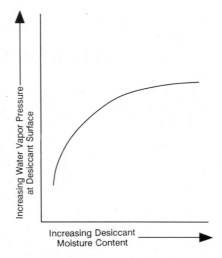

Fig. 1   Desiccant Water Vapor Pressure as Function of Moisture Content
(from *The Dehumidification Handbook* 1982)

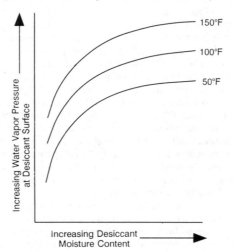

Fig. 2   Desiccant Water Vapor Pressure as Function of Desiccant Moisture Content and Temperature
(from *The Dehumidification Handbook* 1982)

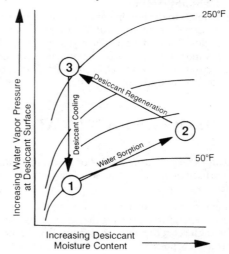

Fig. 3   Desiccant Cycle
(from *The Dehumidification Handbook* 1982)

Table 1   Vapor Pressures of Different Relative Humidities at 70 °F

Relative Humidity at 70°F, %	Dew Point, °F	Vapor Pressure, in. Hg.
10	12	0.07
20	28	0.15
30	37	0.22
40	45	0.30
50	51	0.37
60	55	0.44
70	60	0.52
80	64	0.59
90	67	0.67
100	70	0.74

pressures, the greater the ability of the material to absorb moisture from the air at that moisture content.

The ideal desiccant for a particular application depends on the range of water vapor pressures that are likely to occur in the air, the temperature level of the regeneration heat source, and the moisture sorption and desorption characteristics of the desiccant within those constraints. In commercial practice, however, most desiccants can be made to perform well in a wide variety of operating situations through careful engineering of the mechanical aspects of the dehumidification system. Some of these hardware issues are discussed in Chapter 22 of the 1992 ASHRAE *Handbook—Systems and Equipment*.

TYPES OF DESICCANTS

Desiccants can be solids or liquids and can hold moisture through adsorption or absorption as described earlier. Most absorbents are liquids, and most adsorbents are solids.

Liquid Absorbents

Liquid absorption dehumidification can best be illustrated by comparing it to the operation of an air washer. When air passes through an air washer, its dew point approaches that of the temperature of the water supplied to the machine. More humid air is dehumidified and less humid air is humidified. In a similar manner, a liquid absorption dehumidifier contacts air with a liquid desiccant solution. The liquid has a vapor pressure lower than water at the same temperature, and the air passing over the solution approaches this reduced vapor pressure; it is dehumidified.

The vapor pressure of a liquid absorbtion solution is directly proportional to its temperature and inversely proportional to its concentration. Figure 4 illustrates the effect of increasing desiccant concentration on the water vapor pressure at its surface. The figure shows the vapor pressure of various solutions of water and triethylene glycol, a common commercial desiccant. As the glycol content of the mixture increases, its vapor pressure decreases. This pressure difference allows the glycol solution to absorb moisture from the air whenever the vapor pressure of the air is greater than that of the solution.

From a slightly different perspective, the vapor pressure of a given concentration of absorbent solution approximates the vapor pressure values of a fixed relative humidity line on a psychrometric chart. Higher solution concentrations result in lower equilibrium relative humidities, allowing the absorbent to dry air to lower levels.

Figure 5 illustrates the effect of temperature on the vapor pressure of lithium chloride (LiCl), another liquid desiccant in common use. A solution that is 25% lithium chloride has a vapor pressure of 0.37 in. Hg at a temperature of 70°F. If the same 25% solution is heated to 100°F, its vapor pressure more than doubles to 0.99 in. Hg. This can be expressed another way as the 70°F, 25% solution is in equilibrium with air at a 51°F dew point. The same 25% solution at 100°F is at equilibrium with an airstream at a 79°F dew point. The warmer the desiccant, the less moisture it can attract from the air.

In standard practice, the behavior of a liquid desiccant can be controlled by adjusting its concentration, its temperature, or both. Desiccant temperature is controlled by simple heaters and coolers. Concentration is controlled by heating the desiccant to drive moisture out into a waste airstream or directly to the ambient.

Commercially available liquid desiccants have an especially high water-holding capacity. Each molecule of lithium chloride, for example, can hold two water molecules, even in the dry state. Above two water molecules per molecule of LiCl, the desiccant becomes a liquid and continues to absorb water. If the solution is in equilibrium with air at a 90% rh air moisture condition,

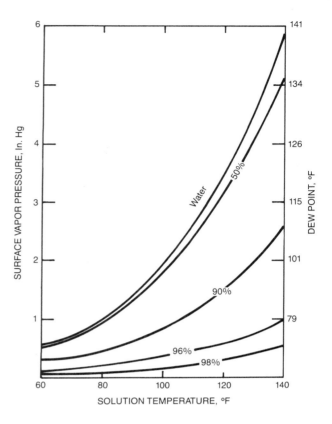

Fig. 4   Surface Vapor Pressure of Water-Triethylene Glycol Solutions
(from *A Guide to Glycols* 1981)

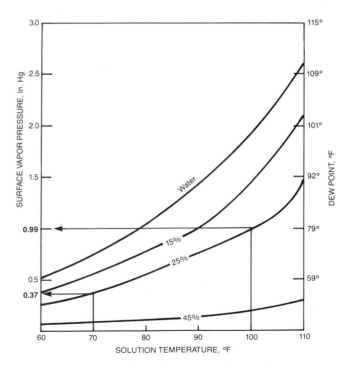

Fig. 5   Surface Vapor Pressure of Water-Lithium Chloride Solutions
(from *Lithium Chloride Technical Data* 1988)

approximately 26 water molecules are attached to each molecule of LiCl. This represents a water absorption of more than 1000% on a dry mass basis.

As a practical matter, however, the absorption process is limited by the surface area of a desiccant exposed to the air being dehumidified and the contact time allowed for the reaction. More surface area and more contact time allows the desiccant to approach its theoretical capacity. Commercial desiccant systems reflect these realities either by spraying the desiccant onto an extended surface much like in a cooling tower, or holding a solution in a rotating extended surface with a large solution capacity.

Solid Adsorbents

Adsorbents are solid materials with a tremendous internal surface area per unit of mass; a single gram can have more than 50,000 ft^2 of surface area. Structurally, adsorbents resemble a rigid sponge, and the surface of the sponge in turn resembles the ocean coastline of a fjord. This analogy indicates the scale of the different surfaces in an adsorbent. The fjords can be compared to the capillaries in the adsorbent. The spaces between the grains of sand on the fjord beaches can be compared to the spaces between the individual molecules of the adsorbent, all of which have the capacity to hold water molecules. The bulk of the adsorbed water is contained by condensation into the capillaries, and the majority of the surface area that attracts individual water molecules is in the crystalline structure of the material itself.

Adsorbents attract moisture because of the electrical field at the desiccant surface. The field is not uniform in either force or charge, so it attracts water molecules that have a net opposite charge from specific sites on the desiccant surface. When the complete surface is covered, the adsorbent can hold still more moisture, as vapor condenses into the first water layer and fills the capillaries throughout the material.

As with liquid absorbents, the ability of an adsorbent to attract moisture depends on how much water is on its surface compared to how much water is in the air. That difference is reflected in the vapor pressure at the surface and in the air.

The capacity of solid adsorbents is generally less than the capacity of liquids. For example, a typical molecular sieve adsorbent will hold 17% of its dry mass in water when the air is at 70°F and 20% rh. In contrast, lithium chloride can hold 130% of its mass at the same temperature and relative humidity. But solid adsorbents have several other favorable characteristics.

For example, molecular sieves continue to adsorb moisture even when they are quite hot, allowing dehumidification of very warm airstreams. Also, several solid adsorbents can be manufactured to precise tolerances, with pore diameters that can be closely controlled. This means they can be tailored to adsorb a molecule of a specific diameter. Water, for example, has an effective molecular diameter of 3.2 nm. A molecular sieve adsorbent with an average pore diameter of 4.0 nm adsorbs water but has almost no capacity for larger molecules, such as organic solvents. This selective adsorption characteristic is useful in many applications. For example, several desiccants with different pore sizes can be combined in series to remove first water and then other specific contaminants from an airstream.

The adsorption behavior of solid adsorbents depends on (1) their total surface area, (2) the total volume of their capillaries, and (3) the range of their capillary diameters. A large surface area gives the adsorbent a larger capacity at low relative humidities. Large capillaries provide a high capacity for condensed water, which gives the adsorbent a higher capacity at high relative humidities. A narrow range of capillary diameters makes an adsorbent more selective in the vapor molecules it can hold; thus, some will fit and others will be too large to pass through the passages in the material.

In designing a desiccant, some trade-offs are necessary. For example, materials with large capillaries necessarily have a smaller surface area per unit of volume than those with smaller capillaries. As a result, adsorbents are sometimes combined to provide a high adsorption capacity across a wide range of operating conditions. Figure 6 illustrates this point. These are all silica gel adsorbents prepared for use in laboratory research. Each has a different internal structure, but since they are all silicas, they have similar surface adsorption characteristics. Gel 1 has large capillaries, making their total volume large, but the total surface area is small. It has a large adsorption capacity at high relative humidities, but adsorbs a small amount at low relative humidities.

In contrast, Gel 8 has a capillary volume one-seventh the size of Gel 1, but a total surface area almost twice as large. This gives it a higher capacity at low relative humidities, but a lower capacity to hold the condensed moisture that occurs at high relative humidities.

The table in Figure 6 illustrates the wide range of performance characteristics possible within a single class of adsorbent. The table shows three noncommercial silica gels. Silica gels and most other adsorbents can be manufactured to provide optimum performance in a specific application, balancing capacity against strength, mass, and other favorable characteristics (MVB Series Engineering Data 1986).

General classes of solid adsorbents include:

- Silica gels
- Zeolites
- Synthetic zeolites (molecular sieves)

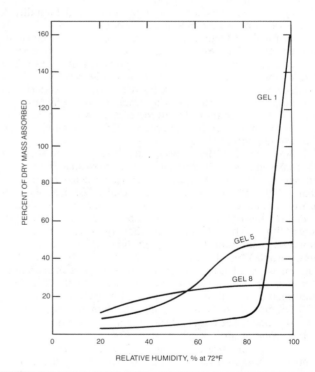

Reference Series Number	Total Surface Area, m^3/g	Average Capillary Diameter, nm	Total Volume of Capillaries, mm^3/g
1	315	21	1700
5	575	3.8	490
8	540	2.2	250

Fig. 6 Adsorption and Structural Characteristics of Some Experimental Silica Gels
(from Oscic *et al.* 1982)

• Activated aluminas
• Carbons
• Synthetic polymers

Silica gels are solid structures formed by condensing soluble silicates from solutions of water or other solvents. They have the advantages of a relatively low cost and relative simplicity of structural customizing. They are available as large spherical beads about 3/16 in. in diameter or as small as grains of a fine powder.

Zeolites are aluminosilicate minerals. They occur in nature and are mined rather than synthesized. Zeolites have a very open crystalline lattice which allows molecules like water vapor to be held inside the crystal itself like an object in a cage. Particular atoms of an aluminosilicate determine the size of the openings between the "bars of the cage," which in turn governs the maximum size of the molecule that can be adsorbed into the structure.

Synthetic zeolites, also called *molecular sieves*, are crystalline aluminosilicates manufactured in a thermal process. Controlling the temperature of the process and composition of the ingredient materials allows close control of the structure and surface characteristics of the adsorbent. At a somewhat higher cost, this provides a much more uniform product than naturally occurring zeolites.

Activated aluminas are oxides and hydrides of aluminum that are also manufactured in thermal processes. Their structural characteristics can be controlled by the gases and the temperature and duration of the thermal process.

Carbons are most frequently used for adsorption of gases other than water vapor, because they often have a greater affinity for the nonpolar molecules typical of organic solvents. Like other adsorbents, carbons have a large internal surface and especially large capillaries. This capillary volume gives them a high capacity to adsorb water vapor at relative humidities from 45 to 100%.

Solid polymers have potential for use as desiccants as well. Long molecules, like those found in polystyrenesulfonic acid sodium salt (PSSASS), are twisted together like the strands of string. The many sodium ions in the long PSSASS molecules each have the potential to bind several water molecules, and the spaces between the packed strings can also contain condensed water, giving the polymer a capacity exceeding that of many other solid adsorbents.

DESICCANT ISOTHERMS

Figure 7 shows a rough comparison of the sorption characteristics of different desiccants. Large variations from these isotherms occur because manufacturers use different methods to optimize the materials for different applications. The suitability of a given desiccant to a particular application is generally governed by the engineering of the mechanical system that presents the material to the airstreams as much as by the characteristics of the material itself.

Brunauer (1945) and Collier (1986, 1988) give details of desiccant equipment design and information about desiccant isotherm characteristics. Brunauer considers five basic isotherm shape types. Each isotherm shape is determined by the dominant sorption mechanisms of the desiccant, giving rise to its specific capacity characteristics at different vapor pressures. Isotherm shape can be important in designing the optimum desiccant for applications where a narrow range of operating conditions can be expected. Collier illustrates how an optimum isotherm shape can be used to ensure a maximum coefficient of performance in one particular air-conditioning desiccant application.

DESICCANT LIFE

The useful life of desiccant materials depends largely on the quantity and type of contamination in the airstreams they dry. In

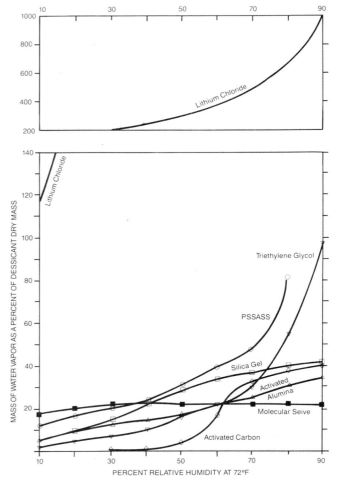

The sources for isotherms presented in the figure include :
PSSASS: Solar Energy Research Insitute Report No. PR-255-3308
Lithium chloride: Cargocaire Engineering Division of Munters USA and Kathabar Systems Division of Sommerset Technologies
Triethylene glycol: Dow Chemical Corporation
Silica gel: Davison Chemical Division of W.R. Grace Co. Inc.
Activated carbon: Calgon Corporation
Activated alumina: LaRoche Chemical Company
Molecular sieve: Davison Chemical Division of W.R. Grace Co.

Fig. 7   Sorption Isotherms of Various Desiccants at 72 °F

commercial equipment, desiccants last between 10,000 and 100,000 h and longer before they need replacement.

Normally, two mechanisms cause the loss of desiccant capacity—change in desiccant sorption characteristics through reactions with contaminants, and loss of effective surface area through clogging or hydrothermal degradation. Liquid absorbents are more susceptible to chemical reaction with airstream contaminants other than water vapor than are solid adsorbents. For example, certain sulfur compounds can react with lithium chloride to form lithium sulfate, which is not a desiccant. If the concentration of such compounds in the airstream were below 10 ppm and the desiccant were in use 24 h a day, the capacity reduction would be approximately 10 to 20% after three years of operation. If the concentration were 30 ppm, the capacity change would occur after one year.

Adsorbents tend to be less reactive chemically and more sensitive to clogging, a function of the type and quantity of particulate material in the airstream. In some situations, certain adsorbents are sensitive to hydrothermal stress, resulting from the thermal expansion and contraction of the desiccant material on rapid changes in desiccant moisture content. For example, silica gel that must move between an airstream above 95% rh at low temperatures and a reactivating airstream at high temperatures six times per hour, 24 h a day can partly fracture; this may mean a 10% reduction in capacity over the course of a year.

In air-conditioning applications, desiccant equipment is designed to minimize the need for desiccant replacement in much the same way that vapor compression cooling systems are designed to avoid the need for compressor replacement. Unlike filters, desiccants are seldom intended to be frequently replaced during normal service in an air-drying application.

COSORPTION OF WATER VAPOR AND INDOOR CONTAMINANTS

Hines *et al.* (1991) have confirmed that many desiccant materials can collect common indoor pollutants at the same time they collect water vapor from ambient air. This characteristic promises to become useful in future air-conditioning systems where the quality of indoor air is especially important.

The behavior of different desiccant and different vapor mixtures is complex but, in general, pollutant absorption reactions can be classified into five categories:

- Humidity-neutral sorption
- Humidity-reduced sorption
- Humidity-enhanced sorption
- Humidity-pollutant displacement
- Desiccant-catalyzed pollutant conversion

An example of humidity-reduced sorption is illustrated by the behavior of water vapor and chloroform on activated carbon. Sorption is humidity-neutral until relative humidity exceeds 45%, when the uptake of chloroform is reduced. The adsorbed water blocks sites that would otherwise attract and hold chloroform. In contrast, water and carbonyl chloride mixtures on activated carbon demonstrate humidity-enhanced sorption. Here, sorption of the pollutant increases at high relative humidities. Hines *et al.* (1991) attribute this phenonmenon to the high water solubility of carbonyl chloride.

REFERENCES

Batelle. 1971. Project No. N-0914-5200-1971. Report 65711-1974. Batelle Memorial Institute, Columbus, OH.

Brunauer, S. 1945. *The adsorption of gases and vapors*, Vol. I. Princeton University Press, Princeton, NJ. This information is quoted and expanded in *The physical chemistry of surfaces*, by Arthur W. Adamson. John Wiley & Sons, New York, 1982.

Buffalo Testing Laboratory. 1974. Report No. 65711-1974.

Collier, R.K. *Advanced desiccant materials assessment*. Research Report 5084-243-1089. Phase I-1986, Phase II-1988. Gas Research Institute, Chicago.

Guide to glycols. 1981. Dow Chemical Corporation, Organic Chemicals Division, Midland, MI.

Harriman, L.G., III. 1990. *The dehumidification handbook*, 2nd ed. Munters Cargocaire, Amesbury, MA.

Hines, A.J., T.K. Ghosh, S.K. Loyalka, and R.C. Warder, Jr. 1991. *Investigation of co-sorption of gases and vapors as a means to enhance indoor air quality*. ASHRAE Research Project 475 RP and Gas Research Institute Project GRI-90/0194. Gas Research Institute, Chicago.

Lithium chloride technical data, Bulletin 151. 1988. Foote Mineral Corporation, Exton, PA.

MVB Series Engineering Data. 1986. Bry-Air Inc., Sunbury, OH.

Oscic, J. and I.L. Cooper. 1982. *Adsorption*. John Wiley & Sons, New York.

SUNY Buffalo School of Medicine. Effects of glycol solution on microsociological growth. Niagrara Blower Report No. 03188.

BIBLIOGRAPHY

Adamson, A.W. 1982. *The physical chemistry of surfaces*. John Wiley & Sons, New York.

Czanderna, A.W. 1988. Polymers as advanced materials for desiccant applications. Research Report NREL/PR-255-3308. National Renewable Energy Laboratory, Golden, CO.

Falcone, J.S., Jr., ed. 1982. Soluble silicates. Symposium Series 194. American Chemical Society, Washington, D.C.

Gas conditioning factbook. 1962. Dow Chemical Company, Midland, MI.

Iler, R.K. *The chemistry of silica*. 1979. John Wiley & Sons, New York.

King, C.J. 1980. *Separation processes*, 2nd ed. McGraw-Hill Book Company, New York.

McCabe, W.L. and J.C. Smith. 1978. *Unit operations of chemical engineering*. McGraw-Hill Book Company, New York.

Perry, R.H. and C.H. Chilton. 1973. *Chemical engineers' handbook*, 5th ed. McGraw-Hill Book Company, New York.

Ruthven, D.M. 1984. *Principles of adsorption and adsorption processes*. John Wiley & Sons, New York.

Strauss, W. 1975. *Industrial gas cleaning*. Pergamon Press, New York.

Valenzuela, D. and A. Myers. 1989. *Adsorption equilibrium data handbook*. Simon and Schuster/Prentice-Hall, Englewood Cliffs, NJ.

CHAPTER 20

THERMAL INSULATION AND VAPOR RETARDERS—FUNDAMENTALS

PROPER design of space heating, air conditioning, refrigeration systems, and other industrial applications requires knowledge of thermal insulations and thermal behavior of building structures. This chapter deals with heat and moisture transfer definitions, fundamentals and properties of thermal insulation materials, heat flow calculations, economic thickness of insulation, and the fundamentals of moisture as it relates to building components and systems.

DEFINITIONS AND SYMBOLS

The following heat and moisture transfer definitions and symbols cover common building industry terms. The units used in the definitions are in I-P units: British thermal units (Btu), hour (h), temperature ($°F$), absolute temperature (R), temperature difference ($°F$), feet (ft), inches (in.), pound (lb), grains (gr), and inches of mercury (in. Hg).

Thermal transmission, heat transfer, or **rate of heat flow**. The flow of heat energy induced by a temperature difference. Heat may be transferred by conduction, mass transfer, convection, and radiation. These can occur separately or in combinations, depending on specific circumstances.

Thermal conductivity, k. The time rate of heat flow through a unit area of homogeneous material in a direction perpendicular to isothermal planes induced by a unit temperature gradient. (ASTM *Standard* C168 defines homogeneity.) Units are $Btu \cdot in/h \cdot ft^2 \cdot °F$ or $Btu/h \cdot ft \cdot °F$ (if thickness is in ft). Thermal conductivity must be evaluated for a specific mean temperature, because in most materials it varies with temperature.

For many thermal insulation materials, heat flows by a combination of modes and may depend on orientation, direction, or both. The measured property of such materials may be called *effective* or *apparent* thermal conductivity. The specific test conditions, *i.e.*, sample thickness, orientation, environment, environmental pressure, surface temperature, mean temperature, and temperature difference, should be reported with the values. With thermal conductivity, the symbol k_{app} is used to note the lack of pure conduction or to state that all values reported are apparent.

Thermal resistivity, R_u. Reciprocal of thermal conductivity. Units are $°F \cdot ft^2 \cdot h/(Btu \cdot in)$.

Thermal conductance, C-factor, C. Time rate of heat flow through a unit area of a body induced by a unit temperature difference between the body surfaces, $Btu/h \cdot ft^2 \cdot °F$.

When the two defined surfaces of mass-type thermal insulation have unequal areas, as in the case of radial heat flow through a curved block or through a pipe covering (see Chapter 3, Table 2) or through materials of nonuniform thickness, an appropriate mean area and mean thickness must be given. Heat flow formulas involving materials that are not uniform slabs must contain *shape factors* to account for the area variation involved.

When heat flow is by conduction alone, thermal conductance of a material may be obtained by dividing the thermal conductivity of the material by its thickness. When several modes of heat transfer are involved, the *apparent* or *effective* thermal conductance may be obtained by dividing the apparent thermal conductivity by the thickness.

Where air circulates within or passes through insulation, as it may with low-density fibrous materials, the effective thermal conductance is affected.

Thermal conductances and thermal resistances of the more common building materials and industrial insulations are tabulated in Chapter 22.

Heat transfer film coefficient (surface coefficient of heat transfer, surface film conductance), h or f. Heat transferred between a surface and a fluid in unit time through unit area induced by unit temperature difference between the surface and the fluid in contact with it, $Btu/h \cdot ft^2 \cdot °F$.

Surface film resistance. The reciprocal of the heat transfer film coefficient, $°F \cdot ft^2 \cdot h/Btu$.

Note 1: The surrounding space must be air or other fluids for radiation or convection or both to take place. If space is evacuated, the only heat flow is by radiation.

Note 2: Subscripts i and o often denote inside and outside surface resistances and conductances, respectively.

Thermal resistance, R. Under steady conditions, the mean temperature difference between two defined surfaces of material or construction that induces unit heat flow through a unit area, $°F \cdot ft^2 \cdot h/Btu$.

Thermal transmittance, U-factor, U. The time rate of heat flow per unit area under steady conditions from the fluid on the warm side of a barrier to the fluid on the cold side, per unit temperature difference between the two fluids. It is evaluated by

The preparation of this chapter is assigned to TC 4.4, Thermal Insulation and Moisture Retarders.

first evaluating the R-value and then computing its reciprocal, U, in Btu/h·ft²·°F.

Note 1: The U-factor is sometimes called the overall coefficient of heat transfer.

Note 2: In building practice, the heat transfer fluid is air. The temperature of the fluid is obtained by averaging its temperature over a finite region of the fluid near the surface involved.

Thermal emittance, ϵ. The ratio of the radiant flux emitted by a body to that which would be emitted by a blackbody at the same temperature and under the same conditions.

Note: Effective emittance E is the combined effect of emittances from the boundary surfaces of an air space, where the boundaries are parallel and of a dimension much larger than the distance between them. Chapter 22 lists values of E for various air spaces.

Surface reflectance, ρ. The fraction of the radiant flux falling on a surface that is reflected by it.

Water vapor permeability, μ. The property of a substance that permits passage of water vapor. When permeability varies with psychrometric conditions, the spot or specific permeability defines the property at a specific condition in gr/h·ft²·(in. Hg/in.), where in. Hg is vapor pressure difference.

Water vapor permeance, M. The rate of water vapor transmission per unit area of a body between two specified parallel surfaces, to the vapor pressure difference between the two surfaces, gr/h·ft²·in. Hg.

The perm is a widely used measure of water vapor transmission rate in gr/h·ft²·in. Hg. Originally, 1 perm was proposed as the maximum allowable permeance for a vapor retarding component to ensure a low probability for moisture problems in building construction. The perm was intended to apply to houses in climates that do not exceed 5000° (heating) degree days (Fahrenheit) of cold weather. Thus the word "perm", as originally defined, has a very specific and limited meaning.

Water vapor resistance, RV. The reciprocal of permeance—signifies a resistance to moisture flow (in. Hg·ft²·h/gr). A *rep* is a unit resistance to water vapor flow, i.e., rep = 1/perm.

THERMAL INSULATION

Thermal insulations are materials or combinations of materials that, when properly applied, retard the flow of heat energy by conductive, convective, and/or radiative transfer modes. These materials can be fibrous, particulate, film or sheet, block or monolithic, open or closed cell, or composites of these materials that can be chemically or mechanically bound or supported.

By retarding heat flow, thermal insulations can serve one or more of the following thermal functions:

1. Conserve energy by reducing heat loss or gain of piping, ducts, vessels, equipment, and structures.
2. Control surface temperatures of equipment and structures for personnel protection and comfort.
3. Help control the temperature of a chemical process, a piece of equipment, or a structure.
4. Prevent vapor condensation on surfaces with a temperature below the dew point of the surrounding atmosphere.
5. Reduce temperature fluctuations within an enclosure when heating or cooling is not needed or available.
6. Reduce temperature variations within a conditioned space for increased personal comfort.
7. Provide fire protection.

Thermal insulation can serve additional functions, although such secondary functions should be consistent with its capabilities and primary purpose. Under certain conditions, insulations may:

1. Add structural strength to a wall, ceiling, or floor section.
2. Provide support for a surface finish.
3. Impede water vapor transmission and air infiltration.
4. Prevent or reduce damage to equipment and structures from exposure to fire and freezing conditions.
5. Reduce noise and vibration.

Thermal insulation is used to control heat flow at all temperatures, the limiting value being its survival temperature.

BASIC MATERIALS

Thermal insulations normally consist of the following basic materials and composites:

- Inorganic, fibrous, or cellular materials such as glass, rock, or slag wool; calcium silicate, bonded perlite, vermiculite, ceramic products; and asbestos. Asbestos has been shown to be a carcinogen. Extreme caution must be used if it is encountered.
- Organic fibrous materials such as cotton, animal hair, wood, pulp, cane, or synthetic fibers, and organic cellular materials such as cork, foamed rubber, polystyrene, polyurethane, and other polymers.
- Metallic or metallized organic reflective membranes. These surfaces must face air, gas-filled, or evacuated spaces to be effective.

PHYSICAL STRUCTURE AND FORM

Mass-type insulation can be cellular, granular, or fibrous solid material to retard heat flow. Reflective insulation consists of smooth-surfaced sheets of metal foil or foil-surfaced material separated by air spaces.

The physical forms of industrial and building insulations include the following:

Loose-fill insulation consists of fibers, powders, granules, or nodules that are usually poured or blown into walls or other spaces.

Insulating cement is a loose material mixed with water or a suitable binder to obtain plasticity and adhesion. It is troweled or blown wet on a surface and dried in place. Both loose-fill and insulating cement are suited for covering irregular spaces.

Flexible and semirigid insulations consist of organic and inorganic materials with and without binders and with varying degrees of compressibility and flexibility. These insulations are generally available as blanket, batt, or felt insulation, and in either sheets or rolls. Coverings and facings may be fastened to one or both sides and serve as reinforcing; air or vapor retarders or both; reflective surfaces; or surface finishes. These coverings include combinations of laminated foil, glass, cloth or plastics and paper, wire mesh, or metal lath. Although standard sizes are generally used, thickness and shape of insulation can be any dimension handled conveniently.

Rigid materials are available in rectangular blocks, boards, or sheets, which are preformed during manufacture to standard lengths, widths, and thicknesses. Insulation for pipes and curved surfaces is supplied in sections or segments, with radii of curvature available to suit all standard sizes of pipe and tubing; it is also supplied in greater diameters.

Reflective materials are available in sheets and rolls of single layer or multilayer construction and in preformed shapes with integral air spaces.

Formed-in-place insulations are available as liquid components or expandable pellets that can be poured, frothed, or sprayed in place to form rigid or semirigid foam insulation. Fibrous materials mixed with liquid binders can also be sprayed in place, and in some products, the binder is also a foam.

Accessory materials for thermal insulation include mechanical and adhesive fasteners, exterior and interior finishes, vapor and air retarder coatings, jackets and weather coatings, sealants, lagging adhesives, membranes, and flashing compounds. ASTM *Standard* C168 defines terms relating to thermal insulating materials.

PROPERTIES

Thermal insulation selection may involve secondary criteria in addition to the primary property of low thermal conductivity. Characteristics such as resiliency or rigidity, acoustical energy absorption, water vapor permeability, airflow resistance, fire hazard and fire resistance, ease of application, applied cost, health and safety aspects, or other parameters may influence the choice among materials that have almost equal thermal performance values.

Thermal

Thermal resistance is a measure of the effectiveness of thermal insulation's ability to retard heat flow. Material with a high thermal resistance (low thermal conductance) is an effective insulator.

Heat transmission in most thermal insulations is accomplished by a combination of gas and solid conduction, radiation, and convection. Heat transfer through materials or systems is controlled by factors such as length of heat flow paths, temperature, temperature difference characteristics of the system, and environmental conditions.

Although heat transmission characteristics are usually determined by measuring thermal conductivity, this property does not strictly apply to thermal insulation. A particular sample of a material has a unique value of thermal conductivity for a particular set of conditions. This value may not be representative of the material at other conditions and should be called *apparent* thermal conductivity. For details, refer to ASTM *Standards* C168, C177, C236, C335, C518, C691, and C976.

Reflective insulations impede radiant heat transfer because the surfaces have high reflectance and low emittance values. (Chapter 22, Table 1, gives typical design values.) To be effective, the reflective face of both single and multiple layer reflective insulations must face an air or evacuated space. Multiple layers of reflective materials and smooth and parallel sealed air spaces increase overall thermal resistance. Air exchange and movement must be inhibited.

Mass-type insulation can be combined with reflective surfaces and air spaces to obtain a higher thermal resistance. However, each design must be evaluated since maximum thermal performance of these systems depends on such factors as condition of the insulation, shape and form of the construction, the means to avoid air leakage and movement, and the condition of the installed reflective surfaces and their aging characteristics.

Design values of effective or apparent thermal conductivity, thermal conductance, and thermal resistance for the most common insulations are listed in Table 4, Chapter 22. These values have been selected as typical and useful for engineering calculations. The manufacturer or test results of the insulation under appropriate conditions can give values for a particular insulation.

Other thermal properties that can be important are specific heat, heat capacity, thermal diffusivity, the coefficient of thermal expansion, and maximum temperature limit. *Heat capacity* is the product of specific heat and mass. *Thermal diffusivity* becomes important for applications where temperature varies with time, since the rate of temperature change within an insulation is proportional to its thermal diffusivity for a given thickness. Chapter 3 covers symbols, definitions, and methods of calculation in steady-state heat transfer.

Mechanical

Some insulations have sufficient structural strength for load bearing. They are used occasionally to support load-bearing floors, form self-supporting partitions, or stiffen structural panels. For such applications, one or more of the following properties of an insulation may be important: strength in compression, tension, shear, impact, flexure, and resistance to vibration. These temperature-dependent mechanical properties vary with basic composition, density, cell size, fiber diameter and orientation, type and amount of binder (if any), and temperature and environmental conditioning.

Health and Safety

Most thermal insulations have good resistance to fire, vermin, rot, objectionable odors, and vapors; some are a potential risk to health and safety. These risks can result from direct exposure to these materials and accessories while they are being stored or transported, during or after installation, or as a result of intervening or indirect actions or events, such as aging, fire, or physical disturbance. The potential health and safety effects of thermal insulation can be considered in two categories: (1) those related to storage, handling, and installation operations and (2) those that occur after installation (such as aging). Potential hazards during manufacture are not discussed. Correct handling, installation, and precautionary measures can reduce or eliminate these problems.

Potential health effects range from temporary irritation to serious changes in body functions. The principal concerns are with insulation containing asbestos. Questions have also been raised about man-made fibers. To date, research is inconclusive as to their potential hazard; however, they can be very annoying in installation, and the use of proper safeguards is desirable. Potential traumatic injury can occur during or from direct contact with materials that are sharp, rough, have protrusions or abrasive surfaces, permit overheating, or transmit electrical energy.

Combustion of insulation materials and accessories may release heat, hazardous gases, fibers, and particulates. Manufacturers' recommendations and applicable government codes and standards (ASTM *Standard* C930) have more details.

Acoustics

Some thermal insulations are used as acoustical control materials, whether or not thermal performance is a design requirement. Acoustical efficiency depends on the physical structure of the material. Materials with open, porous surfaces have sound absorption capability. Those of high density and resilient characteristics can act as vibration insulators; either alone or in combination with other materials, some are an effective barrier to sound transmission. Insulations for sound conditioning include flexible and semirigid, formed-in-place fibrous materials, and rigid fibrous insulation.

Sound absorption insulations are normally installed on interior surfaces or used as interior surfacing materials. Rigid insulations are fabricated into tile or blocks, edge-treated to facilitate mechanical or adhesive application, and prefinished during manufacture. Some insulation units have a natural porous surface, others include mechanical perforations to facilitate the entry of sound waves, and still others use a diaphragm or decorative film surfacing attached only to the edges of the units, which allows the sound waves to reach the fibrous backing by diaphragm action.

Flexible, semirigid, and formed-in-place fibrous materials used for sound absorption are available in a variety of thicknesses and densities that determine their sound absorption characteristics. When density is increased by reducing the thickness of the material, sound absorption is generally reduced; however, as thickness increases, the influence of density is minimal.

Thermal insulations improve sound transmission loss when installed in *discontinuous construction*. A wall of staggered stud construction that uses resilient clips or channels on one side of the stud or resilient insulation boards of special manufacture to prevent acoustical coupling mechanically between the surfaces reduces sound transmission. A sound absorption thermal insulation blanket in a wall cavity reduces sound transmission, depending on the type of construction.

In floor construction, resilient channels or separate floor and ceiling joists form a discontinuous construction; thermal insulation placed within this construction further reduces sound transmission. Sound-deadening boards underlying finish flooring absorb impact sounds and improve the airtightness of the construction, thus reducing airborne sound transmission.

Thermal insulation boards can be placed under mechanical equipment to isolate vibration. The imposed loading and natural resonant frequency of materials are critical for proper design. Since material must deflect properly under load to provide isolation, the system should be neither overloaded nor underloaded.

For further information on sound and vibration control, refer to Chapter 42 of the 1991 ASHRAE *Handbook—HVAC Applications*.

Other Properties

Other properties of insulating materials that can be important, depending on the application, include density, resilience, resistance to settling, permanence, reuse or salvage value, ease of handling, dimensional uniformity and stability, resistance to chemical action and chemical change, resistance to moisture penetration, ease in fabrication, applying of finishing, and sizes and thicknesses obtainable.

HEAT FLOW

FACTORS AFFECTING THERMAL PERFORMANCE

A wide variety of physical, environmental, application, and, in some cases, aging factors affect the thermal performance of insulations.

Thermal conductivity k is a property of a homogeneous material. Building materials, such as lumber, brick, and stone, are usually considered homogeneous. Most thermal insulation, except some reflective types, is porous and consists of combinations of solid matter with small voids.

For most insulating materials, conduction is not the only mode of heat transfer. Consequently, the term *apparent thermal conductivity* describes the heat flow properties of most materials. Some materials with low thermal conductivities are almost purely conductive (silica opacified aerogel, corkboard, etc.).

The apparent thermal conductivity of insulation varies with form and physical structure, environment, and application conditions. Form and physical structure vary with the basic material and manufacturing process. Typical variations include density, cell size, diameter and arrangement of fibers or particles, degree and extent of bonding materials, transparency to thermal radiation, and the type and pressure of gas within the insulation.

Environment and application conditions include mean temperature, temperature gradient, moisture content, air infiltration, orientation, and direction of heat flow. Thermal performance values for insulation materials and systems are usually obtained by standard methods listed in Volume 04.06 of the *Annual Book of ASTM Standards*. The methods apply mainly to laboratory measurements on dried or conditioned samples at specific mean temperatures and temperature gradient conditions. Although the fundamental heat transmission characteristics of a material or

system can be determined accurately, actual performance in a structure may vary from that indicated in the laboratory due to application variations. Field measurement techniques continue to be developed. The design of the envelope, its construction, and the materials used may all affect the test procedure. These factors are detailed in ASTM STP 544-74, STP 660-79, STP 718-80, ASTM STP 789-83, ASTM STP 879-86, STP 885-85, and STP 922-88.

The effective thermal conductivity of some thermal insulation materials varies with density. Figure 1 illustrates this variation at one mean temperature for a number of materials currently used to insulate building envelopes. For most mass insulations, there is a minimum in the respective apparent thermal conductivity versus density. This minimum depends on the type and form of material, temperature, and direction of heat flow. For fibrous materials, the values of density at which the minimum value occurs increase as both the fiber diameter or cell size and the mean temperature increase. These effects are shown in Figures 2 (Lotz 1969) and 3, respectively.

Other factors that affect thermal performance include compaction and settling of insulation, type and amount of binder used, additives that may influence the bond or contact between fibers or particles, and, if used, type and form of radiation transfer inhibitor. In cellular materials, the factors that influence thermal performance and strength properties are the same as those that control the thermal conductivity of the basic structured material: size and shape of the cells, thickness of the cell walls, gas contained in the cells, orientation of the cells, and radiation characteristics of the cell surfaces.

A change in density caused by the degree of compaction of insulation of powders affects the effective thermal conductivity. Insulating concretes made from lightweight aggregates can be produced in a wide range of densities, with corresponding thermal conductances.

Fibrous insulations reach a minimum conductivity when fibers are uniformly spaced and perpendicular to the direction of heat flow. Generally, a decrease in the diameter of the fiber lowers the

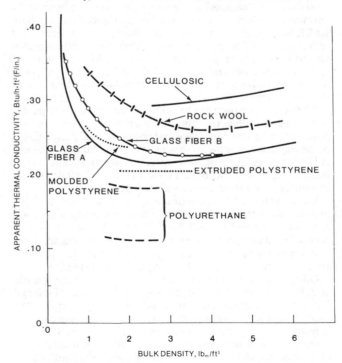

Fig. 1   Apparent Thermal Conductivity versus Density of Several Thermal Insulations Used as Building Insulations

conductivity for the same density (Figure 2). For cellular insulation, a specific combination of cell size, density, and gas composition produces optimum thermal conductivity.

At temperatures below 400 to 570 °F, the greater portion of heat transfer across most insulations occurs by conduction through the air or through other gas in the insulations (Rowley *et al.* 1952, Lander 1955, Simons 1955, Verschoor and Greebler 1952). Consequently, if the gas within the insulation is replaced by another gas with a different thermal conductivity, the apparent thermal conductivity of the insulation is changed by an amount approximately equal to the difference in conductivity of the two gases. For example, replacing air with a fluorinated hydrocarbon gas can lower the apparent thermal conductivity of the insulation by as much as 50%. Fluorocarbon-expanded cellular plastic foams with a high proportion (greater than 90%) of closed cells retain the fluorocarbon within the cells for extended periods. As these products are initially produced, they have apparent thermal conductivities of approximately 0.111 Btu/h·ft^2·(°F/in) at 75 °F when the gas contained has a mean-free-path greater than the dimensions within the cells. However, this value increases with time as atmospheric gases diffuse into the cells and the fluorocarbon gas diffuses out. The rates of diffusion and increase in apparent thermal conductivity depend on several factors, including permeance of the cell walls to the gases involved, age of the foam, temperature, geometry of the insulation (thickness), and surface protection provided. The significance of the surface is apparent when foams are encased in gas-impermeable membranes or some water vapor retarders (Brandreth 1986, Tye 1987).

For polyurethane and polyisocyanurate materials, Brandreth (1986) and Tye (1988) show that the aging process of these materials is reasonably well understood analytically and confirmed experimentally. The dominant parameters for minimum aging are as follows:

* closed-cell content > 90% preferably > 95%
* small uniform cell diameter << 0.04 in.
* small anisotropy in cell structure
* low density (1.8 to 3 lb/ft^3)
* increasing thickness
* high initial pressure of fluorocarbon blowing agent in cell
* a polymer highly resistant to gas diffusion and solubility
* polymer distributed evenly in struts and windows of the cell
* a low aging temperature

Aging may be further reduced, particularly for laminated and spray-applied products with higher density polymer skins, or by well-adhered facings and coverings with low gas and moisture permeance characteristics. An O_2 diffusion rate of less than 0.02 in^3/1000 ft^2·day for 0.001 in. thickness of barrier is one criterion used by some industry organizations manufacturing laminated products. The adhesion of any facing must be continuous, and every effort must be made in the manufacturing process to eliminate or minimize the shear plane layer at the foam/substrate interface (Ostrogorsky and Glicksman 1986).

Closed-cell phenolic-type materials and products, while blown with similar gases, age differently and much more slowly. The reasons for this are under study, but higher material density, smaller, more uniform cell size with more even distribution of polymer in struts and windows, and a basic polymer more resistant to gas diffusion could be contributory causes.

The average distance, or *mean-free-path*, that an enclosed gas molecule travels before striking another gas molecule increases as pressure within an insulation decreases. When the mean-free-path equals the average distance a gas molecule travels before striking a solid part of the insulation, apparent thermal conductivity of the insulation decreases with decreasing pressure. Correspondingly, for materials such as silica gel and fine carbon black, which have an average pore size smaller than the mean-free-path of air at atmospheric pressure, it is possible to attain thermal conductivity values lower than those for still air (Verschoor and Greebler 1952).

For homogeneous, dense materials, the primary mode of heat transfer is conduction. However, as the temperatures and the consequent transparency of materials increase, the heat transmission by thermal radiation and possible convection becomes a greater part of the total heat transferred. The magnitude of radiation and convection transfer depends on temperature difference, direction of heat flow, the nature of materials involved, and geometric considerations. The rate of radiant heat transfer varies in proportion to the fourth power of the absolute temperature.

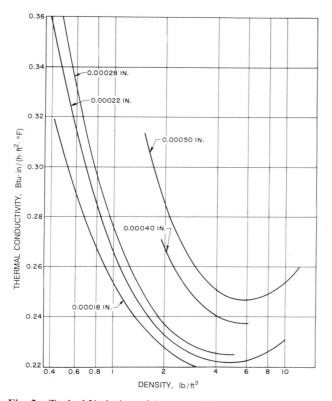

**Fig. 2   Typical Variation of Apparent Thermal Conductivity
with Fiber Diameter and Density
(Heat Flow Perpendicular to Fiber Orientation)**

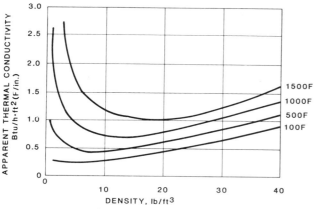

**Fig. 3   Typical Variation of Apparent Thermal Conductivity
with Mean Temperature and Density**

Because of radiation heat transfer in low-density insulation, measured apparent thermal conductivity depends on test thickness. The thickness effect increases the apparent thermal conductivity measured at installed thickness over that commonly determined at 1 in. (Pelanne 1979). From a thermal resistance standpoint, the effect is small, typically less than 10% for most 1- to 6-in. thick insulations.

Environment and Application Conditions

The apparent conductivity of insulating materials generally increases with temperature. Figure 4 shows typical variations with mean temperature. However, some materials, such as fluorocarbon-expanded, closed-cell urethanes, have an inflection in the curve over the temperature range where there is a change of phase of the fluorocarbon from gas to liquid (see Table 4, Chapter 22).

The apparent thermal conductivity of a sample at one mean temperature (average of the two surface temperatures) only applies to the material at the particular thickness tested. Further testing is required to obtain values suitable for all thicknesses. The rate of change of apparent thermal conductivity with temperature and environmental conditions varies with the type and density of material.

Insulating materials that permit a large percentage of heat transfer by radiation, such as low-density fibrous and cellular products, show the greatest change in apparent thermal conductivity with changes of temperature and surrounding surface emittance. The ASTM Standard Test Methods recognize the importance of radiation in heat transmission and require that the surfaces of all

plates be painted or otherwise treated to have a total emittance greater than 0.8 at operating temperature.

The effect of temperature alone on structural integrity is not ordinarily important for most materials in the low-temperature insulation field. Decomposition, excessive linear shrinkage, softening, or some other effects of temperature alone limit the maximum temperature for which a material is suited. At extreme temperatures, both high and low, selecting materials for a specific service becomes more critical and must be based on experience and factual performance data (see Table 4, Chapter 22).

Convection and air infiltration in or through some insulation systems, may increase heat transfer across them. Low-density, loose-fill, large open-cell and fibrous insulations, and poorly designed or installed reflective systems are most susceptible to increased heat transfer by air filtration and forced convection or both. The temperature difference across an insulation, as well as the height, thickness, or width of the insulated space, influences the amount of convection. In some cases, natural convection may be inherent in the systems (Wilkes and Rucker 1983); however, in many cases, the effect of convection in or through insulation can be minimized by careful design of an insulated structure (Donnelly *et al.* 1976).

Gaps between both board- and batt-type insulation can lower insulation effectiveness. Board-type insulation may not be perfectly square, it may be installed improperly, and the surfaces to which it is applied may be uneven. Joints formed in board-type insulation allow it to fit together without air gaps. Batts can be installed in two layers with joints between layers offset and staggered. For example, a 4% void area around batt insulation can produce a 30% loss in effective thermal resistance for an R-19 ceiling application (Verschoor 1977). Similar results have been obtained with different test conditions and for wall configurations (Lewis 1979, Hedlin 1985, Tye *et al.* 1981). Chapter 21 has further details on these effects.

If moisture condenses in the insulation, it may reduce thermal resistance, and, perhaps physically damage the system. The reduction in thermal resistance depends on the material, the moisture content, and its distribution. The thermal characteristics of some insulations may be affected at 0% moisture content (Tobiasson *et al.* 1977, Tye *et al.* 1987).

Verschoor (1985) showed that an insulated residential-type stud wall panel with a poor vapor retarder on the warm side accumulated 0.3 lb of moisture per square foot when exposed to continuously below freezing (steady-state) conditions for 31 days. All of the accumulated moisture was located in the 0.5-in. layer of mineral fiber insulation immediately adjacent to the exterior cold side sheathing and at the interface between the insulation and the sheathing. In a continuation of the test program, when the total exposure period was 60 days, the rate of moisture gain remained constant during the entire period. Subsequently, the same test wall was subjected to diurnal outside temperature cycles through the freezing point. With this exposure, most of the accumulated moisture was found in the sheathing and the bottom of the test wall rather than in the coldest insulation layer. Except for the sheathing, there was negligible change in the overall thermal performance of the wall.

The behavior may be different for closed-cell plastic foam insulations. Field observations of low-temperature insulated tanks have shown that when no air space exists between the cold surface and the insulation, the expected accumulation of condensed moisture did not occur.

Because the thermal conductivities of water and ice are many times greater than that of the gas they displace in insulation, their presence in insulation systems can have a markedly deleterious effect. This is detailed in the section Effect of Moisture on Heat Flow. CIBSE *Guide*, Section A3 (1986) and Chapter 21 cover the thermal properties of building structures affected by moisture.

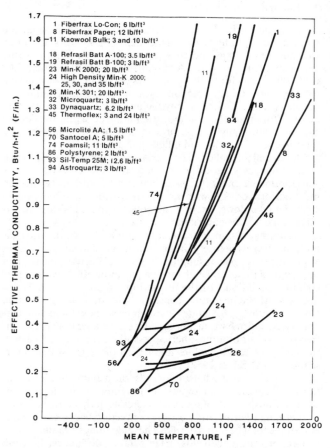

Fig. 4   Apparent Thermal Conductivity versus Mean Temperature for Various Materials (in Air at Atmospheric Pressure)
(Glaser *et al.* 1967, Pelanne 1977)

THERMAL TRANSMITTANCE

The method of calculating an overall coefficient of heat transmission requires knowledge of (1) the apparent thermal conductivity and thickness of homogeneous components, (2) thermal conductance of nonhomogeneous components, (3) surface conductances of both sides of the construction, and (4) conductance of air spaces in the construction. Procedures for calculating thermal conductance and resistance and definitions of heat transfer terms and symbols are included in this chapter and in Chapter 3. In some construction, multidimensional heat flow effects are significant. Parallel heat flow paths of different resistances occur in wood frame house construction, for example. In such cases, the guarded hot box method (ASTM C236) or the calibrated hot box method (ASTM C976) may be used to determine overall thermal transmittance.

Surface Conductance

Surface conductance is the heat transfer to or from the surface by the combined effects of radiation, convection, and conduction. Each of these transport modes can vary independently. Heat transfer by radiation between two surfaces is controlled by the character of the surfaces (emittance and reflectivity), the temperature difference between them, and the solid angle through which they see each other. Heat transfer by convection and conduction is controlled by surface roughness, air movement, and temperature difference between the air and surface. Table 1 illustrates the importance of the effect of temperature of surrounding surfaces on surface heat flux caused by radiation.

In many cases, because the thermal resistance (reciprocal of conductance) of the internal parts of the wall is high compared with the surface resistance, the surface factors are of minor importance. However, surface resistances on a window with a single pane of glass constitute almost the entire resistance and are very important. Raber and Hutchinson (1945) analyzed the factors affecting surface conductance and the difference between surface and air temperatures.

The convective part of the surface conductance is markedly affected by the nature of the air movement on the surface, illustrated by Figure 1, Chapter 22. On smooth surfaces, surface length (Parmelee and Huebscher 1947) also affects the convection part of conductance; the average value decreases as the surface length increases. Moreover, observations (Parmelee and Aubele 1950) of the magnitude of low-temperature radiant energy from outdoor surroundings show that only under certain conditions can outdoors be treated as a blackbody radiating at an effective air temperature. Therefore, selection of surface conductance coefficients for a building becomes a matter of judgment. Surface conductances shown in Table 1 of Chapter 22 are applicable to ordinary building materials. In special cases, where surface conductances become important factors in the overall rates of heat transfer, more accurate coefficients may be required. Principles and data given in Chapter 3 can be applied in such cases.

FACTORS AFFECTING HEAT TRANSFER ACROSS AIR SPACES

The nature of the boundary surfaces, as well as the intervening air space, the orientation of the air space, and the direction of heat flow, affect the transfer of heat across an air space. Air space conductance coefficients represent the total conductance from one surface bounding the air space to the other. The total conductance is the sum of a radiation component and a convection and conduction component. In all cases, the spaces are considered airtight with no through air leakage.

Table 1 Variation in Surface Heat Flux for Vertical Surfaces at 80 °F with Different Temperatures of Surrounding Surface (70 °F Ambient Still Air; 0.83 Emittance)

Surrounding Surface Temperature, °F	Surface Heat Flux, Btu/h · ft²				
	75	70	65	60	50
Convection	6.6	6.6	6.6	6.6	6.6
Radiation	4.4	8.6	12.8	17.0	24.9
Total	11.0	15.2	19.4	23.6	31.5

The radiation portion of the coefficient is affected by the temperature of the two boundary surfaces and by their respective surface properties. For surfaces that can be considered ideal graybodies, the surface properties can be characterized by emissivity. The combined effect of the emittances of the boundary surfaces of an air space is expressed by the effective emittance E of the air space. The radiation component is unaffected by the thickness of the space, its orientation, the direction of heat flow, or the order of emittance (hot or cold surface). The heat transfer by convection and conduction combined is affected markedly by orientation of the air space and the direction of heat flow, by the temperature difference across the space, and, in some cases, by the thickness of the space. It is also slightly affected by the mean temperature of its surfaces. For air spaces in building construction, the radiation and convection-conduction components can vary independently of each other.

Table 2, Chapter 22 lists the thermal resistance values of sealed air spaces of uniform thickness and moderately smooth, plane, parallel surfaces. These data are based on experimental measurements (Robinson et al. 1954). Resistance values for systems with air spaces can be estimated from these results if emissivity values are corrected for field conditions. However, the value of some common composite building insulation systems involving mass-type insulation with a reflective surface in conjunction with an air space may be appreciably lower than the estimated value (Palfey 1980), particularly if the air space is not sealed or of uniform thickness. The thermal resistance values for plane air spaces in Table 2 of Chapter 22 represent typical values; for critical applications, the effectiveness of a particular design should be confirmed by actual test data undertaken by using ASTM hot box methods. This test is especially necessary for constructions combining reflective and nonreflective thermal insulation.

For narrow air spaces, defined as those for which the product of the temperature difference (in degrees Fahrenheit) and the cube of the space thickness (in inches) is less than 3 for heat flow horizontally or downward, or less than 1 for heat flow upward, convection is practically suppressed. The conductance for these spaces is the sum of the radiative heat transfer coefficient and that for heat conduction alone through air. The radiation and conduction component can be computed by the method shown in the footnote to Table 2, Chapter 22. Effects of different mean temperatures, temperature differences, and effective emittances may be found in Table 2 of Chapter 22.

To obtain high thermal resistance with reflective insulation, a series of multiple air layers bounded by reflective surfaces is needed. The total resistance equals the sum of the resistance values across each air space. All air layers or spaces must be sealed because air moving between the layers can increase the heat flow. Depending on the type of reflective insulation, one or both sides may have highly reflective surfaces. Except for thick horizontal air spaces with heat flow down, little is gained thermally by the addition of a second highly reflective surface to the same air space. If an air space has only one reflective surface, the side on which the reflective surface is placed makes no appreciable difference in the rate of heat transfer; however, placing the surface on the warm side minimizes or prevents condensation. Condensation should be prevented from forming on a reflective surface because, apart from other effects, it degrades the reflective properties of the surface. A reflective surface placed on the warm side of an air space usually is not a condensing surface and, therefore, maintains the thermal resistance of the air space and acts as a water vapor retarder if the material and its joints are low in permeance (see discussion on vapor retarders).

The emittance of a surface is the measure of its ability to emit radiant energy and, for the same temperature and wavelength, is equal to its absorptance (ratio of the radiant energy absorbed by a surface to the total radiant energy falling on it). The ratio of the energy reflected by the surface to that falling on it is called reflectance; for an opaque surface, reflectance is equal to one minus the emittance. This emittance varies with surface type and condition and radiation wavelength.

For reflective insulation used with heating, air-conditioning, and refrigeration applications, the emittance value for long-wavelength (infrared) radiation is important, and not the value for the shorter wavelengths of the visible spectrum. Visible brightness is not a true measure of the reflectance for thermal radiation, because the reflectance for light and for long-wavelength radiation is unrelated. Table 3, Chapter 22 lists typical reflectance and emittance values for reflective surfaces and building materials and the corresponding emittance factors for air spaces.

Chemical action, dust or oil accumulations, or the presence of condensation or frost can change a reflective surface enough to reduce its reflectance and increase its emittance. Chemical changes include oxidation, corrosion, or tarnishing caused by air, moisture, wet plaster, or the chemical treatment of wood spacing strips or other adjoining structural members. Surface emittance values should be obtained by tests, such as described in ASTM *Standard* C445. Low-emittance windows have significantly greater thermal resistance, for example.

CALCULATING OVERALL THERMAL RESISTANCES

Using the principles of heat flow presented in Chapter 3, calculating heat flow by the overall thermal resistance method is preferred.

The total resistance to heat flow through building constructions such as flat ceiling, floor, or wall (or curved surface if the curvature is small) is the numerical sum of the resistances (R-values) of all parts of the construction in series.

$$R = R_1 + R_2 + R_3 + R_4 + \ldots + R_n \tag{1}$$

where R_1, R_2, etc., are the individual resistances of the parts, and R is the resistance of the construction from inside surface to outside surface. However, in buildings, to obtain the overall resistance R_T the air film resistances R_1 and R_0 must be added from Table 1, Chapter 22.

$$R_T = R_1 + R + R_0 \tag{2}$$

The U-factor is the reciprocal of R_T.

$$U = 1/R_T \tag{3}$$

With the use of higher values of R_T, the corresponding values of U become very small. This is one reason why it is sometimes preferable to specify resistance rather than transmittance. Also, a whole number is more understandable to an insulation buyer than is a decimal or fraction.

An example of former calculations for a wall with air space construction, consisting of two homogeneous materials of conductivities k_1 and k_2 and thickness x_1 and x_2, respectively, separated by an air space of conductance C, is:

$$R_T = 1/h_i + x_i/k_1 + 1/C + x_2/k_2 + 1/h_o \tag{4}$$

Series and Parallel Heat Flow Paths

In many installations, components are arranged so that heat flows in parallel paths of different conductances. If no heat flows between lateral paths, heat flow in each path may be calculated using Equations (1) and (2). The average transmittance is then:

$$U_{av} = aU_a + bU_b + \ldots + nU_n \tag{5}$$

where $a, b, \ldots, n$ are respective fractions of a typical basic area composed of several different paths with transmittance U_a, $U_b, \ldots, U_n$.

If heat can flow laterally in any continuous layer so that transverse isothermal planes result, total average resistance $R_{T(av)}$ will be the sum of the resistance of the layers between such planes, each layer being calculated by the appropriate Equation (1) or modification of Equation (4), using the resistance values. This is a series combination of layers, of which one (or more) provides parallel paths.

The calculated heat flow, assuming parallel heat flow only, is usually considerably lower than that calculated with the assumption of combined series-parallel heat flow. The actual heat flow will be some value between the two calculated values. In the absence of test values for the combination, an intermediate value should be used; examination of the construction will usually reveal whether a value closer to the higher or lower calculated value should be used. Generally, if the construction contains any highly conductive layer in which lateral conduction is very high compared to heat flow through the wall, a value closer to the series parallel calculation should be used. If, however, there is no layer of high lateral conductance, a value closer to the parallel heat flow calculation should be used.

CALCULATING SURFACE TEMPERATURES

The temperature at any interface can be calculated, since the temperature drop through any component of the wall is proportional to its resistance. Thus, the temperature drop Δt through R_1 in Equation (1) is:

$$\Delta t_1 = R_1 (t_i - t_o)/R_T \tag{6}$$

where t_i and t_o are the indoor and outdoor temperatures, respectively. Hence, the temperature at the interface between R_1 and R_2 is:

$$\Delta t_{1-2} = t_i - t_1 \tag{7}$$

For types of building materials having nonuniform or irregular sections such as hollow clay tile or concrete blocks, it is necessary to use the R-value of the unit as manufactured.

If the resistances of materials in a wall are highly dependent on temperature, the mean temperature must be known to assign the correct value. In such cases, it is perhaps most convenient to use a trial and error procedure for the calculation of the total resistance R_T. First, the mean operating temperature for each layer is estimated, and R-values for the particular materials are selected. The total resistance R_T is then calculated as in Equation (4), and the temperature at each interface is calculated using Equations (6) and (7).

The mean temperature of each component (arithmetic mean of its surface temperatures) can then be used to obtain second generation R-values. This procedure can then be repeated until the R-values have been correctly selected for the resulting mean temperatures. Generally, this can be done in two or three trial calculations.

In many heating and cooling load calculations, it is necessary to determine the inside surface temperature or the temperature of the surfaces within the structure. The resistances through any two paths of heat flow are proportional to the temperature drops through these paths and can be expressed as:

$$R_1/R_2 = (t_i - t_x) / (t_i - t_o) \qquad (8)$$

where

R_1 = resistance from indoor air to any point in structure at which temperature is to be determined
R_2 = overall resistance of wall from indoor air to outdoor air
t_i = indoor air temperature
t_x = temperature to be determined
t_o = outdoor air temperature

HEAT FLOW CALCULATIONS

Equation (9) is used to calculate heat flow through flat surfaces; Equation (10) is used for cylindrical surfaces (Figure 5).

$$q_s = (t_{is} - t_{os})/R \qquad (9)$$

and

$$q_s = \frac{t_{is} - t_{os}}{[r_s \ln(r_1/r_i)]/k_1 + [r_s \ln(r_s/r_1)]/k_2} \qquad (10)$$

where

q_s = rate of heat transfer per unit area of outer surface of insulation
R = surface-to-surface thermal resistance from Equation (1)
k = thermal conductivity of insulation at calculated mean temperature
t_{is} = temperature of inner surface
t_{os} = temperature of outer surface
r_i = inner radius of insulation
r_1, r_2 = outer radius of intermediate layers of insulation
r_s = outer radius of insulation
ln = natural or Napernian logarithm

To calculate heat flow per unit area of pipe surface, use:

$$q_o = q_s(r_s/r_i) \qquad (11)$$

where q_o = rate of heat flow per unit area of pipe surface, Btu/h·ft².

Heat flow per unit length of pipe is the preferred unit. For steady-state conditions, heat flow through each successive material

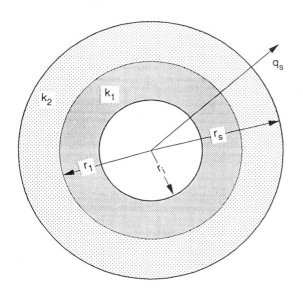

Fig. 5   Heat Flow through Cylindrical Surfaces

is the same. However, the temperature drop through each material is proportional to its thermal resistance. The terms that appear in the denominators of Equations (9) and (10) represent the resistances to heat flow based on log mean thickness. The thermal resistances discussed for flat surfaces and a thermal resistance of a cylindrical insulation should not be confused where the surface areas involved are never equal.

Heat flow is inversely proportional to the sum of the resistances of the system. The various temperature drops in the system are proportional to the resistances.

The assumptions used for calculations of heat flow are usually as follows:

t_{is} = temperature at inner surface of insulation equal to temperature of fluid in pipe or container
t_o = still air ambient temperature
r_i = inner radius of insulation = outside radius of iron pipe
r_s = outer radius of insulation = $r_i + x$
x = equivalent thickness of insulation, based on log mean thickness versus flat insulation thickness equivalent in performance

INSULATION THICKNESS

ECONOMIC THICKNESS: MECHANICAL SYSTEMS

Economics can be used (1) to select the optimum insulation thickness for a specific insulation or (2) to evaluate two or more insulation materials for least cost for a given thermal performance. In either function, economics determines the most cost-effective solution for insulating over a specific period (TIMA 1973, Federal Energy Administration 1976). This solution can be reached by different techniques, but only one solution exists for a given set of economic variables. This section presents the basic concepts used to determine economic insulation thickness for a single-insulation material system.

Greater than optimum insulation thickness may also require more capital investment for structure and piping. However, in some instances, limited energy availability may require more insulation than is normally justified by minimizing insulation and lost energy costs alone.

Determining the most profitable thickness of insulation is difficult. The economics of each plant, including cost of producing energy, cost of insulating, and potential for energy loss, indicate the preferred amount of insulation. Various types of equipment and piping also require different economic thicknesses. This analysis is further complicated because future energy cost and the life of the facility and insulation must also be considered. For every plant, these factors dictate different solutions to the economic analysis.

Economic Analysis

The cost of installed insulation increases with thickness. This incremental cost is for both labor and material. Insulation is often applied in multiple layers because materials are not manufactured in single layers of sufficient thickness and, in many cases, to compensate for expansion and contraction. Figure 6 shows installed costs for a multilayer application. The average slope of the curves increases with the number of layers because labor and material costs increase at a more rapid rate as thickness increases.

Since the optimal economic thickness is the lowest total cost of lost energy and installed insulation over the life of the insulation, these two costs must be compared in similar terms. Either the annual cost of the insulation must be compared to the average annual cost of lost energy, or the cost of the energy lost each year

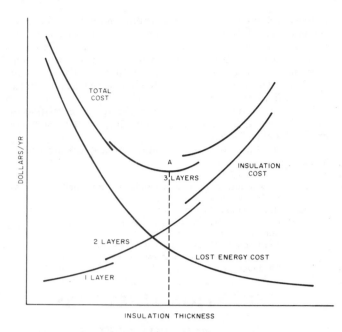

**Fig. 6   Determination of Economic Thickness
of Installed Insulation**

Initially, as insulation is applied, the total cost decreases because the incremental energy savings is greater than the incremental cost of insulation. Additional insulation reduces this cost up to a thickness where the change in the total cost is equal to zero. At this point, $m_t = m_c + m_s = 0$, and no further reduction can be obtained. Beyond this thickness, the incremental insulation costs become greater than the additional energy savings derived by adding another increment of insulation.

Figure 6 shows curves of total annual costs of operation, insulation costs, and lost energy costs. Point A on the total cost curve corresponds to the economic insulation thickness, which, in this example, is in the double-layer range. Viewing the calculated economic thickness as a minimum thickness provides a hedge against unforeseen fuel price increases and conserves energy.

ECONOMIC THICKNESS: BUILDING ENVELOPES

In buildings such as residences and warehouses, the internal energy gains are insignificant compared with the heat losses and gains through the envelope. For these buildings, the heating and cooling requirements are roughly proportional to the difference between the indoor and outdoor temperature. For commercial, industrial, and institutional buildings, internal heat loads can be significant, and the heating and cooling requirements are not directly proportional to the indoor/outdoor temperature difference. In both types of buildings, solar heat can be an important factor and should be evaluated.

Dominant Heat Loss and Gain through Envelope

Thermal insulation is generally installed in building envelope components (*e.g.*, ceilings, walls, and floors) to reduce space heating (and, in some cases, cooling) costs on a long-term basis. Additional benefits may be increased occupant comfort, reduced requirements for heating and cooling system capacity and, perhaps, elimination of condensation on wall surfaces in cold climates. However, since these benefits are difficult to quantify and are likely to be negligible beyond a nominal level of thermal resistance, the following evaluation methods focus only on reductions in energy costs. The economically optimal insulation thickness (best measured in terms of thermal resistance) in an envelope component minimizes total life cycle space heating and cooling costs attributable to that component. Total life cycle costs are the sum of present-value heating and cooling costs over the useful lifetime of the insulation plus the installed cost of the insulation.

Generally, as the thermal resistance of the insulation is increased, space heating costs decrease, while insulation costs increase. As long as the additional reduction in present-value heating plus cooling costs (incremental savings) caused by an increase in insulation thickness exceeds the additional cost of that increased insulation thickness (incremental costs), total life cycle costs are reduced. But, as insulation thickness is increased, the incremental savings per unit change in resistance decreases while incremental insulation costs beyond the first resistance unit typically remain constant or increase. Since start-up costs are attributed to the first resistance unit, this may be the most expensive unit. Beyond a certain resistance level, incremental costs exceed incremental savings, so that such additions increase life cycle costs. At the point where incremental savings equal incremental costs, life cycle costs are minimized. However, if total insulation costs then exceed total insulation benefits, the optimal level of insulation for a given component is no insulation at all. This may occur if the start-up costs are extremely high, as when insulation is installed in a sealed cavity wall.

If the R-value of the insulation used is continuously variable (loose-fill insulation in attics), uniformly small increments of

must be expressed in present dollars and compared with the total cost of the insulation investment. The former method, annualizing the insulation cost and comparing this with the average annual cost of lost energy, is easier to compute.

Life cycle costing spreads the initial cost of the insulation over the number of years the insulation is expected to be in service. The life cycle selected affects the economic thickness of the insulation. Thinner insulation pays back on a short life cycle, while thicker insulation pays back over a longer cycle. An insulation system designed to pay for itself in energy savings in a short payback period that stays in service longer, does not produce the lowest total cost over the service period.

Annualized cost of the installed insulation must also be adjusted for the cost of money, which can be a discount rate including the desired rate of return on the insulation investment. Insulation system maintenance costs should also be included in the annual cost.

Since fuel cost is likely to change during the depreciation period (life of the facility or payback period), the average cost should be estimated and this value used rather than current cost.

Because it lowers energy demand, insulation reduces the size and capital costs of the heating and cooling equipment required for an installation. This capital cost may be annualized by considering the plant depreciation period, cost of money, annual energy output for the plant, and operational expenses.

Total Cost

Since the total annual cost represents the sum of the annual cost of insulation and the annual cost of lost energy, for each incremental increase in insulation thickness, the corresponding change in the total cost is given as:

$$m_t = m_c + m_s$$

where

m_t = incremental change in total annual costs
m_c = incremental change in cost of insulation
m_s = incremental change in cost of lost energy

insulation (R-1 . . .) can be used to determine appropriate optimal thicknesses; or calculus can determine an exact optimum. If the insulation materials used are available only in discreet levels of thermal resistance (R-12, R-18, R-27), the increment used in determining optimal thickness should be based on differences between those levels (R-12 over R-0, R-20 over R-12, R-27 over R-20). Where discrete increments of resistance are used, determining the resistance level for which incremental savings equal incremental costs may not be feasible. In such cases, the selection should be left to the judgment of the analyst based on the level of conservation desired.

If a building envelope component requires structural modifications to accommodate increased insulation thickness, its cost must be included in the installed cost of the additional insulation. Generally, such modifications should only be considered when they are less costly than the use of more efficient (*i.e.*, lower thermal conductivity), but more expensive, insulation materials than those ordinarily used.

Typically, the incremental energy savings and insulation costs differ for each building envelope component; therefore, the optimal insulation level differs for each component in the same building. Less efficient heating plants and higher costs of heating energy necessitate higher optimal insulation levels in each building envelope component. Conversely, more efficient heating equipment lowers the optimal insulation level. The effects of climate, cooling energy costs, and cooling equipment efficiency on optimal insulation levels are less clear and differ widely, depending on overall building design and operational profile.

Computation

Optimal insulation thickness is found by examining the incremental savings and costs of each additional level of insulation included in a building envelope component. The present-value, life cycle savings from each increment are calculated as follows:

$$\Delta \text{LCS}_{ij} = \Delta \text{AHR}_{ij}(P_h)^{n_h}(\text{UPW}_h)^* + \Delta \text{ACR}_{ij}{}^{n_c}(P_c)(\text{UPW}_c)^* \quad (12)$$

where

ΔLCS	=	present-value, life cycle savings in dollars from the i_{th} increment of insulation in the j_{th} component
$\Delta \text{AHR}, \Delta \text{ACR}$	=	changes in annual heating and cooling requirements, respectively, of building due to the i_{th} increment of insulation in the j_{th} component (treating a reduction as positive)
n_h, n_c	=	seasonal conversion efficiencies (or coefficients of performance) of building's heating and cooling equipment, respectively
P_h, P_c	=	current costs per thermal unit for heating and cooling energy, respectively
$\text{UPW}_h^* \text{UPW}_c^*$	=	modified uniform present worth factors for heating and cooling energy types used, respectively

The terms ΔAHR_{ij}, ΔACR_{ij}, n_h, and n_c must be determined through thermal engineering analysis and require consideration of the overall building design, the climate, and the operational profile of the building. In some cases (floors), ΔACR_{ij} from additional insulation may be negative, *i.e.*, additional insulation may increase annual cooling requirements. Careful construction and insulation practices ensure that theoretical savings are actually realized.

The terms UPW_h^* and UPW_c^* relate present-value, life cycle savings to annual savings calculated at current energy costs. They are modified uniform present worth factors in the sense that they incorporate both a discount rate and a rate of energy price increase

over the time horizon considered in the life cycle cost analysis. The UPW* is calculated as follows:

$$\text{UPW}^* = \frac{1+P}{D-P}\left[1 + \left(\frac{1+P}{1+D}\right)^{L}\right], \text{ in which } D \neq P \quad (13)$$

or $\text{UPW}^* = L$, in which $D = P$

where

P = annual rate of energy price increase
D = discount rate
L = useful life of insulation

Since different energy types may have different projected rates of price increase, UPW_h^* and UPW_c^* may differ when different energy types are used for heating and cooling. However, the same discount rate and lifetime should be used to determine the appropriate UPW* factors. Generally, longer lifetimes (30 years) rather than shorter lifetimes (10 years) approach true life cycle cost minimization more closely and avoid the high cost of retrofitting if energy prices increase.

Dominant Internal Heat Loads

In buildings with dominant internal loads, the energy requirements vary so widely that no generalizations can be made regarding insulation. This contrasts with envelope dominated structures, in which more insulation reduces energy consumption (Hart 1981).

In internal load dominated buildings with cooling, for cool climates having as many or more hours below room temperature than above, higher thermal resistance increases cooling energy consumption while reducing heating energy consumption. Therefore, the calculation of economically optimum resistance becomes quite complex and involves multiple measure or hourly methods described in Chapter 28. Spielvogel (1974), Burch and Hunt (1978), and Rudoy (1975) give more details.

Figure 7 shows the results of these calculations for a building in Columbus, Ohio, with 8.2 Btu/h·ft² of internal heat gains that operates 24 h per day (Spielvogel 1974). This solution is not the only one possible but brings forth problems faced by the designer.

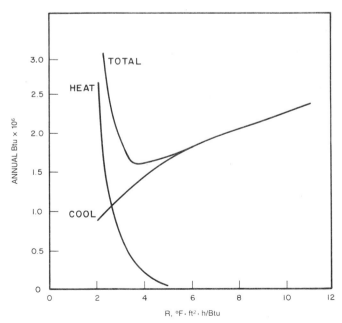

Fig. 7   Example of Optimal Thermal Resistance for Thermally Heavy Building
(Adapted from Spielvogel 1974)

As thermal resistance increases, U-value decreases, annual heating energy decreases, and annual cooling energy increases. The energy optimum exists at Point Y, where the total heating plus cooling energy is minimal. Because the cost of cooling energy differs from the cost of heating energy, the economic optimum will not be the same as the energy optimum.

These results occur in some localities because far more hours per year have outdoor temperatures of 50 to 75 °F than between 75 and 100 °F. At temperatures between 75 and 100 °F low, U-values result in less energy consumption for cooling; however, at temperatures between 50 and 75 °F, low U-values inhibit the flow of internal heat from the building, thereby creating higher cooling loads and higher energy requirements than those in buildings with higher U-values. What might be saved at outdoor temperatures over 75 °F can be more than spent in additional cooling energy at temperatures below 75 °F. Economizer cycles could offset these excess internal gains with ventilation air, however.

Where the hours of use or quantity of internal heat gains vary from room to room, the optimum thermal resistance also varies. For example, in a cold climate, a hotel kitchen requires little or no insulation, since the internal heat is sufficient to heat the space almost all year. In a meeting room adjacent to the kitchen, substantially more insulation will be justified. Thus, the economic thermal resistance of any envelope element, such as a roof, will not be the same throughout the entire building.

This type of analysis must include the level and duration of internal gains and the nature of the energy consumption of the heating and cooling systems. Most buildings need evaluation of walls, roofs, and floors on a room-by-room basis. Computer programs make these more complex analyses possible. Due to the wide diversity of building types, internal gains, system types, and operating conditions, no simple rules can establish U-values for minimum energy consumption.

Effectiveness of Added Insulation

The effectiveness of added insulation varies with many factors, including climate, original insulation level, preparation costs, and predicted life, based on payback calculations. Building codes generally balance life cycle costs between construction, financing, and energy expenditures. Figure 8 shows typical relationships between life cycle costs and energy consumption. Individual points on the curve represent different combinations of ceiling, wall, and floor insulation in R-values and glazing types (single, double, or triple). Since life cycle costs vary not only with construction and energy costs but also with climatic factors, the profiles of this curve vary according to locality. In Figure 8, the optimal condition is attained with R-30 attic insulation,

R-18.7 wall insulation, and double glazing. However, the effectiveness of added insulation can be determined only by analyzing actual conditions.

FUNDAMENTALS OF MOISTURE IN BUILDINGS

PROPERTIES OF WATER VAPOR IN AIR

Water vapor is a gas occupying the same space as the other gases that collectively constitute air and, like them, acts independently. It can migrate with or against airflow and follows its own physical laws of migration. The atmosphere is an air and vapor mixture that must be evaluated for conditions anticipated in building operation. For any given temperature and degree of saturation, water vapor in the air exerts a specific vapor pressure; it diffuses or migrates from points of higher vapor pressure toward points of lower vapor pressure, whether in air or within materials.

Effect of Temperature Change

Changes in mixtures of air and water vapor properties with heating and cooling can be followed using a psychrometric chart, shown in outline in Figure 9. A common indoor condition, 70 °F and 40% rh, is found at point A. Cooling from condition A on the chart to point B is accompanied by increasing relative humidity. At B (45 °F dew point), the relative humidity becomes 100%, and the air/vapor mixture is saturated. On further cooling to 35 °F, the original amount of water vapor is reduced by condensation to condition C, from 0.00633 lb/lb dry air to 0.00427 lb/lb dry air. ABC is the typical process that an air/vapor mixture experiences when it contacts a cool window surface. Cooling from B to C causes visible condensation on the glass. If point C is below 32 °F, the condensate would be frost. For either dew or frost to form, heat must be removed from the water vapor. All conditions on DE are at lower vapor pressure than all points on AB, so moisture will flow from the higher to the lower vapor pressure without necessarily reaching dew points. A common winter process is that shown by DE, where saturated air at 19 °F is heated to 70 °F with a resulting large decrease in relative humidity. This change explains, in part, the greatly reduced relative humidities experienced in buildings in cold weather, when cold outdoor air with high relative humidity enters and is heated.

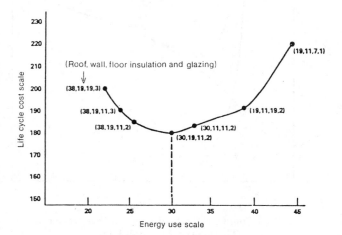

Fig. 8 Typical Life-cycle Cost Relationships

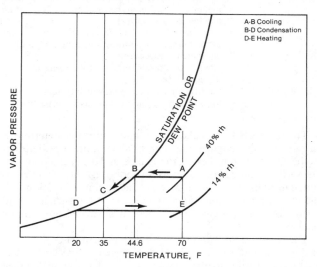

Fig. 9 Typical Cooling and Heating Processes in Buildings

At certain conditions (ice point for ice and water and triple point for ice, water, and vapor), the different physical states can exist together.

Effect of Surface Temperature at or Below Dew Point

The vapor pressure at the condensing surface lowers when the temperature drops below the dew point or frost point, at a window surface, for example. This vapor pressure gradient and convective action in the room moves water vapor continuously to the window surface to be condensed, as long as the room's concentration of water vapor is maintained.

WATER IN BUILDING MATERIALS

Many common building materials have a large number of interstices that may or may not be interconnected. Collectively, they present a large internal surface area that has an affinity for water molecules. Molecular forces of attraction hold water molecules to the surface but decrease very rapidly as the distance of molecular proportions increases. The film thickness and the amount of water held in equilibrium with the surrounding atmosphere is roughly proportional to relative humidity. Surface films of water molecules at low humidities may be only one molecule thick; at moderate humidities, polymolecular films may be established; and at humidities very close to 100%, the films become so thick that small pores can be filled and larger interstices can be partially filled. At saturation conditions, all voids in the material may be filled.

Some materials such as silica gel, alumina, and most natural fibrous materials present large effective internal surfaces to the water molecules, so the amount of water held on the effective surface in these materials can be relatively large, even at moderate humidities. These materials are said to be hygroscopic. Other materials, such as most metals, not penetrated by the water molecules present relatively small internal surfaces and may take only minute quantities of water, except when wetted directly by liquid.

Chapter 19 describes hygroscopic substances with a great affinity for water and their use as dehumidifying agents. Table 2, Chapter 12 of the 1991 ASHRAE *Handbook—HVAC Applications* gives data on the moisture content of various common materials in equilibrium with the atmosphere at various relative humidities.

Wood and other hygroscopic materials change dimensions with variations in moisture content (MC). The MC of wood varies with the relative humidity and temperature of its ambient air. The point at which a material neither gains moisture from the air nor releases moisture to the air is the *equilibrium moisture content* (EMC). The MC at which cell walls are saturated but no free water exists in cell cavities is the *fiber saturation point*. It represents the upper limit for moisture gain from the air as water vapor. It is also the upper limit of swelling; while more water can be absorbed in cell cavities, swelling is not increased. Wood MC is expressed as a percentage of its oven dry weight. The average fiber saturation point for most species is about 30%. The average EMC at 68 °F and 45% rh (heating season indoor conditions) is about 8.5%. At 75 °F and 70% rh (summer conditions), the EMC is about 13%.

The resulting dimensional changes in wood are proportional to change in moisture content, but vary with species and direction of grain. For white oak, as representative of interior trim material, 4.5% MC variation causes 2.5% volumetric shrinkage or swelling. Longitudinal dimensional change in straight-grained wood is insignificant, but it increases with crossgrain and other irregularities. Accordingly, residential wood trusses with top and bottom chords exposed to different temperature and moisture regimes can show measurable seasonal vertical movements (Plewes 1976).

Most plant and animal fibers undergo dimensional changes similar to those noted in wood. Typical values for wood can be found in Table 2; the *Wood Handbook* (FPL 1974) gives more detailed information. Related changes in other materials are not as well documented. However, different expansion rates caused by temperature and moisture changes in different materials used in composite constructions should be considered (Baker 1964, Building Research Station 1979).

Degradation of Materials

Materials exposed to excess moisture may experience reversible performance degradation or irreversible deterioration caused by (1) physical changes (such as masonry spalling on freezing), (2) chemical (steel rusting), or (3) biological (wood decaying) processes. The presence of water may be attributable to condensation, roof and window leaks, capillary flow from the ground, or any combination of these sources.

Wood construction is tolerant of heating season moisture accumulations. If the wetted wood can dry out before the onset of decay-conducive temperatures, moisture levels do not cause glue joints or paint films to fail, and moisture buildup does not occur in assemblies that are difficult to protect.

Decay is a warm weather phenomenon, most active at temperatures between 50 and 100 °F and essentially inactive when the temperature drops below 35 °F. The required wood MC to sustain decay has been estimated at 24 to 31%. Once established, decay fungi produce 0.55 lb of water for every pound of cellulose decomposed. For a sufficient safety margin, the MC should remain at 20% or less (De Groot 1976).

Molds can grow on surfaces of air dry wood (below 20% MC) at relative humidities above 85%. Condensation on wood surfaces cooled by heat radiation to the sky at night is an important moisture source for these fungi (De Groot 1976).

Table 2 Shrinkage Values of Wood, from Green to Oven Dry Moisture Content

Wood Species	Radial %	Tangential %	Volumetric %
Hardwoods			
Birch			
Yellow	7.3	9.5	16.8
Oak			
Northern Red	4.0	8.6	13.7
Southern Red	4.7	11.3	16.1
White	5.6	10.5	16.3
Softwoods			
Cedar			
Northern White	2.2	4.9	7.2
Western Red Cedar	2.4	5.0	6.8
Douglas Fir			
Coast	4.8	7.6	12.4
Fir			
White	3.3	7.0	9.8
Hemlock			
Western	4.2	7.8	12.4
Pine			
Eastern White	2.1	6.1	8.2
Longleaf	5.1	7.5	12.2
Ponderosa	3.9	6.2	9.7
Redwood			
Young Growth	2.2	4.9	7.0
Imported wood			
Lauan	3.8	8.0	—

Note: Value expressed as a percentage of the green dimension (FPL 1974).

Although condensation may have occurred in wall cavities during winter, the moisture content of wood framing members generally falls below the 25 to 30% level in spring because of normal drying, unless water has leaked in around windows and joints in the exterior surface or from an ice dam on the roof. Typical moisture levels and temperatures in roof cavities result in a more decay-conducive environment. Consequently, the incidence of decay and delamination in roof framing or sheathing is higher than that in wall construction, unless the wall is filled with material that dries slowly.

Effect of Moisture on Heat Flow

Pedersen (1990), Pedersen et al. (1991), Kumaran (1991), Hedlin (1983), Langlais et al. (1983), Thomas et al. (1983), Bomberg et al. (1978, 1991), Tobiasson et al. (1979), and Paljak (1973) show that the effect of moisture on heat flow depends on the nature of insulation, the moisture content, and the temperature of the insulation. The moisture may contribute to heat flow in both sensible and latent forms. The first occurs under all temperature conditions. However, the second requires evaporation and condensation and is most active when daily reversals in temperature across the insulation drive moisture back and forth. For example, in a flat roof, during cool nighttime temperatures, moisture migrates upward and condenses in the upper part of the insulation. In the heat of the following day, the temperature is reversed and the moisture is driven downward, transporting heat in the process. The movement depends on the porosity of the insulation. In closed-cell insulations, migration is slow, whereas in porous insulations, it may make a large contribution to energy transfer.

In cold weather, the temperature reversals cease and the driving forces exist only in one direction, i.e., upward in flat roof insulation. In this situation, moisture accumulates in the cold region of the insulation as liquid or frost. Transfer by vapor movement substantially ceases. Epstein et al. (1977) and Larsson et al. (1977) showed a nearly linear increase in energy transfer of about 3 to 5% for each percent increase in moisture content, by volume. For example, insulation with 5% moisture content by volume would have 15 to 25% greater energy transfer than dry insulation. Other field studies by Dechow et al. (1978) and Ovstaas et al. (1983) have shown similar results for insulations installed in below-earth applications, as highways and foundation walls.

In porous insulations, when temperature reversals occur, the effect of moisture follows a different pattern. Hedlin (1988) and Shuman (1980) showed the rate of energy transfer increased sharply as the moisture content increased to about 1%, by volume. The rate of increase diminished rapidly with further increase above that moisture content. The effect was strong; the rate of energy transfer for insulation with 1% moisture content was roughly twice that for dry insulation.

Moisture Effect on Heat Storage

Moisture also affects the thermal storage capacity of hygroscopic building materials considered composites of a given substance and water. At 10% MC, nearly 30% of the heat storage capacity of wood is in the water held in the cell walls. Since the specific heat of wood is a function of its temperature and moisture content (but almost independent of density and species), heat storage calculations must include an estimate of the equilibrium in-service moisture content of wood-building components.

In studies of hourly temperature and moisture content variations in insulated wood frame wall cavities, Duff (1971) showed measurable levels of daily moisture migration across the cavity. With nearly impermeable interior and exterior surfaces, such moisture migration could be viewed as a heat storage rather than a heat flow mechanism. However, the effects of moisture migration within the wall construction on solar heat gains through opaque wall surfaces remain unquantified.

WATER VAPOR MIGRATION

As a constituent part of the atmosphere, water vapor always migrates with airflow. As an independent gas, it also migrates by diffusion through air and materials according to its own pressure differentials. Moisture diffusion in free air is rapid, resulting in minimal vapor pressure differentials between connected spaces and rapid flow to condensing surfaces, such as cold glass or dehumidifier coils. As a result, vapor pressure differentials in buildings and across the building envelope usually are not as strong a driving force as air convection. Conversely, in industrial applications such as cold storage or built-in refrigerators, the primary driving force is diffusion. Therefore, the control of diffusion becomes more important with improved convection and air leakage control in buildings.

Basic Diffusion Equation

The equation used in calculating water vapor diffusion through materials is based on a form of Fick's law:

$$w = -\mu\,(dp/dx) \tag{14}$$

where

w = mass of vapor diffusing through unit area in unit time
p = vapor pressure
x = distance along flow path

Hence,

dp/dx = vapor pressure gradient
μ = permeability

Fick's law closely parallels Fourier's equation for heat flow. The actual diffusion of vapor through a material is complex; the coefficient μ is not simple, but is a function of relative humidity and temperature and may vary along the flow path through the material (Chang and Hutcheon 1956).

However, if μ is taken as the average permeability coefficient applicable to varying conditions along the flow path of length l and is assumed to be independent of vapor pressure and temperatures, Equation (14) can be rewritten as:

$$M_v = \mu A \Theta(\Delta p/l) \tag{15}$$

where

M_v = total mass of vapor transmitted
A = area of cross section of flow path, ft^2
Θ = time during which transmission occurred, h
Δp = difference of vapor pressure between ends of flow path, in. Hg
l = length of flow path (or thickness of specimen), in.
μ = average permeability, gr/h · ft^2 · (in. Hg) (see Table 9, Chapter 22)

Whenever a material of a stated or implied thickness is other than the unit thickness to which μ refers, the permeance coefficient M, which equals μ/l, should be used. The designation M for the symbol of permeance is a convenient substitute for the unit gr/h · ft^2 · in. Hg. The corresponding flow equation is:

$$M_v = M\,A\Theta\,\Delta p \tag{16}$$

where M is the permeance coefficient, gr/h · ft^2 · in. Hg. See Table 9, Chapter 22 for typical permeance and permeability values.

Resistance to vapor flow provided by a sheet or board is the reciprocal of the permeance:

$RV = 1/\mu$ = resistance to water vapor flow, in. Hg · ft^2 · h/gr

The designation *rep* is a convenient substitute for the unit resistance of one in Hg · ft^2 · h/gr.

The overall vapor resistance of an assembly (such as a wall) of materials in series is the sum of the resistances of its component parts:

$$G_t = G_1 + G_2 + G_3 + \ldots + G_n$$

The flow equation using resistance is:

$$M_v = A\Theta\Delta p/G \qquad (17)$$

This simple vapor flow theory, as in the corresponding simple heat flow theory, assumes conditions of unidirectional, steady-state flow. Useful calculations can be made for an assembly or subassembly where the inflow and outflow of vapor are equal (a condition at which no condensation occurs) if a resistance applicable to actual conditions can be assigned to each component. Overall resistances, vapor pressures, and vapor flow can be calculated; in conjunction with thermal calculations, relative humidities can be determined and the imminence of condensation predicted. (See Example 1, Figure 10. Also, see Chapter 19 for use of vapor flow calculations.)

Example 1. A wood frame wall is exposed to indoor conditions of 70 °F and 50% rh with a vapor pressure of 0.37 in. Hg and outdoor conditions of 0 °F and 80% rh with a vapor pressure of 0.03 in. Hg. The wall is painted plaster on gypsum lath on the inside over 2 by 4 studs, with mineral wool fill between studs, 1 in. wood exterior sheathing, paper, and pine lap siding. Check for possible condensation.

To simplify the example, consider the paint, plaster, and lath as a single element, with a vapor resistance $G = 1$ rep and a thermal resistance $R = 0.42$. The exterior sheathing, paper, siding, and paint are another single element for which the vapor resistance is 0.5 rep and $R = 2$. From Table 9, Chapter 22, the value for the resistivity (to vapor flow) or resistance per unit of thickness of mineral wool fill is 0.0086 rep/in. For mineral wool, the thermal resistivity is 3.7 °F·ft²·h/(Btu·in).

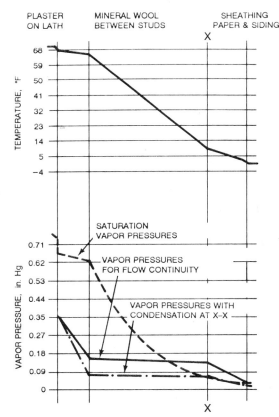

Fig. 10 Temperatures and Vapor Pressures in the Insulated Frame Wall of Example 1

Solution: In this insulated wall, with moderate warm-side relative humidities, condensation is unlikely to occur until the sheathing is reached. To check for condensation on the warm side of the exterior sheathing (designated as plane X-X in Figure 10 for convenient reference), proceed as follows:

According to the method described earlier, calculate the temperature at plane X-X. This is 9 °F. The saturation vapor pressure at this temperature is 0.06 in. Hg. If condensation is imminent or occurs at X-X, the vapor pressure at that point will be 0.06 in. Hg. Calculate the resistance for the portion of the wall from the warm side to X-X, and the vapor flow rate to X-X:

Resistance of wall to X-X = 1 + (0.0086 × 3.625) = 1.03 rep

Vapor pressure drop to X-X = 0.37 − 0.06 = 0.31 in. Hg

Vapor flow to X-X = 0.31/1.03 = 0.30 gr/h·ft²

Calculate the vapor flow rate from X-X to outdoors:

Resistance of wall from X-X to outdoors = 0.5 rep

Vapor pressure drop from X-X to outdoors = 0.06 − 0.03 = 0.03 in. Hg

Vapor flow rate to outdoors = 0.03/0.5 = 0.06 gr/h·ft²

Interpretation of calculations. Continuity of vapor flow is impossible in this example. At the highest vapor pressure permitted by the temperature at X-X in Figure 10, the calculations indicate a greater flow to plane X-X than from plane X-X to outdoors. Condensation is indicated at a rate of 0.30 − 0.06 = 0.24 gr/h·ft².

The designer must decide whether this condensation rate is serious, since it might readily be absorbed by the sheathing without excessive wetting. When the temperature at X-X is below freezing, as in this case, condensation will be frost, which may accumulate until released over a short period with a rise in outdoor temperature.

Condensation may be reduced or avoided if the resistance of the warm side of the wall can be increased so that the flow to X-X is limited to 0.06 gr/h·ft². The resistance required for this is (0.37 − 0.06)/0.06 = 5.2 rep or more. A more resistant paint film on the plaster to increase the paint-plaster-lath resistance to 5.2 rep would also accomplish this.

Determining Critical Temperature Location

If the critical plane for condensation is unknown or if a more informative, graphical representation of the situation throughout the wall is required, temperatures and vapor pressures can be calculated and plotted, as in Figure 10. For continuity of flow, the vapor pressures throughout the wall are calculated in a manner similar to the temperatures; the external vapor pressures are given, and the vapor pressure drops across each element are taken in proportion to resistance to vapor flow.

The curve for saturation vapor pressures at the various temperatures throughout the wall (Figure 10) falls below the curve for vapor pressures with continuity of flow toward the outer portions of the wall. Thus, with the given temperatures and vapor pressures, continuity of flow is not possible, and condensation occurs. Condensation on the sheathing is a definite possibility, and the new vapor pressure curve can be constructed for this condition, as shown. This new curve does not rise above the saturation vapor pressure curve, thereby confirming that the critical plane for condensation was correctly assumed.

With the vapor pressures established, the relative humidities can be found by referring to the saturation vapor pressures. The permeances assigned to the various elements can then be reexamined in the light of the indicated temperatures and relative humidities, and the analysis repeated, if necessary, using more appropriate permeance values. In a more detailed analysis, individual values might be assigned to the elements forming the outer portion of the wall, dealt with here as a composite, homogeneous element.

WATER VAPOR RETARDERS AND AIR BARRIERS

Water vapor retarders and air barriers combine to control the movement of moisture and air. While their functions are different, a single component may serve both functions. The designer should assess the needs for moisture and air movement in the building envelope and provide a system that combines the required vapor retarder and air barrier properties.

VAPOR RETARDER FUNCTIONS AND PROPERTIES

A vapor retarder is a component that retards vapor movement but does not totally prevent its transmission. In building structures, vapor retarders may allow moisture to flow through the system and exit to the outside. However, all cold surfaces that are impervious to moisture flow must be insulated with a system that has a vapor barrier jacket, or with insulation that is impervious to moisture vapor. For example, for pipes carrying cold fluids on a continuous basis, there is no egress; any moisture moving into the system will remain there and eventually result in an unacceptable accumulation.

In HVAC applications, vapor retarders are applied to thermal insulation on tanks, cold pipes, ducts, refrigerated enclosures, and buildings. They retard water vapor diffusion. If conditions are conducive to condensation, water vapor retarders help (1) keep the insulation dry and reduce heat load requirements for the cooling system; (2) prevent structural damage by rot, corrosion, or the expansion of freezing water; and (3) reduce paint problems on exterior wall construction (ASTM 1975).

In addition to vapor permeance, the following properties of vapor retarders are important, depending on the application:

- mechanical strength in tension, shear, impact, and flexure
- adhesion
- elasticity
- thermal stability
- fire and flammability resistance
- inertness to other deteriorating elements
- ease of fabrication, application, and joint sealing

Any vapor retarder's effectiveness depends on its vapor permeance, installation, and location within the insulated section. The vapor retarder is usually located at or near the surface exposed to the higher water vapor pressure. For residences with heating systems, this is usually the winter-warm side of the insulation.

Under conditions of reversible water vapor flow that can occur during temperature cycling of industrial insulations or of special-purpose buildings, the selection and location of water vapor retarders require special study and treatment (Stachelek 1955).

In the construction industry, the common rating of a vapor retarder for use in domestic construction is 1 perm or 1 rep. This stems from the definition of the perm given at the beginning of this chapter. Ratings below this value may be specified by the designer.

Vapor retarder material is usually a thin sheet or coating. However, a construction of several materials, some perhaps of substantial thickness, could also constitute a vapor retarder system.

Korsgaard and Pedersen (1989) describe a vapor retarder composed of a synthetic fabric sandwiched between sheets of plastic film. The fabric wicks free water from the building envelope, while the film retards vapor flow into it.

Water vapor permeances and permeabilities of some vapor retarders and other building materials are presented in Table 9, Chapter 22.

A vapor barrier material is usually sheet metal with soldered seams; heavy foil with wide, sealed overlaps; plastic pipe; or other zero-permeance systems.

AIR BARRIER FUNCTIONS AND PROPERTIES

Wind and stack effects in buildings cause outdoor and indoor air exchange. This results in the exchange of heat and the transfer of moisture. Exfiltrating air carries moisture that may condense in the envelope of the building, causing a variety of moisture problems (Chapter 21). In summer, infiltration of air adds to the sensible cooling load. Further, in warm humid climates, infiltrating air may bring moisture with it and add to the latent cooling load.

The exchange of air may produce beneficial ventilation; however, the balance problems caused by this exchange of air outweigh its usefulness, and buildings should be designed and built to control air exchange through the envelope.

Condensation of moisture in building envelopes can cause serious problems. It may be transported there by diffusion and by exfiltrating air. Exfiltrating air in buildings normally transports far more moisture than does diffusion, as illustrated in the following example: Assume a temperature of 70°F and a relative humidity of 35% (vapor pressure 0.259 in. Hg) on the warm side of a 1 perm vapor retarder 1000 ft² in area. On its cold side, the temperature and relative humidity are assumed to be 0°F and 100% (vapor pressure 0.0377 in. Hg), respectively. The rate of transport through the vapor retarder by diffusion will be:

$$W = (0.259 - 0.0377)\ 1000 \times 1 = 221 \text{ gr/h or } 0.032 \text{ lb/h}$$

Assume that the same vapor retarder has a hole of area A equal to 10 in², i.e., 0.069 ft², in it. This equals 1 in²/100 ft² and corresponds to tight construction. Further, at 70°F, the density of air ρ is 0.075 lb/ft³ and the specific volume of evaporation is 867 ft³/lb. Assume that the pressure drop $p_1 - p_2$ across the hole is 0.02 in. of water. In that case, the volume rate of airflow through an orifice (as discussed in Chapter 32) will be:

$$Q = 60 \times 1097 \times 0.62\ A\ [(p_1 - p_2)/\rho]^{0.5} = 1460 \text{ ft}^3/\text{h}$$

Associated with this air will be $1460 \times 0.35/867 = 0.59$ lb/h of moisture.

Thus, 0.59/0.032 are approximately 18.5 times as much moisture would move through the 10 in² hole by airflow as would move through the 1000 ft² of vapor retarder by diffusion.

Note: The pressure drop across the vapor retarder was assumed to be 0.02 in. of water. This is only an example—pressure drop across the vapor retarder varies with wind and stack forces and with the airtightness of other components of the envelopes.

Air permeates through many building materials. Thus, low air permeability is a requirement for an air barrier; for example, 3.28×10^{-4} ft³/s per square foot under a pressure difference of 1.57 lb/ft².

Overall control of moisture requires both a vapor retarder and an effective air barrier. A vapor retarder also restricts airflow; large amounts of air will pass through it only where it is breached. Thus, a vapor retarder may also be an air barrier, i.e., an air/vapor retarder. In the past, many designs were based on this fact; measures were taken to ensure that the vapor retarder was continuous in order to control airflow through it. This remains a valid approach. Some recent designs treat air barriers and vapor retarders as separate entities, but an "air barrier" should not be placed in a location where it can cause moisture to condense because of its "vapor retarder" properties. For example, an air barrier placed on the cold side of a building envelope may cause condensation, particularly if the vapor retarder is ineffective and the air barrier is impermeable to moisture.

The effectiveness of an air barrier can be greatly reduced if openings, even small ones, exist in it. Such openings can be caused by poor workmanship during application, poorly sealed joints and edges, insufficient coating thickness, improper caulking and flashing, uncompensated thermal expansion, mechanical forces, aging, and other forms of degradation. Common faults or leaks occur at electrical boxes, plumbing penetrations, telephone and television wiring, and other unsealed openings in the structure. In flat roofing, mechanical fasteners are sometimes used to adhere the system to the deck. These often penetrate the air barrier, and the resulting hole may allow air and moisture leakage into the roof.

Because it resists airflow, an air barrier must withstand pressures exerted by chimney (stack) and wind effects, or both, during construction and over the life of the building. The magnitude of the pressure varies depending on the type of building and the sequence of construction. At one extreme, single-family dwellings may be built with the exterior cladding partly or entirely installed and insulation in place before the air barrier is added. In addition, chimney effects in such buildings are small even in cold weather, so stresses on the air barrier during construction are small.

At the other extreme, in tall buildings, wind and chimney-effect forces are much greater than they are on single-family or other low buildings. Thus a fragile, unprotected sheet material used as an air barrier (or vapor retarder) on a tall building will probably be torn by the wind before construction is completed.

In summary, an air barrier must:

- meet air permeability requirements
- be continuous
  - tight joints in the air barrier must be constructible
  - it must accommodate dimensional changes due to temperature or shrinkage without damage to joints or the air barrier material
  - effective bonds in the air barrier must be made at intersections, *e.g.*, wall/roof
- be strong enough to support the stresses applied to it
  - it must not be ruptured or excessively deformed by air pressures due to wind and stack effects
  - where an adhesive is used to complete a joint, it must be designed to withstand forces that might gradually peel it away

Note: A small penetration across an air barrier system may seriously affect its performance. It concentrates the air/vapor flow in such a way that large local deposits of water/ice are possible. Calculations using permeance values are useless when holes are involved.

Classification of Vapor Retarders

Water vapor retarders are classified as rigid, flexible, or coating materials. Rigid retarders include reinforced plastics, aluminum, and stainless steel. These retarders usually are mechanically fastened in place and are vapor-sealed at the joints.

Flexible retarders include metal foils, laminated foil and treated papers, coated felts and papers, and plastic films or sheets. Such retarders are supplied in roll form or as an integral part of a building material (*e.g.*, insulation). Accessory materials are required for sealing joints.

Coating retarders may be semifluid or mastic; paint (arbitrarily called surface coatings); or hot melt, including thermofusible sheet materials. Their basic composition may be asphaltic, resinous, or polymeric, with or without pigments and solvents, as required to meet design conditions. They can be applied by spray, brush, trowel, roller, dip or mop, or in sheet form, depending on the type of coating and surface to which it is applied. Potentially, each of these materials is an air barrier; however, to meet air barrier specifications, it must satisfy the requirements for strength, continuity, and so forth.

Designers have many options. For example, the conditions for control of airflow and moisture movement might be achieved using an interior finish, such as drywall, to provide strength and stiffness, along with a low-permeability coating, such as a vapor retarder paint, to provide the required low level of permeance. In that case, edge sealing would be needed to establish continuity with adjacent air barrier/vapor retarder components.

Other designs may use more than one component. However, (1) any component that qualifies as a vapor retarder also impedes airflow, and is thus subject to pressure differences that it must resist, and (2) any component that impedes airflow also retards vapor movement and delays the exit of moisture to the outdoors. Even if it has a high permeance to water vapor, the vapor retarder may promote condensation or frost formation. The rate of diffusion of water vapor across a building component depends on the product of its permeance and the vapor pressure difference. If the air barrier is situated in a cold part of the envelope, the vapor pressure difference across it may be so small that transmission rates through it are too low to remove vapor supplied from interior sources. This is illustrated in Example 1.

Additional information regarding the control of moisture and airflow through the use of vapor retarders and air barriers may be found in Chapters 22 and 23 and in CSC TEK-AID (1990).

Vapor Diffusion through Vapor Retarders

Foil and foil laminate vapor retarders are used extensively in construction. Foil damages readily when handled; pinholes are easy to spot by holding the foil over a light table in a darkened room, although this is impractical as a field test.

An example of the effect of holes in foil is illustrated by the following tabulation for a typical 0.7-mil (0.0007 in.) foil scrim-kraft laminate.

Condition	Perms
As received	0.02
35 pinholes per square foot	0.04
A few holes larger than pinholes	0.08 to 0.16

A laminate of 0.33-mil thick foil/polyethylene, adhesive/kraft paper has 75% lower permeance (on the average) than a similar laminate made with a latex adhesive, because the polyethylene tends to seal the small holes and cracks in the aluminum foil. Foil 1-mil thick has better corrosion resistance than 0.70-mil foil, and better resistance to handling damage. The improvement is greater than predicted from the 0.30-mil extra thickness.

In any application where high humidity or condensation might be expected, flame resistant foil-kraft laminates must be selected carefully. Some laminates contain a salt flame retardant in the paper that can be highly corrosive to the foil. Many examples have been found where the vapor retarder was corroded to the point of uselessness within 6 months' service.

Most vapor retarders cannot be considered truly homogeneous materials. Doubling the thickness often more than halves the permeance. For example, a vapor retarder mastic has been tested at a dry coating mass of 0.22 lb/ft^2 with a permeance of 1.7 perm—not a good vapor retarder. However, at 0.44 lb/ft^2, the permeance was 0.06 perm, which is an excellent vapor retarder if installed properly. Most vapor retarder coatings have a minimum thickness at which acceptable permeability is attained.

Joints must also be considered. A typical field-applied system, which has a permeance of 0.01 perm in the laboratory, may have an effective overall permeance of 0.05 perm, even with carefully sealed joints. The vapor retarder is the most important component of any low-temperature insulation system. Insulation systems for low-temperature piping and tanks must have a vapor retarder with permeance close to zero, because the vapor pressure difference always drives moisture toward the metal surface.

PERMEANCE AND LABORATORY TESTING

Wet-Cup versus Dry-Cup Methods

The permeance of a specimen is found by sealing it over the top of a cup containing desiccant or water. It is then placed in a controlled atmosphere and weighed periodically. The steady rate of weight gain or loss is normally the water vapor diffusion. When the cup contains a desiccant, the procedure is called the dry-cup method; when it contains water, the wet-cup method (ASTM E-96). Usually, the surrounding atmosphere is held at 50% rh, which provides substantially the same vapor pressure difference in either method. However, the two methods produce very different results, with the wet-cup producing the higher values. The relationship between these values is best understood by referring to Figure 11 at one temperature (isothermal conditions) for a material such as wood.

The vapor permeability varies moderately at low humidities, but it grows at an increasing rate at higher humidities. The dry-cup test of this material, with 0% rh on one side and 50% on the other, exhibits a variation in spot permeability throughout its thickness because of the variation in relative humidity. The average permeability μ, by definition, from Equation (14) is:

$$\int_{p_1}^{p_2} \mu dp / (p_1 - p_2) \tag{18}$$

Since at a fixed temperature a linear relationship vapor pressure and relative humidity follow a linear relationship, this expression corresponds to the mean height of the area under the spot permeability curve, between the appropriate relative humidity limits. The average permeability found for dry-cup conditions should, therefore, have the value μ_1. Similarly, for the wet-cup test between 50 and 100% rh, the value should be μ_2. These values for wood and wood fiber materials are commonly in a ratio of 1:3 or higher (Table 9, Chapter 22).

Other Limits

Average permeability μ for other relative humidities at a particular temperature is the mean height of the area under the curve of spot permeability for the material at that temperature, between the appropriate limits of relative humidity. Only average permeabilities (or permeances) are directly measurable in practical tests. But, if several average permeabilities at different relative humidities are known and can be plotted as in Figure 11, a spot permeability curve can be plotted by trial and error. The average height of this curve between the appropriate limits for each test must

equal the value found in each test. Since each temperature requires separate tests, a large number of permeance cup tests are needed to cover a range of both temperature and humidity conditions.

Temperature Effect

The permeability constants of the least permeable membranes may be highly sensitive to temperature changes, but in porous membranes, the constants are independent of, or only slightly dependent on, temperature. Corrections for temperature have been made with certain materials by the use of an equation based on activation energy (Chang and Hutcheon 1956).

$$\mu = \mu_o \exp(-E/RT) \tag{19}$$

where (in appropriate units)

μ_o = constant at $T = \infty$
R = gas constant
E = activation energy
T = absolute temperature

This equation applies to resins, rubbers, and polymers in which the water molecules enter and move through the molecular structure by activated diffusion. It may serve as an approximate solution in other cases.

Equation (19) can be used to construct spot permeability curves for a variety of temperatures if two tests at different temperatures can be run and from which E, the activation energy, can be evaluated. In this way, permeability of material can be estimated from as few as five single tests at carefully selected conditions.

Limitations of Permeance Projections

Other than direct testing for particular conditions, no method predicts effective permeabilities for materials exposed to both a temperature gradient and a humidity (or vapor pressure) gradient, even though this is the usual situation. Data are available on permeances and permeabilities of various materials, but because the data are for different conditions, the results cannot be compared directly. Often, in fact, they do not cover a range of conditions that would permit construction of basic curves of spot permeability or permeance.

The cup test methods are steady-state procedures that may not represent the behavior or reaction of vapor retarding materials under conditions existing in actual buildings. The tests provide a standard means for comparing the permeability characteristics of materials, but measurement techniques for real, dynamic conditions are needed. Table 9, Chapter 22 indicates that some materials show a great difference in permeance for the wet-cup and dry-cup methods. This difference is even more apparent when resistances are noted.

Details of Standard Tests

Since dry-cup and wet-cup tests are easier to carry out than tests at intermediate humidities, most available data have been obtained in this way. The permeance of a material may be judged adequately, for many purposes, if tested by both dry and wet methods. Any statement regarding permeance of a specimen must include test conditions.

ASTM *Standard* E96, Standard Test Methods for Water Vapor Transmission of Materials, gives procedures for measuring water vapor transmission of materials under a variety of test conditions, using both wet-cup and dry-cup apparatus. ASTM *Standards* E398 and F372 describe convenient apparatus for measuring approximate permeance of sheet materials. ASTM *Standard* C677, Practice for Use of a Standard Reference Sheet for the Measurement of the Time-Averaged Vapor Pressure in a Controlled Humidity Space, covers the use of NBS Standard Reference Materials 707-1 (dry-cup) and 707-2 (wet-cup). ASTM *Standard* C755, Practice for Selection of Vapor Barriers for

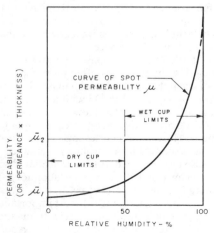

Fig. 11   Relation between Dry- and Wet-Cup Tests and the Spot Permeability for a Material such as Wood

Thermal Insulations, defines water vapor transmission values appropriate for established criteria.

Units and Conversions

There is no general agreement to follow the method described in this chapter and report results as coefficients of permeance (or permeability, where appropriate) or even to use the same basic units. Test data are frequently reported in terms of mass transmitted per (unit time) (unit area). Such data are referred to as water vapor transmission (WVT) data, and values are either high or low, depending on the vapor pressure difference chosen for the test. When this difference is known, WVT data can be converted to permeance by the following equation:

$$\text{Permeance} = \text{WVT}/\Delta p \qquad (20)$$

where

WVT = mass of vapor transmitted, gr/h·ft^2
Δp = vapor pressure difference in test, in. Hg

Table 3 lists common test conditions. Table 9, Chapter 22 presents some data on typical building materials, showing the method and, where applicable, the thickness tested. Note in the table that a paper may be waterproof, *i.e.*, possess water resistance, but still have low water vapor resistance.

PREVENTING SURFACE CONDENSATION

Condensation occurs when water vapor comes in contact with a surface that has a temperature lower than the dew point of that vapor. To prevent condensation from occurring on the warm side of insulated rooms, pipes, ducts, and equipment, insulation should be sufficiently thick to ensure that the insulation surface temperature always exceeds the dew-point temperature.

Figure 12 illustrates the steady-state heat transfer for which the following relationship can be used to calculate the data in Figure 13.

$$q = (t_b - t_p)/R_s = (t_p - t_o)R_i \qquad (21)$$

since

$$R_s/R_i = (t_b - t_p)/(t_p - t_o)$$
$$R_i = L/k$$
$$L = kR_i = kR_s(t_p - t_o)/(t_b - t_p) \qquad (22)$$

where

q = heat flow rate, Btu/h·ft^2
t_b = ambient still air dry-bulb temperature, °F
t_p = dew point, °F
t_o = operating temperature, °F
R_s = surface thermal resistance, 0.62 °F·ft^2·h/Btu
R_i = insulation thermal resistance, L/k, °F·ft^2·h/Btu
L = thickness, in. (After calculating L, Figure 14 can be used to obtain the actual thickness as needed for pipe insulation.)
k = mean thermal conductivity, Btu/h·ft^2(°F/in)

Table 3 Common Test Conditions

Nominal Test Conditions ASTM E96		% Relative Humidity on Two Sides of Specimens	
Procedure[a]	Temperature, °F	In Cup	Outside Cup
A	73.4	0	50
B	73.4	100	50
C	90	0	50
D	90	100	50
E	100	0	90

[a]Data obtained by one procedure cannot be reliably converted to another procedure.

Equivalent thickness is the thickness of insulation on a flat surface that would be required to give the same rate of heat transmission per unit area of outer surface of insulation, as on a cylinder or pipe.

$$L = r_s \ln (r_s/r_i) \qquad (23)$$

where

r_s = outer radius of insulation, in consistent units
r_i = inner radius of insulation, in consistent units

The approximate thickness of insulation L required to prevent condensation on flat metallic surfaces can be obtained from Figure 13, which produces the approximate thermal resistance

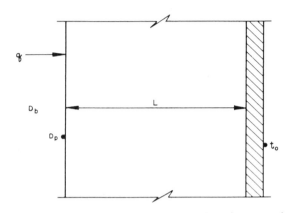

Fig. 12 Steady-State Heat Transfer

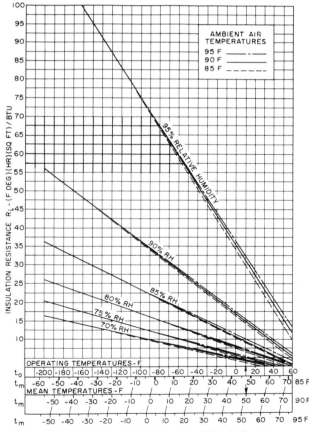

Fig. 13 Thermal Resistance of Insulation to Prevent Surface Condensation

needed to prevent condensation; the formula $L = Rk$ can be used to find thickness. In Figure 13, t_m is the mean temperature for a specific insulating material, and is the arithmetic average of t_b and t_o:

$$t_m = (t_b + t_o)/2$$

A more accurate determination of the insulation thickness L needed to control condensation can be made with the formulas in Figure 12. When pipes are involved, Figure 14 can be applied to relate the thickness of insulation, as determined by either of the methods discussed previously, to the actual thickness needed. By entering Figure 14 with L as the thickness needed on a flat surface (noted as equivalent thickness) and moving vertically until the curve for the pipe size involved is intersected, the actual thickness needed is found by moving horizontally to the Y axis. When choosing pipe insulation, always use the next greater nominal thickness than the thickness determined.

Example 2. A nominal 4-in. pipe operates at 10°F in ambient still-air conditions of 90°F and 90% rh. The mean temperature is 50°F. The insulation specified has a thermal conductivity of 0.28 Btu/h · ft² (°F/in) at the mean temperature, according to manufacturers' tables.

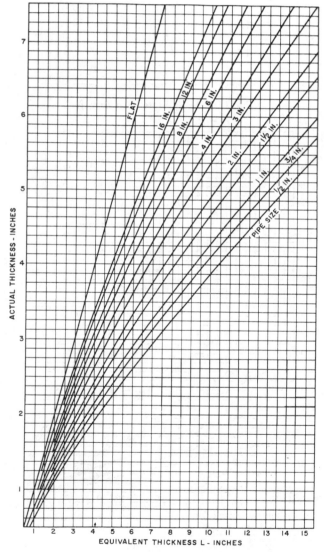

Fig. 14   Conversion of Equivalent Thickness to Actual Thickness for Pipe Insulation

Solution: Using Figure 13, locate the insulation resistance needed to prevent condensation and obtain $R = 14.8$. Since $R = L/k$, the thickness required for a flat surface or the equivalent thickness L for any pipe size is:

$$L = Rk = 14.8 \times 0.28 = 4.14 \text{ in.}$$

To find the actual thickness needed on a 4-in. pipe, read the Equivalent Thickness Scale in Figure 14 from L of 4.1 in. to pipe size of 4 in. and find the actual thickness of 2.9 in. on the ordinate. Select next higher standard thickness of 3 in.

REFERENCES

ASHRAE. 1979. Thermal performance of exterior envelopes of buildings. Proceedings of the First ASHRAE/DOE/BTECC Conference.

ASHRAE. 1982. Thermal performance of exterior envelopes of buildings II. Proceedings of the Second ASHRAE/DOE/ BTECC Conference.

ASHRAE. 1985. Thermal performance of exterior envelopes of buildings III. Proceedings of the Third ASHRAE/DOE/ BTECC Conference.

ASTM. 1980. Standard test methods for water vapor transmission of materials. ASTM *Standard* E96-80.

ASTM. 1980. Thermal insulation performance. STP 718.

ASTM. 1982. Practices for determination of heat gain or loss and the surface temperature of insulated pipe and equipment systems by use of a computer program. ASTM *Standard* C680-82.

ASTM. 1983. Practice for use of a standard reference sheet for the measurement of the time-averaged vapor pressure in a controlled humidity space (Rev. 1983). ASTM *Standard* C677-71.

ASTM. 1983. Test method for water vapor transmission rate of sheet materials using a rapid technique for dynamic measurement. ASTM *Standard* E398-83.

ASTM. 1983. Thermal insulation materials and systems for energy conservation in the 80s. ASTM STP 789.

ASTM. 1984. Standard test method for water vapor transmission rate of flexible barrier materials using an infrared detection technique (Rev. 1984). ASTM *Standard* F372-73.

ASTM. 1985. Standard practice for selection of vapor retarders for thermal insulation. ASTM *Standard* C755-85.

ASTM. 1985. Classification of potential health and safety hazards of thermal insulation materials and accessories during installation. ASTM *Standard* C930-85.

ASTM. 1985. Standard practice for selection of vapor retarders for thermal insulation. ASTM *Standard* C755-85.

ASTM. 1985. Classification of potential health and safety concerns associated with thermal insulation materials and accessories.

ASTM. 1985. Building applications of heat flux transducers. ASTM STP 885.

ASTM. 1988. Thermal insulation—Materials and systems. ASTM STP 922.

Baker, M.C. 1964. Thermal and moisture deformation in building materials. *Canadian Building Digest* 56, National Research Council of Canada, Ottawa.

Bomberg, M. and C. Shirtliffe. 1978. Influence of moisture and moisture gradients on heat transfer through porous building materials. ASTM STP 660:211-33.

Bomberg, M., *et al.* 1991. Moisture research in North America: 25 Years of building science. Lund University, Lund, Sweden.

Brandreth, D.A., ed. 1986. Advances in foam aging—A topic in energy conservation series. Caissa Editions, Yorklyn, DE.

BRS. 1974. Estimation of thermal and moisture movements and stresses, Part 1, 2, and 3. *Building Research Station Digest* 227, 228, and 229, Watford, England.

Burch, D.M. and C.M. Hunt. 1978. Retrofitting an existing wood frame residence to reduce its heating and cooling energy requirements. ASHRAE *Transactions* 84:176.

Chang, S.C. and N.B. Hutcheon. 1956. Dependence of water vapor permeability on temperature and humidity. ASHRAE *Transactions* 62:437.

CIBSE. 1986. Thermal properties of Building Structures in CIBSE *Guide*, Volume A, "Design Data." The Chartered Institution of Building Services Engineers, London, England.

CSC TEK-AID. 1990. Air barriers, Construction specifications. 1 St. Clair Street W., Toronto, M4W 1K6, Canada.

De Groot, R.C. 1976. Wood decay ecosystem in residential construction. Proceedings of XVIth IUFRO World Congress, Vancouver and Oslo.

Dechow, F.J. and K.A. Epstein. 1978. Laboratory and field investigations of moisture absorption and its effect on thermal performance of various insulations. ASTM STP 660-78:234-60.

Donnelly, R.G., V.J. Tennery, D.L. McElroy, T.G. Godfrey, and J.O. Kolb. 1976. Industrial thermal insulation, An assessment. Oak Ridge National Laboratory Report TM-5283 and TM-5515, TID-27120.

Duff, J.E. 1971. The effect of air conditioning on the moisture conditions in wood walls. USDA Forest Service Research Paper SE-78, Madison, WI.

ECON-I. 1973. How to determine economic thickness of thermal insulation. Thermal Insulation Manufacturers Association, Mt. Kisco, NY.

Epstein, K.A. and L.E. Putnam. 1977. Performance criteria for the protected membrane roof system. Proceedings of the Symposium on Roofing Technology, sponsored by the National Bureau of Standards National Roofing Contractors Association.

FEA. 1974. Retrofitting existing housing for energy conservation: An economic analysis. Federal Energy Administration, Washington, D.C.

FEA. 1976. Economic thickness for industrial insulation. GPO No. 41-018-00115-8. Washington, D.C.

FPL. 1974. *Wood handbook*. Agriculture Handbook No. 72, Forest Products Laboratory, Forest Service, U.S. Department of Agriculture.

Glaser, P.E., I.A. Black, R.S. Lindstrom, F.E. Ruccia, and A.E. Wechsler. 1967. Thermal insulation systems—A survey. NASA SP5027.

Hart, G.H. 1981. Heating the perimeter zone of an office building: An analytical study using the proposed ASHRAE comfort standard (55-74R). ASHRAE *Transactions* 87(2):529.

Hedlin, C.P. 1988. Heat flow through a roof insulation having moisture contents between 0 and 1% by volume, in summer. ASHRAE *Transactions* 94(2).

Hedlin, C.P. 1983. Effect of moisture on thermal resistances of some insulations in a flat roof system under field-type conditions. ASTM STP 789:602-25.

Hedlin, C.P. 1985. Effect of insulation joints on heat loss through flat roofs. ASHRAE *Transactions* 91(2B):608-22.

Hite, S.C. and J.L. Dray. 1948. Research in home humidity control. Research Series No. 106, Purdue University, Engineering Experiment Station, West Lafayette, IN.

Korsgard, V. and C.R. Pedersen. 1989. Transient moisture distribution in flat roofs with hygro diode vapor retarder. Thermal Performance of Exterior Envelopes of Buildings IV. ASHRAE/DOE/BTECC/CIBSE Conference Proceedings. Available from ASHRAE.

Kumaran, M.K. 1989. Experimental investigation on simultaneous heat and moisture transport through thermal insulation 1:458-65. Proc. CIB 11th Conference, France.

Lander, R.M. 1955. Gas is an important factor in the thermal conductivity of most insulating materials. ASHRAE *Transactions* 61:151.

Langlais, C., M. Hyrien, and S. Klarsfeld. 1983. Influence of moisture on heat transfer through fibrous insulating materials. ASTM STP 789:563-81.

Larsson, L.E., J. Ondrus, and B.A. Petersson. 1977. The protected membrane roof (PMR)—A study combining field and laboratory tests. Proceedings of the Symposium on Roofing Technology, NBS and NRCA.

Lewis, J.E. 1979. Thermal evaluation of the effects of gaps between adjacent roof insulation panels. *Journal of Thermal Insulation* (October): 80-103.

Lotz, W.A. 1969. Facts about thermal insulation. ASHRAE *Journal* (June):83-84.

Ostrogorsky, A.G. and L.R. Glicksman. 1986. Laboratory tests of effectiveness of diffusion barriers. *Journal of Cellular Plastics* 22:303.

Ovstaas, G., S. Smith, W. Strzepek, and G. Titley. 1983. Thermal performance of various insulations in below-earth-grade perimeter application. ASTM STP 789:435-54.

Palfey, A.J. 1980. Thermal performance of low emittance building sheathing. *Journal of Thermal Insulation* 3.

Paljak, I. 1973. Condensation in slabs of cellular plastics. *Materiaux et Constructions* 6(31):53.

Parmelee, G.V. and W.W. Aubele. 1950. ASHVE Research Report No. 1399. Heat flow through unshaded glass: Design data for load calculations. ASHVE *Transactions* 56:371.

Parmelee, G.V. and R.G. Huebscher. 1947. Forced convection heat transfer from flat surfaces. ASHVE *Research Bulletin* 3:40; also ASHVE *Transactions* 53:245.

Pedersen, C.R. 1990. Combined heat and moisture transfer in building construction. Report No. 214. Thermal Insulation Laboratory, Technical University of Denmark, Lyngby, Denmark.

Pedersen, C.R., T.W. Petrie, G.E. Courville, P.W. Childs, and K.E. Wilkes. 1991. Moisture migration and drying rates for low slope roofs—Preliminary results. Proceedings of the 3rd International Symposium on Roofing Technology. NRCA, Rosemont, IL.

Pelanne, C.M. 1977. Heat flow principles in thermal insulation. *Journal of Thermal Insulation* 1:48.

Pelanne, C.M. 1979. Thermal insulation heat flow measurements: Requirements for implementation. ASHRAE *Journal* 21(3):51.

Plewes, W.G. 1976. Upward deflection of wood trusses in winter. Building Research Note 107, National Research Council of Canada, Ottawa.

Raber, B.F. and F.W. Hutchinson. 1945. Radiation corrections for basic constants used in the design of all types of heating systems. ASHVE *Transactions* 51:213.

Robinson, H.E., F.J. Powlitch, and R.S. Dill. 1954. The thermal insulating value of airspaces. Housing and Home Finance Agency, Housing Research Paper No. 32, U.S. Government Printing Office, Washington, D.C.

Rowley, F.B., R.C. Jordan, C.E. Lund, and R.M. Lander. 1952. Gas is an important factor in the thermal conductivity of most insulating materials. ASHVE *Transactions* 58:15.

Rudoy, W. 1975. Effect of building envelope parameters on annual heating/cooling load. ASHRAE *Journal* 17(7):19.

Shuman, E.C. 1980. Field measurement of heat flux through a roof with saturated thermal insulation and covered with black and white granules. ASTM STP 718:519-39.

Simons, E. 1955. In-place studies of insulated structures. *Refrigerating Engineering* 63:40 and 63:128.

Spielvogel, L.G. 1974. More insulation can increase energy consumption. ASHRAE *Journal* 16(1):61.

Stachelek, S.J. 1955. Vapor barrier requirements for cold storage structures. *Industrial Refrigeration* 34.

Thomas, W.C., G.P. Bal, and R.J. Onega. 1983. Heat and moisture transfer in glass fiber roof-insulating material. ASTM STP 789:582-601.

Tobiasson, W. and J. Richard. 1977. Moisture gain and its thermal consequence for common roof insulations. Proceedings of the 5th Conference on Roofing Technology, NBS and NRCA, Chicago, IL.

Tye, R.P. 1988. Aging of cellular plastics: A comprehensive bibliography. *Journal of Thermal Insulation* 11:196-222.

Verschoor, J.D. 1977. Effectiveness of building insulation applications. USN/CEL Report No. CR78.006—NTIS No. AD-AO53 452/9ST.

Verschoor J.D. 1985. Measurement of water vapor migration and storage in composite building construction. ASHRAE *Transactions* 91(2):390-403.

Verschoor, J.D. and P. Greebler. 1952. Heat transfer by gas conductivity and radiation in fibrous insulations. ASME *Transactions* 74(6):961-68.

Wilkes, K.E. and J.L. Rucker. 1983. Thermal performance of residential attic insulation E. *Energy and Buildings* 5:263-77.

CHAPTER 21

THERMAL INSULATION
AND VAPOR RETARDERS—APPLICATIONS

THERMAL and moisture design and long-term performance of buildings should be considered in the original planning phase. Adequate insulation installed during construction can be much more economical than later installation. Proper selection of thermal insulations and vapor retarders must be based on:

- Thermal efficiency of the material
- Other physical characteristics required by the location of the materials
- Space available
- Compatibility of the materials with adjacent components.

Types of thermal insulation, their properties, economic thickness, vapor transmissions, and vapor retarder are discussed in Chapters 20 and 22. Insulation in various assemblies that can be used interchangeably for a given construction is discussed in this chapter.

GENERAL BUILDING
INSULATION PRACTICE

Wood Frame Construction

Wood frame construction usually includes cavities in walls and ceiling/floor assemblies suitable for insulation. Other horizontal spaces without ceiling finish can also accommodate a fill-type material if it is held in place.

Insulating sheathing placed on exterior walls may also have sufficient structural properties to provide required lateral bracing. Prefinished insulating paneling can be used as an inside finish on exterior walls or on one or both sides of interior partitions; it can also underlay other finishes. A vapor retarder on the winter warm side prevents damaging condensation within the construction.

Structural monolithic insulating decks must be at least 3 in. thick to provide an acceptable overall heat transmission coefficient. Additional thermal insulation can be placed on top of or below the deck. Roof/ceiling construction that does not have a ventilated air space may allow water vapor to condense in it. To control this potentially damaging condensation, a vapor retarder may be needed at or near the inside surface. In any case, roof decks of wood, metal, or preformed units requiring additional high thermal resistance insulation may require a vapor retarder depending on weather conditions.

Conventional attic construction with rafters and ceiling joists can be insulated between framing members with batt, blanket, reflective, or loose-fill insulation. A vapor retarder on the winter warm side of the insulation helps maintain the desired humidity within an interior environment. Batt and blanket insulation with an integral vapor retarder is adequate if properly installed.

Attics are usually vented to limit condensation and to reduce the attic temperature from solar radiation. Figure 1 shows a combination of eave and soffit vents with gable-end vents in an insulated building with knee walls. Roof vents or ridge vents can also be installed to vent attic spaces. Note that the floor joists under the knee wall are blocked to keep cold air from flowing under the finished attic floor.

Preformed metal framing members may be installed in a similar way to wood frame construction, except that physical dimensions of and spacing between members may differ. Preformed insulation material of the proper dimension may be installed if loose-fill or sprayed-in-place insulation is not used.

Existing frame construction can be insulated pneumatically, using suitable loose insulating material that can be applied in this manner. Accessible space can be insulated by manual placement of batt, blanket, or loose-fill material.

In cold climates, the vapor retarder must effectively prevent vapor movement into the roof/ceiling space by diffusion or air flow. Note that air moisture may also enter the roof through openings in the wall, e.g., electrical outlets, finding its way through the plate and into the roof.

Steel Frame Construction

The steel framing usually supports the building. The exterior metal cladding or curtain wall provides no structural support. This type of structure is usually insulated between the frame and cladding with faced blanket or spray insulation as shown in Figure 2. The facing often serves as a combination vapor retarder and interior finish and must be protected against physical damage.

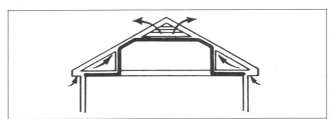

Fig. 1   Location of Insulation and Vents for Finished Attic Spaces

The preparation of this chapter is assigned to TC 4.4, Thermal Insulation and Moisture Retarders.

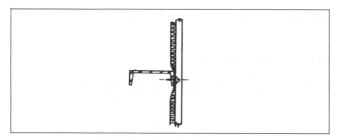

Fig. 2   Steel Frame Building Wall Insulated with Flexible Blanket

Other types of insulation can be used by adding framing or furring to the inside of frame or weather covering. This type of installation should also have an interior vapor retarder. Securing insulation as shown in Figure 2 causes local reduction of thermal resistance. This reduction plus the thermal bridging caused by the screw or bolt penetrating the insulation to the interior skin may result in condensation during cold weather.

Curtain walls may be (1) custom walls for specific projects, where most construction components are developed on a one-time basis; (2) commercial walls, where standard components are adapted to a particular building design; and (3) industrial-type walls, where standard metal sheets are fluted or ribbed to form field-assembled sandwiches to meet job conditions.

Curtain walls may consist of prefabricated panels or sections that can be classed as one of the following general types:

1. To withstand the elements, single-thickness facings are usually inserted in subframing with the windows to form a veneer over separate backup walls. Thermal insulation is integral with, or added to, the exterior of the backup wall and covers the edge of the floor slab or spandrel beam. This method reduces heat loss and keeps floors warm at their exterior edge. Insulating the exterior of the framing reduces thermal movement of the building structure (Figure 3).
2. Sandwich construction or adhesive-bonded panels are generally of three-ply construction: exterior skin, core materials, and interior skin. This type of panel, when manufactured with concrete faces and an insulation core, comes in large sizes and can be attached directly to the exterior of the building frame. When using metal facings, the width of the panels is usually restricted to match that of the standard formed sheet. These panels are installed on the exterior of the building framing. The thermal performance required for a particular installation should be carefully checked because adding insulation later is difficult.
3. Mechanically fastened panels are generally the hollow box type, with the exterior facing nesting over the interior facing. The cavity can be filled with flexible or semirigid insulation (Figure 4). The edges should vent to the exterior of the building. The panels are normally installed in a subframing system.
4. Industrial metal panels have an exterior facing of standard ribbed, corrugated V-beam materials and an interior facing of

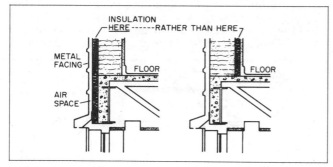

Fig. 3   Proper Placement to Insulate Structural Frame

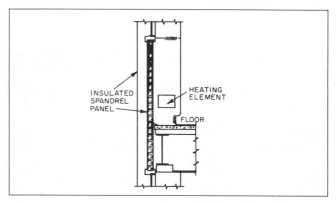

Fig. 4   Insulated Pan-Type Panel

proprietary metal pans or standard corrugated or ribbed sheeting (Figure 5). Insulation varies from semirigid to rigid, depending on the design of the inner surfacing materials. It provides good thermal and moisture control when installed with tight inner surface construction.

All curtain wall panels with insulation in the cavity or as a core should be sufficiently tight at their edges to prevent the entrance of free water or moisture. However, because some moisture may penetrate the vapor retarder and enter the wall, the exterior component should be porous enough to allow moisture vapor and air to escape. Panel edges should not be hermetically sealed. For cold climates, the subframing members or wall mullions should be of noncontinuous construction. The exterior to interior path should have a thermal break or insulated mullion cover on the interior (see Figure 6) to reduce the hazard of condensation.

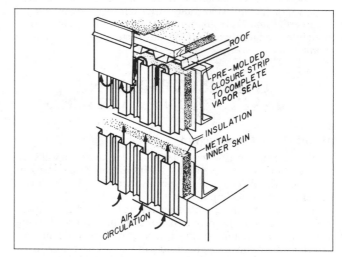

Fig. 5   Vented and Insulated Field-Fabricated Curtain Wall

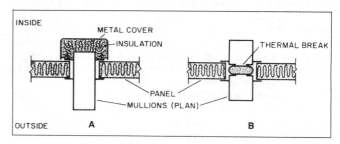

Fig. 6   Methods of Reducing Thermal Conduction through Mullions

Masonry Construction

Because concrete masonry unit walls and masonry cavity walls have hollow vertical cavities, insulating materials can usually be placed within the wall itself. Loose-fill insulation, such as water-repellent perlite and vermiculite, as well as newer foam inserts and foamed-in-place insulations, are commonly used. This insulation method is most effective with low-density masonry units. Rigid insulation can be placed between the wythes in a cavity wall or veneered construction (Figure 7A), and furring on the inside of the wall is still used in many areas (Figure 7B).

The thickness and density of concrete masonry wall construction coupled with its interior air cavities moderate the heat transmission through such an insulated wall construction. The thermal mass or inertia of such heavy construction causes a time lag of heat migration, which may lower the peak gains or losses. Insulation placed on the outside of a concrete masonry wall can also enhance this heat storage, time lag effect and help protect the structure from expansion and contraction caused by temperature extremes.

Resistance to moisture penetration in masonry walls depends on wall construction. For cavity wall construction, metal flashing at each floor and roof intersection collects and diverts moisture out through weep holes in the masonry wall. For through-wall construction, waterproof coatings of portland cement, latex, or oil-base paints are used, as well as epoxy, silicone, and bituminous based coatings. Depending on the severity of the moisture problem, both filler and finish coats can be used. Manufacturers' suggestions should be followed in such applications. Fill coats (fillers or primer-sealers) fill voids in coarse-textured masonry units and help smooth out surface irregularities before application of finish coats. Whatever the construction, tooled concave or V-shaped mortar joints are preferred. These joints impede moisture infiltration because the joint tool compresses the mortar and forces it into holes or voids that may develop when laying the concrete masonry units. They also create a shape that sheds water.

In addition to augmenting effective thermal performance, proper insulation can enhance the characteristic fire resistance and noise-reduction qualities of masonry construction. Also, insulation materials installed during construction and integral with the wall are usually placed by the mason, resulting in reduced construction time and a more economical structure.

Precast or poured-in-place solid concrete walls are insulated similarly to solid masonry construction. The design and method of fabricating the panels dictates the type of insulation selected. Figure 8 details the joint of a typical precast concrete panel with sandwiched insulation.

Floor Systems

Perimeter insulation should be applied vertically on the inside (Figure 9A) or outside (Figure 9B) of foundation walls. Rigid mineral fiber or cellular plastic insulation are used as perimeter outside insulation (Figure 9B). Drainage (e.g., gravel-fill wire mesh) may be required by manufacturers for some perimeter insu-

lations. It may need protection from physical damage if it extends above grade level. Insulation can also be installed horizontally under the slab, from the foundation wall inward for a distance of 2 ft or more, often with vertically placed insulation.

Below-grade, back-filled walls enclosing usable and heated space can be insulated and finished in the same way as solid masonry or concrete walls. Unoccupied and unheated spaces below heated areas, such as crawl spaces, may have insulated walls to maintain crawl space temperatures and minimize the need for underfloor insulation. A ground cover vapor retarder should be used, particularly over moisture-bearing soils. The soil should be graded to drain the crawl space so that surface water will not rise above the vapor retarder. Floor insulation eliminates the need for dampers on crawl space vent openings, but does not allow these spaces to be used for utilities or a heating distribution system unless it is efficiently insulated.

Floor construction with joists can be insulated between the joists (Figure 10). Insulation between joists must have supplemental support. Material with an integral vapor retarder, installed with the vapor retarder up, can be secured by the flange provided for this purpose. Laced wires, barbed-end wires spanning the joist space, wire mesh, screening, or ceiling finish will support loose-fill insulation. Floor construction can also be insulated by placing flexible blanket material over the joists before laying the subfloor.

Another floor insulation method uses double-sided, perforated aluminum foil draped and stapled over floor joists (BRANZ 1983). For about a 4-in. sag depth, laboratory R-values were 5.7 to 14.2 °F·ft²·h/Btu with a mean of 6.8, depending on the humidity above the floor space. Field measurements have shown R-values up to 14 °F·ft²·h/Btu for carpeted floors over good installations with a 4-in. sag.

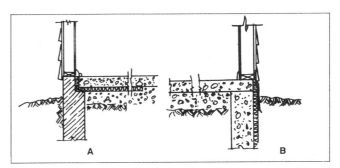

Fig. 8 Plan Section of Insulated Precast Concrete Panel

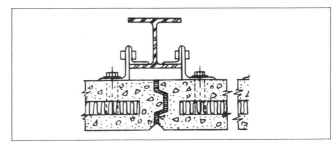

Fig. 9 Perimeter Insulation

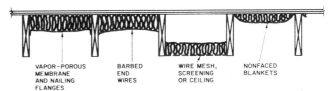

Fig. 10 Insulation Methods for Framed Floors

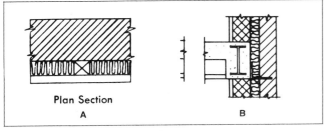

Fig. 7 Section of Insulated Walls

Roof Deck Construction

Almost all structural roof deck construction systems require thermal insulation to economically maintain the design indoor environment. For level or nearly level roofed structures, insulation should be placed on top of the deck, on the outside of the structure. This location moderates the deck temperature, which reduces thermal movement of the deck and the potential for underdeck condensation. Traditionally, the insulation is placed under the roof membrane so that it functions as a base for the built-up roof (BUR) membrane. For a stable base, a good bond must be established between the insulation, through the vapor retarder, to the roof deck, and between the insulation and the BUR. Ineffective bonds that allow the BUR to move in reaction to stresses from temperature changes often cause the roof to fail.

In this configuration, a vapor retarder should be placed between the deck and the insulation when required by interior environmental conditions. An alternate method for obtaining the required vapor retarder for a solid-shaped metal deck is to seal the deck at all overlaps and joints.

Single-ply membrane roofs are frequently installed with ballasted, loose-laid, or mechanically fastened systems. Adhesion to the insulation is usually not desirable.

Insulation can also be placed above the membrane in a Protected Membrane Roof (PMR) system. With this approach, the roof membrane is installed on the deck, where it functions as a waterproofing membrane and vapor retarder. Insulation can be placed above and below membranes to function both as the base for the membrane and the membrane protector, as well as for the insulator. Some insulation can be wetted and then successfully dried; while wet, however, it has a greatly reduced insulation value.

Roof systems designed for continual heavy traffic should have a secondary structure, supported on a curb system, above the water membrane of the roof system.

Insulation for pitched roofs covered with shingles, tile, or similar mechanically fastened watershedding materials should be placed above the ceiling directly on the attic floor. A vapor retarder should be placed on the winter warm side of the insulation, and the attic space must be adequately ventilated.

If the undersurface of the pitched roof is exposed to the interior, the winter warm side vapor retarder and insulation can be applied over the roof framing or decking. The insulation is then covered with a nailing base for the roofing application. Prefinished roof planks with the required thermal resistance can also be placed over the rafters.

Ventilation may be needed between the insulation and the roofing nailing base depending on the influence the permeance of the insulation and vapor retarder has on interior environmental conditions.

If insulation must be placed on a suspended ceiling, the space between the insulation and roof deck must be adequately ventilated for condensation control. Power-driven ventilators are usually necessary for flat decks. The ventilation air must come from the outside. It is important to use effective air and vapor retarders.

INSULATION FIELD PERFORMANCE CHARACTERISTICS

Convection and air infiltration in some insulation systems may increase the heat transfer across them. Low-density loose-fill, large open-cell, and fibrous insulations and poorly designed or installed reflective systems are most susceptible to increased heat transfer caused by natural and forced convection (air infiltration). A change of temperature differential across an insulation, as well as the height, thickness, or width of the insulated space, influences the amount of convection. When an impermeable membrane is applied to one surface or when the cavity is filled with insulation, natural convection is reduced significantly and apparent thermal conductivities measured by standard test methods apply.

The effectiveness of thermal insulation is seriously impaired when it is installed incorrectly. For example, a 4% void area in R-11 $ft^2 \cdot °F \cdot h/Btu$ wall insulation increases heat loss by 15%. A 4% void in R-19 ceiling insulation causes a 50% increase in heat loss. When insulation is properly installed, the wall or ceiling system performs as shown in Table 4, Chapter 22. Verschoor (1977) found that air interchange around thin wall insulation installed vertically with air spaces on both sides increases heat loss by 60%. Other factors, including vibration, temperature cycling, and other mechanical forces, can affect thermal performance by causing settling or other dimensional changes. Section A3 of the CIBSE Guide (1986) gives more information on the effect of moisture on thermal properties of building structures.

MOISTURE IN BUILDINGS

MIGRATION OF WATER VAPOR

Importance of Air Leakage

Vapor retarders have been advocated to avoid condensation problems in buildings, since water vapor migrates by diffusion through the building envelope materials because of a vapor pressure difference. Typical winter indoor vapor pressure at 70°F and 40% rh is about 0.13 psi above outdoor vapor pressure when the temperature is 0°F. Wilson and Garden (1965), Dickens and Hutcheon (1965), Latta (1976), Handegord (1979), Tamura (1975), and Chapter 23 show that air movement carrying the water vapor with it is far more powerful in transporting water vapor within the building envelope. Condensation will occur if the air/vapor mixture is cooled to its dew point and latent heat is removed.

Air passing quickly on a direct path through a wall may not be cooled sufficiently to condense until it mingles with and is diluted by the outdoor air. Air following a longer path on the cold side of the insulation (for example, when it flows down an air space to escape at weep holes) may deposit water on the back of a cold cladding or sheathing.

The relative amounts of water deposited in a wall or roof by vapor diffusion or by a current of air cannot be calculated with certainty. Under one set of conditions, six or seven times as much water can be deposited as a result of air leakage as by vapor diffusion through a wall without a vapor retarder. Under different circumstances, the rates could be 100:1 or higher.

A good quality vapor retarder eliminates condensation by vapor diffusion, but even an excellent vapor retarder is of little benefit if it can be bypassed by a current of air. Therefore, the first defense against condensation is control of humid air movement within a building structure by airtight construction. The section on concealed condensation in this chapter and the section on water vapor retarders in Chapter 20 include further details.

Water entering between two layers with high resistance to vapor diffusion may damage the construction if it does not dry out again. As a result, the double vapor retarder compounds the problem initiated by the defect that allowed the water to enter. Walls and roofs should be designed so that any water that enters can be removed by ventilation or by drainage to the outdoors or to a location capable of withstanding the water. Drainage may not be sufficient in low-slope roofs and is addressed in the section Membrane Roof Systems.

Controlling Air Movement

A controlled amount of ventilation air must be provided to remove moisture, but removal should not be left to uncontrolled air leakage. To control air leakage, either the holes through which the air passes or the driving pressure difference must be eliminated. Since the pressure differences arise from the natural phenomena of wind, air, and temperature differences, they can only be balanced by changing the pressure inside the building. However, the complex controls and equipment needed to sense and balance the pressure difference under constantly changing conditions make this solution impractical. The most practical approach is to minimize the size and number of holes.

It may be simple to design a satisfactory wall or roof construction for clear uninterrupted areas, but intersections with other building components require planning to link the airtightness of a wall section to the airtight glass in a window or an airtight membrane in the roof or ceiling. Leakage paths are often found in the inside of the outer wall where a partition wall joins it. Similar leakage paths can be found in electrical outlet boxes, at openings for piping, up through the hollow core of a partition wall, and through the ceiling, if ceiling airtightness is not carried across the top of the partition. A designer should work closely with construction workers to find practical solutions to construction problems, but in considering alternatives, airtightness of the envelope must be maintained.

Airtightness and Location of Air Barriers

In striving to eliminate air leaks, but realizing that perfection is not attainable, construction should include drainage or ventilation to carry away unwanted water. Ventilation or drainage must go to the outside of the airtight layer of construction or it will increase air leakage of the building. To avoid condensation on the airtight layer, the layer must either be kept above dew point by locating it on the warm side of the insulation or be of adequate permeance to permit vapor to move to the outside.

There must be no gaps between the airtight layer and the insulation, or cold air will penetrate the ventilation or drainage openings into the warm side of the insulation and reduce its effectiveness. Generally, the air barrier is placed on the warm side of, and in contact with, the insulation. The air barrier must be sufficiently strong and well supported to resist wind loads.

Air leakage through the building envelope does not necessarily occur at doors and windows. Tamura and Wilson (1964) showed that leakage from windows and doors in houses constitute only about 20% of the total leakage. Leakage cracks and openings in walls and ceilings, especially at intersections of walls, floors, and ceilings, make a far greater contribution to total leakage. Depending on the particular house, up to 70% of the total leakage openings was in walls, and up to 67% was through the ceiling.

Holes for wiring or plumbing in upper plates of exterior walls and interior partitions may cause about 50% of the total leakage through the ceiling. An equivalent leakage area may develop from shrinkage of upper wall plates after construction. Substantial leakage openings can be caused by wood shrinkage or cantilevering of floor constructions at the foundation wall or second-floor level. Chimney or plumbing stack penetrations and access hatches or fans are also likely leakage locations. In some residential construction, more leakage has been found between a window frame and wall than between a window frame and sash. Unsealed joints in ductwork and around exhaust fan openings are also common leakage areas.

Not all cracks and openings can be sealed in existing houses, nor is absolutely tight construction achieved in new housing. However, an effort should be made to provide as tight an enclosure as possible to reduce leakage and so minimize potential condensation within the envelope. Such measures also reduce energy loss.

Air Leakage in Buildings

The direction and rate of air leakage through cracks and holes depends on the direction and magnitude of air pressure differences created by stack effect and wind, and by the action of chimneys, fans, and blowers. Most significant is the stack effect from the winter temperature difference between inside and outside. Warm inside air, lighter than cold outside air, rises and flows outward through cracks and openings in the upper walls and ceilings. Condensation occurs on contact with cold surfaces below the dew point. The magnitude of the pressures depends on the height of the heated space and the temperature difference, as indicated by Equation (9) and Figures 1 and 2, Chapter 23.

Air pressure differences created by wind may be of the same magnitude as those of stack effect for low-rise buildings in cold climates (see Figure 3, Chapter 23). Wind pressure acts inward over the windward walls and outward over leeward walls and walls parallel to the wind. Although suction pressures are generally created over all flat or sloped roofs, the sustained pressure difference from wind across the ceiling in a ventilated roof space is usually near zero.

Electrically Heated and Fuel-heated Houses

In electrically heated houses or houses without a chimney, the midwinter pattern of pressure differences from average wind and stack effect is likely to be similar to that shown in Figure 4, Chapter 23. Outdoor air leaks inward through windward walls, while room air leaks outward through the ceiling and leeward walls. The pressure acting upward across the ceiling is caused primarily by stack effect; it acts in this direction as long as the inside temperature is higher than the outside. Thus, a constant pressure moves indoor air up through the ceiling.

This upward air pressure difference is counteracted or even reversed by the burner and chimney of a house with a fuel-fired heating system (Tamura and Wilson 1964). The stack effect of an operating chimney is greater than that of the house, because the flue gases are at a much higher temperature. Unless the combustion chamber has a source of outdoor air, the chimney acts like an exhaust fan, lowering the interior air pressure. In this case, the pressure difference across the ceiling is reduced, but the pressure difference acting inward is increased and acts over a larger total area of inlet openings. The total air change rate is thus increased, with most or all of room air exhausting through the chimney with the flue gases rather than into the cold wall and roof spaces.

The furnace chimney represents the most significant difference between electric and fuel-heated houses as far as air leakage and indoor humidity levels are concerned, since it provides an exhaust opening when the burner is off. A fireplace is comparable to a furnace chimney, unless combustion and ventilation air are taken directly from outdoors.

Tolerance of Condensation

As climatic conditions change with the seasons, limited condensation that accumulates at one time can dry out at another. To a lesser extent, diurnal drying can occur from solar heat gain. If the material on which condensation is deposited can temporarily store the quantity of water involved, no harm may result. However, this will not always be the case; each wall or roof should be assessed individually. Refrigerated storage or insulated chilled water pipes, where conditions do not change with the seasons, are examples of situations where condensation does not dry out.

Some materials can temporarily store water without damage. Assume that during one month, 1 lb of water is deposited by diffusion in a single concrete block with an absorptive capacity from dry to saturated of 10 lb. If the block is initially at 40% saturation, this extra water can readily be absorbed by the inner face, raising saturation to only 50%. This quantity of water would probably go undetected.

Wall components with little or no absorptivity, for example metal panels, could collect appreciable quantities of such surface condensation as frost or ice that can be rapidly released with a change in weather conditions. Any water vapor passing through an inadequate or defective vapor retarder in an unventilated flat roof with a highly impermeable exterior membrane will be trapped between the two layers, leading to premature roof failure.

CONDENSATION IN HEATED BUILDINGS

Unwanted moisture accumulation in heated buildings may be caused by either visible or concealed condensation. Each location of condensate requires different control strategies, either active or passive. Active condensation controls attempt to limit, by controlling moisture sources and mechanical equipment, the quantity of moisture present. Passive controls attempt to limit moisture migration, retard unwanted accumulation, and accelerate disposal of condensed moisture with combinations of construction materials and schemes that take advantage of natural forces or mechanisms, such as vapor retarder systems and natural cavity ventilation.

Indoor Humidity Levels

The moisture content of the air in a heated building in winter is higher than that in the outdoors, because moisture is introduced by occupants, processes, or humidification equipment. Actual building humidity depends on the balance between the rate at which it is lost by diffusion, air leakage, ventilation, and condensation. Table 2 in Chapter 20 of the 1992 ASHRAE *Handbook— Systems and Equipment* gives rates at which moisture is supplied by the occupants and their activities. Moisture loss by diffusion depends on the water vapor permeance characteristics of the building envelope materials and the vapor pressure difference across the envelope. Moisture lost through air leakage depends on the air pressure differences and the characteristics of the envelope leakage openings.

Flow rates due to natural ventilation can be estimated from the data in Chapter 23, and flow rates due to mechanical ventilation can be calculated from design data. The accuracy of these calculated estimates depends on whether the number, size, and location of openings are known. Condensation on surfaces exposed to the building interior is normally minimal and can either be eliminated by improved thermal design of the enclosure element or by intentionally lowering indoor humidity.

Lowering indoor humidity to minimize surface condensation on windows is one approach, but increasing the interior surface temperature of the window by multiple glazing may be more effective. Multiple glazing has the added advantages of saving energy, improving occupant comfort, and reducing the possibility of condensate damage to the interior adjacent to the window, *e.g.*, staining of wall, rotting of window sill, and mold growth. Both humidity reduction in the space and additional glazing involve additional cost. Lowering indoor humidity enough to avoid concealed condensation within insulated construction may be difficult or impossible in areas with extremely low dew points. For example, with a sheathing temperature of 5°F, indoor air at 70°F is limited to a relative humidity of 6% if the indoor air comes in contact with the sheathing. However, experience suggests that lowering indoor humidity enough prevents potential condensation damage, even though occasional condensation may still occur.

In houses without chimneys or fireplaces, moisture from normal occupant activity may raise humidities to cause excessive condensation on windows. With fuel-fired heating systems, however, ventilation by the chimney in conjunction with air inlet openings often maintains relative humidity levels below that for window condensation (particularly with multiple glazing), except in extremely cold temperatures.

Control of Excess Indoor Humidity in Houses

The relative humidity maintained in heated buildings is usually determined by the requirements of the occupants, but it may be limited by visible condensation on surfaces exposed to the space. If improved thermal performance of the component is not possible, the removal of moisture sources such as humidifiers or attached greenhouses should be considered. Exposed soil surfaces in crawl spaces or cellars should be covered with vapor retarder membranes held down by sand, gravel, or other suitable means. Houseplants are a source of moisture, as is green or wet wood stored inside for fuel. (See Chapter 20, Table 2, in the 1992 ASHRAE *Handbook—Systems and Equipment*.)

Visible condensation may be removed in several ways. An exhaust fan controlled by a humidistat setting causes dry and cold outdoor air to infiltrate, flushing out moist air. An air-to-air heat recovery device can be included to reduce the heating energy penalty from the exhaust fan. Devices that also recover latent heat may be counterproductive if the moisture is reevaporated into the incoming airstream. Other approaches include the use of mechanical dehumidifiers and insulated vent stacks that go from the living space through the roof. The use of dehumidifiers should be taken as an indicator that greater precautions for moisture control may be required.

Short-term ventilation procedures, such as occasionally opening a window or door, may lower humidity momentarily, but it will rise to its original level soon after the opening is closed. Materials such as wood, fabrics, paper, and rugs absorb water vapor from the air until their moisture content is in equilibrium with the relative humidity. This slow process can remove substantial quantities of water. When relative humidity is reduced, the absorbed moisture slowly evaporates from the material over several days or even weeks. The slow evaporation of moisture accumulated over a summer of high relative humidity is the reason houses have a high relative humidity in the fall.

Windows and doors are not a satisfactory means of ventilation in very cold weather. If there is no other alternative, houses with forced warm air systems can use an outdoor air intake duct with an adjustable damper for continuous ventilation. The chimney acts as an exhaust outlet for room air and combustion products. In this arrangement, the furnace circulating fan should run continuously to control the distribution of outdoor air entering the system. Alternatively, an automatically controlled damper may be installed in the outdoor air duct. The duct should be insulated to avoid condensation on its surface in cold weather. Although this ventilation system can be very effective in reducing humidity, extra energy is needed to heat the additional outdoor air. Also exfiltration increases, and concealed condensation may occur. This alternative should be used only as a last resort to control a serious moisture problem.

VISIBLE CONDENSATION

Figure 11, Chapter 20 in the 1992 ASHRAE *Handbook— Systems and Equipment* shows combinations of indoor relative humidity and outdoor temperature levels at which visible condensation appears on surfaces. Because wall surface temperatures vary with other factors, such as the presence of thermal bridges and circulation or stratification of room air, the values indicated are only an estimate. Condensation will take place at any sufficiently cold spot, and the limit on indoor relative humidity levels for control of visible condensation on a nonhomogeneous wall generally is lower than may be inferred from its average U-factor.

Danger of Surface Condensation

On surfaces that remain only slightly above dew point and where condensation is barely avoided, the average moisture content may still be high enough for noticeable discoloration, mold growth, and dimensional change. Such surfaces are often in poorly ventilated closets and behind furniture against outside walls. The effects of thermal bridges on moisture and dust accumulation are indicated by further discoloration of drywall surfaces over nail heads and framing members. A similar phenomenon may also be observed on the outside surfaces of heavily insulated walls with insufficient thermal break over the studs. As the cavity is better insulated, the exterior wall surface between studs stays colder and has a higher rate of dust accumulation than that directly over studs. The dust marking becomes evident through more rapid discoloration over colder surfaces. Siding may become more vulnerable to mold growth (Neilson 1940). An International Energy Agency (1990) study concluded that the critical monthly mean surface relative humidity is 80% for mold growth on interior building surfaces.

Summer Surface Condensation

Surface condensation on below-grade concrete and masonry walls and floors is a common problem in areas with high summer humidity. Heavyweight construction has a more nearly constant surface temperature close to the soil temperature, while air dew-point temperatures fluctuate from day to day. Mechanical or chemical dehumidification and selective ventilation during periods of low dew point are used to prevent surface condensation. Unless preventive measures are used, surface condensation may deter the successful application of earth-sheltered design concepts based on radiant cooling.

One means of summer surface condensation control is added insulation with a vapor retarder on the warm side, such as that used in the insulation of cold water pipes. Insulation on the outside of basement walls helps control condensation on the interior surface by isolating the walls from colder ground temperatures. Insulation on the room side eliminates visible condensation but can foster concealed summer and winter condensation.

Floor Slab Condensation

Summer surface condensation may also form on concrete floors on grade. Carpeting tends to lower the interior slab surface temperature, increasing the condensation hazard. Dehumidification and ventilation may be sufficient to control odors caused by the moisture; these are generally more objectionable than actual damage to the floor covering.

Entry of ground moisture can be reduced by isolating the slab with the placement of a low-permeability membrane over the soil, beneath the slab, and using coarse gravel to break the capillary moisture rise. Note that application of a membrane is difficult because of the potential damage during construction.

Sealing of floor slabs and basements against the entry of radon should also be considered. Although soil cover sheets are commonly referred to as "vapor retarders," they can also act as a waterproofing membrane when exposed to liquid water.

Control of slab surface temperatures is an important consideration in minimizing the need for mechanical dehumidification, particularly with solar-oriented designs that emphasize the effects of thermal ballast. In localities with severe summer surface condensation problems, lightweight concrete should be considered for floor slabs to increase their insulating value or insulation under the slab.

Preventing Surface Condensation

Low air infiltration or ventilation increases the winter condensation hazard because higher indoor relative humidity levels are sustained. The design expedient controlling condensation by warming cold window surfaces with warm air or radiation may not be compatible with energy conservation objectives.

Multiple window glazing based on the outside design temperature should ensure that windows remain clear most of the time. Based on the U-factors given for the center of glass in Chapter 27, comfortable inside conditions can be maintained without excessive window condensation on double-glazed windows with outside temperatures down to 5 °F and on triple-glazed windows with

outside temperatures down to −28 °F. In both instances, some condensation may appear around the edge of the window perimeter but should disappear with a warming trend. This criterion should be used in deciding whether to install double or triple glazing to maintain the desired humidity, or whether to reduce humidity to avoid condensation during the coldest periods.

CONCEALED CONDENSATION

Condensation Control Strategies

Most buildings are maintained for human occupancy by managing relative humidity in the 30 to 50% range. In parts of healthcare facilities (such as operating and intensive care rooms) and other buildings, higher relative humidities may be required. Concealed condensation control is based largely on passive control measures. Such measures may vary between residential and commercial or institutional buildings because of variations in scale and in construction details. Nevertheless, the basic control strategies are the same.

Since complete control of air and vapor flow is not feasible, moisture accumulating in structural cavities must be disposed of by ventilation and drainage. Disposal provisions must be outside the airtight layer in the construction, or they will increase the total air leakage of the building. This concept is known as the flow-through principle. Furthermore, the air infiltration barrier should be kept warm by placing it on the warm side of the insulation or it must be designed to allow moisture egress. Since the insulation cannot be relied on to be airtight because of its inherent porosity or imperfectly sealed joints, the airtight layer should be placed in contact with the insulation to preclude convection airflow from the room. If so installed, the layer will perform the functions of both air barrier and vapor retarder.

Water vapor and airflow control in a single plane also prevents trapping condensed moisture in a cavity. Water lodged between two impervious layers can rapidly degrade materials.

Control in Residential Construction

Wood frame roofs may be more vulnerable to decay than are frame walls, possibly because of the higher moisture flow rate through defects in the ceiling vapor retarder and insufficient ventilation. Poorly ventilated spaces (behind knee walls, for example) and defects in the ceiling vapor retarder (recessed lights) are typical areas of concern. Particularly troublesome are poorly fitted bathroom exhaust ducts penetrating poorly ventilated flat roof cavities. Damage may also develop in structures with a year-round moisture source, such as swimming pool enclosures, without sufficient protection and ventilation.

An attic space can be ventilated at a sufficient rate to assure satisfactory moisture dissipation. In areas of heavy snowfall, a cold roof surface is also needed to prevent the snow cover from melting on the roof and forming ice dams along the eaves. Reducing the total available vent area during the coldest months may conserve energy without condensation damage, as long as adequate moisture dissipation is still allowed.

The effectiveness of attic ventilation varies with climatic conditions, especially wind. It may be most beneficial in continental climates where, on a sunny day, the daytime outdoor relative humidity decreases with rising temperatures. In coastal climates its effectiveness may be reduced, as experienced in Great Britain. Dickens and Hutcheon (1965) considered attic ventilation a less reliable moisture control measure in Canadian climates than it is in the United States.

In level roofs, free air must circulate through all joist spaces from one eave or soffit to the other. Obstructions such as header beams cannot be effectively bypassed; therefore, vent stacks placed on a raised curb in the middle of the roof should vent the joist spaces. Similar details may also be necessary in construction with soffit vents placed below the plane of the roof, as in mansard roof design. In such cases, the ventilation air must flow down the vertical chase. Since the wind force may not be sufficient to overcome the thermal stack effect, this may lead to stagnation of warm and moist air in the flat roof cavity.

Sloped roofs offer several advantages in residential construction. Water vapor gained in the roof cavity can be dispersed in a larger volume of attic air and may be absorbed in a larger wood sheathing surface. Although the condensation may be localized or distributed, condensed water can cling to the sheathing and drain down to the perimeter soffit, where it may be disposed of by appropriate flashing or other drainage provisions. In flat roofs, condensation forms immediately above leakage openings and usually finds its way back into the house through the same openings.

Cathedral ceilings built with joists are vulnerable to moisture problems similar to those for flat roofs; they can, however, be provided with ridge vents and drained through soffits or eaves. Cathedral ceilings built with plank decking exposed to the interior appear to have a sufficiently high moisture capacity to withstand typical heating season moisture accumulations. In more heavily insulated construction, a superior vapor retarder with a rating not greater than 0.1 perm is necessary between the deck and the insulation. The best construction practices also favor separation between insulation and roofing (sheathing on sleepers) to permit ventilation at that plane.

Moisture dissipation from wood frame wall cavities can be accelerated by solar gain which drives moisture accumulated on the sheathing back into the cavity and out through available vent openings. In severe climatic conditions, such ventilation effects may not be sufficient, and further provisions may be needed for positive drainage at the bottom of the cavity. A single weep hole in each stud space should allow sufficient outward drainage of moisture without promoting excessive circulation of outside air in the wall cavity. At the same time, such weep holes should be large enough to prevent rapid freeze-up and blockage by insects.

Control in Commercial and Institutional Construction

Materials commonly used in commercial and institutional construction tend to be somewhat more moisture-tolerant and decay-resistant than those used in residential construction. In addition, vapor retarders in the walls of commercial buildings are less common. As a result, air leakage through hollow concrete masonry is often greater than in other types of construction. Upward airflow in the cavities is not always adequately blocked, and parallel random leakage paths are found between gypsum wallboard or other finishes and the block face. Because of such conditions, plaster or heavy-textured paints and mastics that fill the pores and holes control airflow better than do sheet materials.

Air can leak through exterior walls where the structural system or services penetrate the air barrier or at joints between dissimilar materials or components. For example, masonry cannot form a tight seal with structural steel columns and beams. To reduce this problem, the structural frame should be inside and separate from the exterior wall. The resulting curtain wall can then incorporate a more continuous air barrier and be protected from the fluctuating weather by insulation applied to the outside. Exterior cladding to control rain penetration is best applied following the weather-tightening system. The interior wythe (masonry course) and air barrier should be accessible for maintenance of the air seal and joints.

The deterioration of exterior structural elements of a building and damage to the interior through condensation from air leakage have an important bearing on the operation and maintenance costs of the building. Improving the airtightness of internal floors and partitions, particularly in high-rise buildings, redistributes the pressure differences caused by stack effect and reduces the pressure difference across the exterior wall on each floor. This approach also improves ventilation and air distribution and reduces the air circulation between occupancies on different floors. It also helps control smoke movement in the case of fire and may enable a more equitable apportioning of energy charges for space heating between individual units in apartment buildings.

Leakage in actual buildings often occurs through holes cut accidentally or deliberately in a reasonably tight membrane or component, such as the penetration of services through specified air or vapor retarders or solid components. Other leakage openings in the exterior envelope result from dimensional changes in improperly placed materials or by inadequate sealants or membranes applied to bridge joints or cracks that will eventually open.

Membrane Roof Systems

Since most membrane roof systems in commercial and institutional construction are highly resistant to vapor leakage, condensation must be prevented when insulation is placed between the heated interior and the roof membrane. In winter, insulation temperatures may be below the dew point of the room air for long periods. Condensed moisture can then saturate insulation, wood decks, and roof sheathing; the subsequent soaking of the insulation can damage interior materials.

Wet insulation in level roof construction is difficult to dry. Even with some slope, drainage is likely to be so slow as to be ineffective. Ventilation to the outside is not effective for drying roof insulation, since forces acting to remove the moisture are small. The vents themselves may present a hazard to the insulation by admitting moisture and drifting snow. Also, water leaks can occur where the vent penetrates the roof, unless they are properly installed. Finally, vents may allow chimney action to carry air upward through openings in the deck and through faults in the vapor retarder. Then as air flows to the outside, further moisture is drawn into the roof with the replacement air; this moisture may condense.

A vapor retarder in a conventional flat roof can trap moisture between the roof itself and the roof membrane. The decision of whether to use a vapor retarder depends on interior humidity and climate. The absence of a vapor retarder allows vapor to enter a roof during the heating season, but it also facilitates the removal of moisture in warm weather. This may not be acceptable in buildings with high humidity, where a large accumulation of frost or liquid condensation results in dripping. Where humidities are lower, the roof system may successfully store moisture through the heating season without problems (Baker 1980).

Regular inspection of the membrane and flashings helps prevent water leakage into the roof. Link (1977), Tobiasson et al. (1977), and Hedlin (1974) describe how infrared scanners or capacitance meters can help detect wet insulation, which can be removed or possibly dried out.

The top layers in protected membrane or inverted roof systems are not waterproof; therefore, insulation is exposed to rainwater. To remain effective, it must be able to resist moisture penetration. Extruded polystyrene board has been used extensively. Insulation moisture contents commonly ranges up to 4 or 5% by volume.

Some insulations are damaged by freezing and thawing, fracturing cell walls, and allowing water into an otherwise low-permeance material. When free moisture is available, the rate of entry increases rapidly as the temperature gradient increases (Hedlin 1977). Even when insulation is immersed in ponded water, moisture absorption through the edges is less than through the upper and lower surfaces, since the temperature gradient is normal to the roof surface.

Protective measures can reduce moisture gains. Roof slope performs much the same function for protected membrane roofs as it does for conventional ones. Covering the bottom surface of the insulation with a low-permeance layer inhibits moisture entry there. The upper surface should be open to the atmosphere so that water can evaporate freely. If it is trapped against the upper surface (*e.g.*, by paving stones), sun heating may drive the water into the insulation.

Where a high thermal resistance is required, roofs may combine conventional and protected membrane systems when they may be applied in two separate lifts. The protected membrane system may be applied to existing conventional roofs to increase the thermal resistance, if the roof structure can support the added weight. This addition keeps the insulation below the roof membrane warmer, so that the danger of moisture condensation entering from below is significantly reduced.

Other Considerations

All previously mentioned considerations require special attention in buildings with high, sustained relative humidity levels. To gain flexible and uninterrupted floor spaces, mechanical service shafts are often located outside the perimeter of the occupied area, allowing unrestricted airflow into exterior wall cavities. Potential problems may be further aggravated by duct leakage. In controlled humidity areas, such as operating rooms, the outflow of moisture to surrounding cavities should be reduced by low-permeance surface finishes such as vinyl fabrics or vapor retarder paints. Complex piping and electrical systems may not permit effective control of air and water vapor flow in the plane of the interior surface. Such interferences require careful placement of air and vapor retarders because they can be rendered ineffective by penetrations of mechanical work. Such penetrations should be kept to a minimum.

Past Practice

While previous methods for determining the need for vapor retarders were based on climatic conditions and geographic location, experience has shown that construction practices, workmanship, and conditions of occupancy can be as significant as climatic conditions in determining the need for vapor retarders.

Average winter temperature and its duration affects condensation. Figure 11 shows the United States divided by -20, 0, and 20°F isotherms of winter design temperature. Zone I roughly includes areas with design temperatures of -20°F or colder; Zone II, design temperatures of 0 to -20°F; and Zone III, those at 0°F and warmer. Within each zone, similar degrees of condensation trouble were expected, and similar corrective measures apply.

Not all buildings require a vapor retarder, but the walls of every dwelling should include a vapor retarder when it includes any material that would be damaged by moisture through rotting, staining of wall or ceiling surface, freeze-thaw action, or through reduction in the thermal resistance of the envelope system. Vapor retarders should be applied in all three condensation zones when the wall U-factor is less than 0.25 Btu/h·ft²·°F and in Zones I and II, to walls of higher transmittance.

The recommended ventilation shown in Table 1 for dwellings is based on insulated ceilings. The net area refers to the total of all openings free of obstruction. For example, louvers with 8-mesh screen require a gross area 2.25 times that listed. In Zone I (Figure 11), a vapor retarder should be included in all ceilings. Also, openings from walls into the attic or around a loose-fitting attic door must be avoided to keep warm, moist air out.

The Housing Development Directorate (1979) developed Table 2, a checklist for diagnosing moisture-related problems in housing in cold climates.

Crawl Spaces

The *Building Foundation Design Handbook* (Labs *et al.* 1988) notes that ground cover membranes that restrict evaporation of soil moisture are the single most important way to prevent condensation and wood decay problems in crawl spaces. The ground

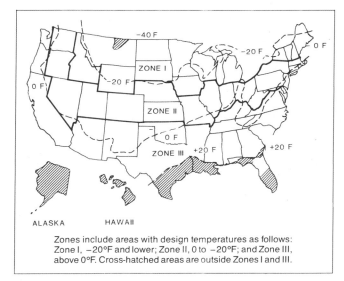

Zones include areas with design temperatures as follows: Zone I, -20°F and lower; Zone II, 0 to -20°F; and Zone III, above 0°F. Cross-hatched areas are outside Zones I and III.

Fig. 11   Condensation Zones in United States

Table 1   Recommended Practice for Loft and Attic Ventilation

Flat Roof—Slope 3:12 or Less

Condensation Zone I[c]: Total net area of ventilation should be 1/300[b] distributed uniformly at the eaves *plus* a vapor retarder in the top story ceiling. Free circulation must be provided through all spaces.

Condensation Zones II and III: Same as for Zone I.

Gable Roof—Slope 3:12 or More

Condensation Zone I: Total net area of at least 2 louvers on opposite sides located near the ridge to be 1/300[b] *plus* a vapor retarder in the top story ceiling.

Condensation Zone II: Same as for Zone I.

Condensation Zone III: Same ventilation as for Zone II. A vapor retarder is not necessary.

Hip Roof

Condensation Zone I: Total net area of ventilation should be 1/300[b] with 1/600[b] distributed uniformly at the eaves *and* 1/600[b] located at the ridge with all spaces interconnected. A vapor retarder should also be used in the top story ceiling.

Condensation Zone II: Same as for Zone I.

Condensation Zone III: Same ventilation as for Zone II. A vapor retarder is not necessary.

Gable or Hip Roof—With Occupancy Contemplated

Condensation Zone I: Total net area of ventilation should be 1/300[b] with 1/600[b] distributed uniformly at the eaves and 1/600[b] located at the ridge with all spaces interconnected. A vapor retarder should also be used on the warm side of the top full story ceiling, the dwarf walls, the sloping part of the roof, and the attic story ceiling.

Condensation Zone II: Same as for Zone I.

Condensation Zone III: Same as for Zone I, except that a vapor retarder is not necessary if insulation is omitted.

Adapted from *Building Foundation Design Handbook* (Labs *et al.* 1988).
[a]In many areas, increased ventilation may be desirable for summer comfort. For winter comfort, insulation should be placed between a living space and a loft or attic ventilated at these rates.
[b]Refers to area enclosed within building lines at eave level.
[c]The zone numbers refer to Figure 11.

Table 2   Checklist for Condensation and Potential Mold Growth

Key
- ● Likely remedial measure
- ○ Possible remedial measure

Symptom columns:
- S1 — Dampness/mold growth in cold weather only
- S2 — Dampness/mold growth in rooms with large areas of exposed walls, roofs, or floors
- S3 — Dampness/mold growth in bedrooms
- S4 — Dampness/mold growth in or behind cupboards or in upper corners of rooms
- S5 — Dampness/mold growth at times of rain or snow, especially with recurrence in summer during periods of rain
- S6 — Moisture staining, tidemarks on ceilings, walls, etc.

Action by Householder / Remedial Measures

	Measure	S1	S2	S3	S4	S5	S6
Action by Householder	Mold Treatment	●	○	○	○		
	Reduce Use of LPG, etc.	○	○	○			
	Adequate Ventilation	●	○	○	●		
	Adequate Heat Input	●	●	●	○		
Remedial Measures	Mold Treatment	●	○	○	○		
Building	Alter Fittings				●		
	Double Glazing		○	○			
	Insulate Floor	○	●				
	Insulate Walls	○	●				
	Insulate Roof	○	●	○			
Ventilation	Cupboard Ventilators			○	●		
	Window Vent (Slot Type)			●			
	Exhaust Vent, Bathroom	○	○	○	○		
	Exhaust Vent, Kitchen	○	○	○	○		
Heating System	Supplement, Hall, Ldg.		●	○			
	Supplement, Bedroom		●	●			
	Supplement, Livingroom		●				
	Maintain	○	○				
						Repair, overhaul, replace as necessary	Repair, overhaul, replace as necessary

Possible Causes or Influences

Symptom	Heating System	Ventilation	Building	Householder
S1. Dampness/mold growth in cold weather only	Low or intermittent heat input leading to inadequate background heating	Heating and ventilation not matched; usually too little ventilation due to absence of natural paths of ventilation	Inappropriate construction for intermittent heating, i.e., slow response heavy weight	Heat input too low, particularly in bedrooms; intermittent heating; drying clothes indoors; use of LPG or wood
S2. Dampness/mold growth in rooms with large areas of exposed walls, roofs, or floors	Low or intermittent heat input leading to inadequate background heating	Heating and ventilation not matched	Exposed or projecting areas of building that are poorly insulated (including stairwells); downstairs bedrooms; large windows, esp. full-height windows	Heat input too low, particularly in bedrooms; uninsulated water pipes; intermittent heating; use of LPG or wood
S3. Dampness/mold growth in bedrooms	Unheated bedrooms; lack of background heat in dwelling generally	Poor ventilation not matched with heating; moist air entering from kitchens and bathrooms	Exposed or projecting bedrooms, poorly insulated, esp. north-facing unheated bedrooms downstairs; bedrooms over unheated spaces, e.g., walkways, garages	Heat input too low, particularly at night; leaving kitchen and bathroom doors open; use of LPG or wood
S4. Dampness/mold growth in or behind cupboards or in upper corners of rooms		Stagnant air pockets; lack of air movement behind furniture; lack of air movement in cupboards, etc.	Built-in cupboards or wardrobes on outside walls	Bad positioning of furniture; overfilling of cupboards; putting clothes away damp
S5. Dampness/mold growth at times of rain or snow, especially with recurrence in summer during periods of rain			Dampness from water penetration; building cracks; defective flashing or pointing; blocked gutters; cracked gutters/rainwater pipes; exposure to driving rain	
S6. Moisture staining, tidemarks on ceilings, walls, etc.	Radiator, boiler, pipe leaks		Dampness from plumbing leaks; cracked water pipes, tanks, etc.	

Reprinted from *Condensation and Mould Growth* (Housing Development Directorate 1979).

cover material should have a perm rating of no more than 1.0 and must be rugged enough to withstand foot and knee traffic. Recommended materials include 6-mil (0.006 in.) polyethylene, 45 or 55-lb smooth roll roofing (federal specification Class A, Type 1), and 45-mil (0.045-in.) thick EPDM membranes. Roll roofing should not be used for very wet soils, because fungi will decompose it over time. Any source of subfloor warmth (heating ducts, furnaces) is likely to seriously increase the evaporation from subfloor soil (Trethowen and Middlemass 1988, Trethowen 1988).

All debris must be removed and the soil leveled before laying the membrane. Edges need only be lapped 4 to 6 in., and no sealing is required. The membrane need not be carried up the face of the wall unless the interior grade is below the outside grade.

A rectangular crawl space requires a minimum of four vents, one on each wall, located no farther than 3 ft from each corner. The vents should be as high on the wall as possible to best capture breezes, and landscaping should be planned to prevent obstruction of the vents. The total free (open) area of all vents should be at least 1/1500 of the floor area. The gross area of various vents can be found by multiplying the 1/1500 area by the factors in Table 3. In the absence of a ground cover, the vent area should be increased to 1/150 of the floor area. Ventilation alone should not be relied on where soils are known to be moist.

The most common vent-related failures include: (1) providing 1/1500 in gross vent area instead of free area, (2) placing vents on only one pair of walls or away from the corners, and (3) obstructing the vents by ducts and plumbing, or by shrubbery, condenser units, porches, and patios. Another common problem is the failure of homeowners to open operable vents.

Adequate space under all beams, pipes, and ducts allows access to the crawl space, especially the perimeter. Adequate space also prevents ventilation being impeded. Codes and standard practice usually call for a minimum of 18 in. between the crawl space floor and the underside of joists, but 24 in. is better to leave space for ducts and plumbing.

CONDENSATION IN COOLED STRUCTURES

Refrigerators, cold pipes, and cold vessels require insulation and should have a vapor retarder placed on the warm side of the insulation or as close as possible to the warm surface of any enclosure. Insulation on a cold pipeline has no opportunity to dry out periodically. Also, vapor should not escape from the cold side of any metal (vaportight) pipe. Such applications require insulating materials that are highly resistant to water vapor, or a highly vapor-resistant jacket or coating whose permeance is not over 0.01 perm. Metal coverings are desirable but difficult to apply.

Similar considerations apply in cold rooms, whether constructed inside a heated building or as a separate building. Cold rooms operating above freezing experience some periods of vapor reversal in winter, but such drying is of little help. Refrigerators lined with cement or other vapor-permeable material allow slight

amounts of vapor to pass, which reduces the accumulation of moisture that penetrates the retarder. However, an adequate warm side retarder not greater than 0.01 perm should be installed.

Air can infiltrate through gaps in joints, tears in the retarder, or cracks where pipes or similar items penetrate the wall. Opening and closing cold room doors also causes a pumping of air, which can increase the infiltration of the moist air.

Summer cooling for comfort in temperate climates does not usually create serious vapor problems in exterior walls and ceilings. Outdoor dew point, especially peak values, may exceed the design dew-point temperatures in common use, but they seldom exceed 75 °F for any prolonged period. Condensation within exterior walls exposed to an indoor temperature of 75 °F is seldom as serious as winter condensation.

In a study of a wood-sided house in Athens, Georgia, Duff (1956) showed under summer cooling conditions that temperatures were lower outside than inside long enough to prevent moisture buildup from damaging the structure. This condition held true regardless of whether or not a vapor retarder was placed near the inner surface. Masonry or brick-veneered structures with an efficient vapor retarder near inner surfaces do have a moisture buildup under summer cooling conditions, however. Observed moisture contents in wood framing exceeds 25 to 30% by dry weight, which is conducive to wood decay and metal corrosion. Mei and Woolrich (1963) reported summer condensation in exterior walls and crawl spaces in some areas. For this reason, in areas of prolonged heavy cooling and shorter heating seasons, tight vapor retarders are discouraged.

A vapor retarder to prevent condensation within exterior walls under summer cooling conditions would normally be located on the outside of the insulation, whereas under winter conditions, it would be located on the inner side. Use of vapor retarders at both locations can restrict not only the entry of moisture into the insulation but the escape of any moisture as well. In dwelling construction, the vapor retarder should be placed to protect for the more serious case of winter condensation; the summer case should be disregarded, even when some condensation is expected. The corresponding situation in cold storage buildings, in which a more serious reversal of vapor flow conditions from winter to summer may occur, poses a wall design problem not yet fully resolved.

Serious wetting within walls can occur in summer under special conditions. The National Research Council of Canada tested the walls of huts of brick masonry, finished inside with furring, insulation, vapor retarder, and plasterboard interior. The walls were opened during a sunny period following rain. Extensive wetting was observed in the insulation, particularly on the back of the vapor retarder. The absorptive brick wall had taken on substantial quantities of water during the rainfall. Subsequent heating by the sun had driven the moisture as vapor into the wall, where it condensed and caused serious wetting. The construction had no protection in the form of parging or paper on the inside of the brick for the furring and insulation. Walls with absorptive exterior coverings capable of absorbing and storing considerable quantities of water during a rain, and providing little resistance to vapor flow into the insulation from outdoors, may experience serious interior wetting by condensation under these unusual conditions. No wetting occurred in a similar construction when a saturated sheathing paper was used between the insulation and the brick. Thus, a moderate vapor flow resistance, such as that provided by parging or a good sheathing paper on the outside of the insulation, can be effective in such cases.

Vapor that diffuses inward under cooling conditions adds to cooling loads. Vapor retarders installed for winter conditions help reduce this. However, if condensation occurs on the back of a vapor retarder installed on the inside of the insulation, the latent heat released may add to the heat gain through the wall.

Table 3 Crawl Space Vent Area Requirements

Obstructions in Vents	No Soil Cover[a]	With Soil Cover[b]
	Multiply Gross Vent Area by[c]	
1/4-in. mesh hardware cloth	10	1.0
1/8-in. mesh screen	12.5	1.25
16-mesh insect screen	20	2.0
Louvers + 1/4-in. hardware cloth	20	2.0
Louvers + 1/8-in. mesh screen	22.5	2.25
Louvers +16-mesh screen	30	3.0

[a] Vent area may be reduced by one-half if the crawl space floor soil is dusty dry.
[b] The indicated amount or four vents, whichever is greater.
[c] As determined by the 1/1500 formula.

AIR-CONDITIONED BUILDINGS IN HUMID CLIMATES

A number of moisture-related problems have been identified in humid and tropical climates (Moore *et al.* 1980). While the principles of moisture flow remain the same, climatic differences present different problems. A humid climate can be defined as one in which one or both of the following conditions occur:

1. A 67 °F or higher wet-bulb temperature for 3500 h or more during the warmest six consecutive months of the year.
2. A 73 °F or higher wet-bulb temperature for 1750 h or more during the warmest six consecutive months of the year.

Figures 12 and 13 show the locations of humid climates satisfying the definition in the United States and the world.

Depending on local experience with moisture problems, a humid-climate design criteria may also be desirable in locations that do not quite meet the foregoing conditions. Fringe areas include some locations in the southeastern United States, notably Alabama, Arkansas, Georgia, Louisiana, Mississippi, North Carolina, South Carolina, and east Texas (see Figure 12). Fringe locations are generally identified by one or both of the following conditions:

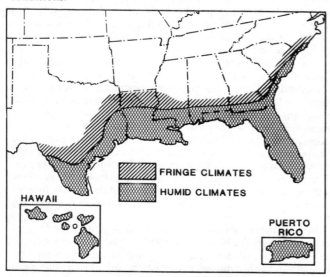

Fig. 12 Humid Climates in Continental United States

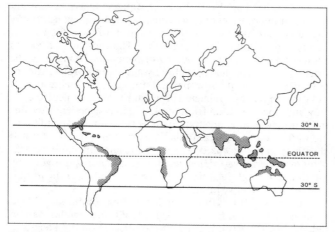

Fig. 13 Humid Climates Generally Occur in Latitudes 30 °S to 30 °N in Low Coastal Regions

1. A 67 °F or higher wet-bulb temperature for 3000 h or more during the warmest six consecutive months of the year.
2. A 73 °F or higher wet-bulb temperature for 1500 h or more during the warmest six consecutive months of the year.

Mold and Mildew

Air-conditioned buildings in humid climates tend to experience growth of mold and mildew which can cause decay of wood. A reliable air-conditioning system with good humidity control will decrease indoor humidity and provide adequate air circulation to reduce indoor mold and mildew.

Design Practices

In high humidity climates, the outdoor to indoor vapor pressure differential can be twice as much as in humidified buildings in cold climates. Thus, the moisture load caused by diffusion through the building envelope may need to be considered, even though the calculation is imprecise. This load is in addition to ventilation, infiltration, and internally generated moisture loads.

Since the latent cooling load on an air-conditioning system in high humidity climates frequently exceeds the sensible load, a system should be capable of handling the latent load without overcooling. Oversized air-conditioning units contribute to the problem of high humidity, since the latent capacity drops off faster than does the sensible capacity.

The most successful approach to solving this problem includes proper sizing of the air-conditioning system and use of reheat. An air-conditioning system capable of operation at a low sensible heat ratio and with multiple capacities may suffice in some cases. The air-conditioning systems commonly used in other climates, *i.e.*, those with limited dehumidification capability, especially under light sensible loads (such as room fan coil units handling the full load), should be avoided. Any air-conditioning system capable of providing comfort conditions within the range shown in the ASHRAE Comfort Standard Chart (Chapter 8), under all sensible loads, will prevent the growth of mold and mildew indoors. All-air systems with reheat generally provide satisfactory conditions.

Ventilating air should be continuously dehumidified since its vapor pressure is usually higher than that of indoor air. Similarly, infiltration should be minimized. Temperature controls for air-conditioning units, especially those handling outside air, should not allow variable temperature supply air unless humidity is also controlled. When the control system permits supply air temperature to increase with a consequent reduction in dehumidification, previously cooled surfaces (such as ceilings, diffusers, or registers) may be at temperatures below the dew point of the supply air, causing condensation.

Unconditioned or unventilated spaces in an air-conditioned building should be conditioned or connected by openings to conditioned spaces to prevent the possibility of mold growth.

In high humidity climates, moisture condenses on the exterior surface of a building because its temperature is frequently lower than the ambient dew point. In roof systems with high relative humidity under the roof, condensation may occur at night on the bottom surface of the roofing, wetting the insulation. This may be mistaken as a leaking roof, even though it may not have rained for days.

If condensation is not reevaporated by wind or solar heat gain, it can be reduced by (1) using higher thermal resistances in the building envelope to raise exterior surface temperatures and (2) avoiding shaded exterior surfaces that do not benefit from the evaporative effects of sun and wind, such as inside corners. Special attention should be focused on building components that have thermal bridges, such as exterior cantilevers, columns, foundations, or window and door frames. While these solutions may not totally eliminate mold and mildew growth, they substantially reduce the potential.

Water vapor diffuses through building materials from the outside into air-conditioned spaces in high humidity climates. Exterior surfaces should have higher vapor resistance than interior surfaces in such climates. Paints and finishes can provide the necessary vapor resistances, with the greater vapor resistance at the outside surface and lower vapor resistances toward the inside.

Where vapor retarders are used, they should be on the outside. Therefore, any water that does enter the construction can flow through to the inside of the building where it can be removed by the air-conditioning system instead of accumulating in the wall or roof construction. Note that this is the reverse of the recommended practice for cold climates.

Insulated cold piping (and sometimes ductwork) should receive special attention when exposed to ambient or nonconditioned air. Because cold piping frequently operates year-round, a constant vapor drive will exist under humid conditions. Even with vapor retarder insulation, jackets, and vapor sealing of joints and fittings, moisture inevitably accumulates in permeable insulations. This moisture not only reduces the thermal resistance of the insulation, it also accelerates condensation on the jacket surface with consequent dripping of water and growth of mold and mildew. Depending on local conditions, these problems can arise in less than 3 years, or they can take as long as 20 or 30 years. Since periodic insulation replacement is the only known solution, the piping installation should be accessible for such replacement and should have a means for draining water. As an alternative, very low permeance-type insulating materials (for example, materials not exceeding 0.1 perms) have been used to extend the life of the system and reduce replacement frequency. The lower the permeance of the insulation material, the longer its life.

Operating and Maintenance Practices

Since the potential for damage to a building and its contents is substantial in an air-conditioned building in humid climates, it is more important to properly operate and maintain the building and its air-conditioning system in humid climates than it is in others. The chilled water supply temperature and flow should be reliable, and multiple chillers and pumps should be considered to assure continuous uninterrupted dehumidification. Raising the chilled water supply temperature to conserve energy should not be attempted under these conditions.

INDUSTRIAL INSULATION PRACTICE

The applications of industrial insulation discussed in this section include: (1) pipes; (2) tanks, vessels, and equipment; (3) ducts; and (4) refrigerated rooms and buildings.

PIPES

Small pipes are insulated with cylindrical half-sections of insulation, with factory-applied jackets that form a hinge-and-lap, or with flexible, closed-cell material. Large pipes can be insulated with flexible material or with curved, flat segmented, or cylindrical half-, third-, or quarter-sections of rigid insulation, particularly where removal for frequent servicing of the pipe is necessary. Fittings (valves, tees, crosses, and elbows) are insulated with preformed fitting insulation, fabricated fitting insulation, individual pieces cut from sectional straight pipe insulation, or insulating cements. Fittings insulation should always be consistent with pipe insulation.

Securing Methods

The method of securement varies with the size of pipe, form, and weight of the insulation, and the type of jacketing, *i.e.*, separate

or factory-applied. Insulation with certain factory-applied jacketing can be secured on small piping by cementing the overlapping jacket. Large piping may require supplemental wiring or banding. Insulation on large piping requiring separate jacketing is wired or banded in place, and the jacket is cemented, wired, or banded, depending on the type. Insulation with factory-applied metal jacketing is secured by specific design of the jacket and its joint closure. The flexible closed-cell materials require no jacket for most applications, and they are applied using specially formulated contact adhesives.

Insulation Finish for Above-Ambient Temperatures

Pipe insulation finishes for indoor use are usually governed by location. If concealed, the simplest finish is a laminate of kraft paper, glass fiber reinforcing, and a vapor retarder. If exposed, the finishes are factory-applied jackets designed to meet fire safety requirements. For maximum fire safety, unusual exposure conditions, or appearance, factory or separately applied metal jackets may be used. Flexible closed-cell material usually is painted where exposed and left unfinished where hidden. An outdoor finish protects the insulation from the weather; chemical exposure, mechanical abuse, and appearance are additional considerations.

Insulation Finish for Below-Ambient Temperatures

Piping at temperatures below ambient is insulated to control heat gain and prevent condensation of moisture from the ambient air. Since piping is an absolute barrier to the passage of water vapor, the outer surface of the insulation must be covered by an impervious membrane or cover, which also helps protect the pipe against corrosion.

Retarder treatment should be recommended by the insulation manufacturer as established by performance testing. To protect the insulation from undue weather exposure, it must be as dry as possible.

Vapor-seal finishes for straight pipe insulation are generally designed to meet operating temperature, fire safety, and appearance requirements. Jacketing commonly consists of various combinations of laminates of paper, aluminum foil, plastic film, and glass fiber reinforcing. An important feature of such jacketing is very low permeance in a relatively thin layer, which provides flexibility for ease of cementing and sealing laps and end-joint strips. This type of jacketing is commonly used indoors without additional treatment. In some cases of subzero operating temperature, multilayer insulation and jacketing may be used. Flexible closed-cell materials must be carefully cemented to avoid openings in the insulations.

Insulation fittings are usually vapor-sealed by field applications of suitable materials and may vary with the type of insulation and operating temperature. For temperatures not below 10°F, the vapor seal can be a lapped spiral wrap of plastic film adhesive tape or a relatively thin coat of vapor-seal mastic. For temperatures below 10°F, common practice is two coats of vapor-seal mastic reinforced with open-weave glass or other fabric. The thickness of the mastic increases with decreasing temperature. With long lines of piping, the insulation should be sealed off every 15 or 20 ft to limit water penetration if vapor-seal damage occurs.

For dual-temperature service, where piping is alternately cold and hot, the vapor-seal finish, including mastics, must withstand pipe movement and exposure temperatures without deterioration. When flexible closed-cell insulation is used, it should be applied slightly compressed to prevent it from being strained when the piping expands.

Outdoor pipe insulation may be vapor-sealed in the same manner as indoor piping, by applying added weather protection jacketing without damage to the retarder and sealing it to keep out water. In some instances, heavy-duty weather and vapor-seal finish may be used. Table 4 shows the minimum spacing allowed to permit proper application and sealing of pipe insulation.

Table 4 Estimated Requirements to Prevent Freezing of Water in Pipes[a]

Nominal Pipe Size, in.	Time to Cool Water to Freezing, h[b]			Water Mass Flow Rate per Unit Length of Exposed Pipe to Prevent Freezing, lb/(h·ft)[b]		
	Insulation Thickness, in.[c]					
	2	3	4	2	3	4
0.5	0.27	0.32	0.36	0.53	0.44	0.40
1	0.61	0.75	0.85	0.70	0.57	0.48
1.5	1.16	1.46	1.69	0.91	0.69	0.59
2	1.66	2.12	2.48	1.04	0.77	0.66
3	2.83	3.71	4.43	1.50	1.03	0.87
4	4.06	5.42	6.54	1.91	1.27	1.00
5	5.45	7.36	8.96	2.41	1.56	1.21
6	6.86	9.36	11.50	2.97	1.89	1.35
8	9.58	13.30	16.50	3.96	2.47	1.73
10	12.50	17.60	22.00	5.80	3.14	2.18
12	15.60	22.00	27.80	8.75	3.91	2.66

[a]Surrounding air temperature = −18°F.
[b]Initial water temperature = 42°F.
[c]Insulation thermal conductivity = 0.30 Btu·in/h·ft²·°F.

Surface Temperature

In elevated temperature applications, the surface temperature of the insulation system should be below that at which personnel coming into contact with the surface could be harmed. In below-ambient temperature applications, the surface temperature of the system should be above dew point to prevent condensation.

A jacket with a reflective surface will have a higher surface temperature for hot applications than for cold temperature applications because the lower emissivity reduces the rate of heat exchange. Therefore, adding a reflective jacket could produce a surface temperature capable of burning personnel on hot applications and causing condensation on cold applications. The jacketing material used also contributes to the relative safety at a given surface temperature. For example, at 175°F, a stainless steel jacket blisters skin more severely than a canvas jacket does. Consult Chapter 22 for surface temperature calculation procedures and surface resistances at several emittance levels.

Insulating Pipes to Prevent Freezing

If the surrounding air temperature remains sufficiently low for an extended period, insulation cannot prevent freezing of still water or of water flowing at a rate insufficient for the available heat content to offset the heat loss. Insulation can only prolong the time required for water to freeze and prevent freezing if water flow is maintained at a sufficient rate; the first section can be used to estimate the thickness of insulation necessary to prevent freezing of still water in pipes; the second section gives the minimum flow of water at an initial temperature of 42°F to prevent the temperature of the pipe from reaching 32°F at the end of the exposed length.

To calculate time required, the following equation can be used:

$$H = B_f(t_i - t_f)/q_o (\pi r_2/6) \qquad (1)$$

with

$$q_o = k \frac{0.5(t_1 + t_f) - t_2}{r_2 \ln (r_2/r_1)} \qquad (2)$$

where

B_f = heat content of water, Btu/lb·°F
= $(\pi r_i^2/a)bc_p$
c_p = specific heat of water, Btu/lb·°F

a = 144 in²/ft²
b = 62.4 lb/ft²
θ = time for water to cool to freezing temperature, h
q_o = heat flow from outer surface of insulation, Btu/h·ft²
r_i = inside radius of pipe, in.
r_1 = outer radius of pipe or inner radius of insulation, in.
r_2 = outer radius of insulation, in.
k = thermal conductivity of insulation, Btu in/h·ft²·°F
t_1 = initial water temperature, °F
t_2 = temperature of outer surface of insulation, °F
t_f = freezing temperature, °F

Refer to Chapter 33 or manufacturers' catalogs and handbooks for pipe capacities.

To calculate water flow, the following equation can be used:

$$W/L = 1/[R_T \ln (O_i/O_o)] \qquad (3)$$

where

W = flow rate required to keep pipe free of ice, lb/h
L = length of exposed pipe, ft
$R_T = R_I + R_f$ = total resistance to heat flow between water and ambient air per foot of pipe
R_I = [ln (D_I/D_p)]/$(2\pi K_I)$ = resistance of thermal insulation per foot of pipe, ft·°F·h/Btu
R_f = $1/(\pi K_W Nu)$ = resistance between water and inner surface of pipe per foot of pipe = 0.23 ft·°F·h/Btu
O_i = $t_i - t_a$, difference between initial water temperature and ambient air temperature, °F
O_o = $(R_T/R_I)(t_f - t_a)$
t_a = ambient air temperature, °F
D_I = outside diameter of insulation, ft
D_p = outside diameter of pipe, ft
K_I = thermal conductivity of insulation, Btu/h·ft·°F
K_W = thermal conductivity of water near freezing = 0.32 Btu/h·ft·°F
Nu = Nusselt number, a constant value of 4.36 for fully developed laminar flow and constant heat flux

When unusual conditions make it impractical to maintain protection with insulation alone, a hot trace pipe or, preferably, electric resistance heating cable is required along the bottom or top of the water pipe. The heating system then supplies the heat lost through the insulation. The insulation thickness is determined by the cost of the heating system, the insulation, and the heat loss.

Underground Pipe Insulation

Both heated and cooled underground piping systems are insulated. Protecting underground insulated piping is more difficult than protecting aboveground piping. Groundwater conditions, including chemical or electrolytic contributions by the soil and the existence of a water pressure, require a special design to protect insulated pipes from corrosion. Walk-through tunnels, conduits, or integral protective coverings are generally provided to protect the pipe and insulation from water. Examples and general design features of conduits and a description of tunnels can be found in Chapter 11 of the 1992 ASHRAE *Handbook—Systems and Equipment*.

Temperatures above ambient. Piping for heated systems in walk-through tunnels is usually covered with sectional insulation and finished with effective mechanical protection such as metal or waterproofing jackets. The use of walk-through tunnels is declining because of cost.

Conduit systems are generally used for underground insulated piping systems. The most successful application is the use of sectional insulation with the conduit sized for drainage and adequate drying of insulation on heated piping in the event of accidental flooding. BRAB Technical Report 66 (1975) gives detailed design criteria for conduit systems. The criteria require that (1) all systems provide for draining and insulation drying, (2) the insulation withstand boiling and drying without physical damage and loss of insulating value, and (3) the conduit casing be watertight in the field. The insulation should be a nonconductor of electricity, vermin-proof, and chemically and dimensionally stable at the operating temperature of the pipe.

BRAB Technical Report 39 (1963) describes suitable laboratory tests for compliance with the criteria just discussed. BRAB Technical Report 47 (1964) describes evaluative tests and field investigations which have shown that calcium silicate is resistant to severe boiling action. Fibrous glass (4 to 7 lb/ft³ density) will not withstand boiling when a conduit becomes flooded, and wet poured-in-place insulations are likely to remain partially wet for their installed life.

The thickness of insulation for underground piping is not determined on the same basis as aboveground piping.

Temperatures below ambient. Integrally protected, insulated piping buried directly in the ground is commonly used for chilled water. However, since no heat is available to drive out moisture, an absolute protective covering against water and insulation with low permeance and water absorption is extremely important. Cellular glass with proper protection has been widely used for this type of application. The acceptance of plastic foams is increasing, but their long-term performance has not yet been established.

Conduit for chilled water piping requires a different approach than for hot piping. Insulation must have low conductivity, and conduit design must use this low conductivity and maintain continuing performance. More recent designs use low-conductivity plastic foam insulation with plastic pipe as the internal water-carrying piping and as the external conduit.

Where the temperature difference between the pipe at 40°F and the soil at 55 to 60°F is small, pipe size, length, and flow rate may economically justify direct burial without insulation. However, good piping protection may be required.

TANKS, VESSELS, AND EQUIPMENT

Flat, curved, and irregular surfaces, such as tanks, vessels, boilers, and chimney connectors, are normally insulated with flat blocks or bevelled lags, curved segments, blanket forms of insulation, or mineral fiber-inorganic binder insulations. On surfaces operating below 185°F, closed cellular insulations are used extensively. Since no general procedure can apply to all materials and conditions, manufacturers' specifications and instructions must be followed for specific applications.

Securing Methods

Insulations are secured in a variety of ways, depending on the form of insulation and contour of the surface to be insulated. On small-diameter, cylindrical vessels, insulation can be secured by banding around the circumference. On larger cylindrical vessels, banding can be supplemented by angle iron support ledges to support the insulation against slippage. Where diameters exceed 10 to 15 ft, slotted angle iron may be run lengthwise on the cylinder, at intervals around the circumference, as securement, and to avoid an excessive length of banding.

On large flat and cylindrical surfaces, banding or wiring can be supplemented by fastening to various welded studs at frequent intervals. On large flat, cylindrical, and spherical surfaces, it is often advantageous to secure the insulation by impaling it on welded studs or pins and fastening it with speed washers. Flexible closed-cell insulations are adhered directly to the surface to be insulated, using a suitable contact adhesive.

Insulation Finish for Above-Ambient Temperatures

For temperatures above ambient, insulation is finished to protect it against mechanical damage and weather, consistent with

acceptable appearance. On smaller equipment indoors, insulation is commonly covered with hexagonal wire mesh tightly stretched and secured. Then a base and hard finish coat of cement is applied. This can be additionally finished by painting.

For the same equipment outdoors, the insulation can be finished with a coat of hard cement, hexagonal mesh properly secured, and a coat of weather-resistant mastic (preferably a breathing type). A variation is to apply only two coats of weather-resistant mastic reinforced with open-mesh glass or other fabric; but this finish is limited to about 300°F operating temperature, as metal expansion can rupture the finish at insulation joints. Larger equipment may be finished indoors and out with suitable sheet metal. Outdoor finish of any type must be properly flashed around penetrations (*e.g.*, manhole access openings, pipe connections, and structural supports) to maintain a weatherproof condition.

Insulation Finish for Below-Ambient Temperatures

For temperatures below ambient, insulation is finished, as required, to prevent condensation and protect against mechanical damage and weather, consistent with acceptable appearance. In accordance with the operating temperature, the finish must retard vapor to avoid moisture entry from surrounding air, which can cause an increase in thermal conductivity of the insulation, deterioration of the insulation, or corrosion of the metal equipment surface. For moderately low temperatures, insulation can be finished with hexagonal mesh properly secured, a coat of hard finish cement, and several coats of suitable paint. Where a greater degree of vapor retarding is required, two coats of mastic reinforced with open-mesh glass or other fabric can be used. For an even greater degree of vapor retarding, two coats of asphalt vapor-retarder mastic reinforced with asphalt-saturated and perforated felt with edges lapped and cemented can be used (see Table 9, Chapter 22). If this type of vapor retarding is used indoors, standards for acceptable appearance and protection from mechanical damage may require the addition of hexagonal mesh, hard finish cement, and finish painting.

For the same equipment outdoors, the insulation can be finished with heavier or additional coats of vapor-retarder mastic reinforced with open-mesh glass or other fabric. For appearance, mastics can be painted, but since they usually contain solvents, they must be properly dried before painting. Paints must be the appropriate type recommended by paint manufacturers. In some cases, the additional finish over the mastic may be suitable sheet metal, applied carefully so as not to damage the vapor retarder (see Table 9, Chapter 22 for material properties).

Whenever vapor retarder finish is applied, all penetrations such as manholes, pipe connections, and structural supports must be properly vapor retarded with mastic or other sealant reinforced with open-mesh glass or other fabric. Equipment must be insulated from structural steel by isolating supports of high compressive strength and reasonably low thermal conductivity, such as hardwood block. The vapor retarding must carry over this insulation from the equipment to the supporting steel to assure proper sealing. If the equipment rests directly on steel supports, the supports must be insulated for some distance from the points of contact. Commonly, insulation and vapor retarder are extended 8 to 10 times the thickness of the insulation applied to the equipment.

For dual-temperature service, where vessels are alternately cold and hot, the materials and a design of vapor retarder finish selected must withstand movement caused by temperature change.

Reflective insulation is commonly applied to large refrigerated tanks. In this system, the external finishing sheets of heavy-gage metal form the vapor retarder, and the joints are sealed to permit contraction and expansion. Valved outlets at the bottom of the tank allow drainage.

DUCTS

The need for duct insulation is influenced by (1) duct location, whether indoors or outdoors; (2) the effect of heat loss or gain on equipment size and operating cost; (3) the need to prevent condensation on low-temperature ducts; (4) the need to control temperature change in long duct lengths; and (5) the need to control noise with interior duct lining.

All ducts exposed to outdoor conditions, as well as those passing through unconditioned spaces, should be insulated. While analyses of temperature change, heat loss or gain, and other factors affecting the economics of thermal insulation are seldom made for residential installations, they are essential for large commercial and industrial projects.

The U-value for uninsulated sheet metal ducts is affected by air velocity, the emittance of the metal, and the shape of the duct. An approximate value of 1.0 Btu/h·ft²·°F may be used. For insulated ducts, U-factors of 0.24 and 0.13 Btu/h·ft²·°F represent 1- and 2-in. thick rigid insulation with a thermal conductivity of 0.27 Btu·in/h·ft²·°F at 75°F mean temperature. A method for determining heat loss or gain for ducts is given in Chapter 32.

Materials for Ducts, Insulations, and Liners

Ducts within buildings can be of insulated sheet metal or fibrous glass, both of which provide combined air barrier, thermal insulation, and sound absorption. Ducts embedding in or below floor slabs may be of compressed fiber, ceramic tile, or other rigid materials.

Duct insulations include semirigid boards and flexible blanket types, composed of organic and inorganic materials in fibrous, cellular, or bonded particle forms. Insulations for exterior surfaces may have attached vapor barriers or facings, or vapor barriers may be applied separately. When applied to the duct interior as a liner, insulation both thermally insulates and absorbs sound. Duct liner insulations have sound-permeable coatings or other treatment on the side facing the airstream to withstand air velocities without deterioration.

Fibrous glass ducts are available in preformed round ducts or in board form for fabrication of rectangular ductwork. Round and rectangular ducts have a minimum density of about 3 lb/ft³, with a thermal conductivity of 0.23 Btu in/h·ft²·°F at 75°F mean temperature. Round ducts with a diameter of up to about 30 in. and boards in various thicknesses and sizes are used to form required sizes of rectangular ducts. Maximum recommended velocity is 2000 fpm, although tests at velocities exceeding 10,000 fpm show no fiber erosion. Primary use is for low-pressure systems tested at 1.5 times the recommended static pressure. Maximum recommended air temperature is 250°F. A complete system provides greater decibel attenuation than is usually provided by standard duct liners, with greater reduction in airborne equipment noise and crosstalk. Higher design velocities are also possible.

To satisfy most building codes, the various duct insulations and fibrous glass duct materials must meet the fire hazard requirements of the National Fire Protection Association's *Standard* 90A (NFPA 1989), to restrict spread of smoke, heat, and fire through duct systems, and to minimize ignition sources; and *Standard* 90B (NFPA 1989), on supply ducts, controls, clearances, heating panels, return ducts, air filters, and heat pumps. Local code authorities should also be consulted.

Where thermally insulated air-conditioning ducts pass through unconditioned spaces, such as attics, the maximum allowable heat flux should be no greater than that required by NFPA *Standard* 90A.

Securing Methods

Exterior duct insulation can be attached with adhesive, with supplemental preattached pins and clips, or with wiring or band-ing. Liners can be attached with adhesive and supplemental pins and clips.

Manufacturers provide information on the construction of fibrous glass duct systems. Preformed round duct for straight runs is combined with fittings fabricated from straight duct. Rectangular ducts and fittings are fabricated by grooving, folding, and taping, with metal accessories such as turning vanes, splitters, and dampers incorporated into the system. When rectangular ducts exceed predetermined dimensions for particular static pressures, ductwork must be reinforced. The Sheet Metal and Air Conditioning Contractors National Association's (SMACNA) *Fibrous Glass Duct Construction Standards* has further information.

Heating Ducts

The effect of duct insulation on residential heating system equipment size can be marginal. However, insulation can reduce operating costs significantly, depending on unit costs for heating and the extent of duct exposed to outside conditions. In addition, duct insulation maintains the supply air temperature, which may keep the entering air within a more comfortable range when it enters the conditioned space.

Vapor retarders are not required on exterior insulation of ducts used for heating alone, but they must be provided for ducts used for alternate heating and cooling.

Cooling Ducts

Insulation can reduce operating costs and cooling equipment size significantly. The advantage of adequate insulation is especially significant in areas with high dry-bulb and dew-point temperatures. Ducts for summer air conditioning are insulated with the same materials as heating ducts. Ducts in any unconditioned space should be insulated and have vapor retarders to prevent condensation. Joints and laps in the vapor retarder must be sealed. Flexible closed-cell insulation does not always need a supplemental vapor retarder, but care must be taken to form vapor-tight seams at joints.

REFRIGERATED ROOMS AND BUILDINGS

Cold storage facilities require the most attention to thermal insulation and air/vapor retarders, since they greatly influence costs. Many refrigerated spaces operate at subzero temperatures (as much as 100°F below outside temperature), which increases the severity of the service condition imposed on the insulation system. The primary source of moisture is airborne water vapor; however, soil moisture must also be considered because whenever moisture freezes (either in the insulation system or in soil under slabs-on-ground), both operating and maintenance costs increase. Control of moisture entry into the insulation is of primary importance in limiting water and ice collection in the structure. The location and integrity of the vapor retarder system is crucial and places a premium on good workmanship.

Interior Finish

The permeance of the interior finish, *e.g.*, paint, should be greater than that of the vapor retarder so that the drywall, etc., can dry out to the direction of the finish and away from the vapor retarder.

The panels in structural panel systems fill in between structural members and, therefore, become the building surfaces. Where panels are used as a building lining, no special surface preparation is needed, assuming the surfaces are sound and reasonably smooth. Sheet metal and other specialty boards (when installed to permit passage of water vapor, if necessary) have been successfully used for walls and ceilings of rooms operating over a wide temperature range.

Table 5 Recommended R-Values for Refrigerated Facilities

Facility	Temperature Range, °F	Recommended R-Values, °F · ft² · h/Btu[a]		
		Floors	Walls/ Suspended Ceilings	Roofs
Cooler	40 to 50	Perimeter insulation only	17 to 30	25 to 35
Chill cooler[b]	28 to 35	15 to 20	25 to 35	30 to 40
Holding freezer	−10 to −20	30 to 35	33 to 50	40 to 50
Blast freezer[c]	−40	35 to 40	50 to 60	50 to 60 (Suspended ceilings)

[a]The above R-values were derived using the economic thickness methods previously described in this chapter. Due to the wide range in the cost of energy and insulation materials on a thermal performance basis, a recommended R-value is a guide. For more exact values, consult a designer, insulation supplier, or both.
[b]If a cooler has the possibility of being converted to a freezer, the owner should consider insulating the facility with the higher R-values from the freezer section.
[c]R-values shown are for a blast freezer built within an unconditioned space. If built within a cooler or freezer, consult a designer and/or insulation supplier.

Insulated structural panels, with metal exteriors and metal or reinforced plastic interior faces, are commonly used for cooler and freezer rooms. These finishes keep moisture out of the insulation, so that only the joints between panels must be vapor-sealed to avoid moisture penetration.

In selecting an interior finish, the following factors should be considered: (1) fire resistance, (2) washdown requirements, (3) mechanical damage, (4) moisture and gas permeance, and (5) applicable codes in food-handling installations. All interior walls of insulated spaces should be protected by bumper guards and curbs whenever there is possibility of damage to the finish.

Thermal Resistance

Generally, the R-values in Table 5 may be used for the different types of facilities noted. The range in R-values results from variation in cost of energy, insulation materials, and climatic conditions. For more exact values, consult a designer and/or insulation supplier. R-values less than those shown should not be used.

LAND TRANSPORT VEHICLES

Insulation installed in transport vehicles should have the same qualities desirable for most other applications, *i.e.*, moderate cost, low density, low thermal conductivity, low moisture permeability and retention, ease of application, uniformity, resistance to breakdown at temperature extremes, and fire resistance. The insulation should resist cracking, crumbling, shifting, and pack-down from shock, vibration, and flexing of the body structure that occurs in transport.

State laws and other factors limit the outside width, height, and length of vehicles. Therefore, the designer must establish an insulation thickness to obtain the optimum cargo space and operating performance. For example, increasing insulation thickness from 3 to 4 in. in a 40-ft long highway trailer decreases cargo space by about 100 ft³, or 5%. A decrease in insulation thickness, assuming other factors remain equal, requires an increase in refrigerating or heating unit capacity. In determining the optimum thickness of insulation, the vehicle body and its refrigeration or heating unit must be considered as one system.

Urethane foam insulations are used increasingly for transport vehicles, although glass fiber, polystyrene, and other materials continue to be used. Urethane can be formed in place between the inner and outer skin of the vehicle, or it can be preformed outside the vehicle.

The total amount of heat transferred from outside to inside a vehicle, excluding heat that enters when the vehicle door is open for loading or unloading of cargo, is composed of (1) heat transferred through insulation, (2) heat transferred through structural members, and (3) heat caused by air and moisture leakage into the vehicle and insulated space.

Since air leakage can significantly affect the overall heat transfer rate of the vehicle body, some buyers specify a maximum rate of air leakage that the body can have with a given air pressure imposed in the cargo space. Further, the buyer often specifies the maximum overall heat transfer rate for the vehicle, usually at conditions of 100°F and 50% rh outside and 0°F inside air temperature. Chapters 28 and 29 of the 1990 ASHRAE *Handbook—Refrigeration* have further information on refrigerated land transport vehicles.

ENVIRONMENTAL SPACES

Atmospheric environmental spaces provide environments of selected temperature, humidity, pressure, and composition. These may be desired singly or in combination, in stabilized degree or in phased fluctuating degree. The designed environment may be required for short- or long-term tests, for continued or periodic production or manufacture, or for short- or long-term storage. Design requirements and specifications for vapor retarder construction, thermal insulation construction, and protective finish of insulation construction should include performance requirements in terms of (1) permissible water vapor permeance; (2) thermal transfer; and (3) physical and structural requirements for protective finish based on environmental conditions, range of temperature exposure, and permeance to water vapor or other gases to which it may be exposed or for which it may be a controlling boundary.

In a tightly designed system, pressure changes caused by temperature change alone can be sufficient to rupture the insulation system. The effect of vacuum during a pulldown can also tear a barrier fastening.

The manufacture of electronic equipment; pharmaceutical and biological chemicals; textiles, paper, and tobaccos; and packing facilities for such items as critical electrical cable, requires refrigeration equipment to maintain the proper environment. The permissible water vapor permeance rates through walls, ceilings, and floors of the manufacturing space must be considered in the design load calculations for humidification equipment.

Controlled atmosphere storages for fruit and other perishable commodities are held at refrigerated temperatures and control the percentage of oxygen, carbon dioxide, and other gases in the air mixture within the space. Permeance of these gases through the internal gas barrier, which is normally also the protective finish of the thermal insulation, as well as the water vapor permeance through the external water vapor retarder, which should not exceed that of the internal gas barrier, should be specified.

General Principles

The insulation principles for refrigerated rooms or buildings given in this chapter apply only to steady-state operation. Environmental facilities used for alternate refrigerated and high temperatures are subjected to reversal of vapor flow and are usually designed with vapor retarders of equal permeance on both sides of the thermal insulation.

In altitude chambers, heavy welded steel shells form an external structure that resists the pressure differences between external atmospheric and internal vacuum conditions.

Environmental conditions at normal ambient temperatures and above, with injected high humidity, present problems of operation similar to those described for tanks and vessels. Interior protective finish on thermal insulation is usually designed with the lowest possible water vapor permeance through the protective material and its sealed joints, or with some provision for draining condensate that can be expected in the thermal insulation construction.

Altitude chambers with prestressed shells are generally designed with the insulation on the inside of the shell. This protects the shell from extremes of temperature and eliminates the effect of the shell's thermal mass on the refrigeration load during a rapid pulldown of temperature.

The insulation design must also consider the heat transfer through sleeves, pipes, conduits, door jambs, observation windows, and other openings, but a general guide to insulation thickness for low temperatures is:

$$x = 0.15k (t_o - t_i) \qquad (4)$$

where

x = insulation thickness, in.
k = conductivity of insulation, Btu·in/h·ft^2·°F
t_o = design ambient temperature, °F
t_i = chamber temperature, °F

The use of Equation (4) results in an outside surface temperature of less than 5°F below ambient.

When the temperature-controlled space involves both heating and refrigeration, the thermal resistance of the structure should be designed to satisfy the most severe need. Thermal insulation should be selected to perform safely at all the temperatures encountered. The effect of thermal mass of insulation and protective finish and the temperature change rate should also be considered.

The thermal insulation construction must prevent the insulation material from (1) settling with ensuing noninsulated voids, (2) deteriorating through thermal shock, and (3) absorbing or retaining flammable liquids or vapors (if present in the environment of the space).

General Installation Technique

The four types of insulation described in this chapter (rigid, loose-fill, flexible, and reflective) can be used in environmental space insulation construction. Requirements of vapor control, temperature range, thermal mass, and effect on temperature pulldown rate are set by the specification of environmental conditions.

Reflective insulations in the form of aluminum alloy sheets with emittances of 0.05 to 0.03, spaced to form the requisite number of dimensioned air spaces, are used extensively in large altitude chambers, low-temperature test rooms, and large all-weather rooms. The inherent low thermal mass of this insulation construction is advantageous for rapid temperature pulldown. Evacuation of air in spaces between reflective insulation sheets must be simultaneous with evacuation of air from the environmental space, either by vent openings through the protective liner of the thermal insulation, or by bleed lines exhausting the thermal areas to the main evacuation line. Bleed lines permit the use of a completely sealed protective liner between the insulation space and the environmental space, with no pressure differential created against the liner. Drainage of the spaces between reflective insulation sheets is also provided. Resistance to fire is afforded by the limited supply of oxygen in the spaces between the reflective insulation sheets.

Wood separator strips between reflective insulation sheets can be employed from the low range of temperature up to 200°F. For temperature ranges of 250 to 700°F, aluminum rods may be used as separators. For temperatures ranging from 700°F to the limits of stainless steel, a separator system of stainless steel rods and reflective insulation sheets of stainless steel may be used. The

protective liner can be heavy-gage aluminum or stainless steel, depending on environmental conditions and temperatures.

Rigid cellular glass insulation can be used in combination with reflective insulation. Use cellular glass in the outer layer of thermal insulation where it does not affect rapid temperature pulldown; use reflective insulation in the inner layer where low thermal mass permits rapid temperature pulldown.

Cellular-glass rigid insulation can be used through a range of high temperatures. It is applied with adhesive or mechanical fasteners and joint sealers designed for the temperatures encountered. The protective finish must be suitable for the same temperatures. The high load-bearing value of cellular-glass permits support of floor constructions. Because the material has low conductivity, it is used to insulate piping in reflective insulation construction.

Mineral fiber insulation in flexible form, often used in reachin and walk-in altitude chambers and low-temperature test facilities, is manufactured in plants and shipped assembled or in sections.

Organic rigid, foamed plastic rigid, semirigid, flexible, and loose-fill insulation, employed in the temperature range permitted by their physical characteristics, must be designed to preclude contraction cracks, settlement, and condensed water vapor collection.

REFERENCES

Baker, M.C. 1980. *Roofs*, Chapters 6 and 8. Multiscience Publications Ltd. Montreal, Canada.

BRAB. 1963. Evaluation of components for underground heat distribution systems. Building Research Advisory Board, Federal Construction Council Technical Report 39, National Academy of Sciences, National Research Council, Publication No. 828. Washington, D.C.

BRAB. 1964. Field evaluation of underground heat distribution systems. Building Research Advisory Board, Federal Construction Council Technical Report 47, National Academy of Sciences, National Research Council, Washington, D.C.

BRAB. 1975. Criteria for underground distribution. Technical Report 66. Standing Committee on Mechanical Engineering of the FCC, BRAB, and NRC.

BRANZ. 1983. Insulating suspended timber frame floors. BRANZ Bulletins 177 and 233. Building Research Association, New Zealand.

CIBSE. 1986. CIBSE *Guide—Volume* A (*Design Data*). Section A3. Chartered Institution of Building Services Engineers, London.

Condensation and Mould Growth. 1979. Domestic Energy Notes 4. Housing Development Directorate, 1 Lambeth Palace Road, London, England SEI 7ER.

Courville, G.E. and A. Tenwolde. 1984. Moisture measurements in buildings. Proceedings of the Workshop on Moisture Control in Buildings. Building Thermal Envelope Coordinating Council of The National Institute of Building Science, Washington, D.C. (September).

Dickens, H.B. and N.B. Hutcheon. 1965. Moisture accumulation in roof spaces under extreme winter conditions. Paper presented at RILEM/CIB symposium, Moisture Problems in Buildings, Helsinki, Finland. NRCC 9132.

Donnelly, R.G., V.J. Tennery, D.L. McElroy, T.G. Godfrey, and J.O. Kolb. 1976. Industrial thermal insulation, An assessment. Oak Ridge National Laboratory Report TM-5283 and TM-5515, TID-27120.

Duff, J.E. 1956. The effect of air conditioning of water vapor permeability on temperature and humidity. ASHRAE *Transactions* 62:437.

Engineering Weather Data. 1978. Departments of Air Force, Army, and Navy, AFM 88-29, TM5-785, NAFAC P-89. Available from Superintendent of Documents, U.S. Government Printing Office, Washington, D.C. 20402, as Stock No. 008-070-00420-8.

Handegord, G.O. 1979. The need for improved air tightness in buildings. Building Research Note 151. National Research Council of Canada, Ottawa.

Hedlin, C.P. 1977. Moisture gains by foam plastic roof insulations under controlled temperature gradients. *Journal of Cellular Plastics* 13(5). NRCC 16317.

Hedlin, C.P. 1974. The use of nuclear moisture density meters to measure moisture in flat roofing insulations. National Research Council of Canada. NRCC 14593.

Hutcheon, N.B. 1963. Requirements for exterior walls. *Canadian Building Digest* 48. National Research Council of Canada.

International Energy Agency. 1990. Vol. 2, Guidelines and practice. *Annex XIV, Condensation and Energy* (August).

Labs, K., J. Carmody, R. Sterling, L. Shen, Y.J. Huang, and D. Parker. 1988. *Building foundation design handbook*. Oak Ridge National Laboratory Report ORNL/sub/86-72143/1.

Latta, J.K. 1976. Vapor barriers: What are they? Are they effective? *Canadian Building Digest* 175.

Link, R.E., Jr. 1977. Airborne thermal infrared and nuclear meter systems for detecting roof moisture. Proceedings of the NBS-NRCA symposium, Roofing Technology.

Mei, H.T. and W.R. Woolrich. 1963. "Condensation problem solutions in the insulation of building in hot climates" and "Soil covers protect basementless houses from wood decay." In *Humidity and Moisture, Measurement and Control of Science and Industry* 2:334, 340. Reinhold Publishing Corp., New York.

Modern Plastics Encyclopedia. 1962. Plastics Catalog Corporation, New York.

Moore, R.J., L.G. Spielvogel, and C.W. Griffin. 1980. Air conditioned buildings in humid climates: Guidelines for design, operation, and maintenance. NTIS AD-A090145. Southern Division, Naval Facilities Engineering Command, Charleston, SC.

NBS. 1951. *District heating handbook*, 3rd ed. National Bureau of Standards, Washington, D.C., 193.

Neilsen, R.A. 1940. Dirt patterns on walls. ASHVE *Transactions* 46:247.

NFPA. 1989. Standard for the installation of air conditioning and ventilating systems. *Standard* 90A-89, National Fire Protection Association, Quincy, MA.

NFPA. 1989. Standard for the installation of warm air heating and air conditioning systems. *Standard* 90B-89. National Fire Protection Association, Quincy, MA.

Proceedings of the Symposium, Roofing Technology. 1977. NBSNRCA.

Romanoff, M. *Corrosion*. 1959. National Bureau of Standards, Washington, D.C.

Tamura, G.T. and A.G. Wilson. 1964. Air leakage and pressure measurements in two occupied houses. ASHRAE *Transactions* 70:110.

Tamura, G.T. 1975. Measurement of air leakage characteristics of house enclosures. ASHRAE *Transactions* 81:202.

Tamura, G.T., G.H. Kuester, and G.O. Handegord. 1974. Condensation problems in flat wood frame roofs. Paper presented at the Second International Symposium, Moisture Problems in Buildings. NRCC 14589. Rotterdam, The Netherlands.

Tobiasson, W., C. Korhonen, and A. van den Berg. 1977. Hand held infrared systems for detecting roof moisture. Proceedings of the NBS-NRCA symposium, Roofing Technology.

Trethowen, H.A. 1988. A survey of subfloor ground evaporation rates. Building Research Association of New Zealand, BRANZ Study Report SR 13.

Trethowen, H.A. and G. Middlemass. 1988. A survey of moisture damage in southern New Zealand buildings. Building Research Association of New Zealand, BRANZ Study Report SR 7.

Verschoor, J.D. 1977. Effectiveness of building insulation applications. USN/CEL Report No. CR78.006-NTIS No. AD-A053 452/9ST.

Verschoor, J.D. 1985. Measurement of water vapor migration and storage in composite building construction. ASHRAE *Transactions* 91(2A):390-403.

Wilson, A.G. and G.K. Garden. 1965. Moisture accumulation in walls due to air leakage. NRCC 9131. Paper presented at RILEM/CIB symposium, Moisture Problems in Buildings, Helsinki, Finland.

Wilson, A.G. 1960. Condensation on inside window surfaces. *Canadian Building Digest* 4. National Research Council of Canada, Ottawa.

THERMAL AND WATER VAPOR TRANSMISSION DATA

THIS chapter presents thermal and water vapor transmission data based on steady-state or equilibrium conditions. Chapter 3 covers heat transfer under transient or changing temperature conditions. Chapter 20 discusses selection of insulation materials and procedures for determining overall thermal resistances by simplified methods.

BUILDING ENVELOPES

Thermal Transmission Data for Building Components

The steady-state thermal resistances (R-values) of building components (walls, floors, windows, roof systems, etc.) can be calculated from the thermal properties of the materials in the component; or the heat flow through the assembled component can be measured directly with laboratory equipment such as the guarded hot box (ASTM *Standard* C 236) or the calibrated hot box (ASTM *Standard* C 976).

Tables 1 through 6 list thermal values, which may be used to calculate thermal resistances of building walls, floors, and ceilings. The values shown in these tables were developed under ideal conditions. In practice, overall thermal performance can be reduced significantly by such factors as improper installation and shrink-

age, settling, or compression of the insulation (Tye and Desjarlais 1983, Tye 1985, 1986).

Most values in these tables were obtained by accepted ASTM test methods described in ASTM *Standards* C 177 and C 518 for materials and ASTM *Standards* C 236 and C 976 for building envelope components. Because commercially available materials vary, not all values apply to specific products. (Previous editions of the handbook can be consulted for data on materials no longer commercially available.)

The most accurate method of determining the overall thermal resistance for a combination of building materials assembled as a building envelope component is to test a representative sample by a hot box method. However, all combinations may not be conveniently or economically tested in this manner. For many simple constructions, calculated R-values agree reasonably well with values determined by hot box measurement.

The performance of materials fabricated in the field is especially subject to the quality of workmanship during construction and installation. Good workmanship becomes increasingly important as the insulation requirement becomes greater. Therefore, some engineers include additional insulation or other safety factors based on experience in their design.

Figure 1 shows how convection affects surface conductance of several materials. Other tests on smooth surfaces show that the average value of the convection part of conductance decreases as the length of the surface increases.

Table 1   Surface Conductances and Resistances for Air

Position of Surface	Direction of Heat Flow	Non-reflective $\epsilon = 0.90$		Reflective $\epsilon = 0.20$		$\epsilon = 0.05$	
		h_i	R	h_i	R	h_i	R
STILL AIR							
Horizontal	Upward	1.63	0.61	0.91	1.10	0.76	1.32
Sloping—45°	Upward	1.60	0.62	0.88	1.14	0.73	1.37
Vertical	Horizontal	1.46	0.68	0.74	1.35	0.59	1.70
Sloping—45°	Downward	1.32	0.76	0.60	1.67	0.45	2.22
Horizontal	Downward	1.08	0.92	0.37	2.70	0.22	4.55
MOVING AIR (Any position)		h_o	R	h_o	R	h_o	R
15-mph Wind (for winter)	Any	6.00	0.17	—	—	—	—
7.5-mph Wind (for summer)	Any	4.00	0.25	—	—	—	—

Notes:
1. Surface conductance h_i and h_o measured in Btu/h·ft²·°F; resistance R in °F·ft²·h/Btu.
2. No surface has both an air space resistance value and a surface resistance value.
3. For ventilated attics or spaces above ceilings under summer conditions (heat flow down), see Table 5.
4. Conductances are for surfaces of the stated emittance facing virtual blackbody surroundings at the same temperature as the ambient air. Values are based on a surface-air temperature difference of 10°F and for surface temperatures of 70°F.
5. See Chapter 3 for more detailed information, especially Tables 5 and 6, and see Figure 1 for additional data.
6. Condensate can have a significant impact on surface emittance (see Table 3).

The preparation of this chapter is assigned to TC 4.4, Thermal Insulation and Moisture Retarders.

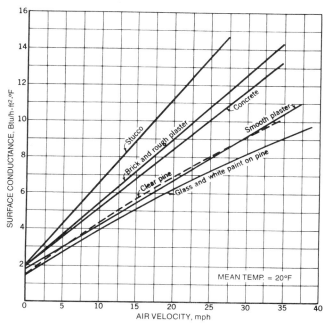

Fig. 1   Surface Conductance for Different 12-Inch-Square Surfaces as Affected by Air Movement

Table 2 Thermal Resistances of Plane Air Spaces[a,b,c], °F·ft²·h/Btu

Position of Air Space	Direction of Heat Flow	Air Space Mean Temp.[d], °F	Temp. Diff.[d], °F	0.5-in. Air Space[c] Effective Emittance ϵ_{eff}[d,e]					0.75-in. Air Space[c] Effective Emittance ϵ_{eff}[d,e]				
				0.03	0.05	0.2	0.5	0.82	0.03	0.05	0.2	0.5	0.82
Horiz. Up ↑		90	10	2.13	2.03	1.51	0.99	0.73	2.34	2.22	1.61	1.04	0.75
		50	30	1.62	1.57	1.29	0.96	0.75	1.71	1.66	1.35	0.99	0.77
		50	10	2.13	2.05	1.60	1.11	0.84	2.30	2.21	1.70	1.16	0.87
		0	20	1.73	1.70	1.45	1.12	0.91	1.83	1.79	1.52	1.16	0.93
		0	10	2.10	2.04	1.70	1.27	1.00	2.23	2.16	1.78	1.31	1.02
		−50	20	1.69	1.66	1.49	1.23	1.04	1.77	1.74	1.55	1.27	1.07
		−50	10	2.04	2.00	1.75	1.40	1.16	2.16	2.11	1.84	1.46	1.20
45° Slope Up ↗		90	10	2.44	2.31	1.65	1.06	0.76	2.96	2.78	1.88	1.15	0.81
		50	30	2.06	1.98	1.56	1.10	0.83	1.99	1.92	1.52	1.08	0.82
		50	10	2.55	2.44	1.83	1.22	0.90	2.90	2.75	2.00	1.29	0.94
		0	20	2.20	2.14	1.76	1.30	1.02	2.13	2.07	1.72	1.28	1.00
		0	10	2.63	2.54	2.03	1.44	1.10	2.72	2.62	2.08	1.47	1.12
		−50	20	2.08	2.04	1.78	1.42	1.17	2.05	2.01	1.76	1.41	1.16
		−50	10	2.62	2.56	2.17	1.66	1.33	2.53	2.47	2.10	1.62	1.30
Vertical Horiz. →		90	10	2.47	2.34	1.67	1.06	0.77	3.50	3.24	2.08	1.22	0.84
		50	30	2.57	2.46	1.84	1.23	0.90	2.91	2.77	2.01	1.30	0.94
		50	10	2.66	2.54	1.88	1.24	0.91	3.70	3.46	2.35	1.43	1.01
		0	20	2.82	2.72	2.14	1.50	1.13	3.14	3.02	2.32	1.58	1.18
		0	10	2.93	2.82	2.20	1.53	1.15	3.77	3.59	2.64	1.73	1.26
		−50	20	2.90	2.82	2.35	1.76	1.39	2.90	2.83	2.36	1.77	1.39
		−50	10	3.20	3.10	2.54	1.87	1.46	3.72	3.60	2.87	2.04	1.56
45° Slope Down ↘		90	10	2.48	2.34	1.67	1.06	0.77	3.53	3.27	2.10	1.22	0.84
		50	30	2.64	2.52	1.87	1.24	0.91	3.43	3.23	2.24	1.39	0.99
		50	10	2.67	2.55	1.89	1.25	0.92	3.81	3.57	2.40	1.45	1.02
		0	20	2.91	2.80	2.19	1.52	1.15	3.75	3.57	2.63	1.72	1.26
		0	10	2.94	2.83	2.21	1.53	1.15	4.12	3.91	2.81	1.80	1.30
		−50	20	3.16	3.07	2.52	1.86	1.45	3.78	3.65	2.90	2.05	1.57
		−50	10	3.26	3.16	2.58	1.89	1.47	4.35	4.18	3.22	2.21	1.66
Horiz. Down ↓		90	10	2.48	2.34	1.67	1.06	0.77	3.55	3.29	2.10	1.22	0.85
		50	30	2.66	2.54	1.88	1.24	0.91	3.77	3.52	2.38	1.44	1.02
		50	10	2.67	2.55	1.89	1.25	0.92	3.84	3.59	2.41	1.45	1.02
		0	20	2.94	2.83	2.20	1.53	1.15	4.18	3.96	2.83	1.81	1.30
		0	10	2.96	2.85	2.22	1.53	1.16	4.25	4.02	2.87	1.82	1.31
		−50	20	3.25	3.15	2.58	1.89	1.47	4.60	4.41	3.36	2.28	1.69
		−50	10	3.28	3.18	2.60	1.90	1.47	4.71	4.51	3.42	2.30	1.71

Position of Air Space	Direction of Heat Flow	Air Space Mean Temp., °F	Temp. Diff., °F	1.5-in. Air Space[c]					3.5-in. Air Space[c]				
Horiz. Up ↑		90	10	2.55	2.41	1.71	1.08	0.77	2.84	2.66	1.83	1.13	0.80
		50	30	1.87	1.81	1.45	1.04	0.80	2.09	2.01	1.58	1.10	0.84
		50	10	2.50	2.40	1.81	1.21	0.89	2.80	2.66	1.95	1.28	0.93
		0	20	2.01	1.95	1.63	1.23	0.97	2.25	2.18	1.79	1.32	1.03
		0	10	2.43	2.35	1.90	1.38	1.06	2.71	2.62	2.07	1.47	1.12
		−50	20	1.94	1.91	1.68	1.36	1.13	2.19	2.14	1.86	1.47	1.20
		−50	10	2.37	2.31	1.99	1.55	1.26	2.65	2.58	2.18	1.67	1.33
45° Slope Up ↗		90	10	2.92	2.73	1.86	1.14	0.80	3.18	2.96	1.97	1.18	0.82
		50	30	2.14	2.06	1.61	1.12	0.84	2.26	2.17	1.67	1.15	0.86
		50	10	2.88	2.74	1.99	1.29	0.94	3.12	2.95	2.10	1.34	0.96
		0	20	2.30	2.23	1.82	1.34	1.04	2.42	2.35	1.90	1.38	1.06
		0	10	2.79	2.69	2.12	1.49	1.13	2.98	2.87	2.23	1.54	1.16
		−50	20	2.22	2.17	1.88	1.49	1.21	2.34	2.29	1.97	1.54	1.25
		−50	10	2.71	2.64	2.23	1.69	1.35	2.87	2.79	2.33	1.75	1.39
Vertical Horiz. →		90	10	3.99	3.66	2.25	1.27	0.87	3.69	3.40	2.15	1.24	0.85
		50	30	2.58	2.46	1.84	1.23	0.90	2.67	2.55	1.89	1.25	0.91
		50	10	3.79	3.55	2.39	1.45	1.02	3.63	3.40	2.32	1.42	1.01
		0	20	2.76	2.66	2.10	1.48	1.12	2.88	2.78	2.17	1.51	1.14
		0	10	3.51	3.35	2.51	1.67	1.23	3.49	3.33	2.50	1.67	1.23
		−50	20	2.64	2.58	2.18	1.66	1.33	2.82	2.75	2.30	1.73	1.37
		−50	10	3.31	3.21	2.62	1.91	1.48	3.40	3.30	2.67	1.94	1.50
45° Slope Down ↘		90	10	5.07	4.55	2.56	1.36	0.91	4.81	4.33	2.49	1.34	0.90
		50	30	3.58	3.36	2.31	1.42	1.00	3.51	3.30	2.28	1.40	1.00
		50	10	5.10	4.66	2.85	1.60	1.09	4.74	4.36	2.73	1.57	1.08
		0	20	3.85	3.66	2.68	1.74	1.27	3.81	3.63	2.66	1.74	1.27
		0	10	4.92	4.62	3.16	1.94	1.37	4.59	4.32	3.02	1.88	1.34
		−50	20	3.62	3.50	2.80	2.01	1.54	3.77	3.64	2.90	2.05	1.57
		−50	10	4.67	4.47	3.40	2.29	1.70	4.50	4.32	3.31	2.25	1.68
Horiz. Down ↓		90	10	6.09	5.35	2.79	1.43	0.94	10.07	8.19	3.41	1.57	1.00
		50	30	6.27	5.63	3.18	1.70	1.14	9.60	8.17	3.86	1.88	1.22
		50	10	6.61	5.90	3.27	1.73	1.15	11.15	9.27	4.09	1.93	1.24
		0	20	7.03	6.43	3.91	2.19	1.49	10.90	9.52	4.87	2.47	1.62
		0	10	7.31	6.66	4.00	2.22	1.51	11.97	10.32	5.08	2.52	1.64
		−50	20	7.73	7.20	4.77	2.85	1.99	11.64	10.49	6.02	3.25	2.18
		−50	10	8.09	7.52	4.91	2.89	2.01	12.98	11.56	6.36	3.34	2.22

[a]See Chapter 20, section Factors Affecting Heat Transfer across Air Spaces. Thermal resistance values were determined from the relation, $R = 1/C$, where $C = h_c + \epsilon_{eff}h_r$, h_c is the conduction-convection coefficient, $\epsilon_{eff}h_r$ is the radiation coefficient $\cong 0.00686 \epsilon_{eff}[(t_m + 460)/100]^3$, and t_m is the mean temperature of the air space. Values for h_c were determined from data developed by Robinson et al. (1954). Equations (5) through (7) in Yarbrough (1983) show the data in Table 2 in analytic form. For extrapolation from Table 2 to air spaces less than 0.5 in. (as in insulating window glass), assume $h_c = 0.159(1 + 0.0016 t_m)/l$ where l is the air space thickness in inches, and h_c is heat transfer through the air space only.

[b]Values are based on data presented by Robinson et al. (1954). (Also see Chapter 3, Tables 3 and 4, and Chapter 39). Values apply for ideal conditions, i.e., air spaces of uniform thickness bounded by plane, smooth, parallel surfaces with no air leakage to or from the space. When accurate values are required, use overall U-factors determined through calibrated hot box (ASTM C 976) or guarded hot box (ASTM C 236) testing. Thermal resistance values for multiple air spaces must be based on careful estimates of mean temperature differences for each air space.

[c]A single resistance value cannot account for multiple air spaces; each air space requires a separate resistance calculation that applies only for the established boundary conditions. Resistances of horizontal spaces with heat flow downward are substantially independent of temperature difference.

[d]Interpolation is permissible for other values of mean temperature, temperature difference, and effective emittance ϵ_{eff}. Interpolation and moderate extrapolation for air spaces greater than 3.5 in. are also permissible.

[e]Effective emittance ϵ_{eff} of the air space is given by $1/\epsilon_{eff} = 1/\epsilon_1 + 1/\epsilon_2 - 1$, where ϵ_1 and ϵ_2 are the emittances of the surfaces of the air space (see Table 3).

Vapor retarders, outlined in Chapters 20 and 21, require special attention. Moisture from condensation or other sources may reduce the thermal resistance of insulation, but the effect of moisture must be determined for each material. For example, some materials with large air spaces are not affected significantly if the moisture content is less than 10% by weight, while the effect of moisture on other materials is approximately linear.

Ideal conditions of components and installations are assumed in calculating overall R-values (*i.e.*, insulating materials are of uniform nominal thickness and thermal resistance, air spaces are of uniform thickness and surface temperature, moisture effects are not involved, and installation details are in accordance with design). The National Bureau of Standards' Building Materials and Structures Report BMS 151 shows that measured values differ from calculated values for certain insulated constructions. For this reason, some engineers decrease the calculated R-values a moderate amount to account for departures of constructions from requirements and practices.

Tables 2 and 3 give values for well-sealed systems constructed with care. Field applications can differ substantially from laboratory test conditions. Air gaps in these insulation systems can seriously degrade thermal performance as a result of air movement due to both natural and forced convection. Sabine *et al.* (1975) found that the tabular values are not necessarily additive for multiple-layer, low-emittance air spaces, and tests on actual constructions should be conducted to accurately determine thermal resistance values.

Values for foil insulation products supplied by manufacturers must also be used with caution because they apply only to systems that are identical to the configuration in which the product was tested. In addition, surface oxidation, dust accumulation, condensation, and other factors that change the condition of the low-emittance surface can reduce the thermal effectiveness of these insulation systems (Moroz 1951, Hooper and Moroz 1952). Deterioration results from contact with several types of solutions, either acidic or basic (*e.g.*, wet cement mortar or the preservatives found in decay-resistant lumber). Polluted environments may cause rapid and severe material degradation. However, site inspections show a predominance of well-preserved installations and only a small number of cases in which rapid and severe deterioration has occurred. An extensive review of the reflective building insulation system performance literature is provided by Goss and Miller (1989).

Table 3   Emittance Values of Various Surfaces and Effective Emittances of Air Spaces[a]

Surface	Average Emittance ϵ	Effective Emittance ϵ_{eff} of Air Space	
		One Surface Emittance ϵ; Other, 0.9	Both Surfaces Emittance ϵ
Aluminum foil, bright	0.05	0.05	0.03
Aluminum foil, with condensate just visible (> 0.7gr/ft²)	0.30[b]	0.29	—
Aluminum foil, with condensate clearly visible (> 2.9 gr/ft²)	0.70[b]	0.65	—
Aluminum sheet	0.12	0.12	0.06
Aluminum coated paper, polished	0.20	0.20	0.11
Steel, galvanized, bright	0.25	0.24	0.15
Aluminum paint	0.50	0.47	0.35
Building materials: wood, paper, masonry, nonmetallic paints	0.90	0.82	0.82
Regular glass	0.84	0.77	0.72

[a]These values apply in the 4 to 40 μm range of the electromagnetic spectrum.

[b]Values are based on data presented by Bassett and Trethowen (1984).

CALCULATING OVERALL THERMAL RESISTANCES

Relatively small conductive elements within an insulating layer or thermal bridges can substantially reduce the average thermal resistance of a component. Examples include wood and metal studs in frame walls, concrete webs in concrete masonry walls, and metal ties or other elements in insulated wall panels. The following examples illustrate how to calculate R-values and U-factors for components containing thermal bridges.

The following conditions are assumed in calculating the design R-values:

- Equilibrium or steady-state heat transfer, disregarding effects of heat storage
- Surrounding surfaces at ambient air temperature
- Exterior wind velocity of 15 mph for winter (surface with $R = 0.17°F \cdot ft^2 \cdot h/Btu$) and 7.5 mph for summer (surface with $R = 0.25°F \cdot ft^2 \cdot h/Btu$)
- Surface emittance of ordinary building materials is 0.90

Wood Frame Walls

The average overall R-values and U-factors of wood frame walls can be calculated by assuming either parallel heat flow paths through areas with different thermal resistances or isothermal planes. Equations (1) through (5) from Chapter 20 are used.

For stud walls 16 in. on center (OC), the fraction of insulated cavity is about 0.75; the fraction of studs, plates, and sills is 0.21; and the fraction of headers is 0.04. For studs 24 in. OC, the respective values are 0.78, 0.18, and 0.04. These fractions contain an allowance for multiple studs, plates, sills, extra framing around windows, headers, and band joists.

Example 1A. Calculate the U-factor of the 2 by 4 stud wall shown in Figure 2. The studs are at 16 in. OC. There is 3.5-in. mineral fiber batt insulation (R-13) in the stud space. The inside finish is 0.5-in. gypsum wallboard; the outside is finished with rigid foam insulating sheathing (R-4) and 0.5-in. by 8-in. wood bevel lapped siding. The insulated cavity occupies approxi-

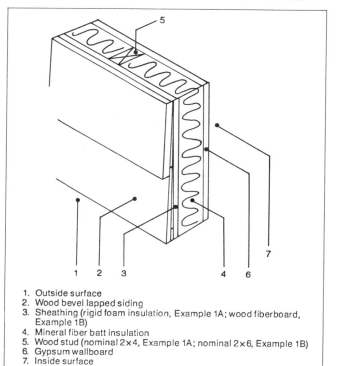

1. Outside surface
2. Wood bevel lapped siding
3. Sheathing (rigid foam insulation, Example 1A; wood fiberboard, Example 1B)
4. Mineral fiber batt insulation
5. Wood stud (nominal 2×4, Example 1A; nominal 2×6, Example 1B)
6. Gypsum wallboard
7. Inside surface

Fig. 2   Insulated Wood Frame Wall (Examples 1A and B)

mately 75% of the transmission area; the studs, plates, and sills occupy 21%; and the headers occupy 4%.

Solution: Obtain the R-values of the various building elements from Tables 1 and 4. Assume the R-value of the wood framing is R-1.25 per inch. Also, assume the headers are solid wood, in this case, and group them with the studs, plates, and sills.

Element	R (Insulated Cavity)	R (Studs, Plates, and Headers)
1. Outside surface, 15 mph wind	0.17	0.17
2. Wood bevel lapped siding	0.81	0.81
3. Rigid foam insulating sheathing	4.0	4.0
4. Mineral fiber batt insulation, 3.5 in.	13.0	—
5. Wood stud, nominal 2 × 4	—	4.38
6. Gypsum wallboard, 0.5 in.	0.45	0.45
7. Inside surface, still air	0.68	0.68
	$R_1 = 19.11$	$R_2 = 10.49$

Since the U-factor is the reciprocal of R-value, $U_1 = 0.052$ and $U_2 = 0.095$ Btu/·h·ft^2·°F.

If the wood framing (thermal bridging) is not included, Equation (3) from Chapter 20 may be used to calculate the U-factor of the wall as follows:

$$U_{av} = U_1 = 1/R_1 = 0.052 \text{ Btu/h·ft}^2 \cdot \text{°F}$$

If the wood framing is accounted for using the parallel flow method, the U-factor of the wall is determined using Equation (5) from Chapter 20 as follows:

$$U_{av} = (0.75 \times 0.052) + (0.25 \times 0.095) = 0.063 \text{ Btu/h·ft}^2 \cdot \text{°F}$$

If the wood framing is included using the isothermal planes method, the U-factor of the wall is determined using Equations (2) and (3) from Chapter 20 as follows:

$$R_{T(av)} = 4.98 + 1/[(0.75/13.0) + (0.25/4.38)] + 1.13$$
$$= 14.82 \text{°F·ft}^2 \cdot \text{h/Btu}$$
$$U_{av} = 0.067 \text{ Btu/h·ft}^2 \cdot \text{°F}$$

For a frame wall with a 24-in. OC stud space, the average overall R-value becomes 15.18°F·ft^2·h/Btu. Similar calculation procedures can be used to evaluate other wall designs.

Example 1B. Calculate the U-factor of a 2 by 6 stud wall, similar to the one considered in Example IA, except that the sheathing is 0.5-in. wood fiberboard and the studs are at 24 in. OC. There is 5.5-in. mineral fiber batt insulation (R-21) in the stud space. Assume the headers are double 2 by 8 framing (with a 0.5-in. air space), with a 2.0-in. air space between the headers and the wallboard.

Solution: Obtain the R-values of the various building elements from Tables 1 and 4. Assume the R-value of the wood framing is 1.25 per inch. In this case, the headers must be treated separately.

Element	R (Insulated Cavity)	R (Studs and Plates)	R (Headers)
1. Outside surface, 15 mph wind	0.17	0.17	0.17
2. Wood bevel lapped siding	0.81	0.81	0.81
3. Wood fiberboard sheathing, 0.5 in.	1.32	1.32	1.32
4. Mineral fiber batt insulation, 5.5 in.	21.0	—	—
5. Wood stud, nominal 2 × 6	—	6.88	—
6. Wood headers, double 2 × 8	—	—	3.75
7. Air space, 0.5 in.	—	—	0.90
8. Air space, 2 in.	—	—	0.90
9. Gypsum wallboard, 0.5 in.	0.45	0.45	0.45
10. Inside surface, still air	0.68	0.68	0.68
	$R_1 = 24.43$	$R_2 = 10.31$	$R_3 = 8.98$

Since U-factor is the reciprocal of R-value, $U_1 = 0.041$, $U_2 = 0.097$, and $U_3 = 0.111$ Btu/h·ft^2·°F.

If the wood framing is accounted for using the parallel flow method, the U-factor of the wall is determined using Equation (5) from Chapter 20 as follows:

$$U_{av} = (0.78 \times 0.041) + (0.18 \times 0.097) + (0.04 \times 0.111)$$
$$= 0.054 \text{ Btu·h·ft}^2 \cdot \text{°F}$$

If the wood framing is included using the isothermal planes method, the U-factor of the wall is determined using Equations (2) and (3) from Chapter 20 as follows:

$$R_{T(av)} = 2.30 + 1/[(0.78/21.0) + (0.18/6.88) + (0.04/5.55)] + 1.13$$
$$= 17.61 \text{°F·ft}^2 \cdot \text{h/Btu}$$
$$U_{av} = 0.057 \text{ Btu/h·ft}^2 \cdot \text{°F}$$

If the headers are insulated with R-10 insulation, the average overall R-value becomes 18.57°F·ft^2·h/Btu.

For a frame wall with a 16-in. OC stud space and uninsulated headers, the average overall R-value becomes 17.05°F·ft^2·h/Btu. If the headers are insulated with R-10 insulation, the average overall R-value becomes 17.93°F·ft^2·h/Btu. Similar calculation procedures can be used to evaluate other wall designs.

Masonry Walls

The average overall R-values of masonry walls can be estimated by assuming a combination of layers in series, one or more of which provides parallel paths. This method is used because heat flows laterally through block face shells so that transverse isothermal planes result. Average total resistance $R_{T(av)}$ is the sum of the resistances of the layers between such planes, each layer calculated as shown in Example 2.

Example 2. Calculate the overall thermal resistance and average U-factor of the 7-5/8-in. thick insulated concrete block wall shown in Figure 3. The two-core block has an average web thickness of 1-in. and a face shell thickness of 1-1/4-in. Overall block dimensions are 7-5/8 by 7-5/8 by 15-5/8 in. Measured thermal resistances of 112 lb/ft^3 concrete and 7 lb/ft^3 expanded perlite insulation are 0.10 and 2.90°F·ft^2·h/Btu per inch, respectively.

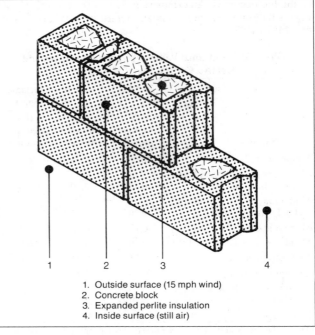

1. Outside surface (15 mph wind)
2. Concrete block
3. Expanded perlite insulation
4. Inside surface (still air)

Fig. 3   Insulated Concrete Block Wall (Example 2)

Solution: The equation used to determine the overall thermal resistance of the insulated concrete block wall is derived from Equations (2) and (5) from Chapter 20 and is given below:

$$R_{T(av)} = R_i + R_f + \left(\frac{a_w}{R_w} + \frac{a_c}{R_c}\right)^{-1} + R_o$$

where

$R_{T(av)}$ = overall thermal resistance based on assumption of isothermal planes
R_i = thermal resistance of inside air surface film (still air)
R_o = thermal resistance of outside air surface film (15 mph wind)
R_f = total thermal resistance of face shells
R_c = thermal resistance of cores between face shells
R_w = thermal resistance of webs between face shells
a_w = fraction of total area transverse to heat flow represented by webs of blocks
a_c = fraction of total area transverse to heat flow represented by cores of blocks

From the information given and the data in Table 1, determine the values needed to compute the overall thermal resistance.

R_i = 0.68
R_o = 0.17
R_f = (2)(1.25)(0.10) = 0.25
R_c = (5.125)(2.90) = 14.86
R_w = (5.125)(0.10) = 0.51
a_w = 3/15.625 = 0.192
a_c = 12.625/15.625 = 0.808

Using the equation given, the overall thermal resistance and average U-factor are calculated as follows:

$$\begin{aligned}
R_{T(av)} &= 0.68 + 0.25 + (0.51)(14.86)/[(0.808)(0.51) \\
&\quad + (0.192)\,(14.86)] + 0.17 \\
&= 0.68 + 0.25 + 2.33 + 0.17 = 3.43\,°F \cdot ft^2 \cdot h/Btu \\
U_{av} &= 1/3.43 = 0.29\ Btu/h \cdot ft^2 \cdot °F
\end{aligned}$$

Based on guarded hot box tests of this wall without mortar joints, Tye and Spinney (1980) measured the average R-value for this insulated concrete block wall as 3.13 °F·ft²·h/Btu.

Assuming parallel heat flow only, the calculated resistance is usually higher than that calculated on the assumption of isothermal planes. The actual resistance generally is some value between the two calculated values. In the absence of test values, examination of the construction usually reveals whether a value closer to the higher or lower calculated R-value should be used. Generally, if the construction contains a layer in which lateral conduction is high compared with transmittance through the construction, the calculation with isothermal planes should be used. If the construction has no layer of high lateral conductance, the parallel heat flow calculation should be used.

Hot box tests of insulated and uninsulated masonry walls constructed with block of conventional configuration show that thermal resistances calculated using the isothermal planes heat flow method agree well with measured values (Van Geem 1985, Valore 1980, Shu *et al.* 1979). Neglecting horizontal mortar joints in conventional block can result in thermal transmittance values up to 16% lower than actual, depending on the density and thermal properties of the masonry, and 1 to 6% lower, depending on the core insulation material (Van Geem 1985, McIntyre 1984). For aerated concrete block walls, other solid masonry, and multicore block walls with full mortar joints, neglecting mortar joints can cause errors in R-values up to 40% (Valore 1988). Horizontal mortar joints usually found in concrete block wall construction are neglected in Example 2.

Panels Containing Metal

Curtain wall constructions often include metallic and other thermal bridges. The thermal resistance of panels can be signifi-cantly reduced by metallic thermal bridges. However, the capacity of the adjacent facing materials to transmit heat transversely to the metal is limited, and some contact resistance between all materials in contact limits the reduction. Contact resistances in building structures are only 0.06 to 0.6 °F·ft²·h/Btu—too small to be of concern in many cases. However, the contact resistances of steel framing members are important. Also, in many cases (as illustrated in Example 3), the area of metal in contact with the facing greatly exceeds the thickness of the metal which mitigates the influence.

Thermal characteristics for panels of sandwich construction can be computed by combining the thermal resistances of the various layers. However, few panels are true sandwich constructions; many have ribs and stiffeners that create complicated heat flow paths. R-values for the assembled sections should be determined on a representative sample by using a hot box method. If the sample is a wall section with air cavities on both sides of fibrous insulation, the sample must be of representative height since convective airflow can contribute significantly to heat flow through the test section. Computer modeling can also be useful, but all heat transfer mechanisms must be considered.

In Example 3, the metal member is only 0.020 in. thick, but it is in contact with adjacent facings over a 1.25 in.-wide area. The steel member is 3.50 in. deep, has a thermal resistance of approximately 0.011°F·ft²·h/Btu, and is virtually isothermal. The calculation involves careful selection of the appropriate thickness for the steel member. If the member is assumed to be 0.020 in. thick, the fact that the flange transmits heat to the adjacent facing is ignored, and the heat flow through the steel is underestimated. If the member is assumed to be 1.25 in. thick, the heat flow through the steel is overestimated. In Example 3, the steel member behaves in much the same way as a rectangular member 1.25 in. thick and 3.50 in. deep with a thermal resistance of 0.69°F·ft²·h/Btu [(1.25/0.020) × 0.011] does. The Building Research Association of New Zealand (BRANZ) commonly uses this approximation.

Example 3. Calculate the C-factor of the insulated steel frame wall shown in Figure 4. Assume that the steel member has an R-value of 0.69 °F·ft²·h/Btu and that the framing behaves as though it occupies approximately 8% of the transmission area.

Solution: Obtain the R-values of the various building elements from Table 4.

Element	R (Insul.)	R (Framing)
1. 0.5-in. gypsum wallboard	0.45	0.45
2. 3.5-in. mineral fiber batt insulation	11	—
3. Steel framing member	—	0.69
4. 0.5-in. gypsum wallboard	0.45	0.45
	R_1 = 11.90	R_2 = 1.59

Therefore, C_1 = 0.084; C_2 = 0.629 Btu/h·ft² °F.

If the steel framing (thermal bridging) is not considered, the C-factor of the wall is calculated using Equation (3) from Chapter 20 as follows:

$$C_{av} = C_1 = 1/R_1 = 0.084\ Btu/h \cdot ft^2 \cdot °F$$

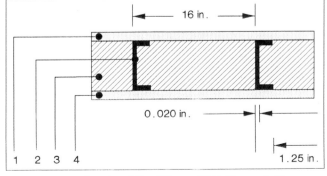

Fig. 4   Insulated Steel Frame Wall (Example 3)

Table 4   Typical Thermal Properties of Common Building and Insulating Materials—Design Values[a]

Description	Density, lb/ft³	Conductivity[b] (k), Btu·in / h·ft²·°F	Conductance (C), Btu / h·ft²·°F	Resistance[c] (R) Per Inch Thickness (1/k), °F·ft²·h / Btu·in	Resistance[c] (R) For Thickness Listed (1/C), °F·ft²·h / Btu	Specific Heat, Btu / lb·°F
BUILDING BOARD						
Asbestos-cement board	120	4.0	—	0.25	—	0.24
Asbestos-cement board ....0.125 in.	120	—	33.00	—	0.03	
Asbestos-cement board ....0.25 in.	120	—	16.50	—	0.06	
Gypsum or plaster board...........0.375 in.	50	—	3.10	—	0.32	0.26
Gypsum or plaster board...........0.5 in.	50	—	2.22	—	0.45	
Gypsum or plaster board...........0.625 in.	50	—	1.78	—	0.56	
Plywood (Douglas Fir)[d]	34	0.80	—	1.25	—	0.29
Plywood (Douglas Fir) ....0.25 in.	34	—	3.20	—	0.31	
Plywood (Douglas Fir) ....0.375 in.	34	—	2.13	—	0.47	
Plywood (Douglas Fir) ....0.5 in.	34	—	1.60	—	0.62	
Plywood (Douglas Fir) ....0.625 in.	34	—	1.29	—	0.77	
Plywood or wood panels ....0.75 in.	34	—	1.07	—	0.93	0.29
Vegetable fiber board						
Sheathing, regular density[e] ....0.5 in.	18	—	0.76	—	1.32	0.31
..............0.78125 in.	18	—	0.49	—	2.06	
Sheathing intermediate density[e] ......0.5 in.	22	—	0.92	—	1.09	0.31
Nail-base sheathing[e] ....0.5 in.	25	—	0.94	—	1.06	0.31
Shingle backer ....0.375 in.	18	—	1.06	—	0.94	0.31
Shingle backer ....0.3125 in.	18	—	1.28	—	0.78	
Sound deadening board ....0.5 in.	15	—	0.74	—	1.35	0.30
Tile and lay-in panels, plain or acoustic	18	0.40	—	2.50	—	0.14
...... 0.5 in.	18	—	0.80	—	1.25	
...... 0.75 in.	18	—	0.53	—	1.89	
Laminated paperboard	30	0.50	—	2.00	—	0.33
Homogeneous board from repulped paper	30	0.50	—	2.00	—	0.28
Hardboard[e]						
Medium density	50	0.73	—	1.37	—	0.31
High density, service-tempered grade and service						
grade	55	0.82	—	1.22	—	0.32
High density, standard-tempered grade	63	1.00	—	1.00	—	0.32
Particleboard[e]						
Low density	37	0.71	—	1.41	—	0.31
Medium density	50	0.94	—	1.06	—	0.31
High density	62.5	1.18	—	0.85	—	0.31
Underlayment ....0.625 in.	40	—	1.22	—	0.82	0.29
Waferboard	37	0.63	—	1.59	—	—
Wood subfloor ....0.75 in.	—	—	1.06	—	0.94	0.33
BUILDING MEMBRANE						
Vapor—permeable felt	—	—	16.70	—	0.06	
Vapor—seal, 2 layers of mopped 15-lb felt	—	—	8.35	—	0.12	
Vapor—seal, plastic film	—	—	—	—	Negl.	
FINISH FLOORING MATERIALS						
Carpet and fibrous pad	—	—	0.48	—	2.08	0.34
Carpet and rubber pad	—	—	0.81	—	1.23	0.33
Cork tile ....0.125 in.	—	—	3.60	—	0.28	0.48
Terrazzo ....1 in.	—	—	12.50	—	0.08	0.19
Tile—asphalt, linoleum, vinyl, rubber	—	—	20.00	—	0.05	0.30
vinyl asbestos						0.24
ceramic						0.19
Wood, hardwood finish ....0.75 in.	—	—	1.47	—	0.68	
INSULATING MATERIALS						
Blanket and Batt[f,g]						
Mineral fiber, fibrous form processed						
from rock, slag, or glass						
approx. 3–4 in.	0.4–2.0	—	0.091	—	11	
approx. 3.5 in.	0.4–2.0	—	0.077	—	13	
approx. 3.5 in.	1.2–1.6	—	0.067	—	15	
approx. 5.5–6.5 in.	0.4–2.0	—	0.053	—	19	
approx. 5.5 in.	0.6–1.0	—	0.048	—	21	
approx. 6–7.5 in.	0.4–2.0	—	0.045	—	22	
approx. 8.25–10 in.	0.4–2.0	—	0.033	—	30	
approx. 10–13 in.	0.4–2.0	—	0.026	—	38	
Board and Slabs						
Cellular glass	8.0	0.33	—	3.03	—	0.18
Glass fiber, organic bonded	4.0–9.0	0.25	—	4.00	—	0.23
Expanded perlite, organic bonded	1.0	0.36	—	2.78	—	0.30
Expanded rubber (rigid)	4.5	0.22	—	4.55	—	0.40
Expanded polystyrene, extruded						
(smooth skin surface) (CFC-12 exp.)	1.8–3.5	0.20	—	5.00	—	0.29
Expanded polystyrene, extruded (smooth skin surface)						
(HCFC-142b exp.)[h]	1.8–3.5	0.20	—	5.00	—	0.29

Table 4 Typical Thermal Properties of Common Building and Insulating Materials—Design Values[a] (Continued)

Description	Density, lb/ft³	Conductivity[b] (k), $\frac{Btu \cdot in}{h \cdot ft^2 \cdot °F}$	Conductance (C), $\frac{Btu}{h \cdot ft^2 \cdot °F}$	Resistance[c] (R) Per Inch Thickness (1/k), $\frac{°F \cdot ft^2 \cdot h}{Btu \cdot in}$	For Thickness Listed (1/C), $\frac{°F \cdot ft^2 \cdot h}{Btu}$	Specific Heat, $\frac{Btu}{lb \cdot °F}$
Expanded polystyrene, molded beads	1.0	0.26	—	3.85	—	—
	1.25	0.25	—	4.00	—	—
	1.5	0.24	—	4.17	—	—
	1.75	0.24	—	4.17	—	—
	2.0	0.23	—	4.35	—	—
Cellular polyurethane/polyisocyanurate[i] (CFC-11 exp.) (unfaced)	1.5	0.16–0.18	—	6.25–5.56	—	0.38
Cellular polyisocyanurate[i] (CFC-11 exp.)(gas-permeable facers)	1.5–2.5	0.16–0.18	—	6.25–5.56	—	0.22
Cellular polyisocyanurate[j] (CFC-11 exp.)(gas-impermeable facers)	2.0	0.14	—	7.04	—	0.22
Cellular phenolic (closed cell)(CFC-11, CFC-113 exp.)	3.0	0.12	—	8.20	—	—
Cellular phenolic (open cell)	1.8–2.2	0.23	—	4.40	—	—
Mineral fiber with resin binder	15.0	0.29	—	3.45	—	0.17
Mineral fiberboard, wet felted						
Core or roof insulation	16–17	0.34	—	2.94	—	—
Acoustical tile	18.0	0.35	—	2.86	—	0.19
Acoustical tile	21.0	0.37	—	2.70	—	—
Mineral fiberboard, wet molded						
Acoustical tile[k]	23.0	0.42	—	2.38	—	0.14
Wood or cane fiberboard						
Acoustical tile[k] ...0.5 in.	—	—	0.80	—	1.25	0.31
Acoustical tile[k] ...0.75 in.	—	—	0.53	—	1.89	—
Interior finish (plank, tile)	15.0	0.35	—	2.86	—	0.32
Cement fiber slabs (shredded wood with Portland cement binder)	25–27.0	0.50–0.53	—	2.0–1.89	—	—
Cement fiber slabs (shredded wood with magnesia oxysulfide binder)	22.0	0.57	—	1.75	—	0.31
Loose Fill						
Cellulosic insulation (milled paper or wood pulp)	2.3–3.2	0.27–0.32	—	3.70–3.13	—	0.33
Perlite, expanded	2.0–4.1	0.27–0.31	—	3.7–3.3	—	0.26
	4.1–7.4	0.31–0.36	—	3.3–2.8	—	—
	7.4–11.0	0.36–0.42	—	2.8–2.4	—	—
Mineral fiber (rock, slag, or glass)[g]						
approx. 3.75–5 in.	0.6–2.0	—	—	—	11.0	0.17
approx. 6.5–8.75 in.	0.6–2.0	—	—	—	19.0	—
approx. 7.5–10 in.	0.6–2.0	—	—	—	22.0	—
approx. 10.25–13.75 in.	0.6–2.0	—	—	—	30.0	—
Mineral fiber (rock, slag, or glass)[g] approx. 3.5 in. (closed sidewall application)	2.0–3.5	—	—	—	12.0–14.0	—
Vermiculite, exfoliated	7.0–8.2	0.47	—	2.13	—	0.32
	4.0–6.0	0.44	—	2.27	—	—
Spray Applied						
Polyurethane foam	1.5–2.5	0.16–0.18	—	6.25–5.56	—	—
Ureaformaldehyde foam	0.7–1.6	0.22–0.28	—	4.55–3.57	—	—
Cellulosic fiber	3.5–6.0	0.29–0.34	—	3.45–2.94	—	—
Glass fiber	3.5–4.5	0.26–0.27	—	3.85–3.70	—	—
METALS (See Chapter 36, Table 3)						
ROOFING						
Asbestos-cement shingles	120	—	4.76	—	0.21	0.24
Asphalt roll roofing	70	—	6.50	—	0.15	0.36
Asphalt shingles	70	—	2.27	—	0.44	0.30
Built-up roofing ...0.375 in.	70	—	3.00	—	0.33	0.35
Slate ...0.5 in.	—	—	20.00	—	0.05	0.30
Wood shingles, plain and plastic film faced	—	—	1.06	—	0.94	0.31
PLASTERING MATERIALS						
Cement plaster, sand aggregate	116	5.0	—	0.20	—	0.20
Sand aggregate ...0.375 in.	—	—	13.3	—	0.08	0.20
Sand aggregate ...0.75 in.	—	—	6.66	—	0.15	0.20
Gypsum plaster:						
Lightweight aggregate ...0.5 in.	45	—	3.12	—	0.32	—
Lightweight aggregate ...0.625 in.	45	—	2.67	—	0.39	—
Lightweight aggregate on metal lath ...0.75 in.	—	—	2.13	—	0.47	—
Perlite aggregate	45	1.5	—	0.67	—	0.32
Sand aggregate	105	5.6	—	0.18	—	0.20
Sand aggregate ...0.5 in.	105	—	11.10	—	0.09	—
Sand aggregate ...0.625 in.	105	—	9.10	—	0.11	—
Sand aggregate on metal lath ...0.75 in.	—	—	7.70	—	0.13	—
Vermiculite aggregate	45	1.7	—	0.59	—	—
MASONRY MATERIALS						
Masonry Units						
Brick, fired clay	150	8.4–10.2	—	0.12–0.10	—	—
	140	7.4–9.0	—	0.14–0.11	—	—
	130	6.4–7.8	—	0.16–0.12	—	—
	120	5.6–6.8	—	0.18–0.15	—	0.19
	110	4.9–5.9	—	0.20–0.17	—	—

Table 4   Typical Thermal Properties of Common Building and Insulating Materials—Design Values[a] (Continued)

Description	Density, lb/ft³	Conductivity[b] (k), Btu·in / h·ft²·°F	Conductance (C), Btu / h·ft²·°F	Resistance[c] (R) Per Inch Thickness (1/k), °F·ft²·h / Btu·in	Resistance[c] (R) For Thickness Listed (1/C), °F·ft²·h / Btu	Specific Heat, Btu / lb·°F
Brick, fired clay continued	100	4.2–5.1	—	0.24–0.20	—	—
	90	3.6–4.3	—	0.28–0.24	—	—
	80	3.0–3.7	—	0.33–0.27	—	—
	70	2.5–3.1	—	0.40–0.33	—	—
Clay tile, hollow						
1 cell deep ....3 in.	—	—	1.25	—	0.80	0.21
1 cell deep ....4 in.	—	—	0.90	—	1.11	—
2 cells deep ....6 in.	—	—	0.66	—	1.52	—
2 cells deep ....8 in.	—	—	0.54	—	1.85	—
2 cells deep ....10 in.	—	—	0.45	—	2.22	—
3 cells deep ....12 in.	—	—	0.40	—	2.50	—
Concrete blocks[i]						
Limestone aggregate						
8 in., 36 lb, 138 lb/ft³ concrete, 2 cores	—	—	—	—	—	—
Same with perlite filled cores	—	—	0.48	—	2.1	—
12 in., 55 lb, 138 lb/ft³ concrete, 2 cores	—	—	—	—	—	—
Same with perlite filled cores	—	—	0.27	—	3.7	—
Normal weight aggregate (sand and gravel)						
8 in., 33-36 lb, 126-136 lb/ft³ concrete, 2 or 3 cores	—	—	0.90–1.03	—	1.11–0.97	0.22
Same with perlite filled cores.................	—	—	0.50	—	2.0	—
Same with verm. filled cores	—	—	0.52–0.73	—	1.92–1.37	—
12 in., 50 lb, 125 lb/ft³ concrete, 2 cores	—	—	0.81	—	1.23	0.22
Medium weight aggregate (combinations of normal weight and lightweight aggregate)						
8 in., 26-29 lb, 97-112 lb/ft³ concrete, 2 or 3 cores..	—	—	0.58–0.78	—	1.71–1.28	—
Same with perlite filled cores	—	—	0.27–0.44	—	3.7–2.3	—
Same with verm. filled cores	—	—	0.30	—	3.3	—
Same with molded EPS (beads) filled cores	—	—	0.32	—	3.2	—
Same with molded EPS inserts in cores	—	—	0.37	—	2.7	—
Lightweight aggregate (expanded shale, clay, slate or slag, pumice)						
6 in., 16-17 lb 85-87 lb/ft³ concrete, 2 or 3 cores...	—	—	0.52–0.61	—	1.93–1.65	—
Same with perlite filled cores	—	—	0.24	—	4.2	—
Same with verm. filled cores	—	—	0.33	—	3.0	—
8 in., 19-22 lb, 72-86 lb/ft³ concrete, ...	—	—	0.32–0.54	—	3.2–1.90	0.21
Same with perlite filled cores	—	—	0.15–0.23	—	6.8–4.4	—
Same with verm. filled cores	—	—	0.19–0.26	—	5.3–3.9	—
Same with molded EPS (beads) filled cores	—	—	0.21	—	4.8	—
Same with UF foam filled cores	—	—	0.22	—	4.5	—
Same with molded EPS inserts in cores	—	—	0.29	—	3.5	—
12 in., 32-36 lb, 80–90 lb/ft³ concrete, 2 or 3 cores...	—	—	0.38–0.44	—	2.6–2.3	—
Same with perlite filled cores	—	—	0.11–0.16	—	9.2–6.3	—
Same with verm. filled cores	—	—	0.17	—	5.8	—
Stone, lime, or sand						
Quartzitic and sandstone	180	72	—	0.01	—	—
	160	43	—	0.02	—	—
	140	24	—	0.04	—	—
	120	13	—	0.08	—	0.19
Calcitic, dolomitic, limestone, marble, and granite..	180	30	—	0.03	—	—
	160	22	—	0.05	—	—
	140	16	—	0.06	—	—
	120	11	—	0.09	—	0.19
	100	8	—	0.13	—	—
Gypsum partition tile						
3 by 12 by 30 in., solid	—	—	0.79	—	1.26	0.19
3 by 12 by 30 in., 4 cells	—	—	0.74	—	1.35	—
4 by 12 by 30 in., 3 cells	—	—	0.60	—	1.67	—
Concretes						
Sand and gravel or stone aggregate concretes (concretes	150	10.0–20.0	—	0.10–0.05	—	—
with more than 50% quartz or quartzite sand have	140	9.0–18.0	—	0.11–0.06	—	0.19–0.24
conductivities in the higher end of the range) ..	130	7.0–13.0	—	0.14–0.08	—	—
Limestone concretes	140	11.1	—	0.09	—	—
	120	7.9	—	0.13	—	—
	100	5.5	—	0.18	—	—
Gypsum-fiber concrete (87.5% gypsum, 12.5% wood chips)	51	1.66	—	0.60	—	0.21
Cement/lime, mortar, and stucco	120	9.7	—	0.10	—	—
	100	6.7	—	0.15	—	—
	80	4.5	—	0.22	—	—
Lightweight aggregate concretes						
Expanded shale, clay, or slate; expanded slags; cinders;	120	6.4–9.1	—	0.16–0.11	—	—
pumice (with density up to 100 lb/ft³); and scoria	100	4.7–6.2	—	0.21–0.16	—	0.20
(sanded concretes have conductivities in the higher	80	3.3–4.1	—	0.30–0.24	—	0.20
end of the range)	60	2.1–2.5	—	0.48–0.40	—	—
	40	1.3	—	0.78	—	—

Table 4 Typical Thermal Properties of Common Building and Insulating Materials—Design Values[a] (*Concluded*)

Description	Density, lb/ft³	Conductivity[b] (k), Btu·in / h·ft²·°F	Conductance (C), Btu / h·ft²·°F	Resistance [c](R) Per Inch Thickness (1/k), °F·ft²·h / Btu·in	Resistance [c](R) For Thickness Listed (1/C), °F·ft²·h / Btu	Specific Heat, Btu / lb·°F
Perlite, vermiculite, and polystyrene beads	50	1.8–1.9	—	0.55–0.53	—	—
	40	1.4–1.5	—	0.71–0.67	—	0.15–0.23
	30	1.1	—	0.91	—	—
	20	0.8	—	1.25	—	—
Foam concretes	120	5.4	—	0.19	—	—
	100	4.1	—	0.24	—	—
	80	3.0	—	0.33	—	—
	70	2.5	—	0.40	—	—
Foam concretes and cellular concretes	60	2.1	—	0.48	—	—
	40	1.4	—	0.71	—	—
	20	0.8	—	1.25	—	—

SIDING MATERIALS (on flat surface)

Shingles

Description	Density, lb/ft³	Conductivity[b] (k)	Conductance (C)	Resistance Per Inch (1/k)	Resistance For Thickness (1/C)	Specific Heat
Asbestos-cement	120	—	4.75	—	0.21	—
Wood, 16 in., 7.5 exposure	—	—	1.15	—	0.87	0.31
Wood, double, 16-in., 12-in. exposure	—	—	0.84	—	1.19	0.28
Wood, plus insul. backer board, 0.3125 in.	—	—	0.71	—	1.40	0.31

Siding

Description	Density	Conductivity	Conductance	Resistance Per Inch	Resistance For Thickness	Specific Heat
Asbestos-cement, 0.25 in., lapped	—	—	4.76	—	0.21	0.24
Asphalt roll siding	—	—	6.50	—	0.15	0.35
Asphalt insulating siding (0.5 in. bed.)	—	—	0.69	—	1.46	0.35
Hardboard siding, 0.4375 in.	—	—	1.49	—	0.67	0.28
Wood, drop, 1 by 8 in.	—	—	1.27	—	0.79	0.28
Wood, bevel, 0.5 by 8 in., lapped	—	—	1.23	—	0.81	0.28
Wood, bevel, 0.75 by 10 in., lapped	—	—	0.95	—	1.05	0.28
Wood, plywood, 0.375 in., lapped	—	—	1.59	—	0.59	0.29
Aluminum or Steel[m], over sheathing						
Hollow-backed	—	—	1.61	—	0.61	0.29
Insulating-board backed nominal 0.375 in.	—	—	0.55	—	1.82	0.32
Insulating-board backed nominal 0.375 in., foil backed	—	—	0.34	—	2.96	—
Architectural (soda-lime float) glass	158	6.9	—	—	—	0.21

WOODS (12% moisture content)[e,n]

Hardwoods

Description	Density	Conductivity	Conductance	Resistance Per Inch	Resistance For Thickness	Specific Heat
						0.39[o]
Oak	41.2–46.8	1.12–1.25	—	0.89–0.80	—	
Birch	42.6–45.4	1.16–1.22	—	0.87–0.82	—	
Maple	39.8–44.0	1.09–1.19	—	0.92–0.84	—	
Ash	38.4–41.9	1.06–1.14	—	0.94–0.88	—	

Softwoods

Description	Density	Conductivity	Conductance	Resistance Per Inch	Resistance For Thickness	Specific Heat
						0.39[o]
Southern Pine	35.6–41.2	1.00–1.12	—	1.00–0.89	—	
Douglas Fir-Larch	33.5–36.3	0.95–1.01	—	1.06–0.99	—	
Southern Cypress	31.4–32.1	0.90–0.92	—	1.11–1.09	—	
Hem-Fir, Spruce-Pine-Fir	24.5–31.4	0.74–0.90	—	1.35–1.11	—	
West Coast Woods, Cedars	21.7–31.4	0.68–0.90	—	1.48–1.11	—	
California Redwood	24.5–28.0	0.74–0.82	—	1.35–1.22	—	

[a]Values are for a mean temperature of 75 °F. Representative values for dry materials are intended as design (not specification) values for materials in normal use. Thermal values of insulating materials may differ from design values depending on their in-situ properties (*e.g.*, density and moisture content, orientation, etc.) and variability experienced during manufacture. For properties of a particular product, use the value supplied by the manufacturer or by unbiased tests.

[b]To obtain thermal conductivities in Btu/h·ft·°F, divide the *k*-factor by 12 in./ft.

[c]Resistance values are the reciprocals of C before rounding off C to two decimal places.

[d]Lewis (1967).

[e]U.S. Department of Agriculture (1974).

[f]Does not include paper backing and facing, if any. Where insulation forms a boundary (reflective or otherwise) of an airspace, see Tables 2 and 3 for the insulating value of an airspace with the appropriate effective emittance and temperature conditions of the space.

[g]Conductivity varies with fiber diameter. (See Chapter 20, Factors Affecting Thermal Performance.) Batt, blanket, and loose-fill mineral fiber insulations are manufactured to achieve specified R-values, the most common of which are listed in the table. Due to differences in manufacturing processes and materials, the product thicknesses, densities, and thermal conductivities vary over considerable ranges for a specified R-value.

[h]This material is relatively new and data are based on limited testing.

[i]For additional information, see Society of Plastics Engineers (SPI) *Bulletin* U108. Values are for aged, unfaced board stock. For change in conductivity with age of expanded polyurethane/polyisocyanurate, see Chapter 20, Factors Affecting Thermal Performance.

[j]Values are for aged products with gas-impermeable facers on the two major surfaces. An aluminum foil facer of 0.001 in. thickness or greater is generally considered impermeable to gases. For change in conductivity with age of expanded polyisocyanurate, see Chapter 20, Factors Affecting Thermal Performance, and SPI *Bulletin* U108.

[k]Insulating values of acoustical tile vary, depending on density of the board and on type, size, and depth of perforations.

[l]Values for fully grouted block may be approximated using values for concrete with a similar unit weight.

[m]Values for metal siding applied over flat surfaces vary widely, depending on amount of ventilation of airspace beneath the siding; whether airspace is reflective of non-reflective; and on thickness, type, and application of insulating backing-board used. Values given are averages for use as design guides, and were obained from several guarded hot box tests (ASTM C236) or calibrated hot box (ASTM C976) on hollow-backed types and types made using backing-boards of wood fiber, foamed plastic, and glass fiber. Departures of ±50% or more from the values given may occur.

[n]See Adams (1971), MacLean (1941), and Wilkes (1979). The conductivity values listed are for heat transfer across the grain. The thermal conductivity of wood varies linearly with the density, and the density ranges listed are those normally found for the wood species given. If the density of the wood species is not known, use the mean conductivity value. For extrapolation to other moisture contents, the following empirical equation developed by Wilkes (1979) may be used:

$$k = 0.1791 + \frac{(1.874 \times 10^{-2} + 5.753 \times 10^{-4}M)\rho}{1 + 0.01M}$$

where ρ is density of the moist wood in lb/ft³, and M is the moisture content in percent.

[o]From Wilkes (1979), an empirical equation for the specific heat of moist wood at 75 °F is as follows:

$$c_p = \frac{(0.299 + 0.01M)}{(1 + 0.01M)} + \Delta c_p$$

where Δc_p accounts for the heat of sorption and is denoted by

$$\Delta c_p = M(1.921 \times 10^{-3} - 3.168 \times 10^{-5}M)$$

where M is the moisture content in percent by mass.

If the steel framing is accounted for using the parallel flow method, the C-factor of the wall is determined using Equation (5) from Chapter 20 as follows:

$$C_{av} = (0.92 \times 0.084) + (0.08 \times 0.629)$$
$$= 0.128 \text{ Btu/h} \cdot \text{ft}^2 \cdot {}^\circ\text{F}$$
$$R_{T(av)} = 7.81 {}^\circ\text{F} \cdot \text{ft}^2 \cdot \text{h/Btu}$$

If the steel framing is included using the isothermal planes method, the C-factor of the wall is determined using Equations (2) and (3) from Chapter 20 as follows:

$$R_{T(av)} = 0.45 + 1/[(0.92/11.00) + (0.08/0.69)] + 0.45$$
$$= 5.91 {}^\circ\text{F} \cdot \text{ft}^2 \cdot \text{h/Btu}$$
$$C_{av} = 0.169 \text{ Btu/h} \cdot \text{ft}^2 \cdot {}^\circ\text{F}$$

For this insulated steel frame wall, Farouk and Larson (1983) measured an average R-value of $6.61 {}^\circ\text{F} \cdot \text{ft}^2 \cdot \text{h/Btu}$.

In ASHRAE/IES *Standard* 90.1-1989, Energy Efficient Design of New Buildings except New Low-Rise Residential Buildings, one method given for determining the thermal resistance of wall assemblies containing metal framing involves using a parallel path correction factor F_c. The F_c values are included in Table 8C-2 of ASHRAE/IES *Standard* 90.1-1989. For 2 by 4 steel framing, 16 in. on center, $F_c = 0.50$. Using the correction factor method, an R-value of $6.40 {}^\circ\text{F} \cdot \text{ft}^2 \cdot \text{h/Btu}$ [0.45 + 11(0.50) + 0.45] is obtained for the wall described in Example 3.

Zone Method of Calculation

For structures with widely spaced metal members of substantial cross-sectional area, calculation by the isothermal planes method can result in thermal resistance values that are too low. For these constructions, the *zone method* can be used. This method involves two separate computations—one for a chosen limited portion, Zone A, containing the highly conductive element; the other for the remaining portion of simpler construction, Zone B. The two computations are then combined using the parallel flow method, and the average transmittance per unit overall area is calculated. The basic laws of heat transfer are applied by adding the area conductances CA of elements in parallel, and adding area resistances R/A of elements in series.

The surface shape of Zone A is determined by the metal element. For a metal beam (see Figure 5), the Zone A surface is a strip of width W that is centered on the beam. For a rod perpendicular to panel surfaces, it is a circle of diameter W. The value of W is calculated from Equation (1), which is empirical. The value of d should not be less than 0.5 in. for still air.

$$W = m + 2d \tag{1}$$

where

- m = width or diameter of metal heat path terminal, in.
- d = distance from panel surface to metal, in.

Generally, the value of W should be calculated using Equation (1) for each end of the metal heat path; the larger value, within the limits of the basic area, should be used as illustrated in Example 4.

Example 4. Calculate transmittance of the roof deck shown in Figure 5. Tee-bars at 24 in. OC support glass fiber form boards, gypsum concrete, and built-up roofing. Conductivities of components are: steel, 314.4 Btu·in/h·ft²·°F; gypsum concrete, 1.66 Btu·in/h·ft²·°F; and glass fiber form board, 0.25 Btu·in/h·ft²·°F. Conductance of built-up roofing is 3.00 Btu/h·ft²·°F.

Solution: The basic area is 2 ft² (24 in. by 12 in.) with a tee-bar (12 in. long) across the middle. This area is divided into Zones A and B.

Zone A is determined from Equation (1) as follows:

Top side $W = m + 2d = 0.625 + (2 \times 1.5) = 3.625$ in.
Bottom side $W = m + 2d = 2.0 + (2 \times 0.5) = 3.0$ in.

Using the larger value of W, the area of Zone A is $(12 \times 3.625)/144 = 0.302$ ft². The area of Zone B is $2.0 - 0.302 = 1.698$ ft².

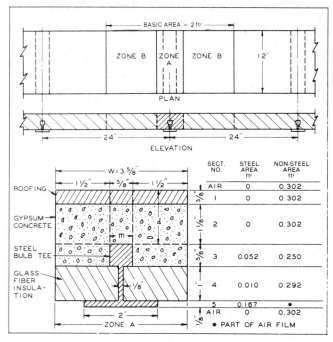

Fig. 5   Gypsum Roof Deck on Bulb Tees (Example 4)

To determine area transmittance for Zone A, divide the structure within the zone into five sections parallel to the top and bottom surfaces (Figure 5). The area conductance CA of each section is calculated by adding the area conductances of its metal and nonmetal paths. Area conductances of the sections are converted to area resistances R/A and added to obtain the total resistance of Zone A.

Section	Area × Conductance	=	CA	$\dfrac{1}{CA} = \dfrac{R}{A}$
Air (outside, 15 mph)	0.302 × 6.00		1.81	0.55
No. 1, Roofing	0.302 × 3.00		0.906	1.10
No. 2, Gypsum concrete	0.302 × 1.66/1.125		0.446	2.24
No. 3, Steel	0.052 × 314.4/0.625		26.2 ⎫	
No. 3, Gypsum concrete	0.250 × 1.66/0.625		0.664 ⎬	0.04
No. 4, Steel	0.010 × 314.4/1.00		3.14 ⎫	
No. 4, Glass fiberboard	0.292 × 0.25/1.00		0.073 ⎬	0.31
No. 5, Steel	0.167 × 314.4/0.125		420.0 ⎭	0.002
Air (inside)	0.302 × 1.63		0.492	2.03
			Total R/A =	6.27

Area transmittance of Zone A = $1/(R/A) = 1/6.27 = 0.159$.

For Zone B, the unit resistances are added and then converted to area transmittance, as shown in the following table.

Section	Resistance, R
Air (outside, 15 mph)	1/6.00 = 0.17
Roofing	1/3.00 = 0.33
Gypsum concrete	1.75/1.66 = 1.05
Glass fiberboard	1.00/0.25 = 4.00
Air (inside)	1/1.63 = 0.61
Total resistance	= 6.16

Since unit transmittance = $1/R = 0.162$, the total area transmittance UA is calculated as follows:

$$\text{Zone B} = 1.698 \times 0.162 = 0.275$$
$$\text{Zone A} = 0.159$$
$$\text{Total area transmittance of basic area} = 0.434$$
$$\text{Transmittance per ft}^2 = 0.434/2.0 = 0.217$$
$$\text{Resistance per ft}^2 = 4.61$$

Overall R-values of 4.57 and $4.85 {}^\circ\text{F} \cdot \text{ft}^2 \cdot \text{h/Btu}$ have been measured in two guarded hot box tests of a similar construction.

When the steel member represents a relatively large proportion of the total heat flow path, as in Example 4, detailed calculations of resistance in sections 3, 4, and 5 of Zone A are unnecessary; if only the steel member is considered, the final result of Example 4 is the same. However, if the heat flow path represented by the steel member is small, as for a tie rod, detailed calculations for sections 3, 4, and 5 are necessary. A panel with an internal metallic structure and bonded on one or both sides to a metal skin or covering presents special problems of lateral heat flow not covered in the zone method.

Ceilings and Roofs

The overall R-value for ceilings of wood frame flat roofs can be calculated using Equations (1) through (5) from Chapter 20. Properties of the materials are found in Tables 1, 2, 3, and 4. The fraction of framing is assumed to be 0.10 for joists at 16 in. OC and 0.07 for joists at 24 in. OC. The calculation procedure is similar to that shown in Example 1. Note that if the ceiling contains plane air spaces (see Table 2), the resistance depends on the direction of heat flow, i.e., whether the calculation is for a winter (heat flow up) or summer (heat flow down) condition.

For ceilings of pitched roofs under winter conditions, calculate the R-value of the ceiling using the procedure for flat roofs. The heat loss from these ceilings can be obtained using a calculated attic temperature (see Chapter 25). Table 5 can be used to determine the effective resistance of the attic space under summer conditions for varying conditions of ventilation air temperature, airflow direction and rates, ceiling resistance, roof or sol-air temperatures, and surface emittances (Joy 1958).

The R-value is the total resistance obtained by adding the ceiling and effective attic resistances. The applicable temperature difference is that difference between room air and sol-air temperatures or between room air and roof temperatures (see Table 5, footnote f). Table 5 can be used for pitched and flat residential roofs over attic spaces. When an attic has a floor, the ceiling resistance should account for the complete ceiling-floor construction.

Windows and Doors

The U-factors given in Table 5 of Chapter 27 are for vertical glazing (e.g., windows, glass in exterior doors, glass doors, and skylights). The values were computed using procedures outlined in Chapter 27. The U-factors in Table 6 are for exterior wood and steel doors. The values given for wood doors were calculated, and those for steel doors were taken from hot box tests (Sabine et al. 1975, Yellott 1965) or from manufacturers' test reports. An outdoor surface conductance of 6.0 Btu/h·ft²·°F was used, and the indoor surface conductance was taken as 1.46 Btu/h·ft²·°F for vertical surfaces with horizontal heat flow. All values given are for exterior doors without glazing. If an exterior door contains glazing, the glazing should be analyzed as a window, as illustrated in Example 5.

Example 5. Determine the U-factor of a fixed wood frame residential window containing double insulating glass with 0.5-in. air space and metal spacer for winter conditions.

Solution: From Chapter 27, Table 5, the U-factor of the center of the glass portion only is 0.49 Btu/h·ft²·°F for glazing 1D6, double glazing, 0.5-in. air space. The wood frame of the window must also be

Table 5 Effective Thermal Resistance of Ventilated Attics[a] (Summer Condition)

PART A. NONREFLECTIVE SURFACES

Ventilation Air Temperature, °F	Sol-Air[f] Temperature, °F	No Ventilation[b]		Natural Ventilation		Power Ventilation[c]					
		Ventilation Rate, cfm/ft²									
		0		0.1[d]		0.5		1.0		1.5	
		Ceiling Resistance R[e], °F·ft²·/Btu									
		10	20	10	20	10	20	10	20	10	20
80	120	1.9	1.9	2.8	3.4	6.3	9.3	9.6	16	11	20
	140	1.9	1.9	2.8	3.5	6.5	10	9.8	17	12	21
	160	1.9	1.9	2.8	3.6	6.7	11	10	18	13	22
90	120	1.9	1.9	2.5	2.8	4.6	6.7	6.1	10	6.9	13
	140	1.9	1.9	2.6	3.1	5.2	7.9	7.6	12	8.6	15
	160	1.9	1.9	2.7	3.4	5.8	9.0	8.5	14	10	17
100	120	1.9	1.9	2.2	2.3	3.3	4.4	4.0	6.0	4.1	6.9
	140	1.9	1.9	2.4	2.7	4.2	6.1	5.8	8.7	6.5	10
	160	1.9	1.9	2.6	3.2	5.0	7.6	7.2	11	8.3	13
PART B. REFLECTIVE SURFACES[g]											
80	120	6.5	6.5	8.1	8.8	13	17	17	25	19	30
	140	6.5	6.5	8.2	9.0	14	18	18	26	20	31
	160	6.5	6.5	8.3	9.2	15	18	19	27	21	32
90	120	6.5	6.5	7.5	8.0	10	13	12	17	13	19
	140	6.5	6.5	7.7	8.3	12	15	14	20	16	22
	160	6.5	6.5	7.9	8.6	13	16	16	22	18	25
100	120	6.5	6.5	7.0	7.4	8.0	10	8.5	12	8.8	12
	140	6.5	6.5	7.3	7.8	10	12	11	15	12	16
	160	6.5	6.5	7.6	8.2	11	14	13	18	15	20

[a]Although the term effective resistance is commonly used when there is attic ventilation, this table includes values for situations with no ventilation. The effective resistance of the attic added to the resistance (1/U) of the ceiling yields the effective resistance of this combination based on sol-air (see Chapter 26) and room temperatures. These values apply to wood frame construction with a roof deck and roofing that has a conductance of 1.0 Btu/h·ft²·°F.
[b]This condition cannot be achieved in the field unless extreme measures are taken to tightly seal the attic.
[c]Based on air discharging outward from attic.
[d]When attic ventilation meets the requirements stated in Chapter 23, 0.1 cfm/ft² is assumed as the natural summer ventilation rate.
[e]When determining ceiling resistance, do not add the effect of a reflective surface facing the attic, as it is accounted for in Table 5, Part B.
[f]Roof surface temperature rather than sol-air temperature (see Chapter 26) can be used if 0.25 is subtracted from the attic resistance shown.
[g]Surfaces with effective emittance ϵ_{eff} = 0.05 between ceiling joists facing attic space.

Table 6   Transmission Coefficients U for Wood and Steel Doors, Btu/h·ft²·°F

Nominal Door Thickness, in.	Description	No Storm Door	Wood Storm Door[c]	Metal Storm Door[d]
Wood Doors[a,b]				
1-3/8	Panel door with 7/16-in. panels[e]	0.57	0.33	0.37
1-3/8	Hollow core flush door	0.47	0.30	0.32
1-3/8	Solid core flush door	0.39	0.26	0.28
1-3/4	Panel door with 7/16-in. panels[e]	0.54	0.32	0.36
1-3/4	Hollow core flush door	0.46	0.29	0.32
1-3/4	Panel door with 1-1/8-in. panels[e]	0.39	0.26	0.28
1-3/4	Solid core flush door	0.40	—	0.26
2-1/4	Solid core flush door	0.27	0.20	0.21
Steel Doors[b]				
1-3/4	Fiberglass or mineral wool core with steel stiffeners, no thermal break[f]	0.60	—	—
1-3/4	Paper honeycomb core without thermal break[f]	0.56	—	—
1-3/4	Solid urethane foam core without thermal break[a]	0.40	—	—
1-3/4	Solid fire rated mineral fiberboard core without thermal break[f]	0.38	—	—
1-3/4	Polystyrene core without thermal break (18 gage commercial steel)[f]	0.35	—	—
1-3/4	Polyurethane core without thermal break (18 gage commercial steel)[f]	0.29	—	—
1-3/4	Polyurethane core without thermal break (24 gage residential steel)[f]	0.29	—	—
1-3/4	Polyurethane core with thermal break and wood perimeter (24 gage residential steel)[f]	0.20	—	—
1-3/4	Solid urethane foam core with thermal break[a]	0.20	—	0.16

Note: All U-factors for exterior doors in this table are for doors with no glazing, except for the storm doors which are in addition to the main exterior door. Any glazing area in exterior doors should be included with the appropriate glass type and analyzed as a window (see Chapter 27). Interpolation and moderate extrapolation are permitted for door thicknesses other than those specified.
[a]Values are based on a nominal 32 by 80 in. door size with no glazing.

[b]Outside air conditions: 15 mph wind speed, 0°F air temperature; inside air conditions: natural convection, 70°F air temperature.
[c]Values for wood storm door are for approximately 50% glass area.
[d]Values for metal storm door are for any percent glass area.
[e]55% panel area.
[f]ASTM C 236 hotbox data on a nominal 3 by 7 ft door size with no glazing.

considered when determining the window U-factor. Referring to Table 5 in Chapter 27, for a fixed wood frame window with a 0.5-in. air space and metal spacer, the U-factor is given as 0.51 Btu/h·ft²·°F.

All R-values are approximate, since a significant portion of the resistance of a window or door is contained in the air film resistances, and some parameters that may have important effects are not considered. For example, the listed U-factors assume the surface temperatures of surrounding bodies are equal to the ambient air temperature. However, the indoor surface of a window or door in an actual installation may be exposed to nearby radiating surfaces, such as radiant heating panels, or opposite walls with much higher or lower temperatures than the indoor air. Air movement across the indoor surface of a window or door, such as that caused by nearby heating and cooling outlet grilles, increases the U-factor; and air movement (wind) across the outdoor surface of a window or door also increases the U-factor.

U_o Concept

In Section 4 of ASHRAE *Standard* 90A-1980, Energy Conservation in New Building Design, requirements are stated in terms of U_o, where U_o is the combined thermal transmittance of the respective areas of gross exterior wall, roof or ceiling or both, and floor assemblies. The U_o equation for a wall is as follows:

$$U_o = (U_{wall} A_{wall} + U_{window} A_{window} + U_{door} A_{door})/A_o \quad (2)$$

where

U_o = average thermal transmittance of gross wall area
A_o = gross area of exterior walls
U_{wall} = thermal transmittance of all elements of opaque wall area
A_{wall} = opaque wall area
U_{window} = thermal transmittance of window area (including frame)

A_{window} = window area (including frame)
U_{door} = thermal transmittance of door area
A_{door} = door area

Where more than one type of wall, window, or door is used, the UA term for that exposure should be expanded into its sub-elements, as shown in Equation (3).

$$\begin{aligned} U_o A_o = {} & U_{wall\,1} A_{wall\,1} + U_{wall\,2} A_{wall\,2} + \cdots + U_{wall\,m} A_{wall\,m} \\ & + U_{window\,1} A_{window\,1} + U_{window\,2} A_{window\,2} + \cdots \\ & + U_{window\,n} A_{window\,n} + U_{door\,1} A_{door\,1} \\ & + U_{door\,2} A_{door\,2} + \cdots + U_{door\,o} A_{door\,o} \end{aligned} \quad (3)$$

Example 6. Calculate U_o for a wall 30 ft by 8 ft, constructed as in Example 1A. The wall contains one window 60 in. by 34 in. and a second window 36 in. by 30 in. Both windows are constructed as in Example 5. The wall also contains a 1.75-in. solid core flush door with a metal storm door 34 in. by 80 in. ($U = 0.26$ Btu/h·ft²·°F from Table 6).

Solution: The U-factors for the wall and windows were obtained in Examples 1A and 5, respectively. The areas of the different components are:

$$A_{window} = [(60 \times 34) + (36 \times 30)]/144 = 21.7 \text{ ft}^2$$
$$A_{door} = (34 \times 80)/144 = 18.9 \text{ ft}^2$$
$$A_{wall} = (30 \times 8) - (21.7 + 18.9) = 199.4 \text{ ft}^2$$

Therefore, the combined thermal transmittance for the wall is:

$$U_o = \frac{(0.063 \times 199.4) + (0.51 \times 21.7) + (0.26 \times 18.9)}{(30 \times 8)}$$
$$= 0.119 \text{ Btu/h·ft}^2\cdot°F$$

Slab-on-Grade and Below-Grade Construction

Heat transfer through basement walls and floors to the ground depends on the following factors: (1) the difference between the air temperature within the room and that of the ground and outside air, (2) the material of the walls or floor, and (3) the thermal conductivity of the surrounding earth. The latter varies with local conditions and is usually unknown. Because of the great thermal inertia of the surrounding soil, ground temperature varies with depth, and there is a substantial time lag between changes in outdoor air temperatures and corresponding changes in ground temperatures. As a result, ground-coupled heat transfer is less amenable to steady-state representation than above-grade building elements. However, several simplified procedures for estimating ground-coupled heat transfer have been developed. These fall into two principal categories: (1) those that reduce the ground heat transfer problem to a closed form solution, and (2) those that use simple regression equations developed from statistically reduced multidimensional transient analyses.

Closed form solutions, including the ASHRAE arc-length procedure discussed in Chapter 25 by Latta and Boileau (1969), generally reduce the problem to one-dimensional, steady-state heat transfer. These procedures use simple, "effective" U-factors or ground temperatures or both. Methods differ in the various parameters averaged or manipulated to obtain these effective values. Closed form solutions provide acceptable results in climates that have a single dominant season, because the dominant season persists long enough to permit a reasonable approximation of steady-state conditions at shallow depths. The large errors (percentage) that are likely during transition seasons should not seriously affect building design decisions, since these heat flows are relatively insignificant when compared with those of the principal season.

The ASHRAE arc-length procedure is a reliable method for wall heat losses in cold winter climates. Chapter 25 discusses a slab-on-grade floor model developed by one study. Although both procedures give results comparable to transient computer solutions for cold climates, their results for warmer U.S. climates differ substantially.

Research conducted by Hougten *et al.* (1942) and Dill *et al.* (1945) indicates a heat flow of approximately 2.0 Btu/h·ft² through an uninsulated concrete basement floor with a temperature difference of 20°F between the basement floor and the air 6 in. above it. A U-factor of 0.10 Btu/h·ft²·°F is sometimes used for concrete basement floors on the ground. For basement walls below grade, the temperature difference for winter design conditions is greater than for the floor. Test results indicate that at the midheight of the below-grade portion of the basement wall, the unit area heat loss is approximately twice that of the floor.

For concrete slab floors in contact with the ground at grade level, tests indicate that for small floor areas (equal to that of a 25 by 25 ft house) the heat loss can be calculated as proportional to the length of exposed edge rather than total area. This amounts to 0.81 Btu/h per linear foot of exposed edge per °F difference between the indoor air temperature and the average outdoor air temperature. This value can be reduced appreciably by installing insulation under the ground slab and along the edge between the floor and abutting walls. In most calculations, if the perimeter loss is calculated accurately, no other floor losses need to be considered. Chapter 25 contains data for load calculations and heat loss values for below-grade walls and floors at different depths.

The second category of simplified procedures uses transient two-dimensional computer models to generate the ground heat transfer data that are then reduced to compact form by regression analysis (see Mitalas 1982 and 1983, Shipp 1983). These are the most accurate procedures available, but the database is very expensive to generate. In addition, these methods are limited to the range of climates and constructions specifically examined. Extrapolating beyond the outer bounds of the regression surfaces can produce significant errors.

Apparent Thermal Conductivity of Soil

Effective or apparent soil thermal conductivity is difficult to estimate precisely and may change substantially in the same soil at different times due to changed moisture conditions and the presence of freezing temperatures in the soil. Figure 6 shows the typical apparent soil thermal conductivity as a function of moisture content for different general types of soil. The figure is based on data presented in Salomone and Marlowe (1989) using envelopes of thermal behavior coupled with field moisture content ranges for different soil types. In Figure 6, the term well-graded applies to granular soils with good representation of all particle sizes from largest to smallest. The term poorly graded refers to granular soils with either a uniform gradation, in which most particles are about the same size, or a skip (or gap) gradation, in which particles of one or more intermediate sizes are not present.

Although thermal conductivity varies greatly over the complete range of possible moisture contents for a soil, this range can be narrowed if it is assumed that the moisture contents of most field soils lie between the "wilting point" of the soil (*i.e.*, the moisture content of a soil below which a plant cannot alleviate its wilting symptoms) and the "field capacity" of the soil (*i.e.*, the moisture

Table 7   Typical Apparent Thermal Conductivity Values for Soils, Btu·in/h·ft²·°F

	Normal Range	Recommended Values for Design[a]	
		Low[b]	High[c]
Sands	4.2 to 17.4	5.4	15.6
Silts	6 to 17.4	11.4	15.6
Clays	6 to 11.4	7.8	10.8
Loams	6 to 17.4	6.6	15.6

[a]Reasonable values for use when no site- or soil-specific data are available.
[b]Moderately conservative values for minimum heat loss through soil (*e.g.*, use in soil heat exchanger or earth-contact cooling calculations). Values are from Salomone and Marlowe (1989).
[c]Moderately conservative values for maximum heat loss through soil (*e.g.*, use in peak winter heat loss calculations). Values are from Salomone and Marlowe (1989).

Table 8   Typical Apparent Thermal Conductivity Values for Rocks, Btu·in/h·ft²·°F

	Normal Range
Pumice, tuff, obsidian	3.6 to 15.6
Basalt	3.6 to 18.0
Shale	6 to 27.6
Granite	12 to 30
Limestone, dolomite, marble	8.4 to 30
Quartzose sandstone	9.6 to 54

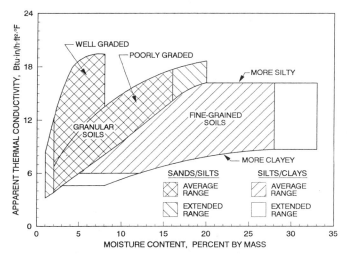

Fig. 6   Trends of Apparent Thermal Conductivity of Moist Soils

Table 9   Typical Water Vapor Permeance and Permeability Values for Common Building Materials[a]

Material	Thickness, in.	Permeance, Perm	Resistance[h] Rep	Permeability, Perm-in.	Resistance/in.[h], Rep/in.
Construction Materials					
Concrete (1:2:4 mix)				3.2	0.31
Brick masonry	4	0.8[f]	1.3		
Concrete block (cored, limestone aggregate)	8	2.4[f]	0.4		
Tile masonry, glazed	4	0.12[f]	8.3		
Asbestos cement board	0.12	4–8[d]	0.1–0.2		
With oil-base finishes		0.3–0.5[d]	2–3		
Plaster on metal lath	0.75	15[f]	0.067		
Plaster on wood lath		11[e]	0.091		
Plaster on plain gypsum lath (with studs)		20[f]	0.050		
Gypsum wall board (plain)	0.375	50[f]	0.020		
Gypsum sheathing (asphalt impregnated)	0.5			20[d]	0.050
Structural insulating board (sheathing quality)				20–50[f]	0.050–0.020
Structural insulating board (interior, uncoated)	0.5	50–90[f]	0.020–0.011		
Hardboard (standard)	0.125	11[f]	0.091		
Hardboard (tempered)	0.125	5[f]	0.2		
Built-up roofing (hot mopped)		0.0			
Wood, sugar pine				0.4–5.4[b]	2.5–0.19
Plywood (douglas fir, exterior glue)	0.25	0.7[f]	1.4		
Plywood (douglas fir, interior glue)	0.25	1.9[f]	0.53		
Acrylic, glass fiber reinforced sheet	0.056	0.12[d]	8.3		
Polyester, glass fiber reinforced sheet	0.048	0.05[d]	20		
Thermal Insulations					
Air (still)				120[f]	0.0083
Cellular glass				0.0[d]	∞
Corkboard				2.1–2.6[d]	0.48–0.38
				9.5[e]	0.11
Mineral wool (unprotected)				116[e]	0.0086
Expanded polyurethane (R-11 blown) board stock				0.4–1.6[d]	2.5–0.62
Expanded polystyrene—extruded				1.2[d]	0.83
Expanded polystyrene—bead				2.0–5.8[d]	0.50–0.17
Phenolic foam (covering removed)				26	0.038
Unicellular synthetic flexible rubber foam				0.02–0.15[d]	50–6.7
Plastic and Metal Foils and Films[c]					
Aluminum foil	0.001	0.0[d]	∞		
Aluminum foil	0.00035	0.05[d]	20		
Polyethylene	0.002	0.16[d]	6.3		3100
Polyethylene	0.004	0.08[d]	12.5		3100
Polyethylene	0.006	0.06[d]	17		3100
Polyethylene	0.008	0.04[d]	25		3100
Polyethylene	0.010	0.03[d]	33		3100
Polyvinylchloride, unplasticized	0.002	0.68[d]	1.5		
Polyvinylchloride, plasticized	0.004	0.8–1.4[d]	1.3–0.72		
Polyester	0.001	0.73[d]	1.4		
Polyester	0.0032	0.23[d]	4.3		
Polyester	0.0076	0.08[d]	12.5		
Cellulose acetate	0.01	4.6[d]	0.2		
Cellulose acetate	0.125	0.32[d]	3.1		

content of a soil that has been thoroughly wetted and then drained until the drainage rate has become negligibly small). After a prolonged dry spell, the moisture will be near the wilting point, and after a rainy period, the soil will have a moisture content near its field capacity. The moisture contents at these limits have been studied by many agricultural researchers, and data for different types of soil are given by Salomone and Marlowe (1989) and Kersten (1949). The shaded areas on Figure 6 approximate (1) the full range of moisture contents for different soil types and (2) a range between average values of each limit.

Table 7 gives a summary of design values for thermal conductivities of the basic soil classes. Table 8 gives ranges of thermal conductivity for some basic classes of rock. The value chosen depends on whether heat transfer is being calculated for minimum heat loss through the soil, as in a ground heat exchange system, or a maxi-

mum value, as in peak winter heat loss calculations for a basement. Hence, a high and a low value are given for each soil class.

As heat flows through the soil, the moisture tends to move away from the source of heat. This moisture migration provides initial mass transport of heat, but it also dries the soil adjacent to the heat source, hence lowering the apparent thermal conductivity in that zone of soil.

The following trends are typical in a soil when other factors are held constant:

1. k increases with moisture content
2. k increases with increasing dry density of a soil
3. k decreases with increasing organic content of a soil
4. k tends to decrease for soils with uniform gradations and rounded soil grains (because the grain-to-grain contacts are reduced)

Table 9 Typical Water Vapor Permeance and Permeability Values for Common Building Materials[a] (Concluded)

Material	Weight, lb/100 ft[2]	Permeance, Perms			Resistance[h] Rep		
		Dry-Cup	Wet-Cup	Other	Dry-Cup	Wet-Cup	Other
Building Paper, Felts, Roofing Papers[g]							
Duplex sheet, asphalt laminated, aluminum foil one side	8.6	0.002	0.176		500	5.8	
Saturated and coated roll roofing	65	0.05	0.24		20	4.2	
Kraft paper and asphalt laminated, reinforced 30-120-30	6.8	0.3	1.8		3.3	0.55	
Blanket thermal insulation backup paper, asphalt coated	6.2	0.4	0.6-4.2		2.5	1.7-0.24	
Asphalt-saturated and coated vapor retarder paper	8.6	0.2-0.3	0.6		5.0-3.3	1.7	
Asphalt-saturated, but not coated, sheathing paper	4.4	3.3	20.2		0.3	0.05	
15-lb asphalt felt	14	1.0	5.6		1.0	0.18	
15-lb tar felt	14	4.0	18.2		0.25	0.055	
Single-kraft, double	3.2	31	42		0.032	0.024	
Liquid-Applied Coating Materials	**Thickness, in.**						
Commercial latex paints (dry film thickness)[i]							
Vapor retarder paint	0.0031			0.45			2.22
Primer-sealer	0.0012			6.28			0.16
Vinyl acetate/acrylic primer	0.002			7.42			0.13
Vinyl-acrylic primer	0.0016			8.62			0.12
Semi-gloss vinyl-acrylic enamel	0.0024			6.61			0.15
Exterior acrylic house and trim	0.0017			5.47			0.18
Paint-2 coats							
Asphalt paint on plywood			0.4			2.5	
Aluminum varnish on wood		0.3-0.5			3.3-2.0		
Enamels on smooth plaster				0.5-1.5			2.0-0.66
Primers and sealers on interior insulation board				0.9-2.1			1.1-0.48
Various primers plus 1 coat flat oil paint on plaster				1.6-3.0			0.63-0.33
Flat paint on interior insulation board				4			0.25
Water emulsion on interior insulation board				30-85			0.03-0.012
	Weight, oz/ft[2]						
Paint-3 coats							
Exterior paint, white lead and oil on wood siding			0.3-1.0			3.3-1.0	
Exterior paint, white lead-zinc oxide and oil on wood			0.9			1.1	
Styrene-butadiene latex coating	2		11			0.09	
Polyvinyl acetate latex coating	4		5.5			0.18	
Chlorosulfonated polyethylene mastic	3.5		1.7			0.59	
	7.0		0.06			16	
Asphalt cutback mastic, 1/16 in., dry			0.14			7.2	
3/16 in., dry			0.0			—	
Hot melt asphalt	2		0.5			2	
	3.5		0.1			10	

[a]This table permits comparisons of materials; but in the selection of vapor retarder materials, exact values for permeance or permeability should be obtained from the manufacturer or from laboratory tests. The values shown indicate variations among mean values for materials that are similar but of different density, orientation, lot, or source. The values should not be used as design or specification data. Values from dry-cup and wet-cup methods were usually obtained from investigations using ASTM E96 and C355; values shown under others were obtained by two-temperature, special cell, and air velocity methods. Permeance, resistance, permeability, and resistance per unit thickness values are given in the following units:

Permeance	Perm	= gr/h · ft^2 · in. Hg
Resistance	Rep	= in. Hg · ft^2 · h/gr
Permeability	Perm-in.	= gr/h · ft^2 · (in. Hg/in.)
Resistance/unit thickness	Rep/in.	= (in. Hg · ft^2 · h/gr)/in.

[b]Depending on construction and direction of vapor flow.

[c]Usually installed as vapor retarders, although sometimes used as exterior finish and elsewhere near cold side, where special considerations are then required for warm side barrier effectiveness.

[d]Dry-cup method.

[e]Wet-cup method.

[f]Other than dry- or wet-cup method.

[g]Low permeance sheets used as vapor retarders. High permeance used elsewhere in construction.

[h]Resistance and resistance/in. values have been calculated as the reciprocal of the permeance and permeability values.

[i]Cast at 10 mils wet film thickness.

5. k of a frozen soil may be higher or lower than that of the same unfrozen soil (because the conductivity of ice is higher than that of water but lower than that of the typical soil grains). Differences in k below moisture contents of 7 to 8% are quite small. At approximately 15% moisture content, differences in k-factors may vary up to 30% from unfrozen values.

When calculating annual energy use, values that represent typical site conditions as they vary during the year should be chosen. In climates where ground freezing is significant, accurate heat transfer simulations should include the effect of the latent heat of fusion of water. The energy released during this phase change significantly retards the progress of the frost front in moist soils.

Water Vapor Transmission Data for Building Components

Table 9 gives typical water vapor permeance and permeability values for common building materials. These values can be used to calculate water vapor flow through building components and assemblies using Equations (14) through (17) in Chapter 20.

Table 10 Typical Thermal Conductivity k for Industrial Insulations at Various Mean Temperatures—Design Values[a]

Material	Max. Temp.,[b] °F	Typical Density, lb/ft³	Typical Conductivity k in Btu·in/h·ft²·°F at Mean Temp., °F													
			−100	−75	−50	−25	0	25	50	75	100	200	300	500	700	900
BLANKETS AND FELTS																
ALUMINOSILICATE FIBER																
7 to 10 μm diameter fiber	1800	4								0.24		0.32	0.54	0.99	1.03	
	2000	6-8								0.25		0.30	0.48	0.78	0.95	
3 μm diameter fiber	2200	4								0.22		0.29	0.45	0.59	0.74	
MINERAL FIBER (Rock, slag, or glass)																
Blanket, metal reinforced	1200	6-12									0.26	0.32	0.39	0.54		
	1000	2.5-6									0.24	0.31	0.40	0.61		
Blanket, flexible, fine-fiber organic bonded	350	<0.75				0.25	0.26	0.28	0.30	0.33	0.36	0.53				
		0.75				0.24	0.25	0.27	0.29	0.32	0.34	0.48				
		1.0				0.23	0.24	0.25	0.27	0.29	0.32	0.43				
		1.5				0.21	0.22	0.23	0.25	0.27	0.28	0.37				
		2.0				0.20	0.21	0.22	0.23	0.25	0.26	0.33				
		3.0				0.19	0.20	0.21	0.22	0.23	0.24	0.31				
Blanket, flexible, textile fiber, organic bonded	350	0.65				0.27	0.28	0.29	0.30	0.31	0.32	0.50	0.68			
		0.75				0.26	0.27	0.28	0.29	0.31	0.32	0.48	0.66			
		1.0				0.24	0.25	0.26	0.27	0.29	0.31	0.45	0.60			
		1.5				0.22	0.23	0.24	0.25	0.27	0.29	0.39	0.51			
		3.0				0.20	0.21	0.22	0.23	0.24	0.25	0.32	0.41			
Felt, semirigid organic bonded	400	3-8						0.24	0.25	0.26	0.27	0.35	0.44			
	850	3	0.16	0.17	0.18	0.19	0.20	0.21	0.22	0.23	0.24	0.35	0.55			
Laminated and felted without binder	1200	7.5											0.35	0.45	0.60	
BLOCKS, BOARDS, AND PIPE INSULATION																
MAGNESIA	600	11-12									0.35	0.38	0.42			
85% CALCIUM SILICATE	1200	11-15									0.38	0.41	0.44	0.52	0.62	0.72
	1800	12-15												0.63	0.74	0.95
CELLULAR GLASS	900	7.8-8.2	0.24	0.25	0.26	0.28	0.29	0.30	0.32	0.33	0.34	0.41	0.49	0.70	1.01	
DIATOMACEOUS SILICA	1600	21-22												0.64	0.68	0.72
	1900	23-25												0.70	0.75	0.80
MINERAL FIBER (Glass)																
Organic bonded, block and boards	400	3-10	0.16	0.17	0.18	0.19	0.20	0.22	0.24	0.25	0.26	0.33	0.40			
Nonpunking binder	1000	3-10									0.26	0.31	0.38	0.52		
Pipe insulation, slag, or glass	350	3-4						0.20	0.21	0.22	0.23	0.24	0.29			
	500	3-10						0.20	0.22	0.24	0.25	0.26	0.33	0.40		
Inorganic bonded block	1000	10-15									0.33	0.38	0.45	0.55		
	1800	15-24									0.32	0.37	0.42	0.52	0.62	0.74
Pipe insulation, slag, or glass	1000	10-15									0.33	0.38	0.45	0.55		
Resin binder		15	0.23	0.24	0.25	0.26	0.28	0.29								
RIGID POLYSTYRENE																
Extruded (CFC-12 exp.) (smooth skin surface)	165	1.8-3.5	0.16	0.16	0.17	0.16	0.17	0.18	0.19	0.20						
Molded beads	165	1	0.17	0.19	0.20	0.21	0.22	0.24	0.25	0.26	0.28					
		1.25	0.17	0.18	0.19	0.20	0.22	0.23	0.24	0.25	0.27					
		1.5	0.16	0.17	0.19	0.20	0.21	0.22	0.23	0.24	0.26					
		1.75	0.16	0.17	0.18	0.19	0.20	0.22	0.23	0.24	0.25					
		2.0	0.15	0.16	0.18	0.19	0.20	0.21	0.22	0.23	0.24					
RIGID POLYURETHANE/POLYISOCYANURATE[c,d]																
Unfaced (CFC-11 exp.)	210	1.5-2.5	0.16	0.17	0.18	0.18	0.18	0.17	0.16	0.16	0.17					
RIGID POLYISOCYANURATE[e]																
Gas-impermeable facers (CFC-11 exp.)	250	2.0						0.12	0.13	0.14	0.15					
RIGID PHENOLIC																
Closed cell (CFC-11, CFC-113 exp.)		3.0						0.11	0.115	0.12	0.125					
RUBBER, Rigid foamed	150	4.5						0.20	0.21	0.22	0.23					
VEGETABLE AND ANIMAL FIBER																
Wool felt (pipe insulation)	180	20						0.28	0.30	0.31	0.33					
INSULATING CEMENTS																
MINERAL FIBER (Rock, slag, or glass)																
With colloidal clay binder	1800	24-30									0.49	0.55	0.61	0.73	0.85	
With hydraulic setting binder	1200	30-40									0.75	0.80	0.85	0.95		
LOOSE FILL																
Cellulose insulation (milled pulverized paper or wood pulp)		2.5-3							0.26	0.27	0.29					
Mineral fiber, slag, rock, or glass		2-5				0.19	0.21	0.23	0.25	0.26	0.28	0.31				
Perlite (expanded)		3-5	0.22	0.24	0.25	0.27	0.28	0.30	0.31	0.33	0.35					
Silica aerogel		7.6				0.13	0.14	0.15	0.16	0.17	0.18					
Vermiculite (expanded)		7-8.2				0.39	0.40	0.42	0.44	0.45	0.47	0.49				
		4-6				0.34	0.35	0.38	0.40	0.42	0.44	0.46				

[a] Representative values for dry materials, which are intended as design (not specification) values for materials in normal use. Insulation materials in actual service may have thermal values that vary from design values depending on their in-situ properties (e.g., density and moisture content). For properties of a particular product, use the value supplied by the manufacturer or by unbiased tests.

[b] These temperatures are generally accepted as maximum. When operating temperature approaches these limits, follow the manufacturers' recommendations.
[c] Some polyurethane foams are formed by means that produce a stable product (with respect to k), but most are blown with refrigerant and will change with time.
[d] See Table 4, footnote i.
[e] See Table 4, footnote j.

MECHANICAL AND INDUSTRIAL SYSTEMS

Thermal Transmission Data

Table 10 lists the thermal conductivities of various materials used as industrial insulations. These values are functions of the arithmetic mean of the temperatures of the inner and outer surfaces for each insulation.

Heat Loss from Pipes and Flat Surfaces

Tables 11A, 11B, and 12 give heat losses from bare steel pipes and flat surfaces and bare copper tubes. These tables were calculated using ASTM *Standard* C 680, Practice for Determination of Heat Gain or Loss and the Surface Temperature of Insulated Pipe and Equipment Systems by the Use of a Computer Program. User inputs for these programs include operating temperature, ambient temperature, pipe size, insulation type, number of insulation layers, and thickness for each layer. A program option allows the user to input a surface coefficient or surface emittance, surface orientation, and wind speed. The computer uses this information to calculate the heat flow and the surface temperature. The programs calculate the surface coefficients if the user has not already supplied them.

The equations used in ASTM C680 are:

$$h_{cv} = C \left(\frac{1}{d}\right)^{0.2} \left(\frac{1}{t_{avg}}\right)^{0.181} \Delta t^{0.266} \sqrt{1 + 1.277 \,(\text{Wind})} \quad (4)$$

where

h_{cv} = convection surface coefficient, Btu/h·ft²·°F
d = diameter for cylinder, in. For flat surfaces and large cylinders ($d > 24$), use $d = 24$.
t_{avg} = average temperature of air film, °F
Δt = surface to air temperature difference, °F
Wind = air speed, mph
C = constant depending on shape and heat flow condition
 = 1.016 for horizontal cylinders
 = 1.235 for longer vertical cylinders
 = 1.394 for vertical plates
 = 1.79 for horizontal plates, warmer than air, facing upward
 = 0.89 for horizontal plates, warmer than air, facing downward
 = 0.89 for horizontal plates, cooler than air, facing upward
 = 1.79 for horizontal plates, cooler than air, facing downward

$$h_{rad} = \frac{\epsilon \times 0.1713 \times 10^{-8}[(t_a + 459.6)^4 - (t_s + 459.6)^4]}{(t_a - t_s)} \quad (5)$$

where

h_{rad} = radiation surface coefficient, Btu/h·ft²·°F
ϵ = surface emittance
t_a = air temperature, °F
t_s = surface temperature, °F

Table 11A   Heat Loss from Bare Steel Pipe to Still Air at 80°F[a], Btu/h·ft

Nominal Pipe Size[b], in.	Pipe Inside Temperature, °F									
	180	280	380	480	580	680	780	880	980	1080
0.50	59.3	147.2	263.2	412.3	600.9	836.8	1128.6	1485.6	1918.0	2436.8
0.75	72.5	180.1	322.6	506.2	739.2	1031.2	1392.9	1836.0	2373.5	3018.8
1.00	88.8	220.8	396.1	622.7	910.9	1272.6	1721.2	2271.5	2939.4	3741.6
1.25	109.7	272.8	490.4	772.3	1131.7	1583.8	2145.6	2835.4	3673.4	4680.9
1.50	123.9	308.5	555.1	875.1	1283.8	1798.3	2438.2	3224.6	4180.5	5330.0
2.00	151.8	378.1	681.4	1076.3	1581.5	2218.9	3012.6	3989.2	5177.2	6606.8
2.50	180.5	450.0	811.9	1284.0	1888.8	2652.6	3604.3	4775.3	6199.5	7912.5
3.00	215.9	538.8	973.5	1541.8	2271.4	3194.0	4344.9	5762.2	7486.9	9562.3
3.50	243.9	609.0	1101.4	1746.1	2574.7	3623.6	4933.0	6546.4	8510.4	10874.3
4.00	271.6	678.6	1228.2	1948.7	2875.9	4050.5	5517.5	7326.0	9528.1	12178.9
4.50	299.2	747.7	1354.4	2150.9	3176.8	4477.7	6103.8	8109.5	10553.2	13496.2
5.00	329.8	824.7	1494.8	2375.4	3510.6	4950.7	6751.3	8972.5	11678.4	14936.3
6.00	387.1	968.7	1757.8	2796.8	4138.0	5841.4	7972.7	10603.1	13808.2	17667.6
7.00	440.5	1102.8	2003.0	3189.9	4723.9	6673.5	9114.2	12127.4	15799.4	20220.8
8.00	493.3	1235.7	2246.1	3580.0	5305.5	7500.0	10248.4	13642.2	17778.2	22758.0
9.00	545.9	1368.1	2488.8	3970.2	5888.7	8331.0	11392.1	15174.5	19787.1	25343.6
10.00	604.3	1514.8	2757.2	4400.7	6530.1	9241.1	12638.6	16835.1	21949.2	28104.9
11.00	656.0	1644.8	2995.5	4783.8	7102.1	10054.9	13756.2	18328.4	23900.3	30606.1
12.00	704.0	1762.3	3203.8	5104.9	7557.3	10661.8	14524.9	19256.7	24967.6	31766.8
14.00	771.0	1934.2	3525.9	5636.0	8373.9	11862.4	16235.5	21635.6	28212.3	36120.3
16.00	872.2	2189.0	3993.2	6387.4	9495.9	13458.0	18424.8	24556.6	32021.1	40990.7
18.00	972.5	2441.7	4456.7	7132.9	10609.4	15041.3	20596.7	27453.2	35795.6	45813.1
20.00	1072.1	2692.4	4916.8	7873.2	11715.1	16613.4	22752.5	30326.8	39537.6	50590.0
24.00	1269.3	3188.9	5828.3	9339.9	13905.5	19726.9	27019.7	36010.1	46930.3	60014.7

Table 11B   Heat Loss from Flat Surfaces to Still Air at 80°F, Btu/h·ft²

	Surface Inside Temperature, °F									
	180	280	380	480	580	680	780	880	980	1080
Vertical surface	212.2	533.1	973.3	1558.6	2321.2	3298.0	4530.1	6062.8	7945.5	10231.5
Horizontal surface										
Facing up	234.7	586.4	1061.1	1683.5	2484.9	3501.9	4775.4	6350.4	8276.3	10606.1
Facing down	183.6	465.3	861.4	1399.6	2112.8	3038.4	4217.8	5696.7	7524.5	9754.7

[a]Calculations from ASTM C680-82; steel: k = 314.4 Btu·in/h·ft²·°F; ϵ = 0.94.

[b]Losses per square foot of pipe for pipes larger than 24 in. can be considered the same as losses per square foot for 24-in. pipe.

Example 7. Compute total annual heat loss from 165 ft of nominal 2-in. bare steel pipe in service 4000 h per year. The pipe is carrying steam at 10 psi and is exposed to an average air temperature of 80°F.

Solution: The pipe temperature is taken as the steam temperature, which is 239.4°F, obtained by interpolation from Steam Tables. By interpolation in Table 11A between 180°F and 280°F, heat loss from a 2-in. pipe is 285.3 Btu/h·ft. Total annual heat loss from the entire line is 285.3 Btu/h·ft × 165 ft × 4000 h = 188 million Btu.

In calculating heat flow, Equations (9) and (10) from Chapter 20 generally are used. For dimensions of standard pipe and fitting sizes, refer to the *Piping Handbook*. For insulation product dimensions, refer to ASTM *Standard* C 585, Recommended Practice for Inner and Outer Diameters of Rigid Thermal Insulation for Nominal Sizes of Pipe and Tubing (NPS) System, or to the insulation manufacturers' literature.

Examples 8 and 9 illustrate how Equations (9) and (10) from Chapter 20 can be used to determine heat loss from both flat and cylindrical surfaces. Figure 7 shows surface resistance as a function of heat transmission for both flat and cylindrical surfaces. The surface emittance is assumed to be 0.85 to 0.90 in still air at 80°F.

Example 8. Compute heat loss from a boiler wall if the interior insulation surface temperature is 1100°F and ambient still air temperature is 80°F. The wall is insulated with 4.5 in. of mineral fiber block and 0.5 in. of mineral fiber insulating and finishing cement.

Solution: Assume that the mean temperature of the mineral fiber block is 700°F, the mean temperature of the insulating cement is 200°F, and the surface resistance R_s is 0.60.

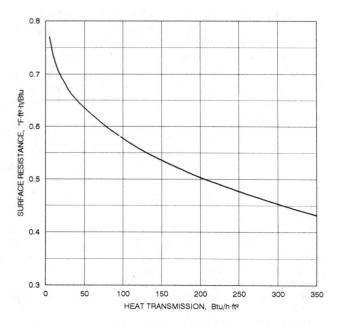

Fig. 7   Surface Resistance as Function of Heat Transmission for Flat Surfaces and Cylindrical Surfaces Greater than 24 Inches in Diameter

Table 12   Heat Loss from Bare Copper Tube to Still Air at 80°F[a], Btu/h·ft

Nominal Tube Size, in.	Tube Inside Temperature, °F								
	120	150	180	210	240	270	300	330	
0.250	7.1	14.1	21.9	30.6	39.9	49.9	60.6	71.9	
0.375	9.1	18.0	28.1	39.1	51.1	63.9	77.6	92.2	
0.500	11.0	21.8	34.0	47.4	61.9	77.5	94.1	111.8	
0.750	14.7	29.1	45.4	63.3	82.7	103.6	126.0	149.8	
1.000	18.3	36.2	56.4	78.7	102.8	128.9	156.7	186.5	
1.250	21.8	43.1	67.2	93.6	122.4	153.4	186.7	222.2	
1.500	25.2	49.8	77.6	108.3	141.5	177.4	216.0	257.1	
2.000	31.8	62.9	98.0	136.7	178.8	224.3	273.1	325.4	
2.500	38.3	75.6	117.9	164.4	215.1	269.8	328.7	391.8	Dull $\epsilon = 0.44$
3.000	44.6	88.1	137.2	191.5	250.5	314.4	383.2	456.9	
3.500	50.8	100.3	156.3	218.0	285.4	358.2	436.7	520.8	
4.000	57.0	112.3	175.0	244.2	319.7	401.4	489.4	583.9	
5.000	69.0	135.9	211.7	295.5	386.9	486.0	592.8	707.6	
6.000	80.7	159.0	247.7	345.7	452.8	568.9	694.2	829.0	
8.000	103.7	204.1	317.8	443.7	581.3	730.7	892.1	1066.0	
10.000	126.1	247.9	386.1	539.1	706.5	888.4	1085.2	1297.4	
12.000	148.0	290.9	453.0	632.5	829.2	1043.1	1274.6	1524.4	
0.250	5.4	10.8	16.9	23.5	30.5	37.9	45.5	53.5	
0.375	6.8	13.7	21.4	29.7	38.6	47.9	57.6	67.6	
0.500	8.2	16.4	25.7	35.7	46.3	57.4	69.1	81.2	
0.750	10.7	21.6	33.8	46.9	60.9	75.6	90.9	106.8	
1.000	13.2	26.5	41.4	57.6	74.7	92.8	111.6	131.2	
1.250	15.5	31.3	48.8	67.8	88.0	109.3	131.6	154.7	
1.500	17.8	35.8	56.0	77.8	100.9	125.3	150.8	177.4	
2.000	22.2	44.6	69.7	96.8	125.7	156.1	187.9	221.1	
2.500	26.4	53.0	82.8	115.1	149.5	185.6	223.5	263.0	Bright $\epsilon = 0.08$
3.000	30.5	61.2	95.6	132.8	172.4	214.2	257.9	303.5	
3.500	34.4	69.1	107.9	150.0	194.8	242.0	291.4	342.9	
4.000	38.3	76.8	120.0	166.8	216.6	269.1	324.1	381.4	
5.000	45.7	91.8	143.4	199.3	258.8	321.6	387.4	456.1	
6.000	53.0	106.3	166.0	230.7	299.7	372.5	448.7	528.3	
8.000	66.8	134.1	209.4	291.1	378.2	470.1	566.5	667.2	
10.000	80.2	160.8	251.0	349.0	453.4	563.7	679.5	800.4	
12.000	93.0	186.5	291.3	404.9	526.1	654.2	788.7	929.3	

[a]Calculations from ASTM C680-82; for copper: $k = 2784$ Btu·in/h·ft^2·°F.

From Table 10, $k_1 = 0.62$ and $k_2 = 0.80$. Using Equation (9) from Chapter 20:

$$q_s = \frac{1100 - 80}{(4.5/0.62) + (0.5/0.80) + 0.60} = \frac{1020}{8.48}$$

$$= 120.2 \text{ Btu/h} \cdot \text{ft}^2$$

As a check, from Figure 7, at 120.2 Btu/h·ft², $R_s = 0.56$. The mean temperature of the mineral fiber block is:

$$4.5/0.62 = 7.26; 7.26/2 = 3.63$$

$$1100 - [(3.63/8.48)(1020)] = 1100 - 437 = 663\,°F$$

and the mean temperature of the insulating cement is:

$$0.5/0.80 = 0.63; 0.63/2 = 0.31; 7.26 + 0.31 = 7.57$$

$$1100 - [(7.57/8.48)(1020)] = 1100 - 911 = 189\,°F$$

From Table 10, at 663 °F, $k_1 = 0.60$; at 189 °F, $k_2 = 0.79$. Using these adjusted values to recalculate q_s:

$$q_s = \frac{1020}{(4.5/0.60) + (0.5/0.79) + 0.56} = \frac{1020}{8.69}$$

$$= 117.4 \text{ Btu/h} \cdot \text{ft}^2$$

From Figure 7, at 117.4 Btu/h·ft², $R_s = 0.56$. The mean temperature of the mineral fiber block is:

$$4.5/0.6 = 7.50; 7.50/2 = 3.75$$

$$1100 - [(3.75/8.69)(1020)] = 1100 - 440 = 660\,°F$$

and the mean temperature of the insulating cement is:

$$0.5/0.79 = 0.63; 0.63/2 = 0.31; 7.50 + 0.31 = 7.81$$

$$1100 - [(7.81/8.69)(1020)] = 1100 - 917 = 183\,°F$$

From Table 10, at 660 °F, $k_1 = 0.60$; at 183 °F, $k_2 = 0.79$. Since R_s, k_1, and k_2 do not change at these values, $q_s = 117.4$ Btu/h·ft².

Example 9. Compute heat loss per square foot of outer surface of insulation if pipe temperature is 1200 °F and ambient still air temperature is 80 °F. The pipe is nominal 6-in. steel pipe, insulated with a nominal 3-in. thick diatomaceous silica as the inner layer and a nominal 2-in. thick calcium silicate as the outer layer.

Solution: From Chapter 42 of the 1992 ASHRAE *Handbook—Equipment*, $r_o = 3.31$ in. A nominal 3-in. thick diatomaceous silica insulation to fit a nominal 6-in. steel pipe is 3.02 in. thick. A nominal 2-in. thick calcium silicate insulation to fit over the 3.02-in. diatomaceous silica is 2.08 in. thick. Therefore, $r_i = 6.33$ in. and $r_s = 8.41$ in.

Assume that the mean temperature of the diatomaceous silica is 600 °F, the mean temperature of the calcium silicate is 250 °F and the surface resistance R_s is 0.50. From Table 10, $k_1 = 0.66$; $k_2 = 0.42$. By Equation (10) from Chapter 20:

$$q_s = \frac{1200 - 80}{[8.41 \ln (6.33/3.31)/0.66] + [8.41 \ln (8.41/6.33)/0.40] + 0.50}$$

$$= \frac{1120}{(5.45/0.66) + (2.39/0.40) + 0.50} = 76.0 \text{ Btu/h} \cdot \text{ft}^2$$

From Figure 7, at 76.0 Btu/h·ft², $R_s = 0.60$. The mean temperature of the diatomaceous silica is:

$$5.45/0.66 = 8.26; 8.26/2 = 4.13$$

$$1200 - [(4.13/14.83)(1120)] = 1200 - 312 = 888\,°F$$

and the mean temperature of the calcium silicate is:

$$2.39/0.40 = 5.98; 5.98/2 = 2.99; 8.26 + 2.99 = 11.25$$

$$1200 - [(11.25/14.83)(1120)] = 1200 - 850 = 350\,°F$$

From Table 10, $k_1 = 0.72$; $k_2 = 0.46$. Recalculating:

$$q_s = \frac{1120}{(5.45/0.72) + (2.39/0.46) + 0.60} = 83.8 \text{ Btu/h} \cdot \text{ft}^2$$

From Figure 7 at 83.8 Btu/h·ft², $R_s = 0.59$. The mean temperature of the diatomaceous silica is:

$$5.45/0.72 = 7.57; 7.57/2 = 3.78$$

$$1200 - [(3.78/13.36)(1120)] = 1200 = 317 = 883\,°F$$

and the mean temperature of the calcium silicate is:

$$2.39/0.46 = 5.20; 5.20/2 = 2.60; 7.57 + 2.60 = 10.17$$

$$1200 - [(10.17/13.36)(1120)] = 1200 - 853 = 347\,°F$$

From Table 10, $k_1 = 0.72$; $k_2 = 0.46$. Recalculating:

$$q_s = \frac{1120}{(5.45/0.72) + (2.39/0.46) + 0.59} = 83.8 \text{ Btu/h} \cdot \text{ft}^2$$

Since R_s, k_1, and k_2 do not change at 83.8 Btu/h·ft², this is q_s. The heat flow per ft² of the inner surface of the insulation is:

$$q_o = q_s (r_s/r_o) = 83.8(8.41/3.31) = 213 \text{ Btu/h} \cdot \text{ft}^2$$

Because trial and error techniques are tedious, the computer programs previously described should be used to estimate heat flows per unit area of flat surfaces or per unit length of piping, and interface temperatures including surface temperatures.

Several methods can be used to determine the most effective thickness of insulation for piping and equipment. Table 13 shows the recommended insulation thicknesses for three different pipe and equipment insulations. Installed cost data can be developed using procedures described by the Federal Energy Administration (1976). Computer programs capable of calculating thickness information are available from several sources. Also, manufacturers of insulations offer computerized analysis programs for designers and owners to evaluate insulation requirements. For more information on determining economic insulation thickness, see Chapter 20.

Chapters 3 and 20 give guidance concerning process control, personnel protection, condensation control, and economics. For specific information on sizes of commercially available pipe insulation, see ASTM *Standard* C585 and consult with the Thermal Insulation Manufacturers Association (TIMA) and its member companies.

CALCULATING HEAT FLOW FOR BURIED PIPELINES

In calculating heat flow to or from buried pipelines, the thermal properties of the soil must be assumed. Table 7 gives the apparent thermal conductivity values of various soil types, and Figure 6 shows the typical trends of apparent soil thermal conductivity with moisture content for various soil types. Table 8 provides ranges of apparent thermal conductivity for various types of rock. Kernsten (1949) also discusses thermal properties of soils. Carslaw and Jaeger (1959) give methods for calculating the heat flow taking place between one or more buried cylinders and the surroundings.

Table 13   Recommended Thicknesses for Pipe and Equipment Insulation

Nominal Pipe Size, in.		MINERAL FIBER (Fiberglass and Rock Wool) Process Temperature, °F										CALCIUM		
		150	250	350	450	550	650	750	850	950	1050	150	250	350
½	Thickness	1	1½	2	2½	3	3½	4	4	4½	5½	1	1½	2
	Heat loss	8	16	24	33	43	54	66	84	100	114	13	24	34
	Surface temperature	72	75	76	78	79	81	82	86	87	87	75	78	80
1	Thickness	1	1½	2	2½	3½	4	4	4½	5	5½	1	2	2½
	Heat loss	11	21	30	41	49	61	79	96	114	135	16	26	38
	Surface temperature	73	76	78	80	79	81	84	86	88	89	76	76	79
1½	Thickness	1	2	2½	3	4	4	4	5½	5½	6	1½	2½	3
	Heat loss	14	22	33	45	54	73	94	103	128	152	17	29	42
	Surface temperature	73	74	77	79	79	82	86	84	88	90	73	75	78
2	Thickness	1½	2	3	3½	4	4	4	5½	6	6	1½	2½	3
	Heat loss	13	25	34	47	61	81	105	114	137	168	19	32	47
	Surface temperature	71	75	75	77	79	83	87	85	87	91	74	76	79
3	Thickness	1½	2½	3½	4	4	4½	4½	6	6½	7	2	3	3½
	Heat loss	16	28	39	54	75	94	122	133	154	184	21	37	54
	Surface temperature	72	74	75	77	81	83	87	86	87	90	73	75	78
4	Thickness	1½	3	4	4	4	5	5½	6	7	7½	2	3	4
	Heat loss	19	29	42	63	88	102	126	152	174	206	25	43	58
	Surface temperature	72	73	74	78	82	86	85	87	88	90	70	76	77
6	Thickness	2	3	4	4	4½	5	5½	6½	7½	8	2	3½	4
	Heat loss	21	38	54	81	104	130	159	181	208	246	33	51	75
	Surface temperature	71	74	75	79	82	84	87	88	89	91	74	75	79
8	Thickness	2	3½	4	4	5	5	5½	7	8	8½	2½	3½	4
	Heat loss	26	42	65	97	116	155	189	204	234	277	35	62	90
	Surface temperature	71	73	76	80	81	86	89	88	89	92	73	76	79
10	Thickness	2	3½	4	4	5	5½	5½	7½	8½	9	2½	4	4
	Heat loss	32	50	77	115	136	170	220	226	259	307	41	66	106
	Surface temperature	72	74	77	81	82	85	90	87	89	91	73	75	80
12	Thickness	2	3½	4	4	5	5½	5½	7½	8½	9½	2½	4	4
	Heat loss	36	57	87	131	154	192	249	253	290	331	47	75	121
	Surface temperature	72	74	77	82	82	86	91	88	89	91	73	76	81
14	Thickness	2	3½	4	4	5	5½	6½	7½	9	9½	2½	4	4
	Heat loss	40	61	94	141	165	206	236	271	297	352	51	81	130
	Surface temperature	72	74	77	82	83	86	87	89	89	91	73	76	81
16	Thickness	2½	3½	4	4	5½	5½	7	8	9	10	3	4	4
	Heat loss	37	68	105	157	171	228	247	284	326	372	50	90	144
	Surface temperature	71	74	78	83	82	87	86	88	89	91	72	76	82
18	Thickness	2½	3½	4	4	5½	5½	7	8	9	10	3	4	4
	Heat loss	41	75	115	173	187	250	270	310	354	404	55	99	159
	Surface temperature	71	74	78	83	83	87	87	88	90	91	73	76	82
20	Thickness	2½	3½	4	4	5½	5½	7	8	9	10	3	4	4
	Heat loss	45	82	126	189	204	272	292	335	383	436	60	108	174
	Surface temperature	71	75	78	83	83	87	87	89	90	92	73	77	82
24	Thickness	2½	4	4	4	5½	6	7½	8	9	10	3	4	4
	Heat loss	53	86	147	221	237	295	320	386	439	498	71	127	203
	Surface temperature	71	74	78	83	83	86	86	89	91	93	73	77	82
30	Thickness	2½	4	4	4	5½	6½	7½	8½	10	10	3	4	4
	Heat loss	65	105	179	268	286	332	383	439	481	591	86	154	247
	Surface temperature	71	74	79	84	84	85	87	89	89	94	73	77	83
36	Thickness	2½	4	4	4	5½	7	8	9	10	10	2½	4	4
	Heat loss	77	123	211	316	335	364	422	486	556	683	119	181	291
	Surface temperature	71	74	79	84	84	84	86	88	90	94	74	77	83
Flat	Thickness	2	3½	4	4½	5½	8½	9½	10	10	10	2½	3½	4
	Heat loss	10	14	20	27	31	27	31	38	47	58	12	20	28
	Surface temperature	72	74	77	80	82	80	82	85	89	93	73	77	81

Consult manufacturer's literature for product temperature limitations. Table is based on typical operating conditions, *e.g.,* 65 °F ambient temperature and 7.5 mph wind speed, and may not represent actual conditions of use. Units for thickness, heat loss, and surface temperature are in inches, Btu/h·ft (Btu/h·ft² for flat surfaces), and °F, respectively.

Table 13   Recommended Thicknesses for Pipe and Equipment Insulation (*Concluded*)

SILICATE							CELLULAR GLASS						
Process Temperature, °F							Process Temperature, °F						
450	550	650	750	850	950	1050	150	250	350	450	550	650	750
2½	3	3½	4	4	4	4	1½	1½	2	2½	3	3½	4
42	53	63	75	90	108	128	9	23	34	48	62	78	92
81	82	83	84	87	91	94	70	76	78	82	83	85	84
3	3½	4	4	4	4	4	1½	2	2½	3	3½	4	4
49	60	72	89	109	130	154	12	25	38	52	68	86	112
80	82	83	86	90	94	98	71	75	77	79	81	83	88
3½	4	4	4	4	5	5	1½	2½	3	4	4	4	4
54	68	86	106	128	139	164	15	28	44	56	79	105	137
80	81	85	88	92	91	94	72	75	77	78	82	87	92
3½	4	4½	5	5½	6	6	1½	2½	3	4	4	4	4½
61	75	90	106	123	142	167	17	31	47	61	84	113	140
81	82	84	85	87	88	91	72	74	77	78	82	86	89
4	4½	5	5½	6	6	6	1½	3	3½	4	4	4½	5
71	87	105	123	143	71	202	22	35	54	75	105	132	161
80	82	84	85	87	90	94	73	74	77	79	84	86	89
4	4½	5	5½	6	6½	7	2	3	4	4	4	4½	5
82	101	121	142	164	187	213	22	41	59	87	122	150	185
81	83	85	87	89	90	92	71	74	76	80	85	87	90
4	4½	5	5½	6	7	8	2	3½	4	4	4½	5½	6
105	129	153	178	205	224	245	30	48	74	111	144	171	212
83	85	87	89	91	91	91	72	74	77	82	85	86	89
4½	5	5	6	7	8	8½	2½	3½	4	4	5	5½	6½
117	144	183	200	220	243	277	30	58	90	134	161	203	238
82	85	89	89	89	90	92	71	74	78	83	84	87	89
4	5	5½	6	7½	8½	9	2½	4	4	4	5½	5½	7
149	168	200	233	243	269	306	37	63	106	159	178	238	264
85	86	88	90	89	89	91	71	74	79	84	84	87	88
4	5	5½	7	8	8½	9½	2½	4	4	4	5½	5½	7½
170	191	266	236	262	300	330	42	71	121	181	201	269	284
86	86	89	88	88	90	91	71	74	79	85	84	90	88
4	5	5½	7	8	9	9½	2½	4	4	4	5½	5½	8
183	205	242	252	262	308	352	47	79	134	199	219	293	293
86	87	89	88	88	89	91	72	74	80	85	85	91	87
4	5½	6½	7½	8	9	10	2½	4	4	4	5½	5½	8
204	211	237	265	307	338	372	53	88	149	222	242	325	322
87	85	86	87	89	90	91	72	75	80	86	86	91	88
4	5½	6½	7½	8½	9	10	2½	4	4	4	5½	5½	8
225	232	259	289	320	367	403	59	96	164	245	266	356	351
87	86	87	87	88	90	91	72	75	80	86	86	92	88
4	5½	6½	7½	8½	9½	10	2½	4	4	4½	5½	5½	8
245	252	281	312	346	381	435	64	105	179	243	289	387	379
87	86	87	88	89	90	92	72	75	81	84	86	92	88
4	5½	6½	7½	8½	9½	10	2½	4	4	5	5½	5½	8
287	293	325	360	397	437	497	76	123	209	260	336	449	436
88	87	88	88	89	90	93	72	75	81	83	87	93	89
4	5½	7	8	9	10	10	2½	4	4	5½	5½	5½	8
349	353	368	409	452	498	589	93	150	254	290	405	542	521
88	87	87	88	89	90	94	72	75	81	82	87	93	90
4	6½	7½	8	9	10	10	2½	4	4	5½	5½	5½	8
410	359	406	475	524	576	681	110	176	229	340	474	635	606
89	84	86	88	89	91	94	73	76	81	82	88	94	90
5½	6½	7½	8½	9½	10	10	2½	4	4	5½	5½	7½	8½
29	33	36	39	43	49	58	11	17	29	31	44	43	50
81	83	84	85	87	89	93	73	76	83	84	90	90	93

REFERENCES

Adams, L. 1971. Supporting cryogenic equipment with wood. *Chemical Engineering* (May):156-58.

Bassett, M.R. and H.A. Trethowen. 1984. Effect of condensation on emittance of reflective insulation. *Journal of Thermal Insulation* 8 (October):127.

Carslaw, H.S. and J.C. Jaeger. 1959. Conduction of heat in solids. Oxford University Press, Amen House, London, England, 449.

Dill, R.S., W.C. Robinson, and H.E. Robinson. 1945. Measurements of heat losses from slab floors. National Bureau of Standards. Building Materials and Structures Report, BMS 103.

Economic thickness for industrial insulation. 1976. GPO No. 41-018-001 15-8, Federal Energy Administration, Washington, D.C.

Farouk, B. and D.C. Larson. 1983. Thermal performance of insulated wall systems with metal studs. Proceedings of the 18th Intersociety Energy Conversion Engineering Conference, Orlando, FL.

Farouki, O.T. 1981. Thermal properties of soil. CRREL *Monograph* 81-1, United States Army Corps of Engineers Cold Regions Research and Engineering Laboratory, December.

Fishenden, M. 1962. Tables of emissivity of surfaces. *International Journal of Heat and Mass Transfer* 5:67-76.

Goss, W.P. and R.G. Miller. 1989. Literature review of measurement and prediction of reflective building insulation system performance: 1900-1989. ASHRAE *Transactions* 95(2).

Hooper, F.C. and W.J. Moroz. 1952. The impact of aging factors on the emissivity of reflective insulations. ASTM *Bulletin* (May):92-95.

Hougten, F.C., S.I. Taimuty, C. Gutberlet, and C.J. Brown. 1942. Heat loss through basement walls and floors. ASHVE *Transactions* 48:369.

Joy, F.A. 1958. Improving attic space insulating values. ASHAE *Transactions* 64:251.

Kersten, M.S. 1949. Thermal properties of soils. University of Minnesota, Engineering Experiment Station Bulletin 28, June.

Latta, J.K. and G.G. Boileau. 1969. Heat losses from house basements. *Canadian Building* 19(10).

Lewis, W.C. 1967. Thermal conductivity of wood-base fiber and particle panel materials. Forest Products Laboratory, Research Paper FPL 77, June.

Lotz, W.A. 1964. Vapor barrier design, neglected key to freezer insulation effectiveness. *Quick Frozen Foods* (November):122.

MacLean, J.D. 1941. Thermal conductivity of wood. ASHVE *Transactions* 47:323.

McElroy, D.L., D.W. Yarbrough, and R.S. Graves. 1987. Thickness and density of loose-fill insulations after installation in residential attics. *Thermal insulation: Materials and systems*. F.J. Powell and S.L. Matthews, eds. ASTM STP 922:423-505.

McIntyre, D.A. 1984. The increase in U-value of a wall caused by mortar joints, ECRC/M1843. The Electricity Council Research Centre, Copenhurst, England, June.

Mitalas, G.P. 1982. Basement heat loss studies at DBR/NRC, NRCC 20416. Division of Building Research, National Research Council of Canada, September.

Mitalas, G.P. 1983. Calculation of basement heat loss. ASHRAE *Transactions* 89(1B):420.

Moroz, W.J. 1951. Aging factors affecting reflective insulations. MS Thesis, University of Toronto, January.

Prangnell, R.D. 1971. The water vapor resistivity of building materials—A literature survey. *Materiaux et Constructions* 4:24 (November).

Robinson, H.E., F.J. Powell, and L.A. Cosgrove. 1957. Thermal resistance of airspaces and fibrous insulations bounded by reflective surfaces. National Bureau of Standards, Building Materials and Structures Report BMS 151.

Robinson, H.E., F.J. Powlitch, and R.S. Dill. 1954. The thermal insulation value of airspaces. Housing and Home Finance Agency, Housing Research Paper No. 32.

Sabine, H.J., M.B. Lacher, D.R. Flynn, and T.L. Quindry. 1975. Acoustical and thermal performance of exterior residential walls, doors and windows. National Bureau of Standards, Building Science Series 77, November.

Salomone, L.A. and J.I. Marlowe. 1989. Soil and rock classification according to thermal conductivity: Design of ground-coupled heat pump systems. EPRI CU-6482, Electric Power Research Institute, August.

Shipp, P.H. 1983. Basement, crawlspace and slab-on-grade thermal performance. Proceedings of the ASHRAE/DOE Conference, Thermal Performance of the Exterior Envelopes of Buildings II, ASHRAE SP 38:160-79.

Shu, L.S., A.E. Fiorato, and J.W. Howanski. 1979. Heat transmission coefficients of concrete block walls with core insulation. Proceedings of the ASHRAE/DOE-ORNL Conference, Thermal Performance of the Exterior Envelopes of Buildings, ASHRAE SP 28:421-35.

Tye, R.P. 1985. Upgrading thermal insulation performance of industrial processes. *Chemical Engineering Progress* (February):30-34.

Tye, R.P. 1986. Effects of product variability on thermal performance of thermal insulation. Proceedings of the First Asian Thermal Properties Conference, Beijing, People's Republic of China.

Tye, R.P. and A.O. Desjarlais. 1983. Factors influencing the thermal performance of thermal insulations for industrial applications. *Thermal insulation, materials, and systems for energy conservation in the '80s*. F.A. Govan, D.M. Greason, and J.D. McAllister, eds. ASTM STP 789:733-48.

Tye, R.P. and S.C. Spinney. 1980. A study of various factors affecting the thermal performance of perlite insulated masonry construction. Dynatech Report No. PII-2. Holometrix, Inc. (formerly Dynatech R/D Company), Cambridge, MA.

USDA. 1974. *Wood handbook*. Wood as an engineering material. Forest Products Laboratory, U.S. Department of Agriculture Handbook No. 72, Tables 3-7 and 4-2, and Figures 3-4 and 3-5.

Valore, R.C. 1980. Calculation of U-values of hollow concrete masonry. American Concrete Institute, *Concrete International* 2(2):40-62.

Valore, R.C. 1988. Thermophysical properties of masonry and its constituents, Parts I and II. International Masonry Institute, Washington, D.C.

Valore, R., A. Tuluca, and A. Caputo. 1988. Assessment of the thermal and physical properties of masonry block products (ORNL/Sub/86-22020/1), September.

Van Geem, M.G. 1985. Thermal transmittance of concrete block walls with core insulation. ASHRAE *Transactions* 91(2).

Wilkes, K.E. 1979. Thermophysical properties data base activities at Owens-Corning Fiberglas. Proceedings of the ASHRAE/DOE-ORNL Conference, Thermal Performance of the Exterior Envelopes of Buildings, ASHRAE SP 28:662-77.

Yarbrough, E.W. 1983. Assessment of reflective insulations for residential and commercial applications (ORNL/TM-8891), October.

Yellott, J.I. 1965. Thermal and mechanical effects of solar radiation on steel doors. ASHRAE *Transactions* 71(2):42.

CHAPTER 23

INFILTRATION AND VENTILATION

OUTDOOR air that flows through a building either intentionally as ventilation air or unintentionally as infiltration (and exfiltration) is important for two reasons. Outdoor air is often used to dilute indoor air contaminants, and the energy associated with heating or cooling this outdoor air is a significant space-conditioning load. The magnitude of these airflow rates should be known at maximum load to properly size equipment and at average conditions, to properly estimate average or seasonal energy consumption. Minimum air exchange rates need to be known to assure proper control of indoor contaminant levels. In large buildings, the effect of infiltration and ventilation on distribution and interzone airflow patterns, which include smoke circulation patterns in the event of fire, should be determined (see Chapter 47 of the 1991 ASHRAE *Handbook—Applications*).

Air exchange between indoors and out is divided into ventilation (intentionally controlled) and infiltration (unintentional and uncontrolled). Ventilation can be natural and forced. Natural ventilation is unpowered airflow through open windows, doors, and other intentional openings in the building envelope. Forced ventilation is intentional, powered air exchange by a fan or blower and intake and/or exhaust vents that are specifically designed and installed for ventilation.

Infiltration is uncontrolled airflow through cracks, interstices, and other unintentional openings. Infiltration, exfiltration, and natural ventilation airflows are caused by pressure differences due to wind, indoor-outdoor temperature differences, and appliance operation.

This chapter focuses on residences and small commercial buildings in which air exchange is due primarily to envelope or shell infiltration. The physical principles are also discussed in relation to large buildings in which air exchange depends more on mechanical ventilation than it does on building envelope performance.

TYPES OF AIR EXCHANGE

Buildings have three different modes of air exchange: (1) forced ventilation, (2) natural ventilation, and (3) infiltration. These modes differ significantly in how they affect energy, air quality, and thermal comfort. They also differ in their ability to maintain over time a desired air exchange rate. The air exchange rate in a building at any given time generally includes all three modes, and they all must be considered even when only one is expected to dominate.

Forced ventilation affords the greatest potential for control of air exchange rate and air distribution within a building through the proper design, installation, operation, and maintenance of the ventilation system. Forced or mechanical ventilation equipment and systems are described in Chapers 1, 2, and 9 of the 1992 ASHRAE *Handbook—Systems and Equipment*. An ideal forced ventilation system has a sufficient ventilation rate to control indoor contaminant levels and, at the same time, avoids overventilation and the associated energy penalty. In addition, it maintains good thermal comfort (see Chapters 8 and 32).

Forced ventilation is generally mandatory in larger buildings, where a minimum amount of outdoor air is required for occupant health and comfort, and where a mechanical exhaust system is advisable or necessary. Forced ventilation has generally not been used in residential and other envelope-dominated structures. However, tighter, more energy-conserving buildings require ventilation systems to assure an adequate amount of outdoor air for maintaining acceptable indoor air quality.

Natural ventilation through intentional openings is caused by pressures from wind and/or indoor-outdoor temperature differences. Airflow through open windows and doors and other design openings can be used to provide adequate ventilation for contaminant dilution and temperature control. Unintentional openings in the building envelope and the associated infiltration can interfere with desired natural ventilation air distribution patterns and lead to larger than design airflow rates. Natural ventilation is sometimes defined to include infiltration, but in this chapter it does not.

Infiltration is the uncontrolled flow of air through unintentional openings driven by wind, temperature difference, and/or appliance-induced presures across the building envelope. Infiltration is least reliable in providing adequate ventilation and distribution, because it depends on weather conditions and the location of unintentional openings. It is the main source of ventilation in envelope-dominated buildings and is also an important factor in mechanically ventilated buildings.

VENTILATION AND THERMAL LOADS

Outdoor air introduced into a building constitutes part of the space-conditioning (heating, cooling, dehumidification) load, which is one reason to limit air exchange rates in buildings to the minimum required. Air exchange typically represents 20 to 40% of the shell-dominated building's thermal load. Chapters 25 and 26 cover thermal loads in more detail.

Air exchange increases a building's thermal load in three ways. First, the incoming air must be heated or cooled from the outdoor air temperature to the indoor air temperature. The rate of energy consumption due to this sensible heating or cooling is given by:

$$q_s = 60\, Q\, \rho\, c_p \Delta t \tag{1}$$

where

q_s = sensible heat load, Btu/h
Q = airflow rate, cfm
ρ = air density, lb_m/ft^3 (about 0.075)
c_p = specific heat of air, Btu/lb·°F (about 0.24)
Δt = indoor-outdoor temperature difference, °F

The preparation of this chapter is assigned to TC 4.3, Ventilation Requirements and Infiltration.

Second, air exchange modifies the moisture content of the air in a building. This is particularly important in the summer in some locations when the humid outdoor air must be dehumidified. In the winter, when the humidity ratio of the indoor air is much lower than that of the outside air, humidification may be needed. The rate of energy consumption associated with these latent loads is given by:

$$q_l = 60 \, Q \, \rho \, h_{fg} \, \Delta W \qquad (2)$$

where

q_l = latent heat loads, Btu/h
h_{fg} = latent heat of vapor at appropriate air temperature, Btu/lb_m (about 1000)
ΔW = humidity ratio of indoor air minus humidity ratio of outdoor air, lb_m water/lb_m dry air

Finally, air exchange can increase a building's thermal loads by decreasing the performance of the envelope insulation system. Air flowing around and through the insulation can increase heat transfer rates above design rates. The effect of such airflow on insulation system performance is difficult to quantify but should be considered. Airflow within the insulation system can also decrease the system's performance due to moisture condensing in and on the insulation.

VENTILATION AND INDOOR AIR QUALITY

Outdoor air requirements for acceptable indoor air quality (IAQ) have long been debated, and different rationales have produced radically different ventilation standards (Grimsrud and Teichman 1989; Janssen 1989; Klauss *et al.* 1970; Yaglou *et al.* 1936, 1937). Historically, the major considerations have included the amount of outdoor air required to control moisture, carbon dioxide (CO_2), odors, and tobacco smoke generated by occupants. These considerations have led to prescriptions of the

Table 1 Indoor Air Pollutants and Sources

Sources	Pollutant Types
OUTDOOR	
Ambient air	SO_2, NO, NO_2, O_3, hydrocarbons, CO, particulates, bioaerosols
Motor vehicles	CO, Pb, hydrocarbons, particulates
Soil	Radon, organics
INDOOR	
Building construction materials	
Concrete, stone	Radon
Particleboard, plywood	Formaldehyde
Insulation	Formaldehyde, fiberglass
Fire retardant	Asbestos
Adhesives	Organics
Paint	Mercury, organics
Building contents	
Heating and cooking combustion appliances	CO, NO, NO_2, formaldehyde, particulates, organics
Furnishings	Organics
Water service; natural gas	Radon
Human occupants	
Metabolic activity	H_2O, CO_2, NH_3, odors
Human activities	
Tobacco smoke	CO, NO_2, organics, particulates, odors
Aerosol spray devices	Fluorocarbons, vinyl chloride, organics
Cleaning and cooking products	Organics, NH_3, odors
Hobbies and crafts	Organics
Damp organic materials, stagnant water	Bioaerosols
Coil drain pans	
Humidifiers	

minimum rate of outside air supply per occupant. More recently, the maintenance of acceptable indoor concentrations of a variety of additional pollutants that are not generated primarily by occupants has been a major concern. The most common pollutants of concern and their sources are presented in Table 1. Additional information on contaminants can be found in Chapter 11; odors are covered in Chapter 12.

Indoor pollutant concentrations depend on the strength of pollutant sources and the total rate of pollutant removal. Pollutant sources include the outdoor air; indoor sources such as occupants, furnishings, and appliances; and the soil adjacent to the building. Pollutant removal processes include ventilation with outside air, local exhaust ventilation, deposition on surfaces, chemical reactions, and air cleaning processes. If general building ventilation is the only significant pollutant removal process, the indoor air is thoroughly mixed, and the pollutant source strength and ventilation rate have been stable for a sufficient period; the steady-state indoor pollutant concentration is given by the equation

$$C_i = C_o + S/Q \qquad (3)$$

where

C_i = steady-state indoor concentration
C_o = outdoor concentration
S = total pollutant source strength
Q = ventilation rate (outdoor air only)

Variations in pollutant source strengths between buildings (and not variation in ventilation rates) is considered the largest cause of the building-to-building variation in the concentrations of pollutants that are not generated by occupants. Turk *et al.* (1989) found that a lack of correlation between average indoor respirable particle concentrations and whole building outdoor ventilation rates indicated that source strength, high outdoor concentrations, building volumes, and other removal processes are important. Because pollutant source strengths are highly variable, maintenance of minimum ventilation rates does not ensure acceptable indoor air quality in all situations. The lack of health-based concentration standards for many indoor air pollutants, primarily because of the lack of health data, makes the specification of minimum ventilation rates even more difficult.

Regardless of the complexities and uncertainties described previously, building designers and operators require guidance on ventilation and indoor air quality.

ASHRAE *Standard* 62-1989 provides guidance on ventilation and indoor air quality in the form of two alternative procedures. In the *ventilation rate procedure*, indoor air quality is assumed to be acceptable if the concentrations of six pollutants in the incoming outdoor air meet the United States national ambient air quality standards and the outside air supply rates meet or exceed values (which vary depending on the type of space) provided in a table. The minimum outside air supply per person for any type of space is 15 cfm. This minimum rate will maintain indoor CO_2 concentrations below 0.1% (1000 parts per million) assuming a typical CO_2 generation rate per occupant (Janssen 1989). This minimum outside air supply rate was based, in part, on research by Berg-Munch *et al.* (1986), which indicated that 15 cfm was required to satisfy the odor perceptions of 80% or more of visitors.

The second alternative in *Standard* 62-1989 is the *indoor air quality procedure*. Under this procedure, any outside air supply rates are acceptable if the indoor concentrations of nine pollutants are maintained below specified values and the air is deemed acceptable via subjective evaluations of odor. If users of the IAQ procedure control pollutant source strengths or use air cleaning or local exhaust ventilation, they may be able to reduce the outside air supply rates below the rates specified in the ventilation rate procedure. Modest energy savings may result as described by Eto and Meyer (1988) and Eto (1990). However, the maximum acceptable CO_2 concentrations of 0.1% in the IAQ procedure effectively limits the minimum ventilation rate to 15 cfm per occupant, unless CO_2 is removed by air cleaning, which is generally considered impractical.

In cases of large contaminant source strengths, impractically high levels of ventilation are required to control contaminant levels, and other methods of control are more effective. Removal or reduction of contaminant sources is a very effective means of control. Spot ventilation, such as range hoods or bathroom exhausts, for controlling a localized source is also effective.

Particles can be removed with various types of air filters. Gaseous contaminants with higher molecular weight can be controlled with activated carbon or alumina pellets impregnated with substances such as potassium permanganate. Chapter 25 of the 1992 ASHRAE *Handbook—Systems and Equipment* has information on air cleaning.

Source control and local exhaust, as opposed to dilution with ventilation air, is the method of choice in industrial environments. Indoor air quality problems and methods of control are covered in the proceedings of annual IAQ conferences sponsored by ASHRAE (1986 through 1992). Industrial ventilation is discussed in Chapters 25 and 27 of the 1991 ASHRAE *Handbook—Applications* and in *Industrial Ventilation: A Manual of Ventilation Practice* (ACGIH 1991).

DRIVING MECHANISMS

Natural ventilation and infiltration are driven by pressure differences across the building envelope caused by wind, air density differences due to temperature differences between indoor and outdoor air (stack effect), and the operation of appliances, such as combustion devices, leaky forced-air thermal distribution systems, and mechanical ventilation systems. The pressure difference at a location depends on the magnitude of these driving mechanisms as well as on the characteristics of the openings in the building envelope, *i.e.*, their locations and the relationship between pressure difference and airflow for each opening.

Pressure differences across the building envelope are based on the requirement that the mass flow of air into the building equals the mass flow out. In general, density differences between indoors and outdoors can be neglected, so the volumetric airflow rate into the building equals the volumetric airflow rate out. Based on this assumption, the envelope pressure differences can be determined; however, such a determination requires a great deal of detailed information that is essentially impossible to obtain.

When wind impinges on a building, it creates a distribution of static pressures on the building's exterior surface, which depends on the wind direction and the location on the building exterior. This pressure distribution is independent of the pressure inside the building p_i. If (1) no other forces act on the building, (2) no indoor-outdoor temperature difference exists, and (3) no appliance forces air through the building, the pressure differences are determined by the interior static pressure, according to:

$$\Delta p = p_o + p_w - p_i \qquad (4)$$

where

Δp = pressure difference between outdoors and indoors at location
p_o = static pressure at reference height in undisturbed flow
p_w = wind pressure at location
p_i = interior pressure at height of location

If no indoor-outdoor temperature difference exists, the interior static pressure p_i decreases linearly with height at a rate dependent on the interior temperature. This rate of pressure decrease equals $-\rho_i g$, where ρ_i is the average interior air density and g is the acceleration of gravity. The interior static pressure equilibrates

to a value such that the total airflow into the building equals the total airflow out of the building. The interior static pressure may be determined by calculating the airflow through each opening as a function of the interior pressure, adding all of these airflow rates together, setting this sum equal to zero, and solving for the interior pressure. However, to solve for the interior pressure in this way, the location of each opening in the building envelope, the value of p_w at each opening, and the relationship between airflow rate and pressure difference for each opening must be known.

When an indoor-outdoor temperature difference exists, it imposes a gradient in the pressure difference. This pressure difference Δp_s is a function of height and temperature difference and may be added to the pressure difference due to wind in Equation (4). The pressure difference is now expressed as:

$$\Delta p = p_o + p_w - p_{i,r} + \Delta p_s \qquad (5)$$

The parameter $p_{i,r}$ is the interior static pressure at some reference height, and this pressure again assumes a value such that the total inflow equals the total outflow. A summation of all the airflows through these openings can be set up, set equal to zero, and solved for the interior pressure at the reference height.

When an appliance such as a combustion device, leaky forced-air distribution system, or a ventilation fan operates, an additional airflow is imposed on the building. The pressure difference is still calculated using Equation (5), but the interior pressure $p_{i,r}$ changes so that the balance between inflow and outflow is maintained. This balance necessarily includes the airflow rate(s) associated with the appliance(s).

To determine the pressure differences across the building envelope and the corresponding air exchange rates, building-specific information about the exterior pressure distribution due to wind and the location and airflow rate/pressure difference relationship for every opening in the building envelope are needed. These inputs are difficult to obtain for any given building, making such a determination unrealistic.

Wind Pressure

Wind pressures are generally positive with respect to the static pressure in the undisturbed airstream on the windward side of a building and negative on the leeward side. Pressures on the other sides are negative or positive, depending on wind angle and building shape. Static pressures over building surfaces are almost proportional to the velocity head of the undisturbed airstream. The wind pressure or velocity head is given by Bernoulli's equation, assuming no height change or head losses:

$$p_v = C_1 C_p \rho v^2/2 \qquad (6)$$

where

p_v = surface pressure relative to static pressure in undisturbed flow, in. of water
ρ = air density, lb_m/ft^3 (about 0.075)
v = wind speed, mph
C_p = surface pressure coefficient, dimensionless
C_1 = unit conversion factor = 0.0129

Therefore, Equation (5) can be rewritten as:

$$\Delta p = p_o + C_1 C_p \rho v^2/2 - p_i \qquad (7)$$

C_p is a function of location on the building envelope and wind direction. Chapter 14 provides additional information on the values of C_p. Although standard conditions are frequently used, the air density and, consequently, the wind pressure can vary for a given wind speed with changes in temperature and/or elevation.

For example, for an elevation rise from sea level to 5000 ft, or an air temperature change from −20 to 70°F, the air density will drop about 20%. If these elevation and temperature changes occur simultaneously, the air density will drop by about 45%. Therefore, the effects of local air density cannot be ignored.

The wind speed incident on a building is generally lower than the average meteorological wind speed for a region; thus, meteorological data usually overestimates wind pressures on a building. Building wind speeds are lower because of the effects of height, terrain, and shielding (Lee *et al.* 1980). The wind speed is zero at the ground surface and increases with height up to an altitude of about 2000 ft above ground level. Meteorological measurements typically are made at a height of 33 ft in open areas. Residential building heights are generally less than 33 ft and are therefore subject to lower wind pressures. Tall buildings are subject to a range in wind speed over the height of the building, exposing the exterior to wind pressures that are both lower and higher than estimates based on Equation (6).

A database of surface pressure coefficients from eight investigators for low-rise buildings and one source for high-rise buildings was developed by Swami and Chandra (1988). The 544 average surface pressure coefficients from the database were used to develop a relationship (with a correlation coefficient 0.80) between wind incident angle, building side ratio, and average surface pressure coefficient:

$$NC_p = \ln [1.248 - 0.703 \sin (a/2) - 1.175 \sin^2 (a)$$
$$+ 0.131 \sin^3 (2aG) + 0.769 \cos (a/2)$$
$$+ 0.07 G^2 \sin^2 (a/2) + 0.717 \cos^2 (a/2)] \qquad (8)$$

where

$\quad NC_p$ = normalized C_p
$\qquad a$ = angle in degrees between wind direction and outward normal of wall under consideration
$\qquad G$ = natural log of ratio of width of wall under consideration to width of adjacent wall

They further noted that the uncertainties in the estimation of site wind speed and the effect of surrounding buildings are likely to be equal to or greater than the uncertainty in estimating C_p from the correlations.

The shielding effects of trees, shrubbery, and other buildings, within several building heights of a particular building, produce large-scale turbulence that not only reduces effective wind speed but also alters wind direction. Thus, meteorological wind speed data must be reduced carefully when applied to low buildings. Chapter 14 provides additional guidance on estimating wind pressures.

The magnitude of the pressure differences found on the surfaces of buildings varies rapidly with time because of turbulent fluctuations in the wind (Grimsrud *et al.* 1979, Etheridge and Nolan 1979). However, the use of average wind pressures to calculate pressure differences is usually sufficient to calculate average infiltration values. In residential buildings, the magnitude of wind pressure differences averaged over 20 min seldom exceeds ±0.02 in. of water under typical conditions. In many cases, the averages are less than ±0.01 in. of water. For tall buildings or buildings completely exposed to open terrain, the pressure on the windward side is much closer to those calculated from average wind speeds for the site (Tamura and Wilson 1968). In the latter cases, for example, if $v = 6.7$ mph, $p_v \cong 0.02$ in. of water; if $v = 15.7$ mph, $p_v \cong 0.12$ in. of water (assuming $C_p = 1$).

Stack Pressures

Temperature differences between indoors and outdoors cause density differences, and therefore pressure differences, that drive

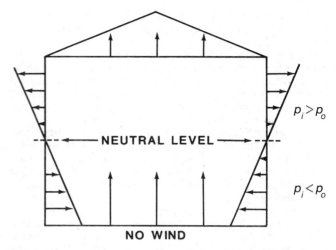

Note: Arrows indicate magnitude and direction of pressure difference.

Fig. 1   Pressure Differences Caused by Stack Effect for Typical Structure (Heating)

infiltration. During the heating season, the warmer inside air rises and flows out of the building near its top. It is replaced by colder outdoor air that enters the building near its base. During the cooling season, the flow directions are reversed and generally lower, because the indoor-outdoor temperature differences are smaller. Qualitatively, the pressure distribution over the building in the heating season due to the stack effect takes the form shown in Figure 1.

The height at which the interior and exterior pressures are equal is called the neutral pressure level (NPL) (Tamura and Wilson 1966, 1967a). Above this point (during the heating season), the interior pressure is greater than the exterior; below this point, the greater exterior pressure causes airflow into the building.

The pressure difference due to the stack effect at height h is:

$$\Delta p_s = C_2 (\rho_o - \rho_i) g (h - h_{NPL})$$
$$= C_2 \rho_i g (h - h_{NPL})(T_i - T_o)/T_o \qquad (9)$$

where

$\quad \Delta p_s$ = pressure difference due to stack effect, in. of water
$\qquad \rho$ = air density, lb_m/ft^3 (about 0.075 for indoor conditions)
$\qquad g$ = gravitational constant, 32.2 ft/s^2
$\qquad h$ = height of observation, ft
$\quad h_{NPL}$ = height of neutral pressure level, ft
$\qquad T$ = average absolute temperature, °R
$\qquad C_2$ = unit conversion factor = 0.00598

Subscripts

$\qquad i$ = inside
$\qquad o$ = outside

Chastain and Colliver (1989) show that the average of the vertical distribution of temperature differences due to stratification rather than the localized temperature difference near the opening is the appropriate temperature difference to use in this equation.

A useful estimate of the magnitude of the stack effect on a building is that the pressure difference induced by the stack effect is 2.7×10^{-5} in. of water/ft^2·°R. This estimate neglects any resistance to airflow within the structure. Therefore, in a one-story house with an 8-ft ceiling, an NPL of one-half the building height, and a temperature difference of 45°R, the stack pressure will be 0.005 in. of water at the ceiling and floor. In a tall building (*e.g.*, 20 stories of 13 ft each) with no internal resistance to airflow, the stack pressure under these same conditions will be 0.18 in. of water. For cold climate conditions, these values are significantly greater due to the increased outdoor air density.

The location of the NPL at zero wind speed is a structure-dependent parameter that depends only on the vertical distribution of openings in the shell, the resistance of the openings to airflow, and the resistance to vertical airflow within the building. If the openings are uniformly distributed vertically, they have the same resistance to airflow, and there is no internal airflow resistance, the NPL is at the midheight of the building (Figure 1). If there is only one opening or an extremely large opening relative to any others, the NPL is at or near the center of this opening. Foster and Down (1987) studied the location of the NPL as it relates to natural ventilation in a building with only two openings. Lee *et al.* (1985) studied, both experimentally and analytically, the characteristics of thermal performance of high-rise buildings using an idealized model building with a number of openings at various locations and temperature distributions. Later Lee *et al.* (1988) showed that the ratio of floor opening area to exterior opening areas had a dominant role on pressure distribution. Chastain and Colliver (1989) investigated the location of the NPL in walls with multiple openings and found that it was independent of the temperature difference and that the resulting mean density difference across the building envelope should be used.

Internal partitions, stairwells, elevator shafts, utility ducts, chimneys, vents, and mechanical supply and exhaust systems complicate the analysis of NPL locations. Chimneys and openings at or above roof height raise the NPL in small buildings. Exhaust systems increase the height of the NPL; outdoor air supply systems lower it.

Available data on the NPL in various kinds of buildings is limited. The NPL in tall buildings varies from 0.3 to 0.7 of total building height (Tamura and Wilson 1966, 1967a). For houses, especially those with chimneys, the NPL is usually above midheight. Operating a combustion heat source with a flue raises the NPL further, sometimes above the ceiling (Shaw and Brown 1982).

Equation (9) provides a maximum stack pressure difference, given no internal airflow resistance. The sum of the pressure differences across the exterior wall at the bottom and top of the building, as calculated by Equation (9), equals the total theoretical draft for the building. The sum of the actual top and bottom pressure differences, divided by the total theoretical draft, equals the *thermal draft coefficient*. The value of the thermal draft coefficient depends on the airflow resistance of the exterior walls relative to the airflow resistance between floors. For a building without internal partitions, the total theoretical draft is achieved across the exterior walls (Figure 2A), and the thermal draft coefficient equals 1. In a building with airtight separations at each floor, each story acts independently, its own stack effect being unaffected by that of any other floor (Figure 2B). The ratio of the actual to the theoretical draft is minimized in this case.

Real multistory buildings are neither open inside (Figure 2A), nor airtight between stories (Figure 2B). Vertical air passages, stairwells, elevators, and other service shafts allow airflow between floors. Figure 2C represents a heated building with uniform openings in the exterior wall, through each floor, and into the vertical shaft at each story. Between floors, the slope of the line representing the inside pressure is the same as that shown in Figure 2A, and the discontinuity at each floor (Figure 2B) represents the pressure difference across it. Total stack effect for the building remains the same, but some of the total pressure difference maintains flow through openings in the floors and vertical shafts. As a result, the pressure difference across the exterior wall at any level is less than it would be with no internal flow resistance.

Maintaining airtightness between floors and from floors to vertical shafts is a means of controlling indoor-outdoor pressure differences, and therefore infiltration. Good separation is also conducive to the proper operation of mechanical ventilation and smoke control systems. Tamura and Wilson (1967b) showed that when vertical shaft leakage is at least two times the envelope leak-

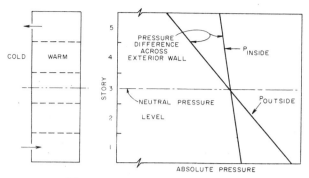

Fig. 2A   Stack Effect in a Building with No Internal Partition

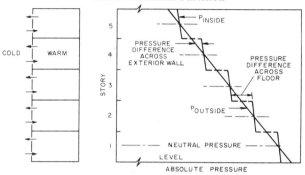

Fig. 2B   Stack Effect in a Building with Airtight Separation of Each Story

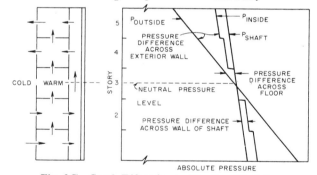

Fig. 2C   Stack Effect for an Idealized Building

age, the thermal draft coefficient is almost 1. Openings in floors are less effective in providing communication between floors; as the building height increases, they become even less effective. Measurements of pressure differences in three tall office buildings by Tamura and Wilson (1967a) indicated that the thermal draft coefficient ranged from 0.8 to 0.9 with the ventilation systems off.

Mechanical Systems

The operation of mechanical equipment, such as ventilation/exhaust systems and vented combustion devices, affects pressure differences across the building envelope. The interior static pressure adjusts such that the airflows through all the openings in the building envelope plus equipment-induced airflows balance to zero. These changes in pressure differences and airflow rates caused by mechanical equipment are unpredictable, unless the location of each opening in the envelope and the relationship between pressure difference and airflow rate for each opening are known. The interaction of mechanical ventilation system operation and envelope airtightness has been discussed for low-rise buildings (Nylund 1980) and for office buildings (Tamura and Wilson 1966, 1967b; Persily and Grot 1985a).

Air exhausted from a building must be balanced by increasing the airflow into the building through other openings. The airflow at some locations changes from outflow to inflow because in the winter (summer), the NPL rises (falls). For supply fans, the situation is reversed, and envelope inflows become outflows. Thus, the effects a mechanical system has on a building must be considered. Depressurization caused by an improperly designed system can increase radon entry rates into a building and interfere with the proper operation of combustion device venting or other exhaust systems. Overpressurization can force moist indoor air through the building envelope and, in cold climates, moisture may condense within the building envelope.

The interaction between mechanical systems and the building envelope also holds for systems serving zones of buildings. The performance of zone-specific exhaust or pressurization systems is affected by the leakage in zone partitions, as well as in exterior walls.

Mechanical systems can also create infiltration-driving forces in single-zone buildings. Specifically, a large fraction of single-family houses with central forced-air duct systems have multiple supply registers, yet only a central return register. In these houses, when internal doors are closed, large positive indoor-outdoor pressure differentials are created for rooms with only supply registers, whereas the return duct tends to depressurize relative to outside. This is caused by the resistance of internal door undercuts to flow from the supply register to the return (Modera *et al.* 1991). The magnitudes of the indoor-outdoor pressure differentials created have been measured to average 0.012 to 0.024 in. of water (Modera *et al.* 1991).

Building envelope airtightness and interzone airflow resistance can also affect the performance of mechanical systems. The actual airflow rate delivered by these systems, particularly ventilation systems, depends on the pressure they work against. This effect is the same as the interaction of a fan with its associated duct work discussed in Chapter 32 of this volume and in Chapter 18 in the 1992 ASHRAE *Handbook—Systems and Equipment.* The building envelope and its leakage can be considered part of the duct work in determining the pressure drop of the system.

Combining Driving Forces

The pressure differences just discussed are considered in combination by adding them together and determining the airflow rate through each opening due to this total pressure difference. Because the airflow rate through these openings is not linearly related to

pressure difference, the driving forces must be combined in this manner, as opposed to adding the airflow rates due to the separate driving forces.

Figure 3 qualitatively shows the addition of driving forces for a building with uniform openings above and below midheight and without significant internal resistance to airflow. The slopes of the pressure lines are functions of the densities of the indoor and outdoor air. In Figure 3A, with inside air warmer than outside and pressure differences caused solely by thermal forces, the NPL is at midheight, with inflow through lower openings and outflow through higher openings. (Direction of flow is always from the higher to the lower pressure region.) A chimney or mechanical exhaust decreases the inside pressure and thus shifts the inside pressure line to the left, raising the NPL; an excess of outdoor supply air over exhaust would lower it. Figure 3B presents qualitative uniform pressure differences caused by wind alone, with the effect on windward and leeward sides equal but opposite. When the temperature difference and wind effects both exist, the pressures due to each are added together to determine the total pressure difference across the building envelope. Figure 3C shows the combination where the wind force of Figure 3B has just balanced the thermal force of Figure 3A, causing no pressure difference at the top windward or bottom leeward side. Total airflow is similar to that with the wind acting alone, but significantly larger than the airflow due only to the stack effect.

The relative importance of the wind and stack pressures in a building depends on building height, internal resistance to vertical airflow, location and flow resistance characteristics of envelope openings, local terrain, and the immediate shielding of the building. The taller the building and the less the internal resistance to airflow, the stronger the stack effect. The more exposed a building, the more susceptible it will be to wind. For any building, there will be ranges of wind speed and temperature difference for which the building's infiltration is dominated by the stack effect, the wind, or a regime in which the driving pressures of both must be considered (Sinden 1978). These factors determine, for specific values of temperature difference and wind speed, in which regime the building's infiltration lies.

The effect of mechanical ventilation on envelope pressure differences is more complex and depends on the direction of the ventilation flow (exhaust or supply) and differences in these ventilation flows among the zones of the building. If mechanically supplied outdoor air is provided uniformly to each story, the change in the exterior wall pressure difference pattern from ther-

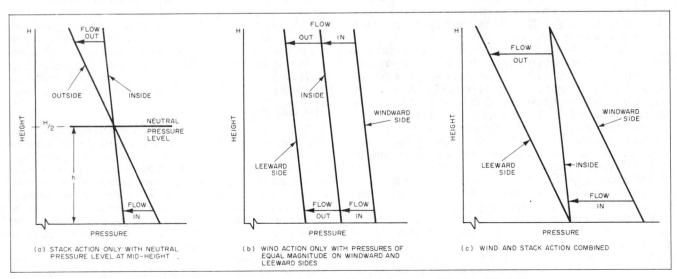

Fig. 3   Distribution of Inside and Outside Pressures over Height of Building

mal pressures is uniform. With a nonuniform supply of outdoor air (for example, to one story only), the extent of pressurization varies from story to story and depends on the internal airflow resistance. Pressurizing all levels uniformly has little effect on the pressure differences across floors and vertical shaft enclosures, but pressurizing individual stories increases the pressure drop across these internal separations. Pressurization of the ground level is often used in tall buildings to reduce the stack pressures across entries.

The pressure fields illustrated in Figure 3 indicate that the wind and stack effects can be combined by simply adding the pressures. Because this pressure addition can cause direction changes in the resulting flow, the flows cannot be combined as simply. An accurate superposition of the two effects requires detailed knowledge of the two pressure fields. Several theoretical models and measurements have been made for combining forced and natural ventilation (Kiel and Wilson 1987, Persily and Norford 1987, Shaw 1987). A simpler model (Sherman 1992) uses the neutral level and wind direction to combine the two effects. For typical buildings, the following equation can be used to combine the stack and wind effects:

$$Q_{ws} = (Q_w^2 + Q_s^2)^{0.5} \qquad (10)$$

where

Q_{ws} = infiltration from both wind and stack effects, cfm
Q_w = infiltration from wind, cfm
Q_s = infiltration from stack effect, cfm

Mechanical ventilation can change the flow through the envelope by affecting the internal pressure. When both supply and exhaust fans are present, the mechanical ventilation can be broken up into the amount of ventilation flow which creates no change in pressure distribution or the "balanced" part:

$$Q_{bal} = \text{minimum of } (Q_{supply}, Q_{exhaust}) \qquad (11)$$

and an "unbalanced" part:

$$Q_{unbal} = \text{maximum of } (Q_{supply}, Q_{exhaust}) - Q_{bal} \qquad (12)$$

(If the building has only supply or exhaust, all the flow is unbalanced.)

Because the balanced part of the flow does not affect the pressure distribution across the envelope, it adds simply to the wind- and stack-driven pressures. The unbalanced part, however, does affect the pressure distribution and, therefore, only part of it contributes to additional ventilation (Sherman 1992) until the fan completely dominates. The efficiency of unbalanced ventilation depends again on the neutral level and wind direction, but it can be assumed to be 50% for typical cases.

The total ventilation then becomes the following:

$$Q = Q_{bal} + \text{Maximum } (Q_{unbal}, Q_{ws} + 0.5\,Q_{unbal}) \qquad (13)$$

Levins (1982) and Kiel and Wilson (1987) further discuss the combination of mechanical ventilation airflow rates with naturally induced infiltration rates.

Shaw and Brown (1982) compared air infiltration in identical homes, with and without a gas furnace with a chimney. Figure 4 presents the effects of exfiltration through the chimney and ceiling with and without the gas furnace; it also shows the impact of the chimney on the NPL.

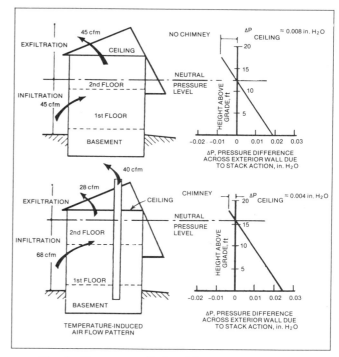

Fig. 4 Temperature-Induced Pressure and Airflow Patterns under Operation of Electric or Gas Furnace for $\Delta t = 50°F$

AIRFLOW THROUGH OPENINGS

The relationship between the airflow through an opening in the building envelope Q and the pressure difference Δp across it is called the leakage function of the opening. The form of the leakage function depends on the geometry of the opening. Background theoretical material relevant to leakage functions may be found in Chapter 2, Hopkins and Hansford (1974), Etheridge (1977), Kronvall (1980a), and Chastain *et al.* (1987).

The relationship describing the airflow through an opening is based on the solution to the Navier-Stokes equation for the idealized case of infinite parallel plates with several simplifying assumptions such as steady, fully developed laminar flow. The most commonly used expression for the airflow rate through an opening is:

$$Q = C_3 C_D A \sqrt{2\,\Delta p/\rho} \qquad (14)$$

where

Q = airflow rate, cfm
C_D = discharge coefficient for opening, dimensionless
A = cross-sectional area of opening, ft^2
ρ = air density, lb/ft^3
Δp = pressure difference across opening, in. of water
C_3 = conversion factor = 776

The discharge coefficient C_D is a dimensionless number that depends on the opening geometry and the Reynolds number of the flow.

Airflow through constant area ducts is well characterized. At sufficiently low Reynolds numbers, the fluid velocity varies only in the direction perpendicular to the flow, and the flow may be visualized as many sheets or laminae flowing parallel to the duct walls. Thus, this type of flow is referred to as laminar. In laminar flow, C_D depends on the square root of the pressure difference;

therefore, Q is proportional to Δp. At large Reynolds numbers, the flow becomes turbulent. The velocity at a given point fluctuates rapidly and at random, even if the net flow rate is constant. In turbulent flow, the discharge coefficient is constant at a fixed Reynolds number and therefore the flow Q is proportional to $\sqrt{\Delta p}$.

The case of fully developed flow impinging on a hole, nozzle, or orifice in a thin plate is also described by Equation (14). Again, for a sufficiently large value of the Reynolds number, the discharge coefficient is constant. The value of C_D for an orifice depends on Reynolds number and the relative areas of the orifice and the duct in which the orifice is placed.

This discussion of laminar and turbulent flow applies to constant area ducts with fully developed flow and orifices in such ducts. The openings in a building envelope are much less uniform in geometry. Generally, the flow never becomes fully developed, thereby preventing the applicability of the simple relations between Q and Δp. Each opening in the building envelope can still be described by Equation (14), where A is an equivalent cross-sectional area, and C_D depends on opening geometry and the pressure difference across it. Equation (15), commonly called the power law equation, is sometimes used instead:

$$Q = c(\Delta p)^n \qquad (15)$$

where

c = flow coefficient, cfm/(in. of water)^n
n = flow exponent, dimensionless

Equation (15) only approximates the relationship between Q and Δp. In fact, the values of c and n depend on the range of Δp over which Equation (15) is applied and do not have a physical interpretation. Honma (1975) measured Q as a function of Δp for several simple openings, and the measured data were fit to Equation (15). The cracks with larger flow resistances, *i.e.*, greater flow lengths or narrower widths, tended to have an exponent n closer to 1 than did gaps with less resistance. For openings in the shell of a building, the value of n depends on the opening geometry, as well as on entrance and exit effects. Additional investigation of pressure/flow data on simple cracks by Chastain *et al.* (1987) further indicated the importance of adequately characterizing the three-dimensional geometry of openings and the entrance and exit effects. A modeling procedure was developed to determine the discharge coefficient in Equation (14) for the complex openings found in the building shell.

The combination of all the openings in a building's envelope produces a relationship between pressure difference and airflow rate for the whole building and is referred to as the air leakage of the building.

The air leakage of a building can be measured (as described in the section on air leakage) and is a physical property of the building envelope that depends on the envelope design, construction, and deterioration over time. A building's air leakage is measured by imposing a uniform pressure difference over the entire building envelope and measuring the airflow rate required to maintain this difference. Such a distribution of envelope pressures never occurs naturally, but it does provide a useful measure of the airtightness of a building.

NATURAL VENTILATION FLOW RATES

Natural ventilation is the intentional flow of outdoor air because of wind and thermal pressures through controllable openings. It can effectively control both temperature and contaminants, particularly in mild climates. Temperature control by natural ventilation is often the only means of providing cooling when mechanical air conditioning is not available. The arrangement, location, and control of ventilation openings should com-

bine the driving forces of wind and temperature to achieve a desired ventilation rate and good distribution of ventilation air through the building.

Natural Ventilation Openings

Natural ventilation openings include: (1) windows, doors, monitor openings, and skylights; (2) roof ventilators; (3) stacks connecting to registers; and (4) specially designed inlet or outlet openings.

Windows transmit light and provide ventilation when open. They may open by sliding vertically or horizontally; by tilting on horizontal pivots at or near the center; or by swinging on pivots at the top, bottom, or side. The type of pivoting used is important for weather protection and airflow rate.

Roof ventilators provide a weatherproof air outlet. Capacity is determined by the ventilator's location on the roof; the resistance the ventilator and its duct work offer to airflow; its ability to use kinetic wind energy to induce flow by centrifugal or ejector action; and the height of the draft.

Natural draft or gravity roof ventilators can be stationary, pivoting, oscillating, or rotating. Selection criteria include ruggedness, corrosion resistance, stormproofing features, dampers and operating mechanisms, noise, cost, and maintenance. Natural ventilators can be supplemented with power-driven supply fans; the motors need only be energized when the natural exhaust capacity is too low. Gravity ventilators can have manual dampers or dampers controlled by thermostat or wind velocity.

A roof ventilator should be positioned so that it receives the full, unrestricted wind. Turbulence created by surrounding obstructions, including higher adjacent buildings, impairs a ventilator's ejector action. The ventilator inlet should be conical or bell mounted to give a high flow coefficient. The opening area at the inlet should be increased if screens, grilles, or other structural members cause flow resistance. Building air inlets at lower levels should be larger than the combined throat areas of all roof ventilators.

Stacks or vertical flues should be located where wind can act on them from any direction. Without wind, stack effect alone removes air from the room with the inlets.

Required Flow

The ventilation airflow rate required to remove a given amount of heat from a building can be calculated from Equation (16) if the quantity of heat to be removed and the indoor-outdoor temperature difference are known.

$$Q = H/c_p \rho (t_i - t_o) \qquad (16)$$

where

Q = airflow rate required to remove heat, cfm
H = heat to be removed, Btu/min
c_p = specific heat of air, Btu/lb$_m \cdot$°F (about 0.24)
ρ = air density, lb$_m$/ft^3 (about 0.075)
$t_i - t_o$ = indoor-outdoor temperature difference, °F

Flow Caused by Wind

Factors that affect the ventilation rate due to wind forces include average speed, prevailing direction, seasonal and daily variation in speed and direction, and local obstructions such as nearby buildings, hills, trees, and shrubbery. Liddament (1988) reviews the relevance of wind pressure as a driving mechanism. A multiflow path simulation model was developed and used to illustrate the effects of wind on air change rate.

Wind speeds are usually lower in summer than in winter; directional frequency is also a function of season. There are relatively few places where speed falls below one-half the average for more than a few hours a month. Therefore, natural ventilation systems are often designed for wind speeds of one-half the seasonal average. Equation (17) shows the quantity of air forced through ventilation inlet openings by wind or determines the proper size of openings to produce given airflow rates:

$$Q = C_4 C_v A V \qquad (17)$$

where

Q = airflow rate, cfm
A = free area of inlet openings, ft^2
V = wind speed, mph
C_v = effectiveness of openings (C_v is assumed to be 0.5 to 0.6 for perpendicular winds and 0.25 to 0.35 for diagonal winds)
C_4 = unit conversion factor = 88.0

Inlets should face directly into the prevailing wind. If they are not advantageously placed, flow will be less than that in the equation; if unusually well placed, flow will be slightly more. Desirable outlet locations are (1) on the leeward side of the building directly opposite the inlet, (2) on the roof, in the low-pressure area caused by a flow discontinuity of the wind, (3) on the side adjacent to the windward face where low-pressure areas occur, (4) in a monitor on the leeward side, (5) in roof ventilators, or (6) by stacks. Chapter 14 gives a general description of the wind pressure distribution on a building. The inlets should be placed in the high-pressure regions; the outlets should be placed in the negative or low-pressure regions.

Flow Caused by Thermal Forces

If building internal resistance is not significant, the flow caused by stack effect can be expressed by:

$$Q = C_5 K A \, [2g\Delta h_{NPL} \, (T_i - T_o)/ \, T_i]^{0.5} \qquad (18)$$

where

Q = airflow rate, cfm
K = discharge coefficient for opening
Δh_{NPL} = height from midpoint of lower opening to NPL, ft
C_5 = unit conversion factor = 60
T_i = indoor temperature, °R
T_o = outdoor temperature, °R

Equation (18) applies when $T_i > T_o$. If $T_i < T_o$, replace T_i in the denominator with T_o, and replace $(T_i - T_o)$ in the numerator with $(T_o - T_i)$. An average temperature for T_i should be used if there is thermal stratification. If the building has more than one opening, the outlet and inlet areas are considered equal. The discharge coefficient K accounts for all viscous effects such as surface drag and interfacial mixing.

Estimation of Δh_{NPL} is difficult. If one window or door represents a large fraction (approximately 90%) of the total opening area in the envelope, the NPL is at the midheight of that aperture, and Δh_{NPL} equals to one-half its height. For this condition, flow through the opening is bidirectional, *i.e.*, air from the warmer side flows through the top of the opening, and air from the colder side flows through the bottom. Interfacial mixing occurs across the counterflow interface, and the orifice coefficient can be calculated (Kiel and Wilson 1986) according to:

$$K = 0.40 + 0.0025 \, |T_i - T_o| \qquad (19)$$

If enough other openings are available, the airflow through the opening will be unidirectional and mixing cannot occur. A discharge coefficient of $K = 0.65$ should then be used. Additional information on stack-driven airflows for natural ventilation can be found in Foster and Down (1987).

Greatest flow per unit area of openings is obtained when inlet and outlet areas are equal; Equations (17) and (18) are based on

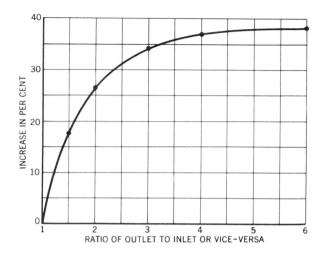

Fig. 5   Increase in Flow Caused by Excess of One Opening over Another

this equality. Increasing the outlet area over inlet area, or vice versa, increases airflow but not in proportion to the added area. When openings are unequal, use the smaller area in the equations and add the increase, as determined from Figure 5.

Natural Ventilation Guidelines

Several general guidelines should be observed in designing for natural ventilation. Some of these may conflict with other climate-responsive strategies (such as orientation and shading devices to minimize solar gain) or other design considerations.

1. In hot, humid climates, maximize air velocities in the occupied zones for bodily cooling. In hot, arid climates, maximize airflow throughout the building for structural cooling, particularly at night when the temperature is low.
2. Take advantage of topography, landscaping, and surrounding buildings to redirect airflow and give maximum exposure to breezes. Use vegetation to funnel breezes and avoid wind dams, which reduce the driving pressure differential around the building. Site objects should not obstruct inlet openings.
3. Shape the building to expose maximum surface area to breezes.
4. Use architectural elements such as wingwalls, parapets, and overhangs to promote airflow into the building interior.
5. The long facade of the building and the majority of the door and window openings should be oriented with respect to the prevailing summer breezes. If there is no prevailing direction, openings should be sufficient to provide ventilation regardless of wind direction.
6. Windows should be located in opposing pressure zones. Two openings on opposite sides of a space increase the ventilation flow. Openings on adjacent sides force air to change direction, providing ventilation to a greater area. The benefits of the window arrangement depend on the outlet location relative to the direction of the inlet airstream.
7. If a room has only one external wall, better airflow is achieved with two widely spaced windows.
8. If the openings are at the same level and near the ceiling, much of the flow may bypass the occupied level and be ineffective in diluting contaminants there.
9. The stack effect requires vertical distance between openings to take advantage of the stack effect; the greater the vertical distance, the greater the ventilation.
10. Openings in the vicinity of the NPL are least effective for thermally induced ventilation. If the building has only one

large opening, the NPL tends to move to that level, which reduces the pressure across the opening.

11. Greatest flow per unit area of total opening is obtained by inlet and outlet openings of nearly equal areas. An inlet window smaller than the outlet creates higher inlet velocities. An outlet smaller than the inlet creates lower but more uniform air speed through the room.
12. Openings with areas much larger than calculated are sometimes desirable when anticipating increased occupancy or very hot weather.
13. Horizontal windows are generally better than square or vertical windows. They produce more airflow over a wider range of wind directions and are most beneficial in locations where prevailing wind patterns shift.
14. Window openings should be accessible to and operable by occupants.
15. Inlet openings should not be obstructed by indoor partitions. Partitions can be placed to split and redirect airflow, but should not restrict flow between the building's inlets and outlets.
16. Vertical airshafts or open staircases can be used to increase and take advantage of stack effects. However, enclosed staircases intended for evacuation during a fire should not be used for ventilation.

INFILTRATION

Although the terms infiltration and air leakage are sometimes used synonymously, they are different, though related, quantities. Infiltration is the rate of uncontrolled air exchange through unintentional openings that occurs under given conditions, while air leakage is a measure of the airtightness of the building shell. The greater the air leakage of a building, the greater its infiltration rate, all else (weather, exposure, and building geometry) being equal. The infiltration rate is the air entering the structure and is equal to the exfiltration rate (air leaving the structure).

Infiltration may be reduced either by reducing the surface pressures driving the flow, or reducing the air leakage of the shell. Surface pressures caused by the wind can be reduced by changing the landscaping in the vicinity of the building (Mattingly and Peters 1977). Stack pressures can be reduced by increasing the airflow resistance between floors and from floors to any vertical shafts within the building, although this is almost exclusively an issue in tall buildings.

The infiltration rate of an individual building depends on weather conditions, equipment operation, and occupant activities. The rate can vary by a factor of five from weather effects alone (Malik 1978). When associating a building with an infiltration rate, it is important to provide the corresponding weather conditions and equipment status, or to describe it as a seasonal or annual average.

Typical infiltration values in housing in North America vary by a factor of about ten, from tight housing with seasonal average air change rates of about 0.2 per hour to housing with air exchange rates as great as 2.0 per hour. Figures 6 and 7 show histograms of infiltration rates measured in two different samples of North American housing (Grimsrud et al. 1982, Grot and Clark 1979). Figure 6 shows the average seasonal infiltration of 312 houses located in different areas in North America. The median infiltration value of this sample is 0.5 air changes per hour (ACH). Figure 7 represents measurements in 266 houses located in 16 cities in the United States. The median value of this sample is 0.90 ACH. The group of houses contained in the Figure 6 sample is biased toward new energy-efficient houses, while the group in Figure 7 represents older, low-income housing in the United States. Additional studies have found average values for houses in regional areas. Palmiter

and Brown (1989) and Parker et al. (1990) found a a heating season average of 0.40 ACH (range: 0.13 to 1.11 ACH) for 134 houses in the Pacific Northwest. In a comparison of 292 houses incorporating energy-efficient features including measures to reduce air infiltration and provide ventilation heat recovery with 331 control houses, Parker et al. (1990) found an average of about 0.25 ACH (range: 0.02 to 1.63 ACH) for the energy-efficient houses versus a rate of 0.49 (range: 0.05 to 1.63 ACH) for the control. Ek et al. (1990) found an average of 0.5 ACH (range: 0.26 to 1.09) for 93 double-wide manufactured homes also in the Pacific Northwest. Canadian housing stock has been characterized by Yuill and Comeau (1989) and Riley (1990). While these do not represent random samples of North American housing, they indicate the distribution of infiltration rates expected in a group of buildings.

Occupancy influences have not been measured directly and vary widely. Desrochers and Scott (1985) estimate they add an average of 0.10 to 0.15 ACH to unoccupied values. Kvisgaard and Collet (1990) found that in 16 Danish dwellings, the users on average provided 63% of the total air change.

Grot and Persily (1986) found that eight recently constructed office buildings had infiltration rates ranging from 0.1 to 0.6 air changes per hour with no outdoor air intake. The infiltration rates of these buildings exhibited varying degrees of weather dependence, generally much lower than that measured in houses.

AIR EXCHANGE MEASUREMENT

The only reliable way to determine the air exchange rate of a building is to measure it. Several tracer gas measurement pro-

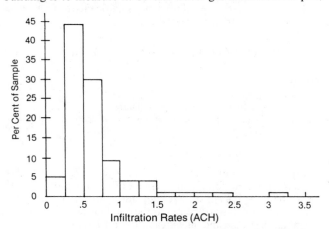

Fig. 6   Histogram of Infiltration Values—New Construction

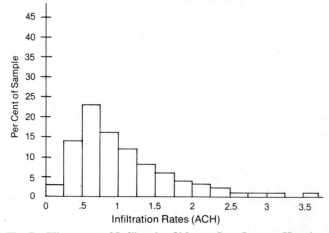

Fig. 7   Histogram of Infiltration Values—Low-Income Housing

cedures exist, all involving an inert or nonreactive gas used to label the indoor air (Hunt 1980, Sherman et al. 1980, Harrje et al. 1981, Lagus and Persily 1985, Dietz et al. 1986, Charlesworth 1988, Persily 1988, Fisk et al. 1989, Lagus 1989, Sherman 1989a, 1989b, Fortmann et al. 1990, Harrje et al. 1990, Persily and Axley 1990, Sherman 1990). The tracer is released into the building in a specified manner, and the concentration of the tracer within the building is monitored and related to the building's air exchange rate. A variety of tracer gases, and associated concentration detection devices, have been used. Desirable qualities of a tracer gas are detectability, nonreactivity, nontoxicity, and relatively low concentration in ambient air (Hunt 1980).

All tracer gas measurement techniques are based on a mass balance of the tracer gas within the building. Assuming the outdoor concentration is zero, this mass balance takes the form:

$$V(dc/d\theta) = F(\theta) - Q(\theta)c(\theta) \qquad (20)$$

where

$$
\begin{aligned}
V &= \text{volume of space being tested, } ft^3 \\
c(\theta) &= \text{tracer gas concentration at time } \theta \\
dc/d\theta &= \text{time rate of change of concentration, } min^{-1} \\
F(\theta) &= \text{tracer gas injection rate at time } \theta, \text{ cfm} \\
Q(\theta) &= \text{airflow rate out of building at time } \theta, \text{ cfm} \\
\theta &= \text{time, min}
\end{aligned}
$$

In Equation (20), density differences between indoor and outdoor air are generally ignored for moderate climates; therefore, Q also refers to the airflow rate into the building. While Q is often referred to as the infiltration rate, any measurement includes both mechanical and natural ventilation in addition to envelope infiltration. The ratio of the air exchange rate Q to the volume being tested V has units of volume/time (often converted to ACH) and is called the air change rate I.

Equation (20) is based on the assumptions that no unknown tracer gas sources exist and that the airflow out of the building is the dominant process removing the tracer gas from the space, i.e., the tracer gas does not react chemically within the space and/or is not absorbed onto interior surfaces. It is also based on the assumption that the tracer gas concentration within the building can be represented by a single value, i.e., the tracer gas concentration is uniformly mixed within the space. Three different tracer gas procedures are used to measure air exchange rates: (1) decay, (2) constant concentration, and (3) constant injection.

Decay

The simplest tracer gas measurement technique is the decay method, which is a standardized procedure (ASTM 1990). In the decay method, a small amount of tracer gas is injected into the space and is allowed to mix with the interior air. After the injection, $F = 0$ and the solution to Equation (20) is:

$$c(\theta) = c_o e^{-I\theta} \qquad (21)$$

where c_o is the concentration at $\theta = 0$.

Equation (21) is generally used to solve for I by measuring the tracer gas concentration periodically during the decay and fitting the data to the log form of Equation (21):

$$\ln c(\theta) = \ln c_o - I\theta \qquad (22)$$

As with all tracer gas techniques, the tracer gas decay method has advantages and disadvantages. The advantages include the fact that Equation (21) is an exact solution to the tracer gas mass balance equation. Also, because logarithms of concentration are taken, only relative concentrations are needed, which can simplify the calibration of the concentration-measuring equipment.

Finally, the tracer gas injection rate need not be measured, although it must be controlled so that the tracer gas concentrations are within the range of the concentration-measuring device. The concentration-measuring equipment can be located on site, or building samples can be collected in suitable containers and analyzed elsewhere.

The most serious problem with the decay technique is imperfect mixing of the tracer gas with the interior air, both at initial injection and during the decay. Equations (20) and (21) employ the assumption that the tracer gas concentration within the building is uniform. If the tracer is not well mixed, this assumption is not appropriate and the determination of I will be subject to errors. It is difficult to estimate the magnitude of the errors due to poor mixing, and little analysis of this problem has been done.

Constant Concentration

In the constant concentration technique, the tracer gas injection rate is adjusted to maintain a constant concentration within the building. If the concentration is truly constant, then Equation (20) reduces to:

$$Q(\theta) = F(\theta)/c \qquad (23)$$

There is less experience with this technique than with the decay procedure, but several applications exist (Kumar et al. 1979, Collet 1981, Bohac et al. 1985, Fortmann 1990).

Because the tracer gas injection is continuous, it requires no initial mixing period. Another advantage is that the tracer concentration in each zone of the building can be separately controlled by injecting into each zone; thus, the amount of outdoor air flowing into each zone can be determined. This procedure has the disadvantage of requiring the measurement of absolute tracer concentrations and injection rates. Also, imperfect mixing of the tracer and the interior air causes a delay in the response of the concentration to changes in the injection rate. This delay in concentration response makes it impossible to keep the concentration constant, and therefore Equation (23) is only an approximation. The magnitude of these errors has not been well examined.

Constant Injection

In the constant injection procedure, the tracer is injected at a constant rate and the solution to Equation (20) becomes:

$$c(\theta) = (F/Q)(1 - e^{-I\theta}) \qquad (24)$$

After sufficient time, the transient term reduces to zero, the concentration attains equilibrium, and Equation (24) reduces to:

$$Q = F/c \qquad (25)$$

This relation is valid only when the air exchange rate is constant; thus, this technique is appropriate for systems at or near equilibrium. It is particularly useful in spaces with mechanical ventilation or with high air exchange rates. Constant injection requires the measurement of absolute concentrations and injection rates.

Dietz et al. (1986) introduced a special case of the constant injection technique. This technique uses permeation tubes as a tracer gas source. The tubes release the tracer at an ideally constant rate into the building being tested. A sampling tube packed with an adsorbent collects the tracer from the interior air at a constant rate by diffusion. After a sampling period of one week or more, the sampler is removed and analyzed to determine the average tracer gas concentration within the building during the sampling period.

Solving Equation (20) for c and taking the time average gives

$$< c > \, = \, < F/Q > \, = \, F < 1/Q > \qquad (26)$$

where $< \ldots >$ denotes time average. (Note that the time average of $dc/d\theta$ is assumed to equal zero.)

Equation (26) shows that the average tracer concentration and the injection rate can be used to calculate the average of the inverse air exchange rate. The average of the inverse is less than the actual average, with the magnitude of this difference depending on the distribution of air exchange rates during the measurement period. Sherman and Wilson (1986) calculated these differences to be about 20% for one-month averaging periods. Differences greater than 30% have been measured when there were large changes in air exchange rate due to occupant airing of houses; errors from 5 to 30% were measured when the variation was due to weather effects (Bohac *et al.* 1987). Longer averaging periods and large changes in air exchange rates during the measurement periods generally lead to larger differences between the average inverse exchange rate and the actual average rate.

Multizone Air Exchange Measurement

Equation (20) is based on the assumption of a single, well-mixed enclosure, and the techniques described are for single-zone measurements. Airflow between internal zones and between the exterior and individual internal zones has lead to the development of multizone measurement techniques (Harrje *et al.* 1985, Sherman and Dickerhoff 1989, Fortmann *et al.* 1990, Harrje *et al.* 1990). This is important when considering the transport of pollutants from one room of a building to another. For a theoretical development, see Sinden (1978b). Multizone measurements will typically use multiple tracer gases for the different zones or the constant concentration technique. A proper error analysis is essential in all multizone flow determination (Charlesworth 1988, D'Ottavio *et al.* 1988).

AIR LEAKAGE

The air leakage of a building characterizes the relationship between the pressure difference across the building envelope and the airflow rate through the envelope (see Airflow through Openings). Building air leakage is a physical property of a building and is determined by its design, construction, seasonal effects, and deterioration over time. Although airtightness is just one factor in determining the air exchange rate of a building, it is useful for comparing buildings to one another or to airtightness standards, for evaluating design and construction quality, and for studying the effectiveness of airtightening retrofits. No simple relationship exists between a building's air leakage and its air exchange rate, but calculation methods do exist (see Calculating Air Exchange).

Measurement

While tracer gas measurement procedures provide building air exchange rates, they are somewhat expensive and time-consuming. In many cases, it is sufficient, or preferable, to measure the air leakage of a building with pressurization testing (Stricker 1975, Tamura 1975, Kronvall 1978, Blomsterberg and Harrje 1979, Gadsby and Harrje 1985). Fan pressurization is relatively quick and inexpensive and characterizes building envelope airtightness independent of weather conditions. In this procedure, a large fan or blower is mounted in a door or window and induces a large and roughly uniform pressure difference across the building shell (CGSB 1986, ASTM 1987). The airflow required to maintain this pressure difference is then measured. The leakier the building, the more airflow is necessary to induce a specific indoor-outdoor pressure difference. The airflow rate is generally measured at a series of pressure differences ranging from about 0.04 to 0.30 in. of water.

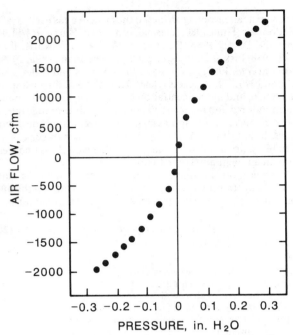

Fig. 8 Airflow Rate versus Pressure Difference Data from Whole House Pressurization Test

The results of a pressurization test, therefore, consist of several combinations of pressure difference and airflow rate data. An example of typical data is shown in Figure 8. These data points characterize the air leakage of a building and are generally converted to a single value that serves as a measure of the building's airtightness. There are several different measures of airtightness, most of which involve fitting the data to a curve in the form of Equation (15), *i.e.*, $Q = c\Delta p^n$. The airtightness ratings are based on airflow rates predicted at particular reference pressures by Equation (15). The basic difference between the different airtightness ratings is the value of the reference pressure.

In some cases, the predicted airflow rate is converted to an equivalent or effective leakage area by rearranging Equation (14) into the following form:

$$L = C_6 \, Q_r \, (\rho/2\Delta p_r)^{0.5}/C_D \qquad (27)$$

where

L = equivalent or effective leakage area, in^2
Δp_r = reference pressure difference, in. of water
Q_r = predicted airflow rate at Δp_r (from curve fit to pressurization test data), cfm
C_D = discharge coefficient
C_6 = unit conversion factor = 0.186

All the openings in the building shell are combined into an overall opening area and discharge coefficient for the building when the equivalent or effective leakage area is calculated. Some users of the leakage area approach set the discharge coefficient equal to 1. Others set $C_D \cong 0.6$, *i.e.*, the discharge coefficient for a sharp-edged orifice. The leakage area of a building is, therefore, the area of an orifice (with an assumed value of C_D) that would produce the same amount of leakage as the building envelope at the reference pressure.

Whether an airtightness rating based on leakage area or a predicted airflow rate is used, either quantity is generally normalized by some factor to account for building size. These normalization factors include floor area, exterior envelope area, and building volume.

With the wide variety of possible approaches to normalization and reference pressure, and the use of the leakage area concept, many different airtightness ratings are being used. Reference pressure differences include 0.016, 0.04, 0.10, 0.20, and 0.30 in. of water. Reference pressures of 0.016 and 0.04 in. of water are advocated because they are closer to the pressures that actually induce air exchange and, therefore, better model the flow characteristics of the openings. While this may be true, they are outside the range of measured values in the test; therefore, the predicted airflow rates at 0.016 or 0.04 in. of water are subject to significant uncertainty. The uncertainty in these predicted airflow rates and the implications for quantifying airtightness are discussed in Persily and Grot (1985b), Chastain (1987), and Modera and Wilson (1990). Round robin tests by Murphy *et al.* (1991) to determine the repeatability and reproducibility of fan pressurization devices found that subtle errors in fan calibration or operator technique are greatly exaggerated when extrapolating the pressure versus flow curve out to 0.016 in. of water with errors as much as ±40% mainly due to the fan calibration errors at low flow.

Some common airtightness ratings include the effective leakage area at 0.016 in. of water assuming $C_D = 1.0$ (Sherman and Grimsrud 1980), the equivalent leakage area at 0.04 in. of water assuming $C_D = 0.611$ (CGSB 1986), and the airflow rate at 0.20 in. of water, divided by the building volume to give units of air changes per hour (Blomsterberg and Harrje 1979).

Leakage areas at one reference pressure can be converted to leakage areas at some other reference pressure according to:

$$L_{r,2} = L_{r,1} \, (C_{D,1}/C_{D,2}) \, (\Delta p_{r,2}/\Delta p_{r,1})^{n-0.5} \qquad (28)$$

where

$L_{r,1}$ = leakage area at reference pressure $\Delta p_{r,1}$, in^2
$L_{r,2}$ = leakage area at reference pressure $\Delta p_{r,2}$, in^2
$C_{D,1}$ = discharge coefficient used to calculate $L_{r,1}$
$C_{D,2}$ = discharge coefficient used to calculate $L_{r,2}$
n = flow exponent from Equation (15)

A leakage area at one reference pressure can be converted to an airflow rate at some other reference pressure according to:

$$Q_{r,2} = C_7 \, C_{D,1} \, L_{r,1} \, (2/\rho)^{0.5} \, (\Delta p_{r,1})^{0.5-n} \, (\Delta p_{r,2})^n \qquad (29)$$

where

$Q_{r,2}$ = airflow rate at reference pressure $\Delta p_{r,2}$, cfm
$L_{r,1}$ = leakage area at reference pressure, in^2
$C_{D,1}$ = discharge coefficient used to calculate $L_{r,1}$
C_7 = conversion factor = 5.39

Finally, a leakage area may be converted to a flow coefficient in Equation (15) according to:

$$c = C_7 \, C_D \, L \, (2/\rho)^{0.5} \, (\Delta p_r)^{0.5-n} \qquad (30)$$

where

c = flow coefficient, cfm / (in. of water)^n
C_D = discharge coefficient used to calculate L
L = leakage area at reference pressure Δp_r
C_7 = conversion factor = 5.39

Equations (28) through (30) require the assumption of a value of n, unless it is reported with the measurement results. When fitting whole building pressurization test data to Equation (15), the value of n generally lies between 0.6 and 0.7. Therefore, using a value of n in this range is reasonable.

Fan pressurization measures a property of a building that ideally varies little with time and weather conditions. In reality, unless the wind and temperature difference during the measurement period are sufficiently mild, the pressure differences they induce during the test will interfere with the test pressures and cause measurement errors. Persily (1982) and Modera and Wilson (1990) have presented experimental studies of the effects of wind speed on pressurization test results. Several experimental studies have also shown variations on the order of 20 to 40% over a year in the measured airtightness in homes (Persily 1982, Kim and Shaw 1986, Warren and Webb 1986).

Several pressurization test results for residential buildings are presented in Figure 9 (Persily 1986). These results are in units of air changes per hour at 0.20 in. of water and show the wide range in airtightness among houses, even houses of identical design. The passive solar and energy-efficient data also show that houses that might be expected to be relatively airtight are not necessarily that tight. The houses in Sweden—which have a residential building airtightness standard of 0.3 air changes per hour at 0.20 in. of water for single-family detached houses (Swedish Building Code 1980)—are exceptionally tight, as are the houses in Canada.

ASHRAE *Standard* 119-1988 establishes air leakage performance levels for residential buildings. These levels are in terms of the normalized leakage L_n:

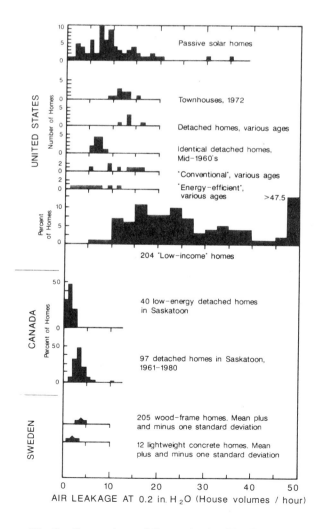

Fig. 9   Comparison of Pressurization Test Results

Table 2 Leakage Classes

Range of Normalized Leakage	Leakage Class
$L_n < 0.10$	A
$0.10 \leqslant L_n < 0.14$	B
$0.14 \leqslant L_n < 0.20$	C
$0.20 \leqslant L_n < 0.28$	D
$0.28 \leqslant L_n < 0.40$	E
$0.40 \leqslant L_n < 0.57$	F
$0.57 \leqslant L_n < 0.80$	G
$0.80 \leqslant L_n < 1.13$	H
$1.13 \leqslant L_n < 1.60$	I
$1.60 \leqslant L_n$	J

$$L_n = C_8 \, (L/A) \, (H/H_o)^{0.3} \qquad (31)$$

where

L = effective leakage area at 0.016 in. of water ($C_D = 1.0$), in^2
A = gross floor area (within exterior walls), ft^2
H = building height, ft
H_o = reference height of one-story building = 8 ft
C_8 = conversion factor = 6940

Table 2 presents the leakage classes of *Standard* 119. The values of L_n in this table correspond approximately to annual average building air exchange rates in units of air changes per hour. The standard specifies appropriate leakage classes for a building based on climate.

Persily and Grot (1986) ran whole building pressurization tests in large office buildings, which showed pressurization airflow rate divided by the building volume is relatively low compared to that of houses. However, if these airflow rates are normalized by building envelope area instead of by volume, the results indicate envelope airtightness levels similar to those in typical American houses.

Air Leakage of Building Components

The fan pressurization procedure discussed earlier enables the measurement of whole building air leakage. The location and size of individual openings in building envelopes are extremely important, as they influence the air infiltration rate of a building as well as the heat and moisture transfer characteristics of the envelope. Additional test procedures exist for pressure testing individual building components such as windows, walls, and doors; they are discussed in ASTM *Standards* E 283 and E 783 for laboratory and field tests, respectively. The following sections discuss component air leakage in both residential and commercial buildings.

Leakage Distribution in Residential Buildings

Dickeroff *et al.* (1982) and Harrje and Born (1982) studied the air leakage of individual building components and systems. The following points summarize the percentages of whole building leakage area associated with various components and systems. The values in parentheses include the range determined for each component, and the mean of the range.

Walls (18 to 50%; 35%). Both interior and exterior walls contribute to the leakage of the structure. Leakage between the sill plate and the foundation, cracks below the bottom of the gypsum wall board, electrical outlets, plumbing penetrations, and leaks into the attic at the top plates of walls all occur. Since interior walls are not filled with insulation, open paths connecting these walls and the attic permit the walls to behave like heat exchanger fins within the conditioned living space of the house.

Ceiling details (3 to 30%; 18%). Leakage across the top ceiling of the heated space is particularly insidious because it reduces the effectiveness of insulation on the attic floor and contributes to infiltration heat loss. Ceiling leakage also reduces the effectiveness of ceiling insulation in buildings without attics. Recessed lighting, plumbing, and electrical penetrations leading to the attic are some particular areas of concern.

Heating system (3 to 28%; 18%). The location of the furnace or ductwork in conditioned or unconditioned spaces, the venting arrangement of a fuel-burning device, and the existence and location of a combustion air supply all affect leakage. Modera *et al.* (1991) and Robison and Lambert (1989), among others, have shown that the variability of leakage in ducts passing through unconditioned spaces is high, the coefficient of variation being on the order of 50%. Field studies have also shown that in-situ repairs can eliminate one-quarter to two-thirds of the observed leakage (Cummings and Tooley 1989, Cummings *et al.* 1990, Robison and Lambert 1989). The 18% contribution of ducts to total-leakage significantly underestimates their impact since during system operation, the pressure differentials across the duct leaks are approximately ten times higher than typical pressures across the envelope leaks (Modera 1989, Modera *et al.* 1991).

Windows and doors (6 to 22%; 15%). More variation is seen in window leakage among window types (*e.g.*, casement versus double-hung) than among new windows of the same type from different manufacturers (Weidt *et al.* 1979). Windows that seal by compressing the weatherstrip (casements, awnings) show significantly lower leakage than windows with sliding seals.

Fireplaces (0 to 30%; 12%). When a fireplace is not in use, poor-fitting dampers allow air to escape. Glass doors reduce excess air while a fire is burning but rarely seal the fireplace structure more tightly than a closed damper does. Chimney caps or fireplace plugs with telltale signs that warn they are in place effectively reduce leakage through a cold fireplace.

Vents in conditioned spaces (2 to 12%; 5%). Exhaust vents in conditioned spaces frequently have no dampers or dampers that do not close properly.

Diffusion through walls (<1%). Diffusion, in comparison to infiltration through holes and other openings in the structure, is not an important flow mechanism. Typical values for the permeability of building materials at 0.02 in. of water produce an air exchange rate of less than 0.01 air changes per hour by wall diffusion in a typical house.

Component leakage areas. Table 3 shows effective leakage areas for a variety of residential building components at 0.016 in. of water with a C_D assumed equal to 1 (Colliver *et al.* 1992). The values in the table present results in terms of leakage area per component. Per unit component means either per component, per unit surface area, or per unit length of crack, whichever is appropriate. These leakage areas may be converted to leakage areas at other reference pressures, airflow rates, or flow coefficients using Equations (28) through (30).

Multifamily Building Leakage

Leakage distribution is particularly important in multifamily apartment buildings. These buildings often cannot be treated as single zones due to the internal resistance between apartments. Moreover, the leakage between apartments varies widely, tending to be small in modern construction, and ranging as high as 60% of the total apartment leakage in turn-of-the-century brick walk-up apartment buildings (Modera *et al.* 1985, Diamond *et al.* 1986). Little data on interzonal leakage has been reported because of the difficulty and expense of these measurements.

Commercial Building Envelope Leakage

The building envelopes of large commercial buildings are often thought to be quite airtight. The National Association of Architectural Metal Manufacturers specifies a maximum leakage per unit of exterior wall area of 0.060 cfm/ft^2 at a pressure difference of 0.30 in. of water exclusive of leakage through operable windows. Tamura and Shaw (1976a) found that air leakage measurements in eight Canadian office buildings with sealed windows, assum-

Table 3 Effective Leakage Areas (Low-Rise Residential Applications)

	Units (see note)	Best Estimate	Mini-mum	Maxi-mum		Units (see note)	Best Estimate	Mini-mum	Maxi-mum
Ceiling					Sole plate, floor/wall, caulked	in^2/lftc	0.4	0.04	0.61
General	in^2/ft^2	0.026	0.011	0.04	Top plate, band joist	in^2/lftc	0.05	0.04	0.19
Drop	in^2/ft^2	0.0027	0.00066	0.003	**Piping/Plumbing/Wiring**				
Chimney	in^2/ea	4.5	3.3	5.6	**penetrations**				
Ceiling penetrations					Uncaulked	in^2/ea	0.9	0.31	3.7
Whole house fans	in^2/ea	3.1	0.25	3.3	Caulked	in^2/ea	0.3	0.16	0.3
Recessed lights	in^2/ea	1.6	0.23	3.3	**Vents**				
Ceiling/Flue vent	in^2/ea	4.8	4.3	4.8	Bathroom with damper closed	in^2/ea	1.6	0.39	3.1
Surface-mounted lights	in^2/ea	0.13			Bathroom with damper open	in^2/ea	3.1	0.95	3.4
Crawl space					Dryer with damper	in^2/ea	0.46	0.45	1.1
General (area for exposed wall)	in^2/ft^2	0.144	0.1	0.24	Dryer without damper	in^2/ea	2.3	1.9	5.3
8 in. by 16 in. vents	in^2/ea	20			Kitchen with damper open	in^2/ea	6.2	2.2	11
Door frame					Kitchen with damper closed	in^2/ea	0.8	0.16	1.1
General	in^2/ea	1.9	0.37	3.9	Kitchen with tight gasket	in^2/ea	0.16		
Masonry, not caulked	in^2/ft^2	0.07	0.024	0.07	**Walls (Exterior)**				
Masonry, caulked	in^2/ft^2	0.014	0.004	0.014	Cast-in-place concrete	in^2/ft^2	0.007	0.0007	0.026
Wood, not caulked	in^2/ft^2	0.024	0.009	0.024	Clay brick cavity wall, finished	in^2/ft^2	0.0098	0.0007	0.033
Wood, caulked	in^2/ft^2	0.004	0.001	0.004	Precast concrete panel	in^2/ft^2	0.017	0.0004	0.024
Trim	in^2/lftc	0.5			Lightweight concrete block,				
Jamb	in^2/lftc	4	3.6	5	unfinished	in^2/ft^2	0.05	0.019	0.058
Threshold	in^2/lftc	1	0.6	12	Lightweight concrete block,				
Doors					painted or stucco	in^2/ft^2	0.016	0.0075	0.016
Attic/crawl space,					Heavyweight concrete block,				
not weatherstripped	in^2/ea	4.6	1.6	5.7	unfinished	in^2/ft^2	0.0036		
Attic/crawl space,					Continuous air infiltration barrier	in^2/ft^2	0.0022	0.0008	0.003
weatherstripped	in^2/ea	2.8	1.2	2.9	Rigid sheathing	in^2/ft^2	0.005	0.0042	0.006
Attic fold down,					**Window framing**				
not weatherstripped	in^2/ea	6.8	3.6	13	Framing, masonry, uncaulked	in^2/ft^2	0.094	0.082	0.148
Attic fold down,					Framing, masonry, caulked	in^2/ft^2	0.019	0.016	0.03
weatherstripped	in^2/ea	3.4	2.2	6.7	Framing, wood, uncaulked	in^2/ft^2	0.025	0.022	0.039
Attic fold down,					Framing, wood, caulked	in^2/ft^2	0.004	0.004	0.007
with insulated box	in^2/ea	0.6			**Windows**				
Attic from unconditioned					Awning not weatherstripped	in^2/ft^2	0.023	0.011	0.035
garage	in^2/ea	0	0	0	Awning with weatherstripping	in^2/ft^2	0.012	0.006	0.017
Double, not weatherstripped	in^2/ft^2	0.16	0.1	0.32	Casement with weatherstripping	in^2/lftc	0.12	0.05	1.5
Double, weatherstripped	in^2/ft^2	0.12	0.04	0.33	Casement without weather-				
Elevator (passenger)	in^2/ea	0.04	0.022	0.054	stripping	in^2/lftc	0.14		
General, average	in^2/lftc	0.16	0.12	0.23	Double horizontal slider				
Interior (pocket, on top floor)	in^2/ea	2.2			without weatherstripping	in^2/lftc	0.56	0.01	1.7
Interior (stairs)	in^2/lftc	0.5	0.13	0.76	Double horizontal slider, wood				
Mail slot	in^2/lftc	2			with weatherstripping	in^2/lftc	0.28	0.076	0.87
Sliding exterior glass patio	in^2/ea	3.4	0.46	9.3	Double horizontal slider, alumi-				
Sliding exterior glass patio	in^2/ft^2	0.079	0.009	0.22	num with weatherstripping	in^2/lftc	0.37	0.29	0.4
Storm (difference between					Double hung without weather-				
with and without)	in^2/ea	0.9	0.46	0.96	stripping	in^2/lftc	1.3	0.44	3.1
Single, not weatherstripped	in^2/ea	3.3	1.9	8.2	Double hung with weather-				
Single, weatherstripped	in^2/ea	1.9	0.6	4.2	stripping	in^2/lftc	0.33	0.1	0.97
Vestibule (subtract per					Double hung without weather-				
each location)	in^2/ea	1.6			stripping, with storm	in^2/lftc	0.5	0.25	0.86
Electrical outlets/Switches					Double hung with weather-				
No gaskets	in^2/ea	0.38	0.08	0.96	stripping, with storm	in^2/lftc	0.4	0.22	0.5
With gaskets	in^2/ea	0.023	0.012	0.54	Double hung with weatherstrip-				
Furnace					ping, with pressurized track	in^2/lftc	0.24	0.2	0.28
Sealed (or no) combustion	in^2/ea	0	0	0	Jalousie	in^2/louver	0.524		
Retention head or stack damper	in^2/ea	4.6	3.1	4.6	Lumped	in^2/lfts	0.24	0.0046	1
Retention head and stack					Single horizontal slider,				
damper	in^2/ea	3.7	2.8	4.6	weatherstripped	in^2/lfts	0.34	0.1	1
Floors over crawl spaces					Single horizontal slider, aluminum	in^2/lfts	0.4	0.14	1
General	in^2/ft^2	0.032	0.006	0.071	Single horizontal slider, wood	in^2/lfts	0.22	0.14	0.5
Without ductwork in crawl					Single horizontal slider, wood clad	in^2/lfts	0.33	0.27	0.41
space	in^2/ft^2	0.0285			Single hung, weatherstripped	in^2/lfts	0.44	0.32	0.63
With ductwork in crawl space	in^2/ft^2	0.0324			Sill	in^2/lftc	0.11	0.071	0.11
Fireplace					Storm inside, heat shrink	in^2/lfts	0.009	0.0046	0.009
With damper closed	in^2/ft^2	0.62	0.14	1.3	Storm inside, rigid sheet with				
With damper open	in^2/ft^2	5.04	2.09	5.47	magnetic seal	in^2/lfts	0.061	0.009	0.12
With glass doors	in^2/ft^2	0.58	0.06	0.58	Storm inside, flexible sheet with				
With insert and damper closed	in^2/ft^2	0.52	0.37	0.66	mechanical seal	in^2/lfts	0.078	0.009	0.42
With insert and damper open	in^2/ft^2	0.94	0.58	1.3	Storm inside, rigid sheet with				
Gas water heater	in^2/ea	3.1	2.3	3.9	mechanical seal	in^2/lfts	0.2	0.023	0.42
Joints					Storm outside, pressurized track	in^2/lftc	0.27		
Ceiling-wall	in^2/lftc	0.76	0.081	1.3	Storm outside, 2 track	in^2/lftc	0.63		
Sole plate, floor/wall,					Storm outside, 3 track	in^2/lftc	1.25		
uncaulked	in^2/lftc	2	0.2	2.8					

Note: Leakage areas are based on values found in the literature. The effective leakage area (in square inches) is based on a pressure difference of 0.016 in. of water and $C_d = 1$.

Abbreviations: ft^2 = gross area in square feet   lftc = lineal feet of crack
                 ea = each   lfts = lineal feet of sash

ing a flow exponent of 0.65 in Equation (15), ranged from 0.120 to 0.480 cfm/ft^2. Other measurements taken by Persily and Grot (1986) in eight U.S. office buildings ranged from 0.213 to 1.028 cfm/ft^2 at 0.30 in. of water. Therefore, office building envelopes are leakier than expected. Typical air leakage values per unit wall area at 0.30 in. of water are 0.10, 0.30, and 0.60 cfm/ft^2 for tight, average, and leaky walls, respectively (Tamura and Shaw 1976a).

Air Leakage through Internal Partitions

In large buildings, the air leakage associated with internal partitions becomes very important. Elevator, stair, and service shaft walls, floors, and other interior partitions are the major separations of concern in these buildings. Their leakage characteristics are needed to determine infiltration through exterior walls and airflow patterns within a building. These internal resistances are also important in the event of a fire to predict smoke movement patterns and evaluate smoke control systems.

Table 4 gives leakage areas (calculated at 0.30 in. of water with $C_D = 0.65$) for different internal partitions of commercial buildings (Klote and Fothergill 1983). Figure 10 presents examples of measured air leakage rates of elevator shaft walls (Tamura and Shaw 1976b), the type of data used to derive the values in Table 4. Chapter 47 of the 1991 ASHRAE *Handbook—Applications* should be consulted for models of performance of smoke control systems and their application.

Table 4 Leakage Areas for Internal Partitions in Commercial Buildings (at 0.3 in. of water and $C_D = 0.65$)

Construction Element	Wall Tightness	Area Ratio
		A/A_w
Stairwell walls	Tight	0.14×10^{-4}
	Average	0.11×10^{-3}
	Loose	0.35×10^{-3}
Elevator shaft walls	Tight	0.18×10^{-4}
	Average	0.84×10^{-3}
	Loose	0.18×10^{-2}
		A/A_f
Floors	Average	0.52×10^{-4}

A = leakage area A_w = wall area A_f = floor area

Fig. 10 Air-Leakage Rates of Elevator Shaft Walls

Leakage openings at the top of elevator shafts are equivalent to orifice areas of 620 to 1550 in^2. Air leakage rates through stair shaft and elevator doors are shown in Figure 11 as a function of average crack width around the door. The leakage areas associated with other openings within commercial buildings are also important for air movement calculations. These include interior doors and partitions, suspended ceilings in buildings where the space above the ceiling is used in the air distribution system, and other components of the air distribution system.

Air Leakage through Exterior Doors

Door infiltration depends on the type of door, room, and building. In residences and small buildings where doors are used infrequently, the air exchange associated with a door can be estimated based on air leakage through cracks between the door and the frame. A frequently opened single door, as in a small retail store, has a much larger amount of airflow. An ASHRAE research program provided data on air leakage characteristics of swinging door entrances (Min 1958, Tamura and Wilson 1966, 1967a) and revolving doors (Schutrum *et al.* 1961). A design chart (Min 1961) based on this report (Schutrum *et al.* 1961) evaluates infiltration through manual and power-operated revolving doors.

CONTROLLING AIR LEAKAGE

New Buildings

It is much easier to build a tight building than to tighten an existing building. Elmroth and Levin (1983), Eyre and Jennings (1983), Marbek (1984), and Nelson *et al.* (1985) provide information and construction details on airtight building design for houses. However, little corresponding information is available for commercial buildings.

A continuous air infiltration barrier is one of the most effective means of reducing air leakage through walls, around window and door frames, and at joints between major building elements. The air infiltration barrier can be installed on the inside of the wall framing, in which case it also usually functions as a vapor retarder, or on the outside of the wall framing, in which case it should have a permeance rating high enough to permit diffusion of water vapor from the wall. For a discussion of moisture transfer in building envelopes, see Chapters 20 and 21.

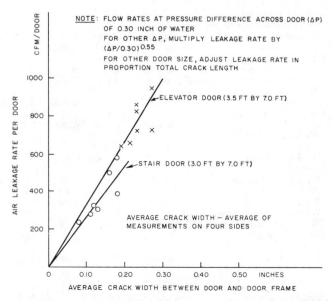

Fig. 11 Air-Leakage Rate of Door versus Average Crack Width

When the air infiltration barrier is also a continuous plastic film vapor retarder, particular care must be taken to ensure its continuity at all wall, floor, and ceiling joints; at window and door frames; and at all penetrations of the air-vapor barrier, such as electrical outlets and switches, plumbing connections, and utility service penetrations. Joints in the air-vapor barrier must be lapped and sealed. Plastic vapor retarders installed in the ceiling should be tightly sealed with the vapor retarder in the outside walls and continuous over the partition walls. A seal at the top of the partition walls prevents leakage into the attic; a plate on top of the studs generally gives a poor seal.

A continuous air-infiltration barrier installed on the outside of wall framing can eliminate many difficult construction details associated with the installation of continuous air-vapor barriers. Interior air-vapor barriers must be lapped and sealed at electrical outlets and switches, at joints between walls and floors and joints between walls and ceilings, and at plumbing connections penetrating the wall's interior finish. The exterior air-infiltration barrier can cover these problem areas with a continuous material. Joints in the air-infiltration barrier should be lapped and sealed or taped. Exterior air-infiltration barriers are generally made of a material stronger than plastic film and are more likely to withstand damage during construction. Sealing the wall against air leakage at the exterior of the insulation also cuts down on convective currents within the wall cavity, allowing insulation to retain more of its effectiveness.

Existing Buildings

The air-leakage sites must first be located to tighten the envelope of an existing building. As discussed earlier, air leakage in buildings is due not only to windows and doors, but to a wide range of unexpected and unobvious construction defects. Many important leakage sites can be very difficult to find. A variety of techniques developed to locate leakage sites are described in ASTM *Standard* E 1186 (1988) and Charlesworth (1988).

Once leakage sites are located, they can be repaired with materials and techniques appropriate to the size and location of the leak. Harrje *et al.* (1979), Diamond *et al.* (1982), and Energy Resource Center (1983) include information on airtightening in existing residential buildings. By using these procedures, the air leakage of residential buildings can be improved dramatically. Depending on the extent of the tightening effort and the experience of those doing the work, residential buildings can be tightened anywhere from 5 to more than 50% (Blomsterberg and Harrje 1979, Harrje and Mills 1980, Jacobson *et al.* 1986, Verschoor and Collins 1986, Giesbrecht and Proskiw 1986). Much less experience is available for airtightening large, commercial buildings, but the same general principles apply (Parekh *et al.* 1991, Persily 1991).

RESIDENTIAL VENTILATION SYSTEMS

Infiltration has traditionally met ventilation requirements for houses on the assumption that the building envelope is leaky enough. Possible difficulties with this approach include low ventilation when natural forces (temperature difference and wind) are weak, unnecessary energy consumption when such forces are strong, drafts in cold climates, lack of control of ventilation rates to meet changing needs, potential for interstitial condensation from exfiltration in cold climates, and no opportunity to recover the energy used to condition the ventilation air. The solution to these concerns is to have a reasonably tight building envelope and a properly designed and operated mechanical ventilation system.

ASHRAE *Standard* 119-1988, and the 1990 National Building Code of Canada encourage the transition to tighter envelope construction. Hamlin (1991) shows a 30% increase in airtightness of tract-built Canadian houses between 1982 and 1989. Also, 82% of the latter houses had natural air change rates below 0.3 ACH in March. Yuill *et al.* (1991) derived a procedure to show the extent to which infiltration contributes to ventilation rate requirements. As a result, the 1990 edition of the NBC of Canada has requirements for mechanical ventilation capability in all new dwelling units.

ASHRAE *Standard* 62-1989 gives ventilation rate requirements for houses, essentially 0.35 ACH with at least 15 cfm per occupant. The Canadian Standards Association *Standard* F 326 expands the requirements for residential mechanical ventilation systems to cover air distribution within the house, thermal comfort, minimum temperatures for equipment and duct work, system controls, pressurization and depressurization of the dwelling, installation requirements, and verification of compliance. Verification can be by design or by test, except that the total rate of outside air delivery must be measured.

Mechanical ventilation is being used in houses, especially in energy-efficient housing demonstration programs (Riley 1990, Palmiter *et al.* 1991). Possible systems can be characterized as local or central, exhaust or supply or balanced, with forced air or radiant/hydronic heating/cooling systems, with or without heat recovery, and with continuous operation or occupant controlled or demand controlled (*i.e.*, by pollutant sensing). Note that not all combinations are viable. Various options are described by Fisk *et al.* (1984), Hekmat *et al.* (1986), Sibbitt and Hamlin (1991), Palmiter *et al.* (1991), and Yuill *et al.* (1991).

The simplest systems use bathroom and kitchen fans to augment infiltration. Noise, installed capacity, life-time under continuous operation, distribution to all rooms (especially bedrooms), and energy-efficiency issues need to be addressed. Many present bath and kitchen fans are ineffective ventilators because of poor installations and designs. However, properly specified and installed exhaust fans can form part of good whole house ventilation systems and are so specified in some Canadian building codes.

Central supply systems use a furnace blower to induce airflow from outdoors and distribute it. However, if the blower is thermostatically controlled, it will operate intermittently and provide little ventilation in mild weather. If the blower operates continuously, cold drafts can be a problem when the furnance is off.

Central exhaust systems use leakage sites and, in some cases, intentional and controllable openings in the building envelope as the supply. Such systems are suitable for retrofit in existing houses. Energy can be recovered from the exhaust airstream with a heat pump to supplement domestic hot water and/or space heating.

For new houses with tightly constructed envelopes, balanced ventilation systems with passive heat recovery (air-to-air heat exchangers or heat recovery ventilators) are appropriate. Nearly equal rates of fan-induced supply and exhaust air flows over a heat exchanger, where heat and sometimes moisture is transferred between the airstreams. This reduces the energy consumption to condition the ventilation air by typically 60 to 80% (Cutter 1987). It also eliminates the thermal comfort problem that would occur if untempered air were introduced directly into the house. Airflow balance, leakage between streams, biological contamination of wet surfaces, and frosting are concerns associated with these devices.

Sibbitt and Hamlin (1991) found several low-cost mechanical systems that met CSA *Standard* F326 requirements, with proper design and commissioning. Palmiter *et al.* (1991) found that mechanical systems provide significant ventilation only when operated continuously. A separate ventilation system (air-to-air heat exchanger or exhaust air heat pump) or integrating an existing system with multispeed fans can best accomplish continuous ventilation. Continuous central supply systems are not recommended for cold climates, because they cause cold air drafts and condensation in the walls.

The type of ventilation system can be selected based on house leakage as defined in Table 2. Balanced air-to-air systems with heat recovery are optimal for tight houses (Leakage Classes A-C). The leakier the house, the larger will be the contribution from infiltration and the heat recovery ventilation will be less effective. In mild climates, these systems can also effectively be used in leakage classes D-F. Central exhaust systems should not be used for leakage classes A-C, unless special provisions are made for air inlets; otherwise, internal pressure will become too low during operation and backdrafting through fossil-fueled appliances could occur. Central exhaust systems are optimal for leakage classes D-F. Ventilation systems are normally not needed for leakage classes G-J, but for those cases in which it is, the central exhaust is usually the best choice.

CALCULATING AIR EXCHANGE

Techniques for calculating building air exchange rates have improved in recent years (Liddament 1983, Liddament and Allen 1983). This section describes several calculation procedures, ranging from simple estimation techniques to more physical models. The air exchange rate of a building cannot be reliably estimated a priori based on the building's construction or age or from a simple visual inspection. Some measurement is necessary, such as a pressurization test of envelope airtightness or a detailed quantification of the leakage sites and their magnitude. As discussed in the section on driving mechanisms, it is straightforward to calculate the air exchange rate of a building given the location and leakage function for every opening in the building envelope and between major building zones, the wind pressure coefficients over the building envelope, and any mechanical ventilation airflow rates. These inputs are generally unavailable for all except very simple structures or extremely well-studied buildings. Therefore, assumptions as to their values must be made. The appropriateness of these assumptions determines the accuracy of predictions of air exchange rates.

Empirical Models

These models of residential infiltration are based on statistical fits of infiltration rate data for specific houses. They use pressurization test results to account for house airtightness and take the form of simple relations between infiltration rate, an airtightness rating, and, in most cases, weather conditions. The models account for envelope infiltration only and do not deal with intentional ventilation. In one approach, the air changes per hour at 0.20 in. of water from a pressurization test is simply divided by a constant approximately equal to 20 (Sherman 1987). This technique does not account for the infiltration driving mechanisms on air exchange. Empirical models that do account for weather effects have been developed by Reeves et al. (1979), Kronvall (1980a), and Shaw (1981). The latter two models account for building air leakage using the values of c and n from Equation (15). The only other inputs required are the wind speed and temperature difference. Such empirical models predict long-term (one week) infiltration rates very well in the houses from which they were developed; they do not, however, work as well in other houses due to the building-specific nature of leakage distribution, wind pressure, and internal partitioning. Persily and Linteris (1983) and Persily (1986) show comparisons between measured and predicted house infiltration rates for these and other models. The average long-term differences between measurements and predictions are generally on the order of 40% for both models, although individual predictions can be off by 100% or more.

Single-Cell Models

Several procedures have been developed to calculate building air exchange rates that are based on physical models of the build-

ing interior as a single zone. These models are only appropriate to buildings with no internal resistance to airflow, and are therefore inappropriate to large, multizone buildings. Models of this type have been developed by the Institute of Gas Technology (IGT), Cole et al. (1980), the Building Research Establishment (Warren and Webb 1980), and the Lawrence Berkeley Laboratory (LBL) (Sherman and Grimsrud 1980). The LBL model has been widely used and serves as the basis of the calculation procedure described in the residential calculation example section that follows. It uses pressurization test results to characterize house air leakage through the effective leakage area at 0.016 in. of water ($C_D = 1$). In addition to the wind speed and temperature difference, the user must input information on distribution of leakage over the building envelope, a shielding parameter, and a local terrain coefficient. The predictive accuracy of this model can be very good ($\pm 7\%$ for weekly, $\pm 20\%$ for short term) when the inputs are well known for the building in question (Sherman and Modera 1986), but the predictions are not as accurate when the inputs are not known. All these single-zone models are sensitive to the values of the inputs, and it is generally quite difficult to determine appropriate values. These models have exhibited average errors on the order of 40% for many measurements on groups of houses and can be off by 100% in individual cases (Persily 1986).

Multicell Models

Multicell models of air exchange treat buildings as a series of interconnected zones and assume that the air within each zone is well mixed. Several such models have been developed by Allard and Herrlin (1989), Etheridge and Alexander (1980), Liddament and Allen (1983), Walton (1984, 1989), Herrlin (1985), and Feustel and Raynor-Hoosen (1990). They are all based on a mass balance for each zone of the building. These mass balances are used to solve for an interior static pressure within the building by requiring that the inflows and outflows for each zone balance to zero. These models require the user to input a location and leakage function for every opening in the building envelope and relevant interior partitions, a value for the wind pressure coefficient C_p at the location of each building envelope leakage site, and any mechanical ventilation airflow rates. This detailed information is difficult to obtain for a building. Wind pressure coefficient data in the literature, air leakage measurement results from the building or its components, and air-leakage data from the literature, can be used. These models not only solve for whole building and individual zone air exchange rates, but also determine airflow rates between zones. These interzone airflow rates are useful for predicting pollutant transport within buildings and smoke movement patterns in the event of a fire. Multizone models have the advantage of providing a physically correct determination of airflow rates, and very complex representations of buildings can be easily modeled on personal computers; however, determining the correct inputs to these models is difficult.

Residential Calculation Example

This section presents a simple, single-zone approach to calculating air infiltration rates in houses based on the LBL model. The approach requires the effective leakage area at 0.016 in. of water, which can be obtained from a whole building pressurization test. If a test value is not available, the data in Table 3 can be used to estimate the leakage area of the building. To obtain the building's total leakage area, multiply the overall dimensions or number of occurrences of each building component by the appropriate table entry. The sum of the resulting products is the total building leakage area.

Table 5 gives the result of an example calculation of the effective leakage area of a residence. Each leakage component is identified in the first column and described in the second. The length, area, or number of components is in the third column. The fourth column contains the leakage area per unit component, from Table 3, and the fifth contains the total leakage area associated with that component. The sum of the terms in the last column is the total leakage area of the building, in this case 131 in^2.

Using the effective leakage area, the airflow rate due to infiltration is calculated according to:

$$Q = L (A\Delta t + Bv^2)^{0.5} \qquad (32)$$

where

Q = airflow rate, cfm
L = effective leakage area, in^2
A = stack coefficient, cfm$^2 \cdot$in$^{-4} \cdot$°F^{-1}
Δt = average indoor-outdoor temperature difference for time interval of calculation, °F
B = wind coefficient, cfm$^2 \cdot$in$^{-4} \cdot$mph^{-2}
v = average wind speed measured at local weather station for time interval of interest, mph

The infiltration rate of the building is obtained by dividing Q by the building volume. Table 6 presents values of A for one-, two-, and three-story houses. The value of the wind coefficient B depends on the local shielding class of the building and building height. Table 7 lists five different shielding classes. Table 8 presents values of B for one-, two-, and three-story houses in shielding classes one through five. In calculating the values in Tables 6 and 8, several assumptions are made regarding input to the LBL model. These include a terrain class of 3 (rural area with scattered obstacles), $R = 0.5$ (half of the building leakage in the walls), and $X = 0$ (equal amounts of leakage in the floor and ceiling). The height of the one-, two-, and three-story buildings are 8, 16, and 24 ft, respectively.

Example 1. Estimate the infiltration at design conditions for a two-story house in Lincoln, NB. The house has an effective leakage area of 77 in^2, a volume of 12,000 ft^3, and is surrounded by a thick hedge (Shielding Class 3).

Solution: The 97.5% design temperature for Lincoln is -2 °F. Assume a design wind speed of 15 mph. Choosing A (0.0313) from Table 6 and B (0.0086) from Table 8, the airflow rate due to infiltration is:

$$Q = 77 [(0.0313 \times 70) + (0.0086 \times 15^2)]^{0.5}$$
$$= 156 \text{ cfm} = 9384 \text{ ft}^3/\text{h}$$

Table 5   Example of Calculation of Building Effective Leakage Area Based on Component Leakage Areas

Component	Description	D_i	L_i	D_iL_i
Sills	Uncaulked	142 ft	0.19 in^2/ft	27.0
Electrical outlets		20	0.08 in^2 ea	1.6
Windows	Sliding	141 ft^2	0.057 in^2/ft^2	11.4
Framing		141 ft^2	0.024 in^2/ft^2	
Exterior doors	Single	62 ft^2	0.11 in^2/ft^2	8.3
Framing		62 ft^2	0.024 in^2/ft^2	
Fireplace	Without damper	1	54.0 in^2 ea	54.0
Penetrations	Pipes	7	0.93 in^2 ea	6.5
Heating ducts	Ducts untaped, in basement	1	22.0 in^2 ea	22.0
			Calculated building leakage area $L_c =$131 in^2	

Table 6   Stack Coefficient A

	House Height (Stories)		
	One	Two	Three
Stack coefficient	0.0156	0.0313	0.0471

The infiltration rate I is equal to Q divided by the building volume:

$$I = (9384 \text{ ft}^3/\text{h}) / (12,000 \text{ ft}^3)$$
$$\text{or } I = 0.78 \text{ ACH}$$

Example 2. Calculate the average infiltration during a one-week period in January for a one-story house in Portland, OR. During this period, the average indoor-outdoor temperature difference is 30 °F, and the average wind speed is 6 mph.

The house has a volume of 9000 ft^3, an effective leakage area of 107 in^2, and is located in an area with buildings and trees within 30 ft in most directions (Shielding Class 4).

Solution: The airflow rate due to infiltration is:

$$Q = 107 [(0.0156 \times 30) + (0.0039 \times 6^2)]^{0.5}$$
$$= 83.5 \text{ cfm} = 5000 \text{ ft}^3/\text{h}$$

The infiltration rate is, therefore:

$$I = (5000 \text{ ft}^3/\text{h}) / (9000 \text{ ft}^3)$$
$$I = 0.56 \text{ ACH}$$

This value of infiltration is an estimate of the average value over the one-week interval for which the weather information was obtained and averaged.

Example 3. Estimate the average infiltration over the heating season in a two-story house with a volume of 11,000 ft^3 and the leakage area calculated in Table 5 (131 in^2). The house is located on a lot with several large trees but no other close buildings (Shielding Class 3). The average wind speed during the heating season is 7 mph, while the average indoor-outdoor temperature difference is 36 °F.

Solution: From Equation (32), the airflow rate due to infiltration is:

$$Q = 131 [(0.0313 \times 36) + (0.0086 \times 7^2)]^{0.5}$$
$$= 163 \text{ cfm} = 9780 \text{ ft}^3/\text{h}$$

The average infiltration is, therefore:

$$I = (9780 \text{ ft}^3/\text{h})/11,000 \text{ ft}^3$$
$$I = 0.89 \text{ ACH}$$

Again, this estimate is valid for the time interval used in computing the average values of the weather variables. Therefore, since the temperature difference and wind speed are values averaged over the entire heating season, the infiltration estimate is valid over the same interval.

Table 7   Local Shielding Classes

Class	Description
1	No obstructions or local shielding
2	Light local shielding; few obstructions, few trees, or small shed
3	Moderate local shielding; some obstructions within two house heights, thick hedge, solid fence, or one neighboring house
4	Heavy shielding; obstructions around most of perimeter, buildings or trees within 30 ft in most directions; typical suburban shielding
5	Very heavy shielding; large obstructions surrounding perimeter within two house heights; typical downtown shielding

Table 8   Wind Coefficient B

Shielding Class	House Height (Stories)		
	One	Two	Three
1	0.0119	0.0157	0.0184
2	0.0092	0.0121	0.0143
3	0.0065	0.0086	0.0101
4	0.0039	0.0051	0.0060
5	0.0012	0.0016	0.0018

REFERENCES

ACGIH. 1991. *Industrial ventilation: A manual of recommended practices*, 21st ed. American Conference of Governmental Industrial Hygienists, Lansing, MI.

AIVC. 1992. AIRBASE Bibliographic Database. International Energy Agency Air Infiltration and Ventilation Centre, Coventry, Great Britain.

Allard, F. and M. Herrlin. 1989. Wind-induced ventilation. ASHRAE *Transactions* 95(2):722-28.

ASHRAE. 1986. Proceedings IAQ '86. Managing Indoor Air for Health and Energy Conservation. Atlanta, GA.

ASHRAE. 1987. Proceedings IAQ '87. Practical Control of Indoor Air Problems. Washington, D.C.

ASHRAE. 1988. Proceedings IAQ '88. Engineering Solutions to Indoor Air Problems. Atlanta, GA.

ASHRAE. 1988a. Air leakage performance for detached single-family residential buildings. *Standard* 119-1988.

ASHRAE. 1989. Ventilation for acceptable indoor air quality. *Standard* 62-1989.

ASHRAE. 1990. Proceedings IAQ '90. Heating Buildings. Washington, D.C.

ASHRAE. 1991. Healthy Buildings. IAQ Conference. Washington, D.C.

ASHRAE. 1992. Proceedings IAQ'92. Environments for People. San Francisco, CA.

ASTM. 1983. Test method for determining air leakage rate by tracer dilution. *Standard* E741. American Society for Testing and Materials, Philadelphia.

ASTM. 1987. Practices for air leakage site detection in building envelopes. *Standard* E-1186. American Society for Testing and Materials, Philadelphia.

ASTM. 1988. Standard test method for determining air leakage rate by fan pressurization. *Standard* E779. American Society for Testing and Materials, Philadelphia.

ASTM. 1991a. Standard test method for determining the rate of air leakage through exterior windows, curtain walls, and doors under specified pressure differences across the specimens. *Standard* E283. American Society for Testing and Materials, Philadelphia.

ASTM. 1991b. Standard test method for field measurement of air leakage through installed exterior windows and doors. *Standard* E783. American Society for Testing and Materials, Philadelphia.

Berg-Munch, B., G. Clausen, and P.O. Fanger. 1986. Ventilation requirements for the control of body odor in spaces occupied by women. *Environmental International* 12(1-4):195.

Blomsterberg, A.K. and D.T. Harrje. 1979. Approaches to evaluation of air infiltration energy losses in buildings. ASHRAE *Transactions* 85(1):797.

Bohac, D., D.T. Harrje, and L.K. Norford. 1985. Constant concentration infiltration measurement technique: An analysis of its accuracy and field measurements, 176. Proceedings of the ASHRAE-DOE-BTECC Conference on the Thermal Performance of the Exterior Envelopes of Buildings III. Clearwater Beach, FL.

Bohac, D., D.T. Harrje, and G.S. Horner. 1987. Field study of constant concentration and PFT infiltration measurements. Proceedings of the 8th IEA Conference of the Air Infiltration and Ventilation Centre, Uberlingen, Germany.

CGSB. 1986. Determination of the airtightness of building envelopes by the fan depressurization method. CGSB *Standard*, 149.10-M86. Canadian General Standards Board, Ottawa.

Charlesworth, P.S. 1988. Chapter 2, Measurement of air exchange rates. In *Air exchange rate and airtightness measurement techniques—An applications guide*. Air Infiltration and Ventilation Centre, Coventry, Great Britain.

Chastain, J.P. 1987. Pressure gradients and the location of the neutral pressure axis for low-rise structures under pure stack conditions. Unpublished M.S. Thesis. University of Kentucky, Lexington.

Chastain, J.P. and D.G. Colliver. 1989. Influence of temperature stratification on pressure differences resulting from the infiltration stack effect. ASHRAE *Transactions* 95(1):256-68.

Chastain, J.P., D.G. Colliver, and P.W. Winner. 1987. Computation of discharge coefficients for laminar flow in rectangular and circular openings. ASHRAE *Transactions* 93(2):2259.

Cole, J.T., T.S. Zawacki, R.H. Elkins, J.W. Zimmer, and R.A. Macriss. 1980. Application of a generalized model of air infiltration to existing homes. ASHRAE *Transactions* 86(2):765.

Collet, P.F. 1981. Continuous measurements of air infiltration in occupied dwellings, 147. Proceedings of the 2nd IEA Conference of the Air Infiltration Centre, Stockholm, Sweden.

Colliver, D.G., W.E. Murphy, and W. Sun. 1992. Evaluation of the techniques for the measurement of air leakage of building components. Final report of ASHRAE Research Project RP-438. University of Kentucky, Lexington.

CSA. 1991. Residential mechanical ventilation systems: CAN/CSA—F326-M91. Toronto.

Cummings, J.B. and J.J. Tooley, Jr. 1989. Infiltration and pressure differences induced by forced air systems in Florida residences. ASHRAE *Transactions* 96(20):551-60.

Cummings, J.B., J.J. Tooley, Jr., and R. Dunsmore. 1990. Impacts of duct leakage on infiltration rates, space conditioning energy use and peak electrical demand in Florida homes. Proceedings of ACEEE Summer Study, Pacific Grove, CA. American Council for an Energy Efficient Economy, Washington, D.C.

Cutter. 1987. Air-to-air heat exchanges. In *Energy design update*. Cutter Information Corporation, Arlington, MA.

Desrochers, D. and A.G. Scott. 1985. Residential ventilation rates and indoor radon daughter levels, 362. Transactions of the APCA Specialty Conference, Indoor Air Quality in Cold Climates: Hazards and Abatement Measures. Ottawa, Ontario.

Diamond, R.C., M.P. Modera, and H.E. Feustel. 1986. Ventilation and occupant behavior in two apartment buildings. Lawrence Berkeley Laboratory Report LBL-21862. Presented at the 7th AIC Conference, Stratford-upon-Avon, UK.

Diamond, R.C., J.B. Dickinson, R.D. Lipschutz, B. O'Regan, and B. Shohl. 1982. The house doctor's manual. Report PUB 3017. Lawrence Berkeley Laboratory, Berkeley, CA.

Dickerhoff, D.J., D.T. Grimsrud, and R.D. Lipschutz. 1982. Component leakage testing in residential buildings. Proceedings of the American Council for an Energy Efficient Economy, 1982 Summer Study, Santa Cruz, CA. Report LBL 14735. Lawrence Berkeley Laboratory, Berkeley, CA.

Dietz, R.N., R.W. Goodrich, E.A. Cote, and R.F. Wieser. 1986. Detailed description and performance of a passive perfluorocarbon tracer system for building ventilation and air exchange measurement. In *Measured air leakage of buildings*, ASTM STP 904, 203. H.R. Trechsel and P.L. Lagus, eds. American Society for Testing and Materials, Philadelphia.

D'Ottavio, T.W., G.I. Senum, and R.N. Dietz. 1988. Error analysis techniques for perfluorocarbon tracer derived multizone ventilation rates. *Building and Environment* 23(40).

Ek, C.W., S.A. Anisko, and G.O. Gregg. 1990. Air leakage tests of manufactured housing in the Northwest United States. In *Air change rate and airtightness in buildings*, ASTM STP 1067. M.H. Sherman, ed. American Society for Testing and Materials, Philadelphia, 152-64.

Elmroth, A. and P. Levin. 1983. Air infiltration control in housing, a guide to international practice. Report D2:1983. Air Infiltration Centre, Swedish Council for Building Research, Stockholm.

Energy Resource Center. 1982. How to house doctor. University of Illinois, Chicago.

Etheridge, D.W. 1977. Crack flow equations and scale effect. *Building and Environment* 12:181.

Etheridge, D.W. and D.K. Alexander. 1980. The British gas multi-cell model for calculating ventilation. ASHRAE *Transactions* 86(2):808.

Etheridge, D.W. and J.A. Nolan. 1979. Ventilation measurements at model scale in a turbulent flow. *Building and Environment* 14(1):53.

Eto, J. 1990. The HVAC costs of increased fresh air ventilation rates in office buildings. Proceedings of Indoor Air '90 4:53-58. International Conference on Indoor Air Quality and Climate, Ottawa, Ontario.

Eto, J. and C. Meyer. 1988. The HVAC costs of increased fresh air ventilation rates in office buildings. ASHRAE *Transactions* 94(2):331-45.

Eyre, D. and D. Jennings. 1983. Air-vapour barriers—A general perspective and guidelines for installation. Energy, Mines, and Resources, Ottawa, Canada.

Feustel, H.E. and A. Raynor-Hoosen, eds. 1990. Fundamentals of the multizone air flow model—COMIS. Technical Note 29. Air Infiltration and Ventilation Centre, Coventry, Great Britain.

Fisk, W.J., R.J. Prill, and O. Steppanen. 1989. A multi-tracer technique for studying rates of ventilation, air distribution patterns and air exchange efficiencies. Proceedings of Conference on Building Systems—Room Air and Air Contaminant Distribution. ASHRAE, Atlanta, 237-40.

Fisk, W.J., R.K. Spencer, D.T. Grimsrud, F.J. Offermann, B. Pedersen, and R. Sextro. 1984. Indoor air quality control techniques: A critical review. Report LBL 16493. Lawrence Berkeley Laboratory, Berkeley, CA.

Fleming, W.S. 1986. Indoor air quality, infiltration and ventilation in residential buildings. Proceedings of IAQ '86, Managing Indoor Air for Health and Energy Conservation. ASHRAE, Atlanta, 192-207.

Fortmann, R.C., N.L. Nagda, and H.E. Rector. 1990. Comparison of methods for the measurement of air change rates and interzonal airflows to two test residences. In *Air change rate and airtightness in buildings*, ASTM STP 1067. M.H. Sherman, ed., American Society for Testing and Materials, Philadelphia, 104-18.

Foster, M.P. and M.J. Down. 1987. Ventilation of livestock buildings by natural convection. *Journal of Agricultural Engineering Research* 37:1.

Gadsby, K.J. and D.T. Harrje. 1985. Fan pressurization of buildings: Standards, calibration and field experience. ASHRAE *Transactions* 91(2):95-104.

Giesbrecht, P. and G. Proskiw. 1986. An evaluation of the effectiveness of air leakage sealing. In *Measured air leakage of buildings*. ASTM STP 904, 312. H.R. Trechsel and P.L. Lagus, eds. American Society for Testing and Materials, Philadelphia.

Grimsrud, D.T. and K.Y. Teichman. 1989. The scientific basis of *Standard* 62-1989. ASHRAE *Journal* 31(10):51-54.

Grimsrud, D.T., M.H. Sherman, and R.C. Sonderegger. 1982. Calculating infiltration: Implications for a construction quality standard, 422. Proceedings of the ASHRAE-DOE Conference on the Thermal Performance of the Exterior Envelope of Buildings II, Las Vegas, NV.

Grimsrud, D.T., M.H. Sherman, R.C. Diamond, P.E. Condon, and A.H. Rosenfeld. 1979. Infiltration-pressurization correlations: Detailed measurements in a California house. ASHRAE *Transactions* 85(1):851.

Grot, R.A. and R.E. Clark. 1979. Air leakage characteristics and weatherization techniques for low-income housing, 178. Proceedings of the ASHRAE-DOE Conference on the Thermal Performance of the Exterior Envelopes of Buildings, Orlando, FL.

Grot, R.A. and A.K. Persily. 1986. Measured air infiltration and ventilation rates in eight large office buildings. In *Measured air leakage of buildings*, ASTM STP 904, 151. H.R. Trechsel and P.L. Lagus, eds. American Society for Testing and Materials, Philadelphia.

Hamlin, T.L. 1991. Ventilation and airtightness in new, detached Canadian housing. ASHRAE *Transactions* 97(2):904-10.

Harrje, D.T. and G.J. Born. 1982. Cataloguing air leakage components in houses. Proceedings of the American Council for an Energy-Efficient Economy, 1982 Summer Study, Santa Cruz, CA.

Harrje, D.T. and K.J. Gadsby. 1986. Indoor air quality—Some residential answers. ASHRAE *Journal* 28(7):32-36.

Harrje, D.T. and T.A. Mills, Jr. 1980. Air infiltration reduction through retrofitting, 89. *Building air change rate and infiltration measurements*, ASTM STP 719. C.M. Hunt, J.C. King, and H.R. Trechsel, eds. American Society for Testing and Materials, Philadelphia.

Harrje, D.T., G.S. Dutt, and J. Beyea. 1979. Locating and eliminating obscure but major energy losses in residential housing. ASHRAE *Transactions* 85(2):521.

Harrje, D.T., R.A. Grot, and D.T. Grimsrud. 1981. Air infiltration site measurement techniques, 113. Proceedings of the 2nd IEA Conference of the Air Infiltration Centre, Stockholm, Sweden.

Harrje, D.T., G.S. Dutt, D.L. Bohac, and K.J. Gadsby. 1985. Documenting air movements and infiltration in multicell buildings using various tracer techniques. ASHRAE *Transactions* 91(2):2012-27.

Harrje, D.T., R.N. Dietz, M. Sherman, D.L. Bohac, T.W. D'Ottavio, and D.J. Dickerhoff. 1990. Tracer gas measurement systems compared in a multifamily building. In *Air change rate and airtightness in buildings*, ASTM STP 1067. M.H. Sherman, ed. American Society for Testing and Materials, Philadelphia, 5-20.

Hekmat, D., H.E. Feustel, and M.P. Modera. 1986. Impacts of ventilation strategies on energy consumption and indoor air quality in single-family residences. *Energy and Buildings* 9(3):239.

Herrlin, M.K. 1985. MOVECOMP: A static-multicell-airflow-model. ASHRAE *Transactions* 91(2B):1989.

Honma, H. 1975. *Ventilation of dwellings and its disturbances*. Faibo Grafiska, Stockholm.

Hopkins, L.P. and B. Hansford. 1974. Air flow through cracks. *Building Service Engineer* 42 (September):123.

Hunt, C.M. 1980. Air infiltration: A review of some existing measurement techniques and data. *Building air change rate and infiltration measurements*. ASTM STP 719, 3. C.M. Hunt, J.C. King, and H.R. Trechsel, eds. American Society for Testing and Materials, Philadelphia.

Jacobson, D.I., G.S. Dutt, and R.H. Socolow. 1986. Pressurization testing, infiltration reduction, and energy savings. In *Measured air leakage of buildings*, ASTM STP 904, 265. H.R. Trechsel and P.L. Lagus, eds. American Society for Testing and Materials, Philadelphia.

Janssen, J.E. 1989. Ventilation for acceptable indoor air quality. ASHRAE *Journal* 31(10):40-48.

Kiel, D.E. and D.J. Wilson. 1986. Gravity driven airflows through open doors, 15.1. Proceedings of the 7th IEA Conference of the Air Infiltration Centre, Stratford-upon-Avon, United Kingdom.

Kiel, D.E. and D.J. Wilson. 1987. Influence of natural ventilation on total building ventilation dominated by strong fan exhaust. ASHRAE *Transactions* 93(2):1286.

Kim, A.K. and C.Y. Shaw. 1986. Seasonal variation in airtightness of two detached houses. In *Measured air leakage of buildings*. ASTM STP 904, 17. H.R. Trechsel and P.L. Lagus, eds. American Society for Testing and Materials, Philadelphia.

Klauss, A.K., R.H. Tull, L.M. Rootsd, and J.R. Pfafflino. 1970. History of the changing concepts in ventilation requirements. ASHRAE *Journal* 12(6):51-55.

Klote, J.H. and J.W. Fothergill, Jr. 1983. *Design of smoke control systems for buildings*. ASHRAE.

Kronvall, J. 1978. Testing of homes for air leakage using a pressure method. ASHRAE *Transactions* 84(1):72.

Kronvall, J. 1980a. Correlating pressurization and infiltration rate data—Tests of an heuristic model. Lund Institute of Technology, Division of Building Technology, Lund, Sweden.

Kronvall, J. 1980b. Air flow in building components, Report TVBH-1002. Lund Institute of Technology, Division of Building Technology, Lund, Sweden.

Kumar, R., A.D. Ireson, and H.W. Orr. 1979. An automated air infiltration measuring system using SF6 tracer gas in constant concentration and decay methods. ASHRAE *Transactions* 85(2):385.

Kvisgaard, B. and P.F. Collet. 1990. The user's influence on air change. In *Air change rate and airtightness in buildings*, ASTM STP 1067. M.H. Sherman, ed. American Society for Testing and Materials, Philadelphia, 67-76.

Lagus, P.L. 1989. Tracer measurement instrumentation suitable for infiltration, air leakage, and air flow pattern characterization. Proceedings of Conference on Building Systems—Room Air and Air Contaminant Distribution. ASHRAE, Atlanta, 97-102.

Lagus, P. and A.K. Persily. 1985. A review of tracer-gas techniques for measuring airflows in buildings. ASHRAE *Transactions* 91(2B):1075.

Lee, B.E., M. Hussain, and B. Soliman. 1980. Predicting natural ventilation forces upon low-rise buildings. ASHRAE *Journal* 22(2):35-39.

Lee, K.H., Y. Lee, and H. Tanaka. 1985. Thermal effect on pressure distribution in simulated high-rise buildings—Experiment and analysis. ASHRAE *Transactions* 91(2):530-44.

Lee, K.H., H. Tanaka, and Y. Lee. 1988. Thermally induced pressure distribution in simulated tall buildings with floor partitions. ASHRAE *Transactions* 94(1):228-42.

Levins, W.P. 1982. Measured effect of forced ventilation on house infiltration rate. Proceedings of the ASHRAE-DOE Conference on the Thermal Performance of the Exterior Envelopes of Buildings II, Las Vegas, NV.

Liddament, M.W. 1983. The Air Infiltration Centre's Program of Model Validation. ASHRAE *Transactions* 89(2):129-45.

Liddament, M.W. 1988. The calculation of wind effect on ventilation. ASHRAE *Transactions* 94(2):1645-60.

Liddament, M. and C. Allen. 1983. The validation and comparison of mathematical models of air infiltration, Technical Note 11. Air Infiltration Centre, Bracknell, Great Britain.

Malik, N. 1978. Field studies of dependence of air infiltration on outside temperature and wind. *Energy and Buildings* 1(3):281.

Marbek Resource Consultants. 1984. Air sealing homes for energy conservation. Energy, Mines and Resources Canada, Buildings Energy Technology Transfer Program, Ottawa.

Mattingly, G.E. and E.F. Peters. 1977. Wind and trees: Air infiltration effects on energy in housing. *Journal of Industrial Aerodynamics* 2(1):1.

Min, T.C. 1958. Winter infiltration through swinging-door entrances in multistory buildings. ASHRAE *Transactions* 64:421.

Min, T.C. 1961. Engineering concept and design of controlling infiltration and traffic through entrances in tall commercial buildings. International Conference on Heating, Ventilating and Air Conditioning, London.

Modera, M.P. 1989. Residential duct system leakage: Magnitude, impacts, and potential for reduction. ASHRAE *Transactions*. 96(2):561-69.

Modera, M.P. and D.J. Wilson. 1990. The effects of wind on residential building leakage measurements. *Air change rate and airtightness in buildings*, ASTM STP 1067. M.H. Sherman, ed. American Society for Testing and Materials, Philadelphia, 132-45; Lawrence Berkeley Laboratory Report LBL-24195.

Modera, M.P., D. Dickerhoff, R. Jansky, and B. Smith. 1991. Improving the energy efficiency of residential air distribution systems in California. Lawrence Berkeley Laboratory Report LBL-30866.

Murphy, W.E., D.G. Colliver, and L.R. Piercy. 1991. Repeatability and reproducibility of fan pressurization devices in measuring building air leakage. ASHRAE *Transactions* 97(2):885-95.

Nelson, B.D., D.A. Robinson, and G.D. Nelson. 1985. Designing the envelope—Guidelines for buildings. Thermal Performance of the Exterior Envelopes of Buildings III. Proceedings of ASHRAE/DOE/BTECC Conference, Florida. ASHRAE SP 49, 1117-24.

NRCC. 1990. National Building Code of Canada, 10th ed. National Research Council of Canada, Ottawa.

Nylund, P.O. 1980. Infiltration and ventilation, Report D22:1980. Swedish Council for Building Research, Stockholm.

Palmiter, L. and I. Brown. 1989. The northwest residential infiltration survey, description and summary of results. Proceedings of the ASHRAE/DOE/BTECC/CIBSE Conference—Thermal Performance of the Exterior Envelopes of Buildings IV, Florida, 445-57.

Palmiter, L., I.A. Brown, and T.C. Bond. 1991. Measured infiltration and ventilation in 472 all-electric homes. ASHRAE *Transactions* 97(2):979-87.

Parekh, A., K. Ruest, and M. Jacobs. 1991. Comparison of airtightness, indoor air quality and power consumption before and after air-sealing of high-rise residential buildings. Proceedings of the 12th AIVC Conference: Air Movement and Ventilation Control within Buildings, Air Infiltration and Ventilation Centre, Coventry, Great Britain.

Parker, D.S. 1990. Air tightness characteristics of electrically heated houses in the residential standards demonstration program. In *Air change rate and airtightness in buildings*. ASTM STP 1067. M.H. Sherman, ed. American Society for Testing and Materials, Philadelphia, 283-93.

Parker, G.B., M. McSorley, and J. Harris. 1990. The northwest residential infiltration survey: A field study of ventilation in new houses in the Pacific northwest. In *Air change rate and airtightness in buildings*, ASTM STP 1067. M.H. Sherman, ed. American Society for Testing and Materials, Philadelphia, 93-103.

Persily, A. 1982. Repeatability and accuracy of pressurization testing, 380. Proceedings of the ASHRAE-DOE Conference, Thermal Performance of the Exterior Envelopes of Buildings II, Las Vegas, NV.

Persily, A.K. 1986. Measurements of air infiltration and airtightness in passive solar homes. In *Measured air leakage of buildings*. ASTM STP 904, 46. H.R. Trechsel and P.L. Lagus, eds. American Society for Testing and Materials, Philadelphia.

Persily, A.K. 1988. Tracer gas techniques for studying building air exchange. Report NBSIR 88-3708. National Bureau of Standards, Gaithersburg, MD.

Persily, A.K. 1991. Design guidelines for thermal envelope integrity in office buildings. Proceedings of the 12th AIVC Conference: Air Movement and Ventilation Control within Buildings, Air Infiltration and Ventilation Centre, Coventry, Great Britain.

Persily, A.K. and J. Axley. 1990. Measuring airflow rates with pulse tracer techniques. In *Air change rate and airtightness in buildings*, ASTM STP 1067. M.H. Sherman, ed. American Society for Testing and Materials, Philadelphia, 31-51.

Persily, A.K. and R.A. Grot. 1985a. The airtightness of office building envelopes, 125. Thermal performance of the exterior envelopes of buildings III. Proceedings of the ASHRAE-DOE-BTECC Conference in Clearwater Beach, FL.

Persily, A.K. and R.A. Grot. 1985b. Accuracy in pressurization data analysis. ASHRAE *Transactions* 91(2B):105.

Persily, A.K. and R.A. Grot. 1986. Pressurization testing of federal buildings. In *Measured air leakage of buildings*. ASTM STP 904, 184. H.R. Trechsel and P.L. Lagus, eds. American Society for Testing and Materials, Philadelphia.

Persily, A.K. and G.T. Linteris. 1983. A comparison of measured and predicted infiltration rates. ASHRAE *Transactions* 89(2):183.

Persily, A.K. and L.K. Norford. 1987. Simultaneous measurements of infiltration and intake in an office building. ASHRAE *Transactions* 93(2):42-56.

Reeves, G., M.F. McBride, and C.F. Sepsy. 1979. Air infiltration model for residences. ASHRAE *Transactions* 85(1):667.

Reinhold, G. and R. Sonderegger. 1983. Component leakage areas in residential buildings. Proceedings of the 4th IEA Conference of the Air Infiltration Centre, Elm, Switzerland. Report LBL 16221. Lawrence Berkeley Laboratory, Berkeley, CA.

Riley, M. 1990. Indoor air quality and energy conservation; The R-2000 home program experience. Indoor Air '90 5:143.

Robison, P.E. and L.A. Lambert. 1989. Field investigation of residential infiltration and heating duct leakage. ASHRAE *Transactions* 95(2):542-50.

Schutrum, L.F., N. Ozisik, C.M. Humphrey, and J.T. Baker. 1961. Air infiltration through revolving doors. ASHRAE *Transactions* 67:488.

Shaw, C.Y. 1981. A correlation between air infiltration and air tightness for a house in a developed residential area. ASHRAE *Transactions* 87(2):333.

Shaw, C.Y. 1987. Methods for estimating air change rates and sizing mechanical ventilation systems for houses. ASHRAE *Transactions* 93(2):2284-2302.

Shaw, C.Y. and W.C. Brown. 1982. Effect of a gas furnace chimney on the air leakage characteristic of a two-story detached house, 12.1. Proceedings of the 3rd IEA Conference of the Air Infiltration Centre, London.

Shaw, C.Y. and G.T. Tamura. 1977. The calculation of air infiltration rates caused by wind and stack action for tall buildings. ASHRAE *Transactions* 83(2):145.

Sherman, M.H. 1987. Estimation of infiltration from leakage and climate indications. *Energy and Buildings* 10(1):81.

Sherman, M.H. 1989a. Uncertainty in airflow calculations using tracer gas measurements. *Building and Environment* 24(4):347-54.

Sherman, M.H. 1989b. On the estimation of multizone ventilation rates from tracer gas measurements. *Building and Environment* 24(4):355-62.

Sherman, M.H. 1990. Tracer gas techniques for measuring ventilation in a single zone. *Building and Environment* 25(4):365-74.

Sherman, M.H. 1992. Superposition in infiltration modeling. *Indoor Air* 2:101-14.

Sherman, M.H. and D. Dickerhoff. 1989. Description of the LBL multitracer measurement system. Proceedings of the ASHRAE/DOE/BTECC/CIBSE Conference—Thermal Performance of the Exterior Envelopes of Buildings IV, 417-32.

Sherman, M.H. and D.T. Grimsrud. 1980. Infiltration-pressurization correlation: Simplified physical modeling. ASHRAE *Transactions* 86(2):778.

Sherman, M.H., D.T. Grimsrud, P.E. Condon, and B.V. Smith. 1980. Air infiltration measurement techniques, 9. Proceedings of the 1st IEA Symposium of the Air Infiltration Centre, London. Report LBL 10705. Lawrence Berkeley Laboratory, Berkeley, CA.

Sherman, M.H. and M.P. Modera. 1986. Comparison of measured and predicted infiltration using the LBL infiltration model. In *Measured air leakage of buildings*. ASTM STP 904, 325. H.R. Trechsel and P.L. Lagus, eds. American Society for Testing and Materials, Philadelphia.

Sherman, M.H. and D.J. Wilson. 1986. Relating actual and effective ventilation in determining indoor air quality. *Buildings and Environment* 21(3/4):135.

Sinden, F.W. 1978a. Wind, temperature and natural ventilation—Theoretical considerations. *Energy and Buildings* 1(3):275.

Sinden, F.W. 1978b. Multi-chamber theory of air infiltration. *Building and Environment* 13:21-28.

Stricker, S. 1975. Measurement of air-tightness of houses. ASHRAE *Transactions* 81(1):148.

Swami, M.V. and S. Chandra. 1988. Correlations for pressure distribution on buildings and calculation of natural-ventilation airflow. ASHRAE *Transactions* 94(1):243-66.

Swedish Building Code. 1980. Thermal insulation and air tightness. SBN 1980.

Tamura, G.T. 1975. Measurement of air leakage characteristics of house enclosures. ASHRAE *Transactions* 81(1):202.

Tamura, G.T. and C.Y. Shaw. 1976a. Studies on exterior wall airtightness and air infiltration of tall buildings. ASHRAE *Transactions* 82(1):122.

Tamura, G.T. and C.Y. Shaw. 1976b. Air leakage data for the design of elevator and stair shaft pressurization system. ASHRAE *Transactions* 82(2):179.

Tamura, G.T. and A.G. Wilson. 1966. Pressure differences for a nine-story building as a result of chimney effect and ventilation system operation. ASHRAE *Transactions* 72(1):180.

Tamura, G.T. and A.G. Wilson. 1967a. Pressure differences caused by chimney effect in three high buildings. ASHRAE *Transactions* 73(2):II.1.1.

Tamura, G.T. and A.G. Wilson. 1967b. Building pressures caused by chimney action and mechanical ventilation. ASHRAE *Transactions* 73(2):II.2.1.

Tamura, G.T. and A.G. Wilson. 1968. Pressure differences caused by wind on two tall buildings. ASHRAE *Transactions* 74(2):170.

Turk, B.T., D.T. Grimrud, J.T. Brown, K.L. Geisling-Sobotka, J. Harrison, and R.J. Prill. 1989. Commercial building ventilation rates and particle concentrations. ASHRAE *Transactions* 95(1):422-33.

Verschoor, J.D. and J.O. Collins. 1986. Demonstration of air leakage reduction program in navy family housing. In *Measured air leakage of buildings*. ASTM STP 904, 294. H.R. Trechsel and P.L. Lagus, eds. American Society for Testing and Materials, Philadelphia.

Walton, G.N. 1984. A computer algorithm for predicting infiltration and interroom airflows. ASHRAE *Transactions* 90(1B):601.

Walton, G.N. 1989. Air flow network models for element-based building air flow modelling. ASHRAE *Transactions* 95(2):611-20.

Warren, P.R. and B.C. Webb. 1980. The relationship between tracer gas and pressurization techniques in dwellings. Proceedings of the 1st IEA Symposium of the Air Infiltration Centre, London.

Warren, P.R. and B.C. Webb. 1986. Ventilation measurements in housing. CIBSE Symposium, Natural Ventilation by Design. Chartered Institution of Building Services Engineers, London.

Weidt, J.L., J. Weidt, and S. Selkowitz. 1979. Field air leakage of newly installed residential windows, 149. Proceedings of the ASHRAE-DOE Conference, Thermal Performance of the Exterior Envelopes of Buildings, Orlando, FL.

Yaglou, C.P. and W.N. Witheridge. 1937. Ventilation requirements. ASHVE *Transactions* 43: 423.

Yaglou, C.P., E.C. Riley, and D.I. Coggins. 1936. Ventilation requirements. ASHVE *Transactions* 42:133.

Yuill, G.K. 1991. The development of a method of determining air change rates in detached dwellings for assessing indoor air quality. ASHRAE *Transactions* 97(2):896-903.

Yuill, G.K. and G.M. Comeau. 1989. Investigation of the indoor air quality, air tightness and air infiltration rates of a random sample of 78 houses in Winnipeg. Proceedings of IAQ '89, The Human Equation—Health and Comfort. ASHRAE, Atlanta, 122-27.

Yuill, G.K. and J.E. Lovatt. 1986. Prediction of pollutant concentration in houses. Proceedings of IAQ '86, Managing Indoor Air for Health and Energy Conservation. ASHRAE, Atlanta, 383-91.

Yuill, G.K., M.R. Jeanson, and C.P. Wray. 1991. Simulated performance of demand-controlled ventilation systems using carbon dioxide as an occupancy indicator. ASHRAE *Transactions* 97(2):963-68.

BIBLIOGRAPHIC DATABASE

A database of bibliographic references which contains abstracts in English of technical papers covering air infiltration in buildings, AIRBASE, has been developed by the Air Infiltration and Ventilation Centre (AIVC 1992). Most of the articles are concerned with the prediction, measurement, and reduction of air infiltration and leakage rates. Selected abstracts are also included for papers relating to: indoor air quality, occupant behavior, thermal comfort, ventilation efficiency, natural and mechanical ventilation, wind pressure and its influence on infiltration, energy saving measures, and moisture and condensation.

CHAPTER 24

WEATHER DATA

THIS chapter presents weather station climatic conditions in the United States, Canada, and other countries. Design temperatures are based on the assumption that the frequency level of a specific temperature over a suitable time period will repeat in the future. The selected summer frequencies of 1%, 2.5%, and 5% and the winter frequencies of 99% and 97.5% enable the engineer to match the risk level desired for the problem at hand. At many locations, meteorological evidence indicates that the temperatures at the 1 and 99% levels may vary in the order of 2 to 4°F in any 15-year period from the previous 15-year period, and even more in any single year from the previous one. The proximity of the 99% level to the median of the annual extreme minimum temperatures indicates that extremely low temperatures occur in rare extended episodes rather than in long-term summations (Ecodyne Cooling Products 1980, Snelling 1985, Crow 1963).

Where stringent requirements for design conditions must be met, a consulting meteorologist with experience in applied climatology can evaluate special conditions, such as the formation of heat sinks in urban areas (heat islands) and the duration of extreme temperatures at project sites located remotely from reporting stations.

Annual degree-day tabulations for seven base temperatures in °F at 2546 stations in the United States can be found in *Climatography of the U.S.* (National Climatic Data Center 1992b). Annual degree-day tabulations for seven base temperatures in °C at 1195 stations in Canada are in *Canadian Climate Normals* (Atmospheric Environment Service 1981).

INTERIOR DESIGN CONDITIONS

ASHRAE *Standard* 55-1992 specifies thermal environmental conditions for the comfort of healthy people in buildings. For other conditions, refer to Chapter 8. Indoor design conditions for specific applications are suggested in Chapters 3 through 7 of the 1991 ASHRAE *Handbook—Applications*. Building codes may refer to the ASHRAE *Standard* 90 series or 100 series. Other legal restrictions on interior temperature set points and/or energy consumption may also apply.

OUTDOOR DESIGN CONDITIONS

The recommended design temperatures presented here are based on data from the National Climatic Data Center of NOAA, U.S. Air Force, U.S. Navy, Canadian Atmospheric Environment Service, and the weather organizations of the countries noted. Ecodyne Cooling Products (1980) gives other tabulations.

Temperatures at these stations, airports, and U.S. Air Force bases are measured each hour by trained observers and are recorded rounded off to the nearest whole degree, Fahrenheit. These recorded values have been statistically analyzed for a recent

5-year period of record and tabulated to the nearest whole degree, Fahrenheit, as shown in Tables 1, 2, and 3.

During the period 1965 through 1981, NOAA digitized only the three-hourly observations for most stations; therefore, the time frame for data in this tabulation is 1950 through 1964, when hourly data were available. The U.S. Air Force data used are for the period of 1957 through 1971. Although published records are for only 3-h intervals, hourly tapes are available. The frequency of occurrence of these recommended design temperatures has been machine tabulated from recorded hourly observations at the stations reporting hourly readings. Tabulated data are obtained from *Engineering Weather Data* (USAF 1978).

To make recommended design temperatures available for locations where hourly readings were not made, but where population concentrations, climatic variation, or geographical representation indicated a need, the data were adjusted by appropriate meteorological techniques from nearby stations reporting hourly data. Methods for interpolation are described in Ecodyne Cooling Products (1980) and Crow (1963).

Winter

Recommended design temperatures are presented in Column 5 of Tables 1, 2, and 3. Two frequency levels are offered for each station representing temperatures that have been equaled or exceeded by 99% or 97.5% of the total hours in the months of December, January, and February (a total of 2160 h) in the Northern Hemisphere and the months of June, July, and August in the Southern Hemisphere (a total of 2208 h). In a normal winter there would be approximately 22 h at or below the 99% value and 54 h at or below the 97.5% value. Column 9, the prevailing wind direction, is the wind direction occurring most frequently and the mean wind speed, which occurs coincidentally with the 97.5% dry-bulb winter design temperature. Column 10 is the median of the annual extreme minimum temperature. Snelling (1985) and Crow (1985) found that the duration in a single episode of extremely low temperatures below the 99% and 97.5% levels can continue for 3 to 5 days.

For Canadian stations, the two design values are based on only the month of January, because the temperature distribution in January in Canada is characteristic of an extremely cold month compared to the temperature distributions in December and February. The Canadian design temperatures are a few degrees lower than those based on three winter months (see Arctic Meteorology Research Group 1960, Boughner 1960).

Summer

Recommended design dry-bulb and wet-bulb temperatures and mean daily range are presented in Columns 6, 7, and 8 of Tables 1, 2, and 3. Column 6 provides dry-bulb temperatures with their corresponding coincident wet-bulb temperatures. In Table 3, the coincident wet-bulb temperatures are not available. The dry-bulb

---

The preparation of this chapter is assigned to TC 4.2, Weather Information.

temperatures represent values that have been equaled or exceeded by 1%, 2.5%, and 5% of the total hours during the months of June through September (a total of 2928 h) in the Northern Hemisphere and the months of December through March in the Southern Hemisphere (a total of 2904 h). The coincident wet-bulb temperature listed with each design dry-bulb temperature is the mean of all wet-bulb temperatures occurring at the specific dry-bulb temperature.

The mean daily range shown in Column 7 of Tables 1 and 2 is the difference between the average daily maximum and average daily minimum temperatures in the warmest month. In Table 3, the daily range is the long-term average. In Tables 1, 2, and 3, wet-bulb temperatures in Column 8 represent values that have been equaled or exceeded by 1%, 2.5%, and 5% of the hours during the summer months. These wet-bulb values were computed independently of the dry-bulb values and are not coincident with the design dry-bulb values in Column 6. Their coincident dry-bulb values have not been tabulated. Ecodyne Cooling Products (1980) has examples of the coincident dry-bulb with the extreme wet-bulb temperature.

In Column 9, the prevailing wind direction is the wind direction occurring most frequently coincident with the 2.5% dry-bulb summer design temperature. Column 10 gives the median of the annual extreme maximum and minimum temperature. In a normal summer, approximately 30 h would be at or above the 1% design value, and 150 h at or above the 5% design value. Snelling (1985) and Crow (1985) found that episodes of temperatures above the 1%, 2.5%, and 5% levels do not continue for more than 24 h.

For Canadian stations, the 1%, 2.5%, and 5% values are based on the month of July only, because the temperature distribution in July in Canada is characteristic of an extremely warm month compared with the temperature distributions of the adjacent months. The Canadian summer design values are a few degrees higher than those based on four summer months (see Arctic Meteorology Research Group 1960, Boughner 1960).

Design temperatures for individual months or other periods are not tabulated here but can be roughly approximated from bin data in USAF (1978). The Atmospheric Environment Service (1883-1987) has detailed monthly temperature percentiles and other data for 1345 Canadian locations. Data summaries for stations throughout the world are also available in Engineering Weather Data (USAF 1978) and Tables of Temperature, Relative Humidity and Precipitation for the World, Parts I-IV (Meteorological Office 1958).

WIND DATA

Winter

The prevailing wind direction is the wind direction occurring most frequently with the 97.5% dry-bulb winter design temperature. The mean wind speed (knots) is the average of those wind speed values occurring coincidentally with the 97.5% dry-bulb winter design temperature (USAF 1978).

At many stations, particularly in and near mountainous terrain, the prevailing wind direction for winter hours at the 97.5% level is strongly influenced by local terrain features. The wind speeds during this local down-valley drainage flow is generally 7 knots or less.

Summer

The prevailing wind direction is the wind direction occurring most frequently with the 2.5% dry-bulb summer design temperature (USAF 1978).

MEDIAN OF ANNUAL EXTREMES

In winter, the median of the annual extreme applies to the minimum temperatures. In summer, the median of the annual extreme applies to the maximum temperatures.

INTERPOLATION BETWEEN STATIONS

Data from many specific weather stations can supply a database for interpolation of expected conditions at nearby locations that lack data. Refer to Crow (1963) for examples of interpolation criteria. The stations included in Tables 1 and 2 were primarily selected because of their status as first-order weather stations having available hourly weather data. Additional locations were selected on the basis of population, climatic variations, or geographical representation. The design temperatures for stations lacking hourly data have been estimated from nearby stations with an hourly database.

The design data in Table 3 were compiled using techniques described by Ecodyne Cooling Products (1980). In *Engineering Weather Data* (USAF 1978), adjustments have been made for longer periods of record.

The statistics in these tables are presented as a guide only, and they should be used with caution for locations other than those measured. The design values at a building site may differ significantly from the reporting location. A meteorologist experienced in climatology can help interpret and extrapolate the data.

WEATHER-ORIENTED DESIGN FACTORS

The usual approach in air-conditioning system design involves computation of peak design load at a specific hour of a design day, using one of the frequency levels of design conditions in Tables 1, 2, or 3. While the values enumerated in this chapter are statistically accurate, certain precautions are recommended concerning their use. These figures are frequency distribution statistics for the ends of the distribution (see Ecodyne Cooling Products 1980, Snelling 1985, Crow 1963).

Design-day hourly profiles of dry-bulb temperatures with coincident wet-bulb temperatures have not been published. These profiles may be estimated by studying local climatology or by examining the daily range and peak design values.

Winter

Minimum temperatures usually occur between 6:00 A.M. and 8:00 A.M. suntime on clear days when the daily range is greatest. For residential applications or other applications where the occupancy is continuous throughout the day, the recommended design temperatures in Column 5 apply. With commercial applications or other applications where occupancy is only during hours near the middle of the day, design temperatures above the recommended minimum may apply. For example, in a supermarket or food store, lower temperatures during unoccupied times increase the shelf life of the merchandise and may reduce the heating plant size, but the recovery to acceptable occupancy levels must be considered in equipment selection and system design. Studies at several stations have found that the duration of extremely cold temperatures can continue below the 99% level for 3 days and below the 97.5% level for 5 days or more (see Ecodyne Cooling Products 1980, Snelling 1985, Crow 1963).

Summer

Maximum temperatures usually occur between 2:00 P.M. and 4:00 P.M. suntime with deviations on cloudy days when the daily range is less. When calculating building cooling loads, it is advisable to determine whether the structure is most sensitive to dry bulb, *i.e.*, extensive exterior exposure, or wet bulb, *i.e.*, outside ventilation. Then the appropriate design dry bulb with its coincident wet bulb from Column 6, or the appropriate design wet bulb from Column 8 with its coincident dry bulb (not shown) may be used. As noted previously, the design dry bulb (Column 6) and design wet bulb (Column 8) temperatures are not coincident. Comparing the coincident wet-bulb temperatures in Column 6 with the design wet-bulb temperatures in Column 8 and considering the daily range in

Column 7 illustrates this phenomenon. Using the design dry-bulb temperatures in Column 6 with the design wet-bulb temperatures in Column 8 will give computed loads significantly greater than actual loads.

Typically, the design dry-bulb temperatures should be used with the coincident wet-bulb temperatures in Column 6 in computing building cooling loads. The design wet-bulb temperatures in Column 8 are primarily intended for evaporative cooling processes, but may also be used for computing ventilation loads. Studies at several stations have found that the duration of extremely hot temperatures does not exceed one day (Ecodyne Cooling Products 1980, Snelling 1985, Crow 1963).

This chapter does not include design dew-point information. But the frequency of occurrence of the humidity ratio (which is obtained from the design dew-point temperature) and the coincident dry-bulb temperature is needed to size desiccant or indirect evaporative cooling systems, particularly those that control humidity.

Note that a dew point calculated from the design dry-bulb and mean coincident wet-bulb temperatures is generally significantly lower than the dew point that corresponds to the same nominal percentile. Thus, desiccant equipment will be significantly undersized if this method of calculating dew points is used for design. This is because the design dry bulb corresponds to extremes of sensible heat, not extremes of humidity ratio. Also, the design wet-bulb temperature cannot be used to estimate design dew points because the corresponding dry-bulb temperature is not included in these tables. Appropriate humidity information for design for various locations is often available from manufacturers of dehumidification and indirect evaporative cooling equipment.

For residential or other applications where the occupancy is continuous throughout the day, the recommended design temperature applies. For commercial applications or other applications where occupancy is only during hours near the middle of the day, design temperatures below the recommended maximum might apply. In some cases, the peak occupancy load occurs before the effect of the outdoor maximum temperature has reached the space by conduction through the building mass. In other cases, the peak occupancy loads may be in months other than the three or four summer months when the maximum outdoor temperature is expected; here design temperatures from other months will apply. Bin data distributions are available in *Engineering Weather Data* (USAF 1978) for three time periods during the day.

Energy consumption of the proposed system is a design concern often solved by intuition, experience, or simple calculation. Current energy programs include occupancy schedules, operating schedules, and weather data varying from hourly to seasonal. These programs may include off-peak design values. They may evaluate system control on days that are characterized as cloudy, small temperature change, windy warm A.M. and cool P.M., fair and warm, and fair and cool. A proposed system design should consider these weather conditions and their effect on temperature control before a final design is chosen and equipment selection is made.

WEATHER YEAR FOR ENERGY CALCULATIONS (WYEC)

Public domain and proprietary energy programs frequently require 8760 h of weather data. Several ASHRAE research projects have quantified weather data for use in energy calculations. Crow (1985) reports a weather year for energy calculations (WYEC) tape at 51 locations with 8760 h of data for each station. The data are in the TRY format, with solar data from 16 solar reporting stations, National Climatic Data Center (1981, 1991), and MacLaren *et al.* (1980). The year is made up of monthly data from the U.S. Weather Service 1440 series tapes selected closest to the long-term mean. Both temperature and solar radiation were examined for correlation and for closeness to the long-term mean. Adjustments were made to secure close compliance to the long-term mean by replac-

ing days in the closest month with days from the same month in other years. At the connections (midnight), the temperatures were adjusted for fit. Erroneous data and atypical conditions were replaced with better data. This tape with accompanying text is available from ASHRAE headquarters.

Degelman (1985) derived bin data from the WYEC tapes for 51 locations in six time periods during the day.

OTHER WEATHER DATA

Sources. Typical Meteorological Year (TMY) tapes were prepared from U.S. Weather Services hourly surface data files for the period 1954 through 1972. A year of 8760 h for each of 249 stations was prepared based on the SOLMET database (National Climatic Data Center 1981). Nine indices (total horizontal radiation, maximum, minimum, and mean of dry bulb and dew point, and the maximum and mean of wind speed) were identified as critical. They were weighted with the solar index as 50%, and the rest at 50%. Typical months were identified by their closeness to long-term cumulative distribution functions.

In the final selection, lengths of hot or cold periods with sunny or cloudy days were used. Discontinuities between months were machine smoothed. The TMY is made up of the typical months selected. Missing and atypical data were not replaced. These tapes are recommended for active solar design problems (Insolation Data Manual 1980). Sandia National Laboratories (1988) include information about the potential yearly variation of the solar resource in various U.S. locations.

The *Climatology of the U.S.: Annual Degree-Days to Selected Bases* from the National Climatic Data Center (1992b) includes United States data for heating and cooling degree-days to seven base temperatures in °F. Similar Canadian data in °C are available through the Canadian Climate Centre, Atmospheric Environment Service.

Investigations at both the U.S. Bureau of Standards and at Lawrence Berkeley Laboratories have evaluated the mean daily threshold value for calculating degree-day deficiencies. Most residential buildings fit energy consumption data best when the base of 57 °F is used. This proven base is the result of the combination of prudent use of insulation and the use of many heat-producing appliances (Nall and Arens 1979). Chapter 28 describes the energy estimating methods using these data.

ASHRAE (1986) contain bin data on magnetic disks for personal computers. The data is derived from WYEC tapes for 51 locations in six time periods during the day. This data does not necessarily represent long-term climatic conditions. The data is primarily for use in comparing simplified energy evaluations with hourly simulated loads, where WYEC tapes were used.

Solar Radiation Data. Approximately 30 stations report solar radiation in the United States. Due to inconsistencies in the calibration of station measuring devices prior to 1975, the data were rehabilitated by the National Climatic Center (National Climatic Data Center 1981). The Canadian Climate Centre has prepared a merged solar and meteorological computer database for 50 stations (MacLaren *et al.* 1980).

Evaporative Cooling Data. Crow (1972) reports on weather data for the design of evaporative cooling systems in the United States and Canada. Frequency of occurrence of median wet-bulb temperatures versus dry-bulb temperature data are available with design recommendations in Chapter 46 of the 1991 ASHRAE *Handbook—Applications*.

Underground Structures. Kusuda (1968) analyzed underground temperature data for a few stations in the contiguous United States. Kusuda (1971) also reported on earth temperatures beneath five different surfaces.

Table 1   Climatic Conditions for the United States

Col. 1	Col. 2 Lat. °'N	Col. 3 Long. °'W	Col. 4 Elev. Feet	Col. 5 Winter,[b] °F Design Dry-Bulb 99% 97.5%	Col. 6 Summer,[c] °F Design Dry-Bulb and Mean Coincident Wet-Bulb 1% 2.5% 5%	Col. 7 Mean Daily Range	Col. 8 Design Wet-Bulb 1% 2.5% 5%	Col. 9 Prevailing Wind Winter Summer Knots[d]	Col. 10 Temp., °F Median of Annual Extr. Max. Min.
ALABAMA									
Alexander City	32 57	85 57	660	18 22	96/77 93/76 91/76	21	79 78 78		
Anniston AP	33 35	85 51	599	18 22	97/77 94/76 92/76	21	79 78 78	SW 5 SW	98.4 12.4
Auburn	32 36	85 30	652	18 22	96/77 93/76 91/76	21	79 78 78		99.8 14.6
Birmingham AP	33 34	86 45	620	17 21	96/74 94/75 92/74	21	78 77 76	NNW 8 WNW	98.5 12.9
Decatur	34 37	86 59	580	11 16	95/75 93/74 91/74	22	78 77 76		
Dothan AP	31 19	85 27	374	23 27	94/76 92/76 91/76	20	80 79 78		
Florence AP	34 48	87 40	581	17 21	97/74 94/74 92/74	22	78 77 76	NW 7 NW	
Gadsden	34 01	86 00	554	16 20	96/75 94/75 92/74	22	78 77 76	NNW 8 WNW	
Huntsville AP	34 42	86 35	606	11 16	95/75 93/74 91/74	23	78 77 76	N 9 SW	
Mobile AP	30 41	88 15	211	25 29	95/77 93/77 91/76	18	80 79 78	N 10 N	
Mobile Co	30 40	88 15	211	25 29	95/77 93/77 91/76	16	80 79 78		97.9 22.3
Montgomery AP	32 23	86 22	169	22 25	96/76 95/76 93/76	21	79 79 78	NW 7 W	98.9 18.2
Selma, Craig AFB	32 20	87 59	166	22 26	97/78 95/77 93/77	21	81 80 79	N 9 SW	100.1 17.6
Talladega	33 27	86 06	565	18 22	97/77 94/76 92/76	21	79 78 78		99.6 11.2
Tuscaloosa AP	33 13	87 37	169	20 23	98/75 96/76 94/76	22	79 78 77	N 5 WNW	
ALASKA									
Anchorage AP	61 10	150 01	114	−23 −18	71/59 68/58 66/56	15	60 59 57	SE 3 WNW	
Barrow	71 18	156 47	31	−45 −41	57/53 53/50 49/47	12	54 50 47	SW 8 SE	
Fairbanks AP	64 49	147 52	436	−51 −47	82/62 78/60 75/59	24	64 62 60	N 5 S	
Juneau AP	58 22	134 35	12	−4 1	74/60 70/58 67/57	15	61 59 58	N 7 N	
Kodiak	57 45	152 29	73	10 13	69/58 65/56 62/55	10	60 58 56	WNW 14 NW	
Nome AP	64 30	165 26	13	−31 −27	66/57 62/55 59/54	10	58 56 55	N 4 W	
ARIZONA									
Douglas AP	31 27	109 36	4098	27 31	98/63 95/63 93/63	31	70 69 68		104.4 14.0
Flagstaff AP	35 08	111 40	7006	−2 2	84/55 82/55 80/54	31	61 60 59	NE 5 SW	90.0 −11.6
Fort Huachuca AP	31 35	110 20	4664	24 28	95/62 92/62 90/62	27	69 68 67	SW 5 W	
Kingman AP	35 12	114 01	3539	18 25	103/65 100/64 97/64	30	70 69 69		
Nogales	31 21	110 55	3800	28 32	99/64 96/64 94/64	31	71 70 69	SW 5 W	
Phoenix AP	33 26	112 01	1112	31 34	109/71 107/71 105/71	27	76 75 75	E 4 W	112.8 26.7
Prescott AP	34 39	112 26	5010	4 9	96/61 94/60 92/60	30	66 65 64		
Tucson AP	32 07	110 56	2558	28 32	104/66 102/66 100/66	26	72 71 71	SE 6 WNW	108.9 0.3
Winslow AP	35 01	110 44	4895	5 10	97/61 95/60 93/60	32	66 65 64	SW 6 WSW	102.7 −0.4
Yuma AP	32 39	114 37	213	36 39	111/72 109/72 107/71	27	79 78 77	NNE 6 WSW	114.8 30.8
ARKANSAS									
Blytheville AFB	35 57	89 57	264	10 15	96/78 94/77 91/76	21	81 80 78	N 8 SSW	
Camden	33 36	92 49	116	18 23	98/76 96/76 94/76	21	80 79 78		
El Dorado AP	33 13	92 49	277	18 23	98/76 96/76 94/76	21	80 79 78	S 6 SE	101.0 13.9
Fayetteville AP	36 00	94 10	1251	7 12	97/72 94/73 92/73	23	77 76 75	NE 9 SSW	99.4 −0.4
Fort Smith AP	35 20	94 22	463	12 17	101/75 98/76 95/76	24	80 79 78	NW 8 SE	101.9 7.0
Hot Springs	34 29	93 06	535	17 23	101/77 97/77 94/77	22	80 79 78	N 8 SW	103.0 10.6
Jonesboro	35 50	90 42	345	10 15	96/78 94/77 91/76	21	81 80 78		101.7 7.3
Little Rock AP	34 44	92 14	257	15 20	99/76 96/77 94/77	22	80 79 78	N 9 SSW	99.0 11.2
Pine Bluff AP	34 18	92 05	241	16 22	100/78 97/77 95/78	22	81 80 80	N 7 SW	102.2 13.1
Texarkana AP	33 27	93 59	389	18 23	98/76 96/77 93/76	21	80 79 78	WNW 9 SSW	104.8 14.0
CALIFORNIA									
Bakersfield AP	35 25	119 03	475	30 32	104/70 101/69 98/68	32	73 71 70	ENE 5 WNW	109.8 25.3
Barstow AP	34 51	116 47	1927	26 29	106/68 104/68 102/67	37	73 71 70	WNW 7 W	110.4 17.4
Blythe AP	33 37	114 43	395	30 33	112/71 110/71 108/70	28	75 75 74		116.8 24.1
Burbank AP	34 12	118 21	775	37 39	95/68 91/68 88/67	25	71 70 69	NW 3 S	
Chico	39 48	121 51	238	28 30	103/69 101/69 98/67	36	71 70 68	NW 5 SSE	109.0 22.6
Concord	37 58	121 59	200	24 27	100/69 97/68 94/67	32	71 70 68	WNW 5 W	
Covina	34 05	117 52	575	32 35	98/69 95/68 92/67	31	73 71 70		
Crescent City AP	41 46	124 12	40	31 33	68/60 65/59 63/58	18	62 60 59		
Downey	33 56	118 08	116	37 40	93/70 89/70 86/69	22	72 71 70		
El Cajon	32 49	116 58	367	42 44	83/69 80/69 78/68	30	71 70 68		
El Centro AP	32 49	115 40	−43	35 38	112/74 110/74 108/74	34	81 80 78	W 6 SE	
Escondido	33 07	117 05	660	39 41	89/68 85/68 82/68	30	71 70 69		
Eureka, Arcata AP	40 59	124 06	218	31 33	68/60 65/59 63/58	11	62 60 59	E 5 NW	75.8 29.7
Fairfield, Travis AFB	38 16	121 56	62	29 32	99/68 95/67 91/66	34	70 68 67	N 5 WSW	
Fresno AP	36 46	119 43	328	28 30	102/70 100/69 97/68	34	72 71 70	E 4 WNW	108.7 25.8
Hamilton AFB	38 04	122 30	3	30 32	89/68 84/66 80/65	28	72 69 67	N 4 SE	
Laguna Beach	33 33	117 47	35	41 43	83/68 80/68 77/67	18	70 69 68		
Livermore	37 42	121 57	545	24 27	100/69 97/68 93/67	24	71 70 68	WNW 4 NW	
Lompoc, Vandenberg AFB	34 43	120 34	368	35 38	75/61 70/61 67/60	20	63 61 60	ESE 5 NW	
Long Beach AP	33 49	118 09	30	41 43	83/68 80/68 77/67	22	70 69 68	NW 4 WNW	

[a]AP, AFB, following the station name designates airport or military airbase temperature observations. Co designates office locations within an urban area that are affected by the surrounding area. Undersigned stations are semirural and may be compared to airport data.

[b]Winter design data are based on the 3-month period, December through February.
[c]Summer design data are based on the 4-month period, June through September.
[d]Mean wind speeds occurring coincidentally with the 99.5% dry-bulb winter design temperature.

Table 1 Climatic Conditions for the United States (*Continued*)

State and Station[a]	Lat. ° 'N	Long. ° 'W	Elev. Feet	Winter,[b] °F Design Dry-Bulb 99%	97.5%	Summer,[c] °F Design Dry-Bulb and Mean Coincident Wet-Bulb 1%	2.5%	5%	Mean Daily Range	Design Wet-Bulb 1%	2.5%	5%	Prevailing Wind Winter Knots[d]	Summer	Temp., °F Median of Annual Extr. Max.	Min.
Los Angeles AP	33 56	118 24	97	41	43	83/68	80/68	77/67	15	70	69	68	E 4	WSW		
Los Angeles Co	34 03	118 14	270	37	40	93/70	89/70	86/69	20	72	71	70	NW 4	NW	98.1	35.9
Merced, Castle AFB	37 23	120 34	188	29	31	102/70	99/69	96/68	36	72	71	70	ESE 4	NW		
Modesto	37 39	121 00	91	28	30	101/69	98/68	95/67	36	71	70	69			105.8	26.2
Monterey	36 36	121 54	39	35	38	75/63	71/61	68/61	20	64	62	61	SE 4	NW		
Napa	38 13	122 17	56	30	32	100/69	96/68	92/67	30	71	69	68			103.1	25.8
Needles AP	34 46	114 37	913	30	33	112/71	110/71	108/70	27	75	75	74			116.4	26.7
Oakland AP	37 49	122 19	5	34	36	85/64	80/63	75/62	19	66	64	63	E 5	WNW	93.0	31.8
Oceanside	33 14	117 25	26	41	43	83/68	80/68	77/67	13	70	69	68				
Ontario	34 03	117 36	952	31	33	102/70	99/69	96/67	36	74	72	71	E 4	WSW		
Oxnard	34 12	119 11	49	34	36	83/66	80/64	77/63	19	70	68	67				
Palmdale AP	34 38	118 06	2542	18	22	103/65	101/65	98/64	35	69	67	66	SW 5	WSW		
Palm Springs	33 49	116 32	411	33	35	112/71	110/70	108/70	35	76	74	73				
Pasadena	34 09	118 09	864	32	35	98/69	95/68	92/67	29	73	71	70			102.8	30.4
Petaluma	38 14	122 38	16	26	29	94/68	90/66	87/65	31	72	70	68			102.0	24.2
Pomona Co	34 03	117 45	934	28	30	102/70	99/69	95/68	36	74	72	71	E 4	W	105.7	26.2
Redding AP	40 31	122 18	495	29	31	105/68	102/67	100/66	32	71	69	68			109.2	26.0
Redlands	34 03	117 11	1318	31	33	102/70	99/69	96/68	33	74	72	71			106.7	27.1
Richmond	37 56	122 21	55	34	36	85/64	80/63	75/62	17	66	64	63				
Riverside, March AFB	33 54	117 15	1532	29	32	100/68	98/68	95/67	37	72	71	70	N 4	NW	107.6	26.6
Sacramento AP	38 31	121 30	17	30	32	101/70	98/70	94/69	36	72	71	70	NNW 6	SW	105.1	27.6
Salinas AP	36 40	121 36	75	30	32	74/61	70/60	67/59	24	62	61	59				
San Bernardino, Norton AFB	34 08	117 16	1125	31	33	102/70	99/69	96/68	38	74	72	71	E 3	W	109.3	25.3
San Diego AP	32 44	117 10	13	42	44	83/69	80/69	78/68	12	71	70	68	NE 3	WNW	91.2	37.4
San Fernando	34 17	118 28	965	37	39	95/68	91/68	88/67	38	71	70	69				
San Francisco AP	37 37	122 23	8	35	38	82/64	77/63	73/62	20	65	64	62	S 5	NW		
San Francisco Co	37 46	122 26	72	38	40	74/63	71/62	69/61	14	64	62	61	W 5	W	91.3	35.9
San Jose AP	37 22	121 56	56	34	36	85/66	81/65	77/64	26	68	67	65	SE 4	NNW	98.6	28.2
San Luis Obispo	35 20	120 43	250	33	35	92/69	88/70	84/69	26	73	71	70	E 4	W	99.8	29.3
Santa Ana AP	33 45	117 52	115	37	39	89/69	85/68	82/68	28	71	70	69	E 3	SW	101.0	29.9
Santa Barbara AP	34 26	119 50	10	34	36	81/67	77/66	75/65	24	68	67	66	NE 3	SW	97.1	31.7
Santa Cruz	36 59	122 01	125	35	38	75/63	71/61	68/61	28	64	62	61			97.5	26.8
Santa Maria AP	34 54	120 27	236	31	33	81/64	76/63	73/62	23	65	64	63	E 4	WNW		
Santa Monica Co	34 01	118 29	64	41	43	83/68	80/68	77/67	16	70	69	68				
Santa Paula	34 21	119 05	263	33	35	90/68	86/67	84/66	36	71	69	68				
Santa Rosa	38 31	122 49	125	27	29	99/68	95/67	91/66	34	70	68	67	N 5	SE	102.5	23.4
Stockton AP	37 54	121 15	22	28	30	100/69	97/68	94/67	37	71	70	68	WNW 4	NW	104.1	24.5
Ukiah	39 09	123 12	623	27	29	99/69	95/68	91/67	40	70	68	67			108.1	21.6
Visalia	36 20	119 18	325	28	30	102/70	100/69	97/68	38	72	71	70			108.4	25.1
Yreka	41 43	122 38	2625	13	17	95/65	92/64	89/63	38	67	65	64			102.8	7.1
Yuba City	39 08	121 36	80	29	31	104/68	101/67	99/66	36	71	69	68				
COLORADO																
Alamosa AP	37 27	105 52	7537	−21	−16	84/57	82/57	80/57	35	62	61	60				
Boulder	40 00	105 16	5445	2	8	93/59	91/59	89/59	27	64	63	62			96.0	−8.4
Colorado Springs AP	38 49	104 43	6145	−3	2	91/58	88/57	86/57	30	63	62	61	N 9	S	92.3	−12.1
Denver AP	39 45	104 52	5283	−5	1	93/59	91/59	89/59	28	64	63	62	S 8	SE	96.8	−10.4
Durango	37 17	107 53	6550	−1	4	89/59	87/59	85/59	30	64	63	62			92.4	−11.2
Fort Collins	40 35	105 05	4999	−10	−4	93/59	91/59	89/59	28	64	63	62			95.2	−18.1
Grand Junction AP	39 07	108 32	4843	2	7	96/59	94/59	92/59	29	64	63	62	ESE 5	WNW	99.9	−3.4
Greeley	40 26	104 38	4648	−11	−5	96/60	94/60	92/60	29	65	64	63				
Lajunta AP	38 03	103 30	4160	−3	3	100/68	98/68	95/67	31	72	70	69	W 8	S		
Leadville	39 15	106 18	10155	−8	−4	84/52	81/51	78/50	30	56	55	54				
Pueblo AP	38 18	104 29	4641	−7	0	97/61	95/61	92/61	31	67	66	65	W 5	SE	100.5	−12.2
Sterling	40 37	103 12	3939	−7	−2	95/62	93/62	90/62	30	67	66	65			100.3	−15.4
Trinidad AP	37 15	104 20	5740	−2	3	93/61	91/61	89/61	32	66	65	64	W 7	WSW	96.8	−10.5
CONNECTICUT																
Bridgeport AP	41 11	73 11	25	6	9	86/73	84/71	81/70	18	75	74	73	NNW 13	WSW		
Hartford, Brainard Field	41 44	72 39	19	3	7	91/74	88/73	85/72	22	77	75	74	N 5	SSW	95.7	−4.4
New Haven AP	41 19	73 55	6	3	7	88/75	84/73	82/72	17	76	75	74	NNE 7	SW	93.0	−0.2
New London	41 21	72 06	59	5	9	88/73	85/72	83/71	16	76	75	74				
Norwalk	41 07	73 25	37	6	9	86/73	84/71	81/70	19	75	74	73				
Norwich	41 32	72 04	20	3	7	89/75	86/73	83/72	18	76	75	74				
Waterbury	41 35	73 04	843	−4	2	88/83	85/71	82/70	21	75	74	72	N 8	SW		
Windsor Locks, Bradley Fld	41 56	72 41	169	0	4	91/74	88/72	85/71	22	76	75	73	N 8	SW		
DELAWARE																
Dover AFB	39 08	75 28	28	11	15	92/75	90/75	87/74	18	79	77	76	W 9	SW	97.0	7.0
Wilmington AP	39 40	75 36	74	10	14	92/74	89/74	87/73	20	77	76	75	WNW 9	WSW	95.4	4.9
DISTRICT OF COLUMBIA																
Andrews AFB	38 5	76 5	279	10	14	92/75	90/74	87/73	18	78	76	75				
Washington, National AP	38 51	77 02	14	14	17	93/75	91/74	89/74	18	78	77	76	WNW 11	S	97.6	7.4

Table 1   Climatic Conditions for the United States (*Continued*)

State and Station[a]	Lat. °'N	Long. °'W	Elev. Feet	Winter °F Design Dry-Bulb 99%	97.5%	Summer °F Design Dry-Bulb and Mean Coincident Wet-Bulb 1%	2.5%	5%	Mean Daily Range	Design Wet-Bulb 1%	2.5%	5%	Prevailing Wind Winter Knots[d]	Summer	Temp. °F Median of Annual Extr. Max.	Min.
FLORIDA																
Belle Glade	26 39	80 39	16	41	44	92/76	91/76	89/76	16	79	78	78			94.7	30.9
Cape Kennedy AP	28 29	80 34	16	35	38	90/78	88/78	87/78	15	80	79	79				
Daytona Beach AP	29 11	81 03	31	32	35	92/78	90/77	88/77	15	80	79	78	NW 8			
E Fort Lauderdale	26 04	80 09	10	42	46	92/78	91/78	90/78	15	80	79	79	NW 9	ESE		
Fort Myers AP	26 35	81 52	15	41	44	93/78	92/78	91/77	18	80	79	79	NNE 7	W	94.9	34.9
Fort Pierce	27 28	80 21	25	38	42	91/78	90/78	89/78	15	80	79	79			96.1	34.0
Gainesville AP	29 41	82 16	152	28	31	95/77	93/77	92/77	18	80	79	78	W 6	W	97.8	23.3
Jacksonville AP	30 30	81 42	26	29	32	96/77	94/77	92/76	19	79	79	78	NW 7	SW	97.5	25.4
Key West AP	24 33	81 45	4	55	57	90/78	90/78	89/78	09	80	79	79	NNE 12	SE	92.0	51.5
Lakeland Co	28 02	81 57	214	39	41	93/76	91/76	89/76	17	79	78	78	NNW 9	SSW		
Miami AP	25 48	80 16	7	44	47	91/77	90/77	89/77	15	79	79	78	NNW 8	SE	92.5	39.0
Miami Beach Co	25 47	80 17	10	45	48	90/77	89/77	88/77	10	79	79	78				
Ocala	29 11	82 08	89	31	34	95/77	93/77	92/76	18	80	79	78			98.6	24.8
Orlando AP	28 33	81 23	100	35	38	94/76	93/76	91/76	17	79	78	78	NNW 9	SSW		
Panama City, Tyndall AFB	30 04	85 35	18	29	33	92/78	90/77	89/77	14	81	80	79	N 8	WSW		
Pensacola Co	30 25	87 13	56	25	29	94/77	93/77	91/77	14	80	79	79	NNE 7	SW	96.3	23.3
St. Augustine	29 58	81 20	10	31	35	92/78	89/78	87/78	16	80	79	79	NW 7	W	97.6	25.8
St. Petersburg	27 46	82 80	35	36	40	92/77	91/77	90/76	16	79	79	78	N 8	W	94.8	35.6
Sanford	28 46	81 17	89	35	38	94/76	93/76	91/76	17	79	78	78				
Sarasota	27 23	82 33	26	39	42	93/77	92/77	90/76	17	79	79	78				
Tallahassee AP	30 23	84 22	55	27	30	94/77	92/76	90/76	19	79	78	78	NW 6	NW	97.6	20.9
Tampa AP	27 58	82 32	19	36	40	92/77	91/77	90/76	17	79	79	78	N 8	W	95.0	31.5
West Palm Beach AP	26 41	80 06	15	41	45	92/78	91/78	90/78	16	80	79	79	NW 9	ESE		
GEORGIA																
Albany, Turner AFB	31 36	84 05	223	25	29	97/77	95/76	93/76	20	80	79	78	N 7	W	100.6	19.9
Americus	32 03	84 14	456	21	25	97/77	94/76	92/75	20	79	78	77			100.4	16.5
Athens	33 57	83 19	802	18	22	94/74	92/74	90/74	21	78	77	76	NW 9	WNW	98.7	13.5
Atlanta AP	33 39	84 26	1010	17	22	94/74	92/74	90/73	19	77	76	75	NW 11	NW	95.7	11.9
Augusta AP	33 22	81 58	145	20	23	97/77	95/76	93/76	19	80	79	78	W 4	WSW	99.0	17.5
Brunswick	31 15	81 29	25	29	32	92/78	89/78	87/78	18	80	79	79			99.3	24.7
Columbus, Lawson AFB	32 31	84 56	242	21	24	95/76	93/76	91/75	21	79	78	77	NW 8	W		
Dalton	34 34	84 57	720	17	22	94/76	93/76	91/76	22	79	78	77				
Dublin	32 20	82 54	215	21	25	96/77	93/76	91/75	20	79	78	77			101.0	16.7
Gainesville	34 11	83 41	50	24	27	96/77	93/77	91/77	20	80	79	78	WNW 7	SW	98.7	21.9
Griffin	33 13	84 16	981	18	22	93/76	90/75	88/74	21	78	77	76				
LaGrange	33 01	85 04	709	19	23	94/76	91/75	89/74	21	78	77	76			97.7	11.9
Macon AP	32 42	83 39	354	21	25	96/77	93/76	91/75	22	79	78	77	NW 8	WNW	99.6	16.9
Marietta, Dobbins AFB	33 55	84 31	1068	17	21	94/74	92/74	90/74	21	78	77	76	NNW 12	NW		
Savannah	32 08	81 12	50	24	27	96/77	93/77	91/77	20	80	79	78	WNW 7	SW	98.7	21.9
Valdosta-Moody AFB	30 58	83 12	233	28	31	96/77	94/77	92/76	20	80	79	78	WNW 6	W		
Waycross	31 15	82 24	148	26	29	96/77	94/77	91/76	20	80	79	78			100.0	19.5
HAWAII																
Hilo AP	19 43	155 05	36	61	62	84/73	83/72	82/72	15	75	74	74	SW 6	NE		
Honolulu AP	21 20	157 55	13	62	63	87/73	86/73	85/72	12	76	75	74	ENE 12	ENE		
Kaneohe Bay MCAS	21 27	157 46	18	65	66	85/75	84/74	83/74	12	76	76	75	NNE 9	NE		
Wahiawa	21 03	158 02	900	58	59	86/73	85/72	84/72	14	75	74	73	WNW 5	E		
IDAHO																
Boise AP	43 34	116 13	2838	3	10	96/65	94/64	91/64	31	68	66	65	SE 6	NW	103.2	0.6
Burley	42 32	113 46	4156	− 3	2	99/62	95/61	92/66	35	64	63	61			98.6	− 8.3
Coeur D'Alene AP	47 46	116 49	2972	− 8	− 1	89/62	86/61	83/60	31	64	63	61			99.9	− 4.5
Idaho Falls AP	43 31	112 04	4741	−11	− 6	89/61	87/61	84/59	38	65	63	61	N 9	S	96.2	−16.0
Lewiston AP	46 23	117 01	1413	− 1	6	96/65	93/64	90/63	32	67	66	64	W 3	WNW	105.9	2.7
Moscow	46 44	116 58	2660	− 7	0	90/63	87/62	84/61	32	65	64	62			98.0	− 5.9
Mountain Home AFB	43 02	115 54	2996	6	12	99/64	97/63	94/62	36	66	65	63	ESE 7	NW	103.2	− 6.5
Pocatello AP	42 55	112 36	4454	− 8	− 1	94/61	91/60	89/59	35	64	63	61	NE 5	W	97.9	−11.4
Twin Falls AP	42 29	114 29	4150	− 3	2	99/62	95/61	92/60	34	64	63	61	SE 6	NW	100.9	− 5.1
ILLINOIS																
Aurora	41 45	88 20	744	− 6	− 1	93/76	91/76	88/75	20	79	78	76			96.7	−13.0
Belleville, Scott AFB	38 33	89 51	453	1	6	94/76	92/76	89/75	21	79	78	76	WNW 8	S		
Bloomington	40 29	88 57	876	− 6	− 2	92/75	90/74	88/73	21	78	76	75			98.4	− 9.6
Carbondale	37 47	89 15	417	2	7	95/77	93/77	90/76	21	80	79	77			100.9	− 0.8
Champaign/Urbana	40 02	88 17	777	− 3	2	95/75	92/74	90/73	21	78	77	75				
Chicago, Midway AP	41 47	87 45	607	− 5	0	94/74	91/73	88/72	20	77	75	74	NW 11	SW		
Chicago, O'Hare AP	41 59	87 54	658	− 8	− 4	91/74	89/74	86/72	20	77	76	74	WNW 9	SW		
Chicago Co	41 53	87 38	590	− 3	2	94/75	91/74	88/73	15	79	77	75			96.1	− 8.3
Danville	40 12	87 36	695	− 4	1	93/75	90/74	88/73	21	78	77	75	W 10	SSW	98.2	− 8.4
Decatur	39 50	88 52	679	− 3	2	94/75	91/74	88/73	21	78	77	75	NW 10	SW	99.0	− 8.1
Dixon	41 50	89 29	696	− 7	− 2	93/75	90/74	88/73	23	78	77	75			97.5	−13.5
Elgin	42 02	88 16	758	− 7	− 2	91/75	88/74	86/73	21	78	77	75				
Freeport	42 18	89 37	780	− 9	− 4	91/74	89/73	87/72	24	77	76	74				
Galesburg	40 56	90 26	764	− 7	− 2	93/75	91/75	88/74	22	78	77	75	WNW 8	SW		

Table 1 Climatic Conditions for the United States (*Continued*)

State and Station[a]	Lat. ° 'N	Long. ° 'W	Elev. Feet	Winter,[b] °F Design Dry-Bulb 99%	97.5%	Summer,[c] °F Design Dry-Bulb and Mean Coincident Wet-Bulb 1%	2.5%	5%	Mean Daily Range	Design Wet-Bulb 1%	2.5%	5%	Prevailing Wind Winter Knots[d]	Summer	Temp., °F Median of Annual Extr. Max.	Min.
Greenville	38 53	89 24	563	− 1	4	94/76	92/75	89/74	21	79	78	76				
Joliet	41 31	88 10	582	− 5	0	93/75	90/74	88/73	20	78	77	75	NW 11	SW		
Kankakee	41 05	87 55	625	− 4	1	93/75	90/74	88/73	21	78	77	75				
La Salle/Peru	41 19	89 06	520	− 7	− 2	93/75	91/75	88/74	22	79	78	76				
Macomb	40 28	90 40	702	− 5	0	95/76	92/76	89/75	22	79	78	76			96.8	− 12.7
Moline AP	41 27	90 31	582	− 9	− 4	93/75	91/75	88/74	23	78	77	75	WNW 8	SW	100.5	− 2.9
Mt Vernon	38 19	88 52	479	0	5	95/76	92/75	89/74	21	79	78	76			98.0	− 10.9
Peoria AP	40 40	89 41	652	− 8	− 4	91/75	89/74	87/73	22	78	76	75	WNW 8	SW	101.1	− 6.7
Quincy AP	39 57	91 12	769	− 2	3	96/76	93/76	90/76	22	80	78	77	NW 11	SSW		
Rantoul, Chanute AFB	40 18	88 08	753	− 4	1	94/75	91/74	89/73	21	78	77	75	W 10	SSW	97.4	− 13.8
Rockford	42 21	89 03	741	− 9	− 4	91/74	89/73	87/72	24	77	76	74			98.1	− 7.2
Springfield AP	39 50	89 40	588	− 3	2	94/75	92/74	89/74	21	79	77	76	NW 10	SW	96.5	− 10.6
Waukegan	42 21	87 53	700	− 6	− 3	92/76	89/74	87/73	21	78	76	75				
INDIANA																
Anderson	40 06	85 37	919	0	6	95/76	92/75	89/74	22	79	78	76	W 9	SW	95.1	− 6.0
Bedford	38 51	86 30	670	0	5	95/76	92/75	89/74	22	79	78	76			97.5	− 4.4
Bloomington	39 08	86 37	847	0	5	95/76	92/75	89/74	22	79	78	76	W 9	SW	97.8	− 4.6
Columbus, Bakalar AFB	39 16	85 54	651	3	7	95/76	92/75	90/74	22	79	78	76	W 9	SW	98.3	− 6.4
Crawfordsville	40 03	86 54	679	− 2	3	94/75	91/74	88/73	22	79	77	76			98.4	− 7.6
Evansville AP	38 03	87 32	381	4	9	95/76	93/75	91/75	22	79	78	77	NW 9	SW	98.2	0.2
Fort Wayne AP	41 00	85 12	791	− 4	1	92/73	89/72	87/72	24	77	75	74	WSW 10	SW		
Goshen AP	41 32	85 48	827	− 3	1	91/73	89/73	86/72	23	77	75	74			96.8	− 10.5
Hobart	41 32	87 15	600	− 4	2	91/73	88/73	85/72	21	77	75	74			98.5	− 8.5
Huntington	40 53	85 30	802	− 4	1	92/73	89/72	87/72	23	77	75	74			96.9	− 8.1
Indianapolis AP	39 44	86 17	792	− 2	2	92/74	90/74	87/73	22	78	76	75	WNW 10	SW	96	− 7
Jeffersonville	38 17	85 45	455	5	10	95/74	93/74	90/74	23	79	77	76			98	2
Kokomo	40 25	86 03	855	− 4	0	91/74	90/73	88/73	22	77	75	74			98.2	− 7.5
Lafayette	40 2	86 5	600	− 3	3	94/74	91/73	88/73	22	78	76	75				
La Porte	41 36	86 43	810	− 3	3	93/74	90/74	87/73	22	78	76	75			98.1	− 10.5
Marion	40 29	85 41	859	− 4	0	91/74	90/73	88/73	23	77	75	74			97.0	− 8.6
Muncie	40 11	85 21	957	− 3	2	92/74	90/73	87/73	22	76	75	74				
Peru, Grissom AFB	40 39	86 09	813	− 6	− 1	90/74	88/73	86/73	22	77	75	74	W 10	SW	94.8	− 8.5
Richmond AP	39 46	84 50	1141	− 2	2	92/74	90/74	87/73	22	78	76	75			97.7	− 6.0
Shelbyville	39 31	85 47	750	− 1	3	93/74	91/74	88/73	22	78	76	75			96.2	− 9.2
South Bend AP	41 42	86 19	773	− 3	1	91/73	89/73	86/72	22	77	75	74	SW 11	SSW	98.3	− 4.9
Terre Haute AP	39 27	87 18	585	− 2	4	95/75	92/74	89/73	22	79	77	76	NNW 7	SSW	95.5	− 11.0
Valparaiso	41 31	87 02	801	− 3	3	93/74	90/74	87/73	22	78	76	75			100.3	− 2.8
Vincennes	38 41	87 32	420	1	6	95/75	92/74	90/73	22	79	77	76				
IOWA																
Ames	42 02	93 48	1099	− 11	− 6	93/75	90/74	87/73	23	78	76	75	NW 9	SSW	97.4	− 17.8
Burlington AP	40 47	91 07	692	− 7	− 3	94/74	91/75	88/73	22	78	77	75	NW 9	S	98.6	− 11.0
Cedar Rapids AP	41 53	91 42	863	− 10	− 5	91/76	88/75	86/74	23	78	77	75			97.7	− 15.6
Clinton	41 50	90 13	595	− 8	− 3	92/75	90/75	87/74	23	78	77	75			97.5	− 13.8
Council Bluffs	41 20	95 49	1210	− 8	− 3	94/76	91/75	88/74	22	78	77	75			98.2	− 14.2
Des Moines AP	41 32	93 39	938	− 10	− 5	94/74	91/74	88/73	23	78	77	75	NW 11	S	95.2	− 15.0
Dubuque	42 24	90 42	1056	− 12	− 7	90/74	88/74	86/72	22	77	75	74	N 10	SSW	98.5	− 19.1
Fort Dodge	42 33	94 11	1162	− 12	− 7	91/74	88/74	86/72	23	77	75	74	NW 11	S	97.4	− 15.2
Iowa City	41 38	91 33	661	− 11	− 6	92/76	89/76	87/74	22	80	78	76	NW 9	SSW	98.4	− 8.8
Keokuk	40 24	91 24	574	− 5	0	95/75	92/75	89/74	22	79	77	76			98.5	− 13.4
Marshalltown	42 04	92 56	898	− 12	− 7	92/76	90/75	88/74	23	78	77	75			96.5	− 21.7
Mason City AP	43 09	93 20	1213	− 15	− 11	90/74	88/74	85/72	24	77	75	74	NW 11	S	98.2	− 14.7
Newton	41 41	93 02	936	− 10	− 5	94/75	91/74	88/73	23	78	77	75			99.1	− 12.0
Ottumwa AP	41 06	92 27	840	− 8	− 4	94/75	91/74	88/73	22	78	77	75	NNW 9	S	99.9	− 17.7
Sioux City AP	42 24	96 23	1095	− 11	− 7	95/74	92/74	89/74	24	78	77	75	NW 9	S	97.7	− 19.8
Waterloo	42 33	92 24	868	− 15	− 10	91/76	89/75	86/74	23	78	77	75				
KANSAS																
Atchison	39 34	95 07	945	− 2	2	96/77	93/76	91/76	23	81	79	77			100.5	− 8.8
Chanute AP	37 40	95 29	981	3	7	100/74	97/74	94/74	23	78	77	76	NNW 11	SSW	102.8	− 2.8
Dodge City AP	37 46	99 58	2582	0	5	100/69	97/69	95/69	25	74	73	71	N 12	SSW	102.9	− 7.0
El Dorado	37 49	96 50	1282	3	7	101/74	98/73	96/73	24	77	76	75			103.5	− 5.0
Emporia	38 20	96 12	1210	1	5	100/74	97/74	94/73	25	78	77	76			102.4	− 6.4
Garden City AP	37 56	100 44	2880	− 1	4	99/69	96/69	94/69	28	74	73	71				
Goodland AP	39 22	101 42	3654	− 5	0	99/66	96/65	93/66	31	71	70	68	WSW 10	S	103.2	− 10.4
Great Bend	38 21	98 52	1889	0	4	101/73	98/73	95/73	28	78	76	75			105.3	− 6.1
Hutchinson AP	38 04	97 52	1542	4	8	102/72	99/72	97/72	28	77	75	74	N 14	S	105.8	− 3.8
Liberal	37 03	100 58	2870	2	7	99/68	96/68	94/68	28	73	72	71			104.5	− 8.6
Manhattan, Ft Riley	39 03	96 46	1065	− 1	3	99/75	95/75	92/74	24	78	77	76	NNE 8	S		
Parsons	37 20	95 31	899	5	9	100/74	97/74	94/74	23	79	77	76	NNW 11	SSW		
Russell AP	38 52	98 49	1866	0	4	101/73	98/73	95/73	29	78	76	75				
Salina	38 48	97 39	1272	0	5	103/74	100/74	97/73	26	78	77	75	N 8	SSW	101.8	− 6.4
Topeka AP	39 04	95 38	877	0	4	99/75	96/75	93/74	24	79	78	76	NNW 10	S	102.5	− 2.8
Wichita AP	37 39	97 25	1321	3	7	101/72	98/73	96/73	23	77	76	75	NNW 12	SSW		

Table 1 Climatic Conditions for the United States (*Continued*)

State and Station[a]	Lat. ° 'N	Long. ° 'W	Elev. Feet	Winter,[b] °F Design Dry-Bulb 99%	97.5%	Summer,[c] °F Design Dry-Bulb and Coincident Wet-Bulb 1%	2.5%	5%	Mean Daily Range	Design Wet-Bulb 1%	2.5%	5%	Prevailing Wind Winter	Knots[d]	Summer	Temp., °F Median of Annual Extr. Max.	Min.
KENTUCKY																	
Ashland	38 33	82 44	546	5	10	94/76	91/74	89/73	22	78	77	75	W	6	SW	97.4	0.8
Bowling Green AP	35 58	86 28	535	4	10	94/77	92/75	89/74	21	79	77	76				99.9	1.2
Corbin AP	36 57	84 06	1174	4	9	94/73	92/73	89/72	23	77	76	75					
Covington AP	39 03	84 40	869	1	6	92/73	90/72	88/72	22	77	75	74	W	9	SW		
Hopkinsville, Ft Campbell	36 40	87 29	571	4	10	94/77	92/75	89/74	21	79	77	76	N	6	W	100.1	−0.4
Lexington AP	38 02	84 36	966	3	8	93/73	91/73	88/72	22	77	76	75	WNW	9	SW	95.3	−0.5
Louisville AP	38 11	85 44	477	5	10	95/74	93/74	90/74	23	79	77	76	NW	8	SW	97.4	1.2
Madisonville	37 19	87 29	439	5	10	96/76	93/75	90/75	22	79	78	77					
Owensboro	37 45	87 10	407	5	10	97/76	94/75	91/75	23	79	78	77	NW	9	SW	98.0	−0.2
Paducah AP	37 04	88 46	413	7	12	98/76	95/75	92/75	20	79	78	77					
LOUISIANA																	
Alexandria AP	31 24	92 18	92	23	27	95/77	94/77	92/77	20	80	79	78	N	7	S	100.1	−5.7
Baton Rouge AP	30 32	91 09	64	25	29	95/77	93/77	92/77	19	80	80	79	ENE	8	W	98.0	21.4
Bogalusa	30 47	89 52	103	24	28	95/77	93/77	92/77	19	80	80	79				99.3	20.2
Houma	29 31	90 40	13	31	35	95/78	93/78	92/77	15	81	80	79				97.2	22.5
Lafayette AP	30 12	92 00	42	26	30	95/78	94/78	92/78	18	81	80	79	N	8	SW	98.2	22.6
Lake Charles AP	30 07	93 13	9	27	31	95/77	93/77	92/77	17	80	79	79	N	9	SSW	99.2	20.5
Minden	32 36	93 18	250	20	25	99/77	96/76	94/76	20	79	79	78				101.7	−4.9
Monroe AP	32 31	92 02	79	20	25	99/77	96/76	94/76	20	79	79	78	N	9	S	101.1	−5.9
Natchitoches	31 46	93 05	130	22	26	97/77	95/77	93/77	20	80	79	78					
New Orleans AP	29 59	90 15	4	29	33	93/78	92/78	90/77	16	81	80	79	NNE	9	SSW	96.3	27.7
Shreveport AP	32 28	93 49	254	20	25	99/77	96/76	94/76	20	79	79	78	N	9	S		
MAINE																	
Augusta AP	44 19	69 48	353	−7	−3	88/73	85/70	82/68	22	74	72	70	NNE	10	WNW		
Bangor, Dow AFB	44 48	68 50	192	−11	−6	86/70	83/68	80/67	22	73	71	69	WNW	7	S		
Caribou AP	46 52	68 01	624	−18	−13	84/69	81/67	78/66	21	71	69	67	WSW	10	SW		
Lewiston	44 02	70 15	200	−7	−2	88/73	85/70	82/68	22	74	72	70				94.0	−13.7
Millinocket AP	45 39	68 42	413	−13	−9	87/69	83/68	80/66	22	72	70	68	WNW	11	WNW	92.4	−23.0
Portland	43 39	70 19	43	−6	−1	87/72	84/71	81/69	22	74	72	70	W	7	S	93.5	−9.9
Waterville	44 32	69 40	302	−8	−4	87/72	84/69	81/68	22	74	72	70					
MARYLAND																	
Baltimore AP	39 11	76 40	148	10	13	94/75	91/75	89/74	21	78	77	76	W	9	WSW		
Baltimore Co	39 20	76 25	20	14	17	92/77	89/76	87/75	17	80	78	76	WNW	9	S	97.9	7.2
Cumberland	39 37	78 46	790	6	10	92/75	89/74	87/74	22	77	76	75	WNW	10	W		
Frederick AP	39 27	77 25	313	8	12	94/76	91/75	88/74	22	78	77	76	N	9	WNW		
Hagerstown	39 42	77 44	704	8	12	94/75	91/74	89/74	22	77	76	75	WNW	10	W		
Salisbury	38 20	75 30	59	12	16	93/75	91/75	88/74	18	79	77	76				96.8	7.4
MASSACHUSETTS																	
Boston AP	42 22	71 02	15	6	9	91/73	88/71	85/70	16	75	74	72	WNW	16	SW	95.7	−1.2
Clinton	42 24	71 41	398	−2	2	90/72	87/71	84/69	17	75	73	72				91.7	−8.5
Fall River	41 43	71 08	190	5	9	87/72	84/71	81/69	18	74	73	72	NW	10	SW	92.1	−1.0
Framingham	42 17	71 25	170	3	6	89/72	86/71	83/69	17	74	73	71				96.0	−7.7
Gloucester	42 35	70 41	10	2	5	89/73	86/71	83/70	15	75	74	72					
Greenfield	42 3	72 4	205	−7	−2	88/72	85/71	82/69	23	74	73	71					
Lawrence	42 42	71 10	57	−6	0	90/73	87/72	84/70	22	76	74	73	NW	8	WSW	95.2	−9.0
Lowell	42 39	71 19	90	−4	1	91/73	88/72	85/70	21	76	74	73				95.1	−8.5
New Bedford	41 41	70 58	79	5	9	85/72	82/71	80/69	19	74	73	72	NW	10	SW	91.4	2.2
Pittsfield AP	42 26	73 18	1194	−8	−3	87/71	84/70	81/68	23	73	72	70	NW	12	SW		
Springfield, Westover AFB	42 12	72 32	245	−5	0	90/72	87/71	84/69	19	75	73	72	N	8	SSW	95.7	−4.7
Taunton	41 54	71 04	20	5	9	89/73	86/71	83/70	18	75	74	73				92.9	−9.8
Worcester AP	42 16	71 52	986	0	4	87/71	84/70	81/68	18	73	72	70	W	14	W		
MICHIGAN																	
Adrian	41 55	84 01	754	−1	3	91/73	88/72	85/71	23	76	75	73				97.2	−7.0
Alpena AP	45 04	83 26	610	−11	−6	89/70	85/70	83/69	27	73	72	70	W	5	SW	93.9	−14.8
Battle Creek AP	42 19	85 15	941	1	5	92/74	88/72	85/70	23	76	74	73	SW	8	SW		
Benton Harbor AP	42 08	86 26	643	1	5	91/72	88/72	85/70	20	75	74	72	SSW	8	WSW		
Detroit	42 25	83 01	619	3	6	91/73	88/72	86/71	20	76	74	73	W	11	SW	95.1	−2.6
Escanaba	45 44	87 05	607	−11	−7	87/70	83/69	80/68	17	73	71	69				88.8	−16.1
Flint AP	42 58	83 44	771	−4	1	90/73	87/72	85/71	25	76	74	72	SW	8	SW	95.3	−9.9
Grand Rapids AP	42 53	85 31	784	1	5	91/72	88/72	85/70	24	75	74	72	WNW	8	WSW	95.4	−5.6
Holland	42 42	86 06	678	2		88/72	86/71	83/70	22	75	73	72				94.1	−6.8
Jackson AP	42 16	84 28	1020	1	5	92/74	88/72	85/70	23	76	74	73				96.5	−7.8
Kalamazoo	42 17	85 36	955	1	5	92/74	88/72	85/70	23	76	74	73				95.9	−6.7
Lansing AP	42 47	84 36	873	−3	1	90/73	87/72	85/70	24	75	74	72	SW	12	W	94.6	−11.0
Marquette Co	46 34	87 24	735	−12	−8	84/70	81/69	77/66	18	72	70	68				94.5	−11.8
Mt Pleasant	43 35	84 46	796	0	4	91/73	87/72	84/71	24	76	74	72				95.4	−11.1
Muskegon AP	43 10	86 14	625	2	6	86/72	84/70	82/70	21	75	73	72	E	8	SW		
Pontiac	42 40	83 25	981	0	4	90/73	87/72	85/71	21	76	74	73				95.0	−6.8

Table 1 Climatic Conditions for the United States (*Continued*)

				Winter,[b] °F		Summer,[c] °F							Prevailing Wind		Temp., °F	
Col. 1	Col. 2	Col. 3	Col. 4	Col. 5		Col. 6			Col. 7	Col. 8			Col. 9		Col. 10	
State and Station[a]	Lat.	Long.	Elev.	Design Dry-Bulb		Design Dry-Bulb and Mean Coincident Wet-Bulb			Mean Daily Range	Design Wet-Bulb			Winter	Summer	Median of Annual Extr.	
	° ′N	° ′W	Feet	99%	97.5%	1%	2.5%	5%		1%	2.5%	5%	Knots[d]		Max.	Min.
Port Huron	42 59	82 25	586	0	4	90/73	87/72	83/71	21	76	74	73	W 8	S	96.1	−7.6
Saginaw AP	43 32	84 05	667	0	4	91/73	87/72	84/71	23	76	74	72	WSW 7	SW		
Sault Ste. Marie AP	46 28	84 22	721	−12	−8	84/70	81/69	77/66	23	72	70	68	E 7	SW	89.8	−21.0
Traverse City AP	44 45	85 35	624	−3	1	89/72	86/71	83/69	22	75	73	71	SSW 9	SW	95.4	−10.7
Ypsilanti	42 14	83 32	716	1	5	92/72	89/71	86/70	22	75	74	72	SW 10	SW		
MINNESOTA																
Albert Lea	43 39	93 21	1220	−17	−12	90/74	87/72	84/71	24	77	75	73				
Alexandria AP	45 52	95 23	1430	−22	−16	91/72	88/72	85/70	24	76	74	72			95.1	−28.0
Bemidji AP	47 31	94 56	1389	−31	−26	88/69	85/69	81/67	24	73	71	69	N 8	S	94.5	−36.9
Brainerd	46 24	94 08	1227	−20	−16	90/73	87/71	84/69	24	75	73	71				
Duluth AP	46 50	92 11	1428	−21	−16	85/70	82/68	79/66	22	72	70	68	WNW 12	WSW	90.9	−27.4
Fairbault	44 18	93 16	940	17	−12	91/74	88/72	85/71	24	77	75	73			95.8	−24.3
Fergus Falls	46 16	96 04	1210	−21	−17	91/72	88/72	85/70	24	76	74	72			96.9	−27.8
International Falls AP	48 34	93 23	1179	−29	−25	85/68	83/68	80/66	26	71	70	68	N 9	S	93.4	−36.5
Mankato	44 09	93 59	1004	−17	−12	91/72	88/72	85/70	24	77	75	73				
Minneapolis/St. Paul AP	44 53	93 13	834	−16	−12	92/75	89/73	86/71	22	77	75	73	NW 8	S	96.5	−22.0
Rochester AP	43 55	92 30	1297	−17	−12	90/74	87/72	84/71	24	77	75	73	NW 9	SSW		
St. Cloud AP	45 35	94 11	1043	−15	−11	91/74	88/72	85/70	24	76	74	72				
Virginia	47 30	92 33	1435	−25	−21	85/69	83/68	80/66	23	71	70	68			92.6	−33.0
Willmar	45 07	95 05	1128	−15	−11	91/74	88/72	85/71	24	76	74	72			96.8	−24.3
Winona	44 03	91 38	652	−14	−10	91/75	88/73	85/72	24	77	75	74				
MISSISSIPPI																
Biloxi, Keesler AFB	30 25	88 55	26	28	31	94/79	92/79	90/78	16	82	81	80	N 8	S	98.0	23.0
Clarksdale	34 12	90 34	178	14	19	96/77	94/77	92/76	21	80	79	78			100.9	13.2
Columbus AFB	33 39	88 27	219	15	20	95/77	93/77	91/76	22	80	79	78	N 7	W	101.6	12.7
Greenville AFB	33 29	90 59	138	15	20	95/77	93/77	91/76	21	80	79	78			99.5	14.9
Greenwood	33 30	90 05	148	15	20	95/77	93/77	91/76	21	80	79	78			100.6	15.3
Hattiesburg	31 16.	89 15	148	24	27	96/78	94/77	92/77	21	81	80	79			99.9	18.2
Jackson AP	32 19	90 05	310	21	25	97/76	95/76	93/76	21	79	78	78	NNW 6	NW	99.8	16.0
Laurel	31 40	89 10	236	24	27	96/78	94/77	92/77	21	81	80	79			99.7	17.8
McComb AP	31 15	90 28	469	21	26	96/77	94/76	92/76	18	80	79	78				
Meridian AP	32 20	88 45	290	19	23	97/77	95/76	93/76	22	80	79	78	N 6	WSW	98.3	15.7
Natchez	31 33	91 23	195	23	27	96/78	94/78	92/77	21	81	80	79			98.4	18.4
Tupelo	34 16	88 46	361	14	19	96/77	94/77	92/76	22	80	79	78			100.7	11.8
Vicksburg Co	32 24	90 47	262	22	26	97/78	95/78	93/77	21	81	80	79			96.9	18.0
MISSOURI																
Cape Girardeau	37 14	89 35	351	8	13	98/76	95/75	92/75	21	79	78	77				
Columbia AP	38 58	92 22	778	−1	4	97/74	94/74	91/73	22	78	77	76	WNW 9	WSW	99.5	−6.2
Farmington AP	37 46	90 24	928	3	8	96/76	93/75	90/74	22	78	77	75			99.9	−2.1
Hannibal	39 42	91 21	489	−2	3	96/76	93/76	90/76	22	80	78	77	NNW 11	SSW	98.4	−7.6
Jefferson City	38 34	92 11	640	2	7	98/75	95/74	92/74	23	78	77	76			101.2	−6.1
Joplin AP	37 09	94 30	980	6	10	100/73	97/73	94/73	24	78	77	76	NNW 12	SSW		
Kansas City AP	39 07	94 35	791	2	6	99/75	96/74	93/74	20	78	77	76	NW 9	S	100.2	−4.3
Kirksville AP	40 06	92 33	964	−5	0	96/74	93/74	90/73	24	78	77	76			98.3	−10.8
Mexico	39 11	91 54	775	−1	4	97/74	94/74	91/73	22	78	77	76			101.2	−8.0
Moberly	39 24	92 26	850	−2	3	97/74	94/74	91/73	23	78	77	76				
Poplar Bluff	36 46	90 25	380	11	16	98/78	95/76	92/76	22	81	79	78				
Rolla	37 59	91 43	1204	3	9	94/77	91/75	89/74	22	78	77	76			99.4	−3.1
St. Joseph AP	39 46	94 55	825	−3	2	96/77	93/76	91/76	23	81	79	77	NNW 9	S	100.6	−8.0
St. Louis AP	38 45	90 23	535	2	6	97/75	94/75	91/74	21	78	77	76	NW 9	WSW		
St. Louis Co	38 39	90 38	462	3	8	98/75	94/75	91/74	18	78	77	76	NW 6	S	99.1	−2.7
Sikeston	36 53	89 36	325	9	15	98/77	95/76	92/75	21	80	78	77				
Sedalia, Whiteman AFB	38 43	93 33	869	−1	4	95/76	92/76	90/75	22	79	78	76	NNW 7	SSW	100.0	−5.1
Sikeston	36 53	89 36	325	9	15	98/77	95/76	92/75	21	80	78	77				
Springfield AP	37 14	93 23	1268	3	9	96/73	93/74	91/74	23	78	77	75	NNW 10	S	97.2	−2.4
MONTANA																
Billings AP	45 48	108 32	3567	−15	−10	94/64	91/64	88/63	31	67	66	64	NE 9	SW	100.5	−19.1
Bozeman	45 47	111 09	4448	−20	−14	90/61	87/60	84/59	32	63	62	60			92.2	−23.2
Butte AP	45 57	112 30	5553	−24	−17	86/58	83/56	80/56	35	60	58	57	S 5	NW	91.8	−26.3
Cut Bank AP	48 37	112 22	3838	−25	−20	88/61	85/61	82/60	35	64	62	61			94.7	−30.9
Glasgow AP	48 25	106 32	2760	−22	−18	92/64	89/63	85/62	29	68	66	64	E 8	S		
Glendive	47 08	104 48	2476	−18	−13	95/66	92/64	89/62	29	69	67	65			103.3	−29.8
Great Falls AP	47 29	111 22	3662	−21	−15	91/60	88/60	85/59	28	64	62	60	SW 7	WSW	98.0	−25.1
Havre	48 34	109 40	2492	−18	−11	94/65	90/64	87/63	33	68	66	65			99.7	−31.3
Helena AP	46 36	112 00	3828	−21	−16	91/60	88/60	85/59	32	64	62	61	N 12	WNW	95.6	−23.7
Kalispell AP	48 18	114 16	2974	−14	−7	91/62	87/61	84/60	34	65	63	62			94.4	−16.8
Lewistown AP	47 04	109 27	4122	−22	−16	90/62	87/61	83/60	30	65	63	62	NW 9	NW	96.2	−27.7
Livingston AP	45 42	110 26	4618	−20	−14	90/61	87/60	84/59	32	63	62	60			97.2	−21.2
Miles City AP	46 26	105 52	2634	−20	−15	98/66	95/66	92/65	30	70	68	67	NW 7	SE	103.6	−27.7
Missoula AP	46 55	114 05	3190	−13	−6	92/62	88/61	85/60	36	65	63	62	ESE 7	NW	98.6	−13.9

Table 1   Climatic Conditions for the United States (*Continued*)

State and Station[a]	Lat. ° 'N	Long. ° 'W	Elev. Feet	Winter,[b] °F Design Dry-Bulb 99%	97.5%	Summer,[c] °F Design Dry-Bulb and Mean Coincident Wet-Bulb 1%	2.5%	5%	Mean Daily Range	Design Wet-Bulb 1%	2.5%	5%	Prevailing Wind Winter	Summer Knots[d]	Temp., °F Median of Annual Extr. Max.	Min.
NEBRASKA																
Beatrice	40 16	96 45	1235	− 5	− 2	99/75	95/74	92/74	24	78	77	76			103.1	− 11.3
Chadron AP	42 50	103 05	3313	− 8	− 3	97/66	94/65	91/65	30	71	69	68				
Columbus	41 28	97 20	1450	− 6	− 2	98/74	95/73	92/73	25	77	76	75				
Fremont	41 26	96 29	1200	− 6	− 2	98/75	95/74	92/74	22	78	77	76				
Grand Island AP	40 59	98 19	1860	− 8	− 3	97/72	94/71	91/71	28	75	74	73	NNW 10	S	103.3	− 14.2
Hastings	40 36	98 26	1954	− 7	− 3	97/72	94/71	91/71	27	75	74	73	NNW 10	S	103.5	− 10.7
Kearney	40 44	99 01	2132	− 9	− 4	96/71	93/70	90/70	28	74	73	72			102.9	− 13.7
Lincoln Co	40 51	96 45	1180	− 5	− 2	99/75	95/74	92/74	24	78	77	76	N 8	S	102.0	− 12.4
McCook	40 12	100 38	2768	− 6	− 2	98/69	95/69	91/69	28	74	72	71				
Norfolk	41 59	97 26	1551	− 8	− 4	97/74	93/74	90/73	30	78	77	75			102.0	− 20.0
North Platte AP	41 08	100 41	2779	− 8	− 4	97/69	94/69	90/69	28	74	72	71	NW 9	SSE	100.8	− 15.8
Omaha AP	41 18	95 54	977	− 8	− 3	94/76	91/75	88/74	22	78	77	75	NW 8	S	100.2	− 13.2
Scottsbluff AP	41 52	103 36	3958	− 8	− 3	95/65	92/65	90/64	31	70	68	67	NW 9	SE	101.6	− 18.9
Sidney AP	41 13	103 06	4399	− 8	− 3	95/65	92/65	90/64	31	70	68	67				
NEVADA																
Carson City	39 10	119 46	4675	4	9	94/60	91/59	89/58	42	63	61	60	SSW 3	WNW	99.2	− 5.0
Elko AP	40 50	115 47	5050	− 8	− 2	94/59	92/59	90/58	42	63	62	60	E 4	SW		
Ely AP	39 17	114 51	6253	− 10	− 4	89/57	87/56	85/55	39	60	59	58	S 9	SSW		
Las Vegas AP	36 05	115 10	2178	25	28	108/66	106/65	104/65	30	71	70	69	ENE 7	SW		
Lovelock AP	40 04	118 33	3903	8	12	98/63	96/63	93/62	42	66	65	64			103.0	− 1.0
Reno AP	39 30	119 47	4404	5	10	95/61	92/60	90/59	45	64	62	61	SSW 3	WNW		
Reno Co	39 30	119 47	4408	6	11	96/61	93/60	91/59	45	64	62	61			98.9	0.2
Tonopah AP	38 04	117 05	5426	5	10	94/60	92/59	90/58	40	64	62	61	N 8	S		
Winnemucca AP	40 54	117 48	4301	− 1	3	96/60	94/60	92/60	42	64	62	61	SE 10	W	100.1	− 8.1
NEW HAMPSHIRE																
Berlin	44 03	71 01	1110	− 14	− 9	87/71	84/69	81/68	22	73	71	70			93.2	− 24.7
Claremont	43 02	72 02	420	− 9	− 4	89/72	86/70	83/69	24	74	73	71				
Concord AP	43 12	71 30	342	− 8	− 3	90/72	87/70	84/69	26	74	73	71	NW 7	SW	94.8	− 16.0
Keene	42 55	72 17	490	− 12	− 7	90/72	87/70	83/69	24	74	73	71			94.6	− 18.9
Laconia	43 03	71 03	505	− 10	− 5	89/72	86/70	83/69	25	74	73	71				
Manchester, Grenier AFB	42 56	71 26	233	− 8	− 3	91/72	88/71	85/70	24	75	74	72	N 11	SW	93.7	− 12.6
Portsmouth, Pease AFB	43 04	70 49	101	− 2	2	89/73	85/71	83/70	22	75	74	72	W 8	W		
NEW JERSEY																
Atlantic City Co	39 23	74 26	11	10	13	92/74	89/74	86/72	18	78	77	75	NW 11	WSW	93.0	7.5
Long Branch	40 19	74 01	15	10	13	93/74	90/73	87/72	18	78	77	75			95.9	4.3
Newark AP	40 42	74 10	7	10	14	94/74	91/73	88/72	20	77	76	75	WNW 11	WSW		
New Brunswick	40 29	74 26	125	6	10	92/74	89/73	86/72	19	77	76	75				
Paterson	40 54	74 09	100	6	10	94/74	91/73	88/72	21	77	76	75				
Phillipsburg	40 41	75 11	180	1	6	92/73	89/72	86/71	21	76	75	74			97.4	− 0.7
Trenton Co	40 13	74 46	56	11	14	91/75	88/74	85/73	19	78	76	75	W 9	SW	96.2	4.2
Vineland	39 29	75 00	112	8	11	91/75	89/74	86/73	19	78	76	75				
NEW MEXICO																
Alamagordo, Holloman AFB	32 51	106 06	4093	14	19	98/64	96/64	94/64	30	69	68	67				
Albuquerque AP	35 03	106 37	5311	12	16	96/61	94/61	92/61	27	66	65	64	N 7	W	98.1	5.1
Artesia	32 46	104 23	3320	13	19	103/67	100/67	97/67	30	72	71	70			105.5	3.7
Carlsbad AP	32 20	104 16	3293	13	19	103/67	100/67	97/67	28	72	71	70	N 6	SSE		
Clovis AP	34 23	103 19	4294	8	13	95/65	93/65	91/65	28	69	68	67			102.0	2.5
Farmington AP	36 44	108 14	5503	1	6	95/63	93/62	91/61	30	67	65	64	ENE 5	SW		
Gallup	35 31	108 47	6465	0	5	90/59	89/58	86/58	32	64	62	61				
Grants	35 10	107 54	6524	− 1	4	89/59	88/58	85/57	32	64	62	61				
Hobbs AP	32 45	103 13	3690	13	18	101/66	99/66	97/66	29	71	70	69				
Las Cruces	32 18	106 55	4544	15	20	99/64	96/64	94/64	30	69	68	67	SE 5	SE		
Los Alamos	35 52	106 19	7410	5	9	89/60	87/60	85/60	32	62	61	60			89.8	− 2.3
Raton AP	36 45	104 30	6373	− 4	1	91/60	89/60	87/60	34	65	64	63				
Roswell, Walker AFB	33 18	104 32	3676	13	18	100/66	98/66	96/66	33	71	70	69	N 6	SSE	103.0	2.7
Santa Fe Co	35 37	106 05	6307	6	10	90/61	88/61	86/61	28	63	62	61			90.1	− 1.2
Silver City AP	32 38	108 10	5442	5	10	95/61	94/60	91/60	30	66	64	63				
Socorro AP	34 03	106 53	4624	13	17	97/62	95/62	93/62	30	67	66	65				
Tucumcari AP	35 11	103 36	4039	8	13	99/66	97/66	95/65	28	70	69	68	NE 8	SW	102.7	1.1

Table 1 Climatic Conditions for the United States (*Continued*)

State and Station[a]	Lat. ° 'N	Long. ° 'W	Elev. Feet	Winter,[b] °F Design Dry-Bulb 99%	97.5%	Summer,[c] °F Design Dry-Bulb and Mean Coincident Wet-Bulb 1%	2.5%	5%	Mean Daily Range	Design Wet-Bulb 1%	2.5%	5%	Prevailing Wind Winter	Knots[d]	Summer	Temp., °F Median of Annual Extr. Max.	Min.
NEW YORK																	
Albany AP	42 45	73 48	275	− 6	− 1	91/73	88/72	85/70	23	75	74	72	WNW	8	S		
Albany Co	42 39	73 45	19	− 4	1	91/73	88/72	85/70	20	75	74	72				95.2	− 11.4
Auburn	42 54	76 32	715	− 3	2	90/73	87/71	84/70	22	75	73	72				92.4	− 9.5
Batavia	43 00	78 11	922	1	5	90/72	87/71	84/70	22	75	73	72				92.2	− 7.5
Binghamton AP	42 13	75 59	1590	− 2	1	86/71	83/69	81/68	20	73	72	70	WSW	10	WSW	92.9	− 9.3
Buffalo AP	42 56	78 44	705	2	6	88/71	85/70	83/69	21	74	73	72	W	10	SW	90.0	− 3.2
Cortland	42 36	76 11	1129	− 5	0	88/71	85/71	82/70	23	74	73	71				93.8	− 11.2
Dunkirk	42 29	79 16	692	4	9	88/73	85/72	83/71	18	75	74	72	SSW	10	WSW		
Elmira AP	42 10	76 54	955	− 4	1	89/71	86/71	83/70	24	74	73	71				96.2	− 6.7
Geneva	42 45	76 54	613	− 3	2	90/73	87/71	84/70	22	75	73	72				96.1	− 6.5
Glens Falls	43 20	73 37	328	− 11	− 5	88/72	85/71	82/69	23	74	73	71	NNW	6	S		
Gloversville	43 02	74 21	760	− 8	− 2	89/72	86/71	83/69	23	75	74	72				93.2	− 14.6
Hornell	42 21	77 42	1325	− 4	0	88/71	85/70	82/69	24	74	73	72					
Ithaca	42 27	76 29	928	− 5	0	88/71	85/71	82/70	24	74	73	71	W	6	SW		
Jamestown	42 07	79 14	1390	− 1	3	88/70	86/70	83/69	20	74	72	71	WSW	9	WSW		
Kingston	41 56	74 00	279	− 3	2	91/73	88/72	85/70	22	76	74	73					
Lockport	43 09	79 15	638	4	7	89/74	86/72	84/71	21	76	74	73	N	9	SW	92.2	− 4.8
Massena AP	44 56	74 51	207	− 13	− 8	86/70	83/69	80/68	20	73	72	70					
Newburgh, Stewart AFB	41 30	74 06	471	− 1	4	90/73	88/72	85/70	21	76	74	73	W	10	W		
NYC-Central Park	40 47	73 58	157	11	15	92/74	89/73	87/72	17	76	75	74				94.9	3.8
NYC-Kennedy AP	40 39	3 47	13	12	15	90/73	87/72	84/71	16	76	75	74	WNW	4	SSW		
NYC-La Guardia AP	40 46	73 54	11	11	15	92/74	89/73	87/72	16	76	75	74	WNW	15	SW		
Niagara Falls AP	43 06	79 57	590	4	7	89/74	86/72	84/71	20	76	74	73	W	9	SW		
Olean	42 14	78 22	2119	− 2	2	87/71	84/71	81/70	23	74	73	71					
Oneonta	42 31	75 04	1775	− 7	− 4	86/71	83/69	80/68	24	73	72	70					
Oswego Co	43 28	76 33	300	1	7	86/73	83/71	80/70	20	75	73	72	E	7	WSW	91.3	− 7.4
Plattsburg AFB	44 39	73 28	235	− 13	− 8	86/70	83/69	80/68	22	73	72	70	NW	6	SE		
Poughkeepsie	41 38	73 55	165	0	6	92/74	89/74	86/72	21	77	75	74	NNE	6	SSW	98.1	− 5.6
Rochester AP	43 07	77 40	547	1	5	91/73	88/71	85/70	22	75	73	72	WSW	11	WSW		
Rome, Griffiss AFB	43 14	75 25	514	− 11	− 5	88/71	85/70	83/69	22	75	73	71	NW	5	W		
Schenectady	42 51	73 57	377	− 4	1	90/73	87/72	84/70	22	75	74	72	WNW	8	S		
Suffolk County AFB	40 51	72 38	67	7	10	86/72	83/71	80/70	16	76	74	73	NW	9	SW		
Syracuse AP	43 07	76 07	410	− 3	2	90/73	87/71	84/70	20	75	73	72	N	7	WNW	93.	− 10.0
Utica	43 09	75 23	714	− 12	− 6	88/73	85/71	82/70	22	75	73	71	NW	12	W		
Watertown	43 59	76 01	325	− 11	− 6	86/73	83/71	81/70	20	75	73	72	E	7	WSW	91.7	− 19.6
NORTH CAROLINA																	
Asheville AP	35 26	82 32	2140	10	14	89/73	87/72	85/71	21	75	74	72	NNW	12	NNW	91.9	5.8
Charlotte AP	35 13	80 56	736	18	22	95/74	93/74	91/74	20	77	76	76	NNW	6	SW	97.8	12.6
Durham	35 52	78 47	434	16	20	94/75	92/75	90/75	20	78	77	76				98.9	9.6
Elizabeth City AP	36 16	76 11	12	12	19	93/78	91/77	89/76	18	80	78	78	NW	8	SW		
Fayetteville, Pope AFB	35 10	79 01	218	17	20	95/76	92/76	90/75	20	79	78	77	N	6	SSW	99.1	13.1
Goldsboro	35 20	77 58	109	18	21	94/77	91/76	89/75	18	79	78	77	N	8	SW	99.8	13.0
Greensboro AP	36 05	79 57	897	14	18	93/74	91/73	89/73	21	77	76	75	NE	7	SW	97.7	9.7
Greenville	35 37	77 25	75	18	21	93/77	91/76	89/75	19	79	78	77					
Henderson	36 22	78 25	480	12	15	95/77	92/76	90/76	20	79	78	77					
Hickory	35 45	81 23	1187	14	18	92/73	90/72	88/72	21	75	74	73				96.5	9.6
Jacksonville	34 50	77 37	95	20	24	92/78	90/78	88/77	18	80	79	78					
Lumberton	34 37	79 04	129	18	21	95/76	92/76	90/75	20	79	78	77					
New Bern AP	35 05	77 03	20	20	24	92/78	90/78	88/77	18	80	79	78				98.2	15.1
Raleigh/Durham AP	35 52	78 47	434	16	20	94/75	92/75	90/75	20	78	77	76	N	7	SW	97.7	12.2
Rocky Mount	35 58	77 48	121	18	21	94/77	91/76	89/75	19	79	78	77					
Wilmington AP	34 16	77 55	28	23	26	93/79	91/78	89/77	18	81	80	79	N	8	SW	96.9	18.2
Winston-Salem AP	36 08	80 13	969	16	20	94/74	91/73	89/73	20	76	75	74	NW	8	WSW		
NORTH DAKOTA																	
Bismarck AP	46 46	100 45	1647	− 23	− 19	95/68	91/68	88/67	27	73	71	70	WNW	7	S	100.3	− 31.5
Devils Lake	48 07	98 54	1450	− 25	− 21	91/69	88/68	85/66	25	73	71	69				97.5	− 30.4
Dickinson AP	46 48	102 48	2585	− 21	− 17	94/68	90/66	87/65	25	71	69	68	WNW	12	SSE	101.0	− 31.3
Fargo AP	46 54	96 48	896	− 22	− 18	92/73	89/71	85/69	25	76	74	72	SSE	11	S	97.3	− 29.7
Grand Forks AP	47 57	97 24	911	− 26	− 22	91/70	87/70	84/68	25	74	72	70	N	8	S	97.6	− 29.0
Jamestown AP	46 55	98 41	1492	− 22	− 18	94/70	90/69	87/68	26	74	74	71				101.3	− 27.9
Minot AP	48 25	101 21	1668	− 24	− 20	92/68	89/67	86/65	25	72	70	68	WSW	10	S		
Williston	48 09	103 35	1876	− 25	− 21	91/68	88/67	85/65	25	72	70	68				99.7	− 32.9

Table 1 Climatic Conditions for the United States (*Continued*)

State and Station[a]	Lat. ° 'N	Long. ° 'W	Elev. Feet	Winter,[b] °F Design Dry-Bulb 99%	97.5%	Summer,[c] °F Design Dry-Bulb and Mean Coincident Wet-Bulb 1%	2.5%	5%	Mean Daily Range	Design Wet-Bulb 1%	2.5%	5%	Prevailing Wind Winter Knots[d]	Summer	Temp., °F Median of Annual Extr. Max.	Min.
OHIO																
Akron, Canton AP	40 55	81 26	1208	1	6	89/72	86/71	84/70	21	75	73	72	SW 9	SW	94.4	−4.6
Ashtabula	41 51	80 48	690	4	9	88/73	85/72	83/71	18	75	74	72				
Athens	39 20	82 06	700	0	6	95/75	92/74	90/73	22	78	76	74				
Bowling Green	41 23	83 38	675	−2	2	92/73	89/73	86/71	23	76	75	73			96.7	−7.3
Cambridge	40 04	81 35	807	1	7	93/75	90/74	87/73	23	78	76	75				
Chillicothe	39 21	83 00	640	0	6	95/75	92/74	90/73	22	78	76	74	W 8	WSW	98.2	−2.1
Cincinnati Co	39 09	84 31	758	1	6	92/73	90/72	88/72	21	77	75	74	W 9	SW	97.2	−0.2
Cleveland AP	41 24	81 51	777	1	5	91/73	88/72	86/71	22	76	74	73	SW 12	N	94.7	−3.1
Columbus AP	40 00	82 53	812	0	5	92/73	90/73	87/72	24	77	75	74	W 8	SSW	96.0	−3.4
Dayton AP	39 54	84 13	1002	−1	4	91/73	89/72	86/71	20	76	75	73	WNW 11	SW	96.6	−4.5
Defiance	41 17	84 23	700	−1	4	94/74	91/73	88/72	24	77	76	74				
Findlay AP	41 01	83 40	804	2	3	92/74	90/74	87/72	24	77	76	74			97.4	−7.4
Fremont	41 20	83 07	600	−3	1	90/73	88/73	85/71	24	76	75	73				
Hamilton	39 24	84 35	650	0	5	92/73	90/72	87/71	22	76	75	73			98.2	−2.8
Lancaster	39 44	82 38	860	0	5	93/74	91/73	88/72	23	77	75	74				
Lima	40 42	84 02	975	−1	4	94/74	91/73	88/72	24	77	76	74	WNW 11	SW	96.0	−6.5
Mansfield AP	40 49	82 31	1295	0	5	90/73	87/72	85/72	22	76	74	73	W 8	SW	93.8	−10.7
Marion	40 36	83 10	920	0	5	93/74	91/73	88/72	23	77	76	74				
Middletown	39 31	84 25	635	0	5	92/73	90/72	87/71	22	76	75	73				
Newark	40 01	82 28	880	−1	5	94/73	92/73	89/72	23	77	75	74	W 8	SSW	95.8	−6.8
Norwalk	41 16	82 37	670	−3	1	90/73	88/73	85/71	22	76	75	73			97.3	−8.3
Portsmouth	38 45	82 55	540	5	10	95/76	92/74	89/73	22	78	77	75	W 8	SW	97.9	1.0
Sandusky Co	41 27	82 43	606	1	6	93/73	91/72	88/71	21	76	74	73			96.7	−1.9
Springfield	39 50	83 50	1052	−1	3	91/74	89/73	87/72	21	77	76	74	W 7	W		
Steubenville	40 23	80 38	992	1	5	89/72	86/71	84/70	22	74	73	72				
Toledo AP	41 36	83 48	669	−3	1	90/73	88/73	85/71	25	76	75	73	WSW 8	SW	95.4	−5.2
Warren	41 20	80 51	928	0	5	89/71	87/71	85/70	23	74	73	71				
Wooster	40 47	81 55	1020	1	6	89/72	86/71	84/70	22	75	73	72			94.0	−7.7
Youngstown AP	41 16	80 40	1178	−1	4	88/71	86/71	84/70	23	74	73	71	SW 10	SW		
Zanesville AP	39 57	81 54	900	1	7	93/75	90/74	87/73	23	78	76	75	W 6	WSW		
OKLAHOMA																
Ada	34 47	96 41	1015	10	14	100/74	97/74	95/74	23	77	76	75				
Altus AFB	34 39	99 16	1378	11	16	102/73	100/73	98/73	25	77	76	75	N 10	S		
Ardmore	34 18	97 01	771	13	17	100/74	98/74	95/74	23	77	77	76				
Bartlesville	36 45	96 00	715	6	10	101/73	98/74	95/74	23	77	77	76				
Chickasha	35 03	97 55	1085	10	14	101/74	98/74	95/74	24	78	77	76				
Enid, Vance AFB	36 21	97 55	1307	9	13	103/74	100/74	97/74	24	79	77	76				
Lawton AP	34 34	98 25	1096	12	16	101/74	99/74	96/74	24	78	77	76				
McAlester	34 50	95 55	776	14	19	99/74	96/74	93/74	23	77	76	75	N 10	S		
Muskogee AP	35 40	95 22	610	10	15	101/74	98/75	95/75	23	79	78	77				
Norman	35 15	97 29	1181	9	13	99/74	96/74	94/74	24	77	76	75	N 10	S		
Oklahoma City AP	35 24	97 36	1285	9	13	100/74	97/74	95/73	23	78	77	76	N 14	SSW		
Ponca City	36 44	97 06	997	5	9	100/74	97/74	94/74	24	77	76	76				
Seminole	35 14	96 40	865	11	15	99/74	96/74	94/73	23	77	76	75				
Stillwater	36 10	97 05	984	8	13	100/74	96/74	93/74	24	77	76	75	N 12	SSW	103.7	1.6
Tulsa AP	36 12	95 54	650	8	13	101/74	98/75	95/75	22	79	78	77	N 11	SSW	107.1	−1.3
Woodward	36 36	99 31	2165	6	10	100/73	97/73	94/73	26	78	76	75				
OREGON																
Albany	44 38	123 07	230	18	22	92/67	89/66	86/65	31	69	67	66			97.5	16.6
Astoria AP	46 09	123 53	8	25	29	75/65	71/62	68/61	16	65	63	62	ESE 7	NNW		
Baker AP	44 50	117 49	3372	−1	6	92/63	89/61	86/60	30	65	63	61			97.5	−6.8
Bend	44 04	121 19	3595	−3	4	90/62	87/60	84/59	33	64	62	60			96.4	−5.8
Corvallis	44 30	123 17	246	18	22	92/67	89/66	86/65	31	69	67	66	N 6	N	98.5	17.1
Eugene AP	44 07	123 13	359	17	22	92/67	89/66	86/65	31	69	67	66	N 7	N		
Grants Pass	42 26	123 19	925	20	24	99/69	96/68	93/67	33	71	69	68	N 5	N	103.6	16.4
Klamath Falls AP	42 09	121 44	4092	4	9	90/61	87/60	84/59	36	63	61	60	N 4	W	96.3	0.9
Medford AP	42 22	122 52	1298	19	23	98/68	94/67	91/66	35	70	68	67	S 4	WMW	103.8	15.0
Pendleton AP	45 41	118 51	1482	−2	5	97/65	93/64	90/62	29	66	65	63	NNW 6	WNW		
Portland AP	45 36	122 36	21	17	23	89/68	85/67	81/65	23	69	67	66	ESE 12	NW	96.6	18.3
Portland Co	45 32	122 40	75	18	24	90/68	86/67	82/65	21	69	67	66			97.6	20.5
Roseburg AP	43 14	123 22	525	18	23	93/67	90/66	87/65	30	69	67	66			99.6	19.5
Salem AP	44 55	123 01	196	18	23	92/68	88/66	84/65	31	69	68	66	N 6	N	98.9	15.9
The Dalles	45 36	121 12	100	13	19	93/69	89/68	85/66	28	70	68	67			105.1	7.9

Table 1 Climatic Conditions for the United States (*Continued*)

Col. 1	Col. 2		Col. 3		Col. 4	Col. 5 Winter,[b] °F		Col. 6 Summer,[c] °F			Col. 7	Col. 8			Col. 9 Prevailing Wind		Col. 10 Temp., °F	
State and Station[a]	Lat.		Long.		Elev.	Design Dry-Bulb		Design Dry-Bulb and Mean Coincident Wet-Bulb			Mean Daily Range	Design Wet-Bulb			Winter	Summer	Median of Annual Extr.	
	° 'N		° 'W		Feet	99%	97.5%	1%	2.5%	5%		1%	2.5%	5%	Knots[d]		Max.	Min.
PENNSYLVANIA																		
Allentown AP	40	39	75	26	387	4	9	92/73	88/72	86/72	22	76	75	73	W 11	SW		
Altoona Co	40	18	78	19	1504	0	5	90/72	87/71	84/70	23	74	73	72	WNW 11	WSW	93.7	− 5.2
Butler	40	52	79	54	1100	1	6	90/73	87/72	85/71	22	75	74	73				
Chambersburg	39	56	77	38	640	4	8	93/75	90/74	87/73	23	77	76	75			97.4	− 0.3
Erie AP	42	05	80	11	731	4	9	88/73	85/72	83/71	18	75	74	72	SSW 0	WSW	91.3	− 2.2
Harrisburg AP	40	12	76	46	308	7	11	94/75	91/74	88/73	21	77	76	75	NW 11	WSW	96.5	3.7
Johnstown	40	19	78	50	2284	− 3	2	86/70	83/70	80/68	23	72	71	70	WNW 8	WSW	96.4	− 1.8
Lancaster	40	07	76	18	403	4	8	93/75	90/74	87/73	22	77	76	75	NW 11	WSW		
Meadville	41	38	80	10	1065	0	4	88/71	85/70	83/69	21	73	72	71			93.2	− 8.5
New Castle	41	01	80	22	825	2	7	91/73	88/72	86/71	23	75	74	73	WSW 10	WSW	94.7	− 6.4
Philadelphia AP	39	53	75	15	5	10	14	93/75	90/74	87/72	21	77	76	75	WNW 10	WSW	96.4	5.9
Pittsburgh AP	40	30	80	13	1137	1	5	89/72	86/71	84/70	22	74	73	72	WSW 10	WSW		
Pittsburgh Co	40	27	80	00	1017	3	7	91/72	88/71	86/70	19	74	73	72			94.6	− 1.1
Reading Co	40	20	75	38	266	9	13	92/73	89/72	86/72	19	76	75	73	W 11	SW	97.0	3.6
Scranton/Wilkes-Barre	41	20	75	44	930	1	5	90/72	87/71	84/70	19	74	73	72	SW 8	WSW	94.8	− 2.2
State College	40	48	77	52	1175	3	7	90/72	87/71	84/70	23	74	73	72	NNW 8	WSW	93.2	− 3.6
Sunbury	40	53	76	46	446	2	7	92/73	89/72	86/70	22	75	74	73				
Uniontown	39	55	79	43	956	5	9	91/74	88/73	85/72	22	76	75	74			93.9	− 2.5
Warren	41	51	79	08	1280	− 2	4	89/71	86/71	83/70	24	74	73	72			93.3	− 10.7
West Chester	39	58	75	38	450	9	13	92/75	89/74	86/72	20	77	76	75				
Williamsport AP	41	15	76	55	524	2	7	92/73	89/72	86/70	23	75	74	73	W 9	WSW	95.5	− 4.6
York	39	55	76	45	390	8	12	94/75	91/74	88/73	22	77	76	75			97.0	− 2.4
RHODE ISLAND																		
Newport	41	30	71	20	10	5	9	88/73	85/72	82/70	16	76	75	73	WNW 10	SW		
Providence AP	41	44	71	26	51	5	9	89/73	86/72	83/70	19	75	74	73	WNW 11	SW	94.6	− 0.5
SOUTH CAROLINA																		
Anderson	34	30	82	43	774	19	23	94/74	92/74	90/74	21	77	76	75			99.5	13.3
Charleston AFB	32	54	80	02	45	24	27	93/78	91/78	89/77	18	81	80	79	NNE 8	SW		
Charleston Co	32	54	79	58	3	25	28	94/78	92/78	90/77	13	81	80	79			97.8	21.4
Columbia AP	33	57	81	07	213	20	24	97/76	95/75	93/75	22	79	78	77	W 6	SW	100.6	16.2
Florence AP	34	11	79	43	147	22	25	94/77	92/77	90/76	21	80	79	78	N 7	SW	99.5	16.5
Georgetown	33	23	79	17	14	23	26	92/79	90/78	88/77	18	81	80	79	N 7	SSW	98.2	19.1
Greenville AP	34	54	82	13	957	18	22	93/74	91/74	89/74	21	77	76	75	NW 8	SW	97.3	12.6
Greenwood	34	10	82	07	620	18	22	95/75	93/74	91/74	21	78	77	76			99.5	14.1
Orangeburg	33	30	80	52	260	20	24	97/76	95/75	93/75	20	79	78	77			101.2	18.0
Rock Hill	34	59	80	58	470	19	23	96/75	94/74	92/74	20	78	77	76				
Spartanburg AP	34	58	82	00	823	18	22	93/74	91/74	89/74	20	77	76	75			99.5	13.9
Sumter, Shaw AFB	33	54	80	22	169	22	25	95/77	92/76	90/75	21	79	78	77	NNE 6	W	100.0	15.4
SOUTH DAKOTA																		
Aberdeen AP	45	27	98	26	1296	− 19	− 15	94/73	91/72	88/70	27	77	75	73	NNW 8	S	102.3	− 28.1
Brookings	44	18	96	48	1637	− 17	− 13	95/73	92/72	89/71	25	77	75	73			97.8	− 26.5
Huron AP	44	23	98	13	1281	− 18	− 14	96/73	93/72	90/71	28	77	75	73	NNW 8	S	101.5	− 25.8
Mitchell	43	41	98	01	1346	− 15	− 10	96/72	93/71	90/70	28	76	75	73			103.0	− 22.7
Pierre AP	44	23	100	17	1742	− 15	− 10	99/71	95/71	92/69	29	75	74	72	NW 11	SSE	105.7	− 20.6
Rapid City AP	44	03	103	04	3162	− 11	− 7	95/66	92/65	89/65	28	71	69	67	NNW 10	SSE	100.9	− 19.0
Sioux Falls AP	43	34	96	44	1418	− 15	− 11	94/73	91/72	88/71	24	76	75	73	NW 8	S		
Watertown AP	44	55	97	09	1738	− 19	− 15	94/73	91/72	88/71	26	76	75	73			97.8	− 26.5
Yankton	42	55	97	23	1302	− 13	− 7	94/73	91/72	88/71	25	77	76	74			100.8	− 19.1
TENNESSEE																		
Athens	35	26	84	35	940	13	18	95/74	92/73	90/73	22	77	76	75				
Bristol-Tri City AP	36	29	82	24	1507	9	14	91/72	89/72	87/71	22	75	73	73	WNW 6	SW		
Chattanooga AP	35	02	85	12	665	13	18	96/75	93/74	91/74	22	78	77	76	NNW 8	WSW	97.2	9.8
Clarksville	36	33	87	22	382	6	12	95/76	93/74	90/74	21	78	77	76			99.8	3.7
Columbia	35	38	87	02	690	10	15	97/75	94/74	91/74	21	78	77	76				
Dyersburg	36	01	89	24	344	10	15	96/78	94/77	91/76	21	81	80	78				
Greeneville	36	04	82	50	1319	11	16	92/73	90/72	88/72	22	76	75	74			95.6	0.8
Jackson AP	35	36	88	55	423	11	16	98/76	95/75	92/75	21	79	78	77			99.2	6.6
Knoxville AP	35	49	83	59	980	13	19	94/74	92/73	90/73	21	77	76	75	NE 8	W	96.0	7.0
Memphis AP	35	03	90	00	258	13	18	98/77	95/76	93/76	21	80	79	78	N 10	SW	97.9	10.4
Murfreesboro	34	55	86	28	600	9	14	97/75	94/74	91/74	22	78	77	76			97.7	4.5
Nashville AP	36	07	86	41	590	9	14	97/75	94/74	91/74	21	78	77	76	NW 8	WSW		
Tullahoma	35	23	86	05	1067	8	13	96/74	93/73	91/73	22	77	76	75	NW 9	WSW	96.7	3.7

Table 1   Climatic Conditions for the United States (*Continued*)

Col. 1	Col. 2		Col. 3		Col. 4	Col. 5 Winter,[b] °F Design Dry-Bulb		Col. 6 Summer,[c] °F Design Dry-Bulb and Mean Coincident Wet-Bulb			Col. 7 Mean Daily Range	Col. 8 Design Wet-Bulb			Col. 9 Prevailing Wind		Col. 10 Temp., °F Median of Annual Extr.	
State and Station[a]	Lat. ° 'N		Long. ° 'W		Elev. Feet	99%	97.5%	Mean 1%	Coincident 2.5%	Wet-Bulb 5%		1%	2.5%	5%	Winter Knots[d]	Summer	Max.	Min.
TEXAS																		
Abilene AP	32	25	99	41	1784	15	20	101/71	99/71	97/71	22	75	74	74	N 12	SSE	103.6	10.4
Alice AP	27	44	98	02	180	31	34	100/78	98/77	95/77	20	82	81	79			104.9	24.8
Amarillo AP	35	14	101	42	3604	6	11	98/67	95/67	93/67	26	71	70	70	N 11	S	100.8	0.9
Austin AP	30	18	97	42	597	24	28	100/74	98/74	97/74	22	78	77	77	N 11	S	101.6	19.7
Bay City	29	00	95	58	50	29	33	96/77	94/77	92/77	16	80	79	79				
Beaumont	29	57	94	01	16	27	31	95/79	93/78	91/78	19	81	80	80			99.7	23.5
Beeville	28	22	97	40	190	30	33	99/78	97/77	95/77	18	82	81	79	N 9	SSE	103.1	22.5
Big Spring AP	32	18	101	27	2598	16	20	100/69	97/69	95/69	26	74	73	72			105.3	10.7
Brownsville AP	25	54	97	26	19	35	39	94/77	93/77	92/77	18	80	79	79	NNW 13	SE	98.1	30.1
Brownwood	31	48	98	57	1386	18	22	101/73	99/73	96/73	22	77	76	75	N 9	S	105.3	13.0
Bryan AP	30	40	96	33	276	24	29	98/76	96/76	94/76	20	79	78	78				
Corpus Christi AP	27	46	97	30	41	31	35	95/78	94/78	92/78	19	80	80	79	N 12	SSE	97.0	27.2
Corsicana	32	05	96	28	425	20	25	100/75	98/75	96/75	21	79	78	77			104.2	15.2
Dallas AP	32	51	96	51	481	18	22	102/75	100/75	97/75	20	78	78	77	N 11	S		
Del Rio, Laughlin AFB	29	22	100	47	1081	26	31	100/73	98/73	97/73	24	79	77	76			103.8	23.0
Denton	33	12	97	06	630	17	22	101/74	99/74	97/74	22	78	77	76			104.5	11.8
Eagle Pass	28	52	100	32	884	27	32	101/73	99/73	98/73	24	78	78	77	NNW 9	ESE	107.7	22.1
El Paso AP	31	48	106	24	3918	20	24	100/64	98/64	96/64	27	69	68	68	N 7	S	103.0	15.7
Fort Worth AP	32	50	97	03	537	17	22	101/74	99/74	97/74	22	78	77	76	NW 11	S	103.2	13.5
Galveston AP	29	18	94	48	7	31	36	90/79	89/79	88/78	10	81	80	80	N 15	S	93.9	27.5
Greenville	33	04	96	03	535	17	22	101/74	99/74	97/74	21	78	77	76			103.6	11.7
Harlingen	26	14	97	39	35	35	39	96/77	94/77	93/77	19	80	79	79	NNW 10	SSE	102.3	29.3
Houston AP	29	58	95	21	96	27	32	96/77	94/77	92/77	18	80	79	79	NNW 11	S		
Houston Co	29	59	95	22	108	28	33	97/77	95/77	93/77	18	80	79	79			99.0	23.5
Huntsville	30	43	95	33	494	22	27	100/75	98/75	96/75	20	78	78	77			100.8	18.7
Killeen, Robert Gray AAF	31	05	97	41	850	20	25	99/73	97/73	95/73	22	77	76	75				
Lamesa	32	42	101	56	2965	13	17	99/69	96/69	94/69	26	73	72	71			105.5	8.9
Laredo AFB	27	32	99	27	512	32	36	102/73	101/73	99/74	23	78	78	77	N 8	SE		
Longview	32	28	94	44	330	19	24	99/76	97/76	95/76	20	80	79	78				
Lubbock AP	33	39	101	49	3254	10	15	98/69	96/69	94/69	26	73	72	71	NNE 10	SSE		
Lufkin AP	31	25	94	48	277	25	29	99/76	97/76	94/76	20	80	79	78	NNW 12	S		
McAllen	26	12	98	13	122	35	39	97/77	95/77	94/77	21	80	79	79				
Midland AP	31	57	102	11	2851	16	21	100/69	98/69	96/69	26	73	72	71	NE 9	SSE	103.6	10.8
Mineral Wells AP	32	47	98	04	930	17	22	101/74	99/74	97/74	22	78	77	76				
Palestine Co	31	47	95	38	600	23	27	100/76	98/76	96/76	20	79	79	78			101.2	16.3
Pampa	35	32	100	59	3250	7	12	99/67	96/67	94/67	26	71	70	70				
Pecos	31	25	103	30	2610	16	21	100/69	98/69	96/69	27	73	72	71				
Plainview	34	11	101	42	3370	8	13	98/68	96/68	94/68	26	72	71	70			102.7	3.1
Port Arthur AP	29	57	94	01	16	27	31	95/79	93/78	91/78	19	81	80	80	N 9	S	97.7	24.0
San Angelo, Goodfellow AFB	31	26	100	24	1877	18	22	101/71	99/71	97/70	24	75	74	73	NNE 10	SSE		
San Antonio AP	29	32	98	28	788	25	30	99/72	97/73	96/73	19	77	76	76	N 8	SSE	101.3	21.1
Sherman, Perrin AFB	33	43	96	40	763	15	20	100/75	98/75	95/74	22	78	77	76	N 10	S	103.0	11.9
Snyder	32	43	100	55	2325	13	18	100/70	98/70	96/70	26	74	73	72				
Temple	31	06	97	21	700	22	27	100/74	99/74	97/74	22	78	77	77				
Tyler AP	32	21	95	16	530	19	24	99/76	97/76	95/76	21	80	79	78	NNE 23	S		
Vernon	34	10	99	18	1212	13	17	102/73	100/73	97/73	24	77	76	75				
Victoria AP	28	51	96	55	104	29	32	98/78	96/77	94/77	18	82	81	79			101.4	23.4
Waco AP	31	37	97	13	500	21	26	101/75	99/75	97/75	22	78	78	77				
Wichita Falls AP	33	58	98	29	994	14	18	103/73	101/73	98/73	24	77	76	75	NNW 12	S		
UTAH																		
Cedar City AP	37	42	113	06	5617	− 2	5	93/60	91/60	89/59	32	65	63	62	SE 5	SW		
Logan	41	45	111	49	4785	− 3	2	93/62	91/61	88/60	33	65	64	63			95.5	− 7.8
Moab	38	36	109	36	3965	6	11	100/60	98/60	96/60	30	65	64	63				
Ogden AP	41	12	112	01	4455	1	5	93/63	91/61	88/61	33	66	65	64	S 6	SW	99.5	− 3.9
Price	39	37	110	50	5580	− 2	5	93/60	91/60	89/59	33	65	63	62				
Provo	40	13	111	43	4448	1	6	98/62	96/62	94/61	32	66	65	64	SE 5	SW		
Richfield	38	46	112	05	5270	− 2	5	93/60	91/60	89/59	34	65	63	62			98.1	− 10.5
St George Co	37	02	113	31	2900	14	21	103/65	101/65	99/64	33	70	68	67			109.3	11.1
Salt Lake City AP	40	46	111	58	4220	3	8	97/62	95/62	92/61	32	66	65	64	SSE 6	N	99.4	− 0.1
Vernal AP	40	27	109	31	5280	− 5	0	91/61	89/60	86/59	32	64	63	62				
VERMONT																		
Barre	44	12	72	31	600	− 16	− 11	84/71	81/69	78/68	23	73	71	70				
Burlington AP	44	28	73	09	332	− 12	− 7	88/72	85/70	82/69	23	74	72	71	E 7	SSW	92.4	− 16.9
Rutland	43	36	72	58	620	− 13	− 8	87/72	84/70	81/69	23	74	72	71			92.5	− 17.5

Table 1 Climatic Conditions for the United States (*Concluded*)

Col. 1	Col. 2		Col. 3		Col. 4	Winter,[b] °F Col. 5		Summer,[c] °F Col. 6			Col. 7	Col. 8			Prevailing Wind Col. 9				Temp., °F Col. 10	
State and Station[a]	Lat.		Long.		Elev.	Design Dry-Bulb		Design Dry-Bulb and Mean Coincident Wet-Bulb			Mean Daily Range	Design Wet-Bulb			Winter		Summer		Median of Annual Extr.	
	° 'N		° 'W		Feet	99%	97.5%	1%	2.5%	5%		1%	2.5%	5%	Knots[d]				Max.	Min.
VIRGINIA																				
Charlottesville	38	02	78	31	870	14	18	94/74	91/74	88/73	23	77	76	75	NE	7	SW		97.4	8.0
Danville AP	36	34	79	20	590	14	16	94/74	92/73	90/73	21	77	76	75					100.1	9.2
Fredericksburg	38	18	77	28	100	10	14	96/76	93/75	90/74	21	78	77	76						
Harrisonburg	38	27	78	54	1370	12	16	93/72	91/72	88/71	23	75	74	73						
Lynchburg AP	37	20	79	12	916	12	16	93/74	90/74	88/73	21	77	76	75	NE	7	SW		97.2	7.6
Norfolk AP	36	54	76	12	22	20	22	93/77	91/76	89/76	18	79	78	77	NW	10	SW		97.2	15.3
Petersburg	37	11	77	31	194	14	17	95/76	92/76	90/75	20	79	78	77						
Richmond AP	37	30	77	20	164	14	17	95/76	92/76	90/75	21	79	78	77	N	6	SW		97.9	9.6
Roanoke AP	37	19	79	58	1193	12	16	93/72	91/72	88/71	23	75	74	73	NW	9	SW			
Staunton	38	16	78	54	1201	12	16	93/72	91/72	88/71	23	75	74	73	NW	9	SW		95.9	2.5
Winchester	39	12	78	10	760	6	10	93/75	90/74	88/74	21	77	76	75					97.3	3.7
WASHINGTON																				
Aberdeen	46	59	123	49	12	25	28	80/65	77/62	73/61	16	65	63	62	ESE	6	NNW		91.9	19.3
Bellingham AP	48	48	122	32	158	10	15	81/67	77/65	74/63	19	68	65	63	NNE	15	WSW		87.4	10.3
Bremerton	47	34	122	40	162	21	25	82/65	78/64	75/62	20	66	64	63	E	8	N			
Ellensburg AP	47	02	120	31	1735	2	6	94/65	91/64	87/62	34	66	65	63						
Everett, Paine AFB	47	55	122	17	596	21	25	80/65	76/64	73/62	20	67	64	63	ESE	6	NNW		84.9	15.2
Kennewick	46	13	119	08	392	5	11	99/68	96/67	92/66	30	70	68	67					103.4	2.0
Longview	46	10	122	56	12	19	24	88/68	85/67	81/65	30	69	67	66	ESE	9	NW		96.0	14.8
Moses Lake, Larson AFB	47	12	119	19	1185	1	7	97/66	94/65	90/63	32	67	66	64	N	8	SSW			
Olympia AP	46	58	122	54	215	16	22	87/66	83/65	79/64	32	67	66	64	NE	4	NE			
Port Angeles	48	07	123	26	99	24	27	72/62	69/61	67/60	18	64	62	61					83.5	19.4
Seattle-Boeing Field	47	32	122	18	23	21	26	84/68	81/66	77/65	24	69	67	65						
Seattle Co	47	39	122	18	20	22	27	85/68	82/66	78/65	19	69	67	65	N	7	N		90.2	22.0
Seattle-Tacoma AP	47	27	122	18	400	21	26	84/65	80/64	76/62	22	66	64	63	E	9	N		90.1	19.9
Spokane AP	47	38	117	31	2357	− 6	2	93/64	90/63	87/62	28	65	64	62	NE	6	SW		98.8	− 4.9
Tacoma, McChord AFB	47	15	122	30	100	19	24	86/66	82/65	79/63	22	68	66	64	S	5	NNE		89.4	18.8
Walla Walla AP	46	06	118	17	1206	0	7	97/67	94/66	90/65	27	69	67	66	W	5	W		103.0	3.8
Wenatchee	47	25	120	19	632	7	11	99/67	96/66	92/64	32	68	67	65					101.1	1.0
Yakima AP	46	34	120	32	1052	− 2	5	96/65	93/65	89/63	36	68	66	65	W	5	NW			
WEST VIRGINIA																				
Beckley	37	47	81	07	2504	− 2	4	83/71	81/69	79/69	22	73	71	70	WNW	9	WNW			
Bluefield AP	37	18	81	13	2867	− 2	4	83/71	81/69	79/69	22	73	71	70						
Charleston AP	38	22	81	36	939	7	11	92/74	90/73	87/72	20	76	75	74	SW	8	SW		97.2	2.9
Clarksburg	39	16	80	21	977	6	10	92/74	90/73	87/72	21	76	75	74						
Elkins AP	38	53	79	51	1948	1	6	86/72	84/70	82/70	22	74	72	71	WNW	9	WNW		90.6	− 7.3
Huntington Co	38	25	82	30	565	5	10	94/76	91/74	89/73	22	78	77	75	W	6	SW		97.1	2.1
Martinsburg AP	39	24	77	59	556	6	10	93/75	90/74	88/74	21	77	76	75	WNW	10	W		99.0	1.1
Morgantown AP	39	39	79	55	1240	4	8	90/74	87/73	85/73	22	76	75	74						
Parkersburg Co	39	16	81	34	615	7	11	93/75	90/74	88/73	21	77	76	75	WSW	7	WSW		95.9	0.7
Wheeling	40	07	80	42	665	1	5	89/72	86/71	84/70	21	74	73	72	WSW	10	WSW		97.5	− 0.6
WISCONSIN																				
Appleton	44	15	88	23	730	− 14	− 9	89/74	86/72	83/71	23	76	74	72					94.6	− 16.2
Ashland	46	34	90	58	650	− 21	− 16	85/70	82/68	79/66	23	72	70	68					94.1	− 26.8
Beloit	42	30	89	02	780	− 7	− 3	92/75	90/75	88/74	24	78	77	75						
Eau Claire AP	44	52	91	29	888	− 15	− 11	92/75	89/73	86/71	23	77	75	73						
Fond Du Lac	43	48	88	27	760	− 12	− 8	89/74	86/72	84/71	23	76	74	72					96.0	− 17.7
Green Bay AP	44	29	88	08	682	− 13	− 9	88/74	85/72	83/71	23	76	74	72	W	8	SW		94.3	− 17.9
La Crosse AP	43	52	91	15	651	− 13	− 9	91/75	88/73	85/72	22	77	75	74	NW	10	S		95.7	− 21.3
Madison AP	43	08	89	20	858	− 11	− 7	91/74	88/73	85/71	22	77	75	73	NW	8	SW		93.6	− 16.8
Manitowoc	44	06	87	41	660	− 11	− 7	89/74	86/72	83/71	21	76	74	72					94.1	− 13.7
Marinette	45	06	87	38	605	− 15	− 11	87/73	84/71	82/70	20	75	73	71					95.9	− 15.8
Milwaukee AP	42	57	87	54	672	− 8	− 4	90/74	87/73	84/71	21	76	74	73	WNW	10	SSW			
Racine	42	43	87	51	730	− 6	− 2	91/75	88/73	85/72	21	77	75	74						
Sheboygan	43	45	87	43	648	− 10	− 6	89/75	86/73	83/72	20	77	75	74					97.0	− 12.4
Stevens Point	44	30	89	34	1079	− 15	− 11	92/75	89/73	86/71	23	77	75	73					95.3	− 24.1
Waukesha	43	01	88	14	860	− 9	− 5	90/74	87/73	84/71	22	76	74	73					95.7	− 14.3
Wausau AP	44	55	89	37	1196	− 16	− 12	91/74	88/72	85/70	23	76	74	72						
WYOMING																				
Casper AP	42	55	106	28	5338	− 11	− 5	92/58	90/57	87/57	31	63	61	60	NE	10	SW		97.3	− 20.9
Cheyenne	41	09	104	49	6126	− 9	− 1	89/58	86/58	84/57	30	63	62	60	N	11	WNW		92.5	− 15.9
Cody AP	44	33	109	04	4990	− 19	− 13	89/60	86/60	83/59	32	64	63	61					97.4	− 21.9
Evanston	41	16	110	57	6780	− 9	− 3	86/55	84/55	82/54	32	59	58	57					89.2	− 21.2
Lander AP	42	49	108	44	5563	− 16	− 11	91/61	88/61	85/60	32	64	63	61	E	5	NW		94.9	− 22.6
Laramie AP	41	19	105	41	7266	− 14	− 6	84/56	81/56	79/55	28	61	60	59						
Newcastle	43	51	104	13	4265	− 17	− 12	91/64	87/63	84/63	30	69	68	66					99.4	− 19.0
Rawlins	41	48	107	12	6740	− 12	− 4	86/57	83/57	81/56	40	62	61	60						
Rock Springs AP	41	36	109	04	6745	− 9	− 3	86/55	84/55	82/54	32	59	58	57	WSW	10	W			
Sheridan AP	44	46	106	58	3964	− 14	− 8	94/62	91/62	88/61	32	66	65	63	NW	7	N		99.8	− 23.6
Torrington	42	05	104	13	4098	− 14	− 8	94/62	91/62	88/61	30	66	65	63					101.1	− 20.7

Table 2 Climatic Conditions for Canada

Col. 1	Col. 2		Col. 3		Col. 4	Winter,[b] °F Col. 5 Design Dry-Bulb		Summer,[c] °F Col. 6 Design Dry-Bulb and Coincident Wet-Bulb			Col. 7 Mean Daily Range	Col. 8 Design Wet-Bulb			Prevailing Wind Col. 9	
State and Station[a]	Lat. ° 'N		Long. ° 'W		Elev. Feet	99%	97.5%	Mean 1%	2.5%	5%		1%	2.5%	5%	Winter	Summer Knots[d]
ALBERTA																
Calgary AP	51	06	114	01	3540	−27	−23	84/63	81/61	79/60	25	65	63	62	NNW 8	SE
Edmonton AP	53	34	113	31	2219	−29	−25	85/66	82/65	79/63	23	68	66	65	E 9	SE
Grande Prairie AP	55	11	118	53	2190	−39	−33	83/64	80/63	78/61	23	66	64	62		
Jasper	52	53	118	04	3480	−31	−26	83/64	80/62	77/61	28	66	64	63		
Lethbridge AP	49	38	112	48	3018	−27	−22	90/65	87/64	84/63	28	68	66	65		
McMurray AP	56	39	111	13	1216	−41	−38	86/67	82/65	79/64	26	69	67	65		
Medicine Hat AP	50	01	110	43	2365	−29	−24	93/66	90/65	87/64	28	70	68	66		
Red Deer AP	52	11	113	54	2965	−31	−26	84/65	81/64	78/62	25	67	66	64		
BRITISH COLUMBIA																
Dawson Creek	55	44	120	11	2164	−37	−33	82/64	79/63	76/61	26	66	64	62		
Fort Nelson AP	58	50	122	35	1230	−43	−40	84/64	81/63	78/62	23	67	65	64		
Kamloops Co	50	43	120	25	1133	−21	−15	94/66	91/65	88/64	29	68	66	65		
Nanaimo	49	11	123	58	230	16	20	83/67	80/65	77/64	21	68	66	65		
New Westminster	49	13	122	54	50	14	18	84/68	81/67	78/66	19	69	68	66		
Penticton AP	49	28	119	36	1121	0	4	92/68	89/67	87/66	31	70	68	67		
Prince George AP	53	53	122	41	2218	−33	−28	84/64	80/62	77/61	26	66	64	62	N 11	N
Prince Rupert Co	54	17	130	23	170	−2	2	64/59	63/57	61/56	12	60	58	57		
Trail	49	08	117	44	1400	−5	0	92/66	89/65	86/64	33	68	67	65		
Vancouver AP	49	11	123	10	16	15	19	79/67	77/66	74/65	17	68	67	66	E 6	WNW
Victoria Co	48	25	123	19	228	20	23	77/64	73/62	70/60	16	64	62	60		
MANITOBA																
Brandon	49	52	99	59	1200	−30	−27	89/72	86/70	83/68	25	74	72	70		
Churchill AP	58	45	94	04	155	−41	−39	81/66	77/64	74/62	18	67	65	63	SE 11	S
Dauphin AP	51	06	100	03	999	−31	−28	87/71	84/70	81/68	23	74	72	70		
Flin Flon	54	46	101	51	1098	−41	−37	84/68	81/66	79/65	19	70	68	67		
Portage La Prairie AP	49	54	98	16	867	−28	−24	88/73	86/72	83/70	22	76	74	71		
The Pas AP	53	58	101	06	894	−37	−33	85/68	82/67	79/66	20	71	69	68	W 8	W
Winnipeg AP	49	54	97	14	786	−30	−27	89/73	86/71	84/70	22	75	73	71	W 8	S
NEW BRUNSWICK																
Campbellton Co	48	00	66	40	25	−18	−14	85/68	82/67	79/66	21	72	70	68		
Chatham AP	47	01	65	27	112	15	−10	89/69	85/68	82/67	22	72	71	69		
Edmundston Co	47	22	68	20	500	−21	−16	87/70	83/68	80/67	21	73	71	69		
Fredericton AP	45	52	66	32	74	−16	−11	89/71	85/69	82/68	23	73	71	70		
Moncton AP	46	07	64	41	248	−12	−8	85/70	82/69	79/67	23	72	71	69		
Saint John AP	45	19	65	53	352	−12	−8	80/67	77/65	75/64	19	70	68	66		
NEWFOUNDLAND																
Corner Brook	48	58	57	57	15	−5	0	76/64	73/63	71/62	17	67	66	65		
Gander AP	48	57	54	34	482	−5	−1	82/66	79/65	77/64	19	69	67	66	WNW 11	SW
Goose Bay AP	53	19	60	25	144	−27	−24	85/66	81/64	77/63	19	68	66	64	N 9	SW
St John's AP	47	37	52	45	463	3	7	77/66	75/65	73/64	18	69	67	66	N 20	WSW
Stephenville AP	48	32	58	33	44	−3	4	76/65	74/64	71/63	14	67	66	65	WNW 10	S
NORTHWEST TERRITORIES																
Fort Smith AP	60	01	111	58	665	−49	−45	85/66	81/64	78/63	24	68	66	65	NW 4	S
Frobisher AP	63	45	68	33	68	−43	−41	66/53	63/51	59/50	14	54	52	51	NNW 9	NW
Inuvik	68	18	133	29	200	−56	−53	79/62	77/60	75/59	21	64	62	61		
Resolute AP	74	43	94	59	209	−50	−47	57/48	54/46	51/45	10	50	48	46		
Yellowknife AP	62	28	114	27	682	−49	−46	79/62	77/61	74/60	16	64	63	62	SSE 7	S
NOVA SCOTIA																
Amherst	45	49	64	13	65	−11	−6	84/69	81/68	79/67	21	72	70	68		
Halifax AP	44	39	63	34	83	1	5	79/66	76/65	74/64	16	69	67	66		
Kentville	45	03	64	36	40	−3	1	85/69	83/68	80/67	22	72	71	69		
New Glasgow	45	37	62	37	317	−9	−5	81/69	79/68	77/67	20	72	70	69		
Sydney AP	46	10	60	03	197	−1	3	82/69	80/68	77/66	19	71	70	68		
Truro Co	45	22	63	16	131	−8	−5	82/70	80/69	78/68	22	73	71	70		
Yarmouth AP	43	50	66	05	136	5	9	74/65	72/64	70/63	15	68	66	65	NW 11	S

Table 2 Climatic Conditions for Canada (*Concluded*)

Col. 1	Col. 2		Col. 3		Col. 4	Winter,[b] °F Col. 5		Summer,[c] °F Col. 6			Col. 7	Col. 8			Prevailing Wind Col. 9	
State and Station[a]	Lat.		Long.		Elev.	Design Dry-Bulb		Design Dry-Bulb and Mean Coincident Wet-Bulb			Mean Daily Range	Design Wet-Bulb			Winter	Summer
	° 'N		° 'W		Feet	99%	97.5%	Mean 1%	Coincident 2.5%	Wet-Bulb 5%		1%	2.5%	5%	Knots[d]	
ONTARIO																
Belleville	44	09	77	24	250	−11	−7	86/73	84/72	82/71	20	75	74	73		
Chatham	42	24	82	12	600	0	3	89/74	87/73	85/72	19	76	75	74		
Cornwall	45	01	74	45	210	−13	−9	89/73	87/72	84/71	21	75	74	72		
Hamilton	43	16	79	54	303	−3	1	88/73	86/72	83/71	21	76	74	73		
Kapuskasing AP	49	25	82	28	752	−31	−28	86/70	83/69	80/67	23	72	70	69		
Kenora AP	49	48	94	22	1345	−32	−28	84/70	82/69	80/68	19	73	71	70		
Kingston	44	16	76	30	300	−11	−7	87/73	84/72	82/71	20	75	74	73		
Kitchener	43	26	80	30	1125	−6	−2	88/73	85/72	83/71	23	75	74	72		
London AP	43	02	81	09	912	−4	0	87/74	85/73	83/72	21	76	74	73		
North Bay AP	46	22	79	25	1210	−22	−18	84/68	81/67	79/66	20	71	70	68		
Oshawa	43	54	78	52	370	−6	−3	88/73	86/72	84/71	20	75	74	73		
Ottawa AP	45	19	75	40	413	−17	−13	90/72	87/71	84/70	21	75	73	72		
Owen Sound	44	34	80	55	597	−6	−2	84/71	82/70	80/69	21	73	72	70		
Peterborough	44	17	78	19	635	−13	−9	87/72	85/71	83/70	21	75	73	72		
St Catharines	43	11	79	14	325	−1	3	87/73	85/72	83/71	20	76	74	73		
Sarnia	42	58	82	22	625	0	3	88/73	86/72	84/71	19	76	74	73		
Sault Ste Marie AP	46	32	84	30	675	−7	−13	85/71	82/69	79/68	22	73	71	70		
Sudbury AP	46	37	80	48	1121	−22	−19	86/69	83/67	81/66	22	72	70	68		
Thunder Bay AP	48	22	89	19	644	−27	−24	85/70	83/68	80/67	24	72	70	68	W 8	W
Timmins AP	48	34	81	22	965	−33	−29	87/69	84/68	81/66	25	72	70	68		
Toronto AP	43	41	79	38	578	−5	−1	90/73	87/72	85/71	20	75	74	73	N 10	SW
Windsor AP	42	16	82	58	637	0	4	90/74	88/73	86/72	20	77	75	74		
PRINCE EDWARD ISLAND																
Charlottetown AP	46	17	63	08	186	−7	−4	80/69	78/68	76/67	16	71	70	68		
Summerside AP	46	26	63	50	78	−8	−4	81/69	79/68	77/67	16	72	70	68		
QUEBEC																
Bagotville AP	48	20	71	00	536	−28	−23	87/70	83/68	80/67	21	72	70	68		
Chicoutimi	48	25	71	05	150	−26	−22	86/70	83/68	80/67	20	72	70	68		
Drummondville	45	53	72	29	270	−18	−14	88/72	85/71	82/69	21	75	73	71		
Granby	45	23	72	42	550	−19	−14	88/72	85/71	83/70	21	75	73	72		
Hull	45	26	75	44	200	−18	−14	90/72	87/71	84/70	21	75	73	72		
Megantic AP	45	35	70	52	1362	−20	−16	86/71	83/70	81/69	20	74	72	71		
Montreal AP	45	28	73	45	98	−16	−10	88/73	85/72	83/71	17	75	74	72		
Quebec AP	46	48	71	23	245	−19	−14	87/72	84/70	81/68	20	74	72	70		
Rimouski	48	27	68	32	117	−16	−12	83/68	79/66	76/65	18	71	69	67		
St Jean	45	18	73	16	129	−15	−11	88/73	86/72	84/71	20	75	74	72		
St Jerome	45	48	74	01	556	−17	−13	88/72	86/71	83/70	23	75	73	72		
Sept. Iles AP	50	13	66	16	190	−26	−21	76/63	73/61	70/60	17	67	65	63		
Shawinigan	46	34	72	43	306	−18	−14	86/72	84/70	82/69	21	74	72	71		
Sherbrooke Co	45	24	71	54	595	−25	−21	86/72	84/71	81/69	20	74	73	71		
Thetford Mines	46	04	71	19	1020	−19	−14	87/71	84/70	81/69	21	74	72	71		
Trois Rivieres	46	21	72	35	50	−17	−13	88/72	85/70	82/69	23	74	72	71		
Val D'or AP	48	03	77	47	1108	−32	−27	85/70	83/68	80/67	22	72	70	68		
Valleyfield	45	16	74	06	150	−14	−10	89/73	86/72	84/71	20	75	74	72		
SASKATCHEWAN																
Estevan AP	49	04	103	00	1884	−30	−25	92/70	89/68	86/67	26	72	70	69		
Moose Jaw AP	50	20	105	33	1857	−29	−25	93/69	89/67	86/66	27	71	69	68		
North Battleford AP	52	46	108	15	1796	−33	−30	87/66	85/66	82/65	23	69	68	66		
Prince Albert AP	53	13	105	41	1414	−42	−35	87/67	84/66	81/65	25	70	68	67		
Regina AP	50	26	104	40	1884	−33	−29	91/69	88/68	84/67	26	72	70	68		
Saskatoon AP	52	10	106	41	1645	−35	−31	89/68	86/66	83/65	26	70	68	67		
Swift Current AP	50	17	107	41	2677	−28	−25	93/68	90/66	87/65	25	70	69	67		
Yorkton AP	51	16	102	28	1653	−35	−30	87/69	84/68	80/66	23	72	70	68		
YUKON TERRITORY																
Whitehorse AP	60	43	135	04	2289	−46	−43	80/59	77/58	74/56	22	61	59	58	NW 5	SE

[a] AP following the station name designates airport temperature observations. Co designates office locations within an urban area that are affected by the surrounding area. Undesignated stations are semirural and may be compared to airport data.

[b] Winter design data are based on the month of January only.

[c] Summer design data are based on the month of July only. See also Boughner (1960).

[d] Mean wind speeds occurring coincidentally with the 99.5% dry-bulb winter design temperature.

Table 3 Climatic Conditions for Other Countries

Col. 1 Country and Station	Col. 2 Lat. °	'N	Long. °	'W	Col. 3 Eleva-tion, ft	Col. 4 Winter, °F Mean of Annual Extremes	99%	97.5%	Col. 5 Design Dry-Bulb 1%	2.5%	5%	Col. 6 Mean Daily Range	Col. 7 Design Wet-Bulb 1%	2.5%	5%	Prevailing Wind Winter (Knots)		Summer
AFGHANISTAN																		
Kabul	34	35N	69	12E	5955	2	6	9	98	96	93	32	66	65	64	N	4	N
ALGERIA																		
Algiers	36	46N	3	03E	194	38	43	45	95	92	89	14	77	76	75			
ARGENTINA																		
Buenos Aires	34	35S	58	29W	89	27	32	34	91	89	86	22	77	76	75	SW	9	NNE
Cordoba	31	22S	64	15W	1388	21	28	32	100	96	93	27	76	75	74			
Tucuman	26	50S	65	10W	1401	24	32	36	102	99	96	23	76	75	74			
AUSTRALIA																		
Adelaide	34	56S	138	35E	140	36	38	40	98	94	91	25	72	70	68	NE	5	NW
Alice Springs	23	48S	133	53E	1795	28	34	37	104	102	100	27	75	74	72	N	6	SE
Brisbane	27	28S	153	02E	137	39	44	47	91	88	86	18	77	76	75	N	7	NNE
Darwin	12	28S	130	51E	88	60	64	66	94	93	91	16	82	81	81	E	10	WNW
Melbourne	37	49S	144	58E	114	31	35	38	95	91	86	21	71	69	68			
Perth	31	57S	115	51E	210	38	40	42	100	96	93	22	76	74	73	N	6	E
Sydney	33	52S	151	12E	138	38	40	42	89	84	80	13	74	73	72	N	8	NE
AUSTRIA																		
Vienna	48	15N	16	22E	644	−2	6	11	88	86	83	16	71	69	67	W	13	SSE
AZORES																		
Lajes	38	45N	27	05W	170	42	46	49	80	78	77	11	73	72	71	W	9	NW
BAHAMAS																		
Nassau	25	05N	77	21W	11	55	61	63	90	89	88	13	80	80	79			
BANGLADESH																		
Chittagong	22	21N	91	50E	87	48	52	54	93	91	89	20	82	81	81			
BELGIUM																		
Brussels	50	48N	4	21E	328	13	15	19	83	79	77	19	70	68	67	NE	8	ENE
BELIZE																		
Belize	17	31N	88	11W	17	55	60	62	90	90	89	13	82	82	81			
BERMUDA																		
Kindley AFB	33	22N	64	41W	129	47	53	55	87	86	85	12	79	78	78	NW	16	S
BOLIVIA																		
La Paz	16	30S	68	09W	12001	28	31	33	71	69	68	24	58	57	56			
BRAZIL																		
Belem	1	27S	48	29W	42	67	70	71	90	89	87	19	80	79	78	SE	5	E
Belo Horizonte	19	56S	43	57W	3002	42	47	50	86	84	83	18	76	75	75			
Brasilia	15	53S	47	56W	3481	46	52	54	86	84	84	23	71	70	70	W	1	E
Campinas	23	01S	47	08W	2169	—	52	54	91	90	88	23	76	75	74	ESE	7	N
Congonhas	23	38S	46	39W	2635	—	46	49	88	86	84	21	73	72	71	S	3	NNW
Cirotoba	25	25S	49	17W	3114	28	34	37	86	84	82	21	75	74	74			
Fortaleza	3	46S	38	33W	89	66	69	70	91	90	89	17	79	78	78			
Galeao	22	50S	43	15W	20	—	61	63	97	95	92	21	80	79	78	NW	2	SSE
Porto Alegre	30	02S	51	13W	33	32	37	40	95	92	89	20	76	76	75			
Recife	8	04S	34	53W	97	67	69	70	88	87	86	10	78	77	77	S	7	ESE
Rio de Janeiro	22	55S	43	12W	201	56	58	60	94	92	90	11	80	79	78	N	5	S
Salvador	13	00S	38	30W	154	65	67	68	88	87	86	12	79	79	78			
Sao Paulo	23	31S	46	37W	2598	36	47	50	89	87	85	14	74	73	72	E	6	NW
BULGARIA																		
Sofia	42	42N	23	20E	1805	−2	3	8	89	86	84	26	71	70	69			
CAMBODIA																		
Phnom Penh	11	33N	104	51E	36	62	66	68	98	96	94	19	83	82	82	N	4	W
CHILE																		
Concepcion	36	47S	73	04W	30	—	41	41	75	73	72	22	64	63	62	S	4	SW
Punta Arenas	53	10S	70	54W	26	22	25	27	68	66	64	14	56	55	54			
Santiago	33	24S	70	47W	1555	27	30	32	90	88	86	37	68	67	66	NE	1	SW
Valparaiso	33	01S	71	38W	135	39	43	46	81	79	77	16	67	66	65			
CHINA																		
Chungking	29	33N	106	33E	755	34	37	39	99	97	95	18	81	80	79			
Shanghai	31	12N	121	26E	23	16	23	26	94	92	90	16	81	81	80	WNW	6	S
COLOMBIA																		
Baranquilla	10	59N	74	48W	44	66	70	72	95	94	93	17	83	82	82			
Bogota	4	36N	74	05W	8406	42	45	46	72	70	69	19	60	59	58	E	8	E
Cali	3	25N	76	30W	3189	53	57	58	84	82	79	15	70	69	68			
Medellin	6	13N	75	36W	4650	48	53	55	87	85	84	25	73	72	72			

Table 3   Climatic Conditions for Other Countries (*Continued*)

Col. 1 Country and Station	Col. 2 Lat. °	′N	Long. °	′W	Col. 3 Elevation, ft	Winter, °F Mean of Annual Extremes	Col. 4 99%	97.5%	Col. 5 Design Dry-Bulb 1%	2.5%	5%	Col. 6 Mean Daily Range	Col. 7 Design Wet-Bulb 1%	2.5%	5%	Prevailing Wind Winter Knots	Summer Knots
COMMONWEALTH OF INDEPENDENT STATES																	
(formerly SOVIET UNION)																	
Alma Ata	43	14N	76	53E	2543	−18	−10	−6	88	86	83	21	69	68	67		
Archangel	64	33N	40	32E	22	−29	−23	−18	75	71	68	13	60	58	57		
Ekaterinburg (Sverdlousk)	56	49N	60	38E	894	−34	−25	−20	80	76	72	16	63	62	60		
Kaliningrad	54	43N	20	30E	23	−3	1	6	83	80	77	17	67	66	65		
Krasnoyarsk	56	01N	92	57E	498	−41	−23	−27	84	80	76	12	64	62	60		
Kiev	50	27N	30	30E	600	−12	−5	1	87	84	81	22	69	68	67		
Kharkov	50	00N	36	14E	472	−19	−10	−3	87	84	82	23	69	68	67		
Minsk	53	54N	27	33E	738	−19	−11	−4	80	77	74	16	67	66	65		
Moscow	55	46N	37	40E	505	−19	−11	−6	84	81	78	21	69	67	65	SW 11	S
Odessa	46	29N	30	44E	214	−1	4	8	87	84	82	14	70	69	68		
Petropavlovsk	52	53N	158	42E	286	−9	−3	0	70	68	65	13	58	57	56		
Rostov on Don	47	13N	39	43E	159	−9	−2	4	90	87	84	20	70	69	68		
Samara	53	11N	50	06E	190	−23	−19	−13	89	85	81	20	69	67	66		
St. Petersburg (Leningrad)	59	56N	30	16E	16	−14	−9	−5	78	75	72	15	65	64	63		
Tashkent	41	20N	69	18E	1569	−4	3	8	95	93	90	29	71	70	69		
Tbilisi	41	43N	44	48E	1325	12	18	22	87	85	83	18	68	67	66		
Vladivostok	43	07N	131	55E	94	−15	−10	−7	80	77	74	11	70	69	68		
Volgograd	48	42N	44	31E	136	−21	−13	−7	93	89	86	19	71	70	69		
CONGO																	
Brazzaville	4	15S	15	15E	1043	54	60	62	93	92	91	21	81	81	80		
CUBA																	
Guantanamo Bay	19	54N	75	09W	21	60	64	66	94	93	92	16	82	81	80	N 6	ESE
Havana	23	08N	82	21W	80	54	59	62	92	91	89	14	81	81	80	N 11	E
CYPRUS																	
Akrotiri	34	36N	32	59E	75	—	41	43	91	90	87	17	78	77	76	NNW 4	SW
Lainaca	34	53N	33	39E	7	—	37	41	93	91	90	21	78	77	76	NW 6	SSW
Paphos	34	43N	32	30E	26	—	42	45	88	86	86	13	79	79	78	NE 7	W
CZECHOSLOVAKIA																	
Prague	50	05N	14	25E	662	3	4	9	88	85	83	16	66	65	64		
DENMARK																	
Copenhagen	55	41N	12	33E	43	11	16	19	79	76	74	17	68	66	64	NE 11	N
DOMINICAN REPUBLIC																	
Santo Domingo	18	29N	69	54W	57	61	63	65	92	90	88	16	81	80	80	NNE 6	SE
EQUADOR																	
Guayaquil	2	10S	79	53W	20	61	64	65	92	91	89	20	80	80	79		
Quito	0	13S	78	32W	9446	30	36	39	73	72	71	32	63	62	62	N 3	N
EGYPT																	
Cairo	29	52N	31	20E	381	39	45	46	102	100	98	26	76	75	74	N 9	NNW
Luxor	25	40N	32	43E	289	—	38	41	109	108	106	31	76	74	73	E 1	N
EL SALVADOR																	
San Salvador	13	42N	89	13W	2238	51	54	56	98	96	95	32	77	76	75	N 7	S
ETHIOPIA																	
Addis Ababa	9	02N	38	45E	7753	35	39	41	84	82	81	28	66	65	64	E 10	S
Asmara	15	17N	38	55E	7628	36	40	42	83	81	80	27	65	64	63	E 9	WNW
FINLAND																	
Helsinki	60	10N	24	57E	30	−11	−7	−1	77	74	72	14	66	65	63	E 4	S
FRANCE																	
Lyon	45	42N	4	47E	938	−1	10	14	91	89	86	23	71	70	69	N 7	S
Marseilles	43	18N	5	23E	246	23	25	28	90	87	84	22	72	71	69	SE 14	W
Nantes	47	15N	1	34W	121	17	22	26	86	83	80	21	70	69	67	NNE 6	E
Nice	43	42N	7	16E	39	31	34	37	87	85	83	15	73	72	72		
Paris	48	49N	2	29E	164	16	22	25	89	86	83	21	70	68	67	NE 7	E
Strasbourg	48	35N	7	46E	465	9	11	16	86	83	80	20	70	69	67		
FRENCH GUIANA																	
Cayenne	4	56N	52	27W	20	69	71	72	92	91	90	17	83	83	82	ENE 5	E
GERMANY																	
Berlin	52	27N	13	18E	187	6	7	12	84	81	78	19	68	67	66	E 6	E
Hamburg	53	33N	9	58E	66	10	12	16	80	76	73	13	68	66	65		
Hannover	52	24N	9	40E	561	7	16	20	82	78	75	17	68	67	65	E 8	E
Mannheim	49	34N	8	28E	359	2	8	11	87	85	82	18	71	69	68	N 5	S
Munich	48	09N	11	34E	1729	−1	5	9	86	83	80	18	68	66	64	S 4	N
GHANA																	
Accra	5	33N	0	12W	88	65	68	69	91	90	89	13	80	79	79	WSW 5	SW
GIBRALTAR																	
Gibraltar	36	09N	5	22W	11	38	42	45	92	89	86	14	76	75	74		
GREECE																	
Athens	37	58N	23	43E	351	29	33	36	96	93	91	18	72	71	71	N 9	NNE
Souda	35	32N	24	09E	479	—	38	41	95	91	90	21	76	74	72	NNW 4	WNW
Thessaloniki	40	37N	22	57E	78	23	28	32	95	93	91	20	77	76	75		

Table 3 Climatic Conditions for Other Countries (*Continued*)

Col. 1 Country and Station	Col. 2 Lat. ° 'N	Col. 2 Long. ° 'W	Col. 3 Elevation, ft	Winter, °F Col. 4 Mean of Annual Extremes	Winter, °F Col. 4 99%	Winter, °F Col. 4 97.5%	Summer, °F Col. 5 Design Dry-Bulb 1%	Summer, °F Col. 5 Design Dry-Bulb 2.5%	Summer, °F Col. 5 Design Dry-Bulb 5%	Summer, °F Col. 6 Mean Daily Range	Summer, °F Col. 7 Design Wet-Bulb 1%	Summer, °F Col. 7 Design Wet-Bulb 2.5%	Summer, °F Col. 7 Design Wet-Bulb 5%	Prevailing Wind Winter Knots	Prevailing Wind Summer Knots
GREENLAND															
Narsarssuaq	61 11N	45 25W	85	−23	−12	−8	66	63	61	20	56	54	52		
GUATEMALA															
Guatemala City	14 37N	90 31W	4855	45	48	51	83	82	81	24	69	68	67	N 9	S
GUYANA															
Georgetown	6 50N	58 12W	6	70	72	73	89	88	87	11	80	79	79		
HAITI															
Port au Prince	18 33N	72 20W	121	63	65	67	97	95	93	20	82	81	80	N 6	ESE
HONDURAS															
Tegucigalpa	14 06N	87 13W	3094	44	47	50	89	87	85	28	73	72	71	N 8	E
HONG KONG															
Hong Kong	22 18N	114 10E	109	43	48	50	92	91	90	10	81	80	80	N 9	W
HUNGARY															
Budapest	47 31N	19 02E	394	8	10	14	90	86	84	21	72	71	70	N 5	S
ICELAND															
Reykjavik	64 08N	21 56E	59	8	14	17	59	58	56	16	54	53	53	E 12	E
INDIA															
Ahmenabad	23 02N	72 35E	163	49	53	56	109	107	105	28	80	79	78		
Bangalore	12 57N	77 37E	3021	53	56	58	96	94	93	26	75	74	74		
Bombay	18 54N	72 49E	37	62	65	67	96	94	92	13	82	81	81	NW	NW
Calcutta	22 32N	88 20E	21	49	52	54	98	97	96	22	83	82	82	N 4	S
Madras	13 04N	80 15E	51	61	64	66	104	102	101	19	84	83	83	W 3	W
Nagpur	21 09N	79 07E	1017	45	51	54	110	108	107	30	79	79	78		
New Delhi	28 35N	77 12E	703	35	39	41	110	107	105	26	83	82	82	N 6	NW
INDONESIA															
Djakarta	6 11S	106 50E	26	69	71	72	90	89	88	14	80	79	78	N 11	N
Kupang	10 10S	123 34E	148	63	66	68	94	93	92	20	81	80	80		
Makassar	5 08S	119 28E	61	64	66	68	90	89	88	17	80	80	79		
Medan	3 35N	98 41E	77	66	69	71	92	91	90	17	81	80	79		
Palembang	3 00S	104 46E	20	67	70	71	92	91	90	17	80	79	79		
Surabaya	7 13S	112 43E	10	64	66	68	91	90	89	18	80	79	79		
IRAN															
Abadan	30 21N	48 16E	7	32	39	41	116	113	110	32	82	81	81	W 6	WNW
Meshed	36 17N	59 36E	3104	3	10	14	99	96	93	29	68	67	66		
Tehran	35 41N	51 25E	4002	15	20	24	102	100	98	27	75	74	73	W 5	SE
IRAQ															
Baghdad	33 20N	44 24E	111	27	32	35	113	111	108	34	73	72	72	WNW 5	WNW
Mosul	36 19N	43 09E	730	23	29	32	114	112	110	40	73	72	72		
IRELAND															
Dublin	53 22N	6 21W	155	19	24	27	74	72	70	16	65	64	62	W 9	SW
Shannon	52 41N	8 55W	8	19	25	28	76	73	71	14	65	64	63	SE 4	W
IRIAN BARAT															
Manokwari	0 52S	134 05E	62	70	71	72	89	88	87	12	82	81	81		
ISRAEL															
Jerusalem	31 47N	35 13E	2485	31	36	38	95	94	92	24	70	69	69	W 12	NW
Tel Aviv	32 06N	34 47E	36	33	39	41	96	93	91	16	74	73	72	N 8	W
ITALY															
Milan	45 27N	9 17E	341	12	18	22	89	87	84	20	76	75	74	W 4	SW
Naples	40 53N	14 18E	220	28	34	36	91	88	86	19	74	73	72	N 6	SSW
Rome	41 48N	12 36E	377	25	30	33	94	92	89	24	74	73	72	E 6	WSW
IVORY COAST															
Abidjan	5 19N	4 01W	65	64	67	69	91	90	88	15	83	82	81	WSW 5	SW
JAMAICA															
Kingston	17 56N	76 47W	46	—	71	72	93	91	89	10	81	80	80	N 10	ESE
Montego	18 30N	77 55W	10	—	68	69	89	89	88	11	80	79	79	ESE 10	ENE
JAPAN															
Fukuoka	33 35N	130 27E	22	26	29	31	92	90	89	20	82	80	79		
Sapporo	43 04N	141 21E	56	−7	1	5	86	83	80	20	76	74	72	SE 3	SE
Tokyo	35 41N	139 46E	19	21	26	28	91	89	87	14	81	80	79	SW 10	S
JOHNSTON ISLAND	16 44N	169 31W	16	—	72	73	87	86	85	6	80	80	79	NE 15	E
JORDAN															
Amman	31 57N	35 57E	2548	29	33	36	97	94	92	25	70	69	68	N 6	NNW
KENYA															
Nairobi	1 16S	36 48E	5971	45	48	50	81	80	78	24	66	65	65	E 13	ENE
KOREA															
Pyongyang	39 02N	125 41E	186	−10	−2	3	89	87	85	21	77	76	76		
Seoul	37 34N	126 58E	285	−1	7	9	91	89	87	16	81	79	78	NW 7	W
LEBANON															
Beirut	33 54N	35 28E	111	40	42	45	93	91	90	15	78	77	76	N 7	SW
LIBERIA															
Monrovia	6 18N	10 48W	75	64	68	69	90	89	88	19	82	82	81	E 3	WSW
LIBYA															
Benghazi	32 06N	20 04E	82	41	46	48	97	94	91	13	77	76	75	SSE 8	S
MADAGASCAR															
Tananarive	18 55S	47 33E	4531	39	43	46	86	84	83	23	73	72	71		

Table 3 Climatic Conditions for Other Countries (*Continued*)

Col. 1 Country and Station	Col. 2 Lat. °	'N	Long. °	'W	Col. 3 Eleva-tion, ft	Winter, °F Mean of Annual Extremes	99%	97.5%	Col. 5 Design Dry-Bulb 1%	2.5%	5%	Col. 6 Mean Daily Range	Col. 7 Design Wet-Bulb 1%	2.5%	5%	Prevailing Wind Winter		Summer
MALAYSIA																		
Kuala Lumpur	3	07N	101	42E	127	67	70	71	94	93	92	20	82	82	81	N	4	W
Penang	5	25N	100	19E	17	69	72	73	93	93	92	18	82	81	80			
MARTINIQUE																		
Fort de France	14	37N	61	05W	13	62	64	66	90	89	88	14	81	81	80			
MEXICO																		
Guadalajara	20	41N	103	20W	5105	35	39	42	93	91	89	29	68	67	66	N	7	W
Merida	20	58N	89	38W	72	56	59	61	97	95	94	21	80	79	77	E	11	E
Mexico City	19	24N	99	12W	7575	33	37	39	83	81	79	25	61	60	59	N	8	N
Monterrey	25	40N	100	18W	1732	31	38	41	98	95	93	20	79	78	77			
Tampico	22	17N	97	52W	79	—	48	50	91	89	89	10	83	81	81	N	11	E
Vera Cruz	19	12N	96	08W	184	55	60	62	91	89	88	12	83	83	82			
MIDWAY ISLAND	28	13N	177	23W	10	—	56	58	87	86	86	9	78	77	76	NNW	9	E
MOROCCO																		
Casablanca	33	35N	7	39W	164	36	40	42	94	90	86	50	73	72	70			
MYANMAR																		
Mandalay	21	59N	96	06E	252	50	54	56	104	102	101	30	81	80	80			
Rangoon	16	47N	96	09E	18	59	62	63	100	98	95	25	83	82	82	W	6	W
NEPAL																		
Katmandu	27	42N	85	12E	4388	30	33	35	89	87	86	25	78	77	76	W	4	NW
NETHERLANDS																		
Amsterdam	52	23N	4	55E	5	17	20	23	79	76	73	10	65	64	63	S	8	E
NEW ZEALAND																		
Auckland	36	51S	174	46E	140	37	40	42	78	77	76	14	67	66	65			
Christ Church	43	32S	172	37E	32	25	28	31	82	79	76	17	68	67	66	W	4	NNW
Wellington	41	17S	174	46E	394	32	35	37	76	74	72	14	66	65	64	NE	6	NNE
NICARAGUA																		
Managua	12	10N	86	15W	135	62	65	67	94	93	92	21	81	80	79	E	9	E
NIGERIA																		
Lagos	6	27N	3	24E	10	67	70	71	92	91	90	12	82	82	81	WSW	8	S
NORWAY																		
Bergen	60	24N	5	19E	141	14	17	20	75	74	73	21	67	66	65			
Oslo	59	56N	10	44E	308	−2	0	4	79	77	74	17	67	66	64	N	10	S
PAKISTAN																		
Karachi	24	48N	66	59E	13	45	49	51	100	98	95	14	82	82	81	N	4	SSW
Lahore	31	35N	74	20E	702	32	35	37	109	107	105	27	83	82	81	NW	3	SE
Peshwar	34	01N	71	35E	1164	31	35	37	109	106	103	29	81	80	79	W	5	NE
PANAMA AND CANAL ZONE																		
Panama City	8	58N	79	33W	21	69	72	73	93	92	91	18	81	81	80			
PAPUA NEW GUINEA																		
Port Moresby	9	29S	147	09E	126	62	67	69	92	91	90	14	80	80	79			
PARAGUAY																		
Ascuncion	25	17S	57	30W	456	35	43	46	100	98	96	24	81	81	80	NE	7	NE
PERU																		
Lima	12	05S	77	03W	394	51	53	55	86	85	84	17	76	75	74	N	10	S
San Juan de Marcona	15	24S	75	10W	197	—	55	57	82	81	79	12	75	73	72	S	10	S
Talara	4	35S	81	15W	282	—	59	61	90	88	86	15	79	78	76	SSE	18	S
PHILIPPINES																		
Manila	14	35N	120	59E	47	69	73	74	94	92	91	20	82	81	81	N	3	ESE
POLAND																		
Krakow	50	04N	19	57E	723	−2	2	6	84	81	78	19	68	67	66			
Warsaw	52	13N	21	02E	394	−3	3	8	84	81	78	19	71	70	68	E	7	SE
PORTUGAL																		
Lisbon	38	43N	9	08W	313	32	37	39	89	86	83	16	69	68	67	ENE	5	N
PUERTO RICO																		
San Juan	18	29N	66	07W	82	65	67	68	89	88	87	11	81	80	79	ENE	10	E
RUMANIA																		
Bucharest	44	25N	26	06E	269	−2	3	8	93	91	89	26	72	71	70			
SAUDI ARABIA																		
Dhahran	26	17N	50	09E	80	39	45	48	111	110	108	32	86	85	84	N	8	N
Jedda	21	28N	39	10E	20	52	57	60	106	103	100	22	85	84	83			
Riyadh	24	39N	46	42E	1938	29	37	40	110	108	106	32	78	.77	76	N	8	N
SENEGAL																		
Dakar	14	42N	17	29W	131	58	61	62	95	93	91	13	81	80	80	N	8	NW
SINGAPORE																		
Singapore	1	18N	103	50E	33	69	71	72	92	91	90	14	82	81	80	N	4	SE
SOMALIA																		
Mogadiscio	2	02N	49	19E	39	67	69	70	91	90	89	12	82	82	81	SSW	16	E
SOUTH AFRICA																		
Cape Town	33	56S	18	29E	55	36	40	42	93	90	86	20	72	71	70			
Johannesburg	26	11S	78	03E	5463	26	31	34	85	83	81	24	70	69	69			
Pretoria	25	45S	28	14E	4491	27	32	35	90	87	85	23	70	69	68	N	4	W
SOUTH YEMEN																		
Aden	12	50N	45	02E	10	63	68	70	102	100	98	11	83	82	82			

Table 3 Climatic Conditions for Other Countries (*Concluded*)

Col. 1 Country and Station	Col. 2 Lat. °	'N	Long. °	'W	Col. 3 Eleva-tion, ft	Winter, °F Col. 4 Mean of Annual Extremes	99%	97.5%	Summer, °F Col. 5 Design Dry-Bulb 1%	2.5%	5%	Col. 6 Mean Daily Range	Col. 7 Design Wet-Bulb 1%	2.5%	5%	Prevailing Wind Winter	Knots	Summer
SPAIN																		
Barcelona	41	24N	2	09E	312	31	33	36	88	86	84	13	75	74	73	N	10	S
Madrid	40	25N	3	41W	2188	22	25	28	93	91	89	25	71	69	67	NNE	5	W
Valencia	39	28N	0	23W	79	31	33	37	92	90	88	14	75	74	73	W	7	ESE
SRI LANKA																		
Colombo	6	54N	79	52E	24	65	69	70	90	89	88	15	81	80	80	W	6	W
SUDAN																		
Khartoum	15	37N	32	33E	1279	47	53	56	109	107	104	30	77	76	75	N	6	NW
SURINAM																		
Paramaribo	5	49N	55	09W	12	66	68	70	93	92	90	18	82	82	81	NE	9	E
SWEDEN																		
Stockholm	59	21N	18	04E	146	3	5	8	78	74	72	15	64	62	60	W	4	S
SWITZERLAND																		
Zurich	47	23N	8	33E	1617	4	9	14	84	81	78	21	68	67	66			
SYRIA																		
Damascus	33	30N	36	20E	2362	25	29	32	102	100	98	35	72	71	70			
TAIWAN																		
Tainan	22	57N	120	12E	70	40	46	49	92	91	90	14	84	83	82	N	10	W
Taipei	25	02N	121	31E	30	41	44	47	94	92	90	16	83	82	81	E	7	E
TANZANIA																		
Dar es Salaam	6	50S	39	18E	47	62	64	65	90	89	88	13	82	81	81			
THAILAND																		
Bangkok	13	44N	100	30E	39	57	61	63	97	95	93	18	82	82	81	N	4	S
TRINIDAD																		
Port of Spain	10	40N	61	31W	67	61	64	66	91	90	89	16	80	80	79			
TUNISIA																		
Tunis	36	47N	10	12E	217	35	39	41	102	99	96	22	77	76	74	W	10	E
TURKEY																		
Adana	36	59N	35	18E	82	25	33	35	100	97	95	22	79	78	77			
Ankara	39	57N	32	53E	2825	2	9	12	94	92	89	28	68	67	66	N	8	W
Istanbul	40	58N	28	50E	59	23	28	30	91	88	86	16	75	74	73	N	10	NE
Izmir	38	26N	27	10E	16	24	27	29	98	96	94	23	75	74	73	NNE	8	N
UNITED KINGDOM																		
Belfast	54	36N	5	55W	24	19	23	26	74	72	69	16	65	64	62			
Birmingham	52	29N	1	56W	535	21	24	27	79	76	73	15	66	64	63			
Cardiff	51	28N	3	10W	203	21	24	27	79	76	73	14	64	63	62			
Edinburgh	55	55N	3	11W	441	22	25	28	73	70	68	13	64	62	61	WSW	6	WSW
Glasgow	55	52N	4	17W	85	17	21	24	74	71	68	13	64	63	61			
London	51	29N	0	00	149	20	24	26	82	79	76	16	68	66	65	W	7	E
URUGUAY																		
Montevideo	34	51S	56	13W	72	34	37	39	90	88	85	21	73	72	71	N	11	NNE
VENEZUELA																		
Caracas	10	30N	66	56W	3418	49	52	54	84	83	81	21	70	69	69	E	8	ENE
Maracaibo	10	39N	71	36W	20	69	72	73	97	96	95	17	84	83	83			
VIETNAM																		
Da Nang	16	04N	108	13E	23	56	60	62	97	95	93	14	86	86	85	NW	5	N
Hanoi	21	02N	105	52E	53	46	50	53	99	97	95	16	85	85	84			
Ho Chi Minh City	10	47N	106	42E	30	62	65	67	93	91	89	16	85	84	83			
YUGOSLAVIA																		
Belgrade	44	48N	20	28E	453	4	9	13	92	89	86	23	74	73	72	ESE	9	SE
ZAIRE																		
Kinshasa	4	20S	15	18E	1066	54	60	62	92	91	90	19	81	80	80	NNW	7	W
Kisangani	0	26S	15	14E	1370	65	67	68	92	91	90	19	81	80	80			

BIBLIOGRAPHY

Arctic Meteorology Research Group. 1960. Temperature and wind frequency tables of North America and Greenland, Vol. 1 and 2. *Meteorology* 24 and 25. McGill University, Quebec, Canada.

Arens, E.A., D.H. Nall, and W.L. Carroll. 1979. The representativeness of TRY data in predicting mean annual heating and cooling requirements. ASHRAE Seminar No. 8, Paper No. 1. Philadelphia.

ASHRAE. 1986. *Bin and degree hour weather data for simplified energy calculations.*

Atmospheric Environment Service. 1981. "Degree days 1951-1980." In *Canadian Climate Normals*, Vol. 4. Canadian Climate Centre, Atmospheric Environment Service, 4905 Dufferin Street, Downsview, Ontario.

Atmospheric Environment Service. 1983-1987. Principal Station Data. PSD 1 to PSD 134. Environment Canada, 4905 Dufferin Street, Downsview, Ontario, M3H 5T4.

Boughner, C.C. 1960. Canadian meteorological memoirs, No. 5. Meteorological Branch, Department of Transport, Toronto. (Now Environment Canada, Ontario.)

Crow, L.W. 1963. Study of weather design conditions. ASHRAE RP 23.

Crow, L.W. 1972. Weather data related to evaporative cooling. ASHRAE *Transactions* 78(1):153-64.

Crow, L.W. 1980. Development of hourly data for weather year for energy calculations (WYEC), including solar data, at 21 weather stations throughout the United States. ASHRAE *Transactions* 87(1):896-906; and *Development of hourly data for weather year for energy calculations* (WYEC), including solar data, at 29 stations throughout the United States and 5 stations in southern Canada. ASHRAE RP 364, Bulletin. November 1983.

Crow, L.W. 1985. Design weather data for the irregular terrain in New Mexico. Seminar No. 2, Paper No. 2. Hawaii.

Ecodyne Cooling Products. 1980. *Weather data handbook for HVAC and cooling equipment design*, 1st ed. McGraw Hill Book Co., New York.

Golden Gate and Southern California Chapters, ASHRAE. 1982. Climatic Data for Region X (Arizona, California, Hawaii, and Nevada), 5th ed. ASHRAE *Special Publication* SPCDX.

Insolation Data Manual—SERI/SP—755-789, October, 1980. Solar Energy Research Institute, Boulder, CO.

Kusuda, T. 1968. Least squares analysis of earth temperature cycles for selected stations in the contiguous United States. National Bureau of Standards Report 9493 (January).

Kusuda, T. 1971. Earth temperatures beneath five different surfaces. National Bureau of Standards Report 10373 (February).

MacLaren, Hooper and Angus, Hay, and Davies. 1980. *Define, develop and establish a merged solar and meteorological computer data base.* Canadian Climate Centre Report 80-8.

Meteorological Office. 1958. Tables of temperature, relative humidity and precipitation for the world, Parts I-IV. M.O. 617 AF. London.

Nall, D.H. and E.A. Arens. 1979. The influence of degree-day base temperature on residential building energy prediction. ASHRAE *Transactions* 85(1):722-35.

National Climatic Data Center. 1981. SOLMET *Users manual.* TD-9734.

National Climatic Data Center. 1992a. *Climatography of the U.S. #81.* Monthly normals of temperature, precipitation and heating and cooling degree-days. Federal Building, Asheville, NC 28801.

National Climatic Data Center. 1992b. *Climatography of the U.S. #81.* Annual degree-days to selected bases (1961-90). Federal Building, Asheville, NC 28801.

Rocky Mountain Chapter ASHRAE. 1976. Climate Data for Air Conditioning Design, Rocky Mountain Chapter Region (Colorado, Wyoming, Montana, and Environs). Also *Supplements* 1978 and 1980.

Sandia National Laboratories. 1988. A comparison of typical meteorological year solar radiation information with the SOLMET data base. SAND87-2379 UC-276. Sandia National Laboratories, Albuquerque, NM.

Snelling, H.J. 1985. Duration study for heating and air-conditioning design temperatures. Seminar No. 2, Paper No. 5. Hawaii.

USAF. 1978. *Engineering weather data.* AFM 88-29. Departments of the Air Force, the Army, and the Navy. Washington, D.C.

CHAPTER 25

RESIDENTIAL COOLING AND HEATING LOAD CALCULATIONS

The procedures described in Chapter 26 may be used to calculate a heating or cooling load for residential buildings. However, this chapter covers the engineering basis of modified residential load calculation procedures for the non-engineer.

RESIDENTIAL FEATURES

The unique features distinguishing residences from other types of buildings, with respect to heating and cooling load calculation and equipment sizing, are:

- Unlike many other structures, residences are usually occupied, and conditioned, for 24 h a day, virtually every day of the cooling and the heating season.
- Residential system loads are primarily imposed by heat loss or gain through structural components and by air leakage or ventilation. Internal loads, particularly those from occupants and lights, are small in comparison to those in commercial or industrial installations.
- Most residences are conditioned as a single zone. Unit capacity cannot be redistributed from one area to another as loads change from hour to hour; however, exceptions do occur.
- Most residential cooling systems use units of relatively small capacity (about 18,000 to 60,000 Btu/h cooling, 60,000 to 110,000 Btu/h heating). Since loads are largely affected by outside conditions and few days each season are design days, only a partial load exists during most of the season; thus, an oversized unit is detrimental to good system performance, especially for cooling in areas of high wet-bulb temperature.
- Dehumidification occurs only during cooling unit operation, and space condition control is usually restricted to use of room thermostats (sensible heat-actuated devices).
- Multifamily living units are similar to single-family, detached houses, but each living unit may not have surfaces exposed in all directions. This impacts load calculation.

Categories

Single-family detached. A house in this category usually has exposed walls in four directions, often more than one story, and a roof. The cooling system is a single-zone, unitary system with a single thermostat. Two-story houses may have separate cooling systems for each floor. The rooms are reasonably open and generally have a centralized air return. In this configuration, air and load from rooms is mixed, and a load-leveling effect, which requires a different distribution of air to each room than a pure commercial system, results. Since the amount of air supplied to each room is based on the load for that room, proper load calculation procedures must be used.

Multifamily buildings. Unlike single-family detached units, multifamily units do not have exposed surfaces facing in all directions. Rather, each unit has only one or two exposed surfaces and possibly a roof. Two exposed walls will be at right angles, and both east and west walls will not be exposed in a given living unit. Each living unit has a single unitary cooling system or a single fan coil unit, and the rooms are relatively open to one another. This configuration does not have the same load-leveling effect as a single-family detached house, and it is not a commercial building. Therefore, a specific load calculation procedure is required.

Other categories. Many buildings do not fall into either of the above categories. Critical to the designation of a single-family detached building is the presence of both east and west walls. Therefore, some multifamily structures should be treated as single-family detached when the exposed surfaces are oriented in a particular way. Examples include duplexes or apartments with exposed (1) east, west, and south, or (2) east, west, and north walls, with or without a roof; and apartments, town houses, or condominiums with only east and west or north and south exposed walls.

COOLING LOAD

COOLING LOAD COMPONENTS

A cooling load calculation determines total sensible cooling load due to heat gain (1) through structural components (walls, floors, and ceilings); (2) through windows; (3) caused by infiltration and ventilation; and (4) due to occupancy. The latent portion of the cooling load is evaluated separately. While the entire structure may be considered a single zone, equipment selection and system design should be based on a room-by-room calculation. To properly design the distribution system, the amount of conditioned air required by each room must be known.

Peak Load Computation

To select a properly sized cooling unit, the peak or maximum load (block load) for each zone must be computed. Since this procedure may vary considerably for different types of buildings, each building type has to be considered; the block load for a single-family detached house with one central system is the sum of all the room loads. If the house has a separate system for each zone, each zone block load (i.e., the sum of the loads for all rooms in each zone) is required. When a house is zoned with one central cooling system, the block load must be computed for the complete house as if it were one zone. In multifamily structures, each living unit has a zone load that equals the sum of the room loads. For apartments with separate systems, the block load for each unit establishes the system size. Apartment buildings with a central

---

The preparation of this chapter is assigned to TC 4.1, Load Calculation Data and Procedures.

cooling system (*i.e.*, a hydronic system with fan coils in each apartment) require a block load calculation for the complete structure to size the central system; each unit load establishes the size of the fan coil and air distribution system for each apartment. One of the methods discussed in Chapter 26 may be used for the block load.

Indoor Temperature Swing

As demonstrated in the section Transfer Function Method in Chapter 26 for hour-by-hour load calculations, allowing for a swing in indoor temperature results in lower peak loads. Since the indoor temperature does swing, such an allowance gives a more reasonable equipment capacity. The tables in this section are based on an assumed indoor temperature swing of not more than 3 °F on a design day, when the residence is conditioned for 24 h per day and the thermostat is set at 75 °F.

Cooling Load Due to Heat Gain through Structure

The sensible cooling load due to heat gains through the walls, floors, and ceilings of each room is calculated using appropriate cooling load temperature differences and U-values for summer conditions (Tables 1 and 2). For ceilings under naturally vented attics or beneath vented flat roofs, use the combined U-value for the roof, vented space, and ceiling. The mass of the walls is a variable in Table 2 and is important in calculating energy use, but it is not used in Table 1 because of the averaging technique required to develop the CLTDs. Values in Tables 1 and 2 assume a dark color, since color is an unpredictable variable in any residence.

Daily range (outdoor temperature swing on a design day) significantly affects the equivalent temperature difference. Tables 1 and 2 list daily temperature ranges classified as high, medium, and low. Tables 1, 2, and 3 in Chapter 24 list outdoor daily ranges for different locations.

Cooling Load Due to Heat Gain through Windows

Direct application of procedures for calculating cooling load due to heat gain for flat glass (discussed in Chapters 26 and 27) results in unrealistically high cooling loads for residential installations. Window glass load factors (GLF) modified for single and multifamily residential cooling load calculations, and including solar heat load plus air-to-air conduction, are given in Tables 3 and 4. Table 5 lists the shading coefficients and U-value used to compile Tables 3 and 4.

In application, the area of each window is multiplied by the appropriate GLF. The effects of permanent outside shading devices should be considered separately in determining the cooling load. Shaded glass is considered the same as north-facing glass. The shade line factor (SLF) is the ratio of the distance a shadow falls beneath the edge of an overhang to the width of the overhang (Table 6). Therefore, assuming the overhang is at the top of the window, the shade line equals the SLF times the overhang width. The shaded and sunlit glass areas may then be computed separately. The tabulated values are the average of the shade line values for 5 h of maximum solar intensity on each wall orientation shown. Northeast- and northwest-facing windows are not effectively protected by roof overhangs; in most cases, they should not be considered as shaded.

Infiltration

Natural air leakage in residential structures is less in summer than in winter, largely because wind velocities are lower in most localities. The data in Tables 7 and 8 showing space air changes per hour (ACH) apply to both single- and multifamily housing, although most of the raw data was for single-family structures (McQuiston 1984). Construction may be defined as follows:

Tight. Good multifamily construction with close-fitting doors, windows, and framing is considered tight. New houses with full

Table 1 CLTD Values for Single-Family Detached Residences[a]

	Design Temperature, °F											
	85		90			95			100		105	110
Daily Temp. Range[b]	L	M	L	M	H	L	M	H	M	H	M	H
All walls and doors												
North	8	3	13	8	3	18	13	8	18	13	18	23
NE and NW	14	9	19	14	9	24	19	14	24	19	24	29
East and West	18	13	23	18	13	28	23	18	28	23	28	33
SE and SW	16	11	21	16	11	26	21	16	26	21	26	31
South	11	6	16	11	6	21	16	11	21	16	21	26
Roofs and ceilings												
Attic or flat built-up	42	37	47	42	37	51	47	42	51	47	51	56
Floors and ceilings												
Under conditioned space, over unconditioned room, over crawl space	9	4	12	9	4	14	12	9	14	12	14	19
Partitions												
Inside or shaded	9	4	12	9	4	14	12	9	14	12	14	19

[a]Cooling load temperature differences (CLTDs) for single-family detached houses, duplexes, or multifamily, with both east and west exposed walls or only north and south exposed walls, °F.
[b]L denotes low daily range, less than 16 °F; M denotes medium daily range, 16 to 25 °F; and H denotes high daily range, greater than 25 °F.

Table 2 CLTD Values for Multifamily Residences[a]

		Design Temperature, °F										
		85		90			95			100		105	110
Daily Temp. Range[b]		L	M	L	M	H	L	M	H	M	H	M	H
Walls and doors													
N	LW	14	11	19	16	12	24	21	17	26	22	27	32
	MW	13	10	18	15	11	23	20	16	25	21	26	31
	HW	9	6	15	11	7	20	16	12	21	17	22	27
NE	LW	23	17	28	22	17	33	27	22	32	26	31	36
	MW	20	15	25	20	16	30	25	21	29	25	29	34
	HW	16	12	21	17	13	26	22	18	26	22	26	31
E	LW	32	27	37	32	27	43	38	32	42	37	42	47
	MW	30	24	34	29	24	40	34	29	39	33	39	44
	HW	23	18	28	23	18	34	29	23	33	28	33	38
SE	LW	31	27	35	31	26	41	37	31	42	37	42	47
	MW	28	22	32	27	22	37	32	27	37	33	38	43
	HW	21	16	26	22	17	32	27	22	31	27	32	37
S	LW	25	22	29	26	22	35	31	26	36	32	37	43
	MW	22	18	26	22	18	31	26	22	31	27	32	38
	HW	16	11	20	16	12	26	21	17	26	21	27	33
SW	LW	39	36	44	40	35	50	46	40	51	47	52	58
	MW	33	29	37	34	29	44	40	35	45	40	46	52
	HW	23	18	28	24	19	36	31	25	35	30	36	42
W	LW	44	41	48	45	40	54	51	46	56	52	57	63
	MW	37	33	41	38	33	46	42	38	48	43	49	55
	HW	26	22	31	27	23	37	32	27	37	32	38	44
NW	LW	33	30	37	34	30	43	39	34	44	40	45	50
	MW	28	25	32	29	24	37	33	29	39	35	40	45
	HW	20	16	25	20	16	31	26	21	31	26	32	37
Roof and ceiling													
Attic or N, S, W		58	53	65	60	55	70	65	60	70	65	72	77
Built-up East		21	18	23	21	18	25	23	21	25	23	25	28
Floors and ceiling													
Under or over unconditioned space, crawl space		9	4	12	9	4	14	12	9	14	12	14	19
Partitions													
Inside or shaded		9	4	12	9	4	14	12	9	14	12	14	19

[a]Cooling load temperature differences (CLTDs) for multifamily low-rise or single-family detached if zoned with separate temperature control for each zone, °F.
[b]L denotes low daily range, less than 16 °F; M denotes medium daily range, 16 to 25 °F; and H denotes high daily range, greater than 25 °F.

Table 3 Window Glass Load Factors (GLF) for Single-Family Detached Residences[a]

Design Temperature, °F	Regular Single Glass						Regular Double Glass						Heat-Absorbing Double Glass						Clear Triple Glass		
	85	90	95	100	105	110	85	90	95	100	105	110	85	90	95	100	105	110	85	90	95
No inside shading																					
North	34	36	41	47	48	50	30	30	34	37	38	41	20	20	23	25	26	28	27	27	30
NE and NW	63	65	70	75	77	83	55	56	59	62	63	66	36	37	39	42	44	44	50	50	53
E and W	88	90	95	100	102	107	77	78	81	84	85	88	51	51	54	56	59	59	70	70	73
SE and SW[b]	79	81	86	91	92	98	69	70	73	76	77	80	45	46	49	51	54	54	62	63	65
South[b]	53	55	60	65	67	72	46	47	50	53	54	57	31	31	34	36	39	39	42	42	45
Horizontal skylight	156	156	161	166	167	171	137	138	140	143	144	147	90	91	93	95	96	98	124	125	127
Draperies, venetian blinds, translucent roller shades fully drawn																					
North	18	19	23	27	29	33	16	16	19	22	23	26	13	14	16	18	19	21	15	16	18
NE and NW	32	33	38	42	43	47	29	30	32	35	36	39	24	24	27	29	29	32	28	28	30
E and W	45	46	50	54	55	59	40	41	44	46	47	50	33	33	36	38	38	41	39	39	41
SE and SW[b]	40	41	46	49	51	55	36	37	39	42	43	46	29	30	32	34	35	37	35	36	38
South[b]	27	28	33	37	38	42	24	25	28	31	31	34	20	21	23	25	26	28	23	24	26
Horizontal skylight	78	79	83	86	87	90	71	71	74	76	77	79	58	59	61	63	63	65	69	69	71
Opaque roller shades, fully drawn																					
North	14	15	20	23	25	29	13	14	17	19	20	23	12	12	15	17	17	20	13	13	15
NE and NW	25	26	31	34	36	40	23	24	27	30	30	33	21	22	24	26	27	29	23	23	26
E and W	34	36	40	44	45	49	32	33	36	38	39	42	29	30	32	34	35	37	32	32	35
SE and SW[b]	31	32	36	40	42	46	29	30	33	35	36	39	26	27	29	31	32	34	29	29	31
South[b]	21	22	27	30	32	36	20	20	23	26	27	30	18	19	21	23	24	26	19	20	22
Horizontal skylight	60	61	64	68	69	72	57	57	60	62	63	65	52	52	55	57	57	59	56	57	59

[a]Glass load factors (GLFs) for single-family detached houses, duplexes, or multi-family, with both east and west exposed walls or only north and south exposed walls, Btu/h·ft^2.

[b]Correct by +30% for latitude of 48° and by −30% for latitude of 32°. Use linear interpolation for latitude from 40 to 48° and from 40 to 32°

To obtain GLF for other combinations of glass and/or inside shading: $GLF_a = (SC_a/SC_t)(GLF_t - U_t D_t) + U_a D_t$, where the subscripts a and t refer to the alternate and table values, respectively. SC_t and U_t are given in Table 5. $D_t = (t_a - 75)$, where $t_a = t_o - (DR/2)$; t_o is the outdoor design temperature and DR is the daily range.

Table 4 Window Glass Load Factors (GLF) for Multifamily Residences

Design Temperature, °F	Regular Single Glass						Regular Double Glass						Heat-Absorbing Double Glass						Clear Triple Glass		
	85	90	95	100	105	110	85	90	95	100	105	110	85	90	95	100	105	110	85	90	95
No inside shading																					
North	40	44	49	54	58	64	34	36	39	42	44	47	23	24	26	29	30	33	30	32	34
NE	88	89	91	95	97	100	78	79	80	83	84	85	52	52	53	55	55	57	71	71	73
East	136	137	139	142	144	147	120	121	122	125	126	127	79	79	81	83	83	84	109	109	111
SE	129	130	134	139	141	144	109	113	116	119	120	122	72	75	77	79	79	81	99	103	105
South[b]	88	91	96	101	105	110	76	78	81	84	86	89	50	52	54	56	58	60	68	70	72
SW	154	159	164	169	174	179	134	137	140	143	145	148	89	91	93	95	97	99	121	123	125
West	174	178	183	188	192	197	151	154	157	160	162	165	100	102	104	106	108	110	137	139	141
NW	123	127	132	137	141	147	107	109	112	115	117	121	71	72	75	77	79	81	96	98	100
Horizontal	249	252	256	261	264	268	218	220	223	226	228	230	144	146	148	150	152	154	198	200	202
Draperies, venetian blinds, translucent roller shades, fully drawn																					
North	21	25	29	33	36	40	18	21	23	26	28	31	15	17	19	21	23	25	17	19	21
NE	43	44	46	50	51	52	39	40	41	44	45	46	33	33	34	36	36	37	39	39	40
East	67	68	70	74	75	76	61	62	63	65	66	67	50	50	51	54	54	55	60	60	61
SE	64	65	69	73	74	77	58	59	61	63	64	66	48	48	50	52	52	54	57	57	59
South[b]	45	48	52	56	59	63	40	42	44	47	49	52	33	34	36	39	40	42	38	40	42
SW	79	83	87	91	94	98	70	72	75	78	80	83	57	59	62	64	66	68	68	69	71
West	89	92	96	100	103	107	79	81	84	86	88	91	65	66	69	71	72	75	76	78	80
NW	63	66	70	74	77	81	56	58	61	63	66	68	46	48	50	52	54	56	54	55	57
Horizontal	126	128	132	135	137	141	113	115	117	120	121	124	93	94	96	98	100	102	110	111	113
Opaque roller shades, fully drawn																					
North	17	21	25	29	32	36	15	17	20	23	25	28	14	15	18	20	22	24	15	16	18
NE	33	34	35	39	40	42	31	32	33	36	35	37	29	28	30	32	32	34	32	31	33
East	51	52	53	57	61	65	48	49	50	53	52	55	45	45	46	48	48	49	49	49	50
SE	49	50	53	57	58	61	46	47	49	52	52	55	42	43	45	47	47	49	46	46	48
South[b]	35	38	42	46	49	53	32	34	37	40	42	42	29	31	33	35	37	39	32	33	35
SW	61	65	69	73	77	81	57	59	62	65	67	70	52	54	56	58	60	62	56	58	60
West	68	71	75	80	83	87	64	66	68	71	73	76	58	60	62	64	66	68	63	64	66
NW	49	52	56	60	63	67	45	47	50	53	55	58	41	43	45	47	49	51	45	46	48
Horizontal	97	99	102	106	108	111	91	93	95	97	99	102	83	85	87	89	90	92	90	92	93

[a]Glass load factors (GLFs) for multifamily low-rise or single-family detached if zoned with separate temperature control for each zone, Btu/h·ft^2.

[b]Correct by +30% for latitude of 48° and by −30% for latitude of 32°. Use linear interpolation for latitude from 40 to 48° and from 40 to 32°.

To obtain GLF for other combinations of glass and/or inside shading, $GLF_a = (SC_a/SC_t)(GLF_t - U_t D_t) + U_a D_t$, where the subscripts a and t refer to the alternate and table values, respectively. SC_t and U_t are given in Table 5. $D_t = (t_a - 75)$, where $t_a = t_o - (DR/2)$; t_o is the outdoor design temperature and DR is the daily range.

Table 5 Shading Coefficients for Residential Windows

Glass Type	None		Inside Shade Drapery, Venetian Blind, or Translucent Roller Shade		Opaque Roller Shade	
	SC	U	SC	U	SC	U
Single	1.00	1.04	0.50	0.81	0.38	0.81
Double	0.88	0.61	0.45	0.55	0.36	0.55
Heat absorbing	0.58	0.45	0.37	0.44	0.33	0.44
Triple	0.80	0.44	0.44	0.40	0.36	0.40

U is in Btu/h·ft^2·°F

Table 6 Shade Line Factors (SLF)

Direction Window Faces	Latitude, Degrees N						
	24	32	36	40	44	48	52
East	0.8	0.8	0.8	0.8	0.8	0.8	0.8
SE	1.8	1.6	1.4	1.3	1.1	1.0	0.9
South	9.2	5.0	3.4	2.6	2.1	1.8	1.5
SW	1.8	1.6	1.4	1.3	1.1	1.0	0.9
West	0.8	0.8	0.8	0.8	0.8	0.8	0.8

Shadow length below the overhang equals the shade line factor times the overhang width. Values are averages for the 5 h of greatest solar intensity on August 1.

vapor retardant, no fireplace, well-fitted windows, weather-stripped doors, one story, and less than 1500 ft^2 floor area fall into this category.

Medium. Medium structures include new, two-story frame houses or one-story houses more than 10 years old with average maintenance, a floor area greater than 1500 ft^2, average fit windows and doors, and a fireplace with damper and glass closure. Below-average multifamily construction falls in this category.

Loose. Loose structures are poorly constructed single- and multifamily residences, with poorly fitted windows and doors. Examples include houses more than 20 years old, of average maintenance, a fireplace without damper or glass closure, or more than an average number of vented appliances. Average mobile homes are in this category.

Ventilation

Residential air-conditioning systems may introduce outdoor air, although it is not a code requirement in most localities. Positive ventilation should be considered, however, if the anticipated infiltration just discussed is less than about 0.5 air changes per hour. When positive means of introducing outdoor air are used, controls, either manual or automatic, should be provided, and an energy recovery device should be considered.

Occupancy

Even though occupant density is low, occupancy loads should be estimated. Sensible heat gain per sedentary occupant is assumed to be 230 Btu/h. To prevent gross oversizing, the number of occupants should not be overestimated. Recent census studies recommend that the total number of occupants be based on two persons for the first bedroom, plus one person for each additional bedroom. The occupancy load should then be distributed equally among the living areas, because the maximum load occurs when most of the residents occupy these areas.

Household Appliances

Appliance loads are concentrated mainly in the kitchen and laundry areas. Based on contemporary living conditions in single-family houses, a sensible load of 1600 Btu/h should be divided between the kitchen and/or laundry and the adjoining room or

Table 7 Air Change Rates as Function of Airtightness

Class	Outdoor Design Temperatures, °F									
	50	40	30	20	10	0	−10	−20	−30	−40
Tight	0.41	0.43	0.45	0.47	0.49	0.51	0.53	0.55	0.57	0.59
Medium	0.69	0.73	0.77	0.81	0.85	0.89	0.93	0.97	1.00	1.05
Loose	1.11	1.15	1.20	1.23	1.27	1.30	1.35	1.40	1.43	1.47

Values for 15 mph wind and indoor temperature of 68°F.

Table 8 Air Change Rates as Function of Outdoor Design Temperatures

Class	Outdoor Design Temperature, °F					
	85	90	95	100	105	110
Tight	0.33	0.34	0.35	0.36	0.37	0.38
Medium	0.46	0.48	0.50	0.52	0.54	0.56
Loose	0.68	0.70	0.72	0.74	0.76	0.78

Values for 7.5 mph wind and indoor temperature of 75°F.

rooms. For multifamily units, the sensible heat gain values should be about 1200 Btu/h. These values assume that the cooking range and clothes drier are vented. Further allowances should be considered when unusual lighting intensities, computers, or other equipment are present.

Air Distribution System—Heat Loss/Gain

Whenever the air distribution system is outside the conditioned space, *i.e.*, in attics, crawl spaces, or other unconditioned spaces, heat loss or gains to the ducts or pipes must be included in the calculated load and should be considered in equipment selection.

Latent Heat Sources

The latent cooling load has three main sources: outdoor air, occupants, and miscellaneous sources, such as cooking, laundry, and bathing. These miscellaneous latent loads are largely covered by outdoor air, because most residences have exhaust fans and clothes driers that vent most of the moisture from these sources. This vent air is accounted for in the infiltration calculation. McQuiston (1984) estimated latent load factors for typical houses located in geographic regions, ranging from very dry to very wet, using the transfer function method (Figure 1). A latent factor (LF) (LF = 1/SHF) of 1.3 or a SHF (SHF = sensible load/total load) of 0.77 matches the performance of typical residential vapor compression cooling systems. Homes in almost all other regions of North America have cooling loads with a SHF greater than 0.77 and latent factors less than 1.3. Figure 1 may be used to estimate the total cooling load by reading LF as a function of the design humidity ratio and airtightness. Then $q_{total} = (LF)q_{sensible}$. If the humidity ratio is less than 0.01, set LF = 1.0.

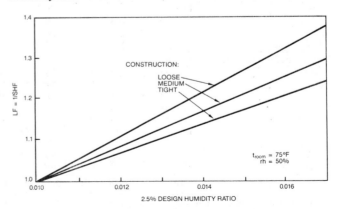

Fig. 1 Effect of Infiltration on Latent Load Factor

SELECTING COOLING EQUIPMENT

Residential cooling equipment that is flexible in application helps match the space SHF to the system. Usually, equipment is designed for a standard operating condition of 80 °F db and 67 °F wb air entering the coil, an airflow of 400 cfm per ton of refrigeration, and a SHF of 0.77 (LF = 1.3). Most regions require a coil SHF greater than 0.77; therefore, a mismatch is not unusual and can be acceptable. A coil designed with a given latent capacity generally transfers more sensible heat in the absence of moisture in the air, but not in the same amount as the design latent heat transfer capacity.

Cooling equipment should be sized on the basis of the estimated sensible cooling load, taking into account the effect of outdoor ambient conditions on the unit performance. The unit should not be oversized more than 25% of the estimated sensible load or the next available size. After selecting a unit, its latent capacity should be checked to ensure that it is equal to or greater than the calculated latent cooling load. The latent capacity of the unit will generally be ample except in very wet climates with excessive infiltration. If the latent cooling capacity is less than the calculated latent load, the unit size must be increased.

The cooling load calculation procedures are summarized in Table 9.

Example 1. A single-family detached house (Figure 2) is located in a southwestern state at 36 °N latitude.

Roof construction. Conventional roof-attic-ceiling combination, vented to remove moisture with 6 in. of fibrous batt insulation and vapor retardant ($U = 0.05$ Btu/h · ft^2 · °F).

Wall construction. Frame with 4 in. face brick, 3.5 in. fibrous batt insulation, 0.75 in. polystyrene sheathing, and 0.5 in. gypsum plaster board ($U = 0.06$ Btu/h · ft^2 · °F).

Floor construction. 4 in. concrete slab on grade.

Fenestration. Clear double glass, 0.125 in. thick, in and out. Assume closed, medium color venetian blinds. The window glass has a 2-ft overhang at the top.

Doors. Solid core flush with all-glass storm doors ($U = 0.32$ Btu/ h · ft^2 · °F).

Outdoor design conditions. Temperature of 96 °F dry bulb with a 24 °F daily range and a humidity ratio of 0.0136 lb vapor/lb dry air (74.6 °F wet bulb).

U-values for all external surfaces are based on a 7.5 mph wind velocity.

Indoor design conditions. Temperature of 75 °F dry bulb and 50% rh.

Occupancy. Four persons, based on two for the master bedroom and one for each additional bedroom. Assign to the living room.

Appliances and lights. Assume 1600 Btu/h for the kitchen, and assign 50% to the living room. Assume 1600 Btu/h for the utility room, and assign 25% to the kitchen and 25% to the storage room.

The conditioning equipment is located in the garage, and the construction of the house is considered average.

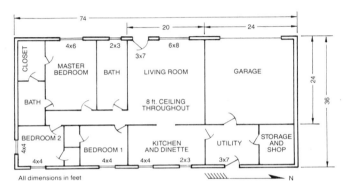

Fig. 2   Floor Plan of Single-Family Detached House

Table 9   Summary of Procedures for Residential Cooling Load Calculations

Load Source	Equation	Tables and Notes
Glass and window areas	$q = (GLF)A$	Glass load factors may be found in Tables 3 and 4 according to window orientation, type of glass, type of interior shading, and outdoor design temperature. The GLF includes effects of both transmission and solar radiation. Glass shaded by overhangs is treated as north glass. Table 6 gives shade line factors.
Doors	$q = U_w A (CLTD)$	Door CLTD are in Tables 1 and 2 according to orientation, outdoor design temperature, and design daily temperature range.
Above-grade exterior walls	$q = U_w A (CLTD)$	Wall CLTD values are in Tables 1 and 2 based on the outdoor design temperature, daily range, and orientation.
Partitions to unconditioned space	$q = U_p A \Delta t$	Where Δt is the temperature difference across the partition.
Ceilings and roofs	$q = U_r A (CLTD)$	Tables 1 and 2 for CLTD, based on outdoor design temperature and daily range.
Exposed floors	$q = U_f A (CLTD)$	Tables 1 and 2 for CLTD, according to outdoor design temperature and daily range.
Infiltration	$q = 1.1 Q\Delta t$ $Q = \text{ACH (room volume)}/60$	Air change rates are given in Tables 7 and 8.
Internal loads— People, appliances, lights	Plan 230 Btu/h per person.	Divide occupants evenly among rooms not used as bedrooms. If number of occupants is not known, assume two people for first bedroom and one person for each additional bedroom.
		The appliance and light load of 1600 Btu/h is divided between the kitchen and adjoining room and the laundry and adjoining room. Use 1200 Btu/h for multifamily units.
Total loads	Total cooling load = LF (Sum of individual sensible cooling load components)	Load factors are from Figure 1 according to outdoor design humidity ratio and airtightness classification.

q = sensible cooling load, Btu/h	Q = volumetric airflow rate, ft^3/min
Δt = design temperature difference between outside and inside air, °F	ACH = air changes per hour, 1/h
A = area of applicable surface, ft^2	GLF = glass load factor, Btu/h · ft^2
U = U-factors for appropriate construction, Btu/h · ft^2 · °F	CLTD = cooling load temperature difference, °F
	LF = latent load multiplier

Find the sensible, latent, and total cooling load; size the cooling unit; and compute the air quantity for each room.

Solution: The cooling load must be made on a room-by-room basis to determine the proper distribution of air. The calculations follow the procedure just outlined.

Walls, roof, windows, and doors. The calculations for the living room and the kitchen, where $q = UA$ (CLTD) for the walls, roof, and door and $q = A$ (GLF) for the windows, are outlined in Table 10. The glass shaded by the overhang is treated as north-facing glass, with the shaded area computed using Table 6.

The internal and infiltration sensible cooling loads are computed as follows:

For the living room:

Infiltration. Using Table 8,

$$Q = \text{ACH (room volume)}/60$$
$$Q = 0.5 \times 3840/60 = 32$$
$$q = 1.1 \, Q(t_o - t_i) = 1.1 \, Q \, \Delta t$$
$$q = 1.1 \times 32(96 - 75) = 740 \text{ Btu/h}$$

Occupants. Assuming 230 Btu/h per person,

$$q = 230 \times \text{(persons)}$$
$$q = 230 \times 4 = 920 \text{ Btu/h}$$

Appliances. Assuming that 50% of the kitchen appliance load is picked up in the living room,

$$q = 0.5 \times \text{(kitchen appliance load)}$$
$$q = 0.5 \times 1600 = 800 \text{ Btu/h}$$

For the kitchen:

Infiltration.

$$Q = 0.5 \times 1920/60 = 16$$
$$q = 1.1 \times 16(96 - 75) = 370 \text{ Btu/h}$$

No occupants.

$$q = 0$$

Appliances. Assuming 25% of the utility appliance load is picked up in the kitchen,

$$q = (1600/2) + (1600/4) = 1200 \text{ Btu/h}$$

Table 10 Transmission Cooling Load for Example 1

Item	Net Area, ft²	GLF Btu h·ft²	U-Value Btu h·ft²·°F	CLTD, °F	Cooling Load, Btu/h	Reference
Living Room						
West wall	91		0.06	24	131	Table 1
Partition (garage)	192		0.07	12	161	Table 1
Roof	480		0.05	48	1152	Table 1
West door	21		0.32	24	161	Table 1
West glass	35	44			1549	Table 3
Shaded glass	13	19			243	Table 3
Kitchen						
East wall	135		0.06	24	194	Table 1
Roof	240		0.05	48	576	Table 1
East glass	14	44			607	Table 3
Shaded glass	11	19			213	Table 3

Table 11 Summary of Sensible Cooling Load Estimate for Example 1

Room	Walls and Doors	Glass	People	Appliances	Infiltration	Total Btu/h	Room cfm
Living room	1605	1792	920	800	740	5857	300
Kitchen	770	820		1200	370	3160	170
Utility and storage	1363			1200	443	3006	150
Bedroom No. 1	582	544			278	1404	75
Bedroom No. 2	664	304			278	1246	70
Master bedroom and bath	1682	816			813	3311	175
Bath	540	276			295	1111	60
Total	7206	4552	920	3200	3217	19095	1000
Duct loss (10%)						1910	
Outdoor ventilation air						1600	
					Total	22605	Btu/h

For the total sensible cooling load for these two rooms and the cooling load for the remaining rooms, see Table 11. At this point, the sensible cooling load for the house is about 19,100 Btu/h. Depending on the design of the air distribution system, heat losses from the supply and return ducts may add to the cooling load. These may be more accurately estimated after designing the system; however, to size the cooling unit, duct losses should be included initially. If all ducts are in the attic space, a duct loss of 10% of the space sensible cooling load is reasonable. For a counterflow system, with ducts below the slab, a 5% loss is more reasonable.

An infiltration rate of 0.5 ACH may not be adequate for good indoor air quality, so some outdoor air should be introduced. This additional cooling load may be estimated in the same way as the infiltration load.

Assume that the entire duct system is in the attic, *i.e.*, the total sensible cooling load with a 10% duct loss is $1.1 \times 19{,}095$ or 21,000 Btu/h. Also, assume that additional outdoor air is needed to assure good indoor air quality, so the total infiltration and outdoor ventilation air is 0.75 ACH. This increases the infiltration rate by 50% or about 1600 Btu/h. The total sensible cooling load is then increased to 22,605 Btu/h (Table 11).

The total cooling load (sensible plus latent) may be estimated by applying the latent factor (LF) from Figure 1. For a design humidity ratio of 0.0136 lb vapor per lb dry air, LF = 1.15 for a house of medium construction. Hence, the total cooling load equals $1.15 \times 22{,}605$ or 26,000 Btu/h.

Most residential unitary equipment operates with a cooling load that has a latent factor of 1.3 or a sensible heat factor of 0.77. Therefore, to ensure that the unit is large enough, it should be selected based on the sensible cooling load and the unit sensible cooling capacity. For this example, the sensible cooling capacity should be equal to or greater than 22,605 Btu/h and the total capacity, about $1.3 \times 22{,}605$ or 29,400 Btu/h. Therefore, a nominal 30,000 Btu/h (2.5 ton) unit is acceptable. On the basis of total capacity, the unit is oversized less than 25%.

The supply air quantities can now be computed. Air should be supplied to each room on the basis of the room sensible cooling load

$$Q_{rm} = Q_{tot} \, (q_{rm}/q)$$

where

Q_{rm} = Q airflow to each room
Q_{tot} = total airflow rate, about 400 cfm per ton of total capacity
q_{rm} = room sensible cooling load
q = total sensible cooling load

Thus:

$$Q_{rm} = (1000/19{,}095) \, q_{rm}$$

If the living space in Example 1 had been a multifamily unit (assume that the north, south, and east walls are not exposed surfaces), the calculation procedure would be the same, except that Table 2 would have been used for the CLTDs and Table 4 for the GLFs. Assumptions regarding infiltration, ventilation, and appliance loads are different for smaller multifamily units.

HEATING LOAD

Calculating a residential heating load involves estimating the maximum probable heat loss of each room or space to be heated and the simultaneous maximum loss for the building, while maintaining a selected indoor air temperature during periods of design outdoor weather conditions. Heat losses are mainly:

- Transmission losses or heat transferred through the confining walls, glass, ceiling, floor, or other surfaces.
- Infiltration losses or energy required to warm outdoor air leaking in through cracks and crevices around doors and windows or through open doors and windows, and through porous building materials.

GENERAL PROCEDURE

To calculate a design heating load, prepare the following information about building design and weather data at design conditions.

1. Select outdoor design weather conditions: temperature, wind direction, and wind speed. Winter climatic data can be found in Chapter 24, or selected weather conditions and temperatures appropriate for the application may be used. Weather station may differ significantly from values in Chapter 24.

2. Select the indoor air temperature to be maintained in each space during design weather conditions.

3. Estimated temperatures in adjacent unheated spaces, attached garages and attics can be at the outdoor ambient temperature.

4. Select or compute heat transfer coefficients for outside walls and glass; for inside walls, nonbasement floors, and ceilings, if these are next to unheated spaces; and for the roof if it is next to heated spaces.

5. Determine the net area of outside wall, glass, and roof next to heated spaces, as well as any cold walls, floors, or ceilings next to unheated spaces. These determinations can be made from building plans or from the actual building, using inside dimensions.

6. Compute heat transmission losses for each kind of wall, glass, floor, ceiling, and roof in the building by multiplying the heat transfer coefficient in each case by the area of the surface and the temperature difference between indoor and outdoor air or adjacent lower temperature spaces.

7. Compute heat losses from basement or grade-level slab floors using the methods in this chapter.

8. Select unit values, and compute the energy associated with infiltration of cold air around outside doors, windows, porous building materials, and other openings. These unit values depend on the kind or width of crack, wind speed, and the temperature difference between indoor and outdoor air. An alternative method is to use air changes (see Chapter 23.)

9. When positive ventilation using outdoor air is provided by an air heating or air-conditioning unit, the energy required to warm the outdoor air to the space temperature must be provided by the unit. The principle for calculation of this load component is identical to that for infiltration. If mechanical exhaust from the space is provided in an amount equal to the outdoor air drawn in by the unit, the unit must also provide for natural infiltration losses. If no mechanical exhaust is used and the outdoor air supply equals or exceeds the amount of natural infiltration that can occur without ventilation, some reduction in infiltration may occur.

10. The sum of the coincidental transmission losses or heat transmitted through the confining walls, floor, ceiling, glass, and other surfaces, plus the energy associated with cold air entering by infiltration or the ventilation air required to replace mechanical exhaust, represents the total heating load.

11. Include the pickup loads that may be required in intermittently heated buildings using night thermostat setback. Pickup loads frequently require an increase in heating equipment capacity to bring the temperature of structure, air, and material contents to the specified temperature. See Figure 9.

12. Use materials and data in Chapter 22, 23, 24, and others as are appropriate to the calculations. See Table 12.

SELECTING HEATING DESIGN CONDITIONS

The ideal solution to a basic heating system design is a plant with a maximum output capacity equal to the heating load that develops with most severe local weather conditions. However, this solution is usually uneconomical. Weather records show that severe weather conditions do not repeat annually. If heating systems were designed for maximum weather conditions,

Table 12 Summary of Loads, Equations, and References for Calculating Design Heating Loads

Heating Load	Equation	Reference, Table, Description
Roofs, ceilings, walls, glass	$q = U A \Delta t$	►Chapter 22, Tables 1, 2, and 4 ►Temperature difference between inside and outside design dry bulbs, Chapter 24. For temperatures in unheated spaces, see Equation (1); for attic temperatures, see Equation (2). ►Area calculated from plans
Walls below grade	$q = U A \Delta t$	►See Table 14 ►Use Figure 6 to assist in determining Δt
Floors Above grade	$q = U A \Delta t$	►For crawl space temperatures, see Equation (4) ►See Table 16
On grade	$q = F_2 P \Delta t$	►See Equation (6) ►Perimeter of slab
Below grade	$q = U A \Delta t$	►Use Figure 6 to assist in determining Δt ►See Table 15
Infiltration and ventilation air Sensible Latent	$q_s = 0.018 \, Q \, \Delta t$ $q_l = 80.7 \, Q \, \Delta W$	►Volume of outdoor air entering building. See Chapter 23 for estimating methods for infiltration. ►Humidity ratio difference, if humidification is to be added.

excess capacity would exist during most of the system's operating life.

In many cases, an occasional failure of a heating plant to maintain a preselected indoor design temperature during brief periods of severe weather is not critical. However, the successful completion of some industrial or commercial processes may depend on close regulation of indoor temperatures. The specific requirements for each building should be carefully evaluated.

Before selecting an outdoor design temperature from Chapter 24, the designer should consider the following for residential buildings:

- Is the type of structure heavy, medium, or light?
- Is the structure insulated?
- Is the structure exposed to high wind?
- Is the load from infiltration or ventilation high?
- Is there more glass area than normal?
- During what part of the day will the structure be used?
- What is the nature of occupancy?
- Will there be long periods of operation at reduced indoor temperature?
- What is the amplitude between local maximum and minimum daily temperatures?
- Are there local conditions that cause significant variation from temperatures reported by the Weather Bureau?
- What auxiliary heating devices will be in the building?

Before selecting an outdoor design temperature, the designer must keep in mind that, if the outdoor to indoor design temperature difference is exceeded, the indoor temperature may fall, depending on the thermal mass of the structure and its contents and on whether or not the internal load was included in calculations and the duration of the cold period.

The effect of wind on the heating requirements of any building should be considered, because:

- Wind movement increases the heat transmission of walls, glass, and roof, affecting poorly insulated walls to a much greater extent than well-insulated walls.
- Wind materially increases the infiltration of cold air through cracks around doors and windows and even through building materials themselves (see Chapter 23).

Theoretically, on a design basis, the most unfavorable combination of temperature and wind speed should be chosen. A building may require more heat on a windy day with a moderately low outdoor temperature than on a quiet day with a much lower outdoor temperature. The worst combination of wind and temperature varies in different buildings, because wind speed has a greater effect on buildings with relatively high infiltration rates. The building heating load may be calculated for several combinations of temperature and wind speed on record and the worst combination selected; however, except for critical applications, designers generally find such a degree of refinement unnecessary. No correlation has been shown between the design temperatures in Chapter 24 and the simultaneous maximum wind speed. If a designer prefers the air change method for computing infiltration rates, such correlation is not important. Designers who use the crack method can use a leakage rate at a wind speed of 15 mph, unless local experience has established that other speeds are more appropriate. Abnormally high wind speeds may have an effect on infiltration and the U-value of the building components (see Chapter 20).

ASHRAE *Standard* 90A-80 suggests 72 °F as the indoor temperature for comfort heating design; however, this may vary depending on building use, type of occupancy, or code requirements. Chapter 8 defines the relation of temperature and human comfort.

ESTIMATING TEMPERATURES IN ADJACENT UNHEATED SPACES

Heat loss from heated rooms to unheated rooms or spaces must be based on the estimated or assumed temperature in such unheated spaces. This temperature will lie in the range between the indoor and outdoor temperatures. If the respective surface areas adjacent to the heated room and exposed to the outdoors are the same and if the heat transfer coefficients are equal, the temperature in the unheated space may be assumed equal to the mean of the indoor and outdoor design temperatures. If, however, the surface areas and coefficients are unequal, the temperature in the unheated space should be estimated by:

$$
\begin{aligned}
t_u = & \ [t_i(A_1U_1 + A_2U_2 + A_3U_3 + \text{etc.}) \\
& + t_o(60\rho c_pQ_o + A_aU_a + A_bU_b + A_cU_c + \text{etc.})] \\
& \div (A_1U_1 + A_2U_2 + A_3U_3 + \text{etc.} \\
& + 60\rho c_pQ_o + A_aU_a + A_bU_b + A_cU_c + \text{etc.})
\end{aligned} \tag{1}
$$

where

ρc_p = density times specific heat of air = 0.018 Btu/ft³·°F for standard air
t_u = temperature in unheated space, °F
t_i = indoor design temperature of heated room, °F
t_o = outdoor design temperature, °F
A_1, A_2, A_3, etc. = areas of surface of unheated space adjacent to heated space, ft²
A_a, A_b, A_c, etc. = areas of surface of unheated space exposed to outdoors, ft²
U_1, U_2, U_3, etc. = heat transfer coefficients of surfaces of A_1, A_2, A_3, etc., Btu/h·ft²·°F
U_a, U_b, U_c, etc. = heat transfer coefficients of surfaces of A_a, A_b, A_c, etc., Btu/h·ft²·°F
Q_o = rate of introduction of outside air into unheated space by infiltration and/or ventilation, cfm

Example 2. Calculate the temperature in an unheated space adjacent to a heated room with surface areas (A_1, A_2, and A_3) of 100, 120, and 140 ft² and overall heat transfer coefficients (U_1, U_2, and U_3) of 0.15, 0.20, and 0.25 Btu/h·ft²·°F, respectively. The surface areas of the unheated space exposed to the outdoors (A_a and A_b) are 100 and 140 ft², respectively, and the corresponding overall heat transfer coefficients are 0.10 and 0.30 Btu/h·ft²·°F. The sixth surface is on the ground and can be neglected for this example, as can the effect of introduction of outdoor air into the unheated space. Assume $t_i = 70$°F and $t_o = -10$°F.

Solution: Substituting into Equation (1)

$$
\begin{aligned}
t_u = & \ [70(100 \times 0.15 + 120 \times 0.20 + 140 \times 0.25) \\
& + (-10)(100 \times 0.10 + 140 \times 0.30)] \\
& \div (100 \times 0.15 + 120 \times 0.20 + 140 \times 0.25 \\
& + 100 \times 0.10 + 140 \times 0.30)
\end{aligned}
$$

$$
t_u = 4660/126 = 37\,°F
$$

Temperatures in unheated spaces with large glass areas and two or more surfaces exposed to the outdoors (such as sleeping porches and sun parlors) are generally assumed to be the same as that of the outdoors.

Attic Temperature

An attic is a space having an average distance of 1 ft or more between a ceiling and the underside of the roof. Estimating attic temperature is a special case of estimating temperature in an adjacent unheated space and can be done using

$$
t_a = \frac{A_cU_ct_c + t_o(\rho c_pA_cQ_c + A_rU_r + A_wU_w + A_gU_g)}{A_c(U_c + \rho c_pQ_c) + A_rU_r + A_wU_w + A_gU_g} \tag{2}
$$

where

ρc_p = air density times specific heat = 0.018 Btu/h·ft³·°F for standard air
t_a = attic temperature, °F
t_c = indoor temperature near top floor ceiling, °F
t_o = outdoor temperature, °F
A_c = area of ceiling, ft²
A_r = area of roof, ft²
A_w = area of net vertical attic wall surface, ft²
A_g = area of attic glass, ft²
U_c = heat transfer coefficient of ceiling, Btu/h·ft²·°F, based on surface conductance of 2.2 Btu/h·ft²·°F (upper surface, see Chapter 22); 2.2 = reciprocal of one-half the air space resistance
U_r = heat transfer coefficient of roof, Btu/h·ft²·°F, based on surface conductance of 2.2 Btu/h·ft²·°F (upper surface, see Chapter 22); 2.2 = reciprocal of one-half the air space resistance
U_w = heat transfer coefficient of vertical wall surface, Btu/h·ft²·°F
U_g = heat transfer coefficient of glass, Btu/h·ft²·°F
Q_c = rate of introduction of outside air into the attic space by ventilation per square foot of ceiling area, cfm/ft²

Example 3. Calculate the temperature in an unheated attic assuming $t_c = 70$°F; $t_o = 10$°F; $A_c = 1000$ ft²; $A_r = 1200$ ft²; $A_w = 100$ ft²; $A_g = 10$ ft²; $U_r = 0.50$ Btu/h·ft²·°F; $U_c = 0.40$ Btu/h·ft²·°F; $U_w = 0.30$ Btu/h·ft²·°F; $U_g = 1.13$ Btu/h·ft²·°F; and $Q_c = 0.5$ cfm/ft².

Solution: Substituting these values into Equation (2):

$$
\begin{aligned}
t_a = & \ [(1000 \times 0.40 \times 70) + 10(1.10 \times 1000 \times 0.5 \\
& + 100 \times 0.30 + 10 \times 1.13)] \\
& \div [1000(0.40 + 1.10 \times 0.5) + 1200 \times 0.50 + 100 \times 0.30 \\
& + 10 \times 1.13]
\end{aligned}
$$

$$
t_a = 39{,}913/1591 = 25.1\,°F
$$

Equation (2) includes the effect of air interchange that would take place through attic vents or louvers intended to preclude attic condensation. Test data from Joy et al. (1956), Joy (1958), and Rowley et al. (1940) indicate that a reduction in the temperature difference between attic air and outside air is linear, with attic ventilation rates between 0 and 0.5 cfm/ft^2 of the ceiling area. When attic ventilation meets the requirements in Chapter 23, 0.5 cfm/ft^2 of the approximate ventilation rate is within design conditions. This reduction in temperature difference affects the overall heat loss of a residence with an insulated ceiling by only 1 or 2%.

Equation (2) also does not consider factors such as heat exchange between chimney and attic or solar radiation to and from the roof. Because of these effects, attic temperatures are frequently higher than values calculated by using the equations. However, Equation (2) can be used to calculate attic temperature, because the resulting error is generally less than that introduced by neglecting the roof and assuming the attic temperature equal to the outdoor air temperature.

When relatively large louvers are installed (customary in southern states), the attic temperature is often assumed as the average between indoor and outdoor air temperatures.

For a shorter, approximate method of calculating heat losses through attics, the combined ceiling and roof coefficient may be used (see Chapter 22).

Ground Temperature

Ground temperatures assumed for estimating basement heat losses will differ for basement floors and walls. The temperatures under floors are generally higher than those adjacent to walls. This is discussed further in the section Basement Design Temperatures.

CALCULATING HEAT LOSS FROM CRAWL SPACES

A crawl space can be considered a half basement. To prevent ground moisture from evaporating and causing a condensation problem, sheets of vapor retardant (e.g., polyethylene film) are used to cover the ground surface (see Chapter 21). Most codes require crawl spaces to be adequately vented all year round. However, venting the crawl space in the heating season causes substantial heat loss through the floor.

The space may be insulated in several ways; the crawl space ceiling (floor above the crawl space) can be insulated, or the perimeter wall can be insulated either on the outside or on the inside of the wall. If the floor above is insulated, the crawl space vents should be kept open, because the temperature of the crawl space is likely to be below the dew point of the indoor space. If the perimeter wall is insulated, the vents should be kept closed in the heating season and open the remainder of the year.

Crawl Space Temperature

The crawl space temperature depends on such factors as venting, heating ducts, and the heating plant. When the crawl space is well ventilated, its temperature is close to that of the ambient air temperature. When the crawl space vent is closed for the heating season, or if the space is used as a plenum (i.e., part of the forced air heating system), the crawl space temperature approaches the indoor conditioned space. In the former case, the floor above the crawl space, the heating duct work, and the utility pipes should be insulated similar to the walls and ceiling of a house.

The following steady-state equation can be used to estimate the temperature of a crawl space.

$$q_f = q_p + q_g + q_a \qquad (3)$$

where

q_f = heat loss through floor into crawl space, Btu/h
q_p = heat loss through foundation walls and sill box from crawl space, Btu/h
q_g = heat loss into ground, Btu/h
q_a = heat loss due to ventilation of crawl space, Btu/h

Latta and Boileau (1969) estimate the air exchange rate for an uninsulated basement at 0.67 air changes per hour under winter conditions. In more detail, Equation (3) can be repeated as

$$U_f A_f(t_i - t_c) = U_p A_p(t_c - t_o) + U_g A_g(t_c - t_g) \\ + 0.67(H_c V_c)(t_c - t_o) \qquad (4)$$

where

t_i = indoor temperature, i.e., air above ceiling of crawl space, °F
t_o = outdoor temperature, °F
t_g = ground temperature (constant), °F
t_c = crawl space temperature, °F
A_f = area of floor above, ft^2
A_p = area of perimeter, exposed foundation wall plus sill box, ft^2
A_g = area of ground below ($A_f = A_g$), ft^2
U_f = average heat transfer coefficient through floor, Btu/h·ft^2·°F
U_g = average heat transfer coefficient through ground (horizontal air film and 10 ft of soil), Btu/h·ft^2·°F
U_p = combined heat transfer coefficient of sill box and foundation wall (both above and below grade), Btu/h·ft^2·°F
V_c = volume of crawl space, ft^3
H_c = volumetric heat capacity of air = 0.018 Btu/ft^3·°F
0.67 = assumed air exchange rate, volume/hour

Example 4. A crawl space of 1200 ft^2, with a 140-ft perimeter, is considered. The construction of the perimeter wall is shown in Figure 3. The indoor, outdoor, and the deep-down ground temperatures are 70, 10, and 50 °F, respectively. Estimate the heat loss and crawl space temperature with and without insulation. The heat transmission coefficient (U-factor) for each component is indicated in Table 13.

Solution: Three cases are examined.

Case A. This base case is a vented and uninsulated crawl space. The crawl space temperature approaches that of the outdoors, 10 °F, and the heat loss is $0.25 \times 1200(70 - 10) = 18,000$ Btu/h.

Case B. The crawl space is vented. The floor above is insulated with an $R = 11$ blanket; no insulation on the perimeter. The temperature of the crawl space approaches that of the outdoors, 10 °F. The heat loss is calculated as

$$q_f - 1200(70 - 10) = 5472 \text{ Btu/h}$$

Table 13   Estimated U-Values for Insulated and Uninsulated Crawl Spaces

Component	Uninsulated Btu/h·°F per ft of Perimeter	Insulated[a] Btu/h·°F per ft of Perimeter
16-in. exposed concrete blocks	0.7	0.18
7.5-in. sill box	0.188	0.071
1st 12-in. block wall below grade	0.355	0.127
2nd 12-in. block wall below grade	0.22	0.14
3rd 12-in. block wall below grade	0.133	0.1
Total for perimeter wall	1.6	0.62

	Btu h·ft^2·°F	Btu h·ft^2·°F
Ground	0.077	0.077
Floor above crawl space	0.25	0.076[a]

[a]Perimeter walls are insulated with $R = 5.4$; the floor is insulated with $R = 11$ blanket or batts.

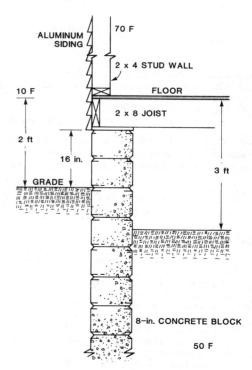

Fig. 3   Uninsulated Crawl Space

Case C. The crawl space is not vented during the heating season. The floor above is not insulated, but the perimeter wall is insulated with $R = 5.4$ down to 3 ft below grade.

$$q_f = 1200 \times 0.25(70 - t_c)$$
$$q_p = 140 \times 0.62(t_c - 10)$$
$$q_g = 1200 \times 0.077(t_c - 50)$$
$$q_a = 1200 \times 3 \times 0.67 \times 0.018(t_c - 10)$$

The crawl space temperature is solved using Equation (4), $t_c = 51.5\,°F$. The heat loss is 5530 Btu/h.

The results show that base case A can potentially lose the most heat. However, when the floor above is insulated, the crawl space must be vented to eliminate any condensation potential, and the heating duct work and utility pipeline in the crawl space must be adequately insulated. When the perimeter is insulated, the vents must be closed during the heating season and opened for the rest of the year; the heating duct work and utility pipeline do not need insulation.

Case	Venting	Insulation	Heat Loss through Floor Above, Btu/h	Temperature of Crawl Space, °F
Base	Yes	None	18,000	10
A	Yes	$R = 11$ on floor above	5,470	10
B	No	$R = 5.4$ on perimeter wall	5,530	51.5

CALCULATING TRANSMISSION HEAT LOSS

Steady-state heat loss by conduction and convection heat transfer through any surface is:

$$q = AU(t_i - t_o) \tag{5}$$

where

q = heat transfer through wall, roof, ceiling, floor, Btu/h
A = area of wall, glass, roof, ceiling, floor, or other exposed surface, ft^2
U = air-to-air heat transfer coefficient, Btu/h·ft^2·°F
t_i = indoor air temperature near surface involved, °F
t_o = outdoor air temperature, or temperature of adjacent unheated space, °F

Example 5. Calculate the transmission loss through an 8-in. brick wall having an area of 150 ft^2, if the indoor temperature t_i is 70°F, and the outdoor temperature t_o is −10°F.

Solution: The overall heat transfer coefficient U of a plain 8-in. brick wall is 0.41 Btu/h·ft^2·°F. Substituting into Equation (5):

$$q = 150 \times 0.41\,[70 - (-10)] = 4920\ \text{Btu/h}$$

Through Ceiling and Roof

Transmission heat loss through top floor ceilings, attics, and roofs may be estimated by either of two methods:

1. Substitute in Equation (5) the ceiling area A, the indoor/outdoor temperature difference $(t_i - t_o)$, and the proper U-value:

   Flat roofs. Use appropriate coefficients in Equation (2), as outlined in Chapter 22, if side walls extend appreciably above the ceiling or the floor below.

   Pitched roofs. Calculate the combined roof and ceiling coefficient by means outlined in Chapter 22.

2. For *pitched roofs*, estimate the attic temperature (based on the indoor and outdoor design temperatures) using Equation (2), and substitute for t_o in Equation (5), obtaining the value of t_a, together with the ceiling area A and the ceiling U-value. Attic temperatures do not need to be calculated for *flat roofs*, as the ceiling-roof heat loss can be determined as suggested in (1) above.

From the Basement

The basement interior is considered conditioned space if a minimum of 10°F is maintained over the heating season. In many instances, the house heating plant, water heater, and heating ducts are in the basement, so it remains at or above 50°F.

The heat transmission from the below-grade portion of the basement wall to the ambient air cannot be estimated by simple, one-dimensional heat conduction. In fact, field measurement of an uninsulated basement by Latta and Boileau (1969) showed that the isotherms near the wall are not parallel lines but closer to radial lines centered at the intersection of the grade line and the wall. Therefore, heat flow paths approximately follow a set of concentric circular patterns (Figure 4).

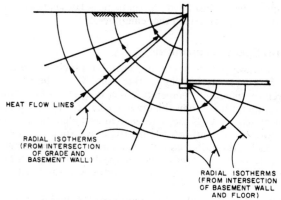

Fig. 4   Heat Flow from Basement

Such heat flow paths are altered when insulation is added to the wall or floor. An extreme case would be no heat loss from the basement wall and floor, i.e., infinite insulation applied to the wall and floor. In this case, the isotherms would be horizontal lines parallel to the grade line, and the heat flow would be vertical. When finite insulation or partial insulation is applied to the wall and floor, the heat flow paths take shapes somewhere between the circular and vertical lines (Figure 5).

Through Basement Walls

Houghten *et al.* (1942) observed nonuniform heat flux across the basement wall with respect to the depth of the wall, because each heat flow path contains a different thermal resistance. For a basement wall that has its top portion exposed to ambient air, heat may be conducted vertically through the concrete wall and dissipated to the ambient from the top portion of the wall (Wang 1979, Bligh *et al.* 1978). Under certain conditions, this vertical heat flux becomes significant and should not be ignored.

Once the heat paths are known or assumed, a steady-state analysis can calculate the overall heat transmission coefficient for each segment of the basement wall. Referring to Figures 4 and 5, the total thermal resistance for each depth increment of the basement wall can be found by summing the thermal resistances along each heat flow path. Based on these resistances, the heat loss at each depth increment can be estimated for a unit temperature difference between the basement and the average mean winter temperature. Table 14 lists such heat loss values at different depths

for an uninsulated and insulated concrete wall (Latta and Boileau 1969). Also listed are the lengths of the heat flow path through the soil (circular path). The conductivity of the soil was assumed to be 9.6 Btu/h·ft²·°F.

Through Basement Floors

The same steady-state design used for the basement wall can be applied to the basement floor, except that the length of the heat flow path is longer (see Figure 4). Thus, the heat loss through the basement floor is much smaller than that from the wall. Therefore, an average value for the heat loss through the basement floor can be multiplied by the floor area to give total heat loss from the floor. Table 15 lists typical values.

Basement Design Temperatures

Although internal design temperature is given by basement air temperature, none of the usual external design air temperatures apply because of the heat capacity of the soil. However, ground surface temperature fluctuates about a mean value by an amplitude *A*, which varies with geographic location and surface cover. Therefore, suitable external design temperatures can be obtained by subtracting *A* for the location from the mean winter air temperature t_a. Values for t_a can be obtained from meteorological records, and *A* can be estimated from the map in Figure 6. This map is part of one prepared by Chang (1958) giving annual ranges in ground temperature at a depth of 4 in.

Example 6. Consider a basement 28 ft wide by 30 ft long sunk 6 ft below grade, with $R = 8.34$ insulation applied to the top 2 ft of the wall below grade. Assume an internal air temperature of 70 °F and an external design temperature ($t_a - A$) of 20 °F.
Solution:

Wall (using Table 14)

First foot below grade	0.093 Btu/h·ft·°F
Second foot below grade	0.079 Btu/h·ft·°F
Third foot below grade	0.155 Btu/h·ft·°F
Fourth foot below grade	0.119 Btu/h·ft·°F
Fifth foot below grade	0.096 Btu/h·ft·°F
Sixth foot below grade	0.079 Btu/h·ft·°F
Total per foot length of wall	0.621 Btu/h·ft·°F
Basement perimeter	2 (28 + 30) = 116 ft
Total wall heat loss	0.62 × 116 = 72 Btu/h·°F

Floor (using Table 4)

Average heat loss per ft²	0.025 Btu/h·°F
Floor area 28 × 30	840 ft²
Total floor heat loss	0.025 × 840 = 21 Btu/h·°F

Total

Total basement heat loss below grade	72 + 21 = 93 Btu/h·°F
Design temperature difference	70 − 20 = 50 °F
Maximum rate of heat loss from below-grade basement	93 × 50 = 4650 Btu/h

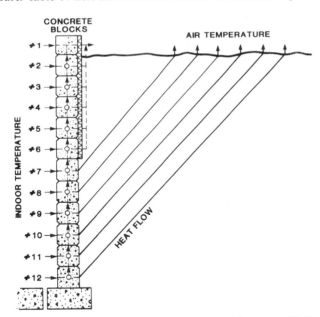

Fig. 5 Heat Flow Path for Partially Insulated Basement Wall

Table 14 Heat Loss Below Grade in Basement Walls

Depth, ft	Path Length through Soil, ft	Heat Loss, Btu/h·ft²·°F							
		Uninsulated		$R = 4.17$		$R = 8.34$		$R = 12.5$	
0 to 1	0.68	0.410		0.152		0.093		0.067	
1 to 2	2.27	0.222	0.632	0.116	0.268	0.079	0.172	0.059	0.126
2 to 3	3.88	0.155	0.787	0.094	0.362	0.068	0.240	0.053	0.179
3 to 4	5.52	0.119	0.906	0.079	0.441	0.060	0.300	0.048	0.227
4 to 5	7.05	0.096	1.002	0.069	0.510	0.053	0.353	0.044	0.271
5 to 6	8.65	0.079	1.081	0.060	0.570	0.048	0.401	0.040	0.311
6 to 7	10.28	0.069	1.150	0.054	0.624	0.044	0.445	0.037	0.348

Table 15 Heat Loss through Basement Floors, Btu/h·ft²·°F

Depth of Foundation Wall Below Grade, ft	Shortest Width of House, ft			
	20	24	28	32
5	0.032	0.029	0.026	0.023
6	0.030	0.027	0.025	0.022
7	0.029	0.026	0.023	0.021

Note: $\Delta F = (t_a - A)$

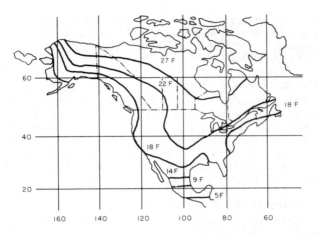

Fig. 6   Lines of Constant Amplitude

If a basement is completely below grade and unheated, its temperature ranges between that in the rooms above and that of the ground. Basement windows lower the basement temperature when it is cold outdoors, and heat given off by the heating plant increases the basement temperature. The exact basement temperature is indeterminate if the basement is not heated. In general, heat from the heating plant sufficiently warms the air near the basement ceiling to make an allowance unnecessary for floor heat loss from rooms located over the basement.

Transient Calculations, Basement Walls

The heat loss from basement walls can be estimated more accurately with a finite element or finite difference computer program by transient simulations (Wang 1979, Bligh *et al.* 1978). The solution is in the form of heat loss over time, which can be converted to an average U-value. This approach also offers the possibility for estimating the depth below grade to which insulation is economical. Direct and indirect evidence of hollow concrete block walls shows a convective path existing within the blocks, vertically along the wall (Harrje *et al.* 1979). Therefore, insulation should be arranged to reduce this convective heat transfer.

Peony *et al.* (1979) showed that the dynamic thermal performance of a masonry wall is better when insulation is placed on the exterior. Moreover, transient simulation showed that insulation is more effective when it is placed on the exterior side of the basement wall. Depending on the exposed portion of the block wall and the temperature difference between indoor and outdoor air, exterior application can be 10 to 20% more efficient than a corresponding interior application. However, such exterior application must be installed properly to maintain its integrity.

Calculating Transmission Heat Loss from Floor Slabs

Concrete slab floors are (1) unheated, relying for warmth on heat delivered above floor level by the heating system, and (2) heated, containing heated pipes or ducts that constitute a radiant slab or portion of it for complete or partial heating of the house.

The perimeter insulation of a slab-on-grade floor is quite important for comfort and energy conservation. In unheated slab floors, the floor edge must be insulated to keep the floor warm. Downdrafts from windows or exposed walls can create pools of chilly air over considerable areas of the floor. In heated slab floors, the floor edge must be insulated to prevent excessive heat loss from the heating pipe or duct embedded in the floor or from the baseboard heater.

Wang (1979) and Bligh *et al.* (1978) found that heat loss from an unheated concrete slab floor is mostly through the perimeter rather than through the floor and into the ground. Total heat loss is more nearly proportional to the length of the perimeter than to

the area of the floor, and it can be estimated by the following equation for both unheated and heated slab floors:

$$q = F_2 P (t_i - t_o) \tag{6}$$

where

> q = heat loss through perimeter, Btu/h
> F_2 = heat loss coefficient per foot of perimeter (see Table 5), Btu/(h·°F·ft)
> P = perimeter or exposed edge of floor, ft
> t_i = indoor temperature, °F. (For the heated slab, t_i is the weighted average heating duct or pipe temperature.)
> t_o = outdoor temperature, °F

Vertical "I"-shape systems are used to insulate slab floor perimeters. In the "I" system, the insulation is placed vertically next to the exposed slab edge, extending downward below grade, as shown in Figure 7.

Breaks or joints must be avoided when the insulation is installed; otherwise, local thermal bridges can be formed, and the overall efficiency of the insulation is reduced.

Transient Calculations, Floor Slabs

Figure 8 shows four basic slab-on-grade constructions analyzed with a finite element computer program by Wang (1979). Figures 8a-c represent unheated slabs; Figure 8d can be considered a heated slab. Each was investigated with and without insulation of $R = 5.4$ under three climate conditions (7433, 5350, and 2950 degree days). Table 16 lists results in terms of heat loss coefficient F_2, based on degree days.

Table 16 shows that the heat loss coefficient F_2 is sensitive to both construction and insulation. The reverse loss, or heat loss into the ground and outward through the edges of the slab and foundation wall, is significant when heating pipes or ducts or when baseboard heaters are placed near the slab perimeters. To prevent reverse loss, the designer may find it advantageous to use perimeter insulation even in warmer climates. For severe winter regions (above 6000 degree days), the insulation value should be increased to $R > 10$.

Figure 8a shows that this construction benefits from the wall insulation between block and brick; the insulation is extended roughly 16 in. below the slab floor. Without this wall insulation, the heat loss coefficient F_2 would be close to that of the 4-in. block wall construction (Figure 8b). Table 16 can be used to estimate F_2 under different degree days of heating season weather.

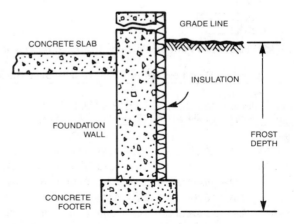

Fig. 7   "I" Shape or Vertical Insulation System

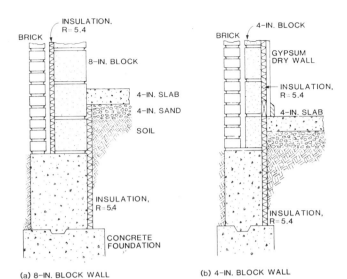

Fig. 8 Slab-on-Grade Foundation Insulation

(a) 8-IN. BLOCK WALL

(b) 4-IN. BLOCK WALL

(c) METAL STUD WALL

(d) CONCRETE WALL

Table 16 Heat Loss Coefficient F_2 of Slab Floor Construction, Btu/h·°F per ft of Perimeter

Construction	Insulation[a]	Degree Days (65°F Base)		
		2950	5350	7433
8-in. block wall, brick facing	Uninsulated	0.62	0.68	0.72
	R = 5.4 from edge to footer	0.48	0.50	0.56
4-in block wall, brick facing	Uninsulated	0.80	0.84	0.93
	R = 5.4 from edge to footer	0.47	0.49	0.54
Metal stud wall, stucco	Uninsulated	1.15	1.20	1.34
	R = 5.4 from edge to footer	0.51	0.53	0.58
Poured concrete wall with duct near perimeter[b]	Uninsulated	1.84	2.12	2.73
	R = 5.4 from edge to footer, 3 ft under floor	0.64	0.72	0.90

[a]R-value units in °F·ft²·h/Btu·in.
[b]Weighted average temperature of the heating duct was assumed at 110°F during the heating season (outdoor air temperature less than 65°F).

CALCULATING INFILTRATION HEAT LOSS

Infiltration of outside air causes both *sensible* and *latent* heat loss. The energy required to raise the temperature of outdoor infiltrating air to indoor air temperature is the sensible component. The energy associated with net loss of moisture from the space is the latent component. Infiltration is discussed in detail in Chapter 23.

Sensible Heat Loss

The energy required to warm outdoor air entering by infiltration to the temperature of the room is given by:

$$q_s = c_p \, Q\rho\,(t_i - t_o) \tag{7}$$

where

q_s = heat flow required to raise temperature of air leaking into building from t_o to t_i, Btu/h
c_p = specific heat of air, Btu/lb·°F
Q = volume of outdoor air entering building, ft³/h
ρ = density of air at temperature t_o, lb/ft³

Using standard air, Q = 0.075 lb/ft³, and c_p = 0.24, Equation (7) reduces to:

$$q_s = 0.018 \, Q(t_i - t_o) \tag{8}$$

The volume Q of outdoor air entering depends on wind speed and direction, width of cracks or size of openings, type of openings, and other factors explained in Chapter 23. The two methods used to obtain the quantity of infiltration air are the *crack length* and the *air change*. Louvers and doors and the direction they face, as well as any other factors that affect infiltration, may need to be considered.

Latent Heat Loss

When moisture must be added to the indoor air to maintain winter comfort conditions, the energy needed to evaporate an amount of water equivalent to what is lost by infiltration (latent component of infiltration heat loss) must be determined. This energy may be calculated by:

$$q_l = Q\rho\,(W_i - W_o)h_{fg} \tag{9}$$

where

q_l = heat flow required to increase moisture content of air leakage into building from W_o to W_i, Btu/h
Q = volume of outdoor air entering building, ft³/h
ρ = density of air at temperature t_i, lb/ft³
W_i = humidity ratio of indoor air, lb/lb$_{dry\ air}$
W_o = humidity ratio of outdoor air, lb/lb$_{dry\ air}$
h_{fg} = latent heat of vapor at t_i, Btu/lb

If the latent heat of vapor h_{fg} is 1076 Btu/lb, and the air density is 0.075 lb/ft³, Equation (7) reduces to:

$$q_l = 80.7 \, Q(W_i - W_o) \tag{10}$$

Crack Length Method

For designers who prefer the crack method, the basis of calculation is: the amount of crack used for computing the infiltration heat loss should not be less than one-half the total length of crack in the outside walls of the room. In a building without partitions, air entering through cracks on the windward side must leave through cracks on the leeward side. Therefore, one-half the total crack for each side and end of the building is used for calculation. In a room with one exposed wall, all the crack is used. With two, three, or four exposed walls, the wall with the crack that will result in the greatest air leakage or at least one-half the total crack is used, whichever is greater.

In residences, total infiltration loss of the house is generally considered equal to the sum of infiltration losses of the various rooms.

But, at any given time, infiltration takes place only on the windward side or sides and not on the leeward. Therefore, for determining total heat requirements of larger buildings, it is more accurate to base total infiltration loss on the wall with the most total crack or at least half the total crack in the building, whichever is greater. When the crack method is used for estimating leakage, rather than Equation (8), the heat loss in terms of the crack length may be expressed as

$$q_s = 0.018BL(t_i - t_o) \qquad (11)$$

and

$$q_l = 80.7BL(W_i - W_o) \qquad (12)$$

where

    B = air leakage for wind velocity and type of window or door crack involved, ft³/h per foot of crack

    L = length of window or door crack to be considered, ft

Air Change Method

Some designers base infiltration on an estimated number of air changes rather than the length of window cracks. The number of air changes given in Chapter 23 should be considered only as a guide. When calculating infiltration losses by the air change method, Equations (8) and (10) can be used by substituting for V the volume of the room multiplied by the number of air changes.

Exposure Factors

Some designers use empirical *exposure* factors to increase calculated heat loss of rooms or spaces on the side(s) of the building exposed to prevailing winds. However, exposure factors are not needed with the method of calculating heat loss described in this chapter. Instead, they may be (1) regarded as safety factors for the rooms or spaces exposed to prevailing winds, to allow for additional capacity for these spaces, or (2) used to balance the radiation, particularly in the case of multistory buildings. Tall buildings may have severe infiltration heat losses, induced by stack effect, that require special analysis. Although a 15% exposure allowance is often assumed, the actual allowance, if any, is largely a matter of experience and judgment; no test data are available from which to develop rules for the many conditions encountered.

PICK-UP LOAD

For intermittently heated buildings, or in the case of night thermostat setback, additional heat is required to raise the temperature of air, building materials, and material contents of a building to the specified indoor temperature. The pick-up load, which is the rate at which this additional heat must be supplied, depends on the heat capacity of the structure and its material contents and the time in which these are to be heated.

Relatively little information on pick-up load exists; however, some early work by Smith (1941, 1942) addressed pick-up loads for buildings heated only occasionally, such as auditoriums and churches. Nelson and MacArthur (1978) studied the relationship between thermostat setback, furnace capacity, and recovery time. Based on this limited information, the following design guidelines are offered.

Because design outdoor temperatures generally provide a substantial margin for outdoor temperatures typically experienced during operating hours, many engineers make no allowance for this additional heat in most buildings. However, if a minimum safety factor is to be used, the additional heat should be computed and allowed for, as conditions require. In the case of intermittently heated buildings, an additional 10% capacity should be provided.

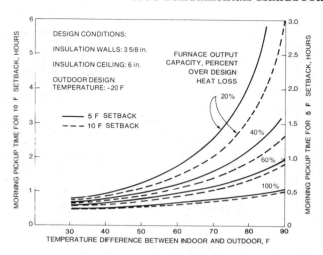

Fig. 9   Operating Time of Furnace Required to Pick Up Space Temperature Following 5 and 10 °F Night Setback for Various Furnace Sizes

In buildings with setback-type thermostats, the furnace must be oversized to allow for reestablishing the space temperature in an acceptable time. The amount of oversizing depends on many factors, such as the amount of setback, inside-to-outside temperature difference, building construction and acceptable pick-up time. Figure 9 indicates this relationship for a particular residence. As a general rule for residences, a 10 °F night setback requires 40% oversizing for acceptable pick-up time and minimum energy requirements (Nelson and MacArthur 1978). For smaller setback, the oversizing can be proportionally less. If daytime as well as night setback is practiced, oversizing of up to 60% is warranted.

REFERENCES

Bligh, T.P., P. Shipp, and G. Meixel. 1978. Energy comparisons and where to insulate earth sheltered buildings and basements. Earth covered settlements, U.S. Department of Energy Conference, Fort Worth, TX.

Chang, J.H. 1958. *Ground temperature*. Bluehill Meteorological Observatory, Harvard University, Cambridge, MA.

Harrje, D.T., G.S. Dutt, and J. Beyea. 1979. Locating and eliminating obscure but major energy losses in residential housing. ASHRAE *Transactions* 85(2).

Houghten, F.C., S.I. Taimuty, C. Gutberlet, and C.J. Brown. 1942. Heat loss through basement walls and floors. ASHVE *Transactions* 48:369.

Joy, F.A. 1958. Improving attic space insulating values. *Heating, Piping and Air Conditioning* 30(1):223.

Joy, F.A., J.J. Zabrony, and S. Bhaduri. 1956. Insulating value of reflective elements in an attic under winter conditions. Pennsylvania State University, University Park, PA.

Latta, J.K. and G.G. Boileau. 1969. Heat losses from house basements. *Canadian Building* 19(10):39.

McQuiston, F.C. 1984. A study and review of existing data to develop a standard methodology for residential heating and cooling load calculations. ASHRAE *Transactions* 90(2A).

McMcQuiston, F.C. and J.D. Spitler. 1992. *Cooling and heating load calculation manual*, 2nd ed. ASHRAE, Atlanta.

Nelson, L.W. and J.W. MacArthur. 1978. Energy savings through thermostat setback. ASHRAE *Transactions* 84(2):3l9-34 and ASHRAE *Journal* 20(9):49-54.

Peony, B.A., F.J. Powell, and D.M. Burch. 1979. Dynamic thermal performance of an experimental masonry building. NBS Report 10 664, National Bureau of Standards, Washington, D.C.

Rowley, F.B., A.B. Algren, and C.E. Lund. 1940. Methods of moisture control and their application to building construction. *Bulletin* No. 17 XLIII(4):28. University of Minnesota Engineering Experiment Station.

Smith, E.G. 1941. Heat requirement of intermittently heated buildings. Texas A&M Engineering Experiment Station Series No. 62 (November). College Station, TX.

Smith, E.G. 1942. A method of compiling tables for intermittent heating. *Heating, Piping, and Air Conditioning* 14(6):386.

Wang, F.S. 1979. Mathematical modeling and computer simulation of insulation systems in below grade applications. ASHRAE/DOE Conference on Thermal Performance of the Exterior Envelopes of Buildings, Orlando, FL.

BIBLIOGRAPHY

Ackridge, J.M. and J.F. Poulos. 1983. The decremented average ground temperature method for predicting the thermal performance of underground walls. ASHRAE *Transactions* 89(2A).

Burch, D.M., D.F. Krintz, and R.S. Spain. 1984. The effect of wall mass on winter heating loads and indoor comfort—An experimental study. ASHRAE *Transactions* 90(1B).

Burch, D.M., B.A. Peavy, and F.J. Powell. 1975. Comparison between measured and computer predicted hourly heating and cooling energy requirements for an instrumented wood-frame townhouse subjected to laboratory tests. ASHRAE *Transactions* 81(2)19-69.

FHA. Minimum property standards. 1978. U.S. Department of Housing and Urban Development, Federal Housing Administration, Washington, D.C.

Kusuda, T. and J.W. Bean. 1984. Simplified methods for determining seasonal heat loss from uninsulated slab-on-grade floors. ASHRAE *Transactions* 90(1B).

Mitalas, G.P. 1983. Calculation of basement heat loss. ASHRAE *Transactions* 89(1B).

Peavy, B.A., D.M. Burch, F.J. Powell, and C.M. Hunt. 1975. Comparison of measured and computer predicted thermal performance of a four-bedroom wood-frame townhouse. NBS Building Science Series 57 (April).

Wilcox, T.N., *et al.* 1954. Analog computer analysis of residential cooling loads. ASHVE *Transactions* 60:505.

Yard, D.C., M. Morton-Gibson, and J.W. Mitchell. 1984. Simplified dimensionless relations for heat loss from basements. ASHRAE *Transactions* 90(1B).

CHAPTER 26

NONRESIDENTIAL AIR-CONDITIONING COOLING AND HEATING LOAD

THIS chapter presents three methods of calculating air-conditioning cooling load for sizing cooling equipment and a general procedure for calculating heating load, for nonresidential applications. In addition, the fundamental principles for calculating heating loads are presented as a counterpart to cooling load calculation. For residential applicaitons, consult Chapter 25. For information on cooling and/or heating equipment energy use, consult Chapter 28.

The heat balance approach is a fundamental concept in calculating cooling loads. While generally cumbersome for widespread or routine use, this underlying concept is the basis for each of the three simplified procedures outlined for varying purposes.

The cooling calculation procedure most closely approximating the heat balance concept is the transfer function method (TFM), first introduced in the 1972 ASHRAE *Handbook of Fundamentals*. This computer-based procedure takes place in two steps, first establishing the heat gain from all sources and then determining the conversion of such heat gain into cooling load. Developed as an hour-by-hour calculation procedure oriented to simulate annual energy use, its normalizing characteristics make it particularly appropriate for that application.

A simplified version of the TFM, which can be used with certain types of buildings for which application data are available, was presented in the 1977 ASHRAE *Handbook of Fundamentals*. This one-step procedure uses cooling load temperature differences (CLTD), solar cooling load factors (SCL), and internal cooling load factors (CLF), to calculate cooling loads as an approximation of the TFM. Where applicable, this method may be suitable for hand calculation use.

An alternative simplification of the heat balance technique uses total equivalent temperature differential values and a system of time-averaging (TETD/TA) to calculate cooling loads. Also a computer-based, two-step procedure (heat gain, then cooling load), first introduced in the 1967 ASHRAE *Handbook of Fundamentals*, this method gives valid broad-range results to experienced users.

COOLING LOAD PRINCIPLES

The variables affecting cooling load calculations are numerous, often difficult to define precisely, and always intricately inter-related. Many cooling load components vary in magnitude over a wide range during a 24-h period. Since these cyclic changes in load components are often not in phase with each other, each must be analyzed to establish the resultant maximum cooling load for a building or zone. A zoned system (a system of conditioning equipment serving several independent areas, each with its own temperature control) need recognize no greater total cooling load capacity than the largest hourly summary of simultaneous zone loads throughout a design day; however, it must handle the peak cooling load for each zone at its individual peak hour. At certain times of the day during the heating or intermediate seasons, some zones may require heating while others require cooling.

Calculation accuracy. The concept of determining the cooling load for a given building must be kept in perspective. A proper cooling load calculation gives values adequate for proper performance. Variation in the heat transmission coefficients of typical building materials and composite assemblies, the differing motivations and skills of those who physically construct the building, and the manner in which the building is actually operated are some of the variables that make a numerically precise calculation impossible. While the designer uses reasonable procedures to account for these factors, the calculation can never be more than a good estimate of the actual cooling load.

Heat flow rates. In air-conditioning design, four related heat flow rates, each of which varies with time, must be differentiated: (1) space heat gain, (2) space cooling load, (3) space heat extraction rate, and (4) cooling coil load.

Space heat gain. This instantaneous rate of heat gain is the rate at which heat enters into and/or is generated within a space at a given instant. Heat gain is classified by (1) the mode in which it enters the space and (2) whether it is a sensible or latent gain.

Mode of entry. The modes of heat gain may be as (1) solar radiation through transparent surfaces; (2) heat conduction through exterior walls and roofs; (3) heat conduction through interior partitions, ceilings, and floors; (4) heat generated within the space by occupants, lights, and appliances; (5) energy transfer as a result of ventilation and infiltration of outdoor air; or (6) miscellaneous heat gains.

Sensible or latent heat. Sensible heat gain is directly added to the conditioned space by conduction, convection, and/or radiation. Latent heat gain occurs when moisture is added to the space (*e.g.*, from vapor emitted by occupants and equipment). To maintain a constant humidity ratio, water vapor must condense on

---

The preparation of this chapter is assigned to TC 4.1, Load Calculation Data and Procedures.

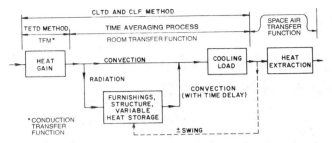

Fig. 1 Origin of Difference between Magnitude of Instantaneous Heat Gain and Instantaneous Cooling Load

cooling apparatus at a rate equal to its rate of addition into the space. The amount of energy required to offset the latent heat gain essentially equals the product of the rate of condensation and the latent heat of condensation. In selecting cooling apparatus, it is necessary to distinguish between sensible and latent heat gain. Every cooling apparatus has a maximum sensible heat removal capacity and a maximum latent heat removal capacity for particular operating conditions.

Space cooling load. This is the rate at which heat must be removed from the space to maintain a constant space air temperature. The sum of all space instantaneous heat gains at any given time does not necessarily (or even frequently) equal the cooling load for the space at that same time.

Radiant heat gain. Space heat gain by radiation is not immediately converted into cooling load. Radiant energy must first be absorbed by the surfaces that enclose the space (walls, floor, and ceiling) and the objects in the space (furniture, etc.). As soon as these surfaces and objects become warmer than the space air, some of their heat is transferred to the air in the space by convection. The composite heat storage capacity of these surfaces and objects determines the rate at which their respective surface temperatures increase for a given radiant input, and thus governs the relationship between the radiant portion of heat gain and its corresponding part of the space cooling load (Figure 1). The thermal storage effect is critically important in differentiating between instantaneous heat gain for a given space and its cooling load for that moment. Predicting the nature and magnitude of this elusive phenomenon in order to estimate a realistic cooling load for a particular combination of circumstances has long been a subject of major interest to design engineers. The bibliography lists some of the early work on the subject.

Space Heat Extraction Rate

The rate at which heat is removed from the conditioned space equals the space cooling load only to the degree that room air temperature is held constant. In conjunction with intermittent operation of the cooling equipment, the control system characteristics usually permit a minor cyclic variation or swing in room temperature. Therefore, a proper simulation of the control system gives a more realistic value of energy removal over a fixed time period than using the values of the space cooling load. This concept is primarily important for estimating energy use over time (see Chapter 28); however, it is not needed to calculate design peak cooling load for equipment selection. Space heat extraction rate calculation is discussed later in this chapter; see also Mitalas (1972).

Cooling Coil Load

The rate at which energy is removed at the cooling coil that serves one or more conditioned spaces equals the sum of the instantaneous space cooling loads (or space heat extraction rate if it is assumed that the space temperature does not vary) for all the spaces served by the coil, plus any external loads. Such external loads include heat gain by the distribution system between the individual spaces and the cooling equipment, and outdoor air heat and moisture introduced into the distribution system through the cooling equipment.

SPACE COOLING LOAD CALCULATION TECHNIQUES

Heat Balance Fundamentals

The calculation of cooling load for a space involves calculating a surface-by-surface conductive, convective, and radiative heat balance for each room surface and a convective heat balance for the room air. Sometimes called "the exact solution," these principles form the foundation for all other methods described in this chapter.

To calculate space cooling load directly by heat balance procedures requires a laborious solution of energy balance equations involving the space air, surrounding walls and windows, infiltration and ventilation air, and internal energy sources. To demonstrate the calculation principle, consider a sample room enclosed by four walls, a ceiling, and a floor, with infiltration air and normal internal energy sources. The calculations that govern energy exchange at each inside surface at a given time are:

$$q_{i,\theta} = [h_{ci}(t_{a,\theta} - t_{i,\theta}) + \sum_{j=1 \neq i}^{m} g_{ij}(t_{j,\theta} - t_{i,\theta})] A_i$$
$$+ RS_{i,\theta} + RL_{i,\theta} + RE_{i,\theta} \text{ for } i = 1, 2, 3, 4, 5, 6 \quad (1)$$

where

m = number of surfaces in room (6 in this case)
$q_{i,\theta}$ = rate of heat conducted into surface i at inside surface at time θ
A_i = area of surface i
h_{ci} = convective heat transfer coefficient at interior surface i
g_{ij} = radiation heat transfer factor between interior surface i and interior surface j
$t_{a,\theta}$ = inside air temperature at time θ
$t_{i,\theta}$ = average temperature of interior surface i at time θ
$RS_{i,\theta}$ = rate of solar energy coming through windows and absorbed by surface i at time θ
$RL_{i,\theta}$ = rate of heat radiated from lights and absorbed by surface i at time θ
$RE_{i,\theta}$ = rate of heat radiated from equipment and occupants and absorbed by surface i at time θ

Conduction transfer functions. The equations governing conduction within the six surfaces cannot be solved independently of Equation (1), since the energy exchanges occurring within the room affect the inside surface conditions, in turn affecting the internal conduction. Consequently, the above mentioned six formulations of Equation (1) must be solved simultaneously with the governing equations of conduction within the six surfaces in order to calculate the space cooling load. Typically, these equations are formulated as conduction transfer functions in the form

$$q_{in,\theta} = \sum_{m=1}^{M} Y_{k,m} t_{o,\theta-m+1} - \sum_{m=1}^{M} Z_{k,m} t_{in,\theta-m+1}$$
$$+ \sum_{m=1}^{m} F_m q_{in,\theta-m} \quad (2)$$

where

in = inside surface subscript
k = order of CTF
m = time index variable
M = number of nonzero CTF values
o = outside surface subscript

t = temperature
θ = time
x = exterior CTF values
Y = cross CTF values
Z = interior CTF values
F_m = flux history coefficients

Space air energy balance. Note that the interior surface temperature, $t_{i,\theta}$ in Equation (1) and $t_{in,\theta}$ in Equation (2), requires simultaneous solution. In addition, Equation (3) representing an energy balance on the space air must also be solved simultaneously

$$Q_{L,\theta} = \left[\sum_{i=1}^{6} h_{ci}(t_{i,\theta} - t_{a,\theta}) \right] A_i + {}_\sigma CV_{L,t}(t_{o,\theta} - t_{a,\theta})$$
$$+ {}_\sigma CV_{v,\theta}(t_{v,\theta} - t_{a,\theta}) + RS_{a,\theta} + RL_{a,\theta} + RE_{a,\theta} \quad (3)$$

where

σ = air density
C = air specific heat
$V_{L,\theta}$ = volume flow rate of outdoor air infiltrating into room at time θ
$t_{o,\theta}$ = outdoor air temperature at time θ
$V_{v,\theta}$ = volume rate of flow of ventilation air at time θ
$t_{v,\theta}$ = ventilation air temperature at time θ
$RS_{a,\theta}$ = rate of solar heat coming through windows and convected into room air at time θ
$RL_{a,\theta}$ = rate of heat from lights convected into room air at time θ
$RE_{a,\theta}$ = rate of heat from equipment and occupants and convected into room air at time θ

Note that the space air temperature is allowed to float. By fixing the space air temperature, the cooling load need not be determined simultaneously.

This rigorous approach to calculating space cooling load is impractical without the speed at which some computations can be done by modern digital computers. Computer programs in use where instantaneous space cooling loads are calculated in this exact manner are primarily oriented to energy use calculations over extended periods (Mitalas and Stephenson 1967, Buchberg 1958, Walton 1982).

The transfer function concept is a simplification to the strict heat balance calculation procedure. In the transfer function concept, Mitalas and Stephenson (1967) used room thermal response factors. In their procedure, room surface temperatures and cooling load were first calculated by the rigorous method just described, for several typical constructions representing offices, schools, and dwellings of heavy, medium, and light construction. In these calculations, components such as solar heat gain, conduction heat gain, or heat gain from the lighting, equipment, and occupants were simulated by pulses of unit strength. The transfer functions were then calculated as numerical constants representing the cooling load corresponding to the input excitation pulses. Once these transfer functions were determined for typical constructions they were assumed independent of input pulses, thus permitting cooling loads to be determined without the more rigorous calculation. Instead, the calculation requires simple multiplication of the transfer functions by a time-series representation of heat gain and subsequent summation of these products, which can be carried out on a small computer. The same transfer function concept can be applied to calculating heat gain components themselves, as explained later.

Total Equivalent Temperature Differential Method

In the total equivalent temperature differential (TETD) method, the response factor technique was used with a number of representative wall and roof assemblies from which data were derived to calculate TETD values as functions of sol-air temperature and maintained room temperature. Various components of space heat gain are calculated using associated TETD values, and the results are added to internal heat gain elements to get an instantaneous total rate of space heat gain. This gain is converted to an instantaneous space cooling load by the time-averaging (TA) technique of averaging the radiant portions of the heat gain load components for the current hour with related values from an appropriate period of immediately preceding hours. This technique provides a rational means to deal quantitatively with the thermal storage phenomenon, but it is best solved by computer because of its complexity.

Transfer Function Method

Although similar in principle to TETD/TA, the transfer function method (TFM) (Mitalas 1972) applies a series of weighting factors, or conduction transfer function (CTF) coefficients to the various exterior opaque surfaces and to differences between sol-air temperature and inside space temperature to determine heat gain with appropriate reflection of thermal inertia of such surfaces. Solar heat gain through glass and various forms of internal heat gain are calculated directly for the load hour of interest. The TFM next applies a second series of weighting factors, or coefficients of room transfer functions (RTF), to heat gain and cooling load values from all load elements having radiant components, to account for the thermal storage effect in converting heat gain to cooling load. Both evaluation series consider data from several previous hours as well as the current hour. RTF coefficients relate specifically to the spatial geometry, configuration, mass, and other characteristics of the space so as to reflect weighted variations in thermal storage effect on a time basis rather than a straight-line average.

Transfer functions. These coefficients relate an output function at a given time to the value of one or more driving functions at a given time and at a set period immediately preceding. The CTF described in this chapter is no different from the thermal response factor used for calculating wall or roof heat conduction, while the RTF is the weighting factor for obtaining cooling load components (ASHRAE 1975). The bibliography lists reports of various experimental work that has validated the predictive accuracy of the TFM. While the TFM is scientifically appropriate and technically sound for a specific cooling load analysis, its computational complexity requires computer use for effective application in a commercial design environment.

CLTD/SCL/CLF Method

ASHRAE has sponsored research to compare the TETD/TA and TFM (Rudoy and Duran 1975). As part of this work, data obtained by using the TFM on a group of applications considered representative were then used to generate cooling load temperature differential (CLTD) data, for direct one-step calculation of cooling load from conduction heat gain through sunlit walls and roofs and conduction through glass exposures (see Bibliography). Cooling load factors (CLF) for similar one-step calculation of solar load through glass and for loads from internal sources were also developed. More recent research (McQuiston 1992) developed an improved factor for solar load through glass, the solar cooling load (SCL) factor, which allows additional influencing parameters to be considered for greater accuracy. CLTDs, SCLs, and CLFs all include the effect of (1) time lag in conductive heat gain through opaque exterior surfaces and (2) time delay by thermal storage in converting radiant heat gain to cooling load. This simplification allows cooling loads to be calculated manually; thus, when data are available and are appropriately used, the results are consistent with those from the TFM, thus making the method popular for instruction.

Application Experience

The CLTD and CLF tables published in previous editions of the Fundamentals volume and in the original *Cooling and Heating Load Calculation Manual* (ASHRAE 1979) are normalized data, based on applications of the original TFM data presented in the 1972 Fundamentals volume. Subsequent studies investigating the effects of 1981 to 1985 RTF data indicated results generally less conservative than those computed with the 1972 data. More recent research, however, suggests otherwise (McQuiston 1992), and the revised values for 1993, including the new SCLs, are currently considered more realistic for design load purposes.

CLTD Data. The originally developed CLTD data were so voluminous that they were first limited to 13 representative flat roof assemblies (with and without ceilings, for 26 total cases) and 7 wall groups (into which 41 different wall assemblies can be categorized). Twenty-four hourly CLTD values were tabulated for each of the 26 roof cases and each of the 7 wall groups, broken down for walls into 8 primary orientations. Adjustments were then required for specific north latitude and month of calculation. Reliability of adjustments was reasonably consistent during summer months but became much less realistic for early and late hours during traditionally noncooling load months.

Solar heat gain data. Solar heat gain through glass required similar data compression to present a corresponding range of conditions. Tables of maximum solar heat gain factors (SHGF) were listed for every 4° of north latitude between 0 and 64°, for each month and by 16 compass directions and horizontal. Cooling load factors (CLF), decimal multipliers for SHGF data, were tabulated for unshaded glass in spaces having carpeted or uncarpeted floors and for inside-shaded glass with any room construction. Unshaded CLFs were presented for each of 24 hours by 8 compass directions plus horizontal, further categorized by light, medium, or heavy room construction. Inside-shaded CLFs disregarded construction mass but included 16 orientations plus horizontal. The product of the selected CLTD and CLF values represented cooling load per unit area as a single process. CLF values published in the *Handbook* were derived for the period May through September as normally the hottest months for load calculation purposes. As with CLTDs, the reliability of CLF data deteriorated rapidly for applications during early and late hours of months considered "noncooling load" periods.

ASHRAE Research. For some space geometries and building constructions, the tabulated CLTD and CLF data published through 1989 were found also to be too restrictive or limited. The weighting factors used to generate these data, based on representative spaces in schools, offices, and dwellings at the time of the original research, did not reflect current design and construction practices. ASHRAE research investigated the sensitivity of the weighting factors to variations in space construction, size, exposure, and related conditions to update the tabular data. However, the investigators discovered that the range and amplitude of this sensitivity was much broader than previously thought, rendering even more impractical the generation of enough tabular material to cover the majority of normal applications. Accordingly, two significant changes in direction have occured:

1. The section describing the CLTD/CLF in the 1985 and 1989 editions of the Fundamentals volume recommended caution in application of this procedure for general practice, and this cautionary notice was also added as an insert to the *Cooling and Heating Load Calculation Manual* (McQuiston and Spitler 1992).
2. The system itself was modified for more specific tabulation of data, abandoning the maximum SHGF concept and incorporating solar cooling load (SCL) factors for estimating cooling load from glass.

The main thrust of recent ASHRAE research has been to update the *Cooling and Heating Load Calculation Manual*. Information from earlier research was used to revise the original factors by incorporating additional parameters, including separating solar load through glass from the CLF category and creating more appropriate SCL factors for that component. Still faced with too much tabular data, information was tabulated only for limited use and representative examples, but it was accompanied by instructions for customizing similar data for specific application; a microcomputer database was also provided to facilitate such calculations. Certain limitations resulting from normalization of data remain, for which anticipated error ranges are listed to aid in evaluating results. The section in this chapter describing the CLTD/SCL/CLF method has incorporated this latest research, but it does not provide the microcomputer program.

TETD/TA Method. Prior to introduction of the CLTD/CLF, most users had turned to computer-based versions of the time-averaging technique, proven successful and practical in ten years of heavy use. Most users, however, recognized the subjectivity of determining the relative percentages of radiant heat in the various heat gain components and selecting the number of hours over which to average such loads—both of which must rely on the individual experience of the user rather than on research or support in the scientific literature. Harris and McQuiston (1988) developed decrement factors and time lag values. In this chapter, these factors have been keyed to typical walls and roofs. All other tabular data pertaining to this method has been deleted, so that since 1989, information has been confined to basic algorithms intended for continued computer applications.

Primary emphasis of future research. Because of the increasing availability and power of microcomputers, emphasis has refocused on the more specific and flexible transfer function method as the fundamental computational technique. Future research will concentrate on refining the weighting factors and broadening the range of circumstances to which the TFM can be applied.

Alternative procedures. The CLTD/SCL/CLF and TETD/TA procedures, tables, and related data will continue to be appropriate and dependable when applied within the limitations identified. Users will likely incorporate newer relationships developed for TFM usage in the development of specific CLTD/SCL/CLF or TETD/TA tabular data, customized for their specialized projects.

INITIAL DESIGN CONSIDERATIONS

To calculate a space cooling load, detailed building design information and weather data at selected design conditions are required. Generally, the following steps should be followed:

Data Assembly

1. **Building characteristics.** Obtain characteristics of the building. Building materials, component size, external surface colors and shape are usually determined from building plans and specifications.
2. **Configuration.** Determine building location, orientation and external shading from building plans and specifications. Shading from adjacent buildings can be determined by a site plan or by visiting the proposed site, but should be carefully evaluated as to its probable permanence before it is included in the calculation. The possibility of abnormally high ground-reflected solar radiation (*i.e.*, from adjacent water, sand, or parking lots), or solar load from adjacent reflective buildings should not be overlooked.
3. **Outdoor design conditions.** Obtain appropriate weather data and select outdoor design conditions. Weather data can be obtained from local weather stations or from the National Climatic Center, Asheville, NC 28801. For outdoor design

conditions for a large number of weather stations, see Chapter 24. Note, however, that the scheduled values for 5%, 2-1/2%, or even 1% design dry-bulb and mean coincident wet-bulb temperatures can vary considerably from data traditionally used in various areas. Use appropriate judgment to ensure that results are consistent with expectations. Also, consider prevailing wind velocity and the relationship of a project site to the selected weather station.

4. **Indoor design conditions**. Select indoor design conditions, such as indoor dry-bulb temperature, indoor wet-bulb temperature, and ventilation rate. Include permissible variations and control limits.

5. **Operating schedules**. Obtain a proposed schedule of lighting, occupants, internal equipment, appliances, and processes that contribute to the internal thermal load. Determine the probability that the cooling equipment will be operated continuously or shut off during unoccupied periods (*e.g.*, nights and/or weekends).

6. **Date and time**. Select the time of day and month to do the cooling load calculation. Frequently, several different times of day and several different months must be analyzed to determine the peak load time. The particular day and month are often dictated by peak solar conditions, as tabulated in Tables 3 through 11 in Chapter 27. For southern exposures in north latitudes above 32° having large fenestration areas, the peak space cooling load usually occurs in December or January. To calculate a space cooling load under these conditions, the warmest temperature for the winter months must be known. These data can be found in the National Climatic Center's Climatic Atlas of the United States.

Use of data. Once the data is assembled, calculate the space cooling load at design conditions, as outlined in the following sections of this chapter.

Additional Considerations

The proper design and sizing of all-air or air-and-water central air-conditioning systems require more than calculation of the cooling load in the space to be conditioned. The type of air-conditioning system, fan energy, fan location, duct heat loss and gain, duct leakage, heat extraction lighting systems, and type of return air all affect system load and component sizing. Adequate system design and component sizing require that system performance be analyzed as a series of psychrometric processes. Chapter 3 of the 1992 ASHRAE *Handbook—Systems and Equipment* describes some elements of this technique in detail, while others are delineated in this chapter.

HEAT GAIN CALCULATION CONCEPTS

Heat Gain through Fenestration Areas

The primary weather-related variable influencing the cooling load for a building is solar radiation. The effect of solar radiation is more pronounced and immediate in its impact on exposed nonopaque surfaces. The calculation of solar heat gain and conductive heat transfer through various glazing materials and associated mounting frames, with or without interior and/or exterior shading devices, is discussed in Chapter 27. This chapter covers the application of such data to the overall heat gain evaluation and the conversion of the calculated heat gain into a composite cooling load for the conditioned space.

Heat Gain through Exterior Surfaces

Heat gain through exterior opaque surfaces is derived from the same elements of solar radiation and thermal gradient as that for fenestration areas. It differs primarily as a function of the mass and nature of the wall or roof construction, since those elements affect the rate of conductive heat transfer through the composite assembly to the interior surface.

Sol-Air Temperature

Sol-air temperature is the temperature of the outdoor air that, in the absence of all radiation changes, gives the same rate of heat entry into the surface as would the combination of incident solar radiation, radiant energy exchange with the sky and other outdoor surroundings, and convective heat exchange with the outdoor air.

Heat flux into exterior sunlit surfaces. The heat balance at a sunlit surface gives the heat flux into the surface q/A as

$$q/A = \alpha I_t + h_o(t_o - t_s) - \epsilon \Delta R \qquad (4)$$

where

α = absorptance of surface for solar radiation
I_t = total solar radiation incident on surface, Btu/(h·ft²)
h_o = coefficient of heat transfer by long-wave radiation and convection at outer surface, Btu/(h·ft²·°F)
t_o = outdoor air temperature, °F
t_s = surface temperature, °F
ϵ = hemispherical emittance of surface
R = difference between long-wave radiation incident on surface from sky and surroundings and radiation emitted by blackbody at outdoor air temperature, Btu/(h·ft²)

Assuming the rate of heat transfer can be expressed in terms of the sol-air temperature t_e

$$q/A = h_o(t_e - t_s) \qquad (5)$$

and from Equations (4) and (5)

$$t_e = t_o + \alpha I_t/h_o - \epsilon \Delta R/h_o \qquad (6)$$

Horizontal surfaces. For horizontal surfaces that receive long-wave radiation from the sky only, an appropriate value of ΔR is about 20 Btu/(h·ft²), so that if $\epsilon = 1$ and $h = 3.0$ Btu/(h·ft²·°F), the long-wave correction term is about -7°F (Bliss 1961).

Vertical surfaces. Because vertical surfaces receive long-wave radiation from the ground and surrounding buildings as well as from the sky, accurate ΔR values are difficult to determine. When solar radiation intensity is high, surfaces of terrestrial objects usually have a higher temperature than the outdoor air; thus, their long-wave radiation compensates to some extent for the sky's low emittance. Therefore, it is common practice to assume $\Delta R = 0$ for vertical surfaces.

Tabulated temperature values. The sol-air temperatures in Table 1 have been calculated based on $\Delta R/h_o$ being -7°F for horizontal surfaces and 0°F for vertical surfaces; total solar intensity values used for the calculations were the same as those used to evaluate the solar heat gain factors (SHGF) for July 21 at 40°N latitude (Chapter 27). These values of I_t incorporate diffuse radiation from a clear sky and ground reflection, but make no allowance for reflection from adjacent walls.

Surface colors. Sol-air temperature values are given for two values of the parameter α/h_o (Table 1); the value of 0.15 is appropriate for a light-colored surface, while 0.30 represents the usual maximum value for this parameter (*i.e.*, for a dark-colored surface, or any surface for which the permanent lightness can not reliably be anticipated).

Air temperature cycle. The air temperature cycle used to calculate the sol-air temperatures is given in Column 2, Table 1. Sol-air temperatures can be adjusted to any other air temperature cycle simply by adding or subtracting the difference between the desired air temperature and the air temperature value given in Column 2.

Table 1   Sol-Air Temperatures t_e for July 21, 40 °N Latitude

$$t_e = t_o + \alpha I_t/h_o/h_o - \epsilon\delta R/h_o$$

Time	Air Temp., t_o	Light Colored Surface, $\alpha/h_o = 0.15$									Time	Air Temp., °F	Dark Colored Surface, $\alpha/h_o = 0.30$								
		N	NE	E	SE	S	SW	W	NW	HOR			N	NE	E	SE	S	SW	W	NW	HOR
1	76	76	76	76	76	76	76	76	76	69	1	76	76	76	76	76	76	76	76	76	69
2	76	76	76	76	76	76	76	76	76	69	2	76	76	76	76	76	76	76	76	76	69
3	75	75	75	75	75	75	75	75	75	68	3	75	75	75	75	75	75	75	75	75	68
4	74	74	74	74	74	74	74	74	74	67	4	74	74	74	74	74	74	74	74	74	67
5	74	74	74	74	74	74	74	74	74	67	5	74	74	75	75	74	74	74	74	74	67
6	74	80	93	95	84	76	76	76	76	72	6	74	85	112	115	94	77	77	77	77	77
7	75	80	99	106	94	78	78	78	78	81	7	75	84	124	136	113	81	81	81	81	94
8	77	81	99	109	101	82	81	81	81	92	8	77	85	121	142	125	86	85	85	85	114
9	80	85	96	109	106	88	85	85	85	102	9	80	90	112	138	131	96	89	89	89	131
10	83	88	91	105	107	95	88	88	.88	111	10	83	94	100	127	131	107	94	94	94	145
11	87	93	93	99	106	102	93	93	93	118	11	87	98	99	111	125	118	100	98	98	156
12	90	96	96	96	102	106	102	96	96	122	12	90	101	101	102	114	123	114	102	101	162
13	93	99	99	99	99	108	112	105	99	124	13	93	104	104	104	106	124	131	117	105	162
14	94	99	99	99	99	106	118	116	102	122	14	94	105	105	105	105	118	142	138	111	156
15	95	100	100	100	100	103	121	124	111	117	15	95	105	104	104	104	111	146	153	127	146
16	94	98	98	98	98	99	118	126	116	109	16	94	102	102	102	102	103	142	159	138	131
17	93	98	96	96	96	96	112	124	117	99	17	93	102	99	99	99	99	131	154	142	112
18	91	97	93	93	93	93	101	112	110	89	18	91	102	94	94	94	94	111	132	129	94
19	87	87	87	87	87	87	87	87	87	80	19	87	87	87	87	87	87	87	88	88	80
20	85	85	85	85	85	85	85	85	85	78	20	85	85	85	85	85	85	85	85	85	78
21	83	83	83	83	83	83	83	83	83	76	21	83	83	83	83	83	83	83	83	83	76
22	81	81	81	81	81	81	81	81	81	74	22	81	81	81	81	81	81	81	81	81	74
23	79	79	79	79	79	79	79	79	79	72	23	79	79	79	79	79	79	79	79	79	72
24	77	77	77	77	77	77	77	77	77	70	24	77	77	77	77	77	77	77	77	77	70
Avg.	83	86	88	90	90	87	90	90	88	90	Avg.	83	89	94	99	97	93	97	99	94	104

Note: Sol-air temperatures are calculated based on $\epsilon\delta/h_o = -7$ °F for horizontal surfaces and 0 °F for vertical surfaces.

Adjustments. Sol-air temperature cycles can be estimated for other dates and latitudes by using the data in Tables 12 through 18, Chapter 27. For any of the times, dates, and wall orientations listed in those tables, the value of I_t is approximately 1.15 × SHGF. However, the 1.15 factor is approximate and only accounts for the solar energy excluded by a single sheet of ordinary window glass. For surfaces with other orientations or slope angles of other than 0°, and for more accurate estimates at incident angles above 50° (particularly critical for southern exposures), the solar intensity can be found by the method outlined in Chapter 27.

Average sol-air temperature. The average daily sol-air temperature t_{ea} can be calculated for any of the situations covered by Tables 12 through 18 of Chapter 27:

$$t_{ea} = t_{oa} + \alpha/[h_o(I_{DT}/24)] - (\epsilon\Delta R/h_o) \qquad (7)$$

where I_{DT} is the sum of two appropriate half-day totals of solar heat gain in Btu/(h·ft²). For example, the average sol-air temperature for a wall facing southeast at 40 °N latitude on August 21 would be

$$t_{ea} = t_{oa} + \alpha/h_o[1.15(956 + 205)/24]$$

The daily solar heat gain of double-strength sheet glass is 956 + 205 Btu/(h·ft²) in a southeast facade at this latitude and date (Table 16, Chapter 27); and $\epsilon\Delta R/h_o$ is assumed to be zero for this vertical surface.

Hourly air temperatures. The hourly air temperatures in Column 2, Table 1 are for a location with a design temperature of 95 °F and a range of 21 °F. To compute corresponding temperatures for other locations, select a suitable design temperature from Column 6, Table 1 of Chapter 24 and note the outdoor daily range (Column 7). For each hour, take the percentage of the daily range indicated in Table 2 of this chapter and subtract from the design temperature.

Table 2   Percentage of Daily Range

Time, h	%	Time, h	%	Time, h	%
1	87	9	71	17	10
2	92	10	56	18	21
3	96	11	39	19	34
4	99	12	23	20	47
5	100	13	11	21	58
6	98	14	3	22	68
7	93	15	0	23	76
8	84	16	3	24	82

Example 1. Air temperature calculation. Calculate the summer dry-bulb temperature at 1200 h for Carson City, Nevada.

Solution: From Table 1, Chapter 24, the daily range is 42 °F and the 1% design dry-bulb temperature is 93 °F. From Table 2, the percentage of the daily range at 1200 hours is 23%. Thus, the dry-bulb temperature at 1200 h = Design dry-bulb − (Daily range × Percentage) = 93.0 − [42(23/100)] = 83.3 °F.

Data limitations. The summer design temperatures in Table 1, Chapter 24, are based on the 4-month period, June through September. (Canadian data are based on July only.) The outdoor daily range is the difference between the average daily maximum and average daily minimum temperatures during the warmest month. More reliable results could be obtained by determining or estimating the shape of the temperature curve for typical hot days at the building site and considering each month separately. Peak cooling load is often determined by solar heat gain through fenestration; this peak may occur in winter months and/or at a time of day when outside air temperature is not at its peak.

Heat Gain through Fenestration

The sections that include Equations (30) through (41) in Chapter 27 describe one method used to calculate space cooling load resulting from heat transfer through fenestration. The solar heat gain profiles listed in Chapter 27 are for fenestration areas with no external shading. The equations for calculating shade angles (Chapter 27) can be used to determine the shape and area of moving shadow falling across a given window from external shading elements during the course of a design day. Thus, a subprofile of heat gain for that window can be created by separating its sunlit and shaded areas for each hour; modifying multipliers for inside shading devices can also be included.

Exterior shading. Nonuniform exterior shading, caused by roof overhangs, side fins, or building projections, require separate hourly calculations for the externally shaded and unshaded areas of the window in question, with the SC still used to account for any internal shading devices. The areas, shaded and unshaded, depend on the location of the shadow line on a surface in the plane of the glass. Sun (1968) developed fundamental algorithms for analysis of shade patterns. The *Cooling and Heating Load Calculation Manual* (McQuiston and Spitler 1992) provides graphical data to facilitate shadow line calculation, and the north exposure SHGF may be taken for shaded glass (with some loss of accuracy at latitudes less than 24° north).

An alternate, more accurate, method suggested by Todorovic and Curcija (1984) first calculates cooling loads as if the external shading were absent, then adjusts (reduces) the result to account for the shading effect. This correction applies a "negative cooling load factor," calculated in much the same way as a conventional cooling load but using the time-varying area of the shaded portion of the glass as the heat gain element. Todorovic (1987) describes the solution of the moving shade line problem in the context of consequent cooling load.

Temperature considerations. To estimate the conduction flow of heat through a fenestration at any time, applicable values of the outdoor and indoor dry-bulb temperatures must be used. Chapter 24 gives design values of summer outdoor dry-bulb temperatures for many locations. These are generally mid-afternoon temperatures; for other times, local weather stations or NOAA can supply temperature data. Winter design temperatures should not be used in Equation (15), since such data are for heating design rather than coincident conduction heat gain with sunlit glass during the heating season.

Heat Gain through Interior Surfaces

Whenever a conditioned space is adjacent to a space with a different temperature, transfer of heat through the separating physical section must be considered. The heat transfer rate is given by

$$q = UA(t_b - t_i) \qquad (8)$$

where

  q = heat transfer rate, Btu/h
  U = coefficient of overall heat transfer between adjacent and conditioned space, Btu/(h·ft²·°F)
  A = area of separating section concerned, ft²
  t_b = average air temperature in adjacent space, °F
  t_i = air temperature in conditioned space, °F

Values of U can be obtained from Chapter 22. Temperature t_b may range widely from that in the conditioned space. The temperature in a kitchen or boiler room, for example, may be as much as 15 to 50°F above the outdoor air temperature. Actual temperatures in adjoining spaces should be measured when possible. Where nothing is known, except that the adjacent space is of conventional construction, contains no heat sources, and itself receives no significant solar heat gain, $t_b - t_i$ may be considered the difference between the outdoor air and conditioned space design dry-bulb temperatures minus 5°F. In some cases, the air

temperature in the adjacent space will correspond to the outdoor air temperature or higher.

Floors. For floors directly in contact with the ground, or over an underground basement that is neither ventilated nor conditioned, heat transfer may be neglected for cooling load estimates.

HEAT SOURCES IN CONDITIONED SPACES

People

Table 3 gives representative rates at which heat and moisture are given off by human beings in different states of activity. Often these sensible and latent heat gains constitute a large fraction of the total load. For short occupancy, the extra heat and moisture brought in by people may be significant. For further details, see Chapter 8.

While Chapter 8 should be referred to for detailed information, Table 3 summarizes practical data representing conditions commonly encountered.

The conversion of sensible heat gain from people to space cooling load is affected by the thermal storage characteristics of that space and is thus subject to application of appropriate room transfer functions (RTF). Latent heat gains are considered instantaneous.

Lighting

Since lighting is often the major space load component, an accurate estimate of the space heat gain it imposes is needed. Calculation of this load component is not straightforward; the rate of heat gain at any given moment can be quite different from the heat equivalent of power supplied instantaneously to those lights.

Only part of the energy from lights is in the form of convective heat, which is picked up instantaneously by the air-conditioning apparatus. The remaining portion is in the form of radiation, which affects the conditioned space only after having been absorbed and rereleased by walls, floors, furniture, etc. This absorbed energy contributes to space cooling load only after a time lag, with some part of such energy still present and reradiating after the lights have been switched off (Figure 2).

There is always significant delay between the time of switching lights on and a point of equilibrium where reradiated light energy equals that being instantaneously stored. Time lag effect must be considered when calculating cooling load, since load felt by the space can be considerably lower than the instantaneous heat gain being generated, and peak load for the space may be affected significantly.

Instantaneous heat gain from lighting. The primary source of heat from lighting comes from light-emitting elements, or lamps, although significant additional heat may be generated from associated appurtenances in the light fixtures that house such

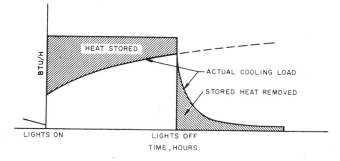

Fig. 2 Thermal Storage Effect in Cooling Load from Lights

lamps. Generally, the instantaneous rate of heat gain from electric lighting may be calculated from

$$q_{el} = 3.41 \, W F_{ul} F_{sa} \tag{9}$$

where

q_{el} = heat gain, Btu/h
W = total light wattage
F_{ul} = lighting use factor
F_{sa} = lighting special allowance factor

The *total light wattage* is obtained from the ratings of all lamps installed, both for general illumination and for display use.

The *use factor* is the ratio of the wattage in use, for the conditions under which the load estimate is being made, to the total installed wattage. For commercial applications such as stores, the use factor would generally be unity.

The *special allowance factor* is for fluorescent fixtures and/or fixtures that are either ventilated or installed so that only part of their heat goes to the conditioned space. For fluorescent fixtures, the special allowance factor accounts primarily for ballast losses, and can be as high as 2.19 for 32-W single lamp high-output fixtures on 277 V circuits. Rapid-start, 40-W lamp fixtures have special allowance factors that vary from a low of 1.18 for two lamps at 277 V to a high of 1.30 for one lamp at 118 V, with a recommended value of 1.20 for general applications. Industrial fixtures other than fluorescent, such as sodium lamps, may have special allowance factors varying from 1.04 to 1.37, depending on the manufacturer, and should be dealt with individually.

For *ventilated* or *recessed* fixtures, manufacturers' or other data must be sought to establish the fraction of the total wattage that may be expected to enter the conditioned space directly (and subject to time lag effect), versus that which must be picked up by return air or in some other appropriate manner.

Light heat components. Cooling load caused by lights recessed into ceiling cavities is made up of two components: one part comes from the light heat directly contributing to the space heat gain, and the other is the light heat released into the above-ceiling cavity, which (if used as a return air plenum) is mostly picked up by the return air that passes over or through the light fixtures. In such

a ceiling return air plenum, this second part of the load (sometimes referred to as heat-to-return) never enters the conditioned space. It does, however, add to the overall load and significantly influences the load calculation.

Even though the total cooling load imposed on the cooling coil from these two components remains the same, the larger the fraction of heat output picked up by the return air, the more the space cooling load is reduced. The minimum required airflow rate for the conditioned space is decreased as the space cooling load becomes less. Supply fan horsepower reduces accordingly, which ultimately results in reduced energy consumption for the system, and, possibly reduced equipment size as well.

For ordinary design load estimation, the heat gain for each component may simply be calculated as a fraction of the total lighting load by using judgment to estimate heat-to-space and heat-to-return percentages (Mitalas and Kimura 1971).

Return air light fixtures. Two generic types of return air light fixture are available—those that allow and those that do not allow return air to flow through the lamp chamber. The first type is sometimes called a heat-of-light fixture. The percentage of light heat released through the plenum side of various ventilated fixtures can be obtained from lighting fixture manufacturers. For representative data, see Nevens *et al.* (1971). Even unventilated fixtures lose some heat to plenum spaces; however, most of the heat ultimately enters the conditioned space from a dead-air plenum or is picked up by return air via ceiling return air openings. The percentage of heat to return air ranges from 40 to 60% for heat-to-return ventilated fixtures or 15 to 25% for unventilated fixtures.

Plenum temperatures. As heat from lighting is picked up by the return air, the temperature differential between the ceiling space and the conditioned space causes part of that heat to flow from the ceiling back to the conditioned space. Return air from the conditioned space can be ducted to capture light heat without passing through a ceiling plenum as such, or the ceiling space can be used as a return air plenum, causing the distribution of light heat to be handled in distinctly different ways. Most plenum temperatures do not rise more than 1 to 3 °F above space temperature, thus generating only a relatively small thermal gradient for heat transfer through plenum surfaces but a relatively large percentage reduction in space cooling load. (Many engineers believe that a

Table 3   Rates of Heat Gain from Occupants of Conditioned Spaces

Degree of Activity		Total Heat, Btu/h Adult Male	Total Heat, Btu/h Adjusted, M/F[a]	Sensible Heat, Btu/h	Latent Heat, Btu/h	% Sensible Heat that is Radiant[b] Low V	% Sensible Heat that is Radiant[b] High V
Seated at theater	Theater, matinee	390	330	225	105		
Seated at theater, night	Theater, night	390	350	245	105	60	27
Seated, very light work	Offices, hotels, apartments	450	400	245	105		
Moderately active office work	Offices, hotels, apartments	475	450	250	105		
Standing, light work; walking	Department store; retail store	550	450	250	105	58	38
Walking, standing	Drug store, bank	550	500	250	105		
Sedentary work	Restaurant[c]	490	550	275	105		
Light bench work	Factory	800	750	275	105		
Moderate dancing	Dance hall	900	850	305	105	49	35
Walking 3 mph; light machine work	Factory	1000	1000	375	105		
Bowling[d]	Bowling alley	1500	1450	580	105		
Heavy work	Factory	1500	1450	580	105	54	19
Heavy machine work; lifting	Factory	1600	1600	635	105		
Athletics	Gymnasium	2000	1800	710	105		

Notes:
1. Tabulated values are based on 75 °F room dry-bulb temperature. For 80 °F room dry bulb, the total heat remains the same, but the sensible heat values should be decreased by approximately 20%, and the latent heat values increased accordingly.
2. Also refer to Table 4, Chapter 8, for additional rates of metabolic heat generation.
3. All values are rounded to nearest 5 Btu/h.

[a]Adjusted heat gain is based on normal percentage of men, women, and children for the application listed, with the postulate that the gain from an adult female is 85%

of that for an adult male, and that the gain from a child is 75% of that for an adult male.
[b]Values approximated from data in Table 6, Chapter 8, where V is air velocity with limits shown in that table.
[c]Adjusted heat gain includes 60 Btu/h for food per individual (30 Btu/h sensible and 30 Btu/h latent).
[d]Figure one person per alley actually bowling, and all others as sitting (400 Btu/h) or standing or walking slowly (550 Btu/h).

major reason for plenum temperatures not becoming more elevated is due to leakage into the plenum from supply air duct work normally concealed there, but consideration of this elusive factor is beyond the scope of this chapter.)

Energy balance. Where the ceiling space is used as a return air plenum, an energy balance requires that the heat picked up from the lights into the return air (1) becomes a part of the cooling load to the return air (represented by a temperature rise of the return air as it passes through the ceiling space), (2) is partially transferred back into the conditioned space through the ceiling material below, and/or (3) may be partially "lost" (from the space) through the floor surfaces above the plenum. In a multistory building, the conditioned space frequently gains heat through its floor from a similar plenum below, offsetting the loss just mentioned. The radiant component of heat leaving the ceiling or floor surface of a plenum is normally so small that all such heat transfer is considered convective for calculation purposes.

Figure 3 shows a schematic diagram of a typical return air plenum. Equations (10) through (14), using the sign convention as shown in Figure 3, represent the heat balance of a return air plenum design for a typical interior room in a multifloor building, as

$$q_1 = U_c A_c (t_p - t_r) \tag{10}$$

$$q_2 = U_f A_f (t_p - t_{fa}) \tag{11}$$

$$q_3 = 1.1 \, Q (t_p - t_r) \tag{12}$$

$$q_{lp} - q_2 - q_1 - q_3 = 0 \tag{13}$$

$$Q = (q_r + q_1)/[1.1 (t_r - t_s)] \tag{14}$$

where

$q_1 =$ heat gain to space from plenum through ceiling, Btu/h
$q_2 =$ heat loss from plenum through floor above, Btu/h
$q_3 =$ heat gain "pickup" by return air, Btu/h
$Q =$ return airflow, cfm
$q_{lp} =$ light heat gain to plenum via return air, Btu/h
$q_{lr} =$ light heat gain to space, Btu/h
$q_f =$ heat gain from plenum below, through floor, Btu/h
$q_w =$ heat gain from exterior wall, Btu/h
$q_r =$ space cooling load, Btu/h, including appropriate treatment of q_{lr}, q_f, and/or q_w
$t_p =$ plenum temperature, °F
$t_r =$ space temperature, °F
$t_{fa} =$ space temperature of floor above, °F
$t_s =$ supply air temperature, °F

By substituting Equations (10), (11), (12), and (14) into heat balance equation (13), t_p can be found as the resultant return air temperature or plenum temperature, by means of a quadratic equation derived from such relationships. The results, although

rigorous and best solved by computer, are important in determining the cooling load, which affects equipment size selection, future energy consumption, and other factors.

Equations (10) through (14) are simplified to illustrate the heat balance relationship. Heat gain into a return air plenum is not limited to the heat of lights alone. Exterior walls directly exposed to the ceiling space will transfer heat directly to or from the return air. For single-story buildings or the top floor of a multistory building, the roof heat gain or loss enters or leaves the ceiling plenum rather than entering or leaving the conditioned space directly. The supply air quantity calculated by Equation (14) is for the conditioned space under consideration only, and is assumed equal to the return air quantity.

The amount of airflow through a return plenum above a conditioned space may not be limited to that supplied into the space under consideration; it will, however, have no noticeable effect on plenum temperature if the surplus comes from an adjacent plenum operating under similar conditions. Where special conditions exist, heat balance Equations (10) through (14) must be modified appropriately. Finally, even though the building's thermal storage has some effect, the amount of heat entering the return air is small and may be considered as convective for calculation purposes.

Power

Instantaneous heat gain from equipment operated by electric motors within a conditioned space is calculated as

$$q_{em} = 2545 \, (P/E_M)(F_{UM})(F_{LM}) \tag{15}$$

where

$q_{em} =$ heat equivalent of equipment operation, Btu/h
$P =$ motor power rating, horsepower
$E_M =$ motor efficiency, as decimal fraction < 1.0
$F_{UM} =$ motor use factor, 1.0 or decimal fraction < 1.0
$F_{LM} =$ motor load factor, 1.0 or decimal fraction < 1.0

The motor use factor may be applied when motor use is known to be intermittent with significant nonuse during all hours of operation (*e.g.,* overhead door operator). For conventional applications, its value would be 1.0.

The motor load factor is the fraction of the rated load being delivered under the conditions of the cooling load estimate. In Equation (15), it is assumed that both the motor and the driven equipment are within the conditioned space. If the motor is outside the space or airstream

$$q_{em} = 2545 \, P(F_{UM})(F_{LM}) \tag{16}$$

When the motor is inside the conditioned space or airstream but the driven machine is outside

$$q_{em} = 2545 \, P[(1.0 - E_M)/E_M (F_{UM})(F_{LM}) \tag{17}$$

Equation (17) also applies to a fan or pump in the conditioned space that exhausts air or pumps fluid outside that space.

Average efficiencies, and related data representative of typical electric motors, generally derived from the lower efficiencies reported by several manufacturers of open, drip-proof motors, are given in Tables 4 and 5. These reports indicate that TEFC (totally enclosed fan-cooled) are slightly more efficient. For speeds lower or higher than those listed, efficiencies may be 1 to 3% lower or higher, depending on the manufacturer. Should actual voltages at motors be appreciably higher or lower than rated nameplate voltage, efficiencies in either case will be lower. If electric motor load is an appreciable portion of cooling load, the motor efficiency should be obtained from the manufacturer. Also, depending on

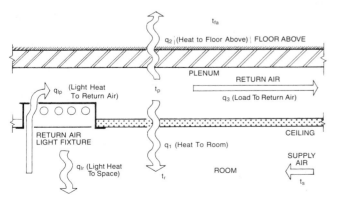

Fig. 3 Heat Balance of Typical Ceiling Return Plenum

Table 4 Heat Gain from Typical Electric Motors

Motor Name-plate or Rated Horse-power	Motor Type	Nominal rpm	Full Load Motor Efficiency, %	Location of Motor and Driven Equipment with Respect to Conditioned Space or Airstream		
				A	B	C
				Motor in, Driven Equipment in, Btu/h	Motor out, Driven Equipment in, Btu/h	Motor in, Driven Equipment out, Btu/h
0.05	Shaded pole	1500	35	360	130	240
0.08	Shaded pole	1500	35	580	200	380
0.125	Shaded pole	1500	35	900	320	590
0.16	Shaded pole	1500	35	1160	400	760
0.25	Split phase	1750	54	1180	640	540
0.33	Split phase	1750	56	1500	840	660
0.50	Split phase	1750	60	2120	1270	850
0.75	3-Phase	1750	72	2650	1900	740
1	3-Phase	1750	75	3390	2550	850
1.5	3-Phase	1750	77	4960	3820	1140
2	3-Phase	1750	79	6440	5090	1350
3	3-Phase	1750	81	9430	7640	1790
5	3-Phase	1750	82	15,500	12,700	2790
7.5	3-Phase	1750	84	22,700	19,100	3640
10	3-Phase	1750	85	29,900	24,500	4490
15	3-Phase	1750	86	44,400	38,200	6210
20	3-Phase	1750	87	58,500	50,900	7610
25	3-Phase	1750	88	72,300	63,600	8680
30	3-Phase	1750	89	85,700	76,300	9440
40	3-Phase	1750	89	114,000	102,000	12,600
50	3-Phase	1750	89	143,000	127,000	15,700
60	3-Phase	1750	89	172,000	153,000	18,900
75	3-Phase	1750	90	212,000	191,000	21,200
100	3-Phase	1750	90	283,000	255,000	28,300
125	3-Phase	1750	90	353,000	318,000	35,300
150	3-Phase	1750	91	420,000	382,000	37,800
200	3-Phase	1750	91	569,000	509,000	50,300
250	3-Phase	1750	91	699,000	636,000	62,900

Table 5 Typical Overload Limits with Standard Motors

Horsepower	0.05-0.25	0.16-0.33	0.67-0.75	1 and up
AC open	1.4	1.35	1.25	1.15
AC TEFC[a] and DC	—	1.0	1.0	1.0

Note: Some shaded pole, capacitor start, and special purpose motors have a service factor varying from 1.0 up to 1.75.
[a]Some totally enclosed fan-cooled (TEFC) motors have a service factor above 1.0.

design, the maximum efficiency might occur anywhere between 75 to 110% of full load; if underloaded or overloaded, the efficiency could vary from the manufacturer's listing.

Overloading or underloading. Heat output of a motor is generally proportional to the motor load, within the overload limits. Because of typically high no-load motor current, fixed losses, and other reasons, FLM is generally assumed to be unity, and no adjustment should be made for underloading or overloading unless the situation is fixed, can be accurately established, and the reduced load efficiency data can be obtained from the motor manufacturer.

Radiation and convection. Unless the manufacturer's technical literature indicates otherwise, the heat gain normally should be equally divided between radiant and convective components for the subsequent cooling load calculations.

Appliances

In a cooling load estimate, heat gain from all appliances—electrical, gas, or steam—should be taken into account. Because

of the variety of appliances, applications, usage schedules, and installations, estimates are very subjective.

Food preparation equipment. This equipment includes the most common types of heat-producing appliances found in the conditioned areas. Marn (1962) found that appliance surfaces contribute most of the heat to commercial kitchens and that when installed under an effective hood, the cooling load is independent of the fuel or energy employed for similar equipment performing the same operations. Marn's tests also indicated that the heat is primarily radiant energy from the appliance surfaces and cooking utensils and that convected and latent heat are negligible.

To establish a heat gain value, actual nameplate energy input ratings should be used with appropriate usage factors, efficiencies, or other judgmental modifiers. Where specific rating data are not available (nameplate missing, equipment not yet purchased, etc.) or as an alternative approach, recommended heat gains tabulated in this chapter for a wide variety of commonly encountered equipment items may be used. In estimating the appliance load, probabilities of simultaneous use and operation for different appliances located in the same space must be considered.

Radiation contributes up to 32% of the maximum hourly heat input to hooded appliances (Talbert *et al.* 1973), suggesting a conservative estimate of heat released into the kitchen, or radiation factor $F_{RA} = 0.32$. Marn (1962) found that radiant heat temperature rises can be substantially reduced by shielding the fronts of cooking appliances. These reductions amounted to 59% with asbestos paper, 61% with glass panels, and 78% using polished aluminum shielding. A floor-slot air curtain in front of the appliances reduced the radiant temperature rise by 15%.

Heat gain from meals. For each meal served the heat transferred to the dining space is approximately 50 Btu/h, of which 75% is sensible and 25% latent.

Heat gain for electric and steam appliances. The maximum hourly input can be estimated as 50% of the total nameplate or catalog input q_i ratings because of diversity of appliance use and the effect of thermostatic controls, giving a usage factor $F_{UA} = 0.50$. Therefore, the maximum hourly heat gain q_m for generic types of electric and steam appliances installed under a hood can be estimated from the following equations

$$q_{sensible} = q_{is}(F_{UA})(F_{RA}) \tag{18}$$

or, for electric or steam,

$$q_{sensible} = 0.16\, q_{is} \tag{19}$$

Heat gain from gas-fired appliances. Direct fuel-fired cooking appliances require more heat input than electric or steam equipment of the same type and size. In the case of gas fuel, the American Gas Association (1948, 1950) established an overall figure of approximately 60% more. Marn (1962) confirmed that where the appliances are installed under an effective hood, only radiant heat adds to the cooling load; convected and latent heat from the cooking process and combustion products are exhausted and do not enter the kitchen. It is therefore necessary to adjust Equation (18) for use with fuel-fired appliances, to compensate for the 60% higher input ratings, since the appliance surface temperatures are the same, and the extra heat input and combustion products are exhausted to outdoors. This correction is made by the introduction of a flue loss factor (FFL) of 1.60 as follows

$$q_{sensible} = [q_{is}(F_{UA})(F_{RA})]/F_{FL} \tag{20}$$

or, for fuel-fired appliances,

$$q_{sensible} = 0.10\, q_{is} \tag{21}$$

Latent heat gain from appliances. Latent heat gain from any kind of appliance is instantaneous, taken as the latent rating for the particular item, conditioned by the usage factor, or

$$q_{latent} = q_{il}(F_{UA}) \qquad (22)$$

Table 6 lists factors for seven typical electrical, steam, or gas appliances; Table 7 lists the relative efficiencies of various types of commercial cooking equipment and gives a representative efficiency ratio for each (Alereza and Breen 1984).

Unhooded equipment. For all cooking appliances not installed under an exhaust hood or directly vent-connected and located in the conditioned area, the heat gain may be estimated as 50% (usage factor = 0.50) of the rated hourly input, regardless of the type of energy or fuel used. On average, 34% of the heat may be assumed to be latent and the remaining 66% sensible heat.

Recommended heat gain values. As an alternative procedure, Table 8 lists recommended rates of heat gain from typical commercial cooking appliances (Alereza and Breen 1984). The data in the "with hood" columns assume installation under a properly designed exhaust hood connected to a mechanical fan exhaust system.

Hospital and laboratory equipment. Hospital and laboratory equipment items are major sources of heat gain in conditioned spaces. Care must be taken in evaluating the probability and duration of simultaneous usage when many components are concentrated in one area, such as laboratory, operating room, etc. Commonly, heat gain from equipment in a laboratory ranges from 15 to 70 Btu/(h·ft²) or, in laboratories with outdoor exposure, as much as four times the heat gain from all other sources combined.

Table 6 Heat Gain Factors of Typical Appliances under Hoods

Appliance	Usage Factor F_{UA}	Radiation Factor F_{RA}	Load Factor $F_L = F_{UA}F_{RA}$ Elec. or Steam	Load Factor $F_L = F_{UA}F_{RA}/F_{FL}$ Gas
Griddle	0.10	0.35	0.04	0.025
Hot-top range				
Without oven	0.79	0.47	0.37	0.231
With oven	0.59	0.48	0.28	0.175
Convection oven	0.14	0.30	0.04	0.025
Broiler	0.55	0.36	0.20	0.125
Charbroiler	0.50	0.15	0.08	0.050
Fryer	0.03	0.30	0.01	0.056
Steam cooker	0.13	0.30	0.04	0.025

Table 7 Efficiency of Commercial Cooking Equipment

Equipment Type	Efficiency, % Electric	Gas
Conventional fryer	78	28
Infrared fryer	72	37
Pressure fryer	83	30
Griddle	62	46
Grooved griddle	71	51
Charbroiler	65	16
Superspeed broiler	37	19
Overfired broiler	52	22
New convection oven	64	46
Convection oven	62	28
Range oven	45	13
Deck oven	55	24
Atmospheric convection steamer	23	13
Pressure steamer	39	19
Jacketed kettle	73	42
Tilting braising pan	79	52
Open burner	73	38
Hot plate	48	23

Table 3 in Chapter 30 of the 1991 ASHRAE *Handbook—Applications* lists heat gain values for various hospital and laboratory equipment.

Office appliances. Electric typewriters, calculators, checkwriters, teletype units, posting machines, etc., can generate 3 to 4 Btu/(h·ft²) for general offices or 6 to 7 Btu/(h·ft²) for purchasing and accounting departments. However, in offices having computer display terminals at most desks, heat gains range up to 15 Btu/(h·ft²) (Table 9).

Computer rooms housing mainframe or minicomputer equipment must be considered individually. Computer manufacturers have data pertaining to various individual components. Additional insight should be sought from data processing managers as to schedules, near-term future planning, etc. Heat gain rates from digital computer equipment range from 75 to 175 Btu/(h·ft²). While the trend in hardware development is toward less heat release on a component basis, the associated miniaturization tends to offset such unitary reduction by a higher concentration of equipment. Chapter 33 of the 1991 ASHRAE *Handbook—Applications* gives further information on the air conditioning of data processing areas.

INFILTRATION AND VENTILATION HEAT GAIN

Ventilation

Outdoor air must be introduced to ventilate conditioned spaces. Chapter 23 suggests minimum outdoor air requirements for representative applications, but the minimum levels are not necessarily adequate for all psychological attitudes and physiological responses. Where maximum economy in space and load is essential, as in submarines or other restricted spaces, as little as 1 cfm of outdoor air per person can be sufficient, provided that recirculated air is adequately decontaminated (Consolazio and Pecora 1947).

Local codes and ordinances frequently specify ventilation requirements for public places and for industrial installations. For example, minimum requirements for safe practice in hospital operating rooms are given in NFPA *Standard* 99 (1990). Although 100% outdoor air is sometimes used in operating rooms, this standard does not require it, and limiting the outdoor air to 6 to 8 changes per hour is finding increasing acceptance.

ASHRAE *Standard* 62 (1989) recommends minimum ventilation rates for most common applications. For general applications, such as offices, 20 cfm per person is suggested.

Ventilation air is normally introduced at the air-conditioning apparatus rather than directly into the conditioned space, and thus becomes a cooling coil load component instead of a space load component. Calculations for estimating this heat gain are discussed later.

Reducing heat gain from outdoor air by using filtered recirculated air in combination with outdoor air should be considered. Recirculated air can also be treated to control odor (see Chapter 12 in this volume and Chapter 50 in the 1991 ASHRAE *Handbook—Applications*). To provide for physiological needs, the outdoor air quantity should never be less than 4 cfm per person.

Infiltration

The principles of estimating infiltration in buildings, with emphasis on the heating season, are discussed in Chapter 23. For the cooling season, infiltration calculations are usually limited to doors and windows. However, in multistory commercial buildings, outdoor air may infiltrate at the top of the building and exfiltrate at the bottom.

Air leakage through doors can be estimated using the information in Chapter 23. Tables 4 and 6, Chapter 23, adjusted for the

Table 8 Recommended Rate of Heat Gain from Restaurant Equipment Located in Air-Conditioned Area

Appliance	Size	Input Rating, Btu/h Maximum	Input Rating, Btu/h Standby	Recommended Rate of Heat Gain,[a] Btu/h Without Hood Sensible	Without Hood Latent	Without Hood Total	With Hood Sensible
Electric, No Hood Required							
Barbeque (pit), per pound of food capacity	80 to 300 lb	136	—	86	50	136	42
Barbeque (pressurized), per pound of food capacity	44 lb	327	—	109	54	163	50
Blender, per quart of capacity	1 to 4 qt	1550	—	1000	520	1520	480
Braising pan, per quart of capacity	108 to 140 qt	360	—	180	95	275	132
Cabinet (large hot holding)	16.2 to 17.3 ft^3	7100	—	610	340	960	290
Cabinet (large hot serving)	37.4 to 406 ft^3	6820	—	610	310	920	280
Cabinet (large proofing)	16 to 17 ft^3	693	—	610	310	920	280
Cabinet (small hot holding)	3.2 to 6.4 ft^3	3070	—	270	140	410	130
Cabinet (very hot holding)	17.3 ft^3	21000	—	1880	960	2830	850
Can opener		580	—	580	—	580	0
Coffee brewer	12 cup/2 brnrs	5660	—	3750	1910	5660	1810
Coffee heater, per boiling burner	1 to 2 brnrs	2290	—	1500	790	2290	720
Coffee heater, per warming burner	1 to 2 brnrs	340	—	230	110	340	110
Coffee/hot water boiling urn, per quart of capacity	11.6 qt	390	—	256	132	388	123
Coffee brewing urn (large), per quart of capacity	23 to 40 qt	2130	—	1420	710	2130	680
Coffee brewing urn (small), per quart of capacity	10.6 qt	1350	—	908	445	1353	416
Cutter (large)	18 in. bowl	2560	—	2560	—	2560	0
Cutter (small)	14 in. bowl	1260	—	1260	—	1260	0
Cutter and mixer (large)	30 to 48 qt	12730	—	12730	—	12730	0
Dishwasher (hood type, chemical sanitizing), per 100 dishes/h	950 to 2000 dishes/h	1300	—	170	370	540	170
Dishwasher (hood type, water sanitizing), per 100 dishes/h	950 to 2000 dishes/h	1300	—	190	420	610	190
Dishwasher (conveyor type, chemical sanitizing), per 100 dishes/h	5000 to 9000 dishes/h	1160	—	140	330	470	150
Dishwasher (conveyor type, water sanitizing), per 100 dishes/h	5000 to 9000 dishes/h	1160	—	150	370	520	170
Display case (refrigerated), per 10 ft^3 of interior	6 to 67 ft^3	1540	—	617	0	617	0
Dough roller (large)	2 rollers	5490	—	5490	—	5490	0
Dough roller (small)	1 roller	1570	—	140	—	140	0
Egg cooker	12 eggs	6140	—	2900	1940	4850	1570
Food processor	2.4 qt	1770	—	1770	—	1770	0
Food warmer (infrared bulb), per lamp	1 to 6 bulbs	850	—	850	—	850	850
Food warmer (shelf type), per square foot of surface	3 to 9 ft^2	930	—	740	190	930	260
Food warmer (infrared tube), per foot of length	39 to 53 in.	990	—	990	—	990	990
Food warmer (well type), per cubic foot of well	0.7 to 2.5 ft^3	3620	—	1200	610	1810	580
Freezer (large)	73 ft^3	4570	—	1840	—	1840	0
Freezer (small)	18 ft^3	2760	—	1090	—	1090	0
Griddle/grill (large), per square foot of cooking surface	4.6 to 11.8 ft^2	9200	—	615	343	958	343
Griddle/grill (small), per square foot of cooking surface	2.2 to 4.5 ft^2	8300	—	545	308	853	298
Hot dog broiler	48 to 56 hot dogs	3960	—	340	170	510	160
Hot plate (double burner, high speed)		16720	—	7810	5430	13240	6240
Hot plate (double burner, stockpot)		13650	—	6380	4440	10820	5080
Hot plate (single burner, high speed)		9550	—	4470	3110	7580	3550
Hot water urn (large), per quart of capacity	56 qt	416	—	161	52	213	68
Hot water urn (small), per quart of capacity	8 qt	738	—	285	95	380	123
Ice maker (large)	220 lb/day	3720	—	9320	—	9320	0
Ice maker (small)	110 lb/day	2560	—	6410	—	6410	0
Microwave oven (heavy duty, commercial)	0.7 ft^3	8970	—	8970	—	8970	0
Microwave oven (residential type)	1 ft^3	2050 to 4780	—	2050 to 4780	—	2050 to 4780	0
Mixer (large), per quart of capacity	81 qt	94	—	94	—	94	0
Mixer (small), per quart of capacity	12 to 76 qt	48	—	48	—	48	0
Press cooker (hamburger)	300 patties/h	7510	—	4950	2560	7510	2390
Refrigerator (large), per 100 ft^3 of interior space	25 to 74 ft^3	753	—	300	—	300	0
Refrigerator (small), per 100 ft^3 of interior space	6 to 25 ft^3	1670	—	665	—	665	0
Rotisserie	300 hamburgers/h	10920	—	7200	3720	10920	3480
Serving cart (hot), per cubic foot of well	1.8 to 3.2 ft^3	2050	—	680	340	1020	328
Serving drawer (large)	252 to 336 dinner rolls	3750	—	480	34	510	150
Serving drawer (small)	84 to 168 dinner rolls	2730	—	340	34	380	110
Skillet (tilting), per quart of capacity	48 to 132 qt	580	—	293	161	454	218
Slicer, per square foot of slicing carriage	0.65 to 0.97 ft^2	680	—	682	—	682	216
Soup cooker, per quart of well	7.4 to 11.6 qt	416	—	142	78	220	68
Steam cooker, per cubic foot of compartment	32 to 64 ft^3	20700	—	1640	1050	2690	784
Steam kettle (large), per quart of capacity	80 to 320 qt	300	—	23	16	39	13
Steam kettle (small), per quart of capacity	24 to 48 qt	840	—	68	45	113	32
Syrup warmer, per quart of capacity	11.6 qt	284	—	94	52	146	45

Table 8 Recommended Rate of Heat Gain from Restaurant Equipment Located in Air-Conditioned Area (*Concluded*)

Appliance	Size	Input Rating, Btu/h		Recommended Rate of Heat Gain,[a] Btu/h			
				Without Hood			With Hood
		Maximum	Standby	Sensible	Latent	Total	Sensible
Toaster (bun toasts on one side only)	1400 buns/h	5120	—	2730	2420	5150	1640
Toaster (large conveyor)	720 slices/h	10920	—	2900	2560	5460	1740
Toaster (small conveyor)	360 slices/h	7170	—	1910	1670	3580	1160
Toaster (large pop-up)	10 slice	18080	—	9590	8500	18080	5800
Toaster (small pop-up)	4 slice	8430	—	4470	3960	8430	2700
Waffle iron	75 in²	5600	—	2390	3210	5600	1770
Electric, Exhaust Hood Required							
Broiler (conveyor infrared), per square foot of cooking area/minute	2 to 102 ft²	19230	—	—	—	—	3840
Broiler (single deck infrared), per square foot of broiling area	2.6 to 9.8 ft²	10870	—	—	—	—	2150
Charbroiler, per square foot of cooking surface	1.5 to 4.6 ft²	7320	—	—	—	—	307
Fryer (deep fat), per pound of fat capacity	15 to 70 lb	1270	—	—	—	—	14
Fryer (pressurized), per pound of fat capacity	13 to 33 lb	1565	—	—	—	—	59
Oven (large convection), per cubic foot of oven space	7 to 19 ft³	4450	—	—	—	—	181
Oven (large deck baking with 537 ft³ decks), per cubic foot of oven space	15 to 46 ft³	1670	—	—	—	—	69
Oven (roasting), per cubic foot of oven space	7.8 to 23 ft³	27350	—	—	—	—	113
Oven (small convection), per cubic foot of oven space	1.4 to 5.3 ft³	10340	—	—	—	—	147
Oven (small deck baking with 272 ft³ decks), per cubic foot of oven space	7.8 to 23 ft³	2760	—	—	—	—	113
Range (burners), per 2 burner section	2 to 10 brnrs	7170	—	—	—	—	2660
Range (hot top/fry top), per square foot of cooking surface	4 to 8 ft²	7260	—	—	—	—	2690
Range (oven section), per cubic foot of oven space	4.2 to 11.3 ft³	3940	—	—	—	—	160
Range (stockpot)	1 burner	18770	—	—	—	—	6990
Gas, No Hood Required							
Broiler, per square foot of broiling area	2.7 ft²	14800	660[b]	5310	2860	8170	1220
Cheese melter, per square foot of cooking surface	2.5 to 5.1 ft²	10300	660[b]	3690	1980	5670	850
Dishwasher (hood type, chemical sanitizing), per 100 dishes/h	950 to 2000 dishes/h	1740	660[b]	510	200	710	230
Dishwasher (hood type, water sanitizing), per 100 dishes/h	950 to 2000 dishes/h	1740	660[b]	570	220	790	250
Dishwasher (conveyor type, chemical sanitizing), per 100 dishes/h	5000 to 9000 dishes/h	1370	660[b]	330	70	400	130
Dishwasher (conveyor type, water sanitizing), per 100 dishes/h	5000 to 9000 dishes/h	1370	660[b]	370	80	450	140
Griddle/grill (large), per square foot of cooking surface	4.6 to 11.8 ft²	17000	330	1140	610	1750	460
Griddle/grill (small), per square foot of cooking surface	2.5 to 4.5 ft²	14400	330	970	510	1480	400
Hot plate	2 burners	19200	1325[b]	11700	3470	15200	3410
Oven (pizza), per square foot of hearth	6.4 to 12.9 ft²	4740	660[b]	623	220	843	85
Gas, Exhaust Hood Required							
Braising pan, per quart of capacity	105 to 140 qt	9840	660[b]	—	—	—	2430
Broiler, per square foot of broiling area	3.7 to 3.9 ft²	21800	530	—	—	—	1800
Broiler (large conveyor, infrared), per square foot of cooking area/minute	2 to 102 ft²	51300	1990	—	—	—	5340
Broiler (standard infrared), per square foot of broiling area	2.4 to 9.4 ft²	1940	530	—	—	—	1600
Charbroiler (large), per square foot of cooking area	4.6 to 11.8 ft²	16500	510	—	—	—	790
Charbroiler (small), per square foot of cooking area	1.3 to 4.5 ft²	19700	510	—	—	—	950
Fryer (deep fat), per pound of fat capacity	11 to 70 lb	2270	660[b]	—	—	—	160
Oven (bake deck), per cubic foot of oven space	5.3 to 16.2 ft³	7670	660[b]	—	—	—	140
Oven (convection), per cubic metre of oven space	7.4 to 19.4 ft³	8670	660[b]	—	—	—	250
Oven (pizza), per square foot of oven hearth	9.3 to 25.8 ft²	7240	660[b]	—	—	—	130
Oven (roasting), per cubic foot of oven space	9 to 28 ft³	4300	660[b]	—	—	—	77
Oven (twin bake deck), per cubic foot of oven space	11 to 22 ft³	4390	660[b]	—	—	—	78
Range (burners), per 2 burner section	2 to 10 brnrs	33600	1325	—	—	—	6590
Range (hot top or fry top), per square foot of cooking surface	3 to 8 ft²	11800	330	—	—	—	3390
Range (large stock pot)	3 burners	100000	1990	—	—	—	19600
Range (small stock pot)	2 burners	40000	1330	—	—	—	7830
Steam							
Compartment steamer, per pound of food capacity/h	46 to 450 lb	280	—	22	14	36	11
Dishwasher (hood type, chemical sanitizing), per 100 dishes/h	950 to 2000 dishes/h	3150	—	880	380	1260	410
Dishwasher (hood type, water sanitizing), per 100 dishes/h	950 to 2000 dishes/h	3150	—	980	420	1400	450
Dishwasher (conveyor, chemical sanitizing), per 100 dishes/h	5000 to 9000 dishes/h	1180	—	140	330	470	150
Dishwasher (conveyor, water sanitizing), per 100 dishes/h	5000 to 9000 dishes/h	1180	—	150	370	520	170
Steam kettle, per quart of capacity	13 to 32 qt	500	—	39	25	64	19

[a] In some cases, heat gain data are given per unit of capacity. In those cases, the heat gain is calculated by: q = (recommended heat gain per unit of capacity) * (capacity)

[b] Standby input rating is given for entire appliance regardless of size.

Table 9   Rate of Heat Gain from Selected Office Equipment

Appliance	Size	Maximum Input Rating, Btu/h	Standby Input Rating, Btu/h	Recommended Rate of Heat Gain, Btu/h
Check processing workstation	12 pockets	16400	8410	8410
Computer devices				
Card puncher		2730 to 6140	2200 to 4800	2200 to 4800
Card reader		7510	5200	5200
Communication/transmission		6140 to 15700	5600 to 9600	5600 to 9600
Disk drives/mass storage		3410 to 34100	3412 to 22420	3412 to 22420
Magnetic ink reader		3280 to 16000	2600 to 14400	2600 to 14400
Microcomputer	16 to 640 KByte[a]	340 to 2050	300 to 1800	300 to 1800
Minicomputer		7500 to 15000	7500 to 15000	7500 to 15000
Optical reader		10240 to 20470	8000 to 17000	8000 to 17000
Plotters		256	128	214
Printers				
Letter quality	30 to 45 char/min	1200	600	1000
Line, high speed	5000 or more lines/min	4300 to 18100	2160 to 9040	2500 to 13000
Line, low speed	300 to 600 lines/min	1540	770	1280
Tape drives		4090 to 22200	3500 to 15000	3500 to 15000
Terminal		310 to 680	270 to 600	270 to 600
Copiers/Duplicators				
Blue print		3930 to 42700	1710 to 17100	3930 to 42700
Copiers (large)	30 to 67[a] copies/min	5800 to 22500	3070	5800 to 22500
Copiers (small)	6 to 30[a] copies/min	1570 to 5800	1020 to 3070	1570 to 5800
Feeder		100	—	100
Microfilm printer		1540	—	1540
Sorter/collator		200 to 2050	—	200 to 2050
Electronic equipment				
Cassette recorders/players		200	—	200
Receiver/tuner		340	—	340
Signal analyzer		90 to 2220	—	90 to 2220
Mailprocessing				
Folding machine		430	—	270
Inserting machine	3600 to 6800 pieces/h	2050 to 11300	—	1330 to 7340
Labeling machine	1500 to 30000 pieces/h	2050 to 22500	—	1330 to 14700
Postage meter		780	—	510
Wordprocessors/Typewriters				
Letter quality printer	30 to 45 char/min	1200	600	1000
Phototypesetter		5890	—	5180
Typewriter		270	—	230
Wordprocessor		340 to 2050	—	300 to 1800
Vending machines				
Cigarette		250	51 to 85	250
Cold food/beverage		3920 to 6550	—	1960 to 3280
Hot beverage		5890	—	2940
Snack		820 to 940	—	820 to 940
Miscellaneous				
Barcode printer		1500	—	1260
Cash registers		200	—	160
Coffee maker	10 cups	5120	—	3580 sensible 1540 latent
Microfiche reader		290	—	290
Microfilm reader		1770	—	1770
Microfilm reader/printer		3920	—	3920
Microwave oven	1 ft^3	2050	—	1360
Paper shredder		850 to 10240	—	680 to 8250
Water cooler	32 qt/h	2390	—	5970

[a]Input is not proportional to capacity.

average wind velocity in the locality, may be used to compute infiltration for windows. In calculating window infiltration for an entire structure, the total window area on all sides of the building is not involved, since wind does not act on all sides simultaneously. In any case, infiltration from all windows in any two adjacent wall exposures should be included. A knowledge of the prevailing wind direction and velocity is helpful in selecting exposures.

When economically feasible, sufficient outdoor air should be introduced as ventilation air through the air-conditioning equipment to maintain a constant outward escape of air, and thus eliminate the infiltration portion of the gain. The pressure maintained must overcome wind pressure through cracks and door openings. When the quantity of outside air introduced through the cooling equipment is not sufficient to maintain the required pressure to eliminate infiltration, the entire infiltration load should be included in the space heat gain calculations.

Standard Air Defined

Since the specific volume of air varies appreciably, calculations will be more accurate when made on the basis of air mass instead of volume. However, volume values are often required for selection of coils, fans, ducts, etc., in which cases volume values based on measurement at standard conditions may be used for accurate results. One standard value is 0.075 lb of dry air/ft^3 (13.33 ft^3/lb of dry air). This density corresponds to about 60°F at saturation, and 69°F dry air (at 14.7 psia). Because air usually passes through the coils, fans, ducts, etc. at a density close to standard, the accuracy desired normally requires no correction. When airflow is to be measured at a particular condition or point, such as at a coil entrance or exit, the corresponding specific volume can be taken from the psychrometric chart and the standard volume multiplied by the ratio of the actual specific volume divided by 13.33.

Example 2. Standard air calculations. Assume that the outdoor air at standard conditions is 1000 cfm. What is the flow rate when the outdoor air is at 95°F dry-bulb and 75°F wet-bulb (14.3 ft^3/lb)? The measured rate at that condition should be 1000(14.3/13.33) = 1070 cfm.

Solution: Air-conditioning design often requires calculation of:

1. Total heat

Total heat gain, in Btu/h, corresponding to the change of a given standard flow rate Q_s through an enthalpy difference Δh

$$\text{Total heat change} = 60 \times 0.075\, Q_s \Delta h$$
$$= 4.5\, Q_s \Delta h \qquad (23)$$

where 60 = min/h and 0.075 = lb of dry air/ft^3.

2. Sensible heat

Sensible heat gain corresponding to the change of dry-bulb temperature Δt for given airflow (standard conditions) Q_s, or sensible heat change q_s, in Btu/h, is

$$q_s = 60 \times 0.075(0.24 + 0.45\, W)\, Q_s \Delta t \qquad (24)$$

where

    0.24 = specific heat of dry air, Btu/(lb·°F)
    W = humidity ratio, lb water/lb dry air
    0.45 = specific heat of water vapor, Btu/(lb·°F)

The specific heats are for a range from about −100 to +200°F. When $W = 0$, the value of $60 \times 0.075 (0.24 + 0.45\, W)$ is 1.08; when $W = 0.01$, the value is 1.10; when $W = 0.02$, the value is 1.12; and when $W = 0.03$, the value is 1.14. Thus, since a value of $W = 0.01$ approximates conditions found in many air-conditioning problems, the sensible heat change (Btu/h) can normally be found as

$$q_s = 1.10\, Q_s \Delta t \qquad (25)$$

3. Latent heat

Latent heat gain corresponding to the change of humidity ratio (ΔW) for given air flow (standard conditions) Q_s is

$$q_l = 60 \times 0.075 \times 1076\, Q_s \Delta W$$
$$= 4840\, Q_s \Delta W \qquad (26)$$

where 1076 is the approximate heat content of 50% rh vapor at 75°F, less the heat content of water at 50°F. The 50% rh at 75°F is a common design condition for the space, and 50°F is normal condensate temperature from cooling and dehumidifying coils.

The constants 4.5, 1.10, and 4840 are useful in air-conditioning calculations at sea level (14.7 psia) and for normal temperatures and moisture ratios. For other conditions, more precise values should be used. For an altitude of 5000 ft (12.2 psia), appropriate values are 3.73, 0.92, and 4020.

Latent Heat Gain from Moisture through Permeable Building Materials

The diffusion of moisture through all common building materials is a natural phenomenon that is always present. Chapters 20 and 21 cover the principles and specific methods used to control moisture. Moisture transfer through walls is often neglected in the usual comfort air-conditioning application, because the actual rate is quite small, and the corresponding latent heat gain is insignificant. The permeability and permeance values for various building materials are given in Table 7, Chapter 22. Vapor retarders are frequently installed to keep moisture transfer to a minimum.

Special conditions. Certain industrial applications call for a low moisture content to be maintained in a conditioned space. In such cases, the latent heat gain accompanying moisture transfer through walls may be greater than any other latent heat gain. This gain is computed by

$$q_m = (M/7000)\, A \Delta p_v (h_g - h_f) \qquad (27)$$

where

    q_m = latent heat gain, Btu/h
    M = permeance of wall assembly in perms, or grains/ (ft^2·h·in. Hg)
    7000 = grains/lb
    A = area of wall surface, ft^2
    Δp_v = vapor pressure difference, in. Hg
    h_g = enthalpy at room conditions, Btu/lb
    h_f = enthalpy of water condensed at cooling coil, Btu/lb
    = 1076 Btu/lb when the room temperature is 75°F and condensate off the coil is 50°F

Heat Gain from Miscellaneous Sources

The calculation of the cooling load is affected by such factors as (1) type of HVAC system, (2) effectiveness of heat exchange surfaces, (3) fan location, (4) duct heat gain or loss, (5) duct leakage, (6) heat-extraction lighting systems, (7) type of return air system, and (8) sequence of controls. System performance needs to be analyzed as a sequence of individual psychrometric processes. The most straightforward method first defines all known (or desired) state points on a psychrometric chart. Next, the actual entering and leaving dry- and wet-bulb conditions are calculated for such components as the cooling and/or heating coils (based on zone or space load), the amount of outside air introduced into the system through the equipment, and the amount of heat gain or loss at various points.

This overall process must verify that the space conditions originally sought can actually be met by the designed system by considering all sensible and latent heat changes to the air as it travels from the space conditions through the return air system and equipment back to the conditioned space. If the design is successful (*i.e.*, within the degree of correctness of the various design assumptions), appropriate equipment components can safely be selected. If not, the designer must judge if the results will be "close enough" to satisfy the needs of the project, or if one or more assumptions and/or design criteria must first be modified and the calculations rerun.

Heat gain from fans. Fans that circulate air through HVAC systems add energy to the system by one or all of the following processess:

- Temperature rise in the airstream from fan inefficiency. Depending on the equipment, fan efficiencies generally range between 50 and 70%, with an average value of 65%. Thus, some 35% of the energy required by the fan appears as instantaneous heat gain to the air being transported.
- Temperature rise in the airstream as a consequence of air static and velocity pressure. The "useful" 65% of the total fan energy that creates pressure to move air spreads out throughout the entire air transport system in the process of conversion to sensible heat. Designers commonly assume that the temperature change equivalent of this heat occurs at a single point in the system, depending on fan location as noted below.
- Temperature rise from heat generated by motor and drive inefficiencies. The relatively small gains from fan motors and drives are normally disregarded unless the motor and/or drive are physically located within the conditioned airstream. Equations (15), (16), and (17) may be used to estimate heat gains from typical motors. Belt drive losses are often estimated as 3% of the motor power rating. Conversion to temperature rise is calculated by Equation (25).

The location of each fan relative to other elements (primarily the cooling coil), and the type of system (*e.g.*, single zone, multizone, double-duct, terminal reheat, VAV) along with the concept of equipment control (space temperature alone, space temperature and relative humidity, etc.) must be known before the analysis can be completed. A fan located upstream of the cooling coil (blowthrough supply fan, return air fan, outside air fan) adds the heat equivalent of its inefficiency to the airstream at that point; thus, a slightly elevated entering dry-bulb temperature to the cooling coil results. A fan located downstream of the cooling coil raises the dry-bulb temperature of air leaving the cooling coil. This rise can be offset by reducing the cooling coil temperature, or alternatively, by increasing airflow across the cooling coil as long as its impact on space conditions is considered.

Duct heat gain and leakage. Unless return air duct systems are extensive or subjected to rigorous conditions, only the heat gained or lost by supply duct systems is significant; it is normally estimated as a percentage of space sensible cooling load (usually about 1%) and applied to the dry-bulb temperature of the air leaving the coil in the form of an equivalent temperature reduction.

Duct leakage. Air leakage out of (or into) duct work can have much greater impact than conventional duct heat gain or loss, but it is normally about the same or less. Outward leakage from supply ducts is a direct loss of cooling and/or dehumidifying capacity and must be offset by increased airflow (sometimes reduced supply air temperatures) unless it enters the conditioned space directly. Inward leakage to return ducts causes temperature and/or humidity variations, but these are often ignored under ordinary circumstances due to the low temperature and pressure differentials involved. Chapter 32 has further details on duct sealing and leakage.

A well-designed and installed duct system should not leak more than 1 to 3% of the total system airflow. All HVAC equipment and volume control units connected into a duct system are usually delivered from manufacturers with allowable leakage not exceeding 1 or 2% of maximum airflow rating. Where duct systems are specified to be sealed and leak tested, both low and medium pressure types can be constructed and required to fall within this range, and designers normally assume this loss to approximate 1% of the space load, handled in a similar manner to that for duct heat gain. Latent heat considerations are frequently ignored.

Poorly designed or installed duct systems can have leakage rates of 10 to 30%. Leakage from low-pressure lighting troffer connections lacking proper taping and sealing can be 35% or more of the terminal air supply. Improperly sealed high-pressure systems can leak as much as 10% or more from the high-pressure side alone, before considering the corresponding low-pressure side of such systems. Such extremes destroy the validity of any load calculation procedures. Although not always affecting overall system loads enough to cause problems, they will, however, always adversely impact required supply air quantities for most air-conditioning systems as outlined in Chapter 3 of the 1992 ASHRAE *Handbook—Applications*. Also, using uninsulated supply duct work running through return air plenums results in high "thermal leakage," thus loss of space cooling capability by the supply air, and potential condensation difficulties during a warm startup.

HEATING LOAD PRINCIPLES

Techniques for estimating design heating load for commercial, institutional, and industrial applications are essentially the same as for those estimating design cooling loads for such uses, except that (1) temperatures outside the conditioned spaces are generally lower than the space temperatures maintained; (2) credit for solar heat gains or for internal heat gains is not included; and (3) the thermal storage effect of building structure or content is ignored. Heat losses (negative heat gains) are thus considered to be instantaneous, heat transfer essentially conductive, and latent heat treated only as a function of replacing space humidity lost to the exterior environment.

Justification of this simplified approach derives from the purpose of a heating load estimate, as identification of "worst case" conditions that can reasonably be anticipated during a heating season. Traditionally this is considered as the load that must be met under design interior and exterior conditions, including infiltration and/or ventilation, but in the absence of solar effect (at night or cloudy winter days) and before the periodic presence of people, lights, and appliances can begin to have an offsetting effect. The primary orientation is thus toward identification of adequately sized heating equipment to handle the normal worst-case condition.

Safety factors and load allowances. Before mechanical cooling of buildings became a usual procedure, buildings included much less insulation, large operable windows, and generally more infiltration-prone assemblies than the energy-efficient and much tighter buildings typical of post-1975 design. Allowances of 10 to 20% of the net calculated heating load for piping losses to unheated spaces, and 10 to 20% more for a warm-up load were common practice, along with occasional other safety factors reflecting the experience and/or concern of the individual designer. Such measures are infrequently used in estimating heating loads for contemporary buildings, with the uncompensated net heating load normally considered as having an adequate margin for error.

Cooling needs during noncooling months. Perimeter spaces exposed to high solar heat gain often justify mechanical cooling during sunlit portions of traditional heating months, as will completely interior spaces with significant internal heat gain. These conditions require special attention by the system designer for proper accommodation as needed, but such spaces can also represent significant heating loads during nonsunlit hours or after periods of nonoccupancy when adjacent spaces have been allowed to cool below interior design temperatures. The loads involved can be estimated conventionally for the purpose of designing the means to offset or to compensate for them and prevent overheating, but they have no direct relationship to design heating loads for the spaces in question.

Other considerations. Calculation of design heating load estimates for this general category of applications has essentially become a subset of the more involved and complex estimation of cooling loads for such spaces. Chapter 28 discusses using the heating load estimate to predict or analyze energy consumption over time. Special provisions to deal with atypical problems are relegated to appropriate chapters in the Systems, Equipment, and Applications volumes.

TRANSFER FUNCTION METHOD CALCULATION PROCEDURE

BASIC COOLING LOAD ANALYSIS

The basic procedures for estimating the maximum design cooling load for a conditioned space were developed when all design calculations were performed manually. For this reason, extensive design analysis was not part of the primary load estimate. Today, with computers used for routine design calculations, the individual load elements may be evaluated more thoroughly and a comprehensive design analysis can be included with the results. The TFM method makes it possible to estimate the cooling load for a conditioned space on an hour-by-hour basis and to predict resultant conditions that can be expected in that space for various system types, control strategies, and operating schedules. The equations and sequence of the TFM Procedure in general are summarized in Table 10.

HEAT GAIN BY CONDUCTION THROUGH EXTERIOR WALLS AND ROOFS

Sensible Heat Gain

The transfer function method (TFM) is particularly well suited for use with a computer. This method is a special case of the calculation of heat flow through building components outlined in Chapter 20. This approach uses (1) sol-air temperature to represent outdoor conditions, and (2) an assumed constant indoor air temperature. Furthermore, both indoor and outdoor surface heat transfer coefficients are assumed constant (Mitalas 1968). Thus, the heat gain through a wall or roof is given by

$$q_{e,\theta} = A\left[\sum_{n=0} b_n(t_{e,\theta-n\delta}) - \sum_{n=1} d_n\{(q_{e,\theta-n\delta})/A\} - t_{rc}\sum_{n=0} c_n\right] \quad (28)$$

where

$q_{e,\theta}$ = heat gain through wall or roof, at calculation hour θ, Btu/h
A = indoor surface area of a wall or roof, ft^2
θ = time, h
δ = time interval, h
n = summation index (each summation has as many terms as there are non-negligible values of coefficients)
$t_{e,\theta-n\delta}$ = sol-air temperature at time $\theta - n\delta$, °F
t_{rc} = constant indoor room temperature, °F
b_n, c_n, d_n = conduction transfer function coefficients

Conduction transfer function coefficients. Conduction transfer function (CTF) coefficients are usually calculated using combined

outdoor heat transfer coefficient h_o = 3.0 Btu/(h·ft^2·°F), indoor coefficient h_i = 1.46 Btu/(h·ft^2·°F), and the wall or roof constructions, as may be appropriate. The use of h_o = 3.0 limits the application of these coefficients to cases with similarly calculated sol-air temperature values. Specific CTF coefficients for different constructions can be calculated using the procedure and computer program outlined in Mitalas and Arseneault (1970) or as discussed by the *Cooling and Heating Load Calculation Manual* (McQuiston and Spitler 1992) and with the microcomputer software issued with that publication.

Representative walls and roofs. Harris and McQuiston (1988) investigated the thermal behavior of approximately 2600 walls and 500 roofs as they influenced transmission of heat gain to conditioned spaces. This work identified 41 representative wall assemblies and 42 roof assemblies with widely varying components, insulating values, and mass, and with the predominant mass concentrated near the inside surface (mass in), outside surface (mass out), or essentially homogeneous (mass integral) with the overall construction. These prototypical assemblies can be used to reflect the overall range of conditions. The CTF and associated data pertaining to these conditions are listed in Tables 11 through 19.

Approximate values of CTF coefficients can be obtained by selecting a set of data from Tables 13 and 14 for a roof construction or Tables 18 and 19 for a wall that is nearly the same as the roof or wall under consideration, and multiplying the bs and cs by the ratio of the U-factor of the roof or wall under consideration over the U-factor of the selected representative roof or wall.

The physical and thermal properties of the various layers that make up roof and wall assemblies are listed in Table 11. Group numbers for various arrangements of layers with differing insulation R value and placement for roofs are listed in Table 12 and those for walls are listed in Tables 15, 16 and 17. Data from these tables identify prototypical roof or wall CTFs and associated data tabulated in Tables 13, 14, 18, and 19.

Example 3. Heat gain through wall. A light-colored wall is constructed of 4 in. heavy concrete, 2 in. insulation (2.0 lb/ft^3, R = 6.667 h·ft^2·°F/Btu), 3/4 in. indoor plaster, and with outdoor and indoor surface resistances of 0.333 and 0.685 h·ft^2·°F/Btu, respectively. There is an air space between the plaster and the insulation. Find the heat gain through 1 ft^2 of the wall area (*i.e.*, A = 1.0 ft^2) with sol-air temperature as listed in Table 1 for July 21, 40° North latitude, West, α/h_0 = 0.15, a room temperature of 75 °F, and assuming that the daily sol-air temperature cycle is repeated on several consecutive days.

Solution: The calculation of heat gain for a particular time requires sol-air temperature values at that and preceding times, as well as the heat flow at preceding times. Heat flow is assumed as zero to start the calculations. The effect of this assumption becomes negligible as the calculation is repeated for successive 24-h cycles.

Sol-Air Temperatures (from Table 1)

Time, h	t_e, °F	Time, h	t_e, °F
1	76	13	110
2	76	14	121
3	75	15	129
4	74	16	131
5	74	17	127
6	75	18	114
7	78	19	87
8	81	20	85
9	85	21	83
10	89	22	81
11	93	23	79
12	97	24	77

and $t_{e,\theta} = t_{e,\theta-24}$ for $\theta > 24$.

Table 10   Summary of TFM Load Calculation Procedures

External Heat Gain

$$t_e = t_o + \alpha I_t / h_o - \epsilon \delta R / h_o \qquad (6)$$

$$t_{ea} = t_{oa} + \alpha / h_o \, (I_{DT}/24) - \epsilon \delta R / h_o \qquad (7)$$

where

t_e = sol-air temperature
t_o = current hour dry-bulb temperature, from design db (Chapter 24, Table 1) adjusted by Table 2 Daily range % values
α = absorptance of surface for solar radiation
α / h_o = surface color factor—0.15 for light colors, 0.30 for dark
I_t = total incident solar load—1.15 (SHGF), with SHGF per Chapter 27, Tables 12 through 18
$\epsilon \delta R / h_o$ = long-wave radiation factor; $-7\,°F$ for horizontal surfaces, $0\,°F$ for vertical
t_{ea} = 24-h average sol-air temperature
t_{oa} = 24-h average dry-bulb temperature
I_{DT} = total daily solar heat gain (Chapter 27, Tables 12 through 18)

Roofs and Walls

$$q_{e,\,\theta} = A \left[\sum_{n=o} b_n (t_{e,\,\theta-\,n\delta}) - \sum_{n=1} d_n [(q_{e,\,\theta-\,n\delta})/A] - t_{rc} \sum_{n=o} c_n \right] \quad (28)$$

where

b and d = conduction transfer coefficients—roof, Table 13; wall, Table 18
c and U_{table} = conduction transfer coefficients—roof, Table 14; wall, Table 19
U_{actual} = design heat transfer coefficient for roof or wall, from Chapter 22, Table 4

Adjust b and c coefficients by ratio U_{actual}/U_{table}.

Roofs

Identify layers of roof construction from Table 11. With R-value of dominant layer, identify R-value Range number R and Roof Group number from Table 12. Proceed to Table 13.

Walls

Identify layers of wall construction from Table 11. With R-value of dominant layer, identify R-value Range number R and Wall Group number from Table 15, 16, or 17. Proceed to Table 14.

θ = hour for which calculation is made
δ = time interval (1 h)
n = number of hours for which b and d values are significant
e = element under analysis, roof or wall assembly
A = area of element under analysis

Glass

$$\text{Convective } q = UA\,(t_o - t_i)$$

$$\text{Solar } q = A\,(SC)(SHGF)$$

U = design heat transfer coefficients, glass—Chapter 27
SC = shading coefficient—Chapter 27
SHGF = solar heat gain factor by orientation, north latitude, hour, and month—Chapter 27, Tables 12 to 18.

Partitions, Ceilings, Floors

$$q = UA\,(t_b - t_i) \qquad (8)$$

t_b = temperature in adjacent space
t_i = inside design temperature in conditioned space

Internal Heat Gain

People

$$q_{sensible} = N \times \text{Sensible heat gain}$$

$$q_{latent} = N \times \text{Latent heat gain}$$

N = number of people in space, from best available source. Sensible and latent heat gain from occupancy—Table 3, or Chapter 8; adjust as required.

Lights

$$q_{el} = 3.41\, W F_{ul} F_{sa} \qquad (9)$$

where

W = watts input from electrical plans or lighting fixture data
F_{ul} = lighting use factor, from Section I, as appropriate
F_{sa} = special allowance factor, from Section I, as appropriate

Power

$$q_p = 2545\, P E_F \qquad (15),\,(16),\,(17)$$

where

P = horsepower rating from electrical plans or manufacturer's data
E_F = efficiency factors and arrangements to suit circumstances

Appliances

where

$$q_{sensible} = q_{is} F_{ua} F_{ra} \qquad (18)$$

or

$$q_{sensible} = (q_{is} F_{ua} F_{ra})/F_{fl} \qquad (20)$$

$$q_{latent} = q_{il} F_{ua} \qquad (22)$$

$q_{is},\, q_{il}$ = sensible and latent heat gain from appliances—Tables 5 to 9, or manufacturer's data (set latent heat = 0, if appliance is under exhaust hood)
$F_{ua},\, F_{ra},\, F_{fl}$ = use factors, radiation factors, flue loss factors from previous section

Ventilation and Infiltration Air

$$q_{sensible} = 1.10Q\,(t_o - t_i) \qquad (25)$$

$$q_{latent} = 4840Q\,(W_o - W_i) \qquad (27)$$

$$q_{total} = 4.5Q\,(H_o - H_i) \qquad (23)$$

Q = ventilation airflow—ASHRAE *Standard* 62; infiltration cfm—Chapter 23
$t_o,\, t_i$ = outside, inside air temperature, °F
$W_o,\, W_i$ = outside, inside air humidity ratio, lb water/lb dry air
$H_o,\, H_i$ = outside, inside air enthalpy, Btu/lb dry air

Cooling Load

Sensible $Q\theta = Q_{tf} + Q_{sc}$

$$Q_{tf} = \sum_{i=1} (v_o q_{\theta,\,i} + v_1 q_{\theta,\,i-\delta} + v_2 q_{\theta,\,i-2\delta} + \ldots)$$
$$- (w_1 Q_{\theta-\delta} + w_2 Q_{\theta-2\delta} + \ldots) \qquad (31)$$

$$Q_{sc} = \sum_{j=1} (q_{c,\,j}) \qquad (33)$$

Q_{tf} = sensible cooling load from heat gain elements having convective and radiant components
v and w = room transfer function coefficients, Tables 24 and 25; select per element type, circulation rate, mass, and/or fixture type
q_θ = each of i heat gain elements having a radiant component; select appropriate fractions for processing, per Tables 24, 25, and 42
δ = time interval (1 h)
Q_{sc} = sensible cooling load from heat gain elements having only convective components
q_c = each of j heat gain elements having only convective component

$$\text{Latent } Q_l = \sum_{n=1} (q_{c,\,n})$$

q_c = each of n latent heat gain elements

Table 11   Thermal Properties and Code Numbers of Layers Used in Wall and Roof Descriptions for Tables 12 and 13

Code Number	Description	Thickness and Thermal Properties					
		L	k	ρ	c_p	R	Mass
A0	Outside surface resistance	0.0	0.0	0.0	0.0	0.33	0.0
A1	1 in. Stucco	0.0833	0.4	116.0	0.20	0.21	9.7
A2	4 in. Face brick	0.333	0.77	125.0	0.22	0.43	41.7
A3	Steel siding	0.005	26.0	480.0	0.10	0.00	2.4
A4	1/2 in. Slag	0.0417	0.11	70.0	0.40	0.38	2.2
A5	Outside surface resistance	0.0	0.0	0.0	0.0	0.33	0.0
A6	Finish	0.0417	0.24	78.0	0.26	0.17	3.3
A7	4 in. Face brick	0.333	0.77	125.0	0.22	0.43	41.7
B1	Air space resistance	0.0	0.0	0.0	0.0	0.91	0.0
B2	1 in. Insulation	0.083	0.025	2.0	0.2	3.33	0.2
B3	2 in. Insulation	0.167	0.025	2.0	0.2	6.67	0.3
B4	3 in. Insulation	0.25	0.025	2.0	0.2	10.00	0.5
B5	1 in. Insulation	0.0833	0.025	5.7	0.2	3.33	0.5
B6	2 in. Insulation	0.167	0.025	5.7	0.2	6.67	1.0
B7	1 in. Wood	0.0833	0.07	37.0	0.6	10.00	3.1
B8	2.5 in. Wood	0.2083	0.07	37.0	0.6	2.98	7.7
B9	4 in. Wood	0.333	0.07	37.0	0.6	4.76	12.3
B10	2 in. Wood	0.167	0.07	37.0	0.6	2.39	6.2
B11	3 in. Wood	0.25	0.07	37.0	0.6	3.57	9.3
B12	3 in. Insulation	0.25	0.025	5.7	0.2	10.00	1.4
B13	4 in. Insulation	0.333	0.025	5.7	0.2	13.33	1.9
B14	5 in. Insulation	0.417	0.025	5.7	0.2	16.67	2.4
B15	6 in. Insulation	0.500	0.025	5.7	0.2	20.00	2.9
B16	0.15 in. Insulation	0.0126	0.025	5.7	0.2	0.50	0.1
B17	0.3 in. Insulation	0.0252	0.025	5.7	0.2	1.00	0.1
B18	0.45 in. Insulation	0.0379	0.025	5.7	0.2	1.50	0.2
B19	0.61 in. Insulation	0.0505	0.025	5.7	0.2	2.00	0.3
B20	0.76 in. Insulation	0.0631	0.025	5.7	0.2	2.50	0.4
B21	1.36 in. Insulation	0.1136	0.025	5.7	0.2	4.50	0.6
B22	1.67 in. Insulation	0.1388	0.025	5.7	0.2	5.50	0.8
B23	2.42 in. Insulation	0.2019	0.025	5.7	0.2	8.00	1.2
B24	2.73 in. Insulation	0.2272	0.025	5.7	0.2	9.00	1.3
B25	3.33 in. Insulation	0.2777	0.025	5.7	0.2	11.00	1.6
B26	3.64 in. Insulation	0.3029	0.025	5.7	0.2	12.00	1.7
B27	4.54 in. Insulation	0.3786	0.025	5.7	0.2	15.00	2.2
C1	4 in. Clay tile	0.333	0.33	70.0	0.2	1.01	23.3
C2	4 in. Lightweight concrete block	0.333	0.22	38.0	0.2	1.51	12.7
C3	4 in. Heavyweight concrete block	0.333	0.47	61.0	0.2	0.71	20.3
C4	4 in. Common brick	0.333	0.42	120.0	0.2	0.79	40.0
C5	4 in. Heavyweight concrete	0.333	1.0	140.0	0.2	0.33	46.7
C6	8 in. Clay tile	0.667	0.33	70.0	0.2	2.00	46.7
C7	8 in. Lightweight concrete block	0.667	0.33	38.0	0.2	2.00	25.3
C8	8 in. Heavyweight concrete block	0.667	0.6	61.0	0.2	1.11	40.7
C9	8 in. Common brick	0.667	0.42	120.0	0.2	1.59	80.0
C10	8 in. Heavyweight concrete	0.667	1.0	140.0	0.2	0.67	93.4
C11	12 in. Heavyweight concrete	1.0	1.0	140.0	0.2	1.00	140.0
C12	2 in. Heavyweight concrete	0.167	1.0	140.0	0.2	0.17	23.3
C13	6 in. Heavyweight concrete	0.5	1.0	140.0	0.2	0.50	70.0
C14	4 in. Lightweight concrete	0.333	0.1	40.0	0.2	3.33	13.3
C15	6 in. Lightweight concrete	0.5	0.1	40.0	0.2	5.00	20.0
C16	8 in. Lightweight concrete	0.667	0.1	40.0	0.2	6.67	26.7
C17	8 in. Lightweight concrete block (filled)	0.667	0.08	18.0	0.2	8.34	12.0
C18	8 in. Heavyweight concrete block (filled)	0.667	0.34	53.0	0.2	1.96	35.4
C19	12 in. Lightweight concrete block (filled)	1.000	0.08	19.0	0.2	12.50	19.0
C20	12 in. Heavyweight concrete block (filled)	1.000	0.39	56.0	0.2	2.56	56.0
E0	Inside surface resistance	0.0	0.0	0.0	0.0	0.69	0.0
E1	3/4 in. Plaster or gypsum	0.0625	0.42	100.0	0.2	0.15	6.3
E2	1/2 in. Slag or stone	0.0417	0.83	55.0	0.40	0.05	2.3
E3	3/8 in. Felt and membrane	0.0313	0.11	70.0	0.40	0.29	2.2
E4	Ceiling air space	0.0	0.0	0.0	0.0	1.00	0.0
E5	Acoustic tile	0.0625	0.035	30.0	0.2	1.79	1.9

L = thickness, ft k = thermal conductivity, Btu/h · ft · °F ρ = density, lb/ft^3
c_p = specific heat, Btu/lb · °F R = thermal resistance, °F · ft^2 · h/Btu   Mass = unit mass, lb/ft^2

Table 12 Roof Group Numbers

Roofs without Suspended Ceilings

Roof Materials[a] No.	Codes	Mass In R-Value Range Numbers[b] 1	2	3	4	5	6	Integral Mass R-Value Range Numbers[b] 1	2	3	4	5	6	Mass Out R-Value Range Numbers[b] 1	2	3	4	5	6
1	B7							1	2	2	4	4							
2	B8							4	5	9	10	18							
3	B9							19	21	27	27	28							
4	C5	6	7	7	10	10		3						6	7	7	10	11	
5	C12	2	2	4	4	5		2						2	3	4	5	5	
6	C13	7	12	13	13	20		5						7	12	13	13	20	
7	C14		4	5	9	9		2	2						4	5	9	9	
8	C15		5	10	18	18	18		4						5	10	10	18	18
9	C16		9	19	20	27	27	9							9	18	20	27	27
10	A3							1	1	1	2	2							
11	Attic							1	2	2	2	4							

Roof Terrace Systems

No.	Codes							Integral Mass 1	2	3	4	5	6	Mass Out 1	2	3	4	5	6
12	C12-C12							4	5	9	9	9		5	5	7	9	9	
13	C12-C5							6	11	12	18	18		7	12	12	12	20	
14	C12-C13							11	20	20	21	27		12	13	21	21	21	
15	C5-C12							5	10	10	17	17		5	10	11	11	18	
16	C5-C5							10	20	20	26	26		10	13	21	21	21	
17	C5-C13							20	27	28	28	35		20	22	22	22	28	
18	C13-C12							10	18	20	20	26		10	13	20	29	21	
19	C13-C5							18	27	27	28	35		20	22	22	28	28	
20	C13-C13							21	29	30	36	36		21	29	30	31	36	

Roofs with Suspended Ceilings

Roof Materials[a] No.	Codes	Mass In R-Value Range Numbers[b] 1	2	3	4	5	6	Integral Mass R-Value Range Numbers[b] 1	2	3	4	5	6	Mass Out R-Value Range Numbers[b] 1	2	3	4	5	6
1	B7								4	5	9	10	10						
2	B8								9	20	21	22	28						
3	B9								20	28	30	37	38						
4	C5	8	15	18	18	23		6						7	7	7	10	10	
5	C12	5	8	13	13	14		3						3	3	4	5		
6	C13		18	24	25	25		11	11						12	13	13	20	
7	C14		4	10	11	18	20		4						4	5	9	9	17
8	C15		10	18	21	21	28		9						9	10	18	19	26
9	C16			20	28	29	36			18						18	26	27	27
10	A3							1	1	2	2	4							
11	Attic																		

Roof Terrace Systems

No.	Codes							Integral Mass 1	2	3	4	5	6	Mass Out 1	2	3	4	5	6
12	C12-C12							6	13	22	22	22		5	5	7	9	9	
13	C12-C5							10	21	23	24	31			12	12	18	20	
14	C12-C13							13	23	24	33	33			13	21	21	21	
15	C5-C12							10	20	22	28	29		10	12	18	18		
16	C5-C5							13	23	32	32	33			20	21	21	21	
17	C5-C13							21	32	34	40				22	22	28	28	
18	C13-C12							12	28	30	31	37			13	20	20	21	
19	C13-C5							21	31	39	40	40			22	22	28	28	
20	C13-C13								39	41	41	42	42		29	30	31	36	37

[a]Blank spaces denote a roof that is not possible with the chosen combinations of parameters. Numbers 12 through 20 are roof terrace systems. First material is outer layer, second material is inner layer. Massive material numbers are: 4, 5, 6, 7, 8, 9, 12, 13, 14, 15, 16, 17, 18, 19, and 20. Nonmassive material numbers are: 1, 2, 3, 10, and 11.

[b]R-Value ranges in $h \cdot ft^2 \cdot °F/Btu$ are:

No.	Range	No.	Range	No.	Range
1	0 to 5	3	10 to 15	5	20 to 25
2	5 to 10	4	15 to 20	6	25 to 30

Table 13 Roof Conduction Transfer Function Coefficients (*b* and *d* Factors)

Roof Group	(Layer Sequence Left to Right = Inside to Outside)		$n=0$	$n=1$	$n=2$	$n=3$	$n=4$	$n=5$	$n=6$
1	Layers E0 A3 B25 E3 E2 A0	b_n	.00487	.03474	.01365	.00036	.00000	.00000	.00000
	Steel deck with 3.33 in. insulation	d_n	1.00000	−.35451	.02267	−.00005	.00000	.00000	.00000
2	Layers E0 A3 B14 E3 E2 A0	b_n	.00056	.01202	.01282	.00143	.00001	.00000	.00000
	Steel deck with 5 in. insulation	d_n	1.00000	−.60064	.08602	−.00135	.00000	.00000	.00000
3	Layers EO E5 E4 C12 E3 E2 A0	b_n	.00613	.03983	.01375	.00025	.00000	.00000	.00000
	2 in. h.w. concrete deck with suspended ceiling	d_n	1.00000	−.75615	.01439	−.00006	.00000	.00000	.00000
4	Layers E0 E1 B15 E4 B7 A0	b_n	.00000	.00065	.00339	.00240	.00029	.00000	.00000
	Attic roof with 6 in. insulation	d_n	1.00000	−1.34658	.59384	−.09295	.00296	−.00001	.00000
5	Layers E0 B14 C12 E3 E2 A0	b_n	.00006	.00256	.00477	.00100	.00002	.00000	.00000
	5 in. insulation with 2 in. h.w. concrete deck	d_n	1.00000	−1.10395	.26169	−.00475	.00002	.00000	.00000
6	Layers E0 C5 B17 E3 E2 A0	b_n	.00290	.03143	.02114	.00120	.00000	.00000	.00000
	4 in. h.w. concrete deck with 0.3 in. insulation	d_n	1.00000	−.97905	.13444	−.00272	.00000	.00000	.00000
7	Layers E0 B22 C12 E3 E2 C12 A0	b_n	.00059	.00867	.00688	.00037	.00000	.00000	.00000
	1.67 in. insulation with 2 in. h.w. concrete RTS	d_n	1.00000	−1.11766	.23731	−.00008	.00000	.00000	.00000
8	Layers E0 B16 C13 E3 E2 A0	b_n	.00098	.01938	.02083	.00219	.00001	.00000	.00000
	0.15 in. insul. with 6 in. h.w. concrete deck	d_n	1.00000	−1.10235	.20750	−.00287	.00000	.00000	.00000
9	Layers E0 E5 E4 B12 C14 E3 E2 A0	b_n	.00000	.00024	.00217	.00251	.00055	.00002	.00000
	3 in. insul. w/4 in. l.w. conc. deck and susp. clg.	d_n	1.00000	−1.40605	.58814	−.09034	.00444	−.00006	.00000
10	Layers E0 E5 E4 C15 B16 E3 E2 A0	b_n	.00000	.00025	.00241	.00303	.00074	.00004	.00000
	6 in. l.w. conc. dk w/0.15 in. ins. and susp. clg.	d_n	1.00000	−1.55701	.73120	−.11774	.00600	−.00008	.00000
11	Layers E0 C5 B15 E3 E2 A0	b_n	.00000	.00013	.00097	.00112	.00020	.00001	.00000
	4 in. h.w. concrete deck with 6 in. insulation	d_n	1.00000	−1.61467	.79142	−.13243	.00611	−.00008	.00000
12	Layers E0 C13 B16 E3 E2 C12 A0	b_n	.00005	.00356	.01058	.00404	.00019	.00000	.00000
	6 in. h.w. deck w/0.15 in. ins. and 2 in. h.w. RTS	d_n	1.00000	−1.59267	.72160	−.08275	.00029	.00000	.00000
13	Layers E0 C13 B6 E3 E2 A0	b_n	.00002	.00136	.00373	.00129	.00006	.00000	.00000
	6 in. h.w. concrete deck with 2 in. insulation	d_n	1.00000	−1.34451	.44285	−.04344	.00016	.00000	.00000
14	Layers E0 E5 E4 C12 B13 E3 E2 A0	b_n	.00000	.00046	.00143	.00057	.00003	.00000	.00000
	2 in. l.w. conc. deck w/4 in. ins. and susp. clg.	d_n	1.00000	−1.33741	.41454	−.03346	.00031	.00000	.00000
15	Layers E0 E5 E4 C5 B6 E3 E2 A0	b_n	.00001	.00066	.00163	.00049	.00002	.00000	.00000
	1 in. insul. w/4 in. h.w. conc. deck and susp. clg.	d_n	1.00000	−1.24348	.28742	−.01274	.00009	.00000	.00000
16	Layers E0 E5 E4 C13 B20 E3 E2 A0	b_n	.00001	.00060	.00197	.00086	.00005	.00000	.00000
	6 in. h.w. deck w/0.76 in. insul. and susp. clg.	d_n	1.00000	−1.39181	.46337	−.04714	.00058	.00000	.00000
17	Layers E0 E5 E4 B15 C14 E3 E2 A0	b_n	.00000	.00001	.00021	.00074	.00053	.00010	.00000
	6 in. insul. w/4 in. l.w. conc. deck and susp. clg.	d_n	1.00000	−1.87317	1.20950	−.32904	.03799	−.00169	.00002
18	Layers E0 C12 B15 E3 E2 C5 A0	b_n	.00000	.00002	.00027	.00052	.00019	.00002	.00000
	2 in. h.w. conc. dk w/6 in. ins. and 2 in. h.w. RTS	d_n	1.00000	−2.10928	1.50843	−.40880	.03249	−.00068	.00000
19	Layers E0 C5 B27 E3 E2 C12 A0	b_n	.00000	.00009	.00073	.00078	.00015	.00000	.00000
	4 in. h.w. deck w/4.54 in. ins. and 2 in. h.w. RTS	d_n	1.00000	−1.82851	1.02856	−.17574	.00556	−.00003	.00000
20	Layers E0 B21 C16 E3 E2 A0	b_n	.00000	.00002	.00044	.00103	.00049	.00005	.00000
	1.36 in. insulation with 8 in. l.w. concrete deck	d_n	1.00000	−1.91999	1.21970	−.30000	.02630	−.00061	.00000
21	Layers E0 C13 B12 E3 E2 C12 A0	b_n	.00000	.00009	.00072	.00077	.00015	.00000	.00000
	6 in. h.w. deck w/3 in. insul. and 2 in. h.w. RTS	d_n	1.00000	−1.84585	1.03238	−.17182	.00617	−.00003	.00000
22	Layers E0 B22 C5 E3 E2 A0	b_n	.00000	.00014	.00100	.00100	.00015	.00000	.00000
	1.67 in. ins. w/4 in. h.w. deck and 6 in. h.w. RTS	d_n	1.00000	−1.79981	.94786	−.13444	.00360	−.00001	.00000
23	Layers E0 E5 E4 C12 B14 E3 E2 C12 A0	b_n	.00000	.00002	.00022	.00031	.00008	.00000	.00000
	Susp. clg, 2 in. h.w. dk, 5 in. ins, 2 in. h.w. RTS	d_n	1.00000	−1.89903	1.13575	−.23586	.01276	−.00015	.00000
24	Layers E0 E5 E4 C5 E3 E2 B6 B1 C12 A0	b_n	.00000	.00008	.00047	.00039	.00006	.00000	.00000
	Susp. clg, 4 in. h.w. dk, 2 in. ins, 2 in. h.w. RTS	d_n	1.00000	−1.73082	.85681	−.11614	.00239	−.00001	.00000
25	Layers E0 E5 E4 C13 B13 E3 E2 A0	b_n	.00000	.00002	.00021	.00031	.00009	.00001	.00000
	6 in. h.w. conc. deck w/4 in. ins. and susp. clg.	d_n	1.00000	−1.63446	.78078	−.14422	.00940	−.00011	.00000
26	Layers E0 E5 E4 B15 C15 E3 E2 A0	b_n	.00000	.00000	.00002	.00014	.00024	.00011	.00002
	6 in. insul. w/6 in. l.w. conc. deck and susp. clg.	d_n	1.00000	−2.29459	1.93694	−.75741	.14252	−.01251	.00046
27	Layers E0 C13 B15 E3 E2 C12 A0	b_n	.00000	.00000	.00007	.00024	.00016	.00003	.00000
	6 in. h.w. deck w/6 in. ins. and 2 in. h.w. RTS	d_n	1.00000	−2.27813	1.82162	−.60696	.07696	−.00246	.00001
28	Layers E0 B9 B14 E3 E2 A0	b_n	.00000	.00000	.00001	.00010	.00017	.00009	.00001
	4 in. wood deck with 5 in. insulation	d_n	1.00000	−2.41915	2.17932	−.93062	.19840	−.02012	.00081
29	Layers E0 E5 E4 C12 B13 E3 E2 C5 A0	b_n	.00000	.00001	.00018	.00026	.00007	.00000	.00000
	Susp. clg, 2 in. h.w. dk, 4 in. ins, 4 in. h.w. RTS	d_n	1.00000	−1.99413	1.20218	−.20898	.01058	−.00010	.00000
30	Layers E0 E5 E4 B9 B6 E3 E2 A0	b_n	.00000	.00000	.00003	.00016	.00018	.00005	.00000
	4 in. wood deck w/2 in. insul. and susp. ceiling	d_n	1.00000	−2.29665	1.86386	−.65738	.10295	−.00631	.00012
31	Layers E0 B27 C13 E3 E2 C13 A0	b_n	.00000	.00000	.00003	.00014	.00014	.00003	.00000
	4.54 in. ins. w/6 in. h.w. deck and 6 in. h.w. RTS	d_n	1.00000	−2.29881	1.85733	−.64691	.10024	−.00593	.00006
32	Layers E0 E5 E4 C5 B20 E3 E2 C13 A0	b_n	.00000	.00002	.00024	.00037	.00011	.00001	.00000
	Susp. clg, 4 in. h.w. dk, 0.76 in. ins, 4 in. h.w. RTS	d_n	1.00000	−2.09344	1.35118	−.26478	.01281	−.00018	.00000
33	Layers E0 E5 E4 C5 B13 E3 E2 C5 A0	b_n	.00000	.00000	.00005	.00013	.00007	.00001	.00000
	Susp. clg, 4 in. h.w. dk, 4 in. ins, 4 in. h.w. RTS	d_n	1.00000	−2.07856	1.33963	−.27670	.02089	−.00058	.00000
34	Layers E0 E5 E4 C13 B23 E3 E2 C5 A0	b_n	.00000	.00000	.00005	.00013	.00007	.00001	.00000
	Susp. clg, 6 in. h.w. dk, 2.42 in. ins, 4 in. h.w. RTS	d_n	1.00000	−2.13236	1.43448	−.32023	.02188	−.00038	.00000

Table 13   Roof Conduction Transfer Function Coefficients (*b* and *d* Factors) (*Concluded*)

Roof Group	(Layer Sequence Left to Right = Inside to Outside)		$n=0$	$n=1$	$n=2$	$n=3$	$n=4$	$n=5$	$n=6$
35	Layers E0 C5 B15 E3 E2 C13 A0	b_n	.00000	.00000	.00002	.00010	.00011	.00003	.00000
	4 in. h.w. deck w/6 in. ins. and 6 in. h.w. RTS	d_n	1.00000	−2.51234	2.25816	−.87306	.14066	−.00785	.00016
36	Layers E0 C13 B27 E3 E2 C13 A0	b_n	.00000	.00000	.00002	.00009	.00011	.00003	.00000
	6 in. h.w. deck w/4.54 in. ins. and 6 in. h.w. RTS	d_n	1.00000	−2.50269	2.23944	−.88012	.15928	−.01176	.00018
37	Layers E0 E5 E4 B15 C13 E3 E2 C13 A0	b_n	.00000	.00000	.00000	.00002	.00005	.00004	.00001
	Susp. clg, 6 in. ins, 6 in. h.w. dk, 6 in. h.w. RTS	d_n	1.00000	−2.75535	2.88190	−1.44618	.36631	−.04636	.00269
38	Layers E0 E5 E4 B9 B15 E3 E2 A0	b_n	.00000	.00000	.00000	.00001	.00003	.00003	.00001
	4 in. wood deck with 6 in. insul. and susp. ceiling	d_n	1.00000	−2.81433	3.05064	−1.62771	.45499	−.06569	.00455
39	Layers E0 E5 E4 C13 B20 E3 E2 C13 A0	b_n	.00000	.00000	.00007	.00019	.00011	.00001	.00000
	Susp. clg, 6 in. h.w. dk, 0.76 in. ins, 6 in. h.w. RTS	d_n	1.00000	−2.30711	1.77588	−.52057	.05597	−.00118	.00001
40	Layers E0 E5 E4 C5 B26 E3 E2 C13 A0	b_n	.00000	.00000	.00002	.00007	.00006	.00001	.00000
	Susp. clg, 4 in. h.w. dk, 3.64 in. ins, 6 in. h.w. RTS	d_n	1.00000	−2.26975	1.68337	−.45628	.04712	−.00180	.00002
41	Layers E0 E5 E4 C13 B6 E3 E2 C13 A0	b_n	.00000	.00000	.00002	.00007	.00006	.00001	.00000
	Susp. clg, 6 in. h.w. deck, 2 in. ins, 6 in. h.w. RTS	d_n	1.00000	−2.35843	1.86626	−.56900	.06466	−.00157	.00001
42	Layers E0 E5 E4 C13 B14 E3 E2 C13 A0	b_n	.00000	.00000	.00000	.00001	.00002	.00001	.00000
	Susp. clg, 6 in. h.w. deck, 5 in. ins, 6 in. h.w. RTS	d_n	1.00000	−2.68628	2.63091	−1.16847	.24692	−.02269	.00062

Table 14   Roof Conduction Transfer Function Coefficients Σc_n, Time Lag, U-Factors, and Decrement Factors

Roof Group		Σc_n	TL, h	U	DF
1	Layers E0 A3 B25 E3 E2 A0	0.05362	1.63	0.080	0.97
2	Layers E0 A3 B14 E3 E2 A0	0.02684	2.43	0.055	0.94
3	Layers EO E5 E4 C12 E3 E2 A0	0.05997	3.39	0.232	0.75
4	Layers E0 E1 B15 E4 B7 A0	0.00673	4.85	0.043	0.82
5	Layers E0 B14 C12 E3 E2 A0	0.00841	4.82	0.055	0.68
6	Layers E0 C5 B17 E3 E2 A0	0.05668	4.57	0.371	0.60
7	Layers E0 B22 C12 E3 E2 C12 A0	0.01652	5.00	0.138	0.56
8	Layers E0 B16 C13 E3 E2 A0	0.04340	5.45	0.424	0.47
9	Layers E0 E5 E4 B12 C14 E3 E2 A0	0.00550	6.32	0.057	0.60
10	Layers E0 E5 E4 C15 B16 E3 E2 A0	0.00647	7.14	0.104	0.49
11	Layers E0 C5 B15 E3 E2 A0	0.00232	7.39	0.046	0.43
12	Layers E0 C13 B16 E3 E2 C12 A0	0.01841	7.08	0.396	0.40
13	Layers E0 C13 B6 E3 E2 A0	0.00645	6.73	0.117	0.33
14	Layers E0 E5 E4 C12 B13 E3 E2 A0	0.00250	7.06	0.057	0.26
15	Layers E0 E5 E4 C5 B6 E3 E2 A0	0.01477	7.16	0.090	0.16
16	Layers E0 E5 E4 C13 B20 E3 E2 A0	0.00349	7.54	0.140	0.15
17	Layers E0 E5 E4 B15 C14 E3 E2 A0	0.00159	8.23	0.036	0.50
18	Layers E0 C12 B15 E3 E2 C5 A0	0.00101	9.21	0.046	0.41
19	Layers E0 C5 B27 E3 E2 C12 A0	0.00176	8.42	0.059	0.37
20	Layers E0 B21 C16 E3 E2 A0	0.00202	8.93	0.080	0.32
21	Layers E0 C13 B12 E3 E2 C12 A0	0.00174	8.93	0.083	0.26
22	Layers E0 B22 C5 E3 E2 C13 A0	0.00222	8.99	0.129	0.20
23	Layers E0 E5 E4 C12 B14 E3 E2 C12 A0	0.00064	9.26	0.047	0.16
24	Layers E0 E5 E4 C5 E3 E2 B6 B1 C12 A0	0.00100	8.84	0.082	0.12
25	Layers E0 E5 E4 C13 B13 E3 E2 A0	0.00063	8.77	0.056	0.09
26	Layers E0 E5 E4 B15 C15 E3 E2 A0	0.00053	10.44	0.034	0.30
27	Layers E0 C13 B15 E3 E2 C12 A0	0.00050	10.48	0.045	0.24
28	Layers E0 B9 B14 E3 E2 A0	0.00038	11.18	0.044	0.19
29	Layers E0 E5 E4 C12 B13 E3 E2 C5 A0	0.00053	10.57	0.056	0.16
30	Layers E0 E5 E4 B9 B6 E3 E2 A0	0.00042	11.22	0.064	0.13
31	Layers E0 B27 C13 E3 E2 C13 A0	0.00034	11.27	0.057	0.12
32	Layers E0 E5 E4 C5 B20 E3 E2 C13 A0	0.00075	11.31	0.133	0.10
33	Layers E0 E5 E4 C5 B13 E3 E2 C5 A0	0.00026	11.47	0.055	0.08
34	Layers E0 E5 E4 C13 B23 E3 E2 C5 A0	0.00026	11.63	0.077	0.06
35	Layers E0 C5 B15 E3 E2 C13 A0	0.00026	12.29	0.045	0.18
36	Layers E0 C13 B27 E3 E2 C13 A0	0.00025	12.67	0.057	0.13
37	Layers E0 E5 E4 B15 C13 E3 E2 C13 A0	0.00012	13.02	0.040	0.11
38	Layers E0 E5 E4 B9 B15 E3 E2 A0	0.00008	13.33	0.035	0.09
39	Layers E0 E5 E4 C13 B20 E3 E2 C13 A0	0.00039	12.23	0.131	0.07
40	Layers E0 E5 E4 C5 B26 E3 E2 C13 A0	0.00016	12.68	0.059	0.06
41	Layers E0 E5 E4 C13 B6 E3 E2 C13 A0	0.00016	12.85	0.085	0.05
42	Layers E0 E5 E4 C13 B14 E3 E2 C13 A0	0.00005	14.17	0.046	0.03

Table 15 Wall Group Numbers, Walls for Mass-In Case—Dominant Wall Material

Combined with Wall Material A1, E1, or Both

R	1	2	3	4	5	6	7	8	9	10	11	12	13	14	15	16	17	18	19	20	21	22	23	24	25		R-Value Ranges, h·ft²·°F/Btu
1	*	*	*	*	*	*	*	*	*	*	*	*	*	*	*	*	2	*	*	*	*	*	*	*	*		1 0.0 – 2.0
2	*	5	*	*	*	*	*	*	*	5	*	*	*	*	11	*	2	6	*	*	*	*	*	*	*		2 2.0 – 2.5
3	*	5	*	*	*	3	*	2	5	6	*	*	5	*	12	18	2	6	*	*	*	*	*	*	*		3 2.5 – 3.0
4	*	5	*	*	*	4	2	2	5	6	*	*	6	12	12	19	2	7	*	*	*	*	*	*	*		4 3.0 – 3.5
5	*	5	*	*	*	4	2	3	6	6	10	4	6	17	12	19	2	7	*	*	*	*	5	*	*		5 3.5 – 4.0
6	*	6	*	*	*	5	2	4	6	6	11	5	10	17	13	19	2	11	*	*	*	*	10	*	16		6 4.0 – 4.75
7	*	6	*	*	*	5	2	4	6	6	11	5	10	18	13	20	2	11	2	*	*	*	10	*	16		7 4.75 – 5.5
8	*	6	*	*	*	5	2	5	10	7	12	5	11	18	13	26	2	12	2	*	*	*	10	*	17		8 5.5 – 6.5
9	*	6	*	*	*	5	4	5	11	7	16	10	11	18	13	20	3	12	4	5	*	*	11	*	18		9 6.5 – 7.75
10	*	6	*	*	*	5	4	5	11	7	17	10	11	18	13	20	3	12	4	9	10	*	11	*	18		10 7.75 – 9.0
11	*	6	*	*	*	5	4	5	11	7	17	10	11	19	13	27	3	12	4	10	15	4	11	*	18		11 9.0 – 10.75
12	*	6	*	*	*	5	4	5	11	11	17	10	11	19	19	27	3	12	4	10	16	4	11	*	24		12 10.75 – 12.75
13	*	10	*	*	*	10	4	5	11	11	17	10	11	19	18	27	4	12	5	11	17	9	12	15	25		13 12.75 – 15.0
14	*	10	*	*	*	10	5	5	11	11	18	11	12	25	19	27	4	12	5	11	17	10	16	16	25		14 15.0 – 17.5
15	*	11	*	*	*	10	5	9	11	11	18	15	16	26	19	28	4	12	5	11	17	10	16	22	25		15 17.5 – 20.0
16	*	11	*	*	*	10	9	9	16	11	18	15	16	26	19	34	4	17	9	16	23	10	16	23	25		16 20.0 – 23.0
17	*	*	*	*	*	*	*	*	*	*	24	16	*	*	*	*	*	*	9	16	24	15	17	24	25		17 23.0 – 27.0

Combined with Wall Material A3 or A6

R	1	2	3	4	5	6	7	8	9	10	11	12	13	14	15	16	17	18	19	20	21	22	23	24	25
1	*	*	*	*	*	*	*	*	*	*	*	*	*	*	*	*	1	*	*	*	*	*	*	*	*
2	*	3	*	*	*	*	*	2	3	5	*	*	*	*	11	*	2	6	*	*	*	*	*	*	*
3	*	5	*	*	*	2	*	2	5	3	*	*	5	*	12	18	2	6	*	*	*	*	*	*	*
4	*	5	*	*	*	3	1	2	5	5	*	*	5	11	12	19	2	7	*	*	*	*	*	*	*
5	*	5	*	*	*	3	2	2	5	5	6	3	5	12	12	19	2	7	*	*	*	*	5	*	*
6	*	6	*	*	*	4	2	2	5	5	10	4	6	12	12	19	2	7	*	*	*	*	5	*	11
7	*	6	*	*	*	5	2	2	6	6	11	5	6	17	13	20	2	7	2	*	*	*	6	*	12
8	*	6	*	*	*	5	2	3	6	6	11	5	6	18	13	20	2	7	2	*	*	*	6	*	17
9	*	6	*	*	*	5	2	3	6	6	11	5	6	18	13	20	2	8	2	5	*	*	10	*	17
10	*	6	*	*	*	5	2	3	6	6	12	5	6	18	14	21	2	12	2	5	10	*	11	*	17
11	*	6	*	*	*	5	2	3	6	6	12	5	6	18	14	21	3	12	4	5	11	4	11	*	18
12	*	6	*	*	*	5	2	3	6	7	12	6	11	19	14	21	3	12	4	10	16	4	11	*	18
13	*	6	*	*	*	5	2	4	6	7	12	10	11	19	14	27	3	12	5	10	17	5	11	10	18
14	*	10	*	*	*	6	4	4	10	7	17	10	11	19	18	27	4	12	5	11	17	9	11	16	18
15	*	10	*	*	*	10	4	4	10	11	17	10	11	25	18	28	4	12	5	11	17	10	11	16	18
16	*	11	*	*	*	10	4	5	11	11	17	10	11	25	18	28	4	12	9	11	18	10	16	17	24
17	*	*	*	*	*	*	*	*	*	*	17	10	*	*	*	*	*	*	9	16	24	11	16	23	25

Combined with Wall Material A2 or A7

R	1	2	3	4	5	6	7	8	9	10	11	12	13	14	15	16	17	18	19	20	21	22	23	24	25
1	*	*	*	*	*	*	*	*	*	*	*	*	*	*	*	*	*	*	*	*	*	*	*	*	*
2	3	*	*	*	*	*	*	*	*	*	11	*	*	*	*	*	*	*	6	*	*	*	*	*	*
3	5	11	*	*	*	*	*	*	6	11	12	*	*	*	*	18	*	6	12	*	*	*	*	*	*
4	5	12	5	*	*	11	*	11	12	12	*	*	12	*	19	26	7	13	*	*	*	*	*	*	*
5	5	12	6	*	*	12	6	12	12	13	*	*	12	24	19	27	7	14	*	*	*	*	*	*	*
6	6	13	6	10	*	13	10	12	12	13	17	11	17	25	20	27	7	18	*	*	*	*	16	*	24
7	6	13	6	11	*	18	11	12	13	13	18	16	17	26	20	28	7	19	11	*	*	*	17	*	25
8	6	13	6	11	*	18	11	12	13	13	24	17	18	26	20	28	12	19	11	*	*	*	17	*	25
9	6	13	6	11	24	18	11	13	18	13	25	17	18	27	20	29	12	19	11	16	*	*	18	*	26
10	6	13	10	16	25	19	11	13	18	13	25	17	18	27	26	35	12	19	11	17	23	*	18	*	26
11	6	14	10	16	32	19	11	13	18	14	25	17	18	33	21	35	12	19	16	23	24	16	18	*	33
12	6	14	10	16	32	19	11	13	18	14	26	18	18	34	27	35	12	19	16	24	31	16	19	*	33
13	6	18	11	16	33	19	12	13	18	18	26	18	18	34	27	36	12	20	17	24	32	17	25	30	33
14	10	18	11	17	33	19	12	13	18	18	26	18	18	34	27	36	12	20	17	24	32	23	25	31	34
15	10	18	11	17	34	19	16	18	18	18	26	24	25	34	27	36	12	20	17	25	33	24	25	32	34
16	11	19	15	23	39	26	16	18	24	19	32	24	25	34	27	36	17	26	23	31	33	24	25	32	34
17	*	*	*	23	39	*	16	*	*	*	33	24	*	35	*	*	*	*	23	32	38	24	25	38	39

Wall Materials Layers (Table 11)

1	A1,A3,A6, or E1
2	A2 or A7
3	B7
4	B10
5	B9
6	C1
7	C2
8	C3
9	C4
10	C5
11	C6
12	C7
13	C8
14	C9
15	C10
16	C11
17	C12
18	C13
19	C14
20	C15
21	C16
22	C17
23	C18
24	C19
25	C20

*Denotes a wall not possible with chosen combination of parameters.

Table 16 Walls for Integral Mass Case—Dominant Wall Material

Combined with Wall Material A1, E1, or Both

R	1	2	3	4	5	6	7	8	9	10	11	12	13	14	15	16	17	18	19	20	21	22	23	24	25
1	1	3	*	*	*	*	*	1	3	3	*	*	*	*	11	*	2	5	*	*	*	*	*	*	*
2	1	3	1	*	*	2	*	2	4	4	*	*	5	*	11	17	2	5	*	*	*	*	*	*	*
3	1	4	1	*	*	2	2	2	4	4	*	*	5	10	12	17	4	5	*	*	*	*	*	*	*
4	1	*	1	*	*	2	2	*	*	*	10	4	5	10	*	17	*	*	*	*	*	*	4	*	*
5	1	*	1	2	*	*	4	*	*	*	10	4	*	10	*	*	*	*	*	*	*	*	4	*	10
6	1	*	1	2	*	*	*	*	*	*	10	4	*	*	*	*	*	*	2	*	*	*	4	*	10
7	1	*	1	2	*	*	*	*	*	*	*	*	*	*	*	*	*	*	2	*	*	*	*	*	10
8	1	*	2	4	10	*	*	*	*	*	*	*	*	*	*	*	*	*	4	4	*	*	*	*	*
9	1	*	2	4	11	*	*	*	*	*	*	*	*	*	*	*	*	*	4	*	*	*	*	*	*
10	1	*	2	4	16	*	*	*	*	*	*	*	*	*	*	*	*	*	*	*	9	*	*	*	*
11	1	*	2	4	16	*	*	*	*	*	*	*	*	*	*	*	*	*	*	*	9	4	*	*	*
12	1	*	2	5	17	*	*	*	*	*	*	*	*	*	*	*	*	*	*	*	*	4	*	*	*
13	2	*	2	5	17	*	*	*	*	*	*	*	*	*	*	*	*	*	*	*	*	*	*	15	*
14	2	*	2	5	17	*	*	*	*	*	*	*	*	*	*	*	*	*	*	*	*	*	*	15	*
15	2	*	2	9	24	*	*	*	*	*	*	*	*	*	*	*	*	*	*	*	*	*	*	*	*
16	2	*	4	9	24	*	*	*	*	*	*	*	*	*	*	*	*	*	*	*	*	*	*	*	*
17	*	*	*	9	24	*	*	*	*	*	*	*	*	*	*	*	*	*	*	*	*	*	*	*	*

	R-Value Ranges, $h \cdot ft^2 \cdot °F/Btu$
1	0.0 – 2.0
2	2.0 – 2.5
3	2.5 – 3.0
4	3.0 – 3.5
5	3.5 – 4.0
6	4.0 – 4.75
7	4.75 – 5.5
8	5.5 – 6.0
9	6.5 – 7.75
10	7.75 – 8.0
11	9.0 – 10.75
12	10.75 – 12.75
13	12.75 – 15.0
14	15.0 – 17.5
15	17.5 – 20.0
16	20.0 – 23.0
17	23.0 – 27.0

Combined with Wall Material A3 or A6

R	1	2	3	4	5	6	7	8	9	10	11	12	13	14	15	16	17	18	19	20	21	22	23	24	25
1	1	3	*	*	*	*	*	1	3	2	*	*	*	*	6	*	1	5	*	*	*	*	*	*	*
2	1	3	1	*	*	2	*	1	3	2	*	*	3	*	6	12	1	5	*	*	*	*	*	*	*
3	1	4	1	*	*	2	1	2	4	4	*	*	3	10	11	12	2	5	*	*	*	*	*	*	*
4	1	*	1	*	*	4	1	*	*	*	5	2	4	10	*	12	*	*	*	*	*	*	4	*	*
5	1	*	1	2	*	*	2	*	*	*	5	2	*	10	*	*	*	*	*	*	*	*	4	*	10
6	1	*	1	2	*	*	*	*	*	*	10	4	*	*	*	*	*	*	2	*	*	*	4	*	10
7	1	*	1	2	*	*	*	*	*	*	*	*	*	*	*	*	*	*	2	*	*	*	*	*	10
8	1	*	1	2	10	*	*	*	*	*	*	*	*	*	*	*	*	*	4	4	*	*	*	*	*
9	1	*	1	4	11	*	*	*	*	*	*	*	*	*	*	*	*	*	4	*	*	*	*	*	*
10	1	*	2	4	16	*	*	*	*	*	*	*	*	*	*	*	*	*	*	*	9	*	*	*	*
11	1	*	2	4	16	*	*	*	*	*	*	*	*	*	*	*	*	*	*	*	9	2	*	*	*
12	1	*	2	4	17	*	*	*	*	*	*	*	*	*	*	*	*	*	*	*	*	4	*	*	*
13	1	*	2	5	17	*	*	*	*	*	*	*	*	*	*	*	*	*	*	*	*	*	*	10	*
14	1	*	2	5	17	*	*	*	*	*	*	*	*	*	*	*	*	*	*	*	*	*	*	15	*
15	1	*	2	5	18	*	*	*	*	*	*	*	*	*	*	*	*	*	*	*	*	*	*	*	*
16	2	*	4	9	24	*	*	*	*	*	*	*	*	*	*	*	*	*	*	*	*	*	*	*	*
17	*	*	*	9	24	*	*	*	*	*	*	*	*	*	*	*	*	*	*	*	*	*	*	*	*

	Wall Materials Layers (Table 11)
1	A1, A3, A6, or E1
2	A2 or A7
3	B7
4	B10
5	B9
6	C1
7	C2
8	C3
9	C4
10	C5
11	C6
12	C7
13	C8
14	C9
15	C10
16	C11
17	C12
18	C13
19	C14
20	C15
21	C16
22	C17
23	C18
24	C19
25	C20

Combined with Wall Material A2 or A7

R	1	2	3	4	5	6	7	8	9	10	11	12	13	14	15	16	17	18	19	20	21	22	23	24	25
1	3	6	*	*	*	*	*	*	*	6	*	*	*	*	*	*	*	3	11	*	*	*	*	*	*
2	3	10	*	*	*	*	*	5	10	10	*	*	*	*	17	24	5	11	*	*	*	*	*	*	*
3	4	10	5	*	*	5	*	5	10	11	*	*	10	*	17	25	5	16	*	*	*	*	*	*	*
4	*	11	5	*	*	10	5	5	11	11	15	10	10	17	18	26	5	17	*	*	*	*	10	*	*
5	*	11	5	10	*	10	5	5	11	11	16	10	16	23	18	26	5	17	*	*	*	*	10	*	*
6	*	11	*	11	*	10	5	5	16	11	17	10	16	24	18	33	5	17	*	*	*	*	16	*	23
7	*	11	*	11	*	10	5	10	16	16	17	10	16	25	25	33	5	17	5	*	*	*	16	*	23
8	*	16	*	*	22	10	9	10	16	11	17	11	16	25	25	34	10	18	9	*	*	*	16	*	24
9	*	16	*	*	23	11	9	10	16	16	24	16	16	26	25	34	10	18	10	15	*	*	17	*	25
10	*	16	*	*	*	15	9	10	16	17	24	15	16	26	26	34	10	18	10	15	22	*	17	*	25
11	*	16	*	*	*	15	10	10	17	16	24	16	17	33	26	35	10	18	10	16	23	10	23	*	25
12	*	16	*	*	*	16	10	10	17	17	24	16	17	33	26	35	10	18	10	16	23	15	23	*	32
13	*	16	*	*	*	16	10	10	17	16	25	17	17	33	26	35	10	24	15	23	24	15	24	23	32
14	*	17	*	*	*	16	10	15	23	17	31	23	24	33	26	40	10	24	15	23	31	16	23	30	32
15	*	17	*	*	*	16	15	15	23	23	31	23	24	38	33	40	10	24	15	24	31	16	23	30	32
16	*	23	*	*	*	22	15	16	24	24	32	23	24	38	33	41	15	25	15	23	32	22	23	31	32
17	*	*	*	*	*	*	15	*	*	*	32	23	*	39	*	*	*	*	22	30	32	23	24	32	38

*Denotes a wall not possible with chosen combination of parameters.

Table 17 Walls for Mass-Out Case—Dominant Wall Material

Combined with Wall Material A1, E1, or Both

R	1	2	3	4	5	6	7	8	9	10	11	12	13	14	15	16	17	18	19	20	21	22	23	24	25
1	*	*	*	*	*	*	*	*	*	*	*	*	*	*	*	*	1	*	*	*	*	*	*	*	*
2	*	3	*	*	*	*	*	2	3	5	*	*	*	*	6	*	1	5	*	*	*	*	*	*	*
3	*	3	*	*	*	2	*	2	4	5	*	*	5	*	11	18	2	5	*	*	*	*	*	*	*
4	*	3	*	*	*	2	2	2	5	5	*	*	5	16	11	18	2	5	*	*	*	*	*	*	*
5	*	3	*	*	*	2	2	2	5	5	10	4	6	17	11	19	2	6	*	*	*	*	5	*	*
6	*	4	*	*	*	4	2	2	5	5	10	4	6	17	11	19	2	6	*	*	*	*	9	*	16
7	*	4	*	*	*	4	2	2	5	6	11	5	10	17	12	19	2	6	2	*	*	*	10	*	16
8	*	5	*	*	*	4	2	2	5	6	11	5	10	18	11	20	2	6	4	*	*	*	10	*	16
9	*	5	*	*	*	4	2	2	5	6	11	5	10	18	11	26	2	6	4	9	*	*	10	*	17
10	*	5	*	*	*	5	2	4	5	6	16	10	10	18	12	26	2	6	4	9	15	*	10	*	17
11	*	5	*	*	*	5	4	4	5	6	16	10	10	18	12	26	2	6	4	10	15	4	11	*	18
12	*	5	*	*	*	5	4	4	10	6	16	10	10	18	12	26	2	10	5	10	16	9	11	*	18
13	*	5	*	*	*	5	4	4	10	10	17	10	11	18	12	26	2	10	5	11	17	9	11	15	24
14	*	5	*	*	*	5	4	4	10	10	17	10	11	24	18	26	2	10	9	15	23	10	16	16	24
15	*	5	*	*	*	9	4	4	10	10	17	10	15	25	18	26	2	10	9	15	23	10	16	22	24
16	*	9	*	*	*	9	9	9	15	10	17	10	15	25	18	33	4	11	9	16	24	15	16	23	24
17	*	*	*	*	*	*	*	*	*	*	23	15	*	*	*	*	*	*	9	22	24	15	16	24	25

R-Value Ranges, h·ft²·°F/Btu

1	0.0 – 2.0
2	2.0 – 2.5
3	2.5 – 3.0
4	3.0 – 3.5
5	3.5 – 4.0
6	4.0 – 4.75
7	4.75 – 5.5
8	5.5 – 6.5
9	6.5 – 7.75
10	7.75 – 9.0
11	9.0 – 10.75
12	10.75 – 12.75
13	12.75 – 15.0
14	15.0 – 17.5
15	17.5 – 20.0
16	20.0 – 23.0
17	23.0 – 27.0

Combined with Wall Material A3 or A6

R	1	2	3	4	5	6	7	8	9	10	11	12	13	14	15	16	17	18	19	20	21	22	23	24	25
1	*	*	*	*	*	*	*	*	*	*	*	*	*	*	*	*	1	*	*	*	*	*	*	*	*
2	*	3	*	*	*	*	*	2	3	2	*	*	*	*	6	*	1	5	*	*	*	*	*	*	*
3	*	3	*	*	*	2	*	2	3	2	*	*	*	*	10	17	1	5	*	*	*	*	*	*	*
4	*	3	*	*	*	2	1	2	4	3	*	*	4	11	11	17	1	5	*	*	*	*	*	*	*
5	*	3	*	*	*	2	2	2	4	3	5	2	5	11	11	18	1	6	*	*	*	*	4	*	*
6	*	3	*	*	*	2	2	2	4	3	10	3	5	12	11	18	2	6	*	*	*	*	5	*	10
7	*	3	*	*	*	2	2	2	5	3	10	4	5	12	11	18	2	6	2	*	*	*	5	*	11
8	*	4	*	*	*	2	2	2	5	3	10	4	5	12	11	18	2	6	2	*	*	*	5	*	12
9	*	4	*	*	*	2	2	2	5	4	11	5	5	17	11	18	2	6	2	5	*	*	6	*	16
10	*	5	*	*	*	2	2	2	5	4	11	5	5	17	11	19	2	6	2	5	10	*	6	*	17
11	*	5	*	*	*	2	2	2	5	4	11	5	5	17	12	19	2	6	4	5	11	4	10	*	17
12	*	5	*	*	*	4	2	2	5	5	11	5	5	17	12	19	2	6	4	10	15	4	10	*	17
13	*	5	*	*	*	4	2	2	5	5	11	5	10	18	12	19	2	10	4	10	16	5	10	10	17
14	*	5	*	*	*	4	2	4	5	5	16	9	10	18	12	25	2	10	4	10	17	9	10	16	17
15	*	5	*	*	*	4	4	4	9	5	16	9	10	18	16	25	2	10	5	11	17	10	10	16	18
16	*	9	*	*	*	4	4	4	9	9	16	10	10	24	17	25	4	10	5	11	17	10	11	17	18
17	*	*	*	*	*	*	*	*	*	*	16	10	*	*	*	*	*	*	9	16	23	10	15	23	24

Wall Materials Layers (Table 11)

1	A1,A3,A6, or E1
2	A2 or A7
3	B7
4	B10
5	B9
6	C1
7	C2
8	C3
9	C4
10	C5
11	C6
12	C7
13	C8
14	C9
15	C10
16	C11
17	C12
18	C13
19	C14
20	C15
21	C16
22	C17
23	C18
24	C19
25	C20

Combined with Wall Material A2 or A7

R	1	2	3	4	5	6	7	8	9	10	11	12	13	14	15	16	17	18	19	20	21	22	23	24	25
1	*	*	*	*	*	*	*	*	*	*	*	*	*	*	*	*	*	*	*	*	*	*	*	*	*
2	3	*	*	*	*	*	*	*	*	*	11	*	*	*	*	*	*	5	*	*	*	*	*	*	*
3	3	10	*	*	*	*	*	5	10	11	*	*	*	*	17	*	5	12	*	*	*	*	*	*	*
4	3	11	5	*	*	10	*	5	11	11	*	*	11	*	18	26	6	12	*	*	*	*	*	*	*
5	3	11	5	*	*	10	5	10	11	11	17	10	11	24	18	26	6	13	*	*	*	*	16	*	23
6	3	11	5	10	*	10	5	10	11	12	17	11	16	25	19	27	6	17	9	*	*	*	16	*	23
7	3	12	5	10	*	10	9	10	11	12	17	15	16	25	19	27	6	17	10	*	*	*	16	*	24
8	4	12	5	10	*	10	10	10	12	12	23	16	17	26	19	27	10	18	10	15	*	*	16	*	25
9	4	12	5	10	23	11	10	10	12	12	23	16	17	26	19	34	10	18	10	16	22	*	17	*	25
10	5	12	5	15	24	11	10	10	16	12	24	16	17	26	19	34	10	18	10	16	23	15	17	*	25
11	5	12	9	15	30	11	10	10	16	12	24	16	17	26	25	34	10	18	10	22	24	15	17	*	32
12	5	12	10	15	31	11	10	10	17	12	24	16	17	26	25	34	10	18	15	23	30	15	23	23	32
13	5	17	10	16	32	11	10	11	17	17	24	16	17	26	25	34	11	18	15	23	30	15	23	23	32
14	5	17	10	16	32	15	10	11	17	17	25	16	17	33	25	34	11	18	15	23	31	22	23	30	32
15	5	17	10	16	32	16	15	15	17	17	25	22	23	33	26	35	11	18	15	23	31	22	23	30	32
16	9	17	15	16	32	16	15	15	23	17	31	22	23	33	26	40	15	24	15	23	32	23	24	31	32
17	*	*	*	22	38	*	15	*	*	*	31	23	*	33	*	*	*	*	22	30	37	23	24	37	38

*Denotes a wall not possible with chosen combination of parameters.

Table 18   Wall Conduction Transfer Function Coefficients (b and d Factors)

Wall Group	(Layer Sequence Left to Right = Inside to Outside)		$n=0$	$n=1$	$n=2$	$n=3$	$n=4$	$n=5$	$n=6$
1	Layers E0 A3 B1 B13 A3 A0	b_n	.00768	.03498	.00719	.00006	.00000	.00000	.00000
	Steel siding with 4 in. insulation	d_n	1.00000	−.24072	.00168	.00000	.00000	.00000	.00000
2	Layers E0 E1 B14 A1 A0 A0	b_n	.00016	.00545	.00961	.00215	.00005	.00000	.00000
	Frame wall with 5 in. insulation	d_n	1.00000	−.93389	.27396	−.02561	.00014	.00000	.00000
3	Layers E0 C3 B5 A6 A0 A0	b_n	.00411	.03230	.01474	.00047	.00000	.00000	.00000
	4 in. h.w. concrete block with 1 in. insulation	d_n	1.00000	−.76963	.04014	−.00042	.00000	.00000	.00000
4	Layers E0 E1 B6 C12 A0 A0	b_n	.00001	.00108	.00384	.00187	.00013	.00000	.00000
	2 in. insulation with 2 in. h.w. concrete	d_n	1.00000	−1.37579	.61544	−.09389	.00221	.00000	.00000
5	Layers E0 A6 B21 C7 A0 A0	b_n	.00008	.00444	.01018	.00296	.00010	.00000	.00000
	1.36 in. insulation with 8 in. l.w. concrete block	d_n	1.00000	−1.16043	.32547	−.02746	.00021	.00000	.00000
6	Layers E0 E1 B2 C5 A1 A0	b_n	.00051	.00938	.01057	.00127	.00001	.00000	.00000
	1 in. insulation with 4 in. h.w. concrete	d_n	1.00000	−1.17580	.30071	−.01561	.00001	.00000	.00000
7	Layers E0 A6 C5 B3 A3 A0	b_n	.00099	.00836	.00361	.00007	.00000	.00000	.00000
	4 in. h.w. concrete with 2 in. insulation	d_n	1.00000	−.93970	.04664	.00000	.00000	.00000	.00000
8	Layers E0 A2 C12 B5 A6 A0	b_n	.00014	.00460	.00733	.00135	.00002	.00000	.00000
	Face brick and 2 in. h.w. concrete with 1 in. insul.	d_n	1.00000	−1.20012	.27937	−.01039	.00005	.00000	.00000
9	Layers E0 A6 B15 B10 A0 A0	b_n	.00000	.00006	.00086	.00146	.00051	.00004	.00000
	6 in. insulation with 2 in. wood	d_n	1.00000	−1.63352	.86971	−.18121	.01445	−.00031	.00000
10	Layers E0 E1 C2 B5 A2 A0	b_n	.00001	.00102	.00441	.00260	.00024	.00000	.00000
	4 in. l.w. conc. block w/1 in. insul. and face brick	d_n	1.00000	−1.66358	.82440	−.11098	.00351	.00000	.00000
11	Layers E0 E1 C8 B6 A1 A0	b_n	.00000	.00061	.00289	.00183	.00018	.00000	.00000
	8 in. h.w. concrete block with 2 in. insulation	d_n	1.00000	−1.52480	.67146	−.09844	.00239	.00000	.00000
12	Layers E0 E1 B1 C10 A1 A0	b_n	.00002	.00198	.00816	.00467	.00044	.00001	.00000
	8 in. h.w. concrete	d_n	1.00000	−1.51658	.64261	−.08382	.00289	−.00001	.00000
13	Layers E0 A2 C5 B19 A6 A0	b_n	.00003	.00203	.00601	.00233	.00013	.00000	.00000
	Face brick and 4 in. h.w. concrete with 0.61 in. ins.	d_n	1.00000	−1.41349	.48697	−.03218	.00057	.00000	.00000
14	Layers E0 A2 A2 B6 A6 A0	b_n	.00000	.00030	.00167	.00123	.00016	.00000	.00000
	Face brick and face brick with 2 in. insulation	d_n	1.00000	−1.52986	.62059	−.06329	.00196	−.00001	.00000
15	Layers E0 A6 C17 B1 A7 A0	b_n	.00000	.00003	.00060	.00145	.00074	.00009	.00000
	8 in. l.w. concrete block (filled) and face brick	d_n	1.00000	−1.99996	1.36804	−.37388	.03885	−.00140	.00002
16	Layers E0 A6 C18 B1 A7 A0	b_n	.00000	.00014	.00169	.00270	.00086	.00006	.00000
	8 in. h.w. concrete block (filled) and face brick	d_n	1.00000	−2.00258	1.32887	−.32486	.02361	−.00052	.00000
17	Layers E0 A2 C2 B15 A0 A0	b_n	.00000	.00000	.00013	.00044	.00030	.00005	.00000
	Face brick and 4 in. l.w. conc. block with 6 in. ins.	d_n	1.00000	−2.00875	1.37120	−.37897	.03962	−.00165	.00002
18	Layers E0 A6 B25 C9 A0 A0	b_n	.00000	.00001	.00026	.00071	.00040	.00005	.00000
	3.33 in. insulation with 8 in. common brick	d_n	1.00000	−1.92906	1.24412	−.33029	.03663	−.00147	.00002
19	Layers E0 C9 B6 A6 A0 A0	b_n	.00000	.00005	.00064	.00099	.00030	.00002	.00000
	8 in. common brick with 2 in. insulation	d_n	1.00000	−1.78165	.96017	−.16904	.00958	−.00016	.00000
20	Layers E0 C11 B19 A6 A0 A0	b_n	.00000	.00012	.00119	.00154	.00038	.00002	.00000
	12 in. h.w. concrete with 0.61 in. insulation	d_n	1.00000	−1.86032	1.05927	−.19508	.01002	−.00016	.00000
21	Layers E0 C11 B6 A1 A0 A0	b_n	.00000	.00001	.00019	.00045	.00022	.00002	.00000
	12 in. h.w. concrete with 2 in. insulation	d_n	1.00000	−2.12812	1.53974	−.45512	.05298	−.00158	.00002
22	Layers E0 C14 B15 A2 A0 A0	b_n	.00000	.00000	.00006	.00026	.00025	.00006	.00000
	4 in. l.w. concrete with 6 in. insul. and face brick	d_n	1.00000	−2.28714	1.85457	−.63564	.08859	−.00463	.00009
23	Layers E0 E1 B15 C7 A2 A0	b_n	.00000	.00000	.00002	.00012	.00019	.00008	.00001
	6 in. insulation with 8 in. l.w. concrete block	d_n	1.00000	−2.54231	2.43767	−1.10744	.24599	−.02510	.00101
24	Layers E0 A6 C20 B1 A7 A0	b_n	.00000	.00000	.00015	.00066	.00062	.00015	.00001
	12 in. h.w. concrete block (filled) and face brick	d_n	1.00000	−2.47997	2.22597	−.87231	.14275	−.00850	.00018
25	Layers E0 A2 C15 B12 A6 A0	b_n	.00000	.00000	.00004	.00019	.00021	.00006	.00001
	Face brick and 6 in. l.w. conc. blk. w/3 in. insul.	d_n	1.00000	−2.28573	1.80756	−.58999	.08155	−.00500	.00013
26	Layers E0 A2 C6 B6 A6 A0	b_n	.00000	.00000	.00010	.00036	.00027	.00005	.00000
	Face brick and 8 in. clay tile with 2 in. insulation	d_n	1.00000	−2.18780	1.60930	−.46185	.05051	−.00218	.00003
27	Layers E0 E1 B14 C11 A1 A0	b_n	.00000	.00000	.00001	.00006	.00011	.00005	.00001
	5 in. insulation with 12 in. h.w. concrete	d_n	1.00000	−2.55944	2.45942	−1.12551	.25621	−.02721	.00107
28	Layers E0 E1 C11 B13 A1 A0	b_n	.00000	.00000	.00002	.00010	.00012	.00004	.00000
	12 in. h.w. concrete with 4 in. insulation	d_n	1.00000	−2.37671	2.04312	−.79860	.14868	−.01231	.00037
29	Layers E0 A2 C11 B5 A6 A0	b_n	.00000	.00000	.00004	.00021	.00021	.00006	.00000
	Face brick and 12 in. h.w. concrete with 1 in. insul.	d_n	1.00000	−2.42903	2.08179	−.75768	.11461	−.00674	.00015
30	Layers E0 E1 B19 C19 A2 A0	b_n	.00000	.00000	.00001	.00006	.00015	.00010	.00002
	0.61 in. ins. w/12 in. l.w. blk. (fld.) and face brick	d_n	1.00000	−2.83632	3.10377	−1.65731	.45360	−.06212	.00393
31	Layers E0 E1 B15 C15 A2 A0	b_n	.00000	.00000	.00000	.00002	.00007	.00006	.00002
	6 in. insul. with 6 in. l.w. conc. and face brick	d_n	1.00000	−2.90291	3.28970	−1.85454	.55033	−.08384	.00599
32	Layers E0 E1 B23 B9 A2 A0	b_n	.00000	.00000	.00000	.00005	.00011	.00007	.00001
	2.42 in. insulation with face brick	d_n	1.00000	−2.82266	3.04536	−1.58410	.41423	−.05186	.00273
33	Layers E0 A2 C6 B15 A6 A0	b_n	.00000	.00000	.00000	.00002	.00006	.00005	.00001
	Face brick and 8 in. clay tile with 6 in. insulation	d_n	1.00000	−2.68945	2.71279	−1.28873	.30051	−.03338	.00175
34	Layers E0 C11 B21 A2 A0 A0	b_n	.00000	.00000	.00003	.00015	.00014	.00003	.00000
	12 in. h.w. conc. with 1.36 in. insul. and face brick	d_n	1.00000	−2.67076	2.58089	−1.07967	.18237	−.01057	.00021

Table 18 Wall Conduction Transfer Function Coefficients (*b* and *d* Factors) (*Concluded*)

Wall Group	(Layer Sequence Left to Right = Inside to Outside)		$n=0$	$n=1$	$n=2$	$n=3$	$n=4$	$n=5$	$n=6$
35	Layers E0 E1 B14 C11 A2 A0	b_n	.00000	.00000	.00000	.00001	.00003	.00003	.00001
	5 in. insul. with 12 in. h.w. conc. and face brick	d_n	1.00000	−2.96850	3.45612	−2.02882	.64302	−.10884	.00906
36	Layers E0 A2 C11 B25 A6 A0	b_n	.00000	.00000	.00000	.00004	.00007	.00004	.00001
	Face brick and 12 in. h.w. conc. with 3.33 in. insul.	d_n	1.00000	−2.55127	2.36600	−.99023	.19505	−.01814	.00075
37	Layers E0 E1 B25 C19 A2 A0	b_n	.00000	.00000	.00000	.00001	.00003	.00003	.00002
	3.33 in. ins. w/12 in. l.w. blk. (fld.) and face brick	d_n	1.00000	−3.17762	4.00458	−2.56328	.89048	−.16764	.01638
38	Layers E0 E1 B15 C20 A2 A0	b_n	.00000	.00000	.00000	.00001	.00002	.00003	.00001
	6 in. ins. w/12 in. h.w. block (fld.) and face brick	d_n	1.00000	−3.14989	3.95116	−2.53790	.89438	−.17209	.01706
39	Layers E0 A2 C16 B14 A6 A0	b_n	.00000	.00000	.00000	.00001	.00002	.00003	.00001
	Face brick and 8 in. l.w. concrete with 5 in. insul.	d_n	1.00000	−2.99386	3.45884	−1.95834	.57704	−.08844	.00687
40	Layers E0 A2 C20 B15 A6 A0	b_n	.00000	.00000	.00000	.00001	.00002	.00003	.00001
	Face brick, 12 in. h.w. block (fld.), 6 in. insul.	d_n	1.00000	−2.97582	3.42244	−1.93318	.56765	−.08568	.00652
41	Layers E0 E1 C11 B14 A2 A0	b_n	.00000	.00000	.00000	.00001	.00002	.00002	.00001
	12 in. h.w. conc. with 5 in. insul. and face brick	d_n	1.00000	−3.08296	3.66615	−2.11991	.62142	−.08917	.00561

Table 19 Wall Conduction Transfer Function Coefficients Σc_n, Time Lag, U-Factors, and Decrement Factors

Wall Group		Σc_n	TL, h	U	DF
1	Layers E0 A3 B1 B13 A3 A0	0.04990	1.30	0.066	0.98
2	Layers E0 E1 B14 A1 A0 A0	0.01743	3.21	0.055	0.91
3	Layers E0 C3 B5 A6 A0 A0	0.05162	3.33	0.191	0.78
4	Layers E0 E1 B6 C12 A0 A0	0.00694	4.76	0.047	0.81
5	Layers E0 A6 B21 C7 A0 A0	0.01776	5.11	0.129	0.64
6	Layers E0 E1 B2 C5 A1 A0	0.02174	5.28	0.199	0.54
7	Layers E0 A6 C5 B3 A3 A0	0.01303	5.14	0.122	0.41
8	Layers E0 A2 C12 B5 A6 A0	0.01345	6.21	0.195	0.35
9	Layers E0 A6 B15 B10 A0 A0	0.00293	7.02	0.042	0.58
10	Layers E0 E1 C2 B5 A2 A0	0.00828	7.05	0.155	0.53
11	Layers E0 E1 C8 B6 A1 A0	0.00552	7.11	0.109	0.37
12	Layers E0 E1 B1 C10 A1 A0	0.01528	7.25	0.339	0.33
13	Layers E0 A2 C5 B19 A6 A0	0.01053	7.17	0.251	0.28
14	Layers E0 A2 A2 B6 A6 A0	0.00337	7.90	0.114	0.22
15	Layers E0 A6 C17 B1 A7 A0	0.00291	8.64	0.092	0.47
16	Layers E0 A6 C18 B1 A7 A0	0.00545	8.91	0.222	0.38
17	Layers E0 A2 C2 B15 A0 A0	0.00093	9.36	0.043	0.30
18	Layers E0 A6 B25 C9 A0 A0	0.00144	9.23	0.072	0.24
19	Layers E0 C9 B6 A6 A0 A0	0.00200	8.97	0.106	0.20
20	Layers E0 C11 B19 A6 A0 A0	0.00326	9.27	0.237	0.16
21	Layers E0 C11 B6 A1 A0 A0	0.00089	10.20	0.112	0.13
22	Layers E0 C14 B15 A2 A0 A0	0.00064	10.36	0.040	0.36
23	Layers E0 E1 B15 C7 A2 A0	0.00042	11.17	0.042	0.28
24	Layers E0 A6 C20 B1 A7 A0	0.00159	11.29	0.196	0.23
25	Layers E0 A2 C15 B12 A6 A0	0.00051	11.44	0.060	0.19
26	Layers E0 A2 C6 B6 A6 A0	0.00078	10.99	0.097	0.15
27	Layers E0 E1 B14 C11 A1 A0	0.00024	11.82	0.052	0.12
28	Layers E0 E1 C11 B13 A1 A0	0.00029	11.40	0.064	0.10
29	Layers E0 A2 C11 B5 A6 A0	0.00052	12.06	0.168	0.08
30	Layers E0 E1 B19 C19 A2 A0	0.00034	12.65	0.062	0.24
31	Layers E0 E1 B15 C15 A2 A0	0.00017	12.97	0.038	0.21
32	Layers E0 E1 B23 B9 A2 A0	0.00025	13.05	0.069	0.16
33	Layers E0 A2 C6 B15 A6 A0	0.00015	12.96	0.042	0.12
34	Layers E0 C11 B21 A2 A0 A0	0.00035	12.85	0.143	0.09
35	Layers E0 E1 B14 C11 A2 A0	0.00009	13.69	0.052	0.08
36	Layers E0 A2 C11 B25 A6 A0	0.00016	12.82	0.073	0.06
37	Layers E0 E1 B25 C19 A2 A0	0.00008	14.70	0.040	0.14
38	Layers E0 E1 B15 C20 A2 A0	0.00008	14.39	0.041	0.12
39	Layers E0 A2 C16 B14 A6 A0	0.00007	14.64	0.040	0.10
40	Layers E0 A2 C20 B15 A6 A0	0.00007	14.38	0.041	0.08
41	Layers E0 E1 C11 B14 A2 A0	0.00005	14.87	0.052	0.06

CTF Coefficients (Tables 11 and 15 through 19)

$$\text{Outside surface resistance} = A0$$
$$\text{4 in. heavyweight concrete} = C5$$
$$\text{2 in. insulation} = B3$$
$$\text{Air space resistance} = B1$$
$$\text{3/4 in. plaster} = E1$$
$$\text{Inside surface resistance} = E0$$

The appropriate arrangement of layers in the wall can be found in Table 17. The dominant wall layer C5 is at the outside surface ("mass out"), and has a Wall Material column number of 10; combined with an E1 layer, this dictates use of the upper array of code numbers for wall assembly groups. Entering this array with an R-value range of 9 ($R = 6.667$), column 10 indicates that Wall Group 6 most nearly represents the wall under consideration.

The CTF coefficients of Wall Group 6 as listed in Table 18 are:

$$b_0 = 0.00051 \qquad d_0 = 1.0000$$
$$b_1 = 0.00938 \qquad d_1 = -1.17580$$
$$b_2 = 0.01057 \qquad d_2 = 0.30071$$
$$b_3 = 0.00127 \qquad d_3 = -0.01561$$
$$b_4 = 0.00001 \qquad d_4 = 0.00001$$
$$b_5 = 0.00000 \qquad d_5 = 0.00000$$
$$b_6 = 0.00000 \qquad d_6 = 0.00000$$

And from Table 19, the U-factor of the wall is 0.199 and $\sum_{n=0} c_n = 0.02174$.

Heat flow calculations. The following format of Equation (28) demonstrates heat flow calculations through the wall:

$$q_{e,\theta}/A = \begin{bmatrix} b_0(t_{e,\theta}) \\ +b_1(t_{e,\theta-\delta}) \\ +b_2(t_{e,\theta-2\delta}) \\ \cdot \\ \cdot \\ \cdot \end{bmatrix} - \begin{bmatrix} d_1[(q_{e,\theta-\delta})/A] \\ +d_2[(q_{e,\theta-2\delta})/A] \\ \cdot \\ \cdot \\ \cdot \end{bmatrix} - \begin{bmatrix} t_{rc}\sum_{n=0} c_n \end{bmatrix}$$

This arrangement indicates that the heat gain through the wall is the sum of three parts:

1. Sum of the products of b coefficients and sol-air temperature values. The current value of this temperature is multiplied by b_0, the sol-air temperature of one step in time earlier is multiplied by b_1, etc.
2. Sum of the products of d coefficients and the previous values of heat gain. Note that the first d used is d_1. Again, the order of values is the same as in the first term, i.e., d_1 is multiplied by the heat gain value that was calculated for the previous step in time, d_2 is multiplied by the value calculated for two steps back in time, etc.
3. A constant, since room air temperature is constant and needs to be calculated only once.

The sequence of calculation using numerical values of this example are then as follows (starting at time $\theta = 1$, expressing heat flux in Btu/(h·ft²), setting $A = 1.0$, and dropping b and d coefficients 4 through 6 as insignificant):

$$q_{e,1} = \begin{bmatrix} +0.00051(76) \\ +0.00938(77) \\ +0.01057(79) \\ +0.00127(81) \end{bmatrix} - \begin{bmatrix} -1.17580(0) \\ +0.30071(0) \\ -0.01561(0) \end{bmatrix} - [0.02174(75)]$$
$$= 0.068$$

$$q_{e,2} = \begin{bmatrix} +0.00051(76) \\ +0.00938(76) \\ +0.01057(77) \\ +0.00127(79) \end{bmatrix} - \begin{bmatrix} -1.17580(0.068) \\ +0.30071(0) \\ -0.01561(0) \end{bmatrix} - (+1.6305)$$
$$= 0.117$$

The values for q_e for this example are given in the summary table.

The convergence of the heat gain values to a periodic steady-state condition is indicated by comparing the average of the last 24 values with the average heat flow. The latter is given by the product of the U-factor and the difference between the average sol-air and room temperature. Thus

$$q_{avg} = 0.199(91.54 - 75.00) = 3.291 \text{ Btu/(h·ft}^2)$$

The average of the last 24 values of heat gain tabulated in the summary table is given by:

$$q_{e,avg} = \frac{\sum_{i=73}^{96} q_{e,i}}{24} = \frac{78.954}{24} = 3.290 \text{ Btu/(h·ft}^2)$$

Summary of Calculations for Example 3

n	$q_{e,n}$	n	$q_{e,n}$	n	$q_{e,n}$	n	$q_{e,n}$
1	0.068	25	3.688	49	3.744	73	3.745
2	0.117	26	3.174	50	3.221	74	3.221
3	0.139	27	2.712	51	2.751	75	2.751
4	0.141	28	2.303	52	2.335	76	2.336
5	0.117	29	1.933	53	1.961	77	1.961
6	0.077	30	1.603	54	1.626	78	1.626
7	0.048	31	1.329	55	1.348	79	1.349
8	0.065	32	1.141	56	1.157	80	1.157
9	0.156	33	1.060	57	1.073	81	1.073
10	0.333	34	1.092	58	1.103	82	1.103
11	0.599	35	1.237	59	1.246	83	1.246
12	0.948	36	1.483	60	1.491	84	1.491
13	1.372	37	1.821	61	1.828	85	1.828
14	1.945	38	2.322	62	2.328	86	2.328
15	2.746	39	3.063	63	3.068	87	3.068
16	3.731	40	3.997	64	4.002	88	4.002
17	4.773	41	4.997	65	5.000	89	5.000
18	5.702	42	5.890	66	5.892	90	5.892
19	6.320	43	6.478	67	6.480	91	6.480
20	6.388	44	6.520	68	6.522	92	6.522
21	5.974	45	6.085	69	6.087	93	6.087
22	5.401	46	5.495	70	5.496	94	5.496
23	4.810	47	4.888	71	4.889	95	4.889
24	4.236	48	4.302	72	4.303	96	4.303

Note: n is in hours and $q_{e,n}$ is in Btu/(h·ft²).

Heat Gain through Interior Partitions, Floors, and Ceilings

Whenever a conditioned space is adjacent to other spaces at different temperatures, the transfer of heat through the partition can be calculated by:

$$q_{p,\theta} A \left[\sum_{n=0} b_n(t_{b,\theta-n\delta}) - \sum_{n=1} d_n \{(q_{p,\theta-n\delta})/A\} - t_{rc}\sum_{n=0} c_n \right] \quad (29)$$

where

A = area, ft²
t_b = air temperature of adjacent space, °F
b, c, d = CTF coefficients derived from Tables 11 through 19, considering partitions as "walls" and floors or ceilings as "roofs"

Heat gain from adjacent spaces. When t_b is constant or at least the variations of t_b are small compared to the difference ($t_b - t_{rc}$), $q_{p,\theta}$ is given by the simple steady-state expression

$$q_{p,\theta} = UA(t_b - t_{rc}) \quad (30)$$

where U = coefficient of overall heat transfer between the adjacent and the conditioned spaces (see Tables 14 or 19, or Chapter 22).

The same expression gives the mean values for $q_{p,\theta}$, when a mean value of t_b is used even though t_b varies. When $q_{p,\theta}$, is relatively small compared to the other room heat gain components, it may be considered constant at its mean value. If this component of heat gain is large, the temperature in the adjacent space should be calculated.

Note the common values $q_{p,\theta}$, A, t_b, and t_{rc} in Equations (29) and (30), illustrating the general functional equivalency of CTF coefficients b, c, and d in dynamic heat transfer over time to the steady-state heat transfer coefficient U, thus setting the rationale for adjustment of tabular CTF values by ratio of U_{actual}/U_{table}.

Conversion of Cooling Load from Heat Gain

The cooling load of a space depends on the magnitude and the nature of the sensible heat gain (*i.e.*, heat conduction through walls, direct and diffuse solar radiation, energy input to lights, etc.) and on the location and mass of room objects that absorb the radiant heat. For example, the cooling load profile resulting from a unit pulse of solar radiation absorbed by window glass is quite different from that absorbed by a floor surface. Thus, each component of the room heat gain gives rise to a distinct component of cooling load, and the sum of these various components at any time is the total cooling load at that time.

Unlike other components, the latent heat gain component of the cooling load may or may not be part of room load depending on the type of air-conditioning system, *i.e.*, ventilation air may be dehumidified at a central location rather than in each room.

Cooling load by room transfer function. Stephenson and Mitalas (1967), Mitalas and Stephenson (1967), and Kimura and Stephenson (1968) related heat gain to the corresponding cooling load by a room transfer function (RTF), which depends on the nature of the heat gain and on the heat storage characteristics of the space (*i.e.*, of the walls, floor, etc., that enclose the space, and of the contents of that space). Where the heat gain q_θ is given at equal time intervals, the corresponding cooling load Q_θ at time θ can be related to the current value of q_θ and the preceding values of cooling load and heat gain by:

$$Q_\theta = \sum_{i=1} (v_o q_\theta + v_1 q_{\theta-d} + v_2 q_{\theta-2\delta} + \ldots)$$
$$- (w_1 Q_{\theta-\delta} + w_2 Q_{\theta-2\delta} + \ldots) \quad (31)$$

where i is taken from 1 to the number of heat gain components and δ = time interval. The terms $v_0, v_1 \ldots, w_1, w_2 \ldots$ are the coefficients of the RTF

$$K_{(z)} = (v_0 + v_1 z^{-1} + v_2 z^{-2} + \ldots)/(1 + w_1 z^{-1} + w_2 z^{-2} + \ldots) \quad (32)$$

which relates the transform of the corresponding parts of the cooling load and of the heat gain. These coefficients depend on (1) the size of the time interval δ between successive values of heat gain and cooling load, (2) the nature of the heat gain (how much is in the form of radiation and where it is absorbed), and (3) on the heat storage capacity of the room and its contents. Therefore, different RTFs are used to convert each distinct heat gain component to cooling load.

While the basic form of Equation (31) anticipates a series of v_n and w_n coefficients, the effect of past v_1 and w_1 is negligible, and data tabulated may generally be used with confidence. A slight inaccuracy does occur in the calculation for the first hour that internal loads begin; up through the second before the hour for which the calculation is made, such load does not exist, and the value generated by the transfer functions is not reached until the end of that hour. The convective component of such load is instantaneous, and the growth of the radiant component (combined with the convective element by the transfer coefficients) as it is absorbed and released by the building mass and contents is realistic throughout the rest of the load period.

Sensitivity of Parameters—Nontypical Applications

The concept of evaluating the thermal storage performance of a given space by means of RTF coefficients is based on the essential similarity of enclosing surfaces, spacial geometry, and related characteristics of that space to corresponding parameters of the space for which the data were calculated. ASHRAE research projects 359-RP [(Chiles and Sowell 1984), (Sowell and Chiles 1984a),

Table 20   Zone Parametric Level Definitions

No.	Parameter	Meaning	Levels (in normal order)
1	ZG	Zone geometry	100 ft × 20 ft, 15 ft × 15 ft 100 ft × 100 ft
2	ZH	Zone height	8 ft, 10 ft, 20 ft
3	NW	No. exterior walls	1, 2, 3, 4, 0
4	IS	Interior shade	100, 50, 0%
5	FN	Furniture	With, Without
6	EC	Exterior wall construction	1, 2, 3, 4 (Table 21)
7	PT	Partition type	5/8 in. gypsum board-air space 5/8 in. gypsum board, 8 in. concrete block
8	ZL	Zone location	Single-story, top floor, bottom floor, mid-floor
9	MF	Mid-floor type	8 in. concrete, 2.5 in. concrete, 1 in. wood
10	ST	Slab type	Mid-floor type, 4 in. slab on 12 in. soil
11	CT	Ceiling type	3/4 in. acoustic tile and air space, w/o ceiling
12	RT	Roof type	1, 2, 3, 4 (Table 23)
13	FC	Floor covering	Carpet with rubber pad, vinyl tile
14	GL	Glass percent	10, 50, 90

Table 21   Exterior Wall Construction Types

Type	Description
1	Outside surface resistance, 1 in. stucco, 1 in. insulation, 3/4 in. plaster or gypsum, inside surface resistance (A0, A1, B1, E1, E0)*
2	Outside surface resistance, 1 in. stucco, 8 in. HW concrete, 3/4 in. plaster or gypsum, inside surface resistance (A0, A1, C10, E1, E0)
3	Outside surface resistance, steel siding, 3 in. insulation, steel siding, inside surface (A0, A3, B12, A3, E0)*
4	Outside surface resistance, 4 in. face brick, 2 in. insulation. 12 in. HW concrete, 3/4 in. plaster or gypsum, inside surface resistance (A0, A2, B3, C11, E1, E0)*

Note: Code letters are defined in Table 11.

Table 22   Floor and Ceiling Types Specified by Zone Location Parameter

Zone Location	Floor	Ceiling
Single story	Slab-on-grade	Roof
Top floor	Mid-floor	Roof
Bottom floor	Slab-on-grade	Mid-floor
Mid-floor	Mid-floor	Mid-floor

Table 23   Roof Construction Types

Type	Description
1	Outside surface resistance, 1/2 in. slag or stone, 3/8 in. felt membrane, 1 in. insulation, steel siding, inside surface resistance (A0, E2, E3, B4, A3, E0)*
2	Outside surface resistance, 1/2 in. slag or stone, 3/8 in. felt membrane, 6 in. LW concrete, inside surface resistance (A0, E2, E3, C15, E0)*
3	Outside surface resistance, 1/2 in. slag or stone, 3/8 in. felt membrane, 2 in. insulation, steel siding, ceiling air space, acoustic tile, inside surface resistance (A0, E2, E3, B6, A3, E4, E5, E0)*
4	Outside surface resistance, 1/2 in. slag or stone, 3/8 in. felt membrane, 8 in. LW concrete, ceiling air space, acoustic tile, inside surface resistance (A0, E2, E3, C16, E4, E5, E0)*

Note: Code letters are defined in Table 11.

(Sowell and Chiles 1984b)], 472-RP [(Harris and McQuiston 1988), (Sowell 1988a), (Sowell 1988b), (Sowell 1988c)], and 626-RP [(Falconer *et al.* 1993), (Spitler and McQuiston 1993), (Spitler *et al.* 1993)] investigated the unexpected sensitivity of such attributes and other counterintuitive phenomena regarding apparent responsiveness of relative masses in the storage and rejection of heat, and identified 14 discrete screening parameters with two to five levels of characterization each (Tables 20 through 23) by which to select representative data and to modify factors appropriately for specific applications. While these selection parameters are arranged so that errors due to deviations are minimal and conservative, careful use is required in situations differing significantly from one or more specific parameters.

Peak heat gain versus peak cooling load. The RTF procedure distributes all heat gained during a 24-h period throughout that period in the conversion to cooling load. Thus, individual heat gain components rarely appear at full value as part of the cooling load unless representing a constant 24-h input (such as a continuously burning light fixture), or in very low mass construction that releases stored radiant heat relatively quickly. This concept is further complicated by the premise of "constant interior space temperature" (*i.e.*, operation of an HVAC system 24 h a day, seven days a week with fixed control settings), which practice is far less prevalent today than in the past. The effect of intermittent system operation is seen primarily during the first hours of operation for a subsequent day, as discussed in the section Heat Extraction Rate, and can impact equipment size selection significantly.

Superposition of load components. Finally, a presupposition of the TFM is that total cooling load for a space can be calculated by simple addition of the individual components. For example, radiation heat transfer from individual walls or roofs is assumed to be independent of the other surfaces, which is slightly incorrect in a theoretical sense. However, extensive ASHRAE sponsored research found that the means for compensation for these limitations fall within the range of acceptable error that must be expected in any estimate of cooling load.

The abovementioned research calculated RTF values for all possible combinations of screening parameter levels for a total of 200,640 individual cases. Access to these data is available electronically by techniques outlined in the Cooling and Heating Load Calculation Manual (McQuiston and Spitler 1992). For illustration of the TFM, a simplified method of RTF selection is presented

in this chapter with RTF coefficients for various types and configurations of room construction and room air circulation rates given in Tables 24 and 25.

Use of Room Transfer Functions

To obtain appropriate room transfer function data for use in Equation (31), (1) select the value of w_1 from Table 24 for the approximate space envelope construction and range of air circulation, and (2) select and/or calculate the values of v_0 and v_1 from

Table 25   Room Transfer Functions: v_0 and v_1 Coefficients

Heat Gain Component	Room Envelope Construction[b]	v_0	v_1 Dimensionless
Solar heat gain through glass[c] with no interior shade; radiant heat from equipment and people	Light	0.224	$1 + w_1 - v_0$
	Medium	0.197	$1 + w_1 - v_0$
	Heavy	0.187	$1 + w_1 - v_0$
Conduction heat gain through exterior walls, roofs, partitions, doors, windows with blinds or drapes	Light	0.703	$1 + w_1 - v_0$
	Medium	0.681	$1 + w_1 - v_0$
	Heavy	0.676	$1 + w_1 - v_0$
Convective heat generated by equipment and people, and from ventilation and infiltration air	Light	1.000	0.0
	Medium	1.000	0.0
	Heavy	1.000	0.0

Heat Gain from Lights[d]				
Furnishings	Air Supply and Return	Type of Light Fixture	v_0	v_1
Heavyweight simple furnishings, no carpet	Low rate; supply and return below ceiling ($V \leqslant 0.5$)[e]	Recessed, not vented	0.450	$1 + w_1 - v_0$
Ordinary furnishings, no carpet	Medium to high rate, supply and return below or through ceiling ($V \geqslant 0.5$)[e]	Recessed, not vented	0.550	$1 + w_1 - v_0$
Ordinary furnishings, with or without carpet on floor	Medium to high rate, or induction unit or fan and coil, supply and return below, or through ceiling, return air plenum ($V \geqslant 0.5$)[e]	Vented	0.650	$1 + w_1 - v_0$
Any type of furniture, with or without carpet	Ducted returns through light fixtures	Vented or free-hanging in airstream with ducted returns	0.750	$1 + w_1 - v_0$

[a]The transfer functions in this table were calculated by procedures outlined in Mitalas and Stephenson (1967) and are acceptable for cases where all heat gain energy eventually appears as cooling load. The computer program used was developed at the National Research Council of Canada, Division of Building Research.

[b]The construction designations denote the following:

Light construction: such as frame exterior wall, 2-in. concrete floor slab, approximately 30 lb of material/ft^2 of floor area.

Medium construction: such as 4-in. concrete exterior wall, 4-in. concrete floor slab, approximately 70 lb of building material/ft^2 of floor area.

Heavy construction: such as 6-in. concrete exterior wall, 6-in. concrete floor slab, approximately 130 lb of building material/ft^2 of floor area.

[c]The coefficients of the transfer function that relate room cooling load to solar heat gain through glass depend on where the solar energy is absorbed. If the window is shaded by an inside blind or curtain, most of the solar energy is absorbed by the shade, and is transferred to the room by convection and long-wave radiation in about the same proportion as the heat gain through walls and roofs; thus the same transfer coefficients apply.

[d]If room supply air is exhausted through the space above the ceiling and lights are recessed, such air removes some heat from the lights that would otherwise have entered the room. This removed light heat is still a load on the cooling plant if the air is recirculated, even though it is not a part of the room heat gain as such. The percent of heat gain appearing in the room depends on the type of lighting fixture, its mounting, and the exhaust airflow.

[e]V is room air supply rate in cfm/ft^2 of floor area.

Table 24   Room Transfer Functions: w Coefficient

	Room Envelope Construction[b]				
Room Air Circulation[a] and S/R Type	2-in. Wood Floor	3-in. Concrete Floor	6-in. Concrete Floor	8-in. Concrete Floor	12-in. Concrete Floor
	Specific Mass per Unit Floor Area, lb/ft^2				
	10	40	75	120	160
Low	−0.88	−0.92	−0.95	−0.97	−0.98
Medium	−0.84	−0.90	−0.94	−0.96	−0.97
High	−0.81	−0.88	−0.93	−0.95	−0.97
Very High	−0.77	−0.85	−0.92	−0.95	−0.97
	−0.73	−0.83	−0.91	−0.94	−0.96

[a]Circulation rate—

Low: Minimum required to cope with cooling load from lights and occupants in interior zone. Supply through floor, wall, or ceiling diffuser. Ceiling space not used for return air, and $h = 0.4$ Btu/h·ft^2·°F (where h = inside surface convection coefficient used in calculation of w_1 value).

Medium: Supply through floor, wall, or ceiling diffuser. Ceiling space not used for return air, and $h = 0.6$ Btu/h·ft^2·°F.

High: Room air circulation induced by primary air of induction unit or by room fan and coil unit. Ceiling space used for return air, and $h = 0.8$ Btu/h·ft^2·°F.

Very high: High room circulation used to minimize temperature gradients in a room. Ceiling space used for return air, and $h = 0.8$ Btu/h·ft^2·°F.

[b]Floor covered with carpet and rubber pad; for a bare floor or if covered with floor tile, take next w_1 value down the column.

Table 25 for the appropriate heat gain component and range of space construction mass.

Example 4. Cooling load due to solar radiation through glass. Consider a room having a 1/2 in. air space double-glazed window (shading coefficient = 0.83) in a multistory office building of heavyweight construction (approximately 120 lb/ft² floor area). The building is located at 40°N latitude, the date is June 21, and the window orientation is NW. The U-factor for the window is 0.56 Btu/(h·ft²·°F). Assume the floor to be carpeted, the air circulation rate "medium" (h_i = 0.6 Btu/(h·ft²·°F), and the ceiling space not ventilated. Calculate the cooling load due to solar radiation through glass. Solar heat gain (SHG) to the room through the window is given as SHG = SHGF × Shading Coefficient = SHGF × 0.83.

Time, h	SHGF	SHG	Time, h	SHGF	SHG
0100	0	0	1300	40	33
0200	0	0	1400	63	52
0300	0	0	1500	114	95
0400	0	0	1600	156	129
0500	1	1	1700	172	143
0600	13	11	1800	143	119
0700	21	17	1900	21	17
0800	27	22	2000	0	0
0900	32	27	2100	0	0
1000	35	29	2200	0	0
1100	38	32	2300	0	0
1200	38	32	2400	0	0
Daily total					753

Note: SHGF from Table 15, Chapter 27.

Solution: The room transfer function coefficients for 120 lb/ft² construction, solar radiation input, medium air circulation rate, and the condition of "no heat loss for the room" are (see Tables 24 and 25):

$$v_0 = 0.187 \qquad w_0 = 1.000 \text{ (in all cases)}$$
$$v_1 = -0.147 \qquad w_1 = -0.960$$

The cooling load component due to solar radiation through glass at any time θ is given by Equation (31). The calculations can be set up as follows:

$$Q_\theta = \begin{bmatrix} v_0(\text{SHG}_\theta) \\ + v_1(\text{SHG}_{\theta-\delta}) \\ - w_1(Q_{\theta-\delta}) \end{bmatrix}$$

As in the earlier heat gain calculation example, the calculation is started by assuming that the previous Qs are zero. Furthermore, in this example, SHG = 0 for θ = 1, 2, 3, and 4; therefore, Qs in Btu/(h·ft²) are:

Q_5	v, w	Hour	SHG	Prev. Qs	Factor
	0.187	5	1		0.187
	−0.147	4	0		0.000
	0.96	4		0	0.000
				Q_5 =	0.187

Q_6	v, w	Hour	SHG	Prev. Qs	Factor
	0.187	6	11		0.187
	−0.147	5	1		−0.147
	0.96	5		0.187	0.180
				Q_6 =	2.090

Values of Q_θ for the remainder of the calculations are listed in the following table. The calculations of Q_θ are terminated at θ = 96 h, because by that time, the effect of the assumed zero initial conditions has decreased to negligible proportions.

Values of Q_θ for Example 4

θ	Q_θ	θ	Q_θ	θ	Q_θ	θ	Q_θ
1 =	0.000	25 =	17.041	49 =	23.440	73 =	25.842
2 =	0.000	26 =	16.359	50 =	22.502	74 =	24.808
3 =	0.000	27 =	15.705	51 =	21.602	75 =	23.816
4 =	0.000	28 =	15.077	52 =	20.738	76 =	22.863
5 =	0.187	29 =	14.661	53 =	20.095	77 =	22.135
6 =	2.090	30 =	15.985	54 =	21.201	78 =	23.160
7 =	3.568	31 =	16.908	55 =	21.915	79 =	23.796
8 =	5.040	32 =	17.847	56 =	22.653	80 =	24.459
9 =	6.653	33 =	18.948	57 =	23.562	81 =	25.296
10 =	7.841	34 =	19.644	58 =	24.074	82 =	25.738
11 =	9.248	35 =	20.579	59 =	24.832	83 =	26.429
12 =	10.158	36 =	21.036	60 =	25.119	84 =	26.652
13 =	11.219	37 =	21.662	61 =	25.581	85 =	27.053
14 =	15.643	38 =	25.669	62 =	29.431	86 =	30.844
15 =	25.138	39 =	34.763	63 =	38.375	87 =	39.731
16 =	34.290	40 =	43.530	64 =	46.998	88 =	48.300
17 =	40.696	41 =	49.567	65 =	52.896	89 =	54.146
18 =	40.300	42 =	48.816	66 =	52.012	90 =	53.212
19 =	24.374	43 =	32.549	67 =	35.618	91 =	36.770
20 =	20.900	44 =	28.748	68 =	31.694	92 =	32.800
21 =	20.064	45 =	27.598	69 =	30.426	93 =	31.488
22 =	19.261	46 =	26.494	70 =	29.230	94 =	30.228
23 =	18.491	47 =	25.434	71 =	28.041	95 =	29.019
24 =	17.751	48 =	24.417	72 =	26.919	96 =	27.858

Note: Values carried to 3 decimals to illustrate degree of convergence.

Cooling Load from Nonradiant Heat Gain

Sensible cooling load from strictly convective heat gain elements is instantaneous, added directly to the results of those gains processed by CTF and RTF coefficients, per the following equation.

$$Q_{sc} = \sum_{j=1} (q_{c,j}) \tag{33}$$

where

Q_{sc} = sensible cooling load from heat gain elements having only convective components

q_c = each of j heat gain elements having only such convective component

Heat Extraction Rate and Room Temperature

Discussion to this point has concentrated on estimating design cooling load for a conditioned space, assuming the maintenance of a constant interior temperature and the hourly total removal of all cooling load entering the space; and allowing the delaying action of building mass and contents to run its course. Certain minor factors have been ignored, such as the relatively indeterminate radiant heat loss to the outside of the building.

The basic principles of the TFM are also useful in estimating dynamic cooling load requirements over an extended period (see Chapter 28). In such cases, however, the goal is no longer to seek the peak load for equipment selection purposes, and the ebb and flow of heat into and out of the building assume much greater importance; thus, any loss back to the environment must be considered. This concept is also critical in predicting temperature swings in the space and the ability of cooling equipment to extract heat when operated in a building with extended off cycles (nights and weekends).

The cooling loads determined by the TFM serve as input data for estimating the resultant room air temperature and the heat extraction rate with a particular type and size of cooling unit, or set of operating conditions, or both. In addition, the characteristics of the cooling unit (*i.e.*, heat extraction rate versus room air temperature), the schedule of operation, and a space air transfer function (SATF) for the room that relates room air temperature and heat extraction rate must also be included to run these calculations.

The heat extraction characteristics of the cooling unit can be approximated by a linear expression of the form

$$ER_\theta = W_\theta + (St_{r\theta}) \qquad (34)$$

where

ER_θ = rate of heat removal from space at time θ
$t_{r\theta}$ = the air temperature in space at time θ
W, S = parameters characterizing performance of specific types of cooling equipment

This linear relationship only holds when $t_{r\theta}$ is within the throttling range of the control system. When $t_{r\theta}$ lies outside of this range, ER_θ has the value of either ER_{max} or ER_{min}, depending on whether the temperature $t_{r\theta}$ is above or below the throttling range. The value of S is the difference $ER_{max} - ER_{min}$ divided by the width of the throttling range, and W_θ is the value ER_θ would have if the straight-line relationship between it and $t_{r\theta}$ held at t_{r1} equals zero. This intercept depends on the set point temperature of the control system, which may be taken as the temperature at the middle of the throttling range. Thus,

$$W_\theta = (ER_{max} + ER_{min})/2 - (St_{r\theta}^*) \qquad (35)$$

where $t_{r\theta}^*$ is the thermostat set point temperature at time θ.

Space Air Transfer Function

The heat extraction rate and the room air temperature are related by the space air transfer function (SATF):

$$\sum_{i=0}^{1} p_i(ER_{\theta-\delta} - Q_{\theta-i\delta}) = \sum_{i=0}^{2} g_i(t_{rc} - t_{r,\theta-i\delta}) \qquad (36)$$

where g_1 and p_1 are the SATF coefficients, and Q is the calculated cooling load for the room at time θ, based on an assumed constant room temperature of t_{rc}. Normalized values of g and p are given in Table 26 for light, medium, and heavy construction.

Thermal conductance to surroundings. In calculating the design cooling load components previously described, it was assumed that all energy transferred into the space eventually appears as space cooling load. However, this is not quite true over an extended period, because a fraction of the input energy can instead be lost back to the surroundings. This fraction F_c depends on the thermal conductance between the space air and the surroundings and can be estimated as

$$F_c = 1 - 0.02 K_\theta \qquad (37)$$

Table 26 Normalized Coefficients of Space Air Transfer Functions[a]

Room Envelope Construction[b]	g_0^*	g_1^* Btu/h · ft² · °F	g_2^*	p_0 Dimensionless	p_1
Light	+1.68	−1.73	+0.05	1.0	−0.82
Medium	+1.81	−1.89	+0.08	1.0	−0.87
Heavy	+1.85	−1.95	+0.10	1.0	−0.93

[a]For simplified procedure for calculating space air transfer function coefficients, see ASHRAE (1975).
[b]The designations Light, Medium, and Heavy denote the same meanings as those footnoted for Table 25.

where K_θ is the unit length conductance between the space air and surroundings, in Btu/(h · ft² · °F), given by

$$K_\theta = (1/L_F)(U_R A_R + U_W A_W + U_{OW} A_{OW} + U_P A_P) \qquad (38)$$

where

L_F = length of space exterior wall, ft
U = U-factor of space enclosure element (subscript R for roof, W for window, OW for outside wall, and P for partition, should such be adjacent to an unconditioned area), Btu/(h · ft² · °F)
A = area of space enclosure element, ft²

The units of K_θ are Btu/(h · ft² · °F). Therefore, if F_c is to be dimensionless, the multiplier is 0.02 h · ft² · °F/Btu.

Adjustment of load components. To adjust the space cooling loads calculated in the previous sections, multiply the value of the following components by the factor F_c from Equation (37):

• Sensible cooling load from heat gain by conduction through exterior roofs and walls
• Sensible cooling load from conduction and solar heat gain through fenestration areas
• Sensible cooling load from heat gain through interior partitions, ceilings, and floors
• Sensible cooling load from radiant portion of heat gain from lights, people, and equipment

Adjustments to g* coefficients. To obtain the SATF coefficients for Equation (36), first select the values of p_0, p_1, g_1^*, g_1^*, and g_2^* from Table 26 for the appropriate space envelope construction. Since the g* coefficients in Table 26 are for a space with zero heat conductance to surrounding spaces and are normalized to a unit floor area, it is necessary to adjust the 0 and 1 values. To get the g_0 and g_1 coefficients for a space with a floor area A, total conductance K_θ [by Equation (38)] between space air surroundings, ventilation rate, and infiltration rate, the relationships are:

$$g_{0,\theta} = g_0^* A + p_0 [K_\theta + 1.10 (V_\theta + VI_\theta)] \qquad (39)$$

$$g_{1\theta} = g_1^* A + p_1 [K_\theta + 1.10 (V_{\theta-\delta} + VI_{\theta-1\delta})] \qquad (40)$$

Note that Equation (40) has no second term when calculating $g_{2,\theta}$, since p_2 has no value.

Heat extraction rate. For either condition (heat loss to surroundings or not, and using the appropriate values of g), Equations (34) and (35) can be solved simultaneously for ER_θ

$$ER_\theta = [W_\theta g_0/(S + g_0)] + [I_\theta S/(S + g_0)] \qquad (41)$$

where

$$I_\theta = t_{rc} \sum_{i=0}^{2} g_{i,\theta} - \sum_{i=1}^{2} g_{i,\theta}(t_{r,\theta-i\delta})$$

$$+ \sum_{i=0}^{1} p_i (Q_{\theta-i\delta}) - \sum_{i=1}^{1} p_i (ER_{\theta-i\delta}) \qquad (42)$$

If the value of ER_θ calculated by Equation (41) is greater than ER_{max}, it is made equal to ER_{max}; if it is less than ER_{min}, it is made equal to ER_{min}. Then $t_{r\theta}$ is calculated from the expression

$$t_{r\theta} = (1/g_{0,\theta})(I_\theta - ER_\theta) \qquad (43)$$

Example 5. Calculation of room air temperature and heat extraction rate. A room is of heavy construction with a floor area of 400 ft². The total room cooling load calculated on the basis of t_{rc} = 70 °F is given as:

θ, h	Q_θ, Btu/h	θ, h	Q_θ, Btu/h	θ, h	Q_θ, Btu/h
1	2200	9	2180	17	7630
2	2030	10	2330	18	6880
3	1850	11	2650	19	5530
4	1730	12	3580	20	4380
5	1680	13	4880	21	3630
6	1750	14	6180	22	3130
7	1880	15	7150	23	2730
8	2030	16	7680	24	2450

The cooling unit has a maximum heat extraction capability of 7500 Btu/h and a minimum of zero. The throttling range is 3 °F wide. Assume no ventilation and no infiltration, and heat loss to the exterior surroundings at the rate of 100 Btu/(h · °F). Calculate room air temperature and heat extraction rate for:

Schedule A. The control thermostat is set at 77 °F from 0700 to 1800 h; during the rest of the time, it is set up to 85 °F.

Schedule B. The control thermostat is set at 77 °F all the time.

Solution:

(a) Space Air Transfer Functions.

The SATF coefficients for a 400-ft^2 room of heavy construction are [from Table 22 and Equations (39) and (40) with V and V_1 dropping out]:

$$g_{0,\theta} = 400(+1.85) + 100(+1.0) = 840.00$$
$$g_{1,\theta} = 400(-1.95) + 100(-0.93) = -873.00$$
$$g_{2,\theta} = 400(+0.10) = 40.00$$
$$\sum_{i=0}^{2} g_i = 7.00$$

(b) Cooling Unit Characteristics.

$$ER_{max} = 7500 \text{ Btu/h}$$
$$ER_{min} = 0$$
$$t_{tr} = 3\,°F \text{ throttling range}$$
$$S = (7500 - 0)/3 = 2500 \text{ Btu/(h · °F)}$$

when $t_{r\theta}^* = 77.0\,°F$,

$$W_\theta = [(7500 - 0)/2] - 2500(77.0) = -188{,}750 \text{ Btu/h}$$

and when $t_{r\theta}^* = 85.0\,°F$,

$$W_\theta = [(7500 - 0)/2] - 2500(85.0) = -208{,}750 \text{ Btu/h}$$

(c) Calculation of ER_θ and $t_{r\theta}$.

Some prior values for ER_θ and $t_{r\theta}$ must be assumed to begin the computation process. The computation is repeated until the results for successive days are the same. At that time, the results are independent of the values assumed initially.

To get the calculation started, assume all previous values of $ER = 0$ and $t_r = 80\,°F$. Thus:

$$I_1 = 70.0(7.0) - \begin{bmatrix} -873\,(80) \\ +40\,(80) \end{bmatrix} + \begin{bmatrix} +1.0\,(2200) \\ -0.93\,(2450) \end{bmatrix}$$

$$- \begin{bmatrix} -0.93\,(0.0) \end{bmatrix}$$

$$= 67{,}052 \text{ Btu/h}$$

$$ER_1 = (-208{,}750)\,840/(2500 + 840)$$
$$+ (67{,}052)\,2500/(2500 + 840)$$
$$= -52{,}500 + 50{,}189 = -2311 \text{ Btu/h}$$

As this is less than ER_{min}, $ER_1 = ER_{min} = 0.0$

and $t_{r1} = (1/840)(67{,}052 - 0.0) = 79.8\,°F$

Table 27 Room Air Temperature and Heat Extraction Rates for Example 6

	Schedule A (Control thermostat set at 77 °F from 0800 to 1800, and at 85 °F at all other times)		Schedule B (Control thermostat set at 77 °F at all times)	
Time, h	Room Air Temperature t_r, °F	Heat Extraction ER, Btu/h	Room Air Temperature t_r, °F	Heat Extraction ER, Btu/h
0100	82.0	0	76.3	1956
0200	81.9	0	76.2	1806
0300	81.8	0	76.2	1649
0400	81.7	0	76.1	1535
0500	81.7	0	76.1	1474
0600	81.8	0	76.1	1505
0700	82.0	0	76.1	1584
0800	77.2	4235	76.2	1680
0900	77.1	4051	76.2	1780
1000	77.1	4025	76.3	1881
1100	77.2	4138	76.3	2113
1200	77.4	4720	76.6	2807
1300	77.7	5602	77.0	3794
1400	78.1	6507	77.4	4799
1500	78.4	7184	77.7	5571
1600	78.6	7500	77.9	6019
1700	78.5	7477	77.9	6032
1800	78.3	6876	77.7	5513
1900	83.8	786	77.3	4528
2000	83.6	268	77.0	3669
2100	83.3	0	76.7	3099
2200	82.8	0	76.6	2710
2300	82.5	0	76.5	2393
2400	82.3	0	76.4	2164
Totals		63369		72061

$$I_2 = 70.0\,(7.0) - \begin{bmatrix} -873\,(79.8) \\ +40\,(80) \end{bmatrix} + \begin{bmatrix} +1.0\,(2030) \\ -0.93\,(2200) \end{bmatrix}$$

$$- \begin{bmatrix} -0.93\,(0.0) \end{bmatrix}$$

$$= 66{,}939 \text{ Btu/h}$$

$$ER_2 = (-208{,}750)\,840/(2500 + 840)$$
$$+ (66{,}939)\,2500/(2500 + 840)$$
$$= -52{,}500 + 50{,}104 = -2396 \text{ Btu/h}$$

As this also is less than ER_{min}, $ER_2 = ER_{min} = 0$ and $t_{r2} = (1/840)$ $(66{,}939 - 0) = 79.7\,°F$, and so on.

The effect of the assumed initial ER_θ and $t_{r\theta}$ values has decreased to negligible proportions by the time $\theta = 145$, i.e., $t_{r145} = t_{r169} = 82.6\,°F$. The complete set of results for operating schedules A and B is given in Table 27.

EXAMPLE COOLING LOAD CALCULATION

Example 6. Cooling load calculation of small office building. A one-story small commercial building (Figure 4) is located in the eastern United States near 40 °N latitude. The adjoining buildings on the north and west are not conditioned, and the air temperature within them is approximately equal to the outdoor air temperature at any time of day.

Note: The small commercial building shown in Figure 4 has been in the ASHRAE literature for several decades to illustrate cooling load procedures. In this example, some materials have been updated to reflect currently available products and associated U-factors, and the calculation month changed to July for better comparison with newer data. Otherwise, all other characteristics of this example remain unchanged.

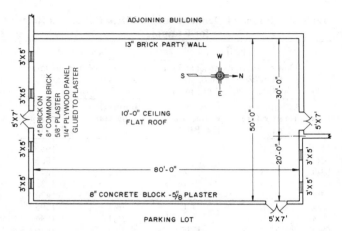

Note: The small commercial building shown in this figure has been in the ASHRAE literature for several decades to illustrate cooling load procedures. In this example, some materials have been updated to reflect currently available products and associated U-factors; the calculation month has been changed to July for better comparison with newer data. Otherwise, all other characteristics of this example remain unchanged.

Fig. 4 Plan of One-Story Office Building

Building Data:

South wall construction. 4-in. light-colored face brick, 8-in. common brick, 0.625-in. plaster, 0.25-in. plywood panel glued on plaster (Summer $U = 0.24$ Btu/h·ft²·°F, or $R = 4.14$).

East wall and outside north wall construction. 8-in. light-colored heavy concrete block, 5/8-in. plaster on walls (Summer $U = 0.48$ Btu/h·ft²·°F, or $R = 2.083$).

West wall and adjoining north party wall construction. 13-in. solid brick (color N/A), no plaster: with U for a 12-in. brick interior wall = 0.26, R for that wall = 1/0.26 = 3.846; subtracting 2 still air film coefficients with $R_{fc} = 0.68$ each leaves $R_b = 2.486$; thus for this wall:

$$R_w = 0.68 + (2.486 \times 13/12) + 0.68 = 4.053$$

and

$$U_w = 1/4.053 = 0.247, \text{ say } U = 0.25 \text{ Btu/(h·ft}^2\text{·°F)}$$

Roof construction. 4-1/2-in. (nominal) flat roof of 2-in. gypsum slab on metal roof deck, 2-in. rigid roof insulation, surfaced with two layers of mopped 15-lb felt vapor-seal built-up roofing having dark-colored gravel surface, and with no false ceiling below underside of roof deck; (Summer $U = 0.09$ Btu/h·ft²·°F, or $R = 11.11$).

Floor construction. 4-in. concrete on ground.

Fenestration. 3 ft by 5 ft nonoperable windows of regular plate glass with light colored venetian blinds (Summer $U = 0.81$ Btu/h·ft²·°F).

Door construction. Light-colored 1.75-in. steel door with solid urethane core and thermal break (Summer $U = 0.19$ Btu/h·ft²·°F or $R = 5.26$ for exterior doors, and $U = 0.18$ Btu/h·ft²·°F or $R = 5.56$ for interior doors).

Front doors. Two 30 in. by 7 ft
Side doors. Two 30 in. by 7 ft
Rear doors. Two 30 in. by 7 ft (interior)

Note: U-factors for all exterior surfaces assume a summer wind velocity of 7.5 mph. Those for party walls and other interior surfaces assume still air.

Summer outdoor design conditions. Dry bulb 94°F, daily range 20°F, wet bulb 77°F: $W_o = 0.0161$ lb vapor/lb dry air; $h_o = 40.3$ Btu/lb dry air.

Winter outdoor design conditions. Dry bulb 10°F.

Summer indoor design conditions. Dry bulb 75°F, wet bulb 62.5°F; $W_i = 0.0092$ lb vapor/lb dry air; $h_i = 28.07$ Btu/lb dry air.

Winter indoor design conditions. Dry bulb 75°F.

Occupancy. 85 office workers from 0800 to 1700 h.

Lights. 17,500 watts, fluorescent, operating from 0800 to 1700 hours daily; along with 4000 watts, tungsten, operated continuously. Lighting fixtures are non-ventilated type.

Power equipment and appliances. For this example, none are assumed.

Ventilation. From ASHRAE *Standard* 62, a ventilation rate of 15 cfm/person is selected as representative of a drugstore or hardware store. With 85 people, the total ventilation air quantity is thus 1275 cfm. Floor area of 4000 ft² with 10 ft ceiling height gives a space volume of 40,000 ft³, corresponding to (1275 cfm×60)/40,000 = 1.91 air changes/h. In practice, ventilation air is normally conditioned to some extent by the air conditioning equipment before being admitted to the conditioned space. However, the variety of such arrangements and the varying impact felt by the load calculation process are not covered by this chapter and should be evaluated as part of a system analysis procedure. For this example, assume the ventilation air is introduced directly into the space and included as part of the space cooling load, but only during scheduled operating hours of the cooling equipment.

Infiltration. Window infiltration is considered zero, since the windows are sealed. Infiltration through wall surfaces is also neglected as insignificant, particularly with plastered interior surfaces. Calculation of door infiltration however, requires some judgement. The pressure of 1.91 air changes/h in the form of positive ventilation could be sufficient to prevent door infiltration, depending on the degree of simultaneous door openings and the wind direction and velocity. For this example, assume that outside and inside doors are frequently opened simultaneously, and that door infiltration should be included as part of the cooling load, estimating 100 ft³ per person per door passage. Further estimating outside door use at 10 persons/h, and inside doors (to unconditioned space, previously estimated to be at ambient temperature and humidity) at 30 persons/h, generates the following infiltration rate:

$$Q_{inf} = 40 \times 100/60 = 67 \text{ cfm}$$

Thermal responsiveness of building and contents. For this example, mass of building construction and contents is "medium."

Conditioning equipment location. Conditioning equipment is in an adjoining structure to the north, thus having no direct impact on heat gain.

Find:

1. Sensible cooling load.
2. Latent cooling load.
3. Total cooling load.
4. Capacity of system to maintain:

   (a) Fixed temperature: 75°F indoor temperature, 24-hour "on" period.
   (b) 2°F throttling range: Indoor temperature in the range 75 to 77°F, 24-hour "on" period.
   (c) 4°F throttling range: Indoor temperature in the range 75 to 79°F, 24-hour "on" period.
   (d) 2°F throttling range: Indoor temperature in the range 75 to 77°F, 10-hour "on" period, 0800 to 1700.
   (e) 4°F throttling range: Indoor temperature in the range 75 to 79°F, 12-hour "on" period, 0600 to 1700.

Solution by Transfer Function Method

1. Daily load cycle: Estimated thermal loads are calculated by the TFM once per hour for a 24-h daily cycle.
2. Hourly heat gain components: The methodology using CTF coefficients is used to calculate heat gain components through walls and roof.
3. Thermal storage: The heat storage effect of the building and contents is accounted for by RTF coefficients.
4. Room temperature and heat extraction: TFM approximates resultant room air temperature and heat extraction rates for a specified schedule of thermostat set-points and/or cooling unit operating periods, by applying SATF coefficients to sensible cooling loads, including consideration for heat loss to surroundings. This process can be used to predict the capability of a particular size and type of cooling equipment, its control, and its operating schedule to maintain room air temperature within a specified range.
5. Summary: The data and summary of results using TFM are tabulated in Table 28. The following describes the calculation procedure used to determine the values for this table:

1. Sensible cooling load

   (a) General

Line 1, Time of day in hours: Various temperatures and heat flow rates were calculated for every hour on the hour, assuming that hourly values are sufficient to define the daily profile.

Line 2, Outside air temperatures: Hourly values derived by the above-mentioned procedure, using the specified maximum dry bulb temperature of 94 °F and daily range of 20 °F.

(b) Solar heat gain factors

Lines 3, 4, 5, and 6, Solar heat gain through opaque surfaces: SHGF values from Table 15, Chapter 27 for July 21 at 40 °N latitude. These values are used to calculate sol-air temperatures of various outside surfaces, and solar heat gain through windows.

Values for June might have been used, since the solar irradiation of horizontal surface (*e.g.*, a roof) is maximum at that time of year and since the heat gain through the roof appears to be the major component of exterior heat gain in this problem. The difference between June and August values is relatively small however, compared to the large percentage increase in solar heat gain through south glass in August versus June at this latitude, thus indicating that August might be the better choice. For this example, data for July were selected as reasonable, and to provide better comparison with the results from other techniques for which tabular data are limited. To determine the month when the maximum building load will occur, the relative loads of various surfaces should first be evaluated and compared for several months.

(c) Sol-air temperatures

Lines 7, 8, 9, and 10, Sol-air temperatures at opaque surfaces: Sol-air temperatures, calculated by Equation (6), of the various opaque surfaces. These values are used in calculations of heat gain through the roof and outside walls.

(d) Instantaneous sensible heat gain

Line 11, Roof heat gain: Instantaneous heat gain through the roof, calculated by CTF coefficients.

From Table 11, the major element of the roof (that layer with the most mass) is the gypsum slab (code number C14). Other elements are the metal deck (A3), rigid insulation (B3), built-up roofing (E3), and gravel surface (E2). Entering Table 12 with these code values, the C14 roof slab designates column 7, and the R-value 11.11 calls for R = 3. From the "mass-in" part of the table and the condition of being "w/o ceiling", the table identifies Roof Group 5 as that whose CTF coefficients will best represent the roof in question.

The CTF coefficients (b, d and $\Sigma_{n=0}c_n$) are then obtained from Tables 13 and 14, by selecting roof group 5 and adjusting the tabulated b_n and $\Sigma_{n=0}c_n$ by the U-factor ratio ($U_{example}/U_{table}$) = 0.09 / 0.055 = 1.636.

The adjusted b_n and $\Sigma_{n=0}c_n$ are:

$$
\begin{aligned}
b_0 &= 0.00006\,(1.636) &&= 0.00010 \\
b_1 &= 0.00256\,(1.636) &&= 0.00419 \\
b_2 &= 0.00477\,(1.636) &&= 0.00780 \\
b_3 &= 0.00100\,(1.636) &&= 0.00164 \\
b_4 &= 0.00002 &&= \text{N/A} \\
b_5 &= 0.00000 &&= \text{N/A} \\
b_6 &= 0.00000 &&= \text{N/A} \\
\sum_{n=0} c_n &= 0.00841\,(1.636) &&= 0.01376
\end{aligned}
$$

The *d* values (used without modification) are:

$$
\begin{aligned}
d_0 &= 1.00000 \\
d_1 &= -1.10395 \\
d_2 &= 0.26169 \\
d_3 &= -0.00475 \\
d_4 &= 0.00002 \\
d_5 &= 0.00000 \\
d_6 &= 0.00000
\end{aligned}
$$

The heat gain through the roof is calculated by Equation (28), using the sol-air temperature cycle given in line 7 and t_{rc} = 75 °F. The calculations are extended for five daily cycles at which time the daily periodic steady state is effectively reached. The last daily cycle is used as the heat gain through the roof. (*Note*: Three daily cycles are sufficiently accurate in this case, but since calculations do not converge for the more massive wall components before the 93rd hour, all calculations are run to hour 120.)

Lines 12, 13, 14, and 15, Wall heat gain. The instantaneous heat gains through the various walls are calculated by the same approach as that used for the roof. The CTF coefficients selected from Tables 11 and 15 to 19 are:

North and East Exterior Walls
Dominant element C8, or col. 13 in Integral Mass table (Table 16);
Interior finish E1;
R value indicating *R* of 2 in Table 16;
Select Wall Group 5 in Tables 18 and 19 for representative factors.

South Wall
Dominant element C9, or col. 14 in Table 16;
Exterior layer A2 or A7;
Interior layer E1 (plywood panel ignored as trivial);
R value indicating *R* of 6;
Select Wall Group 24 for representative factors.

North and West Party Walls
With no specific data for a 13 in. brick wall, use a layer of 8 in. common brick (C9) and a layer of 4 in. face brick (A2 or A7) as an approximation; thus:
Dominant element C9, or column 14 in Table 16;
Exterior layer A2 or A7;
R value indicating *R* of 6;
Select Wall Group 24 in Table 16 for representative factors.

The b_n and $\Sigma_{n=0}c_n$ require multiplication by the U-factor ratio to account for the difference in U-factors. The heat gain is then calculated by Equation (28), using corresponding wall CTF coefficients and sol-air temperatures for south, east, and north walls, and the outside air temperature cycle for north and west party walls.

Lines 16, 17, and 18, Door heat gain: Heat storage of the doors could be assumed negligible, in which case the heat gain would be calculated by Equation (16) as

$$q_{D\theta} = U_D A_D (t_{D\theta} - t_i)$$

where

U_D = 0.19; U-factor of doors (0.18 for interior doors)
A_D = 35; area of a door, ft^2
t_i = 75 °F, inside temperature
$t_{D\theta}$ = outside temperature at door, at time θ

For the door in the north party wall, $t_{D\theta}$ equals outside air temperature. For the doors in east and south walls $t_{D\theta}$ equals the east and south wall sol-air temperatures, respectively.

The foregoing would be a reasonable approach for estimating the minor loads involved. For the purpose of this example however, the relatively brief storage effect of the solid core doors has been considered by use of Equation (28), in accordance with:

Dominant element B7, or column 3 in Table 16;

Interior finish A6;
R value indicating *R* of 8;
Select Wall Group 1 for representative factors.

Lines 19, 20, and 21, Window heat gain: The air to air heat gain (line 19):

$$q_a = U_w A_w (t_{o\theta} - t_i)$$

where

U_w = 0.81; U-factor of window
A_w = 90; area of windows, ft^2
$t_{o\theta}$ = outside air temperature at time θ

The solar radiation heat gain (lines 20 and 21) through south and north windows:

$$Q_r = A_w \times SC \times SHGF_\theta$$

where

$SHGF_\theta$ = Solar heat gain factors given in line 5 for south and line 4 for north.
SC = 0.55; shading coefficient for clear window with light colored curtain or blind.

Lines 22 and 23, Heat gain from tungsten and fluorescent lights: For the gain from lighting, Equation (9) is used with a use factor of unity and special allowance factors of 1.20 for fluorescent lamps and of unity for tungsten lamps. Thus:

Table 28   Tabulation of Data for Example 6

		0100	0200	0300	0400	0500	0600	0700	0800	0900	1000	1100	1200
1	Time, hour	0100	0200	0300	0400	0500	0600	0700	0800	0900	1000	1100	1200
2	Outside air temperature, °F	76	75	74	74	74	74	75	77	79	82	86	89
3	SHGF, Btu/h·ft², Horizontal	0	0	0	0	0	32	88	145	194	231	254	262
4	East	0	0	0	0	1	37	30	28	32	35	37	38
5	West	0	0	0	0	0	11	21	30	52	81	102	109
6	South	0	0	0	0	2	137	204	216	193	146	81	41
7	Sol-air temperature, °F, Horizontal	69	68	67	67	67	77	94	114	130	144	155	161
8	East	76	75	74	74	74	80	80	81	84	87	92	95
9	West	76	75	74	74	74	76	78	82	87	94	101	105
10	South	76	75	74	74	74	95	106	109	108	104	98	95
	Instantaneous Sensible Heat Gain, Btu/h												
11	Roof	5377	3844	2565	1495	599	−140	−555	−208	1218	3715	6989	10682
12	East wall	934	794	664	544	434	337	272	266	292	338	417	532
13	West wall	1466	1454	1424	1379	1321	1255	1181	1103	1024	947	879	824
14	South wall	4390	3723	3107	2542	2029	1579	1471	2235	3679	5277	6646	7551
15	North and east party wall	2498	2535	2534	2497	2429	2335	2218	2086	1943	1797	1656	1529
16	East door (to adjacent building)	19	12	6	1	−3	−5	−4	2	12	26	44	64
17	West door	20	13	6	1	−4	−4	6	20	43	76	118	159
18	South door	20	13	6	1	−4	11	98	176	212	218	200	167
19	Windows, air to air heat gain	117	44	−15	−58	−73	−44	29	160	350	569	816	1050
20	East windows, solar heat gain	0	0	0	0	17	611	495	462	528	578	611	627
21	West windows, solar heat gain	0	0	0	0	0	363	693	990	1716	2673	3366	3597
22	Lights, tungsten (always on)	13640	13640	13640	13640	13640	13640	13640	13640	13640	13640	13640	13640
23	Lights, fluorescent (on-off)	0	0	0	0	0	0	0	71610	71610	71610	71610	71610
24	People	0	0	0	0	0	0	0	21250	21250	21250	21250	21250
25	Infiltration	0	0	0	0	0	0	0	162	354	575	825	1061
26	Ventilation	2244	841	−281	−1122	−1403	−841	561	3086	6732	10940	15708	20196
27	Total instant sensible heat gain	30725	26913	23656	20920	18982	19097	20105	117040	124603	134229	144775	154539
	Latent Heat Gain/Cooling Load, Btu/h												
28	People	0	0	0	0	0	0	0	17000	17000	17000	17000	17000
29	Infiltration	0	0	0	0	0	0	0	2205	2205	2205	2205	2205
30	Ventilation	41963	41963	41963	41963	41963	41963	41963	41963	41963	41963	41963	41963
31	Total latent heat gain/cooling load	41963	41963	41963	41963	41963	41963	41963	61168	61168	61168	61168	61168
32	Sum: sens. + latent heat gain, Btu/h	72688	68876	65619	62883	60945	61060	62068	178208	185771	195397	205943	215707
	Sensible Cooling Load from Convective Heat Gain, Btu/h												
33	Windows, air to air heat gain	117	44	−15	−58	−73	−44	29	160	350	569	816	1050
34	Lights, tungsten (20% convective)	2728	2728	2728	2728	2728	2728	2728	2728	2728	2728	2728	2728
35	Lights, fluorescent (50% conv.)	0	0	0	0	0	0	0	35805	35805	35805	35805	35805
36	People (67% convective)	0	0	0	0	0	0	0	14237	14237	14237	14237	14237
37	Infiltration (100% convective)	0	0	0	0	0	0	0	162	354	575	825	1061
38	Ventilation (100% convective)	2244	841	−281	−1122	−1403	−841	561	3086	6732	10940	15708	20196
	Sensible Cooling Load from Radiant Heat Gain, Btu/h												
39	Lights, tungsten (80% radiant)	10912	10912	10912	10912	10912	10912	10912	10912	10912	10912	10912	10912
40	Lights, fluorescent (50% radiant)	12120	11271	10482	9748	9066	8431	7841	10873	12618	14241	15751	17154
41	People (33% radiant)	2118	1970	1832	1704	1584	1473	1370	2656	2961	3245	3508	3754
	Sensible Cooling Load from Convective and Radiant Heat Gain, Btu/h												
42	From SHG through east windows	85	79	73	68	75	476	406	390	440	480	509	527
43	From SHG through west windows	267	249	231	215	200	433	653	858	1362	2038	2555	2769
44	From roof heat gain	7577	6379	5331	4409	3594	2882	2387	2418	3205	4766	6922	9442
45	From east wall heat gain	963	866	773	683	599	521	464	446	452	472	516	587
46	From west wall heat gain	1362	1361	1347	1322	1287	1244	1194	1140	1084	1027	975	931
47	From south wall heat gain	4984	4488	4015	3567	3146	2761	2605	3046	3973	5040	5989	6651
48	From N. and E. party wall heat gain	2304	2343	2356	2343	2308	2252	2178	2091	1993	1890	1788	1692
49	From east door heat gain	32	26	21	16	13	10	10	13	19	28	40	54
50	From west door heat gain	43	36	30	24	20	18	23	32	46	69	98	127
51	From south door heat gain	46	39	33	27	22	30	89	142	169	176	167	147
52	Total sensible cooling load, Btu/h	47902	43632	39868	36586	34078	33286	33450	91195	99440	109238	119849	129824
53	Sum: sens. + lat. cooling load, Btu/h	89865	85595	81831	78549	76041	75249	75413	152363	160608	170406	181017	190992
	Air Temperature, °F, and Heat Extraction, Btu/h												
54	Total sensible cooling load, Btu/h (LTS)	45482	41375	37758	34612	32232	31516	31749	89212	97208	106702	116999	126680
55	2 °F throttling range: temperature	74.0	74.0	74.0	74.0	74.0	74.0	74.0	75.0	75.0	75.0	75.0	75.0
56	Equipment run 1-24; heat extraction	49487	45493	41936	38803	36358	35399	35314	85674	93257	102010	111520	120541
57	4 °F throttling range: temperature	74.0	74.0	74.0	74.0	74.0	74.0	74.0	75.0	75.0	75.0	75.0	76.0
58	Equipment run 1-24; heat extraction	52695	48849	45392	42318	39866	38753	38445	83238	90364	98444	107245	115651
59	2 °F throttling range: temperature	90	90	91	91	91	91	91	80	79	79	80	80
60	Equipment run 8-17; heat extraction	0	0	0	0	0	0	0	146600	146600	146600	146600	146600
61	4 °F throttling range: temperature	88	88	88	88	88	75	75	76	76	76	76	77
62	Equipment run 6-17; heat extraction	0	0	0	0	0	106261	95365	133756	135291	138399	142778	146600

Table 28   Tabulation of Data for Example 6 (*Concluded*)

	1300	1400	1500	1600	1700	1800	1900	2000	2100	2200	2300	2400	24 h Total	Heat Loss, Btu/h
1	1300	1400	1500	1600	1700	1800	1900	2000	2100	2200	2300	2400	24 h	Heat Loss,
2	91	93	94	93	92	89	87	84	82	80	78	77	Total	Btu/h
3	254	231	194	145	88	32	0	0	0	0	0	0	2150	
4	37	35	32	28	30	37	1	0	0	0	0	0	438	
5	102	81	52	30	21	11	0	0	0	0	0	0	703	
6	37	35	31	26	20	11	0	0	0	0	0	0	1180	
7	160	155	145	130	111	92	80	77	75	73	71	70		
8	97	98	99	97	97	95	87	84	82	80	78	77		
9	106	105	102	98	95	91	87	84	82	80	78	77		
10	97	98	99	97	95	91	87	84	82	80	78	77		

Instantaneous Sensible Heat Gain, Btu/h

	1300	1400	1500	1600	1700	1800	1900	2000	2100	2200	2300	2400	24 h Total	Heat Loss, Btu/h
11	14436	17882	20649	22467	23154	22592	20788	18036	14908	11963	9395	7212	239063	23400
12	685	862	1045	1214	1356	1461	1531	1549	1486	1368	1228	1081	20690	5304
13	789	780	801	851	927	1020	1120	1218	1306	1378	1429	1459	27335	6318
14	7939	8025	8075	8159	8246	8247	8099	7757	7214	6548	5829	5100	133467	23868
15	1427	1358	1331	1351	1417	1525	1667	1831	2001	2165	2308	2421	46859	17306
16	84	100	111	116	115	108	96	81	65	51	38	28	1167	410
17	189	203	200	183	162	139	117	91	70	54	40	29	1931	432
18	147	148	153	156	149	136	116	90	70	54	40	29	2406	432
19	1224	1341	1385	1341	1239	1079	889	700	540	393	277	189	13542	4739
20	611	578	528	462	495	611	17	0	0	0	0	0	7231	*
21	3366	2673	1716	990	693	363	0	0	0	0	0	0	23199	*
22	13640	13640	13640	13640	13640	13640	13640	13640	13640	13640	13640	13640	327360	-13640
23	71610	71610	71610	71610	71610	0	0	0	0	0	0	0	716100	-71610
24	21250	21250	21250	21250	21250	0	0	0	0	0	0	0	212500	-21250
25	1238	1356	1400	1356	1253	0	0	0	0	0	0	0	9580	4791
26	23562	25806	26648	25806	23843	20757	17111	13464	10379	7574	5330	3646	260587	91163
27	162197	167612	170542	170952	169549	71678	65191	58457	51679	45188	39554	34834	2043017	

Latent Heat Gain/Cooling Load, Btu/h

	1300	1400	1500	1600	1700	1800	1900	2000	2100	2200	2300	2400	24 h Total	
28	17000	17000	17000	17000	17000	0	0	0	0	0	0	0	170000	
29	2205	2205	2205	2205	2205	0	0	0	0	0	0	0	22050	
30	41963	41963	41963	41963	41963	41963	41963	41963	41963	41963	41963	41963	1007112	
31	61168	61168	61168	61168	61168	41963	41963	41963	41963	41963	41963	41963	1199162	
32	223365	228780	231710	232120	230717	113641	107154	100420	93642	87151	81517	76797	3242179	

Sensible Cooling Load from Convective Heat Gain, Btu/h

	1300	1400	1500	1600	1700	1800	1900	2000	2100	2200	2300	2400	24 h Total	
33	1224	1341	1385	1341	1239	1079	889	700	540	393	277	189	13542	
34	2728	2728	2728	2728	2728	2728	2728	2728	2728	2728	2728	2728	65472	
35	35805	35805	35805	35805	35805	0	0	0	0	0	0	0	358050	
36	14237	14237	14237	14237	14237	0	0	0	0	0	0	0	142370	
37	1238	1356	1400	1356	1253	0	0	0	0	0	0	0	9580	
38	23562	25806	26648	25806	23843	20757	17111	13464	10379	7574	5330	3646	260587	

Sensible Cooling Load from Radiant Heat Gain, Btu/h

	1300	1400	1500	1600	1700	1800	1900	2000	2100	2200	2300	2400	24 h Total	
39	10912	10912	10912	10912	10912	10912	10912	10912	10912	10912	10912	10912	261888	
40	18460	19674	20803	21853	22830	20158	18747	17434	16214	15079	14024	13042	357914	
41	3982	4194	4391	4575	4746	3523	3276	3047	2834	2635	2451	2279	70108	

Sensible Cooling Load from Convective and Radiant Heat Gain, Btu/h

	1300	1400	1500	1600	1700	1800	1900	2000	2100	2200	2300	2400	24 h Total	Heat Loss, Btu/h
42	523	507	478	437	461	542	143	122	114	106	98	91	7230	
43	2669	2246	1624	1136	924	683	413	384	357	332	309	288	23195	
44	12086	14596	16711	18225	18990	18898	17928	16254	14249	12289	10518	8953	239009	
45	687	808	937	1059	1167	1251	1314	1341	1313	1245	1158	1063	20685	
46	900	886	893	921	967	1028	1095	1164	1227	1282	1324	1351	27312	
47	6979	7104	7203	7321	7439	7496	7448	7261	6926	6492	6007	5498	133439	
48	1611	1551	1520	1520	1553	1617	1708	1816	1933	2049	2155	2243	46814	
49	68	80	89	94	95	92	85	75	65	55	46	38	1164	
50	150	162	163	154	141	128	113	96	81	70	59	50	1933	
51	134	136	140	143	140	131	118	100	86	74	63	54	2406	
52	137955	144129	148067	149623	149470	91023	84028	76898	69958	63315	57459	52425	2042698	178163
53	199123	205297	209235	210791	210638	132986	125991	118861	111921	105278	99422	94388	3241860	71663

Air Temperature, °F, and Heat Extraction, Btu/h

	1300	1400	1500	1600	1700	1800	1900	2000	2100	2200	2300	2400	24 h Total	
54	134552	140509	144275	145693	145432	87215	80409	73476	66754	60326	54673	49830	1974669	
55	75.0	75.0	75.0	75.0	75.0	75.0	75.0	75.0	75.0	74.0	74.0	74.0		
56	128005	133792	137639	139386	139602	88759	82327	75937	69698	63689	58341	53696	1966663	
57	76.0	76.0	76.0	76.0	76.0	75.0	75.0	75.0	75.0	74.0	74.0	74.0		
58	122700	128264	132093	134030	134566	89480	83460	77573	71790	66186	61150	56731	1959283	
59	80	81	81	81	80	89	90	90	90	90	90	90		
60	146600	146600	146600	146600	146600	0	0	0	0	0	0	0	1466000	
61	77	77	78	78	78	86	87	87	87	87	88	88		
62	146600	146600	146600	146600	146600	0	0	0	0	0	0	0	1631450	

$$q_{el\ tung} = 4000 \times 1 \times 1 \times 3.41 = 13,640 \text{ Btu/h, and}$$

$$q_{el\ fluor} = 17,500 \times 1 \times 1.20 \times 3.41 = 71,610 \text{ Btu/h}$$

Line 24, Heat gain from people: Sensible heat gain from occupants, for moderately active office work (Table 3):

$$q_{sp} = \text{(number of people)(sensible heat generated per person)}$$

$$= 85 \times 250 = 21,250 \text{ Btu/h}$$

Lines 25 and 26, Sensible heat gain from infiltration and ventilation: As developed in Building Data, the value used for infiltration is 67 cfm, and that for ventilation, 1275 cfm.

Heat gain from infiltration air is part of the space load, while that from ventilation air normally is not. In this example however, since ventilation is delivered directly to the space rather than through the cooling equipment first, its gain is also included as a direct space load.

Note: Had the ventilation air instead been mixed with return air leaving the occupied space and before entering the cooling equipment, only (4) that portion which passed through the cooling coil untreated due to coil inefficiency (or "Bypass Factor", normally 3 to 5% for a chilled water coil of 6 or more rows and close fin spacing up to 15% or more for refrigerant coils in packaged air-conditioning units), and/or (5) that quantity deliberately bypassed around the coil in response to a "face and bypass" or "conventional multizone" space dry-bulb temperature control scheme, would become a part of the space heat gain as such rather than a part of the cooling coil load directly (see 1991 ASHRAE *Handbook— Applications*).

The sensible loads are determined from Equation (25). At 1600 hours for example, when $t_o = 94\,°F$ and $t_i = 75\,°F$, this generates:

$$q_{si} = 1.1 \, (\text{Infiltration rate, cfm})(t_o - t_i)$$

$$= 1.1 \times 67\,(94 - 75) = 1400 \text{ Btu/h, and}$$

$$q_{sv} = 1.1(\text{Ventilation rate, cfm}) \, (t_o - t_i)$$

$$= 1.1 \times 1275\,(94 - 75) = 26,600 \text{ Btu/h}$$

Line 27, Total instantaneous sensible heat gain: The sum of instantaneous heat gain values listed in lines 11 through 26. All such values take into account the delaying effects of insulation and mass of the elements enclosing the conditioned space on the heat that ultimately enters that space, but before considering the thermal inertia of the overall mass and configuration of the building and contents in delaying conversion of radiant heat gain to space cooling load.

(e) Instantaneous latent heat gain

Line 28, People: The latent heat gain due to people, using Table 3 data:

$$q_{lp} = \text{(number of persons)(latent heat generated by one person)}$$

$$= 85 \times 200 = 17,000 \text{ Btu/h during the occupied period.}$$

Lines 29 and 30, Latent heat gain from infiltration and ventilation: The latent loads are determined from Equation (26). At 1600 hours for example, when $W_o = 0.0161$ and $W_s = 0.0093$, this generates

$$q_{li} = 4840\,(\text{Infiltration rate, cfm}) \, (W_o - W_s)$$

$$= 4840 \times 67\,(0.0161 - 0.0093) = 2,205 \text{ Btu/h, and}$$

$$q_{lv} = 4840\,(\text{Ventilation rate, cfm}) \, (W_o - W_s)$$

$$= 4840 \times 1275\,(0.0161 - 0.0093) = 41,963 \text{ Btu/h}$$

Line 31, Total latent heat gain: The total latent heat gain, *i.e.*, the sum of lines 28, 29, and 30.

Line 32, Sum of instantaneous sensible and latent heat gain: The sum of heat gain values from lines 27 and 31.

(f) Cooling load from convective sensible heat gain components

Lines 33 through 38: Direct inclusion of the convective portions of instantaneous heat gain components listed in lines 19, 25, and 26, and 20%, 50%, and 67% of lines 22, 23, and 24 respectively. These room sensible heat gain components (*i.e.*, loads due to air-to-air heat gain through

windows, tungsten lights, fluorescent lights, infiltration, ventilation, and heat gain due to people by convection, all appear as cooling load without delay. Percentages of heat gain considered corrective are listed in Table 3 and Table 44 under the section describing TETD/TA procedures. Selection of 33% of sensible gain for people as radiant is an approximation for purposes of this example.

(g) Cooling load from radiant sensible heat gain components

Lines 39 through 41: Heat gain data from lights and people (lines 22 through 24) are processed by Equation (31) using RTF coefficients from Tables 24 and 25:

From Table 24, assuming "medium" mass of building and contents, the 75 lb/ft^2 specific mass classification can be considered representative. Assuming a conventional supply diffuser and non-plenum return air arrangement with inside surface coefficient $h = 0.6 \text{ Btu/(h} \cdot \text{ft}^2 \cdot °F)$, or "medium" type indicates a w_1 value of -0.94; except with an uncarpeted floor the next w_1 value down the column is used, or 0.93.

From the lower part of Table 25, assuming ordinary furnishings, no carpet, medium air circulation, supply and return below ceiling, and unvented light fixtures, the v_0 value for lighting is 0.55 and $v_1 = 1 + (-0.93) - 0.55 = -0.48$.

For people, the upper part of Table 25 calls for a v_0 of 1.0 and v_1 of 0 to be applied to convective heat gain (instantaneous conversion to cooling load), and for radiant heat gain a v_0 of 0.197 and $v_1 = 1 + (-0.93) - 0.197 = -0.127$.

Note that the TFM treatment of lighting heat gain is "generic," without individual regard to the differences in radiant/convective percentages of heat gain from incandescent, fluorescent, or other type lamps, and the RTF coefficients are applied to the combined sensible heat gain values. For the purposes of this example, to facilitate comparison with other calculation methods, the values in lines 39 and 40 represent the hourly results of Equation (31) less the amounts of instantaneous cooling load included and indicated on lines 34 and 35.

(h) Cooling load from convective and radiant sensible heat gain components

Lines 42 through 51: Elements of instantaneous heat gain from solar radiation through windows, walls, doors and roof, *i.e.*, sum of values listed in lines 11 to 21, delayed in being felt as cooling load by the space. Data listed in lines 42 through 51 are the results of applying Equation (31) and appropriate RTF coefficients to the heat gain values from lines 11 through 21, without separately considering radiant or convective components. RTF coefficients are taken from Tables 24 and 25 in the manner above described for lighting loads, producing:

From Table 24, $w = -0.93$ in all cases.

From the upper part of Table 25, all cases fall within the second category described, which for "medium" building and contents mass indicates $v_0 = 0.681$ and $v_1 = 1 + (-0.93) - 0.681 = -0.611$.

The heat gain by solar radiation transmitted through windows is included with heat gain through walls and roof because the venetian blind intercepts solar radiation and releases it to the room in a similar way as the heat gain through walls and roof.

Note: If the glass had no internal shading, the solar radiation through windows would have to be treated by a different set of RTF coefficients to account otherwise for thermal storage (see Tables 24 and 25). Translucent draperies fall somewhere between these limits, with assumed linear relationship in the absence of specific research on the subject (see Chapter 27).

Line 52, Total room sensible cooling load: Total sensible cooling load felt by the room, and the design sensible load used as the basis for sizing cooling equipment. This total load is the sum of the values listed in lines 33 through 51. The tiny difference between the 24 hour total of 2,042,698 Btu/h on line 52 and the sum of the 24 hour totals for lines 11 through 26 reflects rounding of values during intermediate computation.

2. Latent Cooling Load

Line 31—The sum of lines 28, 29, and 30: Total Latent Heat Gain is also the Total Latent Cooling Load, as all components occur instantaneously.

3. Total cooling load

Line 53—The sum of lines 52 and 31: Note that the Total Cooling Load for this example problem is the theoretical total for the conditions as defined, and may or may not represent the actual total cooling load imposed upon a system of cooling equipment. An appropriate psychrometric analysis should be performed of supply air, space air, return air, and mixed air (where ventilation air is mixed with return air en route back to the cooling equipment), considering the type of cooling equipment and characteristics of the preferred control scheme. Only an analysis of this type can verify that the design will meet the requirements, and determine whether the actual sensible, latent, and total cooling loads are greater or less than the theoretical values calculated (see 1992 ASHRAE *Handbook—Systems and Applications*).

4. Capacity of system to maintain conditions

(a) *Fixed temperature*: 75 °F indoor temperature, 24 hour "on" period: The basic calculation procedure assumes a fixed indoor temperature, in this case 75 °F; thus the results tabulated in lines 1 through 42 are for this condition.

(b) 2 °F throttling range: Indoor temperature in the range 75 to 77 °F, 24 hour "on" period.

(c) 4 °F throttling range: Indoor temperature in the range 75 to 79 °F, 24 hour "on" period.

(d) 2 °F throttling range: Indoor temperature in the range 75 to 77 °F, 10 hour "on" period, 0800 through 1700.

(e) 4 °F throttling range: Indoor temperature in the range 75 to 79 °F, 12 hour "on" period, 0600 through 1700.

Line 54, Sensible cooling load with loss to surroundings: To be consistent with the concept of heat extraction and resultant space temperatures, certain cooling load elements must be modified to account for heat loss to surroundings. The multiplier $F_c = 0.94362$ was calculated by the process noted for each of the envelope element areas times the respective U-factors, dividing the sum by the building perimeter to develop K_θ, and generating F_c by Equations (37) and (38); then using F_c to reduce the appropriate load elements. The sum of all modified and unmodified load elements is listed on line 53 as the basis for the various heat extraction/space temperature evaluations.

Lines 55 through 62, Air temperatures and heat extraction rates: Heat extraction and indoor air temperatures are based on the normalized SATF coefficients for medium weight construction listed in Table 26 and calculated by use of Equations (39) through (43) in the procedure previously described. The SATF coefficients for this example are thus for hour θ (0800 − 1700):

$$g_{0\theta} = g_0* \text{ (Floor area)} + p_0 [K_\theta \text{ (Perimeter length)} + 1.1(\text{Ventilation and Infiltration)}]$$

$$= (1.81 \times 4000) + 1.0 [(4.86 \times 260) + 1.1(1275 + 67)]$$

$$= 7240 + (1263.4 + 1476.2) = 9980$$

$$g_{1\theta} = g_1* \text{ (Floor area)} + p_1 [K_\theta \text{ (perimeter length)} + 1.1(\text{Ventilation and Infiltration)}]$$

$$= (-1.89 \times 4000)(-0.87)[(4.86 \times 260) + 1.1 (1275 + 67)]$$

$$= -7560 - 0.87 (1263.4 + 1476.2) = -9944$$

$$g_{2\theta} = g_2* \text{ (Floor area)} = 0.08 \times 4000 = 320$$

$$p_0 = 1.0000$$

$$p_1 = -0.87$$

The heat extraction rates and room air temperatures listed in lines 55 through 62 are calculated using these SATF coefficients, the modified total sensible cooling load values listed in line 54, and the specified throttling ranges and "on" and "off" periods.

The maximum sensible heat extraction capacity required to maintain the space temperature at a constant 75 °F can be taken as the design peak value on line 52, or 149,470 Btu/h at 1600 hours.

The maximum sensible heat extraction capacity required to maintain interior temperature within a 75 to 77 °F range is 145,693 Btu/h (hour 1600, line 54), and within a 75 to 79 °F range is 139,602 Btu/h (hour 1700, line 56), assuming continuous operation of cooling equipment.

Comparable maintenance of space temperature ranges during equipment operation hours (limited to 10 hours and 12 hours respectively) requires heat extraction rates of 146,600 Btu/h (hours 0800 to 1700, line 60) and 146,600 Btu/h (hours 1200 to 1700, line 62) respectively. Here ER_{max} needs to be increased if the heat accumulated overnight is to be overcome; but the total daily heat extraction still will be significantly less than for continuous operation.

5. Heating Load

Lines 11 through 19, Heat loss by conduction: The heat loss column lists for each of the building envelope components a single value representing the product of exposed area, U-factor, and the temperature difference between inside design dry bulb and outside design dry bulb temperatures for winter conditions, in an adaptive use of Equation (8). Often, a lower inside design dry bulb temperature is selected for winter conditions than for summer, and, where appropriate, the U-factors are adjusted to reflect different average exterior wind velocities. For this example, the same inside temperatures and U-factors are used year-round.

These results are design heat loss values, which are used to establish a "design heating load" with which to design heating systems and to select properly sized equipment components. When the load calculation is used to analyze energy performance, hourly calculations of heat loss that reflect the profile of outside weather conditions must be run (see Chapter 28).

Lines 20 through 21, Solar heat gain: For *design* heating loss calculations, offsetting values of solar heat gain are routinely ignored at night or during periods of extended cloud cover, and thus not consistently available to assist the installed heating equipment. Designers must, however, consider the higher solar heat gain values that occur during winter months due to low solar angles that often cause peak cooling loads through large areas of exposed glass. Hourly calculations are required for energy use evaluation.

Lines 22, 23, and 24, Internal heat gains: Like solar heat gain, the heat from internal sources requires year-round cooling for completely interior spaces and contributes to unseasonable cooling requirements in conjunction with glass loads on sunny days. For conventional heating load purposes, however, these loads are normally ignored because of their uncertainty during all hours of need and since their full effect does not occur until some number of hours after occupancy begins during intermittent schedules. Heat gain values in this example are given as "negative heat loss" figures, and not routinely included in design heating load summaries.

Lines 25 and 26, Infiltration and ventilation: Values listed for these variables are calculated on the basis of a single "worst case" hour under winter design temperature conditions, adapting Equation (25) in a similar manner to that noted for conduction heat losses.

Humidification: For this example, the issue of maintaining interior humidity levels during winter months has been ignored. While this represents routine practice for most applications in latitudes 35 °N and lower, humidity levels are of major concern in colder climates. (see Chapter 21).

Line 52, Total sensible heat loss: The sum of heat loss values from lines 11 through 19, 25, and 26, and which conventionally represents the design heating load for the building. Internal heat gain figure from lines 22, 23, and 24 are not included in this total.

Line 53, Net sensible heat loss, considering internal heat gains: The heat loss summary value if internal heat gains were to be included in the total, illustrated here only to emphasize the potential significance of such elements and the importance of providing an appropriate means of temperature control for differently affected building areas.

CLTD/SCL/CLF CALCULATION PROCEDURE

To calculate a space cooling load using the CLTD/SCL/CLF convention, the same general procedures outlined for the TFM relative to data assembly and use of data apply. Similarly, the basic heat gain calculation concepts of solar radiation, total heat gain through exterior walls and roofs, heat gain through interior surfaces, and heat gain through infiltration and ventilation are handled in an identical manner.

The CLTD/SCL/CLF method is a one-step, hand calculation procedure, based on the transfer function method (TFM). It may be used to approximate the cooling load corresponding to the first three modes of heat gain (conductive heat gain through surfaces such as windows, walls, and roofs; solar heat gain through fenestrations; and internal heat gain from lights, people, and equipment) and the cooling load from infiltration and ventilation. The acronyms are defined as follows:

CLTD—Cooling Load Temperature Difference
SCL—Solar Cooling Load
CLF—Cooling Load Factor

The following sections give details of how the CLTD/SCL/CLF technique relates to and differs from the TFM. The sources of the space cooling load, forms of equations to use in the calculations, appropriate references, tables, are summarized in Table 29.

SYNTHESIS OF HEAT GAIN AND COOLING LOAD CONVERSION PROCEDURES

Exterior Roofs and Walls

This method was developed by using the TFM to compute one-dimensional transient heat flow through various sunlit roofs and walls. Heat gain was converted to cooling load using the room transfer functions for rooms with light, medium, and heavy thermal characteristics. Variations in the results due to such varying room constructions and other influencing parameters discussed in the TFM description are so large that only one set of factors is presented here for illustration. All calculations for data tabulated were based on the sol-air temperatures in Table 1. The inside air temperature was assumed to be constant at 78 °F (cooling system in operation 24 h/day, seven days a week). The mass of building and contents was "light to medium". For application of CLTD/SCL/CLF techniques, refer to the *Cooling and Heating Load Calculation Manual* (McQuiston and Spitler 1992).

Basic CLTD cooling load for exterior surfaces. The results were generalized to some extent by dividing the cooling load by the U-factor for each roof or wall and are in units of total equivalent cooling load temperature difference (CLTD). This establishes the basic cooling load equation for exterior surfaces as:

$$q = UA\,(\text{CLTD}) \tag{44}$$

where

q = cooling load, Btu/h
U = coefficient of heat transfer, Btu/(h·ft²·°F)

Table 29   Procedure for Calculating Space Design Cooling Load by CLTD/SCL/CLF Method

External Cooling Load

Roofs, walls, and conduction through glass

$$q = UA\,(\text{CLTD}) \tag{44, 55}$$

    U = design heat transfer coefficient for roof or wall from Chapter 22, Table 4; or for glass, Table 5, Chapter 27
    A = area of roof, wall, or glass, calculated from building plans
CLTD = cooling load temperature difference, roof, wall, or glass

Solar load through glass

$$q = A\,(SC)(SCL) \tag{46}$$

    SC = shading coefficient: Chapter 27
    SCL = solar cooling load factor with No Interior Shade, or With Shade Table 36.

Cooling load from partitions, ceilings, floors

$$q = UA\,(t_b - t_{rc}) \tag{8}$$

    U = design heat transfer coefficient for partition, ceiling, or floor, from Chapter 22, Table 4
    A = area of partition, ceiling, or floor, calculated from building plans
    t_b = temperature in adjacent space
    t_{rc} = inside design temperature (constant) in conditioned space

Internal Cooling Load

People

$$q_{sensible} = N\,(\text{Sensible heat gain})\,\text{CLF} \tag{47}$$

$$q_{latent} = N\,(\text{Latent heat gain}) \tag{48}$$

    N = number of people in space, from best available source. Sensible and latent heat gain from occupancy—Table 3, or Chapter 8; adjust as required
    CLF = cooling load factor, by hour of occupancy, Table 37
Note: CLF = 1.0 with high density or 24-h occupancy and/or if cooling off at night or during weekends.

Lights

$$q_{el} = 3.41\,W\,F_{ul}\,F_{sa}\,(\text{CLF}) \tag{9, 49}$$

    W = watts input from electrical plans or lighting fixture data
    F_{ul} = lighting use factor, from Section I, as appropriate
    F_{sa} = special allowance factor, from Section I, as appropriate
    CLF = cooling load factor, by hour of occupancy, Table 38

Note: CLF = 1.0 with 24-h light usage and/or if cooling off at night or during weekends.

Power

$$q_p = 2545\,PE_F\,\text{CLF} \tag{15, 16, 17, 50}$$

    P = horsepower rating from electrical plans or manufacturer's data
    E_F = efficiency factors and arrangements to suit circumstances
    CLF = cooling load factor, by hour of occupancy, Table 37

Note: CLF = 1.0 with 24-h power operation and/or if cooling off at night or during weekends.

Appliances

$$q_{sensible} = q_{is}\,F_{ua}\,F_{ra}\,(\text{CLF}) \tag{18, 50}$$

or

$$q_{sensible} = (q_{is}\,F_{ua}\,F_{ra}\,\text{CLF})/F_{fl} \tag{20, 50}$$

$$q_{latent} = q_{il}\,F_{ua} \tag{22}$$

    q_{is}, q_{il} = sensible and latent heat gain from appliances—Tables 5 through 9, or manufacturer's data
    F_{ua}, F_{ra}, F_{fl} = use factors, radiation factors, flue loss factors from the General Principles section
    CLF = cooling load factor, by scheduled hours and hooded or not; Tables 37 and 39

Note 1: CLF = 1.0 with 24-h appliance operation and/or if cooling off at night or during weekends.
Note 2: Set latent load = 0 if appliance under exhaust hood.

Ventilation and Infiltration Air

$$q_{sensible} = 1.10\,Q\,(t_o - t_i) \tag{25}$$

$$q_{latent} = 4840\,Q\,(W_o - W_i) \tag{27}$$

$$q_{total} = 4.5\,Q\,(H_o - H_i) \tag{23}$$

    Q = ventilation cfm from ASHRAE *Standard* 62; infiltration from Chapter 23
    t_o, t_i = outside, inside air temperature, °F
    W_o, W_i = outside, inside air humidity ratio, lb water/lb dry air
    H_o, H_i = outside, inside air enthalpy, Btu/lb dry air

A = area of surface, ft^2
CLTD = cooling load temperature difference

In developing the method, it was assumed that the heat flow through a similar roof or wall (similar in thermal mass as well as U-factor) can be obtained by multiplying the total CLTDs listed in Tables 30 or 32 by the U-factor of the roof or wall at hand, respectively. The errors introduced by this approach depend on the extent of the differences between the construction in question (components, size, configuration, and general mass of building and contents) and the one used for calculating the CLTDs.

The sol-air temperature value depends on outdoor air temperature as well as the intensity of solar radiation. Consequently, a change in either outdoor air temperature or geographic location changes the sol-air temperature. The CLTD values in the tables were computed for an indoor air temperature of 78 °F, an outdoor maximum temperature of 95 °F, and an outdoor mean temperature of 85 °F, with an outdoor daily range of 21 °F and a solar radiation variation typical of 40 °N latitude on July 21.

The notes associated with Tables 30 and 32 provide descriptions of the conditions under which the CLTD values were calculated. While variations in exterior color and/or outside and inside surface film resistances do have some effect, their impact on roofs or walls of contemporary construction is relatively minor and can be ignored with data that is already normalized for convenience. Variations in inside space temperature or the mean outdoor temperature are of much more significance, and the means of appropriate adjustment are thus outlined. Additional guidance for specific application may be found along with tables for a broad range of latitudes in the *Cooling and Heating Load Calculation Manual* (McQuiston and Spitler 1992).

Space Cooling Load from Fenestration

The basic principles of calculating heat gain from conduction and solar radiation through fenestration are as previously discussed for the TFM.

CLTD Cooling load from conduction. For conduction heat gain, the overall heat transfer coefficient accounts for the heat transfer processes of (1) convection and long-wave radiation exchange outside and inside the conditioned space, and (2) conduction through the fenestration material. To calculate cooling load for this component, the conduction heat gain is treated in a manner similar to that through walls and roofs. The RTF coefficients used to convert the heat gain to cooling load are thus the same as those for walls and roofs. The resulting CLTDs are given in Table 32, again presenting only a single set of factors for all room construction types, neglecting the effects of mass and latitude due to the generally low density and the small magnitude of these components. The CLTDs from Table 32 can also be used for doors with reasonable accuracy. The cooling load from conduction and convection heat gain is calculated by:

$$q_{cond} = UA \, (\text{CLTD}) \qquad (45)$$

where A is the net glass area of the fenestration in square feet. [Note that the equation is identical to Equation (44).]

Solar heat gain. The basic principles of evaluating heat gain from transmitted and absorbed solar energy through fenestration, including the primary terms SHGF and SC, are the same for the CLTD/CLF procedure as previously described for the TFM.

Previous ASHRAE Handbooks tabulated values of maximum solar heat gain factors for sunlit or externally shaded double-strength sheet glass, used as the heat gain input for calculating cooling load factors (CLFs), employing appropriate RTF coefficients as in the TFM discussion. This process, however, introduced new variables into the calculations: (1) the presence or absence of

interior shading devices, which is pivotal, and (2) the construction, furnishings, floor coverings, and relative amounts of fenestration, which are critical when interior shading is absent. Results obtained with this method do not recognize the significant variation of solar cooling load profiles due to different latitudes, months, and other factors. A new term, solar cooling load (SCL), is introduced to more closely approximate cooling loads due to solar radiation transmitted through fenestration.

Cooling load caused by solar radiation through fenestration is calculated by:

$$q_{rad} = A \, (\text{SC})(\text{SCL}) \qquad (46)$$

where

q_{rad} = cooling load caused by solar radiation, Btu/h
A = net glass area of fenestration, ft^2
SC = shading coefficient, for combination of fenestration and shading device, obtained from Chapter 27
SCL = solar cooling load, Btu/(h·ft^2), from Table 36

Total cooling load from fenestration. The total cooling load due to fenestration is the sum of the conductive and radiant components q_{cond} and q_{rad}.

Zone influencing parameters. For purposes of estimating a cooling load, a zone is a particular combination of conditions defining the space under consideration, and which govern the absorption and release of radiant energy. The SCL for a particular zone depends on latitude, direction, nature, and quantity of enclosing surfaces, as well as various internal parameters that influence the SHGF for each glass exposure in that zone.

To determine the most appropriate SCL table for a zone, refer to Tables 35A and 35B, where zone types (A, B, C, or D) are given as functions of some of the more dominant of the 16 zone parameters defined in the TFM discussion. The SCLs for sunlit glass at 40 °N latitude and one month, July, are tabulated in Table 36 for each zone type. SCLs for externally shaded glass may be taken from these tables as those for North exposure, although with some loss of accuracy at latitudes lower than 24 °N. Interpolation between latitudes can be performed with some loss of accuracy. The *Cooling and Heating Load Calculation Manual* (McQuiston and Spitler 1992) includes additional data for multistory buildings and for other latitudes, months, and zone types.

Shading Coefficient

Interior shading. The cooling load from solar radiation must be analyzed for one of two cases: (1) presence of interior shading or (2) absence of interior shading. Blinds (venetian or roller shades) or drapes absorb the solar energy before it can strike the floor or other interior surfaces of the space, which leads to a rapid response in the cooling load due to the low mass of the shading device.

When interior shading is absent, the solar energy is absorbed by the more massive elements of the space, which results in increased delay in such heat gain being converted to cooling load. Many variables, of which the more important are the presence or absence of carpet on the floor, mass of the floor and other surfaces, mass of the contents of the space, amount of glass in the exposed surfaces, presence or absence of a ceiling, the relative size of the space, etc., have influence on this phenomenon.

The composite effect of the various forms of interior shading on solar radiation from glass, relative to unshaded clear double-strength glass, is represented by a shading coefficient (SC) or decimal multiplier, tabulated in Chapter 27 for a wide variety of conditions.

Exterior shading. Where glass is shaded by exterior means of a permanent nature, the hourly mitigating effect of such shading may be estimated by separate evaluations of shaded areas relative to unshaded areas for each situation as previously noted.

Table 30   July Cooling Load Temperature Differences for Calculating Cooling Load from Flat Roofs at 40° North Latitude

Roof No.	1	2	3	4	5	6	7	8	9	10	11	12	13	14	15	16	17	18	19	20	21	22	23	24
1	0	−2	−4	−5	−6	−6	0	13	29	45	60	73	83	88	88	83	73	60	43	26	15	9	5	2
2	2	0	−2	−4	−5	−6	−4	4	17	32	48	62	74	82	86	85	80	70	56	39	25	15	9	5
3	12	8	5	2	0	−2	0	5	13	24	35	47	57	66	72	74	73	67	59	48	38	30	23	17
4	17	11	7	3	1	−1	−3	−3	0	7	17	29	42	54	65	73	77	78	74	67	56	45	34	24
5	21	16	12	8	5	3	1	2	6	12	21	31	41	51	60	66	69	69	65	59	51	42	34	27
8	28	24	21	17	14	12	10	10	12	16	21	28	35	42	48	53	56	57	56	52	48	43	38	33
9	32	26	21	16	13	9	6	4	4	7	12	19	27	36	45	53	59	63	64	63	58	52	45	38
10	37	32	27	23	19	15	12	10	9	10	12	17	23	30	37	44	50	55	57	58	56	52	47	42
13	34	31	28	25	22	20	18	16	16	17	20	24	28	33	38	42	46	48	49	48	46	44	40	37
14	35	32	30	27	25	23	21	20	19	20	22	24	28	32	36	39	42	44	45	45	44	42	40	37

Note 1. Direct application of data
- Dark surface
- Indoor temperature of 78°F
- Outdoor maximum temperature of 95°F with mean temperature of 85°F and daily range of 21°F
- Solar radiation typical of clear day on 21st day of month
- Outside surface film resistance of 0.333 (h·ft^2·°F)/Btu
- With or without suspended ceiling but no ceiling plenum air return systems
- Inside surface resistance of 0.685 (h·ft^2·°F)/Btu

Note 2. Adjustments to table data
- Design temperatures : Corr. CLTD = CLTD + $(78 - t_r)$ + $(t_m - 85)$

  where

  t_r = inside temperature and t_m = mean outdoor temperature

  t_m = maximum outdoor temperature − (daily range)/2

- No adjustment recommended for color
- No adjustment recommended for ventilation of air space above a ceiling

Table 31   Roof Numbers Used in Table 30

Mass Location**	Suspended Ceiling	R-Factor, ft^2·h·°F/Btu	B7, Wood 1 in.	C12, HW Concrete 2 in.	A3, Steel Deck	Attic-Ceiling Combination
Mass inside the insulation	Without	0 to 5	*	2	*	*
		5 to 10	*	2	*	*
		10 to 15	*	4	*	*
		15 to 20	*	4	*	*
		20 to 25	*	5	*	*
		25 to 30	*	*	*	*
	With	0 to 5	*	5	*	*
		5 to 10	*	8	*	*
		10 to 15	*	13	*	*
		15 to 20	*	13	*	*
		20 to 25	*	14	*	*
		25 to 30	*	*	*	*
Mass evenly placed	Without	0 to 5	1	2	1	1
		5 to 10	2	*	1	2
		10 to 15	2	*	1	2
		15 to 20	4	*	2	2
		20 to 25	4	*	2	4
		25 to 30	*	*	*	*
	With	0 to 5	*	3	1	*
		5 to 10	4	*	1	*
		10 to 15	5	*	2	*
		15 to 20	9	*	2	*
		20 to 25	10	*	4	*
		25 to 30	10	*	*	*
Mass outside the insulation	Without	0 to 5	*	2	*	*
		5 to 10	*	3	*	*
		10 to 15	*	4	*	*
		15 to 20	*	5	*	*
		20 to 25	*	5	*	*
		25 to 30	*	*	*	*
	With	0 to 5	*	3	*	*
		5 to 10	*	3	*	*
		10 to 15	*	4	*	*
		15 to 20	*	5	*	*
		20 to 25	*	*	*	*
		25 to 30	*	*	*	*

*Denotes a roof that is not possible with the chosen parameters.   **The 2 in. concrete is considered massive and the others nonmassive.

Table 32 July Cooling Load Temperature Differences for Calculating Cooling Load from Sunlit Walls 40° North Latitude

Wall Number 1

Wall Face	1	2	3	4	5	6	7	8	9	10	11	12	13	14	15	16	17	18	19	20	21	22	23	24
N	1	0	−1	−2	−3	−1	7	11	11	13	17	21	25	27	29	29	28	29	27	17	11	7	5	3
NE	1	0	−1	−2	−3	2	24	42	47	43	35	28	27	28	29	29	27	24	20	14	10	7	5	3
E	1	0	−1	−2	−2	2	28	51	62	64	59	48	36	31	30	30	28	25	20	14	10	7	5	3
SE	1	0	−1	−2	−3	0	15	32	46	55	58	56	49	39	33	31	28	25	20	14	10	7	5	3
S	1	0	−1	−2	−3	−2	0	4	11	21	33	43	50	52	50	44	34	27	20	14	10	7	5	3
SW	2	0	−1	−2	−2	−2	0	4	8	13	17	25	39	53	64	70	69	61	45	24	13	8	5	3
W	2	1	−1	−2	−2	−2	1	4	8	13	17	21	27	42	59	73	80	79	62	32	16	9	6	3
NW	2	0	−1	−2	−2	−2	0	4	8	13	17	21	25	29	38	50	61	64	55	29	15	9	5	3

Wall Number 2

Wall Face	1	2	3	4	5	6	7	8	9	10	11	12	13	14	15	16	17	18	19	20	21	22	23	24
N	5	3	2	0	−1	−2	−1	3	7	9	11	14	18	21	24	26	27	28	28	27	22	17	12	8
NE	5	3	2	0	−1	−2	2	13	26	36	39	37	33	31	29	29	29	28	26	23	18	14	10	7
E	5	3	2	0	−1	−1	2	15	32	47	55	57	52	44	38	34	32	30	27	23	19	14	11	8
SE	5	3	2	0	−1	−2	0	8	20	33	43	50	53	51	45	39	35	31	28	24	19	14	11	8
S	5	3	2	0	−1	−2	−2	−1	2	7	14	24	33	42	47	48	46	40	33	27	21	15	11	8
SW	7	4	2	1	0	−1	−2	0	2	5	9	13	20	30	41	53	61	65	62	53	39	27	17	11
W	8	5	3	1	0	−1	−2	0	2	5	9	13	17	23	33	46	59	69	73	66	50	34	22	14
NW	8	4	2	1	−1	−2	−2	−1	2	5	9	13	17	21	25	32	41	51	57	54	42	29	19	12

Wall Number 3

Wall Face	1	2	3	4	5	6	7	8	9	10	11	12	13	14	15	16	17	18	19	20	21	22	23	24
N	7	5	3	2	1	0	2	5	7	8	11	14	17	20	23	24	25	26	27	24	20	16	13	10
NE	7	5	3	2	0	0	7	17	26	31	33	31	30	29	29	29	29	28	25	22	18	15	12	9
E	7	5	4	2	1	1	8	21	33	42	47	47	44	40	37	35	33	31	28	24	20	16	13	10
SE	8	5	4	2	1	0	4	12	22	32	39	44	46	44	41	38	35	32	29	24	20	16	13	10
S	8	6	4	2	1	0	0	1	4	9	16	24	31	38	41	42	40	36	31	26	22	17	14	11
SW	12	9	6	4	2	1	1	2	4	6	9	14	21	30	40	49	55	57	54	45	36	28	21	16
W	14	10	7	5	3	1	1	2	4	6	9	13	17	24	34	45	56	63	63	54	43	33	25	19
NW	12	8	6	4	2	1	0	2	3	6	9	13	16	20	25	32	40	48	50	44	35	27	21	16

Wall Number 4

Wall Face	1	2	3	4	5	6	7	8	9	10	11	12	13	14	15	16	17	18	19	20	21	22	23	24
N	11	8	6	4	2	0	0	1	3	5	7	10	13	16	19	22	24	26	27	27	26	22	19	15
NE	10	7	5	3	2	0	0	4	12	21	29	32	33	32	31	30	30	29	28	26	23	20	16	13
E	10	8	5	4	2	1	1	5	15	27	38	45	49	47	44	40	37	34	32	29	25	21	17	14
SE	11	8	6	4	2	1	0	2	8	17	27	36	43	46	46	44	41	37	34	30	26	22	18	14
S	11	8	6	4	2	1	0	−1	0	2	6	13	20	28	35	41	43	42	39	35	30	24	19	15
SW	18	13	9	6	3	2	0	0	0	2	5	8	12	18	27	36	46	53	57	57	51	42	33	25
W	21	15	10	7	4	2	1	0	1	2	5	8	11	15	21	30	40	51	60	64	60	50	40	30
NW	18	13	9	6	3	1	0	0	0	2	4	8	11	15	19	23	30	37	45	49	48	41	33	25

Wall Number 5

Wall Face	1	2	3	4	5	6	7	8	9	10	11	12	13	14	15	16	17	18	19	20	21	22	23	24
N	13	10	8	6	5	3	2	3	5	6	8	9	12	14	17	19	21	23	24	24	23	21	18	15
NE	13	10	8	6	5	3	3	7	14	20	25	27	28	28	28	28	28	28	27	26	23	21	18	15
E	14	11	9	7	5	4	4	8	17	26	33	39	40	40	38	37	35	34	32	29	26	23	20	17
SE	14	12	9	7	5	4	3	6	11	18	25	32	37	39	39	38	37	35	33	30	27	24	20	17
S	15	12	9	7	5	4	3	2	3	4	8	13	19	25	31	35	36	36	34	32	28	24	21	18
SW	22	18	14	11	8	6	5	4	4	5	6	9	12	17	25	33	40	46	49	48	44	38	32	26
W	25	20	16	13	10	7	5	4	4	5	7	9	11	15	20	28	37	45	52	54	50	44	37	30
NW	21	17	13	10	8	6	4	3	4	4	6	8	11	14	17	21	27	34	40	42	40	35	30	25

Wall Number 6

Wall Face	1	2	3	4	5	6	7	8	9	10	11	12	13	14	15	16	17	18	19	20	21	22	23	24
N	13	11	9	8	6	5	4	5	6	7	8	10	12	14	16	18	20	21	22	23	21	20	17	15
NE	14	12	10	8	6	5	6	10	15	20	23	25	25	26	26	27	27	27	26	25	23	21	18	16
E	16	13	11	9	7	6	7	11	18	25	31	35	36	36	35	35	34	33	31	29	26	24	21	18
SE	16	14	11	9	8	6	6	8	13	18	24	29	33	35	36	35	34	33	32	29	27	24	21	18
S	16	13	11	9	7	6	5	4	4	6	9	13	18	24	28	31	33	33	31	29	27	24	21	18
SW	23	19	16	14	11	9	7	6	6	7	8	10	13	18	24	31	37	42	44	43	40	35	31	27
W	26	22	18	15	13	10	8	7	7	7	8	10	12	15	20	27	35	42	47	48	45	40	35	30
NW	21	18	15	12	10	8	7	6	6	6	8	9	11	14	16	21	26	32	36	38	36	32	28	25

Table 32 July Cooling Load Temperature Differences for Calculating Cooling Load from Sunlit Walls 40° North Latitude (*Continued*)

Wall Number 7

Wall Face	1	2	3	4	5	6	7	8	9	10	11	12	13	14	15	16	17	18	19	20	21	22	23	24
N	13	12	10	9	7	6	6	7	8	8	9	11	12	14	16	17	18	19	20	20	19	18	16	15
NE	15	13	11	10	9	8	9	13	17	20	22	23	23	24	24	25	25	25	24	23	22	20	18	16
E	17	15	13	12	10	9	11	16	21	26	30	32	32	32	32	32	31	30	29	27	25	23	21	19
SE	17	15	13	12	10	9	9	12	16	21	25	28	31	32	32	32	31	30	29	27	25	23	21	19
S	16	14	13	11	10	8	7	7	7	9	12	15	19	23	26	28	29	29	28	26	24	22	20	18
SW	23	20	18	16	13	12	10	10	10	10	11	12	15	20	25	30	35	38	39	37	34	31	28	25
W	25	22	20	17	15	13	12	11	11	11	12	13	14	17	22	28	34	39	42	41	38	34	31	28
NW	20	18	16	14	12	10	9	9	9	9	10	11	13	15	17	21	26	30	33	33	30	28	25	23

Wall Number 9

Wall Face	1	2	3	4	5	6	7	8	9	10	11	12	13	14	15	16	17	18	19	20	21	22	23	24
N	17	15	13	11	9	7	5	4	4	4	5	7	8	10	12	15	17	19	21	22	23	23	22	20
NE	18	15	13	11	9	7	5	5	6	10	16	20	23	25	26	27	27	28	28	27	26	25	23	20
E	20	17	14	12	10	8	6	5	7	12	19	26	32	36	37	37	37	36	34	33	31	29	26	23
SE	20	17	15	12	10	8	6	5	6	9	13	19	25	31	34	36	37	36	35	34	32	29	26	23
S	21	18	15	12	10	8	6	5	4	3	4	6	10	14	20	25	29	33	34	34	32	30	27	24
SW	31	26	22	18	15	12	9	7	6	5	5	6	8	10	14	19	26	33	39	43	45	44	40	36
W	35	30	25	21	17	14	11	8	7	6	6	7	8	10	12	16	22	30	37	44	48	48	45	41
NW	29	25	21	17	14	11	9	7	5	5	5	6	7	9	11	14	18	22	28	34	37	38	36	33

Wall Number 10

Wall Face	1	2	3	4	5	6	7	8	9	10	11	12	13	14	15	16	17	18	19	20	21	22	23	24
N	17	15	13	11	9	7	6	5	5	5	6	7	8	10	12	14	17	18	20	22	22	22	21	19
NE	18	16	13	11	9	7	6	6	8	12	16	20	22	24	25	26	27	27	27	27	26	24	22	20
E	20	17	15	12	10	8	7	7	10	14	20	26	31	34	35	36	36	35	34	33	31	28	26	23
SE	21	18	15	13	10	8	7	6	7	10	15	20	25	30	33	34	35	35	34	33	31	29	26	23
S	21	18	15	13	11	9	7	5	4	4	5	7	11	15	20	24	28	31	32	32	31	29	26	24
SW	31	27	23	19	16	13	10	8	7	6	6	7	8	11	15	20	26	32	38	41	42	41	38	35
W	34	30	26	22	18	15	12	9	8	7	7	7	8	10	13	17	23	30	37	42	45	45	42	39
NW	28	24	21	18	15	12	10	8	6	6	6	6	8	10	12	14	18	23	28	33	35	36	34	31

Wall Number 11

Wall Face	1	2	3	4	5	6	7	8	9	10	11	12	13	14	15	16	17	18	19	20	21	22	23	24
N	16	14	13	12	10	9	8	7	7	7	8	9	10	11	12	14	15	17	18	19	20	19	18	17
NE	18	17	15	13	12	10	9	9	11	14	17	20	21	22	23	23	24	24	25	25	24	23	21	20
E	21	19	17	16	14	12	11	11	13	17	22	26	29	30	31	31	31	31	31	30	29	27	25	23
SE	21	19	17	16	14	12	11	10	11	14	17	21	24	27	29	30	31	31	30	30	29	27	25	23
S	20	18	16	15	13	11	10	9	8	8	8	10	13	16	19	23	25	27	28	28	27	25	24	22
SW	28	25	23	20	18	16	14	12	11	11	10	11	12	14	17	21	25	30	33	36	36	35	33	30
W	31	28	25	22	20	18	16	14	12	12	11	12	12	13	15	19	23	28	33	37	39	38	36	33
NW	25	23	20	18	16	14	12	11	10	9	9	10	11	12	13	15	18	22	26	29	31	31	29	27

Wall Number 12

Wall Face	1	2	3	4	5	6	7	8	9	10	11	12	13	14	15	16	17	18	19	20	21	22	23	24
N	16	14	13	12	11	10	8	8	8	8	8	9	10	11	12	14	15	16	17	18	19	19	18	17
NE	18	17	15	14	13	11	10	10	12	14	17	19	21	21	22	23	23	24	24	24	23	22	21	20
E	22	20	18	17	15	13	12	12	14	17	21	25	28	29	30	30	30	30	30	29	28	27	25	24
SE	22	20	18	16	15	13	12	11	12	14	17	21	24	26	28	29	30	30	30	29	28	27	25	23
S	20	19	17	15	14	12	11	10	9	9	9	11	13	16	19	22	24	26	26	26	26	25	23	22
SW	27	25	23	21	19	17	15	14	12	12	12	12	12	14	17	20	24	28	32	34	34	34	32	30
W	30	28	25	23	21	19	17	15	14	13	13	13	13	14	16	19	23	27	32	35	37	36	35	33
NW	24	22	20	19	17	15	13	12	11	10	10	11	11	12	13	15	18	21	25	28	29	29	28	26

Table 32 July Cooling Load Temperature Differences for Calculating Cooling Load from Sunlit Walls 40° North Latitude (*Concluded*)

Wall Number 13

Wall Face	1	2	3	4	5	6	7	8	9	10	11	12	13	14	15	16	17	18	19	20	21	22	23	24
N	15	14	13	12	11	10	9	9	9	9	9	10	10	11	12	14	15	16	17	18	18	18	17	16
NE	18	17	16	15	13	12	11	12	13	16	18	19	20	21	21	22	23	23	23	23	23	22	21	20
E	22	20	19	17	16	15	14	14	16	19	22	25	27	28	29	29	29	29	29	28	27	26	25	23
SE	22	20	19	17	16	14	13	13	14	16	18	21	24	26	27	28	28	28	28	28	27	26	24	23
S	20	18	17	16	14	13	12	11	10	10	11	12	14	16	19	21	23	24	25	25	24	23	22	21
SW	26	25	23	21	19	18	16	15	14	13	13	13	14	15	18	21	24	28	30	32	32	31	30	28
W	29	27	25	23	21	19	18	16	15	15	14	14	15	15	17	20	23	27	31	34	34	34	32	31
NW	23	22	20	18	17	15	14	13	12	12	12	12	12	13	14	16	18	21	24	26	27	27	26	25

Wall Number 14

Wall Face	1	2	3	4	5	6	7	8	9	10	11	12	13	14	15	16	17	18	19	20	21	22	23	24
N	15	15	14	13	12	11	10	10	10	10	10	10	10	11	12	13	14	15	15	16	17	17	16	16
NE	19	18	17	16	15	14	13	13	14	15	17	18	19	20	20	21	21	22	22	22	22	22	21	20
E	23	22	21	19	18	17	16	15	16	18	21	23	25	26	27	27	28	28	28	28	27	26	25	24
SE	23	21	20	19	18	16	15	15	15	16	18	20	22	24	25	26	27	27	27	27	26	26	25	24
S	20	19	18	17	16	15	14	13	12	12	12	12	14	15	17	19	21	22	23	23	23	23	22	21
SW	26	25	24	22	21	19	18	17	16	15	15	15	15	16	17	19	22	25	27	29	30	30	29	28
W	29	27	26	24	23	21	20	18	17	16	16	16	16	16	17	19	21	24	27	30	32	32	31	30
NW	23	22	21	19	18	17	16	15	14	13	13	13	13	14	14	15	17	19	21	24	25	25	25	24

Wall Number 15

Wall Face	1	2	3	4	5	6	7	8	9	10	11	12	13	14	15	16	17	18	19	20	21	22	23	24
N	19	18	16	14	12	10	9	7	6	6	6	6	7	8	9	11	13	15	17	19	20	21	21	20
NE	21	19	17	15	13	11	9	8	7	9	11	14	18	20	22	23	25	25	26	26	26	26	25	23
E	25	22	20	17	15	12	10	9	9	10	14	18	23	27	30	32	34	34	34	33	32	31	29	27
SE	25	22	20	17	15	13	11	9	8	8	10	14	18	22	26	30	32	33	34	33	33	31	30	27
S	25	22	20	17	15	13	11	9	7	6	6	6	7	10	13	17	21	25	28	30	30	30	29	27
SW	35	32	28	25	22	18	16	13	11	9	8	8	8	9	11	14	18	23	28	33	37	39	39	37
W	39	35	32	28	24	21	18	15	12	10	9	8	8	9	10	13	16	21	26	32	38	41	42	41
NW	31	28	26	23	20	17	14	12	10	8	7	7	7	8	9	11	13	16	20	25	29	32	33	33

Wall Number 16

Wall Face	1	2	3	4	5	6	7	8	9	10	11	12	13	14	15	16	17	18	19	20	21	22	23	24
N	18	17	16	14	13	11	10	9	8	7	7	7	8	9	10	11	13	14	16	17	18	19	19	19
NE	21	20	18	16	14	13	11	10	10	11	13	15	17	19	21	22	23	24	24	25	25	24	24	23
E	25	23	21	19	17	15	13	11	11	12	15	19	22	26	28	30	31	31	32	32	31	30	29	27
SE	25	23	21	19	17	15	13	11	10	11	12	15	18	21	25	27	29	30	31	31	31	30	29	27
S	24	22	20	18	16	14	12	11	9	8	8	8	9	11	14	17	20	23	25	27	27	27	27	25
SW	33	30	28	25	23	20	18	15	13	12	11	10	10	11	12	15	18	22	27	30	33	35	35	34
W	36	33	31	28	25	22	20	17	15	13	12	11	11	11	12	14	17	20	25	30	34	37	38	37
NW	29	27	25	23	20	18	16	14	12	11	10	9	9	10	11	12	14	16	19	23	27	29	30	30

Note 1. Direct application of data
- Dark surface
- Indoor temperature of 78 °F
- Outdoor maximum temperature of 95 °F with mean temperature of 85 °F and daily range of 21 °F
- Solar radiation typical of clear day on 21st day of month
- Outside surface film resistance of 0.333 (h · ft² · °F)/Btu
- Inside surface resistance of 0.685 (h · ft² · °F)/Btu

Note 2. Adjustments to table data
- Design temperatures

$$\text{Corr. CLTD} = \text{CLTD} + (78 - t_r) + (t_m - 85)$$

where

t_r = inside temperature and

t_m = maximum outdoor temperature − (daily range)/2

- No adjustment recommended for color

Table 33A   Wall Types, Mass Located Inside Insulation, for Use with Table 32

Secondary Material	R-Factor ft²·°F/Btu	A1	A2	B7	B10	B9	C1	C2	C3	C4	C5	C6	C7	C8	C17	C18
Stucco and/or plaster	0.0 to 2.0	*	*	*	*	*	*	*	*	*	*	*	*	*	*	*
	2.0 to 2.5	*	5	*	*	*	*	*	*	*	5	*	*	*	*	*
	2.5 to 3.0	*	5	*	*	*	3	*	2	5	6	*	*	5	*	*
	3.0 to 3.5	*	5	*	*	*	4	2	2	5	6	*	*	6	*	*
	3.5 to 4.0	*	5	*	*	*	4	2	3	6	6	10	4	6	*	5
	4.0 to 4.75	*	6	*	*	*	5	2	4	6	6	11	5	10	*	10
	4.75 to 5.5	*	6	*	*	*	5	2	4	6	6	11	5	10	*	10
	5.5 to 6.5	*	6	*	*	*	5	2	5	10	7	12	5	11	*	10
	6.5 to 7.75	*	6	*	*	*	5	4	5	11	7	16	10	11	*	11
	7.75 to 9.0	*	6	*	*	*	5	4	5	11	7	*	10	11	*	11
	9.0 to 10.75	*	6	*	*	*	5	4	5	11	7	*	10	11	4	11
	10.75 to 12.75	*	6	*	*	*	5	4	5	11	11	*	10	11	4	11
	12.75 to 15.0	*	10	*	*	*	10	4	5	11	11	*	10	11	9	12
	15.0 to 17.5	*	10	*	*	*	10	5	5	11	11	*	11	12	10	16
	17.5 to 20.0	*	11	*	*	*	10	5	9	11	11	*	15	16	10	16
	20.0 to 23.0	*	11	*	*	*	10	9	9	16	11	*	15	16	10	16
	23.0 to 27.0	*	*	*	*	*	*	*	*	*	*	*	16	*	15	*
Steel or other light-weight siding	0.0 to 2.0	*	*	*	*	*	*	*	*	*	*	*	*	*	*	*
	2.0 to 2.5	*	3	*	*	*	*	*	2	3	5	*	*	*	*	*
	2.5 to 3.0	*	5	*	*	*	2	*	2	5	3	*	*	5	*	*
	3.0 to 3.5	*	5	*	*	*	3	1	2	5	5	*	*	5	*	*
	3.5 to 4.0	*	5	*	*	*	3	2	2	5	5	6	3	5	*	5
	4.0 to 4.75	*	6	*	*	*	4	2	2	5	5	10	4	6	*	5
	4.75 to 5.5	*	6	*	*	*	5	2	2	6	6	11	5	6	*	6
	5.5 to 6.5	*	6	*	*	*	5	2	3	6	6	11	5	6	*	6
	6.5 to 7.75	*	6	*	*	*	5	2	3	6	6	11	5	6	*	10
	7.75 to 9.0	*	6	*	*	*	5	2	3	6	6	12	5	6	*	11
	9.0 to 10.75	*	6	*	*	*	5	2	3	6	6	12	5	6	4	11
	10.75 to 12.75	*	6	*	*	*	5	2	3	6	7	12	6	11	4	11
	12.75 to 15.0	*	6	*	*	*	5	2	4	6	7	12	10	11	5	11
	15.0 to 17.5	*	10	*	*	*	6	4	4	10	7	*	10	11	9	11
	17.5 to 20.0	*	10	*	*	*	10	4	4	10	11	*	10	11	10	11
	20.0 to 23.0	*	11	*	*	*	10	4	5	11	11	*	10	11	10	16
	23.0 to 27.0	*	*	*	*	*	*	*	*	*	*	*	10	*	11	16
Face brick	0.0 to 2.0	*	*	*	*	*	*	*	*	*	*	*	*	*	*	*
	2.0 to 2.5	3	*	*	*	*	*	*	*	*	11	*	*	*	*	*
	2.5 to 3.0	5	11	*	*	*	*	*	6	11	12	*	*	*	*	*
	3.0 to 3.5	5	12	5	*	*	11	*	11	12	12	*	*	12	*	*
	3.5 to 4.0	5	12	6	*	*	12	6	12	12	13	*	*	12	*	*
	4.0 to 4.75	6	13	6	10	*	13	10	12	12	13	*	11	*	*	16
	4.75 to 5.5	6	13	6	11	*	*	11	12	13	13	*	16	*	*	*
	5.5 to 6.5	6	13	6	11	*	*	11	12	13	13	*	*	*	*	*
	6.5 to 7.75	6	13	6	11	*	*	11	13	*	13	*	*	*	*	*
	7.75 to 9.0	6	13	10	16	*	*	11	13	*	13	*	*	*	*	*
	9.0 to 10.75	6	14	10	16	*	*	11	13	*	14	*	*	*	16	*
	10.75 to 12.75	6	14	10	16	*	*	11	13	*	14	*	*	*	16	*
	12.75 to 15.0	6	*	11	16	*	*	12	13	*	*	*	*	*	*	*
	15.0 to 17.5	10	*	11	*	*	*	12	13	*	*	*	*	*	*	*
	17.5 to 20.0	10	*	11	*	*	*	16	*	*	*	*	*	*	*	*
	20.0 to 23.0	11	*	15	*	*	*	16	*	*	*	*	*	*	*	*
	23.0 to 27.0	*	*	*	*	*	*	16	*	*	*	*	*	*	*	*

*Denotes a wall that is not possible with the chosen set of parameters.
**See Table 11 for definition of Code letters.

Table 33B Wall Types, Mass Evenly Distributed, for Use with Table 32

Secondary Material	R-Factor ft²·°F/Btu	Principal Wall Material**														
		A1	A2	B7	B10	B9	C1	C2	C3	C4	C5	C6	C7	C8	C17	C18
Stucco and/or plaster	0.0 to 2.0	1	3	*	*	*	*	*	1	3	3	*	*	*	*	*
	2.0 to 2.5	1	3	1	*	*	2	*	2	4	4	*	*	5	*	*
	2.5 to 3.0	1	4	1	*	*	2	2	2	4	4	*	*	5	*	*
	3.0 to 3.5	1	*	1	*	*	2	2	*	*	*	10	4	5	*	4
	3.5 to 4.0	1	*	1	2	*	*	4	*	*	*	10	4	*	*	4
	4.0 to 4.75	1	*	1	2	*	*	*	*	*	*	10	4	*	*	4
	4.75 to 5.5	1	*	1	2	*	*	*	*	*	*	*	*	*	*	*
	5.5 to 6.5	1	*	2	4	10	*	*	*	*	*	*	*	*	*	*
	6.5 to 7.75	1	*	2	4	11	*	*	*	*	*	*	*	*	*	*
	7.75 to 9.0	1	*	2	4	16	*	*	*	*	*	*	*	*	*	*
	9.0 to 10.75	1	*	2	4	16	*	*	*	*	*	*	*	*	4	*
	10.75 to 12.75	1	*	2	5	*	*	*	*	*	*	*	*	*	4	*
	12.75 to 15.0	2	*	2	5	*	*	*	*	*	*	*	*	*	*	*
	15.0 to 17.5	2	*	2	5	*	*	*	*	*	*	*	*	*	*	*
	17.5 to 20.0	2	*	2	9	*	*	*	*	*	*	*	*	*	*	*
	20.0 to 23.0	2	*	4	9	*	*	*	*	*	*	*	*	*	*	*
	23.0 to 27.0	*	*	*	9	*	*	*	*	*	*	*	*	*	*	*
Steel or other light-weight siding	0.0 to 2.0	1	3	*	*	*	*	*	1	3	2	*	*	*	*	*
	2.0 to 2.5	1	3	1	*	*	2	*	1	3	2	*	*	3	*	*
	2.5 to 3.0	1	4	1	*	*	2	1	2	4	4	*	*	3	*	*
	3.0 to 3.5	1	*	1	*	*	4	1	*	*	*	5	2	4	*	4
	3.5 to 4.0	1	*	1	2	*	*	2	*	*	*	5	2	*	*	4
	4.0 to 4.75	1	*	1	2	*	*	*	*	*	*	10	4	*	*	4
	4.75 to 5.5	1	*	1	2	*	*	*	*	*	*	*	*	*	*	*
	5.5 to 6.5	1	*	1	2	10	*	*	*	*	*	*	*	*	*	*
	6.5 to 7.75	1	*	1	4	11	*	*	*	*	*	*	*	*	*	*
	7.75 to 9.0	1	*	2	4	16	*	*	*	*	*	*	*	*	*	*
	9.0 to 10.75	1	*	2	4	16	*	*	*	*	*	*	*	*	2	*
	10.75 to 12.75	1	*	2	4	*	*	*	*	*	*	*	*	*	4	*
	12.75 to 15.0	1	*	2	5	*	*	*	*	*	*	*	*	*	*	*
	15.0 to 17.5	1	*	2	5	*	*	*	*	*	*	*	*	*	*	*
	17.5 to 20.0	1	*	2	5	*	*	*	*	*	*	*	*	*	*	*
	20.0 to 23.0	2	*	4	9	*	*	*	*	*	*	*	*	*	*	*
	23.0 to 27.0	*	*	*	9	*	*	*	*	*	*	*	*	*	*	*
Face brick	0.0 to 2.0	3	6	*	*	*	*	*	*	*	6	*	*	*	*	*
	2.0 to 2.5	3	10	*	*	*	*	*	5	10	10	*	*	*	*	*
	2.5 to 3.0	4	10	5	*	*	5	*	5	10	11	*	*	10	*	*
	3.0 to 3.5	*	11	5	*	*	10	5	5	11	11	15	10	10	*	10
	3.5 to 4.0	*	11	5	10	*	10	5	5	11	11	16	10	16	*	10
	4.0 to 4.75	*	11	*	11	*	10	5	5	16	11	*	10	16	*	16
	4.75 to 5.5	*	11	*	11	*	10	5	10	16	16	*	10	16	*	16
	5.5 to 6.5	*	16	*	*	*	10	9	10	16	11	*	11	16	*	16
	6.5 to 7.75	*	16	*	*	*	11	9	10	16	16	*	16	16	*	*
	7.75 to 9.0	*	16	*	*	*	15	9	10	16	*	*	15	16	*	*
	9.0 to 10.75	*	16	*	*	*	15	10	10	*	16	*	16	*	10	*
	10.75 to 12.75	*	16	*	*	*	16	10	10	*	*	*	16	*	15	*
	12.75 to 15.0	*	16	*	*	*	16	10	10	*	16	*	*	*	15	*
	15.0 to 17.5	*	*	*	*	*	16	10	15	*	*	*	*	*	16	*
	17.5 to 20.0	*	*	*	*	*	16	15	15	*	*	*	*	*	16	*
	20.0 to 23.0	*	*	*	*	*	*	15	16	*	*	*	*	*	*	*
	23.0 to 27.0	*	*	*	*	*	*	15	*	*	*	*	*	*	*	*

*Denotes a wall that is not possible with the chosen set of parameters.
**See Table 11 for definition of Code letters.

Table 33C Wall Types, Mass Located Outside Insulation, for Use with Table 32

Secondary Material	R-Factor ft²·°F/Btu	Principal Wall Material**														
		A1	A2	B7	B10	B9	C1	C2	C3	C4	C5	C6	C7	C8	C17	C18
Stucco and/or plaster	0.0 to 2.0	*	*	*	*	*	*	*	*	*	*	*	*	*	*	*
	2.0 to 2.5	*	3	*	*	*	*	*	2	3	5	*	*	*	*	*
	2.5 to 3.0	*	3	*	*	*	2	*	2	4	5	*	*	5	*	*
	3.0 to 3.5	*	3	*	*	*	2	2	2	5	5	*	*	5	*	*
	3.5 to 4.0	*	3	*	*	*	2	2	2	5	5	10	4	6	*	5
	4.0 to 4.75	*	4	*	*	*	4	2	2	5	5	10	4	6	*	9
	4.75 to 5.5	*	4	*	*	*	4	2	2	5	6	11	5	10	*	10
	5.5 to 6.5	*	5	*	*	*	4	2	2	5	6	11	5	10	*	10
	6.5 to 7.75	*	5	*	*	*	4	2	2	5	6	11	5	10	*	10
	7.75 to 9.0	*	5	*	*	*	5	2	4	5	6	16	10	10	*	10
	9.0 to 10.75	*	5	*	*	*	5	4	4	5	6	16	10	10	4	11
	10.75 to 12.75	*	5	*	*	*	5	4	4	10	6	16	10	10	9	11
	12.75 to 15.0	*	5	*	*	*	5	4	4	10	10	*	10	11	9	11
	15.0 to 17.5	*	5	*	*	*	5	4	4	10	10	*	10	11	10	16
	17.5 to 20.0	*	5	*	*	*	9	4	4	10	10	*	10	15	10	16
	20.0 to 23.0	*	9	*	*	*	9	9	9	15	10	*	10	15	15	16
	23.0 to 27.0	*	*	*	*	*	*	*	*	*	*	*	15	*	15	16
Steel or other light-weight siding	0.0 to 2.0	*	*	*	*	*	*	*	*	*	*	*	*	*	*	*
	2.0 to 2.5	*	3	*	*	*	*	*	2	3	2	*	*	*	*	*
	2.5 to 3.0	*	3	*	*	*	2	*	2	3	2	*	*	*	*	*
	3.0 to 3.5	*	3	*	*	*	2	1	2	4	3	*	*	4	*	*
	3.5 to 4.0	*	3	*	*	*	2	2	2	4	3	5	2	5	*	4
	4.0 to 4.75	*	3	*	*	*	2	2	2	4	3	10	3	5	*	5
	4.75 to 5.5	*	3	*	*	*	2	2	2	5	3	10	4	5	*	5
	5.5 to 6.5	*	4	*	*	*	2	2	2	5	3	10	4	5	*	5
	6.5 to 7.75	*	4	*	*	*	2	2	2	5	4	11	5	5	*	6
	7.75 to 9.0	*	5	*	*	*	2	2	2	5	4	11	5	5	*	6
	9.0 to 10.75	*	5	*	*	*	2	2	2	5	4	11	5	5	4	10
	10.75 to 12.75	*	5	*	*	*	4	2	2	5	5	11	5	5	4	10
	12.75 to 15.0	*	5	*	*	*	4	2	2	5	5	11	5	10	5	10
	15.0 to 17.5	*	5	*	*	*	4	2	4	5	5	16	9	10	9	10
	17.5 to 20.0	*	5	*	*	*	4	4	4	9	5	16	9	10	10	10
	20.0 to 23.0	*	9	*	*	*	4	4	4	9	9	16	10	10	10	11
	23.0 to 27.0	*	*	*	*	*	*	*	*	*	*	*	16	10	*	10	15
Face brick	0.0 to 2.0	*	*	*	*	*	*	*	*	*	*	*	*	*	*	*
	2.0 to 2.5	3	*	*	*	*	*	*	*	*	11	*	*	*	*	*
	2.5 to 3.0	3	10	*	*	*	*	*	5	10	11	*	*	*	*	*
	3.0 to 3.5	3	11	5	*	*	10	*	5	11	11	*	*	11	*	*
	3.5 to 4.0	3	11	5	*	*	10	5	6	11	11	*	*	11	*	*
	4.0 to 4.75	3	11	5	10	*	10	5	10	11	11	*	10	11	*	16
	4.75 to 5.5	3	12	5	10	*	10	9	10	11	12	*	11	16	*	16
	5.5 to 6.5	4	12	5	10	*	10	10	10	12	12	*	15	16	*	16
	6.5 to 7.75	4	12	5	10	*	11	10	10	12	12	*	16	*	*	16
	7.75 to 9.0	5	12	5	15	*	11	10	10	16	12	*	16	*	*	*
	9.0 to 10.75	5	12	9	15	*	11	10	10	16	12	*	16	*	15	*
	10.75 to 12.75	5	12	10	15	*	11	10	10	*	12	*	16	*	15	*
	12.75 to 15.0	5	*	10	16	*	11	10	11	*	*	*	16	*	15	*
	15.0 to 17.5	5	*	10	16	*	15	10	11	*	*	*	16	*	*	*
	17.5 to 20.0	5	*	10	16	*	16	15	15	*	*	*	*	*	*	*
	20.0 to 23.0	9	*	15	16	*	16	15	15	*	*	*	*	*	*	*
	23.0 to 27.0	*	*	*	*	*	15	*	*	*	*	*	*	*	*	*

*Denotes a wall that is not possible with the chosen set of parameters.
**See Table 11 for definition of Code letters.

Example 7. Cooling load from south and west glass. Determine the cooling load caused by glass on the south and west walls of a building at 1200, 1400, and 1600 h in July. The building is located at 40 °N latitude with outside design conditions of 90 °F dry-bulb temperature and a 20 °F daily range. The inside design dry bulb temperature is 78 °F. Assume the room configuration includes two exposed walls, vinyl floor covering, and gypsum partitions, and that the building is a single story. The south glass is insulating type (0.25 in. air space) with an area of 100 ft^2 and no interior shading. The west glass is 7/32 in. single grey-tinted glass with an area of 100 ft^2 and with light-colored venetian blinds.

Solution: By the room configuration described and with inside shading for half the exposed glass area, Table 35B indicates the SCL factors should be selected for a Zone C condition.

Data required for the calculation are as follows:

Variable		South Glass	West Glass
U, Btu/(h·ft^2·°F)*		0.61	0.81
Area A, ft^2		100	100
SC (Chapter 27)		0.82	0.53
SCL (Table 36, Zone C)	1200	79	37
	1400	80	98
	1600	40	153

*U-factors based on previous edition of this Handbook. See Table 5, Chapter 27 for current values.

Table 34 Cooling Load Temperature Differences (CLTD) for Conduction through Glass

Solar Time, h	CLTD, °F	Solar Time, h	CLTD, °F
0100	1	1300	12
0200	0	1400	13
0300	−1	1500	14
0400	−2	1600	14
0500	−2	1700	13
0600	−2	1800	12
0700	−2	1900	10
0800	0	2000	8
0900	2	2100	6
1000	4	2200	4
1100	7	2300	3
1200	9	2400	2

Corrections: The values in the table were calculated for an inside temperature of 78 °F and an outdoor maximum temperature of 95 °F with an outdoor daily range of 21 °F. The table remains approximately correct for other outdoor maximums 93 to 102 °F and other outdoor daily ranges 16 to 34 °F, provided the outdoor daily average temperature remains approximately 85 °F. If the room air temperature is different from 78 °F and/or the outdoor daily average is different from 85 °F, see note 2, Table 32.

The conduction heat gain component of cooling load by Equation (45) is:

Time	CLTD (Table 34)	CLTD Correction	South Glass, Btu/h	West Glass, Btu/h
1200	9	4	244	324
1400	13	8	488	648
1600	14	9	549	729

The correction factor applied to the above CLTDs was −5 °F, computed from the notes of Table 34. Heat gain values are rounded.

The solar heat gain component of cooling load by Equation (46) is:

Time	South Glass SC	SCL	SHG, Btu/h	West Glass SC	SCL	SHG, Btu/h
1200	0.82	79	6478	0.53	37	1961
1400	0.82	70	5740	0.53	98	5194
1600	0.82	40	3280	0.53	153	8109

Table 35A Zone Types for Use with CLF Tables, Interior Rooms

Room Location	Zone Parameters[a] Middle Floor	Ceiling Type	Floor Covering	Zone Type People and Equipment	Lights
Single story	N/A	N/A	Carpet	C	B
	N/A	N/A	Vinyl	D	C
Top floor	2.5 in. Concrete	With	Carpet	D	C
	2.5 in. Concrete	With	Vinyl	D	D
	2.5 in. Concrete	Without	b	D	B
	1 in. Wood	b	b	D	B
Bottom floor	2.5 in. Concrete	With	Carpet	D	C
	2.5 in. Concrete	b	Vinyl	D	D
	2.5 in. Concrete	Without	Carpet	D	D
	1 in. Wood	b	Carpet	D	C
	1 in. Wood	b	Vinyl	D	D
Mid-floor	2.5 in. Concrete	N/A	Carpet	D	C
	2.5 in. Concrete	N/A	Vinyl	D	D
	1 in. Wood	N/A	b	C	B

[a] A total of 14 zone parameters is fully defined in Table 20. Those not shown in this table were selected to achieve an error band of approximately 10%.
[b] The effect of inside shade is negligible in this case.

Table 35B Zone Types for Use with SCL and CLF Tables, Single-Story Building

No. Walls	Floor Covering	Zone Parameters[a] Partition Type	Inside Shade	Zone Type Glass Solar	People and Equipment	Lights	Error Band Plus	Minus
1 or 2	Carpet	Gypsum	b	A	B	B	9	2
1 or 2	Carpet	Concrete block	b	B	C	C	9	0
1 or 2	Vinyl	Gypsum	Full	B	C	C	9	0
1 or 2	Vinyl	Gypsum	Half to None	C	C	C	16	0
1 or 2	Vinyl	Concrete block	Full	C	D	D	8	0
1 or 2	Vinyl	Concrete block	Half to None	D	D	D	10	6
3	Carpet	Gypsum	b	A	B	B	9	2
3	Carpet	Concrete block	Full	A	B	B	9	2
3	Carpet	Concrete block	Half to None	B	B	B	9	0
3	Vinyl	Gypsum	Full	B	C	C	9	0
3	Vinyl	Gypsum	Half to None	C	C	C	16	0
3	Vinyl	Concrete block	Full	B	C	C	9	0
3	Vinyl	Concrete block	Half to None	C	C	C	16	0
4	Carpet	Gypsum	b	A	B	B	6	3
4	Vinyl	Gypsum	Full	B	C	C	11	6
4	Vinyl	Gypsum	Half to None	C	C	C	19	−1

[a] A total of 14 zone parameters is fully defined in Table 20. Those not shown in this table were selected to achieve the minimum error band shown in the righthand column for Solar Cooling Load (SCL). The error band for Lights and People and Equipment is approximately 10%.

[b] The effect of inside shade is negligible in this case.

Table 36 July Solar Cooling Load For Sunlit Glass 40° North Latitude

Zone type A

Glass Face	Hour 1	2	3	4	5	6	7	8	9	10	11	Solar Time 12	13	14	15	16	17	18	19	20	21	22	23	24
N	0	0	0	0	1	25	27	28	32	35	38	40	40	39	36	31	31	36	12	6	3	1	1	0
NE	0	0	0	0	2	85	129	134	112	75	55	48	44	40	37	32	26	18	7	3	2	1	0	0
E	0	0	0	0	2	93	157	185	183	154	106	67	53	45	39	33	26	18	7	3	2	1	0	0
SE	0	0	0	0	1	47	95	131	150	150	131	97	63	49	41	34	27	18	7	3	2	1	0	0
S	0	0	0	0	0	9	17	25	41	64	85	97	96	84	63	42	31	20	8	4	2	1	0	0
SW	0	0	0	0	0	9	17	24	30	35	39	64	101	133	151	152	133	93	35	17	8	4	2	1
W	1	0	0	0	0	9	17	24	30	35	38	40	65	114	158	187	192	156	57	27	13	6	3	2
NW	1	0	0	0	0	9	17	24	30	35	38	40	40	50	84	121	143	130	46	22	11	5	3	1
Hor	0	0	0	0	0	24	69	120	169	211	241	257	259	245	217	176	125	70	29	14	7	3	2	1

Zone type B

Glass Face	Hour 1	2	3	4	5	6	7	8	9	10	11	Solar Time 12	13	14	15	16	17	18	19	20	21	22	23	24
N	2	2	1	1	1	22	23	24	28	32	35	37	38	37	35	32	31	35	16	10	7	5	4	3
NE	2	1	1	1	2	73	109	116	101	73	58	52	48	45	41	36	30	23	13	9	6	5	3	3
E	2	2	1	1	2	80	133	159	162	143	105	74	63	55	48	41	34	25	15	10	7	5	4	3
SE	2	2	1	1	1	40	81	112	131	134	122	96	69	58	49	42	35	26	15	10	8	6	4	3
S	2	2	1	1	1	8	15	21	36	56	74	86	87	79	63	46	37	27	16	11	8	6	4	3
SW	6	5	4	3	2	9	16	22	27	31	36	58	89	117	135	138	126	94	46	31	21	15	11	8
W	8	6	5	4	3	9	16	22	27	31	35	37	59	101	139	166	173	147	66	43	30	21	15	11
NW	6	5	4	3	2	9	16	22	27	31	34	37	37	46	76	108	128	119	51	33	22	16	11	8
Hor	8	6	5	4	3	22	60	104	147	185	214	233	239	232	212	180	137	90	53	37	27	19	14	11

Zone type C

Glass Face	Hour 1	2	3	4	5	6	7	8	9	10	11	Solar Time 12	13	14	15	16	17	18	19	20	21	22	23	24
N	5	5	4	4	4	24	23	24	27	30	33	34	35	34	32	29	29	34	14	10	8	7	6	6
NE	7	6	6	5	6	75	106	107	88	61	49	47	45	43	40	36	31	25	16	13	11	10	9	8
E	9	8	8	7	8	83	130	148	145	124	89	62	56	52	47	43	37	30	20	17	15	13	12	11
SE	9	8	7	6	6	45	82	107	121	121	107	82	59	51	47	42	36	29	19	16	14	13	11	10
S	7	7	6	5	5	12	18	23	36	54	70	79	79	70	54	40	33	26	16	14	13	12	10	8
SW	14	12	11	10	9	15	21	26	29	33	36	57	86	110	124	125	111	80	37	28	23	20	17	15
W	17	15	13	12	11	17	22	27	31	34	36	37	59	98	132	153	156	128	50	35	28	24	21	19
NW	12	11	10	9	8	14	20	25	29	32	34	36	36	44	73	102	118	107	39	26	21	17	15	13
Hor	24	21	19	17	16	34	68	107	144	175	199	212	215	207	189	160	123	83	53	44	38	34	30	27

Zone type D

Glass Face	Hour 1	2	3	4	5	6	7	8	9	10	11	Solar Time 12	13	14	15	16	17	18	19	20	21	22	23	24
N	8	7	6	6	6	21	21	21	24	27	29	31	32	31	30	28	29	32	17	14	12	11	10	9
NE	11	10	9	8	9	63	87	90	77	58	49	48	46	44	42	39	35	29	22	19	17	15	14	12
E	15	13	12	11	11	70	107	123	124	110	85	65	60	57	53	48	43	37	29	25	22	20	18	16
SE	14	13	11	10	10	39	68	90	102	104	95	78	60	55	51	47	42	35	27	24	21	19	17	16
S	11	10	9	8	7	12	17	21	32	46	59	67	69	63	52	41	36	30	22	19	17	15	14	12
SW	21	19	17	15	14	18	22	25	28	31	34	51	74	94	106	109	100	78	45	37	33	29	26	23
W	25	23	20	18	17	21	24	28	30	33	34	35	59	84	112	130	135	116	57	46	39	35	31	28
NW	18	16	15	13	12	17	21	24	27	30	32	33	34	41	64	87	101	94	42	33	29	25	22	20
Hor	37	33	30	27	24	38	64	95	124	150	171	185	191	188	176	156	128	96	72	63	56	50	45	41

Notes:
1. Values are in Btu/h·ft^2.
2. Apply data directly to standard double strength glass with no inside shade.
3. Data applies to 21st day of July.
4. For other types of glass and internal shade, use shading coefficients as multiplier. See text. For externally shaded glass, use north orientation. See text.

The total cooling load due to heat gain through the glass is, therefore:

Time	South Glass, Btu/h	West Glass, Btu/h
1200	6722	2285
1400	6198	5842
1600	3829	8838

HEAT SOURCES WITHIN CONDITIONED SPACE

People

The basic principles of evaluating heat gain and moisture generation from people are the same as those previously described for the TFM. Latent heat gains are considered instantaneous cooling loads.

The total sensible heat gain from people is not converted directly to cooling load. The radiant portion is first absorbed by the surroundings (floor, ceiling, partitions, furniture) then convected to the space at a later time, depending on the thermal characteristics of the room. The radiant portion of the sensible heat gain from people varies widely depending on the circumstances, as indicated by Table 3 and in more detail by Chapter 8. A 70% value was used to generate CLFs for Table 37, which considers the storage effect on this radiant load in its results, plus the 30% convective portion. The instantaneous sensible cooling load is thus:

$$q_s = N(\text{SHG}_p)(\text{CLF}_p) \tag{47}$$

and the latent cooling load is:

$$q_l = N(\text{LHG}_p) \tag{48}$$

where

q_s = sensible cooling load due to people
N = number of people
SHG_p = sensible heat gain per person (Table 3)

Table 37   Cooling Load Factors for People and Unhooded Equipment

	Number of Hours after Entry into Space or Equipment Turned On																							
Hours in Space	1	2	3	4	5	6	7	8	9	10	11	12	13	14	15	16	17	18	19	20	21	22	23	24
Zone Type A																								
2	0.75	0.88	0.18	0.08	0.04	0.02	0.01	0.01	0.01	0.01	0.00	0.00	0.00	0.00	0.00	0.00	0.00	0.00	0.00	0.00	0.00	0.00	0.00	0.00
4	0.75	0.88	0.93	0.95	0.22	0.10	0.05	0.03	0.02	0.02	0.01	0.01	0.01	0.01	0.00	0.00	0.00	0.00	0.00	0.00	0.00	0.00	0.00	0.00
6	0.75	0.88	0.93	0.95	0.97	0.97	0.33	0.11	0.06	0.04	0.03	0.02	0.02	0.01	0.01	0.01	0.01	0.00	0.00	0.00	0.00	0.00	0.00	0.00
8	0.75	0.88	0.93	0.95	0.97	0.97	0.98	0.98	0.24	0.11	0.06	0.04	0.03	0.02	0.02	0.01	0.01	0.01	0.01	0.01	0.00	0.00	0.00	0.00
10	0.75	0.88	0.93	0.95	0.97	0.97	0.98	0.98	0.99	0.99	0.24	0.12	0.07	0.04	0.03	0.02	0.02	0.01	0.01	0.01	0.01	0.01	0.00	0.00
12	0.75	0.88	0.93	0.96	0.97	0.98	0.98	0.98	0.99	0.99	0.99	0.99	0.25	0.12	0.07	0.04	0.03	0.02	0.02	0.02	0.01	0.01	0.01	0.01
14	0.76	0.88	0.93	0.96	0.97	0.98	0.98	0.99	0.99	0.99	0.99	0.99	1.00	1.00	0.25	0.12	0.07	0.05	0.03	0.03	0.02	0.02	0.01	0.01
16	0.76	0.89	0.94	0.96	0.97	0.98	0.98	0.99	0.99	0.99	0.99	0.99	1.00	1.00	1.00	1.00	0.25	0.12	0.07	0.05	0.03	0.03	0.02	0.02
18	0.77	0.89	0.94	0.96	0.97	0.98	0.98	0.99	0.99	0.99	0.99	1.00	1.00	1.00	1.00	1.00	1.00	1.00	0.25	0.12	0.07	0.05	0.03	0.03
Zone Type B																								
2	0.65	0.74	0.16	0.11	0.08	0.06	0.05	0.04	0.03	0.02	0.02	0.01	0.01	0.01	0.01	0.00	0.00	0.00	0.00	0.00	0.00	0.00	0.00	0.00
4	0.65	0.75	0.81	0.85	0.24	0.17	0.13	0.10	0.07	0.06	0.04	0.03	0.03	0.02	0.02	0.01	0.01	0.01	0.01	0.00	0.00	0.00	0.00	0.00
6	0.65	0.75	0.81	0.85	0.89	0.91	0.29	0.20	0.15	0.12	0.09	0.07	0.05	0.04	0.03	0.02	0.02	0.01	0.01	0.01	0.01	0.01	0.00	0.00
8	0.65	0.75	0.81	0.85	0.89	0.91	0.93	0.95	0.31	0.22	0.17	0.13	0.10	0.08	0.06	0.05	0.04	0.03	0.02	0.02	0.01	0.01	0.01	0.01
10	0.65	0.75	0.81	0.85	0.89	0.91	0.93	0.95	0.96	0.97	0.33	0.24	0.18	0.14	0.11	0.08	0.06	0.05	0.04	0.03	0.02	0.02	0.01	0.01
12	0.66	0.76	0.81	0.86	0.89	0.92	0.94	0.95	0.96	0.97	0.98	0.98	0.34	0.24	0.19	0.14	0.11	0.08	0.06	0.05	0.04	0.03	0.02	0.02
14	0.67	0.76	0.82	0.86	0.89	0.92	0.94	0.95	0.96	0.97	0.98	0.98	0.99	0.99	0.35	0.25	0.19	0.15	0.11	0.09	0.07	0.05	0.04	0.03
16	0.69	0.78	0.83	0.87	0.90	0.92	0.94	0.95	0.96	0.97	0.98	0.98	0.99	0.99	0.99	0.99	0.35	0.25	0.19	0.15	0.11	0.09	0.07	0.05
18	0.71	0.80	0.85	0.88	0.91	0.93	0.95	0.96	0.97	0.98	0.98	0.99	0.99	0.99	0.99	0.99	1.00	1.00	0.35	0.25	0.19	0.15	0.11	0.09
Zone Type C																								
2	0.60	0.68	0.14	0.11	0.09	0.07	0.06	0.05	0.04	0.03	0.03	0.02	0.02	0.01	0.01	0.01	0.01	0.01	0.01	0.00	0.00	0.00	0.00	0.00
4	0.60	0.68	0.74	0.79	0.23	0.18	0.14	0.12	0.10	0.08	0.06	0.05	0.04	0.04	0.03	0.02	0.02	0.02	0.01	0.01	0.01	0.01	0.01	0.01
6	0.61	0.69	0.74	0.79	0.83	0.86	0.28	0.22	0.18	0.15	0.12	0.10	0.08	0.07	0.06	0.05	0.04	0.03	0.03	0.02	0.02	0.01	0.01	0.01
8	0.61	0.69	0.75	0.79	0.83	0.86	0.89	0.91	0.32	0.26	0.21	0.17	0.14	0.11	0.09	0.08	0.06	0.05	0.04	0.04	0.03	0.02	0.02	0.02
10	0.62	0.70	0.75	0.80	0.83	0.86	0.89	0.91	0.92	0.94	0.35	0.28	0.23	0.18	0.15	0.12	0.10	0.08	0.07	0.06	0.05	0.04	0.03	0.03
12	0.63	0.71	0.76	0.81	0.84	0.87	0.89	0.91	0.93	0.94	0.95	0.96	0.37	0.29	0.24	0.19	0.16	0.13	0.11	0.09	0.07	0.06	0.05	0.04
14	0.65	0.72	0.77	0.82	0.85	0.88	0.90	0.92	0.93	0.94	0.95	0.96	0.97	0.97	0.38	0.30	0.25	0.20	0.17	0.14	0.11	0.09	0.08	0.06
16	0.68	0.74	0.79	0.83	0.86	0.89	0.91	0.92	0.94	0.95	0.96	0.96	0.97	0.98	0.98	0.98	0.39	0.31	0.25	0.21	0.17	0.14	0.11	0.09
18	0.72	0.78	0.82	0.85	0.88	0.90	0.92	0.93	0.94	0.95	0.96	0.97	0.97	0.98	0.98	0.99	0.99	0.99	0.39	0.31	0.26	0.21	0.17	0.14
Zone Type D																								
2	0.59	0.67	0.13	0.09	0.08	0.06	0.05	0.05	0.04	0.04	0.03	0.03	0.02	0.02	0.02	0.01	0.01	0.01	0.01	0.01	0.01	0.01	0.01	0.00
4	0.60	0.67	0.72	0.76	0.20	0.16	0.13	0.11	0.10	0.08	0.07	0.06	0.05	0.05	0.04	0.03	0.03	0.03	0.02	0.02	0.02	0.01	0.01	0.01
6	0.61	0.68	0.73	0.77	0.80	0.83	0.26	0.20	0.17	0.15	0.13	0.11	0.09	0.08	0.07	0.06	0.05	0.05	0.04	0.03	0.03	0.03	0.02	0.02
8	0.62	0.69	0.74	0.77	0.80	0.83	0.85	0.87	0.30	0.24	0.20	0.17	0.15	0.13	0.11	0.10	0.08	0.07	0.06	0.05	0.05	0.04	0.04	0.03
10	0.63	0.70	0.75	0.78	0.81	0.84	0.86	0.88	0.89	0.91	0.33	0.27	0.22	0.19	0.17	0.14	0.12	0.11	0.09	0.08	0.07	0.06	0.05	0.05
12	0.65	0.71	0.76	0.79	0.82	0.84	0.87	0.88	0.90	0.91	0.92	0.93	0.35	0.29	0.24	0.21	0.18	0.16	0.13	0.12	0.10	0.09	0.08	0.07
14	0.67	0.73	0.78	0.81	0.83	0.86	0.88	0.89	0.91	0.92	0.93	0.94	0.95	0.95	0.37	0.30	0.25	0.22	0.19	0.16	0.14	0.12	0.11	0.09
16	0.70	0.76	0.80	0.83	0.85	0.87	0.89	0.90	0.92	0.93	0.94	0.95	0.95	0.96	0.96	0.97	0.38	0.31	0.26	0.23	0.20	0.17	0.15	0.13
18	0.74	0.80	0.83	0.85	0.87	0.89	0.91	0.92	0.93	0.94	0.95	0.95	0.96	0.97	0.97	0.97	0.98	0.98	0.39	0.32	0.27	0.23	0.20	0.17

Note: See Table 35 for zone type. Data based on a radiative/convective fraction of 0.70/0.30.

CLF_p = cooling load factor for people (Table 37)
q_l = latent cooling load due to people
LHG_p = latent heat gain per person (Table 3)

The CLF for people load is a function of the time such people spend in the conditioned space and the time elapsed since first entering. As defined for estimating cooling load from fenestration, the space under consideration is categorized as a zone, identified in Table 35. The appropriate CLF is selected from Table 37 by zone type, occupancy period, and number of hours after entry.

CLF Usage exceptions. If the space temperature is not maintained constant during the 24-h period, for example, if the cooling system is shut down during the night (night shutdown), a "pulldown load" results because a major part of the stored sensible heat in the structure has not been removed, thus reappearing as cooling load when the system is started the next day. In this case, a CLF of 1.0 should be used.

When there is a *high occupant density*, as in theaters and auditoriums, the quantity of radiation to the walls and room furnishings is proportionally reduced. In these situations, a CLF of 1.0 should also be used.

Example 8. Cooling load from occupants. Estimate the cooling load in a building at 1200, 1400, and 1600 h from four moderately active people occupying an office from 0900 to 1700 h. The office temperature is 78 °F, and the cooling system operates continuously. Assume the conditions of the space as applied to Table 33A; define it as Type D.

Solution: The sensible cooling load is calculated by Equation (47), and the latent cooling load is calculated by Equation (48). The period of occupancy is 8 h. Therefore,

Time	No. of People	Hours in Space	Hours after Entry	Btu/h, each (Table 3) Sen.	Lat.	CLF_p (Table 37) Zone D	Cooling Load Sen., Btu/h	Lat., Btu/h
1200	4	8	3	255	255	0.74	755	1020
1400	4	8	5	255	255	0.80	816	1020
1600	4	8	7	255	255	0.85	867	1020

Lighting

As discussed for the TFM, the cooling load from lighting does not immediately reflect the full energy output of the lights. Kimura and Stephenson (1968), Mitalas and Kimura (1971), and Mitalas (1973) indicated the effect on cooling load of light fixture type, type of air supply and return, space furnishings, and the thermal characteristics of the space. The effect of these influencing parameters have been incorporated in the *Heating and Cooling Load Calculation Manual* (McQuiston and Spitler 1992) into the CLF values for lighting listed in Table 38, and for which selection zones are identified as appropriate by Tables 35. At any time, the space cooling load from lighting can be estimated as:

$$q_{el} = HG_{el}(CLF_{el}) \qquad (49)$$

where

q_{el} = cooling load from lighting, Btu/h
HG_{el} = heat gain from lighting, Btu/h, as $WF_{ul}F_{sa}$ [Equation (9)]
W = total lamp watts
F_{ul} = lighting use factor
F_{sa} = lighting special allowance factor
CLF_{el} = lighting cooling load factor (Table 38)

CLF_{el} data in Table 38 are based on the assumptions that (1) the conditioned space temperature is continuously maintained at a constant value, and (2) the cooling load and power input to the lights eventually become equal if the lights are on for long enough.

Operational exceptions. If the cooling system operates only during occupied hours, the CLF_{el} should be considered 1.0 in lieu of the Table 38 values. Where one portion of the lights serving the space is on one schedule of operation and another portion is on a different schedule, each should be treated separately. Where lights are left on for 24 h a day, the CLF_{el} is 1.0.

Example 9. Cooling load from lighting. Estimate the cooling load in a building at 1200, 1400, and 1600 h from recessed fluorescent lights, turned on at 0800 h and turned off at 1800 h. Lamp wattage is 800 W. The use

Table 38 Cooling Load Factors for Lights

Lights On For	Number of Hours after Lights Turned On																							
	1	2	3	4	5	6	7	8	9	10	11	12	13	14	15	16	17	18	19	20	21	22	23	24
Zone Type A																								
8	0.85	0.92	0.95	0.96	0.97	0.97	0.97	0.98	0.13	0.06	0.04	0.03	0.02	0.02	0.02	0.01	0.01	0.01	0.01	0.01	0.01	0.01	0.01	0.01
10	0.85	0.93	0.95	0.97	0.97	0.97	0.98	0.98	0.98	0.98	0.14	0.07	0.04	0.03	0.02	0.02	0.02	0.02	0.02	0.02	0.01	0.01	0.01	0.01
12	0.86	0.93	0.96	0.97	0.97	0.98	0.98	0.98	0.98	0.98	0.98	0.98	0.14	0.07	0.04	0.03	0.03	0.02	0.02	0.02	0.02	0.02	0.02	0.02
14	0.86	0.93	0.96	0.97	0.98	0.98	0.98	0.98	0.98	0.98	0.99	0.99	0.99	0.99	0.15	0.07	0.05	0.03	0.03	0.03	0.02	0.02	0.02	0.02
16	0.87	0.94	0.96	0.97	0.98	0.98	0.98	0.99	0.99	0.99	0.99	0.99	0.99	0.99	0.99	0.99	0.15	0.08	0.05	0.04	0.03	0.03	0.03	0.02
Zone Type B																								
8	0.75	0.85	0.90	0.93	0.94	0.95	0.95	0.96	0.23	0.12	0.08	0.05	0.04	0.04	0.03	0.03	0.03	0.02	0.02	0.02	0.02	0.02	0.02	0.01
10	0.75	0.86	0.91	0.93	0.94	0.95	0.95	0.96	0.96	0.97	0.24	0.13	0.08	0.06	0.05	0.04	0.04	0.03	0.03	0.03	0.02	0.02	0.02	0.02
12	0.76	0.86	0.91	0.93	0.95	0.95	0.95	0.96	0.96	0.97	0.97	0.97	0.24	0.14	0.09	0.07	0.05	0.05	0.04	0.04	0.03	0.03	0.03	0.03
14	0.76	0.87	0.92	0.94	0.95	0.96	0.96	0.97	0.97	0.97	0.97	0.98	0.98	0.98	0.25	0.14	0.09	0.07	0.06	0.05	0.05	0.04	0.04	0.03
16	0.77	0.88	0.92	0.95	0.96	0.96	0.97	0.97	0.97	0.98	0.98	0.98	0.98	0.98	0.98	0.99	0.25	0.15	0.10	0.07	0.06	0.05	0.05	0.04
Zone Type C																								
8	0.72	0.80	0.84	0.87	0.88	0.89	0.90	0.91	0.23	0.15	0.11	0.09	0.08	0.07	0.07	0.06	0.05	0.05	0.05	0.04	0.04	0.03	0.03	0.03
10	0.73	0.81	0.85	0.87	0.89	0.90	0.91	0.92	0.92	0.93	0.25	0.16	0.13	0.11	0.09	0.08	0.08	0.07	0.06	0.06	0.05	0.05	0.04	0.04
12	0.74	0.82	0.86	0.88	0.90	0.91	0.92	0.92	0.93	0.94	0.94	0.95	0.26	0.18	0.14	0.12	0.10	0.09	0.08	0.08	0.07	0.06	0.06	0.05
14	0.75	0.84	0.87	0.89	0.91	0.92	0.92	0.93	0.94	0.94	0.95	0.95	0.96	0.96	0.27	0.19	0.15	0.13	0.11	0.10	0.09	0.08	0.08	0.07
16	0.77	0.85	0.89	0.91	0.92	0.93	0.93	0.94	0.95	0.95	0.95	0.96	0.96	0.97	0.97	0.97	0.28	0.20	0.16	0.13	0.12	0.11	0.10	0.09
Zone Type D																								
8	0.66	0.72	0.76	0.79	0.81	0.83	0.85	0.86	0.25	0.20	0.17	0.15	0.13	0.12	0.11	0.10	0.09	0.08	0.07	0.06	0.06	0.05	0.04	0.04
10	0.68	0.74	0.77	0.80	0.82	0.84	0.86	0.87	0.88	0.90	0.28	0.23	0.19	0.17	0.15	0.14	0.12	0.11	0.10	0.09	0.08	0.07	0.06	0.06
12	0.70	0.75	0.79	0.81	0.83	0.85	0.87	0.88	0.89	0.90	0.90	0.91	0.92	0.30	0.25	0.21	0.19	0.17	0.15	0.13	0.12	0.11	0.10	0.08
14	0.72	0.77	0.81	0.83	0.85	0.86	0.88	0.89	0.90	0.91	0.92	0.93	0.94	0.94	0.32	0.26	0.23	0.20	0.18	0.16	0.14	0.13	0.12	0.10
16	0.75	0.80	0.83	0.85	0.87	0.88	0.89	0.90	0.91	0.92	0.93	0.94	0.94	0.95	0.96	0.96	0.34	0.28	0.24	0.21	0.19	0.17	0.15	0.14

Note: See Table 35 for zone type. Data based on a radiative/convective fraction of 0.59/0.41.

factor is 1.0, and the special allowance factor is 1.25. The room is an interior type in a one-story building, has tile flooring over a 3-in. concrete floor, and a suspended ceiling. The cooling system runs 24 h/day, including weekends.

Solution: From Table 35B, the room is categorized as Type C for lighting load purposes. Therefore,

Time	Hours in Space	Hours after Entry	Lamp Watts	Heat Gain, Btu/h [Eq. (15)]	CLF$_{el}$ (Table 38) Zone C	Cooling Load, Btu/h [Eq. (42)]
1200	8	3	800	3410	0.85	2489
1400	8	5	800	3410	0.89	2660
1600	8	7	800	3410	0.91	2796

Power and Appliances

Heat gain of power-driven equipment can be estimated by means of Equations (15), (16), or (17) as applicable, or taken directly from Tables 4 and/or 5.

Equations (18) through (21) can be used to estimate heat gain values under various circumstances, and Tables 6 through 9 provide representative data for direct use or as input to the equations. Latent heat gain from appliances is considered instantaneous cooling load and is calculated by Equation (22).

The radiant component of sensible heat gain from power-driven equipment or appliances is delayed in becoming cooling load in the same manner as that of other load categories already discussed. For power-driven equipment, the CLF values tabulated for unhooded equipment (Table 37) are considered appropriate. Tables 37 and 39 tabulate cooling load factors (CLF$_a$) for appliances. Multiplying the sensible portion of heat gain by the appropriate CLF$_a$ will produce the following approximate cooling load values:

$$q = SHG(CLF) \tag{50}$$

Example 10. Appliance cooling load. Determine the cooling load in a building at 1200, 1400, and 1600 h caused by an electric coffee brewer with one brewer and one warmer. The brewer operates continuously from 0900 to 1500 h and does not have an exhaust hood. The room is a "midfloor" type in a multistory building, has carpet over a 3-in. concrete floor, and a suspended ceiling. The cooling system runs 24 h/day, including weekends.

Solution: From Table 7, q_s and q_l for an unhooded, two-burner coffee brewer is 3750 and 1910 Btu/h, respectively (thus 1875 and 955 each burner), and for a coffee heater (per warming burner) is 230 and 110 Btu/h, respectively. The brewer is on for 6 h, and 1200 h is 3 h after the brewer is turned on. From Table 35B, the room is categorized as Type D for equipment load purposes. Therefore,

Table 39 Cooling Load Factors for Hooded Equipment

Hours in Operation	Number of Hours after Equipment Turned On																							
	1	2	3	4	5	6	7	8	9	10	11	12	13	14	15	16	17	18	19	20	21	23	24
Zone Type A																							
2	0.64	0.83	0.26	0.11	0.06	0.03	0.01	0.01	0.01	0.01	0.00	0.00	0.00	0.00	0.00	0.00	0.00	0.00	0.00	0.00	0.00	0.00	0.00
4	0.64	0.83	0.90	0.93	0.31	0.14	0.07	0.04	0.03	0.03	0.01	0.01	0.01	0.01	0.00	0.00	0.00	0.00	0.00	0.00	0.00	0.00	0.00
6	0.64	0.83	0.90	0.93	0.96	0.96	0.33	0.16	0.09	0.06	0.04	0.03	0.03	0.01	0.01	0.01	0.01	0.00	0.00	0.00	0.00	0.00	0.00
8	0.64	0.83	0.90	0.93	0.96	0.96	0.97	0.97	0.34	0.16	0.09	0.06	0.04	0.03	0.03	0.01	0.01	0.01	0.01	0.01	0.00	0.00	0.00
10	0.64	0.83	0.90	0.93	0.96	0.96	0.97	0.97	0.99	0.99	0.34	0.17	0.10	0.06	0.04	0.03	0.03	0.01	0.01	0.01	0.01	0.01	0.00
12	0.64	0.83	0.90	0.94	0.96	0.97	0.97	0.97	0.99	0.99	0.99	0.99	0.36	0.17	0.10	0.06	0.04	0.03	0.03	0.03	0.01	0.01	0.01
14	0.66	0.83	0.90	0.94	0.96	0.97	0.97	0.99	0.99	0.99	0.99	0.99	1.00	1.00	0.36	0.17	0.10	0.07	0.04	0.04	0.03	0.03	0.01
16	0.66	0.84	0.91	0.94	0.96	0.97	0.97	0.99	0.99	0.99	0.99	0.99	1.00	1.00	1.00	1.00	0.36	0.17	0.10	0.07	0.04	0.04	0.03
18	0.67	0.84	0.91	0.94	0.96	0.97	0.97	0.99	0.99	0.99	0.99	1.00	1.00	1.00	1.00	1.00	1.00	1.00	0.36	0.17	0.10	0.07	0.04
Zone Type B																							
2	0.50	0.63	0.23	0.16	0.11	0.09	0.07	0.06	0.04	0.03	0.03	0.01	0.01	0.01	0.01	0.00	0.00	0.00	0.00	0.00	0.00	0.00	0.00
4	0.50	0.64	0.73	0.79	0.34	0.24	0.19	0.14	0.10	0.09	0.06	0.04	0.04	0.03	0.03	0.01	0.01	0.01	0.01	0.00	0.00	0.00	0.00
6	0.50	0.64	0.73	0.79	0.84	0.87	0.41	0.29	0.21	0.17	0.13	0.10	0.07	0.06	0.04	0.03	0.03	0.01	0.01	0.01	0.01	0.01	0.00
8	0.50	0.64	0.73	0.79	0.84	0.87	0.90	0.93	0.44	0.31	0.24	0.19	0.14	0.11	0.09	0.07	0.06	0.04	0.03	0.03	0.01	0.01	0.01
10	0.50	0.64	0.73	0.79	0.84	0.87	0.90	0.93	0.94	0.96	0.47	0.34	0.26	0.20	0.16	0.11	0.09	0.07	0.06	0.04	0.03	0.03	0.01
12	0.51	0.66	0.73	0.80	0.84	0.89	0.91	0.93	0.94	0.96	0.97	0.97	0.49	0.34	0.27	0.20	0.16	0.11	0.09	0.07	0.06	0.04	0.03
14	0.53	0.66	0.74	0.80	0.84	0.89	0.91	0.93	0.94	0.96	0.97	0.97	0.99	0.99	0.50	0.36	0.27	0.21	0.16	0.13	0.10	0.07	0.06
16	0.56	0.69	0.76	0.81	0.86	0.89	0.91	0.93	0.94	0.96	0.97	0.97	0.99	0.99	0.99	0.99	0.50	0.36	0.27	0.21	0.16	0.13	0.10
18	0.59	0.71	0.79	0.83	0.87	0.90	0.93	0.94	0.96	0.97	0.97	0.99	0.99	0.99	0.99	0.99	1.00	1.00	0.50	0.36	0.27	0.21	0.16
Zone Type C																							
2	0.43	0.54	0.20	0.16	0.13	0.10	0.09	0.07	0.06	0.04	0.04	0.03	0.03	0.01	0.01	0.01	0.01	0.01	0.01	0.00	0.00	0.00	0.00
4	0.43	0.54	0.63	0.70	0.33	0.26	0.20	0.17	0.14	0.11	0.09	0.07	0.06	0.06	0.04	0.03	0.03	0.03	0.01	0.01	0.01	0.01	0.01
6	0.44	0.56	0.63	0.70	0.76	0.80	0.40	0.31	0.26	0.21	0.17	0.14	0.11	0.10	0.09	0.07	0.06	0.04	0.04	0.03	0.03	0.01	0.01
8	0.44	0.56	0.64	0.70	0.76	0.80	0.84	0.87	0.46	0.37	0.30	0.24	0.20	0.16	0.13	0.11	0.09	0.07	0.06	0.06	0.04	0.03	0.03
10	0.46	0.57	0.64	0.71	0.76	0.80	0.84	0.87	0.89	0.91	0.50	0.40	0.33	0.26	0.21	0.17	0.14	0.11	0.09	0.07	0.06	0.04
12	0.47	0.59	0.66	0.73	0.77	0.81	0.84	0.87	0.90	0.91	0.93	0.94	0.53	0.41	0.34	0.27	0.23	0.19	0.16	0.13	0.10	0.09	0.07
14	0.50	0.60	0.67	0.74	0.79	0.83	0.86	0.89	0.90	0.91	0.93	0.94	0.96	0.96	0.54	0.43	0.36	0.29	0.24	0.20	0.16	0.13	0.11
16	0.54	0.63	0.70	0.76	0.80	0.84	0.87	0.89	0.91	0.93	0.94	0.94	0.96	0.97	0.97	0.97	0.56	0.44	0.36	0.30	0.24	0.20	0.16
18	0.60	0.69	0.74	0.79	0.83	0.86	0.89	0.90	0.91	0.93	0.94	0.96	0.96	0.97	0.97	0.99	0.99	0.99	0.56	0.44	0.37	0.30	0.24
Zone Type D																							
2	0.41	0.53	0.19	0.13	0.11	0.09	0.07	0.07	0.06	0.06	0.04	0.04	0.03	0.03	0.03	0.01	0.01	0.01	0.01	0.01	0.01	0.01	0.01
4	0.43	0.53	0.60	0.66	0.29	0.23	0.19	0.16	0.14	0.11	0.10	0.09	0.07	0.07	0.06	0.04	0.04	0.04	0.03	0.03	0.03	0.01	0.01
6	0.44	0.54	0.61	0.67	0.71	0.76	0.37	0.29	0.24	0.21	0.19	0.16	0.13	0.11	0.10	0.09	0.07	0.07	0.06	0.04	0.04	0.04	0.03
8	0.46	0.56	0.63	0.67	0.71	0.76	0.79	0.81	0.43	0.34	0.29	0.24	0.21	0.19	0.16	0.14	0.11	0.10	0.09	0.07	0.06	0.06
10	0.47	0.57	0.64	0.69	0.73	0.77	0.80	0.83	0.84	0.87	0.47	0.39	0.31	0.27	0.24	0.20	0.17	0.16	0.13	0.11	0.10	0.09	0.07
12	0.50	0.59	0.66	0.70	0.74	0.77	0.81	0.83	0.86	0.87	0.89	0.90	0.50	0.41	0.34	0.30	0.26	0.23	0.19	0.17	0.14	0.13	0.11
14	0.53	0.61	0.69	0.73	0.76	0.80	0.83	0.84	0.87	0.89	0.90	0.91	0.93	0.93	0.53	0.43	0.36	0.31	0.27	0.23	0.20	0.17	0.16
16	0.57	0.66	0.71	0.76	0.79	0.81	0.84	0.86	0.89	0.90	0.91	0.93	0.93	0.94	0.94	0.96	0.54	0.44	0.37	0.33	0.29	0.24	0.21
18	0.63	0.71	0.76	0.79	0.81	0.84	0.87	0.89	0.90	0.91	0.93	0.93	0.94	0.96	0.96	0.96	0.97	0.97	0.56	0.46	0.39	0.33	0.29

Note: See Table 35 for zone type. Data based on a radiative/convective fraction of 100/0.

26.54

1993 Fundamentals Handbook

Time	q_s Btu/h	Hours in Use	Hours after Start	CLF_a (Table 37)	Cooling Load Sensible, Btu/h	Latent, Btu/h	Total, Btu/h
1200	2105	6	3	0.73	1537	1065	2602
1400	2105	6	5	0.80	1684	1065	2749
1600	2105	6	7	0.26	547	0	547

Total Space Cooling Load

The estimated total space cooling load for a given application is determined by summing the individual components for each hour of interest.

EXAMPLE
COOLING LOAD CALCULATION
USING CLTD/CLF METHOD

Example 11. For this example, the one-story commercial building in Example 6 will be the basis for calculating a cooling load by the CLTD/SCL/CLF method. Refer to Example 6 for the statement of conditions.

Find (for stated design conditions):

1. Sensible cooling load
2. Latent cooling load
3. Total cooling load

Solution: By inspection, the cooling load from the roof can be expected to be the variable making the greatest contribution to the overall cooling load for the building. Therefore, the time of maximum cooling load occurrence will probably be close to the time of maximum CLTD for the roof. The maximum cooling load for the building as a whole can be expected to occur in one of the summer months—June, July, or August. From Table 31, a "mass inside" roof with no ceiling and an R-factor of 11.1 is classified as Type 4. From Table 30, the CLTD for Roof No. 4 at 40°N latitude has a maximum tabulated value of 78°F at 1800 h, but is only somewhat less (73) at 1600 h.

South-facing glass can also be expected to wield considerable influence on the cooling load for this particular building. From Table 35A, a one-story building with three exposed walls, uncarpeted floor, masonry partitions, and fully inside-shaded windows is classified B for solar loads, and C for loads from people, equipment, or lights. Cross-checking the variation of SCLs for glass facing south, in Table 36, the maximum cooling load from these windows would be only slightly more at noon or 1300 h than at 1400 for 40°N latitude. Sometime in the early afternoon seems obvious, but it is necessary to make a quick estimate to establish the peak load hour:

Roof [Equation (44)]

$$U = 0.09 \text{ Btu/(h·ft}^2\text{·°F)} \quad \text{Area} = 4000 \text{ ft}^2$$

Time	1300	1400	1500	1600	1700
CLTD	42	54	65	73	77
C_1	3	3	3	3	
C_2	−1	−1	−1	−1	−1
Corr. CLTD	44	56	67	75	79
Btu/h	15840	20160	24120	27000	28440

$C_1 = (78 - 75) = $ indoor design temperature correction

$C_2 = (94 + 74)/2 - 85 = $ daily average temperature correction

South Glass, Solar [Equation (46)]:

$$SC = 0.55 \quad \text{Area} = 60 \text{ ft}^2$$

Time		1300	1400	1500	1600	1700
SCL at 40°, Btu/h	=	87 2871	79 2607	63 2079	46 1518	37 1221
Roof and south glass, Btu/h	=	18711	22767	26199	28518	29661

Evaluation of the foregoing indicates that 1600 h will be the probable hour of maximum cooling load for this building, considering that although the roof and south glass loads increase another 1143 Btu/h for 1700 h, the trend has slowed and most other load components of significance can be expected to be leveling off or moving toward lower values at that time. In some cases there would be no such clear cut indication, and it would be necessary to estimate the total load for a number of hours, including the potential impact of other significant variables which could exert a determining influence at a different time (such as a major load from appliances, known to occur only in the morning, etc), before selecting the peak load hour for the overall calculation.

Cooling Load from Heat Gain through
Roof, Exposed Walls, and Doors

Such loads are estimated using Equation (44), where the CLTDs are taken from Table 30 after determining appropriate type numbers from Table 31 whose insulation placement, U-factors and general construction are as close to the actual components as possible. Corrections to CLTD values are made in accordance with footnote instructions to Table 30 similar to the above preliminary evaluation. (Note that there are no corrections to CLTD values for building mass variations, per the foregoing discussion, as considered of only limited significance to the overall results. Tabulated data for roofs, walls and doors assume Room Transfer Functions for "light to medium" construction.)

Cooling Load from Heat Gain through Fenestration Areas

The load component from conduction heat gain is calculated using Equation (45), where the CLTD value is taken from Table 34, corrected by −1°F because of a 1°F lower average daily temperature than that for which the table was generated, and +3°F to recognize the 75°F design space temperature. The U-factor of the glass is taken as 0.81 Btu/(h·ft²·°F) for single sheet plate glass, under summer conditions.

The load component from solar heat gain is calculated using Equation (46) as indicated above. A shading coefficient (SC) of 0.55 is used for clear glass with light-colored venetian blinds. SCL values are taken from Table 36 in this chapter, after first identifying the appropriate zone type as B for solar load from Table 35A. Results are tabulated in Table 40.

Cooling Load from Heat Gain through Party Walls

For the north and west party walls, cooling load is calculated using Equation (16) for wall and door areas, using appropriate U-factors from Chapter 22 and the temperature differential existing at 1600 h, 18.4°F. Results are tabulated in Table 41.

Cooling Load from Internal Heat Sources

For the cooling load component from lights, Equation (9) is first used to obtain the heat gain. Assuming a use factor of 1.0, and a special allowance factor of 1.0 for tungsten lamps and 1.20 for fluorescent lamps, these gains are:

$$q_{tung} = 4000 \times 1.0 \times 1.0 \times 3.41 = 13,640 \text{ Btu/h}$$

$$q_{fluor} = 17500 \times 1.0 \times 1.2 \times 3.41 = 71,610 \text{ Btu/h}$$

Since the tungsten lamps are operated continuously, the previously stored radiant heat from this source currently being re-convected to the space equals the rate of new radiant heat from this source being stored, thus the cooling load from this source equals heat gain. The fluorescent lamps however are operated only 10 hours per day, 0800 through hour 1700, and thus contribute radiant heat to cooling load in a cyclic and somewhat delayed manner. From Table 35A, the zone type is identified as C for lighting loads, and from Table 38 a CLF value of 0.92 is obtained for lights which are operated for 10 hours, for a calculation hour 9 hours after the lights have been turned on. The cooling load from fluorescent lights for this estimate is thus:

$$q_{cl\,fluor} = 71,610 \times 0.92 = 65,881 \text{ Btu/h}$$

For people, Table 3 is used to select heat gains for seated occupants doing light office work, as 250 Btu/h per person, sensible, and 200 Btu/h, latent for 75°F space temperature. The CLF for the sensible component is taken from Table 37 as 0.92, for a condition of 10 total hours in a type C space and a load calculation taken 9 h after entry. Cooling load from people is thus estimated at:

$$q_{ps} = 85 \text{ people} \times 250 \times 0.92 = 19,550 \text{ Btu/h}$$

$$q_{pl} = 85 \text{ people} \times 200 = 17,000 \text{ Btu/h}$$

Cooling Load from Power Equipment and Appliances

For this example, none are assumed.

Cooling Load from Infiltration and Ventilation Air

As determined in Example 6, which illustrates the use of the TFM, ventilation for this building is established at 15 cfm/person, or 1275 cfm, and infiltration (through doors) at 67 cfm. For this example, ventilation is assumed to enter directly into the space (as opposed to first passing through the cooling equipment), and thus is included as part of the space cooling load.

The sensible and latent portions of each load component are calculated using Equations (20) and (23), respectively, where at 1600 h: $t_o = 93.4\,°F$; $t_i = 75\,°F$; $W_o = 0.0161$; and $W_i = 0.0093$; thus:

For ventilation:

cfm	Factor	$t_o - t_i$	$W_o - W_i$	Btu/h
1275	1.1	18.4		25,806 Sensible
1275	4840		0.0068	41,963 Latent

For infiltration:

cfm	Factor	$t_o - t_i$	$W_o - W_i$	Btu/h
67	1.1	18.4		1,356 Sensible
67	4840		0.0068	2,205 Latent

Table 40 Solar Cooling Load for Windows, Example 11

Section	Net ft²	SC	CLF Table 36	Cooling Load, Btu/h
South windows	60	0.55	46	1518
North windows	30	0.55	32	528

Table 41 Conduction Cooling Load Summary for Enclosing Surfaces, Example 11

Section	Net ft²	U-Factor, Btu/(h·ft²·°F)	Δt, °F	CLTD, °F	Ref. for CLTD	1600 h Cooling Load, Btu/h
Roof	4000	0.09		73	Table 30 Roof 4	27000
South wall	405	0.24		19	Table 32 Wall 16	1847
East wall	765	0.48		38	Table 32 Wall 10	13954
North exposed wall	170	0.48		16	Table 32 Wall 10	1306
W. and N. party wall	1065	0.25	18.4			4899
Doors in S. Wall	35	0.19		50	Table 32 Wall 2	333
Doors in N. Wall	35	0.18	18.4			116
Doors in E. Wall	35	0.19		36	Table 32 Wall 2	239
South windows	60	0.81		16	Table 34	778
North windows	30	0.81		16	Table 34	389

Summary of Calculations for Example 11

	Dry Bulb, °F	Wet Bulb, °F	Humidity Ratio
Outdoor conditions	93.4	76.8	0.0161
Indoor conditions	75	62.5	0.0093
Difference	19		0.0068

Sensible Cooling Load at 1600 h	Btu/h
Roof and Exposed Walls	
Roof	27,000
South wall	1,847
East wall	13,954
North wall	1,306
South wall doors	333
East wall doors	239
Fenestration Areas	
South windows	2,296
North windows	917
Party Walls	
West and North Walls	4,899
North Wall doors	116
Internal Sources	
People	19,550
Tungsten lights	13,640
Fluorescent lights	65,881
Outside Air	
Infiltration	1,356
Ventilation	25,806
Total	179,140

Latent Cooling Load at 1600 h	Btu/h
People	17,000
Infiltration	2,205
Ventilation	41,963
Total	63,208
Grand Total Load	242,348

Limitations of CLTD/SCL/CLF Methods

The results obtained from using CLTD/CLF data depend on the characteristics of the space and how they vary from those used to generate the weighting factors. Variations can appear in the amplitude and when radiant heat gain components are felt as cooling loads, which affect the hourly cooling loads for the space. Two types of error are possible:

1. The computer software that generated CLTD/SCL/CLF tables uses the TFM to determine cooling loads based on various types of heat gain. The cooling loads for each type of heat gain are normalized appropriately to obtain CLTDs, SCLs, or CLFs. Except, as discussed next, use of the CLTD/SCL/CLF method in conjunction with these tables will yield the same results as the TFM, but only when the same 14 zone parameters are specified.

   Three inherent errors in the TFM are carried through to the CLTD/SCL/CLF data:

   a. Each set of weighting factors or conduction transfer function coefficients are used for a group of walls, roofs, or zones with similar thermal response characteristics. Groups were chosen so that error would be minimal and conservative (Harris and McQuiston 1988, Sowell 1988).

   b. The scheme used for calculating weighting factors is based on 14 discrete parameters applied to a rectangular room. Rarely does a room fit exactly into these parameters. Therefore, engineering judgment must be used to choose the values of the 14 parameters that most closely represent the room for which load calculations are being performed.

Table 42 Potential Errors for Roof and Wall CLTDs in Tables 30 and 32

Roof No.	Error, %		Wall No.	Error, %	
	Plus	Minus		Plus	Minus
1	13	5	1	18	7
2	13	5	2	17	8
3	12	5	3	17	7
4	13	5	4	16	7
5	11	4	5	13	8
6	—	—	6	14	6
7	—	—	7	12	6
8	10	4	—	—	—
9	10	4	9	13	6
10	9	3	10	10	6
11	—	—	11	8	3
12	—	—	12	4	7
13	7	4	13	4	4
14	5	4	14	5	8
15	—	—	15	11	6
16	—	—	16	8	7

Note: Percent error = [(Table Value − TFM Value)/TFM Value] × 100

Deviations of the room from the available levels of the 14 parameters may result in errors that are not easily quantifiable.

c. A fundamental presupposition of the TFM is that total cooling load for a zone can be calculated by simple addition of the individual components. For example, radiation heat transfer from individual walls and roofs is assumed to be independent of the other surfaces. O'Brien (1985) has shown this assumption can cause some error.

2. The printed tables for CLTDs, SCLs, and CLFs have undergone a further grouping procedure. The maximum potential errors due to the second grouping procedure have been analyzed and are tabulated in Tables 35 and 42. These errors are in addition to those inherent in the TFM. However, for usual construction, these errors are modest.

In summary, the CLTD/SCL/CLF method, as with any method, requires engineering judgment in its application. When the method is used in conjunction with custom tables generated by appropriate computer software (McQuiston and Spitler 1992) and for buildings where external shading is not significant, it can be expected to produce results very close to those produced by the TFM. When the printed tables are used, some additional error is introduced. In many cases, the accuracy should be sufficient.

TETD/TA CALCULATION PROCEDURE

To calculate a space cooling load using the TETD/TA convention, the same general procedures for data assembly and precalculation analysis apply as for the TFM. Similarly, the following factors are handled in an identical manner and are not repeated here.

• Basic heat gain calculation concepts of solar radiation (solar and conductive heat gain through fenestration areas, conversion to cooling load)
• Total heat gain through exterior walls and roofs (sol-air temperature, heat gain through exterior surfaces, tabulated temperature values, surface colors, air temperature cycle and adjustments, average sol-air temperature, hourly air temperatures, and data limitations)
• Heat gain through interior surfaces (adjacent spaces, floors)
• Heat gain through infiltration and ventilation

This section describes how the TETD/TA technique differs from the TFM. For sources of the space cooling load, equations, appropriate references, tables, and sources of other information for an overall analysis, see Table 43.

Treatment of Heat Gain and Cooling Load Conversion Procedures

As originally presented in the 1967 and 1972 editions of the *Handbook of Fundamentals*, the TETD/TA method was oriented primarily as a manual procedure. Tables of precalculated time-lags, decrement factors, and total equivalent temperature differential values listed a number of representative wall and roof assemblies, for use in the appropriate heat gain equations. These data were based on a Fourier series solution to the one-dimensional unsteady-state conduction equation for a multiple-component slab, as used to calculate the heat flow through each of the walls and roofs selected for that purpose. All calculations were based on an inside air temperature of 75 °F and a sol-air temperature at the outside equal to those given in Table 1 for horizontal and vertical surfaces of various orientations, at 2-h increments throughout a typical design day, as outlined by Stewart (1948) and Stephenson (1962). Basic equations were also presented to facilitate a computer solution.

Heat gain through walls and roofs. The results of the foregoing calculations were generalized by dividing the derived hourly heat gain values by the U-factor for each typical wall and roof. The quantity obtained from this generalization is called the total equivalent temperature differential (TETD). This establishes the basic heat gain equation for exterior surfaces as:

$$q = UA \, (\text{TETD}) \tag{51}$$

where

q = heat gain, Btu/h
U = coefficient of heat transfer, Btu/(h·ft^2·°F)
A = area of surface, ft^2
TETD = total equivalent temperature differential (as above)

It is assumed that the heat flow through a similar wall or roof (similar in thermal mass as well as U-factor) can be obtained by multiplying the TETDs listed in the appropriate table by the U-factor of the wall or roof of interest. Any errors introduced by this approach depend on the extent of the differences between the construction in question (components, size, color, and configuration) and the one used for calculating the TETDs.

TETD as function of decrement and time lag factors. The heat gain results for representative walls and roofs were also generalized in another way. Effective decrement factors λ and time lags δ were determined for each assembly, such that the equivalent temperature differentials and the corresponding sol-air temperatures are related by:

$$\text{TETD} = t_{ea} - t_i + \lambda(t_{e\delta} - t_{ea}) \tag{52}$$

where

t_{ea} = daily average sol-air temperature, including consideration for surface color
t_i = indoor air temperature
λ = effective decrement factor
$t_{e\delta}$ = sol-air temperature δ hours before the calculation hour for which TETD is intended

This relationship permits the approximate calculation of the heat gain through any of the walls or roofs tabulated, or their near equivalents, for any sol-air temperature cycle.

Manual versus automated calculation. Manual application of the TETD/TA procedure, especially the time-averaging calcu-

Table 43   Summary of TETD/TA Load Calculation Procedures

External Heat Gain

$$t_e = t_o + \alpha I_t/h_o - \epsilon\delta R/h_o \qquad (6)$$

$$t_{ea} = t_{oa} + \alpha/h_o(I_{DT}/24) - \epsilon\delta R/h_o \qquad (7)$$

t_e = sol-air temperature
t_o = current hour dry-bulb temperature, from design db (Chapter 24, Table 1) adjusted by Table 2 daily range % values
α = absorptance of surface for solar radiation
α/h_o = surface color factor = 0.15 for light colors, 0.30 for dark
I_t = total incident solar load = 1.15 (SHGF), with SHGF per Chapter 27, Tables 12 through 18
$\epsilon\delta R/h_o$ = long-wave radiation factor = −7 °F for horizontal surfaces, 0 °F for vertical
t_{ea} = 24-h average sol-air temperature
t_{oa} = 24-h average dry-bulb temperature
I_{DT} = total daily solar heat gain (Chapter 27, Tables 12 through 18)

Roofs and Walls

$$q = UA \text{ (TETD)} \qquad (51)$$

$$\text{TETD} = t_{ea} - t_i + \lambda(t_{e\delta} - t_{ea}) \qquad (52)$$

U = design heat transfer coefficient for roof or wall, from Chapter 22, Table 4
A = area of roof or wall, calculated from building plans
TETD = total equivalent temperature difference, roof or wall
t_i = interior design dry-bulb temperature
λ = decrement factor, from Table 14 or 19
$t_{e\delta}$ = sol-air temperature at time lag δ hours (Table 14 or 19) previous to calculation hour

Roofs

Identify layers of roof construction from Table 11.
With R-value of dominant layer, identify R-value Range number R and Roof Group number from Table 12. From Table 14 obtain decrement factor and time lag data with which to calculate TETD values for each sol-air temperature value by Equation (52). Calculate hourly heat gain with Equation (51).

Walls

Identify layers of wall construction from Table 11.
With R-value of dominant layer, identify R-value Range number R and Wall Group number from Tables 15, 16, or 17.

Glass

$$\text{Convective } q = UA(t_o - t_i)$$

$$\text{Solar } q = A\,(SC)(SHGF)$$

U = design heat transfer coefficients, glass—Chapter 27, Tables 1, 2, 5, and 6
SC = shading coefficient—Chapter 27
SHGF = solar heat gain factor by orientation, north latitude, hour, and month—Chapter 27, Tables 12 through 18

Partitions, Ceilings, Floors

$$q = UA(t_b - t_i) \qquad (8)$$

t_b = temperature in adjacent space
t_i = inside design temperature in conditioned space

Internal Heat Gain

People

$$q_{sensible} = N \text{ (Sensible heat gain)}$$

$$q_{latent} = N \text{ (Latent heat gain)}$$

N = number of people in space, from best available source

Sensible and latent heat gain from occupancy—Table 3, or Chapter 8; adjust as required

Lights

$$q_{el} = 3.41\, W F_{ul} F_{sa} \qquad (9)$$

W = watts input from electrical plans or lighting fixture data
F_{ul} = lighting use factor, from the first section, as appropriate
F_{sa} = special allowance factor, from first section, as appropriate

Power

$$q_p = 2545\, P E_F \qquad (15), (16), (17)$$

P = horsepower rating from electrical plans or manufacturer's data
E_F = efficiency factors and arrangements to suit circumstances

Appliances

$$q_{sensible} = q_{is} F_{ua} F_{ra} \qquad (18)$$

or

$$q_{sensible} = (q_{is} F_{ua} F_{ra})/F_{fl} \qquad (20)$$

$$q_{latent} = q_{il} F_{ua} \qquad (22)$$

q_{is}, q_{il} = sensible and latent heat gain from appliances—Tables 5 to 9, or manufacturer's data (set latent heat = 0 if appliance under exhaust hood)
F_{ua}, F_{ra}, F_{fl} = use factors, radiation factors, flue loss factors

Ventilation and Infiltration Air

$$q_{sensible} = 1.10\, Q(t_o - t_i) \qquad (25)$$

$$q_{latent} = 4840\, Q(W_o - W_i) \qquad (27)$$

$$q_{total} = 4.5\, Q(H_o - H_i) \qquad (23)$$

Q = ventilation from ASHRAE *Standard* 62; infiltration from Chapter 23
t_o, t_i = outside, inside air temperature, °F
W_o, W_i = outside, inside air humidity ratio, lb water/lb dry air
H_o, H_i = outside, inside air enthalpy, Btu/lb dry air

Cooling Load

Sensible

$$q_{sensible} = q_{cf} + q_{arf} + q_c$$

$$q_{cf} = [q_{s,1}(1 - rf_1)] + [q_{s,2}(1 - rf_2)] + \ldots rf_n)]$$

$$q_{arf} = \sum_{\gamma = h_{a+1-\theta}}^{\theta} [(q_{s,1} \times rf_1) + (q_{s,2} \times rf_2) + \ldots rf_n)_\gamma]\,/\,\theta$$

$$q_c = (q_{sc,1} + q_{sc,2} + q_{sc,\beta})$$

$q_{sensible}$ = sensible cooling load, Btu/h
q_{cf} = convective fraction of hourly sensible heat gain (current hour) for n load elements, Btu/h
$q_{s,1}$ = sensible hourly heat gain for load element 1, ... n
rf_1 = radiant fraction (Table 44) of sensible hourly heat gain for load element 1, ... n
q_{arf} = average of radiant fractions of hourly sensible heat gain for n load elements, Btu/h
θ = number of hours over which to average radiant fractions of sensible heat gain
h_a = current hour, 1 to 24, for which cooling load is to be calculated
γ = one of calculation hours, from $h_{a+1-\theta}$ to h_a, for which the radiant fraction of sensible heat gain is to be averaged for each of n load elements
q_c = convective hourly sensible heat gain (current hour) for β load elements having no radiant component, Btu/h

Latent

$$q_{latent} = (q_{l,1} + q_{l,2} + q_{l,\beta})$$

q_{latent} = latent cooling load, Btu/h
q_l = hourly latent heat gain (current hour) for β load elements, Btu/h

lation itself, is tedious in practice. This fact, plus growing interest in the TFM, led to ASHRAE research with the objective of comparing the differences and similarities of the TETD and TFMs.

Later research completed the circle of relationships between the TFM, its subsystem CLTD/CLF, and the TETD/TA techniques for dealing with the conversion of heat gain to cooling load. It also confirmed the logic of maintaining these various approaches to solving the problem, depending on the orientation and needs of the individual user and the means available. Finally, the research showed no further need to continue developing manual TETD/TA procedures. Thus, the tabulated values of TETDs have been eliminated from this Handbook in favor of calculation of TETD values by use of the material in the previous section that discusses the TFM.

U-Factors. The values for TETD, originally tabulated in the 1967 *Handbook of Fundamentals*, were calculated using an outside surface conductance of 3.0 Btu(/h·ft²·°F) and an inside surface conductance of 1.2 Btu/(h·ft²·°F), and thus should most appropriately be used with U-factors based on the same surface conductances. TETD data tabulated in the 1972 *Handbook of Fundamentals* and all data listed in this chapter are based on outside and inside surface conductances of 3.0 and 1.46, respectively. U-factors listed in Tables 14 for roofs and 19 for walls can, however, be used with the 1972 TETD data with negligible error, while calculated TETD values are directly compatible with the Table 14 and 19 U-factors.

Example 12. Wall heat gain by TETD. A wall is constructed of 4 in. heavy concrete, 2 in. insulation (2.0 lb/ft³, $R = 6.667$ h·ft²·°F/Btu), 3/4 in. indoor plaster, and with outdoor and indoor surface resistances of 0.333 and 0.685 h·ft²·°F/Btu, respectively. There is an air space between the plaster and the insulation. The wall faces west, the outside design temperature is 95°F, the outdoor daily range is 21°F, the indoor temperature is 75°F, and the color of the exterior surface is light ($\alpha/h_o = 0.15$). The time is 1400 h on a July day in the central part of the United States (40°N latitude).

Find the heat gain per square foot of wall area.

Solution: Turning first to Table 11, the code numbers for the various layers of the wall described above are:

Outside surface resistance =	A0
4 in. heavyweight concrete =	C5
2 in. insulation =	B3
Air space resistance =	B1
3/4 in. plaster =	E1
Inside surface resistance =	E0

Construction of the wall being "mass out" (as defined in TFM section), Table 17 represents the appropriate arrangement of layers. The dominant wall layer C5 is indicated to have a Wall Material column number of 10, which, combined with an E1 layer, dictates use of the upper array of code numbers for wall assembly "groups." Entering this array with an R value range of 9 ($R = 6.667$) indicates under column 10 that Wall Group 7 is that most nearly representative of the wall under consideration. The appropriate data from Wall Group 7 as listed in Table 19 are:

$$h = 5.14 \quad = \text{time lag, h}$$

$$\lambda = 0.41 \quad = \text{effective decrement factor}$$

$$U = 0.122 \quad = \text{heat transfer coefficient, Btu/(h·ft²·°F)}$$

For this example, the sol-air temperature value for 1400 h t_e, as listed in Table 1, is 121°F, that for 0900 h (5 h earlier) is 85°F, and the daily average is 91°F. Thus, from Equation (52):

$$\text{TETD} = 91 - 75 + [0.41(85 - 91)] = 13.54°F$$

and from Equation (51)

$$q = 0.122 \times 1 \times 13.54 = 1.65 \text{ Btu/(h·ft²)}$$

Roof heat gain by TETD. The procedure for estimating heat gain from an exposed roof assembly is similar to that described for a wall—first identifying the code letters for the various layers from Table 11; identifying the appropriate roof group number from Table 12; reading the time lag, effective decrement factor, and U-factor for the selected roof group from Table 14; calculating the TETD for the hour of interest from these data and reference to Table 1; and then calculating the heat gain by means of Equation (51).

Heat gain from adjacent unconditioned spaces. In a manner similar to that described for the TFM, heat gain from adjacent unconditioned spaces can be estimated in two ways, depending on the thermal storage characteristics of the intervening surface. When storage effect is minor, sufficient accuracy can be obtained by use of Equation (8); otherwise, the appropriate TETD value should be calculated by the manner described for an exposed wall surface and the heat gain calculated by Equation (51).

Instantaneous heat gain from all other sources. Conductive and solar heat gain through fenestration, heat gain from the various internal sources (*e.g.*, people, lighting, power, appliances, etc.), heat gain due to infiltration and ventilation, and latent heat gain from moisture through permeable building surfaces are each calculated in the same manner as described in the TFM section. The basic differences in calculation techniques between TFM and TETD/TA lie in the manner in which the heat gain data are converted to cooling load, as described later.

COOLING LOAD BY TIME AVERAGING

The time-averaging technique for relating instantaneous heat gain to instantaneous cooling load is an approximation of the TFM two-step conversion concept. It recognizes thermal storage by building mass and contents of the radiant portions of heat gain entering a space at any time, with subsequent release of stored heat to the space at some later time. It further recognizes that the cooling load for a space at a given hour is the sum of all convective heat gain and the nonradiant portion of conductive heat gain to that space, plus the amount of previously stored radiant heat gain released back to the space during that same hour.

The effect of room transfer coefficients on hourly heat gain is to generate a load profile that tracks the instantaneous heat gain in amplitude (greater or lower) and delay (negligible for very light structures with a predominance of glass, up to several hours for very massive, monumental construction). Being functions of the mass and configuration of the building and its contents, such coefficients place major emphasis on the immediately preceding hour, and rapidly lessening emphasis on each hour previous to that.

Such TFM-generated cooling load profiles can be closely approximated by averaging the hourly radiant components of heat gain for the previous one to seven or eight hours with those for the current hour, and adding the result to the total convective heat gain for the current hour. As long as results are consistent with results from the more rigorous TFM analysis, those from TETD/TA can be obtained with far less computational effort. The convenient ability to vary the averaging period independently (in recognition of previous experience of the probable thermal performance of an individual building) is also a valuable means of exerting professional judgment on the results.

Success of this approach depends on the accuracy with which the heat gain components are broken down into convective and radiant percentages, as well as on the number of hours used for the averaging period. Weakness of this approach lies in the absence of verified data in the technical literature regarding either determining factor, and the corresponding necessity for experienced judgment by the user.

Heat gain values for either the TFM or TETD/TA method are essentially identical for all load components. Derived cooling load values from properly applied averaging techniques closely track those from the TFM for external heat gain sources. Cooling loads from internal heat gains, however, averaged over the same period as for the external components, normally have peaks that occur more quickly and with greater amplitudes (up to full value of the source heat gain) than those generated by the TFM. This difference is due primarily to the almost constant level of radiant heat input during the occupied periods and the resultant "flattening" of the cooling load curves by the TFM as discussed in the TFM section.

The conservative results obtained from the time-averaging method compared with those of the TFM should be viewed in proper perspective. In the CLTD technique, for example, CFL values profile internal loads as a function of time in the space (up to the hour of interest) versus total time to be in effect during the day, and the tabulated fractional values are used *only* when HVAC equipment is operated 24 h a day and space temperature is not allowed to rise during unoccupied periods; otherwise, internal heat gains are considered instantaneous cooling loads at *full* value. On the other hand, the TFM, while not dealing directly with individual load components when space temperatures are permitted to rise overnight, applies space air transfer functions to estimate resultant increased rates of total sensible heat extraction from that space during periods of equipment operation. Regardless of methodology, good engineering judgment must be applied to predict realistic cooling loads from internal heat gains.

Time-averaging data in this chapter are empirical and offered only as information found dependable in practice by users of the technique. Basic assumptions regarding the percentages of radiant heat gain from various sources are used as default values by the TFM in establishing envelope transfer coefficients and room transfer coefficients. The TETD/TA method requires a specific breakdown by the user to determine what values are to be averaged over time. Table 44 suggests representative percentages for this purpose.

The convective portion of heat gain is treated as instantaneous cooling load. The radiant portion of instantaneous heat gain is considered as reduced or averaged over time by the thermal storage of the building and its contents. For lightweight construction, the instantaneous cooling load may be considered as an average of the radiant instantaneous heat gain over a 1 to 3-h period up to and including the hour of calculation interest, plus the nonradiant component of that hour's heat gain. For very heavy construction, the averaging period for hourly values of radiant instantaneous heat gain may be as long as 6 to 8 h, including the hour of calculation interest. Most users of this technique rarely consider application of an averaging period longer than 5 h, with a general norm of 3 h for contemporary commercial construction.

Table 44 Convective and Radiant Percentages of Total Sensible Heat Gain for Hour Averaging Purposes

Heat Gain Source	Radiant Heat, %	Convective Heat, %
Window solar, no inside shade	100	—
Window solar, with inside shade	58	42
Fluorescent lights	50	50
Incandescent lights	80	20
People	33	67
Transmission, external roof and walls	60	40
Infiltration and ventilation	—	100
Machinery and appliances[a]	20 to 80	80 to 20

[a] The load from machinery or appliances varies, depending on the temperature of the surface. The higher the surface temperature, the greater the percentage of heat gain that is radiant.

The two-step nature of the TETD/TA procedure offers a unique convenience in calculating cooling load through externally shaded fenestration. As described in the beginning of this section, the hourly history of fenestration heat gain as modified by external shading devices is directly usable for averaging purposes. Thus, the engineer has excellent control and can readily use the effect of external shading on cooling load in the conditioned space. Sun (1968) identified convenient algorithms for analysis of moving shade lines on glass from external projections.

EXAMPLE COOLING LOAD CALCULATION USING TETD/TA

Example 13. Cooling load calculation of small office building. For this example, the one-story building used to illustrate the TFM in Example 6 (and indicated in Figure 4) is also used for calculating a cooling load by the TETD/TA method. Refer to Example 6 for the statement of conditions.

Find (for stated design conditions):

1. Sensible cooling load
2. Latent cooling load
3. Total cooling load

Solution: By TETD/Time-averaging method.

1. Daily Load Cycle

The cooling loads are calculated once per hour for a period of time necessary to cover the hour of anticipated peak design load. For the purposes of this example, the full range of loads over a typical 24-h cycle are presented.

2. Hourly Heat Gain Components

Hourly heat gain values for each load component must be calculated for the same range as those for the cooling load, plus as many preceding hours as will be needed for the purposes of time-averaging (in the case of this example, all 24-h values have been calculated). The methodology involving use of time lag, effective decrement factor, and calculated TETD values is used to calculate heat gain components through walls and roof.

3. Thermal Storage

The heat storage effect of the room is accounted for by averaging the radiant elements of heat gain components for the hour in question with those of the immediately previous hours making up the selected averaging period, and combining the result with the convective heat gain elements for the current hour.

4. Summary

The data and summary of results using TETD/TA are tabulated in Table 45. Following the table is a step by step description of the calculation procedure used to determine the values listed.

1. **Sensible Cooling Load**
   (a) General

   Line 1, Time of day in hours. Various temperatures and heat flow rates were calculated for every hour on the hour, assuming that hourly values are sufficient to define the daily profile.
   Line 2, Outside air temperatures. Hourly values were derived by the procedure given previously, using the specified maximum dry-bulb temperature of 94°F and daily range of 20°F.
   (b) Solar Heat Gain Factors

   Lines 3, 4, 5, and 6, Solar heat gain through opaque surfaces. The values in these columns are copies of the SHGF values listed in Table 4 for July 21 and 40°N latitude. These SHGF values are used to calculate sol-air temperatures of various outside surfaces, and solar heat gain through windows.
   The June values might have been used, since the solar irradiation of horizontal surface (i.e., roof) is maximum at that time of year and since the heat gain through the roof appears to be the major component of exterior heat gain in this example problem. The difference between June and August values is relatively small however, compared to the large percentage increase in solar heat gain through south glass in August versus June at this latitude, thus indicating that August might be the better choice. For this example, data for July were selected as reasonable, and to provide better comparison with the results from other techniques for which tabular data are limited. For better assurance of accuracy it is preferable to evaluate

Table 45 Tabulation of Data for Example 13—TETD/TA Method

1	Time, hour	0100	0200	0300	0400	0500	0600	0700	0800	0900	1000	1100	1200
2	Outside air temperature, °F	76	75	74	74	74	74	75	77	79	82	86	89
3	SHGF, Btu/h·ft², Horizontal	0	0	0	0	0	32	88	145	194	231	254	262
4	North	0	0	0	0	1	37	30	28	32	35	37	38
5	South	0	0	0	0	0	11	21	30	52	81	102	109
6	East	0	0	0	0	2	137	204	216	193	146	81	41
7	Sol-air temperature, °F, Horizontal	69	68	67	67	67	77	94	114	130	144	155	161
8	North	76	75	74	74	74	80	80	81	84	87	92	95
9	South	76	75	74	74	74	76	78	82	87	94	101	105
10	East	76	75	74	74	74	95	106	109	108	104	98	95
10a	Calculated TETD, °F, Roof	10	8	7	6	5	5	4	3	3	4	12	24
10b	North wall	10	8	7	6	5	5	4	3	3	3	6	7
10c	South wall	16	15	14	14	13	12	11	11	10	10	10	9
10d	East wall	11	10	9	8	7	7	5	5	4	4	17	24
10e	North and west party wall	10	10	10	9	9	8	8	7	7	6	6	6
10f	North door (to adjacent building)	3	2	1	0	0	0	0	0	1	4	6	10
10g	South door	3	2	1	0	0	0	0	2	6	11	17	24
10h	East door	3	2	1	0	0	0	14	27	33	33	30	24

Instant Sensible Heat Gain, Btu/h

11	Roof	3722	3193	2718	2347	2106	1814	1580	1393	1264	1681	4518	8802
12	North wall	827	710	605	518	454	446	348	305	273	259	556	592
13	South wall	1599	1536	1457	1379	1306	1213	1141	1090	1044	1005	976	968
14	East wall	4326	3801	3323	2938	2648	2614	2170	1979	1832	1774	6246	9099
15	North and west party wall	2686	2740	2724	2652	2532	2380	2221	2079	1954	1848	1768	1747
16	North door (to adjacent building)	19	17	7	1	−3	−5	−3	2	11	26	44	64
17	South door	21	19	8	2	−2	−5	6	20	41	74	117	160
18	East door	21	19	8	2	−2	−4	93	185	219	223	204	166
19	Windows, air to air heat gain	117	44	−15	−58	−73	−44	29	160	350	569	816	1050
20	North windows, solar heat gain	0	0	0	0	17	611	495	462	528	578	611	627
21	South windows, solar heat gain	0	0	0	0	0	363	693	990	1716	2673	3366	3597
22	Lights, tungsten (always on)	13640	13640	13640	13640	13640	13640	13640	13640	13640	13640	13640	13640
23	Lights, fluorescent (on-off)	0	0	0	0	0	0	0	71610	71610	71610	71610	71610
24	People	0	0	0	0	0	0	0	21250	21250	21250	21250	21250
25	Infiltration	0	0	0	0	0	0	0	162	354	575	825	1061
26	Ventilation	2244	841	−281	−1122	−1403	−841	561	3086	6732	10940	15708	20196
27	Total instant sensible heat gain	29222	26560	24194	22299	21220	22182	22974	118413	122818	128725	14 2255	154629

Latent Heat Gain/Cooling Load, Btu/h

28	People	0	0	0	0	0	0	0	17000	17000	17000	17000	17000
29	Infiltration	0	0	0	0	0	0	0	2205	2205	2205	2205	2205
30	Ventilation	41963	41963	41963	41963	41963	41963	41963	41963	41963	419 63	41963	41963
31	Total latent heat gain/cooling load	41963	41963	41963	41963	41963	41963	41963	61168	61168	61168	61168	61168
32	Sum: sensible + latent heat gain, Btu/h	71185	68523	66157	64262	63183	64145	64937	179581	183986	189893	2 03423	215797

Sensible Cooling Load from Convective Heat Gain, Btu/h

33	Windows, air to air heat gain	117	44	−15	−58	−73	−44	29	160	350	569	816	1050
34	Lights, tungsten (20% convective)	2728	2728	2728	2728	2728	2728	2728	2728	2728	2728	2728	272 8
35	Lights, fluorescent (50% conv.)	0	0	0	0	0	0	0	35805	35805	35805	35805	35805
36	People (67% convective)	0	0	0	0	0	0	0	14238	14238	14238	14238	14238
37	Infiltration (100% convective)	0	0	0	0	0	0	0	162	354	575	825	1061
38	Ventilation (100% convective)	2244	841	−281	−1122	−1403	−841	561	3086	6732	10940	15708	20 196

Sensible Cooling Load from Radiant Heat Gain, Btu/h

39	Lights, tungsten (80% radiant)	10912	10912	10912	10912	10912	10912	10912	10912	10912	10912	10 912	10912
40	Lights, fluorescent (50% radiant)	0	0	0	0	0	0	0	7161	14322	21483	28644	35805
41	People (33% Radiant)	0	0	0	0	0	0	0	1403	2805	4208	5610	7013

Sensible Cooling Load from Exposed Surfaces; From Convective Heat Gain, Btu/h

42a	North windows, SHG (42% convective)	0	0	0	0	7	257	208	194	222	243	257	263
43a	South windows, SHG (42% convective)	0	0	0	0	0	152	291	416	721	1123	1414	1511
44a	Roof (40% convective)	1489	1277	1087	939	842	726	632	557	506	672	1807	3521
45a	North wall (40% convective)	331	284	242	207	182	178	139	122	109	104	222	237
46a	South wall (40% convective)	640	614	583	552	522	485	456	436	418	402	390	387
47a	East wall (40% convective)	1730	1520	1329	1175	1059	1046	868	792	733	710	2498	3640
48a	N. and W. party wall (40% conv.)	1074	1096	1090	1061	1013	952	888	832	782	739	707	699
49a	N. door to adj. bldg. (40% conv.)	8	7	3	0	−1	−2	−1	1	4	10	18	26
50a	South door (40% convective)	8	8	3	1	−1	−2	2	8	16	30	47	64
51a	East door (40% convective)	8	8	3	1	−1	−2	37	74	88	89	82	66

Sensible Cooling Load from Exposed Surfaces; From Radiant Heat Gain, Btu/h

42b	SHG at north windows (58% radiant)	0	0	0	0	2	73	130	184	245	310	310	325
43b	SHG at south windows (58% radiant)	0	0	0	0	0	42	122	237	436	746	1095	1432
44b	Roof heat gain (60% radiant)	5056	3532	2515	1959	1691	1461	1268	1109	978	928	1253	2119
45b	North wall heat gain (60% radiant)	742	648	545	439	373	328	285	249	219	196	209	238
46b	South wall heat gain (60% radiant)	904	932	932	910	874	827	780	735	695	659	631	610
47b	East wall heat gain (60% radiant)	3532	3110	2697	2325	2044	1839	1643	1482	1349	1244	1681	2511
48b	N. and W. party wall HG (60% rad.)	1428	1515	1576	1605	1600	1563	1501	1423	1340	1258	1185	1127
49b	North door heat gain (60% radiant)	24	18	13	9	5	2	−1	−1	1	4	9	17
50b	South door heat gain (60% radiant)	26	19	14	9	6	3	1	2	8	16	31	49
51b	East door heat gain (60% radiant)	26	19	14	10	6	3	12	33	59	86	111	120
52	Total sensible cooling load, Btu/h	33027	29132	25990	23662	22387	22686	23491	84540	97175	111027	129 243	147770
53	Sum: sens. + lat. cooling load, Btu/h	74990	71095	67953	65625	64350	64649	65454	145708	158343	172195	1 90411	208938

Table 45 Tabulation of Data for Example 13—TETA/TA Method (*Concluded*)

1	1300	1400	1500	1600	1700	1800	1900	2000	2100	2200	2300	2400	24 h Total	Heat Loss, Btu/h
2	91	93	94	93	92	89	87	84	82	80	78	77	**Total**	**Btu/h**
3	254	231	194	145	88	32	0	0	0	0	0	0	2150	
4	37	35	32	28	30	37	1	0	0	0	0	0	438	
5	102	81	52	30	21	11	0	0	0	0	0	0	703	
6	37	35	31	26	20	11	0	0	0	0	0	0	1180	
7	160	155	145	130	111	92	80	77	75	73	71	70		
8	97	98	99	97	97	95	87	84	82	80	78	77		
9	106	105	102	98	95	91	87	84	82	80	78	77		
10	97	98	99	97	95	91	87	84	82	80	78	77		
10a	37	48	57	64	67	66	62	54	44	31	19	12		
10b	7	9	11	14	16	18	18	19	18	17	17	12		
10c	9	9	9	9	9	10	10	11	13	14	16	16		
10d	27	27	24	20	18	19	20	20	19	18	16	13		
10e	6	5	5	5	5	5	6	6	7	8	9	9		
10f	13	15	17	18	18	17	15	12	10	8	6	4		
10g	28	31	30	27	24	20	17	13	10	8	6	4		
10h	21	21	23	23	22	20	17	13	10	8	6	4		

Instant Sensible Heat Gain, Btu/h

1	1300	1400	1500	1600	1700	1800	1900	2000	2100	2200	2300	2400	24 h Total	Heat Loss, Btu/h
11	13511	17460	20729	23087	24307	23947	22406	19674	15890	11192	6984	4342	238667	23400
12	630	802	962	1184	1358	1488	1529	1561	1490	1462	1404	997	20760	5304
13	930	911	896	889	926	973	1047	1160	1307	1454	1563	1612	28382	6318
14	9969	9958	9110	7627	6870	7300	7487	7630	7311	6767	6033	4990	133802	23868
15	1640	1590	1550	1528	1544	1595	1691	1834	2010	2213	2412	2572	49510	17306
16	84	100	112	117	116	109	96	81	65	51	38	27	1176	410
17	192	206	202	184	161	139	117	90	69	54	41	30	1946	432
18	140	146	153	157	151	137	118	91	70	55	41	30	2423	432
19	1224	1341	1385	1341	1239	1079	889	700	540	393	277	189	13542	4739
20	611	578	528	462	495	611	17	0	0	0	0	0	7231	
21	3366	2673	1716	990	693	363	0	0	0	0	0	0	23199	
22	13640	13640	13640	13640	13640	13640	13640	13640	13640	13640	13640	13640	327360	− 13640
23	71610	71610	71610	71610	71610	0	0	0	0	0	0	0	716100	− 71610
24	21250	21250	21250	21250	21250	0	0	0	0	0	0	0	212500	− 21250
25	1238	1356	1400	1356	1253	0	0	0	0	0	0	0	9580	4791
26	23562	25806	26648	25806	23843	20757	17111	13464	10379	7574	5330	3646	260587	91163
27	163597	169427	171891	171228	169456	72138	66148	59925	52771	44855	37763	32075	2046765	

Latent Heat Gain/Cooling Load, Btu/h

1	1300	1400	1500	1600	1700	1800	1900	2000	2100	2200	2300	2400	24 h Total	
28	17000	17000	17000	17000	17000	0	0	0	0	0	0	0	170000	
29	2205	2205	2205	2205	2205	0	0	0	0	0	0	0	22050	
30	41963	41963	41963	41963	41963	41963	41963	41963	41963	41963	41963	41963	1007112	
31	61168	61168	61168	61168	61168	41963	41963	41963	41963	41963	41963	41963	1199162	
32	224765	230595	233059	232396	230624	114101	108111	101888	94734	86818	79726	74038	3245927	

Sensible Cooling Load from Convective Heat Gain, Btu/h

1	1300	1400	1500	1600	1700	1800	1900	2000	2100	2200	2300	2400	24 h Total	
33	1224	1341	1385	1341	1239	1079	889	700	540	393	277	189	13542	
34	2728	2728	2728	2728	2728	2728	2728	2728	2728	2728	2728	2728	65472	
35	35805	35805	35805	35805	35805	0	0	0	0	0	0	0	358050	
36	14238	14238	14238	14238	14238	0	0	0	0	0	0	0	142380	
37	1238	1356	1400	1356	1253	0	0	0	0	0	0	0	9580	
38	23562	25806	26648	25806	23843	20757	17111	13464	10379	7574	5330	3646	260587	

Sensible Cooling Load from Radiant Heat Gain, Btu/h

1	1300	1400	1500	1600	1700	1800	1900	2000	2100	2200	2300	2400	24 h Total	
39	10912	10912	10912	10912	10912	10912	10912	10912	10912	10912	10912	10912	261888	
40	35805	35805	35805	35805	35805	28644	21483	14322	7161	0	0	0	358050	
41	7013	7013	7013	7013	7013	5610	4208	2805	1403	0	0	0	70130	

Sensible Cooling Load from Exposed Surfaces; From Convective Heat Gain, Btu/h

1	1300	1400	1500	1600	1700	1800	1900	2000	2100	2200	2300	2400	24 h Total	
42a	257	243	222	194	208	257	7	0	0	0	0	0	3039	
43a	1414	1123	721	416	291	152	0	0	0	0	0	0	9745	
44a	5404	6984	8292	9235	9723	9579	8962	7870	6356	4477	2794	1737	95468	
45a	252	321	385	474	543	595	612	624	596	585	562	399	8305	
46a	372	364	358	356	370	389	419	464	523	582	625	645	11352	
47a	3988	3983	3644	3051	2748	2920	2995	3052	2924	2707	2413	1996	53521	
48a	656	636	620	611	618	638	676	734	804	885	965	1029	19805	
49a	34	40	45	47	46	44	38	32	26	20	15	11	471	
50a	77	82	81	74	64	56	47	36	28	22	16	12	779	
51a	56	58	61	63	60	55	47	36	28	22	16	12	967	

Sensible Cooling Load from Exposed Surfaces; From Radiant Heat Gain, Btu/h

1	1300	1400	1500	1600	1700	1800	1900	2000	2100	2200	2300	2400	24 h Total	Heat Loss, Btu/h
42b	343	349	343	325	310	310	245	184	130	73	2	0	4193	
43b	1707	1818	1707	1432	1095	746	436	237	122	42	0	0	13452	
44b	3573	5517	7802	10030	11891	13143	13738	13610	12747	11173	9137	6970	143200	
45b	277	341	425	500	592	695	782	855	891	904	893	830	12456	
46b	591	575	562	551	547	552	568	599	650	713	784	851	17032	
47b	3470	4446	5326	5491	5224	4904	4607	4430	4392	4380	4228	3928	80283	
48b	1075	1031	995	967	942	937	949	983	1041	1121	1219	1325	29706	
49b	27	38	48	57	64	66	67	63	56	49	40	31	706	
50b	70	90	105	113	114	107	96	83	69	56	45	34	1166	
51b	115	106	98	91	90	89	86	79	68	56	45	34	1456	
52	156283	163149	167774	169082	168376	105964	92708	78902	64574	49474	43046	37319	2046781	178163
53	217451	224317	228942	230250	229544	147927	134671	120865	106537	91437	85009	79282	3245943	71663

and compare the relative loads of various surfaces for several months, before making a final determination as to that in which the maximum load will occur.

(c) Sol-Air Temperatures

Lines 7, 8, 9, and 10, Sol-air temperatures at opaque surfaces. Sol-air temperatures were calculated by Equation (6).

(d) Total Equivalent Temperature Differentials

Lines 10a through 10h, Calculated TETD values. Hourly TETD values for each of the expose surfaces, are calculated by Equation (52) to incorporate individual thermal characteristics and orientation.

Line 10a, Roof TETD. Referring to Table 11, the major element of the roof (that with the most mass) is the gypsum slab with code number C14. Other elements are the metal deck (A3), rigid insulation (B3), built-up roofing (E3), and gravel surface (E2). Entering Table 12 with these data, the C14 roof slab with no ceiling and R values of 11.11 calls for an R range of 3. From the "mass-in" part of the upper table these pointers indicate roof group 5 as that whose thermal characteristics will best represent the roof in question.

The time lag and effective decrement factors are then obtained from Table 14, as tabulated for roof group 5. These values are:

Time lag $= 4.82$ h

Effective decrement factor $(\lambda) = 0.68$

The TETD values were then calculated for the roof surface with Equation (52), using the sol-air temperature cycle given in line 7 and $t_i = 75\,°F$.

Lines 10b through 10e, Wall TETD. The TETD values for the various walls were calculated by the same approach as that described for the roof. Time lag and effective decrement factors were selected from Table 19, as:

North and East Exterior Walls

Dominant element C8, from Table 11;
Interior finish E1 from Table 11;
R value indicating R range of 2 from Table 16 (integral mass);
C8 dominant layer indicating Material Layer 13 from Table 16;
From Table 16 select Wall Group 5 from the upper section (combining Layer 13 with E1 finish) as the most representative, and thus obtain the time lag and decrement factors from Table 19 as 5.11 h and 0.64 DF, respectively.

South Wall

Dominant element C9, or Layer 14 in Integral mass table;
Exterior layer A2 or A7;
Interior layer E1 (plywood panel ignored as trivial);
R value indicating R of 6;
Select wall group 24 for representative performance factors of 11.29 h and 0.23 DF.

North and West Party Walls

With no specific data for a 13 in. brick wall, use a layer of 8 in. common brick (C9) and a layer of 4 in. face brick (A2 or A7) as an approximation;
Dominant element C9, or Layer 14 in Integral Mass table;
Exterior layer A2 or A7;
R value indicating R of 6;
  Select wall group 24 for representative factors 11.29 h and 0.23 DF. Calculate TETD values as above.

Lines 10f, 10g, and 10h, Door TETD values. Heat storage of the doors may be assumed negligible, and the heat gain, therefore, is calculated with Equation (8) as:

$$q_{DT} = U_D A_D \, (t_{DT} - t_i)$$

where

$U_D = 0.19$; U-factor of doors (0.18 for interior doors)
$A_D = 35$; area of a door, ft^2
$t_i = 75\,°F$, inside temperature
$t_{DT} =$ outside temperature. For the door in the north party wall, t_{DT} equals outside air temperature. For the doors in east and south walls t_{DT} equals the east and south wall sol-air temperatures, respectively.

While the foregoing calculation would be reasonable in estimating the minor loads involved, for this example, the relatively brief storage effect of the solid core doors has been considered as:

Dominant element B7, or Layer 3 in Integral Mass table;
Interior finish A6;
R-factor indicating R range of 8;

Select wall group 1 for representative time lag of 1.30 h and 0.98 DF, and follow the above procedures to calculate the associated TETD values.

(e) *Instantaneous Sensible Heat Gain*

Line 11, Roof heat gain. Instantaneous heat gain through the roof, calculated with Equation (51) and the TETD values on line 10a.

Lines 12 through 18, Wall and door heat gain. The instantaneous heat gains through the various walls and doors were calculated the same way as heat gain through the roof was calculated. TETD values from lines 10b through 10h were used in Equation (51).

Lines 19, 20, and 21, Window heat gain. The air to air heat gain (line 19) is

$$q_{a-a} = U_w A_w (t_{o\theta} - t_i)$$

where

$U_w = 0.81$; U-factor of window
$A_w = 90$; area of windows
$t_{o\theta} =$ outside air temperature at hour θ

The solar radiation heat gain (Lines 20 and 21) through south and north windows is:

$$q_r = A_w \, SC \, (SHGF)_\theta$$

where

$(SHGF)_\theta =$ Solar heat gain factors given in line 5 for south and line 6 for north
$SC = 0.55$; shading coefficient for clear window with light-colored curtain or blind

Lines 22 and 23, Heat gain from tungsten and fluorescent lights. For the gain from lighting, Equation (9) was used with a use factor of unity, and special allowance factors of 1.20 for the fluorescent lamps and of unity for the tungsten lamps. Thus:

$$q_{el\,tung} = 4000 \times 1 \times 1 \times 3.41 = 13,640 \text{ Btu/h, and}$$

$$q_{el\,fluor} = 17500 \times 1 \times 1.20 \times 3.41 = 71,610 \text{ Btu/h}$$

Line 24, People. Sensible heat gain due to people. For the occupants, the data of Table 3 was used for moderately active office work. Thus:

$$q_p = \text{(Number of people)(Sensible heat generated per person)}$$

$$= 85 \times 250 = 21,250 \text{ Btu/h}$$

Lines 25 and 26, Sensible heat gain from infiltration and ventilation. As developed previously, the value to be used for infiltration was established as 67 cfm, and that for ventilation as 1275 cfm. Heat gain from all air entering as infiltration is routinely part of the space load. In the case of this example (since ventilation is delivered directly to the space, rather than first through the cooling equipment), its gain is also included as a direct space load. (Note: Had the ventilation air instead been mixed with return air after leaving the occupied space and before entering the cooling equipment, only that portion which passed through the cooling coil without being treated by it—as a function of the coil inefficiency or "Bypass Factor," which is normally 3 to 5% for a chilled water coil of 6 or more rows and close fin spacing to 15% or more for refrigerant coils in packaged air-conditioning units—and/or that quantity deliberately bypassed around the coil in response to a "face and bypass" or "conventional multizone" space dry-bulb temperature control scheme, would become a part of the space heat gain as such rather than a part of the cooling coil load per se. While potentially of major significance to the overall design of a cooling system, the detailed treatment of this concept is beyond the scope of this chapter. Refer to the 1992 ASHRAE *Handbook—Systems and Equipment* for assistance.)

The sensible loads are determined from Equation (20). At 1500 hours for example, when $t_o = 94\,°F$ and $t_i = 75\,°F$, this generates:

$$q_{si} = 1.1\,(\text{Infiltration rate, cfm})(t_o - t_i)$$
$$= 1.1 \times 67\,(94 - 75) = 1400\,\text{Btu/h, and}$$

$$q_{sv} = 1.1\,(\text{Ventilation rate, cfm})(t_o - t_i)$$
$$= 1.1 \times 1275\,(94 - 75) = 26{,}600\,\text{Btu/h}$$

Line 27, Total instantaneous sensible heat gain. The sum of sensible heat gain values on lines 11 through 26 for each calculation hour. This represents the total amount of such gain that actually enters the building during each hour, including any delaying effects of the individual surfaces on the passage of heat, but before any consideration of the storage and subsequent release of the radiant components of such heat.

(f) Instantaneous Latent Heat Gain

Line 28, People. The latent heat gain due to people, using Table 3 data

$$= (\text{number of persons})(\text{latent heat generated by one person})$$
$$= 85 \times 200 = 17{,}000\,\text{Btu/h during the occupied period}$$

Lines 29 and 30, Latent heat gain from infiltration and ventilation. The latent loads are determined from Equation (23). At 1500 h for example, when $W_o = 0.0161$ and $W_s = 0.0093$, this generates:

$$q_{li} = 4840\,(\text{Infiltration rate, cfm})(W_o - W_s)$$
$$= 4840 \times 67\,(0.0161 - 0.0093) = 2205\,\text{Btu/h, and}$$

$$q_{lv} = 4840\,(\text{Ventilation rate, cfm})(W_o - W_s)$$
$$= 4840 \times 1275\,(0.0161 - 0.0093) = 41{,}963\,\text{Btu/h}$$

Line 31, Total latent heat gain. The total latent heat gain is the sum of lines 27, 28, and 29.

(g) Total Instantaneous Heat Gain

Line 32, Total instantaneous heat gain. The sum of total instantaneous values on lines 27 and 31, sensible and latent heat gain, respectively. The hourly profile of such a total will normally reach a higher level at an earlier time of day than that of the building total cooling load, although the 24-h totals will be identical.

(h) Cooling Load from Convective Sensible Heat Gain Components

Lines 33 through 38. Direct inclusion of the instantaneous heat gain components listed in Lines 19, 25, and 26, and 20%, 50%, and 67% of lines 22, 23, and 24, respectively. These room sensible heat gain components (*i.e.*, loads due to air-to-air heat gain through windows, tungsten lights, fluorescent lights, people, infiltration, and ventilation) all appear as cooling load without delay. Percentages of sensible heat gain considered convective are taken from Tables 3 and 42. Selection of 67% of sensible gain from people as convective is an approximation for purposes of this example.

(i) Cooling Load Involving Time-Averaging Consideration

Radiant elements of instantaneous heat gain will be felt as cooling load in the space only after having first been absorbed by the mass of building and contents, and later released back into the space as convective heat. This delaying action is approximated by time-averaging, or taking the average of such a heat gain value for the current hour with those from some number of immediately previous hours. An averaging period of about 5 h is used for this example, in which, for example, the value of cooling load for hour 1200 is derived as the average of the radiant fractions of hourly sensible heat gain for hours 1200, 1100, 1000, 0900, and 0800; thus delaying the full impact of such heat gain becoming cooling load for 5 h, and extending the period after the heat gain has ended for some amount of cooling load to be felt by the space.

Line 39, Cooling load from tungsten light sensible heat gain. Although 80% of the sensible heat gain from tungsten lights is radiant heat and subject to the storage/re-release phenomenon, data on line 39 appears as a constant value for every hour. This is due to the constant heat input to the room (line 22), from lights switched on all the time and thus with the radiant heat gain component from prior hours being released as cooling load at the same rate as the absorption by the room of the current hour's radiant component.

Line 40, Cooling load from lighting cycled on and off. Fifty percent (the radiant component) of the fluorescent lighting heat gain from line 23, showing the effect of such gain being processed by time-averaging, as indicated above.

Line 41, Cooling load due to radiant heat gain from people. Of sensible heat generated by people, 33% is dissipated by radiation and felt by the space as cooling load only after having been absorbed by the mass of the building and its contents.

(j) Sensible Cooling Load from Exposed Surfaces

Elements of instantaneous heat gain from solar radiation through windows, walls, doors and roof, *i.e.*, the sum of values listed in lines 11 through 18, 20, and 21, are also delayed in being felt as cooling. The radiant heat gain by solar radiation transmitted through windows is treated the same way as the radiant portion of heat gain through walls and roof surfaces. However, since the windows have inside shading devices, solar radiation is considered reduced to approximately 58% of the solar heat gain through glass because the venetian blind intercepts about 42% of such solar radiation and releases it to the room in a convective form, similar to the treatment of heat gain through walls and roof [see Table 44].

(*Note*: Had there been *no internal shading* of the glass, the solar radiation through windows would have to be treated as 100% radiant, all subject to time-averaging. Translucent draperies fall between these limits, in a linear relationship. Chapter 27 has more specific information on internal shading.)

Lines 42a through 51a, Sensible cooling load from convective heat gain through enclosing surfaces. Data on lines 42b and 43b represent 58% of heat gain values for north and south windows, respectively, form lines 20 and 21, but time-averaged. Data for opaque enclosing surfaces on lines 44b through 51b represent 60% of the corresponding heat gain values on lines 11 through 18, but also time-averaged.

Cooling Load from power equipment and appliances. For this example, none are assumed. Had such loads been involved, with starting or ending periods within the time before the hour of calculation interest that can affect the averaging period, 20 to 80% of the sensible heat gain would have been considered as radiant and subject to time-averaging.

Line 52, Total room sensible cooling load. Total sensible cooling load felt by the room, and the design sensible load which is used as the basis for sizing cooling equipment. It is the sum of the values listed in lines 33 through 51b. The almost exact match between the 24-h total of 2,046,781 Btu/h on line 52 and the sum of the 24-h gain totals on line 27 (differing only by 16 Btu/h), while beyond reasonable limits of arithmetic precision, does verify completeness of the computation.

2. *Latent Cooling Load*

Line 31—The sum of lines 28, 29, and 30. Total latent heat gain is also the total latent cooling load, as all components occur instantaneously.

3. *Total Cooling Load*

Line 52, The sum of lines 52 and 31. The total cooling load for this example problem is the theoretical total for the conditions as defined, and may or may not represent the actual total cooling load imposed upon a system of cooling equipment attempting to maintain the specified space conditions. An appropriate psychrometric analysis of supply air, space air, return air, and mixed air [when ventilation air is mixed with return air enroute back to the cooling equipment] should be performed, in conjunction with proper consideration of the type of cooling equipment and characteristics of the preferred control scheme, in order to verify the ability of the design to meet the requirements, and to determine whether the actual sensible, latent, and total cooling loads are greater or less than the theoretical values calculated. The 1992 ASHRAE *Handbook—Systems and Equipment* has further information.

Comparison of Results

Each of the calculation procedures outlined in this chapter, TFM, CLTD/SCL/CLF, and TETD/TA have used the same building in Examples 6, 11, and 13, respectively. Although widely different in purpose, approach, and mathematical processes, the results have many similarities as illustrated by Figure 5.

Tabular data for hourly total instantaneous sensible heat gain and total sensible cooling load values from Tables 28 and 45 are plotted to compare the two computer-based techniques, TFM and TETD/TA. The curves for heat gain are almost identical. Those for cooling load, however, happen to peak at the same hour, 1600, but with different magnitudes. The TETD/TA cooling load peak has reached almost the peak of its com-

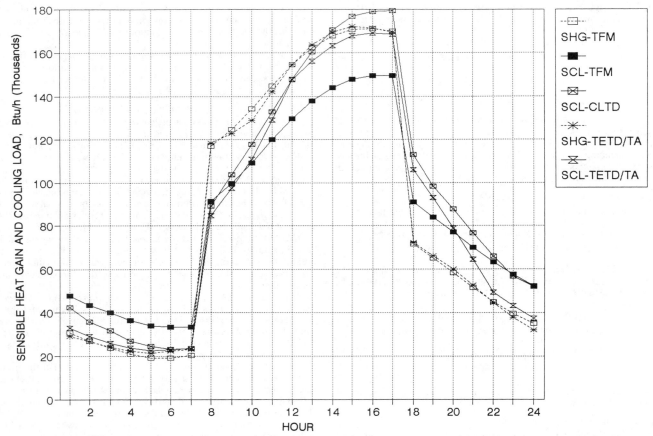

Fig. 5   TFM versus CLTD/SCF/CLF versus TETD/TA Methods of Calculating Sensible Heat Gain and Cooling Load

panion heat gain curve, but one hour later. The TFM heat gain curve reaches a peak at 1600 with a value only 0.5% different from that for TETD/TA, but the TFM cooling load curve peaks at only 87.5% of its heat gain curve. All unoccupied hours show substantially greater TFM cooling loads than for TETD/TA, while 24-h totals vary only by 0.15%.

As a manual procedure, Example 11, illustrating the use of CLTD/SCL/CLF, was carried through for hour 1600 only, in the manner that it would primarily be applied by users. For comparison purposes, it was also calculated for each of the daily 24 h and that cooling load profile plotted on Figure 5. There is no comparable heat gain profile, as this method does not produce such values directly. The curve peaks at 1700 hours, one hour later than the others, but with a total value 19.8% greater than TFM. The profile is somewhat different from and between those for TFM and TETD/TA during unoccupied hours.

Note: The small building used in these examples is more massive than typical for a similar function in post-1990 construction, and it would probably not meet ASHRAE *Standard* 90.1-1989 energy requirements. Calculating the entire building as a single simultaneous load could certainly be questioned, particularly in any larger configuration; thus, it is used here purely to illustrate the techniques discussed.

REFERENCES

Alereza, T. and J.P. Breen, III. 1984. Estimates of recommended heat gain due to commercial appliances and equipment. ASHRAE *Transactions* 90(2): 25-58.

American Gas Association. 1948. A comparison of gas and electric use for commercial cooking. Cleveland, OH.

American Gas Association. 1950. Gas and electric consumption in two college cafeterias. Cleveland, OH.

ASHRAE. 1975. Procedure for determining heating and cooling loads for computerized energy calculations, algorithms for building heat transfer subroutines.

ASHRAE. 1979. *Cooling and heating load calculation manual.*

ASHRAE. 1989. Ventilation for acceptable indoor air quality. ASHRAE *Standard* 62-1989.

Bliss, R.J,V. 1961. Atmospheric radiation near the surface of the ground. Solar Energy 5(3):103.

Buchberg, H. 1958. Cooling load from thermal network solutions ASHAE *Transactions* 64:111.

Chiles, D.C. and E.F. Sowell. 1984. A counter-intuitive effect of mass on zone cooling load response. ASHRAE *Transactions* 91(2A):201-208.

Consolazio, W. and L.J. Pecora. 1947. Minimal replenishment air required for living spaces. ASHVE *Transactions* 53:127.

Falconer, D.R., E.F. Sowell, J.D. Spitler, and B.B. Todorovic. 1993. Electronic tables for the ASHRAE Load Calculation Manual. ASHRAE *Transactions* 99(1).

Harris, S.M. and F.C. McQuiston. 1988. A study to categorize walls and roofs on the basis of thermal response. ASHRAE *Transactions* 94(2): 688-715.

Kimura and Stephenson. 1968. Theoretical study of cooling loads caused by lights. ASHRAE *Transactions* 74(2):189-97.

Marn, W.L. 1962. Commercial gas kitchen ventilation studies. Research Bulletin No. 90 (March). Gas Association Laboratories, Cleveland, Ohio.

McQuiston, F.C. and J.D. Spitler. 1992. *Cooling and heating load calculation manual*, 2nd ed. ASHRAE.

Mitalas, G.P. 1972. Transfer function method of calculating cooling loads, heat extraction rate, and space temperature. ASHRAE *Journal* 14(12):52.

Mitalas, G.P. 1973. Calculating cooling load caused by lights. ASHRAE *Journal* 15(6):7.

Mitalas, G.P. and J.G. Arsenault. 1971. Fortran IV program to calculate Z-transfer functions for the calculation of transient heat transfer through walls and roofs. Proceedings of the conference, Use of Computers for Environmental Engineering Related to Buildings. NBS *Building Science Series* 39 (October). Gaithersburg, MD.

Mitalas, G.P. and K. Kimura. 1971. A calorimeter to determine cooling load caused by lights. ASHRAE *Transactions* 77(2)65.

Mitalas, G.P. and D.G. Stephenson. 1967. Room thermal response factors. ASHRAE *Transactions* 73(2):III.2.1.

Nevins, R.G., H.E. Straub, and H.D. Ball. 1971. Thermal analysis of heat removal troffers. ASHRAE *Transactions* 77(2):58-72.

NFPA. 1987. Health care facilities. National Fire Protection Association *Standard* 99.

Rudoy, W. and F. Duran. 1975. Development of an improved cooling load calculation method. ASHRAE *Transactions* 81(2):19-69.

Sowell, E.F. 1988a. Classification of 200,640 parametric zones for cooling load calculations. ASHRAE *Transactions* 94(2):754-77.

Sowell, E.F. 1988b. Cross-check and modification of the DOE program for calculation of zone weighting factors. ASHRAE *Transactions* 94(2):737-53.

Sowell, E.F. 1988c. Load calculations for 200,640 zones. ASHRAE *Transactions* 94(2):71 6-36.

Sowell, E.F. and D.C. Chiles. 1984a. Characterization of zone dynamic response for CLF/CLTD tables. ASHRAE *Transactions* 91(2A): 162-78.

Sowell, E.F. and D.C. Chiles. 1984b. Zone descriptions and response characterization for CLF/CLTD calculations. ASHRAE *Transactions* 91(2A):179-200.

Spitler, J.D., F.C. McQuiston, and K.L. Lindsey. 1993. The CLTD/SCL/CLF cooling load calculation method. ASHRAE *Transactions* 99(1).

Spitler, J.D. and F.C. McQuiston. 1993. Development of a revised cooling and heating calculation manual. ASHRAE *Transactions* 99(1).

Stephenson, D.G. 1962. Method of determining non-steady-state heat flow through walls and roofs at buildings. *The Journal of the Institution of Heating and Ventilating Engineers* 30:5.

Stephenson, D.G. and G.P. Mitalas. 1967. Cooling load calculation by thermal response factor method. ASHRAE *Transactions* 73(2):III.1.1.

Stewart, J.P. 1948. Solar heat gain through walls and roofs for cooling load calculations. ASHVE *Transactions* 54:361.

Talbert, S.G., L.J. Canigan, and J.A. Eibling. 1973. An experimental study of ventilation requirements of commercial electric kitchens. ASHRAE *Transactions* 79(1):34.

Todorovic, B. 1987. The effect of the changing shade line on the cooling load calculations. ASHRAE videotape "Practical applications for cooling load calculations."

Todorovic, B. and D. Curcija. 1984. Calculative procedure for estimating cooling loads influenced by window shading, using negative cooling load method. ASHRAE *Transactions* 2:662.

BIBLIOGRAPHY

Historical

Alford, J.S., J.E. Ryan, and F.O. Urban. 1939. Effect of heat storage and variation in outdoor temperature and solar intensity on heat transfer through walls. ASHVE *Transactions* 45:387.

Brisken, W.R. and G.E. Reque. 1956. Thermal circuit and analog computer methods, thermal response. ASHAE *Transactions* 62:391.

Buchberg, H. 1955. Electric analog prediction of the thermal behavior of an inhabitable enclosure. ASHAE *Transactions* 61:339-386.

Buffington, D.E. 1975. Heat gain by conduction through exterior walls and roofs—transmission matrix method. ASHRAE *Transactions* 81(2):89.

Headrick, J.B. and D.P. Jordan. 1969. Analog computer simulation of heat gain through a flat composite roof section. ASHRAE *Transactions* 75(2):21.

Houghton, D.G., C. Gutherlet, and A.J. Wahl. 1935. ASHVE Research Report No. 1001—Cooling requirements of single rooms in a modern office building. ASHVE *Transactions* 41:53.

Leopold, C.S. 1947. The mechanism of heat transfer, panel cooling, heat storage. *Refrigerating Engineering* 7:33.

Leopold, C.S. 1948. Hydraulic analogue for the solution of problems of thermal storage, radiation, convection, and conduction. ASHVE *Transactions* 54:3-9.

Livermore, J.N. 1943. Study of actual vs predicted cooling load on an air conditioning system. ASHVE *Transactions* 49:287.

Mackey, C.O. and N.R. Gay, 1949. Heat gains are not cooling loads. ASHVE *Transactions* 55:413.

Mackey, C.O. and N.R. Gay. 1952. Cooling load from sunlit glass. ASHVE *Transactions* 58:321.

Mackey, C.O. and N.R. Gay. 1954. Cooling load from sunlit glass and wall. ASHVE *Transactions* 60:469.

Mackey, C.O. and L.T. Wright, Jr. 1944. Periodic heat flow—homogeneous walls or roofs. ASHVE *Transactions* 50:293.

Mackey, C.O. and L.T. Wright, Jr. 1946. Periodic heat flow—composite walls or roofs. ASHVE *Transactions* 52:283.

Nottage, H.B. and G.V. Parmelee. 1954. Circuit analysis applied to load estimating. ASHVE *Transactions* 60:59.

Nottage, H.B. and G.V. Parmelee. 1955. Circuit analysis applied to load estimating. ASHAE *Transactions* 61(2):125.

Parmelee, G.V., P. Vance, and A.N. Cherry. 1957. Analysls of an airconditioning thermal circuit by an electronic differential analyzer. ASHAE *Transactions* 63:129.

Paschkis, V. 1942. Periodic heat flow in building walls determined by electric analog method. ASHVE *Transactions* 48:75.

Romine, T.B., Jr. 1992. Cooling load calculation, Art or science? ASHRAE *Journal*, 34(1), p. 14.

Sun, T.-Y. 1968. Computer evaluation of the shadow area on a window cast by the adjacent building, ASHRAE *Journal*, September 1968.

Sun, T.-Y. 1968. Shadow area equations for window overhangs and side-fins and their application in computer calculation. ASHRAE *Transactions* 74(1): I-1.1 to I-1.9.

Vild, D.J. 1964. Solar heat gain factors and shading coefficients. ASHRAE *Journal* 6(10):47.

Transfer Function Method

Burch, D.M., B.A. Peavy, and F.J. Powell. 1974. Experimental validation of the NBS load and indoor temperature prediction model. ASHRAE *Transactions* 80(2):291.

Mast, W.D. 1972. Comparison between measured and calculated hour heating and cooling loads for an instrumented building. ASHRAE Symposium Bulletin No. 72-2.

McBridge, M.F., C.D. Jones, W.D. Mast, and C.F. Sepsey. 1975. Field validation test of the hourly load program developed from the ASHRAE algorithms. ASHRAE *Transactions* 1(1):291.

Mitalas, G.P. 1969. An experimental check on the weighting factor method of calculating room cooling load. ASHRAE *Transactions* 75(2):22.

Peavy, B.A., F.J. Powell, and D.M. Burch. 1975. Dynamic thermal performance of an experimental masonry building. NBS *Building Science Series* 45 (July).

CLTD and CLF Data

DeAlbuquerque, A.J. 1972. Equipment loads in laboratories. ASHRAE *Journal* 14(10):59.

Kusuda, T. 1969. Thermal response factors for multilayer structures of various heat conduction systems. ASHRAE *Transactions* 75(1):246

Mitalas, G.P. 1968. Calculation of transient heat flow through walls and roofs. ASHRAE *Transactions* 74(2):182.

Rudoy, W. 1979. Don't turn the tables. ASHRAE *Journal* 21(7):62.

Stephenson, D.G. and G.P. Mitalas. 1971. Calculation of heat conduction transfer functions for multilayer slabs. ASHRAE *Transactions* 77(2):1.17.

Todorovic B. 1982. Cooling load from solar radiation through partially shaded windows, taking heat storage effect into account. ASHRAE *Transactions* 88(2): 924-937.

Todorovic, B. 1984. Distribution of solar energy following its transmittal through window panes. ASHRAE *Transactions* 90(1B): 806-15.

Todorovic, B. 1989. Heat storage in building structure and its effect on cooling load; Heat and mass transfer in building materials and structure. Hemisphere publishing, New York, 603-14.

CHAPTER 27

FENESTRATION

FENESTRATION refers to any glazed aperture in the building envelope. Fenestration components include: (1) glazing material, either glass or plastic; (2) framing, mullions, muntins, and dividers; (3) external shading devices; (4) internal shading devices; and (5) integral (between-glass) shading systems.

Fenestration (1) satisfies human needs for visual communication with the outside world; (2) admits solar radiation to provide supplemental daylight and heat, and, in some cases, outside air; (3) provides egress in low-rise buildings in case of fire or other emergencies; and (4) enhances the exterior and interior appearance of a building.

The designer must consider the following factors when selecting fenestration: (1) *architectural*, by identifying design options and their synthesis to achieve energy conservation, including possible use of both electric illumination and daylight, with controls to reduce electric lighting automatically when daylight is available; (2) *thermal*, by designing for heat losses and/or gains and surface temperatures consistent with occupant comfort and energy conservation; (3) *economic*, by evaluating first costs and life-cycle costs of alternative fenestration designs; and (4) *human need*, by determining the psychological desire or physical need for fenestration and the proper illumination standards for projected use of the space and for occupant comfort and acceptance.

This chapter covers the design and evaluation of fenestration based on thermal analysis and energy conservation. When necessary, the information in this chapter should be modified to follow manufacturers' directions for proper use of their products.

Energy Effects of Fenestration

Fenestration affects building energy use through four basic mechanisms—thermal heat transfer, solar heat gains, air leakage, and daylighting. The net effect of a fenestration system on total building energy requirements, including heating, cooling, and lighting energy, depends on (1) the characteristics and orientation of the fenestration, (2) the weather and solar radiation conditions, and (3) the operation of the building thermal system.

Careful design of the fenestration can minimize energy use. In general, the energy impacts of fenestration can be optimized by (1) using daylight to offset lighting requirements, (2) using passive solar heat gain to offset winter space heating, (3) using glazings with special transmission properties to reject excessive solar heat gain, and (4) specifying insulated glazings and low air leakage fenestration designs.

Fenestration Components

Common glazing materials are glass, plastic, and other materials having good light transmittance. Inward or outward heat flow through fenestration by radiation, conduction, or convection can be controlled by various single or multiple (insulating) glazings, by exterior and/or interior shading, and by spectrally selective tints, films, and/or coatings. Sunlight or skylight that is too bright or glaring may be controlled by tinted glazing and by fixed or adjustable exterior or interior shading or both. Tinted glazing and/or draperies of proper openness may be appropriate to modify an unattractive or distracting view. Privacy with a degree of outward view can be obtained only when the illumination level outside is substantially above that inside. For this purpose, reflective or tinted glazing and/or draperies of the correct shade and openness of weave can be used and are effective in the daytime. For privacy at night, nearly opaque materials, as well as slatted blinds or roller shades, are more effective.

Outdoor sounds can be reduced with an airtight building envelope. Additional noise reduction can be achieved by heavier and/or multiple glazing, well sealed around the edges. While draperies of closely woven fabrics are good sound absorbers and help reduce sound within a room, they do not directly reduce outdoor sound coming through fenestration. Glazing materials must not break when under wind loads and thermal stress; they must also meet building codes. Applicable published standards and manufacturers' recommendations should be followed.

CALCULATING FENESTRATION ENERGY FLOW

Energy flows through fenestration via three physical effects: (1) conductive and convective heat transfer between the outer fenestration surface and the adjacent air caused by the temperature difference between that air and the interior space of the building; (2) net long-wave (above 2500 nm) radiative exchange between the outer fenestration surface and the sky, ground, or adjacent objects; and (3) short-wave (below 2500 nm) solar radiation incident on the window, either directly from the sun after scattering in the atmosphere or after reflection from the ground or adjacent objects. Simplified calculations are based on the observation that the temperatures of the sky, ground, and surrounding objects (and hence their radiant heat transfers) correlate with the exterior air temperature. The radiative interchanges are then approximated by assuming that all the radiating surfaces (including the sky) are at the same temperature as the outdoor air. With this assumption, the basic equation for the instantaneous energy flow q through a fenestration is

The preparation of this chapter is assigned to TC 4.5, Fenestration.

$$q = U_o A_T (t_{out} - t_{in}) + E_t (A_G F_G + A_{Frm} F_{Frm}) \qquad (1)$$

where

U_o = overall coefficient of heat transfer (U-factor)
t_{in} = interior air temperature
t_{out} = exterior air temperature
A_T = total area of fenestration
A_G = glazing (transparent) area
A_{Frm} = frame (opaque) area
F_G = glazing solar heat gain coefficient
F_{Frm} = frame solar heat gain coefficient
E_t = incident total irradiance

The quantities U_o, F_G, and F_{Frm} are frequently considered constants; however, they slowly vary as functions of the environmental variables, most importantly temperatures and wind speed. The solar heat gain coefficients also depend significantly on solar incident angle.

The principal justification for Equation (1) is its simplicity, achieved by collecting all the linked radiative, conductive, and convective energy transfer processes into U and F. These quantities vary slowly because (1) convective heat transfer rates vary as fractional powers of temperature differences or free-stream speeds, (2) variations in temperature due to the weather or climate are small on the absolute temperature scale (°R) that controls radiative heat transfer rates, and (3) fenestration systems always involve at least two thermal resistances in series.

OVERALL COEFFICIENT OF HEAT TRANSFER (U-FACTOR)

In the absence of sunlight, air infiltration, and moisture condensation, the first term in Equation (1) represents the rate of thermal heat transfer through a fenestration system. Most systems consist of transparent multipane glazing units and an opaque sash and frame (hereafter called frame). The glazing unit's heat transfer paths include a one-dimensional center-of-glass contribution and a two-dimensional edge contribution. The frame contribution is primarily two-dimensional.

Consequently, the total rate of heat transfer through a fenestration system can be calculated knowing the separate heat transfer contributions of the center glass, edge glass, and frame. (When present, glazing dividers, such as decorative grilles and muntins, also affect heat transfer, and their contribution must be considered). The overall U-factor is estimated using area-weighted U-factors for each contribution by:

$$U_o = (U_{cg}A_{cg} + U_{eg}A_{eg} + U_fA_f)/A_{pf} \qquad (2)$$

where the subscripts *cg*, *eg*, and *f* refer to the center-of-glass, edge-of-glass, and frame, respectively. A_{pf} is the area of the fenestration product's rough opening in the wall or roof less installation clearances. Where a fenestration product has glazed surfaces in only one direction (typical windows), the sum of the areas equals the projected area. Skylights, greenhouse/garden windows, bay/bow windows, etc., because they extend beyond the plane of the wall/roof, have greater surface area for heat loss than that of a window with a similar glazing option and frame material; consequently, U-factors for such products are greater.

Center-of-glass U-factor. Heat flow across the central glazed portion of a multipane unit must consider both convective and radiative transfer in the gas space. Convective heat transfer is estimated based on high aspect-ratio, natural convection correlations for vertical and inclined air layers (ElSherbiney *et al.* 1982, Shewen 1986, Wright 1991). Radiative heat transfer (ignoring gas absorption) is quantified using a more fundamental approach. Rubin (1982) and Hollands and Wright (1982) devised computational methods to solve the combined heat transfer problem.

Values for U_{cg} at standard indoor and outdoor conditions depend on such glazing construction features as the number of glazing panes, the gas-space dimensions, the orientation relative to vertical, the emittance of each surface, and the composition of the fill gas. Several computer programs can be used to estimate glazing unit heat transfer for a wide range of glazing construction (Arasteh *et al.* 1989, Sullivan and Wright 1987). The National Fenestration Rating Council calls for WINDOW 4.0 from Lawrence Berkeley Laboratory (1992) as a standard calculation method.

Figure 1 shows the effect of gas space width on U_{cg} for various glazing units. U-factors are plotted for air, argon, and krypton fill-gases and for high (uncoated) and low (coated) values of surface emittance. Gas space widths greater than 0.5 in. have no significant effect on U_{cg}, but greater glazing unit thicknesses decrease U_o since the length of the shortest heat flow path through the frame increases. A low-emittance coating combined with krypton gas fill significantly reduces heat transfer in narrow gap-width glazing units. The value of U_{cg} for sloped glazings (with heat flow

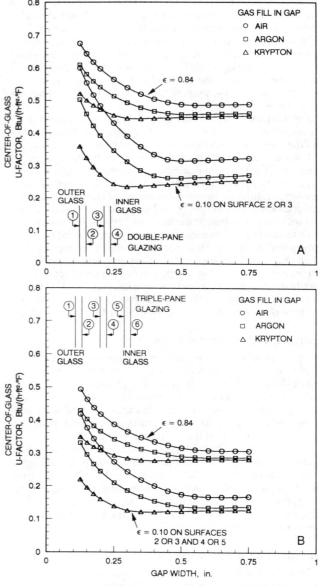

Fig. 1   Center of Glass U-Factor for Vertical Double- and Triple-Pane Glazing Units

up) is greater than that for vertical glazings because both the interior film coefficient and the air space coefficient is larger, particularly with the 0.5 in. gap spacing between glazings.

Edge-of-glass U-factor. Insulating glass units usually have continuous spacer members around the glass perimeter to separate the glazing and provide an edge seal. Aluminum spacers greatly increase conductive heat transfer between the contacted inner and outer glazing, thereby degrading the thermal performance of the glazing unit locally. Laboratory measurements reported by Peterson (1987) showed this conductive region to be limited to a 2.5 inch-wide band around the perimeter of the glazing unit.

Edge-of-glass heat transfer is two-dimensional and requires detailed modeling for accurate determination. Based on detailed two-dimensional modeling, Arasteh (1989) developed the following correlation to calculate the edge-of-glass U-factor as a function of spacer type and center-of-glass U-factor:

$$U_{eg} = A + BU_{cg} + CU_{cg}^2 \qquad (3)$$

where A, B, and C are correlation coefficients, which are listed in Table 1 for metal, insulating (including wood) and fused glass spacers, and a combination of insulating and metal spacers. The correlation constants for the combination of metal and insulated spacers were derived from computer simulations, which showed that 85% of the benefit in triple-glazing is attributable to the insulated spacer.

Approximate edge-of-glass U-factors as a function of the center-of-glass U-factor is shown in Figure 2. The spacer edge is assumed to be even with the line of sight of the glazing. Curves

Table 1   Equation (3) Coefficients for Edge-of-Glass U-Factor

	A	B	C
Metal	0.223	0.842	−0.153
Insulating	0.120	0.682	0.244
Glass	0.158	0.774	0.057
Metal + insulation	0.135	0.706	0.187

Note: A, B, and C have units of $[\text{Btu}/(\text{h} \cdot \text{ft}^2 \cdot {}^\circ\text{F})]^n$, where $n = 1, 0,$ and -1, respectively.

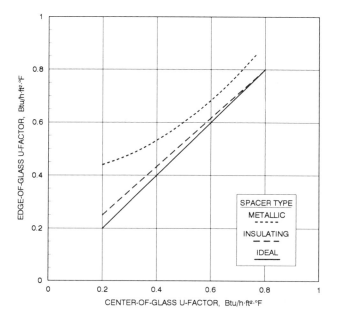

Fig. 2   Relationship between Edge-of-Glass U-Factor and Center-of-Glass U-Factor for Various Spacers

are for aluminum spacers with sealants (metallic) and non-metal (insulating) spacers, including fiberglass, wood, and butyl. Values for glass edges and steel spacers fall between the metallic and insulating spacer curves. This edge effect does not occur with single glazing. For highly insulating glazing, edge heat transfer can significantly increase the overall U-factor. Thus, test data or design-specific computations should account for this effect.

Frame U-factor. Fenestration frame elements consist of all structural members exclusive of the glazing units and include sash, jamb, head, and sill members; meeting rails; mullions; and other glazing dividers. Estimating the rate of heat transfer through the frame is complicated by (1) the variety of fenestration products and frame configurations, (2) the different combinations of materials used for frames, (3) the different sizes available in residential and commercial applications, and to a lesser extent, (4) the glazing unit width and spacer type. Internal dividers or grilles have little impact on the fenestration U-factor, provided there is at least a 1/8-in. gap between the divider and each panel of glass.

Fenestration product type (or operator type) refers to the design configuration (*e.g.*, casement, slider, or fixed window, patio door, skylight, etc.). Materials commonly used for framing members include wood, aluminum, and vinyl. Fiberglass-framed windows are also available. Manufacturers sometimes combine these materials as clad units, (*e.g.*, vinyl-clad aluminum, aluminum-clad wood) to increase durability or improve aesthetics. Aluminum units are available in both solid metal and in "thermally broken" units with plastic inserts that reduce the conductive heat transfer through framing members. Vinyl-framed units may also be structurally reinforced with steel rods and/or may have interior cavities filled with foam insulating materials.

Computer simulations found that frame heat loss in most fenestration is controlled by a single component or controlling resistance, and only changes in this component significantly affect frame heat loss (Enermodal Engineering 1990). For example, the frame U-factor for thermally broken aluminum windows is largely controlled by the depth of the thermal break material in the heat flow direction. For aluminum frames without a thermal break, the inside film coefficient provides most of the resistance to heat flow. For vinyl- and wood-framed fenestrations, the controlling resistance is the shortest distance between the inside and outside surfaces, which usually depends on the thickness of the sealed glazing unit.

Carpenter and McGowan (1993) experimentally validated frame U-factors for a variety of fixed and operable window types, sizes, and materials using computer modelling techniques. Table 2 lists frame U-factors for a variety of frame and spacer materials and glazing unit thicknesses.

Interior and exterior surface coefficients. Part of the overall thermal resistance of a fenestration system is due to the convective and radiative heat transfer between the exposed surfaces and the environment. Surface heat transfer coefficients at the outer and inner glazing surfaces, h_o and h_i, respectively, combine the effects of radiation and convection.

The wind speed and orientation of the building are important in determining h_o. This relationship has long been studied and many correlations have been proposed for h_o as a function of wind speed. However, no universal relationship has been accepted, and limited field measurements at low air speeds by Klems (1989) show significant difference with values used by others.

Convective heat transfer coefficients are usually determined at standard temperature and air velocity conditions on each side. Wind speed can vary from less than 50 fpm for calm weather, free convection conditions, to over 25 mph for storm conditions. A standard value of 5.1 Btu/h · ft² · °F corresponding to a 15 mph is often used to represent winter design conditions. At near-zero

Table 2   Representative Fenestration Frame U-Factors in Btu/(h·ft²·°F)—Vertical Orientation

Frame Material	Type of Spacer	Operable Single[b]	Operable Double[c]	Operable Triple[d]	Fixed Single	Fixed Double	Fixed Triple	Double Door Single	Double Door Double	Double Door Triple	Skylight Single	Skylight Double	Skylight Triple
Aluminum	All	2.18	2.18	2.18	1.78	1.78	1.78	2.24	2.24	2.24	6.80	6.80	6.80
Thermal broken	Metal	0.95	0.95	0.95	1.16	1.16	1.16	1.04	1.04	1.00	6.95	5.11	4.88
aluminum[a]	Insulated	n/a	0.86	0.86	n/a	0.92	0.92	n/a	0.97	0.95	n/a	4.68	4.63
Al-clad wood/	Metal	0.69	0.63	0.58	0.56	0.53	0.49	0.58	0.55	0.51	2.43	2.31	2.27
reinforced vinyl	Insulated	n/a	0.56	0.48	n/a	0.46	0.40	n/a	0.49	0.44	n/a	2.25	2.22
Wood/vinyl	Metal	0.55	0.51	0.48	0.51	0.49	0.48	0.51	0.49	0.48	2.17	2.02	1.99
	Insulated	n/a	0.46	0.39	n/a	0.42	0.37	n/a	0.44	0.40	n/a	1.94	1.87
Insulated	Metal	0.37	0.33	0.30	0.37	0.33	0.30	0.37	0.33	0.30	n/a	n/a	n/a
fiberglass/vinyl	Insulated	n/a	0.28	0.25	n/a	0.28	0.25	n/a	0.28	0.25	n/a	n/a	n/a

[a]Depends strongly on width of thermal break. Value given is for 3/8 in.
[b]Single glazing corresponds to individual glazing unit thickness of 1/8 in (nominal).
[c]Double glazing corresponds to individual glazing unit thickness of 3/4 in (nominal).

[d]Triple glazing corresponds to individual glazing unit thickness of 1 3/8 in (nominal).
n/a Not applicable

Table 3   Indoor Radiation and Convection Coefficient h_i (Still Air Conditions)

Indoor Glass Surface Emittance, e_g						Temp. Diff., °F	Glass Temp., °F	Room Temp. °F
0.05	0.10	0.20	0.40	0.84	0.90			
Indoor Coefficient h_i, Btu/(h·ft²·°F)								
0.45	0.51	0.61	0.81	1.25	1.31	5	65	70
0.53	0.58	0.68	0.88	1.31	1.37	10	60	(winter
0.62	0.67	0.76	0.96	1.38	1.44	20	50	design)
0.68	0.73	0.81	1.01	1.42	1.48	30	40	
0.73	0.77	0.86	1.04	1.44	1.50	40	30	
0.76	0.81	0.90	1.07	1.46	1.51	50	20	
0.79	0.84	0.92	1.10	1.47	1.52	60	10	
0.81	0.88	1.00	1.25	1.79	1.87	60	135	75
0.78	0.84	0.96	1.20	1.73	1.80	50	125	(summer
0.74	0.80	0.91	1.15	1.66	1.73	40	115	design)
0.69	0.75	0.86	1.09	1.59	1.66	30	105	
0.63	0.68	0.79	1.02	1.50	1.57	20	95	
0.53	0.59	0.70	0.91	1.39	1.45	10	85	
0.46	0.51	0.62	0.83	1.30	1.36	5	80	

$h_i = h_c + h_r = A (\Delta t)^{0.25} + [e_g \sigma (T_g^4 - T_i^4)]/(T_g - T_i)$ where $A = 0.27$.

wind speed, h_o varies with outside air and surface temperature, orientation to vertical, and air moisture content.

For natural convection at the inner surface of a vertical window, the inner surface coefficient depends on the indoor air and glass surface temperatures and on the emittance of the glass inner surface. Table 3 shows the variation of h_i for winter ($t_i = 70$°F) and summer ($t_i = 75$°F) design conditions. Designers often use $h_i = 1.46$ Btu/h·ft²·°F, which corresponds to $t_i = 70$°F, glass temperature = 20°F, and uncoated glass with $e_g = 0.84$. For summer conditions, the conventional $h_i = 1.46$ Btu/h·ft² °F corresponds approximately to glass temperature = 90°F and $t_i = 75$°F.

Heat transfer between the glazing surface and its environment is driven not only by the local air temperatures but also by the radiant temperatures to which the surface is exposed. The radiant temperature of the indoor environment is generally assumed to be equal to the indoor air temperature. While this is a safe assumption where a small window is exposed to a large room with surface temperatures equal to the air temperature, it is not valid in rooms where the window is exposed to other large areas of glazing surfaces (e.g., greenhouse, atrium) or to other cooled or heated surfaces (Parmelee 1947).

The radiant temperature of the outdoor environment is frequently assumed to be equal to the outdoor air temperature. This assumption may be in error, since additional radiative heat loss occurs between a fenestration and the clear sky (Duffie and Beck-

man 1980). Therefore, for clear sky conditions, some effective outdoor temperature $t_{o,e}$ should replace t_o, in Equation (1). For methods for determining $t_{o,e}$ see, for example, work by the University of Waterloo (1987). Note that a fully cloudy sky is assumed in ASHRAE design conditions.

Especially for single glass, U-factors depend strongly on interior and exterior film coefficients. The U-factor for single glass (neglecting the glass resistance) is:

$$U_{cg} = h_o h_i/(h_o + h_i) \qquad (4)$$

Values for h_i presented in Table 3 can result in values of U_{cg} for single glass to range from more than 1.4 Btu/h·ft²·°F in winter for an internally coated glass with a low-emittance film to approximately 1.0 Btu/h·ft²·°F in summer for an uncoated clear glass, and 1.2 Btu/h·ft²·°F with bright sunshine.

The air space in an insulating glass panel made up of glass with no reflective coating on the air space surface has a coefficient h_S is given in Table 4 for an effective emittance of 0.72. When a reflective coating is applied to the air space surface, h_S can be selected from Table 4 by first calculating the effective air space emittance E by Equation (5).

$$E = 1/(1/e_o + 1/e_i - 1) \qquad (5)$$

where e_o and e_i are the hemispherical emittances of the two air space surfaces. Hemispherical emittance of ordinary uncoated glass is 0.84 over a wavelength range of 0.4 to 40 μm.

Representative U-factors for fenestration systems. Table 5 lists computed U-factors at winter design conditions for many currently available designs. The table is based on ASHRAE-sponsored research involving laboratory testing and computer simulation of various fenestration products. Values are listed for vertical installation and for skylights that are sloped 20° from the horizontal. When available, test data may provide more accurate results for specific products. However, a wide range of measured U-factors for similar products has been reported (Hogan 1988). Different test methods sometimes give different U-factors (McCabe et al. 1986), but the procedures presented in NFRC Standard 100-91 or CSA Standard A440.2 are generally accepted. Fenestration should be rated in accordance with these standards, while Table 5 can be used as an estimating tool in the early phases of design.

Data are based on center-of-glass and edge-of-glass component U-factors and assume that there are no dividers. If product-specific information is not available from the NFRC or CSA standards programs, the overall U-factors in Table 5 are appropriate

Table 4 Coefficients for Horizontal Heat Flow

Air Space Coefficient h_s, Btu/(h·ft²·°F) Effective Emittance E						Air Temp. Diff., °F	Air Space Temp., °F	Air Space Thickness, °F
0.05	0.10	0.20	0.40	0.72	0.82			
0.36	0.40	0.47	0.61	0.84	0.91	10	10	0.50
0.37	0.41	0.48	0.62	0.85	0.92	30		
0.40	0.43	0.50	0.65	0.87	0.95	50		
0.45	0.48	0.55	0.70	0.93	1.00	70		
0.48	0.52	0.59	0.73	0.96	1.04	90		
0.38	0.42	0.50	0.66	0.92	1.00	10	30	
0.39	0.43	0.51	0.67	0.93	1.01	30		
0.41	0.45	0.54	0.70	0.96	1.04	50		
0.45	0.50	0.58	0.74	1.00	1.08	70		
0.49	0.53	0.61	0.78	1.04	1.12	90		
0.40	0.44	0.53	0.71	1.00	1.10	10	50	
0.41	0.45	0.54	0.72	1.02	1.11	30		
0.43	0.48	0.57	0.75	1.04	1.13	50		
0.46	0.51	0.60	0.78	1.07	1.17	70		
0.50	0.55	0.64	0.82	1.11	1.21	90		
0.43	0.49	0.60	0.83	1.19	1.31	10	90	
0.44	0.50	0.61	0.84	1.21	1.32	30		
0.47	0.53	0.64	0.87	1.24	1.35	50		
0.48	0.54	0.65	0.88	1.25	1.36	70		
0.52	0.58	0.69	0.92	1.29	1.40	90		
0.45	0.51	0.64	0.89	1.30	1.43	10	110	
0.46	0.52	0.65	0.90	1.31	1.44	30		
0.49	0.55	0.68	0.93	1.34	1.47	50		
0.49	0.55	0.68	0.93	1.34	1.47	70		
0.53	0.59	0.72	0.97	1.38	1.51	90		
0.47	0.51	0.58	0.72	0.95	1.02	10	10	0.28–0.38
0.47	0.51	0.58	0.72	0.95	1.02	50		
0.50	0.53	0.60	0.75	0.98	1.05	90		
0.49	0.53	0.61	0.77	1.03	1.11	10	30	
0.49	0.53	0.61	0.77	1.03	1.11	50		
0.52	0.56	0.64	0.80	1.06	1.14	90		
0.51	0.56	0.65	0.83	1.12	1.21	10	50	
0.51	0.56	0.65	0.83	1.12	1.21	50		
0.54	0.58	0.68	0.86	1.15	1.24	90		
0.56	0.61	0.73	0.95	1.32	1.43	10	90	
0.56	0.61	0.73	0.95	1.32	1.43	50		
0.58	0.64	0.76	0.99	1.35	1.47	90		
0.58	0.64	0.77	1.02	1.43	1.55	10	110	
0.58	0.64	0.77	1.02	1.43	1.56	50		
0.61	0.67	0.80	1.05	1.46	1.59	90		
0.69	0.72	0.80	0.94	1.17	1.24	10	10	0.25
0.69	0.72	0.80	0.94	1.17	1.24	90		
0.72	0.76	0.84	1.00	1.26	1.34	10	30	
0.72	0.76	0.84	1.00	1.26	1.34	90		
0.75	0.79	0.88	1.06	1.35	1.45	10	50	
0.75	0.79	0.88	1.07	1.36	1.45	90		
0.81	0.86	0.98	1.20	1.57	1.68	10	90	
0.81	0.86	0.98	1.21	1.57	1.69	90		
0.84	0.90	1.03	1.28	1.69	1.81	10	110	
0.84	0.90	1.03	1.28	1.69	1.82	90		
0.91	0.94	1.01	1.15	1.38	1.45	10	10	0.19
0.91	0.94	1.01	1.16	1.39	1.46	90		
0.94	0.98	1.06	1.22	1.48	1.56	10	30	
0.94	0.98	1.06	1.23	1.49	1.57	90		
0.98	1.03	1.12	1.30	1.59	1.68	10	50	
0.98	1.03	1.12	1.30	1.59	1.68	90		
1.05	1.11	1.23	1.45	1.82	1.93	10	90	
1.05	1.11	1.23	1.46	1.82	1.94	90		
1.09	1.16	1.28	1.54	1.94	2.07	10	110	
1.09	1.16	1.28	1.54	1.95	2.08	90		

Adapted from Building Materials and Structures, Report 151, United States Dept. of Commerce.

for calculating compliance with the requirements of ASHRAE *Standard* 90. However, they apply only to the specific design conditions described in the footnotes in the table, and they are typically used only to determine peak load conditions for sizing heating equipment.

While these U-factors have been determined for winter conditions, they can also be used to estimate heat gain during peak cooling conditions, since conductive gain, which is one of several variables, is usually a small portion of the total heat gain for fenestration in direct sunlight. Glazing designs and framing materials may be compared in choosing a fenestration system that needs a specific winter design U-factor.

The multiple glazing categories are appropriate for sealed glass units and the addition of storm sash to other glazing units. No distinction is made between flat and domed units such as skylights. For acrylic domes, use an average gas-space width to determine the U-factor. Note that skylight U-factors are greater than those of other similar products. While this is partially due to the difference in slope, it is largely because the skylight surface area, which includes the curb, can vary from 13 to 240% greater than the rough opening area depending on the size and mounting method. Unless noted, all multiple-glazed units are filled with dry air. Argon units are filled with 100% argon. U-factors for CO_2 filled units are similar to argon fills. For spaces up to 0.5 in., argon/SF_6 mixtures with up to 70% SF_6 are generally the same as argon fills. Krypton gas can provide U-factors lower than those for argon for glazing spaces less than 0.5 in.

Table 5 provides data for four values of hemispherical emittance and for 0.25 and 0.5 in. gas space width. The emittance of various low-emittance glasses varies considerably between manufacturers and processes. When the emittance is between the listed values, interpolation may be used. When manufacturers' data are not available, assume that glass with a pyrolytic (hard) coating has an emittance of 0.40, and that glass with a sputtered (soft) coating has an emittance of 0.10. Some coated glazings have lower emittance values. Tinted glass does not change the winter U-factor. Also, some reflective glass may have an emittance less than 0.84. Nonlinear interpolation for gas space width should be based on the equation in footnote 5.

Frame type refers to the primary unit. Thus, when storm sash is added over another glazing, use the values for a nonstorm frame. For glazing with a steel frame, use aluminum frame values. For vertical sliders, horizontal sliders, casement, awning, and pivoted and dual-action windows, use the operable category. Fixed windows can also represent glazed wall systems. Double-door U-factors can also represent single glass doors.

The U-factors in Table 5 are based on the definitions of the four product types, nominal frame sizes, and proportion of frame to glass area as shown in Figure 3. The nominal dimensions for skylights (NFRC size AA) correspond to centerline spacings of roof framing members; consequently, the rough opening dimensions are 22.5 in. by 46.5 in. The skylight coefficients assume the presence of an uninsulated 4 in. curb (either integral to the skylight or field installed). Skylights with smaller curbs, insulated curbs, or no curbs would have lower U-factors.

To estimate the overall U-factor of a fenestration product that differs significantly from the assumptions given in Table 5 and/or Figure 3, first determine the percentage area that is frame/sash, center-of-glass, and edge-of-glass (based on a 2.5 in. band around the perimeter of each glazing unit). Next, determine the appropriate component U-factors. These can be taken either from the standard values listed in italics in Table 5 for glass, the values in Table 2 for frames, or from some other source such as test data or computed factors. Finally, multiply the percent area and the component U-factors, and sum these products to obtain the overall U-factor U_o.

Table 5 Overall Coefficients of Heat Transmission of Various Fenestration Products (Btu/h · ft² · °F)

	Glass Only			Aluminum w/o Thermal Break				Aluminum with Thermal Break							
Operator Type	Center Glass	Edge of Glass		Operable	Fixed	Double Door	Sloped Skylight	Operable		Fixed		Double Door		Sloped Skylight	
Spacer Type		Metal	Insul.	All	All	All	All	Metal	Insul.	Metal	Insul.	Metal	Insul.	Metal	Insul.
Glazing ID and Type															
Single Glazing															
1 1/8 in. glass	1.11	1.11	—	1.30	1.17	1.26	1.92	1.07	—	1.11	—	1.10	—	1.93	—
2 1/4 in. acrylic/polycarbonate	0.93	0.93	—	1.15	1.00	1.10	1.72	0.93	—	0.95	—	0.95	—	1.73	—
3 1/8 in. acrylic/polycarbonate	1.02	1.02	—	1.22	1.08	1.18	1.93	1.00	—	1.03	—	1.02	—	1.94	—
4 glass block	0.60														
Double Glazing															
5 1/4 in. air space	0.57	0.65	0.59	0.87	0.69	0.80	1.30	0.67	0.64	0.63	0.60	0.66	0.64	1.13	1.07
6 1/2 in. air space	0.49	0.60	0.51	0.81	0.62	0.74	1.29	0.62	0.58	0.56	0.53	0.59	0.57	1.12	1.06
7 1/4 in. argon space	0.52	0.62	0.54	0.83	0.64	0.76	1.25	0.64	0.60	0.59	0.56	0.62	0.60	1.08	1.02
8 1/2 in. argon space	0.46	0.58	0.49	0.78	0.59	0.71	1.25	0.59	0.56	0.54	0.50	0.57	0.55	1.08	1.02
Double Glazing, E = 0.40 on surface 2 or 3															
9 1/4 in. air space	0.50	0.61	0.52	0.81	0.63	0.74	1.24	0.62	0.59	0.57	0.54	0.60	0.58	1.06	1.00
10 1/2 in. air space	0.40	0.54	0.43	0.74	0.54	0.66	1.23	0.55	0.51	0.49	0.45	0.52	0.50	1.06	0.99
10 1/4 in. argon space	0.43	0.56	0.46	0.76	0.57	0.69	1.17	0.57	0.54	0.51	0.48	0.54	0.52	1.00	0.93
12 1/2 in. argon space	0.36	0.51	0.40	0.71	0.50	0.63	1.18	0.52	0.48	0.45	0.41	0.49	0.46	1.01	0.94
Double Glazing, E = 0.20 on surface 2 or 3															
13 1/4 in. air space	0.46	0.58	0.49	0.78	0.59	0.71	1.20	0.59	0.56	0.54	0.50	0.57	0.55	1.03	0.96
14 1/2 in. air space	0.35	0.50	0.39	0.70	0.50	0.62	1.19	0.52	0.48	0.44	0.40	0.48	0.46	1.02	0.95
15 1/4 in. argon space	0.38	0.52	0.41	0.72	0.52	0.64	1.12	0.54	0.50	0.47	0.43	0.50	0.48	0.95	0.87
16 1/2 in. argon space	0.30	0.46	0.35	0.66	0.45	0.58	1.14	0.48	0.44	0.40	0.36	0.44	0.42	0.97	0.89
Double Glazing, E = 0.10 on surface 2 or 3															
17 1/4 in. air space	0.43	0.56	0.46	0.76	0.57	0.69	1.17	0.57	0.54	0.51	0.48	0.54	0.52	1.00	0.93
18 1/2 in. air space	0.32	0.48	0.36	0.67	0.47	0.59	1.17	0.49	0.45	0.42	0.38	0.46	0.43	1.00	0.93
19 1/4 in. argon space	0.35	0.50	0.39	0.70	0.50	0.62	1.09	0.52	0.48	0.44	0.40	0.48	0.46	0.92	0.84
20 1/2 in. argon space	0.27	0.44	0.32	0.64	0.43	0.55	1.11	0.46	0.42	0.37	0.33	0.42	0.39	0.94	0.86
Triple Glazing															
21 1/4 in. air spaces	0.38	0.52	0.41	0.72	0.52	0.64	1.13	0.54	0.50	0.47	0.43	0.50	0.48	0.93	0.88
22 1/2 in. air spaces	0.31	0.47	0.35	0.67	0.46	0.59	1.10	0.49	0.45	0.41	0.37	0.44	0.42	0.91	0.85
23 1/4 in. argon spaces	0.34	0.49	0.38	0.69	0.49	0.61	1.08	0.51	0.47	0.43	0.40	0.47	0.45	0.89	0.83
24 1/2 in. argon spaces	0.29	0.45	0.34	0.65	0.44	0.57	1.07	0.47	0.43	0.39	0.35	0.43	0.40	0.88	0.82
Triple Glazing, E = 0.40 on surface 2, 3, 4, or 5															
25 1/4 in. air spaces	0.35	0.50	0.39	0.70	0.50	0.62	1.09	0.52	0.48	0.44	0.40	0.47	0.45	0.90	0.84
26 1/2 in. air spaces	0.27	0.44	0.32	0.64	0.43	0.55	1.06	0.46	0.42	0.37	0.33	0.41	0.39	0.87	0.81
27 1/4 in. argon spaces	0.30	0.46	0.35	0.66	0.45	0.58	1.04	0.48	0.44	0.40	0.36	0.43	0.41	0.85	0.79
28 1/2 in. argon spaces	0.24	0.42	0.30	0.61	0.40	0.53	1.03	0.44	0.40	0.35	0.31	0.39	0.36	0.84	0.78
Triple Glazing, E = 0.20 on surface 2, 3 ,4, or 5															
29 1/4 in. air spaces	0.33	0.48	0.37	0.68	0.48	0.60	1.07	0.50	0.46	0.43	0.39	0.46	0.44	0.88	0.82
30 1/2 in. air spaces	0.24	0.42	0.30	0.61	0.40	0.53	1.04	0.44	0.40	0.35	0.31	0.39	0.36	0.85	0.79
31 1/4 in. argon spaces	0.27	0.44	0.32	0.64	0.43	0.55	1.02	0.46	0.42	0.37	0.33	0.41	0.39	0.82	0.76
32 1/2 in. argon spaces	0.21	0.39	0.27	0.59	0.37	0.50	1.01	0.41	0.37	0.32	0.28	0.36	0.34	0.81	0.75
Triple Glazing, E = 0.10 on surface 2, 3, 4, or 5															
33 1/4 in. air spaces	0.32	0.48	0.36	0.67	0.47	0.59	1.05	0.49	0.45	0.42	0.38	0.45	0.43	0.86	0.80
34 1/2 in. air spaces	0.22	0.40	0.28	0.60	0.38	0.51	1.03	0.42	0.38	0.33	0.29	0.37	0.35	0.84	0.78
35 1/4 in. argon spaces	0.26	0.43	0.31	0.63	0.42	0.54	1.00	0.45	0.41	0.36	0.32	0.40	0.38	0.80	0.74
36 1/2 in. argon spaces	0.19	0.38	0.26	0.57	0.36	0.48	0.99	0.40	0.36	0.30	0.26	0.35	0.32	0.79	0.73
Triple Glazing, E = 0.40 on surfaces 3 and 5															
37 1/4 in. air spaces	0.33	0.48	0.37	0.68	0.48	0.60	1.06	0.50	0.46	0.43	0.39	0.46	0.44	0.87	0.81
38 1/2 in. air spaces	0.24	0.42	0.30	0.61	0.40	0.53	1.04	0.44	0.40	0.35	0.31	0.39	0.36	0.85	0.79
39 1/4 in. argon spaces	0.27	0.44	0.32	0.64	0.43	0.55	1.02	0.46	0.42	0.37	0.33	0.41	0.39	0.82	0.76
40 1/2 in. argon spaces	0.21	0.39	0.27	0.59	0.37	0.50	1.00	0.41	0.37	0.32	0.28	0.36	0.34	0.80	0.74
Triple Glazing, E = 0.20 on surfaces 3 and 5															
41 1/4 in. air spaces	0.29	0.45	0.34	0.65	0.44	0.57	1.03	0.47	0.43	0.39	0.35	0.43	0.40	0.84	0.78
42 1/2 in. air spaces	0.20	0.39	0.27	0.58	0.36	0.49	1.01	0.41	0.37	0.31	0.27	0.35	0.33	0.81	0.75
43 1/4 in. argon spaces	0.23	0.41	0.29	0.60	0.39	0.52	0.98	0.43	0.39	0.34	0.30	0.38	0.36	0.79	0.72
44 1/2 in. argon spaces	0.16	0.35	0.24	0.55	0.33	0.46	0.96	0.38	0.34	0.28	0.24	0.32	0.30	0.77	0.71
Triple Glazing, E = 0.10 on surfaces 3 and 5															
45 1/4 in. air spaces	0.27	0.44	0.32	0.64	0.43	0.55	1.02	0.46	0.42	0.37	0.33	0.41	0.39	0.82	0.76
46 1/2 in. air spaces	0.17	0.36	0.24	0.56	0.34	0.47	0.99	0.39	0.34	0.29	0.25	0.33	0.31	0.79	0.73
47 1/4 in. argon spaces	0.21	0.39	0.27	0.59	0.37	0.50	0.95	0.41	0.37	0.32	0.28	0.36	0.34	0.76	0.70
48 1/2 in. argon spaces	0.14	0.34	0.22	0.53	0.31	0.44	0.94	0.36	0.32	0.26	0.22	0.31	0.28	0.75	0.69
Quadruple Glazing, E = 0.10 on surfaces 3 and 5															
49 1/4 in. air spaces	0.23	0.41	0.29	0.60	0.39	0.52	0.97	0.43	0.39	0.34	0.30	0.38	0.36	0.78	0.71
50 1/2 in. air spaces	0.15	0.35	0.23	0.54	0.32	0.45	0.93	0.37	0.33	0.27	0.23	0.31	0.29	0.74	0.68
51 1/4 in. argon spaces	0.17	0.36	0.24	0.56	0.34	0.47	0.92	0.39	0.34	0.29	0.25	0.33	0.31	0.73	0.67
52 1/2 in. argon spaces	0.12	0.32	0.21	0.52	0.29	0.43	0.90	0.35	0.31	0.24	0.20	0.29	0.27	0.71	0.65
53 1/4 in. krypton spaces	0.12	0.32	0.21	0.52	0.29	0.43	0.89	0.35	0.31	0.24	0.20	0.29	0.27	0.69	0.63
54 1/2 in. krypton spaces	0.11	0.31	0.20	0.51	0.29	0.42	0.89	0.34	0.30	0.23	0.19	0.28	0.26	0.69	0.63

Table 5 Overall Coefficients of Heat Transmission of Various Fenestration Products (Btu/h · ft² · °F) (*Concluded*)

Reinforced Vinyl/Aluminum-Clad Wood								Wood Vinyl								Insulated Fiberglass/Vinyl						
Operable		Fixed		Double Door		Sloped Skylight		Operable		Fixed		Double Door		Sloped Skylight		Operable		Fixed		Double Door		ID
Metal	Insul.	Metal	Insul.	Metal	Insul.	Metal	Insul.	Metal	Insul.	Metal	Insul.	Metal	Insul.	Metal	Insul.	Metal	Insul.	Metal	Insul.	Metal	Insul.	
0.98	—	1.05	—	0.99	—	1.50	—	0.94	—	1.04	—	0.98	—	1.47	—	0.86	—	1.02	—	0.93	—	1
0.85	—	0.89	—	0.85	—	1.31	—	0.81	—	0.88	—	0.84	—	1.27	—	0.74	—	0.86	—	0.79	—	2
0.92	—	0.97	—	0.92	—	1.51	—	0.87	—	0.96	—	0.91	—	1.48	—	0.80	—	0.94	—	0.86	—	3
																						4
0.60	0.57	0.58	0.56	0.57	0.55	0.88	0.86	0.56	0.54	0.57	0.56	0.56	0.54	0.85	0.82	0.50	0.47	0.55	0.54	0.52	0.50	5
0.55	0.52	0.51	0.49	0.52	0.49	0.87	0.85	0.51	0.48	0.51	0.49	0.50	0.48	0.84	0.81	0.45	0.42	0.49	0.47	0.46	0.44	6
0.57	0.54	0.54	0.52	0.54	0.52	0.84	0.81	0.53	0.50	0.53	0.51	0.53	0.51	0.80	0.77	0.47	0.44	0.51	0.49	0.48	0.46	7
0.53	0.50	0.49	0.46	0.49	0.47	0.84	0.81	0.49	0.46	0.48	0.46	0.48	0.46	0.80	0.77	0.43	0.40	0.46	0.44	0.44	0.42	8
0.56	0.52	0.52	0.50	0.52	0.50	0.82	0.79	0.52	0.49	0.52	0.50	0.51	0.49	0.79	0.75	0.46	0.43	0.50	0.48	0.47	0.45	9
0.49	0.46	0.44	0.41	0.45	0.43	0.81	0.78	0.45	0.42	0.43	0.41	0.44	0.41	0.78	0.74	0.40	0.36	0.41	0.39	0.40	0.37	10
0.51	0.48	0.46	0.44	0.47	0.45	0.76	0.72	0.47	0.44	0.46	0.43	0.46	0.44	0.72	0.68	0.42	0.38	0.44	0.42	0.42	0.40	11
0.47	0.43	0.40	0.38	0.42	0.40	0.77	0.73	0.43	0.40	0.40	0.37	0.41	0.38	0.73	0.69	0.37	0.34	0.38	0.36	0.37	0.34	12
0.53	0.50	0.49	0.46	0.49	0.47	0.78	0.75	0.49	0.46	0.48	0.46	0.48	0.46	0.75	0.71	0.43	0.40	0.46	0.44	0.44	0.42	13
0.46	0.42	0.39	0.37	0.41	0.39	0.78	0.74	0.42	0.39	0.39	0.36	0.40	0.37	0.74	0.70	0.37	0.33	0.37	0.35	0.36	0.34	14
0.48	0.44	0.42	0.39	0.43	0.41	0.70	0.67	0.44	0.41	0.41	0.39	0.42	0.40	0.67	0.63	0.38	0.35	0.40	0.37	0.38	0.36	15
0.43	0.39	0.35	0.32	0.38	0.35	0.72	0.68	0.39	0.36	0.35	0.32	0.36	0.34	0.69	0.65	0.33	0.30	0.33	0.31	0.33	0.30	16
0.51	0.48	0.46	0.44	0.47	0.45	0.76	0.72	0.47	0.44	0.46	0.43	0.46	0.44	0.72	0.68	0.42	0.38	0.44	0.42	0.42	0.40	17
0.44	0.40	0.37	0.34	0.39	0.37	0.76	0.72	0.40	0.37	0.36	0.34	0.37	0.35	0.72	0.68	0.35	0.31	0.35	0.32	0.34	0.32	18
0.46	0.42	0.39	0.37	0.41	0.39	0.68	0.64	0.42	0.39	0.39	0.36	0.40	0.37	0.64	0.60	0.37	0.33	0.37	0.35	0.36	0.34	19
0.41	0.37	0.32	0.30	0.35	0.33	0.69	0.66	0.37	0.34	0.32	0.29	0.34	0.31	0.66	0.62	0.32	0.28	0.30	0.28	0.30	0.28	20
0.46	0.41	0.41	0.39	0.43	0.40	0.71	0.67	0.43	0.39	0.41	0.38	0.42	0.39	0.67	0.63	0.37	0.34	0.39	0.37	0.38	0.35	21
0.42	0.37	0.35	0.33	0.37	0.35	0.68	0.64	0.38	0.34	0.35	0.32	0.36	0.34	0.65	0.60	0.33	0.29	0.33	0.31	0.32	0.30	22
0.44	0.39	0.38	0.35	0.40	0.37	0.66	0.62	0.40	0.36	0.38	0.35	0.39	0.36	0.63	0.58	0.35	0.31	0.36	0.33	0.35	0.32	23
0.40	0.35	0.34	0.31	0.36	0.33	0.65	0.61	0.37	0.33	0.34	0.31	0.35	0.32	0.62	0.57	0.32	0.28	0.32	0.29	0.31	0.28	24
0.44	0.39	0.39	0.36	0.40	0.38	0.67	0.63	0.41	0.37	0.39	0.36	0.39	0.37	0.64	0.59	0.35	0.32	0.37	0.34	0.35	0.33	25
0.39	0.34	0.32	0.29	0.35	0.32	0.65	0.61	0.36	0.31	0.32	0.29	0.33	0.31	0.61	0.56	0.30	0.27	0.30	0.28	0.30	0.27	26
0.41	0.36	0.35	0.32	0.37	0.34	0.63	0.59	0.38	0.33	0.34	0.31	0.36	0.33	0.59	0.54	0.32	0.29	0.33	0.30	0.32	0.29	27
0.37	0.32	0.30	0.27	0.32	0.29	0.62	0.58	0.34	0.29	0.29	0.26	0.31	0.28	0.58	0.53	0.29	0.25	0.27	0.25	0.27	0.25	28
0.43	0.38	0.37	0.34	0.39	0.36	0.65	0.61	0.40	0.35	0.37	0.34	0.38	0.35	0.62	0.57	0.34	0.31	0.35	0.33	0.34	0.31	29
0.37	0.32	0.30	0.27	0.32	0.29	0.63	0.59	0.34	0.29	0.29	0.26	0.31	0.28	0.59	0.54	0.29	0.25	0.27	0.25	0.27	0.25	30
0.39	0.34	0.32	0.29	0.35	0.32	0.60	0.56	0.36	0.31	0.32	0.29	0.33	0.31	0.57	0.52	0.30	0.27	0.30	0.28	0.30	0.27	31
0.35	0.30	0.27	0.24	0.30	0.27	0.59	0.55	0.32	0.27	0.27	0.24	0.29	0.26	0.56	0.51	0.27	0.23	0.25	0.22	0.25	0.23	32
0.42	0.37	0.36	0.34	0.38	0.35	0.64	0.60	0.39	0.35	0.36	0.33	0.37	0.34	0.60	0.55	0.34	0.30	0.34	0.32	0.33	0.31	33
0.36	0.31	0.28	0.25	0.31	0.28	0.62	0.58	0.33	0.28	0.28	0.25	0.30	0.27	0.58	0.53	0.27	0.24	0.26	0.23	0.26	0.23	34
0.38	0.33	0.31	0.28	0.34	0.31	0.58	0.54	0.35	0.31	0.31	0.28	0.33	0.30	0.55	0.50	0.30	0.26	0.29	0.27	0.29	0.26	35
0.34	0.29	0.25	0.22	0.29	0.26	0.57	0.53	0.31	0.26	0.25	0.22	0.27	0.24	0.54	0.49	0.25	0.22	0.23	0.21	0.24	0.21	36
0.43	0.38	0.37	0.34	0.39	0.36	0.65	0.61	0.40	0.35	0.37	0.34	0.38	0.35	0.61	0.56	0.34	0.31	0.35	0.33	0.34	0.31	37
0.37	0.32	0.30	0.27	0.32	0.29	0.63	0.59	0.34	0.29	0.29	0.26	0.31	0.28	0.59	0.54	0.29	0.25	0.27	0.25	0.27	0.25	38
0.39	0.34	0.32	0.29	0.35	0.32	0.60	0.56	0.36	0.31	0.32	0.29	0.33	0.31	0.57	0.52	0.30	0.27	0.30	0.28	0.30	0.27	39
0.35	0.30	0.27	0.24	0.30	0.27	0.58	0.54	0.32	0.27	0.27	0.24	0.29	0.26	0.55	0.50	0.27	0.23	0.25	0.22	0.25	0.23	40
0.40	0.35	0.34	0.31	0.36	0.33	0.62	0.58	0.37	0.33	0.34	0.31	0.35	0.32	0.58	0.53	0.32	0.28	0.32	0.29	0.31	0.28	41
0.35	0.29	0.26	0.23	0.29	0.26	0.59	0.55	0.31	0.27	0.26	0.23	0.28	0.25	0.56	0.51	0.26	0.23	0.24	0.22	0.24	0.22	42
0.37	0.31	0.29	0.26	0.32	0.29	0.56	0.52	0.33	0.29	0.28	0.25	0.30	0.27	0.53	0.48	0.28	0.24	0.27	0.24	0.27	0.24	43
0.32	0.27	0.23	0.20	0.26	0.23	0.55	0.50	0.29	0.24	0.22	0.19	0.25	0.22	0.51	0.46	0.23	0.20	0.21	0.18	0.22	0.19	44
0.39	0.34	0.32	0.29	0.35	0.32	0.60	0.56	0.36	0.31	0.32	0.29	0.33	0.31	0.57	0.52	0.30	0.27	0.30	0.28	0.30	0.27	45
0.33	0.28	0.24	0.21	0.27	0.24	0.57	0.53	0.29	0.25	0.23	0.20	0.26	0.23	0.54	0.49	0.24	0.21	0.22	0.19	0.22	0.20	46
0.35	0.30	0.27	0.24	0.30	0.27	0.54	0.49	0.32	0.27	0.27	0.24	0.29	0.26	0.50	0.45	0.27	0.23	0.25	0.22	0.25	0.23	47
0.31	0.26	0.21	0.18	0.25	0.22	0.53	0.49	0.27	0.23	0.21	0.18	0.23	0.21	0.49	0.44	0.22	0.19	0.19	0.17	0.20	0.18	48
0.37	0.31	0.29	0.26	0.32	0.29	0.56	0.51	0.33	0.29	0.28	0.25	0.30	0.27	0.52	0.47	0.28	0.24	0.27	0.24	0.27	0.24	49
0.31	0.26	0.22	0.19	0.26	0.23	0.52	0.48	0.28	0.23	0.22	0.19	0.24	0.21	0.48	0.43	0.23	0.19	0.20	0.17	0.21	0.18	50
0.33	0.28	0.24	0.21	0.27	0.24	0.51	0.47	0.29	0.25	0.23	0.20	0.26	0.23	0.48	0.42	0.24	0.21	0.22	0.19	0.22	0.20	51
0.29	0.24	0.19	0.16	0.23	0.20	0.49	0.45	0.26	0.22	0.19	0.16	0.22	0.19	0.46	0.41	0.21	0.18	0.17	0.15	0.19	0.16	52
0.29	0.24	0.19	0.16	0.23	0.20	0.47	0.43	0.26	0.22	0.19	0.16	0.22	0.19	0.44	0.39	0.21	0.18	0.17	0.15	0.19	0.16	53
0.29	0.24	0.18	0.16	0.23	0.20	0.47	0.43	0.25	0.21	0.18	0.15	0.21	0.18	0.44	0.39	0.20	0.17	0.16	0.14	0.18	0.15	54

Notes for Table 5

1. All heat transmission coefficients in this table include film resistances and are based on winter conditions of 0°F outdoor air and 70°F indoor air temperature, with 15 mph outdoor air velocity and zero solar flux. With the exception of single glazing, small changes in the interior and exterior temperatures will not significantly affect overall U-factors. The coefficients are for vertical position except skylight values, which are for 20° from horizontal with heat flow up.

2. Glazing layer surfaces are numbered from the outside to the inside. Double, triple, and quadruple refer to the number of glazing panels. All data are based on 1/8 in. glass, unless otherwise noted. Thermal conductivities are: 0.53 Btu/(h·ft·°F) for glass, and 0.11 Btu/(h·ft·°F) for acrylic and polycarbonate .

3. Standard spacers are metal. Insulation means wood, fiberglass, or butyl. Edge-of-glass effects assumed to extend over the 2-1/2 in. band around perimeter of each glazing unit as seen in Figure 3.

4. Product sizes are described in Figure 3 and frame U-factors are from Table 2.

5. Interpolation procedure for estimating U_{cg} for vertically oriented gas space widths between 0.25 and 0.5 in. The most general equation is:

$$U_t = \left[\frac{1}{U_{1/4}} - \frac{n_g}{h + 48\,k_g} + \frac{n_g}{h + (12\,k_g/t)} \right]^{-1}$$

$$h = -36\,k_g + \left[\frac{24\,n_g\,k_g\,U_{1/4}\,U_{1/2}}{U_{1/4} - U_{1/2}} + 144\,k_g^2 \right]^{0.5}$$

where

n_g = number of gaps (*e.g.*, double-pane windows have one gap)
k_g = gas conductivity Btu/(h·ft·°F)
t = gas space width of window, in.
$U_{1/4}$ = U-factor for identical glazing with 1/4 in. gas spaces
$U_{1/2}$ = U-factor for identical glazing with 1/2 in. gas spaces

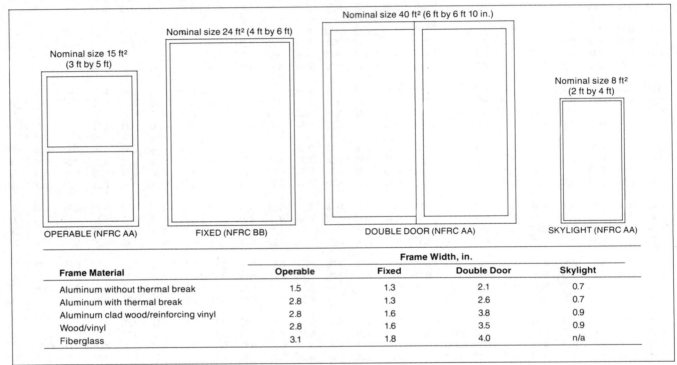

Frame Material	Frame Width, in.			
	Operable	Fixed	Double Door	Skylight
Aluminum without thermal break	1.5	1.3	2.1	0.7
Aluminum with thermal break	2.8	1.3	2.6	0.7
Aluminum clad wood/reinforcing vinyl	2.8	1.6	3.8	0.9
Wood/vinyl	2.8	1.6	3.5	0.9
Fiberglass	3.1	1.8	4.0	n/a

Fig. 3 Standard Fenestration Units

Example 1. Estimate the U-factor for a fixed window (NFRC Size BB) with a reinforced vinyl frame and double-glazing with a sputter-type low-e coating (*e* = 0.10). The gap is 0.5 in. wide and argon-filled and the spacer is metal.

Solution: Locate the glazing system type in the first column of Table 5 (ID = 20). Then, find the appropriate frame type (reinforced vinyl), operator type (fixed), and spacer type (metal). The U-factor listed (in the third column on the facing page) is 0.32.

Example 2. Estimate a representative U-factor for a wood-framed, 38 in. by 82 in. swinging French door with eight 11-in. by 16-in. panes (true divided panels), each consisting of clear double-glazing with a 0.5 in. air space.

Solution: Without more detailed information, assume that the dividers have the same U-factor as the frame, and that the divider edge has the same U-factor as the edge-of-glass. Calculate the center-of-glass, edge-of-glass, and frame areas.

$$A_{cg} = 8\,[(11-5)(16-5)] = 528 \text{ in}^2$$
$$A_{eg} = 8\,(11 \times 16) - 528 = 880 \text{ in}^2$$
$$A_f = (38 \times 82) - 8\,(11 \times 16) = 1708 \text{ in}^2$$

Select the center-of-glass, edge-of-glass, and frame U-factors. These component U-factors are 0.49 and 0.60 (from Table 5, glazing ID = 6,

columns 1 and 2) and 0.49 (from Table 2, wood frame double door, double-glazing) Btu/h·ft²·°F, respectively.

$$U_o = [(0.49 \times 528) + (0.60 \times 880) +$$
$$(0.49 \times 1708)]/(38 \times 82) = 0.52 \text{ Btu/h·ft}^2\cdot°F.$$

Example 3. Estimate the U-factor for a 60 in. by 36 in. aluminum framed garden window (NFRC Size AA) that projects out 12 in. from the wall surface. The glazing is double-glass with a 0.25 in. air space and a metal spacer. All of the glazing is fixed (*i.e.*, nonoperable) and the frame is 1 in. wide. The base is 0.75 in. plywood.

Solution: The U-factor is determined by calculating the heat flow through all five surfaces (front, two sides, top, and bottom) and then dividing by the projected area of the window (*i.e.*, the 60 in. by 36 in. opening in the insulated wall).

The U-factors are 0.57 Btu/h·ft²·°F for the center-of-glass and 0.65 Btu/h·ft²·°F for the edge-of-glass (from Table 5, glazing ID = 5, columns 1 and 2) and 1.78 Btu/h·ft²·°F (from Table 2, aluminum frame, fixed window, double-glazing), respectively.

From Table 4 in Chapter 22, the resistance of 0.75 in. plywood is 0.93 h·ft²·°F/Btu. When the exterior film resistance of 0.17 and interior downward flow film resistance of 0.92 h·ft²·°F/Btu (from Table 1, Chapter 22) are added, the total resistance is 2.02 h·ft²·°F/Btu and the

U-factor is 1/2.02 = 0.50 Btu/h·ft²·°F. (The glazing U-factors in Table 5 already include the film resistances.) Now calculate the areas.

For the front surface:

$$A_{cg} = 53 \times 29 = 1537 \text{ in}^2$$

$$A_{eg} = (58 \times 34) - (53 \times 29) = 435 \text{ in}^2$$

$$A_f = (60 \times 36) - (58 \times 34) = 188 \text{ in}^2$$

For the two sides:

$$A_{cg} = 2(5 \times 29) = 290 \text{ in}^2$$

$$A_{eg} = 2[(10 \times 34) - (5 \times 29)] = 390 \text{ in}^2$$

$$A_f = 2[(12 \times 36) - (10 \times 34)] = 184 \text{ in}^2$$

For the top surface:

$$A_{cg} = 53 \times 5 = 265 \text{ in}^2$$

$$A_{eg} = (58 \times 10) - (53 \times 5) = 315 \text{ in}^2$$

$$A_f = (60 \times 12) - (58 \times 10) = 140 \text{ in}^2$$

For the bottom surface,

$$A_{plywood} = 58 \times 10 = 580 \text{ in}^2$$

$$A_f = (60 \times 12) - (58 \times 10) = 140 \text{ in}^2$$

Then calculate the U-factor as follows:

$$U_o = [0.57(1537 + 290 + 265) + 0.65(435 + 390 + 315)$$
$$+ 1.78(188 + 184 + 140 + 140) + (0.50 \times 580)]/(60 \times 36)$$
$$= 1.57 \text{ Btu/h} \cdot \text{ft}^2 \cdot °\text{F}$$

Note that the U-factor for this garden window is even greater than that for a skylight with the same type of glazing. While both have a surface area that is approximately twice the rough opening area, the skylight curbs are often wood, but the garden window side and top panels are glazed.

Example 4. Find the center-of-glass U-factor for a triple-pane window with 0.375 in. air-filled gaps and a low-e coating (e = 0.2) on the fifth surface.

Solution: From Table 5, for 0.25 in. spacing, $U_{1/4} = 0.33$ and for 0.5 in. spacing, $U_{1/2} = 0.24$. The thermal conductivity of air is 0.014 Btu/h·ft·°F. Using the equation in footnote 5,

$$h = -36 \times 0.014$$
$$+ \left[\frac{24 \times 2 \times 0.014 \times 0.33 \times 0.24}{(0.33 - 0.24)} + 144 \times 0.014^2 \right]^{0.5} = 0.283$$

$$U_{3/8} = \left[\frac{1}{0.33} - \frac{2}{0.283 + 48 \times 0.014} + \frac{2}{0.283 + (12 \times 0.014/0.375)} \right]^{-1}$$
$$= 0.27 \text{ Btu/h} \cdot \text{ft}^2 \cdot °\text{F}$$

Because of the combined effects of solar gain, wind, ambient temperature and cloud cover, fenestration performance is likely to be different from the data in Table 5 when products are installed in buildings. Since these factors influence the rate of heat transfer through fenestration, seasonal or annual energy flows cannot be accurately determined solely on the basis of the design-day U-factors. Where U-factors are to be compared at different wind speeds, Table 6 provides approximate data to convert the overall U-factor data at one specific wind condition to a U-factor at another. Performance data in Table 5 are based on the assumption of an overcast sky, which implies that the effective sky temperature for radiative heat transfer is equal to the outdoor air temperature. This assumption gives reasonable results with 7.5- and 15-mph wind speeds; however, with a clear sky and nearly still wind conditions, significant heat transfer occurs by radiation, and the data in Table 6 may underestimate the actual U-factor.

DETERMINING INCIDENT SOLAR FLUX

Solar Radiation

The flux of solar radiation on a surface normal to the sun's rays beyond the earth's atmosphere at the mean earth-sun distance (92.9 × 10⁶ miles) is defined as the solar constant E_{sc}. The currently accepted value is 433.34 Btu/h·ft² (CIMO 1982). Because the earth's orbit is slightly elliptical, the extraterrestrial radiant flux E_o varies from a maximum of 448.2 Btu/h·ft² on January 3, when the earth is closest to the sun, to a minimum of 419.2 Btu/h·ft² on July 4, when the earth-sun distance reaches its maximum.

The earth's orbital velocity also varies throughout the year, so *apparent solar time*, as determined by a sundial, varies slightly from the *mean time* kept by a clock running at a uniform rate. This variation, called the *equation of time*, is given in Table 7. The sun's position in the sky is determined by *local solar time*, found by adding the equation of time to the local civil time. *Local civil time* is found by adding to or subtracting from *local standard time* (LST) the longitude correction of 4 minutes per degree difference between the local longitude and the longitude of the standard time meridian for that locality. In the United States and Canada, these values are 60° for Atlantic Standard Time (ST); 75° for Eastern ST; 90° for Central ST; 105° for Mountain ST; 120° for Pacific ST, 135° for Yukon ST; 150° for Alaska-Hawaii ST.

Table 6   Glazing U-Factor for Various Wind Speeds

Wind speed, mph		
15	7.5	0
U-Factor, Btu/h·ft²·°F		
0.10	0.10	0.10
0.20	0.20	0.19
0.30	0.29	0.28
0.40	0.38	0.37
0.50	0.47	0.45
0.60	0.56	0.53
0.70	0.65	0.61
0.80	0.74	0.69
0.90	0.83	0.78
1.00	0.92	0.86
1.10	1.01	0.94
1.20	1.10	1.02
1.30	1.19	1.10

Table 7   Extraterrestrial Solar Radiation Intensity and Related Data

	I_o Btu/(h·ft²)	Equation of Time, min.	Declination, degrees	A Btu/(h·ft²)	B (Dimensionless Ratios)	C
Jan	448.8	−11.2	−20.0	390	0.142	0.058
Feb	444.2	−13.9	−10.8	385	0.144	0.060
Mar	437.7	−7.5	0.0	376	0.156	0.071
Apr	429.9	1.1	11.6	360	0.180	0.097
May	423.6	3.3	20.0	350	0.196	0.121
June	420.2	−1.4	23.45	345	0.205	0.134
July	420.3	−6.2	20.6	344	0.207	0.136
Aug	424.1	−2.4	12.3	351	0.201	0.122
Sep	430.7	7.5	0.0	365	0.177	0.092
Oct	437.3	15.4	−10.5	378	0.160	0.073
Nov	445.3	13.8	−19.8	387	0.149	0.063
Dec	449.1	1.6	−23.45	391	0.142	0.057

Note: Data are for 21st day of each month during the base year of 1964.

Equation (6) relates Apparent Solar Time (AST) to Local Standard Time (LST) as follows:

$$AST = LST + ET + 4(LSM - LON) \qquad (6)$$

where

ET = equation of time, minutes of time
LSM = local standard time meridian, ° of arc
LON = local longitude, ° of arc
4 = minutes of time required for 1.0 degree rotation of earth

Because the earth's equatorial plane is tilted at an angle of 23.45° to the orbital plane, the solar declination δ (the angle between the earth-sun line and the equatorial plane) varies throughout the year, as shown in Figure 4 and Table 7. This variation causes the changing seasons, with their unequal periods of daylight and darkness.

The spectral distribution of solar radiation beyond the earth's atmosphere (Figure 5) resembles the radiant energy emitted by a blackbody at about 10,800°F. The invisible ultraviolet region with wavelengths between 0.29 and 0.40 μm contains about 7% of the total energy, while the visible region between 0.4 and 0.7 μm contains 39%, and the near infrared region between 0.7 and 3.5 μm contains the remaining 52%. The peak intensity, about 700 Btu/h·ft²·μm, is reached at about 0.45 μm in the green portion of the visible spectrum.

In passing through the earth's atmosphere, the sun's radiation is reflected, scattered, and absorbed by dust, gas molecules, ozone,

water vapor, and water droplets (clouds). The extent of this depletion at any given time is determined by atmospheric composition and length of the atmospheric path traversed by the sun's rays. This length is expressed in terms of the air mass m, the ratio of the mass of atmosphere in the actual earth-sun path to the mass which would exist if the sun were directly overhead at sea level ($m = 1.0$). For most purposes, the air mass at any time equals the cosecant of the solar altitude, multiplied by the ratio of the existing barometric pressure to standard pressure. Beyond the atmosphere, $m = 0$.

Most ultraviolet solar radiation is absorbed by the ozone in the upper atmosphere, while part of the radiation in the shortwave portion of the spectrum is scattered by air molecules, imparting the blue color to the sky. Water vapor in the lower atmosphere causes the characteristic absorption bands observed in the solar spectrum at sea level (Figure 5). For a solar altitude of 41.8° (air mass = 1.5), the spectrum of the sun's direct radiation on a clear day at sea level shows less than 3% of the total energy in the ultraviolet, 38% in the visible region, and the remaining 59% in the infrared (ASTM 1987). The maximum intensity occurs at 0.50 μm, and virtually no solar energy exists at wavelengths beyond 2.2 μm.

The spectral distributions shown in Part A, Figure 5 for $m = 0$ and $m = 2.0$ are adapted from Moon (1940), whose solar radiation data took no account of monthly variations in intensity caused by changes in the earth-sun distance and by the atmosphere's varying average moisture content. Moon used a solar constant of about 420 Btu/h·ft², while the presently recommended value is 433.3 Btu/h·ft². The solar heat gain factors (SHGF) in Tables 12 through 18 are based on terrestrial measurements (Threlkeld and Jordan 1958) and are not affected directly by the change in the accepted value of the solar constant. However, these calculations may be affected by the accuracy of the calculated solar irradiance incident on the glazing. Iqbal (1983), Galanis and Chatiguy (1986), and Gueymard (1993) discuss the accuracy and universality of the radiation model used to produce these data and propose possible improvements.

For many years the glazing industry has used Moon's spectrum (air mass ≐ 2) to obtain spectrally average optical characteristics of glazings. The National Fenestration Rating Council recommends a spectrum at a standard air mass of 1.5 as described in ASTM *Standard* E 891 (1987a). This difference in reference spectrum may result in slight deviations of the solar heat gain coefficient of spectrally selective glazings, whereas the visible transmittance of all types of glazings should remain unchanged.

Some short-wave radiation scattered by air molecules and dust reaches the earth in the form of diffuse radiation E_d. Since this diffuse radiation comes from all parts of the sky, its intensity is difficult to predict and varies as moisture and dust content of the atmosphere change throughout any given day. On a completely overcast day, the diffuse component accounts for all solar radiation reaching the ground.

Some energy absorbed by carbon dioxide and water vapor in the sky reaches the earth in the form of long-wave atmospheric radiation. Because the apparent emittance of the atmosphere depends primarily on its water vapor content (Parmelee and Aubele 1952, Bliss 1961) and is always less than 1.00, usually long-wave radiant energy flows from terrestrial surfaces.

The total short-wave irradiance E_t reaching a terrestrial surface is the sum of the direct solar radiation E_D, the diffuse sky radiation E_d, and the solar radiation reflected from surrounding surfaces E_r. The irradiance of the direct component is the product of the direct normal irradiation E_{DN} and the cosine of the angle of incidence θ between the incoming solar rays and a line normal (perpendicular) to the surface:

$$E_t = E_{DN} \cos \theta_V + E_d + E_r \qquad (7)$$

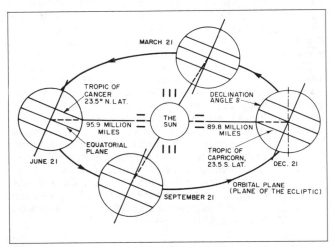

Fig. 4   Motion of Earth around Sun

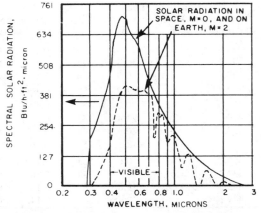

Fig. 5   Extraterrestrial and Terrestrial Solar Radiation

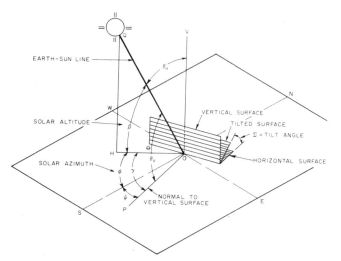

Fig. 6   Solar Angles for Vertical and Horizontal Surfaces

A method of computing all the factors on the right side of Equation (7) follows. Perez *et al.* (1986) give a more detailed model for E_t, which separates the diffuse sky radiation into three components.

Solar Angle Determination

The sun's position in the sky is conveniently expressed in terms of the solar altitude β above the horizontal and the solar azimuth ϕ measured from the south (Figure 6). These angles, in turn, depend on the local latitude L; the solar declination δ, which is a function of the date (Table 7); and the apparent solar time, expressed as the hour angle H, where $H = 0.25$ (number of minutes from local solar noon), in degrees.

Equations (8a) and (8b) relate β and ϕ to the three angles just mentioned:

$$\sin \beta = \cos L \cos \delta \cos H + \sin L \sin \delta \qquad (8a)$$

$$\cos \phi = (\sin \beta \sin L - \sin \delta)/(\cos \beta \cos L) \qquad (8b)$$

Figure 6 shows the solar position angles and incident angles for horizontal and vertical surfaces. Line OQ leads to the sun, the north-south line is NOS, and the east-west line is EOW. Line OV is perpendicular to the horizontal plane in which the solar azimuth, angle HOS, and the surface azimuth, angle POS (ψ) are located. Angle HOP is the surface solar azimuth defined as:

$$\gamma = \phi - \psi \qquad (9)$$

The solar azimuth ϕ is positive for afternoon hours and negative for morning hours. Likewise, surfaces that face west have a positive surface azimuth ψ; those facing east, have a negative surface azimuth (Table 8). If γ is greater than 90° or less than 270°, the surface is in the shade. Table 8 gives values in degrees for the surface azimuth ψ, applicable to the orientations of interest.

The angle of incidence θ for any surface is defined as the angle between the incoming solar rays and a line normal to that surface. For the horizontal surface shown in Figure 6, the incident angle θ_H is QOV; for the vertical surface, the incident angle θ_V is QOP.

**Table 8   Surface Orientations and Azimuths,
Measured from South**

Orientation	N	NE	E	SE	S	SW	W	NW
Surface azimuth, ψ	180°	−135°	−90°	−45°	0	45°	90°	135°

For any surface, the incident angle θ is related to β, γ and the tilt angle of the surface Σ by:

$$\cos \theta = \cos \beta \cos \gamma \sin \Sigma + \sin \beta \cos \Sigma \qquad (10)$$

where Σ = tilt angle of surface from horizontal.
When the surface is horizontal, $\Sigma = 0°$, and

$$\cos \theta_H = \sin \beta \qquad (11a)$$

For a vertical surface, $\Sigma = 90°$, and

$$\cos \theta_V = \cos \beta \cos \gamma \qquad (11b)$$

Example 5. Find the solar azimuth and altitude at 0830 central time on October 21 at 32° north latitude and 95° west longitude.
Solution: Local time is 0830 + 4 (90 − 95) = 0810. The equation of time (Table 7) is +15 min, so apparent solar time (AST) = 0810 + 15 = 0825, or 215 min. before noon, and H = 0.25 × 215 = 53.8°. Table 7 gives the solar declination on October 21 as −10.5°.

Thus, by Equation (8a):

$$\sin \beta = \cos (32) \cos (-10.5) \cos (53.8) + \sin (32) \sin (-10.5) = 0.396$$

$$\beta = 23.3°$$

Using Equation (8b):

$$\cos \phi = \frac{\sin (23.3) \sin (32) - \sin (-10.5)}{\cos (23.3) \cos (32)} = 0.503$$

$$\phi = 59.8°$$

Example 6. For the conditions of Example 5, find the incident angle at a window facing southeast.
Solution: Since the solar azimuth ϕ is to the east (AST < 1200) and the surface azimuth ψ is to the east (Table 8), they are both negative— $\phi = -59.8$, $\psi = -45.0$.

$$\gamma = -59.8 - (-45.0) = -14.8°$$

A negative surface-solar azimuth γ indicates that the sun is east of the normal to the surface. Thus, using Equation (11b)

$$\cos \theta_V = \cos 23.3 \cos (-14.8) = 0.888$$

$$\theta_V = 27.4°$$

Direct Normal Irradiance

At the earth's surface on a clear day, direct normal irradiation, or solar intensity E_{DN} is represented by

$$E_{DN} = A/[\exp (B/\sin \beta)] \qquad (12)$$

where

A = apparent solar irradiation at air mass $m = 0$ (Table 7)
B = atmospheric extinction coefficient (Table 7)

(Values of E_{DN} based on these data are given in Tables 13 through 19 for the daylight hours of the 21st day of each month, for latitudes 16 to 64° North in 8° increments.)

Values of A and B vary during the year because of seasonal changes in the dust and water vapor content of the atmosphere and because of the changing earth-sun distance. Equation (12) does not give the maximum value of E_{DN} that can occur in each month, but yields values that are representative of conditions on cloudless days for a relatively dry and clear atmosphere. For very clear atmospheres, E_{DN} can be 15% higher than indicated by Equation (12), using values of A and B in Table 7.

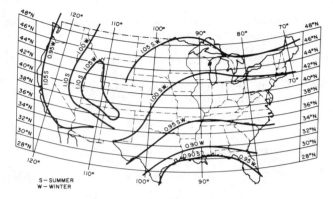

Fig. 7   Estimated Atmospheric Clearness Numbers in United States for Nonindustrial Localities

For locations where clear, dry skies predominate (*e.g.*, at high elevations), or, conversely, where hazy and humid conditions are frequent, values found by using Equation (12) and Table 7 should be multiplied by the Clearness Numbers in Threlkeld and Jordan (1958), reproduced here as Figure 7.

Example 7. Find the direct component of the solar irradiation on a horizontal roof for the conditions of Example 5.
   Solution: From Example 5, $\sin \beta = 0.396$, and from Table 7, $A = 378$ Btu/h·ft^2 and $B = 0.160$. Therefore,

$$E_{DN} = 378/[\exp (0.160/0.396)] = 252 \text{ Btu/h·ft}^2$$

$$E_{DH} = E_{DN} \sin \beta = 252 \times 0.396 = 100 \text{ Btu/h·ft}^2$$

OPTICAL PROPERTIES

Solar radiation (including both direct rays from the sun and diffuse rays from the sky, clouds, and surrounding objects) incident on a fenestration system is partly transmitted and partly reflected by the glazings of that system. An additional fraction is absorbed within the glazings or the coatings on their surfaces. The fraction of incident flux that is reflected is called the reflectance ρ, the fraction absorbed is called the absorptance α, and the fraction transmitted is the transmittance τ. The sum of the transmittance τ, absorptance α, reflectance ρ of a glazing layer is unity:

$$\tau + \alpha + \rho = 1 \tag{13}$$

However, this is complicated by the fact that radiation incident on a surface can have nonconstant distributions over the directions of incidence and over the wavelength (or frequency) scale. Thus, when measuring one or more of the optical properties, the wavelength distribution and direction of incident radiation must be specified.

The emittance of the surface ϵ is the ratio of the emission of thermal radiant flux from a surface to the flux that would be emitted by a blackbody emitter at the same temperature. Given the temperature and spectral emittance of a surface, the emitted irradiance spectrum can be computed.

Angular Dependence

The concept of solid angle is needed to understand angle dependence. A solid angle is a surface formed by all rays joining a point to a closed curve (Figure 8). For a sphere of radius R, the solid angle is the ratio of the projected area A on the sphere to the square of R. A sphere has a solid angle of 4π steradians (4π sr); a hemisphere has a 2π sr solid angle.

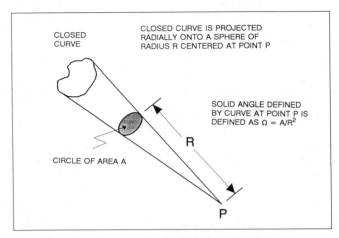

Fig. 8   Geometry for Definition of Solid Angle

Radiation incident on a point in a surface comes to that point from many directions in a conical solid angle. For a cone of half angle α, the solid angle defined by the circular top and point bottom of that cone is given by

$$\Omega = 2\pi (1 - \cos \alpha) \tag{14}$$

In measuring transmittance or reflectance, a sample is illuminated over a specified solid angle. The reflected or transmitted flux is then collected within another solid angle. The size of a conical solid angle and the direction of its axis need to be specified to obtain meaningful results. A conical solid angle is in the shape of a right circular cone.

ASTM *Standards* E 903-82, E 1084-86, E 971-88, and E 972-88 and ASHRAE *Standard* 74-88 on solar optical properties refer to conical-hemispherical measurements because, for most thermal or HVAC calculations, only the total transmitted irradiance due to the direct solar beam is of interest, since all transmitted solar irradiance is considered heat gain.

For complex fenestrations, for daylighting applications, and for some passive solar heating applications, a biconical definition as shown in Figure 9 is needed. This definition is used: (1) to treat sky and direct beam radiation, (2) to handle the directional distribution of the flux entering the room through the window, and (3) to calculate the angle-dependent optical and solar gain properties of multipane window and shade systems.

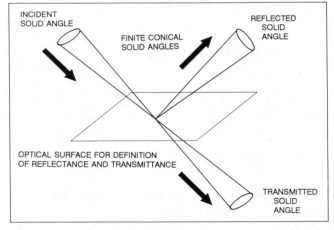

Fig. 9   Geometry for Definition of Biconical Transmittance and Reflectance

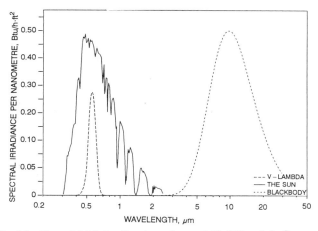

Fig. 10 Terrestrial Solar Spectrum (m = 1.5), V-Lambda Curve, and Blackbody Spectrum of 75 °F

Table 9 Portions of Total Solar Spectral Irradiance Contained in Portions of Visible Spectrum

Wavelength, nm		Percent Irradiance	Percent Illuminance
Start	End		
370	770	54.4	100.0
380	760	52.2	100.0
390	750	50.2	99.9
400	740	47.4	99.9
410	730	44.9	99.8
420	720	41.9	99.8
430	710	39.5	99.8
440	700	36.7	99.8
450	690	35.3	99.5
460	680	31.1	99.1

Note: The integrated total irradiance = 950 W/m² and illuminance = 100 klx (from Figure 10).

Spectral Dependence

Frequency and wavelength are related through the equation

$$\lambda f = c \tag{15}$$

where

   λ = wavelength, m
   f = frequency, Hz
   c = velocity of light = 2.9×10^8 m/s in air at atmospheric pressure

The wavelength dependence of radiometric quantities is denoted with a subscript λ attached to the optical property thus: $\tau_\lambda, \rho_\lambda, \alpha_\lambda,$ and ϵ_λ.

Source spectra. Radiation incident on a surface has a distribution of magnitude or spectrum over a range of wavelengths. For terrestrial applications, the solar spectrum after it has been modified by passage through the earth's atmosphere is more important (Figure 5).

In addition to the radiation incident on and reflected by surfaces, all bodies above absolute zero also *emit* radiation. The hotter they are, the more they emit. Also, as the temperature changes, the spectral distribution of the emitted radiation shifts. Figure 10 shows the emission spectrum of a blackbody at 75 °F. The spectrum is shifted toward the longer wavelengths, and no visible radiation is produced. Also, the total magnitude of the emission is low.

SOLAR-OPTICAL PROPERTIES OF GLAZING

The solar-optical properties of a glazing are the integrated or total transmittance, reflectance, and absorptance of the glazing to incident solar radiation. If the spectral optical properties (τ_λ, ρ_λ, α_λ) of the glazing and the spectral irradiance E_λ incident on the glazing are known, the solar optical properties can be calculated from the following equation where q stands for τ, ρ, or α.

$$q_e = \frac{\int_0^\infty q_\lambda E_\lambda \, d\lambda}{\int_0^\infty E_\lambda \, d\lambda} \tag{16}$$

Many window glazings do not have strong spectral selectivity over the solar spectrum, so their spectral optical properties can be considered constant. Thus, the transmitted spectral irradiance can be determined by multiplying the incident spectral irradiance $E_{t\lambda}$ by the solar transmittance.

With glazings that have nonuniform spectral transmittances τ_λ, the following integral must be evaluated over the solar spectrum.

$$E_t = \int \tau_\lambda E_\lambda \, d\lambda \tag{17}$$

Figure 11 shows the normal incidence spectral transmittance of typical glasses. The approximate transmission of total incident solar radiation for clear float glass at an incidence angle of 0° ranges from 86% for 0.09 in. thicknesses to 84% for 0.12 in. thicknesses to 78% for 0.25 in. thicknesses. Actual transmission varies with the amount of iron or other absorbers in the glass. Bronze, gray, and green heat-absorbing glasses, in 0.25-in. thicknesses, transmit between 45 and 49% of the total incident solar radiation (Rubin 1984).

Almost all architectural glass is opaque to the long-wave radiation emitted by surfaces at temperatures below about 250 °F. This characteristic produces the greenhouse effect, by which solar radiation passing through a window is partially retained inside by the following mechanism. Radiation absorbed by surfaces within the room is emitted as long-wave radiation, and cannot escape directly through the glass since it is opaque to all radiation beyond 4.5 µm. Instead, the radiation from the room surfaces falling on the glass is absorbed and reemitted to both sides as determined by several parameters, such as the inside and outside film coefficients of heat transfer, the surface emissivities, and other glazing properties.

Figure 12 shows how solar-optical properties vary with incident angle of typical uncoated glazing materials. As the incident angle increases from zero, the transmittance diminishes, reflectance increases, and absorptance first increases because of the lengthened optical path and then decreases as more incident radiation is

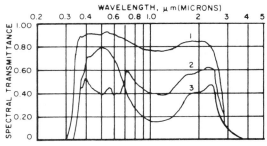

1. 1/8-in. regular sheet
2. 1/4-in. gray heat-absorbing plate/float
3. 1/4-in. green heat-absorbing plate/float

Fig. 11 Spectral Transmittance for Typical Architectural Glass

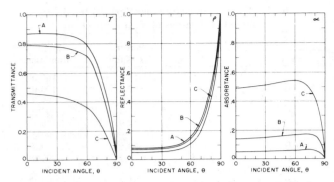

**Fig. 12   Variation with Incident Angle of Solar-Optical
Properties for Double-Strength Sheet, Clear,
and Heat-Absorbing Glass**

reflected. For diffuse radiation, the solar-optical properties are close to the direct radiation properties when $\theta = 60°$. Maximum solar gains through vertical glazing usually occur when the solar altitude is 30 to 35° and the surface-solar azimuth is close to zero.

Thermal Infrared Radiation

All the energy of the radiant flux in the solar spectrum becomes heat when it is absorbed, including that in the visible portion of the spectrum, and all this flux is absorbed eventually. If it is all absorbed inside a building, it all becomes heat gain. If some is reflected by a window or shade as it passes through, then the heat gain inside is reduced.

The Stefan-Boltzmann law predicts the integration of the Planck radiation law for blackbodies over all wavelengths as

$$E_b = \sigma T^4 \qquad (18)$$

If the ratio of the spectral irradiance E_λ to E_b is constant over the spectral range of output, then Equation (18) can be modified to predict the emitted irradiance of real surfaces as

$$E = \epsilon E_b = \epsilon \sigma T^4 \qquad (19)$$

For most windows, only the hemispherical emittance is of interest. As window technology advances, however, angular emittance effects will have to be studied more closely.

Emittance is related to the other optical properties in that a good absorber is a good emitter, and a good reflector is a poor emitter. This is only true at a given wavelength, which is a consequence of Kirchoff's law (Duffie and Beckman 1980):

$$\alpha_\lambda(\theta, \phi) = \epsilon_\lambda(\theta, \phi) \qquad (20)$$

where θ and ϕ are angles defining the directional dependence of the spectral absorptance α_λ and spectral emittance ϵ_λ. A surface appearing to be an excellent reflector in the visible portion of the spectrum may have a high emittance over most of the infrared spectrum, or vice versa.

Spectral Selectivity

Suitable, spectrally selective window glazings can be chosen either to trap solar heat or to block its entry into a building. A glazing with spectral selectivity has a surface with one or more of the optical properties that behaves differently in one part of the spectrum than in another.

The terrestrial spectral distribution of irradiance from the sun and the spectral response of the human eye (the V-lambda curve) is shown in Figure 10. The V-lambda function is used to determine

the visible portion (illuminance) in broadband radiation. Illuminance is measured in lumens and is the visible equivalent of irradiance. The V-lambda curve is nonzero from 370 to 770 nm, but the values outside the range from 430 to 690 nm are less than 1% of the maximum value at 555 nm (Table 9).

The spectral irradiance from 370 to 770 nm is 54.4% of the total solar irradiance of 300 Btu/h·ft^2 shown in Figure 10 or 163 Btu/h·ft^2. This spectral range contains 100% of the light in this spectrum. The spectral irradiance from 450 to 690 nm is 102 Btu/h·ft^2 or 34% of the total. This spectral range contains about 99.4% of the light in the spectrum. Thus, a perfect spectrally selective filter, which rejects all electromagnetic radiation outside the range from 430 to 690 nm and passes all within this range, would pass 99.4% of the light incident on the filter but only 34% of the heat energy. Spectrally selective filters using multilayer dielectric coatings on glass approach this goal. Their transmittances over the visible portion of the spectrum are not 100%, but some are in excess of 80%.

An ideal filter would pass 99.5% of the light (9300 lumens/ft^2) and only 34% of the heat, or 30 W/ft^2. The luminous efficacy of the emerging radiation would be 9300/30 or 310 lm/W, a factor of 2.9 times the 105.5 lm/W luminous efficacy of the whole spectrum shown in Figure 10.

Windows for heating dominated buildings. The best spectral selectivity for cold climates would have a high transmittance over the entire solar spectrum, admitting the most solar radiant heat, and a low emissivity (high reflectivity) over the long-wavelength infrared radiation spectrum, to reflect low-temperature radiant heat from the walls and room furnishings back into the building. Conventional low-emittance (low-e) windows approximate this ideal performance.

Windows for cooling dominated buildings. Table 9 indicates that the visible portion contains from 30 to 50% of the total energy in the solar spectrum. Thus, by eliminating all the solar ultraviolet and infrared radiation transmitted by window glazing, outside the visible spectrum (and even some of the radiation at each end of the visible portion), from 50 to 70% of the solar radiant heat gain could be eliminated with no significant loss of light transmission through the glazing (Figure 13).

Window systems are available with glazing coatings that approximate this ideal spectral selectivity. Most also have low emittance in the long-wavelength, infrared portion of the spectrum. These coatings also block some of the visible spectrum. The effects of this selectivity on the color of the transmitted light and its color-rendering properties must also be considered in the design.

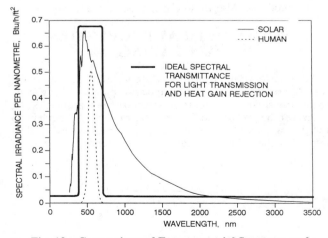

**Fig. 13   Comparison of Extraterrestrial Spectrum and
Human Visibility Function with Ideal Filter
Transmittance Spectrum**

Absorbed Solar Radiation

The absorbed radiation, including ultraviolet, visible, and infrared, is turned to heat inside the absorbing material. In a window, the glazing loses the heat through conduction, convection, and radiation. Some heat goes outside the building, and the remainder adds to the directly transmitted solar heat gain. The size of this inwardly flowing fraction depends on the nature of the air boundary layers adjacent to both sides of the glazing, including any gas trapped between the panes of a multiple-glazed window.

If E_t is the incident solar irradiance (outside), τ_s is the solar transmittance, α_s is the solar absorptance, and N_i is the inwardly flowing fraction of the absorbed irradiance for a single pane of glass, the solar gain q_i (inside) is

$$q_i = E_t (\tau_s + N_i \alpha_s) \qquad (21)$$

The quantity in parentheses is the solar heat gain coefficient (SHGC). It is the fraction of incident irradiance that enters through the glazing as heat gain. It includes both the directly transmitted portion τ_s and the absorbed and reemitted portion $N_i\alpha_s$. The SHGC is needed to determine the solar radiant heat gain from a window. The SHGC should be included, along with window U-factor and other properties, in any description of a window's energy performance.

The total rate of heat flow inward (Vild 1964) by radiation and convection from an unshaded single glazing is

$$q_{RCi} = E_t (\tau_S + N_i \alpha_S) + U (t_o - t_i) \qquad (22)$$

where

N_i = inward-flowing fraction of absorbed radiation
E_t = incident irradiance

For unshaded single glazing, $N_i = U/h_o$, and Equation (22) becomes:

$$q_{RCi} = U \left[\frac{\alpha E_t}{h_o} + t_o - t_i \right] + E_o \tau_s \qquad (23)$$

Shading Coefficient

Tables in this chapter list data that may be used to calculate heat gain through a standard reference glazing system. The shading coefficient is a widely used indicator of solar gain that was devised as a convenient way to convert these numbers to equivalent values for different glazing systems. It is defined as the ratio of the solar gain q_i of the window to that of a standard reference window of single-pane, double-strength clear glass, irradiated in the same way and under the same environmental conditions. It is, therefore, equal to the ratio of solar heat gain coefficients of the two windows

$$SC = \frac{SHGC_{Test}}{SHGC_{Ref}} = \frac{\tau_{Test} + N_{i,\,Test}\,\alpha_{Test}}{\tau_{Ref} + N_{i,\,Ref}\,\alpha_{Ref}} \qquad (24)$$

If the reference window's SHGC is constant, then the SC would differ from the SHGC only by a constant multiplier. However, in general, the SHGC is a function of the direction of incidence of the radiation as well as its spectral distribution. This is true both of the reference window and the test window.

As fenestration systems become more complex, however, the shading coefficient is being replaced by the solar heat gain coefficient (McCluney 1991). For energy analyses, including hourly building performance simulation calculations, the angle-dependent values of the solar heat gain coefficient give better results than the shading coefficient (McCluney 1987).

Data on solar heat gain coefficients are available from some manufacturers of fenestration products. Computer programs are available for calculating solar heat gain factors (incident solar irradiances and solar heat gains per unit area of glazing area) automatically (LBL 1992).

Solar Gain for Complex Multipane Windows

For multiple-pane windows with gas fills other than air and with spectrally selective coatings on one or more of the glazing surfaces, the foregoing equations are inadequate for determining the solar heat gain coefficient, and they do not explicitly include angle-dependent and spectrally selective effects. LBL (1992) has developed a calculation procedure and computer program (WINDOW 4.0) that performs the needed calculations for determining this coefficient for complex windows where the glazings do not exhibit significant scattering effects and where the glazing coatings have known angle-dependent optical properties.

The National Fenestration Rating Council (NFRC) is working to standardize this procedure for use in a national fenestration certification and labeling program. Until such a standard is available, the fenestration manufacturer should be consulted for data on the angle-dependent SHGC of its products.

Frame and Other Nonglazing Elements

Solar radiant heat also enters a building through opaque elements such as the frame and any mullion bars and dividers that are part of the fenestration system. A solar heat gain coefficient can be defined for these opaque elements and used to determine the overall solar gain q_i of the fenestration system as follows

$$q_i = E_t \left(A_g F_g + A_f F_f + \sum_{i=1}^{N} A_i F_i \right) \qquad (25)$$

where

A_g = area of glazing with solar heat gain coefficient F_g
A_f = area of frame with solar heat gain coefficient F_f
A_i = area of the ith additional opaque element having F_i as its solar heat gain coefficient

The overall, area-weighted solar heat gain coefficient $<F>$ can be defined by

$$<F> = \frac{\displaystyle\sum_{i=1}^{M} A_i F_i + A_g F_g + A_f F_f}{\displaystyle\sum_{i=1}^{M} A_i + A_g + A_f} \qquad (26)$$

where M is the total number of different opaque and transparent elements over the total area of the fenestration aperture.

Passive Solar Gain

Energy analysis of a fenestration product should include the value of passive solar gain through the product in winter. As described in Chapter 30 of the 1991 ASHRAE *Handbook— Applications*, the magnitude of this energy gain depends on such variables as latitude and orientation. In some cases, properly designed and operated fenestration allows more energy into the building over a heating season than it loses, thus making it energy contributing rather than energy consuming. Excessive solar gain must be controlled during the cooling season, however.

Solar Gain Rejection

For some buildings in certain climates, preventing solar gain is more important from an energy perspective than improved thermal insulation using multiple panes of glazing. For example, internal load-dominated buildings in cool, clear climates can have substantial daytime solar and internal heat gains. These gains can be rejected by conduction through the building envelope and/or forced ventilation through the HVAC system. Preventing excessive solar gain through the fenestration systems of such buildings is very important. If such a building is occupied and conditioned mostly during daylight hours, improving the U-factor of the windows can have an overall adverse impact on building energy performance, preventing the escape of heat by conduction through the windows, and requiring the HVAC system to remove this heat even when the outside air temperature is substantially below the inside temperature.

AIR MOVEMENT

Infiltration through Fenestration

Air infiltration through fenestration products is significant in determining a building's heating and cooling loads. It should be analyzed separately and not incorporated in a U-factor determination for any specific product. Chapter 23 presents calculation procedures for air infiltration.

Indoor Air Movement

Because supply air grilles are frequently located directly below windows, air sweeps the interior glass surface. Heated supply air should be directed away from the glass to prevent large temperature differences between the center and edges of the glass. These thermal effects must be considered, particularly when annealed glass is used and air is forced over the glass surface during the heating season. Direct flow of heated air over the glass surface can increase the heat transfer coefficient and the temperature difference, causing a substantial increase in heat loss; it may also lead to thermally induced stress and risk of glass breakage.

Systems designed predominantly for cooling lower the glass temperature and rapidly pick up the cooling load. Both tend to improve the comfort condition of the space. However, the air-conditioned space has an increased net heat gain, caused by (1) increase in the shading coefficient (SC) due to delivery of a larger portion of the absorbed heat to the indoor space; (2) increase in the fenestration U-factor, because of the greater convection effect at the indoor surface; and (3) increase in the air-to-air temperature difference, since supply air rather than room air is in contact with the indoor glass surface. The principal increase in heat gain with clear glass is the result of the higher U-factor and the greater air-to-air temperature difference, since supply air, rather than room air, is in contact with the indoor glass surface.

The principal increase in heat gain with clear glass is the result of the higher U-factor and the greater air-to-air temperature difference. McCabe *et al.* (1986) showed that the SC for heat-

absorbing glass is much larger than that for clear glass (Figures 14 and 15).

When supply air flows between glass and an interior shading device, a greater portion of the heat absorbed in a shading device is added to the total heat gain. Pennington and McDuffie (1970) used this arrangement with a medium-colored drape to show an approximate 20% increase over the natural convection condition (Figures 16 and 17).

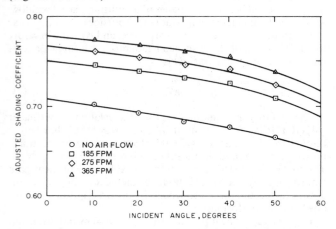

Fig. 15 Theoretically Determined Shading Coefficients for 0.25-in. Green Heat Absorbing Plate Glass without Drape

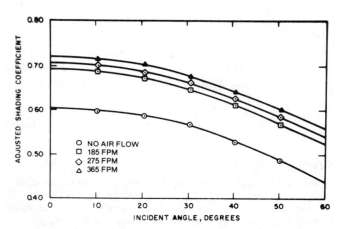

Fig. 16 Theoretically Determined Shading Coefficients for 0.25-in. *Clear* Plate Glass with Drape

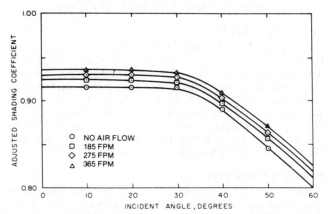

Fig. 14 Theoretically Determined Shading Coefficients for 0.25-in. Clear Plate Glass without Drape

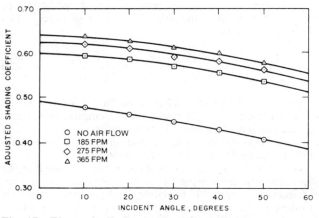

Fig. 17 Theoretically Determined Shading Coefficients for 0.25-in. Green Heat-Absorbing Plate Glass with Drape

Fenestration

Table 10 U-Factors for Summer Conditions, Btu/h·ft²·°F

	Velocity of Air Sweeping Window			Still Air
Type	365 fpm	275 fpm	185 fpm	
CL and CL	0.673	0.658	0.64	0.56
HA and CL	0.673	0.658	0.64	0.56
Refl. and CL	0.378	0.378	0.367	0.34

CL = Clear 0.25-in. float; HA = Heat absorbing 0.25-in. float; Refl = 0.25-in. reflective float.

Insulating glass exhibits the same effects as single glass for the same reasons, but they are less marked. Figure 18 shows theoretically determined SC curves for three types of insulating glass at three different airflow rates and for still air. Figure 19 shows similar curves for glass and drape arrangements, with the supply air delivered between the glass and drape. Pennington *et al.* (1973) experimentally checked the theoretical curves for both single and insulating glass arrangements with good agreement. Table 10 shows theoretically determined U-factors for the three insulating glass types. The values shown for still air conditions were verified experimentally by Pennington *et al.* (1973).

The theoretical determination of the SC, U-factor, and total heat gain when the supply air sweeps the indoor glass surface is similar to that explained in the section Fenestration Heat Balance—Solar Heat Gain. Instead of a combined air space conductance value h_s, convection and radiation effects must be calculated separately for spaces where the surfaces are swept by the supply air. An iterative computer solution is necessary to determine temperatures of the fenestration components.

Equation (27) may be used to calculate heat gain through the center of a single-glass fenestration; convection and radiation components for the interior glass surface are calculated separately.

$$q_A = E_{Tg} + h_c (t_g - t_{aa}) + \sigma e_g (T_g^4 - T_r^4) \qquad (27)$$

where

q_A = total heat gain, Btu/(h·ft²)
E_{Tg} = transmitted solar energy, Btu/(h·ft²)
h_c = inside glass surface convection coefficient = 0.99 + 0.21V, Btu/(h·ft²·°F)
V = air velocity, fps
t_g = temperature of glass, °F
T_g = temperature of glass, °R
t_{aa} = average temperature of air sweeping glass, °F
T_r = room air temperature, °R
σ = Stefan-Boltzmann constant
e_g = hemispherical emittance of room side glass surface

Since the equation has two unknowns, q_A and t_g, a second equation is needed

$$E_{\alpha g} = h_o (t_g - t_{oo}) + h_c (t_g - t_{aa}) + \sigma e_g (T_g^4 - T_r^4)$$

where

$E_{\alpha g}$ = solar energy absorbed in glass, Btu/(h·ft²)
h_o = surface coefficient = 4.0 Btu/(h·ft²·°F)
t_o = outdoor air temperature, °F

The U-factor may be determined by

$$U = 1/(1/h_o + 1/h_i + L/k) \qquad (28)$$

where

h_i = inside glass surface combined radiation and convection coefficient, Btu/(h·ft²·°F)
L/k = glass thickness, in., over thermal conductivity, Btu·in/(h·ft²·°F)

The air-to-air heat gain is expressed by

$$q_{td} = U (t_o - t_{aa}) \qquad (29)$$

The solar heat gain is $q_A - q_{td}$.

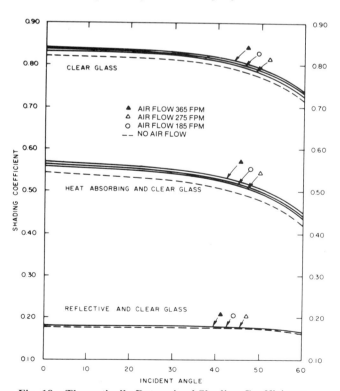

Fig. 18 Theoretically Determined Shading Coefficients for Three Insulating Glass Types

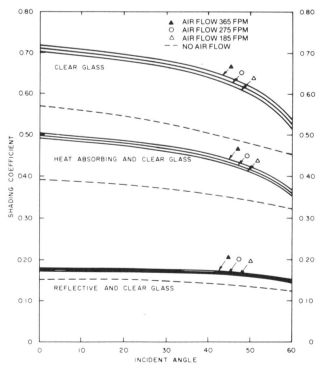

Fig. 19 Theoretically Determined Shading Coefficients for Three Insulating Glass Types with Medium Colored Drapes

DAYLIGHTING FROM FENESTRATION

In the United States, a general rule has been that the fenestration area should be at least 20% of the floor area. In Europe, a similar rule was based on a minimum illumination value on the normal work plane from a standard overcast sky condition.

The secondary visual benefit of fenestration is the amount and quality of light it produces in the work environment. One general rule determined the need for auxiliary electric light by assuming that daylight was adequate for a depth of two and one-half times the height of the window into the room based on a normal sill height. To prevent excessive glare, all fenestration should have sun controls. Variable and removable controls are often more effective in daylight than fixed controls.

For more accurate evaluation of daylight distribution within a space, several prediction tools, such as the *Recommended Practice of Daylighting* (IES 1979), are available. This practice shows a simple way of calculating the daylight distribution on the work plane from windows and skylights with and without controls. Many other daylight prediction tools calculate illuminance from radiant flux transfer or ray tracing.

Any or all of the various daylight prediction tools can be used to compare the relative value of daylight distribution from alternative fenestration systems, but ultimately the designer must evaluate costs and benefits to choose between alternative designs. This may be based on energy use or, more properly, on overall costs and benefits to the client. A simplified prediction tool, BEEM, is available for such analysis (Rundquist 1991). Also, total loss of productivity from an electric brown-out in a space with no natural ventilation or daylight may be as important as the benefits of many energy saving schemes.

SIMPLIFIED METHODS FOR PREDICTING HEAT TRANSFER THROUGH FENESTRATION

Fenestration Heat Balance-Solar Heat Gain

At any instant, the heat balance between a unit area of sunlit single-glazing material and its thermal environment, as shown in Figure 20, is

$$E_t + U(t_o - t_i) = q_R + q_s + q_T + q_{RCo} + q_{RCi} \qquad (30)$$

where q_R, q_s, and q_T represent the heat reflected, stored in the glass, and transmitted, respectively. In general, q_s is relatively small in magnitude, so it is disregarded. The terms q_{RCo} and q_{RCi} are the rates of heat flux outward and inward, respectively, by radiation and convection. The rate of heat rejection to the atmosphere is the sum of the reflected heat q_R and outward radiation convection heat flux q_{RCo}.

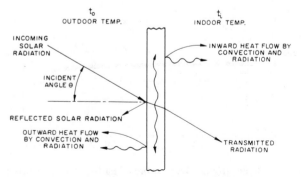

Fig. 20   Instantaneous Heat Balance for Sunlit Glazing Material

The total instantaneous rate of heat gain through the glazing material may be expressed:

Total heat admission through glass	=	Radiation transmitted through glass	+	Heat flow inward by radiation-convection from inner glass surface

$$q_A = E_{D\tau D} + E_{d\tau d} + q_{RCi} \qquad (31)$$

Subscript D denotes direct radiation; d denotes diffuse radiation, and τ indicates the transmitted portion. The solar-optical properties for diffuse radiation can be taken as those for direct radiation where $\theta = 60°$. The following equations consider total solar energy values for simplicity.

This equation may also be written as:

Total heat admission through glass	=	Radiation transmitted through glass	+	Inward flow of absorbed solar radiation
			+	Heat flow due to outdoor-indoor temperature difference

For single glazing,

$$q_A = E_t \tau + N_i(\alpha E_t) + U(t_o - t_i) \qquad (32)$$

For double glazing, the transmittance through both glasses can be calculated by

$$\bar{\tau} = \tau_o \tau_i/(1 - \rho_2 \rho_3) \qquad (33)$$

Since the first two terms of Equation (32) are related to the incident solar radiation, while the third occurs whether or not the sun is shining, Equations (32) and (33) may be combined as

$$q_A = E_t \bar{\tau} + N_{jo}(\alpha E_o) + N_i(\alpha E_i) + U(t_o - t_i) \qquad (34)$$

Total heat admission through glass	=	Solar heat gain	+	Conduction heat gain

These may both be written as

$$q_A = FE_t + U(t_o - t_i) \qquad (35)$$

where

q_A = instantaneous rate of heat admission through fenestration, Btu/h·ft²

F = dimensionless ratio of solar heat gains to incident solar radiation

The solar heat gain coefficient F is a characteristic of each type of fenestration and varies with the incident angle, since transmittance and absorptance of the glazing material depend on θ. For single glazing

$$F = \tau + U\alpha/h_o \qquad (36)$$

For double glazing

$$F = \tau + U\alpha_o/h_o + [(U/h_o) + (U/h_s)]\alpha_i \qquad (37)$$

For shaded glazing materials and other more complex fenestration, F can be calculated for standard conditions, but such values should be verified by solar calorimeter tests.

Because of the constantly growing number of fenestration products and the great variety of combinations of incident angle and solar radiation intensity, the quantities in Equation (37) cannot be estimated in all conditions. Instead, a procedure described in the following section has been developed for estimating solar and total heat gains through fenestration (Vild 1964, Yellott 1966, Pennington *et al.* 1964, Smith and Pennington 1964).

Solar Heat Gain Factors and Shading Coefficients

The reference glazing material for the following calculations is double-strength sheet glass (DSA) with 0.86 transmittance, 0.08 reflectance, and 0.06 absorptance at normal incidence. For other incident angles, Figure 11 gives the solar-optical properties.

The shading coefficient (SC) is the ratio of solar heat gain through a glazing system under a specific set of conditions to solar gain through a single pane of the reference glass under the same conditions. This ratio, a unique characteristic of each type of fenestration, is represented by

$$SC = \frac{\text{Solar heat gain of fenestration}}{\text{Solar heat gain of reference glass}} \quad (38)$$

The SC is a ratio of the solar heat gain coefficient F for the fenestration and F for single-pane, double-strength, clear (DSA) glass, which, for standard summer conditions, is 0.87 for normal incidence.

$$SC = \frac{F \text{ of Fenestration}}{0.87} = 1.15 \,(F \text{ of Fenestration}) \quad (39)$$

Table 11 gives values of SC for frequently used fenestrations.

For this calculation procedure, heat gains through sunlit DSA glass are designated as the solar heat gain factors (SHGF). Tables 12 through 18 give solar intensity and SHGF values calculated for daylight hours of the 21st day of each month. The tables apply solar-optical properties of DSA glass and the incident angles, which apply to 16 orientations for latitudes in 8° increments from 16 to 64° North. For other dates, times, and latitudes, use linear interpolation.

Table 11 Shading Coefficients for Single Glass and Insulating Glass

Type of Glass	Nominal Thickness[a]	Solar Trans.[a]	Shading Coefficient	
			$h_0 = 4.0$	$h_0 = 3.0$
A. Single Glass				
Clear	1/8 in.	0.86	1.00	1.00
	1/4 in.	0.78	0.94	0.95
	3/8 in.	0.72	0.90	0.92
	1/2 in.	0.67	0.87	0.88
Heat absorbing	1/8 in.	0.64	0.83	0.85
	1/4 in.	0.46	0.69	0.73
	3/8 in.	0.33	0.60	0.64
	1/2 in.	0.24	0.53	0.58
B. Insulating Glass				
Clear out, Clear in	1/8 in.[b]	0.71[c]	0.88	0.88
Clear out, Clear in	1/4 in.	0.61	0.81	0.82
Heat absorbing out, Clear in[d]	1/4 in.	0.36	0.55	0.58

Table refers to factory-fabricated units with 3/16, 1/4, or 1/2-in. air space or to prime windows plus storm sash.
[a]Refer to manufacturer's literature for values.
[b]Thickness of each pane of glass, not thickness of assembled unit.
[c]Combined transmittance for assembled unit.
[d]Refers to gray, bronze, and green tinted heat-absorbing float glass.

The SHGF in Tables 12 through 18 do not represent maximum values but the conditions on average cloudless days. For high elevations and very clear atmospheres, the maximum SHGF can be as much as 15% above the tabulated values. For very dusty industrial atmospheres and exceptionally humid locations, they may be from 20 to 30% lower than the tabulated values. The Clearness numbers in Figure 7 can be used for these conditions.

Half-day totals given in Tables 12 through 18 are integrated values using Simpson's rule for integration, with a 10-min. time interval starting at 0100 solar time. Similar integration by hourly periods results in less accurate values, because of the variation of SHGF within this longer time interval. Full-day totals can be estimated from these tables by adding the morning half-day total for the orientation, listed at the top, to the afternoon half-day total for the same orientation, listed at the bottom of the tables.

For any glazing material or fenestration other than double-strength sheet glass, but with a similar angular dependence, the solar heat gain will be:

$$\begin{array}{l} \text{Solar} \\ \text{heat} \\ \text{gain} \end{array} = \begin{array}{l} \text{Shading} \\ \text{coefficient} \\ \text{for fenestration} \end{array} + \begin{array}{l} \text{Solar heat gain factor} \\ \text{for given orientation} \\ \text{and existing conditions} \end{array} \quad (40)$$

The total instantaneous heat gain is

$$q_A = (SC)(SHGF) + U(t_o - t_i) \quad (41)$$

Equation (41) applies at any time of the year. In winter, when outdoor temperature is lower than indoor, the conduction heat flow occurring in the absence of sunshine is negative or outward in summer, conduction heat flow is usually inward.

To estimate the total gain through fenestration at any time, applicable values of outdoor and indoor dry-bulb temperatures must be used. Tables in Chapter 24 give design values of summer outdoor dry-bulb temperatures for many locations. These are generally mid-afternoon temperatures; for other times, local weather stations can supply appropriate temperature data. Winter design temperatures should not be used in Equation (40), since they are usually given for early morning hours before sunrise. The indoor air temperature for air-conditioned spaces is likely to be closer to 75°F than to 80°F (see Chapter 24).

Example 8. Find the total heat gain through the center of a double-glazed window consisting of an outdoor light of 0.25-in. clear glass, 0.5-in. air space, and an indoor light of 0.25-in. clear glass, with a reflective film on the No. 3 surface in a west wall at 40° North latitude, at 1600 on July 21. Outdoor air is 105°F, and indoor air is 75°F. Solar irradiance is 248.3 Btu/(h·ft²).

The spectrally averaged data for the glass are:

Outdoor $\tau = 0.80$; $\rho_1 = 0.07$; $\rho_2 = 0.07$; $\alpha_1 = 0.13$; $\alpha_2 = 0.13$;
$$e_1 = 0.84; e_2 = 0.84$$

Indoor $\tau_i = 0.12$; $\rho_3 = 0.70$; $\rho_4 = 0.45$; $\alpha_3 = 0.18$; $\alpha_4 = 0.81$;
$$e_3 = 0.10; e_4 = 0.84$$

Solution:
Absorption for the glasses

$$\alpha_o = 0.13 + (0.13)\frac{0.80 \times 0.70}{1 - (0.07 \times 0.70)}$$

$$\alpha_i = (0.18)\frac{0.80 \times 0.70}{1 - (0.70 \times 0.70)} = 0.151$$

The solar radiation absorbed

$$\alpha E_o = 248.3 \times 0.207 = 51.4 \text{ Btu/h·ft}^2$$
$$\alpha E_i = 248.3 \times 0.151 = 37.5 \text{ Btu/h·ft}^2$$

Table 12   Solar Intensity (E_{DN}) and Solar Heat Gain Factors (SHGF) for 16° North Latitude

Date	Solar Time	Direct Normal Btu/(h·ft²)	N	NNE	NE	ENE	E	ESE	SE	SSE	S	SSW	SW	WSW	W	WNW	NW	NNW	Hor.	Solar Time
Jan	0700	141	5	6	44	92	124	134	126	96	49	6	5	5	5	5	5	5	14	1700
	0800	262	14	15	55	147	210	240	233	189	114	25	14	14	14	14	14	14	79	1600
	0900	300	21	21	32	122	200	244	251	219	152	58	22	21	21	21	21	21	150	1500
	1000	317	26	26	27	66	150	209	233	223	178	102	31	26	26	26	26	26	203	1400
	1100	325	29	29	29	31	77	148	195	210	194	146	75	31	29	29	29	29	236	1300
	1200	327	30	30	30	30	32	72	139	184	199	184	138	72	32	30	30	30	248	1200
	HALF DAY TOTALS		110	112	196	461	760	1000	1096	1020	781	426	211	127	111	110	110	110	805	
Feb	0700	182	8	17	84	138	169	172	150	103	36	8	8	8	8	8	8	8	25	1700
	0800	273	17	19	96	180	231	247	224	166	77	18	17	17	17	17	17	17	101	1600
	0900	305	23	24	64	153	214	242	233	188	110	30	23	23	23	23	23	23	174	1500
	1000	319	28	29	33	92	161	202	211	188	134	61	30	28	28	28	28	28	229	1400
	1100	326	32	32	32	37	83	136	167	172	149	102	49	33	32	32	32	32	263	1300
	1200	328	33	33	33	33	34	60	107	142	154	142	106	60	34	33	33	33	275	1200
	HALF DAY TOTALS		124	137	321	609	865	1023	1034	885	582	287	174	132	124	124	124	124	930	
Mar	0700	201	11	53	124	172	192	183	145	82	15	10	10	10	10	10	10	10	40	1700
	0800	272	20	50	140	205	239	235	195	123	35	19	19	19	19	19	19	19	120	1600
	0900	299	26	35	109	179	218	225	197	138	57	27	26	26	26	26	26	26	192	1500
	1000	312	31	33	61	120	165	182	172	134	76	34	32	31	31	31	31	31	247	1400
	1100	318	34	35	36	53	87	114	125	116	89	55	36	35	34	34	34	34	280	1300
	1200	320	35	35	36	36	37	47	69	87	93	86	68	47	37	36	36	35	291	1200
	HALF DAY TOTALS		141	226	494	755	928	975	879	643	319	187	153	142	139	139	139	139	1025	
Apr	0600	14	2	8	12	14	14	12	8	2	1	1	1	1	1	1	1	1	1	1800
	0700	197	24	94	153	187	191	167	117	45	14	13	13	13	13	13	13	13	53	1700
	0800	256	27	99	172	216	227	204	150	69	24	22	22	22	22	22	22	22	131	1600
	0900	280	31	79	149	193	208	193	147	77	31	29	29	29	29	29	29	29	197	1500
	1000	293	35	54	102	141	158	151	120	73	37	34	33	33	33	33	33	33	249	1400
	1100	299	38	40	54	72	86	88	78	60	43	38	38	36	36	36	36	37	279	1300
	1200	301	39	39	39	40	40	41	43	45	45	45	43	41	40	39	39	39	289	1200
	HALF DAY TOTALS		179	403	674	859	922	851	653	352	174	159	157	156	155	155	155	156	1057	
May	0600	44	14	30	41	45	43	34	19	4	3	3	3	3	3	3	3	3	5	1800
	0700	193	50	120	168	191	185	150	92	24	16	16	16	16	16	16	16	17	62	1700
	0800	244	52	132	189	218	215	179	115	38	25	24	24	24	24	24	24	25	135	1600
	0900	268	49	116	171	198	197	167	109	45	32	30	30	30	30	30	30	32	197	1500
	1000	280	47	89	130	151	150	126	84	44	37	35	35	35	35	35	35	37	245	1400
	1100	286	47	63	79	87	83	70	52	41	40	39	38	38	38	39	39	41	273	1300
	1200	288	46	46	44	43	42	41	41	41	41	41	41	41	42	43	44	46	282	1200
	HALF DAY TOTALS		283	575	804	916	897	748	493	217	172	167	167	167	167	168	169	176	1058	
Jun	0600	53	20	39	52	55	51	39	20	4	4	4	4	4	4	4	4	4	7	1800
	0700	188	62	128	172	190	179	141	80	20	16	16	16	16	16	16	16	18	64	1700
	0800	238	66	142	194	217	207	167	99	31	25	25	25	25	25	25	25	27	135	1600
	0900	261	63	130	178	198	190	154	93	37	31	31	31	31	31	31	31	33	194	1500
	1000	273	59	104	140	154	145	115	70	39	37	36	36	36	36	36	36	38	241	1400
	1100	279	57	76	90	92	82	63	46	41	40	39	39	39	39	40	41	43	268	1300
	1200	281	57	55	50	45	43	42	41	41	41	41	41	42	42	45	50	55	277	1200
	HALF DAY TOTALS		356	648	850	929	876	700	430	194	174	171	171	171	172	173	176	190	1049	
Jul	0600	41	14	29	39	42	40	31	18	4	3	3	3	3	3	3	3	3	6	1800
	0700	184	51	118	164	185	179	145	88	23	16	16	16	16	16	16	16	17	62	1700
	0800	236	55	132	187	214	210	174	111	37	25	25	25	25	25	25	25	26	133	1600
	0900	259	52	117	170	196	193	163	106	44	32	31	31	31	31	31	31	33	194	1500
	1000	272	50	92	131	151	148	123	81	44	38	36	36	36	36	36	36	38	241	1400
	1100	278	49	66	81	88	83	69	52	42	41	40	39	39	39	40	40	42	269	1300
	1200	279	49	48	46	44	43	42	42	42	42	42	42	43	44	46	48	49	277	1200
	HALF DAY TOTALS		296	580	799	903	878	729	478	215	176	172	171	171	171	172	173	182	1043	
Aug	0600	11	2	7	10	12	12	10	6	2	1	1	1	1	1	1	1	1	1	1800
	0700	180	26	92	145	176	180	156	109	42	15	14	14	14	14	14	14	14	53	1700
	0800	240	30	100	168	209	219	196	143	65	25	23	23	23	23	23	23	23	128	1600
	0900	266	33	82	148	190	203	187	142	74	33	30	30	30	30	30	30	30	193	1500
	1000	279	37	58	104	140	155	147	117	71	39	36	35	35	35	35	35	35	243	1400
	1100	285	40	43	57	75	86	87	76	59	44	40	39	38	38	38	38	39	273	1300
	1200	287	41	41	41	42	42	43	44	45	46	45	44	43	42	41	41	41	282	1200
	HALF DAY TOTALS		191	410	666	837	891	817	624	339	180	167	165	164	163	163	163	164	1033	
Sep	0700	179	12	50	114	158	176	168	133	76	15	11	11	11	11	11	11	11	39	1700
	0800	253	21	49	134	196	227	224	186	119	36	20	20	20	20	20	20	20	116	1600
	0900	281	28	36	106	173	211	217	191	134	57	28	27	27	27	27	27	27	185	1500
	1000	295	32	34	61	118	161	178	168	132	76	35	33	32	32	32	32	32	238	1400
	1100	302	35	36	37	54	86	113	123	114	88	56	38	36	35	35	35	35	271	1300
	1200	304	36	36	37	38	39	49	69	86	93	86	69	48	39	38	37	36	282	1200
	HALF DAY TOTALS		146	226	475	722	885	931	842	622	319	192	159	148	145	144	144	144	991	
Oct	0700	166	8	18	79	128	156	159	139	95	33	9	8	8	8	8	8	8	25	1700
	0800	259	17	20	95	174	223	237	215	159	74	19	17	17	17	17	17	17	99	1600
	0900	292	24	25	65	150	209	235	225	182	106	31	24	24	24	24	24	24	170	1500
	1000	307	29	30	34	92	158	197	205	183	130	60	31	29	29	29	29	29	224	1400
	1100	314	32	32	33	39	83	133	163	167	145	100	49	34	32	32	32	32	258	1300
	1200	316	33	33	33	34	35	60	105	139	150	138	104	60	35	34	33	33	270	1200
	HALF DAY TOTALS		127	141	318	592	836	986	996	852	563	283	175	136	128	127	127	127	911	
Nov	0700	134	5	6	43	89	119	129	120	92	47	6	5	5	5	5	5	5	14	1700
	0800	255	15	15	55	145	206	235	228	185	111	25	15	15	15	15	15	15	78	1600
	0900	295	21	21	33	121	197	241	247	215	150	57	22	21	21	21	21	21	149	1500
	1000	312	26	26	28	67	147	206	230	220	176	100	31	26	26	26	26	26	201	1400
	1100	320	29	29	29	31	77	146	192	207	191	144	74	31	29	29	29	29	234	1300
	1200	322	30	30	30	30	32	72	137	181	196	181	137	72	32	30	30	30	246	1200
	HALF DAY TOTALS		112	113	197	456	749	983	1077	1001	767	420	210	128	112	112	112	112	799	
Dec	0700	118	4	5	30	72	101	112	107	85	48	7	4	4	4	4	4	4	10	1700
	0800	255	13	14	41	132	198	233	231	193	124	33	13	13	13	13	13	13	69	1600
	0900	297	20	20	25	108	191	241	254	227	165	72	21	20	20	20	20	20	138	1500
	1000	315	25	25	26	56	144	208	239	233	192	117	35	25	25	25	25	25	191	1400
	1100	323	28	28	28	29	73	150	202	221	207	161	86	30	28	28	28	28	223	1300
	1200	325	29	29	29	29	30	77	149	197	212	196	149	76	30	29	29	29	234	1200
	HALF DAY TOTALS		104	105	159	402	710	975	1099	1050	836	484	228	125	105	104	104	104	748	
			N	NNW	NW	WNW	W	WSW	SW	SSW	S	SSE	SE	ESE	E	ENE	NE	NNE	Hor.	PM

Notes: 1. Clearness number = 1.00; Ground reflectance = 0.20.   2. Figures shown are for 21st day of each month.

Table 13   Solar Intensity (E_{DN}) and Solar Heat Gain Factors (SHGF) for 24° North Latitude

Date	Solar Time	Direct Normal Btu/(h·ft²)	N	NNE	NE	ENE	E	ESE	SE	SSE	S	SSW	SW	WSW	W	WNW	NW	NNW	Hor.	Solar Time
Jan	0700	71	2	3	21	45	62	67	63	49	25	3	2	2	2	2	2	2	5	1700
	0800	239	12	12	41	128	190	221	218	181	114	28	12	12	12	12	12	12	55	1600
	0900	288	18	18	23	106	190	240	253	227	166	73	19	18	18	18	18	18	121	1500
	1000	308	23	23	24	53	144	211	245	241	200	125	38	24	23	23	23	23	172	1400
	1100	317	26	26	26	27	73	156	211	234	220	173	95	29	26	26	26	26	204	1300
	1200	320	27	27	27	27	29	82	160	210	227	210	160	81	29	27	27	27	214	1200
	HALF DAY TOTALS		95	96	148	372	671	942	1076	1039	840	505	241	120	96	95	95	95	664	
Feb	0700	153	6	12	67	114	141	145	128	90	33	6	6	6	6	6	6	6	17	1700
	0800	262	15	16	80	165	220	240	224	172	89	17	15	15	15	15	15	15	83	1600
	0900	297	21	22	46	138	208	244	243	205	133	42	22	21	21	21	21	21	153	1500
	1000	314	26	26	28	76	157	209	228	213	165	87	28	26	26	26	26	26	205	1400
	1100	321	29	29	29	31	80	148	191	203	185	137	68	31	29	29	29	29	238	1300
	1200	323	30	30	30	30	32	70	134	177	192	177	133	70	32	30	30	30	249	1200
	HALF DAY TOTALS		113	119	257	527	806	1011	1072	965	699	374	200	127	113	113	113	113	820	
Mar	0700	194	11	45	115	164	186	180	145	86	17	10	10	10	10	10	10	10	36	1700
	0800	267	18	35	124	195	234	237	204	138	48	19	18	18	18	18	18	18	112	1600
	0900	295	25	27	85	165	215	232	214	163	82	27	25	25	25	25	25	25	180	1500
	1000	309	30	31	41	103	162	194	195	168	112	47	31	30	30	30	30	30	232	1400
	1100	315	33	33	34	42	85	129	154	155	139	86	43	34	33	33	33	33	264	1300
	1200	317	34	34	34	34	35	56	96	126	137	126	95	56	35	34	34	34	275	1200
	HALF DAY TOTALS		133	189	422	693	906	1011	970	778	458	249	169	139	133	133	133	133	962	
Apr	0600	40	6	21	33	39	39	33	22	7	2	2	2	2	2	2	2	2	4	1800
	0700	203	20	88	151	189	197	176	127	55	15	14	14	14	14	14	14	14	58	1700
	0800	256	24	80	159	209	228	212	164	88	24	22	22	22	22	22	22	22	132	1600
	0900	280	30	54	126	181	208	203	169	105	39	29	28	28	28	28	28	28	195	1500
	1000	292	34	37	75	125	157	165	148	107	56	35	33	33	33	33	33	33	244	1400
	1100	298	36	37	40	59	85	103	106	94	70	45	38	37	36	36	36	36	274	1300
	1200	299	37	37	38	38	39	46	59	70	75	70	58	45	39	38	38	37	283	1200
	HALF DAY TOTALS		168	339	607	826	940	924	773	494	244	180	163	157	155	155	154	154	1048	
May	0600	86	25	57	79	87	84	66	38	8	6	6	6	6	6	6	6	6	13	1800
	0700	203	43	117	171	199	196	163	105	32	17	17	17	17	17	17	17	18	73	1700
	0800	248	38	114	178	214	218	190	132	54	26	25	25	25	25	25	25	26	142	1600
	0900	269	35	88	150	188	198	179	132	66	33	31	31	31	31	31	31	31	201	1500
	1000	280	38	59	103	137	150	141	111	67	39	36	35	35	35	35	35	36	247	1400
	1100	286	40	43	55	72	83	84	75	58	44	40	39	38	38	38	38	39	274	1300
	1200	288	41	41	41	41	42	43	44	46	46	46	44	43	42	41	41	41	282	1200
	HALF DAY TOTALS		238	492	749	909	943	840	614	308	187	176	174	173	172	172	172	175	1089	
Jun	0600	97	36	70	93	101	94	73	39	8	7	7	7	7	7	7	7	8	17	1800
	0700	201	55	127	177	199	192	155	94	26	18	18	18	18	18	18	18	20	77	1700
	0800	242	50	126	184	214	212	179	117	43	27	26	26	26	26	26	26	27	145	1600
	0900	263	43	102	158	189	192	168	116	53	34	32	32	32	32	32	32	33	201	1500
	1000	274	41	72	113	140	146	131	96	55	39	36	36	36	36	36	36	38	245	1400
	1100	279	42	50	65	77	82	77	64	49	42	41	40	39	39	39	40	41	271	1300
	1200	281	43	43	43	43	43	43	43	43	43	43	43	43	43	43	43	43	279	1200
	HALF DAY TOTALS		284	562	802	933	932	797	544	255	187	181	180	179	179	179	180	187	1096	
Jul	0600	81	26	56	76	84	80	63	36	8	6	6	6	6	6	6	6	7	13	1800
	0700	195	45	116	168	194	190	158	101	31	18	18	18	18	18	18	18	19	73	1700
	0800	239	41	115	176	210	213	185	128	52	27	26	26	26	26	26	26	26	141	1600
	0900	261	37	90	150	186	195	175	129	64	34	32	32	32	32	32	32	32	198	1500
	1000	272	39	62	104	137	149	139	108	65	39	37	36	36	36	36	36	37	243	1400
	1100	278	41	44	58	73	83	83	73	57	44	41	40	39	39	39	39	40	270	1300
	1200	280	42	42	42	43	43	44	45	46	46	46	45	43	43	42	42	42	278	1200
	HALF DAY TOTALS		247	498	746	897	925	820	595	300	191	181	178	177	177	177	177	181	1076	
Aug	0600	35	6	20	30	35	35	30	19	6	2	2	2	2	2	2	2	2	4	1800
	0700	186	22	87	144	179	186	165	119	51	16	15	15	15	15	15	15	15	58	1700
	0800	241	26	82	156	203	220	204	157	84	26	24	24	24	24	24	24	24	130	1600
	0900	265	32	57	126	178	202	197	162	101	39	31	30	30	30	30	30	30	191	1500
	1000	278	36	40	78	125	155	161	143	103	55	37	35	35	35	35	35	35	239	1400
	1100	284	38	39	42	61	85	101	104	91	68	46	40	38	37	37	37	37	268	1300
	1200	286	38	39	40	40	41	47	58	69	72	68	58	47	41	40	40	39	277	1200
	HALF DAY TOTALS		179	347	601	806	910	889	740	473	243	186	171	165	164	163	163	162	1028	
Sep	0800	248	19	36	119	185	222	225	194	132	48	20	19	19	19	19	19	19	108	1600
	0900	278	26	28	84	160	207	223	206	158	81	28	26	26	26	26	26	26	174	1500
	1000	292	31	32	42	101	158	188	190	163	110	48	32	31	31	31	31	31	224	1400
	1100	299	34	34	35	43	84	127	151	151	128	86	44	35	34	34	34	34	256	1300
	1200	301	35	35	35	36	37	57	95	124	134	124	94	57	37	36	35	35	266	1200
	HALF DAY TOTALS		139	190	406	661	863	964	927	749	451	251	174	145	139	138	138	138	930	
Oct	0700	138	6	12	62	104	129	133	117	82	31	7	6	6	6	6	6	6	17	1700
	0800	247	16	17	79	159	211	230	214	164	85	17	16	16	16	16	16	16	82	1600
	0900	284	22	23	47	135	202	237	235	198	128	41	23	22	22	22	22	22	150	1500
	1000	301	27	27	29	77	154	204	222	207	160	85	29	27	27	27	27	27	201	1400
	1100	309	30	30	30	33	80	145	186	198	180	133	67	32	30	30	30	30	233	1300
	1200	311	31	31	31	31	33	70	131	173	187	172	130	69	33	31	31	31	244	1200
	HALF DAY TOTALS		116	123	255	512	778	974	1032	929	675	367	200	131	117	116	116	116	804	
Nov	0700	67	2	3	20	43	59	64	60	46	24	3	2	2	2	2	2	2	5	1700
	0800	232	12	13	42	126	186	216	213	177	111	28	12	12	12	12	12	12	55	1600
	0900	282	19	19	23	106	187	236	249	223	163	71	20	19	19	19	19	19	120	1500
	1000	303	23	23	24	53	143	209	241	237	197	123	37	24	23	23	23	23	171	1400
	1100	312	26	26	26	28	73	154	209	230	217	171	93	29	26	26	26	26	202	1300
	1200	315	27	27	27	27	29	81	158	207	224	207	158	80	29	27	27	27	213	1200
	HALF DAY TOTALS		97	97	149	368	661	926	1056	1020	825	497	239	121	98	97	97	97	659	
Dec	0700	30	1	1	7	18	25	28	27	21	12	2	1	1	1	1	1	1	2	1700
	0800	225	10	10	29	112	174	208	209	178	118	35	11	10	10	10	10	10	44	1600
	0900	281	17	17	19	93	180	234	252	231	174	84	18	17	17	17	17	17	107	1500
	1000	304	22	22	22	44	137	209	247	247	209	137	44	22	22	22	22	22	157	1400
	1100	314	25	25	25	26	69	156	216	241	230	183	104	29	25	25	25	25	188	1300
	1200	317	26	26	26	26	27	85	147	219	237	219	167	84	27	26	26	26	199	1200
	HALF DAY TOTALS		88	88	118	313	611	899	1054	1042	868	550	257	117	89	88	88	88	598	
			N	NNW	NW	WNW	W	WSW	SW	SSW	S	SSE	SE	ESE	E	ENE	NE	NNE	Hor.	PM

Notes: 1. Clearness number = 1.00; Ground reflectance = 0.20.   2. Figures shown are for 21st day of each month.

Table 14 Solar Intensity (E_{DN}) and Solar Heat Gain Factors (SHGF) for 32° North Latitude

Date	Solar Time	Direct Normal Btu/(h·ft²)	N	NNE	NE	ENE	E	ESE	SE	SSE	S	SSW	SW	WSW	W	WNW	NW	NNW	Hor.	Solar Time
Jan	0700	1	0	0	0	1	1	1	1	1	0	0	0	0	0	0	0	0	0	1700
	0800	203	9	9	29	105	160	189	189	159	103	28	9	9	9	9	9	9	32	1600
	0900	269	15	15	17	91	175	229	246	225	169	82	17	15	15	15	15	15	88	1500
	1000	295	20	20	20	41	135	209	249	250	212	141	46	20	20	20	20	20	136	1400
	1100	306	23	23	23	24	68	159	221	249	238	191	110	29	23	23	23	23	166	1300
	1200	310	24	24	24	24	25	88	174	228	246	228	174	88	25	24	24	24	176	1200
	HALF DAY TOTALS		79	79	107	284	570	856	1015	1014	853	553	264	112	80	79	79	79	512	
Feb	0700	112	4	7	47	82	102	106	95	67	26	4	4	4	4	4	4	4	9	1700
	0800	245	13	14	65	149	205	228	216	170	95	17	13	13	13	13	13	13	64	1600
	0900	287	19	19	32	122	199	242	248	216	149	55	20	19	19	19	19	19	127	1500
	1000	305	24	24	25	62	151	213	241	232	189	112	31	24	24	24	24	24	176	1400
	1100	314	26	26	26	28	76	156	208	227	212	165	87	28	26	26	26	26	207	1300
	1200	316	27	27	27	27	29	79	155	204	221	204	155	79	29	27	27	27	217	1200
	HALF DAY TOTALS		100	103	201	445	735	978	1080	1010	780	452	228	122	100	100	100	100	691	
Mar	0700	185	10	37	105	153	176	173	142	88	20	9	9	9	9	9	9	9	32	1700
	0800	260	17	25	107	183	227	237	209	150	62	18	17	17	17	17	17	17	100	1600
	0900	290	23	25	64	151	210	237	227	183	107	30	23	23	23	23	23	23	164	1500
	1000	304	28	28	30	87	158	202	215	195	144	70	29	28	28	28	28	28	211	1400
	1100	311	31	31	31	34	82	142	179	188	168	120	59	32	31	31	31	31	242	1300
	1200	313	32	32	32	32	33	66	122	162	176	162	122	66	33	32	32	32	252	1200
	HALF DAY TOTALS		124	162	359	629	875	1033	1041	888	589	326	193	136	125	124	124	124	874	
Apr	0600	66	9	35	54	65	66	56	38	12	4	3	3	3	3	3	3	3	7	1800
	0700	206	17	80	146	188	200	182	136	65	16	14	14	14	14	14	14	14	61	1700
	0800	255	23	61	144	200	227	219	177	107	30	22	22	22	22	22	22	22	129	1600
	0900	278	28	36	103	168	206	212	187	133	58	29	28	28	28	28	28	28	188	1500
	1000	290	32	34	52	108	155	177	172	141	87	39	33	32	32	32	32	32	233	1400
	1100	295	35	35	36	47	83	118	135	132	108	70	40	36	35	35	35	35	262	1300
	1200	297	36	36	36	37	38	53	82	106	115	106	82	53	38	37	36	36	271	1200
	HALF DAY TOTALS		161	296	550	792	952	992	889	645	360	228	177	157	153	152	152	152	1015	
May	0600	119	33	77	108	121	116	94	56	13	8	8	8	8	8	8	8	9	21	1800
	0700	211	36	111	170	202	204	174	118	42	19	18	18	18	18	18	18	19	81	1700
	0800	250	29	94	165	208	220	199	149	73	27	25	25	25	25	25	25	25	146	1600
	0900	269	33	61	128	177	198	190	155	93	37	32	31	31	31	31	31	31	201	1500
	1000	280	36	40	76	121	150	156	138	99	54	37	35	35	35	35	35	35	243	1400
	1100	285	38	39	42	59	83	99	102	90	68	47	40	39	37	37	37	37	269	1300
	1200	286	38	39	40	40	41	47	59	70	74	70	59	47	41	40	40	39	277	1200
	HALF DAY TOTALS		222	438	702	900	985	933	747	447	250	199	183	177	175	174	174	175	1098	
Jun	0600	131	44	92	123	135	127	99	55	12	10	10	10	10	10	10	10	11	28	1800
	0700	210	47	122	176	204	201	168	108	35	20	20	20	20	20	20	20	21	88	1700
	0800	245	36	106	171	208	214	189	135	60	28	27	27	27	27	27	27	27	151	1600
	0900	264	35	74	137	178	193	180	139	77	35	32	32	32	32	32	32	32	204	1500
	1000	274	38	47	86	125	146	145	123	83	45	38	36	36	36	36	36	36	244	1400
	1100	279	40	41	47	64	82	91	89	75	56	43	41	40	39	39	39	39	269	1300
	1200	280	41	41	41	42	42	46	52	58	60	58	52	46	42	42	41	41	276	1200
	HALF DAY TOTALS		261	504	762	935	985	897	678	372	225	197	189	185	184	184	183	186	1122	
Jul	0600	113	34	76	105	117	113	90	53	12	9	9	9	9	9	9	9	9	22	1800
	0700	203	38	111	167	198	198	169	114	41	20	19	19	19	19	19	19	19	81	1700
	0800	241	31	95	163	204	215	194	145	70	28	26	26	26	26	26	26	26	145	1600
	0900	261	34	64	129	175	195	186	150	90	37	32	32	32	32	32	32	32	198	1500
	1000	271	37	42	78	121	148	153	134	96	53	38	36	36	36	36	36	36	240	1400
	1100	277	39	40	43	60	83	98	99	88	66	47	41	40	38	38	38	38	265	1300
	1200	279	40	40	41	41	42	48	58	68	72	68	58	48	42	41	41	40	273	1200
	HALF DAY TOTALS		231	444	701	890	967	912	726	433	248	202	187	182	180	179	179	180	1088	
Aug	0600	59	10	33	50	60	60	51	34	11	4	4	4	4	4	4	4	4	8	1800
	0700	190	19	79	141	179	190	172	128	61	17	15	15	15	15	15	15	15	61	1700
	0800	240	25	63	141	195	219	210	170	102	31	23	23	23	23	23	23	23	128	1600
	0900	263	30	39	104	166	200	206	181	127	57	31	29	29	29	29	29	29	185	1500
	1000	276	34	36	55	109	153	173	167	136	84	40	35	34	34	34	34	34	229	1400
	1100	282	36	37	39	50	84	116	131	127	104	69	41	38	36	36	36	36	256	1300
	1200	284	37	37	37	39	40	54	81	103	111	103	81	54	40	39	37	37	265	1200
	HALF DAY TOTALS		171	303	546	774	922	955	854	618	352	231	184	166	162	161	160	160	999	
Sep	0700	163	10	35	96	139	159	156	128	80	20	10	10	10	10	10	10	10	31	1700
	0800	240	18	26	103	173	215	224	198	143	60	19	18	18	18	18	18	18	96	1600
	0900	272	24	26	64	146	202	227	218	177	105	31	24	24	24	24	24	24	158	1500
	1000	287	29	29	32	86	154	196	208	189	141	70	31	29	29	29	29	29	204	1400
	1100	294	32	32	32	36	81	139	174	182	163	118	59	34	32	32	32	32	234	1300
	1200	296	33	33	33	33	35	66	120	158	171	158	120	66	35	33	33	33	244	1200
	HALF DAY TOTALS		130	164	345	598	831	982	993	852	574	325	197	142	130	129	129	129	845	
Oct	0700	99	4	7	43	74	92	96	85	60	24	5	4	4	4	4	4	4	10	1700
	0800	229	13	15	63	143	195	217	206	162	90	17	13	13	13	13	13	13	63	1600
	0900	273	20	20	33	120	193	234	239	208	144	54	21	20	20	20	20	20	125	1500
	1000	293	24	24	26	62	147	207	234	225	183	109	32	24	24	24	24	24	173	1400
	1100	302	27	27	27	29	76	152	203	221	207	160	85	29	27	27	27	27	203	1300
	1200	304	28	28	28	28	30	78	151	199	215	199	151	78	30	28	28	28	213	1200
	HALF DAY TOTALS		103	106	200	433	708	941	1038	972	753	441	226	125	104	103	103	103	679	
Nov	0700	2	0	0	0	1	1	1	1	1	0	0	0	0	0	0	0	0	0	1700
	0800	196	9	9	29	103	156	184	184	155	100	27	9	9	9	9	9	9	32	1600
	0900	263	16	16	17	90	173	225	241	221	166	80	17	16	16	16	16	16	88	1500
	1000	289	20	20	21	41	134	206	245	246	209	138	45	21	20	20	20	20	136	1400
	1100	301	23	23	23	24	67	157	218	245	234	188	109	29	23	23	23	23	165	1300
	1200	304	24	24	24	24	25	87	171	224	243	224	171	87	25	24	24	24	175	1200
	HALF DAY TOTALS		80	81	108	282	561	841	996	995	838	544	261	113	81	80	80	80	509	
Dec	0800	176	7	7	19	84	135	163	166	143	97	31	7	7	7	7	7	7	22	1600
	0900	257	14	14	15	77	162	218	238	222	171	89	15	14	14	14	14	14	72	1500
	1000	288	18	18	18	34	127	204	246	251	216	148	52	19	18	18	18	18	119	1400
	1100	301	21	21	21	22	63	157	222	252	243	197	116	29	21	21	21	21	148	1300
	1200	304	22	22	22	22	23	89	177	232	252	232	177	89	23	22	22	22	158	1200
	HALF DAY TOTALS		71	71	84	227	500	792	965	986	852	578	275	107	71	71	71	71	440	
			N	NNW	NW	WNW	W	WSW	SW	SSW	S	SSE	SE	ESE	E	ENE	NE	NNE	Hor.	PM

Notes: 1. Clearness number = 1.00; Ground reflectance = 0.20.   2. Figures shown are for 21st day of each month.

Table 15   Solar Intensity (E_{DN}) and Solar Heat Gain Factors (SHGF) for 40° North Latitude

Date	Solar Time	Direct Normal Btu/(h·ft²)	N	NNE	NE	ENE	E	ESE	SE	SSE	S	SSW	SW	WSW	W	WNW	NW	NNW	Hor.	Solar Time
Jan	0800	142	5	5	17	71	111	132	133	114	75	22	6	5	5	5	5	5	14	1600
	0900	239	12	12	13	74	154	205	224	209	160	82	13	12	12	12	12	12	55	1500
	1000	274	16	16	16	31	124	199	241	246	213	146	51	17	16	16	16	16	96	1400
	1100	289	19	19	19	20	61	156	222	252	244	198	118	28	19	19	19	19	124	1300
	1200	294	20	20	20	20	21	90	179	234	254	234	179	90	21	20	20	20	133	1200
	HALF DAY TOTALS		61	61	73	199	452	734	904	932	813	561	273	101	62	61	61	61	354	
Feb	0700	55	2	3	23	40	51	53	47	34	14	2	2	2	2	2	2	2	4	1700
	0800	219	10	11	50	129	183	206	199	160	94	18	10	10	10	10	10	10	43	1600
	0900	271	16	16	22	107	186	234	245	218	157	66	17	16	16	16	16	16	98	1500
	1000	294	21	21	21	49	143	211	246	243	203	129	38	21	21	21	21	21	143	1400
	1100	304	23	23	23	24	71	160	219	244	231	184	103	27	23	23	23	23	171	1300
	1200	307	24	24	24	24	25	86	170	222	241	222	170	86	25	24	24	24	180	1200
	HALF DAY TOTALS		84	86	152	361	648	916	1049	1015	821	508	250	114	85	84	84	84	548	
Mar	0700	171	9	29	93	140	163	161	135	86	22	8	8	8	8	8	8	8	26	1700
	0800	250	16	18	91	169	218	232	211	157	74	17	16	16	16	16	16	16	85	1600
	0900	282	21	22	47	136	203	238	236	198	128	40	22	21	21	21	21	21	143	1500
	1000	297	25	25	27	72	153	207	229	216	171	95	29	25	25	25	25	25	186	1400
	1100	305	28	28	28	30	78	151	198	213	197	150	77	30	28	28	28	28	213	1300
	1200	307	29	29	29	29	31	75	145	191	206	191	145	75	31	29	29	29	223	1200
	HALF DAY TOTALS		114	139	302	563	832	1035	1087	968	694	403	220	132	114	113	113	113	764	
Apr	0600	89	11	46	72	87	88	76	52	18	5	5	5	5	5	5	5	5	11	1800
	0700	206	16	71	140	185	201	186	143	75	16	14	14	14	14	14	14	14	61	1700
	0800	252	22	44	128	190	224	223	188	124	41	22	21	21	21	21	21	21	123	1600
	0900	274	27	29	80	155	202	219	203	156	83	29	27	27	27	27	27	27	177	1500
	1000	286	31	31	37	92	152	187	193	170	121	56	32	31	31	31	31	41	217	1400
	1100	292	33	33	34	39	81	130	160	166	146	102	52	35	33	33	33	33	243	1300
	1200	293	34	34	34	34	36	62	108	142	154	142	108	62	36	34	34	34	252	1200
	HALF DAY TOTALS		154	265	501	758	957	1051	994	782	488	296	199	157	148	147	147	147	957	
May	0500	1	0	1	1	1	1	1	1	0	0	0	0	0	0	0	0	0	0	1900
	0600	144	36	90	128	145	141	115	71	18	10	10	10	10	10	10	10	11	31	1800
	0700	216	28	102	165	202	209	184	131	54	20	19	19	19	19	19	19	19	87	1700
	0800	250	27	73	149	199	220	208	164	93	29	25	25	25	25	25	25	25	146	1600
	0900	267	31	42	105	164	197	200	175	121	53	32	30	30	30	30	30	30	195	1500
	1000	277	34	36	54	105	148	168	163	133	83	40	35	34	34	34	34	34	234	1400
	1100	283	36	36	38	48	81	113	130	127	105	70	42	38	36	36	36	36	257	1300
	1200	284	37	37	37	38	40	54	82	104	113	104	82	54	40	38	37	37	265	1200
	HALF DAY TOTALS		215	404	666	893	1024	1025	881	601	358	247	200	180	176	175	174	175	1083	
Jun	0500	22	10	17	21	22	20	14	6	2	1	1	1	1	1	1	1	2	3	1900
	0600	155	48	104	143	159	151	121	70	17	13	13	13	13	13	13	13	14	40	1800
	0700	216	37	113	172	205	207	178	122	46	22	21	21	21	21	21	21	21	97	1700
	0800	246	30	85	156	201	216	199	152	80	29	27	27	27	27	27	27	27	153	1600
	0900	263	33	51	114	166	192	190	161	105	45	33	32	32	32	32	32	32	201	1500
	1000	272	35	38	63	109	145	158	148	116	69	39	36	35	35	35	35	35	238	1400
	1100	277	38	39	40	52	81	105	116	110	88	60	41	39	38	38	38	38	260	1300
	1200	279	38	38	38	40	41	52	72	89	95	89	72	52	41	40	38	38	267	1200
	HALF DAY TOTALS		253	470	734	941	1038	999	818	523	315	236	204	191	188	187	186	188	1126	
Jul	0500	2	1	2	2	2	2	1	1	0	0	0	0	0	0	0	0	0	0	1900
	0600	138	37	89	125	142	137	112	68	18	11	11	11	11	11	11	11	12	32	1800
	0700	208	30	102	163	198	204	179	127	53	21	20	20	20	20	20	20	20	88	1700
	0800	241	28	75	148	196	216	203	160	90	30	26	26	26	26	26	26	26	145	1600
	0900	259	32	44	106	163	193	196	170	118	52	33	31	31	31	31	31	31	194	1500
	1000	269	35	37	56	106	146	165	159	129	81	41	36	35	35	35	35	35	231	1400
	1100	275	37	38	40	50	81	111	127	123	102	69	43	39	37	37	37	37	254	1300
	1200	276	38	38	38	40	41	55	80	101	109	101	80	55	41	40	38	38	262	1200
	HALF DAY TOTALS		223	411	666	885	1008	1003	858	584	352	248	204	186	181	180	180	181	1076	
Aug	0600	81	12	44	68	81	82	71	48	17	6	5	5	5	5	5	5	5	12	1800
	0700	191	17	71	135	177	191	177	135	70	17	16	16	16	16	16	16	16	62	1700
	0800	237	24	47	126	185	216	214	180	118	41	23	23	23	23	23	23	23	122	1600
	0900	260	28	31	82	153	197	212	196	151	80	31	28	28	28	28	28	28	174	1500
	1000	272	32	33	40	93	150	182	187	165	116	56	34	32	32	32	32	32	214	1400
	1100	278	35	35	36	41	81	128	156	160	141	99	52	37	35	35	35	35	239	1300
	1200	280	35	35	35	36	38	63	106	138	149	138	106	63	38	36	35	35	247	1200
	HALF DAY TOTALS		164	273	498	741	928	1013	956	751	474	296	205	166	157	156	156	156	946	
Sep	0700	149	9	27	84	125	146	144	121	77	21	9	9	9	9	9	9	9	25	1700
	0800	230	17	19	87	160	205	218	199	148	71	18	17	17	17	17	17	17	82	1600
	0900	263	22	23	47	131	194	227	226	190	124	41	23	22	22	22	22	22	138	1500
	1000	280	27	27	28	71	148	200	221	209	165	93	30	27	27	27	27	27	180	1400
	1100	287	29	29	29	31	78	147	192	207	191	146	77	31	29	29	29	29	206	1300
	1200	290	30	30	30	30	32	75	142	185	200	185	142	75	32	30	30	30	215	1200
	HALF DAY TOTALS		119	142	291	534	787	980	1033	925	672	396	222	137	119	118	118	118	738	
Oct	0700	48	2	3	20	36	45	47	42	30	12	2	2	2	2	2	2	2	4	1700
	0800	204	11	12	49	123	173	195	188	151	89	18	11	11	11	11	11	11	43	1600
	0900	257	17	17	23	104	180	225	235	209	151	64	18	17	17	17	17	17	97	1500
	1000	280	21	21	22	50	139	205	238	235	196	125	38	22	21	21	21	21	140	1400
	1100	291	24	24	24	25	71	156	212	236	224	178	101	28	24	24	24	24	168	1300
	1200	294	25	25	25	25	27	85	165	216	234	216	165	85	27	25	25	25	177	1200
	HALF DAY TOTALS		88	89	152	351	623	878	1006	974	791	493	247	117	89	88	88	88	540	
Nov	0800	136	5	5	18	69	108	128	129	110	72	21	6	5	5	5	5	5	14	1600
	0900	232	12	12	13	73	151	201	219	204	156	80	13	12	12	12	12	12	55	1500
	1000	268	16	16	16	31	122	196	237	242	209	143	50	17	16	16	16	16	96	1400
	1100	283	19	19	19	20	61	154	218	248	240	194	116	28	19	19	19	19	123	1300
	1200	288	20	20	20	20	21	89	176	231	250	231	176	89	21	20	20	20	132	1200
	HALF DAY TOTALS		63	63	75	198	445	721	887	914	798	551	269	101	63	63	63	63	354	
Dec	0800	89	3	3	8	41	67	82	84	73	50	17	3	3	3	3	3	3	6	1600
	0900	217	10	10	11	60	135	185	205	194	151	83	13	10	10	10	10	10	39	1500
	1000	261	14	14	14	25	113	188	232	239	210	146	55	15	14	14	14	14	77	1400
	1100	280	17	17	17	17	56	151	217	249	242	198	120	28	17	17	17	17	104	1300
	1200	285	18	18	18	18	19	89	178	233	253	233	178	89	19	18	18	18	113	1200
	HALF DAY TOTALS		52	52	56	146	374	649	822	867	775	557	276	94	53	52	52	52	282	
			N	NNW	NW	WNW	W	WSW	SW	SSW	S	SSE	SE	ESE	E	ENE	NE	NNE	Hor.	PM

Notes: 1. Clearness number = 1.00; Ground reflectance = 0.20.   2. Figures shown are for 21st day of each month.

Table 16 Solar Intensity (E_{DN}) and Solar Heat Gain Factors (SHGF) for 48° North Latitude

Date	Solar Time	Direct Normal Btu/(h·ft²)	N	NNE	NE	ENE	E	ESE	SE	SSE	S	SSW	SW	WSW	W	WNW	NW	NNW	Hor.	Solar Time
Jan	0800	37	1	1	4	18	29	34	35	30	20	6	1	1	1	1	1	1	2	1600
	0900	185	8	8	8	53	118	160	176	166	129	69	10	8	8	8	8	8	25	1500
	1000	239	12	12	12	22	106	175	216	223	195	136	50	12	12	12	12	12	55	1400
	1100	261	14	14	14	15	53	144	208	239	233	190	116	26	14	14	14	14	77	1300
	1200	267	15	15	15	15	16	86	171	226	245	226	171	86	16	15	15	15	85	1200
	HALF DAY TOTALS		43	43	46	117	316	567	729	776	701	512	259	85	43	43	43	43	203	
Feb	0700	4	0	0	1	3	3	3	3	2	1	0	0	0	0	0	0	0	0	1700
	0800	180	8	8	36	103	149	170	166	136	82	17	8	8	8	8	8	8	25	1600
	0900	247	13	13	16	90	168	216	230	209	155	71	14	13	13	13	13	13	66	1500
	1000	275	17	17	17	38	131	203	242	244	207	138	44	18	17	17	17	17	105	1400
	1100	288	19	19	19	20	65	158	221	249	239	192	113	27	19	19	19	19	130	1300
	1200	292	20	20	20	20	22	89	176	231	250	231	176	89	22	20	20	20	138	1200
	HALF DAY TOTALS		68	68	107	274	541	816	968	967	813	531	261	104	68	68	68	68	395	
Mar	0700	153	7	22	80	123	145	145	123	80	23	7	7	7	7	7	7	7	20	1700
	0800	236	14	15	76	154	204	222	206	158	82	15	14	14	14	14	14	14	68	1600
	0900	270	19	19	3	121	193	234	239	207	142	52	20	19	19	19	19	19	118	1500
	1000	287	23	23	24	58	146	208	237	231	189	115	33	23	23	23	23	23	156	1400
	1100	295	25	25	25	26	74	156	210	232	218	172	94	28	25	25	25	25	180	1300
	1200	298	26	26	26	26	27	83	161	211	228	211	161	83	27	26	26	26	188	1200
	HALF DAY TOTALS		100	118	250	494	775	1012	1100	1014	767	465	244	126	101	100	100	100	636	
Apr	0600	108	12	53	86	105	107	93	64	23	6	6	6	6	6	6	6	6	15	1800
	0700	205	15	61	132	180	199	189	148	84	18	14	14	14	14	14	14	14	60	1700
	0800	247	20	32	111	179	219	225	196	138	55	21	20	20	20	20	20	20	114	1600
	0900	268	25	26	60	141	197	223	215	176	106	33	25	25	25	25	25	25	161	1500
	1000	280	28	28	31	77	148	193	209	194	150	80	31	28	28	28	28	28	196	1400
	1100	286	31	31	31	33	78	140	181	193	177	133	69	33	31	31	31	31	218	1300
	1200	288	31	31	31	31	34	71	131	172	186	172	131	71	34	31	31	31	226	1200
	HALF DAY TOTALS		147	242	461	724	957	1098	1081	895	605	370	226	156	141	140	140	140	875	
May	0500	41	17	31	40	42	39	29	14	3	3	3	3	3	3	3	3	3	5	1900
	0600	162	35	97	141	162	160	133	85	24	12	12	12	12	12	12	12	13	40	1800
	0700	219	23	90	158	200	212	191	142	68	21	19	19	19	19	19	19	19	91	1700
	0800	248	26	54	132	190	218	214	178	113	38	25	25	25	25	25	25	25	142	1600
	0900	264	29	32	82	151	194	208	192	147	77	32	29	29	29	29	29	29	185	1500
	1000	274	33	34	39	90	145	178	184	163	116	57	35	33	33	33	33	33	219	1400
	1100	279	35	35	36	40	79	126	155	160	142	101	54	37	35	35	35	35	240	1300
	1200	280	35	35	35	36	38	63	107	139	150	139	107	63	38	36	35	35	247	1200
	HALF DAY TOTALS		215	388	645	893	1065	1114	1007	749	483	316	225	184	174	173	173	174	1045	
Jun	0500	77	35	61	76	80	72	53	24	6	5	5	5	5	5	5	5	8	12	1900
	0600	172	46	110	155	175	169	138	84	22	14	14	14	14	14	14	14	16	51	1800
	0700	220	29	101	165	204	211	187	135	60	23	21	21	21	21	21	21	21	103	1700
	0800	246	29	64	139	191	215	206	168	101	34	27	27	27	27	27	27	27	152	1600
	0900	261	31	36	91	153	190	199	180	133	66	33	31	31	31	31	31	31	193	1500
	1000	269	34	36	45	94	143	169	171	148	101	50	36	34	34	34	34	34	225	1400
	1100	274	36	36	38	44	79	118	142	145	126	88	49	38	36	36	36	36	246	1300
	1200	275	37	37	37	38	40	60	96	124	134	124	96	60	40	38	37	37	252	1200
	HALF DAY TOTALS		257	459	722	955	1095	1102	955	678	436	299	228	197	189	188	188	191	1108	
Jul	0500	43	18	33	42	45	41	30	15	3	3	3	3	3	3	3	3	4	6	1900
	0600	156	37	96	138	159	156	129	82	24	13	13	13	13	13	13	13	14	41	1800
	0700	211	25	90	156	196	207	186	138	66	22	20	20	20	20	20	20	20	92	1700
	0800	240	27	56	132	187	214	209	174	110	38	26	26	26	26	26	26	26	142	1600
	0900	256	30	34	83	149	191	204	187	143	75	33	30	30	30	30	30	30	184	1500
	1000	266	34	35	41	90	143	174	180	158	113	56	36	34	34	34	34	34	217	1400
	1100	271	36	36	37	42	79	124	151	156	138	99	54	38	36	36	36	36	237	1300
	1200	272	36	36	36	37	39	63	104	136	146	136	104	63	39	37	36	36	244	1200
	HALF DAY TOTALS		223	395	646	886	1050	1092	983	730	474	315	229	190	181	179	179	180	1042	
Aug	0600	99	13	51	81	98	100	87	60	22	7	7	7	7	7	7	7	7	16	1800
	0700	190	17	61	128	172	190	179	141	79	19	15	15	15	15	15	15	15	61	1700
	0800	232	22	34	110	174	211	216	188	132	53	23	22	22	22	22	22	22	114	1600
	0900	154	27	28	63	139	192	216	108	169	102	34	27	27	27	27	27	27	159	1500
	1000	266	30	30	33	78	145	188	203	188	144	78	33	30	30	30	30	30	193	1400
	1100	272	32	32	32	36	78	137	175	187	171	129	68	35	32	32	32	32	215	1300
	1200	274	33	33	33	33	36	71	128	167	189	167	128	71	36	33	33	33	223	1200
	HALF DAY TOTALS		157	251	459	709	929	1060	1040	862	587	366	231	165	151	149	149	149	869	
Sep	0700	131	8	21	71	108	128	128	108	71	21	8	7	7	7	7	7	7	20	1700
	0800	215	15	16	72	144	191	207	193	148	77	16	15	15	15	15	15	15	65	1600
	0900	251	20	20	34	116	184	223	227	197	136	52	21	20	20	20	20	20	114	1500
	1000	269	24	24	25	58	141	200	228	221	182	112	34	24	24	24	24	24	151	1400
	1100	278	26	26	26	28	73	151	203	223	210	166	92	29	26	26	26	26	174	1300
	1200	280	27	27	27	27	29	82	156	204	220	204	156	82	29	27	27	27	182	1200
	HALF DAY TOTALS		105	121	240	465	729	953	1040	963	737	453	243	131	106	105	105	105	614	
Oct	0700	4	0	0	2	3	4	4	3	2	1	0	0	0	0	0	0	0	0	1700
	0800	165	8	9	35	96	139	159	155	126	77	16	8	8	8	8	8	8	25	1600
	0900	233	14	14	16	88	161	207	220	199	148	68	15	14	14	14	14	14	66	1500
	1000	262	18	18	18	39	128	196	233	234	199	133	43	18	18	18	18	18	104	1400
	1100	274	20	20	20	21	64	153	213	241	231	186	109	27	20	20	20	20	128	1300
	1200	278	21	21	21	21	23	87	171	223	242	223	171	87	23	21	21	21	136	1200
	HALF DAY TOTALS		71	71	108	266	519	780	925	925	779	513	256	106	72	71	71	71	391	
Nov	0800	36	1	1	4	18	29	34	35	30	20	6	1	1	1	1	1	1	2	1600
	0900	179	8	8	9	52	115	156	171	161	125	67	10	8	8	8	8	8	26	1500
	1000	233	12	12	12	22	104	172	212	218	191	133	49	13	12	12	12	12	55	1400
	1100	255	15	15	15	15	52	142	204	234	228	186	114	26	15	15	15	15	77	1300
	1200	261	15	15	15	15	17	85	168	222	240	222	168	85	17	15	15	15	85	1200
	HALF DAY TOTALS		44	44	47	117	310	555	713	760	686	502	255	85	44	44	44	44	204	
Dec	0900	140	5	5	6	36	86	120	133	127	100	56	8	5	5	5	5	5	13	1500
	1000	214	10	10	10	16	91	156	194	201	179	126	49	10	10	10	10	10	38	1400
	1100	242	12	12	12	13	46	134	195	225	220	180	111	25	12	12	12	12	57	1300
	1200	250	13	13	13	13	14	81	163	215	233	215	168	81	14	13	13	13	65	1200
	HALF DAY TOTALS		33	33	34	73	233	458	610	665	616	468	247	76	34	33	33	33	141	
			N	NNW	NW	WNW	W	WSW	SW	SSW	S	SSE	SE	ESE	E	ENE	NE	NNE	Hor.	PM

Notes: 1. Clearness number = 1.00; Ground reflectance = 0.20. 2. Figures shown are for 21st day of each month.

Table 17 Solar Intensity (E_{DN}) and Solar Heat Gain Factors (SHGF) for 56° North Latitude

Date	Solar Time	Direct Normal Btu/(h·ft²)	N	NNE	NE	ENE	E	ESE	SE	SSE	S	SSW	SW	WSW	W	WNW	NW	NNW	Hor.	Solar Time
Jan	0900	78	3	3	3	21	49	67	74	70	55	30	4	3	3	3	3	3	5	1500
	1000	170	7	7	7	13	74	126	156	162	143	100	38	7	7	7	7	7	21	1400
	1100	207	9	9	9	10	40	116	169	194	190	156	96	21	9	9	9	9	34	1300
	1200	217	10	10	10	10	11	71	144	190	205	190	144	71	11	10	10	10	40	1200
	HALF DAY TOTALS		23	23	24	46	163	343	468	517	487	378	206	61	24	23	23	23	80	
Feb	0800	115	4	4	21	64	95	109	107	88	55	12	4	4	4	4	4	4	10	1600
	0900	203	10	10	11	71	139	183	197	182	136	66	10	10	10	10	10	10	36	1500
	1000	246	13	13	13	28	115	184	223	227	196	133	45	14	13	13	13	13	65	1400
	1100	262	15	15	15	16	57	148	210	239	232	188	112	25	15	15	15	15	84	1300
	1200	267	16	16	16	16	17	86	171	225	244	225	171	86	17	16	16	16	91	1200
	HALF DAY TOTALS		49	50	66	182	409	666	821	846	737	509	253	89	50	49	49	49	241	
Mar	0700	128	6	16	65	101	121	122	105	70	21	6	6	6	6	6	6	6	14	1700
	0800	215	12	13	61	136	185	205	194	152	84	15	12	12	12	12	12	12	49	1600
	0900	253	16	16	23	105	179	224	233	207	148	61	17	16	16	16	16	16	89	1500
	1000	272	19	19	20	46	136	203	238	236	198	128	39	20	19	19	19	19	122	1400
	1100	282	21	21	21	22	68	156	215	241	230	184	106	27	21	21	21	21	142	1300
	1200	284	22	22	22	22	24	86	170	222	241	222	170	86	24	22	22	22	149	1200
	HALF DAY TOTALS		85	97	200	419	699	956	1071	1016	800	502	258	118	86	85	85	85	491	
Apr	0600	122	13	58	95	118	121	107	75	29	7	7	7	7	7	7	7	7	18	1800
	0700	201	15	51	123	173	195	188	152	91	21	14	14	14	14	14	14	14	56	1700
	0800	239	19	23	95	167	211	223	201	148	68	20	19	19	19	19	19	19	101	1600
	0900	260	23	24	44	126	190	223	223	189	126	44	24	23	23	23	23	23	140	1500
	1000	272	26	26	27	63	142	196	220	212	171	102	33	26	26	26	26	26	170	1400
	1100	278	28	28	28	30	74	147	195	213	200	156	86	31	28	28	28	28	189	1300
	1200	280	28	28	28	28	31	79	149	194	210	194	149	79	31	28	28	28	195	1200
	HALF DAY TOTALS		139	226	430	694	951	1132	1147	982	699	437	252	154	132	131	131	131	772	
May	0500	93	36	68	89	95	88	66	33	7	6	6	6	6	6	6	6	7	14	1900
	0600	175	33	99	148	174	173	147	97	31	14	14	14	14	14	14	14	14	48	1800
	0700	219	21	77	149	195	212	197	152	81	22	19	19	19	19	19	19	19	92	1700
	0800	244	25	38	115	179	215	218	189	131	52	25	24	24	24	24	24	24	135	1600
	0900	259	28	30	62	136	189	213	206	168	102	36	28	28	28	28	28	28	171	1500
	1000	268	31	31	33	75	141	185	200	187	145	80	33	31	31	31	31	31	199	1400
	1100	273	32	32	32	35	76	135	174	187	172	131	71	35	32	32	32	32	216	1300
	1200	275	33	33	33	33	36	71	129	168	181	168	129	71	36	33	33	33	222	1200
	HALF DAY TOTALS		222	391	644	906	1112	1202	1120	878	604	392	256	187	172	170	170	173	986	
Jun	0400	21	13	19	22	21	18	11	3	1	1	1	1	1	1	1	2	5	3	2000
	0500	122	53	94	119	126	115	85	40	10	9	9	9	9	9	9	9	12	25	1900
	0600	185	42	111	160	185	182	152	97	30	16	16	16	16	16	16	16	17	62	1800
	0700	222	25	86	156	199	213	195	147	74	24	22	22	22	22	22	22	22	105	1700
	0800	243	27	46	122	181	213	213	181	122	46	27	26	26	26	26	26	26	146	1600
	0900	257	30	32	69	139	187	206	196	156	91	34	30	30	30	30	30	30	181	1500
	1000	265	33	33	36	79	139	178	190	174	132	71	35	33	33	33	33	33	208	1400
	1100	269	34	34	35	38	76	129	164	174	159	119	65	37	34	34	34	34	225	1300
	1200	271	35	35	35	35	38	68	119	155	168	155	119	68	38	35	35	35	231	1200
	HALF DAY TOTALS		275	473	738	989	1162	1207	1082	822	562	376	260	203	190	189	189	196	1070	
Jul	0500	91	37	69	89	95	88	66	33	8	7	7	7	7	7	7	7	8	16	1900
	0600	169	34	98	145	170	170	143	95	31	15	14	14	14	14	14	14	15	50	1800
	0700	212	23	77	147	192	208	193	148	79	23	20	20	20	20	20	20	20	93	1700
	0800	237	26	40	115	177	211	214	185	128	51	26	25	25	25	25	25	25	135	1600
	0900	252	29	31	63	135	186	209	201	164	99	36	29	29	29	29	29	29	171	1500
	1000	261	32	32	34	76	139	181	196	182	142	78	35	32	32	32	32	32	198	1400
	1100	265	33	33	33	37	76	133	171	183	168	128	70	36	33	33	33	33	215	1300
	1200	267	34	34	34	34	37	71	126	164	177	164	126	71	37	34	34	34	221	1200
	HALF DAY TOTALS		231	398	646	901	1097	1180	1096	859	593	390	259	193	179	177	177	180	987	
Aug	0500	1	0	1	1	1	1	1	0	0	0	0	0	0	0	0	0	0	0	1900
	0600	112	14	56	91	111	114	101	71	28	8	8	8	8	8	8	8	8	20	1800
	0700	187	16	51	119	165	186	179	144	86	22	15	15	15	15	15	15	15	58	1700
	0800	225	20	25	94	162	203	214	192	142	66	22	20	20	20	20	20	20	101	1600
	0900	246	25	26	46	124	184	216	215	182	121	44	26	25	25	25	25	25	140	1500
	1000	258	28	28	30	65	139	191	213	204	165	99	34	28	28	28	28	28	169	1400
	1100	264	30	30	30	32	74	143	189	206	193	152	84	33	30	30	30	30	187	1300
	1200	266	30	30	30	30	30	78	153	188	203	188	153	78	33	30	30	30	198	1200
	HALF DAY TOTALS		149	235	429	680	923	1092	1104	946	678	431	256	163	142	140	140	141	771	
Sep	0700	107	6	15	56	87	104	105	90	60	19	6	6	6	6	6	6	6	14	1700
	0800	194	12	14	58	126	171	189	179	140	78	16	12	12	12	12	12	12	48	1600
	0900	233	17	17	24	100	170	211	220	195	140	59	18	17	17	17	17	17	86	1500
	1000	253	20	20	21	46	131	194	227	225	189	123	39	21	20	20	20	20	118	1400
	1100	263	22	22	22	24	67	150	206	230	220	176	103	28	22	22	22	22	137	1300
	1200	266	23	23	23	23	25	85	163	213	231	213	163	85	25	23	23	23	144	1200
	HALF DAY TOTALS		89	99	191	391	652	893	1004	958	761	484	255	121	90	89	89	89	474	
Oct	0800	104	4	5	20	59	87	100	98	81	50	11	4	4	4	4	4	4	10	1600
	0900	193	10	10	11	68	132	173	186	171	129	63	11	10	10	10	10	10	37	1500
	1000	231	14	14	14	28	111	176	213	216	186	127	44	14	14	14	14	14	64	1400
	1100	248	16	16	16	17	56	142	202	229	222	180	108	25	16	16	16	16	84	1300
	1200	253	16	16	16	16	18	83	164	216	234	216	164	83	18	16	16	16	91	1200
	HALF DAY TOTALS		52	52	68	177	390	633	779	804	702	487	246	90	53	52	52	52	240	
Nov	0900	76	3	3	3	21	48	66	72	69	54	29	4	3	3	3	3	3	6	1500
	1000	165	7	7	7	13	72	122	152	157	139	98	37	7	7	7	7	7	21	1400
	1100	201	9	9	9	10	39	113	165	190	186	152	94	21	9	9	9	9	35	1300
	1200	211	10	10	10	10	11	70	140	186	200	186	140	70	11	10	10	10	40	1200
	HALF DAY TOTALS		24	24	24	47	161	336	457	505	475	369	202	61	24	24	24	24	81	
Dec	0900	5	0	0	0	1	3	4	5	5	4	2	0	0	0	0	0	0	0	1500
	1000	113	4	4	4	7	47	82	103	107	96	68	27	4	4	4	4	4	9	1400
	1100	166	6	6	6	7	30	92	135	156	154	127	78	17	6	6	6	6	19	1300
	1200	180	7	7	7	7	8	59	120	159	171	159	120	59	8	7	7	7	23	1200
	HALF DAY TOTALS		14	14	14	20	88	217	311	354	343	277	163	47	15	14	14	14	40	
			N	NNW	NW	WNW	W	WSW	SW	SSW	S	SSE	SE	ESE	E	ENE	NE	NNE	Hor.	PM

Notes: 1. Clearness number = 1.00; Ground reflectance = 0.20.   2. Figures shown are for 21st day of each month.

Table 18 Solar Intensity (E_{DN}) and Solar Heat Gain Factors (SHGF) for 64° North Latitude

Date	Solar Time	Direct Normal Btu/(h·ft²)	N	NNE	NE	ENE	E	ESE	SE	SSE	S	SSW	SW	WSW	W	WNW	NW	NNW	Hor.	Solar Time
Jan	1000	22	1	1	1	1	9	16	20	21	19	13	5	1	1	1	1	1	1	1400
	1100	81	3	3	3	3	15	45	67	77	75	62	38	8	3	3	3	3	6	1300
	1200	100	3	3	3	3	4	33	67	89	96	89	67	33	4	3	3	3	8	1200
	HALF DAY TOTALS		5	5	5	6	25	79	121	142	141	119	75	23	5	5	5	5	11	
Feb	0800	18	1	1	3	10	15	17	17	14	9	2	1	1	1	1	1	1	1	1600
	0900	134	5	5	6	43	89	118	128	119	90	45	6	5	5	5	5	5	13	1500
	1000	190	8	8	8	18	87	144	176	180	157	108	38	9	8	8	8	8	28	1400
	1100	215	10	10	10	11	44	122	177	202	197	160	97	20	10	10	10	10	41	1300
	1200	222	11	11	11	11	12	73	147	194	210	194	147	73	12	11	11	11	45	1200
	HALF DAY TOTALS		29	30	33	89	244	446	578	617	560	411	212	66	30	29	29	29	106	
Mar	0700	95	4	11	47	74	90	91	79	53	17	4	4	4	4	4	4	4	9	1700
	0800	185	9	10	46	113	158	177	170	135	78	14	9	9	9	9	9	9	32	1600
	0900	227	13	13	16	88	159	203	215	194	143	64	14	13	13	13	13	13	59	1500
	1000	249	16	16	16	35	122	190	226	228	194	130	42	16	16	16	16	16	84	1400
	1100	260	17	17	17	18	60	148	209	236	228	184	109	25	17	17	17	17	99	1300
	1200	263	18	18	18	18	19	85	168	221	239	221	168	85	19	18	18	18	105	1200
	HALF DAY TOTALS		68	74	150	334	596	854	984	958	779	504	257	104	68	68	68	68	335	
Apr	0500	27	8	18	24	27	26	20	12	2	1	1	1	1	1	1	1	1	2	1900
	0600	133	12	59	102	127	132	118	84	35	8	8	8	8	8	8	8	8	21	1800
	0700	194	14	41	113	163	189	185	153	96	25	13	13	13	13	13	13	13	51	1700
	0800	228	17	19	79	153	201	217	201	153	79	19	17	17	17	17	17	17	85	1600
	0900	248	21	21	32	111	180	219	225	197	138	55	22	21	21	21	21	21	116	1500
	1000	260	23	23	24	51	134	194	225	221	185	118	38	24	23	23	23	23	140	1400
	1100	266	24	24	24	26	68	148	202	225	214	171	99	29	24	24	24	24	155	1300
	1200	268	25	25	25	25	27	83	159	208	224	208	159	83	27	25	25	25	160	1200
	HALF DAY TOTALS		131	218	410	671	943	1150	1186	1036	763	487	273	149	121	120	120	120	651	
May	0400	51	30	44	51	51	43	28	8	3	3	3	3	3	3	3	3	10	6	2000
	0500	132	48	95	125	135	125	96	50	11	9	9	9	9	9	9	9	11	26	1900
	0600	185	28	97	150	181	183	158	109	40	15	15	15	15	15	15	15	15	55	1800
	0700	218	21	63	138	189	211	201	161	94	24	19	19	19	19	19	19	19	90	1700
	0800	239	23	28	97	167	209	220	198	146	68	25	23	23	23	23	23	23	124	1600
	0900	252	26	27	45	122	183	215	215	184	123	46	27	26	26	26	26	26	152	1500
	1000	261	28	28	30	61	135	188	212	205	167	102	36	28	28	28	28	28	174	1400
	1100	265	30	30	30	32	72	141	188	207	195	154	87	33	30	30	30	30	188	1300
	1200	267	30	30	30	30	33	78	146	189	204	189	146	78	33	30	30	30	192	1200
	HALF DAY TOTALS		247	425	680	950	1177	1291	1218	985	708	465	288	191	169	168	168	176	911	
Jun	0300	21	17	21	22	20	14	6	2	1	1	1	1	1	1	1	2	10	3	2100
	0400	93	53	83	96	94	78	50	14	7	7	7	7	7	7	7	7	21	16	2000
	0500	154	62	114	148	158	145	110	55	14	12	12	12	12	12	12	12	14	39	1900
	0600	194	36	107	162	191	192	163	110	39	18	17	17	17	17	17	17	18	71	1800
	0700	221	24	71	145	193	213	200	158	89	25	22	22	22	22	22	22	22	105	1700
	0800	239	25	33	104	170	208	216	192	139	62	27	25	25	25	25	25	25	137	1600
	0900	251	28	29	51	124	181	210	208	175	115	43	29	28	28	28	28	28	165	1500
	1000	258	30	30	32	65	134	183	204	195	157	94	36	30	30	30	30	30	186	1400
	1100	262	32	32	32	34	72	137	180	196	184	144	82	35	32	32	32	32	199	1300
	1200	263	32	32	32	32	35	76	138	179	193	179	138	76	35	32	32	32	203	1200
	HALF DAY TOTALS		326	538	806	1066	1256	1318	1195	947	679	455	297	212	192	191	192	216	1021	
Jul	0400	53	32	47	55	54	46	29	9	4	4	4	4	4	4	4	4	11	8	2000
	0500	128	49	94	123	133	124	95	50	11	10	10	10	10	10	10	10	11	28	1900
	0600	179	30	96	148	177	180	155	106	39	16	15	15	15	15	15	15	15	57	1800
	0700	211	22	64	137	186	207	197	157	92	25	20	20	20	20	20	20	20	92	1700
	0800	231	24	30	97	165	205	215	193	142	67	26	24	24	24	24	24	24	124	1600
	0900	245	27	28	47	121	180	211	211	179	120	46	28	27	27	27	27	27	152	1500
	1000	253	29	29	31	62	134	185	208	200	164	100	37	29	29	29	29	29	174	1400
	1100	257	31	31	31	33	72	139	185	202	191	151	86	34	31	31	31	31	187	1300
	1200	259	31	31	31	31	34	78	143	185	200	185	143	78	34	31	31	31	192	1200
	HALF DAY TOTALS		258	434	684	946	1163	1269	1193	965	697	462	292	198	177	175	175	185	918	
Aug	0500	29	9	20	27	30	28	22	13	2	2	2	2	2	2	2	2	2	3	1900
	0600	123	13	58	97	121	125	111	80	34	9	9	9	9	9	9	9	9	23	1800
	0700	181	15	42	109	157	180	176	145	92	26	14	14	14	14	14	14	14	53	1700
	0800	214	19	21	78	148	193	208	192	147	76	21	19	19	19	19	19	19	87	1600
	0900	234	22	22	34	109	174	211	217	189	133	55	23	22	22	22	22	22	117	1500
	1000	246	25	25	26	52	131	188	217	214	178	114	39	25	25	25	25	25	140	1400
	1100	252	26	26	26	28	69	144	196	217	207	166	97	31	26	26	26	26	154	1300
	1200	254	27	27	27	27	29	82	155	201	217	201	155	82	29	27	27	27	159	1200
	HALF DAY TOTALS		142	226	410	657	914	1109	1141	997	740	478	275	158	131	130	130	130	656	
Sep	0700	77	4	10	39	62	74	75	65	44	15	4	4	4	4	4	4	4	8	1700
	0800	163	10	10	43	103	143	160	154	123	71	14	10	10	10	10	10	10	31	1600
	0900	206	14	14	17	83	148	189	200	181	133	61	15	14	14	14	14	14	57	1500
	1000	229	16	16	17	35	116	179	213	214	183	123	41	17	16	16	16	16	81	1400
	1100	240	18	18	18	19	59	141	198	224	216	174	104	26	18	18	18	18	96	1300
	1200	244	19	19	19	19	21	82	160	209	227	209	160	82	21	19	19	19	101	1200
	HALF DAY TOTALS		71	77	142	307	547	787	910	891	731	480	249	106	72	71	71	71	324	
Oct	0800	17	1	1	3	10	14	16	16	13	8	2	1	1	1	1	1	1	1	1600
	0900	122	5	5	6	40	82	109	118	110	83	42	6	5	5	5	5	5	13	1500
	1000	176	9	9	9	18	83	135	165	169	147	102	36	9	9	9	9	9	29	1400
	1100	201	11	11	11	11	43	116	167	191	186	152	92	20	11	11	11	11	41	1300
	1200	208	11	11	11	11	13	70	140	184	199	184	140	70	13	11	11	11	46	1200
	HALF DAY TOTALS		31	31	34	86	231	420	542	580	527	388	202	66	32	31	31	31	108	
Nov	1000	23	1	1	1	1	10	17	21	22	20	14	5	1	1	1	1	1	1	1400
	1100	79	3	3	3	3	15	44	65	75	74	61	37	8	3	3	3	3	6	1300
	1200	97	4	4	4	4	4	32	66	87	93	87	66	32	4	4	4	4	8	1200
	HALF DAY TOTALS		5	5	5	6	26	79	120	141	140	117	74	23	6	5	5	5	11	
Dec	1100	4	0	0	0	0	1	2	3	4	4	3	2	0	0	0	0	0	0	1300
	1200	16	0	0	0	0	1	5	11	14	15	14	11	5	1	0	0	0	1	1200
	HALF DAY TOTALS		0	0	0	0	1	5	9	11	11	10	7	3	0	0	0	0	1	

N	NNW	NW	WNW	W	WSW	SW	SSW	S	SSE	SE	ESE	E	ENE	NE	NNE	Hor.	PM

Notes: 1. Clearness number = 1.00; Ground reflectance = 0.20.   2. Figures shown are for 21st day of each month.

Next, the center of glazing U-factor for the fenestration must be found. The effective emittance of the air space is

$$e = \frac{1}{(1/e_2) + (1/e_3) - 1} = \frac{1}{(1/0.84) + (1/0.1) - 1} = 0.098$$

Air space and indoor coefficients are determined by trial and error. Assume the outdoor and indoor glass temperatures are 115 °F and 105 °F, respectively. The mean temperature is 110 °F and the temperature difference 10 °F. From Table 4, $h_s = 0.51$. The indoor coefficient from Table 3 is $h_i = 1.59$ for $e_g = 0.84$. The thermal resistance of 0.25-in. thick glass is 0.035 °F·ft²/Btu. Thus, for an outdoor coefficient of $h_o = 4.0$ Btu/(h·ft²·°F)

$$U = \frac{1}{(1/4.0) + (0.035) + (1/0.51) + 0.035 + (1/1.59)}$$

The inward radiation and convection gain is

$$q_{RCi} = 0.344 \left[\frac{51.4}{4.0} + 37.5 \left(\frac{1}{4.0} + \frac{1}{0.51} \right) + (105 - 75) \right]$$

$$= 43.3 \ \text{Btu/(h·ft}^2)$$

Find the glass temperatures to check on first assumed values of $t_{go} = 115$ °F and $t_{gi} = 105$ °F (inside and outside temperatures, respectively).

$$t_{go} = t_o + (\alpha E_o + E_i - q_{RCi})(1/h_o + R_{go}/2)$$

$$= 105 + (51.4 + 37.5 - 43.3)(1/4.0 + 0.035/2) = 117.2 \ \text{°F}$$

and $t_{gi} = t_i + g_{RCi}(1/h_i + R_{gi}/2)$

$$t_{gi} = 75 + 43.3(1/1.59 + 0.035/2) = 103.0 \ \text{°F}$$

The glass temperatures approximate the assumed temperatures, but repeating the previous calculations for a mean temperature of 110.1 °F and a new temperature difference of 14.2 °F for the space by interpolation results in an $h_s = 0.515$, and an $h_i = 1.57$ and

$$U = 0.345 \ \text{Btu/h·ft}^2\text{·°F}$$

where

$q_{RCi} = 43.1$ Btu/h·ft²
$t_{go} = 116.3$ °F
$t_{gi} = 103.0$ °F

Transmittance through both panes of glass is

$$\tau = (0.80 \times 0.12)/[1 - (0.07 \times 0.70)] = 0.101$$

The F factor for the fenestration is

$$F = 0.101 + \frac{0.345}{4.0}(0.207) + \left(\frac{0.345}{4.0} + \frac{0.345}{0.515} \right)(0.151) = 0.233$$

The shading coefficient is

$$SC = 0.233/0.87 = 0.268$$

The total heat gain is

$$q_A = 0.268 (216) + 0.345(105 - 75) = 68.2 \ \text{Btu/(h·ft}^2)$$

Solar Heat Gain Factors for Ground Reflectances Other Than 0.20

The SHGF tables have been computed using a ground reflectance of 0.20. For other foreground reflectance values, computer calculations can be made. For hand computations, the SHGF can be adjusted for the different SHGF of the new foreground reflectances (see Table 19).

Table 19 Solar Reflectances of Foreground Surfaces

Foreground Surface	Incident Angle					
	20°	30°	40°	50°	60°	70°
New concrete	0.31	0.31	0.32	0.32	0.33	0.34
Old concrete	0.22	0.22	0.22	0.23	0.23	0.25
Bright green grass	0.21	0.22	0.23	0.25	0.28	0.31
Crushed rock	0.20	0.20	0.20	0.20	0.20	0.20
Bitumen and gravel roof	0.14	0.14	0.14	0.14	0.14	0.14
Bituminous parking lot	0.09	0.09	0.10	0.10	0.11	0.12

Adapted from Threlkeld (1962).

Equation (22) can be used to estimate the ground-reflected diffuse radiation falling on a vertical surface. However, the SHGF for a horizontal fenestration is approximately 87% of the total solar radiation falling on a horizontal surface. By using this SHGF in Equation (22) instead of the solar radiation incident on the ground, the ground-reflected contribution to the vertical SHGF will be in the same units. The incident ground-reflected diffuse radiation for the reference reflectance of 0.20 in terms of the SHGF is given by Equation (22)

$$0.87 \ E_{dg} = \text{SHGF} \times 0.20 \times 0.5$$

where 0.87 is the conversion from incident solar radiation to SHGF, SHGF is the value for the horizontal surface, 0.20 is the reference ground reflectance, and 0.5 is the F_{sg} angle factor for vertical surfaces.

The ground component for the new reflectance may be calculated in the same manner, or by means of a ratio of the reflectances and the difference applied to the SHGF for the vertical surface.

Example 9. Find the adjusted SHGF for a foreground having a reflectance of 0.32, for an east-facing window at 1000, July 21, 40° North latitude.
Solution: The SHGF for the horizontal surface, from Table 15, is 231 Btu/(h·ft²). The ground-reflected contribution to the SHGF for the east window, by Equation (22), is

$$0.87 \ E_{dg} = 231 \times 0.20 \times 0.5 = 23.1 \ \text{Btu/h·ft}^2$$

For a foreground reflectance of 0.32, it becomes

$$0.87 \ E_{dg} = 231 \times 0.32 \times 0.5 = 37.0 \ \text{Btu/h·ft}^2$$

The increase in the SHGF for the higher ground reflectance is 37.0 − 23.1 = 13.9 Btu/h·ft². The SHGF, from Table 15 for the east window, is 146 Btu/h·ft²; and, for the 0.32 ground reflectance, the adjusted SHGF becomes 146 + 13.9 = 159.9 Btu/h·ft².

Computer Calculation of Solar Heat Gain Factors

The following equations can be used to generate SHGF, where all angles are in degrees. The solar azimuth θ and the surface azimuth ψ are measured in degrees from south; angles to the east of south are negative, and angles to the west of south are positive.

Variables

H = hour angle
L = latitude
δ = declination
β = solar altitude
ϕ = solar azimuth
ψ = surface azimuth
γ = surface-solar azimuth
Σ = surface tilt
θ = incident angle
A = apparent solar constant
B = atmospheric extinction coefficient
C = sky diffuse factor
E_{DN} = direct normal irradiance

E_D = direct irradiance
Y = ratio of vertical/horizontal sky diffuse
E_{ds} = diffuse sky irradiance
ρ_g = ground reflectance
E_{dg} = diffuse ground reflected irradiance
E_d = diffuse irradiance
t_j = transmission coefficients for glass
a_j = absorption coefficients for glass
N_i = heat transfer factor, inward flow fraction
τ_D = transmittance of DSA glass
α_D = absorptance of DSA glass

Values of A, B, and C are given in Table 7 for the 21st day of each month. Values for other dates can be obtained by interpolation. The transmission and absorption coefficients for DSA glass are given in Table 20. The absorption or transmission of direct solar radiation incident at an angle θ is

$$\alpha_D = \sum_{J=0}^{5} \alpha_j \cos^j \theta \tag{42}$$

$$\tau_D = \sum_{j=0}^{5} t_j \cos^j \theta \tag{43}$$

Note that this calculation procedure, using the coefficients found in Table 20, gives a normal transmittance for DSA glass of 0.88, which is slightly higher than values used elsewhere in this chapter.

Tables 12 through 18 are based on a ground reflectance ρ_g of 0.2 and a heat transfer factor, inward flow fraction N_i of 0.267.

Hour angle, degrees	$H = 0.25$ (minutes of time from local solar noon)	
Solar altitude β	$\sin \beta = \cos L \cos \delta \cos H + \sin L \sin \delta$	
Solar Azimuth ϕ	$\cos \phi = \dfrac{\sin \beta \sin L - \sin \beta}{\cos \beta \cos L}$	
Surface-solar azimuth	$\gamma = \phi - \psi$	
Incident angle θ	$\cos \theta = \cos \beta \cos \gamma \sin \Sigma + \sin \beta \cos \Sigma$	
Direct normal irradiance E_{DN}	$E_{DN} = A \exp(-B/\sin \beta)$	
Direct irradiance E_D	$E_D = E_{DN} \cos \theta$ if $\cos \theta > 0$	
	$E_D = 0$ otherwise	
Ratio of sky diffuse on vertical surface to sky diffuse on horizontal surface Y	If $\cos \theta > -0.2$ $Y = 0.55 + 0.437 \cos \theta + 0.313 \cos^2 \theta$ otherwise, $Y = 0.45$	
Diffuse irradiance E_d	$E_d = E_{ds} + E_{dg}$	

where

Vertical surfaces	$E_{ds} = CY E_{DN}$
Surfaces other than vertical	$E_{ds} = CE_{DN}(1 + \cos \Sigma/2)$
and	$E_{dg} = E_{DN}(C + \sin \beta)\rho_g(1 - \cos \Sigma)/2$

Solar heat gain

Transmitted component $= E_D \sum_{j=0}^{+5} t_j \cos^j \theta + E_d^2 \sum_{j=0}^{5} t_j/(j + 2)$

Absorbed components $= E_D \sum_{j=0}^{5} a_j \cos^j \theta + E_d^2 \sum_{j=0}^{5} a_j/(j + 2)$

Solar heat gain factor = Energy transmitted
+ N_i (Energy absorbed)

Table 20 Coefficients for DSA Glass for Calculation of Transmittance and Absorptance

i	a_j	t_j
0	0.01154	-0.00885
1	0.77674	2.71235
2	-3.94657	-0.62062
3	8.57881	-7.07329
4	-8.38135	9.75995
5	3.01188	-3.89922

Values of SC given in Table 11 are based on an incident solar radiation equivalent to a SHGF of 216 Btu/(h·ft^2). This value corresponds approximately with the SHGF for July 21 through September 21 for a west-oriented window at 1600 sun time for north latitudes from 16 to 40°. The angles of incidence for these times and latitudes are 36, 32, and 30° for July 21, August 21, and September 21, respectively. These low-incidence angles permit use of normal incidence solar properties of the fenestration without adjustment for angular effects. Figure 11 shows how the solar optical properties of glass varies with angle of incidence.

The values of SC given in Table 11 are based on still air (natural convection) at the inner surface of the fenestration and 7.5 mph wind at the outer surface. For these conditions, h_i is about 1.46 Btu/(h·ft^2·°F) (Table 3 gives specific values), and h_o is 4.0 Btu/(h·ft^2·°F), so the inward flowing fraction of absorbed heat for single glazing is

$$N_i = h_i/(h_i + h_o) = 1.46/5.46 = 0.267$$

Note: Resistance of the glass plate is neglected.

For other indoor and outdoor conditions, the surface coefficients vary over a considerable range, and N_i will also change. Figure 21 shows the variation in SC caused by these changes in surface coefficients. For comparison, Table 11 also lists SC for $h_o = 3.0$. The designer should be aware of the possible variations upward and downward in the SC for any fenestration when the surface conditions vary from still air and 7.5 mph wind at the inner and outer surfaces, respectively.

The SC for any fenestration will rise above the tabulated values when the inner surface coefficient is increased because of forced airflow over that surface and when the outer surface coefficient is decreased by lower wind velocity. Conversely, the SC will fall somewhat below the tabulated values when the air at the inner surface is still and the wind velocity rises above 7.5 mph.

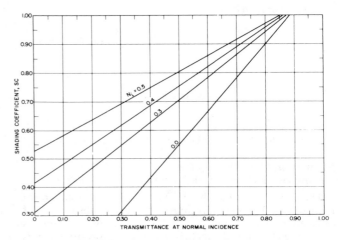

Fig. 21 Shading Coefficient versus Transmittance

The SC and SHGC given in Table 11 and Figure 21 are for uncoated single glass, in which the reflectivity ranges downward from 0.08 for double-strength sheet to 0.05 for 0.25-in. heat-absorbing plate or float. If solar-reflective films or coatings are used, the absorptance is reduced for any given transmittance value. Hence, less absorbed heat flows inward, and the SC is reduced below the tabulated values. For coated single glass with higher reflectivities, SC is given as a function of transmittance and reflectance in Figure 22.

For double and triple glazing with low-emittance coatings, the SC varies as a function of glazing transmittance, absorptance, and the glazing surface to which the low emittance coating is applied. For summer conditions, SC is given as a function of transmittance and absorptance in Figure 23 for double glazing with low-emittance coatings on the No. 2 surface (measured from the outside) and in Figure 24, for triple glazing with low emittance coatings on the No. 4 surface. Similarly, under winter conditions, SC is given in Figure 25 for double glazing with coatings on the No. 3 surface and in Figure 26, for triple glazings, again on No. 3 surfaces. In all figures, the transmittance and absorptance values are for the low-emittance coating and glass or plastic substrate only (the SC values are for the complete glazing system). The remaining light(s) of glass making up the window unit are assumed clear and 0.125 in. thick.

Low-emittance coatings used in windows with bronze, gray, or blue-green glass produce different shading coefficients. Moderate changes in design temperature conditions, gap widths (assumed to be 0.5 in.), exterior and interior film coefficients, and the overall U-factor do not significantly change the SC for the given transmittance, absorptance, and emittance. SC for the actual coating emittance can be found in interpolating between the emittance values of 0 and 0.3 (Rubin 1984, Selkowitz 1979).

Shading coefficients for uncoated single-glazing materials (plastic sheet as well as glass) with transmittances other than those cited in Table 21 can be estimated from Figure 21, since SC is virtually a linear function of the normal incidence transmittance. The tabulated values can be interpolated for insulating glass when the inner light is clear sheet or plate glass. When heat-absorbing glass is used in double glazing in cooling dominated climates, it should be installed in the outer light, so that the absorbed heat can be more readily dissipated to the atmosphere.

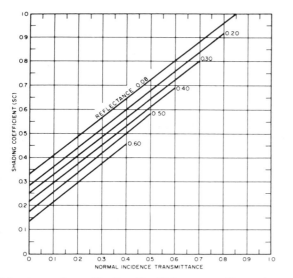

Fig. 22   Approximate Shading Coefficient versus Transmittance and Reflectance for Coated Single Glass

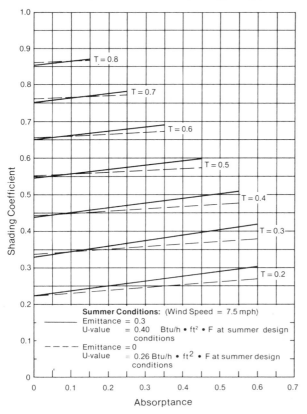

Fig. 23   Shading Coefficient for Double Glazing with Low Emittance Coating on Surface 2

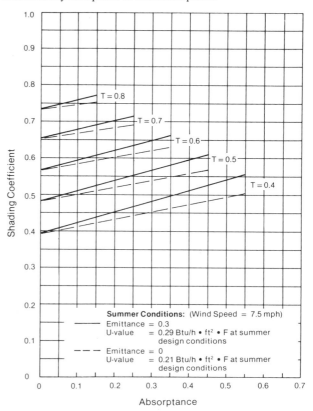

Fig. 24   Shading Coefficient for Triple Glazing (Glass-Plastic-Glass) with Low-Emittance Coating on Surface 4

When indoor shading devices are used with glass fenestration, small differences in glass properties become unimportant. Consequently, the wide range of glasses listed in Table 11, Part A, can be expanded into four classes; the insulating glasses in Table 11, Part B, into two classes.

Airflow Windows

If properly managed, airflow between panes of a double-glazed window can improve fenestration performance. In normal use, a venetian blind is located between the glazing layers. Ventilation air from the room enters the double-glazed cavity, flows over the blind, and is, in some designs, exhausted from the building or returned through the ducts to the central HVAC system. In cold weather, the window acts as a heat exchanger when sunlit, so that the inner glass temperature nearly equals the room air temperature and improves thermal comfort.

The apparent conductance across the inner glazing is very low, but this is misleading since additional heat is lost to the outdoors from the moving airstream in the window cavity. During sunny winter days, the blind acts as a solar air collector; heat removed by the moving air can be used elsewhere in the building. In the summer, the window can have a very low shading coefficient if the blinds are appropriately placed, since the majority of solar gains are removed from the window. These systems can control window heat transfer under many different operating conditions. Sodergren and Bostrom (1971) and Brandle and Boehm (1982) give details on airflow or exhaust windows.

Domed Skylights

Solar and total heat gains for domed skylights can be determined by the same procedure used for windows. The SHGF values for such calculations should be consistent with the dome orientations. For horizontal roofs, Tables 12 through 18 give approxi-

mate SHGF values. For sloping roofs, an approximate SHGF can be found from SHGF $= E_t/1.15$, where E_t = total solar irradiation on the sloping surface. Table 21 gives SC for plastic domed skylights. Manufacturers' literature has further details.

Glass Block Walls

Glass block provides light transmission in exterior walls when optical clarity for view is not desired. Table 22 describes the glass block patterns discussed in the following text and gives shading

Table 21   Shading Coefficients for Domed Horizontal Skylights

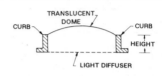

Dome	Light Diffuser (Translucent)	Curb Height, in.	Width to Height Ratio	Shading Coefficient
Clear	Yes	0	∞	0.61
$\tau = 0.86$	$\tau = 0.58$	9	5	0.58
		18	2.5	0.50
Clear		0	∞	0.99
$\tau = 0.86$	None	9	5	0.88
		18	2.5	0.80
Translucent	None	0	∞	0.57
$\tau = 0.52$		18	2.5	0.46
Translucent	None	0	∞	0.34
$\tau = 0.27$		9	5	0.30
		18	2.5	0.28

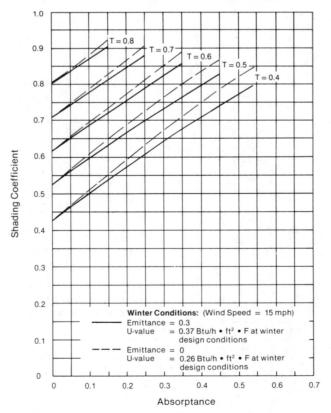

Fig. 25   Shading Coefficient for Double Glazing with Low-Emittance Coating on Surface 3

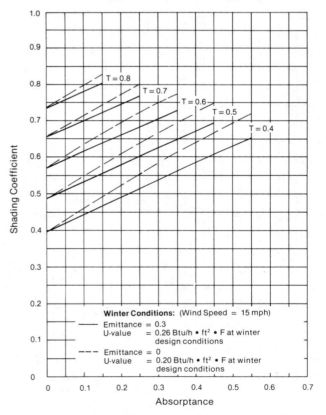

Fig. 26   Shading Coefficient for Triple Glazing (Glass-Plastic-Glass) with Low-Emittance Coating on Surface 3

Table 22 Shading Coefficients and U-Factors for Standard Hollow Glass Block Wall Panels

Type of Glass Block[a]	Description of Glass Block	Shading Coefficient[b]		U-Factor,[e] Btu/h·ft²·°F
		Panels[c] in Sun	Panels[d] in Shade (N, NW, W, SW)	
Type I	Glass colorless or aqua A, D: Smooth B, C: Smooth or wide ribs, or flutes horizontal or vertical, or shallow configuration E: None	0.65	0.40	0.51
Type IA	Same as Type I except ceramic enamel on A	0.27	0.20	0.51
Type II	Same as Type I except glass fiber screen partition E	0.44	0.34	0.48
Type III	Glass colorless or aqua A, D: Narrow vertical ribs or flutes. B, C: Horizontal light-diffusing prisms, or horizontal light-directing prisms E: Glass fiber screen	0.33	0.27	0.48
Type IIIA	Same as Type III except E: Glass fiber screen with green ceramic spray coating or glass fiber screen and gray glass or glass fiber screen with light-selecting prisms	0.25	0.18	0.48
Type IV	Same as Type I except reflective oxide coating on A	0.16	0.12	0.51

[a]All values are for 7 3/4 by 7 3/4 by 3 7/8 in. block, set in light-colored mortar. For 11 3/4 by 11 3/4 by 3 7/8 in. block, increase coefficients by 15%, and for 5 3/4 by 5 3/4 by 3 7/8 in. block reduce coefficents by 15%.
[c]Shading coefficients are applied to heat gain factors for one hour earlier than the load calculation time to allow for heat storage in the panel.
[c]Shading coefficients are for peak load condition, but provide a close approximation for other conditions.
[d]For NE, E, and SE panels in shade, add 50% to the values listed for panels in the shade.
[e]Values shown are the same for all size block.

coefficients to be applied to solar heat gain factors from Tables 12 through 18 so that instantaneous solar heat gains can be calculated.

Convection and low-temperature radiation heat gain for all hollow glass block panels fall within a narrow range. Differences in shading coefficients are largely the result of differences in the transmittance of the glass blocks for solar radiation. Shading coefficients for any particular glass block pattern vary depending on orientation and time of day. The SC for western exposures in the morning (in the shade) is depressed because of the heat storage within the block, whereas the SC for eastern exposures in the afternoon (in the shade) is elevated as the stored heat is dissipated. Time lag effects from heat storage are estimated by using SHGFs and air-to-air temperature differences for one hour earlier than the time for which the load calculation is made.

Calorimeter tests of Type 1A glass block showed little difference in solar heat gains between glass block with either black or white ceramic enamel on the exterior of the block. Because white and black ceramic enamel surfaces represent the two extremes for reflecting or absorbing solar energy, glass block with enamel surfaces of other colors should have shading coefficients between these values.

GLAZING MATERIALS

PLASTIC MATERIALS FOR GLAZING

Generally, the factors outlined for glass apply also to glazing materials such as acrylic or other plastic panels. If the solar transmittance, absorptance, and reflectance are known, a shading coefficient can be calculated in the same way it is for glass. These properties can be obtained from the manufacturer or determined by simple laboratory tests. Both ASHRAE and ASTM have developed standards for performing these tests.

Table 23 lists solar optical properties of typical plastic materials. Plastic panels are available with transmittance values from 10 to

92% and reflectance values from 4 to over 60%. Emittances for acrylic, polystyrene, and polycarbonate glazing all range between 0.88 and 0.92. Plastics are available in translucent and transparent form; some are corrugated. Many tints (colors) are also available.

Figure 17 can be used to obtain an approximate SC for single panels of low reflectivity. Where shading is involved, the SC for a glass of similar solar optical properties can be used. The plastic manufacturer has the engineering data.

In selecting plastic panels for glazing, possible deterioration from the sun, expansion and contraction because of temperature extremes, and possible damage from abrasion are concerns.

Table 23 Solar Optical Properties and Shading Coefficients of Transparent Plastic Sheeting
(Burkhardt 1975, 1976)

Type of Plastic	Transmittance		SC
	Visible	Solar	
Acrylic			
Clear	0.92	0.85	0.98
Gray tint	0.16	0.27	0.52
"	0.33	0.41	0.63
"	0.45	0.55	0.74
"	0.59	0.62	0.80
"	0.76	0.74	0.89
Bronze tint	0.10	0.20	0.46
"	0.27	0.35	0.58
"	0.49	0.56	0.75
"	0.61	0.62	0.80
"	0.75	0.75	0.90
Reflective[a]	0.14	0.12	0.21
Polycarbonate			
Clear, 1/8-in.	0.88	0.82	0.98
Gray, 1/8-in.	0.50	0.57	0.74
Bronze, 1/8-in.	0.50	0.57	0.74

[a]Aluminum metallized polyester film on plastic.

CHOOSING GLASS FOR BUILDINGS

The choice of glass as a glazing material, should be based on (1) light transmittance and daylight use, (2) occupant comfort, (3) sound reduction, (4) strength (deflection under load), (5)safety (security), (6) life cycle costs, and (7) aesthetics. In many respects, these factors and the thermal performance of glass are related.

Light transmittance and daylight use. When daylight is to be the primary lighting system, the minimum expected daylight in the building must be calculated for the building performance cycle and integrated into the lighting calculations. IES (1979) gives daylight design and calculation procedures. In some glazing applications, such as artists' studios and showrooms, maximum transmittance may be required for adequate daylighting of the interior. Regular clear glass, produced by float, plate, or sheet process, may be the logical choice.

When daylight is a supplementary light source, the electric lighting can be designed independently of the daylight system. But adequate switching must be included in the electric distribution to substitute available daylight for electric lighting by automatic or prescribed manual control. Photosensitive controls automatically adjust shading devices to provide uniform illumination and reduce energy consumption. Manual control is less effective.

Buildings with large areas of glass usually have insulating glass units with clear, tinted, or reflective-film. The tinted and reflecting units reduce the brightness contrast between windows and other room surfaces and provide a relatively glare-free environment for most daylight conditions.

Tables 11 and 24 list typical solar energy transmittances and daylight transmittances for various glass types. Manufacturers' literature have more appropriate values.

The color of glass chosen for a building depends largely on where and how it is used. For commercial building lobbies, showroom windows, and other areas where maximum visibility from exterior to interior is required, regular clear glass is generally best. For other glass areas, a tinted glass may best complement the interior colors. Bronze, gray, and reflective-film glasses also give some privacy to building occupants during daylight hours. Patterned, etched, or sandblasted glass that diffuses lighting is available.

Occupant comfort and acceptance. Tinted and reflective film glasses reduce summer heat gain as well as winter heat loss. Insulating glass units, especially reflective-film types, also reduce radiation heat transfer between building occupants and window areas, thereby providing greater year-round comfort. ASHRAE *Standard* 55 requires that the variations in radiant temperature be limited to achieve a minimum level of occupant comfort. For buildings with sliding glass doors or equivalent size glazing areas, the net effect is to require double-glazing in all climates where the

winter design temperature is less than 45 °F. Drafts from cold glass surfaces are also reduced in cold weather, and higher relative humidities can be maintained without condensation. Such high performance glasses often reduce cooling load 15 to 30% below that with clear glass (Rudoy and Duran 1975). Clear insulating glass with light-colored continuous interior shading, or particularly with exterior shading, can reduce fenestration cooling load substantially.

Sound reduction. Proper acoustical treatment of exterior walls can decrease noise levels in certain areas. The airtightness of a wall is the primary factor to consider in reducing sound transmission from the exterior. Once walls and windows are tight, the choice of glass and draperies becomes important. Draperies do not prevent sound from coming through the fenestration; they act as an absorber for sound that does penetrate. Table 25 lists average sound transmission losses for various types of glass. These averages apply for the frequency range of 125 to 4000 Hz and were determined by tests based on ASTM *Standard* E 90.1. Chapter 42 of the 1991 ASHRAE *Handbook—Applications* has further details.

Strength and safety. In addition to its thermal, visual, and aesthetic functions, glass for building exteriors must also structurally perform. Wind loads are specified in most building codes, and these requirements may be adequate for many structures. However, detailed wind tunnel tests should be run for tall or unusually shaped buildings and for buildings where the surroundings create unusual wind patterns. Alternative loading conditions should be chosen in developing design specifications when wind load tests suggest wind loading conditions different from those shown in Figure 27. All manufacturers do not agree on the values presented here; for specific information, consult the glass manufacturer.

Figure 27 shows the relationship between size, thickness, and strength of annealed glass using a design factor of 2.5 (average breaking pressure divided by 2.5). The graphs are based on actual tests to destruction, in which the glass was exposed to uniform loads increased in increments and held static for one minute. Figure 31 applies to rectangular lights of glass, with length-to-width ratios of not more than 5:1, firmly supported on all four edges. To determine the uniform load of nonannealed glass held on four sides, use the appropriate multiplying factor listed in Table 26.

Thermal expansion and contraction of glass can cause breakage or produce openings in the glazing system causing greater air infiltration. Tinted and reflective-type glasses are especially vulnerable to thermal stress. The manufacturer should provide information on thermal stress performance.

Building codes may require glass in certain positions to perform with certain breakage characteristics, which can be satisfied by tempered, laminated, or wired glass. In this case, glass should meet *Federal Standard* 16 CRF 1201 or other appropriate breakage performance requirements.

Table 24   Daylight Transmittance for Various Types of Glass

Type of Glass	Visible Transmittance
1/8-in. regular sheet or float glass	0.86 to 0.91
1/8-in. gray sheet	0.31 to 0.71
3/16-in. gray sheet	0.61
7/32-in. gray sheet	0.14 to 0.56
1/4-in. gray sheet	0.52
1/4-in. green/float glass	0.75
1/4-in. gray plate glass	0.44
1/4-in. bronze plate glass	0.49
1/2-in. gray plate glass	0.21
1/2-in. bronze plate glass	0.25
Coated glasses (single, laminated, insulating)	0.07 to 0.50

Table 25   Sound Transmittance Loss for Various Types of Glass

Type of Glass	Sound Transmittance Loss, db
1/8-in. double-strength sheet glass	24
1/4-in. plate or float glass	27
1/2-in. plate glass	32
3/4-in. plate glass	35
1-in. plate glass	36
1/4-in. laminated glass (9/20-in. plastic interlayer)	30
1-in. insulating glass	32
1/2-in. laminated glass (9/20-in. plastic interlayer)	34
Insulating glass, 6-in. air space, 1/4-in. plate or float glass	40

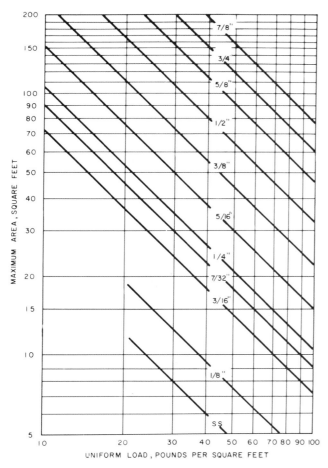

Fig. 27   Design Thickness for Annealed Glass

Table 26   Multiplier for Other than Annealed Glass

Type of Glass	Multiplier	Type of Glass	Multiplier
Tempered	4.0	Rough plate	1.0
Heat strengthened	2.0	Laminated	0.6
Double glazed	1.5	Wired	0.5

Note: Sandblasting significantly weakens glass and can reduce its strength by 60% or more. Sandblasted glass should not be used for exterior glazing.

Life-cycle costs. Alternative building shells should be compared to assure satisfactory energy use and total energy budget compliance, if required. ASHRAE *Standard* 90.1-1989 should be used as a starting point. A life-cycle cost model should be developed for each system considered, see Chapter 33 of the 1991 ASHRAE *Handbook—Applications*.

SHADING

EXTERIOR SHADING

The most effective way to reduce the solar load on fenestration is to intercept direct radiation from the sun before it reaches the glass. Windows fully shaded from the outside will reduce solar heat gain as much as 80%. In one way or another, fenestration can be shaded by roof overhangs, vertical and horizontal architectural projections, awnings, heavily proportioned exterior louvers, insect or shading screens, patterned screens having a weave designed for sunlight interception, or sun screens of narrow fixed louvers. In all exterior shading structures, the air must move freely to carry away heat absorbed by the shading and glazing materials. See manufacturer's instructions regarding proper installation for achieving expected performance by providing suitable free convection ventilation between shading and glazing. Also, consider the geometry of the structures relative to changing sun position to determine the times and quantities of direct sunlight penetration. Detailed discussions of the effectiveness of various outside shading devices are given in Pennington (1968), Yellott (1972), and Ewing and Yellott (1976).

Louvers and Sunshades

The ability of horizontal panels or louvers to intercept the direct component of solar radiation depends on their geometry and the profile or shadow-line angle Ω (Figure 27), defined as the angular difference between a horizontal plane and a plane tilted about a horizontal axis in the plane of the fenestration until it includes the sun. The profile angle can be calculated by

$$\tan \Omega = \tan \beta / \cos \gamma \qquad (44)$$

For slat-type sunshades, the transmitted solar radiation consists of straight-through and transmitted-through components. When the profile angle Ω is above the cutoff angle (see Figure 28), straight-through transmission of direct radiation is completely eliminated, but the transmitted diffuse and the reflected-through components remain. Their magnitude depends largely on the reflectance of the sunshade surfaces and of exterior objects.

Narrow horizontal louvers fabricated in conventional width-spacing ratios and framed as window screens, retain their shading characteristics, while gaining in effective transparency (view) by eliminating the coarse striation pattern of wide louvers. Table 27 gives shading coefficients (SC) for several types of louvered sun screens. Commercially available sun screens completely exclude direct solar radiation when the profile angle exceeds approximately 26° (Groups 1, 2, and 5) or 40° (Groups 3, 4, and 6). Group designations are defined in the table footnote.

Table 27   Shading Coefficients for Louvered Sun Screens

Profile Angle	Group 1 Transmittance	SC	Group 2 Transmittance	SC	Group 3 Transmittance	SC	Group 4 Transmittance	SC	Group 5 Transmittance	SC	Group 6 Transmittance	SC
10°	0.23	0.35	0.25	0.33	0.40	0.51	0.48	0.59	0.15	0.27	0.26	0.45
20°	0.06	0.17	0.14	0.23	0.32	0.42	0.39	0.50	0.04	0.11	0.20	0.35
30°	0.04	0.15	0.12	0.21	0.21	0.31	0.28	0.38	0.03	0.10	0.13	0.26
≥ 40°	0.04	0.15	0.11	0.20	0.07	0.18	0.20	0.30	0.03	0.10	0.04	0.13

Group 1. Black, width over spacing ratio 1.15/1; 23 louvers/in. Group 2. Light color; high reflectance, otherwise same as Group 1. Group 3. Black or dark color; w/s ratio 0.85/1; 17 louvers/in. Group 4. Light color or unpainted aluminum; high reflectance; otherwise same as Group 3. Group 5. Same as Group 1, except two lights of 0.25 in. clear glass with 0.5 in. air space. Group 6. Same as Group 3, except two lights of 0.25 in. clear glass with 0.5 in. air space. U-value = 0.85 Btu/h·ft²·°F for all groups when used with single glazing.

Roof Overhangs; Horizontal and Vertical Projections

In the northern hemisphere, horizontal projections can considerably reduce solar heat gain on south, southeast, and southwest exposures during late spring, summer, and early fall. On east and west exposures during the entire year, and on southerly exposures in winter, the solar altitude is generally so low that to be effective horizontal projections must be excessively long.

The shadow width S_W and shadow height S_H (Figure 30), produced by the vertical and horizontal projections (P_V and P_W),

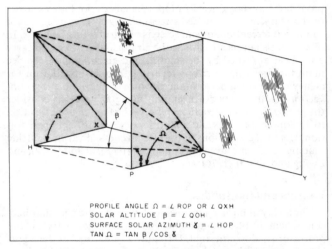

PROFILE ANGLE Ω = $\angle$ ROP OR $\angle$ QXH
SOLAR ALTITUDE β = $\angle$ QOH
SURFACE SOLAR AZIMUTH γ = $\angle$ HOP
TAN Ω = TAN β / COS γ

Fig. 28   Profile Angle for South-Facing Slat-Type Sunshades

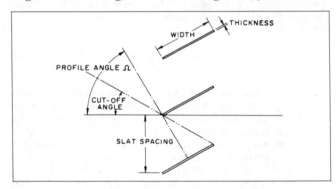

Fig. 29   Geometry of Slat-Type Sunshades

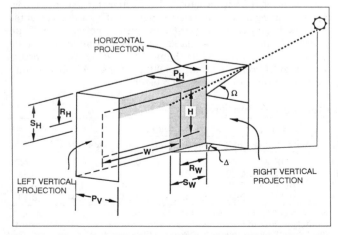

Fig. 30   Vertical and Horizontal Projections and Related Profile Angles for Vertical Surface Containing Fenestration System

respectively, can be calculated using the solar surface azimuth γ and the horizontal profile angle Ω determined by Equation (44).

$$S_W = P_V \,|\tan \gamma| \qquad (45)$$

$$S_H = P_H \tan \Omega \qquad (46)$$

Note: When the solar surface azimuth γ is greater than 90° and less than 270°, the window is completely in the shade; thus, $S_W = W + R_W$ and $A_{SL} = 0$.

The sunlit A_{SL} and shaded A_{SH} area of the window is variable during the day and can be calculated for each moment using the following relations (see Figure 30)

$$A_{SL} = [W - (S_W - R_W)] [H - S_H - R_H] \qquad (47)$$

$$A_{SH} = A - A_{SL} \qquad (48)$$

where A is total window area.

Example 10   A window in the southwest wall of a building at 40°N latitude at 1500 on July 21 is 34.5 in. wide and 58.5 in. high. The depth of the horizontal and vertical projections are 6 in., located 3 in. beyond the edges of the window.
Part A. Find the sunlit and shaded area of the window.
Part B. Find the depth of projections needed to fully shade the window.
Solution
Part A. The wall azimuth ψ for a southwest wall is +45° (Table 8). The solar azimuth ϕ can be calculated using Equation (8b). At 1500, $H = 0.25 \times 180 = 45°$; from Table 7, for July 21, $\delta = 20.6°$.
Find the solar altitude β using Equation (8a)

$$\sin \beta = \cos (40)\cos (20.6)\cos (45) + \sin (40)\sin (20.6)$$
$$\beta = 47.2°$$

Find the solar azimuth ϕ using Equation (8b)

$$\cos \phi = [\sin (47.2)\sin (40) - \sin (20.6)]/[\cos (47.2)\cos (40)]$$
$$\phi = 76.7°$$

Thus, $\gamma = 76.7 - 45 = 31.7°$

Using Equation (45), the width of the vertical projection shadow is

$$S_W = 6 \,|\tan 31.7| = 3.71 \text{ in.}$$

Using Equation (44), the profile angle for the horizontal projection is

$$\tan \Omega = \tan (47.2)/\cos (31.7)$$
$$\Omega = 51.8°$$

Using Equation (46), the height of the horizontal projection shadow is

$$S_H = 6 \tan 51.8 = 7.62 \text{ in.}$$

Using Equations (47) and (48), the sunlit and shaded area of the window are now

$$A_{SL} = [34.5 - (3.71 - 3)] [58.5 - (7.62 - 3)]/144 = 12.6 \text{ ft}^2$$
$$A_{sh} = (34.5 \times 58.5/144) - 12.6 = 1.4 \text{ ft}^2$$

Part B. The shadow length necessary to fully shade the given window $S_{H(fs)}$ and $S_{W(fs)}$ from the horizontal and vertical projection are given by (see Figure 30)

$$S_{H(fs)} = 58.5 + 3 = 61.5 \text{ in.}$$

$$S_{W(fs)} = 34.5 + 3 = 37.5 \text{ in.}$$

Thus, using Equations (45) and (46)

$$P_{H(fs)} = 61.5 \cot(51.8) = 48.4 \text{ in.}$$

$$P_{W(fs)} = 37.5 \,|\cot(31.7)| = 60.7 \text{ in.}$$

For this example, since both the horizontal and vertical projections do not need to fully shade the window, a horizontal projection of 48.4 in. is satisfactory. Also, to accurately analyze the influence of external shadowing by projections, an hour-by-hour calculation must be performed over the periods of the year for which shadowing is desired.

Partially Shaded Windows

All solar heat gain data in this section are based on the sunlit area of the glass itself. For actual windows, the sunlit glass area is likely to be significantly less than the total window area, and the effect of external shadowing as well as the effect of the opaque areas must be considered.

Besides the horizontal and vertical projections, external shade can be produced by the mullions and the transom. The shaded area varies continuously throughout the day, but it can be estimated readily by treating the mullions and transom as vertical and horizontal projections and applying Equation (46).

The solar heat gain factors (defined in the section Simplified Methods for Predicting Heat Transfer through Fenestration) through a partially shaded window, can be estimated using

$$SHGF = E_D (A_{SL}/A) + E_d$$

The diffuse component E_d can be found using the SHGF values (Tables 12 through 18) for the nearest facade not receiving direct sunlight. The direct component of solar radiation can be estimated by subtracting the diffuse component E_d from the SHGF values (Tables 12 through 18) for an unshaded window.

The solar heat gain for the entire window is then

$$\text{Solar heat gain} = (SC) \times (SHGF) \times (A) \qquad (50)$$

Example 11. Find the solar heat gain through the window given in Example 10 assuming that the window consists of 0.25 in. clear plate glass.

Solution: At 1500, the SHGF for an unshaded window in a southwest wall is 170 Btu/(h·ft²) of which the diffuse component is 31 Btu/(h·ft²).

$$SHGF = \frac{12.64}{14.02}(139 + 31) = 156.3 \text{ Btu/(h·ft}^2)$$

The SC for 0.25 in. plate glass (assuming in this example that one-half the absorbed energy is dissipated to the room air) is 0.97; thus, the solar heat gain through the window is

$$\text{Solar heat gain} = 0.97 \times 156.3 \times 14.02 = 2125.6 \text{ Btu/h}$$

Equations for Computer Calculations of External Shadowing of Inclined Surfaces

Incidence angle: $\theta = \cos^{-1}(\cos \beta \cos \gamma \sin \Sigma + \sin \beta \cos \Sigma)$

Vertical surface $\qquad \theta_V = \cos^{-1}(\cos \beta \cos \gamma)$

Horizontal surface $\theta_H = \cos^{-1}(\sin \beta)$

Vertical projection profile angle:

$$\Delta = \tan^{-1}\left(\frac{\sin \gamma \cos \beta}{\cos \theta}\right); |2| < 90°$$

For $\theta > 90°$ $A_{SL} = 0$ and $A_{SH} = A$

Vertical surface $\qquad \Delta_V = \tan^{-1}(\gamma); |\gamma| < 90°$ and $|\gamma| > 270°$

Horizontal surface $\quad \Delta_H = \tan^{-1}\left(\frac{\sin \gamma}{\tan \beta}\right);$ for all γ

Horizontal projection profile:

Angle: $\Omega = \tan^{-1}\left(\frac{\sin \beta \sin \Sigma - \cos \beta \cos \gamma \cos \Sigma}{\cos \theta}\right); |\theta| < 90°$

Vertical surface $\qquad \Omega_V = \tan^{-1}\left(\frac{\tan \beta}{\cos \gamma}\right); |\gamma| < 90°$

and $|\gamma| > 270°$

Horizontal surface $\quad \Omega_H = \tan^{-1}\left(-\frac{\cos \gamma}{\tan \beta}\right);$

$$90° < |\gamma| < 270°$$

Length of shadow from vertical projection:

$$S_W = P_V |\tan \Delta|$$

Length of shadow from horizontal projection:

$$S_H = P_H |\tan \Omega|$$

Sunlit area of the window:

$$A_{SL} = [W - (S_W - R_W)] [H - (S_H - R_H)]$$

Shaded area of the window:

$$A_{SH} = A - A_{SL}$$

where

ϕ = solar azimuth
β = solar altitude
γ = solar surface azimuth
Σ = surface tilt angle
P_V = vertical projection depth
P_H = horizontal projection depth
W = window width
H = window height
R_W = width of opaque surface between window and vertical projection
R_H = height of opaque surface between window and horizontal projection
A = total projected area of the window
θ = angle of incidence
Ω = horizontal projection profile angle
Δ = vertical projection profile angle

INDOOR SHADING DEVICES

Venetian Blinds and Roller Shades

Most fenestration has some type of internal shading to provide privacy and aesthetic effects, as well as to give varying degrees of sun control (Ozisik and Schutrum 1960). The SC values for typical internal shading are given in Tables 28, 29, and 30. The effectiveness of any internal shading device depends on its ability to reflect incoming solar radiation back through the fenestration before it can be absorbed and converted into heat within the building. Table 31 lists approximate values of solar-optical properties for the typical indoor shading devices described in Tables 28, 29, and 30.

Table 28   Shading Coefficients for Single Glass with Indoor Shading by Venetian Blinds or Roller Shades

Type of Glass	Nominal Thickness[a], in.	Solar Transmittance[b]	Venetian Blinds		Roller Shade		
					Opaque		Translucent
			Medium	Light	Dark	White	Light
Clear	3/32[c]	0.87 to 0.80	0.74[d] (0.63)[e]	0.67[d] (0.58)[e]	0.81	0.39	0.44
Clear	1/4 to 1/2	0.80 to 0.71					
Clear pattern	1/8 to 1/2	0.87 to 0.79					
Heat-absorbing pattern	1/8	—					
Tinted	3/16, 7/32	0.74, 0.71					
Heat-absorbing[f]	3/16, 1/4	0.46					
Heat-absorbing pattern	3/16, 1/4	—	0.57	0.53	0.45	0.30	0.36
Tinted	1/8, 7/32	0.59, 0.45					
Heat-absorbing or pattern	—	0.44 to 0.30	0.54	0.52	0.40	0.28	0.32
Heat-absorbing[f]	3/8	0.34					
Heat-absorbing or pattern	—	0.29 to 0.15 0.24	0.42	0.40	0.36	0.28	0.31
Reflective coated glass	S.C. = 0.30[g]		0.25	0.23			
	= 0.40		0.33	0.29			
	= 0.50		0.42	0.38			
	= 0.60		0.50	0.44			

[a]Refer to manufacturers' literature for values.
[b]For vertical blinds with opaque white and beige louvers in the tightly closed position, SC is 0.25 and 0.29 when used with glass of 0.71 to 0.80 transmittance.
[c]Typical residential glass thickness.
[d]From Van Dyck and Konen (1982), for 45° open venetian blinds, 35° solar incidence, and 35° profile angle.

[e]Values for closed venetian blinds. Use these values only when operation is automated for solar gain reduction (as opposed to daylight use).
[f]Refers to gray, bronze, and green tinted heat-absorbing glass.
[g]SC for glass with no shading device.

Table 29   Shading Coefficients for Insulating Glass with Indoor Shading by Venetian Blinds or Roller Shades

Type of Glass	Nominal Thickness, Each Light	Solar Transmittance[a]		Venetian Blinds[b]		Roller Shade		
		Outer Pane	Inner Pane			Opaque		Translucent
				Medium	Light	Dark	White	Light
Clear out Clear in	3/32, 1/8 in.	0.87	0.87	0.62[e] (0.63)[d]	0.58[c] (0.58)[d]	0.71	0.35	0.40
Clear in	1/4 in.	0.80	0.80					
Heat-absorbing[e] out Clear in	1/4 in.	0.46	0.80	0.39	0.36	0.40	0.22	0.30
Reflective coated glass	SC = 0.20[f]			0.19	0.18			
	= 0.30			0.27	0.26			
	= 0.40			0.34	0.33			

Table refers to factory-fabricated units with 3/16, 1/4, or 1/2-in. air space, or to prime windows plus storm windows.

[a]Refer to manufacturers' literature for exact values.
[b]For vertical blinds with opaque white or beige louvers, tightly closed, SC is approximately the same as for opaque white roller shades.

[c]From Van Dyck and Konen (1982), for 45° open venetian blinds, 35° solar incidence, and 35° profile angle.
[d]Values for closed venetian blinds. Use these values only when operation is automated for solar gain reduction (as opposed to daylight use).
[e]Refers to bronze, or green tinted, heat-absorbing glass.
[f]SC for glass with no shading device.

Table 30   Shading Coefficients for Double Glazing with Between-Glass Shading

Type of Glass	Nominal Thickness, Each Pane	Solar Transmittance[a]		Description of Air Space	Venetian Blinds		Louvered Sun Screen
		Outer Pane	Inner Pane		Light	Medium	
Clear out, Clear in	3/32, 1/8 in.	0.87	0.87	Shade in contact with glass or shade separated from glass by air space.	0.33	0.36	0.43
Clear out, Clear in	1/4 in.	0.80	0.80	Shade in contact with glass-voids filled with plastic.	—	—	0.49
Heat-absorbing[b] out, Clear in				Shade in contact with glass or shade separated from glass by air space.	0.28	0.30	0.37
	1/4 in.	0.46	0.80	Shade in contact with glass-voids filled with plastic.	—	—	0.41

[a]Refer to manufacturers' literature for exact values.
[b]Refers to grey, bronze and green tinted heat-absorbing glass.

Table 31 Properties of Representative Indoor Shading Devices Shown in Tables 28, 29, and 30

Indoor Shade	Solar-Optical Properties (Normal Incidence)		
	Trans.	Reflect.	Absorp.
Venetian blinds[a] (ratio of slat width to slat spacing 1.2, slat angle 45°)			
Light colored slat	0.05	0.55	0.40
Medium colored slat	0.05	0.35	0.60
Vertical blinds			
White louvers	0.00	0.77	0.23
Roller shades			
Light shades (translucent)	0.25	0.60	0.15
White shade (opaque)	0.00	0.65	0.35
Dark colored shade (opaque)	0.00	0.20	0.80

[a]Values in this table and preceding tables are based on horizontal venetian blinds. However, tests show that these values can be used for vertical blinds with good accuracy.

The values in Table 28 apply both to sunlit fenestration and fenestration on the shaded side of the building. The values are similar in both cases because shades generally are open on shaded exposures. The tabulated values apply specifically to horizontal venetian blinds but are usable for vertical blinds when adjusted so that no direct solar radiation can enter through them.

Table 29 gives SCs for venetian blinds and roller shades used with insulating glass. The first row applies to windows in which both lights are high-transmittance glass; the second applies when the outer light is heat-absorbing and the inner light is clear glass.

Because of the wide variety of glass available, the manufacturer should be consulted for specific data. The SC with no interior shading is included as a reference point for each classification. Note that the energy benefit of a shade decreases as the SC of the unshaded glass decreases. Similarly, the flexibility of the fenestration system decreases as the SC decreases, due to the low transmittances and the inability of the occupant to change this factor.

Draperies

Draperies reduce heating and cooling loads, depending on the type and the use by the occupant. Rudoy and Duran (1975) found annual reductions between 5 and 20%.

The solar optical properties of drapery fabrics can be determined accurately by laboratory tests (Yellott 1963), and manufacturers can usually supply solar transmittance and reflectance values of their products. In addition to these properties, the openness factor (ratio of the open area between the fibers to the total area of the fabric) is a useful property that can be measured exactly (Keyes 1967, Pennington and Moore 1967). It can also be estimated by inspection, since the human eye can readily distinguish between tightly woven fabrics that permit little direct radiation to pass between the fibers and loosely woven fabrics which allow the sun's rays to pass freely.

Drapery fabrics can be classified in terms of their solar-optical properties as having specific values of fabric transmittance and reflectance. Fabric reflectance is the major factor in determining the ability of a fabric to reduce solar heat gain. Based on their appearance, draperies can also be classified by yarn color as dark, medium, and light; and by weave as closed, semiopen, and open. The apparent color of a fabric is determined by the reflectance of the yarn itself. The figure in Table 32 shows yarn reflectance. Figure 31 classifies drapery fabrics into nine types, rated by openness and yarn reflectances.

Figure 31, with the aid of Table 32, guides in estimating the probable SC for a fabric-glass combination when the solar-optical properties are unknown. Whenever possible, fabric reflectance and transmittance values should be obtained from the manufacturer, which permits more accurate SC estimates to be made. Visual

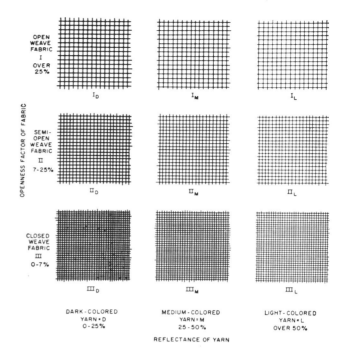

Note: Classes may be approximated by eye. With closed fabrics, no objects are visible through the material, but large light or dark areas may show. Semiopen fabrics do not permit details to be seen, and large objects are clearly defined. Open fabrics allow details to be seen, and the general view is relatively clear with no confusion of vision. The yarn color or shade of light or dark may be observed to determine whether the fabric is light, medium, or dark.

Fig. 31 Classification of Drapery Fabrics

estimations of openness and yarn reflectance, interpreted through Table 32, are valuable in judging the effectiveness of drapes for (1) protection from excessive radiant energy from either sunlight or sun-heated glass, (2) brightness control, (3) providing either outward view or privacy, and (4) sound control.

Table 32 applies to glass and a single drape hung with 100% fullness (drapery width is twice the width of the draped area). If the drapery is hung flat, like a window shade, a different SC applies; with a low transmittance and high reflectance, the SC is appreciably lower. As an extreme example, a flat opaque drapery having an aluminized or similar coating with reflectance of 0.80, in combination with 0.25-in. clear glass, has an SC of 0.18, as compared with 0.32 that can be extrapolated from Table 32 for this material in draped form. Pennington and Moore (1967) explain the effect of folding drapery materials to provide 100% fullness, and describe a method for calculating SC when materials are used flat.

Example 12. A drape with 100% fullness, having a fabric transmittance of 0.20 and a fabric reflectance of 0.40, is used with 0.25-in. glass. What SC should be used?

Solution: From the figure in Table 32, the 0.20 and 0.40 intersection is nearest line F. Table 32 assigns SC 0.55 to column F for 0.25-in. clear single glass. Interpolate if necessary; see notes for other uses.

Example 13. For the same drapery as in Example 12, the incident angle for which the SC is desired is 30°. What SC should be used?

Solution: Add 5% to the value found in Example 12. Thus,

$$SC = (1 + 0.05)\,0.55 = 0.58$$

Example 14. Determine the Fabric Designator for a fabric having an openness factor of 0.10 and a yarn reflectance of 0.60.

Solution: On the figure in Table 32, these lines intersect in the area of Designator II_L. Refer also to Figure 31. Fabric is semiopen and light in color. Additional information: probable fabric reflectance is 0.50, and fabric transmittance is 0.35.

Table 32   Shading Coefficients for Single and Insulating Glass with Draperies

Glazing	Glass Trans- mission	Glass Alone SC (No Drapes)	A	B	C	D	E	F	G	H	I	J
Single glass												
1/8 in. Clear	0.86	1.00	0.87	0.82	0.74	0.69	0.64	0.59	0.53	0.48	0.42	0.37
1/4 in. Clear	0.80	0.95	0.80	0.75	0.70	0.65	0.60	0.55	0.50	0.45	0.40	0.35
1/2 in. Clear	0.71	0.88	0.74	0.70	0.66	0.61	0.56	0.52	0.48	0.43	0.39	0.35
1/4 in. Heat absorbing	0.46	0.67	0.57	0.54	0.52	0.49	0.46	0.44	0.41	0.38	0.36	0.33
1/2 in. Heat absorbing	0.24	0.50	0.43	0.42	0.40	0.39	0.38	0.36	0.34	0.33	0.32	0.30
Reflective coated	—	0.60	0.57	0.54	0.51	0.49	0.46	0.43	0.41	0.38	0.36	0.33
(see manufacturers' literature	—	0.50	0.46	0.44	0.42	0.41	0.39	0.38	0.36	0.34	0.33	0.31
for exact values)	—	0.40	0.36	0.35	0.34	0.33	0.32	0.30	0.29	0.28	0.27	0.26
	—	0.30	0.25	0.24	0.24	0.23	0.23	0.23	0.22	0.21	0.21	0.20
Insulating glass, 1/4-in. air space												
(1/8 in. out and 1/8 in. in)	0.76	0.89	0.75	0.71	0.65	0.63	0.57	0.53	0.48	0.45	0.38	0.36
Insulating glass 1/2-in. air space												
clear out and clear in	0.64	0.83	0.66	0.62	0.58	0.56	0.52	0.48	0.45	0.42	0.37	0.35
Heat absorbing out and clear in	0.37	0.55	0.49	0.47	0.45	0.43	0.41	0.39	0.37	0.35	0.33	0.32
Reflective coated	—	0.40	0.38	0.37	0.37	0.36	0.34	0.32	0.31	0.29	0.28	0.28
(see manufacturers' literature for	—	0.30	0.29	0.28	0.27	0.27	0.26	0.26	0.25	0.25	0.24	0.24
exact values)	—	0.20	0.19	0.19	0.18	0.18	0.17	0.17	0.16	0.16	0.15	0.15

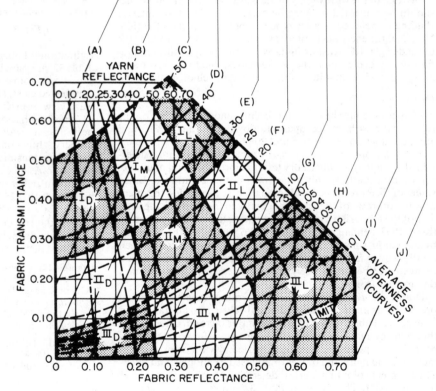

Shading Coefficient Index Letter

Notes:
1. Shading coefficients are for draped fabrics.
2. Other properties are for fabrics in flat orientation.
3. Use fabric reflectance and transmittance to obtain accurate shading coefficients.
4. Use openness and yarn reflectance or openness and fabric reflectance to obtain the various environmental characteristics, or to obtain approximate shading coefficients.

Classification of Fabrics

   I = Open weave
  II = Semiopen weave
 III = Closed weave

  D = Dark color
  M = Medium color
  L = Light color

To obtain fabric designator (III_L, I_M, etc.). Using either (1) Fabric Transmittance and Fabric Reflectance coordinates, or (2) Openness and Yarn Reflectance coordinates, find a point on the chart and note the designator for that area. If properties are not known, the classification may be approximated by eye as described in the note in Figure 31. Table 33 lists appropriate classifications for various applications.

To obtain shading coefficient (SC). (1) Locate drapery fabric as a point using its known properties, or approximate using its fabric classification designator. For accuracy, use fabric transmittance and fabric reflectance; (2) follow diagonal SC lines to lettered columns in the table. Find SC on line with glazing used. For example, SC is 0.45 for 0.25-in. clear single glass with III_L drapery (Column H).

Note: Shading coefficients are for 45° incident angle. For 30° or less, add 5% to the number found in the table.

VISUAL AND THERMAL CONTROLS

The ideal fenestration system permits optimum light, heat, ventilation, and visibility; minimizes moisture and sound transfer between the exterior and the interior; and produces a satisfactory physiological and psychological environment. The controls of an optimum system react to varying climatological and occupant demands. Fixed controls may have operations or cost advantages or both but do not react to physical and psychological variations. Variable controls are, therefore, more effective in energy conservation and environmental satisfaction.

Operational Effectiveness of Shading Devices

Shading devices vary in their operational effectiveness. Some devices such as overhangs and tinted glazings do not require operation, have long life expectancies, and do not degrade significantly over their effective life. Other types of shading devices, especially operable interior shades, may have reduced effectiveness due to less than optimal operation and degradation of effectiveness over time. It is important to evaluate operational effectiveness when considering the actual heat rejection potential of shading devices.

The performance of shading devices for the reduction of peak cooling loads and annual energy use should account for operational effectiveness or reliability in actual operation. Passive devices, such as architectural elements and glazing tinting, are considered 100% effective in operation. Glazing coatings and adherent films may degrade over time. Shade screens are removable and may be assumed to be operated seasonally but in any given population of users some will remain in place all year long and some will not be installed or removed at optimum times. Automated shading devices controlled for optimum thermal operation are considered more effective than manual devices, but controls require ongoing maintenance. Automated shading devices may also be operated for nonthermal purposes such as glare and daylighting optimization which may reduce thermal effectiveness. Manually operated devices will be subject to wide variation in use effectiveness, and this diversity in effective use should be considered when evaluating performance.

Indoor Shading Devices

Thermal comfort of occupants within the glazed space is paramount. Other factors (see Table 32 and Figure 31) include the following.

Radiant energy protection. Unshaded windows become sources of radiant heat by transmitting short-wave solar radiation and by emitting long-wave radiation to dissipate some of the absorbed solar energy. In winter, glass temperatures usually fall below room air temperature, which may produce thermal discomfort to occupants near the fenestration. In summer, individuals seated near the unshaded window may experience discomfort from both direct solar rays and long-wave radiation emitted by sunheated glass. In winter, loss of heat by radiation to cold glass can also cause discomfort. Tightly woven, highly reflective drapes minimize such discomfort; drapes with high openness factors are less effective because they permit short-wave and long-wave radiation to pass more freely. Light-colored shading devices with maximum total surface usually provide the best protection, since they absorb less heat and tend to lose heat readily by convection to the conditioned air.

Outward vision. Outward vision is normally desirable in both business and living spaces. Open-weave, dark-colored fabrics of uniform pattern permit maximum outward vision, while uneven pattern weaves reduce the ability to see out. A semi-open weave modifies the view without completely obscuring the outdoors. Tightly woven fabrics block off outward vision completely.

Privacy. Venetian blinds, either vertical or horizontal, can be adjusted and, when completely closed, afford full privacy. When draperies are closed, the degree of privacy is determined by their color and tightness of weave, and the source of the principal illumination. To obscure the view so completely that not even shadows or silhouettes can be detected, fully opaque materials are used.

Brightness control. Visual comfort is essential in many occupied areas, and freedom from glare is an important factor in performing tasks. Direct sunlight must not strike the eye, and reflected sunlight from bright or shiny surfaces is equally disturbing and even disabling. A tightly woven white fabric with high solar transmittance attains such brilliance when illuminated by direct sunshine that, by contrast with its surroundings, it creates excessive glare. Off-white colors should be used so their surface brightness is not too great. Venetian blinds permit considerable light to enter by inter-reflection between slats. When two shading devices are used, the one on the inside (away from the window) should be darker and more open. With this arrangement, the inside can be used to control brightness for the other shading devices and, when used alone, reduce brightness while still permitting some view of the outside.

Table 33   Summary of Environmental Control Capabilities of Draperies

Item	Designator (Table 32 and Figure 31)								
	I_D	I_M	I_L	II_D	II_M	II_L	III_D	III_M	III_L
1. Protection from direct solar radiation and long-wave radiation to or from window areas	Fair	Fair	Fair	Fair	Good	Good	Fair	Good	Good
2. Effectiveness in allowing outward vision through fenestration	Good	Good	Fair	Fair	Fair	Some	None	None	None
3. Effectiveness in attaining privacy (limiting inward vision from outside)	None	None	Poor[a] Good[a]	Poor	Fair	Fair[a] Good[a]	Good[b]	Good[b]	Good[b]
4. Protection against excessive brightness and glare from sunshine and external objects	Mild	Mild	Mild[c] Poor[c]	Good	Good	Good[c] Poor[c]	Good	Good	Good[c] Poor[c]
5. Effectiveness in modifying unattractive or distracting view out of window	Little	Little	Some	Some	Good	Good	Blocks	Blocks	Blocks

[a]Good when bright illumination is on the viewing side.
[b]To obscure view completely, material must be completely opaque.

[c]Poor rating applies to white fabric in direct sunlight. Use off-white color to avoid excessive transmitted light.

View modification. When the view is unattractive or distracting, draperies modify the view to some degree, depending on the fabric weave and color (summarized in Table 33). Thus, the window remains an effective connection to the outside.

Sound control. Indoor shading devices, particularly draperies, can absorb some of the sounds originating within the room, but have little or no effect in preventing outdoor sounds from entering. For excessive internally generated sound, the usual remedy is to apply acoustical treatment to the ceiling and other room surfaces. While these materials can be effective in controlling sound, they are often located on the two horizontal surfaces (ceiling and floor) and leave the opposing vertical surfaces of glass and bare wall to reflect sound. The noise reduction coefficient (NRC = average absorptance coefficient at four frequencies) for venetian blinds is about 0.10, as compared to 0.02 for glass and 0.03 for plaster. For drapery fabrics at 100% fullness, NRC ranges from 0.10 to 0.65, depending on the tightness of weave. Class III (tightly woven) fabrics have NRC values of 0.35 to 0.65. Figure 32 shows the relation between NRC and openness factor for fabrics of normal weight.

Example 15. To select a drapery fabric, consider the five environmental factors listed in Table 31. Choose a fabric designator which has suitable performance for all the factors important to the case being considered. If this is not possible, make compromises resulting in an acceptable designator. Determine from Table 32 if the SC for the chosen designator is satisfactory. Specific cases follow:

1. Where modification of a distracting view is necessary, but a degree of outward vision is needed and an SC of 0.50 with 0.25-in. gray glass is satisfactory (see Table 33, Item 5), select II_M or II_L; Item 2, select II_M. The SC for II_M on Table 33 is approximately 0.46, therefore satisfactory.
2. Where protection from radiation is paramount and minimum SC is necessary (see Table 33, Item 1), select a closed weave, III_M or III_L. Since the SC for III_L is lowest (see Figure 32), choose III_L.
3. When good outward vision is desired, together with some reduction in brightness, choose I_M or I_D (Table 33).

Double Drapery

Double draperies (two sets of draperies covering the same area) have a light, open weave on the window side of the fenestration for outward vision and daylight when desired, and a heavy, close weave or opaque drapery on the room side to block out sunlight and provide privacy when desired. When properly selected and used, double draperies provide a reduced U-factor and a lowered SC.

The reduced U-factor results principally from adding a semi-closed air space to the barrier. A U-factor of about 0.57 Btu/h·ft²·°F is achieved using double draperies with single glass, and about 0.37 Btu/h·ft²·°F with insulating glass.

To most effectively reduce solar heat gain, the drapery exposed to sunlight should have high reflectance and low transmittance. The light, open-weave drapery should be opened when the heavy drapery is closed to prevent entry of sunlight. The open weave drapery works better on the room side, since this arrangement improves the SC and increases the U-factor when both draperies are closed.

Properly used double draperies give (1) extreme flexibility of vision and light intensity, (2) a lowered U-factor and SC, and (3) an improved comfort condition, since the room-side drapery is more nearly at room temperature. Table 33 gives characteristics of individual draperies. For large areas, the SC should be calculated in detail to determine the cooling load.

REFERENCES

ANSI/ASHRAE. 1988. Method of measuring solar-optical properties of materials. *Standard* 74-88.

Arasteh, D. 1989. An analysis of edge heat transfer in residential windows. Proceedings of ASHRAE/DOE/BTECC Conference, Thermal Performance of the Exterior Envelopes of Buildings IV, Orlando, FL, 376-87.

ASHRAE. 1988. Method of measuring solar-optical properties of materials. ANSI/ASHRAE *Standard* 74-1988.

ASHRAE. 1989. Energy efficient design of new buildings except low-rise residential buildings. ASHRAE *Standard* 90.1-1989.

ASTM. 1982. Standard test method for solar absorptance, reflectance, and transmittance of materials using integrating spheres. *Standard* E 903-82. American Society for Testing and Materials, Philadelphia, PA.

ASTM. 1986. Standard test method for solar transmittance (terrestrial) of sheet materials using sunlight. *Standard* E1084-86. American Society for Testing and Materials, Philadelphia, PA.

ASTM. 1987a. Standard tables for terrestrial direct normal solar spectral irradiance for air mass 1.5. *Standard* E891-87.

ASTM. 1987b. Standard tables for terrestrial solar spectral irradiance at air mass 1.5 for a 37° tilted surface. *Standard* E892-87.

ASTM. 1988a. Standard practice for calculation of photometric transmittance and reflectance of materials to solar radiation. *Standard* E971-88. American Society for Testing and Materials, Philadelphia, PA.

ASTM. 1988b. Standard test method for solar photometric transmittance of sheet materials using sunlight. *Standard* E972-88. American Society for Testing and Materials, Philadelphia, PA.

Bird, R.E. and R.L. Hulstrom. 1982. Terrestrial solar spectral data sets. SERI/TR-642-1149. Solar Energy Research Institute.

Bliss, R.W. 1961. Atmospheric radiation near the surface of the ground *Solar Energy* 5(3):103.

Brandle, K. and R.F. Boehm. 1982. Air flow windows: Performance and applications. Proceedings of Thermal Performance of the Exterior Envelopes of Buildings II, ASHRAE/DOE Conference, ASHRAE SP38, December.

Carpenter, S. and A. McGowan. 1993. Effect of framing systems on the thermal performance of windows. ASHRAE *Transactions* 99(1).

CSA. 1991. Windows/User selection guide to CSA standards. CAN/CSA A440-M90/A440.1-M90. Canadian Standards Association, Rexdale, Ontario.

Duffie, J.A. and W.A. Beckman. 1980. *Solar engineering of thermal processes.* John Wiley & Sons, Inc., New York.

ElSherbiny, S.M., *et al.* 1982. Heat transfer by natural convection across vertical and inclined air layers. *Journal of Heat Transfer* 104:96-102.

Enermodal Engineering Limited. 1990. FRAME/VISION Window performance modelling and sensitivity analysis. Institute for Research in Construction, National Research Council of Canada, Ottawa.

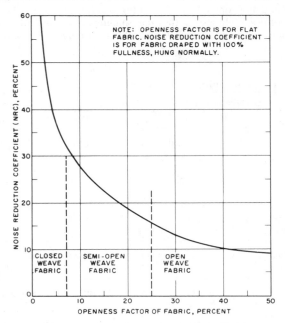

Fig. 32   Noise Reduction Coefficient versus Openess Factor

Ewing, W.B. and J.I. Yellott. 1976. Energy conservation through the use of exterior shading of fenestration. ASHRAE *Transactions* 82(1):703-33.

Federal Standard 16 CRF 1201. Safety standard for architectural glazing materials.

Galanis, N. and R. Chatiguy. 1986. A critical review of the ASHRAE solar radiation model. ASHRAE *Transactions* 92(1).

Gates, D.M. 1966. Spectral distribution of solar radiation at the earth's surface. *Science* 151(2):3710.

Gueymard, C.A. 1987. An anisotropic solar irradiance model for tilted surfaces and its comparison with selected engineering algorithms. *Solar Energy* 38:367-86. Erratum, *Solar Energy* 40:175 (1988).

Gueymard, C.A. 1993. Critical analysis and performance asessment of clear sky solar irradiance models using theoretical and measured data. *Solar Energy* (in press).

Hickey, J.R., *et al.* 1981. Observations of the solar constant and its variations: Emphasis on Nimbus 7 results. Symposium on the Solar Constant and the Special Distribution of Solar Irradiance, IAMAP 3rd Scientific Assembly, Hamburg, Germany.

Hogan, J.F. 1988. A summary of tested glazing U-values and the case for an industry wide testing program. ASHRAE *Transactions* 94(2).

Hollands, K.G.T. and J.L. Wright. 1982. Heat loss coefficients and effective $\tau\alpha$ products for flat plate collectors with diathermous covers. *Solar Energy* 30:211-16.

IES. 1979. Recommended practice of daylighting, IES RP-5. Illuminating Engineering Society of North America, New York.

Iqbal, M. 1983. *An introduction to solar radiation.* Academic Press, Toronto.

Keyes, M.W. 1967. Analysis and rating of drapery materials used for indoor shading. ASHRAE *Transactions* 73(1):8.4.1.

Klems, J.H. 1989. U-values, solar heat gain, and thermal performance: Recent studies using the MoWitt. ASHRAE *Transactions* 95(1).

McCabe, M.E., *et al.* 1986. U-value measurements for windows and movable insulations from hot box tests in two commercial laboratories. ASHRAE *Transactions* 92(1).

McCluney, R. 1987. Determining solar radiant heat gain of fenestration systems. *Passive Solar Journal* 4(4):439-87.

McCluney, R. 1991. The death of the shading coefficient? ASHRAE *Journal* 33(3):36-45.

Moon, P. 1940. Proposed standard solar radiation curves for engineering use. *Journal of the Franklin Institute* 11:583.

NFRC. 1991. *Standard* 100-91.

Ozisik, N. and L.F. Schutrum. 1960. Solar heat gain factors for windows with drapes. ASHRAE *Transactions* 66:228.

Parmelee, G.V. and W.W. Aubele. 1952. Radiant energy emission of atmosphere and ground. ASHRAE *Transactions* 58:85.

Parmelee, G.V. and R.G. Huebscher. 1947. Forced convection heat transfer from flat surfaces. ASHVE *Transactions* 245-84.

Pennington, C.W. 1968. How louvered sun screens cut cooling, heating loads. *Heating, Piping, and Air Conditioning*, December.

Pennington, C.W., *et al.* 1964. Experimental analysis of solar heat gain through insulating glass with indoor shading. ASHRAE *Journal* 2:27.

Pennington, C.W. and D.E. McDuffie, Jr. 1970. Effect of inner surface air velocity and temperature upon heat gain and loss through glass fenestration. ASHRAE *Transactions* 76:190.

Pennington, C.W. and G.L. Moore. 1967. Measurement and application of solar properties of drapery shading materials. ASHRAE *Transactions* 73(1):8.3.1.

Pennington, C.W., C. Morrison, and R. Pena. 1973. Effect of inner surface air velocity and temperatures upon heat loss and gain through insulating glass. ASHRAE *Transactions* 79(2).

Perez, R., *et al.* 1986. An anisotropic hourly diffuse radiation model for sloping surfaces—Description, performance validation, and site dependency evaluation. *Solar Energy* 36:481-98.

Peterson, C.O., Jr. 1987. How is low-E performance criteria determined? *Glass Digest* 1:70-76.

Rubin, M. 1982. Solar optical properties of windows. *Energy Research* 6:122-33.

Rubin, M. 1982. Calculating heat transfer through windows. Energy Research 6:341-49.

Rubin, M. 1984. Optical constants and bulk optical properties of soda lime silica glasses for windows. Report No. 13572, Applied Sciences Division, Lawrence Berkeley Laboratory, Berkeley, CA, June.

Rudoy, W. and F. Duran. 1975. Effect of building envelope parameters on annual heating/cooling load. ASHRAE *Journal* 7:19.

Rundquist, R.A. 1991. Calculation procedure for daylighting and fenestration effects on energy and peak demand. ASHRAE *Transactions* 97(2).

Selkowitz, S.E. 1979. Thermal performance of insulating windows. ASHRAE *Transactions* 85(2):669-85.

Shewen, E.C. 1986. A Peltier-effect technique for natural convection heat flux measurement applied to the rectangular open cavity. Ph.D. thesis, Department of Mechanical Engineering, University of Waterloo, Ontario.

Smith, W.A. and C.W. Pennington. 1964. Shading coefficients for glass block panels. ASHRAE *Journal* 5(December):31.

Sodergren, D. and T. Bostrom. 1971. Ventilating with the exhaust air window. ASHRAE *Journal* 13(4):51.

Sullivan, H.F. and J.L. Wright. 1987. Recent improvements and sensitivity of the VISION glazing system thermal analysis program. Proceedings of the 12th Passive Solar Conference, ASES/SESCI, Portland, OR, 145-49.

Threlkeld, J.L. and R.C. Jordan. 1958. Direct solar radiation available on clear days. ASHRAE *Transactions* 64:45.

University of Waterloo. 1987. Vision: A computer program to evaluate the thermal performance of innovative glazing systems. University of Waterloo, Ontario.

Vild, D.J. 1964. Solar heat gain factors and shading coefficients. ASHRAE *Journal* 10:47.

Wright, J.L. 1991. Correlations for quantifying convective heat transfer between window glazings. Communication prepared for ASHRAE SPC-142 (Standard method for determining and expressing the heat transfer and total optical properties of fenestration products).

Yellott, J.I. 1963. Selective reflectance—A new approach to solar heat control. ASHRAE *Transactions* 69:418.

Yellott, J.I. 1966. Shading coefficients and sun-control capability of single glazing. ASHRAE *Transactions* 72(1):72.

Yellott, J.I. 1972. Effect of louvered sun screens upon fenestration heat loss. ASHRAE *Transactions* 78(1):199-204.

BIBLIOGRAPHY

ASTM. 1970. Recommended practice for laboratory measurements of airborne sound transmission loss of building partitions. ASTM *Standard* E90-70.

Brambley, M.R. and S.S. Penner. 1979. Fenestration devices for energy conservation I. Energy savings during the cooling season. Energy, February.

Burkhardt, W.C. 1975. Solar optical properties of gray and brown solar control series transparent acrylic sheet. ASHRAE *Transactions* 81(1):384-97.

Burkhardt, W.C. 1976. Acrylic plastic glazing; properties, characteristics and engineering data. ASHRAE *Transactions* 82(1):683.

Collins, B.L. 1975. Windows and people: A literature survey, psychological reaction with and without windows. National Bureau of Standards, Building Science Series 70.

Energy, Mines and Resources Canada. 1987. Enermodal Engineering Ltd.—The effect of frame design on window heat loss phase 1. Report prepared for Renewable Energy Branch, Ottawa.

Johnson, B. 1985. Heat transfer through windows. Swedish Council for Building Research, Stockholm.

Lawrence Berkeley Laboratory. 1987. *Window 3.0 users guide.* Windows and Daylighting Group, Lawrence Berkeley Laboratory, Berkeley, CA.

McCluney, W.R. 1968. Radiometry and photometry. *American Journal of Physics* 36:977-79.

Meyer-Arendt, J.R. 1968. Radiometry and photometry: Units and conversion Factors. *Applied Optics* 7:2081-84.

Moore, G.L. and C.W. Pennington. 1967. Measurement and application of solar properties of drapery shading materials. ASHRAE *Transactions* 73(1):3.1-15.

Nicodemus, F.E. 1963. Radiance. *American Journal of Physics* 31:368.

Nicodemus, F.E., *et al.* 1976-84. Self-study manual on optical radiation measurements, Part I—Concepts. NBS Technical Notes 910-1, 910-2, 910-3, and 910-4. National Bureau of Standards (now National Institute of Standards and Technology).

Nicodemus, F.E., *et al.* 1977. Geometrical considerations and nomenclature for reflectance. NBS Monograph 160, NBS (now NIST), October.

Pierpoint, W. and J. Hopkins. 1982. The derivation of a new area source equation. Presented at the Annual Conference of the Illuminating Engineering Society, Atlanta, August.

Threlkeld, J.L. 1962. *Thermal environmental engineering.* Prentice Hall, New York, 321.

Van Dyke, R.L. and T.P. Konen. 1982. Energy conservation through interior shading of windows: An analysis, test and evaluation of reflec-tive venetian blinds. LBL-14369, Lawrence Berkeley Laboratory, March.

Yellott, J.I. 1965. Drapery fabrics and their effectiveness in sun control. ASHRAE *Transactions* 71(1):260-72.

CHAPTER 28

ENERGY ESTIMATING METHODS

THE energy requirements and fuel consumption of HVAC systems have a direct impact on the cost of operating a building and an indirect impact on the environment. This chapter discusses methods for estimating energy use as a guide in design, for standards compliance, and for economic optimization. The general requirements of an energy analysis are reviewed; then the various analysis methods are considered in order of increasing complexity—degree-day methods, bin methods, correlation methods, and transient simulation. The chapter concludes with a discussion of factors unique to solar energy systems.

These energy estimating methods can provide quantitative energy and cost comparisons among design alternatives. A large number of uncontrolled and unknown factors generally preclude, except in cases where extraordinarily detailed and careful measurements and observations have been made, the use of such methods for the precise calculation of absolute energy consumption. Thus, a careful measurement of energy use is the standard against which energy calculation results should be referenced. In no case should these methods be used to predict future utility bills.

GENERAL CONSIDERATIONS

PURPOSES

A primary objective of building energy analysis is economic—to determine which of the available options has the lowest total cost. Chapter 33 of the 1991 ASHRAE *Handbook—Applications* outlines how different analysis techniques can be used to optimize system selection and provides tables and data to assist in these analyses. Software programs are available to perform these calculations.

Many areas of engineering use life-cycle cost (LCC) optimization techniques in design. Unfortunately, LCC optimization has been inhibited by the complexity of buildings, which makes an analysis complex and costly, and by the uniqueness of each building, which makes it difficult to spread the cost of the analysis over several units. These barriers are being overcome, largely due to the availability of computer programs and inexpensive computers.

Many factors related to building energy system design are not considered in the LCC of a building. One example is the thermal comfort of the occupants, which affects productivity that could be accounted for if the relationship between the two were accu-

The preparation of this chapter is assigned to TC 4.7, Energy Calculations.

rately known. A small loss in productivity due to reduced comfort can quickly offset an energy cost savings. Therefore, the designer must treat comfort conditions as constraints on a design rather than as variables to be included in the optimization.

Environmental impacts and other societal concerns also depend on building energy system design. However, they are difficult to quantify in an LCC optimization unless some authority provides a constraint or price signal. If the cost of equipment and energy that the designer plans to use reflects the total cost to society of producing them, then the optimum building design will be an optimum for society as well as for the building owner. If these costs fail to account for some societal costs, they are not considered, and the optimum design chosen for the building owner may not be an optimum for society.

A few building energy codes and standards allow the use of an energy analysis to demonstrate compliance with the energy performance goals in the code. This use of energy analysis programs may be more prevalent than actual comparative energy studies.

COMMON FACTORS

Although the procedures for estimating energy requirements vary considerably in their degree of complexity, they all have three common elements, *i.e.*, the calculation of (1) space load, (2) secondary equipment load, and (3) primary equipment energy requirements. Here, secondary refers to equipment that distributes the heating, cooling, or ventilating medium to conditioned spaces, while primary refers to central plant equipment that converts fuel or electric energy to heating or cooling effect. A major distinction is made between steady-state methods (based on degree days or temperature bins) and dynamic methods (*e.g.*, based on transfer functions).

The space load calculations determine the amounts of energy that must be added to or extracted from a space to maintain thermal comfort. The simplest procedures assume that the energy required to maintain comfort is only a function of the outdoor dry-bulb temperature. More detailed methods consider solar effects, internal gains, heat storage in the walls and interiors, and the effects of wind on both building envelope heat transfer and infiltration. These calculations are similar to the heating and cooling load calculations used to size equipment, but they are not the same. Energy calculations are based on average usage schedules and typical weather conditions rather than maximum usage schedules and worst case weather. The most sophisticated procedures are currently based on hourly profiles for climatic conditions and operational characteristics for a number of typical days of the year or on 8760 h of operation per year.

The second step translates the space load into a load on the secondary system equipment. This can be a simple estimate of duct or piping losses or gains, or a complex hour-by-hour simulation of an air system, such as variable air volume with outdoor air cooling. This step must include the calculation of all forms of energy required by the secondary system, *i.e.*, electrical energy to operate fans and/or pumps, as well as energy in the form of heated or chilled water.

The third step calculates the fuel and energy required by the primary equipment to meet these loads and peak demand on the utility system. It considers equipment efficiencies and part-load characteristics. It is often necessary to keep track of the different forms of energy, such as electrical, natural gas, or oil. In some cases, where calculations are required to assure compliance with codes or standards, these energies must be converted to source energy or resource consumed, as opposed to energy delivered to the building boundary.

Often energy calculations lead to an economic analysis to establish the cost-effectiveness of conservation measures (ASHRAE 1989). Thus, thorough energy analysis provides intermediate data, such as time of energy usage and maximum demand, so that utility charges can be accurately estimated. Although not part of the energy calculations, estimated capital equipment costs should be included in such an analysis.

Complex and often unexpected interactions can occur between the systems or between various modes of heat transfer. For example, radiant heating panels impact the space loads by raising the mean radiant temperature in the space (Howell 1990). As a result, the air temperature can be lowered while maintaining comfort. Compared to a conventional heated air system, radiant panels create a greater temperature difference from the inside surface to the outside air. Thus, conduction losses through the walls and roof increase because the inside surface temperatures are greater. At the same time, the heating load due to infiltration or ventilation decreases because of the reduced indoor air to outdoor air temperature difference and a reduced stack effect. The infiltration rate may also decrease because the reduced air temperature difference reduces the stack effect.

CHOOSING AN ANALYSIS METHOD

The most important step in selecting an energy analysis method is to match the method capabilities with project requirements. The method must be capable of evaluating all design options with sufficient accuracy to make correct choices. The following factors apply generally (Sonderegger 1985):
- **Accuracy.** The method should be sufficiently accurate to allow correct choices. Because of the many parameters involved in energy estimation, absolutely accurate energy prediction is not possible (Waltz 1992).
- **Sensitivity.** The method should be sensitive to the design options being considered. The difference in energy use between two choices should be accurate.
- **Versatility.** The method should allow the analysis of all options under consideration. When different methods must be used to consider different options, an accurate estimate of the differential energy use cannot be made.
- **Speed and cost.** The total time (gathering data, preparing input, calculations, and analysis of output) to make an analysis should be appropriate to the potential benefits gained. With greater speed, more options can be considered in a given time. The cost of analysis is largely determined by the total time of analysis.
- **Reproducibility.** The method should not allow so many vaguely defined choices that different analysts would get completely different results (Corson 1992).

- **Ease of use.** This impacts both on the economics of analysis (speed) and the reproducibility of results.

Since almost all manual methods have been implemented on a computer, the selection of an energy analysis method is the selection of a computer program. The ASHRAE Journal's 1990 HVAC&R Software directory lists suppliers of various programs.

Selecting Energy Analysis Computer Programs

The selection of a building energy analysis program depends on its application, the number of times it will be used, the experience of the user, and the hardware available to run it. The first of the criteria to be considered must be the capability of the program to deal with the application. For example, if the effect of a shading device is to be analyzed on a building that will also be shaded by other buildings part of the time, the capability of analyzing detached shading is an absolute requirement, regardless of any other factors.

Today, all well-known programs run on microcomputers. Some programs require very large memory and hard disk capacities or need a very fast computer. However, the cost of the computer facilities and the software itself are typically a small part of carrying out a building energy analysis. The major costs are the cost of the effort involved in learning to use the program and the cost of the effort involved in using it. Major issues that influence the cost of learning a program include: (1) complexity of the input procedures, (2) quality of the user's manual, and (3) availability of a good support system to answer questions. As the user becomes more experienced, the cost of learning the program becomes less important. However, the requirement for a complex set of input data, will continue to consume the time of even an experienced user.

The complexity of input is largely influenced by the availability of default values for the input variables. Default values can be used as a simple set of input data when detail is not needed or when the building design is very conventional; but additional complexity can be supplied when needed. Secondary defaults, which can be supplied by the user, are also useful in the same way. Some programs allow the user to specify a level of detail. Then the program requests only the information appropriate to that level of detail, using default values for all others.

The quality of the output is another factor to consider. Reports should be easy to read and uncluttered. The titles and headings should be unambiguous. Units should be stated explicitly. The users' manual should explain the meanings of the data presented. Graphic output can be very helpful. In most cases, simple summaries of overall results are the most useful, but very detailed output is needed for certain studies and also for debugging program input during the early stages of an energy analysis project.

Before making a final decision, the user should obtain and review the users' manuals for the most suitable programs and, if possible, should obtain and run demonstration versions of the programs. During this last part of the selection process, the user should test the support from the software supplier. The availability of training should be considered when choosing a more complex program.

The availability of weather data and/or the availability of a weather data processing subroutine or program is a major feature of a program. Some programs include subroutine or supplementary programs that allow the user to create a weather file for any site for which weather data is available. Programs that do not have this capability must have weather files for various sites created by the program supplier. In that case, the available weather data and the terms on which the supplier will create new weather data files must be checked.

Auxiliary capabilities, such as economic analysis and design calculations, are a final concern in selecting a program. Economic analysis may include only the ability to calculate annual energy bills from utility rates, or it might extend to calculations, or even to LCC optimization. An integrated program may save time, because some input data will have been entered already for other purposes.

The results of computer calculations should be accepted with caution as the software vendor does not accept responsibility for the correctness of calculations or the use of the program. A manual calculation should be run to develop a good understanding of the underlying physical processes and building behavior. In addition, the user should (1) review the computer program documentation to determine what calculation procedures are used, (2) compare the results with manual calculations and measured data, and (3) conduct sample tests to confirm that the program delivers acceptable results.

Since energy analysis computer programs take different approaches in their method of calculation, significant differences in results are found (Spielvogel 1977, Corson 1992).

STEADY-STATE METHODS

Degree-day methods are the simplest methods for energy analysis and are appropriate if the building use and the efficiency of the HVAC equipment are constant. Where efficiency or conditions of use vary with outdoor temperature, the consumption can be calculated for different values of the outdoor temperature and multiplied by the corresponding number of hours; this approach is used in various bin methods. When the indoor temperature is allowed to fluctuate or when interior gains vary, models other than simple steady-state models must be used.

Even in an age when computers can easily calculate the energy consumption of a building, the concepts of degree days and balance point temperature remain valuable tools. The severity of a climate can be characterized concisely in terms of degree days. Also, the degree-day method and its generalizations can provide a simple estimate of annual loads, which can be accurate if the indoor temperature and internal gains are relatively constant and if the heating or cooling systems are to operate for a complete season. For these reasons, basic steady-state methods continue to be important.

BALANCE POINT TEMPERATURE AND DEGREE DAYS

The balance point temperature t_{bal} of a building is defined as that value of the outdoor temperature t_o at which, for the specified value of the interior temperature t_i, the total heat loss q_{gain} is equal to the heat gain from sun, occupants, lights, and so forth.

$$q_{gain} = K_{tot} (t_i - t_{bal}) \qquad (1)$$

where K_{tot} is the total heat loss coefficient of the building in Btu/h · °F. For any of the steady-state methods described in this section, heat gains must be the average for the period in question, not for the peak values. In particular, solar radiation must be based on averages, not peak values. The balance point temperature is therefore obtained as

$$t_{bal} = t_i - \frac{q_{gain}}{K_{tot}} \qquad (2)$$

Heating is needed only when t_o drops below t_{bal}. The rate of energy consumption of the heating system is

$$q_h = \frac{K_{tot}}{\eta_h} [t_{bal} - t_o (\tau)]^+ \qquad (3)$$

where η_h is the efficiency of the heating system, also designated on an annual basis as the annual fuel use efficiency (AFUE) and τ is time. If t_{bal}, K_{tot}, and η_h are constant, the annual heating consumption can be written as an integral

$$Q_{h, yr} = \frac{K_{tot}}{\eta_h} \int [t_{bal} - t_o (\tau)]^+ \, d\tau \qquad (4)$$

where the plus sign above the bracket indicates that only positive values are to be counted. This integral of the temperature difference conveniently summarizes the effect of outdoor temperatures on a building. In practice, it is approximated by summing averages over short time intervals (daily or hourly); the result is termed *degree days* or *degree hours*.

If daily average values of outdoor temperature are used for evaluating the integral, the degree days for heating $DD_h(t_{bal})$ are obtained as

$$DD_h (t_{bal}) = (1 \text{ day}) \sum_{\text{days}} (t_{bal} - t_o)^+ \qquad (5)$$

with dimensions of °F · days. Here the summation is to extend over the entire year or over the heating season. It is a function of t_{bal}, reflecting the roles of t_i, heat gain, and loss coefficient. The balance point temperature t_{bal} is also known as the base of the degree days. In terms of degree days, the annual heating consumption is

$$Q_{h, yr} = \frac{K_{tot}}{\eta_h} DD_h (t_{bal}) \qquad (6)$$

Heating degree days or degree hours for a balance point temperature of 65 °F have been widely tabulated based on the observation that this has represented average conditions in typical buildings in the past. The 65 °F base is assumed whenever t_{bal} is not indicated explicitly. The extension of degree-day data to different bases will be discussed later.

Cooling degree days can be calculated using an expression analogous to heating degree days

$$DD_c (t_{bal}) = (1 \text{ day}) \sum_{\text{days}} (t_o - t_{bal})^+ \qquad (7)$$

While the definition of the balance point temperature is the same as that for heating, in a given building its numerical value for cooling is generally different from that for heating because q_i, K_{tot}, and t_i can be different. According to Claridge *et al.* (1987), t_{bal} can include both solar and internal gains as well as losses to the ground.

In some cases, heating and cooling degree days for shorter time periods than a year are desirable, so that listings of monthly values are common.

Figure 1(a) shows how the heating degree-days vary with t_{bal} for a particular site, *i.e.*, New York. This plot is obtained by evaluating Equation (5) with data for the number of hours per year during which t_o is within 5 °F temperature intervals centered at 72 °F, 67 °F, 62 °F, . . . , −8 °F. The data for the number of hours in each interval, or bin, are included as labels in this plot. Analogous curves, without these labels, are shown in Figure 1(b) for three other sites: Houston, Washington, and Denver.

If the annual average of t_o is known, the cooling degree days to any base below 72 + 2.5 °F can also be found. But calculating cooling energy consumption using degree days is more difficult than heating. For cooling, the equation analogous to Equation (6) is

$$Q_{c, yr} = \frac{K_{tot}}{\eta_c} DD_c (t_{bal}) \qquad (8)$$

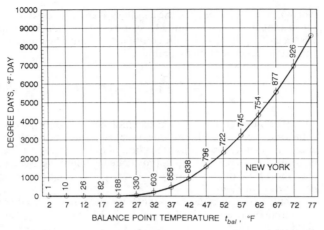

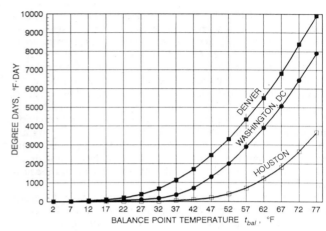

a. For New York, with labels = hours/year when t_o is in 5°F bin centered at the point (interpreting x-axis as t_o values)

b. For Denver, Houston, and Washington, D.C., degree days only.

Fig. 1   Annual Heating Days $DD_h(t_{bal})$ as Function of t_{bal}

for a building whose K_{tot} does not change. That assumption is generally acceptable during the heating season, when windows are closed and the air exchange rate is fairly constant. However, during the intermediate or cooling season, heat gains can be eliminated, and the onset of mechanical cooling can be postponed by opening windows or increasing the ventilation. (In buildings with mechanical ventilation, this is called the *economizer* mode.) Mechanical air conditioning is needed only when the outdoor temperature extends beyond the threshold t_{max}. This threshold is given by an equation analogous to Equation (2), with the replacement of the closed window heat transmission coefficient K_{tot} by its value K_{max} for open windows

$$t_{max} = t_i - \frac{q_{gain}}{K_{max}} \tag{9}$$

K_{max} varies considerably with wind speed; but for simple cases, assume a constant value. The resulting sensible cooling load is shown schematically in Figure 2 as a function of t_o. The solid line is the load with open windows or increased ventilation; the dashed line shows the load if K_{tot} were kept constant. The annual cooling load for this mode can be calculated by breaking the area under the solid line into a rectangle and a triangle, or

$$Q_c = K_{tot}[DD_c(t_{max}) + (t_{max} - t_{bal})N_{max}] \tag{10}$$

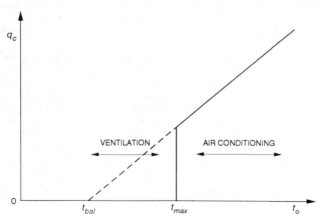

Fig. 2   Cooling Load as Function of Outdoor Temperature t_o
(Below t_{max}, cooling can be avoided by using ventilation.)

where $DD_c(t_{max})$ are the cooling degree days for base t_{max}, and N_{max} is the number of days during the season when t_o rises above t_{max}. This is merely a schematic model of air conditioning. In practice, heat gains and ventilation rates vary, as does the occupant behavior in using the windows and the air conditioner. Also, in commercial buildings with economizers, the extra fan energy for increased ventilation must be added to the calculations. Finally, air-conditioning systems are often turned off during unoccupied periods. Therefore, cooling degree hours better represents the period when equipment is operating than cooling degree days, because degree days assume uninterrupted equipment operation as long as there is a cooling load.

Latent loads can form an appreciable part of a building's cooling load. The degree-day method can be used to estimate the latent load during the cooling season on a monthly basis by adding the following term to Equation (10):

$$Q_{latent} = \dot{m}h_{fg}(W_o - W_i) \tag{11}$$

where

q_{latent} = monthly latent cooling load
$\dot{m}$ = monthly infiltration (total airflow)
h_{fg} = heat of vaporization of water
W_o = outdoor humidity ratio (monthly averaged)
W_i = indoor humidity ratio (monthly averaged)

The degree-day method assumes that t_{bal} is constant, which is not well satisfied in practice. Solar gains are zero at night, and internal gains tend to be highest during the evening. The pattern for a typical house is shown in Figure 3. As long as t_o always stays below t_{bal}, the variations average out without changing the consumption. But for the situation in Figure 3, t_o rises above t_{bal} from shortly after 10:00 A.M. to 10:00 P.M.; the consequences for energy consumption depend on the thermal inertia and on the control of the HVAC system. If this building had low inertia, and if temperature control were critical, heating would be needed at night and cooling during the day. In practice, this effect is reduced by thermal inertia and by the deadband of the thermostat which allows t_i to float.

The closer t_o is to t_{bal}, the greater the uncertainty. If the occupants keep the windows closed during mild weather, t_i will rise above the set point. If they open the windows, the potential benefit of heat gains is reduced. In either case, the true values of t_{bal} become uncertain. Therefore, the degree-day method, like any

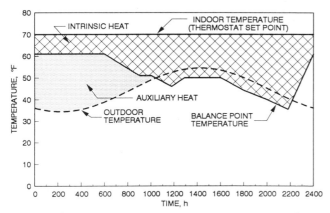

**Fig. 3 Variation of Balance Point Temperature and
Internal Gains for Typical House**
(Nisson and Dutt 1985)

steady-state method, is unreliable for estimating the consumption during mild weather. In fact, the consumption becomes most sensitive to occupant behavior and cannot be predicted with certainty.

Despite these problems, the degree-day method (using an appropriate base temperature) can give remarkably accurate results for the annual heating energy of single-zone buildings dominated by gains through the walls and roof and/or ventilation. Typical buildings have time constants that are about one day, and a building's thermal inertia essentially averages over the diurnal variations, especially if t_i is allowed to float. Furthermore, the energy consumption in mild weather is small; hence a relatively large error here has only a small effect on the total for the season.

Seasonal Efficiency

The seasonal efficiency of heating equipment η_h depends on such factors as steady-state efficiency, sizing, cycling effects, and energy conservation devices. Sometimes it is much lower than and other times it is comparable to steady-state efficiency. Alereza and Kusuda (1982) developed expressions to estimate the seasonal efficiency for a variety of furnaces, if information on rated input and output is available. These expressions correlate seasonal efficiency with variables determined by using the equipment simulation capabilities of a large hourly simulation program and typical equipment performance curves supplied by the National Institute of Standards and Technology.

$$\bar{\eta} = \eta_{ss} \, CF_{pl}/(1 + \alpha_D) \tag{12}$$

where

η_{ss} = steady-state efficiency (rated output/input)
α_D = fraction of heat loss from ducts

The term CF_{pl} is a characteristic of the part-load efficiency of the heating equipment, which may be calculated as follows:

Gas-forced air furnaces

With pilot

$$CF_{pl} = 0.6328 + 0.5738 \, RLC - 0.3323 \, (RLC)^2$$

With intermittent ignition

$$CF_{pl} = 0.7791 + 0.1983 \, RLC - 0.0711 \, (RLC)^2$$

With intermittent ignition and loose stack damper

$$CF_{pl} = 0.9276 + 0.0732 \, RLC - 0.0284 \, (RLC)^2$$

Oil furnaces without stack damper

$$CF_{pl} = 0.7092 + 0.6515 \, RLC - 0.4711 \, (RLC)^2$$

Resistance electric furnaces

$$CF_{pl} = 1.0$$

These equations are based on many annual simulations for the equipment in which RLC is defined as follows:

$$RLC = BLC \, (t_{bal} - t_{od})(1 + \alpha_D)/CHT$$

where

t_{od} = outside design temperature
CHT = rated output of equipment

Seasonal efficiency is also discussed by Chi and Kelly (1978), Parker *et al.* (1980), and Mitchell (1983).

VARIABLE BASE DEGREE DAYS

The calculation of Q_h from degree days $DD_h \, (t_{bal})$ depends on the value of t_{bal}. This value varies widely from one building to another because of widely differing personal preferences for thermostat settings and setbacks and because of different building characteristics. In response to the fuel crises of the 1970s, heat transmission coefficients have been reduced, and thermostat setback has become common. At the same time, the energy use by appliances has increased. These trends all reduce t_{bal} (Fels and Goldberg 1986). Hence, in general, degree days with the traditional base 65 °F are not to be used.

Many formulas have been proposed for estimating the degree days relative to an arbitrary base when detailed data are not available. The basic idea is to assume a typical probability distribution of temperature data, characterized by its average $\bar{t}_o$ and by its standard deviation σ. Erbs *et al.* (1983) developed a model that needs as input only the average $\bar{t}_o$ for each month of the year. The standard deviations σ_m for each month are then estimated from the correlation

$$\sigma_m = 3.54 - 0.0290 \, \bar{t}_o + 0.0664 \, \sigma_{yr} \tag{13a}$$

This is a dimensional equation with t and σ in °F; σ_{yr} is the standard deviation of the monthly average temperatures about the annual average $\bar{t}_{o,\,yr}$

$$\sigma_{yr} = \sqrt{\frac{1}{12} \sum_1^{12} (\bar{t}_o - \bar{t}_{o,\,yr})^2} \tag{13b}$$

To obtain a simple expression for the degree days, a normalized temperature variable θ is defined as

$$\theta = \frac{t_{bal} - \bar{t}_o}{\sigma_m \sqrt{N}} \tag{14}$$

with N = number of days in the month (N has units of day/month and θ has units of $\sqrt{\text{month/day}}$). While temperature distributions can be different from month to month and location to location, most of this variability can be accounted for by the average and the standard deviation of $\bar{t}_o$. Being centered around $\bar{t}_o$ and scaled by σ_m, the quantity θ eliminates these effects. In terms of θ, the monthly heating degree days for any location are well approximated by

$$DD_h\,(t_{bal}) \;=\; \sigma_m\,N^{3/2}\left[\frac{\theta}{2} + \frac{\ln\,(e^{-a\theta} + e^{a\theta})}{2a}\right] \qquad (15)$$

where $a = 1.698\sqrt{\text{day/mo.}}$

For nine locations spanning most climatic zones of the United States, Erbs et al. (1983) verified that the annual heating degree days can be estimated with a maximum error of 315 °F-days if this equation is used for each month. For cooling degree days, the largest error is 270 °F-days. Such errors are quite acceptable, representing less than 5% of the total.

Example 1. Find the monthly heating degree-days for New York City, using the model of Erbs et al. (1983), given monthly averages of t_o as reproduced in column 2 of Table 1. Base the degree days on a balance temperature of 60 °F.

The results are included in Table 1. Column 2 lists the given values of monthly average outdoor temperature, and N is the number of days in the month. Intermediate quantities are shown in columns 4 and 5, and $t_{o,\,yr}$ and σ_{yr} are shown at the bottom. Column 6 shows the monthly and annual results.

Table 2 contains degree-day data for several sites and monthly averaged outdoor temperatures needed for the algorithm used in the example. More complete tabulations of the latter are contained in Cinquemani et al. (1978) and in local climatological data summaries available from the National Climatic Data Center, Asheville, NC. Monthly degree-day data to the bases of 50 °F, 55 °F, 60 °F, 65 °F, and 70 °F, as well as other climatic infor-

mation for 209 U.S. and 14 Canadian cities, may be found in Appendix 3 to Balcomb et al. (1982).

BIN METHOD

For many applications, the degree-day method should not be used, even with the variable base method, because the heat loss coefficient K_{tot}, the efficiency η_h of the HVAC system, or the balance point temperature may not be sufficiently constant. The efficiency of a heat pump, for example, varies strongly with outdoor temperature; or the efficiency of the HVAC equipment may be affected indirectly by t_o when the efficiency varies with the load, a common situation for boilers and chillers. Furthermore, in most commercial buildings, the occupancy has a pronounced pattern, which affects heat gain, indoor temperature, and ventilation rate.

In such cases, a steady-state calculation can yield good results for the annual energy consumption, if different temperature intervals and time periods are evaluated separately. This approach is known as the bin method, because the consumption is calculated for several values of the outdoor temperature t_o and multiplied by the number of hours N_{bin} in the temperature interval (bin) centered around that temperature

$$Q_{bin} = N_{bin}\,\frac{K_{tot}}{\eta_h}\,[t_{bal} - t_o]^+ \qquad (16)$$

The superscript plus sign indicates that only positive values are counted; no heating is needed when t_o is above t_{bal}. This equation is evaluated for each bin, and the total consumption is the sum of the Q_{bin} over all bins.

In the United States, the necessary data are widely available (ASHRAE 1984, USAF 1978). The bins are usually in 5 °F increments and are often collected in three daily 8-h shifts. Mean coincident wet-bulb temperature data (for each dry-bulb bin) are used to calculate latent cooling loads from infiltration and ventilation. The bin method considers both occupied and unoccupied building conditions and gives credit for internal loads by adjusting the balance point. For example, a calculation could be performed for 42 °F outdoors (representing all occurrences from 39.5 to 44.5 °F) and with building operation during the midnight to 0800 shift. Since there are 23 5 °F bins between −10 and 105 °F and 3 8-h shifts, 69 separate operating points are calculated. For many applications, the number of calculations can be reduced. A residential heat pump (heating mode), for example, could be calculated for just the bins below 65 °F without the three-shift breakdown. The data included in Table 3 are samples of annual totals for a few sites, but ASHRAE (1984) and USAF (1978) include monthly data and data further separated into time intervals during the day.

Table 1 Degree-Day Calculation from Monthly Averaged Data

Month	$\bar{t}_o$, °F	N, day/mo.	σ_m, °F	σ, $\sqrt{\text{mo./day}}$	$DD_h(t_{bal})$, °F·day
January	32.2	31	3.65	1.37	864
Feburary	33.4	28	3.62	1.39	746
March	41.1	31	3.40	1.00	592
April	52.1	30	3.08	0.47	265
May	62.3	31	2.79	−0.15	67
June	71.6	30	2.52	−0.84	7
July	76.6	31	2.38	−1.26	2
August	74.9	31	2.41	−1.11	3
September	68.4	30	2.61	−0.59	16
October	58.7	31	2.88	0.08	123
November	46.4	30	3.22	0.72	391
December	35.5	31	3.56	1.24	762
$t_{o,\,yr}$	54.4			Sum	3837
σ_{yr}	15.8				

Note: Use Equation (15) to calculate $DD_h\,(t_{bal})$

Table 2 Degree-Day and Monthly Average Temperatures for Various Locations

Site	Variable Base Heating Degree-Day, °F·days[a]					Monthly Average Outdoor Temperature, °F[b]											
	65	60	55	50	45	Jan	Feb	Mar	Apr	May	Jun	Jul	Aug	Sep	Oct	Nov	Dec
Los Angeles, CA	1245	522	158	26	0	54.5	55.6	56.5	58.8	61.9	64.5	68.5	69.6	68.7	65.2	60.5	56.9
Denver, CO	6016	4723	3601	2653	1852	29.9	32.8	37.0	47.5	57.0	66.0	73.0	71.6	62.8	52.0	39.4	32.6
Miami, FL	206	54	8	0	0	67.2	67.8	71.3	75.0	78.0	81.0	82.3	82.9	81.7	77.8	72.2	68.3
Chicago, IL	6127	4952	3912	2998	2219	24.3	27.4	36.8	49.9	60.0	70.5	74.7	73.7	65.9	55.4	40.4	28.5
Albuquerque, NM	4292	3234	2330	1557	963	35.2	40.0	45.8	55.8	65.3	74.6	78.7	76.6	70.1	58.2	44.5	36.2
New York, NY	4909	3787	2806	1980	1311	32.2	33.4	41.1	52.1	62.3	71.6	76.6	74.9	68.4	58.7	47.4	35.5
Bismarck, ND	9044	7656	6425	5326	4374	8.2	13.5	25.1	43.0	54.4	63.8	70.8	69.2	57.5	46.8	28.9	15.6
Nashville, TN	3696	2758	1964	1338	852	38.3	41.0	48.7	60.1	68.5	76.6	79.6	78.5	72.0	60.9	48.4	40.4
Dallas/Ft. Worth, TX	2290	1544	949	526	250	45.4	49.4	55.8	66.4	73.8	81.6	85.7	85.8	78.2	68.0	55.9	48.2
Seattle, WA	4727	3269	2091	1194	602	39.7	43.5	45.5	50.4	56.5	61.3	65.7	64.9	60.6	54.2	45.7	42.0

[a]Source: NOAA (1973)   [b]Source: Cinquemani et al. (1978)

Example 2. Estimate the energy requirements for a residence in Washington, D.C., with a design heat loss of 40,000 Btu/h at 53 °F temperature difference. Assume a 3-ton heat pump, with characteristics given in Figure 4 and Table 4.

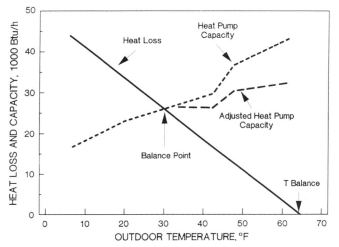

Fig. 4   Heat Pump Capacity versus Building Load

Solution: The design heat loss is based on no internal heat generation. To compute the heat pump system energy input, compute the net heat requirement of the space, *i.e.*, envelope loss minus internal heat generation. For this example, assume an average internal heat generation of 4280 Btu/h and a room temperature of 70 °F. The net heat loss per degree and the heating/cooling balance temperature may be computed:

$$(HL/\Delta t) = (40,000/53) = 755 \text{ Btu/h} \cdot {}^\circ F$$

$$t_{balance} = 70 - (4280/755) = 64.3 \,{}^\circ F$$

Table 4 is then computed resulting in 9578 kWh.

Figure 4 shows the relationship between heat loss and the integrated heat pump capacity. The balance point occurs at the temperature where the heat loss and steady-state capacity are equal. At lower temperatures, supplemental heat is required, and at higher temperatures the heat pump cycles on and off.

To properly predict the energy use of the heat pump system, actual dynamic performance data, including cycling losses, should be used. For the heat pump system, performance data should include the effects of cycling above the building balance temperature, frosting, defrosting, and the effect of auxiliaries. These losses may be significant, depending on the size of the heat pump compared to the building heat loss and the control methodology of the heat pump.

Table 3   Sample Annual Bin Data

														Bin												
City	100/ 104	95/ 99	90/ 94	85/ 89	80/ 84	75/ 79	70/ 74	65/ 69	60/ 64	55/ 59	50/ 54	45/ 49	40/ 44	35/ 39	30/ 34	25/ 29	20/ 24	15/ 19	10/ 14	5/ 9	0/ 4	−5/ 1	−10/ 6	−15/ 11	−20/ 16	−25/ 21
Albuquerque	1	54	191	348	511	617	789	785	816	676	637	720	678	676	560	406	180	101	31	3						
Bismarck		11	68	173	252	320	450	590	625	550	583	506	624	539	626	596	424	399	391	306	364	144	131	43	42	3
Chicago			97	222	362	512	805	667	615	622	585	577	636	720	957	511	354	243	125	66	58	6				
Dallas/Ft. Worth	27	210	351	527	804	1100	947	705	826	761	615	615	523	364	289	57	29									
Denver		3	118	235	348	390	472	697	699	762	783	718	665	758	713	565	399	164	106	65	80	22				
Los Angeles	8	.8	9	17	53	194	632	1583	234	2055	1181	394	74	4												
Miami			45	864	1900	2561	1605	871	442	222	105	77	36	12												
Nashville		7	137	407	616	756	1100	866	706	692	650	670	720	582	342	280	107	71	29							
New York City		5	26	170	383	664	820	941	763	699	593	690	765	858	648	377	212	99	20	5						
Seattle			16	62	139	256	450	769	1353	1436	1461	1413	915	358	51	43	15	1								

Table 4   Calculation of Annual Heating Energy Consumption for Example 2

Climate		House				Heat Pump						Supplemental	
A	B	C	D	E	F	G	H	I	J	K	L	M	N
Temp. Bin, °F	Temp. Diff., $t_{bal} - t_{bin}$	Weather Data Bin, h	Heat Loss Rate, 1000 Btu/h	Heat Pump Integrated Heating Capacity, 1000 Btu/h	Cycling Capacity Adjustment Factor[a]	Adjusted Heat Pump Capacity, 1000 Btu/h[b]	Rated Electric Input, kW	Operating Time Fraction[c]	Heat Pump Supplied Heating, 10^6 Btu[d]	Seasonal Heat Pump Electric Consumption, kWh[e]	Space Load, 10^6 Btu[f]	Supplemental Heating Required, kWh[g]	Total Electric Energy Consumption[h]
62	2.3	740	1.8	44.3	0.760	33.7	3.77	0.05	1.30	146	1.30	—	146
57	7.3	673	5.5	41.8	0.783	32.7	3.67	0.17	3.72	417	3.72	—	417
52	12.3	690	9.3	39.3	0.809	31.8	3.56	0.29	6.42	719	6.42	—	719
47	17.3	684	13.1	36.8	0.839	30.9	3.46	0.42	8.95	1002	8.95	—	1002
42	22.3	790	16.9	29.9	0.891	26.6	3.23	0.63	13.31	1614	13.31	—	1614
37	27.3	744	20.6	28.3	0.932	26.4	3.15	0.78	15.35	1833	15.35	—	1833
32	32.3	542	24.4	26.6	0.979	26.0	3.07	0.94	13.22	1559	13.22	—	1559
27	37.3	254	28.2	25.0	1.000	25.0	3.00	1.00	6.35	762	7.16	236	998
22	42.3	138	31.9	23.4	1.000	23.4	2.92	1.00	3.23	403	4.41	345	748
17	47.3	54	35.7	21.8	1.000	21.8	2.84	1.00	1.18	153	1.93	220	373
12	52.3	17	39.5	19.3	1.000	19.3	2.74	1.00	0.33	47	0.67	101	147
7	57.3	2	43.3	16.8	1.000	16.8	2.63	1.00	0.03	5	0.09	16	21
2	62.3	0	47.0	14.3	1.000	—	—	—	—	—	—	—	—
								TOTALS:	73.39	8660	76.52	917	9578

[a] Cycling Capacity Adjustment Factor $= 1 - C_d(1 - x)$ when $C_d =$ degradation coefficient, default $= 0.25$, unless part load factor is known, and $x =$ building heat loss/unit capacity at temperature bin. Cycling capacity $= 1$ at the balance point and below.
[b] Col. G = Col. E × Col. F
[c] Operating Time Factor equals the smaller of 1 or Col. D/Col. G

[d] Col. J = (Col. I × Col. G × Col. C)/1000
[e] Col. K = Col. I × Col. H × Col. C
[f] Col. L = Col. C × Col. D/1000
[g] Col. M = Col. L − Col. J
[h] Col. N = Col. K + Col. M

The Department of Energy has adopted test procedures (10 CFR 430, Appendix M, USGPO) to determine the effect of dynamic operations. The bin method uses these results for a specific heat pump to adjust the integrated capacity for the effect of part-load operation. Figure 4 compares the adjusted heat pump capacity to the building heat loss. This type of curve must be developed for each model heat pump as applied to an individual profile. The heat pump cycles on and off above the balance point temperature to meet the house load. This cycling can reduce performance depending on the part-load factor at a given temperature. The cycling capacity adjustment factors (ANSI/ASHRAE *Standard* 116-1983) used in this example to account for cycling degradation can be calculated from the equation in footnote a of Table 4.

Frosting and the necessary defrost cycle can reduce performance over steady-state conditions that do not include frosting. The effects of frosting and defrosting are already integrated into many (but not all) manufacturers' performance data. The example problem assumes that the manufacturers' data already accounts for the frosting/defrosting losses (as indicated by the characteristic notch of the capacity curve in Figure 4) and shows how to adjust an integrated performance curve for cycling losses. For those cases where the manufacturers' data on heating capacities do not include the effect of defrost cycles in the temperature range where they occur, steady-state, instant system capacities at a given outdoor temperature should be multiplied by a factor X_{int} to obtain the integrated heating capacity to be used in the annual energy calculation. The factor X_{int} is given by

$$X_{int} = 1.0 \text{ for } t_o \geqslant 45\,°\text{F}$$
$$= 1.0 - (0.02)(45 - t_o) \text{ for } 40\,°\text{F} \leqslant t_o < 45\,°\text{F}$$
$$= 0.95 - 0.00125\, t_o \text{ for } t_o < 40\,°\text{F}$$

where t_o is the outdoor dry-bulb temperature (Didion and Kelly 1979).

The total annual heating energy consumption used by the heat pump in this example is computed in Table 4 using bin weather data. Column F contains the capacity adjustment factor used to account for losses due to cycling.

Degree-Day Data from Bin Data

To calculate degree days from hourly bin data such as those in ASHRAE (1984), the base or balance temperature t_{bal} must first be determined. When t_{bal} is known, the following summation can be used on any time scale, either monthly or annually, or for several periods of a day on either a monthly or annual basis.

$$\text{DD}_h(t_{bal}) = \frac{1}{24}\left(\sum_{bins}(t_{bal} - t_{bin})^+ N_{bin}\right) \quad (17)$$

where

t_{bin} = temperature at center of bin
N_{bin} = number of hours in bin corresponding to t_{bin}

Cooling degree days are calculated analogously from

$$\text{DD}_c(t_{bal}) = \frac{1}{24}\left(\sum_{bins}(t_{bin} - t_{bal})^+ N_{bin}\right) \quad (18)$$

This method generally produces degree-day values slightly higher than published values from NOAA or the NCDC. This small but systematic deviation can be suppressed by ignoring degree days during the swing seasons when totals are less than a minimum, *e.g.*, 50 to 100 °F-days per month. During such periods, thermal storage of daytime heat often makes nighttime heating unnecessary.

Modified Bin Method

Various refinements, such as the seasonal variation of solar gains, can be included in a bin calculation. If a separate calculation is done for each month, Q_{gain} could be based on the average solar heat gain of the month. The diurnal variation of solar gains can be accounted for by calculating the average solar gain for each of the hourly time periods of the bin method. If such a detailed calculation of solar gains is considered unnecessary, assume a linear correlation of monthly average solar heat gains with monthly average outdoor temperature t_o.

This procedure is illustrated in Figure 5(a) for the simplest case—a building with constant operating conditions every hour of the day and every day of the year (*i.e.*, t_i, K_{tot}, and nonsolar heat gains $q_{non-sol}$ are all constant). In this case, it suffices to calculate the daily average heating load q_{heat} for the peak winter day and the daily average cooling load q_{cool} on the peak summer day

$$q_{heat} = K_{tot}(t_i - t_{peak,\ winter}) - q_{non-sol} - q_{sol,\ winter} \quad (19)$$

and

$$q_{cool} = K_{tot}(t_i - t_{peak,\ summer}) - q_{non-sol} - q_{sol,\ summer} \quad (20)$$

where $t_{peak,\ winter}$ and $t_{peak,\ summer}$ are the design winter and summer outdoor temperatures and $q_{sol,\ winter}$ and $q_{sol,\ summer}$ are the corresponding daily average solar gains. The solar gains in Equations (19) and (20) are totals for all orientations of windows, roofs, and walls. Solar gains are calculated for each building element for summer and winter days and then summed to form the building total for the summer and winter endpoints of the load line in Figure 5 [see Knebel (1983) for details]. In the northern hemisphere, the summer month is usually taken as July and the winter month as January.

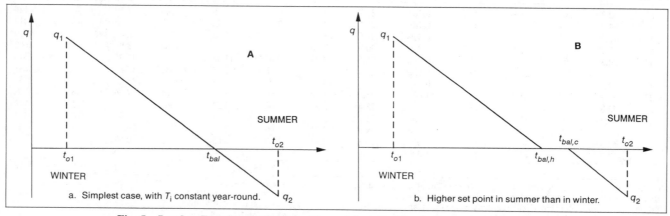

Fig. 5   Load as Function of Outdoor Temperature T_o (Ventilative cooling not shown.)

In equation form, the solar flux ordinate for fenestration in Figure 5 is found from

$$q_{sol, \, summer} = \frac{\sum_{i=1}^{N_{exp}} \text{MSHGF}_i \cdot \text{AG}_i \cdot \text{SC}_i \cdot \text{CLFTOT}_i \cdot \text{FPS}}{\theta A_f} \qquad (21)$$

where

N_{exp} = number of building surfaces (exposures)
MSHGF_i = maximum solar heat gain factor for exposure i
AG_i = glazing area for exposure i
SC_i = shading coefficient of exposure i
CLFTOT_i = daily total cooling load factor for exposure i
FPS = fraction of possible sunshine for summer design month
A_f = floor area [the modified bin method algorithm of Knebel (1983) is executed on a per unit floor area basis]
θ = operating time (h/day) of HVAC system in summer

Note: The methodology for calculating the daily-average hourly solar heat gain is taken from the section Air-Conditioning Cooling Load in Chapter 26 of the 1989 ASHRAE Handbook—Fundamentals; the maximum solar gain factors and cooling load factors used in this and related calculations can be obtained from that source. However, the methodology has been significantly improved; for the revised procedure, consult McQuiston and Spitler (1992).

The solar component of the winter load line endpoint is found from a similar equation, except that the corresponding winter values for all terms are used. Also, the value of θ is 24 h for the winter endpoint.

The total summer and winter load values are plotted versus the corresponding outdoor temperatures. Connecting these two points by a straight line, a simple linear approximation for the load as a function of outdoor temperature t_o is obtained. The balance point temperature t_{bal} of Equation (2) is the value of t_o where this line crosses the abscissa.

This approach does not estimate solar-related loads well for all climates (Vadon et al. 1991), since the endpoints of the solar load lines are defined by the infrequently occurring maximum and minimum bin temperatures. Improvements are possible if solar gains are related to a different outdoor dry-bulb temperature statistic, the annual standard deviation σ_t of hourly data about the annual mean.

The solar flux on any vertical surface of a building is given by the following set of equations. The solar flux is linear with bin temperature (Figure 5), but the slope and intercept are not determined by peak design conditions. The solar flux I in any bin is

$$I = mt_{bin} + b \qquad (22)$$

The slope m and intercept b of this linear expression are given by

$$m = \frac{A}{\sqrt{\sigma_t}} + B \qquad (23)$$

$$b = \frac{C}{\sqrt{\sigma_t}} + D \qquad (24)$$

The standard deviation of hourly temperature is found from

$$\sigma_t = \sqrt{\frac{1}{8759} \sum_{i=1}^{N} N_{bin, \, i} \, (t_i - \bar{t}_{o, \, yr})^2} \qquad (25)$$

where N is the number of bins.

The coefficients in this equation depend on the surface azimuth angle

$$A = 10.63 - 4.980 \cos(a_w - 1)$$
$$B = -1.887 - 0.9754 \cos(a_w - 0.5)$$
$$C = -837.73 - 398.39 \cos(a_w - 1)$$
$$D = 223.29 + 100.53 \cos(a_w - 0.5) \qquad (26)$$

in which the wall azimuth a_w is expressed in radians, with south zero and west positive. Note that the preceding equations are dimensional, with solar flux being in $\text{Btu/h} \cdot \text{ft}^2$.

Equations (22) through (26) are used as follows. The modified bin method load line endpoints are dependent on the solar gain terms as described in the previous section. The term $\text{MSHGF}_i \cdot \text{FPS}$ is replaced by $0.87I$, I being found from Equation (22). The constant 0.87 accounts for the basis of SHGF being single-strength glass (see Chapter 26).

After dividing by the equipment efficiency (using the proper part-load value for each bin), the annual heating energy is found by summing over all temperature bins below the balance point, each bin weighted by the corresponding number of hours. The same can be done for cooling above the balance point temperature, although ventilative cooling should be taken into account according to Figure 2.

It is customary to set the thermostat a few degrees higher in summer than in winter. This has the effect of displacing the cooling characteristic line to the right as shown in Figure 5(b). Evaluating q_{cool} for the higher summer value of t_i, one finds by how much the cooling part of the line is shifted to the right.

As a further refinement, a separate calculation of the solar gain for the transition season, when t_o is close to the values of t_{bal}, could be performed. Then the corrected values of the points ($t_{bal, \, heat}$ and $t_{bal, \, cool}$) where the lines intersect the $q = o$ axis are calculated with the mid-season solar gain. Now the slopes for heating and for cooling can be different.

Finally, the correlations shown in Figure 5(b) can be developed separately for different times of the day and for different occupancy conditions, for example, 1200 to 1600 on Saturdays. Using bin data for the corresponding periods, the calculation can thus account for the operating schedules of commercial buildings.

The modified bin method has the advantage of allowing the use of diversified (part-load) rather than peak load values to establish the load as a function of outdoor dry-bulb temperature. This method also allows both secondary and primary (plant) HVAC equipment effects to be included in the energy calculation. The modified bin method permits the user to predict more accurately effects such as reheat and heat recovery that can only be guessed at with the degree-day or conventional bin methods.

In the modified bin method, average solar gain profiles, average equipment and lighting use profiles, and cooling load temperature difference (CLTD) values are used to characterize time-dependent diversified loads. An improved method is described in Claridge et al. (1992a, 1992b). The CLTDs approximate the transient effects of building mass. Time dependencies resulting from scheduling are averaged over a selected period, or multiple calculation periods are established. The duration of a calculation period determines the number of bin hours included. Normally, two calculation periods representing occupied and unoccupied hours are sufficient, although any number can be used. The method can be further refined by making calculations on a monthly, not annual, basis.

As described earlier, loads resulting from solar gains through glazings are calculated by determining a weighted-average solar load for a summer and for a winter day (each being of average cloudiness and having average solar conditions), and then establishing a linear relationship of this solar load as a function of outdoor temperature. When outdoor ambient temperature exceeds room temperature, the effect of heat transfer through opaque, sunlit roofs and walls is included in the transmission load by averaging the CLTDs from Chapter 26 and applying corrections for deviations from the CLTD-reference outdoor-to-indoor temperature difference.

Once a total load profile is determined as a function of outdoor ambient temperature for occupied and unoccupied periods, the performance of the HVAC system is computed based on the heating and cooling coil loads. Next, the annual energy consumption at the coils is determined using bin-hour weather data. Finally, the annual plant energy consumption is calculated using boiler and chiller part-load performance models. Example 3 demonstrates the first step of this method (diversified coil loads) as applied to a variable air volume system; note that it could be applied to any type of system.

Example 3. Given the following information, determine the annual energy consumption for an office building with the simplified floor plan shown in Figure 6.

1. Building type: office, single-story

2. Location: Washington, D.C., near Andrews Air Force Base

3. Summer design load: 515,000 Btu/h

4. Summer outside design conditions: 90°F db, 74°F wb
   Fraction possible sunshine (FPS) = 0.64

5. Inside design conditions (occupied hours): 75°F db, 50% rh. This is chosen as constant throughout the year to simplify the example. Different set points can be used for the heating and cooling season. During unoccupied hours, a 55°F setback is used, and the cooling coils are shut off.

6. Winter design load: 285,200 Btu/h. These design loads were calculated using standard ASHRAE heating load methods (Chapter 25).

7. Winter outside design conditions: 14°F db (Chapter 24)
   FPS = 0.44

8. *Physical data*

   Gross floor area: 15,000 ft²
   Net conditioned area: 12,000 ft²
   Perimeter depth: 12 ft
   Interior space area: 6576 ft²
   Perimeter space area: 5424 ft²

   Surface data

Direction	Wall Area, ft²	Glass Area, ft²
N	900	900
E	600	600
S	900	900
W	600	600

   Wall: $U = 0.0684$ Btu/h·ft²·°F (group G)
   Glazing: $UG = 0.97$ Btu/h·ft·°F
   Glazing: $SC = 0.55$
   Roof: $U = 0.0807$ Btu/h·ft²·°F (type 3)

9. *Internal loads and schedules*

   Lights   2.5 W/ft²
                   0700 to 1900, Mon-Fri, except holidays
                   0.25 W/ft² otherwise
                   Utilization factor (UF) = 0.8 (constant)
   Equipment   1.0 W/ft²
                   0700 to 1900, Mon-Fri, except holidays
                   0 W/ft² other times
   People   1 person/100 ft²
                   0700 to 1900, Mon-Fri, except holidays
                   0 person/ft² otherwise
                   $Q_s = 255$ Btu/h·person
                   $Q_L = 255$ Btu/h·person

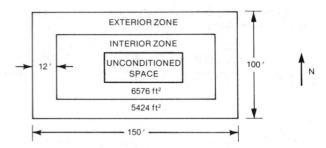

Fig. 6   Floor Plan of Building for Example 3

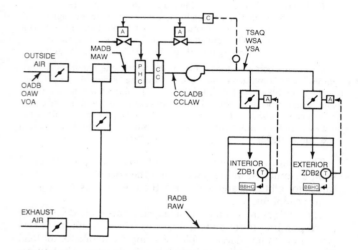

Fig. 7   Variable Air Volume System with Baseboard Heating (Example 3)

10. Ventilation: 0.2 cfm/ft²
                       0.05 cfm/ft² (unoccupied period)

11. *Air distribution system data* (Figure 7)

    Variable air volume in both interior and exterior zones, with baseboard heat in perimeter space
    Design supply air temperature: 55°F
    (includes a 2°F drawthrough fan temperature rise)
    Design supply air: 20,000 cfm
    Design power: 17.3 hp

The design load was calculated using the CLTD/CLF cooling load method described in Chapter 26. A construction of 87 lb/ft² (medium weight) was assumed. CLFs for this construction weight, assuming interior shading (SC = 0.55), were used to calculate the solar gain through the glazings for July. CLTDs for Group G walls and a Type 3 roof (with suspended ceiling) were used to calculate heat gains through opaque envelope surfaces. Corrections were made for outside and inside temperature conditions. A corrected CLTD for conduction through glass was used. The internal heat gains listed in item 9 result in the design supply airflow rate and summer design load at the cooling coil.

Solution: Calculate the zone load, HVAC system coil loads, and primary equipment energy input at all outdoor air temperature bins.

1. *Sub-building or zone selection*. This building has four distinct perimeter zones and an interior zone. If the perimeter spaces are physically separated from the interior zone, each of the perimeter and the interior zones will exhibit a different load profile. In this case, the building should be subdivided into five separate zones for analysis—the four perimeter zones and the interior zone. However, to simplify this example, the building will be subdivided into two zones—one perimeter zone and one interior zone.

2. *Perimeter zone diversified loads.*
(a) Solar heat gain through glass—July

$$QSOL,Jul = \frac{\sum\limits_{i=1}^{N_{exp}} (MSHGF_i \cdot AG_i \cdot SC_i \cdot CLFTOT_i \cdot FPS)}{\theta A_F}$$

where

$QSOL,Jul$ = averaged solar contribution for July
$MSHGF_i$ = maximum solar heat gain factor for orientation i for July at specified latitude
AG_i = glass area for exposure i
SC_i = shading coefficient of glass for exposure i
$CLFTOT_i$ = 24-h sum of CLF for orientation i
FPS = fraction of possible sunshine for July
θ = runtime of air-conditioning system, h (12 h)
A_F = building conditioned floor area, ft^2

Exposure	MSHGF $\times$	AG $\times$	SC $\times$	CLFTOT $\times$	FPS	= Numerator
N	38	900	0.55	11.57	0.64	139,284
E	216	600	0.55	5.46	0.64	249,080
S	109	900	0.55	6.43	0.64	222,035
W	216	600	0.55	5.46	0.64	249,080
					TOTAL =	859,479 Btu/day

$QSOL,Jul = 859,479/(12 \times 12,000) = 5.969$ Btu/h·ft^2

(b) Solar heat gain through glass—January

$$QSOL,Jan = \frac{\sum\limits_{i=1}^{N_{exp}} (MSHGF_i \cdot AG_i \cdot SC_i \cdot CLFTOT_i \cdot FPS)}{24\, A_F}$$

where

$MSHGF_i$ = maximum solar heat gain factor for orientation i for January at specified latitude
FPS = fraction of possible sunshine in January

Exposure	MSHGF $\times$	AG $\times$	SC $\times$	CLFTOT $\times$	PPS	= Numerator
N	20	900	0.55	11.57	0.44	50,399
E	154	600	0.55	5.46	0.44	122,090
S	254	900	0.55	6.43	0.44	355,715
W	154	600	0.55	5.46	0.44	122,090
					TOTAL =	650,294 Btu/day

$QSOL,Jan = 650,294/(24 \times 12,000) = 2.258$ Btu/h·ft^2

(c) Linearize solar contribution from glass with outside air temperature.

QSOL may be expressed as:

$$QSOL = M(t - t_{ph}) + QSOL,Jan$$

where

t = summer bin temperature
t_{ph} = winter peak bin temperature
t_{pc} = summer peak bin temperature (Table 8)

$$M = \frac{QSOL,Jul - QSOL,Jan}{t_{pc} - t_{ph}}$$

$$= \frac{5.969 - 2.258}{97 - 2} = 0.03906$$

$$QSOL = 0.03906(t - 2) + 2.258$$
$$= 0.03906\, t + 2.1799 \text{ (Btu/h·ft}^2)$$

(d) Transmission solar loads

Compute solar component through opaque walls in July.

Table 5 CLTD Solar Component 24-h Average (CLTDS) for Group G Walls and Type 3 Roof at 40°N Latitude

Direction	July	January
N	5	0
NE	11	1
E	15	6
SE	14	15
S	10	21
SW	14	15
W	15	6
NW	11	1
Roof	22	3

	Color Correction Multipliers K		
	D	**M**	**L**
Wall	1	0.83	0.65
Roof	1	0.75	0.50

$$QTS,Jul = \frac{\sum\limits_{i=1}^{N_{surf}} (A_i U_i\, CLTDS_i\, K_i\, FPS)}{A_F}$$

where

QTS = solar contribution to transmission load
U_i = overall heat transmission coefficients for walls
$CLTDS_i$ = 24-h averaged solar component of CLTD, Table 5
K_i = color correction factor

July Surface	Area $\times$	U $\times$	CLTDS$_i$ $\times$	K_i $\times$	FPS	= Numerator
N	900	0.0684	5	1.0	0.64 =	196.99
E	600	0.0684	15	1.0	0.64 =	393.99
S	900	0.0684	10	1.0	0.64 =	393.99
W	600	0.0684	15	1.0	0.64 =	393.99
ROOF	5424	0.0807	22	1.0	0.64 =	6163.00
					TOTAL =	7541.96 Btu/h

$QTS,Jul = 7541.96/12,000 = 0.6285$ Btu/h·ft^2

Compute solar component through opaque walls in January

Surface	Area $\times$	U	$\times$ CLTDS$_i$ $\times$	K_i $\times$	FPS	= Numerator
N	900	0.0684	0	1.0	0.44 =	0.00
E	600	0.0684	6	1.0	0.44 =	108.34
S	900	0.0684	21	1.0	0.44 =	568.81
W	600	0.0684	6	1.0	0.44 =	108.34
ROOF	5424	0.0807	3	1.0	0.44 =	577.80
					TOTAL =	1363.29 Btu/h

$QTS, Jan = 1363.29/12,000 = 0.1136$ Btu/h·ft^2

Linearize solar contribution *from opaque walls* with outside air temperature:

QTS may be expressed as:

$$QTS = M(t - t_{ph}) + QTS,Jan$$
$$M = (QTS,Jul - QTS,Jan)/(t_{pc} - t_{ph})$$
$$= (0.6285 - 0.1136)/(97 - 2) = 0.00542$$
$$QTS = 0.00542(t - 2) + 0.1136$$
$$= 0.00542\, t + 0.1028 \text{ (Btu/h·ft}^2)$$

Note: This contribution to the diversified load is much less than the solar contribution through glass. Where there are large expanses of glass, ignore this term.

(e) Transmission conduction load

Compute conduction component through opaque walls, roof, and glass:

$$QT = \frac{\sum\limits_{i=1}^{N_{surf}} A_i U_i (t - t_i)}{A_F}$$

where

QT = conduction component of load
A_i = surface area for surface i
U_i = overall U-value for surface i
t = bin temperature
t_i = inside air temperature

Surface	Area ×	U	=	AU
N-WALL	900	0.0684	=	61.56
E-WALL	600	0.0684	=	41.04
S-WALL	900	0.0684	=	61.56
W-WALL	600	0.0684	=	41.04
ROOF	5424	0.0807	=	437.72
GLASS	3000	0.97	=	2910.00
		TOTAL	=	3552.92 Btu/h·°F

$$QT = 3552.92 \, (t - t_i)/12{,}000$$

$$= 0.2961 \, (t - t_i)$$

With t_i set to 75 °F, this becomes

$$QT = 0.2961t - 22.208 \; (\text{Btu/h·ft}^2)$$

When the space is unoccupied, the night setback temperature is 55 °F. Therefore, during this period, the conduction through walls, roof, and glass is:

$$QT = 0.2961t - 16.28 \; (\text{Btu/h·ft}^2)$$

(f) Perimeter zone internal loads

$$\text{Internal Loads} = \frac{\text{Use Factor} \cdot \text{Maximum Load} \cdot \text{Heat Factor}}{A_F}$$

where

Use Factor = average usage during occupied or unoccupied periods, expressed as fraction of maximum load
Maximum Load = connected load for lighting and other internal equipment or, for people, maximum number of occupants
Heat Factor = factor for converting unit of load to Btu/h/(3.41 Btu/h = 1 W)

Period			Load, Btu/h·ft²
Occupied period (0700 to 1900)			
Lights	2.5 × 3.41 × 0.8 × 5424/12,000 =		3.08
Equipment	1.0 × 3.41 × 5424/12,000 =		1.54
Occupants, sensible	(255/100) (5424/12,000) =		1.15
Total sensible		=	5.77
Occupants, latent	(255/100) (5424/12,000) =		1.15
Unoccupied period (1900 to 0700)			
Lights	0.25 × 3.41 × 0.8 × 5424/12,000 =		0.31
Equipment			0.00
Occupants			0.00
Total sensible			0.31

(g) Total perimeter zone loads. The total space load for the perimeter zone is the sum of contributions from all the components at a given outdoor temperature. Therefore, the load profile can be expressed as a sum of linear equations described above:

	Occupied		*Unoccupied*
QSOL	= 0.03906 t + 2.1799	QSOL	= 0.03906 t + 2.1799
QTS	= 0.00542 t + 0.1028	QTS	= 0.00542 t + 0.1028
QT	= 0.2961 t − 22.208	QT	= 0.2961 t − 16.28
QI	= 5.77	QI	= 0.31
QTOT	= 0.3406 t − 14.1553	QTOT	= 0.3406 t − 13.6873

3. *Interior zone diversified loads*

(a) Solar heat gain through glass—none.
(b) Transmission solar loads component. Compute solar component through opaque roof. This zone has no exterior walls.

$$QTS = \frac{A_R U_R \cdot \text{CLTDS} \cdot K \cdot \text{FPS}}{A_F}$$

Month	Surface	Area ×	U	× CLTDS ×	K ×	FPS
July	Roof	6576	0.0807	22	1	0.64 = 7472

QTS,Jul = 7472/12,000 = 0.622

Month	Surface	Area ×	U	× CLTDS ×	K ×	FPS
Jan	Roof	6576	0.0807	3	1	0.44 = 700.5

QTS,Jan = 700.5/12,000 = 0.0584

Linearize solar contribution from the roof:

$$QTS = 0.00593 \, t + 0.0465 \; \text{Btu/h·ft}^2$$

(c) Compute transmission load due to conduction from roof

$$QT = \frac{6576 \times 0.0807}{12{,}000} (t - t_i) = 0.04422 \, (t - t_i)$$

For the occupied period ($t_i = 75$ °F) and the unoccupied period ($t_i = 55$ °F). The transmission profile becomes

$$QT = 0.04422 \, t_o - 3.3165 \; (\text{occupied})$$

$$QT = 0.04422 t_o - 2.4321 \; (\text{unoccupied})$$

(d) Interior Loads

Period			Load, Btu/h·ft²
Occupied period (0700 to 1900)			
Lights	(2.5 × 3.74) (3.41 × 0.8 × 6576/12,000) =		3.81
Equipment	1 × 3.41 × 6576/12,000 =		1.87
Occupants, sensible	(255/100) (6576/12,000) =		1.40
Total sensible			7.01
Occupants, latent	255/100 × 6576/12,000		1.40
Unoccupied period (1900 to 0700)			
Lights	0.25 (3.41 × 0.8 × 6576/12,000) =		0.37
Equipment		=	0.00
Occupants			0.00
Total sensible		=	0.37

(e) Total interior zone loads
Compute the total space load for the interior zone by summing the individual components. The load profile can be expressed as a sum of linear equations.

	Occupied		*Unoccupied*
QTS	= 0.00593 t + 0.0465	QTS	= 0.00593 t + 0.0465
QT	= 0.04422 t − 3.3165	QT	= 0.04422 t − 2.4321
QI	= 7.01	QI	= 0.37
QTOT$_{occ}$	= 0.0502 t + 3.740	QTOT$_{unocc}$	= 0.0502 t − 2.015

where t is the outside air temperature.

4. *Determine HVAC system performance*
The following calculation steps lead to completion of noted columns in Table 6.

Column A. Obtain bin temperature weather data (ASHRAE 1984 or USAF 1978).
Column B. Compute outside air humidity ratio OAW from outside air dry-bulb and mean coincident wet-bulb bin data using a psychrometric chart or relationships in Chapter 6.

Table 6 Coil Loads Summary (Example 3)

A OADB	MCWB	B OAW	C VSA	D MADB	E CCLADB	F PHLDB	G RAW	H MAW	I CCLAW	J QPHC	K QCS	L QCL	M QCT	N QHC
						Occupied Period								
97	76	.01450	1.250	78.52	53.00	78.52	.00812	.00914	.00770	0.00	35.09	8.724	43.81	0.00
92	74	.01400	1.161	77.93	53.00	77.93	.00815	.00916	.00770	0.00	31.84	8.209	40.05	0.00
87	72	.01340	1.072	77.24	53.00	77.24	.00819	.00916	.00770	0.00	28.59	7.592	36.18	0.00
82	70	.01300	0.983	76.42	53.00	76.42	.00824	.00920	.00770	0.00	25.34	7.162	32.50	0.00
77	68	.01270	0.895	75.45	53.00	75.45	.00829	.00928	.00770	0.00	22.09	6.820	28.91	0.00
72	66	.01230	0.806	74.26	53.00	74.26	.00835	.00933	.00770	0.00	18.84	6.370	25.21	0.00
67	61	.01010	0.717	72.77	53.00	72.77	.00843	.00890	.00770	0.00	15.59	4.162	19.75	0.00
62	56	.00820	0.628	70.86	53.00	70.86	.00854	.00843	.00770	0.00	12.34	2.222	14.56	0.00
57	51	.00660	0.539	68.32	53.00	68.32	.00868	.00791	.00770	0.00	9.09	0.540	9.63	0.00
52	46	.00520	0.450	64.79	53.00	64.79	.00785	.00668	.00668	0.00	5.84	0.000	5.84	0.00
47	42	.00450	0.368	59.77	53.00	59.77	.00714	.00571	.00571	0.00	2.74	0.000	2.74	0.13
42	37	.00350	0.356	56.47	53.00	56.47	.00615	.00466	.00466	0.00	1.36	0.000	1.36	1.83
37	33	.00310	0.345	52.96	53.00	53.00	.00575	.00422	.00422	0.02	0.00	0.000	0.00	3.54
32	29	.00270	0.333	49.21	53.00	53.00	.00535	.00376	.00376	1.39	0.00	0.000	0.00	5.24
27	24	.00200	0.322	45.19	53.00	53.00	.00465	.00300	.00300	2.77	0.00	0.000	0.00	6.94
22	19	.00150	0.311	40.87	53.00	53.00	.00414	.00245	.00245	4.14	0.00	0.000	0.00	8.65
17	15	.00130	0.299	36.23	53.00	53.00	.00394	.00218	.00218	5.52	0.00	0.000	0.00	10.35
12	10	.00090	0.288	31.22	53.00	53.00	.00355	.00171	.00171	6.90	0.00	0.000	0.00	12.05
7	6	.00090	0.276	25.79	53.00	53.00	.00354	.00163	.00163	8.27	0.00	0.000	0.00	13.76
2	2	.00090	0.265	19.90	53.00	53.00	.00354	.00155	.00155	9.65	0.00	0.000	0.00	15.46
						Unoccupied Period								
52														0.00
47														0.00
42														0.09
37														2.23
32														4.46
27					**All values in columns B through M = 0**									6.69
22														8.92
17														11.14
12														13.37
7														15.60
2														17.83

Variable Description and Units for Columns A through N

A	Outside air bin dry-bulb temperature	°F		H	Mixed air humidity ratio	lb_w/lb_a
B	Outside air humidity ratio of water vapor to dry air	lb_w/lb_a		I	Cooling coil leaving air humidity ratio	lb_w/lb_a
C	Supply air volume	cfm/ft^2		J	Preheat coil load	Btu/h · ft^2
D	Mixed air dry-bulb temperature	°F		K	Cooling coil sensible load	Btu/h · ft^2
E	Cooling coil leaving air dry-bulb temperature	°F		L	Cooling coil latent load	Btu/h · ft^2
F	Preheat coil leaving air dry-bulb temperature	°F		M	Cooling coil total load	Btu/h · ft^2
G	Return air humidity ratio	lb_w/lb_a		N	Baseboard heating load	Btu/h · ft^2

$$OAW = f(OADB, MCWB)$$

Column C. For a VAV system operating with a constant supply air temperature, determine the required air volume based on the load profiles developed earlier. The supply air quantity VSA is calculated as a function of outside air for a particular design supply air temperature difference Δt.

$$\Delta t = 75 - 55 = 20\,°F \text{ (occupied period)}$$

Perimeter zone:

$$VSA = \frac{Load}{60\,\rho\,c_p\,\Delta t} = \frac{Load}{1.1\,\Delta t}$$

$$VSA = (0.3406\,t - 14.1553)/(1.1 \times 20) = 0.0155t - 0.6434 \text{ (cfm/ft}^2)$$

If VSA less than 0.2 cfm/ft^2 (5424/12000) = 0.0904, set VSA = 0.0904

Interior zone:

$$VSA = (0.0502\,t + 3.74)/(1.1 \times 20)$$
$$= 0.0023\,t + 0.1700 \text{ cfm/ft}^2$$

If VSA is less than 0.2 cfm/ft^2 (6576/12000) = 0.1096, set VSA = 0.1096

Total supply air volume:

$$VSA = VSA \text{ (perimeter)} + VSA \text{ (interior)}$$

During unoccupied periods, the system is assumed to be off. Therefore, VSA = 0.

Column D. Compute mixed air dry-bulb temperature (MADB, Col. D) from air energy balance on the outside air mixing box.

$$MADB = RADB + (VO/VSA)(OADB - RADB)$$

where

RADB = return air dry-bulb temperature; assumed constant at 75°F during occupied periods, 55°F during unoccupied periods

VO = outside air volume; assumed constant 0.20 cfm/ft^2 during occupied hours, 0 otherwise

VSA = total supply air volume (Col. C)

OADB = outside air dry-bulb temperature; bin value (Col. A), same as t above *thus*

$$MADB = 75 + (0.20/VSA)(OADB - 75) \text{ (occupied)}$$
$$= 55 \text{ (unoccupied)}$$

Column E. Compute cooling coil leaving air dry-bulb temperature CCLADB as the required supply air temperature minus the fan temperature rise (assumed constant at 2°F).

$$CCLADB = 55 - 2 = 53°F$$

More accurate models would calculate the fan temperature rise at part-load conditions.

Column F. Compute preheat coil leaving air dry-bulb temperature. If MADB less than CCLADB, set PHLADB = CCLADB; otherwise set PHLADB = MADB.

Column G. Compute return air humidity ratio RAW, which is essentially the cooling coil leaving air humidity ratio plus any latent heat gain from the spaces.

$$RAW = QL/(4840 \text{ VSA}) + CCLAW$$

where

QL = total latent heat gain; assumed constant at 2.55 Btu/h·ft^2 because of people

CCLAW = cooling coil leaving air humidity ratio; (Col. I) assumed constant at 0.0077 lb$_w$/lb$_a$ when the coil is wet for the coil model used in this example. When the coil is not operating, the humidity ratio will not have this value, and an iteration is performed to determine the true system humidity ratios.

Thus $$RAW = 0.0077 + 0.00053/VSA$$

Column H. Compute mixed air humidity ratio MAW from an energy balance on the outside air mixing box.

$$MAW = RAW + (VO/VSA)(OAW - RAW)$$

Column I. Cooling coil leaving air humidity ratio discussed under G.

Column J. Compute preheat coil load.

$$QPHC = (1.1 \text{ VSA})(PHLADB - MADB)$$

Column K. Compute cooling coil sensible load.

$$QCS = (1.1 \text{ VSA})(MADB - CCLADB)$$

Column L. Compute cooling coil latent load.

$$QCL = (4840 \text{ VSA})(MAW - CCLAW)$$

Column M. Compute cooling coil total load.

$$QCT = QCS + QCL$$

Column N. Compute baseboard heating load QHC. If a heating load exists (the net load on the space is negative), the supply air volume is set to minimum value. Therefore, baseboard heating must account for the overcooling because of this supply air as well as the sensible load on the space.

Thus, for the *occupied period*:

Perimeter zone

$$QHC = -(0.3406 \, t - 14.1553) + 1.1 \times 0.0904 \times 20$$
$$= 16.14 - 0.3406 \, t$$

Interior zone

$$QHC = -(0.0502 \, t + 3.74) + 1.1 \times 0.1096 \times 20$$
$$= -1.3228 - 0.0502 \, t$$

Total

$$QHC = QHC \text{ (perimeter)} + QHC \text{(interior)}$$

For the unoccupied period during which $t_i = 55°F$, the supply air is zero. For the perimeter zone, an outside air load term is added to account for infiltration (0.05 cfm/ft^2):

Perimeter zone

$$QHC = 0.3406 \, t - 13.6829 + 1.1 \times 0.05 \times (t - t_o)$$
$$= 0.3956 \, t - 16.7079$$

Interior zone

$$QHC = 0.0502 \, t - 2.015$$

If, for either zone, QHC < 0, set QHC = 0.

Table 6 summarizes the coil load calculation for both occupied and unoccupied periods. Annual energy consumption can be found from the loads in Table 6 by applying fan, chiller, and boiler part-load characteristics. An example of the calculation for this problem is contained on page 28.17 in the 1989 ASHRAE *Handbook—Fundamentals* (I-P edition).

OTHER ENERGY CONSIDERATIONS

Energy Conservation Effects

Studies of the energy conservation effects from wintertime thermostat setback, summertime thermostat setup, and night setback for electric, gas, or oil furnaces and heat pumps are listed in the bibliography. Predicted energy savings up to 50% for combined setbacks are reported and are shown to be a strong function of outdoor design temperatures, number of degree days, residential construction, and type of heating unit. Typical annual heating energy savings for a 4°F winter setback range between 14 and 20%; typical annual heating energy savings for a 10°F winter nighttime setback for an 8 h/day range, between 7 and 12%. Typical seasonal energy savings for residential cooling units for a 3°F temperature setup in summer are between 15 and 30%.

Shavit (1975) reports savings from increased relative humidity during winter operation. Savings that might be realized from electric ignition and/or automatic vent dampers on gas and oil furnaces are reported by AGA (1977), Zabinski and Loverme (1974), and Macriss and Elkins (1976). Annual energy savings reported for a combination of an intermittent ignition device and an automatic vent-damper range between −16 and +40%, with a mean reduction in gas consumption of 10% (AGA 1977).

Threlkeld and Jordan (1954), Yellott and Ewing (1976), Beckman *et al.* (1977), and Yellott (1972) report savings during the cooling season from applying sunscreens and/or attic ventilation. Sunscreens yield cooling energy savings ranging from negligible to 25% for typical residential applications, while attic ventilation has resulted in cooling energy savings ranging from negligible to 9% for specific applications.

Thermal Mass Effects

Burch *et al.* (1984a, 1984b), Robertson and Christian (1985), Childs *et al.* (1983), and Arumi-Noe (1984) in studies in Maryland and New Mexico show that the massiveness of an exterior wall reduces the heating and cooling requirements of buildings, provided the room air temperature floats above the thermostat set point in heating and below the thermostat set point in the cooling season. Robertson and Christian (1985) found 2 to 4% reductions. The floating temperatures occur more frequently in mild climates and during the spring, summer, and fall. In periods where the room air temperature did not float (during the winter), steady-state analysis accurately calculated the net energy transfer in the 20 by 20-ft test buildings (U-values and outdoor temperatures averaged over several days) during the winter. Steady-state analysis underestimated the amount of energy required by the lightweight buildings during spring, summer, and fall. The energy requirement of the heavyweight test buildings could be more accurately quantified by steady-state analysis during all four seasons.

Burch *et al.* (1986) modeled the annual space heating and cooling loads of a typical house operated with fixed thermostat settings and found interior mass features (partition walls and interior furnishings) had only a small effect on the annual space heating and cooling loads, except when the house operated near its balance-point temperature.

Buildings with low balance-point temperature (with large solar or internal gains) experience more room air temperature floating in winter. This sensitivity of energy use to the mass of building components could cause lightweight buildings to use more energy than that predicted by a steady-state analysis for the winter.

Location of the mass in relation to the conditioned space and the insulation has a large effect on the deviation between measured energy use and steady-state analysis. In heavyweight buildings where mass was outside the insulation, the measured energy use closely matched that predicted by steady-state analysis. However, when the insulation was outside the mass, the energy use was greater than that predicted by steady-state analysis. Although Christian and Downing (1986) developed simplified procedures to estimate the effect of thermal mass in reducing heating and cooling loads in various climates, more research is needed to develop generalized thermal mass credits for load reduction.

CORRELATION METHODS

Even with simplified multiple-measure methods, manual estimation of the energy requirements of intermediate (20,000 to 50,000 ft^2) or large (> 50,000 ft^2) commercial or institutional buildings can be complex and time-consuming. Properly using the bin method for large buildings normally requires subdividing the building into several thermal zones and calculating loads for each. The loads can then be used to perform a system simulation to obtain equipment loads, which in turn are used with the equipment characteristics to obtain the equipment energy requirements. Since these calculations must be performed at each temperature bin point and the calculations may have to be repeated for alternative designs, the total effort required to perform a comparative energy analysis may be great.

One way of simplifying the analysis is to develop correlations that relate energy requirements to various input parameters. Correlations are typically developed by performing many detailed analyses and using statistical means to reduce the results to a simple form. That form is typically an equation, easy to use in a calculator or small computer program, or a graph that is convenient to use manually and provides quick insight into the basic input/output relationships of the energy requirements.

Sud *et al.* (1979) indicate that graphical procedures simplify annual energy requirements estimates for intermediate and large commercial and institutional buildings. The information required can provide relatively reliable results without laborious manual calculations. This method is based on three sets of graphs that correspond to the three elements of any energy estimating method: (1) space load, (2) secondary equipment load, and (3) primary equipment energy requirements. A separate set of graphs is required for each weather zone, building type, and size range and HVAC system type. Parameters used in the graphs include: (1) high or low minimum ventilation rates for three types of economizer cycles, (2) high or low interior loads with combinations of high or low thermostat settings, and (3) continuous or setback HVAC system operation. Plant operation, and thus equipment energy requirements, are characterized by an annual equivalent coefficient of performance for generic heating and cooling equipment. The sum of HVAC equipment, auxiliary equipment and lighting, and internal load energy use determines the total annual building energy use.

Two other correlation methods, *f*-Chart for active solar systems and SLR for passive space heating, are described later. Although the name indicates a graphic method, note that *f*-Chart also provides equations for the correlations.

The accuracy of correlation methods depends on the size and accuracy of the database and the appropriateness of the statistical means used to develop the correlation. A database generated from measured data can lead to accurate correlations (Lachal 1992). The key to the proper use of a correlation is that the case being studied matches the cases used in developing the database. Inputs to the correlation (the independent variables) indicate the factors that are considered to have significant impact on energy consumption. The use of a correlation is invalid when either an input parameter is used beyond its valid range (corresponding to extrapolation rather than interpolation), or when some important feature of the building/system is not included in the available inputs to the correlation. When designers wish to analyze such features, they must find another method of analysis. The most flexible methods are detailed dynamic simulations.

DYNAMIC METHODS

Because of the wide-ranging and constantly changing internal and external factors that determine thermal loads on commercial and industrial buildings, frequent evaluation is needed to obtain reasonably accurate estimates of annual energy consumption. High peak internal heat loads in many commercial buildings necessitate a review of these load patterns hourly and for each different type of day during the year.

To apply these methods, a mathematical model of the building and its energy systems must represent (1) the thermal behavior of the building structure (the loads model), (2) the thermodynamic behavior of the air-conditioning delivery system (the secondary systems model), and (3) a mathematical relationship for load versus energy requirements of the primary energy conversion equipment (the plant model). Usually, each model is formulated so that input quantities allow calculation of output quantities. For example, weather and internal load information are inputs to the loads model, allowing calculation of sensible loads on the occupied spaces in the building. Most often, several thermally different zones within the building must be modeled rather than one representing the building as a whole.

The secondary system model is comprised of conservation equations (mass and energy) for each component of the actual system, *e.g.*, heating coils, cooling and dehumidifying coils, mixing boxes, and fans. These can usually be arranged to allow calculation of the required rate of energy delivery to the heating and cooling coils for a given space sensible load, previously calculated by the loads model. In an actual simulation, there may be several secondary system models, each representing a physical delivery system serving a group of zones.

The plant model relates the required input energy rate, *e.g.*, gas and/or electrical power, to the required energy rates of several secondary systems. Typically, a central plant model includes such relationships for chillers, boilers, cooling towers, and so forth.

Figure 8 shows the basic function of each major model element on an input and/or output basis. The control interaction path shows that capacity limits and/or control characteristics of the air-conditioning system can affect the space load. Therefore, any

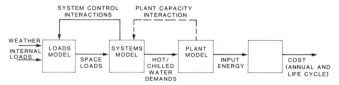

Fig. 8 Flow Diagram for Calculating Hourly Heating Cooling Loads Using ASHRAE Algorithms

variation in space temperature allowed by the system causes the space sensible load to vary. Capacity limits and control characteristics of the central plant can cause variation in secondary system performance, thus affecting loads as shown by the dashed lines in Figure 8.

Figure 8 shows an economic model that calculates energy costs (and sometimes capital costs) based on the estimated required input energy. Thus, the simulation model calculates energy usage and cost for any given input weather and internal loads. By applying this model (*i.e.*, determining output for given inputs) at each hour (or other suitable interval), the hour-by-hour energy consumption and cost can be determined. Maintaining running sums of these quantities yields monthly or annual energy usage and costs.

These models only compare design alternatives; a large number of uncontrolled and unknown factors usually precludes the use of such models for accurate prediction of utility bills. The bibliography lists several models. Generally, the loads model tends to be the most complex and time-consuming; the central plan model, the least.

CALCULATING INSTANTANEOUS SPACE SENSIBLE LOAD

Calculating instantaneous space sensible load is a key step in any building energy simulation. The *heat balance method* and the *weighting factor method* are two widely used methods for these calculations.

The *instantaneous space sensible load* is the rate of heat flow into the space air mass. This quantity, sometimes called the cooling load, differs from heat gain in that the latter usually contains a radiative component that passes entirely through the air mass and is absorbed by other bounding surfaces. Instantaneous space sensible load is entirely convective; even loads from internal equipment, lights, and occupants enter the air by convection from the surface of such objects or by convection from room surfaces that have absorbed the radiant component of energy emitted from these sources. However, some adjustment must be made when radiant cooling and heating systems are evaluated because a portion of the space load is offset directly by radiant transfer without convective transfer to the air mass.

For equilibrium conditions, the instantaneous space sensible load must be matched by the heat removal rate of the conditioning equipment. Any imbalance in these rates changes the energy stored in the air mass. Customarily, however, the thermal mass (heat capacity) of the air itself is ignored in the analysis, so that the air is always assumed to be in thermal equilibrium. Under these assumptions, the instantaneous space sensible load and the rate of heat removal are equal in magnitude and opposite in sign.

In sensible load calculations, the cooling situation is often the reference frame. If there is heat *loss*, only the algebraic signs change. Therefore, discussion of the heating situation is omitted when developing mathematical models that work equally well for either situation.

Weighting and response factors and conduction transfer functions all stem from the same mathematical treatment (discreet convolution), but each represent a different concept. *Weighting factors*, also called room response factors, are used in the weighting factor method to calculate space instantaneous sensible load components from corresponding heat gain components. *Response factor* and *conduction transfer functions*, on the other hand, refer to conductive heat transfer (*e.g.*, through walls and floors) and are used to calculate the transmission component of heat gain (see Chapter 26).

Note that both the weighting factor method and the heat balance method conduction transfer functions (or their equiva-

lents) are used to calculate transmission heat gain or loss. The principal difference is in the methods used to calculate the subsequent internal heat transfers to the room. Experience with both methods has indicated largely the same results, provided the weighting factors are determined for the specific building under analysis.

Heat Balance Method

The heat balance method for calculating net space sensible loads is the more fundamental of the two methods. Its development relies on the first law of thermodynamics (conservation of energy) and the principles of matrix algebra. Since it requires fewer assumptions than the weighting factor method, it is also more flexible. However, the heat balance method requires more calculations at each point in the simulation process, using more computer time. The weighting factors used in the weighting factor method are determined by the heat balance equations in a form similar to the one presented here. And, although not necessary, linearization is commonly used to simplify the radiative transfer formulation.

The heat balance method allows the net instantaneous sensible heating and/or cooling load to be calculated on the space air mass. Generally, a heat balance equation is written for each enclosing surface, plus one for room air. This set of equations can then be solved for the unknown surface and air temperatures. Once these temperatures are known, they can be used to calculate the convective heat flow to or from the space air mass.

To write heat balance equations, consider a space within the building under analysis as an enclosure bounded by a number of discrete surfaces (walls, floors, windows, and ceiling). At any time θ, each of these surfaces is assumed to be at some uniform temperature $t_{i,\theta}$. Also, the space air mass is assumed to be of uniform temperature $t_{a,\theta}$. At any plain boundary, the flux entering the boundary must equal the flux leaving the boundary. Thus, at the inside surface of any room wall, the heat flow into the surface—because of convection from room air, radiation from interior sources such as lights and people, and the net radiant interchange between the wall and all other surfaces in the room—is exactly balanced by the conductive flux leaving the surface to penetrate the solid.

When radiant cooling and heating systems are evaluated, the radiant source should be identified as a room surface. The calculation procedure considers the radiant source in the heat balance analysis. Therefore, this method is preferred over the weighting factor method evaluating radiant systems.

Similarly, for the outside surface of an exterior wall, the conductive flux leaving the surface to penetrate the solid toward the room is exactly balanced by absorbed solar radiation, net longwave radiant flux from the surroundings, and convective flux from the outdoor air. A heat balance on the room air volume requires the air-conditioning system to remove the net heat added to the volume by *convection* from interior surfaces and interior heat sources (lights, people) and by mass transfer due to infiltration.

Thus, at the *ith* surface at time θ, the energy (per unit area) balance, is

$$q_{i,\theta} = h_i(t_{a,\theta} - t_{i,\theta}) + \sum_{k=1}^{n} g_{i,k}(t_{a,\theta} - t_{i,\theta}) + R_i \quad (27)$$

where the first term represents heat gain due to convection from the room air, the summation represents radiative interchange (longwave), and the last term R_i represents all other radiant energy absorbed by the surface. The last term includes shortwave radiation from lights, window solar gain, and the radiant component of occupant and equipment loads.

The left side of Equation (27) $q_{i,\theta}$ represents the *net* transfer to the surface from the air and surroundings. It must be exactly matched by the heat conducted into the solid surface. The latter can be represented in terms of historical values of surface temperatures and heat flux using conduction transfer functions (ASHRAE 1976). [Response factor and conduction transfer functions are *closely* related, elegant solutions to the transient one-dimensional, multilayered-slab heated conduction problem. Mitalas and Stephenson (1967), Mitalas (1968), and Kusuda (1969) give several theoretically equivalent solutions, and at least two forms appear in Chapters 22 and 26.]

$$q_{i,t} = \sum_{j=0}^{m} X_j t_{i,\theta-j} - \sum_{j=0}^{m} Y_j t_{oi,\theta-j} + CR_i q_{i,\theta-j}$$

$$= X_o t_{i,\theta} + \sum_{j=1}^{m} X_j t_{i,\theta-j} - \sum_{j=1}^{m} Y_j t_{oi,\theta-j} + CR_i q_{i,\theta-j} \quad (28)$$

By substituting Equation (28) into Equation (27), $q_{i,t}$ can be eliminated, and after rearrangement

$$\left(X_o + h_i + \sum_{k=1}^{m} g_{i,k} \right) t_{i,\theta} - \sum_{k=1}^{m} g_{i,k} t_{k,\theta} - h_i t_{a,\theta} =$$

$$- \sum_{j=1}^{m} X_j t_{i,\theta-j} + \sum_{j=0}^{m} Y_j t_{oi,\theta-j} + CR_i q_{i,\theta-1} + R_i \quad (29)$$

In addition to the n equation forms of Equation (29), a heat balance equation also can be written for the space air. Assuming negligible heat capacity of this air relative to the more massive building elements, the sum of all heat flow to it must be zero, yielding

$$\sum_{j=1}^{n} S_j h_j (t_{j,\theta} - t_{a,\theta}) + m_1 c_p (t_{oa,\theta} - t_{a,\theta})$$

$$+ QI_\theta + QS_\theta = 0 \quad (30)$$

Here, the summation represents convective transfer from the enclosing surfaces, second-term infiltration QI_θ, convective transfer from internal objects, and QS_θ, the rate of energy addition and/or removal by the space-conditioning equipment. This equation can be rearranged to give

$$\sum_{j=1}^{n} S_j h_j (t_{j,\theta} - t_{a,\theta}) - m_1 c_p t_{a,\theta} =$$

$$m_1 c_p t_{oa,\theta} - QI_\theta - QS_\theta \quad (31)$$

Equations (29) and (31) together represent a set of $n + 1$ equations in $n + 1$ unknown temperatures. The equations are arranged so that all unknowns are on the left, and all known quantities are on the right. This facilitates rewriting the problem to be solved in vector-matrix notation as

$$A t_\theta = B \quad (32)$$

where A is an $n + 1$ by $n + 1$ matrix, and B is a $(n + 1)$ vector. By convention, the first n elements of the t_θ vector are the surface temperatures, while the $n + 1$ element is the air temperature.

The elements of A and B can be determined from the expanded equations given previously. The first n diagonal elements of A are

$$a_{ii} = X_o + h_i + \sum_{k=1}^{n} q_{i,k}; \; i = 1, n \quad (33)$$

while the last diagonal element is

$$a_{n+1, n+1} = - \left(\sum_{k=1}^{n} S_k h_k + m_1 c_p \right) \quad (34)$$

and off-diagonal elements are

$$a_{i,j} = - g_{i,j}; \; i, j = 1, n \quad (35)$$

$$a_{i,n+1} = - h_i; \; i = 1, n \quad (36)$$

$$a_{n+1,j} = S_j h_j; \; j = 1, n \quad (37)$$

The elements of B are the terms on the right of Equations (29) and (31).

Through these equations, the heat transfer at any surface is coupled to heat transfer at all other surfaces and to heat transfer to or from the room air. The heat transfer at an inside wall surface, for example, is coupled to the heat transfer at the outside surface through conduction transfer functions for the wall. It is coupled to the heat transfer at all other inside surfaces through equations governing intersurface radiant exchange, and it is coupled to the room air volume through the inside surface convection coefficient. The radiant and convective fluxes from solar effects and the effects of interior energy sources can be calculated independently, so that an equation for the flux at any surface in terms of the surface temperature, the temperature of all other surfaces, and the room air temperature can be written (assume that all *past* temperatures and fluxes are known).

The heat balance method requires that Equation (28) be solved for surface and air temperatures at each point, allowing subsequent calculation of sensible load. The solution can be derived by various mathematical methods, some more efficient than others. As discussed by Sowell and Walton (1980), computational speed can be increased by using the special properties of the coefficient matrix A.

Regardless of the solution method, the procedure is to compute updated values of the right vector B and then solve for t_θ. Note that B involves (1) historical values of all surface temperatures, (2) historical values of the outside surface temperatures, (3) the previous value of heat flux at each surface, (4) outside air temperature, (5) the absorbed shortwave radiant flux at each surface, (6) the internal load, and (7) the supply heating and/or cooling effect.

Historical data are developed and retained as the solution progresses, starting with some assumed initial conditions and past performance. Internal loads are usually calculated from schedules of occupancy, lighting, and equipment usage. The radiant absorption at each surface is usually approximated by uniformly distributing window solar load and known radiant components of lighting, equipment, and occupants to each unit area of walls, ceiling, and floor.

Outside surface temperatures can be found by writing a heat balance at each outside surface. This produces an equation similar to Equation (27), equating heat flux into the solid material to the algebraic sum of convection from outdoor air, longwave radiant absorption from the sky and surroundings, and shortwave solar absorption. The resulting equations, one for each outside surface, can be rearranged to give an expression for the outside surface temperatures, shown in Equation (38).

$$t_{oi,\theta} = (q_{si,\theta} + q_{li,\theta} + h_o t_{oa,\theta} + \sum_{j=0}^{m} Y_i t_{i,\theta-j}$$

$$- \sum_{j=1}^{m} Z_j t_{oi,\theta-j} + CR q_{oi,\theta-1})/(h_o + Z_o) \quad (38)$$

Here, $q_{si,t}$ and $q_{li,t}$ represent the net short- and longwave radiant flux at the ith outside surface at time t, while h_o is the outside film coefficient. It is important to note that the right side of Equation (38) contains the *inside* surface temperature at time θ; thus this set of equations must be solved simultaneously with Equation (32). However, since this would double the size of the overall heat balance problem, significantly increasing computer memory requirements and solution time, a simplified assumption is often made. For example, some algorithms simply use the outside surface temperature from the *previous* time step when evaluating the right side of Equation (28). Resulting errors are thought to be small for thermally massive elements. For lightweight walls, however, the equations should probably be solved simultaneously.

The supply heating and/or cooling term in Equation (31) QS_θ presents a similar dilemma. Because of the thermostat and the nature of the air-conditioning process, QS_θ depends on room temperature $t_{a\theta}$, which is one of the unknowns in Equation (32). This requires some type of simultaneous solution procedure for Equation (32) and the space control characteristic, e.g., Figure 9. Iteration is usually employed to find a space temperature and supply heating and/or cooling amount that simultaneously satisfies the heat balance equations and the control characteristics.

In principle, the method described above extends directly to multiple spaces. An equation such as Equation (32) can be written for each space; this recognizes that the outside of partitions are in fact the inside surfaces of adjacent spaces. In practice, however, the size of the coefficient array required for solving the simultaneous equations becomes prohibitively large, and the solution time excessive. For this reason, many programs solve only one space at a time and assume that the adjacent space temperatures are either the same as the space in question or some assigned, constant value. Other approaches may remove this limitation (Walton 1980).

Weighting Factor Method

The weighting factor method of calculating instantaneous space sensible load is a simple but flexible technique that accounts for

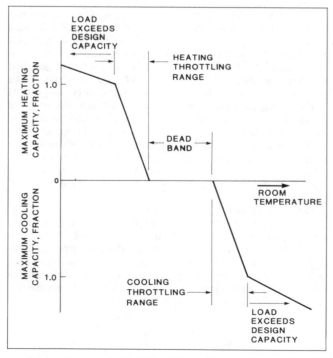

Fig. 9 Thermostat Model of Proportional Control with Deadband and Dual Throttling Ranges

the important parameters affecting building energy flow. It represents a compromise between simpler methods, such as a steady-state calculation that ignores the ability of building mass to store energy, and more complex methods, such as complete energy balance calculations. With this method, space heat gains at constant space temperature are determined from a physical description of the building, the ambient weather conditions, and internal load profiles. Along with the characteristics and availability of heating and cooling systems for the building, these space heat gains are used to calculate air temperatures and heat extraction rates. This discussion is in terms of heat gains, cooling loads, and heat extraction rates. Heat losses, heating loads, and heat addition rates are merely different terms for the same quantities when the direction of heat flow is reversed, *i.e.*, when the quantity has the opposite algebraic sign.

The weighting factors represent Z-transfer functions. The Z-transform is a method for solving differential equations that is used with discrete data. Two groups of weighting factors are used: heat gain and air temperature. Heat gain weighting factors represent transfer functions that relate space cooling load to instantaneous heat gains. A set of weighting factors is calculated for each group of heat sources that differ significantly in (1) the relative amounts of energy appearing as convection to the air versus radiation, and (2) in the distribution of radiant energy intensities on different surfaces.

Air temperature weighting factors represent a transfer function that relates room air temperature to the net energy load of the room. The weighting factors for a particular heat source are determined by introducing a unit pulse of energy from that source into the room's network. The network is a set of equations (described in the previous section) that represents a heat balance for the room. At each time step (1-h intervals), including the initial introduction, the energy flow to the room air represents the amount of the pulse that becomes a cooling load. Thus a long sequence of cooling loads can be generated from which the weighting factors are calculated. Similarly, a unit pulse change in room air temperature can be used to produce a sequence of cooling loads.

A two-step process is used to determine the air temperature and heat extraction rate of a room or building zone for a given set of conditions. First, the room's air temperature is assumed to be fixed at some reference value. This reference temperature is usually chosen as the average air temperature expected for the room over the simulation period. Instantaneous heat gains are calculated based on this constant air temperature. Various types of heat gains are considered. Some, such as solar energy entering through windows or energy from lighting, people, or equipment, are independent of the reference temperature. Others, such as conduction through walls, depend directly on the reference temperature.

A space sensible cooling load for the room, defined as the rate at which energy must be removed from the room to maintain the reference value of the air temperature, is calculated for each type of instantaneous heat gain. The cooling load generally differs from the instantaneous heat gain because some of the energy from the heat gain can be absorbed by walls or furniture and stored for later release to the air. At hour t, the calculation employs present and past values of the instantaneous heat gain (q_θ, $q_{\theta-1}$), past values of the cooling load ($Q_{\theta-1}$, $Q_{\theta-2}, \ldots$), and the heat gain weighting factors (v_0, v_1, $v_2, \ldots w_1$, $w_2, \ldots$) for the type of heat gain under consideration. Thus, for each type of heat gain q_θ, the cooling load Q_θ is calculated as:

$$Q_\theta = v_0 q_\theta + v_1 q_{\theta-1} + \ldots - w_1 Q_{\theta-1} - w_2 Q_{\theta-2} \ldots \quad (39)$$

The heat gain weighting factors incorporate information about the room and the heat gain source. They are a set of parameters that quantitatively determine how much of the energy entering a room is stored and how rapidly the stored energy is released during later hours. Mathematically, the weighting factors are parameters in a Z-transfer function relating the heat gain to the cooling load (ASHRAE 1976).

These weighting factors differ for different heat gain sources because the relative amounts of convective and radiative energy leaving various sources differ, and because the distribution of radiative energy can differ. The heat gain weighting factors also differ for different rooms because room construction influences the amount of incoming energy stored by walls or furniture and the rate at which it is released. After the first step, the cooling loads from various heat gains are added to give a total cooling load for the room.

In the second step, the total cooling load is used—along with information on the heating, ventilating, and air-conditioning system attached to the room, and a set of *air temperature weighting factors*—to calculate the actual heat extraction rate and air temperature. The actual heat extraction rate differs from the cooling load because, in practice, the air temperature can vary from the reference value used to calculate the cooling load or because of HVAC system characteristics. The deviation of the air temperature from the reference value at hour θ, t_θ, is calculated as:

$$t_\theta = 1/g_o [(Q_\theta - ER_\theta) + P_1 (Q_{\theta-1} - ER_{\theta-1}) +$$
$$P_2 (Q_{\theta-2} - ER_{\theta-2}) + \ldots - g_1 t_{\theta-1} - g_2 t_{\theta-2} - \ldots] \quad (40)$$

where ER_θ is the energy removal rate of the HVAC system at hour θ, and $g_0, g_1, g_2, \ldots, P_1, P_2, \ldots$ are the air temperature weighting factors, which incorporate information about the room, particularly the thermal coupling between the air and the storage capacity of the massive elements.

Tables of values of weighting factors for typical building rooms are presented in Chapter 26. One of the three groups of weighting factors, for light, medium, and heavy construction rooms, can be used to approximate the behavior of any room. Some automated simulation techniques allow weighting factors to be calculated specifically for the building under consideration (LBL 1981). This option improves the accuracy of the calculated results, particularly for a building with an unconventional design.

Two general assumptions are made in the weighting factor method. First, the processes modeled are linear. This assumption is necessary because heat gains from various sources are calculated independently and summed to obtain the overall result (*i.e.*, the superposition principle is employed). Therefore, nonlinear processes, such as radiation or natural convection, must be approximated linearly. This assumption does not represent a significant limitation, because these processes can be linearly approximated with sufficient accuracy for most calculations. The second assumption is that system properties influencing the weighting factors are constant, *i.e.*, they are not functions of time. This assumption is necessary because only one set of weighting factors is used during the entire simulation period. This assumption can limit the use of weighting factors in situations where important room properties vary during the calculation. Two examples are the distribution of solar radiation incident on the interior walls of a room that can vary hourly, and inside surface heat transfer coefficients.

When the weighting factor method is used, a combined radiative-convective heat transfer coefficient is used as the inside surface heat transfer coefficient. This value is assumed constant even though in a real room the radiant heat transferred from a surface depends on the temperature of other room surfaces (not on the room air temperature) and the combined heat transfer coefficient is not constant. Under these circumstances, an average value of the property must be used to determine the weighting factors. Cumali *et al.* (1979) have investigated extensions to the weighting factor method to eliminate this limitation.

SIMULATING SECONDARY SYSTEMS

Secondary system models mathematically relate the rate of heating and/or cooling energy delivery, *e.g.*, hot and/or chilled water, to space sensible loads. Usually, such models are formulated to receive the space sensible loads as input and allow calculation of heating and/or cooling rates at the coils (also see the section Overall Modeling Strategies).

Fundamental Relations for Moist Air

The underlying principles for secondary system models are primarily those for mass and energy balances on moist air described in Chapter 6. For example, Equation (45) from Chapter 6 gives the relationship between space sensible heat gain, space moisture addition, supply airflow rate, and air enthalpy change from supply to exhaust conditions. Similarly, Equation (43) then relates input and output enthalpies and flow rates of mixing boxes, and Equation (36) the heating requirements at a coil to its inlet and outlet enthalpies. Equations (37) and (38) relate air mass flows, enthalpies, and moisture-removal rates for cooling-dehumidifying coils. [These particular equations must be augmented, however, with some type of model representing moisture removal and sensible cooling capability of the physical cooling coil, as discussed in Chapter 5, Equations (65) through (67).]

Since secondary systems are composed of mixing boxes, cooling coils, heating coils, and fans interconnected in various ways, mathematical models can be developed by assembling the applicable equations. For example, all mass and energy balance equations for a variable air volume terminal reheat system can be developed into an algorithm in which airflow rate, cooling coil energy rate, and reheating energy rate are calculated, given a particular required space sensible load on one or more service zones.

Models for Fans

A complete secondary system also requires a mathematical model for supply and return fans for two reasons: the required fan power must be calculated, since fan energy usage is an important factor and most, if not all, of the fan power is ultimately degraded to heat in the airstream, adding to cooling coil loads or reducing heating coil loads, depending on the fan location.

Computation of fan power is usually based on a characteristic curve, giving fraction of rated input power versus fraction of rated volume. This curve is determined principally by the method used to control air volume (see Figure 10). These curves can be determined experimentally or obtained from the manufacturer, although those shown in the figure are often accurate enough for energy analysis. Rated power can be calculated from rated volumetric flow rate, pressure rise, and efficiency as

$$\text{Fan horsepower} = \frac{Q \Delta p}{6370 \, \eta_F \, \eta_M} \quad (41)$$

where

Q = air volumetric flow, cfm
Δp = fan total pressure rise, in. of water
η_F = fan efficiency
η_M = motor efficiency

Most models assume that some fraction of the fan motor power goes directly into heating (enthalpy rise) of the air as it passes through the fan. The fraction is 1.0 if the fan motor is mounted in the airstream, and 1.0 minus motor efficiency η_m if mounted outside. The increase in air enthalpy occurs immediately downstream of the fan.

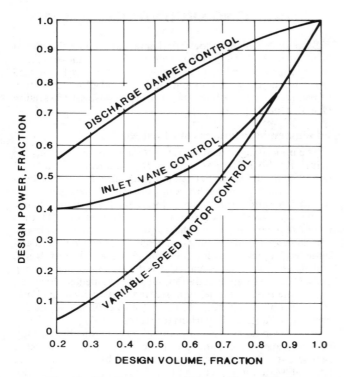

Fig. 10   Fan Power versus Volume Characteristics

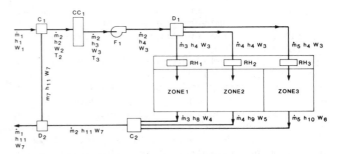

**Fig. 11   Schematic of Variable Air Volume System
Serving Three Zones**

Modeling of System Controls

From a mathematical viewpoint, system control devices represent equations that must be satisfied at each point during the simulation and equations that model other system components. For example, the room thermostat can be represented as a function relating heating and cooling delivery rate to space temperature, as shown in Figure 9. Similarly, cooling coil reset controls can be modeled as a relationship between outside or zone temperature and coil discharge temperature. An accurate secondary system model must ensure that all controls are properly represented and that the governing equations are satisfied at each simulation time step. This usually creates a need for iteration or, alternately, for use of values from an earlier solution point.

The controls on space temperature affect the interaction between loads calculation and secondary system simulation (see Figure 8). A realistic model might require a dead band in space temperature (Figure 9), in which no heating or cooling is called for; within this range, the true space sensible load is zero, and the true space temperature must be adjusted accordingly. If the thermostat has proportional control between zero and full capacity, the space temperature will rise in proportion to the load during cooling and fall similarly during heating. Capacity to heat or cool also varies with space temperature after the control device has reached its maximum because capacity is proportional to the *difference* between supply and space temperatures. Failure to properly model these phenomena results in overestimating required energy.

Integration of System Models

To demonstrate the approach to system modeling, a variable air volume (VAV) system serving three zones is shown in Figure 11. For simplicity, the following assumptions are made: (1) the space temperatures, sensible loads, and latent loads have been previously determined by the heat balance method; (2) the loads are within the capacities of the respective heating and/or cooling equipment; (3) the fan is a variable volume unit with the fraction of nominal power expressible by a second-order polynomial in fraction of nominal volume; (4) the cooling coil discharge temperature is scheduled linearly with the temperature of the warmest zone; and (5) the outside air quantity is fixed.

In this example, the physical quantities used are mass flow rate, enthalpy, and absolute humidity. Temperature, volumetric flow rate, and other measures of air moisture are viewed as auxiliary quantities that can be computed using the psychrometric relationships from Chapter 6. However, many computer programs are based on volume flow rate, temperature, and absolute humidity, treating enthalpy and mass flow rate as auxiliary variables. The latter approach uses more familiar quantities, but tends to make the conservation equations more complicated. For example, at a mixing box, mass, not volume, is conserved. However, if density changes and the effect of moisture on enthalpy are ignored, the conservation equations can be written in terms of volumetric flow and temperature. Many analysts take this approach and accept the small attendant errors.

The schematic diagram is first labeled for mass flow rates, enthalpies, humidity ratios, and temperatures. In doing this, the number of unique variables can be minimized by observing that enthalpy and humidity do not change across a distributor, humidity does not change across the fan, and so on. Heat *added* to the space is positive and heat *removed* from the space is negative. A data table, showing the quantities assumed, is helpful.

These preliminary steps are followed by constructing a calculation sequence or algorithm that shows exactly how the input data are transformed into the output data. Usually, the best place to start is at a point where the most information is available, in this case, at the outside air mixing box M_1. Unknown quantities are simply assumed. Mass, energy, and moisture balances are then used to calculate subsequent quantities around the flow circuit. Occasionally one or more quantities must be assumed to proceed; then the calculations are repeated until the calculated quantities agree with the assumed quantities.

Algorithm for VAV System Simulation

1. Assume m_7, W_7, and h_{11}.
2. Compute $m_2 = m_1 + m_7$
$$h_2 = (m_1 h_1 + m_7 h_{11})/m_2$$
$$W_2 = (m_1 W_1 + m_7 W_7)/m_2$$
3. Compute T_2 using Equation (29), Chapter 6.
4. Determine cooling coil leaving conditions:
  (a) Find maximum zone temperature.
  (b) From cooling coil control schedule, determine desired leaving temperature T_{cc}.
  (c) If $T_{cc} < T_2$
Then: set $T_3 = T_{cc}$
Else: set $T_3 = T_2$
  (d) Compute h_3 and W_3, using h_2, W_2, T_3, and a suitable model for coil moisture removal, with Equation (29), Chapter 6.

5. Compute fan leaving conditions h_4.
   (a) Fan power: $P_F = P_{Fn} F_F(m_2/m_{2n})$,
where n represents nominal quantities, and F_F is the fan characteristic, *e.g.*, Figure 10. Note that computing volume ratio would be more accurate.
   (b) Leaving enthalpy: $h_4 = h_3 + F_{Fa}P_F/m_2$,
where F_{Fa} is the fraction of fan power entering the airstream.

6. Determine air mass flow rate, supply enthalpy, return humidity ratio, and reheat coil energy rate for each zone. Repeat (a) through (f) for each zone:
   (a) Assume zone mass flow m_z (minimum or last value)
   (b) Compute zone humidity $W_z = W_3 + Q_{zl}/(m_z h_{zwv})$, where h_{zwv} is the enthalpy of vaporization at temperature T_z
   (c) Compute zone enthalpy h_z from Equation (29), Chapter 6
   (d) If $Q_{zs} + Q_{zl} \ll \underline{\quad} m_{z,\,min} (h_z - h_4)$
      Then: $m_2 = m_{z,\,min}$
              $h_{z,\,sup} = h_z - (Q_{zs} + Q_{zl})/m_z$
      Else: $h_{z,\,sup} = h_4$
              $m_z = (Q_{zs} + Q_{zl})/(h_z - h_{z,\,sup})$
   (e) Compute reheat required
              $Q_{RH,\,z} = m_z (h_{z,\,sup} - h_4)$
   (f) If m_z differs significantly from that used in 6a, repeat from 6b

7. Compute $m_2 = m_3 + m_4 + m_5$
             $h_{11} = (m_3 h_8 + m_4 h_9 + m_5 h_{10})/m_2$
             $W_7 = (m_3 W_4 + m_4 W_5 + m_5 W_6)/m_2$

8. Compute $m_7 = m_2 - m_1$

9. If any of the values m_7, W_9, h_{11} is significantly different from values assumed in step 1, repeat all calculations from steps 2 through 8.

10. Compute cooling coil load

$$Q_{cc} = m_2 [(h_2 - h_3) - (W_2 - W_3) h_{w2}]$$

where h_{wz} is the enthalpy of *liquid* water at temperature of leaving air.

This basic algorithm for simulation of a VAV system might be used in conjunction with a heat balance-type load calculation. For a weighting factor approach, it would have to be modified to allow variation in the zone temperatures and consequently in readjustment of zone loads. It should also be enhanced to allow for possible limits on reheat temperature and/or cooling coil limits, zone humidity limits, outside air control schemes (economizers), and/or heat-recovery devices, zone exhaust, return air fan, heat gain in the return air path because of lights, the presence of baseboard-type heaters, and more realistic control profiles. Most current building energy programs incorporate these and other features as user options, as well as algorithms for other types of systems.

SIMULATING PRIMARY ENERGY CONVERSION SYSTEMS

Primary energy conversion equipment (plant equipment) is simulated to calculate the energy used and associated cost required to meet the heating, cooling, and/or electrical loads of a building. Manufacturers specify the performance of equipment for the design operating conditions. Temperatures and flow rates other than the design values affect both the COP and capacity of the chiller. Also, operating the chiller at less than full capacity affects the COP. Some parameters can affect the performance of the machine significantly on an hour-by-hour basis; others may have very small effects. Modeling the equipment accurately requires hourly determination of the important parameters.

Performance Correction Curves

Off-design performance data are much harder to obtain; in some cases, the manufacturer does not know the off-design performance. For this reason, equipment simulation algorithms should include general correction functions to modify the design data for off-design performance. If the off-design performance is known, the algorithm should allow this to override the general correction functions.

The functions can take any appropriate form, including polynomials, exponentials, Fourier series, and so forth. Most computer algorithms use polynomials, usually third order or less. The necessary degree of the polynomial depends on the behavior of the equipment. It is also possible to use tables of data; however, this method may require excessive data input and computer memory.

Boiler Algorithms

The hourly performance of a boiler depends on boiler load, boiler characteristics and, to a lesser extent, on ambient air temperature. Because the ambient temperature dependence is small (usually $< \pm 2\%$), it is often ignored.

Electric boilers. Assuming the electric resistance heating losses of the loads supplying the boiler are negligible, the only losses affecting its efficiency are the radiative and/or conductive losses through the shell. If the ambient temperature is relatively constant, the loss through the shell can be considered constant, and the electrical consumption is

$$E = \text{Load} + \text{Loss} \tag{42}$$

If the boiler experiences wide swings in environmental temperature, the loss term must take this into account.

Fossil fuel boilers. Fossil fuel-fired boilers are more complex to model because the combustion efficiency usually varies with the load. If the combustion efficiency is constant, the electric boiler Equation (42) can be used. Nonconstant combustion efficiency requires one of the following models:

Fuel consumption calculated as a function of efficiency (heat energy output and/or fuel energy input). Load divided by the capacity defines the part-load ratio

$$\text{PLR} = \text{Load}/\text{Capacity} \tag{43}$$

The efficiency for part-load ratios less than 1.0 (PLR = 1.0 corresponds to the design output) is the design efficiency corrected by a function of the part-load ratio

$$\text{Eff} = \text{Eff}_{design} f(\text{PLR}) \tag{44}$$

The fuel consumed is then

$$F = \text{Load}/\text{Eff} \tag{45}$$

Since the efficiency at PLR = 0 must be zero, the correction function must have a form similar to curve 1 in Figure 12.

Some boilers with dampers or other means of maintaining an almost constant air/fuel ratio for varying loads may have an almost constant efficiency correction curve until very low loads are encountered. At very low loads, the efficiency curve will bend sharply to zero (curve 2 in Figure 12). This kind of curve cannot be modeled accurately with a low-order polynomial. Therefore, the form presented next should be used instead.

Fuel consumption calculated as fraction of full-load fuel consumption. This model takes the form

$$F = F_{design} f(\text{PLR}) \tag{46}$$

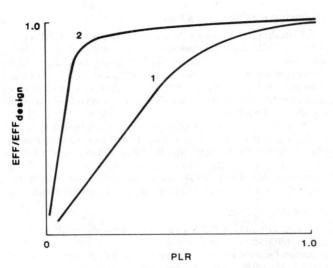

Fig. 12 Efficiency Model for Fossil Fuel Boiler

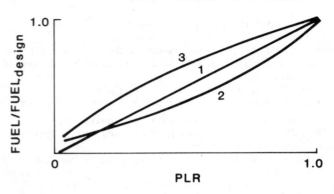

Fig. 13 Fuel Consumption Model for Fossil Fuel Boiler

Figure 13 indicates the possible shapes of the correction factor. Curve 1 is for a boiler with a constant efficiency (not possible in reality); curve 2 is for a boiler that is most efficient toward the middle of its operating range; and curve 3 is for a boiler that is most efficient at full load. For many boilers, the function f (PLR) is consistent and can be approximated easily with a low-order polynomial. For this reason, this model, rather than the efficiency model described earlier, should be used.

If a fossil boiler has draft fans, fuel pumps, etc., the electrical consumption can be modeled either as a constant (no dependence on load) or as a variable (changes with the load). If the electrical consumption varies with the load, it can be modeled in a fashion analogous to Equation (44).

Chillers

The capacity of a chiller usually depends on the condenser and chilled water temperatures. The design capacity is usually rated at a chilled water temperature of 44 °F and an entering condenser water temperature of 85 °F. Water flow rates are usually 2.5 gpm/ton and 3.0 gpm/ton, respectively. For entering condenser water temperatures less than 85 °F and for chilled water temperatures greater than 44 °F, the capacity is greater than the design capacity. The converse is also true. The hourly operating capacity must be calculated to determine the number of chiller units needed to meet the cooling load (in multiple-unit installations), and to calculate the space conditions or cooling overloads.

Assuming the water flow rates remain constant, the capacity adjustment factor can be expressed as a biquadratic polynomial function of the condenser and chilled water temperatures. The hourly available capacity is then

$$C_{avail} = C_{nom} f\ (t_{cold}, t_{cond}) \tag{47}$$

The power consumption of a chiller also depends on the chilled and condenser water temperatures. The load also affects the performance. Dependence on load can be modeled by either of the methods described for boilers, although the second method is used here for f_1. The dependence on temperatures can be modeled as a biquadratic function or as two quadratic functions. A biquadratic function f_2 is used here. The power consumed is expressed as

$$P = P_{design} f_1\ (t_{cold}, t_{cond}) f_2\ (PLR) \tag{48}$$

When modeling an absorption chiller, an equation of the same form as Equation (48) can be used to calculate thermal energy consumption. An absorption machine's electrical consumption is usually modeled as constant, although variable consumption can be modeled by using an equation of the same form as Equation (46).

When simulating a double-bundle chiller operating in the heat-recovery mode, the capacity and energy consumption are determined by the leaving condenser water temperature rather than the entering condenser water temperature. During times of heat recovery operation, the function f_1 in Equation (48) should be replaced by a function that takes the (constant) higher leaving condenser water temperature into account.

When simulating chillers with air-cooled condensers, Equation (48) can still be used to calculate the energy consumption; however, function f_1 should take the ambient air temperature into account rather than the entering condenser water temperature.

Modeling False Loading and Cycling Effects in Chillers

If a compression chiller uses hot gas bypass or some other means to maintain an adequate refrigerant flow through the compressor during low load conditions, the compressor operates under a load larger than the external load on the machine. This additional load, called a false load, causes the compressor to consume as much electricity as if the entire load were external to the chiller. The *minimum unloading ratio* (MUR) is the smallest part-load ratio at which the compressor can operate without false loading. When the chiller is operating at less load, the compressor is operating at MUR. PLR in Equation (48) should be replaced by MUR during periods of false loading to simulate this effect.

If a chiller cycles on and off at low loads with no false loading, the function f_2 in Equation (48) usually includes this behavior, so that f_2 is then valid for

$$0 \leqslant PLR \leqslant 1.0$$

However, if the chiller first false loads and then starts cycling on and off as the load decreases, the function f_2 will not be smooth and may need to be divided into discrete intervals of PLR. Another way to approximate this behavior without breaking the range of PLR into discrete intervals is to assume that the energy consumed while cycling is proportional to the fraction of the hour the machine is cycled on.

Cooling Towers

A cooling tower is simulated to calculate the leaving water temperature for use in the chiller algorithms and to calculate the electrical consumption of tower fans. The performance of a tower depends on the wet-bulb temperature, the temperature drop of the water flowing through the tower (the range), and how closely the water temperature approaches the wet-bulb temperature (the approach). Performance charts relating the wet-bulb, range, and approach to a fourth quantity defined as the *rating factor* are available. The rating factor, when coupled with the tower water flow rate, can be used to determine the number of tower cells and airflow rates needed to produce the desired exit water temperature. Additionally, the same set of relations can be used to calculate the floating temperature that occurs when all cells running at the maximum airflow rate cannot produce the desired exit temperature.

OVERALL MODELING STRATEGIES

In developing a simulation model for building energy prediction, two basic issues must be considered—modeling of components or subsystems, and the overall modeling strategy. Component or subsystem models are merely mathematical equations relating the physical variable that must be continuously satisfied. Overall modeling strategy refers to the *sequence* and *procedures* used to solve these equations, which have important implications, including accuracy of the results and computer resources required to achieve the results. These are discussed briefly in this section.

Virtually all building energy programs in use today employ the same overall modeling strategy. In this strategy, the loads model is executed for every space for every hour of the simulation period. This is followed by employing the systems model for every secondary system, one at a time, for every hour of the simulation. Finally, the plant simulation model is executed again for the entire period. Each sequential execution processes the *fixed* output of the preceding step.

Because of this loads-systems-plants sequence, certain phenomena cannot be modeled precisely. For example, if the heat balance method for computing loads is used, and some component in the system simulation model cannot meet the load, the program can only report the current load. In actuality, the space temperature should readjust until the load matches the equipment capacity, but this cannot be modeled since the loads have been precalculated and fixed. If the weighting factor method is used for loads, this problem is partially overcome, since loads are continually readjusted during the system simulation execution. However, the weighting factor technique is based on linear mathematics, and wide departures of room temperatures from those used during the loads program execution can introduce errors.

A similar problem arises in plant simulation. For example, in an actual building, as the load on the central plant varies, so does the supply chilled water temperature, which in turn affects the capacity of the secondary system equipment. In the extreme case, when the central plant becomes overloaded, space temperatures should rise to reduce the load. However, in modeling strategies used in most current programs, this cannot occur; thus, only the overload condition can be reported. These are some of the penalties associated with decoupling of the loads, systems, and plants models.

An alternative overall modeling strategy, in which all calculations are carried out at each time step, is conceivable. Here the loads, systems, and plant models are viewed together as a single set of equations to be solved simultaneously at each point in time. The unmet loads and imbalances discussed earlier cannot arise; conditions at the plant are immediately reflected to the secondary system and then to the load model, forcing them to readjust to the instantaneous conditions throughout the building. The results of this modeling strategy are superior to those currently available, although the magnitude and importance of the improvement are difficult to judge.

The principal disadvantage of the alternative approach, and the reason that it is not widely used, is that it is more demanding of computing resources. For example, memory requirements tend to be somewhat proportional to the number of surfaces used to model the individual spaces in the building. For example, if 25 spaces are modeled with the alternative strategy, 10 to 15 times as much memory is required. Also, solution time tends to increase at least linearly with the number of simultaneous equations to be solved, *i.e.*, it takes longer to solve 10 spaces together than to solve them one at a time. Conversely, if these difficulties could be overcome, perhaps by the use of more efficient algorithms in the models themselves, the approach might be feasible.

Another issue related to overall modeling strategy, and one that also affects component models, is the *timewise resolution* sought from the model. Practically all models in use today use one hour as the time-step, which precludes any information on phenomena occurring in a shorter time span. Miller (1980) found that the dynamics of control of components may have at least minor effects on predicted energy use.

SMALL TIME-STEP SIMULATION

The nature of a building's design, mechanical system, or control strategy affects the level of simulation needed. For most analyses, an hour-by-hour, quasi-steady state simulation is adequate. However, a closed loop control or real-time performance testing and evaluation of controls may require advanced simulation techniques that involve small time steps. Closed-loop control regulates the output of a "plant" or physical process by comparing it with the desired output response. The difference (error) is then used to control the process close to a predefined (although possible time-varying) set point.

Simulations can model the performance of energy-saving application programs or algorithms for Energy Management and Control Systems (ECMS). The application algorithm then controls the modeled system, and any problems can be corrected before the algorithm is implemented on a real system. The simulated system can be part of a main program that calls the application algorithm and runs at computer speed, or it may reside on a separate computer and run in "real" time. The latter process is often called *emulation* rather than simulation.

A simulation that considers system dynamics is often required to evaluate sophisticated control strategies. If local loop control dynamics are unlikely to significantly impact the energy consumed, a minute-by-minute simulation is probably sufficient. If local loop control dynamics are important, a second-by-second simulation may be needed. Simulation programs reported by Kelly *et al.* (1984), Shavit (1977), Hacker *et al.* (1984, 1985), and Benton *et al.* (1982) have the ability, to varying degrees, of modeling dynamic building control strategies and the interactions between the building shell, the HVAC system, and the controls.

Performance Testing and Evaluation

The control algorithms provided by most EMCS manufacturers or suppliers are usually proprietary. The source code may be available for purchase on a site-specific basis, but it usually does not help in evaluating how well such algorithms actually work or how much energy they save. A control algorithm can be evaluated only by:

• Gathering data on actual buildings controlled by the EMCS
• Testing an EMCS connected to actual HVAC equipment in a laboratory
• Using a building/HVAC system emulator

The emulator, the only practical alternative for testing, consists of a special computer connected to the sensor input and command outputs of an EMCS. No actual HVAC equipment or building is used. Instead, a computer program simulates the response of building systems (including the equipment, building space, and building envelope) to the control strategies and commands implemented by the EMCS. Thus the EMCS can be tested with any type of building, HVAC system, and/or climatic conditions that the emulator can simulate. The emulator keeps track of the energy consumed by the hypothetical building, so that a quantitative measure of the performance of the EMCS system and the supervisory control algorithm is obtained. The results can then be compared with those obtained with different EMCS systems or with a standard set of EMCS application algorithms. May (1985) and Wise (1984) describe two prototypes that demonstrate EMCS testing with an emulator.

SOLAR ENERGY ESTIMATES

Solar heating and/or cooling can be incorporated into buildings through active or passive systems. Active solar systems use either liquid or air as the collector fluid. A complete system includes solar collectors, energy storage devices, and pumps or fans for transferring energy to storage or to the load. The load can be space cooling, heating, or hot water. Although it is technically possible to construct a solar heating and cooling system to supply 100% of the design load, that system would be oversized for seasonal operation and, generally, uneconomical. The size of the solar system, and thus its ability to meet the load, is determined by life-cycle cost analysis that weighs the cost of energy saved against the amortized solar system cost.

Active solar energy systems have been combined with heat pumps for water and/or space heating. The most economical arrangement in residential heating is a solar system in parallel with a heat pump, which supplies auxiliary energy when the solar source is not available. For domestic water systems requiring high water temperatures, a heat pump placed in series with the solar storage tank may be advantageous. Freeman et al. (1979) and Morehouse and Hughes (1979) present information on performance and estimated energy savings for solar heat pump systems.

Passive hybrid solar systems incorporate solar collection, storage, and distribution into the architectural design of the building and make minimal or no use of fans to deliver the collected energy to the structure. The concepts involved in passive hybrid solar heating, cooling, and lighting include the building envelope and its orientation, thermal storage mass, and window configuration and design. DOE (1980), LBL (1981), Mazria (1979), and ASHRAE (1984b) give estimates of energy savings resulting from the application of passive solar design concepts. Hybrid systems combine elements of both active and passive systems.

Methods used to determine solar heating and/or cooling energy requirements for both active and passive/hybrid systems are described by Feldman and Merriam (1979) and Hunn et al. (1987). Descriptions of public and private domain methods are included. An overview of the simulation techniques suitable for active heating and cooling systems analysis, and for passive/hybrid heating, cooling, and lighting analysis follows.

SIMPLIFIED METHODS

Simplified analysis methods have advantages of computational speed, low cost, rapid turnaround (especially important during iterative design phases), and the ability to be used by persons with little technical experience. Disadvantages include limited flexibility for design optimization, lack of control over assumptions, and a limited selection of systems that can be analyzed. Thus, if the system application, configuration, or load characteristics under consideration are significantly nonstandard, a detailed computer simulation may be required to achieve accurate results. This section describes the f-Chart method for active solar heating and the Solar Load Ratio method for passive solar heating (see Dickinson and Cheremisinoff (1980), Lunde (1980), and Klein and Beckman (1979)).

ACTIVE HEATING/COOLING SYSTEMS

Beckman et al. (1977) developed the f-Chart method using an hourly simulation program (Klein et al. 1976) to evaluate space heating and service water heating systems in many climates and conditions. The results of these analyses correlate the fraction of the heat load f met by the solar energy system. The correlations give the fraction f of the monthly heating load (for space heating and hot water) supplied by solar energy as a function of collector characteristics, heating loads, and weather. The standard error of the differences between detailed simulations in 14 locations in the United States and the f-Chart predictions was about 2.5%. Correlations also agree within the accuracy of measurements of long-term system performance data. Beckman et al. (1977, 1981), and Duffie and Beckman (1980) discuss the method in detail.

The f-Chart method requires the following data:

- Monthly average daily radiation on a horizontal surface
- Monthly average ambient temperatures
- Collector thermal performance curve slope and intercept from standard collector tests, i.e., $F_R U_L$ and $F_R(\tau\alpha)_n$ (see ASHRAE Standard 93 and Chapter 34, 1992 ASHRAE Handbook—Systems and Equipment)
- Monthly space and water heating loads

Standard Systems

The f-Chart assumes several standard systems and applies only to these liquid configurations. The standard *liquid system* uses water, an antifreeze solution, or air as the heat transfer fluid in the collector loop and water as the storage medium (Figure 14). Energy is stored in the form of sensible heat in a water tank. A water-to-air heat exchanger transfers heat from the storage tank to the building. A liquid-to-liquid heat exchanger transfers energy from the main storage tank to a domestic hot water preheat tank, which in turn supplies solar heated water to a conventional water heater. A conventional furnace or heat pump is used to meet the space heating load when the energy in the storage tank is depleted.

Figure 15 shows the assumed configuration for a solar *air heating system* with a pebble-bed storage unit. Energy for domestic hot water is provided by heat exchange from the air leaving the collector to a domestic water preheat tank as in the liquid system. The hot water is further heated, if necessary, by a conventional water heater. During summer operation, a seasonal, manually operated storage bypass damper is used to avoid heat loss from the hot bed into the building.

The standard solar *domestic water heating system* collector heats either air or liquid. Collected energy is transferred by a heat exchanger to a domestic water preheat tank that supplies solar-heated water to a convectional water heater. The water is further heated to the desired temperature by conventional fuel if necessary.

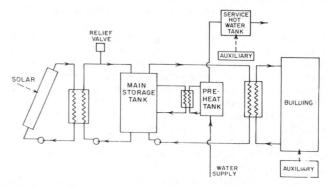

Fig. 14 Schematic Diagram of Liquid-Based Solar Heating System
Adapted from Beckman et al. (1977)

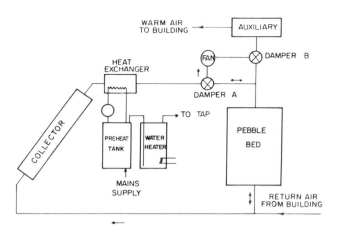

Fig. 15   Schematic Diagram of Solar Air Heating System
Adapted from Beckman *et al.* (1977)

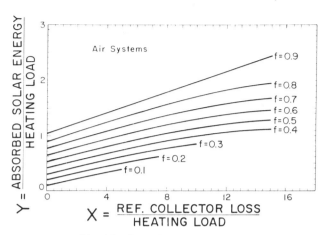

Fig. 16   Chart for Air System
Adapted from Beckman *et al.* (1977)

f-Chart Method

Computer simulations correlate dimensionless variables and the long-term performance of the systems. The fraction of the monthly space and water heating loads supplied by solar energy *f* is empirically related to two dimensionless groups. The first dimensionless group is collector loss; the second is collector gain. These groups are defined as:

$$X = F_R U_L (F_r/F_R)(t_{ref} - \bar{t}_a) \Delta\theta A_c/L \qquad (49)$$

$$Y = F_R (\tau\alpha)_n (F_r/F_R) (\overline{\tau\alpha})/(\tau\alpha)_n \bar{H}_T N A_c/L \qquad (50)$$

where

A_c = area of solar collector, ft^2
F_r = collector-heat exchanger efficiency factor
F_R = collector efficiency factor
U_L = collector overall energy loss coefficient, Btu/h·ft^2·°F
$\Delta\theta$ = total number of seconds (or hours) in month, depending on units of U_L
$\bar{t}_a$ = monthly average ambient temperature, °F
L = monthly total heating load for space heating and hot water, Btu
$\bar{H}_T$ = monthly averaged, daily radiation incident on collector surface per unit area, Btu/day·ft^2
N = number of days in month
$(\overline{\tau\alpha})$ = monthly average transmittance-absorptance product
t_{ref} = reference temperature, 212°F

$F_R U_L$ and $F_R (\tau\alpha)_n$ are obtained from collector test results. The ratios F_r/F_R and $(\overline{\tau\alpha})/(\tau\alpha)_n$ are calculated using methods given by Beckman *et al.* (1977). The value of $\bar{t}_a$ is obtained from meteorological records for the month and location desired. $\bar{H}_T$ is found from the monthly average, daily radiation on a horizontal surface by the methods in Chapter 34 of the 1992 ASHRAE *Handbook—Systems and Equipment* or in Duffie and Beckman (1980). The monthly load L can be determined by any appropriate load estimating method, including analytical techniques or measurements. Values of the collector area A_c are selected for the calculations. Thus, all the terms in these equations can be determined from available information.

Transmittance of the transparent collector cover system τ and the absorptance of the collector plate α depend on the angle at which solar radiation is incident on the collector surface. Collector tests are usually carried out with the radiation incident on the collector in a nearly perpendicular direction. Thus, the value of $F_R (\tau\alpha)$ determined from collector tests ordinarily corresponds to the transmittance and absorptance values for radiation at normal incidence. Depending on the collector orientation and the time of year,

the monthly averaged values of the transmittance and absorptance can be significantly lower. The *f*-Chart method requires a knowledge of the ratio of the monthly average to normal incidence transmittance-absorptance.

The *f*-Chart method for liquid systems is similar to that for air systems. The fraction of the monthly total heating load supplied by the solar air heating system is correlated with the dimensionless groups X and Y, as shown in Figure 16. To determine the fraction of the heating load supplied by solar energy for a month, values of X and Y are calculated for the collector and heating load in question. The value of f is determined at the intersection of X and Y on the *f*-Chart, or from the following equivalent equations.

Air system: $f = 1.04 Y - 0.065 X - 0.159 Y^2$
$$+ 0.00187 X^2 - 0.0095 Y^3 \qquad (51)$$

Liquid system: $f = 1.029 Y - 0.065 X - 0.245 Y^2$
$$+ 0.0018 X^2 + 0.025 Y^3 \qquad (52)$$

This is done for each month of the year. The solar energy contribution for the month is the product of f and the total heating load L for the month. Finally, the fraction of the annual heating load supplied by solar energy F is the sum of the monthly solar energy contributions divided by the annual load:

$$F = \Sigma f L/\Sigma L$$

Example 4. Calculating the heating performance of a residence, assume that a solar heating system is to be designed for Madison, WI, with two-cover collectors facing south, inclined 58° with respect to horizontal. The air heating collectors have the characteristics: $F_R U_L = 0.50$ Btu/h·ft^2·°F and $F_R (\tau\alpha)_n = 0.49$. The total space and water heating load for January is calculated to be 34.1×10^6 Btu, and the solar radiation incident on the plane of the collector is calculated to be 1.16×10^3 Btu/day·ft^2. Determine the fraction of the load supplied by solar energy with a system having a collector area of 538.2 ft^2.

Solution: For air systems, there is no heat exchanger penalty factor and $F_r/F_R = 1$. The value of $(\overline{\tau\alpha})/(\tau\alpha)_n$ is 0.94 for a two-cover collector in January. Therefore, the values of X and Y are

$X = (0.50 \text{ Btu/h·ft}^2\text{·°F}) (212°\text{F} - 19.4°\text{F}) (31 \text{ days})$
$(24 \text{ h/day}) (538.2 \text{ ft}^2)/(3.41 \times 10^7 \text{ Btu}) = 1.13$

$Y = (0.49) (1) (0.94) (1.16 \times 10^3 \text{ Btu/day·ft}^2)$
$(31 \text{ days}) (538.2 \text{ ft}^2)/(34.1 \times 10^6 \text{ Btu}) = 0.26$

Then the fraction of the energy f supplied for January from Figure 16 is 0.19. The total solar energy supplied by this system in January is

$$fL = 0.19 \times 34.1 \text{ MMBtu} = 6.4 \times 10^6 \text{ Btu}$$

The annual system performance is obtained by summing the energy quantities for all months. The result is that 37% of the annual load is supplied by solar energy.

The collector heat removal factor F_R that appears in X and Y is a function of the collector fluid flow rate. Because of the higher cost of power for flowing fluid through air collectors than through liquid collectors, the capacitance rate used in air heaters is ordinarily much lower than that in liquid heaters. As a result, air heaters generally have a lower value of F_R. Values of F_R corresponding to the expected airflow rate in the collector must be used in calculating X and Y.

An increase in airflow rate tends to improve collector performance by increasing F_R, but it tends to decrease system performance by reducing the degree of thermal stratification in the pebble bed (or water storage tank). The f-Chart for air systems is based on a collector airflow rate of 2 scfm/ft^2 of collector area. The performance of systems with different collector airflow rates can be estimated by using the appropriate values of F_R in both X and Y. A further modification to the value of X is required to account for the change in degree of stratification in the pebble bed.

The performance of air systems is less sensitive to storage capacity than that of liquid systems for two reasons: (1) air systems can operate with air delivered directly to the building in which the storage component is not used, and (2) pebble beds are highly stratified and additional capacity is effectively added to the cold end of the bed, which is seldom heated and cooled to the same extent as the hot end. The f-Chart for air systems is for a nominal storage capacity. The performance of systems with other storage capacities can be determined by modifying the dimensionless group X as described in Beckman et al. (1977).

With modification, f-Charts can be used to estimate the performance of a solar water heating system operating in the range of 120 to 160°F. The main water supply temperature, and the minimum acceptable hot water temperature (i.e., desired delivery temperature) both affect the performance of solar water heating systems. The dimensionless group X, which is related to collector energy losses, can be redefined to include these effects. If monthly values of X are multiplied by a correction factor, the f-Chart for liquid-based solar space and water heating systems can be used to estimate monthly values of f for water heating systems. Experiments and analysis show that the load profile for a well-designed system has little effect on long-term performance. Although the f-Chart was originally developed for two-tank systems, it may be applied to single- and double-tank domestic hot water systems with and without collector tank heat exchangers.

For industrial process heating, absorption air conditioning, or other processes for which the delivery temperature is outside the normal f-Chart range, modified f-Charts are applicable (Klein et al. 1976). The concept underlying these charts is that of solar usability, which is the fraction of the total solar energy that is useful in the given process. This fraction depends on the required delivery temperature as well as collector characteristics and solar radiation. The procedure allows the energy delivered to be calculated in a manner similar to that for f-Charts. An example of the application of this method to solar-assisted heat pumps is presented in Svard et al. (1981).

Other Active System Methods

The *relative areas method*, based on correlations of the f-Chart method, predicts annual rather than monthly active heating system performance (Barley and Winn 1978). An hourly simulation program has been used to develop the *monthly solar-load*

ratio (SLR) *method*, another simplified procedure for residential systems (Dickinson and Cheremisinoff 1980). Based on hour-by-hour simulations, a method was devised to estimate system performance based on monthly values of horizontal solar radiation and heating degree days. This SLR method has also been extended to nonresidential buildings for a range of design water temperatures (Dickinson and Cheremisinoff 1980, Schnurr et al. 1981).

Passive Heating Systems

The most widely accepted simplified passive space heating design tool is the solar-load ratio method (DOE 1980, 1982, ASHRAE 1984b). It can be applied manually, although like the f-Chart, it is available on microcomputer software. The SLR method for passive systems is based on correlating results of multiple hour-by-hour computer simulations, the algorithms of which have been validated against test cell data for generic passive heating system types: direct gain, thermal storage wall, and attached sunspace. Monthly and annual performance, as expressed by the auxiliary heating requirement, is predicted by this method. The method applies to single-zone, envelope-dominated buildings. A simplified, annual-basis distillation of SLR results, the *load collector ratio* (LCR) *method*, and several simple-to-use rules have grown out of the SLR method. Several hand-held calculator and microcomputer programs have been written using the methodology (Nordham 1981).

The SLR method uses a single dimensionless correlating parameter (SLR), which Balcomb et al. (1982) define as a particular ratio of solar energy gains to building heating load.

$$\text{SLR} = \frac{\text{Solar energy absorbed}}{\text{Building heating load}} \tag{53}$$

A correlation period of 1 month is used; thus the quantities in the SLR are calculated for a 1-month period.

The parameter that is correlated to the SLR, the *solar savings fraction* (SSF), is defined as

$$\text{SSF} = 1 - \frac{\text{Auxiliary heat}}{\text{Net reference load}} \tag{54}$$

The SSF measures energy savings expected from the passive solar building, relative to a reference nonpassive solar building.

In this equation, the net reference load is equal to the degree-day load of the nonsolar elements of the building as

$$\text{Net reference load} = (\text{NLC})(\text{DD}) \tag{55}$$

where NLC is the net load coefficient. This term is a modified UA coefficient computed by leaving out the solar elements of the building. The nominal units are Btu/°F·day. The quantity DD is the degree-days computed for an appropriate base temperature. A building energy analysis based on the SLR correlations begins with a calculation of the monthly SSF values. The monthly auxiliary heating requirements are then calculated by

$$\text{Auxiliary heat} = (\text{NLC})(\text{DD})(1 - \text{SSF}) \tag{56}$$

The annual auxiliary heat is the sum of the monthly values.

By definition, SSF is the fraction of the degree-day load of the nonsolar portions of the building met by the solar element. If the solar elements of the building (south-facing walls and window) were replaced by other elements so that the net annual flow of heat through these elements was zero, the annual heat consumption of the building would be the net reference load. The savings achieved by the solar elements would therefore be the net reference load minus the auxiliary heat, or

$$\text{Solar savings } = \text{(NLC) (DD) (SSF)} \qquad (57)$$

Although simple, in many situations and climates, Equation (57) is only approximately true because a normal solar-facing wall, with a normal complement of opaque walls and windows, has a near zero effect over the entire heating season. In any case, the auxiliary heat estimate is the primary result and does not depend on this assumption.

The hour-by-hour simulations used as the basis for the SLR correlations are done with a detailed model of the building in which all the design parameters are specified. The only parameter that remains a variable is the solar collector area expressed as the load collector ratio (LCR)

$$\text{LCR } = \frac{\text{Net load coefficient}}{\text{Projected collector area}} \qquad (58)$$

Performance variations are estimated from the correlations, which allow the user to account directly for thermostat set point, internal heat generation, glazing orientation, and configuration, shading, and other solar radiation modifiers. Major solar system characteristics are accounted for by selecting one of 94 reference designs. Other design parameters, such as thermal storage thickness and conductivity, and the spacing between glazings, are included in a series of sensitivity calculations obtained using the hour-by-hour simulations. These results are generally presented in graphic form so that the designer can see the effect of changing a particular design parameter.

Solar radiation correlations for the collector area have been determined using hour-by-hour simulations and Typical Meteorological Year weather data. These correlations are expressed as ratios of incident-to-horizontal radiation, transmitted-to-incident, and absorbed-to-transmitted, as a function of the latitude minus midmonth solar declination and the atmospheric clearness index K_T.

The performance predictions of the SLR method have been compared to predictions made by the detailed hour-by-hour simulations for a variety of climates in the United States. The standard error in the prediction of the annual SSF, compared to the hour-by-hour simulation, is typically 2 to 4%.

An annual solar savings fraction calculation involves summing the results of 12 monthly calculations. For a particular city, the resulting SSF depends only on LCR (the ratio of the net load coefficient to projected collector area), the system type, and the temperature base used in calculating degree days. Thus, tables that relate SSF to LCR for the various systems and for various degree-day base temperatures may be generated for a particular city. Such tables are easier for hand analysis than are the SLR correlations.

Annual SSF versus LCR tables have been developed for 209 locations in the United States and 14 cities in southern Canada for 94 reference designs and 12 base temperatures (ASHRAE 1984).

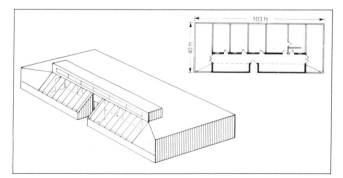

Fig. 17 Floor Plan and Perspective Drawing for Commercial Building in Example 5

Example 5. Consider a small office building located in Denver, CO, with 3000 ft² of usable space and a sunspace entry foyer that faces due south; the projected collector area A_p is 420 ft². A sketch and preliminary plan are shown in Figure 17. Distribution of solar heat to the offices is primarily by convection through the doorways from the sunspace. The principle thermal mass is in the common wall that separates the sunspace from the offices and in the sunspace floor. This example is abstracted from the detailed version given in Balcomb *et al.* (1982). Even though lighting and cooling are likely to have the greatest energy costs for this building, heating is a significant energy item and should be addressed by a design that integrates passive solar heating, cooling, and lighting strategies.

Solution: Calculations of the *net load coefficient* (NLC) are shown in Table 7.

$$\text{NLC } = 24 \times 525 = 12{,}600 \text{ Btu/°F·day}$$

and the total load coefficient (TLC) includes the solar aperture

$$\text{TLC } = 24 \times 676 = 16{,}294 \text{ Btu/°F·day}$$

The load collector ratio (LCR) is

$$\begin{aligned} \text{LCR } &= \text{NLC}/A_p \\ &= 12{,}600/420 = 30 \text{ Btu/ft}^2 \cdot \text{°F·day} \end{aligned}$$

Suppose the daily internal heat is 130,000 Btu/day, and the average thermostat setting is 68 °F. Then

$$T_{base} = 68 - 130{,}000/16{,}294 = 60 \text{°F}$$

(Note that solar gains to the space are not included in the internal gain term as they customarily are for nonsolar buildings.)

In this example, the solar system is type SSD1, defined in ASHRAE (1984) and Balcomb *et al.* (1982). It is a semiclosed sunspace with a 12-in. masonry common wall between it and the heated space (offices). The aperture is double-glazed, with a 50° tilt and no night insulation. To achieve a projected area of 420 ft², a sloped glazed area of 420/sin 50° = 548 ft² is required.

The SLR correlation for solar system type SSD1 is shown in Figure 18. Values of the absorbed solar energy (S) and heating degree-days (DD) to base temperature 60 °F are determined monthly. For S, the solar radiation correlation presented in ASHRAE (1984) for tabulated Denver, CO, weather data are used. For January, the horizontal surface incident radiation is 840 Btu/ft²·day. The 60 °F base degree-days are 933. Using the tabulated incident-to-absorbed coefficients found in ASHRAE (1984), S = 43,681 Btu/ft². Thus S/DD = 46.8 Btu/ft²·°F·day. From Figure 19 at an LCR = 30 Btu/ft²·°F·day, the SSF = 0.51. Therefore, for January

$$\begin{aligned} \text{Net reference load } &= (933)(12{,}600) = 11.76 \times 10^6 \text{ Btu} \\ \text{Solar savings } &= (11.76 \times 10^6)(0.51) = 6.00 \times 10^6 \text{ Btu} \\ \text{Auxiliary heat } &= (11.76 \times 10^6) - (6.00 \times 10^6) = 5.76 \times 10^6 \text{ Btu} \end{aligned}$$

Repeating this calculation for each month and adding the results for the year yields an annual auxiliary heat of 19.94 × 10⁶ Btu.

Table 7   Estimated Heat Load Coefficients for Example Building

	Area A, ft²	U-Value, Btu/h·ft²·°F	UA, Btu/h·°F
Opaque wall	2000	0.04	80
Ceiling	3000	0.03	90
Floor (over crawl space)	3000	0.04	120
Windows (E, W, N)	100	0.55	55
		Subtotal	345
Infiltration			180
		Subtotal	525
Sunspace (treated as unheated space			151
		Total	676

ASHRAE procedures are approximated with V = volume, ft³; c = heat capacity of Denver air, Btu/ft³·°F; ACH = air changes/h; equivalent UA for infiltration = $VcACH$. In this case, V = 24,000 ft³, c = 0.015 Btu/ft³·°F, and ACH = 0.5.

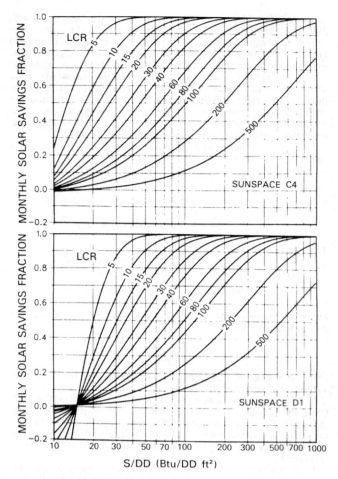

Fig. 18 Monthly SSF versus Monthly S/DD for Various LCR Values

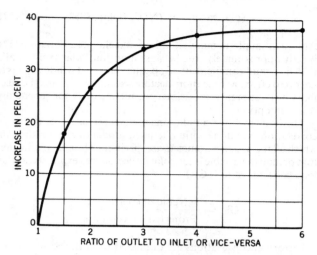

Fig. 19 Increase in Flow Caused by Excess of One Opening over Another

Other Passive Heating System Methods

The concept of usability has been applied to passive buildings. In this approach, the energy requirements of a zero- and infinite-capacity building are calculated. The amount of solar energy that enters the building and exceeds the instantaneous load of the zero-capacity building is then calculated. This excess energy must be dumped in the zero-capacity building, but it can be stored to offset heating loads in a finite-capacity building. Methods are provided to interpolate between the zero- and infinite-capacity limits for finite-capacity buildings. Equations and graphs for direct gain and collector-storage wall systems are given in Monsen *et al.* (1981, 1982).

PASSIVE/HYBRID COOLING SYSTEMS

Passive hybrid cooling systems use natural environmental conditions to cool the conditioned space. Mechanical power only transports air throughout the building; it does not directly cool the air, such as in a conventional air-conditioning system. In this section, cooling by mechanically induced ventilation, natural ventilation, increased air motion, direct and indirect evaporative coolers, radiative heat transfer, and the reduction of lighting loads through daylighting are discussed.

Ventilative Cooling

A building's occupants can be cooled directly by ventilation (physiological cooling) or indirectly by reducing the mean radi-

ant temperature of the building surfaces to which they are exposed (structural cooling). Physiological cooling can be accomplished by (1) the exchange of interior air with exterior air (air exchange method), *e.g.*, opening windows, thereby removing heat and moisture, and (2) direct cooling of occupants by air motion across the body (air motion method), thereby increasing convective and evaporative cooling. The latter can be called task cooling in that the occupant is cooled locally. Both physiological and structural cooling offer the potential for significant energy savings to the extent that they can be substituted for mechanical cooling. These methods use ambient air as a natural cooling sink and rely on either natural convection or forced convection to move the air through the building.

The portion of the cooling season amenable to ventilative cooling is limited by the climatic conditions at the site. For example, when exterior dry-bulb temperatures are above interior temperatures or when exterior humidity is above acceptable comfort levels, then admission of outside air is detrimental rather than helpful. However, interior air motion can be used even when outdoor conditions are above the comfort limits. ASHRAE *Standard* 55-1981 places the upper limit of the summer comfort zone at 79°F for air at a velocity less than or equal to 50 fpm, but permits extending the limit to 82°F if the air velocity is 160 fpm. Rohles *et al.* (1983) indicate that the summer comfort zone may be extended to as high as 85°F if the mean air speed is 200 fpm.

All ventilative cooling methods require air movement induced by either (1) wind, (2) thermal differences (stack effect), (3) mechanical devices (fan or blower), or (4) some combination of the three. Therefore, the volumetric airflow through the building space must be estimated before the cooling effect is determined. But equations for estimating airflows are based on crude models and further refinements and extensions are still needed.

The seasonal energy savings resulting from ventilative cooling are obtained from integration of the hourly cooling effect, which is straightforward for physiological cooling methods. However, the calculation of structural cooling is more complicated because dynamic thermal mass effects and the coupling of mass to the ventilation air must be included. Simplified methods are not yet available.

Cooling by Air Exchange

Traditionally known as ventilation, the air exchange method replaces warm air from an interior space by cooler outside air. The amount of cooling is computed by the same equation as that used for calculating infiltration losses. The airflow volume for forced ventilation systems may be estimated from the fan or blower specifications. The airflow rate for wind-driven ventilation can be estimated by

$$Q = 88\, C_v VA \qquad (59)$$

where

Q = airflow, cfm
A = free area of inlet openings, ft^2
V = wind velocity, mph
C_v = effectiveness of opening = 0.5 to 0.6 for wind perpendicular to opening and 0.25 to 0.35 for wind diagonal to opening

Chandra *et al.* (1983) give more precise values for C_v for windows on opposite and adjacent walls and for several configurations of interior partitions. The local wind speed V may be estimated from monthly average (or monthly average hourly) data, which may be modified to account for characteristics of the local terrain as described by Vickery (1981).

Airflow volume can also be expressed in terms of external pressures exerted by airflow around a building. Vickery (1981) presents a method to determine this pressure distribution. For buoyancy-driven flow, the airflow rate is expressed by the stack effect, which is modified by a coefficient that accounts for the outlet-to-inlet size ratio.

If no significant internal building resistance exists, and assuming indoor and outdoor temperatures are close to 80°F, the flow due to stack effect is

$$Q = CA\sqrt{H\,(T_i - T_o)/T_i} \qquad (60)$$

where

Q = airflow, cfm
A = free area of inlets or outlets (assumed equal), ft^2
H = height from inlets to outlets, ft
T_i = average temperature of indoor air in height, °R
T_o = temperature of outdoor air, °R
C = constant of proportionality, including a value of 65% for effectiveness of openings. This should be reduced to 50% if conditions are not favorable ($C = 7.2$).

Equation (60) applies when $T_i > T_o$.

Greatest flow per unit area of openings is obtained when inlets and outlets are equal; Equations (59) and (60) are based on their equality. Increasing the outlet area over inlet area, or vice versa, will increase airflow but not in proportion to the added area. When openings are unequal, use the smaller area in the equations and add the increase as determined from Figure 19.

Equation (60) requires the height from the inlet to outlet, as well as the average temperature of the interior air at this height. Because the stratification in the space depends on many interrelated factors, it is difficult to predict the temperature at this height. Therefore, this temperature is typically estimated as a constant value obtained from an iterative calculation of the energy balance in the space.

Finally, to calculate the cooling effect of the air exchange method, the estimated hourly cooling rates are integrated over the cooling season by using the bin method with concurrent local wind velocities and/or dry- and wet-bulb temperatures. Temperature bins during occupied hours, unoccupied hours, or both are used, depending on the ventilation strategy. Fan power requirements, if any, are deducted to obtain a net cooling effect or to determine a COP for cooling.

Cooling by Air Motion

In this strategy, the amount of cooling depends directly on the air speed achieved in the space. Acceptable comfort can be maintained above the customary comfort zone for air speeds exceeding 50 fpm. Generally, for speeds above 30 fpm, most people will perceive a 15 fpm increase in air speed to be equal to a 1°F decrease in temperature. The permissible upper limits for dry-bulb temperatures should be determined by using the Fanger comfort charts in Chapter 8 that allow for varying conditions of activity, clothing, relative humidity, and air speed.

Cooling airflows can be induced by natural or forced convection. Outside air can be introduced into a space through openings, or fans or blowers can move the air when the exterior air cannot enter because it is too warm or humid. The average air velocity in the space is approximated by the total flow rate divided by the cross-sectional area.

The interior air motion caused by ceiling fans varies as a function of fan position, power, speed, blade size, and the number of fans within the space. Moreover, air speeds within a space can vary drastically at different distances from the fan. Rohles (1983) found that at the level of seated occupants, air speeds of 50 to 200 fpm occurred throughout a 144-ft^2 space with a 52-in. diameter fan. Figure 20 shows typical air speed distribution. Thus, air speeds necessary to extend the upper end of the comfort zone are easily attainable by using approximately one ceiling fan in an average size room in a residential structure.

Airflow rate for blower-, wind-, or buoyancy-driven flows can be determined from the section on air exchange. To calculate the cooling effect of the air motion method, the seasonal cooling energy required at standard comfort zone conditions (no air motion) is determined first using the bin method. Then the cooling required at each temperature bin is calculated with the air motion using the concurrent local air speed and wet- and dry-bulb temperature conditions in the space. These bin data are integrated over the cooling season for occupied and unoccupied hours, with a deduction taken for the required fan energy. The difference between the cooling energy requirements with and without air motion is the net cooling accomplished with the air motion method. The fan energy required to create the air motion, if any, can be used to calculate a COP for cooling.

Evaporative Cooling

Evaporative cooling can be accomplished by direct or indirect systems, or a multistage combination of the two. In a direct system, water evaporates directly into the supply air, resulting in cooling, but increasing the moisture content of the supply air. Indirect evaporative cooling evaporates water into a secondary airstream that exchanges heat sensibly through a heat exchanger, with the primary (supply) airstream. In either case or combination of the two approaches, the amount of cooling depends on the psychrometric conditions of the outside air. Typically, only sensible cool-

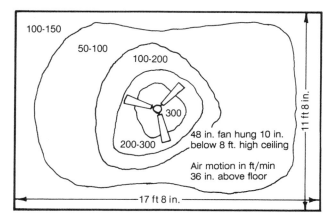

Fig. 20   Approximate Air Speeds from Ceiling Fan

ing is accomplished by these evaporative processes. For a detailed discussion on evaporative cooling, see Chapter 19 of the 1992 ASHRAE *Handbook—Systems and Equipment*.

The performance of evaporative cooling systems is usually determined using a bin method. For each dry-bulb temperature bin, calculate the supply air temperature approach to the mean coincident wet-bulb temperature for the direct evaporative process. For the bin under consideration, the sensible cooling provided by the evaporative system, for sea level air, is given by

$$q_c = 1.10 \, Q \, (t_{space} - t_{sa}) \tag{61}$$

where

q_c = cooling rate or space sensible load, Btu/h
Q = airflow rate, cfm
t_{sa} = temperature of air entering space, °F
t_{space} = temperature of space, °F

The system is assumed to be sized so that the cooling capacity meets the design sensible load. If the cooling capacity is insufficient, either auxiliary cooling is required or a higher room air temperature, consistent with the cooling capacity-to-load ratio, should be specified for the bin under consideration.

Next, the fan and pump requirements are determined. Fan power is given by

$$P = 0.12 \, Q \Delta p / \eta \tag{62}$$

where

P = fan power, W
Δp = static pressure difference, in. of water
η = combined fan and motor efficiency

Add to this an estimate of the pumping power to obtain the total electrical input to the evaporative cooling unit. The system COP becomes the ratio of the cooling capacity to the electrical input.

Finally, the energy savings rate, compared to a conventional vapor compression refrigeration system, is the difference between the electrical input to the conventional system (cooling capacity/COP for conventional system) and the electrical input to the evaporative system. The energy savings for the bin being considered is the difference in electrical input power multiplied by the number of operating hours for this bin. This calculation is repeated for each bin, and the energy savings, generally expressed in kWh, is summed for the cooling season.

Direct Evaporative Cooling

Residential cooling loads in arid or semiarid regions usually use direct evaporative coolers rather than indirect systems. These direct, open systems are usually sized to meet the full sensible load using once-through airflow (no recirculation of return air). In such cases, the monthly or seasonal energy savings compared to conventional vapor compression cooling can be estimated using the bin method.

Example 6. Calculate the cooling energy savings of a direct evaporative system in a residence in Salt Lake City, UT, for which the bin and load data shown in Table 8 are specified. Assume a 3-ton design sensible cooling load, and that for outside air temperatures < 70°F, ventilation is sufficient to meet this load. The space temperature is to be maintained at 78°F.

Solution: For an assumed evaporative heat exchange effectiveness of 0.85, the supply air temperature for the first bin is

$$t_{sa} = 95 - 0.85 \, (95 - 62) = 67°F$$

Thus, the design airflow rate is

$$Q = \frac{(3 \text{ tons}) \, (12{,}000 \text{ Btu/h} \cdot \text{ton})}{(78 - 67°F) \, (1.10 \text{ Btu/h} \cdot \text{cfm} \cdot °F)} = 2975 \text{ cfm}$$

Table 8   Direct Evaporative Cooling of Residence in Salt Lake City, Utah

Temperature Bin	1	2	3
Dry-bulb temperature range, °F	91 to 100	81 to 90	71 to 80
Average db temperature/Mean coincident wb temperature, °F/°F	95/62	85/57	75/54
Hours of operation at this bin	121	428	471
Sensible cooling load, kW	10.57	7.05	3.52
Sensible cooling provided by evaporative unit, kW	10.57	7.05	3.52
Fan power[a], kW	0.17	0.11	0.06
Pumping power, kW	0	0	0
Total electrical power input, kW	0.17	0.11	0.06
Evaporative system COP	62.2	62.2	62.2
Conventional system power input, kW	4.23	2.82	1.41
Energy savings, kWh	491	1160	635

[a]Equivalent fan power consistent with a constant evaporative unit COP.

at standard summer conditions. Therefore, the evaporative cooling provided for the first bin is by Equation (61)

$$q_c = (1.10) \, (2975)(78 - 67) = 36{,}000 \text{ Btu/ or } 10.57 \text{ kW}$$

The fan power from Equation (62) is

$$P = 0.12 \times 2975 \times 0.17/0.35 = 173 \text{ W}$$

where a pressure drop of 0.17 in. of water is assumed for the evaporative unit. A combined fan and motor efficiency of 0.35, typical for small units, is assumed. If the water to the wetted pad is supplied by line pressure, the pumping power is zero. For the other bins, the evaporative cooling capacity increases while the load decreases. Thus, the unit will cycle to meet the load. The fan power for Bins 2 and 3 has been reduced, consistent with the assumption of a constant COP for the evaporative unit.

For a conventional vapor compression system with an EER = 8.5 Btu/Wh (COP = 2.5) for sensible cooling, assumed to include all appropriate parasitic power, the total electrical input for the cooling load is

$$P = \frac{10.57}{2.5} = 4.23 \text{ kW}$$

Thus, the energy savings for this bin are

$$ES = (4.23 - 0.17 \text{ kW}) \, (121\text{h}) = 491 \text{ kWh}$$

The calculation is repeated (see Table 8) for Bins 2 and 3, resulting in a seasonal energy savings of 2286 kWh.

Indirect Evaporative Cooling

Indirect cooling, or a combination of both, is customarily used with the central fan system because of humidity control requirements in commercial buildings. With indirect evaporative cooling, the wet-side (secondary) air can consist of full outside air or full return air, or any proportion of the two. The bin method illustrated next is applicable to these configurations; however, the details must be modified as appropriate.

If a precise hourly analysis is required, or unusual control strategies or system configurations are used, a system simulation should be applied. The bin method approach requires the calculation of the space load, the return and outside air loads, and the performance and operating power of the indirect evaporative system at each of several dry-bulb temperature bins at the above supply air temperature.

For each dry-bulb temperature bin, first determine the space sensible load using the bin or other suitable method. Next, determine the HVAC system sensible load (cooling coil sensible load) by adding the return air, outside air, and duct sensible loads to the space sensible load. To do this, the supply airflow rate at design conditions (Bin No. 1) is needed. This is calculated from Equation (61) as

$$Q = \frac{q_c}{1.10\,(t_{space} - t_{sa})} \quad (63)$$

at sea level for fixed space conditions. If the space temperature floats appreciably, the space load and supply airflow rate calculation must take this into account. For a VAV system, this quantity is reduced at the other bins in proportion to the reduced space sensible load.

The return air load (temperature rise) from lighting is calculated from an energy balance on the airflow through the light fixtures as

$$\Delta t = \frac{q_{lighting}}{1.10\,Q_{return\,air}} \quad (64)$$

Similarly, the return air duct gain is estimated. Then the temperature rise of the return air resulting from the introduction of minimum outdoor air (at the outdoor air conditions of the bin under consideration) is determined. The sum of these return air temperature rises, when added to the space air temperature, gives the mixed air temperature. Finally, an energy balance on the mixed air will give the mixed air load. When added to the space sensible load, this gives

Mixed air sensible load (Btu/h) $= 1.10\,Q_{sa}\,(t_{mixed\,air} - t_{space})$ (65)

However, when the outdoor air temperature is lower than the return air temperature and the economizer is in operation, the system sensible load can be calculated as

System sensible load (Btu/h) $= 1.10\,Q_{sa}\,(t_{oa} - t_{sa})$ (66)

plus any fan and/or duct heat gain.

Next, determine the cooling capacity of the evaporative unit. Based on the supply airflow rate through each dry-side core of the evaporative unit, the indirect evaporative heat exchanger effectiveness is assumed to represent the combined wet- and dry-side heat exchange processes. If dictated by the equipment configuration, a two-step heat exchange process can be represented. The temperature drop of the supply air across the dry side of the heat exchanger is then calculated from

$$\Delta t_{sa} = \epsilon\,(t_{dry\text{-}side\,inlet} - t_{wet\text{-}side\,wb}) \quad (67)$$

where ϵ is the effectiveness, $T_{dry\text{-}side\,inlet}$ is the mixed air dry-bulb temperature (when the evaporative unit is located downstream of the economizer mixing box), and $t_{wet\text{-}side\,wb}$ is the wet-bulb temperature of the outside air at mean coincident wet-bulb conditions for the bin (when the wet-side is provided with all outside air). The evaporative cooling capacity is then

$$Q_c = 1.10\,Q_{sa}\,\Delta t_{sa} \quad (68)$$

To determine the operating power, the fan and pump power must be combined. The fan power for each side of the heat exchanger may be calculated by Equation (62). The pressure drop is obtained from the manufacturer's specifications. The pump power is then estimated and added to the fan power to obtain the total electrical input to the system. The system COP is the ratio of cooling capacity to electrical input.

Finally, the energy savings rate, compared to a conventional vapor compression refrigeration system, is the difference between the electrical input of the conventional system and the electrical input of the evaporative system. When these rates are multiplied by the number of hours of operation for each bin and summed for all bins, the seasonal energy savings are obtained.

Example 7. Calculate the cooling energy savings of an indirect evaporative VAV system for a 100-ton design cooling load in a 26,700 ft^2 commercial building in Sacramento, CA, for which the bin data in Table 9 are specified. A two-stage process, indirect evaporative precooling followed by a conventional cooling coil, will be used (see Figure 21). The space air temperature will be maintained at 78 °F and the supply air at 60 °F.

Solution: For the first bin, the space sensible load is assumed specified as 65 tons. Therefore, the supply airflow rate to meet this load is

$$Q_{sa} = \frac{65 \times 12,000}{1.10\,(78 - 60)} = 39,4000 \text{ cfm}$$

Table 9 Indirect Evaporative Cooling of Commercial Building in Sacramento, California

Temperature Bin	1	2	3	4	5
Dry-bulb temperature range, °F	100–104	90–99	80–89	70–79	60–69
Average db temperature/ Mean coincident wb temperature, °F/°F	102/71	95/68	85/64	75/60	65/56
Hours of operation at this bin	11	74	234	407	941
Mixed air temperature, °F	86.5	85.7	83.5	75.0[a]	65.0[a]
Supply airflow, scfm	39,400	37,400	34,600	28,000	24,000
System sensible cooling load, kW	336.6	309.8	262.0	150.4[b]	51.9[b]
Evaporative system cooling Δt, °F	8.1	9.4	10.7	9.0	5.7
Evaporative supply air temperature, °F	78.4	76.3	72.8	66.0	59.3
Evaporative cooling capacity, kW	102.9	113.3	119.4	81.2	44.1
Evaporative system fraction of total load	0.31	0.37	0.46	0.54	0.85
Evaporative system power input, kW	5.0	4.8	3.7	3.4	3.1
Evaporative system COP	20.6	23.6	32.3	23.9	14.2
Conventional system COP	3.5	3.5	3.5	3.5	3.5
Conventional system power input, kW	29.4	32.4	34.1	23.2	12.6
Energy savings, kWh	268	2040	7117	8059	8940

[a]Full economizer operation
[b]System outside air (OA) load plus fan and/or duct heat gain

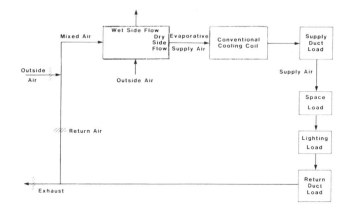

Fig. 21 Indirect Evaporative Cooling System Schematic

The temperature rise resulting from lighting (assume 1.8 W/ft^2) is calculated as

$$\Delta t = \frac{1.8 \times 26,700 \times 3.412}{1.10 \times 39,400} = 3.8\,°F$$

A temperature rise of 2 °F from duct heat gain is assumed. At an assumed minimum outside air volume of 15% at this bin, energy and mass balances on the mixing of outside and return air result in a mixed air-to-outside air temperature rise of 2.7 °F. Therefore, the mixed air temperature is 78 + 3.8 + 2.0 + 2.7 = 86.5 °F. Thus, the mixed sensible air load is

$$Q_{ma} = 39,400 \times 1.10\,(86.5 - 78.0) = 368,400\,Btu/h = 30.7\,tons$$

and the system sensible load is 65.0 + 30.7 = 95.7 tons.

If the manufacturer of the evaporative cooling unit has specified a heat exchange effectiveness of 52% at this airflow through each core, then the supply air temperature drop through the unit is

$$\Delta t_{sa} = 0.52\,(86.5 - 71.0) = 8.1\,°F$$

The evaporative cooling capacity is then

$$Q_c = 1.10 \times 39,400 \times 8.1 = 351,100\,Btu/h\ or\ 102.9\,kW$$

Assuming a specification of a 0.38 in. of water pressure drop across the dry side of the heat exchanger for these flow conditions, 2 hp blower for the wet side, and 75% fan efficiency for each side, the fan power for the dry side is

$$P_{dry\,side} = \frac{0.12 \times 39,400 \times 0.38}{0.75} = 2396\,W$$

and for the wet side

$$P_{wet} = \frac{(2hp)\,(746W/hp)}{0.75} = 1989\,W$$

If the pumping power is 600 W, the total electrical input is then 2396 + 1989 + 600 = 4985 W = 5.0 kW. Therefore, the system COP is

$$COP = \frac{102.9}{5.0} = 20.6$$

The energy consumed at this bin for a conventional cooling system, assuming the COP = 3.5, is 102.9/3.5 = 29.4 kW. The energy savings for all hours in this bin are

$$ES = (29.4\,kW - 5.0\,kW)\,(11) = 268\,kWh$$

The calculation is repeated for bins 2 through 5 (see Table 8), and assuming the conventional system uses an economizer cycle (100% OA) for bins 4 and 5, the seasonal energy savings are 26,424 kWh.

Radiative Cooling

Radiative cooling is a natural heat loss mechanism responsible for the formation of dew, frost, and ground fog. Because of its commonly observed effectiveness at night, it is sometimes termed *nocturnal radiation*, although the process continues to operate throughout the day. Thermal infrared radiation plays a role in determining the surface temperature of a building wall or roof, and it has been treated in an approximate manner using the sol-air temperature concept. Radiative cooling of window and skylight surfaces can be significant, especially under winter conditions when the dew-point temperature is low.

The most useful parameter for characterizing the radiative heat transfer between horizontal nonspectral emitting surfaces and the sky is the *sky temperature* T_{sky}. If S designates the total downcoming radiant heat flux emitted by the atmosphere, then T_{sky} is defined as:

$$T^4_{sky} = S/\sigma \qquad (69)$$

where $\sigma = 0.1713 \times 10^{-8}$ Btu/(h·ft^2·°R^4).

The sky radiance is treated as if it originates from a blackbody emitter of temperature T_{sky}. The *net radiative cooling rate* R_{net} of a horizontal surface with absolute temperature T_{rad} and a nonspectral emittance ϵ is then

$$R_{net} = \epsilon\sigma\,(T^4_{rad} - T^4_{sky}) \qquad (70)$$

Values of ϵ for most nonmetallic construction materials are usually about 0.9.

Radiative building cooling has not been fully developed. Design methods and performance data compiled by Hay and Yellott (1969) and Marlatt et al. (1984) are available for residential roof pond systems that use a sealed volume of water covered by sliding insulation panels as the combined rooftop radiator and thermal storage element. Other conceptual radiative cooling designs have been proposed, but more developmental work is required (Givoni 1981, Mitchell and Biggs 1979).

The sky temperature is a function of atmospheric water vapor; the amount of cloud cover and air temperature with the lowest sky temperatures occur under an arid, cloudless sky. The monthly average sky temperature depression, which is the average of the difference between the ambient air temperature and the sky temperature, typically lies between 10 and 43 °F throughout the continental United States. Martin and Berdahl (1984) have calculated this quantity using hourly weather data from 193 sites. They suggest that the sky temperature should be less than 61 °F to achieve reasonable cooling in July (Figure 22). In regions where sky temperatures fall below 61 °F for 40% or more of the month, all nighttime hours are effectively available for radiative cooling systems.

Clark (1981) modeled a horizontal radiator at various surface temperatures in convective contact with outdoor air for 77 U.S. locaions. Average monthly cooling rates for a surface temperature of 76 °F are plotted in Figure 23. If effective steps are taken to reduce the surface convection coefficient by modifying the radiator geometry or an infrared-transparent glazing, it may be possible to improve performance beyond these values.

Design Methods for Daylighting Systems

Estimates of the energy conservation effects of automatic lighting control systems that respond to daylight can be made in conjunction with single-measure cooling energy estimates. Light levels at various locations within the room for both uniformly overcast and sunny days in different seasons can be calculated with procedures recommended by the Illumination Engineering Society (IES 1987) or nomographs (IES 1979, Sain et al. 1983). For conditions

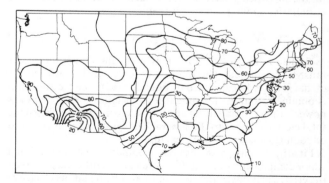

Fig. 22   Percentage of Monthly Hours when Sky Temperature Falls below 61 °F
Adapted from Martin and Berdahl (1984)

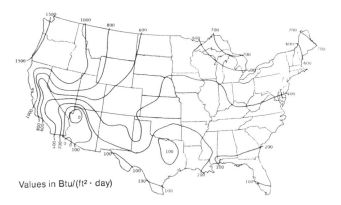

Fig. 23   July Nocturnal Net Radiative Cooling Rate from Horizontal Dry Surface at 76 °F
Adapted from Clark (1981)

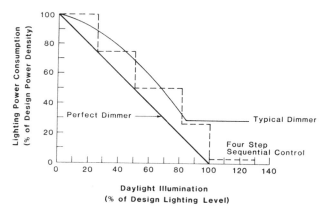

Fig. 24   Control Profile for Perfect Dimmer, Typical Dimmer, and Four-Step Sequential Control
Adapted from Gillette (1983)

in which the daylight level within the space is greater than the design illumination level, electric lights are assumed to be off. Based on these calculations and with knowledge of the distribution of cloudy and clear days, the reduction in hours of artificial lighting and the resultant reduction in lighting energy consumption can be estimated. Effects of solar gain in the space can be calculated using conventional methods.

Reduction of annual lighting energy reduces the annual cooling requirement and increases the annual heating requirement of the building. The simplest method for calculating the reduction in cooling energy is to estimate the coefficient of performance of the cooling system and the number of cooling months during the year. The total reduction from daylighting is equal to

$$E_T = (1 + \frac{M_c}{12 \, COP_c}) \, Q_L \qquad (71)$$

where

E_T = annual total energy savings from daylighting
M_c = number of cooling months in year
COP_c = cooling system annual coefficient of performance
Q_L = annual lighting energy savings from daylighting

A more sophisticated method by Robbins (1983) estimates the fraction of the year in which a room is daylit, using a set of coefficients for a given city, orientation, and standard work year. Estimates of cooling energy reduction can be made as shown in Equation (71).

Energy savings estimates from daylighting can be made using somewhat more complex models than the ones just described. Most of these start with the daylighting calculation methodology, otherwise known as the lumen method, adopted by IES (1979) for the estimate of illumination within a room from transmitted daylight. Several programs are available to calculate daylight levels for different room geometries under uniformly clear and cloudy skies (Bryan and Clear 1982). The increased complexity of these methods incorporates seasonal effects of daylight availability and heating and cooling requirements and the effect of lighting control schemes. Lighting and resultant electric lighting savings calculations are made for each month and are incorporated into monthly energy calculations.

After the daylighting contribution in the space has been estimated for a particular exterior condition, the response of the automatic lighting control system is calculated. Automatic lighting controls fall into three categories: simple on-off, sequential stepping of lamps, and continuous dimming (Treado and Kusuda 1980, Gillette 1983). For simple on-off control, artificial lighting is considered to be on, unless the daylighting illumination is greater than the design illumination level within the space.

Calculations for the other two-control schemes use the fact that illumination is additive, and therefore, the supplementary lighting required is merely the daylighting illumination subtracted from the design illumination level. The percentage of supplemental lighting is the supplemental illumination divided by the design illumination (IES 1979). Typical relationships for percentage input power to percentage supplemental lighting are shown in Figure 24 for a continuous dimming system and a four-step sequential control (Gillette 1983). The percentage of input power is multiplied by the design uniform lighting power density in the daylit zone to determine the actual power density for a daylit condition.

Calculations should be performed for numerous solar and sky conditions to determine the daylighting potential over the course of the year, typically on a monthly basis for both clear and cloudy days. Daylight levels can be compared to the actual lighting requirement of the space and the lighting control profile to obtain the actual lighting power density. Monthly weather data, such as that in the *Local Climatological Data* series from NOAA, can be used to determine the number of clear and cloudy days. Lighting power density for clear and cloudy days can be prorated according to these weather data.

Calculations can be made each month to determine the cooling energy decrease and heating increase resulting from daylighting. These calculations can be made in conjunction with a monthly bin method calculation into which the effects of cooling and heating system performance factors are incorporated. This method allows calculations for each month to prorate heating and cooling conditions with daylit and nondaylit periods. Robbins (1983) and Rundquist (1982) have developed methods to reduce the calculations required to estimate daylighting savings.

SIMULATION METHODS

A number of computer programs are available for active and passive solar system analysis. These range from full building energy analysis programs with passive solar capabilities to programs for active solar energy systems, but with some passive and/or hybrid capability. For passive and/or hybrid solar applications in which the details of energy transfer between walls and rooms are required, or for multizone applications in which HVAC system simulation is important, building energy analysis programs are more appropriate. For active solar system studies, or for passive applications for which the nonpassive elements are relatively unimportant, solar simulation programs are appropriate.

Detailed simulation methods can be used for design and energy analysis studies, and in some cases, for HVAC system trade-off analyses. Advantages are the ability to simulate accurately the details of complex systems, the incorporation of a large variety of system types, and flexibility in doing optimization studies. Disadvantages include the complexity of program input, long running times, high costs, accessibility of computing facilities, and requirements for learning how to use a complex program to perform accurate simulations.

Solar heating and/or cooling simulation programs can be purchased, leased, timeshared, or batch processed. Some programs are proprietary, in which case the user has limited access to program documentation and assumptions, and limited opportunity to change the program to suit special needs. A brief description of the major, accepted simulation programs is provided here. Feldman and Merriam (1979) and Kreider and Kreith (1981) provide more detail on program capabilities.

Building Energy Analysis Computer Programs

Detailed building energy analysis programs, such as those described in U.S. Army (1979) and DOE (1981), were intended for simulation of large, multizone buildings and their HVAC systems. However, they may also be used for active and passive hybrid solar energy analysis. Representative of hour-by-hour simulators, both of these contain active solar system simulators for liquid heating and/or cooling systems; the simulator in LBL (1981) also includes air simulation capabilities and contains many component models (Roschke *et al.* 1978). Using these components, the user can assemble a system to suit special requirements or select a preassembled, standard liquid-to-air system from among several in the program library. To save time, most users use a preassembled system. Both commercial and residential building systems, including space heating and cooling, process hot water, and combined solar heat systems, are modeled.

These detailed building energy analysis programs can also be used for passive-hybrid system analysis. They include detailed evaluations of solar and other thermal energy exchanges inside the structure. Transfer function techniques are used to simulate the dynamics of heat flow through walls, roofs, and other envelope structures. In the program described in U.S. Army (1979), radiation exchange between surfaces and convection between walls and room air are included. An energy balance is used to compute surface and room temperatures. Another approach to calculating loads is used in the program described in LBL (1981). Here the loads are calculated assuming a constant temperature in the conditioned space. The thermal storage effect is simulated by applying weighting factors to internal and solar gains; custom weighting factors may be generated for the building being analyzed. The effect of assuming a constant space temperature is corrected by a perturbation technique in the next stage of the program when the heating and cooling system is simulated, although the load predictions of these two programs agree closely when custom weighting factors are used. The approach that uses an energy balance on each space can simulate temperatures more accurately in spaces where conditions fluctuate widely under the influence of solar gains. While these programs provide a detailed picture of thermal energy flows and the interaction between the heating and cooling load and equipment, they include simplifications that may not fully account for all significant energy terms.

Both programs are flexible and can accommodate passive components such as direct gain and storage walls. External shading devices and the transmittance of the glazing are accurately computed. An accurate simulation of a passive structure can be made. One disadvantage of these programs may be the large amount of input information required to simulate a simple passive structure. Many applications require a significant level of detail and may require an analyst with training or experience in computer simulation.

Solar Heating and Cooling Simulation Methods

The transient solar system simulation program described by Klein *et al.* (1976) has been developed for analyzing active solar heating and cooling systems. It is a modular program, with each module representing a specific physical component. The standard library contains component models for collectors, storage devices, heat exchangers, heat pumps, controllers, and so forth. The building load components include walls, roofs, and windows that can be used to build up to a complete structure and can also be a component for modeling internal rooms. Passive elements, such as direct gain and storage walls, are also part of the library. The components needed for simulating active and passive systems are provided.

The components are connected in a manner analogous to the actual system, which allows any solar system configuration to be specified and modeled. Passive elements can be connected into the system in the same way as active elements. The building loads calculations are less sophisticated than those of comprehensive building energy analysis programs.

Another active solar system program described by Arumi-Noe and Northrop (1979) simulates system performance for a representative day each month. It is not as flexible as full building energy analysis computer programs. Nor is it modular, and the user lacks the freedom to build simulation models of nonstandard solar systems. It is more flexible than the *f*-Chart method, since it includes several types of solar systems that are not included in the *f*-Chart. This program includes tracking and concentrating collector solar cooling systems, in addition to conventional heating systems on which the *f*-Chart is based.

Other special-purpose computer programs, both proprietary and public domain, are available for active solar system simulation. For descriptions of these and other programs, as well as accessibility, see Feldman and Merriam (1979) and Kreider and Kreith (1981).

Thermal network programs, which are less sophisticated and comprehensive than building energy analysis programs, may be more suitable for simulating passive-hybrid solar buildings. Representative programs are presented in McFarland (1978) and Berkeley Solar Group (1982). These programs were intended for residential and light commercial HVAC systems limited to a furnace or central air conditioner. They treat the building as having one or two thermal zones. The basis of the calculation procedure is a thermal network that connects all elements with thermal resistances. Energy balances are performed on each mass element, and the resulting equations are integrated over time. The programs readily accommodate common passive components such as direct gain and storage walls, sunspaces, underfloor rock beds, and natural ventilation. A minimum of input data is required.

Another hour-by-hour analysis tool for passive systems is a modular program that uses a thermal network, finite-difference solution procedure (Arumi-Noe and Northrup 1979). Graphical output routines are included in the program. Although developed primarily for residential passive heating applications, this program can treat multizone buildings and has limited passive cooling and daylighting capabilities. However, HVAC system simulation capabilities are limited.

Simplified passive solar analysis programs, based on fixed and limited system configurations and a small number of thermal nodes, have been developed. Examples are described in Nordham (1981), Milne and Yoshikawa (1979), and Kohler and Sullivan (1980). Because of the simplified algorithms employed, the results of these programs are not likely to be as accurate as detailed simulations are. However, they require fewer inputs and computer resources.

Because of their modularity, building energy analysis computer programs and thermal network programs can, in principle, simulate passive-hybrid cooling options based on ventilation, evaporation, and radiation. Most of the algorithms in the available programs that represent passive cooling processes are fairly crude. For example, a number of the widely used programs include first-order air-exchange natural ventilation models; others include evaporative cooler models as well (Berkeley Solar Group 1982). None of these as yet includes roof pond, central system evaporative, or radiative cooling models that are coupled to the space air temperatures and/or include thermal cool storage effects. Several detailed models of wet and dry roof ponds, which include coupling and storage effects, have been developed (Loxsom *et al.* 1983); however, none is well documented or in widespread use.

Daylighting Simulation Methods

Simulation of daylighting is important to evaluate (1) the reduction of electricity used for lighting, and (2) associated thermal effects, which can impact the heating and cooling loads of a building. The thermal impacts of daylighting vary considerably, depending on electric lighting power density and control, window size, transmittance, placement, and climate (Place *et al.* 1987, Johnson *et al.* 1984, Arasteh *et al.* 1985). Adverse cooling impacts due to excessive solar gain can be limited by a daylighting system that prevents acceptance of more daylight than is useful for offsetting electric lighting. Careful design can reduce both cooling energy consumption and peak cooling demand. Adverse heating impacts can be minimized by using insulating windows and avoiding glazing areas that are larger than necessary.

Availability of daylight has not been measured in enough locations over sufficient time to be directly applied in most simulations. Because of the different proportions of solar heat and light with variations in sun altitude and atmospheric conditions, undifferentiated solar radiation data cannot be used directly in daylighting simulations either. Nevertheless, Littlefair (1985) found that this proportion, the luminous efficacy, which usually varies from about 70 to 130 lm/W, can be applied to the direct and diffuse components of solar radiation to estimate the light available under given conditions. For tables of available daylight, refer to Robbins (1986).

Controls affect the efficiency of the daylighting system, especially its impact on heating and cooling in a building. Aperture controls, such as drapes, blinds, insulating shutters, etc., moderate the amount of light entering an aperture and/or contribute to the insulating value of the glazing. They maximize useful solar heat gains and reduce thermal losses through the glazing. In cooling situations, they can permit just enough light to assure proper visibility in the space. This technique can assure lower cooling loads and limit peak demand. Aperture controls can also ameliorate daylight discomfort glare and prevent penetration of beam radiation onto the workplane.

Electric lighting controls adjust electric light levels based on the quantity of daylight available. Lighting controls can be of either the dimming or switching type. With a switching system, lights near windows or skylights are switched off when daylight in these areas exceeds some threshold value. With dimming systems, the light levels are sensed by a photoelectric cell that dims the electric light in proportion to the amount of available daylight. The mechanism that controls the electric lights can be either automatic or manually operated or a combination of the two. Crisp (1981) found that while occupants may switch on lights when daylight is insufficient, they do not ordinarily switch them off when daylight is adequate.

In simulating daylighting, a comprehensive daylighting system should be integrated with other elements of the building design, especially HVAC systems, the building structure, roof and wall sections, and interior design. A variety of daylighting simulation techniques and products are available, ranging from tables and nomographs to computerized methods (Wilde 1985). Many tech-

niques develop information on illumination and/or electric lighting response to be passed on to a more general building energy analysis procedure.

Cole *et al.* (1983) describe the Illuminating Engineering Society's (IES) daylighting calculation procedure (the lumen method), which may be used in an hour-by-hour program to estimate daylighting levels in a room. The electric lighting power density is then calculated each hour as a function of the supplementary illumination required. This function can be modified to reflect various lighting control techniques described earlier. Although the IES procedures have very specific limitations on the configuration of a room and its fenestration in the calculation assumptions, the calculation is simple and can be repeated without significant expense in computer time.

Another approach documented by Gillette (1983) and Jurovics (1979) uses explicit, but simplified geometric calculations to estimate the hour-by-hour illumination at a specific point in a room. The most advanced method using this approach has achieved good approximations of illumination levels compared with test models (Gillette 1983). Supplemental illumination requirements for each hour are then compared with the lighting control function to determine the lighting power required.

A third approach uses data from a separate illumination calculation computer model (Dilaura and Hauser 1979) or instrumented physical models to develop simple linear models of illumination level in a room (Davenport and Nall 1983). Statistical analysis is used on resulting data to develop a regression model of the illumination level in a space as a function of exterior solar conditions. The linear statistical model is then used to estimate daylighting illumination in the space. The control system response can be precalculated to develop a profile of lighting power density against specific levels of daylight illumination in the space. Switching schemes can be input as a step function with the discontinuities at the illumination levels that initiate switching.

Daylighting simulation methods have been combined in various ways with hourly building energy analysis programs. One approach described by Winkelmann and Selkowitz (1986) and Herron *et al.* (1981) integrates routines that determine daylight illuminance and simulate lighting and aperture controls directly in the loads calculation. In this approach, the illuminance calculation may be based on a geometric calculation such as that described earlier, or it may be calculated from empirically determined factors that correlate daylight illuminance with transmitted solar gain from room apertures. Another approach is to use an illuminance calculation program as a preprocessor to determine effective electric lighting schedules for use in a separate loads program (Treado *et al.* 1986). With either approach, the hourly reduction in electric lighting power due to daylighting is used in the thermal calculation, so that the impact of daylighting on heating and cooling loads, electricity consumption and demand, and HVAC equipment sizing can be determined.

REFERENCES

AGA. 1977. Space heating system efficiency improvement program. 1976/77 Heating Season Status Report. American Gas Association, Cleveland, OH.

Alereza, T. and T. Kusuda. 1982. Development of equipment seasonal performance models for simplified energy analysis methods. ASHRAE *Transactions* 88(2):249-62.

Arasteh, D., *et al.* 1985. Energy performance and savings potentials with skylights. ASHRAE *Transactions* 91:1A.

Arumi-Noe, F. 1984. DEROB simulation of the NBS test buildings. ASHRAE *Transactions* 90:2.

Arumi-Noe, F. and D.O. Northrup. 1979. A field validation of the thermal performance of a passively heated building as simulated by the DEROB system. *Energy in Buildings* 2:1.

ASHRAE. 1976. Energy calculations 1—Procedures for determining heating and cooling loads for computerizing energy calculations. Algorithms for Building Heat Transfer Subroutines.

ASHRAE's Simplified energy calculations. 1983. ASHRAE *Bulletin.*

ASHRAE. 1984a. Bin Weather Data, ASHRAE Research Project 385. Larry Degelman, Principal Investigator; bin data are available on computer diskette for 58 sites, based on Weather Year for Energy Calculations (WYEC) data.

ASHRAE. 1984b. *Passive solar heating analysis: A design manual.*

ASHRAE. 1986. Bin and degree-hour weather data for simplified energy calculations. Computer data available from ASHRAE.

ASHRAE. 1989. Energy efficient design of new buildings except low-rise residential buildings. ASHRAE *Standard* 90.1.

ASHRAE. 1990. ASHRAE Journal's HVAC&R Software Directory.

Balcomb, J.D., R.W. Jones, R.D. McFarland, and W.O. Wray. 1982. Expanding the SLR method. *Passive Solar Journal* 1:2.

Barley, C.D. and C.B. Winn. 1978. Optimal sizing of solar collectors by the method of relative areas. *Solar Energy* 21:4.

Beckman, W.A., S.A. Klein, and J.A. Duffie. 1977. *Solar heating design by the f-Chart method.* John Wiley & Sons, New York.

Beckman, W.A., S.A. Klein, and J.A. Duffie. 1981. Performance predictions for solar heating systems. *Solar energy handbook,* J.F. Kreider and F. Kreith, eds., McGraw Hill, New York.

Benton, R., J.W. MacArthur, J.K. Mahesh, and J.P. Cockroit. 1982. Generalized modeling and simulation software tools for building systems. ASHRAE *Transactions* 88:2.

Berkeley Solar Group. 1982. CALPAS3 User's Manual. Berkeley, CA.

Bryan, H.J. and R.D. Clear. 1982. Microlite I, A microcomputer program for daylighting design. Proceedings of the Seventh National Passive Solar Conference, Knoxville, TN.

Burch, D.M., K.L. Davis, and S.A. Malcolm. 1984a. The effect of wall mass on the summer space cooling of six test buildings. ASHRAE *Transactions* 90:24.

Burch, D.M., D.F. Krintz, and R.F. Spain. 1984b. The effect of wall mass on winter heating loads and indoor comfort—An experimental study. ASHRAE *Transactions* 90(1B):94-121.

Burch, D.M., G.N. Walton, K. Cavanaugh, and B.A. Licitra. 1986. Effect of interior mass surfaces on the space heating and cooling loads of a single-family residence. Proceedings of the ASHRAE/DOE/ BTECC Conference on Thermal Performance of the Exterior Envelopes of Buildings III, ASHRAE SP49.

CALPAS3 *User's Manual.* Berkeley Solar Group, Berkeley, CA.

Chandra, S., P.W. Fairey, and M. Houston. 1983. *A handbook for designing naturally ventilated buildings.* FSEC-CR83EA, Florida Solar Energy Center, September.

Chi, J. and G.E. Kelly. 1978. A method for estimating the seasonal performance of residential gas and oil-fired heating systems. ASHRAE *Transactions* 84(1):405.

Childs, K.W., G.E. Courville, and E.L. Bales. 1983. Thermal mass assessment. ORNL/CON-97, Oak Ridge National Laboratory, September.

Christian, J. and D. Downing. 1986. A statistically balanced parametric study for building simulation. Proceedings of the ASHRAE/DOE/ BTECC Conference on Thermal Performance on the Exterior Envelopes of Buildings III, ASHRAE SP49.

Cinquemani, V., J.R. Owenby, and R.G. Baldwin. 1978. Input data for solar systems. U.S. Department of Energy Report No. E(49-26)1041.

Claridge, D.E., M. Krarti, and M. Bida. 1987. A validation study of variable-base degree-day cooling calculations. ASHRAE *Transactions* 93(2):90-104.

Claridge, D.E., R. Balasubramanya, L.K. Norford, and J.F. Kreider. 1992b. A multiclimate comparison of the improved TC 4.7 simplified energy analysis procedure with DOE-2. ASHRAE *Transactions* 98(1).

Claridge, D.E., L.K. Norford, and R. Balasubramanya. 1992a. A thermal mass treatment for the TC 4.7 simplified energy analysis procedure. ASHRAE *Transactions* 98(1).

Clark, G. 1981. Passive/hybrid comfort cooling by thermal radiation. Proceedings of the International Passive and Hybrid Cooling Conference, American section of the International Solar Energy Society, Miami Beach, FL.

Cole, R.J., L.W. Harder, and A.S. Ostevik. 1983. Modeling energy savings from daylighting strategies. Proceedings of the 1983 International Daylighting Conference, Phoenix, AZ.

Corson, G.C. 1992. Input-output sensitivity of building energy simulations. ASHRAE *Transactions* 98.

Crisp, V. 1981. The energy implications of flexible lighting controls. Building Research Establishment Report R6/81, London.

Cumali, Z.O., *et al.* 1979. Passive solar calculation methods. U.S. Department of Energy Report, Contract No. EMxC5221.

Davenport, M.R. and D.H. Nall. 1983. The correlation of measured interior daylight with calculated solar gains. Proceedings of the 1983 International Daylighting Conference, Phoenix, AZ.

Degelmann, L.O. and G. Andrade. 1986. A bibliography of available computer programs in the area of heating, ventilating, air conditioning, and refrigeration. ASHRAE, Atlanta, GA.

Dickinson, W.C. and P.N. Cheremisinoff, eds. 1980. *Solar energy technology handbook,* Part B: Application, systems design and economics. Marcel Dekker, Inc., New York.

Didion, D.A. and G.E. Kelly. 1979. New testing and rating procedures for seasonal performance of heat pumps. ASHRAE *Journal* (September).

Dilaura, D.L. and G.A. Hauser. 1979. On calculating the effects of daylighting in interior spaces. Journal of the Illuminating Engineering Society, October.

DOE. 1980. *Passive solar design handbook.* DOE/CS-0127/2, II. U.S. Department of Energy.

Duffie, J.A. and W.A. Beckman. 1980a. *Solar engineering of thermal processes.* John Wiley and Sons, New York.

Duffie, J.A. and W.A. Beckman. 1980b. *Solar thermal energy processes.* Wiley Interscience, New York.

EPRI. 1980. Project RP 1351 Final Report, Electric Power Research Institute.

Erbs, D.G., S.A. Klein, and W.A Beckman. 1983. Estimation of degree days and ambient temperature bin data from monthly-average temperatures. ASHRAE *Journal* 25(6):60.

Feldman, S.J. and R.L. Merriam. 1979. Building energy analysis computer programs with solar heating and cooling system capabilities. Arthur D. Little, Inc. Report to the Electric Power Research Institute No. EPRIER-1146 (August).

Fels, M.F. and M. Goldberg. 1986. Refraction of PRISM results in components of saved energy. *Energy and Buildings* 9:169.

Freeman, T.L., J.W. Mitchell, and T.E. Audit. 1979. Performance of combined solar-heat pump systems. *Solar Energy* 22:2.

Gillette, G. 1983. A daylighting model for building energy simulation. NBS Building Science Series 152, USGPO, Washington, D.C.

Givoni, B. 1981. Experimental studies on radiant and evaporative cooling of roofs. Proceedings of the International Passive and Hybrid Cooling Conference, American Section of the International Solar Energy Society, Miami Beach, FL.

Hacker, R.J., J.W. Mitchell, and W.A. Beckman. 1984. HVAC system dynamics and energy use in existing buildings. ASHRAE *Transactions* 90:2.

Hacker, R.J., J.W. Mitchell, and W.A. Beckman. 1985. HVAC system dynamics and energy use in buildings—Part II. ASHRAE *Transactions* 91:1.

Hay, H.R. and J.I. Yellott. 1969. Natural air conditioning with roof ponds and movable insulation. ASHRAE *Transactions* 75(1):165-77.

Herron, D., G. Walton, and L. Lawrie. 1981. Building loads analysis and systems thermodynamics (BLAST) program users manual, Vol. 1 supplement, Version 3.0, U.S. Army report CERL-TR-E-171, March.

Howell, R.H. and S. Suryanarayana. 1990. Sizing of radiant heating systems: Part I and Part II. ASHRAE *Transactions* 96.

IES. 1979. Recommended practice of daylighting. RP-5-79. Prepared by the IES Daylighting Committee, Lighting Design and Applications, New York, February.

IES. 1987. *Lighting handbook,* Application Volume. Illumination Engineering Society of North America, New York.

Johnson, R., *et al.* 1984. Glazing energy performance and design optimization with daylighting. *Energy and Buildings* 6:3.

Jurovics, S. 1979. Solar radiation data, natural lighting and building energy minimization. ASHRAE *Transactions* 85:2.

Kelly, G.E., C. Park, D.R. Clark, and W.B. May, Jr. 1984. HVACSIM—A dynamic building/HVAC/control system simulation program. Proceedings of the workshop on HVAC controls modeling and simulation, Georgia Institute of Technology, Atlanta.

Klein, S.A. and W.A. Beckman. 1979. A general design method for closed-loop solar energy systems. *Solar Energy* 22(3):269-82.

Klein, S.A., W.A. Beckman, and J.A. Duffie. 1976. TRNSYS—A transient simulation program. ASHRAE *Transactions* 82(1):623-33.

Knebel, D.E. 1983. Simplified energy analysis using the modified bin method. ASHRAE, Atlanta, GA.

Knebel, D. and S. Silver. 1985. Upgraded documentation of the TC 4.7 simplified energy analysis procedure. ASHRAE *Transactions* 92(2).

Kohler, J.T. and P.W. Sullivan. 1980. Thermal modeling of passive solar buildings on a programmable calculator. ASHRAE *Transactions* 86:2.

Kreider, J.E. and F. Kreith. 1981. *Solar energy handbook.* McGraw Hill, New York.

Kusuda, T. 1969. Thermal response factors for multi-layer structures of various heat conduction systems. ASHRAE *Transactions* 75.

LBL. 1981. DOE-2 Reference Manual Version 2.1A. Los Alamos Scientific Laboratory, Report LA-7689-M, Version 2.1A. Report LBL-8706 Rev. 2, Lawrence Berkeley Laboratory, May.

Lachel, B., W.U. Weber, and O. Guisan. 1992. Simplified methods for the thermal analysis of multifamily and administrative buildings. ASHRAE *Transactions* 98.

Littlefair, P.J. 1985. The luminous efficacy of daylight: A review. *Lighting Research and Technology* 17.

Loxsom, F., *et al.* 1983. Validation of a thermal simulation of a roofpond building. Trinity University Report to the U.S. Department of Energy, DE-AC03-79CS53020.

Lunde, P.J. 1980. *Thermal engineering.* John Wiley and Sons, New York.

Macriss, R.A. and R.H. Elkins. 1976. Standing pilot gas consumption. ASHRAE *Journal* 18(6):54-57.

Marlatt, W., C. Murray, and S. Squire. 1984. Roofpond systems energy technology engineering center. Rockwell International, Report No. ETEC6, April.

Martin, M. and P. Berdahl. 1984. Characteristics of infrared sky radiation in the United States. *Solar Energy* 33(3/4):321-36.

May, W.B. 1985. HVAC emulation and on-line testing of EMCS systems. International Symposium on Recent Advances in Control and Operation of Building HVAC Systems, Norwegian Institute of Technology, Trondheim, Norway.

Mazria, E. 1979. *The passive solar energy book.* Rodale Press, Emmaus, PA.

McFarland, R.D. 1978. PASOLE, A general simulation program for passive solar energy. Los Alamos National Laboratory, LA7433-MS, October.

McQuiston, F.C. and J.D. Spitler. 1992. *Cooling and heating load calculation manual*, 2nd ed. ASHRAE, Atlanta.

Miller, D.E. 1980. The impact of HVAC process dynamics on energy use. ASHRAE *Transactions* 86(2):535-56.

Milne, M. and S. Yoshikawa. 1979. Solar-5—An interactive computer-aided passive solar building design system. Proceedings of the Third National Passive Solar Conference, San Jose, CA.

Mitalas, G.P. 1968. Calculation of transient heat flow through walls and roofs. ASHRAE *Transactions* 74(2):182-88.

Mitalas, G.P. and D.G. Stephenson. 1967. Room thermal response factors. ASHRAE *Transactions* 73:III, 2.1—III 2.10.

Mitchell, D. and K.L. Biggs. 1979. Radiative cooling of buildings at night. *Applied Energy* 5:263-75.

Mitchell, J.W. 1983. *Energy engineering.* John Wiley and Sons, Inc., New York.

Monsen, W.A., S.A. Klein, and W.A. Beckman. 1981. Prediction of direct gain solar heating system performance. *Solar Energy* 27(2):143-47.

Monsen, W.A., S.A. Klein, and W.A. Beckman. 1982. The un-utilizability design method for collector-storage walls. *Solar Energy* 29(5):421-29.

Morehouse, J.H. and P.J. Hughes. 1979. Residential solar-heat pump systems: Thermal and economic performance. Paper 79-WA/SOL-25, ASME Winter Annual Meeting, New York, December.

NOAA. 1973. Degree-days to selected bases. U.S. National Climatic Data Center, Asheville, NC.

Nordham, D. 1981. Microcomputer methods for solar design and analysis. Solar Energy Research Institute, SERI-SP-722-1127, February.

Parker, W.H., G.E. Kelly, and D. Didion. 1980. A method for testing, rating, and estimating the heating seasonal performance of heat pumps. National Bureau of Standards, NBSIR 80-2002, April.

Place, J.W., *et al.* 1987. The impact of glazing orientation and tilt on the energy performance of roof apertures. ASHRAE *Transactions* 93:1A.

Robbins, C.L. 1983. A simplified method for predicting energy savings attributed to daylighting. Proceedings of the 1983 International Daylighting Conference, Phoenix, AZ.

Robbins, C.L. 1986. *Daylighting.* Van Nostrand Reinhold, New York.

Robertson, D.K. and J.E. Christian. 1985. Comparisons of four computer models with experimental data from test buildings in northern New Mexico. ASHRAE *Transactions* 91:2.

Rohles, F.H., S.A. Konz, and B.W. Jones. 1983. Ceiling fans as extenders of the summer comfort envelope. ASHRAE *Transactions* 89(1):245-62.

Roschke, M.A., B.D. Hunn, and S.C. Diamond. 1978. A component based simulator for solar systems. Proceedings of the conference on systems simulation and economic analysis for solar heating and cooling, San Diego, CA, June.

Rundquist, R.A. 1982. Daylighting energy savings algorithms. ASHRAE *Transactions* 88(1):343-76.

Rundquist, R.A. 1991. Calculation procedure for daylighting and fenestration effects on energy and peak demand. ASHRAE *Transactions* 91(1).

Sain, A.G., P.G. Rockwell, and J.E. Davy. 1983. Energy nomographs as a design tool for daylighting. Proceedings of the 1983 International Daylighting Conference, Phoenix, AZ.

Schnurr, N.M., B.D. Hunn, and K.D. Williamson. 1981. The solar load ratio method applied to commercial buildings active solar system sizing. Proceedings of the ASME Solar Energy Division Third Annual Conference on System Simulation, Economic Analysis/Solar Heating and Cooling Operational Results, Reno, NV, May.

Shavit, G. 1975. Humidity control and energy conservation. ASHRAE *Transactions* 81(1):701-15.

Shavit, G. 1977. Energy conservation and fan systems: Floating space temperature. ASHRAE *Journal* 19(10):29-34.

Sonderegger, R.C. 1985. Thermal modeling of buildings as a design tool. Proceedings of CHMA 2000, Vol 1.

Sowell, E.F. and G.N. Walton. 1980. Efficient calculation of zone loads. ASHRAE *Transactions* 86(1):49-72.

Spielvogel, L.G. 1977. Comparison of energy analysis computer programs. ASHRAE *Transactions* 83(2):293-99.

Sud, I., *et al.* 1979. Development of the Duke University building energy analysis method (DUBEAM) and generation of plots for North Carolina. NTIS:NCEI-0009, November.

Svard, C.D., J.W. Mitchell, and W.A. Beckman. 1981. Design procedure and application of solar-assisted series heat pump systems. *Journal of Solar Energy Engineering* 103(5):135.

Threlkeld, J.L. and R.C. Jordon. 1954. Availability and utilization of solar energy. Part III: Design and economics of solar energy heat pump systems. ASHVE *Transactions* 60:212-38.

Treado, S. and T. Kusuda. 1980. Daylighting, Window Management Systems and Lighting Controls. NBSIR80-2147 NTIS, Springfield, VA.

Treado, S., D. Holland, and W. Remmart. 1986. Building energy analysis with BLAST and CEL-1, National Bureau of Standards Report NBSIR 86-3256, February.

U.S. Air Force. 1978. Engineering weather data. Air Force Manual AFM 88-29, U.S. Government Printing Office, Washington, D.C.

U.S. Army. 1979. BLAST, The Building Loads Analysis and System Thermodynamics Program—Users Manual, Vol. 1. U.S. Army Construction Engineering Research Laboratory Report E-153, June.

U.S. Department of Commerce. 1973. Degree days to selected bases for first-order type stations. National Climatic Center, Asheville, NC.

U.S. Department of Energy. 1979. Energy Budget Levels Selection. DOE/CS-0119, November.

U.S. Department of Energy. 1980 and 1982. Passive Solar Design Handbook. Vols. 2 and 3, Passive solar design analysis. DOE Reports/CS-0127/2 and CS-0127/3. January, July.

Vadon, M., J.F. Kreider, and L.K. Norford. 1991. Improvement to the solar calculations in the modified bin method. ASHRAE *Transactions* 97(1).

Vickery, B.J. 1981. The use of the wind tunnel in the analysis of naturally ventilated structures. Proceedings of the International Passive and Hybrid Cooling Conference, American section of the Solar Energy Society, Miami Beach, FL.

Walton, G.N. 1980. A new algorithm for radiant interchange in room loads calculations. ASHRAE *Transactions* 86(2):190-208.

Waltz, J.P. 1992. Practical experience in achieving high levels of accuracy in energy simulations of existing buildings. ASHRAE *Transactions* 98.

Wilde, M. 1985. Daylighting design tool survey. Lawrence Berkeley Laboratory Report LBL-21054, June.

Winkelmann, F.C. and S. Selkowitz. 1986. Daylighting simulation in the DOE-2 building energy analysis program. *Energy and Buildings* 6:3.

Wise, B.B. 1984. Energy monitoring and control system software testing device. Proceedings of the Ninth Energy Management and Control Society Conference, Tampa, FL, November.

Yellott, J.I. 1972. Effect of louvered sun screens upon fenestration heat loss. ASHRAE *Transactions* 78:1.

Yellott, J.I. and W.B. Ewing. 1976. Energy conservation-exterior shading of fenestration techniques. ASHRAE *Journal* 18(7):23-30.

Zabinski, M.P. and L. Loverme. 1974. Fuel consumption in residential heating at various thermostat settings. ASHRAE *Journal* 16(12):67.

BIBLIOGRAPHY

Alereza, T. and R.I. Hossli. 1979. A simplified method of calculating heat loss and solar gain through residential windows during the heating season. ASHRAE *Transactions* 85:1.

Beard, W.L. 1977. A method for calculating energy consumption of a room air conditioner. ASHRAE *Transactions* 83:1.

Boyd, R.L. 1959. Annual energy use for electric space heating. ASHRAE *Transactions* 65.

Cane, R.L.D. 1979. A "modified" bin method for estimating annual heating requirements of air source heat pumps. ASHRAE *Journal* 21:9.

Effect of reducing excess firing rate on the seasonal efficiency of 26 Boston oil-fired heating systems. 1976. Conference on Efficiency in HVAC Equipment and Components, II, Proceedings 81 and 375. Purdue University, April.

Ellison, R.D. 1977. The effects of reduced indoor temperature and night setback on energy consumption of residential heat pumps. ASHRAE *Journal* 19(2):21-25.

Kusuda, T., I. Sud, and T. Alereza. 1981. Comparison of DOE-2 generated residential design energy budgets with those calculated by the degree day and bin methods. ASHRAE *Transactions* 87(1):491-506.

Nall, D.H. and E.A. Arens. 1979. The influence of degree day base temperature on residential building energy prediction. ASHRAE *Transactions* 85:1.

Quentzel, D. 1976. Night-time thermostat set back: Fuel savings in residential heating. ASHRAE *Journal* 18(3):39-43.

Sherwood, L. 1979. Total energy use of home heating systems. ASHRAE *Transactions* 85:1.

Solar design handbook. 1980. Solar Energy Research Institute Manual, SERI/SPb308, May.

CHAPTER 29

COOLING AND FREEZING TIMES OF FOODS

THIS chapter reviews selected procedures available for estimating heat conduction that occurs during the cooling, freezing, and thawing of foods. These procedures may be classified as *semitheoretical* or *theoretical*. Semitheoretical methods combine experimental data and heat balance equations; theoretical methods rely solely on mathematical solutions. Examples in the chapter use selected procedures to estimate cooling or freezing time.

THEORETICAL METHODS

Most theoretical models use heat conduction equations similar to Fourier's equation. Simple geometrical shapes can be solved analytically, but complex shapes require numerical solutions. This chapter classifies theoretical methods as (1) those assuming that freezing or defrosting completes at a single temperature and (2) those assuming that the phase change completes over a range of temperatures.

Phase Change at Single Temperature

Plank's formula (1941) is widely used because of its mathematical simplicity. It can only be used to estimate freezing or defrosting time, and it gives no information on transient temperature distribution. It was derived analytically by assuming conditions B2, I1, P2, and M1 as defined in Table 1. In addition, the surrounding medium is assumed to be in a steady-state condition. Plank's formula, Equation (1), applies to samples with infinite slab, infinite cylinder, spherical, or rectangular parallelepiped configurations.

$$\theta_f = \frac{L_\rho}{|t_f - t_a|} \, (Pd/h + Rd^2/k) \tag{1}$$

To use Plank's equation for freezing, k and ρ should be the values of frozen food, since heat is removed through a frozen layer. In thawing, the property values of unfrozen food should be used. The following table shows values of P and R for the shapes indicated.

Shapes	P	R
Slab	1/2	1/8
Cylinder	1/4	1/16
Sphere	1/6	1/24
Rectangular parallelepiped[a]	See Figure 1	See Figure 1

[a]Figure 1 shows P and R values for food of dimensions d by $\beta_1 d$ by $\beta_2 d$. In this figure, β_1 and β_2 are interchangeable, since P and R curves are symmetric with respect to the 45° diagonal line.

Ede (1949) prepared charts to simplify the estimation of freezing time of food of brick-shape (rectangular parallelepiped).

Even though condition I1 was assumed, the formula is applicable to condition I2, as Plank originally suggested. For this application, latent heat of the phase change in the equation is replaced with total heat required for completing freezing or thawing as used by Earle and Fleming (1967) and Nagaoka *et al.* (1955). Scott and Hayakawa (1977) prepared a set of charts for the determination of phase change time by using Plank's formula.

Table 1 Classification of Restrictive Assumptions Imposed for Deriving Analytical or Numerical Solutions of Heat Balance Equations

Code No.	Condition	Assumption
B1	Boundary conditions	Given surface temperature of sample object or infinitely large Biot number.
B2		Convective heat exchange between surrounding medium and surface of sample or finite Biot number.
B3		Radiative heat exchange between radiative heat source and surface of sample object.
B4		Prescribed heat flux on surface.
B5		Moisture loss or gain on surface.
I1	Initial condition	Constant initial temperature of sample object, identical to a phase change temperature or identical to upper or lower limit of phase change temperatures.
I2		Constant initial temperature of sample object, different from phase change temperatures.
I3		Location variable initial temperatures, different from phase change temperatures.
D1	Density of sample object	No difference in density between frozen and unfrozen samples. Temperature-independent density.
D2		Difference in density between frozen and unfrozen samples. Temperature-independent density.
D3		Similar to D2. Temperature-dependent density.
P1	Other physical properties	Temperature-independent properties. Properties of frozen sample different from those of unfrozen sample.
P2		Temperature-dependent properties.
M1	Movement of unfrozen mass	No convective nor diffusional movement of unfrozen mass of sample.
M2		Convective movement of unfrozen mass or solute owing to volumetric change.
M3		Diffusional movement of unfrozen mass or solute owing to its concentration gradient.
M4		Both of M2 and M3.

The preparation of this chapter was assigned to TC 11.9, which is presently inactive.

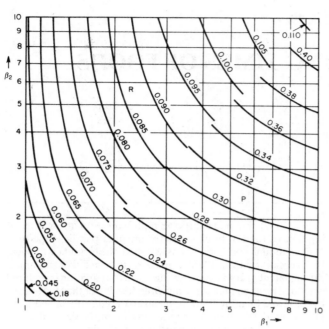

Fig. 1 P and R Values for Brick Shape
(Ede 1949)

ing and the unfrozen region during thawing) and in region 2 (the unfrozen region during freezing and the frozen region during thawing).

$$t_1 = t_s + (t_{fp} - t_s)\, x_c/X_c \tag{2}$$

$$t_2 = t_{fp} + (t_o - t_{fp}) \left[2 \frac{x_g - X_g}{\delta - X_g} - \left(\frac{x_g - X_g}{\delta - X_g} \right)^2 \right] \tag{3}$$

where $X_g = 2\,(\rho_1/\rho_2)\,\gamma\,\sqrt{\alpha, \theta}$

$$\delta = bX_g \qquad b = \left(\frac{9}{4} + \frac{3\,(\alpha_2/\alpha_1)}{(\rho_1/\rho_2)^2\gamma^2} \right)^{0.5} - 0.5 \tag{4}$$

and where

$$X_c = 2\,\gamma\,\sqrt{\alpha_1\theta} \qquad \gamma^2 = \frac{b_1 - \sqrt{b_1^2 - 4a\,S_T^2}}{2a} \tag{5}$$

$$b_1 = 2S_T[2 + S_T + \phi(k_2/k_1)(\rho_1/\rho_2)\,S_T/(\alpha_2/\alpha_1)] + 4/3\,(\phi k_2 S_T/k_1)^2/(\alpha_2/\alpha_1)$$

$$a = (2 + S_T)[2 + S_T + 2S_T\phi k_2\rho_1\alpha_1/(k_1 P_2\alpha_2)]$$

$$X_c - X_g = 2\,\gamma\,\sqrt{\alpha_1\theta}\,(1 - 2\,\rho_1/\rho_2)$$

In Equation (5), $X_c - X_g$ is either a positive value representing a change in the location of free surface caused by expansion or a negative value caused by shrinkage.

Talmon and Davis (1981) estimated freezing and thawing times of an infinite slab by assuming B2, I2, D2, P1, and M1 and by using a modified isotherm migration method. The estimated freezing or thawing times agreed with published results based on a finite difference procedure.

Charm *et al.* (1972) estimated transient state temperatures of food in an infinite slab or cylindrical shape by numerically solving the heat balance equation that was applicable to each of the imaginarily subdivided layers of the food. The estimation was made by assuming conditions B2, I2, D1, P1, and M1.

Tao (1967, 1968) prepared useful formulas and charts for estimating the phase change time of a body in the shape of an infinite cylinder, infinite slab, or sphere. For this preparation, Fourier's equation for heat conduction was solved numerically by applying a finite difference technique and assuming conditions B2, I1, D1, P1, and M1. The phase change time of a sphere can be computed using Equation (6).

$$\theta_f = 1/6 + f/(3\mathrm{Bi}) + \eta_1 + \eta_2/\mathrm{Bi} - \eta_3 e^{-1/\mathrm{Bi}} \tag{6a}$$

where

$$\eta_1 = 0.096\,[1 - \exp(5.49\,\gamma^3 - 6.62\,\gamma^2 - 4.59\,\gamma)] + 0.176/\mathrm{Bi} \tag{6b}$$

$$\eta_2 = \gamma\,[\,0.024 + 0.179\,\exp(-2.19\,\gamma)] \tag{6c}$$

$$\eta_3 = 0.051\,[1 - \exp(-2.52\,\gamma)] + 0.104\,\gamma \tag{6d}$$

Kern (1977) examined heat transfer in an infinite slab by assuming conditions B2, I1, D1, P1, and M1. He then derived simple analytical expressions for estimating the upper and lower bounds of true phase change times.

Voller and Cross (1981) examined heat conduction in an infinite cylinder undergoing freezing by assuming B1, I2, D1, P1, and M1. They derived the following simple formula for estimating the freezing time.

Chung and Yeh (1975) considered heat conduction in a semiinfinite body. They assumed conditions B2, B3 (simultaneous convection and radiation on the surface of the sample), I1, D1, P1, and M1. They reduced Fourier's partial differential equations by applying Biot's variational method or Goodman's integral method. Even though the ordinary differential equations are not suitable for deriving analytical expressions, they may be easily solved by using any proper numerical integration procedure.

Yuen (1980) obtained approximate, analytical solutions for locating a melting front and temperature distribution in a semi-infinite body by assuming B1, B2, or B4, as well as I2, D1, P1, and M1. He also assumed that the product of specific heat and density did not change with the phase conversion. He used Goodman's integral method to obtain these formulas. The formulas obtained by assuming B1 and B4 are closed form are relatively simple and could be used if his assumed conditions are met. The formula for assumption B2 is in the form of a nonlinear ordinary differential equation that should be solved numerically. Lunardini (1983a, 1983b) obtained approximate, analytical solutions in a closed form for freezing or thawing a semiinfinite body. His assumptions are identical to Yuen's, except for a change in the product of specific heat and density with the phase change. Lunardini observed that results obtained from his solutions agreed with available exact or approximate solutions.

The initial temperature distribution in some sample bodies is not uniform. Therefore, Lunardini (1983a) obtained an analytical solution for estimating transient state temperature distribution in a semiinfinite body. The solution represented the initial temperature distribution as the linear function of the location variable (assumed as an I3 condition). The other conditions assumed were B1, D2, P1, and M1.

Lunardini (1983b) considered the expansion of water when it freezes, to derive an approximate, analytical solution for freezing a semiinfinite body by using assumptions B1, I2, D2, P1, and M1. He used two location variables, x_1 and x_2, to solve the problem. The origin of x_1 is on the current exposed surface, which expands with freezing and shrinks with thawing; x_2 begins on the original surface. The following equations can be used to estimate the temperature distribution in region 1 (the frozen region during freez-

$$\text{Fo}_f = (0.14 + 0.085 \, Y_o) + (0.252 - 0.0025 \, Y_o) \, Ld \qquad (7)$$

Equation (7) estimated the freezing time with an error of less than 2%, compared to the time estimated by numerical solutions when $0 \leqslant Y_o \leqslant 2$ and $2 \leqslant Ld \leqslant 50$.

Hill and Kucera (1983) obtained an approximate analytical formula for estimating freezing time of a spherical body by assuming B2, I1, D2, P1, and M1.

Rathjen and Jiji (1970) derived approximate analytical formulas to estimate transient state heat transfer and to locate the phase change front in a two-dimensional corner by assuming conditions B1, I1, D1, P1, and M1. They used a superhyperbola to approximate a curve representing the front location. The locations determined by the derived analytical solutions correlated with those determined by solving the original heat conduction equation using a finite difference technique. They also prepared a series of charts for predicting the locations.

Most foods are opaque to thermal radiative heat transfer. However, internal radiative heat transfer contributes significantly to freezing and thawing of some foods such as clear gels or clear solutions, if they are in transparent containers. Chan *et al.* (1986) simulated freezing or thawing of a transparent, semiinfinite body under conditions B1, I1, D2, P1, and M1. They solved the model for several limiting cases, including cases of freezing and thawing by internal radiative transfer with no internal conduction.

A difficulty in solving a heat conduction equation with phase conversion is estimating the location of the freezing or defrosting front as a function of time and temperature distribution. This can be overcome by using enthalpy as a dependent variable, in addition to temperature (enthalpy method). For example, one-dimensional heat conduction with cartesian coordinates becomes:

$$\rho \, \frac{\delta H}{\delta \theta} = \frac{\delta}{\delta x} \left(k \frac{\delta t}{\delta x} \right) \qquad (8)$$

Since enthalpy H depends only on temperature, the above equation can be solved by a proper numerical method. The freezing or thawing front can then be located from estimated enthalpy or temperature distribution.

Shamsunder and Sparrow (1975) used the enthalpy method to examine freezing a square column by assuming B2, I1, D1, P1, and M1. The enthalpy method can be applied to cases of freezing over a range of temperatures.

Several researchers have developed a method for evaluating the flux of unfrozen liquid in a body undergoing freezing or thawing. Although most of these methods are for analyzing the flux in nonfoods, they also apply to food freezing.

Shamsunder and Sparrow (1976) modified their previous equation based on the application of the enthalpy method by including an additional term representing changes in enthalpy caused by the movement of unfrozen fluid. This modified equation was used to estimate the formation of cavity in a eutectic fluid undergoing a phase change by assuming conditions B2, I1, D2, P1, and M4.

Tarnawski (1976) investigated heat and moisture transfer in slab-shaped food by assuming conditions B2, I3, D2, P2, and M3. He represented the transfer problems with two simultaneous partial differential equations — Fourier's heat conduction equation and a moisture diffusion equation. He assumed that internal moisture transfer was caused by a gradient in moisture transfer potential, although the volumetric change in the food undergoing freezing or thawing could be one of the major factors responsible for the moisture transfer.

Cho and Sunderland (1974) derived an exact analytical solution for heat conduction in a semiinfinite body by assuming conditions B1, I2, D2, P2, and M2. They assumed that thermal conductivities of frozen and unfrozen material changed linearly with temperature, while condition P1 was applicable to the specific heat.

Grange *et al.* (1976) examined freezing of salt solution in a body whose overall configuration was assumed to be an infinite slab. They also assumed conditions B1, I2, D2, P1, and M4. Their work is unique in that they considered the convective movement of unfrozen water as well as the diffusional movement of the solute. Wollhoever *et al.* (1985) examined theoretically the diffusional transfer of NaCl (condition M3) in an aqueous solution occupied in a semiinfinite space. Assumed conditions for this examination are B1, D2, and P1.

Phase Change over Range of Temperatures

Because most foods do not complete freezing or thawing at one fixed temperature but gradually complete the phase change over a range of temperatures, the mathematical procedures in this section represent heat transfer in food.

Keller and Ballard (1956) estimated transient state temperature distributions in concentrated orange juice by assuming conditions B2, I2, D1, P1, and M1. Schmidt's graphical method was used for this estimation. Using effective thermal conductivity and effective specific heat, which included latent heat of phase change, simplified the computations required.

Tien and Geiger (1967) examined heat conduction in a semiinfinite body by assuming conditions B1, I1, D1, P1, and M1. Because they assumed a phase change over a range of temperatures, a freezing front was replaced with a phase change zone in which there was a mixture of frozen and unfrozen materials. Goodman's integral method was applied to obtain an approximate analytical solution. Tien and Geiger (1968) obtained another set of solutions by assuming time-variable surface temperatures and conditions similar to those in their previous paper. Tien and Koump (1968) reported similar analysis of heat transfer in an infinite slab.

Hayakawa and Bakal (1973) examined heat conduction in an infinite slab and found it necessary to consider the several different combinations of frozen, phase changing and/or unfrozen zones in the slab for the complete evaluation of heat conduction. For each combination of these zones, they derived a set of approximate analytical formulas by applying Goodman's integral technique and assuming conditions B2, I2, D1, P1, and M1. There was agreement between temperatures predicted by most of the analytical formulas and those determined experimentally.

The results of similar analyses were presented by Muehlbauer *et al.* (1973). They simplified the analyses by using apparent specific heat values in which latent heat was included. With this inclusion, the numerical analysis of the heat conduction equation was simplified because there was no need to determine a freezing front.

Tien and Koump (1969) evaluated heat transfer in a semiinfinite body when density changes during freezing or thawing. For this evaluation, they derived approximate analytical formulas by assuming conditions B1, I1, D2, P1, and M2; these are the only available formulas for predicting the influence of the density change on heat conduction, although several researchers have examined this influence by numerically solving the heat and mass balance equations. The application of these formulas for food freezing has been discussed by Hayakawa and Bakal (1973).

Joshi and Tao (1974), as well as Katayama and Hattori (1975), used procedures similar to the enthalpy model technique described by Shamsunder and Sparrow (1975). They obtained solutions by using finite difference techniques. Joshi and Tao (1974) studied heat conduction in lean meat, the overall configuration of which was approximated with an infinite cylinder, by assuming conditions B1, I2, D1, P2, and M1. Shamsunder and Sparrow (1975) presented brief discussions on a mathematical model and then examined heat conduction in sands saturated with water by assuming conditions B1, I2, D1, P1, and M1. The sands were assumed to be filled either in a semiinfinite space or in an infinite space around a circular pipe. For the former configuration, they also examined the mass flux of unfrozen water by assuming condition D2. This procedure can be applied for freezing porous food.

Mascheroni and Calvelo (1982) developed a method for predicting the freezing time of an infinite plate by assuming B2, I2, D3, P2, and M1. They computed a freezing time by adding precooling, phase conversion, and tempering times. The first and third times were estimated by using analytical solutions of a simple heat conduction equation without phase conversion, the second by numerically solving a nonlinear heat conduction equation. They assumed (1) that precooling terminated when temperature on the midplane between the geometrical center and surface of the plate reached the initial freezing point, (2) that the phase conversion ended when the temperature at the geometrical center reached the initial freezing point, and (3) that the tempering completed when the thermal center reached 14 or 0°F. They estimated the phase change time by using the following equation:

$$\theta_f = \frac{\rho_w f_{rw} L_w f_{ri} a^2}{(t_{sh} - t_a)k} \left(k/ha + 1/2\right) \tag{9}$$

where thermal conductivity of food k is evaluated at the mid-temperature between the initial freezing point and cooling medium temperature.

De Michelis and Calvelo (1982) developed a method for predicting freezing time of an infinite plate where two surfaces were subjected to two different surface heat conductances and/or medium temperatures. Their assumptions were identical to those of Mascheroni and Calvelo (1982).

Cleland and Earle (1979) analyzed heat conduction in food approximated by an infinite cylinder, sphere, or rectangular parallelepiped by assuming B2, I2, D3, P2, and M1 and by using thermophysical property values of the Karlsruhe test substance that has properties similar to lean beef. From this analysis, they obtained the following Plank's-type formula.

$$\theta_f = \frac{\Delta H_v}{t_{sh} - t_a} \left(P_2 d/h + R_2 d^2/k_s\right) \tag{10}$$

The values of modified Plank's constants P_2 and R_2 can be estimated by following the two-step calculation below:

Step 1. Calculate P_1 and R_1 from the original Plank's constants, P and R.

$$P_1 = P\,[1.026 + 0.5808\,Pk + Ste\,(0.2296\,Pk$$
$$+ 0.0182/Bl + 0.1050)] \tag{11}$$

$$R_1 = R\,[1.202 + Ste\,(0.410\,Pk + 0.7336)] \tag{12}$$

Step 2. Calculate P_2 and R_2 for a respective configuration. For infinite plate:

$$P_2 = P_1 \text{ and } R_2 = R_1 \tag{13}$$

For infinite cylinder and sphere:

$$P_2 = P_1 + 0.1278P \tag{14}$$

$$R_2 = R_1 - 0.1888R \tag{15}$$

For rectangular brick:

$$P_2 = P_1 + P\,[0.1136 + Ste\,(5.766P - 1.242)] \tag{16}$$

$$R_2 = R_1 + R\,[0.7344 + Ste\,(49.89R - 2.900)] \tag{17}$$

The above equations are applicable within the following ranges of parametric values:

$$0.155 \leqslant Ste \leqslant 0.34 \qquad 0 \leqslant Pk \leqslant 0.55$$
$$0.2 \leqslant Bi \leqslant 20 \text{ for plate}$$
$$0.5 \leqslant Bi \leqslant 4.5 \text{ for cylinder and sphere}$$
$$0.5 \leqslant Bi \leqslant 22 \text{ for brick}$$
$$1 \leqslant \beta_1 \leqslant 4 \text{ for brick}$$
$$1 \leqslant \beta_2 \leqslant 4 \text{ for brick}$$

According to Cleland and Earle (1979), differences between freezing time estimated by their formula and determined experimentally were less than 10%. Equation (10) should be limited to substances with properties similar to lean beef.

Hayakawa et al. (1983) developed a computer program to simulate two-dimensional heat conduction in anisotropic food of an arbitrary shape by modifying a program developed by Comini and Del Giudice (1976). They assumed conditions B2, B3, B4, B5, I2, D3, P2, and M1. For this development, the following formulas were used to estimate temperature-dependent thermophysical property values, because it was more convenient to use the formulas than tabulated data, and the influence of empirical constants included in the formulas on freezing or thawing rate could be examined.

Density:

$$\rho = f_l \text{ for } t \geqslant t_{sh} \tag{18}$$

$$\rho = \rho_\rho + S_{ds}(t_{sh} - t) + (\rho_l - \rho_r)(t_{sw} - t_{sh})/(t_{sw} - t)$$
$$\text{for } t < t_{sh} \tag{19}$$

Apparent specific heat:

$$C = C_l \text{ for } t \geqslant t_{sh} \tag{20}$$

$$C = C_r + E/(t_{sw} - t)^{n_c} \text{ for } t < t_{sh} \tag{21}$$

Anisotropic thermal conductivity:

$$k_x = k_{lx} + S_{klx}(t - t_{sh}) \text{ for } t \geqslant t_{sh} \tag{22}$$

$$k_y = k_{ly} + S_{kly}(t - t_{sh}) \text{ for } t \geqslant t_{sh} \tag{23}$$

$$k_x = k_{rx} + S_{ksx}(t_{sh} - t)$$
$$+ (k_{lx} - k_{rx})(t_{sw} - t_{sh})/(t_{sw} - t) \text{ for } t < t_{sh} \tag{24}$$

$$k_y = \gamma k_x \text{ for } t < t_{sh} \tag{25}$$

where

$$k_{ly} = \gamma k_{lx} \tag{26}$$

Note that Equation (24) can be used to estimate y-directional or isotropic thermal conductivity if the empirical parameters included in this equation are replaced with theoretical parameters. For example, to estimate y-directional conductivity, replace k_{rx}, S_{ksx}, and k_{lx} with k_{ry}, S_{ksy}, and k_{ly}, respectively. Because of Equation (26), $k_{ry} = \gamma k_{rx}$, $S_{ksy} = \gamma S_{ksx}$, and $k_{ly} = \gamma k_{lx}$.

The latent heat of the phase conversion is included in Equation (21), which was obtained by slightly modifying Schwartzberg's formula (Schwartzberg 1977). Equation (24) is based on the direct application of Schwartzberg's formula for estimating thermal conductivity at temperatures below the highest freezing point t_{sh}. According to available data, the ratio of thermal conductivities parallel and perpendicular to muscle fibers is almost constant when $t \leqslant t_{sh}$; therefore, Equation (25) is used.

Hayakawa *et al.* (1983) examined the influence of all independent parameters on freezing food thermophysically approximated with an infinite rectangular column or finite cylinder. For this examination, they used the maximum range of values of empirical constants in Equations (18) through (26) and also of operational parameters for commercial freezing such as freezing medium temperature, initial and final food temperatures, surface heat conductance, rate for moisture loss, and radiative heatsink temperature.

To simplify computations, the following dimensionless expressions were used.

Dimensionless density:

$$Pd = \begin{cases} Rd + 1 \text{ for } U \geqslant U_{sh} & (27) \\ 1 + Md(U_{sh} - U) + Rd/(D_a - UD_b) \\ \text{for } U \geqslant U_{sh} & (28) \end{cases}$$

Dimensionless apparent specific heat:

$$Pc = \begin{cases} Rc + 1 \text{ for } U \geqslant U_{sh} & (29) \\ 1 + C_{eh}/(D_a - UD_b)^{n_c} \text{ for } U < U_{sh} & (30) \end{cases}$$

Dimensionless thermal conductivity:

$$P_{kx} = \begin{cases} 1 + R_{kx} + M_{klx}(U - U_{sh}) \text{ for } U \geqslant U_{sh} & (31) \\ 1 + L_{kx}U - M_{klx} + R_{kx}/(D_a - UD_b) \\ \text{for } U < U_{sh} & (32) \end{cases}$$

$$P_{ky} = \begin{cases} 1 + R_{ky} + M_{kly}(U - U_{sh}) \text{ for } U \geqslant U_{sh} & (33) \\ \gamma P_{kx} \text{ for } U < U_{sh} & (34) \end{cases}$$

According to statistical analysis of the computed results, the following parameters influenced dimensionless freezing time Fo_f significantly for both configurations: U_o (initial food temperature), U_a (freezing medium temperature), U_f (final food temperature), U_{sh} (initial freezing point of food), C_{eh} [constant in Equation (30)], Bi_u (parameter applicable to the three sides of an infinite rectangular column or applicable to the top and side surfaces of a finite cylinder), Bi_n (parameter applicable to remaining surfaces of both configurations), and Sf (shape factor). For rectangular food, Sf is the ratio of two sides; for cylindrical food, it is height/radius.

The freezing time was not influenced significantly by the radiative source temperature, absorbance of radiative energy, temperature dependence of thermal conductivity, anisotropy in thermal conductivity, temperature dependence of density, or rate of surface moisture loss or gain.

Through further analysis of computational results, Hayakawa *et al.* (1983) obtained the following formulas for estimating freezing time.

Rectangular food:

$$\begin{aligned} \ln(Fo_f) &= 2.454463 + 0.032643\,X_o - 0.240336\,X_a \\ &\quad - 0.257882\,X_{sh} + 0.070507\,X_f + 0.292865\,X_{eh} \\ &\quad + 0.830827\,X_{iu} + 0.257515\,X_{in} + 0.070274\,X_{sf} \\ &\quad - 0.012814\,X_o^2 - 0.051904\,X_a^2 - 0.078139\,X_{sh}^2 \\ &\quad - 0.056448\,X_{eh}^2 + 0.067824\,X_{iu}^2 - 0.028723\,X_{in}^2 \\ &\quad - 0.068525\,X_{sf}^2 - 0.030382\,X_a X_f + 0.013005\,X_f X_{eh} \\ &\quad + 0.021406\,X_{in}X_{sf} + 0.096347\,X_{iu}X_{in} - 0.015276\,X_{sh}^3 \\ &\quad - 0.004532\,X_{eh}^3 - 0.010683\,X_{iu}^3 - 0.006472\,X_{in}^3 \\ &\quad + 0.0016285\,X_{sf}^3 \end{aligned} \qquad (35)$$

Table 2 Minimum, Frequent, and Maximum Values of Significant Parameters

Parameter	Minimum	Frequent	Maximum	Variables in Eqs. (35) and (36)
U_o	0.813	1.000	1.46	X_o
C_{eh}	19.4	157.14	769.0	X_{eh}
U_{sh}	$0.07665 \log C_{eh} + 0.58879$	$0.04064 \log C_{eh} + 0.70165$	0.829	X_{sh}
U_f	0.20833	$0.70498 U_{sh} - 0.16228$	$1.63645 U_{sh} - 0.65197$	X_f
U_a[a]	$-0.99902 U_f + 0.041459$	$-0.87688 U_f + 0.13828$	0.83333	X_a
$1/Bi_n$[b]	0.008226	1.43	146.512	X_{iu}
$1/Bi_n$[b]		when $0.008226 \leqslant 1/Bi_u \leqslant 1.43$		X_{in}
	$0.0526053/Bi_u + 0.0077933$	$0.9061257/Bi_u + 0.134240$	$8.42908/Bi_u + 1.248744$	
$1/Bi_n$[b]		when $1.43 < 1/Bi_u \leqslant 146.512$		X_{in}
	$0.0057302/Bi_u + 0.074825$	$0.098703/Bi_u + 1.28886$	$0.918168/Bi_u + 11.98935$	
$1/Bi_u$[c]	0.0126	1.021	109.88	X_{iu}
$1/Bi_n$[c]		when $0.0126 \leqslant 1/Bi_u \leqslant 1.021$		X_{iu}
	$0.046311/Bi_u + 0.012016$	$0.796807/Bi_u + 0.207460$	$7.417691/Bi_u + 1.92654$	
$1/Bi_n$[c]		when $1.021 < 1/Bi \leqslant 109.88$		X_{in}
	$0.0057478/Bi_u + 0.053432$	$0.0991098/Bi_u + 0.919809$	$0.922110/Bi_u + 8.55853$	
Sf[b]	0.5	4	16	X_{sf}
Sf[c]	0.125	2.4	5	X_{sf}

[a]Negative U_a is used for transformation to the design variable X_a [b]Rectangle [c]Cylinder

Cylindrical food:

$$
\begin{aligned}
\ln (\text{Fo}_f) =\ & 2.298986 + 0.035752\,X_o - 0.229786\,X_a \\
& - 0.258789\,X_{sh} + 0.078752\,X_f + 0.291006\,X_{eh} \\
& + 0.777841\,X_{iu} + 0.070700\,X_{in} + 0.137491\,X_{sf} \\
& - 0.035143\,X_a{}^2 - 0.068954\,X_{sh}{}^2 + 0.011703\,X_f{}^2 \\
& - 0.038801\,X_{eh}{}^2 + 0.137763\,X_{iu}{}^2 - 0.129691\,X_{sf}{}^2 \\
& - 0.028044\,X_a X_f + 0.010418\,X_f X_{eh} - 0.01739\,X_f X_{iu} \\
& + 0.037177\,X_{iu}X_{in} + 0.028813\,X_{iu}X_{sf} \\
& - 0.036392\,X_{in}X_{sf} - 0.016211\,X_{sh}{}^3 - 0.006843\,X_{iu}{}^3 \\
& - 0.004018\,X_{in}{}^3 - 0.037937\,X_{sf}{}^3 \qquad\qquad (36)
\end{aligned}
$$

Equations (35) and (36) only apply when the parametric values are within their respective minimum and maximum values shown in Table 2. Since these equations were obtained through the regression analysis of computational results obtained by applying a statistical procedure, they use variables transformed from the significant parameters by using the following equation:

$$
X_x = G_1 \ln (\pi_x + G_3) + G_3 \qquad\qquad (37a)
$$

where π_x is any significant parameter, and X_x is a transformed variable.

Constants G_1, G_2, and G_3 in Equation (37) can be estimated from the minimum, frequent, and maximum values shown in Table 2 by using the following equation:

$$
\begin{aligned}
G_1 &= 5.6568/\ln[(\pi_{x,\,max} + G_2)/(\pi_{x,\,min} + G_2)] \\
G_2 &= (\pi_{x,\,max}\pi_{x,\,min} - \pi_{x,\,frq}{}^2)/(2\pi_{x,\,frq} - \pi_{x,\,min} - \pi_{x,\,max}) \\
G_3 &= -G_f \ln(\pi_{x,\,frq} + G_2) \qquad\qquad (37b)
\end{aligned}
$$

It is useful to know the magnitude of error in estimated freezing time when there are errors in the significant variable values. Table 3 shows the results of error analyses performed by using Equations (35) and (36) at the frequently observed or practiced levels of significant variables. Case 1 is an analysis performed by assuming 1.8 °F error in initial and final food temperatures, 3.6 °F error in a freezing medium temperature, 5% error in specific heat variables, 10% error in thermal diffusivity, 10% error in Biot numbers, and 0.04-in. error in food dimensions. With these reasonable errors, the maximum errors are less than 60%—one-third of which can be attributed to errors in specific heat parameters. Case 2 results were obtained by reducing the errors of operational temperatures, specific heat parameters, and thermal diffusivity to one-fifth of those for Case 1, which is possible when data on these properties are accurate. Reduction of the other errors is almost impossible. Even with errors of small magnitude, maximum errors are less than 25%. Any error caused by the *method* used to estimate freezing time would be additional to those stated above.

Table 3   Relative Errors of Estimated Freezing Times[a]

Source of Errors	Rectangle		Cylinder	
	Case 1[b]	Case 2[c]	Case 1	Case 2
Operational temperature	10.9 (%)	2.2 (%)	7.7 (%)	1.5 (%)
Food dimension	7.6	7.6	5.1	5.1
Thermophysical property	30.4	6.1	30.4	6.1
Boundary property	8.4	8.4	5.6	5.6
Total	57.3	24.3	48.6	18.3

[a]Relative error $= 100 \dfrac{\text{Error in estimated freezing time caused by source error}}{\text{Estimated freezing time}}$

[b]Error analysis based on large source errors.

[c]Error analysis based on small errors.

Reprinted with permission from *Journal of Food Science*, 1982, 48 (6):1841-1848. Copyright © by Institute of Food Technologists.

Table 4 (Succar and Hayakawa 1983) shows parametric values useful in applying Equations (35) and (36). The freezing times estimated by these equations fluctuate with an increase in Bi for some combinations of parameters instead of monotonic reduction, and care should be exercised in their application.

A computational method for estimating freezing time cannot determine thawing time when physical properties vary with temperature. Therefore, Nonino and Hayakawa (1986) developed formulas for estimating the thawing time of food thermophysically approximated by an infinitely long rectangular column or a finite cylinder. The conditions assumed for this development are identical to those applicable to Equations (35) and (36). The overall approach to this development is similar to the approach of Hayakawa *et al.* (1983). Dimensionless groups influencing thawing time significantly were identical to those included in Equations (35) and (36).

Succar and Hayakawa (1984) developed a formula for estimating freezing time of an infinite slab whose surfaces were subjected to convective and radiative heat exchange with a surrounding environment (conditions B2 and B3). The convective surface heat conductance and radiative adsorbance of one surface was assumed to be different from respective physical parameters of another side (nonsymmetric freezing). Additional assumed conditions are I2, D3, P2, and M1.

The developed formula is as follows:

$$
\begin{aligned}
\ln (\text{Fo}_1) =\ & 2.5949 + 0.0222\,X_o - 0.2117\,X_a - 0.2181\,X_{sh} \\
& + 0.05161\,X_f + 0.249\,X_{eh} + 0.770\,X_{iu} + 0.2411\,X_{in} \\
& - 0.3756\,X_f{}^2 + 0.0507\,X_{in}{}^2 + 0.083\,X_{in}X_{iu} \\
& - 0.0092\,X_{sh}{}^3 - 0.0107\,X_{iu}{}^3 - 0.0044\,X_{sh}{}^4 + 0.0334\,X_f{}^4 \\
& - 0.0028\,X_a{}^4 - 0.026\,X_{eh}{}^4 - 0.0010\,X_{in}{}^4 \qquad\qquad (38)
\end{aligned}
$$

where $\text{Fo}_1 = \alpha t/d^2$.

The values of X_o, X_a, X_{sh}, X_f, and X_{eh} may be estimated by using values given in Table 2 and Equation (37a). The values of G_1, G_2, and G_3 in Equation (37a) were estimated by using the set of equations obtained from Equation (37b) by replacing the constant of the first equation, 5.6568, with 3.76374. The values of X_{iu} and X_{in} are estimated by:

$$
X_{iu} = 0.72720 \ln (1/\text{Bi}_u + 0.005845) - 0.26307 \qquad (39)
$$

$$
X_{in} = 1.54552 \ln \left[\frac{1/\text{Bi}_n + 0.05664/\text{Bi}_u + 0.0084}{0.96275/\text{Bi}_u + 0.14263} \right]
$$
for $0.0082 \leqslant 1/\text{Bi}_u + < 1.43$ (40)

$$
X_{in} = 1.54552 \ln \left[\frac{1/\text{Bi}_n + 0.006168/\text{Bi}_u + 0.08054}{0.10487/\text{Bi}_u + 1.369396} \right]
$$
for $1.43 \leqslant 1/\text{Bi}_n + < 146.4$ (41)

Equation (38) may be used to estimate nonsymmetric freezing time of an infinite slab. This equation was obtained by assuming a frequently observed level of radiative heat adsorbance (0.8), since the radiative surface heat transfer did not influence the freezing time significantly, according to a parametric analysis. The whole thickness of a slab d was used as the characteristic dimension to calculate Bi_n and Bi_u values included in Equations (39) through (41).

Comini *et al.* (1974) used apparent specific heat, which included the latent heat of phase change, and showed numerical procedures for obtaining finite element solutions by assuming B2, B3, I3, D3, P2, and M1. Specific heat is a strong function of temperature, and its value suddenly increases to a large value at the initial freezing point t_{sh}. Therefore, the error caused by the contribution of latent heat to heat conduction can be overlooked if a functional

Table 4 Selected Empirical Parameters Required to Use Equations (35) and (36)[a]

Apparent Specific Heat Below t_{sh}

Food	Moisture Content (Mass Fraction)	t_{sh}, °F	C_1, Btu/(lb·°F)	E, Btu·°F$^{(n_c-1)}$/lb	C_r, Btu/(lb·°F)	n_c
Cellulose gel[b]	0.77	30.92	0.874	43.94	0.339	1.696
Cellulose gel + NaCl	0.77	30.29	0.879	54.96	0.265	1.597
Beef meat (lean)	0.74	30.22	0.834	91.13	0.447	1.968
$t_{sh} \geqslant t \geqslant -40$ °F	0.70	30.18	0.810	61.70	0.368	1.717
	0.63	28.83	0.774	132.39	0.416	1.981
	0.57	28.36	0.740	51.69	0.235	1.497
	0.45	24.64	0.700	50.42	0.136	1.416
Fish meat (lean)	0.82	30.56	0.915	88.68	0.447	1.999
$t_{sh} \geqslant t \geqslant -40$ °F	0.75	30.20	0.874	60.15	0.282	1.628
	0.66	28.49	0.817	144.54	0.387	1.934
	0.57	26.67	0.776	171.11	0.373	1.933
Fresh fruit and vegetables	0.96	31.30	0.967	53.12	0.365	1.928
$t_{sh} \geqslant t \geqslant -40$ °F	0.87	29.52	0.931	144.22	0.416	1.896
	0.75	26.26	0.862	181.95	0.466	1.750
	0.61	19.44	0.771	245.59	0.327	1.675
Sucrose solution	0.96	31.62	0.977	28.71	0.428	1.903
$t_{sh} \geqslant t \geqslant -6.5$ °F	0.61	24.15	0.781	224.20	0.145	1.799

Thermal Conductivity

Food	Moisture Content (Mass Fraction)	t_{sh}, °F	k_{lx} or k_{ly}, Btu/(h·ft·°F)	k_{rx} or k_{ry}, Btu/(h·ft·°F)	S_{ksx} or S_{ksy}, Btu/(h·ft·°F)
Cellulose gel[b]	0.77	30.92	0.283	0.912	0.002481
Beef meat (lean) k_x[c]	0.75	30.22	0.276	0.811	0.002455
Beef meat (lean) k_y[c]	0.74	30.22	0.276	0.623	0.001999
Cod fish	0.82	30.56	0.302	0.753	0.003870
Cod and haddock	0.75	30.20	0.302	0.896	0.004871
Asparagus	0.93	30.74	0.306	0.751	0.005362
Strawberries	0.89	30.38	0.312	1.127	0.004685
Carrots	0.88	30.02	0.289	0.742	0.007104
Cherries	0.87	29.48	0.306	0.993	0.006222
Peas	0.76	28.76	0.272	1.146	0.003761
Plums	0.76	27.86	0.295	1.234	0.000472

Density

Food	Moisture Content (Mass Fraction)	t_{sh}, °F	ρ_1, lb/ft^3	ρ_r, lb/ft^3	S_{ds}, lb/(ft^3·°F)
Cellulose gel[b]	0.77	30.92	62.8	58.6	0.005348
Beef meat (lean)	0.70	30.18	66.8	63.2	0.000194
	0.63	28.83	67.1	63.6	0.007030
	0.57	26.67	67.4	64.0	0.011521
	0.45	24.64	68.0	66.0	0.003829
Fish meat (lean)	0.82	30.56	66.2	61.5	− 0.007061
	0.75	30.20	66.8	62.9	0.001477
	0.66	28.49	67.1	63.4	0.006936
	0.57	26.67	67.4	64.3	0.008452
Fresh fruit and vegetables	0.96	31.30	54.9	51.0	0.004675
	0.87	29.52	58.9	58.9	− 0.003742
	0.75	26.26	68.7	64.3	0.003742
	0.61	19.44	76.6	71.5	0.007797
Sucrose solution	0.96	31.62	62.4	57.5	0.001963
	0.61	24.15	63.0	59.7	0.009486

[a]Reproduced with slight modifications from Lebens. Wiss. U. Technol., Vol. 16, pp. 326-31, 1983, by the publisher.
[b]Crystallized cellulose, Tylose MH 1000, American Hoechst Corp.
[c]k_x = parallel to meat fibers, k_y = perpendicular to meat fibers.

relationship between apparent specific heat and temperature is used directly. To eliminate this error, Comini et al. (1974) estimated apparent specific heat values from enthalpy and temperature distributions around a nodal point.

Although the numerical procedures described are for three-dimensional heat conduction, the sample solutions are for bodies of relatively simple shapes. Comini and Del Giudice (1976) applied similar computerized numerical procedures to estimate temperature distribution in biological tissues during cryosurgery. Rebellato et al. (1978) used the same procedures to analyze two-dimensional heat conduction in lamb carcasses and beef sides during airblast freezing.

Fang et al. (1984) analyzed transient state heat transfer in an aqueous binary solution, occupying a semiinfinite space and undergoing freezing by assuming conditions B2, I2, D3, D2, and M1. They assumed no mass transfer of solute (M1) because the mass diffusivity was much smaller than the thermal diffusivity.

O'Callaghan et al. (1982) and Levin (1980, 1981) estimated solute transport in a solution undergoing freezing by assuming a constant rate of freezing or thermal equilibrium of a sample body with a surrounding medium. Because excessive transport of solute in food results in poor food quality, methods presented in these papers help to evaluate methods for postfreezing handling. However, since most of these researchers assumed a known, constant freezing rate or steady-state temperature distribution in food, these methods cannot be used to predict freezing time.

SEMITHEORETICAL METHODS

Hung and Thompson (1983) obtained Plank-like formulas through the regression analysis of freezing data, which were experimentally collected by using a Karlsruhe test substance (whose thermophysical properties were similar to lean beef). They found that their formulas accurately estimated the freezing times of lean beef, mashed potato, minced carp meat, and ground beef. However, their formulas should be applied carefully to food with properties considerably different from lean beef. Mott (1964) developed a method for predicting freezing times of several different foods. Bailey and James (1974) prepared a chart for estimating the thawing time of frozen pork legs.

Cleland and Earle (1982) modified Equation (10) as follows:

$$\theta_f = \frac{\Delta H_v}{(t_{sh} - t_2) W} (P_2 d/h + R_2 d^2/k_s) \qquad (42)$$

where W is an equivalent heat transfer dimension.

The W of a rectangular brick, d by $d\beta_1$ by $d\beta_2$, is estimated by:

$$W = 1 + W_1 + W_2 \qquad (43)$$

where

$$W_i = \left(\frac{Bl}{Bl+2} \right) \left(\frac{5}{8\beta_i{}^3} \right) \left(\frac{2}{Bl+2} \right) \left(\frac{2}{\beta_i (\beta_i + 1)} \right) \qquad (44)$$

where $i = 1$ or 2 and $Bl = hd/k$.

In many situations, freezing medium temperatures t_a and surface heat transfer conductances h change with time during freezing processes. Because most methods readily adoptable for hand-calculations assume constant t_a and h, Loeffen et al. (1981) developed two methods applicable to variable medium temperature and conductance; one is based on the numerical integration of Plank's equation for variable t_a and h, and the other is described later.

One freezing process was divided into proper time intervals $\Delta\theta$, during each of which t_a and h were assumed to be constant. By using Equation (10), a freezing time θ_f was estimated for conditions in each time interval. The freezing time for variable environmental conditions was estimated by assuming freezing was completed when the sum of fractions $\Delta\theta/\theta_f$ reached unity. The results obtained by both methods correlated with those predicted by Cleland and Earle's computer program (Cleland and Earle 1979).

EMPIRICAL FORMULAS FOR ESTIMATING COOLING TIME

Although the major emphasis of this chapter is on methods for estimating freezing or thawing time, precooling time is often an important consideration. Therefore, this section covers empirical formulas unique to food processing engineers.

Empirical formulas containing two experimental constants, f and j, are used to predict food temperatures when food is exposed to constant surrounding temperature. These constants were introduced by Ball (1932) to predict temperatures of canned foods during heat processing. Values for f and j represent slope and intercept coefficients, respectively, for a linear portion of a temperature history curve. This curve is obtained by computer analysis or by plotting temperature differences between food and its

surrounding medium against a time variable on semilog paper. The f value represents the time required for a linear portion of the history curve to pass one log cycle. The j value is obtained by taking the ratio between the phantom and real temperature differences at zero time. The former temperature difference is obtained by extrapolating the linear portion to a zero time. Formulas containing f and j values can predict a temperature history curve of any food without restriction on shape or heat transfer, if the f and j values are experimentally determined for each food and heating or cooling treatment; the f and j values are independent of an initial temperature difference between food and medium if food is heated by conduction only.

A semilogarithmic temperature history curve always consists of one curvilinear portion and one or more linear portions, except when $j = 1.0$. The following are empirical formulas for estimating food temperatures in both portions (Hayakawa 1982).

Initial curvilinear segment $(0 \leqslant \theta \leqslant \theta_v)$

1. $0.001 \leqslant j \leqslant 0.4$

$$\Delta t = 10\Delta t - (\theta/B)^{1/u} \qquad (45)$$

$$\theta_v = f(0.3913 - 0.3737 \log j) \qquad (46)$$

$$u = (\theta_v/f - \log j) / (\theta_v/f) \qquad (47)$$

$$B = \theta_v(\theta_v/f - \log j)^{-u} \qquad (48)$$

2. $0.4 \leqslant j < 1$

$$\Delta t = \Delta t_o{}^{\cos(B\theta + \pi/4)} \qquad (49)$$

$$\theta_v = 0.9 f(1 - j) \qquad (50)$$

$$B = 1/\theta_v \left[\tan^{-1} \left(\frac{\log \Delta t_o}{\log (j\Delta T_o) - \theta_v/f} \right) - \pi/4 \right] \qquad (51)$$

3. $1 < j$

$$\Delta t = \Delta t_o{}^{\cos(B\theta)} \qquad (52)$$

$$\begin{aligned}
\theta_v &= 0.7 f(j - 1); \quad 1 < j \leqslant 5.8 \\
&= 1.54 f \log (j/1.8); \quad 5.8 < j
\end{aligned} \qquad (53)$$

$$B = \frac{\cos^{-1}}{\theta_v} \left(\frac{\log (j\Delta t_o) - \theta_v/f}{\log \Delta t_o} \right) \qquad (54)$$

Linear portion $(\theta_v \leqslant \theta)$

$$\Delta t = j\Delta t_o \cdot 10^{-\theta/f} \qquad (55)$$

When a dimensionless temperature ratio is used for the semilogarithmic plot, it is likely that $\Delta t_o \leqslant 1$ or $\log (j\Delta t_o) \leqslant \theta_v/f$. In this case, an initial temperature difference Δt_o should be replaced with a hypothetical temperature difference Δt_{oh}, which satisfies:

$$0.2 \leqslant \log (j\Delta t_{oh}) - \theta_v/f \leqslant 1 \qquad (56)$$

With this temperature difference, a hypothetical temperature difference Δt_h is estimated by using the appropriate equations among Equations (45) through (54). This hypothetical difference is then converted to an actual temperature difference applicable to Δt_o by:

$$\Delta t = \Delta t_h \Delta t_o/\Delta t_{oh} \qquad (57)$$

Half cooling time, value Z, has been used to estimate a cooling time of fresh fruits and vegetables, as described in Chapter 11 of the 1990 ASHRAE *Handbook—Refrigeration*. This Z value is particularly useful when $j = 1$. Otherwise, the temperature response of food cannot be estimated. The following relationship among f, j, and Z values exists if Z is located in the linear portion of a semilogarithmic curve.

$$Z = f \log (z j) \qquad (58a)$$

If it is in the curvilinear portion, the proper curvilinear formulas should be used. For example, if $0.001 \leqslant j \leqslant 0.4$:

$$\log Z = \log B - 0.52140 \ U \qquad (58b)$$

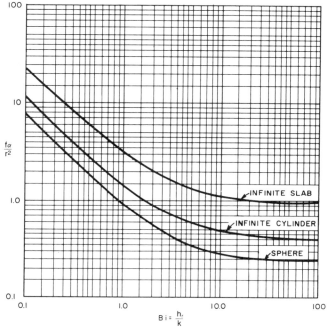

Fig. 2   Relationship between Dimensionless Parameter $f\alpha/r^2$ and Biot Number hr/k for Infinite Slab, Infinite Cylinder, or Sphere
(Pflug *et al.* 1965)

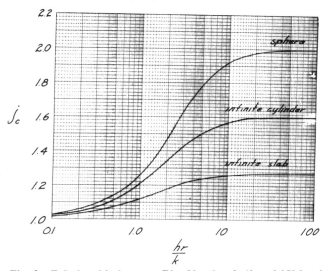

Fig. 3   Relationship between Biot Number hr/k and j Value j_c at Thermal Center of Infinite Slab, Infinite Cylinder, or Sphere
(Pflug *et al.* 1965)

Pflug *et al.* (1965) developed charts for estimating f and j values for conduction heating of food thermophysically approximated with either an infinite slab, infinite cylinder, or sphere. For this development, they used analytical solutions derived by assuming a uniform initial temperature distribution in the body, a constant surrounding medium temperature, convective heat exchange on the surface, constant thermophysical properties, and constant surface heat conductance. Figure 2 is the chart for estimating a dimensionless f value. Figures 3, 4, and 5 are for estimating a j value at the geometrical center of food, a j value for computing volumetric average temperature of food, and a j value on the food surface, respectively. Since f value is location-independent, only one figure estimates this value. In some cases, one side of a slab is insulated, while another side is exposed to a convective medium. In this case, one whole thickness of the slab should be used as the r value for estimating the f and j values using the charts.

The f and j values of the following composite shapes can be estimated using the charts of Pflug *et al.* (1965): (1) infinite rec-

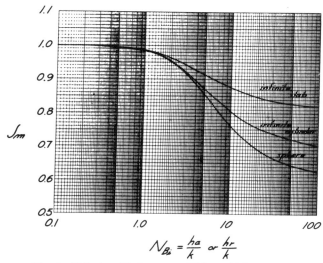

Fig. 4   Relationship between j Value of Mass Average Temperature j_m and Biot Number hr/k for Infinite Slab, Infinite Cylinder, or Sphere
(Pflug *et al.* 1965)

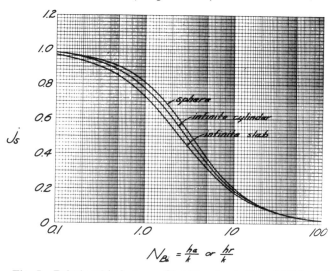

Fig. 5   Relationship between Biot Number hr/k and j Values j_s on Surface of Infinite Slab, Infinite Cylinder, or Sphere
(Pflug *et al.* 1965)

tangular column, (2) finite cylinder, and (3) rectangular parallelepiped (brick).

Each of these shapes can be perceived as volumetric elements produced by intersecting two infinite slabs of proper thicknesses (rectangular column), one infinite slab and one infinite cylinder (finite cylinder), or three infinite slabs of proper thicknesses at right angles to each other (rectangular parallelepiped). The f and j of a composite body are estimated by:

$$1/f_{composite} = \sum_i (1/f_i) \qquad (59)$$

$$j_{composite} = \pi_i j_i \qquad (60)$$

where subscript i represents f or j values of corresponding infinite slab or infinite cylinder.

The values of the Biot number of corresponding, infinite bodies are all likely to be different, since the characteristic dimensions r of the infinite bodies are different in most cases, although the surface heat conductance h is fixed.

Example 1. Determine the time for the center of a thin slab of meat to reach 40 °F, if the slab is thin enough to simulate an infinite slab for which all heat is transferred through the two large parallel faces. The following information is known: $t_a = 35$ °F; $t_o = 80$ °F; $t = 40$ °F; $h = 4.0$ Btu/ (h·ft^2·°F); $k = 0.35$ Btu/(h·ft·°F); $r = 0.167$ ft; $c_p = 0.86$ Btu/ (lb·°F); and $\rho = 64.9$ lb/ft^3.

Solution:

$$Bi = 4.0 \times 0.167/0.35 = 1.91$$

$$\alpha = 0.35/(0.86 \times 64.9) = 6.27 \times 10^{-3} \text{ ft}^2/\text{h}$$

$f\alpha/r^2 = 2.1$ for infinite slab of Figure 2

Therefore:

$$f = 2.1 \times 0.167^2/6.27 \times 10^{-3} \times 9.34 \text{ h}$$

From Figure 3, $j_c = 1.17$.
By using Equation (55):

$$40 - 35 = 1.17 (80 - 35) 10^{-\theta/9.34}$$

Solve for θ: $\theta = 9.55$ h
Since the equation is used for the linear portion, check that the estimated θ is in this portion. From Equation (53):

$$\theta_v = 0.7 \times 9.34 (1.17 - 1) = 1.11 \text{ h}$$

Since the solution is greater than θ_v, it is a correct estimation.

Example 2. Consider a package of meat, 0.166 ft by 0.332 ft by 1.0 ft, being cooled under the same conditions as the slab in Example 1. Determine the time at which the center of the package reaches 39 °F.

Solution: The package can be considered the intersection of three infinite slabs—0.166 ft, 0.332 ft, and 1.0 ft thick—all mutually perpendicular. Therefore, the value of the Biot number applicable to each slab can be estimated by using 0.083 ft, 0.166 ft, or 0.5 ft as the characteristic dimension. Using these values in Figures 2 and 3, the f and j_c values are obtained for each slab component, as shown below:

Slab	Dimen., ft	Bi	$f\alpha/r^2$	f, h	j_c
1	0.083	0.948	3.4	3.73	1.116
2	0.166	1.90	2.1	9.23	1.17
3	0.5	5.71	1.3	51.8	1.24

By using Equations (59) and (60):

$$1/f = 1/3.73 + 9.23 + 51.8; f = 2.53 \text{ h}$$

$$j_c = 1.116 \times 1.17 \times 1.24 = 1.62$$

From Equation (24):

$$39 - 35 = 1.62 (80 - 35) 10^{-\theta/2.53}$$

$$\theta = 3.19 \text{ h}$$

From Equation (52): $\theta_v = 1.1$ h. Since $\theta > \theta_v$, the above estimation is correct.

Smith *et al.* (1968) developed a useful method for estimating the temperature of irregularly shaped food by using a shape factor called the *geometry index G*. This index is defined as:

$$G = 1/4 + 3/(8A_1^2) + 3/(8A_2^2) \qquad (61)$$

The A_1 and A_2 of Equation (61) are related to the cross-sectional areas of the food, as shown below.

$$A_1 = \text{area } 1/(\pi r^2); A_2 = \text{area } 2/(\pi r^2) \qquad (62)$$

In the above equation, r is the characteristic dimension of the food, which is the shortest distance between the thermal center of food and its surface exposed to a heat exchange medium. Then, area 1 is the minimum cross-sectional area containing the radial line segment that defines the characteristic dimension. The cross-sectional area orthogonal to area 1 is area 2.

The value of M_1^2, which is related to f, is determined by entering the geometry index and inverse of the Biot number into Smith *et al.* (1968) (Figure 6). Based on the alignment chart of the observed relationship, the following formulas are used to estimate

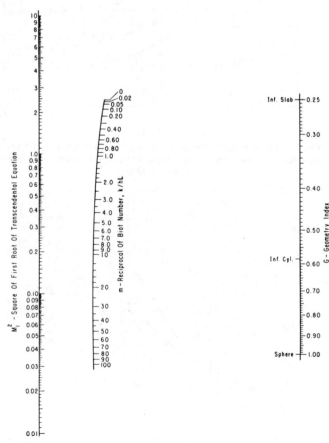

Fig. 6 Nomograph for Estimating M^2 Value from Reciprocal Biot Number and Smith's Geometry Index
(Smith 1966)

the f and j values estimating the mass average temperature of food.

$$f = 2.303 \, r^2/(M_1^2 \alpha) \tag{63}$$

$$j_{G_m} = 0.892 \, \exp^{(-0.0388M_1^2)} \tag{64}$$

The use of j_{Gm} produces a maximum error of 4% in the estimated average temperature, according to Smith *et al.* (1968).

Smith *et al.* (1968) observed that the dimensionless location ζ, where the temperature response was identical to the mass average temperature of an object, could be estimated by:

$$\zeta = G^{0.14} - 0.25 \tag{65}$$

Because $G = 1.0$ for a sphere, $\zeta = 0.75$. Smith *et al.* (1968) proposed the following method for estimating the mass average temperature of an irregularly shaped object.

1. Enter G and $1/Bi$ of an irregularly shaped object into Figure 6 to find M_1^2 value.
2. Align this M_1^2 with 1.0 on the geometry index, and find a modified $1/Bi$ applicable to the sphere.
3. Estimate a mass average temperature from the thermal response at a radial, dimensionless location of 0.75 of a sphere subjected to cooling or heating with a modified Bi found in Step 2.

Since the same M_1^2 is applicable to the sphere, Equation (63) can be used to estimate the f value. The value of j_{Gs} for the sphere should be estimated by:

$$j_{Gs} = j_c \sin(0.75 \, \beta_1)/(0.75 \, \beta_1) \tag{66}$$

where β_1 is the first positive root of Equation (67).

$$Bi = 1 - \beta_1 (\cot \beta_1) \tag{67}$$

and where j_c is the central j value of the sphere.

The estimated j_{Gs} values are shown in Table 5 for selected Bi values.

Example 3. A piece of ham is to be cooled in a 30 °F airblast cooler from 160 to 50 °F. The overall dimension of the ham is 0.335 ft by 0.541 ft by 0.915 ft. Estimate the cooling time.

Given property values are: $C = 0.893$ Btu/(lb·°F), $h = 8.45$ Btu/(h·ft²·°F), $k = 0.219$ Btu/(h·ft²·°F), $\rho = 67.4$ lb/ft³.

Solution: Assume that the thermal center is at the geometrical center of the ham. Since the characteristic dimension r is the shortest distance between the thermal center and exposed surface, it is equal to 0.335/2 = 0.167 ft. The overall dimension of cross section for area 1 is 0.335 ft by 0.541 ft. Assume that this cross section consists of two half circles of 0.335 ft diameters and of a 0.335 ft by (0.541 − 0.335) ft rectangle. Therefore:

$$\text{area } 1 = 2 \times 0.167^2 \, \pi/2 + (0.541 - 0.335) \, 0.335 = 0.157 \text{ ft}^2$$

The cross section for area 2 is orthogonal to that of area 1. Therefore, the overall dimension of this cross section is 0.335 ft by 0.915 ft. Similarly, it can be assumed that:

$$\text{area } 2 = 2 \times 0.167^2 \pi/2 + (0.915 - 0.335) \, 0.335 = 0.282 \text{ ft}^2$$

Therefore, from Equation (63), $A_1 = 0.157/(0.167^2\pi) = 1.79$ and $A_2 = 0.282/(0.167^2\pi) = 3.21$. By using Equation (61):

$$G = 1/4 + 3/(8 \times 1.79^2) + 3/(8 \times 3.21^2) = 0.403$$

The reciprocal of Bi is: $0.219/(8.45 \times 0.167) = 0.155$. From Figure 6: $M_1^2 = 3.0$. From the definition of thermal diffusivity:

$$\alpha = 0.219/(0.893 \times 67.4) = 0.00364 \text{ ft}^2/\text{h}$$

From Equations (63) and (64):

$$f = 2.303 \times 0.167^2/(3.0 \times 0.00364) = 5.88 \text{ h}$$

$$j_{Gm} = 0.892 \, \exp(-0.0388 \times 3.0) = 0.794$$

By using Equation (55):

$$50 - 30 = 0.794 (160 - 30) \, 10^{-\theta/5.88}$$

$$\theta = 4.19 \text{ h}$$

From Equation (50):

$$\theta_v = 0.9 \times 5.88 \, (1 - 0.794) = 1.09 \text{ h}$$

The solution is $\theta > \theta_v$.

Smith *et al.* (1968) recommend a second method to solve the same problem. By aligning $G = 1.0$ (sphere) and $M_1^2 = 3.0$, $1/Bi = 0.8$ or $Bi = 1.25$. From Table 5: $j_{Gs} = 0.993$.

Since the f value is identical:

$$50 - 30 = 0.9932 (160 - 30) \, 10^{-\theta/5.88}$$

$$\theta = 4.76 \text{ h}$$

From Equation (50), $\theta_v = 0.036$ h. The above estimation is correct; there is less than 15% difference between the two estimated times.

Computerizing the method of Smith *et al.* (1968) is difficult, since Figure 6 is required. Therefore, Hayakawa and Villalobos (1988) obtained the following regression formulas to estimate M_1^2 values as functions of geometry index G and Biot number Bi.

When Bi = ∞

$$\ln M_1^2 = 2.2893825 + 0.35330539 \, Y_g - 3.8044156 \, Y_g^2$$
$$- 9.6821811 \, Y_g^3 - 12.0321827 \, Y_g^4$$
$$- 7.1542411 \, Y_g^5 - 1.6301018 \, Y_g^6 \tag{68}$$

where $Y_g = \ln G$.

Table 5   Value of j at Mass Average Temperature Location in Sphere for Selected Biot Numbers

Bi	j_{Gs}	Bi	j_{Gs}	Bi	j_{Gs}	Bi	j_{Gs}	Bi	j_{Gs}
0.0	1.00000	0.09	1.00150	0.9	1.00017	1.9	0.97439	10	0.76839
0.005	1.00009	0.10	1.00164	1.0	0.99849	2.0	0.97105	11	0.75609
0.01	1.00019	0.15	1.00229	1.1	0.99656	2.5	0.95344	16	0.71375
0.02	1.00037	0.2	1.00283	1.2	0.99440	3	0.93520	21	0.68922
0.03	1.00054	0.3	1.00358	1.3	0.99203	4	0.89991	31	0.66212
0.04	1.00071	0.4	1.00391	1.4	0.98946	5	0.86851	41	0.64765
0.05	1.00088	0.5	1.00386	1.5	0.98672	6	0.84150	51	0.63861
0.06	1.00104	0.6	1.00343	1.6	0.98383	7	0.81859	61	0.63251
0.07	1.00120	0.7	1.00266	1.7	0.98080	8	0.79915	81	0.62468
0.08	1.00135	0.8	1.00156	1.8	0.97765	9	0.78261	101	0.61989

When Bi $< \infty$

$$\ln M_1{}^2 = 0.92083090 + 0.83409615\, Y_g - 0.78765739\, Y_b$$
$$- 0.04821784\, Y_g\, Y_b - 0.0408987\, Y_g{}^2$$
$$- 0.10045526\, Y_b + 0.01521388\, Y_g{}^3$$
$$+ 0.00119941\, Y_g\, Y_b{}^3 + 0.00129982\, Y_b{}^4 \quad (69)$$

where $Y_g = \ln G$ and $Y_b = \ln(1/\text{Bi})$.

An $M_1{}^2$ value estimated by Equations (68) or (69) may be used to estimate f and j_{Gm} values for calculating the mass average temperature of an irregular body by using Equations (63) and (64). Equation (64) may be replaced by more accurate equations [Equations (70) and (71)], since j_{Gm} is equal to a spherical j at dimensionless location 0.75 and a plate $-j$ at dimensionless location 0.57359, according to Equation (65).

$$j_{Gm} = 2 \sin M_1 \cdot \cos(0.57359\, M_1)/(M_1 + \sin M_1 \cos M_1)$$
$$\text{when } 0.01 \leqslant M_1{}^2 \leqslant 0.03 \quad (70)$$

$$j_{Gm} = \frac{2\,(\sin M_1 - M_1 \cos M_1)\sin(0.75\, M_1)}{0.75\, M_1\,(M_1 - \sin M_1 \cos M_1)}$$
$$\text{when } 0.03 \leqslant M_1{}^2 \quad (71)$$

M_1, the square root of $M_1{}^2$, should be used in Equations (70) and (71).

Surface heat conductances applicable to the two sides of an infinite slab can be different from each other. Uno and Hayakawa's charts (Uno and Hayakawa 1979) determine the location of thermal center and f and j values for these cases.

All of the methods described above are applicable only when the temperature of heat exchange medium is constant. Uno and Hayakawa (1979) estimate the temperatures of food exposed to time variable medium temperatures when its f and j values are known.

SAMPLE PROBLEMS FOR ESTIMATING FREEZING TIME

To provide better understanding of these procedures, four sample problems are presented. The first three show combined application of Cleland and Earle's procedure (1979) and empirical formulas for estimating the thermophysical properties of food. The fourth shows the application of the procedure developed by Hayakawa *et al.* (1983).

Example 4. Spherical carp fish balls 0.1 ft diameter are to be frozen in an IQF airblast freezer. The initial temperature of the balls is 62.7 °F. The freezer air temperature is -12.9 °F, and the coefficient of surface heat transfer is 12 Btu/(h·ft²·°F). Calculate the time required for the center temperature of the fish balls to reach 14.1 °F.

Solution: The thermophysical property values of sample food are estimated first by using the empirical formulas for this estimation. Because no information is available on fat content and other ingredients in the food, its thermophysical properties are approximated with those of lean fish.

The enthalpy can be calculated by integrating Equations (20) and (21), as shown below:

$$\Delta H = \int_{t_f}^{t_o} C\,dt$$
$$= C_l\,(t_o - t_{sh}) + C_r\,(t_{sh} - t_f)$$
$$+ \frac{E}{n_c - 1}\left[(t_{sw} - t_{sh})^{1-n_c} - (t_{sw} - t_f)^{1-n_c}\right] \quad (72)$$

By assuming that the mass fraction of water in the food is 0.82, the following constants are needed to use the above equation from Table 4:

$$t_{sh} = 30.56;\ C_l = 0.915;\ E = 88.68;\ C_r = 0.447;\ n_c = 1.999$$

The units of the above values are omitted, since they are given in Table 4. From the conditions given in the problem:

$$t_f = 14.1\,°\text{F and } t_{sh} = 30.6\,°\text{F}$$

Because of the definition of t_{sw}:

$$t_{sw} = 32.0\,°\text{F}$$

By substituting these values into Equation (72):

$$\Delta H = 89.2 \text{ Btu/lb}$$

To convert the above enthalpy to the volumetric value, the density of food is estimated by using Equation (19). Since the heat is removed through the frozen layer of food, the density is estimated at t_f. From Table 4:

$$\rho_r = 61.5;\ \rho_l = 66.2;\ S_{ds} = 0.00706$$

Substituting the above values into Equation (19):

$$\rho = 61.9 \text{ lb/ft}^3$$

The volumetric enthalpy is then:

$$\Delta H_v = 89.2 \times 61.9 = 5520 \text{ Btu/ft}^2$$

The specific heat of unfrozen food C_l is 0.915. Since the density of the food ρ_l is 66.2 from Table 4, the volumetric specific heat C_{lv} is:

$$C_{lv} = 0.915 \times 66.2 = 60.57 \text{ Btu/(ft}^3\cdot°\text{F)}$$

The specific heat of frozen food C_s at t_f is estimated by Equation (21). When the constants given in Table 4 are substituted into this equation, $C_s = 0.836$ Btu/(lb·°F). Since the density at t_f is 61.9 lb/ft³, the volumetric specific heat C_{sv} is 51.7 Btu/(ft³·°F).

The thermal conductivity of the sample food is approximated by that of codfish, as listed in Table 4. From the table:

$$t_{sh} = 30.56;\ S_{ksx} = 0.00387;\ k_{lx} = 0.302;\ k_{rx} = 0.753$$

The thermal conductivity at t_f is estimated by substituting the above listed values into Equation (24):

$$k_s = 0.782 \text{ Btu/(h}\cdot\text{ft}\cdot°\text{F)}$$

The value of Ste is then determined by using its definition as:

$$\text{Ste} = C_{sv}\,(t_{sh} - t_a)\,/\Delta H_v = 51.7\,[30.56 - (-12.9)]/5520 = 0.407$$

The value of Pk is calculated also as:

$$\text{Pk} = C_{lv}\,(t_o - t_{sh})/\Delta H_v = 60.57\,(62.7 - 30.56)/5520 = 0.352$$

By using the definition of Bl:

$$Bl = hd/k = 12 \times 0.1/0.782 = 1.53$$

where d = fish ball diameter, ft.

For a spherical shape, Plank's constants are $P = 1/16$ and $R = 1/24$. Therefore, using Equations (11), (12), (14), and (15), $P_2 = 0.233$ and $R_2 = 0.0750$. Finally, using Equation (10):

$$\theta_f = \frac{5520}{30.56 - (-12.9)}\left(0.233\,\frac{0.1}{12} + 0.0750\,\frac{0.1^2}{0.782}\right) = 0.368\text{h}$$

Example 5. Whole fish are to be hung on a rack and frozen in an airblast tunnel. The shape of the fish can be approximated by a cylinder 0.165 ft diameter by 0.1 ft long and, for calculation purposes, can be considered an infinite cylinder. The balance of the conditions and thermal properties are the same as in Example 4. As before, calculate the time for the center temperature to reach 14.1 °F.

Solution: All physical property values and dimensionless, parametric values are identical to those estimated in Example 4, except the Bl value. Since $d = 0.165$ ft:

$$Bl = 12 \times 0.165/0.782 = 2.53$$

For an infinite cylinder, P and R values are equal to 1/4 and 1/16, respectively. The values of P_2 and R_2 are estimated, as described before, by using Equations (11), (12), (14), and (15), as given below:

$$P_2 = 0.347 \qquad R_2 = 0.103$$

Therefore, by using Equation (10):

$$\theta_f = \frac{5520}{30.6 - (-12.9)} \left(0.359 \frac{0.165}{12} + 0.103 \frac{0.165^2}{0.782} \right) = 1.125 \text{ h}$$

Example 6. Packaged lean beef of rectangular brick shape with a thickness of 0.13 ft, a width of 0.39 ft, and a length of 0.52 ft is to be frozen in a blast freezer. The initial temperature of the meat is 44.7 °F, the freezer air temperature is −21.9 °F, and the coefficient of surface heat transfer is 7.04 Btu/(h · ft² · °F). Calculate the time for the average temperature of the package to reach 14.1 °F.

Solution: From the data given in the problem, $t_o = 44.7$ °F; $t_a = 21.9$ °F; $t_f = 14.1$ °F; $h = 7.04$ Btu/(h · ft² · °F).

Assume that the mass fraction of moisture is 0.74. As in Example 4, constants required to estimate the enthalpy or specific heat of lean beef are obtained.

$$t_{sh} = 30.22; \; C_l = 0.834; \; E = 91.13; \; C_r = 0.447; \; n_c = 1.968$$

By using Equation (72), the mass-based enthalpy is obtained:

$$\Delta H = 92.9 \text{ Btu/lb}$$

The following are constants required to estimate the density of food. Use mass fraction of moisture = 0.70 (closest value from Table 4).

$$\rho_r = 63.2; \; \rho_l = 66.8; \; S_{ds} = 0.000194$$

By substituting the above values into Equation (19), the density at 14.1 °F is determined.

$$\rho = 63.56 \text{ lb/ft}^3$$

Therefore, the volumetric enthalpy ΔH_v is:

$$\Delta H_v = \rho \, \Delta H = 5905 \text{ Btu/ft}^3$$

The volumetric specific heat is:

$$c_{lv} = 55.7 \text{ Btu/(ft}^3 \cdot \text{°F)}$$

The value of C_s at t_f is estimated by Equation (21):

$$C_s = 64.07 \text{ Btu/(ft}^3 \cdot \text{°F)}$$

Therefore, the volumetric specific heat C_{sv} is:

$$C_{sv} = 64.07 \text{ Btu/(ft}^3 \cdot \text{°F)}$$

The thermal conductivity at t_f is estimated by Equation (24).

$$k_s = 0.6207 \text{ Btu/(h · ft · °F)}$$

The values of Ste and Pk are estimated by using their definitions:

$$Pk = 0.137 \text{ and } Ste = 0.566$$

The value of Bl is 1.48.

Since the shortest side dimension is 0.13 ft, the dimensionless side lengths are: $\beta_1 = 0.52/0.13 = 4$; $\beta_2 = 0.39/0.13 = 3$.

From Figure 1, $P = 0.315$ and $R = 0.089$.

Note that β_1 and β_2 are interchangeable in Figure 1. Therefore, to obtain P, $\beta_1 = 4$ and $\beta_2 = 3$ are entered into the figure. To obtain R, these values of β_1 and β_2 are interchanged. By using Equations (11), (12), (16), and (17), the values of P_2 and R_2 are equal to 0.512 and 0.310, respectively. Finally:

$$\theta_f = \frac{5960}{30.22 - (-21.9)} \left(0.512 \times 0.13/7.04 + \frac{0.310 \times 0.13^2}{0.6207} \right) = 2.04 \text{ h}$$

Example 7. A sample of cherry juice, mass fraction of moisture 0.87, in a 1-ft diameter by 1.5-ft height container is frozen in a −29.2 °F airblast freezer. Estimate the freezing time of the juice. Assume that the freezing completes when the thermal center temperature reaches 14 °F, that the initial temperature of juice is 41 °F, and that the surface overall heat conductances applicable to the top and side surfaces and applicable to the bottom surface are equal to 5.28 and 0.88 Btu/(h · ft² · °F), respectively.

Solution: Table 4 contains the following physical properties—

$t_{sh} = 29.5$ °F	$k_{rx} = 0.993$ Btu/(h · ft · °F)
$\rho_r = 58.9$ lb/ft³	$E = 144.22$ Btu · °F$^{(n_c - 2)}$/lb
$C_r = 0.416$ Btu/(lb · °F)	$n_c = 1.896$

Since the radius of the container is 0.5 ft, $r = 0.5$. The values of dimensionless parameters, required to use Equation (36), are estimated by using the definitions given in the Symbols section at the end of the chapter.

$U_o = 0.9375$	$C_{eh} = 152.5$
$U_{sh} = 0.8042$	$U_f = 0.625$
$-U_a = -0.1250$	$1/Bi_u = 0.3818$
$1/Bi_n = 2.2907$	$Sf = 3.00$

The values of transformed variables used in Equation (36) are estimated by Equations (37a) and (37b) and the minimum, frequent, and maximum values listed in Table 2. For example, the minimum, frequent, and maximum values of $-U_a$ are:

Minimum: $-0.99902 \times 0.625 + 0.041459 = -0.582929$

Frequent: $-0.87688 \times 0.625 + 0.13828 = -0.40977$

Maximum: 0.8333

By entering these values into Equation (37b):

$$G_2 = 0.610953; \; G_1 = 1.43490; \; G_3 = 2.3009$$

When the above estimated G_1, G_2, and G_3 values are entered with the $-U_a$ value of -0.1250 into Equation (37a), $X_a = 1.2654$.

The estimated values of all transformed variables are:

$X_o = -0.69472$	$X_{eh} = -0.5656$
$X_{sh} = 1.30876$	$X_f = 2.44967$
$X_a = 1.26543$	$X_{iu} = -0.597423$
$X_{in} = 1.8799$	$X_{sf} = 0.687036$

By substituting the above values into Equation (36) ln Fo$_f = 1.075$. Since Fo$_f = a_r t_f / l^2$, estimate α_r:

$$\alpha_r = k_{rx}/(C_r \rho_r) = 0.993/(0.416 \times 58.9) = 0.0405 \text{ ft}^2/\text{h}$$

Therefore:

$$\theta_f = e^{3801} \times 0.5^2/(0.0405 \times 10^{-6}) = 24.54 \text{ h}$$

Approximately 0.75 day is required to freeze the sample.

SYMBOLS

A_1, A_2 = dimensionless cross-sectional areas defined by Eq. (62); used to estimate the geometry index of Smith *et al.* (1968) by Eq. (61)

a = parameter defined by one of Eq. (5)

B = parameter used in Hayakawa's empirical formulas (1983); estimated by Eqs. (48), (51), or (54)

Bi = hr/k; Biot number

Bl = hd/k; used in Eqs. (11) and (44)

Bi_n = dimensionless parameter $h_n r/k_{rx}$, applicable to one side of an infinite rectangular column or one end of a finite cylinder. Symbol r is radius for a cylinder and normally a smaller side dimension for a rectangle in the case of isotropic food. In the case of anisotropic, rectangular food, it is the y-directional side dimension. Note that Bi_n and Bi_u for rectangular food are not Biot numbers, since r is not the shortest dimension between the thermal center and exposed surface of the food. Symbol k_{rx} is a reference k_x normally perpendicular to tissue fibers for a cylinder and parallel to tissue fibers for a rectangle when food is anisotropic.

Bi_n = $h_n d/k$ in Eqs. (40) and (41). For an infinite slab, Bi_n is related to a slab-surface with a lower surface heat conductance h_n.

Bi_u = dimensionless parameter h_r/k_{rx}, applicable to three sides of an infinite rectangular column or to one end and side of a finite cylinder

Bi_u = $h_u d/k$ in Eqs. (39), (40), and (41). For an infinite slab, Bi_u is related to a slab surface with a higher surface heat conductance h_u.

b, b_1 = parameters estimated by Eqs. (4) and (5), respectively

C = specific heat (energy per unit mass per unit temperature difference); if it is affixed with subscript v, it is per unit volume basis

C_{eh} = $E/[C_r (t_{sw} - t_{sh})^{n_c - 1}]$

D_a = $U_{sw}(U_{sw} - U_{sh})$

D_b = D_a/U_{sw}

d = diameter of sphere or infinite cylinder or whole thickness of infinite slab

E = empirical parameter for estimating apparent specific heat

Fo = Fourier modulus, $\alpha_r \theta/r^2$

Fo_1 = $\alpha_r \theta/d^2$

f = slope index of semilogarithmic temperature history curve

F_{ri} = fraction of ice, mass of ice/total mass of water, in food at $(t_a + t_x)/2$

f_{rw} = fraction of water in food, mass of water/mass of food

G = Smith *et al.* (1968) geometry index; see Eq. (61)

G_1, G_2, G_3 = constants estimated by Eq. (37b); used to transform a dimensionless parameter to a corresponding design variable; see Eq. (37a)

h = coefficient of convective surface heat transfer

H = enthalpy of food

j = intercept coefficient of semilogarithmic temperature history curve of food

k = thermal conductivity of food

k_{lx} = thermal conductivity of unfrozen food in x-direction at t_{sh}

k_{lx} = thermal conductivity of unfrozen food in y-direction at t_{sh}

L = latent heat of freezing

Ld = $L/[C(t_{fp} - t_a)]$

L_{kx} = $M_{kx} U_{sh}$

L_w = latent heat of freezing of water

Md = $S_{ds}(t_{or} - t_{ar})/\rho_r$

M_{kx} = $S_{ksx}(t_{sh} - t_{ar})/k_{rx}$

M_{klx} = $S_{klx}(t_{or} - t_{ar})/k_{rx}$

M_{kly} = $S_{kly}(t_{or} - t_{ar})/k_{rx}$

M_1^2 = exponential constant determined by Smith *et al.* (1968) chart

n_c = empirical constant for estimating apparent specific heat of food

P = constant in Plank's formula, Eq. (1)

Pc = dimensionless specific heat, which is the ratio of dimensional specific heat at any temperature and reference specific heat

Pd = dimensionless density, which is the ratio of dimensional density at any temperature and reference density

Pk_x = dimensionless thermal conductivity, which is the ratio of dimensional thermal conductivity at any temperature and the reference thermal conductivity in x-direction; used in Eqs. (31) through (34)

Pk = $C_{lv}(t_o - t_{sh})/\Delta H_v$; Plank number, used in Eqs. (11) and (12)

P_1, P_2 = Cleland and Earle's parameters (1979)

R = Plank's constant; see Eq. (1)

R_c = C_l/C_r

R_d = ρ_l/ρ_r

R_{kx} = $k_{lx}/k_{rx} - 1$

R_{ky} = $k_{ly}/k_{rx} - 1$

R_1, R_2 = Cleland and Earle's parameters (1979)

r = characteristic dimension of food; equal to the shortest distance between the thermal center of food and its exposed surface. For example, r = radius for a sphere or cylinder and r = half thickness of an infinite slab with both surfaces exposed to a heat exchange medium.

S = linear rate of change in thermal conductivity or density of food with change in its temperature.

Sf = shape factor; for rectangular column, it is x-directional side dimension/y-directional side dimension; for cylindrical food, it is height/radius.

S_T = $C_l(t_{fp} - t_s)/L$; used in Eq. (5)

Ste = $C_{sv}(t_l - t_a)/\Delta H_v$; Stefan number

t = dimensional temperature

U = $(t - t_{ar})/(t_{or} - t_{ar})$; dimensionless temperature; reference cooling medium temperature t_{ar} and reference initial food temperature t_{or} are equal to $-40°$ and $46.4°$F, respectively

u = empirical constant; value is estimated by Eq. (47)

W = equivalent heat transfer dimension

W_1, W_2, W_i = parameters defined by Eq. (44)

X = location of freezing or defrosting front measured from an exposed surface; used in Eqs. (2) through (5); also used as a design variable transformed from a dimensionless parameter in Eqs. (35) and (36)

x = location variable; measured from an exposed surface; used in Eqs. (2) through (5) and Eqs. (38) through (41)

x = location variable; for a finite cylinder, it is the radial directional variable; for a rectangular column, it is the direction of tissue fibers or the direction of a larger thermal conductivity

X_x = variable corresponding to any dimensionless parameter

Y_b = ln (1/Bi)

Y_g = ln G

Y_o = $(t_o - t_{fp})/(t_{fp} - t_a)$

α = thermal diffusivity; α_r is reference thermal diffusivity, equal to $k_r/(C_r \rho)$.

β_1, β_2 = ratio of a side dimension of a brick and its smallest side dimension; β_1 also is the first positive root of Eq. (67)

γ = constant estimated by Eq. (5); in Eqs. (25) and (34), it is the ratio of thermal conductivities in the y and x directions for $t \times t_{sh}$

γ = in Eq. (6b) through (6d), $\gamma = C(t_{fp} - t_a)/L$

δ = variable defined by Eq. (4)

ζ = dimensionless location defined by Eq. (65)

ΔH = enthalpy difference, usually per unit mass basis; however, if it is affixed with subscript v, it is per unit volume basis

Δt = temperature difference

Z = half cooling time

θ = dimensional time

π_x = any dimensionless parameter

ρ = density of food

ρ_w = density of water

ϕ = $(t_o - t_{fp})/(t_{fp} - t_s)$

η_1, η_2, η_3 = parameters defined by Eqs. (6b), (6c), and (6d), respectively

Subscripts

a = surrounding heat exchange medium

ar = reference surrounding medium temperature

c = distance measured from current position of exposed surface [Eqs. (2) through (5)].

c = central position (Figure 3)

c = value related to specific heat [Eqs. (29) and (30)]

d = value related to density [Eqs. (19), (27), and (28)]

eh = value related to C_{eh}

f = final food temperature or freezing time

fp = freezing point when freezing or defrosting completes at one single temperature

frq = frequently observed value

Gm = related to average j value, estimated from geometry index

Gs = related to equivalent, spherical j value

g = distance measured from the original position of exposed surface [Eqs. (2) through (5)]

h = hypothetical

i = summation or multiplication index

in = value related to Bi_n

iu = value related to Bi_u

l = unfrozen

m = average (Figure 4)

k = value related to thermal conductivity [Eqs. (22), (23), and (31) through (34)]

max = maximum value

min = minimum value

o = initial or value related to U_o

r = reference value

s = frozen or surface

sh = value related to initial freezing temperature of food t_{sh}

sw = value related to freezing point of water

sf = value related to Sf

u = per unit volume basis (C_{lv}, C_{sv}, ΔH_v) or length of curvilinear portion of semilogarithmic temperature history curve [Eqs. (45) through (55)]

x, y = values related to x and y directions, respectively

1 = frozen region in case of freezing. Defrosted region in case of defrosting [Eqs. (2) through (5)]

2 = unfrozen region in case of freezing; frozen region in case of defrosting [Eqs. (2) through (5)]

REFERENCES

Bailey, C. and S.J. James. 1974. Predicting thawing time of frozen pork legs. ASHRAE *Journal* 16(3):68.

Ball, C.O. 1932. Thermal process time for canned food. *Bulletin of the National Research Council of the National Academy of Science* 7, Part I, No. 37, Washington, D.C.

Chan, S.H., D.H. Cho, and G. Kocamustafaogullari. 1986. Melting and solidification with internal radiative transfer—A generalized phase change model. *International Journal of Heat and Mass Transfer* 26:621.

Charm, S.E., D.H. Brand, and D.W. Baker. 1972. A simple method for estimating freezing and thawing times of cylinders and slabs. ASHRAE *Journal* 14(11):39.

Cho, S.H. and J.E. Sunderland. 1974. Phase change problems with temperature-dependent thermal conductivity. *Journal of Heat Transfer*, ASME *Transactions*, Series C 96:214.

Chung, B.T.F. and L.T. Yeh. 1975. Solidification and melting of materials subjected to convection and radiation. *Journal of Spacecraft and Rockets* 12:329.

Cleland, A.C. and R.L. Earle. 1979. Prediction of freezing times for foods in rectangular packages. *Journal of Food Science* 44:964.

Cleland, A.C. and R.L. Earle. 1982. Freezing time prediction for foods—A simplified procedure. *Revue Internationale du Froid* 5:134.

Cleland, A.C. and R.L. Earle. 1979. A comparison of methods for predicting the freezing times of cylindrical and spherical foodstuffs. *Journal of Food Science* 44:958.

Comini, G. and S. Del Giudice. 1976. Thermal aspect of cryosurgery. *Journal of Heat Transfer* 98:54.

Comini, G., S. Del Giudice, R.W. Lewis, and O.C. Zienkiewicz. 1974. Finite element solution of non-linear heat conduction problems with special reference to phase change. *International Journal of Numerical Method Engineering* 8:613.

De Michelis, A. and A. Calvelo. 1982. Mathematical models for nonsymmetric freezing of beef. *Journal of Food Science* 47:1211.

Earle, R.L. and A.K. Fleming. 1967. Cooling and freezing of lamb and mutton carcasses, I. Cooling and freezing rates in legs. *Food Technology* 21:79.

Ede, A. 1949. The calculation of the rate of freezing and thawing of food stuffs. *Modern Refrigeration* 52:52.

Fang, L.J., F.B. Cheung, J.H. Linehan, and D.R. Pedersen. 1984. Selective freezing of a dilute salt solution on a cold ice surface. *Journal of Heat Transfer* 106:385.

Grange, B.W., R. Viskanta, and W.H. Stevenson. 1976. Diffusion of heat and solute during freezing of salt solutions. *International Journal of Heat Mass Transfer* 19:373.

Hayakawa, K. 1982. Empirical formulae for estimating nonlinear survivor curves of thermallly vulnerable factors. *Canadian Institute of Food Science Technology Journal* 15:116.

Hayakawa, K. and A. Bakal. 1973. Formulas for predicting transient temperatures in food during freezing or thawing. AICHE *Symposium Series* 69(132):14.

Hayakawa, K. and G. Villalobos. 1988. Unpublished results. Formulas for estimating Smith *et al.* parameters to determine the mass average temperature of irregularly shaped bodies.

Hayakawa, K., C. Nonino, and J. Succar. 1983. Two dimensional heat conduction in food undergoing freezing: predicting freezing time of rectangular or finitely cylindrical food. *Journal of Food Science* 48:1841.

Hayakawa, K., C. Nonino, J. Succar; G. Comini, and S. Del Giudice. 1983. Two dimensional heat conduction in food undergoing freezing: Development of computerized model. *Journal of Food Science* 48:1849.

Hill, J.M. and A. Kucera. 1983. Freezing a saturated liquid inside a sphere. *International Journal of Heat Mass Transfer* 26:1631.

Hung, Y.C. and D.R. Thompson. 1983. Freezing time prediction for slab shape foodstuffs by an improved analytical method. *Journal of Food Science* 48:555.

Joshi, C. and L.C. Tao. 1974. A numerical method of simulating the axisymmetrical freezing of food systems. *Journal of Food Science* 39:623.

Katayama, K. and M. Hattori. 1975. Research on heat conducting with freezing, I. Numerical method on Stefan's problem. JSME *Bulletin* 18:41.

Keller, G.J. and J.H. Ballard. 1956. Predicting temperature changes in frozen liquids. *Industrial Engineering Chemistry* 48:188.

Kern, J. 1977. A simple and apparently safe solution to the general Stefan problem. *International Journal of Heat Mass Transfer* 20:475.

Levin, R.L. 1980. Generalized analytical solution for the freezing of supercooled aqueous solution in a finite domain. *International Journal of Heat Mass Transfer* 23:951.

Levin, R.L. 1981. The freezing of finite domain aqueous solution: solute redistribution. *International Journal of Heat Mass Transfer* 24:1443.

Loeffen, M.P.F., R.L. Earle, and A.C. Cleland. 1981. Two simple methods for predicting food freezing times with time-variable boundary conditions. *Journal of Food Science* 46:1032.

Lunardini, V.J. 1983. Approximate solution to conduction freezing with density variation. *Journal of Energy Resources Technology* 105:43.

Lunardini, V.J. 1983a. Freezing and thawing: Heat balance integral approximations. *Journal of Energy Resources Technology* 105:30.

Lundardini, V.J. 1983b. Freezing of a semi-infinite medium with initial temperature gradient. *Journal of Energy Resources Technology* 106:103.

Mascheroni, R.H. and A. Calvelo. 1982. A simplified method for freezing time calculation in foods. *Journal of Food Science* 47:1201.

Mott, L.F. 1964. The prediction of product freezing time. *Australian Refrigerating, Air Conditioning and Heating* 18(2):16.

Muehlbauer, J.C., J.D. Hatcher, D.W. Lyons, and J.E. Sunderland. 1973. Transient heat transfer analysis of alloy solidification. ASME Paper No. 73-HT-4.

Nagaoka, J., S. Takagi, and S. Hotani. 1955. Experiments on the freezing of fish in an air-blast freezer. Proceedings of the 9th International Congress of Refrigeration, Paris, 4:105.

Nonino, C. and K. Hayakawa. 1986. Thawing time of frozen food of a rectangular or finitely cylindrical shape. *Journal of Food Science* 51:116.

O'Callaghan, M.G., E.G. Carvalho, and C.E. Huggins. 1982a. An analysis of heat and solute transport during solidification of an aqueous binary solution-I. Basal plane region. *International Journal of Heat Mass Transfer* 25:553.

O'Callaghan, M.G., E.G. Carvalho, and C.E. Huggins. 1982b. An analysis of the heat and solute transport during solidification of an aqueous binary solution-II. Dendrite tip region. *International Journal of Heat Mass Transfer* 25:563.

Pflug, I.J., J.L Blaisdell, and I.J. Kopelman. 1965. Developing temperature-time curves for objects that can be approximated by a sphere, infinite plate or infinite cylinder. ASHRAE *Transactions* 71(1):238.

Rathjen, K.A. and L.M. Jiji. 1970. Meat conduction with melting or freezing in a corner. ASME Paper No. 70-HT-Q.

Rebellato, L., S. Del Giudice, and G. Comini. 1978. Finite element analysis of freezing processes in foodstuffs. *Journal of Food Science* 43:239.

Schwartzberg, H.G. 1977. Effective heat capacities for the freezing and thawing of food. *Refrigeration Science and Technology, Freezing, Frozen Storage and Freeze-Drying*, International Institute of Refrigeration, Commissions C1 and C2, 303.

Shamsunder, N. and E.M. Sparrow. 1975. Analysis of multidimensional conduction phase change via the enthalpy model. *Journal of Heat Transfer*, ASME *Transactions*, Series C 97:333.

Shamsunder, N. and E.M. Sparrow. 1976. Effect of density change on multi-dimensional conduction phase change *Journal of Heat Transfer*, ASME *Transactions*, Series C 98:550.

Smith, R.E. 1966. Analysis of transient heat transfer from anomalous shape with heterogeneous properties. Ph.D. thesis, Oklahoma State University, Stillwater, OK.

Smith, R.E., G.L. Nelson, and R.L. Henrickson. 1968. Applications of geometry analysis of anomalous shapes to problems in transient heat transfer. *Transactions of the ASAE* 11(2):296.

Succar, J. and K. Hayakawa. 1983. Empirical formulae for predicting thermal physical properties of food at freezing or defrosting temperatures. Lebensm.-Wiss. U.-Technol. 16:326.

Succar, J. and K. Hayakawa. 1984. Parametric analysis for predicting freezing time of infinitely slab-shaped food. *Journal of Food Science* 49:468.

Talmon, Y. and H.T. Davis. 1981. Analysis of propagation of freezing and thawing fronts. *Journal of Food Science* 46:1478.

Tao, L.C. 1967. Generalized numerical solutions of freezing a saturated liquid in cylinders and spheres. AICHE *Journal* 13:165.

Tao, L.C. 1968. Generalized solution of freezing a saturated liquid in a convex container. AICHE *Journal* 14:720.

Tarnawski, W. 1976. Mathematical model of frozen consumption products. *International Journal of Heat Mass Transfer* 19:15.

Tien, R.H. and G.E. Geiger. 1967. A heat transfer analysis of the solidification of a binary eutectic system. *Journal of Heat Transfer*, ASME *Transactions*, Series C 69:230.

Tien, R.H. and V. Koump. 1968. Undirectional solidification of a slab-variable surface temperature. AIME *Transactions* 242:283.

Tien, R.H. and V. Koump. 1969. Effect of density change on the solidification of alloys. ASME Paper No. 69-HT-45.

Uno, J. and K. Hayakawa. 1979. Nonsymmetric heat conduction in an infinite slab. *Journal of Food Science* 44:396.

Voller, V.R. and M. Cross. 1981. Estimating the solidification/melting times of cylindrically symmetric regions. *International Journal of Heat Mass Transfer* 24:1457.

Wallhoever, K., C. Koerber, M.W. Scheive, and U. Hartmann. 1985. Undirectional freezing of binary aqueous solutions: An Analysis of transient diffusion of heat and mass. *International Journal of Heat and Mass Transfer* 28:761.

Yuen, W.W. 1980. Application of the heat balance integral to melting problems with internal subcooling. *International Journal of Heat Mass Transfer* 23:1157.

CHAPTER 30

THERMAL PROPERTIES OF FOODS

THIS chapter summarizes thermal process (heating, cooling, freezing, and thawing) calculations for foods and food substances. The material will help engineers and designers applying these processes to foods in industry, teaching, and research. The references listed at the end of the chapter give more details on errors, accuracy, and methods for obtaining data.

The variable composition and structure of foods greatly affects their thermal properties. In addition, their chemical and physical (and hence thermal) properties change with time, temperature, and other ambient conditions. If the food is a living commodity, such as a stored fruit or vegetable, it generates heat; uses atmospheric oxygen; and gives off carbon dioxide, water vapor, and other gases. All these processes affect ambient surroundings, quality, and storage life.

The composition of meat (fat, lean, bone, moisture, muscle, bone size and shape, and fiber direction) varies depending on the species, age, feeding, and slaughter and postslaughter conditions of the animal. Above freezing, a temperature change can affect the conductivity caused by changes in the fat; below freezing, the conductivity changes rapidly until nearly all the water is frozen. After thawing and refreezing, or a period of above-freezing storage, the conductivity may change because of water loss and other changes.

COMMODITY REFRIGERATION LOAD DATA

Water content. Although not a thermal property, water content significantly influences all thermal properties. Values of specific heat and latent heat of fusion (Table 1) are calculated directly from the water content.

Values for percent of water content (Table 1) are the average for the commodity. For fruits and vegetables, water content varies with the stage of development or maturity when harvested, cultivar, growing conditions, and amount of moisture lost after harvest. Values normally apply to mature products shortly after harvest. For fresh meat, water content values are for time of slaughter or after the usual aging period. In cured or processed products, water content depends on the particular process or product.

Some foods are hygroscopic (gaining moisture during storage), but most lose water that cannot be recovered. Values for specific heat and latent heat of fusion in Table 1 apply only for the corresponding water content given.

Freezing point. Freezing point values in Table 1 are based on experiments where the product was cooled slowly until freezing occurred. Sensitive temperature indicators, such as thermocouples, were inserted into the product, and a sudden temperature rise, caused by ice formation, indicated the beginning of freezing. For fruits and vegetables, Table 1 lists the highest temperature at which

specimens froze; they can be seriously damaged by freezing. For other foods, average freezing temperature is shown, largely because initial freezing temperatures are unavailable. Freezing points of foods vary with composition. All apples in a given lot, for example, do not freeze at the same temperature.

As a food product is frozen, ice crystals form, concentrating the remaining unfrozen portion and lowering the freezing temperature. Foods high in sugar content or packed in high syrup concentrations (e.g., dried fruits) may never be completely frozen—even at frozen food storage temperatures. Also, foods normally low in moisture, (e.g., dried beans and peas) do not freeze at these temperatures.

Specific heat. Siebel's formulas can be used to calculate listed specific heat values for above and below freezing (Siebel 1892):

$$c_p = 0.00800a + 0.200 \tag{1}$$

$$c_p = 0.00300a + 0.200 \tag{2}$$

where c_p is specific heat in Btu/lb·°F; a is percent water content; and 0.200 is an arbitrary base, assumed to represent the specific heat of the solid constituents.

Specific heat is a function of temperature. Siebel's formulas are based on the specific heat of water and ice at 32°F of 1.00 and 0.500 Btu/lb·°F, respectively. Specific heat values in Table 1 are for 32°F. In nonfrozen foods, specific heat becomes slightly lower as temperatures rise from 32 to 68°F. In frozen foods, there is a large change in specific heat as temperatures decrease. For example, frozen beef at −40°F has a specific heat of 0.50 Btu/lb·°F, as compared with 0.24 Btu/lb·°F at −166°F (Dickerson 1968). As stated previously, variability in composition is a major characteristic of food substances; this is true for those constituents, particularly water content, that affect specific heat.

In calculating specific heat of frozen foods, it is assumed that all the water is frozen to ice and that the specific heat involved is that of ice. However, this assumption is not entirely correct. As indicated, freezing of most foods is a gradual process, occurring over a wide temperature range. Specific heat values are subject to error; because freezing may be incomplete, the specimen may not be ice but a mixture of frozen and unfrozen constituents (Dickerson 1968).

Latent heat of fusion. Latent heat of fusion is useful in determining the amount of refrigeration required to freeze a product. As with specific heat, the values listed are subject to error because they do not consider any chemical composition other than water content. They are simply the product of the heat of fusion of water (143.59 Btu/lb) and the water content expressed in decimal form. Their use is limited to practical applications, and these values are subject to some discrepancy, depending on product composition.

The following examples show that errors inherent in calculating heat of fusion of foods are not significantly large.

Example 1. Assume a food of 75% moisture content, initially at 32°F, is cooled to 12°F; latent heat of fusion of water, 143.59 Btu/lb; specific heat of ice, 0.50 Btu/lb·°F; specific heat of water, 1.01 Btu/lb·°F; specific heat of nonaqueous 25% of the food [Equation (2)], 0.200 Btu/lb·°F. Assum-

The preparation of this chapter was assigned to TC 11.9, which is presently inactive.

30.1

Table 1 Thermal and Related Properties of Food and Food Materials

Food or Food Material	Water Content, %(mass)[a]	Highest Freezing Point, °F[b]	Specific Heat[c] Above Freezing, Btu/lb·°F	Specific Heat[c] Below Freezing, Btu/lb·°F	Latent Heat of Fusion,[d] Btu/lb
Vegetables					
Artichokes,					
Globe	84	30	0.902	0.453	121
Jerusalem	80	28	0.878	0.441	115
Asparagus	93	31	0.956	0.480	134
Beans, Snap	89	31	0.932	0.468	128
Beans, Lima	67	31	0.801	0.401	96
Beans, Dried	11	—	0.465	0.233	16
Beets, Roots	88	30	0.926	0.465	127
Broccoli	90	31	0.938	0.471	130
Brussels sprouts	85	31	0.908	0.456	122
Cabbage, Late	92	30	0.950	0.477	132
Carrots, Roots	88	29	0.926	0.465	127
Cauliflower	92	31	0.950	0.477	132
Celeriac	88	30	0.926	0.465	127
Celery	94	31	0.926	0.483	135
Collards	87	31	0.920	0.462	125
Corn, Sweet	74	31	0.843	0.423	107
Cucumbers	96	31	0.974	0.489	138
Eggplant	93	31	0.956	0.480	134
Endive (Escarole)	93	32	0.956	0.480	134
Garlic	61	31	0.765	0.383	88
Ginger, Rhizomes	87	—	0.920	0.462	125
Horseradish	75	29	0.849	0.426	108
Kale	87	31	0.920	0.462	125
Kohlrabi	90	30	0.938	0.471	130
Leeks	85	31	0.908	0.456	122
Lettuce	95	32	0.968	0.486	137
Mushrooms	91	30	0.944	0.474	131
Okra	90	29	0.938	0.471	130
Onions, Green	89	30	0.932	0.468	128
Onions, Dry	88	31	0.926	0.465	127
Parsley	85	30	0.908	0.456	122
Parsnips	79	30	0.872	0.438	114
Peas, Green	74	31	0.792	0.423	107
Peas, Dried	12	—	0.471	0.236	17
Peppers, Dried	12	—	0.471	0.236	17
Peppers, Sweet	92	31	0.950	0.477	132
Potatoes, Early	81	31	0.884	0.444	117
Potatoes, Main crop	78	31	0.866	0.435	112
Potatoes, Sweet	69	30	0.813	0.407	99
Yams	74	—	0.843	0.423	107
Pumpkins	91	31	0.944	0.474	131
Radishes	95	31	0.968	0.486	137
Rhubarb	95	30	0.968	0.486	137
Rutabagas	89	30	0.932	0.468	128
Salsify	79	30	0.872	0.438	114
Spinach	93	31	0.956	0.480	134
Squash, Summer	94	31	0.926	0.483	135
Squash, Winter	85	31	0.908	0.456	122
Tomatoes,					
Mature green	93	31	0.956	0.480	134
Tomatoes, Ripe	94	31	0.962	0.483	135
Turnip greens	90	32	0.968	0.471	130
Turnip	92	30	0.950	0.477	132
Watercress	93	31	0.956	0.480	134
Fruits					
Apples, Fresh	84	30	0.902	0.453	121
Apples, Dried	24	—	0.543	0.272	35
Apricots	85	30	0.908	0.456	122
Avocados	65	31	0.789	0.395	94
Bananas	75	31	0.849	0.426	108
Blackberries	85	31	0.908	0.456	122
Blueberries	82	29	0.890	0.447	118
Cantaloupes	92	30	0.950	0.477	132
Cherries, Sour	84	29	0.902	0.453	121

Food or Food Material	Water Content, %(mass)[a]	Highest Freezing Point, °F[b]	Specific Heat[c] Above Freezing, Btu/lb·°F	Specific Heat[c] Below Freezing, Btu/lb·°F	Latent Heat of Fusion,[d] Btu/lb
Cherries, Sweet	80	29	0.878	0.441	115
Cranberries	87	30	0.920	0.462	125
Currants	85	30	0.908	0.456	122
Dates, Cured	20	4	0.519	0.260	29
Figs, Fresh	78	28	0.866	0.435	112
Figs, Dried	23	—	0.537	0.269	33
Gooseberries	89	30	0.932	0.468	128
Grapefruit	89	30	0.932	0.468	128
Grapes, American	82	29	0.890	0.447	118
Grapes, Vinifera	82	28	0.890	0.447	118
Lemons	89	29	0.932	0.468	128
Limes	86	29	0.914	0.459	124
Mangoes	81	30	0.884	0.444	117
Melons, Casaba	93	30	0.956	0.480	134
Melons, Crenshaw	93	30	0.956	0.480	134
Melons, Honeydew	93	30	0.956	0.480	134
Melons, Persian	93	31	0.956	0.480	134
Melons, Watermelon	93	31	0.956	0.480	134
Nectarines	82	30	0.890	0.447	118
Olives	75	29	0.849	0.426	108
Oranges	87	31	0.920	0.462	125
Peaches, Fresh	89	30	0.932	0.468	128
Peaches, Dried	25	—	0.549	0.275	36
Pears	83	29	0.896	0.450	120
Persimmons	78	28	0.866	0.435	112
Pineapples	85	30	0.908	0.456	122
Plums	86	31	0.914	0.459	124
Pomegranates	82	27	0.890	0.447	118
Prunes	28	—	0.567	0.284	40
Quinces	85	28	0.908	0.456	122
Raisins	18	—	0.507	0.254	26
Raspberries	81	31	0.884	0.444	117
Strawberries	90	31	0.938	0.471	130
Tangerines	87	30	0.920	0.462	125
Whole Fish					
Haddock—Cod	78	28	0.866	0.435	112
Halibut	75	28	0.849	0.426	108
Herring, Kippered	70	28	0.819	0.441	101
Herring, Smoked	64	28	0.783	0.392	92
Menhaden	62	28	0.771	0.386	89
Salmon	64	28	0.783	0.392	92
Tuna	70	28	0.819	0.411	101
Filets or Steaks					
Haddock-Cod-Perch	80	28	0.878	0.441	115
Hake—Whiting	82	28	0.890	0.447	118
Pollock	79	28	0.872	0.438	114
Mackerel	57	28	0.741	0.371	82
Shellfish					
Scallop, Meat	80	28	0.878	0.441	115
Shrimp	83	28	0.896	0.450	120
Lobster, American	79	28	0.872	0.438	114
Oysters—Clams,					
Meat and liquor	87	28	0.920	0.462	125
Oyster in Shell	80	27	0.878	0.441	115
Beef					
Carcass (60% Lean)	49	29	0.693	0.347	71
Carcass (54% Lean)	45	28	0.669	0.335	65
Sirloin, Retail cut	56	—	0.735	0.368	81
Round, Retail cut	67	—	0.801	0.401	96
Dried, Chipped	48	—	0.687	0.344	69
Liver	70	29	0.819	0.411	101
Veal, Carcass (81% Lean)	66	—	0.795	0.398	95
Pork					
Bacon	19	—	0.513 27	0.257	
Ham, Light cure	57	—	0.741	0.371	82
Ham, Country cure	42	—	0.651	0.326	60

Table 1 Thermal and Related Properties of Food and Food Materials (*Concluded*)

Food or Food Material	Water Content, %(mass)[a]	Highest Freezing Point, °F[b]	Specific Heat[c] Above Freezing, Btu/lb·°F	Specific Heat[c] Below Freezing, Btu/lb·°F	Latent Heat of Fusion,[d] Btu/lb	Food or Food Material	Water Content, %(mass)[a]	Highest Freezing Point, °F[b]	Specific Heat[c] Above Freezing, Btu/lb·°F	Specific Heat[c] Below Freezing, Btu/lb·°F	Latent Heat of Fusion,[d] Btu/lb
Carcass (47% Lean)	37	—	0.621	0.311	53	Dried (Nonfat)	3	—	0.417	0.209	4
Bellies (33% Lean)	30	—	0.579	0.290	43	Fluid (3.7% Fat)	87	31	0.920	0.462	125
Backfat (100% Fat)	8	—	0.447	0.224	12	Fluid (Skim)	91	—	0.944	0.474	131
Shoulder (67% Lean)	49	28	0.693	0.347	71	Whey, Dried	5	—	0.425	0.215	7
Ham (74% Lean)	56	29	0.735	0.368	81	**Poultry Products**					
Sausage,						Eggs,					
Links or bulk	38	—	0.627	0.314	55	Whole (Fresh)	74	31	0.843	0.423	107
Country style, Smoked	50	25	0.699	0.350	72	Whites	88	31	0.926	0.465	127
Frankfurters	56	29	0.735	0.368	81	Yolks	51	31	0.705	0.353	73
Polish style	54	—	0.723	0.362	78	Yolks (Sugared)	51	25	0.705	0.353	73
Lamb						Yolks (Salted)	50	1	0.699	0.350	72
Composite of cuts						Dried (Whole)	4	—	0.425	0.212	6
(67% Lean)	61	29	0.765	0.383	88	Dried (White)	9	—	0.453	0.227	13
Leg (83% Lean)	65	—	0.789	0.395	94	Chicken	74	27	0.843	0.423	107
Dairy Products						Turkey	64	—	0.783	0.392	92
Butter	16	—	0.495	0.248	23	Duck	69	—	0.513	0.407	99
Cheese,						**Miscellaneous**					
Camembert	52	—	0.711	0.356	75	Honey	17	—	0.501	0.251	25
Cheddar	37	9	0.621	0.311	53	Maple syrup	33	—	0.597	0.299	48
Cottage (Uncreamed)	79	30	0.872	0.438	114	Popcorn, Unpopped	10	—	0.459	0.230	14
Cream	51	—	0.705	0.353	73	Yeast, Bakers,					
Limburger	45	19	0.669	0.335	65	Compressed	71	—	0.835	0.414	102
Roquefort	40	3	0.639	0.320	58	**Candy**					
Swiss	39	14	0.633	0.317	56	Milk chocolate	1	—	0.405	0.203	1
Processed American	40	20	0.639	0.320	58	Peanut brittle	2	—	0.411	0.206	3
Cream,						Fudge, Vanilla	10	—	0.459	0.230	14
Half and half	80	—	0.878	0.441	115	Marshmallows	17	—	0.501	0.257	25
Table	72	28	0.831	0.417	106	**Nuts, Shelled**					
Whipping, Heavy	57	—	0.741	0.371	82	Peanuts (With skins)	6	—	0.435	0.218	9
Ice cream (10% Fat)	63	22	0.777	0.389	91	Peanuts (With					
Milk,						skins, roasted)	2	—	0.411	0.206	3
Canned, Condensed,						Pecans	3	—	0.417	0.209	4
Sweetened	27	5	0.561	0.281	39	Almonds	5	—	0.429	0.215	7
Evaporated,						Walnuts, English	4	—	0.423	0.212	6
Unsweetened	74	29	0.843	0.423	107	Filberts	6	—	0.435	0.218	9
Dried (Whole)	2	—	0.411	0.206	3						

[a]Water contents of fruits and vegetables are from Lutz and Hardenburg (1968), except for Jerusalem artichokes; dried beans; and peas, yams, dried apples, figs, peaches, prunes, and raisins; the latter are from Watt and Merrill (1963). Water content of meats, dairy and poultry products, miscellaneous candy, and nuts are also from Watt and Merrill (1963); water content of eggs (yolks, salted) and fish are from ASHRAE (1972, 1974, 1978)

[b]Freezing points of fruits and vegetables are from Whiteman (1957), and average freezing oints of other foods are from ASHRAE (1972, 1974, 1978).
[c]Specific heat was calculated from Siebel's formulas (1892).
[d]Latent heat of fusion was obtained by multiplying water content expressed in decimal form by 144, the heat of fusion of water in Btu/lb.

ing only 90% of the water present freezes, a good assumption for many foods is:

(a) Latent heat of freezing:
$143.59 \times 0.75 \times 0.9$ = 96.92 Btu/lb
(b) Sensible heat to cool the ice:
$(0.75 \times 0.9 \times 0.50) \times 20$ = 6.75 Btu/lb
(c) Sensible heat to cool unfrozen water:
$(0.75 \times 0.1 \times 1.01) \times 20$ = 1.52 Btu/lb
(d) Sensible heat to cool nonaqueous material:
$(0.25 \times 0.200) \times 20$ = 1.00 Btu/lb
Total = 106.19 Btu/lb

Calculated from the freezing of water only, $(0.75 \times 144 = 108.0$ Btu/lb) for an error of $100 (108.0 - 106.19)/106.19 = 1.7\%$.

Example 2. Assume a food of 50% water content. Calculating as above:

(a) 64.62 Btu/lb
(b) 4.50 Btu/lb
(c) 1.01 Btu/lb
(d) 2.00 Btu/lb

Total 72.13 Btu/lb

Using heat of fusion of water gives only 71.80 Btu/lb for an error of 0.5%. In some foods with low water content, very little water will be frozen at -4°F. This causes a large error if the 90% freezing assumption is maintained.

For computing cooling or freezing loads of frozen products, specific heat below freezing and latent heat of fusion are interdependent and should be used together. For many food products, both freezing and temperature change of frozen material take place simultaneously. This occurs over a temperature range, narrow for foods with low solid matter content, wide for those with high solid matter content (see Enthalpy, Specific Heat, and Thermal Diffusivity section).

In Table 1, no latent heat of fusion values or specific heat values below freezing are given for products so low in moisture that the water in them does not freeze.

Heat of respiration. All living food products respire. In respiration, a sugar, usually glucose, combines with oxygen by a step process involving enzymes. A simplified formula for the process is:

$$C_6H_{12}O_6 + 6O_2 \rightarrow 6CO_2 + 6H_2O$$

$$+ \text{ energy (heat and energy of ATP)}$$

Table 2 Heat of Respiration of Fresh Fruits and Vegetables Held at Various Temperatures[a]

Commodity	| Heat of Respiration, Btu/day per Ton of Produce						Reference
	32°F	41°F	50°F	59°F	68°F	77°F	
Apples, Y, Transparent	1513	2665	—	7889	12,392	—	Wright et al. (1954)
Apples, Delicious	757	1117	—	—	—	—	Lutz and Hardenburg (1968)
Apples, Golden Delicious	793	1189	—	—	—	—	Lutz and Hardenburg (1968)
Apples, Jonathan	865	1295	—	—	—	—	Lutz and Hardenburg (1968)
Apples, McIntosh	793	1189	—	—	—	—	Lutz and Hardenburg (1968)
Apples, Early cultivars	720–1369	1153–2342	3062–4503	3962–6844	4323–9005	—	IIR (1967)
Apples, Late cultivars	396–793	1008–1549	1513–2306	2053–4323	3242–5403	—	IIR (1967)
Apples, Average of many cultivars	505–901	1117–1585	—	2990–6808	3711–7709	—	Lutz and Hardenburg (1968)
Apricots	1153–1261	1405–1982	2449–4143	4683–7565	6484–11,527	—	Lutz and Hardenburg (1968)
Artichokes, Globe	5007–9907	7025–13,220	1203–21,649	1704–31,951	3004–51,403	—	Sastry et al. (1978), Rappaport and Watuda (1958)
Asparagus	6015–17,651	12,032–30,043	23,630–67,146	35,086–72,152	60,121–110,228	—	Sastry et al. (1978), Lipton (1957)
Avocados	*[b]	*[b]	—	13,616–34,581	16,246–76,439	—	Lutz and Hardenburg (1968), Biale (1960)
Bananas,							
Green	*[b]	*[b]	□[b]	4431–7626	6484–11,527	—	IIR (1967)
Ripening	*[b]	*[b]	□[b]	6484–9726	7204–18,011	—	IIR (1967)
Beans, Lima							
Unshelled	2306–6628	4323–7925	—	22,046–27,449	29,250–39,480	—	Lutz and Hardenburg (1968), Tewfik and Scott (1954)
Shelled	3890–7709	6412–13,436	—	—	46,577–59,509	—	Lutz and Hardenburgh (1968), Tewfik and Scott (1954)
Beans, Snap	*[b]	7529–7709	12,032–12,824	18,731–20,533	26,044–28,673	—	Ryall and Lipton (1972), Watuda and Morris (1966)
Beets, Red, Roots	1189–1585	2017–2089	2594–2990	3711–5115	—	—	Ryall and Lipton (1972), Smith (1957)
Berries,							
Blackberries	3458–5043	6304–10,086	11,527–20,893	15,489–32,060	28,818–43,227	—	IIR (1967)
Blueberries	505–2306	2017–2702	—	7529–13,616	11,419–19,236	—	Lutz and Hardenburg (1968)
Cranberries	*[b]	901–1008	—	—	2413–3999	—	Lutz and Hardenburg (1968) Anderson et al. (1963)
Gooseberries	1513–1909	2702–2990	—	4791–7096	—	—	Lutz and Hardenburg (1968), Smith (1966)
Raspberries	3890–5512	6808–8501	6124–12,248	18,119–22,334	25,215–54,033	—	Lutz and Hardenburg (1968), IIR (1967), Haller et al. (1941)
Strawberries	2702–3890	3602–7313	10,807–20,893	15,634–20,317	22,514–43,154	37,247–46,468	Lutz and Hardenburg (1968), IIR (1967), Maxie et al. (1959)
Broccoli, Sprouting	4107–4719	7601–35,226	—	38,256–74,890	61,274–75,106	85,805–123,376	Lutz and Hardenburg (1968), Morris (1947), Scholz et al. (1963)
Brussels Sprouts	3386–5295	7096–10,698	13,904–18,623	21,037–23,523	19,848–41,894	—	Sastry et al. (1978), Smith (1957)
Cabbage,							
Penn State	865	2089–2234	—	4935–6988	—	—	Van den Berg and Lentz (1972)
White, Winter	1081–1801	1621–3062	2702–3962	4323–5944	7925–9006	—	IIR (1967)
White, Spring	2089–2990	3890–4719	6412–7313	11,815–12,609	—	—	Sastry et al. (1978), Smith (1957)
Red, Early	1693–2161	3423–3783	5224–61,238	8105–9366	12,248–12,608	—	IIR (1967)
Savoy	3422–4683	5584–6484	11,527–13,509	19,272–21,794	28,818–32,420	—	IIR (1967)
Carrots, Roots,							
Imperator, Texas	3386	4323	6916	8718	15,526	—	Scholz et al. (1963)
Main Crop, U.K.	757–1513	1296–2666	2161–3423	6448–14,589 at 65°F	—	—	Smith (1957)
Nantes, Can.[d]	684	1477	—	4755–6232	—	—	Van den Berg and Lentz (1972)
Cauliflower,							
Texas	3926	4503	7456	10,158	17,687	—	Scholz et al. (1963)
U.K.	1693–5295	4323–6015	9006–10,734	14,841–18,047	—	—	Smith (1957)
Celery,							
N.Y., White	1585	2413	—	82,148	14,229	—	Lutz and Hardenburg (1968)
U.K.	1117–1585	2017–2810	4323–6015	8609–9221 at 65°F	—	—	Smith (1957)
Utah, Can.[e]	1117	1982	—	6556	—	—	Van den Berg and Lentz (1972)
Cherries,							
Sour	296–2918	2810–2918	—	6015–11,022	8609–11,022	11,708–15,634	Lutz and Hardenburg (1968), Hawkins (1929)
Sweet	901–1189	2089–3098	—	5512–9907	6196–7025	—	Lutz and Hardenburg (1968), Micke et al. (1965)
Corn, Sweet with husk,							
Texas	9366	17,111	24,676	35,878	63,543	89,695	Scholz et al. (1963)
Cucumbers, Calif.	*[b]	*[b]	5079–6376	5295–7313	6844–10,591	—	Eaks and Morris (1956)

Table 2   Heat of Respiration of Fresh Fruits and Vegetables Held at Various Temperatures[a] (*Continued*)

Commodity	Heat of Respiration, Btu/day per Ton of Produce						Reference
	32 °F	41 °F	50 °F	59 °F	68 °F	77 °F	
Figs, Mission	—	2413–2918	4863–5079	10,807–13,940	12,536–20,929	18,731–20,929	Lutz and Hardenburg (1968), Claypool and Ozbek (1952)
Garlic	648–2413	1296–2125	2017–2125	2413–6015	2197–3999	—	Sastry et al. (1978), Mann and Lewis (1956)
Grapes, Labrusca, Concord	612	1189	—	3494	7204	8501	Lutz and Hardenburg (1968), Lutz (1938)
Grapes, Vinifera, Emperor	288–505	684–1296	1801	2197–2594	—	5512–6628	Lutz and Hardenburg (1968), Pentzer et al. (1933)
Grapes, Vinifera, Thompson Seedless	432	1045	1693	—	—	—	Wright et al. (1954)
Grapes, Vinifera, Ohanez	288	720	2	—	—	—	Wright et al. (1954)
Grapefruit, Calif. Marsh	*[b]	*[b]	*[b]	2594	3890	4791	Haller et al. (1945)
Grapefruit, Florida	*[b]	b	*[b]	2810	3494	4214	Haller et al. (1945)
Horseradish	1801	2377	5800	7204	9834	—	Sastry et al. (1978)
Kiwi fruit	616	1455	2889	—	3858–4254	—	Saravacos and Pilsworth (1965)
Kohlrabi	2197	3602	6916	10,807	—	—	Sastry et al. (1978)
Leeks	2089–3062	4323–6412	11,815–15,021	18,227–25,756	—	—	Sastry et al. (1978), Smith (1957)
Lemons, Calif., Eureka	*[b]	*[b]	*[b]	3494	5007	5727	Haller (1945)
Lettuce, Head Calif.	2017–3711	2918–4395	6015–8826	8501–9006	13,220	—	Sastry et al. (1978)
Lettuce, Head Texas	2306	2918	4791	7925	12,536	181 at 180 °F	Watt and Merrill (1963), Lutz and Hardenburg, (1968)
Lettuce, Leaf, Texas	5079	6448	8681	13,869	22,118	32,275	Scholz et al. (1963)
Lettuce, Romaine, Texas	—	4575	7817	9762	15,093	23,883	Scholz et al. (1963)
Limes, Persian	*[b]	*[b]	576–1261	1296–2306	1513–4107	3314–10,014	Lutz and Hardenburg (1968)
Mangos	*[b]	*[b]	—	9907	16,534–33,356	26,441	Lutz and Hardenburg (1968), Gore (1911), Karmarkar and Joshe (1941)
Melons, Cantaloupes	*[b]	1909–2197	3423	7420–8501	9834–14,229	13,725–15,741	Lutz and Hardenburg (1968), Sastry et al. (1978), Scholz et al. (1963)
Melons, Honeydew	—	*[b]	1765	2594–3494	4395–5259	5800–7601	Lutz and Hardenburg (1968), Scholz et al. (1963), Pratt and Morris (1958)
Melons, Watermelon	*[b]	*[b]	1657	—	3818–5512	—	Lutz and Hardenburg (1968), Scholz et al. (1963)
Mint [m]	1769–3306	6614	16,754–20,061	23,148–29,981	36,595–50,041	56,655–69,883	Hruschka and Want (1979)
Mushrooms	6196–9618	15,634	—	—	58,104–69,738	—	Lutz and Hardenburg (1968) Smith (1964)
Nuts (kind not specified)	181	360	720	720	1081	—	IIR (1967)
Okra, Clemson	*[b]	76,043	19,236	32,132	57,527	76,040 at 85 °F	Scholz et al. (1963)
Onions, Dry, Autumn Spice[f]	505–684	793–1477	—	2089–5548	—	—	Van den Berg and Lentz (1972)
Onions, Dry, White Bermuda	648	757	1585	2449	3711	6196 at 80 °F	Scholz et al. (1963)
Onions, Green, N.J.	2306–4899	3819–15,021	7961–12,968	14,553–21,434	17,205–34,225	21,541–46,217	Lutz and Hardenburg (1968)
Olives, Manzanillo	*[b]	*[b]	—	4791–8609	8501–10,807	9006–13,436	Maxie et al. (1959)
Oranges, Florida	684	1405	2702	4611	6628	7817 at 80 °F	Haller (1945)
Oranges, Calif., W. Navel	*[b]	1405	2990	5007	6015	7997	Haller (1945)
Oranges, Calif., Valencia	*[b]	1008	2594	2810	3890	4611	Haller (1945)
Papayas	*[b]	*[b]	2485	3314–4791	—	8609–21,613	Pantastico (1974), Jones (1942)
Parsley[m]	7277–10,140	14,549–18,738	28,879–36,155	31,746–49,163	43,208–56,216	67,902–75,174	Hruschka and Want (1979)
Parsnips, U.K.	2558–3423	1946–3854	4503–5800	7096–9438	—	—	Smith (1957)
Parsnips, Canada, Hollow Crown[g]	793–1801	1369–3386	—	4755–10,195	—	—	Van den Berg and Lentz (1972)
Peaches, Elberta	829	1441	3458	7565	13,509	19,812 at 80 °F	Haller et al. (1932)
Peaches, Several cultivars	901–1405	1405–2017	—	7313–9330	13,040–22,549	17,939–26,837	Lutz and Hardenburg (1968)
Peanuts, Cured[h]	3 at 85 °F	—	—	—	—	51 at 85 °F	Thompson et al. (1951)
Peanuts, Not cured, Virginia Bunch[i]	—	—	—	—	—	3120 at 85 °F	Schenk (1959, 1961)
Peanuts, Dixie Spanish	—	—	—	—	—	1823 at 85 °F	Schenk (1959, 1961)
Pears, Bartlett	684–1513	1117–2197	—	3314–13,220	6628–15,417	—	Lutz and Hardenburg (1968)
Pears, Late ripening	576–793	1296–3062	1729–4143	6124–9366	7204–16,210	—	IIR (1967)
Pears, Early ripening	576–1081	1621–3423	2161–4683	7565–11,887	8645–19,812	—	IIR (1967)
Peas, Green-in-Pod	6700–10,302	12,139–16,822	—	39,372–44,595	54,105–79,645	75,646–83,067	Lutz and Hardenburg (1968), Tewfik and Scott (1954)
Peas, Shelled	10,410–16,642	17,435–21,444	—	—	76,871–10,893	—	Lutz and Hardenburg (1968), Tewfik and Scott (1954)

Table 2 Heat of Respiration of Fresh Fruits and Vegetables Held at Various Temperatures[a] (Concluded)

Commodity	Heat of Respiration, Btu/day per Ton of Produce						Reference
	32 °F	41 °F	50 °F	59 °F	68 °F	77 °F	
Peppers, Sweet	*[b]	*[b]	3170	5043	9654	—	Scholz et al. (1963)
Persimmons	—	1296	—	2594–3098	4395–5295	6412–8826	Lutz and Hardenburg (1968), Gore (1911)
Pineapple, Mature green	*[b]	*[b]	1225	2846	5331	7817 at 80 °F	Scholz et al. (1963)
Pineapple, Ripening	*[b]	*[b]	1657	3999	8790	13,797	Scholz et al. (1963)
Plums, Wickson	432–648	865–1982	1981–2522	2630–2737	3962–5727	6160–15,634	Claypool and Allen (1951)
Potatoes, Calif. White Rose, Immature	*[b]	2594–	3098–4611	3098–6808	3999–9932	—	Sastry et al. (1978)
Potatoes, Calif. White Rose, Mature	*[b]	1296–1513	1467–2197	1467–2594	1467–3494	—	Sastry et al. (1978)
Potatoes, Calif. White Rose, Very Mature[j]	*[b]	1117–1513	1513	1513–2197	2017–2630	—	Sastry et al. (1978)
Potatoes, Katahdin, Can.[k]	*[b]	865–936	—	1729–2234	—	—	Van den Berg and Lentz (1972)
Potatoes, Kennebec	*[b]	793–936	—	936–1982	—	—	Van den Berg and Lentz (1972)
Radishes, with Tops	3206–3818	4214–4611	6808–8105	15,417–17,146	27,341–30,043	34,869–42,470	Lutz and Hardenburg (1968)
Radishes, Topped	1189–1296	1693–1801	3314–3494	6124–7204	10,519–10,807	14,841–16,751	Lutz and Hardenburg (1968)
Rhubarb, Topped	1801–2918	2413–3999	—	6808–10,014	8826–12,536	—	Hruschka (1966)
Rutabaga, Laurentian, Can.[l]	432–612	1045–1124	—	2342–3458	—	—	Van den Berg and Lentz (1972)
Spinach, Texas	—	10,122	24,387	39,409	50,683	—	Scholz et al. (1963)
Spinach, U.K., Summer	2558–4719	6015–7096	12,896–16,534	—	40,777–47,657 at 65 °F	—	Smith (1957)
Spinach, U.K., Winter	3854–5584	6448–13,869	15,021–22,766	—	42,938–53,673 at 65 °F	—	Smith (1957)
Squash, Summer, Yellow, Straight-neck	□[b]	□[b]	7709–8105	16,534–20,028	18,731–21,434	—	Lutz and Hardenburg (1968)
Squash, Winter Butternut	*[b]	*[b]	—	—	—	16,318–26,908	Lutz and Hardenburg (1968)
Sweet Potatoes, Cured, Puerto Rico	*[b]	*[b]	□[b]	3530–4863	—	—	Lewis and Morris (1956)
Sweet Potatoes, Cured, Yellow Jersey	*[b]	*[b]	□[b]	4863–5079	—	—	Lewis and Morris (1956)
Sweet Potatoes, Noncured	*[b]	*[b]	*[b]	6304	—	11,923–16,138	Lutz and Hardenburg (1968)
Tomatoes, Texas, Mature Green	*[b]	*[b]	*[b]	4503	7637	9402 at 80 °F	Scholz et al. (1963)
Tomatoes, Texas, Ripening	*[b]	*[b]	*[b]	5872	8933	10,627 at 80 °F	Scholz et al. (1963)
Tomatoes, Calif. Mature Green	*[b]	*[b]	*[b]	—	5295–7709	6592–10,591	Workman and Pratt (1957)
Turnip, Roots	1909	2089–2197	—	4719–5295	5295–5512	—	Lutz and Hardenburg (1968)
Watercress[m]	3306	9920	20,061–26,674	29,981–43,208	66,576–76,719	76,720–96,561	Hruschka and Want (1979)

[a]Column headings indicate temperatures at which respiration rates were determined, within 2 °F, except where the actual temperatures are given.

[b]The symbol * denotes a chilling temperature. The symbol □ denotes the temperature is borderline, not damaging to some cultivars, if exposure is short.

[c]Rates are for 30 to 60 days and 60 to 120 days storage, the longer storage having the higher rate, except at 32 °F, where they were the same.

[d]Rates are for 30 to 60 days and 120 to 180 days storage, respiration increasing with time only at 59 °F.

[e]Rates are for 30 to 60 days storage.

[f]Rates are for 30 to 60 days and 120 to 180 days storage; rates increased with time at all temperatures as dormancy was lost.

[g]Rates are for 30 to 60 days and 120 to 180 days; rates increased with time at all temperatures.

[h]Shelled peanuts with about 7% moisture. Respiration after 60 h curing was almost negligible, even at 85 °F.

[i]Respiration for freshly dug peanuts, not cured, with about 35 to 40% moisture. During curing, peanuts in the shell were dried to about 5 to 6% moisture, and in roasting are dried further to about 2% moisture.

[j]Harvested 141 days after planting (Morris 1952).

[k]Rates are for 30 to 60 days and 120 to 180 days with rate declining with time at 41 °F but increasing at 59 °F as sprouting started.

[l]Rates are for 30 to 60 days and 120 to 180 days; rates increased with time, especially at 59 °F where sprouting occurred.

[m]Rates are for 1 day after harvest.

The end products are CO_2, H_2, and energy in the forms of heat and adenosine triphosphate (ATP) that the cell uses for growth and development. In most stored plant products, little cell development takes place, and the greater part of respiration energy is in the form of heat, which must be taken into account when cooling and storing these living products.

As in all chemical reactions, temperature is important to the rate of respiration. An 18 °F rise in temperature causes respiration rates to double or triple in the range of about 32 to 86 °F. Higher temperatures usually retard respiration.

Fruits, vegetables, flowers, bulbs, florists' greens, and nursery stock are storage commodities with significant heats of respiration. Dry plant products, such as seeds and nuts, have very low respiration rates. Young, actively growing tissues, such as asparagus, broccoli, and spinach, have high rates, as do immature seeds such as green peas and sweet corn. Fast-developing fruits such as strawberries, raspberries, and blackberries have much higher respiration rates than do fruits that are slow to develop, such as apples, grapes, and citrus fruits (Table 2).

Almost all commodities in Table 2 have a low and a high value for heat of respiration at each temperature. When no range is given, the value is an average for the specified temperature and may be an average of the respiration rates for many days.

Most vegetables, other than root crops, have a high initial rate for the first day or two after harvest. Within a few days, they quickly lower to the equilibrium rate (Ryall and Lipton 1972). Asparagus is a good example of this. The first day, heat of respiration at 32 °F is 17,652 Btu/ton·day. Within three days, it is down to 8682, and in 16 days to 6160 (Table 3). Sweet corn in the husk at 32 °F produces 11,312 Btu/ton·day the first day, and in four days, decreases to 6772 Btu/ton·day. Onions are an exception; they increase in respiration with time as the bulbs lose dormancy. The same is true with garlic (Table 3).

Fruits are different from most vegetables. Those that do not ripen in storage, such as citrus fruits and grapes, have fairly constant respiration rates. Those that ripen in storage (*e.g.*, apples, peaches, and avocados) increase in respiration rate. If a fruit can be held at 32 °F, as most apples can, respiration rarely increases, since no ripening takes place. But if these fruits are held at higher temperatures (50 or 60 °F), respiration increases and then decreases (see Apples, Table 3). Soft fruits, such as blueberries, figs, and strawberries, show a decrease in respiration at 32 °F with time. If they become infected with decay organisms, however, respiration increases.

To use Table 2, select the lower value when estimating respiration heat at the equilibrium state for storage; use the higher value if calculating the heat load for the first day or two, as for precooling and short-distance transport. If the storage temperature is 32 or 40 °F, respiration increase in fruits caused by ripening is slight. In fruits that must be held at 50 °F or higher, such as mangos, avocados, or bananas, ripening occurs, and the higher rates should be used. Vegetables that lose dormancy in storage, such as onions,

garlic, and cabbage, can increase in heat production after long storage.

Not all variations in respiration heat can be attributed to change in rate with time. Broccoli with many flower heads respires faster than if it is mostly stem tissue. Immature fruits and vegetables usually respire faster than more mature specimens (see Potatoes, Table 2). Usually, early, fast-growing cultivars of fruits and vegetables have higher respiration rates than later, slower-developing types.

To obtain the values in Tables 2 and 3, the heat of respiration was assumed to be derived from glucose oxidation in the reaction given earlier. One mole of glucose (180 lb) is oxidized by six moles of oxygen to produce six moles of CO_2 (264 lb). Glucose oxidation produces 1,214,994 Btu/mol. Therefore, one pound of CO_2 represents:

$$1,214,994/264 = 4602 \text{ Btu}$$

If substrates other than glucose are oxidized in the respiration process, the heat produced amounts to less when organic acids are the substrate, and considerably more when fats are used.

Transpiration of Fruits and Vegetables

Transpiration of moisture from fresh fruits and vegetables is a mass transfer process in which water vapor moves from the surface of the commodity to the surrounding air. Water, the most abundant constituent of these products, exists as a continuous liquid phase within the fruit or vegetable. Terms used to describe transpiration are defined here.

The *transpiration coefficient* of a fruit or vegetable is the mass of moisture transpired per unit mass of commodity, per unit environmental water vapor pressure deficit per unit time. In some

Table 3   Change in Respiration Rates with Time

Commodity	Days in Storage	Heat of Respiration, Btu/day per Ton of Produce 32 °F	41 °F	Reference	Commodity	Days in Storage	Heat of Respiration, Btu/day per Ton of Produce 32 °F	41 °F	Reference
Apples, Grimes	7	648	2882 at 50 °F	Harding (1929)	Garlic	10	865	1982	Mann and Lewis (1956)
	30	648	3854			30	1333	3314	
	80	648	2413			180	3098	7277	
Artichokes, Globe	1	9907	13,220	Rappaport and Watada (1957)	Lettuce, Great Lakes	1	3747	4395	Pratt *et al.* (1954)
	4	5512	7709			5	1982	33	
	16	3314	5727			10	1765	3314	
Asparagus, Martha Washington	1	17,652	2316	Lipton (1957)	Olives, Manzanillo	1	—	8610 at 60 °F	Maxie *et al.* (1960)
	3	8682	14,337			5	—	6376	
	16	6160	6629			10	—	4864	
Beans, Lima, in Pod	2	6593	7925	Tewfik and Scott (1954)	Onions, Red	1	360	—	Karmarkar and Joshe (1941)
	4	4431	6376			30	541	—	
	6	3890	5836			120	720	—	
Blueberries, Blue Crop	1	1585	—	Hardenburg (1966)	Plums, Wickson	2	432	865	Claypool and Allen (1951)
	2	584	—			6	432	1549	
	3	1261	—			18	648	1982	
Broccoli, Waltham 29	1	—	16,102	Rappaport and Watada (1953)	Potatoes	2	—	1333	Morris (1959)
	4	—	9690			6	—	1765	
	8	—	7277			10	—	1549	
Corn, Sweet, in Husk	1	11,312	—	Scholz *et al.* (1963)	Strawberries, Shasta	1	3873	6305	Maxie *et al.* (1959)
	2	8106	—			2	2918	6772	
	4	6772	—			5	2918	7277	
Figs, Mission	1	2882	—	Claypool and Ozbek (1952)	Tomatoes, Pearson, Mature green	5	—	7060 at 70 °F	Workman and Pratt (1957)
	2	2630	—			15	—	6160	
	12	2630	—			20	—	5295	

cases, the transpiration coefficient is expressed per unit surface area of commodity, rather than per unit mass.

Transpiration rate of a fruit or vegetable is the mass of moisture transpired per unit mass of commodity per unit time. This rate is sometimes expressed per unit commodity surface area. The transpiration rate of a commodity varies with environmental conditions, among other factors.

The rate of transpiration in fruits and vegetables affects product quality. Moisture transpires continuously from fruits and vegetables during handling and storage when subjected to a water vapor pressure deficit. Some moisture loss is inevitable and can be tolerated. However, under many conditions, the loss may be sufficient to cause the commodity to shrivel. The resulting loss in mass not only affects appearance, texture, and flavor, but also reduces the salable mass.

All fruits and vegetables do not lose moisture at the same rate when held under the same conditions. Improving present systems to reduce moisture loss from fresh produce maintains the initial harvest time quality longer. The transpiration rate of stored produce must be known to establish the ideal cold storage conditions for fresh fruits and vegetables.

Figure 1 shows some of the factors that affect transpiration rates or transpiration behavior of stored fruits and vegetables and some phenomena associated with transpiration.

Table 4 lists values for the transpiration coefficients of the investigated fruits and vegetables. These values have either been obtained directly or calculated from actual data presented in Sastry *et al.* (1978).

Because of the many factors, the variables involved, and the methods used to obtain the data, not all the values in Table 4 are reliable; they are to be used primarily as a guide or as a comparative indication of transpiration rates among certain commodities obtained from the literature.

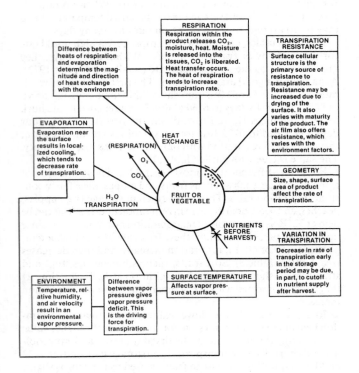

Fig. 1   Transpiration of Fruits and Vegetables; Related Factors and Phenomena

Table 4   Transpiration Coefficients of Certain Fruits and Vegetables

Product and Variety	Transpiration Coefficient, ppm/(h·in. Hg)[a]	Product and Variety	Transpiration Coefficient, ppm/(h·in. Hg)[a]	Product and Variety	Transpiration Coefficient, ppm/(h·in. Hg)[a]
Apples		*Leeks*		*Pears*	
Johnathan	430	Musselburgh	12,600	Passe Crane	974
Golden Delicious	710			Beurre Clairgeau	986
Bramley's Seedling	510	*Lemons*			
		Eureka		*Plums*	
Brussels Sprouts		Dark green	2760	Victoria	
Unspecified	40,100	Yellow	1700	Unripe	2410
				Ripe	1400
Cabbage		*Lettuce*		Wickson	1510
Penn State Ballhead		Unrivalled	106,000		
Trimmed	3300			*Potatoes*	
Untrimmed	4920	*Onions*		Manona	
Mammoth		Autumn Spice		Mature	304
Trimmed	2920	Uncured	1170	Kennebec	
		Cured	535	Uncured	2080
Carrots		Sweet White Spanish		Cured	730
Nantes	20,000	Cured	1500	Sebago	
Chantenay	21,500			Uncured	1920
		Oranges		Cured	462
Celery		Valencia	710		
Unspecified varieties	25,400	Navel	1270	*Rutabagas*	
				Laurentian	5710
Grapefruit		*Parsnips*			
Unspecified varieties	380	Hollow Crown	23,500		
March	670			*Tomatoes*	
		Peaches		Marglobe	864
Grapes		Redhaven		Eurocross BB	1410
Emperor	960	Hard mature	11,200		
Cardinal	1220	Soft mature	12,400		
Thompson	2480	Elberta	3330		

[a]lb of water transpired/(million lb of produce · h · in. Hg)

ENTHALPY, SPECIFIC HEAT, AND THERMAL DIFFUSIVITY

Frozen Foods

The data shown in Table 1 consist of latent heat, freezing point, and specific heat above and below freezing. This table is used to perform refrigeration load calculations by (1) using specific heat above freezing to compute the sensible heat removed during cooling from the starting temperature to freezing point; (2) using latent heat to compute the heat removed during freezing; and (3) using specific heat below freezing to compute the sensible heat removed during cooling from the freezing point to the final frozen storage temperature. This procedure assumes the latent heat of fusion is removed at constant temperature (the freezing point).

Foods do not freeze at constant temperature, however. Beef begins to freeze at about 30°F, but 25% of the latent heat still must be removed at 25°F (Riedel 1957). In haddock muscle, some water is bound to proteins and does not freeze at −40°F (Charm and Moody 1966). In reviewing problems of measuring specific heat and latent heat of foods, Woolrich (1966) noted that fresh beef is not completely frozen at −80°F, and orange juice is not completely frozen at −139°F.

The process of food freezing was explained by Staph and Woolrich (1951). When food begins to freeze, the concentration of food solids is increased in the remaining unfrozen water, establishing a lower freezing point for additional phase change. With additional freezing, there is a gradual depression of the freezing point until all freezable water is frozen. Measurement of heat removed during freezing includes both latent and sensible heat. Consequently, the generally accepted criterion of latent heat (change of phase at constant temperature) cannot be rigorously applied to the process of freezing foods (Woolrich 1966). Neither the concept of specific heat nor thermal diffusivity can be applied because there is, as yet, no way to separate the specific heat component from the latent heat component in a food-freezing process.

Table 1 gives conservative results (*i.e.*, the computed refrigeration load is always somewhat greater than that required). The error diminishes for lower frozen storage temperatures. Conversely, if the product is to be cooled only slightly below the initial freezing temperature, using Table 1 can produce significant errors because only a small fraction of the latent heat is actually removed from the product. Results from Table 1 assume all latent heat is removed.

When the product is cooled only slightly below the initial freezing temperature, or where greater accuracy is desired for any final frozen storage temperature, the physical quantity employed in frozen food calculations is the total heat content or enthalpy (Woolrich 1966, Mannheim *et al.* 1955, Short and Staph 1951, Bartlett 1944). This approach has the additional advantage of yielding estimates of percent of water unfrozen.

When enthalpy is tabulated as a function of temperature, a base temperature must be identified. This base is the temperature at which enthalpy is arbitrarily designated as zero. In this chapter, zero enthalpy is at −40°F.

Enthalpy of Meats, Fruits, Vegetables, and Eggs

Enthalpy of some frozen foods and percent (by mass) of unfrozen water in the food is presented in Table 5. The following equation may be used to calculate the heat transferred from the food.

$$Q = W(h_2 - h_1) \qquad (3)$$

where

Q = total heat transferred, Btu
W = mass, lb
h = enthalpy, Btu/lb

Example 3. A quantity of beef (300 lb) is to be frozen to a temperature of −5°F. Initial temperature of the beef is 50°F, and moisture content is 74.5%. How much heat must be removed, and what is the total mass of unfrozen water at −5°F?

Solution: The heat removed in cooling from 50 to 32°F is calculated from temperatures and specific heat c_p. From Table 5, specific heat is 0.84 Btu/lb·°F, so from 50 to 32°F:

$$Q = 300 \times 0.84 \, (50-32) = 4536 \text{ Btu}$$

From Table 5:

$$h_2 \text{ at } 32°F = 131 \text{ Btu/lb}; \; h_1 \text{ at } -5°F = 18 \text{ Btu/lb}$$

Therefore, from 32°F to −5°F:

$$Q = W(h_2 - h_1) = 300 \, (131-18) = 33,900 \text{ lb}$$

The total heat to be removed is:

$$Q_{total} = 4536 + 33,900 = 38,436 \text{ Btu}$$

The amount of unfrozen water at −5°F is taken from Table 5 as 12%. The total mass of unfrozen water is:

$$300 \times 0.745 \times 0.12 = 27 \text{ lb}$$

Reidel (1957) developed graphs of enthalpy versus water content and percent water frozen versus temperature in beef. Dickerson (1968) combined these two graphs (Figure 2). Below −40°F, beef is nearly frozen, and the specific heat values shown in Table 6 (Moline *et al.* 1961) may be used for calculations.

Working with twelve fruit juices, Riedel (1951) showed that enthalpy values for fruit juices are represented by Figure 3 within about 2%.

The effect of soluble solids in the juice on the temperature range over which latent heat is removed is apparent in Figure 3. For example, a vertical line at a water content of 95% shows the heat to be removed during freezing. For this high water content, freezing starts at a temperature close to 32°F, and the main part

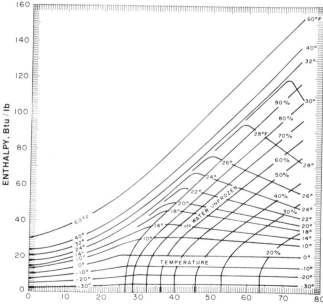

Fig. 2 Enthalpy of Beef
(Adapted from Riedel 1957)

Table 5   Enthalpy of Frozen Foods[a]

Product	Water Content, % (mass)	Mean Specific Heat[b] (40 to 90°F), Btu/lb·°F		-40	-20	-10	-5	0	5	10	15	18	20	22	24	26	28	30	32
Fruits and Vegetables																			
Applesauce	82.8	0.89	Enthalpy, Btu/lb	0	11	17	21	25	30	36	43	49	56	61	71	84	114	145	147
			% water unfrozen[c]	—	5	7	9	11	14	17	20	25	28	33	41	52	76	100	—
Asparagus, Peeled	92.6	0.95	Enthalpy, Btu/lb	0	8	14	16	19	22	26	30	34	37	40	44	51	63	101	162
			% water unfrozen	—	—	—	—	4	5	6	7	9	10	12	16	20	28	55	100
Bilberries	85.1	0.90	Enthalpy, Btu/lb	0	10	15	18	22	25	30	37	41	45	50	56	67	87	149	151
			% water unfrozen	—	—	5	6	7	9	11	14	17	19	22	27	35	50	100	—
Carrots	87.5	0.93	Enthalpy, Btu/lb	0	10	15	18	22	26	31	37	41	45	50	57	68	88	152	154
			% water unfrozen	—	—	5	6	7	9	11	14	17	19	22	27	35	50	100	—
Cucumbers	95.4	0.96	Enthalpy, Btu/lb	0	8	13	16	18	21	24	27	30	32	35	38	43	52	78	167
			% water unfrozen	—	—	—	—	—	—	—	6	7	8	9	12	18	36	100	
Onions	85.5	0.91	Enthalpy, Btu/lb	0	10	16	20	24	28	34	40	46	52	57	66	79	105	149	151
			% water unfrozen	—	5	7	8	9	12	15	18	21	24	28	35	45	65	100	—
Peaches, Without stones	85.1	0.90	Enthalpy, Btu/lb	0	10	16	20	24	28	34	42	47	53	59	67	81	108	148	150
			% water unfrozen	—	5	7	8	10	12	15	18	22	26	30	37	48	69	100	—
Pears, Bartlett	83.8	0.89	Enthalpy, Btu/lb	0	10	17	21	25	29	35	42	47	53	59	69	83	111	146	148
			% water unfrozen	—	6	8	9	10	12	15	19	23	27	31	38	49	72	100	—
Plums, Without stones	80.3	0.87	Enthalpy, Btu/lb	0	12	19	24	28	33	40	50	57	64	73	85	113	139	141	143
			% water unfrozen	—	8	11	13	16	18	22	28	34	38	46	55	71	100	—	—
Raspberries	82.7	0.89	Enthalpy, Btu/lb	0	10	16	19	22	26	31	38	42	46	52	59	71	92	146	148
			% water unfrozen	—	4	6	7	8	9	12	15	18	21	24	30	39	56	100	—
Spinach	90.2	0.93	Enthalpy, Btu/lb	0	8	14	16	19	22	26	29	32	35	38	42	48	59	93	158
			% water unfrozen	—	—	—	—	—	—	5	7	9	10	11	14	18	25	50	100
Strawberries	89.3	0.94	Enthalpy, Btu/lb	0	9	15	18	21	25	29	34	39	41	45	51	60	77	127	158
			% water unfrozen	—	—	—	5	6	7	8	10	13	15	18	21	28	40	79	100
Sweet Cherries, Without stones	77.0	0.86	Enthalpy, Btu/lb	0	12	20	24	29	35	42	51	59	67	76	89	110	134	136	138
			% water unfrozen	—	9	12	14	17	20	25	32	38	43	50	62	80	100	—	—
Tall peas	75.8	0.85	Enthalpy, Btu/lb	0	10	17	21	25	30	36	43	49	54	61	70	86	114	137	139
			% water unfrozen	—	6	8	10	12	15	18	22	27	30	37	44	57	82	100	—
Tomato pulp	92.9	0.96	Enthalpy, Btu/lb	0	10	14	17	20	23	27	32	36	39	42	47	54	68	112	163
			% water unfrozen	—	—	—	—	—	5	6	8	10	12	14	18	22	31	62	100
Eggs																			
Egg white	86.5	0.91	Enthalpy, Btu/lb	0	9	14	16	19	22	25	29	31	33	36	40	45	55	87	151
			% water unfrozen	—	—	—	—	—	—	—	—	10	12	13	14	17	22	48	100
Egg yolk	50.0	0.74	Enthalpy, Btu/lb	0	9	14	16	19	22	25	29	31	33	35	38	42	47	65	98
			% water unfrozen	—	—	—	—	—	—	—	—	—	—	20	23	27	32	66	100
Egg yolk	40.0	0.68	Enthalpy, Btu/lb	0	9	14	17	20	23	26	31	33	35	38	41	46	53	76	82
			% water unfrozen	20	—	—	—	24	—	27	—	30	—	34	38	43	54	89	100
Whole egg, With shell[d]	66.4	0.79	Enthalpy, Btu/lb	0	9	13	15	18	20	23	27	29	31	34	37	41	49	73	121
Fish and Meat																			
Cod	80.3	0.88	Enthalpy, Btu/lb	0	10	15	18	21	24	28	33	36	39	43	48	56	73	123	139
			% water unfrozen	10	10	10	11	12	13	14	16	18	20	22	26	32	45	88	100
Haddock	83.6	0.89	Enthalpy, Btu/lb	0	9	15	18	21	24	28	33	36	39	43	48	56	73	127	145
			% water unfrozen	8	8	9	9	10	11	12	14	15	17	19	23	29	42	86	100
Perch	79.1	0.86	Enthalpy, Btu/lb	0	9	14	17	20	23	27	32	35	38	42	46	53	68	117	137
			% water unfrozen	10	10	11	11	12	13	14	16	17	19	21	24	30	41	83	100
Beef, Lean, Fresh[e]	74.5	0.84	Enthalpy, Btu/lb	0	9	15	18	21	24	27	32	35	38	42	48	57	74	119	131
			% water unfrozen	10	10	11	12	12	13	15	18	20	22	24	28	37	48	92	100
Beef, Lean, Dried	26.1	0.59	Enthalpy, Btu/lb	0	9	14	17	20	24	28	31	—	33	—	36	—	38	—	40
			% water unfrozen	96	96	96	97	98	99	100	—	—	—	—	—	—	—	—	—
Bread																			
White bread	37.3	0.62	Enthalpy, Btu/lb	0	9	13	15	18	21	26	34	40	45	51	55	56	57	58	59
Whole wheat bread	42.4	0.64	Enthalpy, Btu/lb	0	9	13	15	18	22	27	36	43	48	55	62	67	68	69	70

[a]Above -40°F. Adapted from Dickerson (1968) and Riedel (1951, 1956, 1957, 1959).
[b]Temperature range limited to 32 to 68°F for meats, and 68 to 104°F for egg yolk.
[c]Total mass of unfrozen water = (Total mass of food) (% water content/100) (% water unfrozen/100).

[d]Calculated for a mass composition of 58% white (86.5% water) and 32% yolk (50% water).
[e]Data for chicken, veal, and venison nearly matched the data for beef of the same water content (Riedel 1957).

Table 6 Specific Heat of Beef

Product	Temperature, °F	Specific Heat, Btu/lb · °F
Beef, Chuck	− 40	0.49
	− 60	0.44
	−110	0.35
	−170	0.24

From Moline *et al.* (1961).

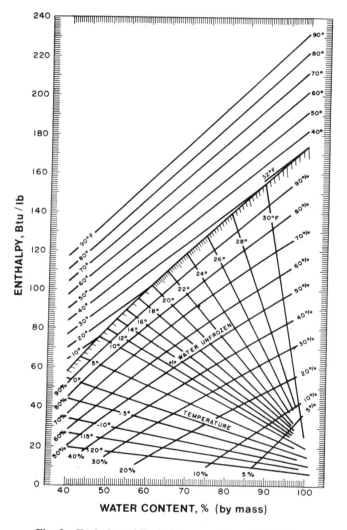

WATER CONTENT, % (by mass)

Fig. 3 Enthalpy of Fruit Juices and Vegetable Juices
(Adapted from Riedel 1957)

of the freezing process occurs over a narrow temperature range. Conversely, for 60% water content, freezing starts at a much lower temperature, and appreciable freezing occurs over a greater temperature range due to the lower water content; however, the energy associated with freezing is significantly less.

Figure 3 does not give enthalpy of whole fruits and vegetables because of the effect of solids (fibers, rind, membranes, stems, and seeds) that are not a part of the juice and do not participate in the change of phase. From data collected on 16 fruits and vegetables, Riedel (1951) made the following correlation between enthalpy of fruits and vegetables, enthalpy of juice (Figure 3), and percent solids-not-juice (solids other than those present in the juice).

$$\Delta h = (1 - X_{snj}/100)\, \Delta h_j + 0.29(X_{snj}/100)\, \Delta t \qquad (4)$$

where

Δh = enthalpy difference, Btu/lb
X_{snj} = solids-not-juice, percent by mass
Δh_j = enthalpy difference of juice from Figure 3, Btu/lb
Δt = temperature difference, °F

Equation (4) and Figure 3 have been used to calculate enthalpy of the fruits and vegetables investigated, with the results shown in Table 5. Except for plums and onions, enthalpy of fruits and vegetables (Table 5) differs from Riedel's measured values by less than 5% of enthalpy at 32 °F. Maximum deviation for plums and onions is 12%.

Using measured enthalpy values for egg white, yolk, and shell, Riedel (1957) calculated enthalpy of whole egg with shell. These calculated values are shown in Table 5. Also in Table 5 are enthalpy values of frozen egg products.

Fresh Foods

Specific heat. Above freezing, specific heat can be used instead of enthalpy. Specific heat of fruits, vegetables, juices, meats, breads, and eggs is shown in Table 5; selected data are shown as a function of water content in Figure 4. For juices, variation of specific heat with water content is uniform (Figure 4) and closely matches the equation:

$$c_p = 0.40 + 0.006\,[\%\text{ water (mass)}] \qquad (5)$$

where c_p = specific heat at constant pressure, Btu/lb · °F.

When data are needed for juices not in Figure 4, Equation (5) yields reasonably accurate specific heat values.

For other foods, the correlation with Equation (5) is not as precise due to the distorting effects of the solids-not-juice content (fats, fibers, rind, membranes, stems, and seeds). However, deviations are not great and, when data are lacking, Equation (5) can be used to estimate specific heat.

Data for meats (Figure 4) were obtained in the temperature range of 32 to 68 °F; caution is recommended in extrapolating the data into the range of 68 to 120 °F, particularly for meats with high fat content. For beef fat, the change in enthalpy from 69 to 120 °F is more than double the change in sensible heat due to specific heat alone (Riedel 1955); the additional heat is required to melt the fat.

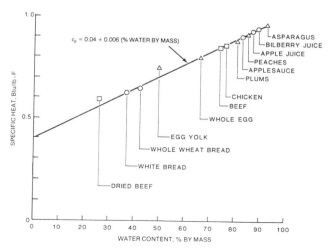

Fig. 4 Specific Heat of Food in Temperature Range of 40 to 90 °F (32 to 68 °F for Meats, 68 to 104 °F for Egg Yolk)

At temperatures between -40 and $32\,°F$, the specific heat of beef fat is constant at $0.40\ Btu/lb\cdot°F$ (Riedel 1955). Above $120\,°F$, it is constant at $0.43\ Btu/lb\cdot°F$. Between 32 and $120\,°F$, the energy required for a unit change in temperature is highly variable, because beef fat components change state at different temperatures. Similar data are available for 26 other fats and oils (Riedel 1955).

For similar reasons, the data for egg yolk should not be extrapolated into the temperature range of 32 to $68\,°F$. Riedel (1957) gave data for egg yolk in this temperature range.

Thermal diffusivity. For nonsteady-state heat transfer, the important property is thermal diffusivity α defined by the Fourier equation:

$$\delta t/\delta\tau = \alpha\,[(\delta^2 t/\delta x^2) + (\delta^2 t/\delta y^2) + (\delta^2 t/\delta z^2)] \qquad (6)$$

where

$$\begin{aligned}
x, y, z &= \text{rectangular coordinates, ft}\\
t &= \text{temperature, }°F\\
\tau &= \text{time, h}\\
\alpha &= \text{thermal diffusivity, ft}^2/\text{h}
\end{aligned}$$

Equation (6) has been solved for numerous conditions and graphical solutions are available (Heisler 1947, Schneider 1963, Smith *et al.* 1967, 1968, Dickerson 1972). Use of Equation (6), however, is limited to temperatures above freezing; and, from a mathematically rigorous point of view, it is restricted to homogeneous, isotropic substances. Generally, foods are nonhomogeneous and anisotropic; however, Equation (6) has been used to find the thermal behavior of foods (Smith *et al.* 1968, Dickerson 1972, Gane 1936, Slavicek *et al.* 1962).

Thermal diffusivity can also be defined as:

$$\alpha = k/\rho\,c_p \qquad (7)$$

where

$$\begin{aligned}
\alpha &= \text{thermal diffusivity, ft}^2/\text{h}\\
k &= \text{thermal conductivity, Btu/h}\cdot\text{ft}\cdot°F\\
\rho &= \text{density, lb/ft}^3\\
c_p &= \text{specific heat at constant pressure, Btu/lb}\cdot°F
\end{aligned}$$

The denominator on the right-hand side of Equation (7) denotes heat absorbing ability; the numerator represents the ability of the material to conduct heat through itself. Schneider (1955) interpreted thermal diffusivity in terms of heating time:

In a transient heating process the thermal capacity of the conducting material dictates the quantity of heat absorbed and the thermal conductivity of the conducting material sets the rate of this heat addition. The reciprocal of the diffusivity, $1/\alpha = h/ft^2$, is a measure of the time required to heat this material to some required temperature level and evidently this time is directly proportional to the square of the conducting path length.

Thermal diffusivity data for foods are scarce; but reasonable estimates can be obtained from Equation (7), using values of thermal conductivity, specific heat, and density. A few experimental values are available (Table 7). Like other thermal properties of foods, thermal diffusivity strongly depends on water content, as shown by Riedel's correlation (Reidel 1969):

$$\alpha = 0.00341 + (\alpha_w - 0.00341)\,[\%\ \text{water (mass)}/100] \qquad (8)$$

where

$$\begin{aligned}
\alpha &= \text{thermal diffusivity, ft}^2/\text{h}\\
\alpha_w &= \text{thermal diffusivity of water at desired temperature, ft}^2/\text{h}
\end{aligned}$$

When data are lacking, Equation (8) can be used to estimate thermal diffusivity of foods, but its use must be limited to water contents above 40% by mass. For water contents below 40%, there is

some question whether the water represents the continuous phase. If another constituent (such as fat—usually a dispersed phase) replaces the water as the continuous phase, there would be an abrupt drop in thermal conductivity and thermal diffusivity.

Meats. Equation (8) applies to foods with variable amounts of water, fat, and fiber; if the water content is reduced, it is assumed that the water is replaced with fat. Consequently, the thermal diffusivity of $0.00341\ ft^2/h$, obtained from Equation (8) for zero water content, is reasonably close to a computed value for fat [from Equation (7)]. Therefore, Equation (8) is applicable only where a reduction in water content is compensated for by an increase in fat content. Equation (8) cannot, for example, be used for freeze-dried samples, where a reduction in water content is compensated for by an increase in air content.

When meat samples are heated to temperatures in the range of 100 to $150\,°F$, juices sometimes exude. Heat transfer rates can be altered significantly, depending on whether the exuded juices remain in the sample and contribute to heat transfer, or drain away from the meat and lower thermal conductivity. The data in Table 7 for the range of 100 to $150\,°F$ applies only where juices remain in the meat sample.

Fruits. Most data for fruits (Table 7) apply to the flesh of the fruit; for products such as apples, cherries, and peaches, the data can be used for the whole fruit (with rind) because a very thin rind neither improves nor inhibits heat transfer.

Conversely, the relatively thick rind of grapefruits, oranges, and lemons has a pronounced effect on heat transfer rates, even though thermal diffusivity of the rind is about the same as that of the juice vesicle (Bennett *et al.* 1970). Although it may seem that these equal thermal diffusivities should result in equal heat transfer rates through both components, the problem arises because the thermal conductivity of the rind is only half that of the juice vesicle (Bennett *et al.* 1964). The rind has a spongy layer of loosely arranged cells with many gas-filled intercellular spaces, producing an insulating effect. Because of these spaces, the rind density is only 0.5 that of the juice vesicle (Bennett *et al.* 1970). The lower density compensates for the lower thermal conductivity, leaving thermal diffusivity essentially unchanged [Equation (7)].

Food Container Materials

In calculating heat transfer in foods, it may be necessary to consider the effect of the food container. Thermal properties of some food container materials are given in Table 8; the data for plastics apply only for monolayer materials and not for plastic laminates. Thin plastic films can have a thermal resistance much greater than indicated by their thermal conductivities is because of imperfect contact between laminated films and gas bubble buildup on the hydrophobic surface of the plastic.

In much the same way as a thick rind insulates the flesh of fruit, a food container may insulate food. Thermal diffusivities and conductivities of glass and stainless steel are greater than those of foods (Tables 7 and 9). Consequently, these two materials do not delay heat transfer in foods. However, if there is no intimate contact between the food and the container, the resulting void space adds resistance to heat transfer that may be greater than the resistance caused by the container.

THERMAL CONDUCTIVITY

Heat conduction (as opposed to convection and radiation) can be described as the transfer of heat associated with motion of the particles (molecules, atoms, electrons) of a substance without appreciable displacement or flow of those particles. This mode of heat transfer depends on a property of substances called the *coefficient of thermal conductivity*, which is the quantity of heat that flows in unit time through a plate of unit thickness and unit area with unit temperature difference between its faces.

Table 7   Thermal Diffusivity of Some Foods

Product	Water Content, % (mass)	Fat Content, % (mass)	Apparent Density, lb/ft³	Temperature, °F	Thermal Diffusivity, ft²/h	Reference
Cakes						
Angel food	36	—	9.2	73	0.0099	Sweat (1985)
Applesauce	24	—	18.7	73	0.0045	Sweat (1985)
Carrot	22	—	20.0	73	0.0045	Sweat (1985)
Chocolate	32	—	21.2	73	0.0047	Sweat (1985)
Pound	23	—	30.0	73	0.0048	Sweat (1985)
Yellow	25	—	18.7	73	0.0046	Sweat (1985)
White	32	—	27.8	73	0.0040	Sweat (1985)
Fruits and Vegetables						
Apple, Whole, Red Delicious[a]	85	—	52.4	32 to 86	0.0053	Bennett *et al.* (1969)
Apple, Dried	42	—	53.4	73	0.0037	Sweat (1985)
Applesauce	37	—	—	41	0.0041	Riedel (1969)
	37	—	—	149	0.0043	Riedel (1969)
	80	—	—	41	0.0047	Riedel (1969)
	80	—	—	149	0.0054	Riedel (1969)
Apricots, Dried	44	—	82.6	73	0.0044	Sweat (1985)
Bananas, Flesh	76	—	—	41	0.0046	Riedel (1969)
	76	—	—	149	0.0055	Riedel (1969)
Cherries, Flesh[b]	—	—	65.5	32 to 86	0.0051	Parker and Stout (1967)
Dates	35	—	82.3	73	0.0040	Sweat (1985)
Figs	40	—	77.4	73	0.0037	Sweat (1985)
Jam, Strawberry	41	—	81.7	68	0.0045	Sweat (1985)
Jelly, Grape	42	—	82.4	68	0.0047	Sweat (1985)
Peaches[b]	—	—	59.9	36 to 90	0.0054	Bennett (1963)
Peaches, Dried	43	—	78.6	73	0.0046	Sweat (1985)
Potatoes, Whole	—	—	64.9 to 66.8	32 to 158	0.0052	Minh *et al.* (1969) Mathews and Hall (1968)
Potatoes, Mashed, Cooked	78	—	—	41	0.0048	Riedel (1969)
	78	—	—	149	0.0056	Riedel (1969)
Prunes	43	—	76.1	73	0.0046	Sweat (1985)
Raisins	32	—	86.1	73	0.0041	Sweat (1985)
Strawberries, Flesh	92	—	—	41	0.0049	Riedel (1969)
Sugar beets	—	—	—	32 to 140	0.0049	Slavicek (1962)
Meats						
Codfish	81	—	—	41	0.0047	Riedel (1969)
	81	—	—	149	0.0055	Riedel (1969)
Halibut[c]	76	1	66.8	104 to 149	0.0057	Dickerson and Read (1975)
Beef, Chuck[d]	66	16	66.2	104 to 149	0.0048	Dickerson and Read (1975)
Beef, Round[d]	71	4	68.0	104 to 149	0.0052	Dickerson and Read (1975)
Beef, Tongue[d]	68	13	66.2	104 to 149	0.0051	Dickerson and Read (1975)
Beefstick	37	—	65.5	68	0.0042	Sweat (1985)
Bologna	65	—	62.4	68	0.0050	Sweat (1985)
Corned beef	65	—	—	41	0.0044	Riedel (1969)
	65	—	—	149	0.0051	Riedel (1969)
Ham, Country	72	—	64.3	68	0.0053	Sweat (1985)
Ham, Smoked	64	—	—	41	0.0046	Riedel (1969)
Ham, Smoked[d]	64	14	68.0	104 to 149	0.0053	Dickerson and Read (1975)
Pepperoni	32	—	66.1	68	0.0036	Sweat (1985)
Salami	36	—	59.9	68	0.0050	Sweat (1985)
Water	—	—	—	86	0.0057	Dickerson (1968)
	—	—	—	149	0.0062	Dickerson (1968)

[a] Data are applicable only to raw whole apple.
[b] Freshly harvested.
[c] Stored frozen and thawed prior to test.
[d] Data are applicable only where the juices exuded during heating remain in the food samples.

Table 8   Thermal Properties of Food Container Materials (32 to 175 °F)

Product	Thermal Conductivity, Btu/h·ft·°F	Specific Heat, Btu/lb·°F	Apparent Density, lb/ft³	Thermal Diffusivity[a] $\alpha = k/\rho c_p$, ft²/h	References
Stainless steel (302), Type 18-8 Austenitic	9.17	0.118	494	0.157	ASM (1948)
Glass, Borosilicate	0.65	0.20	140	0.023	Lange (1961)
Nylon, Type 6/6[b]	0.14	0.40	70	0.0050	Breskin Pub. (1963)
Polyethylene, High density[b]	0.28	0.55	60	0.0085	Breskin Pub. (1963)
Polyethylene, Low density[b]	0.19	0.55	58	0.0060	Breskin Pub. (1963)
Polypropylene[b]	0.068	0.46	57	0.0026	Breskin Pub. (1963)
Polytetrafluoroethylene[b]	0.15	0.25	130	0.0046	Breskin Pub. (1963)

[a] To obtain in²/min, multiply by 2.4.
[b] Data applicable only to monolayer materials.

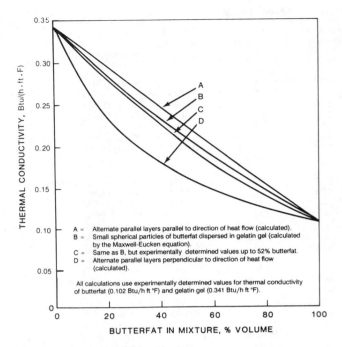

**Fig. 5   Thermal Conductivity of Gelatin Gel [6% (mass)]—
Butter Fat Mixtures at 39°F**

Thermal conductivity depends on many factors, including the kind of substance (metal, dielectric, crystalline, amorphous, solid, liquid, gas); composition (impurities, mixtures); structure and structural orientation; temperature; and pressure. Thus, accurate measurement and prediction of thermal conductivity may be difficult for many foods, although a large amount of experimental data are available and estimations or predictions sufficiently accurate for practical purposes can be made for many food materials.

Thermal Conductivity of Mixtures

Thermal conductivity for mixtures of 6% gelatin gel and butter fat arranged in several ways are shown in Figure 5, which was adapted from a more detailed study done by Lentz (1961). Since the ratio of thermal conductivities of butter fat and 6% gelatin gel is roughly the same as for fat and lean portions of meat, the curves indicate the variability to be expected in the thermal conductivity of meat. Conductivity is maximum for layers arranged parallel to the direction of heat flow (Curve A) and minimum for perpendicular layers (Curve D); other arrangements fall in between. Curves B (calculated) and C (experimental) give values for small spherical butter fat particles dispersed in the gel. The calculated curve was obtained using Eucken's adaptation (Eucken 1940) of Maxwell's equation (Maxwell 1904) for conductivity of a mixture composed of small spheres of one substance dispersed in another:

$$k = k_c \frac{1 - [1 - a(k_d/k_c)]b}{1 + (a - 1)b} \qquad (9)$$

where

k = conductivity of mixture
k_c = conductivity of continuous phase
k_d = conductivity of dispersed phase

$a = 3k_c/(2k_c + k_d)$
$b = V_d/(V_c + V_d)$
V_d = volume of dispersed phase
V_c = volume of continuous phase

The derivation of this equation assumes the dispersed particles are sufficiently separated for their effects in disturbing heat flow to be independent of each other. The units in Equation (9) can be any units as long as the thermal conductivity units are all the same and the volume units are all the same.

Effect of Temperature

Temperature effects on thermal conductivity of a number of meats and fats, as well as gelatin gel and ice, in the range of −13 to 50°F are shown in Figure 6 (see Table 9 for the composition of sample materials), adapted from Lentz (1961). Rapid change of phase and the possibility of subcooling and metastable states make values in the 14 to 32°F range more difficult to measure and less reliable than at other temperatures. The effect of meat fiber or structural orientation in the frozen state appears to be 10 to 20%. Thermal conductivity of fats is affected only slightly by temperature in the range studied.

Thermal Conductivity and Water Content

Figure 7 shows the relation between thermal conductivity and moisture content for a wide range of food materials, based on a study using the Maxwell-Eucken equation (Van den Berg and Lentz 1975). The effect of temperature is also included. The curves represent data on sugar solutions and fruit juices (Riedel 1949), milk and evaporated milk (Leidenfrost 1959), butter fat and gelatin gels (Lentz 1961, Cherneeva 1956), fats (Lentz 1961, Cherneeva 1956), and meats (Lentz 1961, Cherneeva 1956, Hill *et al.* 1967, Miller and Sutherland 1963), with an accuracy of ±10% if an allowance of ±7%

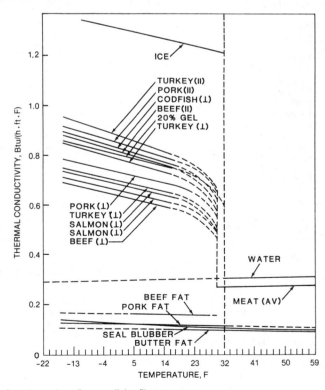

∥ indicates heat flow parallel to fiber structure
⊥ indicates heat flow perpendicular to fiber structure

**Fig. 6   Thermal Conductivity of Meats, Fats, Gelatin Gel,
and Water between −13 and 50°F**

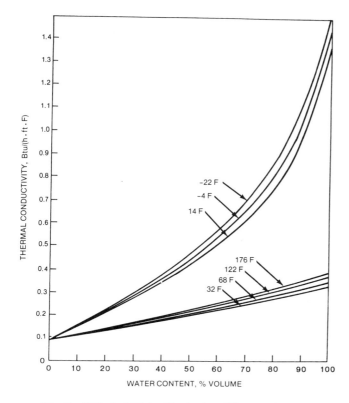

Fig. 7   Effect of Water Content and Temperature on Thermal Conductivity of Food Materials

is made for fiber direction in frozen meats. The allowance is positive for heat flow parallel to the fiber structure; negative for flow across the structure.

The study on which Figure 7 is based differs from studies done by Lentz (1961), Riedel (1949), Leidenfrost (1959), Spells (1960-61), and Sweat (1974), in that water content was calculated on a volumetric rather than gravimetric basis, and a wider range of food materials and temperatures is covered. In addition, only materials not containing appreciable air were included. For the above-freezing temperatures, the curves follow the Maxwell-Eucken equation, assuming water to be the continuous phase, and the nonaqueous part the dispersed phase. For below-freezing temperatures, the curves best fitting the experimental data (Figure 7) represent an average of the results calculated, assuming that ice is the dispersed phase in one instance and the nonaqueous part, plus unfrozen water, is the dispersed phase in the other instance.

The Maxwell-Eucken equation predicts thermal conductivity from water content, even though the assumptions on which it is based do not appear to be closely met, because:

1. The thermal conductivity of water is high compared to that of the nonaqueous components, especially at below-freezing temperatures.

2. The thermal conductivity of the nonaqueous components falls within a relatively narrow range—most organic substances between 0.081 to 0.12 Btu/h·ft·°F with most fats in the 0.092 to 0.10 range.

3. The mass densities of fats, proteins, and carbohydrates fall within narrow ranges that are relatively close together [values used in the study were 57.5, 84.4, and 96.9 lb/ft³ for fat, protein, and carbohydrate, respectively (Van den Berg and Lentz 1975)].

Tabular Data

Table 9 lists thermal conductivities for many foods (Qashou *et al.* 1972). Data in the table have been averaged, interpolated, extrapolated, selected, or rounded off from the original research data. Tables 7 and 9 also include ASHRAE research data on low and intermediate moisture foods (Sweat 1985).

SURFACE HEAT TRANSFER COEFFICIENT

The surface heat transfer coefficient, while not a thermal property of a food or any other material, is needed to design heat transfer equipment for foods where convection is involved. Newton's cooling law in one dimension defines it as

$$q = h(t_s - \theta_\infty) = -k \left(\frac{\partial t}{\partial \mathbf{n}} \right), \text{ for time} > 0 \qquad (10)$$

where

q = heat flux
h = heat transfer coefficient
t_s = surface temperature
t_∞ = ambient temperature
k = thermal conductivity
$\mathbf{n}$ = normal surface vector

Using a linear boundary condition helped mathematicians solve the heat conduction equations but required h to be determined by experiments. Some efforts have theoretically determined h for simple geometries and laminar flow by solving the Navier-Stokes energy and continuity equations simultaneously.

The results showed that factors affecting the surface heat transfer coefficient are not within the object being cooled, frozen, or thawed, but are related to the boundary layer that forms outside the body. This boundary layer is affected by the dynamics of the surrounding fluids and the conditions at the surface of the body. Surface roughness, packaging, etc., influence the flow regime around the body, thus affecting the surface heat transfer coefficient.

Because the heat transfer coefficient depends on the velocity of the fluid to which heat is transferred, experimenters represent their results in the correlation of the Nusselt number as a function of Reynolds and Prandtl numbers. These Nu-Re correlations predict theoretical values accurately for simple geometries such as flat plates.

Theoretical derivations, as well as laboratory experiments, showed that h is a local phenomenon and not a constant—even in the simplest geometries and flow patterns. The value of h used in the Nu-Re correlations is the area averaged value of the local heat transfer coefficient.

Surface Heat Transfer Coefficient Table

The data were organized in Table 10 to be comprehensive yet compact enough for simple use. The first two columns of the table describe the product or geometrical shape used in the experiment. Columns 3 through 7 describe the experimental conditions used to determine h. Column 8 describes the types of heat transfer present during the experiment. Columns 9 through 12 describe the Nusselt number correlation, the error associated with h, the range of the Reynolds number or other parameter where the correlation was determined, and the value of the surface heat transfer coefficient. The last three columns contain the classification, the reference, and comments. An explanation of each column follows.

1. *Product.* The product name is entered in alphabetical order.

2. *Shape and characteristic length.* The shape assumed for the calculation of h is given; otherwise, it is the characteristic

(*Text continues on page 30.22*)

Table 9 Thermal Conductivity of Food and Food Materials

Food or Food Material[a]	Temp., °F	Water Content, % (mass)	k^b, Btu/h·ft·°F	Rating,[c] A, I, U	Reference[d]	Remarks—Composition[e]
Grains, Cereals, and Seeds						
Corn, Yellow	90	0.9	0.081	U	Kazarian (1962)	0.75 d
		14.7	0.092			0.75 d
		30.2	0.099			0.68 d
Flax seed	90		0.066	U	Griffiths and Hickman (1951)	0.66 d
Oats, White English	81	12.7	0.075	U	Oxley (1944)	
Sorghum	41	13	0.076	U	Miller (1963)	Hybrid Rs610 grain
		22	0.087			
Wheat, No. 1 Northern Hard Spring	93	2	0.078	A	Moote (1953)	Values taken from plot of series of values given by authors
		7	0.086		Babbitt (1945)	
		10	0.090			
		14	0.097			
Wheat, Soft White Winter	88	5	0.070	U	Kazarian (1962)	Values taken from plot of series of values given by author; 0.78 d
		10	0.075			
		15	0.079			
Fruits, Vegetables, and Byproducts						
Beans, Runner	48		0.225	U	Smith *et al.* (1952)	0.75 d; machine sliced, scalded, packed in slab
Broccoli	21		0.223	I	Smith *et al.* (1952)	0.56 d; heads cut and scalded
Carrots	3		0.387	I	Smith *et al.* (1952)	0.6 d; scraped, sliced and scalded
Carrots, Puree	18		0.728	U	Smith *et al.* (1952)	0.89 d; slab
Potatoes, Mashed	9		0.630	I	Smith *et al.* (1952)	0.97 d; tightly packed slab
Potato salad	36		0.277	I	Dickerson and Read (1968)	1.01 d
Apple juice	68	87	0.323	A	Riedel (1949)	Refractive index at 70°F = 1.35
	176		0.365			
	68	70	0.291			Refractive index at 70°F = 1.38
	176		0.325			
	68	36	0.225			Refractive index at 70°F = 1.45
	176		0.251			
Apple	46		0.242	A	Gane (1936)	Tasmanian French Crab apple, whole fruit; 5 oz
Apple, Dried	73	41.6	0.127	A	Sweat (1985)	0.86 d
Apricots, Dried	73	43.6	0.217	A	Sweat (1985)	1.32 d
Black currants	1		0.179	I	Smith *et al.* (1952)	0.64 d
Dates	73	34.5	0.195	A	Sweat (1985)	1.32 d
Figs	73	40.4	0.179	A	Sweat (1985)	1.24 d
Gooseberries	5		0.160	I	Smith *et al.* (1952)	0.58 d; mixed sizes
Grapefruit, Juice vesicle	86		0.268	A	Bennett *et al.* (1964)	Marsh, seedless; 786 p
Grapefruit, Rind	86		0.137	A	Bennett *et al.* (1964)	Marsh, seedless; 812 p
Grape, Green, Juice	68	89	0.328	A	Riedel (1949)	Refractive index at 70°F = 1.35
	176		0.369			
	68	68	0.287			Refractive index at 70°F = 1.38
	176		0.320			
	68	37	0.229			Refractive index at 70°F = 1.45
	176		0.254			
	77		0.254	A	Turrell and Perry (1957)	Eureka
Jam, Strawberry	68	41.0	0.224	A	Sweat (1985)	1.31 d
Jelly, Grape	68	42.0	0.226	A	Sweat (1985)	1.32 d
Orange, Juice vesicle	86		0.251	A	Bennett *et al.* (1964)	Valencia; 104 p
Orange, Rind	86		0.103	A	Bennett *et al.* (1964)	Valencia; 108 p
Peaches, Dried	73	43.4	0.209	A	Sweat (1985)	1.26 d
Pear, Juice	68	85	0.318	A	Riedel (1949)	Refractive index at 70°F = 1.36
	176		0.363			
	68	60	0.275			Refractive index at 70°F = 1.40
	176		0.307			
	68	39	0.232			Refractive index at 70°F = 1.44
	176		0.258			
Plums	3		0.143	I	Smith *et al.* (1952)	0.61d; 1.5 in. diam., 2 in. long
Prunes	73	42.9	0.217	A	Sweat (1985)	1.22 d
Raisins	73	32.2	0.194	A	Sweat (1985)	1.38 d
Strawberries	7		0.636	I	Smith *et al.* (1952)	Mixed sizes, 0.80 d slab
	5		0.555			Mixed sizes in 57% sucrose syrup, slab
Meat and Animal Byproducts						
Beef brain	95	77.7	0.287	A	Poppendick *et al.* (1966)	12% fat; 10.3% protein; 1.04 d
Beef fat	95	0.0	0.110	A	Poppendick *et al.* (1966)	Melted 100% fat; 0.81 d
	95	20	0.133			0.86 d
Beef fat⊥[a]	36	9	0.125	A	Lentz (1961)	89% fat
	16		0.166			
Beef kidney	95	76.4	0.303	A	Poppendick *et al.* (1966)	8.3% fat, 15.3% protein; 1.02 d
Beef liver	95	72	0.282	A	Poppendick *et al.* (1966)	7.2% fat; 20.6% protein

Table 9 Thermal Conductivity of Food and Food Materials (*Continued*)

Food or Food Material[a]	Temp., °F	Water Content, % (mass)	k^b, Btu/h·ft·°F	Rating,[c] A, I, U	Reference[d]	Remarks—Composition[e]
Beef, Lean = [a]	37 5	75	0.292 0.821	A	Lentz (1961)	Sirloin; 0.9% fat
Beef, Lean = [a]	68 5	79	0.249 0.826	A	Hill *et al.* (1967)	1.4% fat
Beef, Lean = [a]	43 5	76.5	0.231 0.786	A	Hill *et al.* (1967), Hill (1966)	2.4% fat
Beef, Lean⊥[a]	68 5	79	0.277 0.780	A	Hill *et al.* (1967)	Inside round; 0.8% fat
Beef, Lean⊥[a]	43 5	76	0.237 0.659	A	Hill *et al.* (1967), Hill (1966)	3% fat
Beef, Lean⊥[a]	37 5	74	0.272 0.647	A	Lentz (1961)	Flank; 3 to 4% fat
Beef, Ground	43 39 43 37	67 62 55 53	0.235 0.237 0.203 0.210	A	Qashou *et al.* (1970)	12.3% fat; 0.95 *d* 16.8% fat; 0.98 *d* 18% fat; 0.93 *d* 22.2% fat; 0.95 *d*
Beefstick	68	36.6	0.172	A	Sweat (1985)	1.05 *d*
Bologna	68	64.7	0.243	A	Sweat (1985)	1.00 *d*
Dog food	73	30.6	0.184	A	Sweat (1985)	1.24 *d*
Cat food	73	39.7	0.188	A	Sweat (1985)	1.14 *d*
Ham, Country	68	71.8	0.277	A	Sweat (1985)	1.03 *d*
Horse meat⊥[a]	86	70	0.266	I	Griffiths and Cole (1948)	Lean
Lamb⊥[a]	68 5	72	0.264 0.647	A	Hill *et al.* (1967)	8.7% fat
Lamb = [a]	68 5	71	0.231 0.734	A	Hill *et al.* (1967)	9.6% fat
Pepperoni	68	32.0	0.148	A	Sweat (1985)	1.06 *d*
Pork Fat	37 5	6	0.124 0.126	A	Lentz (1961)	93% fat
Pork, Lean = [a]	39 5	72	0.276 0.861	A	Lentz (1961)	6.1% fat
Pork, Lean = [a]	68 9	76 76	0.262 0.821	A	Hill *et al.* (1967)	6.7% fat
Pork, Lean⊥[a]	39 5	72	0.264 0.745	A	Lentz (1961)	6.1% fat
Pork, Lean⊥[a]	68 7	76	0.292 0.751	A	Hill *et al.* (1967)	6.7% fat
Salami	68	35.6	0.180	A	Sweat (1985)	0.96 *d*
Sausage	77	68	0.243	A	Woodams (1965), Nowrey and Woodams (1968)	Mixture of beef and pork; 16.1% fat; 12.2% protein
	77	62	0.223	A	Woodams (1965), Nowrey and Woodams (1968)	Mixture of beef and pork; 24.1% fat; 10.3% protein
Veal⊥[a]	68 5	75	0.272 0.797	A	Hill *et al.* (1967)	2.1% fat
Veal = [a]	82 5	75	0.257 0.844	A	Hill *et al.* (1967)	2.1% fat
Poultry and Eggs						
Chicken, Breast⊥[a]	68	69-75	0.238	I	Walters and May (1963)	0.6% fat
Chicken, Breast with skin	68	58-74	0.212	I	Walters and May (1963)	0-30% fat
Egg White	97	88	0.322	I	Spells (1960-61), Spells (1958)	
Egg, Whole	18		0.555	I	Smith *et al.* (1952)	0.98 *d*
Egg Yolk	88	50.6	0.243	A	Poppendick *et al.* (1966)	32.7% fat; 16.7% protein; 1.02 *d*
Turkey, Breast⊥[a]	37 5	74	0.287 0.797	A	Lentz (1961)	2.1% fat
Turkey, Leg⊥[a]	39 5	74	0.287 0.711	A	Lentz (1961)	3.4% fat
Turkey, Breast = ⊥[a]	37 5	74	0.290 0.884	A	Lentz (1961)	2.1% fat
Fish and Sea Products						
Fish, Cod⊥[a]	37 5	83	0.309 0.844	A	Lentz (1961)	0.1% fat
Fish, Cod	34 5		0.324 0.977	I	Long (1955), Jason and Long (1955)	
Fish, Herring	−2		0.462	I	Smith *et al.* (1952)	0.91*d*; whole and gutted
Fish, Salmon⊥[a]	37 5	67	0.307 0.717	A	Lentz (1961)	12% fat; *Salmo salar* from Gaspe penisula
Fish, Salmon⊥[a]	41 5	73	0.288 0.653	A	Lentz (1961)	5.4% fat, *Oncorhynchus tchawytscha* from British Columbia
Seal bubber⊥[a]	41 5	4.3 2.19	0.114	A	Lentz (1961)	95% fat
Whale bubber⊥[a]	64		0.121	I	Griffiths and Cole (1948)	1.04 *d*
Whale mat	90 16		0.375 0.832	I	Griffiths and Hickman (1951)	1.07 *d*
	10		0.740	I	Smith *et al.* (1952)	0.51% fat; 1.00 *d*

Table 9 Thermal Conductivity of Food and Food Materials (*Continued*)

Food or Food Material[a]	Temp., °F	Water Content, % (mass)	k^b, Btu/h·ft·°F	Rating,[c] A, I, U	Reference[d]	Remarks—Composition[e]
Dairy Products						
Butterfat	43	0.6	0.100	A	Lentz (1961)	
	5		0.103			
Butter	39		0.114	I	Hooper and Chang (1952)	
Buttermilk	68	89	0.329	A	Riedel (1949)	0.35% fat
Milk, Whole	82	90	0.335	A	Leidenfrost (1959)	3% fat
	36	83	0.302	A	Riedel (1949)	3.6% fat
	68		0.318			
	122		0.339			
	176		0.355			
Milk, Skimmed	36	90	0.311	A	Riedel (1949)	0.1% fat
	68		0.327			
	122		0.350			
	176		0.367			
Milk, Evaporated	36	72	0.281	A	Riedel (1949)	4.8% fat
	68		0.291			
	122		0.313			
	176		0.327			
Milk, Evaporated	36	62	0.264	A	Riedel (1949)	6.4% fat
	68		0.273			
	122		0.295			
	176		0.307			
Milk, Evaporated	73	67	0.272	A	Leidenfrost (1959)	10% fat
	106		0.291			
	140		0.298			
	174		0.305			
Milk, Evaporated	79	50	0.187	A	Leidenfrost (1959)	15% fat
	104		0.197			
	138		0.206			
	174		0.210			
Whey	36	90	0.312	A	Riedel (1949)	No fat
	68		0.328			
	122		0.364			
	176		0.370			
Sugar, Starch, Bakery Products, and Derivatives						
Sugar beet juice	77	79	0.318	A	Khelemskii and Zhadan (1964)	
		82	0.329			
Sucrose solution	32	90	0.309	A	Riedel (1949)	Cane or beet sugar solution
	68		0.327			
	122		0.351			
	176		0.368			
	32	80	0.291			
	68		0.309			
	122		0.331			
	176		0.347			
	32	70	0.273			
	68		0.290			
	122		0.310			
	176		0.325			
	32	60	0.256			
	68		0.272			
	122		0.290			
	176		0.303			
	32	50	0.239			
	68		0.253			
	122	94–80	0.270			
	176		0.283			
	32	40	0.221			
	68		0.233			
	122		0.251			
	176		0.262			
Glucose solution	36	89	0.311	A	Riedel (1949)	
	68		0.327			
	122		0.347			
	176		0.369			
	36	80	0.294			
	68		0.309			
	122		0.330			
	176		0.346			
	36	70	0.276			
	68		0.291			
	122		0.311			
	176		0.326			
	36	60	0.258			
	68		0.272			
	122		0.289			
	176		0.306			

Table 9 Thermal Conductivity of Food and Food Materials (*Concluded*)

Food or Food Material[a]	Temp., °F	Water Content, % (mass)	k[b], Btu/h·ft·°F	Rating,[c] A, I, U	Reference[d]	Remarks—Composition[e]
Corn Syrup	77		0.325	I	Metzner and Friend (1959)	1.16d
			0.280			1.31d
			0.270			1.34d
Molasses syrup	86	23	0.200	U	Popov and Terentiev (1966)	
Angel food cake	73	36.1	0.0572	A	Sweat (1985)	0.15d, Porosity—88%
Applesauce cake	73	23.7	0.0457	A	Sweat (1985)	0.30d, Porosity—78%
Carrot cake	73	21.6	0.0485	A	Sweat (1985)	0.32d, Porosity—75%
Chocolate cake	73	31.9	0.0613	A	Sweat (1985)	0.34d, Porosity—74%
Pound cake	73	22.7	0.0757	A	Sweat (1985)	0.48d, Porosity—58%
Yellow cake	73	25.1	0.0473	A	Sweat (1985)	0.30d, Porosity—78%
White cake	73	32.3	0.0636	A	Sweat (1985)	0.45d, Porosity—62%
Fats, Oils, Gums, and Extracts						
Gelatin gel	41	94–80	0.302	A	Lentz (1961)	Conductivity did not vary with concentration in range tested (6,12,20%)
	5	94	1.24			6% gelatin concentration
	5	88	1.12			12% gelatin concentration
	5	80	0.815			20% gelatin concentration
Margarine	41		0.135	I	Hooper and Chang (1952)	1.00d
Almond oil	39		0.102	U	Wachsmuth (1892)	0.92d
Cod liver oil	95		0.098	I	Spells (1960-61) Spells (1958)	
Lemon oil	43		0.090	U	Weber (1880)	0.82d
Mustard oil	77		0.098	U	Weber (1886)	1.02d
Nutmeg oil	39		0.090	U	Wachsmuth (1892)	0.94d
Olive oil	45		0.101	U	Weber (1880)	0.91d
Olive oil	90		0.097	U	Kaye and Higgins (1928)	0.91d
	149		0.096			
	304		0.092			
	365		0.090			
Peanut oil	39		0.097	U	Wachsmuth (1892)	0.92d
Peanut oil	77		0.098	I	Woodams (1965)	
Rapeseed oil	68		0.092	U	Kondrat'ev (1950)	0.91d
Sesame oil	39		0.102	U	Wachsmuth (1892)	0.92d

		Pressure, psia				
Freeze-Dried Foods						
Apple	95	0.00039	0.0090	A	Harper (1960, 1962)	Delicious; 88% porosity; 5.1 tortuosity factor; measured in air
		0.00306	0.0107			
		0.0271	0.0163			
		0.418	0.0234			
Peach	95	0.00087	0.0095	A	Harper (1960, 1962)	Clingstone; 91% porosity; 4.1 tortuosity factor; measured in air
		0.00311	0.0107			
		0.0271	0.0161			
		0.387	0.0237			
		7.4	0.0249			
Pears	95	0.00031	0.0107	A	Harper (1960, 1962)	97% porosity; measured in nitrogen
		0.00282	0.0120			
		0.0271	0.0177			
		0.311	0.0242			
		10.1	0.0261			
Beef = [a]	95	0.00021	0.0221	I	Harper (1960, 1962)	Lean; 64% porosity; 4.4 tortuosity factor, measured in air
		0.00329	0.0238			
		0.0345	0.0307			
		0.392	0.0358			
		14.7	0.0377			
Egg albumin gel	106	14.7	0.0227	U	Saravacos and Pilsworth (1965)	2% water content; measured in air
Egg albumin gel	106	0.00064	0.0075	U	Saravacos and Pilsworth (1965)	Measured in air
Turkey = [a]		0.00077	0.0166	U	Triebes and King (1966)	Cooked white meat; 68 to 72% porosity; measured in air
		0.00218	0.0256			
		0.0676	0.0408			
		0.31	0.0498			
		14.3	0.0536			
Turkey⊥[a]		0.00081	0.0098	U	Triebes and King (1966)	Cooked white meat; 68 to 72% porosity; measured in air
		0.0027	0.0101			
		0.019	0.0128			
		0.18	0.0241			
		12.7	0.0339			
Potato starch gel		0.00062	0.0053	U	Saravacos and Pilsworth (1965)	Measured in air
		0.026	0.0083			
		0.32	0.0168			
		14.9	0.0227			

[a]The symbol ⊥ indicates heat flow perpendicular to the grain or structure; the symbol = indicates heat flow parallel to the grain of structure.

[b]The symbol k is used for thermal conductivity.

[c]The rating *A* indicates data considered reliable by reviewers from TC 11.9, within practical limits. The rating *I* indicates data that appears to be reliable within practical limits, but for which background information supplied (*e.g.*, composition factors or conditions of measurement) was insufficient for a firm assessment. *U* indicates data unrated because it did not fall within the area of competence of any reviewer of TC 11.9.

[d]References quoted are those on which given data are based, although actual values in this table may have been averaged, interpolated, extrapolated, selected, or rounded off.

[e]This column includes density (*d*, in lb/ft^3), pressure (*P*, in psia), fat content (%), and other details.

Table 10 Surface Heat Transfer Coefficients for Food Products

Column 1 Product	2 Shape Length, in.[a]	3 Transfer Medium	4 Δt or Temp. t of Medium, °F	5 Velocity of Medium, ft/s	6 Exp. Method[b]	7 Temperature Measurement[c]	8 Type of Heat Transfer[d]	9 Nu–Re Correlation, if Any[e]
Apple Jonathan	Spherical 2.0 2.3 2.4	Air	$t = 81$	0.0 1.3 3.0 6.7 17.0 0.0 1.3 3.0 6.7 17.0 0.0 1.3 3.0 6.7 17.0	TI	24 ga. T	FC RAD EVAP	N/A
Apple Red Delicious	2.5 2.8 3.0 2.2 2.8 3.0	Air Water	$\Delta t = 41$ $t = 31$ $\Delta t = 46$ $t = 32$	4.9 15.0 4.9 15.0 0.0 4.9 9.8 15.0 .90	TI	24 ga. T	FC RAD EVAP	N/A
Beef Carcasses	142.2 lb 187.4 lb	Air	$t = -3$	5.9 0.98	TF	Heat Flux Probe	FC RAD EVAP	N/A
Eggs Jifujitori	1.3	Air	$\Delta t = 81$	6.6 – 26	TB	T 0.25D from surface	FC RAD	$Nu = 0.46\,Re^{0.56}$
Eggs Leghorn	1.7	Air	$\Delta t = 81$	6.6 – 26	TB	T 0.25D from surface	FC RAD	$Nu = 0.71\,Re^{0.55}$
Fish Pike Perch Sheatfish	Not given	Air	N/A	3.2 – 22	NS	Not specified	FC RAD EVAP	$Nu = 4.5\,Re^{0.28}$
Hams Boneless Processed	G* 0.4-0.45	Air	$\Delta t = 132$ $t = 150$	N/A	TI	36 ga. T	FC RAD	$Nu = 0.329\,Re^{0.564}$
Hams Processed	N/A	Air	$t = -10$ $t = -55$ $t = -60$ $t = -70$ $t = -80$	2.0	TI	36 ga. T at center and every 1/2 in. along axis	FC	N/A
Meat	Slabs Thickness = 0.91 in.	Air	$t = 32$	1.8 4.6 12.0	TN	N/A	FC RAD EVAP	N/A
Oranges Grapefruit Tangelos Bulk packed	Spheroids 2.3 3.1 2.1	Air	$\Delta t = 70$ to 56 $t = 16$	0.36–1.1	TC	36 ga. T every 0.25 in. along diameter	FC RAD EVAP	$Nu = 5.05\,Re^{0.333}$
Oranges Grapefruit Bulk packed	Spheroids 3.0 4.2	Air	$\Delta t = 59$ $t = 32$	0.17–6.7	TN	36 ga. thermocouples	FC RAD EVAP	$Nu = 1.17\,Re^{0.529}$

Table 10 Surface Heat Transfer Coefficients for Food Products (*Continued*)

Column 10	11	12	13	14	15
Error of Correlation Coefficient	Reynolds Number Range[g]	h, Btu/h·ft²·°F	Classifi-cation[i]	Reference	Comments
N/A	N/A	2.0 3.0 4.8 8.0 9.4 2.0 3.0 4.9 7.9 9.6 2.0 2.8 4.6 6.9 8.9	220	Kopelman *et al.* (1966)	N/A indicates that data were not reported in original article.
N/A		4.8 10.0 2.5 6.5 1.8 4.0 5.8 6.1 16.0 14.0 9.8	110	Nicholas *et al.* (1964)	Thermocouples at center of fruit
± 0.058 Btu/h·ft²·°F ± 0.014 Btu/h·ft²·°F	N/A	3.8 1.8	120	Fedorov *et al.* (1972)	*For size indication
± 1.0%	6000–15,000	N/A	112	Chuma *et al.* (1970)	5 points in correlation
±1.0%	8000–25,000	N/A	112	Chuma *et al.* (1970)	5 points in correlation
±10%	5000–35,000	N/A	022	Khatchaturov (1958)	32 points in correlation
N/A	1000–86,000	N/A	222	Clary *et al.* (1968)	*Geometrical factor for shrink-fitted plastic bag $G = 1/4 + 3/(8A^2) + 3/(8B^2)$ $A = a/Z, B = b/Z$ A = characteristic length = 0.5 min. distance ⊥ to airflow a = minor axis b = major axis Correlation included 18 points Recalculated with minimum distance ⊥ to airflow Calculated Nu with 0.5 char. length
Max coefficient of variation = 12%	N/A	3.6 3.6 3.5 3.5 3.2	220	Van den Berg and Lentz (1957)	38 points in total; values are averages
N/A	N/A	1.9 3.5 6.2	200	Radford *et al.* (1976)	
r = 0.586	35000–135,000	*11.7 (See comments)	111	Bennett *et al.* (1966)	Bins 42 by 42 by 158 in. 36 points in correlation. Random packaging. Interstitial velocity. *Average for oranges
r = 0.996	180–18,000	N/A	222	Baird and Gaffney (1976)	20 points in correlation Bed depth 26 in.

Table 10   Surface Heat Transfer Coefficients for Food Products (*Continued*)

Column 1	2	3	4	5	6	7	8	9
Product	Shape Length, in.[a]	Transfer Medium	Δt or Temp. t of Medium, °F	Velocity of Medium, ft/s	Exp. Method[b]	Temperature Measurement[c]	Type of Heat Transfer[d]	Nu–Re Correlation, if Any[e]
Peas Fluidized bed	Spherical N/A	Air	$t = -15$ to -35	4.9–24 ±1.0	TB	4 sets of resistance thermometers	FC RAD EVAP	$Nu = 3.5 \times 10^{-4}\ Re$
Peas Bulk packed	Spherical N/A	Air	$t = -15$ to -35	4.9–24 ±1.0	TB	4 sets of resistance thermometers	FC RAD EVAP	$Nu = 0.016\ Re^{0.95}$
Potatoes Pungo Bulk packed	Ellipsoids N/A	Air	$t = 40$	2.2 4.0 4.5 5.7	TI	36 ga. T	FC RAD EVAP	$Nu = 0.364\ Re^{0.558}$ $Pr^{1/3}$

Poultry
 	
  Chickens
  Turkey
	2.6 to
	20.8 lb
	
	See comments
		*
		
		See comments
			$\Delta t = 32$	**
				
				See comments
					TC	30 ga. T	FC
							RAD
								N/A

Soybeans	Spherical 2.6	Air	N/A	22	TS	N/A	FC	$Nu = 1.07\ Re^{0.64}$
Karlsruhe substance	1-dimensional block 3.0	Air	$\Delta t = 96$ $t = 100$	N/A	SS	T	FC	N/A
Milk container	Cylindrical 2.8×3.9 2.8×5.9 2.8×9.8	Air	$\Delta t = 9.5$	N/A	TB	30 ga. T	RAD NC	$Nu = 0.75\ Gr^{-0.264}$

[a]Characteristic length is used in Reynolds number and illustrated in the Comments column where appropriate.

[b]Experimental methods are coded as follows:

TI—transient; infinite series heat transfer solution is simplified to one term

TB—transient; solution of convection boundary condition assuming uniform product temperature

TC—transient; heat flux quantified using calorimetry

TF—transient; monitoring of heat flux with heat flux meter

TN—transient; heat conduction equation solved numerically

NS—experimental method not specified

[c]Thermocouple wire gage if given; Thermocouple type; T is copper constantan.

[d]FC: forced convection; RAD: radiation effects included; EVAP: evaporation included; NC: natural convection.

[e]Nu = Nusselt number, Re = Reynolds number, Gr = Grashoff number, Pr = Prandtl number

product shape. The characteristic length given is the dimension used to calculate the Nusselt number. If another dimension was used for the Reynolds number, it is shown in the Comments column.

3. *Transfer medium.* The fluid used for forced convection is listed. The correlations that did not include a Prandtl number should be used with fluids similar to the experimental conditions, since significant variations in the calculation of *h* may be encountered if different fluids are used. If the correlation used the Prandtl number, the use of other transfer media can be accounted for by the correlation when the Prandtl dimensionless number is within the same order of magnitude.

4. *Temperature difference* (Δt) or *temperature of medium.* The surface heat transfer coefficient exhibits a weak dependence on temperature; therefore, for accurate determinations, the temperature of the body and the medium should be known. The user can also compare the experiment with design conditions.

5. *Velocity of medium.* This column lists either the velocity range or the velocity corresponding to the surface heat transfer coefficient. The location at which the velocity was measured is explained in the Comments column.

6. *Experimental method.* Each data set was typified according to the experimental methods used. The methods of estimating the surface heat transfer coefficient fall mainly into three categories: (1) the steady-state measurement of surface temperature for a given heat dissipation of the body, (2) the measurement of the transient temperature as the body cools or heats, and (3) the measurement of heat flux at the surface

of the body. The code that indicates the type of experimental method used is explained in the footnotes to the table. For more details, refer to Arce and Sweat (1980).

7. *Temperature measurement.* Size, type, and location of temperature sensors are described. This information can help to evaluate the accuracy and relevancy of the temperature measurements.

8. *Type of heat transfer.* Especially with forced convection cooling, freezing, or heating of agricultural products, two other kinds of heat transfer occur: evaporative cooling and radiative transfer. Evaporative cooling is inherent with any object containing moisture. Radiative cooling is experienced especially by single standing objects cooled in an environment substantially cooler than their surface temperature. Some experimenters corrected their measurements for these types of transfer, but where not corrected, the types included in the correlation are listed in this column.

9. *Nu-Re correlation.* The Nusselt number as a function of Reynolds number, Prandtl number, and any other variables, such as the geometry factor *G* are listed in this column. All the correlations listed are of the form $Nu = c_1\ Re^{c2}\ Pr^{1/3}$, where the Nusselt number hD/k characterizes the heat transfer and contains the surface heat transfer coeffficient *h*, the characteristic length *D*, and the thermal conductivity *k*, of the fluid at the surface temperature conditions. The Reynolds number $DV\rho/\mu$, characterizes the dynamic properties of the fluid and has the same characteristic dimension *D*, the velocity of the fluid *V*, and the density ρ and viscosity μ of the fluid at the average temperature conditions. The Prandtl number

Table 10   Surface Heat Transfer Coefficients for Food Products (*Concluded*)

Column 10	11	12	13	14	15
Error of Correlation Coefficient	Reynolds Number Range[g]	h, Btu/h·ft²·°F	Classification[i]	Reference	Comments
N/A	1000–4000	N/A	111	Kelly (1965)	Fluidized bed 2 in. deep
N/A	1000–6000	N/A	111	Kelly (1965)	
r = 0.866 Top of bin	3000–9000	*2.5 3.4 3.6 4.3	121	Minh *et al.* (1969)	Use intersticial velocity to calculate Re. Bin is 30 by 20 by 9 in. *Each h value is an average of 3 replications with airflow from top to bottom.
±6% Based on water freezing	N/A	74.0 to 83.3	120	Lentz (1969)	Cry-o-vac packaged *CaCl₂ brine 26% wt **moderately agitated Chickens 2.4 to 6.4 lb Turkeys 11.9 to 21 lb
N/A	1200–4600	N/A	222	Otten (1974)	8 points in correlation Bed depth 1.3 in.
±20%	N/A	2.9	220	Cleland and Earle (1976)	Packed in tin foil and brown paper
σ = 5.5	Gr = 10⁶ to 5 × 10⁷	N/A	222	Leichter *et al.* (1976)	Emissivity = 0.7 300 points in correlation L = characteristic length All cylinders 2.8 in. diam.

[f]Errors are given as perent of measured value. Correlation coefficients are given as r values

[g]Characteristic length is given in column 2, free stream velocity is used, unless specified otherwise in column 15.

[h]N/A indicates that hs was not reported in original article.

[i]The three values in the classification represent the method used to obtain h, the accuracy reported, and the location of the point or Nu-Re correlation relative to the rest of the data obtained. For method used: 0 = not reported, 1 = less accurate, 2 = most accurate methods. For accuracy: 0 = not reported, 1 = less accurate, 2 = most accurate. For comparison with other data: 0 = no comparison possible, 1 = substantially different from other data, and 2 = close to other published data.

$c_p\mu/k$ characterizes the heat transfer properties of the fluid at average temperature conditions.

10. *Error or correlation coefficient.* Measurement errors are estimated in this column. All the entries do not use the same basis; some researchers measured the error as the correlation coefficient of a least squares fit to the experimental data, while others measured it comparing their results to some known correlation or the mean of their experiment. The correlation coefficient r is a relative measurement and should be used to compare only experiments performed under the same conditions; correlation coefficients higher than 0.95 represent a close fit to the experimental data even though smaller values for r may have even better fits to the data for other conditions.

11. *Reynolds number range.* Reynolds numbers indicate the range where the Nusselt-Reynolds correlation is valid, to avoid extrapolation. The range of other variables, if used, is also reported, such as the geometry factor G.

12. *Surface heat transfer coefficient h.* Values obtained in the literature are listed in this column. Whenever possible, all available data were listed, except where an overwhelming amount of data made it impractical. In such a case, a representative sample covering the complete range of experimental conditions has been included. The surface heat transfer coefficients are listed in Btu/h·ft²·°F. The values for h include other types of heat transfer, such as evaporative cooling, which are listed in separate columns. For each h reported in this column, there should be a corresponding set of values for velocity, characteristic length, and temperature at which the h value was obtained. The table leaves blanks below values

in some columns, meaning that the value of h is calculated keeping that variable constant.

13. *Classification.* Classification of the data obtained from the references is listed to help the designer choose an accurate value for the surface heat transfer coefficient. The figures compare the theoretical method used, accuracy, and overall placement of the results relative to the rest of the data reported. Therefore, three values were needed: (a) for the theoretical methods used, (b) for the relative accuracy, and (c) for the agreement with the rest of the data reported.

   (a) *Theoretical method.* This value shows the inherent accuracy of the method used to calculate h. For instance, a value for h calculated using the truncated infinite series solution should be more accurate than one calculated using some type of calorimetric determination. This classification does not evaluate the experimental setup nor the analysis of the data. The number 2 indicates that the method used was of higher accuracy than for 1, and 0 indicates that the method used was not reported.

   (b) *Relative error.* The error was evaluated qualitatively with the following parameters: the error reported (either as a correlation coefficient, a standard deviation, or a percentage deviation from the mean); the number of data points per log cycle of Reynolds numbers; and any other indication of accuracy available such as actual measurements of an assumed constant. The cutoff points between 1 and 2 were 0.8 for the correlation coefficient, an error of 40% or larger for the experiment, or five or less data points per cycle of Nu-Re correlation reported. The number 2 indicates the most accuracy, 1 indicates less accuracy, and 0 indicates no error data.

   (c) *General agreement of the data.* The Nu-Re correlations were plotted, and the relative location of the curves was noted for the classification. The slope of the curve was emphasized more than its location in the plane, since Reynolds numbers may have been calculated using different values of velocity and characteristic length, depending on where these were measured. The number 2 indicates the closest agreement with other similar data, 1 indicates less agreement with other data, and 0 indicates that the data were not plotted; therefore, no comparison could be made.

14. *Reference.* The article from which the data was obtained is listed.

15. *Comments.* This column lists any additional information.

How to Use Tables

Users should obtain *h* values from the Nu-Re correlations in the tables, where available, rather than selecting *h* values directly from column 12. The *h* values are applicable only for one set of experimental conditions, whereas the Nu-Re correlations are more generally applicable. The design conditions may vary from the experimental conditions in the article, provided the Reynolds number is within the same range.

The amount of data in the tables is limited when the almost infinite number of products, product form and shape, packaging condition, and process condition are considered. Therefore, the Nu-Re correlations have much greater utility in estimating *h* values. Techniques for dealing with irregular shapes may be used in conjunction with the Nu-Re correlations.

Considering the great variety of geometrical shapes and arrangements possible, some experimenters have attempted to categorize all the shapes with some type of parameter. For instance, Smith *et al.* (1967) use a geometry index *G* to characterize ellipsoids or shapes simulated by ellipsoids. Only the overall shape is important and any agricultural product can be approximated with an ellipsoid; therefore, one correlation covers all possible shapes.

The following guidelines are important in choosing the Nu-Re correlation: (1) try to use a Nu-Re correlation for a Reynolds number in the same range as is being designed for; (2) avoid extrapolations; (3) use data for the same heat transfer medium with the temperature and temperature difference similar to the design conditions; and (4) consider the same types of heat transfer that the design includes, *i.e.*, evaporative, radiative, etc. If, after looking at these factors, the choice is narrowed down to a few correlations, use the classification number to decide which one is best. The classification reflects the accuracy and the method used in the experiment.

Once the Nu-Re correlation has been chosen, use the same dimensions and velocities defined in the article. In case of doubt, check to see if the Nusselt number obtained is in the range of the experiment. If there are still doubts, consult the original source. Since the Nu-Re correlations are logarithmic, any small errors mean a large change in *h*. Special emphasis should be given to the use of characteristic lengths and free stream, or interstitial velocity should be emphasized in calculating Reynolds and Nusselt numbers.

If data from column 12 are used, the conditions of the design must equal those in the experimental procedure, since the value of *h* is valid only for one set of conditions. If all conditions are matched, *h* values from the table would be the most accurate, since a generalization was made to fit all the data in one Nu-Re correlation. However, if this match is not exact, a Nu-Re correlation should be used to calculate *h*; data not reported in correlation form can be used to check the estimate.

Many foods not packaged or packaged in permeable materials lose moisture during cooling. The effects of packaging on the value of surface heat transfer coefficient are not well documented.

REFERENCES

Anderson, R.E., R.E. Hardenburg, and H.C. Vaught. 1963. Controlled atmosphere storage studies with cranberries. *Proceedings of the American Society for Horticultural Science* 83:416.

Arce, J.A. and V.E. Sweat. 1980. Survey of published heat transfer coefficients encountered in food processes. ASHRAE *Transactions* 86(2):235-60.

ASHRAE. 1972. *Fundamentals Volume*, Chapter 30, Table 4, p. 572.

ASHRAE. 1974. *Applications Volume*, Chapter 29, Table 3, p. 29.10.

ASHRAE. 1978. *Applications Volume*, Chapter 40, Table 7, p. 40.5.

ASM. 1948. *Metals handbook.* American Society for Metals. (Now ASM International).

Babbitt, J.D. 1945. The thermal properties of wheat in bulk. *Canadian Journal Research* 23F:338.

Baird, C.D. and J.J. Gaffney. 1976. A numerical procedure for calculating heat transfer in bulk loads of fruits or vegetables. ASHRAE *Transactions* 82:525-35.

Bartlett, L.H. 1944. A thermodynamic examination of the "latent heat" of food. *Refrigerating Engineering* 47:377.

Bennett, A.H. 1963. Thermal characteristics of peaches as related to hydrocooling. USDA Agricultural Marketing Service, Technical Bulletin No. 1292, U.S. Government Printing Office, Washington, D.C.

Bennett, A.H., W.G. Chace, Jr., and R.H. Cubbedge. 1964. Thermal conductivity of Valencia orange and Marsh grapefruit rind and juice vesicles. ASHRAE *Transactions* 70:256.

Bennett, A.H., J. Soule, and G.E. Yost. 1966. Temperature response of Florida citrus to forced-air precooling. ASHRAE *Journal* 8(4):48-54.

Bennett, A.H., W.G. Chace, Jr., and R.H. Cubbedge. 1969. Heat transfer properties and characteristics of Appalachian area, Red Delicious apples. ASHRAE *Transactions* 75(2):133.

Bennett, A.H., W.G. Chace, Jr., and R.H. Cubbedge. 1970. Thermal properties and heat transfer characteristics of marsh grapefruit. USDA Agricultural Research Service, Technical Bulletin No. 1413, U.S. Government Printing Office, Washington, D.C.

Biale, J.B. 1960. Respiration of fruits. *Encyclopedia of Plant Physiology* 12:536.

Breskin Pub. 1963. *Modern plastics encyclopedia.* Breskin Publications, New York.

Charm, S.E. and P. Moody. 1966. Bound water in haddock muscle. ASHRAE *Journal* 8(4):39.

Cherneeva, L.I. 1956. Study of the thermal properties of foodstuffs. *Report of VNIKHI*, Gostorgisdat, Moscow, 16.

Chuma, Y., S. Murata, and S. Uchita. 1970. Determination of heat transfer coefficients of farm products by transient method using lead model. *Journal of the Society of Agricultural Machinery*, Japan 31(4):298-302.

Clary, B.L., G.L. Nelson, and R.E. Smith. 1968. Heat transfer from hams during freezing by low temperature air. *Transactions of the ASAE* 11:496-99.

Claypool, L.L. and F.W. Allen. 1951. The influence of temperature and oxygen level on the respiration and ripening of Wickson plums. *Hilgardea* 21:129.

Claypool, L.L. and S. Ozbek 1952. Some influences of temperature and carbon dioxide on the respiration and storage life of the Mission Fig. *Proceedings of the American Society for Horticultural Science* 60:226.

Cleland, A.C. and R.L. Earle. 1976. A new method for prediction of surface heat transfer coefficients in freezing. *Bulletin de L'Institut International du Froid* 1:361-68.

Dickerson, R.W., Jr. 1968. Thermal properties of food. *The freezing preservation of foods*, 4th ed., Vol. 2, AVI Publishing Co., Westport, CT.

Dickerson, R.W. 1972. Computing heating and cooling rates of foods. Symposium on Prediction of Cooling/Freezing Times for Food Products. ASHRAE *Bulletin* No-72-3, 5.

Dickerson, R.W., Jr. and R.B. Read, Jr. 1968. Calculation and measurement of heat transfer in foods. *Food Technology* 22(December):37.

Dickerson, R.W., Jr. and R.B. Read, Jr. 1975. Thermal diffusivity of meats. ASHRAE *Transactions* 81(1):356.

Eaks, J.L. and L.L. Morris. 1956. Respiration of cucumber fruits associated with physiological injury at chilling temperatures. *Plant Physiology* 31:308.

Eucken, A. 1940. Allgemeine Gesetzmassigkeiten fur das Warmeleitvermogen verschiedener Stoffarten und Aggregatzustande. *Forschung auf dem Gebiete des Ingenieurwesens, Ausgabe A* 11(1):6.

Fedorov, V.G., D.N. Il'Inskiy, O.A. Gerashchenko, and L.D. Andreyeva. 1972. Heat transfer accompanying the cooling and freezing of meat carcasses. *Heat Transfer—Soviet Research* 4:55-59.

Gane, R. 1936. The thermal conductivity of the tissue of fruits. *Annual Report*, Food Investigation Board, Great Britain, 211.

Gerhardt, F., H. English, and E. Smith. 1942. Respiration, internal atmosphere, and moisture studies of sweet cherries during storage. Proceedings of the American Society for Horticultural Science 41:119.

Gore, H.C. 1911. Studies on fruit respiration. USDA *Bur. Chem Bulletin*, Vol. 142.

Griffiths, E. and D.H. Cole. 1948. Thermal properties of meat. *Society of Chemical Industry Journal*, London, England 67:33.

Griffiths, E. and M.J. Hickman. 1951. *The thermal conductivity of some nonmetallic materials.* Institute of Mechanical Engineers, London, England, 289.

Haller, M.H., P.L. Harding, J.M. Lutz, and D.H. Rose. 1932. The respiration of some fruits in relation to temperature. Proceedings of the American Society for Horticultural Science 28:583.

Haller, M.H., D.H. Rose, and P.L. Harding. 1941. Studies on the respiration of strawberry and raspberry fruits. USDA *Circular*, Vol. 613.

Haller, M. H. *et al.* 1945. Respiration of citrus fruits after harvest. *Journal of Agricultural Research* 71(8):327. Also in 1974 ASHRAE *Handbook*, Table 1, 32.4.

Hardenburg, R.E. 1966. Unpublished data.

Harding, P.L. 1929. Respiration studies of Grimes apples under various controlled temperatures. Proceedings of the American Society for Horticultural Science 26:319.

Harper, J.C. 1960. Microwave spectra and physical characteristics of fruit and animal products relative to freeze-dehydration. Report No. 6, Army Quarter Master Food and Container Institute for the Armed Forces, ASTIA AD 255 818, 16.

Harper, J.C. 1962. Transport properties of gases in porous media at reduced pressures with reference to freeze-drying. *American Institute of Chemical Engineering Journal* (8)3:298.

Harris, S. and B. McDonald. 1975. Physical data for Kiwi fruit (actinidia chinensis). *New Zealand Journal of Science* 18:307.

Hawkins, L.A. 1929. Governing factors in transportation of perishable commodities. *Refrigerating Engineering* 18:130.

Heisler, M.P. 1947. Temperature charts for induction and constant temperature heating. ASME *Transactions* 69:227.

Hill, J.E. 1966. *The thermal conductivity of beef.* Georgia Institute of Technology, Atlanta, GA, 49.

Hill, J.E., J.D. Leitman, and J.E. Sunderland. 1967. Thermal conductivity of various meats. *Food Technology* 21(8):91.

Hooper, F.C. and S.C. Chang. 1952. Development of the thermal conductivity probe. *Heating, Piping and Air Conditioning*, ASHVE 24(10):125.

Hruschka, H.W. 1966. Storage and shelf life of packaged rhubarb. USDA *Marketing Research Report*, 771.

Hruschka, H.W. and Chien Yi Want. 1979. Storage and shelf life of packaged watercress, parsley, and mint. USDA *Marketing Research Report* 1102.

International Institute of Refrigeration. 1967. *Recommended conditions for the cold Storage of perishable produce*, 2nd ed., Paris, France.

Jason, A.C. and R.A.K. Long. 1955. The specific heat and thermal conductivity of fish muscle. Proceedings of the 9th International Congress of Refrigeration, Paris, France 1:2160.

Jones, W.W. 1942. Respiration and chemical changes of papaya fruit in relation to temperature. *Plant Physiology* 17:481.

Karmarkar, D.V. and B.M. Joshe. 1941a. Respiration of onions. *Indian Journal of Agricultural Science* 11:82.

Karmarka, D.V. and B.M. Joshe. 1941b. Respiration studies on the Alphonse mango. *Indian Journal of Agricultural Science* 11:993.

Kaye, G.W.C. and W.F. Higgins. 1928. The thermal conductivities of certain liquids. Proceedings of Royal Society of London A117:459.

Kazarian, E.A. 1962. *Thermal properties of grain*. Michigan State University, East Lansing, MI, 74.

Kelly, M.J. 1965. Heat transfer in fluidized beds. *Duchema Monographien* 56:119.

Khatchaturov, A.B. 1958. Thermal processes during air-blast freezing of fish. *Bulletin of the I.I.R.* 2:365-78.

Khelemskii, M.Z. and V.Z. Zhadan. 1964. Thermal conductivity of normal beet juice. *Sakharnaya Promyshlennost* 10:11.

Kondrat'ev, G.M. 1950. Application of the theory of regular cooling of a two-component sphere to the determination of heat conductivity of poor heat conductors (method, sphere in a sphere). *Otdelenie Tekhnicheskikh Nauk, Isvestiya Akademii Nauk*, USSR 4(April):536.

Kopelman, I., J.L. Blaisdell, and I.J. Pflug. 1966. Influence of fruit size and coolant velocity of Jonathan apples in water and air. ASHRAE *Transactions* 72(1):209-16.

Lange, N.A. 1961. *Handbook of chemistry*, 10th ed. McGraw-Hill, New York, 834.

Leichter, S., S. Mizrahi, and I.J. Kopelman. 1976. Effect of vapor condensation on rate of warming up of refrigerated products exposed to humid atmosphere: Application to the prediction of fluid milk shelf life. *Journal of Food Science* 41:1214-18.

Leidenfrost, W. 1959. Measurements on the thermal conductivity of milk. ASME Symposium on Thermophysical Properties, Purdue University, 291.

Lentz, C.P. 1961. Thermal conductivity of meats, fats, gelatin gels, and ice. *Food Technology* 15(5):243.

Lentz, C.P. 1969. Calorimetric study of immersion freezing of poultry. *Journal of the Canadian Institute of Food Technology* 3(2):132-36.

Lewis, D.A. and L.L. Morris. 1956. Effects of chilling storage on respiration and deterioration of several sweet potato varieties. Proceedings of the American Society for Horticultural Science 68:421.

Lipton, W.J. 1957. *Physiological changes in harvested asparagus (Asparagus officinales) as related to temperature*. University of California at Davis.

Long, R.A.K. 1955. Some thermodynamic properties of fish and their effect on the rate of freezing. *Journal of the Science of Food and Agriculture* 6:621.

Lutz, J.M. 1938. Factors influencing the quality of American grapes in storage. USDA *Technical Bulletin*, Vol. 606.

Lutz, J.M. and R.E. Hardenburg. 1968. The commercial storage of fruits, vegetables, and florists and nursery stocks. USDA *Handbook* 66.

Mann, L.K. and D.A. Lewis. 1956. Rest and dormancy in garlic. *Hilgardia* 26:161.

Mannheim, H.C., M.P. Steinberg and A.I. Nelson. 1955. Determinations of enthalpies involved in food freezing. *Food Technology* 9:556.

Mathews, F.V., Jr. and C.W. Hall. 1968. Method of finite differences used to relate changes in thermal and physical properties of potatoes. ASAE *Transactions* 11(4):558.

Maxie, E.C., F.G. Mitchell and A. Greathead. 1959. Studies on strawberry quality. *California Agriculture* 13(2):11, 16.

Maxie, E.C., P.B. Catlin and H.T. Hartmann. 1960. Respiration and ripening of olive fruits. Proceedings of the American Society for Horticultural Science 75:275.

Maxwell, J.C. 1904. *A treatise on electricity and magnetism*, 3rd ed., Vol. 1. The Clarendon Press, Oxford, England, 440.

Metzner, A.B. and P.S. Friend. 1959. Heat transfer to turbulent non-Newtonian fluids. *Industrial and Engineering Chemistry* 51:879.

Micke, W.C., F.G. Mitchell and E.C. Maxie. 1965. Handling sweet cherries for fresh shipment. *California Agriculture* 19(4):12.

Miller, C.F. 1963. *Thermal conductivity and specific heat of sorghum grain*. Texas Agricultural and Mechanical College, College Station, TX, 79.

Miller, H.L. and J.E. Sunderland. 1963. Thermal conductivity of beef. *Food Technology* 17(4):124.

Minh, T.V., J.S. Perry and A.H. Bennett. 1969. Forced-air precooling of white potatoes in bulk. ASHRAE *Transactions* 75(2):148-50.

Moline, S.W., J.A. Sawdye, J.A. Short and A.P. Rinfret. 1961. Thermal properties of foods at low temperatures. *Food Technology* 15:228.

Moote, I. 1953. The effect of moisture on the thermal properties of wheat. *Canadian Journal Technology*, Vol. 31, No. 2 and 3, February/March, 57.

Morris, L.L. 1947. A study of broccoli deterioration. *Ice and Refrigeration* 113(5):41.

Morris, L.L. 1952. Unpublished data.

Nicholas R.C., K.E.H. Motawi, and J.L. Blaisdell. 1964. Cooling rate of individual fruit in air and water. *Quarterly Bulletin*, Michigan State University Agricultural Experiment Station 47(1):51-64.

Nowrey, J.E. and E.E. Woodams. 1968. Thermal conductivity of a vegetable oil-in-water emulsion. *Journal of Chemical and Engineering Data* 13(3):297.

Otten, L. 1974. Thermal parameters of agricultural materials and food products. *Bulletin I.I.F./I.I.R.* 3:191-99.

Oxley, T.A. 1944. The properties of grain in bulk; III—The thermal conductivity of wheat, maize and oats. *Society of Chemical Industry Journal*, London, England 63:53.

Pantastico, E.B. 1974. Handling and utilization of tropical and subtropical fruits and vegetables. *Postharvest Physiology*, Table 16.2, AVI Publishing Co., Inc., Westport, CT, 323.

Parker, R.E. and B.A. Stout. 1967. Thermal properties of tart cherries. ASAE *Transactions* 10(4):489.

Pentzer, W.T., C.E. Asbury and K.C. Hamner. 1933. The effect of sulfur dioxide fumigation on the respiration of Emperor grapes. *Proceedings of the American Society for Horticultural Science* 30:258.

Perry, R.L., F.M. Turrell, and S.W. Austin. 1964. Thermal diffusivity of citrus fruits. *Heat Transfer, Thermodynamics, and Education*, H.A. Johnson, ed., McGraw-Hill, New York, 242.

Popov, V.D. and Y.A. Terentiev. 1966. Thermal properties of highly viscous fluids and coarsely dispersed media. *Teplofizicheskie Svoistva Veshchestv, Akademiya Nauk, Ukrainskoi SSSR, Respublikanskii Sbornik* 18:76.

Poppendick, H.F., *et al.* 1965-66. Annual report on thermal and electrical conductivities of biological fluids and tissues. ONR *Contract* 4095 (00), A-2, GLR-43 Geoscience Ltd., 39.

Pratt, H.K. and L.L. Morris. 1958. Some physiological aspects of vegetable and fruit handling. *Food Technology in Australia* 10:407.

Pratt, H.K., L.L. Morris, and C.L. Tucker. 1954. Temperature and lettuce deterioration. *Proceedings of the Conference on Transportation of Perishables*, University of California at Davis, 77.

Qashou, M.S., G. Nix, R.I. Vachon, and G.W. Lowery. 1970. Thermal conductivity values for ground beef and chuck. *Food Technology* 23(4):189.

Qashou, M.S., R.I. Vachon, and Y.S. Touloukian. 1972. Thermal conductivities of foods. ASHRAE *Transactions* 78(1):165-83.

Radford, R.D., L.S. Herbert, and D.A. Lorett. 1976. Chilling of meat—A mathematical model for heat and mass transfer. *Bulletin de L'Institut du Froid* 1:323:30.

Rappaport, L. and A.E. Watada. 1953. Unpublished data.

Rappaport, L. and A.E. Watada. 1958. Effect of temperature on artichoke quality. *Proceedings of the Conference on Transportation of Perishables*, University of California at Davis, 142.

Riedel, L. 1949. Warmeleitfahigkeitsmessungen an Zukerlosungen, Fruchtsaften, und Milch. *Chemie-Ingenieur-Technik* 21(17-18):340.

Riedel, L. 1951. The refrigerating effect required to freeze fruits and vegetables. *Refrigerating Engineering* 59:670.

Riedel, L. 1955. Calorimetric investigations of the melting of fats and oils. *Fette, Seifen, Anstrichmittel* 57(10):771.

Riedel, L. 1956. Calorimetric investigations of the freezing of fish meat. *Daltetechnik* 8(12):374.

Riedel, L. 1957a. Calorimetric investigations of freezing of egg white and yolk. *Kaltetechnik* 9(11):342.

Riedel, L. 1957b. Calorimetric investigations of the meat freezing process. *Kaltetechnik* 9(2):38.

Riedel, L. 1959. Calorimetric investigations of the freezing of white bread and other flour products. *Kaltetechnik* 11(2):41.

Riedel, L. 1969. Measurements of thermal diffusivity on foodstuffs rich in water. *Kaltetechnik-Klimatisierung* 21(11):315.

Ryall, A.L. and W.J. Lipton. 1972. Vegetables as living products. Respiration and heat production. *Transportation and Storage of Fruits and Vegetables*, Vol. 1, AVI Publishing Co., Westport, CT, 5.

Saravacos, G.D. and M.N. Pilsworth. 1965. Thermal conductivity of freeze-dried model food gels. *Journal of Food Science* 30:773.

Sastry, S.K., C.D. Baird, and D.E. Buffington. 1978. Transpiration rates of certain fruits and vegetables. ASHRAE *Transactions* 84(1).

Schenk, R.U. 1959. Respiration of peanut fruit during curing. *Proceedings of the Association of Southern Agricultural Workers* 56:228.

Schenk, R.U. 1961. Development of the Peanut Fruit. Georgia Agricultural Experiment Station Bulletin N.S., Vol. 22.

Schneider, P.J. 1955. *Conduction heat transfer*. Addison-Wesley Publishing Co., Reading, MA, 15.

Schneider, P.J. 1963. *Temperature response charts*. John Wiley and Sons, New York.

Scholz, E.W., H.B. Johnson, and W.R. Buford. 1963. Heat evolution rates of some Texas-grown fruits and vegetables. *Rio Grande Valley Horticultural Society Journal* 17:170.

Short, B.F. and H.E. Staph. 1951. The energy content of foods. *Ice and Refrigeration* 121(11):23.

Siebel, J.E. 1892. Specific heat of various products. *Ice and Refrigeration*, April, 256.

Slavicek, E., K. Handa and M. Kminek. 1962. Measurements of the thermal diffusivity of sugar beets. *Cukrovarnicke Listy*, Vol. 78, Czechoslovakia, 116.

Smith, F.G., A.J. Ede, and R. Gane. 1952. The thermal conductivity of frozen foodstuffs. *Modern Refrigeration* 55:254.

Smith, R.E., G.L. Nelson and R.L. Henrickson. 1967. Analyses on transient heat transfer from anomalous shapes. ASAE *Transactions* 10(2):236.

Smith, R.E., G.L. Nelson and R.L. Henrickson. 1968. Applications of geometry analysis of anomalous shapes to problems in transient heat transfer from anomalous shapes. ASAE *Transactions* 10(2):296.

Smith, W.H. 1957. The production of carbon dioxide and metabolic heat by horticultural produce. *Modern Refrigeration* 60:493.

Smith, W.H. 1963-64. The storage of mushrooms. *Ditton and Covent Garden Garden Laboratories Annual Report*, Great Britain Agricultural Research Council, 18.

Smith, W.H. 1965-66. The storage of gooseberries. *Ditton and Covent Garden Laboratories Annual Report*, Great Britain Agricultural Research Council, 13.

Spells, K.E. 1958. The thermal conductivities of some biological fluids. Flying Personnel Research Committee, Institute of Aviation Medicine, Royal Air Force, Farnborough, England, FPRC-1071 AD 229 167, 8.

Spells, K.E. 1960-61. The thermal conductivities of some biological fluids. *Physics in Medicine and Biology* 5:139.

Staph, H.E. and W.R. Woolrich. 1951. Specific and latent heats of foods in the freezing zone. *Refrigerating Engineering* 59:1086.

Sweat, V.E. 1974. Modelling the conductivity of meats. Paper presented at the 1974 meeting of the American Society of Agricultural Engineers.

Sweat, V.E. 1985. Thermal properties of low- and intermediate-moisture food. ASHRAE *Transactions* 91(2):369-89.

Tewfik, S. and L.E. Scott. 1954. Respiration of vegetables as affected by postharvest treatment. *Journal of Agricultural and Food Chemistry* 2:415.

Thompson, H., S.R. Cecil, and J.G. Woodroof. 1951. Storage of edible peanuts. Georgia Agricultural Experiment Station Bulletin, Vol. 268.

Triebes, T.A. and C.J. King. 1966. Factors influencing the rate of heat conduction in freeze-drying. *I and EC Process Design and Development* 5(4):430.

Turrell, F.M. and R.L. Perry. 1957. Specific heat and heat conductivity of citrus fruit. *Proceedings of the American Society for Horticultural Science* 70:261.

Van den Berg, L. and C.P. Lentz. 1957. Factors affecting freezing rates of poultry immersed in liquid. *Food Technology* 11(7):377-80.

Van den Berg, L. and C.P. Lentz. 1972. Respiratory heat production of vegetables during refrigerated storage. *Journal of the American Society for Horticultural Science* 97:431.

Van den Berg, L. and C.P. Lentz. 1975. Effect of composition on thermal conductivity of fresh and frozen foods. *Canadian Institute of Food Science and Technology Journal* 8(2):79.

Wachsmuth, R., XIII. 1892. Untersuchungen auf dem Gebiet der inneren Warmeleitung. *Annalen der Physik* 3(48):158.

Walters, R.E. and K.N. May. 1963. Thermal conductivity and density of chicken breast muscle and skin. *Food Technology* 17(June):130.

Watada, A.E. and L.L. Morris. 1966. Effect of chilling and nonchilling temperatures on snap bean fruits. Proceedings of the American Society for Horticultural Science 89(368).

Watt, B.K. and A.L. Merrill. 1963. Composition of foods. USDA *Handbook* 8.

Weber, H.F., VIII. 1880. Untersuchungen Uber die Warmeleitung in Flussigkeiten. *Annalen der Physik* 10(3):304.

Weber, H.F. 1886. The thermal conductivity of drop forming liquids. *Exner's Reportorium* 22:116.

Whiteman, T.M. 1957. Freezing points of fruits, vegetables and florists stocks. USDA *Marketing Research Report* 196.

Woodams, E.E. 1965. *Thermal conductivity of fluid foods*. Cornell University, Ithaca, NY, 96.

Woolrich, W.R. 1966. Specific and latent heat of foods in the freezing zone. ASHRAE *Journal* 8(4):43.

Workman, M. and H.K. Pratt. 1957. Studies on the physiology of tomato fruits; II, Ethylene production at 20°C as related to respiration, ripening and date of harvest. *Plant Physiology* 32:330.

Wright, R.C., D.H. Rose, and T.H. Whiteman. 1954. The commercial storage of fruits, vegetables, and florists and nursery stocks. USDA *Handbook* 66.

CHAPTER 31

SPACE AIR DIFFUSION

TERMINOLOGY

THE following definitions refer to air diffusion equipment and methods.

Aspect ratio The ratio of length to width of an opening or core of a grille.

Axial flow jet A stream of air whose motion is approximately symmetrical along a line, although some spreading and drop or rise can occur from diffusion and buoyancy effects.

Coefficient of discharge Ratio of area at vena contracta to area of opening.

Core area The total plane area of that portion of a grille, included within lines tangent to the outer edges of the outer openings through which air can pass.

Damper A device used to vary the volume of air passing through a confined cross section by varying the cross-sectional area.

Diffuser An outlet discharging supply air in various directions and planes.

Diffusion Distribution of air within a space by an outlet discharging supply air in various directions and planes.

Drop The vertical distance that the lower edge of the horizontally projected airstream drops between the outlet and the end of its throw.

Effective area The net area of an outlet or inlet device through which air can pass, equal to the free area times the coefficient of discharge.

Entrainment The movement of room air into the jet caused by the airstream discharged from the outlet (secondary air motion).

Entrainment ratio The total air divided by the air discharged from the outlet.

Envelope The outer boundary of an airstream moving at a perceptible velocity.

Exhaust opening or inlet Any opening through which air is removed from a space.

Free area The total minimum area of the openings in the air outlet or inlet through which air can pass.

Grille A covering for any opening through which air passes.

Induction Room air drawn into an outlet by the primary airstream (aspiration).

Isothermal jets Air jets with the same temperature as its surrounding air.

Lower zone Room volume below the stratification level created by displacement ventilation.

Nonisothermal jets Air jets with a initial temperature different from its surrounding air.

Outlet velocity The average velocity of air emerging from the outlet, measured in the plane of the opening.

Primary air The air delivered to the outlet by the supply duct.

Radius of diffusion The horizontal axial distance an airstream travels after leaving an air outlet before the maximum stream velocity is reduced to a specified terminal level, e.g., 200, 150, 100 fpm.

Register A grille equipped with a damper or control valve.

Spread The divergence of the airstream in a horizontal and/or vertical plane after it leaves the outlet.

Supply opening or outlet Any opening through which supply air is delivered into a ventilated space being heated, cooled, humidified, or dehumidified. Supply outlets are classified according to their location in a room as sidewall, ceiling, baseboard, or floor outlets. However, since numerous designs exist, they are more accurately described by their construction features. (See Chapter 17 of the 1992 ASHRAE *Handbook—Systems and Equipment*.)

Temperature differential Temperature difference between primary and room air.

Terminal velocity The maximum airstream velocity at the end of the throw.

Throw The horizontal or vertical axial distance an airstream travels after leaving an air outlet before the maximum stream velocity is reduced to a specified terminal velocity, e.g., 100, 150, 200 fpm. Data for the throw of a jet from various outlets are generally given by each manufacturer for isothermal jet conditions and without boundary walls interfering with the jet.

Total air The mixture of discharged air and entrained air.

Vane A thin plate in the opening of a grille.

Vane ratio The ratio of depth of vane to minimum width between two adjacent vanes.

Upper zone Room volume above the stratification level created by the displacement ventilation.

METHODS OF ROOM AIR DISTRIBUTION

Room air distribution systems can be classified as mixing, displacement, and local systems.

Mixing Systems

Conditioned air is normally supplied to air outlets at velocities much greater than those acceptable in the occupied zone. Conditioned air temperature may be above, below, or equal to the air temperature in the occupied zone, depending on the heating/cooling load. The diffuser jets mix with the ambient room air by entrainment, which reduces the air velocity and equalizes the air temperature. The occupied zone is either ventilated by the decayed air jet directly or by the reverse flow created by the jets.

Mixing air distribution creates relatively uniform air velocity, temperature, humidity, and air quality conditions in the occupied zone.

The preparation of this chapter is assigned to TC 5.3, Room Air Distribution.

Typical outlets and their performance. Straub *et al.* (1956, 1957) classified outlets into five groups:

Group A. Outlets mounted in or near the ceiling that discharge air horizontally.

Group B. Outlets mounted in or near the floor that discharge air vertically in a nonspreading jet.

Group C. Outlets mounted in or near the floor that discharge air in a vertical spreading jet.

Group D. Outlets mounted in or near the floor that discharge air horizontally.

Group E. Outlets mounted in or near the ceiling that project primary air vertically.

Analysis of outlet performance was based on primary air pattern, total air pattern, stagnant air layer, natural convection currents, return air pattern, and room air motion.

Figures 1 through 5 show the room air motion characteristics of the five outlet groups; exterior walls are depicted by heavy lines.

The principles of air diffusion emphasized by these tests are as follows:

1. The primary air (shown by clear envelopes in Figures 1 through 5) from the outlet down to a velocity of about 150 fpm can be treated analytically. The heating or cooling load has a strong effect on the characteristics of the primary air.

2. The total air, shown by diagonally lined envelopes in Figures 1 through 5, is influenced by the primary air and is of relatively high velocity (but less than 150 fpm), with air temperatures generally within 1°F of room temperature. The total air is also influenced by the environment and drops during cooling or rises during heating; it is not subject to precise analytical treatment.

3. Natural convection currents form a stagnant zone from the ceiling down during cooling, and from the floor up during heating. This zone forms below the terminal point of the total air during heating and above the terminal point during cooling.

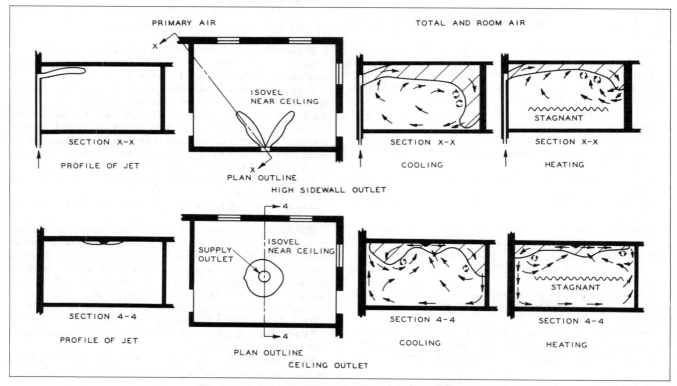

Fig. 1   Air Motion Characteristics of Group A Outlets
(Straub *et al.* 1956)

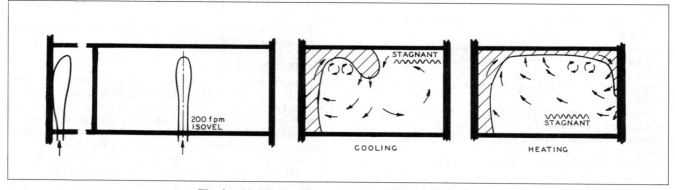

Fig. 2   Air Motion Characteristics of Group B Outlets
(Straub *et al.* 1956)

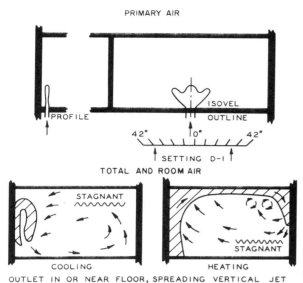

Fig. 3 Air Motion Characteristics of Group C Outlets
(Straub *et al.* 1956)

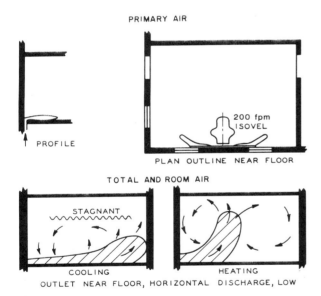

Fig. 4 Air Motion Characteristics of Group D Outlets
(Straub *et al.* 1956)

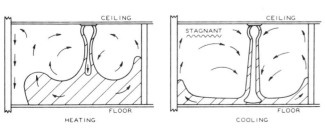

Fig. 5 Air Motion Characteristics of Group E Outlets
(Straub *et al.* 1956)

Since this zone results from natural convection currents, the air velocities within it are usually low (approximately 20 fpm), and the air stratifies in layers of increasing temperatures. The concept of a stagnant zone is important in properly applying and selecting outlets, since it considers the natural convection currents from warm and cold surfaces and internal loads.

4. A return inlet affects the room air motion only within its immediate vicinity. The intake should be located in the stagnant zone to return the warmest room air during cooling or the coolest room air during heating. The importance of the location depends on the relative size of the stagnant zone that results from various types of outlets.

5. The general room air motion (shown by clear areas in Figures 1 through 5) is a gentle drift toward the total air. Room conditions are maintained by the entrainment of the room air into the total airstream. The room air motion between the stagnant zone and the total air is relatively slow and uniform. The highest air motion occurs in and near the total airstreams.

Group A outlets. This group includes high sidewall grilles, sidewall diffusers, ceiling diffusers, linear ceiling diffusers, and similar outlets. High sidewall grilles and ceiling diffusers are illustrated in Figure 1.

The primary air envelopes (isovels) show a horizontal, two-jet pattern for the high sidewall and a 360° diffusion pattern for the ceiling outlet. Although variation of vane settings might cause a discharge in one, two, or three jets in the case of the sidewall outlet, or have a smaller diffusion angle for the ceiling outlet, the general effect in each is the same.

During cooling, the total air drops into the occupied zone at some distance from the outlet, which depends on air quantity, supply velocity, temperature differential between supply and room air, deflection setting, ceiling effect, and type of loading within the space. Analytical methods of relating some of these factors are presented in the section Room Air Distribution Theory.

The cooling diagram for the high sidewall outlet shows an overthrow condition, causing the total air to drop along the opposite wall and flow slowly for some distance across the floor. This condition should be avoided, since the total air region becomes one of high velocity accompanied by temperature depressions. Velocities of about 100 to 150 fpm may be found near the wall but will dissipate within about 4 in. of the wall.

The cooling diagram of the ceiling outlet shows that the total air movement is counteracted by the rising natural convection currents on the heated wall and, therefore, drops before reaching the wall. On the other hand, the total air reaches the inside wall and descends for some distance along it. With this type of outlet, temperature variations within the room are minimized, with minimal stagnant volume. Since the maximum velocity and the maximum temperature variation occur within and near the total air envelope, the drop region becomes important because it is an area with high effective draft temperature θ. Consequently, it is necessary to know how far the air drops before velocities and temperatures reach acceptable limits.

Since these outlets discharge horizontally near the ceiling, the warmest air in the room is mixed immediately with the cool primary air far above the occupied zone. Therefore, the outlets are capable of handling relatively large quantities of air at large temperature differentials.

During heating, warm supply air introduced at the ceiling can cause stratification in the space if there is insufficient induction of room air at the outlet. Selecting diffusers properly, limiting the room supply temperature differential, and maintaining air supply rates at a level high enough to ensure air mixing by induction provide adequate air diffusion and minimize stratification.

Several building codes and ASHRAE *Standard* 90.1 require sufficient insulation in exterior walls, so most perimeter spaces can be heated effectively by ceiling air diffusion systems. Interior spaces, which generally have only cooling demand conditions, seldom require long-term heating and are seldom a design problem.

Flow rate and velocity for both heating and cooling are the same for outlet types in Figure 1. The heating diagram for the sidewall unit shows that, under these conditions, the total air does not descend along the wall. Consequently, higher velocities might be beneficial in eliminating the stagnant zone, since high velocity causes some warm air to reach floor level and counteract stratification of the stagnant region.

The heating diagram for the ceiling outlet shows the effect of the natural convection currents that produce a larger throw toward the cold exposed wall. The velocity of the total air toward the exposed wall complements the natural convection currents. However, the warm total air loses its downward momentum at its terminal point, and buoyancy forces cause it to rise toward the primary air. Although these forces are complementary, the heating effect of the total air replaces the cool natural convection currents with warm total air.

Group B outlets. This group includes floor registers, baseboard units, low sidewall units, linear-type grilles in the floor or windowsill, and similar outlets. Figure 2 illustrates a floor outlet adjacent to an inside wall.

Since these outlets have no deflecting vanes, the primary air is discharged in a single vertical jet. When the total air strikes the ceiling, it fans out in all directions from the point of contact and, during cooling, follows the ceiling for some distance before dropping toward the occupied zone. During heating, the total airflow follows the ceiling across the room, then descends partway down the exterior wall.

The cooling diagram shows that a stagnant zone forms outside the total air region above its terminal point. Below the stagnant zone, air temperature is uniform, effecting complete cooling. Also, space below the terminal point of the total air is cooled satisfactorily. For example, if total airflow is projected upward for 8 ft, the region from this level down to the floor will be cooled satisfactorily. This, however, does not apply to an extremely large space. Judgment to determine the acceptable size of the space outside the total air is needed. A distance of 15 to 20 ft between the drop region and the exposed wall is a conservative design value.

A comparison of Figures 1 and 2 for heating shows that the stagnant region is smaller for Group B than Group A outlets because the air entrained in the immediate vicinity of the outlet is taken mainly from the stagnant region, which is the coolest air in the room. This results in greater temperature equalization and less buoyancy in the total air than would occur with Group A outlets.

While the temperature gradients for both outlet groups are about the same, the stagnant layer for Group B is lower than that for Group A.

Group C outlets. This group includes floor diffusers, sidewall diffusers, linear-type diffusers, and other outlets installed in the floor or windowsill (Figure 3).

Although outlets of this group are related to Group B, they are characterized by wide-spreading jets and diffusing action. Total air and room air characteristics are similar to those of Group B, although the stagnant zone formed is larger during cooling and smaller during heating. Diffusion of the primary air usually causes the total air to fold back on the primary and total air during cooling, instead of following the ceiling. This diffusing action of the outlets makes it more difficult to project the cool air, but it also provides a greater area for induction of room air. This action is beneficial during heating, since the induced air comes from the lower regions of the room.

Group D outlets. This group includes baseboard and low sidewall registers and similar outlets (Figure 4), which discharge the primary air in single or multiple jets. However, since the air is

discharged horizontally across the floor, the total air, during cooling, remains near the floor, and a large stagnant zone forms in the entire upper region of the room.

During heating, the total air rises toward the ceiling because of the buoyant effect of warm air. The temperature variations are uniform, except in the total air region.

Group E outlets. This group includes ceiling diffusers, linear-type grilles, sidewall diffusers and grilles, and similar outlets, mounted or designed for vertical downward air projection. Figure 5 shows the heating and cooling diagrams for this ceiling diffuser.

During cooling, the total air projects to and follows the floor, producing a stagnant region near the ceiling. During heating, the total airflow reaches the floor and folds back toward the ceiling. If projected air does not reach the floor, a stagnant zone results.

General Factors Affecting Outlet Performance

Effect of vanes. Vanes affect grille performance if their depth is at least equal to the distance between the vanes. If the vane ratio is less than unity, effective control of the airstream discharged from the grille by means of the vanes is impossible. Increasing the vane ratio above two has little or no effect, so vane ratios should be between one and two.

A grille discharging air uniformly forward (vanes in straight position) has a spread of 14 to 24°, depending on the type of outlet, duct approach, and discharge velocity. Turning the vanes influences the direction and throw of the discharged airstream.

A grille with diverging vanes (vertical vanes with uniformly increasing angular deflection from the centerline to a maximum at each end of 45°) has a spread of about 60° and reduces the throw considerably. With increasing divergence, the quantity of air discharged by a grille for a given upstream total pressure decreases.

A grille with converging vanes (vertical vanes with uniformly decreasing angular deflection from the centerline) has a slightly higher throw than a grille with straight vanes, but the spread is approximately the same for both settings. The airstream converges slightly for a short distance in front of the outlet and then spreads more rapidly than air discharged from a grille with straight vanes.

In addition to vertical vanes that normally spread the air horizontally, horizontal vanes may spread the air vertically. However, spreading the air vertically risks hitting beams or other obstructions or blowing primary air into the occupied zone at excessive velocities. On the other hand, vertical deflection may increase adherence to the ceiling and reduce the drop.

Effect of beamed ceilings and obstructions. In spaces with exposed beams, the outlets should be located below the bottom of the lowest beam level, preferably low enough to employ an upward or arched air path. The air path should be arched sufficiently to miss the beams and prevent the primary or induced airstream from striking furniture and obstacles and producing objectionable drafts (Wilson 1970). Obstructions influence airflow patterns and can reduce air distribution efficiency. Obstructions can reduce jet throw, increase air velocities in portions of the occupied zone, and create stagnant zones.

Variable air volume systems (VAV). The design of air distribution systems is usually based on the full load (heating/cooling). When only a partial load exists, VAV systems reduce the supply airflow, which, in turn, reduces the air velocity at the outlet. This decrease in outlet air velocity reduces the throw, which may cause a cold draft in the occupied zone or a poorly ventilated (stagnant) zone. Therefore, different operation modes of the system (airflow and initial temperature difference) should be considered in designing a VAV system air distribution (Zhivov 1990).

Displacement Ventilation

Conditioned air with a temperature slightly lower than the desired room air temperature in the occupied zone is supplied from air outlets at low air velocities (100 fpm or less). The outlets are

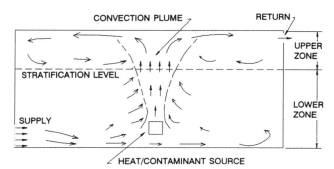

Fig. 6 Schematic of Displacement Ventilation

located at or near the floor level, and the supply air is directly introduced to the occupied zone. Returns are located at or close to the ceiling through which the warm room air is exhausted from the room. The supply air is spread over the floor and then rises as it is heated by the heat sources in the occupied zone. Heat sources (*e.g.*, person, computer) in the occupied zone create upward convective flows in the form of thermal plumes; these remove heat and contaminants which are less dense than the air from the occupied zone (Figure 6).

The air volume in the plumes increases as they rise because the plumes entrain ambient air. A stratification level exists where the airflow rate in the plumes equals the supply airflow rate. Two distinct zones are thus formed within the room, one lower zone below the stratification level and with no recirculation flow (close to displacement flow), and one upper zone, with recirculation flow (Figure 6). The height of the lower zone depends on the supply airflow rate and characteristics of heat sources and their distribution across the floor area. In a properly designed displacement ventilation system, the height of the lower zone is above the occupied zone so that the occupied zone can be ventilated effectively. For this type of system to function properly, a stable vertically stratified temperature field is essential.

In contrast to mixing ventilation, displacement ventilation is designed to minimize mixing of air within the occupied zone. The objective of the displacement ventilation is to create close to supply air conditions in the occupied zone. This type of ventilation was originally used in industrial buildings as an effective method for removing contaminants in the occupied zone. It is now also used for ventilating and cooling office buildings. However, local discomfort due to draft and vertical temperature gradient may be critical (Melikov and Nielsen 1989). Sandberg and Blomqvist (1989) suggest that the maximum convective cooling load in office buildings with displacement ventilation not exceed about 8 Btu/ft$^2 \cdot$h so that the maximum vertical temperature gradient in the occupied zone will not be larger than 5°F. This is equivalent to 1 cfm/ft^2 at a maximum cooling differential of 7.5°F. Kegel (1989) and Svensson (1989) suggest somewhat higher cooling load limits of 10 to 13 Btu/ft$^2 \cdot$h.

One way of increasing the cooling capacity of displacement ventilation systems is to recirculate some of the room air in the occupied zone through an induction circuit, *i.e.*, the room air is induced into the supply air and is mixed before discharge through the low-velocity air terminal device into the room. This reduces the room air temperature gradient for a given cooling load, thus allowing a higher limit of up to 16 Btu/ft$^2 \cdot$h (Jackman 1991).

Air diffusers with a large outlet area are used to supply air with low velocities. Displacement ventilation has been compared with conventional mixing-type ventilation (Svensson 1989, Seppanen *et al.* 1989, Stymne 1991). Design guidelines for displacement ventilation can be found in Scaret (1985), Jackman (1990, 1991), and Shilkrot and Zhivov (1992).

Unidirectional Airflow Ventilation

Air is either (1) supplied from the ceiling and exhausted through the floor or vice versa, or (2) supplied through the wall and exhausted through returns at the opposite wall. The outlets are uniformly distributed over the ceiling, floor, or wall to provide a low turbulent "plug"-type flow across the entire room. This type of system is mainly used for ventilating clean rooms, in which the main objective is to remove contaminant particles within the room. Details about clean room ventilation are described by ASHRAE (1991). Unidirectional flow ventilation is also used in certain areas, such as computer rooms, paint booths, etc.

Localized Ventilation

Air is supplied locally for occupied regions, such as desks in offices, seats in theaters and cinemas, or working places in industrial buildings. Conditioned air is supplied towards the breathing zone of the occupants to create comfortable conditions and/or to reduce the concentration of pollutants. Several special air diffusers are available. Figure 7 shows one arrangement, with diffusers placed on the desks in front of the occupants. In theaters and cinemas, air can be supplied through the grille or perforated panels in the back (Figure 8A), through the legs (Figures 8B and D) of the theater chairs, or through air diffusers installed in the floor under the chairs (Figure 8C). A scheme of air distribution through the back of the chair is presented on Figure 8E (Scheunemann 1989, Rowlinson and Croom 1987).

METHODS OF EVALUATION

Standards for Satisfactory Conditions

The object of air diffusion in warm air heating, ventilating, and air-conditioning systems is to create the proper combination of temperature, humidity, and air motion in the occupied zone of the conditioned room—from the floor to 6 ft above floor level (Miller 1989). To obtain comfort conditions within this zone, standard limits have been established as acceptable effective draft temperature, which combines the effects of air temperature, air motion, and relative humidity in terms of their physiological effects on a human body. Variation from accepted standards causes occupant discomfort. Lack of uniform conditions within the space or excessive fluctuation of conditions in the same part of the space also produce discomfort. Such discomfort can arise because of (1)

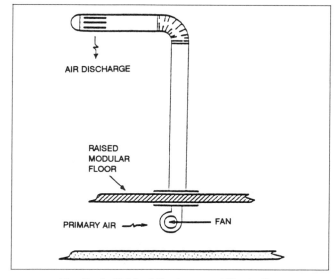

Fig. 7 Individual Environment Controller

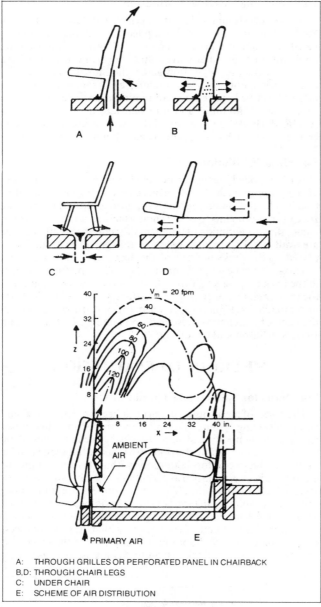

Fig. 8   Local Air Supply in Theaters

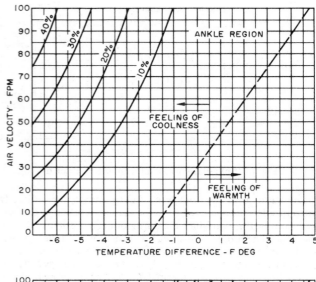

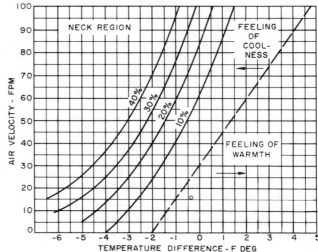

Fig. 9   Percentage of Occupants Objecting to Drafts in
Air-Conditioned Rooms

$$\theta = t_x - t_c - 0.07\,(V_x - 30) \qquad (1)$$

where

t_o = local airstream dry-bulb temperature, °F
t_c = average room dry-bulb temperature, °F
V_x = local airstream velocity, fpm

Equation (1) accounts for the feeling of coolness produced by air motion and is used to establish the neutral line in Figure 9. In summer, the local airstream temperature t_x is below the control temperature. Hence, both temperature and velocity terms are negative when velocity V_x is greater than 30 fpm and they both add to the feeling of coolness. If, in winter, t_x is above the control temperature, any air velocity above 30 fpm subtracts from the feeling of warmth produced by t_x. Therefore, it is usually possible to have zero difference in effective temperature between location, x, and the control point in winter, but not in summer.

Houghten *et al.* (1938) presented data that make it possible to statistically interpret the percentage of room occupants that will object to a given draft condition. Figure 6 presents the data in the form used by Koestel and Tuve (1955). The data show that a person tolerates higher velocities and lower temperatures at ankle

excessive room air temperature variations (horizontally, vertically, or both), (2) excessive air motion (draft), (3) failure to deliver or distribute air according to the load requirements at different locations,(4) or overly rapid fluctuation of room temperature.

Koestel and Tuve (1955) and Reinmann *et al.* (1959) studied the effect of air motion on comfort and defined draft as any localized feeling of coolness or warmth of any portion of the body, due to both air movement and air temperature, with humidity and radiation considered constant. The warmth or coolness of a draft was measured above or below a controlled room condition of 76°F dry-bulb at the center of the room, 30 in. above the floor, with air moving at about 30 fpm.

To define the difference in effective draft temperature θ between any point in the occupied zone and the control condition, the investigators used the following equation proposed by Rydberg and Norback (1949) and modified by Straub in discussion of a paper by Koestel and Tuve (1955):

A: THROUGH GRILLES OR PERFORATED PANEL IN CHAIRBACK
B,D: THROUGH CHAIR LEGS
C: UNDER CHAIR
E: SCHEME OF AIR DISTRIBUTION

level than at neck level. Because of this, conditions in the zone extending from approximately 30 to 60 in. above the floor are more critical than conditions nearer the floor.

Room air velocities less than 50 fpm are generally preferred; however, Figure 6 shows that even higher velocities may be acceptable to some occupants. ASHRAE *Standard* 55-1992 recommends elevated air speeds at elevated air temperatures. No minimum air speeds are recommended for comfort, although air speeds below 20 fpm are usually imperceptible.

Figure 6 also shows that up to 20% of occupants will not accept an ankle-to-sitting-level gradient of about 4 °F. Poorly designed or operated systems in a heating mode can create this condition, which emphasizes the importance of proper selection and operation of perimeter systems. The section Typical Outlets and Their Performance describes possible regions of high room air velocities caused by various outlets; the section Outlet Location and Selection describes how to evaluate acceptable air diffusion.

Air Diffusion Performance Index (ADPI)

Comfort criteria. A high percentage of people are comfortable in sedentary (office) occupations where the effective draft temperature θ, as defined in Equation (1), is between -3 and 2 °F and the air velocity is less than 70 fpm. If several measurements of air velocity and air temperature were made throughout the occupied zone of an office, the ADPI would be defined as the percentage of locations where measurements that meet the previous specifications on effective draft temperature and air velocity were taken. If the ADPI is maximum (approaching 100%), the most desirable conditions are achieved (Miller and Nevins 1969, 1970, 1972, 1974; Miller 1971; Miller and Nash 1971; Nevins and Ward 1968; Nevins and Miller 1972).

The ADPI is based only on air velocity and effective draft temperature, a combination of local temperature differences from the room average, and is not directly related to the level of dry-bulb temperature or relative humidity. These and similar effects, such as mean radiant temperature, must be accounted for separately according to ASHRAE recommendations.

The ADPI is a measure of cooling mode conditions. Heating conditions can be evaluated using ASHRAE *Standard* 55-1992 guidelines or the ISO *Standard* 7730-84.

The following cooling zone design criteria for various air diffusion devices maximize the ADPI and comfort. These criteria also account for airflow rate, outlet size, manufacturer's design qualities, and dimensions of the room for which the system is designed.

Jet Throw

The throw of a jet is the distance from the diffuser to a point where the maximum velocity in the stream cross section has been reduced to a selected terminal velocity. For all diffusers, the terminal velocity V_T was selected as 50 fpm, except in the case of ceiling slot diffusers, where the terminal velocity was selected as 100 fpm. Each manufacturer gives data for the throw of a jet from various diffusers for isothermal conditions and without a boundary wall interfering with the jet. Throw data certified under ASHRAE *Standard* 70-72 must be taken under isothermal conditions; throw data not certified by this standard may be isothermal or not, as the manufacturer chooses.

The throw distance of a jet is denoted by T_V, where subscript V indicates the terminal velocity for which the throw is given. The characteristic room length is the distance from the diffuser to the nearest boundary wall in the principle horizontal direction of the airflow. However, where air injected into the room does not impinge on a wall surface but mixes with air from a neighboring diffuser, the characteristic length L is one-half the distance between diffusers, plus the distance the mixed jet travels downward to reach the occupied zone. Table 1 summarizes definitions of characteristic length for various diffusers.

Table 1 Characteristic Room Length for Several Diffusers

Diffuser Type	Characteristic Length L
High sidewall grille	Distance to wall perpendicular to jet
Circular ceiling diffuser	Distance to closest wall or intersecting air jet
Sill grille	Length of room in direction of jet flow
Ceiling slot diffuser	Distance to wall or midplane between outlets
Light troffer diffusers	Distance to midplane between outlets, plus distance from ceiling to top of occupied zone
Perforated, louvered ceiling diffusers	Distance to wall or midplane between outlets

The midplane between diffusers also can be considered the module line when diffusers serve equal modules throughout a space, and a characteristic length consideration can be based on module dimensions d.

Load Considerations

These recommendations cover cooling loads of up to 80 Btu/h·ft² of floor surface. The loading is distributed uniformly over the floor up to about 7 Btu/h·ft², lighting contributes about 10 Btu/h·ft², and the remainder is supplied by a concentrated load against one wall that simulates a business machine or a large sun-loaded window. Over this range of data, the maximum ADPI condition is lower for the highest loads; however, the optimum design condition changes only slightly with the load.

Design Conditions

The quantity of air must be known from other design specifications. If it is not known, the solution must be obtained by trial and error.

The devices for which data were obtained are (1) high sidewall grilles, (2) sill grilles, (3) two- and four-slot ceiling diffusers, (4) cone-type circular ceiling diffusers, (5) light troffer diffusers, and (6) square-faced perforated and louvered ceiling diffusers. Table 2 summarizes the results of the recommendations on values of

Table 2 Air Diffusion Performance Index (ADPI) Selection Guide

Terminal Device	Room Load, Btu/(h·ft²)	T_{50}/L for Max. ADPI	Maximum ADPI	For ADPI Greater Than	Range of $T_{0.25}/L$
High sidewall grilles	80	1.8	68	—	—
	60	1.8	72	70	1.5–2.2
	40	1.6	78	70	1.2–2.3
	20	1.5	85	80	1.0–1.9
Circular ceiling diffusers	80	0.8	76	70	0.7–1.3
	60	0.8	83	80	0.7–1.2
	40	0.8	88	80	0.5–1.5
	20	0.8	93	90	0.7–1.3
Sill grille straight vanes	80	1.7	61	60	1.5–1.7
	60	1.7	72	70	1.4–1.7
	40	1.3	86	80	1.2–1.8
	20	0.9	95	90	0.8–1.3
Sill grille spread vanes	80	0.7	94	90	0.8–1.5
	60	0.7	94	80	0.6–1.7
	40	0.7	94	—	—
	20	0.7	94	—	—
Ceiling slot diffusers (for T_{100}/L)	80	0.3*	85	80	0.3–0.7
	60	0.3*	88	80	0.3–0.8
	40	0.3*	91	80	0.3–1.1
	20	0.3*	92	80	0.3–1.5
Light troffer diffusers	60	2.5	86	80	<3.8
	40	1.0	92	90	<3.0
	20	1.0	95	90	<4.5
Perforated and louvered ceiling diffusers	11–51	2.0	96	90	1.4–2.7
				80	1.0–3.4

T_V/L by giving the value of T_V/L where the ADPI is a maximum for various loads, as well as a range of values of T_V/L where ADPI is above a minimum specified value.

PRINCIPLES OF ROOM AIR DISTRIBUTION

Air Jets

As a rule, air supplied into rooms through the various types of outlets (e.g., grille-like, ceiling diffusers, perforated panels) is distributed by turbulent air jets. These air jets are the primary factor affecting room air motion; for further information on the relationship between the air jet and the occupied zone, see Baturin (1972), Christianson (1989), and Murakami (1992). If the air jet is not obstructed by walls, ceiling, or other obstructions, it is considered a *free jet*. If the air jet is attached to a surface, it is an *attached air jet*.

Characteristics of the air jet in a room might be influenced by reverse flows, created by the same jet entraining the ambient air. This air jet is called a *confined jet*. If the temperature of the supplied air is equal to the temperature of the ambient room air, the air jet is called an *isothermal jet*. A jet with an initial temperature different from the temperature of the ambient air is called a *nonisothermal jet*. The air temperature differential between supplied and ambient room air generates thermal forces in jets, affecting (1) the trajectory of the jet, (2) the location at which the jet attaches and separates from the ceiling/floor, and (3) the throw of the jet. The significance of such effects depends on the ratio between the thermal buoyancy and inertial forces (characterized by the Archimedes number Ar).

Depending on the diffuser type, air jets can be classified as follows:

- *Compact* air jets are formed by cylindrical tubes, nozzles, square or rectangular openings with small aspect ratio, unshaded or shaded by perforated plates, grilles etc. Compact air jets are three-dimensional and axisymmetric at least on some distance from the diffuser opening. The maximum velocity in the cross section of the compact jet is on the axis.
- *Linear* air jets are formed by slots or rectangular openings with a large aspect ratio. The jet flows are approximately two-dimensional. Air velocity is symmetric in the plane at which air velocities in the cross section are maximum. At some distance from the diffuser, linear air jets tend to transform into compact jets.
- *Radial* air jets are formed when the ceiling cylindrical air diffusers with flat discs or multidiffusers direct the air horizontally in all directions.
- *Conical* air jets are formed by cone-type or regulated multidiffuser ceiling-mounted air distribution devices. They have an axis of symmetry. The air flows parallel to the conical surface (the angle at the top of the cone is 120°) with the maximum velocities in the cross section perpendicular to the axis.
- *Incomplete radial* jets are formed by outlets with grilles having diverging vanes and a forced angle of expansion. At a distance, this jet tends to transform into a compact one.
- *Swirling* jets are forced by diffusers with vortex-forming devices. These devices create rotation, which has besides the axial component of velocity vectors, tangential and radial ones. Depending on the type of air diffuser, swirling jets can be compact, conical, or radial.

Isothermal Free Jet

The shape of jets at a short distance from the outlet face is very similar whether the outlet is round, rectangular, grille-like, or a perforated panel. The jet discharged from a round opening forms an expanding cone; jets from rectangular outlets rapidly pass from a rectangular to an elliptical cross-sectional shape and then to a circular shape, at a rate depending primarily on the aspect ratio and jet width. Even for wide-angle grilles and annular outlets, the similarities permit the same performance analysis for both.

For many conditions of jet discharge, it is possible to analyze jet performance and determine (1) the angle of divergence of the jet boundary, (2) the velocity patterns along the jet axis, (3) the velocity profile at any cross section in the zone of maximum engineering importance, and (4) the entrainment ratios in the same zone (Tuve 1953).

Using the data in this section, the following must be considered:

1. Since the method of finding the jet velocities is based on several approximations, the two recommended equations must be used cautiously for extreme axial and radial distances.
2. The characteristics of the low-velocity regions of ventilating jets are not well understood, and the effects at various Reynolds numbers are not fully known for axial or radial jets.
3. The quantitative treatment of the forces governing room air diffusion problems is limited, and nonisothermal conditions involving buoyant forces are more difficult to predict.
4. Most investigations have addressed free jets, whereas airstreams in practical room air diffusion are not free streams but are influenced by walls, ceilings, floors, and other obstructions.

Angle of Divergence

The angle of divergence is well-defined near the outlet face, but the boundary contours are billowy and easily affected by external influences. Here, as in the room, air movement has local eddies, vortices, and surges. The internal forces governing this air motion are extremely delicate (Nottage *et al.* 1952).

Measured angles of divergence (spread) for discharge into large open spaces usually range from 20 to 24° with an average of 22°. Coalescing jets for closely spaced multiple outlets expand at smaller angles, averaging 18°, and jets discharging into relatively small spaces show even smaller angles of expansion (McElroy 1943). In cases where the outlet area is small compared to the dimensions of the space normal to the jet, the jet may be considered free as long as

$$X \leqslant 1.5\sqrt{A_R} \tag{2}$$

where

X = distance from face of outlet, ft
A_R = cross-sectional area of confined space normal to jet, ft^2

Jet Expansion Zones

The full length of an air jet (compact, radial, linear, or conical), in terms of the maximum or centerline velocity and temperature differential at the cross section, can be divided into four zones:

Zone 1. A core zone; a short zone, extending about four diameters or widths from the outlet face, in which the maximum velocity (temperature) of the airstream remains practically unchanged.
Zone 2. A transition zone, the length of which depends on the type of diffuser, diffuser's aspect ratio, initial airflow turbulence, and so forth.
Zone 3. A zone of fully established turbulent flow, which may be 25 to 100 equivalent air diffuser diameters (widths for slot-type air diffusers).
Zone 4. A zone of diffuser jet degradation, where the maximum air velocity and temperature decreases rapidly. The distance to this zone and its length depends on the velocities and turbulence characteristics of the ambient air. In a few diameters, the air velocity becomes less than 50 fpm. Although this zone was studied by several researchers (Madison *et al.* 1946, Weinhold 1969), its characteristics are still not well understood.

Zone 3 is of major engineering importance because, in most cases, the diffuser jet enters the occupied area within this zone.

Centerline Velocities in Zones 1 and 2

In Zone 1, the ratio V_x/V_o is constant and equal to the ratio of the center velocity of the jet at the start of expansion to the average velocity. The ratio V_x/V_o varies from approximately 1.0 for rounded entrance nozzles to about 1.2 for straight pipe discharges; it has much higher values for diverging discharge outlets. Experimental evidence indicates that in Zone 2

$$V_x/V_o = \sqrt{K'H_o/X} \tag{3}$$

where

V_x = centerline velocity, fpm
V_o = $V_c/(C_d R_{fa})$ = average initial velocity at discharge from open-ended duct or across contracted stream at vena contracta of orifice or multiple-opening outlet, fpm
V_c = nominal velocity of discharge based on core area, fpm
C_d = coefficient of discharge (usually between 0.65 and 0.90)
R_{fa} = ratio of free area to gross (core) area
H_o = width of jet at outlet or at vena contracta, ft
K' = centerline velocity constant depending on outlet type and discharge pattern (see Table 3)
X = distance to centerline velocity V_x, ft

Aspect ratio (Tuve 1953) and turbulence (Nottage *et al.* 1952) primarily affect the centerline velocities in Zones 1 and 2. Aspect ratio has little effect on the terminal zone of the jet when H_o is greater than 4 in. This is particularly true of nonisothermal jets. When H_o is very small, it is possible for the induced air to penetrate the core of the jet and, thus, reduce the centerline velocities. The difference in performance between the radial-type outlet with a small H_o and the axial-type outlet with a large H_o shows the importance of the thickness of the jets.

When air is discharged from relatively large perforated panels, the constant velocity core formed by the coalescence of the individual jets extends a considerable distance from the panel face. In Zone 1, when the ratio (Distance from Panel)/$\sqrt{\text{Panel Area}}$ is less than 5, Equation (4)

$$V_x = 1.2 V_o \sqrt{C_d R_{fa}} \tag{4}$$

should be used for estimating centerline velocities (Koestel *et al.* 1949).

Centerline Velocity in Zone 3

In Zone 3, maximum or centerline velocities of straight flow isothermal jets can be determined with accuracy from the equations

$$\frac{V_x}{V_o} = \frac{K D_o}{X} = \frac{K'\sqrt{A_o}}{X} \tag{5}$$

$$V_x = \frac{K' V_o \sqrt{A_o}}{X} = \frac{K' Q}{X\sqrt{A_o}} \tag{6}$$

$$V_x = \frac{K' Q}{X\sqrt{A_c C_d R_{fa}}} \tag{7}$$

where

K = proportionality constant, with $K' = 1.13 K$
D_o = effective or equivalent diameter of stream at discharge from open-end duct or at contracted section, ft
A_o = $A_c C_d R_{fa}$ = effective area of stream at discharge from open-end duct or at contracted section, ft^2
A_c = measured gross (core) area of outlet, ft^2
Q = discharge from outlet, ft^3/min

Since A_o equals the effective area of the stream, the flow area for commercial registers and diffusers, according to ASHRAE *Standard* 70-72, can be used in Equation (5) with the appropriate value of K and K'.

Equation (5) is nondimensional and requires only that consistent units be used. Values of K and K' are listed in Table 4 (Tuve 1953, Koestel *et al.* 1950).

In multiple-opening outlets and annular ring outlets, the streams coalesce into a solid jet before actual jet expansion takes place. This coalescence affects the proportionality constants K or K' and accounts for some divergence in reported values for similar outlets.

For perforated panels of relatively large size, the values of K and K' given in Table 4 apply only when the ratio (Distance from Panel)/$\sqrt{\text{Panel Area}}$ is larger than 5 (see section Centerline Velocities in Zones 1 and 2).

Low-velocity test results, in the range $V_x < 150$ fpm, indicate that normal values of K and K' should be reduced about 20% for $V_x = 50$ fpm, as used later in Equation (10) for throw.

Table 3 Recommended Values of Centerline Velocity Constant K or K′ for Commercial Supply Outlets

Outlet Type	Discharge Pattern	Area A_o	K	K′
High sidewall grilles	0° deflection	Flow area	5.0	5.7
	Wide deflection	Flow area	3.7	4.2
High sidewall linear	Core less than 4 in. high	Flow area	3.9	4.4
	Core more than 4 in. high	Flow area	4.4	5.0
Low sidewall	Up and on wall, no spread	Free area	4.4	4.5
	Wide spread	Free area	2.6	3.0
Baseboard	Up and on wall, no spread	Free area	3.9	4.0
	Wide spread	Free area	1.8	2.0
Floor	No spread	Free area	4.1	4.7
	Wide spread	Free area	1.4	1.6
Ceiling circular directional	360° horizontal	Flow area	1.0	1.1
	4 way—little spread	Flow area	3.3	3.8
Ceiling linear	One way—horizontal along ceiling	Flow area	4.8	5.5

[a]These values are representative for the commercial type of outlet and discharge pattern given (Straub *et al.* 1956, 1957, Air Diffusion Council 1984).

Table 4 Recommended Values of Centerline Velocity Constant for Standard Openings

	K		K′	
	$V_o =$ 500 to	$V_o =$ 2000 to	$V_o =$ 500 to	$V_o =$ 2000 to
Type of Outlet	1000 fpm	10,000 fpm	1000 fpm	10,000 fpm
Free openings				
Round or square	5.0	6.2	5.7	7.0
Rectangular, large aspect ratio (<40)	4.3	5.3	4.9	6.0
Annular slots axial or radial[a]	—	—	3.9	4.8
Grilles and grids				
Free area 40% or more	4.1	5.0	4.7	5.7
Perforated panels				
Free area 3 to 5%	2.7	3.3	3.0	3.7
Free area 10 to 20%	3.5	4.3	4.0	4.9

[a]For radial slots, use X/H instead of $X/\sqrt{A}$. H is the height or width of the slot.
Note: K and K' are indexes of loss in axial kinetic energy. Interpolate as required. Departures from maximum value indicate losses in first and second zones when compared with the jet from a rounded-entrance, circular nozzle.

Determining Centerline Velocities

To correlate data from all four zones, centerline velocity ratios are plotted against distance from the outlet in Figure 9 according to Equation (3). A nomogram for calculating the parameters x/A_o and V_x/V_c from $X/\sqrt{A_c}$ and V_x/V_o through R_{fa} and C_d is given in the same illustration.

The variation of the centerline velocity ratio with distance from the outlet or, more properly, from start of jet expansion for Zones 1 and 2 is also shown in Figure 9. V_x/V_o is plotted against X/H_o and, for a range of aspect ratios, against $X/\sqrt{A}$ for the single value of $K' = 7.0$. Values of V_x/V_o for other values of K' can be obtained by direct proportioning of $\sqrt{K'}$ to $\sqrt{7.0}$.

The following example, which is solved in Figure 10, illustrates the use of Figure 9.

Example 1. A grille has a core area of 12 in. by 18.75 in., $R_{fa} = 0.90$, $C_d = 0.80$, and $K' = 5.0$. Find V_c (velocity through core area) when V_x is 50 fpm for a throw of 50 ft ($X = 50$).

Solution:

$$A_c = (12 \times 18.75)/144 = 1.56 \text{ ft}^2$$
$$X/\sqrt{A_c} = 50/1.25 = 40$$
$$A_o = 1.56 \times 0.80 \times 0.90 = 1.123 \text{ ft}^2$$
$$X/A_o = 50/1.06 = 47.2$$
$$V_x/V_o = K'\sqrt{A_o}/X = 5\sqrt{1.123}/50 = 0.106$$
$$V_x/V_c = V_x/V_o(C_d R_{fa}) = 0.106/(0.80 \times 0.90) = 0.147$$

For $V_x = 50$ fpm, $V_c = 50/0.147 = 340$ fpm
The quantity of air discharged is then

$$Q = V_c A_c = 340 \times 1.56 = 530 \text{ cfm}$$

Throw

Equation (8) can be transposed to determine the throw X of an outlet, if the discharge volume and the center velocity are known.

$$X = \frac{K'}{V_x} \frac{Q}{\sqrt{A_c C_d R_{fa}}} \tag{8}$$

Or, if $Z = \sqrt{C_d R_{fa}}$, then

$$X = \frac{K'}{V_x} \frac{Q}{Z\sqrt{A_c}} \tag{9}$$

The maximum throw L is usually defined as the distance from the outlet face where the centerline velocity is 50 fpm. Therefore, for $V_x = 50$ fpm

$$L = X = \frac{K'}{50} \frac{Q}{Z\sqrt{A_c}} \tag{10}$$

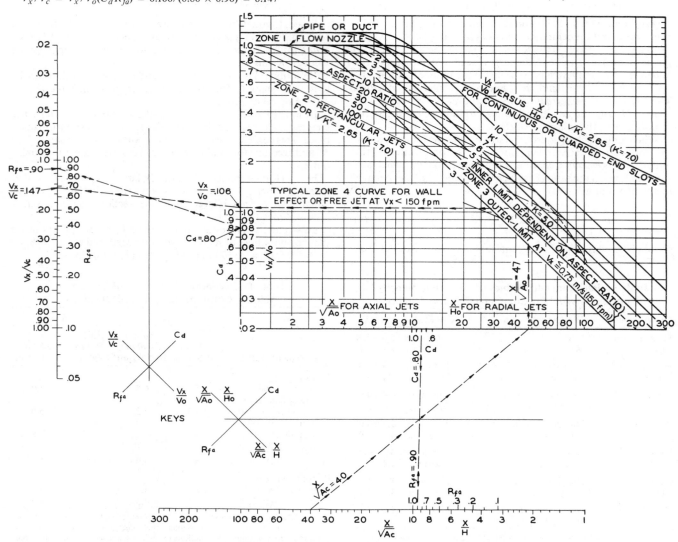

Fig. 10   Chart for Determining Centerline Velocities of Axial and Radial Jets

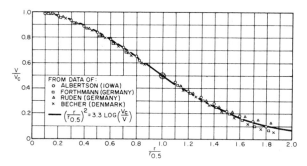

Fig. 11 Cross-Sectional Velocity Profiles for Straight-Flow Turbulent Jets

Any other terminal centerline velocity could be inserted in Equation (10) for V_x.

Velocity Profiles of Jets

In Zone 3 of both axial and radial jets, the velocity distribution may be expressed by a single curve (Figure 11) in terms of dimensionless coordinates; this same curve can be used as a good approximation for adjacent portions of Zones 2 and 4. Temperature and density differences have little effect on cross-sectional velocity profiles.

Velocity distribution in Zone 3 can be expressed by the Gauss error function or probability curve, which is approximated by Equation (11)

$$\left(\frac{r}{r_{0.5V}}\right)^2 = 3.3 \log \frac{V_x}{V} \tag{11}$$

where

r = radial distance of point under consideration from centerline of jet
$r_{0.5V}$ = radial distance in same cross-sectional plane from axis to point where velocity is one-half centerline velocity, *i.e.*, $V = 0.5\,V_x$
V_x = centerline velocity in same cross-sectional plane
V = actual velocity at point being considered

Experiments show that the conical angle for $r_{0.5V}$ is approximately one-half the total angle of divergence of a jet. The velocity profile curve for one-half a straight-flow turbulent jet (the other half being a symmetrical duplicate) is shown in Figure 11. For multiple-opening outlets, such as grilles or perforated panels, the velocity profiles are similar, but the angles of divergence are smaller.

Entrainment Ratios

Equations for the entrainment of circular jets and of jets from long slots follow.

For third-zone expansion of circular jets

$$\frac{Q_x}{Q_o} = \frac{2\,X}{K'\sqrt{A_o}} \tag{12}$$

By substituting from Equation (5)

$$\frac{Q_x}{Q_o} = 2\,\frac{V_o}{V_x} \tag{13}$$

For a long slot

$$\frac{Q_x}{Q_o} = \frac{\sqrt{2}}{K'}\sqrt{\frac{X}{H_o}} \tag{14}$$

or

$$\frac{Q_x}{Q_o} = \sqrt{2}\,\frac{V_o}{V_x} \tag{15}$$

where

Q_x = total volume flow rate at distance X from face of outlet, ft³/min
Q_o = discharge from outlet, ft³/min
X = distance from face of outlet, ft
K' = proportionality constant
A_o = effective area of stream at discharge from open-end duct or at contracted section, ft²
H_o = width of slot, ft

Entrainment ratios are important to determine total air movement at a given distance from an outlet. With a given outlet, the entrainment ratio is proportional to the distance X [Equation (12)] or proportional to the square root of the distance X [Equation (14)] from the outlet. Equations (13) and (15) show that, for a fixed centerline velocity, the entrainment ratio is proportional to the outlet velocity. Equations (13) and (15) also show that, at the same centerline and outlet velocity, a circular jet has greater entrainment and total air movement than a long slot. Comparing Equations (12) and (14), the long slot should have a greater rate of induction. The entrainment ratio at a given distance is less with a large K' than with a small K'.

Isothermal Radial Flow Jets

In a radial jet, as with an axial jet, the cross-sectional area at any distance from the outlet varies as the square of this distance. Centerline velocity gradients and cross-sectional velocity profiles are similar to those of Zone 3 of axial jets, and the angles of divergence are about the same.

Jets from ceiling plaques have the same form as one-half a free radial jet. The jet is wider and longer than a free jet, with the maximum velocity close to the surface. Koestel (1957) provides an equation for radial flow outlets.

Nonisothermal Free Jet

When the temperature of introduced air is different from the room air temperature, the behavior of the diffuser air jet is affected by the thermal buoyancy due to air density difference. The trajectory of a nonisothermal jet introduced horizontally is determined by the Archimedes number (Baturin 1972):

$$\text{Ar} = \frac{gL_o\,(t_o - t_s)}{V_o^2\,T_s} \tag{16}$$

where

g = gravitational acceleration rate, ft/min²
L_o = length scale of diffuser outlet equal to hydraulic diameter of outlet, ft
t_o = initial temperature of jet, °F
t_s = temperature of surrounding air, °F
V_s = initial air velocity of jet, fpm
T_s = room air temperature, K

The paths assumed by horizontally projected heated and chilled jets influenced by buoyant forces are significant in heating and cooling with wall outlets. Koestel's equation (1955) describes the behavior of these jets.

Helander *et al.* (1948, 1953, 1954, 1957), Yen *et al.* (1956), and Knaak (1957) developed equations for outlet characteristics that affect the downthrow of heated air. Koestel (1954, 1955) developed equations for temperatures and velocities in heated and chilled jets.

Surface Jets (Wall and Ceiling)

Jets discharging parallel to a surface with one edge of the outlet coinciding with the surface take the form of one-half of an axial jet discharging from an outlet twice as large, similar to radial jets from ceiling plaques. Entrainment takes place almost only along the surface of a half cone, and the maximum velocity remains close to the surface (Tuve 1953).

Values of K and K' are approximately those for a free jet multiplied by $\sqrt{2}$; *i.e.*, the normal maximum of 7.0 for K' for free jets becomes 9.9 for a similar jet adjacent to and discharged parallel to a surface, and $X/\sqrt{2A_o}$ should replace $X\sqrt{A_o}$ in Figure 9.

When a jet is discharged parallel to but at some distance from a solid surface (wall, ceiling, or floor), its expansion in the direction of the surface is reduced, and entrained air must be obtained by recirculation from the jet instead of from ambient air (McElroy 1943, Nottage *et al.* 1952, Zhang *et al.* 1990). The restriction of the solid surface to the jet entrainment induces the Coanda effect, which makes the jet attach to a surface after it leaves the diffuser outlet for a short distance. The jet then remains attached to the surface for some distance before separating from the surface again.

In nonisothermal cases, the trajectory of the jet is determined by the balance between the thermal buoyancy and the Coanda effect which depends on the jet momentum and the distance between the jet exit and the solid surface. The behavior of such nonisothermal surface jets have been studied by Kirkpatrick *et al.* (1991), Wilson *et al.* (1970), Oakes (1987), and Zhang *et al.* (1990), each addressing different factors. A more systematic study of these types of jets in room ventilation flows is needed to provide reliable guidelines for designing air diffusion systems.

Multiple Jets

Twin parallel air jets act independently until they interfere. The point of interference and its distance from the outlets varies with the distance between the outlets. From the outlets to the point of interference, the maximum velocity, as for a single jet, is on the centerline of each jet. Then, the velocity on a line midway between and parallel to the two jet centerlines increases until it equals the jet centerline velocity. From this point, a maximum velocity of the combined jet stream is on the midway line, and the profile seems to emanate from a single outlet of twice the area of one of the two outlets.

Koestel and Austin (1956) determined the spacing between outlets for noninterference between the jets. For a K value of 6.5, the outlets should be placed three to eight diameters apart, with V_o values from 500 to 1500 fpm.

Airflow in Occupied Zone in Mixing-Type Systems

Laboratory experiments on jets usually involve recirculated air with negligible resistance to flow on the return path of the jet air. Experiments in mine tunnels of small cross-sectional areas, where considerable resistance meets the return flow of jet air to outlets, show that expansion of the jet terminates abruptly at a distance independent of velocity of discharge and is only slightly affected by the size of the outlet. These distances are determined primarily by size and length of the return path. In a long 5 ft by 6 ft tunnel, a jet may not travel more than 25 ft; with a relatively large section (25 ft by 60 ft), it may travel more than 250 ft. McElroy (1943) provides data on this phase of jet expansion.

Zhang *et al.* (1990) found that air velocity in the occupied zone increased with the diffuser air velocity for the same ventilation rate. Therefore, the supply air velocity should be designed high enough to maintain the jet traveling in the desired direction to ensure good mixing before it reaches the occupied zone. Excessively high outlet air velocity would induce high air velocity in the occupied zone and result in thermal discomfort.

Turbulence Production and Transport

The turbulence within a room is mainly produced at the diffuser jet region by interaction of the supply air with the room air and with the solid surfaces (walls or ceiling) in the vicinity. It is then transported to other parts of the room, including the occupied zone (Zhang *et al.* 1991). Meanwhile, the turbulence is also damped by viscous effect. Air in the occupied zone usually contains very small amounts of turbulent kinetic energy as compared to that in the jet region. Since turbulence may cause thermal discomfort (Fanger *et al.* 1989), air diffusion systems should be designed so that the primary mixing between the introduced air and the room air occurs away from occupied regions.

SYSTEM FACTORS

Noise

The noise generated by diffusers transmits to the occupied space directly and cannot be attenuated. Therefore, the diffusion system design should meet the criteria of sound level as specified in Chapter 42 of the 1991 ASHRAE *Handbook—Applications*.

Duct Arrangements

Duct approaches to diffuser outlets. The manner in which the airstream approaches the diffuser outlet is important. To obtain correct air diffusion, the velocity of the airstream must be as uniform as possible over the entire connection to the duct and must be perpendicular to the outlet face. Effects of improper duct approach generally cannot be corrected by the diffuser.

If the system is designed carefully, a wall grille installed at the end of a horizontal duct and a ceiling outlet at the end of a vertical duct receive the air perpendicularly and at uniform velocity over the entire duct cross section. However, few outlets are installed in this way. Most sidewall outlets are installed either at the end of vertical ducts or in the side of horizontal ducts, and most ceiling outlets are attached either directly to the bottom of horizontal ducts or to special vertical takeoff ducts that connect the outlet with the horizontal duct. In all these cases, special devices for directing and equalizing the airflow are necessary for proper direction and diffusion of the air.

The influence of the duct approach on outlet performance has been investigated for vertical stack heads with plain openings (Nelson *et al.* 1940) or equipped with grilles (Nelson *et al.* 1942) and side outlets on horizontal ducts (Nelson and Smedberg 1943). In tests conducted with the stack heads, splitters or guide vanes in the elbows at the top of the vertical stacks are needed, regardless of the shape of the elbows (whether rounded, square, or expanding types). Cushion chambers at the top of the stack heads are not beneficial. Figure 12 shows the direction of flow, diffusion, and velocity (measured 12 in. from the opening) of the air for various stack heads tested, expanding from a 14 in. by 6 in. stack to a 14 in. by 9 in. opening, without a grille. The air velocity for each was 500 fpm in the stack below the elbow, but the direction of flow and the diffusion pattern indicate performance obtained with nonexpanding elbows of similar shapes for velocities from 200 to 400 fpm.

In tests conducted with 3 in. by 10 in., 4 in. by 9 in., and 6 in. by 6 in. side outlets in a 6-in. by 20-in. horizontal duct at duct velocities of 200 to 1400 fpm in the horizontal duct section, multiple curved deflectors produced the best flow characteristics. Vertical guide strips in the outlet were not as effective as curved deflectors. A single scoop-type deflector at the outlet did not improve the flow pattern obtained from a plain outlet and, therefore, was not desirable.

Return and exhaust openings. The selection of return and exhaust openings depends on (1) velocity in the occupied zone near the openings, (2) permissible pressure drop through the openings, and (3) noise.

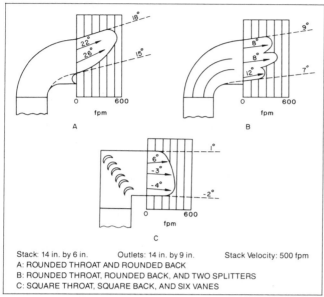

Stack: 14 in. by 6 in.   Outlets: 14 in. by 9 in.   Stack Velocity: 500 fpm
A: ROUNDED THROAT AND ROUNDED BACK
B: ROUNDED THROAT, ROUNDED BACK, AND TWO SPLITTERS
C: SQUARE THROAT, SQUARE BACK, AND SIX VANES

Fig. 12   Outlet Velocity and Air Direction Diagrams for Stack Heads with Expanding Outlets

Velocity. Airflow patterns as well as room air movement are not influenced by the location of the return and exhaust outlets at a distance more than one characteristic length of the return or exhaust opening (*e.g.*, square root of the opening area). Air handled by the opening approaches the opening from all directions, and its velocity decreases rapidly as the distance from the opening increases. Therefore, drafty conditions rarely occur near return openings. Table 4 shows recommended return opening face velocities.

Permissible pressure drop. Permissible pressure drop depends on the choice of the designer. Proper pressure drop allowances should be made for control or directive devices.

Noise. The problem of return opening noise is the same as that for supply outlets. In computing resultant room noise levels from the operation of an air-conditioning system, the return opening must be included as part of the total grille area.

Return and Exhaust Opening Location

The openings should be located to minimize short-circuiting of supply air. If air is supplied by the jets attached to the ceiling, exhaust openings should be located between the jets or at the other side of the room away from the supply air jets. In rooms with temperature stratification along its height, such as in foundries, computer rooms, theaters, bars, kitchens, dining rooms, and club rooms, exhaust openings should be located near the ceiling to collect warm air, odors, and fumes.

For industrial rooms with gas release, selection of exhaust opening locations depends on the specific weight of the released gases and their temperature; the locations should be specified for each application.

Exhaust outlets located in walls and doors, depending on their elevation, have the characteristics of either floor or ceiling returns. In large buildings with many small rooms, return air may be brought through door grilles or door undercuts into the corridors and then to a common return or exhaust. The pressure drop through door returns should not be excessive, or the air diffusion to the room may be seriously unbalanced by opening or closing the doors. Outward leakage through doors or windows cannot be counted on for dependable results.

System Balancing

In designing a system, ducts and diffusers should be sized so that the supply of air is distributed properly. However, for flexibility, use standard sizes and allow for future redistribution; the system, as designed, may not be self-balancing. Chapter 34 of the 1991 ASHRAE *Handbook—Applications* describes the procedures used to balance air distribution systems.

SYSTEM DESIGN

Design Procedure

1. Determine the air volume requirements and room size.
2. Select the tentative diffuser type and location within room.
3. Determine the room's characteristic length L (Table 1).
4. Select the recommended T_V/L ratio from Table 2.
5. Calculate the throw distance T_V by multiplying the recommended T_V/L ratio from Table 2 by the room length L.
6. Locate the appropriate outlet size from the manufacturer's catalog.
7. Ensure that this outlet meets other imposed specifications, such as noise and static pressure.

Example 2.
Specifications:

Room size	20 ft by 12 ft with 9-ft ceiling
Type device	High sidewall grille, located at the center of 12-ft endwall, 9 in. from ceiling
Loading	Uniform, 10 Btu/h·ft² or 2400 Btu/h
Air volume	1 cfm/ft² or 240 cfm for the one outlet

Data required:

Characteristic length	L = 20 ft (Length of room: Table 1)
Recommended	T_V/L = 1.5 (Table 2)
Throw to 50 fpm	T_{50} = 1.5 × 20 = 30 ft

Refer to the manufacturer's catalog for a size that gives this isothermal throw to 50 fpm. Manufacturer recommends the following sizes, when vanes are straight, discharging 240 cfm: 16 in. by 4 in., 12 in. by 5 in., or 10 in. by 6 in.

Outlet Location and Type Selection

No criteria have been established for choosing among the six types of outlets to obtain an optimum ADPI. All outlets tested, when used according to these recommendations, can have ADPI values that are satisfactory (greater than 90% for loads less than 40 Btu/h·ft²).

The design of an air distribution and air diffusion system is influenced by the same factors as those influencing the design of an air-conditioning plant—building's use, size, and construction type. Location and selection of the supply outlets is further influenced by the interior design of the building, local sources of heat gain or loss, and outlet performance and design.

Local sources of heat gain or loss promote convection currents or cause stratification; they may, therefore, determine both the type and location of the supply outlets. Outlets should be located to neutralize any undesirable convection currents set up by a concentrated load. If a concentrated heat source is located at the occupancy level of the room, the heating effect can be counteracted (1) by directing cool air toward the heat source or (2) by locating an exhaust or return grille adjacent to the heat source. The second method is more economical for cooling applications, since heat is withdrawn at its source rather than dissipated into the conditioned space. Where lighting loads are heavy (5 W/ft²) and ceilings relatively high (above 15 ft), outlets should be located below the lighting load, and the stratified warm air should be removed by an exhaust or return fan. An exhaust fan is recommended if the wet-bulb temperature of the air is above that of the outdoors; a return fan is recommended if the wet-bulb temperature is below this temperature. These methods reduce the requirements for supply air. Enclosed lights are more economical than exposed lights, since a considerable portion of the energy is radiant.

Based on the analysis of the outlet performance tests conducted by Straub *et al.* (1956, 1957), the following are selection considerations for outlet types in Groups A to E:

Group A outlets. Outlets mounted in or near the ceiling with horizontal air discharge should not be used with temperature differentials exceeding 25 °F during heating. Consequently, such outlets should be used for heating buildings located in regions where winter heating is only a minor problem and, in northern latitudes, solely for interior spaces. However, these outlets are particularly suited for cooling and can be used with high airflow rates and large temperature differentials. They are usually selected for their cooling characteristics.

The performance of these outlets is affected by various factors. Vane deflection settings reduce throw and drop by changing air from a single straight jet to a wide-spreading or fanned out jet. Accordingly, a sidewall outlet with 0° deflection has a longer throw and a greater drop than a ceiling diffuser with a single 360° angle of deflection. Sidewall grilles and similar outlets with other deflection settings may have performance characteristics between these two extremes.

Wide deflection settings also cause a ceiling effect, which increases the throw and decreases the drop. To prevent smudging, the total air should be directed away from the ceiling, but this rarely is practicable, except for very high ceilings. For optimum air diffusion in areas without high ceilings, total air should scrub the ceiling surface.

Drop increases and throw decreases with larger temperature differentials. For constant temperature differential, airflow rate affects drop more than velocity. Therefore, to avoid drop, several small outlets instead of one large outlet may be better in a room .

With the data in the section Isothermal Jets, the throw may be selected for a portion of the distance between the outlet and wall or, preferably, for the entire distance. For outlets in opposite walls, the throw should be one-half the distance between the walls. Following the above recommendations, the air drops before striking the opposite wall or the opposing airstream. To counteract specific sources of heat gain or to provide higher air motion in rooms with high ceilings, it may be necessary to select a longer throw. In no case should the drop exceed the distance from the outlet to the 6-ft level.

To maintain maximum ventilation effectiveness with ceiling diffusers, throws should be kept as long as possible. With VAV designs, some overthrow at maximum design volumes will be desirable—the highest induction can be maintained at reduced flows. Adequate induction by a ceiling-mounted diffuser prevents short-circuiting of unmixed supply air between supply outlet and ceiling-mounted returns.

Group B outlets. In selecting these outlets, it is important to provide enough throw to project the air high enough for proper cooling in the occupied zone. An increase of supply air velocity improves air diffusion during both heating and cooling. Also, during heating and cooling, a terminal velocity of about 150 fpm is found at the same distance from the floor. Therefore, outlets should be selected from the data given in the Isothermal Jets section, with throw based on a terminal velocity of 150 fpm.

With outlets installed near the exposed wall, the primary air is drawn toward the wall, resulting in a wall effect similar to the ceiling effect for ceiling outlets. This scrubbing of the wall increases heat gain or loss. To reduce scrubbing, outlets should be installed some distance from the wall, or the supply air should be deflected at an angle away from the wall. However, the distance should not be too large, nor the angle too wide, to prevent the air from dropping into the occupied zone before it reaches maximum projection. A distance of 6 in. and an angle of 15° is satisfactory.

These outlets do not counteract natural convection currents, unless sufficient outlets are installed around the perimeter of the space—preferably in locations of greatest heat gain or loss (under

windows). The effect of drapes and blinds must be considered with outlets installed near windows. If installed correctly, outlets of this type handle large airflow rates with uniform air motion and temperatures.

Group C outlets. These outlets can be used for heating, even with severe heat load conditions. Higher supply velocities produce better room air diffusion than lower velocities, but velocity is not critical in selecting these units for heating.

To achieve the required projection for cooling, the outlets should be used with temperature differentials of less than 15 °F. With higher temperature differentials, supply air velocity is not sufficient to project the total air up to the desired level.

The outlets have been used successfully for residential heating, but they may also offer a solution for applications where heating requirements are severe and cooling requirements are moderate. For throw, refer to the Isothermal Jets section.

Group D outlets. This type of outlet directs high-velocity total air into the occupied zone, and, therefore, is not recommended for comfort, particularly for summer cooling. If used for heating, outlet velocities should not be higher than 300 fpm, so that air velocities in the occupied zone will not be excessive. These outlets have been applied successfully to process installations where controlled air velocities are desired.

Group E outlets. The heating and cooling diagrams for these outlets show different throws that become critical in selecting and applying these outlets. Since the total air enters the occupied zone for both cooling and heating, outlets are used for either cooling or heating—seldom for both.

During cooling, temperature differential, supply air velocity, and airflow rate have considerable influence on projection. Therefore, low values of each should be selected.

During heating, it is important to select the correct supply air velocity to project the warm air into the occupied zone. Temperature differential is also critical, because a small temperature differential reduces variation of the throw during the cyclic operation of the supply air temperature. Vane setting for deflection is as important here as it is for Group B and C outlets.

Investigations by Nevins and Ward (1968) and Miller and Nevins (1969) in full-scale interior test rooms indicate that air temperatures and velocities throughout a room cooled by a ventilating ceiling are a linear function of room load (heat load per unit area) and are not affected significantly by variations in ceiling type, total air temperature differential, or air volume flow rate. Higher room loading produces wider room air temperature variations and higher velocities, which decrease performance.

These studies also found no appreciable difference in the performance of air diffusing ceilings and circular ceiling diffusers for lower room loads (20 Btu/h · ft²). For higher room loads (80 Btu/h · ft²), an air-diffusing ceiling system has only slightly larger vertical temperature variations and slightly lower room air velocities than a ceiling diffuser system.

When the ventilating ceiling is used at exterior exposures, the additional load at the perimeter must be considered. During heating operation, the designer must provide for the cold wall effect, as with any ceiling supply diffusion system. The sound generated by the air supply device must also be considered in total system analysis to ensure that room sound levels do not exceed the design criteria.

Return Air Design for Optimum Performance

An HVAC system operating in the cooling mode performs best when generated heat is removed at its source rather than distributed throughout the conditioned space. Heat from solar and miscellaneous loads such as machinery and floor or desk-mounted lamps is difficult to remove at the source. However, return air flowing over ceiling-mounted lighting fixtures keeps most of that heat from being distributed into the conditioned space. Besides

Table 5 Recommended Return Inlet Face Velocities

Inlet Location	Velocity over Gross Area, fpm
Above occupied zone	800 up
Within occupied zone, not near seats	600 to 800
Within occupied zone, near seats	400 to 600
Door or wall louvers	200 to 300
Through undercut area of doors	200 to 300

increasing HVAC system efficiency, return air lighting fixtures improve light output and extend the lamp life. The manufacturers of fixtures, ceiling grids, and grilles give performance information (airflow rate, pressure drop, and heat removal rate) of their product. Ball *et al.* (1971) found that the heat removal performance of return air fixtures covers a narrow range.

With a suspended ceiling, low operating static pressure across the ceiling must be maintained. Failure to do so can result in return air being forced around the edges of the ceiling panels or, in some cases, through the ceiling panels. The result is often a soiled ceiling and a mechanical system that is choked for return air. To avoid this, the static pressure difference across the ceiling should be as low as possible. If necessary, slotted tees or grilles can be used with return air fixtures to obtain the specified pressure drop. A maximum pressure drop of 0.02 to 0.03 in. of water is acceptable under most conditions.

At the typical air supply rates found in office interior zone spaces (usually less than 1.5 cfm/ft^2) and with adequate induction at the supply diffusers, the location of the return diffuser has no effect on air patterns in the space. For most office spaces, it is only necessary that sufficient return outlets be provided to maintain inlet velocities within recommendations (see Table 5).

In spaces expected to operate in a cooling mode most of the time, returning the warmest air in the space can effectively reduce energy costs and increase circulation in the space. This is especially true in climates where economizer systems operate for long periods during the year. In very high ceiling spaces, with atria, skylights, or large vertical glass surfaces, and where the highest areas are unoccupied, air stratification may be used as an energy-saving measure, by locating returns near the occupied zone.

Recommendations for Ceiling-Mounted Air Diffuser Systems

For the best thermal comfort conditions and highest ventilation effectiveness in an occupied space (*i.e.*, office or retail store), the entire system performance of air diffusers should be considered. This is particularly true for open spaces, where airstreams from diffusers may interact with each other, and for perimeter spaces, where airstreams from diffusers interact with hot or cold perimeter walls. While throw data for an individual diffuser is used in system design, an air diffuser system should maintain a high quality of air diffusion in the occupied space. This is achieved with low temperature variation, good air mixing, and no objectionable drafts in the occupied space (typically 6 in. to 6 ft. above the floor).

Adequate ventilation requires that the selected diffusers effectively mix (by induction) the total air in the room with the supplied conditioned air, which is assumed to contain adequate ventilation air.

Interior spaces. An interior space is conditioned exclusively for cooling loads, except after unoccupied periods when the space may have cooled to below a comfortable temperature. Tests by Miller *et al.* (1970, 1971, 1979) and Hart and Int-Hout (1981) suggest that the air diffusion performance index (ADPI) can be improved by moving diffusers closer together (*i.e.*, specifying more diffusers for a given space and air quantity) and by limiting the value of the supply air/room air temperature difference. In a given system of diffusers, these studies found an optimum operating

range of air volumes at a given thermal load. The operating load varies with diffuser design, ceiling height, thermal load, and diffuser orientation. This information can be obtained by constructing a mock-up representing the proposed building space, with several alternatives tested for ADPI values, in accordance with ASHRAE *Standard* 113-90. Usually, the diffuser manufacturer has performed these tests and can provide the best choice of design options for a particular building. For a VAV system, the diffuser spacing selection should not be based on maximum or design air volumes but rather on the air volume range where the system is expected to operate most of the time. For VAV applications, Miller (1979) recommends that the designer consider the expected variation range in the outlet air volume to ensure that ADPI values remain above a specified minimum.

Perimeter spaces. All-air mechanical systems that handle both heating and cooling thermal loads are commonly used in modern office buildings instead of using baseboards for heating and forced air for cooling. State energy codes (most based on ASHRAE *Standard* 90) require that commercial buildings have exterior walls that meet minimum thermal performance criteria for a particular location. Typically, walls of new buildings have design heat losses as low as 200 to 300 Btu/h per linear foot of wall.

A successful all-air heating-cooling mechanical system requires the designer to consider several design variables that have been the subject of research by Hart and Int-Hout (1980), Lorch and Straub (1983), and Rousseau (1983). The most important design variables include:

- Supply air/room air temperature difference
- Diffuser type and design
- Design heating and cooling loads
- Supply air volumes
- Distance between diffusers and perimeter wall
- Direction of air throw (toward wall, away from wall, or both)
- Ceiling height
- Desired air diffusion performance criteria

The diffuser manufacturer is best able to recommend the use of equipment.

For an office environment in cooling mode, the design goal should be an ADPI greater than 80. The ADPI should not be used as a measure of performance for heating conditions. In both cases, ASHRAE *Standard* 55-1992 recommends that the maximum temperature gradient should not exceed 5 °F (where temperature gradient refers to the difference between temperatures at any two points). Linear diffusers placed parallel to the perimeter wall perform well. For year-round operation, linear diffusers with two-way throw (*i.e.*, that throw air both toward and away from the perimeter wall) work best. Lorch and Straub (1983) reported optimum performance with a diffuser that throws warm air toward the perimeter wall under heating season conditions and chilled air in both directions under cooling season conditions. All researchers found less than optimum performance with high discharge temperatures (greater than 15 °F above ambient), both with one-way throw of air away from a cold wall and with one-way throw with chilled air toward the perimeter wall. Under heating season conditions, the supply air temperature must be limited to avoid excessive thermal stratification. To resolve any uncertainty about performance, a mock-up should be constructed with provisions for a cold wall; several variations of the design should be tested. As a result, the best diffuser wall spacing and supply air volumes can be selected. The ADPI or room temperature gradients or both, measured in accordance with ASHRAE *Standard* 113-90, can help gage system performance.

The following principles provide the best air diffusion quality and minimum energy use:

- For cooling load conditions, return air should exhaust from a location that takes advantage of any thermal stratification design. In many cases, this should be a high point to take advantage of rising warm air. Cooling supply air should be introduced as close to the heat sources as possible. Alternately, stratification designs may condition only part of the total space. In these cases, conditioned air is supplied and exhausted as close to the occupants as possible. In either case, comfort zone temperature gradients should be maintained within 5°F.

• For heating load conditions, thermal stratification should be discouraged. Heat should be introduced at points low in the large space. Ceiling-mounted fans may reduce stratification.

REFERENCES

ASHRAE. 1972. Method of testing for rating the air flow performance of outlets and inlets. ASHRAE *Standard* 70-72.

ASHRAE. 1981. Thermal environmental conditions for human occupancy. ASHRAE *Standard* 55-81.

ASHRAE. 1989. Building systems: Room air and air contaminant distribution. Christianson, L.L., ed.

ASHRAE. 1990. Method of testing for room air diffusion. ASHRAE *Standard* 113-1990.

ASHRAE. 1992. Thermal environmental conditions for human occupancy. ASHRAE *Standard* 55-92.

Ball, H.D., R.G. Nevins, and H.E. Straub. 1971. Thermal analysis of heat removal troffers. ASHRAE *Transactions* 77(2).

Baturin, V.V. 1972. *Fundamentals of industrial ventilation,* 3rd England Edition. Translated by O.M. Blunn. Pergamon Press, New York.

Christianson, L.L. 1989. Building systems: Room air and air containment distribution. ASHRAE, Atlanta.

Fanger, P.O., A.K. Melikov, H. Hanzawa, and J. Ring. 1988. Air turbulence and sensation of draft. *Energy and Buildings* 12:21-39.

Hart, G.H. and D. Int-Hout. 1980. The performance of a continuous linear diffuser in the perimeter zone of an office environment. ASHRAE *Transactions* 86(2).

Hart, G.H. and D. Int-Hout. 1981. The performance of a continuous linear diffuser in the interior zone of an open office environment. ASHRAE *Transactions* 87(2).

Helander, L. and C.V. Jakowatz. 1948. Downward projection of heated air. ASHVE Research Report No. 1327. ASHVE *Transactions* 54:71.

Helander, L., S.M. Yen, and R.E. Crank. 1953. Maximum downward travel of heated jets from standard long radius ASME nozzles. ASHVE Research Report No. 1475. ASHVE *Transactions* 59:241.

Helander, L., S.M. Yen, and L.B. Knee. 1954. Characteristics of downward jets of heated air from a vertical delivery discharge unit heater. ASHVE Research Report No. 1511. ASHVE *Transactions* 60:359.

Helander, L., S.M. Yen, and W. Tripp. 1957. Outlet characteristics that affect the downthrow of heated air jets. ASHAE Research Report No. 1601. ASHAE *Transactions* 63:255.

Houghten, F.C., C. Gutberlet, and E. Witkowski. 1938. Draft temperatures and velocities in relation to skin temperatures and feelings of warmth. ASHVE *Transactions* 44:289.

ISO. 1984. Moderate thermal environments—Determination of the PMV and PPD indices and specification of the conditions for thermal comfort. ISO *Standard* 7730.

Int-Hout, D. 1981. Measurement of room air diffusion in actual office environments to predict occupant thermal comfort. ASHRAE *Transactions* 87(2).

Jackman, P.J. 1991. Displacement ventilation. CIBSE National Conference, University of Kent, Canterbury.

Jackman, P.J. and P.A. Appleby. 1990. Displacement flow ventilation. BSRIA Project Report. The Building Services Research and Information Association, Old Bracknell Lane, Berkshire, RG12 4AH, UK.

Kegel, B. and U.W. Schulz. 1989. Displacement ventilation for office buildings. Proceedings 10th AIVC Conference, Helsinki. Air Infiltration and Ventilation Center, University of Warwick Science Park, Barclays Venture Centre, Sir William Lyons Road, Coventry CV4 7EZ, U.K.

Kirkpatrick, A., T. Malmstrom, P. Miller, and V. Hassani. 1991. Use of low temperature air for cooling of buildings. Proceedings Building Simulation. Nice, France.

Knaak, R. 1957. Velocities and temperatures on axis of downward heated jet from 4-inch long-radius ASME nozzle. ASHAE Research Report No. 1619. ASHAE *Transactions* 63:527.

Koestel, A. 1954. Computing temperatures and velocities in vertical jets of hot or cold air. ASHVE Research Report No. 1512. ASHVE *Transactions* 60:385.

Koestel, A. 1955. Paths of horizontally projected heated and chilled air jets. ASHAE Research Report No. 1534. ASHAE *Transactions* 61:213.

Koestel, A. 1957. Jet velocities from radial flow outlets. ASHAE Research Report No. 1618. ASHAE *Transactions* 63:505.

Koestel, A. and J.B. Austin, Jr. 1956. Air velocities in two parallel ventilating jets. ASHAE Research Report No. 1580. ASHAE *Transactions* 62:425.

Koestel, A. and G.L. Tuve. 1955. Performance and evaluation of room air distribution systems. ASHRAE Research Report No. 1553. ASHRAE *Transactions* 61:533.

Koestel, A., P. Hermann, and G.L. Tuve. 1949. Air streams from perforated panels. ASHVE Research Report No. 1366. ASHVE *Transactions* 55:283.

Koestel, A., P. Hermann, and G.L. Tuve. 1950. Comparative study of ventilating jets from various types of outlets. ASHVE Research Report No. 1404. ASHVE *Transactions* 56:459.

Lorch, F.A. and H.E. Straub. 1983. Performance of overhead slot diffusers with simulated heating and cooling conditions. ASHRAE *Transactions* 89(1).

Madison, R.D. and W.R. Elliot. 1946. Throw of air from slots and jets. *Heating, Piping, and Air Conditioning* 11:108.

McElroy, G.E. 1943. Air flow at discharge of fan-pipe lines in mines. U.S. Bureau of Mines Report of Investigations, 19.

Melikov, A.K. and J.B. Nielsen. 1989. Local thermal discomfort due to draft and vertical temperature difference in rooms with displacement ventilation. ASHRAE *Transactions* 95(2):1050-57.

Miller, P.L. 1971. Room air distribution performance of four selected outlets. ASHRAE *Transactions* 77(2):194.

Miller, P.L. 1979. Design of room air diffusion systems using the air diffusion performance index (ADPI). ASHRAE *Journal* 10:85.

Miller, P.L. 1989. Descriptive methods. Building Systems: Room Air and Air Contaminant Distribution. ASHRAE, Atlanta, GA.

Miller, P.L. and R.T. Nash. 1971. A further analysis of room air distribution performance. ASHRAE *Transactions* 77(2):205.

Miller, P.L. and R.G. Nevins. 1969. Room air distribution with an air distributing ceiling—Part II. ASHRAE *Transactions* 75:118.

Miller, P.L. and R.G. Nevins. 1970. Room air distribution performance of ventilating ceilings and cone-type circular ceiling diffusers. ASHRAE *Transactions* 76(1):186.

Miller, P.L. and R.G. Nevins. 1972. An analysis of the performance of room air distribution systems. ASHRAE *Transactions* 78(2):191.

Miller, P.L. and R.G. Nevins. 1974. Room air distribution—An ASHRAE engineering practice monograph. ASHRAE *Journal* 1:92.

Murakami, S. 1992. New scales for ventilation efficiency and their application based on numerical simulation of room airflow. International Symposium on Room Air Convection and Ventilation Effectiveness. Tokyo, Japan.

Nelson, D.W. and G.E. Smedberg. 1943. Performance of side outlets on horizontal ducts. ASHVE Research Report No. 1226. ASHVE *Transactions* 49:58.

Nelson, D.W., H. Krans, and A.F. Tuthill. 1940. The performance of stack heads. ASHVE Research Report No. 1155. ASHVE *Transactions* 46:205.

Nelson, D.W., D.H. Lamb, and G.E. Smedberg. 1942. Performance of stack heads equipped with grilles. ASHVE Research Report No. 1206. ASHVE *Transactions* 48:279.

Nevins, R.G. and P.L. Miller. 1972. Analysis, evaluation and comparison of room air distribution performance. ASHRAE *Transactions* 78(2):235.

Nevins, R.G. and E.D. Ward. 1968. Room air distribution with an air distributing ceiling. ASHRAE *Transactions* 74:VI.2.1.

Nottage, H.B., J.G. Slaby, and W.P. Gojsza. 1952. Isothermal ventilation—jet fundamentals. ASHVE Research Report No. 1443. ASHVE *Transactions* 58:107.

Nottage, H.B., J.G. Slaby, and W.P. Gojsza. 1952. Outlet turbulence intensity as a factor in isothermal-jet flow. ASHVE Research Report No. 1458. ASHVE *Transactions* 58:343.

Oakes, W.C. 1987. Experimental investigation of Coanda jet. M.S. Thesis, Michigan State University, East Lansing, MI.

Reinmann, J.J., A. Koestel, and G.L. Tuve. 1959. Evaluation of three room air distribution systems for summer cooling. ASHRAE Research Report No. 1697. ASHRAE *Transactions* 65:717.

Rousseau, W.H. 1983. Perimeter air diffusion performance index tests for heating with a ceiling slot diffuser. ASHRAE *Transactions* 89(1).

Rowlinson, D. and D. Croom. 1987. Supply characteristics of the floor mounted diffusers. Air distribution in ventilated spaces. ROOMVENT-87. Stockholm, Sweden.

Rydberg, J. and P. Norback. 1949. Air distribution and draft. ASHVE Research Report No. 1362. ASHVE *Transactions* 55:225.

Sandberg, M. and C. Blomqvist. 1989. Displacement ventilation in office rooms. ASHRAE *Transactions* 95(2):1041-49.

Scaret, E. 1985. Ventilation by displacement—Characterization and design implications. Elsevier Science Publishers.

Scheunemann, K.H. 1989. Local ventilating and air conditioning in industrial and culture buildings. The 2nd World Congress on Heating, Ventilating, Refrigerating and Air Conditioning, CLIMA 2000. Thermal comfort air quality and design parameters, 391-97.

Seppanen, O.A., W.J. Fisk., J. Eto, and D.T. Grimsrud. 1989. Comparison of conventional mixing and displacement air-conditioning and ventilating systems in U.S. commercial buildings. ASHRAE *Transactions* 95(2):1028-40.

Shilkrot, E. and A. Zhivov. 1992. Room ventilation with designed vertical air temperature stratification. Roomvent 92, Proceedings 3rd International Conference on Engineering Aero- and Thermodynamics of Ventilated Rooms.

Straub, H.E. and M.M. Chen. 1957. Distribution of air within a room for year-round air conditioning—Part II. University of Illinois Engineering Experiment Bulletin No. 442.

Straub, H.E., S.F. Gilman, and S. Konzo. 1956. Distribution of air within a room for year-round air conditioning—Part I. University of Illinois Engineering Experiment Station Bulletin No. 435.

Stymne, H., M. Sandberg, and M. Mattsson. 1991. Dispersion pattern of contaminants in a displacement ventilation room-implications for demand control. Proceedings, 12th Air Movement and Ventilation Control Within Buildings.

Svensson, A.G.L. 1989. Nordic experiences of displacement ventilation systems. ASHRAE *Transactions* 95(2):1013-17.

Tuve, G.L. 1953. Air velocities in ventilating jets. ASHVE Research Report No. 1476. ASHVE *Transactions* 59:261.

Weinhold, K., R. Dannecker, U. Schwiegk. 1969. Über auslengungsverfahren von lüftungsdecken. Luft-und Kältetechnick 2:78-84.

Wilson, J.D., M.L. Esmay, and S. Persson. 1970. Wall-jet velocity and temperature profiles resulting from a ventilation inlet. ASAE *Transactions*.

Yen, S.M., L. Helander, and L.B. Knee. 1956. Characteristics of downward jets from a vertical discharge unit heater. ASHAE Research Report No. 1562. ASHAE *Transactions* 62:123.

Zhang, J.S., L.L. Christianson, and G.L. Riskowski. 1990. Regional airflow characteristics in a mechanically ventilated room under nonisothermal conditions. ASHRAE *Transactions* 96(1):751-59.

Zhang, J.S., L.L. Christianson, G.J. Wu, and G.L. Riskowski. 1992. Detailed measurements of room air distribution for evaluating numerical simulation models. ASHRAE *Transactions* 98(1):58-65.

Zhivov A. 1990. Variable-Air volume ventilation systems for industrial buildings. ASHRAE *Transactions* 96(2).

BIBLIOGRAPHY

Davies, E.L. 1930. Measurement of the flow of air through registers and grilles. ASHVE Research Report No. 857. ASHVE *Transactions* 36:201.

Davies, E.L. 1931. Measurement of the flow of air through registers and grilles. ASHVE Research Report No. 911. ASHVE *Transactions* 37:619.

Davies, E.L. 1933. Measurement of the flow of air through registers and grilles. ASHVE Research Report No. 966. ASHVE *Transactions* 39:373.

Elrod, H.G., Jr. 1954. Computation charts and theory for rectangular and circular jets. ASHVE Research Report No. 1514. ASHVE *Transactions* 60:431.

Gilman, S.R., H.E. Straub, A.E. Hershey, and R.B. Engdahl. 1953. Room air distribution research for year-round air conditioning, Part I—Supply outlets at one high sidewall location. ASHVE Research Report No. 1471. ASHVE *Transactions* 59:151.

Greene, A.M., Jr. and M.H. Dean. 1938. The flow of air through exhaust grilles. ASHVE Research Report No. 1092. ASHVE *Transactions* 44:387.

Grimitlyn, M. 1970. Zurluftverteilung in raumen. *Luft-und Kältetechnick* 5.

Koestel, A. and G.L. Tuve. 1948. The discharge of air from a long slot. ASHVE Research Report No. 1328. ASHVE *Transactions* 54:87.

Koestel, A. and C.Y. Young. 1951. The control of airstreams from a long slot. ASHVE Research Report No. 1429. ASHVE *Transactions* 57:407.

Koestel, A., P. Hermann, and G.L. Tuve. 1949. Airstreams from perforated panels. ASHVE Research Report No. 1366. ASHVE *Transactions* 55:283.

Koestel, A., P. Hermann and G.L. Tuve. 1950. Comparative study of ventilating jets from various types of outlets. ASHVE Research Report No. 1404. ASHVE *Transactions* 56:459.

Kratz, A.P., A.E. Hershey, and R.B. Engdahl. 1940. Development of instruments for the study of air distribution in rooms. ASHVE Research Report No. 1165. ASHVE *Transactions* 46:351.

Nelson, D.W. and G.E. Smedberg. 1943. Performance of side outlets on horizontal ducts. ASHVE Research Report No. 1226. ASHVE *Transactions* 49:58.

Nelson, D.W. and D.J. Stewart. 1938. Air distribution from side wall outlets. ASHVE Research Report No. 1076. ASHVE *Transactions* 44:77.

Nielsen, R.A. 1940. Dirt patterns on walls. ASHVE Research Report No. 1158. ASHVE *Transactions* 46:247.

Nottage, H.B. 1949. Turbulence—A fundamental frontier in air distribution. ASHVE Research Report No. 1360. ASHVE *Transactions* 55:193.

Nottage, H.B. 1950. A simple heated-thermocouple anemometer. ASHVE Research Report No. 1402. ASHVE *Transactions* 56:431.

Nottage, H.B., J.G. Slaby, and W.P. Gojsza. 1952a. A smoke-filament technique for experimental research in room air distribution. ASHVE Research Report No. 1461. ASHVE *Transactions* 58:399.

Nottage, H.B., J.G. Slaby, and W.P. Gojsza. 1952b. A V-wire direction probe. ASHVE Research Report No. 1441. ASHVE *Transactions* 58:79.

Nottage, H.B., J.G. Slaby, and W.P. Gojsza. 1952. Exploration of a chilled jet. ASHVE Research Report No. 1459. ASHVE *Transactions* 58:357.

Poz, M.Y. 1991. Theoretical investigation and practical applications of nonisothrmal jets for the rooms ventilating. Current East/West HVAC Developments. IEI/CIBSE/ABOK Joint Conference, Dublin.

Shepelev, I. 1978. *Airdynamics of the air flows in the premises*. Stroiizdat, Moscow.

Shilkrot, E. 1974. About simulation of radiant-convective heat exchange in the premises with the natural ventilation. Proceedings of the Central Research Institute for Industrial Buildings, 37.

Shilkrot, E. 1986. Evaluating of the designing loads for the heating and ventilating systems of the premises using the method of the heat balances in its zones. Proceedings of the Central Research Institute for Industrial Buildings. TsNIIpromzdanii, Moscow.

Stewart, D.J. and G.F. Drake. 1937. The noise characteristics of air supply outlets. ASHVE Research Report No. 1051. ASHVE *Transactions* 43:81.

Straub, H.E. and S.F. Gilman. 1954. Room air distribution research for year-round air conditioning, Part II—Supply outlets at three floor locations. ASHVE Research Report No. 1504. ASHVE *Transactions* 60:249.

Tasker, C. 1948. ASHVE research in air distribution and air duct friction. *Heating, Piping, and Air Conditioning* 4:125.

Tuve, G.L. 1953. Air velocities in ventilating jets. ASHVE Research Report No. 1476. ASHVE *Transactions* 59:261.

Tuve, G.L. and G.B. Priester. 1944. Control of airstreams in large spaces. ASHVE Research Report No. 1248. ASHVE *Transactions* 50:153.

Tuve, G.L. and D.K. Wright, Jr. 1940. Air flow measurements at intake and discharge openings and grilles. ASHVE Research Report No. 1162. ASHVE *Transactions* 46:313.

Tuve, G.L., G.B. Priester, and D.K. Wright, Jr. 1942. Entrainment and jet-pump action of airstreams. ASHVE Research Report No. 1204. ASHVE *Transactions* 48:241.

Tuve, G.L., D.K. Wright, Jr., and L.J. Seigel. 1939. The use of air velocity meters. ASHVE Research Report No. 1140. ASHVE *Transactions* 45:645.

DUCT DESIGN

COMMERCIAL or industrial air duct system design must consider (1) space availability, (2) space air diffusion, (3) noise levels, (4) duct leakage, (5) duct heat gains and losses, (6) balancing, (7) fire and smoke control, (8) initial investment cost, and (9) system operating cost.

Deficiencies in duct design can result in systems that operate incorrectly or are expensive to own and operate. Poor air distribution can cause discomfort; lack of sound attenuators may permit objectionable noise levels. Poorly designed sections of ductwork can result in unbalanced systems. Faulty duct construction or lack of duct sealing produces inadequate airflow rates at the terminals. Proper duct insulation eliminates the problem caused by excessive heat gain or loss.

In this chapter, system design considerations and the calculation of a system's frictional and dynamic resistance to airflow are considered. Chapter 16 of the 1992 ASHRAE *Handbook—Systems and Equipment* examines duct construction and presents construction standards for residential, commercial, and industrial heating, ventilating, air-conditioning, and exhaust systems.

BERNOULLI EQUATION

Bernoulli's equation can be developed by equating the forces on an element of a stream tube in a frictionless fluid flow to the rate of momentum change. On integrating this relationship for steady flow, the following expression (Osborne 1966) results:

$$\frac{v^2}{2g_c} + \int \frac{dP}{\rho} + \frac{gz}{g_c} = \text{constant, ft-lb}_f/\text{lb}_m \quad (1)$$

where

v = streamline (local) velocity, ft/s
g_c = dimensional constant, 32.2 $(\text{lb}_m\text{-ft})/(\text{lb}_f\text{-s}^2)$
P = absolute pressure, lb_f/ft^2
ρ = density, lb_m/ft^3
g = acceleration due to gravity, ft/s^2
z = elevation, ft

Assuming constant fluid density within the system, Equation (1) reduces to:

$$\frac{v^2}{2g_c} + \frac{P}{\rho} + \frac{gz}{g_c} = \text{constant, ft-lb}_f/\text{lb}_m \quad (2)$$

Although Equation (2) was derived for steady flow along a stream tube of an ideal frictionless flow, it can be extended to

analyze flow through ducts in real systems. In terms of pressure, the relationship for fluid resistance between two sections is:

$$\frac{\rho_1 V_1^2}{2g_c} + P_1 + \gamma z_1 = \frac{\rho_2 V_2^2}{2g_c} + P_2 + \gamma z_2 + \Delta p, \frac{\text{ft-lb}_f}{\text{ft}^3} \quad (3)$$

where

V = average duct velocity, ft/s
Δp = total pressure loss due to friction and dynamic losses between stations 1 and 2, lb_f/ft^2
γ = $\rho g/g_c$ specific weight, lb_f/ft^3

In Equation (3), V (section average velocity) replaces v (streamline velocity) because experimentally determined loss coefficients allow for errors in calculating $\rho v^2/2g_c$ (velocity pressure) across streamlines.

Add to, and subtract the quantities p_{z1} and p_{z2} from each side of Equation (3), respectively, where p_{z1} and p_{z2} are the values of atmospheric air at heights z_1 and z_2. Thus:

$$\frac{\rho_1 V_1^2}{2g_c} + P_1 + (p_{z1} - p_{z1}) + \gamma z_1 = \frac{\rho_2 V_2^2}{2g_c} + P_2$$
$$+ (p_{z2} - p_{z2}) + \gamma z_2 + \Delta p \quad (4)$$

The atmospheric pressure at any elevation (p_{z1} and p_{z2}) expressed in terms of the atmospheric pressure at the same datum elevation (p_a) is given by:

$$p_{z1} = p_a - (g/g_c) \rho_a z_1 \quad (5)$$

$$p_{z2} = p_a - (g/g_c) \rho_a z_2 \quad (6)$$

Substituting Equations (5) and (6) into Equation (4) and simplifying yields the total pressure change between stations 1 and 2. Air densities ρ_1 and ρ_2 are assumed equal at heights z_1 and z_2, thus $\rho_1 = \rho_2 = \rho$.

$$\Delta p_{1-2} = \left(p_{s,1} + \frac{\rho V_1^2}{2g_c}\right) - \left(p_{s,2} + \frac{\rho V_2^2}{2g_c}\right)$$
$$= \Delta p - \frac{g}{g_c}(\rho_a - \rho)(z_2 - z_1) \quad (7)$$

where

$p_{s,1}$ = static pressure, gage at elevation z_1 ($P - p_{z1}$), lb_f/ft^2
$p_{s,2}$ = static pressure, gage at elevation z_2 ($P - p_{z2}$), lb_f/ft^2
V_1 = average velocity at station 1, ft/s
V_2 = average velocity at station 2, ft/s
ρ_a = density of ambient air, lb_m/ft^3

The preparation of this chapter is assigned to TC 5.2, Duct Design.

ρ = density of air or gas within duct, lb_m/ft^3

Δp_{1-2} = total pressure change between stations 1 and 2, lb_f/ft^2

Δp = total pressure loss due to friction and dynamic losses between stations 1 and 2, lb_f/ft^2

HEAD AND PRESSURE

The terms *head* and *pressure* are often used interchangeably; however, head is the height of a fluid column supported by fluid flow, while pressure is the normal force per unit area. For liquids, it is convenient to measure the head in terms of the flowing fluid. With a gas or air, however, it is customary to measure pressure on a column of liquid.

Static Pressure

The term $(pg_c/\rho g)$ is static head; p is static pressure.

Velocity Pressure

The term $(V^2/2g)$ refers to velocity head, and the term $(\rho V^2/2g_c)$ to velocity pressure. Although velocity head is independent of fluid density, velocity pressure calculated by Equation (8) is not.

$$p_v = \rho (V/1097)^2 \qquad (8)$$

where

p_v = velocity pressure, in. of water

V = fluid mean velocity, fpm

For air at standard conditions (0.075 lb_m/ft^3), Equation (8) becomes:

$$p_v = (V/4005)^2 \qquad (9)$$

Velocity is calculated by Equation (10) or (11).

$$V = 144Q/A \qquad (10)$$

where

Q = airflow rate, cfm

A = cross-sectional area of duct, in^2

$$V = Q/A \qquad (11)$$

where

A = cross-sectional area of duct, ft^2

Total Pressure

Total pressure is the sum of static pressure and velocity pressure:

$$p_t = p_s + \rho (V/1097)^2 \qquad (12)$$

or

$$p_t = p_s + p_v \qquad (13)$$

where

p_t = total pressure, in. of water

p_s = static pressure, in. of water

Instruments

The range, precision, and limitations of instruments for measuring pressure and velocity are discussed in Chapter 13. The manometer is a simple and useful means for measuring partial vacuum and low pressure. A primary instrument, it is often used as a standard for calibrating other instruments. The static, velocity, and total pressures in a duct system relative to the atmospheric pressure are measured with a pitot tube connected to a manometer. Pitot tube construction and locations for traversing round and rectangular ducts are presented in Chapter 13.

SYSTEM ANALYSIS

The friction and dynamic losses for each section of a duct system are calculated by Equation (14).

$$\Delta p_{t_i} = \Delta p_{f_i} + \sum_{j=1}^{m} \Delta p_{ij} + \sum_{k=1}^{n} \Delta p_{ik} - \sum_{r=1}^{\lambda} p_{se_{ir}},$$

$$i = 1, 2, \ldots, n_{up} + n_{dn} \qquad (14)$$

where

Δp_{t_i} = net total pressure change for *i*-section, in. of water

Δp_{f_i} = pressure loss due to friction for *i*-section, in. of water

Δp_{ij} = total pressure loss due to *j*-fittings, including fan system effect (FSE), for *i*-section, in. of water

Δp_{ik} = pressure loss due to *k*-equipment for *i*-section, in. of water

$p_{se_{ir}}$ = stack effect due to *r*-stacks for *i*-section, in. of water

m = number of fittings within *i*-section

n = number of equipment within *i*-section

n_{dn} = number of duct sections downstream of fan (supply air subsystems)

n_{up} = number of duct sections upstream of fan (exhaust/return air subsystems)

λ = number of stacks within *i*-section

From Equation (7), the stack effect for each nonhorizontal duct with a density other than ambient air is determined by Equation (15).

$$p_{se} = 0.192 (\rho_a - \rho)(z_2 - z_1) \qquad (15)$$

where

p_{se} = stack effect, in. of water

z_1 and z_2 = elevation from datum in direction of airflow (Figure 1), ft

ρ_a = density of ambient air, lb_m/ft^3

ρ = density of air or gas within duct, lb_m/ft^3

The elevation/density pressure due to stack effect (p_{se}) can be positive or negative depending on whether it assists or resists airflow. As shown by Figure 1, an imaginary fan helps to explain buoyancy effects within systems. For flow upward and $\rho > \rho_a$ (Figure 1A) and flow downward and $\rho < \rho_a$ (Figure 1B), the excess pressure is negative and resists flow, thus increasing the fan total pressure requirement. On the other hand, for flow upward and $\rho < \rho_a$ (Figure 1C) and flow downward and $\rho > \rho_a$ (Figure 1D), the excess pressure is positive and assists flow, thus decreasing the fan total pressure requirement. To determine the fan total pressure requirement for a system, use Equation (16).

$$P_t = \sum_{i \in F_{up}} \Delta p_{t_i} + \sum_{i \in F_{dn}} \Delta p_{t_i}, i = 1, 2, \ldots, n_{up} + n_{dn} \qquad (16)$$

where F_{up} and F_{dn} are sets of duct sections upstream and downstream of a fan, and the symbol $\in$ ties the duct section into system paths from the exhaust/return air terminals to the supply terminals. Figure 2 illustrates the use of Equation (16). This system has three supply and two return terminals consisting of nine sections connected in six paths: 1-3-4-9-7-5, 1-3-4-9-7-6, 1-3-4-9-8, 2-4-9-7-5, 2-4-9-7-6, and 2-4-9-8. Sections 1 and 3 are unequal area; thus, they are assigned separate numbers in accordance with the rules for identifying sections (see the HVAC Duct Design Procedures section, step 4). To determine the fan pressure requirement, the following six equations, derived from Equation (16), are applied. These equations must be satisfied to attain pressure balancing for design airflow. Relying entirely on dampers is not economical and may create objectionable flow-generated noise.

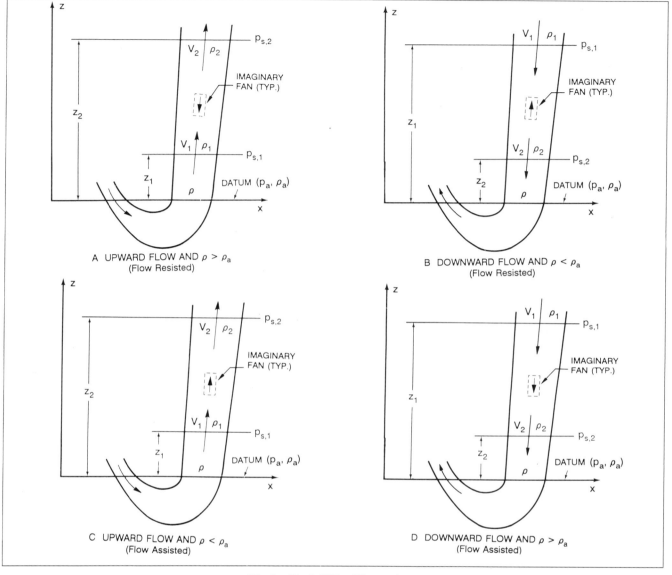

Fig. 1   Stack Effect Illustrations

$$P_t = \Delta p_1 + \Delta p_3 + \Delta p_4 + \Delta p_9 + \Delta p_7 + \Delta p_5$$
$$P_t = \Delta p_1 + \Delta p_3 + \Delta p_4 + \Delta p_9 + \Delta p_7 + \Delta p_6$$
$$P_t = \Delta p_1 + \Delta p_3 + \Delta p_4 + \Delta p_9 + \Delta p_8$$
$$P_t = \Delta p_2 + \Delta p_4 + \Delta p_9 + \Delta p_7 + \Delta p_5$$
$$P_t = \Delta p_2 + \Delta p_4 + \Delta p_9 + \Delta p_7 + \Delta p_6$$
$$P_t = \Delta p_2 + \Delta p_4 + \Delta p_9 + \Delta p_8$$

(17)

where

P_t = fan total pressure, in. of water

Pressure Changes in System

Figure 3 shows total and static pressure changes in a fan/duct system consisting of a fan with both supply and return air ductwork. Also shown are the total and static pressure gradients referenced to atmospheric pressure.

For all constant-area sections, the total and static pressure losses are equal. At the diverging transitions, velocity pressure decreases, absolute total pressure decreases, and absolute static pressure can increase. The static pressure increase at these sections is known as *static regain*.

At the converging transitions, velocity pressure increases in the direction of airflow, and the absolute total and absolute static pressures decrease.

At the exit, the total pressure loss depends on the shape of the fitting and the flow characteristics. Exit loss coefficients C_o can be greater than, less than, or equal to one. The total and static pressure grade lines for the various coefficients are shown in Figure 3. Note that for a loss coefficient less than one, static pressure upstream of the exit is less than atmospheric pressure (negative). The static pressure just upstream of the discharge fitting can be calculated by subtracting the upstream velocity pressure from the upstream total pressure.

At section 1, the total pressure loss depends on the shape of the entry. The total pressure immediately downstream of the entrance equals the difference between the upstream pressure, which is zero (atmospheric pressure), and the loss through the fitting. The static pressure of the ambient air is zero; several diameters downstream, static pressure is negative, algebraically equal to the total pressure (negative) and the velocity pressure (always positive).

System resistance to airflow is noted by the total pressure grade line in Figure 3. Sections 3 and 4 include fan system effect pressure losses. To obtain the fan static pressure requirement for fan selection where the fan total pressure is known, use:

$$P_s = P_t - p_{v,o} \qquad (18)$$

where

> P_s = fan static pressure, in. of water
> P_t = fan total pressure, in. of water
> $p_{v,o}$ = fan outlet velocity pressure, in. of water

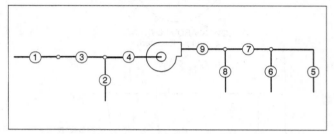

Fig. 2   Illustrative 6-Path, 9-Section System

FLUID RESISTANCE

Duct system losses are the irreversible transformation of mechanical energy into heat. The two types of losses are (1) frictional losses and (2) dynamic losses.

FRICTIONAL LOSSES

Frictional losses are due to fluid viscosity and are a result of momentum exchange between molecules in laminar flow and between particles moving at different velocities in turbulent flow. Frictional losses occur along the entire duct length.

Darcy, Colebrook, and Altshul Equations

For fluid flow in conduits, friction loss can be calculated by the Darcy equation:

$$\Delta p_f = f(12L/D_h)\, \rho \,(V/1097)^2 \qquad (19)$$

where

> Δp_f = friction losses in terms of total pressure, in. of water
> f = friction factor, dimensionless
> L = duct length, ft
> D_h = hydraulic diameter [Equation (24)], in.
> V = velocity, fpm
> ρ = density, lb_m/ft^3

Within the region of laminar flow (Reynolds numbers less than 2000), the friction factor is a function of Reynolds number only.

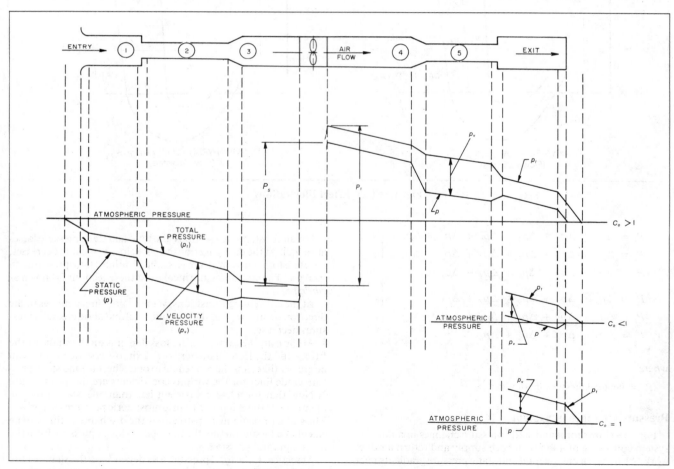

Fig. 3   Pressure Changes During Flow in Ducts

For turbulent flow, the friction factor depends on Reynolds number, duct surface roughness, and internal protuberances such as joints. The traditional Moody chart depicts the behavior for round passages. For hydraulically smooth ducts, the friction factor again depends only on Reynolds number, but the dependence is markedly different from that for laminar flow. In general, for nonsmooth surfaces, the friction factor depends on roughness and Reynolds number; however, for a particular level of roughness beyond a sufficiently large Reynolds number, the friction factor becomes independent of Reynolds number, a flow condition considered as fully rough. Between the bounding limits of hydraulically smooth behavior and fully rough behavior, is a transitional roughness zone where the friction factor depends on both roughness and Reynolds number. In this transitionally rough, turbulent zone, where most cases of airflow occur in air-conditioning applications, the friction factor f is calculated by Colebrook's equation (1938-39). Since this equation cannot be solved explicitly for f, use iterative techniques (Behls 1971).

$$\frac{1}{f^{0.5}} = -2\log\left[\frac{12\epsilon}{3.7 D_h} + \frac{2.51}{\text{Re}\, f^{0.5}}\right] \quad (20)$$

where

ϵ = material absolute roughness factor, ft
Re = Reynolds number

A simplified formula for calculating friction factor, developed by Altshul (1975) and modified by Tsal, is

$$f' = 0.11\left(\frac{12\epsilon}{D_h} + \frac{68}{\text{Re}}\right)^{0.25}$$

If $f' \geq 0.018$: $f = f'$

If $f' < 0.018$: $f = 0.85 f' + 0.0028 \quad (21)$

Friction factors obtained from Altshul's modified equation are within 1.6% of those obtained by Colebrook's equation.

Reynolds number (Re) may be calculated by using Equation (22).

$$\text{Re} = \frac{D_h V}{720\,\nu} \quad (22)$$

where ν = kinematic viscosity, ft^2/s.

For standard air, Re can be calculated by

$$\text{Re} = 8.56\, D_h V \quad (23)$$

Roughness Factors (ϵ)

The ϵ-values listed in Table 1 are recommended for use with the Colebrook or Altshul-Tsal equation. These values should be interpreted as representing a combination of material, duct construction, joint type, and joint spacing (Griggs and Khodabakhsh-Sharifabad 1992). Roughness factors for other materials are presented in Idelchik *et al.* (1986). Idelchik summarizes roughness factors for 80 materials including metal tubes; conduits made from concrete and cement; and wood, plywood, and glass tubes.

Swim (1978) conducted tests on duct liners of varying densities, surface treatments, transverse joints (workmanship), and methods of attachment to sheet metal ducts. As a result of these tests, Swim recommends for design 0.015 ft for spray-coated liners and 0.005 ft for liners with a facing material cemented onto the air side. In both cases, the roughness factor includes the resistance offered by mechanical fasteners and assumes good joints. Liners cut too long

and fastened to the duct cause much more loss than a liner cut too short; therefore, any fabrication error in liner length should be on the short side. Liner density does not significantly influence flow resistance.

Manufacturers' data indicate that the absolute roughness for fully extended nonmetallic flexible ducts ranges from 0.0035 to 0.015 ft. For fully extended flexible metallic ducts, absolute roughness ranges from 0.0004 to 0.007 ft. This range covers flexible duct with the supporting wire exposed to flow or covered by the material. Figure 4 provides a pressure drop correction factor for straight flexible duct when less than fully extended.

Table 1   Duct Roughness Factors

Duct Material	Roughness Category	Absolute Roughness ϵ, ft
Uncoated carbon steel, clean (Moody 1944) (0.00015 ft)	Smooth	0.0001
PVC plastic pipe (Swim 1982) (0.00003 – 0.00015 ft)		
Aluminum (Hutchinson 1953) (0.000015 – 0.0002 ft)		
Galvanized steel, longitudinal seams, 4-ft joints (Griggs *et al.* 1987) (0.00016 – 0.00032 ft)	Medium smooth	0.0003
Galvanized steel, continuously rolled, spiral seams, 10-ft joints (Jones 1979) (0.0002 – 0.0004 ft)		
Galvanized steel, spiral seam with 1, 2, and 3 ribs, 12-ft joints (Griggs *et al.* 1987) (0.00029 – 0.00038 ft)		
Galvanized steel, longitudinal seams, 2.5-ft joints (Wright 1945) (0.0005 ft)	Average	0.0005
Fibrous glass duct, rigid	Medium	0.003
Fibrous glass duct liner, air side with facing material (Swim 1978) (0.005 ft)	Rough	
Fibrous glass duct liner, air side spray coated (Swim 1978) (0.015 ft)	Rough	0.01
Flexible duct, metallic (0.004 – 0.007 ft when fully extended)		
Flexible duct, all types of fabric and wire (0.0035 – 0.015 ft when fully extended)		
Concrete (Moody 1944) (0.001 – 0.01 ft)		

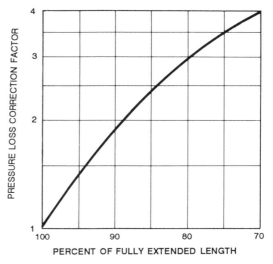

Fig. 4   Correction Factor for Unextended Flexible Duct

Friction Chart

Fluid resistance caused by friction in round ducts can be determined by the Friction Chart (Figure 5). This chart is based on standard air flowing through round galvanized ducts with beaded slip couplings on 48-in. centers, equivalent to an absolute roughness of 0.0003 ft.

Changes in barometric pressure, temperature, and humidity affect air density, air viscosity, and Reynolds number. No corrections to the Friction Chart are needed for (1) duct materials with a medium smooth roughness factor, (2) temperature variations in the order of $\pm 30\,°F$ from $70\,°F$, (3) elevations to 1500 ft, and (4) duct pressures from -20 in. of water to $+20$ in. of water relative to the ambient pressure. These individual variations in temperature, elevation, and duct pressure result in duct losses within $\pm 5\%$ of the standard air friction chart.

For duct materials other than those categorized as medium smooth in Table 1, and for variations in temperature, barometric pressure (elevation), and duct pressures (outside the range listed), calculate the pressure loss in a duct due to friction by the Altshul-Tsal and Darcy equations [(21) and (19), respectively].

Noncircular Ducts

A momentum analysis can relate average wall shear stress to pressure drop per unit length for fully developed turbulent flow in a passage of arbitrary shape but of uniform longitudinal cross-sectional area. Combining the result with the definition of the Darcy friction factor leads to Equation (24), with the ratio $4A/P$ defined as hydraulic diameter:

$$D_h = 4A/P \tag{24}$$

where

D_h = hydraulic diameter, in.
A = duct area, in^2
P = perimeter of cross section, in.

While the hydraulic diameter is often used to correlate noncircular data, exact solutions for laminar flow in noncircular passages show that such practice causes some inconsistencies. No exact solutions exist for turbulent flow. Tests over a limited range of turbulent flow indicated that fluid resistance is the same for equal lengths of duct for equal mean velocities of flow if the ducts have the same ratio of cross-sectional area to perimeter. From a series of experiments using round, square, and rectangular ducts having essentially the same hydraulic diameter, Huebscher (1948) found that each, for most purposes, had the same flow resistance at equal mean velocities. Tests by Griggs and Khodabakhsh-Sharifabad (1992) also indicated that experimental rectangular duct data for airflow over the range typical of HVAC systems can be correlated satisfactorily using Equation (20) together with hydraulic diameter, particularly when a realistic experimental uncertainty is accepted. These tests support using hydraulic diameter to correlate noncircular duct data.

Rectangular ducts. Huebscher developed the relationship between rectangular and round ducts that is used to determine size equivalency based on equal flow, resistance, and length. This relationship, Equation (25), is the basis for Table 2.

$$D_e = 1.30 \frac{(ab)^{0.625}}{(a+b)^{0.250}} \tag{25}$$

where

D_e = circular equivalent of rectangular duct for equal length, fluid resistance, and airflow, in.
a = length of one side of duct, in.
b = length of adjacent side of duct, in.

To size rectangular ducts, determine the circular duct diameter by any design method, and use Table 2 to select the equivalent duct size as a function of aspect ratio. Equations (21) or (20) and (19) must be used to determine pressure loss.

Flat oval ducts. To convert round ducts to spiral flat oval sizes, use Table 3. Table 3 is based on Equation (26) (Heyt and Diaz 1975), the circular equivalent of a flat oval duct for equal airflow, resistance, and length. Equations (21) or (20) and (19) must be used to determine frictional pressure loss.

$$D_e = \frac{1.55 \, A^{0.625}}{P^{0.250}} \tag{26}$$

where A is the cross-sectional area of flat oval duct defined as:

$$A = (\pi b^2/4) + b(a-b) \tag{27}$$

and the perimeter P is calculated by:

$$P = \pi b + 2(a-b) \tag{28}$$

where

P = perimeter of flat oval duct, in.
a = major dimension of flat oval duct, in.
b = minor dimension of flat oval duct, in.

DYNAMIC LOSSES

Dynamic losses result from flow disturbances caused by fittings that change the airflow path's direction and/or area. These fittings include entries, exits, transitions, and junctions. Idelchik (1986) discusses parameters affecting fluid resistance of fittings and presents loss coefficients in three forms: tables, curves, and equations.

Local Loss Coefficients

The following dimensionless coefficient is used for fluid resistance, since this coefficient has the same value in dynamically similar streams, *i.e.*, streams with geometrically similar stretches, equal values of Reynolds number, and equal values of other criteria necessary for dynamic similarity. The fluid resistance coefficient represents the ratio of total pressure loss to velocity pressure at the referenced cross section.

$$C = \frac{\Delta p_j}{\rho(V/1097)^2} = \frac{\Delta p_j}{p_v} \tag{29}$$

where

C = local loss coefficient, dimensionless
Δp_j = fitting total pressure loss, in. of water
ρ = density, lb$_m$/ft^3
V = velocity, fpm
p_v = velocity pressure, in. of water

Dynamic losses occur along a duct length and cannot be separated from frictional losses. For ease of calculation, dynamic losses are assumed to be concentrated at a section (local) and to exclude friction. Frictional losses must be considered only for relatively long fittings. Generally, fitting friction losses are accounted for by measuring duct lengths from the centerline of one fitting to that of the next fitting. For fittings closely coupled (less than six hydraulic diameters apart), the flow pattern entering subsequent fittings differs from the flow pattern used to determine loss coefficients. Adequate data for these situations are unavailable.

For all fittings, except junctions, calculate the total pressure loss Δp_j at a section by:

$$\Delta p_j = C_o \, p_{v,o} \tag{30}$$

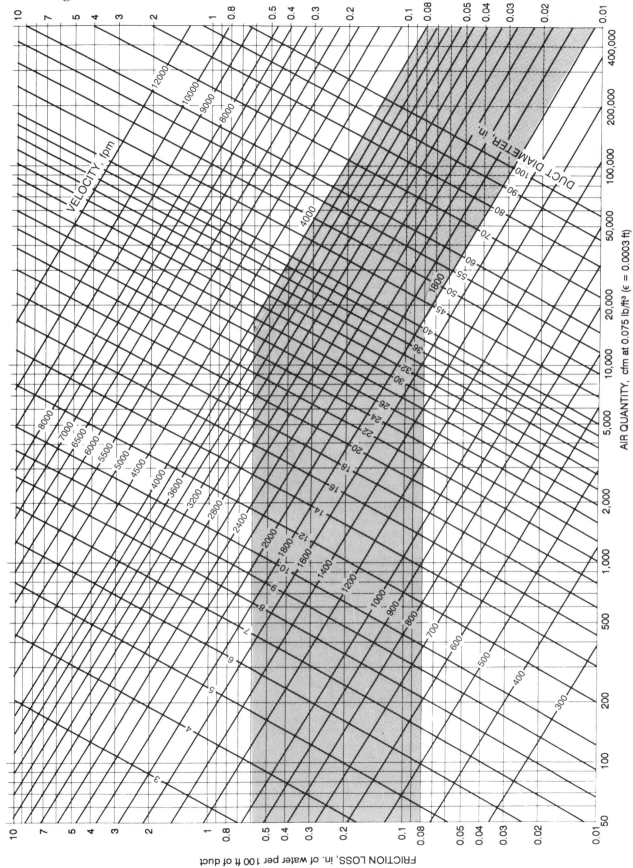

Fig. 5 Friction Chart

Table 2 Equivalent Rectangular Duct Dimension

Duct Diameter, in.	Rectangular Size, in.	Aspect Ratio														
		1.00	1.25	1.50	1.75	2.00	2.25	2.50	2.75	3.00	3.50	4.00	5.00	6.00	7.00	8.00
6	Width	—	6													
	Height	—	5													
7	Width	6	8													
	Height	6	6													
8	Width	7	9	9	11											
	Height	7	7	6	6											
9	Width	8	9	11	11	12	14									
	Height	8	7	7	6	6	6									
10	Width	9	10	12	12	14	14	15	17							
	Height	9	8	8	7	7	6	6	6							
11	Width	10	11	12	14	14	16	18	17	18	21					
	Height	10	9	8	8	7	7	7	6	6	6					
12	Width	11	13	14	14	16	16	18	19	21	21	24				
	Height	11	10	9	8	8	7	7	7	7	6	6				
13	Width	12	14	15	16	18	18	20	19	21	25	24	30			
	Height	12	11	10	9	9	8	8	7	7	7	6	6			
14	Width	13	14	17	18	18	20	20	22	24	25	28	30	36		
	Height	13	11	11	10	9	9	8	8	8	7	7	6	6		
15	Width	14	15	17	18	20	20	23	25	24	28	28	35	36	42	
	Height	14	12	11	10	10	9	9	9	8	8	7	7	6	6	
16	Width	15	16	18	19	20	23	23	25	27	28	32	35	42	42	48
	Height	15	13	12	11	10	10	9	9	9	8	8	7	7	6	6
17	Width	16	18	20	21	22	25	25	28	27	32	32	35	42	49	48
	Height	16	14	13	12	11	11	10	10	9	9	8	7	7	7	6
18	Width	16	19	21	23	24	25	28	28	30	32	36	40	42	49	56
	Height	16	15	14	13	12	11	11	10	10	9	9	8	7	7	7
19	Width	17	20	21	23	24	27	28	30	30	35	36	40	48	49	56
	Height	17	16	14	13	12	12	11	11	10	10	9	8	8	7	7
20	Width	18	20	23	25	26	27	30	30	33	35	40	45	48	56	56
	Height	18	16	15	14	13	12	12	11	11	10	10	9	8	8	7
21	Width	19	21	24	26	28	29	30	33	33	39	40	45	54	56	64
	Height	19	17	16	15	14	13	12	12	11	11	10	9	9	8	8
22	Width	20	23	26	26	28	32	33	36	36	39	44	50	54	56	64
	Height	20	18	17	15	14	14	13	13	12	11	11	10	9	8	8
23	Width	21	24	26	28	30	32	35	36	39	42	44	50	54	63	64
	Height	21	19	17	16	15	14	14	13	13	12	11	10	9	9	8
24	Width	22	25	27	30	32	34	35	39	39	42	48	55	60	63	72
	Height	22	20	18	17	16	15	14	14	13	12	12	11	10	9	9
25	Width	23	25	29	30	32	36	38	39	42	46	48	55	60	70	72
	Height	23	20	19	17	16	16	15	14	14	13	12	11	10	10	9
26	Width	24	26	30	32	34	36	38	41	42	46	52	55	66	70	72
	Height	24	21	20	18	17	16	15	15	14	13	13	11	11	10	9
27	Width	25	28	30	33	36	38	40	41	45	49	52	60	66	70	80
	Height	25	22	20	19	18	17	16	15	15	14	13	12	11	10	10
28	Width	26	29	32	35	36	38	43	44	45	49	56	60	66	77	80
	Height	26	23	21	20	18	17	17	16	15	14	14	12	11	11	10
29	Width	27	30	33	35	38	41	43	44	48	53	56	65	72	77	88
	Height	27	24	22	20	19	18	17	16	16	15	14	13	12	11	11
30	Width	27	31	35	37	40	43	45	47	48	53	60	65	72	77	88
	Height	27	25	23	21	20	19	18	17	16	15	15	13	12	11	11
31	Width	28	31	35	39	40	43	45	50	51	56	60	70	78	84	88
	Height	28	25	23	22	20	19	18	18	17	16	15	14	13	12	11
32	Width	29	33	36	39	42	45	48	50	54	56	60	70	78	84	96
	Height	29	26	24	22	21	20	19	18	18	16	15	14	13	12	12
33	Width	30	34	38	40	44	47	50	52	54	60	64	75	78	91	96
	Height	30	27	25	23	22	21	20	19	18	17	16	15	13	13	12
34	Width	31	35	39	42	44	47	50	52	57	60	64	75	84	91	96
	Height	31	28	26	24	22	21	20	19	19	17	16	15	14	13	12
35	Width	32	36	39	42	46	50	53	55	57	63	68	75	84	91	104
	Height	32	29	26	24	23	22	21	20	19	18	17	15	14	13	13
36	Width	33	36	41	44	48	50	53	55	60	63	68	80	90	98	104
	Height	33	29	27	25	24	22	21	20	20	18	17	16	15	14	13
38	Width	35	39	44	47	50	54	58	61	63	67	72	85	96	105	112
	Height	35	31	29	27	25	24	23	22	21	19	18	17	16	15	14

*Shaded area not recommended.

Table 2 Equivalent Rectangular Duct Dimension (*Continued*)

Duct Diameter, in.	Rectangular Size, in.	Aspect Ratio														
		1.00	1.25	1.50	1.75	2.00	2.25	2.50	2.75	3.00	3.50	4.00	5.00	6.00	7.00	8.00
40	Width	37	41	45	49	52	56	60	63	66	70	76	90	96	105	120
	Height	37	33	30	28	26	25	24	23	22	20	19	18	16	15	15
42	Width	38	43	48	51	56	59	63	66	69	74	80	90	102	112	120
	Height	38	34	32	29	28	26	25	24	23	21	20	18	17	16	15
44	Width	40	45	50	54	58	61	65	69	72	81	84	95	108	119	128
	Height	40	36	33	31	29	27	26	25	24	23	21	19	18	17	16
46	Width	42	48	53	56	60	65	68	72	75	84	88	100	114	126	136
	Height	42	38	35	32	30	29	27	26	25	24	22	20	19	18	17
48	Width	44	49	54	60	62	68	70	74	78	88	92	105	120	126	136
	Height	44	39	36	34	31	30	28	27	26	25	23	21	20	18	17
50	Width	46	51	57	61	66	70	75	77	81	91	96	.110	120	133	144
	Height	46	41	38	35	33	31	30	28	27	26	24	22	20	19	18
52	Width	48	54	59	63	68	72	78	83	84	95	100	115	126	140	152
	Height	48	43	39	36	34	32	31	30	28	27	25	23	21	20	19
54	Width	49	55	62	67	70	77	80	85	90	98	104	120	132	147	160
	Height	49	44	41	38	35	34	32	31	30	28	26	24	22	21	20
56	Width	51	58	63	68	74	79	83	88	93	102	108	125	138	147	160
	Height	51	46	42	39	37	35	33	32	31	29	27	25	23	21	20
58	Width	53	60	66	70	76	81	85	91	96	105	112	130	144	154	168
	Height	53	48	44	40	38	36	34	33	32	30	28	26	24	22	21
60	Width	55	61	68	74	78	83	90	94	99	109	116	130	144	161	
	Height	55	49	45	42	39	37	36	34	33	31	29	26	24	23	
62	Width	57	64	71	75	82	88	93	96	102	112	120	135	150	168	
	Height	57	51	47	43	41	39	37	35	34	32	30	27	25	24	
64	Width	59	65	72	79	84	90	95	99	105	116	124	140	156		
	Height	59	52	48	45	42	40	38	36	35	33	31	28	26		
66	Width	60	68	75	81	86	92	98	105	108	119	128	145	162		
	Height	60	54	50	46	43	41	39	38	36	34	32	29	27		
68	Width	62	70	77	82	90	95	100	107	111	123	132	150	168		
	Height	62	56	51	47	45	42	40	39	37	35	33	30	28		
70	Width	64	71	80	86	92	99	105	110	114	126	136	155			
	Height	64	57	53	49	46	44	42	40	38	36	34	31			
72	Width	66	74	81	88	94	101	108	113	117	130	140	160			
	Height	66	59	54	50	47	45	43	41	39	37	35	32			
74	Width	68	76	84	91	98	104	110	116	123	133	144	165			
	Height	68	61	56	52	49	46	44	42	41	38	36	33			
76	Width	70	78	86	93	100	106	113	118	126	137	148	165			
	Height	70	62	57	53	50	47	45	43	42	39	37	33			
78	Width	71	80	89	95	102	110	115	121	129	140	152				
	Height	71	64	59	54	51	49	46	44	43	40	38				
80	Width	73	83	90	98	104	113	118	124	132	144	156				
	Height	73	66	60	56	52	50	47	45	44	41	39				
82	Width	75	84	93	100	108	115	123	129	135	147	160				
	Height	75	67	62	57	54	51	49	47	45	42	40				
84	Width	77	86	95	103	110	117	125	132	138	151	164				
	Height	77	69	63	59	55	52	50	48	46	43	41				
86	Width	79	88	98	105	112	119	128	135	141	154	168				
	Height	79	70	65	60	56	53	51	49	47	44	42				
88	Width	80	90	99	107	116	124	130	138	144	158					
	Height	80	72	66	61	58	55	52	50	48	45					
90	Width	82	93	102	110	118	126	133	140	147	161					
	Height	82	74	68	63	59	56	53	51	49	46					
92	Width	84	94	104	112	120	128	138	143	150	165					
	Height	84	75	69	64	60	57	55	52	50	47					
94	Width	86	96	107	116	124	131	140	146	153	168					
	Height	86	77	71	66	62	58	56	53	51	48					
96	Width	88	99	108	117	126	135	143	151	159						
	Height	88	79	72	67	63	60	57	55	53						
98	Width	90	100	111	119	128	137	145	154	162						
	Height	90	80	74	68	64	61	58	56	54						
100	Width	91	103	113	123	132	140	148	157	165						
	Height	91	82	75	70	66	62	59	57	55						
102	Width	93	105	116	124	134	142	153	160	168						
	Height	93	84	77	71	67	63	61	58	56						
104	Width	95	106	117	128	136	146	155	162							
	Height	95	85	78	73	68	65	62	59							

*Shaded area not recommended.

Table 2 Equivalent Rectangular Duct Dimension (*Concluded*)

Duct Diameter, in.	Rectangular Size, in.	Aspect Ratio														
		1.00	1.25	1.50	1.75	2.00	2.25	2.50	2.75	3.00	3.50	4.00	5.00	6.00	7.00	8.00
106	Width	97	109	120	130	140	149	158	165							
	Height	97	87	80	74	70	66	63	60							
108	Width	99	110	122	131	142	151	160	168							
	Height	99	88	81	75	71	67	64	61							
110	Width	101	113	125	135	144	153	163								
	Height	101	90	83	77	72	68	65								
112	Width	102	115	126	137	146	158	165								
	Height	102	92	84	78	73	70	66								
114	Width	104	116	129	140	150	160									
	Height	·104	93	86	80	75	71									
116	Width	106	119	131	142	152	162									
	Height	106	95	87	81	76	72									
118	Width	108	121	134	144	154	164									
	Height	108	97	89	82	77	73									
120	Width	110	123	135	147	158										
	Height	110	98	90	84	79										

*Shaded area not recommended.

Table 3 Equivalent Spiral Flat Oval Duct Dimensions

Duct Diameter, in.	3	4	5	6	7	8	9	10	11	12	14	16	18	20	22	24
5	8															
5.5	9	7														
6	11	9														
6.5	12	10	8													
7	15	12	10	8												
7.5	19	13	—	9												
8	22	15	11	—												
8.5		18	13	11	10											
9		20	14	12	—	10										
9.5		21	18	14	12	—										
10			19	15	13	11										
10.5			21	17	15	13	12									
11				19	16	14	—	12								
11.5				20	18	16	14	—								
12				23	20	17	15	13								
12.5				25	21	—	—	15	14							
13				28	23	19	17	16	—	14						
13.5				30	—	21	18	—	16	—						
14				33	—	22	20	18	17	15						
14.5				36	—	24	22	19	—	17						
15				39	—	27	23	21	19	18						
16				45	—	30	—	24	22	20	17					
17				52	—	35	—	27	24	21	19					
18				59	—	39	—	30	—	25	22	19				
19						46	—	34	—	28	23	21				
20						50	—	38	—	31	27	24	21			
21						58	—	43	—	34	28	25	23			
22						65	—	48	—	37	31	29	26			
23						71	—	52	—	42	34	30	27			
24						77	—	57	—	45	38	33	29	26		
25								63	—	50	41	36	32	29		
26								70	—	56	45	38	34	31		
27								76	—	59	49	41	37	34		
28										65	52	46	40	36		
29										72	58	49	43	39	35	
30										78	61	54	46	40	38	
31										81	67	57	49	44	39	37
32											71	60	53	47	42	40
33											77	66	56	51	46	41
34												69	59	55	47	44
35												76	65	58	50	46
36												79	68	61	53	49
37													71	64	57	52
38													78	67	60	55
40														77	69	62
42															75	68
44															82	74

where the subscript *o* is the cross section at which the velocity pressure is referenced. The dynamic loss is based on the actual velocity in the duct, not the velocity in an equivalent noncircular duct. For the cross section to reference a fitting loss coefficient, refer to step 4 of the HVAC Duct Design Procedures section. Where necessary (unequal area fittings), convert a loss coefficient from section *o* to section *i* by Equation (31), where *V* is the velocity at the respective sections.

$$C_i = \frac{C_o}{(V_i/V_o)^2} \tag{31}$$

For converging and diverging flow junctions, total pressure losses through the straight (main) section are calculated as:

$$\Delta p_j = C_{c,s}\, p_{v,c} \tag{32}$$

For total pressure losses through the branch section:

$$\Delta p_j = C_{c,b}\, p_{v,c} \tag{33}$$

where $p_{v,c}$ is the velocity pressure at the common section *c*, and $C_{c,s}$ and $C_{c,b}$ are losses for the straight (main) and branch flow paths, respectively, each referenced to the velocity pressure at section *c*. To convert junction local loss coefficients referenced to straight and branch velocity pressures, use Equation (34).

$$C_i = \frac{C_{c,i}}{(V_i/V_c)^2} \tag{34}$$

where

C_i = local loss coefficient referenced to section being calculated (see subscripts), dimensionless

$C_{c,i}$ = straight ($C_{c,s}$) or branch ($C_{c,b}$) local loss coefficient referenced to dynamic pressure at common section, dimensionless

V_i = velocity at section to which C_i is being referenced, fpm

V_c = velocity at common section, fpm

Subscripts:

b = branch
s = straight (main) section
c = common section

The junction of two parallel streams moving at different velocities is characterized by turbulent mixing of the streams, accompanied by pressure losses. In the course of this mixing, an exchange of momentum takes place between the particles moving at different velocities, finally resulting in the equalization of the velocity distributions in the common stream. The jet with higher velocity loses a part of its kinetic energy by transmitting it to the slower moving jet. The loss in total pressure before and after mixing is always large and positive for the higher velocity jet and increases with an increase in the amount of energy transmitted to the lower velocity jet. Consequently, the local loss coefficient, defined by Equation (29), will always be positive. The energy stored in the lower velocity jet increases as a result of mixing. The loss in total pressure and the local loss coefficient can, therefore, also have negative values for the lower velocity jet (Idelchik 1986).

Duct Fitting Database

A duct fitting database, developed by ASHRAE (1993), which includes 228 round and rectangular fittings with the provision to include flat oval fittings is available from ASHRAE in electronic form with the capability to be linked to duct design programs.

The fittings are numbered (coded) as shown in Table 4. Entries and converging junctions are only in the exhaust/return portion of systems. Exits and diverging junctions are only in supply systems. Equal-area elbows, obstructions, and duct-mounted equipment are common to both supply and exhaust systems. Transitions and unequal-area elbows can be either supply or exhaust fittings. Fitting ED5-1 (see Fitting Loss Coefficients section) is an **E**xhaust fitting with a round shape (**D**iameter). The number 5 indicates that the fitting is a junction, and 1 is its sequential number. Fittings SR3-1 and ER3-1 are **S**upply and **E**xhaust fittings, respectively. The R indicates that the fitting is **R**ectangular, and 3 identifies the fitting as an elbow. Note that the cross-sectional area at sections 0 and 1 are not equal. Otherwise, the elbow would be a **C**ommon fitting such as CR3-6. Additional fittings are reproduced in the Fitting Loss Coefficients section to support the example design problems. These fittings are identified by Tables 10 (Example 3) and 12 (Example 4).

Table 4 Duct Fitting Codes

Fitting Function	Geometry	Category	Sequential Number
S: **S**upply	D: round (**D**iameter)	1. Entries	1,2,3 . . . n
		2. Exits	
E: **E**xhaust/ Return	R: **R**ectangular	3. Elbows	
		4. Transitions	
C: **C**ommon (supply and return)	O: flat **O**val	5. Junctions	
		6. Obstructions	
		7. Fan and system interactions	
		8. Duct-mounted equipment	
		9. Dampers	
		10. Hoods	
		11. Straight duct	

DUCTWORK SECTIONAL LOSSES

Darcy-Weisbach Equation

Total pressure loss in a duct section is calculated by combining Equations (19) and (29) in terms of Δp, where ΣC is the summation of local loss coefficients within the duct section. Each fitting loss coefficient must be referenced to that section's velocity pressure.

$$\Delta p = \left(\frac{12fL}{D_h} + \Sigma C\right) \rho \, (V/1097)^2 \qquad (35)$$

FAN-SYSTEM INTERFACE

Fan Inlet and Outlet Conditions

Fan performance data measured in the field may show lower performance capacity than manufacturers' ratings. The most common causes of deficient performance of the fan/system combination are improper outlet connections, nonuniform inlet flow, and swirl at the fan inlet. These conditions alter the aerodynamic characteristics of the fan so that its full flow potential is not realized. One bad connection can reduce fan performance far below its rating. No data has been published that accounts for the effects of fan inlet and outlet flexible vibration connectors.

Normally, a fan is tested with open inlets and a section of straight duct attached to the outlet (ASHRAE 1985). This setup results in uniform flow into the fan and efficient static pressure recovery on the fan outlet. If good inlet and outlet conditions are not provided in the actual installation, the performance of the fan suffers. To select and apply the fan properly, these effects must be considered, and the pressure requirements of the fan, as calculated by standard duct design procedures, must be increased.

Figure 6 illustrates deficient fan/system performance. The system pressure losses have been determined accurately and a fan selected for operation at Point 1. However, no allowance has been made for the effect of system connections to the fan on fan performance. To compensate, a fan system effect must be added to the calculated system pressure losses to determine the actual system curve. The point of intersection between the fan performance curve and the actual system curve is Point 4. The actual flow volume is, therefore, deficient by the difference from 1 to 4. To achieve design flow volume, a fan system effect pressure loss

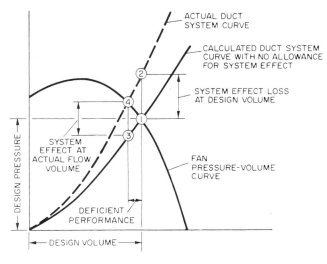

Fig. 6 Deficient System Performance with System Effect Ignored

equal to the pressure difference between Points 1 and 2 should be added to the calculated system pressure losses and the fan selected to operate at Point 2.

Fan System Effect Coefficients

The system effect concept was formulated by Farquhar (1973) and Meyer (1973); the magnitudes of the system effect, called *system effect factors*, were determined experimentally in AMCA's laboratory (Brown 1973, Clarke *et al.* 1978), and published in their *Fan Application Manual, Publication* 201 (1990a). The system effect factors, converted to system effect coefficients, are in the ASHRAE Duct Fitting Database (1993) for both centrifugal and axial fans. Fan system effect coefficients are only an approximation, however. Fans of different types and even fans of the same type, but supplied by different manufacturers, do not necessarily react to a system in the same way. Therefore, judgment based on experience must be applied to any design.

Fan outlet conditions. Fans intended primarily for duct systems are usually tested with an outlet duct in place (ASHRAE 1985b). Figure 7 shows the changes in velocity profiles at various distances from the fan outlet. For 100% recovery, the duct, including transition, must meet the requirements for 100% effective duct length [L_e (Figure 7)] which is calculated as follows:

$V_o > 2500$ fpm:

$$L_e = \frac{V_o A_o^{0.5}}{10,600} \qquad (36)$$

$V_o \leqslant 2500$ fpm:

$$L_e = \frac{A_o^{0.5}}{4.3} \qquad (37)$$

where

V_o = duct velocity, fpm
L_e = effective duct length, ft
A_o = duct area, in^2

As illustrated by Fitting SR7-1 in the Fitting Loss Coefficients section, centrifugal fans should not abruptly discharge to the atmosphere. A diffuser design should be selected from Fitting SR7-2 or SR7-3 (ASHRAE 1993).

Fan inlet conditions. For rated performance, air must enter the fan uniformly over the inlet area in an axial direction without prerotation. Nonuniform flow into the inlet is the most common cause of reduced fan performance. Such inlet conditions are not equivalent to a simple increase in the system resistance; therefore, they cannot be treated as a percentage decrease in the flow and pressure from the fan. A poor inlet condition results in an entirely new fan performance. Many poor inlet conditions affect the fan more at near-free delivery conditions than at peak pressure, so there is a continually varying difference between these two points. An elbow at the fan inlet, for example Fitting ED7-2, causes turbulence and uneven flow into the fan impeller. The losses due to the fan system effect can be eliminated by including an adquate length of straight duct between the elbow and the fan inlet.

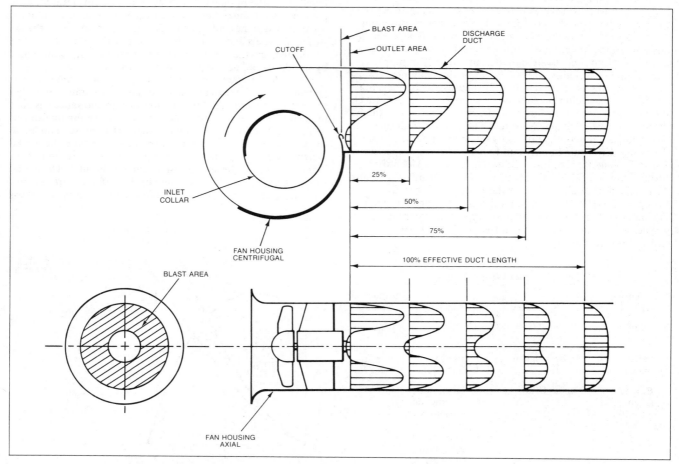

Fig. 7   Establishment of Uniform Velocity Profile in Straight Fan Outlet Duct
(Adapted by permission from AMCA Publication 201)

The ideal inlet condition allows air to enter axially and uniformly without spin. A spin in the same direction as the impeller rotation reduces the pressure-volume curve by an amount dependent on the intensity of the vortex. A counterrotating vortex at the inlet slightly increases the pressure-volume curve, but the power is increased substantially.

Inlet spin may arise from a great variety of approach conditions, and sometimes the cause is not obvious. Inlet spin can be avoided by providing an adequate length of straight duct between the elbow and the fan inlet. Figure 8 illustrates some common duct connections that cause inlet spin and includes recommendations for correcting spin.

Fans within plenums and cabinets or next to walls should be located so that air may flow unobstructed into the inlets. Fan performance is reduced if the space between the fan inlet and the enclosure is too restrictive. The system effect coefficients for fans in an enclosure or adjacent to walls are listed under Fitting ED7-1. The manner in which the airstream enters an enclosure in relation to the fan inlets also affects fan performance. Plenum or enclosure inlets or walls that are not symmetrical with the fan inlets cause uneven flow and/or inlet spin.

TAB Considerations

Fan system effects (FSEs) are not only to be used in conjunction with the system resistance characteristics in the fan selection process, but are also applied in the calculations of the results of testing, adjusting, and balancing (TAB) field tests to allow direct comparison to design calculations and/or fan performance data. Fan inlet swirl and the effect of poor fan inlet and outlet ductwork connections on system performance cannot be measured directly. Poor inlet flow patterns affect fan performance within the impeller wheel (centrifugal fan) or wheel rotor impeller (axial fan), while the fan outlet system effect is flow instability and turbulence within the fan discharge ductwork.

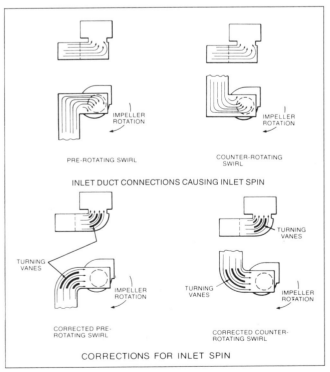

Fig. 8   Inlet Duct Connections Causing Inlet Spin and Corrections for Inlet Spin
(Adapted by permission from AMCA Publication 201)

The static pressure at the fan inlet and the static pressure at the fan outlet may be measured directly in some systems. In most cases, static pressure measurements for use in determining fan total (or static) pressure will not be made directly at the fan inlet and outlet, but at locations a relatively short distance from the fan inlet and downstream from the fan outlet. To calculate fan total pressure for this case from field measurements, use Equation (38), where $\Delta p_{x\text{-}y}$ is the summation of calculated total pressure losses between the fan inlet and outlet sections noted. If necessary, use Equation (18) to calculate fan static pressure knowing fan total pressure. For locating measurement planes and calculation procedures, consult AMCA Publication 203 (1990b).

$$P_t = (p_{s,\,5} + p_{v,\,5}) + \Delta p_{2\text{-}5} + \text{FSE}_2$$
$$+ (p_{s,\,4} + p_{v,\,4}) + \Delta p_{4\text{-}1} + \text{FSE}_1 + \text{FSE}_{1,\,s} \qquad (38)$$

where

P_t = fan total pressure, in. of water
p_s = static pressure, in. of water
p_v = velocity pressure, in. of water
FSE = fan system effect, in. of water
$\Delta p_{x\text{-}y}$ = summarization of total pressure losses between planes x and y, in. of water

Subscripts:

1 = fan inlet
2 = fan outlet
4 = plane of static pressure measurement upstream of fan
5 = plane of static pressure measurement downstream of fan
s = swirl

DUCT DESIGN CONSIDERATIONS

Space Pressure Relationships

Space pressure is determined by fan location and duct system arrangement. For example, a supply fan that pumps air into a space increases space pressure; an exhaust fan reduces space pressure. If both supply and exhaust fans are used, space pressure depends on the relative capacity of the fans. Space pressure is positive if supply exceeds exhaust; negative, if exhaust exceeds supply (Osborne 1966). System pressure variations due to wind can be minimized or eliminated by careful selection of intake air and exhaust vent locations (Chapter 14).

Fire and Smoke Control

Since duct systems can convey smoke, hot gases, and fire from one area to another and can accelerate a fire within the system, fire protection is an essential part of air-conditioning and ventilation system design. Generally, fire safety codes require compliance with the standards of national organizations. NFPA *Standard* 90A examines fire safety requirements for (1) ducts, connectors, and appurtenances; (2) plenums and corridors; (3) air outlets, air inlets, and fresh air intakes; (4) air filters; (5) fans; (6) electric wiring and equipment; (7) air cooling and heating equipment; (8) building construction, including protection of penetrations; and (9) controls, including smoke control.

Fire safety codes often refer to the testing and labeling practices of nationally recognized laboratories, such as Factory Mutual and Underwriters Laboratories (UL). The *Building Materials Directory* compiled by UL lists fire and smoke dampers that have been tested and meet the requirements of UL *Standards* 555 (1990) and 555S (1983). This directory also summarizes maximum allowable sizes for individual dampers and assemblies of these dampers. Fire dampers are 1.5 or 3-h fire-rated. Smoke dampers are classified by (1) temperature degradation [ambient air or high temperature (250 °F minimum)] and (2) leakage at 1 in. of water and 4 in. of water pressure difference (8 in. of water and 12 in. of water classification optional). Smoke dampers are tested under conditions of maximum airflow. UL's *Fire Resistance Directory* lists the fire resistance of floor/roof and ceiling assemblies with and without ceiling fire dampers.

For a more detailed presentation of fire protection, see Chapter 47 of the 1991 ASHRAE *Handbook—Applications* and the NFPA *Fire Protection Handbook* (1991).

Duct Insulation

All new construction (except low-rise residential buildings) air-handling ducts and plenums installed as part of an HVAC air distribution system should be thermally insulated in accordance with Section 9.4 of ASHRAE *Standard* 90.1 (1989). Additional insulation, vapor retarders, or both may be required to limit vapor transmission and condensation. Accessible ducts, plenums, and enclosures located in high-rise residential buildings (ASHRAE *Standard* 100.2), commercial buildings (ASHRAE *Standard* 100.3), industrial buildings (ASHRAE *Standard* 100.4), institutional buildings (ASHRAE *Standard* 100.5), and public assembly buildings (ASHRAE *Standard* 100.6) should also be thermally insulated in accordance with ASHRAE *Standard* 90.1.

Duct heat gains or losses must be known to calculate supply air quantities, supply air temperatures, and coil loads (see Chapter 26 of this volume and Chapter 2 of the 1992 ASHRAE *Handbook—Systems and Equipment*). To estimate duct heat transfer and entering or leaving air temperatures, use Equations (39) through (41).

$$Q_l = \frac{UPL}{12} \left[\left(\frac{t_e + t_l}{2} \right) - t_a \right] \tag{39}$$

$$t_e = \frac{t_l (y + 1) - 2t_a}{(y - 1)} \tag{40}$$

$$t_l = \frac{t_e (y - 1) + 2t_a}{(y + 1)} \tag{41}$$

where

y = 2.4$AV\rho/UPL$ for rectangular ducts
y = 0.6$DV\rho/UL$ for round ducts
A = cross-sectional area of duct, in²
V = average velocity, fpm
D = diameter of duct, in.
L = duct length, ft
Q_l = heat loss/gain through duct walls, Btu/h (negative for heat gain)
U = overall heat transfer coefficient of duct wall, Btu/(h·ft²·°F)
P = perimeter of bare or insulated duct, in.
ρ = density, lb$_m$/ft³
t_e = temperature of air entering duct, °F
t_l = temperature of air leaving duct, °F
t_a = temperature of air surrounding duct, °F

Use Figure 9A to determine U-factors for insulated and uninsulated ducts. Lauvray (1978) has shown the effects of (1) compressing insulation wrapped externally on sheet metal ducts and (2) insulated flexible ducts with air porous liners. For a 2-in. thick, 0.75 lb/ft³ fibrous glass blanket compressed 50% during installation, the heat transfer rate increases approximately 20% (see Figure 9A). Pervious flexible duct liners also influence heat transfer significantly (see Figure 9B). At 2500 fpm, the pervious liner U-value is 0.33 Btu/(h·ft²·°F); for an impervious liner, 0.19 Btu/(h·ft²·°F).

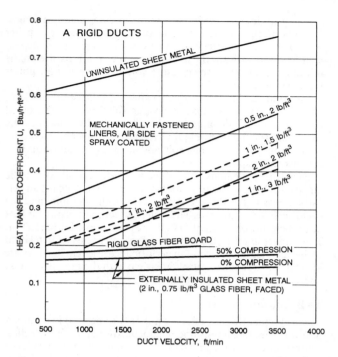

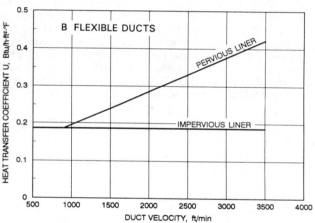

Fig. 9   Duct Heat Transfer Coefficients

Example 1. A 65-ft length of 24 in. by 36 in. uninsulated sheet metal duct, freely suspended, conveys heated air through a space maintained above freezing at 40 °F. Based on heat loss calculations for the heated zone, 17,200 cfm of standard air at a supply air temperature of 122 °F is required. The duct is connected directly to the heated zone. Determine the air temperature entering the duct and the duct heat loss.

Solution: Calculate duct velocity using Equation (10):

$$V = \frac{(144)\,(17{,}200 \text{ cfm})}{(24 \text{ in.})\,(36 \text{ in.})} = 2900 \text{ fpm}$$

Calculate entering air temperature using Equation (40):

U = 0.73 Btu/(h·ft²·°F) (from Figure 9A)

P = 2(24 in. + 36 in.) = 120 in.

$$y = \frac{(2.4)\,(24 \text{ in.})\,(36 \text{ in.})\,(2900 \text{ fpm})\,(0.075 \text{ lb}_m/\text{ft}^3)}{(0.73 \text{ Btu/h·ft}^2\text{·°F})\,(120 \text{ in.})\,(65 \text{ ft})} = 79.2$$

$$t_e = \frac{122\,°F\,(79.2 + 1) - (2 \times 40\,°F)}{(79.2 - 1)} = 124.1\,°F$$

Calculate duct heat loss using Equation (39):

$$Q_l = \frac{[0.73\ \text{Btu}/(\text{h}\cdot\text{ft}^2\cdot°F)]\ (120\ \text{in.})\ (65\ \text{ft})}{12}$$

$$\times \left[\frac{124.1\,°F + 122\,°F}{2} - 40\,°F \right]$$

$$= 39{,}200\ \text{Btu/h}$$

Example 2. Same as Example 1, except the duct is insulated externally with 2 in. thick fibrous glass with a density of 0.75 lb/ft³. The insulation is wrapped with 0% compression.

Solution: All values, except U and P, remain the same as Example 1. From Figure 9A, $U = 0.15\ \text{Btu}/(\text{h}\cdot\text{ft}^2\cdot°F)$ at 2900 fpm. $P = 136$ in. Therefore:

$$y = 441$$
$$t_e = 122.4\,°F$$
$$Q_l = 9083\ \text{Btu/h}$$

Insulating this duct reduces heat loss to 20% of the uninsulated duct.

Duct System Leakage

Leakage in all unsealed ducts varies considerably with the fabricating machinery used, the methods for assembly, and installation workmanship. For sealed ducts, a wide variety of sealing methods and products exists. Each has a relatively short shelf life, and no documented research has identified the in-service aging characteristics of sealant applications. Many sealants contain volatile solvents that evaporate and introduce shrinkage and curing factors. Surface cleanliness and sealant application in relation to air pressure direction (infiltration and exfiltration) are other variables. With the exception of pressure-sensitive adhesive tapes, no standard tests exist to evaluate performance and grade sealing products. A variety of sealed and unsealed duct leakage tests (AISI/SMACNA 1972, ASHRAE/SMACNA/TIMA 1985, ASHRAE 1988) have confirmed that longitudinal seam, transverse joint, and assembled duct leakage can be represented by Eqation (42), and that for the same construction leakage is not significantly different in the negative and positive modes. A range of leakage rates for longitudinal seams commonly used in the construction of metal ducts is presented in Table 5. Longitudinal seam leakage for metal ducts is about 10 to 15% of total duct leakage.

$$Q = C\,\Delta p_s{}^N \tag{42}$$

where

Q = duct leakage rate, cfm
C = constant reflecting area characteristics of leakage path
Δp_s = static pressure differential from duct interior to exterior
N = exponent relating turbulent or laminar flow in leakage path

Table 5   Unsealed Longitudinal Seam Leakage, Metal Ducts

Type of Duct/Seam	Leakage[a], cfm/ft (Seam Length)	
	Range	Average
Rectangular		
Pittsburgh lock	0.01 to 0.56	0.16
Button punch snaplock	0.01 to 0.16	0.08
Round		
Snaplock	0.04 to 0.14	0.11
Grooved	0.11 to 0.18	0.12

[a]Leakage rate is at 1 in. of water static pressure.

Analysis of the AISI/ASHRAE/SMACNA/TIMA data resulted in the categorization of duct systems into a *leakage class* C_L based on Equation (43), where the exponent N is assumed to be 0.65. A selected series of leakage classes based on Equation (43) is shown in Figure 10.

$$C_L = Q/\Delta p_s{}^{0.65} \tag{43}$$

where

Q = leakage rate, cfm/100 ft² (surface area)
C_L = leakage class, cfm per 100 ft² duct surface at 1 in. of water static pressure

Table 6 is a summary of the leakage class attainable for quality duct construction and sealing practices. Connections of ducts to grilles, diffusers, and registers are not represented in the test data. The designer is responsible for assigning acceptable leakage rates. Although leakage as a percentage of fan flow rate is an important evaluation criterion (see Table 7), designers should first become familiar with the leakage rates from selected construction detail. This knowledge allows the designer to analyze both first cost and life cycle cost of a duct system so that the owner may benefit. In performing an analysis, the designer should independently account for air leakage in casings and frames of equipment in the duct system. Casings or volume-controlling air terminal units may leak 2 to 5% of their maximum flow. The effects of such leakage should be anticipated, if allowed, and the ductwork should not be expected to compensate for equipment leakage. Allowable leakage should be controlled consistent with airflow tolerances at the air terminals (see Chapter 34 of the 1991 ASHRAE *Handbook— Applications*). A leakage class of 3 is attainable for all duct systems by careful selection of joints and sealing methods and by good workmanship. Where less leakage is required, designers should understand that less experienced contractors may have difficulty meeting their requirements. Zero leakage is not a practical objective, except in critical situations such as nuclear safety-related applications. For additional discussion of leakage analysis, consult the HVAC *Air Duct Leakage Test Manual* (SMACNA 1985).

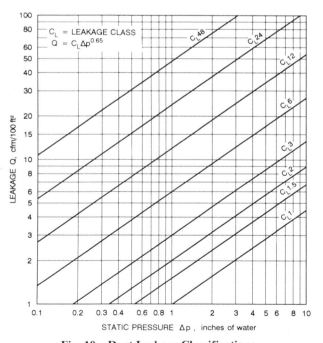

Fig. 10   Duct Leakage Classifications
[Adapted with permission from
HVAC *Air Duct Leakage Test Manual* (SMACNA 1985)]

Energy loss via duct leakage is a serious concern. It is also of special importance in systems serving clean rooms, solar energy air collectors, industrial processes, and moisture removal applications. Sealing criteria appropriate for each special application should be specified by the designer. All new construction (except low-rise residential buildings) ductwork designed to operate at static pressures in excess of 3 in. of water should be, or may be required by code to be leak tested in accordance with Section 9.4 of ASHRAE *Standard* 90.1 (1989). Allowable leakage shall also be in compliance with ASHRAE *Standard* 90.1. Pressure-sensitive tape shall not be used as the primary sealant where such ducts are designed to operate at static pressures of 1 in. of water or greater. Soldered or welded duct construction is necessary where sealants are not suitable.

Shaft and compartment pressure changes affect duct leakage and are important to health and safety in the design and operation of contaminant and smoke control systems. Airflow around buildings, building component leakage, and the distribution of inside and outside pressures over the height of a building, including shafts, are discussed in Chapters 14 and 22. Smoke control system design is covered in Chapter 47 of the 1991 ASHRAE *Handbook—Applications* and in Klote and Milke (1992).

System Component Design Velocities

Table 8 summarizes face velocities for HVAC components in built-up systems. In most cases, the values are abstracted from pertinent chapters in the 1992 ASHRAE *Handbook—Systems and Equipment*; final selection of the components should be based on data in these chapters or from manufacturers.

Louvers require special treatment since the blade shapes, angles, and spacing cause significant variations in louver-free area and performance (pressure drop and water penetration). Selection and analysis should be based on test data obtained in accordance with AMCA *Standard* 500 (1989). This standard presents not only the

pressure drop and water penetration test procedures, but a uniform method for calculating louver-free area. Tests are conducted on a 48-in. square louver with the frame mounted flush in the wall. For the water penetration tests, the rainfall is 4 in/h, no wind, and the water flow down the wall is 0.25 gpm per linear foot of louver width.

Use Figure 11 for preliminary sizing of air intake and exhaust louvers. For air quantities greater than 7000 cfm per louver, the air intake gross louver openings are based on 400 fpm; for exhaust louvers, 500 fpm is used for air quantities of 5000 cfm per louver and greater. For air quantities less than these, refer to Figure 11.

Table 7 Leakage as Percentage of Airflow[a,b]

Leakage Class	System cfm/ft^2 Duct Surface[c]	Static Pressure, in. of water					
		0.5	1	2	3	4	6
48	2	15	24	38	49	59	77
	2.5	12	19	30	39	47	62
	3	10	16	25	33	39	51
	4	7.7	12	19	25	30	38
	5	6.1	9.6	15	20	24	31
24	2	7.7	12	19	25	30	38
	2.5	6.1	9.6	15	20	24	31
	3	5.1	8.0	13	16	20	26
	4	3.8	6.0	9.4	12	15	19
	5	3.1	4.8	7.5	9.8	12	15
12	2	3.8	6	9.4	12	15	19
	2.5	3.1	4.8	7.5	9.8	12	15
	3	2.6	4.0	6.3	8.2	9.8	13
	4	1.9	3.0	4.7	6.1	7.4	9.6
	5	1.5	2.4	3.8	4.9	5.9	7.7
6	2	1.9	3	4.7	6.1	7.4	9.6
	2.5	1.5	2.4	3.8	4.9	5.9	7.7
	3	1.3	2.0	3.1	4.1	4.9	6.4
	4	1.0	1.5	2.4	3.1	3.7	4.8
	5	0.8	1.2	1.9	2.4	3.0	3.8
3	2	1.0	1.5	2.4	3.1	3.7	4.8
	2.5	0.8	1.2	1.9	2.4	3.0	3.8
	3	0.6	1.0	1.6	2.0	2.5	3.2
	4	0.5	0.8	1.3	1.6	2.0	2.6
	5	0.4	0.6	0.9	1.2	1.5	1.9
2	2	0.5	1.0	1.4	1.7	2.0	2.4
	2.5	0.4	0.8	1.1	1.4	1.6	2.0
	3	0.3	0.7	0.9	1.2	1.3	1.6
	4	0.3	0.5	0.7	0.9	1.0	1.2
	5	0.2	0.4	0.6	0.7	0.8	1.0
1.5	2	0.5	0.8	1.1	1.3	1.5	1.8
	2.5	0.4	0.6	0.8	1.0	1.2	1.5
	3	0.4	0.5	0.7	0.9	1.0	1.2
	4	0.3	0.4	0.5	0.6	0.8	0.9
	5	0.2	0.3	0.4	0.5	0.6	0.7
1	2	0.4	0.5	0.7	0.9	1.0	1.2
	2.5	0.3	0.4	0.6	0.7	0.8	1.0
	3	0.2	0.3	0.5	0.6	0.7	0.8
	4	0.2	0.3	0.4	0.4	0.5	0.6
	5	0.1	0.2	0.3	0.3	0.4	0.5

[a]Adapted with permission from HVAC *Air Duct Leakage Test Manual* (SMACNA 1985, Appendix A).
[b]Percentage applies to the airflow entering a single section of duct operating at an assumed pressure equal to the average of the upstream and downstream pressures. When several duct pressure classifications occur in a system, ductwork in each pressure class should be evaluated independently to arrive at an aggregate leakage for the system.
[c]The ratios in this column are typical of fan volume flow rate divided by total system surface. Portions of the systems may vary from these averages.

Table 6 Duct Leakage Classification[a]

Type Duct	Predicted Leakage Class C_L	
	Sealed[b, c]	Unsealed[c]
Metal (flexible excluded)		
Round and flat oval	3	30 (6 to 70)
Rectangular		
≤ 2 in. of water (both positive and negative pressures)	12	48 (12 to 110)
> 2 and ≤ 10 in. of water (both positive and negative pressures)	6	48 (12 to 110)
Flexible		
Metal, Aluminum	8	30 (12 to 54)
Nonmetal	12	30 (4 to 54)
Fibrous glass		
Rectangular	6	NA
Round	3	NA

[a]The *leakage classes* listed in this table are averages based on tests conducted by AISI/SMACNA (1972), ASHRAE/SMACNA/TIMA (1985), and ASHRAE (1988). Leakage classes listed are not necessarily recommendations on allowable leakage. The designer should determine allowable leakage and specify acceptable duct leakage classifications.
[b]The *leakage classes* listed in the sealed category are based on the assumptions that for metal ducts, all transverse joints, seams, and openings in the ductwall are sealed at pressures over 3 in. of water, that transverse joints and longitudinal seams are sealed at 2 and 3 in. of water, and that transverse joints are sealed below 2 in. of water. Lower leakage classes are obtained by careful selection of joints and sealing methods.
[c]*Leakage classes* assigned anticipate about 25 joints per 100 linear foot of duct. For systems with a high fitting to straight duct ratio, greater leakage occurs in both the sealed and unsealed conditions.

Table 8 Typical Design Velocities for HVAC Components

Duct Element	Face Velocity, fpm
LOUVERS[a]	
Intake	
7000 cfm and greater	400
Less than 7000 cfm	See Fig. 11
Exhaust	
5000 cfm and greater	500
Less than 5000 cfm	See Fig. 11
FILTERS[b]	
Panel filters	
Viscous impingement	200 to 800
Dry-type, extended-surface	
Flat (low efficiency)	Duct Velocity
Pleated media (intermediate efficiency)	Up to 750
HEPA	250
Renewable media filters	
Moving-curtain viscous impingement	500
Moving-curtain dry-media	200
Electronic air cleaners	
Ionizing type	150 to 350
HEATING COILS[c]	
Steam and hot water	500 to 1000
	200 min., 1500 max.
Electric	
Open wire	Refer to mfg. data
Finned tubular	Refer to mfg. data
DEHUMIDIFYING COILS[d]	400 to 500
AIR WASHERS[e]	
Spray type	300 to 600
Cell type	Refer to mfg. data
High-velocity, Spray type	1200 to 1800

[a] Based on assumptions presented in text.
[b] Abstracted from Chapter 25, 1992 ASHRAE *Handbook—Systems and Equipment.*
[c] Abstracted from Chapter 24, 1992 ASHRAE *Handbook—Systems and Equipment.*
[d] Abstracted from Chapter 21, 1992 ASHRAE *Handbook—Systems and Equipment.*
[e] Abstracted from Chapter 19, 1992 ASHRAE *Handbook—Systems and Equipment.*

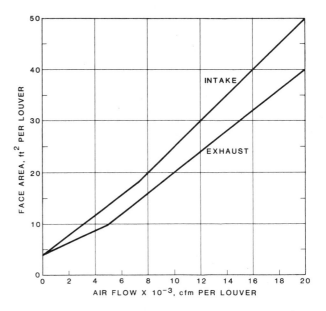

Parameters Used to Establish Figure	Intake Louver	Exhaust Louver
Minimum free area (48-in. square test section), %	45	45
Water penetration, oz/(ft² · 0.25 h)	Negligible (less than 0.2)	NA
Maximum static pressure drop, in. of water	0.15	0.25

Fig. 11   Criteria for Louver Sizing

These criteria are presented on a per louver basis (*i.e.*, each louver in a bank of louvers) to include each louver frame. Representative production-run louvers were used in establishing Figure 11, and all data used in that analysis are based on AMCA standard tests. For louvers larger than 16 ft², the free areas are greater than 45%, while for louvers less than 16 ft², the free areas are less than 45%. Unless specific louver data are analyzed, no louver should have a face area less than 4 ft². If debris collection on the screen of an intake louver is possible, or if louvers are located at grade with adjacent pedestrian traffic, louver face velocity should not exceed 100 fpm.

System and Duct Noise

The major sources of noise from air-conditioning systems are diffusers, grilles, fans, ducts, fittings, and vibrations. Chapter 42 of the 1991 ASHRAE *Handbook—Applications* discusses sound control for each of these sources. Sound control for terminal devices consists of selecting devices that meet the design goal under all operating conditions and installing them properly so that no additional sound is generated. The sound power output of a fan is determined by the type of fan, airflow, and pressure. Sound control in the duct system requires proper duct layout, sizing, and provision for installing duct attenuators, if required. The noise generated by a system increases with both duct velocity and system pressure. Chapter 42 of the 1991 ASHRAE *Handbook—Applications* presents methods for calculating required sound attenuation.

Testing and Balancing

Each air duct system should be tested, adjusted, and balanced. Detailed procedures are given in Chapter 34 of the 1991 ASHRAE *Handbook—Applications*. To properly determine fan total (or static) pressure from field measurements taking into account fan system effect, refer to the Fan-System Interface section. Equation (38) allows direct comparison of system resistance to design calculations and/or fan performance data. It is important that the system effect magnitudes be known prior to testing. If necessary, use Equation (18) to calculate fan static pressure knowing fan total pressure [Equation (38)]. For TAB calculation procedures of numerous fan/system configurations encountered in the field, refer to AMCA Publication 203 (1990b).

DUCT DESIGN METHODS

Duct design methods for HVAC systems, and exhaust systems conveying vapors, gases, and smoke are *equal friction, static regain,* and the *T-method*. The Industrial Exhaust System Duct Design section presents the design criteria and procedures for exhaust systems conveying particulates. Equal friction and static regain are nonoptimizing methods, while the T-method is a practical optimization method introduced by Tsal (1988).

To assure that system designs are acoustically acceptable, noise generation should be analyzed and sound attenuators and/or acoustically lined duct provided where necessary. Dampers must be installed throughout systems designed by equal friction, static regain, and the T-method, since inaccuracies are introduced into these design methods by duct size round-off and the effect on the total pressure loss calculations by close-coupled fittings.

Equal Friction Method

In the equal friction method, ducts are sized for a constant pressure loss per unit length. The shaded area of the friction chart (Figure 5) is the suggested range of friction rate and air velocity. When energy cost is high and installed duct work cost is low, a low friction rate design is more economical. For low energy cost and high duct cost, a higher friction rate is more economical. After initial sizing, calculate the total pressure loss for all duct sections, then resize sections to balance pressure losses at each junction.

Static Regain Method

The objective of the static regain method is to obtain the same static pressure at diverging flow junctions by changing downstream duct sizes. This design objective can be developed by rearranging Equation (7) and setting p_2 equal to p_1 (neglect stack effect term). Thus

$$p_1 - p_2 = \Delta p - \left[\frac{\rho V_1^2}{2g_c} - \frac{\rho V_2^2}{2g_c} \right] \quad (44)$$

and

$$\Delta p = \frac{\rho V_1^2}{2g_c} - \frac{\rho V_2^2}{2g_c} \quad (45)$$

where Δp is the total pressure loss from upstream of junction 1 to upstream of junction 2, or the terminal of section 2. Since Δp can be calculated, the immediate downstream duct size that satisfies Equation (45) is determined by iteration. Equation (45) can be solved only when the downstream velocity is reduced, which may require the downstream duct size to be greater than the upstream duct. If this solution is undesirable, Equation (44) may be used by maintaining the same duct size downstream as upstream to solve for the static pressure at the next junction.

To start the design of a system, a maximum velocity is selected for the root section (duct section upstream and/or downstream of a fan). In Figure 13, section 6 is the root for the return air subsystem. Section 19 is the root for the supply air subsystem. The shaded area on the friction chart (Figure 5) is the suggested range of air velocity. When energy cost is high and installed ductwork cost is low, a lower initial velocity is more economical. For low energy cost and high duct cost, a higher velocity is more economical. All other sections, except terminal sections, are sized iteratively by Equation (45). In Figure 13, terminal sections are 1, 2, 4, 7, 8, 11, 12, 15, and 16. Knowing the terminal static pressure requirements, Equation (44) is used to calculate the duct size of terminal sections. If the terminal is an exit fitting rather than a register, diffuser, or terminal box, the static pressure at the exit of the terminal section is zero.

The classical static regain method (Carrier 1960, Chun-Lun 1983, Shataloff 1966) is based on Equation (46), where R is the static pressure regain factor, and Δp_r is the static pressure regain between junctions.

$$\Delta p_r = R \left[\frac{\rho V_1^2}{2g_c} - \frac{\rho V_2^2}{2g_c} \right] \quad (46)$$

Typically R-factors ranging from 0.5 to 0.95 have been used. Tsal and Behls (1988) show that this uncertainty results because the

splitting of mass at junctions and the dynamic (fitting) losses between junctions are ignored. The classical static regain method using an R-factor should not be used since R is not predictable.

T-Method, Optimization

T-method optimization (Tsal *et al.* 1988) is a procedure based on the same tee-staging idea as dynamic programming (Bellman 1957). However, phase level vector tracing is eliminated by optimizing locally at each stage. This modification reduces the number of calculations, but requires iteration. Usually, three iterations are sufficient.

Optimization basis. The objective function, Equation (47), includes both initial system cost and the present worth of energy. Hours of operation, annual escalation and interest rates, and amortization period are also required for optimization.

$$E = E_p \, (\text{PWEF}) + E_s \quad (47)$$

where

$\quad E$ = present worth owning and operating cost
$\quad E_p$ = first year energy cost
$\quad E_s$ = initial cost
PWEF = present worth escalation factor (Smith 1968), dimensionless

Energy cost is determined by

$$E_p = Q_f \left[\frac{1.176 \times 10^4 \, (E_d + E_c \, T)}{\eta_f \, \eta_e} \right] P_t \quad (48)$$

where

$\quad Q_f$ = fan airflow rate, cfm
$\quad E_c$ = unit energy cost, cost/kWh
$\quad E_d$ = energy demand cost, cost/kW
$\quad T$ = system operating time, h/year
$\quad P_t$ = fan total pressure, in. of water
$\quad \eta_f$ = fan total efficiency, decimal
$\quad \eta_e$ = motor-drive efficiency, decimal

Energy cost depends on both applicable energy rates E_c and demand cost E_d. Since the difference between the fan pressures for an optimized and nonoptimized system is a small part of demand, it is usually neglected. Initial cost includes ducts and HVAC equipment, which is primarily the central handling unit. The cost of duct systems is given by the following equations:

Round $\qquad\qquad E_s = S_d \, \pi D L / 12 \qquad\qquad (49)$

Rectangular $\quad E_s = 2 \, S_d \, (H + W) \, L / 12 \qquad (50)$

where

$\quad S_d$ = unit ductwork cost/ft^2 (including material and labor)
$\quad H$ = duct height, in.
$\quad W$ = duct width, in.
$\quad L$ = duct length, ft

The cost of space required by ducts and equipment is another important factor of duct optimization. Including this cost reduces the size of ducts, thereby increasing energy consumption. Since the space available for ductwork is usually not used for anything else, its cost is ignored.

Both electrical energy rates and duct work costs vary widely. For example, electric energy costs vary by a factor of 9 to 1 between New York and Seattle (DOE 1984). Black iron rectangular ductwork can cost about 3.9 times that of spiral ductwork (Wendes 1989). Combining these ratios yields a factor of 34 to 1 based on locale and type of ductwork. Therefore, a great potential exists for reducing duct system life-cycle cost due to energy and ductwork cost variations.

The following constraints are necessary for duct optimization (Tsal and Adler 1987):

- *Continuity.* For each node, the flow in equals the flow out.
- *Pressure balancing.* The total pressure loss in each path must equal the fan total pressure; or, in effect, at any junction, the total pressure loss for all paths is the same.
- *Nominal duct size.* Ducts are constructed in discrete, nominal sizes. Each diameter of a round duct or height and width of a rectangular duct is rounded to the nearest increment, usually 1 or 2 in. If a lower nominal size is selected, the initial cost decreases, but the pressure loss increases and may exceed the fan pressure. If the higher nominal size is selected, the opposite is true—the initial cost increases, but the section pressure loss decreases. However, this lower pressure at one section may allow smaller ducts to be selected for sections that follow. Therefore, optimization must consider size rounding.
- *Air velocity restriction.* The maximum allowable velocity is an acoustic limitation (ductwork regenerated noise).
- *Construction restriction.* Architectural limits may restrict duct sizes. If air velocity or construction constraints are violated during an iteration, a duct size must be calculated. The pressure loss calculated for this preselected duct size is considered a fixed loss.

Calculation procedure. The T-method comprises the following major procedures:

- *System condensing.* This procedure condenses a multiple-section duct system into a single imaginary duct section with identical hydraulic characteristics and the same owning cost as the entire system. By Tsal *et al.* [1988, Equation (1.41)], two or more converging or diverging sections and the common section at a junction can be replaced by one condensed section. By applying Tsal's equation from junction to junction in the direction to the root section (fan), the entire supply and return systems can be condensed into one section (a single resistance).
- *Fan selection.* From the condensed system, the ideal optimum fan total pressure P_t^{opt} is calculated and used to select a fan. If a fan with a different pressure is selected, its pressure (P^{opt}) is considered optimum.
- *System expansion.* The expansion process distributes the available fan pressure P^{opt} throughout the system. Unlike the condensing procedure, the expansion procedure starts at the root section and continues in the direction of the terminals.

Economic analysis. Tsal *et al.* (1988) describe the calculation procedure and include an economic analysis of the T-method.

T-Method, Simulation

T-method simulation, also developed by Tsal (1990), determines the flow in each duct section of an existing system with a known operating fan performance curve. The simulation version of T-method converges very efficiently. Usually three iterations are sufficient to obtain a solution with a high degree of accuracy.

Calculation procedure. The simulation version of T-method includes the following major procedures:

- *System condensing.* This procedure condenses a branched tee system into a single imaginary duct section with identical hydraulic characteristics. Two or more converging or diverging sections and the common section at a junction can be replaced by one condensed section [by Equation (18) in Tsal *et al.* (1990)]. By applying Tsal's equation from junction to junction in the direction to the root section (fan), the entire system, including supply and return subsystems, can be condensed into one imaginary section (a single resistance).
- *Fan operating point.* This step determines the system flow and pressure by locating the intersection of the fan performance and

system curves, where the system curve is represented by the imaginary section from the previous step.
- *System expansion.* Knowing system flow and pressure, the previously condensed imaginary duct section is expanded into the original system with flow distributed in accordance to the ratio of pressure losses calculated in the system condensing step.

Simulation applications. The need for duct system simulation appears in many HVAC problems. In addition to the following concerns that can be clarified by simulation, T-method is an excellent design tool for simulating the flow distribution within a system with various modes of operation.

- Flow distribution in a variable air volume (VAV) system due to terminal box flow diversity
- Airflow redistribution due to HVAC system additions and/or modifications
- System airflow analysis for partially occupied buildings
- Necessity to replace fans and/or motors when retrofitting an air distribution system
- Multiple-fan system operating condition when one or more fans shut down
- Pressure differences between adjacent confined spaces within a nuclear facility when a design basis accident (DBA) occurs (Farajian *et al.* 1992)
- Smoke control system performance during a fire, when certain fire/smoke dampers close and others remain open

HVAC DUCT DESIGN PROCEDURES

The general procedure for HVAC system duct design is as follows:

1. Study the building plans and arrange the supply and return outlets to provide proper distribution of air within each space. Adjust calculated air quantities for duct heat gains or losses and duct leakage. Also, adjust the supply, return, and/or exhaust air quantities to meet space pressurization requirements.
2. Select outlet sizes from manufacturers' data (see Chaper 31).
3. Sketch the duct system, connecting supply outlets and return intakes with the air-handling units/air conditioners. Space allocated for supply and return ducts often dictates system layout and ductwork shape. Use round ducts whenever feasible.
4. Divide the system into sections and number each section. A duct system should be divided at all points where flow, size, or shape changes. Assign fittings to the section toward the supply and return (or exhaust) terminals. The following examples are for the fittings identified for Example 3 (Figure 12), and system section numbers assigned (Figure 13). For converging flow fitting 3, assign the straight-through flow to section 1 (toward terminal 1), and the branch to section 2 (toward terminal 4). For diverging flow fitting 24, assign the straight-through flow to section 13 (toward terminals 26 and 29) and the branch to section 10 (toward terminals 43 and 44). For transition fitting 11, assign the fitting to upstream section 4 [toward terminal 9 (intake louver)]. For fitting 20, assign the unequal area elbow to downstream section 9 (toward diffusers 43 and 44). The fan outlet diffuser, fitting 42, is assigned to section 19 (again, toward the supply duct terminals).
5. Size ducts by the selected design method. Calculate system total pressure loss; then select the fan (refer to Chapter 18 in the 1992 ASHRAE *Handbook—Systems and Equipment*).
6. Lay out the system in detail. If duct routing and fittings vary significantly from the original design, recalculate the pressure losses. Reselect the fan if necessary.
7. Resize duct sections to approximately balance pressures at each junction.

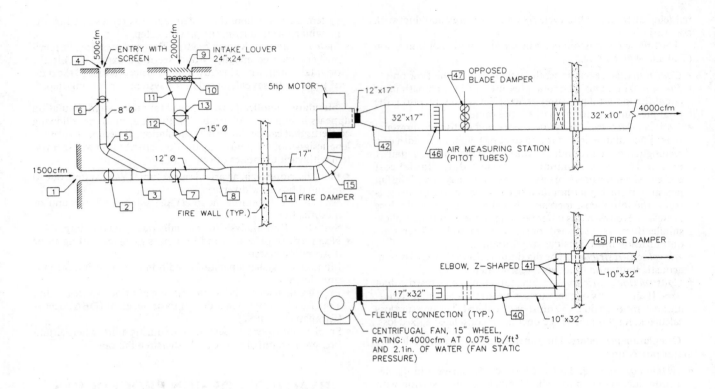

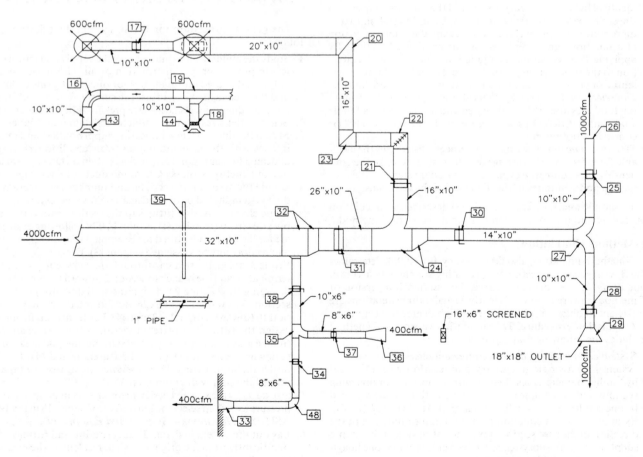

Fig. 12   Example 3 Schematic

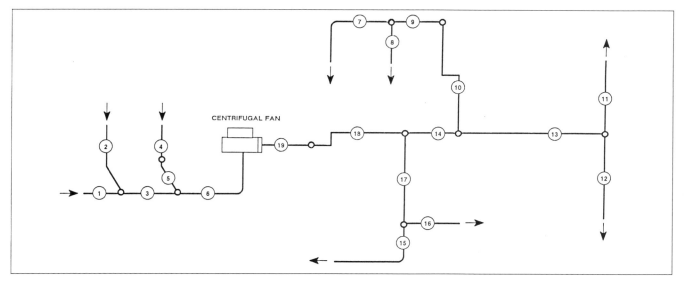

Fig. 13   Example 3—System Schematic with Section Numbers

8. Analyze the design for objectionable noise levels and specify sound attenuators as necessary. Refer to the section System and Duct Noise.

Example 3. For the system illustrated by Figure 12, size the ductwork by the equal friction method, and pressure balance the system by changing duct sizes (use 1-in. increments). Determine the system resistance and total pressure unbalance at the junctions. The airflow quantities are actual values adjusted for heat gains or losses, and ductwork is sealed (assume no leakage), galvanized steel ducts with transverse joints on 4-ft centers (ϵ = 0.0003 ft). Air is at standard conditions (0.075 lb_m/ft^3 density).

Since the primary purpose of Figure 12 is to illustrate calculation procedures, its duct layout is not typical of any real duct system. The layout includes fittings from the local loss coefficient tables, with emphasis on converging and diverging tees and various types of entries and discharges. The supply system is constructed of rectangular ductwork; the return system, round ductwork.

Solution: See Figure 13 for section numbers assigned to the system. The duct sections are sized within the suggested range of friction rate shown on the Friction Chart (Figure 5). Tables 9 and 10 give the total pressure loss calculations and the supporting summary of loss coefficients by sections. The straight duct friction factor and pressure loss were calculated by Equations (19) and (20). The fitting loss coefficients are from the ASHRAE Duct Fitting Database (1993). Loss coefficients were calculated automatically by the database program (not by manual interpolation). The pressure loss values in Table 9 for the diffusers (fittings 43 and 44), the louver (fitting 9), and the air-measuring station (fitting 46) are manufacturers' data.

The pressure unbalance at the junctions may be noted by referring to Figure 14, the total pressure grade line for the system. The system resistance P_t is 2.63 in. of water. Noise levels and the need for duct silencers were not evaluated. To calculate the fan static pressure, use Equation (18):

$$P_s = 2.63 - 0.50 = 2.1 \text{ in. of water}$$

where 0.50 in. of water is the fan outlet velocity pressure.

INDUSTRIAL EXHAUST SYSTEM DUCT DESIGN

Chapter 27 of the 1991 ASHRAE *Handbook—Applications* discusses design criteria, including hood design, for industrial exhaust systems. Exhaust systems conveying vapors, gases, and smoke can be designed by equal friction, static regain, or T-method. Systems conveying particulates are designed by the constant velocity method at duct velocities adequate to convey particles to

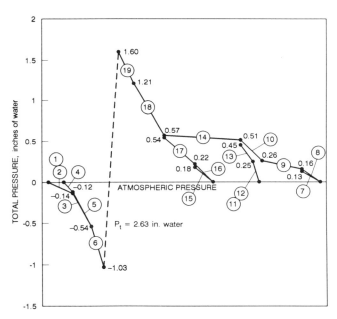

Fig. 14   Total Pressure Grade Line for Example 3

the system air cleaner. For contaminant transport velocities, see Table 2, Chapter 27 of the 1991 ASHRAE *Handbook—Applications*.

Two pressure-balancing methods can be considered when designing industrial exhaust systems. One method uses balancing devices (*e.g.*, dampers, blast gates) to obtain design airflow through each hood. The other approach balances systems by adding resistance to ductwork sections, *i.e.*, changing duct size, selecting different fittings, and increasing airflow. This self-balancing method is preferred, especially for systems conveying abrasive materials. Where potentially explosive or radioactive materials are conveyed, the prebalanced system is mandatory, since contaminants could accumulate at the balancing devices. To balance systems by increasing airflow, use Equation (51), which assumes that all ductwork has the same diameter and that fitting loss coefficients, including main and branch tee coefficients, are constant.

Table 9   Example 3—Total Pressure Loss Calculations by Sections

Duct Section[a]	Fitting No.[b]	Duct Element	Airflow, cfm	Duct Size (Equivalent Round)	Velocity, fpm	Velocity Pressure, in. of water	Duct Length[c], ft	Summary of Fitting Loss Coefficients[d]	Duct Pressure Loss/100 ft, in. of water[e]	Total Pressure Loss, in. of water	Section Pressure Loss, in. of water
1	—	Duct	1500	12 in. ϕ	1910	—	15	—	0.40	0.06	
	—	Fittings	1500	—	1910	0.23	—	0.33	—	0.08	0.14
2	—	Duct	500	8 in. ϕ	1432	—	60	—	0.39	0.23	
	—	Fittings	500	—	1432	0.13	—	−0.71	—	−0.09	0.14
3	—	Duct	2000	12 in. ϕ	2546	—	20	—	0.67	0.13	
	—	Fittings	2000	—	2546	0.40	—	0.67	—	0.27	0.40
4	—	Duct	2000	24 in. × 24 in. (26.2)	500	—	5	—	0.01	0.00	
	—	Fittings	2000	—	500	0.02	—	1.11	—	0.02	
	9	Louver	2000	24 in. × 24 in.	—	—	—	—	—	0.10[f]	0.12
5	—	Duct	2000	15 in. ϕ	1630	—	55	—	0.23	0.13	
	—	Fittings	2000	—	1630	0.17	—	1.73	—	0.29	0.42
6	—	Duct	4000	17 in. ϕ	2538	—	30	—	0.45	0.14	
	—	Fittings	4000	—	2538	0.40	—	0.87	—	0.35	0.49
7	—	Duct	600	10 in. × 10 in. (10.9)	864	—	14	—	0.11	0.02	
	—	Fittings	600	—	864	0.05	—	0.26	—	0.01	
	43	Diffuser	600	10 in. × 10 in.	—	—	—	—	—	0.10[f]	0.13
8	—	Duct	600	10 in. × 10 in. (10.9)	864	—	4	—	0.11	0.00	
	—	Fittings	600	—	864	0.05	—	1.25	—	0.06	
	44	Diffuser	600	10 in. × 10 in.	—	—	—	—	—	0.10[f]	0.16
9	—	Duct	1200	20 in. × 10 in. (15.2)	864	—	25	—	0.08	0.02	
	—	Fittings	1200	—	864	0.05	—	1.67	—	0.08	0.10
10	—	Duct	1200	16 in. × 10 in. (13.7)	1080	—	45	—	0.13	0.06	
	—	Fittings	1200	—	1080	0.07	—	2.66	—	0.19	0.25
11	—	Duct	1000	10 in. × 10 in. (10.9)	1440	—	10	—	0.29	0.03	
	—	Fittings	1000	—	1440	0.13	—	1.68	—	0.22	0.25
12	—	Duct	1000	10 in. × 10 in. (10.9)	1440	—	22	—	0.29	0.06	
	—	Fittings	1000	—	1440	0.13	—	1.45	—	0.19	0.25
13	—	Duct	2000	14 in. × 10 in. (12.9)	2057	—	35	—	0.47	0.16	
	—	Fittings	2000	—	2057	0.26	—	0.16	—	0.04	0.20
14	—	Duct	3200	26 in. × 10 in. (17.1)	1772	—	15	—	0.27	0.04	
	—	Fittings	3200	—	1772	0.20	—	0.12	—	0.02	0.06
15	—	Duct	400	8 in. × 6 in. (7.6)	1200	—	40	—	0.32	0.13	
	—	Fittings	400	—	1200	0.09	—	0.58	—	0.05	0.18
16	—	Duct	400	8 in. × 6 in. (7.6)	1200	—	20	—	0.32	0.06	
	—	Fittings	400	—	1200	0.09	—	1.74	—	0.16	0.22
17	—	Duct	800	10 in. × 6 in. (8.4)	1920	—	22	—	0.70	0.15	
	—	Fittings	800	—	1920	0.23	—	0.76	—	0.17	0.32
18	—	Duct	4000	32 in. × 10 in. (18.8)	1800	—	23	—	0.25	0.06	
	—	Fittings	4000	—	1800	0.20	—	2.91	—	0.58	0.64
19	—	Duct	4000	32 in. × 17 in. (25.2)	1059	—	12	—	0.06	0.01	
	—	Fittings	4000	—	1059	0.07	—	4.71	—	0.33	
	46	Air measuring station	4000	—	—	—	—	—	—	0.05[f]	0.39

[a] See Figure 13.
[b] See Figure 12.
[c] Length of ducts are to fitting centerlines.
[d] See Table 10.
[e] Duct pressure based on a 0.0003 ft absolute roughness factor.
[f] Pressure drop based on manufacturers' data.

Table 10   Example 3—Loss Coefficient Summary by Sections

Duct Section	Fitting Number	Type of Fitting	ASHRAE Fitting No.[a]	Parameters	Loss Coefficient
1	1	Entry	ED1-3	$r/D = 0.2$	0.03
	2	Damper	CD9-1	$\theta = 0°$	0.19
	3	Wye (30°), Main	ED5-1[b]	$A_s/A_c = 1.0$, $A_b/A_c = 0.444$, $Q_s/Q_c = 0.75$	0.11 (C_s)
		Summation of Section 1 loss coefficients			0.33
2	4	Entry	ED1-1[b]	$L = 0$, $t = 0.064$ in. (16 gage)	0.50
	4	Screen	CD6-1[b]	$n = 0.70$, $A_1/A_o = 1$	0.58
	5	Elbow	CD3-6	$60°$, $r/D = 1.5$, pleated	0.27
	6	Damper	CD9-1	$\theta = 0°$	0.19
	3	Wye (30°), Branch	ED5-1[b]	$A_s/A_c = 1.0$, $A_b/A_c = 0.444$, $Q_b/Q_c = 0.25$	−2.25 (C_b)
		Summation of Section 2 loss coefficients			−0.71
3	7	Damper	CD9-1	$\theta = 0°$	0.19
	8	Wye (45°), Main	ED5-2	$Q_s/Q_c = 0.5$, $A_s/A_c = 0.498$, $A_b/A_c = 0.779$	0.48 (C_s)
		Summation of Section 3 loss coefficients			0.67

[a] From Duct Fitting Database (ASHRAE 1993).
[b] Duct Fitting Database (ASHRAE 1993) data for this fitting reproduced in the Fitting Loss Coefficients section.

Table 10 Example 3—Loss Coefficient Summary by Sections (*Concluded*)

Duct Section	Fitting Number	Type of Fitting	ASHRAE Fitting No.[a]	Parameters	Loss Coefficient
4	10	Damper	CR9-4[b]	$\theta = 0°$, 5 blades (opposed), $L/R = 1.25$	0.52
	11	Transition	ER4-3	$L = 30$ in., $A_o/A_1 = 3.26$, $\theta = 17°$	0.59
		Summation of Section 4 loss coefficients			1.11
5	12	Elbow	CD3-17	45°, mitered	0.34
	13	Damper	CD9-1	$\theta = 0°$	0.19
	8	Wye (45°), Branch	ED5-2	$Q_b/Q_c = 0.5$, $A_s/A_c = 0.498$, $A_b/A_c = 0.779$	1.20 (C_b)
		Summation of Section 5 loss coefficients			1.73
6	14	Fire damper	CD9-3[b]	Curtain type, Type C	0.12
	15	Elbow	CD3-9	90°, 5 gore, $r/D = 1.5$	0.15
	—	Fan and system interaction	ED7-2[b]	90° elbow, 5 gore, $r/D = 1.5$, $L = 34$ in.	0.60
		Summation of Section 6 loss coefficients			0.87
7	16	Elbow	CR3-3	90°, $r/W = 0.70$, 1 splitter vane	0.14
	17	Damper	CR9-1	$\theta = 0°$, $H/W = 1.0$	0.08
	19	Tee, Main	SR5-13	$Q_s/Q_c = 0.5$, $A_s/A_c = 0.50$	0.04 (C_s)
		Summation of Section 7 loss coefficients			0.26
8	19	Tee, Branch	SR5-13	$Q_b/Q_c = 0.5$, $A_b/A_c = 0.50$	0.73 (C_b)
	18	Damper	CR9-4[b]	$\theta = 0°$, 3 blades (opposed), $L/R = 0.75$	0.52
		Summation of Section 8 loss coefficients			1.25
9	20	Elbow	SR3-1[b]	90°, mitered, $H/W_1 = 0.625$, $W_o/W_1 = 1.25$	1.67
		Summation of Section 9 loss coefficients			1.67
10	21	Damper	CR9-1	$\theta = 0°$, $H/W = 0.625$	0.08
	22	Elbow	CR3-10	90°, single-thickness vanes, design 2	0.12
	23	Elbow	CR3-6[b]	$\theta = 90°$, mitered, $H/W = 0.625$	1.25
	24	Tee, Branch	SR5-1[b]	$r/W_b = 1.0$, $Q_b/Q_c = 0.375$, $A_s/A_c = 0.538$, $A_b/A_c = 0.615$	1.21 (C_b)
		Summation of Section 10 loss coefficients			2.66
11	25	Damper	CR9-1	$\theta = 0°$, $H/W = 1.0$	0.08
	26	Exit	SR2-1	$H/W = 1.0$, Re $= 122,500$	1.00
	27	Wye, Dovetail	SR5-14	$r/W_c = 1.5$, $Q_{b1}/Q_c = 0.5$, $A_{b1}/A_c = 0.714$	0.60 (C_b)
		Summation of Section 11 loss coefficients			1.68
12	28	Damper	CR9-1	$\theta = 0°$, $H/W = 1.0$	0.08
	29	Exit	SR2-5[b]	$\theta = 19°$, $A_1/A_o = 3.24$, Re $= 130,000$	0.77
	27	Wye, Dovetail	SR5-14	$r/W_c = 1.5$, $Q_{b2}/Q_c = 0.5$, $A_{b2}/A_c = 0.714$	0.60 (C_b)
		Summation of Section 12 loss coefficients			1.45
13	30	Damper	CR9-1	$\theta = 0°$, $H/W = 0.71$	0.08
	24	Tee, Main	SR5-1[b]	$r/W_b = 1.0$, $Q_s/Q_c = 0.625$, $A_s/A_c = 0.538$, $A_b/A_c = 0.615$	0.08 (C_s)
		Summation of Section 13 loss coefficients			0.16
14	31	Damper	CR9-1	$\theta = 0°$, $H/W = 0.38$	0.08
	32	Tee, Main	SR5-13	$Q_s/Q_c = 0.8$, $A_s/A_c = 0.813$	0.04 (C_s)
		Summation of Section 14 loss coefficients			0.12
15	48	Elbow	CR3-1	$\theta = 90°$, $r/W = 1.5$, $H/W = 0.75$	0.19
	33	Exit	SR2-6	$L = 18$ in., $D_h = 6.86$	0.28
	34	Damper	CR9-1	$\theta = 0°$, $H/W = 0.75$	0.08
	35	Tee, Main	SR5-1[b]	$r/W_b = 1.0$, $Q_s/Q_c = 0.5$, $A_s/A_c = 0.80$, $A_b/A_c = 0.80$	0.03 (C_s)
		Summation of Section 15 loss coefficients			0.58
16	36	Exit	SR2-3	$\theta = 20°$, $A_1/A_o = 2.0$, Re $= 70,000$	0.63
	36	Screen	CR6-1[b]	$n = 0.8$, $A_1/A_o = 2.0$	0.08
	37	Damper	CR9-1	$\theta = 0°$, $H/W = 0.75$	0.08
	35	Tee, Branch	SR5-1[b]	$r/W_b = 1.0$, $Q_b/Q_c = 0.5$, $A_s/A_c = 0.80$, $A_b/A_c = 0.80$	0.95 (C_b)
		Summation of Section 16 loss coefficients			1.74
17	38	Damper	CR9-1	$\theta = 0°$, $H/W = 0.6$	0.08
	32	Tee, Branch	SR5-13	$Q_b/Q_c = 0.2$, $A_b/A_c = 0.187$	0.68 (C_b)
		Summation of Section 17 loss coefficients			0.76
18	39	Obstruction, Pipe	CR6-4	Re $= 15,000$, $y = 0$, $d = 1$ in., $S_m/A_o = 0.1$, $y/H = 0$	0.17
	40	Transition	SR4-1[b]	$\theta = 22°$, $A_o/A_1 = 0.588$, $L = 18$ in.	0.04
	41	Elbows, Z-shaped	CR3-17[b]	$L = 42$ in., $L/W = 4.2$, $H/W = 3.2$, Re $= 240,000$	2.51
	45	Fire damper	CR9-6	Curtain type, Type B	0.19
		Summation of Section 18 loss coefficients			2.91
19	42	Diffuser, Fan	SR7-17[b]	$\theta_1 = 28°$, $L = 40$ in., $A_o/A_1 = 2.67$	4.19 (C_1)
	47	Damper	CR9-4[b]	$\theta = 0°$, 8 blades (opposed), $L/R = 1.39$	0.52
		Summation of Section 19 loss coefficients			4.71

[a]From Duct Fitting Database (ASHRAE 1993).
[b]Duct Fitting Database (ASHRAE 1993) data for this fitting reproduced in the Fitting Loss Coefficients section.

$$Q_c = Q_d \, (P_h/P_l)^{0.5} \qquad (51)$$

where

Q_d = total airflow rate through low-resistance duct run, cfm
P_h = absolute value of pressure loss in high-resistance ductwork section(s), in. of water
P_l = absolute value of pressure loss in low-resistance ductwork section(s), in. of water
Q_c = airflow rate required to increase P_l to P_h, cfm

For systems conveying particulates, use elbows with a large centerline radius-to-diameter ratio (r/D) greater than 1.5 whenever possible. If the r/D is 1.5 or less, abrasion in dust-handling systems can reduce the life of elbows. Elbows are often made of seven or more gores, especially in large diameters. For converging flow fittings, a 30° entry angle is recommended to minimize energy losses and abrasion in dust-handling systems. For the entry loss coefficients of hoods and equipment for specific operations, refer to Chapter 27 of the 1991 ASHRAE *Handbook—Applications* and to ACGIH (1992).

Example 4. For the metalworking exhaust system in Figures 15 and 16, size the ductwork and calculate the fan static pressure requirement for an industrial exhaust designed to convey granular materials. Pressure balance the system by changing duct sizes and adjusting airflow rates. The minimum particulate transport velocity for the chipping and grinding table ducts (Sections 1 and 5, Figure 16) are 4000 fpm. For the ducts associated with the grinder wheels (Sections 2, 3, 4, and 5), the minimum duct velocity is 4500 fpm. Ductwork is galvanized steel, with the absolute roughness being 0.0003 ft. Assume standard air and that duct and fittings are available in the following sizes: 3-in. through 9.5 diameters in 0.5-in. increments, 10-in. through 37-in. diameters in 1-in. diameter increments, or 38-in. through 90-in. diameters in 2-in. diameter increments.

The building is one story, and the design wind velocity is 20 mph. For the stack, use Design J shown in Figure 15 of Chapter 14 for complete rain protection. The stack height, determined by calculations from Chapter 14, is 16 ft above the roof. This stack height is based on minimized stack downwash; therefore, the stack discharge velocity must exceed 1.5 times the design wind velocity.

Solution: For the contaminated ducts upstream of the collector, initial duct sizes and transport velocities are summarized below. The 4474-fpm velocity in Sections 2 and 3 are acceptable since the transport velocity is not significantly lower than 4500 fpm. For the next available duct size (4.5-in. diameter), the duct velocity is 5523 fpm, significantly greater than 4500 fpm.

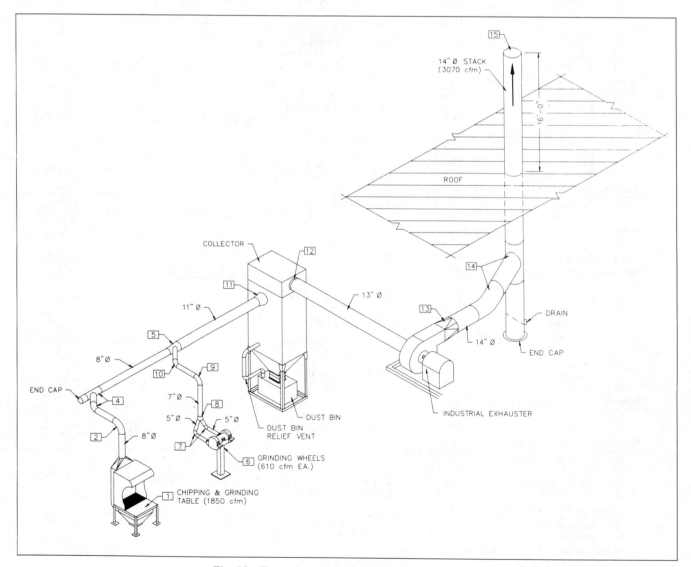

Fig. 15   Example 4—Metalworking Exhaust System

Duct Section	Design Airflow, cfm	Transport Velocity, fpm	Duct Diameter, in.	Duct Velocity, fpm
1	1800	4000	9	4074
2,3	610 each	4500	5	4474
4	1220	4500	7	4565
5	3020	4500	11	4576

The following tabulation summarizes design calculations up through the junction after Sections 1 and 4.

Design No.	D_1, in.	Δp_1, in. of water	Δp_{2+4}, in. of water	Imbalance, $\Delta p_1 - \Delta p_{2+4}$
1	9	1.46	3.09	−1.63
2	8.5	2.00	3.08	−1.08
3	8	2.79	3.00	−0.21
4	7.5	3.92	2.88	+1.04

$Q_1 = 1800$ cfm
$Q_2 = 610$ cfm; $D_2 = 5$ in. dia.
$Q_3 = 610$ cfm; $D_3 = 5$ in. dia.
$Q_4 = 1220$ cfm; $D_4 = 7$ in. dia.

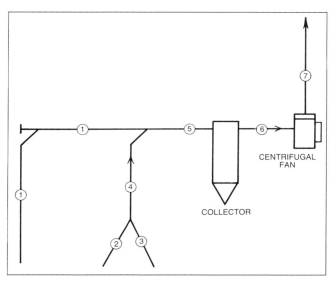

Fig. 16   Example 4—System Schematic with Section Numbers

For the intial design, Design 1, the imbalance between Section 1 and Section 2 (or 3) is 1.63 in. of water, with Section 1 requiring additional resistance. Decreasing Section 1 duct diameter by 0.5-in. increments results in the least imbalance, 0.21 in. of water, when the duct diameter is 8 in. (Design 3). Since Section 1 requires additional resistance, estimate the new airflow rate using Equation (51):

$$Q_{c,1} = (1800)(3.00/2.79)^{0.5} = 1870 \text{ cfm}$$

At 1870 cfm flow in Section 1, 0.13 in. of water imbalance remains at the junction of Sections 1 and 4. At 1850 cfm, balance is attained. The duct between the collector and the fan inlet is 13 in. round to match the fan inlet (12.75 in. diameter). To minimize downwash, the stack discharge velocity must exceed 2640 fpm, 1.5 times the design wind velocity (20 mph) as stated in the problem definition. Therefore, the stack is 14 in. round, and the stack discharge velocity is 2872 fpm.

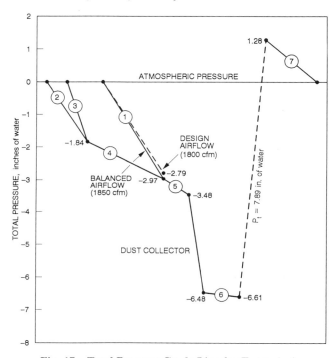

Fig. 17   Total Pressure Grade Line for Example 4

Table 11   Example 4—Total Pressure Loss Calculations by Sections

Duct Section[a]	Duct Element	Airflow, cfm	Duct Size	Velocity, fpm	Velocity Pressure, in. of water	Duct Length[b], ft	Summary of Fitting Loss Coefficients[c]	Duct Pressure Loss/100 ft, in. of water[d]	Total Pressure Loss, in. of water	Section Pressure Loss, in. of water
1	Duct	1850	8 in. ϕ	5300	—	22.5	—	4.63	1.04	
	Fittings	1850	—	5300	1.75	—	1.10	—	1.93	2.97
2,3	Duct	610	5 in. ϕ	4474	—	9	—	5.94	0.53	
	Fittings	610	—	4474	1.24	—	1.06	—	1.31	1.84
4	Duct	1220	7 in. ϕ	4565	—	11.5	—	4.08	0.47	
	Fittings	1220	—	4565	1.30	—	0.51	—	0.66	1.13
5	Duct	3070	11 in. ϕ	4652	—	8.5	—	2.44	0.21	
	Fittings	3070	—	4652	1.35	—	0.22	—	0.30	0.51
—	Collector[e], Fabric	3070	—	—	—	—	—	—	3.0	3.0
6	Duct	3070	13 in. ϕ	3331	—	12	—	1.05	0.13	
	Fittings	3070	—	3331	0.69	—	0.00	—	0.00	0.13
7	Duct	3070	14 in. ϕ	2872	—	29	—	0.72	0.21	
	Fittings	3070	—	2872	0.51	—	2.09	—	1.07	1.28

[a]See Figure 16.
[b]Length of ducts are to fitting centerlines.
[c]See Table 12.
[d]Duct pressure based on a 0.0003 ft absolute roughness factor.

[e]Collector manufacturers set the fabric bag cleaning mechanism to actuate at a pressure difference of 3.0 in. of water between the inlet and outlet plenums. The pressure difference across the clean media is approximately 1.5 in. of water.

Table 12   Example 4—Loss Coefficient Summary by Sections

Duct Section	Fitting Number	Type of Fitting	ASHRAE Fitting No.[a]	Parameters	Loss Coefficient
1	1	Hood[b]	—	Hood face area: 3 ft by 4 ft	0.25
	2	Elbow	CD3-10	$90°$, 7 gore, $r/D = 2.5$	0.11
	4	Capped wye (45°), with 45° elbow	ED5-7	$A_b/A_c = 1$	0.64 (C_b)
	5	Wye (30°), Main	ED5-1	$Q_s/Q_c = 0.60$, $A_s/A_c = 0.529$, $A_b/A_c = 0.405$	0.10 (C_s)
		Summation of Section 1 loss coefficients			1.10
2,3	6	Hood[c]	—	Type hood: For double wheels, Dia. = 22 in. each, Wheel width = 4 in. each, Type takeoff: Tapered	0.40
	7	Elbow	CD3-12	$90°$, 3 gore, $r/D = 1.5$	0.34
	8	Symmetrical wye (60°)	ED5-10	$Q_b/Q_c = 0.5$, $A_b/A_c = 0.510$	0.32 (C_b)
		Summation of Sections 2 and 3 loss coefficients			1.06
4	9	Elbow	CD3-10	$90°$, 7 gore, $r/D = 2.5$	0.11
	10	Elbow	CD3-13	$60°$, 3 gore, $r/D = 1.5$	0.19
	5	Wye (30°), Branch	ED5-1	$Q_b/Q_c = 0.40$, $A_s/A_c = 0.529$, $A_b/A_c = 0.405$	0.21 (C_b)
		Summation of Section 4 loss coefficients			0.51
5	11	Exit, conical diffuser to collector	ED2-1	$L = 24$ in., $L/D_o = 2.18$, $A_1/A_o \approx 16$	0.22
		Summation of Section 5 loss coefficients			0.22
6	12	Entry, Bellmouth from collector	ER2-1	$r/D_1 = 0.20$	0.00 (C_1)
		Summation of Section 6 loss coefficients			0.00
7	13	Diffuser, Fan outlet	SR7-17	Fan outlet size[d]: 10.125 in. by 12.125 in., $A_o/A_1 = 1.596$ (assume 14 in. by 14 in. outlet rather than 16 in. round), $L = 18$ in.	0.45 (C_o)
	14	Capped wye (45°), with 45° elbow	ED5-7	$A_b/A_c = 1$	0.64 (C_b)
	15	Stackhead	SD2-6	$D_e/D = 1$	1.0
		Summation of Section 7 loss coefficients			2.09

[a] From Duct Fitting Database (ASHRAE 1993).
[b] From *Industrial Ventilation* (ACGIH 1992, Figure VS-80-19).
[c] From *Industrial Ventilation* (ACGIH 1992, Figure VS-80-11).

[d] Fan specified: Industrial exhauster for granular materials: 21-in. wheel diameter, 12.75-in. inlet diameter, 10.125 in. by 12.125 in. outlet, 7.5-hp motor.

Table 11 summarizes the system losses by sections. The straight duct friction factor and pressure loss were calculated by Equations (19) and (20). Table 12 lists fitting loss coefficients and input parameters necessary to determine the loss coefficients. The fitting loss coefficients are from the ASHRAE Duct Fitting Database (1993). The fitting loss coefficient tables are included in the Fitting Coefficients section for illustration but can not be obtained exactly by manual interpolation since the coefficients were calculated by the duct fitting database algorithms (more significant figures). For a pressure grade line of the system, see Figure 17. The fan total pressure, calculated by Equation (16), is 7.89 in. of water. To calculate the fan static pressure, use Equation (18):

$$P_s = 7.89 - 0.81 = 7.1 \text{ in. of water}$$

where 0.81 in. of water is the fan outlet velocity pressure. The fan airflow rate is 3070 cfm and its outlet area is 0.853 ft^2 (10.125 in. by 12.125 in.). Therefore, the fan outlet velocity is 3600 fpm.

The hood suction for the chipping and grinding table hood is 2.2 in. of water, calculated by Equation (19) from Chapter 27 of the 1991 ASHRAE *Handbook—Applications* [$HS = (1 + 0.25) (1.74) = 2.2$ in. of water, where 0.25 is the hood entry loss coefficient C_o, and 1.74 is the duct velocity pressure P_v a few diameters downstream from the hood]. Similarly, the hood suction for each of the grinder wheels is 1.7 in. of water:

$$HS_{2,3} = (1 + 0.4) (1.24) = 1.7 \text{ in. of water}$$

where 0.4 is the hood entry loss coefficient, and 1.24 in. of water is the duct velocity pressure.

REFERENCES

ACGIH. 1992. *Industrial ventilation—A manual of recommended practices*, 19th ed. American Conference of Governmental Industrial Hygienists, Lansing, MI.

AISI/SMACNA. 1972. Measurement and analysis of leakage rates from seams and joints of air handling systems.

Altshul, A.D., L.C. Zhivotovckiy, and L.P. Ivanov. 1987. *Hydraulics and aerodynamics*. Stroisdat Publishing House, Moscow.

AMCA. 1989. Test methods for louvers, dampers and shutters. *Standard 500*. Air Movement and Control Association, Arlington Heights, IL.

AMCA. 1990a. Fans and systems. Publication 201. Air Movement and Control Association, Arlington Heights, IL.

AMCA. 1990b. Field performance measurement of fan systems. Publication 203. Air Movement and Control Association, Arlington Heights, IL.

ASHRAE. 1984. Energy conservation in existing facilities—Industrial. *Standard* 100.4-1984.

ASHRAE. 1985a. Energy conservation in existing buildings—Commercial. *Standard* 100.3-1985.

ASHRAE. 1985b. Laboratory methods of testing fans for rating. ANSI/ASHRAE *Standard* 51-1985.

ASHRAE. 1988. Duct leakage: Measurement, analysis and prediction models. ASHRAE Research Project 447.

ASHRAE. 1989. Energy efficient design of new buildings except new low-rise residential buildings. ASHRAE/IES *Standard* 90.1-1989.

ASHRAE. 1991. Energy conservation in existing buildings—High-rise residential. *Standard* 100.2-1991.

ASHRAE. 1991a. Energy conservation in existing buildings—Institutional. *Standard* 100.5-1991.

ASHRAE. 1991b. Energy conservation in existing buildings—Public assembly. *Standard* 100.6-1991.

ASHRAE. 1993. Duct fitting database.

ASHRAE/SMACNA/TIMA. 1985. Investigation of duct leakage. ASHRAE Research Project 308.

Behls, H.F. 1971. Computerized calculation of duct friction. *Building Science Series* 39. U.S. National Bureau of Standards, October, 363.

Bellman, R.E. 1957. *Dynamic programming*. Princeton University Press, New York.

Brown, R.B. 1973. Experimental determinations of fan system effect factors. In *Fans and systems*, ASHRAE Symposium Bulletin LO-73-1, Louisville, KY (June).

Brown, E.J. and J.R. Fellows. 1957. Pressure losses and flow characteristics of multiple leaf dampers. *Heating, Piping and Air Conditioning*, August:119.

Carrier Corporation. 1960. System design manual—Part 2, Air distribution. Air duct design, Chapter 2, 17-63, Syracuse, NY.

Chun-Lun, S. 1983. Simplified static-regain duct design procedure. ASHRAE *Transactions* 89(2A):78.

Clark, M.S., J.T. Barnhart, F.J. Bubsey, and E. Neitzel. 1978. The effects of system connections on fan performance. ASHRAE *Transactions* 84(2):227.

Colebrook, C.F. 1938-39. Turbulent flow in pipes, with particular reference to the transition region between the smooth and rough pipe laws. *Journal of the Institution of Civil Engineers*, London 11:133.

DOE. 1984. *Electric power annual*. DOE/EIA-0348/84. Energy Information Administration, Washington, D.C.

Farajian, T., G. Grewal, and R.J. Tsal. 1992. Post-accident air leakage analysis in a nuclear facility via T-method airflow simulation. 22nd DOE/NRC Nuclear Air Cleaning and Treatment Conference, Denver, October.

Farquhar, H.F. 1973. System effect values for fans. In *Fans and systems*, ASHRAE Symposium Bulletin LO-73-1, Louisville, KY (June).

Griggs, E.I. and F. Khodabakhsh-Sharifabad. 1992. Flow characteristics in rectangular ducts. ASHRAE *Transactions* 92(1).

Griggs, E.I., W.B. Swim, and G.H. Henderson. 1987. Resistance to flow of round galvanized ducts. ASHRAE *Transactions* 93(1):3-16.

Heyt, J.W. and M.J. Diaz. 1975. Pressure drop in flat-oval spiral air duct. ASHRAE *Transactions* 81(2):221-32.

Horowitz, E. and S. Sahni. 1976. *Fundamentals of data structures*. Computer Science Press, New York.

Huebscher, R.G. 1948. Friction equivalents for round, square and rectangular ducts. ASHVE *Transactions* 54:101-44.

Hutchinson, F.W. 1953. Friction losses in round aluminum ducts. ASHVE *Transactions* 59:127-38.

Idelchik, I.E., G.R. Malyavskaya, O.G. Martynenko, and E. Fried. 1986. *Handbook of hydraulic resistance*, 2nd ed. Hemisphere Publishing Corp., subsidiary of Harper & Row, New York.

Jones, C.D. 1979. Friction factor and roughness of United Sheet Metal Company spiral duct. United Sheet Metal, Division of United McGill Corp., Westerville, OH (August). Based on data contained in Friction loss tests, United Sheet Metal Company Spiral Duct, Ohio State University Engineering Experiment Station, File No. T-1011, September, 1958.

Klote, J.H. and J. Milke. 1992. *Design of smoke management systems*. ASHRAE, Atlanta.

Lauvray, T.L. 1978. Experimental heat transmission coefficients for operating air duct systems. ASHRAE *Journal* (June):69.

Locklin, D.W. 1950. Energy losses in 90 degree duct elbows. ASHVE *Transactions* 56:479.

Meyer, M.L. 1973. A new concept: The fan system effect factor. In *Fans and systems*, ASHRAE Symposium Bulletin LO-73-1, Louisville, KY (June).

Moody, L.F. 1944. Friction factors for pipe flow. ASME *Transactions* 66:671.

NFPA. 1991. *Fire protection handbook*, 17th ed. National Fire Protection Association, Quincy, MA.

Osborne, W.C. 1966. *Fans*. Pergamon Press Ltd., London.

Sepsy, C.F. and D.B. Pies. 1973. An experimental study of the pressure losses in converging flow fittings used in exhaust systems. Prepared by Ohio State University for National Institute for Occupational Safety and Health, Document PB-221-130.

Shataloff, N.S. 1966. Static-regain method of design for ducts. ASHRAE *Journal* (May):43.

SMACNA. 1985. *HVAC air duct leakage manual*. Sheet Metal and Air Conditioning Contractors' National Association, Chantilly, VA.

SMACNA. 1990. *HVAC systems duct design*. Sheet Metal and Air-Conditioning Contractors' National Association, Chantilly, VA.

Smith, G.W. 1968. *Engineering economy: Analysis of capital expenditures*. The Iowa State University Press, Ames, IA.

Swim, W.B. 1978. Flow losses in rectangular ducts lined with fiberglass. ASHRAE *Transactions* 84(2):216.

Swim, W.B. 1982. Friction factor and roughness for airflow in plastic pipe. ASHRAE *Transactions* 88(1):269.

Trane Company, The. 1965. *Air conditioning manual* 298-99, LaCrosse, WI.

Tsal, R.J. and M.S. Adler. 1987. Evaluation of numerical methods for ductwork and pipeline optimization. ASHRAE *Transactions* 93(1):17-34.

Tsal, R.J. and H.F. Behls. 1988. Fallacy of the static regain duct design method. ASHRAE *Transactions* 94(2):76-89.

Tsal, R.J., H.F. Behls, and R. Mangel. 1988. T-method duct design, part I: Optimization theory, part II: Calculation procedure and economic analysis. ASHRAE *Transactions* 94(2):90-111.

Tsal, R.J., H.F. Behls, and R. Mangel. 1990. T-method duct design, part III: Simulation. ASHRAE *Transactions* 96(2).

UL. Published annually. *Fire resistance directory*. Underwriters Laboratories, Inc., Northbrook, IL.

UL. 1983. Leakage rated dampers for use in smoke control systems. UL *Standard* 555S-83. Underwriters Laboratories, Inc., Northbrook, IL.

UL. 1990. Fire dampers. UL *Standard* 555-90. Underwriters Laboratories, Inc., Northbrook, IL.

UMC. 1983. Capped exhaust fittings. Laboratory Report No. SRN1083, United McGill Corporation, Westerville, OH.

UMC. 1985a. An experimental study of the performance characteristics of 90 degree elbows. Laboratory Report No. SRF1285, United McGill Corporation, Westerville, OH.

UMC. 1985b. Converging flow fitting test program. Laboratory Report No. SRF785E, United McGill Corporation, Westerville, OH.

UMC. 1986. An experimental evaluation of the performance characteristics of various cybermation fittings. Laboratory Report No. SRF386, United McGill Corporation, Westerville, OH.

Wendes, H.C. 1989. *Sheet metal estimating*. Wendes Mechanical Consulting Services, Inc., Elk Grove Village, IL.

Wright, D.K., Jr. 1945. A new friction chart for round ducts. ASHVE *Transactions* 51:303-16.

BIBLIOGRAPHY

SMACNA. 1987. Duct research destroys design myths. Videotape (VHS). Sheet Metal and Air Conditioning Contractors' National Association, Chantilly, VA.

FITTING LOSS COEFFICIENTS

ROUND FITTINGS

CD3-10 Elbow, 7 Gore, 90 Degree, $r/D = 2.5$
(UMC 1985a, Report SRF1285)

D, in.	3	6	9	12	15	18	27	60
C_o	0.16	0.12	0.10	0.08	0.07	0.06	0.05	0.03

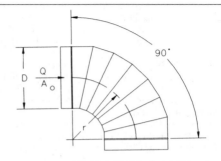

CD3-12 Elbow, 3 Gore, 90 Degree, $r/D = 0.75$ to 2.0
(Locklin 1950, Figure 10)

r/D	0.75	1.00	1.50	2.00
C_o	0.54	0.42	0.34	0.33

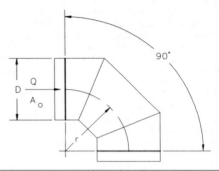

CD3-13 Elbow, 3 Gore, 60 Degree, $r/D = 1.5$
(UMC 1985a, Report SRF1285; Idelchik *et al.* 1986, Diagram 6-1)

D, in.	3	6	9	12	15	18	21	24	27	30	60
C_o	0.40	0.21	0.16	0.14	0.12	0.12	0.11	0.10	0.09	0.09	0.09

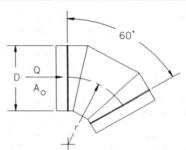

CD6-1 Screen (Only) (Idelchik *et al.* 1986, Diagrams 8-6 and 3-15)

	C_o Values												
	n												
A_1/A_o	0.30	0.35	0.40	0.45	0.50	0.55	0.60	0.65	0.70	0.75	0.80	0.90	1.00
0.2	155.00	102.50	75.00	55.00	41.25	31.50	24.25	18.75	14.50	11.00	8.00	3.50	0.00
0.3	68.89	45.56	33.33	24.44	18.33	14.00	10.78	8.33	6.44	4.89	3.56	1.56	0.00
0.4	38.75	25.63	18.75	13.75	10.31	7.88	6.06	4.69	3.63	2.75	2.00	0.88	0.00
0.5	24.80	16.40	12.00	8.80	6.60	5.04	3.88	3.00	2.32	1.76	1.28	0.56	0.00
0.6	17.22	11.39	8.33	6.11	4.58	3.50	2.69	2.08	1.61	1.22	0.89	0.39	0.00
0.7	12.65	8.37	6.12	4.49	3.37	2.57	1.98	1.53	1.18	0.90	0.65	0.29	0.00
0.8	9.69	6.40	4.69	3.44	2.58	1.97	1.52	1.17	0.91	0.69	0.50	0.22	0.00
0.9	7.65	5.06	3.70	2.72	2.04	1.56	1.20	0.93	0.72	0.54	0.40	0.17	0.00
1.0	6.20	4.10	3.00	2.20	1.65	1.26	0.97	0.75	0.58	0.44	0.32	0.14	0.00
1.2	4.31	2.85	2.08	1.53	1.15	0.88	0.67	0.36	0.40	0.31	0.22	0.10	0.00
1.4	3.16	2.09	1.53	1.12	0.84	0.64	0.49	0.38	0.30	0.22	0.16	0.07	0.00
1.6	2.42	1.60	1.17	0.86	0.64	0.49	0.38	0.29	0.23	0.17	0.13	0.05	0.00
1.8	1.91	1.27	0.93	0.68	0.51	0.39	0.30	0.23	0.18	0.14	0.10	0.04	0.00
2.0	1.55	1.03	0.75	0.55	0.41	0.32	0.24	0.19	0.15	0.11	0.08	0.04	0.00
2.5	0.99	0.66	0.48	0.35	0.26	0.20	0.16	0.12	0.09	0.07	0.05	0.02	0.00
3.0	0.69	0.46	0.33	0.24	0.18	0.14	0.11	0.08	0.06	0.05	0.04	0.02	0.00
4.0	0.39	0.26	0.19	0.14	0.10	0.08	0.06	0.05	0.04	0.03	0.02	0.01	0.00
6.0	0.17	0.11	0.08	0.06	0.05	0.04	0.03	0.02	0.02	0.01	0.01	0.00	0.00

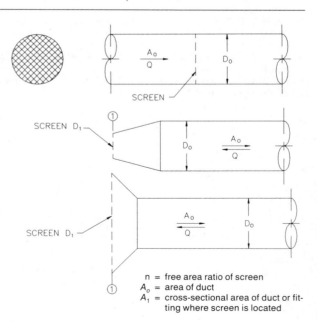

n = free area ratio of screen
A_o = area of duct
A_1 = cross-sectional area of duct or fitting where screen is located

CD9-3 Fire Damper, Curtain Type, Type C

$C_o = 0.12$

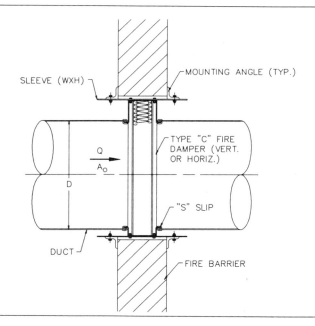

ED1-1 Duct Mounted in Wall (Idelchik *et al.* 1986, Diagram 3-1)

	C_o Values								
	L/D								
t/D	0.000	0.002	0.010	0.050	0.100	0.200	0.300	0.500	10.000
0.00	0.50	0.57	0.68	0.80	0.86	0.92	0.97	1.00	1.00
0.02	0.50	0.51	0.52	0.55	0.60	0.66	0.69	0.72	0.72
0.05	0.50	0.50	0.50	0.50	0.50	0.50	0.50	0.50	0.50
10.00	0.50	0.50	0.50	0.50	0.50	0.50	0.50	0.50	0.50

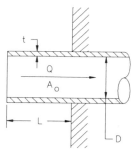

ED2-1 Conical Diffuser, Round to Plenum, Exhaust/Return Systems (Idelchik *et al.* 1986, Diagram 5-8)

	C_o Values										
	L/D_o										
A_1/A_o	0.5	1.0	2.0	3.0	4.0	5.0	6.0	8.0	10.0	12.0	14.0
1.5	0.03	0.02	0.03	0.03	0.04	0.05	0.06	0.08	0.10	0.11	0.13
2.0	0.08	0.06	0.04	0.04	0.04	0.05	0.05	0.06	0.08	0.09	0.10
2.5	0.13	0.09	0.06	0.06	0.06	0.06	0.06	0.06	0.07	0.08	0.09
3.0	0.17	0.12	0.09	0.07	0.07	0.06	0.06	0.07	0.07	0.08	0.08
4.0	0.23	0.17	0.12	0.10	0.09	0.08	0.08	0.08	0.08	0.08	0.08
6.0	0.30	0.22	0.16	0.13	0.12	0.10	0.10	0.09	0.09	0.09	0.08
8.0	0.34	0.26	0.18	0.15	0.13	0.12	0.11	0.10	0.09	0.09	0.09
10.0	0.36	0.28	0.20	0.16	0.14	0.13	0.12	0.11	0.10	0.09	0.09
14.0	0.39	0.30	0.22	0.18	0.16	0.14	0.13	0.12	0.10	0.10	0.10
20.0	0.41	0.32	0.24	0.20	0.17	0.15	0.14	0.12	0.11	0.11	0.10

	Optimum Angle, θ										
A_1/A_o	0.5	1.0	2.0	3.0	4.0	5.0	6.0	8.0	10.0	12.0	14.0
1.5	34	20	13	9	7	6	4	3	2	2	2
2.0	42	28	17	12	10	9	8	6	5	4	3
2.5	50	32	20	15	12	11	10	8	7	6	5
3.0	54	34	22	17	14	12	11	10	8	8	6
4.0	58	40	26	20	16	14	13	12	10	10	9
6.0	62	42	28	22	19	16	15	12	11	10	9
8.0	64	44	30	24	20	18	16	13	12	11	10
10.0	66	46	30	24	22	19	17	14	12	11	10
14.0	66	48	32	26	22	19	17	14	13	11	11
20.0	68	48	32	26	22	20	18	15	13	12	11

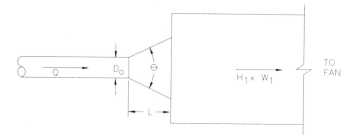

ED4-1 Transition, Round to Round, Exhaust/Return Systems
(Idelchik *et al.* 1986, Diagrams 5-2 and 5-22)

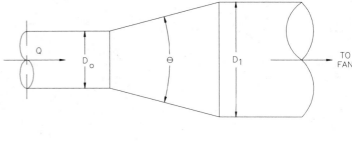

A_o/A_1 < or > 1

					C_o Values					
A_o/A_1	$\theta=10$	15	20	30	45	60	90	120	150	180
0.06	0.21	0.29	0.38	0.60	0.84	0.88	0.88	0.88	0.88	0.88
0.10	0.21	0.28	0.38	0.59	0.76	0.80	0.83	0.84	0.83	0.83
0.25	0.16	0.22	0.30	0.46	0.61	0.68	0.64	0.63	0.62	0.62
0.50	0.11	0.13	0.19	0.32	0.33	0.33	0.32	0.31	0.30	0.30
1.00	0.00	0.00	0.00	0.00	0.00	0.00	0.00	0.00	0.00	0.00
2.00	0.20	0.20	0.20	0.20	0.22	0.24	0.48	0.72	0.96	1.04
4.00	0.80	0.64	0.64	0.64	0.88	1.12	2.72	4.32	5.60	6.56
6.00	1.80	1.44	1.44	1.44	1.98	2.52	6.48	10.10	13.00	15.10
10.00	5.00	5.00	5.00	5.00	6.50	8.00	19.00	29.00	37.00	43.00

ED5-1 Wye, 30 Degree, Converging (Sepsy and Pies 1973)

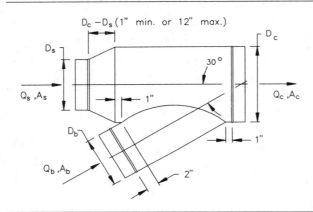 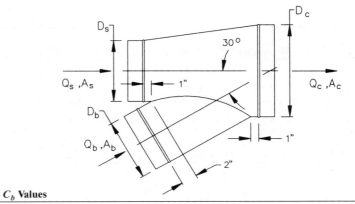

					C_b Values					
					Q_b/Q_c					
A_s/A_c	A_b/A_c	0.1	0.2	0.3	0.4	0.5	0.6	0.7	0.8	0.9
0.2	0.2	−24.17	−3.78	−0.60	0.30	0.64	0.77	0.83	0.88	0.98
	0.3	−55.88	−9.77	−2.57	−0.50	0.25	0.55	0.67	0.70	0.71
	0.4	−99.93	−17.94	−5.13	−1.45	−0.11	0.42	0.62	0.68	0.68
	0.5	−156.51	−28.40	−8.37	−2.62	−0.52	0.30	0.62	0.71	0.69
	0.6	−225.62	−41.13	−12.30	−4.01	−0.99	0.20	0.66	0.78	0.75
	0.7	−307.26	−56.14	−16.90	−5.61	−1.51	0.11	0.73	0.90	0.86
	0.8	−401.44	−73.44	−22.18	−7.44	−2.08	0.04	0.84	1.06	1.01
	0.9	−508.15	−93.02	−28.15	−9.49	−2.71	−0.03	0.99	1.27	1.20
	1.0	−627.39	−114.89	−34.80	−11.77	−3.39	−0.08	1.18	1.52	1.43
0.3	0.2	−13.97	−1.77	0.08	0.59	0.77	0.84	0.88	0.92	1.06
	0.3	−33.06	−5.33	−1.09	0.10	0.51	0.66	0.71	0.72	0.74
	0.4	−59.43	−10.08	−2.52	−0.41	0.32	0.59	0.67	0.68	0.66
	0.5	−93.24	−16.11	−4.30	−1.00	0.14	0.56	0.69	0.70	0.66
	0.6	−134.51	−23.45	−6.44	−1.68	−0.03	0.57	0.76	0.77	0.70
	0.7	−183.25	−32.08	−8.93	−2.45	−0.21	0.61	0.87	0.88	0.79
	0.8	−239.47	−42.01	−11.77	−3.32	−0.38	0.69	1.02	1.03	0.91
	0.9	−303.16	−53.25	−14.97	−4.27	−0.56	0.80	1.21	1.23	1.07
	1.0	−374.32	−65.79	−18.53	−5.32	−0.73	0.94	1.45	1.47	1.27
0.4	0.2	−9.20	−0.85	0.39	0.71	0.82	0.87	0.90	0.94	1.09
	0.3	−22.31	−3.24	−0.38	0.39	0.64	0.73	0.76	0.78	0.85
	0.4	−40.52	−6.48	−1.37	0.02	0.48	0.64	0.67	0.66	0.65
	0.5	−63.71	−10.50	−2.50	−0.33	0.40	0.63	0.69	0.67	0.63
	0.6	−92.00	−15.37	−3.84	−0.71	0.33	0.67	0.75	0.71	0.65
	0.7	−125.40	−21.08	−5.40	−1.13	0.28	0.75	0.85	0.80	0.70
	0.8	−163.90	−27.65	−7.16	−1.59	0.25	0.86	1.00	0.93	0.80
	0.9	−207.52	−35.07	−9.14	−2.09	0.25	1.02	1.18	1.10	0.93
	1.0	−256.25	−43.35	−11.33	−2.63	0.26	1.21	1.42	1.31	1.09

ED5-1 Wye, 30 Degree, Converging (Sepsy and Pies 1973) (*Continued*)

		C_b **Values** (*Continued*)								
					Q_b/Q_c					
A_s/A_c	A_b/A_c	0.1	0.2	0.3	0.4	0.5	0.6	0.7	0.8	0.9
0.5	0.2	−6.62	−0.36	0.54	0.77	0.85	0.88	0.90	0.95	1.11
	0.3	−16.42	−2.11	−0.01	0.54	0.72	0.78	0.80	0.83	0.96
	0.4	−30.26	−4.59	−0.79	0.22	0.54	0.64	0.66	0.64	0.64
	0.5	−47.68	−7.55	−1.61	−0.02	0.48	0.63	0.65	0.62	0.59
	0.6	−68.93	−11.13	−2.56	−0.28	0.45	0.67	0.69	0.65	0.58
	0.7	−94.00	−15.31	−3.65	−0.55	0.44	0.74	0.77	0.71	0.61
	0.8	−122.90	−20.12	−4.88	−0.83	0.46	0.85	0.90	0.81	0.68
	0.9	−155.63	−25.54	−6.25	−1.12	0.51	1.00	1.06	0.94	0.77
	1.0	−192.18	−31.58	−7.77	−1.43	0.59	1.19	1.26	1.12	0.90
0.6	0.2	−5.12	−0.10	0.62	0.79	0.85	0.87	0.90	0.95	1.11
	0.3	−13.00	−1.49	0.18	0.61	0.75	0.79	0.82	0.86	1.02
	0.4	−24.31	−3.55	−0.50	0.30	0.55	0.62	0.63	0.62	0.63
	0.5	−38.41	−5.94	−1.16	0.09	0.48	0.59	0.60	0.57	0.55
	0.6	−55.58	−8.80	−1.92	−0.12	0.45	0.61	0.62	0.57	0.52
	0.7	−75.83	−12.16	−2.79	−0.33	0.44	0.66	0.67	0.60	0.52
	0.8	−99.17	−16.00	−3.76	−0.54	0.46	0.74	0.76	0.67	0.56
	0.9	−125.60	−20.33	−4.83	−0.76	0.51	0.86	0.88	0.77	0.62
	1.0	−155.12	−25.14	−6.02	−0.99	0.58	1.02	1.04	0.90	0.71
0.7	0.2	−4.24	0.05	0.65	0.80	0.85	0.87	0.89	0.94	1.12
	0.3	−11.00	−1.15	0.27	0.63	0.75	0.79	0.82	0.87	1.06
	0.4	−20.82	−3.00	−0.38	0.31	0.52	0.59	0.60	0.59	0.61
	0.5	−32.99	−5.09	−0.98	0.10	0.43	0.53	0.54	0.52	0.51
	0.6	−47.78	−7.58	−1.67	−0.11	0.38	0.52	0.53	0.49	0.45
	0.7	−65.22	−10.50	−2.44	−0.32	0.34	0.53	0.54	0.49	0.43
	0.8	−85.32	−13.83	−3.30	−0.53	0.33	0.58	0.59	0.52	0.43
	0.9	−108.07	−17.58	−4.26	−0.75	0.34	0.66	0.67	0.58	0.46
	1.0	−133.48	−21.76	−5.30	−0.97	0.38	0.76	0.78	0.67	0.51
0.8	0.2	−3.75	0.11	0.65	0.79	0.84	0.86	0.88	0.94	1.12
	0.3	−9.88	−0.99	0.29	0.63	0.74	0.78	0.81	0.87	1.09
	0.4	−18.88	−2.75	−0.36	0.28	0.48	0.55	0.56	0.57	0.61
	0.5	−29.98	−4.71	−0.96	0.04	0.36	0.46	0.47	0.46	0.47
	0.6	−43.46	−7.05	−1.64	−0.20	0.26	0.41	0.43	0.41	0.39
	0.7	−59.34	−9.77	−2.40	−0.44	0.19	0.38	0.41	0.38	0.34
	0.8	−77.64	−12.88	−3.26	−0.69	0.13	0.38	0.42	0.37	0.31
	0.9	−98.35	−16.38	−4.20	−0.95	0.09	0.40	0.45	0.39	0.30
	1.0	−121.48	−20.27	−5.24	−1.23	0.06	0.45	0.51	0.43	0.31
0.9	0.2	−3.52	0.12	0.64	0.78	0.82	0.85	0.88	0.93	1.12
	0.3	−9.34	−0.95	0.28	0.60	0.71	0.76	0.80	0.87	1.10
	0.4	−17.96	−2.70	−0.40	0.22	0.43	0.50	0.53	0.54	0.60
	0.5	−28.58	−4.65	−1.05	−0.07	0.26	0.37	0.40	0.41	0.42
	0.6	−41.45	−6.97	−1.77	−0.35	0.12	0.28	0.32	0.32	0.32
	0.7	−56.61	−9.66	−2.58	−0.65	0.00	0.21	0.27	0.26	0.24
	0.8	−74.08	−12.74	−3.49	−0.97	−0.12	0.16	0.23	0.22	0.18
	0.9	−93.84	−16.21	−4.50	−1.30	−0.23	0.13	0.21	0.19	0.14
	1.0	−115.92	−20.06	−5.61	−1.66	−0.34	0.11	0.21	0.18	0.11
1.0	0.2	−3.48	0.10	0.62	0.76	0.81	0.84	0.87	0.92	1.11
	0.3	−9.22	−1.00	0.23	0.56	0.68	0.74	0.78	0.86	1.11
	0.4	−17.76	−2.79	−0.50	0.14	0.37	0.45	0.49	0.52	0.60
	0.5	−28.31	−4.82	−1.21	−0.20	0.15	0.28	0.33	0.35	0.38
	0.6	−41.06	−7.21	−2.01	−0.55	−0.04	0.15	0.22	0.23	0.25
	0.7	−56.09	−9.99	−2.91	−0.92	−0.23	0.03	0.12	0.14	0.15
	0.8	−73.39	−13.17	−3.92	−1.32	−0.41	−0.07	0.04	0.06	0.06
	0.9	−92.98	−16.75	−5.04	−1.75	−0.60	−0.17	−0.03	−0.01	−0.02
	1.0	−114.85	−20.74	−6.28	−2.21	−0.79	−0.26	−0.09	−0.07	−0.09

		C_s **Values**									
					Q_s/Q_c					
A_s/A_c	A_b/A_c	0.1	0.2	0.3	0.4	0.5	0.6	0.7	0.8	0.9
0.2	0.2	−16.02	−3.15	−0.80	0.04	0.45	0.69	0.86	0.99	1.10
	0.3	−11.65	−1.94	−0.26	0.32	0.60	0.77	0.90	1.01	1.10
	0.4	−8.56	−1.20	0.05	0.47	0.68	0.82	0.92	1.02	1.11
	0.5	−6.41	−0.71	0.25	0.57	0.73	0.84	0.93	1.02	1.11
	0.6	−4.85	−0.36	0.38	0.63	0.76	0.86	0.94	1.02	1.11
	0.7	−3.68	−0.10	0.48	0.68	0.79	0.87	0.95	1.03	1.11
	0.8	−2.77	0.10	0.56	0.71	0.81	0.88	0.95	1.03	1.11
	0.9	−2.04	0.26	0.62	0.74	0.82	0.89	0.95	1.03	1.11
	1.0	−1.45	0.38	0.66	0.76	0.83	0.89	0.96	1.03	1.11

ED5-1 Wye, 30 Degree, Converging (Sepsy and Pies 1973) (*Concluded*)

C_s **Values (*Concluded*)**

A_s/A_c	A_b/A_c	0.1	0.2	0.3	0.4	0.5	0.6	0.7	0.8	0.9
0.3	0.2	−36.37	−7.59	−2.48	−0.79	−0.06	0.29	0.47	0.57	0.61
	0.3	−26.79	−5.07	−1.42	−0.27	0.21	0.42	0.53	0.59	0.61
	0.4	−19.94	−3.49	−0.80	0.02	0.35	0.49	0.56	0.60	0.62
	0.5	−15.18	−2.44	−0.41	0.20	0.43	0.54	0.58	0.61	0.62
	0.6	−11.73	−1.70	−0.13	0.32	0.49	0.56	0.60	0.61	0.62
	0.7	−9.13	−1.14	0.07	0.41	0.53	0.58	0.60	0.61	0.62
	0.8	−7.11	−0.72	0.23	0.48	0.57	0.60	0.61	0.62	0.62
	0.9	−5.49	−0.38	0.35	0.53	0.59	0.61	0.62	0.62	0.62
	1.0	−4.17	−0.11	0.45	0.58	0.61	0.62	0.62	0.62	0.62
0.4	0.2	−64.82	−13.76	−4.74	−1.81	−0.59	−0.02	0.24	0.36	0.39
	0.3	−47.92	−9.38	−2.93	−0.94	−0.16	0.19	0.34	0.39	0.40
	0.4	−35.81	−6.62	−1.88	−0.46	0.07	0.30	0.38	0.41	0.40
	0.5	−27.39	−4.78	−1.20	−0.16	0.22	0.36	0.41	0.42	0.41
	0.6	−21.28	−3.48	−0.73	0.04	0.31	0.41	0.43	0.43	0.41
	0.7	−16.68	−2.51	−0.38	0.20	0.38	0.44	0.45	0.43	0.41
	0.8	−13.10	−1.77	−0.12	0.31	0.44	0.46	0.46	0.44	0.41
	0.9	−10.24	−1.18	0.09	0.40	0.48	0.48	0.46	0.44	0.41
	1.0	−7.90	−0.69	0.26	0.47	0.51	0.50	0.47	0.44	0.41
0.5	0.2	−101.39	−21.64	−7.61	−3.07	−1.19	−0.34	0.05	0.22	0.26
	0.3	−75.05	−14.87	−4.83	−1.75	−0.54	−0.03	0.19	0.26	0.27
	0.4	−56.18	−10.59	−3.21	−1.02	−0.20	0.13	0.26	0.29	0.27
	0.5	−43.04	−7.74	−2.16	−0.56	0.02	0.23	0.30	0.30	0.27
	0.6	−33.51	−5.72	−1.43	−0.24	0.16	0.30	0.33	0.31	0.28
	0.7	−26.34	−4.22	−0.90	−0.01	0.27	0.35	0.35	0.32	0.28
	0.8	−20.75	−3.06	−0.49	0.16	0.35	0.39	0.37	0.33	0.28
	0.9	−16.29	−2.14	−0.17	0.30	0.41	0.41	0.38	0.33	0.28
	1.0	−12.64	−1.39	0.10	0.41	0.46	0.44	0.39	0.33	0.28
0.6	0.2	−146.06	−31.26	−11.09	−4.56	−1.89	−0.68	−0.12	0.10	0.16
	0.3	−108.19	−21.55	−7.12	−2.69	−0.97	−0.24	0.07	0.17	0.17
	0.4	−81.04	−15.40	−4.80	−1.65	−0.48	−0.01	0.17	0.20	0.18
	0.5	−62.13	−11.31	−3.30	−0.99	−0.17	0.13	0.22	0.22	0.18
	0.6	−48.43	−8.41	−2.25	−0.54	0.03	0.22	0.26	0.24	0.18
	0.7	−38.10	−6.25	−1.49	−0.22	0.18	0.29	0.29	0.25	0.19
	0.8	−30.07	−4.59	−0.90	0.03	0.30	0.34	0.31	0.25	0.19
	0.9	−23.64	−3.27	−0.44	0.23	0.39	0.38	0.33	0.26	0.19
	1.0	−18.39	−2.20	−0.06	0.39	0.46	0.42	0.34	0.27	0.19
0.7	0.2	−198.85	−42.62	−15.17	−6.31	−2.68	−1.04	−0.29	0.01	0.08
	0.3	−147.33	−29.41	−9.78	−3.77	−1.44	−0.45	−0.04	0.10	0.10
	0.4	−110.40	−21.07	−6.64	−2.36	−0.77	−0.14	0.09	0.15	0.11
	0.5	−84.67	−15.50	−4.60	−1.48	−0.36	0.05	0.17	0.17	0.11
	0.6	−66.02	−11.56	−3.19	−0.86	−0.08	0.18	0.23	0.19	0.12
	0.7	−51.97	−8.63	−2.15	−0.42	0.12	0.27	0.27	0.20	0.12
	0.8	−41.04	−6.37	−1.35	−0.08	0.27	0.34	0.29	0.21	0.12
	0.9	−32.30	−4.58	−0.72	0.19	0.39	0.39	0.32	0.22	0.12
	1.0	−25.16	−3.12	−0.21	0.40	0.49	0.43	0.33	0.23	0.13
0.8	0.2	−259.75	−55.70	−19.86	−8.29	−3.56	−1.43	−0.46	−0.06	0.03
	0.3	−192.48	−38.47	−12.84	−4.99	−1.95	−0.66	−0.12	0.05	0.05
	0.4	−144.25	−27.58	−8.74	−3.16	−1.09	−0.26	0.05	0.11	0.06
	0.5	−110.65	−20.32	−6.08	−2.00	−0.55	−0.01	0.15	0.15	0.07
	0.6	−86.30	−15.17	−4.24	−1.20	−0.19	0.15	0.22	0.17	0.08
	0.7	−67.95	−11.34	−2.88	−0.62	0.08	0.27	0.27	0.19	0.08
	0.8	−53.67	−8.40	−1.84	−0.18	0.28	0.36	0.30	0.20	0.08
	0.9	−42.26	−6.05	−1.02	0.16	0.44	0.43	0.33	0.21	0.08
	1.0	−32.93	−4.15	−0.35	0.44	0.56	0.49	0.36	0.22	0.09
0.9	0.2	−328.76	−70.51	−25.16	−10.53	−4.54	−1.84	−0.62	−0.12	0.00
	0.3	−243.63	−48.72	−16.28	−6.35	−2.50	−0.87	−0.20	0.03	0.03
	0.4	−182.60	−34.94	−11.09	−4.03	−1.41	−0.37	0.02	0.10	0.04
	0.5	−140.07	−25.75	−7.74	−2.57	−0.74	−0.06	0.15	0.14	0.05
	0.6	−109.25	−19.24	−5.40	−1.56	−0.28	0.15	0.23	0.17	0.05
	0.7	−86.04	−14.40	−3.68	−0.83	0.06	0.30	0.30	0.20	0.06
	0.8	−67.96	−10.66	−2.37	−0.27	0.31	0.41	0.34	0.21	0.06
	0.9	−53.52	−7.70	−1.33	0.17	0.51	0.50	0.38	0.22	0.06
	1.0	−41.71	−5.29	−0.49	0.52	0.67	0.57	0.41	0.23	0.07
1.0	0.2	−405.88	−87.06	−31.07	−13.01	−5.62	−2.29	−0.77	−0.16	−0.02
	0.3	−300.78	−60.15	−20.11	−7.85	−3.10	−1.09	−0.26	0.02	0.02
	0.4	−225.44	−43.14	−13.70	−4.99	−1.76	−0.47	0.01	0.11	0.04
	0.5	−172.93	−31.80	−9.56	−3.18	−0.92	−0.09	0.17	0.17	0.05
	0.6	−134.89	−23.76	−6.68	−1.94	−0.35	0.17	0.28	0.20	0.06
	0.7	−106.23	−17.78	−4.56	−1.04	0.06	0.36	0.35	0.23	0.06
	0.8	−83.92	−13.18	−2.93	−0.35	0.37	0.50	0.41	0.25	0.06
	0.9	−66.08	−9.52	−1.65	0.19	0.62	0.61	0.46	0.26	0.07
	1.0	−51.51	−6.54	−0.61	0.63	0.81	0.70	0.49	0.28	0.07

ED5-7 Capped Wye, Branch with 45-Degree Elbow, Branch 90 Degrees to Main, Converging (UMC 1983, Report SRN1083)

A_b/A_c	0.1	0.2	0.3	0.4	0.5	0.6	0.7	0.8	0.9	1.0
C_b	1.26	1.07	0.94	0.86	0.81	0.76	0.71	0.67	0.64	0.64

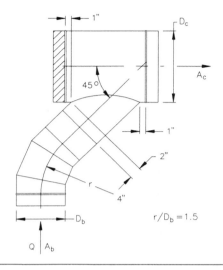

ED5-10 Symmetrical Wye, 60 Degree, $D_{b1} >$ or $= D_{b2}$, Converging (UMC 1985b, Report SRF785E)

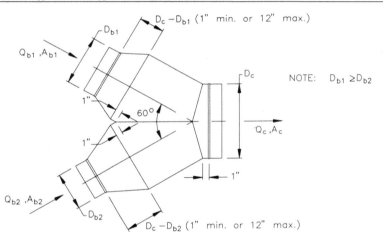

C_{b1} Values

A_{b1}/A_c	A_{b2}/A_c	0.1	0.2	0.3	Q_{b1}/Q_c 0.4	0.5	0.6	0.7	0.8	0.9
0.2	0.2	−11.95	−1.89	−0.09	0.41	0.62	0.74	0.80	0.80	0.79
	0.3	−11.95	−1.89	−0.09	0.41	0.62	0.74	0.80	0.80	0.79
0.3	0.2	−45.45	−9.39	−2.44	−0.41	0.33	0.68	0.89	1.03	1.13
	0.3	−16.88	−2.92	−0.09	0.59	0.86	1.02	1.09	1.10	1.08
0.4	0.2	−72.04	−14.00	−4.26	−1.24	−0.10	0.33	0.50	0.57	0.63
	0.3	−52.95	−9.91	−2.86	−0.69	0.07	0.30	0.40	0.49	0.62
	0.4	−28.86	−6.22	−2.15	−0.57	0.19	0.55	0.72	0.79	0.85
0.5	0.2	−126.04	−23.80	−7.44	−2.64	−0.85	−0.13	0.16	0.26	0.28
	0.3	−91.07	−16.91	−5.16	−1.73	−0.46	0.04	0.23	0.29	0.28
	0.4	−56.41	−10.07	−2.90	−0.82	−0.07	0.21	0.30	0.31	0.29
	0.5	−30.58	−5.23	−1.06	0.00	0.32	0.43	0.47	0.47	0.41
0.6	0.2	−209.81	−39.31	−12.13	−4.35	−1.54	−0.40	0.06	0.22	0.23
	0.3	−147.43	−27.69	−8.75	−3.20	−1.13	−0.29	0.05	0.17	0.18
	0.4	−85.06	−16.07	−5.38	−2.04	−0.71	−0.17	0.04	0.12	0.13
	0.5	−58.22	−11.03	−3.84	−1.49	−0.50	−0.09	0.07	0.11	0.12
	0.6	−40.57	−7.86	−2.60	−0.99	−0.26	0.00	0.14	0.21	0.25
0.7	0.2	−291.57	−54.52	−17.03	−6.21	−2.27	−0.68	−0.04	0.19	0.21
	0.3	−197.37	−38.02	−12.54	−4.92	−2.01	−0.76	−0.22	0.01	0.08
	0.4	−102.97	−21.41	−8.05	−3.64	−1.75	−0.84	−0.40	−0.17	−0.05
	0.5	−65.15	−14.75	−6.16	−3.07	−1.61	−0.85	−0.44	−0.22	−0.09
	0.6	−48.24	−11.70	−4.97	−2.59	−1.40	−0.76	−0.37	−0.15	−0.03
	0.7	−73.02	−16.68	−6.90	−3.29	−1.61	−0.80	−0.29	0.02	0.22
0.8	0.2	−373.33	−69.73	−21.93	−8.08	−3.00	−0.95	−0.13	0.15	0.20
	0.3	−247.31	−48.35	−16.32	−6.65	−2.89	−1.24	−0.49	−0.15	−0.02

ED5-10 Symmetrical Wye, 60 Degree, $D_{b1} <$ or $= D_{b2}$, Converging (UMC 1985b, Report SRF785E) *(Concluded)*

C_{b1} Values *(Concluded)*

A_{b1}/A_c	A_{b2}/A_c	0.1	0.2	0.3	Q_{b1}/Q_c 0.4	0.5	0.6	0.7	0.8	0.9
	0.4	−120.88	−26.76	−10.71	−5.24	−2.78	−1.52	−0.84	−0.45	−0.24
	0.5	−72.08	−18.46	−8.48	−4.65	−2.71	−1.61	−0.95	−0.55	−0.31
	0.6	−55.91	−15.54	−7.35	−4.20	−2.54	−1.53	−0.89	−0.51	−0.30
	0.7	−80.68	−20.52	−9.27	−4.90	−2.75	−1.56	−0.80	−0.34	−0.06
	0.8	−105.46	−25.49	−11.19	−5.59	−2.96	−1.60	−0.72	−0.18	0.19
0.9	0.2	−479.24	−89.56	−28.39	−10.59	−4.04	−1.41	−0.36	0.01	0.09
	0.3	−305.31	−61.27	−21.50	−9.28	−4.39	−2.16	−1.07	−0.54	−0.29
	0.4	−131.17	−32.88	−14.60	−7.98	−4.74	−2.91	−1.79	−1.10	−0.68
	0.5	−67.90	−22.76	−12.17	−7.53	−4.89	−3.19	−2.05	−1.30	−0.81
	0.6	−68.95	−23.08	−12.11	−7.45	−4.84	−3.15	−2.01	−1.26	−0.79
	0.7	−90.48	−27.35	−13.58	−7.95	−4.97	−3.16	−1.96	−1.17	−0.65
	0.8	−112.02	−31.63	−15.05	−8.44	−5.11	−3.18	−1.90	−1.07	−0.51
	0.9	−130.32	−35.19	−16.07	−8.70	−5.18	−3.19	−1.88	−1.08	−0.53
1.0	0.2	−585.16	−109.39	−34.85	−13.11	−5.09	−1.86	−0.59	−0.13	−0.01
	0.3	−363.31	−74.20	−26.68	−11.91	−5.90	−3.08	−1.66	−0.94	−0.56
	0.4	−141.46	−39.00	−18.50	−10.71	−6.71	−4.29	−2.74	−1.74	−1.12
	0.5	−63.71	−27.06	−15.85	−10.41	−7.07	−4.77	−3.16	−2.05	−1.31
	0.6	−81.99	−30.62	−16.87	−10.70	−7.13	−4.77	−3.13	−2.02	−1.28
	0.7	−100.28	−34.19	−17.89	−11.00	−7.19	−4.76	−3.11	−1.99	−1.24
	0.8	−118.58	−37.76	−18.91	−11.29	−7.26	−4.76	−3.09	−1.96	−1.20
	0.9	−136.88	−41.32	−19.93	−11.55	−7.32	−4.77	−3.07	−1.98	−1.23
	1.0	−155.18	−44.89	−20.95	−11.80	−7.39	−4.78	−3.05	−1.99	−1.25

C_{b2} Values

A_{b1}/A_c	A_{b2}/A_c	0.1	0.2	0.3	Q_{b2}/Q_c 0.4	0.5	0.6	0.7	0.8	0.9
0.2	0.2	−11.95	−1.89	−0.09	0.41	0.62	0.74	0.80	0.80	0.79
	0.3	−11.95	−1.89	−0.09	0.41	0.62	0.74	0.80	0.80	0.79
0.3	0.2	−8.24	−1.18	0.05	0.42	0.61	0.73	0.78	0.77	0.76
	0.3	−16.88	−2.92	−0.09	0.59	0.86	1.02	1.09	1.10	1.08
0.4	0.2	−6.95	−1.00	0.16	0.53	0.67	0.71	0.72	0.72	0.71
	0.3	−16.21	−2.90	−0.44	0.40	0.79	0.98	1.05	1.06	1.05
	0.4	−28.86	−6.22	−2.15	−0.57	0.19	0.55	0.72	0.79	0.85
0.5	0.2	−4.82	−0.01	0.56	0.71	0.82	0.89	0.92	0.90	0.89
	0.3	−12.27	−1.17	0.44	0.88	1.11	1.25	1.29	1.25	1.23
	0.4	−20.76	−2.93	−0.21	0.48	0.73	0.84	0.88	0.87	0.82
	0.5	−30.58	−5.23	−1.06	0.00	0.32	0.43	0.47	0.47	0.41
0.6	0.2	−3.68	0.07	0.77	0.98	1.06	1.08	1.08	1.06	1.04
	0.3	−9.06	−0.55	0.86	1.27	1.42	1.48	1.49	1.46	1.42
	0.4	−17.62	−2.12	0.06	0.60	0.83	0.95	0.98	0.95	0.91
	0.5	−28.00	−4.26	−0.99	−0.16	0.20	0.39	0.45	0.41	0.38
	0.6	−40.57	−7.86	−2.60	−0.99	−0.26	0.00	0.14	0.21	0.25
0.7	0.2	−5.44	−0.40	0.55	0.86	0.98	1.02	1.04	1.03	1.02
	0.3	−9.36	−0.77	0.73	1.20	1.39	1.47	1.49	1.47	1.44
	0.4	−19.57	−3.09	−0.44	0.36	0.71	0.89	0.97	0.98	0.97
	0.5	−31.88	−6.02	−1.90	−0.63	−0.05	0.26	0.40	0.44	0.46
	0.6	−46.44	−9.82	−3.47	−1.41	−0.48	−0.04	0.21	0.36	0.45
	0.7	−73.02	−16.68	−6.90	−3.29	−1.61	−0.80	−0.29	0.02	0.22
0.8	0.2	−7.21	−0.87	0.33	0.73	0.90	0.97	1.00	1.00	0.99
	0.3	−9.67	−0.99	0.60	1.13	1.36	1.45	1.49	1.48	1.46
	0.4	−21.53	−4.06	−0.93	0.11	0.59	0.83	0.96	1.01	1.03
	0.5	−35.77	−7.77	−2.82	−1.09	−0.29	0.13	0.35	0.48	0.55
	0.6	−52.32	−11.78	−4.34	−1.83	−0.70	−0.09	0.28	0.51	0.65
	0.7	−78.89	−18.64	−7.76	−3.71	−1.83	−0.85	−0.22	0.16	0.42
	0.8	−105.46	−25.49	−11.19	−5.59	−2.96	−1.60	−0.72	−0.18	0.19
0.9	0.2	−4.98	−0.34	0.54	0.85	0.97	1.03	1.04	1.03	1.01
	0.3	−9.97	−1.21	0.48	1.06	1.32	1.44	1.49	1.49	1.48
	0.4	−23.54	−4.98	−1.39	−0.12	0.47	0.78	0.95	1.04	1.09
	0.5	−40.14	−9.57	−3.69	−1.56	−0.55	−0.01	0.31	0.51	0.63
	0.6	−58.25	−14.28	−5.64	−2.53	−1.08	−0.30	0.18	0.49	0.70
	0.7	−84.09	−21.02	−8.91	−4.38	−2.22	−1.04	−0.31	0.15	0.46
	0.8	−109.92	−27.77	−12.18	−6.22	−3.35	−1.79	−0.81	−0.19	0.23
	0.9	−130.32	−35.19	−16.07	−8.70	−5.18	−3.19	−1.88	−1.08	−0.53
1.0	0.2	−2.75	0.19	0.76	0.96	1.05	1.08	1.08	1.06	1.04
	0.3	−10.28	−1.43	0.35	0.99	1.29	1.43	1.49	1.50	1.50
	0.4	−25.56	−5.89	−1.86	−0.36	0.35	0.72	0.93	1.07	1.15
	0.5	−44.52	−11.37	−4.56	−2.02	−0.81	−0.14	0.27	0.54	0.72
	0.6	−64.19	−16.77	−6.94	−3.24	−1.47	−0.50	0.09	0.48	0.74
	0.7	−89.28	−23.41	−10.05	−5.05	−2.61	−1.24	−0.40	0.14	0.50
	0.8	−114.38	−30.04	−13.16	−6.86	−3.75	−1.97	−0.89	−0.20	0.27
	0.9	−134.78	−37.47	−17.06	−9.33	−5.57	−3.38	−1.97	−1.09	−0.49
	1.0	−155.18	−44.89	−20.95	−11.80	−7.39	−4.78	−3.05	−1.99	−1.25

ED7-1 Centrifugal Fan Located in Plenum or Cabinet (AMCA 1990a, Figure 9-11)

L/D_o	0.30	0.40	0.50	0.75
C_o	0.80	0.53	0.40	0.22

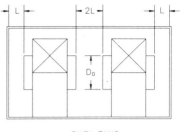
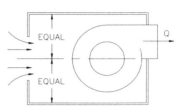

DWDI FANS

ED7-2 Fan Inlet, Centrifugal, SWSI, with 4-Gore Elbow (AMCA 1990a, Figure 9-5)

	C_o Values			
		L/D_o		
r/D_o	0.0	2.0	5.0	10.0
0.50	1.80	1.00	0.53	0.53
0.75	1.40	0.80	0.40	0.40
1.00	1.20	0.67	0.33	0.33
1.50	1.10	0.60	0.33	0.33
2.00	1.00	0.53	0.33	0.33
3.00	0.67	0.40	0.22	0.22

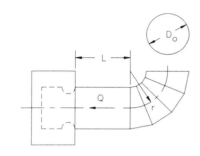

SD2-6 Stackhead (Idelchik *et al.* 1986, Diagram 11-23)

D_e/D	0.3	0.4	0.5	0.6	0.7	0.8	0.9	1.0
C_o	129.63	41.02	16.80	8.10	4.37	2.56	1.60	1.00

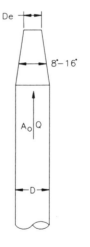
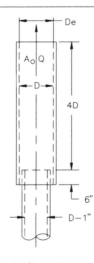

$De/D < 1.0$ $De/D = 1$

SD4-1 Transition, Round to Round, Supply Air Systems (Idelchik *et al.* 1986, Diagrams 5-2 and 5-22)

	C_o Values									
				θ						
A_o/A_1	10	15	20	30	45	60	90	120	150	180
0.10	0.05	0.05	0.05	0.05	0.07	0.08	0.19	0.29	0.37	0.43
0.17	0.05	0.04	0.04	0.04	0.06	0.07	0.18	0.28	0.36	0.42
0.25	0.05	0.04	0.04	0.04	0.06	0.07	0.17	0.27	0.35	0.41
0.50	0.05	0.05	0.05	0.05	0.06	0.06	0.12	0.18	0.24	0.26
1.00	0.00	0.00	0.00	0.00	0.00	0.00	0.00	0.00	0.00	0.00
2.00	0.44	0.52	0.76	1.28	1.32	1.32	1.28	1.24	1.20	1.20
4.00	2.56	3.52	4.80	7.36	9.76	10.88	10.24	10.08	9.92	9.92
10.00	21.00	28.00	38.00	59.00	76.00	80.00	83.00	84.00	83.00	83.00
16.00	53.76	74.24	97.28	153.60	215.04	225.28	225.28	225.28	225.28	225.28

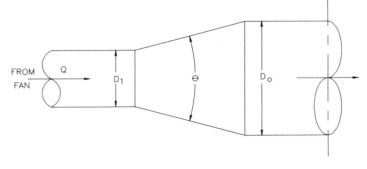

$A_o/A_1 \ < \ \text{or} \ > 1$

SD5-2 Wye, 45 Degree, Conical Branch Tapped into Body, Diverging (UMC 1986, Report SRF386)

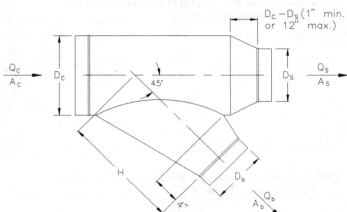

$$H = \frac{D_c}{2 \times \sin 45^\circ} + \frac{D_b}{2 \times \tan 45^\circ} + 4$$

C_b Values

A_b/A_c	Q_b/Q_c								
	0.1	0.2	0.3	0.4	0.5	0.6	0.7	0.8	0.9
0.1	0.15	0.05							
0.2	1.90	0.15	0.04	0.05					
0.3	5.77	0.75	0.15	0.05	0.04	0.05			
0.4	11.74	1.90	0.49	0.15	0.06	0.04	0.04	0.05	
0.5	19.78	3.57	1.08	0.38	0.15	0.07	0.04	0.04	0.04
0.6	29.90	5.77	1.90	0.75	0.32	0.15	0.07	0.05	0.04
0.7	42.09	8.50	2.96	1.26	0.59	0.29	0.15	0.08	0.05
0.8	56.35	11.74	4.25	1.90	0.94	0.49	0.26	0.15	0.09
0.9	72.68	15.50	5.77	2.67	1.38	0.75	0.43	0.25	0.15

C_s Values

A_s/A_c	Q_s/Q_c								
	0.1	0.2	0.3	0.4	0.5	0.6	0.7	0.8	0.9
0.1	0.13	0.16							
0.2	0.20	0.13	0.15	0.16	0.28				
0.3	0.90	0.13	0.13	0.14	0.15	0.16	0.20		
0.4	2.88	0.20	0.14	0.13	0.14	0.15	0.15	0.16	0.34
0.5	6.25	0.37	0.17	0.14	0.13	0.14	0.14	0.15	0.15
0.6	11.88	0.90	0.20	0.13	0.14	0.13	0.14	0.14	0.15
0.7	18.62	1.71	0.33	0.18	0.16	0.14	0.13	0.15	0.14
0.8	26.88	2.88	0.50	0.20	0.15	0.14	0.13	0.13	0.14
0.9	36.45	4.46	0.90	0.30	0.19	0.16	0.15	0.14	0.13

RECTANGULAR FITTINGS

CR3-6 Elbow, Mitered (Idelchik _et al._ 1986, Diagram 6-5)

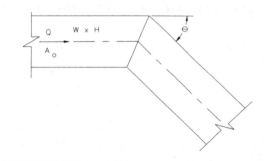

C_o Values

θ	H/W										
	0.25	0.50	0.75	1.00	1.50	2.00	3.00	4.00	5.00	6.00	8.00
20	0.08	0.08	0.08	0.07	0.07	0.07	0.06	0.06	0.05	0.05	0.05
30	0.18	0.17	0.17	0.16	0.15	0.15	0.13	0.13	0.12	0.12	0.11
45	0.38	0.37	0.36	0.34	0.33	0.31	0.28	0.27	0.26	0.25	0.24
60	0.60	0.59	0.57	0.55	0.52	0.49	0.46	0.43	0.41	0.39	0.38
75	0.89	0.87	0.84	0.81	0.77	0.73	0.67	0.63	0.61	0.58	0.57
90	1.30	1.27	1.23	1.18	1.13	1.07	0.98	0.92	0.89	0.85	0.83

CR3-17 Elbow, Z-Shaped (Idelchik _et al._ 1986, Diagram 6-11)

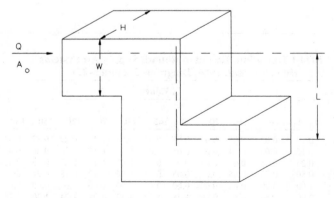

$C_o = K_r C_p$

where K_r = Reynolds number correction factor

C_p Values

H/W	L/W													
	0.0	0.4	0.6	0.8	1.0	1.2	1.4	1.6	1.8	2.0	4.0	8.0	10.0	100.0
0.25	0.00	0.68	0.99	1.77	2.89	3.97	4.41	4.60	4.64	4.60	3.39	3.03	2.70	2.53
0.50	0.00	0.66	0.96	1.72	2.81	3.86	4.29	4.47	4.52	4.47	3.30	2.94	2.62	2.46
0.75	0.00	0.64	0.94	1.67	2.74	3.75	4.17	4.35	4.39	4.35	3.20	2.86	2.55	2.39
1.00	0.00	0.62	0.90	1.61	2.63	3.61	4.01	4.18	4.22	4.18	3.08	2.75	2.45	2.30
1.50	0.00	0.59	0.86	1.53	2.50	3.43	3.81	3.97	4.01	3.97	2.93	2.61	2.33	2.19
2.00	0.00	0.56	0.81	1.45	2.37	3.25	3.61	3.76	3.80	3.76	2.77	2.48	2.21	2.07
3.00	0.00	0.51	0.75	1.34	2.18	3.00	3.33	3.47	3.50	3.47	2.56	2.28	2.03	1.91
4.00	0.00	0.48	0.70	1.26	2.05	2.82	3.13	3.26	3.29	3.26	2.40	2.15	1.91	1.79
6.00	0.00	0.45	0.65	1.16	1.89	2.60	2.89	3.01	3.04	3.01	2.22	1.98	1.76	1.66
8.00	0.00	0.43	0.63	1.13	1.84	2.53	2.81	2.93	2.95	2.93	2.16	1.93	1.72	1.61

Reynolds No. Correction Factor, K_r

Re/1000	10	20	30	40	60	80	100	140	500
K_r	1.40	1.26	1.19	1.14	1.09	1.06	1.04	1.00	1.00

CR6-1 Screen (Only) (Idelchik *et al.* 1986, Diagrams 8-6 and 3-15)

C_o Values

A_1/A_o	0.30	0.35	0.40	0.45	0.50	0.55	0.60	0.65	0.70	0.75	0.80	0.90	1.00
0.2	155.00	102.50	75.00	55.00	41.25	31.50	24.25	18.75	14.50	11.00	8.00	3.50	0.00
0.3	68.89	45.56	33.33	24.44	18.33	14.00	10.78	8.33	6.44	4.89	3.56	1.56	0.00
0.4	38.75	25.63	18.75	13.75	10.31	7.88	6.06	4.69	3.63	2.75	2.00	0.88	0.00
0.5	24.80	16.40	12.00	8.80	6.60	5.04	3.88	3.00	2.32	1.76	1.28	0.56	0.00
0.6	17.22	11.39	8.33	6.11	4.58	3.50	2.69	2.08	1.61	1.22	0.89	0.39	0.00
0.7	12.65	8.37	6.12	4.49	3.37	2.57	1.98	1.53	1.18	0.90	0.65	0.29	0.00
0.8	9.69	6.40	4.69	3.44	2.58	1.97	1.52	1.17	0.91	0.69	0.50	0.22	0.00
0.9	7.65	5.06	3.70	2.72	2.04	1.56	1.20	0.93	0.72	0.54	0.40	0.17	0.00
1.0	6.20	4.10	3.00	2.20	1.65	1.26	0.97	0.75	0.58	0.44	0.32	0.14	0.00
1.2	4.31	2.85	2.08	1.53	1.15	0.88	0.67	0.36	0.40	0.31	0.22	0.10	0.00
1.4	3.16	2.09	1.53	1.12	0.84	0.64	0.49	0.38	0.30	0.22	0.16	0.07	0.00
1.6	2.42	1.60	1.17	0.86	0.64	0.49	0.38	0.29	0.23	0.17	0.13	0.05	0.00
1.8	1.91	1.27	0.93	0.68	0.51	0.39	0.30	0.23	0.18	0.14	0.10	0.04	0.00
2.0	1.55	1.03	0.75	0.55	0.41	0.32	0.24	0.19	0.15	0.11	0.08	0.04	0.00
2.5	0.99	0.66	0.48	0.35	0.26	0.20	0.16	0.12	0.09	0.07	0.05	0.02	0.00
3.0	0.69	0.46	0.33	0.24	0.18	0.14	0.11	0.08	0.06	0.05	0.04	0.02	0.00
4.0	0.39	0.26	0.19	0.14	0.10	0.08	0.06	0.05	0.04	0.03	0.02	0.01	0.00
6.0	0.17	0.11	0.08	0.06	0.05	0.04	0.03	0.02	0.02	0.01	0.01	0.00	0.00

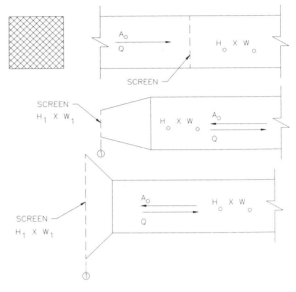

n = free area ratio of screen
A_o = area of duct
A_1 = cross-sectional area of duct of fitting
where screen is located

CR9-4 Damper, Opposed Blades (Brown and Fellows 1957)

C_o Values

L/R	0	10	20	30	40	50	60	70	80
0.3	0.52	0.79	1.91	3.77	8.55	19.46	70.12	295.21	807.23
0.4	0.52	0.85	2.07	4.61	10.42	26.73	92.90	346.25	926.34
0.5	0.52	0.93	2.25	5.44	12.29	33.99	118.91	393.36	1045.44
0.6	0.52	1.00	2.46	5.99	14.15	41.26	143.69	440.25	1163.09
0.8	0.52	1.08	2.66	6.96	18.18	56.47	193.92	520.27	1324.85
1.0	0.52	1.17	2.91	7.31	20.25	71.68	245.45	576.00	1521.00
1.5	0.52	1.38	3.16	9.51	27.56	107.41	361.00	717.05	1804.40

(column heading θ)

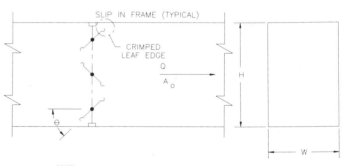

$$L/R = \frac{N\,W}{2(H + W)}$$

where

N = number of damper blades
W = duct dimension parallel to blade axis, in.
H = duct height, in.
L = sum of damper blade lengths, in.
R = perimeter of duct, in.

ER2-1 Bellmouth, Plenum to Round, Exhaust/Return Systems (Idelchik *et al.* 1986, Diagram 4-9)

C_o Values

A_o/A_1	0.00	0.01	0.02	0.03	0.04	0.05	0.06	0.08	0.10	0.12	0.16	0.20	10.00
1.5	0.22	0.20	0.15	0.14	0.12	0.10	0.09	0.07	0.05	0.04	0.03	0.01	0.01
2.0	0.13	0.11	0.08	0.08	0.07	0.06	0.05	0.04	0.03	0.02	0.02	0.01	0.01
2.5	0.08	0.07	0.05	0.05	0.04	0.04	0.03	0.02	0.02	0.01	0.01	0.00	0.00
3.0	0.06	0.05	0.04	0.03	0.03	0.02	0.02	0.02	0.01	0.01	0.01	0.00	0.00
4.0	0.03	0.03	0.02	0.02	0.02	0.01	0.01	0.01	0.01	0.01	0.00	0.00	0.00
8.0	0.01	0.01	0.01	0.00	0.00	0.00	0.00	0.00	0.00	0.00	0.00	0.00	0.00

(column heading r/D_1)

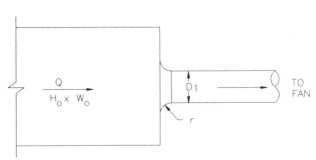

ER3-1 Elbow, 90 Degree, Variable Inlet/Outlet Areas, Exhaust/Return Systems (Idelchik *et al.* 1986, Diagram 6-4)

	C_o Values						
	W_1/W_o						
H/W_o	0.6	0.8	1.0	1.2	1.4	1.6	2.0
0.25	1.76	1.43	1.24	1.14	1.09	1.06	1.06
1.00	1.70	1.36	1.15	1.02	0.95	0.90	0.84
4.00	1.46	1.10	0.90	0.81	0.76	0.72	0.66
100.00	1.50	1.04	0.79	0.69	0.63	0.60	0.55

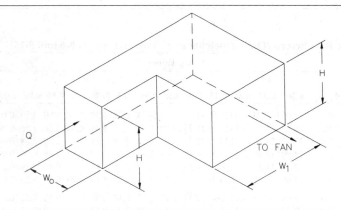

ER4-1 Transition, Rectangular, Two Sides Parallel, Symmetrical, Exhaust/Return Systems (Idelchik *et al.* 1986, Diagram 5-5)

	C_o Values									
	θ									
A_o/A_1	10	15	20	30	45	60	90	120	150	180
0.06	0.26	0.27	0.40	0.56	0.71	0.86	1.00	0.99	0.98	0.98
0.10	0.24	0.26	0.36	0.53	0.69	0.82	0.93	0.93	0.92	0.91
0.25	0.17	0.19	0.22	0.42	0.60	0.68	0.70	0.69	0.67	0.66
0.50	0.14	0.13	0.15	0.24	0.35	0.37	0.38	0.37	0.36	0.35
1.00	0.00	0.00	0.00	0.00	0.00	0.00	0.00	0.00	0.00	0.00
2.00	0.23	0.20	0.20	0.20	0.24	0.28	0.54	0.78	1.02	1.09
4.00	0.81	0.64	0.64	0.64	0.88	1.12	2.78	4.38	5.65	6.60
6.00	1.82	1.44	1.44	1.44	1.98	2.53	6.56	10.20	13.00	15.20
10.00	5.03	5.00	5.00	5.00	6.50	8.02	19.10	29.10	37.10	43.10

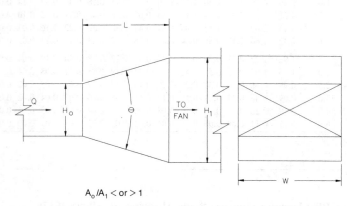

$A_o/A_1 <$ or > 1

SR2-5 Pyramidal Diffuser, Free Discharge (Idelchik *et al.* 1986, Diagram 11-5)

		C_o Values								
		θ								
A_1/A_o	Re/1000	8	10	14	20	30	45	60	90	120
1	50	0.00	0.00	0.00	0.00	0.00	0.00	0.00	0.00	0.00
	100	0.00	0.00	0.00	0.00	0.00	0.00	0.00	0.00	0.00
	200	0.00	0.00	0.00	0.00	0.00	0.00	0.00	0.00	0.00
	400	0.00	0.00	0.00	0.00	0.00	0.00	0.00	0.00	0.00
	2000	0.00	0.00	0.00	0.00	0.00	0.00	0.00	0.00	0.00
2	50	0.65	0.68	0.74	0.82	0.92	1.05	1.10	1.08	1.08
	100	0.61	0.66	0.73	0.81	0.90	1.04	1.09	1.08	1.08
	200	0.57	0.61	0.70	0.79	0.89	1.04	1.09	1.08	1.08
	400	0.50	0.56	0.64	0.76	0.88	1.02	1.07	1.08	1.08
	2000	0.50	0.56	0.64	0.76	0.88	1.02	1.07	1.08	1.08
4	50	0.53	0.60	0.69	0.78	0.90	1.02	1.07	1.09	1.09
	100	0.49	0.55	0.66	0.78	0.90	1.02	1.07	1.09	1.09
	200	0.42	0.50	0.62	0.74	0.87	1.00	1.06	1.08	1.08
	400	0.36	0.44	0.56	0.70	0.84	0.99	1.06	1.08	1.08
	2000	0.36	0.44	0.56	0.70	0.84	0.99	1.06	1.08	1.08
6	50	0.50	0.57	0.66	0.77	0.91	1.02	1.07	1.08	1.08
	100	0.47	0.54	0.63	0.76	0.98	1.02	1.07	1.08	1.08
	200	0.42	0.48	0.60	0.73	0.88	1.00	1.06	1.08	1.08
	400	0.34	0.44	0.56	0.73	0.86	0.98	1.06	1.08	1.08
	2000	0.34	0.44	0.56	0.73	0.86	0.98	1.06	1.08	1.08
10	50	0.45	0.53	0.64	0.74	0.85	0.97	1.10	1.12	1.12
	100	0.40	0.48	0.62	0.73	0.85	0.97	1.10	1.12	1.12
	200	0.34	0.44	0.56	0.69	0.82	0.95	1.10	1.11	1.11
	400	0.28	0.40	0.55	0.67	0.80	0.93	1.09	1.11	1.11
	2000	0.28	0.40	0.55	0.67	0.80	0.93	1.09	1.11	1.11

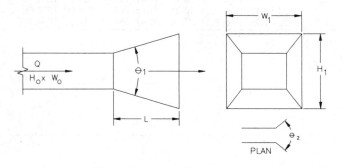

θ is larger of θ_1 and θ_2

SR3-1 Elbow, 90 Degree, Variable Inlet/Outlet Areas, Supply Air Systems (Idelchik _et al._ 1986, Diagram 6-4)

			C_o Values				
			W_o/W_1				
H/W_1	0.6	0.8	1.0	1.2	1.4	1.6	2.0
0.25	0.63	0.92	1.24	1.64	2.14	2.71	4.24
1.00	0.61	0.87	1.15	1.47	1.86	2.30	3.36
4.00	0.53	0.70	0.90	1.17	1.49	1.84	2.64
100.00	0.54	0.67	0.79	0.99	1.23	1.54	2.20

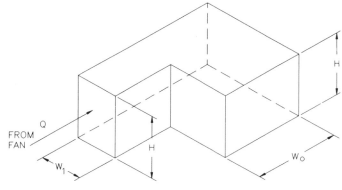

SR4-1 Transition, Rectangular, Two Sides Parallel, Symmetrical, Supply Air Systems (Idelchik _et al._ 1986, Diagram 5-5)

					C_o Values					
					θ					
A_o/A_1	10	15	20	30	45	60	90	120	150	180
0.10	0.05	0.05	0.05	0.05	0.07	0.08	0.19	0.29	0.37	0.43
0.17	0.05	0.04	0.04	0.04	0.05	0.07	0.18	0.28	0.36	0.42
0.25	0.05	0.04	0.04	0.04	0.06	0.07	0.17	0.27	0.35	0.41
0.50	0.06	0.05	0.05	0.05	0.06	0.07	0.14	0.20	0.26	0.27
1.00	0.00	0.00	0.00	0.00	0.00	0.00	0.00	0.00	0.00	1.00
2.00	0.56	0.52	0.60	0.96	1.40	1.48	1.52	1.48	1.44	1.40
4.00	2.72	3.04	3.52	6.72	9.60	10.88	11.20	11.04	10.72	10.56
10.00	24.00	26.00	36.00	53.00	69.00	82.00	93.00	93.00	92.00	91.00
16.00	66.56	69.12	102.40	143.36	181.76	220.16	256.00	253.44	250.88	250.88

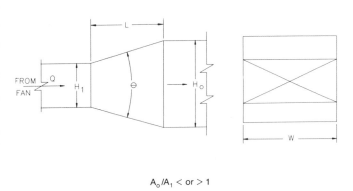

$A_o/A_1 < \text{or} > 1$

SR5-1 Smooth Wye of Type $A_s + A_b > \text{or} = A_c$, Branch 90° to Main, Diverging (Idelchik _et al._ 1986, Diagram 7-21)

					C_b Values					
					Q_b/Q_c					
A_s/A_c	A_b/A_c	0.1	0.2	0.3	0.4	0.5	0.6	0.7	0.8	0.9
0.50	0.25	3.44	0.70	0.30	0.20	0.17	0.16	0.16	0.17	0.18
	0.50	11.00	2.37	1.06	0.64	0.52	0.47	0.47	0.47	0.48
	1.00	60.00	13.00	4.78	2.06	0.96	0.47	0.31	0.27	0.26
0.75	0.25	2.19	0.55	0.35	0.31	0.33	0.35	0.36	0.37	0.39
	0.50	13.00	2.50	0.89	0.47	0.34	0.31	0.32	0.36	0.43
	1.00	70.00	15.00	5.67	2.62	1.36	0.78	0.53	0.41	0.36
1.00	0.25	3.44	0.78	0.42	0.33	0.30	0.31	0.40	0.42	0.46
	0.50	15.50	3.00	1.11	0.62	0.48	0.42	0.40	0.42	0.46
	1.00	67.00	13.75	5.11	2.31	1.28	0.81	0.59	0.47	0.46

					C_s Values					
					Q_s/Q_c					
A_s/A_c	A_b/A_c	0.1	0.2	0.3	0.4	0.5	0.6	0.7	0.8	0.9
0.50	0.25	8.75	1.62	0.50	0.17	0.05	0.00	−0.02	−0.02	0.00
	0.50	7.50	1.12	0.25	0.06	0.05	0.09	0.14	0.19	0.22
	1.00	5.00	0.62	0.17	0.08	0.08	0.09	0.12	0.15	0.19
0.75	0.25	19.13	3.38	1.00	0.28	0.05	−0.02	−0.02	0.00	0.06
	0.50	20.81	3.23	0.75	0.14	−0.02	−0.05	−0.05	−0.02	0.03
	1.00	16.88	2.81	0.63	0.11	−0.02	−0.05	0.01	0.00	0.07
1.00	0.25	46.00	9.50	3.22	1.31	0.52	0.14	−0.02	−0.05	−0.01
	0.50	35.00	6.75	2.11	0.75	0.24	0.00	−0.10	−0.09	−0.04
	1.00	38.00	7.50	2.44	0.81	0.24	−0.03	−0.08	−0.06	−0.02

$A_s = A_b \geq A_c$
$r/W_b = 1.0$

$A_s = A_b \geq A_c$
$r/W_b = 1.0$

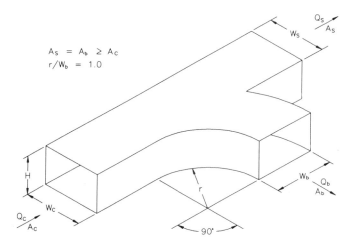

SR7-1 Fan, Centrifugal, without Outlet Diffuser, Free Discharge (AMCA 1990a, Figure 8-3)

A_b/A_o	0.4	0.5	0.6	0.7	0.8	0.9	1.0
C_o	2.00	2.00	1.00	0.80	0.47	0.22	0.00

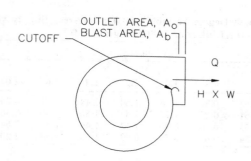

SR7-2 Plane Asymmetric Diffuser at Centrifugal Fan Outlet, Free Discharge (Idelchik *et al.* 1986, Diagram 11-11)

	C_o Values					
	A_1/A_o					
θ	1.5	2.0	2.5	3.0	3.5	4.0
10	0.51	0.34	0.25	0.21	0.18	0.17
15	0.54	0.36	0.27	0.24	0.22	0.20
20	0.55	0.38	0.31	0.27	0.25	0.24
25	0.59	0.43	0.37	0.35	0.33	0.33
30	0.63	0.50	0.46	0.44	0.43	0.42
35	0.65	0.56	0.53	0.52	0.51	0.50

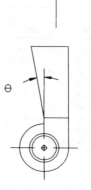
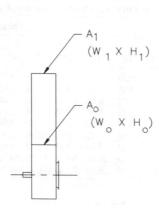

SR7-17 Pyramidal Diffuser at Centrifugal Fan Outlet with Ductwork (Idelchik *et al.* 1986, Diagram 5-16)

	C_1 Values					
	A_o/A_1					
θ	1.5	2.0	2.5	3.0	3.5	4.0
10	0.10	0.18	0.21	0.23	0.24	0.25
15	0.23	0.33	0.38	0.40	0.42	0.44
20	0.31	0.43	0.48	0.53	0.56	0.58
25	0.36	0.49	0.55	0.58	0.62	0.64
30	0.42	0.53	0.59	0.64	0.67	0.69

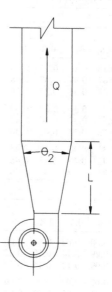
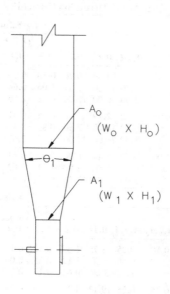

θ is larger of θ_1 and θ_2

PIPE SIZING

THIS chapter includes tables and charts to size piping for various fluid flow systems. Further details on specific piping systems can be found in appropriate chapters of the ASHRAE Handbook series.

PRESSURE DROP EQUATIONS

Pressure drop caused by fluid friction in fully developed flows of all "well behaved" (Newtonian) fluids is described by the Darcy-Weisbach equation

$$\Delta p = f (L/D)(\rho/g_c)(V^2/2) \qquad (1)$$

where

Δp = pressure drop, lb_f/ft^2
f = friction factor, dimensionless (from Moody chart in Chapter 2)
L = length of pipe, ft
D = internal diameter of pipe, ft
ρ = fluid density at mean temperature, lb_m/ft^3
V = average velocity, ft/s
g_c = units conversion factor, $ft \cdot lb_m/lb_f \cdot s^2$

This equation is often presented in head or specific energy form as

$$\Delta h = (\Delta p/\rho)(g_c/g) = f(L/D)(V^2/2g) \qquad (2)$$

where

Δh = head loss, ft
g = acceleration of gravity, ft/s^2

In this form, the density of the fluid does not appear explicitly (although it is in the Reynolds number, which influences f).

The friction factor f is a function of the pipe roughness ϵ, inside diameter D, and a dimensionless parameter, the Reynolds number

$$Re = DV\rho/\mu \qquad (3)$$

where

ϵ = absolute roughness of pipe wall, ft
μ = dynamic viscosity of fluid, $lb_m/ft \cdot s$

The friction factor is frequently presented on a Moody chart (Chapter 2), giving f as a function of Re with ϵ/D as a parameter.

A useful fit of smooth and rough pipe data for the usual turbulent flow regime is the Colebrook equation.

$$1/\sqrt{f} = 1.74 - 2 \log [2\epsilon/D + 18.7/Re\sqrt{f}]$$

Another form of this equation appears in Chapter 2, but the two are equivalent. The present form is more useful in showing behavior at limiting cases—as ϵ/D approaches 0 (smooth limit), the $18.7/Re \sqrt{f}$ term dominates; at high ϵ/D and Re (fully rough limit), the $2\epsilon/D$ term dominates.

The equation is implicit in f; f appears on both sides, so a value for f is usually obtained iteratively.

A less widely used alternative to the Darcy-Weisbach formulation for calculating pressure drop is the Hazen-Williams equation, expressed as

$$\Delta p = 3.022 L(V/C)^{1.852} (1/D^{1.167})(\rho g/g_c) \qquad (4)$$

or

$$\Delta h = 3.022 L(V/C)^{1.852} (1/D^{1.167}) \qquad (5)$$

where C = roughness factor.

Typical values of C are 150 for plastic pipe and copper tubing, 140 for new steel pipe, down to 100 and below for badly corroded or very rough pipe.

Valve and Fitting Losses

Valves and fittings cause pressure losses greater than those caused by the pipe alone. One formulation expresses losses as

$$\Delta p = k(\rho/g_c)(V^2/2) \text{ or } \Delta h = k(V^2/2g) \qquad (6)$$

where k = geometry and size dependent loss coefficient (Tables 1, 2, and 3).

Example 1. Determine the pressure drop for 60°F water flowing at 4 ft/s through a nominal 1-in., 90° screwed ell.
 Solution: From Table 1, the k for a 1-in., 90° screwed ell is 1.5. $\Delta p = 1.5 \times 62.4/32.2 \times 4^2/2 = 23.3 \text{ lb}/ft^2$ or 0.16 psi.

The loss coefficient for valves appears in another form as C_v, a dimensional coefficient expressing the flow through a valve at a specified pressure drop.

$$Q = C_v\sqrt{\Delta p} \qquad (7)$$

where C_v = volume flow, gal/min for a 1-psi pressure drop.

Example 2. Determine the volume flow through a valve with C_v = 10 for an allowable pressure drop of 5 psi.
 Solution: $Q = 10\sqrt{5} = 22.4$ gpm.

Alternative formulations express fitting losses in terms of equivalent lengths of straight pipe (Tables 4 and 5, Figure 4). Pressure loss data for fittings are also presented in Idelchik (1986).

The preparation of this chapter is assigned to TC 6.1, Hydronic and Steam Equipment and Systems.

Table 1   k Values—Screwed Pipe Fittings

Nominal Pipe Dia., in.	90° Ell Reg.	90° Ell Long	45° Ell	Return Bend	Tee-Line	Tee-Branch	Globe Valve	Gate Valve	Angle Valve	Swing Check Valve	Bell Mouth Inlet	Square Inlet	Projected Inlet
3/8	2.5	—	0.38	2.5	0.90	2.7	20	0.40	—	8.0	0.05	0.5	1.0
1/2	2.1	—	0.37	2.1	0.90	2.4	14	0.33	—	5.5	0.05	0.5	1.0
3/4	1.7	0.92	0.35	1.7	0.90	2.1	10	0.28	6.1	3.7	0.05	0.5	1.0
1	1.5	0.78	0.34	1.5	0.90	1.8	9	0.24	4.6	3.0	0.05	0.5	1.0
1-1/4	1.3	0.65	0.33	1.3	0.90	1.7	8.5	0.22	3.6	2.7	0.05	0.5	1.0
1-1/2	1.2	0.54	0.32	1.2	0.90	1.6	8	0.19	2.9	2.5	0.05	0.5	1.0
2	1.0	0.42	0.31	1.0	0.90	1.4	7	0.17	2.1	2.3	0.05	0.5	1.0
2-1/2	0.85	0.35	0.30	0.85	0.90	1.3	6.5	0.16	1.6	2.2	0.05	0.5	1.0
3	0.80	0.31	0.29	0.80	0.90	1.2	6	0.14	1.3	2.1	0.05	0.5	1.0
4	0.70	0.24	0.28	0.70	0.90	1.1	5.7	0.12	1.0	2.0	0.05	0.5	1.0

Source: *Engineering Data Book*, Hydraulic Institute, 1979.

Table 2   k Values—Flanged Welded Pipe Fittings

Nominal Pipe Dia., in.	90° Ell Reg.	90° Ell Long	45° Ell Long	Return Bend Reg.	Return Bend Long	Tee-Line	Tee-Branch	Globe Valve	Gate Valve	Angle Valve	Swing Check Valve
1	0.43	0.41	0.22	0.43	0.43	0.26	1.0	13	—	4.8	2.0
1-1/4	0.41	0.37	0.22	0.41	0.38	0.25	0.95	12	—	3.7	2.0
1-1/2	0.40	0.35	0.21	0.40	0.35	0.23	0.90	10	—	3.0	2.0
2	0.38	0.30	0.20	0.38	0.30	0.20	0.84	9	0.34	2.5	2.0
2-1/2	0.35	0.28	0.19	0.35	0.27	0.18	0.79	8	0.27	2.3	2.0
3	0.34	0.25	0.18	0.34	0.25	0.17	0.76	7	0.22	2.2	2.0
4	0.31	0.22	0.18	0.31	0.22	0.15	0.70	6.5	0.16	2.1	2.0
6	0.29	0.18	0.17	0.29	0.18	0.12	0.62	6	0.10	2.1	2.0
8	0.27	0.16	0.17	0.27	0.15	0.10	0.58	5.7	0.08	2.1	2.0
10	0.25	0.14	0.16	0.25	0.14	0.09	0.53	5.7	0.06	2.1	2.0
12	0.24	0.13	0.16	0.24	0.13	0.08	0.50	5.7	0.05	2.1	2.0

Source: *Engineering Data Book*, Hydraulic Institute, 1979.

Table 3   Approximate Range of Variation for k Factors

90° Elbow	Regular screwed	±20% above 2 in.	Tee	Screwed, line or branch	±25%
		±40% below 2 in.		Flanged, line or branch	±35%
	Long radius screwed	±25%	Globe valve	Screwed	±25%
	Regular flanged	±35%		Flanged	±25%
	Long radius flanged	±30%	Gate valve	Screwed	±25%
45° Elbow	Regular screwed	±10%		Flanged	±50%
	Long radius flanged	±10%	Angle valve	Screwed	±20%
				Flanged	±50%
Return bend (180°)	Regular screwed	±25%	Check valve	Screwed	±50%
	Regular flanged	±35%		Flanged	+200%
	Long radius flanged	±30%			−80%

Source: *Engineering Data Handbook*, Hydraulic Institute, 1979.

Table 4   Equivalent Length in Feet of Pipe for 90° Elbows

Velocity, ft/s	Pipe Size														
	1/2	3/4	1	1-1/4	1-1/2	2	2-1/2	3	3-1/2	4	5	6	8	10	12
1	1.2	1.7	2.2	3.0	3.5	4.5	5.4	6.7	7.7	8.6	10.5	12.2	15.4	18.7	22.2
2	1.4	1.9	2.5	3.3	3.9	5.1	6.0	7.5	8.6	9.5	11.7	13.7	17.3	20.8	24.8
3	1.5	2.0	2.7	3.6	4.2	5.4	6.4	8.0	9.2	10.2	12.5	14.6	18.4	22.3	26.5
4	1.5	2.1	2.8	3.7	4.4	5.6	6.7	8.3	9.6	10.6	13.1	15.2	19.2	23.2	27.6
5	1.6	2.2	2.9	3.9	4.5	5.9	7.0	8.7	10.0	11.1	13.6	15.8	19.8	24.2	28.8
6	1.7	2.3	3.0	4.0	4.7	6.0	7.2	8.9	10.3	11.4	14.0	16.3	20.5	24.9	29.6
7	1.7	2.3	3.0	4.1	4.8	6.2	7.4	9.1	10.5	11.7	14.3	16.7	21.0	25.5	30.3
8	1.7	2.4	3.1	4.2	4.9	6.3	7.5	9.3	10.8	11.9	14.6	17.1	21.5	26.1	31.0
9	1.8	2.4	3.2	4.3	5.0	6.4	7.7	9.5	11.0	12.2	14.9	17.4	21.9	26.6	31.6
10	1.8	2.5	3.2	4.3	5.1	6.5	7.8	9.7	11.2	12.4	15.2	17.7	22.2	27.0	32.0

Table 5   Iron and Copper Elbow Equivalents[a]

Fitting	Iron Pipe	Copper Tubing
Elbow, 90°	1.0	1.0
Elbow, 45°	0.7	0.7
Elbow, 90° long turn	0.5	0.5
Elbow, welded, 90°	0.5	0.5
Reduced coupling	0.4	0.4
Open return bend	1.0	1.0
Angle radiator valve	2.0	3.0
Radiator or convector	3.0	4.0
Boiler or heater	3.0	4.0
Open gate valve	0.5	0.7
Open globe valve	12.0	17.0

[a]See Table 4 for equivalent length of one elbow.
Source: Giesecke (1926) and Giesecke and Badgett (1931, 1932).

Calculating Pressure Losses

The most common engineering design flow loss calculation selects a pipe size for the desired total flow rate and available or allowable pressure drop.

Since either formulation of fitting losses requires a known diameter, pipe size must be selected before calculating the detailed influence of fittings. A frequently used rule of thumb assumes that the design length of pipe is 50 to 100% longer than actual to account for fitting losses. After a pipe diameter has been selected on this basis, the influence of each fitting can be evaluated.

WATER PIPING

FLOW RATE LIMITATIONS

Stewart and Dona (1987) surveyed the literature relating to water flow rate limitations. This section briefly reviews some of the findings. Noise, erosion, and installation and operating costs all limit the maximum and minimum velocities in piping systems. If piping sizes are too small, noise and erosion levels and pumping costs can be unfavorable; if piping sizes are too large, installation costs are excessive. Therefore, pipe sizes are chosen to minimize initial cost while avoiding the undesirable effects of high velocities.

A variety of upper limits of water velocity and/or pressure drop in piping and piping systems is used. One recommendation places a velocity limit of 4 ft/s for 2-in. pipe and smaller, and a pressure drop limit of 4 ft of water/100 ft for piping over 2 in. Others are based on the type of service (Table 6) or the annual operating hours (Table 7). These limitations are imposed either to control the levels of pipe and valve noise, erosion, and water hammer pressure or for economic reasons. Carrier (1960) recommends that the velocity not exceed 15 ft/s in any case.

Velocity-dependent noise in piping and piping systems results from any or all of four sources: turbulence, cavitation, release of entrained air, and water hammer. In investigations of flow-related noise, Marseille (1965), Ball and Webster (1976), and Rogers (1953, 1954, 1956) report that velocities on the order of 10 to 17 ft/s lie within the range of allowable noise levels for residential and commercial buildings. The experiments showed considerable variation in the noise levels obtained for a specified velocity. Generally, systems with longer length pipe and with more numerous fittings and valves were noisier. In addition, sound measurements were taken under widely differing conditions; for example, some tests used plastic-covered pipe, while others did not. Thus, no detailed correlations relating sound level to flow velocity in generalized systems are available.

The noise generated by fluid flow in a pipe system increases sharply if cavitation or the release of entrained air occurs. Usually

Table 6   Water Velocities Based on Type of Service

Type of Service	Velocity, ft/s	Reference
General service	4 to 10	a, b, c
City water	3 to 7	a, b
	2 to 5	c
Boiler feed	6 to 15	a, c
Pump suction and drain lines	4 to 7	a, b

[a]Crane Co. 1976. Flow of fluids through valves, fittings, and pipe. Technical Paper 410.
[b]*System Design Manual.* 1960. Carrier Air Conditioning Co., Syracuse, NY.
[c]*Piping Design and Engineering.* 1951. Grinnell Company, Inc., Cranston, RI.

Table 7   Maximum Water Velocity to Minimize Erosion

Normal Operation, h/yr	Water Velocity, ft/s
1500	15
2000	14
3000	13
4000	12
6000	10

Source: *System Design Manual*, Carrier Air Conditioning Co., 1960.

the combination of a high water velocity with a change in flow direction or a decrease in the cross section of a pipe causing a sudden pressure drop is necessary to cause cavitation. Ball and Webster (1976) found that at their maximum velocity of 42 ft/s, cavitation in straight 3/8 and 1/2 in. pipe did not occur; using the apparatus with two elbows, cold water velocities up to 21 ft/s did not cause cavitation. Cavitation did occur in orifices of 1:8 area ratio (orifice flow area is 1/8 of pipe flow area) at 5 ft/s and at 10 ft/s for 1:4 area ratio orifices (Rogers 1954).

Some data are available for predicting hydrodynamic noise generated by control valves. The Instrument Society of America (1985) compiled prediction correlations in an effort to develop control valves for reduced noise levels. The correlation to predict hydrodynamic (liquid) noise from control valves is

$$SL = 10 \log C_v + 20 \log \Delta p - 30 \log t + 5 \qquad (8)$$

where

SL = sound level, dB
C_v = flow coefficient, $Q/(sg/\Delta p)^{0.5}$, gpm/(psi)$^{0.5}$
Q = flow rate, gpm
sg = specific gravity of liquid
Δp = pressure drop across valve, psi
t = downstream pipe wall thickness, in.

Air entrained in water usually has a higher partial pressure than the water. Even when flow rates are small enough to avoid cavitation, the release of entrained air may create noise. Every effort should be made to vent the piping system or otherwise remove entrained air.

Erosion

Erosion in piping systems is caused by water bubbles, sand, or other solid matter impinging on the inner surface of the pipe. Generally, at velocities lower than 100 ft/s, erosion is not significant as long as there is no cavitation. When solid matter is entrained in the fluid at high velocities, erosion occurs rapidly, especially in bends. Thus, high velocities should not be used in systems where sand or other solids are present or where slurries are transported.

Allowances for Aging

With age, the internal surfaces of pipes become increasingly rough, which reduces the available flow with a fixed pressure supply. However, designing with excessive age allowances may result in oversized piping. Age-related decreases in capacity depend on the type of water, type of pipe material, temperature of water, and type of system (open or closed) and include:

- Sliming (biological growth or deposited soil on the pipe walls) occurs mainly in unchlorinated, raw water systems.
- Caking of calcareous salts occurs in hard water, *i.e.*, water bearing calcium salts, and increases with increases in the temperature of the water.
- Corrosion (incrustations of ferrous and ferric hydroxide on the pipe walls) occurs in metal pipe in soft water. Since oxygen is necessary for corrosion to take place, significantly more corrosion takes place in open systems.

Allowances for expected decreases in capacity are sometimes treated as a specific amount (percentage). Dawson and Bowman (1933) added an allowance of 15% friction loss to new pipe (equivalent to an 8% decrease in capacity). Hennington *et al.* (1981) increased the friction loss by 15 to 20% for closed piping systems and 75 to 90% for open systems. Carrier (1960) indicates a factor of approximately 1.75 between friction factors for closed and open systems.

Obrecht and Pourbaix (1967) differentiated between the corrosive potential of different metals in potable water systems and concluded that iron is the most severely attacked, then galvanized steel, lead, copper, and finally copper alloys (*i.e.*, brass). Hunter (1941) and Freeman (1941) showed the same trend. After four years of cold and hot water use, copper pipe had a capacity loss of 25 to 65%. Aged ferrous pipe has a capacity loss of 40 to 80%. Smith (1983) recommended increasing the design discharge by 1.55 for uncoated cast iron, 1.08 for iron and steel, and 1.06 for cement or concrete.

The Plastic Pipe Institute (1971) found that corrosion is not a problem in plastic pipe, the capacity of plastic pipe used in Europe and the United States remaining essentially the same after 30 years in use.

Extensive age-related flow data are available for use with the Hazen-Williams empirical equation. Difficulties arise in its application, however, because the original Hazen-Williams roughness coefficients are valid only for the specific pipe diameters, water velocities, and water viscosities used in the original experiments. Thus, when the *C*s are extended to different diameters, velocities, and/or water viscosities, errors of up to about 50% in pipe capacity can occur (Williams and Hazen 1933, Sanks 1978).

Water Hammer

When any moving fluid (not just water) is abruptly stopped, as when a valve closes suddenly, large pressures can develop. While detailed analysis requires knowledge of the elastic properties of the pipe and the flow-time history, the limiting case of rigid pipe and instantaneous closure is simple to calculate. Under these conditions

$$p_h = \rho c_s V/g_c \qquad (9)$$

where

p_h = pressure rise caused by water hammer, lb_f/ft^2
ρ = fluid density, lb_m/ft^3
c_s = velocity of sound in fluid, ft/s
V = fluid flow velocity, ft/s

The c_s for water is 4720 ft/s, although the elasticity of the pipe reduces the effective value.

Example 3. What is the maximum pressure rise if water flowing at 10 ft/s is stopped instantaneously?

Solution: $p_h = 62.4 \times 4720 \times 10/32.2 = 91468 \ lb/ft^2$
$\qquad\quad = 635 \ psi$

Other Considerations

Not discussed in detail in this chapter, but of potentially great importance, are a number of physical and chemical considerations: pipe and fitting design, materials, and joining methods must be appropriate for working pressures and temperatures encountered, as well as being suitably resistant to chemical attack by the fluid.

Other Piping Materials and Fluids

For fluids not included in this chapter or for piping materials of different dimensions, manufacturers' literature frequently supplies pressure drop charts. The Darcy-Weisbach equation and the Moody chart or the Colebrook equation can be used as an alternative to pressure drop charts or tables.

HOT AND CHILLED WATER PIPE SIZING

The Darcy-Weisbach equation with friction factors from the Moody chart or Colebrook equation (or, alternatively, the Hazen-Williams equation) is fundamental to calculating pressure drop in hot and chilled water piping; however, charts calculated from these equations (such as Figures 1, 2, and 3) provide easy determination of pressure drops for specific fluids and pipe standards. In addition, tables of pressure drops can be found in Hydraulic Institute (1979) and Crane Co. (1976).

Most tables and charts for water are calculated for properties at 60°F. Using these for hot water introduces some error, although the answers are conservative; *i.e.*, cold water calculations overstate the pressure drop for hot water. Using 60°F water charts for 200°F water should not result in errors in Δp exceeding 20%.

Range of Usage of Pressure Drop Charts

General design range. The general range of pipe friction loss used for design of hydronic systems is between 1 and 4 ft/100 ft. A value of 2.5 ft/100 ft represents the *mean* to which most systems are designed. Wider ranges may be used in specific designs, if certain precautions are taken.

Piping noise. Closed loop hydronic system piping is generally sized below certain arbitrary upper limits, such as a velocity limit of 4 ft/s for 2-in. pipe and under, and a pressure drop limit of 4 ft per 100 ft for piping over 2 in. in diameter. Velocities in excess of 4 ft/s can be used in piping of larger size. This limitation is generally accepted, although it is based on relatively inconclusive experience with noise in piping. Water *velocity noise* is not caused by water but by free air, sharp pressure drops, turbulence, or a combination of these, which in turn cause cavitation or flashing of water into steam. Therefore, higher velocities may be used if proper precautions are taken to eliminate air and turbulence.

Air Separation

Because piping noise can be caused by free air, the hydronic system must be equipped with air separation devices to minimize the amount of entrained air in the piping circuit. Air should be vented at the highest point of the system.

In the absence of such venting, air can be entrained in the water and carried to separation units at flow velocities of 1.5 to 2 ft/s or more in pipe sizes of 2 in. and under. Minimum velocities of 2 ft/s are therefore recommended. For pipe sizes 2 in. and over, minimum velocities corresponding to a head loss of 0.75 ft/100 ft are normally used. Maintenance of minimum velocities is particularly important in the upper floors of high-rise buildings when the air tends to come out of solution because of reduced pressures. Higher velocities should be used in *down-comer* return mains feeding into air separation units located in the basement.

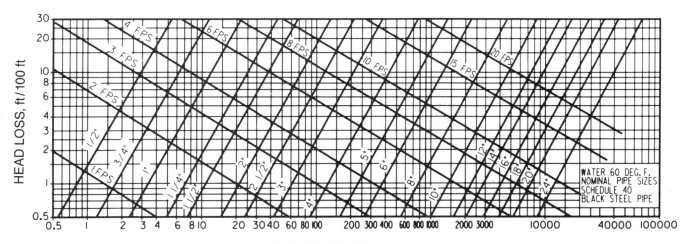

Fig. 1   Friction Loss for Water in Commercial Steel Pipe (Schedule 40)

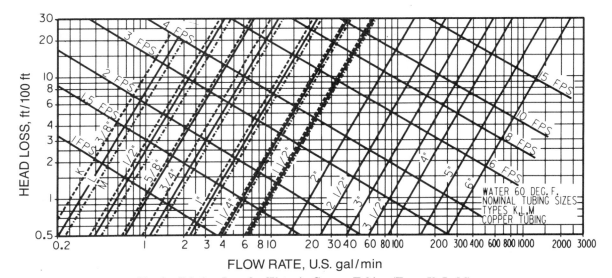

Fig. 2   Friction Loss for Water in Copper Tubing (Types K, L, M)

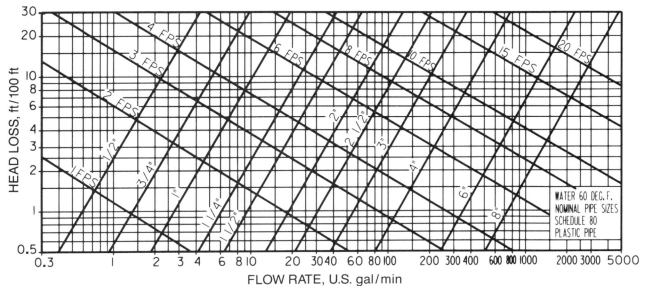

Fig. 3   Friction Loss for Water in Plastic Pipe (Schedule 80)

Example 4. Determine pipe size for circuit requiring 20 gpm flow.

Solution: Enter Figure 1 at 20 gpm, read up to pipe size within normal design range, select 1-1/2 in. Velocity is 3.1 ft/s, which is between 2 and 4. Pressure loss is 2.9 ft/100 ft, which is between 1 and 4 ft/100 ft.

Valve and Fitting Pressure Drop

Valve and fitting pressure drop can be listed in elbow equivalents, with an elbow being equivalent to a length of straight pipe. Table 4 lists equivalent lengths of 90° elbows; Table 5 lists elbow equivalents for valves and fittings for iron and copper.

Example 5. Determine equivalent feet of pipe for a 4-in. open gate valve at a flow velocity of approximately 4 ft/s.

Solution: From Table 4, at 4 ft/s, each equivalent elbow is equal to 10.6 ft of 4-in. pipe. From Table 5, the 4-in. gate valve is equal to 0.5 elbows. The actual equivalent pipe length (added to measure circuit length for pressure drop determination) will be 10.6 × 0.5, or 5.3 equivalent feet of 4-in. pipe.

Tee fitting pressure drop. Pressure drop through pipe tees varies with flow through the branch. Figure 4 illustrates pressure drops for nominal 1-in. tees of equal inlet and outlet sizes and for the flow patterns illustrated. Idelchik (1986) also presents data for threaded tees.

Different investigators present tee loss data in different forms, and it is sometimes difficult to reconcile results from several sources. As an estimate of the upper limit to tee losses, a pressure

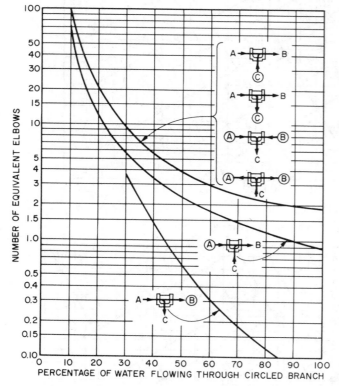

Notes: 1. The chart is based on straight tees, *i.e.*, branches A, B, and C are the same size.
2. Pressure loss in desired circuit is obtained by selecting proper curve according to illustrations, determining the flow at the circled branch, and multiplying the pressure loss for the same size elbow at the flow rate in the circled branch by the equivalent elbows indicated.
3. When the size of an outlet is reduced, the equivalent elbows shown in the chart do not apply. Therefore, the maximum loss for any circuit for any flow will not exceed 2 elbow equivalents at the maximum flow (gpm) occurring in any branch of the tee.
4. The top curve of chart is average of 4 curves, one for each circuit shown.

Fig. 4   Elbow Equivalents of Tees at Various Flow Conditions
(Giesecke and Badgett 1931, 1932)

or head loss coefficient of 1.0 may be assumed for entering and leaving flows (*i.e.*, $\Delta p = 1.0\,\rho V_{in}^2/2 + 1.0\,\rho V_{out}^2/2$).

Example 6. Determine the pressure or head losses for a 1-in. (all openings) threaded pipe tee flowing 25% to the side branch, 75% through. The entering flow is 10 gpm (3.71 ft/s).

Solution: From Figure 4, bottom curve, the number of equivalent elbows for the through-flow is 0.15 elbows; the through-flow is 7.5 gpm (2.78 ft/s); and the pressure loss is based on the exit flow rate. Table 4 gives the equivalent length of a 1-in. elbow at 3 ft/s as 2.7 ft. Using Equations (1) and (2) (friction factor $f = 0.0290$, diameter = 0.0874 ft):

$$\Delta p = (0.15)(0.0290)(2.7/0.0874)(62.4)(2.78^2)/[(2)(32.2)]$$

$$\Delta p = 1.01\ \text{lb/ft}^2 = 0.00699\ \text{lb/in}^2\ \text{for pressure drop, or}$$

$$\Delta h = (0.15)(0.0290)(2.7/0.0874)(2.78^2)/[(2)(32.2)]$$

$$\Delta h = 0.0161\ \text{ft for head loss}$$

From Figure 4, top curve, the number of equivalent elbows for the branch flow of 25% is 13 elbows; the branch flow is 2.5 gpm (0.93 ft/s); and the pressure loss is based on the exit flow rate. Table 4 gives the equivalent of a 1-in. elbow at 1 ft/s as 2.2 ft. Using Equations (1) and (2) (friction factor $f = 0.0350$, diameter = 0.0874 ft):

$$\Delta p = (13)(0.0350)(2.2/0.0874)(62.4)(0.93^2)/[(2)(32.2)]$$

$$\Delta p = 9.60\ \text{lb/ft}^2 = 0.0667\ \text{lb/in}^2\ \text{for pressure drop, or}$$

$$\Delta h = (13)(0.0350)(2.2/0.0874)(0.93^2)/[(2)(32.2)]$$

$$\Delta h = 0.154\ \text{ft for head loss}$$

HOT AND COLD
SERVICE WATER PIPING

Before any part of a water piping system can be sized, the probable rate of flow in any particular section of piping should be determined. The rate of flow in the service line, risers, and main branches, however, is rarely equal to the sum of the rates of flow of all connected fixtures. In fact, the probability that every fixture in a large group will be in use at the same time is so remote that designing the piping to take care of such simultaneous flow is unnecessarily expensive.

The demand load in building water supply systems cannot be determined exactly and is not readily standardized. The two main problems to be considered are (1) the satisfactory supply of water for a given fixture and (2) the number of fixtures that are assumed to be in use at the same time.

The minimum satisfactory flow depends largely on the living standard and professional needs of the consumer, family size, garden requirements, and similar factors. Depending on these factors, the per capita water consumption for domestic use usually varies between 20 and 80 gal per day. Type of dwelling also has considerable influence on water consumption.

Conclusive water consumption data for housing projects in the United States have not been gathered. The daily per capita water consumption in housing projects apparently falls between the consumption in apartment houses and single dwellings at the same geographical location. A daily per capita water consumption of 70 gal can be used as a safe design figure for housing projects, as is indicated by New York City housing project records.

Table 8 gives the rate of flow desirable for many common fixtures and the average pressure necessary to give this rate of flow. The pressure varies with fixture design. Generally, the lower the quality of the faucet, the greater the pressure required.

In estimating the load, the rate of flow is frequently computed in *fixture units*, which are relative indicators of flow. Table 9 gives the demand weights in terms of fixture units for different plumbing fixtures under several conditions of service, and Figure 5 gives the estimated demand in gallons per minute corresponding to any total number of *fixture units*. Figures 6 and 7 provide more accurate estimates at the lower end of the scale.

The estimated demand load for fixtures used intermittently on any supply pipe can be obtained by multiplying the number of each kind of fixture supplied through that pipe by its weight from Table 9, adding the products, and then referring to the appropriate curve of Figures 5, 6, or 7 to find the demand corresponding to the total fixture units. In using this method, note that the demand for fixture or supply outlets other than those listed in the table of fixture units is not yet included in the estimate. The de-

Table 8 Proper Flow and Pressure Required during Flow for Different Fixtures

Fixture	Flow Pressure[a]	Flow, gpm
Ordinary basin faucet	8	3.0
Self-closing basin faucet	12	2.5
Sink faucet—3/8 in.	10	4.5
Sink faucet—1/2 in.	5	4.5
Dishwasher	15–25	—[b]
Bathtub faucet	5	6.0
Laundry tube cock—1/4 in.	5	5.0
Shower	12	3–10
Ball cock for closet	15	3.0
Flush valve for closet	10–20	15–40[c]
Flush valve for urinal	15	15.0
Garden hose, 50 ft, and sill cock	30	5.0

[a]Flow pressure is the pressure (psig) in the pipe at the entrance to the particular fixture considered.
[b]Varies; see manufacturers' data.
[c]Wide range due to variation in design and type of flush valve closets.

Table 9 Demand Weights of Fixtures in Fixture Units[a]

Fixture or Group[b]	Occupancy	Type of Supply Control	Weight in Fixture Units[c]
Water closet	Public	Flush valve	10
Water closet	Public	Flush tank	5
Pedestal urinal	Public	Flush valve	10
Stall or wall urinal	Public	Flush valve	5
Stall or wall urinal	Public	Flush tank	3
Lavatory	Public	Faucet	2
Bathtub	Public	Faucet	4
Shower head	Public	Mixing valve	4
Service sink	Office, etc	Faucet	3
Kitchen sink	Hotel or restaurant	Faucet	4
Water closet	Private	Flush valve	6
Water closet	Private	Flush tank	3
Lavatory	Private	Faucet	1
Bathtub	Private	Faucet	2
Shower head	Private	Mixing valve	2
Bathroom group	Private	Flush valve for closet	8
Bathroom group	Private	Flush tank for closet	6
Separate shower	Private	Mixing valve	2
Kitchen sink	Private	Faucet	2
Laundry trays (1 to 3)	Private	Faucet	3
Combination fixture	Private	Faucet	3

Note: See Hunter (1941).
[a]For supply outlets likely to impose continuous demands, estimate continuous supply separately, and add to total demand for fixtures.
[b]For fixtures not listed, weights may be assumed by comparing the fixture to a listed one using water in similar quantities and at similar rates.
[c]The given weights are for total demand. For fixtures with both hot and cold water supplies, the weights for maximum separate demands can be assumed to be 75% of the listed demand for the supply.

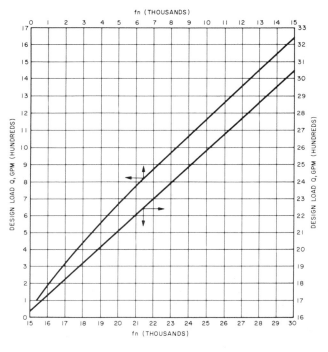

Fig. 5 Design Load versus Fixture Units, Mixed System, High Part of Curve
(Hunter 1941)

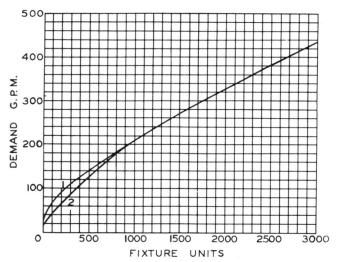

No. 1 for system predominantly for flush valves.
No. 2 for system predominantly for flush tanks.

Fig. 6 Estimate Curves for Demand Load
(Hunter 1941)

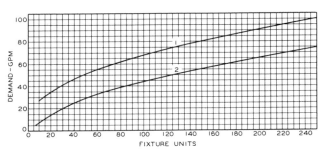

Fig. 7 Section of Figure 6 on Enlarged Scale

mands for outlets (such as hose connections and air-conditioning apparatus) that are likely to impose continuous demand during heavy use of the weighted fixtures, should be estimated separately and added to demand for fixtures used intermittently to estimate total demand.

The Hunter curves in Figures 5, 6, and 7 are based on use patterns in residential buildings and can be erroneous for other usages such as sports arenas. Williams (1976) discusses the Hunter assumptions and presents an analysis using alternative assumptions.

So far, the information presented shows the *design rate of flow* to be determined in any particular section of piping. The next step is to determine the *size* of piping. As water flows through a pipe, the pressure continually decreases along the pipe due to loss of energy from friction. The problem is then to ascertain the minimum pressure in the street main and the minimum pressure required to operate the topmost fixture. (A pressure of 15 psig may be ample for most flush valves, but reference should be made to the manufacturers' requirements. Some fixtures require a pressure up to 25 psig. A minimum of 8 psig should be allowed for other fixtures.) The pressure differential overcomes pressure losses in the distributing system and the difference in elevation between the water main and the highest fixture.

The pressure loss, in psi, resulting from the difference in elevation between the street main and the highest fixture, can be obtained by multiplying the difference in elevation in feet by the conversion factor 0.434.

Pressure losses in the distributing system consist of pressure losses in the piping itself, plus the pressure losses in the pipe fittings, valves, and the water meter, if any. Approximate design pressure losses and flow limits for disk-type meters for various rates of flow are given in Figure 8. Water authorities in many localities require compound meters for greater accuracy with varying flow; consult the local utility. Design data for compound meters differ from the data in Figure 8. Manufacturers give data on exact pressure losses and capacities.

Figure 9 shows the variation of pressure loss with rate of flow for various faucets and cocks. The water demand for hose bibbs or other large-demand fixtures taken off the building main frequently results in inadequate water supply to the upper floor of a building. This condition can be prevented by sizing the distribution system so that the pressure drops from the street main to all fixtures are the same. An ample building main (not less than 1 in. where possible) should be maintained until all branches to hose bibbs have been connected. Where the street main pressure is excessive and a pressure-reducing valve is used to prevent water hammer or excessive pressure at the fixtures, the hose bibbs should be connected ahead of the reducing valve.

The principles involved in sizing upfeed and downfeed systems are the same. In the downfeed system, however, the difference in elevation between the overhead supply mains and the fixtures provides the pressure required to overcome pipe friction. Since friction pressure loss and height pressure loss are not additive, as in an upfeed system, smaller pipes may be used with a downfeed system.

Plastic Pipe

The maximum safe water velocity in a thermoplastic piping system (5 ft/s) depends on the specific details of the system and on the operating conditions. Higher velocities can be used in cases where the operating characteristics of valves and pumps are known so that sudden changes in flow velocity can be controlled. The total pressure in the system at any time (operating plus surge of water hammer) should not exceed 150% of the pressure rating of the system.

Procedure for Sizing Cold Water Systems

The recommended procedure for sizing piping systems is outlined below.

1. Sketch the main lines, risers, and branches, and indicate the fixtures to be served. Indicate the rate of flow of each fixture.
2. Using Table 9, compute the demand weights of the fixtures in fixture units.
3. Determine the total demand in fixture units and, using Figures 5, 6, or 7, find the expected demand in gal/min.
4. Determine the equivalent length of pipe in the main lines, risers, and branches. Since the sizes of the pipes are not known, the exact equivalent length of various fittings cannot be made. Add the equivalent lengths, starting at the street main and proceeding along the service line, the main line of the building, and up the riser to the top fixture of the up served.
5. Determine the average minimum pressure in the street main and the minimum pressure required for the operation of the topmost fixture. This should be 8 to 25 psi.

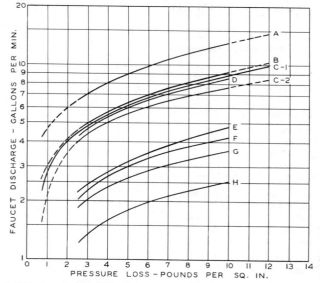

A. 1/2-in. laundry bibb (old style)
B. Laundry compression faucet
C-1. 1/2-in. compression sink faucet (mfr. 1)
C-2. 1/2-in. compression sink faucet (mfr. 2)
D. Combination compression bathtub faucets (both open)
E. Combination compression sink faucet
F. Basin faucet
G. Spring self-closing faucet
H. Slow self-closing faucet
(Dashed lines indicate recommended extrapolation)

**Fig. 9   Variation of Pressure Loss with Rate of Flow
for Various Faucets and Cocks**

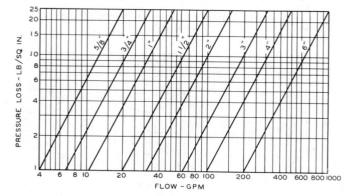

Fig. 8   Pressure Losses in Disk-Type Water Meters

6. Using Equation (1), calculate the approximate design value of the average pressure drop per 100 ft of pipe in equivalent length determined in paragraph 4.

$$\Delta p = (p_s - 0.434 H - p_f - p_m) \, 100/L \qquad (10)$$

where

Δp = average pressure loss per 100 ft of equivalent length of pipe, psi
p_s = pressure in street main, psig
p_f = minimum pressure required to operate topmost fixture, psig
p_m = pressure drop through water meter, psi
H = height of highest fixture above street main, ft
L = equivalent length determined in paragraph 4, ft

If the system is downfeed supply from a gravity tank, height of water in the tank, converted to psi by multiplying by 0.434, replaces the street main pressure, and the term $0.434 H$ is added instead of subtracted in calculating p. In this case, H is the vertical distance of the fixture below the bottom of the tank.

7. From the expected rate of flow, as determined in paragraph 3, and the value of p, calculated in paragraph 6, choose the sizes of pipe from Figures 1, 2, or 3.

Example 7. Assume a minimum street main pressure of 55 psig; a height of topmost fixture (a urinal with flush valve) above street main of 50 ft; a developed pipe length from water main to highest fixture of 100 ft; a total load on the system of 50 fixture units; and that the water closets are flush valve operated. Find the required size of supply main.

Solution: From Figure 7 the estimated peak demand is 51 gpm. From Table 8, the minimum pressure required to operate the topmost fixture is 15 psig. For a trial computation, choose the 1-1/2-in. meter. From Figure 8, the pressure drop through a 1-1/2-in. disk-type meter for a flow of 51 gpm is 6.5 psi.

The pressure drop available for overcoming friction in pipes and fittings is $55 - (0.434 \times 50 + 15 + 6.5) = 12$ psi.

At this point, estimate the equivalent pipe length of the fittings on the direct line from the street main to the highest fixture. The exact equivalent length of the various fittings cannot be determined since the pipe sizes of the building main, riser, and branch leading to the highest fixture are not yet known, but a first approximation is necessary to tentatively select pipe sizes. If the computed pipe sizes differ from those used in determining the equivalent length of pipe fittings, a recalculation will be necessary, using the computed pipe sizes for the fittings. For this example, assume that the total equivalent length of the pipe fittings is 50 ft.

The permissible pressure loss per 100 ft of equivalent pipe is $12 \times 100/(100 + 50) = 8$ psi. A 2-in. building main is adequate.

The sizing of the *branches* of the building main, the *risers*, and *fixture branches* follows these principles. For example, assume that one of the branches of the building main carries the cold water supply for 3 water closets, 2 bathtubs, and 3 lavatories. Using the permissible pressure loss of 8 psi per 100 ft, the size of branch (determined from Table 9 and Figures 1 and 7) is found to be 1-1/2 in. Items included in the computation of pipe size are given as follows:

Fixtures No. and Type	Fixture Units (Table 7 and Note C)	Demand (Figure 7)	Pipe Size (Figure 1)
3 flush valves	$3 \times 6 = 18$		
2 bath tubs	$0.75 \times 2 \times 2 = 3$		
3 lavatories	$0.75 \times 3 \times 1 = 2.25$		
Total	$= 23.25$	38 gpm	1-1/2 in.

Table 10 is a guide to minimum pipe sizing where flush valves are used.

Velocities exceeding 10 ft/s cause undesirable noise in the piping system. This usually governs the size of larger pipes in the system while, in small pipe sizes, the friction loss usually governs the selection because the velocity is low compared with friction loss. Velocity is the governing factor in downfeed systems, where friction loss is usually neglected. Velocity in branches leading to pump suctions should not exceed 5 ft/s.

Table 10 Allowable Number of 1-in. Flush Valves Served by Various Sizes of Water Pipe[a]

Pipe Size, in.	No. 1-in. Flush Valves
1-1/4	1
1-1/2	2–4
2	5–12
2-1/2	13–25
3	26–40
4	41–100

[a]Two 3/4-in. flush valves are assumed equal to one 1-in. flush valve, but can be served by a 1-in. pipe. Water pipe sizing must consider demand factor, available pressure, and length of run.

If the street pressure is too low to adequately supply upper-floor fixtures, the pressure must be increased. Constant or variable speed booster pumps, alone or in conjunction with gravity supply tanks, or hydropneumatic systems may be used.

Flow control valves for individual fixtures under varying pressure conditions automatically adjust the flow at the fixture to a predetermined quantity. These valves allow the designer to (1) limit the flow at the individual outlet to the minimum suitable for the purpose, (2) hold the total demand for the system more closely to the required minimum, and (3) design the piping system as accurately as is practicable for the requirements.

STEAM PIPING

Pressure losses in steam piping for flows of dry or nearly dry steam are governed by the equations in the Pressure Drop Equations section. This section incorporates these principles with other information specific to steam systems.

Pipe Sizes

Determining pipe sizes for a given load in steam heating depends on the following factors: (1) the initial pressure and the total pressure drop that can be allowed between the source of supply and at the end of the return system; (2) the maximum velocity of steam allowable for quiet and dependable operation of the system, taking into consideration the direction of condensate flow; and (3) the equivalent length of the run from the boiler or source of steam supply to the farthest heating unit.

Initial Pressure and Pressure Drop

Table 11 lists pressure drops commonly used with corresponding initial steam pressures for sizing steam piping.

Several factors, such as initial pressure and pressure required at the end of the line, should be considered, but it is most important that (1) the total pressure drop does not exceed the initial gage pressure of the system, and in practice it should never exceed one-

Table 11 Pressure Drops Used for Sizing Steam Pipe[a]

Initial Steam Pressure, psig	Pressure Drop per 100 ft	Total Pressure Drop in Steam Supply Piping
Vacuum return	2 to 4 oz/in^2	1 to 2 psi
0	0.5 oz/in^2	1 oz/in^2
1	2 oz/in^2	1 to 4 oz/in^2
2	2 oz/in^2	8 oz/in^2
5	4 oz/in^2	1.5 psi
10	8 oz/in^2	3 psi
15	1 psi	4 psi
30	2 psi	5 to 10 psi
50	2 to 5 psi	10 to 15 psi
100	2 to 5 psi	15 to 25 psi
150	2 to 10 psi	25 to 30 psi

[a]Equipment, control valves, and so forth must be selected based on delivered pressures.

Table 12   Comparative Capacity of Steam Lines at Various Pitches for Steam and Condensate Flowing in Opposite Directions

Pitch of Pipe, in/10 ft	Nominal Pipe Diameter, in.									
	3/4		1		1-1/4		1-1/2		2	
	Capacity	Maximum Velocity	Capacity	Maximum Velocity	Capacity	Maximum Velocity	Capacity	Maximum Velocity	Capacity	Maximum Velocity
1/4	3.2	8	6.8	9	11.8	11	19.8	12	42.9	15
1/2	4.1	11	9.0	12	15.9	14	25.9	16	54.0	18
1	5.7	13	11.7	15	19.9	17	33.0	19	68.8	24
1-1/2	6.4	14	12.8	17	24.6	20	37.4	22	83.3	27
2	7.1	16	14.8	19	27.0	22	42.0	24	92.9	30
3	8.3	17	17.3	22	31.3	25	46.8	26	99.6	32
4	9.9	22	19.2	24	33.4	26	50.8	28	102.4	32
5	10.5	22	20.5	25	38.5	31	59.2	33	115.0	33

Source: Laschober *et al.* (1966).   Velocity in ft/s; capacity in lb/h.

half of the initial gage pressure; (2) the pressure drop is not great enough to cause excessive velocities; (3) a constant initial pressure should be maintained, except on systems specially designed for varying initial pressures, such as subatmospheric pressure, which normally operate under controlled partial vacuums; and (4) the rise in water from pressure drop does not exceed the difference in level, for gravity return systems, between the lowest point on the steam main, the heating units or the dry return, and the boiler water line.

Maximum Velocity

For quiet operation, steam velocity should be 8000 to 12,000 fpm, with a maximum of 15,000 fpm. The lower the velocity, the quieter the system. When the condensate must flow against the steam, even in limited quantity, the velocity of the steam must not exceed limits above which (1) the disturbance between the steam and the counterflowing water may produce objectionable sound, such as water hammer, or (2) result in the retention of water in certain parts of the system until the steam flow is reduced sufficiently to permit the water to pass. The velocity at which these disturbances take place is a function of (1) pipe size, whether the pipe runs horizontally or vertically; (2) pitch of the pipe if it runs horizontally; (3) the quantity of condensate flowing against the steam; and (4) freedom of the piping from water pockets that, under certain conditions, act as a restriction in pipe size. Table 12 lists maximum capacities for various size steam lines.

Equivalent Length of Run

All tables for the flow of steam in pipes, based on pressure drop, must allow for pipe friction, as well as for the resistance of fittings and valves. These resistances are generally stated in terms of straight pipe; *i.e.*, a certain fitting produces a drop in pressure equivalent to the stated number of feet of straight run of the same size of pipe. Table 13 gives the number of feet of straight pipe usually allowed for the more common types of fittings and valves. In all pipe-sizing tables in this chapter, the *length of run* refers to the *equivalent length of run* as distinguished from the *actual length* of pipe. A common sizing method is to assume the length of run and to check this assumption after pipes are sized. For this purpose, the length of run is usually assumed to be double the actual length of pipe.

Example 8. Determine the length in feet of pipe to be added to actual length of run illustrated.

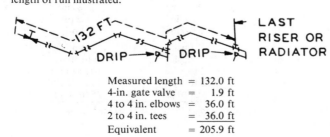

Measured length	= 132.0 ft
4-in. gate valve	= 1.9 ft
4 to 4 in. elbows	= 36.0 ft
2 to 4 in. tees	= 36.0 ft
Equivalent	= 205.9 ft

LOW-PRESSURE STEAM PIPING

Figure 10 is the basic chart for determining the flow rate and velocity of steam in Schedule 40 pipe for various values of pressure drop per 100 ft, based on zero psig saturated steam. Using the multiplier charts, Figure 10 can be used at all saturation pressures between 0 and 200 psig (see Example 10).

Values in Table 14 (taken from the basic chart) provide a more rapid means of selecting pipe sizes for the various pressure drops listed and for systems operated at 3.5 and 12 psig. The flow rates shown for 3.5 psig can be used for saturated pressures from 1 to 6 psig, and those shown for 12 psig can be used for saturated pressures from 8 to 16 psig with an error not exceeding 8%.

Both Figure 10 and Table 14 can be used where the flow of condensate does not inhibit the flow of steam. Columns *B* and *C* of Table 15 are used in cases where steam and condensate flow in opposite directions as in risers or runouts that are not dripped. Columns *D, E,* and *F* are for one-pipe systems and include risers, radiator valves and vertical connections, and radiator and riser runout sizes, all of which are based on the critical velocities of the steam to permit the counterflow of condensate without noise.

Return piping can be sized using Table 16, where pipe capacities for wet, dry, and vacuum return lines are shown for several values of pressure drop per 100 ft of equivalent length.

Table 13   Equivalent Length of Fittings to Be Added to Pipe Run

Nominal Pipe Diameter, in.	Length to Be Added to Run, ft				
	Standard Elbow	Side Outlet Tee[b]	Gate Valve[a]	Globe Valve[a]	Angle Valve[a]
1/2	1.3	3	0.3	14	7
3/4	1.8	4	0.4	18	10
1	2.2	5	0.5	23	12
1-1/4	3.0	6	0.6	29	15
1-1/2	3.5	7	0.8	34	18
2	4.3	8	1.0	46	22
2-1/2	5.0	11	1.1	54	27
3	6.5	13	1.4	66	34
3-1/2	8	15	1.6	80	40
4	9	18	1.9	92	45
5	11	22	2.2	112	56
6	13	27	2.8	136	67
8	17	35	3.7	180	92
10	21	45	4.6	230	112
12	27	53	5.5	270	132
14	30	63	6.4	310	152

[a]Valve in full open position.
[b]Values apply only to a tee used to divert the flow in the main to the last riser.

Table 14 Flow Rate of Steam in Schedule 40 Pipe

Nominal Pipe Size, in.	1/16 psi (1 oz) Sat. Press., psig 3.5	1/16 psi (1 oz) Sat. Press., psig 12	1/8 psi (2 oz) Sat. Press., psig 3.5	1/8 psi (2 oz) Sat. Press., psig 12	1/4 psi (4 oz) Sat. Press., psig 3.5	1/4 psi (4 oz) Sat. Press., psig 12	1/2 psi (8 oz) Sat. Press., psig 3.5	1/2 psi (8 oz) Sat. Press., psig 12	3/4 psi (12 oz) Sat. Press., psig 3.5	3/4 psi (12 oz) Sat. Press., psig 12	1 psi Sat. Press., psig 3.5	1 psi Sat. Press., psig 12	2 psi Sat. Press., psig 3.5	2 psi Sat. Press., psig 12
3/4	9	11	14	16	20	24	29	35	36	43	42	50	60	73
1	17	21	26	31	37	46	54	66	68	82	81	95	114	137
1-1/4	36	45	53	66	78	96	111	138	140	170	162	200	232	280
1-1/2	56	70	84	100	120	147	174	210	218	260	246	304	360	430
2	108	134	162	194	234	285	336	410	420	510	480	590	710	850
2-1/2	174	215	258	310	378	460	540	660	680	820	780	950	1150	1370
3	318	380	465	550	660	810	960	1160	1190	1430	1380	1670	1950	2400
3-1/2	462	550	670	800	990	1218	1410	1700	1740	2100	2000	2420	2950	3450
4	640	800	950	1160	1410	1690	1980	2400	2450	3000	2880	3460	4200	4900
5	1200	1430	1680	2100	2440	3000	3570	4250	4380	5250	5100	6100	7500	8600
6	1920	2300	2820	3350	3960	4850	5700	7000	7200	8600	8400	10,000	11,900	14,200
8	3900	4800	5570	7000	8100	10,000	11,400	14,300	14,500	17,700	16,500	20,500	24,000	29,500
10	7200	8800	10,200	12,600	15,000	18,200	21,000	26,000	26,200	32,000	30,000	37,000	42,700	52,000
12	11,400	13,700	16,500	19,500	23,400	28,400	33,000	40,000	41,000	49,500	48,000	57,500	67,800	81,000

Notes:
1. Flow rate is in lb/h at initial saturation pressures of 3.5 and 12 psig. Flow is based on Moody friction factor, where the flow of condensate does not inhibit the flow of steam.

2. The flow rates at 3.5 psig cover saturated pressure from 1 to 6 psig, and the rates at 12 psig cover saturated pressure from 8 to 16 psig with an error not exceeding 8%.
3. The steam velocities corresponding to the flow rates given in this table can be found from the basic chart and velocity multiplier chart, Figure 11.

Table 15 Steam Pipe Capacities for Low-Pressure Systems

	Capacity, lb/h				
	Two-Pipe System		One-Pipe Systems		
Nominal Pipe Size, in.	Condensate Flowing against Steam Vertical	Condensate Flowing against Steam Horizontal	Supply Risers Up-feed	Radiator Valves and Vertical Connections	Radiator and Riser Runouts
A	Bª	Cᵇ	Dᶜ	E	Fᵇ
3/4	8	7	6	—	7
1	14	14	11	7	7
1-1/4	31	27	20	16	16
1-1/2	48	42	38	23	16
2	97	93	72	42	23
2-1/2	159	132	116	—	42
3	282	200	200	—	65
3-1/2	387	288	286	—	119
4	511	425	380	—	186
5	1050	788	—	—	278
6	1800	1400	—	—	545
8	3750	3000	—	—	—
10	7000	5700	—	—	—
12	11,500	9500	—	—	—
16	22,000	19,000	—	—	—

Notes:
1. For one- or two-pipe systems in which condensate flows against the steam flow.
2. Steam at an average pressure of 1 psig is used as a basis of calculating capacities.

ªDo not use Column *B* for pressure drops of less than 1/16 psi per 100 ft of equivalent run. Use Figure 10 or Table 13 instead.
ᵇPitch of horizontal runouts to risers and radiators should be not less than 0.5 in/ft. Where this pitch cannot be obtained, runouts over 8 ft in length should be one pipe size larger than that called for in this table.
ᶜDo not use Column *D* for pressure drops of less than 1/24 psi per 100 ft of equivalent run, except on sizes 3 in. and over. Use Figure 10 or Table 13 instead.

Example 9. What pressure drop should be used for the steam piping of a system if the measured length of the longest run is 500 ft, and the initial pressure must not exceed 2 psig?

Solution: It is assumed, if the measured length of the longest run is 500 ft, that when the allowance for fittings is added, the equivalent length of run does not exceed 1000 ft. Then, with the pressure drop not over one-half of the initial pressure, the drop could be 1 psi or less. With a pressure drop of 1 psi and a length of run of 1000 ft, the drop per 100 ft would be 0.1 psi; if the total drop were 0.5 psi, the drop per 100 ft would be 0.05 psi. In both cases, the pipe could be sized for a desired capacity according to the 0.1 and 0.05 pressure drop lines in Figure 10.

On completion of the sizing, the drop could be checked by taking the longest line and actually calculating the equivalent length of run from the pipe sizes determined. If the calculated drop is less than that assumed, the pipe size is adequate; if it is more, probably an unusual number of fittings is involved, and either the lines must be straightened or the column for the next lower drop must be used and the lines resized. Ordinarily, resizing is unnecessary.

HIGH-PRESSURE STEAM PIPING

Many heating systems for large industrial buildings use high-pressure steam (15 to 150 psig). These systems usually have unit heaters or large built-up fan units with blast heating coils. Temperatures are controlled by a modulating or throttling thermostatic valve or by face or bypass dampers controlled by the room air temperature, fan inlet, or fan outlet.

Figures 10a through 10d present charts for sizing steam piping for systems of 30, 50, 100, and 150 psig at various pressure drops. These charts are based on the Moody friction factor, which considers the Reynolds number and the roughness of the internal pipe surfaces, and contains the same information as the basic chart (Figure 10) but in a more convenient form.

Use of Basic and Velocity Multiplier Charts

Example 10.

Given:
  Flow rate = 6700 lb/h
  Initial steam pressure = 100 psig
  Pressure drop = 11 psi/100 ft
Find:
  Size of Schedule 40 pipe required
  Velocity of steam in pipe

Solution: The following steps are illustrated by the broken line on Figures 10 and 11.

1. Enter Figure 10 at a flow rate of 6700 lb/h and move vertically to the horizontal line at 100 psig.

2. Follow along inclined multiplier line (upward and to the left) to horizontal 0 psig line. The equivalent weight flow at 0 psig is about 2500 lb/h.

3. Follow the 2500 lb/h line vertically until it intersects the horizontal line at 11 psi per 100 ft pressure drop. Nominal pipe size is 2.5 in. The equivalent steam velocity at 0 psig is about 32,700 fpm.

4. To find the steam velocity at 100 psig, locate the value of 32,700 fpm on the ordinate of the velocity multiplier chart (Figure 11) at 0 psig.

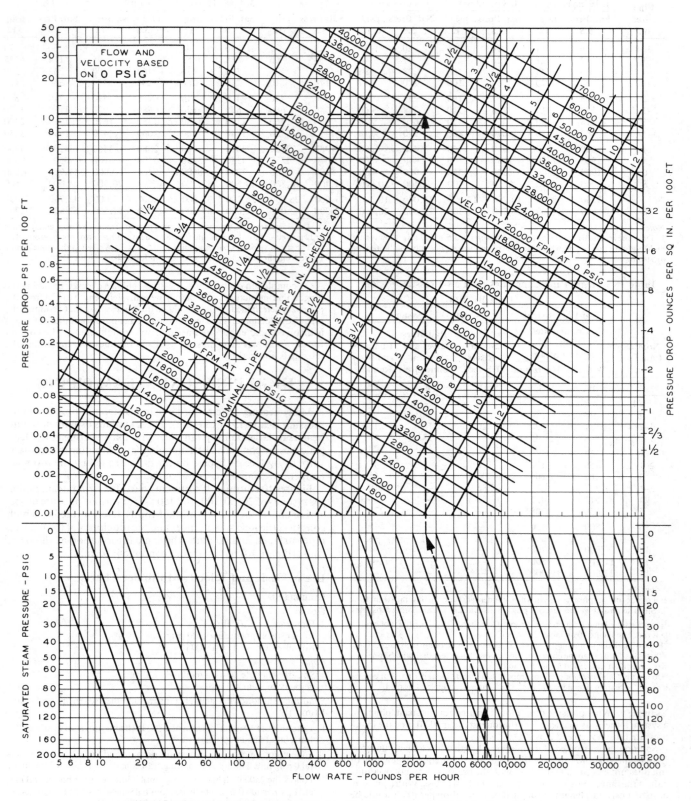

Notes: Based on Moody Friction Factor where flow of condensate does not inhibit the flow of steam.
See Figure 11 for obtaining flow rates and velocities of all saturation pressures between 0 and 200 psig; see also Examples 9 and 10.

Fig. 10   Flow Rate and Velocity of Steam in Schedule 40 Pipe Saturation Pressure of 0 psig

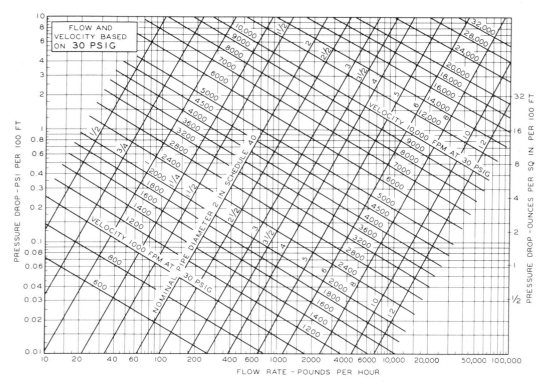

Notes: Based on Moody Friction Factor where flow of condensate does not inhibit the flow of steam.
May be used for steam pressures from 23 to 37 psig with an error not exceeding 9%.

Fig. 10a   Flow Rate and Velocity of Steam in Schedule 40 Pipe at Saturation Pressure of 30 psig

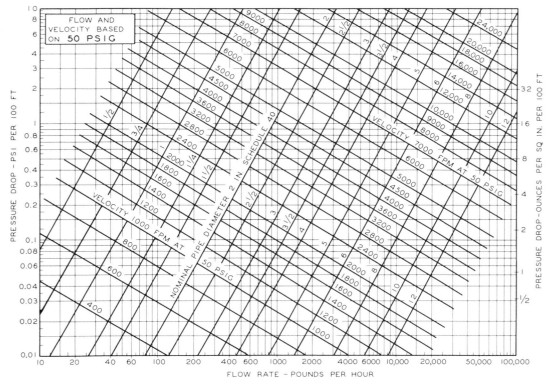

Notes: Based on Moody Friction Factor where flow of condensate does not inhibit the flow of steam.
May be used for steam pressures from 40 to 60 psig with an error not exceeding 8%.

Fig. 10b   Flow Rate and Velocity of Steam in Schedule 40 Pipe at Saturation Pressure of 50 psig

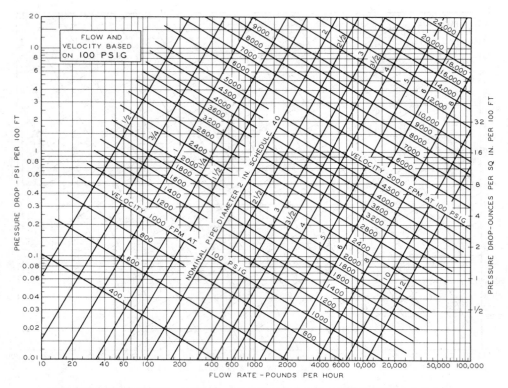

Notes: Based on Moody Friction Factor where flow of condensate does not inhibit the flow of steam.
May be used for steam pressures from 85 to 120 psig with an error not exceeding 8%.

Fig. 10c   Flow Rate and Velocity of Steam in Schedule 40 Pipe at Saturation Pressure of 100 psig

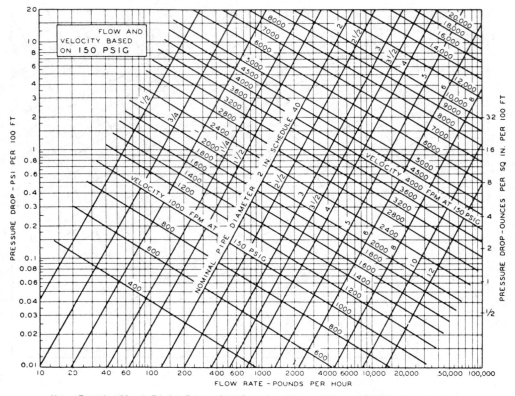

Notes: Based on Moody Friction Factor where flow of condensate does not inhibit the flow of steam.
May be used for steam pressures from 127 to 180 psig with an error not exceeding 8%.

Fig. 10d   Flow Rate and Velocity of Steam in Schedule 40 Pipe at Saturation Pressure of 150 psig

Table 16 Return Main and Riser Capacities for Low-Pressure Systems, lb/h

Pipe Size, in.	1/32 psi or 1/2 oz Drop per 100 ft			1/24 psi or 2/3 oz Drop per 100 ft			1/16 psi or 1 oz Drop per 100 ft			1/8 psi or 2 oz Drop per 100 ft			1/4 psi or 4 oz Drop per 100 ft			1/2 psi or 8 oz Drop per 100 ft		
	Wet	Dry	Vac.	Wet	Dry	Vac.	Wet	Dry	Vac.	Wet	Dry	Vac.	Wet	Dry	Vac.	Wet	Dry	Vac.
G	H	I	J	K	L	M	N	O	P	Q	R	S	T	U	V	W	X	Y
Return Main																		
3/4	—	—	—	—	—	42	—	—	100	—	—	142	—	—	200	—	—	283
1	125	62	—	145	71	143	175	80	175	250	103	249	350	115	350	—	—	494
1-1/4	213	130	—	248	149	244	300	168	300	425	217	426	600	241	600	—	—	848
1-1/2	338	206	—	393	236	388	475	265	475	675	340	674	950	378	950	—	—	1340
2	700	470	—	810	535	815	1000	575	1000	1400	740	1420	2000	825	2000	—	—	2830
2-1/2	1180	760	—	1580	868	1360	1680	950	1680	2350	1230	2380	3350	1360	3350	—	—	4730
3	1880	1460	—	2130	1560	2180	2680	1750	2680	3750	2250	3800	5350	2500	5350	—	—	7560
3-1/2	2750	1970	—	3300	2200	3250	4000	2500	4000	5500	3230	5680	8000	3580	8000	—	—	11 300
4	3880	2930	—	4580	3350	4500	5500	3750	5500	7750	4830	7810	11,000	5380	11,000	—	—	15,500
5	—	—	—	—	—	7880	—	—	9680	—	—	13,700	—	—	19,400	—	—	27,300
6	—	—	—	—	—	12,600	—	—	15,500	—	—	22,000	—	—	31,000	—	—	43,800
Riser																		
3/4	—	48	—	—	48	143	—	48	175	—	48	249	—	48	350	—	—	494
1	—	113	—	—	113	244	—	113	300	—	113	426	—	113	600	—	—	848
1-1/4	—	248	—	—	248	388	—	248	475	—	248	674	—	248	950	—	—	1340
1-1/2	—	375	—	—	375	815	—	375	1000	—	375	1420	—	375	2000	—	—	2830
2	—	750	—	—	750	1360	—	750	1680	—	750	2380	—	750	3350	—	—	4730
2-1/2	—	—	—	—	—	2180	—	—	2680	—	—	3800	—	—	5350	—	—	7560
3	—	—	—	—	—	3250	—	—	4000	—	—	5680	—	—	8000	—	—	11,300
3-1/2	—	—	—	—	—	4480	—	—	5500	—	—	7810	—	—	11,000	—	—	15,500
4	—	—	—	—	—	7880	—	—	9680	—	—	13,700	—	—	19,400	—	—	27,300
5	—	—	—	—	—	12,600	—	—	15,500	—	—	22,000	—	—	31,000	—	—	43,800

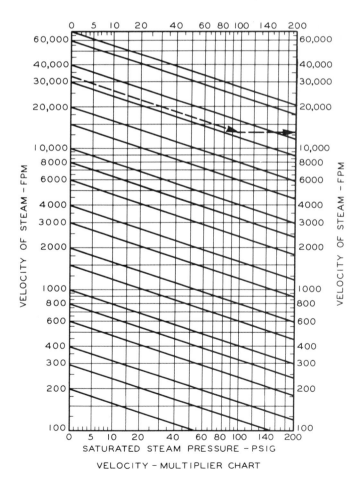

Fig. 11 Velocity Multiplier Chart for Figure 10

5. Move along the inclined multiplier line (downward and to the right) until it intersects the vertical 100 psig pressure line. The velocity as read from the right (or left) scale is about 13,000 fpm.

Note: Steps (1) through (5) would be rearranged or reversed if different data were given.

TWO-PIPE STEAM SYSTEMS

Condensate from steam heating devices is generally saturated liquid or slightly subcooled liquid. The amount of subcooling is a function of the size of the heating device in comparison to the heating load, the control system for the heating device, and the size of the trap and return piping. In properly designed systems, the subcooling is usually no greater than a few degrees. This condensate then passes through a steam trap where the pressure is reduced and some of the liquid may flash into vapor. The amount of flashing depends on the overall pressure drop through the trap, the amount of subcooling in the steam heating device, and the type and condition of the trap.

The process of the fluid passing through the steam trap is a throttling or constant enthalpy process. The resulting fluid on the downstream side of the trap can be a mixture of saturated liquid and vapor. The overall pressure drop and resulting flash vapor are the physical characteristics that differentiate condensate pipe sizing from other types of pipe sizing where only a single phase exists. For this reason, it is important to understand the condition of the condensate when it enters the return line from the trap.

The condition of the condensate downstream of the trap can be expressed by the quality x, defined by:

$$x = \frac{m_v}{m_l + m_v} \tag{11}$$

where

m_v = mass of saturated vapor in condensate
m_l = mass of saturated liquid in condensate

Likewise, the volume fraction V_c of the vapor in the condensate is expressed by:

$$V_c = \frac{V_v}{V_l + V_v} \qquad (12)$$

where

V_v = volume of saturated vapor in condensate
V_l = volume of saturated liquid in condensate

The quality and the volume fraction of the condensate downstream of the trap can be estimated from Equations (13) and (14), respectively.

$$x = \frac{h_1 - h_{f_2}}{h_{g_2} - h_{f_2}} \qquad (13)$$

$$V_c = \frac{x V_{g_2}}{V_{f_2}(1 - x) + x V_{g_2}} \qquad (14)$$

where

h_1 = enthalpy of liquid condensate entering trap evaluated at supply pressure for saturated condensate or at saturation pressure corresponding to temperature of subcooled liquid condensate
h_{f_2} = enthalpy of saturated liquid at return or downstream pressure of trap
h_{g_2} = enthalpy of saturated vapor at return or downstream pressure of trap
V_{f_2} = specific volume of saturated liquid at return or downstream pressure of trap
V_{g_2} = specific volume of saturated vapor at return or downstream pressure of trap

Table 17 presents some values for quality and volume fraction for typical supply and return pressures in heating and ventilating systems. Note that the percent of vapor on a mass basis x is small, while the percent of vapor on a volume basis V_c is very large. This indicates that the return pipe cross section is predominantly occupied by vapor. Figure 12 is a working chart to determine the quality of the condensate entering the return line from the trap for various combinations of supply and return pressures. If the liquid is subcooled entering the trap, the saturation pressure corresponding to the liquid temperature should be used for the supply or upstream pressure. Typical pressures in the return line are given in Table 18.

Table 17   Flash Steam from Steam Trap on Pressure Drop

Supply Pressure, psig	Return Pressure, psig	x Fraction Vapor Mass Basis	V_c Fraction Vapor Volume Basis
5	0	0.016	0.962
15	0	0.040	0.985
30	0	0.065	0.991
50	0	0.090	0.994
100	0	0.133	0.996
150	0	0.164	0.997
100	15	0.096	0.989
150	15	0.128	0.992

Table 18   Estimated Return Line Pressures

Pressure Drop, psi/100 ft	Pressure in Return Line, psig	
	30 psig Supply	150 psig Supply
1/8	1/2	1–1/4
1/4	1	2–1/2
1/2	2	5
3/4	3	7–1/2
1	4	10
2	—	20

Condensate return systems can be either *wet* or *dry*, *open* or *closed*. In *wet return systems*, the return pipe contains only the liquid phase and no flash vapor. This occurs when the condensate entering the trap is sufficiently subcooled or when the liquid and vapor are separated and the wet return line is kept below the boiler water line so that liquid only flows through the line. In *dry return systems*, the condensate piping contains both saturated liquid and saturated vapor. This is typical of most condensate return systems. An *open return system* is vented to the atmosphere, and the condensate line is essentially at atmospheric pressure. The driving force for returning the condensate is gravitational acceleration, and the return line must be sloped. In the *closed return system*, the pressure in the condensate return line is above or below atmospheric pressure and is not vented to the atmosphere. The driving force for returning the condensate is a pressure loss along the return line. These types of systems are illustrated in Figure 13.

For the *wet-closed return* where the condensate return line is horizontal, the Darcy-Weisbach Equation (1) for closed conduit flow can be used (Howell 1985). The section Pressure Drop Equa-

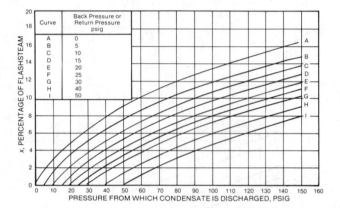

Fig. 12   Working Chart for Determining Percent of Flash Steam (Quality)

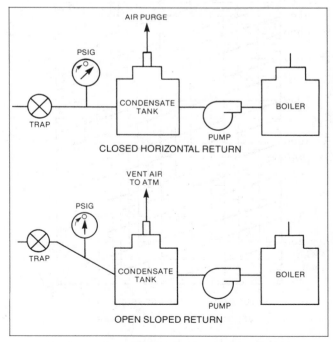

Fig. 13   Types of Condensate Return Lines

tions describes how to apply this equation to full-closed conduits. This equation can be applied to horizontal pipes completely filled with liquid or with vapor. A design pressure loss per unit length ($\Delta p/L$) is usually assumed and then, knowing the design mass flow rate required for the heating device, the required pipe size can be determined by trial and error. Often, a chart such as Figure 1 (where f is assumed constant) can be used to avoid the trial and error solution. Typical design pressure losses per unit length ($\Delta p/L$) are in the range of 1/16 to 1 psi/100 ft. The velocity for liquid condensate systems should not exceed 150 ft/min; for vapor systems, it should not exceed 7000 ft/min.

For the *wet-open return system* with sloped return pipes, the open channel Manning Equation (15) can be used.

$$Q = 1.49 \, A r^{2/3} S^{1/2}/n \qquad (15)$$

where

Q = volumetric flow rate, ft³/s
A = cross-sectional area of conduit, ft²
r = hydraulic radius of conduit, ft
n = coefficient of roughness (usually = 0.012)
S = slope of conduit, ft/ft

For the wet return, the pipe would be full of liquid. Howell (1985) showed that an equivalency between horizontal pipe flows and sloped pipe flows can be made by assuming that the increase in head due to the slope is equal to the friction loss of the fluid in the pipe. Table 19 gives this equivalency between sloped and horizontal flows. For a wet-open system with a sloped return pipe, the slope used yields the pressure loss from Table 19. Then, the same procedure as for the wet-closed return described earlier can be used to size the sloped condensate return. A constant or appropriate slope in the installed system is difficult to ensure, however.

Table 19 Equivalency between Sloped and Horizontal Flows

Pipe Slope in 10 ft	Pressure Loss, psi/100 ft
1/2 in.	0.180
1 in.	0.361
1–1/2 in.	0.540
2 in.	0.722
3 in.	1.084
4 in.	1.440
5 in.	1.805

When a dry return is *open* and the pipe is sloped, a reasonable procedure can be used to size the return pipe (Howell 1985). When the pipe contains both liquid and vapor, it would run at least at a depth of flow of 0.3 times the pipe diameter. For this conservative situation, the pipe slope should be selected as 1 in. in 10 ft; then the equivalent pressure loss (0.36 psi/100 ft) from Table 19 should be selected for sizing the return pipe at the given heater condensate flow rate. The same procedure described for the wet-closed return is used to size the pipe.

For a *dry-closed* return line with horizontal pipe, which is common, Howell (1985) developed another procedure. Here the *condensate is assumed all vapor*, since V_c is large (0.96 to 0.99). Equation (1) is used to determine the velocity of the vapor in the pipe when the Moody friction factor is used for f. This trial and error procedure starts with a specific $\Delta p/L$ and pipe diameter D, and assumes a value for f. The velocity can then be found from Equation (1). A Reynolds number is found for this velocity and pipe diameter. For the relative roughness and Reynolds number, the friction factor is found from the Moody chart in Chapter 2. If it does not agree with the assumed value, a new f is tried until they do agree.

Once the velocity is determined, the continuity equation is used to find the quantity of vapor moving through the pipe of diameter

Table 20 Flow Rate for Dry-Closed Returns

Pipe Dia.	Supply Pressure = 5 psig Return Pressure = 0 psig			Supply Pressure = 15 psig Return Pressure = 0 psig			Supply Pressure = 30 psig Return Pressure = 0 psig			Supply Pressure = 50 psig Return Pressure = 0 psig		
D, in.	1/16	1/4	1	1/16	1/4	1	1/16	1/4	1	1/16	1/4	1
	Flow Rate, lb/h											
1/2	240	520	1100	95	210	450	60	130	274	42	92	200
3/4	510	1120	2400	210	450	950	130	280	590	91	200	420
1	1000	2150	4540	400	860	1820	250	530	1120	180	380	800
1–1/4	2100	4500	9500	840	1800	3800	520	1110	2340	370	800	1680
1–1/2	3170	6780	14,200	1270	2720	5700	780	1670	3510	560	1200	2520
2	6240	13,300	a	2500	5320	a	1540	3270	a	1110	2350	a
2–1/2	10,000	21,300	a	4030	8520	a	2480	5250	a	1780	3780	a
3	18,000	38,000	a	7200	15,200	a	4440	9360	a	3190	6730	a
4	37,200	78,000	a	14,900	31,300	a	9180	19,200	a	6660	13,800	a
6	110,500	a	a	44,300	a	a	27,300	a	a	19,600	a	a
8	228,600	a	a	91,700	a	a	56,400	a	a	40,500	a	a

Pipe Dia.	Supply Pressure = 100 psig Return Pressure = 0 psig			Supply Pressure = 150 psig Return Pressure = 0 psig			Supply Pressure = 100 psig Return Pressure = 15 psig			Supply Pressure = 150 psig Return Pressure = 15 psig		
D, in.	1/16	1/4	1	1/16	1/4	1	1/16	1/4	1	1/16	1/4	1
	Flow Rate, lb/h											
1/2	28	62	133	23	51	109	56	120	260	43	93	200
3/4	62	134	290	50	110	230	120	260	560	93	200	420
1	120	260	544	100	210	450	240	500	1060	180	390	800
1–1/4	250	540	1130	200	440	930	500	1060	2200	380	800	1680
1–1/2	380	810	1700	310	660	1400	750	1600	3320	570	1210	2500
2	750	1590	a	610	1300	a	1470	3100	6450	1120	2350	4900
2–1/2	1200	2550	a	980	2100	a	2370	5000	10,300	1800	3780	7800
3	2160	4550	a	1760	3710	a	4230	8860	a	3200	6710	a
4	4460	9340	a	3640	7630	a	8730	18,200	a	6620	13,800	a
6	13,200	a	a	10,800	a	a	25,900	53,600	a	19,600	40,600	a
8	27,400	a	a	22,400	a	a	53,400	110,300	a	40,500	83,600	a

ᵃFor these sizes and pressure losses, the velocity is above 7000 fpm. Select another combination of size and pressure loss.

D. This vapor flow is then divided by the quality x in the condensate line to determine the total (vapor and liquid) condensate flowing in the return pipe. This method assumes that the condensate flows at a uniform average density through the pipe.

This procedure was used to develop several tables that can be used for sizing the condensate return line. For a given supply pressure to the trap and a return line pressure along with an assumed pressure drop per 100 ft ($\Delta p/L$), and knowing the condensate flow rate for the heating device, the proper pipe diameter can be selected from Table 20.

Example 11. A condensate return system has the steam supply at 30 psig, and the return line is nonvented and at 0 psig. The return line is to be horizontal and is to have the capacity for returning 2000 lb/h of condensate. What must the size of the return line be?

Solution: Since the system will be throttling the condensate from 30 psig to 0 psig, there will be flash steam (assuming no subcooling), and the system will be a dry-closed return with a horizontal pipe. The data in Table 19 can be used. A pressure drop of 1/4 psi per 100 ft is selected. In Table 20, for a 30 psig supply and a 0 psig return for $\Delta p/L = 1/4$, a pipe size for the return line of 2 in. is selected.

Example 12. A condensate return system has the steam supply at 100 psig, and the return line is nonvented and at 0 psig. The return line is horizontal and must have a capacity of 2500 lb/h. What size pipe is required?

Solution: Since the system will be throttling non-subcooled condensate from 100 psig to 0 psig, there will be flash steam, and the system will be a dry-closed return with horizontal pipe. Selecting a pressure drop of 1 psi/100 ft from Table 20 a nonrecommended situation. Select a pressure drop of 1/4 psi/100 ft, and a 2-1/2 in. pipe can be used for this system.

Example 13. A condensate return system is vented and passes subcooled liquid to the return line. The liquid in the return is at 180 °F, and the flow rate must be 1000 lb/h. The return line will be installed with a slope of 1-1/2 in. per 10 ft. What size return line is needed?

Solution: This system is a wet-open condensate return. From Table 19, the equivalent pressure loss for a slope of 1-1/2 in. in 10 ft is 0.54 psi/100 ft. Converting the 1000 lb/h to volume flow (gal/min) yields:

$$1000 \text{ lb/h} \times 0.01651 \text{ ft}^3/\text{lb} \times 7.48 \text{ gal/ft}^3/(60 \text{ m/h}) = 2.06 \text{ gpm}$$

Converting the pressure drop to head loss yields

$$\frac{\Delta h}{100 \text{ ft}} = 0.54 \times \frac{\text{psi}}{100 \text{ ft}} \times 2.307 \frac{\text{ft water}}{\text{psi}} = 1.24 \frac{\text{ft}}{100 \text{ ft}}$$

$$\Delta h/100 \text{ ft} = 0.54 \text{ psi}/100 \text{ ft} \times 2.307 \text{ ft water/psi}$$
$$= 1.24 \text{ ft}/100 \text{ ft of pipe}$$

Figure 1 shows a 1 in. schedule 40 pipe for $\Delta h/100$ ft = 1.24 at 2.06 gpm.

ONE-PIPE STEAM SYSTEMS

Gravity one-pipe air-vent systems, in which the equivalent length of run does not exceed 200 ft, should be sized using Tables 14, 15, and 16 and Figure 10 as follows:

1. For the *steam main* and *dripped runouts to risers*, where the steam and condensate flow in the same direction, use 1/16 psi drop per 100 ft (Table 14).
2. Where the *riser runouts are not dripped* and the steam and condensate flow in opposite directions, and for the *radiator runouts* with the same condition, use Table 15, Column *F*.
3. For *up-feed steam risers* carrying condensate back from the radiators, use Table 15, Column *D*.
4. For *down-feed systems*, the *main risers* of which do not carry any radiator condensate, use Table 15, Column *B*.
5. For the *radiator valve* size and the *stub connection*, use Table 15, Column *E*.
6. For the *dry-return main*, use Table 16, Column *O*.
7. For the *wet-return main*, use Table 16, Column *N*.

Notes on Gravity One-Pipe Air Vent Systems

1. Pitch of mains should not be less than 0.25 in. in 10 ft.
2. Pitch of horizontal runouts to risers and radiators should not be less than 0.5 in/ft. Where this pitch cannot be obtained, runouts over 8 ft in length should be one size larger than specified in the table.
3. In general, it is not desirable for a main to be less than 2 in. The diameter of the far end of the supply main should not be less than one-half its diameter at its largest part.
4. Supply mains, runouts to risers, or risers should be dripped where necessary.
5. Where supply mains are decreased in size they should be dripped or provided with eccentric couplings, flush on the bottom.

Example 14. Size the one-pipe gravity steam system shown in Figure 14, assuming that this is all there is to the system or that the riser and main shown involve the longest run on the system.

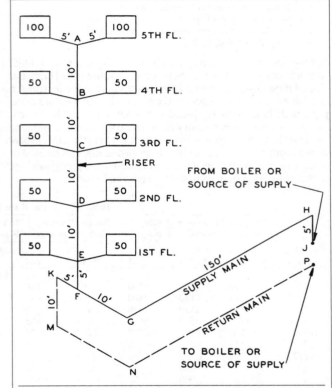

Part of System	Section of Pipe	Radiation Supplied EDR, ft²	Theoretical Pipe Size, in.	Practical Pipe Size, in.
Branches to radiators	—	100	2	2
Branches to radiators	—	50	1-1/4	1-1/4
Riser	A to B	200	2	2
Riser	B to C	300	2-1/2	2-1/2
Riser	C to D	400	2-1/2	2-1/2
Riser	D to E	500	3	3
Riser	E to F	600	3	3
Runout to riser	F to G	600	3	3
Supply main	G to H	600	3	3
Branch to supply main	H to J	600	2-1/2	3
Dry return main	F to K	600	1-1/4	2
Wet return main	K to M	600	1	2
Wet return main	M to N	600	1	2
Wet return main	N to P	600	1	2

Note: EDR = Equivalent direct radiation

Fig. 14   Riser, Supply Main, and Return Main of One-Pipe System

Solution: The total length of run shown is 215 ft. If the equivalent length of the run is double this, it will be 430 ft, and with a total drop of 1/4 psi, the drop per 100 ft will be slightly less than 1/16 psi. In this case, if 1/24 psi is used, the theoretical changes in Table 13 will result. However, these theoretical sizes should be modified by using a wet return not less than 2 in. The main supply, *G-H*, if it is from the uptake of a boiler, should be the full size of the main, or 3 in. The *K-M* portion of the main should be 2 in. if the wet return is also 2 in.

On systems exceeding an equivalent length of 200 ft, total drop should not be over 0.25 psi. The return piping sizes should correspond with the drop used on the steam side of the system. Thus, where a 1/24 psi per 100 ft drop is used, the steam main and dripped runouts are sized from Figure 10, radiator runouts and undripped riser runouts from Table 15, Column *F*. Up-feed risers are sized from Column *D*. The main riser on a downfeed system is sized from Figure 10. (Note that if Column *E* is used, the drop exceeds 1/24 psi per 100 ft.) The dry return is sized from Table 16, Column *L*, and the wet return from Column *K*.

With a 1/32 psi per 100 ft drop, the sizing is the same as for 1/24 psi, except that the steam mean and dripped runouts are sized from Figure 10, the main riser on a down-feed system from Figure 10, the dry return from Column *I*, and the wet return from Column *H*.

GAS PIPE SIZING

Piping for gas appliances should be of adequate size and installed so that it provides a supply of gas sufficient to meet the maximum demand without undue loss of pressure between the point of supply (the meter) and the appliance. The size of gas pipe required depends on (1) maximum gas consumption to be provided, (2) length of pipe and number of fittings, (3) allowable loss in pressure from the outlet of the meter to the appliance, and (4) specific gravity of the gas.

Gas consumption in ft^3/h is obtained by dividing the Btu input rate at which the appliance is operated by the average Btu heating value per ft^3 of the gas. Insufficient gas flow from excessive pressure losses in gas supply lines can cause inefficient operation of gas-fired appliances and sometimes create hazardous operations. Gas-fired appliances are normally equipped with a data plate giving information on maximum gas flow requirements or Btu input as well as inlet gas pressure requirements. The gas utility in the area of installation can give the gas pressure available at the utilities gas meter. Using the information, the required size of gas piping can be calculated for satisfactory operation of the appliance(s).

Table 21 gives pipe capacities for gas flow for up to 200 ft of pipe based on a specific gravity of 0.60. Capacities for pressures less than 1.5 psig may also be determined by the following equation from NFPA *Standard* 54-1988:

$$Q = 2313 d^{2.623} (\Delta p/CL)^{0.541} \qquad (16)$$

where

Q = flow rate at 60°F and 30 in. Hg, ft^3/h
d = inside diameter of pipe, in.
Δp = pressure drop, in. of water
C = factor for viscosity, density, and temperature
= $0.00354(t + 460)s^{0.848}\mu^{0.152}$
t = temperature, °F
s = ratio of density of gas to density of air at 60°F and 30 in. Hg
μ = viscosity of gas, centipoise (0.012 for natural gas, 0.008 for propane)
L = pipe length, ft

Gas service in buildings is generally delivered in the "low-pressure" range of 7 in of water. The maximum pressure drop allowable in piping systems at this pressure is generally 0.5 in. of water, but is subject to regulation by local building, plumbing, and gas appliance codes (see also the *National Fuel Gas Code*, NFPA 54).

Where large quantities of gas are required or where long lengths of pipe are used, as in industrial buildings, low-pressure limitations result in large pipe sizes. Local codes may allow and local gas companies may deliver gas at higher pressures (*e.g.*, 2, 5, or 10 psig). Under these conditions, an allowable pressure drop of 10% of the initial pressure is used, and pipe sizes can be reduced significantly. Gas pressure regulators at the appliance must be specified to accommodate higher inlet pressures. NFPA (1988) provides information on pipe sizing for various inlet pressures and pressure drops at higher pressures.

More complete information on gas piping can be found in the *Gas Engineers' Handbook* (1970).

FUEL OIL PIPE SIZING

The pipe used to convey fuel oil to oil-fired appliances must be large enough to maintain low pump suction pressure and, in the case of circulating loop systems, to prevent overpressure at the burner oil pump inlet. Pipe materials must be compatible with the fuel and must be carefully assembled to eliminate all leaks. Leaks in suction lines cause pumping problems that result in unreliable burner operation. Leaks in pressurized lines create fire hazards. Cast-iron or aluminum fittings and pipe are unacceptable. Pipe joint compounds must be selected carefully.

Oil pump suction lines should be sized so that at maximum suction line flow conditions, the maximum vacuum will not exceed 10 in. Hg for distillate grade fuels and 15 in. Hg for residual oils. Oil supply lines to burner oil pumps should not be pressurized by circulating loop systems or aboveground oil storage tanks to more than 5 psi, or pump shaft seals may fail. A typical oil circulating loop system is shown in Figure 15.

In assembling long fuel pipe lines, care should be taken to avoid air pockets. On overhead circulating loops, the line should vent air at all high points. Oil supply loops for one or more burners should

Table 21 Maximum Capacity of Gas Pipe in Cubic Feet per Hour

Nominal Iron Pipe Size, in.	Internal Diameter, in.	Length of Pipe, ft													
		10	20	30	40	50	60	70	80	90	100	125	150	175	200
1/4	0.364	32	22	18	15	14	12	11	11	10	9	8	8	7	6
3/8	0.493	72	49	40	34	30	27	25	23	22	21	18	17	15	14
1/2	0.622	132	92	73	63	56	50	46	43	40	38	34	31	28	26
3/4	0.824	278	190	152	130	115	105	96	90	84	79	72	64	59	55
1	1.049	520	350	285	245	215	195	180	170	160	150	130	120	110	100
1-1/4	1.380	1050	730	590	500	440	400	370	350	320	305	275	250	225	210
1-1/2	1.610	1600	1100	890	760	670	610	560	530	490	460	410	380	350	320
2	2.067	3050	2100	1650	1450	1270	1150	1050	990	930	870	780	710	650	610
2-1/2	2.469	4800	3300	2700	2300	2000	1850	1700	1600	1500	1400	1250	1130	1050	980
3	3.068	8500	5900	4700	4100	3600	3250	3000	2800	2600	2500	2200	2000	1850	1700
4	4.026	17,500	12,000	9700	8300	7400	6800	6200	5800	5400	5100	4500	4100	3800	3500

Notes: 1. Capacity is in cubic feet per hour at gas pressures of 0.5 psig or less and a pressure drop of 0.5 in. of water; Specific gravity = 0.60.

2. Copyright by the American Gas Association and the National Fire Protection Association. Used by permission of the copyright holder.

be the continuous circulation type, with excess fuel returned to the storage tank. Dead-ended pressurized loops can be used, but air or vapor venting is more problematic.

Where valves are used, select ball or gate valves. Globe valves are not recommended because of their high pressure drop characteristics.

Oil lines should be tested after installation, particularly if they are buried, enclosed, or otherwise inaccessible. Failure to perform this test is a frequent cause of later operating difficulties. A suction line can be hydrostatically tested at 1.5 times its maximum operating pressure or at a vacuum of not less than 20 in. Hg. Pressure or vacuum tests should continue for at least 60 min. If there is no noticeable drop in the initial test pressure, the lines can be considered tight.

Pipe Sizes for Heavy Oil

Tables 22 and 23 give recommended pipe sizes for handling No. 5 and No. 6 oils (residual grades) and No. 1 and No. 2 oils (distillate grades), respectively.

Storage tanks and piping and pumping facilities for delivering the oil from the tank to the burner are important considerations in the design of an industrial oil-burning system. The construction and location of the tank and oil piping are usually subject to local regulations and the National Fire Protection Association's *Flammable and Combustible Fuels Code* and *Standard* 31.

Table 22 Recommended Nominal Size for Fuel Oil Suction Lines from Tank to Pump (Residual Grades, Numbers 5 and 6)

Pumping Rate, gal/h	Length of Run in Feet at Maximum Suction Lift = 15 ft									
	25	50	75	100	125	150	175	200	250	300
10	1-1/2	1-1/2	1-1/2	1-1/2	1-1/2	1-1/2	2	2	2-1/2	2-1/2
40	1-1/2	1-1/2	1-1/2	2	2	2-1/2	2-1/2	2-1/2	2-1/3	3
70	1-1/2	2	2	2	2	2-1/2	2-1/2	2-1/2	3	3
100	2	2	2	2-1/2	2-1/2	3	3	3	3	3
130	2	2	2-1/2	2-1/2	2-1/2	3	3	3	3	4
160	2	2	2-1/2	2-1/2	2-1/2	3	3	3	4	4
190	2	2-1/2	2-1/2	2-1/2	3	3	3	4	4	4
220	2-1/2	2-1/2	2-1/2	3	3	3	4	4	4	4

Notes: 1. Pipe sizes smaller than 1 in. IPS are not recommended for use with residual grade fuel oils.
2. Lines conveying fuel oil from pump discharge port to burners and tank return may be reduced by 1 or 2 sizes, depending on piping length and pressure losses.

Table 23 Recommended Nominal Size for Fuel Oil Suction Lines from Tank to Pump (Distillate Grade Numbers 1 and 2)

Pumping Rate, gal/h	Length of Run in Feet at Maximum Suction Lift = 10 ft									
	25	50	75	100	125	150	175	200	250	300
10	1/2	1/2	1/2	1/2	1/2	1/2	1/2	3/4	3/4	1
40	1/2	1/2	1/2	1/2	1/2	3/4	3/4	3/4	3/4	1
70	1/2	1/2	3/4	3/4	3/4	3/4	3/4	1	1	1
100	1/2	3/4	3/4	3/4	3/4	1	1	1	1	1-1/4
130	1/2	3/4	3/4	1	1	1	1	1	1-1/4	1-1/4
160	3/4	3/4	3/4	1	1	1	1	1-1/4	1-1/4	1-1/4
190	3/4	3/4	1	1	1	1	1-1/4	1-1/4	1-1/4	2
220	3/4	1	1	1	1	1-1/4	1-1/4	1-1/4	1-1/4	2

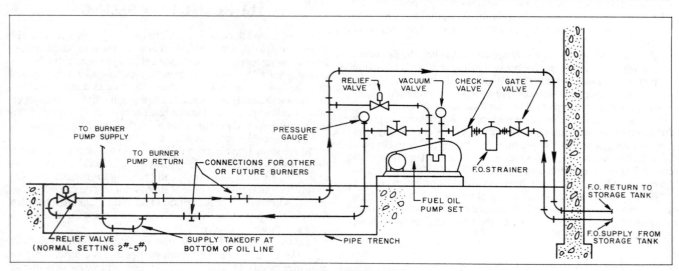

Fig. 15   Typical Oil Circulating Loop

REFERENCES

Ball, E.F. and C.J.D. Webster. 1976. Some measurements of water-flow noise in copper and ABS pipes with various flow velocities. *The Building Services Engineer* 44(2):33.

Crane Co. 1976. Flow of fluids through valves, fittings and pipe. Technical Paper No. 410.

Dawson, F.M. and J.S. Bowman. 1933. Interior water supply piping for residential buildings. University of Wisconsin Experiment Station, No. 77.

Borg-Warner Corp. 1955. *Equipment manual*, Section B-120. York Division, Borg-Warner Corp., York, PA.

Freeman, J.R. 1941. *Experiments upon the flow of water in pipes*. ASME, New York.

Gas engineers' handbook. 1970. The Industrial Press, New York.

Giesecke, F.E. 1926. Friction of water elbows. ASHVE *Transactions* 32:303.

Giesecke, F.E. and W.H. Badgett. 1931. Friction heads in one-inch standard cast-iron tees. ASHVE *Transactions* 37:395.

Giesecke, F.E. and W.H. Badgett. 1932a. Loss of head in copper pipe and fittings. ASHVE *Transactions* 38:529.

Giesecke, F.E. and W.H. Badgett. 1932b. Supplementary friction heads in one-inch cast-iron tees. ASHVE *Transactions* 38:111.

HDR design guide. 1981. Hennington, Durham and Richardson, Omaha, NE.

Howell, R.H. 1985. Evaluation of sizing methods for steam condensate systems. ASHRAE *Transactions* 91(1).

Hunter, R.B. 1940. Methods of estimating loads in plumbing systems. National Bureau of Standards Building Materials and Structures, Report BMS 65.

Hunter, R.B. 1941. Water distributing systems for buildings. National Bureau of Standards Report BMS 79.

Idelchik, I.E. 1986. *Handbook of hydraulic resistance*. Hemisphere Publishing Corporation, New York.

ISA. 1985. Flow equations for sizing control valves. *Standard* S75.01-85.

Laschober, R.R., G.Y. Anderson, and D.G. Barbee. 1966. Counterflow of steam and condensate in slightly pitched pipes. ASHRAE *Transactions* 72(1):157.

Manas, V.T. 1957. *National plumbing code handbook*. McGraw-Hill Book Co., Inc., New York.

Marseille, B. 1965. Noise transmission in piping. *Heating and Ventilating Engineering* (June):674.

NFPA. 1987. Standard for the Installation of oil burning equipment. NFPA *Standard* 31-87.

NFPA. 1988. National fuel gas code. NFPA *Standard* 54-1988. National Fire Protection Association, Quincy, MA. Also ANSI *Standard* Z223.1-1988.

NFPA. 1990. *Flammable and combustible liquids code handbook*, 4th ed. National Fire Protection Association, Quincy, MA.

Obrecht, M.F. and M. Pourbaix. 1967. Corrosion of metals in potable water systems. AWWA 59:977.

Piping design and engineering. 1951. Grinnell Company, Inc. Cranston, RI.

Rogers, W.L. 1953. Experimental approaches to the study of noise and noise transmission in piping systems. ASHVE *Transactions* 59:347-60.

Rogers, W.L. 1954. Sound-pressure levels and frequencies produced by flow of water through pipe and fittings. ASHRAE *Transactions* 60:411-30.

Rogers, W.L. 1956. Noise production and damping in water piping. ASHAE *Transactions* 62:39.

Sanks, R.L. 1978. *Water treatment plant design for the practicing engineer*. Ann Arbor Science Publishers, MI.

Smith, T. 1983. Reducing corrosion in heating plants with special reference to design considerations. *Anti-Corrosion Methods and Materials* 30 (October):4.

Water flow characteristics of thermoplastic pipe. 1971. Plastic Pipe Institute, New York.

Williams, G.J. 1976. The Hunter curves revisited. *Heating/Piping/Air Conditioning* (November):67.

Williams, G.S. and A. Hazen. 1933. *Hydraulic tables*. John Wiley & Sons, New York.

CHAPTER 34

ABBREVIATIONS AND SYMBOLS

THIS chapter contains information about abbreviations and symbols for heating, ventilating, refrigerating, and air-conditioning engineers.

Abbreviations are shortened forms of names and expressions used in drawings, texts, and computer programs. This chapter discusses conventional English language abbreviations that may be different in other languages. A *letter symbol* represents a quantity or a unit, not its name, and is independent of language. Because of this, use of a letter symbol is preferred over abbreviations for unit or quantity terms. Letter symbols necessary for individual chapters are defined in the chapters where they occur.

Abbreviations are never used when a mathematical sign is involved, such as the equality sign ($=$) or division sign ($/$), except in computer programming, where the abbreviation takes on the function of a letter symbol. Mathematical operations are performed only with symbols. Abbreviations should be used only where necessary to save time and space; avoid their usage in documents circulated in foreign countries.

Graphical symbols in this chapter are easy to draw and recognize and were selected to save engineering drafting time. Symbols of piping, ductwork, fittings, and in-line accessories can be used on scale drawings and diagrams.

Identifying piping by legend and color promotes greater safety and lessens the chance of error in emergencies. Piping identification is now required throughout the United States by the Occupational Safety and Health Act (OSHA) for some industries and by many federal, state, and local codes.

ABBREVIATIONS FOR TEXTS, DRAWINGS, AND COMPUTER PROGRAMS

Abbreviations for text and drawings have been compiled from *Abbreviations for Use on Drawings and in Text*, ANSI Y1.1-72. Table 1 gives some of these abbreviations, as well as others commonly found on mechanical drawings and abbreviations (symbols) used in computer programming.

Abbreviations specific to a single subject are defined in the chapters in which they appear. Additional abbreviations used on drawings can be found in the Graphical Symbols section of this chapter.

The abbreviations (symbols) used for computer programming for the heating, ventilating, refrigerating, and air-conditioning industries have been developed by ASHRAE Technical Committee 1.5, Computer Applications. These symbols identify computer variables, subprograms, subroutines, and functions commonly applied in the industry. Use of these symbols enhances comprehension of the program listings and provides a clearly defined nomenclature in applicable computer programs.

Certain programming languages differentiate between *real numbers* (numbers with decimals) and *integers* (numbers without decimals). This is done by reserving certain initial letters of a variable for integer numbers. For instance, in Fortran, any variable beginning with the letters H through N is defined by the computer as an integer. Many of the symbols listed in this chapter begin with these letters and, in order to make them real numbers, must be prefixed with a noninteger letter. Thus, HP would become XHP if the programmer wanted to define horsepower as a decimal value.

Many symbols have two or more options listed. The longest abbreviation is the preferred one and should be used if possible. However, it is sometimes necessary to shorten the symbol to further identify the variable. For instance, the area of a wall cannot be defined as WALLAREA because most computer languages restrict the number of letters in a variable name. Therefore, a shorter variable symbol is applied, and WALLAREA becomes WALLA or WAREA.

In Table 1, the same symbol is sometimes used for different terms. This liberty is taken because it is highly unlikely that the two terms would be used in the same program. If such were the case, one of the terms would require a suffix or prefix to differentiate it from the other.

LETTER SYMBOLS

Letter symbols include symbols for physical quantities (quantity symbols) and symbols for the units in which these quantities are measured (unit symbols). *Quantity symbols*, such as *I* for electric current, are listed in this chapter and are printed in italic type. A *unit symbol* is a letter or group of letters such as mm for millimetre, or a special sign such as ° for degrees, and is printed in Roman type. Subscripts and superscripts are governed by the same principles. Letter symbols are restricted mainly to the English and Greek alphabets.

Quantity symbols may be used in mathematical expressions in any way consistent with good mathematical usage. The product of two quantities, *a* and *b*, is indicated by *ab*. The quotient is a/b, or ab^{-1}. To avoid misinterpretation, parentheses must be used if more than one slash ($/$) is employed in an algebraic term, *e.g.*, ($a/b/c$ or $a/b/c$), but not $a/b/c$.

Subscripts and superscripts, or several of them separated by commas, may be attached to a single basic letter (kernel), but not to other subscripts or superscripts. A symbol that has been modified by a superscript should be enclosed in parentheses before an exponent is added $(X_a)^3$. Symbols can also have alphanumerical marks such as ' (prime), + (plus), and * (asterisk).

More detailed information on the general principles of letter symbol standardization can be found in *IEEE Standard Letter Symbols for Units of Measurement* (ANSI/IEEE *Standard* 260-78) and *Glossary of Terms Concerning Letter Symbols* (ASME Y10.1-72), the source of the above paragraphs.

The letter symbols have, in general, been taken from the following American National Standards Institute (ANSI) standards:

Y10.3M-84   *Letter Symbols for Mechanics and Time-Related Phenomena*

The preparation of this chapter is assigned to TC 1.6, Terminology

Table 1 Abbreviations for Text, Drawings, and Computer Programming

Term	Text	Drawings	Program
above finished floor	—	AFF	—
absolute	abs	ABS	ABS
accumulat(e, -or)	acc	ACCUM	ACCUM
air condition(-ing, -ed)	—	AIR COND	—
air-conditioning unit(s)	—	ACU	ACU
air-handling unit	—	AHU	AHU
air horsepower	ahp	AHP	AHP
alteration	altrn	ALTRN	—
alternating current	ac	AC	AC
altitude	alt	ALT	ALT
ambient	amb	AMB	AMB
American National Standards Institute[1]	ANSI	ANSI	—
American wire gage	AWG	AWG	—
ampere (amp, amps)	Amp	AMP	AMP, AMPS
angle	—	—	ANG
angle of incidence	—	—	ANGI
apparatus dew point	adp	ADP	ADP
approximate	approx	APPROX	—
area	—	—	A
atmosphere	atm	ATM	—
average	avg	AVG	AVG
azimuth	az	AZ	AZ
azimuth, solar	—	—	SAZ
azimuth, wall	—	—	WAZ
barometer(-tric)	baro	BARO	—
bill of material	b/m	BOM	—
boiling point	bp	BP	—
brake horsepower	bhp	BHP	BHP
Brown & Sharpe wire gage	B&S	B&S	—
British thermal unit	Btu	BTU	BTU
Btu per hour (thousand)	Mbh	MBH	MBH
center to center	c to c	C TO C	—
circuit	ckt	CKT	CKT
clockwise	cw	CW	—
coefficient	coef	COEF	COEF
coefficient, valve flow	C_v	C_v	C_v
coil	—	—	COIL
compressor	cprsr	CMPR	CMPR
condens(-er, -ing, -ation)	cond	COND	COND
conductance	—	—	C
conductivity	cndct	CNDCT	K
conductors, number of (3)	3/c	3/c	—
contact factor	—	—	CF
cooling load	clg load	CLG LOAD	CLOAD
counterclockwise	ccw	CCW	—
cubic feet	cu ft	CU FT	CUFT, CFT
cubic inch	cu in	CU IN	CUIN, CIN
cubic feet per minute	cfm	CFM	CFM
cfm, standard conditions	scfm	SCFM	SCFM
cu ft per sec, standard	scfs	SCFS	SCFS
Daylight Saving Time	DST	DST	DST
decibel	dB	DB	DB
degree	deg or 1	DEG or 1	DEG
density	dens	DENS	RHO
depth or deep	dp	DP	DPTH
dew-point temperature	dpt	DPT	DPT
diameter	dia	DIA	DIA
diameter, inside	id	ID	ID
diameter, outside	od	OD	OD
difference or delta	diff	DIFF	D, DELTA
diffuse radiation	—	—	DFRAD
direct current	dc	DC	DC
direct radiation	dir radn	DIR RADN	DIRAD
dry	—	—	DRY
dry-bulb temperature	dbt	DBT	DB, DBT
effectiveness	—	—	EFT
effective temperature[2]	ET*	ET*	ET
efficiency	eff	EFF	EFF
efficiency, fin	—	—	FEFF
efficiency, surface	—	—	SEFF
electromotive force	emf	EMF	—
elevation	el	EL	ELEV
entering	entr	ENT	ENT
entering water temperature	ENT	ENT	ENT
entering air temperature	EAT	EAT	EAT
enthalpy	—	—	H
entropy	—	—	S
equivalent direct radiation	edr	EDR	—
equivalent feet	eqiv ft	EQIV FT	EQFT
equivalent inches	eqiv in	EQIV IN	EQIN
evaporat(-e, -ing, -ed, -or)	evap	EVAP	EVAP
expansion	exp	EXP	XPAN
face area	fa	FA	FA
face to face	f to f	F to F	—
face velocity	fvel	FVEL	FV
factor, correction	—	—	CFAC, CFACT
factor, friction	—	—	FFACT, FF
Fahrenheit	F	F	F
fan	—	—	FAN
feet per minute	fpm	FPM	FPM
feet per second	fps	FPS	FPS
film coefficient, inside	—	—	FI, HI
film coefficient, outside	—	—	FO, HO
flow rate, air	—	—	QAR, QAIR
flow rate, fluid	—	—	QFL
flow rate, gas	—	—	QGA, QGAS
foot or feet	ft	FT	FT
foot pound	ft lb	FT LB	—
freezing point	fp	FP	FP
frequency	Hz	HZ	—
gage or gauge	ga	GA	GA, GAGE
gallons	gal	GAL	GAL
gallons per hour	gph	GPH	GPH
gph, standard	std gph	SGPH	SGPH
gallons per day	gpd	GPD	GPD
grains	gr	GR	GR
gravitational constant	g	G	G
greatest temp difference	gtd	GTD	GTD
head	hd	HD	HD
heat	—	—	HT
heater	—	—	HTR
heat gain	HG	HG	HG, HEATG
heat gain, latent	LHG	LHG	HGL
heat gain, sensible	SHG	SHG	HGS
heat loss	—	—	HL, HEATL
heat transfer	—	—	Q
heat transfer coefficient	U	U	U
height	hgt	HGT	HGT, HT
high-pressure steam	hps	HPS	HPS
high-temperature hot water	hthw	HTHW	HTHW
horsepower	hp	HP	HP
hour(s)	hr	HR	HR
humidity, relative	rh	RH	RH
humidity ratio	W	W	W
incident angle	—	—	INANG
indicated horsepower	ihp	IHP	—
International Pipe Std	IPS	IPS	—
iron pipe size	ips	IPS	—
kelvin	K	K	K
kilowatt	kW	KW	KW
kilowatt hour	kWh	KWH	KWH
latent heat	LH	LH	LH, LHEAT
least mean temp difference[4]	lmtd	LMTD	LMTD
least temp difference[4]	ltd	LTD	LTD
leaving air temperature	lat	LAT	LAT
leaving water temperature	lwt	LWT	LWT
length	lg	LG	LG, L
linear feet	lin ft	LF	LF
liquid	liq	LIQ	LIQ
logarithm (natural)	ln	LN	LN
logarithm to base 10	log	LOG	LOG
low-pressure steam	lps	LPS	LPS

Term			
low-temp. hot water	lthw	LTHW	LTHW
Mach number	Mach	MACH	—
mass flow rate	mfr	MFR	MFR
maximum	max	MAX	MAX
mean effective temp.	met	MET	MET
mean temp. difference	mtd	MTD	MTD
medium pressure steam	mps	MPS	MPS
medium temp. hot water	mthw	MTHW	MTHW
mercury	HG	HG	HG
miles per hour	mph	MPH	MPH
minimum	min	MIN	MIN
noise criteria	NC	NC	—
normally open	n o	N O	—
normally closed	n c	N C	—
not applicable	n/a	N/A	—
not in contract	n i c	N I C	—
not to scale	—	N T S	—
number	no	NO	N, NO
number of circuits	—	—	NC
number of tubes	—	—	NT
ounce	oz	OZ	OZ
outside air	oa	OA	OA
parts per million	ppm	PPM	PPM
percent	%	%	PCT
phase(electrical)	ph	PH	—
pipe	—	—	PIPE
pounds	lbs	LBS	LBS
pounds per square foot	psf	PSF	PSF
psf absolute	psfa	PSFA	PSFA
psf gage	psfg	PSFG	PSFG
pounds per square inch	psi	PSI	PSI
psi absolute	psia	PSIA	PSIA
psi gage	psig	PSIG	PSIG
pressure	—	PRESS	PRES, P
pressure, barometric	baro pr	BARO PR	BP
critical pressure	—	—	CRIP
pressure, dynamic (velocity)	vp	VP	VP
pressure drop or difference	pd	PD	PD, DELTP
pressure, static	sp	SP	SP
pressure, vapor	vap pr	VAP PR	VAP
primary	pri	PRI	PRIM
quart	qt	QT	QT
radian	—	—	RAD
radiat(-e, -or)	—	RAD	—
radiation	—	RADN	RAD
radius	—	—	R
rankine	R	R	R
receiver	rcvr	RCVR	REC
recirculate	recirc	RECIRC	RCIR, RECIR
refrigerant (12, 22, etc.)	R12, R22	R12, R22	R12, R22
relative humidity	rh	RH	RH
resist(-ance, -ivity, -or)	res	RES	RES
return air	ra	RA	RA
revolutions	rev	REV	REV
revolutions per minute	rpm	RPM	RPM
revolutions per second	rps	RPS	RPS
roughness	rgh	RGH	RGH, E
safety factor	sf	SF	SF
saturation	sat	SAT	SAT
saybolt seconds furol	ssf	SSF	SSF
saybolt seconds universal	ssu	SSU	SSU
sea level	sl	SL	SE
second	sec	SEC	SEC
sensible heat	SH	SH	SH
sensible heat gain	SHG	SHG	SHG
sensible heat ratio	SHR	SHR	SHR
shading coefficient	—	—	SC
shaft horsepower	sft hp	SFT HP	SHP
solar	—	—	SOL
specification	spec	SPEC	—
specific gravity	SG	SG	—
specific heat	SP HT	SP HT	C
sp ht at constant pressure	—	—	CP
sp ht at constant volume	—	—	CV
specific volume	sp vol	SP VOL	V, CVOL
square	sq	SQ	SQ
standard	std	STD	STD
standard time meridian	—	—	STM
static pressure	SP	SP	SP
suction	suct	SUCT	SUCT, SUC
summ(-er, -ary, -ation)	—	—	SUM
supply	sply	SPLY	SUP, SPLY
supply air	SA	SA	SA
surface	—	—	SUR, S
surface, dry	—	—	SURD
surface, wet	—	—	SURW
system	—	—	SYS
tabulat(-e, -ion)	tab	TAB	TAB
tee	—	—	TEE
temperature	temp	TEMP	T, TEMP
temperature difference	TD	TD	TD, TDIF
temperature entering	TE	TE	TE, TENT
temperature leaving	TL	TL	TL, TLEA
thermal conductivity	K	K	K
thermal expansion coef	—	—	TXPC
thermal resistance	R	R	RES, R
thermocouple	tc	TC	TC, TCPL
thermostat	T STAT	T STAT	T STAT
thick(-ness)	thkns	THKNS	THK
thousand circular mils	Mcm	MCM	MCM
thousand cubic feet	Mcf	MCF	MCF
thousand foot lbs	kip ft	KIP FT	KIPFT
thousand pounds	kip	KIP	KIP
time	t	T	T
ton	—	—	TON
tons of refrigeration	tons	TONS	TONS
total	—	—	TOT
total heat	tot ht	TOT HT	—
transmissivity	—	—	TAU
U factor	—	—	U
unit	—	—	UNIT
vacuum	vac	VAC	VAC
valve	v	V	VLV
vapor proof	vap prf	VAP PRF	—
variable	var	VAR	VAR
variable air volume	vav	VAV	VAV
velocity	vel	VEL	VEL, V
velocity, wind	w vel	W VEL	W VEL
ventilation, vent	vent	VENT	VENT
vertical	vert	VERT	VERT
viscosity	visc	VISC	MU, VISC
volt	V	V	E
volt ampere	VA	VA	VA
volume	vol	VOL	VOL
volumetric flow rate	—	—	VFR
wall	—	—	W, WAL
water	—	—	WTR
watt	W	W	WAT, W
watthour	Wh	WH	WHR
weight	wt	WT	WT
wet bulb	WB	WB	WB
wet-bulb temperature	WBT	WBT	WBT
width	—	—	WI
wind	—	—	WD
wind direction	wdir	WDIR	WDIR
wind pressure	wpr	WPR	WP, WPRES
yard	yd	YD	YD
year	yr	YR	YR
zone	z	Z	Z, ZN

[1] Abbreviations of most proper names use capital letters in both text and drawings.

[2] The asterisk (*) is used with ET*, effective temperature, as in Chapter 8 of this volume.

[3] These are surface heat transfer coefficients.

[4] Letter L also used for *Logarithm of* these temperature differences in computer programming.

| Y10.4-82 | *Letter Symbols for Heat and Thermodynamics* (Reaffirmed 1988) |
| Y10.2-58 | *Letter Symbols for Hydraulics* |

260-78   *Letter Symbols for Units of Measurement*, ANSI/ IEEE Standard.

When standard symbols in related fields are needed, reference to the following ASME standards is recommended:

Y10.2-58   *Letter Symbols for Hydraulics*
Y10.11-84   *Letter Symbols and Abbreviations for Quantities Used in Acoustics*
Y10.12-55   *Letter Symbols for Chemical Engineering* (Reaffirmed 1988)

Other symbols chosen by an author for a physical magnitude not appearing in any standard list should be ones that do not already have different meanings in the field of the text.

LETTER SYMBOLS

Symbol	Description of Item	Typical Units
A	area	ft^2
a	acoustic velocity	fps or fpm
b	breadth or width	ft
B	barometric pressure	psia or in. Hg
c	concentration	lb/ft^3, mol/ft^3
c	specific heat	Btu/lb·°F
c_p	specific heat at constant pressure	Btu/lb·°F
c_v	specific heat at constant volume	Btu/lb·°F
C	coefficient	—
C	thermal conductance	Btu/h·ft^2·°F
C	fluid capacity rate	Btu/h·°F
C_L	loss coefficient	—
C_P	coefficient of performance	—
d	prefix meaning differential	—
D or d	diameter	ft
d_e	equivalent or hydraulic diameter	ft
D_v	mass diffusivity	ft^2/s
e	base of natural logarithms	—
E	energy	Btu
E	electrical potential	V
f	film conductance (alternate for h)	Btu/h·ft^2·°F
f	frequency	Hz
f_D	friction factor, D'Arcy-Weisbach formulation	—
f_F	friction factor, Fanning formulation	—
F	force	lb
F_{ij}	angle factor (radiation)	—
g	gravitational acceleration	ft/s^2
G	mass velocity	lb/h·ft^2
h	heat transfer coefficient	Btu/h·ft^2·°F
h	hydraulic head	ft
h_D	mass transfer coefficient	lb/h·ft^2· lb per ft^3
h	specific enthalpy	Btu/lb
h_a	enthalpy of dry air	Btu/lb
h_s	enthalpy of moist air at saturation	Btu/lb
H	total enthalpy	Btu
I	electric current	A
J	mechanical equivalent of heat	ft·lb/Btu
k	thermal conductivity	Btu/h·ft·°F
k (or γ)	ratio of specific heats, c_p/c_v	—
K	proportionality constant	—
K_D	mass transfer coefficient	lb/h·ft^2
l or L	length	ft
L_w	sound power	dB
L_p	sound pressure	dB
m or M	mass	lb
M	molecular weight	lb/lb mol
N	rate of rotation	rpm
N or n	number in general	—
p or P	pressure	psi
p_a	partial pressure of dry air	psi
p_w	partial pressure of water vapor in moist air	psi
p_s	vapor pressure of water in saturated moist air	psi
P	power	hp,watts

Symbol	Description of Item	Typical Units
q	time rate of heat transfer	Btu/h
Q	volumetric flow rate	cfm
Q	total heat transfer	Btu
r	radius	ft
r or R	thermal resistance	ft^2·h·°F/Btu
R	gas constant	ft·lb/lb·°R
s	specific entropy	Btu/lb·°R
S	total entropy	Btu/°R
t	temperature	°F
T	absolute temperature	°R
Δt_m or ΔT_m	mean temperature difference	°F
u	specific internal energy	Btu/lb
U	total internal energy	Btu
U	overall heat transfer coefficient	Btu/h·ft^2·°F
v	specific volume	ft^3/lb
V	total volume	ft^3
V	linear velocity	fps
w	mass rate of flow	lb/h
W	weight	lb
W	work	ft-lb
W	humidity ratio of moist air	lb (water)/ lb (dry air)
W_s	humidity ratio of moist air at saturation	lb (water)/ lb (dry air)
x	mole fraction	—
x	quality, mass fraction of vapor	—
x,y,z	lengths along principal coordinate axes	ft
Z	figure of merit	—
α	absolute Seebeck coefficient	V/°C
α	absorptivity, absorptance radiation	—
α	linear coefficient of thermal expansion	per °F
α	thermal diffusivity	ft^2/h
β	volume coefficient of thermal expansion	per °F
γ (or k)	ratio of specific heats, c_p/c_v	—
γ	specific weight	lb/ft^3
Δ	difference between values	—
ϵ	emissivity, emittance (radiation)	—
θ	time	s, h
η	efficiency or effectiveness	—
λ	wavelength	nm
μ	degree of saturation	—
μ	dynamic viscosity	lb/ft·h
ν	kinematic viscosity	ft^2/h
ρ	density	lb/ft^3
ρ	volume resistivity	ohm-cm
ρ	reflectivity, reflectance (radiation)	—
σ	Stefan-Boltzmann constant	Btu/h·ft^2·°R^4
σ	surface tension	lb/ft
τ	stress	lb/ft^2
τ	time	s, h
τ	transmissivity, transmittance (radiation)	—
ϕ	relative humidity	—

DIMENSIONLESS NUMBERS

Fo	Fourier number	$\alpha\tau/L^2$
Gz	Graetz number	wc_p/kL
Gr	Grashof number	$L^3\rho^2\beta g(\Delta t)/\mu^2$
j_D	Colburn mass transfer	$Sh/ReSc^{1/3}$
j_H	Colburn heat transfer	$Nu/RePr^{1/3}$
Le	Lewis number	α/D_v
M	Mach number	V/a
Nu	Nusselt number	hD/k
Pe	Peclet number	GDc_p/k
Pr	Prandtl number	$c_p\mu/k$
Re	Reynolds number	$\rho VD/\mu$
Sc	Schmidt number	$\mu/\rho D_v$
Sh	Sherwood number	$h_D L/D_v$
St	Stanton number	h/Gc_p
Str	Strouhal number	fd/V

MATHEMATICAL SYMBOLS

equal to	$=$
not equal to	$\neq$
approximately equal to	$\approx$
greater than	$>$
less than	$<$
greater than or equal to	$\geqslant$
less than or equal to	$\leqslant$
plus	$+$
minus	$-$
plus or minus	$\pm$
a multiplied by b	$ab, a \cdot b, a \times b$
a divided by b	$\dfrac{a}{b}, a/b, ab^{-1}$
ratio of the circumference of a circle to its diameter	π
a raised to the power n	a_n
square root of a	$\sqrt{a}, a^{0.5}$
infinity	∞
percent	$\%$
summation of	Σ
natural log	$\ln$
logarithm to base 10	$\log$

SUBSCRIPTS

These are to be affixed to the appropriate symbols. Several subscripts may be used together to denote combinations of various states, points, or paths. Often the subscript indicates that a particular property is to be kept constant in a process.

$a,b,\ldots$	referring to different phases, states or physical conditions of a substance, or to different substances
a	air
a	ambient
b	barometric (pressure)
c	referring to critical state or critical value
c	convection
db	dry bulb
dp	dew point
e	base of natural logarithms
f	referring to saturated liquid
f	film
fg	referring to evaporation or condensation
F	friction
g	referring to saturated vapor
h	referring to change of phase in evaporation
H	water vapor
i	referring to saturated solid
i	internal
if	referring to change of phase in melting
ig	referring to change of phase in sublimation
k	kinetic
L	latent
m	mean value
M	molar basis
o	referring to initial or standard states or conditions
p	referring to constant pressure conditions or processes
p	potential
r	refrigerant
r	radiant or radiation
s	referring to moist air at saturation
s	sensible
s	referring to isentropic conditions or processes
s	static (pressure)
s	surface
t	total (pressure)
T	referring to isothermal conditions or processes
v	referring to constant volume conditions or processes
v	vapor
v	velocity (pressure)
w	wall
w	water
wb	wet bulb
$1,2,\ldots$	different points in a process, or different instants of time

GRAPHICAL SYMBOLS FOR DRAWINGS

Graphical symbols have been extracted from (1) *Graphic Symbols for Pipe Fittings, Valves, and Piping* [ANSI 232.2.3-1949 (Reaffirmed 1953)]; (2) *Graphic Symbols for Heating, Ventilating, and Air Conditioning* [ASME Y32.2.4-49 (Reaffirmed 1984)]; (3) *American Standard Abbreviations for Use on Drawings and in Text* (ASME Y1.1-72); and (4) *Graphic Symbols for Plumbing Fixtures for Diagrams Used in Architectural and Building Construction* (ASME Y32.4-77).

Some of these symbols have been modified and others have been added to reflect current practice. Symbols and quotations are used with permission of the publisher, The American Society of Mechanical Engineers.

Piping

Heating

High Pressure Steam	——— HPS ———
Medium Pressure Steam	——— MPS ———
Low Pressure Steam	——— LPS ———
High Pressure Condensate	——— HPC ———
Medium Pressure Condensate	——— MPC ———
Low Pressure Condensate	——— LPC ———
Boiler Blow Down	——— BBD ———
Pumped Condensate	——— PC ———
Vacuum Pump Discharge	——— VPD ———
Makeup Water	——— MU ———
Atmospheric Vent	——— ATV ———
Fuel Oil Discharge	——— FOD ———
Fuel Oil Gage	——— FOG ———
Fuel Oil Suction	——— FOS ———
Fuel Oil Return	——— FOR ———
Fuel Oil Tank Vent	——— FOV ———
Low Temperature Hot Water Supply	——— HWS ———
Medium Temperature Hot Water Supply	——— MTWS ———
High Temperature Hot Water Supply	——— HTWS ———
Low Temperature Hot Water Return	——— HWR ———
Medium Temperature Hot Water Return	——— MTWR ———
High Temperature Hot Water Return	——— HTWR ———
Compressed Air	——— A ———
Vacuum (Air)	——— VAC ———
Existing Piping	——— (NAME)E ———
Pipe to be Removed	—X—X— (NAME) —X—X—

Air Conditioning and Refrigeration

Refrigerant Discharge	——— RD ———
Refrigerant Suction	——— RS ———
Brine Supply	——— B ———
Brine Return	——— BR ———
Condenser Water Supply	——— C ———
Condenser Water Return	——— CR ———
Chilled Water Supply	——— CWS ———
Chilled Water Return	——— CWR ———
Fill Line	——— FILL ———
Humidification Line	——— H ———
Drain	——— D ———
Hot/Chilled Water Supply	——— HCS ———
Hot/Chilled Water Return	——— HCR ———
Refrigerant Liquid	——— RL ———
Heat Pump Water Supply	——— HPWS ———
Heat Pump Water Return	——— HPWR ———

Plumbing

Sanitary Drain above Floor or Grade	——— SAN ———
Sanitary Drain below Floor or Grade	— — — SAN — — —
Storm Drain above Floor or Grade	——— ST ———
Storm Drain below Floor or Grade	— — — ST — — —

Condensate Drain above Floor or grade
Condensate Drain below Floor or grade
Vent
Cold Water
Hot Water
Hot Water Return
Gas
Acid Waste
Drinking Water Supply
Drinking Water Return
Vacuum (Air)
Compressed Air
Chemical Supply Pipes[a]
Floor Drain
Funnel Drain, open

Fire Safety Devices[b]

Signal Initiating Detectors

Heat (Thermal)   Gas

Smoke   Flame

Valves

Valves for Selective Actuators

Air Line

Ball

Butterfly

Diaphragm

Gate

Gate, Angle

Globe

Globe, Angle

Plug Valve

Three Way

---

[a] See Piping Identification in this chapter.
[b] Refer to *Fire Protection Symbols for Architectural and Engineering Drawings* (NFPA *Standard* 172 P-86) for additional symbols.

Valve Actuators

Manual
  Non-Rising Stem

  Outside Stem & Yoke

  Lever

  Gear

Electric
  Motor

  Solenoid

Pneumatic
  Motor

  Diaphragm

Valves, Special Duty

Check, Swing Gate

Check, Spring

Control, Electric-Pneumatic

Control, Pneumatic-Electric

Hose End Drain

Lock Shield

Needle

Pressure Reducing
(number and specify)

Quick Opening

Quick Closing, Fusible
Link

Relief (R) or Safety (S)

Solenoid

Square Head Cock

Unclassified (number
and specify)

Fittings

The following fittings are shown with screwed connections. The symbol for the body of a fitting is the same for all types of connections, unless otherwise specified. The types of connections are often specified for a range of pipe sizes, but are shown with the fitting symbol where required. For example, an elbow would be:

| Flanged | Screwed | Belt & Spigot |
| Welded[a] | Soldered | Solvent Cement |

Fitting	Symbol
Bushing	
Cap	
Connection, Bottom	
Connection, Top	
Coupling (Joint)	
Cross	
Elbow, 90°	
Elbow, 45°	
Elbow, Turned Up	
Elbow, Turned Down	
Elbow, Reducing, Show Sizes	
Elbow, Base	
Elbow, Long Radius	
Elbow, Double Branch	
Elbow, Side Outlet, Outlet Up	
Elbow, Side Outlet, Outlet Down	
Lateral	

[a] Includes fusion, specify type.

Fitting	Symbol
Reducer, Concentric	
Reducer, Eccentric Straight Invert	
Reducer, Eccentric Straight Crown	
Tee	
Tee, Outlet Up	
Tee, Outlet Down	
Tee, Reducing (Show Sizes)	
Tee, Side Outlet, Outlet Up	
Tee, Side Outlet, Outlet Down	
Tee, Single Sweep	
Union, Screwed	
Union, Flanged	

Piping Specialties

Air Vent, Automatic	
Air Vent, Manual	
Air Separator	
Alignment Guide	
Anchor, Intermediate	
Anchor, Main	
Ball Joint	
Expansion Joint	
Expansion Loop	
Flexible Connector	
Flowmeter, Orifice	

Flowmeter, Venturi	VFM-I
Flow Switch	FS
Hanger, Rod	H
Hanger, Spring	H
Heat Exchanger, Liquid	
Heat Transfer Surface (indicate type)	RAD-I
Pitch of Pipe, Rise (R) Drop (D)	R
Pressure Gauge and Cock	
Pressure Switch	PS
Pump (indicate use)	CW-I
Pump Suction Diffuser	PSD
Spool Piece, Flanged	
Strainer	
Strainer, Blow Off	
Strainer, Duplex	
Tank (indicate use)	FO
Thermometer	
Thermometer Well, only	TW
Thermostat, Electric	T
Thermostat, Pneumatic	T
Thermostat, Self-Contained	F&T
Traps, Steam (indicate type)	
Unit Heater (indicate type)	UH

Air Moving Devices and Components

Fans (indicate use)[a]

Axial Flow	R 1,2
Centrifugal	S 1,2

Propeller	E 1,2
Roof Ventilator, Intake	SRV-I
Roof Ventilator, Exhaust	ERV-I
Roof Ventilator, Louvered	

Ductwork[b]

Direction of Flow	
Duct Size, first figure is side shown	12/20
Duct Section, Positive Pressure, first figure is top	20/12
Duct Section, Negative Pressure	20/12
Change of Elevation Rise (R) Drop (D)	R
Access Doors, Vertical or Horizontal	AD 10/10
Acoustical Lining (insulation)	
Cowl, (Gooseneck) and Flashing	
Flexible Connection	
Flexible Duct	
Sound Attenuator	SA
Terminal Unit, Mixing	H C TU M-I
Terminal Unit, Reheat	TU RH-I
Terminal Unit, Variable Volume	TU VAV-I
Transition[c]	20/10 15/8
Turning Vanes	
Detectors, Fire and/or Smoke	

---

[a]Units of measurement are not shown herein, but should be shown on drawings. The first of the two dimensions on ducts indicates the side of the duct showing; on duct sections, the top; on grilles and registers, the horizontal edge.
[b]Adapted from SMACNA, Symbols for Ventilation and Air Conditioning Figure 4.2. *HVAC Duct System Design.*
[c]Indicate Flat on Bottom or Top (FOB or FOT) if applicable.

Dampers

Back Draft Damper

BDD

Pneumatic Operated Damper

POD

Electric Operated Damper

EOD

Fire Damper and Sleeve
(provide access door)

Vertical Position

FD

AD

Horizontal Position

FD

AD

Manual Volume

VD

Manual Splitter

S

Smoke Damper
(provide access door)

SD

AD

Standard Branch, Supply
or Return, No Splitter

S   R

Heater, Duct, Electric

Grilles, Register and Diffusers[a]

Exhaust Grille or Register

±6   20/12
700

Supply Grille or Register

SG 20/12
700

Grille or Register, Ceiling

CG 20/20
700

Heat Stop for
Fire Rated Ceiling

Louver and Screen

40/36 L
700

Louver, Door or Wall

20/12 L
200

Door Grille

DG   12 X 6

Undercut Door

UC 1/2"
100

Ceiling Diffuser, Rectangular

CD   300
20/12   300

Ceiling Diffuser, Round

CD 20 NECK
1000

Diffuser, Linear

48/3   300

Diffuser and Light Fixture
Combination

100
100

Transfer Grille Assembly

TG
1000   24/12

Refrigeration

Compressors

Centrifugal

Reciprocating

Rotary

Rotary Screw

Condensers

Air Cooled

Evaporative

Water Cooled,
(specify type)

Condensing Units

Air Cooled[b]

RS   RL

Water Cooled[b]

S   RL
W

Condenser-Evaporator
(Cascade System)

L.S.
COND

H.S.
EVAP

Cooling Towers

Cooling Tower

Spray Pond

---

[a] Show volumetric flow rate at each device.
[b] L = Liquid being cooled, RL = Refrigerant liquid, RS = Refrigerant suction.

Evaporators[b]
  Finned Coil

  Forced Convection

  Immersion Cooling Unit

  Plate Coil

  Pipe Coil[c]

Liquid Chillers
  (Chillers only)
  Direct Expansion[a]

  Flooded[a]

  Tank, Closed

  Tank, Open

Chilling Units
  Absorption

  Centrifugal

  Reciprocating

  Rotary Screw

Controls

Refrigerant Controls
  Capillary Tube

  Expansion Valve, Hand

  Expansion Valve, Automatic

  Expansion Valve, Thermostatic

  Float Valve, High Side

  Float Valve, Low Side

[a]L = Liquid being cooled, RL = Refrigerant liquid, RS = Refrigerant suction.
[b]Specify manifolding.
[c]Frequently used diagrammatically as evaporator and/or condenser with label indicating name and type.

Thermal Bulb

Solenoid Valve

Constant Pressure Valve, Suction

Evaporator Pressure Regulating
  Valve, Thermostatic, Throttling-
  Type

Evaporator Pressure Regulating
  Valve, Thermostatic, Snap-
  Action Type

Evaporator Pressure-Regulating
  Valve, Throttling-Type, Evaporator Side

Compressor Suction Valve, Pres-
  sure-Limiting, Throttling-Type,
  Compressor Side

Thermo-Suction Valve

Snap-Action Valve

Refrigerant Reversing Valve

*Temperature or Temperature-
Actuated Electrical or Flow Controls*
  Thermostat, Self-Contained

  Thermostat, Remote Bulb

*Pressure of Pressure-Actuated
Electrical or Flow Controls*
  Pressure Switch

  Pressure Switch, Dual
  (High-Low)

  Pressure Switch,
    Differential Oil Pressure

  Valve, Automatic Reducing

  Valve, Automatic Bypass

Valve, Pressure-Reducing

Valve, Condenser Water Regulating

Auxiliary Equipment
Refrigerant

Filter

Strainer

Filter and Drier

Scale Trap

Drier

Vibration Absorber

Heat Exchanger

Oil Separator

Sight Glass

Fusible Plug

Rupture Disc

Receiver, High Pressure, Horizontal

Receiver, High Pressure, Vertical

Receiver, Low Pressure

Intercooler

Intercooler/Desuperheater

Energy Recovery Equipment

Condenser, Double Bundle

Air to Air Energy Recovery
Rotary Heat Wheel

Coil Loop

Heat Pipe

Fixed Plate

Plate Fin, Cross Flow

Power Sources

Motor, Electric (number for identification of description in specifications)

Engine (indicate fuel)

Gas Turbine

Steam Turbine

Steam Turbine, Condensing

Electrical Equipment[a]

Symbols for electrical equipment shown on mechanical drawings are usually geometric figures with an appropriate name or abbreviation, with details described in the specifications. The following are some common examples.[b]

Motor Control — MC

Disconnect Switch, Unfused — DS

Disconnect Switch, Fused — DSF

Time Clock — TC

Automatic Filter Panel — AFP

Lighting Panel — LP

Power Panel — PP

[a] See *Graphic Electrical Symbols for Air-Conditioning and Refrigeration Equipment* (ARI *Standard* 130-82) for preferred symbols of common electrical parts.
[b] Number each symbol if more than one; see *Graphic Symbols for Plumbing Fixtures for Diagrams Used in Architecture and Building Construction* (ANSI Y32.4-77).

PIPING SYSTEM IDENTIFICATION

The material in piping systems is identified to promote greater safety and lessen the chances of error, confusion or inaction in times of emergency. Primary identification should be by means of a lettered legend naming the material conveyed by the piping. In addition to, but not instead of lettered identification, color can be used to identify the hazards or use of the material.

The data has been extracted from the *Scheme for the Identification of Piping Systems,* ASME A13.1-81, with the permission of the publisher, The American Society of Mechanical Engineers.

DEFINITIONS

Piping Systems

Piping systems include pipes of any kind, fittings, valves, and pipe coverings. Supports, brackets, and other accessories are not included. Pipes are defined as conduits for the transport of gases, liquids, semiliquids, or fine particulate dust.

Materials Inherently Hazardous to Life and Property

Flammable or Explosive. Materials that are easily ignited, including materials known as fire producers or explosives.

Chemically Active or Toxic. Materials that are corrosive or are in themselves toxic or productive of poisonous gases.

At Temperatures or Pressures. Materials that, when released from the piping, cause a sudden outburst with the potential for inflicting injury or property damage by burns, impingement, or flashing to vapor state.

Radioactive. Materials that emit ionizing radiation.

Table 2   Examples of Legends

HOT WATER
AIR 100 PSIG
H.P. RETURN
STEAM 100 PSIG

Table 3   Classification of Hazardous Materials and Designation of Colors[a]

Classification	Color Field	Color of Letters for Legend
Materials Inherently Hazardous		
Flammable or Explosive	Yellow	Black
Chemically Active or Toxic	Yellow	Black
Extreme Temperatures or Pressures	Yellow	Black
Radioactive[b]	Purple	Yellow
Materials of Inherently Low Hazard		
Liquid or Liquid Admixture[c]	Green	Black
Gas or Gaseous Admixture	Blue	White
Fire Quenching Materials		
Water, Foam, CO$_2$, Halon, etc.	Red	White

[a] When the color scheme above is used, the colors should be as recommended in ANSI Z-53.1 latest revision, *Safety Color Code for Marking Physical Hazards.*
[b] Previously specified radioactive markers using yellow or purple are acceptable if already installed and/or until existing supplies are depleted, subject to applicable Federal regulations.
[c] Markers with black letters on a green color field are acceptable if already installed and/or until existing supplies are depleted.

Table 4   Size of Legend Letters

Outside Diameter of Pipe or Covering, in.	Length of Color Field A, in.	Size of Letters B, in.
¾ to 1¼	8	½
1½ to 2	8	¾
2½ to 6	12	1¼
8 to 10	24	2½
over 10	32	3½

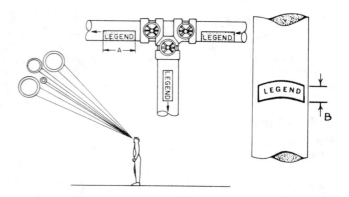

Fig. 1   Visibility of Pipe Markings

Materials of Inherently Low Hazard

All materials that are not hazardous by nature, and are near enough to ambient pressure and temperature that people working on systems carrying these materials run little risk through their release.

Fire Quenching Materials

This classification includes sprinkler systems and other piped fire fighting or fire protection equipment. This includes water (for fire fighting), chemical foam, CO$_2$, Halon, and so forth.

METHOD OF IDENTIFICATION

Legend

The legend is the primary and explicit identification of content. Positive identification of the content of the piping system is by lettered legend giving the name of the contents, in full or abbreviated form, as shown in Table 2. Arrows are used to indicate the direction of flow. Use the legend to identify contents exactly and to provide temperature, pressure and other details necessary to identify the hazard.

The legend shall be brief, informative, pointed, and simple. Legends should be applied close to valves and adjacent to changes in direction, branches, where pipes pass through walls or floors and as frequently as needed along straight runs to provide clear and positive identification. Identification may be applied by stenciling, tape or markers (see Figure 1). The number and location of identification markers on a particular piping system is based on judgment.

Color

Colors listed in Table 3 are used to identify the characteristic properties of the contents. Color can be shown on or contiguous to the piping by any physical means, but it should be used in combination with a legend. Color can be used in continuous total length coverage or in intermittent displays.

Visibility

Pipe markings should be highly visible. If pipe lines are above the normal line of vision, the lettering is placed below the horizontal centerline of the pipe (Figure 1).

Type and Size of Letters

Provide the maximum contrast between color field and legend (Table 3). Table 4 shows the size of letters recommended. Use of standard size letters of 1/2 in. or larger is recommended. For identifying materials in pipes of less than 3/4 in. in diameter and for valve and fitting identification, use a permanently legible tag.

Unusual or Extreme Situations

When the piping layout occurs in or creates an area of limited accessibility or is extremely complex, other identification techniques may be required. While a certain amount of imagination may be needed, the designer should always clearly identify the hazard and use the recommended color and legend guidelines.

CHAPTER 35

UNITS AND CONVERSIONS

Table 1 Conversions to SI Units

Multiply	By	To Obtain	Multiply	By	To Obtain
acre	0.4047	ha	$in \cdot lb_f$ (torque or moment)	113	$mN \cdot m$
bar	*100	kPa	in^2	645	mm^2
barrel (42 U.S. gal, petroleum)	159	L	in^3 (volume)	16.4	mL
	0.159	m^3	in^3/min (SCIM)	0.273	mL/s
Btu (International Table)	1.055	kJ	in^3 (section modulus)	16 400	mm^3
Btu/ft^3	37.3	kJ/m^3; J/L	in^4 (section moment)	416 200	mm^4
Btu/gal	0.279	kJ/L	km/h	0.278	m/s
$Btu \cdot ft/h \cdot ft^2 \cdot °F$	1.731	$W/(m \cdot K)$	kWh	*3.60	MJ
$Btu \cdot in/(h \cdot ft^2 \cdot °F)$ (thermal conductivity, k)	0.1442	$W/(m \cdot K)$	kW/1000 cfm	2.12	J/L
Btu/h	0.2931	W	kilopond (kg force)	9.81	N
Btu/ft^2	11.36	kJ/m^2	kip (1000 lb_f)	4.45	kN
$Btu/(h \cdot ft^2)$	3.155	W/m^2	kip/in^2 (ksi)	6.895	MPa
$Btu/(h \cdot ft^2 \cdot °F)$ (overall heat trans. coefficient, U)	5.678	$W/(m^2 \cdot K)$	litre	*0.001	m^3
Btu/lb	2.326	kJ/kg	met	*58.15	W/m^2
$Btu/(lb \cdot °F)$ (specific heat, c_p)	4.184	$kJ/(kg \cdot K)$	micron (μm) of mercury (60°F)	133	mPa
bushel	0.03524	m^3	mile	1.61	km
calorie, gram	4.187	J	mile, nautical	1.85	km
calorie, kilogram; kilocalorie	4.187	kJ	mph	1.61	km/h
centipoise, dynamic viscosity, μ	*1.00	$mPa \cdot s$	mph	0.447	m/s
centistokes, kinematic viscosity, ν	*1.00	mm^2/s	millibar	*0.100	kPa
clo	0.155	$m^2 \cdot K/W$	mm of mercury (60°F)	0.133	kPa
$dyne/cm^2$	*0.100	Pa	mm of water (60°F)	9.80	Pa
EDR hot water (150 Btu/h)	44.0	W	ounce (mass, avoirdupois)	28.35	g
EDR steam (240 Btu/h)	70.3	W	ounce (force or thrust)	0.278	N
EER	0.293	COP	ounce (liquid, U.S.)	29.6	mL
ft	*0.3048	m	ounce inch (torque, moment)	7.06	$mN \cdot m$
ft	*304.8	mm	ounce (avoirdupois) per gallon	7.49	g/L
ft/min, fpm	*0.00508	m/s	perm (permeance)	57.45	$ng/(s \cdot m^2 \cdot Pa)$
ft/s, fps	*0.3048	m/s	perm inch (permeability)	1.46	$ng/(s \cdot m \cdot Pa)$
ft of water	2.99	kPa	pint (liquid, U.S.)	473	mL
ft of water per 100 ft pipe	0.0981	kPa/m	pound		
ft^2	0.09290	m^2	lb (mass)	0.4536	kg
$ft^2 \cdot h \cdot °F/Btu$ (thermal resistance, R)	0.176	$m^2 \cdot K/W$	lb (mass)	453.6	g
ft^2/s, kinematic viscosity, ν	92 900	mm^2/s	lb_f (force or thrust)	4.45	N
ft^3	28.32	L	lb/ft (uniform load)	1.49	kg/m
ft^3	0.02832	m^3	$lb_m/(ft \cdot h)$ (dynamic viscosity, μ)	0.413	$mPa \cdot s$
ft^3/h, cfh	7.866	mL/s	$lb_f/(ft \cdot s)$ (dynamic viscosity, μ)	1490	$mPa \cdot s$
ft^3/min, cfm	0.4719	L/s	$lb_f \cdot s/ft^2$ (dynamic viscosity, μ)	47.88	$Pa \cdot s$
ft^3/s, cfs	28.32	L/s	lb/h	0.126	g/s
$ft \cdot lb_f$ (torque or moment)	1.36	$N \cdot m$	lb/min	0.00756	kg/s
$ft \cdot lb_f$ (work)	1.36	J	lb/h [steam at 212°F (100°C)]	0.284	kW
$ft \cdot lb_f/lb$ (specific energy)	2.99	J/kg	lb_f/ft^2	47.9	Pa
$ft \cdot lb_f/min$ (power)	0.0226	W	lb/ft^2	4.88	kg/m^2
footcandle	1.076	lx	lb/ft^3 (density, ρ)	16.0	kg/m^3
gallon (U.S., *231 in^3)	3.7854	L	lb/gallon	120	kg/m^3
gph	1.05	mL/s	ppm (by mass)	*1.00	mg/kg
gpm	0.0631	L/s	psi	6.895	kPa
gpm/ft^2	0.6791	$L/(s \cdot m^2)$	quad	1.055	EJ
gpm/ton refrigeration	0.0179	mL/J	quart (liquid U.S.)	0.946	L
grain (1/7000 lb)	0.0648	g	square (100 ft^2)	9.29	m^2
gr/gal	17.1	mg/L	tablespoon (approximately)	15	mL
gr/lb	0.143	g/kg	teaspoon (approximately)	5	mL
horsepower (boiler)	9.81	kW	therm (U.S.)	105.5	MJ
horsepower (550 $ft-lb_f/s$)	0.746	kW	ton, long (2240 lb)	1.016	Mg
inch	*25.4	mm	ton, short (2000 lb)	0.907	Mg
in. of mercury (60°F)	3.377	kPa	ton, refrigeration (12 000 Btu/h)	3.52	kW
in. of water (60°F)	248.8	Pa	torr (1 mm Hg at 0°C)	133	Pa
in/100 ft, thermal expansion	0.833	mm/m	watt per square foot	10.8	W/m^2
			yd	*0.9144	m
			yd^2	0.836	m^2
			yd^3	0.7646	m^3

To Obtain	By	Divide

To Obtain	By	Divide

Note: Units are U.S. values unless noted otherwise.
*Conversion factor is exact.

The preparation of this chapter is assigned to the TC 1.6, Terminology.

Table 2   Conversion Factors

Pressure psi	in. of water (60 °F)	in. Hg (32 °F)	mm Hg atmosphere	mm Hg (32 °F)	bar	kgf/cm²	pascal
1	= 27.708	= 2.0360	= 0.068046	= 51.715	= 0.068948	= 0.07030696	= 6894.8
0.036091	1	0.073483	2.4559×10^{-3}	1.8665	2.4884×10^{-3}	2.537×10^{-3}	248.84
0.491154	13.609	1	0.033421	25.400	0.033864	0.034532	3386.4
14.6960	407.19	29.921	1	760.0	1.01325*	1.03323	101,325*
0.0193368	0.53578	0.03937	0.00131579	1	0.0013332	0.0013595	133.32
14.5038	401.86	29.530	0.98692	750.062	1	1.01972	10^5*
14.223	394.1	28.959	0.96784	735.559	0.98066	1	98,066.5*
1.45038×10^{-4}	4.0186×10^{-3}	2.953×10^{-4}	9.8692×10^{-6}	0.00750	10^{-5}	1.0192×10^{-5}	1

Mass

lb (avoir.)	grain	ounce (avoir.)	kg
1	= 7000*	= 16*	= 0.45359
1.4286×10^{-4}	1	2.2857×10^{-3}	6.4800×10^{-5}
0.06250	437.5*	1	0.028350
2.20462	15,432	35.274	1

Volume

cubic inch	cubic foot	gallon	litre	cubic metre (m³)
1	= 5.787×10^{-4}	= 4.329×10^{-3}	= 0.0163871	= 1.63871×10^{-5}
1728*	1	7.48055	28.317	0.028317
231.0*	0.13368	1	3.7854	0.0037854
61.02374	0.035315	0.264173	1	0.001*
61,023.74	35.315	264.173	1000	1

Energy

Btu	ft·lb	calorie (cal)	joule (J)	watt-second (w·s)
1	= 778.17	= 251.9957	= 1055.056	= 1055.056
1.2851×10^{-3}	1	0.32383	1.355818	1.355818
3.9683×10^{-3}	3.08803	1	4.1868*	4.1868*
9.4782×10^{-4}	0.73756	0.23885	1	1

Density

lb/ft³	lb/gal	g/cm³	kg/m³ (g/L)
1	= 0.133680	= 0.016018	= 16.018463
7.48055	1	0.119827	119.827
62.4280	8.34538	1	1000
0.0624280	0.008345	0.001	1

Specific Volume

ft³/lb	gal/lb	cm³/g	m³/kg (L/g)
1	= 7.48055	= 62.4280	= 0.0624280
0.133680	1	8.34538	0.008345
0.016018	0.119827	1	0.001
16.018463	119.827	1000	1

Specific Heat or Entropy

Btu/lb·°F	cal/(g·K)	kJ/(kg·K)
1	= 1	= 4.1868*
1.0	1	4.1868*
0.23885	0.23885	1

Enthalpy

Btu/lb	cal/g	J/g
1	= 0.55556	= 2.326*
1.8*	1	4.1868*
0.42992	0.23885	1

Thermal Conductivity

Btu/h·ft·°F	cal/(s·cm·°C)	W/(m·K)
1	= 4.1338×10^{-3}	= 1.7307
241.91	1	418.68*
0.57779	2.3885×10^{-3}	1

Viscosity (absolute)   1 poise = 1 dyne-sec/cm² = 0.1 Pa·s = 1 g/(cm·s)

poise	lb_f·s/ft²	lb_f·h/ft²	kg/(m·s)	N·s/m²	lb_m/(ft·s)
1	= 2.0885×10^{-3}	= 5.8014×10^{-7}	= 0.1	= 0.1	= 6.71955×10^{-2}
478.8026	1	2.7778×10^{-4}	47.88026	47.88026	32.17405
1,723,689	3600	1	172,369	172,369	115,827
10	0.020885	5.8014×10^{-6}	1	1	0.0671955
14.8819	3.1081×10^{-2}	8.6336×10^{-6}	1.4882	1.4882	1

Coefficient of Heat Transfer

Btu/h·ft²·°F	cal/(s·cm²·°C)	kcal/(h·m²·°C)	W/(m²·K)
1	= 1.3562×10^{-4}	= 4.8824	= 5.6783
7373.5	1	36,000	41,868*
0.2048	2.7778×10^{-5}	1	1.1630*
0.1761	2.3885×10^{-5}	0.8598	1

Note: The Btu, calorie, and kilocalorie are based on the International Table.

CHAPTER 36

PHYSICAL PROPERTIES OF MATERIALS

VALUES in the following tables are in consistent units to assist the engineer looking for approximate values. For data on refrigerants, see Chapter 16; for secondary coolants, see Chapter 18. Chapter 22 gives more information on the values for materials used in building construction and insulation. Many properties vary with temperature, material density, and composition. The references document the source of the values and provide more detail or values for materials not listed here. The preparation of this chapter is assigned to TC 1.3, Heat Transfer and Fluid Flow.

Table 1   Properties of Vapor

	Molecular Mass	Normal Boiling Point, °F	Critical Temp., °F	Critical Pressure, psia	Density, lb/ft³	Specific Heat, Btu/lb·°F	Thermal Conductivity, Btu/h·ft·°F	Viscosity, lb/ft·h
Alcohol, Ethyl	46.07ᵃ	173.3ᵃ	469.6ᵇ	927.3ᵇ		0.362ʲ	0.0073ᵃ	0.0343ʲ (60)
Alcohol, Methyl	32.04ᵃ	148.9ᵃ	464.0ᵇ	1157ᵇ		0.322ʲ	0.0174ʳ	0.0358ʲ (30)
Ammonia	17.03ᵃ	−28ᵃ	270.3ᵇ	1639ᵇ	0.0482ᵇ	0.525ᵃᵃ	0.0128ᵇ	0.0225ᵃᵃ
Argon	39.948ᵃ	−302.3ᵃ	−187.6ᵇ	705ᵇ	0.1114ᵇ	0.125ᶜ	0.0094ᵃ	0.0507ᵃ
Acetylene	26.04ᵃ	−118.5ᵃ	96.8ᵇ	911ᵇ	0.0732ᵇ	0.377ᵃ	0.0108ᵇ	0.0226ᵃ
Benzene	78.11ᵃ	176.2ᵃ	553.1ᵈ	714.2ᵈ	0.167ᵉ (176)	0.31ᵉ (176)	0.0041ᵉ	0.017ᵃ
Bromine	159.82ᵃ	137.8ᵃ	591.8ᵈ	1499ᵈ	0.38ᶠ (138)	0.055ᶠ (212)	0.0035ᵃ	0.041ᵃ
Butane	58.12ᵃ	31.1ᵃ	305.6ᵈ	550.7ᵈ	0.168ᵍ	0.377ᵃᵃ	0.0079ᵃ	0.017ᵃ
Carbon dioxide	44.01ᵃ	−109.3ᵃ	87.9ᵈ	1071ᵈ	0.123ᵍ	0.20ᵍ	0.0084ᵃ	0.033ʰ
Carbon disulfide	76.13ʰ	115.2ʰ	534ʰ	1046ʰ		0.1431ᵖ (80)		
Carbon monoxide	28.01ᵃ	−312.7ᵃ	−220.4ᵈ	507ᵈ	0.078ᵈ	0.25ᶠ	0.0133ᵃ	0.040ᵃ
Carbon tetrachloride	153.84ʰ	169.8ʰ	541.8ʰ	661ʰ		0.206�q (80)	0.0375ʲ	
Chlorine	70.91ᵃ	−30.3ᵃ	291.2ᵈ	1118ᵈ	0.201ᵈ	0.117ᵃ	0.0046ᵃ	0.030ᵃ
Chloroform	119.39ʰ	143.1ʰ	506.1ʰ	794ʰ		0.126ʲ	0.0081ʳ	0.038ʲ
Ethyl chloride	64.52ʰ	54.2ʰ	369.0ʰ	764ʰ	0.1793ᵇ	0.426ʳ	0.00504ᵈ	0.0378q
Ethylene	28.03ʰ	−154.6ʰ	49.9ʰ	742ʰ	0.0783ʰ	0.352ᵃᵃ	0.0102ᵃᵃ	0.0231ᵃᵃ
Ethyl ether	74.12ʰ	94.4ʰ	378.8ʰ	523ʰ		0.589ʰ (95)	0.0273q	
Fluorine	38.00ʰ	−304.5ʰ	−200.5ʰ	808ʰ	0.1022ᵇ	0.194ʲ	0.0147ʲ	0.089ʲ
Helium	4.0026ᵃ	−452.1ⁱ	−450.2ʰ	33.21ⁱ	0.0111ⁱ	1.241ᵃᵃ	0.0823ᵃᵃ	0.0452ᵃᵃ
Hydrogen	2.0159ᵃ	−423.0ⁱ	−399.9ⁱ	190.8ⁱ	0.00562ⁱ	3.40ʲ	0.0972ᵃᵃ	0.0203ᵃᵃ
Hydrogen chloride	36.461ᵃ	−120.8ᵃ	124.5ᵈ	1198ᵈ	0.1024ᵇ	0.191ʲ	0.00757ʲ	0.0321ʲ
Hydrogen sulfide	34.080ᵃ	−77.3ᵃ	212.7ᵈ	1307ᵈ	0.0961ᵇ	0.238ʲ	0.00751ʲ	0.0281ʲ
Heptane (m)	100.21ᵃ	209.2ᵃ	512.2ᵇ	394ᵇ	0.21ᵏ	0.476ʲ	0.0107ʲ	0.0168ʲ
Hexane (m)	86.18ᵃ	154ᵃ	454.5ᵈ	440ᵈ	0.21ᵏ	0.449ʲ	0.00971ʲ	0.0182ʲ
Isobutane	58.12ᶠ	−10.9ᵃ	275.0ʲ	529.1ʲ	0.154ˢ (70)	0.376ᵃᵃ	0.0081ᵃᵃ	0.0168ᵃᵃ
Methyl chloride	50.49ᵃ	−11.6ᵃ	289.6ʲ	968.5ᵇ	0.1440ᵈ	0.184ᵃᵃ	0.0054ᵃᵃ	0.0244ᵃᵃ
Methane	16.04ᵃ	−263.2ᵃ	−115.18ʲ	673.1ᵇ	0.0448ᵇ	0.520ᵃᵃ	0.0178ᵃᵃ	0.0250ᵃᵃ
Napthalene	128.19ᵃ	−360.4ᵃ	876.2ʲ	576.1ʲ		0.313q (77)		
Neon	20.183ᵃ	−412.6ᵃ	−379.7ʲ	391.3ʲ		0.246ᵃᵃ	0.0268ᵃᵃ	0.0718ᵃᵃ
Nitric oxide	30.01ᵃ	−241.6ᵃ	−135.2ʲ	949.4ʲ		0.238ʲ		0.0712ʲ
Nitrogen	28.01ᵃ	−320.4ᵃ	−232.4ʲ	492.3ᵇ		0.248ʲ	0.0138ᵃᵃ	0.0402ᵃᵃ
Nitrous oxide	44.01ᵃ	−127.3ᵃ	97.5ʲ	1049.3ʲ		0.203ʲ	0.01001ʲ (80.3)	0.0543ʲ
Nitrogen tetroxide	92.02ᵃ		316.8ʲ	1469.6ʲ		0.201ᵖ (80)	0.0232ʳ (131)	
Oxygen	32.00ᵃ	−297.4ᵃ	181.1ʲ	736.3ʲ		0.218ʲ	0.0141ᵃᵃ	0.0462ᵃᵃ
n-Pentane	72.53ᵃ		385.9ʲ	489.5ʲ		0.400ᵃ (80)	0.00877ʲ (80.3)	0.0282ʲ
Phenol	74.11ᵇ	358.5ᵇ	786ᵇ	889ᵇ	0.16ᵏ	0.34ᵏ	0.0099ᵏ	0.029ᵏ
Propane	44.09ᵃ	−43.73ᵍ	206.3ᵍ	617.4ᵍ	0.126ᵍ	0.3753ʲ (40)	0.0087ʲ	0.0179ʲ
Propylene	42.08ᵇ	−53.86ˡ	197.2ˡ	670.3ˡ	0.120ˡ	0.349ᵃᵃ	0.0081ᵃᵃ	0.0195ᵃᵃ
Sulfur dioxide	64.06ᵇ	14.0ᵇ	315ᵇ	1142ᵇ	0.183ᵇ	0.145ˡ	0.0049ʲ	0.0281ʲ
Water vapor	18.02ᵇ	212.0ᵐ	705.47ᵐ	3208.2ᵐ	0.0373ᵐ	0.489ᵃᵃ	0.0143ᵐ	0.0293ᵃᵃ

Notes:
1. Properties at 14.696 psia and 32 °F, or the saturation temperature if higher than 32 °F, unless otherwise noted in parentheses.
2. Superscript letters indicate data source from references.

Table 2 Properties of Liquids

Name or Description	Normal Boiling Point, °F at 14.696 psia	Enthalpy of Vaporization, Btu/lb	Specific Heat, c_p Btu/lb·°F	Temp., °F	Viscosity lb/h·ft	Temp., °F	Enthalpy of Fusion, Btu/lb	Mass Density lb/ft³	Temp., °F	Thermal Conductivity Btu/h·ft·°F	Temp., °F	Vapor Pressure mm of Hg	Temp., °F	Freezing Point, °F
Acetic acid	245.3[a]	174.1[b]	0.522[b]	79–203	2.956[f]	68	84.0[b]	65.49[a]	68	0.099[b]	68	400[a]	210	61.9[a]
Acetone	133.2[a]	228.9[b]	0.514[b]	37–73	0.801[f]	68	42.1[b]	49.4[a]	68	0.102[b]	86	400[a]	103	−139.6[a]
Allyl alcohol	206.6[a]	294.1[b]	0.655[b]	70–205	3.298[f]	68		53.31[a]	68	0.104[b]	77–86	400[a]	176	−200.2[a]
n-Amyl alcohol	280.6[i]	216.3[b]			9.686[f]	73.4	48.0[b]	51.06[f]	59	0.094[b]	86	100[a]	186	−110.2[a]
Ammonia	−28[a]	583.2[b]	1.099[b]	32	0.643[f]	−28.3	142.9[b]	43.50[b]	−50	0.29[b]	5–86	400[a]	−49.7	−107.9[a]
Alcohol-ethyl	173.3[a]	367.5[b]	0.680[b]	32–208	2.889[f]	68	46.4[b]	49.27[a]	68	0.105[b]	68	100[a]	94.8	−179.1[a]
Alcohol-methyl	148.9[a]	473.0[b]	0.601[b]	59–68	1.434[f]	68	42.7[a]	49.40[a]	68	0.124[b]	68	100[a]	70.2	−144.0[a]
Aniline	363.8[a]	186.6[b]	0.512[b]	46–180	10.806[f]	68	48.8[b]	63.77[a]	68	0.100[b]	32–68	10[a]	156.9	20.84[a]
Benzene	176.2[a]	169.4[b]	0.412[b]	68	1.58[a]	68	54.2[h]	54.9[d]	68	0.085[h]	68	75[d]	68	42[a]
Bromine	137.8[a]	79.4[d]	0.107[f]	68	2.39[a]	68	28.5[d]	194.7[f]	68	0.070[a]	77[a]	165[d]	68	19[a]
n-Butyl alcohol	243.5[a]	254.3[h]	0.563[f]	68	7.13[f]	68	53.9[b]	50.6[a]	68	0.089[h]	68	5[d]	68	−130[a]
n-Butyric acid	326.3[a]	217.0[h]	0.515[f]	68	3.73[a]	68	54.1[a]	60.2[a]	68	0.094[h]	54	0.7[d]	68	20[a]
Calcium chloride brine (20% by wt)			0.744[i]	68	4.8[i]	68		73.8[i]	68	0.332[i]	68			2[i]
Carbon disulfide	115.3[a]	148.8[h]	0.240[i]	68	0.88[a]	68	24.8[d]	78.9[d]	68	0.093[b]	86	295[d]	68	−168[a]
Carbon tetrachloride	170.2[a]	83.7[h]	0.201[f]	68	2.34[a]	68	12.8[d]	99.5[d]	68	0.062[j]	68	87[d]	68	−9[a]
Chloroform	142.3[v]	106[v]	0.234[v]	68	1.36[v]	68		92.96[v]	68	0.075[v]	68	160[v]	68	−81.8[v]
Decane-n	345.2[b]		0.50[b]	68			86.9[b]	45.6[b]	68	0.086[b]	68	1.3[b]	68	−21.5[b]
Ethyl ether	94.06[v]	151[v]	0.541[v]	68	0.56[v]	68	42.4[v]	44.61[v]	68	0.081[b]	68	440[v]	68	−177.3[v]
Ethyl acetate	170.8[v]	183.8[v]	0.468[v]	68	1.09[v]	68	51.2[b]	52.3[v]	68	0.101[b]	68	72[b]	68	−116.3[v]
Ethyl chloride	54.2[j]	165.9[f] [68]	0.368[f]	32			29.68[a]	56.05[a]	68	0.179[f]	33.6	400[y]	53.1	−213.5[a]
Ethyl iodide	162.1[a]	82.1[f] [160]	0.368[f]	32	0.0239[f]	68		120.85[a]	68	0.214[f]	86	100[y]	64.4	−162.4*
Ethylene bromide	268.8[a]	99.2[f] [210]	0.174[f]	68	0.0694[f]	68	24.82[a]	136.05[a]	68			10[y]	65.5	49.2[a]
Ethylene chloride	182.3[a]	153.4[f] [308]	0.301[f]	68	0.0338[f]	68	38.02[a]	77.10[a]	68			60[y]	64.6	−31.64[a]
Ethylene glycol	388.4[a]	344.0[f] [651]					77.86[a]	69.22[a]	68	0.100[f]	68	1[y]	128	12.7[a]
Formic acid	213.3[a]	215.8[f] [420]	0.526[f]	68	0.0719[f]	68	118.89[a]	76.16[a]	68	0.104[a]	33	40[y]	75.2	47.1[a]
Glycerine (glycerol)	359 (20 mm)				43.1[f]	68		78.72[a]	68	0.113[a]	68	1[a]	125.5	68[a]
Heptane	209.2[a]	138[f]	0.532[j]	68	0.990[a]	68	60.4[b]	42.7[a]	68	0.0741[j]	68	35.5[y]	68	−132[a]
Hexane	154[a]	145[f]	0.538[j]	68	0.775[d]	68	65.0[b]	41.1[a]	68	0.0720[j]	68	120.0[y]	68	−139[a]
Hydrogen chloride	−120.8[a]	191[f]					23.6[f]	74.6[d]	b.p.					−174.6[a]
Isobutyl alcohol	226.4[a]	249[f]	0.116[f]	68	9.45[f]	68		50.0[f]	68	0.082[f]	68	9.7[y]	68	−162.4[a]
Kerosine	400560[b]		0.50[n]	68	6.0[b]	68		51.2[a]	68	0.086[n]	68			
Linseed oil					104[b]	68		58[d]	68					
Methyl acetate	134.6[a]	177[f]	0.468[f]	68	0.940[f]	68		60.6[a]	68	0.093[f]	68	169.8[y]	68	−11†[a]
Methyl iodide	108.5[a]	82.6[f]			1.21[f]	68		142[a]	68			320[y]	68	−144.6[a]
Naphthalene	411.4[a]	136[f]	0.402[f]	m.p.	2.18[b]	m.p.	64.9[b]	60.9[y]	m.p.					−87.7[a]
Nitric acid	186.8[v]	270[v]	0.42[v]	68	2.2[k]	68	71.5[v]	94.45[v]	68	0.16[v]	68	2.18[b]	68	176.4[a]
Nitrobenzene	411.6[b]	142[b]	0.348[b]	68	5.20[b]	68	40.28[v]	75.2[b]	68	0.96[b]	68	1.77[v]	68	42.9[v]
Octane	258.3[b]	131.7[b]	0.51[b]	68	1.36[b]	68	77.70[b]	43.9[b]	68	0.084[b]	68	< 0.01[b]	68	42.3[b]
Petroleum		98–165[w]	0.4–0.6[w]	68	19–2900[w]	68		40–66[w]	68			0.42[b]	68	−69.7[b]
n-Petane	96.8[a]	153.6[h]	0.558[h]	68	0.546[d]	68	50.1[h]	39.1[a]	68	0.066[h]	68	425[d]	68	−201.5[a]
Propionic acid	286.0[a]	177.8[f]	0.473[h]	68	2.666[a]	68		61.9[a]	68	0.100*	54	3[d]	68	−5.4[a]
Sodium chloride brine 20% by wt.	220.8[a]		0.745[x]	68	3.80[x]	68		71.8[x]	68	0.337[x]	68	0.57[x]	68	2.6[x]
10% by wt.	215.5[a]		0.865[x]	68	2.85[x]	68		66.9[x]	68	0.343[x]	68	0.65[x]	68	20.6[x]
Sodium hydroxide and water 15% by wt.	215.0[v]		0.864[b]	68				72.4[b]	68					−5.8[b]
Sulfuric acid and water 100% by wt.	550.0[v]		0.335[b]	68	53[b]	68		114.4[v]	68			< 0.01[b]	68	50.9[b]
95% by wt.	575.0[v]		0.35[v]	68	52[v]	68		114.6[v]	68			< 0.01[v]	68	−18[v]
90% by wt.	500.0[v]		0.39[v]	68	60[v]	68		113.4[v]	68	0.22[b]	68	< 0.01[v]	68	15.0[v]
Toluene (C$_6$H$_5$CH$_3$)	231[b]	156[b]	0.404[v]	68	1.42[v]	68	30.9[b]	54.1[b]	68	0.090[b]	68	0.88[b]	68	−139[b]
Turpentine	303[a]	123[v]	0.42[b]	68	1.32[b]	68		53.9[b]	68	0.073[b]	68			
Water	212[m]	970.3[m]	0.999[m]	68	2.39[m]	68	143.5[b]	62.32[m]	68	0.348[m]	68	0.339[m]	68	32.018[m]
Xylene C$_6$H$_4$(CH$_3$)$_2$ Ortho	291[b]	149[b]	0.411[b]	68	2.01[b]	68	55.1[b]	55.0[b]	68	0.90[b]	68	0.196[b]	68	−13[b]
Meta	283[b]	147[b]	0.400[b]	68	1.52[b]	68	46.9[b]	54.1[b]	68	0.90[b]	68	0.218[b]	68	−53[b]
Para	281[b]	146[b]	0.393[b]	68	1.62[b]	68	69.3[b]	53.8[b]	68			0.227[b]	68	+56[b]
Zinc sulfate and water 10% by wt.			0.90[b]	68	3.80[a]	68		69.2[r]	68	0.337[a]	68			29.7[a]
1% by wt.			0.80[b]	68	2.54[a]	68		63.0[r]	68	0.346[a]	68			31.7[a]

†Approximate solidification temperature.
*Data source not determined.

Note:
Superscript letters indicate data source from References.

Table 3 Properties of Solids

Material Description	Specific Heat, Btu/lb·°F	Density, lb/ft³	Thermal Conductivity, Btu/h·ft·°F	Emissivity Ratio	Surface Condition
Aluminum (alloy 1100)	0.214[b]	171[u]	128[u]	0.09[n] 0.20[n]	commercial sheet heavily oxidized
Aluminum bronze (76% Cu, 22% Zn, 2% Al)	0.09[u]	517[u]	58[u]		
Alundum (aluminum oxide)	0.186[b]				
Asbestos: fiber	0.25[b]	150[u]	0.097[u]		
insulation	0.20[t]	36[b]	0.092[b]	0.93[b]	"paper"
Ashes, wood	0.20[t]	40[b]	0.041[b] (122)		
Asphalt	0.22[b]	132[b]	0.43[b]		
Bakelite	0.35[b]	81[u]	9.7[u]		
Bell metal	0.086[t] (122)				
Bismuth tin	0.040*		37.6*		
Brick, building	0.2[b]	123[u]	0.4[b]	0.93*	
Brass: red (85% Cu, 15% Zn)	0.09[u]	548[u]	87[u]	0.030[b]	highly polished
yellow (65% Cu, 35% Zn)	0.09[u]	519[u]	69[u]	0.033[b]	highly polished
Bronze	0.104[t]	530[t]	17[d] (32)		
Cadmium	0.055[a]	540[f]	53.7[b]	0.02[d]	
Carbon (gas retort)	0.17[a]		0.20[b] (2)	0.81[a]	
Cardboard			0.04[b]		
Cellulose	0.32[b]	3.4[t]	0.033[t]		
Cement (Portland clinker)	0.16[b]	120[i]	0.017[i]		
Chalk	0.215[t]	143[t]	0.48*	0.34*	about 250°F
Charcoal (wood)	0.20[t]	15[a]	0.03[a] (392)		
Chrome brick	0.17[b]	200[b]	0.67[b]		
Clay	0.22[b]	63[t]			
Coal	0.3[b]	90[t]	0.098[f] (32)		
Coal Tars	0.35[b] (104)	75[b]	0.07[b]		
Coke (petroleum, powdered)	0.36[b] (752)	62[b]	0.55[b] (752)		
Concrete (stone)	0.156[b] (392)	144[b]	0.54[b]		
Copper (electrolytic)	0.092[u]	556[u]	227[u]	0.072[n]	commercial, shiny
Cork (granulated)	0.485[t]	5.4[t]	0.028[t] (23)		
Cotton (fiber)	0.319[u]	95[u]	0.024[u]		
Cryolite (AlF₃ · 3NaF)	0.253[b]	181[b]			
Diamond	0.147[b]	151[t]	27[t]		
Earth (dry and packed)		95[t]	0.037*	0.41*	
Felt		20.6[b]	0.03[b]		
Fireclay Brick	0.198[b] (212)	112[t]	0.58[b] (392)	0.75[n]	at 1832°F
Fluorspar (CaF₂)	0.21[b]	199[v]	0.63[v]		
German Silver (nickel silver)	0.09[u]	545[u]	19[u]	0.135[n]	polished
Glass: crown (soda-lime)	0.18[b]	154[u]	0.59[t] (200)	0.94[n]	smooth
flint (lead)	0.117[b]	267[u]	0.79[r]		
pyrex	0.20[b]	139[t]	0.59[t] (200)		
"wool"	0.157[b]	3.25[t]	0.022[t]		
Gold	0.0312[u]	1208[u]	172[t]	0.02[n]	highly polished
Graphite: powder	0.165*		0.106*		
"Karbate" (impervious)	0.16[u]	117[u]	75[u]	0.75[n]	
Gypsum	0.259[b]	78[b]	0.25[b]	0.903[b]	on a smooth plate
Hemp (fiber)	0.323[u]	93[u]			
Ice: 32°F	0.487[t]	57.5[b]	1.3[b]	0.95*	
−4°F	0.465[t]		1.41*		
Iron: cast	0.12[v] (212)	450[b]	27.6[b] (129)	0.435[b]	freshly turned
wrought		485[b]	34.9[b]	0.94[b]	dull, oxidized
Lead	0.0309[u]	707[u]	20.1[u]	0.28[n]	gray, oxidized
Leather (sole)		62.4[b]	0.092[b]		
Limestone	0.217[b]	103[b]	0.54[b]	0.36* to 0.90	at 145 to 380°F
Linen			0.05[b]		
Litharge (lead monoxide)	0.055[b]	490[b]			
Magnesia: powdered	0.234[b](212)	49.7[b]	0.35[b] (117)		
light carbonate		13[b]	0.034[b]		
Magnesite brick	0.222[b] (212)	158[b]	2.2[b] (400)		
Magnesium	0.241[b]	108[u]	91[u]	0.55[n]	oxidized
Marble	0.21[b]	162[b]	1.5[b]	0.931[b]	light gray, polished
Nickel, polished	0.105[u]	555[u]	34.4[u]	0.045[n]	electroplated
Paints: White lacquer				0.80[n]	
White enamel				0.91[n]	on rough plate
Black lacquer				0.80[n]	
Black shellac		63[u]	0.15[u]	0.91[n]	"matte" finish
Flat black lacquer				0.96[n]	
Aluminum lacquer				0.39[n]	on rough plate

Notes:
1. Values are for room temperature unless otherwise noted in parentheses.

2. Superscript letters indicate data source from References.
*Data source not determined.

Table 3 Properties of Solids (*Concluded*)

Material Description	Specific Heat, Btu/lb·°F	Density, lb/ft³	Thermal Conductivity, Btu/h·ft·°F	Emissivity Ratio	Emissivity Surface Condition
Paper	0.32*	58[b]	0.075[b]	0.92[b]	pasted on tinned plate
Paraffin	0.69[t]	56[t]	0.14[b] (32)		
Plaster		132[b]	0.43[b] (167)	0.91[b]	rough
Platinum	0.032[u]	1340[u]	39.9[u]	0.054[b]	polished
Porcelain	0.18	162[u]	1.3[u]	0.92[b]	glazed
Pyrites (copper)	0.131[b]	262[b]			
Pyrites (iron)	0.136[b] (156)	310[v]			
Rock Salt	0.219[u]	136[u]			
Rubber:					
Vulcanized (soft)	0.48*	68.6[t]	0.08[t]	0.86[b]	rough
(hard)		74.3[t]	0.092[t]	0.95[b]	glossy
Sand	0.191[b]	94.6[b]	0.19[b]		
Sawdust		12[b]	0.03[b]		
Silica	0.316[b]	140[v]	0.83[t] (200)		
Silver	0.0560[u]	654[u]	245[u]	0.02[n]	polished and at 440°F
Snow (freshly fallen)		7[t]	0.34[t]		
(at 32°F)		31[t]	1.3[t]		
Steel (mild)	0.12[b]	489[b]	26.2[b]	0.12[n]	cleaned
Stone (quarried)	0.2[b]	95[t]			
Tar: pitch	0.59[v]	67[u]	0.51[v]		
bituminous		75[t]	0.41[u]		
Tin	0.0556[u]	455[u]	37.5[u]	0.06[h]	bright and at 122°F
Tungsten	0.032[u]	1210[u]	116[u]	0.032[n]	filament at 80°F
Wood: Hardwoods—	0.45/0.65[b]	23/70[z]	0.065/0.148[z]		
Ash, white		43[z]	0.0992[z]		
Elm, American		36[z]	0.0884[z]		
Hickory		50[z]			
Mahogany		34[u]	0.075[u]		
Maple, sugar		45[z]	0.108[z]		
Oak, white	0.570[b]	47[z]	0.102[z]	0.90[n]	planed
Walnut, black		39[z]			
Softwoods—	See Table 4,	22/46[z]	0.061/0.093[z]		
Fir, white	Chapter 22	27[z]	0.068[z]		
Pine, white		27[z]	0.063[z]		
Spruce		26[z]	0.065[z]		
Wool: Fiber	0.325[u]	82[u]			
Fabric		6.9/20.6[u]	0.021/0.037[u]		
Zinc: Cast	0.092[u]	445[u]	65[u]	0.05[n]	polished
Hot-rolled	0.094[b]	445[b]	62[b]		
Galvanizing				0.23[n]	fairly bright

Notes:
1. Values are for room temperature unless otherwise noted in parentheses.

2. Superscript letters indicate data source from References.
*Data source unknown.

NOTES FOR TABLES 1, 2, AND 3

[a] *Handbook of chemistry & physics*, 63rd ed. 1982-83. Chemical Rubber Publishing Co., Cleveland, OH.

[b] Perry, R.H. *Chemical engineers' handbook*, 2nd ed., 1941, 5th ed., 1973. McGraw-Hill Book Co., Inc., New York.

[c] *Tables of thermodynamic and transport properties of air, argon, carbon dioxide, carbon monoxide, hydrogen, nitrogen, oxygen and steam.* 1960. Pergamon Press, Elmsford, NY.

[d] *American institute of physics handbook*, 3rd ed. 1972. McGraw-Hill Book Co., Inc., New York.

[e] Organick and Studhalter. 1948. *Thermodynamic properties of benzene. Chemical Engineering Progress* (November):847.

[f] Lange. 1972. *Handbook of chemistry*, rev. 12th ed. McGraw-Hill Co., Inc., New York.

[g] ASHRAE. 1969. *Thermodynamic properties of refrigerants.*

[h] Reid and Sherwood. 1969. *The properties of gases and liquids*, 2nd ed. McGraw-Hill Book Co., Inc., New York.

[i] Chapter 17, 1993 ASHRAE *Handbook—Fundamentals.*

[j] *T.P.R.C. data book.* 1966. Thermophysical Properties Research Center, W. Lafayette, IN.

[k] Estimated.

[l] Canjar, L.N., M. Goldman, and H. Marchman. 1951. Thermodynamic properties of propylene. *Industrial and Engineering Chemistry* (May):1183.

[m] *ASME steam tables.* 1967. American Society of Mechanical Engineers, New York.

[n] McAdams, W.H. 1954. *Heat transmission*, 3rd ed. McGraw-Hill Book Co., Inc., New York.

[o] Stull, D.R. 1947. Vapor pressure of pure substances (organic compounds). *Industrial and Engineering Chemistry* (April):517.

[p] *JANAF thermochemical tables.* 1965. PB 168 370, National Technical Information Service, Springfield, VA.

[q] *Physical properties of chemical compounds.* 1955-61. American Chemical Society, Washington, D.C.

[r] *International critical tables of numerical data.* 1928. National Research Council of USA, McGraw-Hill Book Co., Inc., New York.

[s] *Matheson gas data book*, 4th ed. 1966. Matheson Company, Inc., East Rutherford, NJ.

[t] Baumeister and Marks. 1967. *Standard handbook for mechanical engineers.* McGraw-Hill Book Co., Inc., New York.

[u] Miner and Seastone. *Handbook of engineering materials.* John Wiley and Sons, New York.

[v] Kirk and Othmer. 1966. *Encyclopedia of Chemical Technology.* Interscience Division, John Wiley and Sons, New York.

[w] Gouse and Stevens. 1960. *Chemical technology of petroleum*, 3rd ed. McGraw-Hill Book Co., Inc., New York.

[x] *Saline water conversion engineering data book.* 1955. M.W. Kellogg Co. for U.S. Department of Interior.

[y] Timmermans, J. *Physicochemical constants of pure organic compounds*, 2nd ed. American Elsevier, New York.

[z] *Wood Handbook.* 1955. Handbook No. 72, Forest Products Laboratory, U.S. Department of Agriculture.

[aa] ASHRAE. 1976. *Thermophysical properties of refrigerants.*

ENVIRONMENTAL HEALTH

An HVAC practitioner's goal is to provide comfortable and healthy environments for clients. This chapter provides the practitioner with an introduction to the field of environmental health as it pertains to buildings. In many cases, architectural, construction, cleaning, maintenance, and other materials and activities that affect the environment are outside the control of the HVAC designer. But whenever possible, the designer should encourage features and decisions that create a healthy building environment.

EPIDEMIOLOGY AND BIOSTATISTICS

Epidemiology is the study of distributions and determinants of disease. It is the application of quantitative methods to evaluate diseases or conditions of interest. The subjects may be humans, animals, or even buildings. Epidemiology is traditionally subdivided into observational and analytical components. It may be primarily descriptive, or it may attempt to identify causal associations. Some classical criteria for the determination of causal relationships in epidemiology are consistency, temporality, plausibility, specificity, strength of the association, and dose-response relationships.

Observational studies are generally performed by defining some group of interest because of a specific exposure or risk factor. A control group is selected on the basis of similar criteria, but without the factor of interest. Observations conducted at one point in time are considered cross-sectional studies. On the other hand, a group may be defined by some criteria at a specific time, for example, all employees who worked in a certain building for at least one month in 1972. They may then be followed over time, leading to an observational cohort study.

Analytical studies may be either experimental or case-control studies. In experimental studies, individuals are selectively exposed to specific agents or conditions. Such studies are generally performed with the consent of the participants, unless the conditions are part of their usual working conditions and known to be harmless. Sometimes exposures cannot be controlled on an individual basis, and the intervention must be applied to entire groups. Control groups must be observed in parallel. Case-control studies are conducted by identifying individuals with the condition of interest and comparing risk factors between these people and individuals without that condition.

Measurement

All factors of interest must be measured in an unbiased fashion to avoid a subconscious influence of the investigator. The method of measurement should be repeatable and the technique meaningful.

Statistical methods for data analysis follow standard procedures. Tests of hypotheses are performed at a specific probability; they must have adequate power, i.e., if a sample size or a measurement difference between the factors or groups is too small, a statistically significant association may not be found even if one is present.

Results obtained in a specific situation, i.e., in a sample of exposed individuals, may be generalized to others only if they share the same characteristics. For example, it may not be legitimate to assume that all individuals have the same tolerance of thermal conditions irrespective of their heritage. Therefore, the results of studies and groups must be evaluated as they apply to a specific situation. Results obtained from a specific problem may not apply to another problem.

TOXICOLOGY

Toxicology is the study of poisons. Most researchers hold that essentially all substances may function as poisons and that only smaller or larger doses prevent them from becoming harmful. Of fundamental importance is defining which component of the chemical structure predicts the harmful effect. The second issue is defining the dose-response relationships. The definition of dose may refer to delivered dose (exposure that is presented to the lungs) or absorbed dose (the dose that is actually absorbed through the lungs into the body and available for metabolism). Measures of exposure may be quite distinct from measures of effect, because of internal dose modifiers; for example, the delayed metabolism of some poisons because of lack of enzymes to degrade them. In addition, the mathematical characteristics of a dose may vary, depending on whether a peak dose, a geometric or arithmetic mean dose, or an integral under the dose curve is used.

Because humans often can not be exposed in experimental conditions, most toxicological literature is based on animal studies. Recent studies suggest that it is not easy to extrapolate between dose level effects from animals to man. Isolated animal systems, such as homogenized rat livers, purified enzyme systems, or other isolated living tissues, may be used to study the impact of chemicals.

INDUSTRIAL HYGIENE

Industrial hygiene is the science of exposure characterization. It involves measurement, interpretation of the measurements, and setting standards. Measurement methods are specified for various environmental characteristics by several organizations. For example, in the United States the National Institute for Occupational Safety and Health (NIOSH) specifies acceptable measurement techniques (P & CAM methods), as does the American Conference of Governmental Industrial Hygienists (ACGIH). The Occupational Safety and Health Administration (OSHA) may specify which method should be used for some standards.

The preparation of this chapter is assigned to the Environmental Health Committee.

Standards

Standards are set by a variety of agencies. The ACGIH reviews data on a yearly basis and publishes a Documentation of Recommended Threshold Limit Values (TLV), levels that are frequently agreed on as reasonably safe. NIOSH has developed criteria documents for specific standards, generally recommending exposure limits below those of the ACGIH. OSHA sets Permissible Exposure Limits (PEL), which are the only enforceable standards in the United States. However, OSHA limits were developed for 8-h exposures in industrial settings and may exceed levels acceptable to occupants in office, residential, and other spaces for various reasons.

Sampling Strategies

Exposure to something may be measured for various purposes. Ascertaining average exposures in specific areas requires a sampling strategy, *i.e.*, the selection of representative samples. On the other hand, sampling for worst-case exposure requires identifying the potentially highest exposures and measuring those.

In short-term exposures likely to cause health effects and work processes likely to lead to short-term overexposures, sampling periods may have to be short or measurements performed with continuous logging equipment. For example, noise-induced high-frequency hearing loss from explosions may be missed if noise exposures are integrated over 8-h periods. On the other hand, short of thermal burns, peak radiant temperatures may not lead to harmful effects if they are kept relatively short.

Generally, if a random exposure is within 40% of a specific danger level, overexposures are likely. The larger the number of samples and the smaller the standard deviations or the lower the geometric mean of exposures, the less likely overexposures are to occur.

AIR QUALITY

Airborne materials include gases and particulates that may be generated by occupants and their activities in a space. Airborne materials may (1) occur from outgassing and/or shedding of building materials and systems, (2) originate in outside air, and/or (3) be from building operating and maintenance programs and procedures. Airborne particles include bioaerosols, asbestos, man-made mineral fibers, and silica, and gaseous contaminants include radon and soil gases as well as volatile organic compounds.

BIOAEROSOLS

Bioaerosols are airborne microbiological particulate matter derived from viruses, bacteria, fungi, protozoa, mites, pollen, and their cellular or cell mass components. Bioaerosols are everywhere in indoor and outdoor environments.

In the past, much attention was given to saprophytic fungi (nutrients that are derived from dead organic material), probably because of readily available sampling and analytical methods for this type of microorganism. Viruses, rickettsia, chlamydia, protozoa, and many pathogenic fungi and bacteria are more difficult to culture, and air sampling methodology for these agents is less well known (ASTM 1990).

Sources

Microorganisms break down the complex molecules found in dead organic materials and recycle minerals and macromolecules to simple substances such as carbon dioxide, water, and nitrates. These components are then used by green plants. Thus, the presence of saprophytic bacteria and fungi in the soil and in the atmosphere is a normal occurrence. For example, the spores of

Cladosporium, a fungus commonly found on dead vegetation, are almost always found in outdoor air. *Cladosporium* spores are also found in indoor air, depending on the amount of outdoor air that infiltrates into interior spaces or is brought into the HVAC system. Bacteria that are saprophytic on human skin (*Staphylococcus*) and viruses (Influenza A) that are obligate parasites in the human respiratory tract, are shed from humans and are thus commonly present in indoor environments.

Although microorganisms are normally present in indoor environments, the presence of abundant moisture and nutrients in interior niches amplifies the growth of some microbial agents to the extent that the interior environment is microbiologically atypical. Thus, certain types of humidifiers, water spray systems, and wet porous surfaces can be reservoirs and sites for growth of fungi, bacteria (including actinomycetes), protozoa, or even nematodes (Strindehag *et al.* 1988, Arnow *et al.* 1978, Morey *et al.* 1986, Morey and Jenkins 1989). Excessive air moisture (Brundrett and Onions 1980) and floods (Hodgson *et al.* 1985) may cause the proliferation of microorganisms indoors. Turbulence associated with the start-up of air-handling unit plenums may also elevate concentrations of bacteria and fungi in occupied spaces (Yoshizawa *et al.* 1987).

Health Effects

The presence of microorganisms in indoor environments may cause infective and/or allergic building-related illnesses (Morey and Feeley 1988, Burge 1989). Some microorganisms under certain conditions may produce volatile chemicals (Hyppel 1984) that are malodorous or irritative (Holmberg 1987), thus contributing to the development of what is called the sick building syndrome.

Legionellosis refers to two distinct types of illnesses—Legionnaires' disease and Pontiac fever. Legionnaires' disease is a multisystem disease, including pneumonia, with a fatality rate of about 15%. Hospitalization and appropriate antibiotic therapy are required for effective treatment. *Legionella pneumophila*, a recently discovered bacterium (the genus name *Legionella* is derived from the American Legion Convention in Philadelphia in 1976), is the microorganism closely associated with infectious disease in buildings. Cooling towers, evaporative condensers, and domestic water service systems (particularly hot water) all provide water and nutrients for amplification of microorganisms such as *Legionella*. Growth of microbial populations to excessive concentrations is generally associated with inadequate preventative maintenance of these systems. Pontiac fever is milder and non-pneumonic. Recovery usually occurs without hospitalization or use of antibiotics.

There are 31 species and 50 serogroups of *Legionella* identified to date. These gram-negative bacteria are naturally occurring and associated with nonsaline surface waters. Many of these species have not been associated with human disease; however, the *Legionella pneumophila* serogroup 1 is most frequently isolated from nature and most frequently associated with disease. Because *Legionella* requires special nutrients for growth and does not produce resistant endospores, this microorganism is difficult to recover from the air. Collection and analysis of water samples is usually relied on for the detection of *Legionella*, *e.g.*, from cooling towers. Although some strains of the bacterium have not been associated with human disease, probably because of their rarity, all *Legionella* strains should be considered potentially pathogenic.

Outbreaks of infective illness in the indoor air may be caused by other types of microorganisms, such as viruses. For example, most of the passengers in an airline cabin developed influenza following exposure to one acutely ill person (Moser *et al.* 1979). In this case, the plane had been parked on a runway for several hours with the ventilation system turned off.

...smosis, an infective illness caused by the fungus *H...* *capsulatum*, has occurred (very rarely) as a building-r... ...ss among individuals involved in the removal of bat d... abandoned buildings (Bartlett *et al.* 1982) and among c... cleaners. Presumably asexual spores (conidia) from t... were inhaled by workers who removed droppings v... quate respiratory protection.

...anisms may cause building-related illness in indoor e... ...ts by affecting the immune system. Thus, allergic re... illness may develop due to the inhalation of particulat... ...aining microorganisms or their components, such as spor... ...zymes, and cell wall fragments. Numerous cases of allergic respiratory illness (humidifier fever, hypersensitivity pneumonitis) report affected people manifesting acute symptoms such as malaise, fever, chills, shortness of breath, and coughing (Hodgson and Morey 1989, Edwards 1980, Morey 1988). In buildings, these illnesses may occur as a response to microbiological contaminants originating from HVAC system components, such as humidifiers and water spray systems, or other mechanical components that have been damaged by chronic water exposure (Hodgson *et al.* 1985, 1987). Affected individuals usually experience relief only after having left the building for an extended period in contrast to occupants with sick building syndrome, where relief is relatively rapid.

Microbiological contamination is reported to be important in about 5% of the indoor air quality evaluations carried out by NIOSH (Wallingford and Carpenter 1986). This is probably an understatement because neither microorganisms nor the possible importance of microbial volatiles, endotoxins, mycotoxins, nonviable microbial particulates, and viruses may have been investigated. Several recent studies suggest that microorganisms in indoor air quality issues are more important than previously suspected (Burge *et al.* 1987, Brundage *et al.* 1988).

Sampling

The principles of sampling and analysis for microorganisms are reviewed in the ACGIH's, *Air Sampling Instruments Manual* (Chatigny 1983). The ACGIH Bioaerosols Committee has developed assessment guidelines (ACGIH 1989) for the collection of microbial particulates.

Preassessment. Sampling for microorganisms should be undertaken when medical evidence indicates the occurrence of diseases such as humidifier fever, hypersensitivity pneumonitis, allergic asthma, and allergic rhinitis. A walk-through examination of the indoor environment for visual detection of possible microbial reservoirs and amplification sites is recommended before sampling. If a microbial reservoir/amplifier is visually identified, it is useful to obtain a bulk or source sample from the reservoir or amplifier. Also, removal of clearly identified reservoirs and amplifiers are preferable to complicated and costly air sampling procedures.

Air sampling. The same principles that affect the collection of an inert particulate aerosol also govern air sampling for microorganisms (ACGIH 1989). However, it is essential that the viability and/or antigenicity of the microbial particulate be protected. In general, culture plate impactors, including multiple- and single-stage sieve devices as well as slit-to-agar samplers, are most useful in office environments where low concentrations of fungi and bacteria are expected. However, not all microbes will grow on the same media; thus, liquid impingement subculturing may be more suitable. Filter cassette samplers are useful for microorganisms or components of microorganisms (*e.g.*, endotoxin) that are not affected by desiccation.

Data interpretation. Rank order assessment is a method used to interpret air sampling data for saprophytic microorganisms (ACGIH 1989). Individual taxa are listed in descending order of abundance for a complainant indoor site and for one or more con-

Table 1 Example Case of Airborne Fungi in Building and in its Outdoor Air

Location	cfu/m³	Rank Order Taxa
Outdoors Complaint	210	Cladosporium>Fusarium>Epicoccum>Aspergillus
Office #1 Complaint	2500	Tritirachium > Aspergillus > Cladosporium
Office #2	3000	Tritirachium > Aspergillus > Cladosporium

1. cfu/m³ = Colony-forming units per cubic metre of air.
2. Culture media was malt extract agar (ACGIH 1989).

trol locations. The predominance of one or more taxa in the complainant site, but not in the control sites, suggests the presence of an amplifier (for that taxa). In the example in Table 1, Tritirachium and Aspergillus species (fungi) were the predominant taxa present in complainant locations in an office building, where Cladosporium and Fusarium dominated outdoor collections. In this case, Tritirachium and Aspergillus were being amplified in the building.

Guidelines

At present, numerical guidelines for bioaerosol exposure in indoor environments are not available for the following reasons (Morey 1990):

- Incomplete data on concentrations and types of microbial particulate indoors, especially as affected by geographical, seasonal, and type-of-building parameters
- Absence of data relating bioaerosol exposure to building-related illness
- Enormous variability in kinds of microbial particulate including viable cells, dead spores, toxins, antigens, and viruses
- Large variation in human susceptibility to microbial particulate, making estimates of health risk difficult

However, even in the absence of numerical guidelines, bioaerosol sampling data can be interpreted based on several considerations:

- Rank order assessment of the kinds (genera species) of microbial agents present in complainant and control locations (ACGIH 1989)
- Medical or laboratory evidence that a building-related illness is caused by a microorganism (ACGIH 1989)
- Indoor/outdoor concentration ratios for various saprophytic microbial agents (Morey 1989, 1990)

For a microorganism to cause a building-related illness, it must be transported in sufficient dose to the breathing zone of a susceptible occupant. Thus, the concepts of reservoir, amplifier, and disseminator need to be understood and considered in interpreting data.

ASBESTOS

The term asbestos refers to a group of naturally occurring silicates that occur in fiber bundles with unusual tensile strength and fire resistance. They are found in two main forms—serpentine, containing only chrysotile, and amphiboles, containing amosite, crocidolite, tremolite, anthrophyllite, and actinolite. Exposure can occur in four settings—production (mining and milling), materials production (insulation, brake linings, etc.), construction, and removal.

Health Effects

Groups of health effects are widely recognized. All clearly follow a dose-response pattern, *i.e.*, the more intense and the longer the exposure, the greater the likelihood of the disease.

Asbestosis. Asbestosis refers to a chronic disease of the lung tissue, leading to scarring, and in end stages, respiratory failure because of inadequate ability to transport oxygen across the lung tissue. This disease is associated not only with disability, but it also gives rise to elevated lung cancer rates. Exposure levels have been reduced substantially in the last 30 years, so new cases of this disease are seen only rarely in developed countries.

Lung cancer. Lung cancer is a chronic, progressive, generally incurable disease; less than 13% of individuals with the disease survive after 5 years. Many risk factors are recognized, although over 85% of cases are attributed to cigarette smoking. Occupational exposure to asbestos is associated with elevated rates of lung cancer in occupations and industries where asbestos is found. Some industries and trades with previously uncontrolled exposures had a 10 to 20-fold higher rate of lung cancer than those found in the general population; more recently, some building trades with regular asbestos exposure have only a 50% elevation. Where adequate protection exists, no excess is seen.

Mesothelioma. Asbestos exposure is associated with a rare tumor of the pleura (lining of the lung) and peritoneum known as mesothelioma. It is a sentinel event for asbestos exposure, *i.e.*, if the disease is seen, asbestos exposure is almost certainly to have occurred, and other people in the same group may be at risk. The disease is generally progressive, leading to death within a year.

There is some controversy about the relative ability of various forms of asbestos to cause disease. Some researchers find that chrysotile, uncontaminated with amphiboles, is not associated with mesothelioma.

Other diseases. A variety of other diseases have been associated with asbestos exposure, including cancers of the larynx and gastrointestinal tract (esophagus, stomach, colon, and rectum). This relationship to other diseases has not been clearly established, as each has other risk factors and is not strongly associated with asbestos exposure.

Sampling and Assessment

Exposure assessment follows several guidelines, including an inspection procedure and sample collection. Environmental monitoring involves both visual assessment and, where appropriate, bulk sampling, including core drilling, and air sampling.

Asbestos is identified through phase contrast or transmission electron microscopy after collection on filters, or through polarized light microscopy.

Guidelines

Asbestos use in the United States is strictly regulated through federal agencies. OSHA has jurisdiction in the workplace, with an overall permissible exposure limit set the same for both general industry and construction. The Environmental Protection Agency (EPA) regulates asbestos use and exposure in a variety of other settings, including schools and residential environments. Use guidelines are provided, including a document on safe handling and abatement/remediation.

MAN-MADE MINERAL FIBERS

Man-made mineral fibers (MMMF), also known as man-made vitreous fibers (MMVF), are synthetic, amorphous, noncrystalline vitreous structures. They are used primarily as reinforcement, textiles, and insulation materials. Their specific composition varies, some being made from naturally occurring products, such as rock wool from rocks, others from slags. Table 2 presents an overview of their general structure. Even though their source material defines their chemical composition, they have similar physical characteristics.

Since they are amorphous, they do not split lengthwise, as do crystalline fibers such as asbestos. They break across their diameter, leading to progressively shorter pieces. Under the in-

Table 2   Types of Man-Made Mineral Fibers

Type	Source Material
Continuous fibers	Glass
Insulation wool	Glass wool
	Rock wool
	Slag wool
Special purpose fibers	Glass microfiber
Refractory fibers	Ceramic
	Others

fluence of high temperatures, refractory fibers may change their chemical/physical characteristics to crystalline silica.

Health Effects

Three broad categories of possible health effects from man-made mineral fibers are possible, although they are not all likely.

Dermatitis. Fiberglass dermatitis is well recognized. Itching and erythema (reddening of skin) after dermal exposure among installers is common. Two mechanisms have been demonstrated. Most cases are due to simple irritation; rarely are true allergies (usually from the coatings applied to the fibers for binding etc.) seen. The diagnosis is easily made if complaints stop after showering. Allergic contact dermatitis requires patch-testing for diagnosis. No long-term health effects have been described.

Microbial contamination. The potential for microbial contamination of porous materials (duct liners and filters) with molds and spores has been widely discussed. Some conditions (moisture, nutrients, temperature) can amplify microbial populations. Two consequences are recognized. First, exposure to specific microbial agents may sensitize some individuals. While individuals (with atopy) are more likely to become sensitized, many nonatopic individuals suffer similar problems. Second, microbial agents may give off specific products, such as aldehydes or complex elaborated toxins, which have more complex health effects.

The relationship between filtration, water incursion, and microbial growth are not clear. Some researchers consider inadequate filtration a primary cause of contamination.

Lung disease. Several surveys have examined the relationship between MMMF exposure and interstitial lung disease, the kind of disease associated with asbestos exposure. Weill *et al.* (1983) and Hughes *et al.* (1993) found no such effects. Kilburn *et al.* (1992) measured some radiographic abnormalities, but these involved persons also exposed to asbestos and welding fumes, both associated with lung disease. At present, there is no evidence that MMMF causes interstitial lung disease.

Cancer and mesothelioma. Man-made mineral fibers have been suspected of leading to elevated cancer rates (WHO 1988). Several mortality studies, in the United States and Europe, have suggested a mild increase, ranging from 10 to over 50%, among MMMF workers. However, in these studies, other factors (asbestos and smoking) were present and probably account for the apparent increase in death rate. When smoking habits were examined, it was found that MMMF workers in the study smoked more than the general control population. When investigators statistically controlled for the effect of smoking, the excess disappeared. Another factor is the asbestos fibers found in the autopsies of the MMMF workers. This led to the conclusion that the excess death rate might also be due to asbestos fibers, which were usually used in the same plants.

One consideration is that most MMMF are simply too large to reach the lungs, as they are filtered out in the upper airways. However, because the man-made mineral fiber can break into shorter pieces, the smaller fragments may reach the airways and lung tissue. Further, refractory fibers fired at a high temperature may have relatively small aerodynamic diameters, enabling them to reach and lodge in the lungs, and not to decay as rapidly. Still, at present there is no clear evidence that any forms of MMMF are carcinogenic in humans. However, the International Agency for Research on Cancer and the EPA have classified mineral wool and rock wool as possible human carcinogens. This classification appeared before a case-control study which suggested that lung cancer associated with rock wool and mineral wool was more strongly associated with smoking than with occupational exposure to MMMF.

Exposure

Some data exist on exposure likely to occur in buildings. Background levels are almost uniformly below 0.1 fibers per millilitre (f/mL). Substantial exposures occur primarily in production facilities. Studies on fiberglass batt application measured exposure in the range of 0.4 to 1.0 f/mL. Installation of blown insulation into attics developed exposures of 1.5 to 2.0 f/mL; installing acoustic ceiling tile 0.0028 f/mL; and installing duct insulation 0.05-0.6 f/mL. Further exposure assessment of MMMF installation is under consideration.

Man-made mineral fibers can be measured gravimetrically or counted visually using light or electron microscopy.

Guidelines and Environmental Criteria

OSHA regulates man-made mineral fibers as nuisance dust, at 5 mg/m^3 for general industry and 15 mg/m^3 for construction, specifying gravimetric measurement methods as adequate.

NIOSH proposes a recommended exposure level of 3 f/mL for respirable fibers and a level of 5 mg/m^3 for total fibrous glass.

The ACGIH recommends 10 mg/m^3. Sweden has set a level of 1 f/mL for respirable fibers, and Denmark has set a level of 2 f/mL.

SILICA

Silica consists of silicon dioxide, which is abundant in the earth's crust and is used in the manufacture of glass, refractories, abrasives, buffing and scouring compounds, and lubricants. Its biological effects depend on the grade of the silica. Increasing temperatures in the processing of silica leads to its refinement and increased grade. Higher grades are progressively tridymite, critoblite, and quartz. Diatomaceous earth is considered relatively inert.

Health Effects

Silica is associated with three forms of silicosis—acute, accelerated, and chronic (nodular). They are characterized by an immunologic action of silica in the lungs, leading to progressive tissue destruction and immunological abnormalities. All three are irreversible diseases, although chronic silicosis generally does not progress to cause symptoms. Accelerated and acute silicosis do progress and may be fatal.

Guidelines

Standards have been established separately by OSHA and MSHA. Environmental criteria in fibers per millilitre include:

	OSHA	ACGIH
Amorphous silica	5	10
Diatomaceous earth	5	10
Crystalline		
Cristobalite	0.05	
Quartz	0.05	0.1
Tridymite	0.05	0.05
Tripoli	0.05	0.1

Sampling may occur on particulate filters. Quantification is through low-temperature ashing or X-ray diffraction spectrometry.

RADON AND OTHER SOIL GASES

Radon is a naturally occurring, chemically inert, odorless, tasteless radioactive gas. It is produced from the radioactive decay of radium, which is formed, through several intermediate steps, from the decay of uranium. Since radium and uranium are ubiquitous elements in rock and soil, radon is widely found in the natural environment. Because Radon-222, the decay product of Radium-226, is not chemically bound or attached to other materials, it can move through very small spaces, such as those between particles of soil and rock, and enter indoor environments (Nazaroff *et al.* 1988, Tanner 1980). Additional, but secondary, sources of indoor radon include groundwater and radium-containing building materials.

Radon gas enters buildings through cracks or openings such as sewer pipe and sump pump openings, cracks in concrete, and wall-floor joints. The amount of radon entering and subsequent indoor concentration distribution depends on several factors, including the concentration of radium in the surrounding soil or rock, the soil porosity and permeability, and the air pressure differential between the building and the soil or between various indoor spaces which may result from the stack effect, operation of exhaust fans, or operation (or lack of) HVAC equipment.

Average concentrations of radon in homes in the United States is about 55 Bq/m^3 (1.5 pCi/L), with a log normal concentration distribution. Although approximately 93% of homes have concentrations three times below this average (150 Bq/m^3)(4 pCi/L), homes with higher concentrations are not uncommon, and some have been found with concentrations exceeding 4000 Bq/m^3 (110 Pci/L). While several sources of radon may contribute to average radon levels in homes, pressure-driven flow of soil gas constitutes the principal source. With the exception of a few homes with private water wells, where radon in water can be important, soil gas constitutes the only significant source in homes with elevated concentrations (Nazaroff 1987, Akerblom 1984, DSMA 1983).

Health Effects

The radioactive decay of radon produces a series of radioactive isotopes of polonium, bismuth, and lead. Unlike their chemically inert radon parent, these progeny are chemically active and can adhere to dust particles or other surfaces and can, on inhalation, also attach to the airways of the lung. Two of these progeny are alpha emitters, and the passage of these particles through the tissue lining the lung can cause lung cancer (Samet 1989). Thus, the health effects associated with radon are mainly due to exposures to these radon decay products, and the amount of risk is assumed to be directly related to the total exposure. Studies of uranium and other underground mines form the principal basis for knowledge about health risks due to radon. These studies have also identified synergistic effects between smoking and radon progeny exposure, so that smokers will have a greater risk of lung cancer than non-smokers if both are exposed to the same radon concentration.

Measurements

Indoor concentrations of radon can vary hourly, daily, and seasonally, in some cases by as much as a factor of 10 to 20 on a daily basis (Scott 1989, Turk 1990). For example, measurements made during a mild spring may underestimate the annual average level because of ventilation from open windows or the operation of HVAC equipment (*e.g.*, economizer operation, which increases outdoor air ventilation rates). Similarly, indoor levels during a cold winter may be higher than average because the building is sealed and the outdoor air ventilation rates are minimized. Thus, longer-term measurements (6 months to 1 year) made during normal use generally provide more reliable estimates of the average indoor concentration.

Two techniques widely used for homeowners' measurements are the charcoal canister (short-term) and the alpha track (long-term) methods. Generally, short-term measurements should only be used as a screening technique to see if a long-term measurement is necessary. The great uncertainties in measurement accuracy (which, for concentrations in the range found in most homes, is approximately 50%), as well as the natural variability of radon concentrations should be considered.

Controls

Exposure to indoor radon may be reduced by (1) preventing radon entry, or (2) removing or diluting radon or radon decay products after entry. The first approach is more desirable, especially if it does not require large maintenance or energy expenditures. The most effective measures are generally those that limit soil gas entry into the building (Turk 1991). One of the most common techniques is subslab depressurization, in which the pressure below the floor slab is reduced so that air between the house substructure and the soil tends to flow out of rather than into the building. Other techniques, such as basement pressurization, have been attempted with some success.

Although sealing various parts of the building substructure has been attempted, it is often not completely effective by itself, since some openings in the building shell may not be accessible or new openings can develop with time. Because furnaces or HVAC systems can contribute to the depressurization of buildings, the location and tightness of supply and return ducts are important, especially in new construction.

Soil-Gas Transportation of Other Pollutants

Evidence from some studies suggest that contaminants other than radon may enter buildings from the surrounding soil, although the extent of such occurrences is unknown. For example, methane migrating away from landfills through unsaturated soils has reached explosive levels in nearby buildings. In other instances, potentially toxic and carcinogenic, volatile organic compounds, present in soil gas as contaminants, have been transported into buildings. These compounds may pose health risks in indoor environments, depending on their concentrations and the length of exposure. Some specific examples include the entry of chlorinated solvents into houses near landfills (Wood and Porter 1987, Hodgson *et al.* 1988, Garbesi *et al.* 1989), the entry of gasoline vapors from leaking underground fuel tanks into buildings (Kullman and Hill 1990), and the detection of pesticides in houses in which the underlying soil had been treated (Livingston and Jones 1981, Wright and Leidy 1982).

Any volatile organic compounds present in soil gases in the vicinity of buildings may enter those buildings under certain, as yet undefined, soil and building conditions. Therefore, their presence, or possible presence, should be considered in the design, construction, and operation of buildings—particularly in the vicinity of inactive landfills or hazardous waste sites. Buildings in these areas should be made (1) to minimize substructure leakage, to prevent substructures from being depressurized by the operation of HVAC systems, and (2) to seal duct systems in crawl spaces or beneath slab foundations against the possible intrusion of gases. Techniques installed to prevent entry of radon gas will also prevent entry of other soil gases.

VOLATILE ORGANIC COMPOUNDS

Volatile organic compounds (VOCs) are air pollutants found in all nonindustrial environments. After ventilation, VOCs are probably the first concern when diagnosing an IAQ problem (Morey and Singh 1991). VOCs are organic compounds with vapor pressures greater than about 10^{-3} to 10^{-4} mm Hg (torr). This

Table 3 Classification of Indoor Organic Pollutants (WHO 1989)

Description	Abbreviation	Boiling Point Range[a], °F
Very volatile (gaseous) organic compounds[b]	VVOC	< 32 to 120-212
Volatile organic compounds	VOC	120-212 to 460-500
Semivolatile organics (pesticides, polynuclear aromatic compounds, plasticizers)	SVOC	460-500 to 720-750

[a]Polar compounds and higher molecular weight VOCs appear at the higher end of each boiling-point range.
[b]Alcohols, aldehydes, olefins, terpenes, aliphatic aromatic hydrocarbons, chlorinated hydrocarbons.

includes 4 to 16 carbon alkanes, chlorinated hydrocarbons, alcohols, aldehydes, ketones, esters, terpenes, ethers, aromatic hydrocarbons (such as benzene and toluene), and heterocyclics. The entire range of VOCs has been categorized as indicated in Table 3 (WHO 1989). No sharp limits exist between the categories, which were defined by boiling-point ranges.

Even in geographical areas with significant outdoor pollution, indoor total VOC concentrations 2 to 10 times higher have been measured (Wallace 1987). In new office buildings, the total VOC concentration at the time of initial occupancy can be 50 to 100 times that present in outdoor air. The variety of VOCs found in indoor air is almost always greater than that in outdoor air, largely because of the large number of possible sources in indoor environments. Although VOCs in indoor, nonindustrial environments are generally present at concentrations considered to be elevated with respect to the outdoor air, the absolute concentrations of specific VOCs is almost always 100 to 1000 times less than that considered significant in industrial workplaces. However, standards for the industrial workplace are higher than would be appropriate for the general population, which includes the elderly, children, and people who are more sensitive to VOCs than the "average" industrial worker.

Sources

Studies on the sources of VOCs in nonindustrial indoor environments are confounded by the variable nature of emissions from potential sources (Berglund 1988). Emissions from some sources, such as building materials, are continuous and regular, whereas emissions from other sources (*e.g.*, paints used in renovation work) are continuous, but irregular in nature. Emission of VOCs from yet other sources are intermittent and regular (*e.g.*, VOCs in combustion products from gas stoves or cleaning products), or intermittent and irregular (*e.g.*, VOCs from carpet shampoos) (Morey and Singh 1991).

While building and furniture materials are known to emit VOCs (Molhave 1986), ventilation may also transport outdoor pollutants to the indoor environment; so the ventilation system itself may be a source of VOCs (Molhave and Thorsen 1990). Human activity, including maintenance, cleaning, and cooking, is another potential source, as are human metabolism and smoking. Additional sources include liquid process copy machines, printing machines, glue, spray cans, cosmetics, and so forth (Miksch 1982). Yet other sources include chloroform in water; tetrachloroethylene and 1,1,1-trichloroethane from cleaning solvents; and methylene chloride from paint strippers, fresheners, cleaners, and polishers. Formaldehyde, a major VOC, has many sources, but pressed wood products appear to be the most significant.

Normally 50 to 300 volatile organic compounds are found in air samples from most nonindustrial environments using standard sorbent sampling techniques. Each compound seldom exceeds a concentration of about 50 $\mu g/m^3$, 100 to 1000 times lower than relevant occupational threshold limit values (TLVs) (ACGIH 1988). An upper extreme average total concentration of all volatile organic compounds in normally occupied houses is approximately 20 mg/m^3.

Health Effects

Adverse health effects potentially caused by VOCs in nonindustrial indoor environments fall into three categories: (1) irritant effects, including perception of unpleasant odors and mucous membrane irritation; (2) systemic effects, such as fatigue and difficulty concentrating; and (3) toxic chronic effects, such as carcinogenicity (Girman 1989).

The chronic adverse health effects due to VOC exposure are of concern because some VOCs commonly found in indoor air are human (benzene) or animal (chloroform, trichloroethylene, carbon tetrachloride, p-di-chlorobenzene) carcinogens. Some other VOCs are also genotoxic. Theoretical risk assessment studies indicate that chronic exposure risk due to VOCs in residential indoor air is greater than that associated with exposure to VOCs in the outdoor air or in drinking water (McCann 1987, Tancrede 1987).

A biological model for acute human response to low levels of VOCs indoors is based on three mechanisms—sensory perception of the environment, weak inflammatory reactions, and environmental stress reaction (Molhave 1991). The observations are based on limited investigations, but they do, however, indicate that indoor VOCs may cause discomfort due to odors, irritative symptoms in the eyes, nose, and throat, and headaches at levels well below TLVs. The syndrome may include other related effects, such as decreased productivity and performance, although such effects have not yet been positively identified.

Molhave (1991) tentatively concluded that no irritation should be expected as a result of exposure to TVOC (total VOC) below about 0.2 mg/m^3 for normally encountered indoor air. Outdoor air concentrations are about 0.1 mg/m^3 or below. At concentrations higher than about 3 mg/m^3, complaints occurred in all investigated buildings where occupants experienced symptoms. In controlled exposure experiments, odors are significant at 3 mg/m^3. At 5 mg/m^3, objective effects were seen in addition to the subjective irritation. Exposures for 50 min to 8 mg/m^3 of synthetic mixtures of 20 VOCs lead to significant irritation of mucous membranes in the eyes, nose, and throat.

Sampling

Existing sampling methods for VOCs in industrial workplaces are not readily adaptable to nonindustrial indoor studies. Methods developed for industrial workplaces are often bulky, noisy, and validated for about 0.1 the applicable threshold limit value (TLV). Because concentrations of VOCs found in nonindustrial indoor air are usually 100 or more times lower than industrial TLVs (Lewis 1988), often more sensitive sampling and analytical methods are necessary. Proprietary sorbents and activated charcoal, combined with analysis involving gas chromatography and mass spectrometry, are extensively used for VOC sampling and analysis (Hodgson *et al.* 1989).

Variables such as building age and outdoor air ventilation must be considered when interpreting VOC sampling results. In new office buildings, the indoor/outdoor concentration ratio of total VOCs may be great (Sheldon 1988a, 1988b). With adequate outdoor air ventilation, these ratios often fall to less than 5 to 1 after 4 or 5 months of aging. In older buildings with continuous, regular, and irregular emission sources, indoor/outdoor ratios of total VOCs may vary from nearly 1 when maximum amounts of outdoor air are being used in HVAC systems to greater than 10 during winter and summer months when minimum amounts of outdoor air are being used (Morey and Jenkins 1989, Morey and Singh 1991).

Reducing VOC Exposures

Approaches to reducing indoor exposures to VOCs include the following:

- Use low-emitting products indoors.
- Substitute products where possible to reduce emissions.
- Use products according to manufacturers' directions.
- Increase general ventilation, although this may not be energy-efficient and may not be effective for some sources, such as building materials, where increasing the ventilation six times only decreases the VOC by 50%. For wet process photocopiers (this can be the major source in a building and account for 80 to 90% of TVOC), the effect of ventilation is directly proportional, *i.e.*, doubling the ventilation, halves the VOCs.
- Use volatile organic products when occupant density is lowest.
- Install local ventilation, *i.e.*, local exhaust ventilation, near photocopiers, printers, and other point sources.
- Store organics in well-ventilated spaces.
- Do not store large quantities of products containing volatile organics, especially those used infrequently.

THERMAL ENVIRONMENT

The thermal environment affects human health in that it affects body temperatures, both internally and externally (of the skin). For the normal, healthy adult, internal or core body temperatures are very stable, with variations seldom exceeding 1°F. The internal temperature of a resting adult, measured orally, averages about 98.6°F; measured rectally, it is about 1°F higher (Guyton 1972). The temperature of the core is carefully modulated by an elaborate physiological control system. In contrast, the temperature of the skin is basically unregulated and experiences variations from about 88 to 96°F in normal environments and activities.

Range of Healthy Living Conditions

The environmental conditions for thermal comfort are those that minimize the physiological effort of internal temperature regulation. For a resting person wearing trousers and a long-sleeved shirt, comfort corresponds to a still air environment of about 75°F. A zone of comfort extends about 3°F above and below this optimum level. Physiological efforts to increase heat loss and maintain internal temperatures include increased blood flow (vasodilation) to the skin and sweating. Behavioral efforts that might be initiated by the individual to help in this regard include reducing clothing, reducing activity, and seeking cooler conditions. Physiological and behavioral efforts to reduce and balance heat loss include decreasing blood flow (vasoconstriction) to the periphery, increased metabolism, increased clothing and activity, and/or finding shelter from the cold. Involuntary increases in metabolism are achieved by increasing muscle tone that may eventually result in shivering. Some of the human responses to the thermal environment are summarized in Figure 1.

Cardiovascular and other diseases and the inevitable processes of aging can reduce the capacity or ability of physiological systems to balance heat losses with production. Thus, such persons are less able to deal with thermal challenges and deviations from comfort conditions. Metabolism decreases with aging. This decline in average heat production comes from a declining basal metabolism together with decreased physical activity. The metabolism at 80 years of age is about 20% less than that at 20 years. Therefore, in a given environment, an older person is likely to have a lower core and skin temperature. As a result, persons in their eighties prefer to have the air temperature about 3°F warmer than persons in their twenties. Older people may have reduced capacity to sweat and to increase and decrease skin blood flow and are, therefore, likely to experience greater strain in warm and hot conditions and are less able to prevent the fall of body temperature in cool and cold conditions.

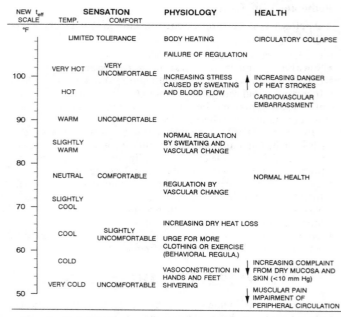

Fig. 1   Related Human Sensory, Physiological, and Health Responses for Prolonged Exposure

Hypothermia

Hypothermia refers to conditions where body temperatures are below the normal range. A temperature decrease reduces the chemical activity of the cells and metabolism. Reduced activity of the nervous system causes sleepiness, irritability, and affects judgement. While mild hypothermia may not cause disease, it impairs the activity and mobility of the immune system, thereby increasing the susceptibility to infections and colds. Thermoregulation is impaired below core temperatures of 94 °F. Consciousness is lost at about 92 °F, and thermoregulation fails completely at 86 °F. Cardiac arrhythmia can develop at temperatures around 82 °F and may lead to fibrillation and death.

In very cold conditions, the skin and peripheral areas of the body can freeze (frostbite). If the tissue is thawed immediately, no permanent damage may result. However, prolonged freezing can permanently damage local circulation and cause tissue loss.

Hyperthermia

Hyperthermia refers to the condition where body temperatures are above normal. A body temperature increase of 4 °F above the normal range does not generally impair function. For example, it is not unusual for runners to have rectal temperatures of 104 °F after a long race.

An elevated body temperature increases metabolism. However, central nervous system function deteriorates at temperatures above 106 to 108 °F and convulsions may occur. At this level, cells may be damaged. This is particularly dangerous for the brain, since lost neurons are not replaced. Thermoregulatory functions of sweating and vasodilation cease at about 110 °F, after which body temperatures may rise very rapidly if external cooling is not imposed.

Seasonal Patterns

Ordinary seasonal changes in temperate climates affect the prevalence of illness. Most acute and several chronic diseases vary in frequency or severity with time of year; some are present only in certain seasons. Minor respiratory infections, such as colds and sore throats, occur mainly in fall and winter; more serious infections, such as pneumonia, have a somewhat shorter season in

winter. Intestinal infections, such as dysentery and typhoid fever, are more prevalent in summer. This is also true of encephalitis, endemic typhus, and other diseases transmitted by insects, since insects are active in warm temperatures only.

Many correlations exist between seasonal illnesses and weather (Hope-Simpson 1958, Hemmes *et al.* 1960). These correlations do not establish a causal relationship. Certain aspects of health, such as mortality and heat stress in heat waves, and the effects of hot and cold extremes on specific diseases appear to have a strong physiological basis directly linked to outdoor temperature. Therefore, the effects of outdoor climate should be considered separately from those of indoor climate.

Increased Deaths in Heat Waves

The role of temperature extremes in producing discomfort, incapacity, and death has been studied extensively (Katayama 1970). Physiological studies have focused on military duty or similar work situations (Liethead 1964). The importance of temperature stress, affecting both the sick and well is not sufficiently appreciated. A study of weekly deaths over many years in large U.S. cities by Collins and Lehmann (1953) demonstrates the impact of heat waves in producing conspicuous periods of excess deaths. Heat waves had the same amplitude as lesser influenza epidemics, but were of shorter duration with 7 days as compared to 28 to 42 for influenza.

In an epidemiological study of the effects of an Illinois heat wave in July 1966, daily weather data were related to death rates in various age groups. Concurrent disease (if any), cause of death, sex, age, and race were compared to a similar period in 1965, a normal year for July temperatures. Three hot spells were identified; the first peaked on July 5, the second on July 12, and the last on July 18. Mortality increased 33% during the first spell, 36% during the second, and 10% during the last. Of the population at risk, people over 65 suffered the greatest mortality, attributed largely to lesions in the vascular system of the brain and to other circulatory diseases. This circulatory impairment led to failure of the physiological thermoregulatory system. Females suffered greater losses than males, and whites more than blacks. Studies of heat-related deaths using data collected in the 1939, 1955, and 1963 Los Angeles heat waves showed similar results (Oeschliand Buechley 1970). For an extensive review of heat-related death, refer to Ellis (1972).

Hardy (1971) showed the relationship of health data to comfort on a psychrometric diagram (Figure 2). The diagram contains ASHRAE effective temperature ET* lines and lines of constant skin moisture level or skin wettedness w. Skin wettedness is defined as that fraction of the skin covered with water to account for the observed evaporation rate. The ET* lines are loci of constant physiological strain, and also correspond to constant levels of thermal discomfort—slightly uncomfortable, comfortable, and very comfortable (Gonzalez *et al.* 1978). Skin wettedness, an indicator of strain (Berglund *et al.* 1977, Berglund and Cunningham 1986) and the fraction of the skin wet with perspiration, is fairly constant along an ET* line. Numerically ET* is the equivalent temperature at 50% rh that produces the strain and discomfort of the actual condition. The comfort range is between an ET* of 72 and 80 °F, depending on clothing. In this region, skin wettedness is less than 0.2. Heat strokes generally occur when ET* exceeds 93 °F (Bridger and Helfand 1966). Thus, the ET* line of 95 °F is considered dangerous. At this point, skin wettedness is 0.4 or higher. The black dots in Figure 2 correspond to heat stroke deaths of healthy male U.S. soldiers assigned to sedentary duties in midwestern army camp offices (Shickele 1947).

From previous discussion, it is expected that older persons respond less well to thermal challenges than do healthy soldiers. This was apparently the case in the Illinois heat wave study mentioned earlier, where the first wave with a 33% increase in death

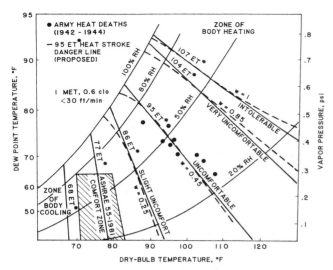

Fig. 2 Isotherms for Comfort, Discomfort, Physiological Strain, Effective Temperature (ET*) and Heat Stroke Danger Threshold

rate and an ET* of 85 °F affected mainly the over 65-year-old group. In the second wave, with an ET* of 93 °F, the observed increase in death rate was 36%. The studies suggest that the "danger line" represents a threshold of significant risk for young healthy people, and that the danger tends to move to lower values of ET* with increasing age.

Effects of Environment on Specific Diseases

Cardiovascular diseases are largely responsible for excess death during heat waves. For example, Burch and DePasquale (1962) found that heart disease cases in which decompensation is present are extremely sensitive to high temperatures and particularly to moist heat. However, both cold and hot temperature extremes are associated with increased coronary heart disease deaths and anginal symptoms (Teng and Heyer 1955).

Both acute and chronic respiratory diseases frequently increase and become more severe during extreme cold weather. No increase in these diseases has been noted with extreme heat. Additional studies of hospital admissions for acute respiratory illness show a negative correlation between acute respiratory illness and temperature after removal of seasonal trend (Holland 1961).

The symptoms of patients with chronic respiratory diseases (bronchitis, emphysema) increase in colder weather. This is thought to be caused by reflex constriction of the bronchi, which adds to the obstruction already present.

Greenburg (1964) revealed evidence of cold sensitivity in asthmatics; emergency room treatments for asthma increased abruptly in local hospitals with severe early autumn cold spells. Later cold waves with even lower temperatures produced no such effects, and years without early extreme cold had no asthma epidemics of this type. The resuspension of allergens caused by resumed indoor heating may account for this phenomenon.

Patients with cystic fibrosis are extremely sensitive to heat because their diminished sweat gland function greatly decreases their ability to cope with increased temperature (Kessler and Anderson 1951).

Itching and chapping of the skin, an annoying affliction that can become severe and disabling, is influenced by: (1) atmospheric factors, particularly cold and dry weather, (2) frequent washing or wetting of skin, and (3) low indoor humidities. Although it is usually a winter (cold climate) illness in the general population, cases of itching and chapping can be caused by excessive summer air conditioning (Susskind 1965, Gaul 1952).

People suffering from chronic illness (heart disease) or serious acute illnesses that require hospitalization are subjected to less temperature stress. Katayama (1970) found that countries with the most carefully regulated indoor climates (Scandinavian countries and the United States) have had only small seasonal fluctuations in recent decades, while countries that use less adequate space heating and cooling exhibit greater seasonal swings in mortality. For example, mandatory air-conditioning in homes for the aged in the southwest has virtually eliminated previously observed mortality increases in heat waves.

Although summer cooling reduces heat stress by removing both sensible and latent heat from the occupied space, winter heating has a mixed effect. It reduces cold stress, but it frequently contributes to greatly reduced relative humidity. Low humidity can contribute to dehydration as well as discomfort and can cause injury to the skin, eyes, nose, throat, and mucous membranes. These dry tissues may be less resistant to infection. Animal experiments also show that infection rates increase with low levels of either ventilation or humidity (Schulman 1962).

In various tests conducted under identical conditions except humidity level, mechanical humidification raised the relative humidity in one space above that in the matched space; no humidified room had a level greater than 50% rh (Green 1979 and 1982, Gelperin 1973, Serati 1969). In each investigation, the humidified rooms showed a reduction in absenteeism and upper respiratory infection—49% reduction in kindergarten children, 6 and 18% in office workers, and 8 and 18% in army recruits. Since occupants in each pair of spaces were subject to the same outdoor conditions and the same indoor air temperature, reductions were attributed to differences in humidity or a related factor (*e.g.*, reduced dust levels and coughing). Therefore, while low humidity does not have a direct pathological effect, it is a contributing factor. A more direct effect of low humidity has been indicated among users of contact lenses on long airline flights in cabins at low humidity. Here, dehydration of the eyes has been blamed for causing irritation and corneal edema or even ulceration of the corneal epithelium (Laviana *et al.* 1988).

Injury from Hot and Cold Surfaces

The skin has cold, warm, and pain sensors to feed back thermal information about surface contacts. When the skin temperature rises above 113 °F or falls below about 59 °F, sensations from the skin's warm and cold receptors are replaced by those from the pain receptors to warn of thermal injury to the tissue (Guyton 1969). The temperature of the skin depends on the temperature of the contact surface, its conductivity, and the contact time. Table 4 gives approximate temperature limits to avoid pain and injury when contacting three classes of conductors for various contact times (CEN).

RADIO FREQUENCY RADIATION

Just as the body absorbs infrared and light energy, which can affect thermal balance, it can also absorb other longer wavelength electromagnetic radiation. For comparison, visible light has

Table 4 Approximate Surface Temperature Limits to Avoid Pain and Injury

Material	Contact Time				
	1 s	10 s	1 min	10 min	8 h
Metal, water	149 °F	133 °F	124 °F	118 °F	109 °F
Glass, concrete	176 °F	151 °F	129 °F	118 °F	109 °F
Wood	248 °F	190 °F	140 °F	118 °F	109 °F

wavelengths in the range 0.4 to 0.7 μm and infrared from 0.7 to 10 μm, while K and X band radar is 12 and 28.6 mm. The wavelength of electromagnetic waves of a typical microwave oven is 120 mm. Infrared and visible light are absorbed within 1 mm of the surface (ASHRAE 1985). The heat of the absorbed radiation raises the skin temperature and, if sufficient, is detected by the skin's thermoreceptors, warning the person of the possible thermal danger. With increasing wavelength, the radiation penetrates deeper into the body. The energy can thus be deposited well beneath the skin's thermoreceptors making the person less able or slower to detect and be warned of the radiation (Justesen *et al.* 1982). Physiologically, these longer waves only heat the tissue and, because the heat may be deeper and less detectable, the maximum power density of such waves in occupied areas is regulated (ANSI) (Figure 3). The maximum permitted power densities are less than half of sensory threshold values.

ELECTRICAL HAZARDS

Electrical current can cause burns, neural disturbances, and cardiac fibrillation (Billings 1975). The threshold of perception is about (5 mA) for direct current, with a feeling of warmth at the contact site. The threshold is 1 mA for alternating current, which causes a tingling sensation.

The resistance of the current pathway through the body is the resistance of the skin and body core. However, the core is basically a saline volume conductor with very little resistance. Thus, the pathway resistance is essentially the resistance of the skin, which is greatly affected by the skin's hydration and sweating activity. Dry skin has a resistance of about 300 kΩ, but the resistance of wet and/or sweating skin is only about 2000 Ω. Therefore, using Ohm's law, the voltage to produce the current for threshold perception is:

Skin	Current	
	AC	**DC**
Dry	300 V	1500 V
Wet or sweating	2 V	10 V

The AC currents necessary to cause pain and involuntary muscle contraction are 3 and 16 mA. The dangerous aspect of alter-

nating electrical current is its ability to cause cardiac arrest by ventricular fibrillation. If a weak alternating current (100 mA for 2 s) passes through the heart (as it would in going from hand to foot), the current can force the heart muscle to fibrillate and lose the rhythmic contractions of the ventricles necessary to pump blood. Unconsciousness and death will soon follow if medical aid can not rapidly restore normal rhythm.

VIBRATION

Vibration in a building originates from both outside and inside the building. Sources outside a building include blasting operations, road traffic, overhead aircraft, underground railways, earth movements, and weather conditions. Sources inside a building include doors closing, foot traffic, moving machinery, elevators, HVAC systems, and other building services. Vibration is an omnipresent, integral part of the built environment. The effects of the vibration on building occupants depend on whether it is perceived by those persons and factors related to the building, the location of the building, the activities of the occupants in the building, and the perceived source and magnitude of the vibration. Factors influencing the acceptability of building vibration are presented in Figure 4.

Whatever the source of vibration, a person will perceive its effect by hearing it, seeing it, or feeling it. It is the combination of these perceptions that will determine human response. Components concerned with hearing and seeing are part of the visual environment of a room and can be assessed as such. The perception of mechanical vibration by feeling it is generally through the cutaneous and kinesthetic senses at high frequencies, and through the vestibular and visceral senses at low frequencies. Because of this and the nature of vibration sources and building responses, it is convenient to consider building vibration in two categories—low-frequency vibrations less than 1 Hz and high-frequency vibrations of 1 to 80 Hz.

Measurement and Assessment

Human response to vibration depends on the vibration of the body. The main vibrational characteristics are vibration level, frequency, axis (and area of the body), and exposure time. A root-mean-square (rms) averaging procedure (over the time of interest)

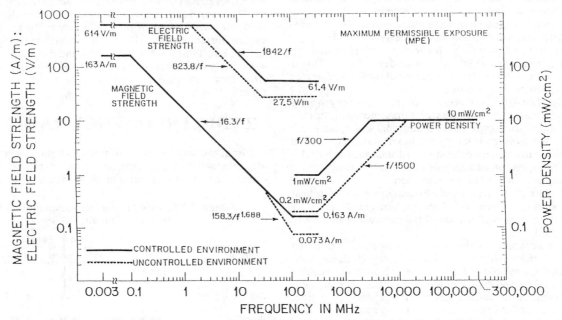

Fig. 3   Maximum Permissible Levels of Radio Frequency Radiation for Human Exposure

is often used to represent vibration acceleration (ft/s^2 rms). Vibration frequency is measured in cycles per second (Hz), and the vibration axis is usually considered in three orthogonal, human-centered translational directions (up-and-down, side-to-side, and fore-and-aft). Although the coordinate system is centered inside the body, in practice, vibration is measured at the human surface and measurements are directly compared with relevant limit values or other data concerning human response.

Rotational motions of a building in roll, pitch, and yaw are usually about an axis of rotation some distance from the building occupants. For most purposes, these motions can be considered as the translational motions of the person. For example, a roll motion in a building about an axis of rotation some distance from a seated person will have a similar effect as side-to-side translational motions of that person, etc.

Most methods assess building vibrations with rms averaging and frequency analysis. However, human response is related to the time-varying characteristics of vibration as well. For example, many stimuli are transient, such as those caused by a train passing a building. The vibration event builds to a peak followed by a decay in level over a total period of about 10 s. The nature of the time varying-event and the number of occasions it occurs during a day are important factors that might be overlooked if data are treated as steady-state and continuous.

Standard Limits

Low-frequency motion (< 1 Hz). The most commonly experienced form of slow vibration in buildings is building sway. This motion can be alarming to occupants if there is fear of building damage or injury. While occupants of two-story wood frame

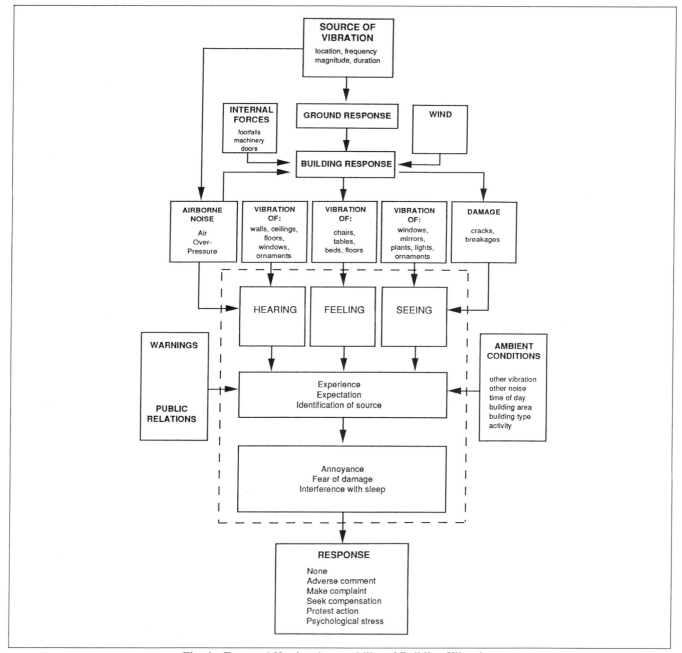

Fig. 4   Factors Affecting Acceptability of Building Vibration

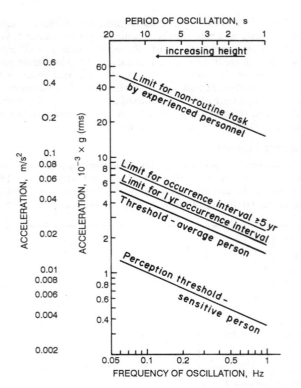

Fig. 5 Acceleration Perception Thresholds and Acceptability Limits for Horizontal Oscillations

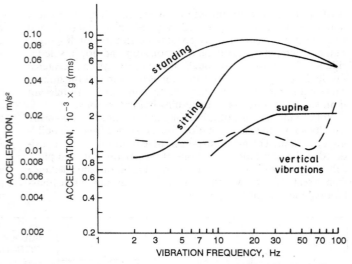

Fig. 6 Median Perception Thresholds to Horizontal (Solid Lines) and Vertical (Dashed Line) Vibrations

houses accept occasional creaks and motion from wind storms or a passing heavy vehicle, such events are not as accepted by occupants of high-rise buildings. Detected motion in tall buildings can cause discomfort and alarm. The perception thresholds of normal sensitive humans to low-frequency horizontal motion are given in Figure 5 (ISO 1984, Chen 1972). The frequency range is from 0.06 to 1 Hz or, conversely, for oscillations with periods of 1 to 17 s. The natural frequency of sway of the Empire State Building in New York City, for example, has a period of 8.3 s (Davenport 1987). The thresholds are expressed in terms of relative acceleration which is the actual acceleration divided by the standard acceleration of gravity ($g = 32.2$ ft/s^2). The perception threshold to sway in terms of building accelerations decreases with increasing frequency and ranges from 5 to 2 milli-g (mg).

For tall buildings, the highest horizontal accelerations generally occur near the top at the building's natural frequency of oscillation. Other parts of the building may have high accelerations at multiples of the natural frequency. Tall buildings always oscillate at their natural frequency, but the deflection is small and the motion undetectable. In general, short buildings have a higher natural frequency of vibration than taller ones. However, strong wind forces energize the oscillation and increase the horizontal deflection, speed, and accelerations of the structure.

Current recommendations state that building motions are not to produce alarm and adverse comment from more than 2% of the building's occupants (ISO 1984). The level of alarm depends on the interval between events. If noticeable building sway occurs for at least 10 min at intervals of 5 years or more, the acceptable acceleration limit is higher than if this sway occurs annually (Figure 5). For annual intervals, the acceptable limit is only slightly above the normal person's threshold of perception. Motion at the 5-year limit level is estimated to cause 12% to complain if it occurred annually. The recommended limits are for purely horizontal motion; rotational oscillations, wind noise, and/or visual cues of the building's motion exaggerate the sensation of

motion and, for such factors, the acceleration limit would be lower.

The upper line in Figure 5 is intended for offshore fixed structures such as oil drilling platforms. The line indicates the level of horizontal acceleration above which routine tasks by experienced personnel would be difficult to accomplish on the structure.

High-frequency motion (1 to 80 Hz). Higher frequency vibrations in buildings are caused by machinery, elevators, foot traffic, fans, pumps, and HVAC equipment. Further, the steel structures of modern buildings are good transmitters of high-frequency vibrations. The sensitivity to these higher frequency vibrations is indicated in Figure 6 (Parsons and Griffin 1988). Displayed are the median perception thresholds to vertical and horizontal vibrations in the 2 to 100-Hz frequency range. The average perception threshold for vibrations of this type is from 1 to 9 mg, depending on the frequency and on whether the person is standing, sitting, or lying down.

People detect horizontal vibrations at lower acceleration levels when lying down than when standing. However, a soft bed decouples and isolates a person fairly well from the vibrations of the structure. The threshold to vertical vibrations is nearly constant at approximately 1.2 mg for both sitting and standing positions from 2 to 100 Hz. This agrees with earlier observations by Reiher and Meister (1931).

Many building spaces with critical work areas (surgery, precision laboratory work) are considered unacceptable if vibration is perceived by the occupants. In other situations and activities, perceived vibration may be acceptable. Parsons and Griffin (1988) found that accelerations twice the threshold level would be unacceptable to occupants in their homes.

A method of assessing vibrational acceptability in buildings is to compare the vibration with perception threshold values (Table 5).

Table 5 Acceptable to Threshold Vibration Level Ratios

Place	Time	Continuous or Intermittent Vibration	Impulse or Transient Vibration Several Times per Day
Critical work areas	Day/night	1	1
Residential	Day	2 to 4	30 to 90
	Night	1.4	1.4 to 20
Office	Day/night	4	60 to 128
Workshop	Day/night	8	90 to 128

Note: The ratios for continuous or intermittent vibration and repeated impulse shock are in the range of 0.7 to 1.0 for hospital operating theaters (room) and critical working areas. In other situations, impulse shock can generally be much higher than when the vibration is more continuous.

COMBUSTION-GENERATED CONTAMINANTS

In addition to the water vapor and carbon dioxide produced during combustion processes, air contaminants are generated which include varying amounts of carbon monoxide, nitrogen oxides, sulfur oxides, respirable particles, and polycyclic aromatic compounds.

The term polycyclic aromatic compounds (PAC) is generally used to refer to polycyclic aromatic hydrocarbons (PAH); nitrogen-, sulfur-, and oxygen-heterocyclic analogues of PAH; and other related PAH derivatives. Depending on their molecular weights and vapor pressures, PACs are distributed between vapor and particulate phases. Because PACs are generally produced by combustion, the particulate compounds are found exclusively in the respirable fraction (*i.e.*, < 2.5 μm aerodynamic diameter).

For indoor air in residences and nonindustrial buildings, the principal sources of combustion-generated contaminants include tobacco smoking (cigarettes, pipes, and cigars), wood-burning stoves and fireplaces, unvented space heaters and gas ranges, and flue gas spillage and backdrafting from improperly vented gas- or oil-fired heaters. Infiltration of combustion contaminants generated outdoors into indoor air can also be a significant source of such contaminants indoors.

Environmental tobacco smoke (ETS) consists of exhaled mainstream smoke from the smoker and sidestream smoke which is emitted from the smoldering tobacco. ETS consists of between 70 and 90% sidestream smoke. More than 4000 compounds have been identified in laboratory-based studies, including many known human toxic and carcinogenic compounds such as carbon monoxide, ammonia, formaldehyde, nicotine, tobacco-specific nitrosamines, benzo(a)pyrene, benzene, cadmium, nickel, and aromatic amines. Many of these toxic constituents are more concentrated in sidestream than in mainstream smoke (Glantz and Parmley 1991). In studies conducted in residences and office buildings with tobacco smoking, ETS was a substantial source of many gas and particulate PACs (Offermann *et al.* 1991).

Unvented or improperly vented combustion of natural gas or propane can be a substantial indoor source of nitrogen dioxide and carbon monoxide. When gas flames burn hot and blue, they are fuel lean and produce a minimum of carbon monoxide and a maximum of nitrogen dioxide. When gas flames burn cool and yellow, they are fuel rich and produce a maximum of carbon monoxide and a minimum of nitrogen dioxide. Thus, unvented or improperly vented gas appliances are principally a source of nitrogen dioxide when properly adjusted and principally a source of carbon monoxide when the gas/air ratio is rich, either because of flue failure or improper adjustment. Under fuel-rich conditions, substantial amounts of aldehydes and formaldehyde can also be produced.

Wood-burning fireplaces that are properly vented and have adequate amounts of combustion air normally have sufficient draft to contain most of the combustion contaminants generated within the fireplace from where they are vented up the chimney. Wood-burning stoves can leak combustion contaminants into the indoor air during normal operation and especially during wood loading and/or stoking times. Traynor (1985) reported elevated levels of carbon monoxide, nitric oxide, nitrogen dioxide, and sulfur dioxide in three residences with wood stoves. Offermann *et al.* (1991) identified wood smoke as a likely source of three gas-phase PACs: biphenyl, acenaphthylene, and anthracene.

Laboratory studies of both unvented kerosene and gas space heaters indicate that they can emit significant quantities of carbon monoxide, nitric oxide, nitrogen dioxide, formaldehyde, and respirable particles (Apte and Traynor 1986). Kerosene heaters using sulfur-containing fuel (*e.g.*, grade 2K kerosene) can also generate significant amounts of sulfur dioxide.

Health Effects

The health effects attributed to carbon monoxide exposure range from neuromotor effects, such as decreased attention span and reaction time, to headaches, nausea, and extreme drowsiness, to death by asphyxiation. Carbon monoxide's toxicity is related to the fact that it rapidly and preferentially binds with hemoglobin in blood, thus restricting the hemoglobin's ability to transfer oxygen to body tissues. The United States Occupational Safety and Health Administration (OSHA) recommends a permissible exposure level of 35 ppm. The National Ambient Air Quality Standards (NAAQS), set by the Environmental Protection Agency and designed to protect sensitive individuals, recommend a maximum 8-h average carbon monoxide concentration of 9 ppm for outdoor air. ASHRAE *Standard* 62-1989, Ventilation for Acceptable Indoor Air Quality, requires that the concentrations of air contaminants in the outside air used for building ventilation not exceed the maximum NAAQS concentrations.

Exposure to nitrogen dioxide (Melial *et al.* 1977) have been related to decreased lung function and increased respiratory disease in children living in homes with gas cooking ranges. While the evidence collected to date is suggestive of such a correlation because of confounding factors related to socioeconomics, the correlation is not definitive. OSHA's recommended permissible 8-h exposure level is 1 ppm, and NAAQS recommends a maximum annual average nitrogen dioxide concentration of 0.05 ppm for outdoor air.

Sulfur dioxide is a well-established respiratory irritant, and levels of sulfur dioxide in homes using unvented kerosene heaters occasionally may be sufficient to induce asthmatic attacks in individuals with hypersensitive airways (Cooper and Alberti 1984). OSHA's recommended permissible 8-h exposure level is 2 ppm, and NAAQS recommends a maximum annual average sulfur dioxide concentration of 0.03 ppm for outdoor air.

Polycyclic aromatic compounds generated by combustion processes include many PAH, nitro-PAH and aza-arenes which have been shown to be carcinogenic in animals (NAS 1983). Other PACs are biologically active as tumor promoters and/or cocarcinogens. High exposures to PAH and aza-arenes have recently been reported for a population in China with very high lung cancer rates (Mumford *et al.* 1987). There are no OSHA or NAAQS exposure guidelines for PACs. The estimated cancer risk for exposure to benzo(a)pyrene, a classical indicator compound for PAH, is 1 per million, for a lifetime exposure to 0.3 ng/m³ (Offermann *et al.* 1991).

ETS, which is a source of many PACs, has been shown to be causally associated with lung cancer (NRC 1986, U.S. Department of Health and Human Services 1986) and cardiovascular disease (Taylor *et al.* 1992) in adults, and respiratory infections, asthma, middle ear effusion (NRC 1986, U.S. Department of Health and Human Services 1986), and low birth weight in children (Martin and Bracken 1986). The U.S. Environmental Protection Agency is considering classifying ETS as a known human carcinogen (EPA 1992), while the National Institute for Occupational Safety and Health has determined that ETS is potentially carcinogenic to occupationally exposed workers (NIOSH 1991). ETS is also a cause of sensory irritation and annoyance (odors and eye irritation).

Carbon monoxide concentrations may be determined directly with real time instrumentation using either portable infrared spectrometers or electrochemical detectors. Nitrogen oxides and sulfur oxides may be measured using passive or active sorbent collection methods followed by spectrophotometric analysis. PACs may be measured in the sub-ng/m³ range through collection of large air samples using medium volume samplers (20 to 40 litres per minute for 12 to 24 h). A sampler developed specifically for measuring indoor concentrations of PACs (Loiselle *et al.* 1991) uses a fluorocarbon resin-impregnated glass fiber filter followed by a cartridge containing XAD-4 sorbent resin. The samples must be extracted in a laboratory and analyzed using gas chromatography and mass spectrometry for the volatile fraction and high-performance liquid chromatography with UV fluorescence for the nonvolatile fraction.

REFERENCES

ACGIH. 1988. TLV's, Threshold limit values and biological exposure indices for 1988-1989. American Conference of Government Industrial Hygienists, Cincinnati, OH.

ACGIH. 1989. Guidelines for the assessment of bioaerosols in the indoor environment. American Conference of Government Industrial Hygienists, Cincinnati, OH.

Akerblom, G., P. Andersson, and B. Clavensjo. 1984. Soil gas radon—A source for indoor radon daughters. *Radiation Protection Dosimetry*, 7, 49.

ANSI. 1991. Safety level with respect to human exposure to radio frequency electromagnetic radiation of 3KH-300GH *Standard* C95.1-1991. American National Standards Institute, New York.

Apte, M.G. and G. W. Traynor. 1986. Comparison of pollutant emission rates from unvented kerosene and gas space heaters. Proceedings of IAQ'86: Managing Indoor Air for Health and Energy Conservation. ASHRAE, Atlanta, 405-416.

Arnow, P.M., J.N. Fink, D.P. Schlueter, J.J. Barboriak, G. Mallison, S.I. Said, S. Martin, G.F. Unger, G.T. Scanlon, and V.P. Kurup. 1978. *American Journal of Medicine*, 64, 237.

ASTM. 1990. *Biological contaminants in indoor environments*. Morey, P., J. Feeley, Sr., and J. Otten, eds. STP 1071. American Society for Testing Materials, Philadelphia.

Bartlett, P.C., L.A. Vonbehren, R.P. Tewari, R.J. Martin, L. Eagleton, M.J. Isaac, and P.S. Kulkarni. 1982. *American Journal of Public Health*, 72, 1369.

Berglund, B., I. Johansson, and T. Lindvall. 1988. *Healthy Buildings* 88, 3:299-309. Stockholm, Sweden.

Berglund, L.G. 1991. Societal and environmental aspects of tall buildings. ASHRAE *Transactions* 97(1).

Berglund, L.G. and D. Cunningham. 1986. Parameters of human discomfort in warm environments. ASHRAE *Transactions* 92(2).

Berglund, L.G. and R.R. Gonzalez. 1977. Evaporation of sweat from sedentary man in humid environments. *Journal of Applied Physiology, Respiratory, Environmental and Exercise Physiology*. 42(5):767-72.

Billings, C.E. 1975. Electrical shock. In *Textbook of medicine*. Saunders, Philadelphia, 72-73.

Bridger, C.A. and L.A. Helfand. 1968. Mortality from heat during July 1966 in Illinois. *International Journal of Biometeorology* 12:51.

Brundage, J.F., R.M. Scott, W.M. Lodnar, D.W. Smith, and R.N. Miller. 1988. *JAMA* 259:2108.

Brundrett, G. and A.H.S. Onions. 1980. *J. Consumer Stud. Home Economics*, 4, 311.

Burge, H. 1989. *Occupational medicine: State of the art reviews*. Cone, J.E. and M.J. Hodgson, eds. 4:713-21. Hanley and Belfus, Inc., Philadelphia.

Burge, S., A. Hedge, S. Wilson, J.H. Bass, and A. Robertson. 1987. *Annals of Occupational Hygiene* 31:493.

Burch, G.E. and N.P. DePasquale. 1962. *Hot climates, Man and his heart*. Charles C. Thomas, Springfield, IL.

CEN/TC 114 N 122 D/E. Surface temperatures of touchable parts, A draft proposal.

Chatigny, M. 1983. *Air sampling instruments*, 6th ed. American Conference of Governmental Industrial Hygienists, E2-E9. Cincinnati, OH.

Chen, P.W. and L.E. Robertson. 1972. Human perception thresholds of horizontal motion. ASCE *Journal* Structure Division, August.

Collins, S.D. and J. Lehmann. 1953. Excess deaths from influenza and pneumonia and from important chronic diseases during epidemic periods, 1918-51. Public Health Monograph 20.10, Public Health Service Publication No. 213.

Cooper, K.K. and R. R. Alberti. 1984. Effect of kerosene heater emissions on indoor air quality and pulmonary function. *American Review of Respiratory Disease* 129:629-31.

Davenport, A.G. 1988. "The response of supertall buildings to wind." In *Second century of the skyscraper*. Van Nostrand Reinhold, New York, 705-26.

DSMA Atcon, Ltd. 1983. Review of existing instrumentation and evaluation of possibilities for research and development of instrumentation to determine future levels at a proposed building site. Report INFO-0096, Atomic Energy Control Board, Ottawa, Canada.

Edwards, J.H. 1980. Microbial and immunological investigations and remedial action after an outbreak of humidifier fever. *British Journal of Industrial Medicine*, 37, 55-62.

Ellis, F.P. 1972. Mortality from heat illness and heat aggravated illness in the United States. *Environmental Research* 5.

EPA. 1989. Radon and radon reduction technology, Report EPA-600/9-89/006a, 1:4-15.

Feeley, J.C. 1988. *Architectural design and indoor microbial pollution*, R.B. Kundsin, ed. Oxford University Press, New York, 218-27.

Garbesi, K. and R.G. Sextro. 1989. Modeling and field evidence of pressure-driven entry of soil gas into a home through permeable below-grade walls. *Environment Science Technology* 23:1481-87.

Gaul, L.E. and G.B. Underwood. 1952. Relation of dew point and barometric pressure to chapping of normal skin. *Journal of Investigative Dermatology* 19:9.

Gelperin, A. 1973. Humidification and upper respiratory infection incidence. *Heating, Piping and Air Conditioning* 45:77.

Girman, J.R. 1989. *Occupational medicine: State of the art review*. Cone, J.E. and M.J. Hodgson, eds. 4:695-712. Hanley and Belfus, Inc., Philadelphia.

Glantz, S.A. and W.W. Parmley. 1991. Passive smoking and heart disease epidemiology, physiology, and biochemistry. *Circulation* 83:633-42.

Gonzalez, R.R., L.G. Berglund, and A.P. Gagge. 1978. Indices of thermoregulatory strain for moderate exercise. *Journal of Applied Physiology:* Respiratory Environmental and Exercise Physiology. 44(6):889-99.

Green, G.H. 1979. The effect of indoor relative humidity on colds. ASHRAE *Transactions* 85(1).

Green, G.H. 1982. The positive and negative effects of building humidification. ASHRAE *Transactions* 88(1):1049.

Greenberg, L. 1964. Asthma and temperature change. *Archives of Environmental Health* 8:642.

Guyton, A.C. 1968. *Textbook of medical physiology*. Saunders, Philadelphia.

Hardy, J.D. 1971. Thermal comfort and health. ASHRAE *Journal* 13:43.

Hemmes, J., K.C. Winkler, and S.M. Kool. 1960. Virus survival as a seasonal factor in influenza and poliomyelitis. *Nature* 188:430.

Hodgson, A.T., K. Garbesi, R.G. Sextro, and J.M. Daisey. 1988. Transport of volatile organic compounds from soil into a residential basement. Paper No. 88-95B.1, Proceedings of the 81st Annual Meeting of the Air Pollution Control Association, June. LBL Report No. 25267.

Hodgson, A.T. and J.R. Girman. 1989. Design and protocol for monitoring indoor air quality. ASTM STP 1002. Nagda, N.L. and J.P. Harber, eds. American Society for Testing and Materials, Philadelphia, 244-56.

Hodgson, M.J., P.R. Morey, J.S. Simon, T.D. Waters, and J.N. Fink. 1987. *American Journal of Epidemiology*, 125, 631.

Hodgson, M.J., P.R. Morey, M. Attfield, W. Sorenson, J.N. Fink, W.W. Rhodes, and G.S. Visvesvara. 1985. *Archives of Environmental Health*, 40, 96.

Holland, W.W. 1961. Influence of the weather on respiratory and heart disease. *Lancet* 2:338.

Holmberg, K. 1987. Proceedings of the 4th International Conference on Indoor Air Quality and Climate, Berlin 1:637-42.

Hope-Simpson, R.E. 1958. The epidemiology of non-infectious diseases. *Royal Society of Health Journal* 78:593.

Hughes, J.M., R.M. Jones, H.W. Glindmeyer, Y.Y. Hammad, and H. Weill. 1993. Follow up study of MMMF workers. *British Journal of Industrial Medicine* 50.

Hyppel, A. 1984. Proceedings of the 3rd International Conference on Indoor Air Quality and Climate, Stockholm, Sweden, 3:443-47.

ISO. 1984. Draft of ISO 6897, Guide to the evaluation of responses of occupants of fixed structures especially buildings and off-shore structures, to low frequency horizontal motion 0.063 to 1 Hz. British Standards Institution, London.

ISO. 1989. ISO 2631-2, Evaluation of human exposure to whole body vibration—Part 2, Continuous and shock-induced vibration in buildings 1-80. British Standards Institution, London.

Jordan, W.S. 1967. Etiology of the acute infectious respiratory diseases. *Archives of Environmental Health* 14:730.

Justesen, D.R., E.R. Adair, J.C. Stevens and V. Bruce-Wolfe. 1982. A comparative study of human sensory thresholds: 2450 MHz Microwaves vs. far-infrared radiation. *Bioelectromagnetics* 3:117-25.

Katayama, K. and M. Momiyana-Sakamoto. 1970. A biometeorological study of mortality from stroke and heart diseases. *Meteorological Geophysics* 21:127.

Kessler, W.R. and W.R. Anderson. 1951. Heat prostration in fibrocystic disease of pancreas and other conditions. *Pediatrics* 8:648.

Kilburn, K.H., D. Powers, and R.H. Warshaw. 1992. Pulmonary effects of exposure to fine fiberglass: Irregular opacities and small airways obstruction. *British Journal of Industrial Medicine* 49:714-20.

Kullman, G.J. and R.A. Hill 1990. Indoor air quality affected by abandoned gasoline tanks. *Applied Occupational Environmental Hygiene* 5:36-37.

Laviana, J.E., F.H. Rohles, and P.E. Bullock. 1988. Humidity, comfort and contact lenses. ASHRAE *Transactions* 94(1).

Lewis, R.G. and L.A. Wallace. 1988. ASTM *Standardization News* 16:40.

Livingston, J.M. and C.R. Jones. 1981. Living area contamination by chlordane used for termite treatment. *Bulletin of Environmental Contaminant Toxicology* 27:406-11.

Loiselle, S.A., F.J. Offermann, and A.T. Hodgson. 1991. Development of an indoor air sampler for polycyclic aromatic compounds. *Indoor Air* 2:191-210.

Martin, T.R. and M.B. Braken. 1986. Association of low birth weight passive smoke and exposure in pregnancy. *American Journal of Epidemiology* 124:633-42.

McCann, J., L. Horn, J. Girman, and A.V. Nero. 1987. Short-term bioassays in the analysis of complex mixtures. Sandhu, S.S., D.M. De Marini, J.J. Mass, eds. Plenum Press, New York 5:325-54.

Melial, R.J.W. *et al.* 1977. Association between gas cooking and respiratory disease in children. *British Medial Journal* 2:149-52.

Miksch, R.R., C.D. Hollowell, and H.E. Schmidt. 1982. Trace organic chemical contaminants in office spaces. *Atmospheric Environment* 8:129-37.

Molhave, L. 1986. Indoor air quality in relation to sensory irritation due to volatile organic compounds. ASHRAE *Transactions* 92(1A):306-16.

Molhave, L. 1991. Volatile organic compounds, indoor air quality and health. *Indoor Air* 1(4):357-76.

Molhave, L. and M. Thorsen. 1990. A model for investigations of ventilation systems as sources for volatile organic compounds in indoor climate. *Atmospheric Environment.* 25A:241-49.

Morey, P.R. 1988. *Architectural design and indoor microbial pollution.* R.B. Kundsin, ed. Oxford University Press, New York, 40-80.

Morey, P.R. 1990. The practitioner's approach to indoor air quality investigations. Proceedings of the Indoor Air Quality International Symposium, American Industrial Hygiene Association, Akron, OH.

Morey, P.R. and J.C. Feeley, Sr. 1988. ASTM *Standardization News*, 16, 54.

Morey, P.R., M.J. Hodgson, W.G. Sorenson, G.J. Kullman, W.W. Rhodes, G.S. Visvesvara. 1986. ASHRAE *Transactions*, 93, 399.

Morey, P.R. and B.A. Jerkins. 1989. Proceedings of IAQ '89, The Human Equation: Health and Comfort. ASHRAE, Atlanta, 67-71.

Morey, P.R. and J. Singh. 1991. "Indoor air quality in non-industrial occupational environments." In *Patty's industrial hygiene and toxicology*, 4th ed. 1(1).

Moser, M.R., T.R. Bender, H.S. Margolis, G.R. Noble, A.P. Kendal, and D.G. Ritter. 1979. *American Journal of Epidemiology* 110(1).

Mumford, J.L., X.Z. He, R.S. Chapman, S.R. Cao, D.B. Harris, K.M. Li, Y.L. Xian, W.Z. Jiang, C.W. Xu, J.C. Chang, W.E. Wilson, and M. Cooke. 1987. Lung cancer and indoor air pollution in Xuan Wei, China. *Science* 235:217-20.

NAS. 1983. Polycyclic aromatic hydrocarbons: Evaluation of sources and effects. National Academy Press, Washington, D.C.

Nazaroff, W.W., S.R. Lewis, S.M. Doyle, B.A. Moed, and A.V. Nero. 1987. Experiments on pollutant transport from soil into residential basements by pressure-driven air flow. *Environment Science and Technology* 21:459.

Nazaroff, W.W., B.A. Moed, and R.G. Sextro. 1988. Soil as a source of indoor radon: Generation, migration and entry. In *Radon and its decay products in indoor air*, W.W. Nazaroff and A.V. Nero, eds. Wiley, New York, 57-112.

NIOSH. 1991. Current Intelligence Bulletin 54: Environmental tobacco smoke in the workplace; Lung cancer and other health effects. DHHS (NIOSH) Publication No. 91-108. U.S. Department of Health and Human Services, Public Health Service, Centers for Disease Control,

National Institute for Occupational Safety and Health, Cincinnati, OH.

NRC. 1986. Environmental tobacco smoke: Measuring exposures and assessing health effects. National Research Council. National Academy Press, Washington, D.C.

Oeschli, F.W. and R.W. Beuchley. 1970. Excess mortality associated with three Los Angeles September hot spells. *Environmental Research* 3:277.

Offermann, F.J., S.A. Loiselle, A.T. Hodgson, L.A. Gundel, and J.M. Daisey. 1991. A pilot study to measure indoor concentrations and emission rates of polycyclic aromatic hydrocarbons. *Indoor Air* 4:497-512.

Parsons, K.C. and M.J. Griffin. 1988. Whole-body vibration perception thresholds. *Journal of Sound and Vibration* 121(2):237-58.

Pellizzari, E.D. 1987. The influence of personal activities on exposure to volatile organic compounds. Proceedings of the 4th International Conference on Indoor Air Quality and Climate, Berlin 1:117-21.

Reiher, H. and F.J. Meister. 1931. Translation of Report No. F-Ts-616-RE 1946. Forshung VDI. 2,381-386, "The sensitivities of the human body to vibrations." Headquarters Air Material Command, Wright Field, Dayton, OH.

Ritzel, G. 1968. Sozialmedizinische erhegungen zur pathagenese und prophylaxe von erkaltungskrankheiten. *Zeitschrift feur Praeventivmedizin* 11:16.

Samet, J.M. 1989. Radon and lung cancer. *Journal of the National Cancer Institute.* 81:145.

Schulman, J.H. and E.M. Kilbourne. 1962. Airborne transmission of influenza virus infection in mice. *Nature* 195:1129.

Serati, A. and M. Wuthrich. 1969. Luftfreughtigkeit und saison Krankenheit. *Schweizerische Medizinische Wochenscrift* 99:46.

Sheldon, L., H. Zelon, J. Sickles, C. Easton, T. Hartwell, and L. Wallace. 1986. *Indoor air quality in public buildings*, Vol. II. EPA/600/S6-88/009b, Research Triangle Park, NC.

Sheldon, L., R.W. Handy, T. Hartwell, R.W. Whitmore, H. Zelon, and E.D. Pellizzari. 1988. *Indoor air quality in public buildings*, Vol. I. EPA/600/S6-88/009a, Washington, D.C.

Shickele, E. 1947. Environment and fatal heat stroke. *Military Surgeon* 100:235.

Stingdehag, O., I. Josefsson, and E. Hennington. 1988. Healthy Buildings '88, 3:611-20. Stockholm, Sweden.

Susskind, R.R. and M. Ishihara. 1965. The effects of wetting on cutaneous vulnerability. *Archives of Environmental Health* 11:529.

Tancrede, M., R. Wilson, L. Ziese, and E.A.C. Crouch. 1987. *Atmospheric Environment* 21:2187.

Tanner, A.B. 1980. "Radon migration in the ground: A supplementary review." In *Natural radiation environment* III., U.S. Department of Commerce, NTIS, Springfield, VA, 5.

Taylor, A.E., D.C. Johnson, and H. Kazemi. 1992. Environmental tobacco smoke and cardiovascular disease: A position paper from the council on cardiopulmonary and critical care. American Heart Association, Dallas, TX.

Teng, H.C. and H.E. Heyer, eds. 1955. The relationship between sudden changes in the weather and acute myocardial infarction. *American Heart Journal* 49:9.

Traynor, G.W. 1985. Indoor air pollution due to emissions from wood burning stoves. LBL-17854. Lawrence Berkeley Laboratory, Berkeley, CA.

Turk, B.H., R.J. Prill, W.J. Fisk, D.T. Grimsrud, and R.G. Sextro. 1990. Effectiveness of radon control techniques in 15 homes. *Journal of the Air and Waste Management Association* 41:723-34.

Turk, B.H., R.J. Prill, D.T. Grimsrud, B.A. Moed, and R.G. Sextro. 1990. Characterizing the occurrence, sources and variability of radon in Pacific Northwest homes. *Journal of the Air and Waste Management Association* 40:498-506.

U.S. Department of Health and Human Services. 1986. The health consequences of involuntary smoking. A report of the surgeon general. DHHS Publication No. (PHS) 87-8398. U.S. Department of Health and Human Services, Public Health Services, Office of the Assistant Secretary for Health, Office of Smoking and Health.

EPA. 1992. Respiratory health effects of passive smoking: Lung cancer and other disorders, review draft. Environmental Protection Agency. EPA/600-6-90/006B. Office of Research and Development, Washington, D.C.

Wallace, L.A. 1987. The total exposure assessment methodology TEAM. Study: Summary and analysis, Vol. 1. Office of Research and Development, U.S. Environmental Protection Agency.

Wallace, L.A. and C.A. Clayton. 1987. Volatile organic compounds in 600 U.S. houses: Major sources of personal exposure. Proceedings of the 4th International Conference on Indoor Air Quality and Climate, Berlin 1:183-87.

Wallingford, K.M. and J. Carpenter. 1986. Proceedings of IAQ '86, Managing indoor air for health and energy conservation, ASHRAE, Atlanta, 448-53.

Weill, H., J.M. Hughes, Y.Y. Hammad, H.W. Glindmeyer, G. Sharon, and R.N. Jones. 1983. Respiratory health in workers exposed to man-made vitreous fibers. *American Review of Respiratory Disease* 128:104-12.

WHO. 1988. IARC Monographs on the evaluation of carcinogenic risks to humans: Man-made mineral fibers and radon, Vol. 43. WHO Regional Office for Europe, Copenhagen, Denmark.

WHO. 1989. Indoor air quality: Organic pollutants. Report on a WHO-meeting, Euro Report and Studies III, WHO Regional Office for Europe, Copenhagen, Denmark.

Wright, C.G. and R.B. Leidy. 1982. Chlorodane and heptachlor in the ambient air of houses treated for termites. *Bulletin of Environmental Contaminant Toxicology* 28:617-23.

Wood, J.A. and M.L. Porter. 1987. Hazardous pollutants in Class II landfills. *Journal of the Air Pollution Control Association* 37:609-15.

Yoshizawa, S., F. Surgawa, S. Ozawo, Y. Kohsaka, and A. Matsumae. 1987. Proceedings of the 4th International Conference on Indoor Air Quality and Climate, Berlin 1:627-31.

CHAPTER 38

CODES AND STANDARDS

THE Codes and Standards listed in Table 1 represent practices, methods, or standards published by the organizations indicated. They are valuable guides for the practicing engineer in determining test methods, ratings, performance requirements, and limits applying to the equipment used in heating, refrigerating, ventilating, and air conditioning. *Copies can usually be obtained from the organization listed in the Publisher column.* These listings represent the most recent information available at the time of publication.

Table 1 Codes and Standards Published by Various Societies and Associations

Subject	Title	Publisher	Reference
Air Conditioners	Room Air Conditioners	CSA	C22.2 No. 117-1970
Rooms	Room Air Conditioners	AHAM	ANSI/AHAM (RA C-1)
	Method of Testing for Rating Room Air Conditioners and Packaged Terminal Air Conditioners	ASHRAE	ANSI/ASHRAE 16-1983 (RA 88)
	Method of Testing for Rating Room Air Conditioners and Packaged Terminal Air Conditioner Heating Capacity	ASHRAE	ANSI/ASHRAE 58-1986 (RA 90)
	Methods of Testing for Rating Room Fan-Coil Air Conditioners	ASHRAE	ANSI/ASHRAE 79-1984 (RA 91)
	Commercial and Residential Central Air Conditioners	CSA	C22.2 No. 119-M1985
	Performance Standard for Room Air Conditioners	CSA	CAN/CSA-C368.1-M90
	Room Air Conditioners (1982)	UL	ANSI/UL 484-1986
Packaged Terminal	Packaged Terminal Air Conditioners	ARI	ARI 310-90
	Packaged Terminal Heat Pumps	ARI	ARI 380-90
Transport	Air Conditioning of Aircraft Cargo (1978)	SAE	SAE AIR806A
	Nomenclature, Aircraft Air-Conditioning Equipment (1978)	SAE	SAE ARP147C
Unitary	Load Calculation for Commercial Summer and Winter Air Conditioning, 4th ed. (1988)	ACCA	ACCA Manual N
	Application of Sound Rated Outdoor Unitary Equipment	ARI	ARI 275-84
	Commercial and Industrial Unitary Air-Conditioning Equipment	ARI	ANSI/ARI 360-86
	Sound Rating of Outdoor Unitary Equipment	ARI	ARI 270-84
	Unitary Air-Conditioning and Air-Source Heat Pump Equipment	ARI	ANSI/ARI 210/240-89
	Methods of Testing for Rating Heat Operated Unitary Air-Conditioning Equipment for Cooling	ASHRAE	ANSI/ASHRAE 40-1986 (RA 92)
	Methods of Testing for Rating Unitary Air-Conditioning and Heat Pump Equipment	ASHRAE	ANSI/ASHRAE 37-1988
	Methods of Testing for Seasonal Efficiency of Unitary Air Conditioners and Heat Pumps	ASHRAE	ANSI/ASHRAE 116-1983
	Performance Standard for Split-System Air Conditioners and Heat Pumps	CSA	CAN/CSA-C273.3-M91
	Performance Standard for Single Package Central Air Conditioners and Heat Pumps	CSA	CAN/CSA-C656-M92
	Air Conditioners, Central Cooling (1982)	UL	ANSI/UL 465-1984
Air Conditioning	Commercial Low Pressure, Low Velocity Duct System Design	ACCA	Manual Q
	Duct Design for Residential Buildings	ACCA	ACCA Manual D
	Load Calculation for Residential Winter and Summer Air Conditioning, 7th ed. (1986)	ACCA	ACCA Manual J
	Gas-Fired Absorption Summer Air Conditioning Appliances (with 1982 addenda)	AGA	ANSI Z21.40.1-1981
	Heating and Cooling Equipment	UL	ANSI/UL 1995-1992
	Environmental System Technology (1984)	NEBB	NEBB
	Automotive Air-Conditioning Hose (1989)	SAE	ANSI/SAE J51 MAY89
	HVAC Systems—Applications, 1st ed. (1986)	SMACNA	SMACNA
	HVAC Systems—Duct Design (1990)	SMACNA	SMACNA
	Installation Standards for Residential Heating and Air Conditioning Systems (1988)	SMACNA	SMACNA
Transport	Air Conditioning Equipment, General Requirements for Subsonic Airplanes (1961)	SAE	SAE ARP85D
	General Requirements for Helicopter Air Conditioning (1970)	SAE	SAE ARP292B
	Testing of Commercial Airplane Environmental Control Systems (1973)	SAE	SAE ARP217B
Unitary	Method of Rating Computer and Data Processing Room Unitary Air Conditioners	ASHRAE	ANSI/ASHRAE 127-1988
	Method of Rating Unitary Spot Air Conditioners	ASHRAE	ANSI/ASHRAE 128-1988
Air Curtains	Air Curtains for Entranceways in Food Establishments	NSF	NSF-20
	Air Distribution Basics for Residential and Small Commercial Buildings	ACCA	ACCA Manual T
	Residential Equipment Selection	ACCA	ACCA Manual S

Table 1   Codes and Standards Published by Various Societies and Associations (*Continued*)

Subject	Title	Publisher	Reference
Air Curtains (continued)	Flexible Duct Performance and Installation Standards	ADC	ADC
	Laboratory Certification Manual	ADC	ADC 1062:LCM-83
	Test Code for Grilles, Registers and Diffusers	ADC	ADC 1062:GRD-84
	Metric Units and Conversion Factors	AMCA	AMCA 99-0100-76
	Test Methods for Air Curtain Units	AMCA	AMCA 220-91
	Air Volume Terminals	ARI	ANSI/ARI 880-89
	Method of Testing for Rating the Performance of Air Outlets and Inlets	ASHRAE	ANSI/ASHRAE 70-1991
	Standard Methods for Laboratory Air Flow Measurement	ASHRAE	ANSI/ASHRAE 41.2-1987 (RA 92)
	Rating the Performance of Residential Mechanical Ventilating Equipment	CSA	CAN/CSA-C260-M90
	Residential Air Exhaust Equipment (1e)	CSA	C260.2-1976
	High Temperature Pneumatic Duct Systems for Aircraft (1981)	SAE	ANSI/SAE ARP699D
Air Ducts and Fittings	Commercial Low Pressure, Low Velocity Duct Systems	ACCA	Manual Q
	Duct Design for Residential Winter and Summer Air Conditioning	ACCA	Manual D
	Flexible Air Duct Test Code	ADC	ADC FD-72R1-1979
	Pipes, Ducts and Fittings for Residential Type Air Conditioning Systems	CSA	B228.1-1968
	Installation of Air Conditioning and Ventilating Systems (1989)	NFPA	ANSI/NFPA 90A-1989
	Installation of Warm Air Heating and Air-Conditioning Systems (1989)	NFPA	ANSI/NFPA 90B-1989
	Ducted Electric Heat Guide for Air Handling Systems (1971)	SMACNA	SMACNA
	HVAC Air Duct Leakage Test Manual (1985)	SMACNA	SMACNA
	HVAC Duct Construction Standards—Metal and Flexible, 1st ed. (1985)	SMACNA	SMACNA
	HVAC Duct Systems Inspection Guide (1989)	SMACNA	SMACNA
	Rectangular Industrial Duct Construction (1980)	SMACNA	SMACNA
	Round Industrial Duct Construction (1977)	SMACNA	SMACNA
	Thermoplastic Duct (PVC) Construction Manual (Rev. A, 1974)	SMACNA	SMACNA
	Closure Systems for Use with Rigid Air Ducts and Air Connectors (1991)	UL	UL 181A
	Factory-Made Air Ducts and Air Connectors (1990)	UL	UL 181
	Marine Rigid and Flexible Air Ducting (1986)	UL	ANSI/UL 1136-1986
Air Filters	Method for Measuring Performance of Portable Household Electrical Cord-Connected Room Air Cleaners	AHAM	ANSI/AHAM AC-1
	Commercial and Industrial Air Filter Equipment	ARI	ARI 850-84
	Residential Air Filter Equipment	ARI	ARI 680-86
	Gravimetric and Dust Spot Procedures for Testing Air Cleaning Devices Used in General Ventilation for Removing Particulate Matter	ASHRAE	ANSI/ASHRAE 52.1-1992
	Method for Sodium Flame Test for Air Filters	BSI	BS 3928
	Methods of Test for Atmospheric Dust Spot Efficiency and Synthetic Dust Weight Arrestance	BSI	BS 6540 Part 1
	Electrostatic Air Cleaners (1989)	UL	ANSI/UL 867-1988
	High Efficiency, Particulate, Air Filter Units (1990)	UL	ANSI/UL 586-1990
	Test Performance of Air Filter Units (1987)	UL	ANSI/UL 900-1987
Air-Handling Units	Commercial Low Pressure, Low Velocity Duct Systems	ACCA	Manual Q
	Duct Design for Residential Winter and Summer Air Conditioning	ACCA	Manual D
	Central Station Air-Handling Units	ARI	ANSI/ARI 430-89
Air Leakage	Air Leakage Performance for Detached Single-Family Residential Buildings	ASHRAE	ANSI/ASHRAE 119-1988
Boilers	A Guide to Clean and Efficient Operation of Coal Stoker-Fired Boilers	ABMA	ABMA
	Boiler Water Limits and Steam Purity Recommendations for Watertube Boilers	ABMA	ABMA
	Boiler Water Requirements and Associated Steam Purity—Commercial Boilers	ABMA	ABMA
	Fluidized Bed Combustion Guidelines	ABMA	ABMA
	Guidelines for Industrial Boiler Performance Improvement	ABMA	ABMA
	Lexicon Boiler and Auxiliary Equipment	ABMA	ABMA
	Matrix of Recommended Quality Control Requirements	ABMA	ABMA
	Operation and Maintenance Safety Manual	ABMA	ABMA
	Recommended Design Guidelines for Stoker Firing of Bituminous Coals	ABMA	ABMA
	(Selected) Summary of Codes and Standards of the Boiler Industry	ABMA	ABMA
	Thermal Shock Damage to Hot Water Boilers as a Result of Energy Conservation Measures	ABMA	ABMA
	Commercial Applications Systems and Equipment	ACCA	Manual CS
	Boiler and Pressure Vessel Code (11 sections) (1989)	ASME	ASME
	Boiler, Pressure Vessel, and Pressure Piping Code	CSA	B51-M1986
	Heating, Water Supply, and Power Boilers—Electric (1991)	UL	ANSI/UL834-1991
Cast-Iron	Ratings for Cast-Iron and Steel Boilers (1992)	HYDI	IBR
	Testing and Rating Heating Boilers (1989)	HYDI	IBR
Gas or Oil	Gas-Fired Low-Pressure Steam and Hot Water Boilers	AGA	ANSI Z21.13-1987; Z21.13a-1989

Table 1 Codes and Standards Published by Various Societies and Associations (*Continued*)

Subject	Title	Publisher	Reference
Boilers (continued)	Gas Utilization Equipment in Large Boilers (with 1972 and 1976 addenda; R-1983, 1989)	AGA	ANSI Z83.3-1971
	Control and Safety Devices for Automatically Fired Boilers	ASME	ANSI/ASME CSD.1-1988
	Oil-Fired Steam and Hot-Water Boilers for Residential Use	CSA	B140.7.1-1976
	Oil-Fired Steam and Hot-Water Boilers for Commercial and Industrial Use	CSA	B140.7.2-1967 (R 1991)
	Explosion Prevention of Fuel Oil and Natural Gas-Fired Single-Burner Boiler-Furnaces (1987)	NFPA	ANSI/NFPA 85A-1987
	Prevention of Furnace Explosions/Implosions in Multiple Burner Boiler-Furnaces (1991)	NFPA	NFPA 85C-1991
	Commercial-Industrial Gas Heating Equipment (1973)	UL	UL 795
	Oil Fired Boiler Assemblies (1990)	UL	ANSI/UL 726-1990
Building Codes	ASTM Standards Used in Building Codes	ASTM	ASTM BOCA
	National Building Code, 11th ed. (1990)	BOCA	BOCA BOCA
	National Property Maintenance Code, 3rd ed. (1990)	BOCA	BOCA CABO
	One- and Two-Family Dwelling Code (1992)	CABO	CABO
	Model Energy Code (1992)	CABO	CABO
	Uniform Building Code (1991)	ICBO	ICBO
	Uniform Building Code Standards (1991)	ICBO	ICBO
	Directory of Building Codes and Regulations (1993 ed.)	NCSBCS	NCSBCS
	Standard Building Code (1991)	SBCCI	SBCCI
Mechanical	Safety Code for Elevators and Escalators (plus two yearly supplements)	ASME	ANSI/ASME A 17.1-1990
	BOCA National Mechanical Code, 7th ed. (1990)	BOCA	BOCA
	Uniform Mechanical Code (1991) (with Uniform Mechanical Code Standards)	ICBO/ IAPMO	ICBO/IAPMO
	Standard Gas Code (1991 ed. with 1992 revisions)	SBCCI	SBCCI
	Standard Mechanical Code (1991 ed. with 1992 revisions)	SBCCI	SBCCI
Burners	Guidelines for Burner Adjustments of Commercial Oil-Fired Boilers	ABMA	ABMA
	Domestic Gas Conversion Burners	AGA	ANSI Z21.17-1984; Z21.17a-1990
	Installation of Domestic Gas Conversion Burners	AGA	ANSI Z21.8-1984; Z21.8a-1990; Z21.176-1990
	General Requirements for Oil Burning Equipment	CSA	CAN/CSA-B140.0-M87 (R 1991)
	Installation Code for Oil Burning Equipment	CSA	CAN/CSA-B139-M91
	Oil Burners; Atomizing Type	CSA	CAN/CSA-B140.2.1-M90
	Pressure Atomizing Oil Burner Nozzles	CSA	B140.2.2-1971 (R 1991)
	Replacement Burners and Replacement Combustion Heads for Residential Oil Burners	CSA	B140.2.3-M1981 (R1991)
	Supplement No. 1 to B139-1976, Installation Code for Oil Burning Equipment	CSA	B139S1-1982
	Vaporizing-Type Oil Burners	CSA	B140.1-1966 (R 1991)
	Commercial-Industrial Gas Heating Equipment (1973)	UL	UL 795
	Oil Burners (1989)	UL	ANSI/UL 296-1988
Capillary Tubes	Capillary Tubes Method of Testing Flow Capacity of Refrigerant	ASHRAE	ANSI/ASHRAE 28-1988
Chillers	Methods of Testing Liquid Chilling Packages	ASHRAE	ASHRAE 30-1978
	Absorption Water-Chilling Packages	ARI	ARI 560-82
	Centrifugal or Rotary Screw Water-Chilling Packages	ARI	ARI 550-90
	Reciprocating Water-Chilling Packages	ARI	ARI 590-86
Chimneys	Design and Construction of Masonry Chimneys and Fireplaces	CSA	CAN/CSA-A405-M87
	Chimneys, Fireplaces, and Vents, and Solid Fuel Burning Appliances	NFPA	ANSI/NFPA 211-1988
	Chimneys, Factory-Built, Medium Heat Appliance (1992)	UL	ANSI/UL 959-1992
	Chimneys, Factory-Built, Residential Type and Building Heating Appliances (1989)	UL	ANSI/UL 103-1988
Cleanrooms	Procedural Standards for Certified Testing of Cleanrooms (1988)	NEBB	NEBB-1988
Coils	Forced-Circulation Air-Cooling and Air-Heating Coils	ARI	ARI 410-91
	Methods of Testing Forced Circulation Air Cooling and Air Heating Coils	ASHRAE	ASHRAE 33-1978
Comfort Conditions	Thermal Environmental Conditions for Human Occupancy	ASHRAE	ANSI/ASHRAE 55-1992
Compressors	Compressors and Exhausters (reaffirmed 1986)	ASME	ANSI/ASME PTC 10-1965 (R 1986)
	Displacement Compressors, Vacuum Pumps and Blowers	ASME	ANSI/ASME PTC9-1974 (R 1992)
	Safety Standard for Air Compressor Systems	ASME	ANSI/ASME B19.1-1990
	Safety Standard for Compressors for Process Industries	ASME	ASME/ANSI B19.3-1991
	Compressed Air and Gas Handbook, 5th ed. (1988)	CAGI	CAGI
Refrigeration	Ammonia Compressor Units	ARI	ANSI/ARI 510-87
	Method for Presentation of Compressor Performance Data	ARI	ARI 540-91

Table 1   Codes and Standards Published by Various Societies and Associations (*Continued*)

Subject	Title	Publisher	Reference
Compressors (*continued*)	Positive Displacement Refrigerant Compressors and Condensing Units	ARI	ANSI/ARI 520-90
	Method for Presentation of Compressor Performance Data	ARI	ARI 540-91
	Methods of Testing for Rating Positive Displacement Refrigerant Compressors	ASHRAE	ASHRAE 23-1978
	Hermetic Refrigerant Motor-Compressors	CSA	CAN/CSA-C22.2 No. 140.2-M91
	Hermetic Refrigerant Motor-Compressors (1991)	UL	ANSI/UL 984
Computers	Protection of Electronic Computer/Data Processing Equipment	NFPA	ANSI/NFPA 75-1989
Condensers	Commercial Applications Systems and Equipment (for equipment selection only)	ACCA	Manual CS
	Remote Mechanical Draft Air-Cooled Refrigerant Condensers	ARI	ARI 460-87
	Water-Cooled Refrigerant Condensers, Remote Type	ARI	ARI 450-87
	Methods of Testing for Rating Remote Mechanical-Draft Air-Cooled Refrigerant Condensers	ASHRAE	ASHRAE 20-1970
	Methods of Testing Remote Mechanical-Draft Evaporative Refrigerant Condensers	ASHRAE	ANSI/ASHRAE 64-1989
	Methods of Testing for Rating Water-Cooled Refrigerant Condensers	ASHRAE	ANSI/ASHRAE 22-1992
	Addendum I, Standards for Steam Surface Condensers (1989)	HEI	HEI
	Standards for Steam Surface Condensers, 8th ed. (1984)	HEI	HEI
Condensing Units	Commercial Applications Systems and Equipment	ACCA	Manual CS
	Residential Equipment Selection	ACCA	Manual S
	Commercial and Industrial Unitary Air-Conditioning Condensing Units	ARI	ARI 365-87
	Methods of Testing for Rating Positive Displacement Condensing Units	ASHRAE	ASHRAE 14-80
	Heating and Cooling Equipment	CSA	CAN/CSA C22.2 No. 236-M90
	Refrigeration and Air-Conditioning Condensing and Compressor Units (1987)	UL	ANSI/UL303-1988
Contactors	Definite Purpose Contactors for Limited Duty	ARI	ANSI/ARI 790-86
	Definite Purpose Magnetic Contactors	ARI	ANSI/ARI 780-86
Controls	Quick-Disconnect Devices for Use with Gas Fuel	AGA	ANSI Z21.41-1989; Z21.41a-1990
	Energy Management Control Systems Instrumentation	ASHRAE	ANSI/ASHRAE 114-1986
	Temperature-Indicating and Regulating Equipment	CSA	C22.2 No. 24-1987
	Control Centers for Changing-Message Type Electric Signals	UL	UL 1433
	Industrial Control Equipment (1989)	UL	ANSI/UL 508-1988
	Limit Controls (1989)	UL	ANSI/UL 353-1988
	Primary Safety Controls for Gas- and Oil-Fired Appliances (1985)	UL	ANSI/UL 372-1985
	Solid State Controls for Appliances (1987)	UL	ANSI/UL244A-1987
	Temperature-Indicating and Regulating Equipment (1988)	UL	ANSI/UL 873-1987
	Test for Safety-Related Controls Employing Solid-State Devices (1991)	UL	UL 991
Commercial and Industrial	General Standards for Industrial Control and Systems	NEMA	NEMA ICS 1-1988
	Industrial Control Devices, Controllers and Assemblies	NEMA	NEMA ICS 2-1988
	Instructions for the Handling, Installation, Operation and Maintenance of Motor Control Centers	NEMA	NEMA ICS 2.3-1983 (R 1990)
	Maintenance of Motor Controllers after a Fault Condition	NEMA	NEMA ICS 2.2-1983 (1988)
	Preventive Maintenance of Industrial Control and Systems Equipment	NEMA	NEMA ICS 1.3-1986
Residential	Automatic Gas Ignition Systems and Components	AGA	ANSI Z21.20-1989; Z21.20a-1991; Z21.20b-1992
	Gas Appliance Pressure Regulators	AGA	ANSI Z21.18-1987; Z21.18-1989
	Gas Appliance Thermostats	AGA	ANSI Z21.23-1989; Z21.23a-1991
	Manually Operated Gas Valves for Appliances, Appliance Connector Valves and Hose End Valves	AGA	ANSI Z21.15-1992
	Manually-Operated Piezo Electric Spark Gas Ignition Systems and Components	AGA	ANSI Z21.77-1989
	Hot Water Immersion Controls	NEMA	NEMA DC-12-1985 (R 1991)
	Line Voltage Integrally-Mounted Thermostats for Electric Heaters	NEMA	NEMA DC 13-1979 (R 1985)
	Quick Connect Terminals	NEMA	ANSI/NEMA DC 2-1982 (R 1988)
	Residential Controls—Class 2 Transformers	NEMA	NEMA DC 20-1986
	Residential Controls—Surface Type Controls for Electric Storage Water Heaters	NEMA	NEMA DC 5-1989
	Safety Guidelines for the Application, Installation, and Maintenance of Solid State Controls	NEMA	NEMA ICS 1.1-1984 (R 1988)
	Temperature Limit Controls for Electric Baseboard Heaters	NEMA	NEMA DC 10-1083 (R 1989)
	Wall-Mounted Room Thermostats	NEMA	NEMA DC 3-1989
	Warm Air Limit and Fan Controls	NEMA	NEMA DC 4-1986
	Electrical Quick-Connect Terminals (1991)	UL	UL 310

Table 1 Codes and Standards Published by Various Societies and Associations (*Continued*)

Subject	Title	Publisher	Reference
Coolers			
Air	Unit Coolers for Refrigeration	ARI	ANSI/ARI 420-89
	Methods of Testing Forced Convection and Natural Convection Air Coolers for Refrigeration	ASHRAE	ANSI/ASHRAE 25-1990
	Milk Coolers	CSA	C22.2 No. 132-1973
	Commercial Bulk Milk Dispensing Equipment	NSF	NSF 20
Bottled Beverage	Methods of Testing and Rating Bottled and Canned Beverage Vendors and Coolers	ASHRAE	ANSI/ASHRAE 32-1986 (RA 90)
	Refrigerated Vending Machines (1989)	UL	ANSI/UL 541-1988
Drinking Water	Application and Installation of Drinking Water Coolers	ARI	ANSI/ARI 1020-84
	Drinking Fountains and Self-Contained, Mechanically Refrigerated Drinking Water Coolers	ARI/ANSI	ANSI/ARI 1010-84
	Methods of Testing for Rating Drinking-Water Coolers with Self-Contained Mechanical Refrigeration Systems	ASHRAE	ANSI/ASHRAE 18-1987 (RA 91)
	Drinking Water Coolers and Beverage Dispensers	CSA	C22.2 No. 91-1971 (R 1981)
	General Standard on Refrigeration Equipment	CSA	CAN/CSA-C22.2 No. 120-M91
	Drinking Water Coolers (1987)	UL	ANSI/UL 399-1986
	Manual Food and Beverage Dispensing Equipment	NSF	NSF 18
Liquid	Refrigerant-Cooled Liquid Coolers, Remote Type	ARI	ANSI/ARI 480-87
	Methods of Testing for Rating Liquid Coolers	ASHRAE	ANSI/ASHRAE 24-1989
Cooling Towers	Commercial Applications Systems and Equipment	ACCA	Manual CS
	Atmospheric Water Cooling Equipment	ASME	ANSI/ASME PTC 23-1986 (R 1991)
	Water-Cooling Towers	NFPA	ANSI/NFPA 214-1988
	Acceptance Test Code for Spray Cooling Systems (1985)	CTI	CTI ATC-133-1985
	Acceptance Test Code for Water Cooling Towers: Mechanical Draft, Natural Draft Fan Assisted Types, Evaluation of Results, and Thermal Testing of Wet/Dry Cooling Towers (1990)	CTI	CTI ATC-105-1990
	Certification Standard for Commercial Water Cooling Towers (1991)	CTI	CTI STD-201-1991
	Code for Measurement of Sound from Water Cooling Towers	CTI	CTI ATC-128-1981
	Fiberglass-Reinforced Plastic Panels for Application on Industrial Water Cooling Towers	CTI	CTI STD-131-1986
	Nomenclature for Industrial Water-Cooling Towers	CTI	CTI NCL-109-1983
Dehumidifiers	Commercial Applications Systems and Equipment	ACCA	Manual CS
	Dehumidifiers	AHAM	ANSI/AHAM DH 1
	Dehumidifiers	CSA	C22.2 No. 92-1971
	Dehumidifiers (1987)	UL	ANSI/UL 474-1987
Desiccants	Method of Testing Desiccants for Refrigerant Drying	ASHRAE	ANSI/ASHRAE 35-1992
Driers	Method of Testing Liquid Line Refrigerant Driers	ASHRAE	ANSI/ASHRAE 63.1-1988
	Liquid Line Driers	ARI	ANSI/ARI 710-86
Electrical	Voltage Ratings for Electrical Power Systems and Equipment	ANSI	ANSI C84.1-1989
	Canadian Electrical Code, Part 1 (16th ed.)	CSA	C22.1-1990
	Application Guide for Ground Fault Interrupters	NEMA	NEMA 280-1990
	Application Guide for Ground Fault Protective Devices for Equipment	NEMA	NEMA PB 2.2-1988
	Enclosures for Electric Equipment	NEMA	NEMA 250-1991
	Enclosures for Industrial Control and Systems	NEMA	NEMA ICS 6-1988
	General Requirements for Wiring Devices	NEMA	NEMA WD 1-1983 (R 1989)
	Low Voltage Cartridge Fuses	NEMA	NEMA FU 1-1986
	Molded Case Circuit Breakers	NEMA	NEMA AB 1-1986
	Terminal Blocks for Industrial Use	NEMA	NEMA ICS 4-1983 (R 1988)
	National Electric Code (1990)	NFPA	ANSI/NFPA 70-1990
	Compatibility of Electrical Connectors and Wiring (1988)	SAE	SAE AIR1329 A
	Manufacturers' Identification of Electrical Connector Contacts, Terminals and Splices (1982)	SAE	SAE AIR1351 A
	Class T Fuses (1992)	UL	ANSI/UL 198H-1987
	Enclosures for Electrical Equipment (1992)	UL	UL 50
	Fuseholders (1987)	UL	ANSI/UL 512-1986
	High-Interrupting-Capacity Fuses, Current-Limiting Types (1990)	UL	ANSI/UL 198C-1986
	Molded-Case Circuit Breakers and Circuit Breaker Enclosures (1992)	UL	UL 489
	Terminal Blocks (1991)	UL	UL 1059
	Thermal Cutoffs for Use in Appliances and Components (1992)	UL	ANSI/UL 1020-1986
Energy	Air Conditioning and Refrigerating Equipment Nameplate Voltages	ARI	ARI 110-90
	Energy Conservation in Existing Buildings—Commercial	ASHRAE	ANSI/ASHRAE/IES 100.3-1985
	Energy Conservation in Existing Buildings—High Rise Residential	ASHRAE	ANSI/ASHRAE/IES 100.2-1991
	Energy Conservation in Existing Buildings—Institutional	ASHRAE	ANSI/ASHRAE/IES 100.5-1991
	Energy Conservation in Existing Buildings—Public Assembly	ASHRAE	ANSI/ASHRAE/IES 100.6-1991
	Energy Conservation in Existing Facilities—Industrial	ASHRAE	ANSI/ASHRAE/IES 100.4-1984
	Energy Conservation in New Building Design—Residential only	ASHRAE	ANSI/ASHRAE/IES 90A-1980

Table 1   Codes and Standards Published by Various Societies and Associations (*Continued*)

Subject	Title	Publisher	Reference
Energy (continued)	Energy Efficient Design of New Buildings Except Low Rise Residential Buildings	ASHRAE	ASHRAE/IES 90.1-1989
	Model Energy Code (MEC) (1992)	CABO	BOCA/ICBO/SBCCI
	Uniform Solar Energy Code (1991)	IAPMO	IAPMO
	Energy Directory	NCSBCS	NCSBCS
	Energy Management Guide for the Selection and Use of Polyphase Motors	NEMA	NEMA MG 10-1983 (R 1988)
	Energy Management Guide for the Selection and Use of Single Phase Motors	NEMA	MEA MG 11-1977 (R 1987)
	Total Energy Management Handbook, 3rd ed.	NEMA	NEMA 05101-1986
	Energy Conservation Guidelines (1984)	SMACNA	SMACNA
	Energy Recovery Equipment and Systems, Air-to-Air (1991)	SMACNA	SMACNA
	Retrofit of Building Energy Systems and Processes (1982)	SMACNA	SMACNA
	Energy Management Equipment (1991)	UL	ANSI/UL 916-1987
Exhaust Systems	Commercial Low Pressure, Low Velocity Duct Systems	ACCA	Manual Q
	Fundamentals Governing the Design and Operation of Local Exhaust Systems	ANSI	ANSI/AIHA Z9.2-1979
	Open-Surface Tanks—Ventilation and Operation	ANSI	ANSI/AIHA Z9.1-1991
	Safety Code for Design, Construction, and Ventilation of Spray Finishing Operations (reaffirmed 1971)	ANSI	ANSI/AIHA Z9.3-1985
	Ventilation and Safe Practices of Abrasives Blasting Operations	ANSI	ANSI/AIHA Z9.4-1985
	Method of Testing Performance of Laboratory Fume Hoods	ASHRAE	ANSI/ASHRAE 110-1985
	Compressors and Exhausters	ASME	ANSI/ASME PTC 10-1974 (R 1986)
	Mechanical Flue-Gas Exhausters	CSA	CAN 3-B255-M81
	Installation of Blower and Exhaust Systems for Dust, Stock, Vapor Removal or Conveying (1990)	NFPA	ANSI/NFPA 91-1990
	Draft Equipment (1973)	UL	UL 378
Expansion Valves	Thermostatic Refrigerant Expansion Valves	ARI	ANSI/ARI 750-87
	Method of Testing for Capacity Rating of Thermostatic Refrigerant Expansion Valves	ASHRAE	ANSI/ASHRAE 17-1986 (R 1990)
Fan Coil Units	Room Fan-Coil Air Conditioners	ARI	ARI 440-89
	Methods of Testing for Rating Room Fan-Coil Air Conditioners	ASHRAE	ANSI/ASHRAE 79-1984 (R 1991)
	Fan Coil Units and Room Fan Heater Units (1986)	UL	ANSI/UL 883-1986
Fans	Commercial Low Pressure, Low Velocity Duct Systems	ACCA	Manual Q
	Duct Design for Residential Winter and Summer Air Conditioning	ACCA	Manual D
	Designation for Rotation and Discharge of Centrifugal Fans	AMCA	AMCA 99-2406-83
	Drive Arrangements for Centrifugal Fans	AMCA	AMCA 99-2404-78
	Drive Arrangements for Tubular Centrifugal Fans	AMCA	AMCA 99-2410-82
	Inlet Box Positions for Centrifugal Fans	AMCA	AMCA 99-2405-83
	Laboratory Methods of Testing Fans for Rating	AMCA	ANSI/AMCA 210-85
	Motor Positions for Belt or Chain Drive Centrifugal Fans	AMCA	AMCA 99-2407-66
	Site Performance Test Standard Power Plant and Industrial Fans	AMCA	AMCA 803-87
	Standards Handbook	AMCA	AMCA 99-86
	Fans and Blowers	ARI	ARI 670-90
	Laboratory Methods of Testing Fans for Rating	ASHRAE	ANSI/ASHRAE 51-1985 ANSI/AMCA 210-85
	Methods of Testing Dynamic Characteristics of Propeller Fans— Aerodynamically Excited Fan Vibrations and Critical Speeds	ASHRAE	ANSI/ASHRAE 87.1-1992
	Fans	ASME	ANSI/ASME PTC 11-1984 (R 1990)
	Fans and Ventilators	CSA	C22.2 No. 113-M1984
	Performance of Ventilating Fans for Use in Livestock and Poultry Buildings	CSA	CAN/CSA C320-M86
	Rating the Performance of Residential Mechanical Ventilating Equipment	CSA	CAN/CSA C260-M90
	Electric Fans (1991)	UL	ANSI/UL 507
Ceiling	AC Electric Fans and Regulators	ANSI	ANSI-IEC Pub. 385
Filters	Flow-Capacity Rating and Application of Suction-Line Filters and Filter Driers	ARI	ANSI/ARI 730-86
	Grease Extractors for Exhaust Ducts (1981)	UL	ANSI/UL 710-1990
	Grease Filters for Exhaust Ducts (1979)	UL	UL 1046
Fire Dampers	Fire Dampers (1990)	UL	ANSI/UL 555-1989
Fireplaces	Factory-Built Fireplaces (1988)	UL	ANSI/UL 127-1992
Fire Protection	Standard Method for Fire Tests of Building Construction and Materials	ASTM	ASTM E 119-88
	Test Method for Surface Burning Characteristics of Building Materials	ASTM/ NFPA	ASTM E 84-89a; NFPA 255-1984

Table 1   Codes and Standards Published by Various Societies and Associations (*Continued*)

Subject	Title	Publisher	Reference
Fire Protection (continued)	BOCA National Fire Prevention Code, 8th ed. (1990)	BOCA	BOCA
	1992 Supplement to the Uniform Fire Codes	ICBO/IFCI	ICBO/IFCI
	Uniform Fire Code (1991)	IFCI	IFCI
	Uniform Fire Code Standards (1991)	IFCI	IFCI
	Interconnection Circuitry of Non-Coded Remote Station Protective Signalling Systems	NEMA	NEMA SB 3-1969 (R 1989)
	Fire Doors and Windows	NFPA	ANSI/NFPA 80-1990
	Fire Prevention Code	NFPA	ANSI/NFPA 1-1987
	Fire Protection Handbook, 17th ed.	NFPA	NFPA
	Flammable and Combustible Liquids Code	NFPA	ANSI/NFPA 30-1990
	Life Safety	NFPA	ANSI/NFPA 101-1991
	National Fire Codes (issued annually)	NFPA	NFPA
	Smoke Control Systems	NFPA	NFPA 92A-1988
	Standard Method of Fire Tests of Door Assemblies	NFPA	ANSI/NFPA 252-1990
	Standard Fire Prevention Code (1991 ed. with 1992 revisions)	SBCCI	SBCCI
	Fire Tests of Building Construction and Materials (1992)	UL	UL 263
	Heat Responsive Links for Fire Protection Service (1987)	UL	ANSI/UL 33-1987
Fireplace Stoves	Fireplace Stoves (1988)	UL	ANSI/UL 737-1988
Flow Capacity	Method of Testing Flow Capacity of Suction Line Filters and Filter Driers	ASHRAE	ANSI/ASHRAE 78-1985 (RA 90)
Freezers Household	Household Refrigerators, Combination Refrigerator-Freezers, and Household Freezers	AHAM	ANSI/AHAM; HRF 1
	Capacity Measurement and Energy Consumption Test Methods for Refrigerators, Combination Refrigerator-Freezers, and Freezers	CSA	CAN/CSA-C300-M91
	General Standard on Refrigeration Equipment	CSA	CAN/CSA-C22-2 No. 120-M91
	Household Refrigerators and Freezers	CSA	C22.2 No. 63-M1987
	Household Refrigerators and Freezers (1983)	UL	ANSI/UL 250-1991
Commercial	Dispensing Freezers	NSF	ANSI/NSF-6-1989
	Food Service Refrigerators and Storage Freezers	NSF	NSF-7
	Commercial Ice Cream Makers (1986)	UL	ANSI/UL 621-1985
	Commercial Refrigerators and Freezers (1992)	UL	ANSI/UL 471-1992
	Ice Makers (1992)	UL	ANSI/UL 563-1991
Furnaces	Commercial Applications Systems and Equipment	ACCA	Manual CS
	Residential Equipment Selection	ACCA	Manual S
	Direct Vent Central Furnaces	AGA	ANSI Z21.64-1990
	Gas-Fired Central Furnaces (except Direct Vent)	AGA	ANSI Z21.47-1990; Z21.47a-1990
	Gas-Fired Duct Furnaces	AGA	ANSI Z83.9-1990
	Gas-Fired Gravity and Fan Type Direct Vent Wall Furnaces	AGA	ANSI Z21.44-1991
	Gas-Fired Gravity and Fan Type Floor Furnaces	AGA	ANSI Z21.48-1989; Z21.48a-1990; Z21.48b-1991
	Gas-Fired Gravity and Fan Type Vented Wall Furnaces	AGA	ANSI Z21.49-1989; Z21.49a-1990; Z21.49b-1991
	Methods of Testing for Heating Seasonal Efficiency of Central Furnaces and Boilers	ASHRAE	ANSI/ASHRAE 103-1988
	Installation Code for Solid-Fuel-Burning Appliances and Equipment	CSA	CAN/CSA-B365-M91
	Solid Fuel-Fired Central Heating Appliances	CSA	CAN/CSA-B366.1-M91
	Electric Central Warm-Air Furnaces	CSA	C22.2 No.23-1980
	Heating and Cooling Equipment	CSA	CAN/CSA C22.2 No.236-M90
	Oil Burning Stoves and Water Heaters	CSA	B140.3-1962 (R 1991)
	Oil-Fired Warm Air Furnaces	CSA	B140.4-1974 (R 1991)
	Installation of Oil Burning Equipment	NFPA	NFPA 31-1987
	Standard Gas Code (1991 ed. with 1992 revisions)	SBCCI	SBCCI
	Standard Mechanical Code (1991 ed. with 1992 revisions)	SBCCI	SBCCI
	Commercial-Industrial Gas Heating Equipment (1973)	UL	UL 795
	Oil-Fired Central Furnaces (1986)	UL	UL 727-1986
	Oil-Fired Floor Furnaces (1987)	UL	ANSI/UL 729-1987
	Oil-Fired Wall Furnaces (1987)	UL	ANSI/UL 730-1986
	Residential Gas Detectors (1991)	UL	UL 1484
	Single and Multiple Station Carbon Monoxide Detectors (1992)	UL	UL 2034
	Solid-Fuel and Combination-Fuel Central and Supplementary Furnaces (1991)	UL	ANSI/UL 391-1991
Heat Exchangers	Remote Mechanical-Draft Evaporative Refrigerant Condensers	ARI	ANSI/ARI 490-89
	Method of Testing Air-to-Air Heat Exchangers	ASHRAE	ANSI/ASHRAE 84-1992

Table 1   Codes and Standards Published by Various Societies and Associations (*Continued*)

Subject	Title	Publisher	Reference
Heat Exchangers (continued)	Standard Methods of Test for Rating the Performance of Heat-Recovery Ventilators	CSA	CAN/CSA-C439-88
	Standards for Power Plant Heat Exchangers, 2nd ed. (1990)	HEI	HEI
	Standards of Tubular Exchanger Manufacturers Association, 7th ed. (1988)	TEMA	TEMA
Heat Meters	Method of Testing Thermal Energy Heat Meters for Liquid Streams in HVAC Systems	ASHRAE	ANSI/ASHRAE 125-1992
Heat Pumps	Commercial Applications Systems and Equipment	ACCA	Manual CS
	Heat Pump Systems: Principles and Applications (Commercial and Residence)	ACCA	Manual H
	Residential Equipment Selection	ACCA	Manual S
	Commercial and Industrial Unitary Heat Pump Equipment	ARI	ANSI/ARI 340-86
	Ground Water-Source Heat Pumps	ARI	ARI 325-85
	Water-Source Heat Pumps	ARI	ANSI/ARI 320-86
	Methods of Testing for Rating Unitary Air-Conditioning and Heat Pump Equipment	ASHRAE	ANSI/ASHRAE 37-1988
	Add-on Heat Pumps	CSA	C22.2 No. 186.2-M1980
	Central Forced Air Unitary Heat Pumps with or without Electric Resistance Heat	CSA	C22.2 No. 186.1-M1980
	Heating and Cooling Equipment	CSA	CAN/CSA C22.2 No. 236-M90
	Installation Requirements for Air-to-Air Heat Pumps	CSA	C273.5-1980 (R 1991)
	Performance Standard for Split System Central Air Conditioners and Heat Pumps	CSA	CAN/CSA-C273.3-M91
	Heat Pumps (1985)	UL	ANSI/UL 559-1985
Heat Recovery	Gas Turbine Heat Recovery Steam Generators	ASME	ANSI/ASME PTC 4.4-1981 (R 1992)
	Energy Recovery Equipment and Systems, Air-to-Air (1991)	SMACNA	SMACNA
Heaters	Direct Gas-Fired Make-Up Air Heaters	AGA	ANSI Z83.4-1991
	Direct Gas-Fired Door Heaters	AGA	ANSI Z83.17-1990; Z83.17a-1991
	Direct Gas-Fired Industrial Air Heaters	AGA	ANSI Z83.18-1990; Z83.18a-1991
	Gas-Fired Construction Heaters	AGA	ANSI Z83.7-1990; Z83.7a-1991
	Gas-Fired Infrared Heaters	AGA	ANSI Z83.6-1990
	Gas-Fired Pool Heaters	AGA	ANSI Z21.56-1991
	Gas-Fired Room Heaters, Vol. I, Vented Room Heaters (with 1989 and 1990 addenda)	AGA	ANSI Z21.11.1-1988
	Gas-Fired Room Heaters, Vol. II, Unvented Room Heaters (with 1990 addenda)	AGA	ANSI Z21.11.2-1989; Z21.11.2a-1990; Z21.11.2b-1991
	Gas-Fired Unvented Commercial and Industrial Heaters (with 1984 and 1989 addenda)	AGA	ANSI Z83.16-1982; Z83.16a-1984; Z83.16b-1989
	Desuperheater/Water Heaters	ARI	ARI 470-87
	Air Heaters	ASME	ANSI/ASME PTC 4.3-1968 (R 1991)
	Space Heaters for Use with Solid Fuels	CSA	B366.2 M1984
	Standards for Closed Feedwater Heaters, 4th ed. (1984)	HEI	HEI
	Fuel-Fired Heaters—Air Heating—for Construction and Industrial Machinery (1989)	SAE	SAE J1024 MAY89
	Motor Vehicle Heater Test Procedure (1982)	SAE	SAE J638 JUN82
	Electric Air Heaters (1980)	UL	ANSI/UL 1025-1988
	Electric Central Air Heating Equipment (1986)	UL	ANSI/UL 1096-1985
	Electric Dry Bath Heaters	UL	UL 875
	Electric Heaters for Use in Hazardous (Classified) Locations (1985)	UL	ANSI/UL 823-1990
	Electric Heating Appliances (1987)	UL	ANSI/UL 499-1987
	Electric Oil Heaters (1990)	UL	ANSI/UL 574-1990
	Fixed and Location Dedicated Electric Room Heaters (1992)	UL	UL 2021 (1992)
	Gas Heating Equipment, Commercial-Industrial (1973)	UL	UL 795
	Movable and Wall- or Ceiling-Hung Electric Room Heaters (1992)	UL	UL 1278 (1992)
	Oil-Fired Air Heaters and Direct-Fired Heaters (1975)	UL	UL 733
	Oil-Fired Room Heaters (1973)	UL	UL 896
	Solid Fuel-Type Room Heaters (1988)	UL	ANSI/UL 1482-1988
	Unvented Kerosene-Fired Room Heaters and Portable Heaters (1982)	UL	UL 647
Heating	Commercial Applications Systems and Equipment	ACCA	Manual CS
	Residential Equipment Selection	ACCA	Manual S
	Determining the Required Capacity of Residential Space Heating and Cooling Appliances	CSA	CAN/CSA-F280-M90
	Automatic Flue-Pipe Dampers for Use with Oil-Fired Appliances	CSA	B140.14-M1979 (R 1991)
	Electric Duct Heaters	CSA	C22.2 No. 155-M1986
	Heater Elements	CSA	C22.2 No.72-M1984 (R 1992)
	Oil-Fired Service Water Heaters and Swimming Pool Heaters	CSA	B140.12-1976 (R 1991)

Table 1 Codes and Standards Published by Various Societies and Associations (*Continued*)

Subject	Title	Publisher	Reference
Heating (continued)	Performance Requirements for Electric Heating Line-Voltage Wall Thermostats	CSA	C273.4-M1978
	Performance Standard for Residential Electrical Baseboard Heaters	CSA	C273.2-1971
	Portable Industrial Oil-Fired Heaters	CSA	B140.8-1967 (R 1991)
	Portable Kerosene-Fired Heaters	CSA	CAN 3-B140.9.3 M86
	Electric Air Heaters	CSA	C22.2 No.46-M1988 (R 1991)
	Advanced Installation Guide for Hydronic Heating Systems, 1991	HYDI	IBR 250
	Comfort Conditioning Heat Loss Calculation Guide	HYDI	IBR H-21 1984, IBR H-22 (1989)
	Installation Guide for Residential Hydronic Heating Systems, 6th ed. (1988)	HYDI	IBR 200
	Environmental System Technology (1984)	NEBB	NEBB
	Installation and Operation of Pulverized Fuel Systems	NFPA	ANSI/NFPA 85F-1988
	Aircraft Electrical Heating Systems (1965) (reaffirmed 1983)	SAE	SAE AIR860
	HVAC Systems—Applications, 1st ed. (1986)	SMACNA	SMACNA
	Installation Standards for Residential Heating and Air Conditioning Systems (1988)	SMACNA	SMACNA
	Electric Baseboard Heating Equipment (1987)	UL	ANSI/UL 1042-1986
	Electric Central Air Heating Equipment (1986)	UL	ANSI/UL 1096-1985
	Heating and Cooling Equipment	UL	ANSI/UL 1995-1992
Humidifiers	Appliance Humidifiers	AHAM	ANSI/AHAM HU 1
	Central System Humidifiers	ARI	ANSI/ARI 610-89
	Self-Contained Humidifiers	ARI	ANSI/ARI 620-89
	Humidifiers and Evaporative Coolers	CSA	C22.2 No. 104-M1983
	Humidifiers (1987)	UL	ANSI/UL 998-1985
Ice Makers	Automatic Commercial Ice Makers	ARI	ARI 810-91
	Ice Storage Bins	ARI	ANSI/ARI 820-88
	Methods of Testing Automatic Ice Makers	ASHRAE	ANSI/ASHRAE 29-1988
	Ice-Making Machines	CSA	C22.2 No. 133-1964 (R 1981)
	Automatic Ice-Making Equipment	NSF	NSF-12
	Ice Makers (1984)	UL	ANSI/UL 563-1985
Incinerators	Incinerator Performance	CSA	Z103-1976
	Incinerators, Waste and Linen Handling Systems and Equipment	NFPA	ANSI/NFPA 82-1990
	Residential Incinerators (1973)	UL	UL 791
Induction Units	Room Air-Induction Units	ARI	ANSI/ARI 445-87
	Frame Assignments for Alternating Current Integral-Horsepower Induction Motors	NEMA	NEMA MG 13-1984 (R 1990)
Industrial Duct	Rectangular Industrial Duct Construction (1980)	SMACNA	SMACNA
	Round Industrial Duct Construction (1977)	SMACNA	SMACNA
Insulation	Specification for Adhesives for Duct Thermal Insulation	ASTM	ASTM C916
	Specification for Thermal and Acoustical Insulation (Mineral Fiber, Duct Lining Material)	ASTM	ASTM C1071-86
	Test Method for Steady-State Heat Flux Measurements and Thermal Transmission Properties by Means of the Guarded Hot Plate Apparatus	ASTM	ASTM C177-85
	Test Method for Steady-State Heat Flux Measurements and Thermal Transmission Properties by Means of the Heat Flow Meter Apparatus	ASTM	ASTM C518-85
	Test Method for Steady-State Heat Transfer Properties of Horizontal Pipe Insulations	ASTM	ASTM C335-89
	Test Method for Steady-State and Thermal Performance of Building Assemblies by Means of a Guarded Hot Box	ASTM	ASTM C236-89
	Thermal Insulation, Mineral Fibre, for Buildings	CSA	A101-M1983
	National Commercial and Industrial Insulation Standards	MICA	MICA 1988
Louvers	Test Method for Louvers, Dampers, and Shutters	AMCA	AMCA 500-89
Lubricants	Method of Testing the Floc Point of Refrigeration Grade Oils	ASHRAE	ANSI/ASHRAE 86-1983
	Practice for Calculating Viscosity Index from Kinematic Viscosity at 40 and 100°C	ASTM	ASTM D2270-91
	Practice for Conversion of Kinematic Viscosity to Saybolt Universal Viscosity or to Saybolt Furol Viscosity	ASTM	ASTM D2161-87
	Method for Estimation of Molecular Weight of Petroleum Oils from Viscosity Measurements	ASTM	ASTM D2502-87
	Method for Separation of Representative Aromatics and Nonaromatics Fractions of High-Boiling Oils by Elution Chromatography	ASTM	ASTM D2549-91
	Classification for Viscosity System for Industrial Fluid Lubricants	ASTM	ASTM D2422-86
	Test Method for Carbon-Type Composition of Insulating Oils of Petroleum Origin	ASTM	ASTM D2140-86

Table 1   Codes and Standards Published by Various Societies and Associations (*Continued*)

Subject	Title	Publisher	Reference
Lubricants (continued)	Test Method for Dielectric Breakdown Voltage of Insulating Liquids Using Disk Electrodes	ASTM	ASTMD877-87
	Test Method for Dielectric Breakdown Voltage of Insulating Oils of Petroleum Origin Using VDE Electrodes	ASTM	ASTM D1816-84a (90)
	Test Method for Mean Molecular Weight of Mineral Insulating Oils by the Cryoscopic Method	ASTM	ASTM D2224-78 (1983)
	Test Method for Molecular Weight of Hydrocarbons by Thermoelectric Measurement of Vapor Pressure	ASTM	ASTM D2503-82 (1987)
	Test Methods for Pour Point of Petroleum Oils	ASTM	ASTM D97-87
	Semiconductor Graphite	NEMA	NEMA CB 4-1989
Measurements	A Standard Calorimeter Test Method for Flow Measurement of a Volatile Refrigerant	ASHRAE	ANSI/ASHRAE 41.9-1988
	Engineering Analysis of Experimental Data	ASHRAE	ASHRAE Guideline 2-1986 (RA 90)
	Procedure for Bench Calibration of Tank Level Gaging Tapes and Sounding Rules	ASME	ANSI MC88.2-1974 (R 1987)
	Standard Method for Measurement of Flow of Gas	ASHRAE	ANSI/ASHRAE 41.7-1984 (RA 91)
	Standard Method for Measurement of Proportion of Oil in Liquid Refrigerant	ASHRAE	ANSI/ASHRAE 41.4-1984
	Standard Method for Temperature Measurement	ASHRAE	ANSI/ASHRAE 41.1-1986 (RA 91)
	Standard Method for Pressure Measurement	ASHRAE	ANSI/ASHRAE 41.3-1989
	Standard Methods of Measurement of Flow of Liquids in Pipes Using Orifice Flowmeters	ASHRAE	ANSI/ASHRAE 41.8-1989
	Standard Method for Measurement of Moist Air Properties	ASHRAE	ANSI/ASHRAE 41.6-1982
	Standard Methods of Measuring and Expressing Building Energy Performance	ASHRAE	ANSI/ASHRAE 105-1984 (RA 90)
	Glossary of Terms Used in the Measurement of Fluid Flow in Pipes	ASME	ANSI/ASME MFC-1M-1991
	Guide for Dynamic Calibration of Pressure Transducers	ASME	ANSI MC88-1-1972 (R 1987)
	Measurement of Fluid Flow in Pipes Using Orifice, Nozzle, and Venturi	ASME	ASME MFC-3M-1989
	Measurement of Fluid Flow in Pipes Using Vortex Flow Meters	ASME	ASME/ANSI MFC-6M-1987
	Measurement of Gas Flow by Means of Critical Flow Venturi Nozzles	ASME	ASME/ANSI MFC-7M-1987
	Measurement of Gas Flow by Turbine Meters	ASME	ANSI/ASME MFC-4M-1986 (R 1990)
	Measurement of Industrial Sound	ASME	ANSI/ASME PTC 36-1985
	Measurement of Liquid Flow in Closed Conduits Using Transit-Time Ultrasonic Flowmeters	ASME	ANSI/ASME MFC-5M-1985 (R 1989)
	Measurement of Rotary Speed	ASME	ANSI/ASME PTC 19.13-1961 (R 1986)
	Measurement Uncertainty	ASME	ANSI/ASME PTC 19.1-1985 (R 1990)
	Measurement Uncertainty for Fluid Flow in Closed Conduits	ASME	ANSI/ASMEMFC-2M-1983 (R 1988)
	Pressure Measurement	ASME	ASME/ANSI PTC 19.2-1987
	Temperature Measurement	ANSI	ANSI/ASME PTC 19.3-1974 (R 1986)
Mobile Homes and Recreational Vehicles	Load Calculation for Residential Winter and Summer Air Conditioning	ACCA	Manual J
	Recreational Vehicle Cooking Gas Appliances (with 1989 addenda)	AGA	ANSI Z21.57-1990; Z21.57a-1991
	Mobile Homes	CSA	CAN/CSA-Z240 MH Series-M1992
	Mobile Home Parks	CSA	Z240.7.1-1972
	Oil-Fired Warm-Air Heating Appliances for Mobile Housing and Recreational Vehicles	CSA	B140.10-1974 (R 1991)
	Recreational Vehicle Parks	CSA	Z240.7.2-1972
	Recreational Vehicles	CSA	CAN/CSA-Z240 RV Series-M92
	Gas Supply Connectors for Manufactured Homes	IAPMO	IAPMO TSC 9-1992
	Manufactured Home Installations	NCSBCS	ANSI A225.1-93
	Recreational Vehicles	NFPA	NFPA 501C-1990
	Plumbing System Components for Manufactured Homes and Recreational Vehicles	NSF	NSF-24
	Gas Burning Heating Appliances for Mobile Homes and Recreational Vehicles (1965)	UL	UL 307B
	Gas-Fired Cooking Appliances for Recreational Vehicles (1976)	UL	UL 1075
	Liquid-Fuel-Burning Heating Appliances for Mobile Homes and Recreational Vehicles (1990)	UL	ANSI/UL 307A-1989
	Low Voltage Lighting Fixtures for Use in Recreational Vehicles (1990)	UL	UL 234
	Roof Jacks for Manufactured Homes and Recreational Vehicles (1990)	UL	UL 311
	Roof Trusses for Mobile Homes (1990)	UL	UL 1298
	Shear Resistance Tests for Ceiling Boards for Mobile Homes (1990)	UL	UL 1296
Motors and Generators	Steam Generating Units	ASME	ANSI/ASME PTC 4.1-1964 (R 1991)
	Testing of Nuclear Air-Treatment Systems	ASME	ANSI/ASME N510-1989
	Nuclear Power Plant Air Cleaning Units and Components	ASME	ANSI/ASME N509-1989
	Energy Efficiency Test Methods for Three-Phase Induction Motors (Efficiency Quoting Method and Permissible Efficiency Tolerance)	CSA	C390-M1985
	Motors and Generators	CSA	C22.2 No. 100-M1985

Table 1   Codes and Standards Published by Various Societies and Associations (*Continued*)

Subject	Title	Publisher	Reference
Motors and Generators (continued)	Guide for the Development of Metric Standards for Motors and Generators	NEMA	NEMA 10407-1980
	Motion/Position Control Motors and Controls	NEMA	NEMA MG 7-1987
	Motors and Generators	NEMA	NEMA MG 1-1987
	Renewal Parts for Motors and Generators—Performance, Selection, and Maintenance	NEMA	NEMA MG 7-1987
	Electric Motors (1989)	UL	ANSI/UL 1004-1988
	Electric Motors and Generators for Use in Hazardous (Classified) Locations (1989)	UL	UL 674
	Impedance Protected Motors (1990)	UL	ANSI/UL 519-1989
	Thermal Protectors for Motors (1986)	UL	ANSI/UL 547-1991
Outlets and Inlets	Method of Testing for Rating the Performance of Air Outlets and Inlets	ASHRAE	ANSI/ASHRAE 70-1991
Pipe, Tubing, and Fittings	Power Piping	ASME	ANSI/ASME B31.1-1989
	Refrigeration Piping	ASME	ASME/ANSI B31.5-1987
	Scheme for the Identification of Piping Systems	ASME	ANSI/ASME A13.1-1981 (R 1985)
	Specification for Acrylonitrile-Butadiene-Styrene (ABS) Plastic Pipe, Schedules 40 and 80	ASTM	ASTM D1527-89
	Specification for Polyethylene (PE) Plastic Pipe, Schedule 40	ASTM	ASTM D2104-90
	Specification for Polyvinyl Chloride (PVC) Plastic Pipe, Schedules 40, 80, and 120	ASTM	ASTM D1785-91
	Specification for Seamless Copper Pipe, Standard Sizes	ASTM	ASTM B42-89
	Specification for Seamless Copper Tube for Air Conditioning and Refrigeration Field Service	ASTM	ASTM B280-88
	Specification for Welded Copper and Copper Alloy Tube for Air Conditioning and Refrigeration Service	ASTM	ASTM B640-90
	Standards of the Expansion Joint Manufacturers Association, Inc., 5th ed. (1980 with 1985 addenda)	EJMA	EJMA
	Corrugated Polyolefin Coilable Plastic Utilities Duct	NEMA	NEMA TC 5-1990
	Corrugated Polyvinyl-Chloride (PVC) Coilable Plastic Utilities Duct	NEMA	NEMA TC 12-1990
	Electrical Nonmetallic Tubing (ENT)	NEMA	NEMA TC 13-1990
	Electrical Plastic Tubing (EPT) and Conduit Schedule EPC-40 and EPC-80	NEMA	NEMA TC 2-1990
	Extra-Strength PVC Plastic Utilities Duct for Underground Installation	NEMA	NEMA TC 8-1990
	Filament Wound Reinforced Thermosetting Resin Conduit and Fittings	NEMA	NEMA TC 14-1984 (R 1986)
	Fittings, Cast Metal Boxes, and Conduit Bodies for Conduit and Cable Assemblies	NEMA	NEMA FB 1-1988
	Fittings for ABS and PVC Plastic Utilities Duct for Underground Installation	NEMA	NEMA TC 9-1990
	Polyvinyl-Chloride (PVC) Externally Coated Galvanized Rigid Steel Conduit and Intermediate Metal Conduit	NEMA	NEMA RN 1-1989
	PVC and ABS Plastic Utilities Duct for Underground Installations	NEMA	NEMA TC 6-1990
	Smooth Wall Coilable Polyethylene Electrical Plastic Duct	NEMA	NEMA TC 7-1990
	National Fuel Gas Code	NFPA/ AGA/ANSI	ANSI/NFPA 54-1988/ ANSI Z223.1-1988
	Plastics Piping Components and Related Materials	NSF	NSF-14
	Refrigeration Tube Fittings (1977)	SAE	ANSI/SAE J513 OCT77
	Seismic Restraint Manual Guidelines for Mechanical Systems (1991)	SMACNA	SMACNA
	Rubber Gasketed Fittings for Fire Protection Service (1982)	UL	UL 213
	Tube Fittings for Flammable and Combustible Fluids, Refrigeration Service and Marine Use (1978)	UL	UL 109
Plumbing	BOCA National Plumbing Code, 8th ed. (1990)	BOCA	BOCA
	Uniform Plumbing Code (1991) (with IAPMO Installation Standards)	IAPMO	IAPMO
	National Standard Plumbing Code (NSPC)	NAPHCC	NSPC 1993
	Standard Plumbing Code (1991 ed. with 1992 revisions)	SBCCI	SBCCI
Pumps	Centrifugal Pumps	ASME	ASME PTC 8.2-1990
	Displacement Compressors, Vacuum Pumps and Blowers	ASME	ANSI/ASME PTC 9-1974 (R 1992)
	Liquid Pumps	CSA	CAN/CSA C.22.2 No.108-M89
	Performance Standard for Liquid Ring Vacuum Pumps, 1st ed. (1987)	HEI	HEI
	Centrifugal Pump Test Standard (1988)	HI	HI
	Hydraulic Institute Engineering Data Book, 2nd ed. (1991)	HI	HI
	Hydraulic Institute Standards, 14th ed. (1983)	HI	HI
	Vertical Pump Test Standard (1988)	HI	HI
	Circulation System Components for Swimming Pools, Spas, or Hot Tubs	NSF	NSF-50
	Electric Swimming Pool Pumps, Filters and Chlorinators (1986)	UL	ANSI/UL 1081-1985
	Motor-Operated Water Pumps (1991)	UL	ANSI/UL 778-1991
	Pumps for Oil-Burning Appliances (1986)	UL	ANSI/UL 343-1985

Table 1   Codes and Standards Published by Various Societies and Associations (*Continued*)

Subject	Title	Publisher	Reference
Radiation	Ratings for Baseboard and Fin-Tube Radiation (1992)	HYDI	IBR
	Testing and Rating Code for Baseboard Radiation, 6th ed. (1990)	HYDI	IBR
	Testing and Rating Code for Finned-Tube Commercial Radiation (1990)	HYDI	IBR
Receivers	Refrigerant Liquid Receivers	ARI	ANSI/ARI 495-85
Refrigerant-Containing Components	Refrigerant-Containing Components for Use in Electrical Equipment	CSA	C22.2 No.140.3-M1987
	Refrigerant-Containing Components and Accessories, Non-Electrical (1986)	UL	ANSI/UL 207-1986
Refrigerants	Performance of Refrigerant Recovery, Recycling, and/or Reclaim Equipment	ARI	ARI 740-91
	Specifications for Fluorocarbon Refrigerants	ARI	ANSI/ARI 700-88
	Methods of Testing Discharge Line Refrigerant-Oil Separators	ASHRAE	ANSI/ASHRAE 69-1990
	Number Designation and Safety Classification of Refrigerants	ASHRAE	ANSI/ASHRAE 34-1992
	Reducing Emission of Fully Halogenated Chlorofluorocarbon (CFC) Refrigerants in Refrigeration and Air-Conditioning Equipment and Applications	ASHRAE	ASHRAE Guideline 3-1990
	Refrigeration Oil Description	ASHRAE	ANSI/ASHRAE 99-1981 (RA 87)
	Sealed Glass Tube Method to Test the Chemical Stability of Material for Use Within Refrigerant Systems	ASHRAE	ANSI/ASHRAE 97-1983 (RA 89)
	Refrigerant Recovery/Recycling Equipment (1989)	UL	ANSI/UL 1963-1991
Refrigeration	Capacity Measurement of Field Erected Compression-Type Refrigeration and Air-Conditioning Systems	ASHRAE	ANSI/ASHRAE 83-1985
	Safety Code for Mechanical Refrigeration	ASHRAE	ANSI/ASHRAE 15-1992
	General Standard on Refrigeration Equipment	CSA	CAN/CSA-C22.2 No.120-M91
	Equipment, Design and Installation of Ammonia Mechanical Refrigeration Systems	IIAR	ANSI/IIAR 2-1984
	Refrigerated Medical Equipment (1978)	UL	UL 416
Refrigeration Systems			
Steam Jet	Ejectors	ASME	ASME PTC 24-1976 (R 1982)
	Standards for Steam Jet Vacuum Systems, 4th ed. (1988)	HEI	HEI
Transport	Mechanical Transport Refrigeration Units	ARI	ARI 1110-83
	Mechanical Refrigeration Installations on Shipboard	ASHRAE	ANSI/ASHRAE 26-1978 (RA 85)
	General Requirements for Application of Vapor Cycle Refrigeration Systems for Aircraft (1973) (reaffirmed 1983)	SAE	SAE ARP731A
	Safety Practices for Mechanical Vapor Compression Refrigeration Equipment or Systems Used to Cool Passenger Compartment of Motor Vehicles (1987)	SAE	SAE J639 JAN87
Refrigerators	Method of Testing Open Refrigerators for Food Stores	ASHRAE	ANSI/ASHRAE 72-1983
	Methods of Testing Closed Refrigerators	ASHRAE	ANSI/ASHRAE 117-1992
Commercial	Food Carts	NSF	NSF 59
	Food Service Equipment	NSF	NSF-2
	Food Service Refrigerators and Storage Freezers	NSF	NSF 7
	Soda Fountain and Luncheonette Equipment	NSF	NSF 1
	Commercial Refrigerators and Freezers (1992)	UL	ANSI/UL 471-1991
	Refrigerating Units (1989)	UL	ANSI/UL 427-1989
	Refrigeration Unit Coolers (1980)	UL	ANSI/UL 412-1984
Household	Refrigerators Using Gas Fuel	AGA	ANSI Z21.19-1990
	Household Refrigerators and Household Freezers	AHAM	AHAM HRF 1
	Capacity Measurement and Energy Consumption Test Methods for Refrigerators, Combination Refrigerator-Freezers, and Freezers	CSA	CAN/CSA C300-M89
	Household Refrigerators and Freezers (1983)	UL	ANSI/UL 250-1984
Roof Ventilators	Commercial Low Pressure, Low Velocity Duct Systems	ACCA	Manual Q
	Power Ventilators (1984)	UL	ANSI/UL 705-1984
Solar Equipment	Method of Measuring Solar-Optical Properties of Materials	ASHRAE	ANSI/ASHRAE 74-1988
	Method of Testing to Determine the Thermal Performance of Flat-Plate Solar Collectors Containing a Boiling Liquid Materials	ASHRAE	ANSI/ASHRAE 109-1986 (RA 90)
	Method of Testing to Determine the Thermal Performance of Solar Collectors	ASHRAE	ANSI/ASHRAE 93-1986 (RA 91)
	Methods of Testing to Determine the Thermal Performance of Solar Domestic Water Heating Systems	ASHRAE	ASHRAE 95-1981 (RA 87)
	Methods of Testing to Determine the Thermal Performance of Unglazed Flat-Plate Liquid-Type Solar Collectors	ASHRAE	ANSI/ASHRAE 96-1980 (RA 90)
Solenoid Valves	Solenoid Valves for Liquid Flow Use with Volatile Refrigerants and Water	ARI	ARI 760-87
Sound Measurement	Measurement of Sound from Boiler Units, Bottom-Supported Shop or Field Erected, 3rd ed.	ABMA	ABMA
	Methods for Calculating Fan Sound Ratings from Laboratory Test Data	AMCA	AMCA Standard 301-90

Table 1 Codes and Standards Published by Various Societies and Associations (*Continued*)

Subject	Title	Publisher	Reference
Sound Measurement (continued)	Laboratory Method of Testing—In-Duct Sound Power Measurement Procedure for Fans	AMCA	ANSI/AMCA 330-86
	Reverberant Room Method for Sound Testing of Fans	AMCA	AMCA 300-85
	Application of Sound Rated Outdoor Unitary Equipment	ARI	ARI 275-84
	Method of Measuring Machinery Sound within Equipment Rooms	ARI	ARI 575-87
	Method of Measuring Sound and Vibration of Refrigerant Compressors	ARI	ANSI/ARI 530-89
	Rating the Sound Levels and Transmission Loss of Package Terminal Equipment	ARI	ANSI/ARI 300-88
	Sound Rating of Large Outdoor Refrigerating and Air-Conditioning Equipment	ARI	ARI 370-86
	Sound Rating of Non-Ducted Indoor Air-Conditioning Equipment	ARI	ARI 350-86
	Sound Rating of Outdoor Unitary Equipment	ARI	ARI 270-84
	Guidelines for the Use of Sound Power Standards and for the Preparation of Noise Test Codes	ASA	ASA 94; ANSI S12.30-1990
	Method for the Calibration of Microphones (reaffirmed 1986)	ASA	ANSI S1.10-1966 (R 1986)
	Specification for Sound Level Meters (reaffirmed 1986)	ASA	ASA 47; ANSI S1.4-1983; ANSI S1.4A-1985
	Measurement of Industrial Sound	ASME	ASME/ANSI PTC 36-1985
	Procedural Standards for Measuring Sound and Vibration	NEBB	NEBB-1977
	Sound and Vibration in Environmental Systems	NEBB	NEBB-1977
	Sound Level Prediction for Installed Rotating Electrical Machines	NEMA	NEMA MG 3-1974 (R 1989)
Space Heaters	Electric Air Heaters	CSA	C22.2 No. 46-M1988
	Electric Air Heaters (1980)	UL	ANSI/UL 1025-1980
	Fixed and Location-Dedicated Electric Room Heaters (1992)	UL	UL 2021
	Movable and Wall- or Ceiling-Hung Electric Room Heaters (1992)	UL	UL 1278
Symbols	Graphic Electrical Symbols for Air-Conditioning and Refrigeration Equipment	ARI	ARI 130-88
	Graphic Symbols for Electrical and Electronic Diagrams	IEEE	ANSI/IEEE 315-1975
	Graphic Symbols for Heating, Ventilating, and Air Conditioning	ASME	ANSI/ASME Y32.2.4-1949 (R 1984)
	Graphic Symbols for Pipe Fittings, Valves and Piping	ASME	ANSI/ASME Y32.2.3-1949 (R 1988)
	Graphic Symbols for Plumbing Fixtures for Diagrams used in Architecture and Building Construction	ASME	ANSI/ASME Y32.4-1977 (R 1987)
	Symbols for Mechanical and Acoustical Elements as used in Schematic Diagrams	ASME	ANSI/ASME Y32.18-1972 (R 1985)
Testing and Balancing	Site Performance Test Standard-Power Plant and Industrial Fans	AMCA	AMCA 803-87
	Procedural Standards for Certified Testing of Cleanrooms (1988)	NEBB	NEBB-1988
	Procedural Standards for Testing, Adjusting, Balancing of Environmental Systems, 5th ed. (1991)	NEBB	NEBB-1991
	HVAC Systems—Testing, Adjusting and Balancing (1983)	SMACNA	SMACNA
Terminals, Wiring	Quick Connect Terminals	NEMA	NEMA DC 2-1982 (R 1988)
	Electrical Quick-Connect Terminals (1991)	UL	UL 310
	Equipment Wiring Terminals for Use with Aluminum and/or Copper Conductors (1991)	UL	ANSI/UL 486E-1987
	Splicing Wire Connectors (1992)	UL	ANSI/UL 486C-1990
	Wire Connectors and Terminal Lugs for Use with Copper Conductors (1992)	UL	ANSI/UL 486A-1990
	Wire Connectors for Use with Aluminum Conductors (1992)	UL	ANSI/UL 486B-1990
Thermal Storage	Commissioning of HVAC Systems	ASHRAE	ASHRAE Guideline 1-1989
	Metering and Testing Active Sensible Thermal Energy Storage Devices Based on Thermal Performance	ASHRAE	ANSI/ASHRAE 94.3-1986 (RA 90)
	Method of Testing Active Latent Heat Storage Devices Based on Thermal Performance	ASHRAE	ANSI/ASHRAE 94.1-1985 (RA 91)
	Methods of Testing Thermal Storage Devices with Electrical Input and Thermal Output Based on Thermal Performance	ASHRAE	ANSI/ASHRAE 94.2-1981 (RA 89)
	Practices for Measurement, Testing and Balancing of Building Heating, Ventilation, Air-Conditioning, and Refrigeration Systems	ASHRAE	ANSI/ASHRAE 111-1988
Turbines	Land Based Steam Turbine Generator Sets	NEMA	NEMA SM 24-1991
	Steam Turbines for Mechanical Drive Service	NEMA	NEMA SM23-1991
Unit Heaters	Gas Unit Heaters	AGA	ANSI Z83.8-1990; Z83.8a-1990
	Oil-Fired Unit Heaters (1988)	UL	ANSI/UL 731-1987
Valves	Automatic Gas Valves for Gas Appliances	AGA	ANSI Z21.21-1987; Z21.21a-1989
	Manually Operated Gas Valves for Appliances, Appliance Connection Valves, and Hose End Valves	AGA	ANSI Z21.15-1992
	Relief Valves and Automatic Gas Shutoff Devices for Hot Water Supply Systems	AGA	ANSI Z21.22-1986; Z21.22a-1990

Table 1 Codes and Standards Published by Various Societies and Associations (*Continued*)

Subject	Title	Publisher	Reference
Valves (continued)	Refrigerant Access Valves and Hose Connectors	ARI	ANSI/ARI 720-88
	Refrigerant Pressure Regulating Valves	ARI	ARI 770-84
	Solenoid Valves for Use with Volatile Refrigerants and Water	ARI	ARI 760-87
	Thermostatic Refrigerant Expansion Valves	ARI	ANSI/ARI 750-84
	Methods of Testing Nonelectric, Nonpneumatic Thermostatic Radiator Valves	ASHRAE	ANSI/ASHRAE 102-1983 (RA 89)
	Face-to-Face and End-to-End Dimensions of Valves	ASME	ASME/ANSI B16.10-1986
	Large Metallic Valves for Gas Distribution (Manually Operated, NPS-2 1/2 to 12, 125 psig Maximum)	ASME	ANSI/ASME B16.38-1985
	Manually Operated Metallic Gas Valves for Use in Gas Piping Systems up to 125 psig	ASME	ANSI B16.33-1990
	Manually Operated Thermoplastic Gas Shutoffs and Valves in Gas Distribution Systems	ASME	ANSI/ASME B16.40-1985
	Safety and Relief Valves	ASME	ANSI/ASME PTC25.3-1988
	Valves—Flanged Threaded, and Welding End	ASME	ANSI/ASME B16.34-1988
	Electrically Operated Valves (1982)	UL	ANSI-UL 429-1988
	Pressure Regulating Valves for LP-Gas (1985)	UL	ANSI/UL 144-1985
	Safety Relief Valves for Anhydrous Ammonia and LP-Gas (1984)	UL	UL 132
	Valves for Anhydrous Ammonia and LP-Gas (Other than Safety Relief) (1980)	UL	UL 125
	Valves for Flammable Fluids (1980)	UL	UL 842
Vending Machines	Methods of Testing Pre-Mix and Post-Mix Soft Drink Vending and Dispensing Equipment	ASHRAE	ANSI/ASHRAE 91-1976 (RA 91)
	Vending Machines	CSA	CAN/CSA-C22.2 No.128-M90
	Vending Machines for Food and Beverages	NSF	NSF-25
	Refrigerated Vending Machines (1989)	UL	ANSI/UL 541-1988
Vent Dampers	Automatic Vent Damper Devices for Use with Gas-Fired Appliances	AGA	ANSI Z21.66-1988; Z21.66a-1991; Z21.66b-1991
	Vent or Chimney Connector Dampers for Oil-Fired Appliances (1988)	UL	ANSI/UL 17-1988
Venting	Draft Hoods	AGA	ANSI Z21.12-1990
	National Fuel Gas Code	AGA	ANSI Z223.1-1992
	Chimneys, Fireplaces, Vents and Solid Fuel Burning Appliances	NFPA	ANSI/NFPA 211-1988
	Explosion Prevention Systems	NFPA	ANSI/NFPA 69-1992
	Guide for Steel Stack Design and Construction (1983)	SMACNA	SMACNA
	Draft Equipment (1973)	UL	UL 378
	Gas Vents (1991)	UL	ANSI/UL 441-1991
	Low-Temperature Venting Systems, Type L (1986)	UL	ANSI/UL 641-1985
Ventilation	Commercial Low Pressure, Low Velocity Duct Systems	ACCA	Manual Q
	Industrial Ventilation (1992)	ACGIH	ACGIH
	Method of Testing for Room Air Diffusion	ASHRAE	ANSI/ASHRAE 113-1990
	Ventilation for Acceptable Indoor Air Quality	ASHRAE	ANSI/ASHRAE 62-1989
	Residential Mechanical Ventilation Requirements	CSA	F326.1-M1989
	Residential Mechanical Ventilation System Requirements	CSA	F326.2-M1989
	Verification of the Performance of Residential Mechanical Ventilation Systems	CSA	F326.3-M1990
	Ventilation Directory	NCSBCS	NCSBCS
	Parking Structures; Repair Garages	NFPA	ANSI/NFPA 88A-1991; 88B-1991
	Removal of Smoke and Grease-Laden Vapors from Commercial Cooking Equipment	NFPA	ANSI/NFPA 96-1987
	Food Service Equipment	NSF	NSF-2
	Class II (Laminar Flow) Biohazard Cabinetry	NSF	NSF-49
Water Heaters	Gas Water Heaters, Vol. I, Storage Water Heaters with Input Ratings of 75,000 Btu per Hour or Less	AGA	ANSI Z21.10.1-1990; Z21.10.1a-1991
	Gas Water Heaters, Vol. III, Storage, with Input Ratings Above 75,000 Btu per Hour, Circulating and Instantaneous Water Heaters	AGA	ANSI Z21.10.3-1990; Z21.10.3a-1990
	Methods of Testing to Determine the Thermal Performance of Solar Domestic Water Heating Systems	ASHRAE	ANSI/ASHRAE 95-1981 (RA 87)
	Methods of Testing for Rating Combination Space-Heating and Water-Heating Appliances	ASHRAE	ANSI/ASHRAE 124-1991
	Construction and Test of Electric Storage-Tank Water Heaters	CSA	CAN/CSA-C22.2 No. 110-M90
	CSA Standards on Performance of Electric Storage Tank Water Heaters	CSA	CAN/CSA-C191-series-M90
	Oil Burning Stoves and Water Heaters	CSA	B140.3-1962 (R 1991)
	Oil-Fired Service Water Heaters and Swimming Pool Heaters	CSA	B140.12-1976 (R 1991)
	Systems Construction and Test of Electric Storage-Tank Water Heaters	CSA	CAN/CSA C22.2 No. 110-M90
	Hot Water Generating and Heat Recovery Equipment	NSF	NSF-5
	Commercial-Industrial Gas Heating Equipment (1973)	UL	UL 795
	Electric Booster and Commercial Storage Tank Water Heaters (1988)	UL	ANSI/UL 1453-1987

Table 1   Codes and Standards Published by Various Societies and Associations (*Concluded*)

Subject	Title	Publisher	Reference
Water Heaters *(continued)*	Household Electric Storage Tank Water Heaters (1989)	UL	ANSI/UL 174-1989
	Oil-Fired Storage Tank Water Heaters (1988)	UL	ANSI/UL 732-1987
Woodburning *Appliances*	Method of Testing for Performance Rating of Woodburning Appliances	ASHRAE	ANSI/ASHRAE 106-1984
	Installation Code for Solid Fuel Burning Appliances and Equipment	CSA	CAN/CSA-B365-M91
	Solid-Fuel-Fired Central Heating Appliances	CSA	CAN/CSA-B366.1-M91
	Space Heaters for Use with Solid Fuels	CSA	B366.2-M1984
	Chimneys, Fireplaces, Vents and Solid Fuel Burning Appliances	NFPA	ANSI/NFPA 211-1988
	Commercial Cooking and Hot Food Storage Equipment	NSF	NSF-4
	Solid Fuel Type Room Heaters (1988)	UL	ANSI/UL 1482-1988

ABBREVIATIONS AND ADDRESSES

ABMA	American Boiler Manufacturers Association, 950 N. Glebe Road, Suite 160, Arlington, VA 22203
ACCA	Air Conditioning Contractors of America, 1513 16th Street, NW, Washington, D.C. 20036
ACGIH	American Conference of Governmental Industrial Hygienists, 6500 Glenway Avenue, Building D-7, Cincinnati, OH 45211
ADC	Air Diffusion Council, Suite 200, 111 E. Wacker Drive, Chicago, IL 60601
AGA	American Gas Association, 1515 Wilson Boulevard, Arlington, VA 22209
AHAM	Association of Home Appliance Manufacturers, 20 N. Wacker Drive, Chicago, IL 60606
AIHA	American Industrial Hygiene Association, 345 White Pond Drive, Akron, OH 44320
AMCA	Air Movement and Control Association, Inc., 30 W. University Drive, Arlington Heights, IL 60004-1893
ANSI	American National Standards Institute, 11 West 42nd Street, New York, NY 10036
ARI	Air-Conditioning and Refrigeration Institute, 4301 North Fairfax Drive, Suite 425, Arlington, VA 22203
ASA	Acoustical Society of America, 335 E. 45 Street, New York, NY 10017-3483
ASHRAE	American Society of Heating, Refrigerating and Air-Conditioning Engineers, Inc., 1791 Tullie Circle, NE, Atlanta, GA 30329
ASME	The American Society of Mechanical Engineers, 345 E. 47 Street, New York, NY 10017
	For ordering publications: ASME Marketing Department, Box 2350, Fairfield, NJ 07007-2350
ASTM	American Society for Testing and Materials, 1916 Race Street, Philadelphia, PA 19103
BOCA	Building Officials and Code Administrators International, Inc., 4051 W. Flossmoor Road, Country Club Hills, IL 60478-5795
BSI	British Standards Institution, 2 Park Street, London, W1A 2BS, England
CABO	Council of American Building Officials, 5203 Leesburg Pike, Suite 708, Falls Church, VA 22041
CAGI	Compressed Air and Gas Institute, 1300 Sumner Avenue, Cleveland, OH 44115
CSA	Canadian Standards Association, 178 Rexdale Boulevard, Rexdale, Ontario M9W 1R3, Canada
CTI	Cooling Tower Institute, P.O. Box 73383, Houston, TX 77273
EJMA	Expansion Joint Manufacturers Association, Inc., 25 N. Broadway, Tarrytown, NY 10591
HEI	Heat Exchange Institute, 1621 Euclid Avenue, Keith Building, Suite 1230, Cleveland, OH 44115
HI	Hydraulic Institute, 9 Sylvan Way, Suite 360, Parsippany, NJ 07054-3802
HYDI	Hydronics Institute, 35 Russo Place, Berkeley Heights, NJ 07922
IAPMO	International Association of Plumbing and Mechanical Officials, 20001 Walnut Drive South, Walnut, CA 91789-2825
ICBO	International Conference of Building Officials, 5360 S. Workman Mill Road, Whittier, CA 90601
IFCI	International Fire Code Institute, 5360 S. Workman Mill Road, Whittier, CA 90601
IIAR	International Institute of Ammonia Refrigeration, 111 East Wacker Drive, Chicago, IL 60601
MICA	Midwest Insulation Contractors Association, 2017 South 139th Circle, Omaha, NE 68144
NCSBCS	National Conference of States on Building Codes and Standards, 505 Huntmar Park Drive, Suite 210, Herndon, VA 22070
NEBB	National Environmental Balancing Bureau, 1385 Piccard Drive, Rockville, MD 20850
NEMA	National Electrical Manufacturers Association, 2101 L Street, NW, Suite 300, Washington, D.C. 20037
NFPA	National Fire Protection Association, 1 Batterymarch Park, P.O. Box 9101, Quincy, MA 02269-9101
NSF International	National Sanitation Foundation, P.O. Box 130140, Ann Arbor, MI 48113-0140
SAE	Society of Automotive Engineers, 400 Commonwealth Drive, Warrendale, PA 15096
SBCCI	Southern Building Code Congress International, Inc., 900 Montclair Road, Birmingham, AL 35213-1206
SMACNA	Sheet Metal and Air Conditioning Contractors' National Association, 4201 Lafayette Center Drive, Chantilly, VA 22021
TEMA	Tubular Exchanger Manufacturers Association, Inc., 25 N. Broadway, Tarrytown, NY 10591
UL	Underwriters Laboratories Inc., 333 Pfingsten Road, Northbrook, IL 60062-2096

ADDITIONS AND CORRECTIONS

This section supplements the current handbooks and notes technical errors found in the series. Occasional typographical errors and nonstandard symbol labels will be corrected in future volumes. The authors and editor encourage you to notify them if you find other technical errors. Please send corrections to: Handbook Editor, ASHRAE, 1791 Tullie Circle NE, Atlanta, GA 30329.

1991 HVAC Applications

p. 4.7, Equation (1). The equation should read:

$$w_p = \frac{A(95 + 0.425v)}{Y}[p_w - p_a] \qquad (1)$$

p. 7.11, 1st column. Revise the paragraph after item "4. Zoning" in the Design Criteria section to read:

Air-cleaning requirements are taken from Table 1 for operating rooms. A recovery lounge need not be considered a sensitive area. The bacteria concern is the same as for an acute-care facility. The minimum ventilation rates, desired pressure relationships, desired relative humidity, and design temperature ranges are similar to the requirements for hospitals shown in Table 3.

p. 21.9, 1st column, 5th line up. The equation for infiltration heat loss should read:

$$q_i = 0.03\ V\ N\ (t_i - t_o)$$

p. 30.25, 1st column. The reference by Diamond and Avery (1986) is a Los Alamos Scientific Laboratory technical applications manual.

p. 36.6, 1st column, 13th line up. Correct reference to read "(Arkin and Shitzer 1979)." Also correct spelling in Bibliography section on p. 36.16.

p. 39.7, 2nd column, 2nd line up. Change "(see Figure 20)" to "(see Figures 18 and 19)."

1992 HVAC Systems and Equipment

p. 2.2, 2nd column, line 3. Change to read: (see section on dehumidification).

p. 2.3, 1st column. In the Humidification section, revise the first sentence of numbered paragraph 2 to read:

2. The use of compressed air to force water through a nozzle into the airstream is essentially a constant wet-bulb (adiabatic) process.

p. 2.3, Figure 5. Change the caption to read:

Fig. 5   Humidification

p. 3.5, Equation (3). Delete subscripts on the numbers.

p. 3.5, Equation (4). The equation only applies at or near sea level.

p. 5.5, left column, lines 7 and 9. Change "mbh values" to read "36×10^3 to 240×10^3 Btu/h and 240×10^3 to 1.5×10^6 Btu/h, respectively."

p. 6.2, 1st column, Equation (2). The equation should read:

$$q_r = \sigma F_e (T_p^4 - T_r^4) \qquad (2)$$

Also, the units for σ are in $Btu/(h \cdot ft^2 \cdot °F^4)$.

p. 6.3, 2nd column, 2nd paragraph. The first sentence should read:

Convection in a panel system is a function of the panel surface temperature and the temperature of the airstream immediately adjacent to the panel.

p. 6.3, 2nd column. In Equations (6) and (7):

D_e = equivalent diameter of panel
   = 4 × (panel area)/(perimeter)

p. 6.4, 1st column, last paragraph. The first sentence should read:

Figures 7 and 8 may be used to estimate the combined heat transfer from typical ceiling and floor panels.

p. 6.6, Figure 9. Add the following notes to the figure.

Graph based on data from Wilkes and Petersen (1938) for heat removal by natural convection with D_e = 5.5 ft, 2 in. to panel and room air temperature = AUST.

p. 8.1, 2nd column, line 13. Change "barriers" to read "retarders."

p. 9.5, 1st column. At equation for Q add note "at sea level."

p. 9.12, 1st column, near bottom. At equation for Q_r add note "at sea level."

p. 10.6, 1st column. Rewrite item 6 to read:

6. Take off all branch lines from the top of the steam mains, preferably. . .

p. 10.7, 2nd column, line 1. Delete "minimum."

p. 11.16, 2nd column. In the line below Equation (25) the variable should read t_{la}.

p. 11.5, Figure 4. The arrow showing the direction of flow from the chillers should point up toward the flow meter, not down. Also, the direction of flow shown on the check valves between the central plant circulating pumps and the chillers should be reversed.

p. 12.11, 1st column. Second paragraph should read

Control valves may be sized on the basis of the valve coefficient C_v as follows:

$$Q = C_v \sqrt{\Delta p/s_f}$$

where

Q = flow rate, gpm
C_v = valve coefficient
Δp = pressure drop across valve, psi
s_f = specific gravity of fluid

p. 19.4, 1st column, 5th paragraph. The pressure range should read "0.2 to 2 in. of water," not "50 to 500 Pa."

p. 19.4, 2nd column, 8th line up. Change "3 °C" to read "5 °F."

p. 19.4, 2nd column, last paragraph. Change "19 °C" to read "66 °F." Change "0.06 kW/kW" to read "0.22 kW/ton." Change "23 °C" to read "74 °F." Change "0.23 kW/kW" to read "0.81 kW/ton."

p. 19.5, 1st column. The first word should read "comparison," not "son."

p. 20.1, 1st column, 6th paragraph. Delete reference to "punched cards."

p. 21.11, 2nd column, Equation (28a). Change "h_{ab}" to read "h_{a1}."

p. 35.28, 1st column. The m in the first term of Equation (22) is misplaced. It should be an exponent.

p. 35.29, 2nd column, Equation (30). The equation should read:

$$W_{pi} = \mu_i u_i^2/g \qquad (30)$$

p. 35.29, 2nd column, Equation (34b). Add = sign so it reads:

$$a = \sqrt{-(\partial p/\partial v)_s} = \sqrt{n_s pv} \qquad (34b)$$

p. 35.30, 1st column, Equation (36a). The equation should read:

$$\Omega = g W_p/a_i^2 = \mu(\Sigma\, u_i^2/a_i^2) \qquad (36a)$$

p. 36.16, 1st column, 3rd paragraph. In the next to the last sentence, change "brake power" to read "brake horsepower."

p. 36.16, 2nd column, 2nd paragraph up. In the sixth line, change "R-12" to read "R-22."

p. 36.18, 1st column. Refer to ANSI/ASHRAE Standard 15-1992, which is the current Safety Code for Mechanical Refrigeration.

p. 37.8, Figure 17. The piping in this figure is improperly routed because it does not keep the condenser water and chilled water circuits positively separated. It will be corrected in the next revision of the chapter.

p. 37.10, Table 1. Revise as follows:

4.	Lubricate	Delete Q under V-Belt Drives
		Add Q under Fan Shaft Bearings
15.	Clean	Delete R under Bleed Rate
		Delete W under Flow Control Valves
		Add R under Flow Control Valves
		Add W under Suction Screen
16.	Repaint	Delete R under Fill
		Delete R under Bleed Rate
		Add R under Casing
		Add R under Drive Shaft
17.	Completely open and close	Delete S under Drive Shaft
		Add S under Flow Control Valve
18.	Make sure vents are open	Delete M under Motor
		Add M under Gear Reducer

p. 39.3, Table 3. Revise the 2nd part of the table, which lists how to correct flow, head, and power when the impeller diameter varies, as follows:

To correct for flow, multiply by (New diameter/Old diameter)^3.

To correct for head, multiply by (New diameter/Old diameter)^2.

To correct for power, multiply by (New diameter/Old diameter)^5.

p. 42.10, Equation (8). Change constant in the equation to read "4000," not "4."

p. 43.9, Figure 21. Switch upstream and downstream pressure captions. Flow should be from right to left.

p. 43.10, 1st column. The third paragraph should read:

The area of an orifice is changed either by (1) moving a piston or cup across a shear plate, or (2) by increasing pressure (to squeeze the rubber) in a rubber or a grommet type valve.

COMPOSITE INDEX

ASHRAE HANDBOOK SERIES

This index covers the current Handbook volumes published by ASHRAE. Listings from each volume are identified as follows:

F = 1993 Fundamentals
S = 1992 Systems and Equipment
A = 1991 HVAC Applications
R = 1990 Refrigeration
E = 1988 Equipment

The index is alphabetized in a *word-by-word* format; for example, *air diffusers* is listed before *aircraft* and *heat flow* is listed before *heaters*.

Note that the code for a volume includes the chapter number followed by a decimal point and the page number(s) within the chapter. For example, F32.4 means the information may be found in the Fundamentals volume, chapter 32, page 4.